TECHNIQUES IN THE BEHAVIORAL AND NEURAL SCIENCES

VOLUME 15

HANDBOOK OF STRESS AND THE BRAIN

Part 1: The Neurobiology of Stress

Previously published in TECHNIQUES IN THE BEHAVIORAL AND NEURAL SCIENCES

Volume 1: Feeding and Drinking, by F. Toates and N.E. Rowland (Eds.), 1987, ISBN 0-444-80895-7

Volume 2: Distribution-free Statistics: Application-oriented Approach, by J. Krauth, 1988, ISBN 0-444-80934-1, Paperback ISBN 0-444-80988-0

Volume 3: Molecular Neuroanatomy, by F.W. Van Leeuwen, R.M. Buijs, C.W. Pool and O. Pach (Eds.), 1989, ISBN 0-444-81014-5, Paperback ISBN 0-444-81016-1

Volume 4: Manual of Microsurgery on the Laboratory Rat, Part 1, by J.J. van Dongen, R. Remie, J.W. Rensema and G.H.J. van Wunnik (Eds.), 1990, ISBN 0-444-81138-9, Paperback ISBN 0-444-81139-7

Volume 5: Digital Biosignal Processing, by R. Weitkunat (Ed.), 1991, ISBN 0-444-81140-0, Paperback ISBN 0-444-98144-7

Volume 6: Experimental Analysis of Behavior, by I.H. Iversen and K.A. Lattal (Eds.), 1991, Part 1, ISBN 0-444-81251-2, Paperback ISBN 0-444-89160-9, Part 2, ISBN 0-444-89194-3, Paperback ISBN 0-444-89195-1

Volume 7: Microdialysis in the Neurosciences, by T.E. Robinson and J.B. Justice, Jr. (Eds.), 1991, ISBN 0-444-81194-X, Paperback ISBN 0-444-89375-X

Volume 8: Techniques for the Genetic Analysis of Brain and Behavior, by D. Goldowitz, D. Wahlsten and R.E. Wimer (Eds.), 1992, ISBN 0-444-81249-0, Paperback ISBN 0-444-89682-1

Volume 9: Research Designs and Methods in Psychiatry, by M. Fava and J.F. Rosenbaum (Eds.), 1992, ISBN 0-444-89595-7, Paperback ISBN 0-444-89594-9

Volume 10: Methods in Behavioral Pharmacology, by F. van Haaren (Ed.), 1993, ISBN 0-444-81444-2, Paperback ISBN 0-444-81445-0

Volume 11: Methods in Neurotransmitter and Neuropeptide Research, by S.H. Parvez (Eds.), 1993, Part 1, ISBN 0-444-81369-1, Paperback ISBN 0-444-81674-7, Part 2, ISBN 0-444-81368-3, Paperback ISBN 0-444-81675-5

Volume 12: Neglected Factors in Pharmacology and Neuroscience Research, by V. Claassen (Ed.), 1994, ISBN 0-444-81871-5, Paperback ISBN 0-444-81907-X

Volume 13: Handbook of Molecular–Genetic Techniques for Brain and Behavior Research, by W.E. Crusio and R.T. Gerlai (Eds.), 1999, ISBN 0-444-50239-4

Volume 14: Experimental Design. A Handbook and Dictionary for Medical and Behavioral Research, by J. Krauth (Ed.), 2000, ISBN 0-444-50637-3, Paperback ISBN 0-444-50638-1

Cover image by Tim Teebken/Getty Images.

TECHNIQUES IN THE BEHAVIORAL AND NEURAL SCIENCES

Series Editor

J.P. HUSTON
Düsseldorf

VOLUME 15

HANDBOOK OF STRESS AND THE BRAIN

Part 1: The Neurobiology of Stress

Edited by

T. STECKLER

Johnson & Johnson Pharmaceutical Research & Development,
A Division of Janssen Pharmaceutica N.V., Turnhoutseweg 30, 2340 Beerse, Belgium

N.H. KALIN

Department of Psychiatry and Health Emotions Research Institute, University of Wisconsin Medical School,
6001 Research Park Boulevard, Madison, WI 53719-1176, USA

J.M.H.M. REUL

Henry Wellcome Laboratories for Integrative Neuroscience and Endocrinology,
The Dorothy Hodgkin Building, University of Bristol, Whitson Street, Bristol BS1 3NY, UK

ELSEVIER

AMSTERDAM – BOSTON – HEIDELBERG – LONDON – NEW YORK – OXFORD
PARIS – SAN DIEGO – SAN FRANCISCO – SINGAPORE – SYDNEY – TOKYO
2005

ELSEVIER B.V.
Radarweg 29
P.O. Box 211
1043 NX Amsterdam
The Netherlands

ELSEVIER Inc.
525 B Street, Suite 1900
San Diego
CA 92101-4495
USA

ELSEVIER Ltd
The Boulevard
Langford Lane
Kidlington
Oxford OX5 1GB, UK

ELSEVIER Ltd
84 Theobalds Road
London WC1X 8RR
UK

First edition 2005

Library of Congress Cataloging in Publication Data
A catalog record is available from the Library of Congress.

British Library Cataloguing in Publication Data
A catalogue record is available from the British Library.

Handbook of stress and the brain. - (Techniques in the behavioral and neural sciences; v. 15)
1. Brain - Effect of stress on
I. Steckler, T. II. Kalin, N.H. III. Reul, J.M.H.M.
612.8′2

ISBN: 0-444-51173-3 (Part 1)
ISBN: 0-444-51823-1 (Part 2)
ISBN: 0-444-51822-3 (Volume 15 Two-Part Set)
Series ISSN: 0921-0709

∞ The paper used in this publication meets the requirements of ANSI/NISO Z39.48-1992 (Permanence of Paper).

Printed in The Netherlands.

List of Contributors, Part 1

K.-B. Abel, Division of Endocrinology, Children's Hospital Boston, Harvard Medical School, 300 Longwood Avenue, Boston, MA 02115, USA

D. Adams, Laboratory of Molecular Psychiatry, Departments of Psychiatry and Pharmacology, Yale University School of Medicine, 34 Park Street, New Haven, CT 06508, USA

I. Akirav, Department of Psychology *and*, The Brain and Behavior Research Center, University of Haifa, Haifa 31905, Israel

O.F.X. Almeida, Max Planck Institute of Psychiatry, Kraepelinstrasse 2–10, D-80804 Munich, Germany

B. Bali, Laboratory of Molecular Neuroendocrinology, Institute of Experimental Medicine, Hungarian Academy of Science, Szigony u. 43, Budapest H-1083, Hungary

C.W. Berridge, Departments of Psychology and Psychiatry, University of Wisconsin, 1202 W. Johnson Street, Madison, WI 53706, USA

J.J. Cerqueira, Life and Health Science Research Institute, Health Science School, University of Minho, Campus de Gualtar, 4710-057 Braga, Portugal

K.C. Chambers, Department of Psychology, University of Southern California, Seely G. Mudd Bldg. # 501, Los Angeles, CA 90089-1061, USA

G.P. Chrousos, Pediatric and Reproductive Endocrinology Branch, National Institute of Child Health and Human Development, National Institutes of Health, Building 10 Room 9D42, 10 Center Drive MSC 1583, Bethesda, MD 20892-1583, USA

O. Cohen, Departments of Biological Chemistry and Psychology, The Hebrew University of Jerusalem, Jerusalem, Israel

W.E. Cullinan, Department of Biomedical Sciences, Marquette University, Milwaukee, WI 53233, USA

B. Czéh, Clinical Neurobiology Laboratory, German Primate Center, Kellnerweg 4, 37077 Gottingen, Germany

F.M. Dautzenberg, Johnson & Johnson Pharmaceutical Research & Development, A Division of Janssen Pharmaceutica NV, Turnhoutseweg 30, 2340 Beerse, Belgium

E.R. De Kloet, Division of Medical Pharmacology, LACDS-LUMC, University of Leiden, P.O. Box 9502, 2300 RA Leiden, The Netherlands

A.J. Douglas, Section of Biomedical Sciences, DBCLS, University of Edinburgh, Hugh Robson Building, George Square, Edinburgh EH8 9XD, UK

G. Drolet, Centre de Recherche en Neurosciences, CHUL, RC-9800, 2705 Boulevard Laurier, Ste-Foy G1V 4G2 QC, Canada

S.K. Droste, Henry Wellcome Laboratories for Integrative Neuroscience and Endocrinology, The Dorothy Hodgkin Building, University of Bristol, Whitson Street, Bristol, BS1 3NY, UK

J. Du, Laboratory of Molecular Pathophysiology, National Institute of Mental Health, Building 49, Room B1EE16, 49 Convent Drive, Bethesda, MD 20892-4405, USA

R.S. Duman, Laboratory of Molecular Psychiatry, Departments of Psychiatry and Pharmacology, Yale University School of Medicine, 34 Park Street, New Haven, CT 06508, USA

Y. Dwivedi, Psychiatric Institute, Department of Psychiatry, University of Illinois at Chicago, 1601 W. Taylor St, Chicago, IL 60612, USA

W.C. Engeland, Departments of Surgery and Neuroscience, University of Minnesota, Mayo Mail Code 120, 420 Delaware St SE, Minneapolis, MN 55455, USA

N. Farzad, Laboratory of Molecular Pathophysiology, National Institute of Mental Health, Building 49, Room B1EE16, 49 Convent Drive, Bethesda, MD 20892-4405, USA

H. Figueiredo, Department of Psychiatry, University of Cincinnati, 231 Albert Sabin Way, Cincinnati, OH 45267-0559, USA

E. Fuchs, Clinical Neurobiology Laboratory, German Primate Center, Kellnerweg 4, 37077 Gottingen, Germany

A.J. Fulford, Department of Anatomy, University of Bristol, Southwell Street, Bristol, BS2 8EJ, UK

D. Glick, Department of Biological Chemistry, The Hebrew University of Jerusalem, Jerusalem, Israel

E. Gould, Department of Psychology, Princeton University, Princeton, NJ 08544, USA

T.D. Gould, Laboratory of Molecular Pathophysiology, National Institute of Mental Health, Building 49, Room B1EE16, 49 Convent Drive, Bethesda, MD 20892-4405, USA

A. Gratton, McGill University, Douglas Hospital Research Center, Montreal, H4H 1R3 QC, Canada

N.A. Gray, Laboratory of Molecular Pathophysiology, National Institute of Mental Health, Building 49, Room B1EE16, 49 Convent Drive, Bethesda, MD 20892-4405, USA

M.S. Harbuz, University Research Centre for Neuroendocrinology, Bristol Royal Infirmary, Marlborough Street, Bristol, BS2 8HW, UK

U.L. Hayes, University of Massachusetts, Amherst, MA, USA

A.L.O. Hebb, Department of Pharmacology, Faculty of Medicine, Dalhousie University, Sir Charles Tupper Medical Building, 5850 College Street, Halifax, NS B3H 1X5, Canada

S.C. Heinrichs, Boston College, Department of Psychology, McGuinn Hall, 140 Commonwealth Avenue, Chestnut Hill, MA 02467, USA

J.P. Herman, Department of Psychiatry, University of Cincinnati, 231 Albert Sabin Way, Cincinnati, OH 45267-0559, USA

M.C. Holmes, Endocrinology Unit, Molecular Medicine Centre, Edinburgh University, Western General Hospital, Edinburgh EH4 2XU, UK

S.Y.T. Hsu, Department of Obstetrics and Gynecology, Division of Reproductive Biology, Stanford University School of Medicine, Pasteur Drive, Room A344E, Stanford, CA 94305-5317, USA

C.D. Ingram, Psychobiology Research Group, School of Neurology, Neurobiology and Psychiatry, University of Newcastle, Royal Victoria Infirmary, Newcastle NE1 4LP, UK

M. Joels, Swammerdam Institute for Life Sciences, Section Neurobiology, University of Amsterdam, Kruislaan 320, 1098 SM Amsterdam, The Netherlands

N.H. Kalin, Department of Psychiatry and Psychology, University of Wisconsin, 6001 Research Park Blvd., Madison, WI 53719, USA

A.M. Karssen, Department of Psychiatry and Behavioral Sciences, School of Medicine, Stanford University, MSL5, P124, 1201 Welch Road, Palo Alto, CA 94304-5485, USA

T. Kino, Pediatric and Reproductive Endocrinology Branch, National Institute of Child Health and Human Development, National Institutes of Health, Building 10 Room 9D42, 10 Center Drive MSC 1583, Bethesda, MD 20892-1583, USA

G.F. Koob, Department of Neuropharmacology, The Scripps Research Institute, La Jolla, CA 92037, USA

K.J. Kovács, Laboratory of Molecular Neuroendocrinology, Institute of Experimental Medicine, Hungarian Academy of Science, Szigony u. 43, Budapest H-1083, Hungary

H.J. Krugers, Swammerdam Institute for Life Sciences, Section Neurobiology, University of Amsterdam, Kruislaan 320, 1098 SM Amsterdam, The Netherlands

S. Laforest, Centre de Recherche en Neurosciences, CHUL, RC-9800, 2705 Boulevard Laurier, Ste-Foy G1V 4G2 QC, Canada

M. Le Moal, INSERM U588, Institut François Magendie, 1 rue Camille St Saëns, 33077 Bordeaux, France

V. Lemaire, INSERM U588, Institut François Magendie, 1 rue Camille St Saëns, 33077 Bordeaux, France

S. Levine, Department of Psychiatry, Center for Neuroscience, University of California at Davis, Davis, CA 95616, USA

A.C.E. Linthorst, Henry Wellcome Laboratories for Integrative Neuroscience and Endocrinology, University of Bristol, The Dorothy Hodgkin Building, Whitson Street, Bristol BS1 3NY, UK

J. Lu, Max Planck Institute of Psychiatry, Kraepelinstrasse 2–10, D-80804 Munich, Germany

S.J. Lupien, Laboratory of Human Psychoneuroendocrine Research, Douglas Hospital Research Center, 6875 Bldg. Lasalle, Verdun, H4H-1RS QC, Canada

F.S. Maheu, Department of Psychology, University of Montreal, 6875 Bld Lasalle, Montreal, H4H 1R3 QC, Canada

J.A. Majzoub, Division of Endocrinology, Children's Hospital Boston, Harvard Medical School, 300 Longwood Avenue, Boston, MA 02115, USA

H.K. Manji, Laboratory of Molecular Pathophysiology, National Institute of Mental Health, Building 49, Room B1EE16, 49 Convent Drive, Bethesda, MD 20892-4405, USA

C.A. Marsden, School of Biomedical Sciences, Institute of Neuroscience, University of Nottingham Medical School, Queen's Medical Centre, Nottingham NG7 2UH, UK

O.C. Meijer, Division of Medical Pharmacology/LACDR-LUMC, Leiden/Amsterdam Center for Drug Research, Leiden University Medical Center, P.O. Box 9502, 2300 RA Leiden, The Netherlands

I.H. Miklós, Laboratory of Molecular Neuroendocrinology, Institute of Experimental Medicine, Hungarian Academy of Science, Szigony u. 43, Budapest H-1083, Hungary

N.K. Mueller, Department of Psychiatry, University of Cincinnati, 231 Albert Sabin Way, Cincinnati, OH 45267-0559, USA

Z. Némethy, Max Planck Institute of Psychiatry, Kraepelinstrasse 2–10, D-80804 Munich, Germany

M. Páez-Pereda, Department of Endocrinology, Max Planck Institute of Psychiatry, Kraepelinstr 10, 80804 Munich, Germany

G.N. Pandey, Department of Psychiatry, Psychiatric Institute, University of Illinois at Chicago, 1601 W. Taylor St, Chicago, IL 60612, USA

J.M. Pego, Life and Health Science Research Institute, Health Science School, University of Minho, Campus de Gualtar, 4710-057 Braga, Portugal

J.D. Peters, Department of Psychology, Princeton University, Princeton, NJ 08544, USA

P.V. Piazza, INSERM U-588, Université de Bordeaux 2, Institut François Magendie, 1 Rue Camille Saint-Saëns, 33077 Bordeaux Cedex, France

J. Prickaerts, Johnson & Johnson Pharmaceutical Research & Development, a Division of Janssen Pharmaceutica NV, Turnhoutseweg 30, B-2340 Beerse, Belgium

J.M.H.M. Reul, Henry Wellcome Laboratories for Integrative Neuroscience and Endocrinology, The Dorothy Hodgkin Building, University of Bristol, Whitson Street, Bristol, BS1 3NY, UK

G. Richter-Levin, Department of Psychology, The Brain and Behavior Research Center, University of Haifa, Haifa 31905, Israel

M.A. Riva, Department of Pharmacological Sciences, Center for Neuropharmacology, University of Milan, Via Balzaretti 9, 20133 Milan, Italy

P.H. Roseboom, Department of Psychiatry and Pharmacology, 6001 Research Park Blvd., Madison, WI 53719, USA

R. Rupprecht, Department of Psychiatry, Nussbaumstr. 7, 80336 Munich, Germany

M. Schmidt, Division of Medical Pharmacology/LACDR-LUMC, University of Leiden, P.O. Box 9502, 2300 RA Leiden, The Netherlands

J.R. Seckl, Endocrinology Unit, Molecular Medicine Centre, Edinburgh University, Western General Hospital, Edinburgh EH4 2XU, UK

B.B. Simen, Laboratory of Molecular Psychiatry, Departments of Psychiatry and Pharmacology, Yale University School of Medicine, 34 Park Street, New Haven, CT 06508, USA

H. Soreq, Department of Biological Chemistry, The Hebrew University of Jerusalem, Jerusalem, Israel

N. Sousa, Life and Health Science Research Institute, Health Science School, University of Minho, Campus de Gualtar, 4710-057 Braga, Portugal

G.K. Stalla, Department of Endocrinology, Max Planck Institute of Psychiatry, Kraepelinstr. 10, 80804 Munich, Germany

S. Stanford, Department of Pharmacology, University College London, Gower Street, London WC1E 6BT, UK

T. Steckler, Johnson & Johnson Pharmaceutical Research & Development, A Division of Janssen Pharmaceutica N.V., Turnhoutseweg 30, 2340 Beerse, Belgium

R.M. Sullivan, Centre de Recherche Fernand-Séguin, Université de Montréal, Montreal, H4H 1R3 QC, Canada

Y.M. Ulrich-Lai, Department of Psychiatry, Albert Sabin Way, University of Cincinnati, Cincinnati, OH 45267-0559, USA

R.J. Valentino, The Children's Hospital of Philadelphia, 402C Abramson Pediatric Research Center, 34th and Civic Center Blvd, Philadelphia, PA 19104, USA

E.J. Van Bockstaele, Department of Pathology, Anatomy and Cell Biology, Thomas Jefferson University, 1020 Locust St, Philadelphia, PA 19107, USA

J.M. Verkuyl, Swammerdam Institute for Life Sciences, Section Neurobiology, University of Amsterdam, Kruislaan 320, 1098 SM Amsterdam, The Netherlands

N. Weekes, Department of Psychology, Pomona College, 550 N. Harvard Avenue, Claremont, CA 91711, USA

J.L.W. Yau, Endocrinology Unit, Molecular Medicine Centre, Edinburgh University, Western General Hospital, Edinburgh EH4 2XU, UK

R. Yirmiya, Department of Psychology, The Hebrew University of Jerusalem, Jerusalem, Israel

P.-X. Yuan, Laboratory of Molecular Pathophysiology, National Institute of Mental Health, Building 49, Room B1EE16, 49 Convent Drive, Bethesda, MD 20892-4405, USA

R. Zhou, Laboratory of Molecular Pathophysiology, National Institute of Mental Health, Building 49, Room B1EE16, 49 Convent Drive, Bethesda, MD 20892-4405, USA

E.P. Zorrilla, Department of Neuropharmacology, The Scripps Research Institute, CVN-7, La Jolla, CA 92037, USA

Preface

Stress is a phenomenon being all around us, but seemingly being too well known and too little understood at the same time, despite the fact that the field has advanced enormously over recent years. We have learned that stress can shape various types of behaviour in the individual long after exposure to the stressor itself has terminated. Exposure to a stressful stimulus during the perinatal period, for example, can have long-term consequences over weeks and months, well into adulthood. This is accompanied by a variety of characteristic neurochemical, endocrine and anatomical changes in the brain, leading, for example, to changes in neural plasticity and cognitive function, motivation and emotionality.

We have started to discover the differentiated effects of various stressors in the brain and how expression of a wide variety of gene products will be altered in the CNS as a function of the type and duration of the stressor. Activity in higher brain areas in turn will shape the response to acute and chronic stress and there are intricate interactions with, for example, immune functions. Cytokines will access the brain and affect its function at various levels.

It has become increasingly clear that stress serves as one of the main triggers for psychiatric and non-psychiatric disorders, including depression, anxiety, psychosis, drug abuse and dementia. Recognizing these intricate relationships has initiated a wealth of research into the development of novel animal models and novel treatment strategies aiming at influencing stress responsivity in patients suffering from these diseases.

Moreover, novel technologies, such as molecular techniques, including gene targeting methods and DNA microarray methods start to unravel the cellular events taking place as a consequence of stress and facilitate the understanding of how stress affects the brain.

Thus, the topic of stress, the brain and behaviour gains increasing relevance, both from a basic scientific and clinical perspective, and spans a wide field of expertise, ranging from the molecular approach to in-depth behavioural testing and clinical investigation. This book aims at bringing these disciplines together to provide an update of the field and an outlook to the future. We think these are exciting times in a rapidly developing area of science and hope that the reader will find it both useful as an introductory text as well as a detailed reference book. The Handbook of Stress and the Brain is presented in two parts, i.e. Part 1: The Neurobiology of Stress, and Part 2: Stress: Integrative and Clinical Aspects.

This part, Part 1, addresses basic aspects of the neurobiology of the stress response including the involvement of neuropeptide, neuroendocrine and neurotransmitter systems, and its corollaries regarding gene expression and behavioural processes such as cognition, motivation and emotionality.

Thomas Steckler
Ned Kalin
Hans Reul

A Memorial for David de Wied (1925–2004)

It is almost an eerie coincidence that this volume, dedicated to the subjects of Stress and Behavior, should be published at a time when the field has lost one of its giants and the man whose work has inspired much of what is written here. On February 21, 2004 Professor David de Wied died. David had just celebrated his 79th birthday. For me not only did the field lose one of its founding fathers but I lost a dear friend. Professor de Wied was born on January 12, 1925. His life prior to embarking on his professional career was marked by a period of several years when he went underground and was in hiding during the German occupation of Holland. Following the war he decided to attend the University of Groningen to study medicine. This involved a tremendous effort since he had lost many precious academic years. He did receive his medical degree in 1955. I shall not document the details of his remarkable academic achievements. These are presented in detail in a volume dedicated to David on his 75th birthday (Smelik and Witter, 2000) and more recently by de Kloet (2004). In my chapter on the history of stress research I devoted several pages to David de Wied and his importance to the field, but his impact on the field was of such significance that it is worth repeating.

David was in every sense a pioneer and a visionary. I have often wondered how one defines a visionary. Perhaps the critical dimension is the ability to see relationships between events that are not immediately apparent to normal mortals. He is best known for his formulation of the "neuropeptide concept" although throughout his career he made many other major contributions. In its simplicity the neuropeptide concept postulated that there were peptides produced in the brain and pituitary that directly influenced brain function, and of particular importance, behavior. The field of hormones and behavior at the time when David began to work on the effects of peptides on behavior was almost exclusively dedicated to studying the effects of gonadal steroids on sexual behavior. There were a few scattered reports of effects of thyroid and adrenal steroid compounds but they had little impact. The demonstration that neuropeptides could influence complex behavioral processes such as learning and memory was indeed revolutionary and met with a great deal of skepticism when first introduced. However, the skeptics were silenced when he continued to demonstrate the powerful influence of these molecules on behavior. It was primarily based on this work that fundamental behavioral processes were integrated into the general rubric of neuroendocrinology and new dimensions of the effects of the hormones of the hypothalamic–pituitary–adrenal axis on behavior were introduced. The neuropeptide concept pre-dated the characterization of the "releasing hormones" synthesized in the hypothalamus. That these hormones have been shown to have a profound influence on behavior is one of the legacies of David's work.

In 1963 he became Professor of Medical Pharmacology at the University of Utrecht which in 1968 became the Rudolf Magnus Institute of Pharmacology in honor of the Dutch pharmacologist Rudolf Magnus. This institute rapidly became the Mecca for the study of hormones and behavior. It was *the* place to visit and study if your field of interest encompassed neuroendocrinology and behavior. Investigators came from every part of the world to study at the institute. On the numerous occasions that I lectured in the institute I was always prepared to be challenged by David and his students. The discussions were vigorous, animated and sometimes heated, but always stimulating and provocative.

David's legacy extends well beyond his scientific contributions. There are multitudes of Ph.D. and post-doctoral students as well as collaborators who are indebted to him. They were privileged to share his scientific rigor, and perhaps of more importance, his unique intellect. On the occasion of his 75th birthday celebrations the room was filled with many of these students and colleagues. What was impressive is that many of them are now the current leaders in the field.

Although he is best known for his life as a scientist there was much more to the man. He had other passions that made up his life. He was an avid art collector and loved music. Until his death he continued to play the violin and take music lessons. We shared a common love for both music and art. One of our secret ambitions was to perform the famous tenor and baritone duet from Bizet's "The Pearl Fisher." There were two problems, first neither of us could qualify as a tenor and second we really did not sing very well. This did not prevent us from singing opera at any occasion whether it be a dinner at a congress or a party in Hungary.

David was a complex man. He received numerous prestigious honors throughout his career and yet in his later years he did not feel he had achieved the recognition he deserved. Perhaps the problem with being the originator of a concept that gains universal acceptance is that its origin is often forgotten. David had the most unusual sense of humor I have encountered. On one occasion my youngest daughter spent a weekend with us and David and his wife Lie on the Italian Riviera. She spent the first day terrified of him until she realized that he was indeed one of the funniest people she had ever met. She spent the next few days in almost constant laughter.

It was my dream and hope that David and I would have a grand celebration for our joint 80th birthdays in 2005. We were both overjoyed when we experienced the millennium. That dream has now been shattered. I will continue to revere and respect him for all the contributions he has made to biology and to the quality of all our lives. We shared many adventures, many avid scientific discussions and the pleasure of watching the growth of our science. These memories are always present and it was indeed a privilege to have shared these with him over many years.

Seymour Levine
Center for Neuroscience
University of California, Davis

References

De Kloet, E.R. (2004) In honour of David de Wied. Psychoneuroendocrinology (in press).

Smelik, P.G. and Witter, A. (2000) David de Wied a biographical sketch. In: David de Wied (Honorary Editor), Neuropeptides, Basics and Perpectives. Elsevier, Amsterdam.

Contents, Part 1

SECTION 1

Concepts of Stress

T. Steckler, N.H. Kalin and J.M.H.M. Reul (Eds.)
Handbook of Stress and the Brain, Vol. 15
ISBN 0-444-51173-3

CHAPTER 1.1

Stress: an historical perspective

Seymour Levine

Department of Psychiatry, Center for Neuroscience, University of California, Davis, California 95616, USA

Abstract: Some of the major landmarks in the history of neuroendocrinology, glucocorticoid physiology and psychoneuroendocrinology are discussed in this chapter. The primary emphasis is on the evolution of the major theories and their experimental underpinnings on the regulation of the hypothalamic–pituitary–adrenal (HPA) axis. Initially an attempt was made to deal with the issues concerning the definitions of stress. The origins of the stress concept, the neural control of the pituitary, the history of the search for corticotrophin-releasing factor (CRF), and the developments that resulted in shaping the current views of the action of the adrenal hormones are elaborated. Further, the role of environment and behavior on the regulation of the HPA axis and the effects of specific neuropeptides on behavior were also covered. The purpose of this chapter is to provide a perspective on the major events that were crucial in the history of stress research that shaped the directions of the field.

"The past is never dead. It's not even the past"
William Faulkner

Introduction

During the course of my career, which now spans over five decades, I could not begin to count the number of conferences, workshops, and symposia related to stress that I have attended. I will not attempt to describe the number of times during these meeting that at least one, if not several, of the participants had championed the notion that we discard the concept for a more precise definition. The absolute failure of these attempts is attested to by my most recent visit to one of my favorite biomedical computer searches. As of this moment Pub-Med listed 209,744 references that in one way or another had some reference to the term stress. It would be difficult to predict what this number will be at the time this chapter is published. These staggering figures are at best an underestimate since computer searches rarely go beyond the late 1970s and publications in this field began long before this time. Further, the particular search that was used lists mostly articles and ignores the extensive list of books dedicated to this subject. What is also evident from the information obtained from the computer is that the number of publications is accelerating. Over 60,000 papers have been published since the beginning of the new millennium. A close examination of only a small sample of these references made it abundantly clear that the term stress was used in so many different ways that it would be necessary to determine, for each article listed, the precise manner the term was used and in what context. It would be further difficult to specify all the different disciplines that have in some way found the concept of stress useful, though each discipline will define stress in its own idiosyncratic manner.

Defining stress

After the completion of my last effort to define stress (Levine and Ursin, 1991), I made myself the promise that I would never again engage in what I consider a futile exercise. One of the more recent definitions was presented by McEwen (2000). "Stress may be defined as a real or interpreted threat to the physiological

Tel.: +1530 752 1887; Fax: +1530 757 8827;
E-mail: slevine@ucdavis.edu

or psychological integrity of an individual that results in physiological and/or behavioral responses. In biomedicine, stress often refers to situations in which adrenal glucocorticoids (GCs) and catecholamines are elevated because of an experience." Chrousos and Gold (1992) state "we define stress as state of disharmony, or threatened homeostasis. The adaptive response can be specific or can be generalized and non specific." At the core of these definitions is the concept of homeostasis. Thus, some disturbance of homeostasis results in a cascade of physiological and/or behavioral responses that presumably are required to reinstate the ideal homeostatic balance.

However, these definitions as well as most others are problematic. As stated by Levine and Ursin (1991) "The major problem with the concept of stress is that we are confronted with a composite, multidimensional concept. All existing definitions include some components. We can identify three main subclasses. These subclasses can be identified as the input (stress stimuli), the processing systems, including the subjective experience of stress and the output (stress responses). One basic difficulty is that these subclasses interact. The essential picture we want to convey is one of a complex system with feedback and control loops, no less but no more complicated than any other of the body's self-regulated systems. This system affects many other biological processes and may function as a common alarm and drive system, whenever there is a real or apparent challenge to the self-regulating systems of the organism."

Steptoe (2000) has suggested that "the effects of stress are manifest in four distinct domains; physiology, behavior, subjective experience, and cognitive function. The physiological effects of stress include alterations in neuroendocrine, autonomic nervous system and immune function." If one were to isolate only the physiological effects of stress, the history of three major areas of investigation would need to be covered and this would not include the relationships or lack of relationship among these systems. This historical perspective will focus primarily on the neuroendocrine system and more specifically with the hypothalamic–pituitary–adrenal (HPA) axis. For this discussion we will use the more traditional designation HPA, although more recently we have seen a trend to label the axis as LHPA. The "L" stands for limbic system and is intended to indicate that the regulation of the hormonal cascade caused by exposure to stress involves extra-hypothalamic structures. The aim of this chapter is to provide a glimpse into the historical events that have provided the framework, the insights, and the theories that in many ways still guide our current research.

Insofar as several papers will appear in subsequent volumes that deal extensively with many different aspects of the HPA axis, I will provide only a very simple description of some of the key elements in the neuroendocrine cascade that results in an increased secretion of adrenocorticotropic hormone (ACTH), and that ultimately results in elevations of the levels of GCs. Some environmental event which involves either physical demands or is psychologically challenging or a combination of both, induces an increase in the release of corticotrophin-releasing factor (CRF) and arginine vasopressin (AVP) into the portal circulation. CRF/AVP then activates the corticotrophs in the pituitary to release ACTH into the general circulation. ACTH acts upon the adrenal cortex to induce synthesis and increased secretion of the GCs. Under normal circumstances these elevated levels of GC activate the GC receptors which serve to terminate the release of CRF and ACTH, thus returning the organism to its basal state.

Traditionally, a discourse on the history of this or any other field would trace the evolution of the critical components along some time dimension. If I were to take this approach the result would simply be a compilation of the information contained in many other sources. Since history does unfold through time it is difficult to avoid using a temporal framework. However, I intend to trace this history as it unfolded in my academic lifetime and how it influenced and shaped the field as we know it today. This will be an historical perspective, but the perspective will be autobiographical.

Not so ancient history

As was stated earlier I will not attempt to present all the details of the variety of thought and experimental evidence that eventually leads us to the 21st century views of stress. The earliest references to the concept of homeostasis come from the Greek philosophers and physicians, in particular, Hippocrates. This is best summarized by Chrousos et al. (1988). What is important is that the concepts of harmony and disharmony (homeostasis?) of man and animals with both the external and internal environment have long been a concern of serious thinkers. The concept of stress was originally taken from the dynamics of physics to describe the relationship between stress and strain in an elastic body. "The term stress is applied to the mutual actions which take place across any section of a body to which a system of forces is applied. The term strain is applied to any changes occurring in the dimension or shape of a body when forces are applied" (Duncan and Starling, 1959).

However, our current interest in and views on the physiology and psychology of stress are of more recent vintage and can be traced primarily to the contributions of Walter B. Cannon (1914, 1915, 1932) and Hans Selye (1950, 1956). The overarching principles that emerged from the studies by Cannon and Selye were: (1) there was a physiology that was specific to stress and (2) that an integral part of this physiology was related to some function of the adrenal. Although both emphasized the role of the adrenal, there was a clear distinction between them. (1) Cannon focused primarily on the sympathetic nervous system, including the adrenal medulla, and the role of the adrenal medullary hormones, epinephrine also called adrenaline and noradrenaline, in the response to emergency situations. Selye, in contrast, emphasized the hormones of the adrenal cortex, primarily the GCs. (2) Canon was describing the responses to an acute threat, whereas Selye was concerned with the adaptation of the organism to chronic challenges. (3) For Canon, stress was defined in terms of the stimulus required to elicit these responses. Selye described a triad of responses that he hypothesized constituted stress: hypertrophy of the adrenal, stomach ulceration, and involution of the thymus gland. This triad of responses implicated the endocrine, autonomic, and immune systems. Stress was defined in terms of the response. Selye in fact turned the original physical definition of stress on its head. Instead of strain being produced by stress, he did not use the term strain, and assumed that stress was produced by stimuli he referred to as stressors. There is some controversy over who first used the term stress in a biological context. Although it is a common belief that Selye was responsible for introducing the term (Medvei, 1982), Sapolsky (1994) contends that it was indeed Cannon (1914) who was responsible. However, Selye was clearly responsible for popularizing the concept and bringing it to the attention of the biomedical community and the general public.

The history of science is replete with examples of serendipity. Sapolsky (1994), in what I consider one of the best written popular books on stress, describes the origins of Selye's observations as follows: "To be only a bit facetious, stress physiology exists only because this man (Selye) was both a very insightful scientist and somewhat inept at handling rats." Originally Selye was attempting to discover the function of some extract of ovarian tissue. However, "he attempted to inject his rats daily, but apparently with not a great display of dexterity. Selye would try to inject the rats, miss them, drop them, spend half the morning chasing the rats around the room or visa versa, and so on. At the end of a number of months of this, Selye examined the rats and discovered something extraordinary: the rats had peptic ulcers, greatly enlarged adrenals, and shrunken immune tissues. He was delighted; he had discovered the effects of the mysterious ovarian extract." However, following several subsequent presumably control experiments the physiological manifestations of these procedures continued to be evident. This led Selye (1936) to postulate that the response to stress was nonspecific. Thus, a wide array of stimuli (stressors) resulted in a similar set or responses and that eventually exposure to stress will result in illness. Selye (1949) defined the General Adaptation Syndrome (GAS) as the "Physiological mechanism which raises the resistance to damage as such." The GAS consisted of a three-stage reaction to stress, an alarm reaction, a stage of resistance, and a final stage of exhaustion. It remains one of the more robust theories in the stress field, although it was in fact incorrect in many of its assumptions. I should state that I come to praise Selye not to bury him. The impact of Selye cannot be measured in the specific details of his work and theories. He pioneered and launched a field of investigation that has had an enormous influence on biology and medicine and is still growing at an exponential rate.

There were a number of developments in the field that followed shortly after Selye's early publications that cast doubt on some of the fundamental aspects of Selye's theories and led to a decline in the interest in GC physiology. One of Selye's claims was that the excess secretion of adrenocortical hormone would cause arthritis, allergies, and collagen-related disorders. However, Hench et al. (1949) demonstrated that the GCs had profound anti-inflammatory activity and could be used therapeutically to treat some of these pathologies. One of the coauthors on this paper was Kendall, who was awarded the Nobel Prize for determining the structure of the steroid hormones. These findings were paradoxical to what Selye had proposed. That there is a relationship between stress and illness is now extensively documented, as is the relationship between stress and GCs. What is not clear is the relationship between the GCs and illness.

The history of science is a history of great ideas and advances in technologies. The techniques that were available to the stress researcher during the embryonic phase of the field were at best crude and often laborious. There were no known methods for directly measuring the levels of circulating hormones. The physiology of the stress response was based initially on changes in weight of the adrenal and counts made of the circulating lymphocytes and eosinophils in blood. The first direct biochemical measure was the depletion of ascorbic acid in the adrenal. It had been observed that, following an injection of ACTH or stress, there was a drop in the content of adrenal ascorbic acid within the adrenal. This observation was to be the basis for an ACTH bioassay developed by Sayers (1950). There were many limitations to these indices of adrenocortical activity. At best they were crude and only by inference could they be related to the output of the GCs. It was difficult to obtain a time course since only the measures of eosinophils did not require sacrificing the animals, and therefore the dynamics of the adrenal response were difficult to ascertain. It was not until the 1950s (Silber et al., 1958) that the first direct measurement of the adrenal steroids was available. However, despite these limitations, many of the initial observations, hypotheses, and theories have held up remarkably well, which is a testimony to the brilliance and insightfulness of the founders of this field.

Although the Selye's definition of stress was to dominate the thinking and direction of stress research for many years, there were other groups of investigators that had a very different perspective. In the late 1940s and early 1950s, one of the bastions of biological psychiatry was located at the Michael Reese Hospital in Chicago. This group had undertaken a large study of normal human subjects who were volunteers in the US Army undergoing airborne training. The definition of stress used in the context of this study was "Any stimulus may in principle arouse an anxiety response because of the particular meaning of threat it may have acquired for that particular individual. However, we distinguish a class of stimuli which are *more likely* to produce disturbances in most individuals. The term stress is applied to this class of conditions, thus, we can conceive of a continuum of stimuli differing in meaning to the organism and in their anxiety-producing consequences. At one end are such stimuli or cues, often highly symbolic, which have meaning only to a single or limited number of persons and which to the observer appear as innocuous or trivial. At the other end are such stimuli, *here called stress*, which by their explicit threat to vital functioning and their intensity are likely to overload the capacity of most organisms coping mechanisms.

Anxiety has been defined in terms of an affective response; stress is the stimulus condition likely to arouse such responses. Ultimately, we can truly speak of a *stress situation* only when a given response occurs, but for schematic purposes as well as consistency with common usage, we may use the term stress to designate certain kinds of stimulating conditions without regard to response." (Basowitz et al., 1955). This definition stands in direct contrast to the one proposed by Selye. In this instance stress is defined almost exclusively in terms of the stimulus with expectations that these events will evoke some response. For Selye, stress is only stress if the triumvirate of physiological responses is elicited. Thus far none of the definitions of stress have focused on or even mentioned the brain although inherent in the psychological approach to stress was the notion that stimuli could be interpreted differentially based on a host of experiential factors, which implies some process by which the stimuli are evaluated.

The brain: the birth of neuroendocrinology

In 1948 Geoffrey W. Harris published a paper entitled "Neural Control of the Pituitary Gland." This was followed in 1955 by an expanded monograph with the same title (Harris, 1955). At the time this book was published, Harris was the Fitzmary Professor of Physiology at the Institute of Psychiatry, Maudsley Hospital, London. He was to become the Professor of Anatomy at Oxford University until his very untimely death at the age of 57. I was privileged to have been a student of Harris from 1960 to 1962 and continued my relationship with him until his death. I cannot recall any time that was more exhilarating in the course of my professional career. Every discovery was major advance, every important scientist at the time came through the laboratory, and it was a new way to look at the world. Geoffrey Harris was many things to me. He was my hero, my mentor, my friend, and my squash instructor. An excellent description of him as a scientist and as a man is contained in the book by Nicholas Wade "The Nobel Duel" (1981).

At the time the "Neural Control" book was published, I was a young faculty member in a new research unit that has just been opened in the psychiatry department at Ohio State University. I had been there only a few years when I encountered the Harris volume. For me this was a revelation, a new approach, whereby there was at last a way to link behavior with endocrinology. I applied for a postdoctoral position in his laboratory in London, was accepted, and packed my wife, three very small children, and one large dog and off we went. We had numerous adventures along the way, but the fateful day arrived in March 1960 when I first encountered the laboratory. After meandering through the grounds of the Maudsley Hospital I finally came upon two rather shabby looking temporary structures that was the Laboratory of Neuroendocrinology. I was both surprised and depressed. My vision based upon my image of the greatness of the man in charge led me to expect a physical plant that would be commensurate with his stature. One could not swing a fat cat or even a skinny one in the laboratory that was assigned to me. To add further to my depression, Harris wanted me to work on a problem related to sexual differentiation, a far cry from my passion for stress and anxiety. As I discovered years later Geoffrey had little regard for psychology and/or psychologists. The only useful activity for an experimental psychologist was the study of sex behavior, and since I had been the first postdoctoral applicant with behavioral credentials it was obvious that I knew how to examine sex behavior. After several months watching rats with a well-worn description of rat sexual behavior written by Frank Beach, I did learn how. In the end what began as a nightmare turned out to be a fulfillment of a dream far beyond my expectations and was to shape the remainder of my scientific career.

It would be erroneous to imply that neuroendocrinology was born with the publication of Harris' extremely influential paper and monograph. There were numerous suggestions and hypotheses that postulated a relationship between the brain and the endocrine system. Harris did not readily embrace the concept of neuroendocrinology. For him the brain was an endocrine organ and therefore neuroendocrinology was a redundancy. Hippocrates in his discourse on the glands anticipated this perspective many centuries ago when he wrote "The flesh of the glands is different from the rest of the body, being spongy and full of veins; they are found in moist parts of the body where they receive humidity... and the brain is a gland as well as the mammae" (Medvei, 1982). In 1936 Francis Marshall concluded in his discussion of periodicity in reproduction that "the primary periodicity is a function of the gonad, the anterior pituitary, acting as a regulator, and the internal rhythms adjusted to the environment, by the latter acting on the pituitary, partly or entirely, through the intermediation of the nervous system" (Marshall, 1936). The role of the brain in the regulation of reproductive processes was critical in the development of neuroendocrinology. Implicating the central nervous system (CNS) with regards to the regulation of the adrenal cortex did not occur until several years later.

One of the early models concerning the regulation of adrenocorticotropin (ACTH) secretion and the secretion of GCs from the adrenal proposed that ACTH secretion is regulated by the systemic blood levels of epinephrine. Epinephrine or adrenaline, as it was then commonly called, was proposed as the corticotrophin-releasing factor (CRF). The principle proponent of this view was Long (1947, 1952).

In response to stress, the increased secretion of adrenaline by the adrenal medulla acts directly on the pituitary gland to induce the secretion of ACTH. an alternative view of the regulation of ACTH was presented by Sayers (1950). He proposed that ACTH secretion is regulated by the blood levels of GCs. Under conditions of stress the peripheral tissues utilize these hormones more rapidly. The blood content of the GCs falls rapidly as a consequence, and this in turn directly stimulates anterior pituitary to secrete increased amounts of ACTH. According to this view, the anterior pituitary and the adrenal cortex form a self-regulating system, the balance of which is broken mainly by variations in the activity of the peripheral tissue.

The position held by Long and coworkers would not withstand the scrutiny of subsequent experimental examination. There was no question that injections of adrenaline could stimulate ACTH secretion. It did not appear, however, that it was absolutely essential. Amongst the experiments that challenged the adrenaline hypothesis, perhaps the most damaging was an experiment conducted by Marthe Vogt (1952), who also stands as one of my heroes. She determined that enucleating the adrenal medulla reduced circulating levels of adrenaline to close to zero. However, in response to emotional stress there was a fall in adrenal ascorbic acid in the adrenal demedullated rat that followed an identical time course to those of intact animals. Similar results were reported by Hodges (1953), effectively dismissing the hypothesis that adrenaline was the CRF.

Regulation of ACTH secretion based upon the level of circulating adrenal hormones was to suffer a similar fate. Although the evidence that supported Sayers indicated that blood levels of the adrenal hormones do in part regulate the output of ACTH, there were many facts that could not be explained by this theory. One example is that this theory could not explain why adrenal atrophy occurs when the pituitary is transplanted to a site remote from the sella turcica. As long as the pituitary was properly vascularized it should maintain a normal functional adrenal cortex.

The history of stress research in some ways reads like a chamber of horrors. Animals were subjected to a wide variety of stress inducing stimuli that were to say the least massively invasive. They were boiled, frozen, surgically insulted, subjected to electric shock, injected with toxic agents, irritants, hemorrhaged, and this does not come close to an exhaustive list. That the peripheral responses to this wide variety of stimuli were similar supported Selye's view of the nonspecificity of the stress response.

Studies were beginning to appear that challenged the view that all stress was equal, and at the same time implicated the CNS as critical for the regulation of the pituitary. Fortier (1951) transplanted pituitary tissue to the anterior chamber of eye. This was a favorite transplantation site for the obvious reason that this area is heavily endowed with a rich blood supply, though I have commented that the real purpose was to better visualize the pituitary. Fortier exposed the animal with the transplanted pituitary to several stress-inducing stimuli. He divided them into two groups: (a) systemic stress which included injection of adrenaline or histamine, exposure to cold and (b) neurotropic stress, which included exposure to loud sounds and immobilization. Fortier reported that systemic stress was able to elicit an ACTH response in animals bearing a transplanted pituitary, but that in the case of neurotropic stress there was no indication of ACTH release. Fortier argued that the so-called neurotropic stress required the mediation of the CNS in order to release ACTH, whereas the response to systemic stress may be mediated by other blood-borne factors. There were other studies that suggested that stress-inducing stimuli could be differentiated. Dallman (1979) summarized a number of experiments and classified the different stimuli into those in which the adrenal response was blocked by the pretreatment with a single dose of GCs (GC sensitive) and those that still elicited a response after pretreatment with large doses of GCs (GC insensitive). Feldman (Feldman et al., 1970; Feldman, 1985) reported that in rats with hypothalamic islands, the GC response to peripheral neural stimulation was inhibited. The responses to systemic stress remained intact. That there are differences in the types of stress has been demonstrated by at least three distinctly different procedures, transplants, GC inhibition, and hypothalamic deafferentation. Recent methodological advances have made it possible to examine the neural pathways in the CNS in response to different stimuli that induce increased secretion of corticosterone and

ACTH. Sawchenko (1991) has described at least five different pathways that converge on the PVN and stimulate the release of CRF. Herman and Cullinan (1997) have presented the most recent version of the stress dichotomy. They state "Stressors involving an immediate physiologic threat ('systemic stressors') are relayed directly to the PVN, probably via brainstem catcholaminergic projections. By contrast, stressors requiring interpretation by higher brain structures ('processive stressors') appear to be channeled through limbic forebrain structures."

At the core of Harris' theory was the hypothesis that the pituitary gland was regulated by blood-borne chemical substances that were transported via the portal vessels from the hypothalamus to the pituitary. The growth of neuroendocrinology was slowed down initially by the description, by Popa and Fielding (1930), that the blood flowed from the pituitary to the hypothalamus. If the blood flow was from south to north, then the question of how the hypothalamic hormones could be transported to the pituitary gland was an enigma. Subsequently, it was demonstrated (Wisloski, 1937; Harris, 1947) that the blood flow through the portal vessels could indeed flow from the hypothalamus to the pituitary, thus making it possible for the brain hormones to directly communicate with the pituitary.

Harris postulated a specific releasing factor for each of the pituitary hormones. The term "releasing factor," to describe the substances in the hypothalamus that could induce secretion of the pituitary hormones, had been already proposed by Schally. Harris used two different approaches to prove that the regulation of the pituitary was a consequence of hormones being transported via the portal circulation. These were (1) pituitary transplants and (2) pituitary stalk sections. That endocrine organs could be removed from their original anatomical location and remain viable had history in experimental endocrinology, dating back to the work of John Hunter (1728–1793). Harris utilized these techniques to demonstrate that in order for the pituitary to function normally the communication between the hypothalamus and the pituitary via the portal vessels had to be intact. When the pituitary was transplanted to a site remote from the hypothalamus (Harris and Jacobsohn, 1952), as long as it was adequately supplied with blood, it remained viable though largely nonfunctional. The experiment by Fortier (1951) described previously is but one example of the importance of the transplant experiments in proving the neuroendocrine hypothesis. A more direct approach intended to show the importance of the portal system was that of sectioning the portal vessels. Under these conditions the pituitary remained in situ, but was deprived of the blood flow from the hypothalamus. One of the goals I had set for myself when I went to the Harris laboratory was to learn the technique of sectioning the portal vessels. It did not take long to realize that the surgical skills involved were far beyond my abilities. Harris' surgical skills were legendary. In the early stalk-section experiments although there was unambiguous evidence that the ACTH secretion was seriously impaired, after a relatively short period of time the pituitary once again became functional. The reasons for this appeared to be the capacity for the portal vessels to show remarkable regeneration. If regeneration was prevented by placing a wax paper barrier between the cut ends of the stalk (Harris, 1950), then the release of ACTH was continually impaired. I continue, even though many years have elapsed, to be impressed by the simplicity and elegance of these studies. By using basic physiological logic and with some surgical magic, the major tenants of neuroendocrinology were established.

In addition to the experiments which disrupted the blood-borne communication between the hypothalamus and the pituitary, there were other lines of investigation that presented convincing evidence that the brain was directly involved in regulating ACTH. (1) *Hypothalamic lesions:* Numerous experiments (Porter, 1953; McCann, 1953) were reported using animals bearing lesions in the hypothalamus. These studies indicated that hypothalamic lesions prevented the ACTH response to most stressors. Many of these lesion studies failed to support the conclusions of Fortier that the response to systemic stress did not require CNS involvement. Lesions which obliterated the hypothalamus prevented the release of ACTH to all stimuli. I use the term obliterate to indicate the state of the art for making lesions. Most of the early lesions studies usually produced large, massive lesions making it difficult to specify which if any of the specific nuclei in the hypothalamus was responsible for the secretion of the putative CRF. (2) *Electrical stimulation of the hypothalamus:* On my first tour of the laboratory in London I was given the "grand" tour. I was ushered into a large room which contained a large circular arena. Seated in this arena was a rabbit with something implanted in its skull. The outer part of this contraption appeared to be something resembling an electrical coil. I was informed that this apparatus (I use this term kindly) was presenting electrical stimulation directly to the hypothalamus on the assumption that the electrical pulses would elicit a response from the hypothalamus that would result in the release of the tropic hormones for the pituitary. Despite its Rube Goldberg appearance this procedure produced some of the most convincing data that the hypothalamus was intimately involved in the regulation of pituitary hormones in general and ACTH specifically (deGroot and Harris, 1950).

The CRF quest

Given this background and the growing acceptance of the role of the brain in regulating ACTH, it is not surprising that during the period of time when neuroendocrinology was blossoming there should also be a growing interest in attempting to find the chemical substance in the hypothalamus that was responsible for activating ACTH. The names most associated with the first attempts to describe CRF were Roger Guillemin and Andrew Schally, who were awarded the Noble Prize in Medicine in 1977 for their identification of TRH and LH–RH. The relationship between these two men on the way to Stockholm is legendary and well documented in Nicholas Wade's book "The Nobel Duel." In addition to describing much of the persona and background of Guillemin and Schally and the intensity of the animus between them, it is perhaps the best written and most entertaining history of neuroendocrinology.

Since most of the evidence indicated that the communication between the hypothalamus and pituitary was by some blood-borne factor, it should be possible to find the compound and characterize its structure. Previously it was mentioned that the first proposed CRF was adrenaline secreted by the adrenal medulla. The search for the new secretagogue responsible for activating ACTH focused on the hypothalamus. During the early 1950s several investigators independently reported that the extracts of hypothalamic tissue could stimulate the release of ACTH from anterior pituitary cell in vitro. Ironically, although the initial attempts to isolate a releasing factor centered on CRF, it proved to be one of the most difficult and elusive of all of the major hypophysiotropic hormones in the CNS. The pitfalls on the way to the CRF were many. There were problems with obtaining sufficient hypothalamic tissue needed to extract and analyze adequate quantities of material. My introduction to Guillemin was in 1961 during my first visit to Paris. I traveled to Paris with a fellow postdoc from London, who had been an associate of Guillemin at Baylor. Guillemin had recently arrived in Paris and had established his laboratory at the very prestigious College de France. At first glance he had achieved the prominence and all that went with it that a young scientist could ever hope for. He lived in part of a lovely château with beautiful grounds on the outskirts of Paris. Unlike the laboratory in London, this lab was sparkling and replete with all the latest technology, but what was most impressive was the very large freezer which contained thousands and thousands of sheep hypothalami that Guillemin had obtained from the slaughter houses of France.

There were numerous problems that were encountered in attempting to isolate and characterize CRF. Hypothalamic extracts often contained other, weaker secretagogues of ACTH such as vasopressin, catecholamines, etc. Another major obstacle was the claim (McCann and Brobeck, 1954) that in fact the CRF was in reality vasopressin and there was little need to try to identify a novel substance. The similarities in the size of ACTH (39 amino acids) and what ultimately turned out to be the size of CRF (41 amino acids) was not conducive to easy separation of the peptides by the existing methods. It also turned out to be the case that the structure of CRF was much more complex than other hypophysiotropic hormones. Thyrotrophic-releasing hormone (TRH), for example, contains only three amino acids. Concentrating on the identification of CRF proved to be a quagmire in the hormone-releasing enterprise. In his conclusions to the chapter describing the quest for CRF, Wade wrote "To this day, CRF has not been found, and may not even exist as such."

However, at almost the identical time this statement was uttered, Vale et al. (1981) at the Salk Institute reported the isolation, characteristics, synthesis, and in vitro and in vivo biological activity of a 41-amino acid ovine hypothalamic CRF. This was a testimony once again to perseverance, but also to the major advances in the available technology, amongst them being the development of the radio-immunoassay (RIA) for peptide hormones, advances in molecular biology, ion exchange, and liquid chromatographic techniques. In one way the quest for CRF was finally over and yet in another way it has just begun. Several years after presenting the structure of CRF, Vale reported the cloning of the CRF1 receptor and further the identification of additional CRF2 receptors which has resulted in the discovery of at least three new and distinct CRF-like peptides – Urocortin 1, 2, or stress copin-related peptide and Urocortin 3 also known as stresscopin (Hsu and Hsueh, 2001). The functional significance of these peptides is only beginning to be described (Dautzenberg and Hauger, 2002).

CRF, however, has proven to be a remarkable molecule. Not only is it found in abundance in the paraventricular nuclei (PVN) of the hypothalamus, but it is widely distributed throughout the brain and the periphery. CRF is also involved in the regulation of a broad range of physiological and psychological processes. Amongst these is the modulation of the autonomic system, gastrointestinal activity, behavior, and immune function. CRF has also been implicated in a variety of psychiatric and other neurological disorders. Thus, what began as the search for a hypothalamic hypophysiotropic hormone may eventually extend into many areas of pathophysiology that are not immediately associated with the regulation of the pituitary–adrenal axis.

I stand in awe of many of the intellectual giants that were involved in the ontogeny of neuroendocrinology. Although the specifics of the hypotheses and theories may not have been entirely correct, they each in some way contained elements that have proven to be prophetic. Peripheral adrenaline was not the CRF, but the catecholamines are clearly part of the neuroendocrine cascade that activates the release of the peripheral stress hormones. Harris postulated a specific releasing factor for each of the pituitary hormones. We now know that there are hypothalamic inhibiting as well as releasing hormones. Vasopressin may not have been the CRF but it is an active cosecretagogue.

Once it became doctrine that the regulation of the pituitary tropic hormones was via some blood-borne substance synthesized in the hypothalamus and released into the portal circulation, there were other investigators who attempted to characterize the structure of these hormones. However, only Guillemin and Schally were willing to take the big gamble and devote all of their time and resources that resulted in the identification of the first releasing factors and ultimately to the discovery of CRF and the CRF-like peptides.

GCs

We have thus far discussed some of the milestones that were crucial in the establishment of neuroendocrinology as a discipline. However, when we trace the history of stress research it began with the adrenal cortex, and although the peptides in the brain appear to have multiple functions, at least one of the major functions of CRF is to regulate the secretion of ACTH and GCs. Conversely, the adrenal hormones in turn regulate the synthesis of CRF by the way of a negative-feedback mechanism. Any basic textbook of endocrinology presents a detailed description of most, if not all, of the known physiological actions of the GCs and the dire consequences of either under or overproduction of these hormones. Previously we discussed the developments that focused on the response of the adrenal as a central component of the stress response. However, Sapolsky et al. (2000) state that "Few contemporary endocrinologists view the GC actions as part of a coherent physiological picture, or see the need to. Today the focus is on the molecular and cell biology of GC action, e.g., GC receptors as ligand-activated transcription factors or GC-induced apoptosis in lymphocytes."

What has been, and still is to some extent, one of the important questions that plague the stress researcher is what the adaptive significance of increased GC secretion is? It is not difficult to understand the importance of GC when the organism is confronted with acute life-threatening and physically invasive events. Under these circumstances the HPA axis is an exquisite adaptive mechanism. However, in our contemporary view of stress there is as much, if not more focus, on psychological and social stress as on the physical.

The question of how GCs influence the stress response continues to be an issue that has still not been completely resolved. Historically there have been several critical hypotheses that have guided the way we view this problem. Originally Selye speculated that GCs facilitate or mediate the ongoing stress response. One of the pioneers in GC physiology was Dwight Ingle. Ingle was an important member of the group at the Mayo clinic and contributed the bioassay that was used to test the purified fractions of the adrenal cortical hormones that were eventually characterized by this group. He also was amongst the first to demonstrate negative feedback when he observed that administration of adrenal cortical extracts or purified GCs caused atrophy of the adrenal gland that could be reversed by simultaneous administration of pituitary extracts. One of his major contributions was to propose a new and unique function for the GCs. The assumption that prevailed in the stress literature was that the damaging effects of stress were due to hypersecretion of the adrenal cortex. Ingle (1952) showed that the characteristic damaging effects of stress persisted when the GCs were supplied to the adrenalectomized animals at a constant, but not excessive rate of administration. From these observations he deduced that the role of the GCs in stress appears to be due to a subtle "permissive" or supporting role rather than as the primary mediator of the stress reaction. The most recent definition of the permissive action is "permissive actions are exerted by GCs present before the stressor and prime the defense mechanisms by which an organism responds to stress. Their consequences are first manifested during the initial stress response and occur whether or not there is a stress-induced increase in GC concentrations" (Sapolsky et al., 2000). Thus the GCs were now relegated to an essential, but a more supportive role.

Around the same time a Dutch physiologist/pharmacologist, Marius Tausk, proposed another role for the GCs. He suggested that GCs exerted a suppressive action. However, Tausk's view of GC actions did not receive much attention due to the fact that it was originally published in the house organ of Organon Pharmaceuticals. In a now well-quoted metaphor he viewed stress as a fire and that the function of the GCs was to limit the water damage created by the fire fighters. This view of the role of GCs in the stress response remained obscure until many years later when Alan Munck wrote a classic paper that independently proposed that GCs role in the stress response was to exert a suppressive action. Munck et al. (1984) wrote "We propose that: (a) the physiological function of stress induced increases in GC levels is to protect not against the source of stress itself, but against the normal defense reactions that are activated by stress; and (b) the GCs accomplish this function by turning off those defense reactions, thus preventing them from overshooting and themselves threatening homeostasis." Thus we begin to see an integrated set of functions that are all part of the complex actions of the GCs in the stress response. Recently, Sapolsky et al. (2000) have attempted once again to answer the question "How do glucocorticoids influence the stress response." In this chapter the previously proposed actions of the GCs, stimulatory, permissive, and suppressive, are integrated along with another newly proposed function of the GCs, "preparative."

It has been over 30 years since McEwen et al. (1968) injected adrenalectomized rats with radioactive GCs and observed that the GC was retained in high levels by the hippocampus as well as other regions of the limbic forebrain. Although it is beyond the scope of this chapter to cover the molecular biology of the GCs, one of the most influential developments in our attempts to understand the physiology of the GCs was the discovery of the GC receptors, their anatomical distributions, structure, and functions. Steroid hormones are a privileged class of molecules that have direct access to the brain and the brain is a target organ for these hormones. One of the surprises occasioned by the McEwen et al.'s results was the hippocampal localization of the GC. There was little evidence at the time that the hippocampus was associated with neuroendocrine function. The presence of GC receptors in areas of the CNS other than the hypothalamic and preoptic areas, which were known to have direct neuroendocrine functions, indicated that the effects of the GC on the CNS may extend beyond its presumed role in the regulation of the stress response and that GC may influence those functions, spatial learning and memory, that have been ascribed to the hippocampus. Since the original observation there has been considerable progress in GC receptor research. A further significant development was the findings of Reul and deKloet (1985) that there were two distinct types of GC receptors. Based primarily on studies which examined the cytosol-binding properties of these receptors, there appeared a high-affinity receptor which was occupied by low levels of GCs and a low-affinity receptor which required levels of GCs that are present only at the peak of the circadian rhythm or when elevated due to stress. Originally these two receptors were designated as type 1 and type 2, but in recent years type 1 is referred to more frequently as mineralo-corticoids (MR) and type 2 as gluco-corticoids (GR). Since McEwen used tracer amounts of radiolabeled GC, the receptors that were found in the hippocampus were most likely the high affinity MRs and not GR. Although both types of receptors can be found in the hippocampus, the MR is more localized anatomically whereas the GR is more widely distributed throughout the brain. The development of specific steroid agonists and antagonists firmly established that the receptors located in the hippocampus were indeed MR. The cloning of these receptors demonstrated that each form was the product of a distinct gene. de Kloet et al. (1998) present a table indicating the "milestones" in GC receptor research. Although their last entry is 1996, I doubt that we have seen the end of new contributions to this field.

What makes the discovery of two distinct GC receptors in the brain important is that it created a different approach to our view of the actions of GCs. de Kloet et al.'s (1998) hypothesis is that the "tonic influences of corticosterone are exerted via hippocampal MRs, while the additional occupancy of GRs with higher levels of corticosterone mediate feedback actions aimed at restoring disturbances in homeostasis" (de Kloet and Reul, 1987). de Kloet et al. (1998) further posited two modes of negative feedback which are implied in his hypothesis. (1) "proactive" negative feedback mediated via the MR maintains normal variations due to circadian rhythms and (2) "reactive" feedback operating via the GR serves to inhibit further secretion of ACTH and facilitate the return to basal levels. Well before the discovery of the GC receptors there was considerable work that described the different aspects of GC negative feedback (Jones et al., 1972; Dallman and Jones, 1973; Dallman, 1979) and postulated that reactive negative feedback occurred in three different modes, fast, intermediate, and slow. According to Dallman (2000) "fast effects occurs in milliseconds and must be exerted at the cell membrane, 'intermediate feedback has its onset at about 30 min after a pulse or continuous exposure to steroid and last about for a period of hours', slow feedback is found in conditions in which supraphysiological levels of exogenous GC have been provided for days or weeks."

The unique anatomical distribution of the MRs has caused the field to examine a much broader range of possible effects of GCs on brain function including cognition. One example is the studies that suggest that the ratio of MR/GR appear to be involved in depression (Young et al., 2003). As we begin to determine the many ways that the GCs influence the brain, these hormones have and will continue to move beyond their simple designation as stress hormones. The history of the stress and the HPA system is, when thought of in real time, relatively recent. It should not come as surprise that many of the individuals that have been mentioned as historical figures are now the leaders in the field. Notably amongst these are Mary Dallman, Bruce McEwen, E. Ron de Kloet, and Wylie Vale.

Psychoneuroendocrinology

In 1970 I attended the inaugural meeting in New York City of a newly formed society, which had been named the International Society of Psychoneuroendocrinology. The name was indeed daunting, but there was a group of people who believed that the relationship between psychology, psychiatry, and endocrinology had advanced to the point where a new forum for the dissemination of information regarding these relationships was necessary. Psychoneuroendocrinology is a broad-based discipline that examines the relationship between brain, behavior, and numerous endocrine and immune systems. The psychobiology of the HPA axis is an integral part of psychoneuroendocrinology. Once again it is almost impossible to determine any event or group of studies that could be considered the catalyst for including psychological factors within the rubric of HPA physiology.

John Mason (1968, 1975) pointed out that the most potent stimuli for activating the pituitary–adrenocortical system were psychological. Mason's position was a direct attack on doctrine presented by Selye. Mason emphasized the crucial role for what he called "the psychological approaches involved in the emotional or arousal reactions to threatening or unpleasant factors in the life situation as a whole." He argued that the so-called non-specificity of the endocrine responses to stress occurs because of the emotional component surrounding the experience associated with exposure to stress-inducing stimuli. Thus, the nonspecific responses described by Selye are primarily behavioral or psychological in nature and "the interpretive processes underlying the nonspecific bodily responses probably involves a higher level of CNS function than was previously realized." For Mason there was only one type of stress, that involving some emotional component. This doctrine has become known as the "Mason Principle." Hennessy and Levine (1979) hypothesized that the HPA axis was a sensitive indicator of emotional arousal and therefore its response was a reflection of heightened emotional arousal. These views certainly reflect a parochial approach to the psychobiology of stress. We now know there are many different pathways that can activate the CRF neuron. However, the hypotheses presented by Mason and others were very influential in the growth of HPA psychoneuroendocrinology. Once it was accepted that the psychological factors play an important role in the activation of the HPA axis, it became theoretically possible that the psychological processes can also modulate the HPA response. The psychobiology of stress has many faces and one cannot presume to cover all of the clinical and experimental data that are germane to this topic. The focus in the discussion will emphasize: (1) some of the psychological factors that modulate the HPA response, and (2) the effects of some of the hormones that comprise the HPA axis on behavior.

Early experiences

That the psychological factors were important in the expression of the HPA response to stress long predated the views expressed by Mason. In 1952 I went as a postdoctoral fellow to the Institute of Psychiatry and Psychosomatic Research at the Michael Reese Hospital in Chicago. Although my espoused purpose was to be trained as a clinical psychologist, the fellowship turned out to be very different than I had bargained for. For the first time I was exposed to the area of stress and also to the endocrinology of the HPA axis. It was during that period we conducted the initial experiment (Levine et al., 1956) that demonstrated long-term behavioral consequences as a result of early postnatal manipulations. This first experiment examined the effects of postnatal exposure to electric shock as a model for early exposure to trauma on a behavior presumed to be a measure of anxiety, the conditioned avoidance response. In addition to pups that were exposed to a brief electric shock (3 min daily starting on day 1 of life) we included two control groups. One group was removed from the dam placed in the shock apparatus for the identical period of time (3 min), but not shocked. The second control group remained in the nest totally undisturbed until weaning. The results were completely paradoxical to what we had predicted. The group that was most emotionally disturbed was the totally undisturbed (nonhandled) pups. Thus, the postnatal experience of being removed and separated briefly from the dam, with or without shock, appeared to permanently modify the behavioral responses to subsequent traumatic events well into adulthood. The first indication that brief separations from the dam which became known as early handling also influenced the HPA system was a study (Levine, 1957) that exposed early handled and nonhandled adults to an injection of glucose and examined adrenal weight 24 h after the injection. Although basal adrenal weights did not differentiate between the two early experience groups, adrenal weight was significantly increased only in the non-handled animals that had received the 20% glucose injection. These results were surprising since adrenal hypertrophy following stress usually takes considerably more time to develop than the 24 h that elapsed between the injection and removal of the adrenal.

Between 1957 and 1960 we conducted a series of experiments that indicated these early experiences also altered the developmental trajectory of the HPA axis. These data will be discussed in the chapter (Vazquez and Levine, this volume) on the development of the HPA axis.

These initial studies did strongly suggest that early experiences have a long-term influence on the activity of the adrenal and the mechanisms that regulate this organ system. Research in this area has continued for four decades and although many other biological processes have been investigated, the original findings have held up remarkably well. It was not until almost ten years had passed that the effects of early experience the HPA axis were revisited. The impetus was the development of a reliable biochemical assay for examining the levels of circulating corticosterone. This fluoremetric assay made it possible to examine the dynamics of the release of corticosterone and to obtain repeated measures on the same animal. In the years following the introduction of the fluoremetric assay, which was modified to assay corticosterone in small quantities of plasma (Glick et al., 1964), numerous studies emanating from many laboratories studied adrenocortical activity in animals subjected to different early experiences using a multitude of different paradigms. In our laboratory we published numerous studies on the effects of early handling in rats and mice. The effects of early handling on the dynamics of the adrenal response to stress have proved to be an extremely robust phenomenon. The initial experiments (Levine et al., 1967) demonstrated that the corticosterone response following exposure to an open field was significantly reduced in early handled animals. This study was conducted in collaboration with Victor Denenberg whose contributions were vital to the growth of developmental psychobiology. These findings have been robust and reproduced under a variety of different conditions (Meaney et al., 1993). The adrenocortical responses to novelty, restraint, shock, conditioned taste aversion, etc. all have been shown to be significantly reduced in early handled animals. Further, early handled animals appear to have a more efficient negative feedback regulation. Early handling also modifies the response to neonatal malnutrition (Wiener and Levine, 1978) and fetal alcohol syndrome (Weinberg et al., 1995).

It has always been assumed that the effects of early handling were to modify those aspects of the CNS that regulated the response of the HPA axis. There were numerous reasons for this assumption. The evidence indicated that there were no differences between handled and nonhandled animals in basal levels of corticosterone and ACTH. No differences were observed between the different early experience groups on adrenal sensitivity to ACTH, pituitary response to exogenous CRF, or clearance of corticosterone and ACTH. Adult early handled and nonhandled rats have similar levels of corticosterone-binding globulin. Insofar as none of the indices of the peripheral aspects of the HPA axis differ as a consequence of early handling, the origins of the differences on the response to stress must be a function of some change in programming of the central regulatory mechanisms. At least two components of the central regulatory mechanisms have been reported to differ between the early experience groups. Plotsky and Meaney (1993) reported that the resting levels of CRF mRNA in the PVN were significantly higher in nonhandled compared to handled adult animals. Median eminence levels of CRF and AVP are also higher in nonhandled rats. These differences in CRF gene expression and protein levels are apparent under resting conditions, although there are no differences in circulating corticosterone levels. More recently we have shown that increased CRF gene expression following restraint is much more rapid in nonhandled rats (Gordon and Levine 1999). Early handling also markedly changes the GR receptor density in the hippocampus of adult rats. In general, early handled animals have increased hippocampal GR sites and also show an increased gene expression for the GR receptor. The MR receptors do not appear to change with early handling. Bhatnagar et al. (1996) suggest that "it seems that the increase in GR sites is a critical feature of neonatal handling on HPA function. This increased receptor density appears to increase the sensitivity of the hippocampus to circulating GCs, enhancing the efficacy of negative feedback inhibition on HPA activity, and serving to reduce post-stress secretion of ACTH and GC in handled animals."

Although there have been several hypotheses proposed to account for the effectiveness of early handling, one hypothesis that seems to have received the most attention and support from the existing data is the "maternal mediation" hypothesis (Smotherman and Bell, 1980). Intuitively, it was difficult to understand why the seemingly innocuous manipulation of very brief bouts of maternal separation could have such permanent and pervasive influence on behavior and the HPA axis. The first suggestion that the effects of early handling could be maternally mediated was an experiment by Denenberg and Whimbey (1963). These investigators reared pups with mothers who had no prior early interventions and compared them to offspring of mothers who had been handled as infants. Elaborate cross-fostering procedures were included in an attempt to tease out postnatal from prenatal effects. The results indicated that, on some measures of emotionality, the early experience of the mother significantly altered the behavioral outcome measure. On other indices of emotionality the effects were a result of the interaction between pre- and postnatal influences. Levine (1967) demonstrated that the maternal early experience could influence the adrenocortical response of the pups. The corticosterone response to novelty of pups reared by handled dams was reduced compared to pups reared by nonhandled dams. Handling the pups resulted in a reduction of the response when reared by a nonhandled dam. In contrast, handled pups from handled dams did not differ from their nonhandled counterparts if the dam had been handled as a pup. There is a direct evidence that handling altered maternal behavior (Smotherman et al., 1977). They observed maternal behavior in dams when reunited with pups that were handled or shocked. Several aspects of maternal behavior were intensified when the treated pups were returned to the dams. Clearly, manipulation of the pup altered and increased dam–pup interactions.

For those of us who have had the opportunity to observe the field over an extended period of time, the cyclic quality of science is apparent. Questions that were posed decades ago are revisited often with new approaches, new techniques, and insights. Recently, Meaney and coworkers (Liu et al., 1997; Francis et al., 1999) have used a naturalistic approach to the question of maternal mediation. They observed maternal behavior in undisturbed females. They reported that variations in maternal behavior influence the development of behavioral and endocrine responses to stress in the offspring. In particular, increased licking/grooming and arched back nursing are correlated with a reduction in HPA activity and less-fearful behavior in the offspring. They further demonstrated that variations in maternal care serve as the basis for nongenomic behavioral transmission of individual differences in stress reactivity.

In recent years new paradigms have been introduced in an attempt to investigate the effects of adverse early experiences on the neurobiology of stress. Several laboratories have begun to explore the consequences of more prolonged periods of maternal separation (Plotsky and Meaney 1993; Patchev et al., 1997). Although there are only a limited number of published papers using these paradigms, it does appear as though these longer periods of separation can reverse the effects of brief separations (early handling). In some instances these animals as adults exhibit hyperreactive HPA activity in contrast to the well-established hyporeactivity that results from early handling. Although the ultimate outcome of these postnatal manipulations depends upon a number of different variables, there is little question that one major source of individual differences in the neuroendocrine regulation of the HPA axis is based on the environment during development and that these effects are pervasive and difficult to reverse.

Stress and coping

One issue that was confronted early in this chapter was the definition of stress. There was no attempt to arrive at a consensus definition. Also discussed was the pervasive issue that all stress-inducing events do not fall into simple distinct categories and that there are different neural pathways that are activated in response to different types of stress-inducing events. What is now abundantly clear is that events that can be viewed as purely psychological, and do not involve any immediate threat to homeostasis, are potent activators of the HPA axis. With the advent of a simple noninvasive procedure for measuring cortisol in saliva, the literature demonstrating that the activation of the HPA axis in response to psychological stimuli has been extensive (Kirschbaum and Hellhammer, 1994). Further, there is evidence that in animals and man the anticipation of an event is as potent an activator of the HPA axis as the event itself. Phobic patients show the highest elevation of cortisol on the day prior to being exposed to the phobic stimulus (Wiedenfeld et al., 1990). What has also emerged from studies in this area is that there are large individual differences in response to these stimuli. Although some of these individual differences may be attributable to early experiences, there are other behavioral processes that modulate the responses to psychological stimuli.

There is abundant evidence that unambiguously supports the hypothesis that psychological factors are important in determining the endocrine responses to stress. The field is indebted to Weiss (1972) for his innovative research on stress and coping. What was demonstrated in these studies was that a physiological response (stomach ulceration), in response to a well-defined stimulus (electric shock), can be modified: if the animal is permitted to exert *control* which regulates in some manner the duration and/or intensity of the shock, and/or is presented with information concerning the onset or offset of the shock, *predictability*, or is given information concerning the efficacy of the response, *feedback*. Weiss further postulated that the amount of stress an animal actually experiences when exposed to noxious stimuli depends on two variables: the number of coping attempts (responses) and the amount of relevant feedback these responses produce.

Some of these psychological principles are directly applicable to the regulation of the HPA axis. Evidence that control is a potent modulator of HPA activity is found in studies using a variety of species. Coover et al. (1973) examined corticosterone levels during active avoidance in rats. Plasma samples were obtained following the first training session during which time the rats received shock on the majority of the trials, after the seventh training session when the animals have achieved asymptotic performance, and ten days later. There was a decline in plasma corticosterone levels from the first to the seventh training session, which was attributed to the absence of shock. However, as the training continued there was a further decline in corticosterone levels, although performance of the avoidance task did not differ from the early training. This decline was interpreted as evidence for the effects of control and predictability on the response of the HPA axis. Other studies (Weinberg and Levine 1977) reported similar findings. Davis et al. (1977) found declines in adrenocortical activity using a lever-press escape paradigm that permitted escape but not avoidance of the aversive stimulus. The avoidance component of the shuttle box task appears to be less important than the ability to make an active response (control) that terminates the noxious event. Dess et al. (1983) examined the issues of control and predictability in dogs. The results revealed that control reduced the cortisol response to shock, whereas predictability in the absence of control has no discernable immediate effect. Increased circulating levels of cortisol induced by shock occurred whether or not the shock was predictable. Hanson et al. (1976) presented a clear demonstration that monkeys who could control the duration of a noxious sound reduce the cortisol response to these loud aversive noise levels. These are but a few of the many examples of importance of control in modulating the GC and presumably the neural components of the HPA axis.

The influence of predictability on HPA activity is more problematic. Davis and Levine (1982) and Dess et al. (1983) failed to show that predictability in the absence of control exerted any effect on the HPA axis. However, these investigators pointed out the interaction between control and predictability. Whereas prediction may occur with or without control, control in the absence of predictability is not necessarily true. The very act of making a stimulus (or its offset) response contingent dictates that the stimulus will also be predictable.

As we have documented the absence of control results in an exacerbated GC response. There is further evidence that the loss of control can induce increased HPA activity. In the study by Coover et al. (1973), a procedure was introduced that prevented the animal from making the now well-learned avoidance response. During this "forced extinction" period a locked door was placed between the compartments. Although the conditioned stimulus was presented and shock was omitted, corticosterone levels were again elevated compared to the plasma levels of corticosterone during avoidance conditioning. These data were seen to indicate that preventing the rat from making its response-contingent avoidance response represented a loss of control. Another example of the effects of loss of control is the rise in corticosterone following extinction of an appetitive response. Rats trained to press a lever for water or food on a continuous reinforcement schedule show an elevation of corticosterone levels as a consequence of reinforcement being withdrawn during extinction (Coover et al., 1971). These data were interpreted as suggesting that "frustration," defined as the absence of reinforcement occurring in a context where reinforcement is expected, does result in activation of the HPA axis. Frustration, however, can also be viewed as loss of control. This notion is supported by a study by Davis et al. (1976). Extinction of an instrumental response can be achieved in several ways. The traditional procedure is to permit the animal to respond and omit the reward. Another extinction paradigm is to permit the subject to continue to respond and receive reward, but to make obtaining the reinforcement no longer response contingent. Under this extinction procedure no change in corticosterone levels occurs.

The concept of loss of control implies that there is an accompanying loss of predictability. This would suggest that HPA activation would also be observed under circumstances where predictability of obtaining reward is altered from high to low predictability, and conversely that a shift from low to high predictability should result in a reduction of arousal and therefore a decrease in HPA activity. Goldman et al. (1973) trained rats to bar press for water on either a continuous reinforcement or a variable interval schedule. When the variable interval-trained rats were shifted to continuous reinforcement their corticosterone levels decreased. In contrast, when rats are shifted from a predictable schedule of reward to a more unpredictable schedule, corticosterone levels invariably elevate (Levine et al., 1972). That the HPA axis is bidirectional has been well established. The presentation of food or water to deprived rats results in a decline in the levels of ACTH and GC (Gray et al., 1978; Romero et al., 1995). Perhaps even more impressive is that simply presenting cues signaling the occurrence of food and water also produce declines in circulating levels of GCs (Levine and Coover, 1976; Coover et al., 1977).

Is there an overall set of assumptions that can account for the data just presented? Several attempts have been made to provide a theoretical basis to account for the activation and inhibition of the HPA axis (Hennessy and Levine, 1979; Levine and Ursin, 1991). The underlying principle that pervades most psychobiological approaches to stress invokes cognitive processes. The primary cognitive operation is one of the comparison between the immediate external event and some cognitive representation based on prior experiences. When discrepancies occur between the event and the cognitive representation, arousal is increased. The neurophysiological basis for this concept is derived from Sokolov's (1963) theory of the orienting response. Insofar as stress and arousal appear to be analogous concepts (Hennessey and Levine, 1979), those events that serve to increase arousal should activate the HPA axis and those, which would reduce arousal, should result in inhibition. Thus, such notions as uncertainty, expectancies, response outcomes, etc. have all been invoked to deal with the manner in which psychological events can influence the HPA system. What is clear is that the psychological processes can exert profound influences on the magnitude and direction of the responses of the HPA axis. Over the years there has been a concerted effort to delineate the neurobiology of affect and emotions. It should be noted that the CNS structures that have been implicated are critical in the regulation of fear and anxiety. The amygdala and other limbic system structures have been shown to be activated by the psychological stress.

What has been presented in this section is far from an exhaustive view of the psychobiology of stress. There is an extensive literature that social isolation can exacerbate and that social support can reduce the responses to stress (Levine, 1993). Dysregulation of the HPA axis has been demonstrated to occur in a number of mental diseases. One of the earliest examples was the studies by Sachar (1980) that patients with depression were hypercortisolemic. Caroll et al. (1976) reported deficiencies in negative-feedback regulation in some cases of depression using the dexamethasone suppression test. Throughout this paper the major focus has been on the activation of the HPA axis and the factors that modulate this activation. In recent years there has been an ever-growing body of evidence that stress can result in a persistent hypocortisolemia (Yehuda, 1998; Gunnar and Vazquez, 2001). The mechanisms that are involved in the persistent down regulation of the HPA axis are still unknown.

The neuropeptide concept

No discourse on the history of stress research would be even mildly comprehensive if the pioneering work of David deWied were not included. I met David when we were still young Turks trying to convince the world that our then somewhat offbeat research findings were of some importance in the complex universe of stress. deWied and I are of the same generation (our birthdays are 12 days apart). Whereas I was engaged in research that was attempting to delineate some of the factors that determine individual differences in the HPA response, his mission was to show that the peptides of the pituitary were active molecules that influenced behavior not through their regulation of some remote hormone but acted directly on the neural substrates that regulate specific behavioral outcomes. His early work was concerned with the pituitary response to stress. However, critical to the development of his future interests was the time he spent with I. Arthur Mirsky in Pittsburgh in 1957–1958. It was during that time de Wied collaborated with R.E. Miller, an experimental psychologist. These studies demonstrated that the effects of ACTH on shuttle-box performance were not a function of the steroidogenic action of ACTH, and in fact ACTH and prednisone had paradoxical effects in the shuttle box. That the effect of ACTH was not mediated by the adrenal cortex suggested that the brain was another target organ for ACTH. There was emerging a body of work around this time that suggested that ACTH may have behavioral effects that could be isolated from the behavioral consequences of the GCs. Much of this work came from the group of researchers at the University of Pecs in Hungary which included Endroczi, Koryani, and Bohus. Bela Bohus later left Hungary and joined with deWied's group in Utrecht. He went on to become a Professor at Groningen and an influential figure in the area of stress and behavior.

Upon his return to the Netherlands, de Wied and a very notable cohort of coworkers concentrated their research efforts on demonstrating what is often referred to as the "Neuropeptide Concept" (for a review, see de Wied, 1990). The grand hypothesis of this work was that peptides emanating from the pituitary had actions on the brain that were a consequence of these peptides acting directly on the neural substrates that regulates specific behaviors. Under deWied's guidance, numerous ingenious experiments were conducted. In his laboratory in Groningen they were able to successfully surgically remove either the posterior or anterior pituitary. This was accomplished by a laboratory technician J. Melchior. As far as I can determine he may have been the only man on earth who successfully performed this surgery. However, the ability to isolate either lobe of the pituitary made it possible to examine the role of specific anterior or posterior pituitary hormones. He discovered that not only ACTH but also vasopressin had observable effects on learning and memory. In his laboratory in Utrecht, where he had assumed the professorship of pharmacology in 1963, he demonstrated that the complete sequence of these peptides was not required for the behavioral effects to be manifested. He initiated a structure–activity program that eventually showed that the complete sequence of the peptide was not essential and that fragments of the ACTH molecule have profound behavioral effects in the absence of any influence on the secretion of GCs. The shortest of these fragments contained only four amino acids ACTH 4–7, though the most behaviorally potent fragment was ACTH 4–10. I could present a litany of experiments which were conducted not only on ACTH fragments but also on fragments of vasopressin. Over the years deWied has convincingly made the case for the neuropeptide concept. There is now complete acceptance of this doctrine to the point where its origin has almost been forgotten. Peptides derived from the gut, fat cells, pituitary, and brain have all been shown to exert their effects on specific behaviors. deWied's accomplishments extend well beyond the limits of this discussion. What makes him an important figure in the history of stress is that he was able to "push the envelope" and demonstrate that the so-called pituitary stress hormones can act directly on the brain to influence behaviors relevant to the stress response.

Conclusions

While writing this brief history of research on stress I have been constantly aware of some of the sage comments regarding history. Santayana wrote "history is a pack of lies, about events that never occurred, written by people who were never there." In my defense of this particular history I can say that the events described did happen and that I was, in some instances, there when the events unfolded. I can also say with pride that I was acquainted with most of the names that are mentioned in this chapter with of course some notable exceptions. However, what is also true is that it is impossible to present an historical account of anything that is not biased by the historian. Thus the history of this field, as presented, is what I believe to have been the critical milestones, the important figures, and the concepts that altered the course of how we understand stress.

The most glaring omission in this history is the absence of a discussion concerning the impact of molecular biology on stress research. The year 2003 is the 50th anniversary of the Watson and Crick's classic paper on the structure of DNA. Since that time the face of biology has been dramatically transformed. It is difficult to communicate to the student of today what research in biology was like without the vast array of techniques that have become commonplace. Much of what has been discovered through molecular biology was beyond the wildest dreams of those of us who began their careers prior to the molecular biology revolution. Much of what we have learned about neural circuitry, receptors, gene expression, and genetic regulation of the biological systems involved in the stress response has come about through the use of the tools provided by molecular biology. Many of the chapters that will follow will focus on the molecular biology of stress. What I have described is my perspective on the milestones in thinking and experimentation that dictated the directions of the field. The techniques provided by the molecular biologist have built upon this foundation. The early stress researchers were not unlike the early explorers. They used the scientific equivalent of wooden sailing ships to make major discoveries. We are no longer earth bound and are light-years way from where we started. But clearly we have not solved the mysteries that surround the exquisite adaptive capacity of living organisms and the consequences of the failure of the adaptive functions. In 1960 I attended a lecture by the eminent physiologist Sir A.S. Parkes. This lecture occurred at the time the Russians had launched the first space craft. He commented that for him inner space was as exciting and engrossing as outer space. I have had the privilege and joy of having observed and participated in the most exciting period in the history of biology. I suspect that if I were able to see the face of stress research 50 years from now it would be as unrecognizable to me as the contemporary face of biology would be to Charles Darwin.

Acknowledgments

I would like to express my appreciation to the National Institutes of Health, in particular NICHD and NIMH for the many years of support that enabled me to pursue much of the research contained in the chapter. I would like to thank Drs. F. Robert Brush, Robert Murison, and Juan Lopez for their advice and critical comments. Finally, my deepest gratitude to the postdoctoral fellows, graduate, and undergraduate students who made my life as a scientist far more gratifying and rewarding and who contributed immeasurably to the research in my laboratories at Stanford University and the University of Delaware. Many of these students are now the leaders in the field.

References

Basowitz, H., Persky, H., Korchin, S. and Grinker, R. (1955) Anxiety and Stress. McGraw-Hill, New York.

Bhatnagar, S. and Meaney, M.J. (1995) Hypothalamic–pituitary–adrenal function in chronic intermittently cold-stressed neonatally handled and non handled rats. J. Neuroendocrinol, 7: 97–108.

Bhatnagar, S., Shanks, N., Plotsky, P.M. and Meaney, M.J. (1996) Hypothalamic–pituitary–adrenal responses in neonatally handled and nonhandled rats: differences in facilitatory and inhibitory neural pathways. In: McCarthy, R., Agulara, G., Sabba, E. and Kvetnansky, R. (Eds.), Stress; Molecular and Neurobiological Advances. Gordon and Breach, New York, pp. 1–24.

Cannon, W.B. (1914) The interrelations of emotions as suggested by recent physiological researches. Am. J. Psychology, 25: 256.

Cannon, W.B. (1915) Bodily Changes in Pain, Hunger, Fear and Rage. Applegate and Co., New York, London.

Cannon, W.B. (1932) Wisdom of the Body. Norton, New York.

Carroll, B.J., Curtis, G.C. and Mendels, J. (1976) Neuroendocrine regulation in depression: II. Discrimination of depressed from nondepressed patients. Arch. Gen. Psychiatry, 33: 1039–1051.

Chrousos, G.P. and Gold, P.W. (1992) The concepts of stress systems disorders: overview of physical and behavioral homeostasis. JAMA, 267: 1244–1252.

Chrousos, G., Loriaux, L.D. and Gold, P.W. (1988) The concept of stress and its historical development. Adv. Exp. Med. Biol., 245: 3–7.

Coover, G.D., Goldman, L. and Levine, S. (1971) Plasma corticosterone increases produced by extinction of operant behavior in rats. Physiol. Behav., 6: 261–263.

Coover, G.D., Ursin, H. and Levine, S. (1973) Plasma-corticosterone levels during active-avoidance learning in rats. J. Comp. Physiol. Psychol., 82: 170–174.

Coover, G.D., Sutton, B.R., and Heybach, J.P (1977) Conditioning decreases in plasma corticosterone levels in rats by pairing stimuli with daily feedings. J. Comp. Physiol. Psychol., 91(4): 716–726.

Dallman, M.F. (1979) Adrenal feedback on stress induced corticoliberin (CRF) and corticotropin (ACTH). In: Jones, M.T., Gillham, B., Dallman, M.F. and Cahttopadhyay, S. (Eds.), Interactions within the Brain–Pituitary–Adrenocortical System. Academic Press, London, pp. 149–162.

Dallman, M.F. (2000) Glucocorticoid negative feedback. In: Fink, G. (Ed.), Encyclopedia of Stress, Vol. 2. Academic Press, San Diego, pp. 224–228.

Dallman, M.F. Jones, M.T. (1973) Corticosteroid feedback control of stress-induced ACTH secretion. In: Brodish, A. and Redgate, W.S. (Eds.), Brain–Pituitary–Adrenal Interrelationships. S. Karger, Basel, pp. 176–196.

Dautzenberg, F.M. and Hauger, R.L. (2002) The CRF peptide family and their receptors: yet more partners discovered. Trends Pharmacol. Sci., 23: 71–77.

Davis, H. and Levine, S. (1982) Predictability, control, and the pituitary–adrenal axis. J. Comp. Physiol. Psych., 96: 393–404.

Davis, H., Memmot, J., MacFadden, L. and Levine, S. (1976) Pituitary–adrenal activity under different appetitive extinction procedures. Physiol. Beh., 17: 687–690.

Davis, H., Porter, J.W., Livingston, J., Hermann, T., MacFadden, L. and Levine, S. (1977) Pituitary–adrenal activity and lever press escape behavior. Physiol. Beh., 17: 280–284.

deGroot, J. and Harris, G.W. (1950) Hypothalamic control of the anterior pituitary and blood lymphocytes. J. Physiol., 111: 335–356.

de Kloet, E.R. and Reul, J.M. (1987) Feedback action and tonic influence of corticosteroids on brain function: a concept arising form the heterogeneity of brain receptor systems. Psychoneuroendocrinology, 12: 83–105.

de Kloet, E.R., Vreugdenhil, E., Oitzl, M. and Joels, M. (1998) Brain glucocorticoid receptors in health and disease. Endocr. Rev., 19: 269–301.

Denenberg, V.H. and Whimbey, A.E. (1963) Behavior of adult rats is modified by the experiences the mother had as infants. Science, 142: 1192–1193.

Dess, N.K., Linwick, D., Patterson, J., Overmier, J.B. and Levine, S. (1983) Immediate and proactive effects of controllability and predictability on plasma cortisol responses to shocks in dogs. Behav. Neurosci., 97: 1005–1016.

de Wied, D. (1990) Effects of peptides hormones on behavior. In: de Wied, D. (Honorary Ed.), Neuropeptides: Basic and Perpectives. Elsevier, Amsterdam, pp. 1–35.

Duncan, J. and Starling, S.G. (1959) A Textbook of Physics, Macmillan, London.

Feldman, S. (1985) Neural pathways mediating adrenocortical responses. Fed. Proc., 44: 169–175.

Feldman, S.C., Conforti, N., Chowers, I. and Davidson, J.M. (1970) Pituitary–adrenal activation in rats with medial basal hypothalamic islands. Acta Endocrinol., 63(3): 405–414.

Fortier, C. (1951) Dual control of adrenocorticotropin release. Endocrinology, 49: 782–788.

Francis, D.D., Champagne, F.A., Liu, D. and Meaney, M.J. (1999) Maternal care, gene expression, and the development of individual differences in stress reactivity. Ann. N.Y.. Acad Sc., 896: 66–84.

Glick, D., von Redlich, D. and Levine, S. (1963) Flouremetric determination of corticosterone and cortisol in 0.02–0.05 milliliters of plasma or submilligrams of adrenal tissue. Endocrinology, 74: 653–655.

Goldman, L., Coover, G.D. and Levine, S. (1973) Bidirectional effects of reinforcement shifts on pituitary adrenal activity. Physiol. Behav., 10: 209–214.

Gordon, M.K. and Levine, S. (1999) Behavioral and neuroregulatory patterns in rats that experienced maternal separation as pups. Soc. Neurosci. Abstracts, 270.

Gray, G.D., Bergfors, A.M., Levin, R. and Levine, S. (1978) Comparison of the effects of restricted morning or evening water intake on the adrenocortical activity in female rats. Neuroendocrinology, 25: 236–246.

Gunnar, M.R. and Vazquez, D.M. (2001) Low cortisol and a flattening of daytime rhythm: potential indices of risk in human development. Dev. Psychopathol., 3(3): 515–538.

Hanson, J.D., Larson, M.E. and Snowdon, C.T. (1976) The effects of control over high intensity noise on plasma cortisol levels in rhesus monkeys. Beh. Biol., 16: 333–340.

Harris, G.W. (1947) The blood-vessels of the rabbit's pituitary gland and the significance of the pars and zona tuberalis. J. Anat. Lond., 81: 343–351.

Harris, G.W. (1948) Neural control of the pituitary gland. Physiol. Rev., 28: 139–179.

Harris, G.W. (1950) Oestrus rhythm, pseudopregnancy, and the pituitary stalk in the rat. J. Physiol., 111: 347–360.

Harris, G.W. (1955) Neural Control of the Pituitary Gland. Edward Arnold LTD, London.

Harris, G.W. and Jacobsohn, D. (1952) Functional grafts of the anterior pituitary gland. Proc. Roy. Soc. B., 139: 263–276.

Hench, P.S., Kendall, E.C., Slocumb, C.H. and Polley, H.F. (1949) The effect of a hormone of the adrenal cortex (17-hydroxy-11-dehydrocorticosterone); compound E and of pituitary adrenocorticotropic hormone on rheumatoid arthritis. Proc. Mayo Clinic, 24: 181–197.

Hennessy, J.W. and Levine, S. (1979) Stress, arousal and the pituitary–adrenal system: a psychoendocrine model. In: Sprague, J.M. and Epstein, A.S. (Eds.), Progress in Psychobiology and Physiological Psychology. Academic Press, New York, pp. 133–178.

Herman, J.P. and Cullinan, W.E. (1997) Neurocircuitry of stress, central control of the hypothalamic–pituitary–adrenal axis. Trends Neurosci., 20: 78–84.

Hodges, J.R. (1953) The function of adrenaline in the production of pituitary adrenocorticotropic activity. J. Endocrinol., 9: 342–350.

Hsu, S.Y. and Hsueh, A.J. (2001) Human stresscopin and stress copin-related peptide are selective ligands for the type 2 corticotropin-releasing hormone receptor. Nat. Med., 7: 605–611.

Ingle, D.J. (1952) The role of the adrenal cortex in homeostasis. J. Endocrinol., 8: 23–37.

Jones, M.T., Brush, F.R. and Neame, R.L.B. (1972) Characteristics of fast feedback control of corticotrophin release by glucocorticoids. J. Endocrinol., 55: 489–497.

Kirschbaum, C. and Hellhammer, D.H. (1994) Salivary cortisol on psychoendocrine research: recent developments and applications. Psychoneuroendocrinology, 19(4): 313–333.

Levine, S. (1957) Infantile experience and resistance to stress. Science, 126: 405

Levine, S. (1967) Maternal and environmental influences on adrenocortical responses to stress in weanling rats. Science, 156: 258–260.

Levine, S. (1993) The influence of social factors on the response to stress. Psychotherapy and Psychosomatics, 60: 33–38.

Levine, S. and Coover, G.D. (1976) Environmental control of suppression of the pituitary–adrenal system. Physiol. Behav., 17: 35–37.

Levine, S. and Ursin, H. (1991) What is stress? In: Brown, M.R. and Koob, G.F. (Eds.), Stress Neurobiology and Neuroendocrinology. Rivier Marcel Dekker, Inc, New York, pp. 3–21.

Levine, S., Chevalier, J.A. and Korchin, S.J. (1956) The effects of early handling and shock on later avoidance behavior. J. Pers., 24: 475–493.

Levine, S., Haltmeyer, G.C., Karas, G.G. and Denenberg, V.H. (1967) Physiological and behavioral effects of infantile stimulation. Physiol. Behav., 2: 55–59.

Levine, S., Goldman, L. and Coover, G.D. (1972) Expectancy and the pituitary–adrenal system. In: Porter, R. and Knight, J. (Eds.), Physiology, Emotion and Psychosomatic Illness. Ciba Foundation Symposium, Elsevier, Amsterdam, pp. 281–291.

Liu, D., Diorio, J., Tannenbaum, B., Caldji, C., Francis, D., Freedman, A., Sharma, S., Pearson, D., Plotsky, P.M. and Meaney, M.J. (1997) Maternal care, hippocampal glucocorticoid receptors, and hypothalamic–pituitary–adrenal responses to stress. Science, 277: 1659–1662.

Long, C.N.H. (1947) The conditions associated with the secretion of the adrenal cortex. Fed. Proc., 6: 461–471.

Long, C.N.H. (1952) Regulation of A.C.T.H. Secretion. Rec. Prog. in Horm. Res., 7: 75–97.

Marshall, F.H.A. (1936) Sexual periodicity and the causes which determine it. Phil. Trans. R. Soc. London, 226: 423–456.

Mason, J.W. (1968) A review of psychoendocrine research on the pituitary adrenal system. Psychosom.Med., 30: 576–607.

Mason, J.W. (1975) A historical view of the stress field. J. Hum. Stress, 16: 11–22.

McCann, S.M. (1953) Effect of hypothalamic lesions on the adrenocortical response to stress. Amer. J. Physiol., 175: 13–20.

McCann, S. and Brobeck, J.R. (1954) Evidence for the role of the supraoptochypophyseal system in the regulation of adrenocorticotrophin secretion. Proc. Soc. Experimental Biol. and Med., 87: 318–324.

McEwen, B. (2000) Stress, definition and concepts of. In: Fink, G. (Ed.), Encyclopedia of Stress, Vol. 3. Academic Press, San Diego, pp. 508–509.

McEwen, B.s., Weiss, J. and Schwartz, L. (1968) Selective retention of corticosterone by limbic system structures in the rat brain. Nature, 220: 911–912.

Meaney, M.J., Aitken, D.H., Viau, V., Sharma, S. and Sarrieau, A. (1993) Individual differences in the hypothalamic–pituitary–adrenal stress response and the hypothalamic CRF system. In: Tache, Y. and Rivier, C. (Eds.), Corticotropin-Releasing Factor and Cytokines: The Role of the Stress Responses. New York Academy of Science, New York, pp. 70–85.

Medvei, V.C. (1982) A History of Endocrinology. MTP Press, Lancaster.

Munck, A.U., Guyre, P.M. and Holbrook, N.J. (1984) Physiological functions of glucocorticoids in stress and their relation to pharmacological actions. Endocr. Rev, 5: 25–44.

Patchev, V.K., Montkowski, A., Rouskova, D., Koranyi, L., Holsboer, F. and Almeida, O.F. (1997) Neonatal treatment of rats with the neuroactive steroid tetrahydrodeoxycorticosterone (THDOC) abolishes the behavioral and neuroendocrine consequences of adverse early life events. J Clin. Invest., 99: 962–966.

Plotsky, P.M. and Meaney, M.J. (1993) Early, postnatal experience alters hypothalamic corticotropin-releasing factor (CRF) mRNA, median eminence CRF content and stress-induced release in adult rats. Brain Res. Mol. Brain Res., 18: 195–200.

Popa, G.T. and Fielding, U. (1930) A portal circulation from the pituitary to the hypothalamus. J. Anat., 65: 227–232.

Porter, R.W. (1953) Hypothalamic involvement in the pituitary adrenocortical response to stress stimuli. Amer. J. Physiol., 172: 515–519.

Reul, J.M. and de Kloet, E.R. (1985) Two receptor systems for corticosterone in rat brain: microdistribution and differential occupation. Endocrinology, 117: 2505–2511.

Romero, L.M., Levine, S., and Sapolsky, R. (1995) Adrenocorticotropin secretagog release: stimulaion by frustration and paradoxically by reward presentation. Brain Res. 676(1): 151–156.

Sachar, E.J. (1980) Hormonal changes in stress and mental illness. In: Krieger, D.T. and Hughes, J.C. (Eds.), Neuroendocrinology. Sinauer Associates, Sunderland, pp. 177–184.

Sapolsky, R.M. (1994) Why Zebras Don't Get Ulcers. W.H. Freeman and Co., New York.

Sapolsky, R.M., Romero, L.M. and Munck, A.U. (2000) How do glucocorticoids influence stress responses? Integrating permissive, suppressive, stimulatory, and preparatory actions. Endrocr. Rev., 21: 55–89.

Sawchenko, P.E. (1991) The final commom path: issues concerning the organization of central mechanisms controlling corticotropin secretion. In: Brown, M.R., Koob, G.F. and Rivier, C. (Eds.), Stress: Neurobiology and Neuroendocrinology. Marcel Dekker, Inc, New York, pp. 55–72.

Sayers, G. (1950) The adrenal cortex and homeostasis. Physiol. Rev., 30: 241–320.

Selye, H. (1936) A syndrome produced by diverse nocuous agents. Nature, 38: 32.

Selye, H. (1949) General adaptation syndrome. In: Textbook of Endocrinology, 2nd ed. Acta Endocrinol., Montreal, pp. 837–839.

Selye, H. (1956) The Stress of Life. McGraw-Hill, New York.

Silber, R.H., Busch, R.D. and Oslaps, R. (1958) Practical procedure for the estimation of corticosterone and hydrocortisone. Clin. Chem., 4: 278–284.

Smotherman, W.P. and Bell, R.W. (1980) Maternal mediation of experiences. In: Smotherman, W.P. and Bell, R.W. (Eds.), Maternal Influences and Early Behavior. Spectrum Publications, New York, pp. 201–210.

Smotherman, W.P., Brown, C.P. and Levine, S. (1977) Maternal responsiveness following differential pup treatment and mother–pup interactions. Dev. Psychobiol., 10: 242–253.

Sokolov, E.N. (1963) Perception and the Conditioned Reflex. Pergamon, Oxford.

Steptoe, A. (2000) Stress effects, overview. In: Fink, G. (Ed.), Encyclopedia of Stress, Vol. 3. Academic Press, San Diego, pp. 510–511.

Vale, W.S., Speiss, J., Rivier, C. and Rivier, J. (1981) Characterization of a 41-residue ovine hypothalamic peptide that stimulates the secretions of corticotropin and beta-endorphin. Science, 213(4514): 1394–1397.

Vogt, M. (1952) Plasma adrenaline and release of A.C.T.H. in normal and demedullated rats. J. Physiol., 118: 588–594.

Wade, N. (1981) The Nobel Duel; Two Scientists 21-year Race to Win the Worlds most Coveted Research Prize. Anchor Press, Doubleday, Garden City.

Weinberg, J. and Levine, S. (1977) Early handling influences on behavioral and physiological responses during active avoidance. Dev. Psychobiol., 10: 1661–1669.

Weinberg, J., Kim, C.K. and Yu, W. (1995) Early handling can attenuate adverse effects of fetal alcohol exposure. Alcohol, 12: 317–327.

Weiss, J.M. (1972) Influences of psychological variables on stress induced ulcers. In: Porter, R. and Knight, J. (Eds.), Physiology, Emotions and Psychosomatic Illness. Ciba Foundation Symposium, Elsevier, Amsterdam, pp. 253–264.

Wiedenfeld, S.S., O'Leary, A., Bandura, A., Brown, S., Levine, S. and Raska, K. (1990) Impact of self-efficacy in coping with stressors on components of the immune system. J. Pers. Psychol., 59: 1082–1094.

Wiener, S.G. and Levine, S. (1978) Perinatal malnutrition and early handling: interactive effects on the development of the pituitary–adrenal system. Dev. Psychobiol., 11: 251–259.

Wislocki, G.B. (1937) The vascular supply of the hypophysis cerebri of the cat. Anat. Rec., 69: 361–387.

Yehuda, R. (1998) Psychoneuroendocrinology of post-traumatic stress disorder. Psychiatr. Clin. North Am., 21: 359–379.

Young, E.A., Lopez, J.F., Murphy-Wienberg, V., Watson, S.J. and Akil, H. (2003) Mineralocorticoid receptor function in major depression. Arch. Gen. Psychiatry, 60: 24–28.

T. Steckler, N.H. Kalin and J.M.H.M. Reul (Eds.)
Handbook of Stress and the Brain, Vol. 15
ISBN 0-444-51173-3

CHAPTER 1.2

The neuropsychology of stress

Thomas Steckler

Johnson & Johnson Pharmaceutical Research & Development, A Division of Janssen Pharmaceutica N.V., Turnhoutseweg 30, 2340 Beerse, Belgium

Abstract: This chapter focuses on the psychological processes which govern the stress response. After an introduction aiming to define the terminology used in the area of stress research and some of the related concepts, such as emotions, the stress response will be discussed within the framework of cognitive functions, including learning theory. It will be demonstrated how situational appraisal and anticipation, predictability and controllability, and differences in coping style will affect stress responsivity. In the final parts of the chapter, these concepts will be related to the behavioural inhibition theory as defined by Gray and McNaughton (*The Neuropsychology of Anxiety*, 2nd ed. Oxford University Press, Oxford, 2000), and how this could be mediated by various areas in the brain.

Stressors and the stress response

Stress can be defined as any challenge to homeostasis of an individuum that requires an adaptive response of that individuum (Newport and Nemeroff, 2002). Conceptually, stress consists of three components, that is the input of a stimulus, the evaluation of this information, and a response output.

Aversive exteroceptive (e.g., electric shock, cold, social dominance in animals, but also several psychosocial aversive situations in humans, such as public speech) or interoceptive (e.g., pain) stimuli are referred to as stressors. A stressor can be defined as a change in the environment that is sensed by an organism, is aversive and potentially harmful to that organism and elicits acute and/or chronic responses (Ottenweller, 2000).

A stress response in turn consists of a complex pattern of physiological, behavioural, cognitive, and/or emotional components. Physiological processes can acutely comprise of, for example, piloerection, increases in heart rate, modulation of intestinal motility, activation of the hypothalamic–pituitary–adrenal (HPA) axis,[1] activation of peripheral catecholaminergic systems, and of various neurotransmitter changes in the brain. More chronically, physiological stress responses can consist of the development of gastric ulcers, chronic changes in HPA axis and neurotransmitter activity, hypertrophy of the adrenal cortex, atrophy of the thymus, and loss of body weight, amongst other effects.

At the behavioural level, exposure to stress has been reported to lead to a decrease in food and water intake (Pare, 1964), to inhibit exploratory activity (Weiss et al., 1980), to suppress appetitively motivated responses (Annau and Kamin, 1961), to increase anxiety-related behaviour (File, 1980), and to enhance or to impair both aversive and appetitive learning under certain conditions (Overmier and Seligman, 1967; Rosellini, 1978; Shors, 2001).

Fax: +32 11 460 6121; E-mail: Tsteckle@PRDBE.jnj.com

[1]One of the main players regulating HPA axis activity is corticotropin-releasing factor (CRF). CRF is released from the parvocellular part of the hypothalamic paraventricular nucleus (PVN) into portal vessels, subsequently activating the HPA axis by stimulation of release of adrenocorticotropic hormone (ACTH) from the anterior pituitary. ACTH in turn triggers the release of glucocorticoids (corticosterone or cortisol, respectively) from the adrenal cortex.

In other words, there is a prioritisation of stress responses over other types of behaviour, and the stressor-induced behavioural responses serve the goal to reduce or eliminate the negative effects of a stressor.

Types of stressors

A stressor can be any unpleasant intrusion to the external or internal environment (e.g., a social encounter, noise, an electric footshock, exposure to extreme temperatures, or an infection), or a withdrawal from the environment (e.g., starvation, social isolation, or separation of an infant from its mother). The stressor can be presented once only for a short time (e.g., administration of a single footshock of a few milliseconds duration), for longer times (e.g., exposure to cold temperatures for several hours), repeatedly (e.g., exposure to a series of repeated footshocks spaced over time, or repeated social defeat by a dominant subject, with the two subjects being fully separated in-between), or chronically/enduring (e.g., exposure to cold temperatures over days, or constant exposure of a defeated subject to the dominant subject). Moreover, some internal stimuli, such as anxiety and fear, can constitute a component of a stressor (e.g., in case of a patient suffering from an anxiety disorder) but, as will be discussed below, can also be part of a stress response (Young and Liberzon, 2002).

A stress response (e.g., an increase in HPA-axis activity) can be induced in a relatively simple, reflex-like manner, in which case it does not necessarily require an evaluation of the situation by the subject. Alternatively, a stress response can also entail inputs from higher brain areas, i.e., a stress response can involve the evaluation of the stressor as being stressful. In particular, so-called psychological stress (for example, novelty stress and social defeat) has been suggested to activate these higher systems (Herman et al., 1996; Herman and Cullinan, 1997).

Psychological stress can be defined as involving a reaction to an aversive stimulus in an individual's *external* environment (Kollack-Walker et al., 2000), and has been viewed as an asymmetry between the motivational systems of reward and punishment (Walker, 1987). This already implies that cognitive functions play an important role in the processing of this type of stress-related information.

However, as mentioned above, not every stressor may involve cognitive processing and a stress response can be induced directly by the exposure to an aversive stimulus. This second type of stressors has been called a physical stressor and can be defined as disturbing an individual's *internal* milieu, leading to activation of regulatory mechanisms that serve to restore homeostasis (Kollack-Walker et al., 2000). Such stressors would include, for example, starvation, noise, cold exposure, or haemorrhage.

Besides covering different baskets with stressors, which involve cognitive processes to different degrees, the distinction between psychological and physical stressors gains relevance by the finding that these two types of stressors seem to activate different parts of the brain. Physical stress seems to be relayed directly to the PVN of the hypothalamus, part of the HPA axis, by ascending viscero- and somatosensory pathways (Sawchenko and Swanson, 1983; Kovacs and Makara, 1990; Sawchenko et al., 1996), rather than to higher brain areas. Psychological stress, on the other hand, seems to involve a number of higher brain areas, including various neocortical areas, the hippocampus, and the septal complex, presumably because it needs to be identified and evaluated first, and the PVN serves as relay point between the higher brain areas and the HPA axis.

What exactly constitutes a physical stressor rather than a psychological stressor can, however, sometimes be a matter of debate. Although Kollack-Walker and colleagues (2000) provide a clear definition, which unambiguously distinguishes between stressors based on their primary location, i.e., whether they originate internally or externally of the subject, it is obvious that most stressors must be seen as compound stressors, affecting the subject both from its external and internal milieu. Swim stress, for example, includes both an external component (e.g., a situation where the subject faces the threat of drowning) and an internal component (e.g., hypothermia due to swimming in cold water). Therefore, it could be argued that it may be more appropriate to see the two definitions of physical and psychological stressors as two extremes of a continuous scale, with most stressors laying in-between. According to this view, it is the relative balance between psychological and physical components that varies between stressors. Indeed, some stressors, originally considered as physical stressors (e.g., swim stress), are now increasingly considered psychological stress. Thus, there is some intuitive blurring of the dichotomy and there appears to be a gradual difference rather than an absolute distinction. On the functional neuroanatomical level, however, it becomes increasingly clear that different stressors activate different pathways, and it has been suggested that one way to distinguish between psychological and physical stressors is by directly looking at the pathways activated by these stressors, i.e., the brain itself would be the best tool to categorise complex stressors (Dayas et al., 2001).

Inherent to the concept of psychological and physical stressors is the fact that the subject is exposed to (external or internal) *aversive* stimuli. To fully appreciate the role of stress systems, it is however important to realise that not only aversive stimuli, but also reinforcement and withdrawal from an appetitive situation affect the activity of the HPA axis (withdrawal from reward may of course be considered as representing an aversive stimulus in its own right). Thus, reinforcement has been shown to decrease plasma corticosterone level in rats trained on an operant task, while extinction, i.e., operant responding followed by withdrawal of the expected reward, increases plasma corticosterone level (Coe et al., 1983; De Boer et al., 1990).

In the following parts of this chapter, I will focus on the processes which take place during the evaluation of the information provided by a stressor. Before going into further detail, it is however important to first clarify a concept central to this discussion, that is the question of what is meant when talking about an emotion.

Emotions related to stress exposure

The stress-induced behavioural changes mentioned above can be, but do not have to be, part of what is considered an emotional response. James (1890) suggested that emotion-provoking stimuli induce bodily changes, and that the feeling of these changes would be the emotion. Others extended this view and associated emotions with changes in the brain and the body, i.e., considered emotions to consist of behavioural, autonomic, and endocrine responses, which can differ according to the nature of the emotion. For example, there is evidence that negative emotions increase heart rate and change skin temperature more than pleasant emotions (Ekman et al., 1983). Behaviourally, emotions can be viewed as central states elicited by reinforcing stimuli (Plutchik, 1967). Especially secondary punishing stimuli have been associated with negative emotions such as fear (Mowrer, 1960), which can be viewed as a state anticipating primary punishment. Rolls (1999) defined emotions as 'a state elicited by rewards or punishers, leading to changes in rewards or punishers' (p. 60). He considers emotions as a cognitive process which results in a decoded signal that an environmental event (at the time of presentation or as a remembrance) is reinforcing or punishing, together with the affective state produced as a result. Within that definition, an affective state differs from an emotion in that the former on its own lacks an external sensory input and the cognitive decoding, i.e., there is no present or past environmental event towards which the affective state is directed, which would then lead to a goal-directed behaviour (Rolls, 1999).

It is clear from these definitions that emotions form an integral part of the response to a (psychological) stressor. Functionally, it has been suggested that the emotional responses are of relevance for coping reactions to short-term aversive events (a topic which will be covered in more detail below) and for the initiation of fight–flight–freezing reactions, which serve the subject to deal effectively with a source of danger and to return into a state of safety. Interestingly, there seems to be a more rapid target detection for emotional stimuli such as fear-related (but also positive emotional) stimuli in humans (Ohman et al., 2001), and such target stimuli seem to be processed even though the stimuli are not perceived, as can be evidenced by alterations in skin conductance responses (Esteves et al., 1994), i.e., these emotionally relevant stimuli will even be detected under conditions of limited attentional resources (pre-attentive processing), which makes sense in that those stimuli to which a value has been ascribed, e.g., a stressor, will be of high relevance to the individuum.

The stress response – a cognitive view

It is obvious from the points raised above that cognitive processes play an important role for the adequate reaction to stress exposure. The importance of such processes appears evident by the simple notion that subjects should learn to repeat responses that lead to reward or prevent punishment, and should learn to inhibit responses that prevent or truncate rewards or lead to punishment, i.e., it is *learned* that the appropriate action should prevent the (re-)occurrence of a certain stressor. As an extreme example, a subject should remember the appearance, smell, sound, and environmental location of a predator to predict its next occurrence to maximize the likelihood for survival. Under daily laboratory conditions, a rat will use related information about its cage mates to adhere to hierarchical orders for food and water access.

This concept is not new, but the relevance of cognitive processes for the response to a stressor has been noted as early as 1926 by Freud in the psychoanalytical theory of defence mechanisms, suggesting that denial and intellectualisation are fundamental ways to reduce anxiety. Thus, changes in cognitive function can result in changes in the way a stressor is perceived, remembered, or how a stressful situation is solved.

The concept of defensive distance

In order to cope with a stressful situation, a subject can have two major strategies: it can try to avoid the stressor (defensive avoidance) or to approach the stressor (defensive approach), the latter of which includes risk assessment and behavioural inhibition. It has been proposed that the relative importance of the two competing types of behaviour (avoidance and approach) will depend on the so-called defensive distance between the subject and the stressor (Blanchard and Blanchard, 1990). Moreover, the *speed* with which such conflict can be resolved in an approach–avoidance situation seems to be related to genetic factors and underlying stress responsivity. For example, Tryon Maze Bright rats have been reported to be faster in speed of conflict resolution than Tryon Maze Dull rats in an operant-conditioned punishment paradigm, and Roman Low Avoidance rats were faster than Roman High Avoidance rats (Koene and Vossen, 1991). In a runway situation, Wistar Kyoto rats were faster in solving a conflict than randomly bred Wistar Wu rats and Brown Norway rats were faster than Wistar Wu rats (Koene and Vossen, 1991).

The *value* of the defensive distance is composed of the objective physical or temporal distance between subject and stressor, and the subjective assessment of threat originating from the stressor (Blanchard and Blanchard, 1990). This in turn implies that in the absence of a stressor, there should be no behavioural inhibition. In the presence of a moderate stressor, there may be behavioural inhibition of non-defensive behaviour (e.g., a decrease in food and water intake, inhibition of exploratory activity, and suppression of appetitively motivated responses), but not of defensive behaviour (fight or flight; risk assessment), while under conditions of high stress, both defensive and non-defensive types of behaviour would be inhibited (freezing).

In this respect, it is interesting to note that Takahashi (1996) suggested that glucocorticoids, which are released under stressful conditions, may also influence freezing during development by actions at the septohippocampal level, thereby modulating the individual's levels of stress-induced arousal and attention to threat. This is an interesting idea in the context of Gray's theory on behavioural inhibition, which will be discussed below.

Appraisal and anticipation

In order to choose between approach and avoidance most effectively, the subject must be able to appraise the situation it has to deal with. The concept of defensive distance suggests that the subject is able to appraise its position relative to the stressor, i.e., there is a quantitative aspect. However, a subject must also be able to appraise a stressor in a qualitative way, i.e., whether a given stimulus is a severe stressor, a mild stressor, or no stressor at all. In 1968, Mason suggested that the appraisal of a stimulus determines whether it is perceived as a stressor, i.e., that the subject must discriminate between threatening and non-threatening situations. In other words, the meaning of a stimulus is, at least in part, determined by the representations held by the subject, which can be innate or acquired. This can be exemplified by the condition of novelty stress. Exposure to novelty alters both HPA axis and behavioural activities (Lemaire et al., 1999; Hall et al., 2000). Novelty, however, is not an inherent characteristic of a stimulus, but an attribution given to the stimulus by the subject.

In addition to the appraisal of a situation, it is important to acknowledge that a subject's responses are not only driven by reactive demands, but also by anticipation. Arthur (1987), for example, demonstrated that anticipation of an aversive stimulus, such as an electric shock, may result in an even greater stress response in rats (as measured in activation of the HPA-axis) than confrontation with the aversive stimulus itself. Similar findings were reported by others and extended to other measures, including anticipatory changes in pain threshold (Przewlocka et al., 1990; Yamamotova et al., 2000). Likewise, anticipatory stress responses can be observed in humans, for example, when facing a public speech, and this stress response can be altered by manipulation of cognitive processes (Rohrmann et al., 1999): the degree to which individuals are stressed anticipating public speech (as measured by HPA-axis activation, changes in heart rate, subjective arousal, and state anxiety) depends on manipulative feedback (reassuring or arousing) given during that period (Rohrmann et al., 1999), but also on the level of control that subjects experience over the situation (see below).

Stress coping

It is evident that cognitive appraisal of a situation will determine the strategy chosen by an individual to deal with, or to *cope* with a stressor. In this context, it is common knowledge derived from everyday experience, but also from sophisticated and well-controlled studies, that there are very clear individual differences in stress coping. As already pointed out, these differences between individuals seem to be in part based on genetic factors (Koene and Vossen, 1991), which interact with developmental and situational influences (see below).

The importance of the concept of coping is highlighted by the fact that the ability to cope is the integral part of some definitions of stress. Lazarus (1966), for example, conceptualised stress as the interaction between the demands of the situation and the individuals ability to cope. Hence, coping is an active response to resolve a stressful situation. It encompasses cognitive and other behavioural efforts to reduce or adapt to a stressor.

In principle, a subject can appraise two components of a situation: First, it can assess whether a situation is irrelevant, appetitive, or aversive, as was discussed above (so-called primary appraisals). Secondly, it can make an assessment of the relevance and availability of its own coping strategies (so-called secondary appraisals). The success of a given coping strategy is situation specific and its appropriateness will, at least in part, depend on defensive distance, as the defensive distance will determine whether approach or avoidance are more likely to be successful.

Differences in stress coping have been implicated in stress resiliance and hence different vulnerability to psychiatric disease. The constitutional vulnerability of individuals to stress is named diathesis. As already mentioned, such individual differences are based on genetic factors, physiological developmental factors, and lifetime experience. We have already seen that different strains of rats exhibit different speeds of conflict resolution. Another example derives from the simple reaction to novelty, which has been shown to differ between inbred strains of rats, suggesting a genetic base (Gilad and Shiller, 1989).

Conversely, differences in reaction to novelty can also be observed within a population of rats (Lemaire et al., 1999; Piazza et al., 1991), suggesting epigenetic factors also play a role. The importance of epigenetic factors is also illustrated by the findings that prenatally stressed animals are more impaired in coping with stress compared to non-stressed controls (Weinstock, 1997), and that rat pups, which differ in the level of maternal care (maternal licking and grooming) while being nursed, will show differences in the response to novelty stress later in life (Liu et al., 1997). Interestingly, these high- and low-reacting rats do not only differ in behavioural and neuroendocrine measures, but also in neuroanatomical features, and show a reduced neurogenesis in the dentate gyrus of the hippocampus (Lemaire et al., 1999).

As an interim summary, we can conclude so far that different stressors will lead to different patterns of brain activation. Some, but not all, stressors will involve a cognitive component, including appraisal and anticipation of a situation, leading to an emotional response and a variety of different coping strategies, which are under control of genetic and epigenetic factors. From that, it can easily be seen that behavioural and affective changes can even precede the presentation of a primary stressor and are under control of the laws of classical- and operant-conditioning processes, where stimuli presented to the subject have been associated with the stressor before.

Learning processes modulating stress responsivity

Next, the role of learning processes in the modulation of stress responsivity will be discussed in more detail. Associating a given stimulus with a stressor is equivalent to meaning that a certain stimulus will have acquired the qualities of a secondary stressor (or a conditioned stimulus). Once the subject has made such an association, it can respond in a way to the aversive situation similar to a response that has led to a successful outcome in the past, i.e., it can elicit a prepared response (Phillips, 1989).

One model which would allow a subject to predict the occurrence of a stressor would be related to classical conditioning procedures, whereby the subject has to associate the stressor with a discrete cue, which could be simple (such as a tone or a light) or complex (such as the environmental context in which this stressor appears). Such classical conditioning procedures are frequently used in animals and have also been reported in humans (Watson and Rayner, 1920). The conditional stimulus in turn can be associated with an emotional tone, such as fear.

Alternatively, the subject could learn an instrumental or so-called operant response, i.e., the subject acquires a behavioural pattern that has the capacity to alter the frequency of the subject's exposure to certain events.

Within this framework, two types of learning have been distinguished, namely escape and avoidance learning, essential forms of behavioural reactions in stressful environments. In escape learning, the aversive stimulus occurs conspicuously before each response, while in avoidance learning, the stressor, if learning is successful, is rarely, if ever, seen. These avoidance responses can be very persistent under conditions of extinction, i.e., in the absence of the primary stressor, and well-learned avoidance responses may be sustained for some time purely as an automatic habit (Mackintosh, 1983). By analogy, presentation of a secondary stressor, followed by presentation of a primary stressor and escape, is likely to generate anxiety, while absence of the primary stressor due to avoidance after presentation of the secondary stressor, should result in diminished anxiety and merely lead to a habitual response.

It is of course also possible to extinguish an avoidance response. Interestingly, extinction of avoidance learning has been suggested to be under control of the activity of the HPA axis. More specifically, extinction of passive avoidance has been suggested to be modulated via activation of the mineralocorticoid receptor (MR), while extinction of active avoidance has been suggested to be modulated via activation of the glucocorticoid receptor (GR)[2] (Korte, 2001).

Such learned responses can be assessed in the animal in procedures such as conditioned suppression or conflict procedures, conditioned emotional response (CER), or fear-potentiated startle. Conditioned suppression or conflict procedures entail the suppression of a consummatory or exploratory behaviour in the presence of a stimulus (the conditioned stimulus) previously associated with an aversive stimulus (the unconditioned stimulus) and could, for example, involve the suppression of a lick for water or a freezing response (the latter being mediated by the periaqueductal grey; Graeff, 1994). These behavioural responses can occur within parts of a second, faster than any neuroendocrine changes, but are still under neuroendocrine control. The freezing response, for example, already takes place

---

[2]Two corticosteroid receptors have been identified in the brain, the MR and the GR, which differ in expression patterns and binding properties (GR binds corticosteroids with a 10-fold lower affinity than MRs). It has been suggested that MRs are involved in the maintenance of the activity of the stress system, while GRs, in conjunction with MRs, may mediate the recovery from stress (Reul and De Kloet, 1985; De Kloet et al., 1998).

before activation of the HPA axis, but has been suggested to be acutely modulated by MR activation (Korte, 2001). Korte argues further that the first reaction of an animal (such as a rat or a mouse) visiting a dangerous location for which an aversive memory has been formed will be conditioned freezing behaviour, modulated by a permissive role of the MR. Although it is evident that the other factors will contribute to such response as well, this example again nicely illustrates the interrelationship between behavioural processes and stress systems, which goes both ways. If the environment turns out to be safe, extinction of passive and active avoidance will take place.

In contrast, a CER often involves suppression of an instrumental response, for example, of lever pressing for food, upon presentation of an aversive CS. In fear-potentiated startle (mediated by the pontine reticular nucleus; Davis, 1992b), the magnitude of the startle response is increased in the presence of an aversive CS, the state retrieved by the CS is fear. Fear-potentiated startle has been suggested to be a sensitive measure of anticipatory anxiety (Davis et al., 1993).

The examples given above highlight the importance of first-order conditioning processes in the response of a subject to a stressor, but this can of course also be extended to second-order conditioning. For example, a rat might have learned to stop responding in the presence of a CS (e.g., a light) in a conditioned suppression paradigm in the absence of an UCS. If the CS is now preceded by another neutral stimulus (e.g., a tone), the presentation of the second stimulus can eventually also lead to suppression of the ongoing behaviour after a few pairings. The relevance of second-order conditioning lays in the fact that it helps to explain how anxiety can generalise from one anxiogenic event to a number of other, more or less closely related, originally neutral stimuli.

However, conditioning is not restricted to behavioural responses, but extends to autonomic and neuroendocrine functions as well. For example, it has been shown that increases in plasma corticosterone levels can be conditioned to stimuli associated with a poison (Adler, 1976). Conversely, decreases in plasma corticosterone levels have been conditioned to stimuli associated with daily feeding and drinking (Coover et al., 1977). Thus, our stress systems are capable of learning and we can adapt – or maladapt – to a stressful situation, with all the consequences associated with stress-related psychiatric disorders.

In the context of learning theory, adaptation and maladaptation are better referred to as habituation and sensitization. Habituation, an active-learning process not to respond to irrelevant situations, governs the decline of the stress response when the subject is repeatedly exposed to the stressor. During habituation, the orienting response towards the stressor is diminished. This can be seen, for example, by the fact that a novel stimulus, which induces a stress response upon initial exposure, becomes a habituated stimulus after repeated exposure. Behaviourally, such stimulus will elicit exploration, which is reduced upon repeated exposure.

Sensitization refers to an exaggeration of the stress response and is seen, for example, if animals are exposed to a mild stressor (e.g., a novel stimulus) after exposure to a strong aversive stimulus. Rats pre-exposed to inescapable footshock, for example, have been demonstrated to display progressive and long-lasting increases in anxiety-related behaviour (Van Dijken et al., 1992), and to develop autonomic (Bruijnzeel et al., 2001) and neuroendocrine changes (Van Dijken et al., 1993), which last for weeks. Likewise, social defeat of one rat by another rat can induce long-lasting behavioural changes in the defeated animal, such as an increase in immobility (Koolhaas et al., 1990). During this stage, the orienting response towards the stressor would be enhanced.

At the neuroendocrine level, it has been shown that a single exposure to electric footshock can induce vasopressin (AVP) levels in corticotropin-releasing factor (CRF) terminals in the median eminence for at least 11 days (Schmidt et al., 1996), and it has been suggested that the sensitized response is due to enhanced AVP release, which potentiates the effect of CRF on ACTH release (Van Dijken et al., 1993).

Sensitization has also been reported in humans, leading to greater hormonal stress response over time and an increase in baseline cortisol levels (Young and Akil, 1985; Dallman, 1993). Moreover, it has been shown that pre-exposure to stress sensitizes the release of noradrenaline at various brain levels

(Tsuda et al., 1986; Pacac et al., 1995), which in turn can activate the HPA axis by stimulation of CRF release by parvocellular PVN neurons (Plotsky et al., 1989; Plotsky, 1991; Pacak et al., 1995). More recently, Bruijnzeel et al. (1999) demonstrated long-term activation of brain areas involved in the mediation of anxiety-related behaviour, autonomic, and neuroendocrine responses for at least two weeks in the pre-shocked rats, as measured by an upregulation of Fos-immunoreactivity.

These alterations are not non-specific changes, but seem to affect certain stages of information processing more than others. Shors (2001), for example, has shown that exposure to a stressor enhances the formation of new associations rather than affecting retention or performance of the conditioned response. In this study, acute exposure to intermittent tail shock or acute inescapable swim stress, but not inescapable noise stress or the unconditioned stimulus of periorbital eyelid stimulation, enhanced classical eyeblink conditioning in male rats when trained 24 h after stress cessation (Shors, 2001). This facilitating effect of the stressor occurred within 30 min of stressor cessation, but did not occur once the response had been acquired (Shors, 2001). Thus, the response to stress will be shaped by prior learning experience, but at the same time the ability to acquire new information will also be affected by the stressor. However, it remains to be shown that this is a direct effect on memory and not due to changes in other types of behaviour.

Arousal and attention

One non-mnemonic type of behaviour which is affected by stress is arousal. It is a *condition sine qua non* that subjects ought to be alerted to orient, process, and respond adequately to a stressful stimulus. An increase in arousal prepares the subject to change behavioural responses within parts of a second to alter between approach and avoidance. As already discussed, this processing of emotionally relevant stimuli is not only rapid, but can occur in a pre-attentive state, i.e., without conscious perception (Esteves et al., 1994).

Arousal is therefore important in face of threatening situations, as it alters the stimulus-processing capacities of the individuum, it lowers the sensory threshold, and allows more rapid processing of sensory stimuli (Lynn, 1966). Various neurotransmitter systems have been implicated in the modulation of arousal. The noradrenergic projections of the locus coeruleus to higher brain areas, including various cortical areas, the hippocampus, and the basal forebrain, seem to play an important role in the mediation of arousal (Berridge, 2004). This is of particular interest as exactly these projections are also activated under stressful conditions, suggesting that this could be one mechanism via which the organism ensures proper information processing to relevant stimuli.

However, too much arousal can also be detrimental: according to the arousal hypothesis of Yerkes and Dodson (1908), arousal will affect performance in a bell-shaped manner, i.e., medium levels of stress will lead to increased arousal and improve stress reactivity, while performance is lower under conditions of low stress and low arousal. This can explain the improved stimulus-processing capacity under conditions of low to medium stress level. However, when arousal is increasing further, e.g., due to high stress levels, performance is likely to drop again. This could be one mechanism explaining the sometimes controversial findings of cognition enhancing or impairing effects of stress exposure.

Let me illustrate this point with a few examples derived from the animal and the human literature: we have already seen in one of the studies discussed above that an anticipatory stress response can be manipulated externally by giving arousing feedback to humans facing public speech (Rohrmann et al., 1999; see above). Thus, the level of arousal can determine the degree of stress exposure. The same holds true for non-human animals. In the experiment by Shors (2001; see above), for example, an interaction between arousal and stress could also perfectly well explain the results obtained, as stress exposure would have led to an increase in arousal, which could have led to improved learning, i.e., the effects of the stressor on conditioning could have been indirect, rather than on learning per se. This then would suggest that arousal influences the formation of new memories.

Another example clearly demonstrating that arousal processes affect not only the sensory–perceptual systems, but also the cognitive systems involved (Pribram and McGinnes, 1975) derives from a classical study by Schachter and Singer (1962). In this study, human volunteers were pharmacologically aroused with adrenaline and placed in a stressful situation, which led to both subjective and objective evidence of emotional states. In contrast, pharmacological treatment alone did not alter emotional states. Likewise, stress-induced emotional changes were less pronounced when the subjects were not pharmacologically aroused. Thus, the combination between an increase in arousal and the exposure to stress had synergistic effects, leading to a higher appraisal of a situation as being stressful.

According to Easterbrook (1959), a stress-induced increase in arousal serves the purpose to narrow *attention* to central details of a situation, leading to better recall of details central to the event. This is in agreement with Mandler (1984), showing that memory for the emotionally arousing event is enhanced (presumably up to a certain stress level and hence a certain level of arousal, as the descending part of the bell-shaped curve will be reached at a certain point).

Narrowing of attention towards relevant stimuli seems to be one of the prerequisites for enhanced mnemonic processing under stressful conditions and attention must be regarded as another preparatory response altered in such situation: an increase in attention leads to increased scanning of the environment for relevant stimuli and/or display of risk assessment behaviours, which would enhance acquisition of information associated with the stressor. Moreover, it is under these conditions of increased attention to a stressor that there may also be an increasing retrieval of memories associated with the stressor. Under conditions of sensitization, attention/orientation towards the stressor is enhanced, while habituation will lead into reduced attention/orientation to the stressor, as mentioned above.

As with arousal and the appraisal of a signal as being stressful (the subject becoming aroused by a stressor and the level of arousal in turn will determine the appraisal of the stressor), there are intricate relationships between stress and attention. In humans, an increase of attention to bodily sensations has been shown to exacerbate anxiety and physiological reactivity (Epstein et al., 1978; Pennebaker, 1982). However, not only will an appropriate level of attention be required for detection of a stressor, but the stress systems also influence attentional mechanisms. Hypocortisolaemic patients, for example, have been reported to be impaired in their ability to recognize and integrate sensory information, although they can detect sensory stimuli at lower levels than control subjects (Henkin, 1970). Wolkowitz et al. (1990) suggested that the glucocorticoids affect selective attention by raising arousal to sub-optimal level. Thus, the hormones of the HPA axis seem to play an important role in the direct or indirect modulation of attention.

Stress predictability, control, and helplessness

From the preceding paragraphs it is evident that appraisal and anticipation are important psychological processes involved in the processing of stressful information. In procedures such as conditioned suppression or conflict procedures, CER, or fear-potentiated startle, the subject is able to make a prediction about the likelihood for the occurrence of a stressor (the unconditioned stimulus) by presentation of a warning stimulus (the conditioned stimulus). The behavioural response to the warning stimulus is an anticipatory response to prevent occurrence of the primary stressor. This is called behavioural control. On the other hand, the subject can just have the sense that it is or is not in control of an aversive situation, which is called perceived control. Under this condition, the subject has the perception that the stressor is potentially controllable by making the correct response. Behavioural and perceived control are two different phenomena. For example, a subject may perceive itself out of control when, in fact, objective behavioural control is given. As such, perceived control is the appraisal the subject gives to its ability to control the situation.

While behavioural control should allow a rat (or in fact any other subject) to perform appropriate in tasks such as CER (we do not really know about an animal's degree of perceived control), lack of control should have opposite effects. Therefore, it can be expected that lack of control has anxiogenic-like effects and, for example, potentiates fear responses. Exposure of rats to a yoked control design of shock exposure, where half of the animals have control over the shock (i.e., they are able to make a behavioural response that terminates the shock) and the other half has no control (i.e., animals receive identical shocks, but are unable to control it, irrespective of responses made), leads to increased freezing in the rats that previously had no control when they are tested for conditioned fear 24 h later (Maier, 1990). Along similar lines, it has been shown that the level of plasma corticosterone and the degree of fear conditioning are higher after exposure to an inescapable stressor than following exposure to an escapable and, hence, better controllable, but otherwise comparable, stressor (Mineka et al., 1984). This goes as far that the degree of control over a stressor has been shown to be inversely related to levels of plasma corticosterone (Sandi et al., 1992), and it has been suggested that in particular in those situations where the future demands are uncertain, there is a strong activation of the HPA axis (Toates, 1995).

Other behavioural and physiological responses are also affected by controllability of the stressful situation. Seligman (1968), for example, showed that shock produced more extensive gastric ulcerations and suppression of food-rewarded lever presses in rats if the shock was unpredictable rather than signalled. Moreover, lack of stress control interfered with subsequent learning (Seligman, 1975). From this one would assume that the ability to control a situation is the preferred state over an uncontrollable situation. Indeed, rats will choose signalled and predictable over unsignalled and unpredictable shocks (Lockard, 1963; Badia et al., 1979). One reason could be that the perception of a stimulus as a stressor will be stronger when it is uncontrollable or unpredictable (Weiss, 1970). Thus, the ability of a subject to predict the occurrence of a stressor and hence its ability to control its own environment has a major impact on the appraisal of the stressor by the subject.

How does the degree to which a stressful situation can be controlled by an individual affect cognitive function? In the previous paragraphs it was shown that the learning processes are influenced by the degree to which a stressor can be controlled: if subjects are pre-exposed to unpredictable shock, they show an impairment in learning a subsequent escape response, a phenomenon called *learned helplessness* (Seligman, 1975). Helplessness refers to the inability of a subject to control a (stressful) situation even when it has the opportunity to do so. A number of concepts have been proposed to explain this phenomenon, ranging from general emotional exhaustion leading to inhibition of active learning, over adoption of inappropriate, passive, response habits, to more cognitive explanations, suggesting that the subjects learn to be helpless and give up trying. The latter concept again relates to the idea that the attribution given to a stimulus will interfere with subsequent performance. This could then be viewed as a special case of classically conditioned learned irrelevance (Gray and McNaughton, 2000).

The degree to which the aversive situation can be controlled by the subject will also determine whether the subject regards its coping skills as sufficient. This serves as a model to explain why some subjects may perceive a situation as stressful, while others may not. Being in control of a stressful situation entails that a subject can learn that the consequence of its action is termination of an aversive situation, and this knowledge should reduce an individual's stress response. For example, it has been shown that monkeys being able to terminate loud noise have lower plasma cortisol level than yoked monkeys exposed to the same noise, but having no control (Hanson et al., 1976). Comparable effects of control on stress coping have been reported in rats (Overmier and Seligman, 1967; Weiss, 1968; Rosellini, 1978).

Thus, exposure to uncontrollable stress not only results in higher glucocorticoid plasma levels, but also in behavioural alterations, such as impaired learning of a controllable avoidance response (Seligman, 1975).

Gray's behavioural inhibition system

Having discussed some of the cognitive processes which play a role in the modulation of stress responsivity, the next question would be whether one can fit this into a contemporary psychological theory. Gray (1976) proposed a behavioural inhibition system, which controls ongoing behaviour whenever the subject wants to achieve a goal which requires to move towards a source of danger, i.e., when there are conflicting aims. Under these conditions, there is an increase in vigilance (the subject orients towards threat, i.e., attention is directed towards the stressor) and in arousal, which can be induced by the presentation of a stressor, such as by secondary (conditioned) punishing or painful stimuli, and by secondary frustrative stimuli, such as failure or loss of reward, but also by innate aversive stimuli, such as novelty or uncertainty.

Important to this concept is that the subject has to experience conflict to activate the behavioural inhibition system, i.e., it is not sufficient to just present an aversive stimulus to the subject, but the subject has to know of a variety of stimuli and contingencies between them as to engender a conflict between mutually incompatible appetitive and aversive goals.

The behavioural inhibition system has been proposed to oppose, to a certain degree, the so-called fight–flight system, a defence system which serves the purpose to remove the subject from a source of danger (Gray and McNaughton, 2000). Thus, activation of the fight–flight system leads to avoidance of anxiety-arousing stimuli and their consequences and, hence, should ultimately result into a reduction of exposure to a stressful stimulus, i.e., activation of the fight–flight system should direct attention away from the stressor. In contrast, activation of the behavioural inhibition system could lead to even enhanced stress exposure.

Gray and McNaughton (2000) propose a third system, which deals with approach in appetitive conditions, the behavioural approach system. However, for the purpose of the present review, the focus will be on the behavioural inhibition and the fight–flight systems. According to Gray and McNaughton, the distinction between these two systems is not only behaviourally, but also neuroanatomically relevant, as will be discussed below.

Brain regions involved in the processing of the stress response

A hierarchical defence system has been proposed by Graeff (1994), where progressively more brain areas will be involved with progressively more complex aspects of defence against stressful stimuli (also discussed in Gray and McNaughton, 2000). The hierarchical defence system proposes that the periaqueductal grey at the lowest level coordinates undirected escape, resulting in freezing, flight or fight. The medial hypothalamus at the next level coordinates directed escape. The amygdala coordinates simple active avoidance, and the cingulate cortex at the highest level coordinates more complex active avoidance. This has been extended more recently by Gray and McNaughton (2000), suggesting that the periaqueductal grey, the medial hypothalamus, the amygdala and the anterior cingulate cortex are involved in defensive avoidance, while the septo-hippocampal system and the posterior cingulate are more relevant for defensive approach, including risk assessment and behavioural inhibition (Fig. 1).

Gray and McNaughton (2000) further propose that the posterior cingulate cortex plays a role in the generation of innate anxiety plans, while the prefrontal cortex is involved in working memory processes and motor programming functions required for acquired anxiety plans.

Under certain conditions, automated responding will be advantageous over more slow cognitive processing of stimuli. Noradrenaline release increases in the prefrontal cortex under stressful conditions (Feenstra, 2000). This increase in noradrenergic activity in the prefrontal cortex may lead to activation of postsynaptic α_1 adrenoceptors, which causes a decline in prefrontal functioning. An inhibition of prefrontal cortex functioning under stressful conditions in turn may lead to more rapid habitual subcortical modes of responding, adding to survival (Arnsten, 1998). Indeed, stress can impair working memory and this effect can be attenuated by infusion of an α_1 adrenoceptor antagonist directly into the prefrontal cortex (Birnbaum et al., 1999), suggesting that it is the stress-induced release of noradrenaline at this site which favours rapid automatic processing over slower declarative processes. Moreover, changes in prefrontal dopaminergic activity have been demonstrated following exposure to stress. Activation of the prefrontal dopaminergic system has been shown to occur preferentially after exposure to mild stressors and has been suggested to be of relevance for experiencing anxiety states, possibly in an attempt to cope with the stressful situation (Deutch and Roth, 1990).

Central to the behavioural defence theory by Gray and McNaughton (2000) is the septohippocampal

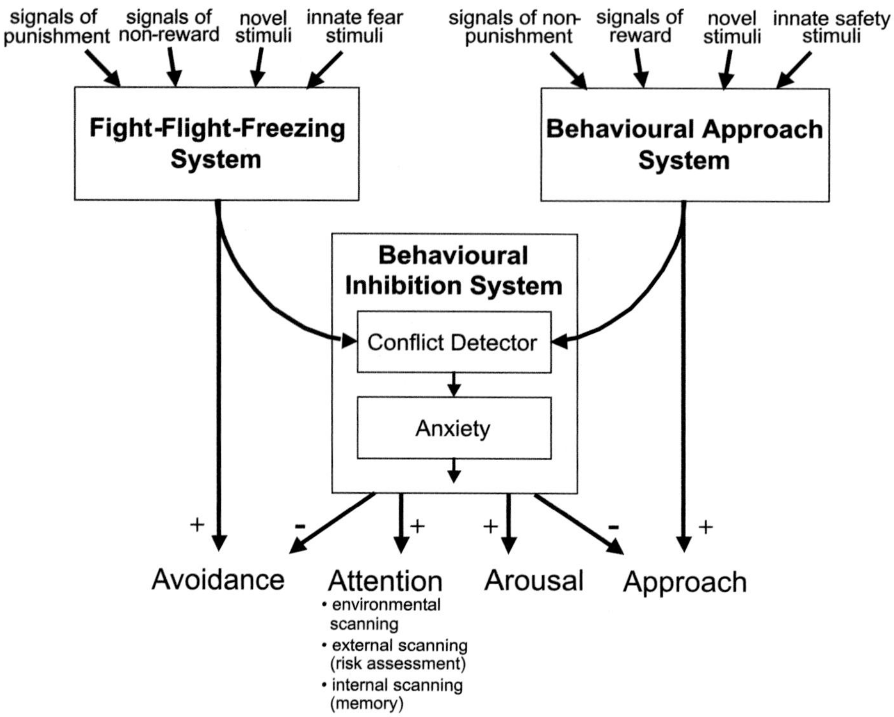

Fig. 1. The behavioural inhibition system (modified after Gray and McNaughton, 2000): the behavioural inhibition system can be activated by simultaneous activation of the fight–flight–freezing system and the behavioural approach system, leading to conflict. The consequence will be an increase in arousal and attention (scanning of the environment, risk assessment, and retrieval of relevant memories) and concurrent inhibition of simple approach and simple avoidance behaviour.

system, which according to these authors forms an essential part of the behavioural defence system. In this context, the septohippocampal system should act to detect conflict between concurrently available goals and to resolve conflict between these goals by increasing bias towards affectively negative information. Under normal conditions, the system has been suggested to act as a comparator and to play a role in interrupting ongoing behaviour to allow information gathering, which in the animal is seen as risk assessment. As mentioned before, Takahashi (1996) suggested that glucocorticoids may influence the development of the freezing response by actions at the septohippocampal level. Given that, according to Gray and McNaughton, the septohippocampal system is central to behavioural inhibition, the ideas of Takahashi may also have implications for the manifestation of the behavioural inhibition system of Gray and McNaughton: it is tempting to speculate that glucocorticoid action during development may affect the behavioural inhibition system over and above just freezing by influencing the development of the septohippocampal system, thereby modulating the individual's levels of stress-induced arousal and attention to threat.

Both cholinergic and monoaminergic mechanisms have been suggested to play important roles in the modulation of stressor-induced increases of arousal and attention at hippocampal level: various stressors have been reported to increase hippocampal acetylcholine release (Dudar et al., 1979; Gilad et al., 1985; Imperato et al., 1991; Acquas et al., 1996; Day et al., 1998). The hippocampal cholinergic system has been suggested to be involved in the mechanisms underlying the arousal that is associated with fear and anxiety-provoking stimuli (Hess and Blozovski, 1987; Smythe et al., 1992) and one of the possible roles of the cholinergic system could be to ensure that the animal is appropriately responsive to its environment, being able to monitor and amend its behaviour in an appropriate manner when exposed to a fearful or anxiety-provoking stimulus (Bhatnagar et al., 1997).

Likewise, stress exposure increases noradrenergic and serotonergic activity at the hippocampal level (Abercombie et al., 1988; Vahabzadeh and Fillenz, 1994; Linthorst et al., 2000). These stress-induced increases in hippocampal monoamine release have been suggested to facilitate information processing by hippocampal circuits and hence bias towards affectively negative information (Gray and McNaughton, 2000).

Another brain area closely involved in the processing of stress-related information and in the modulation of an individual's responses to stress is the amygdala. Within the framework proposed by Gray and McNaughton (2000), the projection from the septohippocampal system to the amygdala is viewed as being crucial for generating arousal and the autonomic components of anxiety, which is part of the stress response.

For example, the amygdala plays an important role in the conditioning of emotional responses. Amygdaloid activation has been demonstrated using fMRI during fear stimuli and fear conditioning (LaBar et al., 1998; Morris et al., 1999). Information from the thalamus is relayed in the lateral nucleus of the amygdala to the central nucleus of the amygdala (CeA). There are both direct and indirect (via the basolateral amygdaloid nucleus) projections from the lateral nucleus of the amygdala to the CeA. In that function, the lateral nucleus has been suggested to represent the gateway into the amygdala (Le Doux, 1996). The basolateral amygdala in turn seems to control the activity of the central nucleus of the amygdala (CeA), which then affects hypothalamic and brainstem function (Davis, 1992a; Le Doux, 2000), leading to increases of adrenaline and plasma corticosterone (Roozendaal et al., 1991, 1997). The CeA also affects the periaqueductal grey and the pontine reticular nucleus, thereby modulating freezing and fear-potentiated startle, respectively. More specifically, it has been suggested that the CeA modulates stimulus-specific fear (related to specific tones, visual signals, etc.), while more complex information such as, for example, provided by exposure to a threatening environment for several minutes, may activate another structure, the bed nucleus of the stria terminalis (Davis, 1998). The basolateral amygdaloid nucleus, on the other hand, has been viewed as being responsible for emotional classical conditioning and seems to be critical for the memory-enhancing effects of stress hormones such as adrenaline and glucocorticoids (McGaugh et al., 2000; Prickaerts and Steckler, 2004).

Cardinal et al. (2002) suggested that the basolateral amygdala is of particular relevance for the re-evaluation of the motivational significance of a stimulus. These authors have argued that the basolateral amygdala will be activated under conditions when a subject has to assess the affective value of a stressor predicted by a conditioned stimulus. The basolateral nucleus has also been suggested to provide information about affect to the frontal and cingulate cortices, and to the striatum, and in return to receive projections providing information about current plans from cortical areas (Gray and McNaughton, 2000).

This learned information of the affective value of a stimulus from the basolateral nucleus to the striatum seems to be relayed to the core of the nucleus accumbens, where its effects may be potentiated by dopamine (Cardinal et al., 2002). Conversely, unconditioned aversive stimuli have been reported to increase dopamine in the shell of the accumbens but not in the core (Deutch and Cameron, 1992), opening the possibility that dopamine plays a similar role in the accumbens shell for unconditioned stimuli as it does in the core for conditioned stimuli.

More specifically, the accumbens shell has been suggested to be a high-level control system which allows the subject to switch between behavioural patterns based on primary motivational states, such as to override feeding and instead to engage in a more appropriate behaviour, such as fleeing, freezing, or fighting, in the face of a stressor such as a predator (Kelley, 1999). The CRF and the noradrenergic projections innervating the shell of the nucleus accumbens have been proposed to signal this stress-related information to the accumbens (Kelley, 1999).

But how does this information from higher brain areas affect HPA axis activity? Earlier on, it has been mentioned that the hypothalamic PVN serves as relay point between the higher brain areas and the HPA axis (the limbic input into the parvocellular PVN is extensively reviewed by Herman et al., 2004). It has been proposed that the indirect projection from the hippocampus to the PVN (from the subiculum via the BNST and GABAergic relay neurons in the hypothalamus) communicates learned information, including spatial information to the PVN and hence to the HPA axis, while the amygdala relays input to the PVN on perceived danger (Herman et al., 2004). Furthermore, and as pointed out by these authors, there are interconnections between amygdala and hippocampus, which may subserve complementary roles in assessing the relative safety of an environment, which can then be integrated into an appropriate activity state of the HPA axis.

Conclusion

We have seen that the effects of aversive stimuli do not only depend on the nature of the stimulus, but also on the circumstances in which the stimulus is presented, the individual perception, the past experience, and cognitive abilities of the individuum. These processes are mediated by a network of brain regions comprising of the hippocampal formation and amygdala, prefrontal and cingulate areas, basal ganglia, hypothalamic areas, periaqueductal grey, and other subcortical structures. An impaired capacity to discriminate between relevant (threatening) and irrelevant (non-threatening) information, be it genetically or environmentally determined, will increase reactivity to stress.

This cognitive model of stress helps to explain the susceptibility of individuals suffering from certain psychiatric disorders, which will be discussed in more detail in other parts of this book: impaired ability to discriminate threatening and non-threatening stimuli leads to increased reactivity to stress, which in turn can lead to more rapid symptom development of, for example, depression or psychosis. For example, it has been suggested that depressed patients have a diminished behavioural approach system in conjunction with a high activation of the behavioural inhibition system, a pattern which remains stable over time and clinical state (Meyer et al., 2001; Kasch et al., 2002). This in turn leads to lower responsiveness to reward and higher responsiveness to punishment, i.e., stressful situations. Consequently, therapeutic approaches which improve stress coping, be it pharmacologically or by other means, should be beneficial in these conditions.

References

Abercombie, E.D., Keller, R.W., Jr. and Zigmand, M.J. (1988) Characterization of hippocampal noradrenaline release as measured by microdialysis perfusion: pharmacological and behavioral studies. Neuroscience, 27: 897–904.

Acquas, E., Wilson, C. and Fibiger, H.C. (1996) Conditioned and unconditioned stimuli increase frontal cortical and hippocampal acetylcholine release: effects of novelty, habituation, and fear. J. Neurosci., 16: 3089–3096.

Adler, R. (1976) Conditioned adrenocortical steroid elevations in the rat. J. Comp. Physiol. Psychol., 90: 1156–1163.

Annau, Z. and Kamin, L.J. (1961) The CER as a function of the intensity of the U.S. J. Comp. Physiol. Psychol., 54: 428–432.

Arnsten, A.F.T. (1998) Catecholamine modulation of prefrontal cortical cognitive function. Trends Cognit. Sci., 2: 436–447.

Arthur, A.Z. (1987) Stress as a state of anticipatory vigilance. Percept. Motor Skills, 64: 75–85.

Badia, P., Harsh, J. and Abbott, B. (1979) Choosing between predictable and unpredictable shock conditions: data and theory. Psychol. Bull., 86: 1107–1131.

Berridge, C.W. (2005) The locus coeruleus-noradrenergic system as a modulator of arousal state and state-dependent behavioural processes. In: Steckler, T., Kalin, N. and Reul, J.M.H.M. (Eds.), Handbook of Stress and the Brain, Part 1, Elsevier, Amsterdam, pp. 437–464.

Bhatnagar, S., Costall, B. and Smythe, J.W. (1997) Hippocampal cholinergic blockade enhances hypothalamic–pituitary–adrenal responses to stress. Brain Res., 766: 244–248.

Birnbaum, S., Gobeske, K.T., Auerbach, J., Taylor, J.R. and Arnsten, A.F. (1999) A role for norepinephrine in stress-induced cognitive deficits: alpha-1-adrenoceptor mediation in the prefrontal cortex. Biol. Psychiatry, 46: 1266–1274.

Blanchard, D.C. and Blanchard, R.J. (1990) An etho-experimental analysis of defense, fear and anxiety. In: McNaughton, N. and Andrews, G. (Eds.), Anxiety. Otago University Press, Dunedin, pp. 124–133.

Bruijnzeel, A.W., Stam, R., Croiset, G. and Wiegant, V.M. (2001) Long-term sensitization of cardiovascular stress responses after a single stressful experience. Physiol. Behav., 73: 81–86.

Brujnzeel, A.W., Stam, R., Compaan, J.C., Croiset, G., Akkermans, L.M.A., Olivier, B. and Wiegant, V.M. (1999) Long-term sensitization of Fos-responsivity in the rat central nervous system after a single stressful experience. Brain Res., 819: 15–22.

Cardinal, R.N., Parkinson, J.A., Hall, J. and Everitt, B.J. (2002) Emotion and motivation: the role of the amygdala, ventral striatum, and prefrontal cortex. Neurosci. Biobehav. Rev., 26: 321–352.

Coe, C.L., Stanton, M.E. and Levine, S. (1983) Adrenal responses to reinforcement and extinction: role of expectancy versus instrumental responding. Behav. Neurosci., 97: 654–657.

Coover, B.D., Sutton, B.R. and Heybach, J.P. (1977) Conditioning decreases plasma corticosterone levels in rats by pairing stimuli with daily feedings. J. Comp. Physiol. Psychol., 91: 716–726.

Dallman, M.F. (1993) Stress update: adaptation of the hypothalamic–pituitary–adrenal axis to chronic stress. Trends Endocrinol. Metab., 4: 62069.

Davis, M. (1992a) The role of the amygdala in conditioned fear. In: Aggleton, J.P. (Ed.), The Amygdala: Neurobiological Aspects of Emotion, Memory, and Mental Dysfunction. Wiley, New York, pp. 255–306.

Davis, M. (1992b) The role of the amygdala in fear-potentiated startle: implications for animal models of anxiety. Trends Pharmacol. Sci., 13: 35–41.

Davis, M. (1998) Are different parts of the extended amygdala involved in fear versus anxiety? Biol. Psychiatry, 44: 1239–1247.

Davis, M., Falls, W.A., Champeau, S. and Kim, M. (1993) Fear-potentiated startle: a neural and pharmacological analysis. Behav. Brain Res., 58: 175–198.

Day, J.C., Koehl, M., Deroche, V., Le Moal, M. and Maccari, S. (1998) Prenatal stress enhances stress- and corticotropin-releasing factor-induced stimulation of hippocampal acetylcholine release. J. Neurochem., 18: 1886–1892.

Dayas, C.V., Buller, K.M., Crane, J.W. and Day, T.A. (2001) Stressor categorization: acute physical and psychological stressors elicit distinctive recruitment patterns in the amygdala and medullary noradrenergic cell groups. Eur. J. Neurosci., 14: 1143–1152.

De Boer, S.F., De Beun, R., Slangen, J.L. and Van der Gugten, J. (1990) Dynamics of plasma catecholamine and corticosterone concentrations during reinforced and extinguished operant behaviour in rats. Physiol. Behav., 47: 691–698.

De Kloet, E.R., Vreugdenhil, E., Oitzl, M.S. and Joels, M. (1998) Brain corticosteroid receptor balance in health and disease. Endocrine Rev., 66: 183–201.

Deutch, A.Y. and Roth, R.H. (1990) The determinants of stress-induced activation of the prefrontal cortical dopamine system. Prog. Brain Res., 85: 367–403.

Deutch, A.Y. and Cameron, D.S. (1992) Pharmacological characterization of dopamine systems in the nucleus accumbens core and shell. Neuroscience, 46: 49–56.

Dudar, J.D., Whishaw, I.Q. and Szerb, J.C. (1979) The role of the septal nuclei in the release of acetylcholine from the hippocampus of freely moving rats during sensory stimulation and running. Neuropharmacology, 18: 673–678.

Easterbrook, J.A. (1959) The effect of emotion on cue utilization and the organization of behavior. Pychol. Rev., 66: 183–201.

Ekman, P., Levenson, R.W. and Friesen, W.V. (1983) Autonomic nervous system activity distinguishes emotions. Science, 221: 1208–1210.

Epstein, S., Rosenthal, S. and Szpiler, J. (1978) The influence of attention upon arousal, habituation, and reactivity to a noxious stimulus. J. Res. Pers., 12: 30–40.

Esteves, F., Parra, C., Dimberg, U. and Ohman, A. (1994) Nonconscious associative learning: Pavlovian conditioning of skin conductance responses to masked fear-relevant facial stimuli. Psychophysiology, 31: 375–385.

Feenstra, M.G. (2000) Dopamine and noradrenaline release in the prefrontal cortex in relation to unconditioned and conditioned stress and reward. Prog. Brain Res., 126: 133–163.

File, S.E. (1980) The use of social interaction as a method for detecting anxiolytic activity of chlordiazepoxide-like drugs. J. Neurosci. Meth., 2: 219–238.

Freud, S. (1926) Hemmung, Symptom und Angst. Internationaler Pychoanalytischer Verlag, Leipzig.

Gilad, G.M. and Shiller, I. (1989) Differences in open-field behavior and in learning tasks between two rat strains differing in their reactivity to stressors. Behav. Brain Res., 32: 89–93.

Gilad, G.M., Mahon, B.D., Finkelstein, Y., Koffler, B. and Gilad, V.H. (1985) Stress induced activation of the hippocampal cholinergic system and the pituitary adrenocortical axis. Brain Res., 347: 404–408.

Graeff, F.G. (1994) Neuroanatomy and neurotransmitter regulation of defensive behaviors and related emotions in mammals. Brazil. J. Med. Biol. Res., 27: 67–70.

Gray, J.A. (1976) The behavioural inhibition system: a possible substrate for anxiety. In: Feldman, M.P. and Broadhurst, A.M. (Eds.), Theoretical and Experimental Bases of Behaviour Modification. Wiley, London, pp. 3–41.

Gray, J.A. and McNaughton, N. (2000) The Neuropsychology of Anxiety, 2nd ed. Oxford University Press, Oxford.

Hall, F.S., Huang, S., Fong, G.W., Sundstrom, J.M. and Pert, A. (2000) Differential basis of strain and rearing effects on open-field behavior in Fawn Hooded and Wistar rats. Physiol. Behav., 71: 525–532.

Hanson, J.D., Larson, M.E. and Snowdon, C.T. (1976) The effects of control over high intensity noise on plasma cortisol levels in rhesus monkeys. Behav. Biol., 16: 333–340.

Henkin, R.I. (1970) The effects of corticosteroids and ACTH on sensory systems. Prog. Brain Res., 32: 270–294.

Herman, J.P. and Cullinan, W.E. (1997) Neurocircuitry of stress: central control of the hypothalamo–pituitary–adrenocortical axis. Trends Neurosci., 20: 78–84.

Herman, J.P., Prewitt, C.M. and Cullinan, W.E. (1996) Neuronal circuit regulation of the hypothalamo–pituitary–adrenocortical stress axis. Crit. Rev. Neurobiol., 10: 371–394.

Herman, J.P., Mueller, N.K., Figueiredo, H. and Cullinan, W.E. (2005) Neurocircuit regulation of the hypothalamo-pituitary-adrenocortical stress response – an overview. In: Steckler, T., Kalin, N. and Reul, J.M.H.M. (Eds.), Handbook of Stress and the Brain, Part 1, Elsevier, Amsterdam, pp. 405–418.

Hess, C. and Blozovski, D. (1987) Hippocampal muscarinic cholinergic mediation of spontaneous alternation and fear in the developing rat. Behav. Brain Res., 24: 203–214.

Imperato, A., Puglisi-Allegra, S., Casolini, P. and Angelucci, L. (1991) Changes in brain dopamine and acetylcholine release during and following stress are independent of pituitary–adrenocortical axis. Brain Res., 538: 111–117.

James, W. (1890) The Principles of Psychology. Holt, New York.

Kasch, K.L., Rottenberg, J., Arnow, B.A. and Gotlib, I.H. (2002) Behavioral activation and inhibition systems and the severity and course of depression. J. Abnorm. Psychol., 111: 589–597.

Kelley, A.E. (1999) Neural integrative activities of nucleus accumbens subregions in relation to learning and motivation. Psychobiology, 27: 198–213.

Koene, P. and Vossen, J.M. (1991) Strain differences in rats with respect to speed of conflict resolution. Behav. Genet., 21: 21–33.

Kollack-Walker, S., Day, H.E.W. and Akil, H. (2000) Central stress neurocircuits. In: Fink, G. (Ed. in chief), Encyclopedia of Stress, Vol. 1. Academic Press, San Diego, pp. 414–423.

Koolhaas, J.M., Hermann, P.M., Kempermann, C., Bohus, B., Van den Hoofddakker, R.H. and Beersma, D.G.M. (1990) Single social defeat in male rats induces a gradual but long-lasting behavioural change: a model of depression? Neurosci. Res. Comm., 7: 35–41.

Korte, S.M. (2001) Corticosteroids in relation to fear, anxiety and psychopathology. Neurosci. Biobehav. Rev., 25: 117–142.

Kovacs, K.J. and Makara, G.B. (1990) Partial deafferentiation of the hypothalamic paraventricular nucleus: effect on the stress or adrenalectomy-induced ACTH secretion. Neuroendocrinol. Lett., 12: 383–389.

LaBar, K.S., Gatenby, J.C., Gore, J.C., LeDoux, J.E. and Phelps, E.A. (1998) Human amygdala activation during conditioned fear acquisition and extinction: a mixed trial fMRI study. Neuron, 20: 937–945.

Lazarus, R.S. (1966) Psychological Stress and the Coping Process. McGraw-Hill, New York.

Le Doux, J.E. (1996) The Emotional Brain. Simon & Schuster, New York.

Le Doux, J.E. (2000) Emotion circuits in the brain. Ann. Rev. Neurosci., 23: 155–184.

Lemaire, V., Aurousseau, C., Le Moal, M. and Abrous, D.N. (1999) Behavioural trait of reactivity to novelty is related to hippocampal neurogenesis. Eur. J. Neurosci., 11: 4006–4014.

Linthorst, A.C., Flachskamm, C., Barden, N., Holsboer, F. and Reul, J.M.H.M. (2000) Glucocorticoid receptor impairment alters CNS responses to a psychological stressor: an in vivo microdialysis study in transgenic mice. Eur. J. Neurosci., 12: 283–291.

Liu, D., Diorio, J., Tannenbaum, B., Caldji, C., Francis, D., Freedman, A., Sharma, S., Pearson, D., Plotsky, P.M. and Meaney, M.J. (1997) Maternal care, hippocampal glucocorticoid receptors, and hypothalamic–pituitary–adrenal responses to stress. Science, 277: 1659–1662.

Lockart, J.A. (1963) Choice of a warning signal or no warning signal in an unavoidable shock situation. J. Comp. Physiol. Psychol., 56: 526–530.

Lynn, R. (1966) Attention, Arousal and the Orientation Reaction. Pergamon, Oxford.

Mackintosh, N.J. (1983) Conditioning and Associative Learning. Clarenndon, Oxford.

Maier, S.F. (1990) The role of fear in mediating the shuttle escape learning deficit produced by inescapable shock. JEP ABP, 16: 137–150.

Mandler, G. (1984) Mind and Body, the Psychology of Emotion and Stress. W.W. Norton & Company, New York.

Mason, J.W. (1968) A review of psychoendocrine research on the pituitary–adrenal cortical system. Psychosom. Med., 30: 576–607.

McGaugh, J.L., Ferry, B., Vazdarjanova, A. and Roozendaal, B. (2000) Amygdala: role in modulation of memory storage. In: Aggleton, J.P. (Ed.), The Amygdala: A Functional Analysis, 2nd ed. Oxford University Press, New York, pp. 391–423.

Meyer, B., Johnson, S.L. and Winters, R. (2001) Responsiveness to threat and incentive in bipolar disorder: relations of the BIS/BAS scales with symptoms. J. Psychopathol. Behav. Assess., 23: 133–143.

Mineka, S., Cook, M. and Miller, S. (1984) Fear conditioned with escapable and inescapable shock: the effects of a feedback stimulus. J. Exp. Psychol., 10: 307–323.

Morris, J.S., Ohman, A. and Dolan, R.J. (1999) A subcortical pathway to the right amygdala mediating "unseen" fear. Proc. Natl. Acad. Sci. USA, 96: 1680–1685.

Mowrer, O.H. (1960) Learning Theory and Behavior. Wiley, New York.

Newport, D.J. and Nemeroff, C.B. (2002) Stress. In: (Ed. in chief), Encyclopedia of the Human Brain, Vol. 4. Elsevier, pp. 449–462.

Ohman, A., Flykt, A. and Esteves, F. (2001) Emotion drives attention: detecting the snake in the grass. J. Exp. Psychol., 130: 466–478.

Ottenweller, J.E. (2000) Animal models (nonprimate) for human stress. In: Fink, G. (Ed. in chief), Encyclopedia of Stress, Vol. 1. Academic Press, San Diego, pp. 200–205.

Overmier, J.B. and Seligman, M.E.P. (1967) Effects of inescapable shock upon subsequent escape and avoidance responding. J. Comp. Physiol. Psychol., 63: 28–33.

Pacac, K., Palkovits, M., Kopin, I.J. and Goldstein, D.S. (1995) Stress-induced norepinephrine release in the hypothalamic paraventricular nucleus and pituitary–adrenocortical and sympathoadrenal activity: in vivo microdialysis studies. Front. Neuroendocrinol., 16: 89–150.

Pare, W.P. (1964) The effects of chronic environmental stress and stomach ulceration, adrenal function and consummatory behavior in the rat. J. Psychol., 57: 143–151.

Pennebaker, J.W. (1982) The Psychology of Physical Symptoms. Springer, New York.

Phillips, K. (1989) Psychophysiological consequences of behavioural choice in aversive situations. In: Steptoe, A. and Appels, A. (Eds.), Stress, Personal Control and Health. John Wiley, Chichester, pp. 239–256.

Piazza, P.V., Maccari, S., Deminiere, J.M., Le Moal, M., Mormede, P. and Simon, H. (1991) Corticosterone levels determine individual vulnerability to amphetamine self-administration. Proc. Natl. Acad. Sci. USA, 88: 2088–2092.

Plotsky, P.M. (1991) Pathways to the secretion of adrenocorticotropin: a view from the portal. J. Neuroendocrinol., 3: 1–9.

Plotsky, P.M., Cunningham, E.T. and Widmaier, E.P. (1989) Catecholaminergic modulation of corticotropin-releasing factor and adrenocorticotropin secretion. Endocr. Rev., 10: 437–458.

Plutchik, R. (1962) The Emotions: Facts, Theories and a New Model. Random House, New York.

Pribram, K.H. and McGuinnes, D. (1975) Arousal, activation and effort in the control of attention. Psychol. Rev., 182: 116–145.

Prickaerts, J. and Steckler, T. (2005) Effects of glucocorticoids on emotion and cognitive processes in animals. In: Steckler, T., Kalin, N. and Reul, J.M.H.M. (Eds.), Handbook of Stress and the Brain, Part 1, Elsevier, Amsterdam, pp. 359–386.

Przewlocka, B., Sumova, A. and Lason, W. (1990) The influence of conditioned fear-induced stress on the opioid systems in the rat. Pharmacol. Biochem. Behav., 37: 661–666.

Reul, J.M.H.M. and De Kloet, E.R. (1985) Two receptor systems for corticosterone in the rat brain: microdistribution and differential occupation. Endocrinology, 117: 2505–2512.

Rohrmann, S., Henning, J. and Netter, P. (1999) Changing psychobiological stress reactions by manipulating cognitive processes. Int. J. Psychophysiol., 33: 149–161.

Rolls, E.T. (1999) The Brain and Emotion. Oxford University Press, Oxford.

Roozendaal, B., Koolhaas, J.M. and Bohus, B. (1991) Attenuated cardiovascular, neuroendocrine, and behavioral responses after a single footshock in central amygdaloid lesioned rats. Phyisol. Behav., 50: 777–781.

Roozendaal, B., Koolhaas, J.M. and Bohus, B. (1997) The role of the central amygdala in stress and adaptation. Acta Physiol. Scand. Suppl., 640: 51–54.

Rosellini, R.A. (1978) Inescapable shock interferes with the acquisition of an appetitive operant. Anim. Learn. Behav., 6: 155–159.

Sandi, C., Borrell, J. and Guaza, C. (1992) Behavioral, neuroendocrine, and immunological outcomes of escapable and inescapable shocks. Physiol. Behav., 51: 651–656.

Sawchenko, P.E. and Swanson, L.W. (1983) The organization of forebrain afferents to the paraventricular and supraoptic nuclei of the rat. J. Comp. Neurol., 218: 121–144.

Sawchenko, P.E., Brown, E.R., Chan, R.K.W., Ericsson, A., Li, H.Y., Ronald, B.L. and Kovacs, K.J. (1996) The paraventricular nucleus of hypothalamus and the functional neuroanatomy of visceromotor responses to stress. Prog. Brain Res., 107: 201–222.

Schachter, S. and Singer, J.E. (1962) Cognitive, social and physiological determinants of emotional state. Psychol. Rev., 69: 379–399.

Schmidt, E.D., Binnekade, R., Janszen, A.W.J.W. and Tilders, F.J.H. (1996) Short stressor induced long-lasting increases in vasopressin stores in hypothalamic corticotropin-releasing hormone (CRH) neurons in adult rats. J. Neuroendocrinol., 8: 703–712.

Seligman, M. (1968) Chronic fear produced by unpredictable electric shock. J. Comp. Physiol. Psychol., 66: 402–411.

Seligman, M. (1975) Helplessness. W.H. Freeman, San Francisco.

Shors, T.J. (2001) Acute stress rapidly and persistently enhances memory formation in the male rat. Neurobiol. Learn. Mem., 75: 10–29.

Smythe, J.W., Colom, L.V. and Bland, B.H. (1992) The extrinsic modulation of hippocampal theta depends on the coactivation of cholinergic and GABA-ergic medial septal inputs. Neurosci. Biobehav. Rev., 16: 289–308.

Takahashi, L.K. (1996) Glucocorticoids and the hippocampus. Developmental interactions facilitating the expression of behavioural inhibition. Mol. Neurobiol., 13: 213–226.

Toates, F. (1995) Stress: Conceptual and Biological Aspects. Wiley, Chichester.

Tsuda, A., Tanaka, M., Ida, Y., Tsujimaru, S., Ushijima, I. and Nagasaki, N. (1986) Effects of preshock experience on enhancement of rat brain noradrenaline turnover induced by psychological stress. Pharmacol. Biochem. Behav., 24: 115–119.

Vahabzadeh, A. and Fillens, M. (1994) Comparison of stress-induced changes in noradrenergic and serotonergic neurons in the rat hippocampus using microdialysis. Eur. J. Neurosci., 6, 1205–1212.

Van Dijken, H.H., Van der Heyden, J.A.M., Mos, J. and Tilders, F.J.H. (1992) Inescapable footshocks induce progressive and long-lasting behavioural changes in male rats. Physiol. Behav., 51: 787–794.

Van Dijken, H.H., De Groeij, D.C.E., Sutano, W., Mos, J., De Kloet, E.R. and Tilders, F.J.H. (1993) Short inescapable stress produces long-lasting changes in the brain–pituitary–adrenal axis of adult male rats. Neuroendocrinology, 58: 57–64.

Walker, S. (1987) Animal Learning: An Introduction. Routledge and Kegan Paul, London.

Watson, J.B. and Rayner, R. (1920) Conditioned emotional reactions. J. Exp. Psychol., 3: 1–14.

Weinstock, M. (1997) Does prenatal stress impair coping and regulation of hypothalamic–pituitary–adrenal axis? Neurosci. Biobehav. Rev., 21: 1–10.

Weiss, J.M. (1968) Effects of coping responses on stress. J. Comp. Physiol. Psychol., 65: 251–260.

Weiss, J.M. (1970) Somatic effects of predictable and unpredictable shock. Psychosom. Med., 32: 397–408.

Weiss, J.M., Baley, W.H., Pohorecky, L.A., Koreniowski, D. and Grillone, G. (1980) Stress-induced depression of motor activity correlates with regional changes in brain norepinephrine but not in dopamine. Neurochem. Res., 5: 9–22.

Wolkowitz, O.M., Reus, V.I., Weingartner, H., Thompson, K., Breier, A., Doran, A., Rubinow, D. and Pickar, D. (1990) Cognitive effects of corticosteroids. Am. J. Psychiatry, 147: 1297–1303.

Yamamotova, A., Starec, M., Holecek, V., Racek, J., Trefil, L., Raskova, H. and Rokyta, R. (2000) Anticipation of acute stress in isoprenaline-sensitive and -resistant rats: strain and gender differences. Pharmacol. Toxicol., 87: 161–168.

Yerkes, R.M. and Dodson, J.D. (1908) The relation of strength of stimulus to rapidity of habit-information. J. Comp. Neurol. Psychol., 18: 459–482.

Young, E.A. and Akil, H. (1985) CRF stimulation of ACTH/β-endorphine release: effect of acute and chronic stress. Endocrinology, 117: 23–30.

Young, E.A. and Liberzon, I. (2002) Stress and anxiety disorders. In: Pfaff, D.W., Arnold, A.P., Etgen, A.M., Fahrbach, S.E. and Rubin, R.T. (Eds.), Hormones, Brain and Behavior, Vol. 5. Academic Press, Amsterdam, pp. 443–465.

T. Steckler, N.H. Kalin and J.M.H.M. Reul (Eds.)
Handbook of Stress and the Brain, Vol. 15
ISBN 0-444-51173-3

CHAPTER 1.3

An introduction to the HPA axis

Allison J. Fulford[1,*] and Michael S. Harbuz[2]

[1]*Department of Anatomy, University of Bristol, Southwell Street, Bristol, BS2 8EJ, UK*
[2]*University Research Centre for Neuroendocrinology, Bristol Royal Infirmary, Marlborough Street, Bristol, BS2 8HW, UK*

Abstract: Integrity of the hypothalamo–pituitary–adrenal (HPA) axis is essential to survival of vertebrate species. This neuroendocrine axis functions to coordinate neural, endocrine and immune responses to diverse stressful stimuli that threaten homeostasis. The final products of activation of the HPA axis are the glucocorticoids that exert widespread effects on body functions, including cellular metabolism and immune function. Inappropriate secretion of endogenous glucocorticoids is potentially damaging and may predispose to disease. Homeostatic regulation of the HPA axis is complex and involves coordination of multiple systems of the body, in part mediated by the bi-directional communication network between the brain, endocrine and immune systems. Health and integrity of the individual relies on the appropriate integration of stress signals, including pro-inflammatory messages, generated at central and peripheral sites. Functional balance between pro- and anti-inflammatory mediators is fundamental to the appropriate control of the HPA axis and the prevention of dysregulation in its activity, a characteristic of numerous stress-related disorders including chronic inflammatory disease.

Introduction

The body's ability to adapt to external and internal factors that challenge the self-regulation of biological systems, or homeostasis, is essential to survival. An inappropriate response to any factor that impinges on homeostasis may result in a stress response. Stress is a term, originally defined by Hans *Selye*, to describe the pathophysiological state associated with specific physiological changes that could be induced by diverse physical and psychological stimuli (Selye, 1936). If exposure to a stressful stimulus persists or is intensified, the consequences for the animal may be severe, leading to disease or even death. The research pursued by Selye in the 1940s lead to further characterisation of the 'general adaptation syndrome' and the proposal that the pituitary gland was critical for the control of endocrine secretions, including those of the adrenal glands (Selye, 1946). He proposed that the adrenocorticotrophic hormone (ACTH), released from the pituitary, was associated with increased production of glucocorticoids from the adrenal glands that exerted profound and widespread effects on metabolism and lymphoid organ activity. These seminal research findings laid the foundations for our understanding of how neuroendocrine systems respond to stress. In addition to activation of the pituitary–adrenalcortical system, endocrine responses to stress include activation of the sympathoadreno–medullary system. Sympathetic nerves innervate the adrenal glands directly. In response to stress sympathetic activation enhances secretion of the adrenal catecholamines, principally the hormone adrenaline (epinephrine), which is important in the regulation of the autonomic response to stress.

Thus the stress response is a complex phenomenon encompassing autonomic, physical and behavioural changes in addition to neuroendocrine changes. For appropriate adaptation to an acute stress challenge, a coordinated response involving widespread systems of the body is required. Changes are necessary to

*Corresponding author. Tel.: +44 117 928 8692;
Fax: +44 117 925 4794; E-mail: A.J.Fulford@bris.ac.uk

promote arousal and attention to salient stimuli. Non-essential vegetative behaviours are inhibited, such as feeding and sexual behaviour, so that energy can be conserved for the 'fight–flight' response. The latter term, coined by Cannon (1939), describes those non-specific rapid autonomic and physiological changes necessary for mounting an acute stress response, including an increase in heart rate, blood pressure, respiration rate and liver glycogenolysis.

Individuals vary in their response to stress and several factors contribute to this inherent variability. Genetic and environmental factors exert significant influence over human stress responsiveness. Human studies suggest that the genetic variables contribute strongly to an individual's ability to respond to a stressful stimulus. However, it is now widely appreciated that environmental influences, especially during development, exert significant effects on the neural pathways controlling emotional responses and behaviour (see Sanchez et al., 2001; Lapiz et al., 2003). The contribution of both genes and environment shape the neural mechanisms subserving stress responses and in doing so, govern the vulnerability of an individual to stress.

The hypothalamo–pituitary axis

The endocrine system encompasses the pituitary gland, the peripheral endocrine organs and the hormones the two produce. The pituitary gland, or hypophysis, is responsible for the production of a range of hormones, which exert strong regulatory control over a wide range of bodily functions, including behaviour, growth and development, metabolism, salt and water balance, reproduction and immunity. Harris (1948) first proposed the functional link between the brain and the adenohypophysis (anterior pituitary). He suggested that the hypothalamus was responsible for the release of factors into the hypophysial blood that could directly influence the control of adenohypophysial hormone secretion. Indeed, the pituitary gland is subject to afferent control, via the actions of specific hypothalamic peptide 'releasing' or 'inhibiting factors' that act on the discrete populations of pituitary cells to regulate synthesis and release of pituitary hormones. Thus the hypothalamus represents the neural control centre whereby the brain can coordinate endocrine activity.

Stress influences the neuroendocrine regulation of a number of pituitary hormones including *ACTH*, prolactin, growth hormone, luteinising hormone, thyrotrophin, vasopressin and oxytocin. The neuroendocrine regulation of stress is a major focus of research interest, since dysregulated hormone activity may contribute to major life illnesses including cancer, metabolic, cardiovascular, autoimmune and psychiatric disorders. This chapter aims to provide an overview of the *hypothalamo–pituitary–adrenal* (*HPA*)*-axis* and its regulation. In addition, we will describe the significance of the changes in HPA-axis regulation associated with acute and chronic stress and the importance of the bi-directional communication network between the HPA axis and the immune system.

The hypothalamo–pituitary–adrenal axis

In vertebrates, appropriate functioning of the HPA axis is absolutely vital for species survival. Upon release into the hypophysial blood supply, the hypothalamic-releasing factors, *corticotrophin-releasing factor* (*CRF*) and the nonapeptide, *arginine vasopressin* (*AVP*), are transported to the adenohypophysis where they activate pituitary corticotrophs to synthesise and release ACTH into the general circulation (see Fig. 1). A corticotrophin-releasing factor, which was able to stimulate secretion of ACTH, was first identified in the 1950s but was not characterised until many years later as the 41 amino acid residue CRF (Vale et al., 1981). In the blood ACTH passes to the adrenal glands where it binds to receptors on cells of the zona fasciculata of the adrenal cortex to promote the conversion of cholesterol esters into free cholesterol and stimulate the steroidogenic pathway. The rapid enzymatic conversion of the precursor cholesterol into steroid intermediates ultimately results in the formation of the end product of the steroid pathway, the *glucocorticoids* (cortisol in humans and most mammals, corticosterone in rodents) (see James and Few, 1985). Only small amounts of these hormones are stored in the adrenal gland, therefore HPA activation results in rapid secretion of nascent glucocorticoids

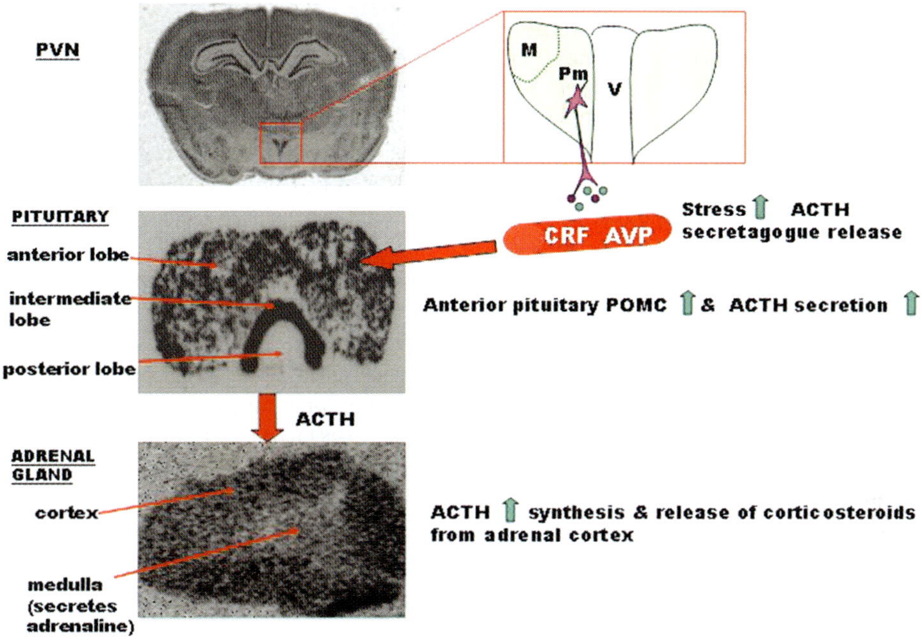

Fig. 1. Diagramatic representation of the hypothalamo–pituitary–adrenal axis in the rat. Following stimulation by a range of stressful stimuli, CRF- and AVP-containing parvocellular neurones of the medial PVN release their contents into the hypophysial portal blood. Once transported to the adenohypophysis, CRF and AVP increase synthesis of the ACTH precursor, POMC, which is cleaved into bioactive ACTH at the pituitary corticotrophs. ACTH is released from the pituitary and circulates to the adrenal glands, where it promotes synthesis and release of the glucocorticoids from the adrenal cortex. M, magnocellular division; Pm, medial parvocellular division; V, 3rd ventricle.

into the systemic circulation. In a normal human, cortisol secretion rates (8–25 mg/day) and plasma concentration (40–180 ng/ml) are maintained within close limits, although the plasma concentration will vary depending on the time of day, and in women, on the stage of the menstrual cycle. In conditions associated with chronic ACTH secretion, cortisol release may increase several fold resulting in secretion rates of up to 200–250 mg/day. In the rat, corticosterone is the principal glucocorticoid product and a small amount of this steroid is also secreted in humans (1–4 mg/day). Circulating ACTH is the major factor regulating glucocorticoid release; however, additional hormones from the adrenal medulla are involved but to a far lesser extent (Ehrhart-Bornstein et al., 2000).

The role of the effector glucocorticoids is to promote homeostatic adaptation to stress and this is achieved through catabolic actions that mobilise energy resources necessary for appropriate adaptive responses. The secretion of glucocorticoids during adversity promotes survival and the integrity of the HPA axis is critical, since homeostatic dysregulation may culminate in immunosuppression, neuroendocrine/autonomic dysfunction and tissue atrophy (McEwen and Stellar, 1993).

The parvocellular paraventricular nucleus is the apex of the HPA axis

The peptides, CRF and AVP are synthesised in the tuberoinfundibular *parvocellular cells* of the *paraventricular nucleus* (*PVN*) that evoke release of ACTH via their synergistic actions on pituitary corticotrophs. The axon terminals of the parvocellular neurons terminate in the external zone of the median eminence adjacent to the capillaries of the hypophysial portal blood supply where they secrete their contents into the portal blood. Parvocellular AVP, in contrast to magnocellular AVP, is involved with regulation of pituitary ACTH release and does not contribute to osmotic balance regulation.

CRF is the principal ACTH secretagogue, whereas AVP colocalises with CRF in approximately 50% of the CRF-containing neurones of resting animals and humans (Whitnall, 1993). The two peptides act synergistically on ACTH secretion in vitro (Gillies et al., 1982) and in vivo (Rivier and Vale, 1983); however, AVP alone has little ACTH secretagogue activity. In addition to evoking release of ACTH, CRF induces transcription of *pro-opiomelanocortin* (*POMC*) mRNA, the ACTH precursor protein (Lightman and Young, 1988). CRF is thought to be the only hypothalamic-releasing factor that can induce POMC gene expression. Thus, stress-induced HPA axis activation is highly reliant on neuroendocrine CRF. A population of parvocellular CRF-containing neurones project to extrahypothalamic sites including limbic nuclei and the brainstem (Sawchenko, 1987a). Therefore, in addition to coordinating the pituitary–adrenal system, CRF is directly involved in the orchestration of robust autonomic and behavioural responses to stress.

During activation of the HPA axis, the synthesis and secretion of both secretagogues is increased leading to a direct increase in ACTH and glucocorticoid secretion. Thus, expression of CRF mRNA and AVP mRNA in the PVN is increased and POMC mRNA expression is increased in the adenohypophysis (Antoni, 1986; Harbuz and Lightman, 1992). The importance of dual peptide control of ACTH release by the pituitary corticotroph is not fully understood. Although CRF is the principal and most potent ACTH secretagogue, AVP appears to be involved in the regulation of stress-induced ACTH release (Scott and Dinan, 1998). Evidence suggests that during chronic stress, the CRF:AVP ratio may increase, possibly due to differential sensitivity of the secretagogues to negative-feedback regulation (Scott and Dinan, 1998). In addition to AVP, various neuropeptides are colocalised within the parvocellular CRF neurones including enkephalin, neurotensin, cholecystokinin, vasoactive intestinal peptide and galanin (Palkovits, 1988). In some cases CRF-containing neurones express the inhibitory amino acid neurotransmitter, γ-aminobutyric acid (GABA), instead of AVP (Meister et al., 1988). The coexistence of these peptides or transmitters with CRF provides a mechanism for subtle regulation of ACTH release.

Pulsatility in the HPA axis

Across a variety of species, glucocorticoid secretion varies markedly throughout the day in a pulsatile fashion and is subject to *circadian regulation* (Kaneko et al., 1981). ACTH is also secreted in a pulsatile manner. Thus, circulating levels of glucocorticoids closely mirror the pulsatility exhibited by the ACTH release. Pulsatile control of HPA-axis hormone secretion facilitates the exquisitely sensitive and dynamic relationship between brain, adenohypophysis and adrenal glands. Peaks of glucocorticoid secretion are typically seen 15–30 min after an ACTH pulse in normal human subjects. A major pulse of ACTH occurs in the early hours of the morning. Following this major pulse, release of ACTH and glucocorticoids is stimulated by further pulses throughout the day, approximately once per hour. Precise information regarding the temporal profile of the peaks is subject to the resolution of sampling methodologies employed (Gudmundsson and Carnes, 1997). Following each brief pulse, rising glucocorticoid levels stimulate negative-feedback loops to inhibit further ACTH release. However, circulating glucocorticoid levels gradually decline to the set-point level so that activation of the HPA axis is stimulated resulting in an additional pulse. This profile of episodic hormone release allows precise control over the HPA axis under normal conditions. Pulsatile release also ensures that receptor down-regulation is prevented, as would most likely happen in the face of continuous exposure to an endogenous agonist. It is important to point out that ACTH does not always stimulate glucocorticoid secretion, although concordance is approximately 80% (see Gudmundsson and Carnes, 1997). The origins of pulsatility in the HPA axis are largely unknown; however, it has been suggested that the pulsatile ACTH release may originate at the level of the pituitary corticotroph, rather than at the level of hypothalamic CRF and AVP release (Gambacciani et al., 1987). Concerted action of the corticotroph population leading to a pulsatile burst would presumably require coordinated signal transfer by efficient autocrine, paracrine or juxtacrine actions.

The PVN subparaventricular zone receives dense afferent input consisting of vasopressin-containing neurones originating in the suprachiasmatic nucleus (SCN) (Buijs et al., 1998). The dorsal and lateral compartments of the parvocellular PVN also receive direct inputs from the SCN. The SCN is involved in the control of circadian rhythms of the body; however, it may also serve a role in the regulation of stress responses. The circadian regulation of the HPA axis is subserved by the SCN innervation of the PVN region (Kalsbeek et al., 1996). Studies in animals have demonstrated how activation of the SCN can stimulate evening secretion of ACTH (Cascio et al., 1987).

There is some evidence for gender differences in the pulse pattern of ACTH secretion; however, there are inconsistencies in the human data, with reports of an increased number of pulses, but similar pulse amplitude, in males than females (Horrocks et al., 1990). In another study, males also differed in the pattern of their ACTH pulsatility, however, in this case the amplitude of the pulses was greater with the pulse frequency unchanged (Roelfsema et al., 1993). Abnormal *pulsatility of the HPA axis* or subtle alterations in the frequency or magnitude of the ACTH signal could have significant consequences for feedback regulation of the HPA axis. Such a mechanism may contribute to the apparent disturbances in neuroendocrine parameters characteristic of certain psychiatric conditions, such as major depression. Chronological decline in HPA-axis regulation has also been suggested to contribute to age-related illnesses; however, there is little evidence in direct support of this.

Negative-feedback regulation of the HPA axis

The HPA axis functions as a closed-loop system involving tight *negative-feedback control* mediated by the glucocorticoids exerting multiple regulatory actions. Autoregulation of the HPA axis is essential for ensuring that the stress response is terminated, preventing excessive activation in order for restoration of internal homeostasis. Regulatory feedback occurs at several sites and involves both rapid and delayed feedback in humans and rats (Keller-Wood and Dallman, 1984; Krishnan et al., 1991; Young and Vasquez, 1996). Rapid feedback occurs immediately following a rise in circulating glucocorticoids and lasts from 5 to 15 min, whereas delayed feedback emerges 1–2 h later, can persist for up to 4 h and is dependent on the glucocorticoid level. In the case of prolonged activation of the HPA axis, delayed feedback may continue for up to 24 h. This temporal profile suggests that delayed feedback relies on genomic actions of glucocorticoid receptors, whereas rapid feedback is presumably a consequence of non-genomic actions of glucocorticoids (see De Kloet et al., 1998).

Rapid feedback is exerted primarily via an inhibitory action of glucocorticoids on the synthesis and release of ACTH at the hypothalamic level, by decreasing mRNA expression for CRF and AVP. Delayed feedback is also manifested at the level of the adenohypophysis where glucocorticoids decrease mRNA expression level of the ACTH precursor protein, pro-opiomelanocortin (POMC) (see Harbuz and Lightman, 1992). In addition, glucocorticoids can act centrally, at the hypothalamus and higher centres, principally the hippocampus, to exert delayed negative-feedback inhibition and thereby prevent continued activation of the HPA axis.

The actions of the corticosteroids are mediated primarily through specific nuclear receptors of which there are two subtypes, the *mineralocorticoid receptor* (*MR*) and *glucocorticoid receptor* (*GR*). The steroid receptors are located intracellularly in the cytoplasm and bind steroids that can freely diffuse across the plasma membrane. Once bound, the receptor–ligand complex translocates to the nucleus and interacts with palindromic hormone response elements on the DNA molecule. Thus, activated steroid receptors function as transcription factors and influence transcription of target genes, ultimately leading to changes in protein synthesis.

The MR subtype has high and equal affinity for aldosterone, *corticosterone*, *cortisol* and deoxycorticosterone, but lower affinity for the synthetic glucocorticoid, dexamethasone. The GR subtype can be distinguished by the following affinity profile:

dexamethasone > cortisol > corticosterone > deoxycorticosterone > aldosterone. Although, GR has lower affinity for the glucocorticoids than MR, its lack of affinity for aldosterone means that this subtype is effectively glucocorticoid selective. The two receptors are found densely expressed in the central nervous system, however, the distribution of both is generally quite distinct with overlap in a few areas. The distribution of MR is more restricted, being found in the hippocampus and sensory and motor nuclei outside the hypothalamus (Arriza et al., 1988; Reul et al., 2000). GRs, however, are more widely localised in the hypothalamic PVN, the brainstem catecholaminergic cell groups, amygdala and hippocampus, in addition to the pituitary gland (Fuxe et al., 1985). Research suggests that MR may be more important in regulating the basal expression of the ACTH secretagogues, CRF and AVP, at the nadir of diurnal ACTH secretion, and in the regulation of the peak ACTH release (Dallman et al., 1989). GR may be more critical for the termination of the HPA-axis response to stress. The *hippocampus* is an important component of the negative-feedback regulation of the neuroendocrine stress response. Both subtypes of corticosteroid receptor are expressed by hippocampal neurones. Lesions interrupting signals from the hippocampus to the PVN are associated with increased basal levels of circulating glucocorticoids and enhanced responsiveness to stress, highlighting the importance of this structure in the feedback regulation of the HPA axis (Herman et al., 1992).

The importance of the *adrenal glands* to normal bodily function can be readily demonstrated by studying the effect of their removal. Excision of the adrenal glands is associated with enlargement of the thymus and impairments in the stress response. Adrenalectomised rats have excessive secretion of ACTH and enhanced expression of the POMC precursor gene in the adenohypophysis (Jingami et al., 1985; Marti et al., 1999). Adrenalectomy also increases immunoreactivity and mRNA expression for the ACTH secretagogues in the parvocellular neurones of the PVN (Wolfson et al., 1985; Sawchenko, 1987b). Interestingly, adrenalectomy appears to selectively affect those CRF/AVP-containing parvocellular neurones involved in the regulation of ACTH secretion. Replacement with exogenous steroids in drinking water, food or implanted pellets normalises the ACTH secretion in rats (Akana et al., 1988), confirming the importance of glucocorticoids in the regulation of the HPA axis. However, adrenalectomised rats do not have maximal increases in basal plasma ACTH levels as evidenced by the fact that novel acute stress can still stimulate further ACTH release (Akana et al., 1988). This implicates an additional mechanism, aside from glucocorticoid feedback, in restraining HPA-axis activation and providing a tonic inhibitory input to the axis.

Afferent regulation of the HPA axis

The parvocellular neurones are subject to close regulation from diverse afferent inputs. These neurones process excitatory and inhibitory inputs and function to coordinate secretion of CRF and AVP, thereby controlling the extent of ACTH stimulation of glucocorticoid secretion. Many brain regions are involved in the integration of responses to fear or stressful stimuli, including hypothalamic, septohippocampal and amygdaloid nuclei, cingulate and prefrontal cortex, brainstem catecholamine cell groups (A2/C2 cell bodies in the nucleus of the solitary tract (NTS); A1/C1 cell bodies in the ventrolateral medulla; A6 cell bodies in the locus coeruleus) and the dorsal raphe nucleus (see Pacak and Palkovits, 2001). Indeed, extensive neuroanatomical studies analysing the connections of the parvocellular PVN identify links with neural pathways concerned with homeostatic adaptation, cognition and affective behaviour (Silverman et al., 1981; Sawchenko and Swanson, 1983; Cunningham and Sawchenko, 1988). Parvocellular PVN afferents implicated in preserving homeostasis arise in the brainstem, hypothalamus and basal forebrain (see Fig. 2). *Noradrenergic* inputs, arising from the lower brainstem, innervate the medial parvocellular PVN and convey stress-related inputs to the parvocellular PVN for the coordination of endocrine, autonomic and behavioural output responses. These include strong visceral afferent inputs transmitting information directly from the NTS or via relays in the ventrolateral medulla that convey sensory stimuli (Sawchenko and Swanson, 1981; Cunningham and Sawchenko, 1988; Sawchenko et al., 2000).

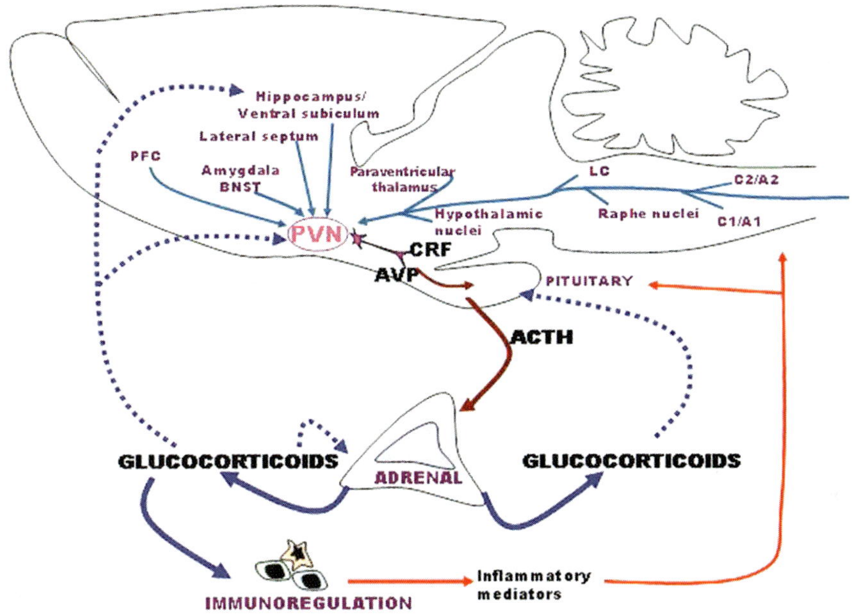

Fig. 2. Schematic representation of the neuroendocrine–immune axis. Activation of the HPA axis results in secretion of immunoregulatory glucocorticoids. Immune cells respond to stimulation by the release of inflammatory mediators that modulate local inflammation and can communicate to the brain and pituitary. Regulation of HPA-axis activity involves integration at the hypothalamic PVN of diverse afferent inputs from other hypothalamic areas, limbic nuclei and brainstem nuclei. Solid lines represent stimulatory inputs; dashed lines represent inhibitory feedback loops.

Noradrenaline is vital for the response to some stressors and hypothalamo–pituitary–adrenal axis activation. It stimulates CRF-containing neurones in vitro and in vivo, providing supporting evidence for direct connections with the hypophysiotropic CRF-containing neurones (Sawchenko and Swanson, 1982; see Pacak and Palkovits, 2001). More recent research suggests that the majority of ascending noradrenergic afferents may relay in the hypothalamus with local excitatory *glutamatergic neurones* that innervate the PVN (Daftary et al., 2000). There are also reciprocal connections from the CRF neurones projecting to the brainstem locus coeruleus noradrenergic cell groups (Valentino et al., 1983; see Koob, 1999). In addition to these projection neurones, short-loop feedback mechanisms allow for close autoregulation of both CRF and noradrenergic neurones of the hypothalamus (Calogero et al., 1988).

The importance of intact catecholaminergic inputs to the PVN for stress responses have been confirmed by selective lesioning studies of the PVN. Injection of the catecholaminergic neuronal toxin, 6-hydroxydopamine, completely prevents the glucocorticoid response to conditioned fear or immobilisation stress (see Van de Kar and Blair, 1999). In contrast, ascending noradrenergic neurones do not appear to regulate AVP expression, suggesting that the two ACTH secretagogues are subject to differential regulation. The occurrence of dual mechanisms for maintenance of ACTH release from the adenohypophysis and their differential regulation highlights the complexity of the HPA axis and demonstrates the essence of its physiological importance under different stress conditions, as will be discussed later.

The immediate vicinity of the PVN, but not the PVN itself, receives input from brainstem cholinergic (Ohmori et al., 1995) and serotonergic (Liposits et al., 1987) nuclei that are concerned with the control of arousal and wakefulness. Both ascending pathways are thought to mediate excitatory effects over HPA-axis drive. The PVN also receives afferents via the fimbria fornix and angular bundle that originate in the ventral hippocampus. These are important for the tonic regulation of the HPA axis and termination of the stress response is thought to involve substance P arising in the arcuate nucleus (Nussdorfer and Malendowicz, 1998; Jessop et al., 2000b). Dense projections from the ventral subiculum directly innervate the subparaventricular zone, proximal to the PVN, and in addition transmit signals via the bed nucleus of the stria terminalis (BNST) and preoptic area to the medial PVN (Cullinan et al., 1993). These inputs from the *ventral subiculum* provide a strong influence over the PVN. The inputs via the medial preoptic area or BNST appear to exert both excitatory and inhibitory regulation over HPA-axis drive (Herman et al., 1994). These functional differences are probably due to discrete topographical organisation of inputs to the PVN from subdivisions of the nuclei.

Neuronal afferents from the prefrontal cortex, lateral septum and paraventricular thalamus also terminate in the peri-PVN zone, providing an additional rich source of projections from limbic structures (see Herman et al., 2002). Furthermore, the *amygdala* can exert a stimulatory effect on the HPA axis, via some direct input to the peri-PVN zone from the medial nucleus and indirect input via the BNST and preoptic nucleus (Canteras et al., 1995). The paraventricular thalamus has links with the suprachiasmatic nucleus (SCN), which also innervates the peri-PVN zone, and these connections are thought to regulate the circadian rhythm of the HPA axis (Watts et al., 1987).

Local interneurones in the vicinity of the PVN appear to directly modulate outgoing signals from the PVN, permitting fine integration of HPA-axis stimuli. The majority of these local circuit neurones express glutamic acid decarboxylase, the rate-limiting enzyme in the synthesis of the major inhibitory amino acid neurotransmitter, GABA (see Herman et al., 2002). Local excitatory inputs derived from glutamatergic neurones in the vicinity of the PVN may also contribute to its regulation, providing a balance between proximal inhibitory and excitatory influences that gate HPA-axis drive. In summary, in addition to the major inputs direct to the medial PVN, e.g. from brainstem noradrenergic afferents, the PVN is subject to complex regulation by a rich supply of limbic inputs which largely relay with local circuit neurones just proximal to the PVN. This additional level of regulation allows for concerted integration of HPA-axis activity so that diverse stimuli are prioritised and can be responded to with appropriate intensity and urgency. Thus the complexity of HPA-axis input regulation allows for hierarchical integration of various stimuli, whether cognitive or physiological stimuli.

Immune–HPA axis interactions

The interaction between the *immune system* and HPA axis has been appreciated for many years. The most widely recognised effect of the communication between these two systems is demonstrated by the actions of the glucocorticoids, the end products of activation of the HPA axis. Glucocorticoids have potent anti-inflammatory effects with extensive effects on the immune system, including actions on every population of immune cell. Glucocorticoid effects are largely inhibitory, and encompass effects of cell growth, proliferation and differentiation, leukocyte trafficking, cytokine and eicosanoid production, antibody formation and cell death including receptor-mediated apoptosis (Munck and Naray Fejes-Toth, 1992; see Webster et al., 2002). The profound actions of the glucocorticoids require tight regulation, hence the importance of intact negative-feedback control of the HPA axis. Unchecked glucocorticoid secretion, due to prolonged inflammation, would have pathological consequences for host immunity, leading to hypercortisolaemia, immunosuppression and increased susceptibility to disease and infection (Munck and Naray Fejes-Toth, 1994). Integrity of the *immune–neuroendocrine interactions* is critically linked to physiological well being, since the ability of immune- or endocrine-derived cytokines to stimulate activation of the HPA axis represents a powerful defence mechanism against chronic inflammatory disease.

Since the late 1980s, understanding of the extent of interaction between the immune and neuroendocrine systems has greatly increased. We now appreciate that the communication network between these two systems is bi-directional in nature (Blalock and Smith, 1985; see Besedovsky and del Rey, 2002) and encompasses a diverse collection of common chemical mediators including peptides, cytokines and neurotransmitters that modulate activity of both systems by receptor-mediated actions (see Gaillard, 2001). *Cytokines* are polypeptides that are synthesised by immune cells and are key regulators of inflammatory responses. Cytokines are also synthesised by neuroendocrine tissues (Koenig, 1991) and by higher centres of the brain (e.g. hippocampus) where they contribute to local inflammation. Such complex interrelationships provide an important mechanism by which the immune system can modulate the activity of the HPA axis in order to preserve immune homeostasis and limit stress responses.

The brain can influence the immune system by stimulation of the sympathetic nervous system that innervates the lymphoid organs. Thus, in conditions of stress, the brain can stimulate immune cell function by coordinated activation of the HPA axis and sympathetic nervous system. The neuroendocrine system synthesises hormones and neuropeptides that can influence immune function. These factors appear to act as *paracrine* or *autocrine mediators*, providing a system of discrete local regulation over inflammatory reactions. Many of the mediators are also synthesised locally by immune cells and their levels may be increased following immune activation (Weigent and Blalock, 1997). The immune–endocrine mediators are often characterised by having pleiotropic effects, including stimulatory or inhibitory actions, dependent on their relative local concentration or activation state of the immune system. Immune-derived peptides and hormones are believed to contribute to local inflammatory control mechanisms. Evidence suggests that sites of inflammation are associated with increased local concentrations of immune-derived pro-inflammatory peptides, such as CRF and substance P (Jessop et al., 2001; Jessop, 2002). Levels of immune-derived opioid peptides, such as enkephalins and dynorphin, are also expressed at high levels during *inflammation*, although their actions are considered to be predominantly anti-inflammatory (Cabot et al., 2001).

A feature of the immune-derived peptide or hormone mediators and their receptors is that, although many appear to be largely similar to the classical brain or endocrine equivalents, in many cases the immune-expressed receptor may be a truncated version or express different pharmacological characteristics (Sharp et al., 1998; Sharp, 2003; see Fulford and Jessop, 2001). The biologically active immune-derived peptides or hormones may also vary from the brain or endocrine equivalent. For example, immune cells express ACTH identical to that found in the pituitary in addition to truncated variants of ACTH (Smith et al., 1990). This highlights the potential importance of the immune-derived peptides as possessing functional significance. Furthermore, immune-derived peptides or hormones are expressed at very low levels suggesting roles as paracrine or autocrine regulators (Karalis et al., 1991; Aird et al., 1993; Sharp et al., 1998).

POMC was one of the first endocrine proteins found expressed in the immune system. POMC, the precursor hormone for ACTH, is also cleaved into other biologically active peptides including the potent *opioid peptide*, β-endorphin. Both ACTH and β-endorphin are expressed in immune cells and are thought to contribute to local immunoregulatory control. As relatively low levels of these endogenous peptides and hormones are synthesised and expressed in the immune system, it seems unlikely that these would be able to interact directly with the endocrine system. Peptides are subject to rapid metabolism and so would be unlikely to remain biologically active following transportation in the circulation. However, in conditions associated with very high secretion of immune-derived peptide or hormone mediators, it is possible that these may be able to modulate peripheral endocrine secretions.

Stress-induced changes in the HPA axis

We will now concentrate on the response of the HPA axis to the effects of (1) *acute stress* and (2) *chronic or repeated stress*. The majority of work underlying our understanding of stress mechanisms has arisen from preclinical research in animals despite the obvious limitations in extrapolating findings from animals to humans. However, these studies have undoubtedly made a major contribution to our current understanding of the mechanisms governing activation of the HPA axis in mammals. A wide range of behavioural paradigms have been developed in animals that closely correlate stress in humans

(Van de Kar and Blair, 1999; Pacak and Palkovits, 2001; Willner and Mitchell, 2002).

HPA-axis responsiveness to acute stress

Systemic versus neurogenic stress

A wide range of acute stressors have been used in the study of HPA-axis regulation in animals. These stressors can be classified as either *systemic or neurogenic stressors*. Systemic stressors include physical stressors such as cold, ether, hypertonic saline challenge, insulin-induced hypoglycaemia, immune challenge such as by cytokine or endotoxin injection, formalin injection or surgical stress. Neurogenic stressors encompass those stressful stimuli bearing predominantly an emotional or psychological component. These include footshock, conditioned fear paradigms, forced swimming or restraint/immobilisation stress. Neurogenic stressors involve a strong somatosensory stimulus that requires cognitive or emotional interpretation.

Exposure to these acute stress challenges results in enhanced secretion of ACTH and glucocorticoids, demonstrating acute activation of the HPA axis. The HPA axis responds to the intensity of the individual stressor, so that repeated or intensified stress results in enhanced secretion of the stress hormones. However, it is not possible to distinguish different stressors on the basis of simple measurement of circulating ACTH and glucocorticoid levels as these are common to all types of acute mild stress (Kant et al., 1982). However, studies at the central level have been important in the identification of stressor-specific neurocircuitry and neuroendocrine responses (see Pacak and Palkovits, 2001).

Acute stress causes a short-lasting and rapid increase in CRF immunoreactivity in the median eminence (Buckingham, 1979). This is followed by increased synthesis of hypothalamic CRF, demonstrated using the technique of in situ hybridisation histochemistry to detect changes in mRNA expression (Lightman and Young, 1988). A range of acute stress paradigms have been demonstrated to increase CRF mRNA including footshock, insulin-induced hypoglycaemia, restraint and forced swim stress (see Lightman and Harbuz, 1993). However, the physical stress of acute cold exposure does not alter CRF mRNA (Harbuz and Lightman, 1989). Acute stressors including cold, footshock, restraint and osmotic stress also increase expression of POMC mRNA in the anterior pituitary (Lightman and Young, 1988; Harbuz and Lightman, 1989; Wu and Childs, 1991).

Differences between systemic and neurogenic stressors have been observed at the level of the hypothalamus. It appears that neurogenic stressors, including restraint or forced swim stress, only activate CRF mRNA, whereas physical stressors including osmotic stress or naloxone-precipitated opiate withdrawal increase *proenkephalin A* mRNA, in addition to CRF mRNA (Harbuz et al., 1991, 1994b). Neurogenic stressors that involve some physical component, e.g. footshock stress, also increase expression of proenkephalin A mRNA in the hypothalamus (Harbuz and Lightman, 1989). The stimulus-specific neuronal activation following stress has been further investigated by mapping stress-induced activation of immediate early gene products, such as c-fos (Chan et al., 1993). Immobilisation/restraint stress, a neurogenic stressor with a physical component, has been shown to increase immunostaining for fos in the hypothalamus, specifically in the medial and dorsal parvocellular PVN where most neurones are CRF positive (Kononen et al., 1992). c-Fos mRNA expression can be seen 30 min after the start of physical restraint; however, maximal fos expression is observed 90 min after stress onset (see Pacak and Palkovits, 2001). Acute footshock stress induces a similar pattern of fos activation to that seen following acute immobilisation stress. This is true for the main subdivisions of the PVN including the parvocellular CRF-containing neurones. The wider neurocircuitry recruited in the processing of the footshock stress is essentially similar to the c-fos response observed following neurogenic stimuli (Sawchenko et al., 2000). These observations are suggestive of the existence of a neurogenic-stress circuit that becomes activated by phenotypically similar, acute stress challenges.

Systemic stressors, which typically contain a strong physical component, appear to evoke a different pattern of fos activation to that shown for the neurogenic stressors. Whereas in the PVN the parvocellular CRF neurones are activated, the wider circuitry involved in the processing of systemic stressors is contained within subcortical structures, a number of which are specifically involved in the integration of systemic stressors (Gaillet et al., 1991). Extensive research, using double-label immunohistochemical staining for tyrosine hydroxylase and c-fos, has confirmed that in the case of both systemic and neurogenic stress, brainstem catecholaminergic cell groups are strongly activated. These neurones project directly to the PVN (Sawchenko and Swanson, 1982) and are likely to contribute to the robust neurohormonal responses seen following exposure to both types of stressors.

Immunological stress

Of the diverse group of cytokines produced by the immune system, the effects of the inflammatory cytokines, *interleukin-1* (IL-1), *interleukin-6* (IL-6) and *tumour-necrosis factor-α* (TNF-α), have been the best characterised in terms of their ability to stimulate the HPA axis (see Turnbull and Rivier, 1999). HPA-axis activation can be induced by the cytokines alone or by concerted actions between them. These cytokines are able to act at the hypothalamus, pituitary and adrenal cortex to increase glucocorticoid secretion and so to suppress further immune/inflammatory reactions.

A single, cytokine injection (either central or peripheral) has been shown to cause potent activation of the HPA axis in rats and mice. For example, fos-immunostaining studies have shown how systemic injection of IL-1 causes activation of the PVN in an identical way to that seen for neurogenic stressors (Ericsson et al., 1994). In addition, ascending aminergic pathways appear to be involved in the neurohumoral response to IL-1 (Chuluyan et al., 1992), suggesting that for all types of acute stressors studied, activation of the HPA axis is strongly dependent on ascending aminergic inputs to the PVN.

Many studies investigating the effect of cytokines on the HPA axis have studied the effect of injection of the endotoxin, *lipopolysaccharide* (*LPS*), a constituent of the bacterial cell wall (Tilders et al., 1994; Quan et al., 1998). Injection of this endotoxin causes the release of a wide spectrum of cytokines, the exact profile of which is dependent on the concentration of endotoxin injected. The acute response involves predominantly IL-1 (principally, IL-1β), IL-6 and TNF-α (Perlstein et al., 1993). If these cytokines are injected separately they also provoke a robust activation of the HPA axis, which confirms that during an acute inflammatory episode, cytokine-mediated responses involve potent activation of the neuroendocrine stress axis (Sapolsky et al., 1987; Rivest et al., 1995; Turnbull and Rivier, 1999).

In the hypothalamus, IL-1β, TNFα and IL-6 can all increase CRF production (Besedovsky et al., 1991). In the case of IL-1β, activation of the HPA axis occurs at the hypothalamic level and is not due to activation of pituitary corticotrophs at physiological concentrations of the cytokines. Pleiotropic cytokines, including IL-1 and IL-6, can act directly at the level of the adenohypophysis to promote differentiation and activation of corticotrophs at higher doses (see Buckingham et al., 1996; Navarra et al., 1997). This effect involves stimulation of POMC expression and hence ACTH formation. These in vitro studies suggest that the cytokines caused a slow release of ACTH. Evidence suggests that the folliculostellate cells of the pituitary may mediate the actions of cytokines on the corticotroph population, since the former are macrophage-like and likely to respond to cytokine stimulation. Indeed, the folliculostellate cells also synthesise cytokines that are able to contribute to local activation of the HPA axis at the level of the pituitary (Gloddek et al., 2001). If cytokines are administered in vivo, plasma ACTH levels rise rapidly demonstrating an acute activation of the HPA axis. Evidence suggests that the effects of peripherally administered cytokines are acting principally at the level of the hypothalamus via actions on the parvocellular CRF-containing neurones of the PVN. These effects can be blocked by central, but not peripheral, administration of an IL-1 receptor antagonist (Habu et al., 1998).

The exact mechanism whereby cytokines are able to stimulate the HPA axis in vivo is incompletely understood; however, a number of potential sites of action have been proposed. Firstly, the presence of cytokine receptors in the brain and hypothalamus suggests that cytokines may be able to act locally to induce activation of hypophysiotropic parvocellular

PVN neurones (Farrar et al., 1987; Katsuura et al., 1988). A problem concerning the mechanism of action of cytokines on central activation of the HPA axis centres on the ability of peripheral cytokines to gain access to the brain. These large polypeptides are unable to traverse the blood–brain barrier alone; however, there is evidence for carrier-mediated transport mechanisms. Blood-borne cytokines may gain access to the brain via saturable transport systems. These have been described for a number of cytokines including IL-1α, IL-1β, IL-6 and TNFα, but not IL-2 (Banks et al., 1995). Circulating cytokines may be able to penetrate the brain at the level of the *circumventricular organs* including the organum vasculosum of the lamina terminalis of the hypothalamus (OVLT), median eminence, the subfornical organ, choroid plexus and area postrema (Saper and Breder, 1994). These structures line the cerebral ventricles, where the blood–brain barrier is weak or absent, and therefore present sites whereby peripheral mediators may gain access to the CNS and stimulate the HPA axis. The circumventricular organs may alternatively represent sites where peripheral cytokine messages are communicated to the hypothalamus. Potential candidates for the signal communication molecule involved in the transduction of blood-borne cytokine effects to the brain include neurotransmitters, most notably 5-HT, prostaglandins, brain cytokines or nitric oxide (Van Dam et al., 1993; Rivier, 1995). Although, the circumventricular organs and transport carriers of the blood–brain barrier provide mechanisms whereby cytokines can gain access to the brain, these are unlikely to accumulate cytokines in the concentrations required to produce the profound behavioural and physiological effects typically seen following peripheral cytokine administration (Watkins et al., 1995).

Cytokines also evoke pain pathways leading to activation of somatosensory networks in the brain that reflexly activate the HPA axis (Dantzer, 2001). Cytokine activation may additionally induce profound systemic effects including hypotension or hypoglycaemia that may activate vagal reflexes. In addition, secretion of inflammatory mediators will also influence C-fibre activity that will signal to the spinal cord and the brain. Systemic infection may also activate resident microglia, monocytes or macrophages present in the CNS. These may also secrete IL-1β or TNFα in response to infection that can signal to the HPA axis and cause direct stimulation of ACTH secretion (Hopkins and Rothwell, 1995). Of these mechanisms, stimulation of vagal afferent nerves is considered to be of major importance. Studies examining the effect of subdiaphragmatic vagotomy in rats have described complete abrogation of many of the effects of peripheral cytokines in the brain (see Maier and Watkins, 1998). This neural pathway may provide a vital link for the communication of peripheral immune signals to the CNS.

The bi-directional communication network regulating the immune–neuroendocrine interface provides a dynamic link by which stress can impact on host immunity. The outcome of exposure to a stressor will depend on the interplay between psychological, neuroendocrine, behavioural and immunological factors. Thus, one must consider the response to stress as a phenomenon governed by adaptations in multiple systems of the body. From this holistic standpoint it becomes clear that in some clinical disorders, such as major depression, linked to stress, the breadth of symptoms displayed in some patients is consistent with dysregulation in neurobehavioural, immunological and neuroendocrine mechanisms. Now the pathways of communication between these physiological systems have been revealed, we will be better able to understand the possible mechanisms contributing to the aetiology of other diseases bearing a strong cognitive or emotional component.

HPA-axis responsiveness to repeated stress

Regardless of the exact phenotypic profile of responses to the various stimuli that elicit an acute stress response, each is characterised by return of HPA-axis activity to baseline once the stressful stimulus is removed or is diminished. Studies of the impact of repeated acute stress, continued for several days, have been undertaken to establish the effect of long-term stress. However, models of repeated stress are generally poor correlates of long-term, persistent stress as repeated exposure to the same stress is often associated with habituation of the HPA-axis response and attenuation of the neurohormonal stress response. Repeated footshock or restraint stress is associated with increased plasma glucocorticoid levels that remain above baseline for several days and then return to control levels (Kant et al., 1985). Similarly, plasma ACTH levels generally return to basal levels following repeated stress. Attenuation in the afferent regulatory control of the HPA axis is thought to underlie this habituation of the HPA axis with repeated stress. Interestingly, habituation to one type of acute stress appears to be specific as cross-tolerance to other types of acute stress does not generally occur (Spencer and McEwen, 1990). This phenomenon may indicate differences in the neural processing of the various types of acute stressors.

A problem with the use of experimental repeated stress paradigms is that they bear little physiological relevance to disorders associated with chronic dysfunction of the stress axis. To better understand the aetiology and progression of human disorders involving long-term changes in the activity of the HPA axis, a more successful strategy is to adopt chronic models with improved validity that provide better insight into the mechanisms governing adaptation to long-term stress.

HPA-axis responsiveness to chronic stress

Studies of sustained chronic stress differ significantly from the effects of repeated stress, since a feature of the former condition is persistent elevated levels of circulating glucocorticoids. *Inflammatory diseases* may be considered as disorders associated with chronic stress as they are typically characterised by high circulating levels of glucocorticoids. The mechanistic changes that confer long-term upregulation of activity of the HPA axis and continuous secretion of stress hormones remain largely unknown; however, preclinical studies have furthered our understanding of the adaptations contributing to a state of chronic stress. It is apparent that the mechanisms essential to the maintenance of HPA-axis integrity, including negative-feedback control, have become dysregulated in the chronic inflammatory condition.

Experimental models of chronic inflammatory stress

A particularly well-characterised animal model of inflammatory stress is *adjuvant-induced arthritis* (AA) in the rat, which is a T lymphocyte-dependent chronic inflammatory disease. AA has been used as a model with relevance for certain clinical inflammatory conditions including pain and *rheumatoid arthritis*. The arthritis can be induced in susceptible strains of rats following an intradermal injection of an oil suspension of ground, heat-killed *Mycobacterium butyricum* (10 mg/ml) into the base of the tail. Specific strains of rat will develop hindpaw inflammation within 12–14 days and other limbs are additionally affected by day 21 post-injection (Rook et al., 1994). The neuroendocrine effects of long-term inflammation in AA rats have been extensively studied. The objective of such studies has been to identify whether HPA-axis dysregulation is aetiologically relevant and essential for disease progression.

The AA rat is characterised by similar pituitary and adrenal changes to that seen following acute and repeated stress, including raised circulating levels of ACTH and glucocorticoids and increased expression of POMC mRNA in the adenohypophysis (Harbuz et al., 1992). There is an apparent defect in the circadian regulation of the HPA axis resulting in the consistently high secretion rate of glucocorticoids that is elevated in the early hours of the morning, a time normally representing the nadir of the daily cycle (Sarlis et al., 1992). In AA, at the level of the hypothalamus, there is a paradoxical decrease in CRF mRNA in the parvocellular PVN and reduced release of CRF into the hypophysial blood (Harbuz et al., 1992). This effect is not entirely due to enhanced glucocorticoid feedback regulation of the CRF neurones, and the exact inhibitory mechanism responsible for the arthritis-induced CRF hypofunction is not completely understood although it may involve substance P (Jessop et al., 2000b). Time-course studies have identified that the reduction in CRF mRNA is apparent when the first signs of inflammation appear (about day 11) and the maximal reduction in CRF is observed when inflammation is most severe (around day 21) (Harbuz et al., 1994a). Interestingly, the inhibition of parvocellular PVN CRF neurones is also seen in other chronic immunological disease models in rodents, including the preclinical model for multiple sclerosis, *experimental allergic encephalomyelitis* (EAE) (Harbuz et al., 1993), and systemic lupus erythematosus (Shanks et al., 1997). This possibly also applies in human conditions (Harbuz, 2002).

In contrast to the effect on CRF neurones, AA is associated with increased expression of AVP mRNA in the parvocellular PVN and AVP release into the portal circulation, indicating that the dominant ACTH secretagogue during chronic inflammatory stress is AVP (Chowdrey et al., 1995). However, how the increased activity of parvocellular AVP neurones contributes to increased synthesis of precursor POMC by the pituitary corticotrophs is unknown, although CRF may act in a permissive role. Evidently, there is marked derangement in the regulation of the HPA axis in association with chronic stress of immunological origin. The transition from the dominance of CRF in the stimulation of ACTH secretion to a major role for AVP has also been associated with repeated stress paradigms. Chronic exposure to immobilisation or other psychological stressors increases the proportion of parvocellular CRF PVN neurones that contain AVP and an increased ratio of AVP to CRF levels in the zona externa of the median eminence (de Goeij et al., 1991). It has been widely suggested that AVP of parvocellular origin is most important for maintaining HPA-axis responsiveness under conditions associated with defective CRF function, like chronic stress.

The characteristic reduction in parvocellular PVN CRF neuronal activity associated with chronic immune-mediated stress impairs the ability of the HPA axis to respond to certain types of acute novel stress. Specifically, AA is associated with a blunted glucocorticoid response to psychological and physical stressors (Aguilera et al., 1997; Windle et al., 2001). In contrast, in AA, acute immunological stressors, such as LPS injection, elicit a robust neuroendocrine stress response that is equivalent to that seen in non-arthritic control rats (Harbuz et al., 1999). In this case, CRF neurones are activated following stimulation with peripheral endotoxin injection, albeit to a lesser extent than in non-arthritic controls. Clearly, there is differential feedback regulation of CRF neurones mediating ACTH release in response to immunological versus other physical or psychological stressors, presumably reflecting the importance of responding to immune-mediated stimuli that directly threatens host survival. Of additional interest is the observation that, in contrast to males, AA female rats are unable to mount a robust corticosterone response to acute endotoxin treatment, possibly relating to significantly higher basal glucocorticoid secretion rates and impaired adrenal responsiveness (Harbuz et al., 1999).

Studies in the late 1980s suggested that susceptibility to autoimmune disease may be linked to a defect in CRF regulation at the level of the PVN and a subsequent inability to mount an HPA-axis response (MacPhee et al., 1989; Sternberg et al., 1989). This inability to damp down the endogenous immune response could thus precipitate autoimmunity. Although a compelling hypothesis, subsequent studies have noted a number of exceptions to this. It is now believed that, although the HPA axis has a major role to play in determining severity of disease, susceptibility is more likely to reflect the balance of pro- and anti-inflammatory factors. Both neuroendocrine and immune factors have been implicated although the exact relationship remains to be determined (Harbuz, 2002).

HPA-axis activity and clinical inflammatory disease

An inability to respond appropriately to novel stressful stimuli will influence host integrity and have serious implications for the long-term health of the individual. There are clear correlations between preclinical findings and clinical data as patients with rheumatoid arthritis (RA) experience defective glucocorticoid responses to the stress of surgery (Chikanza et al., 1992), although this is a contentious issue. A widely held hypothesis is that defective regulation of the HPA axis and the associated excessive secretion of powerful glucocorticoids will cause prolonged immunosuppression and dysregulation of immune cells, ultimately predisposing to autoimmune disease. However, alterations in HPA-axis activity in patients with RA, for example, are not reliably discernible. There is clear involvement of glucocorticoids in the disease process since treatment of RA patients with a cortisol synthesis inhibitor, metyrapone, profoundly worsens symptoms of inflammation (Saldanha et al., 1986). This observation suggests that the HPA axis exerts inhibitory control over the disease process in RA and thereby regulates disease severity. Whether derangements in HPA-axis integrity contribute to disease progression in RA, however, is incompletely understood. There is some evidence in favour of aberrant HPA-axis activity in patients with chronic inflammatory conditions (Dekkers et al., 2000). Recent unpublished observations in our laboratory suggest that there may be sub-populations of patients in RA with altered glucocorticoid regulation. These sub-populations may explain some of the discrepancies in the literature. In one study RA patients did not show deficits in a CRF challenge test, indicating that pituitary ACTH and cortisol secretion were largely unaffected by the disease process. However, other studies have identified increased circulating levels of ACTH in RA, without change in plasma cortisol levels (see Morand and Leech, 2001). Evidence is in favour of an underactive HPA axis in RA, since the absence of elevated plasma glucocorticoids during persistent inflammation points towards a state of *adrenal hyporesponsiveness*. Importantly, studies of disease aetiology in man may be complicated by subject medication, such as prostaglandin synthesis inhibitors (Hall et al., 1994). Clearly, it is essential that adaptations to *chronic inflammatory stress* be considered in light of pre-existing therapeutic treatment since such adaptive changes can occur secondary to drug intervention. Additionally, in the case of clinical inflammatory disease, it is difficult to discern whether the HPA-axis dysfunction is of primary aetiological importance or is a state marker of the disease.

Even during chronic stress, it is vital that the HPA-axis response to acute stressors is maintained. Pre-existing inflammatory disease in man may have serious consequences for the ability to appropriately respond to novel stressful situations. Indeed, a failure to respond to acute activation of the HPA axis may have implications for *stress coping* in individuals. Individuals vary in their ability to cope with stressful situations. In addition to genetic factors, a number of external factors influence propensity to stress, including childhood trauma, other early environmental factors, major life events or infections (Chrousos, 1998; Sanchez et al., 2001). These can influence the development or adaptation of stress responses, in many cases exerting long-lasting effects. For example, in man, traumatic events may promote the onset of disease or exacerbate existing conditions, including RA (Marcenaro et al., 1999) and mutliple sclerosis (Mohr et al., 2000). Evidently, the factors regulating responses to stress are highly complex and changes in basal HPA activity alone cannot explain the phenomena of disease onset, progression and outcome. However, further advances through clinical and preclinical research will improve our understanding of the mechanisms driving dysregulation in the stress system and the consequences for disease.

Endogenous opioids and integrated stress responses

In addition to CRF and noradrenaline, there are a number of other neuromediators strongly implicated in the regulation of responses to stress. The opioid peptides and more recently, certain *opioid-like peptides*, have been identified as specific peptide transmitter molecules that potently modulate both the HPA axis and the immune system and are therefore important regulators of the dynamic communication network between the two.

Evidence that opioids are important regulators of the HPA axis has come from research in rats showing that acute administration of *morphine* or related opioid agonists induces activation of the HPA axis and increased levels of plasma ACTH and corticosterone (Ignar and Kuhn, 1990; Martinez et al., 1990; see Pechnick, 1993). Other opiate agonists, including kappa- and delta-opiate receptor ligands also stimulate the axis (Gonzalvez et al., 1991; Laorden and Milanes, 2000), suggesting that all major classes of opioid are able to stimulate this stress pathway. The effects of the opioids appear to involve stimulation of CRF-containing neurones by either a direct or indirect mechanism of activation (Martinez-Pinero et al., 1994b). In addition to exogenous opiate drugs, endogenous opioid peptides are also able to activate the HPA axis when administered into the CNS. Certain acute stressors are able to stimulate expression of mRNA for the opioid peptide,

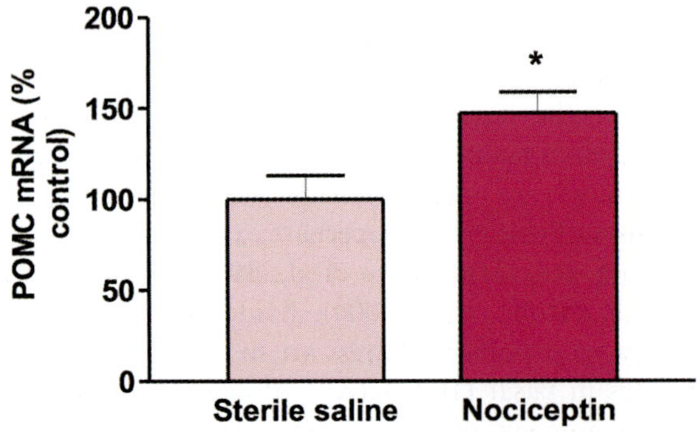

Fig. 3. The opioid-like peptide, nociceptin, induces activation of the HPA axis in rats when administered by intracerebroventricular (i.c.v.) injection. Graph shows the effect of nociceptin (1 μg/rat in 5 μl sterile saline) on mRNA expression of the ACTH precursor, POMC, in the adenohypophysis of male Sprague–Dawley rats 4 h following i.c.v. injection. Nociceptin caused a significant increase (approximately 50%) in expression of POMC mRNA that is associated with HPA-axis activation.

enkephalin, in the PVN, suggesting that this endogenous opioid is important in regulating the central control of adaptations to stress (Harbuz et al., 1994b). The ACTH precursor, POMC, is also subject to cleavage into the endogenous opioid, β-endorphin. Thus, in response to CRF stimulation during stress, synthesis and release of pituitary β-endorphin is increased (Young et al., 1986). It is of interest to note that the endogenous μ-opiate peptides, the endomorphins, are unable to stimulate the HPA axis or modulate the HPA response to morphine suggesting that other μ-opiate receptor subtypes may be responsible for the effects of morphine in vivo (Coventry et al., 2001).

Recently, the opioid-like peptide, *nociceptin*, has also been demonstrated to activate the HPA axis (Devine et al., 2001; Fulford et al., 2002) by stimulation of CRF expression in the parvocellular PVN leading to increased pituitary POMC mRNA expression (see Fig. 3). Nociceptin is an endogenous opioid-like molecule that is widely distributed throughout the CNS and is involved in regulation of phenomena strongly linked to stress responding, including pain modulation, motivation, anxiety state, autonomic and endocrine control and feeding (Calo et al., 2000). Some evidence supports a role for nociceptin in the regulation of the opioidergic mechanisms (see Harrison et al., 1998); thus the homologous nociceptin system may represent a system for fine-tuning opioidergic responses to stress. Clearly, further preclinical and clinical research is necessary to understand the exact role served by both the opioids and opioid-like mediators in the coordinated response to acute and chronic stress. Dysregulation of endogenous opioids may contribute significantly to the aetiology of human disorders of stress adaptation, such as affective disorders and anxiety. Furthermore, stress is widely recognised as a factor contributing to drug misuse, and observations of chronic morphine-induced tolerance of HPA-axis activity (Buckingham and Cooper, 1984; Martinez-Pinero et al., 1994a) provides strong evidence in support of adaptations in opioid–HPA axis interactions bearing clinical significance.

Peripheral peptides and immune regulation

Outside the CNS, a major site for localisation of the endogenous opioids, e.g. endomorphins (Jessop et al., 2000a), and the opioid-like peptide, nociceptin (Pampusch et al., 2000), is the immune system. This is not an unusual finding as many neuropeptides, including those regulating the HPA axis, are also synthesised and released from immune cells. CRF, AVP, substance P and POMC-derived peptides (ACTH, α-melanocyte-stimulating hormone and β-endorphin) are expressed by a wide variety of immunocyte populations including leukocytes, macrophages and monocytes, in addition to the resident cells of the lymphoid organs, the spleen and thymus (see Fulford and Jessop, 2001). These peptides appear to function as local immune regulators or modulators of inflammation as immune cells also express receptors that bind the endogenous peptides (see Jessop, 2002; Walker, 2003). These immune-derived peptides exert potent and complex effects over very low-concentration ranges (10^{-10} to 10^{-14} M) demonstrating the existence of high-affinity receptors mediating discrete paracrine/autocrine functions of these endogenous ligands (see Sharp et al., 1998). In many cases, the immune-derived population of peptides and receptors are identical to those found in the CNS and endocrine system; however, immune cells may also generate unique peptide variants or receptor proteins, with subtle structural differences or ligand-binding characteristics. Such heterogeneity points toward unique functional properties of *immunoneuropeptides* that control immune homeostasis. Indeed, molecular techniques, such as the use of antisense oligonucleotides, have been important in the functional characterisation of immune-derived peptides (Fulford et al., 2000).

In response to stimulation immune cells rapidly increase the production of the peptides, which are then readily released into the extracellular environment to modulate local inflammatory processes. In this way, immune-derived peptides can contribute significantly to peripheral inflammation and the regulation of stress responses. A number of the immune-derived peptides exert potent pro-inflammatory actions, including CRF and substance P (see Jessop et al., 2001; Santoni et al., 2002); however, other peptides exert anti-inflammatory effects including the enkephalins (Fulford et al., 2000) and endomorphins (Khalil et al., 1999). Recently, the opioid-like peptide, nociceptin and its receptor have been found in mammalian immune cells (Pampusch et al., 2000). This system also appears to play a significant role in the modulation of immune cell proliferation (Peluso et al., 2001). It is likely that the peripheral inflammation is regulated through dynamic interactions between the sympathetic nervous system, HPA axis and immune system with a balance required between pro- and anti-inflammatory mediators.

Experimental studies in rodents have revealed that in conditions associated with chronic inflammatory stress, such as AA, the production of certain peptides, such as the opioids, is markedly increased at peripheral sites of inflammation, in addition to increased immunoreactivity in lymphoid organs (*spleen and thymus*) (Jessop et al., 1995). It has been proposed that the enhanced levels of the opioids, met-enkephalin and dynorphin, may act to inhibit further release of nociceptive mediators, such as substance P, from peripheral sensory nerve endings (see Machelska, 2003). These peptides may also be co-released by sensory afferent neurones and sympathetic nerve endings to modulate inflammation. Opioids are expressed by a range of immunocytes, including macrophages and lymphocytes, that can migrate to sites of inflammation to modulate local immune responses (Smith, 2003). By such mechanisms, peripheral opioids may function to limit the damage caused by an inflammatory response. Taken together these findings support the contention that peripheral, immune-derived opioids are significant contributors to the body's response to inflammatory stress that may actually function to modulate disease progression. A condition associated with chronic inflammatory disease may be associated with an imbalance between pro- and anti-inflammatory mechanisms, with dominance of proinflammatory mediators prolonging inflammation. The physiological relevance of immune-derived neuropeptides is only now being recognised, indeed it is possible that these will represent targets for therapeutic intervention in inflammatory states.

Conclusions

The HPA axis is subject to dynamic regulation to ensure activity is maintained within close limits to prevent extreme fluctuations in background activity. Stress impacts at all levels of the HPA axis and prompts adaptive responses commensurate with the type, intensity and duration of the stimulus. Individual characteristics including sex, age, health, childhood experiences and propensity to stress are additional factors that shape the stress response. Stress, in essence, is a defensive response, that if unchecked may lead to damaging effects by predisposing to illness or disease. Our understanding of factors that contribute to the *integration of stress* responses continues apace. Adaptation of the HPA axis to stress relies on complex interplay between multiple body systems, encompassing changes in emotional and cognitive behaviour, autonomic outflow, endocrine and immune function. Maintenance of the balance between pro- and anti-inflammatory responses is the key to safe-guarding homeostasis and the prevention of inflammation. Advances in molecular technologies combined with integrated approaches to the study of human pathophysiology promise enhanced understanding of the mechanisms underlying stress adaptations and the factors conferring susceptibility or resilience to stress.

Abbreviations

AA adjuvant-induced arthritis
ACTH adrenocorticotropic hormone

AVP arginine vasopressin
CRF corticotropic-releasing factor
GABA gamma-aminobutyric acid
HPA hypothalamo–pituitary–adrenal axis
POMC pro-opiomelanocortin
PVN paraventricular nucleus

References

Aguilera, G., Jessop, D.S., Harbuz, M.S., Kiss, A. and Lightman, S.L. (1997) Biphasic regulation of hypothalamic–pituitary corticotropin releasing hormone receptors during development of adjuvant-induced arthritis in the rat. J. Endocrinol., 153: 185–191.

Aird, F., Clevenger, C.V., Prystowsky, M.B. and Redei, E. (1993) Corticotropin-releasing factor mRNA in rat thymus and spleen. Proc. Natl. Acad. Sci. USA, 90: 7104–7108.

Akana, S.F., Jacobson, L., Cascio, C.S., Shinsako, J. and Dallman, M.F. (1988) Constant corticosterone replacement normalises basal adrenocorticotropin (ACTH) but permits sustained ACTH hypersecretion after stress in adrenalectomised rats. Endocrinology, 122: 1337–1342.

Antoni, F.A. (1986) Hypothalamic control of adrenocorticotropin secretion: advances since the discovery of 41-residue corticotropin-releasing factor. Endocrinol. Rev., 7: 351–378.

Arriza, J.L., Simmerly, R.B., Swanson, L.W. and Evans, R.M. (1988) The neuronal mineralocorticoid receptor as a mediator of the glucocorticoid response. Neuron, 1: 887–900.

Banks, W.A., Kastin, A.J. and Broadwell, R.D. (1995) Passage of cytokines across the blood brain barrier. Neuroimmunomodulation, 2: 241–248.

Besedovsky, H.O. and del Rey, A. (2002) Introduction: immune-neuroendocrine network. Front. Horm. Res., 29: 1–14.

Besedovsky, H.O., del Rey, A., Sorkin, E., Da Prada, M., Burri, R. and Honegger, C. (1991) Cytokines as modulators of the hypothalamus–pituitary–adrenal axis. J. Steroid Biochem. Mol. Biol., 40: 613–618.

Blalock, J.E. and Smith, E.M. (1985) A complete regulatory loop between the immune and neuroendocrine systems. Fed. Proc., 44: 108–111.

Buckingham, J.C. (1979) The influence of corticosteroids on the secretion of corticotropin and its hypothalamic releasing hormone. J. Physiol., 286: 331–342.

Buckingham, J.C. and Cooper, T.A. (1984) Differences in hypothalamo–pituitary–adrenocortical activity in the rat after acute anf prolonged treatment with morphine. Neuroendocrinol., 38: 411–417.

Buckingham, J.C., Loxley, H.D., Christian, H.C. and Philip, J.G. (1996) Activation of the HPA axis by immune insults: roles and interactions of cytokines, eicosanoids and glucocorticoids. Pharmacol. Biochem. Behav., 54: 285–298.

Buijs, R.M., Hermes, M.H., van der Woude, T.P., van Heerikhuize, J.J. and Shinn, S. (1998) The suprachiasmatic nucleus-paraventricular nucleus interactions: a bridge to the neuroendocrine and autonomic nervous system. Prog. Brain Res., 119: 365–382.

Cabot, P.J., Carter, L., Schafer, M. and Stein, C. (2001) Methionine-enkephalin and dynorphin A release from immune cells and control of inflammatory pain. Pain, 93: 207–212.

Calo, G., Guerrini, R., Rizzi, A., Salvadori, S. and Regoli, D. (2000) Pharmacology of nociceptin and its receptor: a novel therapeutic target. Br. J. Pharmacol., 129: 1261–1283.

Calogero, A., Galluci, W.T., Gold, P.W. and Chrousos, G.P. (1988) Multiple regulatory feedback loops on hypothalamic corticotropin releasing hormone secretion. J. Clin. Invest., 82: 767–774.

Cannon, W.B. (1939) The Wisdom of the Body. W.W. Norton, New York.

Canteras, N.S., Simerley, R.B. and Swanson, L.W. (1995) Organisation of projections from the medial nucleus of the amygdala: a PHAL study in the rat. J. Comp. Neurol., 360: 213–245.

Cascio, C.S., Shinsako, J. and Dallman, M.F. (1987) The suprachiasmatic nuclei stimulate evening ACTH secretion in the rat. Brain Res., 423: 173–178.

Chan, R.K., Brown, E.R., Ericsson, A., Kovacs, K.J. and Sawchenko, P.E. (1993) A comparison of two immediate early genes, c-fos and NGFI-B, as markers for functional activation in stress-related neuroendocrine circuitry. J. Neurosci., 13: 5126–5138.

Chikanza, I.C., Petrou, P., Kinglsey, G., Chrousos, G.P. and Panayi, G.S. (1992) Defective hypothalamic response to immune and inflammatory stimuli in patients with rheumatoid arthritis. Arthritis. Rheum., 35: 1281–1288.

Chowdrey, H.S., Larsen, P.J., Harbuz, M.S., Jessop, D.S., Aguilera, G., Eckland, D.J.A. and Lightman, S.L. (1995) Evidence for arginine vasopressin as the primary activator of the HPA axis during adjuvant-induced arthritis. Br. J. Pharmacol., 116: 2417–2424.

Chrousos, G.P. (1998) Stressors, stress and neuroendocrine integration of the adaptive response: The 1997 Hans Selye Memorial Lecture. Ann. N.Y. Acad. Sci., 851: 311–335.

Chuluyan, H.E., Saphier, D., Rohn, W.M. and Dunn, A.J. (1992) Noradrenergic innervation of the hypothalamus participates in adrenocortical responses to interleukin-1. Neuroendocrinology, 56: 106–111.

Coventry, T.L., Jessop, D.S., Finn, D.P., Crabb, M.D., Kinoshita, H. and Harbuz, M.S. (2001) Endomorphins and activation of the hypothalamo–pituitary–adrenal axis. J. Endocrinol., 169: 185–193.

Cullinan, W.E., Herman, J.P. and Watson, S.J. (1993) Ventral subicular interaction with the hypothalamic paraventricular nucleus: evidence for a relay in the bed nucleus of the stria terminalis. J. Comp. Neurol., 332: 1–20.

Cunningham, E.T. and Sawchenko, P.E. (1988) Anatomical specificity of noradrenergic inputs to the paraventricular and supraoptic nuclei of the rat hypothalamus. J. Comp. Neurol., 274: 60–76.

Daftary, S.S., Boudaba, C. and Tasker, J.G. (2000) Noradrenergic regulation of parvocellular neurones in the rat hypothalamic paraventricular nucleus. Neuroscience, 96: 742–751.

Dallman, M.F., Levin, N., Cascio, C.S., Akana, S.F., Jacobson, L. and Kuhn, R.W. (1989) Pharmacological evidence that the inhibition of diurnal adrenocorticotropin secretion by corticosteroids is mediated via type I corticosterone-preferring receptors. Endocrinology, 124: 2844–2850.

Dantzer, R. (2001) Cytokine-induced sickness behaviour: mechanisms and implications. Ann. N.Y. Acad. Sci., 933: 222–234.

de Goeji, D.C.E., Kvetnansky, R., Whitnall, M.H., Jesova, D., Berkenbosch, F. and Tilders, F.J.H. (1991) Repeated stress-induced activation of corticotropin-releasing factor neurones enhances vasopressin stores and colocalisation with corticotropin-releasing factor in the median eminence of rats. Neuroendocrinol., 53: 150–159.

Dekkers, J.C., Geenen, R., Godaert, G.L., van Doornen, L.J. and Bijlsma, J.W. (2000) Diurnal rhythm of salivary cortisol levels in patients with recent onset rheumatoid arthritis. Arthritis Rheum., 43: 465–467.

De Kloet, E.R., Vreugdenhil, M.S., Oitzl, M.S. and Joels, M. (1998) Brain corticosteroid receptor balance in health and disease. Endocr. Rev., 19: 269–301.

Devine, D.P., Watson, S.J. and Akil, H. (2001) Nociceptin/orphaninFQ regulates neuroendocrine function of the limbic–hypothalamic–pituitary–adrenal axis. Neuroscience, 102: 541–553.

Ehrhart-Bornstein, M., Haidan, A., Alesci, S. and Bornstein, S.R. (2000) Neurotransmitters and neuropeptides in the differential regulation of steroidogenesis in adrenocortical-chromaffin co-cultures. Endocr. Res., 26: 833–842.

Ericsson, A., Kovacs, K.J. and Sawchenko, P.E. (1994) A functional anatomical analysis of central pathways subserving the effects of interleukin-1 on stress-related neuroendocrine neurones. J. Neurosci., 14: 897–913.

Farrar, W.L., Kilian, P.L., Ruff, M.R., Hill, J.M. and Pert, C.B. (1987) Visualisation and characterisation of interleukin-1 receptors in brain. J. Immunol., 139: 459–463.

Fulford, A.J. and Jessop, D.S. (2001) Neuropeptides in the immune system: mediators of the stress response. In: McCarthy, R., Aguilera, G., Sabban, E. and Kvetnansky, R. (Eds.), Stress: Neural, Endocrine and Molecular Studies. Taylor & Francis Ltd., London.

Fulford, A.J., Harbuz, M.S. and Jessop, D.S. (2000) Antisense inhibition of pro-opiomelanocortin and proenkephalin A messenger RNA translation alters rat immune cell function in vitro. J. Neuroimmunol., 106: 6–13.

Fulford, A.J., Harbuz, M.S., McLeod, J.D., Purcell, W.M. and Waits, P.S. (2002) Central nociceptin increases corticotropin-releasing factor MRNA expression in the hypothalamic paraventricular nucleus and corticosterone release in the rat. The Pharmacologist, 44(2) suppl. 1: A101.

Fuxe, K., Wikstrom, A.C., Okret, S., Agnati, L.F., Harfstrand, A., Yu, Z.-Y., Granholm, L., Zoli, M., Vale, W. and Gustaffson, I.-A. (1985) Mapping of glucocorticoid receptor immunoreactive neurones in the rat tel- and diencephalon using a monoclonal antibody against rat liver glucocorticoid receptor. Endocrinol., 117: 1803–1812.

Gaillard, R.C. (2001) Interaction between the hypothalamo–pituitary–adrenal axis and the immunological system. Ann. Endocrinol., 62: 155–163.

Gaillet, S., Lachuer, J., Malaval, F., Assenmacher, I. and Szafararczyk, A. (1991) The involvement of noradrenergic ascending pathways in the stress-induced activation of ACTH and corticosterone secretions is dependent on the nature of stressors. Exp. Brain Res. 87: 173–180.

Gambacciani, M., Liu, J.H., Swartz, L.H., Tuaros, V.S., Rasmussen, D.D. and Yen, S. (1987) Intrinsic pulsatility of ACTH release from the human pituitary in vitro. Clin. Endocrinol., 26: 557–563.

Gillies, G.E., Linton, E.A. and Lowry, P.J. (1982) Corticotropin-releasing activity of the new CRF is potentiated several times by vasopressin. Nature, 299: 355–357.

Gloddek, J., Lohrer, P., Stalla, J., Arzt, E., Stalla, G.K. and Renner, U. (2001) The intrapituitary stimulatory effect of lipopolysaccharide on ACTH secretion is mediated by paracrine acting IL-6. Exp. Clin. Endocrinol. Diab., 109: 410–415.

Gonzalvez, M.L., Milanes, M.V. and Vargas, M.L. (1991) Effects of acute and chronic administration of mu- and delta-opioid agonists on the hypothalamic–pituitary–adrenocortical (HPA) axis in the rat. Eur. J. Pharmacol., 200: 155–158.

Gudmundsson, A. and Carnes, M. (1997) Pulsatile adrenocorticotropic hormone: an overview. Biol. Psychiat., 41: 342–365.

Habu, S., Watanobe, H., Yasujima, M. and Suda, T. (1998) Different roles of brain interleukin-1 in the adrenocorticotropin response to central versus peripheral administration of lipopolysaccharide in the rat. Cytokine, 10: 390–394.

Hall, J., Morand, E.F., Medbak, S., Zaman, M., Perry, L., Goulding, N.J., Maddison, P.J. and O'Hare, J.P. (1994) Abnormal hypothalamic–pituitary–adrenal axis function in rheumatoid arthritis. Effects of nonsteroidal antiinflammatory drugs and water immersion. Arthritis Rheum., 37: 1132–1137.

Harbuz, M.S. (2002) Neuroendocrinology of autoimmunity. In: Clow, A. and Hucklebridge, F. (Eds.), Neurobiology of the Immune System, International Review of Neurobiology, Vol. 52. Elsevier Science, San Diego, USA.

Harbuz, M.S. and Lightman, S.L. (1989) Responses of hypothalamic and pituitary mRNA to physical and psychological stress in the rat. J. Endocrinol., 122: 705–711.

Harbuz, M.S. and Lightman, S.L. (1992) Stress and the hypothalamo–pituitary adrenal axis: acute, chronic and immunological activation. J. Endocrinol., 134: 327–339.

Harbuz, M.S., Russell, J.A., Sumner, B.E., Kawata, M. and Lightman, S.L. (1991) Rapid changes in the content of proenkephalin A and corticotrophin releasing hormone mRNAs in the paraventricular nucleus during morphine withdrawal in urethane-anaesthetised rats. Brain Res. Mol. Brain Res., 9: 285–291.

Harbuz, M.S., Rees, R.G., Eckland, D., Jessop, D.S., Brewerton, D. and Lightman, S.L. (1992) Paradoxical responses of hypothalamic CRF mRNA and CRF-41 peptide and adenohypophyseal POMC mRNA during chronic inflammatory stress. Endocrinology, 130: 1394–1400.

Harbuz, M.S., Leonard, J.P., Lightman, S.L. and Cuzner, M.L. (1993) Changes in hypothalamic corticotrophin releasing factor (CRF) and pituitary proopiomelanocortin (POMC) messenger RNA during the course of experimental allergic encephalomyelitis (EAE). J. Neuroimmunol., 45: 127–132.

Harbuz, M.S., Chover-Gonzalez, A.J., Biswas, S., Lightman, S.L. and Chowdrey, H.S. (1994a) Role of central catecholamines in the modulation of corticotrophin-releasing factor mRNA during adjuvant-induced arthritis in the rat. Br. J. Rheumatol., 33: 205–209.

Harbuz, M.S., Jessop, D.S., Lightman, S.L. and Chowdrey, H.S. (1994b) The effects of restraint or hypertonic saline stress on corticotrophin-releasing factor, arginine vasopressin, and proenkephalin A mRNAs in the CFY, Sprague-Dawley and Wistar strains of rat. Brain Res., 667: 6–12.

Harbuz, M.S., Rooney, C., Jones, M. and Ingram, C.D. (1999) Hypothalamo–pituitary–adrenal responses to lipopolysaccharide in male and female rats with adjuvant-induced arthritis. Brain Behav. Immunity, 13: 335–347.

Harris, G. (1948) Neural control of the pituitary gland. Physiol. Rev., 28: 139–179.

Harrison, L.M., Kastin, A.J. and Zadina, J.E. (1998) Opiate tolerance and dependence: receptors, G-proteins and antiopiates. Peptides, 19: 1603–1630.

Herman, J.P., Cullinan, W.E., Young, E.A., Akil, H. and Watson, S.J. (1992) Selective forebrain fibre tract lesions implicate ventral hippocampal structures in tonic regulation of paraventricular nucleus corticotropin-releasing hormone (CRH) and arginine vasopressin (AVP) mRNA expression. Brain Res., 592: 228–238.

Herman, J.P., Cullinan, W.E. and Watson, S.J. (1994) Involvement of the bed nucleus of the stria terminalis in tonic regulation of paraventricular hypothalamic CRH and AVP mRNA expression. J. Neuroendocrinol., 6: 433–442.

Herman, J.P., Tasker, J.G., Ziegler, D.R. and Cullinan, W.E. (2002) Local circuit regulation of paraventricular nucleus stress integration. Glutamate-GABA connections. Pharmacol. Biochem. Behav., 71: 457–468.

Hopkins, S.J. and Rothwell, N.J. (1995) Cytokines and the nervous system: I. Expression and recognition. Trends Neurosci., 18: 83–88.

Horrocks, P.M., Jones, A.F. and Ratcliffe, W.A. (1990) Patterns of ACTH and cortisol pulsatility over twenty-four hours in normal males and females. Clin. Endocrinol., 32: 127–134.

Ignar, D.M. and Kuhn, C.M. (1990) Effects of specific Mu and Kappa opiate tolerance and abstinence on hypothalamo–pituitary–adrenal axis secretion in the rat. J. Pharmacol. Exp. Ther., 255: 1287–1295.

James, V.H. and Few, J.D. (1985) Adrenocorticosteroids: chemistry, synthesis and disturbances in disease. Clin. Endocrinol. Metab., 14: 867–892.

Jessop, D.S. (2002) Neuropeptides in the immune system: functional roles in health and disease. Front. Horm. Res., 29: 50–68.

Jessop, D.S., Renshaw, D., Lightman, S.L. and Harbuz, M.S. (1995) Changes in ACTH and β-endorphin immunoreactivity in immune tissues during a chronic inflammatory stress are not correlated with changes in corticotropin-releasing hormone and arginine vasopressin. J. Neuroimmunol., 60: 29–35.

Jessop, D.S., Major, G.N., Coventry, T.L., Kaye, S.J., Fulford, A.J., Harbuz, M.S. and De Bree, F.M. (2000a) Novel opioid peptides endomorphin-1 and endomorphin-2 are present in mammalian immune tissues. J. Neuroimmunol., 106: 53–59.

Jessop, D.S., Renshaw, D., Larsen, P.J., Chowdrey, H.S. and Harbuz, M.S. (2000b) Substance P is involved in terminating the hypothalamo–pituitary–adrenal axis response to acute stress through centrally located neurokinin-1 receptors. Stress, 3: 209–220.

Jessop, D.S., Harbuz, M.S. and Lightman, S.L. (2001) CRH in chronic inflammatory stress. Peptides, 22: 803–807.

Jingami, H., Matsukura, S., Numa, S. and Imura, H. (1985) Effects of adrenalectomy and dexamethasone administration on the level of prepro-corticotropin-releasing factor messenger ribonucleic acid in the hypothalamus and adrenocorticotropin/β-lipotropin precursor mRNA in the pituitary in rats. Endocrinology, 117: 1314–1320.

Kalsbeek, A., van Heerikhuize, J.J., Wortel, J. and Buijs, R.M. (1996) A diurnal rhythm of stimulatory input to the hypothalamo–pituitary–adrenal system as revealed by timed intrahypothalamic administration of the vasopressin V1 antagonist. J. Neurosci., 16: 5555–5565.

Kaneko, M., Kaneko, K., Shinsako, J. and Dallman, M.F. (1981) Adrenal sensitivity to adrenocorticotropin varies diurnally. Endocrinology, 109: 70–75.

Kant, G.J., Meyerhoff, J.L., Bunnell, B.N. and Lennox, R.H. (1982) Cyclic AMP and cyclic GMP response to stress in brain and pituitary: stress elevates pituitary cyclic AMP. Pharmacol. Biochem. Behav., 17: 1967–1072.

Kant, G.J., Eggleston, T., Landman-Roberts, L., Kenion, C.C., Driver, G.C. and Meyerhoff, J.L. (1985) Habituation to repeated stress is stressor specific. Pharmacol. Biochem. Behav., 22: 631–634.

Karalis, K., Sano, H., Redwine, J., Listwak, S., Wilder, R.L. and Chrousos, C.P. (1991) Autocrine or paracrine inflammatory actions of corticotropin releasing hormone in vivo. Science, 254: 421–423.

Katsuura, G., Gottshall, P.E. and Arimura, A. (1988) Identification of a high affinity receptor for interleukin-1 beta in rat brain. Biochem. Biophys. Res. Commun., 156: 61–67.

Keller-Wood, M.E. and Dallman, M.F. (1984) Corticosteroid inhibition of ACTH secretion. Endocrine Rev., 5: 1–24.

Khalil, Z., Sanderson, K., Modig, M. and Nyberg, F. (1999) Modulation of peripheral inflammation by locally administered endomorphin-1. Inflamm. Res., 48: 550–556.

Koenig, J.I. (1991) Presence of cytokines in the hypothalamo–pituitary axis. Prog. Neuroendocr. Immunol., 48: 160–166.

Kononen, J., Honkaniemi, J., Alho, H., Koistinaho, J., Iadarola, M. and Pelto-Huikko, M. (1992) Fos-like immunoreactivity in the rat hypothalamic–pituitary axis after immobilisation stress. Endocrinology, 130: 3041–3047.

Koob, G.F. (1999) Corticotropin-releasing factor, norepineprhine and stress. Biol. Psychiat., 46: 1167–1180.

Krishnan, K.R., Ritchie, J.C., Manepalli, A.N., Saunders, W., Li, S.W., Venkataraman, S. and Nemeroff, C.B. and Carroll, B.J. (1991) Fast feedback regulation of ACTH by cortisol. Prog. Neuropsychopharmacol. Biol. Psychiat., 15: 523–529.

Laorden, M.L. and Milanes, M.V. (2000) Effects of U-50,488H and U-50,488H withdrawal on catecholaminergic neurones of the rat hypothalamus. Life Sci., 66: 803–815.

Lapiz, M.D., Fulford, A.J., Muchimapura, S., Mason, R., Parker, T. and Marsden, C.A. (2003) Influence of postweaning social isolation in the rat on brain development, conditioned behaviour and neurotransmission. Neurosci. Behav. Physiol., 33: 13–29.

Lightman, S.L. and Young, W.S. (1988) Corticotrophin-releasing factor, vasopressin and proopiomelanocortin mRNA responses to stress and opiates in the rat. J. Physiol., 403: 511–523.

Lightman, S.L. and Harbuz, M.S. (1993) Expression of corticotropin-releasing factor mRNA in response to stress. Ciba Found Symp., 172: 173–187.

Liposits, Z., Phelix, C. and Paull, W.K. (1987) Synaptic interaction of serotonergic axons and corticotropin releasing factor (CRF) synthesising neurones in the hypothalamic paraventricular nucleus of the rat: a light and electron micorscopic immunocytochemical study. Histochem., 86: 541–549.

Machelska, H. (2003) Functional evidence of pain control by the immune system. Adv. Exp. Med. Biol., 521: 88–97.

MacPhee, I.A., Antoni, F.A. and Mason, D.W. (1989) Spontaneous recovery of rats from experimental allergic encephalomyelitis is dependent on regulation of the immune system by endogenous adrenal corticosteroids. J. Exp. Med., 169: 431–435.

Maier, S.F. and Watkins, L.R. (1998) Cytokines for psychologists: implications of bi-directional immune-to-brain communication for understanding behaviour, mood and cognition. Psychol. Rev., 105: 83–107.

Marcenaro, M., Prete, C., Badini, A., Sulli, A., Magi, E. and Cutolo, M. (1999) Rheumatoid arthritis, personality, stress response style and coping with illness. A preliminary survey. Ann. N.Y. Acad. Sci., 876: 419–425.

Marti, O., Harbuz, M.S., Andres, R., Lightman, S.L. and Armario, A. (1999) Activation of the hypothalamo–pituitary–adrenal axis in adrenalectomised rats: potentiation by chronic stress. Brain Res., 821: 1–7.

Martinez, J.A., Vargas, M.L., Fuente, T., de Rio-Garcia, J. and Milanes, M.V. (1990) Plasma β-endorphin and cortisol levels in morphine-tolerant rats and in naloxone-induced withdrawal. Eur. J. Pharmacol., 182: 117–123.

Martinez-Pinero, M.G., Milanes, M.V. and Vargas, M.L. (1994a) Modulation by catecholamine of hypothalamus–pituitary–adrenocortical (HPA) axis activity in morphine-tolerance and withdrawal. Gen. Pharmacol., 25: 187–192.

Martinez-Pinero, M.G., Milanes, M.V., Alcaraz, C. and Vargas, M.L. (1994b) Catecholaminergic mediation of morphine-induced activation of pituitary–adrenocortical axis in the rat: implication of alpha- and beta-adrenoceptors. Brain Res., 668: 122–128.

McEwen, B.S. and Stellar, E. (1993) Stress and the individual. Mechanisms leading to disease. Arch. Intern. Med., 153: 2093–2101.

Meister, B., Hokfelt, T., Geffard, M. and Oertel, W. (1988) Glutamic acid decarboxylase- and GABA-like immunoreactivities in corticotropin-releasing factor containing parvocellular neurones of the hypothalamic paraventricular nucleus. Neuroendocrinology, 48: 516–526.

Mohr, D.C., Goodkin, D.E., Bacchetti, P., Boudewyn, A.C., Hunag, L., Marrietta, P., Cheuk, W. and Dee, B. (2000) Psychological stress and the subsequent appearance of new brain MRI lesions in MS. Neurology, 55: 55–61.

Morand, E.F. and Leech, M. (2001) Hypothalamic–pituitary–adrenal axis regulation of inflammation in rheumatoid arthritis. Immunol. Cell Biol., 79: 395–399.

Munck, A. and Naray Fejes-Toth, A. (1992) The ups and downs of glucocorticoid physiology. Permissive and suppressive effects revisited. Mol. Cell Endocrinol., 90: C1–C4.

Munck, A. and Naray Fejes-Toth, A. (1994) Glucocorticoids and stress: permissive and suppressive actions. Ann. N.Y. Acad. Sci., 746: 115–130.

Navarra, P., Schettini, G. and Grossman, A.B. (1997) Responses of the stress axis to immunological challenge: the role of eicosanoids and cytokines. In: Buckingham, J.C., Gillies, G.E. and Cowell, A.-M. (Eds.), Stress, Stress Hormones and the Immune System. Wiley, Chichester.

Nussdorfer, G.G. and Malendowicz, L.K. (1998) Role of tachykinins in the regulation of the hypothalamo–pituitary–adrenal axis. Peptides, 19: 949–968.

Ohmori, N., Itoi, K., Tozawa, F., Sakai, Y., Sakai, K., Horiba, N., Demura, H. and Suda, T. (1995) Effect of acetylcholine on corticotropin-releasing factor gene expression in the hypothalamic paraventricular nucleus of conscious rats. Endocrinology, 136: 4858–4863.

Pacak, K. and Palkovits, M. (2001) Stressor specificity of central neuroendocrine responses: implications for stress-related disorders. Endocrine Rev., 22: 502–548.

Palkovits, M. (1988) Distribution of neuropeptides in the brain: a review of biochemical and immunohistochemical studies. In: Negro-Vilar, A. and Conn, P.M. (Eds.), Peptide Hormones: Effects and Mechanisms of Action, Vol. 1, Neural Distribution of Brain Peptides and Changes during Aging. CRC, Boca Raton, FL, pp. 3–67.

Pampusch, M.S., Serie, J.R., Osinski, M.A., Seybold, V.S., Murtaugh, M.P. and Brown, D.R. (2000) Expression of nociceptin/OFQ receptor and prepro-nociceptin/OFQ in lymphoid tissues. Peptides, 21: 1865–1870.

Pechnick, R.N. (1993) Effect of opioids on the hypothalamo–pituitary–adrenal axis. Ann. Rev. Pharmacol. Toxicol., 32: 353–382.

Peluso, J., Gaveriaux-Ruff, C., Matthes, H.W., Filliol, D. and Kieffer, B.L. (2001) Orphanin FQ/nociceptin binds to functionally coupled ORL1 receptors on human immune cell lines and alters peripheral blood mononuclear cell proliferation. Brain Res. Bull., 54: 655–660.

Perlstein, R.S., Whitnall, M.H., Abrams, J.S. and Moougey, E.H. (1993) Synergistic roles of interleukin-6, interleukin-1 and tumor necrosis factor in adrenocorticotropin response to bacterial lipopolysaccharide in vivo. Endocrinology, 132: 946–963.

Quan, N., Whiteside, M. and Herkemham, M. (1998) Time course and localisation patterns of interleukin-1b messenger RNA expression in brain and pituitary after peripheral administration of lipopolysaccharide. Neuroscience, 83: 281–293.

Reul, J.M., Gesing, A., Droste, S., Stec, I.S., Weber, A., Bachmann, C., Bilang-Bleuel, A., Holsboer, F. and Linthorst, A.C. (2000) The brain mineralocorticoid receptor: greedy for ligand, mysterious in function. Eur. J. Pharmacol., 405: 235–249.

Rivest, S., Laflamme, N. and Nappi, R. (1995) Immune challenge and immobilisation stress induce transcription of the gene encoding the CRF receptor in selective nuclei of the rat hypothalamus. J. Neurosci., 15: 2680–2695.

Rivier, C. (1995) Influence of immune signals on the hypothalamic–pituitary axis in the rodent. Front. Neuroendocrinol., 16: 151–162.

Rivier, C. and Vale, W. (1983) Interaction of corticotropin-releasing factor and arginine vasopressin on adrenocorticotrophin secretion in vivo. Endocrinol., 13: 939–942.

Roelfsema, F., van den Berg, G., Frolich, M., Veldhuis, J.D., van Eijk, A., Buurman, M.M. and Etman, B.H. (1993) Sex-dependent alteration in cortisol response to endogenous adrenocorticotropin. J. Clin. Endocrinol. Metab., 59: 1159–1163.

Rook, G.A.W., Hernandez-Pando, R. and Lightman, S.L. (1994) Hormones, peripherally acting prohormones and regulation of thw Th1/Th2 balance. Immunol. Today, 15: 301–303.

Saldanha, C., Tougas, G. and Grace, E. (1986) Evidence for anti-inflammatory effect of normal circulating plasma cortisol. Clin. Exp. Rheumatol., 4: 365–366.

Sanchez, M.M., Ladd, C.O. and Plotsky, P.M. (2001) Early adverse experience as a developmental risk factor for later psychopathology: evidence from rodent and primate models. Dev. Psychopathol., 13: 419–449.

Santoni, G., Amantini, C., Lucciarini, R., Pompei, P., Perfumi, M., Nabissi, M., Morrone, S. and Piccoli, M. (2002) Expression of substance P and its neurokinin-1 receptor on thymocytes: functional relevance in the regulation of thymocyte apoptosis and proliferation. Neuroimmunomod., 10: 232–246.

Saper, C.B. and Breder, C.D. (1994) The neurologic basis of fever. New Eng. J. Med., 330: 1880–1886.

Sapolsky, R., Rivier, C.L., Yamamoto, G., Plotsky, P. and Vale, W. (1987) Interleukin-1 stimulates the secretion of hypothalamic corticotropin-releasing factor. Science, 238: 522–524.

Sarlis, N.J., Chowdrey, H.S., Stephanou, A. and Lightman, S.L. (1992) Activation of the hypothalamo–pituitary–adrenal axis and loss of diurnal rhythmicity during adjuvant induced arthritis in the rat. Endocrinology, 130: 1775–1779.

Sawchenko, P.E. (1987a) Evidence for differential regulation of corticotropin releasing factor and vasopressin immunoreactivities in parvocellular neurosecretory and autonomic-related projections of the paraventricular nucleus. Brain Res., 437: 253–263.

Sawchenko, P.E. (1987b) Adrenalectomy-induced enhancement of CRF and vasopressin immunoreactivity in parvocellular neurosecretory neurones: anatomic, peptide and steroid specificity. J. Neurosci., 7: 1093–1106.

Sawchenko, P.E. and Swanson, L.W. (1981) Central noradrenergic pathways for the integration of hypothalamic neuroendocrine and autonomic responses. Science, 214: 685–687.

Sawchenko, P.E. and Swanson, L.W. (1982) The organisation of noradrenergic pathways from the brainstem to the paraventricular and supraoptic nuclei in the rat. Brain Res. Rev., 4: 275–325.

Sawchenko, P.E. and Swanson, L.W. (1983) The organisation of the forebrain afferents to the paraventricular and supraoptic nuclei of the rat. J. Comp. Neurol., 218: 121–144.

Sawchenko, P.E., Li, H.-Y. and Ericsson, A. (2000) Circuits and mechanisms governing hypothalamic responses to stress: a tale of two paradigms. Prog. Brain Res., 122: 61–78.

Scott, L.V. and Dinan, T.G. (1998) Vasopressin and the regulation of hypothalamic–pituitary–adrenal axis function: implications for the pathophysiology of depression. Life Sci., 62: 1985–1998.

Selye, H. (1936) A syndrome produced by diverse nocuous agents. Nature, 138: 32.

Selye, H. (1946) The general adaptation syndrome and the diseases of adaptation. J. Clin. Endocrinol., 6: 117–230.

Shanks, N., Moore, P.M., Perks, P. and Lightman, S.L. (1997) Endocrine correlates of murine systemic lupus erythematosus in the MRL lpr/lpr model. Ann. N.Y. Acad. Sci., 823: 252–255.

Sharp, B.M. (2003) Opioid receptor expression and intracellular signalling by cells involved in host defense and immunity. Adv. Exp. Med. Biol., 521: 98–105.

Sharp, B.M., Roy, S. and Bidlack, J.M. (1998) Evidence for opioid receptors on cells involved in host defense and the immune system. J. Neuroimmunol., 83: 45–56.

Silverman, A.-J., Hoffman, D.L. and Zimmerman, E.A. (1981) The descending afferent connections of the paraventricular nucleus of the hypothalamus. Brain Res. Bull., 6: 47–61.

Smith, E.M. (2003) Opioid peptides in immune cells. Adv. Exp. Med. Biol., 521: 51–68.

Smith, E.M., Galin, F.S., LeBoeuf, R.D., Coppenhaver, D.H., Harbour, D.V. and Blalock, J.E. (1990) Nucleotide and amino acid sequence of lymphocyte-derived corticotropin: endotoxin induction of a truncated peptide. Proc. Natl. Acad. Sci. USA, 87: 1057–1060.

Spencer, R.L. and McEwen, B.S. (1990) Adaptation of the hypothalamo–pituitary–adrenal axis to chronic ethanol stress. Neuroendocrinol., 52: 481–487.

Sternberg, E.M., Young, W.S., Bernardini, R., Calogero, A.E., Chrousos, G.P., Gold, P.W. and Wilder, R.L. (1989) A central nervous system defect in biosynthesis of corticotropin-releasing hormone is associated with the susceptibility in streptococcal cell wall-induced arthritis in Lewis rats. Proc. Natl. Acad. Sci. USA, 86: 4771–4775.

Tilders, F.J.H., Derijk, R.H., Van Dam, A.M., Vincent, V.A.M., Schotanus, K. and Persoons, J.H.A. (1994) Activation of the hypothalamus–pituitary–adrenal axis by bacterial endotoxins-routes and intermediates. Psychoneuroendocrinology, 19: 209–232.

Turnbull, A.V. and Rivier, C.L. (1999) Regulation of hypothalamo–pituitary–adrenal axis by cytokines: actions and mechanisms of action. Physiol. Rev., 79: 1–71.

Vale, W., Spiess, J., Rivier, C. and Rivier, J. (1981) Characterisation of a 41-residue ovine hypothalamic peptide that stimulates secretion of corticotropin and β-endorphin. Science, 213: 1394–1397.

Valentino, R.J., Foote, S.L. and Aston-Jones, G. (1983) Corticotropin-releasing hormone activates noradrenergic neurones of the locus coeruleus. Brain Res., 270: 363–367.

Van Dam, A.M., Brown, M., Man-A-Hing, W. and Barkenbosch, F. (1993) Immunocytochemical detection of prostaglandin E2 in microvasculature and in neurones of rat brain after administration of bacterial endotoxin. Brain Res., 613: 331–336.

Van de Kar, L.D. and Blair, M.L. (1999) Forebrain pathways mediating stress-induced hormone secretion. Front. Neuroendocrinol., 20: 1–48.

Waits, P.S., Harbuz, M.S., Purcell, W.M., McLeod, J.D. and Fulford, A.J. (2003) Central nociceptin/orphaninFQ activates the hypothalamo–pituitary–adrenal axis by stimulation of hypothalamic corticotropin releasing factor neurones in conscious rats. J. Neuroendocrinol. (submitted).

Walker, J.S. (2003) Antiinflammatory effects of opioids. Adv. Exp. Med. Biol., 521: 148–160.

Watkins, L.R., Maier, S.F. and Goehler, L.E. (1995) Cytokine-to-brain communication: a review and analysis of alternative mechanisms. Life Sci., 57: 1011–1026.

Watts, A.G., Swanson, L.W. and Sanchez-Watts, G. (1987) Efferent projections of the suprachiasmatic nucleus: studies using anterograde transport of *Phaseolus vulgaris*-leucoagglutinin in the rat. J. Comp. Neurol., 258: 204–229.

Webster, J.I., Tonelli, L. and Sternberg, E.M. (2002) Neuroendocrine regulation of immunity. Ann. Rev. Immunol., 20: 125–163.

Weigent, D.A. and Blalock, J.E. (1997) Production of peptide hormones and neurotransmitters by the immune system. Chem. Immunol., 69: 1–30.

Whitnall, M.H. (1993) Regulation of the hypothalamic corticotrophin-releasing hormone neurosecretory system. Prog. Neurobiol., 40573–40629.

Willner, P. and Mitchell, P.J. (2002) The validity of animal models of predisposition to depression. Behav. Pharmacol., 13: 169–188.

Windle, R.J., Wood, S.A., Kershaw, Y.M., Lightman, S.L., Ingram, C.D. and Harbuz, M.S. (2001) Increased corticosterone pulse frequency during adjuvant-induced arthritis and its relationship to alterations in stress responsiveness. J. Neuroendocrinol., 13: 905–911.

Wolfson, B., Manning, R.W., Davis, L.G., Arentzen, R. and Baldrino, F.J. (1985) Co-localisation of corticotropin-releasing factor and vasopressin mRNA in neurones after adrenalcctomy. Nature, 315: 59–61.

Wu, P. and Childs, G.V. (1991) Changes in rat pituitary POMC mRNA after exposure to cold or a novel environment, detected by in situ hybridisation. J. Histochem. Cytochem., 39: 843–852.

Young, E.A. and Vasquez, D. (1996) Hypercortisolemia, hippocampal glucocorticoid receptors, and fast feedback. Mol. Psychiat., 1: 149–159.

Young, E.A., Lewis, J. and Akil, H. (1986) The preferential release of beta-endorphin from the anterior pituitary lobe by corticotropin releasing factor (CRF). Peptides, 7: 603–607.

T. Steckler, N.H. Kalin and J.M.H.M. Reul (Eds.)
Handbook of Stress and the Brain, Vol. 15
ISBN 0-444-51173-3

CHAPTER 1.4

Hormones of the pituitary

Marcelo Páez-Pereda* and Günter K. Stalla

Department of Endocrinology, Max Planck Institute of Psychiatry, Kraepelinstr. 10, 80804 Munich, Germany

Abstract: The pituitary gland is an important interface between the brain and the peripheral systems. Through the secretion of peptide hormones, the pituitary regulates the activity of the endocrine and immune systems and it controls many metabolic functions. The biosynthesis of pituitary hormones is under the control of hypothalamic factors. These factors as well as pituitary hormones are produced under different physiological and pathological conditions and have the effect of bringing back the organism to a normal homeostatic state. Under stress, corticotrophin-releasing factor stimulates the production of adrenocorticotrophic hormone in the pituitary and this, in turn, induces the release of glucocorticoids in the adrenal glands. In parallel, prolactin and growth hormone are also produced in the pituitary in response to stress. All these hormones have an impact on the endocrine and immune systems. The immune system, on the other hand, signals the brain and the pituitary through the production of cytokines. Cytokines control the production of pituitary hormones through the release of hypothalamic factors as well as acting directly on pituitary cells. All these regulatory mechanisms together coordinate the production of pituitary hormones in response to stress and immune challenge and ensure the maintenance of homeostasis.

Introduction

The pituitary gland is a major control point for the proper function of the endocrine system. It has a small size and it is located at a midline depression of the sphenoid bone, the sella turcica (Fig. 1). Hormones released from the anterior pituitary gland target and control the function of other systemic endocrine organs. Six main hormones are produced by the adenohypophysis: adrenocorticotrophic hormone (ACTH), growth hormone (GH), prolactin (PRL), follicle-stimulating hormone (FSH), luteinizing hormone (LH), and thyroid-stimulating hormone (TSH) (Fig. 1). ACTH is a product of proopiomelanocortin (POMC), together with b-lipotropic hormone (b-LPH), endorphins, encephalin, corticotrophin-like immunoreactive peptide (CLIP), and α-MSH. ACTH stimulates glucocorticoid production from the adrenal cortex and, therefore, it is the main mediator of the stress response (McCann et al., 2000; Pacak and Palkovits, 2001) (Fig. 2). PRL has important role in the initiation and maintenance of lactation and it also plays a role in the integration of the endocrine and immune response to stress (Bole-Feysot et al., 1998). GH promotes growth of the skeleton and soft tissues and has important metabolic effects. Its effects are mediated directly through GH receptors or indirectly by inducing insulin-like growth factor (IGF-I) synthesis. FSH and LH are collectively referred to as gonadotrophins (Themmen and Huhtaniemi, 2000). FSH promotes follicular growth in the ovaries and spermatogenesis in the testes. TSH is important for the physiological growth and function of the thyroid gland. Besides the hormone-producing/endocrine cells, the anterior pituitary also contains the folliculostellate cells, which comprise 3–5% of all adenohypophyseal cells (Allaerts et al., 1990). These cells are the source of growth factors and inflammatory cytokines, therefore suggesting an important role in the paracrine regulation of hormone secretion (Renner et al., 1996) related to the immune response.

*Corresponding author. Tel.: +49-89-30622272;
Fax: +49-89-30622605; E-mail: marcelo@mpipsykl.mpg.de

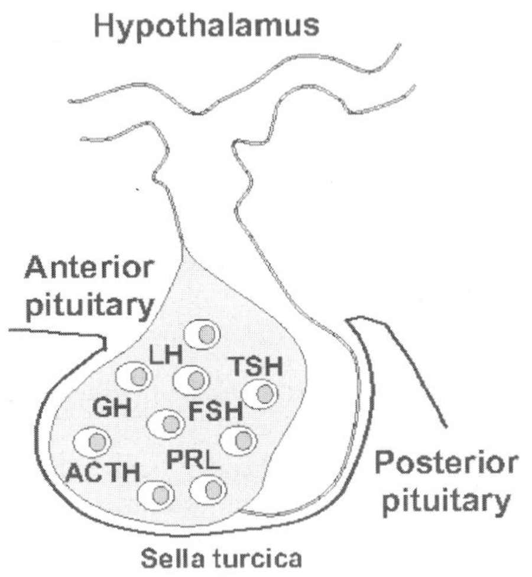

Fig. 1. Schematic diagram of the pituitary gland. Hypothalamic factors reach the pituitary gland through the portal system and stimulate the production of the anterior pituitary hormones. Different anterior pituitary cell types produce different hormones.

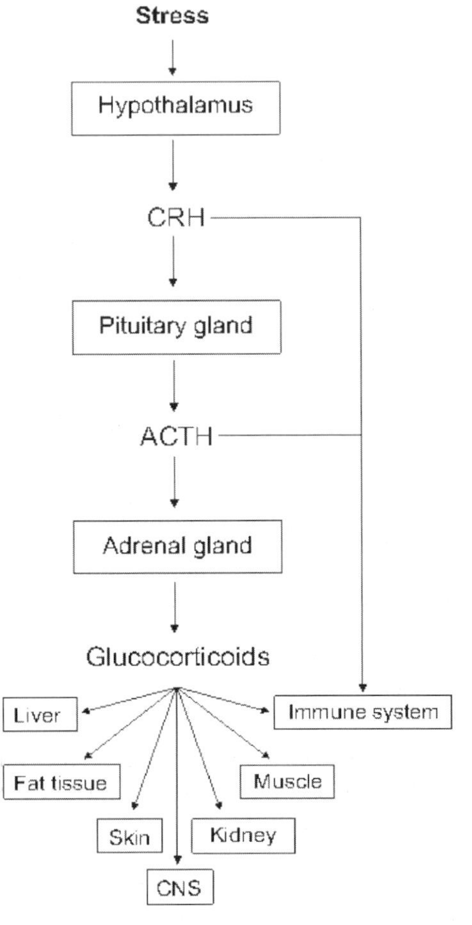

Fig. 2. Diagram of the regulation of the hypothalamus–pituitary–adrenal axis and its effects on target organs. CRF, ACTH, and glucocorticoids act directly on the immune cells and modulate their functions.

Regulation of the pituitary function

The pituitary gland is under the continuous control of peptides, proteins, and other substances derived from the hypothalamus and the periphery. The stimulatory or inhibitory effect of these substances is mediated by binding to cell surface or nuclear receptors.

Corticotroph cells are modulated by the hypothalamic peptides corticotrophin-releasing factor (CRF) and arginine vasopressin (AVP) to increase the synthesis and secretion of ACTH. CRF is a potent mediator of endocrine responses to stress in corticotroph cells (Fig. 2). There are two types of CRF receptors: type I and type II, both of them are expressed in ACTH-secreting cells of the human anterior pituitary (Reul and Holsboer, 2002). However, in rodents, only the CRF type I is expressed in the pituitary (Van Pett et al., 2000; Reul and Holsboer, 2002). CRF as well as CRF receptors are important for the stimulation of ACTH production under stress, but the maintenance of the basal levels of ACTH production depend on AVP and its receptor, V1b (De Keyzer et al., 1997). CRFR1 is highly expressed in the anterior pituitary, neocortex, hippocampus, amygdala, and cerebellum, and activation of this receptor stimulates adenylate cyclase. While CRF receptors activate the cAMP pathway and calcium channels, the AVP receptor, V1b, activates the inositol phosphate pathway and calcium channels as well (Boutillier et al., 1995; Autelitano and Cohen, 1999; Reul and Holsboer, 2002). In mice lacking CRFR1, the stress-induced release of ACTH and corticosterone is reduced (Timpl et al., 1998). Therefore, the CRFR1 plays a key role in the control of ACTH production.

Other important regulators of corticotroph cell physiology are cytokines (Renner et al., 1996; Ray and Melmed, 1997; Arzt et al., 1999). The activation of the HPA axis during infection includes the activation of ACTH production by interleukin-1 (IL-1) (Renner et al., 1996; Ray and Melmed, 1997; Arzt et al., 1999). IL-1 and its receptors are not only expressed in normal but also in tumor ACTH-secreting cells (Arzt et al., 1999) providing the basis for an autocrine loop to stimulate ACTH secretion. Leukemia inhibitory factor (LIF) also plays an important role in ACTH regulation in normal and tumor ACTH-secreting cells (Bousquet et al., 1999, 2000; Auernhammer and Melmed, 2000).

Beside cytokines, other factors regulate the production of ACTH as well. The extracellular matrix (ECM) conveys signals through membrane receptors called integrins producing changes in cell morphology, proliferation, differentiation, and apoptosis. The ECM plays an important role in pituitary physiology and tumorogenesis. ACTH production is inhibited by fibronectin, laminin, and collagen I at the level of POMC gene transcription (Kuchenbauer et al., 2001). ACTH-secreting cell proliferation is stimulated by collagen IV and fibronectin, but inhibited by collagen I and laminin (Kuchenbauer et al., 2001). The production of reactive oxygen species mediates the effects of laminin and collagen IV. On the other hand, the effect of fibronectin is mediated by the β1-integrin and Rho activation (Kuchenbauer et al., 2001).

The cannabinoid receptor I (CB1) was detected in acromegaly associated pituitary adenomas, Cushing's adenomas, and prolactinomas, whereas faint or no expression was found in nonfunctioning pituitary adenomas (Pagotto et al., 2001). In Cushing's adenomas, the CB1 agonist WIN 55,212–2 does not modify basal ACTH secretion. However, CRF and WIN 55,212–2 produce a synergistic stimulation of ACTH secretion (Pagotto et al., 2001).

Molecular mechanisms that control pituitary hormone production

ACTH biosynthesis is coordinately controlled by different transcription factors at the level of the POMC promotor (Therrien and Drouin 1991; Boutillier et al., 1995; Philips et al., 1997; Bousquet et al., 2000).

CRF induces the expression of the nuclear receptors Nur77 and Nurr1 and stimulates the transcriptional activity at the NurRE site in the POMC promotor, leading to an increase in ACTH production. This effect is mediated by the production of cAMP, the activation of protein kinase A (PKA), and calcium-dependent and -independent mechanisms. The induction of Nur77 and Nurr1 by calcium pathways depends on L-type calcium channels and Ca^{2+}-calmodulin-dependent kinase II (CAMKII) activity (Kovalovsky et al., 2002). In parallel, calcium-independent pathways are activated, which are mediated in part by MAPK activation (Kovalovsky et al., 2002). This activation of MAPK is mediated by B-Raf (Kovalovsky et al., 2002). All these signaling pathways can be blocked by small molecules. The inhibition of these signaling pathways results in a reduced transcriptional activity of Nur leading to reduced POMC transcription and ACTH production. The substances that have been proved to inhibit the signaling pathways that control ACTH biosynthesis include L-type calcium channel blockers such as nifedipine, PKA inhibitors such as H89, CAMK inhibitors such as KN62, adenylate cyclase inhibitors such as MDL12330A, and MEK inhibitors such as PD98059 and UO126 (Kovalovsky et al., 2002). These drugs showed different degrees of inhibition of the POMC promotor.

We have recently demonstrated that retinoic acid inhibits the transcriptional activity of AP-1 and Nur in ACTH-secreting cells (Paez-Pereda et al., 2001). This inhibition results in reduced secretion of ACTH in ACTH-secreting tumor cells. However, normal ACTH-secreting cells and other normal pituitary cells are not affected by retinoic acid, demonstrating a specific effect of this drug on tumor cells probably related to the degree of differentiation of normal versus tumor cells (Paez-Pereda et al., 2001). Moreover, COUP-TF1 was able to inhibit the inhibitory effects of retinoic acid in ACTH-secreting tumor cells, which is in line with a hypothetical role for COUP-TF1 in the differentiation of ACTH-secreting cells and in the control of POMC transcription (Kliewer et al., 1992; Wu et al., 1997; Paez-Pereda et al., 2001).

Neuroendocrine–immune interactions

Effects of pituitary hormones on the immune system

Adaptive changes in response to stress are coordinated by the pituitary gland by detecting neuroendocrine signals from the brain and producing a concerted reaction of target organs through the secretion of different specific hormones. This adaptive mechanism is mainly mediated by the hypothalamus–pituitary–adrenal axis (HPA) resulting in the release of glucocorticoids. CRF itself has direct action on monocytes. CRF has been shown to play a modulatory role in monocyte activation through the regulation of the inflammatory cytokine IL-1 and its antagonist IL-1 receptor antagonist IL-1Ra (Paez-Pereda et al., 1995). These direct effects of CRF on immune cells depend on the previous activation state of the immune system and prevent an overproduction of IL-1 after bacterial infection (Paez-Pereda et al., 1995). Besides CRF, ACTH and β-endorphin produced in the pituitary gland can also act directly on immune cells. ACTH inhibits the expression of IL-1ra, whereas β-endorphin produces the opposite effects (Kovalovsky et al., 1999). The different effects of ACTH and β-endorphin could account for their differential contribution to the inflammatory response: while ACTH contributes to the glucocorticoid overall control of the inflammatory response, β-endorphin exerts an inhibitory tone on the resting IL-1 system. IL-1ra is essential in setting the level of monocyte and inflammatory response. Therefore, its differential regulation by the HPA-axis hormones contributes to regulating the inflammatory response mediated by IL-1 (Kovalovsky et al., 1999).

Glucocorticoids induce apoptosis of CD4+ CD8+ thymocytes, which are the precursors of the CD4+ helper and CD8+ cytotoxic T cells. Glucocorticoids have also a clear immunosuppressive effect exerted primarily by the inhibition of lymphocyte activation through the inhibition of the production of inflammatory cytokines, such as interleukins, tumor necrosis factor-α (TNF-α), and γ-interferon (IFN-γ). Lipopolysaccharide (LPS)-stimulated IL-1 and IL-1Ra production are inhibited by glucocorticoids. Glucocorticoids inhibit IL-1β-stimulated IL-1Ra mRNA expression and protein production. Both IFN-γ and IL-4 reverse the inhibitory effect of glucocorticoids on IL-1Ra expression and secretion (Kovalovsky et al., 1998). The differential regulation and involvement of IL-1 in the expression of IL-1Ra by IFN-γ, IL-4, and glucocorticoids sets the level of monocyte responsiveness during the Th1 or Th2 responses (Kovalovsky et al., 1998).

Altogether, the actions of CRF, ACTH, and glucocorticoids on monocyte activation and inflammatory cytokine production indicate that the whole HPA axis has a modulatory effect, which prevents an overreaction of the immune system during bacterial infection.

In addition to the HPA activation during immune challenge, there is a simultaneous production of immunoprotective factors by the pituitary gland. These factors include PRL, GH, and TSH, and their respective hormonal axes.

Similarly, the adrenergic response to stress also produces an inhibition of antibody production and cell-mediated immunity. Some of the pituitary hormones contribute to compensate these immunosuppressive effects, although they are not obligate immunoregulatory factors. Mice with specific defects in PRL, GH, TSH, or their receptors show impairment in the development of some immune cell populations, but all other aspects of humoral and cell-mediated immunity are normal (Bouchard et al., 1999; Hall et al., 2002; Pichurin et al., 2004).

Among the pituitary hormones, not only ACTH mediates the stress response. Prolactin is produced in large amounts during stress and it has two main mechanisms of action on the immune system. On the one hand, PRL inhibits the production of corticosteroids, which could prevent their immunosuppressive effects. On the other hand, PRL acts directly on lymphoid tissues, such as the thymus, to prevent the deleterious effects of corticosteroid. Therefore, PRL promotes survival of thymus-derived cells. GH is also produced during stress and it activates the production of IGF-I. Both factors have inhibitory effects on glucocorticoid actions, such as inhibition of T-cell proliferation. Certain types of stress can also stimulate the production of thyroid hormones, which have immunostimulatory properties as well.

Effects of the immune system on the endocrine system

To date it is well known that the interaction between the immune system and the neuroendocrine system is necessary for the organism to overcome pathological events like inflammation or infection in an optimal manner (Besedovsky and del Rey, 1996). It has become evident that cytokines produced by the immune cells affect and modify the hormone secretion of the endocrine cells (Spangelo and Gorospe, 1995). In turn, hormones modulate the function of immune cells and influence cytokine production (Besedovsky and Del Rey, 1996). In these interactive processes the activation of the HPA axis by cytokines is important since it results in increased levels of anti-inflammatory-acting glucocorticoids which prevent overshooting reactions (septic shock) of the activated immune system (Chrousos, 1995).

The pituitary gland constitutes an integration point among neuroendocrine and immune signals. Corticotroph cells are modulated by the hypothalamic peptides, such as CRF and AVP, to increase the secretion of ACTH. ACTH, in turn, induces the secretion of glucocorticoids from the adrenal cortex to finally shut down the HPA-axis activity. The pituitary corticotroph activity is also modulated by several cytokines (Ray and Melmed, 1997; Arzt et al., 1999) and secretagogues such as the pituitary adenylate cyclase-activating polypeptide, vasointestinal peptide (VIP), epinephrine, and angiotensin II, which also stimulate POMC expression (Yohiaki et al., 1997).

Cytokines integrate the neuroimmune signaling

Interleukins are major mediators in immune regulatory networks, but also participate in the functional regulation of nonimmune cell types (Besedovsky and del Rey, 1992; Arzt et al., 1999). Interleukins play a central role in immune–endocrine interactions affecting the function of endocrine cells in different manners at the level of the hypothalamus, the pituitary, and the peripheral endocrine glands (Besedovsky and del Rey, 1992; Spangelo and Gorospe, 1995). It has been shown that the pituitary cells not only express receptors for IL-1, IL-2, and IL-6 (Arzt et al., 1999) but are also able to produce these cytokines (Vankelecom et al., 1989; Koenig et al., 1990; Spangelo et al., 1990; Arzt et al., 1992; Sauer et al., 1998). The intrinsic interleukin production would allow the cytokines to act in an auto- or paracrine manner on endocrine cells (Renner et al., 1996; Arzt et al., 1999). In the normal anterior pituitary, IL-6 is produced by nonhormone-producing folliculostellate (FS) cells (Vankelekom et al., 1989; Carmeliet et al., 1991; Allaerts et al., 1997; Renner et al., 1998). In human pituitary adenomas, in which FS cells are mostly absent (Marin et al., 1992), considerable amounts of IL-6 have been shown to be produced by the tumor cells (Jones et al., 1991, 1994; Tsagarakis et al., 1992). Therefore, hormone-producing pituitary cells also have the ability to produce IL-6. Cytokines like IL-1, IL-2, IL-6, and TNF-α have regulatory effects on ACTH, PRL, LH, FSH, GH, and TSH secretion, both in vivo and in vitro (Spangelo and Gorospe, 1995; Arzt et al., 1999). It has been shown that IL-6 increases ACTH levels in vivo and that this effect is mediated by CRF produced in the hypothalamus (Naitoh et al., 1988). However, IL-6 also directly affects ACTH release from hemi-pituitaries of normal rats and anterior pituitary cell lines in vitro (Fukata et al., 1989; Lyson and McCann, 1991) through intrapituitary IL-6 receptors (Ohmichi et al., 1992). In addition to its effect on hormone production, IL-6 has been shown to regulate anterior pituitary cell proliferation (Arzt et al., 1993; Sawada et al., 1995; Renner et al., 1997), intrapituitary vascular endothelial growth factor (VEGF) production (Gloddek et al., 1999), and the c-fos protooncogene expression in human anterior pituitary adenomas (Páez-Pereda et al., 1996). Therefore, IL-6 also has the ability to regulate tumor growth and vascularization.

IL-6 stimulates ACTH release and POMC expression in human corticotroph pituitary adenomas in vitro. IL-6 induced a rapid 3- to 30-fold increase in ACTH secretion and, therefore, seems to stimulate ACTH release in corticotroph adenoma cells more potently than CRF (Stalla et al., 1994). IL-6 may be an important regulator of not only normal, but also particularly adenomatous corticotroph pituitary cell function.

There is a considerable amount of evidence that IL-6 is a potent regulator of ACTH in vivo and in vitro. Therefore, IL-6 can act at the hypothalamic level as well as directly on pituitary cells. IL-6 stimulates the release of ACTH in normal rats. Suboptimal amounts of IL-1 and IL-6 synergize to induce an early (30–60 min) ACTH response in mice. In agreement with this synergism, it has been shown that an anti-IL-6 antibody blocked the IL-1-induced increase in plasma ACTH in mice. IL-6 induced an increase in ACTH levels at 4 h postinjection. IL-6 enhances ACTH release from rat hemipituitary glands and stimulates the release of ACTH from AtT-20 cells. Recombinant IL-6 has been shown to activate the HPA axis in humans; at different doses IL-6 treatment of cancer patients induces an increase in both ACTH and cortisol plasma levels. IL-6 stimulates both ACTH secretion and POMC gene expression in corticotroph adenoma cell cultures (Paez-Pereda et al., 2000b). This demonstration of the stimulatory action of IL-6 on human corticotroph adenoma cell function provides further evidence for a direct action of IL-6 on corticotroph pituitary cells.

In mouse pituitary primary cell cultures, a cytokine from the family of IL-6 called leukemia inhibitory factor (LIF) potently stimulates ACTH secretion (Akita et al., 1995). LIF stimulates the secretion of ACTH and the expression of POMC message in AtT-20 cells (Ray et al., 1996). In addition, LIF potentiates the stimulatory action of CRF on ACTH secretion (Akita et al., 1996a). The functional importance of LIF action on ACTH secretion is further underlined by the studies in LIF gene KO mice, in which a defect in the activation of the HPA axis was observed. ACTH levels are diminished after fasting in the KO animals and replacement of LIF restores the HPA response (Akita et al., 1996b). LIF KO mice show attenuated ACTH responses after immobilization stress (Chesnokova et al., 1998).

Another cytokine of the IL-6 family of gp-130-binding cytokines, oncostatin M (OSM), stimulates ACTH in AtT-20 cells and primary human fetal pituitary cultures (Shimon et al., 1997; Kim et al., 2000). In AtT-20 cells IL-11, which also binds gp-130, stimulates ACTH secretion and POMC expression (Auernhammer and Melmed, 1999). IL-11 simulates the expression of the cytoplasmic-signaling protein SOCS3 in AtT-20 cells and overexpression of SOCS3 causes a significant inhibition of IL-11-induced ACTH secretion (Auernhammer and Melmed, 1999) and also LIF activation of corticotrophs, indicating that the intracellular gp130 negative feedback mediated by SOCS3 is functional in corticotroph cells (Auernhammer and Melmed, 1999).

LIF rapidly induces tyrosyl phosphorylation of STAT1 and STAT3 in AtT-20 cells (Ray et al., 1996). By using two dominant negative forms of STAT3, it has been shown that the LIF action on corticotroph cells depends on STAT3 (Bousquet and Melmed, 1999). Using progressive 5′-deletions of POMC promotor, it has been demonstrated that the LIF-responsive POMC promotor region contains two juxtaposed sequences similar to STAT3 DNA-binding motif. POMC gene expression through a STAT3-dependent mechanism provides new mechanistic insights for HPA-axis stimulation by gp130 cytokines. This provides the molecular basis for CRF and LIF synergism in ACTH secretion and may be an important mechanism for the stimulation of ACTH following inflammatory and stress-derived STAT3-inducing cytokines. Interestingly, LIF did not induce Nurr1 or Nur77 mRNA, which mediate the CRF-stimulated POMC transcription in AtT-20 cells and none of the gp130 cytokines (LIF, IL-6, IL-11, or CNTF) did activate Nur transactivation in AtT-20 corticotrophs, as does IL-1 (Kovalovsky and Arzt, unpublished observations). Thus, gp130 cytokines activate the ACTH biosynthesis in the corticotrophs through a novel mechanism independent from Nur factors.

IL-1 is critically involved in the inflammatory process, and has been shown to affect hormone secretion, acting at different levels of the HPA axis (Ray and Melmed, 1997; Arzt et al., 1999). IL-1-binding sites and receptors have been characterized both in hypothalamic and pituitary glands (Marquette et al., 1995), and IL-1 stimulates CRF secretion and mRNA synthesis in the rat hypothalamus (Tsagarakis et al., 1989). The IL-1 receptor (IL-1R) type I and the IL-1R mRNA were characterized in normal pituitary cells and in the corticotroph cell line AtT-20 (De Souza et al., 1989; Haour et al., 1990). IL-1 stimulates the secretion of ACTH and b-endorphin, and induces POMC mRNA expression in the pituitary in vivo and in vitro (Ruzicka and Akil, 1995; Parsadaniantz et al., 1997).

Chronic infusion of IL-1 results in a persistent elevation of ACTH levels in rats (Sweep et al., 1992), and IL-1β stimulates ACTH secretion from human pituitary Cushing's tumors in vitro (Malarkey and Zvara, 1989). Mutual modulation among cytokines and hypothalamic peptides were described. IL-1β, IL-6, and TNF-α stimulate the POMC promotor and potentiate the stimulatory effect of CRF in AtT-20 cells (Katahira et al., 1998). CRF pretreatment of rats sensitizes the pituitary to IL-1-dependent ACTH secretion (Payne et al., 1994). This effect is also observed in vitro, as pretreatment of AtT-20 cells with CRF increases IL-1R density (Webster et al., 1991). IL-1 stimulates β-endorphin secretion in AtT-20 pituitary cells and potentiates β-endorphin secretion stimulated by CRF and VIP (Fagarnasan et al., 1989).

Not only hypothalamic factors and cytokines derived from the immune cells or from pituitary cells themselves activate the production of pituitary hormones. LPS, a component of the outer layer of the cell wall of gram-negative bacteria, is a potent activator of the immune system and it can also activate directly the pituitary cells. In monocytes, LPS induces the subsequent release of TNF-α, IL-1 and IL-6 from macrophages and monocytes (van Deventer et al., 1990). LPS interacts with the Toll-like receptor-4 (Tlr4), a member of the Toll protein family (Medzhitov et al., 1997). In immune cells the Toll-4 receptor induces the activation of nuclear factor-κ B (NF-κB) (McDonald et al., 1997), but the preceding signaling cascade is largely unknown; so far only the participation of mitogen-activated protein kinase (MAPK), kinase-3 (MKK3), and p38 (Nick et al., 1999) or p42/44 (Zhang et al., 1997) has been shown. In the pituitary, LPS can activate the intrapituitary IL-6 production both in vivo and in vitro. Tlr4 is expressed in folliculo-stellate cells of the pituitary (Lohrer et al., 2000). Also CD14, which binds LPS, is expressed in pituitary cells. Thus, both CD14 and Tlr4, which are essential tools for the action of LPS, are present in FS cells. LPS increases phosphoinositol turnover in TtT/GF cells and induces phosphorylation of p38α mitogen-activated protein kinase and the inhibitor (IκB) of nuclear factor-κB (Lohrer et al., 2000). Nuclear factor-κB transcriptional activity is also activated by LPS in TtT/GF cells (Lohrer et al., 2000). SB203580, a specific inhibitor of p38 MAPK phosphorylation (Nick et al., 1999), almost completely suppressed IL-6 production in response to LPS. IL-6 stimulation by LPS was also blocked by dexamethasone. Glucocorticoid receptors represent suppressive counterparts of NF-κB and might act as repressors at the level of IL-6 transcription (McKay and Cidlowski, 1999).

Concluding remarks

During stress produced by different circumstances including immune challenge, hypothalamic factors are released to regulate the production of pituitary hormones. Cytokines, neurotransmitters, and immunomodulatory substances produced during stress by different organs and cells also act directly on pituitary cells. As a result, the production of pituitary hormones during stress and immune challenge integrates the signals from the brain and from the immune system. Pituitary hormones, in turn, control the activity of endocrine glands and also regulate the activation state of the immune system. These bidirectional interactions guarantee the coordination between the functions of the brain, the endocrine, and the immune system under physiological conditions or under stress and adapt the metabolic state of the organism to different situations.

Abbreviations

CRF	corticotropin-releasing factor
ACTH	adrenocorticotrophin
PRL	prolactin
GH	growth hormone
LH	luteinizing hormone
FSH	follicle-stimulating hormone
TSH	thyroid-stimulating hormone
LPH	lipotrophin
CLIP	corticotrophin-like immunoreactive peptide
MSH	melanocyte-stimulating hormone
IGF	insulin-like growth factor
AVP	arginine vassopressin
HPA	hipothalamus–pituitary–adrenal
IL	interleukin
ECM	extracellular matrix
MAPK	mitogen-activated protein kinase

PKA	protein kinase A
VIP	vasointestinal peptide
FS	folliculo-stellate
VEGF	vascular endothelial growth factor
IFN	interferon
LIF	leukemia inhibitory peptide
LPS	lipopolysaccharide

References

Akita, S., Webster, J., Ren, S.G., Takino, H., Said, J., Zand, O. and Melmed, S. (1995) Human and murine pituitary expression of leukemia inhibitory factor. Novel intrapituitary regulation of adrenocorticotropin hormone synthesis and secretion. J. Clin. Invest., 95: 1288–1298.

Akita, S., Conn, P.M. and Melmed, S. (1996a) Leukemia inhibitory factor (LIF) induces acute adrenocorticotrophic hormone (ACTH) secretion in fetal rhesus macaque primates: a novel dynamic test of pituitary function. J. Clin. Endocrinol. Metab., 81: 4170–4178.

Akita, S., Malkin, J. and Melmed, S. (1996b) Disrupted murine leukemia inhibitory factor (LIF) gene attenuates adrenocorticotropic hormone (ACTH) secretion. Endocrinology, 137: 3140–3143.

Allaerts, W., Carmeliet, P. and Denef, C. (1990) New perspectives in the function of pituitary folliculo-stellate cells. Mol. Cell. Endocrinol., 71: 73–81.

Allaerts, W., Jeucken, P.H., Debets, R., Hoefakker, S., Claassen, E. and Drexhage, H.A. (1997) Heterogeneity of pituitary folliculo-stellate cells: implications for interleukin-6 production and accessory function in vitro. J. Neuroendocrinol., 9: 43–53.

Arzt, E., Stelzer, G., Renner, U., Lange, M., Muller, O.A. and Stalla, G.K. (1992) Interleukin-2 and interleukin-2 receptor expression in human corticotrophic adenoma and murine pituitary cell cultures. J. Clin. Invest., 90: 1944–1951.

Arzt, E., Buric, R., Stelzer, G., Stalla, J., Sauer, J., Renner, U. and Stalla, G.K. (1993) Interleukin involvement in anterior pituitary cell growth regulation: effects of interleukin-2 and interleukin-6. Endocrinology, 132: 459–467.

Arzt, E., Paez-Pereda, M., Perez-Castro, C., Pagotto, U., Renner, U. and Stalla, G.K. (1999) Pathophysiological role of the cytokine network in the anterior pituitary gland. Front. Neuroendocrinol., 20: 71–95.

Auernhammer, C.J. and Melmed, S. (1999) IL-11 stimulates proopiomelanocortin gene expression and adrenocorticotropin secretion in corticotroph cells: evidence for a redundant cytokine network in the hypothalamo–pituitary–adrenal axis. Endocrinology, 140: 1559–1566.

Auernhammer, C.J. and Melmed, S. (2000) Leukemia-inhibitory factor-neuroimmune modulator of endocrine function. Endocr. Rev., 21: 313–345.

Autelitano, D.J. and Cohen, D.R. (1996) CRF stimulates expression of multiple fos and jun related genes in the AtT-20 corticotroph cell. Mol. Cell. Endocrinol., 119: 25–35.

Besedovsky, H. and del Rey, A. (1992) Immune–endocrine circuits: integrative role of cytokines. Front. Neuroendocrinol., 13: 61–94.

Besedovsky, H.O. and del Rey, A. (1996) Immune–neuroendocrine interactions: facts and hypotheses. Endocr. Rev., 17: 64–102.

Bole-Feysot, C., Goffin, V., Edery, M., Binart, N. and Kelly, P.A. (1998) Prolactin (PRL) and its receptor: actions, signal transduction pathways and phenotypes observed in PRL receptor knockout mice. Endocr. Rev., 19: 225–268.

Bouchard, B., Ormandy, C.J., Di Santo, J.P. and Kelly, P.A. (1999) Immune system development and function in prolactin receptor-deficient mice. J. Immunol., 163: 576–582.

Bousquet, C. and Melmed, S. (1999) Critical role for STAT3 in murine pituitary adrenocorticotropin hormone leukemia inhibitory factor signaling. J. Biol. Chem., 274: 10723–10730.

Bousquet, C., Susini, C. and Melmed, S. (1999) Inhibitory roles for SHP-1 and SOCS-3 following pituitary proopiomelanocortin induction by leukemia inhibitory factor. J. Clin. Invest., 104: 1277–1285.

Bousquet, C., Zatelli, M.C. and Melmed, S. (2000) Direct regulation of pituitary proopiomelanocortin by STAT3 provides a novel mechanism for immuno-neuroendocrine interfacing. J. Clin. Invest., 106: 1417–1425.

Boutillier, A.L., Monnier, D., Lorang, D., Lundblad, J.R., Roberts, J.L. and Loefler, J.P. (1995) Corticotropin-releasing hormone stimulates proopiomelanocortin transcription by cFos-dependent and -independent pathways: characterization of an AP-1 site in exon 1. Mol. Endocrinol., 9: 745–755.

Carmeliet, P., Vankelecom, H., van Damme, J., Billiau, A. and Denef, C. (1991) Release of IL-6 from anterior pituitary cell aggregates: developmental pattern and modulation by glucocorticoids and forskolin. *Neuroendocrinology*, 53: 29–34.

Chesnokova, V., Auernhammer, C.J. and Melmed, S. (1998) Murine leukemia inhibitory factor gene disruption attenuates the hypothalamo–pituitary–adrenal axis stress response. Endocrinology, 139: 2209–2216.

Chrousos, G.P. (1995) The hypothalamic–pituitary–adrenal axis and immune-mediated inflammation. New Engl. J. Med., 332: 1351–1362.

De Keyzer, Y., Rene, P., Lenne, F., Auzan, C., Clauser, E. and Bertagna, X. (1997) V3 vasopressin receptor and corticotropic phenotype in pituitary and nonpituitary tumours. Horm. Res., 47: 259–262.

De Souza, E.B., Webster, E.L., Grigoriadis, D.E. and Tracey, D.E. (1989) Corticotropin-releasing factor (CRF) and interleukin-1 (IL-1) receptors in the brain–pituitary–immune axis. Psychopharmacol. Bull., 25: 299–305.

Fagarnasan, M.O., Eskay, R. and Axelrod, J. (1989) Interleukin-1 potentiates the secretion of β-endorphin induced by secretagogues in a mouse pituitary cell line (AtT-20). Proc. Natl. Acad. Sci. USA, 86: 2070–2073.

Fukata, J., Usui, T., Maitoh, Y., Nakai, Y. and Imura, H. (1989) Effects of recombinant human interleukin-1α, 1β and 6 on ACTH synthesis and release in the mouse pituitary tumor cell line AtT-20. J. Endocrinol., 122: 33–39.

Gloddek, J., Pagotto, U., Paez-Pereda, M., Arzt, E., Stalla, G.K. and Renner, U. (1999) Pituitary adenylate cyclase-activating polypeptide, interleukin-6 and glucocorticoids regulate the release of vascular endothelial growth factor in pituitary folliculostellate cells. J. Endocrinol., 160: 483–490.

Hall, M.A., Bartke, A. and Martinko, J.M. (2002) Humoral immune response in mice over-expressing or deficient in growth hormone. Exp. Biol. Med., 227: 535–544.

Haour, F., Ban, E.M., Milon, G.M., Baran, D. and Fillion, G.M. (1990) Brain interleukin-1 receptors: characterization and modulation after lipopolysaccharide injection. Prog. Neuroimmunoendocrinol., 3: 196–204.

Jones, T.H., Justice, S., Price, A. and Chapman, K. (1991) IL-6 secreting human pituitary adenomas in vitro. J. Clin. Endocrinol. Metab., 73: 207–209.

Jones, T.H., Daniels, A., James, R., Justice, S., McCorkle, R., Price, A., Kendall-Taylor, P. and Weetman, A. (1994) Production of bioactive and immunoreactive interleukin-6 and expression of IL-6 messenger RNA by human pituitary adenomas. J. Clin. Endocrinol. Metab., 78: 180–187.

Katahira, M., Iwasaki, Y., Aoki, Y., Oiso, Y. and Saito, H. (1998) Cytokine regulation of the rat proopiomelanocortin gene expression in AtT-20 cells. *Endocrinology*, 139: 2414–2422.

Kim, D.S., Choi, H.S., Park, Y.S. and Kim, T.W. (2000) Effects of oncostatin M on hormone release of rat pituitary cells in primary culture. J. Korean Med. Sci., 15: 323–326.

Kliewer, S.A., Kazuhiko, U., Heyman, R.A., Mangelsdorf, D.J., Dyck, J.A. and Evans, R.M. (1992) Retinoid X receptor-COUP-TF interactions modulate retinoic acid signalling. Proc. Natl. Acad. Sci. USA, 89: 1448–1452.

Koenig, J.I., Snow, K., Clark, B.D., Toni, R., Cannon, J.G., Shaw, A.R., Dinarello, C.A., Reichlin, S., Lee, S.L. and Lechan, R.M. (1990) Intrinsic pituitary interleukin-1β is induced by bacterial lipopolysaccharide. Endocrinology, 126: 3053–3058.

Kovalovsky, D., Paez-Pereda, M., Sauer, J., Perez-Castro, C., Nahmod, V.E., Stalla, G.K., Holsboer, F. and Arzt, E. (1998) The Th1 and Th2 cytokines IFN-γ and IL-4 antagonize the inhibition of monocyte IL-1 receptor antagonist by glucocorticoids: involvement of IL-1. Eur. J. Immunol., 28: 2075–2085.

Kovalovsky, D., Paez-Pereda, M., Stalla, G.K., Holsboer, F. and Arzt, E. (1999) Differential regulation of interleukin-1 receptor antagonist by proopiomelanocortin peptides adrenocorticotropic hormone and beta-endorphin. Neuroimmunomodulation, 6: 367–372.

Kovalovsky, D., Refojo, D., Liberman, A.C., Hochbaum, D., Paez-Pereda, M., Coso, O.A., Stalla, G.K., Holsboer, F. and Arzt, E. (2002) Activation and induction of NUR77/NURR1 in corticotrophs by CRH/cAMP: involvement of calcium, protein kinase A, and MAPK pathways. Mol. Endocrinol., 16: 1638–1651.

Kuchenbauer, F., Hopfner, U., Stalla, J., Arzt, E., Stalla, G.K. and Paez-Pereda, M. (2001) Extracellular matrix components regulate ACTH production and proliferation in corticotroph tumor cells. Mol. Cell. Endocrinol., 175: 141–148.

Lohrer, P., Gloddek, J., Nagashima, A.C., Korali, Z., Hopfner, U., Paez-Pereda, M., Arzt, E., Stalla, G.K. and Renner, U. (2000) Lipopolysaccharide directly stimulates the intrapituitary interleukin-6 production by folliculostellate cells via specific receptors and the p38α mitogen-activated protein kinase/nuclear factor-kappaB pathway. Endocrinology, 141: 4457–4465.

Lyson, K. and McCann, S. (1991) The effect of interleukin-6 on pituitary hormone release in vivo and in vitro. Neuroendocrinology, 54: 262–266.

Malarkey, W.B. and Zvara, B.J. (1989) Interleukin-1b and other cytokines stimulate adrenocorticotropin release from cultured pituitary cells of patients with Cushing's disease. J. Clin. Endocrinol., 69: 196–199.

Marin, F., Kovacs, K., Stefaneanu, L., Horvath, E. and Cheng, Z. (1992) S-100 protein immunopositivity in human nontumorous hypophyses and pituitary adenomas. Endocr. Pathol., 3: 28–38.

Marquette, C., Van Dam, A.-M., Ban, E., Lanièce, P., Crumeyrolle-Arias, M., Fillion, G., Berkenbosch, F. and Haour, F. (1995) Rat interleukin-1b binding sites in the rat hypothalamus and pituitary gland. Neuroendocrinol., 62: 362–369.

McCann, S.M., Kimura, M., Karanth, S., Yu, W.H., Mastronardi, C.A. and Rettori, V. (2000) The mechanism of action of cytokines to control the release of hypothalamic and pituitary hormones in infection. Ann. N.Y. Acad. Sci., 917: 4–18.

McDonald, P.P., Bald, A. and Cassatella, M.A. (1997) Activation of the NF-kB pathway by inflammatory stimuli in human neutrophils. Blood, 89: 3421–3433.

McKay, L.I. and Cidlowski, J.A. (1999) Molecular control of immune/inflammatory responses: interactions between nuclear factor-kB and steroid receptor-signaling pathways. Endocr. Rev., 20: 435–459.

Medzhitov, R., Preston-Hurlburt, P. and Janeway, Jr., C.A. (1997) A human homologue of the Drosophila Toll protein signals activation of adaptive immunity. *Nature*, 388: 394–397.

Naitoh, Y., Fukata, J., Tominaga, T., Nakai, Y., Tamai, S., Mori, K. and Imura, H. (1988) Interleukin-6 stimulates the secretion of adrenocorticotropic hormone in conscious, freely-moving rats. Biochem. Biophys. Res. Commun., 155: 1459–1463.

Nick, J.A., Avdi, N.J., Young, S.K., Lehman, L.A., McDonald, P.P., Frasch, S.C., Billstrom, M.A., Henson, P.M., Johnson, G.L. and Worthen, G.S. (1999) Selective activation and functional significance of p38α mitogen-activated protein kinase in lipopolysaccharide-stimulated neutrophils. J. Clin. Invest., 103: 851–858.

Ohmichi, M., Hirota, K., Koike, K., Kurachi, K., Ohtsuka, S., Matsusaki, N., Yamaguchi, M., Miyake, A. and Tanizawa, O. (1992) Binding sites for interleukin-6 in the anterior pituitary gland. Neuroendocrinology, 55: 199–203.

Pacak, K. and Palkovits, M. (2001) Stressor specificity of central neuroendocrine responses: implications for stress-related disorders. Endocr. Rev., 22: 502–548.

Paez-Pereda, M., Sauer, J., Perez-Castro, C., Finkielman, S., Stalla, G.K., Holsboer, F. and Arzt, E. (1995) Corticotropin-releasing hormone differentially modulates the interleukin-1 system according to the level of monocyte activation by endotoxin. Endocrinology, 136: 5504–5510.

Páez-Pereda, M., Goldberg, V., Chervín, A., Carrizo, G., Molina, A., Andrada, J., Sauer, J., Renner, U., Stalla, G.K. and Arzt, E. (1996) Interleukin2 (IL-2) and IL-6 regulate c-fos protooncogene expression in human pituitary adenoma explants. Mol. Cell. Endocrinol., 124: 33–42.

Paez-Pereda, M., Ledda, M.F., Goldberg, V., Chervin, A., Carrizo, G., Molina, H., Muller, A., Renner, U., Podhajcer, O., Arzt, E. and Stalla, G.K. (2000a) High levels of matrix metalloproteinases regulate proliferation and hormone secretion in pituitary cells. J. Clin. Endocrinol. Metab., 85: 263–269.

Paez-Pereda, M.P., Lohrer, P., Kovalovsky, D., Perez-Castro, C., Goldberg, V., Losa, M., Chervin, A., Berner, S., Molina, H., Stalla, G.K., Renner, U. and Arzt, E. (2000b) Interleukin-6 is inhibited by glucocorticoids and stimulates ACTH secretion and POMC expression in human corticotroph pituitary adenomas. Exp. Clin. Endocrinol. Diabetes, 108: 202–207.

Paez-Pereda, M., Kovalovsky, D., Hopfner, U., Theodoropoulou, M., Pagotto, U., Uhl, E., Losa, M., Stalla, J., Grubler, Y., Missale, C., Arzt, E. and Stalla, G.K. (2001) Retinoic acid prevents experimental Cushing syndrome. J. Clin. Invest., 108: 1123–1131.

Pagotto, U., Marsicano, G., Fezza, F., Theodoropoulou, M., Grubler, Y., Stalla, J., Arzberger, T., Milone, A., Losa, M., Di Marzo, V., Lutz, B. and Stalla, G.K. (2001) Normal human pituitary gland and pituitary adenomas express cannabinoid receptor type 1 and synthesize endogenous cannabinoids: first evidence for a direct role of cannabinoids on hormone modulation at the human pituitary level. J. Clin. Endocrinol. Metab., 86: 2687–2696.

Parsadaniantz, S.M., Batsché, E., Gegout-Pottie, P., Terlain, B., Gillet, P., Netter, P. and Kerdelhué, B. (1997) Effects of continuous infusion of Interleukin-1β on corticotropin-releasing hormone (CRH), CRH receptors, proopiomelanocortin gene expression and secretion of corticotropin, b-endorphin and corticosterone. Neuroendocrinology, 65: 53–63.

Payne, L.C., Weigent, D.A. and Blalock, J.E. (1994) Induction of pituitary sensitivity to interleukin-1: a new function for corticotropin-releasing hormone. Biochem. Bioph. Res. Commun., 198: 485–449.

Perlstein, R.S., Mougey, E.H., Jackson, W.E. and Neta, R. (1991) Interleukin-1 and Interleukin-6 act synergistically to stimulate the release of adrenocorticotropic hormone in vivo. Lymphokine Cytok. Res., 10: 141–146.

Philips, A., Lesage, S., Gingras, R., Maira, M.H., Gauthier, Y., Hugo, P. and Drouin, J. (1997) Novel dimeric Nur77 signalling mechanism in endocrine and lymphoid cells. Mol. Cell. Biol., 17: 5946–5951.

Pichurin, P.N., Pichurina, O., Marians, R.C., Chen, C.R., Davies, T.F., Rapoport, B. and McLachlan, S.M. (2004) Thyrotropin receptor knockout mice: studies on immunological tolerance to a major thyroid autoantigen. Endocrinology, 145: 1294–1301.

Raetz, C.R.H. (1990) Biochemistry of endotoxins. Annu. Rev. Biochem., 59: 129–170.

Ray, D. and Melmed, S. (1997) Pituitary cytokine and growth factor expression and action. Endocr. Rev., 18: 206–228.

Ray, D.W., Ren, S.G. and Melmed, S. (1996) Leukemia inhibitory factor (LIF) stimulates proopiomelanocortin (POMC) expression in a corticotroph cell line. Role of STAT pathway. J. Clin. Invest., 97: 1852–1859.

Renner, U., Pagotto, U., Arzt, E. and Stalla, G.K. (1996) Autocrine and paracrine roles of polypeptide growth factors, cytokines and vasogenic substances in normal and tumorous pituitary function and growth: a review. Eur. J. Endocrinol., 135: 515–532.

Renner, U., Gloddek, J., Arzt, E., Inoue, K. and Stalla, G.K. (1997) Interleukin-6 is an autocrine growth factor for folliculostellate-like TtT/GF mouse pituitary tumor cells. Exp. Clin. Endocrinol. Diabetes, 105: 345–352.

Renner, U., Gloddek, J., Paez-Pereda, M., Arzt, E. and Stalla, G.K. (1998) Regulation and role of intrapituitary IL-6 production by folliculostellate cells. Domest. Anim. Endocrinol., 15: 353–362.

Reul, J.M. and Holsboer, F. (2002) Corticotropin-releasing factor receptors 1 and 2 in anxiety and depression. Curr. Opin. Pharmacol., 2: 23–33.

Rietschel, E.T. and Brade, H. (1992) Bacterial endotoxins. Sci. Am., 267: 26–31.

Ruzicka, B.B. and Akil, H. (1995) Differential cellular regulation of proopiomelanocortin by interleukin-1 beta and corticotropin-releasing hormone. Neuroendocrinology, 61: 136–151.

Sauer, J., Renner, U., Hopfner, U., Lange, M., Müller, A., Strasburger, C.J., Pagotto, U., Arzt, E. and Stalla, G.K. (1998) Interleukin-1β enhances interleukin-1 receptor antagonist content in human somatotroph adenoma cell cultures. J. Clin. Endocrinol. Metab., 83: 2429–2434.

Sawada, T., Koike, K., Kanda, Y., Ikesami, H., Jikihara, T., Maeda, T., Osako, Y., Hirota, K. and Miyake, A. (1995) IL-6 stimulates cell proliferation of rat anterior pituitary clonal cell lines in vitro. J. Endocrinol. Invest., 18: 83–90.

Shimon, I., Yan, X., Ray, D.W. and Melmed, S. (1997) Cytokine-dependent gp130 receptor subunit regulates human fetal pituitary adrenocorticotropin hormone and growth hormone secretion. J. Clin. Invest., 100: 357–363.

Spangelo, B.L. and Gorospe, W.C. (1995) Role of the cytokines in the neuroendocrine–immune system axis. Front Neuroendocrinol., 16: 1–22.

Spangelo, B.L., Isakson, P.C. and MacLeod, R.M. (1990) Production of interleukin-6 by anterior pituitary cells is stimulated by increased intracellular adenosine 3′,5′-monophosphate and vasoactive intestinal peptide. Endocrinology, 127: 403–409.

Stalla, G.K., Brockmeier, S.J., Renner, U., Newton, C., Buchfelder, M., Stalla, J. and Müller, O.A. (1994) Octreotide exerts different effects in vivo and in vitro in Cushing's disease. Eur. J. Endocrinol., 130: 125–131.

Sweep, C.G.J., Van der Meer, M.J.M., Hermus, A.R.M.M., Smals, A.G.H., Van der Meer, J.W.M., Pesman, G.J., Willemsen, S.J., Benraad, T.J. and Kloppenborg, P.W.C. (1992) Chronic stimulation of the pituitary–adrenal axis in rats by interleukin-1b infusion: in vivo and in vitro studies. Endocrinology, 130: 1153–1164.

Themmen, A.P.N. and Huhtaniemi, I.T. (2000) Mutations of gonadotropins and gonadotropin receptors: elucidating the physiology and pathophysiology of pituitary–gonadal function. Endocr. Rev., 21: 551–583.

Therrien, M. and Drouin, J. (1991) Pituitary pro-opiomelanocortin gene expression requires synergistic interactions of several regulatory elements. Mol. Cell. Biol., 11: 3492–3503.

Timpl, P., Spanagel, R., Sillaber, I., Kresse, A., Reul, J.M., Stalla, G.K., Blanquet, V., Steckler, T., Holsboer, F. and Wurst, W. (1998) Impaired stress response and reduced anxiety in mice lacking a functional corticotropin-releasing hormone receptor. Nat. Genet., 19: 162–166.

Tsagarakis, S., Gillies, G., Rees, L.H., Besser, M. and Grossman, A. (1989) Interleukin-1 directly stimulates the release of corticotrophin releasing factor from rat hypothalamus. Neuroendocrinology, 49: 98–101.

Tsagarakis, S., Kontogeorgeos, G., Giannou, P., Thalassinos, N., Woollley, J., Besser, G.M. and Grossman, A. (1992) Interleukin-6, a growth promoting cytokine, is present in human pituitary adenomas: an immuno-cytochemical study. Clin. Endocrinol., 37: 163–167.

van Deventer, S.J.H., Buller, H.R., ten Cate, J.W., Aarden, L.A., Hack, C.E. and Sturk, A. (1990) Experimental endotoxemia in humans: analysis of cytokine release and coagulation, fibrinolytic, and complement pathways. Blood, 76: 2520–2526.

Vankelecom, H., Carmeliet, P., Van Damme, J., Billau, A. and Denef, C. (1989) Production of interleukin-6 by folliculostellate cells of the anterior pituitary gland in a histotipic cell aggregate culture system. Neuroendocrinology, 49: 102–106.

Van Pett, K., Viau, V., Bittencourt, J.C., Chan, R.K.W., Li, H.-Y., Arias, C., Prins, G.S., Perrin, M., Vale, W. and Sawchenko, P.E. (2000) Distribution of mRNAs encoding CRF receptors in brain and pituitary of rat and mouse. J. Comp. Neurol., 428: 191–212.

Webster, E.L., Tracey, D.E. and De Souza, E.B. (1991) Upregulation of IL-1 receptors in mouse AtT-20 pituitary tumors cells following treatment with corticotropin-releasing factor. Endocrinology, 129: 2796–2798.

Wu, Q., Li, Y., Liu, R., Agadir, A., Lee, M.O., Liu, Y. and Zhang, X. (1997) Modulation of retinoic acid sensitivity in lung cancer cells through dynamic balance of orphan receptors nur77 and COUP-TF and their heterodimerization. EMBO J., 16: 1656–1669.

Yohiaki, A., Yasumasa, I., Masahito, K., Yutaka, O. and Hidehiko, S. (1997) Regulation of rat proopiomelanocortin gene expression in AtT-20 cells. I: Effects of the common secretagogues. Endocrinology, 138: 1923–1929.

Zhang, C., Baumgartner, R.A., Yamada, K. and Beaven, M.A. (1997) MAPk regulates production of TNF-α and release of arachidonic acid in mast cells. Indication of communication between p38 and p42 MAPks. J. Biol. Chem., 272: 13397–13402.

T. Steckler, N.H. Kalin and J.M.H.M. Reul (Eds.)
Handbook of Stress and the Brain, Vol. 15
ISBN 0-444-51173-3

CHAPTER 1.5

Molecular biology of the HPA axis

Kirsten-Berit Abel and Joseph A. Majzoub*

Division of Endocrinology, Children's Hospital Boston, Harvard Medical School, 300 Longwood Avenue, Boston, MA 02115, USA

Abstract: The hypothalamic-pituitary-adrenal (HPA) axis is a complex endocrine system. Its main role is maintainance of homeostasis following stress. Corticotropin releasing factor (CRF) is the major regulator of the HPA axis, controling the release of pituitary adrenocorticotropin (ACTH). CRF is cosynthesized with vasopressin (AVP) in neurons of the hypothalamic paraventricular nucleus and released into the portal0hypophyseal blood circulation, from whrer it is transported to the anterior pituitary gland. The binding of CRF to its specific receptor (CRF type 1 receptor) on pituitary corticotrophic cells results in stimulation of proopiomelanocortin (POMC) mRNA synthesis and ACTH secretion. ACTH and other biologically active peptides (N-terminal glycopeptide, γ-melanotropin, joining peptide, α-melanotropin, CLIP, β-lipotropin, β-melanotropin and β-endorphin) are generated by posttranslational cleavage of their precursor peptide POMC. POMC gene transcription is induced by CRF-stimulated elevation in cyclic adenosine 3',5'-momophosphate (cAMP). Several mediators of CRF-induced POMC regulation and corticotroph differentiation have been identified, such as Nur factors, TPIT, leukaemia-inhibitory factor and interleukin 6. Upon stimulation, ACTH is released into the systemic circulation in a circadian rhythm, orchestrated by the suprachiasmatic nucleus of the hypothalamus. ACTH stimulates steroidogenesis (mineralocorticoids, glucocorticoid and, in humans, adrenal androgens) via the G_s-coupled melanocortin 2 receptor (MC2R). ACTH, via cAMP, promotes the synthesis of steroidogenic acute regulatory protein (StAR), which in turn mediates the uptake of cholesterol into the mitochondria of adrenocortical cells. This is followed by coordinated action of cytochrome P 450 enzymes resulting in biosynthesis and secretion of adrenal steroids. Glucocorticoids in particular have wide-ranging effects on many organ systems, as well as a including negative feedback action at the pituitary and hypothalamic level.

Introduction

The hypothalamic–pituitary–adrenal (HPA) axis is a complex endocrine system, which functions to maintain homeostasis following stress. The normal stress response consists of a range of physiologic- and behavioral-adaptive changes designed to return the organism toward homeostatic equilibrium. The response includes heightened arousal, increased attention and cognition, decreased appetite and sexual arousal, and increased pain tolerance. In addition, an adaptive stress response leads to elevated cardiovascular and respiratory rate, activation of catecholamine and glucose production, and both activation and inhibition of immune-mediated inflammation. The HPA system contributes to this homeostasis at the endocrine, autonomic, and behavioral levels. The HPA axis not only plays a critical role in an organism's physiology, but also can contribute pathologically to several human diseases, including Cushing's syndrome, Addison's disease, and stress-related and affective diseases such as major depression (Sapolsky et al., 2000). In this introductory chapter, we will provide an overview of gene regulation within the HPA axis, and demonstrate how molecular biological tools reveal important insights into the regulation of this axis. With the recent elucidation of the sequence of the human genome, the links between the structure, function, and pathology of genes within the HPA axis will become better established. Hopefully, this will lead to a better understanding of diseases caused by

*Corresponding author. Tel.: +1617-355-6421;
Fax: +1617-734-0062; E-mail: joseph.majzoub@tch.harvard.edu

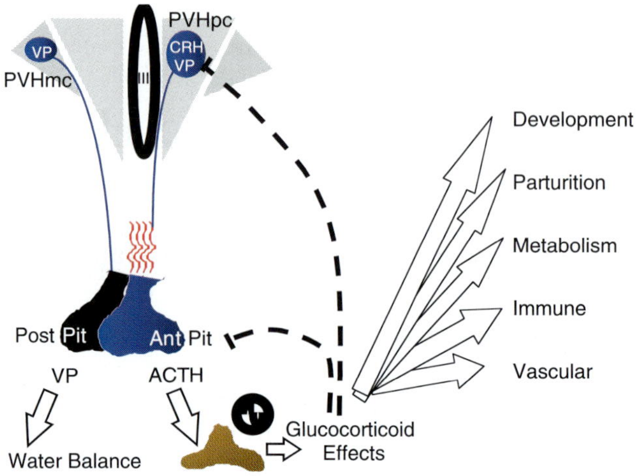

Fig. 1. Components of the HPA axis. CRF and vasopressin (AVP) are synthesized in the paraventricular nucleus of the hypothalamus (PVH), parvocellular (pc) division, whose axons terminate in the median eminence, from which the portal hypophyseal blood system (wavy lines) carries the neuropeptide to the anterior pituitary (Ant Pit). In addition, AVP is made in the magnocellular (mc) division of the PVH, whose axons terminate in the posterior pituitary (Post Pit). From anterior pituitary corticotrophs, adrenocorticotropic hormone (ACTH) is released into the systemic circulation with a circadian rhythm (circle with sine wave). ACTH stimulates glucocorticoid release. Glucocorticoid has multiple actions in development, parturition (in humans), metabolism, immune action, and vascular tone maintenance.

pathology in this axis, and to better diagnosis and therapy of these diseases.

Overview of the components of the HPA axis

Figure 1 depicts a broad overview of the components of the HPA axis. It consists of three interacting stages: hypothalamus, pituitary, and adrenal. These three stages are arranged in series, with each successive stage amplifying the signal from the preceding stage and transmitting this signal to a progressively wider range of targets. Thus, corticotropin-releasing factor (CRF, also known as corticotrophin-releasing hormone, CRH) and vasopressin (AVP) are cosynthesized and released from parvocellular neurons within the paraventricular nuclei into the portal–hypophyseal blood circulation, which carries them to the anterior pituitary gland. There, the two neuropeptides interact with their cognate receptors (CRF type 1 and AVP V1b [or V3], respectively) expressed on corticotrophs, to stimulate the synthesis of proopiomelanocortin (POMC) and the release of adrenocorticotrophic hormone (ACTH). ACTH is synthesized and released with a circadian rhythm into the systemic circulation, where it encounters its receptor, melanocortin receptor type 2 (MCR2) on cells of the adrenal cortex, to stimulate the synthesis and release of all of the adrenal steroids: aldosterone, cortisol (in humans, and corticosterone in rodents), and adrenal androgens (only in humans and higher primates).

Cortisol and corticosterone have wide-ranging effects on many organ systems. As discussed subsequently, using mice with targeted deletion of individual genes, which encode components of the HPA axis, investigators have provided strong evidence for many of the functions of glucocorticoid hormones. One of their most important functions is in fetal tissue and organ maturation. This is perhaps most clearly demonstrated in the lung, where a lack of glucocorticoid action, due to induced mutations in the gene encoding either glucocorticoid receptor (Cole et al., 1995), CRF (Muglia et al., 1995), or the CRF type 1 receptor (Smith et al., 1998), leads to lung dysplasia and neonatal death from pulmonary insufficiency. In humans, cortisol, working in concert with placental CRF, is likely involved in a feed-forward loop that is responsible both for fetal organ maturation and the timing of the onset of labor and delivery (Robinson et al., 1988; McLean et al., 1995; Karalis et al., 1996; Majzoub and Karalis, 1999). A major postnatal action of cortisol is to stimulate gluconeogenesis and glycogenolysis as a counter-regulatory response to hypoglycemia (Cryer, 1993). Although cortisol clearly has potent immunosuppressive effects when used as a pharmacological agent, a role for it in the physiological regulation of immune function is not as clear. Nevertheless, adrenalectomized rats (Kapcala et al., 1995) and CRF knockout mice (Karalis et al., 1997) lacking glucocorticoid action are more sensitive to the lethal effects of endotoxin. Lastly, glucocorticoid hormones have been implicated in regulating vascular tone and reactivity to vasoactive substances. This may be explained by the ability of glucocorticoids to suppress the expression of endothelial nitric oxide synthase, such that glucocorticoid deficiency causes increased vascular nitric oxide levels, vascular dilation, and hypotension (Wallerath et al., 1999).

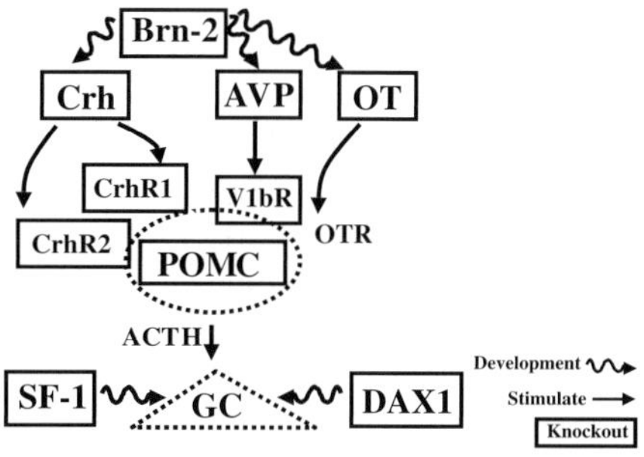

Fig. 2. Targeted gene deletions in the HPA axis. The genes of the HPA axis in which targeted deletion (or knockout) have been made are enclosed in rectangles. Arrowheads attached to straight bodies depict pathways of stimulation, whereas those attached to wavy bodies depict pathways of development. Brn-2, Brain 2; CRF, corticotropin-releasing factor; CRFR1, CRFR2, CRF type 1 and type 2 receptors; DAX1, dosage-sensitive sex reversal, adrenal hypoplasia, located on the X chromosome; GC, gluocorticoid; OTR, oxytocin receptor; OT, oxytocin; POMC, proopiomelanocortin; SF-1, steroidogenic factor 1.

Genetically engineered mouse models – a powerful tool of molecular biology

The mouse has long been used for genetic studies, resulting in new insights into cancer biology, immunology, and development. The acquisition of novel molecular biological techniques, such as gene targeting, have turned the mouse into the most powerful mammalian model system. The ability to generate mice with genetic mutations or complete deletions ("knockout") of specific genes through homologous recombination in embryonic stem (ES) cells has been an important tool to study the physiological role of genes in vivo (Melton, 1994) and their relation to human diseases (Majzoub and Muglia, 1996). Knockout mutations are loss of function mutations and they offer the possibility to study single-gene functions by deleting the gene of interest. Using this technology, the role of a gene throughout embryonic development and, in adult animals, its physiological role and possible relation to certain human diseases can be studied. These animal models, however, have their limitations. It is likely that, in mice which are deficient for a specific gene product during their entire fetal and postnatal life, compensatory mechanisms might further alter the phenotype. So, the phenotype of a genetically modified animal does not necessarily reveal the function of the gene, but might show the reaction of the organism to the alteration.

Conditional gene-targeting methods, either time-specific using the tetracycline-controlled transactivator (tTA) system (Gossen et al., 1995) or cell-specific using a bacteriophage-derived cre/loxP recombination system (Sternberg and Hamilton, 1981; Hoess et al., 1982), can address these issues.

Transgenic mice, created by injection of foreign DNA into a zygote pronucleus, are usually gain-of-function mutants. It is also possible to generate transgenic mice expressing antisense RNA. The antisense RNA is thought to form RNA:RNA hybrids with endogenous mRNA, which results in reduced gene function, either by impaired translation or decreased stability of the mRNA (Munir et al., 1990; Pepin et al., 1992). Similarly, antisense oligonucleotides can be directly applied to or injected within cells, in an attempt to "knockdown" gene expression (Reul et al., 1997). Finally, a recent method using RNA interference (RNAi) is capable of blocking the expression of specific genes (Bhargava et al., 2002). The creation of animals with inappropriate expression or lack of central components of the HPA system have contributed enormously to better understand the physiology and pathophysiology of this system. The genes within the HPA axis, which have been targeted to create knockout mice, are shown in Fig. 2. All of these are complete, nonconditional knockouts thus far, and include targeted deletions of the gene encoding Brain 2 (Schonemann et al., 1995), involved in differentiation of the paraventricular nucleus; CRF (Muglia et al., 1995); AVP (F. Grant and J. Majzoub, unpublished); CRF type 1 receptor (Smith et al., 1998; Timpl et al., 1998); CRF type 2 receptor (Bale et al., 2000; Coste et al., 2000; Kishimoto et al., 2000); V1b (or V3) receptor (Wersinger et al., 2002); proopiomelanocortin (Yaswen et al., 1999); steroidogenic factor 1 (SF-1) (Luo et al., 1994), required for adrenal gland differentiation; and DAX-1 (Yu et al., 1998), also required for adrenal gland development. In some cases mice with multiple-gene deficiencies have been created (Babu, 2002, Rivier et al., 2003).

Corticotropin-releasing factor and its gene

Acute physical and psychological stress activates the HPA axis and increases plasma levels of corticotropin (ACTH) and cortisol. CRF is the major constituent of the HPA axis and regulates both basal- and stress-induced release of pituitary ACTH. CRF is synthesized in parvocellular neurons of the paraventricular nucleus (PVN) of the hypothalamus. In the mouse, using a sensitive reverse transcription polymerase chain reaction (RT-PCR) method specific for CRF mRNA, CRF mRNA was detected, as expected, in the cerebral cortex and hypothalamus. Hybridization to radiolabeled CRF sequences revealed additional positive signals in the anterior pituitary, adrenal glands, testis, ovary, gut, heart, and lung. Although CRF peptide and mRNA have been found in several extra hypothalamic regions of the brain and a wide range of peripheral tissues (Sasaki et al., 1987; Karalis et al., 1991; Muglia et al., 1994), its main source is the hypothalamus. CRF is a 41-amino-acid peptide and as with other neuropeptides it is synthesized as a larger precursor, from which it is cleaved at flanking basic amino acids. In 1981 the chemical structure of CRF was first identified in ovine hypothalamic tissue (Vale et al., 1981), and the biologically active peptide was synthesized. Meanwhile, the mouse CRF gene has been cloned and sequenced by different groups, using molecular biological tools. For example, a Balb/c genomic library in lambda EMBL3 SP6/T7 phages was screened with a random–primed radiolabeled fragment from the coding region of the rat CRF cDNA (Frim et al., 1990) to isolate the mouse CRF gene. There exists a considerable sequence homology among species. The percent identity of the mouse compared to human or ovine CRF sequences from nucleotide −336 to −446 is 75% and 72%, respectively, whereas over a 90% identity of sequences from −336 to +1 can be found (Muglia et al., 1994). Three other neuropeptides, which share a 45% sequence homology with CRF, have been discovered, named urocortin I, urocortin II, and urocortin III (Vaughan et al., 1995; Lewis et al., 2001; Reyes et al., 2001).

The single human CRF gene is located on chromosome 8 (Arbiser et al., 1988) and the mouse CRF gene is mapped to chromosome 3 (Muglia et al., 1994). The basic structure of the CRF gene consists of a 5′ regulatory/promoter region, and two exons with an intervening intron. Exon 1 remains untranslated, whereas exon 2 contains the entire coding region for the CRF precursor peptide. Within the 5′ regulatory region several *cis* elements could be identified. These include TATA elements, CAAT boxes, asymmetric and consensus cAMP (cyclic adenosine 3′,5′-monophosphate)-responsive elements (CREs), CACCC boxes, a polypurine site, AP-1 sites, estrogen-responsive element (ERE), and glucocorticoid-responsive element (GRE) (Nichols et al., 1992; Vamvakopoulos and Chrousos, 1993; Wolfl, 1999). In the intron a stretch of 21 nucleotides, which is over 90% conserved among species, has been detected, and corresponds to the restrictive element-1 (RE-1) (Kraner et al., 1992; Mori et al., 1992; Schoenherr et al., 1996; Seth and Majzoub, 2001).

Several neurotransmitter, such as catecholamines, acetylcholine, serotonin, etc., or cytokines and glucocorticoid hormones alter CRF gene expression and ACTH secretion (Assenmacher et al., 1987; Berkenbosch et al., 1987; Sapolsky et al., 1987). A number of these neurotransmitters and polypeptide hormones seem to induce CRF gene expression by utilizing cAMP or 1,2-diacylglycerol as second messengers. In vitro studies have revealed a core sequence for a cAMP-response element (CRE, 5′-TGACGTCA-3′) at −221 nucleotides relative to the cap site of the CRF gene (Spengler et al., 1992; Mugele et al., 1993; Guardiola-Diaz et al., 1994; Wolfl, 1999). A rise in intracellular cAMP leads to activation of protein kinase A (PKA) and subsequently to phosphorylation of cellular proteins, including cAMP response element-binding protein (CREB). Phosphorylation of CREB is followed by association with CREB-binding protein (CBP). This CREB/CBP complex binds to the CRE and TATA box regions of the CRF promoter, thus resulting in an activation of gene transcription (Wolfl et al., 1999). 1,2-Diacylglycerol and intracellular calcium (Ca^{2+}), via the protein kinase C (PKC) pathway, increase CRF gene expression. Phorbolesters, such as 12-*O*-tetradecanyoyl-phorbol-13-acetate (TPA), which bind to PKC and activate it directly, have been shown to elevate CRF mRNA expression in cell-culture studies (Adler et al., 1992). Several partial TPA-response elements (TREs, 5′-TGAGTACA-3′) have been identified in the CRF gene (Van et al., 1990), but their involvement in CRF gene regulation is not yet well understood. The CRE and TRE regions differ merely by one basepair. In vitro studies have detected that deletion of the CRE abolished the positive effect of TPA on CRF gene expression (Van, 1993). In addition, there is evidence that modification of c-fos and c-jun is involved in the PKC-dependent increase in CRF expression (Van, 1993).

As mentioned above, a highly conserved element, RE-1, has been detected in the intron of the CRF gene. The high degree of conservation among different species might reflect the importance of this sequence in CRF gene regulation. Binding of the repressor element-silencing transcription factor (REST), which is absent in neural cells, to RE-1 results in repression of transcriptional activity. It has been shown that in nonneural L6 cells, which contain high levels of REST, CRF transcription is repressed and that this repression is relieved by mutation of RE-1. In neural PC12 cells, which express CRF, but no REST, addition of REST leads to repression of CRF transcription. If RE-1 has been deleted or mutated, CRF expression is not repressed (Seth and Majzoub, 2001).

Other biogenic amines, such as acetylcholine and serotonin as well as several cytokines, such as IL-1, IL-2, IL-6, stimulate CRF release from the hypothalamus (Hillhouse and Milton, 1989; Hillhouse and Reichlin, 1990; Hu et al., 1992; Lyson and McCann, 1993).

POMC gene and POMC-derived peptides

The POMC gene encodes a variety of peptides, including N-terminal glycopeptide, γ-melanotropin (MSH), joining peptide, α-melanotropin, ACTH, CLIP, β-lipotropin, β-melanotropin, and β-endorphin. The human POMC gene is located on chromosome 2p23q and consists of a promoter of at least 400 basepairs (bp) at the 5′-end followed by three exons and two introns. Exon 1 is not translated and there is less than 50% homology between the human and other mammalian POMC genes. Except for the signal peptide and 18 amino acids of the amino-terminal glycopeptide, which are located on exon 2, the majority of the POMC precursor peptide is encoded by exon 3 (Eberwine and Roberts, 1983). The biologically active peptides mentioned above are generated by posttranslational cleavage of their precursor POMC by trypsin-like prohormone convertase endopeptidase enzymes. These enzymes cleave the precursor on the C-terminal side of regions of two or more basic amino acid residues, which are subsequently removed by carbopeptidase activity. Then, some POMC proteolytic products are amidated at their C-terminus, or they undergo N-terminal acetylation. Proteolytic processing of the human POMC is mediated by either two structurally related enzymes: prohormone convertase 1 (PC1) or prohormones convertase 2 (PC2) (Seidah et al., 1990). These enzymes belong to a seven-member family of subtilisin/kexin-like mammalian proteinases (Seidah and Chretien, 1997) and are distributed within endocrine cells and neurons. PC1 is mainly distributed in the anterior pituitary, the intermediate lobe pituitary cells of rodents, and the supraoptic nucleus of the hypothalamus, whereas PC2 is absent from anterior pituitary, but expressed in rodent intermediate lobe and several sites within the central nervous system (Korner et al., 1991).

Regulation of POMC gene expression and ACTH secretion

CRF is the major positive regulator of POMC transcription. After CRF is released into the hypophysial portal blood system, it is carried to the anterior pituitary and binds to its specific CRF type 1 receptor (CRFR1) located on corticotrophs. Receptor binding stimulates POMC mRNA synthesis and the release of ACTH. The CRFR1 belongs to a family of G_s-protein-coupled receptors. After the ligand has bound to its receptor, adenylyl cyclase is activated, which leads to an increase in cytosolic cAMP and activation of protein kinase A increasing the influx of calcium through L-type Ca^{2+} channels and other yet unknown effects. This results in secretion of ACTH (Aguilera et al., 1983). Whereas CRFR1 is predominantly expressed in the corticotrophs of the anterior pituitary, the type 2 receptor (CRFR2) is more widely distributed throughout the brain and periphery (Lovenberg et al., 1995). The first one mediates the actions of the HPA axis, but also anxiety-related behavior (Skutella et al., 1998; Liebsch et al., 1999). The latter receptor is mainly involved in the regulation of feeding behavior and cardiovascular functions (Spina et al., 1996; Coste et al., 2000).

Although POMC transcription is induced by CRF-stimulated elevation in cAMP, a classical CRE of the TGACGTCA type has not been detected, using heterologous POMC promoter reporter fusion gene constructs (Jin et al., 1994). Other mediators of CRF and cAMP stimulation of POMC in corticotrophs are Nur factors, which are orphan nuclear receptors that belong to the Nur family of transcription factors (Enmark and Gustafsson, 1996; Murphy and Conneely, 1997; Philips et al., 1997a,b; Maira et al., 1999). Nur 77, Nur 1, and NOR-1 possess a highly conserved DNA-binding domain, so they can act with the same DNA sequences in promoters, although independent and specific actions are known for these factors. In the corticotroph POMC promoter, two binding sites for Nur 77 and Nur 1 have been described: the NGFI-B response element (NBRE) and the Nur response element (NurE) (Philips et al., 1997a,b; Maira et al., 1999).

Furthermore, CRF induces c-fos via cAMP and Ca^{2+}-dependent mechanisms. Since it had been shown that overexpression of c-fos in corticotrophs increase POMC transcription (Boutillier et al., 1991), CRF-induced POMC gene expression appears to be mediated to some extent via c-fos.

Recently, a novel regulator of corticotroph differentiation and POMC regulation, TPIT, has been discovered (Lamolet et al., 2001). TPIT is a member of the T-box family of transcription factors and is crucial for cell-specific expression of the POMC gene. TPIT activates POMC gene transcription in cooperation with the homeoprotein Pitx1, and both factors bind DNA within the TPIT/Pitx regulatory element (TPIT/Pitx-RE) (Lamolet et al., 2001). It has been shown that mutation in the TPIT-binding site reduces POMC promoter activity (Lamolet et al., 2001). Furthermore, *TPIT* gene mutations were found in patients with isolated ACTH deficiency, indicating the importance of TPIT for POMC lineage differentiation (Pulichino et al., 2003).

Another factor, which is involved in CRF-induced POMC expression in the anterior pituitary is the cytokine leukemia-inhibitory factor (LIF). Its biological functions are mediated through its binding to a high-affinity cell-surface LIF–receptor complex, which is formed by a low-affinity LIF-binding subunit (LIF-R) and a gp130 subunit (Gearing et al., 1991). In vitro and in vivo experiments revealed a stimulatory effect of LIF on both POMC gene expression and ACTH secretion (Akita et al., 1995; Ray et al., 1996; Chesnokova, 1998).

Following inflammatory stress, cytokines that stimulate POMC expression and ACTH secretion antagonize their peripheral proinflammatory action by stimulation of the HPA axis, which results in immunosuppression. In experiments with CRF-deficient mice it could be demonstrated that interleukin 6 (IL-6) can activate the HPA axis in a CRF-independent manner (Bethin et al., 2000).

Another regulator of the HPA axis is AVP. AVP by itself has only a mild effect on ACTH secretion, but it potentiates the effect of CRF on ACTH release in the anterior pituitary (Whitnall, 1993), whereas it has no synergistic effect with CRF on POMC gene expression (Levin et al., 1989). Besides its role as a companion to CRF-induced ACTH secretion, AVP is released in response to a variety of stimuli, e.g., osmotic stress, hypovolemia, or hypotension. AVP is mostly found in magnocellular neurons of the supraoptic nucleus (SCN) and PVN and is transported via fibers to the internal zone of the median eminence to the posterior pituitary. On the other hand, AVP is colocalized with CRF in parvocellular neurons of the PVN, released into the hypophyseal portal vein system, and regulates pituitary ACTH secretion. In the anterior pituitary the effects of AVP are mediated via $V1_b$ (V3)-receptors, which are linked to G_q-proteins that activate the plasma-membrane-bound enzyme phospholipase C followed by activation of the 1,4,5-triphosphate (IP3) and diacylglycerol (DAG) pathway. AVP seems to be involved in regulation of pituitary ACTH during chronic stress. In chronic psychosocial stress a shift in the hypothalamic CRF/AVP signal in favor of AVP can be observed (Scott and Dinan, 1998). If mice are subjected to repeated immobilization stress, hypothalamic CRF levels are decreased, despite increased pituitary ACTH output in response to acute stress. This fact implies a role for AVP and indeed elevated AVP mRNA levels in the hypothalamus and increased pituitary $V1_b$-receptor mRNA can be found, whereas a down regulation of CRF receptors occurs during chronic stress (Scott and Dinan, 1998). There is evidence that AVP plays a role in modulating negative-feedback mechanisms as a number of chronic stress paradigms have shown. After a novel superimposed stress, pituitary ACTH response is unchanged or enhanced despite an elevation of corticosterone levels (Scott and Dinan, 1998). In support of this assumption, it is interesting that the distribution of glucocorticoid receptors on CRF- and AVP-secreting neurons differ.

Despite the above data, which suggest that AVP and CRF function synergistically to control the pituitary release of ACTH, using AVP-deficient mice with targeted deletion of *Avp*, it has been shown that AVP is not needed for a normal adrenal response to severe, acute stress (Grant, F. and Majzoub, J., unpublished). Furthermore, under basal conditions or with mild stressors encountered in day-to-day activities, AVP acts as a tonic inhibitor of pituitary–adrenal activity in a CRF-independent manner. Indeed, in the absence of AVP in the SCN, CRF in the PVN is not upregulated, and the circadian cycle of the HPA axis remains intact (Grant, F. and Majzoub, J., unpublished). Thus, rather than being a coactivator of the adrenal response to acute stress, the primary role of AVP in this setting is to inhibit the activity of the adrenal axis. It still may be that, under conditions of more prolonged chronic stress, or with different types of stressors, AVP has an activating function in the HPA axis.

CRF-deficient mice have taught us important lessons about POMC gene expression and ACTH release during adrenal insufficiency. Considering the poor adrenal response to stress and adrenocortical atrophy in CRF knockout mice, it is surprising that they have normal POMC immunoreactivity and normal plasma ACTH levels (Muglia et al., 1995, 1997). Being glucocorticoid deficient, one would expect upregulated ACTH levels in these mice. This raises the question, whether CRF might be crucial for augmented ACTH release into the circulation, but not essential for POMC gene transcription. The latter is supported by the fact that POMC gene expression is increased after adrenalectomy in normal mice. In contrast, ACTH does not increase after adrenalectomy in CRF knockout mice, but rises after CRF administration equally to levels seen in adrenalectomized wild-type mice (Muglia et al., 2000).

Action of ACTH on the adrenal cortex

The primary action of ACTH on the adrenal cortex is to increase cortisol, corticosterone in rodents, synthesis, and secretion. ACTH stimulates steroidogenesis via a seven-transmembrane G_s-coupled receptor, melanocortin 2 receptor (MC2R), which interacts with adenylyl cyclase to form the second messenger cAMP. In situ hybridization experiments revealed the presence of MC2R mRNA in zona glomerulosa cells, zona fasciculata cells, and a weaker signal in the zona reticularis of the adrenal cortex of rhesus monkeys (Mountjoy et al., 1992). The different enzymes involved in the synthesis of glucocorticoid and mineralocorticoid hormones and sex steroids are depicted in Fig. 3. It is important to note that there exists no 11-β-hydroxylase in rodents, which therefore lack a fully developed zona reticularis. In adrenocortical cells ACTH stimulates lipoprotein receptors, resulting in increased receptor-mediated endocytosis of lipoproteins from the plasma to intracellular lipid droplets. Within the lipid droplets ACTH regulates hydrolysis of cholesterol esters by activation of cholesterol esterases or suppression of cholesterol acetyltransferase, through cAMP-dependent protein kinase (Pedersen and Brownie, 1980). Stimulation of the steroid-producing cells in the adrenals with ACTH produces an increase in steroid hormone synthesis within minutes. The rate-limiting step in this acute synthesis is

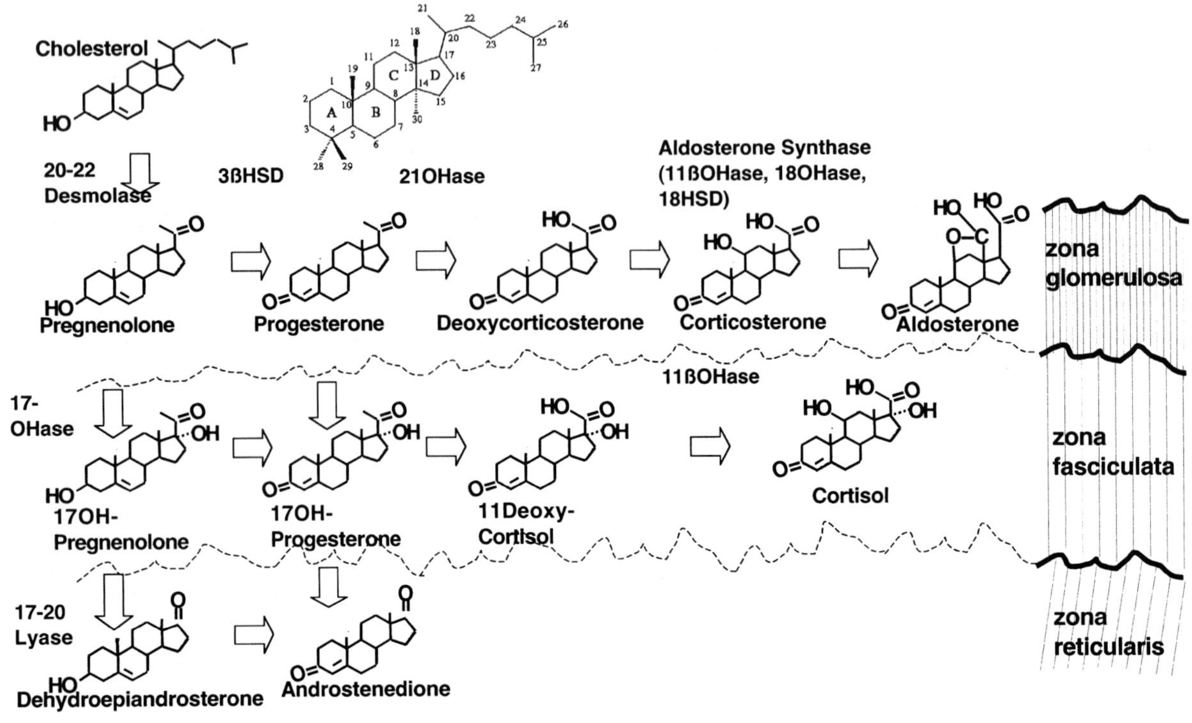

Fig. 3. Adrenal steroidogenesis. After translocation of cholesterol from the outer mitochondrial membrane to the inner mitochondrial membrane, aldosterone, cortisol, and adrenal androgens are synthesized in a zone-specific manner by various enzymes. 3βHSD, 3β-hydroxysteroid dehydrogenase; 21OHase, 21-hydroxylase; 11βOHase, 11-β-hydroxylase; 18OHase, 18-hydroxylase; 17OHase, 17-hydroxylase.

the translocation of cholesterol from the outer mitochondrial membrane to the inner mitochondrial membrane (Privalle et al., 1983). The P450 side-chain-cleavage enzyme system, which is located on the inner mitochondrial membrane, then catalyzes the cleavage of cholesterol to form pregnolone, another rate-limiting step in steroid hormone synthesis. The fact that the acute steroidogenic response could be blocked within minutes by treatment with inhibitors of protein synthesis (i.e., cyclohexemide), implied a model, in which ACTH, cAMP-mediated, promotes the de novo synthesis of a short-acting protein, supporting the trans-mitochondrial translocation of cholesterol. The responsible protein has be identified as steroidogenic acute regulatory protein (StAR) (Clark et al., 1994; Sugawara, 1995, #103).

ACTH exerts not only an acute effect on adrenal steroid hormone synthesis, but also a long-term effect by increasing the transcription of genes coding for steroid hormone enzymes. ACTH has a positive regulatory effect on its own receptor and on the cAMP response to binding of ACTH to the receptor (Saez et al., 1989). With sustained ACTH stimulation, however, receptor downregulation occurs, but it is a less physiologically relevant effect.

ACTH has often been regarded to promote mitogenesis in adrenals. The adrenal gland certainly needs constant stimuli from POMC-derived peptides, since both hypophysectomy (Vinson et al., 1985) as well as suppression of POMC peptide secretion by treatment with dexamethasone (Brandsome, 1968; Wright et al., 1974) result in rapid adrenal atrophy. Whether ACTH is the responsible peptide is, however, questionable. On the one hand, in vitro studies have revealed anti-mitogenic effects of ACTH. On the other hand, in vivo immunoneutralization of ACTH increases mitogenesis, despite significantly downregulated corticosteroid levels (Rao et al., 1978; Estivariz et al., 1982). This observation implies that decreased corticosteroid levels lead to elevated POMC expression and POMC-derived mitogens. There is evidence that the mitogenic POMC peptides reside in the pro-gamma-MSH fragment. N-terminal peptides derived from this fragment not containing the gamma-MSH sequence have mitogenic effects on isolated adrenal cells (Estivariz et al., 1982), whereas pro-gamma-MSH itself does not own mitogenic properties. Therefore, it is to assume that pro-gamma-MSH undergoes a postsecretional cleavage at or near the adrenal gland (Estivariz et al., 1982; Lowry et al., 1983). A strong candidate for this cleavage is a serine protease, called adrenal secretory protease (AsP), which has been cloned and characterized recently (Bicknell et al., 2001).

Negative feedback by glucocorticoid hormones

Glucocorticoid feedback occurs at the pituitary and hypothalamic level and possibly at higher centers. Glucocorticoid hormones exert their inhibitory effects on CRF gene expression and ACTH secretion by utilizing several mechanisms, and the precise picture is yet to be determined. One mechanism, by which glucocorticoids repress CRF transcription, suggests interference with CREB binding to the CRE of the CRF promoter by inhibition of CREB phosphorylation (Legradi et al., 1997). There is also evidence that glucocorticoid hormones directly bind to a functionally defined GRE in the CRF promoter (Malkoski et al., 1997). Another mechanism of negative regulation of CRF gene expression might be through chromatin remodeling, which could involve histone acetylation, deacetylation, dephosphorylation, or DNA methylation (Ito et al., 2000; Banks et al., 2001; Sheldon et al., 2001; Thomassin et al., 2001). Glucocorticoids exert their repressive effects not only at the hypothalamic level of the HPA axis, but also at the pituitary level. Glucocorticoid feedback appears to consist of different phases. An initial fast negative effect, which occurs within seconds to minutes due to inhibition of CRF and ACTH release and not due to an effect on biosynthesis of these peptides. Later, downregulated CRF and POMC gene expression lead to decreased hormone synthesis. This delayed form of negative feedback usually occurs after more than 24 h and persist for several days (Keller-Wood and Dallman, 1984).

Adrenal insufficiency physiologically results in elevated CRF, AVP, POMC mRNA, and ACTH levels. Replacement of corticosterone, but not aldosterone, in adrenalectomized wild-type mice prevents this rise, indicating that glucocorticoids, but not mineralocorticoids play an important role in feedback mechanisms (Muglia et al., 2000).

Circadian rhythmicity of the HPA axis

A circadian timing system seems to be universal in biology. In order to maintain a normal pattern of daily glucocorticoid production in higher mammals, circadian rhythmicity of the HPA axis activity is required. The supraoptic nucleus (SCN) orchestrates circadian rhythmicity consistent with the fact that animals with lesions in the SCN lose their circadian pattern of locomotor, adrenocortical, and sleep–wake modulation (Ibuka and Kawamura, 1975). In humans, as well as in mice and rats, peak adrenal glucocorticoid production occurs just before the active period, which corresponds to the light phase of a light/dark cycle for humans and the dark phase for rodents. Genetically engineered CRF-deficient mice have been very helpful to establish the need for CRF to maintain normal diurnal corticosterone production and to determine physiological manifestations of impaired diurnal glucocorticoid production. Studies on these mice have shown that CRF is required for the normal amplitude of glucocorticoid peak before the onset of the active period, since there was an absent circadian rise in blood corticosterone in males and a markedly impaired in females (Muglia et al., 1997). Interestingly, the diurnal pattern could be restored by constant infusion of CRF indicating that the diurnal variations in CRF secretion is not required to drive rhythmic adrenal glucocorticoid secretion. Although AVP might be the factor which drives circadian rhythmicity in CRF KO mice, deletion of the AVP gene does not abolish the diurnal rhythm in HPA-axis activity (Grant, F. and Majzoub, J., unpublished).

CRF and behavior

In addition to being a central regulator of the hormonal stress response, CRF is also widely believed to mediate stress-induced behaviors, implying a broader, integrative role for the hormone in the psychological stress response. Global (Timpl et al., 1998) and brain-specific (Muller et al., 2003) models of CRF type 1 receptor deficiency indicate that these behaviors may be mediated via the CRF type 1 receptor. However, it has been demonstrated that mice lacking the CRF gene exhibit normal stress-induced behavior that is specifically blocked by a CRF type 1 receptor antagonist (Weninger et al., 1999). The other known mammalian ligand for CRF type 1 receptors is urocortin. Normal and CRF-deficient mice have an identical distribution of urocortin mRNA, which is confined to the region of the Edinger–Westphal nucleus, and is absent from regions known to mediate stress-related behaviors. Since the Edinger–Westphal nucleus is not known to project to any brain regions believed to play a role in anxiety-like behavior, an entirely different pathway must be postulated for urocortin in the Edinger–Westphal nucleus to mediate these behaviors in CRF-deficient mice. Alternatively, an unidentified CRF-like molecule other than CRF or urocortin, acting through the CRF receptors in brain regions believed to mediate stress-induced behaviors, may mediate the behavioral response to stress, either alone or in concert with CRF.

Genetic models of adrenal insufficiency – insights into adrenal development and function

As mentioned before, StAR is a key protein in adrenal steroidogenesis, and the lack of functional StAR causes congenital lipoid adrenal hyperplasia (Lin et al., 1995). In StAR knockout (StAR KO) mice the adrenal glands are severely affected with abnormal cellular architecture and abundant lipoid deposits (Caron et al., 1997b). Disruption of Cyp11a1, the gene, which encodes the side-cleavage enzyme, also leads to lipid accumulation in steroidogenic tissue (Hu et al., 2002).

Another important factor in the regulation of steroid hormone synthesis is the orphan nuclear receptor steroidogenic factor 1 (SF-1). Targeted disruption of this gene has shown its necessity for adrenal and gonadal organogenesis (Luo et al., 1994). SF-1 KO mice lack adrenals, are therefore absolutely glucocorticoid and mineralcorticoid deficient, and show female external genitalia, regardless of their chromosomal sex (Luo et al., 1994). They also lack follicle-stimulating hormone FSH and luteinizing hormone in the anterior pituitary, but all other anterior pituitary hormones are normal except for POMC, which is in the presence of glucocorticoid hormone-deficiency elevated. SF-1 regulates all major enzymes in the steroidogenic pathway, such as P450 side-chain-cleavage enzyme, 17-α-hydroxylase, 11-β-hydroxylase, and 21-hydroxylase, by binding to a DNA element within the proximal promoter in their encoding genes (Lala et al., 1992; Lynch et al., 1993; Wilson et al., 1993; Clemens et al., 1994). Additionally, SF-1 has been shown to regulate the transcription of both the ACTH receptor (Cammas et al., 1997; Marchal et al., 1998) and StAR (Sugawara et al., 1996, 1997; Caron et al., 1997a). SF-1 KO mice die shortly after birth with symptoms of dehydration. They can, however, be rescued with glucocorticoid and mineralocorticoid treatment. Lung development appears to be normal in these mice, implying maternal compensation of glucocorticoid deficiency during fetal life.

Fetal/maternal interaction and HPA-axis regulation have been studied in CRF KO mice. The absence of fetal CRF causes poor adrenal growth, decreased corticosterone secretion, and impairment of pulmonary development (Muglia et al., 1995). Unlike CRF KO fetuses from CRF KO mothers, which die of pulmonary distress, CRF KO from heterozygous mothers are viable. A circadian evening increase in corticosterone levels in heterozygous mothers result in a significant evening rise in fetal corticosterone levels, not seen in CRF KO fetuses from CRF KO mothers. Maternally derived corticosterone seems to be crucial for fetal lung development, since pups of all genotypes, i.e., knockout, heterozygous, and wild-type, die immediately after birth in respiratory distress (Venihaki et al., 2000). Substitution of corticosterone in the drinking water of pregnant CRF KO mice leads to normal viability of all genotypes. CRF KO mice have glucocorticoid, but no mineralocorticoid deficiency, so they do not show any symptoms of dehydration during postnatal development, as SF-1 KO mice do. Interestingly, another genetic mouse model of absolute glucocorticoid and mineralocorticoid deficiency, POMC knockout mice (POMC KO), shows normal postnatal viability (Yaswen et al., 1999). Serum corticosterone and aldosterone were undetectable, using radioimmunoassay, and macroscopically no adrenal glands were discernable. Histological examination showed areas of tissue reminiscent of rudimentary adrenal cortex and medulla. Given their normal viability one has to assume that either of these mice have very low glucocorticoid and mineralocorticoid levels, which cannot be measured with the RIA. Or, other, precursor, molecules compensate for glucocorticoid and mineralocorticoid function.

References

Adler, G.K., Rosen, L.B., Fiandaca, M.J. and Majzoub, J.A. (1992) Protein kinase-C activation increases the quantity and poly(A) tail length of corticotropin-releasing hormone messenger RNA in NPLC cells. Mol. Endocrinol., 6: 476–484.

Aguilera, G., Harwood, J.P., Wilson, J.X., Morell, J., Brown, J.H. and Catt, K.J. (1983) Mechanisms of action of corticotropin-releasing factor and other regulators of corticotropin release in rat pituitary cells. J. Biol. Chem., 258: 8039–8045.

Akita, S., Webster, J., Ren, S.G., Takino, H., Said, J., Zand, O. and Melmed, S. (1995) Human and murine pituitary expression of leukemia inhibitory factor. Novel intrapituitary regulation of adrenocorticotropin hormone synthesis and secretion. J. Clin. Invest., 95: 1288–1298.

Arbiser, J.L., Morton, C.C., Bruns, G.A. and Majzoub, J.A. (1988) Human corticotropin releasing hormone gene is located on the long arm of chromosome 8. Cytogenet. Cell Genet., 47: 113–116.

Assenmacher, I., Szafarczyk, A., Alonso, G., Ixart, G. and Barbanel, G. (1987) Physiology of neural pathways affecting CRH secretion. Ann. N.Y. Acad. Sci., 512: 149–161.

Babu, P.S., Bavers, D.L., Beuschlein, F., Shah, S., Jeffs, B., Jameson, J.L., and Hammer, G.D. (2002) Interaction between *Dax-1* and steroidogenic factor-1 in vivo: increased adrenal responsiveness to ACTH in the absence of *Dax-1*. Endocrinology, 143: 665–673.

Bale, T.L., Contarino, A., Smith, G.W., Chan, R., Gold, L.H., Sawchenko, P.E., Koob, G.F., Vale, W.W. and Lee, K.F. (2000) Mice deficient for corticotropin-releasing hormone receptor-2 display anxiety-like behaviour and are hypersensitive to stress. Nat. Genet., 24: 410–414.

Banks, G.C., Deterding, L.J., Tomer, K.B. and Archer, T.K. (2001) Hormone-mediated dephosphorylation of specific histone H1 isoforms. J. Biol. Chem., 276: 36467–36473.

Berkenbosch, F., van Oers, J., del Rey, A., Tilders, F. and Besedovsky, H. (1987) Corticotropin-releasing factor-producing neurons in the rat activated by interleukin-1. Science, 238: 524–526.

Bethin, K.E., Vogt, S.K. and Muglia, L.J. (2000) Interleukin-6 is an essential, corticotropin-releasing hormone-independent stimulator of the adrenal axis during immune system activation. Proc. Natl. Acad. Sci. USA, 97: 9317–9322.

Bhargava, A., Mathias, R.S., McCormick, J.A., Dallman, M.F. and Pearce, D. (2002) Glucocorticoids prolong Ca(2+) transients in hippocampal-derived H19-7 neurons by repressing the plasma membrane Ca(2+)-ATPase-1. Mol. Endocrinol., 16: 1629–1637.

Bicknell, A., Lomthaisong, K., Woods, R., Hutchinson, E., Bennett, H., Gladwell, R. and Lowry, P. (2001) Characterization of a serine protease that cleaves pro-gamma-melanotropin at the adrenal to stimulate growth. Cell, 105: 903–912.

Boutillier, A.L., Sassone-Corsi, P. and Loeffler, J.P. (1991) The protooncogene c-fos is induced by corticotropin-releasing factor and stimulates proopiomelanocortin gene transcription in pituitary cells. Mol. Endocrinol., 5: 1301–1310.

Brandsome, E. (1968) Regulation of adrenal growth. Differences in the effects of ACTH in normal and dexamethasone-suppressed guinea pigs. Endocrinology, 83: 956–964.

Cammas, F.M., Pullinger, G.D., Barker, S. and Clark, A.J. (1997) The mouse adrenocorticotropin receptor gene: cloning and characterization of its promoter and evidence for a role for the orphan nuclear receptor steroidogenic factor 1. Mol. Endocrinol., 11: 867–876.

Caron, K.M., Ikeda, Y., Soo, S.C., Stocco, D.M., Parker, K.L. and Clark, B.J. (1997a) Characterization of the promoter region of the mouse gene encoding the steroidogenic acute regulatory protein. Mol. Endocrinol., 11: 138–147.

Caron, K.M., Soo, S.C., Wetsel, W.C., Stocco, D.M., Clark, B.J. and Parker, K.L. (1997b) Targeted disruption of the mouse gene encoding steroidogenic acute regulatory protein provides insights into congenital lipoid adrenal hyperplasia. Proc. Natl. Acad. Sci. USA, 94: 11540–11545.

Chesnokova, V., Auernhammer, C.J., Melmed, S. (1998) Murine leukaemia inhibitory factor gene disruption attenuates the hypothalamo-pituitary-adrenal axis stress response. Endocrinology, 139: 2209–2216.

Clark, B.J., Wells, J., King, S.R. and Stocco, D.M. (1994) The purification, cloning and expression of a novel luteinizing hormone-induced mitochondrial protein in MA-10 mouse Leydig tumor cells. Characterization of the steroidogenic acute regulatory protein (StAR). J. Biol. Chem., 269: 28314–28322.

Clemens, J.W., Lala, D.S., Parker, K.L. and Richards, J.S. (1994) Steroidogenic factor-1 binding and transcriptional activity of the cholesterol side-chain cleavage promoter in rat granulosa cells. Endocrinology, 134: 1499–1508.

Cole, T.J., Blendy, J.A., Monaghan, A.P., Krieglstein, K., Schmid, W., Aguzzi, A., Fantuzzi, G., Hummler, E., Unsicker, K. and Schutz, G. (1995) Targeted disruption of the glucocorticoid receptor gene blocks adrenergic chromaffin cell development and severely retards lung maturation. Genes Dev., 9: 1608–1621.

Coste, S.C., Kesterson, R.A., Heldwein, K.A., Stevens, S.L., Heard, A.D., Hollis, J.H., Murray, S.E., Hill, J.K., Pantely, G.A., Hohimer, A.R., Hatton, D.C., Phillips, T.J., Finn, D.A., Low, M.J., Rittenberg, M.B., Stenzel, P. and Stenzel-Poore, M.P. (2000) Abnormal adaptations to stress and impaired cardiovascular function in mice lacking corticotropin-releasing hormone receptor-2. Nat. Genet., 24: 403–409.

Cryer, P.E. (1993) Glucose counterregulation: prevention and correction of hypoglycemia in humans. Am. J. Physiol., 264: E149–E155.

Eberwine, J.H. and Roberts, J.L. (1983) Analysis of pro-opiomelancortin gene structure and function. DNA, 2: 1–8.

Enmark, E. and Gustafsson, J.A. (1996) Orphan nuclear receptors – the first eight years. Mol. Endocrinol., 10: 1293–1307.

Estivariz, F., Iturriza, F., Hope, J., McClean, C. and Lowry, P. (1982) Stimulation of adrenal mitogenesis by N-terminal pro-opiocortin peptide. Nature, 297: 419–422.

Frim, D.M., Robinson, B.G., Pasieka, K.B. and Majzoub, J.A. (1990) Differential regulation of corticotropin-releasing hormone mRNA in rat brain. Am. J. Physiol., 258: E686–E692.

Gearing, D.P., Thut, C.J., VandeBos, T., Gimpel, S.D., Delaney, P.B., King, J., Price, V., Cosman, D. and Beckmann, M.P. (1991) Leukemia inhibitory factor receptor is structurally related to the IL-6 signal transducer, gp130. EMBO J., 10: 2839–2848.

Gossen, M., Freundlieb, S., Bender, G., Muller, G., Hillen, W. and Bujard, H. (1995) Transcriptional activation by tetracyclines in mammalian cells. Science, 268: 1766–1769.

Guardiola-Diaz, H.M., Boswell, C. and Seasholtz, A.F. (1994) The cAMP-responsive element in the corticotropin-releasing hormone gene mediates transcriptional regulation by depolarization. J. Biol. Chem., 269: 14784–14791.

Hillhouse, E.W. and Milton, N.G. (1989) Effect of acetylcholine and 5-hydroxytryptamine on the secretion of corticotrophin-releasing factor-41 and arginine vasopressin from the rat hypothalamus in vitro. J. Endocrinol., 122: 713–718.

Hillhouse, E. and Reichlin, S. (1990) Acetylcholine stimulates the secretion of corticotropin-releasing factor from primary dissociated cell cultures of the rat telencephalon and diencephalon. Brain Res., 506: 9–13.

Hoess, R.H., Ziese, M. and Sternberg, N. (1982) P1 site-specific recombination: nucleotide sequence of the recombining sites. Proc. Natl. Acad. Sci. USA, 79: 3398–3402.

Hu, S.B., Tannahill, L.A. and Lightman, S.L. (1992) Interleukin-1 beta induces corticotropin-releasing factor-41 release from cultured hypothalamic cells through protein kinase C and cAMP-dependent protein kinase pathways. J. Neuroimmunol., 40: 49–55.

Hu, M.C., Hsu, N.C., El Hadj, N.B., Pai, C.I., Chu, H.P., Wang, C.K. and Chung, B.C. (2002) Steroid deficiency syndromes in mice with targeted disruption of Cyp11a1. Mol. Endocrinol., 16: 1943–1950.

Ibuka, N. and Kawamura, H. (1975) Loss of circadian rhythm in sleep-wakefulness cycle in the rat by suprachiasmatic nucleus lesions. Brain Res., 96: 76–81.

Ito, K., Barnes, P.J. and Adcock, I.M. (2000) Glucocorticoid receptor recruitment of histone deacetylase 2 inhibits interleukin-1beta-induced histone H4 acetylation on lysines 8 and 12. Mol. Cell. Biol., 20: 6891–6903.

Jin, W.D., Boutillier, A.L., Glucksman, M.J., Salton, S.R., Loeffler, J.P. and Roberts, J.L. (1994) Characterization of a corticotropin-releasing hormone-responsive element in the rat proopiomelanocortin gene promoter and molecular cloning of its binding protein. Mol. Endocrinol., 8: 1377–1388.

Kapcala, L.P., Chautard, T. and Eskay, R.L. (1995) The protective role of the hypothalamic–pituitary–adrenal axis against lethality produced by immune, infectious, and inflammatory stress. Ann. N.Y. Acad. Sci., 771: 419–437.

Karalis, K., Sano, H., Redwine, J., Listwak, S., Wilder, R.L. and Chrousos, G.P. (1991) Autocrine or paracrine inflammatory actions of corticotropin-releasing hormone in vivo. Science, 254: 421–423.

Karalis, K., Goodwin, G. and Majzoub, J.A. (1996) Cortisol blockade of progesterone: a possible molecular mechanism involved in the initiation of human labor. Nat. Med., 2: 556–560.

Karalis, K., Muglia, L.J., Bae, D., Hilderbrand, H. and Majzoub, J.A. (1997) CRH and the immune system. J. Neuroimmunol., 72: 131–136.

Keller-Wood, M.E. and Dallman, M.F. (1984) Corticosteroid inhibition of ACTH secretion. Endocr. Rev., 5: 1–24.

Kishimoto, T., Radulovic, J., Radulovic, M., Lin, C.R., Schrick, C., Hooshmand, F., Hermanson, O., Rosenfeld, M.G. and Spiess, J. (2000) Deletion of crhr2 reveals an anxiolytic role for corticotropin-releasing hormone receptor-2. Nat. Genet., 24: 415–419.

Korner, J., Chun, J., Harter, D. and Axel, R. (1991) Isolation and functional expression of a mammalian prohormone processing enzyme, murine prohormone convertase 1. Proc. Natl. Acad. Sci. USA, 88: 6834–6838.

Kraner, S.D., Chong, J.A., Tsay, H.J. and Mandel, G. (1992) Silencing the type II sodium channel gene: a model for neural-specific gene regulation. Neuron, 9: 37–44.

Lala, D.S., Rice, D.A. and Parker, K.L. (1992) Steroidogenic factor I, a key regulator of steroidogenic enzyme expression, is the mouse homolog of fushi tarazu-factor I. Mol. Endocrinol., 6: 1249–1258.

Lamolet, B., Pulichino, A.M., Lamonerie, T., Gauthier, Y., Brue, T., Enjalbert, A. and Drouin, J. (2001) A pituitary cell-restricted T box factor, Tpit, activates POMC transcription in cooperation with Pitx homeoproteins. Cell, 104: 849–859.

Legradi, G., Holzer, D., Kapcala, L.P. and Lechan, R.M. (1997) Glucocorticoids inhibit stress-induced phosphorylation of CREB in corticotropin-releasing hormone neurons of the hypothalamic paraventricular nucleus. Neuroendocrinology, 66: 86–97.

Levin, N., Blum, M. and Roberts, J.L. (1989) Modulation of basal and corticotropin-releasing factor-stimulated proopiomelanocortin gene expression by vasopressin in rat anterior pituitary. Endocrinology, 125: 2957–2966.

Lewis, K., Li, C., Perrin, M.H., Blount, A., Kunitake, K., Donaldson, C., Vaughan, J., Reyes, T.M., Gulyas, J., Fischer, W., Bilezikjian, L., Rivier, J., Sawchenko, P.E. and Vale, W.W. (2001) Identification of urocortin III, an additional member of the corticotropin-releasing factor (CRF) family with high affinity for the CRF2 receptor. Proc. Natl. Acad. Sci. USA, 98: 7570–7575.

Liebsch, G., Landgraf, R., Engelmann, M., Lorscher, P. and Holsboer, F. (1999) Differential behavioural effects of chronic infusion of CRH 1 and CRH 2 receptor antisense oligonucleotides into the rat brain. J. Psychiatr. Res., 33: 153–163.

Lin, D., Sugawara, T., Strauss, J.F., 3rd, Clark, B.J., Stocco, D.M., Saenger, P., Rogol, A. and Miller, W.L. (1995) Role of steroidogenic acute regulatory protein in adrenal and gonadal steroidogenesis. Science, 267: 1828–1831.

Lovenberg, T.W., Chalmers, D.T., Liu, C. and De Souza, E.B. (1995) CRF2 alpha and CRF2 beta receptor mRNAs are differentially distributed between the rat central nervous system and peripheral tissues. Endocrinology, 136: 4139–4142.

Lowry, P., Silas, N., McClean, C., Linton, E. and Estivariz, F. (1983) Pro-gamma-melanocyte-stimulating hormone cleavage in the adrenal gland undergoing compensatory growth. Nature, 306: 70–73.

Luo, X., Ikeda, Y. and Parker, K.L. (1994) A cell-specific nuclear receptor is essential for adrenal and gonadal development and sexual differentiation. Cell, 77: 481–490.

Lynch, J.P., Lala, D.S., Peluso, J.J., Luo, W., Parker, K.L. and White, B.A. (1993) Steroidogenic factor 1, an orphan nuclear receptor, regulates the expression of the rat aromatase gene in gonadal tissues. Mol. Endocrinol., 7: 776–786.

Lyson, K. and McCann, S.M. (1993) Alpha-melanocyte-stimulating hormone abolishes IL-1- and IL-6-induced corticotropin-releasing factor release from the hypothalamus in vitro. Neuroendocrinology, 58: 191–195.

Maira, M., Martens, C., Philips, A. and Drouin, J. (1999) Heterodimerization between members of the Nur subfamily of orphan nuclear receptors as a novel mechanism for gene activation. Mol. Cell. Biol., 19: 7549–7557.

Majzoub, J.A. and Muglia, L.J. (1996) Knockout mice. N. Engl. J. Med., 334: 904–907.

Majzoub, J.A. and Karalis, K.P. (1999) Placental corticotropin-releasing hormone: function and regulation. Am. J. Obstet. Gynecol., 180: S242–S246.

Malkoski, S.P., Handanos, C.M. and Dorin, R.I. (1997) Localization of a negative glucocorticoid response element of the human corticotropin releasing hormone gene. Mol. Cell. Endocrinol., 127: 189–199.

Marchal, R., Naville, D., Durand, P., Begeot, M. and Penhoat, A. (1998) A steroidogenic factor-1 binding element is essential for basal human ACTH receptor gene transcription. Biochem. Biophys. Res. Commun., 247: 28–32.

McLean, M., Bisits, A., Davies, J., Woods, R., Lowry, P. and Smith, R. (1995) A placental clock controlling the length of human pregnancy. Nat. Med., 1: 460–463.

Melton, D.W. (1994) Gene targeting in the mouse. Bioessays, 16: 633–638.

Mori, N., Schoenherr, C., Vandenbergh, D.J. and Anderson, D.J. (1992) A common silencer element in the SCG10 and type II Na+ channel genes binds a factor present in nonneuronal cells but not in neuronal cells. Neuron, 9: 45–54.

Mountjoy, K.G., Robbins, L.S., Mortrud, M.T. and Cone, R.D. (1992) The cloning of a family of genes that encode the melanocortin receptors. Science, 257: 1248–1251.

Mugele, K., Kugler, H. and Spiess, J. (1993) Immortalization of a fetal rat brain cell line that expresses corticotropin-releasing factor mRNA. DNA Cell Biol., 12: 119–126.

Muglia, L.J., Jenkins, N.A., Gilbert, D.J., Copeland, N.G. and Majzoub, J.A. (1994) Expression of the mouse corticotropin-releasing hormone gene in vivo and targeted inactivation in embryonic stem cells. J. Clin. Invest., 93: 2066–2072.

Muglia, L., Jacobson, L., Dikkes, P. and Majzoub, J.A. (1995) Corticotropin-releasing hormone deficiency reveals major fetal but not adult glucocorticoid need. Nature, 373: 427–432.

Muglia, L.J., Jacobson, L., Weninger, S.C., Luedke, C.E., Bae, D.S., Jeong, K.H. and Majzoub, J.A. (1997) Impaired diurnal adrenal rhythmicity restored by constant infusion of corticotropin-releasing hormone in corticotropin-releasing hormone-deficient mice. J. Clin. Invest., 99: 2923–2929.

Muglia, L.J., Jacobson, L., Luedke, C., Vogt, S.K., Schaefer, M.L., Dikkes, P., Fukuda, S., Sakai, Y., Suda, T. and Majzoub, J.A. (2000) Corticotropin-releasing hormone links pituitary adrenocorticotropin gene expression and release during adrenal insufficiency. J. Clin. Invest., 105: 1269–1277.

Muller, M.B., Zimmermann, S., Sillaber, I., Hagemeyer, T.P., Deussing, J.M., Timpl, P., Kormann, M.S., Droste, S.K., Kuhn, R., Reul, J.M., Holsboer, F. and Wurst, W. (2003)

Limbic corticotropin-releasing hormone receptor 1 mediates anxiety-related behavior and hormonal adaptation to stress. Nat. Neurosci., 6: 1100–1107.

Munir, M.I., Rossiter, B.J. and Caskey, C.T. (1990) Antisense RNA production in transgenic mice. Somat. Cell. Mol. Genet., 16: 383–394.

Murphy, E.P. and Conneely, O.M. (1997) Neuroendocrine regulation of the hypothalamic pituitary adrenal axis by the nurr1/nur77 subfamily of nuclear receptors. Mol. Endocrinol., 11: 39–47.

Nichols, M., Weih, F., Schmid, W., DeVack, C., Kowenz-Leutz, E., Luckow, B., Boshart, M. and Schutz, G. (1992) Phosphorylation of CREB affects its binding to high and low affinity sites: implications for cAMP induced gene transcription. EMBO J., 11: 3337–3346.

Pedersen, R.C. and Brownie, A.C. (1980) Adrenocortical response to corticotropin is potentiated by part of the amino-terminal region of pro-corticotropin/endorphin. Proc. Natl. Acad. Sci. USA, 77: 2239–2243.

Pepin, M.C., Pothier, F. and Barden, N. (1992) Impaired type II glucocorticoid-receptor function in mice bearing antisense RNA transgene. Nature, 355: 725–728.

Philips, A., Lesage, S., Gingras, R., Maira, M.H., Gauthier, Y., Hugo, P. and Drouin, J. (1997a) Novel dimeric Nur77 signaling mechanism in endocrine and lymphoid cells. Mol. Cell. Biol., 17: 5946–5951.

Philips, A., Maira, M., Mullick, A., Chamberland, M., Lesage, S., Hugo, P. and Drouin, J. (1997b) Antagonism between Nur77 and glucocorticoid receptor for control of transcription. Mol. Cell. Biol., 17: 5952–5959.

Privalle, C.T., Crivello, J.F. and Jefcoate, C.R. (1983) Regulation of intramitochondrial cholesterol transfer to side-chain cleavage cytochrome P-450 in rat adrenal gland. Proc. Natl. Acad. Sci. USA, 80: 702–706.

Pulichino, A.M., Vallette-Kasic, S., Tsai, J.P., Couture, C., Gauthier, Y. and Drouin, J. (2003) Tpit determines alternate fates during pituitary cell differentiation. Genes Dev., 17: 738–747.

Rao, A., Long, J. and Ramachandran, J. (1978) Effects of antiserum to adrenocorticotrophin on adrenal growth and function. Endocrinology, 102: 371–378.

Ray, D.W., Ren, S.G. and Melmed, S. (1996) Leukemia inhibitory factor (LIF) stimulates proopiomelanocortin (POMC) expression in a corticotroph cell line. Role of STAT pathway. J. Clin. Invest., 97: 1852–1859.

Reul, J.M., Probst, J.C., Skutella, T., Hirschmann, M., Stec, I.S., Montkowski, A., Landgraf, R. and Holsboer, F. (1997) Increased stress-induced adrenocorticotropin response after long-term intracerebroventricular treatment of rats with antisense mineralocorticoid receptor oligodeoxynucleotides. Neuroendocrinology, 65: 189–199.

Reyes, T.M., Lewis, K., Perrin, M.H., Kunitake, K.S., Vaughan, J., Arias, C.A., Hogenesch, J.B., Gulyas, J., Rivier, J., Vale, W.W. and Sawchenko, P.E. (2001) Urocortin II: a member of the corticotropin-releasing factor (CRF) neuropeptide family that is selectively bound by type 2 CRF receptors. Proc. Natl. Acad. Sci. USA, 98: 2843–2848.

Rivier, C.L., Grigoriadis, D.E. and Rivier, J.E. (2003) Role of corticotropin-releasing factor receptors type 1 and 2 in modulating the rat adrenocorticotropin response to stressors. Endocrinology, 144: 2396–2403.

Robinson, B.G., Emanuel, R.L., Frim, D.M. and Majzoub, J.A. (1988) Glucocorticoid stimulates expression of corticotropin-releasing hormone gene in human placenta. Proc. Natl. Acad. Sci. USA, 85: 5244–5248.

Saez, J.M., Begeot, M. and Durand, P. (1989) [ACTH receptors]. Ann. Endocrinol. (Paris), 50: 409–417.

Sapolsky, R., Rivier, C., Yamamoto, G., Plotsky, P. and Vale, W. (1987) Interleukin-1 stimulates the secretion of hypothalamic corticotropin-releasing factor. Science, 238: 522–524.

Sapolsky, R.M., Romero, L.M. and Munck, A.U. (2000) How do glucocorticoids influence stress responses? Integrating permissive, suppressive, stimulatory, and preparative actions. Endocr. Rev., 21: 55–89.

Sasaki, A., Sato, S., Murakami, O., Go, M., Inoue, M., Shimizu, Y., Hanew, K., Andoh, N., Sato, I., Sasano, N. and Yoshinaga, K. (1987) Immunoreactive corticotropin-releasing hormone present in human plasma may be derived from both hypothalamic and extrahypothalamic sources. J. Clin. Endocrinol. Metab., 65: 176–182.

Schoenherr, C.J., Paquette, A.J. and Anderson, D.J. (1996) Identification of potential target genes for the neuron-restrictive silencer factor. Proc. Natl. Acad. Sci. USA, 93: 9881–9886.

Schonemann, M.D., Ryan, A.K., McEvilly, R.J., O'Connell, S.M., Arias, C.A., Kalla, K.A., Li, P., Sawchenko, P.E. and Rosenfeld, M.G. (1995) Development and survival of the endocrine hypothalamus and posterior pituitary gland requires the neuronal POU domain factor Brn-2. Genes Dev., 9: 3122–3135.

Scott, L.V. and Dinan, T.G. (1998) Vasopressin and the regulation of hypothalamic–pituitary–adrenal axis function: implications for the pathophysiology of depression. Life Sci., 62: 1985–1998.

Seidah, N.G. and Chretien, M. (1997) Eukaryotic protein processing: endoproteolysis of precursor proteins. Curr. Opin. Biotechnol., 8: 602–607.

Seidah, N.G., Gaspar, L., Mion, P., Marcinkiewicz, M., Mbikay, M. and Chretien, M. (1990) cDNA sequence of two distinct pituitary proteins homologous to Kex2 and furin gene products: tissue-specific mRNAs encoding candidates for pro-hormone processing proteinases. DNA Cell Biol., 9: 415–424.

Seth, K.A. and Majzoub, J.A. (2001) Repressor element silencing transcription factor/neuron-restrictive silencing factor (REST/NRSF) can act as an enhancer as well as a repressor of corticotropin-releasing hormone gene transcription. J. Biol. Chem., 276: 13917–13923.

Sheldon, L.A., Becker, M. and Smith, C.L. (2001) Steroid hormone receptor-mediated histone deacetylation and transcription at the mouse mammary tumor virus promoter. J. Biol. Chem., 276: 32423–32426.

Skutella, T., Probst, J.C., Renner, U., Holsboer, F. and Behl, C. (1998) Corticotropin-releasing hormone receptor (type I) antisense targeting reduces anxiety. Neuroscience, 85: 795–805.

Smith, G.W., Aubry, J.M., Dellu, F., Contarino, A., Bilezikjian, L.M., Gold, L.H., Chen, R., Marchuk, Y., Hauser, C., Bentley, C.A., Sawchenko, P.E., Koob, G.F., Vale, W. and Lee, K.F. (1998) Corticotropin releasing factor receptor 1-deficient mice display decreased anxiety, impaired stress response and aberrant neuroendocrine development. Neuron, 20: 1093–1102.

Spengler, D., Rupprecht, R., Van, L.P. and Holsboer, F. (1992) Identification and characterization of a 3′,5′-cyclic adenosine monophosphate-responsive element in the human corticotropin-releasing hormone gene promoter. Mol. Endocrinol., 6: 1931–1941.

Spina, M., Merlo-Pich, E., Chan, R.K., Basso, A.M., Rivier, J., Vale, W. and Koob, G.F. (1996) Appetite-suppressing effects of urocortin, a CRF-related neuropeptide. Science, 273: 1561–1564.

Sternberg, N. and Hamilton, D. (1981) Bacteriophage P1 site-specific recombination. I. Recombination between loxP sites. J. Mol. Biol., 150: 467–486.

Sugawara, T., Lin, D., Holt, J.A., Martin, K.O., Javitt, N.B., Miller, W.L., Strauss, J.F. 3rd (1995) Structure of the human steroidogenic acute regulatory protein (StAR) gene: StAR stimulates mitochondria cholesterol 27-hydroxylase activity. Biochemistry, 34: 12506–12512.

Sugawara, T., Holt, J.A., Kiriakidou, M. and Strauss, J.F., 3rd (1996) Steroidogenic factor 1-dependent promoter activity of the human steroidogenic acute regulatory protein (StAR) gene. Biochemistry, 35: 9052–9059.

Sugawara, T., Kiriakidou, M., McAllister, J.M., Kallen, C.B. and Strauss, J.F., 3rd (1997) Multiple steroidogenic factor 1 binding elements in the human steroidogenic acute regulatory protein gene 5′-flanking region are required for maximal promoter activity and cyclic AMP responsiveness. Biochemistry, 36: 7249–7255.

Thomassin, H., Flavin, M., Espinas, M.L. and Grange, T. (2001) Glucocorticoid-induced DNA demethylation and gene memory during development. EMBO J., 20: 1974–1983.

Timpl, P., Spanagel, R., Sillaber, I., Kresse, A., Reul, J.M., Stalla, G.K., Blanquet, V., Steckler, T., Holsboer, F. and Wurst, W. (1998) Impaired stress response and reduced anxiety in mice lacking a functional corticotropin-releasing hormone receptor 1. Nat. Genet., 19: 162–166.

Vale, W., Spiess, J., Rivier, C. and Rivier, J. (1981) Characterization of a 41-residue ovine hypothalamic peptide that stimulates secretion of corticotropin and beta-endorphin. Science, 213: 1394–1397.

Vamvakopoulos, N.C. and Chrousos, G.P. (1993) Regulated activity of the distal promoter-like element of the human corticotropin-releasing hormone gene and secondary structural features of its corresponding transcripts. Mol. Cell. Endocrinol., 94: 73–78.

Van, L.P. (1993) Phorbolester stimulates the activity of human corticotropin-releasing hormone gene promoter via 3′,5′-cyclic adenosine monophosphate response element in transiently transfected chicken macrophages. Endocrinology, 132: 30–34.

Van, L.P., Spengler, D.H. and Holsboer, F. (1990) Glucocorticoid repression of 3′,5′-cyclic-adenosine monophosphate-dependent human corticotropin-releasing-hormone gene promoter activity in a transfected mouse anterior pituitary cell line. Endocrinology, 127: 1412–1418.

Vaughan, J., Donaldson, C., Bittencourt, J., Perrin, M.H., Lewis, K., Sutton, S., Chan, R., Turnbull, A.V., Lovejoy, D., Rivier, C. Rivier, J., Sawchenko, P.E. and Vale, W. (1995) Urocortin, a mammalian neuropeptide related to fish urotensin I and to corticotropin-releasing factor. Nature, 378: 287–292.

Venihaki, M., Carrigan, A., Dikkes, P. and Majzoub, J.A. (2000) Circadian rise in maternal glucocorticoid prevents pulmonary dysplasia in fetal mice with adrenal insufficiency. Proc. Natl. Acad. Sci. USA, 97: 7336–7341.

Vinson, G., Hinson, J. and Raven, P. (1985) The relationship between tissue preparation and function: methods for the study of control of aldosterone secretion: a review. Cell Biochem. Funct., 3: 235–253.

Wallerath, T., Witte, K., Schafer, S.C., Schwarz, P.M., Prellwitz, W., Wohlfart, P., Kleinert, H., Lehr, H.A., Lemmer, B. and Forstermann, U. (1999) Down-regulation of the expression of endothelial NO synthase is likely to contribute to glucocorticoid-mediated hypertension. Proc. Natl. Acad. Sci. USA, 96: 13357–13362.

Weninger, S.C., Dunn, A.J., Muglia, L.J., Dikkes, P., Miczek, K.A., Swiergiel, A.H., Berridge, C.W. and Majzoub, J.A. (1999) Stress-induced behaviors require the corticotropin-releasing hormone (CRH) receptor, but not CRH. Proc. Natl. Acad. Sci. USA, 96: 8283–8288.

Wersinger, S.R., Ginns, E.I., O'Carroll, A.M., Lolait, S.J. and Young, W.S., 3rd (2002) Vasopressin V1b receptor knockout reduces aggressive behavior in male mice. Mol. Psychiatry, 7: 975–984.

Whitnall, M.H. (1993) Regulation of the hypothalamic corticotropin-releasing hormone neurosecretory system. Prog. Neurobiol., 40: 573–629.

Wilson, T.E., Fahrner, T.J. and Milbrandt, J. (1993) The orphan receptors NGFI-B and steroidogenic factor 1 establish monomer binding as a third paradigm of nuclear receptor–DNA interaction. Mol. Cell. Biol., 13: 5794–5804.

Wolfl, S., Martinez, C. and Majzoub, J.A. (1999) Inducible binding of cyclic adenosine 3′,5′-monophosphate (cAMP)-responsive element binding protein (CREB) to a cAMP-responsive promoter in vivo. Mol. Endocrinol., 13: 659–669.

Wright, N., Appelton, D. and Morley, A. (1974) Effect of dexamethasone on cell population kinetics in the adrenal cortex of the prepubertal male rat. J. Endocrinol., 62: 527–536.

Yaswen, L., Diehl, N., Brennan, M.B. and Hochgeschwender, U. (1999) Obesity in the mouse model of pro-opiomelanocortin deficiency responds to peripheral melanocortin. Nat. Med., 5: 1066–1070.

Yu, R.N., Ito, M., Saunders, T.L., Camper, S.A. and Jameson, J.L. (1998) Role of Ahch in gonadal development and gametogenesis. Nat. Genet., 20: 353–357.

T. Steckler, N.H. Kalin and J.M.H.M. Reul (Eds.)
Handbook of Stress and the Brain, Vol. 15
ISBN 0-444-51173-3

CHAPTER 1.6

The hypothalamic–pituitary–adrenal axis as a dynamically organized system: lessons from exercising mice

Johannes M.H.M. Reul* and Susanne K. Droste

The Henry Wellcome Laboratories for Integrative Neuroscience and Endocrinology, The Dorothy Hodgkin Building, University of Bristol, Whitson Street, Bristol BS1 3NY, UK

Abstract: The hypothalamic–pituitary–adrenal (HPA) axis is a neuroendocrine system which is vital for the organism to meet and adapt to stressful challenges and physical demands. A complex network of stimulatory and inhibitory influences from neural and endocrine origin governs the activity of the HPA axis. The regulatory setpoints (e.g. responsiveness to secretagogues, negative-feedback sensitivity, limbic and frontal cortical inhibitory and stimulatory control) within this network are not static but, on the contrary, are dynamically adjustable in response to changes in environmental conditions such as stressful events. Thus, the HPA axis does not only respond to stressful events by an acute surge in glucocorticoid hormone production, but also by adjusting its regulatory setpoints in order to adapt to the altering conditions.

In recent work on voluntarily exercising mice, we obtained new insights into the regulatory control mechanisms of the HPA axis. Voluntary exercise is established by providing mice with a running wheel in their home cage. They run exuberantly (~7 km/day), mainly during the first half of the dark phase. Voluntary exercise has been shown to act beneficially on the organism (either animal or human) precipitating effects at the cellular, system and behavioural level. Here, we account of the "lessons" we learned from studying the HPA axis in this animal model. We found that the HPA axis of control mice is asymmetric at least at the level of the adrenal gland (i.e. the left adrenal is larger than the right one) and, possibly, at the level of the central nervous system as well. Moreover, this asymmetry is most likely associated with the asymmetry in the sympathoadrenomedullary system. Strikingly, in exercised mice, this asymmetry was abolished due to a selective potentiation of the right sympathoadrenomedullary branch and growth of the right adrenal gland. The impact of these effects of long-term voluntary exercise on regulatory components of the HPA axis and its responsiveness to different types of stressors is reviewed and discussed.

Introduction

Glucocorticoid hormones, the end product of the hypothalamic–pituitary–adrenal (HPA) axis, play a pivotal role in numerous processes in the body. Classically, by enhancing gluconeogenesis and lipolysis, glucocorticoids support energy metabolism at times of increased energy expenditure, such as during or after stress but also during enhanced physical activity. The secretion of glucocorticoid hormones is particularly increased after a stressful event, this secretion being superimposed on the diurnal rhythm of adrenocortical output of glucocorticoids. The diurnal variation in glucocorticoid secretion usually parallels the diurnal cycle in physical activity, inherently suggesting that the diurnal rise in glucocorticoid levels is sufficient to cover the (metabolic) need for this hormone. Moreover, as physical activity goes up, e.g. due to

*Corresponding author. Tel.: +44 117 33 13137;
Fax: +44 117 33 13 139; E-mail: hans.reul@bristol.ac.uk

exercise, then also glucocorticoid hormone levels rise in parallel (see below).

Glucocorticoid hormones also strongly influence immune function in that they accelerate the immune response (Wiegers et al., 1995, 2000, 2001; Wiegers and Reul, 1998; see also Part II: Chapter 2.3 by Wiegers et al., in this volume). In the central nervous system, by acting on a plethora of mechanisms on the molecular (see Chapter 1.5) and cellular (see Chapter 5.4) levels in both neurons and glial cells, glucocorticoids influence the activity of neural circuits throughout the CNS culminating in changes in neuroendocrine, autonomic and behavioural output. The changes in emotional and cognitive behaviour (see Chapters 3.6 and 3.7) are thought to subserve largely the adaptation of the organism to stressful and possibly also ongoing (see below) events.

The HPA axis is not a static neuroendocrine system but, on the contrary, it is very dynamically organised. Previously, we have pointed out that the regulatory setpoints of the HPA axis are adjusted rather quickly after a subject has experienced a psychologically stressful event (Reul et al., 2000; Gesing et al., 2001). The goal of such an adjustment is possibly to maintain normal baseline activity and stress responsivity in the HPA axis and/or to modify stress responsiveness to specific stressors. Numerous studies have also shown adjustments in the HPA axis after long-term adverse conditions, such as chronic stress (Young et al., 1990; Bhatnagar and Dallman, 1998; Bhatnagar et al., 2000) and perinatal manipulations (Levine, 1957, 1967; Meaney et al., 1989; Vazquez et al., 1996; Vallée et al., 1997; Shanks et al., 2000), after mutations in the genome (Stenzel-Poore et al., 1994; Barden et al., 1997; Smith et al., 1998; Timpl et al., 1998; Bale et al., 2000; Coste et al., 2000; Kishimoto et al., 2000; Reul and Holsboer, 2002), and during aging (which by some may also be seen as an adverse condition (...)) (Sapolsky et al., 1983; Reul et al., 1988, 1991; Patacchioli et al., 1989; Rothuizen et al., 1993). Surprisingly, until now only relatively little attention has been payed to positive events or circumstances (such as the impact of more beneficial-holding conditions (e.g. housing in an enriched environment, access to a running wheel)) on the HPA axis of experimental animals. We have recently reported (Droste et al., 2003) that allowing mice to voluntarily exercise in a running wheel results in major changes in their HPA axis, involving changes at the molecular, cellular, organ and system level. These studies have clearly demonstrated the enormous dynamic range of adjustments that the HPA axis can be subjected to by the rather subtle – non-stressful – modification of the housing condition of a mouse. Moreover, these studies revealed novel HPA axis regulatory mechanisms operating during non-adverse conditions, among others denoting a greater than until now thought role of the sympatho adrenomedullary system in HPA axis regulation and providing clues about a bodily asymmetry in HPA axis control. The aim of this chapter is to review our studies in the context of their relevance for expanding our general understanding of the regulation and function of the HPA axis, and their relevance for the understanding the etiology of stress-related disorders.

The HPA axis proper

The activity of the HPA axis is primarily controlled by parvocellular neurons located in the hypothalamic paraventricular nucleus (PVN). These neurons produce corticotropin-releasing factor (CRF) and sometimes also vasopressin (AVP) or other neuropeptides. The neuropeptides are released in the external zone of the median eminence where they diffuse into the portal bloodstream to be transported to the anterior lobe of the pituitary. Here, CRF and vasopressin in a synergistic manner stimulate the synthesis and endocrine secretion of adrenocorticotropic hormone (ACTH) (see also Chapters 1.3 and 1.4). Other secretagogues, such as Angiotensin II and oxytocin, also stimulate ACTH production, but to a much lesser extent. ACTH in turn stimulates the production and secretion of glucocorticoid hormones from the adrenal cortex. In rodents, such as rats and mice, and birds the main glucocorticoid hormone is corticosterone, whereas in humans, dogs, cats, pigs, sheep and guinea pigs the main glucocorticoid is cortisol. The elevated levels of glucocorticoid hormones exert negative feedback on the parvocellular neurons in the PVN and corticotrophic cells in the anterior pituitary to inhibit the production and secretion of CRF (and, to a lesser extent, AVP) and ACTH, respectively, in order to restrain the activity of the HPA axis.

Corticosteroid receptors

Glucocorticoid hormones exert their action in the brain via mineralocorticoid (MRs) and glucocorticoid receptors (GRs) (for reviews, see Dallman et al., 1987; De Kloet and Reul, 1987; De Kloet, 1991; De Kloet et al., 1998; Reul et al., 2000). The issue that MRs besides responding to the mineralocorticoid hormones, aldosterone and deoxycorticosterone also respond to natural glucocorticoids such as corticosterone and cortisol will not be elaborated here as it does not concern the topic of this chapter and it has been dealt with in other reviews (Edwards et al., 1988; Funder et al., 1988; De Kloet, 1991; Seckl et al., 2002). Now almost 20 years ago, de Kloet and Reul conceptualised a novel framework for glucocorticoid action in the brain and pituitary based on the considerable difference in hormone-binding affinity between MR and GR and, therefore, consequently the considerable difference in receptor occupancy under varying conditions (Reul and De Kloet, 1985; De Kloet and Reul, 1987). Owing to their very high affinity for binding natural glucocorticoids (K_d: 0.1–0.5 nM at 4°C), MRs are under early morning baseline conditions (in rats and mice) already extensively occupied ($>$70–80%) by circulating corticosterone (Reul and De Kloet, 1985; Reul et al., 1987b; for a review on MR, see Reul et al., 2000). Indeed, based on in vivo microdialysis experiments of Linthorst and Reul, at this time of the day, free corticosterone levels in the brain are estimated to be approximately 0.5 nM (Linthorst et al., 1995, 1997; Reul et al., 2000).

It should be noted that the synthetic glucocorticoid dexamethasone, at least in vitro, presents also a high-binding affinity for MRs. However, although the on-rate of dexamethasone to MR is comparable to that of corticosterone and aldosterone, yet, in contrast to that of corticosterone or aldosterone, the off-rate is extremely high ($\sim$20-fold higher than that of corticosterone and aldosterone), possibly explaining why the dexamethasone–MR complex is unstable and cannot be activated (Reul et al., 2000). These observations explain why in early studies ^{3}H-dexamethasone was not retained in hippocampal pyramidal and granular neurons (De Kloet et al., 1975; Warembourg, 1975), this in addition to recent observations that dexamethasone poorly penetrates the brain as it is a substrate for the multidrug resistance (mdr) pump (Meijer et al., 1998). Furthermore, the lack of stability of the dexamethasone–MR complex very likely explains the poor mineralocorticoid activity of dexamethasone in mineralocorticoid target tissues, such as the kidney and the colon. We have proposed that due to its unstable interaction with MRs, dexamethasone might even hamper the action of endogenous mineralocorticoids and glucocorticoids via MRs, thereby acting as an MR antagonist (Reul et al., 2000). Such MR antagonist activity of dexamethasone has been indeed reported (Bohus and De Kloet, 1981; Hoefnagels and Kloppenborg, 1983).

Glucocorticoid receptors displaying a considerably lower affinity for natural glucocorticoids (K_d: 3–5 nM at 4°C) are hardly occupied ($\leq$10%) under early morning baseline conditions. Occupancy of GRs rises as soon as glucocorticoid levels start to rise due to the increased circadian drive when time is moving to the dark, i.e. active, phase of the day for the rat or mouse (Reul and De Kloet, 1985; Reul et al., 1987b, 2000). After exposure of the animal to strong stressors such as forced swimming or restraint, GR occupancy is even higher than that at the diurnal peak of the circadian rhythm. Based on the differential occupancy pattern of MRs and GRs, the concept was formulated that apparently MRs mediate a tonic action of glucocorticoids whereas GRs are the receptors responsible for mediating the negative-feedback action of elevated levels of glucocorticoids as occurring after stress and during the activity phase (Reul and De Kloet, 1985; De Kloet and Reul, 1987; Reul et al., 2000).

Afferent control of the HPA axis

Preface

To understand its physiological implications, the concept needs to be considered in the context of the neuroanatomical localisation of MRs and GRs.

Whereas MRs are most richly located in the hippocampus and to a much lesser extent in the lateral septum, central nucleus of the amygdala and the brainstem motor nuclei, GRs can be found in virtually all neurons (and glia) of the central nervous system with particularly high concentrations present in the PVN, neocortex and hippocampus (Fuxe et al., 1985; Reul and De Kloet, 1985, 1986; Reul et al., 1987b).

Hippocampus

In view of the virtually exclusive localisation of MR in the hippocampus and its role in mediating tonic influences of corticosterone on various HPA axis-related parameters (e.g. CRF and AVP expression in the PVN, early morning baseline ACTH and corticosterone levels), excitability of pyramical neurons and other parameters (see De Kloet and Reul, 1987; De Kloet, 1991; Joels and De Kloet, 1992; De Kloet et al., 1998), it was proposed that hippocampal MRs were involved in modulating the level of excitatory output of the hippocampus. Amongst other, this tonic excitatory output feeding into GABAergic interneurons located in the bed nucleus of the stria terminalis (BNST)/lateral septum area and known to inhibit PVN parvocellular neurons (Herman et al., 1995; Herman and Cullinan, 1997; see also Herman et al., in this volume) thereby would exert a tonic inhibitory influence on HPA activity. Potentially, the hippocampus can also affect HPA axis activity via its modulatory influence on autonomic output (Van den Berg et al., 1990, 1994) (see below) and its projections to the amygdala and frontal cortex (see below). It has been reported that in mice hippocampal MRs, but not GRs, are asymmetrically distributed with higher levels in the right hippocampus than in the left one (Neveu et al., 1998). The significance of the asymmetric distribution of hippocampal MRs for HPA axis regulation, autonomic output and other MR-regulated brain functions needs to be clarified.

The localisation of GR throughout the brain and in the anterior pituitary is congruous to its role in regulating metabolism and negative-feedback action. GRs located in the PVN and anterior pituitary mediate negative feedback of elevated glucocorticoid levels directly to the core structures of the HPA axis. GRs inhibit the synthesis and release of CRF and, to a lesser extent, AVP in parvocellular neurons of the PVN and the synthesis of proopiomelanocortin mRNA (POMC mRNA, the precursor molecule of ACTH) and release of ACTH from the corticotrophic cells in the anterior pituitary.

In the hippocampus, GRs are substantially expressed in pyramidal (primarily CA1) and granular dentate gyrus neurons (Fuxe et al., 1985; Reul and De Kloet, 1986; Van Eekelen et al., 1988). Nevertheless, it seems that the GRs in this higher limbic brain structure with regard to exerting negative feedback on the HPA axis appear to be acting secondary to those in the PVN and anterior pituitary (Van Haarst et al., 1997). Reports suggesting a potent role of the hippocampus in the negative feedback control of the HPA axis have unfortunately mainly based their argument on indirect evidence. Therefore, until now, the exact role of hippocampal GRs in the regulation of HPA activity has not been exactly clarified yet. It seems clear, however, that GRs in the hippocampus play a prominent role in the behavioural adaptation to stressful events, as for instance in the forced swim test, also called the Porsolt swim test (Jefferys et al., 1985; De Kloet et al., 1986; Korte et al., 1996a) (J.M.H.M. Reul and S. Ulbricht, unpublished observations).

Central nucleus of the amygdala

The central nucleus of the amygdala exerts via its projection to the PVN, and possibly also indirectly via a projection to the BNST, a stimulatory influence on the activity of the HPA axis under baseline conditions and after chronic stress (Beaulieu et al., 1986; Roozendaal et al., 1991; Van der Kar et al., 1991; Goldstein et al., 1996; Bhatnagar and Dallman, 1998). Both MRs and GRs are expressed in neurons of this nucleus, whereas GRs are also expressed in other nuclei of the amygdaloid complex (Reul and De Kloet, 1986). However, the role of amygdaloid MRs and GRs in HPA-axis regulation is still unknown. Interestingly, in contrast to their effect in the PVN, GRs exert a stimulatory effect on CRF expression in the central amygdaloid nucleus (and BNST)

(Schulkin et al., 1998), but its significance regarding the HPA axis is also unknown.

The medial prefrontal cortex

Another important region of the brain playing an important role in the regulation of the HPA axis is the medial prefrontal cortex (for review, see Sullivan and Gratton, 2002). The role of the prefrontal cortex generally in stress-related processes is rather complex, as for instance the ventral region of the medial prefrontal cortex seems to act stimulatory and/or facilitatory on the HPA axis, whereas the dorsomedial region is thought to be responsible for inhibitory influences, at least partly through modulation of infralimbic outputs. Diorio et al. (1993) showed that GRs in the prefrontal cortex mediate negative-feedback effects of glucocorticoids on restraint stress-induced increases in circulating ACTH and corticosterone levels. In this study, baseline HPA hormone levels were not altered. Substantial amounts of GRs and, to a lesser extent, MRs have been found in the prefrontal cortex of rats, dogs and primates (Meaney and Aitken, 1985; Reul et al., 1990; Cintra et al., 1994; Sanchez et al., 2000). Interestingly, recent observations indicate that the influence of the prefrontal cortex on stress responses of the HPA axis and autonomic nervous system is lateralized. It was observed that the ibotenic acid lesions in the right or bilateral medial prefrontal cortex of rats affect baseline and stress-induced corticosterone levels and stress ulcer development, but not lesions in the left medial prefrontal cortex (Sullivan and Gratton, 1999). Thus, with regard to the influence of the frontal cortex on the HPA axis there seems to be a clearcut right-sided bias.

The ventral medial prefrontal cortex most likely influences the activity of the PVN mainly indirectly as there are only few direct projections. However, regions in the direct vicinity of the PVN are heavily innervated as well as the brainstem (e.g. locus coeruleus, nucleus tractus solitarius (NTS), pontine (raphe nuclei) and limbic nuclei (Amygdala, BNST), which are known to directly affect the PVN activity (Terreberry and Neafsey, 1983, 1987; Hurley et al., 1991; Takagishi and Chiba, 1991; Jodo et al., 1998). Thus, the ventral region of the medial prefrontal cortex, which is also been called the visceral motor cortex (Cechetto and Saper, 1990), is supremely positioned to modulate and coordinate neuroendocrine (and autonomic) responses to stressful and emotional stimuli.

However, conversely, many of the mentioned nuclei have ascending projections to the medial prefrontal cortex thereby modulating its function. In addition, the hippocampus projects via the fimbria-fornix to the ventral medial prefrontal cortex, very likely modulating its output. Hence, the sketched network underscores the complexity of the afferent control of the HPA axis.

Dynamic changes in HPA axis control due to acute stress

The organisation of the HPA axis is not only complex in terms of space but also in terms of time. However, until recently, available evidence in the literature suggests that the overall control mechanisms of the HPA axis are mainly sensitive to long-term manipulations (i.e. time range: weeks to months). With overall control mechanism here is meant those driving and feedback mechanisms controlling the baseline activity and stress responsiveness of the HPA axis. The long-term manipulations observed to alter control processes within the HPA axis include aging (Meaney et al., 1987; Reul et al., 1988), perinatal manipulations (e.g. prenatal immune challenge, prenatal stress, maternal deprivation) (Levine, 1957, 1967; Meaney et al., 1985, 1987, 1989; Reul et al., 1994b; Vallée et al., 1997), pharmacological treatments (e.g. antidepressant treatments) (Brady et al., 1991; Reul et al., 1993, 1994a) and endocrine extirpations (e.g. adrenalectomy) (Reul et al., 1987a, 1988; Spencer et al., 1990).

Recently, we discovered that control mechanisms governing the HPA axis can be adjusted within hours after a stimulus. We found that hippocampal MRs are relatively quickly upregulated (i.e. within 8 h) by acute psychological stressors such as forced swimming and exposure to a novel environment (Gesing et al., 2001). A physical stressor such as cold exposure was ineffective. Interestingly, the effect of stress on MR levels could be blocked by pretreatment with a

CRF receptor antagonist and mimicked by intracerebroventricular injection of CRF. The effect of stress on MR levels was also observed in the amygdala and neocortex and lasted for up to 24–48 h (Gesing et al., 2001). Importantly, using a challenge test with the selective MR antagonist RU 28318, we observed that the rise in MR levels was associated with a stronger MR-mediated inhibitory tonus on the activity of the HPA axis. Thus, setpoints governing baseline activity and stress responsiveness of the HPA axis are rapidly adjustable allowing the system to adapt quickly to novel conditions.

The exact functional significance of the transient rises in MR after psychological stress needs to be discerned. Possibly, the rises in MR levels are instigated to balance the impact of putative increases in HPA-driving processes, this in order to maintain "normal" HPA physiology. Given that MRs are known to be involved in several processes at the cellular and behavioural level, the stress-induced increases in MR levels are also expected to impact on these levels. Thus, given that MRs in the dentate gyrus granular neurons exert anti-apoptotic effects (Sloviter et al., 1989; Hassan et al., 1997; Almeida et al., 2000), an increased protection of these neurons after stress by elevated MR levels may be expected. Furthermore, in view of the involvement of hippocampal MRs in anxiety-related behaviour (Korte et al., 1996b; Smythe et al., 1997), increases in anxiety after the stressful experience may be anticipated. Increases in anxiety have indeed been observed after exposure to stressful stimuli (Owens and Nemeroff, 1991; Gutman et al., 2003). These MR-associated cellular, neuroendocrine and behavioural changes make sense from a physiological perspective as they act protectively, thereby maintaining a state of well-being and increasing the chances of survival for the organism.

Voluntary exercise exercises the HPA axis

Preface

Until recently, the HPA axis has virtually always been studied under "normal" baseline conditions (in adulthood or during development) or during and after acutely or chronically stressful conditions inflicted upon animals (and sometimes upon humans). Such conditions also include disease states and aging, the latter not necessarily needing to be stressful. A problem is, however, defining what is "normal" (with respect to both animals and humans) and how to keep animals under "normal" conditions. Without even attempting to try and define what is normal and what are normal conditions (it would blast the scope of this book), we just wish to bring out that conditions can be altered in such a way that they are regarded as being "positive" to animals.

Such positive effects have been observed in several animal species after enriching their environment with toys, burying material, etc. However, these studies have been focussing mainly on behavioural changes (Paylor et al., 1992; Chapillon et al., 1999; Williams et al., 2001) and, more recently, on neurogenesis (Kempermann et al., 1997; van Praag et al., 1999b) and spontaneous apoptosis (Young et al., 1999) in the adult dentate gyrus. In general, it seems though that relatively little attention has been payed to changes in stress-related systems such as the HPA axis. In view of the observation that mice held in an enriched environment show reduced anxiety-related behaviour (Chapillon et al., 1999), it would be apt to have a closer look at the HPA axis of such animals. Given that it has been recently reported that the presence of a running wheel as part of an enriched environment, at least regarding its effects on neurogenesis (van Praag et al., 1999b), seems to be the critical object, we decided a few years ago to conduct a detailed study on the changes in the HPA axis of mice allowed free access to a running wheel in their home cage (i.e. the voluntary exercise model). Before we report on our recent results regarding the effects of voluntary exercise on the mouse HPA axis, we provide briefly some background information on the voluntary exercise model.

Cellular and physiological correlates of voluntary exercise

It is now widely accepted that the physical exercise has positive effects on a variety of biological systems, such as body composition, the cardiovascular system, the immune system and also the brain. Regarding body composition, the amount of peritoneal and perirenal adipose tissue, which is indicative of the risk for cardiovascular pathology, is decreased in subjects regularly performing physical activity (Friedman et al., 1997; Lambert and Jonsdottir, 1998). A decreased heart rate, an enhanced oxidative capacity and a decreased blood pressure have been observed under resting conditions in exercised rats (Kramer et al., 2000) and humans (Gielen et al., 2001). Moderate training intensity also has been shown to enhance immune system function and to increase resistance to infections (Jonsdottir, 2000; Pedersen and Hoffman-Goetz, 2000).

It has been shown in rodents that, at the level of the brain, voluntary exercise results in an enhanced performance in spatial learning and memory tasks (van Praag et al., 1999a). Moreover, voluntary exercise evokes in rats and mice increases in neurogenesis in the dentate gyrus of the hippocampus which is thought to be the result of an enhanced action of growth factors (e.g. IGF-1, BDNF) in the brain (van Praag et al., 1999b; Bilang-Bleuel et al., 2000; Carro et al., 2000; Russo-Neustadt et al., 2000; Trejo et al., 2001). In contrast to these stimulatory effects of exercise, a decrease in neurogenesis has been observed after exposing rats or mice to psychological stressors such as forced swimming or predators (van Praag et al., 1999b; Bilang-Bleuel et al., 2000). Thus, it appears that the regular physical exercise has effects on various biological systems which generally can be valued as positive. It has been suggested that, animals, including humans, show improved coping with stressful events after regular performance of moderate physical exercise (Steptoe et al., 1989; Byrne and Byrne, 1993; Salmon, 2001).

The above-mentioned form of regular exercise is not to be mixed up with the high-demand endurance training (e.g. marathon running) in humans and forced exercise in rodents. Voluntary exercise yields many positive biological effects, whereas endurance training, due to its excessive (eccentric) physical demand, has been found to cause injuries (Warren and Stiehl, 1999; Proske and Morgan, 2001), reproductive disturbances (Chen and Brzyski, 1999; Warren and Stiehl, 1999), impaired immunity (Nieman, 2000) and accelerated wear of the movement apparatus (Warren and Stiehl, 1999; Feasson et al., 2002). Moreover, endurance training in humans has been shown to elicit chronic stress-like changes in the HPA axis (Villaneueva et al., 1986; Luger et al., 1987; Duclos et al., 1998, 2001). In forced exercised rats, blunted ACTH and unchanged corticosterone responses to footshock and forced swimming have been found (Watanabe et al., 1991, 1992; Dishman et al., 1998). Furthermore, substantially enlarged adrenal medullas (in man called the so-called "sports adrenal medulla") have been observed in both man and animals after high-intensity exhaustive exercise (Man: Kjær, 1998) and forced treadmill or swim exercise (Rats: Stallknecht et al., 1990; Schmidt et al., 1992).

HPA axis changes after long-term voluntary exercise: introduction

We were prompted to investigate whether voluntary exercise would also affect the HPA axis, this in view of the many reported positive effects at the cellular, physiological and behavioural level. Studying the HPA axis is particularly apt as this is a neuroendocrine system which is highly involved in the coping response to metabolic and stressful challenges (for review, see Dallman et al., 1993; De Kloet et al., 1998; Reul et al., 2000).

In our study published recently (Droste et al., 2003), male C57BL/6N mice were allowed to run voluntarily for a period of 4 weeks in a running wheel provided in their home cage. They ran approximately 7 km/day and this virtually exclusively during the first half of their active period, i.e. the night phase (Fig. 1A). As can be taken from Fig. 1B, the mean maximal distance accomplished by the animals per day was established within a few days. Figure 1C shows that the running capacity among mice can vary, but the variance in running within each individual animal is strikingly small. All in all, these observations are in agreement with other publications (Festing, 1977; Harri et al., 1999; Lancel et al., 2003). Mice run voluntarily when offered a running wheel. Thus, it seems that it complies with a natural urge of the animals (Brant and Kavanau, 1964, 1995) increasing physical fitness (Goodrick, 1978) and helping to control body weight (Leshner, 1971; Goodrick, 1978). Importantly, wheel running is not regarded as a form of stereotypic behaviour (Harri et al., 1999) because it is not expressed at the cost of

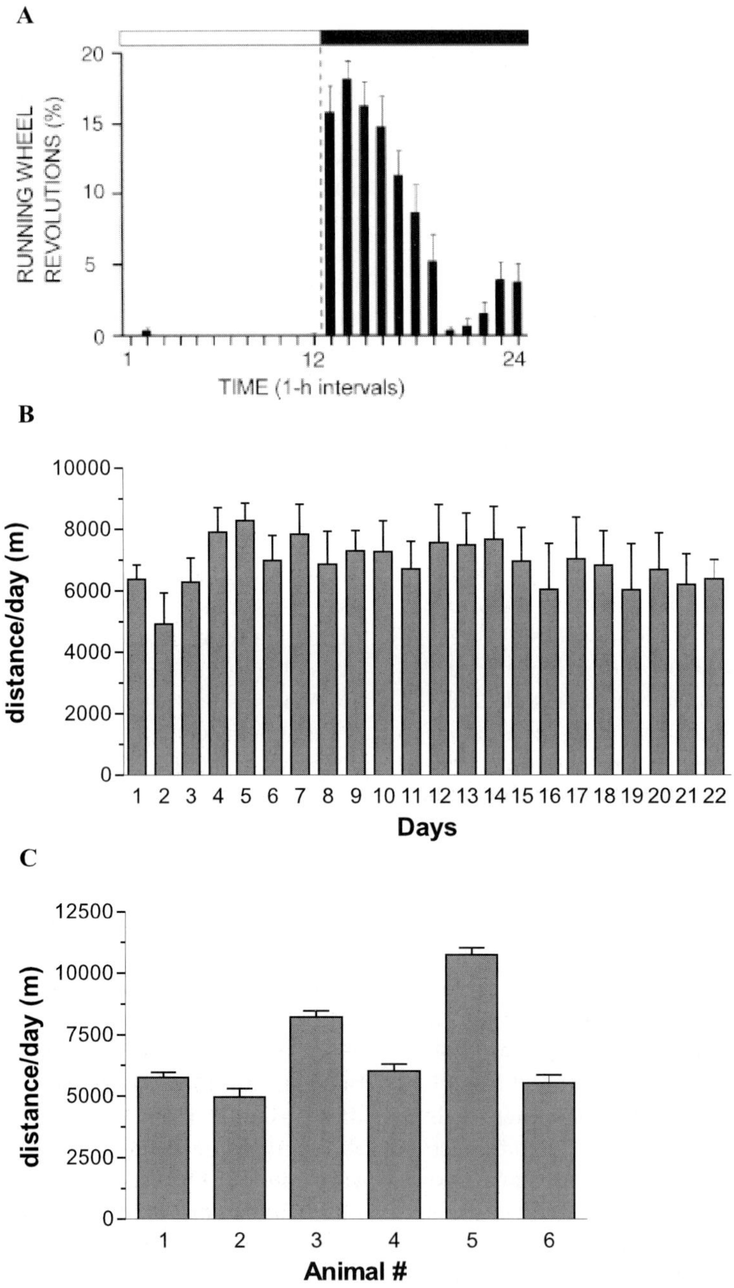

Fig. 1. Running performance in the voluntary exercise model. Male C57BL/6N mice were housed individually and were provided with a running wheel (diameter, 14 cm) in their cage. The number of revolutions was recorded and thereof the distance run calculated. Figure 1A shows that mice run almost exclusively during the first half of the dark phase (the open and closed bar indicate the light and dark period, respectively). Figure 1B shows the establishment of the running performance during the first 22 days of running. It is clear that the full running performance is established within 4–5 days. Figure 1C shows the running performance of each individual mouse averaged over the 22-day period. Apparently, there is an inter-individual variance in running performance, but the inter-day variance within each animal is rather limited. Data presented in Figure 1A were taken with permission of the European Journal of Neuroscience (Publisher: Blackwell) from Lancel et al. (2003).

resting behaviour as is the case with other reported locomotor stereotypes (Cooper and Nicol, 1991, 1996).

HPA axis changes after long-term voluntary exercise: physical changes

Long-term voluntary exercise resulted in remarkable physical changes in the mice. Although body weight of exercised animals did not change, the abdominal fat mass was largely reduced (Droste et al., 2003), for which the reduction in weight was presumably compensated for by increases in muscle substance, heart weight and blood volume. Thymus weights were reduced, whilst total adrenal weight (i.e. the two added together) was increased, suggesting enhanced glucocorticoid action over an extended period of time. Increased adrenal weights after exercise is a long-known finding (Ingle, 1938; Riss et al., 1959; Kjær, 1992). Using histological methods, however, we could demonstrate that the changes that had occurred in the adrenal glands were quite comprehensive. Similar to reports in the literature (Idelman, 1978; Coleman et al., 1998), we observed that the left adrenal gland is bigger than the right one. We could show that this is due to both a larger adrenal medulla and a larger adrenal cortex (Droste et al., 2003). Strikingly, after 4 weeks of voluntary exercise, the difference in size of the left versus the right adrenal gland was abolished. This was solely due to an enlargement of the right adrenal gland and, more specifically, mainly due to a substantial enlargement of the adrenal cortex and, to a lesser extent, to that of the adrenal medulla (Droste et al., 2003). An observation underscoring the tremendous impact of the voluntary exercise on the organism was that the overall shape of the adrenal gland was altered in that it had turned from an elongated form into a spherical organ (S.K. Droste, S. Ulbricht and J.M.H.M. Reul, unpublished observation). This observation also highlights that apparently a complex nervous/endocrine organ, such as the adrenal gland, is capable of significant organ restructuring as part of the complex of adaptational actions of the organism to cope with the demands pressed upon it by the high physical activity.

HPA axis changes after long-term voluntary exercise: hormone secretion and the sympathoadrenomedullary system

When considering the baseline HPA hormone secretion over the diurnal cycle, the most striking change in the exercised mice was a highly increased corticosterone level at the time of lights off (i.e. 18:00 h), thus at the start of the active phase (Droste et al., 2003). Actually, it appeared that the rise in glucocorticoid levels preceded the onset of the running behaviour (see Fig. 1A). Thus, the rise in glucocorticoids can be regarded as anticipatory, an observation which has been seen already several decades ago in other experimental paradigms (Levine et al., 1972; Goldman et al., 1973). In the exercised mice, the rise in corticosterone levels is most likely anticipatory to support the metabolic demand of the enhanced physical activity.

The enhanced corticosterone secretion in the exercised mice at the crest of the diurnal cycle occurred in the absence of a concomitant enhancement in ACTH secretion. This change in adrenocortical sensitivity to ACTH is most likely associated with changes in the sympathoadrenomedullary system of the exercised mice. For an elaborate review about the role of the sympathoadrenomedullary system in adrenal function, see the chapter of W. Engeland in this volume. As mentioned above already, exercised animals do show enlargements of the right adrenal medulla. Moreover, we also discovered distinct changes in mRNA expression of tyrosine hydroxylase (TH; the rate-limiting enzyme in the catecholamine synthesis) in the adrenal medulla of exercised mice (Droste et al., 2003). The level of expression of this enzyme in the adrenal medulla is regarded as an indicator for the overall activity of the sympathoadrenomedullary system (Hamelink et al., 2002). However, first of all, we also found, in parallel to the asymmetry in size of the adrenal medullas in control mice, an asymmetry in the adrenomedullar TH mRNA expression in the control animals, i.e. the left medulla expressing higher levels than the right one (Fig. 2). Strikingly, after exercise the adrenomedullar TH mRNA expression had risen selectively in the right medulla, thereby abolishing the asymmetry in TH mRNA levels seen in the control mice (Fig. 2; Droste et al., 2003).

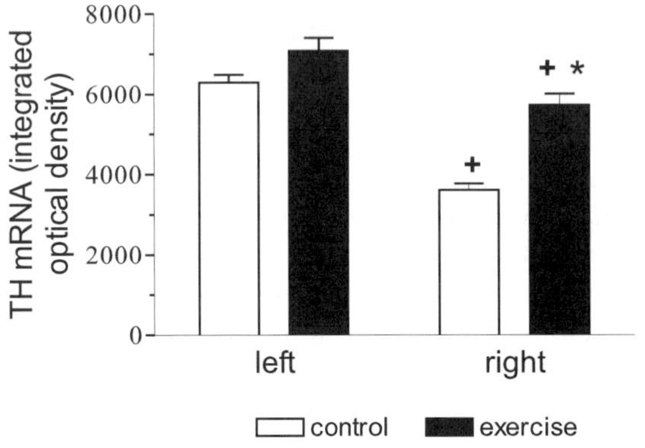

Fig. 2. Changes in tyrosine hydroxylase (TH) mRNA levels in the adrenal medulla of control and (4 weeks) exercising mice ($n = 10$ for both groups). TH mRNA was detected by in situ hybridization histochemistry and autoradiograms were analyzed by computerized image analysis. TH mRNA levels are expressed as integrated optical density (i.e. nett grey values × square pixels/1000). Data are presented as means ± SEM. Exercise evoked an overall increase in TH mRNA expression in the adrenal medulla (analysis of variance (ANOVA): effect of exercise: $F(1, 36) = 34.9$, $P < 0.0001$). In addition, overall higher TH mRNA levels in the left adrenal medulla than in the right one were found (ANOVA: left vs. right: $F(1, 36) = 67.6$, $P < 0.0001$). *, significant difference between exercise and control within the left or right adrenal medulla (post-hoc tests with contrasts); $^{+}$, significant difference between left and right adrenal medulla within the same treatment group (post-hoc tests with contrasts). Data were taken from Droste et al. (2003) with permission of *The Endocrine Society, Copyright 2003.*

How do these observations fit together? First, sympathoadrenomedullary activity in control mice seems to be biased to the left side of the body. This observation is in line with the concept of a predominant involvement of the right brain hemisphere in the control of sympathetic activity and the higher levels of noradrenaline found in this side of the brain (Wittling et al., 1998; Wittling, 2001). However, until now asymmetry in the autonomic nervous system has been hardly studied and one should be careful drawing far-reaching conclusions. Nevertheless, our finding on the left–right difference in adrenomedullar size strengthens the concept on sympathetic asymmetry. Second, the increased adrenocortical sensitivity in exercised mice at the start of the active, i.e. running, phase may be explained by an increased sympathoadrenomedullary activity in these animals. However, such an increased sympathoadrenomedullary activity is, for reasons presently unknown, virtually exclusively restricted to the right branch of the sympathoadrenomedullary input. Given that neural inputs to the adrenal glands are important growth-determining factors for the adrenal cortices (Engeland and Dallman, 1975, 1976; Dallman et al., 1976), the selective enlargement of the right adrenal cortex in the exercising mice may be due to an intensified right-sided sympathoadrenomedullary input. Thus, the enhanced glucocorticoid secretion at the start of the nocturnal phase appears to be due to an increased right-sided sympathoadrenomedullary activity in combination with an increased adrenocortical capacity. Nevertheless, despite the attractiveness of these suppositions, the exact causalities between the depicted interrelationships need to be further clarified in future research.

The sympathoadrenomedullary changes in the exercised mice were found to be of relevance for the distinct responses of the HPA axis to stressful challenges seen in these animals (Droste et al., 2003). We observed that the plasma corticosterone responses, but not ACTH responses, were enhanced in exercising mice in response to stressors demanding or involving physical activity such as forced swimming or restraint. Previous reports have shown enhanced sympathoadrenomedullary activation in response to stressors demanding physical activity (Sothmann et al., 1996; Koolhaas et al., 1997; Kjær, 1998) and to psychosocial stress (Sinyor et al., 1983). Our observation in exercising mice is in line with these reports as these mice show an increased sympathoadrenomedullary capacity as well as an enlarged (right) adrenal cortex.

Thus, exercised mice respond to stressors containing a significant physical component with amplified glucocorticoid responses, most likely because their HPA and sympathoadrenomedullary axes have adapted to meet the enhanced metabolic demand during running. However, the exercised animals react in a novel environment with different HPA hormone responses. Overall, the plasma ACTH responses in the exercised mice were markedly lower than those in the sedentary mice, suggesting strongly that the exposure to novelty has lower impact in these animals. This observation dovetails with observations that exercised mice (Binder et al., 2004) and humans (Steptoe et al., 1989; Byrne and Byrne, 1993; Salmon, 2001) show reduced anxiety. It also corresponds with the observed reduced expression of CRF mRNA in the PVN of exercised mice (Droste et al., 2003). Presently, the reason for the reduced CRF mRNA expression awaits further clarification, but could involve increased GR-mediated negative-feedback signalling as a result of the increased glucocorticoid levels during the first half of the dark phase and increased afferent inhibitory signals from as yet unknown sources.

The plasma corticosterone response to novelty in our exercising mice depended decisively on the presence of a functional running wheel in the new cage (Droste et al., 2003). In the absence of the running wheel, the exercising animals produced the same glucocorticoid levels as the control mice which can be explained by the adrenocortical hyperresponsiveness generally observed in the exercising mice. However, if the exercising animals had access to a running wheel in the new cage, then the corticosterone responses were much lower. The animals indeed used the running wheel for a significant amount of the exposure time. Reversely, when the running wheel was made dysfunctional by blocking the turning mechanism, then plasma ACTH and corticosterone reponses of the exercised mice to the novelty stimulus was indistinguishable from that of the control animals (Droste et al., 2003). Thus, the use of the running wheel may be regarded as displacement behaviour, resulting in a further reduction of the emotional impact of the novel environment. A similar effect on corticosterone secretion has been observed when rats were given access to a running wheel after being exposed to a footshock paradigm (Starzec et al., 1983). The exact mechanism by which the glucocorticoid response is restrained under these conditions is presently unknown, but may involve reduced sympathoadrenomedullary outflow and local adrenomedullary inhibitory mechanisms.

Summing up, long-term voluntary exercise elicits multiple changes in the HPA axis of mice which involves major alterations in the sympathoadrenomedullary system and central afferent control mechanisms. These changes are part of a network of physiological and behavioural changes effectuated in the exercised mice. The results obtained with this model underscore the power of adaptive mechanisms capable of vastly altering the phenotype of an organism.

Concluding remarks

In this chapter, it was our aim to point out that the HPA axis is a neuroendocrine system which is capable of adjusting its regulatory setpoints in a very dynamic fashion. It can do so irrespective of whether the adaptive measures are in response to adverse or positive challenges. As illustrated by our recent publication (Droste et al., 2003), long-term voluntary exercise has a huge impact on the HPA axis. In view of the increased amplitude in the diurnal cycle of plasma corticosterone, the increased glucocorticoid response to stressors entailing physical activity, the decreased responsiveness in HPA hormones to emotional stimuli, it is evident that the voluntary exercise increases the dynamic range of the HPA axis.

Changes in the sympathoadrenomedullary system of the exercising mice turned out to be a principal factor substantially altering the responsiveness of the HPA axis in terms of glucocorticoid responses. The observations underscore that the final glucocorticoid output is actually steered by two pathways, being the neuroendocrine hypothalamic–hypophyseal pathway and the neural sympathoadrenomedullary pathway. We emphasize this view here, because it seems that the evidently prominent role of the sympathoadrenomedullary system in the control of adrenocortical glucocorticoid secretion has been, to put it mildly, somewhat neglected over the past 25 years. Importantly, these pathways do not interact only at the adrenal gland level, but more so at multiple levels including various levels within the central nervous system in addition to the adrenal gland and, possibly, the pituitary gland.

One of the most striking observations in the exercised mice was the phenomenon that the increase in adrenal weight was solely the result of an enlargement of the right adrenal cortex and, to a lesser extent, right adrenal medulla. This enlargement appeared to be the result of a selective rise in the activity of the right branch of the sympathoadrenomedullary system; the reason for this selectivity is currently unknown and awaits further investigations.

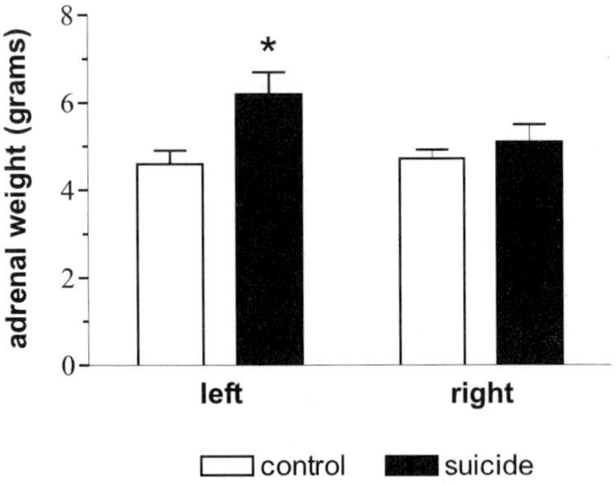

Fig. 3. Left and right adrenal weights of suicide victims and controls. Data are presented as means ± SEM. *, significant increase as compared to control ($p = 0.0030$, $t = 3.22$, df = 31; two-tailed t-test after analysis of covariance (ANCOVA)). There was no significant difference between suicides and controls in mean age, body weight, body height, of sex distribution. Data were taken from Szigethy et al. (1994) with permission of the Society of Biological Psychiatry (Publisher: Elsevier Science Inc.).

The enlargement of the right adrenal gland abolished the asymmetry between the left and right adrenal gland seen in the control mice, in which the left adrenal is bigger than the right one. The asymmetry in the adrenal glands and the role of the sympathoadrenomedullary system therein may also be of pathophysiological interest in view of a post-mortem study on adrenal weights in suicide victims (Szigethy et al., 1994). It was reported that control subjects showed no difference in weight between the left and the right adrenal gland, whereas suicide victims showed a selectively increased weight of the left adrenal gland (Fig. 3; Szigethy et al., 1994). This increase appeared to be most likely due to an enlargement of the adrenal cortex. Clearly, it is impossible at this stage to make direct comparisons between the (patho)-physiological corollaries leading to adrenal gland asymmetries in control mice versus suicide victims, because the mechanisms underlying the asymmetries have not been clarified but also mere species differences could be playing a role. Nevertheless, the observed asymmetry in adrenal weight in the suicide victims is certainly of interest as it is associated with a pathological state. Speculatively, the asymmetry may be the result of an imbalanced sympathoadrenomedullary system. This may not even be that speculative given that vegetative instability has often been reported in subjects with affective disorders (Steptoe et al., 1989; Byrne and Byrne, 1993; Salmon, 2001). Thus, the observation on voluntary exercise abolishing adrenal and sympathoadrenomedullary asymmetry can be regarded as an anti-stress phenomenon, similar to other reported anti-stress phenomena such as the effects of exercise on neurogenesis (van Praag et al., 1999b; Bilang-Bleuel et al., 2000; Carro et al., 2000; Russo-Neustadt et al., 2000; Trejo et al., 2001), sleep (Lancel et al., 2003) and anxiety-related behaviour (Steptoe et al., 1989; Byrne and Byrne, 1993; Salmon, 2001) (E. Binder, S.K. Droste, F. Ohl, and J.M.H.M. Reul, unpublished observations). Therefore, the asymmetry in the HPA axis and sympathoadrenomedullary system, including the central nervous system mechanisms controlling these neuroendocrine and neural output systems, should be the subject of future investigations.

Acknowledgments

The authors are greatly indebted to Ms. Sabine Ulbricht who, due to her tremendous technical skills, contributed significantly to the mouse exercise project reviewed in this chapter.

References

Almeida, O.F.X., Conde, G.L., Crochemore, C., Demeneix, B.A., Fischer, D., Hassan, A.H.S., Meyer, M., Holsboer, F. and Michaelidis, T.M. (2000) Subtle shifts in the ratio between pro- and antiapoptotic molecules after activation of corticosteroid receptors decide neuronal fate. FASEB Journal, 14: 779–790.

Bale, T.L., Contarino, A., Smith, G.W., Chan, R., Gold, L.H., Sawchenko, P.E., Koob, G.F., Vale, W.W. and Lee, K.-F. (2000) Mice deficient for corticotropin-releasing hormone receptor-2 display anxiety-like behaviour and are hypersensitive to stress. Nature Genet., 24: 410–414.

Barden, N., Stec, I.S.M., Montkowski, A., Holsboer, F. and Reul, J.M.H.M. (1997) Endocrine profile and neuroendocrine challenge tests in transgenic mice expressing antisense RNA against the glucocorticoid receptor. Neuroendocrinol., 66: 212–220.

Beaulieu, S., Di Paolo, T. and Barden, N. (1986) Control of ACTH secretion by the central nucleus of the amygdala: implication of the serotoninergic system and its relevance to the glucocorticoid delayed negative feedback mechanism. Neuroendocrinol., 44: 247–254.

Bhatnagar, S. and Dallman, M. (1998) Neuroanatomical basis for facilitation of hypothalamic–pituitary–adrenal responses to a novel stressor after chronic stress. Neuroscience, 84: 1025–1039.

Bhatnagar, S., Viau, V., Chu, A., Soriano, L., Meijer, O.C. and Dallman, M.F. (2000) A cholecystokinin-mediated pathway to the paraventricular thalamus is recruited in chronically stressed rats and regulates hypothalamic–pituitary–adrenal function. J. Neurosci., 20: 5564–5573.

Bilang-Bleuel, A., Droste, S., Gesing, A., Rech, J., Linthorst, A.C.E. and Reul, J.M.H.M. (2000) Impact of stress and voluntary exercise on neurogenesis in the adult hippocampus: quantitative analysis by detection of Ki-67. Proc. 30th Ann. Meeting Soc. Neurosci., 26: 1534.

Binder, E., Droste, S.K., Ohl, F., Reul, J.M.H.M. (2004) Regular voluntary exercise reduces anxiety-related behavior and impulsivity in mice. Behav. Brain Res., in press.

Bohus, B. and De Kloet, E.R. (1981) Adrenal steroids and extinction behavior: antagonism by progesterone, deoxycorticosterone and dexamethasone of a specific effect of corticosterone. Life Sci., 28: 433–440.

Brady, L.S., Whitfield, Jr., H.J., Fox, R.J., Gold, P.W. and Herkenham, M. (1991) Long-term antidepressant administration alters corticotropin-releasing hormone, tyrosine hydroxylase, and mineralocorticoid receptor gene expression in rat brain. Therapeutic implications. J. Clin. Invest., 87: 831–837.

Brant, D.H. and Kavanau, J.L. (1964) Unrewarded exploration and learning of complex mazes by wild and domestic mice. Nature, 204: 267–269.

Brant, D.H. and Kavanau, J.L. (1995) Exploration and movement patterns of the Canyon mouse *Peromyscus crinitus*. Ecology, 46: 452–461.

Byrne, A. and Byrne, D.G. (1993) The effect of exercise on depression, anxiety and other mood states: a review. J. Psychosom. Res., 37: 565–574.

Carro, E., Nuñez, A., Busiguina, S. and Torres-Aleman, I. (2000) Circulating insulin-like growth factor I mediates effects of exercise on the brain. J. Neurosci., 20: 2926–2933.

Cechetto, D.F. and Saper, C.B. (1990) Role of the cerebral cortex in autonomic function. In: Loewy, A.D. and Spyer, K.M. (Eds.), Central Regulation of Autonomic Functions. Oxford University Press, Oxford, pp. 208–223.

Chapillon, P., Manneche, C., Belzung, C. and Caston, J. (1999) Rearing environmental enrichment in two inbred strains of mice: 1. Effects on emotional reactivity. Behav. Genet., 29: 41–46.

Chen, E.C. and Brzyski, R.G. (1999) Exercise and reproductive dysfunction. Fertil. Steril., 71: 1–6.

Cintra, A., Zoli, M., Rosén, L., Agnati, L.F., Okret, S., Wikström, A.-C., Gustafsson, J.-A. and Fuxe, K. (1994) Mapping and computer assisted morphometry and microdensitometry of glucocorticoid receptor immunoreactive neurons and glial cells in the rat central nervous system. Neuroscience, 62: 843–897.

Coleman, M.A., Garland, T., Jr., Marler, C.A., Newton, S.S., Swallow, J.G. and Carter, P.A. (1998) Glucocorticoid response to forced exercise in laboratory house mice (*Mus domesticus*). Physiol. Behav., 63: 279–285.

Cooper, J.J. and Nicol, C.J. (1991) Stereotypic behavior affects environmental preferences in bank voles, *Chlethrionomys glareolus*. Anim. Behav., 41: 971–977.

Cooper, J.J. and Nicol, C.J. (1996) Stereotypic behavior in wild caught and laboratory bred bank voles (*Clethrionomus glareolus*). Anim. Welfare, 5: 245–257.

Coste, S.C., Kesterson, R.A., Heldwein, K.A., Stevens, S.L., Heard, A.D., Hollis, J.H., Murray, S.E., Hill, J.K., Pantely, G.A., Hohimer, A.R., Hatton, D.C., Phillips, T.J., Finn, D.A., Low, M.J., Rittenberg, M.B., Stenzel, P. and Stenzel-Poore, M.P. (2000) Abnormal adaptions to stress and impaired cardiovascular function in mice lacking corticotropin-releasing hormone receptor-2. Nature Genet., 24: 403–409.

Dallman, M.F., Engeland, W.C. and Shinsako, J. (1976) Compensatory adrenal growth: a neurally mediated reflex. Am. J. Physiol., 231: 408–414.

Dallman, M.F., Akana, S.F., Cascio, C.S., Darlington, D.N., Jacobson, L. and Levin, N. (1987) Regulation of ACTH secretion: variations on a theme of B. Rec. Prog. Horm. Res., 43: 113–173.

Dallman, M.F., Strack, A.M., Akana, S.F., Bradbury, M.J., Hanson, E.S., Scribner, K.A. and Smith, M. (1993) Feast and famine: critical role of glucocorticoids with insulin in daily energy flow. Front. Neuroendocrinol., 14: 303–347.

De Kloet, E.R. (1991) Brain corticosteroid receptor balance and homeostatic control. Front. Neuroendocrinol., 12: 95–164.

De Kloet, E.R. and Reul, J.M.H.M. (1987) Feedback action and tonic influence of corticosteroids on brain function: a concept arising from the heterogeneity of brain receptor systems. Psychoneuroendocrinol., 12: 83–105.

De Kloet, E.R., Wallach, G. and Mcewen, B. (1975) Differences in corticosterone and dexamethasone binding to rat brain and pituitary. Endocrinology, 96: 598–609.

De Kloet, E.R., Reul, J.M.H.M., De Ronde, F.S.W. and Veldhuis, H.D. (1986) Brain corticosteroid receptor systems: heterogeneity, function and plasticity. Central Actions of ACTH and Related Peptides Fidia Research Series Symposia in Neuroscience IV. Liviana Press, Padova, pp. 115–129.

De Kloet, E.R., Vreugdenhil, E., Oitzl, M.S. and Joëls, M. (1998) Brain corticosteroid receptor balance in health and disease. Endocr. Rev., 19: 269–301.

Diorio, D., Viau, V. and Meaney, M.J. (1993) The role of the medial prefrontal cortex (cingulate gyrus) in the regulation of hypothalamic–pituitary–adrenal responses to stress. J. Neurosci., 13: 3839–3847.

Dishman, R.K., Bunnell, B.N., Youngstedt, S.D., Yoo, H.S., Mougey, E.H. and Meyerhoff, J.L. (1998) Activity wheel running blunts increased plasma adrenocorticotrophin (ACTH) after footshock and cage-switch stress. Physiol. Behav., 63: 911–917.

Droste, S.K., Gesing, A., Ulbricht, S., Muller, M.B., Linthorst, A.C. and Reul, J.M.H.M. (2003) Effects of long-term voluntary exercise on the mouse hypothalamic–pituitary–adrenocortical axis. Endocrinology, 144: 3012–3023.

Duclos, M., Corcuff, J.B., Arsac, L., Moreau-Gaudry, F., Rashedi, M., Roger, P., Tabarin, A. and Manier, G. (1998) Corticotroph axis sensitivity after exercise in endurance-trained athletes. Clin. Endocrinol., 48: 493–501.

Duclos, M., Corcuff, J.-B., Pehourcq, F. and Tabarin, A. (2001) Decreased pituitary sensitivity to glucocorticoids in endurance-trained men. Eur. J. Endocrinol., 144: 363–368.

Edwards, C.R.W., Burt, D., McIntyre, M.A., De Kloet, E.R., Stewart, P.M., Brett, L., Sutanto, W.S. and Monder, C. (1988) Localisation of 11β-hydroxysteroid dehydrogenase-tissue specific protector of the mineralocorticoid receptor. Lancet, 2: 986–989.

Engeland, W.C. and Dallman, M.F. (1975) Compensatory adrenal growth is neurally mediated. Neuroendocrinol., 19: 352–362.

Engeland, W.C. and Dallman, M.F. (1976) Neural mediation of compensatory adrenal growth. Endocrinology, 99: 1659–1662.

Feasson, L., Stockholm, D., Freyssenet, D., Richard, I., Duguez, S., Beckmann, J.S. and Denis, C. (2002) Molecular adaptations of neuromuscular disease-associated proteins in response to eccentric exercise in human akeletal muscle. J. Physiol., 543: 297–306.

Festing, M.F.W. (1977) Wheel activity in 26 strains of mouse. Lab. Animals, 11: 257–258.

Friedman, J.E., Ferrara, C.M., Aulak, K.S., Hatzoglou, M., McCune, S.A., Park, S. and Sherman, W.M. (1997) Exercise training down-regulates ob gene expression in the genetically obese SHHF/Mcc-fa(cp) rat. Horm. Metab. Res., 29: 214–219.

Funder, J.W., Pearce, P.T., Smith, R. and Smith, A.I. (1988) Mineralocorticoid action: target tissue specifity is enzyme, not receptor, mediated. Science, 242: 583–585.

Fuxe, K., Wikström, A.-C., Okret, S., Agnati, L.F., Härfstrand, A., Yu, Z.-Y., Granholm, L., Zoli, M., Vale, W. and Gustafsson, J.-A. (1985) Mapping of glucocorticoid receptor immunoreactive neurons in the rat Tel and diencephalon using a monoclonal antibody against rat liver glucocorticoid receptor. Endocrinol., 177: 1803–1812.

Gesing, A., Bilang-Bleuel, A., Droste, S.K., Linthorst, A.C.E., Holsboer, F. and Reul, J.M.H.M. (2001) Psychological stress increases hippocampal mineralocorticoid receptor levels: involvement of corticotropin-releasing hormone. J. Neurosci., 21: 4822–4829.

Gielen, S., Schuler, G. and Hambrecht, R. (2001) Exercise training in coronary artery disease and coronary vasomotion. Circulation, 103: e1–e6.

Goldman, L., Coover, G.O. and Levine, S. (1973) Bidirectional effects of reinforcement shifts on pituitary adrenal activity. Physiol. Behav., 10: 209–214.

Goldstein, L.E., Rasmusson, A.M., Bunney, B.S. and Roth, R.H. (1996) Role of the amygdala in the coordination of behavioral, neuroendocrine, and prefrontal cortical monoamine responses to psychological stress in the rat. The Journal of Neuroscience, 16: 4787–4798.

Goodrick, C.L. (1978) Effect of voluntary wheel exercise on food intake, water intake, and body weight for C57BL/6J mice and mutations which differ in maximal body weight. Physiol. Behav., 21: 345–351.

Gutman, D.A., Owens, M.J., Skelton, K.H., Thrivikraman, K.V. and Nemeroff, C.B. (2003) The corticotropin-releasing $factor_1$ receptor antagonist R121919 attenuates the behavioral and endocrine responses to stress. J. Pharmacol. Exp. Ther., 304: 874–880.

Hamelink, C., Tjurmina, O., Damadzic, R., Young, W.S., Weihe, E., Lee, H.W. and Eiden, L.E. (2002) Pituitary adenylate cyclase-activating polypeptide is a sympathoadrenal neurotransmitter involved in catecholamine regulation and glucohomeostasis. Proc. Natl. Acad. Sci. USA, 99: 461–466.

Harri, M., Lindblom, J., Malinen, H., Hyttinen, M., Lapveteläinen, T., Eskola, S. and Helminen, H.J. (1999) Effect of access to a running wheel on behavior of C57BL/6J mice. Laboratory Animal Science, 49: 401–405.

Hassan, A.H.S., Von Rosenstiel, P., Patchev, V.K., Holsboer, F. and Almeida, O.F.X. (1997) Exacerbation of apoptosis in the dentate gyrus of the aged rat by dexamethasone and the protective role of corticosterone. Exp. Neurol., 140: 43–52.

Herman, J.P. and Cullinan, W.E. (1997) Neurocircuitry of stress: central control of the hypothalamo–pituitary–adrenocortical axis. Trends Neurosci., 20: 78–84.

Herman, J.P., Cullinan, W.E., Morano, M.I., Akil, H. and Watson, S.J. (1995) Contribution of the ventral subiculum to inhibitory regulation of the hypothalamo–pituitary–adrenocortical axis. J. Neuroendocrinol., 7: 475–482.

Hoefnagels, W.H.L. and Kloppenborg, P.W.C. (1983) Antimineralocorticoid effects of dexamethasone in subjects treated with glycyrrhetinic acid. J. Hypertens., 1(Suppl. 2): 313–315.

Hurley, K.M., Herbert, H., Moga, M.M. and Saper, C.B. (1991) Efferent projections of the infralimbic cortex of the rat. J. Comp. Neurol., 308: 249–276.

Idelman, S. (1978) The structure of the mammalian adrenal cortex. In: Jones, I.C. and Henderson, I.W. (Eds.), General, Comparative and Clinical Endocrinology of the Adrenal Cortex. Academic Press, New York, pp. 1–180.

Ingle, D.J. (1938) The time for the occurrence of cortico-adrenal hypertrophy in rats during continued work. Am. J. Physiol., 124: 627–630.

Jefferys, D., Boublik, J. and Funder, J.W. (1985) A kappa-selective opioidergic pathway is involved in the reversal of a behavioural effect of adrenalectomy. Eur. J. Pharmacol., 107: 331–335.

Jodo, E., Chiang, C. and Aston-Jones, G. (1998) Potent excitatory influence of prefrontal cortex activity on noradrenergic locus coeruleus neurons. Neuroscience, 83: 63–79.

Joels, M. and De Kloet, E.R. (1992) Control of neuronal excitability by corticosteroid hormones. Trends Neurosci., 15: 25–30.

Jonsdottir, I.H. (2000) Neuropeptides and their interaction with exercise and immune function. Immunol. Cell Biol., 78: 562–570.

Kempermann, G., Kuhn, H.G. and Gage, F.H. (1997) More hippocampal neurons in adult mice living in an enriched environment. Nature, 386: 493–495.

Kishimoto, T., Radulovic, J., Radulovic, M., Lin, C.R., Schrick, C., Hooshmaand, F., Hermanson, O., Rosenfeld, M.G. and Spiess, J. (2000) Deletion of Crhr2 reveals an anxiolytic role for corticotropin-releasing hormone receptor-2. Nature Genet., 24: 415–419.

Kjær, M. (1992) Regulation of hormonal and metabolic responses during exercise in humans. Exerc. Sport Sci. Rev., 20: 161–184.

Kjær, M. (1998) Adrenal medulla and exercise training. Eur. J. Appl. Physiol., 77: 195–199.

Koolhaas, J.M., De Boer, S.F., de Ruiter, A.J.H., Meerlo, P. and Sgoifo, A. (1997) Social stress in rats and mice. Acta Physiol. Scand., 161(Suppl. 640): 69–72.

Korte, S.M., De Kloet, E.R., Buwalda, B., Bouman, S.D. and Bohus, B. (1996a) Antisense to the glucocorticoid receptor in hippocampal dentate gyrus reduces immobility in forced swim test. Eur. J. Pharmacol., 301: 19–25.

Korte, S.M., Kortebouws, G.A.H., Koob, G.F., Dekloet, E.R. and Bohus, B. (1996b) Mineralocorticoid and glucocorticoid receptor antagonists in animal models of anxiety. Pharmacol. Biochem. Behav., 54: 261–267.

Kramer, J.M., Plowey, E.D., Beatty, J.A., Little, H.R. and Waldrop, T.G. (2000) Hypothalamus, hypertension, and exercise. Brain Res. Bull., 53: 77–85.

Lambert, G.W. and Jonsdottir, I.H. (1998) Influence of voluntary exercise on hypothalamic norepinephrine. J. Appl. Physiol., 85: 962–966.

Lancel, M., Droste, S.K., Sommer, S. and Reul, J.M.H.M. (2003) Influence of regular voluntary exercise on spontaneous and social stress-affected sleep in mice. Eur. J. Neurosci., 17: 2171–2179.

Leshner, A.I. (1971) The adrenals and the regulatory nature of running wheel activity. Physiol. Behav., 6: 551–558.

Levine, S. (1957) Infantile experience and resistance to physiological stress. Science, 126: 405.

Levine, S. (1967) Maternal and environmental influences on the adrenocortical response to stress in weanling rats. Science, 156: 258–260.

Levine, S., Goldman, L. and Coover, G.D. (1972) Expectancy and the pituitary–adrenal system. Ciba Found. Symp., 8: 281–291.

Linthorst, A.C.E., Flachskamm, C., Müller-Preuss, P., Holsboer, F. and Reul, J.M.H.M. (1995) Effect of bacterial endotoxin and interleukin-1 beta on hippocampal serotonergic neurotransmission, behavioral activity, and free corticosterone levels: an in vivo microdialysis study. J. Neurosci., 15: 2920–2934.

Linthorst, A.C.E., Flachskamm, C., Hopkins, S.J., Hoadley, M.E., Labeur, M.S., Holsboer, F. and Reul, J.M.H.M. (1997) Long-term intracerebroventricular infusion of corticotropin-releasing hormone alters neuroendocrine, neurochemical, autonomic, behavioral and cytokine responses to a systemic inflammatory challenge. J. Neurosci., 17: 4448–4460.

Luger, A., Deuster, P.A., Kyle, S.B., Gallucci, W.T., Montgomery, L.C., Gold, P.W., Loriaux, D.L. and Chrousos, G.P. (1987) Acute hypothalamic–pituitary–adrenal response to the stress of treadmill exercise. N. Engl. J. Med., 316: 1309–1315.

Meaney, M.J. and Aitken, D.H. (1985) (3H)Dexamethasone binding in rat frontal cortex. Brain Res., 328: 176–180.

Meaney, M.J., Aitken, D.H., Bodnoff, S.R., Iny, L.J. and Tatarewicz, J.E. (1985) Early postnatal handling alters glucocorticoid receptor concentrations in selected brain regions. Behav. Neurosci., 99: 765–770.

Meaney, M.J., Aitken, D.H. and Sapolsky, R.M. (1987) Thyroid hormones influence the development of hippocampal glucocorticoid receptors in the rat: a mechanism for the effects of postnatal handling on the development of the adrenocortical stress response. Neuroendocrinol., 45: 278–283.

Meaney, M.J., Aitken, D.H., Viau, V., Sharma, S. and Sarrieau, A. (1989) Neonatal handling alters adrenocortical negative feedback sensitivity and hippocampal type II glucocorticoid receptor binding in the rat. Neuroendocrinol., 50: 597–604.

Meijer, O.C., de Lange, E.C., Breimer, D.D., de Boer, A.G., Workel, J.O. and De Kloet, E.R. (1998) Penetration of dexamethasone into brain glucocorticoid targets is enhanced in mdr1A P-glycoprotein knockout mice. Endocrinology, 139: 1789–1793.

Neveu, P.J., Liege, S. and Sarrieau, A. (1998) Asymmetrical distribution of hippocampal mineralocorticoid receptors depends on lateralization in mice. Neuroimmunomodulation, 5: 16–21.

Nieman, D.C. (2000) Exercise effects on systemic immunity. Immunol. Cell Biol., 78: 496–501.

Owens, M.J. and Nemeroff, C.B. (1991) Physiology and pharmacology of corticotropin-releasing factor. Pharmacol. Rev., 43: 425–473.

Patacchioli, F.R., Amenta, F., Ramacci, M.T., Taglialatela, G., Maccari, S. and Angelucci, L. (1989) Acetyl-L-carnitine reduces the age-dependent loss of glucocorticoid receptors in the rat hippocampus – an autoradiographic study. J. Neurosci. Res., 23: 462–466.

Paylor, R., Morrison, S.K., Rudy, J.W., Waltrip, L.T. and Wehner, J.M. (1992) Brief exposure to an enriched environment improves performance on the morris water task and increases hippocampal cytosolic protein kinase C activity in young rats. Behavioural Brain Research, 52: 49–59.

Pedersen, B.K. and Hoffman-Goetz, L. (2000) Exercise and the immune, system: regulation, integration, and adaptation. Physiol. Rev., 80: 1055–1081.

Proske, U. and Morgan, D.L. (2001) Muscle damage from eccentric exercise: mechanism, mechanical signs, adaptation and clinical application. J. Physiol., 537: 333–345.

Reul, J.M.H.M. and De Kloet, E.R. (1985) Two receptor systems for corticosterone in rat brain: microdistribution and differential occupation. Endocrinology, 117: 2505–2512.

Reul, J.M.H.M. and De Kloet, E.R. (1986) Anatomical resolution of two types of corticosterone receptor sites in rat brain with in vitro autoradiography and computerized image analysis. J. Steroid Biochem., 24: 269–272.

Reul, J.M.H.M. and Holsboer, F. (2002) Corticotropin-releasing factor receptors 1 and 2 in anxiety and depression. Curr. Opin. Pharmacol., 2: 23–33.

Reul, J.M.H.M., Van den Bosch, F.R. and De Kloet, E.R. (1987a) Differential response of type I and type II corticosteroid receptors to changes in plasma steroid level and circadian rhythmicity. Neuroendocrinol., 45: 407–412.

Reul, J.M.H.M., Van den Bosch, F.R. and De Kloet, E.R. (1987b) Relative occupation of type-I and type-II corticosteroid receptors in rat brain following stress and dexamethasone treatment: functional implications. J. Endocrinol., 115: 459–467.

Reul, J.M.H.M., Tonnaer, J.A.D.M. and De Kloet, E.R. (1988) Neurotrophic ACTH analogue promotes plasticity of type I corticosteroid receptor in brain of senescent male rats. Neurobiol. Aging, 9: 253–260.

Reul, J.M.H.M., De Kloet, E.R., Van Sluijs, F.J., Rijnberk, A. and Rothuizen, J. (1990) Binding characteristics of mineralocorticoid and glucocorticoid receptors in dog brain and pituitary. Endocrinology, 127: 907–915.

Reul, J.M.H.M., Rothuizen, J. and De Kloet, E.R. (1991) Age-related changes in the dog hypothalamic–pituitary–adrenocortical system: neuroendocrine activity and corticosteroid receptors. J. Steroid Biochem., 40: 63–69.

Reul, J.M.H.M., Stec, I., Söder, M. and Holsboer, F. (1993) Chronic treatment of rats with the antidepressant amitriptyline attenuates the activity of the hypothalamic–pituitary–adrenocortical system. Endocrinology, 133: 312–320.

Reul, J.M.H.M., Labeur, M.S., Grigoriadis, D.E., De Souza, E.B. and Holsboer, F. (1994a) Hypothalamic–pituitary–adrenocortical axis changes in the rat after long-term treatment with the reversible monoamine oxidase-A inhibitor moclobemide. Neuroendocrinol., 60: 509–519.

Reul, J.M.H.M., Stec, I., Wiegers, G.J., Labeur, M.S., Linthorst, A.C.E., Arzt, E. and Holsboer, F. (1994b) Prenatal immune challenge alters the hypothalamic–pituitary–adrenocortical axis in adult rats. J. Clin. Invest., 93: 2600–2607.

Reul, J.M.H.M., Gesing, A., Droste, S.K., Stec, I.S.M., Weber, A., Bachmann, C.G., Bilang-Bleuel, A., Holsboer, F. and Linthorst, A.C.E. (2000) The brain mineralocorticoid receptor: greedy for ligand, mysterious in function. Eur. J. Pharmacol., 405: 235–249.

Riss, W., Burstein, S.D., Johnson, R.W. and Lutz, A. (1959) Morphologic correlates of endocrine and running activity. J. Comp. Physiol. Psychol., 52: 618–620.

Roozendaal, B., Koolhaas, J.M. and Bohus, B. (1991) Attenuated cardiovascular, neuroendocrine, and behavioral responses after a single footshock in central amygdaloid lesioned male rats. Physiol. Behav., 50: 771–775.

Rothuizen, J., Reul, J.M.H.M., Van Slujs, F.J., Mol, J.A., Rijnberk, A. and De Kloet, E.R. (1993) Increased Neuroendocrine reactivity and decreased brain mineralocorticoid receptor-binding capacity in aged dogs. Endocrinology, 132: 161–168.

Russo-Neustadt, A.A., Beard, R.C., Huang, Y.M. and Cotman, C.W. (2000) Physical activity and antidepressant treatment potentiate the expression of specific brain-derived neurotrophic factor transcripts in the rat hippocampus. Neuroscience, 101: 305–312.

Salmon, P. (2001) Effects of physical exercise on anxiety, depression, and sensitivity to stress: a unifying theory. Clin. Psychol. Rev., 21: 33–61.

Sanchez, M.M., Young, L.J., Plotsky, P.M. and Insel, T.R. (2000) Distribution of corticosteroid receptors in the rhesus brain: relative absence of glucocorticoid receptors in the hippocampal formation. J. Neurosci., 20: 4657–4668.

Sapolsky, R.M., Krey, L.C. and Mcewen, B.S. (1983) The adrenocortical stress-response in the aged male rat: impairment of recovery from stress. Experimental Gerontology, 18: 55–64.

Schmidt, K.N., Gosselin, L.E. and Stanley, W.C. (1992) Endurance exercise training causes adrenal medullary hypetrophy in young and old Fischer 344 rats. Horm. Metab. Res., 24: 511–515.

Schulkin, J., Gold, P.W. and Mcewen, B.S. (1998) Induction of corticotropin-releasing hormone gene expression by glucocorticoids: implication for understanding the states of fear and anxiety and allostatic load. Psychoneuroendocrinol., 23: 219–243.

Seckl, J.R., Yau, J. and Holmes, M. (2002) 11Beta-hydroxysteroid dehydrogenases: a novel control of glucocorticoid action in the brain. Endocr. Res., 28: 701–707.

Shanks, N., Windle, R.J., Perks, P.A., Harbuz, M.S., Jessop, D.S., Ingram, C.D. and Lightman, S.L. (2000) Early-life exposure to endotoxin alters hypothalamic–pituitary–adrenal function and predisposition to inflammation. Proc. Natl. Acad. Sci. USA, 97: 5645–5650.

Sinyor, D., Schwartz, S.G., Peronnet, F., Brisson, G. and Seraganian, P. (1983) Aerobic fitness level and reactivity to psychosocial stress: physiological, biochemical, and subjective measures. Psychosom. Med., 45: 205–217.

Sloviter, R.S., Valiquette, G., Abrams, G.M., Ronk, E.C., Sollas, A.L., Paul, L.A. and Neubort, S. (1989) Selective loss of hippocampal granule cells in the mature rat brain after adrenalectomy. Science, 243: 535–538.

Smith, G.W., Aubry, J.-M., Dellu, F., Contarino, A., Bilezikijan, L.M., Gold, L.H., Chen, R., Marchuk, Y., Hauser, C., Bentley, C.A., Sawchenko, P.E., Koob, G.F., Vale, W. and Lee, K.-F. (1998) Corticotropin releasing factor receptor 1-deficient mice display decreased anxiety, impaired stress response, and aberrant neuroendocrine development. Neuron, 20: 1093–1102.

Smythe, J.W., Murphy, D., Timothy, C. and Costall, B. (1997) Hippocampal mineralocorticoid, but not glucocorticoid, receptors modulate anxiety-like behavior in rats. Pharmacol. Biochem. Behav., 56: 507–513.

Sothmann, M.S., Buckmworth, J., Claytor, R.P., Cox, R.H., White-Welkey, J.E. and Dishman, R.K. (1996) Exercise training and the cross-stressor adaptation hypothesis. Exerc. Sport Sci. Rev., 24: 267–287.

Spencer, R.L., Young, E.A., Choo, P.H. and McEwen, B.S. (1990) Adrenal steroid type-I and type-II receptor binding – estimates of in vivo receptor number, occupancy, and activation with varying level of steroid. Brain Res., 514: 37–48.

Stallknecht, B., Kjær, M., Ploug, T., Maroun, L., Ohkuwa, T., Vinten, J., Mikines, K.J. and Galbo, H. (1990) Diminished epinephrine response to hypoglycemia despite enlarged adrenal medulla in trained rats. Am. J. Physiol., 259: R998–R1003.

Starzec, J.J., Berger, D.F. and Hesse, R. (1983) Effects of stress and exercise on plasma corticosterone, plasma cholesterol and aortic cholesterol levels in rats. Psychosom. Med., 45: 219–226.

Stenzel-Poore, M.P., Heinrichs, S.C., Rivest, S., Koob, G.F. and Vale, W.W. (1994) Overproduction of corticotropin-releasing factor in transgenic mice: a genetic model of anxiogenic behavior. J. Neurosci., 14: 2579–2584.

Steptoe, A., Edwards, S., Moses, J. and Mathews, A. (1989) The effects of exercise training on mood and perceived coping ability in anxious adults from the general population. J. Psychosom. Res., 33: 537–547.

Sullivan, R.M. and Gratton, A. (1999) Lateralized effects of medial prefrontal cortex lesions on neuroendocrine and autonomic stress responses in rats. J. Neurosci., 19: 2834–2840.

Sullivan, R.M. and Gratton, A. (2002) Prefrontal cortical regulation of hypothalamic–pituitary–adrenal function in the rat and implications for psychopathology: side matters. Psychoneuroendocrinol., 27: 99–114.

Szigethy, E., Conwell, Y., Forbes, N.T., Cox, C. and Caine, E.D. (1994) Adrenal weight and morphology in victims of completed suicide. Biol. Psychiatry, 36: 374–380.

Takagishi, M. and Chiba, T. (1991) Efferent projections of the infralimbic (area 25) region of the medial prefrontal cortex in the rat: an anterograde tracer PHA-L study. Brain Res., 566: 26–39.

Terreberry, R.R. and Neafsey, E.J. (1983) Rat medial frontal cortex: a visceral motor region with a direct projection to the solitary nucleus. Brain Res., 278: 245–249.

Terreberry, R.R. and Neafsey, E.J. (1987) The rat medial frontal cortex projects directly to autonomic regions of the brainstem. Brain Res. Bull., 19: 639–649.

Timpl, P., Spanagel, R., Sillaber, I., Kresse, A., Reul, J.M.H.M., Stalla, G.K., Blanquet, V., Steckler, T., Holsboer, F. and Wurst, W. (1998) Impaired stress response and reduced anxiety in mice lacking a functional corticotropin-releasing hormone receptor 1. Nature Genet., 19: 162–166.

Trejo, J.L., Carro, E. and Torres-Aleman, I. (2001) Circulating insulin-like growth factor I mediates exercise-induced increases in the number of new neurons in the adult hippocampus. J. Neurosci., 21: 1628–1634.

Vallée, M., Mayo, W., Dellu, F., Le Moal, M., Simon, H. and Maccari, S. (1997) Prenatal stress induces high anxiety and postnatal handling induces low anxiety in adult offspring: correlation with stress-induced corticosterone secretion. J. Neurosci., 17: 2626–2636.

Van den Berg, D.T.W.M., De Kloet, E.R., Van Dijken, H.H. and De Jong, W. (1990) Differential central effects of mineralocorticoid and glucocorticoid agonists and antagonists on blood pressure. Endocrinology, 126: 118–124.

Van den Berg, D.T.W.M., De Kloet, E.R. and De Jong, W. (1994) Central effects of mineralocorticoid antagonist RU-28318 on blood pressure of DOCA-salt hypertensive rats. Am. J. Physiol., 267: E927–E933.

Van der Kar, L.D., Piechowski, R.A., Rittenhouse, P.A. and Gray, T.S. (1991) Amygdaloid lesions: differential effect on conditioned stress and immobilization-induced increases in corticosterone and renin secretion. Neuroendocrinol., 54: 89–95.

Van Eekelen, J.A.M., Jiang, W., De Kloet, E.R. and Bohn, M.C. (1988) Distribution of the mineralocorticoid and the glucocorticoid receptor mRNAs in the rat hippocampus. J. Neurosci. Res., 21: 88–94.

Van Haarst, A.D., Oitzl, M.S. and De Kloet, E.R. (1997) Facilitation of feedback inhibition through blockade of glucocorticoid receptors in the hippocampus. Neurochem. Res., 22: 1323–1328.

van Praag, H., Christie, B.R., Sejnowski, T.J. and Gage, F.H. (1999a) Running enhances neurogenesis, learning and long-term potentiation in mice. Proc. Natl. Acad. Sci. USA, 96: 13427–13431.

van Praag, H., Kempermann, G. and Gage, F.H. (1999b) Running increases cell proliferation and neurogenesis in the adult mouse dentate gyrus. Nature Neurosci., 2: 266–270.

Vazquez, D.M., Van Oers, H., Levine, S. and Akil, H. (1996) Regulation of glucocorticoid and mineralocorticoid receptor mRNAs in the hippocampus of the maternally deprived infant rat. Brain Res., 731: 79–90.

Villaneueva, A.L., Schlosser, C., Hopper, B., Liu, J.H., Hoffman, D.I. and Rebar, R.W. (1986) Increased cortisol production in women runners. J. Clin. Endo. Metab., 63: 133–136.

Warembourg, M. (1975) Radioautographic study of the rat brain and pituitary after injection of 3H dexamethasone. Cell Tissue Res., 161: 183–191.

Warren, M.P. and Stiehl, A.L. (1999) Exercise and female adolescents: effects on the reproductive and skeletal systems. J. Am. Med. Womens Assoc., 54: 115–120.

Watanabe, T., Morimoto, A., Sakata, Y., Wada, M. and Murakami, N. (1991) The effect of chronic exercise on the pituitary–adrenocortical response in conscious rats. J. Physiol., 439: 691–699.

Watanabe, T., Morimoto, A., Sakata, Y., Tan, N., Morimoto, K. and Murakami, N. (1992) Running training attenuates the ACTH responses in rats to swimming and cage-switch stress. J. Appl. Physiol., 73: 2452–2456.

Wiegers, G.J. and Reul, J.M.H.M. (1998) Induction of cytokine receptors by glucocorticoids: functional and pathological significance. Trends Pharmacol. Sci., 19: 317–321.

Wiegers, G.J., Labeur, M.S., Stec, I.E.M., Klinkert, W.E.F., Holsboer, F. and Reul, J.M.H.M. (1995) Glucocorticoids accelerate anti-T cell receptor-induced T cell growth. J. Immunol., 155: 1893–1902.

Wiegers, G.J., Stec, I.E.M., Klinkert, W.E.F. and Reul, J.M.H.M. (2000) Glucocorticoids regulate TCR-induced elevation of CD4: functional implications. J. Immunol., 164: 6213–6220.

Wiegers, G.J., Stec, I.E.M., Klinkert, W.E., Linthorst, A.C.E., and Reul, J.M.H.M., (2001) Bidirectional effects of corticosterone on splenic T-cell activation: critical role of cell density and culture time. Neuroendocrinol., 73: 139–148.

Williams, B.M., Luo, Y., Ward, C., Redd, K., Gibson, R., Kuczaj, S.A. and McCoy, J.G. (2001) Environmental enrichment: effects on spatial memory and hippocampal CREB immunoreactivity. Physiol. Behav., 73: 649–658.

Wittling, W. (2001) Brain asymmetry in the control of the stress response. In: Gainotti, G. (Ed.), Emotional Behavior and its Disorders. Elsevier Science B.V., Amsterdam, pp. 207–233.

Wittling, W., Block, A., Schweiger, E. and Genzel, S. (1998) Hemisphere asymmetry in sympathetic control of the human myocardium. Brain Cogn., 38: 17–35.

Young, E.A., Akana, S. and Dallman, M.F. (1990) Decreased sensitivity to glucocorticoid fast feedback in chronically stressed rats. Neuroendocrinol., 51: 536–542.

Young, D., Lawlor, P.A., Leone, P., Dragunow, M. and During, M.J. (1999) Environmental enrichment inhibits spontaneous apoptosis, prevents seizures and is neuroprotective. Nature Med., 5: 448–453.

SECTION 2

Hypothalamic Hormones Involved in Stress Responsivity

T. Steckler, N.H. Kalin and J.M.H.M. Reul (Eds.)
Handbook of Stress and the Brain, Vol. 15
ISBN 0-444-51173-3

CHAPTER 2.1

Novel CRF family peptides and their receptors: an evolutionary analysis

Sheau Yu Teddy Hsu*

Division of Reproductive Biology, Department of Obstetrics and Gynecology, Stanford University School of Medicine, 300 Pasteur Drive, Room A344E, Stanford, California 94305-5317, USA

Abstract: During the last decade, the availability of genomic databases has provided a resource where biological questions can be addressed in an unprecedented manner. Based on searches of homologous sequences, paralogous[†] genes from individual species or orthologous[†] genes from different species can be identified. The function of paralogous or orthologous genes can be inferred or predicted based on their degree of similarity to known genes, thereby augmenting our understanding of the gene functions that coevolved during evolution.

Recent studies on human genomic sequences have led to the discovery of novel type 2 corticotropin-releasing factor (CRF) receptor-selective agonists that are related to CRF by structural and functional characteristics. In addition, analysis of vertebrate genomes showed that the CRF peptide family in mammals includes four distinct genes, CRF, Urocortin 1, Urocortin 3/stresscopin (SCP), and Urocortin 2/stresscopin-related peptide (SRP). Phylogenetic analysis suggested that the origin of each of these peptides predated the separation of tetrapods and teleosts. It is likely that CRF family genes in modern vertebrates evolved from an ancestor gene that gave rise to the CRF/Urocortin 1 and Urocortin 3(SCP)/Urocortin 2(SRP) branches through a gene duplication event. These two ancestor genes then gave rise to additional paralogs through a second round of gene duplication. Each of these four genes is tightly conserved ranging from the >96% identity found for CRF to the >55% for Urocortin 3/SCP, thus suggesting that these peptides played essential signaling roles over the 550 million years of vertebrate evolution.

The finding of type 2 CRF receptor-selective Urocortin 3/SCP and Urocortin 2/SRP not only provided an opportunity to understand the physiology regulated by these novel ligands but also that of related peptides. The present review focuses on the recent findings of selective CRF receptor ligands, the evolution of signaling molecules associated with the CRF pathway as well as the implications of a complete inventory of CRF family ligands, receptors, and binding proteins in genomes of different organisms. In addition, new findings on CRF receptor subtype-dependent functions derived from studies using Urocortin 3/SCP and Urocortin 2/SRP are discussed. Instead of the traditional analysis of single-gene function in endocrine research, the complete assembly of CRF-associated signaling molecules throughout evolution can provide an integrated view for understanding the physiology and pathophysiology of all CRF family peptides and their receptors, thereby providing new therapeutic approaches for the pathology associated with stress.

Introduction

Over the past decade, whole-genome sequencing projects for numerous organisms have been completed. Thus far, the completed genomes of eukaryotic model organisms include that of baker's yeast, *Saccharomyces cerevisiae*, the nematode *Caenorhabditis elegans*, the fruit fly *Drosophila melanogaster*, two species of puffer fish (*Fugu rubripes* and *Tetraodon nigroviridis*), mouse, and human. In addition, genome sequencing for dozens of other species are ongoing. The availability of these genome sequences provides an unprecedented opportunity to discover novel signaling molecules that have

*E-mail: teddyhsu@stanford.edu
[†]paralogous: genes evolved from gene duplication in a species; orthologous: genes evolved from speciation.

coevolved with diverse polypeptide hormones that were previously discovered using traditional approaches.

During the 1950s, Guillemin and Rosenberg, and Saffran and Schally observed independently the presence of a factor in hypothalamic extracts (termed corticotropin-releasing factor (CRF) that could stimulate the release of adrenocorticotropic hormone (ACTH) from anterior pituitary cells in vitro (Guillemin, 1967; Peterson and Guillemin, 1974; Saffran and Schally, 1977; Furutani et al., 1983; Shibahara et al., 1983). Later, purification of this factor using ion exchange and high-performance liquid chromatography led to the isolation, synthesis, and characterization of a 41-amino acid hypothalamic ovine CRF (Spiess et al., 1981; Vale et al., 1981). This original work paved the way for numerous studies into the role of CRF as the predominant hypothalamic neuropeptide regulating adrenal glucocorticoid release via pituitary ACTH release, thus establishing the importance of the hypothalamus–pituitary–adrenal (HPA) axis (Contarino et al., 1999; Dunn and Swiergiel, 1999; Swiergiel and Dunn, 1999; Turnbull et al., 1999; Muglia et al., 2000; Muller et al., 2000).

The characterization of CRF was followed in the 1980s by the identification of the related peptide ligands urotensin I from teleost and sauvagine from amphibian, and in 1995, Urocortin 1 from mammals (Vaughan et al., 1995). Urotensin-I, originally characterized by Lederis and his associates in 1982 (Ichikawa et al., 1982; Lederis et al., 1982, 1983), is a structural homolog of CRF and shares similar biological properties. However, piscine urotensin-I, produced in the caudal neurosecretory system (urophysis), stimulates the release of glucocorticoids directly from the interrenal gland, apparently bypassing the pituitary. Another related peptide, sauvagine, isolated from the serous glands of frog, *Phyllomedusa sauvagei*, is 50% identical to CRF/urotensin I and could be important for defense in some anurans. Because Urocortin 1 possesses approximately 50–60% sequence identity with urotensin-I, it has been proposed that Urocortin 1 and its anuran homolog, sauvagine, represent an orthologous peptide to the piscine urotensin-I in tetrapod lineage (Vaughan et al., 1995). The structural organization of all these CRF family genes is similar and all prepropeptides could be divided into five functional segments: a signal peptide for secretion, an N-terminal prepropeptide, conserved proteolytic sites, an alpha-helical-forming bioactive peptide, and a C-terminal amidation donor residue. Studies on the solution structure of CRF using proton nuclear magnetic resonance spectroscopy suggest that CRF and related peptides comprise an N-terminal coil connected to an extended helix (Lau et al., 1983). In addition to similar secondary structures, all family proteins appear to share similar posttranslational modification features required for the generation of functional mature peptides; all mature peptides have a free N-terminus and are amidated at the C-terminus.

The biological actions of CRF and Urocortin 1 are mediated via binding to two G protein-coupled receptors (GPCRs), type 1 and type 2 CRF receptors (CRF1 and CRF2) found throughout the CNS, and periphery (Potter et al., 1991, 1994; Chen et al., 1993; Vita et al., 1993; Suman-Chauhan et al., 1999). The two receptors are closely related and are members of the family of "brain-gut" neuropeptide receptors that include receptors for calcitonin, CGRP, vasoactive intestinal peptide, parathyroid hormone, pituitary adenylate cyclase-activating peptide, growth hormone-releasing factor, CRF, and secretin. These receptors belong to a subset of GPCRs, referred to as the class B family that also includes the receptors for other brain-gut neuropeptides found in vertebrates and invertebrates. Each of the two CRF receptor genes give rise to multiple-splicing variants (type I: α, β, γ; type II: α, β) with distinct expression profiles and possibly alternative physiological function. Similar to a number of other type B GPCRs, ligand binding led to the activation of adenyl cyclase and cAMP production by CRF receptors. However, coupling of the CRF receptors to other G proteins has been reported (Dautzenberg and Hauger, 2002).

In addition to CRF and Urocortin 1, recent studies based on structural and posttranslational characteristics as well as comparative genomic approaches have led to the identification of two mammalian CRF/Urocortin 1-like peptides, Urocortin 3/stresscopin (SCP) (Hsu and Hsueh, 2001; Reyes et al., 2001) and Urocortin 2/stresscopin-related peptide (SRP) (Hsu and Hsueh, 2001; Lewis et al., 2001), with limited sequence relatedness to other family peptides. Phylogenetic analysis and functional characterization studies showed that these two novel peptides represent a distant evolutionary branch in the evolution of CRF family peptides and they emerged as early as CRF and Urocortin 1 during vertebrate evolution. Unlike known CRF family peptides, which activate both CRF1 and CRF2, Urocortin 3/SCP and Urocortin 2/SRP are selective type 2 CRF receptor agonists. In addition to having an important role in the regulation of the HPA axis, CRF receptors are directly involved in the regulation of diverse functions in the cardiovascular, endocrine, and immune systems. Due to the receptor selectivity of Urocortin 3/SCP and Urocortin 2/SRP, these peptides could be important in CRF2-mediated stress-coping responses for homeostasis maintenance in different tissues, thus giving the CRF receptor system a high flexibility and dynamic role in the adaptation to environmental challenges (Bale et al., 2000; Kishimoto et al., 2000; Hsu and Hsueh, 2001; Aggelidou et al., 2002). Because abnormal signaling by CRF receptors may contribute to the pathophysiology of stress-related disorders, abnormal regulation of Urocortin 3/SCP and Urocortin 2/SRP could be involved in the regulation of anxiety, depression, and cardiac and inflammatory disorders.

Before genome sequencing, it was usually difficult to understand complex hormonal systems in an integrated manner, and scientists frequently resorted to the speculation that unexplained experimental results are due to the actions of unknown genes. With a complete inventory of CRF family peptides and their receptors, scientists will be able to move from a search approach in their investigations to a reconstructionistic mode.

Evolution of the CRF-associated signaling molecules

Evolution of the CRF family peptides – two independent evolutionary branches in vertebrates

Four core genes in mammals and other vertebrates

Recent analysis of genome sequences from both vertebrates and invertebrates provided a complete repertoire of CRF family peptides. Studies based on both pairwise sequence comparison and phylogenetic profiling showed that there are a total of four unique CRF-related genes, CRF, Urocortin 1, Urocortin 3/SCP, and Urocortin 2/SRP, in the genomes of different mammals (Fig. 1). Although it is impossible to exclude the possibility that there is an additional paralog that evolved at an accelerated pace in the mammalian genomes, thus avoiding the detection by the computational method, integrated analysis on the evolution of CRF-related peptides and their receptors indicated that such possibility is negligible. As will be discussed later, the sequences of both the CRF family ligands and CRF receptors are highly conserved and can be traced to invertebrates. For genes with such high conservation, the existence of additional paralogs with a similar function is unlikely.

The four CRF family genes exhibited conservation in both sequence and function throughout the vertebrate lineage (Hsu and Hsueh, 2001; Lewis et al., 2001; Reyes et al., 2001). Among these family members, CRF has been isolated from different mammals, anurans, and fish (Lovejoy and Balment, 1999). The highly conserved CRF appeared to play a central role in the regulation of the HPA axis and to function as an important neuropeptide in the brains of all species studied. Studies on mice deficient for CRF showed that the stress-related activation of the HPA axis was absent in the mutant mice and these mice displayed increased locomotor activity (Dunn and Swiergiel, 1999; Swiergiel and Dunn, 1999) (Table 1). Urocortin 1 has also been cloned from multiple vertebrates including mouse, human, and sheep (Vaughan et al., 1995; Lovejoy and Balment, 1999). Earlier studies on the Urocortin 1 expression pattern in the brain suggested that the Urocortin 1 could be involved in the regulation of feeding, anxiety, and auditory processing. Indeed, studies on mice deficient for the Urocortin 1 gene showed that mutant mice exhibit heightened anxiety-like behaviors in the elevated plus maze and open-field tests (Vetter et al., 2002; Wang et al., 2002). In addition, these mutant mice display an impaired acoustic startle response at the level of the inner ear, suggesting that Urocortin 1 is involved in the normal development of cochlear sensory-cell function and hearing (Vetter et al., 2002). The effect of Urocortin 1

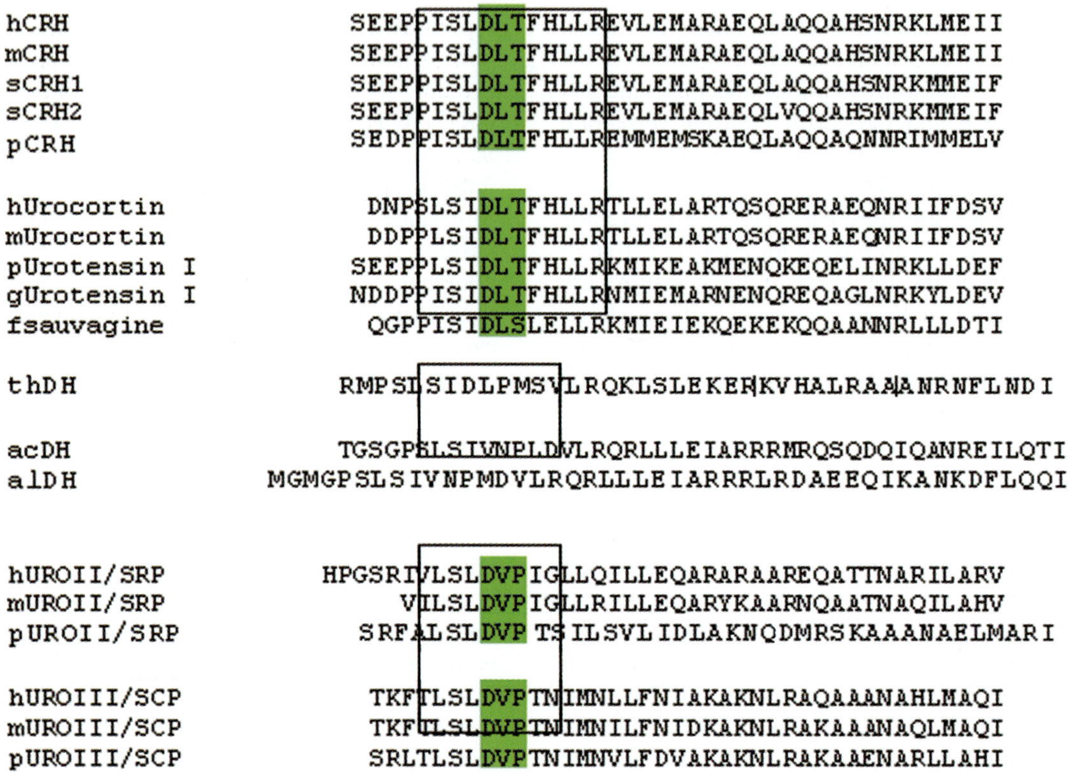

Fig. 1. Sequence alignment of vertebrate and invertebrate CRF/diuretic hormone family peptides. There are a total of four unique CRF-related genes, CRF, Urocortin 1, Urocortin 3/SCP, and Urocortin 2/SRP, in the genomes of different mammals. The insect diuretic hormone showed similarity to CRF family peptides in both primary and secondary structures. The structural determining motif between the activation and binding domain of each subgroup of peptides are enclosed by lines. h: human; m: mouse; p: puffer fish (*Fugu rubripes*); s: sucker (*Catostomus commersoni*); g: goldfish (*Carassius auratus*); f: frog (*P. sauvagei*); th: tobacco hornworm (*Manduca sexta*); ac: american cockroach (*Periplaneta Americana*); al: African migratory locust (*Locusta migratoria*).

on the acoustic startle response could be mediated through the Urocortin 1-expressing neuron projections from the region of the Edinger-Westphal nucleus (Vetter et al., 2002; Wang et al., 2002). However, in comparison with CRF1- and CRF2-deficient mice, Urocortin 1-deficient mice appeared to exhibit minimal alteration in anxiety-like behavior as well as autonomic regulation in response to stress. In contrast, sauvagine and urotensin I are found only in anuran and teleosts, respectively. Sauvagine and urotensin I are clearly related in structure to the CRF/Urocortin 1, but the peptides are found exclusively in specific tissues. The former was from the serosal gland of the skin of *P. sauvagei*, a frog native to Central and South America (Montecucchi and Henschen, 1981; Negri et al., 1983), whereas the latter was purified from the urophysis of two teleost fish, the white sucker *C. commersoni* (Lederis et al., 1982) and the carp *Cyprinus carpio* (Ichikawa et al., 1982). Because these two peptides exhibit pharmacological and sequence characteristics similar to CRF, they likely derived from additional lineage-specific gene duplication (Fig. 2), one duplication in anuran to produce sauvagine and an independent duplication in teleost to generate urotensin I. Subsequent divergence of the promoter region of these newly duplicated genes then gave rise to the distinct expression profile of these paralogous genes.

Unlike the better characterized family members that clustered in a branch of the evolutionary

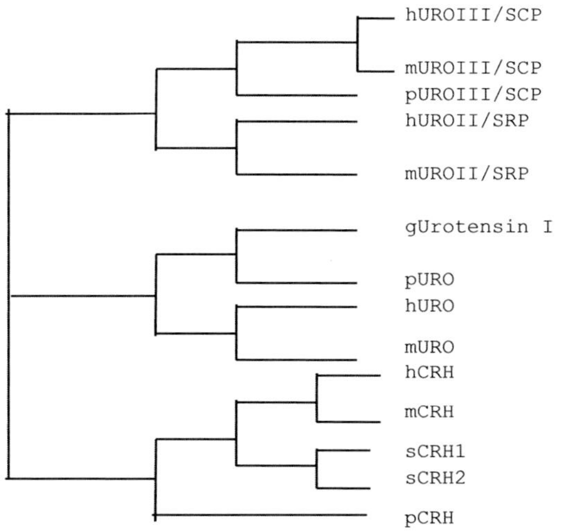

Fig. 2. Evolution of the CRF/diuretic hormone family peptides. The four vertebrate family peptides likely derived from a single ancestor gene that also gave rise to diuretic hormones in insects through gene duplication. A second round of gene duplication of the CRF/Urocortin 1 and Urocortin 3(SCP)/Urocortin 2(SRP) ancestors generated the four core ligand genes in vertebrates. h: human; m: mouse; p: puffer fish (*F. rubripes*); s: sucker (*C. commersoni*); g: goldfish (*C. auratus*).

dendrogram, the newly identified Urocortin 3/SCP and Urocortin 2/SRP form a separate branch, suggesting that these genes emerged as early as the CRF/Urocortin 1 branch. Similar to CRF and Urocortin 1, studies on Urocortin 3/SCP have shown that this gene is expressed in specific areas of the brain (Li et al., 2002). Urocortin 3/SCP-positive neurons were found predominately within the hypothalamus and medial amygdala. In the hypothalamus, Urocortin 3/SCP neurons were observed in the median preoptic nucleus and in the rostral perifornical area lateral to the paraventricular nucleus (Li et al., 2002). In the periphery, it can be detected in multiple tissues including the cardiovascular system, the gastrointestinal tract, and the skin (Hsu and Hsueh, 2001; Lewis et al., 2001). As many of these tissues are known to express high levels of CRF2, Urocortin 3/SCP is likely to be an endogenous ligand for CRF2 in these areas and play a role in mediating physiological functions, including food intake, gastrointestinal mobility, and neuroendocrine regulation (Li et al., 2002). In contrast, Urocortin 2/SRP is expressed in the rodent central nervous system, including stress-related cell groups in the hypothalamus and brainstem, and peripheral tissues (Lewis et al., 2001; Reyes et al., 2001). Behaviorally, central Urocortin 2/SRP attenuates night time feeding, but with a time course distinct from that seen in response to CRF. In contrast to CRF, central Urocortin 2/SRP failed to increase gross motor activity suggesting that Urocortin 2/SRP is involved in central autonomic and appetitive control, but not in generalized behavioral activation (Hsu and Hsueh, 2001; Reyes et al., 2001; Valdez et al., 2002). In addition, studies on the pufferfish Urocortin 3/SCP peptide suggested that the selective activation of CRF2 by Urocortin 3/SCP and Urocortin 3/SRP is evolutionarily conserved and that physiology processes regulated by these two peptides are essential for stress regulation and adaptation in all vertebrates (Hsu and Hsueh, 2001).

Tracing of the CRF family peptides to the common ancestors of Insects and vertebrates

In addition to vertebrate CRF family peptides, studies on insect diuretic hormones have shown that this group of highly conserved hormones exhibit sequence and structural similarity to vertebrate CRF family peptides. Diuretic hormones are essential for fluid secretion regulation by Malpighian tubules in insects and have been isolated from insects, including the tobacco hornworm, moth, and cockroach. The diuretic hormone peptides range from 40 to 47 amino acids and exhibit >18% sequence identity to vertebrate CRF family peptides. Based on GenBank sequence search, diuretic hormone homologs from the *Drosophila* and the *Anopheles* mosquito have also been deduced (Adams et al., 2000; Riehle et al., 2002). Although there is a possibility that the relationship between vertebrate CRF family peptides and insect diuretic hormones is a result of convergent sequence evolution, comparative analysis has shown that the ancestors of diuretic hormone and the diuretic hormone receptor likely coevolved to give rise to diuretic hormone/diuretic hormone receptor signaling in insects and the CRF receptor signaling in vertebrates. Assuming the diuretic hormone and CRF family peptides followed the same rate of evolution, all vertebrate CRF family peptides will have a similarity greater than that between the insect diuretic hormone and the mammalian CRF family peptides.

CRF/Urocortin 1 and Urocortin 3(SCP)/ Urocortin 2(SRP) as two separate evolutionary branches with unique functional characteristics

Although CRF/Urocortin 1 and Urocortin 3(SCP)/ Urocortin 2(SRP) have been characterized as family peptides for CRF receptors, the presence of multiple ligands for the two CRF receptor subtypes in a given species raises the question as to how the corresponding genes coevolved. Because we can trace the origin of both the ligand and the receptor to invertebrates, the common theme of the coevolution of ligands and receptors suggests that CRF family peptides evolved from a single ancestral gene and all family peptides in vertebrates could be identified without ambiguity using the current analytical tools (Darlison and Richter, 1999).

When all four groups of orthologous peptides from different vertebrates are compared with each other, it is clear that CRF/Urocortin 1 and Urocortin 3(SCP)/ Urocortin 2(SRP) have unique group-specific primary sequences and structural characteristics (Figs. 1 and 2). Because homologous genes for each of the four unique mammalian peptides, CRF, Urocortin 1, Urocortin 3/SCP, and Urocortin 2/SRP, can be found in species from human to teleost, the orthologous relationship of each gene among different vertebrates can be established unambiguously. As mentioned earlier, functional characterization indicates that the evolution of Urocortin 3(SCP)/Urocortin 2(SRP) is as ancient as CRF and Urocortin 1 and that the four core CRF family genes in modern vertebrates evolved from two sequential gene duplications of an ancient CRF-like gene (Fig. 2). The first duplication resulted in a CRF-like gene and a Urocortin 3/SCP-like gene, which subsequently evolved into four distinct genes and each of these paralogous genes persisted during evolution.

Evolution of the CRF receptor family

Although less information is available regarding the sequences of CRF receptors in different vertebrates, two subtypes of CRF receptors, CRF1 and CRF2, have been identified in human, rat, mouse, *Xenopus*, and multiple species of fish (Lovenberg et al., 1995; Liaw et al., 1996; Pohl et al., 2001). Sequence analysis showed that CRF receptors are highly conserved and the mammalian and fish CRF receptors exhibited a >50% sequence identity. Earlier studies have shown that although CRF and Urocortin 1 are capable of interacting with both Type I and Type II receptors from different species, CRF appears to have a greater potency on CRF1 whereas Urocortin 1 exhibited equivalent potency on both types of receptors. Based on anatomical distribution and the characteristics of ligand–receptor interactions, Urocortin 1 has been postulated to be an endogenous ligand for CRF2 and CRF mainly activates CRF1 to stimulate ACTH release and, consequently, glucocorticoid production from the adrenal cortex (Contarino et al., 1999; Turnbull et al., 1999; Bale et al., 2000; Coste et al., 2000; Kishimoto et al., 2000).

Expression analyses in different vertebrates have shown that CRF1 is the major form of the CRF receptor expressed in the pituitary gland, whereas CRF2 is expressed in diverse tissues. In addition to differences in distribution between CRF1 and CRF2, there exists a distinct pattern of distribution between the receptor splicing variants (Valdenaire et al., 1997; Palchaudhuri et al., 1999; Suman-Chauhan et al., 1999; Pisarchik and Slominski, 2001, 2002). The CRF2α isoforms primarily expressed within the CNS, where CRF2β isoforms are found both centrally and peripherally. Within the brain, CRF2α is the predominant isoform, whereas the CRF2β is localized primarily to nonneuronal structures. Peripherally, the highest levels of CRF2β mRNA were found in heart and skeletal muscle, with lower levels detected in lung and intestine. The data suggest that different receptor variants could subserve specific physiological roles both centrally and peripherally (Palchaudhuri et al., 1999; Pisarchik and Slominski, 2001, 2002).

The phenotypes of mice deficient in either CRF1 or CRF2 demonstrate the critical role these receptors play (Table 1). CRF1-mutant mice have an impaired stress response and display decreased anxiety-like behavior (Smith et al., 1998; Timpl et al., 1998), whereas CRF2-mutant mice are hypersensitive to

Table 1. Mice deficient for CRF, Urocortin 1, CRF1, CRF2, or CRF-BP show different phenotypes in stress-related responses

Phenotypes	Anxiety and hypothalamic–pituitary–adrenal (HPA) axis	Stress-related behavioral responses	Unique phenotype
Gene			
CRF	Absence of stress-related activation of the HPA axis	Increased locomotor activity	
Urocortin 1	Normal stress-related activation of the HPA axis	Heightened anxiety-like behaviors	Impaired acoustic startle response
CRF1	Impaired HPA axis	Decreased anxiety-like behavior	
CRF2	Heightened stress-related activation of the HPA axis	Increased anxiety-like behavior; hypersensitive to stress	
CRF-BP	Normal HPA axis function	Increased anxiogenic-like behavior	
CRF1 + CRF2	Greater impairment of the HPA axis response as compared to the single receptor knockout	Similar to CRF1-deficient mice but shows sexual dichotomy	

stress and display increased anxiety-like behavior under select conditions (Bale et al., 2000; Coste et al., 2000; Kishimoto et al., 2000). Therefore, it appears that CRF2 mediates a central anxiolytic response, opposing the anxiogenic effect of CRF mediated by CRF1. However, double-mutant mice deficient in both CRF1 and CRF2 displayed an even greater impairment of their HPA axis response to stress than that of the CRF1-mutant mice, suggesting that both CRF1 and CRF2 have critical roles in the regulation of the HPA axis and the maintenance of homeostasis in response to stress (Smith et al., 1998; Timpl et al., 1998; Contarino et al., 1999; Turnbull et al., 1999; Bale et al., 2000; Coste et al., 2000; Kishimoto et al., 2000). In addition, studies on mice deficient in both CRF1 and CRF2 showed that the role of these receptors is gender specific. Although the female double-mutant mice displayed anxiolytic-like behavior, the male double-mutant mice showed significantly more anxiety-like behavior.

CRF receptors from different species share a characteristics long N-terminal extracellular domain containing a number of cysteines that are important in ligand binding. Studies on nonmammalian species have identified two CRF receptor homologs closely related to the mammalian CRF1 and CRF2, respectively, from *Xenopus laevis* (Dautzenberg et al., 1997) and chun salmon, *Oncorhynchus keta* (Pohl et al., 2001). Like the mammalian receptors, both types of receptors from nonmammalian vertebrates do not differentiate the CRF branch of ligands; all CRF-like peptides including CRF, urotensin I, sauvagine, and Urocortin 1 can activate both nonmammalian receptors (Dautzenberg et al., 1997; Pohl et al., 2001). Similar to *X. laevis* and chun salmon, studies of the puffer fish genome showed that *F. rubripes* encodes two CRF receptor homologs (type 1 CRF receptor: SINFRUP00000079341; type 2 CRF receptor: CAAB01000850.1) with greater than 81% and 80% similarity to human CRF1 and CRF2, respectively (Fig. 3). In addition, studies on CRF receptors in a diploid catfish, *Ameiurus nebulosus*, have also been reported (Arai et al., 2001). Unlike all other species studied, the *Ameiurus* catfish encodes two distinct mammalian CRF1 homologs and a single CRF2 ortholog (Arai et al., 2001). The first catfish CRF1 is highly expressed in the brain and its distribution pattern correlates with that of mammalian CRF1, whereas the second CRF1 homolog (CRF3) is mainly expressed in the pituitary gland, urophysis, and brain. In contrast, the catfish CRF2 is most abundantly expressed in the atrium of the heart, reminiscent to that of mammalian CRF2. Therefore, there has been extensive conservation in the expression of the two types of CRF receptors in different organs during evolution. Because the two CRF1 homologs (CRF1 and CRF3) found in catfish showed greater similarity to each other than other teleos CRF receptor homologs, the additional CRF1-like gene (CRF3) found in the catfish

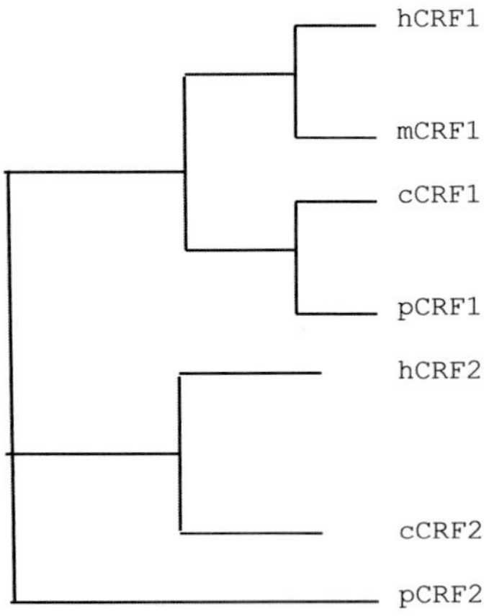

Fig. 3. Two distinct CRF receptors are conserved from pufferfish *F. rubripes* to human. Similar to other vertebrates, pufferfish encode two CRF receptors (type 1: SINFRUP00000079341; type 2: CAAB01000850.1) homologous to the mammalian CRF1 and CRF2, respectively. Human and fish CRF1 are clustered in one branch separate from the human and fish CRF2 homologs. h: human; m: mouse; p: puffer fish (*F. rubripes*); c: catfish (*A. nebulosus*).

likely derived from an additional gene duplication of the ancestral CRF1 gene. Future studies on the functional characteristics of teleost Urocortin 3/SCP and Urocortin 2/SRP peptides and their native receptors should reveal whether the nonmammalian CRF receptors could differentiate Urocortin 3/SCP and Urocortin 2/SRP as found in mammalian receptors.

As described earlier, the presence of close sequence homologs in insects provides support for understanding the evolution of the CRF family peptides and the coevolved CRF receptor signaling system. Similar to the vertebrate CRF family peptides, insect diuretic hormones mediate the action through a type B GPCR homologous to the mammalian CRF receptors (Reagan, 1994, 1995). The orthologous relationship between the vertebrate CRF receptors and the insect diuretic hormone receptors is obvious because the insect diuretic hormone receptors have the closest sequence homology to CRF receptors as compared to all known GPCRs. In insects, the diuretic hormone receptors activated adenylyl cyclase and the protein kinase A-dependent pathway in target tissues, the Malpighian tubules (Reagan, 1994, 1995, 1996; Wiehart et al., 2002). Of interest, functional studies using the Malpighian tubules have actually shown that vertebrate peptides including sauvagine and CRF have a significant effect on cAMP production in insect cells (Audsley et al., 1995, 1997). Analysis of recently completed insect genomes showed that the *Anopheles* mosquito and *Drosophila* encode orthologs of the diuretic hormone and the diuretic hormone receptors that have been identified in other insects. The *Anopheles* mosquito and *Drosophila* encoded one (agCP6328) and two (CG8422-PA and CGG12370-PA) diuretic hormone receptor homologs, respectively, sharing a >60% amino acid identity with known cockroach and house cricket diuretic hormone receptors. Important, the two *Drosophila* diuretic hormone receptors showed distinct similarity to mammalian CRF1 (51% similarity) and CRF2 (57% similarity), respectively.

These data provide a clear evolutionary trail for the origin of the CRF receptor signaling system from invertebrates to vertebrates. They also indicate that the divergence of the two CRF receptor subtypes may have taken place prior to the emergence of vertebrates. Because the diuretic hormone receptor represents one of the few type B GPCRs found in insects, CRF receptors, and diuretic hormone receptor likely derived from a common ancestral gene in invertebrates and represents one of the earliest forms of type B GPCR signaling.

Evolution of the CRF-binding protein (CRF-BP)

The biological activity of CRF and Urocortin 1 has been shown to be modulated by a secreted glycoprotein, the CRF-BP (Potter et al., 1991; Behan et al., 1995; Seasholtz et al., 2002). CRF-BP binding to CRF and Urocortin 1 leads to the sequestration of bioactive peptides and the inhibition of CRF-induced ACTH release by pituitary cells in vitro (Behan et al., 1995, 1996b; Cortright et al., 1995). Studies on mice deficient in CRF-BP showed that mutant mice exhibit a normal increase in ACTH and corticosterone after restraint stress; however, mutant mice exhibit a significant increase in anxiogenic-like behavior and a significant reduction of body weight in males (Karolyi et al., 1999). The increased anorectic and anxiogenic-like behavior likely is a result of increased "free" CRF and/or Urocortin 1 levels in the brain

(Seasholtz et al., 2002). Thus, CRF-BP plays an important role in maintaining appropriate levels of CRF and Urocortin 1 in the central nervous system (Karolyi et al., 1999; Seasholtz et al., 2002). Of interest, studies have shown that CRF-BP does not show appreciable interaction with Urocortin 3/SCP and Urocortin 2/SRP (Lewis et al., 2001), suggesting that the interaction between CRF-BP and the ligands diverged for peptides in the CRF/Urocortin 1 and Urocortin 3(SCP)/Urocortin 2(SRP) branches. The lack of constraint from CRF-BP could be important for the full action of Urocortin 3(SCP)/Urocortin 2 (SRP) during stress regulation.

CRF-BP has been characterized from diverse vertebrates including human, mouse, sheep, rat, and *Xenopus* (Behan et al., 1993, 1995, 1996a; Valverde et al., 2001). Like CRF receptors, CRF-BPs from different vertebrates are highly conserved. For example, *Xenopus* and human CRF-BPs share >78% identity. A search of GenBank sequences showed that CRF-BP is a single copy gene in mammals and is unique among known genes. Analyses of GenBank databases also indicated that, unlike mammalian genomes, the pufferfish encodes two copies of the CRF-BP homologs (SINFRUP00000086841 and SINFRUP00000069650) showing a >70% similarity to mammalian CRF-BP. Because the two pufferfish CRF-BPs showed a divergence similar to that between mammalian and piscine CRF-BPs, they likely derived from a gene duplication event early in teleost evolution. The pufferfish CRF-BP homologs share the ten cysteine residues that have been shown to be essential for forming five disulfide bridges and maintaining the binding activity of mammalian CRF-BP (Fischer et al., 1994). Great conservation of sequence and structural characteristics in the piscine CRF-BP homologs suggest that these polypeptides could have a functional characteristic similar to the CRF-BP in mammals wherein they function as a binding protein for urotensin I and CRF, but not the piscine Urocortin 3/SCP and Urocortin 2/SRP.

Functional evolution of the CRF receptor signaling pathway

The broad and overlapping distribution of CRF family peptides and their receptors have confounded our view of CRF signaling in different tissues. Thus, it seems appropriate to investigate the origin of CRF function with a view toward identifying the divergent nature of this signaling pathway. In view of the fact that CRF receptors and their ligands show higher conservation when compared to most other peptide hormone/GPCR signaling systems during the last one billion years of evolution, this system likely maintained many similar functional features in different lineages. As illustrated by the study of the insect diuretic hormone system, the CRF receptor originated as a paracrine signaling system important for osmoregulation. In lower vertebrates, CRF receptor signaling pathway assumed additional functions. The hypothalamic CRF control of the pituitary gland in teleosts has been well established. In addition to the hypothalamic–hypophyseal system, fish possess a lineage-specific neurosecretory organ, urophysis, which secretes urotensin I. In fish, both the pituitary CRF and the urophysis urotensin I have been implicated in the regulation of osmoregulation, ionic balance, and vasodilation (Mainoya and Bern, 1982; Bern et al., 1985; Lenz et al., 1985). Although data is not available on hormones and receptors in primitive chordates, the presence of multiple CRF family ligands and receptors in modern vertebrates suggests that gene duplications and the subsequent divergence of the regulatory mechanism of these paralogous genes provide an advantage as vertebrates' niches evolved during evolution.

Other than having a role in osmoregulation and vascular homeostasis, CRF family peptides play major roles in the stress responses of teleosts. Similar to that observed in the mammalian system, handling and confinement stress increased the plasma cortisol and POMC peptides within minutes. CRF and urotensin I have been reported to coexist in the fish hypothalamus and both exert ACTH-stimulating action on the pituitary, reminiscent to that found in mammals. Interestingly, CRF has been shown to regulate metamorphosis in response to pond drying in some amphibian species (Denver, 1997; Boorse and Denver, 2002), whereas CRF-BP is closely regulated by the thyroid hormone during tail resorption in the frog (Brown et al., 1996). Though the tripeptide thyrotropin-releasing hormone is active on TSH secretion in adults, CRF seems to assume the thyrotropin-releasing hormone activity in amphibian larvae, suggesting that the CRF function could include the thyroid axes for coordinating the metamorphosis program during the transition from an aquatic environment to land. Although this role may seem to be unrelated, it is consistent with the role of CRF family peptides as osmoregulators and as a stress transducer between the environment and the physiological responses of an organism (Denver, 1997; Boorse and Denver, 2002). Thus, CRF family peptides are phylogenetically ancient developmental signaling molecule that allows developing organisms to coordinate physiological responses in a changing environment (Denver, 1997, 1999; Boorse and Denver, 2002). In addition, the presence of sauvagine and urotensin I in specific secretory tissues of lower vertebrates as paralogs of CRF illustrates the advantage of divergent evolution even though the number of signaling molecular modules remains similar.

Recent comparative genomic analyses of gene families across different species have provided insight into the evolution and associated adaptation of different hormonal regulatory circuits (Sherwood et al., 2000; Leo et al., 2002; Riehle et al., 2002). It has been hypothesized that diverse hormones and receptors coevolved during evolution and the vertebrate endocrine mechanism rooted in invertebrate paracrine signaling system. Although experimental evidence is required for the functional assignment of genes isolated from different species, the conservation of gene function in different phylogenies and networks has provided a solid foundation for translating gene function from one species to another. The identification of the two evolutionarily ancient CRF family peptides, Urocortin 3/SCP and Urocortin 2/SRP, in different vertebrates provides critical elements for the future characterization of the physiology associated with CRF receptors and the better characterized CRF and Urocortin 1.

New findings derived from studies of the CRF2-selective ligands, Urocortin 3/SCP and Urocortin 2/SRP

As CRF and Urocortin 1 possess high affinity for CRF1 and CRF2 as well as the CRF-BP at physiological concentrations, in vivo physiological studies with these peptides produce many confounding observations. As mentioned earlier, recent characterization of CRF1 and CRF2 in mutant mice have revealed a far-reaching complexity and physiological importance for the CRF family peptides (Dautzenberg and Hauger, 2002). It has become clear that adaptive responses induced by stressors are mediated by the autonomic nervous system and two interrelated and somewhat antagonistic CRF receptor pathways (Steckler, 2004). CRF1 is important for the initial flight and fight response whereas CRF2 could be essential for the delayed stress-coping responses prompted by the initial stress inducer. The cloning of Urocortin 3/SCP and Urocortin 2/SRP genes clarified many misconceptions and provided a unique opportunity in the study of stress regulation by CRF receptors. The ancient origin of these CRF2-selective ligands indicates that the prevalent view of stress regulation by the CRF unintentionally tilted to the side of stress induction and much less on the equally important stress-coping responses (Steckler, 2004). Although it is expected that future studies on Urocortin 3/SCP and Urocortin 2/SRP may reveal an undescribed stress-coping network in humans and improve the understanding of stress regulation and environmental adaptation in general, recent studies using these two peptides have unambiguously dissected the critical CRF receptor subtypes involved in a variety of CRF-dependent functions (Steckler, 2004). These studied are summarized in the following section (Table 2).

CRF1-dependent functions

ACTH release from the pituitary

Earlier studies on CRF receptor expression and pharmacological analogs suggested that CRF is the main mediator for pituitary ACTH release through the CRF1 and that CRF/CRF1 signaling is important for proper function of the HPA. Consistent with this hypothesis, deletion of CRF or CRF1 led to altered HPA axis responses in mutant mice (Swiergiel and Dunn, 1999; Turnbull et al., 1999; Muglia et al., 2000) (Table 1). Studies on the effect of Urocortin 3/SCP and Urocortin 2/SRP on pituitary cells indicated that these CRF2-selective

Table 2. CRF receptor subtype-dependent functions

CRF1-dependent functions	CRF2-dependent functions
1. ACTH release from pituitary	1. Regulation of gastric mobility and food intake
2. Protection of hippocampal neurons from excitotoxic insults	2. Regulation of vasculature homeostasis
3. Regulation of locomotor activation	
4. Regulation of colon mobility	
5. Regulation of myometrial relaxation/contractility	

agonists have negligible stimulation on the secretion of ACTH from anterior pituitary both in vivo and in vitro (Hsu and Hsueh, 2001; Lewis et al., 2001; Reyes et al., 2001), confirming the selective property of these ligands in vivo and that CRF1 is the main mediator of pituitary ACTH release.

Regulation of locomotor activation

It was thought that CRF1 and CRF2 mediate CRF actions in the central nervous system and that both could be involved in locomotor behaviors. Studies on CRF1 null mice showed that the CRF treatment resulted in increased levels of locomotion in wild type mice, whereas no change was observed in CRF1 null mice (Contarino et al., 2000). Likewise, mice deficient for CRF exhibited altered locomotor activity (Dunn and Swiergiel, 1999; Swiergiel and Dunn, 1999). This hypothesis was supported by studies showing that, in contrast to CRF, central administration of Urocortin 2/SRP failed to increase gross motor activity (Reyes et al., 2001). In fact, Urocortin 2/SRP actually mildly suppressed locomotor activity during the inactive phase (Reyes et al., 2001; Valdez et al., 2002). Therefore, the activational effects of CRF on locomotor activity observed in many earlier studies are mediated by the CRF1.

Protection of hippocampal neurons from excitotoxic insults

Hippocampus neurons are vulnerable to damage during disease and stress conditions, including cerebral ischemia and anxiety disorders. Although both CRF1 and CRF2 are present in the hippocampal regions of mammalian brain, CRF and Urocortin 1 protect hippocampal neurons from insults whereas Urocortin 2/SRP is ineffective (Pedersen et al., 2002). These studies clearly demonstrated that the neuroprotective effect of CRF and urocortin on the hippocampal neurons to oxidative and excitotoxic insults is mediated by CRF1 (Pedersen et al., 2002).

Regulation of colon mobility

Peripheral CRF has been shown to stimulate colonic motor function and inhibit gastric emptying in different animals. Earlier studies have shown that CRF-related peptides exert similar activity on gastrointestinal function (Miampamba et al., 2002; Million et al., 2002; Saunders et al., 2002); however, it was not clear which type(s) of CRF receptor is important for colonic response. Studies using ovine CRF, a preferential CRF1 agonist, suggested that CRF1 could be important for the regulation of colon transit. In support of this hypothesis, colonic response was shown to be dose dependently blocked by the selective CRF1 receptor antagonists, NBI-27914 and CP-154,526, but not the CRF2-selective peptide anti-sauvagine-30 (Martinez et al., 2002). To identify the exact receptor subtypes responsible for the regulation of colonic motor function, the effects of Urocortin 2/SRP on colon transit have been investigated in conscious rats. It was shown that Urocortin 2/SRP did not influence colonic transit whereas CRF stimulated distal colonic transit motor activity. Thus, the stimulation of colonic propulsive activity involves CRF1 and the nonselective ligands, CRF and Urocortin 1 (Miampamba et al., 2002; Saunders et al., 2002).

Regulation of myometrial relaxation/contractility

It has been well established that CRF modulates myometrial relaxation/contractility and stimulates the nitric oxide system (Rivier, 2001). CRF caused increased basal or atrial natriuretic peptide-stimulated cGMP production and this mechanism may be operational during human pregnancy. Studies using receptor-selective antagonists have shown that treatment of the nonselective antagonists, astressin and antalarmin, but not the CRF2-selective antagonist antisauvagine 30, blocks the CRF-stimulated cGMP production (Aggelidou et al., 2002). Consistent with this observation, incubation of myometrial cells with CRF, but not Urocortin 3/SCP or Urocortin 2/SRP, induced mRNA and protein expression of different nitric oxide synthase isoforms (Aggelidou et al., 2002), indicating that during pregnancy signaling of CRF1, but not CRF2, contributes to the maintenance of myometrial quiescence (Aggelidou et al., 2002).

CRF2-dependent functions

Regulation of gastric mobility and food intake

Peripheral treatment with CRF/Urocortin 1 peptides inhibits gastric emptying and food intake in different mammals. Unlike colon motor activity that is mainly mediated by CRF1, studies have shown that the nonselective antagonist astressin B and the selective CRF2 antagonist antisauvagine-30 dose dependently antagonized CRF- or Urocortin 1-induced delayed gastric emptying action, whereas the selective CRF1 antagonists, NBI-27914 and CP-154,526, had no effect (Wang et al., 2001; Chen et al., 2002). Likewise, Urocortin 1-induced hypophagia was partially antagonized by antisauvagine-30 whereas the CRF1 antagonists, CP-154,526 and DMP904, had no effect (Wang et al., 2001). These data suggest that peripheral CRF/Urocortin 1-induced suppression of gastric emptying involves primarily CRF2 whereas the action on feeding is only partly mediated by CRF2 (Wang et al., 2001). Consistent with these data, studies on CRF1-deficient mice show that CRF decreased food intake equally in both wild type and CRF1-deficient mice (Contarino et al., 2000). Studies using Urocortin 3/SCP and Urocortin 2/SRP have shown that both peptides dose dependently inhibited gastric emptying and food intake (Hsu and Hsueh, 2001; Lewis et al., 2001; Reyes et al., 2001; Miampamba et al., 2002; Million et al., 2002; Saunders et al., 2002), reinforcing the hypothesis that peripheral effects of CRF family peptides on gastric mobility and appetite suppression are mediated through CRF2. Therefore, peripheral CRF family peptides induce opposite actions on upper and lower gut transit through different CRF receptor subtypes: the activation of CRF1 stimulates colonic propulsive activity whereas the activation of CRF2 inhibits gastric emptying (Martinez et al., 2002).

Regulation of vasculature homeostasis

Since its discovery two decades ago, potent CRF/Urocortin 1 effects on the vascular system have been consistently observed. The diverse cardiovascular effects included coronary vasodilation, enhanced cardiac contractility and heart rate, and the protection of cardiomyocytes from ischemia simulated by glucose deprivation, acidosis, or hypoxia reoxygenation injury (Terui et al., 2001; Aggelidou et al., 2002; Brar et al., 2002; Gordon et al., 2002; Schulman et al., 2002). In addition, it has been suggested that CRF1 mediates CRF-induced blood pressure elevation, whereas peripheral CRF2 mediates the hypotensive effect of systemically administered CRF and Urocortin 1. Because CRF2 has been shown to be the main receptor expressed in different chambers of the heart and vascular system, the CRF family peptides likely exert their effects through CRF2 in an endocrine or paracrine manner (Kimura et al., 2002). Furthermore, it has been shown that Urocortin 1 effects on ventricular performance, bioenergetics, and cell survival are not secondary to the inotropic effects of Urocortin 1, and the survival effect of CRF family peptides is likely direct (Scarabelli et al., 2002). Because Urocortin 1 expression has also been detected in the heart, it has been suggested that Urocortin 1 and CRF2 could form a peripheral cardiac CRF system important in mediating the adaptive responses of the heart to stress (Coste et al., 2002). Studies using Urocortin 3/SCP and Urocortin 2/SRP demonstrated that these peptides also have potent suppressive effects on paw edema formation (Hsu and Hsueh, 2001), suggesting a role in vascular homeostasis. Recent studied have also demonstrated that Urocortin 3/SCP and Urocortin 2/SRP could protect cardiomyocytes from injuries (Chanalaris et al., 2003; Brar et al., 2004), but would not provoke the unwanted fight or flight response mediated by the pituitary CRF1. Therefore, Urocortin 3/SCP and Urocortin 2/SRP represent novel cardioprotective agents of therapeutic use in treating the damaging effects of cardiac ischaemia and associated diseases (Latchman, 2002).

Conclusions and future direction

Traditional methods in genetics and molecular biology focused on the characterization of individual genes and their products, whereas the new genomic approach takes advantage of our knowledge concerning the totality of genes in diverse organisms. With the completion of a number of genomes of different species, it becomes feasible to identify orthologous genes belonging to the CRF-signaling pathway throughout eukaryotic evolution. One approach to understanding the role of different family peptides is to consider the basic functions of individual genes in the context of evolution. Current evidence suggests that the origin of CRF-related peptides and their receptors predates insects and vertebrates, and CRF/Urocortin 1 and Urocortin 3 (SCP)/Urocortin 2(SRP) represent the two parallel branches of functional paralogs (Steckler, 2004). Therefore, future integration of the current knowledge on the CRF physiology with that of Urocortin 3/SCP and Urocortin 2/SRP is essential for the full understanding of stress regulation and the environmental adaptation in vertebrates.

The array of effects triggered by CRF/Urocortin 1 span almost all systems in the body, including nervous, endocrine, vasculature, muscular, bone, and immune. The complexity resembles programs that coordinate different tissues during development and these interacting ligands and receptors could play roles in a circuit that ultimately coordinates stress adaptation and homeostasis in different tissues in response to environmental changes. In the short time since their discovery, Urocortin 3/SCP and Urocortin 2/SRP have facilitated the experimental differentiation of a number of CRF receptor subtype-dependent functions (Steckler, 2004). This information serves as a first step for understanding the physiological pathways and functional properties of Urocortin 3 (SCP)/Urocortin 2(SRP)/CRF2 signaling. However, the exact CRF receptor(s) involved in a variety of CRF/urocortin responses remains to be studied (Bamberger and Bamberger, 2000; Baigent, 2001; Cullen et al., 2001; Schulman et al., 2002). Future studies using Urocortin 3/SCP and Urocortin 2/SRP will reveal the exact receptor type(s) important for the majority of CRF receptor-mediated responses and allow improved pharmacological treatments for diseases associated with abnormalities in CRF receptor action (Steckler, 2004). In addition, future generations of knockout mice in which Urocortin 3/SCP or Urocortin 2/SRP is disrupted will add valuable insight to this research. Likewise, studies on the integrated expression of these ligands and receptors in a tissue-specific and time-coordinated manner are important for the full understanding of this GPCR-signaling pathway. Only then can we gain a complete understanding of the functional overlap in these closely related ligands and receptors.

Acknowledgments

We thank Caren Spencer for editorial assistance and Dr. Aaron Hsueh for comments on the manuscript.

References

Adams, M.D., Celniker, S.E., Holt, R.A., Evans, C.A., Gocayne, J.D., Amanatides, P.G., Scherer, S.E., Li, P.W., Hoskins, R.A., Galle, R.F., George, R.A., Lewis, S.E., Richards, S., Ashburner, M., Henderson, S.N., Sutton, G.G., Wortman, J.R., Yandell, M.D., Zhang, Q., Chen, L.X., Brandon, R.C., Rogers, Y.H., Blazej, R.G., Champe, M., Pfeiffer, B.D., Wan, K.H., Doyle, C., Baxter, E.G., Helt, G., Nelson, C.R., Gabor, G.L., Abril, J.F., Agbayani, A., An, H.J., Andrews-Pfannkoch, C., Baldwin, D., Ballew, R.M., Basu, A., Baxendale, J., Bayraktaroglu, L., Beasley, E.M., Beeson, K.Y., Benos, P.V., Berman, B.P., Bhandari, D., Bolshakov, S., Borkova, D., Botchan, M.R., Bouck, J., Brokstein, P., Brottier, P., Burtis, K.C., Busam, D.A., Butler, H., Cadieu, E., Center, A., Chandra, I., Cherry, J.M., Cawley, S., Dahlke, C., Davenport, L.B., Davies, P., de Pablos, B., Delcher, A., Deng, Z., Mays, A.D., Dew, I., Dietz, S.M., Dodson, K., Doup, L.E., Downes, M., Dugan-Rocha, S., Dunkov, B.C., Dunn, P., Durbin, K.J., Evangelista, C.C., Ferraz, C., Ferriera, S., Fleischmann, W., Fosler, C., Gabrielian, A.E., Garg, N.S., Gelbart, W.M., Glasser, K., Glodek, A., Gong, F., Gorrell, J.H., Gu, Z., Guan, P., Harris, M., Harris, N.L., Harvey, D., Heiman, T.J., Hernandez, J.R., Houck, J., Hostin, D., Houston, K.A., Howland, T.J., Wei, M.H., Ibegwam, C., et al. (2000) The genome sequence of Drosophila melanogaster. Science, 287: 2185–2195.

Aggelidou, E., Hillhouse, E.W. and Grammatopoulos, D.K. (2002) Up-regulation of nitric oxide synthase and modulation of the guanylate cyclase activity by corticotropin-releasing hormone but not urocortin II or urocortin III in cultured human pregnant myometrial cells. Proc. Natl. Acad. Sci. USA, 99: 3300–3305.

Arai, M., Assil, I.Q. and Abou-Samra, A.B. (2001) Characterization of three corticotropin-releasing factor receptors in catfish: a novel third receptor is predominantly expressed in pituitary and urophysis. Endocrinology, 142: 446–454.

Audsley, N., Goldsworthy, G.J. and Coast, G.M. (1997) Circulating levels of Locusta diuretic hormone: the effects of feeding. Peptides, 18: 59–65.

Audsley, N., Kay, I., Hayes, T.K. and Coast, G.M. (1995) Cross reactivity studies of CRF-related peptides on insect malpighian tubules. Comp. Biochem. Physiol. A Physiol., 110: 87–93.

Baigent, S.M. (2001) Peripheral Corticotropin-releasing hormone and urocortin in the control of the immune response. Peptides, 22: 809–820.

Bale, T.L., Contarino, A., Smith, G.W., Chan, R., Gold, L.H., Sawchenko, P.E., Koob, G.F., Vale, W.W. and Lee, K.F. (2000) mice deficient for corticotropin-releasing hormone receptor-2 display anxiety-like behaviour and are hypersensitive to stress. Nat. Genet., 24: 410–414.

Bamberger, C.M. and Bamberger, A.M. (2000) The peripheral CRH/urocortin system. Ann. N. Y. Acad. Sci., 917: 290–296.

Behan, D.P., Cepoi, D., Fischer, W.H., Park, M., Sutton, S., Lowry, P.J. and Vale, W.W. (1996a) Characterization of a sheep brain corticotropin releasing factor binding protein (Brain Research 709 (1996) 265–274) (BRES 11959). Brain Res., 732: 267.

Behan, D.P., De Souza, E.B., Lowry, P.J., Potter, E., Sawchenko, P. and Vale, W.W. (1995) Corticotropin releasing factor (CRF) binding protein: a novel regulator of CRF and related peptides. Front. Neuroendocrinol., 16: 362–382.

Behan, D.P., De Souza, E.B., Potter, E., Sawchenko, P., Lowry, P.J. and Vale, W.W. (1996b) Modulatory actions of corticotropin-releasing factor-binding protein. Ann. N. Y. Acad. Sci., 780: 81–95.

Behan, D.P., Potter, E., Lewis, K.A., Jenkins, N.A., Copeland, N., Lowry, P.J. and Vale, W.W. (1993) Cloning and structure of the human corticotrophin releasing factor-binding protein gene (CRHBP). Genomics, 16: 63–68.

Bern, H.A., Pearson, D., Larson, B.A. and Nishioka, R.S. (1985) Neurohormones from fish tails: the caudal neurosecretory system. I. "Urophysiology" and the caudal neurosecretory system of fishes. Recent Prog. Horm. Res., 41: 533–552.

Boorse, G.C. and Denver, R.J. (2002) Acceleration of ambystoma tigrinum metamorphosis by corticotropin-releasing hormone. J. Exp. Zool., 293: 94–98.

Brar, B.K., Jonassen, A.K., Egorina, E.M., Chen, A., Negro, A., Perrin, M.H., Mjos, O.D., Latchman, D.S., Lee, K.F. and Vale, W. (2004) Urocortin-II and urocortin-III are cardioprotective against ischemia reperfusion injury: an essential endogenous cardioprotective role for corticotropin releasing factor receptor type 2 in the murine heart. Endocrinology, 145: 24–35; discussion 21–3.

Brar, B.K., Stephanou, A., Knight, R. and Latchman, D.S. (2002) Activation of protein kinase B/Akt by urocortin is essential for its ability to protect cardiac cells against hypoxia/reoxygenation-induced cell death. J. Mol. Cell Cardiol., 34: 483–492.

Brown, D.D., Wang, Z., Furlow, J.D., Kanamori, A., Schwartzman, R.A., Remo, B.F. and Pinder, A. (1996) The thyroid hormone-induced tail resorption program during Xenopus laevis metamorphosis. Proc. Natl. Acad. Sci. USA, 93: 1924–1929.

Chanalaris, A., Lawrence, K.M., Stephanou, A., Knight, R.D., Hsu, S.Y., Hsueh, A.J. and Latchman, D.S. (2003) Protective effects of the urocortin homologues stresscopin (SCP) and stresscopin-related peptide (SRP) against hypoxia/reoxygenation injury in rat neonatal cardiomyocytes. J. Mol. Cell Cardiol., 35: 1295–1305.

Chen, C.Y., Million, M., Adelson, D.W., Martinez, V., Rivier, J. and Tache, Y. (2002) Intracisternal urocortin inhibits vagally stimulated gastric motility in rats: role of CRF(2). Br. J. Pharmacol., 136: 237–247.

Chen, R., Lewis, K.A., Perrin, M.H. and Vale, W.W. (1993) Expression cloning of a human corticotropin-releasing-factor receptor. Proc. Natl. Acad. Sci. USA, 90: 8967–8971.

Contarino, A., Dellu, F., Koob, G.F., Smith, G.W., Lee, K.F., Vale, W. and Gold, L.H. (1999) Reduced anxiety-like and cognitive performance in mice lacking the corticotropin-releasing factor receptor 1. Brain Res., 835: 1–9.

Contarino, A., Dellu, F., Koob, G.F., Smith, G.W., Lee, K.F., Vale, W.W. and Gold, L.H. (2000) Dissociation of locomotor activation and suppression of food intake induced by CRF in CRFR1-deficient mice. Endocrinology, 141: 2698–2702.

Cortright, D.N., Nicoletti, A. and Seasholtz, A.F. (1995) Molecular and biochemical characterization of the mouse brain corticotropin-releasing hormone-binding protein. Mol. Cell Endocrinol., 111: 147–157.

Coste, S.C., Kesterson, R.A., Heldwein, K.A., Stevens, S.L., Heard, A.D., Hollis, J.H., Murray, S.E., Hill, J.K., Pantely, G.A., Hohimer, A.R., Hatton, D.C., Phillips, T.J., Finn, D.A., Low, M.J., Rittenberg, M.B., Stenzel, P. and Stenzel-Poore, M.P. (2000) Abnormal adaptations to stress and impaired cardiovascular function in mice lacking corticotropin-releasing hormone receptor-2. Nat. Genet., 24: 403–409.

Coste, S.C., Quintos, R.F. and Stenzel-Poore, M.P. (2002) Corticotropin-releasing hormone-related peptides and receptors: emergent regulators of cardiovascular adaptations to stress. Trends Cardiovasc. Med., 12: 176–182.

Cullen, M.J., Ling, N., Foster, A.C. and Pelleymounter, M.A. (2001) Urocortin, corticotropin releasing factor-2 receptors and energy balance. Endocrinology, 142: 992–999.

Darlison, M.G. and Richter, D. (1999) Multiple genes for neuropeptides and their receptors: co-evolution and physiology. Trends Neurosci., 22: 81–88.

Dautzenberg, F.M., Dietrich, K., Palchaudhuri, M.R. and Spiess, J. (1997) Identification of two corticotropin-releasing factor receptors from Xenopus laevis with high ligand selectivity: unusual pharmacology of the type 1 receptor. J. Neurochem., 69: 1640–1649.

Dautzenberg, F.M. and Hauger, R.L. (2002) The CRF peptide family and their receptors: yet more partners discovered. Trends Pharmacol. Sci., 23: 71–77.

Denver, R.J. (1997) Environmental stress as a developmental cue: corticotropin-releasing hormone is a proximate mediator of adaptive phenotypic plasticity in amphibian metamorphosis. Horm. Behav., 31: 169–179.

Denver, R.J. (1999) Evolution of the corticotropin-releasing hormone signaling system and its role in stress-induced phenotypic plasticity. Ann. N. Y. Acad. Sci., 897: 46–53.

Dunn, A.J. and Swiergiel, A.H. (1999) Behavioral responses to stress are intact in CRF-deficient mice. Brain Res., 845: 14–20.

Fischer, W.H., Behan, D.P., Park, M., Potter, E., Lowry, P.J. and Vale, W. (1994) Assignment of disulfide bonds in corticotropin-releasing factor-binding protein. J. Biol. Chem., 269: 4313–4316.

Furutani, Y., Morimoto, Y., Shibahara, S., Noda, M., Takahashi, H., Hirose, T., Asai, M., Inayama, S., Hayashida, H., Miyata, T. and Numa, S. (1983) Cloning and sequence analysis of cDNA for ovine corticotropin-releasing factor precursor. Nature, 301: 537–540.

Gordon, J.M., Dusting, G.J., Woodman, O.L. and Ritchie, R.H. (2002) Cardioprotective action of the CRF peptide urocortin against simulated ischemia in adult rat cardiomyocytes. Am. J. Physiol. Heart Circ. Physiol.

Guillemin, R. (1967) The Adenohypophysis and its hypothalamic control. Annu. Rev. Physiol., 29: 313–348.

Hsu, S.Y. and Hsueh, A.J. (2001) Human stresscopin and stresscopin-related peptide are selective ligands for the type 2 corticotropin-releasing hormone receptor. Nat. Med., 7: 605–611.

Ichikawa, T., McMaster, D., Lederis, K. and Kobayashi, H. (1982) Isolation and amino acid sequence of urotensin I, a vasoactive and ACTH-releasing neuropeptide, from the carp (Cyprinus carpio) urophysis. Peptides, 3: 859–867.

Karolyi, I.J., Burrows, H.L., Ramesh, T.M., Nakajima, M., Lesh, J.S., Seong, E., Camper, S.A. and Seasholtz, A.F. (1999) Altered anxiety and weight gain in corticotropin-releasing hormone-binding protein-deficient mice. Proc. Natl. Acad. Sci. USA, 96: 11595–11600.

Kimura, Y., Takahashi, K., Totsune, K., Muramatsu, Y., Kaneko, C., Darnel, A.D., Suzuki, T., Ebina, M., Nukiwa, T. and Sasano, H. (2002) Expression of urocortin and corticotropin-releasing factor receptor subtypes in the human heart. J. Clin. Endocrinol. Metab., 87: 340–346.

Kishimoto, T., Radulovic, J., Radulovic, M., Lin, C.R., Schrick, C., Hooshmand, F., Hermanson, O., Rosenfeld, M.G. and Spiess, J. (2000) Deletion of crhr2 reveals an anxiolytic role for corticotropin-releasing hormone receptor-2. Nat. Genet., 24: 415–419.

Latchman, D.S. (2002) Urocortin. Int. J. Biochem. Cell Biol., 34: 907–910.

Lau, S.H., Rivier, J., Vale, W., Kaiser, E.T. and Kezdy, F.J. (1983) Surface properties of an amphiphilic peptide hormone and of its analog: corticotropin-releasing factor and sauvagine. Proc. Natl. Acad. Sci. USA, 80: 7070–7074.

Lederis, K., Letter, A., McMaster, D., Ichikawa, T., MacCannell, K.L., Kobayashi, Y., Rivier, J., Rivier, C., Vale, W. and Fryer, J. (1983) Isolation, analysis of structure, synthesis, and biological actions of urotensin I neuropeptides. Can. J. Biochem. Cell Biol., 61: 602–614.

Lederis, K., Vale, W., Rivier, J., MacCannell, K.L., McMaster, D., Kobayashi, Y., Suess, U. and Lawrence, J. (1982) Urotensin I – a novel CRF-like peptide in catostomus commersoni urophysis. Proc. West Pharmacol. Soc., 25: 223–227.

Lenz, H.J., Fisher, L.A., Vale, W.W. and Brown, M.R. (1985) Corticotropin-releasing factor, sauvagine, and urotensin I: effects on blood flow. Am. J. Physiol., 249: R85–90.

Leo, C.P., Hsu, S.Y. and Hsueh, A.J. (2002) Hormonal genomics. Endocr. Rev., 23: 369–381.

Lewis, K., Li, C., Perrin, M.H., Blount, A., Kunitake, K., Donaldson, C., Vaughan, J., Reyes, T.M., Gulyas, J., Fischer, W., Bilezikjian, L., Rivier, J., Sawchenko, P.E. and Vale, W.W. (2001) Identification of urocortin III, an additional member of the corticotropin-releasing factor (CRF) family with high affinity for the CRF2 receptor. Proc. Natl. Acad. Sci. USA, 98: 7570–7575.

Li, C., Vaughan, J., Sawchenko, P.E. and Vale, W.W. (2002) Urocortin III-immunoreactive projections in rat brain: partial overlap with sites of type 2 corticotrophin-releasing factor receptor expression. J. Neurosci., 22: 991–1001.

Liaw, C.W., Lovenberg, T.W., Barry, G., Oltersdorf, T., Grigoriadis, D.E. and de Souza, E.B. (1996) Cloning and characterization of the human corticotropin-releasing factor-2 receptor complementary deoxyribonucleic acid. Endocrinology, 137: 72–77.

Lovejoy, D.A. and Balment, R.J. (1999) Evolution and physiology of the corticotropin-releasing factor (CRF) family of neuropeptides in vertebrates. Gen. Comp. Endocrinol., 115: 1–22.

Lovenberg, T.W., Liaw, C.W., Grigoriadis, D.E., Clevenger, W., Chalmers, D.T., De Souza, E.B. and Oltersdorf, T. (1995) Cloning and characterization of a functionally distinct corticotropin-releasing factor receptor subtype from rat brain. Proc. Natl. Acad. Sci. USA, 92: 836–840.

Mainoya, J.R. and Bern, H.A. (1982) Effects of teleost urotensins on intestinal absorption of water and nacl in tilapia, sarotherodon mossambicus, adapted to fresh water or seawater. Gen. Comp. Endocrinol., 47: 54–58.

Martinez, V., Wang, L., Rivier, J.E., Vale, W. and Tache, Y. (2002) Differential actions of peripheral corticotropin-releasing factor (CRF), Urocortin II, and Urocortin III on gastric emptying and colonic transit in mice: role of CRF receptor subtypes 1 and 2. J. Pharmacol. Exp. Ther., 301: 611–617.

Miampamba, M., Maillot, C., Million, M. and Tache, Y. (2002) Peripheral CRF activates myenteric neurons in the proximal colon through CRF(1) receptor in conscious rats. Am. J. Physiol. Gastrointest. Liver Physiol., 282: G857–865.

Million, M., Maillot, C., Saunders, P., Rivier, J., Vale, W. and Tache, Y. (2002) Human Urocortin II, a new CRF-related peptide, displays selective CRF(2)-mediated action on gastric transit in rats. Am. J. Physiol. Gastrointest. Liver Physiol., 282: G34–40.

Montecucchi, P.C. and Henschen, A. (1981) Amino acid composition and sequence analysis of sauvagine, a new active peptide from the skin of Phyllomedusa sauvagei. Int. J. Pept. Protein Res., 18: 113–120.

Muglia, L.J., Bethin, K.E., Jacobson, L., Vogt, S.K. and Majzoub, J.A. (2000) Pituitary-adrenal axis regulation in CRH-deficient mice. Endocr. Res., 26: 1057–1066.

Muller, M.B., Keck, M.E., Zimmermann, S., Holsboer, F. and Wurst, W. (2000) Disruption of feeding behavior in CRH receptor 1-deficient mice is dependent on glucocorticoids. Neuroreport, 11: 1963–1966.

Negri, L., Melchiorri, P., Montecucchi, P.C., Henschen, A., D'Urso, R., Iellamo, R. and Falaschi, P. (1983) Sauvagine induces release of adrenocorticotropin, beta-endorphin and corticosterone in rats. Pharmacol. Res. Commun., 15: 427–438.

Palchaudhuri, M.R., Hauger, R.L., Wille, S., Fuchs, E. and Dautzenberg, F.M. (1999) Isolation and pharmacological characterization of two functional splice variants of corticotropin-releasing factor type 2 receptor from tupaia belangeri. J. Neuroendocrinol., 11: 419–428.

Pedersen, W.A., Wan, R., Zhang, P. and Mattson, M.P. (2002) Urocortin, but not urocortin II, protects cultured hippocampal neurons from oxidative and excitotoxic cell death via corticotropin-releasing hormone receptor type I. J. Neurosci., 22: 404–412.

Peterson, R.E. and Guillemin, R. (1974) The hormones of the hypothalamus. Am. J. Med., 57: 591–600.

Pisarchik, A. and Slominski, A. (2002) Corticotropin releasing factor receptor type 1: molecular cloning and investigation of alternative splicing in the hamster skin. J. Invest. Dermatol., 118: 1065–1072.

Pisarchik, A. and Slominski, A.T. (2001) Alternative splicing of CRH-R1 receptors in human and mouse skin: identification of new variants and their differential expression. Faseb. J., 15: 2754–2756.

Pohl, S., Darlison, M.G., Clarke, W.C., Lederis, K. and Richter, D. (2001) Cloning and functional pharmacology of two corticotropin-releasing factor receptors from a teleost fish. Eur. J. Pharmacol., 430: 193–202.

Potter, E., Behan, D.P., Fischer, W.H., Linton, E.A., Lowry, P.J. and Vale, W.W. (1991) Cloning and characterization of the cDNAs for human and rat corticotropin releasing factor-binding proteins. Nature, 349: 423–426.

Potter, E., Sutton, S., Donaldson, C., Chen, R., Perrin, M., Lewis, K., Sawchenko, P.E. and Vale, W. (1994) Distribution of corticotropin-releasing factor receptor mRNA expression in the rat brain and pituitary. Proc. Natl. Acad. Sci. USA, 91: 8777–8781.

Reagan, J.D. (1994) Expression cloning of an insect diuretic hormone receptor. A member of the calcitonin/secretin receptor family. J. Biol. Chem., 269: 9–12.

Reagan, J.D. (1995) Functional expression of a diuretic hormone receptor in baculovirus-infected insect cells: evidence suggesting that the N-terminal region of diuretic hormone is associated with receptor activation. Insect Biochem. Mol. Biol., 25: 535–539.

Reagan, J.D. (1996) Molecular cloning and function expression of a diuretic hormone receptor from the house cricket, Acheta domesticus. Insect. Biochem. Mol. Biol., 26: 1–6.

Reyes, T.M., Lewis, K., Perrin, M.H., Kunitake, K.S., Vaughan, J., Arias, C.A., Hogenesch, J.B., Gulyas, J., Rivier, J., Vale, W.W. and Sawchenko, P.E. (2001) Urocortin II: a member of the corticotropin-releasing factor (CRF) neuropeptide family that is selectively bound by type 2 CRF receptors. Proc. Natl. Acad. Sci. USA, 98: 2843–2848.

Riehle, M.A., Garczynski, S.F., Crim, J.W., Hill, C.A. and Brown, M.R. (2002) Neuropeptides and peptide hormones in Anopheles gambiae. Science, 298: 172–175.

Rivier, C. (2001) Role of gaseous neurotransmitters in the hypothalamic-pituitary-adrenal axis. Ann. N. Y. Acad. Sci., 933: 254–264.

Saffran, M. and Schally, A.V. (1977) The status of the corticotropin releasing factor (CRF). Neuroendocrinology, 24: 359–375.

Saunders, P.R., Maillot, C., Million, M. and Tache, Y. (2002) Peripheral corticotropin-releasing factor induces diarrhea in rats: role of CRF1 receptor in fecal watery excretion. Eur. J. Pharmacol., 435: 231–235.

Scarabelli, T.M., Pasini, E., Stephanou, A., Comini, L., Curello, S., Raddino, R., Ferrari, R., Knight, R. and Latchman, D.S. (2002) Urocortin promotes hemodynamic and bioenergetic recovery and improves cell survival in the isolated rat heart exposed to ischemia/reperfusion. J. Am. Coll. Cardiol., 40: 155–161.

Schulman, D., Latchman, D.S. and Yellon, D.M. (2002) Urocortin protects the heart from reperfusion injury via upregulation of p42/p44 MAPK signaling pathway. Am. J. Physiol. Heart Circ. Physiol., 283: H1481–1488.

Seasholtz, A.F., Valverde, R.A. and Denver, R.J. (2002) Corticotropin-releasing hormone-binding protein: biochemistry and function from fishes to mammals. J. Endocrinol., 175: 89–97.

Sherwood, N.M., Krueckl, S.L. and McRory, J.E. (2000) The origin and function of the pituitary adenylate cyclase-activating polypeptide (PACAP)/glucagon superfamily. Endocr. Rev., 21: 619–670.

Shibahara, S., Morimoto, Y., Furutani, Y., Notake, M., Takahashi, H., Shimizu, S., Horikawa, S. and Numa, S. (1983) Isolation and sequence analysis of the human corticotropin-releasing factor precursor gene. Embo. J., 2: 775–779.

Smith, G.W., Aubry, J.M., Dellu, F., Contarino, A., Bilezikjian, L.M., Gold, L.H., Chen, R., Marchuk, Y., Hauser, C., Bentley, C.A., Sawchenko, P.E., Koob, G.F., Vale, W. and Lee, K.F. (1998) Corticotropin releasing factor receptor 1-deficient mice display decreased anxiety, impaired stress response, and aberrant neuroendocrine development. Neuron, 20: 1093–1102.

Spiess, J., Rivier, J., Rivier, C. and Vale, W. (1981) Primary structure of corticotropin-releasing factor from ovine hypothalamus. Proc. Natl. Acad. Sci. USA, 78: 6517–6521.

Steckler, T. (2005). CRF antagonists as novel treatment strategies for stress-related disorders. In: Steckler, T., Kalin, N. and Reul, J.M.H.M. (Eds.), Handbook of Stress and the Brain, Part 1, Elsevier, Amsterdam, pp. 371–400.

Suman-Chauhan, N., Carnell, P., Franks, R., Webdale, L., Gee, N.S., McNulty, S., Rossant, C.J., Van Leeuwen, D., MacKenzie, R. and Hall, M.D. (1999) Expression and characterisation of human and rat CRF2alpha receptors. Eur. J. Pharmacol., 379: 219–227.

Swiergiel, A.H. and Dunn, A.J. (1999) CRF-deficient mice respond like wild-type mice to hypophagic stimuli. Pharmacol. Biochem. Behav., 64: 59–64.

Terui, K., Higashiyama, A., Horiba, N., Furukawa, K.I., Motomura, S. and Suda, T. (2001) Coronary vasodilation and positive inotropism by urocortin in the isolated rat heart. J. Endocrinol., 169: 177–183.

Timpl, P., Spanagel, R., Sillaber, I., Kresse, A., Reul, J.M., Stalla, G.K., Blanquet, V., Steckler, T., Holsboer, F. and Wurst, W. (1998) Impaired stress response and reduced anxiety in mice lacking a functional corticotropin-releasing hormone receptor 1. Nat. Genet., 19: 162–166.

Turnbull, A.V., Smith, G.W., Lee, S., Vale, W.W., Lee, K.F. and Rivier, C. (1999) CRF type I receptor-deficient mice exhibit a pronounced pituitary-adrenal response to local inflammation. Endocrinology, 140: 1013–1017.

Valdenaire, O., Giller, T., Breu, V., Gottowik, J. and Kilpatrick, G. (1997) A new functional isoform of the human CRF2 receptor for corticotropin-releasing factor. Biochim. Biophys. Acta, 1352: 129–132.

Valdez, G.R., Inoue, K., Koob, G.F., Rivier, J., Vale, W. and Zorrilla, E.P. (2002) Human urocortin II: mild locomotor suppressive and delayed anxiolytic-like effects of a novel corticotropin-releasing factor related peptide. Brain Res., 943: 142–150.

Vale, W., Spiess, J., Rivier, C. and Rivier, J. (1981) Characterization of a 41-residue ovine hypothalamic peptide that stimulates secretion of corticotropin and beta-endorphin. Science, 213: 1394–1397.

Valverde, R.A., Seasholtz, A.F., Cortright, D.N. and Denver, R.J. (2001) Biochemical characterization and expression analysis of the Xenopus laevis corticotropin-releasing hormone binding protein. Mol. Cell Endocrinol., 173: 29–40.

Vaughan, J., Donaldson, C., Bittencourt, J., Perrin, M.H., Lewis, K., Sutton, S., Chan, R., Turnbull, A.V., Lovejoy, D., Rivier, C., Rivier, J., Sawenchko, P.E. and Vale, W. (1995) Urocortin, a mammalian neuropeptide related to fish urotensin I and to corticotropin-releasing factor. Nature, 378: 287–292.

Vetter, D.E., Li, C., Zhao, L., Contarino, A., Liberman, M.C., Smith, G.W., Marchuk, Y., Koob, G.F., Heinemann, S.F., Vale, W. and Lee, K.F. (2002) Urocortin-deficient mice show hearing impairment and increased anxiety-like behavior. Nat. Genet, 31: 363–369.

Vita, N., Laurent, P., Lefort, S., Chalon, P., Dumont, X., Kaghad, M., Gully, D., Le Fur, G., Ferrara, P. and Caput, D. (1993) Cloning and expression of a complementary DNA encoding a high affinity human neurotensin receptor. FEBS Lett., 317: 139–142.

Wang, L., Martinez, V., Rivier, J.E. and Tache, Y. (2001) Peripheral urocortin inhibits gastric emptying and food intake in mice: differential role of CRF receptor 2. Am. J. Physiol. Regul. Integr. Comp. Physiol., 281: R1401–1410.

Wang, X., Su, H., Copenhagen, L.D., Vaishnav, S., Pieri, F., Shope, C.D., Brownell, W.E., De Biasi, M., Paylor, R. and Bradley, A. (2002) Urocortin-deficient mice display normal stress-induced anxiety behavior and autonomic control but an impaired acoustic startle response. Mol. Cell Biol., 22: 6605–6610.

Wiehart, U.I., Nicolson, S.W., Eigenheer, R.A. and Schooley, D.A. (2002) Antagonistic control of fluid secretion by the Malpighian tubules of Tenebrio molitor: effects of diuretic and antidiuretic peptides and their second messengers. J. Exp. Biol., 205: 493–501.

T. Steckler, N.H. Kalin and J.M.H.M. Reul (Eds.)
Handbook of Stress and the Brain, Vol. 15
ISBN 0-444-51173-3

CHAPTER 2.2

Molecular regulation of the CRF system

P.H. Roseboom[1,3,*], N.H. Kalin[1,2], T. Steckler[4] and F.M. Dautzenberg[4,†]

[1]Department of Psychiatry, University of Wisconsin, 6001 Research Park Blvd, Madison, WI 53719, USA
[2]Department of Psychology, University of Wisconsin, 1202 West Johnson Street, Madison, WI 53706, USA
[3]Department of Pharmacology, University of Wisconsin, 1300 University Avenue, Madison, WI 53706, USA
[4]Johnson and Johnson Pharmaceutical Research & Development, Janssen Pharmaceutica N.V., Turnhoutseweg 30, 2340 Beerse, Belgium

Abstract: The corticotropin-releasing factor (CRF) system, including the urocortin peptides, is a key regulator of how the body responds to stress. A large amount of preclinical data in rodents and nonhuman primates implicates the CRF system in mediating the various physiological and psychological responses to stress. Importantly, alterations in the CRF system are associated with depression and anxiety disorders in humans. The goal of this chapter is to review what is known about the molecular mechanisms that regulate the activity of the CRF system. The role of the CRF system in stress-induced psychopathology is initially reviewed. Stress-induced molecular changes that are associated with activation of the CRF system are then described, along with the effects of manipulations that mimic or block the effects of stress. CRF receptor regulation is outlined in detail including an overview of recent data implicating the role of G-protein receptor kinase 3 in the phosphorylation and desensitization of the CRF_1 receptor. The limited data on the regulation of the CRF_2 receptors is also described. Finally, preliminary data from the use of microarrays and gene chips aimed at identifying stress-induced changes in gene expression that are CRF receptor dependent or independent will be described. A detailed understanding of the molecular mechanisms that mediate the stress-induced changes in the CRF system will enable identification of novel targets for the treatment and prevention of stress-related disorders.

The CRF system and stress-induced psychopathology

The corticotropin-releasing factor (CRF) system is one of the major peptide systems implicated in regulating the stress response. CRF is a 41-amino acid peptide that was originally discovered as a novel hypothalamic factor controlling the release of pituitary proopiomelanocortin peptides (Guillemin and Rosenberg, 1955; Saffran and Schally, 1956; Vale et al., 1981). It has since been found to play an important role in coordinating various components of the stress response (Dunn and Berridge, 1990; De Souza, 1995; Kalin, 1997; Koob and Heinrichs, 1999). Numerous clinical and preclinical reports indicate that the CRF system mediates behavioral, autonomic, neuroendocrine, and immune responses to stress. Furthermore, alterations in the CRF system are often associated with stress-related psychopathology, such as depression and anxiety disorders (Mitchell, 1998; Arborelius et al., 1999).

More recently, additional components of the CRF system have been identified. CRF and the related endogenous peptide agonists urocortin 1 (Vaughan et al., 1995), urocortin 2 (Reyes et al., 2001), and urocortin 3 (Lewis et al., 2001) bind to the two cloned CRF receptors, designated CRF_1 and CRF_2 (Chen et al., 1993; Perrin et al., 1995). A CRF-binding protein (CRF-BP) that putatively buffers the effects

*Corresponding author. E-mail: roseboom@wisc.edu
†E-mail: fdautzen@prdbe.jnj.com

of endogenous CRF system ligands has also been identified (Potter et al., 1991). Several isoforms of the receptors have been cloned including $CRF_{2(a)}$, $CRF_{2(b)}$, and $CRF_{2(c)}$ receptors (Lovenberg et al., 1995a; Kostich et al., 1998; Hauger et al., 2003b). $CRF_{2(a)}$ and $CRF_{2(c)}$ are the subtypes that predominate in the brain (Lovenberg et al., 1995b; Kostich et al., 1998). Urocortin 1, urocortin 2, and urocortin 3 are likely important in mediating stress-induced behavior because studies have noted that CRF knockout mice exhibit normal stress-induced behavioral responses that can be blocked by CRF receptor antagonism (Weninger et al., 1999).

There is a great deal of clinical and preclinical evidence implicating the CRF system in stress-induced psychopathology, such as depression and anxiety disorders (for review Mitchell, 1998; Arborelius et al., 1999). Elevated levels of CRF in the cerebrospinal fluid of depressed patients have been reported (Nemeroff et al., 1984), and increased numbers of CRF-expressing neurons in the paraventricular nucleus of the hypothalamus (PVN) of depressed patients have been observed (Raadsheer et al., 1994). In addition, altered levels of CRF binding sites in the brains of suicide victims have been reported (Nemeroff, 1988). Recent efforts are directed toward developing CRF_1 antagonists for treating stress-related psychiatric disorders (McCarthy et al., 1999; Steckler, 2005). The first open label trial of such a compound in depressed patients revealed significant reductions in depression and anxiety after CRF_1 antagonist treatment (Zobel et al., 2000).

In addition to clinical data, there are numerous preclinical studies implicating the CRF system in stress-induced psychopathology. Early studies administering CRF intraventricularly to rodents and primates reported that this peptide caused marked behavioral effects that mimicked fear-related and depressive responses (Kalin et al., 1983; Sherman and Kalin, 1986; Dunn and Berridge, 1990; Koob et al., 1993). Infusion of nonselective CRF antagonists such as alpha-helical CRF_{9-41} blocked the effects of stress on behavior (Kalin, 1985; Koob and Heinrichs, 1999). CRF-overexpressing mice demonstrate behavioral phenotypes that are consistent with increased levels of stress (Stenzel-Poore et al., 1992, 1994; van der Meer et al., 2001; van Gaalen et al., 2002), while CRF_1 receptor knockout mice display responses indicative of decreased anxiety-like behavior (Smith et al., 1998; Timpl et al., 1998; Contarino et al., 1999). Data from recent site-specific antagonist administration and antisense oligonucleotide studies suggest that the $CRF_{2(a)}$ receptor has a role similar to that of the CRF_1 receptor in mediating fearful behaviors (Ho et al., 2001; Bakshi et al., 2002). It should also be noted that the CRF and urocortin peptides mimic the anorectic effects of stress most likely through activation of the CRF_2 receptor (Krahn et al., 1986; Arase et al., 1988; Spina, 1996).

Results from three reports of CRF_2 receptor knockout mice (Bale et al., 2000; Coste et al., 2000; Kishimoto et al., 2000) are inconsistent. Two of these studies suggest that the behavioral profile of CRF_2 receptor knockouts increases stress-like responding as evidenced by a decrease in open arm entries in the elevated plus maze (Bale et al., 2000; Kishimoto et al., 2000). However, other aspects of the behavioral profile indicate either no alteration of stress-related responding (Bale et al., 2000; Coste et al., 2000), or a decrease in anxiety-like behaviors (Kishimoto et al., 2000).

There has been one study characterizing mice deficient in both CRF_1 and CRF_2 receptors (Bale et al., 2002). These mice display a greater impairment of the hypothalamic–pituitary–adrenal (HPA) axis response to stress than CRF_1 knockout mice. The behavioral effects described are complex and sexually dichotomous, with female double-mutant mice displaying anxiolytic-like behavior and male double-mutant mice demonstrating significantly greater anxiety-like behaviors compared to females. In addition, this study hypothesized that the CRF_2 receptor genotype of the mothers had a significant effect on the anxiety-like behaviors of the male pups regardless of the pup's genotype. Male pups born from dams that have one (heterozygous) or both copies (homozygous) of the CRF_2 receptor gene knocked out display significantly more anxiety-like behavior (Bale et al., 2002). Nevertheless in the light of the rather inconsistent results obtained with the CRF_2 receptor knockout mice, data obtained from transgenic animals should be treated with caution and might not reflect the situation in animals that carried the gene through their entire development. Small molecule CRF_1 and CRF_2 receptor-specific antagonists might better address questions concerning the involvement of both receptors in the development of stress-related and depressive disorders. In addition, studies have shown that the central distribution of CRF_2 receptors in humans and primates is much broader than in rodents (Kostich et al., 1998; Palchaudhuri et al., 1999; Sanchez et al., 1999) suggesting a more prominent role for CRF_2 receptors in mediating behavior in primates and humans.

In summary, there is a great deal of preclinical data indicating the importance of the various components of the CRF system in mediating the various physiological and behavioral responses to stress. A detailed understanding of the roles played by each of the two known CRF receptors is emerging, and evidence is accumulating that implicates the CRF system in stress-induced psychopathologies, such as depression and anxiety disorders.

Stress-mediated activation of the CRF system

The data above indicate the necessity for understanding how stress exerts its effects on the CRF system. There is considerable evidence that stress activates the hypothalamic CRF system, resulting in regulation of the HPA axis. In response to an acute stress, CRF that is produced in the PVN is secreted into the median eminence, enters the portal vessels, and is transported to CRF_1 receptor binding sites in the anterior pituitary. Binding of CRF to the CRF_1 receptor and subsequent receptor activation leads to increased intracellular cAMP levels within pituitary corticotrophs. This results in immediate release of adrenocorticotropic hormone (ACTH) into the general circulation, and also a long-term increase in ACTH synthesis. The circulating ACTH then binds to its specific receptor, the melanocortin 2 receptor, in the adrenal cortex and thereby stimulates the release of glucocorticoids. Glucocorticoids (i.e. cortisol) distribute throughout the body and produce physiological changes to enable the body to acutely respond to the stressful situation. In addition, via a negative receptor feedback loop to the brain, glucocorticoids inhibit further activation of the HPA system (Kaplan, 1992).

Activation of the hypothalamic CRF system is evidenced by an increase in CRF gene expression in the PVN. A wide variety of acute psychological, physical, physiological, or immunological stressors can produce an increase in CRF mRNA levels in the PVN (for review Bakshi and Kalin, 2000). Outside the hypothalamus, the components of the CRF system are widely expressed throughout the brain. In some instances, the extrahypothalamic CRF system is regulated differently from the hypothalamic CRF system. For example, stress-induced release of CRF from the amygdala appears to occur under a more limited set of circumstances compared to stress-induced release of CRF from the PVN (Bakshi and Kalin, 2000). Extrahypothalamic CRF is thought to play an important role in a variety of psychological and physiological responses to stress. The main question is how exposure to acute and repeated stress affects the expression of CRF and other components of the CRF system.

Acute stress

In rats, restraint stress for 1 h or less or ethanol withdrawal have been shown to increase the release of CRF peptide and the expression of CRF mRNA in the amygdala (Kalin et al., 1994; Pich et al., 1995; Hsu et al., 1998; Merali et al., 1998) and in various thalamic nuclei (Hsu et al., 2001). This stress-induced increase in CRF peptide release occurs immediately with onset of restraint (Pich et al., 1995; Merali et al., 1998) and mRNA changes are seen as early as 1 h after cessation of stress (Hsu et al., 1998). Three hours of restraint stress also increases rat urocortin 1 mRNA in the Edinger-Westphal nucleus (Weninger et al., 2000). CRF_1 receptor mRNA is increased in the PVN following a variety of acute stressors (Makino et al., 1995; Aguilera et al., 1997; Bonaz and Rivest, 1998). Restraint stress increases mRNA expression for CRF-BP in the pituitary and amygdala (Lombardo et al., 2001), and food deprivation increases CRF-BP mRNA in the amygdala and hypothalamic regions (Timofeeva et al., 1999).

Repeated stress

The effects of repeated stress on expression of CRF mRNA have been most extensively studied. Repeated stress (restraint or immobilization) produces elevation in basal levels of CRF mRNA levels in the PVN (Mamalaki et al., 1992; Bartanusz et al., 1993). Although acute stress elevates CRF mRNA in the amygdala, repeated stress has not been found to alter basal levels of amygdalar CRF mRNA (Mamalaki et al., 1992). However, chronic social stress produces an elevation of CRF mRNA in this brain area, but decreases CRF mRNA levels in the PVN (Albeck et al., 1997). In the pituitary, two weeks of restraint stress decreased CRF_1 receptor mRNA (Makino et al., 1995). However, other groups were unable to replicate this effect (Aguilera et al., 1997). In certain strains of mice, cortical levels of CRF_1 receptor mRNA increase in response to repeated restraint stress (Giardino et al., 1996). In rats, however, CRF_1 receptor mRNA levels in several extrahypothalamic regions including the cortex, amygdala, and hippocampus appear not to be affected by repeated restraint stress (Makino et al., 1995; Iredale et al., 1996). However, when the repeated stress paradigm consists of a variable, unpredictable, multimodal regimen, a significant reduction in CRF_1 receptor gene expression occurs in the cortex, and a significant increase of transcript is seen in the hippocampus (Iredale et al., 1996).

Only one study reported the effects of repeated immobilization stress on $CRF_{2(a)}$ receptor gene expression and showed a small (12%) decrease in $CRF_{2(a)}$ receptor mRNA in the ventromedial hypothalamus with stress (Makino et al., 1999).

Effects of repeated stress on response to acute stress

Repeated exposure of rats to stress prevents the acute stress-induced increase of CRF mRNA in the amygdala when the acute stressor is identical to the repeated stressor (Mamalaki et al., 1992). Similarly, the acute stress-induced increase in CRF-BP mRNA in the amygdala is blocked by prior exposure to repeated restraint stress (Lombardo et al., 2001). The acute stress-induced increase in CRF mRNA in the PVN is significantly dampened by prior exposure to repeated restraint (Ma et al., 1997, 1999). However, the acute response is still present if the challenge is with a heterotypic stressor (Ma et al., 1999). There is also evidence for some involvement of the arginine vasopressin system in regulating the activity of the HPA axis in response to repeated stress. This system controls how the HPA axis responds to a homotypic stressor; however, the CRF system controls the HPA axis response to a heterotypic stressor (Bakshi and Kalin, 2000).

Developmental stress

A large body of literature exists indicating that separating both animals and humans from their mothers is a significant stressor that can negatively impact the emotional development of the infant (Bowlby, 1973; Carlson and Earls, 1997). In rat pups, during the stress hyporesponsive period (postnatal days 3–14), an intense stressor, such as 24 h of maternal separation, is necessary to activate the HPA axis. This results in decreased expression of CRF mRNA in the PVN (Smith et al., 1997; van Oers et al., 1998; Dent et al., 2000). In contrast, restraint stress of 30 min or less given to postnatal day 12 pups immediately following 24 h of maternal deprivation results in an increase in CRF mRNA in the PVN. This increase was seen 15 min after the cessation of the stress compared to nonrestrained controls (Dent et al., 2000). CRF mRNA levels are also increased in the amygdala of pups that have been exposed to a maternal deprivation/cold stress exposure (Hatalski et al., 1998). The stress of maternal deprivation results in heightened responsiveness of the CRF system to stressors that occur after the stress hyporesponsive period. For example, rat pups that are subjected to maternal separation on postnatal day 3 demonstrate an increase in CRF mRNA expression in the PVN relative to nondeprived rats that are stressed two weeks later (van Oers et al., 1998).

Alterations in the responsiveness of the CRF system following maternal deprivation of rat pups is maintained into adulthood (Anisman et al., 1998; Francis et al., 1999b; Heim and Nemeroff, 1999). Interestingly, the direction of the change in responsiveness is dependent upon the length of maternal separation. Short bouts of separation from the mother (3–15 min per bout once a day for 2 weeks) results in a profile in adulthood that is indicative of lower levels of anxiety. These rats also have decreased basal levels of hypothalamic CRF mRNA and median eminence CRF peptide as adults compared to control rats (Plotsky and Meaney, 1993). In contrast, when separation from the mother is extended to 3 h or more, the pups develop into adults displaying increased stress-like responding. This change is associated with increased CRF gene expression (Plotsky and Meaney, 1993; Rots et al., 1996; Wigger and Neumann, 1999).

The intense stress of an endotoxin insult to rat pups elevates basal levels of CRF gene expression and enhances stress-induced HPA axis response in adulthood (Shanks et al., 1995). Similar findings have been reported in nonhuman primates. CSF levels of CRF are basally and persistently elevated in adult macaques whose mothers have been exposed to 3 months of variable foraging demand compared to macaques whose mothers faced a predictable foraging demand (Coplan et al., 1996, 2000). This condition results in greater uncertainty in food availability for mother and infant and likely alters mother–infant interactions. In addition, macaques raised by variable foraging demand-exposed mothers develop impaired affiliative social behaviors in adulthood (Andrews and Rosenblum, 1994). In rodents it has been shown that mothering styles can be passed on to offspring by nongenomic modes of transmission. For example, it has been shown that pups will develop the mothering style of the mother who raises them regardless of whether she is their biological mother (Francis et al., 1999a). A similar pattern of behavioral transmission may occur in primates. In rhesus monkeys there is a relationship between birth order and cortisol concentration with an early position in the birth order associated with higher basal cortisol concentrations (Kalin et al., 1998).

In summary, it has been proposed that the perinatal environment and mothering style can shape the offspring's stress-coping system that persists into adulthood and in part this is related to alterations in the CRF system. It is likely that, at some level, the setting of the stress-coping system involves the CRF system. An intriguing suggestion from this theory is that stress in the early days of life has a profound impact on how an individual responds to stress for the remainder of its life (Anisman et al., 1998; Francis et al., 1999b; Heim and Nemeroff, 1999). It is possible that stressors occurring early in life may have a more profound impact on the stress–response profile than any effects of acute or prolonged stress occurring in adulthood. Therefore, it is important to continue to identify how, on a molecular level, developmental stress affects the CRF family of genes (Bakshi and Kalin, 2000). This may ultimately lead to the identification of novel drug targets for treatment in adulthood of developmental stress-induced pathologies such as depression and anxiety disorders.

Effects on the CRF system of hormonal and pharmacological manipulations that either mimic or block the effects of stress

Mimicking the effects of stress

It is known that stress results in release of CRF and activation of the HPA axis, ultimately leading to the release of cortisol or corticosterone into the circulation. Therefore, it is possible to pharmacologically mimic effects of stress by administering either CRF or glucocorticoids. Glucocorticoids affect various components of the CRF system. For example, corticosterone administration decreases CRF mRNA levels in the PVN (Makino et al., 1995). Extensive studies that have been performed on this topic are covered in greater detail in Kino and Chrousos, 2005.

Though expression of brain CRF receptors does not readily change in response to various manipulations, administration of high intracerebral doses of CRF can downregulate brain CRF receptor expression. Four daily intracisternal injections of CRF (5 nmol/day) decreased CRF receptor binding in the amygdala (Hauger et al., 1993). Downregulation of CRF_1 receptors was also reported in rat pups 4 h after administration of 0.75 nmol CRF into the left lateral cerebral ventricle, with a decrease of 21% in the frontal cortex (Brunson et al., 2002). In addition, this 0.75 nmol CRF treatment also caused persistent increases in CRF_1 mRNA levels in the frontal cortex and hippocampal CA3, but not in the CA1 or amygdala (Brunson et al., 2002). A CRF-induced change in the level of CRF_2 receptor mRNA or CRF_2 receptor binding was not detected.

Blocking the effects of stress

Treatment with various psychopharmacologic agents alters the dysregulation of the HPA axis that is observed in some depressed patients. It is possible that some antidepressant and anxiolytic agents produce their effects through actions on the CRF system. In rodents, acute and chronic treatment with these agents has been shown to affect the CRF system. For example, both acute and chronic alprazolam administration decreased CRF concentrations in the locus coeruleus (Skelton et al., 2000). In the same study, alprazolam also decreased CRF mRNA expression in the central nucleus of the amygdala, and decreased expression of CRF_1 receptors and mRNA levels in the basolateral amygdala. In contrast, alprazolam resulted in increased $CRF_{2(a)}$ receptor binding in the lateral septum and ventromedial hypothalamus. Urocortin 1 mRNA expression was also increased in the Edinger-Westphal nucleus that supplies preganglionic parasympathetic nerve fibers to the eye (Skelton et al., 2000). The authors of this study postulated that two separate CRF systems exist that could coordinately and inversely be regulated by stress and other manipulations.

There have been several studies reporting on the effects of antidepressant treatment on the CRF system. In depressed humans successful antidepressant therapy is associated with decreased cortisol concentrations as well as reductions in CSF–CRF concentrations (Heuser et al., 1998). Some preclinical studies show that long-term antidepressant treatment (4–8 weeks) decreases CRF mRNA levels in the PVN (Brady et al., 1991, 1992; Fadda et al., 1995; Aubry et al., 1999). However, other studies have failed to demonstrate a decrease in CRF mRNA in the PVN following similar antidepressant treatments (Jensen et al., 1999; Stout et al., 2002). Finally, four weeks of treatment with amitriptyline has been shown to decrease the levels of CRF_1 receptor mRNA in the rat amygdala (Aubry et al., 1999).

At least one study has described how the effects of long-term antidepressant treatment influence the effects of stress on the CRF system. Four weeks of treatment with either the dual serotonin/noradrenaline reuptake inhibitor venlafaxine or the monoamine oxidase inhibitor tranylcypromine inhibited both acute and chronic stress-induced increases in CRF mRNA levels in the rat PVN (Stout et al., 2002). This study suggests that antidepressants may decrease the sensitivity of CRF-expressing neurons in the PVN to the effects of stress.

Molecular mechanisms underlying effects of stress on expression of the CRF family of genes

The above discussions summarized the effects of stress on the CRF system. To understand at a molecular level how stress alters CRF gene expression, we must first understand the promoter regions that control expression from each of the genes in the CRF family.

CRF

Stress-related increases in CRF gene transcription are thought to be mediated via activation of a variety of different kinases resulting in phosphorylation of cAMP-response element binding protein (CREB) (Chen et al., 2001). The promoter for the CRF genehas been extensively characterized. The cAMP-response element (CRE) that is present in the promoter region of the CRF gene at −224 nucleotides affects basal promoter activity as well as cAMP-dependent, 12-*O*-tetradecanoylphorbol-13-acetate-dependent, and depolarization-dependent transcriptional activation of the CRF promoter (Spengler et al., 1992; Mugele et al., 1993; Guardiola-Diaz et al., 1994). The CRE binds the transcription factor CREB that is activated via phosphorylation.

Glucocorticoids appear to regulate CRF expression through direct inhibition of gene transcription (Herman et al., 1992). Following binding of ligand, glucocorticoid receptors translocate to the nucleus and bind to a negative glucocorticoid response element (nGRE) at bases −279 to −249 that inhibits CRF gene transcription (Malkoski et al., 1997; Malkoski and Dorin, 1999). Interestingly, mutating or eliminating this nGRE from the promoter sequence diminishes cAMP-dependent induction of gene transcription, indicating that there is recruitment of proteins capable of binding to elements within the nGRE that contribute to cAMP-inducible expression. Interestingly, the nGRE site contains an AP-1 site that binds c-*fos* and c-*jun* transcription factors, both of which are known to be stimulated by activation of the cAMP pathway. Mutations within this nGRE that disrupted either GR or AP-1 binding activity abolished glucocorticoid-dependent repression of gene expression. In summary, there are elements within the nGRE that contribute to the glucocorticoid-dependent repression of gene expression and cAMP-dependent stimulation of gene expression (Malkoski and Dorin, 1999). This complex activity of the nGRE within the CRF gene is not novel, and several examples of coregulation between GR and AP-1 proteins have been reported (Diamond et al., 1990; Zhang et al., 1991).

Finally the human cell line BE(2)-M17 expresses CRF protein following differentiation with retinoic acid. In this cell line, the POU transcription factor Brn-2 is required for retinoic acid-induced expression of the CRF gene. However, overexpression of Brn-2 in the absence of retinoic acid is not sufficient to induce CRF mRNA expression (Ramkumar and Adler, 1999). This study suggests that Brn-2 is important in mediating retinoic acid-induced CRF gene expression. Brn-2 is expressed in the parvocellular neurons of the PVN that synthesize CRF, and the CRF promoter contains multiple binding sites for Brn-2 (Li et al., 1993). However, the involvement of this transcription factor in vivo in the regulation of CRF expression remains to be demonstrated. When Brn-2 was overexpressed in vivo in the mouse PVN using an adenoviral vector delivery strategy, no effect on CRF gene expression was demonstrated (Wong and Murphy, 2003).

Urocortin 1

The urocortin 1 gene, like the CRF_1 gene, contains two exons, with exon 2 containing the entire coding region of the urocortin 1 mRNA (Park et al., 2000). Similar to the CRF gene, the urocortin 1 promoter and 5′-upstream region contains CRE elements and a Brn-2 binding site (Zhao et al., 1998). This CRE element seems to be important for basal transcriptional activity and protein kinase A (PKA)-mediated responses (Zhao et al., 1998).

Urocortins 2 and 3

Recently the urocortin 2 promoter region was identified (Chen et al., 2003). In contrast to CRF and urocortin 1, glucocorticoids appear to play a dominant role in the regulation of urocortin 2 gene expression as its 5′-flanking region is clustered with 14 GRE (Chen et al., 2003). Transient transfection experiments have shown that the urocortin 2 promoter is driven by glucocorticoid activation (Nanda et al., 2002; Chen et al., 2003). Additionally results from transfection experiments show that steroid hormone antagonists block the actions of glucocorticoids, and prevent dexamethasone induction of urocortin 2 gene expression, suggesting a direct glucocorticoid receptor-mediated effect (Chen et al., 2003). These results are consistent with in vivo mouse studies demonstrating dexamethasone-induced increases in urocortin 2 gene expression in the hypothalamus and brain stem (Chen et al., 2003).

In contrast to the urocortin 2 gene, characterization of the promoter region of the urocortin 3 gene has not been reported.

CRF-binding protein

The CRF-binding protein (CRF-BP) is thought to buffer the actions of CRF at the synapse (Kemp et al., 1998). This protein preferentially binds CRF and urocortin 1 but displays no affinity for urocortin 2 or 3 (Dautzenberg and Hauger, 2002). As also observed for its natural ligands CRF and urocortin 1, CRF-BP gene expression may be regulated by nuclear factors binding to a CRE element in the 5′-upstream promoter region, as elevation of intracellular cAMP concentrations increase CRF-BP gene transcription (Kemp et al., 1998; McClennen and Seasholtz, 1999). Furthermore, glucocorticoids repress the transcription of the CRF-BP gene (McClennen and Seasholtz, 1999). Besides CRE and GRE elements, acute phase response elements have been located in the promoter region of the CRF-BP gene, one of which has been shown to promote binding of NF-κB (Kemp et al., 1998).

CRF_1 and CRF_2 receptors

CRF_1 and CRF_2 receptors are members of the class B subfamily of G protein-coupled receptors (GPCRs) (Dautzenberg and Hauger, 2002), which are characterized by genes with several introns, in contrast to the majority of members of the GPCR superfamily (Gentles and Karlin, 1999).

The genomic organization of the rat and human CRF_1 gene has been reported (Tsai-Morris et al., 1996; Sakai et al., 1998). Both genes span more than 36 (rat) and 20 kilobases (human), respectively, with similar organization and location of exons and exon/intron boundaries (Sakai et al., 1998). The rat gene contains 13 exons and 12 introns (Tsai-Morris et al., 1996), whereas the human gene contains 14 exons and 13 introns (Sakai et al., 1998). The additional 87 bp exon (number 5 in the human gene) gives rise to a splice variant with a 29 amino acid insertion in the first intracellular domain of the receptor (Chen et al., 1993). It is important to note that in humans a large variety of splice variants have been observed (Dautzenberg et al., 2001b), which arise from exon-skipping and encode truncated versions of this receptor. Most of these variants have not been found in other species. None of these splice variants have been shown to be functional, or to act as dominant-negative receptors, reducing the action of the normal CRF_1 receptor in vitro or in vivo. Thus, the International Union of Pharmacology subcommittee on the nomenclature of CRF receptors has recently discarded and refused to number these variants, as has been done with the functional CRF_2 receptor variants $CRF_{2(a)}$, $CRF_{2(b)}$, and $CRF_{2(c)}$ (Hauger et al., 2003b). A short segment of the rat CRF_1 gene promoter has been obtained with the cloning of the gene and several consensus binding sites for transcriptional factors have been determined (Tsai-Morris et al., 1996). However, it is not known which nuclear factors bind to the promoter region of the CRF_1 gene and regulate its transcription.

The gene encoding the CRF_2 receptor is more complex than that of the CRF_1 receptor. The human CRF_2 gene is more than 50 kb in size and spans 15 exons and 14 introns (Dautzenberg et al., 2000a). The first four exons encode the N-terminal segments of the three human CRF_2 receptor splice variants, with exons 1 and 2 encoding the $CRF_{2(b)}$ variant, followed by exon 3 encoding the $CRF_{2(c)}$ variant, and finally exon 4, which gives rise to the $CRF_{2(a)}$ variant (Dautzenberg et al., 2000a; Fig. 1). The two $CRF_{2(b)}$-specific exons are separated from each other by a 10.5 kb intron (Fig. 1). It was recently speculated that the three CRF_2 receptor splice variants may not be generated by excision of exons from the pre-mRNA but by alternate usage of multiple promoter sites (Nanda et al., 2001; Catalano et al., 2003). Indeed, as exons 1, 3, and 4 encode the start methionine triplet ATG for each of the three splice variants, it is likely that these variants arise from alternate promoter usage (Catalano et al., 2003). First crude mapping of the putative promoter regions revealed various consensus sites for tissue-specific nuclear binding factors (Catalano et al., 2003). For the $CRF_{2(b)}$ promoter region, consensus sequences for vascular factors have been found, whereas the $CRF_{2(c)}$ promoter contains pituitary specific nuclear factor consensus sequences (Catalano et al., 2003). However, it is important to note that in contrast to rodents that show tissue-specific expression of the $CRF_{2(a)}$ and

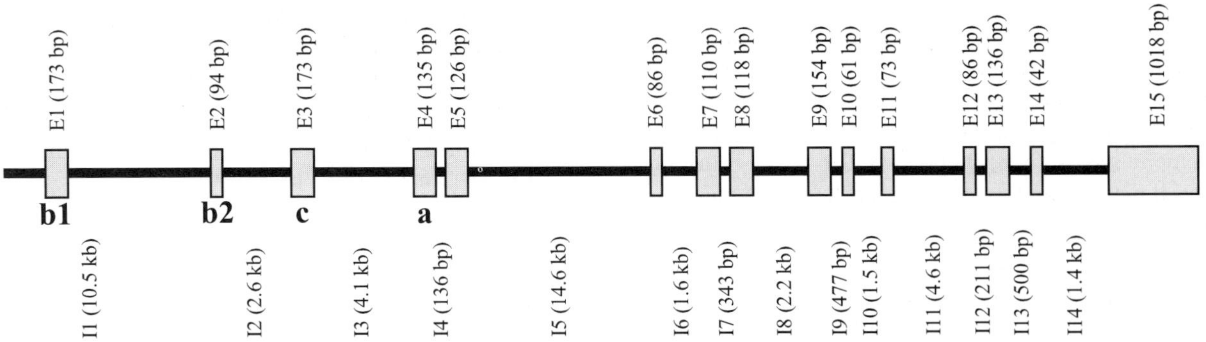

Fig. 1. Schematic representation of the gene encoding the human CRF_2 receptor. The exons (E1–15) are represented as grey boxes and their sizes are indicated above. The introns (I1–14) are drawn as lines. Exons 1 and 2 encode the N-terminal residues of the human $CRF_{2(b)}$ splice variants and are separated by a large 10.5 kb intron. Exon 3, which is specific for the $CRF_{2(c)}$ variant is downstream of the $CRF_{2(b)}$-specific exons, whereas exon 4, which encodes the $CRF_{2(a)}$ variant, is separated from exon 3 by a 4.1 kb intron. The first exon common for all CRF_2 receptor variants is only separated from the $CRF_{2(a)}$-specific exon by a very small intron of 136 bp in size.

$CRF_{2(b)}$ receptors, in humans all three variants are coexpressed in several tissues. Additionally, the $CRF_{2(a)}$ variant dominates in all human tissues that have been analyzed, including the heart and pituitary (Valdenaire et al., 1997; Kostich et al., 1998). Thus, the factors regulating the generation of either splice variants are by far more complex than described by Catalano and colleagues. Nevertheless, it is clear that the expression of the CRF_2 gene is very tightly regulated. To date only a few cell lines have been reported to express very low levels of CRF_2 receptors, including the A-431 epidermoid (Kiang et al., 1998), the A7R5 aortic (Kageyama et al., 2000), and the AR-5 hippocampal cells (Sheriff et al., 2001). Interestingly, primary hippocampal neurons, when brought into tissue culture, seem to downregulate their CRF_2 receptors (Pedersen et al., 2002). It is tempting to speculate that either a silencer region within the CRF_2 gene may be activated when these cells are taken out of their natural context, or that in vivo neighboring neurons may secrete factors that are required for active transcription of the CRF_2 gene and that these factors are missing in the primary cultures. Silencer regions have recently been identified in the CRF_2 gene (Nanda et al., 2001).

CRF receptor signaling

CRF receptors belong to the superfamily of G-protein-coupled receptors that are characterized by the presence of seven transmembrane domains. The CRF receptors belong to the class B subfamily, which includes receptors with small peptide ligands like growth hormone-releasing factor, calcitonin, vasoactive intestinal peptide, parathyroid hormone, and others (Segre and Goldring, 1993). The two CRF receptor subtypes share 70% identity in their amino acid sequence (Spiess et al., 1998). Agonist binding to either CRF receptor causes activation of adenylate cyclase (AC) through the stimulatory G-protein (G_s) and thereby results in an increase in the production of cAMP, which can then bind to the regulatory subunit of PKA. The cAMP-bound regulatory subunit dissociates from the catalytic subunit, thereby activating it and resulting in phosphorylation of a wide variety of proteins (McKnight et al., 1988). One of these proteins is the transcription factor CREB; phosphorylation of CREB converts it into a powerful activator of gene transcription (Gonzalez and Montminy, 1989).

In vitro studies demonstrate that CRF receptors are positively coupled to AC. CRF stimulation of AC activity has been demonstrated in pituitary cell culture (Labrie et al., 1982; Litvin et al., 1984) and CRF treatment of homogenates from a wide variety of rat brain regions increases AC activity (Wynn et al., 1984; Chen et al., 1986; Battaglia et al., 1987). In addition, CRF stimulation of a variety of CRF_1 receptor-expressing neuronal-like cell lines elevates AC activity (Dieterich and DeSouza, 1996; Hogg et al., 1996; Iredale et al., 1996; Hauger et al., 1997; Schoeffter et al., 1999; Dautzenberg et al., 2000b; Roseboom et al., 2001b). Lastly, CRF treatment of a variety of cell lines transfected with vectors that express CRF_1 receptors under control of a constitutively active promoter elevates intracellular cAMP levels (Chen et al., 1993; Xiong et al., 1995; Dieterich et al., 1996; Suman-Chauhan et al., 1999). Similar results have also been reported for the $CRF_{2(a)}$ receptor (Lovenberg et al., 1995a; Suman-Chauhan et al., 1999).

In contrast to the extensive data on CRF-stimulated AC, relatively less has been done to characterize the effects of CRF on PKA activity. CRF receptor activation increases PKA activity in a pituitary cell culture (Aguilera et al., 1983; Litvin et al., 1984). In addition, many of the effects of CRF on cell physiology are blocked by inhibitors of PKA activity, providing evidence that CRF receptor activation increases PKA activity in vitro. For example, activation of both CRF_1 receptor and $CRF_{2(a)}$ receptor in transfected cells results in CREB phosphorylation that is blocked by the PKA inhibitor H89 (Rossant et al., 1999). In addition to the cAMP pathway, other second messenger pathways including MAP kinase, calcium, and phospholipase C have been implicated in the actions of CRF (Rossant et al., 1999).

A recent study identified a variety of G-proteins that couple CRF receptors to various intracellular signaling pathways in the mouse hippocampus (Blank et al., 2003). The G-proteins and intracellular pathways that were activated depended on the strain of mouse used. In BALB/c mice, CRF stimulation coupled CRF receptors to $G_{q/11}$ and was associated with protein kinase C (PKC) activation. In C57BL/6N mice, CRF stimulation coupled CRF receptors to G_s, $G_{q/11}$, and G_i and was associated with PKA activation. Application of CRF to mouse hippocampal slices from both strains of mice increased neuronal activity based on intracellular recording data. As expected, this CRF-dependent increase in BALB/c mice was blocked by an inhibitor of PKC, bisindoylmaleimide, and in C57BL/6N mice by an inhibitor of PKA, H-89. This study indicates that CRF receptors can couple to multiple G-proteins and intracellular pathways and varies with the strain of mouse that is studied.

G protein receptor kinases (GRKs) and CRF_1 receptor desensitization

Sustained exposure of the anterior pituitary and the brain to stress or high levels of administered CRF decreased CRF binding sites (Hauger et al., 1993; Fuchs and Flugge, 1995; Brunson et al., 2002), desensitized CRF-stimulated cAMP accumulation, and decreased ACTH release in corticotrophs (Hauger and Dautzenberg, 1999; Aguilera et al., 2001). Since alterations of signaling properties in vivo are impossible to measure, cellular systems are required for measuring long-term effects of CRF treatment. Several pituitary- and brain-derived cell lines expressing CRF_1 receptors endogenously have been utilized to mimic the in vivo situation and have been compared to standard cell culture lines that recombinantly express CRF_1 receptors, such as HEK293, COS, and Ltk cells.

Likewise, CRF treatment of anterior pituitary cells or AtT20 cells causes large reductions in CRF-stimulated ACTH release resulting from downregulation of CRF_1 receptors and desensitization of CRF_1 receptor-stimulated cAMP accumulation (Reisine and Hoffman, 1983; Hoffman et al., 1985). In principle, CRF_1 receptor desensitization can be subdivided into two groups: homologous and heterologous (see following section). During homologous desensitization, only agonist-activated GPCRs desensitize, mainly due to phosphorylation through specific GRKs, followed by binding of cytosolic proteins called arrestins. Arrestins then facilitate sequestration of the respective GPCRs and result in receptor internalization (Ferguson, 2001). In an environment favoring GRK action (i.e. preincubation with physiological or pharmacological levels of CRF), a subsequent challenge with CRF revealed a marked decrease in CRF-stimulated cAMP accumulation in human retinoblastoma Y79 and human neuroblastoma IMR-32 cells (Hauger et al., 1997; Aguilera et al., 2001; Dautzenberg et al., 2001a; Roseboom et al., 2001b). Homologous CRF_1 receptor desensitization occurs in neuron-like cells exposed to physiological concentrations of CRF without any changes in CRF_1 mRNA levels, even after 24 h of continuous agonist exposure (Hauger et al., 1997; Aguilera et al., 2001; Roseboom et al., 2001b). However, CRF_1 receptor mRNA expression decreases in anterior pituitary cells exposed to CRF for several hours (Pozzoli et al., 1996).

Homologous CRF_1 receptor desensitization can be markedly inhibited in Y79 cells after a large reduction in GRK3 expression is induced by uptake of a GRK3 antisense oligonucleotide or transfection of a GRK3 antisense cDNA construct (Dautzenberg et al., 2001a). These findings suggest that GRK3-mediated phosphorylation contributes to the homologous desensitization of brain CRF_1 receptors. However, a high degree of CRF_1 receptor phosphorylation was detected in COS cells expressing an epitope-tagged CRF_1 receptor after exposure to CRF (Hauger et al., 2000). Because COS cells express GRK2, but not GRK3 (Menard et al., 1997), GRK2 appears to be capable of desensitizing CRF_1 receptors in this cell system. Homologous desensitization of CRF_1 receptors was significantly less in transfected HEK293 and Ltk cells when compared to native cells endogenously expressing CRF_1 receptors (Dieterich et al., 1996; Hauger et al., 1997). It is possible that receptor reserve present in the transfected HEK293 and Ltk cells may compensate for agonist-mediated receptor desensitization (Fig. 2). Indeed, preincubation of transfected HEK293 cells with pharmacological levels of CRF strongly elevates basal cAMP levels but only minimally reduces CRF-stimulated cAMP levels (Fig. 2). In contrast, CRF challenge of Y79 retinoblastoma cells that endogenously express CRF_1 receptors only minimally elevates basal cAMP levels, but CRF-stimulated cAMP production is strongly reduced (Fig. 2). This is similar to results reported for the IMR-32 cell

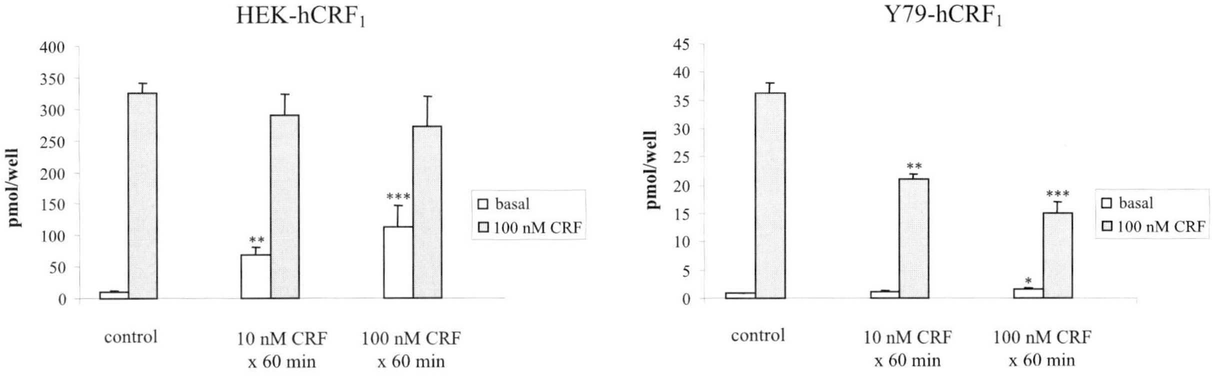

Fig. 2. Desensitization of cAMP responses in HEK293 cells recombinantly expressing human CRF$_1$ receptors (HEK-hCRF$_1$) vs. Y79 retinoblastoma cells endogenously expressing the human CRF$_1$ receptor (Y79-hCRF$_1$). Both cell lines were exposed to a physiological CRF concentration (10 nM), extensively washed with > 1000-fold cellular volume of physiological buffers and then restimulated with a maximally active CRF concentration (100 nM). In HEK-hCRF$_1$ cells, despite a 10-fold increase in basal cAMP levels, only small reductions (up to 17%) of CRF-mediated cAMP production were observed, likely due to a large reserve of spare receptors. In contrast, strong reductions of CRF-mediated cAMP accumulation (up to 60% loss of second messenger responsiveness), accompanied by only minimal alterations of basal cAMP levels, were observed in Y79-hCRF$_1$ cells after CRF preincubation. Statistical significance: $*p < 0.05$ vs. control; $**p < 0.01$ vs. control; $***p < 0.001$ vs. control.

line, a line that also endogenously expresses CRF$_1$ receptors (Roseboom et al., 2001b). Studies aimed at understanding the physiological regulation of receptors should consider using endogenously expressing cell lines versus experiments utilizing transfected cells that express super-physiological levels of receptor.

Heterologous regulation of CRF$_1$ receptors by PKC

In contrast to homologous desensitization, heterologous mechanisms can attenuate receptor responsiveness independent of agonist binding. Heterologous desensitization normally occurs when a receptor unbound by agonist is phosphorylated via a specific second messenger kinase that has been activated via another GPCR (Penn and Benovic, 1998). Since CRF$_1$ and CRF$_2$ receptors mainly couple to the stimulatory G_s protein (Hauger et al., 2003b), it was originally expected that PKA may also be involved in the heterologous desensitization of CRF receptors as demonstrated for other G_s-coupled GPCRs (Penn and Benovic, 1998; Kohout and Lefkowitz, 2003). However, both CRF$_1$ and CRF$_2$ receptors lack the classical PKA consensus sequence Arg-Arg-X-*Ser* (where X can represent any amino acid). Indeed, our laboratories have not found an involvement of PKA in either homologous or heterologous CRF receptor desensitization (Hauger et al., 1997, 2000; Dautzenberg et al., 2001a; Roseboom et al., 2001b).

Several potential phosphorylation sites Arg/Lys-X-*Ser*-X-Arg for PKC have been identified in intracellular domains, especially the C-terminus of all vertebrate CRF$_1$ and CRF$_2$ receptors (Dautzenberg and Hauger, 2002). It has recently been shown that PKC activation potently desensitizes the CRF$_1$ receptor, even in the absence of its ligand, whereas PKC inhibitors reduce the magnitude of CRF receptor desensitization (Hauger et al., 2003a). Furthermore, in transiently transfected COS cells, PKC activation results in CRF$_1$ receptor phosphorylation (Hauger et al., 2003a), suggesting that PKC-mediated CRF$_1$ receptor phosphorylation may be involved in physiological regulation of CRF$_1$ receptors. Interestingly, in the recombinant setting, a high degree of basal CRF$_1$ receptor phosphorylation was observed that could be decreased with specific PKC inhibitors (Hauger et al., 2003a; Fig. 3). It remains to be determined if this high degree of basal CRF$_1$ receptor phosphorylation is restricted to the artificial overexpression system.

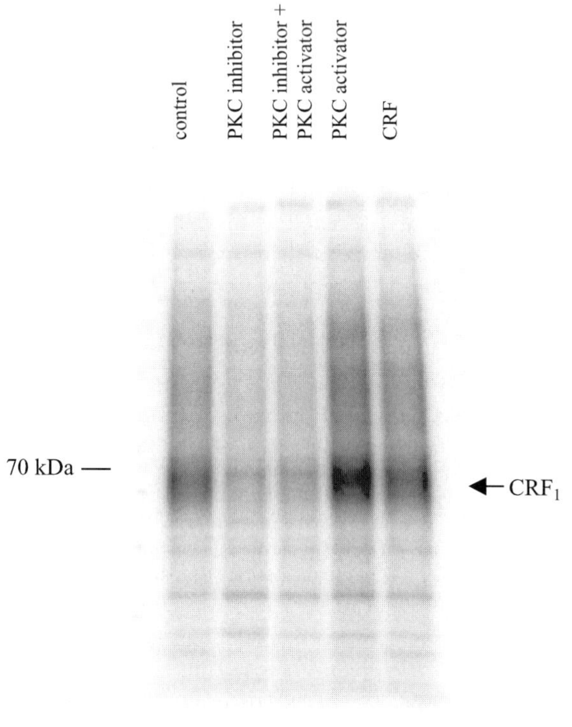

Fig. 3. PKC- and CRF-mediated phosphorylation of the human CRF_1 receptor. The ^{32}P-labeled human epitope-tagged CRF_1 receptor was exposed to a PKC activator or to 1 μM CRF in the presence or absence of PKC inhibitors for 5 min, followed by an immunoprecipitation step and SDS-PAGE. The ~66 kDa band representing the phosphorylated CRF_1-receptor was ~2.5-fold greater in CRF- and PKC-activated cells. In contrast PKC inhibitors blocked the basal phosphorylation >2-fold.

Ligand-dependent CRF receptor adaptations

CRF and urocortin 1 bind to the CRF_1 receptor with similar affinity, and both peptides stimulate cAMP production with a similar potency in cells expressing endogenous or recombinant CRF_1 receptors (Hauger et al., 2003b). In addition, both ligands desensitize second messenger responses of the CRF_1 receptor in Y79 cells with a similar potency and time course (Dautzenberg et al., 2002). However, while CRF promotes CRF_1 receptor phosphorylation (Hauger et al., 2000), this has yet to be examined for urocortin 1. Interestingly, CRF rapidly upregulates GRK3 mRNA, the main GRK capable of phosphorylating and desensitizing the CRF_1 receptor (Dautzenberg et al., 2001a). Urocortin 1 exposure has no effect on GRK3 mRNA levels (Dautzenberg et al., 2002). In agreement with its potency to upregulate GRK3 mRNA levels, CRF exposure to Y79 cells also increases GRK3 protein levels (Fig. 4), albeit only during long-term exposure (Dautzenberg et al., 2002). It can be speculated that elevated GRK3 levels promote a faster desensitization and internalization of CRF_1 receptors. Based on the data above, it appears that long-term elevated levels of CRF or urocortin 1 produce differential CRF receptor adaptation.

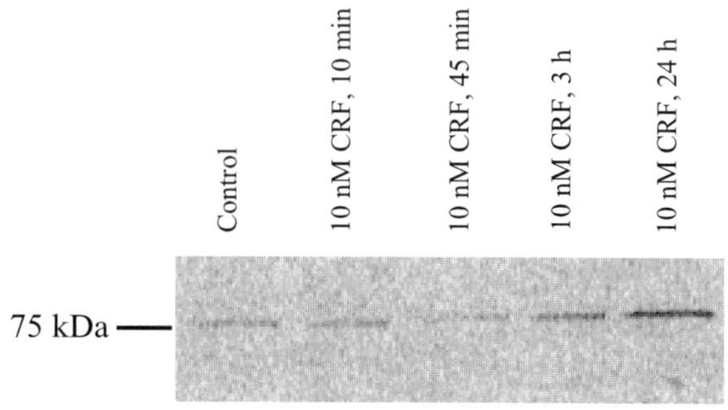

Fig. 4. Western blot probed with a mouse GRK3-specific monoclonal antibody. Y79 cells were exposed to physiological CRF concentrations (10 nM) for up to 24 h. GRK3 protein levels increased more than 3-fold (~350%) compared to control cells not exposed to CRF.

Regulation of CRF_2 receptors

Stress in the form of an inflammatory response can downregulate CRF_2 receptor mRNA expression in the mouse heart (Heldwein et al., 1997), and left ventricular hypertrophy can increase urocortin 1 mRNA expression and downregulate $CRF_{2(b)}$ receptor mRNA expression in the human heart (Nishikimi et al., 2000). When aortic smooth muscle A7R5 cells or ventricular myocytes are exposed to urocortin 1 or CRF, a dose-dependent reduction in CRF_2 receptor mRNA expression occurs (Kageyama et al., 1999; Coste et al., 2001). However, the homologous desensitization of CRF_2 receptors has not been investigated to date, mainly due to the lack of cell lines expressing significant amounts of endogenous CRF_2 receptors.

A few cell lines have been reported to endogenously express CRF_2 receptor mRNA, among those are the above mentioned A7R5 rat aortic

(Kageyama et al., 2000), the A-431 human epidermoid (Kiang et al., 1998), and the AR-5 rat amygdalar (Kasckow et al., 1999) cell lines. The A7R5 cell line expresses the $CRF_{2(b)}$ receptors that increase cAMP following CRF treatment. This cAMP increase is blocked by the CRF receptor antagonist Astressin (Kageyama et al., 2002). In the A-431 cell line, CRF receptor activation alters intracellular Ca^{2+} levels (Kiang, 1994), activates PKC (Kiang et al., 1994), and phosphorylates phospholipase C-gamma (Kiang et al., 1998). The AR-5 amygdalar cells express the $CRF_{2(a)}$ receptor and CRF treatment increases cAMP in these cells. The cells also express neuropeptide Y (NPY) receptors and NPY treatment can decrease CRF-induced increases in cAMP (Sheriff et al., 2001). This latter finding is of interest because CRF and NPY can produce opposite effects on anxiety-like behaviors when injected into the amygdala, with CRF producing anxiogenic responses (Liang and Lee, 1988) and NPY producing anxiolytic responses (Heilig et al., 1993). The results from the AR-5 cells led the authors to suggest that the opposite effects of CRF and NPY on anxiety-like behaviors may be mediated in part by the opposing actions of these two peptides on cAMP accumulation in the amygdala (Sheriff et al., 2001). Additionally, nicotinic agonists can increase CRF mRNA levels in AR-5 cells (Kasckow et al., 1999), which is interesting because acetylcholine induces CRF release from the rat amygdala and hypothalamus, possibly through nicotinic receptors (Raber et al., 1995). Overall, however, the usefulness of these cell lines is limited by the relatively low second messenger responses seen upon receptor stimulation. Furthermore, CRF_2 receptor mRNA and protein levels seem to decrease rapidly when primary cells are established from tissues shown to express CRF_2 mRNA in situ (Pedersen et al., 2002).

Microarray studies to identify genes affected by the CRF system

Great advances have been made in the areas of gene chip and laser capture technology, where the expression patterns of thousands of genes can be investigated within single cells. However, these relatively new techniques also present challenges in data analysis and interpretation.

DNA microarray chips are templates of genes of interest made of cDNA clones or oligonucleotides. Using this technology, small amounts of tissue are required, for example, a few cells which can be microdissected out of a brain area of interest using laser capture methodology (Simone et al., 1998). The importance of this technology in understanding basic mechanisms is illustrated by the estimation that 50–60% of known genes are expressed in the brain (Sandberg et al., 2000), an organ that is believed to exhibit the greatest complexity of gene expression throughout the body (Colantuoni et al., 2000).

Studies using microarray technology to understand how stress affects gene expression through CRF system activation and how these responses may be altered in pathological conditions are scarce. Recently, changes in the expression of genes that are involved in intracellular calcium signaling, neurogenesis, and myelination have been reported in several brain regions in transgenic mice overexpressing CRF (Peeters et al., 2002), suggesting that adaptive mechanisms exist to compensate for life-long exposure to elevated levels of CRF. Both down-regulation of enhancers of glucocorticoid receptor signaling (11ß-HSD1) and upregulation of repressors of this signaling pathway (FK506 binding protein 5) were observed in transgenic mice, suggesting adaptation to the six-fold increase in plasma glucocorticoid levels seen in these CRF overexpressing mice. At a cellular level, stress and increased HPA axis activity have been shown to decrease neurogenesis (Gould et al., 2000). This is consistent with the reduction in neurogenesis in CRF overexpressing mice. In addition to an overactive HPA axis, inhibited neurogenesis may be relevant to the development of mood disorders (Yuan et al., 2003). In summary, these microarray experiments in transgenic mice identify the compensatory changes that take place in the brain in response to elevation of CRF and corticosterone from an early age, and may identify those changes that underlie stress-induced psychopathology.

A recent study compared the mRNA expression patterns from the whole brains of mice lacking a functional CRF_1 receptor to that of mice that had received 40 mg/kg of the CRF_1 antagonist R121919 administered orally for 1 or 7 days. Importantly, the alterations in gene expression seen in the knockout mice were mimicked by 7-day treatment with the

CRF_1 antagonist (Landgrebe et al., 2002). There was a strain difference, with the antagonist having a bigger effect in wildtype mice of a 129SvJ background than in wildtype mice of a 129Ola/CD1 background. Therefore, for certain genetic backgrounds, administration of CRF_1 receptor antagonists can mimic the changes seen in CRF_1 receptor knockout mice.

In studying the effects of acute CRF exposure and of blockade of the CRF_1 receptor with R121919 in AtT-20 cell cultures, microarray analysis of 7256 genes revealed altered gene expression in about 90 genes that was attenuated by the antagonist (Peeters et al., 2002). Known targets of CRF_1 signaling that were altered included immediate early genes such as Jun/B, Nurr1, and Nurr77. For Nurr1, it has been shown that CRF signaling leads to induction of mRNA through PKA- and calcium calmodulin kinase II-dependent mechanisms, while Nurr77 transactivation is regulated through a MAPK-dependent pathway (Kovalovsky et al., 2002). Moreover, several previously unknown targets involved in this signaling cascade were identified and subsequently confirmed by quantitative real time-polymerase chain reaction (qRT-PCR), demonstrating the usefulness of microarrays to accelerate the discovery of the function of unknown genes.

Another microarray study using Clontech nylon filter arrays focusing on the rat septal brain region revealed that 19 genes with altered expression in response to restraint stress (Roseboom et al., 2001a). Each of these genes could potentially mediate adaptive or maladaptive responses to stress and could serve as novel targets for therapeutic intervention. In this study Pat Roseboom and Ned Kalin also identified 16 genes with altered expression in response to treatment with the non-selective CRF receptor antagonist [D-Phe^{12}, Nle^{21,38}, C$^{\alpha}$-MeLeu^{37}]$CRF_{12–41}$ (D-Phe). These changes occurred in the absence of stress, demonstrating intrinsic effects of D-Phe. Some of these changes may be the result of antagonism of baseline CRF neurotransmission, whereas others may be the result of D-Phe acting through as yet unidentified pathways. Most importantly, we determined that D-Phe blocked the effects of stress on 13 of 19 genes that were altered in response to restraint (Roseboom et al., 2001a). Ned Kalin and Pat Roseboom are in the process of confirming these changes with qRT-PCR, and so far have confirmed the D-Phe blockade of the stress-induced increase in somatostatin mRNA expression. This study demonstrates that stress-induced activation of CRF receptors results in downstream effects in a variety of biochemical pathways. Future studies will determine the extent to which these effects are physiologically relevant. We also established that stress induces changes in 6 genes through mechanisms other than CRF receptor activation, for example an increase in α-calcium/calmodulin-dependent protein kinase II. Although gene expression changes need to be confirmed by qRT-PCR and in situ hybridization, the present results illustrate novel pathways that may be involved in mediating stress effects. In addition, these newly implicated genes could lead to novel targets for the development of therapeutic agents for the treatment of stress-related disorders.

Future directions

There is a great deal of preclinical and clinical data associating alterations in the CRF system with stress-induced psychopathologies, such as depression and anxiety disorders. Ultimately, it is important to identify the molecular mechanisms that account for these alterations in the CRF system. Several questions remain to be addressed. As with many other genes, polymorphisms in the genes of the CRF system may alter the body's response to stressful stimuli and have a great impact on the susceptibility to stress-induced pathology. So far, there is no report on polymorphisms within the coding regions of the various genes of the CRF system. In addition, polymorphisms may exist in the promoter and 5′-upstream regions, as well as in the large introns that may up- or downregulate gene transcription and/or mRNA translation. It is important to note that the promoter and 5′-upstream regions, especially in the CRF receptor genes and the genes encoding urocortin 2 and urocortin 3, are either poorly characterized or not identified yet. Thus, extensive studies unraveling their structure and identifying transcription factor interactions may facilitate understanding the complex nature of CRF system regulation. Drugs that influence these promoter regions either directly, or indirectly by altering the actions of promoter-specific transcription factors, represent a novel approach to correcting or preventing stress-induced psychopathology.

Current studies utilizing gene-chip technology may identify novel biochemical pathways mediating the CRF receptor-dependent effects of stress on the brain. Identifying alterations in these pathways in the brains of animals that show a pathological response to stress may lead to novel drug targets for the development of "anti-stress" medications that could significantly improve quality of life and productivity for a large fraction of the human population.

Abbreviations

AC	adenylate cyclase
ACTH	adrenocorticotropic hormone
AP-1	activator protein-1
CRF	corticotropin-releasing factor
CRE	cAMP-response element
CREB	cAMP-response element binding protein
CRF-BP	corticotropin-releasing factor binding protein
CRF_1	CRF_1 receptor
CRF_2	CRF_2 receptor
D-Phe	[D-Phe^{12}, $Nle^{21,38}$, C^{α}-$MeLeu^{37}$] $CRF_{12\text{-}41}$
GR	glucocorticoid receptor
GRE	glucocorticoid response element
GRK	G protein-coupled receptor kinase
GPCRs	G-protein-coupled receptors
G_s	stimulatory G-protein
HPA	hypothalamic–pituitary–adrenal
qRT-PCR	quantitative real time-polymerase chain reaction
nGRE	negative glucocorticoid response element
NPY	nuropeptide Y
PKA	cAMP-dependent protein kinase
PKC	protein kinase C
PVN	paraventricular nucleus of the hypothalamus

Acknowledgments

This work was supported by NIH grant MH40855 and the UW HealthEmotions Research Institute (Madison, WI). The authors would like to thank Stephany G. Jones and Kim A. Jochman for advice and encouragement during the preparation of this review. Ned H. Kalin and Patrick H. Roseboom have equity positions in Promoter Neurosciences, LLC (Madison, WI), and have applied for patents through the Wisconsin Alumni Research Foundation (WARF, Madison, WI) related to promoter-based treatments to alter CRF regulation. In addition to other pharmaceutical companies, Ned H. Kalin is also a scientific consultant to Neurocrine Biosciences (San Diego, CA). Ned H. Kalin and Patrick H. Roseboom had a contractual arrangement with Johnson and Johnson to investigate the effects of CRF receptor antagonists in cell culture.

References

Aguilera, G., Harwood, J.P., Wilson, J.X., Morell, J., Brown, J.H. and Catt, K.J. (1983) Mechanisms of action of corticotropin-releasing factor and other regulators of corticotropin release in rat pituitary cells. J. Biol. Chem., 258: 8039–8045.

Aguilera, G., Jessop, D.S., Harbuz, M.S., Kiss, A. and Lightman, S.L. (1997) Differential regulation of hypothalamic pituitary corticotropin releasing hormone receptors during development of adjuvant-induced arthritis in the rat. J. Endocrinol., 153: 185–191.

Aguilera, G., Rabadan-Diehl, C. and Nikodemova, M. (2001) regulation of pituitary corticotropin releasing hormone receptors. Peptides, 22: 769–774.

Albeck, D.S., McKittrick, C.R., Blanchard, D.C., Blanchard, R.J., Nikulina, J., McEwen, B.S. and Sakai, R.R. (1997) Chronic social stress alters levels of corticotropin-releasing factor and arginine vasopressin mRNA in rat brain. J. Neurosci., 17: 4895–4903.

Andrews, M.W. and Rosenblum, L.A. (1994) The development of affiliative and agonistic social patterns in differentially reared monkeys. Child Dev., 65: 1398–1404.

Anisman, H., Zaharia, M.D., Meaney, M.J. and Merali, Z. (1998) Do early-life events permanently alter behavioral and hormonal responses to stressors? Int. J. Dev. Neurosci., 16: 149–164.

Arase, K., York, D.A., Shimizu, H., Shargill, N. and Bray, G.A. (1988) Effects of corticotropin-releasing factor on food intake and brown adipose tissue thermogenesis in rats. Am. J. Physiol., 255: E255–259.

Arborelius, L., Owens, M.J., Plotsky, P.M. and Nemeroff, C.B. (1999) The role of corticotropin-releasing factor in depression and anxiety disorders. J. Endocrinol., 160: 1–12.

Aubry, J.M., Pozzoli, G. and Vale, W.W. (1999) Chronic treatment with the antidepressant amitriptyline decreases

CRF-R1 receptor mRNA levels in the rat amygdala. Neurosci. Lett., 266: 197–200.

Bakshi, V.P. and Kalin, N.H. (2000) Corticotropin-releasing hormone and animal models of anxiety: gene-environment interactions. Biol. Psychiatry, 48: 1175–1198.

Bakshi, V.P., Smith-Roe, S., Newman, S.M., Grigoriadis, D.E. and Kalin, N.H. (2002) Reduction of stress-induced behavior by antagonism of corticotropin-releasing hormone 2 (CRH_2) receptors in lateral septum or CRH_1 receptors in amygdala. J. Neurosci., 22: 2926–2935.

Bale, T.L., Contarino, A., Smith, G.W., Chan, R., Gold, L.H., Sawchenko, P.E., Koob, G.F., Vale, W.W. and Lee, K.F. (2000) Mice deficient for corticotropin-releasing hormone receptor-2 display anxiety-like behaviour and are hypersensitive to stress. Nat. Genet., 24: 410–414.

Bale, T.L., Picetti, R., Contarino, A., Koob, G.F., Vale, W.W. and Lee, K.F. (2002) Mice deficient for both corticotropin-releasing factor receptor 1 (CRFR1) and CRFR2 have an impaired stress response and display sexually dichotomous anxiety-like behavior. J. Neurosci., 22: 193–199.

Bartanusz, V., Jezova, D., Bertini, L.T., Tilders, F.J., Aubry, J.M. and Kiss, J.Z. (1993) Stress-induced increase in vasopressin and corticotropin-releasing factor expression in hypophysiotrophic paraventricular neurons. Endocrinology, 132: 895–902.

Battaglia, G., Webster, E.L. and DeSouza, E.B. (1987) Characterization of corticotropin-releasing factor receptor-mediated adenylate cyclase activity in the rat central nervous system. Synapse, 1: SYN1.

Blank, T., Nijholt, I., Grammatopoulos, D.K., Randeva, H.S., Hillhouse, E.W. and Spiess, J. (2003) Corticotropin-releasing factor receptors couple to multiple G-proteins to activate diverse intracellular signaling pathways in mouse hippocampus: role in neuronal excitability and associative learning. J. Neurosci., 23: 700–707.

Bonaz, B. and Rivest, S. (1998) Effect of a chronic stress on CRF neuronal activity and expression of its type 1 receptor in the rat brain. Am. J. Physiol., 275: R1438–R1449.

Bowlby, J. (1973) Attachment and Loss, Vol. II: Separation, Basic Books, New York.

Brady, L.S., Gold, P.W., Herkenham, M., Lynn, A.B. and Whitfield, H.J., Jr. (1992) The antidepressants fluoxetine, idazoxan and phenelzine alter corticotropin-releasing hormone and tyrosine hydroxylase mRNA levels in rat brain: therapeutic implications. Brain Res., 572: 117–125.

Brady, L.S., Whitfield, H.J., Jr., Fox, R.J., Gold, P.W. and Herkenham, M. (1991) Long-term antidepressant administration alters corticotropin-releasing hormone, tyrosine hydroxylase, and mineralocorticoid receptor gene expression in rat brain. Therapeutic implications. J. Clin. Invest., 87: 831–837.

Brunson, K.L., Grigoriadis, D.E., Lorang, M.T. and Baram, T.Z. (2002) Corticotropin-Releasing hormone (CRH) downregulates the function of its receptor (CRF_1) and induces CRF1 expression in hippocampal and cortical regions of the immature rat brain. Exp. Neurol., 176: 75–86.

Carlson, M. and Earls, F. (1997) Psychological and neuroendocrinological sequelae of early social deprivation in institutionalized children in Romania. Ann. N. Y. Acad. Sci., 807: 419–428.

Catalano, R.D., Kyriakou, T., Chen, J., Easton, A. and Hillhouse, E.W. (2003) Regulation of corticotropin-releasing hormone type 2 receptors by multiple promoters and alternative splicing: identification of multiple splice variants. Mol. Endocrinol., 17: 395–410.

Chen, A., Vaughan, J. and Vale, W.W. (2003) Glucocorticoids regulate the expression of the mouse urocortin II gene: a putative connection between the corticotropin releasing factor receptor pathways. Mol. Endocrinol., 17: 1622–1639.

Chen, F.M., Bilezikjian, L.M., Perrin, M.H., Rivier, J. and Vale, W. (1986) Corticotropin releasing factor receptor-mediated stimulation of adenylate cyclase activity in the rat brain. Brain Res., 381: 49–57.

Chen, R., Lewis, K.A., Perrin, M.H. and Vale, W.W. (1993) Expression cloning of a human corticotropin-releasing-factor receptor. Proc. Natl. Acad. Sci. USA, 90: 8967–8971.

Chen, Y., Hatalski, C.G., Brunson, K.L. and Baram, T.Z. (2001) Rapid phosphorylation of the CRE binding protein precedes stress-induced activation of the corticotropin releasing hormone gene in medial parvocellular hypothalamic neurons of the immature rat. Brain Res. Mol. Brain Res., 96: 39–49.

Colantuoni, C., Purcell, A.E., Bouton, C.M. and Pevsner, J. (2000) high throughput analysis of gene expression in the human brain. J. Neurosci. Res., 59: 1–10.

Contarino, A., Dellu, F., Koob, G.F., Smith, G.W., Lee, K.F., Vale, W. and Gold, L.H. (1999) Reduced anxiety-like and cognitive performance in mice lacking the corticotropin-releasing factor receptor 1. Brain Res., 835: 1–9.

Coplan, J.D., Andrews, M.W., Rosenblum, L.A., Owens, M.J., Friedman, S., Gorman, J.M. and Nemeroff, C.B. (1996) Persistent elevations of cerebrospinal fluid concentrations of corticotropin-releasing factor in adult nonhuman primates exposed to early-life stressors: implications for the pathophysiology of mood and anxiety disorders. Proc. Natl. Acad. Sci. USA, 93: 1619–1623.

Coplan, J.D., Smith, E.L., Trost, R.C., Scharf, B.A., Altemus, M., Bjornson, L., Owens, M.J., Gorman, J.M., Nemeroff, C.B. and Rosenblum, L.A. (2000) Growth hormone response to clonidine in adversely reared young adult primates: relationship to serial cerebrospinal fluid corticotropin-releasing factor concentrations. Psychiatry Res., 95: 93–102.

Coste, S.C., Heldwein, K.A., Stevens, S.L., Tobar-Dupres, E. and Stenzel-Poore, M.P. (2001) IL-1alpha and TNFalpha down-regulate CRH receptor-2 mRNA expression in the mouse heart. Endocrinology, 142: 3537–3545.

Coste, S.C., Kesterson, R.A., Heldwein, K.A., Stevens, S.L., Heard, A.D., Hollis, J.H., Murray, S.E., Hill, J.K., Pantely,

G.A., Hohimer, A.R., Hatton, D.C., Phillips, T.J., Finn, D.A., Low, M.J., Rittenberg, M.B., Stenzel, P. and Stenzel-Poore, M.P. (2000) Abnormal adaptations to stress and impaired cardiovascular function in mice lacking corticotropin-releasing hormone receptor-2. Nat. Genet., 24: 403–409.

Dautzenberg, F.M. and Hauger, R.L. (2002) The CRF peptide family and their receptors: yet more partners discovered. Trends Pharmacol. Sci., 23: 71–77.

Dautzenberg, F.M., Braun, S. and Hauger, R.L. (2001a) GRK3 mediates homologous desensitization of CRF_1 receptors: a potential mechanism regulating stress adaptation. Am. J. Physiol. Regul. Integr. Comp. Physiol., 280: R935–946.

Dautzenberg, F.M., Higelin, J. and Teichert, U. (2000b) Functional characterization of corticotropin-releasing factor type 1 receptor endogenously expressed in human embryonic kidney 293 cells. Eur. J. Pharmacol., 390: 51–59.

Dautzenberg, F.M., Huber, G., Higelin, J., Py-Lang, G. and Kilpatrick, G.J. (2000a) Evidence for the abundant expression of arginine 185 containing human $CRF_{(2alpha)}$ receptors and the role of position 185 for receptor-ligand selectivity. Neuropharmacology, 39: 1368–1376.

Dautzenberg, F.M., Kilpatrick, G.J., Hauger, R.L. and Moreau, J. (2001b) Molecular biology of the CRH receptors – in the mood. Peptides, 22: 753–760.

Dautzenberg, F.M., Wille, S., Braun, S. and Hauger, R.L. (2002) GRK3 regulation during CRF- and urocortin-induced CRF_1 receptor desensitization. Biochem. Biophys. Res. Commun., 298: 303–308.

De Souza, E.B. (1995) Corticotropin-releasing factor receptors: physiology, pharmacology, biochemistry and role in central nervous system and immune disorders. Psychoneuroendocrinology, 20: 789–819.

Dent, G.W., Smith, M.A. and Levine, S. (2000) Rapid induction of corticotropin-releasing hormone gene transcription in the paraventricular nucleus of the developing rat. Endocrinology, 141: 1593–1598.

Diamond, M.I., Miner, J.N., Yoshinaga, S.K. and Yamamoto, K.R. (1990) Transcription factor interactions: selectors of positive or negative regulation from a single DNA element. Science, 249: 1266–1272.

Dieterich, K.D. and DeSouza, E.B. (1996) Functional corticotropin-releasing factor receptors in human neuroblastoma cells. Brain Res., 733: 113–118.

Dieterich, K.D., Grigoriadis, D.E. and De Souza, E.B. (1996) Homologous desensitization of human corticotropin-releasing $factor_1$ receptor in stable transfected mouse fibroblast cells. Brain Res., 710: 287–292.

Dunn, A.J. and Berridge, C.W. (1990) Physiological and behavioral responses to corticotropin-releasing factor administration: is CRF a mediator of anxiety or stress responses? Brain Res. Brain Res. Rev., 15: 71–100.

Fadda, P., Pani, L., Porcella, A. and Fratta, W. (1995) Chronic imipramine, L-sulpiride and mianserin decrease corticotropin releasing factor levels in the rat brain. Neurosci. Lett., 192: 121–123.

Ferguson, S.S. (2001) Evolving concepts in G protein-coupled receptor endocytosis: the role in receptor desensitization and signaling. Pharmacol. Rev., 53: 1–24.

Francis, D., Diorio, J., Liu, D. and Meaney, M.J. (1999a) nongenomic transmission across generations of maternal behavior and stress responses in the rat. Science, 286: 1155–1158.

Francis, D.D., Caldji, C., Champagne, F., Plotsky, P.M. and Meaney, M.J. (1999b) The role of corticotropin-releasing factor – norepinephrine systems in mediating the effects of early experience on the development of behavioral and endocrine responses to stress. Biol. Psychiatry, 46: 1153–1166.

Fuchs, E. and Flugge, G. (1995) Modulation of binding sites for corticotropin-releasing hormone by chronic psychosocial stress. Psychoneuroendocrinology, 20: 33–51.

Gentles, A.J. and Karlin, S. (1999) Why are human G-protein-coupled receptors predominantly intronless? Trends Genet., 15: 47–49.

Giardino, L., Puglisi-Allegra, S. and Ceccatelli, S. (1996) CRH-R1 mRNA expression in two strains of inbred mice and its regulation after repeated restraint stress. Brain Res. Mol. Brain Res., 40: 310–314.

Gonzalez, G.A. and Montminy, M.R. (1989) Cyclic AMP stimulates somatostatin gene transcription by phosphorylation of CREB at serine 133. Cell, 59: 675–680.

Gould, E., Tanapat, P., Rydel, T. and Hastings, N. (2000) Regulation of hippocampal neurogenesis in adulthood. Biol. Psychiatry, 48: 715–720.

Guardiola-Diaz, H.M., Boswell, C. and Seasholtz, A.F. (1994) The cAMP-responsive element in the corticotropin-releasing hormone gene mediates transcriptional regulation by depolarization. J. Biol. Chem., 269: 14784–14791.

Guillemin, R. and Rosenberg, B. (1955) Humoral hypothalamic control of anterior pituitary: a study with combined tissue cultures. Endocrinology, 57: 599–607.

Hatalski, C.G., Guirguis, C. and Baram, T.Z. (1998) Corticotropin releasing factor mRNA expression in the hypothalamic paraventricular nucleus and the central nucleus of the amygdala is modulated by repeated acute stress in the immature rat. J. Neuroendocrinol., 10: 663–669.

Hauger, R.L. and Dautzenberg, F.M. (1999) Regulation of the stress response by corticotropin-releasing factor receptors. In: Conn, P.M. and Freeman, M.E. (Eds.), Neuroendocrinology in Physiology and Medicine, Humana Press, Totowa, NJ, pp. 261–286.

Hauger, R.L., Dautzenberg, F.M., Flaccus, A., Liepold, T. and Spiess, J. (1997) Regulation of corticotropin-releasing factor receptor function in human Y-79 retinoblastoma cells: rapid and reversible homologous desensitization but prolonged recovery. J. Neurochem., 68: 2308–2316.

Hauger, R.L., Grigoriadis, D.E., Dallman, M.F., Plotsky, P.M., Vale, W.W. and Dautzenberg, F.M. (2003b) International union of pharmacology. XXXVI. Current status of the nomenclature for receptors for corticotropin-releasing factor and their ligands. Pharmacol. Rev., 55: 21–26.

Hauger, R.L., Irwin, M.R., Lorang, M., Aguilera, G. and Brown, M.R. (1993) High intracerebral levels of CRH result in CRH receptor downregulation in the amygdala and neuroimmune desensitization. Brain Res., 616: 283–292.

Hauger, R.L., Olivares-Reyes, J.A., Braun, S.D., Catt, K.J. and Dautzenberg, F.M. (2003a) Mediation of CRF_1 receptor phosphorylation and desensitization by protein kinase C: a possible role in stress adaptation. J. Pharmacol. Exp. Ther., 306: 794–803.

Hauger, R.L., Smith, R.D., Braun, S., Dautzenberg, F.M. and Catt, K.J. (2000) Rapid agonist-induced phosphorylation of the human CRF receptor, type 1: a potential mechanism for homologous desensitization. Biochem. Biophys. Res. Commun., 268: 572–576.

Heilig, M., McLeod, S., Brot, M., Heinrichs, S.C., Menzaghi, F., Koob, G.F. and Britton, K.T. (1993) Anxiolytic-like action of neuropeptide Y: mediation by Y1 receptors in amygdala, and dissociation from food intake effects. Neuropsychopharmacology, 8: 357–363.

Heim, C. and Nemeroff, C.B. (1999) The impact of early adverse experiences on brain systems involved in the pathophysiology of anxiety and affective disorders. Biol. Psychiatry, 46: 1509–1522.

Heldwein, K.A., Duncan, J.E., Stenzel, P., Rittenberg, M.B. and Stenzel-Poore, M.P. (1997) Endotoxin regulates corticotropin-releasing hormone receptor 2 in heart and skeletal muscle. Mol. Cell Endocrinol., 131: 167–172.

Herman, J.P., Schafer, M.K., Thompson, R.C. and Watson, S.J. (1992) Rapid regulation of corticotropin-releasing hormone gene transcription in vivo. Mol. Endocrinol., 6: 1061–1069.

Heuser, I., Bissette, G., Dettling, M., Schweiger, U., Gotthardt, U., Schmider, J., Lammers, C.H., Nemeroff, C.B. and Holsboer, F. (1998) Cerebrospinal fluid concentrations of corticotropin-releasing hormone, vasopressin, and somatostatin in depressed patients and healthy controls: response to amitriptyline treatment. Depress Anxiety, 8: 71–79.

Ho, S.P., Takahashi, L.K., Livanov, V., Spencer, K., Lesher, T., Maciag, C., Smith, M.A., Rohrbach, K.W., Hartig, P.R. and Arneric, S.P. (2001) Attenuation of fear conditioning by antisense inhibition of brain corticotropin releasing factor-2 receptor. Brain Res. Mol. Brain Res., 89: 29–40.

Hoffman, A.R., Ceda, G. and Reisine, T.D. (1985) Corticotropin-releasing factor desensitization of adrenocorticotropic hormone release is augmented by arginine vasopressin. J. Neurosci., 5: 234–242.

Hogg, J.E., Myers, J. and Hutson, P.H. (1996) The human neuroblastoma cell line, IMR-32, expresses functional corticotropin-releasing factor receptors. Eur. J. Pharmacol., 312: 257–261.

Hsu, D.T., Chen, F.L., Takahashi, L.K. and Kalin, N.H. (1998) Rapid stress-induced elevations in corticotropin-releasing hormone mRNA in rat central amygdala nucleus and hypothalamic paraventricular nucleus: an in situ hybridization Analysis. Brain Res., 788: 305–310.

Hsu, D.T., Lombardo, K.A., Herringa, R.J., Bakshi, V.P., Roseboom, P.H. and Kalin, N.H. (2001) Corticotropin-releasing hormone messenger RNA distribution and stress-induced activation in the thalamus. Neuroscience, 105: 911–921.

Iredale, P.A., Terwilliger, R., Widnell, K.L., Nestler, E.J. and Duman, R.S. (1996) Differential regulation of corticotropin-releasing factor (1) receptor expression by stress and agonist treatments in brain and cultured cells. Mol. Pharmacol., 50: 1103–1110.

Jensen, J.B., Jessop, D.S., Harbuz, M.S., Mork, A., Sanchez, C. and Mikkelsen, J.D. (1999) Acute and long-term treatments with the selective serotonin reuptake inhibitor citalopram modulate the HPA axis activity at different levels in male rats. J. Neuroendocrinol., 11: 465–471.

Kageyama, K., Bradbury, M.J., Zhao, L., Blount, A.L. and Vale, W.W. (1999) Urocortin messenger ribonucleic acid: tissue distribution in the rat and regulation in thymus by lipopolysaccharide and glucocorticoids. Endocrinology, 140: 5651–5658.

Kageyama, K., Gaudriault, G.E., Bradbury, M.J. and Vale, W.W. (2000) Regulation of corticotropin-releasing factor receptor type 2 beta messenger ribonucleic acid in the rat cardiovascular system by urocortin, glucocorticoids, and cytokines. Endocrinology, 141: 2285–2293.

Kageyama, K., Gaudriault, G.E., Suda, T. and Vale, W.W. (2002) Regulation of corticotropin-releasing factor receptor type 2beta mRNA via cyclic AMP pathway in A7r5 aortic smooth muscle cells. Cell Signal, 15: 17–25.

Kalin, N.H. (1985) Behavioral effects of ovine corticotropin-releasing factor administered to rhesus monkeys. Fed. Proc., 44: 249–253.

Kalin, N.H. (1997) The neurobiology of fear. Sci. Am, Mysteries of the Mind: 76–83.

Kalin, N.H., Shelton, S.E., Kraemer, G.W. and McKinney, W.T. (1983) Corticotropin-Releasing factor administered intraventricularly to rhesus monkeys. Peptides, 4: 217–220.

Kalin, N.H., Shelton, S.E., Rickman, M. and Davidson, R.J. (1998) Individual differences in freezing and cortisol in infant and mother rhesus monkeys. Behav. Neurosci., 112: 251–254.

Kalin, N.H., Takahashi, L.K. and Chen, F.L. (1994) Restraint stress increases corticotropin-releasing hormone mRNA content in the amygdala and paraventricular nucleus. Brain Res., 656: 182–186.

Kaplan, N.M. (1992) The adrenal glands. In: Griffin, J.E. and Ojeda, S.R. (Eds.), Textbook of Endocrine Physiology, Oxford University Press, New York, pp. 247–275.

Kasckow, J.W., Regmi, A., Sheriff, S., Mulchahey, J. and Geracioti, T.D., Jr. (1999) Regulation of corticotropin-releasing factor messenger RNA by nicotine in an immortalized amygdalar cell line. Life. Sci., 65: 2709–2714.

Kemp, C.F., Woods, R.J. and Lowry, P.J. (1998) The corticotrophin-releasing factor-binding protein: an act of several parts. Peptides, 19: 1119–1128.

Kiang, J.G. (1994) Corticotropin-releasing factor increases [Ca2+]i via receptor-mediated Ca2+ channels in human epidermoid A-431 cells. Eur. J. Pharmacol., 267: 135–142.

Kiang, J.G., Ding, X.Z., Gist, I.D., Jones, R.R. and Tsokos, G.C. (1998) Corticotropin-releasing factor induces phosphorylation of phospholipase C-gamma at tyrosine residues via its receptor 2beta in human epidermoid A-431 cells. Eur. J. Pharmacol., 363: 203–210.

Kiang, J.G., Wang, X. and McClain, D.E. (1994) Corticotropin-releasing factor increases protein kinase C activity by elevating membrane-bound alpha and beta isoforms. Chin J Physiol., 37: 105–110.

Kino, T. and Chrousos, G.P. (2005) Glucocorticoid effects on gene expression. In: Steckler, T., Kalin, N. and Reul, J.M.H.M. (Eds.), Handbook of Stress and the Brain, Part 1, Elsevier, Amsterdam, pp. 295–312.

Kishimoto, T., Radulovic, J., Radulovic, M., Lin, C.R., Schrick, C., Hooshmand, F. Hermanson, O., Rosenfeld, M.G. and Spiess, J. (2000) Deletion of crhr2 reveals an anxiolytic role for corticotropin-releasing hormone receptor-2. Nat. Genet, 24: 415–419.

Kohout, T.A. and Lefkowitz, R.J. (2003) Regulation of G protein-coupled receptor kinases and arrestins during receptor desensitization. Mol. Pharmacol., 63: 9–18.

Koob, G.F. and Heinrichs, S.C. (1999) A role for corticotropin releasing factor and urocortin in behavioral responses to stressors. Brain Res, 848: 141–152.

Koob, G.F., Heinrichs, S.C., Pich, E.M., Menzaghi, F., Baldwin, H., Miczek, K. and Britton, K.T. (1993) The role of corticotropin-releasing factor in behavioural responses to stress. Ciba. Found Symp., 172: 277–289.

Kostich, W.A., Chen, A., Sperle, K. and Largent, B.L. (1998) Molecular identification and analysis of a novel human corticotropin-releasing factor (CRF) receptor: the CRF_{2gamma} receptor. Mol. Endocrinol, 12: 1077–1085.

Kovalovsky, D., Refojo, D., Liberman, A.C., Hochbaum, D., Pereda, M.P., Coso, O.A., Stalla, G.K., Holsboer, F. and Arzt, E. (2002) Activation and induction of NUR77/NURR1 in corticotrophs by CRH/cAMP: involvement of calcium, protein kinase A, and MAPK pathways. Mol. Endocrinol, 16: 1638–1651.

Krahn, D.D., Gosnell, B.A., Grace, M. and Levine, A.S. (1986) CRF antagonist partially reverses CRF- and stress-induced effects on feeding. Brain Res. Bull., 17: 285–289.

Labrie, F., Veilleux, R. and LeFevre, G. (1982) Corticotropin-releasing factor stimulates accumulation of adenosine $3' \times 5'$-monophasphate in rat pituitary corticotrophs. Science, 216: 1007–1008.

Landgrebe, J., Wurst, W. and Welzl, G. (2002) Permutation-validated principal components analysis of microarray data. Genome. Biol., 3: research0019.0011–0019.0011.

Lewis, K., Li, C., Perrin, M.H., Blount, A., Kunitake, K., Donaldson, C., Vaughan, J., Reyes, T.M., Gulyas, J., Fischer, W., Bilezikjian, L., Rivier, J., Sawchenko, P.E. and Vale, W.W. (2001) Identification of urocortin III, an additional member of the corticotropin-releasing factor (CRF) family with high affinity for the CRF_2 receptor. Proc. Natl. Acad. Sci. USA, 98: 7570–7575.

Li, P., He, X., Gerrero, M.R., Mok, M., Aggarwal, A. and Rosenfeld, M.G. (1993) Spacing and orientation of bipartite DNA-binding motifs as potential functional determinants for POU domain factors. Genes Dev., 7: 2483–2496.

Liang, K.C. and Lee, E.H.Y. (1988) Intra-amygdala injections of corticotropin releasing factor facilitate inhibitory avoidance learning and reduce exploratory behavior in rats. Psychopharmacology (Berl). 96: 232–236.

Litvin, Y., PasMantier, R., Fleischer, N. and Erlichman, J. (1984) Hormonal activation of the cAMP-dependent protein kinases in AtT20 cells. J. Biol. Chem., 259: 10296–10302.

Lombardo, K.A., Herringa, R.J., Balachandran, J.S., Hsu, D.T., Bakshi, V.P., Roseboom, P.H. and Kalin, N.H. (2001) Effects of acute and repeated restraint stress on corticotropin-releasing hormone binding protein mRNA in rat amygdala and dorsal hippocampus. Neurosci. Lett., 302: 81–84.

Lovenberg, T.W., Chalmers, D.T., Liu, C. and De Souza, E.B. (1995b) CRF_{2alpha} and CRF_{2beta} receptor mRNAs are differentially distributed between the rat central nervous system and peripheral tissues. Endocrinology, 136: 4139–4142.

Lovenberg, T.W., Liaw, C.W., Grigoriadis, D.E., Clevenger, W., Chalmers, D.T., DeSouza, E.B. and Oltersdorf, T. (1995a) Cloning and characterization of a functionally distinct corticotropin-releasing factor receptor subtype from rat brain. Proc. Natl. Acad. Sci. USA, 92: 836–840.

Ma, X., Levy, A. and Lightman, S. (1997) Emergence of an isolated arginine vasopressin (AVP) response to stress after repeated restraint: a study of both AVP and corticotropin-releasing hormone messenger ribonucleic acid (RNA) and heteronuclear RNA. Endocrinology, 138: 4351–4357.

Ma, X.M., Lightman, S.L. and Aguilera, G. (1999) Vasopressin and corticotropin-releasing hormone gene responses to novel stress in rats adapted to repeated restraint. Endocrinology, 140: 3623–3632.

Makino, S., Asaba, K., Nishiyama, M. and Hashimoto, K. (1999) Decreased type 2 corticotropin-releasing hormone receptor mRNA expression in the ventromedial hypothalamus during repeated immobilization stress. Neuroendocrinology, 70: 160–167.

Makino, S., Schulkin, J., Smith, M., Pacak, K., Palkovits, M. and Gold, P. (1995) Regulation of corticotropin-releasing hormone receptor messenger ribonucleic acid in the rat brain and pituitary by glucocorticoids and stress. Endocrinology, 136: 4517–4525.

Malkoski, S.P. and Dorin, R.I. (1999) Composite glucocorticoid regulation at a functionally defined negative glucocorticoid response element of the human corticotropin-releasing hormone gene. Mol. Endocrinol., 13: 1629–1644.

Malkoski, S.P., Handanos, C.M. and Dorin, R.I. (1997) Localization of a negative glucocorticoid response element of the human corticotropin releasing hormone gene. Mol. Cell Endocrinol., 127: 189–199.

Mamalaki, E., Kvetnansky, R., Brady, L.S., Gold, P.W. and Herkenham, M. (1992) Repeated immobilization stress alters tyrosine hydroxylase, corticotropin-releasing hormone and corticosteroid receptor messenger ribonucleic acid levels in rat brain. J. Neuroendocrinol., 4: 689–699.

McCarthy, J.R., Heinrichs, S.C. and Grigoriadis, D.E. (1999) Recent advances with the CRF_1 receptor: design of small molecule inhibitors, receptor subtypes and clinical indications. Curr. Pharm. Des., 5: 289–315.

McClennen, S.J. and Seasholtz, A.F. (1999) Transcriptional regulation of corticotropin-releasing hormone-binding protein gene expression in astrocyte cultures. Endocrinology, 140: 4095–4103.

McKnight, G.S., Clegg, C.H., Uhler, M.D., Chrivia, J.C., Cadd, G.G., Correll, L.A. and Otten, A.D. (1988) Analysis of the cAMP-Dependent protein kinase system using molecular genetic approaches. Recent Prog. Horm. Res., 44: 307–335.

Menard, L., Ferguson, S.S., Zhang, J., Lin, F.T., Lefkowitz, R.J., Caron, M.G. and Barak, L.S. (1997) Synergistic regulation of beta2-adrenergic receptor sequestration: intracellular complement of beta-adrenergic receptor kinase and beta-arrestin determine kinetics of internalization. Mol. Pharmacol., 51: 800–808.

Merali, Z., McIntosh, J., Kent, P., Michaud, D. and Anisman, H. (1998) Aversive and appetitive events evoke the release of corticotropin-releasing hormone and bombesin-like peptides at the central nucleus of the amygdala. J. Neurosci., 18: 4758–4766.

Mitchell, A.J. (1998) The role of corticotropin releasing factor in depressive illness: a critical review. Neurosci. Biobehav Rev., 22: 635–651.

Mugele, K., Kulger, H.I. and Spiess, J. (1993) Immortalization of a fetal rat brain cell line that expresses corticotropin-releasing factor mRNA. DNA Cell Biol., 12: 119–126.

Nanda, S.A., Jones, S.G., Kalin, N.H. and Roseboom, P.H. (2002) Regulation of human urocortin II promoter activity. Soc. Neurosci. Abstract, 28: 769.762.

Nanda, S.A., Roseboom, P.H., Nash, G.A., Digre, J.R. and Kalin, N.H. (2001) Regulation of the human corticotropin-releasing $hormone_2$ receptor promoter activity. Soc. Neurosci. Abstract, 27: 914.913.

Nemeroff, C.B. (1988) The role of corticotropin-releasing factor in the pathogenesis of major depression. Pharmacopsychiatry, 21: 76–82.

Nemeroff, C.B., Widerlov, E., Bissette, G., Walleus, H., Karlsson, I., Eklund, K., Kilts, C.D., Loosen, P.T. and Vale, W. (1984) Elevated concentrations of CSF corticotropin-releasing factor-like immunoreactivity in depressed patients. Science, 226: 1342–1344.

Nishikimi, T., Miyata, A., Horio, T., Yoshihara, F., Nagaya, N., Takishita, S., Yutani, C., Matsuo, H., Matsuoka, H. and Kangawa, K. (2000) Urocortin, a member of the corticotropin-releasing factor family, in normal and diseased heart. Am. J. Physiol. Heart Circ. Physiol., 279: H3031–H3039.

Palchaudhuri, M.R., Hauger, R.L., Wille, S., Fuchs, E. and Dautzenberg, F.M. (1999) Isolation and pharmacological characterization of two functional splice variants of corticotropin-releasing factor type 2 receptor from tupaia belangeri. J. Neuroendocrinol., 11: 419–428.

Park, J.H., Lee, Y.J., Na, S.Y. and Kim, K.L. (2000) Genomic organization and tissue-specific expression of rat urocortin. Neurosci. Lett., 292: 45–48.

Pedersen, W.A., Wan, R., Zhang, P. and Mattson, M.P. (2002) Urocortin, but not urocortin II, protects cultured hippocampal neurons from oxidative and excitotoxic cell death via corticotropin-releasing hormone receptor type I. J. Neurosci., 22: 404–412.

Peeters, P., Moechars, D., Grohlmann, H., Swagemakers, S., Kass, S., Langlois, X., Stenzel-Poore, M., Bakker, M. and Steckler, T. (2002) Gene expression profiling of CRF signaling. Soc. Neurosci. Abstract, 28: 223.222.

Penn, R.B. and Benovic, J.L. (1998) Mechanisms of receptor regulation. In: Conn, P.M. (Ed.), Handbook of physiology, section 7: Endocrinology, American Physiological Society, Bethesda, MD, pp. 125–164.

Perrin, M., Donaldson, C., Chen, R., Blount, A., Berggren, T., Bilezikjian, L., Sawchenko, P. and Vale, W. (1995) Identification of a second corticotropin-releasing factor receptor gene and characterization of a cDNA expressed in heart. Proc. Natl. Acad. Sci. USA, 92: 2969–2973.

Pich, E.M., Lorang, M., Yeganeh, M., Rodriguez de Fonseca, F., Raber, J., Koob, G.F. and Weiss, F. (1995) Increase of extracellular corticotropin-releasing factor-like immunoreactivity levels in the amygdala of awake rats during restraint stress and ethanol withdrawal as measured by microdialysis. J. Neurosci., 15: 5439–5447.

Plotsky, P.M. and Meaney, M.J. (1993) Early, postnatal experience alters hypothalamic corticotropin-releasing factor (CRF) mRNA, median eminence CRF content and stress-induced release in adult rats. Brain Res. Mol. Brain Res., 18: 195–200.

Potter, E., Behan, D.P., Fischer, W.H., Linton, E.A., Lowry, P.J. and Vale, W.W. (1991) Cloning and characterization of the cDNAs for human and rat corticotropin releasing factor-binding proteins. Nature, 349: 423–425.

Pozzoli, G., Bilezikjian, L.M., Perrin, M.H., Blount, A.L. and Vale, W.W. (1996) Corticotropin-Releasing factor (CRF) and glucocorticoids modulate the expression of type 1 CRF receptor messenger ribonucleic acid in rat anterior pituitary cell cultures. Endocrinology, 137: 65–71.

Raadsheer, F., Hoogendijk, W., Stam, F., Tilders, F. and Swaab, D. (1994) Increased numbers of corticotropin-releasing hormone expressing neurons in the hypothalamic paraventricular nucleus of depressed patients. Neuroendocirnology, 60: 436–444.

Raber, J., Koob, G.F. and Bloom, F.E. (1995) Interleukin-2 (IL-2) induces corticotropin-releasing factor (CRF) release from the amygdala and involves a nitric oxide-mediated signaling; comparison with the hypothalamic response. J. Pharmacol. Exp. Ther., 272: 815–824.

Ramkumar, T. and Adler, S. (1999) A Requirement for the POU transcription factor, Brn-2, in corticotropin-releasing hormone expression in a neuronal Cell Line. Mol. Endocrinol, 13: 1237–1248.

Reisine, T. and Hoffman, A. (1983) Desensitization of corticotropin-releasing factor receptors. Biochem. Biophys. Res. Commun., 111: 919–925.

Reyes, T.M., Lewis, K., Perrin, M.H., Kunitake, K.S., Vaughan, J., Arias, C.A., Hogenesch, J.B., Gulyas, J., Rivier, J., Vale, W.W. and Sawchenko, P.E. (2001) Urocortin II: a member of the corticotropin-releasing factor (CRF) neuropeptide family that is selectively bound by type 2 CRF receptors. Proc. Natl. Acad. Sci. USA, 98: 2843–2848.

Roseboom, P.H., Bakshi, V.P., Herringa, R.J. and Kalin, N.H. (2001a) Microarray analysis of restraint stress-induced gene expression changes in the rat septum. Soc. Neurosci. Abstract, 27: 739.733.

Roseboom, P.H., Urben, C.M. and Kalin, N.H. (2001b) Persistent corticotropin-releasing factor1 receptor desensitization and downregulation in the human neuroblastoma cell line IMR-32. Brain Res. Mol. Brain Res., 92: 115–127.

Rossant, C.J., Pinnock, R.D., Hughes, J., Hall, M.D. and McNulty, S. (1999) Corticotropin-releasing factor type 1 and type 2alpha receptors regulate phosphorylation of calcium/cyclic adenosine 3′,5′- monophosphate response element-binding protein and activation of p42/p44 mitogen-activated protein kinase. Endocrinology, 140: 1525–1536.

Rots, N.Y., deJong, J., Workel, J.O., Levine, S., Cools, A.R. and DeKloet, E.R. (1996) Neonatal maternally deprived rats have as adults elevated basal pituitary-adrenal activity and enhanced susceptibility to apomorphine. J. Neuroendocrinol., 8: 501–506.

Saffran, M. and Schally, A.V. (1956) The release of corticotrophin by anterior pituitary tissue in vitro. Can. J. Biochem. Physiol., 33: 408–415.

Sakai, K., Yamada, M., Horiba, N., Wakui, M., Demura, H. and Suda, T. (1998) The genomic organization of the human corticotropin-releasing factor type-1 receptor. Gene, 219: 125–130.

Sanchez, M.M., Young, L.J., Plotsky, P.M. and Insel, T.R. (1999) Autoradiographic and in situ hybridization localization of corticotropin-releasing factor 1 and 2 receptors in nonhuman primate brain. J. Comp. Neurol., 408: 365–377.

Sandberg, J.A., Parker, V.P., Blanchard, K.S., Sweedler, D., Powell, J.A., Kachensky, A., Bellon, L., Usman, N., Rossing, T., Borden, E. and Blatt, L.M. (2000) Pharmacokinetics and tolerability of an antiangiogenic ribozyme (ANGIOZYME) in healthy volunteers. J. Clin. Pharmacol., 40: 1462–1469.

Schoeffter, P., Feuerbach, D., Bobirnac, I., Gazi, L. and Longato, R. (1999) Functional, endogenously expressed corticotropin-releasing factor receptor type 1 (CRF_1) and CRF1 receptor mRNA expression in human neuroblastoma SH-SY5Y cells. Fundam. Clin. Pharmacol., 13: 484–489.

Segre, G.V. and Goldring, S.R. (1993) Trends Endocrin. Metab., 4: 309–314.

Shanks, N., Larocque, S. and Meaney, M.J. (1995) Neonatal endotoxin exposure alters the development of the hypothalamic–pituitary–adrenal axis: early illness and later responsivity to stress. J. Neurosci., 15: 376–384.

Sheriff, S., Dautzenberg, F.M., Mulchahey, J.J., Pisarska, M., Hauger, R.L., Chance, W.T., Balasubramaniam, A. and Kasckow, J.W. (2001) Interaction of neuropeptide Y and corticotropin-releasing factor signaling pathways in AR-5 amygdalar cells. Peptides, 22: 2083–2089.

Sherman, J.E. and Kalin, N.H. (1986) Corticotropin releasing hormone: effects on stress-related behavior in rats. In: Moody, T. (Ed.) Neural. and Endocrine Peptides and Receptors, Plenum Publ Co., New York, pp. 195–204.

Simone, N.L., Bonner, R.F., Gillespie, J.W., Emmert-Buck, M.R. and Liotta, L.A. (1998) Laser-capture microdissection: opening the microscopic frontier to molecular analysis. Trends Genet, 14: 272–276.

Skelton, K.H., Nemeroff, C.B., Knight, D.L. and Owens, M.J. (2000) Chronic administration of the triazolobenzodiazepine alprazolam produces opposite effects on corticotropin-releasing factor and urocortin neuronal systems. J. Neurosci., 20: 1240–1248.

Smith, G.W., Aubry, J.M., Dellu, F., Contarino, A., Bilezikjian, L.M., Gold, L.H., Chen, R., Marchuk, Y., Hauser, C., Bentley, C.A., Sawchenko, P.E., Koob, G.F., Vale, W. and Lee, K.F. (1998) Corticotropin releasing factor receptor 1-deficient mice display decreased anxiety, impaired stress response, and aberrant neuroendocrine development. Neuron, 20: 1093–1102.

Smith, M.A., Kim, S.Y., VanOers, H.J.J. and Levine, S. (1997) Maternal deprivation and stress induce immediate early genes in the infant rat brain. Endocrinology, 138: 4622–4628.

Spengler, D., Rupprecht, R., Van, L.P. and Holsboer, F. (1992) Identification and characterization of a 3′, 5′-cyclic adenosine monophosphate-responsive element in the human corticotropin-releasing gene promoter. Mol. Endocrinol., 6: 1931–1941.

Spiess, J., Dautzenberg, F.M., Sydow, S., Hauger, R.L., Ruhmann, A. and Radulowic, J. (1998) Molecular properties of the CRF receptor. Trends Endocrin Metab., 9: 140–145.

Spina, M. (1996) Appetite-suppressing effects of urocortin, a CRF-related neuropeptide. Science, 273: 1561–1564.

Steckler, T. (2005) CRF antagonists as novel treatment strategies for stress-related disorders. In: Steckler, T., Kalin, N. and Reul, J.M.H.M. (Eds.), Handbook of Stress and the Brain, Part 1, Elsevier, Amsterdam, pp. 371–400.

Stenzel-Poore, M.P., Cameron, V.A., Vaughan, J., Sawchenko, P.E. and Vale, W. (1992) Development of cushing's syndrome in corticotropin-releasing factor transgenic mice. Endocrinology, 130: 3378–3386.

Stenzel-Poore, M.P., Heinrichs, S.C., Rivest, S., Koob, G.F. and Vale, W.W. (1994) Overproduction of corticotropin-releasing factor in transgenic mice: a genetic model of anxiogenic behavior. J. Neurosci., 14: 2579–2584.

Stout, S.C., Owens, M.J. and Nemeroff, C.B. (2002) Regulation of corticotropin-releasing factor neuronal systems and hypothalamic-pituitary-adrenal axis activity by stress and chronic antidepressant treatment. J. Pharmacol. Exp. Ther., 300: 1085–1092.

Suman-Chauhan, N., Carnell, P., Franks, R., Webdale, L., Gee, N.S., McNulty, S., Rossant, C.J., Van Leeuwen, D., MacKenzie, R. and Hall, M.D. (1999) Expression and characterisation of human and rat CRF_{2alpha} receptors. Eur. J. Pharmacol., 379: 219–227.

Timofeeva, E., Deshaies, Y., Picard, F. and Richard, D. (1999) Corticotropin-releasing hormone-binding protein in brain and pituitary of food-deprived obese (*fa/fa*) zucker rats. Am. J. Physiol., 277: R1749–R1759.

Timpl, P., Spanagel, R., Sillaber, I., Kresse, A., Reul, J.M., Stalla, G.K., Blanquet, V., Steckler, T., Holsboer, F. and Wurst, W. (1998) Impaired stress response and reduced anxiety in mice lacking a functional corticotropin-releasing hormone receptor 1. Nat. Genet., 19: 162–166.

Tsai-Morris, C.H., Buczko, E., Geng, Y., Gamoa-Pinto, A. and Dufau, M.L. (1996) The genomic structure of the rat corticotropin releasing factor receptor. J. Biol. Chem., 271: 14519–14525.

Valdenaire, O., Giller, T., Breu, V., Gottowik, J. and Kilpatrick, G. (1997) A new functional isoform of the human CRF2 receptor for corticotropin-releasing factor. Biochim. Biophys. Acta., 1352: 129–132.

Vale, W., Spiess, J., Rivier, C. and Rivier, J. (1981) Characterization of a 41-residue ovine hypothalamic peptide that stimulates secretion of corticotropin and beta-endorphin. Science, 213: 1394–1397.

van der Meer, M., Baumans, V., Olivier, B. and van Zutphen, B.L. (2001) Impact of transgenic procedures on behavioral and physiological responses in postweaning mice. Physiol. Behav., 73: 133–143.

van Gaalen, M.M., Stenzel-Poore, M.P., Holsboer, F. and Steckler, T. (2002) Effects of transgenic overproduction of CRH on anxiety-like behaviour. Eur. J. Neurosci., 15: 2007–2015.

van Oers, H.J., de Kloet, E.R. and Levine, S. (1998) Early vs. late maternal deprivation differentially alters the endocrine and hypothalamic responses to stress. Brain Res. Dev. Brain Res., 111: 245–252.

Vaughan, J., Donaldson, C., Bittencourt, J., Perrin, M.H., Lewis, K., Sutton, S., Chan, R., Turnbull, A.V., Lovejoy, D., Rivier, C. et al. (1995) Urocortin, a mammalian neuropeptide related to fish urotensin I and to corticotropin-releasing factor. Nature, 378: 287–292.

Weninger, S.C., Dunn, A.J., Muglia, L.J., Dikkes, P., Miczek, K.A., Swiergiel, A.H., Berridge, C.W. and Majzoub, J.A. (1999) Stress-induced behaviors require the corticotropin-releasing hormone (CRH) receptor, but not CRH. Proc. Natl. Acad. Sci. USA, 96: 8283–8288.

Weninger, S.C., Peters, L.L. and Majzoub, J.A. (2000) Urocortin expression in the edinger-westphal nucleus is up-regulated by stress and corticotropin-releasing hormone deficiency. Endocrinology, 141: 256–263.

Wigger, A. and Neumann, I.D. (1999) Periodic maternal deprivation induces gender-dependent alterations in behavioral and neuroendocrine responses to emotional stress in adult rats. Physiol. Behav., 66: 293–302.

Wong, L.F. and Murphy, D. (2003) Adenoviral-mediated overexpression of Brn2 in the rat paraventricular nucleus: no effect on vasopressin or corticotrophin releasing factor RNA levels. Mol. Cell Endocrinol., 200: 165–175.

Wynn, P.C., Hauger, R.L., Holmes, M.C., Millan, M.A., Catt, K.J. and Aguilera, G. (1984) Brain and pituitary receptors for corticotropin releasing factor: localization and differential regulation after adrenalectomy. Peptides, 5: 1077–1084.

Xiong, Y., Xie, L.Y. and Abou-Samra, A.B. (1995) Signaling properties of mouse and human corticotropin-releasing factor (CRF) receptors: decreased coupling efficiency of human type II CRF receptor. Endocrinology, 136: 1828–1834.

Yuan, P.X., Zhou, R., Farzad, N., Gray, N.A., Du, J. and Manji, H.K. (2005) Enhancing resilience to stress: the role of signaling cascades. In: Steckler, T., Kalin, N. and Reul, J.M.H.M. (Eds.), Handbook of Stress and the Brain, Part 1, Elsevier, Amsterdam, pp. 751–772.

Zhang, X.K., Dong, J.M. and Chiu, J.F. (1991) Regulation of alpha-fetoprotein gene expression by antagonism between AP-1 and the glucocorticoid receptor at their overlapping binding site. J. Biol. Chem., 266: 8248–8254.

Zhao, L., Donaldson, C.J., Smith, G.W. and Vale, W.W. (1998) The structures of the mouse and human urocortin genes (Ucn and UCN). Genomics, 50: 23–33.

Zobel, A.W., Nickel, T., Kunzel, H.E., Ackl, N., Sonntag, A., Ising, M. and Holsboer, F. (2000) Effects of the high-affinity corticotropin-releasing hormone receptor 1 antagonist R121919 in major depression: the first 20 patients treated. J. Psychiatr. Res., 34: 171–181.

T. Steckler, N.H. Kalin and J.M.H.M. Reul (Eds.)
Handbook of Stress and the Brain, Vol. 15
ISBN 0-444-51173-3

CHAPTER 2.3

Behavioral consequences of altered corticotropin-releasing factor activation in brain: a functionalist view of affective neuroscience

Stephen C. Heinrichs*

Boston College, Department of Psychology, McGuinn Hall, 140 Commonwealth Avenue, Chestnut Hill, MA 02467, USA

Abstract: Organisms exposed to challenging stimuli that alter the status quo inside or outside of the body are required for survival purposes to generate appropriate coping responses which counteract the departure from homeostasis. Identification of an executive control mechanism within the brain capable of coordinating the multitude of endocrine, physiological and functional coping responses has high utility for discovery of efficacious pharmacological tools in stressor-exposed organisms under normal or pathological conditions. One such mechanism is the corticotropin-releasing factor (CRF)/urocortin family of neuropeptides and receptors which constitutes an affective regulatory system, so called due to the integral role this neuropeptide system plays in controlling core components of affect such as arousal, emotionality, and aversive processes. In particular, available evidence from pharmacological intervention in multiple species and phenotyping of mutant mice is that CRF/urocortin systems mediate negatively valenced activating, avoidant, and distressing components of affective responses. It is suggested that affective regulation is exerted by CRF/urocortin systems within the brain based upon the sensitivity of local brain sites to CRF/urocortin ligand administration and the necessary but insufficient nature of the hypothalamo–pituitary–adrenocortical activation following stressor exposure. Moreover, these same stress neuropeptides may constitute a mechanism for learning to avoid noxious stimuli by forming so-called emotional memories. In additional to presentation of experimental evidence relevant for understanding the role of brain CRF/urocortin from a functional perspective, a conceptual framework is provided for extrapolation of animal model findings to humans and for viewing CRF/urocortin activation as a continuum measure which links normal and pathological states.

Introduction

The close of two complete decades of research into the biology of the corticotropin-releasing factor (CRF) family of neuropeptides seems an opportune moment to ask a simple but beguiling question: what is the ultimate function of these brain signalling systems? This question is typically addressed by citation of the mass of descriptive and experimental studies performed in animals that support the role of CRF as a mediator of interoceptive coping reactions to environmental or physiological challenges. However, the advent of small molecule ligands for CRF receptors suitable for administration in human beings begs the question of what utility, if any, antistress pharmacology via CRF manipulations in animals provides in the context of human biology and psychopathology. The present chapter tackles the pressing task of predicting the consequences of brain CRF system activation or deactivation in man based upon the adaptive utility of CRF systems and

*Tel.: 617-552-0852; Fax: 617-552-0523;
E-mail: stephen.heinrichs@bc.edu

the homologous and analogous characteristics of CRF system activation in multiple species. The functionalist thesis most strongly supported by the available experimental evidence is that activating, avoidant and learning/memory functions of CRF systems have evolved in order to serve affective regulatory functions in man and animals.

One means of assessing the ultimate functions of CRF systems is to identify cross-species homologies in the comparative biology of this peptide, that is, the extent to which characteristics of a contemporary organism have been inherited from ancestors. The term "homology" is used here to denote evidence of a shared underlying mechanism for a particular behavioral function that is expressed in a variety of species but derived from a single evolutionary point of origin. Comparative studies of CRF system biology in animals employing genetic and neurobiological techniques are thus suited to address the issue of homology. For example, the earliest known functions of CRF family peptides are related to osmoregulation and diuresis which are exerted by all variants of the urotensin-I family of CRF-like peptides (Lovejoy and Balment, 1999). Recently evolved vertebrate species exhibit physiological effects of CRF activation related to sympathetic activation and reproduction (Lovejoy and Balment, 1999). In particular, CRF expression in amniotes is associated with descending autonomic nerves that innervate smooth and cardiac muscle and various visceral glands and organs (Lovejoy and Balment, 1999). This cluster of consequences of brain CRF activation can be conceptualized as an action set designed to accomplish important biological functions such as avoiding and escaping life-threatening events. From the perspective of ultimate functionality, such affect programs can be viewed as evolutionarily derived mechanisms for successfully passing genes onto coming generations (Lohman et al., 2000).

Brain CRF systems can be hypothesized to modulate three specific aspects of affective programming: (1) the activation of negatively valenced action sets related to avoidance and escape tendencies, (2) the amplitude of activation as measured by motor output or subjective arousal, and (3) the acquisition of new information necessary for optimal performance in learning and memory tasks. Affective valence can be defined as a bipolar scale defining a continuum of subjective feeling states from the positive extreme of pleasantness to the negative extreme of unpleasantness (Bradley and Lang, 2000). Energized brain CRF systems would be expected to excite unpleasant states associated with the emotional labels anger, sadness, or fear (Panksepp, 1998). It should be noted that the neural substrates of reward which presumably mediate the pleasantly valenced affect programs are well characterized (Rolls, 1999) but beyond the scope of this chapter. Activation is similarly construed as a measure of the intensity of affect program expression ranging from an unaroused state to high arousal (Bradley and Lang, 2000). Energized brain CRF systems would be expected to induce features of high arousal such as excitement and motor activity. These orthogonal valence/arousal scales of affect programming have been proposed as primary motivational systems of the brain in multiple species (Rolls, 1999; Russell, 2003). Finally, brain plasticity required for learning and memory functions necessary to accomplish classical conditioning and working memory are hardwired aspects of brain affect programming (LeDoux, 2000). In particular, fear conditioning in response to noxious stimulus exposure is reported to promote emotional memories which are particularly vivid and enduring (LeDoux, 2000). The remaining sections of this review are organized to present the thesis that CRF systems mediate the valence, intensity, and learned flexibility of affect programs.

Relevance of CRF systems for affective neuroscience

Environmental change invokes an integrated state of endocrine, autonomic, and behavioral activation that is critically dependent on CRF substrates within the central nervous system. A contemporary understanding of the role of classic corticotropin secretagogues has emerged in which functional targets for peptide hormones such as CRF exist in brain sites unrelated to neuroendocrine hypophysiotropic circuits and even entirely outside of the brain. Thus, direct neurobiological actions of CRF acting as a synaptic transmitter at specialized extra-hypothalamic receptors must be coordinated with the capacity of CRF to stimulate secretion of pituitary hormones.

Factors that mobilize brain CRF systems appear to have one feature in common, the ability to disturb homeostasis. For example, demands on the organism may be induced either internally or externally by exposure to physical trauma, infection, or social conflict. Coping responses to such deflections in steady state that include sympathetic nervous system activation, promotion of negative energy balance and augmentation of vigilance and emotionality appear to be CRF dependent. Comparative pharmacology of CRF receptor agonists suggests that CRF, urocortin 1, sauvagine, and urotensin consistently mimic, and peptide CRF receptor antagonists consistently lessen, functional consequences of stressor exposure. Together with the development of novel nonpeptide CRF receptor antagonists, a growing number of CRF receptor selective ligands are available to elucidate the neurobiology and physiological role of CRF systems.

The classification of CRF as a peptide with activating, arousing, and anxiogenic properties is based on a broad and comprehensive in vivo testing battery whose dependent measures reflect changes in locomotion, choice accuracy, energy balance, species-typical social interactions, and many other constructs (Koob, 1999; Steckler and Holsboer, 1999). CRF administrated intracerebroventricularly results in behavioral, physiological, and autonomic responses that are similar to those observed when animals are exposed to a stressor. The striking behavioral effects include decreased food consumption, altered locomotor activity, diminished sexual behavior, sleep disruption, and anxiogenic-like increases in emotionality (see also chapter by Zorrilla and Koob, this volume). These functional responses to pharmacological doses of exogenous CRF are analogous to those produced in animals by exposure to a species-typical, naturalistic stressor such as social conflict (Meerlo et al., 1996; Buwalda et al., 1997). This evidence implicates CRF as a physiological mediator of stress responses or stress-induced psychopathology.

Since the frank and nonspecific behavioral actions of exogenous CRF receptor agonists are sufficiently robust to mask more subtle functional effects that would otherwise emerge (see section of Energy Balance and Reward), a more meaningful classification of animal models that reveal the true physiological significance of CRF neurobiology may be reflected in behavioral actions of CRF receptor antagonists. Indeed, complete reversal of the actions of exogenous CRF is accomplished by coadministration of a competitive receptor antagonist such as α-helical CRF (9–41) in pharmacological competition studies irrespective of the nature of the testing protocol (Dunn and Berridge, 1990). Accordingly, dependent measure selection ought not be based on efficacy of CRF itself or the results of CRF receptor antagonist competition studies but rather on testing contexts that reveal intrinsic actions of the CRF receptor antagonist alone.

This strategy has been applied extensively in studies employing numerous dependent measures used to reveal antistress actions of CRF receptor antagonists following stressor pretreatment. Thus, central administration of a peptide CRF receptor antagonist, or alternatively CRF/urocortin immunoneutralization, blocks or attenuates the behavioral responses to centrally administered CRF and also reverses the effects of a variety of stressors including restraint, foot shock, hemorrhage, insulin-induced hypoglycemia, cholecystokinin administration, and ether exposure (Ida et al., 2000; Ohata et al., 2000; Heinrichs and De Souza, 2001). CRF receptor antagonists reverse not only decreases but also increases in behavior associated with stress, suggesting that the effects of CRF receptor antagonists are stress dependent and not due to nonspecific changes in behavior. The bulk of these studies suggest that efficacy of CRF receptor antagonists can be generalized to many stress situations and is reproducible (Table 1). Current studies in the field are now largely focused on the real world relevance and exact nature of stressors employed in experiments designed to reveal the true physiological role of CRF in biological systems.

Neuroendocrine and neurotransmitter interactions

Hypophysiotropic CRF neurons which mediate ACTH release (see chapter by Fulford and Harbuz, this volume) do not act autonomously but rather are reciprocally regulated by a wide variety of neurochemically and functionally distinct brain peptides and neurotransmitters (Grossman et al., 1993; Menzaghi et al., 1993). CRF neurons within hypothalamus

Table 1. Testing contexts sensitive to the behavioral actions of CRF system activation by receptor agonist administration or CRF system inactivation by competitive receptor antagonist administration

Test	Ligand and behavioral action	References
Evidence for mediation of negatively valenced affective events by CRF		
Elevated Plus-Maze	CRF suppresses exploration and CRF receptor antagonist reverses stress, drug and genotypically induced suppression of exploration	Griebel et al., 1998; Heinrichs et al., 2002
Startle reactivity	CRF facilitates startle amplitude and CRF receptor antagonist blocks fear-potentiated startle	Swerdlow et al., 1989; Arborelius et al., 2000
Defensive burying	CRF enhances defensive burying and CRF receptor antagonist reduces defensive burying	Korte et al., 1994; Basso et al., 1999
Defensive withdrawal	CRF enhances and CRF receptor antagonist attenuates defensive withdrawal	Arborelius et al., 2000; Heinrichs and Joppa, 2001
Evidence for mediation of activation by CRF		
Locomotor activity	CRF enhances locomotion and CRF receptor antagonist reduces stress and drug-induced locomotion	Sarnyai et al., 1992; Menzaghi et al., 1994
Sleep latency and architecture	CRF promotes wakefullness and CRF receptor antagonists delay spontaneous awakening	Ehlers et al., 1986a; Chang and Opp, 1999; Lancel et al., 2002
Food intake	CRF decreases food intake and CRF receptor antagonist reverses stress and drug-induced anorexia	Bell et al., 1998; Jones et al., 1999
Evidence for mediation of learning and memory performance by CRF		
Taste/place conditioning	CRF produces aversion and CRF receptor antagonist weakens drug-induced place aversion	Heinrichs et al., 1991; Benoit et al., 2000a
Amphetamine-induced stereotypy	CRF enhances sensitization and CRF receptor antagonist attenuates stress-induced sensitization	Cador et al., 1989; Cole et al., 1990b
Conditioned emotional response	CRF induces conditioned fear and CRF receptor antagonist blocks acquisition of conditioned emotional response	Cole et al., 1987
Conditioned freezing	CRF facilitates acquisition of immobility and CRF receptor antagonist blocks stress-induced facilitation of immobility	Heinrichs et al., 1997b; Bakshi et al., 1999

apparently interact with peptides such as vasopressin, endorphins, and neuropeptide Y as well as gamma-amino-butyric acid (GABA), catecholamine, noradrenergic, and indolamine systems to modulate adrenocorticotropin secretion (Plotsky et al., 1989). When relevant tissues are isolated and examined using in vitro techniques, many of these interactions persist suggesting neuronal codistribution or mediation by short synaptic circuits. This multiplicity of afferent input attests to the need for precise homeostatic control of the HPA axis and suggests a role for CRF in particular as an integrative neuropeptide which consolidates relevant input from dispersed sources.

The hypothesis that CRF systems constitute the final common pathway for mediation of hypothalmo–pituitary–adrenal (HPA) tone has been tested experimentally (Cole et al., 1990a). Using the technique of immunoneutralization in which the pituitary actions of CRF in portal blood are limited by prior introduction of a CRF antibody that sequesters and inactivates the peptide, numerous studies have shown that HPA-activating properties of footshock, social defeat, and pharmacological stressors, such as cocaine and cholecystokinin, are CRF dependent. Moreover, while other ACTH secretagogues, such as vasopressin exert synergistic activation of HPA tone in the presence of CRF by virtue of clearly defined codistribution and copackaging of these hypophysiotropic factors in the paraventricular nucleus and median eminence, these actions appear to be mediated predominantly or exclusively via increased release of endogenous CRF (Brunani et al., 1995).

Several extrapituitary effects of other neurotransmitters and nonCRF neuropeptide systems including feeding suppression, increased emotionality, and fever induction also appear to the CRF dependent (Gray, 1993). For instance, the anxiogenic-like and anorexic actions of serotonin agonists, such as fenfluramine as well as cholecystokinin, caffeine, and estradiol, are blunted or reversed by reduction in CRF tone accomplished by CRF immunoneutralization or central administration of a CRF receptor antagonist. Similarly, anxiogenic-like behavior produced by central administration of cholecystokinin octapeptide is reversed by concurrent administration of a CRF receptor antagonist or CRF antiserum in a dose-dependent manner. In addition, behavioral despair, anorexic actions, and antinociceptive effects of cytokines such as interleukin-1 appear to be CRF dependent. These results suggest that a possible countermeasure for departure from behavioral homeostasis is to normalize CRF tone regardless of whether CRF is directly or indirectly involved in the behavioral dysregulation.

CRF-binding protein

The majority of late gestational maternal plasma CRF is bound to a high affinity CRF-binding protein (CRF-BP) which neutralizes the receptor agonist's ACTH-releasing properties (Lowry et al., 1996). Thus, maternal plasma CRF-BP levels determines the amount of "free" CRF that will bind to pituitary CRF receptors and thereby modulate the activity of the pituitary–adrenocortical axis during late human pregnancy. Many workers have now demonstrated that CRF is substantially elevated during the third trimester of human pregnancy and that this process is likely to participate in a cascade of events which eventually leads to parturition (Behan et al., 1993). This beneficial biological action of CRF is presumably exerted without undesirable Cushingoid-like symptoms of pituitary–adrenal overactivation due to the simultaneous, buffering presence of CRF-BP.

In contrast to humans, all other species examined to date do not express CRF-BP in the liver and plasma. The predominant tissues expressing CRF-BP in all species are the brain and pituitary gland. With respect to the central nervous system and the role of CRF-BP, it has been demonstrated, by immunohistochemistry and in situ hybridization techniques, that CRF-BP is expressed in various areas of rat brain including the cerebral cortex, amygdala, hippocampus as well as sensory relay nuclei associated with the auditory, olfactory, vestibular, and trigeminal systems (Potter et al., 1992). Of note, there are brain areas that are enriched with CRF and CRF-BP but contain low densities of receptors and conversely, other brain areas that are enriched with receptors and devoid of CRF-BP. Thus, the differential distribution of brain CRF-BP and CRF receptors presents multiple distinct sites of interaction with CRF (Behan et al., 1993).

A number of observations suggest that there is a membrane associated form of the CRF-BP within the brain (Behan et al., 1995a). Using a ligand immunoradiometric assay to detect the CRF-BP, a specific CRF-BP-like activity could be solubilized from sheep, rat, and human brain membranes. Further purification of crude rat and sheep brain membranes by sucrose density gradient centrifugation and detergent solubilization localized CRF-BP activity to the plasma membrane fraction. The detergent solubilized rat and human brain membrane preparations have similar pharmacological profiles to the recombinant protein present in human plasma. A large proportion of CRF in normal human brain may be complexed to the CRF-BP and thus unavailable for actions at the receptor. It is likely that the interaction between CRF and membrane-associated CRF-BP in brain is important in maintaining synaptic CRF concentrations either by presynaptic uptake or by modulating the quantity of neuropeptide that activates CRF receptors at the membrane interface (Turnbull and Rivier, 1997).

The identification of a membrane-associated form of the CRF-BP in brain with kinetic and pharmacological characteristics comparable to those previously observed for human plasma CRF-BP provides a novel target to modulate endogenous CRF levels in select brain areas enriched in CRF-BP. Selective peptide ligands that dissociate CRF from the CRF-BP, termed CRF-BP ligand inhibitors, mimic a number of behavioral effects of CRF including food intake suppression (Heinrichs et al., 1999; Bjenning and Rimvall, 2000) and locomotor activation (Heinrichs and Joppa, 2001). CRF-BP ligand inhibitors also normalize the reduced levels of unbound CRF in postmortem cerebral cortex of Alzheimer's patients to those seen in age-matched controls (Behan et al., 1995b) and exert significant cognitive-enhancing properties in animal models of learning and memory such as the Morris water maze and visual discrimination tests without producing any overt anxiogenic actions characteristic of CRF receptor agonists (Behan et al., 1995a; Radulovic et al., 2000; Zorrilla et al., 2001). Thus, CRF-BP represents a novel target for the symptomatic treatment of cognitive deficits associated with neurodegenerative dementia (see section on Learning and Memory Modulation) (Fig. 1).

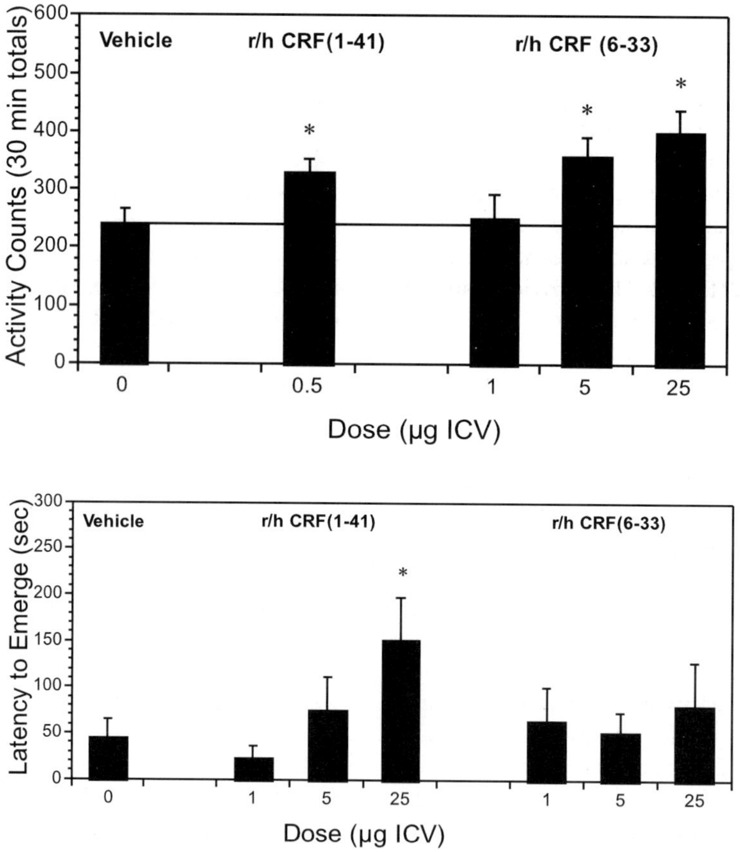

Fig. 1. Limited and selective pharmacological actions of a CRF-binding protein ligand inhibitor peptide relative to CRF itself, a full postsynaptic receptor agonist peptide. Photocell activity counts (mean ± SEM) measured cumulatively over 30 min (top panel) following administration of vehicle, a 0.5 μg dose of r/h CRF or 1, 5, or 25 μg doses of r/h CRF (6–33). Latency to emerge (mean ± SEM) from an enclosure placed within an open field (bottom panel) following administration of vehicle, 1, 5, or 25 μg doses of r/h CRF (1–41) or 1, 5, or 25 μg doses of r/h CRF (6-33) in the Defensive Withdrawal test. Modified from Heinrichs and Joppa, 2001.

Modulation of avoidance, approach, and arousal by stress and CRF systems

The following three sections review evidence in favor of the view that brain CRF system activation promotes negatively valenced avoidance behaviors, counteracts positively valenced consummatory behaviors, and exerts neural and behavioral activation. An assortment of animal models sensitive to these processes as well as the bidirectional effects of CRF receptor agonists and antagonists in these testing contexts is provided in Table 1.

Anxiogenesis, dispair and aversion: avoidance behavior and CRF

Anxiogenesis

Compelling evidence of a role for brain CRF-like neuropeptides in behavioral responses to stressors comes from the demonstration of antistress actions of central administration of CRF receptor antagonists. Evidence from studies employing competitive CRF receptor antagonist peptides, such as α-helical CRF (9–41) and D-Phe CRF (12–41), provides strong support for the hypothesis that brain CRF systems play a role in mediating behavioral responses to stress (Akwa et al., 1999; Bakshi and Kalin, 2000). These two neuropeptide antagonists have high affinity for both CRF_1 and CRF_2 receptors and have been shown in rats to reverse the decrease in food intake induced by stress and to attenuate stress-induced fighting, suggesting that both the suppression and the activation of behaviors associated with stressors may involve endogenous CRF systems. Subsequent studies have shown that the peptide CRF antagonists are very effective in reversing the decrease in exploration of the open arms of an elevated plus-maze caused by exposure to a variety of stressors including restraint, swim stress, ethanol withdrawal, and social stress (Heinrichs et al., 1994). The multiple actions of CRF on behavioral functions can now be examined using new, high affinity, small molecule receptor antagonists that exhibit central activity upon systemic application (see chapter by Steckler, this volume).

The nonpeptide CRF_1 receptor selective antagonist, CP-154,526, is reported to exert anxiolytic-like activity in the elevated plus-maze test in rats (Chen et al., 1996). Other studies have compared the behavioral effects of the CP-154,526 with those of diazepam and the 5-HT_{1A} receptor partial agonist, buspirone, in classical animal models of anxiety (Griebel et al., 1998). Unlike diazepam and buspirone and as expected given negative data with peptide CRF receptor antagonists, CP-154,526 was devoid of significant activity in conflict tests (punished lever pressing and punished drinking tests in rats). In a mouse defense test battery that has been validated for the screening of anxiolytic drugs, diazepam attenuated all defensive reactions of mice confronted with a rat stimulus (i.e. flight, risk assessment, and defensive attack) or with a situation associated with this threat (i.e. contextual defense). Buspirone reduced defensive attack and contextual defense, while CP-154,526 affected all defensive behaviors, with the exception of one risk assessment measure (Griebel et al., 1998). Thus, in mice the anxiolytic-like efficacy of CP-154,526 is superior to that of the atypical anxiolytic buspirone but is weaker than that of diazepam in terms of the magnitude of the effects and the number of indices of anxiety affected (Griebel et al., 1998). Anxiolytic-like efficacy of CRF_1 antagonists is also reported using the fear-potentiated startle and the pentobarbital-induced hypnosis tests (Steckler and Holsboer, 1999). Recent studies in monkeys also indicate that the selective CRF_1 antagonist, antalarmin, administered orally significantly inhibited a repertoire of behaviors associated with anxiety and fear induced by an intense social stressor (Habib et al., 2000).

Dispair

CRF systems are conceptualized as mediators of such basic emotional constructs as dispair (Panksepp, 1998; Buck, 1999). For instance, potential antidepressant-like effects of the small molecule CRF_1 receptor antagonist, CP-154,526, have been studied using the learned helplessness procedure, a putative model of depression with documented sensitivity to antidepressant drugs (Chen et al., 1997). Rats were exposed to a series of inescapable foot shocks on three consecutive days and tested in a shock-escape procedure on the fourth day. Animals trained to exhibit behavioral despair performed poorly in the shock-escape test compared with control animals not receiving inescapable shocks. Systemically administered CP-154,526 dose dependently reversed the escape deficit when administered 1 h prior to the test session, but had no effect on the performance of control rats not receiving prior exposure to inescapable stress (Chen et al., 1997). Likewise, in another experimental model of human depression, the mouse tail-suspension test, subcutaneous injection of CP-154,526 alleviated depression-like behavior (immobility) induced by consensus interferon-α (IFN-α) or sumiferon, a natural IFN-α (Yamano et al., 2000). Consensus IFN-α is effective in treating chronic hepatitis C, however, psychiatric side effects including depression are common and treatable by antidepressants. These data support evidence implicating CRF_1 receptors in the pathophysiology of depression and suggest potential therapeutic efficacy of small molecule CRF_1 receptor antagonists in the treatment of affective disorders.

A number of observations have suggested that CRF functions abnormally in depressed patients (Hartline et al., 1996). Many patients with major depression are hypercortisolemic and exhibit an abnormal dexamethasone suppression test. Given the primary role of CRF in stimulating pituitary–adrenocortical secretion, the hypothesis has been put forth that hypersecretion of CRF in brain might underlie the hypercortisolemia and symptomatology seen in major depression. The cerebrospinal fluid (CSF) concentration of CRF is significantly elevated in depressed patients, and a significant positive correlation is observed between CRF concentrations in the CSF and the degree of insensitivity to dexamethasone suppression of plasma cortisol in depressed individuals. Furthermore, the observation of a decrease in CRF binding sites in the frontal cerebral cortex of suicide victims compared to controls is consistent with the hypothesis that CRF is hypersecreted in major depression (Nemeroff, 2000). The increased CSF concentrations of CRF seen in depressed individuals are decreased by following treatment with electroconvulsive therapy and this normalization correlates well with improvement.

In view of the data suggesting a central role for CRF in depression and consistent with the above discussion of CRF-dependent neurotransmitter interactions (see section on Neuroendocrine and Neurotransmitter Interactions and the chapter by Linthorst, this volume), the hypothesis has been put forth that antidepressants may produce their therapeutic effects, in part, by decreasing CRF secretion (Cizza et al., 1995). An increase in CRF binding sites, presumably to compensate for chronic suppression of CRF secretion, is observed in some brain regions such as the brain stem in rats treated chronically with tricyclic antidepressants such as imipramine. CRF is postulated to have an important role in the locus coeruleus of the brain stem, the primary source of noradrenergic innervation of the forebrain (Koob, 1999). The locus coeruleus receives a rich CRF innervation, contains a moderate density of CRF receptors, and is markedly activated following central administration of CRF. Furthermore, coerulear concentrations of CRF are selectively increased following application of acute or chronic stressors. Given the major involvement of the brain noradrenergic system in depression and activation of noradrenergic neurons in this brain region by CRF, it is possible that antidepressants may function by suppressing CRF secretion in the locus coeruleus, resulting in the observed increase in brain stem CRF binding sites.

The role that has been proposed for CRF in major depressive disorders along with preclinical data in rats demonstrating anxiogenic-like behavioral effects of CRF administration have led to the suggestion that CRF may also be involved in anxiety-related disorders (Owens and Nemeroff, 1993). A role for CRF in panic disorder has been suggested by observations of blunted ACTH responses to intravenously administered CRF in panic disorder patients when compared to control subjects. The blunted ACTH response to CRF in panic disorder patients most likely reflects a process pituitary occurring at or above the, possibly resulting in excess secretion of endogenous CRF.

In view of the evidence described above suggesting that hypersecretion of CRF may underlie some of the symptomatology seen in affective disorders and anxiety-related disorders, it stands to reason that CRF receptor antagonists may be useful in the treatment of these disorders (Holsboer, 1999). Thus, a CRF antagonist may be a useful antidepressant, anxiolytic, or antistress treatment. While major advances have been made in the design of metabolically stable peptide analogs, the problem of achieving access to the brain following systemic administration and oral bioavailability with such compounds is only now being solved (see chapter by Steckler, this volume). Moreover, the well validated antistress actions of peptide CRF receptor antagonists in counteracting the anxiogenic behavior produced by acute exposure to a variety of individual stressors beg the question of long-term efficacy of CRF antagonists in attenuating anxiogenic behavior arising from repeated, long-term exposure to multiple stressors using a chronic unpredictable stress procedure. Such exposure persistently activates the HPA axis and has been advanced by some investigators as an animal model of depression.

Aversion

Postprandial administration in the rat of a wide variety of drugs, peptides, and toxins suppresses future consumption of a meal of previously unfamiliar but otherwise attractive saccharin-flavored solution. Since the intensity of this conditioned flavor aversion in the rat is sensitive to plasma stress hormone levels, a series of studies has examined the effects of CRF peptides on flavor conditioning. In two-bottle water versus saccharin choice tests, an intracerebroventricular dose of CRF (5 µg, but not 0.5 µg) abolished saccharin intake following two saccharin/CRF pairings (Heinrichs et al., 1991). Hence, exogenous CRF is capable of inducing flavor aversion in a dose-dependent manner. Further, direct neurotropic actions of CRF probably subserve its aversive effect since dexamethasone pretreatment weakened but did not prevent CRF-induced conditioned taste avoidance (Heinrichs et al., 1991). These results suggest CRF receptor agonist treatment can exert aversive effects that are reflected in taste avoidance.

Another series of experiments compared the conitioned aversive consequences of central administration of CRF/urocortin at doses that produced comparable decrements in food intake (Benoit et al., 2000). In particular, 1.0 µg urocortin 1 and 2.0 µg CRF administered intracerebroventricularly produced similar reductions in food intake whereas CRF but not urocortin 1 promoted robust and reliable taste aversion learning (Benoit et al., 2000). It is concluded that urocortin 1, at doses that reduce food intake to levels like that observed after administration of CRF, do not produce similarly aversive consequences. Available evidence suggests that affective taste reactivity patterns in multiple species reflect a core hedonic process of palatability or affect, rather than being an ingestion measure, consummatory behavior measure, or a sensory reflex measure (Berridge, 2000).

Energy balance and reward: modulation of approach behavior by CRF

Energy balance

Energy balance regulation encompassing ingestive behavior and coordination of humoral, gastrointestinal, and metabolic responses to nutritional status represents a specific and well-documented example of a finely tuned homeostatic system which is quite sensitive to environmental change. Dynamic control over the amount and type of foodstuff consumed is present in infancy in both man and animal and matures during weaning. Indeed, the simultaneous emergence of adult glucose sensitivity and receptivity of the HPA axis to stress activation suggests that adaptive mechanisms, which react to stressors metamorphose from neonatal to adult life, are intricately related to metabolic needs of an organism (Widmaier, 1990). Accordingly, given the seminal role of CRF in initiating the active "fight or flight" components of the response to stressors, a complementary intrinsic role of CRF in coordinating passive sequelae of stress exposure such as inhibition of intestinal motility and extinguished appetite would not be surprising (Heinrichs and Richard, 1999).

Considerable evidence suggests a role for endogenous brain CRF systems in appetite regulation, energy balance, and in the etiology of eating disorders (Glowa et al., 1992; Dagnault et al., 1993). Food intake is diminished by administration of CRF agonists or treatments that elevate endogenous CRF levels, such as stress, tumor induction, or appetite-suppressing drugs. It is noteworthy that CRF treatment induces, concurrently with a reduction in food intake, an increase in the activity of the sympathetic nervous system. This finding suggests that the anorectic effect of CRF may be mediated, as are its thermogenic effects, by central control over the autonomic nervous system (Rothwell, 1990). Interestingly, there are few reports of intrinsic effects of CRF receptor antagonists on energy balance. In particular, central administration of the CRF receptor antagonist, α-helical CRF (9-41), does not alter intake in nondeprived or food-deprived subjects at doses that potentiate appetite induced by neuropeptide Y and attenuate stress-induced appetite suppression. These clues point to a physiological role for CRF in the induction of negative energy balance not at steady state, but under conditions of exaggerated hunger/weight gain which may be counteracted by anorexic and sympathomimetic effects of activated CRF systems. Indeed, brain CRF content is dependent on feeding/weight status in animal models of dysregulated energy balance, such as the Zucker obese rat, tumor-bearing cachexia, chronic exercise, and in the context of drug- or stress-induced changes in appetite (Heinrichs and Richard, 1999).

The first indications that CRF might play a role in the regulation of energy balance, body weight, and obesity arose indirectly from experiments which assessed the effects of surgical adrenalectomy in genetically or surgically obese rodents. Bilateral adrenalectomy activates hypothalamic CRF systems and prevents, attenuates, or reverses the normally high rates of energy deposition in genetically obese Zucker rats, ob/ob mice, and db/db mice and in rodents made obese by chemical or surgical lesions of the hypothalamus (Rothwell, 1990). Moreover, excessive food intake is also suppressed by adrenalectomy. However, this anorexic effect of adrenalectomy accounts for only a portion of the reduced weight gain since comparable levels of food restriction in intact rats do not prevent the development of obesity. Measurement of energy balance and oxygen consumption as indices of metabolic rate have revealed that adrenalectomy suppresses the very high efficiency of weight gain in obese rats and increases metabolic rate. One feature of the primary etiology of obesity in these animal models, insufficient heat production within sympathetically enervated brown adipose tissue, is normalized by adrenalectomy such that the nonshivering thermogenic response to food is increased by adrenalectomy and restored to values seen in lean animals. The effects of adrenalectomy on energy balance and thermogenesis are not restricted to genetically obese rodents and have been reported to inhibit the development of obesity following electrolytic lesions of the ventromedial and paraventricular hypothalamic nuclei. Particularly important is the finding that overeating and overweight induced pharmacologically by chronic administration of neuropeptide Y is also reversed by adrenalectomy. Taken together, these generalized energy balance restorative effects of adrenalectomy suggest the beneficial efficacy of CRF activation in human forms of obesity.

Several lines of research suggest that an overactive endogenous neuropeptide Y (NPY) system may contribute to overeating and weight gain, while anorexic and cachexic properties of CRF may act to restore energy balance (Beck et al., 1990; Brady et al., 1990; Dryden et al., 1993; Jeanrenaud, 1994). The hypothesis that endogenous CRF has an inhibitory action on food intake has been tested using brain microinjections of α-helical CRF (9–41). Pretreatment with the CRF antagonist either intracerebroventricularly or directly into the paraventricular nucleus of the hypothalamus (PVN) enhances the ability of NPY administered into the same locus to stimulate feeding. Enhancement of NPY orexigenic effects were also observed two weeks after immunotargeted impairment of CRF neurons in the PVN using local administration of a monoclonal antibody against CRF and toxins. These results are in agreement with previous reports of the antistress effect of α-helical CRF (9–41) in reversing the anorexia produced by restraint stress. In addition, food intake occurring in response to a physiological stressor such as nutritional imbalance may be constrained by anorexic actions of endogenous CRF systems. Thus, CRF systems may serve to limit food intake when an element of risk intrudes upon established feeding patterns of animals forced by biological need to consume novel foodstuffs or to consume food under stressful conditions. In particular, appetite that is exaggerated pharmacologically by treatments such as NPY may be kept in check by concurrent activation of endogenous CRF.

CRF receptor antagonists may have utility in the context of eating disorders (Krahn and Gosnell, 1989). Anorexia and bulimia nervosa are eating disorders characterized by psychological pathologies such as stress-related alterations in food intake as well as physiological irregularities such as delayed gastric emptying (Holt et al., 1981; Inui et al., 1995; Asakawa et al., 2000). Interestingly, central administration of mixed CRF receptor antagonists results in normalization of stress-induced anorexia (Contarino et al., 1999b) and gastric stasis (Taché et al., 1999) and the CRF_1 receptor antagonist, CRA 1000, prevented emotional stress-induced inhibition of food intake (Hotta et al., 1999). Comorbidity of eating disorders and depression (Wiederman and Pryor, 2000) may favor efficacy of CRF_1 receptor antagonist drugs in eating disorders accompanied by affective psychopathology. Moreover, recent studies showed that central administration of CRF and the more potent urocortin 1 peptide suppressed food intake in rodents (Adinoff et al., 1996) and this action was prevented by intracerebroventricular administration of the antisauvagine-30 suggesting a role of CRF_2 receptors in the anorexic syndrome induced by CRF (Pelleymounter et al., 2000).

Reward

The behavioral profile of CRF in mediating anxiogenic-like and aversive responses to stress may be particularly relevant for drug dependence and withdrawal states (Vargas et al., 1992). While neurochemical adaptations to chronic drug use almost certainly occur within brain pathways responsible for the acute reinforcing actions of drugs, separate brain systems may coordinate the generalized anxiogenic-like and aversive behavioral responses which accompany chronic use and abstinence from drugs of abuse including cocaine, ethanol, and morphine (Sarnyai et al., 2001). For example, motivational measures of ethanol withdrawal have suggested a possible role for central nervous system CRF in alcohol dependence. Ethanol injected acutely can reverse the anxiogenic-like effects of intracerebroventricular administration of CRF and rats withdrawn from chronic ethanol show a stress-like response on the elevated plus maze which is reversed by intracerebroventricular administration of α-helical CRF (9–41). Much smaller amounts of the CRF receptor antagonist, when injected into the amygdala, are effective in reversing the stress-like effects of ethanol on the plus maze. Moreover, intraamygdala administration of the CRF receptor antagonist also exerts anti-aversive properties in attenuating the strength of place aversion learning or operant response disruption produced by naloxone-precipitated withdrawal in morphine dependent subjects. One possible interpretation of the anti-anxiogenic and anti-aversive effects of the CRF antagonist in ethanol and opiate-dependent groups is that the persistent drug presence gives rise to an opponent process involving activated CRF systems which is counterbalanced in the steady state but which is unmasked during withdrawal. This generalized involvement of CRF systems in drug-related negative motivational states is consistent with the comprehensive role of CRF in mediating the emotional response to stressors.

Brain CRF also appears to have a role in the interaction of stressors with psychostimulant drug responses. Acute cocaine administration activates the HPA axis and produces a reduction in CRF immunoreactivity within CRF-containing hypothalamic and limbic nuclei suggesting that brain CRF is released in response to systemic cocaine injection (Sarnyai et al., 2001). Chronic cocaine administration in rats produces a downregulation in CRF receptors characteristic of increased CRF release in the forebrain areas such as nucleus accumbens and frontal cortex that have been implicated in the reinforcing effects of cocaine; these cocaine-induced alterations in CRF receptors were reversed by selective ablation of dopamine neurons with 6-hydroxydopamine treatment. Activation of brain CRF may also be involved in the behavioral cross-sensitization between stress and psychostimulant drug exposure since the CRF receptor antagonist α-helical CRF (9–41) administered centrally prevents the development of restraint stress-induced sensitization to subsequent amphetamine exposure (Cole et al., 1990b).

There are reports that corticosterone stimulation and locomotor activation produced by cocaine infusion are blocked by coadministration of a competitive CRF receptor antagonist or CRF antiserum in a dose-dependent manner (Sarnyai et al., 2001). Studies on individual differences in the vulnerability to acquire amphetamine self-administration suggest a positive relationship between the plasma levels of corticosterone and subsequent drug responses (Piazza and Le Moal, 1996). Similarly, plasma corticosterone is positively correlated with cocaine intake following a stress-induced increase in drug responses and in animals selectively bred for high or low catecholamine responses to stress. While the behavioral responses to stress are largely independent of HPA tone, this HPA activation often parallels brain CRF activation. Consistent with this hypothesis, an in vivo microdialysis study has shown that acute cocaine administration increases the release of CRF in the central nucleus of the amygdala (Richter et al., 1995).

An increase in reinforcement efficacy of abused drugs as measured in conditioning and self-administration contexts is accompanied by persistent or exaggerated stress-like responses in animal models, which may reflect the nervousness and restlessness reported by drug-abusing populations in the clinic (Koob and Le Moal, 2001). CRF receptor antagonists seem to be most effective in modifying the acquisition of novel behaviors as opposed to the performance of well-learned behaviors. A growing body of literature distinguishes "first time" from habitual drug users in terms of the underlying adaptive processes that alter behavioral and physiological responses to a single drug over time. One can then hypothesize that vulnerability to addiction at the time of initial exposure to the drug is determined in part by individual differences in reactivity to stress. Accordingly, it would be interesting to examine the ability of a CRF receptor antagonist to alter the acquisition of drug self-administration (Goeders and Guerin, 2000).

In particular, CRF_1 receptors have been implicated in the withdrawal and relapse syndromes for various drugs of abuse (Abbott et al., 2000; Iredale et al., 2000). Administration of CP-154,526 prior to naltrexone or naloxone significantly decreased many of the behavioral signs of opiate withdrawal while central administration of a CRF_2 receptor, receptor antagonist, anti-sauvagine-30, had no effect (Abbott et al., 2000; Iredale et al., 2000). Anti-stress efficacy of CP-154,526 has also been examined in a paradigm of stress-induced relapse to drug seeking in cocaine- and heroin-trained rats (Erb et al., 1998; Abbott et al., 2000). Rats were first trained to self-administer heroin or cocaine and then responding for intravenous administration of drug solution was extinguished by substitution of saline. A footshock stressor reliably reinstated extinguished cocaine- and heroin-taking behavior and pretreatment with CP-154,526 significantly attenuated the reinstatement effect of the stressor in both heroin- and cocaine-trained rats (Erb et al., 1998). CP-154,526 also completely blocked the relapse to opiate dependence in the 28-day extinction of morphine-conditioned place preference which is used to test for anti-craving activity (Abbott et al., 2000). These results highlight an important role for the CRF system working through CRF_1 receptors in the expression of drug withdrawal symptoms and vulnerability to stress-induced relapse (Erb et al., 2000).

Waking and locomotor activity: arousal and CRF

Waking

Electrophysiologically, CRF/urocortin have excitatory properties. CRF and urocortin 1 injected intracerebroventricularly in doses of 0.01–0.10 μg produce electroencephalographic activation characteristic of arousal (Ehlers et al., 1983; Slawecki et al., 1999), and at higher doses CRF produces seizure-like activity (Ehlers et al., 1983). In particular, CRF administration produced in one report decreases in slow-wave sleep concomitant with significant decreases in spectral power in lower (1–6 Hz) frequencies and increases in high (32–64 Hz) frequencies (Ehlers et al., 1986b). At sufficiently high doses urocortin 1, like CRF, elicits limbic seizures, an effect that appears to be mediated by CRF_1 receptors (Baram et al., 1999; Brunson et al., 2001). The relationship between CRF/urocortin levels and seizure incidence is reciprocal since limbic seizure kindling results in increased levels of CRF and CRF-binding protein in hippocampus (Smith et al., 1997).

Brain CRF systems appear to mediate arousal processes including regulation of the sleep–wake cycle (Opp, 1995). The stimulus for activation of limbic, hypothalamic, and brain CRF circuits is any perceived change in the external environment or physiological homeostasis for which arousal is an appropriate coping response. For instance, immobilization restraint applied at the beginning of the light period, but not the dark period, increased waking and reduced rapid eye movement sleep without dramatically altering slow-wave sleep and central administration of a CRF receptor antagonist blocked the increase in waking after physical restraint (Chang and Opp, 2002). CRF-induced changes in sleep architecture appear to be primarily centrally mediated since systemic administration of CRF has little or no effect on the duration of slow-wave or rapid eye movement sleep epochs (Born et al., 1989). Additional support for the physiological relevance of endogenous CRF system activation in the regulation of wakefullness is provided by the finding that central administration of a CRF receptor antagonist reduces the time spent awake (Opp, 1995). In particular, intracerebroventricular injection of a 6.5 nmol dose of α-helical CRF (9–41) at the onset of the dark phase of the circadian cycle significantly increased nonrapid eye movement sleep over the subsequent 24 h (Opp, 1995). This finding provides the rationale for administration of a CRF receptor antagonist as a pharmacological means of diminishing wakefullness.

Locomotor activity

In non-stressed animals under low arousal conditions, CRF and urocortin 1 administered intracerebroventricularly produce a dose-dependent behavioral activation that includes increases in locomotor activity, rearing and grooming when rats are tested in a familiar environment (Sutton et al., 1982; Britton et al., 1986; Spina et al., 2000). This activation is not observed following systemic administration of CRF and is not blocked by hypophysectomy or pretreatment with dexamethasone, suggesting that this effect of CRF is mediated by actions in the central nervous system independent of the pituitary–adrenal axis (Eaves et al., 1985; Swerdlow and Koob, 1985). Moreover, the neural substrates for the locomotor-activating effects of centrally administered CRF are separate from those circuits which mediate the activating effects of psychostimulant drugs such as caffeine and amphetamine (Swerdlow and Koob, 1985).

When animals are exposed to a more stressful environment, the profile of the behavioral activation produced by exogenously administered CRF and urocortin 1 changes to reflect behavioral inhibition. The same intracerebroventricular doses of peptide that produce marked behavioral activation in a familiar environment, produce behavioral suppression in a novel, presumably stressful environment. Rodents pretreated with CRF show decreases in behavior in an open field (Takahashi et al., 1989), with or without food availability (Britton et al., 1982), decreased exploration in a multi-compartment chamber (Berridge and Dunn, 1986), and decreased exploration in an elevated plus-maze (Baldwin et al., 1991). Thus, both increases and decreases in brain excitability and locomotor activation can be induced by administration of CRF receptor agonists in a dose-dependent fashion. This same continuum of arousal from activation to deactivation is conceptualized as an indivisible component of the mechanism for the psychological construction of human emotion (Russell, 2003).

Learning and memory modulation: fine-tuning affect

Stressor exposure in the scientific literature and popular press typically connotes an impending decline into failing health, dysfunctionality, and psychiatric illness. While such untoward consequences were collectively categorized by Selye as "distress" in his book, "The Stress of Life" (Selye, 1976), an antithetical and in some sense paradoxical enhancement of performance and improvement in well being also associated with stressor exposure was identified and labeled "eustress". Although pathological consequences undeniably result from uncontrollable or long-term stress (see chapter by Kovacs, this volume), it should be borne in mind that the departure from homeostasis under normal conditions is brief and has adaptive value. Salutary properties of CRF system activation have been identified experimentally using low doses of receptor agonist under minimal intensity stimulation conditions in learning and memory-related tasks (Heinrichs et al., 1997b). In parallel fashion, exposure to a maternal separation procedure which activates brain CRF during development is reported to enhance selective attention and improve learning performance in

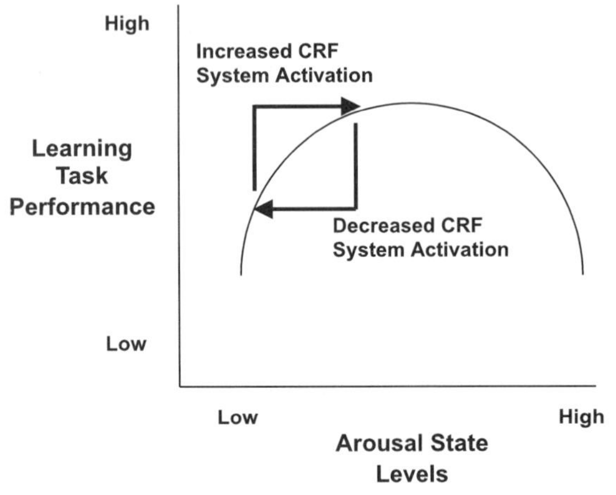

Fig. 2. Schematic diagram relating arousal levels and performance in a working memory task. Arousal attenuating manipulations, such as administration of a CRF receptor antagonist, are conceptualized to damage successful performance in keeping with an underlying inverted U-shaped function. Arousal enhancing administration of low doses of a CRF receptor agonist or a CRF-binding protein ligand inhibitor are hypothesized to optimize performance from a low arousal baseline.

adulthood (Lehmann et al., 2000). These results suggest that environmental and pharmacological manipulations that alter learning and memory performance do so in part by modulating CRF system activation levels (Fig. 2).

Several lines of evidence support the present identification of a physiological role for CRF systems in information processing functions of the central nervous system. First, steady-state levels of endogenous CRF family neuropeptide receptor agonists appear sufficient to modulate learning and memory functions since pharmacological dissociation of CRF and a related neuropeptide, urocortin 1, from their binding protein in brain enhances performance in appetitively and aversively motivated memory tasks (Behan et al., 1995a; Heinrichs et al., 1997a; Eckart et al., 1999; Liang et al., 2001; Weiss et al., 2001). Second, central administration of CRF exerts electrophysiological and neurochemical activation (but see also Rebaudo et al., 2001) of hippocampal circuits relevant for learning and memory processes in several species (Bonaz and Rivest, 1998; Wang et al., 1998; Fuchs et al., 2001). For example, several comprehensive papers from Spiess and colleagues (Blank et al., 2002, 2003a,b) document that intrahippocampal administration of r/h CRF induces neural excitability via several different signaling cascades together with an accompanying increase in hippocampus-dependent fear conditioning and the pharmacological reversibility of these effects using competitive peptide and nonpeptide CRF receptor antagonists. Finally, the brain and cortical CRF levels are significantly reduced in patients with both mild and severe dementia such that cerebrospinal fluid levels of CRF correlate with the degree of cognitive impairment in dementia sufferers (De Souza et al., 1986). Thus, CRF decrements may serve as a potential neurochemical marker of early dementia and possibly early Alzheimer's Disease.

One report has validated the utility of the sauvagine analog, antisauvagine-30, a peptide ligand with preferential affinity for CRF_2 receptors, in an information processing context (Eckart et al., 1999). In particular, results indicate region specific modulation of learning/anxiety through differential mediation by CRF_1 versus CRF_2 receptors. Injection of CRF into the dorsal hippocampus before training enhanced learning of fear conditioning through CRF_1 receptors as demonstrated by the finding that this effect is prevented by the local injection of the nonselective CRF receptor antagonist, astressin, but not by antisauvagine-30 into the dorsal hippocampus (Eckart et al., 1999). In contrast, injection of CRF into the lateral intermediate septum impaired learning of an aversive stimulus through CRF_2 receptors, as demonstrated by the ability of antisauvagine-30 to block this effect (Eckart et al., 1999). Note that the involvement of CRF systems in information processing is in keeping with the hypothesized role of CRF neurobiological derangement in dementia (Contarino et al., 1999b).

Behavioral phenotype of CRF overexpressing mice

CRF overproduction has been hypothesized to be involved in a number of stress-related psychiatric anxiety and affective disorders (Groenink et al., 2002). A transgenic mouse model of CRF overproduction has been developed (Heinrichs et al., 1996) using a CRF transgene composed of rat genomic CRF gene and 3′ and 5′ substitutions. The CRF transgenic mice exhibit endocrine abnormalities including elevations of ACTH and corticosterone, but also enhanced reactivity to novelty and an anxiogenic-like response on the Elevated Plus-Maze. These behavioral effects are reversed by central administration of α-helical CRF (9–41). Thus, CRF-transgenic mice provide a valuable tool for investigating the overproduction of CRF not only in the HPA axis but extrahypothalamic systems as well (Groenink et al., 2002). This model lends support to the hypothesis that extrahypothalamic CRF plays an important role in behavioral responses to stressors. Moreover, the persistent anxiogenic state exhibited spontaneously by CRF overexpressing mice provides a natural complement to the endocrine and behavioral hyperreactivity following long-term, chronic stressor application, but one in which the disturbances in homeostasis can be attributed largely and selectively to CRF dysregulation.

One likely consequence of hyperarousal engendered by chronic CRF exposure in transgenic overexpressing mice is a decrement in focused attention or information processing tasks (Valentino et al., 1993). In order to determine the consequences of CRF overproduction on behavioral plasticity, the learning and memory capacities of CRF-transgenic mice have been tested in forced alternation water T-maze, Morris water maze, and 5-choice serial reaction-time tasks (Heinrichs et al., 1996; van Gaalen et al., 2003). In T-maze testing, littermate control mice reached a criterion of 70% correct responses after five days of trials while the performance of transgenic subjects was still random after the same training. In Morris maze testing, control subjects reached the submerged platform significantly faster after three days of trials while the performance of transgenic mice was unimproved over the same period. Pretest administration of the benzodiazepine anxiolytic, chlordiazepoxide (10 mg/kg), prior to retention training produced a significant improvement in Morris maze performance in transgenic mice when compared to vehicle-treated transgenic littermates. These results suggest that CRF overexpressing mice exhibit a profound learning deficit and that memory plasticity can be restored by anxiolytic pretreatment. Thus, constitutive overabundance of brain CRF may produce hyperemotionality that interferes with learned behaviors.

Profound but reversible performance deficits of CRF-transgenic mice in tests of emotionality and learning plasticity suggest the utility of this animal model in studying both anxiety and learning disorders of clinical populations with an inferred etiology of CRF excess (Stratakis and Chrousos, 1995; Dirks et al., 2002). One inherent advantage of this approach is the ability to evaluate restorative properties of CRF receptor antagonists in functionally impaired CRF-transgenic mice relative to null effects in naive control subjects that do not typically exhibit intrinsic actions of CRF receptor antagonist administration. A follow-up series of experiments applying peptide CRF receptor antagonists in CRF-transgenic mice performing learning acquisition may therefore alter behavior by modifying the output of brain CRF systems themselves or by modifying the output of neurochemically distinct systems such as GABAergic or cholinergic pathways, which are functionally colocalized with CRF in local brain sites and are validated modulators of learning/memory capacity in their own right.

A role for CRF_1 receptors in mediating emotionality produced by stress and CRF receptor agonists has been supported by antistress and calming effects of CRF_1 receptor knockdown (Heinrichs et al., 1997a), knockout (Smith et al., 1998; Timpl et al., 1998), and pharmacological blockade (Schulz et al., 1996). Thus, one study was designed to evaluate the ability of a selective CRF_1 receptor antagonist, R121919, to alter the anxiogenic-like phenotype of CRF-transgenic mice. For example, locomotor performance of CRF-transgenic mice can be tested using the Two-Compartment model of anxiety which has been used extensively to phenotype CRF mutant mice (Contarino et al., 1999a; Bale et al., 2000). The anxiolytic-like action of R121919 was documented using the Two-Compartment test (Fig. 3). This result is also consistent with that reported for administration of another CRF_1 receptor antagonist, CP-154,526, in inbred BALB/c mice tested in the Two-Compartment task (Griebel et al., 1998). Additionally, administration of R121919 effectively promoted efficacy of sex hormone-inducted mating receptivity in CRF-transgenic mice (Fig. 3). It is important to note that the actions of R121919 in disinhibiting motor exploration of the Two-Compartment apparatus and increasing mounting

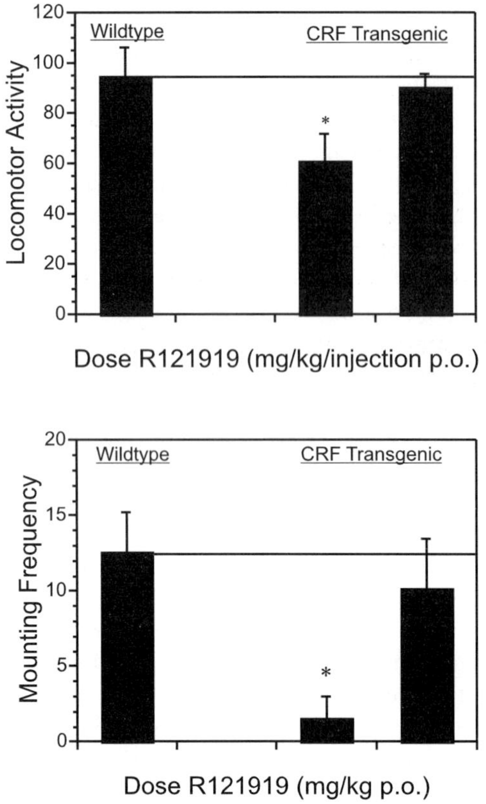

Fig. 3. CRF Type 1 receptor antagonist attenuated the anxiogenic-like phenotype of CRF-transgenic mice and diminished sexual receptivity of female CRF-transgenic mice. Motor activity (mean ± SEM) measured over 10 min in wildtype control or CRF-transgenic mice administered subchronic 0 or 10 mg/kg p.o. × 3 doses of R121919 in the Two-Compartment model of anxiety (top panel). Frequency of mounting (mean ± SEM) of experimental female wildtype and CRF-transgenic mice treated with a 0 or 20 mg/kg p.o. doses of R121919 by male conspecifics during a 20 min sexual receptivity test (bottom panel). $*p < 0.05$ relative to vehicle-treated, wildtype controls.

frequency both reflect increases in behavioral output. Accordingly, the present functional effects of R121919 cannot be attributed to a nonspecific sedative-like action that would constitute one likely behavioral toxicity of CRF receptor antagonist drugs. Instead, anxiolytic-like efficacy of R121919 in the CRF overexpressor mouse model suggests the potential utility of R121919 in anxiety related-disorders with an etiology of CRF excess.

The ability of a repeated 10 mg/kg × 3 dose of the CRF receptor antagonist administered in the Two-Compartment test to normalize the behavior of CRF-transgenic mice is consistent with the reported sensitivity of this exploratory suppression measure to overactivation of endogenous CRF systems (Heinrichs et al., 1997b). Previous studies have shown that environmental novelty effectively inhibits exploratory behavior of CRF-transgenic mice and that habituation of the novelty stressor or central administration of a peptide CRF receptor antagonist both nullify this phenotypic difference relative to wildtype control mice (Stenzel-Poore et al., 1994). While the Two-Compartment task was chosen for characterization of CRF receptor antagonist efficacy due to the species-typical nature of this task, a recent report (Weninger et al., 1999) documents anxiolytic-like efficacy of CP-154,526 in CRF knockout and wildtype (C57BL/6 x 129SVJ) mice tested in a shock-induced model of conditioned freezing. Finally, it is important to mention that a growing literature describes similar disinhibitory properties of CRF_1 receptor antagonists in rats tested in comparable fashion (McCarthy et al., 1999). Thus, the present results are consistent with available literature in proposing that exploratory behavior in mice is sensitive to CRF system dysregulation and that inhibitory effects of exposure to novel or aversive environments can be attenuated by CRF_1 receptor blockade.

Diminished sexual receptivity among female CRF-transgenic mice can be viewed as one maladaptive consequence of the anxiogenic-like behavioral profile induced by CRF overabundance (Heinrichs et al., 1997b). However, this behavioral deficit can be reversed pharmacologically by administration of a serotonin receptor antagonist and appears to affect only female, but not male, CRF-transgenic mice (Heinrichs et al., 1997b). Thus, available data do not support a conclusion of global behavioral disruption in CRF-transgenic mice but rather a profile of transient impairment in particular exploratory, information processing and social contexts. Validation of the pharmacology of CRF ligands in numerous murine models will provide new tools both for mutant phenotyping (Crawley, 2000) and for criticially testing the cross-species generality of CRF receptor agonist and antagonist actions that have been researched predominantly in the rat (Dunn and Berridge, 1990).

Unmet challenges for future research

The wish list for future research in the field of CRF neurobiology commonly includes a request for receptor selective ligands with which to probe physiological relevance of this stress neuropeptide system. However, the evidence described in the present review linking CRF activation with both increased arousal as well as increased emotionality demands a more theory driven investigation of the extent to which a homeostatic neuropeptide described primarily using basic research in animal species can participate in core affect and emotional regulation. Some theorists conceptualize a dual role in emotional regulation both for reward systems in mediating approach as well as punishment-related systems that facilitate avoidance and withdrawal (Rolls, 1999); certainly the known functions of CRF are consistent with the later affective state. Other investigators posit a more modular role for CRF in subserving specific affect programs labeled as "fear" or "aversion" and orchestrated by a separate executive process (Panksepp, 1998). Whatever the resolution of this distinction, it seems reasonable to consider CRF neurobiology from a contemporary perspective that acknowledges both the evolutionarily ancient visceral and physiological functions of CRF systems described in animals as well as the potential role of brain CRF in regulating the appetites, feelings, and cognitions of human beings.

It is important to note that the consistency of the body of literature related to the behavioral effects of CRF systems is blemished by surprising discrepancies between pharmacological studies performed in rats versus the phenotype of CRF mutant mice. For example, several reports describe inconsequential effects of CRF peptide knockout on a host of behavioral indices of activation, emotionality, and stress-reactivity. Similarly, the CRF Type 2 receptor knockout mice exhibit paradoxical anxiogenesis and hyperreactivity to stressor exposure (Bale et al., 2000). Thus, it would be desirable to evolve future studies of CRF biology away from purely descriptive exercises in favor of more integrative accounts of CRF functions in the brain. For example, one set of investigators cautions against unquestioned acceptance of the face validity of results from CRF-mutant mice and instead proposes that the state of CRF overexpression, for example, is a good model for life-long compensatory mechanisms for CRF excess (Groenink et al., 2003). Thus, as suggested by the use of the term "functionalist" in the title of this review, one productive avenue for future research would be to determine the causes and effects of mental states characterized by CRF system activation.

In summary, CRF in the central nervous system appears to have activating properties on behavior and to coordinate behavioral responses to stress. CRF receptor antagonists reverse changes in behavior associated with exposure to a wide variety of stressors and in a wide variety of experimental contexts thus suggesting that the physiological role of CRF is stress dependent and not intrinsic to a given behavioral response. Further, other neurotransmitter and neuropeptide systems which mimic features of the stress response such as ACTH release, thermogenesis, and emotionality appear to do so via a CRF-dependent mechanism. Hence, consistent with the dual role of other hypothalamic releasing factors in integrating hormonal and neural mechanisms by acting both as secretagogues for anterior pituitary hormones and as extrapituitary peptide neurotransmitters, CRF may coordinate coping responses to stress at several bodily levels and in the fashion of a final common effector pathway. Moreover, dysfunction in such a fundamental homeostatic system may be the key to a variety of pathophysiological conditions including mental disorders.

Acknowledgments

I am endebted to Wylie Vale for provision of the CRF-transgenic mice and thank Ric Alvarez for skillful and diligent maintenance of transgenic mouse stock and Helen Min for expert technical assistance.

References

Abbott, C.R., Edwards, C.M.B., Sunter, D., Kim, M., Dakin, C.L., Murphy, K.G., Abusnana, S., Taheri, S., Rossi, M. and Bloom, S.R. (2000) Cocaine- and amphetamine-regulated transcript (CART), Glucagon-like peptide-1 (GLP-1) and corticotrophin releasing factor (CRF) Reduce feeding via melanocortin 3 and 4 receptor independent pathways. Journal of Endocrinology, 164(Suppl.): 183.

Adinoff, B., Anton, R., Linnoila, M., Guidotti, A., Nemeroff, C.B. and Bissette, G. (1996) Cerebrospinal fluid concentrations of corticotropin-releasing hormone (CRH) and Diazepam-binding inhibitor (DBI) during alcohol withdrawal and abstinence. Neuropsychopharmacology, 15(3): 288–295.

Akwa, Y., Purdy, R.H., Koob, G.F. and Britton, K.T. (1999) The amygdala mediates the anxiolytic-like effect of the neurosteroid allopregnanolone in rat. Behavioural Brain Research, 106(1–2): 119–125.

Arborelius, L., Skelton, K.H., Thrivikraman, K.V., Plotsky, P.M., Schulz, D.W. and Owens, M.J. (2000) Chronic administration of the selective corticotropin-releasing factor 1 receptor antagonist CP-154,526: behavioral, endocrine and neurochemical effects in the rat. JPET, 294(2): 588–597.

Asakawa, A., Inui, A., Iwamoto, M., Yonenaga, C., Yuzuriha, H., Nagata, T., Kaga, T., Ueno, N. and Kasuga, M. (2000) Intraperitoneal injection of urocortin markedly reduces food intake and body weight and is closely related to gastric emptying in lean and ob/ob mice. Gastroenterology, 118(4 Suppl. 2 Part 1): A69.

Bakshi, V.P. and Kalin, N.H. (2000) Corticotropin-releasing hormone and animal models of anxiety: Gene-environment interactions. Biological Psychiatry, 48: 1175–1198.

Bakshi, V.P., Smith-Roe, S., Yang, L.W. and Kalin, N.H. (1999) Antagonism of CRF receptors within the lateral septum decreases stress-induced behavioral inhibition in rats. Society for Neuroscience Abstracts, 25(1–2): 62.

Baldwin, H. A., Rassnick, S., Rivier, J., Koob, G.F. and Britton, K.T. (1991) CRF Antagonist reverses the "Anxiogenic" response to ethanol withdrawal in the rat. Psychopharmacology, 103: 227–232.

Bale, T.L., Contarino, A., Smith, G.W., Chan, R., Gold, L.H., Sawchenko, P.E., Koob, G.F., Vale, W.W. and Lee, K.-F. (2000) Mice deficient for corticotropin-releasing hormone receptor-2 display anxiety-like behaviour and are hypersensitive to stress. Nature Genetics, 24(4): 410–414.

Baram, T.Z., Mitchell, W.G., Brunson, K. and Haden, E. (1999) Infantile spasms: Hypothesis-driven therapy and pilot human infant experiments using corticotropin-releasing hormone receptor antagonists. Developmental Neuroscience, 21(3–5): 281–289.

Basso, A.M., Spina, M., Rivier, J., Vale, W. and Koob, G.F. (1999) Corticotropin-releasing factor antagonist attenuates the "Anxiogenic-like" effect in the defensive burying paradigm but not in the elevated plus-maze following chronic cocaine in rats. Psychopharmacology, 145(1): 21–30.

Beck, B., Burlet, A., Nicolas, J.-P. and Burlet, C. (1990) Hypothalamic neuropeptide Y (NPY) in obese zucker rats: implications in feeding and sexual behaviors. Physiol. Behav., 47: 449–453.

Behan, D.P., De Souza, E.B., Lowry, P.J., Potter, E., Sawchenko, P. and Vale, W.W. (1995a) Corticotropin releasing factor (CRF) binding protein: A novel regulator of CRF and related peptides. Frontiers in Neuroendocrinology, 16: 362–382.

Behan, D.P., Heinrichs, S.C., Troncoso, J.C., Liu, X.J., Kawas, C.H., Ling, N. and De Souza, E.B. (1995b) Displacement of corticotropin releasing factor from its binding protein as a possible treatment for Alzheimer's disease. Nature, 378(6554): 284–287.

Behan, D.P., Potter, E., Lewis, K.A., Jenkins, N.A., Copeland, N., Lowry, P.J. and Vale, W. (1993) Cloning and structure of the human corticotrophin releasing factor-binding protein gene (CRHBP). Genomics, 16: 63–68.

Bell, S.M., Reynolds, J.G., Thiele, T.E., Gan, J., Figlewicz, D.P. and Woods, S.C. (1998) Effects of third intracerebroventricular injections of corticotropin-releasing factor (CRF) on ethanol drinking and food intake. Psychopharm., 139: 128–135.

Benoit, S.C., Thiele, T.E., Heinrichs, S.C., Rushing, P.A., Blake, K.A. and Seeley, R.J. (2000) Comparison of central administration of corticotropin-releasing hormone and urocortin on food intake, conditioned taste aversion, and c-Fos expression. Peptides, 21(3): 345–351.

Berridge, K.C. (2000) Measuring hedonic impact in animals and infants: microstructure of affective taste reactivity patterns. Neurosci. Biobehav. Rev., 24(2): 173–198.

Berridge, C.W. and Dunn, A.J. (1986) Corticotropin-releasing factor elicits naloxone sensitive stress-like alterations in exploratory behavior in mice. Regulatory Peptides, 16: 83–93.

Bjenning, C.A. and Rimvall, K. (2000) CRF and CRF (6-33) depress food but not water intake in the obese Zucker rat. International Journal of Obesity, 24(Suppl. 2): S140–S141.

Blank, T., Nijholt, I., Eckart, K. and Spiess, J. (2002) Priming of long-term potentiation in mouse hippocampus by corticotropin-releasing factor and acute stress: implications for hippocampus-dependent learning. J. Neurosci., 22(9): 3788–3794.

Blank, T., Nijholt, I., Grammatopoulos, D.K., Randeva, H.S., Hillhouse, E.W. and Spiess, J. (2003a) Corticotropin-releasing factor receptors couple to multiple G-proteins to activate diverse intracellular signaling pathways in mouse hippocampus: role in neuronal excitability and associative learning. J Neurosci., 23(2): 700–707.

Blank, T., Nijholt, I., Vollstaedt, S. and Spiess, J. (2003b) The corticotropin-releasing factor receptor 1 antagonist CP-154,526 reverses stress-induced learning deficits in mice. Behav. Brain. Res., 138(2): 207–213.

Bonaz, B. and Rivest, S. (1998) Effect of a chronic stress on CRF neuronal activity and expression of its type 1 receptor in the rat brain. American Journal of Physiology, 275(5 PART 2): R1438–R1449.

Born, J., Spath-Schwalbe, E., Schwakenhofer, H., Kern, W. and Fehm, H.L. (1989) Influences of corticotropin-releasing hormone, adrenocorticotropin, and cortisol on sleep in normal man. Journal of Clinical Endocrinology and Metabolism, 68(5): 904–911.

Bradley, M.M. and Lang, P.J. (2000) Measuring emotion: behavior, feeling, and physiology. In: Lane, R.D. and Nadel, L. (Eds.), Cognitive Neuroscience of Emotion. Oxford University Press, New York.

Brady, L.S., Smith, M.A., Gold, P.W. and Herkenham, M. (1990) Altered expression of hypothalamic neuropeptide mRNAs in food-restricted and food-deprived rats. Neuroendocrinology, 52: 441–447.

Britton, D.R., Koob, G.F., Rivier, J. and Vale, W. (1982) Intraventricular corticotropin-releasing factor enhances behavioral effects of novelty. Life Sci., 31: 363–367.

Britton, K.T., Lee, G., Vale, W., Rivier, J. and Koob, G.F. (1986) Corticotropin releasing factor (CRF) receptor antagonist blocks activating and 'anxiogenic' actions of CRF in the rat. Brain Res., 369(1–2): 303–306.

Brunani, A., Invitti, C., Dubini, A., Piccoletti, R., Bendinelli, P., Maroni, P., Pezzoli, G., Ramella, G., Calogero, A. and Cavagnini, F. (1995) Cerebrospinal gluid and plasma concentrations of SRIH, beta-endorphin, CRF, NPY and GHRH in obese and normal weight subjects. International Journal of Obesity, 19: 17–21.

Brunson, K.L., Eghbal-Ahmadi, M. and Baram, T.Z. (2001) How do the many etiologies of West syndrome lead to excitability and seizures? The corticotropin releasing hormone excess hypothesis. Brain & Development, 23(7): 533–538.

Buck, R. (1999) The Biological Affects: A typology. Psychological Review, 106(2): 301–336.

Buwalda, B., de Boer, S.F., Van Kalkeren, A.A. and Koolhaas, J.M. (1997) Physiological and behavioral effects of chronic intracerebroventricular infusion of corticotropin-releasing factor in the rat. Psychoneuroendocrinology, 22(5): 297–309.

Cador, M., Cole, B.J., Koob, G.F., Stinus, L. and Le, M.M. (1989) Central administration of CRF produces sensitization of amphetamine-induced behavior. Soc. Neurosci. Abstr., 15(252): 252–252.

Chang, F.-C. and Opp, M.R. (1999) Pituitary CRH receptor blockade reduces waking in the rat. Physiol. Behav., 67(5): 691–696.

Chang, F.C. and Opp, M.R. (2002) Role of corticotropin-releasing hormone in stressor-induced alterations of sleep in rat. Am. J. Physiol. Regul. Integr. Comp. Physiol., 283(2): R400–407.

Chen, Y.L., Braselton, J., Collins, J., Costello, D., Corman, M., Dunaiskis, A., Faraci, S., Heym, J., Mansbach, R., Schmidt, A., Schulz, D., Seeger, T., Sprouse, J., Tingley, D. and Winston, E. (1996) CP-154-526: A potent and selective non-peptide. 212 ACS National Meeting, Orlando FL, American Chemical Society.

Chen, Y.L., Mansbach, R.S., Winter, S.M., Brooks, E., Collins, J., Corman, M.L., Dunaiskis, A.R., Faraci, W.S., Gallaschun, R.J., Schmidt, A. and Schulz, D.W. (1997) Synthesis and oral efficacy of a 4-(Butylethylamino)pyrrolo [2,3-d]pyrimidine: a centrally active corticotropin-releasing factor1 receptor antagonist. Journal of Medicinal Chemistry, 40(11): 1749–1752.

Cizza, G., Gold, P.W. and Chrousos, G.P. (1995) Aging is associated in the 344/N Fischer rat with decreased stress responsivity of central and peripheral catecholaminergic systems and impairment of the hypothalamic-pituitary-adrenal Axis. Ann. N. Y. Acad. Sci., 771: 491–511.

Cole, B.J., Britton, K.T. and Koob, G.F. (1987) Central administration of alpha-helical corticotropin-releasing factor attenuates the acquisition of a conditioned emotional response. Neuroscience Abstracts, 13: 427.

Cole, B.J., Cador, M., Stinus, L., Rivier, C., Rivier, J., Vale, W., LeMoal, M. and Koob, G.F. (1990a) Critical role of the hypothalamic pituitary adrenal axis in amphetamine-induced sensitization of behavior. Life Sci., 47: 1715–1720.

Cole, B.J., Cador, M., Stinus, L., Rivier, J., Vale, W. and Koob, G.F. (1990b) Central administration of a CRF antagonist blocks the development of stress-induced behavioral sensitization. Brain Res., 512: 343–346.

Contarino, A., Dellu, F., Koob, G.F. and Gold, L.H. (1999) Reduced anxiety-like and cognitive performance in mice. lacking the corticotropin-releasing factor receptor 1. Brain Res., 835: 1–9.

Contarino, A., Heinrichs, S.C. and Gold, L.H. (1999b) Understanding CRF neurobiology: contribution from mutant mice. Neuropeptides, 33: 1–12.

Crawley, J.N. (2000) What's wrong with my mouse? – Behavioral phenotyping of trangenic and knockout mice. Wiley-Liss, New York.

Dagnault, A., Ouerghi, D. and Richard, D. (1993) Treatment with alpha-helical-CRF(9-41) prevents the anorectic effect of 17-beta-estradiol. Brain Research Bulletin, 32(6): 689–692.

De Souza, E., Whitehouse, P.J., Kuhar, M.J., Price, D.L. and Vale, W. (1986) Reciprocal changes in corticotropin-releasing factor (CRF)-like Alzheimer's disease. Nature, 319(593): 593–595.

Dirks, A., Groenink, L., Schipholt, M.I., van der Gugten, J., Hijzen, T.H., Geyer, M.A. and Olivier, B. (2002) Reduced startle reactivity and plasticity in transgenic mice over-expressing corticotropin-releasing hormone. Biol. Psychiatry, 51(7): 583–590.

Dryden, S., McCarthy, H.D., Malabu, U.H., Ware, M. and Williams, G. (1993) Increased neuropeptide Y concentrations in specific hypothalamic nuclei of the rat following treatment with methysergide: evidence that NPY may mediate serotonin's effects on food intake. Peptides, 14: 791–796.

Dunn, A.J. and Berridge, C.W. (1990) Physiological and behavioral responses to corticotropin-releasing factor administration: is CRF a mediator of anxiety of stress responses? Brain Research Reviews, 15: 71–100.

Eaves, M., Thatcher, B.K., Rivier, J., Vale, W. and Koob, G.F. (1985) Effects of corticotropin releasing factor on locomotor activity in hypophysectomized rats. Peptides, 6(5): 923–926.

Eckart, K., Radulovic, J., Radulovic, M., Jahn, O., Blank, T., Stiedl, O. and Spiess, J. (1999) Actions of CRF and its analogs. Curr. Med. Chem., 6(11): 1035–1053.

Ehlers, C.L., Henriksen, S.J., Wang, M., Rivier, J., Vale, W. and Bloom, F.E. (1983) Corticotropin releasing factor produces increases in brain excitability and convulsive seizures in the rat. Brain Res., 278(332): 332–336.

Ehlers, C.L., Reed, T.K. and Henriksen, S.J. (1986) Effects of corticotropin-releasing factor and growth hormone-releasing factor on sleep and activity in rats. Neuroendocrinology, 42(6): 467–474.

Erb, S., Hitchcott, P.K., Rajabi, H., Mueller, D., Shaham, Y. and Stewart, J. (2000) Alpha-2 adrenergic receptor agonists block stress-induced reinstatement of cocaine seeking. Neuropsychopharmacology, 23(2): 138–150.

Erb, S., Shaham, Y. and Stewart, J. (1998) The role of corticotropin-releasing factor and corticosterone in stress- and cocaine-induced relapse to cocaine seeking in rats. J. Neurosci., 14: 5529–5536.

Fuchs, E., Flügge, G., Ohl, F., Lucassen, P., Vollmann-Honsdorf, G.K. and Michaelis, T.M. (2001) Psychosocial stress, glucocorticoids, and the structural alterations in the tree shrew hippocampus. Physiol. Behav., 73: 285–291.

Glowa, J.R., Barrett, J.E., Russell, J. and Gold, P.W. (1992) Effects of corticotropin releasing hormone on appetitive behaviors. Peptides, 13: 609–621.

Goeders, N.E. and Guerin, G.F. (2000) Effects of the CRH receptor antagonist CP-154,526 on intravenous cocaine self-administration in rats. Neuropsychopharmacology, 23(5): 577–586.

Gray, T.S. (1993) Amygdaloid CRF Pathways: Role in autonomic, neuroendocrine, and behavioral responses to stress. Annals of the New York Academy of Sciences, 697: 53–60.

Griebel, G., Perrault, G. and Sanger, D.J. (1998) Characterization of the behavioral profile of the non-peptide CRF receptor antagonist CP-154,526 in anxiety models in rodents. Psychopharmacology, 138(1): 55–66.

Groenink, L., Dirks, A., Verdouw, P.M., Schipholt, M., Veening, J.G., van der Gugten, J. and Olivier, B. (2002) HPA axis dysregulation in mice overexpressing corticotropin releasing hormone. Biol. Psychiatry, 51(11): 875–881.

Groenink, L., Pattij, T., De Jongh, R., Van der Gugten, J., Oosting, R.S., Dirks, A. and Olivier, B. (2003) 5-HT(1A) Receptor knockout mice and mice overexpressing cortico-tropin-releasing hormone in models of anxiety. Eur. J. Pharmacol., 463(1–3): 185–197.

Grossman, A., Costa, A., Navarra, P. and Tsagarakis, S. (1993) The regulation of hypothalamic corticotropin-releasing factor release: in vitro studies. In: Chadwick, D.J., Marsh, J. and Ackrill, K. (Eds.), Corticotropin-releasing Factor. Wiley, Chichester, pp. 129–150.

Habib, K.E., Rice, K.C., Chrousos, G.P. and Gold, P.W. (2000) CRH-1 Receptor antagonism exerts antidepressant- and anxiolytic-like effects in attenuating stress ulcerogenesis. FASEB Journal, 14(4): A338.

Hartline, K.M., Owens, M.J. and Nemeroff, C.B. (1996) Postmortem and cerebrospinal fluid studies of corticotro-pin-releasing factor in humans. Annals of the New York Academy of Sciences, 780: 96–105.

Heinrichs, S.C., Britton, K.T. and Koob, G.F. (1991) Both conditioned taste preference and aversion induced by corticotropin-releasing factor. Pharmacol. Biochem. Behav., 40(4): 717–721.

Heinrichs, S.C. and De Souza, E.B. (2001) Corticotropin-releasing factor in brain: executive gating of neuroendocrine and functional outflow. In: McEwen. B.S. (Ed.), Handbook of Physiology. Volume 7: Coping with the Environment: Neural and Endocrine Mechanisms. Oxford University Press, New York, pp. 125–137.

Heinrichs, S.C., De Souza, E.B., Schulteis, G., Lapsansky, J.L. and Grigoriadis, D.E. (2002) Brain penetrance, receptor occupancy and antistress in vivo efficacy of a small molecule corticotropin releasing factor type I receptor selective antagonist. Neuropsychopharmacology, 27(2): 194–202.

Heinrichs, S.C. and Joppa, M. (2001) Dissociation of arousal-like from anxiogenic-like actions of brain corticotropin-releasing factor receptor ligands in rats. Behav. Brain Res., 122(1): 43–50.

Heinrichs, S.C., Lapsansky, J., Lovenberg, T.W., De Souza, E.B. and Chalmers, D.T. (1997a) Corticotropin-releasing factor CRF1, but not CRF2, receptors mediate anxiogenic-like behavior. Regulatory Peptides, 71: 15–21.

Heinrichs, S.C., Li, D.L., Katner, J. and Ivengar, S. (1999) Effects of corticotropin-releasing factor (CRF) and the CRF binding-protein ligand inhibitor, CRF(6-33), on acute food intake and body temperature in rodents. Society for Neuroscience Abstracts, 25(1–2): 1119.

Heinrichs, S.C., Menzaghi, F., Merlo Pich, E., Baldwin, H.A., Rassnick, S., Britton, K.T. and Koob, G.F. (1994) Anti-stress action of a corticotropin-releasing factor antagonist on behavioral reactivity to stressors of varying type and intensity. Neuropsychopharmacology, 11(3): 179–186.

Heinrichs, S.C. and Richard, D. (1999) The role of cortico-tropin-releasing factor and urocortin in the modulation of ingestive behavior. Neuropeptides, 33(5): 350–359.

Heinrichs, S.C., Stenzel-Poore, M.P., Gold, L.H., Battenberg, E., Bloom, F.E., Koob, G.F., Vale, W.W. and Merlo Pich, E. (1996) Learning deficits in transgenic mice with central overexpression of corticotropin-releasing factor. Neuroscience, 74(2): 303–311.

Heinrichs, S.C., Vale, E.A., Lapsansky, J., Behan, D.P., McClure, L.V., Ling, N., De Souza, E.B. and Schulteis, G. (1997b) Enhancement of performance in multiple learning tasks by corticotropin-releasing factor-binding protein ligand inhibitors. Peptides, 18(5): 711–716.

Holsboer, F. (1999) The rationale for corticotropin-releasing hormone receptor (CRH-R) antagonists to treat depression and anxiety. Journal of Psychiatric Research, 33(3): 181–214.

Holt, S., Ford, M.J., Grant, S. and Heading, R.C. (1981) Abnormal gastric emptying in primary anorexia nervosa. British Journal of Psychiatry, 139: 550–552.

Hotta, M., Shabasaki, T., Arai, K. and Demura, H. (1999) Corticotropin-releasing factor receptor type 1 mediates emotional stress-induced inhibition of food intake and behavioral changes in rats. Brain Research, 823: 221–225.

Ida, T., Nakahara, K., Kuroiwa, T., Fukui, K., Nakazato, M., Murakami, T. and Murakami, N. (2000) Both corticotropin releasing factor and neuropeptide Y are involved in the effect of orexin (hypocretin) on the food intake in rats. Neuroscience Letters, 293(2): 119–122.

Inui, A., Okano, H., Miyamoto, M., Aoyama, N., Uemoto, M., Baba, S. and Kasuga, M. (1995) Delayed gastric emptying in bulimic patients. Lancet, 436: 1240.

Iredale, P.A., Alvaro, J.D., Lee, Y., Terwilliger, R., Chen, Y.L. and Duman, R.S. (2000) Role of corticotropin-releasing factor receptor-1 in opiate withdrawal. Journal of Neurochemistry, 74(1): 199–208.

Jeanrenaud, B. (1994) Central nervous system and peripheral abnormalities: clues to the understanding of obesity and NIDDM. Diabetologia 37(Suppl. 2): S169–S178.

Jones, D.N.C., Kortekaas, R., Hatcher, P.D., Middlemiss, D.N., White, A. and Hagan, J.J. (1999) Influence of peptide CRF receptor antagonists upon the behavioural effects of human/rat CRF. European Journal of Pharmacology, 373(2–3): 141–145.

Koob, G.F. (1999) Corticotropin-releasing Factor, norepinephrine, and stress. Biological Psychiatry, 46(9): 1167–1180.

Koob, G.F. and Le Moal, M. (2001) Drug addiction, dysregulation of reward, and allostasis. Neuropsychopharmacology, 24(2): 97–129.

Korte, S.M., Korte-Bouws, G.A.H., Gohus, B. and Koob, G.F. (1994) Effect of corticotropin-releasing factor antagonist on behavioral and neuroendocrine responses during exposure to defensive burying paradigm in rats. Physiology & Behavior, 56: 115–120.

Krahn, D.D. and Gosnell, B.A. (1989) Corticotropin-releasing hormone: possible role in eating disorders. Psychiatric Medicine, 7: 235–245.

Lancel, M., Muller-Preuss, P., Wigger, A., Landgraf, R. and Holsboer, F. (2002) The CRH1 receptor antagonist R121919 attenuates stress-elicited sleep disturbances in rats, particularly in those with high innate anxiety. J. Psychiat. Res., 36(4): 197–208.

LeDoux, J. (2000) Cognitive-emotional interactions: listen to the brain. In: Lane, R.D. and Nadel, L. (Eds.), Cognitive Neuroscience of Emotion. Oxford University Press, New York.

Lehmann, J., Stohr, T. and Feldon, J. (2000) Long-term effects of prenatal stress experience and postnatal maternal separation on emotionality and attentional processes. Behavioural Brain Research, 107(1–2): 133–144.

Liang, K.C., Chen, H.C. and Chen, D.Y. (2001) Posttraining infusion of norepinephrine and corticotropin releasing factor into the bed nucleus of the stria terminalis enhanced retention in an inhibitory avoidance task. Chin. J. Physiol., 44(1): 33–43.

Lohman, T.G., Caballero, B., Himes, J.H., Davis, C.E., Stewart, D., Houtkooper, L., Going, S.B., Hunsberger, S., Weber, J.L., Reid, R. and Stephenson, L. (2000) Estimation of body fat from anthropometry and bioelectrical impedance in Native American children. International Journal of Obesity, 24(8): 982–988.

Lovejoy, D.A. and Balment, R.J. (1999) Evolution and physiology of the corticotropin-releasing factor (CRF) family of neuropeptides in vertebrates. General and Comparative Endocrinology, 115(1): 1–22.

Lowry, P.J., Woods, R.J. and Baigent, S. (1996) Corticotropin releasing factor and its binding protein. Pharmacol. Biochem. Behav., 54(1): 305–308.

Meerlo, P., De Boer, S.F., Koolhaas, J.M., Daan, S. and Van Den Hoofdakker, R.H. (1996) Changes in daily rhythms of body temperature and activity after a single social defeat in rats. Physiology & Behavior, 59(4/5): 735–739.

Menzaghi, F., Heinrichs, S.C., Merlo Pich, E., Tilders, F.J.H., and Koob, G.F., (1993) Functional impairment of hypothalamic corticotropin-releasing factor neurons with immunotargeted toxins enhances food intake induced by neuropeptide Y. Brain Research, 618: 76–82.

Menzaghi, F., Howard, R.L., Heinrichs, S.C., Vale, W., Rivier, J. and Koob, G.F. (1994) Characterization of a novel and potent corticotropin-releasing factor antagonist in rats. JPET, 269(2): 564–572.

Nemeroff, C. (2000) Clinical studies on the role of CRF in mood and anxiety disorders. Neuropsychopharmacology, 23(S2): S5.

Ohata, H., Suzuki, K., Oki, Y. and Shibasaki T. (2000) Urocortin in the ventromedial hypothalamic nucleus acts as an inhibitor of feeding behavior in rats. Brain Res., 861(1): 1–7.

Opp, M.R. (1995) Corticotropin-releasing hormone involvement in stressor-induced alterations in sleep and in the regulation of waking. Advances in Neuroimmunology, 5: 127–143.

Owens, M.J. and Nemeroff, C.B. (1993) The role of corticotropin-releasing factor in the pathophysiology of affective and anxiety disorders: laboratory and clinical studies. In: Chadwick, D.J. Marsh, J. and Ackrill, K. (Eds.)

Corticotropin-releasing Factor. John Wiley and Sons, Chichester, pp. 296–316.
Panksepp, J. (1998) Affective Neuroscience. Oxford University Press, New York.
Pelleymounter, M.A., Joppa, M., Carmouche, M., Cullen, M.J., Brown, B., Murphy, B., Grigoriadis, D.E., Ling, N. and Foster, A.C. (2000) Role of corticotropin-releasing factor (CRF) receptors in the anorexic syndrome induced by CRF. JPET, 293(3): 799–806.
Piazza, P.V. and Le Moal, M. (1996) Pathophysiological basis of vulnerability to drug abuse: interaction between stress, glucocorticoids, and dopaminergic neurons. Ann. Rev. Pharmacol. Toxicol, 36: 359–378.
Plotsky, P.M., Cunningham, E.T. and Widmaier, E.P. (1989) Catecholaminergic modulation of corticotropin-releasing factor and adrenocorticotropin secretion. Endocrinological Reviews, 10: 437–458.
Potter, E., Behan, D.P., Linton, E.A., Lowry, P.J., Sawchenko, P.E. and Vale, W.W. (1992) The central distribution of a corticotropin-releasing factor (CRF)-binding protein predicts multiple sites and modes of interaction with CRF. Proceedings of the National Academy of Sciences (USA), 89: 4192–4196.
Radulovic, J., Fischer, A., Katerkamp, U. and Spiess, J. (2000) Role of regional neurotransmitter receptors in corticotropin-releasing factor (CRF)-mediated modulation of fear conditioning. Neuropharmacology, 39(4): 707–710.
Rebaudo, R., Melani, R., Balestrino, M. and Izvarina, N. (2001) Electrophysiological effects of sustained delivery of CRF and its receptor agonists in hippocampal slices. Brain Res, 922(1): 112–117.
Richter, R.M., Merlo Pich, E., Koob, G.F. and Weiss, F. (1995) Sensitization of cocaine-stimulated increase in extracellular levels of corticotropin-releasing factor from the rat amygdala after repeated administration as determined by intracranial microdialysis. Neuroscience Letters, 187: 169–172.
Rolls, E.T. (1999) The Brain and Emotion. Oxford University Press, Oxford.
Rothwell, N.J. (1990) Central effects of CRF on metabolism and energy balance. Neuroscience and Biobehavioral Reviews, 14: 263–271.
Russell, J.A. (2003) Core affect and the psychological construction of emotion. Psychological Review, 110(1): 145–172.
Sarnyai, Z., Höhn, J., Szabó, G. and Penke, B. (1992) Critical role of endogenous corticotropin-releasing factor (CRF) in the mediation of the behavioral action of cocaine in rats. Life Sciences, 51: 2019–2024.
Sarnyai, Z., Shaham, Y. and Heinrichs, S.C. (2001) The role of corticotropin-releasing factor in drug addiction. Pharmacological Reviews, 53(2): 209–243.
Schulz, D.W., Mansbach, R.S., Sprouse, J., Braselton, J.P., Collins, J., Dunaiskis, C.M.A., Faraci, S., Schmidt, A.W., Seeger, T., Seymour, P., Tingley III, F.D., Winston, E.N., Chen, Y.L. and Heym, J. (1996) CP-154,526: A potent and selective nonpeptide antagonist of corticotropin releasing factor receptors. Proceedings of the National Academy of Sciences (USA), 93: 10477–10482.
Selye, H. (1976) The Stress of Life. McGraw-Hill, New York.
Slawecki, C.J., Somes, C., Rivier, J.E. and Ehlers, C.L. (1999) Neurophysiological effects of intracerebroventricular administration of urocortin. Peptides (New York), 20(2): 211–218.
Smith, G.W., Aubry, J.-M., Dellu, F., Contarino, A., Bilezikjian, L.M., Gold, L.H., Chen, R., Marchuk, Y., Hauser, C., Bentley, C.A., Sawchenko, P.E., Koob, G.F., Vale, W. and Lee, K.-F. (1998) Corticotropin releasing factor receptor 1-deficient mice display decreased anxiety, impaired stress response, and aberrant neuroendocrine development. Neuron, 20: 1093–1102.
Smith, M.A., Weiss, S.R.B., Berry, R.L., Zhang, L.-X., Clark, M., Massenburg, G. and Post, R.M. (1997) Amygdala-kindled seizures increase the expression of corticotropin-releasing factor (CRF) and CRF-binding protein in GABAergic interneurons of the dentate hilus. Brain Res., 745: 248–256.
Spina, M.G., Basso, A.M., Zorrilla, E.P., Heyser, C.J., Rivier, J., Vale, W., Merlo-Pich, E. and Koob, G.F. (2000) Behavioral effects of central administration of the novel CRF antagonist astressin in rats. Neuropsychopharmacology, 22(3): 230–239.
Steckler, T. and Holsboer, F. (1999) Corticotropin-releasing hormone receptor subtypes and emotion. Biological Psychiatry, 46(11): 1480–1508.
Stratakis, C.A. and Chrousos, G.P. (1995) Neuroendocrinology and pathophysiology of the stress system. Annals of the New York Academy of Sciences, 771: 1–18.
Sutton, R. E., Koob, G.F., Le Moal, M., Rivier, J. and Vale, W. (1982) Corticotropin-releasing factor (CRF) produces behavioral activation in rats. Nature, 297: 331–333.
Swerdlow, N.R., Britton, K.T. and Koob, G.F. (1989) Potentiation of acoustic startle by corticotropin-releasing factor (CRF) and by fear are both reversed by a-helical CRF (9-41). Neuropsychopharmacology, 2: 285–292.
Swerdlow, N.R. and Koob, G.F. (1985) Separate neural substrates of the locomotor-activating properties of amphetamine, caffeine and corticotropin releasing factor (CRF) in the rat. Pharmacology Biochemistry and Behavior, 23: 303–307.
Taché, Y., Martinez, V., Million, M. and Rivier, J. (1999) Corticotropin-releasing factor and the brain-gut motor response to stress. Canadian Journal of Gastroenterology, 13(Suppl. A): 18A–25A.
Takahashi, L.K., Kalin, N.H., Vandenburgt, J.A. and Sherman, J.E. (1989) Corticotropin-releasing factor modulates defensive-withdrawal and exploratory behavior in rats. Behavioral Neuroscience, 103: 648–654.
Timpl, P., Spanagel, R., Sillaber, I., Kress, A., Reul, J.M., Stalla, G.K., Blanquet, V., Steckler, T., Holsboer, F. and Wurst, W. (1998) Impaired stress response and reduced

anxiety in mice lacking a functional corticotropin-releasing hormone receptor. Nature Genetics, 19(2): 162–166.

Turnbull, A.V. and Rivier, C. (1997) Corticotropin-releasing factor (CRF) and endocrine responses to stress: CRF receptors, binding protein, and related peptides. Proceedings of the Society for Experimental Biology and Medicine, 215: 1–10.

Valentino, R.J., Foote, S.L. and Page, M.E. (1993) The locus coeruleus as a site for integrating corticotropin-releasing factor and noradrenergic mediation of stress responses. Annals of the New York Academy of Sciences, 697: 173–188.

van Gaalen, M.M., Stenzel-Poore, M., Holsboer, F. and Steckler, T. (2003) Reduced attention in mice overproducing corticotropin-releasing hormone. Behav. Brain Res., 142(1–2): 69–79.

Vargas, M.A., Bissette, G., Owens, M.J., Ehlers, C.L. and Nemeroff, C.B. (1992) Effects of chronic ethanol and benzodiazepine treatment and withdrawal on corticotropin-releasing factor neural systems. Annals of the New York Academy of Sciences, 654: 145–152.

Wang, H.L., Wayner, M.J., Chai, C.Y. and Lee, E.H. (1998) Corticotrophin-releasing factor produces a long-lasting enhancement of synaptic efficacy in the hippocampus. Eur. J. Neurosci., 10(11): 3428–3437.

Weiss, F., Ciccocioppo, R., Parsons, L.H., Katner, S., Liu, X., Zorrilla, E.P., Valdez, G.R., Ben-Shahar, O., Angeletti, S. and Richter, R.R. (2001) Compulsive Drug-seeking behavior and relapse. Neuroadaptation, stress, and conditioning factors. Ann. N. Y. Acad. Sci., 937: 1–26.

Weninger, S.C., Dunn, A.J., Muglia, L.J., Dikkes, P., Miczek, K.A., Swiergiel, A.H., Berridge, C.W. and Majzoub, J.A. (1999a) Stress-induced behaviors require the corticotropin-releasing hormone (CRH) receptor, but not CRH. Proceedings of the National Academy of Sciences of the United States of America, 96(14): 8283–8288.

Weninger, S.C., Muglia, L.J., Jacobson, L. and Majzoub, J.A. (1999b) CRH-deficient mice have a normal anorectic response to chronic stress. Regulatory Peptides, 84(1–3): 69–74.

Widmaier, E.P. (1990) Changes in responsiveness of the hypothalamic-pituitary-adrenal axis to 2-deoxy-D-glucose in developing rats. Endocrinology, 126(6): 3116–3123.

Wiederman, M.W. and Pryor, T.L. (2000) Body dissatisfaction, bulimia, and depression among women: the mediating role of drive for thinness. Int. J. Eat., Dis., 27(1): 90–95.

Yamano, M., Yuki, H., Yasuda, S. and Miyata, K. (2000) Corticotropin-releasing hormone1 receptors mediate consensus interferon-a YM643-induced depression-like behavior in mice. JPET, 292: 181–187.

Zorrilla, E.P., Schulteis, G., Ling, N., Koob, G.F. and De Souza, E.B. (2001) Performance-enhancing effects of CRF-BP ligand inhibitors. Neuroreport, 12(6): 1231–1234.

T. Steckler, N.H. Kalin and J.M.H.M. Reul (Eds.)
Handbook of Stress and the Brain, Vol. 15
ISBN 0-444-51173-3

CHAPTER 2.4

The roles of urocortins 1, 2, and 3 in the brain

Eric P. Zorrilla* and George F. Koob

Department of Neuropharmacology, The Scripps Research Institute, 1055N. Torrey Pines Rd., La Jolla, CA 92037, USA

Abstract: The neuropeptides urocortin 1, urocortin 2, and urocortin 3 share a family resemblance. Each shows some sequence identity with the stress-related neuropeptide corticotropin-releasing factor (CRF), high affinity for the CRF_2 receptor, predominantly subcortical distributions and anorectic properties when administered centrally. In substantive ways, however, the peptides differ from one another. Urocortin 1 differs pharmacologically, as it also exhibits affinity for the CRF_1 receptor and CRF-binding protein. Brain distributions also differ. Urocortin 1 shows predominant correspondence with hindbrain CRF_1 and CRF_2 receptors and a much more limited correspondence with forebrain CRF receptors. In contrast, urocortin 3 shows almost exclusive correspondence with forebrain CRF_2 receptors unique from those innervated by urocortin 1 projections. Urocortin 2 is synthesized in yet another set of hypothalamic and brainstem nuclei, distinct from those that prominently express urocortin 1 and urocortin 3. Brain urocortins may be involved in the regulation of diverse functions, including: osmotic balance; activity of the somatotropic and pituitary–gonadal neuroendocrine axes; modulation of arousal, attention, anxiety-like behavior, and learning and memory; control of autonomic functions, including cardiovascular and gastrointestinal responses; and the regulation of energy balance, both in terms of food intake and energy expenditure. Commonly described as being stress mediators for their structural resemblance to CRF, available evidence suggests that the type 2 urocortins, urocortin 2 and urocortin 3, do not share the activating and anxiogenic-like properties of stress- or CRF-induced CRF_1 receptor activation. The endogenous role of each urocortin in the autonomic/energy balance components of stress-related responses remains to be determined.

Introduction

Since the isolation of corticotropin-releasing factor (CRF) in 1981 (Vale et al., 1981), three additional mammalian CRF-like neuropeptides have been identified. The first of these – urocortin 1 – was identified in 1995 and was named for its similar primary structure and bioactivity to suckerfish urotensin and CRF (Vaughan et al., 1995). Because urocortin 1 exhibited greater pharmacological and putative in vivo functional activity at the type 2 CRF receptor (CRF_2) than did CRF, it was hypothesized to be a natural CRF_2 receptor ligand (Vaughan et al., 1995; Spina et al., 1996). Urocortin 2 (Hsu and Hsueh, 2001; Reyes et al., 2001) and urocortin 3 (Hsu and Hsueh, 2001; Lewis et al., 2001), subsequently identified in 2001, were recognized for their structural relation to CRF and urocortin 1, but named urocortins due to their predominant affinity for the CRF_2 receptor. Although they share moderate sequence identity, all mammalian CRF-related peptides appear to be phylogenetically distinct from one another based on the heterogeneous pattern of their conserved residues (Hsu and Hsueh, 2001; Lewis et al., 2001). Furthermore, each peptide's expression is regulated through a different gene and exhibits a unique neuroanatomical distribution with little overlap. Consequently, despite a structural and pharmacological family resemblance, the natural functions of CRF-like peptides in vivo might differ

*Corresponding author. E-mail: ezorrilla@scripps.edu

quite significantly. Toward beginning to understand the central roles of urocortins 1, 2, and 3, we review the following relevant areas of research: (1) pharmacology of the urocortins, (2) central distribution of the urocortins and their putative targets, (3) effects of central administration of the urocortins, (4) factors that regulate central activity of the urocortins, and (5) effects of treatments that specifically reduce central transmission of the urocortins.

Pharmacology of the urocortins

CRF_1 and CRF_2 receptors

Two class B, G-protein-coupled receptor families encoded by separate genes (CRF_1 and CRF_2) are putative natural targets of mammalian CRF-related peptides (Perrin and Vale, 1999). CRF is a preferential CRF_1 receptor agonist, exhibiting only low-to-moderate affinity for the CRF_2 receptor depending on the ortholog's species (Vaughan et al., 1995; Dautzenberg et al., 2001). For example, ovine CRF, the most preferential of the known natural CRF_1 receptor ligands, has almost 200-fold greater affinity for the CRF_1 than CRF_2 receptor (Behan et al., 1996a). In contrast, all urocortins bind appreciably to the CRF_2 receptor, though they differ from one another in their selectivity.

Urocortin 1 is the least selective urocortin, binding and activating both known CRF receptors with similarly high (subnanomolar) potency (Vaughan et al., 1995; Lewis et al., 2001; Reyes et al., 2001). In contrast, urocortin 2 (an N-terminally shortened sequence of stresscopin-related peptide; Hsu and Hsueh, 2001) and urocortin 3 (an N-terminally shortened sequence of stresscopin; Hsu and Hsueh, 2001) are both selective agonists for the CRF_2 receptor. Each of these "type 2 urocortins" shows comparable, subnanomolar functional activity at both rodent CRF_2 receptor splice variants ($CRF_{2(a)}$ and $CRF_{2(b)}$) (Lewis et al., 2001; Reyes et al., 2001). Urocortin 2 shows approximately 1000-fold more functional selectivity for the CRF_2 receptor than does urocortin 1 (Reyes et al., 2001). Urocortin 2 does not induce adenylate cyclase in primary anterior pituitary cell cultures (a putative CRF_1-mediated endpoint) until concentrations of 100–1000 nM are reached (Lewis et al., 2001). Urocortin 3 is the most selective urocortin, not showing CRF_1-like agonism in this bioassay up through the 3 μM concentration (Lewis et al., 2001).

Possible unidentified urocortin receptor

Species differences in the binding affinity of urocortin 3 orthologs has led to speculation that urocortin 3 also may be an agonist for an unknown CRF/urocortin receptor (Dautzenberg and Hauger, 2002). Despite sharing 89% sequence identity with murine urocortin 3, human urocortin 3 shows only moderate binding affinity for the CRF_2 receptor (K_i = 10–25 nM for human urocortin 3 versus 0.5–2.5 nM for murine urocortin 3; Lewis et al., 2001). Also, despite their similar functional activity in stable CRF_2-expressing Chinese hamster ovary cell transfectants, human urocortin 3 exhibits at least one-log order less functional potency than murine urocortin 3 in a natural $CRF_{2(b)}$-expressing cell line (Lewis et al., 2001). The existence, let alone the possible identity (Arai et al., 2001), of this hypothesized receptor remains uncertain.

Interactions with the CRF-binding protein

The biological actions of CRF are modulated by a CRF-binding protein (CRF-BP), an evolutionarily conserved 37-kDa secreted glycoprotein (Orth and Mount, 1987; Behan et al., 1989). The CRF-BP binds CRF with equal or greater affinity than do CRF receptors (Sutton et al., 1995). Importantly, the structural requirements for binding to the CRF receptors and the CRF-BP differ (Sutton et al., 1995; Jahn et al., 2001).

The CRF-BP has been suggested to curtail the effects of CRF by sequestering circulating ligand and facilitating subsequent enzymatic degradation, thereby limiting its bioavailability to the receptor. This hypothesis is supported by in vitro and in vivo studies (Behan et al., 1995a,b; Peto et al., 1999; Seasholtz et al., 2001; Zorrilla et al., 2001). In contrast, it also has been postulated that the CRF-BP, like other binding proteins (Ferry et al., 1999), may enhance the effects of its bound ligand, by shielding CRF from metabolic degradation during diffusion to membrane-bound CRF receptors

(Kemp et al., 1998). A more recent hypothesis is that the CRF-BP also has signaling properties independent of known CRF receptors, some of which may depend on ligand/CRF-BP interactions (Chan et al., 2000). Thus, the degree to which the urocortins interact with the CRF-BP could have biological relevance, either via a direct CRF-BP-dependent signaling pathway (Chan et al., 2000) or via an indirect agonism achieved through competitive liberation of bound CRF (Behan et al., 1996b). Conversely, the degree to which the urocortins themselves are sequestered by the CRF-BP under basal conditions would constitute a physiologically relevant reservoir that could be released by provocation of another CRF-BP ligand (such as CRF).

Urocortin 1, like CRF, binds potently to the CRF-BP (Vaughan et al., 1995) and is believed to be the predominant natural ligand for the CRF-BP in ovine brain (Baigent and Lowry, 2000). Also, urocortin 1 dissociates from the CRF-BP approximately twice as slowly as does CRF (Henriot et al., 1999), potentially increasing the physiologic significance of urocortin 1/CRF-BP interactions. Neither urocortin 2 nor urocortin 3 bind to the human CRF-BP (Lewis et al., 2001).

Summary

Pharmacologically, urocortin 1 feasibly could exert its effects directly via either known mammalian CRF receptor or via its interactions with the CRF-BP. In contrast, physiologic effects of the type 2 urocortins, urocortin 2 and urocortin 3, could be mediated by the CRF_2 receptor. The existence of other binding sites for neurotransmission, in particular for human urocortin 3, is speculated.

Central distribution of the urocortins and CRF receptors

Urocortin 1

Despite an extensive peripheral distribution (Bittencourt et al., 1999; Harada et al., 1999; Kageyama et al., 1999; Baigent and Lowry, 2000; Kozicz and Arimura, 2002), urocortin 1 has a restricted, subcortical, predominantly caudal, distribution in the brain. The major locus of urocortin 1 synthesis in brain, as inferred from the presence of urocortin 1 mRNA and both basal and colchicine-enhanced urocortin 1-like immunoreactivity (LI), is the Edinger-Westphal nucleus (E-W). Secondary, validated sites of urocortin 1 synthesis include the lateral superior olive, the supraoptic nucleus (SON), the lateral hypothalamic area and several brain stem and spinal cord motoneuron nuclei. Tertiary, putative sites of urocortin 1 synthesis include the mammillary nucleus of the hypothalamus, sphenoid nucleus, substantia nigra, tegmentum, periaqueductal gray (PAG), raphe, and vestibular nucleus (Kozicz et al., 1998; Yamamoto et al., 1998; Bittencourt et al., 1999).

The prominent synthesis and expression of urocortin 1 in the E-W is well-conserved across rats, sheep, humans, and frogs (Vaughan et al., 1995; Kozicz et al., 1998, 2002; Cepoi et al., 1999; Iino et al., 1999). The E-W is a dorsal midbrain structure well-characterized for its role in oculomotor and pupillary control via parasympathetic preganglionic projections to the ciliary ganglion (Trimarchi, 1992; Gamlin, 1999). The pattern of descending urocortin 1-LI-positive fibers, presumed to originate from the E-W, corresponds to a likely role for urocortin in these functions (Bittencourt et al., 1999). Based on immunohistochemical and neuroanatomical findings, additional roles for the E-W have been hypothesized, but not yet substantiated, including the regulation of behavioral responses to stressors (Jansen et al., 1998; Weninger et al., 1999a), temperature homeostasis (Parver 1991; Smith et al., 1998b; Bachtell, 2003), nociception (Innis and Aghajanian, 1986; Koyama et al., 2000), motor control (Roste and Dietrichs, 1988), vestibular function (Kaufman et al., 1992), and the effects of and motivation to consume alcohol (Bachtell et al., 2002, 2003b). Additional projections of urocortin 1-LI-positive fibers in the rat, putatively from the E-W, may support these hypothesized functions. *Descending* urocortin 1-LI-positive fibers of possible E-W origin are observed in (1) *midbrain*: substantia nigra, periaqueductal gray, interpenduncular nucleus, and red nucleus; (2) *caudal midbrain/rostral pons*: dorsal raphe nucleus, ventral tegmentum, basilar pontine nuclei, and parabrachial nucleus; (3) *medulla*: recognized E-W projection fields, including the facial, lateral reticular, and spinal trigeminal nuclei, inferior olive, and the dorsal column nuclei, as well as targets without known E-W innervation, including the nucleus of the solitary tract (NTS) and area postrema; (4) *cerebellum*: notably in the flocculus and paraflocculus of the cerebellar cortex as well as deep cerebellar and vestibular nuclei, regions in which urocortin 1 also is synthesized (Swinny et al., 2002); and (5) *spinal cord*: throughout the spinal gray and, less so, in the dorsal and ventral horns. The most prominent *ascending* urocortin 1-LI-positive projection from the E-W origin targets the septal/preoptic region, robustly to the lateral septum (LS) and somewhat less so to the bed nucleus of the stria terminalis (BNST), globus pallidus, and medial septal/diagonal band complex. Other ascending urocortin 1-LI-positive fibers innervate the hypothalamus, including the lateral and anterior hypothalamic areas, suprachiasmatic nucleus, dorsomedial nucleus, medial preoptic area, and parvocellular aspects of the paraventricular nucleus (PVN), the thalamus and the rostral periacqueductal gray.

A second caudal site of urocortin 1 synthesis is the lateral superior olive, best known for its role in auditory processing. Urocortin 1-LI-positive fibers exit this structure to follow the olivocochlear projection and target the cochlear and vestibular nuclei as well as the inferior colliculus (Bittencourt et al., 1999). The most posterior sites of urocortin 1 synthesis are a distributed group of motoneuron nuclei in the brain stem and ventral horns of the spinal cord (Bittencourt et al., 1999).

The most prominent forebrain site of urocortin 1 synthesis is the SON, an integrative hypothalamic region of magnocellular, neuroendocrine cells best known for its role in osmoregulation. Accordingly, appreciable urocortin 1-LI-positive fibers have been observed in the internal, but not external, layer of the median eminence, where they are joined by projections from magnocellular neurons of the PVN. Thereafter, urocortin-LI-positive axons terminate in the peripheral posterior, but not anterior, pituitary, though not in the same abundance as oxytocin or vasopressin. In addition to the SON and magnocellular PVN, the ventral lateral hypothalamic area exhibits moderate synthesis and expression of urocortin 1.

Although minor urocortin 1-positive projections to the central and posterior cortical amygdala nuclei exist, urocortin 1-LI generally is scarce or absent in many regions in which its paralog, CRF, is prominent, including the external layer of the median eminence, the hypophysiotropic, dorsal medial parvocellular subdivision of the PVN, basal ganglia, amygdala, hippocampus, locus coeruleus (LC), and cerebral cortex (Kozicz et al., 1998; Bittencourt et al., 1999; Iino et al., 1999; Morin et al., 1999).

The cross-reactivity of antibodies used for immunohistochemistry confounded the initial neuroanatomical mapping of urocortin I. Regions that expressed putative urotensin-like immunoreactivity including the ventrolateral (as opposed to dorsomedial) LS, the ventromedial hypothalamus (VMH), and the medial amygdala (Vaughan et al., 1995) subsequently were determined not to express urocortin 1-LI once more specific antibodies were applied (Bittencourt et al., 1999). Many of these regions now are believed to express urocortin 3 (see below).

Urocortin 2

In addition to marked expression in peripheral tissues, including heart, posterior pituitary, adrenals, and circulating blood cells (Hsu and Hsueh, 2001), urocortin 2 exhibits a consistent, but restricted, subcortical expression in rat and mouse brain (Reyes et al., 2001). Like urocortin 1, urocortin 2 mRNA is localized in the SON and magnocellular subdivision of the PVN as well as brainstem motoneurons and the ventral horns of the spinal cord. Unlike urocortin 1, urocortin 2 has a marked expression in the arcuate nucleus of the hypothalamus and the LC. The projection targets of urocortin 2-positive neurons are unknown. Non-neuronal urocortin 2 expression has been localized in the meninges, but not in glial cells.

Urocortin 3

In addition to marked expression in the periphery, including the gastrointestinal tract, muscle, thyroid, adrenals, pancreas, heart, and spleen, urocortin 3 exhibits a predominantly rostral, subcortical distribution in brain (Hsu and Hsueh, 2001; Lewis et al., 2001; Li et al., 2002). The three prominent sites of forebrain urocortin 3 synthesis are: (1) the median preoptic nucleus of the hypothalamus, (2) a hypothalamic region bordered laterally by the fornix and medially by the PVN that extends rostrally into the posterior BNST, and (3) the dorsal medial amygdala. Less prominent forebrain sites of urocortin 3 mRNA expression include the dorsomedial hypothalamus, both magnocellular and parvocellular components of the PVN, a region dorsal to the SON, and the posterior cortical and amygdalohippocampal transition areas of the amygdala. Urocortin 3-LI-positive fibers are found in every prominent site of urocortin 3 mRNA synthesis. In addition, urocortin 3-LI-positive fibers of unknown origin project heavily to the VMH and arcuate nucleus. In the posterior hypothalamus, urocortin 3-LI-positive fibers project to the ventral premammillary nucleus. Urocortin 3 fibers are much scarcer in the SON, PVN, and anterior, dorsomedial, and lateral areas of the hypothalamus and are seen in the internal, but not external, zone of the median eminence (Li et al., 2002).

Urocortin 3 fibers of unknown origin also are abundant in forebrain limbic structures, including the LS, posterior BNST, and the medial amygdala. The distribution of urocortin 3 in the LS differs from that of urocortin 1 with the latter innervating ventral and intermediate aspects of the structure and the former innervating dorsal aspects. Other forebrain sites with scattered urocortin 3 fibers include the basomedial and posterior cortical nuclei of the amygdala and the ventral hippocampus. In mid/hindbrain, urocortin 3 cell bodies are found in the auditory complex, notably in the superior paraolivary nucleus. Scattered urocortin 3 fibers are evident caudally in the periacqueductal gray, superior and inferior colliculi, and the ventral lateral lemniscus.

Concordance of urocortin expression with CRF receptor expression

Figure 1 summarizes the prominent sites of synthesis for urocortins 1, 2, and 3. Table 1 recapitulates the prominent sites of urocortin synthesis and also reviews prominent projection fields (as known for urocortin 1 and urocortin 3) in relation to known sites of CRF receptor expression in the rat. Inspection of Table 1 reveals that urocortin 1 is a candidate endogenous ligand for many hindbrain CRF receptors, including CRF_1 receptors in visual, somatosensory, auditory, vestibular, motor, tegmental, parabrachial, pontine, median raphe and cerebellar nuclei and CRF_2 receptors in the solitary tract nucleus, area postrema, and raphe complex (especially its dorsal nucleus). In the forebrain, urocortin 1 is a candidate ligand for a considerably more limited distribution of CRF receptors, including CRF_1 receptors in the medial septum, amygdalohippocampal transition area, rostral BNST, globus pallidus, substantia nigra, thalamus, and selected hypothalamic regions, including the suprachiasmatic, dorsomedial and mammilary nuclei. Urocortin 1 also is a candidate ligand for CRF_2 receptors in the medial lateral septum.

No major sites of urocortin 2 synthesis concord with known sites of prominent CRF_2 receptor expression, although scattered cells in the arcuate and supraoptic nuclei of the hypothalamus, where urocortin 2 mRNA is prominent, exhibit CRF_2 mRNA expression. A clearer understanding of likely targets for urocortin 2 neurons awaits mapping of their projection fields.

Urocortin 3 is a leading candidate ligand for major distributions of forebrain CRF_2 receptors whereas urocortin 1 is a leading candidate ligand for many hindbrain CRF_1 and CRF_2 receptors. These include intermediate and lateral divisions of the lateral septum, medial and posterior cortical amygdala nuclei, the posterior BNST, and the ventromedial and medial preoptic nuclei of the hypothalamus. Urocortin 3 expression only overlaps significantly with hindbrain CRF_2 expression in specific auditory nuclei.

Effects of central administration of the urocortins

Exogenous administration of urocortins may result in a spatially, temporally, and quantitatively different pattern of activation than that resulting from physiologic activation of urocortin neurocircuitry. Still, central administration of the urocortins sheds light on their *possible* endogenous roles by identifying functions that might be regulated by CRF family peptides.

It should also be noted that some regulatory functions initially attributed to CRF based on exogenous administration studies ultimately may be shown to be urocortin mediated. For example, the relative lack of a behavioral phenotype of CRF-deficient mice has raised uncertainty about which CRF-related peptides mediate the behavioral responses to stress (Dunn and Swiergiel, 1999; Swiergiel and Dunn, 1999; Weninger et al., 1999a; Weninger et al., 1999b). In this regard, we review information obtained from intracerebroventricular and site-specific administration of both CRF and the urocortins.

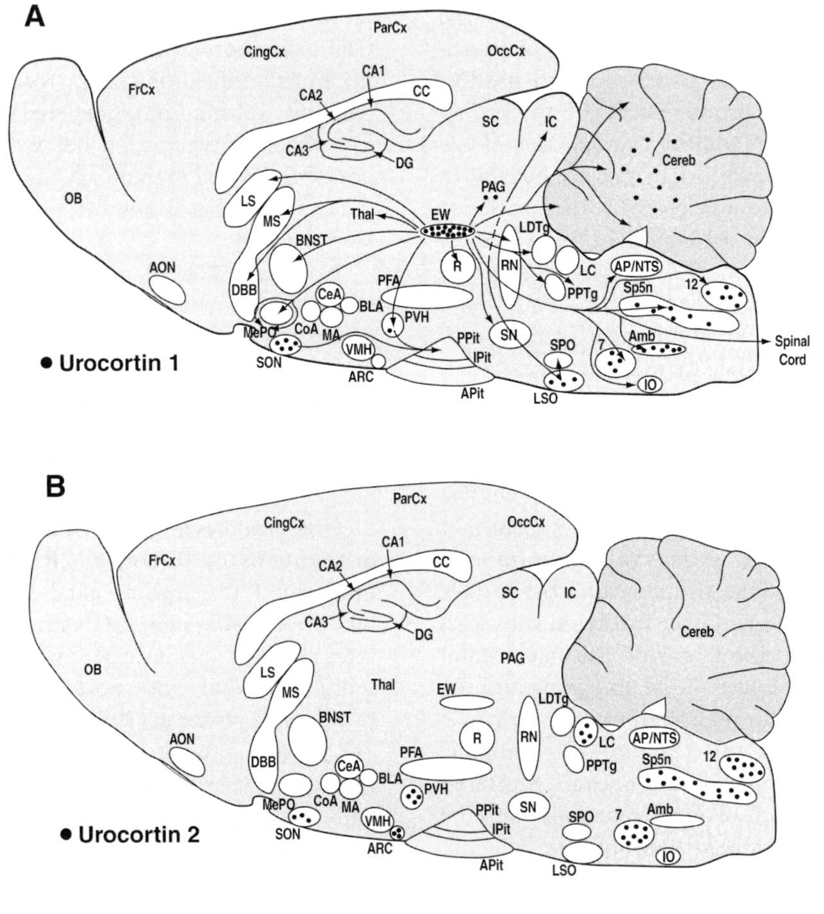

Fig. 1. Distribution of (A) urocortin 1, (B) urocortin 2, and (C) urocortin 3 mRNA in a sagittal section of the rodent brain. The presented mRNA distribution (small circles) and putative immunoreactive projections (lines with arrows) are based on in situ hybridization and immunohistochemical studies reported in Vaughan et al., 1995; Kozicz et al., 1998; Yamamoto et al., 1998; Bittencourt et al., 1999; Morin et al., 1999; Lewis et al., 2001; Reyes et al., 2001; Li et al., 2002. The drawn sagittal sections are only 2-dimensional schematic representations and, therefore, cannot be neuroanatomically exact. 7, facial nucleus; 12, hypoglossal nucleus; Amb, ambiguus nucleus; AON, anterior olfactory nucleus; AP/NTS, area postrema/nucleus tractus solitarus; Apit, anterior pituitary; ARC, arcuate nucleus; BNST, bed nucleus of the strict terminals; EW, Edinger-Westphal nucleus; BLA, basolateral amygdala; CA1-3, fields CA1-3 of Ammon's horn, CC, corpus callosum; CeA, central nucleus of the amygdala; Cereb, cerebellum; CingCx, cingulate cortex; CoA, cortical nucleus of the amygdala; DBB, diagonal band of Broca; DG, dentate gyrus; FrCx, frontal cortex; IC, inferior colliculi; IO, inferior olive; IPit, intermediate pituitary; LC, locus coeruleus; LDTg, laterodorsal tegmental nucleus; LS, lateral septum; LSO, lateral superior olive; MA, medial nucleus of the amygdala; MePO, median preoptic area; MS, medial septum; PVH, paraventricular nucleus of the hypothalamus; OB, olfactory bulb; OccCx, occipital cortex; PAG, periaquaductal gray; ParCx, parietal cortex; PFA, perifornical area; VMH, ventromedial nucleus of the hypothalamus, PPit, posterior pituitary; PPTg, pedunculopontine tegmental nucleus; R, red nucleus; RN, raphe nuclei; SC, superior colliculi; SN, substantia nigra; SON, supraoptic nucleus; SP5n, spinal trigeminus nucleus; SPO, superior paraolivary nucleus; Thal, thalamus. Adapted with permission from Reul and Holsboer, 2002.

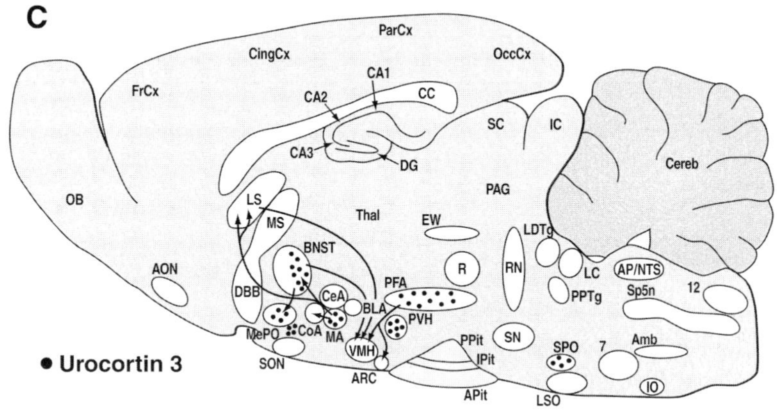

Fig. 1. Continued.

Neuroendocrine effects

Pituitary–adrenal activation

CRF was identified for its hypophysiotropic effects following its release into the portal blood from the median eminence of the hypothalamus (Spiess et al., 1981; Vale et al., 1981). CRF activates the hypothalamic–pituitary–adrenal (HPA) axis not only following systemic administration (Rivier et al., 1982), but also following i.c.v. administration, as observed in amphibians (Moore and Miller, 1984), birds (Furuse et al., 1997), rats (Ono et al., 1984), mice (Rosenthal and Morley, 1989), hamsters (Seifritz et al., 1998), sheep (Donald et al., 1983), dogs (Inoue et al., 1989), pigs (Johnson et al., 1994), and primates (Kalin et al., 1983). Peripheral administration of anti-CRF antisera blocks i.c.v. CRF-induced HPA activation, suggesting that the HPA-activating effects of both i.c.v. and i.v. CRF administration involve stimulation of anterior pituitary corticotrophs (Bruhn et al., 1986; Irwin et al., 1990). Likewise, HPA-activating effects of stress are blocked by peripheral administration of CRF antisera (Brown et al., 1984; Ono et al., 1985). The i.c.v. dose of CRF that stimulates the HPA axis exceeds the minimum dose that is behaviorally active (Veldhuis and de Wied, 1984), indicating that the behavioral and neuroendocrine effects of i.c.v. CRF are dissociable.

Like CRF, i.c.v. urocortin 1 increases circulating HPA hormone levels in diverse species (Smagin et al., 1998; Grill et al., 2000; Parrott et al., 2000. Weisinger et al., 2000; Whitley et al., 2000; Cullen et al., 2001; Jones et al., 2002). In fact, i.c.v. urocortin 1 has shown greater potency than CRF in several studies. It is doubtful, however, that urocortin 1 is a physiologic regulator of the HPA axis. Unlike CRF-deficient mice (Venihaki and Majzoub, 2002), urocortin 1-deficient mice exhibit normal basal and stress-induced HPA hormone levels (Vetter et al., 2002; Wang et al., 2002). Similarly, unlike CRF antisera, peripheral administration of specific urocortin 1 antisera does not modify basal, stress-induced or adrenalectomy-induced ACTH levels (Masuzawa et al., 1999; Turnbull et al., 1999). Finally, unlike the distribution of CRF, urocortin 1-immunoreactive fibers are scarce in the paraventricular nucleus of the hypothalamus and the external layer of the median eminence under basal conditions (Hara et al., 1997a,b; Kozicz et al., 1998). Thus, brain-derived CRF, but likely not urocortin 1, is an endogenous, primary regulator of HPA activity.

Osmoregulatory axis

An osmoregulatory role for urocortin 1 has been hypothesized because of the rich distribution of urocortin 1 fibers projecting from magnocellular neurons of the SON and PVN to the posterior pituitary. Possibly consistent with such a role, i.c.v. administration of urocortin 1 reportedly reduced plasma arginine vasopressin levels (Kakiya et al., 1998),

Table 1. Relation of prominent urocortin projection fields/expression to CRF receptor expression

Region	Urocortin 1	Urocortin 2	Urocortin 3	CRF_1	CRF_2
I. Forebrain					
Septum					
Lateral	++/+++	–	++++	–/+	++++
Medial/diagonal band complex	+/++	–	–/+	+++	–
Amygdala					
Central nucleus	+/++	–	–	–/+	–
Medial nucleus	+	–	+++	+/++	++
Cortical nuclei	+	–	+++	++	+++
Amygdalohippocampal area	++	–	++	++	–
Bed nucleus of the stria terminalis					
Rostral	++	–	+	++	–
Posterior	+	–	+++	+++	++
Globus pallidus	+/++	–	–	++	–
Substantia nigra	++	–	–	++/+++	–
Thalamus	++	–	+/++	+/++	–
Hypothalamus					
Supraoptic nucleus	+	+++	–/+	++	+
Arcuate nucleus	–/+	+++	++	++	+
Ventromedial nucleus	–/+	–	++++	–	+++
Lateral hypothalamus	++	–	+	+	+
Perifornical region	–/+	–	++++		?
Lateral preoptic area	++	–	+	+	+
Anterior hypothalamus	++/+++	–	+	+	+
Suprachiasmatic nucleus	+++	–	+	++	–
Dorsomedial nucleus	++	–	+/++	+++	+
Medial preoptic nucleus	+/++	–	+++	+/++	+/++
Paraventricular nucleus	+/++	+++	+/++	–/+	–/+
Mammillary nuclei	++	–	++	+/++	–/+
II. Brainstem					
Visual nuclei (superior colliculus, anterior pretectal nucleus)	+++	–	+	+/++	–
Somatosensory nuclei (dorsal column)	++	–	–	++++	–/+
Auditory nuclei (cochlear nuclei, inferior colliculus, lateral superior olive)	++++	–	+/++	++/+++	–/+
Vestibular nuclei	++++	–	–	++	–
Visceral nuclei					
Solitary tract nucleus	++	–	–	–/+	++
Area postrema	++	–	–	–	++
Parabrachial nucleus	+/++	–	–	++/+++	–
Motor nuclei (oculomotor, facial, hypoglossal)	++/+++	+++	–	+/++	–
Periaqueductal gray	++/+++	–	+/++	+	+
Tegmental nuclei	++/+++	–	–	++/+++	–
Red nucleus	+++	–	–	+++	–
Edinger-Westphal nucleus	++++	–	–		
Pontine nuclei	++	–	–	++/+++	–
Raphe nuclei			–		
Dorsal	++/+++	–	–/+	+	+++
Median	+/++	–	–	++	++
Locus coeruleus	+/++	+++	–	–/+	–
III. Cerebellum					
Deep nuclei	++	–	–	+++	–
Cortex	+/++	–	–	++/+++	–

(+) Indicates expression of peptide immunoreactivity or mRNA; increasing numbers of plus signs semi-quantitatively indicate greater expression.
(–) Not detected.
(/) Indicates expression intermediate to those indicated.

whereas i.c.v. administration of CRF has no such effect (Fisher and Brown, 1984; Kakiya et al., 1998). The prominent synthesis of urocortin 2 in magnocellular SON and PVN neurons also is consistent with a possible osmoregulatory role of urocortin 2. Interestingly, salt loading, dehydration, and hypophysectomy increased urocortin in magnocellular SON and PVN neurons (Hara et al., 1997a,b, 2000), whereas food deprivation decreased urocortin-LI in the SON (Hara et al., 1997b). The antibody's specificity for urocortin 1 as opposed to urocortin 2 is unknown, but specific increases in magnocellular SON urocortin 1 mRNA expression also have been observed following salt loading (Imaki et al., 2001). Altogether, the findings support a potential endogenous role for urocortins in the regulation of salt/water balance and/or nutrient homeostasis.

Somatotropic axis

Stress increases hypothalamic somatostatin release (Laczi et al., 1994a,b) and decreases plasma growth hormone (GH) levels (Rivier and Vale, 1985). Intracerebroventricular administration of CRF mimics the effects of stress on somatostatin (Mitsugi et al., 1990) and GH activity (Ono et al., 1984; Rivier and Vale, 1984). Moreover, stress-induced reductions in GH levels are reversed by CRF antagonists, indicating a role for CRF-like peptides in the effects of stress. Central, but not peripheral, administration of anti-CRF antisera also block stress-induced inhibition of GH secretion, supporting a specific endogenous role for CRF (Rivier and Vale, 1985). The possible mediating role of urocortin 1 in stress-induced inhibition of the somatotropic axis also has been examined recently.

Unlike the effects of stress and CRF, acute urocortin 1 (i.c.v.) administration increased GH levels in ovariectomized gilts (Whitley et al., 2000) and did not alter GH levels in male pigs (Parrott et al., 2000). Chronic urocortin 1 also increased GH levels in ovariectomized ewes (Holmberg et al., 2001). Thus, it appears unlikely that urocortin 1 mediates the inhibitory effects of stress on the neuroendocrine growth axis. Rather, urocortin 1 may serve to promote GH secretion under as yet unspecified conditions.

Pituitary–gonadal axes

Stress also is recognized to suppress circulating leutenizing-hormone (LH) levels in a CRF-antagonist reversible fashion (Rivier et al., 1986). Acute administration of high doses of CRF (i.c.v.) (Petraglia et al., 1986; Akema et al., 1996; Roozendaal et al., 1996; Chiba et al., 1997) and chronic i.c.v. CRF administration (Miskowiak et al., 1986) similarly reduce LH secretion, supporting a hypothesized role for CRF in the anti-sexual effects of stress.

The effects of urocortin 1 on plasma LH levels also were examined recently. Acute i.c.v. urocortin 1 administration decreased LH levels in ovariectomized gilts (Whitley et al., 2000), but not ewes (Holmberg et al., 2001). In fact, repeated urocortin 1 administration transiently increased circulating LH levels in ewes (Holmberg et al., 2001). Thus, urocortin 1 is not a well-conserved mediator of stress-induced inhibition of the pituitary–gonadal axes, but it remains possible that endogenous urocortin 1 modulates LH secretion under some conditions.

Summary

Urocortin 1 likely does not share the mediating role of CRF in stress-induced activation of the HPA axis or stress-induced inhibition of the somatotropic or pituitary–gonadal axes. However, urocortin 1 or urocortin 2 may have an osmoregulatory role that includes suppression of plasma arginine vasopressin levels. Urocortin 1 also may subtly modulate LH and GH release.

Electrophysiologic effects

CRF generally has excitatory properties on neurons (Aldenhoff et al., 1983; Siggins et al., 1985), although local application can be inhibitory in some regions (e.g., septum, thalamus, dorsal raphe; Siggins et al., 1985; Kirby et al., 2000). Following i.c.v. administration, CRF induces an "arousal-like" electrophysiologic profile (Ehlers et al., 1983; Ehlers, 1986).

At higher doses, i.c.v. CRF induces seizures (Ehlers et al., 1983; Weiss et al., 1986; Marrosu et al., 1988; Baram and Schultz, 1991). The excitatory/proconvulsant effects of i.c.v. CRF are mediated by the CRF_1 receptor (Schulz et al., 1996; Baram et al., 1997; Okuyama et al., 1999).

Urocortin 1, like CRF, stimulates the central nervous system (CNS) following i.c.v. administration. Despite the similarly high affinity of urocortin 1 for the CRF_1 receptor, however, urocortin 1 is less efficacious than CRF in both its arousal-like (Slawecki et al., 1999) and proconvulsant electrophysiologic properties (Baram et al., 1997). The electrophysiologic properties of the type 2 urocortins are not known.

Behavioral activation

Corresponding to the neurophysiologic effects of CRF, behavioral activation in familiar environments was one of the earliest observed consequences of i.c.v. CRF administration in the rat (Sutton et al., 1982; Koob et al., 1984). This activation consists largely of locomotor behavior and grooming (Koob and Thatcher-Britton, 1985; Dunn et al., 1987; Sherman and Kalin, 1987). The arousing and anxiogenic-like effects of CRF are at least partly dissociable from one another (Cole and Koob, 1988; Heinrichs and Joppa, 2001). Behavioral arousal following i.c.v. CRF administration is a well-conserved behavioral response, also having been observed in primates (Kalin et al., 1983; Winslow et al., 1989), pigs (Johnson et al., 1994; Salak-Johnson et al., 1997; Parrott and Vellucci, 2000; Parrott et al., 2000), rabbits (Opp et al., 1989), mice (Dunn and Berridge, 1987; Dunn et al., 1987), birds (Ohgushi et al., 2001; Zhang et al., 2001), fish (Clements et al., 2002), and amphibians (Lowry et al., 1990, 2001; Lowry and Moore, 1991).

In familiar environments, i.c.v. urocortin 1 induces behavioral arousal in the rat, including grooming, with similar potency to, but less efficacy than, CRF (Spina et al., 1996; Jones et al., 1998). In contrast, both i.c.v. urocortin 2 (Valdez et al., 2002) and urocortin 3 (Valdez et al., 2003) mildly suppress locomotor activity. These findings are consistent with the observation that the locomotor activating effects of i.c.v. CRF, like its excitatory properties, are mediated by CRF_1 receptors (Contarino et al., 2000; Zorrilla et al., 2002b), perhaps in the ventral forebrain (Tazi et al., 1987). That is, the diminished efficacy of urocortin 1 to increase behavioral activity may reflect opposing actions of CRF_1 and CRF_2 receptors.

Summary

Given circuit-appropriate access to their cognate receptors, endogenous CRF_1 (i.e., CRF or urocortin 1) and CRF_2 receptor agonists (i.e., urocortins 1, 2, or 3) might subserve arousing and dearousing functions, respectively.

Anxiety-like behavior

Intracerebroventricular administration of CRF increases anxiety-like behavior in multiple animal models of anxiety. In rodents, i.c.v. CRF reduces exploration of novel and/or exposed spaces in the elevated plus maze (EPM; Baldwin et al., 1991; Momose et al., 1999), defensive withdrawal (Takahashi et al., 1989; Yang et al., 1990), open field (Sutton et al., 1982; Koob and Thatcher-Britton, 1985), and multi-compartment tests (Berridge and Dunn, 1989). CRF also selectively suppresses punished responding during operant conflict tests, an anxiogenic-like effect (Britton et al., 1985; Koob and Thatcher-Britton, 1985; De Boer et al., 1992). Other anxiety-like signs of behavioral inhibition following i.c.v. CRF administration include its potentiation of conditioned response suppression (Cole and Koob, 1988; Gupta and Brush, 1998), conditioned immobility (Heinrichs and Joppa, 2001), and freezing behavior (Sherman and Kalin, 1988; Pelleymounter et al., 2000). Likewise, CRF inhibits isolation-induced ultrasonic vocalizing by neonatal rodents (Insel and Harbaugh, 1989; Hennessy et al., 1992; Harvey and Hennessy, 1995; Dirks et al., 2002) and reduces active social interaction in adult rodents (Dunn and File, 1987; Mele et al., 1987; Elkabir et al., 1990). Importantly, the anxiety-like effects of i.c.v. CRF are not limited to behavioral inhibition or promotion of inactivity. Intracerebroventricular CRF potentiates *active* behavioral responses to novelty (Britton et al., 1982), potentiates the acoustic startle response (Swerdlow et al., 1986), and increases conditioned defensive burying in habituated rats (Diamant et al., 1992a). Intracerebral administration of CRF has revealed candidate sites of anxiogenic-like action that include the BNST (Lee and Davis, 1997), amygdala (Liang and Lee, 1988), septum (Radulovic et al., 1999; Kask et al., 2001), dorsal periacqueductal gray (Martins et al., 1997), and LC (Butler et al., 1990). The anxiogenic-like effects of i.c.v. CRF result from a direct neurotropic, CRF_1-mediated action independent of its endocrine effects (Koob and Heinrichs, 1999) and also have been observed in rabbits (Tarjan et al., 1991), primates (Kalin et al., 1983; Kalin et al., 1989; Strome et al., 2002), and birds (Barrett et al., 1989; Zhang and Barrett, 1990).

Following i.c.v. administration, urocortin 1 generally shares the anxiogenic-like properties of i.c.v. CRF (Moreau et al., 1997), albeit with a slightly delayed onset of action and subtle test-specific differences in potency and efficacy (Spina et al., 1996, 2002; Jones et al., 1998). Like CRF, urocortin 1 (i.c.v.) produces anxiogenic-like effects in rodent tests predicated on exploration of novel and/or exposed spaces, including the open field (Moreau et al., 1997; Zorrilla et al., 2001), EPM (Moreau et al., 1997; Jones et al., 1998; Spina et al., 2002), defensive withdrawal (Spina et al., 2002), and light/dark box tests (Moreau et al., 1997). Unlike CRF, however, i.c.v. urocortin 1 does not potentiate the acoustic startle response (Jones et al., 1998) and does not preferentially reduce punished responding in the Geller-Seifter food/shock conflict test (Spina et al., 2002). Microinfusion of urocortin 1 into the basolateral amygdala leads to anxiogenic-like behavior in the social interaction test (Sajdyk et al., 1999; Sajdyk and Gehlert, 2000), but other aforementioned CRF-responsive candidate sites of action have yet to be probed with urocortin 1.

Quite unlike the CRF_1 receptor agonists, i.c.v. administration of the type 2 urocortins did not have anxiogenic-like effects in the rat (Valdez et al., 2002, 2003). Rather, although i.c.v. urocortin 2 did not acutely affect exploration in the EPM, it showed *anxiolytic-like* effects under delayed, higher baseline anxiety test conditions (Valdez et al., 2002). Urocortin 3 potently and acutely increased open arm exploration and reduced risk assessment in the EPM (Valdez et al., 2003). In contrast to its reported anxiolytic-like effects in the rat, i.c.v. urocortin 2 potently produced acute anxiogenic-like effects in the mouse EPM (Pelleymounter et al., 2002). Separate studies also suggest that the anxiety-related effects of CRF_2 activation may be site-dependent, further complicating interpretation of the anxiety-related functions of the urocortins (see Valdez et al., 2002; 2003, for a discussion). Thus, firm conclusions about the anxiety-related effects of central administration of the type 2 urocortins cannot yet be drawn.

Summary

Both CRF_1 receptor agonists – CRF and urocortin 1 – have well-conserved anxiogenic-like properties following i.c.v. administration. Urocortin 1 has a delayed onset of action and does not share all anxiogenic-like properties of its paralog, CRF. Candidate sites of action for i.c.v. CRF and urocortin include the BNST, septum, amygdala, periacqueductal gray, and locus coeruleus. The anxiety-related properties of the type 2 urocortins are not yet clear, as they may vary according to time, site of action, or species.

Learning and memory

Possibly related to the regulation of anxiety, antagonist studies have suggested a role for CRF_1 receptors and septal CRF_2 receptors in fear conditioning (Deak et al., 1999; Radulovic et al., 1999; Kikusui et al., 2000). The neuroanatomical concordance of urocortins with medial septal CRF_1 receptors (urocortin 1) and lateral septal CRF_2 receptors (urocortin 1 and urocortin 3) make it possible that urocortins play an endogenous role in fear conditioning. In addition to these specific effects on learning about threatening stimuli, urocortin 1 also may play a more general role in stress-related information processing. Central administration of both CRF and urocortin 1 exert biphasic effects on the acquisition and consolidation of information, with the direction of effect (enhancement or impairment) depending on factors such as dose and training conditions. In animals trained under either aversive or non-aversive conditions (e.g., appetitively motivated visual discrimination task; Heinrichs et al., 1997), i.c.v. CRF and urocortin 1 facilitate certain forms of learning and memory, often with an inverted U-shaped dose–response function. This dose–response profile is characteristic of arousal-modulating substances.

For example, urocortin 1 facilitated acquisition, but not consolidation, of place memory in the Morris maze. At the same time, urocortin 1 enhanced consolidation of passive avoidance learning, a form of stimulus–response learning (Zorrilla et al., 2002a). This profile of cognitive enhancement is consistent with increased activity of the medial septal area (MSA), the primary subcortical, cholinergic projection to the hippocampus, and cingulate cortex. The MSA, composed of the vertical limb of the diagonal band and the medial septum, subserves arousal and attention. Increased arousal resulting from activation of the MSA is hypothesized to facilitate processing of emotional, novel, or highly salient stimuli with the subsequent initiation of adaptive responses. The MSA is rich in both urocortin 1 and CRF_1 expression, with CRF_1 receptors colocalizing heavily with both cholinergic and catecholaminergic neurons (Sauvage and Steckler, 2001). Thus, endogenous urocortin 1 may help prime attention for and processing of adaptively relevant stimuli (see Zorrilla et al., 2002a, for a discussion). A role for endogenous CRF receptor transmission in learning and memory is further suggested by the observation that pretreatment with a CRF receptor antagonist prior to meeting an unfamiliar conspecific impairs subsequent social recognition (Heinrichs, 2003).

Energy balance

Energy expenditure

Almost since its discovery, it has been recognized that i.c.v. CRF promotes negative energy balance, both by reducing food intake and by increasing energy expenditure. The increased metabolism achieved by CRF occurs through activation of diverse behavioral, thermogenic, and other autonomic pathways. Independent of its effects on behavioral activation (Overton and Fisher, 1989; Diamant et al., 1992b), i.c.v. administration of CRF elevated mean arterial pressure (Fisher and Brown, 1984; Lenz et al., 1987; Grosskreutz and Brody, 1988; Overton and Fisher, 1989; Richter and Mulvany, 1995), elevated heart rate (Fisher and Brown, 1984; Lenz et al., 1987; Grosskreutz and Brody, 1988; Overton and Fisher, 1989; Diamant and de Wied, 1991; Korte et al., 1992; Diamant et al., 1992b; Richter and Mulvany, 1995; Nijsen et al., 2000), increased plasma catecholamine levels (Fisher and Brown, 1984; Lenz et al., 1987; Overton and Fisher, 1989; Irwin et al., 1992; Nijsen et al., 2000), increased firing of sympathetic nerves to brown adipose tissue (Holt and York, 1989), increased brown adipose tissue thermogenesis (LeFeuvre et al., 1987; Arase et al., 1988, 1989a,b), increased core body temperature (Diamant and de Wied, 1991; Buwalda et al., 1997; Linthorst et al., 1997), and increased resting whole body oxygen consumption (VO_2) (Rothwell et al., 1991). Consistent with these findings, i.c.v. CRF reduced body weight gain even after reductions in food intake were accounted for by pair-feeding (Rohner-Jeanrenaud et al., 1989; Hotta et al., 1991; Cullen et al., 2001). CRF antagonists attenuated CRF-induced increases in blood pressure and heart rate (Brown et al., 1986), with more recent studies indicating a specific role for CRF_1 receptors (Nijsen et al., 2000). It also has been proposed that VMH CRF_2-bearing neurons may modulate autonomic outflow (De Fanti and Martinez, 2002).

Although it has been suggested that i.c.v. urocortin 1 does not share the metabolic effects of i.c.v. CRF (Cullen et al., 2001), studies have observed that i.c.v. urocortin 1 increases mean arterial pressure (Spina et al., 1996), increases colonic body temperature via activation of the sympathetic nervous system (De Fanti and Martinez, 2002), and increases whole body oxygen consumption as measured by indirect calorimetry (De Fanti and Martinez, 2002). Intra-PVN urocortin 1 administration also increased plasma leptin levels as well as uncoupling protein-1 mRNA in brown adipose tissue (Kotz et al., 2002). Thus, it remains possible that endogenous urocortin 1 has catabolic-like effects in the regulation of energy balance, although the brain and receptor substrates for such an effect remain unclear.

Food intake

In addition to the potential effects of urocortin 1 on energy expenditure, i.c.v. administration of urocortins has potent anorectic effects. The joint observations that i.c.v. urocortin 1 and urotensin 1 more potently bind the CRF_2 receptor and induce anorexia than does CRF led to the hypothesis that brain CRF_2 receptor activation reduces food intake (Spina et al., 1996). Supporting this hypothesis, i.c.v. administration of antisauvagine-30, a preferential CRF_2 receptor antagonist (Ruhmann et al., 1998), and antisense knockdown of CRF_2 receptor expression attenuate i.c.v. CRF- and urocortin 1-induced anorexia (Smagin et al., 1998; Cullen et al., 2001; Pelleymounter et al., 2002). Similarly, CRF_2 knockout mice exhibit abbreviated urocortin 1-induced anorexia (Coste et al., 2000). Finally, VMH CRF_2 mRNA levels vary inversely with appetite. Hyperphagic states or stimuli, including obesity, diabetes, and food-deprivation, are associated with downregulated VMH CRF_2 mRNA levels in the rat (Richard et al., 1996; Timofeeva and Richard, 1997; Makino et al., 1998). Conversely, i.c.v. leptin upregulates VMH CRF_2 mRNA levels (Nishiyama et al., 1999). Collectively, findings spurred interest in the role of urocortins and the CRF_2 receptor in the physiology of feeding.

Very recently, i.c.v. urocortin 2 was observed to potently suppress food intake in nondeprived rats, most prominently during peak nocturnal feeding 3–6 h post-injection (Reyes et al., 2001; Inoue et al., 2003). Rats ate smaller, shorter meals more slowly, without altering their frequency of initiating meals (Inoue et al., 2003), an effect consistent with facilitation of satiation. Unlike CRF_1 agonists, i.c.v. urocortin 2 did not produce malaise, arousal, or anxiety-like effects at anorectic doses in the rat (Reyes et al., 2001; Valdez et al., 2002; Inoue et al., 2003). Thus, selective CRF_2 agonists produce satiation-like effects in the rat. The degree to which endogenous brain urocortins produce these effects under physiologic conditions remains unclear. Appetite-regulating regions under intense scrutiny include the ventromedial (co-occurrence of urocortin 3 terminals and CRF_2 receptor expression) and arcuate (urocortin 3 projections and urocortin 2 synthesis) nuclei of the hypothalamus, the lateral septum (urocortin 1 and urocortin 3 terminals corresponding with CRF_2 receptor expression), the medial amygdala (urocortin 3 terminals corresponding with CRF_2 receptor expression), and visceral nuclei in the hindbrain (urocortin 1 expression corresponding with CRF_2 receptor expression in the solitary tract nucleus and area postrema).

In addition to the hypothesized anorectic role of urocortins via the CRF_2 receptor, brain urocortin 1 may reduce food intake via a CRF_1-dependent mechanism. Brain CRF_1 receptor activation suppresses feeding through a different mechanism than CRF_2 receptor activation. In rats, i.c.v. CRF_1 agonists elicited abbreviated, short-onset anorexia; CRF_2 agonists provoked delayed-onset anorexia; and nonselective CRF agonists produced an additive effect (prolonged, short-onset anorexia; Reyes et al., 2001). Similarly, nonselective CRF agonists yielded prolonged, short-onset anorexia in wildtype mice; abbreviated, short-onset anorexia in CRF_2 null mice; and delayed-onset anorexia in CRF_1 knockouts (see Inoue et al., 2003). Possible loci for putative endogenous urocortin 1-CRF_1 mediated anorexia include the dorsomedial nucleus of the hypothalamus and the parabrachial nucleus, both of which contain moderate urocortin 1-LI and prominent CRF_1 receptor expression.

Summary

Both CRF_1 receptor agonists – CRF and urocortin 1 – appear to increase indices of energy expenditure following i.c.v. administration, but any catabolic-like effects of the type 2 urocortins remain unknown. Several lines of evidence indicate that both CRF_1- and CRF_2-mediated anorexia exist, and food intake is potently suppressed by i.c.v. administration of urocortins 1 and 2. The microstructural and neuroactivational consequences of i.c.v. urocortin 2 administration resemble those of recognized satiating appetite suppressants. The concordant distribution of urocortin fibers with their putative cognate receptors in caudal hindbrain, hypothalamus, and feeding-regulatory limbic regions further suggests an endogenous role of urocortin signaling in body weight homeostasis.

Gastrocolonic motility

Stressors inhibit gastric motility in a CRF antagonist-reversible fashion (Taché et al., 2001; Million et al., 2002). Conversely, through vagal mechanisms, central urocortin 1 administration reduced antral gastric motility, inhibited high-amplitude gastric contractions, shifted duodenal activity from fasted to fed motor patterns, and delayed gastric emptying (Kihara et al., 2001; Wang et al., 2001; Chen et al., 2002). Gastric stasis induced by urocortin 1 is believed to be CRF_2-mediated as it was reversed by nonselective CRF antagonists, but not by NBI 27914, a selective CRF_1 antagonist (Kihara et al., 2001; Chen et al., 2002). Thus, findings indicate that urocortin 1 may interact with CRF_2 receptors in the solitary tract nucleus to reduce gastric motility under stress and possibly other conditions.

In contrast to the inhibition of gastric motility, many stressors stimulate colonic motor function, reflected as increased colonic motility, decreased colonic transit time, elevated fecal output and watery diarrhea. Central administration of CRF_1-binding (urocortin 1), but not CRF_2-binding (urocortin 2) urocortins, mimicked these effects (Taché et al., 2001, 2002; Million et al., 2002). Conversely, selective CRF_1, but not CRF_2, antagonists attenuated stress-induced stimulation of colonic motility (Taché et al., 2001, 2002). Candidate brain substrates for CRF_1-mediated increases in colonic motor function include the PVN of the hypothalamus and the LC/Barrington nuclei complex, regions that express both CRF and urocortin 1, but the endogenous role of urocortin 1 in physiologic regulation of colonic motility remains unproven.

Regulation of central urocortin activity

Remarkably little is known about factors that regulate synthesis and secretion of the urocortins. As discussed, stimuli relevant to osmotic balance alter urocortin 1 mRNA and urocortin-LI in magnocellular neurons of the SON and, possibly, PVN of the hypothalamus. Urocortin 1 mRNA expression in the E-W, the principal neural site of urocortin 1 synthesis, was upregulated by 3 h restraint stress (Weninger et al., 2000), an effect that could be blocked by chronic treatment with stress-like corticosterone levels. Also consistent with the hypothesis that urocortin 1-expressing neurons are stress responsive, stress-induced Fos expression in the E-W was associated with colocalization of urocortin 1- and cocaine/amphetamine-regulated transcript-LI (Kozicz, 2003). On the other hand, two antistress-like conditions –chronic administration of the benzodiazepine anxiolytic alprazolam and CRF-gene deletion – were also associated with upregulated urocortin 1 mRNA expression in the E-W (Skelton et al., 2000; Weninger et al., 2000). Interestingly, CRF_2 null mice (Bale et al., 2000), but neither CRF_1 (Smith et al., 1998a; Timpl et al., 1998) nor conjoint CRF_1 and CRF_2 null mice (Bale et al., 2002), also exhibited elevated urocortin 1 mRNA levels in the E-W, suggesting CRF_1 receptors may serve a permissive role in upregulation of urocortin 1 synthesis. Perifornical urocortin 3 mRNA levels were dramatically increased in CRF_2 null and conjoint CRF_1 and CRF_2 null mutant mice (Bale et al., 2002), possibly consistent with the absence of an auto-regulatory negative feedback mechanism.

Effects of specifically reducing central urocortin activity

Many studies have examined the effects of increasing or reducing CRF receptor transmission, thereby circumstantially shedding light on possible urocortin functions. Very few studies, however, have specifically opposed endogenous urocortin activity through immunoneutralization or molecular knockdown/knockout methods. As discussed previously, immunoneutralization of urocortin 1 did not attenuate the adrenocorticotropic hormone (ACTH) response to footshock, suggesting a lack of involvement of urocortin 1 in stress-induced HPA activation (Turnbull et al., 1999). Similarly, immunoneutralization of urocortin 1 did not modulate adrenalectomy-induced disinhibition of ACTH levels (Masuzawa et al., 1999). With respect to feeding, the anorectic effects of CRF and leptin were not attenuated by i.c.v. anti-urocortin 1 antiserum administration (Okamoto et al., 2001). However, intra-VMH administration of an anti-urocortin 1 antiserum reportedly increased food and water intake (Ohata et al., 2000). The study was performed prior to the discoveries of urocortin 2 and urocortin 3, so the specificity of the putative urocortin 1 antisera is not clear. Nonetheless, the findings are consistent with a hypothesized endogenous anti-consummatory role of urocortins in the VMH. Urocortin 1 antisera (i.c.v.) also did not alter gastroduodenal motility in fed or fasted states, suggesting that any endogenous role for urocortin 1 in regulating gastric motor function is situation specific (e.g., under stress; Kihara et al., 2001).

Two independent groups have generated urocortin 1-deficient mice (Vetter et al., 2002; Wang et al., 2002), with conflicting results regarding phenotypes. One group observed that the urocortin 1-deficient mice exhibited increased anxiogenic-like behavior, reflected as decreased exploration of the open arms of the EPM and center of an open field (Vetter et al., 2002). These mice also showed impaired auditory function, reflected as increased thresholds for auditory brainstem responses during adulthood and diminished amplitude distortion product otoacoustic emissions, consistent with impaired outer hair-cell function of the organ of Corti (Vetter et al., 2002). Mutant mice also exhibited downregulated CRF_2 mRNA expression in the lateral septum and decreased CRF mRNA levels in the rostral BNST. Moreover, urocortin 1 null mice had normal body weight and both 24 h deprivation-induced and nondeprived total daily food intake.

However, the other set of urocortin 1-deficient mice exhibited normal anxiety-like behavior, as measured in the open field, EPM, and light/dark box tests (Wang et al., 2002). These urocortin 1-null mutant mice reportedly were *more* sensitive to low intensity sounds, as demonstrated by reduced startle thresholds and heightened startle responses. However, mutant mice exhibited reduced maximal startle amplitudes. Findings were not interpreted as reflecting a hearing defect, because these urocortin 1-null mice exhibited normal multiwave response and interpeak latency at the range of 70 to 90-dB sound stimuli in auditory brainstem response testing as well as normal prepulse inhibition with low intensity prepulses. Rather, the findings were interpreted as a nonperceptual alteration in acoustic startle response neurocircuitry. Transgenics exhibited normal autonomic responses to stress, as demonstrated by restraint-induced changes in heart rate and plasma catecholamine levels. Thus, phenotypes differed with respect to reported changes in anxiety-like behavior, hearing, and startle responding. In sole agreement, both urocortin 1-deficient transgenics exhibited normal basal and restraint stress-induced HPA hormone levels. The main conclusion from the urocortin 1 knockout studies may be that urocortin 1 is not *required* for stress-induced anxiety-like, autonomic or endocrine responses, nor for maintenance of normal body weight or food intake. Because of the redundancy of peptide systems, however, the functional significance of urocortin 1 in the intact mammal remains less clear.

Summary and conclusions

The neuropeptides urocortin 1, urocortin 2, and urocortin 3 share a structural and pharmacological family resemblance. Each urocortin shares high affinity for the CRF_2 receptor, mild-to-moderate sequence identity with the stress-related neuropeptide CRF, predominantly subcortical distributions and anorectic properties when administered centrally. In many significant ways, however, the peptides differ from one another. Urocortin 1 differs pharmacologically, as it also exhibits affinity for the CRF_1 receptor and CRF-BP. Brain distributions differ as well. Urocortin 1 shows predominant correspondence with hindbrain CRF_1 and CRF_2 receptors and a much more limited correspondence with forebrain CRF receptors. In contrast, urocortin 3 shows almost exclusive correspondence with forebrain CRF_2 receptors unique from those innervated by urocortin 1 projections. Urocortin 2 is synthesized in yet another discrete group of hypothalamic and brainstem nuclei, largely distinct from those that prominently express urocortin 1 and urocortin 3. Neuroanatomical and neuropharmacologic studies suggest that urocortins may be involved in the regulation of diverse functions, including osmotic balance; activity of the somatotropic and pituitary–gonadal neuroendocrine axes; modulation of arousal, attention, anxiety-like behavior, and learning and memory; control of autonomic functions, including cardiovascular and gastrointestinal responses; and the regulation of energy balance, both in terms of food intake and energy expenditure. Much less is currently known about urocortin 2 and urocortin 3 than about urocortin 1.

Although they are commonly described as being mediators of the organism stress response because of their structural resemblance to CRF, very little evidence directly and specifically supports the hypothesis that any of the urocortins is an integral part of either the activating or anxiogenic responses to stressors. No studies to date have shown that stress modifies the synthesis or secretion of urocortin 2 or urocortin 3, and both stress-like and antistress-like treatments have been shown to upregulate urocortin 1 synthesis. Moreover, unlike CRF- and CRF receptor-deficient mice, urocortin 1-deficient mice did not consistently exhibit a striking stress-related phenotype. Urocortins, likely through activation of the CRF_2 receptor, may be more involved in autonomic-energy balance responses that comprise a part of the global organism response to stress. Clearly, more remains to be learned about these putatively stress-related neuropeptides.

Abbreviations

ACTH	adrenocorticotropic hormone
BNST	bed nucleus of the stria terminalis
CRF	corticotropin-releasing factor
CRF-BP	corticotropin-releasing factor binding protein
CRF_1	type 1 corticotropin-releasing factor receptor
CRF_2	type 2 corticotropin-releasing factor receptor
dB	decibel
E-W	Edinger-Westphal nucleus
EPM	elevated plus maze
GH	growth hormone
HPA	hypothalamic–pituitary–adrenal axis
i.c.v.	intracerebroventricular
i.v.	intravenous
LC	locus coeruleus
LH	leutenizing hormone
LI	like immunoreactivity
mRNA	messenger ribonucleic acid
nM	nanomolar
PVN	paraventricular nucleus
SON	supraoptic nucleus
VMH	ventromedial hypothalamus

Acknowledgments

The authors were supported by National Institutes of Health grants DK26741 and DK64871 from the National Institute of Diabetes and Digestive and Kidney Diseases. The authors gratefully recognize Mike Arends for editorial assistance. This is the publication number 15746-NP from The Scripps Research Institute.

References

Akema, T., Chiba, A., Shinozaki, R., Oshida, M., Kimura, F. and Toyoda, J. (1996) Acute immobilization stress and intraventricular injection of CRF suppress naloxone-induced LH release in ovariectomized estrogen-primed rats. J. Neuroendocrinol., 8: 647–652.

Aldenhoff, J.B., Gruol, D.L., Rivier, J., Vale, W. and Siggins, G.R. (1983) Corticotropin releasing factor decreases postburst hyperpolarizations and excites hippocampal neurons. Science, 221: 875–877.

Arai, M., Assil, I.Q. and Abou-Samra, A.B. (2001) Characterization of three corticotropin-releasing factor receptors in catfish: a novel third receptor is predominantly expressed in pituitary and urophysis. Endocrinology, 142: 446–454.

Arase, K., Shargill, N.S. and Bray, G.A. (1989a) Effects of corticotropin releasing factor on genetically obese (fatty) rats. Physiol. Behav., 45: 565–570.

Arase, K., Shargill, N.S. and Bray, G.A. (1989b) Effects of intraventricular infusion of corticotropin-releasing factor on VMH-lesioned obese rats. Am. J. Physiol., 256: R751–R756.

Arase, K., York, D.A., Shimizu, H., Shargill, N. and Bray, G.A. (1988) Effects of corticotropin-releasing factor on food intake and brown adipose tissue thermogenesis in rats. Am. J. Physiol., 255: E255–E259.

Bachtell, R.K., Tsivkovskaia, N.O. and Ryabinin, A.E. (2002) Alcohol-induced c-Fos expression in the Edinger-Westphal nucleus: pharmacological and signal transduction mechanisms. J. Pharmacol. Exp. Ther., 302: 516–524.

Bachtell, R.K., Tsivkovskaia, N.O. and Ryabinin, A.E. (2003a) Identification of temperature-sensitive neural circuits in mice using c-Fos expression mapping. Brain Res., 960: 157–164.

Bachtell, R.K., Weitemier, A.Z., Galvan-Rosas, A., Tsivkovskaia, N.O., Risinger, F.O., Phillips, T.J., Grahame, N.J. and Ryabinin, A.E. (2003b) The Edinger-Westphal-lateral septum urocortin pathway and its relationship to alcohol consumption. J. Neurosci., 23: 2477–2487.

Baigent, S.M. and Lowry, P.J. (2000) Urocortin is the principal ligand for the corticotrophin-releasing factor binding protein in the ovine brain with no evidence for a sauvagine-like peptide. J. Mol. Endocrinol., 24: 53–63.

Baldwin, H.A., Rassnick, S., Rivier, J., Koob, G.F. and Britton, K.T. (1991) CRF antagonist reverses the "anxiogenic" response to ethanol withdrawal in the rat. Psychopharmacology, 103: 227–232.

Bale, T.L., Contarino, A., Smith, G.W., Chan, R., Gold, L.H., Sawchenko, P.E., Koob, G.F., Vale, W.W. and Lee, K.F. (2000) Mice deficient for corticotropin-releasing hormone receptor-2 display anxiety-like behaviour and are hypersensitive to stress. Nat. Genet., 24: 410–414.

Bale, T.L., Picetti, R., Contarino, A., Koob, G.F., Vale, W.W. and Lee, K.F. (2002) Mice deficient for both corticotropin-releasing factor receptor 1 (CRFR1) and CRFR2 have an impaired stress response and display sexually dichotomous anxiety-like behavior. J. Neurosci., 22: 193–199.

Baram, T.Z., Chalmers, D.T., Chen, C., Koutsoukos, Y. and De Souza, E.B. (1997) The CRF1 receptor mediates the excitatory actions of corticotropin releasing factor (CRF) in the developing rat brain: in vivo evidence using a novel, selective, non-peptide CRF receptor antagonist. Brain Res., 770: 89–95.

Baram, T.Z. and Schultz, L. (1991) Corticotropin-releasing hormone is a rapid and potent convulsant in the infant rat. Brain Res. Dev. Brain Res., 61: 97–101.

Barrett, J.E., Zhang, L., Ahlers, S.T. and Wojnicki, F.H. (1989) Acute and chronic effects of corticotropin-releasing factor on schedule-controlled responding and neurochemistry of pigeons. J. Pharmacol. Exp. Ther., 250: 788–794.

Behan, D.P., De Souza, E.B., Lowry, P.J., Potter, E., Sawchenko, P. and Vale, W.W. (1995a) Corticotropin releasing factor (CRF) binding protein: a novel regulator of CRF and related peptides. Front. Neuroendocrinol., 16: 362–382.

Behan, D.P., Grigoriadis, D.E., Lovenberg, T., Chalmers, D., Heinrichs, S., Liaw, C. and De Souza, E.B. (1996a) Neurobiology of corticotropin releasing factor (CRF) receptors and CRF-binding protein: implications for the treatment of CNS disorders. Mol. Psychiatry, 1: 265–277.

Behan, D.P., Heinrichs, S.C., Troncoso, J.C., Liu, X.J., Kawas, C.H., Ling, N. and De Souza, E.B. (1995b) Displacement of corticotropin releasing factor from its binding protein as a possible treatment for Alzheimer's disease. Nature, 378: 284–287.

Behan, D.P., Khongsaly, O., Ling, N. and De Souza, E.B. (1996b) Urocortin interaction with corticotropin-releasing factor (CRF) binding protein (CRF-BP): a novel mechanism for elevating "free" CRF levels in human brain. Brain Res., 725: 263–267.

Behan, D.P., Linton, E.A. and Lowry, P.J. (1989) Isolation of the human plasma corticotrophin-releasing factor-binding protein. J. Endocrinol., 122: 23–31.

Berridge, C.W. and Dunn, A.J. (1989) CRF and restraint-stress decrease exploratory behavior in hypophysectomized mice. Pharmacol. Biochem. Behav., 34: 517–519.

Bittencourt, J.C., Vaughan, J., Arias, C., Rissman, R.A., Vale, W.W. and Sawchenko, P.E. (1999) Urocortin expression in rat brain: evidence against a pervasive relationship of urocortin-containing projections with targets bearing type 2 CRF receptors. J. Comp. Neurol., 415: 285–312.

Britton, D.R., Koob, G.F., Rivier, J. and Vale, W. (1982) Intraventricular corticotropin-releasing factor enhances behavioral effects of novelty. Life Sci., 31: 363–367.

Britton, K.T., Morgan, J., Rivier, J., Vale, W. and Koob, G.F. (1985) Chlordiazepoxide attenuates response suppression induced by corticotropin-releasing factor in the conflict test. Psychopharmacology, 86: 170–174.

Brown, M.R., Gray, T.S. and Fisher, L.A. (1986) Corticotropin-releasing factor receptor antagonist: effects on the autonomic nervous system and cardiovascular function. Regul. Pept., 16: 321–329.

Brown, M.R., Rivier, C. and Vale, W. (1984) Central nervous system regulation of adrenocorticotropin secretion: role of somatostatins. Endocrinology, 114: 1546–1549.

Bruhn, T.O., Sutton, S.W., Plotsky, P.M. and Vale, W.W. (1986) Central administration of corticotropin-releasing factor modulates oxytocin secretion in the rat. Endocrinology, 119: 1558–1563.

Butler, P.D., Weiss, J.M., Stout, J.C. and Nemeroff, C.B. (1990) Corticotropin-releasing factor produces fear-enhancing and behavioral activating effects following infusion into the locus coeruleus. J. Neurosci., 10: 176–183.

Buwalda, B., De Boer, S.F., Van Kalkeren, A.A. and Koolhaas, J.M. (1997) Physiological and behavioral effects of chronic intracerebroventricular infusion of corticotropin-releasing factor in the rat. Psychoneuroendocrinology, 22: 297–309.

Cepoi, D., Sutton, S., Arias, C., Sawchenko, P. and Vale, W.W. (1999) Ovine genomic urocortin: cloning, pharmacologic characterization, and distribution of central mRNA. Brain Res. Mol. Brain Res., 68: 109–118.

Chan, R.K., Vale, W.W. and Sawchenko, P.E. (2000) Paradoxical activational effects of a corticotropin-releasing factor- binding protein "ligand inhibitor" in rat Brain. *Neuroscience*, 101: 115–129.

Chen, C.Y., Million, M., Adelson, D.W., Martinez, V., Rivier, J. and Taché, Y. (2002) Intracisternal urocortin inhibits vagally stimulated gastric motility in rats: role of CRF(2). Br. J. Pharmacol., 136: 237–247.

Chiba, A., Akema, T., Nagami, Y., Kimura, F. and Toyoda, J. (1997) Differential inhibition of NMDA- and naloxone-induced LH release by NMDA receptor antagonist and CRH in ovariectomized estrogen-primed rats. Neuroendocrinology, 65: 141–146.

Clements, S., Schreck, C.B., Larsen, D.A. and Dickhoff, W.W. (2002) Central administration of corticotropin-releasing hormone stimulates locomotor activity in juvenile chinook salmon (Oncorhynchus tshawytscha). Gen. Comp., 125: 319–327.

Cole, B.J. and Koob, G.F. (1988) Propranolol antagonizes the enhanced conditioned fear produced by corticotropin releasing factor. J. Pharmacol. Exp. Ther., 247: 902–910.

Contarino, A., Dellu, F., Koob, G.F., Smith, G.W., Lee, K.F., Vale, W.W. and Gold, L.H. (2000) Dissociation of locomotor activation and suppression of food intake induced by CRF in CRFR1-deficient mice. Endocrinology, 141: 2698–2702.

Coste, S.C., Kesterson, R.A., Heldwein, K.A., Stevens, S.L., Heard, A.D., Hollis, J.H., Murray, S.E., Hill, J.K., Pantely, G.A., Hohimer, A.R., Hatton, D.C., Phillips, T.J., Finn, D.A., Low, M.J., Rittenberg, M.B., Stenzel, P. and Stenzel-Poore, M.P. (2000) Abnormal adaptations to stress and impaired cardiovascular function in mice lacking corticotropin-releasing hormone receptor-2. Nat. Genet., 24: 403–409.

Cullen, M.J., Ling, N., Foster, A.C. and Pelleymounter, M.A. (2001) Urocortin, corticotropin releasing factor-2 receptors and energy balance. Endocrinology, 142: 992–999.

Dautzenberg, F.M. and Hauger, R.L. (2002) The CRF peptide family and their receptors: yet more partners discovered. Trends Pharmacol. Sci., 23: 71–77.

Dautzenberg, F.M., Py-Lang, G., Higelin, J., Fischer, C., Wright, M.B. and Huber, G. (2001) Different binding modes of amphibian and human corticotropin-releasing factor type 1 and type 2 receptors: evidence for evolutionary differences. J. Pharmacol. Exp. Ther., 296: 113–120.

De Boer, S.F., Katz, J.L. and Valentino, R.J. (1992) Common mechanisms underlying the proconflict effects of corticotropin-releasing factor, a benzodiazepine inverse agonist and electric foot-shock. J. Pharmacol. Exp. Ther., 262: 335–342.

De Fanti, B.A. and Martinez, J.A. (2002) Central urocortin activation of sympathetic-regulated energy metabolism in wistar rats. Brain Res., 930: 37–41.

Deak, T., Nguyen, K.T., Ehrlich, A.L., Watkins, L.R., Spencer, R.L., Maier, S.F., Licinio, J., Wong, M.L., Chrousos, G.P., Webster, E. and Gold, P.W. (1999) The impact of the nonpeptide corticotropin-releasing Hormone antagonist antalarmin on behavioral and endocrine responses to stress. Endocrinology, 140: 79–86.

Diamant, M., Croiset, G. and de Wied, D. (1992a) The Effect of corticotropin-releasing factor (CRF) on Autonomic and behavioral responses during shock-prod burying test in rats. Peptides, 13: 1149–1158.

Diamant, M. and de Wied, D. (1991) Autonomic and behavioral effects of centrally administered corticotropin-releasing factor in rats. Endocrinology, 129: 446–454.

Diamant, M., Kashtanov, S.I., Fodor, M. and de Wied, D. (1992b) Corticotropin-releasing factor induces differential behavioral and cardiovascular effects after intracerebroventricular and lateral hypothalamic/perifornical injections in rats. Neuroendocrinology, 56: 750–760.

Dirks, A., Fish, E.W., Kikusui, T., van der Gugten, J., Groenink, L., Olivier, B. and Miczek, K.A. (2002) Effects of corticotropin-releasing Hormone on distress vocalizations and locomotion in maternally separated mouse pups. Pharmacol. Biochem. Behav., 72: 993–999.

Donald, R.A., Redekopp, C., Cameron, V., Nicholls, M.G., Bolton, J., Livesey, J., Espiner, E.A., Rivier, J. and Vale, W. (1983) The Hormonal actions of corticotropin-releasing factor in sheep: effect of intravenous and intracerebroventricular injection. Endocrinology, 113: 866–870.

Dunn, A.J. and Berridge, C.W. (1987) Corticotropin-releasing factor administration elicits a stress-like activation of cerebral catecholaminergic systems. Pharmacol. Biochem. Behav., 27: 685–691.

Dunn, A.J., Berridge, C.W., Lai, Y.I. and Yachabach, T.L. (1987) CRF-induced excessive grooming behavior in rats and mice. Peptides, 8: 841–844.

Dunn, A.J. and File, S.E. (1987) Corticotropin-releasing factor has an anxiogenic action in the social interaction test. Horm. Behav., 21: 193–202.

Dunn, A.J. and Swiergiel, A.H. (1999) Behavioral responses to stress are intact in CRF-deficient mice. Brain Res., 845: 14–20.

Ehlers, C.L. (1986) EEG stability following corticotropin releasing factor in rats. Psychoneuroendocrinology, 11: 121–125.

Ehlers, C.L., Henriksen, S.J., Wang, M., Rivier, J., Vale, W. and Bloom, F.E. (1983) Corticotropin releasing factor produces increases in brain excitability and convulsive seizures in rats. Brain Res., 278: 332–336.

Elkabir, D.R., Wyatt, M.E., Vellucci, S.V. and Herbert, J. (1990) The effects of separate or combined infusions of corticotrophin-releasing factor and vasopressin either intraventricularly or into the amygdala on aggressive and investigative behaviour in the rat. Regul. Pept., 28: 199–214.

Ferry, R.J., Jr., Cerri, R.W. and Cohen, P. (1999) Insulin-like growth factor binding proteins: new proteins, new functions. Horm. Res., 51: 53–67.

Fisher, L.A. and Brown, M.R. (1984) Corticotropin-releasing factor and angiotensin II: comparison of CNS actions to influence neuroendocrine and cardiovascular function. Brain Res., 296: 41–47.

Furuse, M., Matsumoto, M., Saito, N., Sugahara, K. and Hasegawa, S. (1997) The central corticotropin-releasing factor and glucagon-like peptide-1 in food intake of the neonatal chick. Eur. J. Pharmacol., 339: 211–214.

Gamlin, P.D. (1999) Subcortical neural circuits for ocular accommodation and vergence in primates. Ophthalmic Physiol. Opt., 19: 81–89.

Grill, H.J., Markison, S., Ginsberg, A. and Kaplan, J.M. (2000) Long-term effects on feeding and body weight after stimulation of forebrain or hindbrain CRH receptors with urocortin. Brain Res., 867: 19–28.

Grosskreutz, C.L. and Brody, M.J. (1988) Regional hemodynamic responses to central administration of corticotropin-releasing factor (CRF). Brain Res., 442: 363–367.

Gupta, P. and Brush, F.R. (1998) Differential behavioral and endocrinological effects of corticotropin-releasing hormone (CRH) in the Syracuse high- and low-avoidance rats. Horm. Behav., 34: 262–267.

Hara, Y., Ueta, Y., Isse, T., Kabashima, N., Shibuya, I., Hattori, Y. and Yamashita, H. (1997a) Increase of urocortin-like immunoreactivity in the rat hypothalamo-neurohypophysial system after salt loading and hypophysectomy. Neurosci. Lett., 227: 127–130.

Hara, Y., Ueta, Y., Isse, T., Kabashima, N., Shibuya, I., Hattori, Y. and Yamashita, H. (1997b) Increase of urocortin-like immunoreactivity in the rat supraoptic nucleus after dehydration but not food deprivation. Neurosci. Lett., 229: 65–68.

Hara, Y., Ueta, Y., Isse, T., Serino, R., Shibuya, I., Hattori, Y. and Yamashita, H. (2000) Increase of urocortin-like immunoreactivity in the supraoptic nucleus of Dahl rats given a high salt diet. Neurosci. Lett., 279: 17–20.

Harada, S., Imaki, T., Naruse, M., Chikada, N., Nakajima, K. and Demura, H. (1999) Urocortin mRNA is expressed in the enteric nervous system of the rat. Neurosci. Lett., 267: 125–128.

Harvey, A.T. and Hennessy, M.B. (1995) Corticotropin-releasing factor modulation of the ultrasonic vocalization rate of isolated rat pups. Brain Res. Dev. Brain Res., 87: 125–134.

Heinrichs, S.C. (2003) Modulation of social learning in rats by brain corticotropin-releasing factor. Brain Res., 994: 107–114.

Heinrichs, S.C. and Joppa, M. (2001) Dissociation of arousal-like from anxiogenic-like actions of brain corticotropin-releasing factor receptor ligands in rats. Behav. Brain Res., 122: 43–50.

Heinrichs, S.C., Vale, E.A., Lapsansky, J., Behan, D.P., McClure, L.V., Ling, N., De Souza, E.B. and Schulteis, G. (1997) Enhancement of performance in multiple learning tasks by corticotropin-releasing factor-binding protein ligand inhibitors. Peptides, 18: 711–716.

Hennessy, M.B., O'Neil, D.R., Becker, L.A., Jenkins, R., Williams, M.T. and Davis, H.N. (1992) Effects of centrally administered corticotropin-releasing factor (CRF) and alpha-helical CRF on the vocalizations of isolated Guinea pig pups. Pharmacol. Biochem. Behav., 43: 37–43.

Henriot, S., Dautzenberg, F.M. and Kilpatrick, G.J. (1999) Urocortin: slower dissociation than corticotropin releasing factor from the CRF binding protein. Eur. J. Pharmacol., 376: 321–324.

Holmberg, B.J., Morrison, C.D. and Keisler, D.H. (2001) Endocrine responses of ovariectomized ewes to i.c.v. infusion of urocortin. J. Endocrinol., 171: 517–524.

Holt, S.J. and York, D.A. (1989) The effects of adrenalectomy, corticotropin releasing factor and vasopressin on the sympathetic firing rate of nerves to interscapular brown adipose tissue in the Zucker rat. Physiol. Behav., 45: 1123–1129.

Hotta, M., Shibasaki, T., Yamauchi, N., Ohno, H., Benoit, R., Ling, N. and Demura, H. (1991) The effects of chronic central administration of corticotropin-releasing factor on food intake, body weight, and hypothalamic-pituitary-adrenocortical hormones. Life Sci., 48: 1483–1491.

Hsu, S.Y. and Hsueh, A.J. (2001) Human stresscopin and stresscopin-related peptide are selective ligands for the type 2 corticotropin-releasing hormone receptor. Nat. Med., 7: 605–611.

Iino, K., Sasano, H., Oki, Y., Andoh, N., Shin, R.W., Kitamoto, T., Takahashi, K., Suzuki, H., Tezuka, F., Yoshimi, T. and Nagura, H. (1999) Urocortin expression in the human central nervous system. Clin. Endocrinol. (Oxf), 50: 107–114.

Imaki, T., Katsumata, H., Miyata, M., Naruse, M., Imaki, J. and Minami, S. (2001) Expression of corticotropin releasing factor (CRF), urocortin and CRF type 1 receptors in hypothalamic-hypophyseal systems under osmotic stimulation. J. Neuroendocrinol., 13: 328–338.

Innis, R.B. and Aghajanian, G.K. (1986) Cholecystokinin-containing and nociceptive neurons in rat edinger-westphal nucleus. Brain Res., 363: 230–238.

Inoue, K., Valdez, G.R., Reyes, T.M., Reinhardt, L.E., Tabarin, A., Rivier, J., Vale, W.W., Sawchenko, P.E., Koob, G.F. and Zorrilla, E.P. (2003) Human urocortin II, a selective agonist for the type 2 corticotropin-releasing factor receptor, decreases feeding and drinking in the rat. J. Pharmacol. Exp. Ther., 305: 385–393.

Inoue, T., Inui, A., Okita, M., Sakatani, N., Oya, M., Morioka, H., Mizuno, N., Oimomi, M. and Baba, S. (1989) Effect of neuropeptide Y on the hypothalamic-pituitary-adrenal axis in the dog. Life Sci., 44: 1043–1051.

Insel, T.R. and Harbaugh, C.R. (1989) Central administration of corticotropin releasing factor alters rat pup isolation calls. Pharmacol. Biochem. Behav., 32: 197–201.

Irwin, M., Hauger, R. and Brown, M. (1992) Central corticotropin-releasing hormone activates the sympathetic nervous system and reduces immune function: increased responsivity of the aged rat. Endocrinology, 131: 1047–1053.

Irwin, M., Vale, W. and Rivier, C. (1990) Central corticotropin-releasing factor mediates the suppressive effect of stress on natural killer cytotoxicity. Endocrinology, 126: 2837–2844.

Jahn, O., Eckart, K., Sydow, S., Hofmann, B.A. and Spiess, J. (2001) Pharmacological characterization of recombinant rat corticotropin releasing factor binding protein using different sauvagine analogs. Peptides, 22: 47–56.

Jansen, A.S., Farkas, E., Mac Sams, J. and Loewy, A.D. (1998) Local connections between the columns of the periaqueductal gray matter: a case for intrinsic neuromodulation. Brain Res., 784: 329–336.

Johnson, R.W., von Borell, E.H., Anderson, L.L., Kojic, L.D. and Cunnick, J.E. (1994) Intracerebroventricular injection of corticotropin-releasing hormone in the pig: acute effects on behavior, adrenocorticotropin secretion, and immune suppression. Endocrinology, 135: 642–648.

Jones, D.N., Kortekaas, R., Slade, P.D., Middlemiss, D.N. and Hagan, J.J. (1998) The behavioural effects of corticotropin-releasing factor-related peptides in rats. Psychopharmacology, 138: 124–132.

Jones, J.E., Pick, R.R., Davenport, M.D., Keene, A.C., Corp, E.S. and Wade, G.N. (2002) Disinhibition of female sexual behavior by a CRH receptor antagonist in syrian hamsters. Am. J. Physiol. Regul. Integr. Comp. Physiol., 283: R591–R597.

Kageyama, K., Bradbury, M.J., Zhao, L., Blount, A.L. and Vale, W.W. (1999) urocortin messenger ribonucleic acid: tissue distribution in the rat and regulation in thymus by lipopolysaccharide and glucocorticoids. Endocrinology, 140: 5651–5658.

Kakiya, S., Yokoi, H., Arima, H., Iwasaki, Y., Oki, Y. and Oiso, Y. (1998) Central administration of urocortin inhibits vasopressin release in conscious rats. Neurosci. Lett., 248: 144–146.

Kalin, N.H., Shelton, S.E. and Barksdale, C.M. (1989) Behavioral and physiologic effects of crh administered to infant primates undergoing maternal separation. Neuropsychopharmacology, 2: 97–104.

Kalin, N.H., Shelton, S.E., Kraemer, G.W. and McKinney, W.T. (1983) Corticotropin-releasing factor administered intraventricularly to rhesus monkeys. Peptides, 4: 217–220.

Kask, A., Nguyen, H.P., Pabst, R. and Von Horsten, S. (2001) Neuropeptide Y Y1 receptor-mediated anxiolysis in the dorsocaudal lateral septum: functional antagonism of corticotropin-releasing hormone-induced anxiety. Neuroscience, 104: 799–806.

Kaufman, G.D., Anderson, J.H. and Beitz, A.J. (1992) Fos-defined activity in rat brainstem following centripetal acceleration. J. Neurosci., 12: 4489–4500.

Kemp, C.F., Woods, R.J. and Lowry, P.J. (1998) The corticotrophin-releasing factor-binding protein: an act of several parts. Peptides, 19: 1119–1128.

Kihara, N., Fujimura, M., Yamamoto, I., Itoh, E., Inui, A. and Fujimiya, M. (2001) Effects of central and peripheral urocortin on fed and fasted gastroduodenal motor activity in conscious rats. Am. J. Physiol. Gastrointest. Liver Physiol., 280: G406–G419.

Kikusui, T., Takeuchi, Y. and Mori, Y. (2000) Involvement of corticotropin-releasing factor in the retrieval process of fear-conditioned ultrasonic vocalization in rats. Physiol. Behav., 71: 323–328.

Kirby, L.G., Rice, K.C. and Valentino, R.J. (2000) Effects of corticotropin-releasing factor on neuronal activity in the serotonergic dorsal raphe nucleus. Neuropsychopharmacology, 22: 148–162.

Koob, G.F. and Heinrichs, S.C. (1999) A role for corticotropin releasing factor and urocortin in behavioral responses to stressors. Brain Res., 848: 141–152.

Koob, G.F., Swerdlow, N., Seeligson, M., Eaves, M., Sutton, R., Rivier, J. and Vale, W. (1984) Effects of alpha-flupenthixol and naloxone on CRF-induced locomotor activation. Neuroendocrinology, 39: 459–464.

Koob, G.F. and Thatcher-Britton, K. (1985) Stimulant and anxiogenic effects of corticotropin releasing factor. Prog. Clin. Biol. Res., 192: 499–506.

Korte, S.M., Eisinga, W., Timmerman, W., Nyakas, C. and Bohus, B. (1992) Behavioral and cardiac responses after intracerebroventricular corticotropin-releasing hormone (CRH) administration: role of adrenal cortical hormones. Horm. Behav., 26: 375–384.

Kotz, C.M., Wang, C., Levine, A.S. and Billington, C.J. (2002) Urocortin in the hypothalamic PVN increases leptin and affects uncoupling proteins-1 and -3 in rats. Am. J. Physiol. Regul. Integr. Comp. Physiol., 282: R546–R551.

Koyama, N., Nishio, T. and Yokota, T. (2000) Non-serotonergic midbrain neurons are involved in picrotoxin-induced analgesia. an immunohistochemical study in the rat. Neurosci. Lett., 291: 147–150.

Kozicz, T. (2003) Neurons colocalizing urocortin and cocaine and amphetamine-regulated transcript immunoreactivities are induced by acute lipopolysaccharide stress in the Edinger-Westphal nucleus in the rat. Neuroscience, 116: 315–320.

Kozicz, T. and Arimura, A. (2002) Distribution of urocortin in the rat's gastrointestinal tract and its colocalization with tyrosine hydroxylase. Peptides, 23: 515–521.

Kozicz, T., Arimura, A., Maderdrut, J.L. and Lazar, G. (2002) Distribution of urocortin-like immunoreactivity in the central nervous system of the frog Rana esculenta. J. Comp. Neurol., 453: 185–198.

Kozicz, T., Yanaihara, H. and Arimura, A. (1998) Distribution of urocortin-like immunoreactivity in the central nervous system of the rat. J. Comp. Neurol., 391: 1–10.

Laczi, F., Ivanyi, T., Sarnyai, Z., Vecsernyes, M., Lengyel, G., Szabo, G., Biro, E., Gardi, J., Julesz, J. and Telegdy, G. (1994a) The role of central corticoliberin in the ether stress-induced secretion of neurohypophyseal hormones and corticosterone in the rat. Neuropeptides, 26: 33–37.

Laczi, F., Szabo, G., Vecsernyes, M., Biro, E., Gardi, J., Julesz, J. and Telegdy, G. (1994b) The role of central corticoliberin in the hyperosmosis-induced secretion of neurohypophysial hormones and corticosterone in the rat. Neuropeptides, 27: 15–18.

Lee, Y. and Davis, M. (1997) Role of the hippocampus, the bed nucleus of the stria terminalis, and the amygdala in the excitatory effect of corticotropin-releasing hormone on the acoustic startle reflex. J. Neurosci., 17: 6434–6446.

LeFeuvre, R.A., Rothwell, N.J. and Stock, M.J. (1987) Activation of brown fat thermogenesis in response to central injection of corticotropin releasing hormone in the rat. Neuropharmacology, 26: 1217–1221.

Lenz, H.J., Raedler, A., Greten, H. and Brown, M.R. (1987) CRF initiates biological actions within the brain that are observed in response to stress. Am. J. Physiol., 252: R34–R39.

Lewis, K., Li, C., Perrin, M.H., Blount, A., Kunitake, K., Donaldson, C., Vaughan, J., Reyes, T.M., Gulyas, J., Fischer, W., Bilezikjian, L., Rivier, J., Sawchenko, P.E. and Vale, W.W. (2001) Identification of urocortin III, an additional member of the corticotropin-releasing factor (CRF) family with high affinity for the CRF2 receptor. Proc. Natl. Acad. Sci. USA, 98: 7570–7575.

Li, C., Vaughan, J., Sawchenko, P.E. and Vale, W.W. (2002) Urocortin III-immunoreactive projections in rat brain: partial overlap with sites of type 2 corticotrophin-releasing factor receptor expression. J. Neurosci., 22: 991–1001.

Liang, K.C. and Lee, E.H. (1988) Intra-amygdala injections of corticotropin releasing factor facilitate inhibitory avoidance learning and reduce exploratory behavior in rats. Psychopharmacology, 96: 232–236.

Linthorst, A.C., Flachskamm, C., Hopkins, S.J., Hoadley, M.E., Labeur, M.S., Holsboer, F. and Reul, J.M. (1997) Long-term intracerebroventricular infusion of corticotropin-releasing hormone alters neuroendocrine, neurochemical, autonomic, behavioral, and cytokine responses to a systemic inflammatory challenge. J. Neurosci., 17: 4448–4460.

Lowry, C.A., Burke, K.A., Renner, K.J., Moore, F.L. and Orchinik, M. (2001) Rapid changes in monoamine levels following administration of corticotropin-releasing factor or corticosterone are localized in the dorsomedial hypothalamus. Horm. Behav., 39: 195–205.

Lowry, C.A., Deviche, P. and Moore, F.L. (1990) Effects of corticotropin-releasing factor (CRF) and opiates on amphibian locomotion. Brain Res., 513: 94–100.

Lowry, C.A. and Moore, F.L. (1991) Corticotropin-releasing factor (CRF) antagonist suppresses stress-induced locomotor activity in an amphibian. Horm. Behav., 25: 84–96.

Makino, S., Nishiyama, M., Asaba, K., Gold, P.W. and Hashimoto, K. (1998) Altered expression of type 2 CRH receptor mRNA in the VMH by glucocorticoids and starvation. Am. J. Physiol., 275: R1138–R1145.

Marrosu, F., Fratta, W., Carcangiu, P., Giagheddu, M. and Gessa, G.L. (1988) Localized epileptiform activity induced by murine CRF in rats. Epilepsia, 29: 369–373.

Martins, A.P., Marras, R.A. and Guimaraes, F.S. (1997) Anxiogenic effect of corticotropin-releasing hormone in the dorsal periaqueductal grey. Neuroreport, 8: 3601–3604.

Masuzawa, M., Oki, Y., Ozawa, M., Watanabe, F. and Yoshimi, T. (1999) Corticotropin-releasing factor but not urocortin is involved in adrenalectomy-induced adrenocorticotropin release. J. Neuroendocrinol., 11: 71–74.

Mele, A., Cabib, S., Oliverio, A., Melchiorri, P. and Puglisi-Allegra, S. (1987) Effects of corticotropin releasing factor and sauvagine on social behavior of isolated mice. Peptides, 8: 935–938.

Million, M., Maillot, C., Saunders, P., Rivier, J., Vale, W. and Taché, Y. (2002) Human urocortin II, a new CRF-related peptide, displays selective CRF(2)-mediated action on gastric transit in rats. Am. J. Physiol. Gastrointest. Liver Physiol., 282: G34–G40.

Miskowiak, B., Janecki, A., Jakubowiak, A. and Limanowski, A. (1986) Reproductive functions in the adult male rats after prolonged intraventricular administration of corticotropin-releasing factor (CRF). Exp. Clin. Endocrinol., 88: 25–30.

Mitsugi, N., Arita, J. and Kimura, F. (1990) Effects of intracerebroventricular administration of growth hormone-releasing factor and corticotropin-releasing factor on somatostatin secretion into rat hypophysial portal blood. Neuroendocrinology, 51: 93–96.

Momose, K., Inui, A., Asakawa, A., Ueno, N., Nakajima, M., Fujimiya, M. and Kasuga, M. (1999) Intracerebroventricularly administered corticotropin-releasing factor inhibits food intake and produces anxiety-like behaviour at very low doses in mice. Diabetes Obes. Metab., 1: 281–284.

Moore, F.L. and Miller, L.J. (1984) Stress-induced inhibition of sexual behavior: corticosterone inhibits courtship behaviors of a male amphibian (Taricha granulosa). Horm .Behav., 18: 400–410.

Moreau, J.L., Kilpatrick, G. and Jenck, F. (1997) Urocortin, a novel neuropeptide with anxiogenic-like properties. Neuroreport, 8: 1697–1701.

Morin, S.M., Ling, N., Liu, X.J., Kahl, S.D. and Gehlert, D.R. (1999) Differential distribution of urocortin- and corticotropin-releasing factor-like immunoreactivities in the rat brain. Neuroscience, 92: 281–291.

Nijsen, M.J., Croiset, G., Stam, R., Bruijnzeel, A., Diamant, M., de Wied, D. and Wiegant, V.M. (2000) The role of the CRH type 1 receptor in autonomic responses to corticotropin-releasing hormone in the rat. Neuropsychopharmacology, 22: 388–399.

Nishiyama, M., Makino, S., Asaba, K. and Hashimoto, K. (1999) Leptin effects on the expression of type-2 CRH receptor mRNA in the ventromedial hypothalamus in the rat. J. Neuroendocrinol., 11: 307–314.

Ohata, H., Suzuki, K., Oki, Y. and Shibasaki, T. (2000) Urocortin in the ventromedial hypothalamic nucleus acts as an inhibitor of feeding behavior in rats. Brain Res., 861: 1–7.

Ohgushi, A., Bungo, T., Shimojo, M., Masuda, Y., Denbow, D.M. and Furuse, M. (2001) Relationships between feeding and locomotion behaviors after central administration of CRF in chicks. Physiol. Behav., 72: 287–289.

Okamoto, S., Kimura, K. and Saito, M. (2001) Anorectic effect of leptin is mediated by hypothalamic corticotropin-releasing hormone, but not by urocortin, in rats. Neurosci. Lett., 307: 179–182.

Okuyama, S., Chaki, S., Kawashima, N., Suzuki, Y., Ogawa, S., Nakazato, A., Kumagai, T., Okubo, T. and Tomisawa, K. (1999) Receptor binding, behavioral, and electrophysiological profiles of nonpeptide corticotropin-releasing factor subtype 1 receptor antagonists CRA1000 and CRA1001. J. Pharmacol. Exp. Ther., 289: 926–935.

Ono, N., Lumpkin, M.D., Samson, W.K., McDonald, J.K. and McCann, S.M. (1984) Intrahypothalamic action of corticotrophin-releasing factor (CRF) to inhibit growth hormone and LH release in the rat. Life Sci., 35: 1117–1123.

Ono, N., Samson, W.K., McDonald, J.K., Lumpkin, M.D., Bedran de Castro, J.C. and McCann, S.M. (1985) Effects of intravenous and intraventricular injection of antisera directed against corticotropin-releasing factor on the secretion of anterior pituitary hormones. Proc. Natl. Acad. Sci. USA, 82: 7787–7790.

Opp, M., Obal, F., Jr. and Krueger, J.M. (1989) Corticotropin-releasing factor attenuates interleukin 1-induced sleep and fever in rabbits. Am. J. Physiol., 257: R528–R535.

Orth, D.N. and Mount, C.D. (1987) Specific high-affinity binding protein for human corticotropin-releasing hormone in normal human plasma. Biochem. Biophys. Res. Commun., 143: 411–417.

Overton, J.M. and Fisher, L.A. (1989) Central nervous system actions of corticotropin-releasing factor on cardiovascular function in the absence of locomotor activity. Regul. Pept., 25: 315–324.

Parrott, R.F. and Vellucci, S.V. (2000) Behaviour of pigs given corticotrophin-releasing hormone in combination with flumazenil or diazepam. Pharmacol. Biochem. Behav., 67: 465–471.

Parrott, R.F., Vellucci, S.V. and Goode, J.A. (2000) Behavioral and hormonal effects of centrally injected "anxiogenic" neuropeptides in growing pigs. Pharmacol. Biochem. Behav., 65: 123–129.

Parver, L.M. (1991) Temperature modulating action of choroidal blood flow. Eye, 5 (Pt 2): 181–185.

Pelleymounter, M.A., Joppa, M., Carmouche, M., Cullen, M.J., Brown, B., Murphy, B., Grigoriadis, D.E., Ling, N. and Foster, A.C. (2000) Role of corticotropin-releasing factor (CRF) receptors in the anorexic syndrome induced by CRF. J. Pharmacol. Exp. Ther., 293: 799–806.

Pelleymounter, M.A., Joppa, M., Ling, N. and Foster, A.C. (2002) pharmacological evidence supporting a role for central corticotropin-releasing factor(2) receptors in behavioral, but not endocrine, response to environmental stress. J. Pharmacol. Exp. Ther., 302: 145–152.

Perrin, M.H. and Vale, W.W. (1999) Corticotropin releasing factor receptors and their ligand family. Ann. N. Y. Acad. Sci., 885: 312–328.

Peto, C.A., Arias, C., Vale, W.W. and Sawchenko, P.E. (1999) Ultrastructural localization of the corticotropin-releasing factor-binding protein in rat brain and pituitary. J. Comp. Neurol., 413: 241–254.

Petraglia, F., Vale, W. and Rivier, C. (1986) Opioids act centrally to modulate stress-induced decrease in luteinizing hormone in the rat. Endocrinology, 119: 2445–2450.

Radulovic, J., Ruhmann, A., Liepold, T. and Spiess, J. (1999) Modulation of learning and anxiety by corticotropin-releasing factor (CRF) and stress: differential roles of CRF receptors 1 and 2. J. Neurosci., 19: 5016–5025.

Reul, J.M. and Holsboer, F. (2002) Corticotropin-releasing factor receptors 1 and 2 in anxiety and depression. Curr. Opin. Pharmacol., 2: 23–33.

Reyes, T.M., Lewis, K., Perrin, M.H., Kunitake, K.S., Vaughan, J., Arias, C.A., Hogenesch, J.B., Gulyas, J., Rivier, J., Vale, W.W. and Sawchenko, P.E. (2001) Urocortin II: a member of the corticotropin-releasing factor (CRF) neuropeptide family that is selectively bound by type 2 CRF receptors. Proc. Natl. Acad. Sci. USA, 98: 2843–2848.

Richard, D., Rivest, R., Naimi, N., Timofeeva, E. and Rivest, S. (1996) Expression of corticotropin-releasing factor and its receptors in the brain of lean and obese Zucker rats. Endocrinology, 137: 4786–4795.

Richter, R.M. and Mulvany, M.J. (1995) Comparison of hCRF and oCRF effects on cardiovascular responses after central, peripheral and in vitro application. Peptides, 16: 843–849.

Rivier, C., Brownstein, M., Spiess, J., Rivier, J. and Vale, W. (1982) In vivo corticotropin-releasing factor-induced secretion of adrenocorticotropin, beta-endorphin, and corticosterone. Endocrinology, 110: 272–278.

Rivier, C., Rivier, J. and Vale, W. (1986) Stress-induced inhibition of reproductive functions: role of endogenous corticotropin-releasing factor. Science, 231: 607–609.

Rivier, C. and Vale, W. (1984) Corticotropin-releasing factor (CRF) acts centrally to inhibit growth hormone secretion in the rat. Endocrinology, 114: 2409–2411.

Rivier, C. and Vale, W. (1985) Involvement of corticotropin-releasing factor and somatostatin in stress-induced inhibition of growth hormone secretion in the rat. Endocrinology, 117: 2478–2482.

Rohner-Jeanrenaud, F., Walker, C.D., Greco-Perotto, R. and Jeanrenaud, B. (1989) Central corticotropin-releasing factor administration prevents the excessive body weight gain of genetically obese (fa/fa) rats. Endocrinology, 124: 733–739.

Roozendaal, M.M., Swarts, J.J., Wolbers, W.B., Threels, A., Wiegant, V.M. and Mattheij, J.A. (1996) Effect of CRH on the preovulatory LH and FSH surge in the cyclic rat: a role for arginine vasopressin? J. Neuroendocrinol., 8: 765–770.

Rosenthal, M.J. and Morley, J.E. (1989) Corticotropin releasing factor (CRF) and age-related differences in behavior of mice. Neurobiol. Aging, 10: 167–171.

Roste, G.K. and Dietrichs, E. (1988) Cerebellar cortical and nuclear afferents from the Edinger-Westphal nucleus in the cat. Anat. Embryol., 178: 59–65.

Rothwell, N.J., Hardwick, A., LeFeuvre, R.A., Crosby, S.R. and White, A. (1991) Central actions of CRF on thermogenesis are mediated by pro-opiomelanocortin products. Brain Res., 541: 89–92.

Ruhmann, A., Bonk, I., Lin, C.R., Rosenfeld, M.G. and Spiess, J. (1998) Structural requirements for peptidic antagonists of the corticotropin- releasing factor receptor (CRFR): development of CRFR2beta-selective antisauvagine-30. Proc. Natl. Acad. Sci. USA, 95: 15264–15269.

Sajdyk, T.J. and Gehlert, D.R. (2000) Astressin, a corticotropin releasing factor antagonist, reverses the anxiogenic effects of urocortin when administered into the basolateral amygdala. Brain Res., 877: 226–234.

Sajdyk, T.J., Schober, D.A., Gehlert, D.R. and Shekhar, A. (1999) Role of corticotropin-releasing factor and urocortin within the basolateral amygdala of rats in anxiety and panic responses. Behav. Brain Res., 100: 207–215.

Salak-Johnson, J.L., McGlone, J.J., Whisnant, C.S., Norman, R.L. and Kraeling, R.R. (1997) Intracerebroventricular porcine corticotropin-releasing hormone and cortisol effects on pig immune measures and behavior. Physiol. Behav., 61: 15–23.

Sauvage, M. and Steckler, T. (2001) Detection of corticotropin-releasing hormone receptor 1 immunoreactivity in cholinergic, dopaminergic and noradrenergic neurons of the murine basal forebrain and brainstem nuclei – potential implication for arousal and attention. Neuroscience, 104: 643–652.

Schulz, D.W., Mansbach, R.S., Sprouse, J., Braselton, J.P., Collins, J., Corman, M., Dunaiskis, A., Faraci, S., Schmidt, A.W., Seeger, T., Seymour, P., Tingley, F.D., 3rd, Winston, E.N., Chen, Y.L. and Heym, J. (1996) CP-154,526: a potent and selective nonpeptide antagonist of corticotropin releasing factor receptors. Proc. Natl. Acad. Sci. USA, 93: 10477–10482.

Seasholtz, A.F., Burrows, H.L., Karolyi, I.J. and Camper, S.A. (2001) Mouse models of altered CRH-binding protein expression. Peptides, 22: 743–751.

Seifritz, E., Klemfuss, H., Montes, J.M., Britton, K.T. and Ehlers, C.L. (1998) Effects of corticotropin-releasing factor on circadian locomotor rhythm in the golden hamster. Pharmacol. Biochem. Behav., 60: 855–862.

Sherman, J.E. and Kalin, N.H. (1987) The effects of ICV-CRH on novelty-induced behavior. Pharmacol. Biochem. Behav., 26: 699–703.

Sherman, J.E. and Kalin, N.H. (1988) ICV-CRH alters stress-induced freezing behavior without affecting pain sensitivity. Pharmacol. Biochem. Behav., 30: 801–807.

Siggins, G.R., Gruol, D., Aldenhoff, J. and Pittman, Q. (1985) Electrophysiological actions of corticotropin-releasing factor in the central nervous system. Fed. Proc., 44: 237–242.

Skelton, K.H., Nemeroff, C.B., Knight, D.L. and Owens, M.J. (2000) Chronic administration of the triazolobenzodiazepine alprazolam produces opposite effects on corticotropin-releasing factor and urocortin neuronal systems. J. Neurosci., 20: 1240–1248.

Slawecki, C.J., Somes, C., Rivier, J.E. and Ehlers, C.L. (1999) Neurophysiological effects of intracerebroventricular administration of urocortin. Peptides, 20: 211–218.

Smagin, G.N., Howell, L.A., Ryan, D.H., De Souza, E.B. and Harris, R.B. (1998) The role of CRF2 receptors in corticotropin-releasing factor- and urocortin-induced anorexia. Neuroreport, 9: 1601–1606.

Smith, G.W., Aubry, J.M., Dellu, F., Contarino, A., Bilezikjian, L.M., Gold, L.H., Chen, R., Marchuk, Y., Hauser, C., Bentley, C.A., Sawchenko, P.E., Koob, G.F., Vale, W. and Lee, K.F. (1998a) Corticotropin releasing factor receptor 1-deficient mice display decreased anxiety, impaired stress response, and aberrant neuroendocrine development. Neuron, 20: 1093–1102.

Smith, J.E., Jansen, A.S., Gilbey, M.P. and Loewy, A.D. (1998b) CNS cell groups projecting to sympathetic outflow of tail artery: neural circuits involved in heat loss in the rat. Brain Res., 786: 153–164.

Spiess, J., Rivier, J., Rivier, C. and Vale, W. (1981) Primary structure of corticotropin-releasing factor from ovine hypothalamus. Proc. Natl. Acad. Sci. USA, 78: 6517–6521.

Spina, M., Merlo-Pich, E., Chan, R.K., Basso, A.M., Rivier, J., Vale, W. and Koob, G.F. (1996) Appetite-suppressing effects of urocortin, a CRF-related neuropeptide. Science, 273: 1561–1564.

Spina, M.G., Merlo-Pich, E., Akwa, Y., Balducci, C., Basso, A.M., Zorrilla, E.P., Britton, K.T., Rivier, J., Vale, W.W. and Koob, G.F. (2002) Time-dependent induction of anxiogenic-like effects after central infusion of urocortin or corticotropin-releasing factor in the rat. Psychopharmacology, 160: 113–121.

Strome, E.M., Wheler, G.H., Higley, J.D., Loriaux, D.L., Suomi, S.J. and Doudet, D.J. (2002) Intracerebroventricular corticotropin-releasing factor increases limbic glucose metabolism and has social context-dependent behavioral effects in nonhuman primates. Proc. Natl. Acad. Sci. USA, 99: 15749–15754.

Sutton, R.E., Koob, G.F., Le Moal, M., Rivier, J. and Vale, W. (1982) Corticotropin releasing factor produces behavioural activation in rats. Nature, 297: 331–333.

Sutton, S.W., Behan, D.P., Lahrichi, S.L., Kaiser, R., Corrigan, A., Lowry, P., Potter, E., Perrin, M.H., Rivier, J. and Vale, W.W. (1995) Ligand requirements of the human corticotropin-releasing factor-binding protein. Endocrinology, 136: 1097–1102.

Swerdlow, N.R., Geyer, M.A., Vale, W.W. and Koob, G.F. (1986) Corticotropin-releasing factor potentiates acoustic startle in rats: blockade by chlordiazepoxide. Psychopharmacology, 88: 147–152.

Swiergiel, A.H. and Dunn, A.J. (1999) CRF-deficient mice respond like wild-type mice to hypophagic stimuli. Pharmacol. Biochem. Behav., 64: 59–64.

Swinny, J., Kalicharan, D., Gramsbergen, A. and van der Want, J. (2002) The localisation of urocortin in the adult rat cerebellum: a light and electron microscopic study. Neuroscience, 114: 891–903.

Taché, Y., Martinez, V., Million, M. and Maillot, C. (2002) Role of corticotropin-releasing factor receptor subtype 1 in stress-related functional colonic alterations: implications in irritable bowel syndrome. Eur. J. Surg. Suppl., 587: 168–170.

Taché, Y., Martinez, V., Million, M. and Wang, L. (2001) Stress and the gastrointestinal tract III. stress-related alterations of gut motor function: role of brain corticotropin-releasing factor receptors. Am. J. Physiol. Gastrointest. Liver Physiol., 280: G173–G177.

Takahashi, L.K., Kalin, N.H., Vanden Burgt, J.A. and Sherman, J.E. (1989) Corticotropin-releasing factor modulates defensive-withdrawal and exploratory behavior in rats. Behav. Neurosci., 103: 648–654.

Tarjan, E., Denton, D.A. and Weisinger, R.S. (1991) Corticotropin-releasing factor enhances sodium and water intake/excretion in rabbits. Brain Res., 542: 219–224.

Tazi, A., Swerdlow, N.R., LeMoal, M., Rivier, J., Vale, W. and Koob, G.F. (1987) Behavioral activation by CRF: evidence for the involvement of the ventral forebrain. Life Sci., 41: 41–49.

Timofeeva, E. and Richard, D. (1997) Functional activation of CRH neurons and expression of the genes encoding CRH and its receptors in food-deprived lean (Fa/?) and obese (fa/fa) Zucker rats. Neuroendocrinology, 66: 327–340.

Trimarchi, F. (1992) Neuro-ophthalmology. Curr. Opin. Neurol. Neurosurg., 5: 740–743.

Turnbull, A.V., Vaughan, J., Rivier, J.E., Vale, W.W. and Rivier, C. (1999) Urocortin is not a significant regulator of intermittent electrofootshock-induced adrenocorticotropin secretion in the intact male rat. Endocrinology, 140: 71–78.

Valdez, G.R., Inoue, K., Koob, G.F., Rivier, J., Vale, W. and Zorrilla, E.P. (2002) Human urocortin II: mild locomotor suppressive and delayed anxiolytic-like effects of a novel corticotropin-releasing factor related peptide. Brain Res., 943: 142–150.

Valdez, G.R., Zorrilla, E.P., Rivier, J., Vale, W.W. and Koob, G.F. (2003) Locomotor suppressive and anxiolytic-like effects of urocortin 3, a highly selective type 2 cortictropin-releasing factor agonist. Brain Res., 980: 206–212.

Vale, W., Spiess, J., Rivier, C. and Rivier, J. (1981) Characterization of a 41-residue ovine hypothalamic peptide that stimulates secretion of corticotropin and beta-endorphin. Science, 213: 1394–1397.

Vaughan, J., Donaldson, C., Bittencourt, J., Perrin, M.H., Lewis, K., Sutton, S., Chan, R., Turnbull, A.V., Lovejoy, D., Rivier, C., Sawchenko, P.E. and Vale, W. (1995) Urocortin, a mammalian neuropeptide related to fish urotensin I and to corticotropin-releasing factor. Nature, 378: 287–292.

Veldhuis, H.D. and de Wied, D. (1984) Differential behavioral actions of corticotropin-releasing factor (CRF). Pharmacol. Biochem. Behav., 21: 707–713.

Venihaki, M. and Majzoub, J. (2002) Lessons from CRH knockout mice. Neuropeptides, 36: 96–102.

Vetter, D.E., Li, C., Zhao, L., Contarino, A., Liberman, M.C., Smith, G.W., Marchuk, Y., Koob, G.F., Heinemann, S.F., Vale, W. and Lee, K.F. (2002) Urocortin-deficient mice show hearing impairment and increased anxiety-like behavior. Nat. Genet., 31: 363–369.

Wang, L., Martinez, V., Rivier, J.E. and Taché, Y. (2001) Peripheral urocortin inhibits gastric emptying and food intake in mice: differential role of CRF receptor 2. Am. J. Physiol. Regul. Integr. Comp. Physiol., 281: R1401–R1410.

Wang, X., Su, H., Copenhagen, L.D., Vaishnav, S., Pieri, F., Shope, C.D., Brownell, W.E., De Biasi, M., Paylor, R. and Bradley, A. (2002) Urocortin-deficient mice display normal stress-induced anxiety behavior and autonomic control but an impaired acoustic startle response. Mol. Cell. Biol., 22: 6605–6610.

Weisinger, R.S., Blair-West, J.R., Burns, P., Denton, D.A., McKinley, M.J., Purcell, B., Vale, W., Rivier, J. and Sunagawa, K. (2000) The inhibitory effect of hormones associated with stress on Na appetite of sheep. Proc. Natl. Acad. Sci. USA, 97: 2922–2927.

Weiss, S.R., Post, R.M., Gold, P.W., Chrousos, G., Sullivan, T.L., Walker, D. and Pert, A. (1986) CRF-induced seizures and behavior: interaction with amygdala kindling. Brain Res., 372: 345–351.

Weninger, S.C., Dunn, A.J., Muglia, L.J., Dikkes, P., Miczek, K.A., Swiergiel, A.H., Berridge, C.W. and Majzoub, J.A. (1999a) Stress-induced behaviors require the corticotropin-releasing hormone (CRH) receptor, but not CRH. Proc. Natl. Acad. Sci. USA, 96: 8283–8288.

Weninger, S.C., Muglia, L.J., Jacobson, L. and Majzoub, J.A. (1999b) CRH-deficient mice have a normal anorectic response to chronic stress. Regul. Pept., 84: 69–74.

Weninger, S.C., Peters, L.L. and Majzoub, J.A. (2000) Urocortin expression in the Edinger-Westphal nucleus is up-regulated by stress and corticotropin-releasing hormone deficiency. Endocrinology, 141: 256–263.

Whitley, N.C., Barb, C.R., Kraeling, R.R., Barrett, J.B., Rampacek, G.B., Carroll, J.A. and Keisler, D.H. (2000) Feed intake and serum GH, LH and cortisol in gilts after intracerebroventricular or intravenous injection of urocortin. Domest. Anim. Endocrinol., 19: 209–221.

Winslow, J.T., Newman, J.D. and Insel, T.R. (1989) CRH and alpha-helical-CRH modulate behavioral measures of arousal in monkeys. Pharmacol. Biochem. Behav., 32: 919–926.

Yamamoto, H., Maeda, T., Fujimura, M. and Fujimiya, M. (1998) Urocortin-like immunoreactivity in the substantia nigra, ventral tegmental area and Edinger-Westphal nucleus of rat. Neurosci. Lett., 243: 21–24.

Yang, X.M., Gorman, A.L. and Dunn, A.J. (1990) The involvement of central noradrenergic systems and corticotropin-releasing factor in defensive-withdrawal behavior in rats. J. Pharmacol. Exp. Ther., 255: 1064–1070.

Zhang, L. and Barrett, J.E. (1990) Interactions of corticotropin-releasing factor with antidepressant and anxiolytic drugs: behavioral studies with pigeons. Biol. Psychiatry, 27: 953–967.

Zhang, R., Nakanishi, T., Ohgushi, A., Ando, R., Yoshimatsu, T., Denbow, D.M. and Furuse, M. (2001) Interaction of corticotropin-releasing factor and glucagon-like peptide-1 on behaviors in chicks. Eur. J. Pharmacol., 430: 73–78.

Zorrilla, E.P., Schulteis, G., Ling, N., Koob, G.F. and De Souza, E.B. (2001) Performance-enhancing effects of CRF-BP ligand inhibitors. Neuroreport, 12: 1231–1234.

Zorrilla, E.P., Schulteis, G., Ormsby, A., Klaassen, A., Ling, N., McCarthy, J.R., Koob, G.F. and De Souza, E.B. (2002a) Urocortin shares the memory modulating effects of corticotropin-releasing factor (CRF): mediation by CRF1 receptors. Brain Res., 952: 200–210.

Zorrilla, E.P., Valdez, G.R., Nozulak, J., Koob, G.F. and Markou, A. (2002b) Effects of antalarmin, a CRF type 1 receptor antagonist, on anxiety-like behavior and motor activation in the rat. Brain Res., 952: 188–199.

T. Steckler, N.H. Kalin and J.M.H.M. Reul (Eds.)
Handbook of Stress and the Brain, Vol. 15
ISBN 0-444-51173-3

CHAPTER 2.5

Vasopressin and oxytocin

A.J. Douglas*

Division of Biomedical Sciences, SBCLS, University of Edinburgh, Hugh Robson Building, George Square, Edinburgh EH8 9XD, UK

Abstract: Vasopressin and oxytocin are neuroendocrine peptide hormones that are synthesised in the hypothalamus and released into the peripheral circulation and into the brain in response to stress, and they regulate stress responses and anxiety behaviours. Stress responses comprise corticotropin-releasing factor (CRF) secretion at the median eminence that drives peripheral adrenocorticotropin (ACTH) secretion from the anterior pituitary and consequently glucocorticoid secretion from the adrenal cortex. Vasopressin is a key co-secretagogue for ACTH, synergizing with CRF, and vasopressin synthesis in the paraventricular nucleus (PVN) and secretion at the median eminence are increased during stress and further enhanced with chronic stress: it has an important role in maintaining stress responsiveness. Centrally, vasopressin inhibits secretory stress responses and increases anxiety. Oxytocin may play a minor role as a secretagogue for ACTH but has more defined roles within the brain, inhibiting both stress-induced ACTH secretion and anxiety. Peripheral oxytocin secretion also increases with stress in some species. Vasopressin and oxytocin mechanisms both respond to glucocorticoid feedback and adapt with ageing and through reproduction (pregnancy, parturition and lactation): their contribution to stress responsiveness in such physiological conditions is discussed.

Introduction

Vasopressin and oxytocin are related neuroendocrine nine amino acid peptide hormones that are integrally involved in stress responses. However, this role was discovered relatively late; oxytocin was originally described as uterotonic and to mediate the milk-ejection reflex, while vasopressin was initially described as antidiuretic and as a vasoconstrictor of vascular smooth muscle.

In the paraventricular nucleus (PVN), corticotropin-releasing factor (CRF) neurons respond to inputs from stress-processing brain regions and mediate the body's response to stress by inducing secretion of a cascade of hormones: these allow the body to compensate and thus cope with stress. Thus, CRF neurons secrete CRF into the hypophysial portal blood, which acts on corticotroph cells in the anterior pituitary to induce adrenocorticotropin (ACTH) secretion into the peripheral circulation that then causes glucocorticoid (specifically: corticosterone in the rat, cortisol in the human) secretion from the adrenal cortex. Glucocorticoids negatively feedback to regulate this hypothalamo–pituitary–adrenal (HPA) axis as well as mobilizing the body's compensatory mechanisms. These stress hormones are secreted by the body in response to a variety of external stimuli, which comprise physical stressors, including immobilization and pain; emotional stressors, e.g. fear, novel environment and social defeat; or a combination of these influences, e.g. forced swimming and restraint (Dallman et al., 2000).

Vasopressin and oxytocin are two of the key factors, other than CRF, that are involved in stress responses. Their role comprises release and action at several levels. They are secreted peripherally in response to some stressors but their key roles are: (i) at the anterior pituitary – where they have

*Tel.: +44 (0) 131 650 3274; Fax: +44 (0) 131 650 2872; E-mail: alison.j.douglas@ed.ac.uk

synergistic secretagogue effects on corticotrophs (Aguilera and Rabadan-Diehl, 2000a), and (ii) centrally in the brain – where they modulate PVN CRF neuron responses and the perception of stress and anxiety (Landgraf, 2001). Oxytocin was initially described as a stress hormone in the early 1980s, after discovering stress-induced increases in peripheral oxytocin secretion. Although this was before vasopressin's role was fully recognised, vasopressin is now accepted to play a central role in mediating anterior pituitary secretory responses to stress.

These peptides are synthesised in neurons of distinct and limited regions of the hypothalamus and their axons project widely around the brain and to the posterior pituitary. Vasopressin and oxytocin are found in separate, but immediately adjacent, populations of neurons in the medial/ventral hypothalamus, predominantly the PVN and supraoptic nucleus (SON). One subpopulation of parvocellular vasopressin neurons of the PVN projects to the median eminence, where vasopressin is released into the hypophysial portal blood as a secretagogue for corticotroph cells. Other parvocellular vasopressin and oxytocin neurons project centrally, within the hypothalamus and to the limbic system (i.e. hippocampus, amygdala, septum, bed nucleus of the stria terminalis, BNST) and brainstem, where they can act as neurotransmitters/neuromodulators. The magnocellular neurons in the other subpopulations of the PVN and in the SON all project to the posterior pituitary and the hormones are stored in nerve terminals, from where they are secreted into the peripheral circulation. Additionally, vasopressin and oxytocin are released from the magnocellular neuron dendrites and/or cell bodies (somato-dendritic release) into the extracellular space in the PVN and SON, from where they may influence local neurons (Fig. 1; see also Leng, 2000; Renaud, 2000).

Vasopressin and oxytocin act via membrane receptors that are of the seven transmembrane domain receptor family, which are G-protein coupled (Chini and Fanelli, 2000). Three vasopressin receptor types are currently described: V1a, V1b and V2; of which V1b is expressed in the hypothalamus, limbic regions and brainstem and in the anterior pituitary, and mediates stress responses. The V1a receptor is

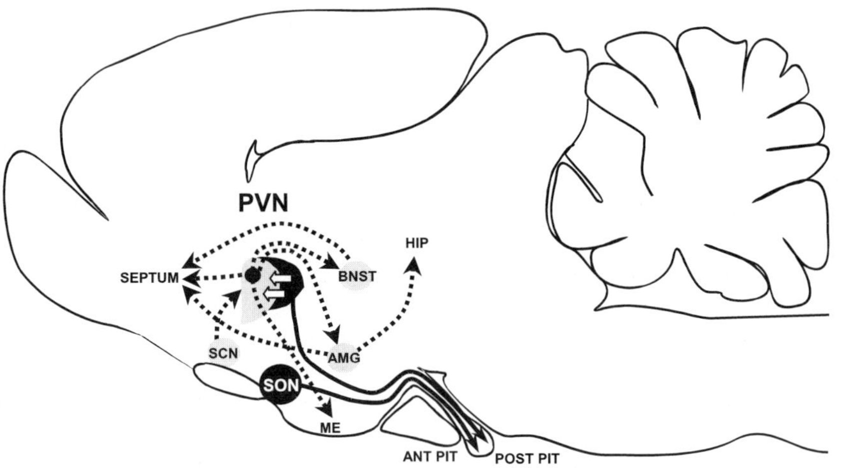

Fig. 1. Vasopressin and oxytocin pathways involved in stress responses. Schematic sagittal section of a rat brain showing sources (parvocellular: grey areas, and magnocellular: black areas) and main output pathways (arrows) of vasopressin and oxytocin neurons. Vasopressin neurons are primarily located in the PVN and SON, but are also found in the medial amygdala (AMG), BNST and suprachiasmatic nucleus (SCN). Oxytocin neurons are located mainly in the PVN and SON. Parvocellular PVN vasopressin neurons project to the median eminence (ME) to release vasopressin into the hypophysial portal blood, which then acts on corticotroph cells in the anterior pituitary (ANT PIT). The intranuclear neural pathways include vasopressin and oxytocin projections to the septum, BNST, AMG and the hippocampus (HIP). Magnocellular vasopressin and oxytocin neurons project to the posterior pituitary (POST PIT), secreting directly into the peripheral circulation; they also release vasopressin and oxytocin, respectively, within the PVN and SON which diffuses within the nucleus to act on parvocellular cells ($\Leftarrow$).

localised in the liver, kidney and vascular smooth muscle, as well as many brain areas, including the hypothalamus and limbic system, where it is thought to mediate behavioural and cognitive effects of vasopressin. The V2 receptor is predominantly expressed in the kidney and mediates the antidiuretic effect of vasopressin. Oxytocin receptor is also expressed in the above brain regions and is generally accepted to be the same as that expressed peripherally, for example, in the uterus and mammary gland. V1b and oxytocin receptors are linked to phospholipase C and upon activation generate diacylglycerol and IP3, which increase intracellular calcium and activate protein kinase C, leading to the cell responses, such as secretion and/or induction of gene expression. Vasopressin and oxytocin have distinct and separate effects on stress processing and stress responses.

Vasopressin

Parvocellular vasopressin neurons in the PVN that project to the median eminence are the main source of vasopressin that mediates HPA axis secretory responses to stress. Other populations of vasopressin cells in the amygdala and BNST project within the hypothalamus, limbic system and brainstem, mediating behaviours such as aggression, while those in the suprachiasmatic nucleus project within the hypothalamus to participate in circadian timing. The expression of vasopressin in these neurons has been extensively investigated under a variety of stress conditions; changes in expression have been used as an effective marker for identifying the involvement and activation of vasopressin neurons during stress.

Stress and vasopressin expression

Vasopressin expression is present in both CRF and non-CRF parvocellular cells in the PVN (Walker et al., 2001a,b) and, while expression in parvocellular neurons is low under basal conditions, it increases in response to stress. Thus acute stress, applied in the short term (e.g. ether, restraint, footshock), is generally agreed to rapidly increase vasopressin expression in rat PVN neurons, measured as either the immediate early transcript-heterogeneous nuclear (hn) RNA (peak at 1–2 h post-stress) or as messenger (m) RNA (Lightman and Young, 1987, 1988; Bartanusz et al., 1993a,b; Sawchenko et al., 1993; Ma et al., 1997a,b; Ma and Aguilera, 1999a; Aguilera and Rabadan-Diehl, 2000a), although in some species vasopressin mRNA is not found in parvocellular neurons and does not respond to stress (e.g. pig, Vellucci and Parrott, 1997). CRF hnRNA expression responses to stress are extremely fast (peak at 5–15 min) and vasopressin hnRNA expression responses can be equally fast, but after some other stressors are more delayed (Kovacs and Sawchenko, 1996a; Ma et al., 1997a,b; Aguilera and Rabadan-Diehl, 2000a). Thus, vasopressin transcription, and subsequently peptide synthesis, increases in PVN neurons during stress, indicating that vasopressin plays an important role. The immediate early genes (also known as transcription factors) *c-fos* and NGFI-B are also rapidly expressed in parvocellular PVN neurons after stress and they are co-localised with vasopressin (Chan et al., 1993; da Costa et al., 1996,1997; Kovacs and Sawchenko, 1996a,b; Krukoff et al., 1997; Kc et al., 2002). Since the vasopressin gene promoter region is reported to contain both AP-1 binding sites (for Fos protein) and NGFI-B response elements, both of these transcription factors have been implicated in inducing the vasopressin RNA expression responses to stress (Chan et al., 1993), and are used as neuroanatomical markers for neuron activation. Upon removal of an animal from acute exposure to a stressor, *c-fos* and NGFI-B gene expression in the PVN neurons decreases again to pre-stress levels.

Glucocorticoid feedback on vasopressin expression

Glucocorticoids inhibit vasopressin gene expression and thus restore baseline mRNA levels following a stress response (Herman, 1995; Herman et al., 1995). Direct glucocorticoid negative feedback at the transcriptional level of the vasopressin promoter has been recently reported (Pearce et al., 1998; Kim et al., 2001). Glucocorticoid type II receptors are expressed in the parvocellular PVN neurons and, after coupling to these, the glucocorticoid–receptor complex inhibits vasopressin transcription by binding to a glucocorticoid regulatory element in the vasopressin promoter region (Uht et al., 1988; Burke et al., 1997). Additionally, evidence supports the binding of the glucocorticoid–receptor complex to Jun (co-transcription factor with Fos) that would repress AP-1 activity and thus attenuate vasopressin gene transcription responses (Diamond et al., 1990; Schule et al., 1990; Stauber et al., 1990; Yang-Yen et al., 1990; Unlap and Jope, 1994; Kovacs et al., 2000). Loss of glucocorticoid negative feedback, e.g. after adrenalectomy, which removes the source of glucocorticoids, increases vasopressin mRNA expression and levels do not return to baseline following the end of exposure to a stressor (Kiss et al., 1984; Sawchenko et al., 1984a; Jingami et al., 1985; Wolfson et al., 1985; Kovacs et al., 1986; Young et al., 1986a,b; Kovacs and Mezey, 1987; Swanson and Simmons, 1989; Viau et al., 1999). When prolonged elevated levels of glucocorticoid are present (such as with chronic or repeated stress) vasopressin expression also increases, adapting to the new conditions and contributing to the maintenance of stress responsiveness.

Chronic stress and vasopressin expression

With chronic (i.e. maintained stress exposure) or repeated stress (e.g. daily repetitions of exposure to the same stressor, such as restraint) or long-term activation of the HPA axis, which are associated with hypercortisolemia, there is an upregulation of the vasopressin system (Bertini and Kiss, 1991; de Groeij et al., 1992; Bartanusz et al., 1993a,b; Aguilera and Rabadan-Diehl, 2000a). Basal vasopressin expression is elevated while basal CRF expression is reduced under these conditions (Herman, 1995; Ma et al., 1997c; Kovacs et al., 2000), and co-expression of vasopressin with CRF in parvocellular PVN neurons increases (Lightman and Young, 1988; Herman et al., 1992, 1995; Whitnall, 1993; Makino et al., 1995a,b; Herman, 1995; Pinnock and Herbert, 2001). Additionally, vasopressin expression responses to repeated stress are maintained or even increased (Aguilera and Rabadan-Diehl, 2000a), in contrast to CRF expression responses to repeated stress that are attenuated.

The differential between CRF and vasopressin expression responses is mediated by the action of glucocorticoids, which selectively enhance vasopressin responses (Kovacs et al., 2000) since the negative feedback from glucocorticoids on vasopressin gene expression becomes lost, enabling continuing elevated vasopressin expression (Pinnock and Herbert, 2001). Together these findings indicate an important role for vasopressin in the brain adaptation to chronic stress (Aguilera and Rabadan-Diehl, 2000a) and reveal that vasopressin is an important variable susceptible to modulation under physiological or disease circumstances (Kovacs et al., 2000). Vasopressin expression is also implicated in the generation of the circadian profile of glucocorticoid secretion.

Vasopressin and circadian corticosterone secretion

Vasopressin that is synthesised in neurons of the suprachiasmatic nuclei is involved in the generation of circadian rhythms (Larsen et al., 1994; Kalsbeek et al., 1996; Vellucci and Parrott, 1997; Murphy et al., 1998). A decrease in vasopressin expression is necessary for the normal circadian rise in plasma corticosterone and lesions of the suprachiasmatic nucleus increase corticosterone secretion (Buijs et al., 1993a; Kalsbeek et al., 1996). These effects may be mediated by the projections from the suprachiasmatic nucleus to the hypothalamic stress regions (PVN, Buijs et al., 1993b; Kalsbeek et al., 2002). Glucocorticoid feedback to the suprachiasmatic nucleus regulates a diurnal pattern of expression of vasopressin (Kalsbeek et al., 1996); thus, vasopressin from the suprachiasmatic nucleus inhibits the HPA axis in a circadian pattern. On the other hand, exposure to stress increases vasopressin release in the suprachiasmatic nucleus (Engelmann et al., 1998), thus, vasopressin may also have local actions within the nucleus on other factors mediating circadian timing and stress may interfere with circadian rhythms via a vasopressinergic mechanism.

Vasopressin expression in magnocellular neurons

Vasopressin mRNA is abundant in magnocellular neurons under basal conditions and vasopressin in these cells is constitutively synthesised, stored and released as a function of body osmoregulation.

However, these neurons additionally respond to stress. Exposure to selected stressors, such as haemorrhage, restraint and forced swimming, increases Fos expression and vasopressin expression in magnocellular neurons in rats and pigs (Shen et al., 1992; Roberts et al., 1993; Vellucci and Parrott, 1997; Wotjak et al., 2001), indicating a role in responses to stress for magnocellular vasopressin release, either from the neuron dendrites/cell bodies centrally or from nerve terminals into the periphery.

Thus, vasopressin expression in parvocellular and magnocellular hypothalamic neurons is responsive to stress exposure and adapts during chronic stress conditions, indicating that vasopressin is synthesised and secreted during these circumstances and consequently that it plays a key role in stress responses under a variety of physiological conditions, particularly in conditions of chronic stress.

Vasopressin actions in the anterior pituitary

At the same time as inducing increased vasopressin gene expression in PVN neurons, stress increases vasopressin release in the external zone of the median eminence into the hypophysial portal capillaries that acts on corticotroph cells of the anterior pituitary to synergize with CRF in stimulating the secretion of ACTH (Gillies et al., 1982; Rivier et al., 1984; Antoni, 1993; Wotjak et al., 2002). Vasopressin alone is a weak secretagogue, but acts via V1b receptors to enhance protein kinase C-mediated potentiation of CRF-stimulated (via CRF receptor 1) cAMP production and post-cAMP-dependent mechanisms in the corticotroph cells (Bilezikjian et al., 1987; Carvallo and Aguilera, 1989; Rene and De Keyzer, 2002). There is some evidence that vasopressin released from PVN and SON magnocellular neurons can also facilitate ACTH secretion (Antoni, 1993; Whitnall, 1993; Wotjak et al., 1996a; Klavdieva, 1996; Aguilera and Rabadan-Diehl, 2000a), reaching the anterior pituitary via short portal vessels, although recent evidence appears to rule this out (Wotjak et al., 2002). However, in general the PVN responds to all stresses (e.g. Bartanusz et al., 1993a,b; Sawchenko et al., 1993; Ma et al., 1997c; Ma and Aguilera, 1999b) and to glucocorticoid feedback (Herman, 1995; Herman et al., 1995; Burke et al., 1997; Pearce et al., 1998; Kim et al., 2001), while increased vasopressin peripheral secretion from magnocellular neurons in response to stress is not consistently shown, or decreases (Lang et al., 1983; Kasting, 1988; Onaka and Yagi, 1993; Nishioka et al., 1998; Wotjak et al., 1998a). The importance of vasopressin in synergizing with CRF to drive ACTH and glucocorticoid secretion is further supported by reports that the Brattleboro rat (vasopressin-deficient) has a blunted corticosterone response to stressors (Scaccianoce et al., 1991), and reduction of vasopressin expression in adult rats using antisense oligonucleotides decreases the secretory responses of the HPA axis (Neumann, 2000). Mice lacking CRF receptor 1 have significantly elevated vasopressin immunoreactivity in the median eminence; even so the secretory responses to stress remain severely impaired (Muller et al., 2000) so vasopressin alone cannot maintain normal HPA axis secretory responses. On the other hand, in conditions of chronic/repeated stress, enhanced vasopressin secretion and action at the anterior pituitary is an essential component of stress adaptation.

Chronic stress and vasopressin action in the anterior pituitary

Vasopressin action at the anterior pituitary plays a particularly central role in adaptation to chronic stress conditions. Anterior pituitary vasopressin receptor V1b mRNA expression increases during chronic stress (Sawchenko et al., 1993; Aguilera and Rabadan-Diehl, 2000a) and glucocorticoids potentiate vasopressin-stimulated IP3 generation during receptor activation (Rabadan-Diehl and Aguilera, 1998), thus, vasopressin signalling capacity becomes enhanced and there is potentiation of the ACTH secretory response to CRF. The pattern of vasopressin release into the portal blood is a key determinant of V1b receptor expression in the anterior pituitary (Aguilera and Rabadan-Diehl, 2000a) and since vasopressin secretion increases in chronic stress it presumably mediates the increased vasopressin receptor expression (Aguilera, 1994). Additionally, vasopressin attenuates glucocorticoid negative feedback at the pituitary (Webster and Cidlowski, 1999) and increases the responsiveness of corticotroph cells to CRF by upregulating CRF receptor 1 (Jia et al., 1992). So, vasopressin and glucocorticoids act synergistically to enhance the vasopressin receptor effectiveness in conditions of chronic stress. Together these mechanisms support vasopressin's role in maintaining stress responsiveness during chronic stress and/or high glucocorticoid conditions. Thus vasopressin, as well as compensating for chronic stress centrally with increased synthesis, also appears to control the capacity of the corticotrophs to respond to stimulation of the HPA axis during chronic stress, ensuring continued HPA axis responsiveness, which is essential for the body's stress-coping mechanisms.

Stress and vasopressin release and action in the brain

As well as its key role at the anterior pituitary, vasopressin is released centrally in stressful conditions into the extracellular space from the dendrites and/or cell bodies of magnocellular cells in the PVN and SON, and as a neurotransmitter/neuromodulator in various hypothalamic, limbic and brainstem regions.

PVN

Vasopressin is released within the PVN from magnocellular neuron dendrites and cell bodies and acts on adjacent PVN cells in an autocrine/paracrine way. Various stimuli increase vasopressin release within the nucleus, e.g. an hyperosmotic stimulus, dehydration, haemorrhage (see Ludwig, 1998, for review) and stress. Stressors, such as forced swimming and social defeat, increase vasopressin release in the PVN in both male and female rats (Wotjak et al., 1996b, 1998a; Engelmann et al., 2000, 2001; Wigger and Neumann, 2002), but some stressors, such as novel cage, have no effect (Wotjak et al., 1995, 1996b, 1998a,b). In contrast to vasopressin action in the anterior pituitary, vasopressin released within the PVN is inhibitory to HPA axis activity (Plotsky et al., 1984; Wotjak et al., 1996b) and thus reduces ACTH secretion. Vasopressin administration intracerebroventricularly (icv) inhibits CRF secretion into the hypophysial portal blood (Plotsky et al., 1984) indicating a inhibition of CRF cells, and combined vasopressin and CRF receptor antagonist infused into the PVN reduces basal and social defeat-stimulated ACTH secretion (Wotjak et al., 1996b), thus vasopressin action in the PVN appears to act by inhibiting CRF neuron responsiveness to stress. This role for vasopressin is separate from any anterior pituitary or peripheral action, and release is dissociated between the different sites (Wotjak et al., 1998a). The effects are not mutually exclusive and it might be hypothesised that there is a simultaneous mechanism stimulating release of ACTH, while also preventing further ongoing stimulation of the HPA axis to limit the secretory responses to stress.

SON

Vasopressin is also released within SON from magnocellular neuron dendrites and cell bodies in response to stressors, such as forced swimming, but novel cage has no effect (Wotjak et al., 1996b, 1998a; Engelmann et al., 2000, 2001). Vasopressin release in the SON during stress is also inhibitory to ACTH secretion (Wotjak et al., 2002), and it may do this by diffusing into the cerebrospinal fluid and to the PVN, supplementing vasopressin release in the PVN. Thus, the action of vasopressin in the SON and PVN during stress is an important component of the neuronal network that balances appropriate HPA axis responses to stressor exposure.

Other brain areas

Apart from the PVN and SON, vasopressin acts in the limbic system during stress exposure. The limbic regions are important areas mediating emotionality and stress coping, which are at least partly regulated by vasopressin (Koolhaas et al., 1998). The source of vasopressin released in the limbic system is neurons in the BNST and medial amygdala (Caffe et al., 1987; Viau et al., 2001) and these respond to stress. Vasopressin release in the septum increases upon exposure to stress and modulates the enhanced anxiety-related behaviours and active behavioural coping responses (Ebner et al., 1999; Engelmann et al., 2000). Selective downregulation of V1b receptor binding in the septum, using antisense oligonucleotides or vasopressin V1b receptor antagonist administration, reduces anxiety-related behaviours measured on the elevated plus maze (Landgraf et al., 1995; Liebsch et al., 1996; Engelmann et al., 2000); thus vasopressin acts in the septum to promote anxiety. Vasopressin release is also increased in the amygdala during forced swimming and vasopressin receptor antagonist increases struggling time and reduces floating time, thus vasopressin also promotes passive coping strategies (Ebner et al., 2002). In contrast, an emotional stressor such as social defeat does not alter vasopressin release in the septum (Ebner et al., 2000). It seems that there is a balance between vasopressin effects in the septum and amygdala that determines behavioural responses to stress exposure. Enhanced vasopressinergic mechanisms also mediate anxiety-related behaviours in rats selectively bred for high anxiety by Landgraf and colleagues (Liebsch et al., 1998; Landgraf et al., 1999, for further discussion see below).

The role of vasopressin in the limbic regions is dependent upon the rat gender and whether they are aggressive or not (Koolhaas et al., 1998); non-aggressive males (and females) have more vasopressin in the septum and cope with stress/anxiety in a more passive than active way (Koolhaas et al., 1998; Everts and Koolhaas, 1999), presumably due to differences in sex steroid concentrations that vary greatly between aggressive and non-aggressive males (as between males and females), and are responsible for the stress-coping differences as well as HPA axis function (Koolhaas et al., 1998; Viau, 2002).

Stress and peripheral vasopressin secretion

Although a role for vasopressin in the control of ACTH secretion was proposed in the early 1980s, vasopressin was originally not recognised as a stress hormone since peripheral secretion was not reported to increase during exposure to a variety of stressors, and the potential effects of central vasopressin were not recognised (Lang et al., 1983). More recently, subtle differences in peripheral vasopressin concentration have been reported in response to exposure to some selected stressors, reflecting altered secretion from the nerve terminals of magnocellular neurons in the posterior pituitary. Hyperosmolality or hypovolemia/hypotension normally increase vasopressin secretion but fear or novel stimuli inhibit vasopressin secretion in the rat (Onaka and Yagi, 1993) and in female, but not male, mice exposure to novel environment or swim stress both decrease plasma vasopressin concentrations (Muller et al., 2000). Forced swimming or shaker stresses do not increase plasma vasopressin in rats (Lang et al., 1983; Verbalis et al., 1986; Kasting, 1988; Wotjak et al., 1996b, 1998a; Nishioka et al., 1998). However, forced swimming increases vasopressin mRNA in magnocellular cells and local somato-dendritic vasopressin release in the PVN (Wotjak et al., 1998b, 2001), so peripheral vasopressin secretion and central vasopressin release and synthesis are dissociated in stress conditions. Furthermore, forced swimming increases plasma osmolality, which would be expected to change vasopressin secretion but does not (Wotjak et al., 1998a). The role of change/decrease in peripheral vasopressin secretion during stress is not known, but would reduce the vasopressin-related osmoregulatory and vasoconstrictor effects at that time. However, unlike the above stressors, conditions such as depression are associated with increased vasopressin activity and secretion.

Vasopressin and depression

Depression is associated with hypersecretion of the HPA axis (van Bardeleben and Holsboer, 1991) and reduced glucocorticoid feedback mechanisms (Holsboer, 1995). Basal vasopressin secretion into the blood has been reported to be elevated in humans suffering from major depression (van Londen et al., 1997) and in post-mortem brains from depressed patients the synthetic activity of vasopressin cells is increased, accompanied by increased numbers of vasopressin neurons (Purba et al., 1996). These reports, together with the finding that vasopressin receptor antagonist prevents the CRF-induced ACTH secretion in the dexamethasone/CRF test (non-responsiveness correlates with depression), indicate a possible role for vasopressin in such affective disorders and in the hyperresponsiveness of the HPA axis.

Anxiety

Anxiety is one of the key symptoms associated with depression. Landgraf and colleagues have developed rat lines selectively bred for high and low anxiety (Liebsch et al., 1998; Landgraf et al., 1999); high-anxiety rats exhibit greater ACTH and corticosterone secretory responses to emotional stressors (elevated plus maze, on which measurements of anxiety are made as a definition of the rat line) than low-anxiety or normal rats. The high-anxiety rat lines show increased basal vasopressin synthesis and release within the PVN compared to the low-anxiety rats, and treatment with a vasopressin type 1 receptor antagonist abolishes the difference in secretory responses to exogenous CRF in dexamethasone-pretreated rats (Keck et al., 2002). Thus, vasopressin-mediated effects are involved in the HPA axis disturbance of the high-anxiety rat lines, and seems parallel to findings in depressed humans. However, basal vasopressin secretion into the peripheral circulation is not different between the rat lines, and vasopressin secretion is not induced with stress exposure, suggesting that portal or central vasopressin release underlies its role, rather than any other peripheral effect.

Effect of physiological circumstances on vasopressin responses to stress

During aging and reproduction there are progressive alterations in the HPA axis and its responses to stress. With increasing age there are similar symptoms to chronic stress in the HPA axis and hypothalamo–neurohypophysial systems, which become disinhibited, showing hypercortisolemia, hypervasopressinemia and non-suppression to dexamethasone in basal conditions (see Holsboer, 1995). During the reproductive cycle of the female generally the opposite occurs, and the HPA axis and the hypothalamo–neurohypophysial system become hyporesponsive to stress during pregnancy and lactation, a phenomenon that may contribute to prevention of overexposure of glucocorticoids to the offspring, and thus to reduced risk of depression and cardiovascular disease in the offspring in adulthood (Weinstock, 1997). Changes in vasopressin appear to, at least partly, underlie the HPA axis plasticity under these conditions.

Aging

In humans and rodents the HPA axis alters profoundly with increasing age (for review see Sapolsky, 1991; Seeman and Robbins, 1994): HPA axis basal secretion is elevated and responses to stress are enhanced and prolonged, with an age-related decrease in glucocorticoid negative feedback (van Eekelen et al., 1991; Meaney et al., 1995; Keck et al., 2000) – although in aged animals HPA axis changes are strain-dependent. In parallel, there are age-related changes in the hypothalamo–neurohypophysial system (Miller, 1987; Johnson et al., 1994). Vasopressin synthetic and secretory activity in the brain increases in humans and rats, and accompanies an increase in colocalisation of vasopressin with CRF in parvocellular PVN neurons (Hoogendijk et al., 1985; Lucassen et al., 1993; Raadsheer et al., 1993; van der Woude et al., 1995; Hatzinger et al., 2000). There is enhanced vasopressin release from hypothalami of aged rats in vitro (Scaccianoce et al., 1990), and there is enhanced signalling in the anterior pituitary since vasopressin receptor antagonist prevents CRF-induced ACTH secretion in old but not young rats, indicating that the increased vasopressin activity is required for HPA secretion (Keck et al., 2000). Within the PVN basal somato-dendritic vasopressin release increases approximately twofold in aged rats and release in response to forced swimming is blunted (Keck et al., 2000). Thus, there is adaptation of the effect of central vasopressin with age that leads to increased HPA axis activity (Keck et al., 2000). Reduced corticosterone receptor function in the hippocampus and hypothalamus has been proposed to cause disinhibition of vasopressin neurons with aging (van Bardeleben and Holsboer, 1991; Hatzinger et al., 2000). Basal vasopressin secretion into the blood increases with age (Miller, 1987; Terwel et al., 1992; Johnson et al., 1994; van Londen et al., 1997; Keck et al., 2000), although some have reported no change or decreased vasopressin levels in rats (Aravich and Sladek, 1987; Silverman et al., 1990; Geelen and Corman, 1992). Thus in this respect, as with the other parameters, the vasopressin system seems to respond similarly in aging as with chronic stress.

Pregnancy and lactation

In late pregnancy and lactation in the female rat the HPA axis is hyporesponsive to a variety of emotional and physical stressors (Lightman, 1993; Walker, 1995; Neumann et al., 1998; Johnstone et al., 2000), similar to other species (Carter and Altemus, 1997; Douglas et al., 2001a,b; Heinrichs et al., 2002). In pregnancy there is reduced secretagogue expression, thus basal CRF and vasopressin mRNA expression in parvocellular PVN neurons decrease and mRNA expression responses to stress are attenuated (Johnstone et al., 2000; Russell et al., 2000; da Costa et al., 2001). Increased glucocorticoid negative feedback may partly explain the hyporesponsiveness since there is increased expression of 11beta hydroxysteroid dehydrogenase type I (a corticosteroid reductase enzyme) in the PVN and anterior pituitary of pregnant rats (Johnstone et al., 2000), which would increase their exposure to glucocorticoids and thus may explain the reduced vasopressin expression. Effects of vasopressin on stress-induced ACTH secretion are attenuated in late pregnancy, and vasopressin V1b mRNA expression is reduced in the anterior pituitary (Ma et al., 2001), indicating a weaker action of vasopressin on the corticotrophs. However, the stress-induced responses of vasopressin release within the PVN and peripheral vasopressin secretion are not different in late pregnancy compared to virgins (Douglas et al., 1998; Neumann et al., 1998; Wigger and Neumann, 2002). Interestingly though, rats bred for high anxiety have reduced vasopressin peripheral secretion in pregnancy in response to forced swimming compared to rats bred for low anxiety, although the HPA axis secretory hyporesponsiveness of pregnancy is retained (Neumann et al., 1998).

Following birth, in lactation, there is an increase in CRF and vasopressin co-localisation in parvocellular PVN neurons. However, unlike pregnancy plasma corticosterone concentration is tonically elevated, and basal vasopressin mRNA expression and vasopressin peptide in the parvocellular PVN are increased in lactation (Walker et al., 2001a,b) and vasopressin mRNA expression increases after stress (da Costa et al., 2001; Walker et al., 2001a) in lactation. This is in contrast to CRF mRNA expression that continues to decrease after birth (Walker et al., 2001a,b). Thus, there appears to be opposing regulating of CRF and vasopressin genes during lactation and the upregulation of vasopressin expression may partly compensate for the reduced CRF synthesis (Walker et al., 2001a,b). In the anterior pituitary vasopressin V1b receptor binding decreases in lactation (Aguilera and Rabadan-Diehl, 2000b), correlating with the lower ACTH response at this time to stress, even though exogenous vasopressin is more effective at stimulating ACTH secretion (Toufexis et al., 1999; Walker et al., 2001a).

Summary

Vasopressin regulates HPA axis responsiveness by acting at the anterior pituitary to enhance corticotroph responses to CRF and centrally in the brain to attenuate PVN neuron responses. Vasopressin's role in promoting the body response to stress becomes enhanced upon chronic or repeated stress exposure, compensating for an apparently less prominent role of CRF in those conditions. As such vasopressin plays an important role in conditions such as anxiety, depression and aging. Peripheral vasopressin secretion decreases or remains unchanged with exposure to stressors: it may play a role in decreasing blood pressure in stressful situations.

Oxytocin

Although oxytocin is a structurally similar hormone to vasopressin and is generally synthesised and secreted from adjacent neurons in the same hypothalamic regions, it plays a distinct role in relation to stress responsiveness. Initially peripheral oxytocin secretion was observed to increase after exposure to certain stressors (Lang et al., 1983), and subsequently it was reported that oxytocin action in the brain has anti-anxiety and anti-stress effects. Oxytocin is also involved in maternal aggression, as part of the range of maternal behaviours expressed by the postpartum mother.

Stress and oxytocin expression

As with vasopressin, stress stimulates oxytocin expression in the hypothalamus of the rat and, thus, oxytocin mRNA expression increases after forced swimming in magnocellular neurons of the PVN and SON, indicating activation of the neurons and a role for oxytocin release (Callahan et al., 1992; Wotjak et al., 2001). Fos expression also increases in magnocellular oxytocin neurons in the PVN and SON after stress (Miyata et al., 1995), indicating their synaptic activation and further confirming their involvement in and/or responses to stress. Oxytocin mRNA expression in parvocellular PVN neurons does not evidently increase with stress in the rat (Wotjak et al., 2001) and where reported, oxytocin expression does not increase with stress in the PVN and SON of other species (e.g. the pig, Vellucci and Parrott, 1998). As well as being co-expressed with vasopressin, in the rat CRF is co-localised with oxytocin in a subset of magnocellular PVN neurons (Sawchenko et al., 1984b; Levis and Sawchenko, 1993), and if secreted into the peripheral circulation or released from dendrites within the nuclei, CRF could directly influence oxytocin or vasopressin peripheral or central secretion (Bruhn et al., 1986; Lightman, 1992) as another dimension to the neuroendocrine control of stress responses. Additionally, oxytocin neuron activity and secretion are sensitive to glucocorticoids and are regulated in part by glucocorticoid feedback since the neurons co-express glucocorticoid receptors (Jirikowski et al., 1993) and the oxytocin gene has a glucocorticoid response element (Mohr and Schmitz, 1991). Oxytocin expression in magnocellular neurons and its regulation by glucocorticoids during stress imply that peripheral and/or central oxytocin secretion and action are a normal aspect of stress responsiveness.

Stress and peripheral oxytocin secretion

The immediate effect of stress exposure is an increase in oxytocin secretion into the blood circulation. Stressors such as restraint (Gibbs, 1984; Miyata et al., 1995; Johnstone et al., 2000), immobilisation (Lang et al., 1983; Lightman, 1992), foot shock (Crine and Buijs, 1987), forced swimming (Lang et al., 1983; Gibbs, 1984; Kasting, 1988; Neumann et al., 1998; Wotjak et al., 1998a) and shaker stress (Hashiguchi et al., 1997; Nishioka et al., 1998) all increase oxytocin secretion in parallel with the ACTH secretory responses in the rat. However, some stressors (especially those described as 'emotional') are not effective at driving peripheral oxytocin secretion, for example, social defeat in male rats (Engelmann et al., 1999). Interestingly, central action of CRF increases plasma oxytocin concentration (Bruhn et al., 1986; Lightman, 1992), indicating that centrally acting CRF may be involved in the oxytocin secretory responses to stress. However, in mice deficient in CRF receptor 1 oxytocin secretory responses to forced swimming or social defeat are the same as in wild type CRF receptor 1 expressing mice, despite strong attenuation of ACTH and corticosterone responses (Muller et al., 2000). Thus CRF must act via another receptor to mediate oxytocin responses, and/or another factor stimulates the oxytocin responses. Since in the same animals vasopressin secretory responses to stress were severely attenuated it is apparent that vasopressin and oxytocin responses are independently regulated and influence the HPA axis via different mechanisms in the male mouse. In other species, whether peripheral oxytocin secretion increases in response to stress is less clear. In humans oxytocin concentration in the periphery does not reliably change with stress exposure (Kalin et al., 1985; Uvnas-Moberg, 1998a; Light et al., 2000; Altemus et al., 2001; Turner et al., 2002).

The role of stress-induced oxytocin secretion

The role for peripheral secretion of oxytocin in response to stress has not been established. One possibility is a role for oxytocin as a secretagogue for ACTH secretion. Oxytocin release into pituitary portal vessels may arise from magnocellular neuron axons or terminals and axons and their collaterals pass through and/or terminate in the median eminence (Brownstein et al., 1980; Buma and Nieuwenhuys, 1987). Oxytocin secreted from the posterior pituitary may influence the anterior pituitary corticotrophs since short portal vessels have been described between the posterior and anterior lobes. It has been shown that peripheral oxytocin can induce ACTH secretion (Watanabe et al., 1989; Petersson et al., 1999), acting synergistically with CRF, as vasopressin does. Oxytocin receptors are expressed in the anterior pituitary at low levels (Adan et al., 1995) and since oxytocin receptor expression is enhanced by estrogen administration (Quinones-Jenab et al., 1997), oxytocin may have a sexually dimorphic effect on HPA axis secretory responses to stress. However, physiological conditions which exhibit high peripheral oxytocin secretion do not necessarily coincide with high ACTH concentrations (e.g. parturition and lactation).

Other proposed roles for oxytocin include induction of a prolonged decrease in corticosterone secretion (Petersson et al., 1999), acting at the adrenal level (Stachowiak et al., 1995). Oxytocin also transiently increases blood pressure which is reversible by oxytocin antagonist (Saameli, 1968; Petersson et al., 1999), but in contrast may induce a long-term decrease in blood pressure, although this is not reversible by oxytocin antagonist (Petersson et al., 1999, Petersson, 2002). Oxytocin and oxytocin receptors are expressed in the heart (Gutkowska et al., 1997; Jankowski et al., 1998) and oxytocin may regulate atrial natriuretic peptide release (Haanwinckel et al., 1995; Gutkowska et al., 1997), which decreases blood pressure and cardiac output, and thus may explain the above effect of oxytocin on blood pressure. Oxytocin effects on glucocorticoid secretion and on blood pressure may help to restore HPA axis secretion to basal levels following stress, and its tendency to increase nociceptive threshold (Petersson et al., 1996) may reduce painful stress signals. Although oxytocin also has a role in mediating immune signalling (for overview see Geenen et al., 1998), or natriuresis (Verbalis et al., 1991), its involvement in stress-induced changes is not presently apparent.

Effect of oxytocin action in the brain on secretory responses to stress

On the other hand, oxytocin does have a defined physiological role mediating stress responses by acting centrally. Oxytocin does not readily cross the blood brain barrier so peripheral secretion is unlikely to be the source of oxytocin that moderates HPA axis responses: the oxytocin probably comes from the magnocellular oxytocin neurons within the PVN. Oxytocin is released from dendrites and/or cell bodies of magnocellular neurons into the extracellular space, from where it can act locally on adjacent neurons and synaptic contacts have been demonstrated between magnocellular oxytocin neurons and CRF neurons in the PVN (Hisano et al., 1992), so oxytocin actions in the PVN to moderate HPA axis responses may be paracrine or transsynaptic. Both physical and threatening stressors increase oxytocin release within the PVN and SON in parallel with increased peripheral secretion (Hashiguchi et al., 1997; Nishioka et al., 1998; Wotjak et al., 1998a; Wigger and Neumann, 2002). However, social defeat increases oxytocin release in the SON, but not PVN (Wotjak et al., 1996b; Engelmann et al., 1999), despite having no effect on peripheral oxytocin secretion.

Oxytocin has been described as an anti-stress factor since oxytocin antagonist given centrally to rats (but not mice, Douglas et al., 2001b) causes an increase in basal plasma ACTH and enhances ACTH secretory responses to a variety of stressors, an effect that is gender independent (Neumann et al., 2000a,b, 2002). Evidently, the oxytocin released within the PVN inhibits HPA axis secretion and central administration of exogenous oxytocin attenuates HPA axis responses to stress (Windle et al., 1997a,b). However, as well as effects within the PVN, the oxytocin may be acting in other brain regions that express oxytocin receptor (e.g. limbic regions – Brinton et al., 1984; Insel, 1990; Broad et al., 1999). Oxytocin antagonist effects on basal and stress-induced ACTH secretion appear to be dependent upon the stressor and upon the specific brain region and can be bimodal, for example in the PVN oxytocin antagonist decreases stress-stimulated ACTH secretion but given in the septum ACTH secretory responses are increased (Neumann et al., 2000a; Neumann, 2002). Oxytocin receptor is expressed in many brain regions, including those rich in glucocorticoid receptors and involved in HPA axis regulation (Brinton et al., 1984; Tribollet et al., 1988; Ur and Grossman, 1994; Feldman and Weidenfeld, 1995) and oxytocin receptor binding in limbic brain regions is increased by glucocorticoids, dexamethasone and prolonged stress (Patchev et al., 1993; Liberzon and Young, 1997). With repeated stress exposure (prolonged elevated glucocorticoid levels), oxytocin release in the SON has been shown to increase during each bout of forced swimming, coupled to peripheral secretion (Wotjak et al., 1998a); thus oxytocin may, in combination with vasopressin, maintain HPA axis secretory responses to chronic stress by preventing prolonged activation of CRF neurons. Reduction in glucocorticoids, by adrenalectomy, decreases oxytocin release in the PVN in response to forced swimming in male rats, presumably contributing to the large increase in plasma ACTH concentration (Neumann, 2002). Replacement of glucocorticoids partly prevented the effect (Torner et al., 2000), thus the presence of corticosterone seems necessary for the intra-hypothalamic oxytocin release during stress.

Effect of oxytocin action in the brain on behavioural responses to stress

In addition to oxytocin's role in inhibiting HPA axis secretory responses, it has also been reported to have anti-anxiety effects in the rat and mouse when given centrally, or even peripherally (King et al., 1985; Arletti and Bertolini, 1987; Roozendaal et al., 1992; Uvnas-Moberg et al., 1994; McCarthy et al., 1996; Windle et al., 1997a,b; Uvnas-Moberg, 1998b; Lightman et al., 2001). However, in virgin and male rats oxytocin antagonist given centrally does not affect anxiety-related behaviours on the elevated plus maze (Neumann et al., 2000b,c). Anxiety and related behaviours such as coping, fear and social behaviours, including social defeat and maternal behaviour, are mediated by limbic structures and oxytocin, like vasopressin (Koolhaas et al., 1998), is involved in mediating anxiety-related behaviours here under the influence of glucocorticoids (Patchev et al., 1993; Liberzon et al., 1994; Liberzon and Young, 1997). Oxytocin release can be measured in the limbic regions, although the limited amounts released make accurate quantification difficult. Oxytocin release increases in the septum in male rats exposed to social defeat but the role of oxytocin there is uncertain: oxytocin antagonist had no effect on freezing (defensive) or investigating (explorative/activity) behaviours during the social encounter (Ebner et al., 2000).

Oxytocin and anxiety – effect of oxytocin deficiency

The necessity for oxytocin in controlling anxiety in these regions is revealed by transgenic mouse studies, where knockout or inactivation of the oxytocin gene impairs social recognition, while having no effect on other memory functions such as spatial memory. This indicates a specific effect of oxytocin on social memory and is mediated in the medial amygdala (Ferguson et al., 2001), the deficit being restored by central oxytocin administration. However, although the mice deficient in oxytocin exhibit normal maternal care, the pups vocalise less during maternal separation (Winslow et al., 2000) and as adults are reported to have altered aggression (DeVries et al., 1997; Winslow et al., 2000; Winslow and Insel, 2002) and to be less fearful (Winslow et al., 2000) compared to wild type, oxytocin-expressing mice.

Oxytocin receptor and inheritance of stress and anxiety

Further evidence highlighting a key role for oxytocin in stress and anxiety proposed a non-genomic mechanism of inheritance of stress responsivity. A variety of experimental approaches have demonstrated a link between oxytocin receptor expression in the limbic system in neonates and their aggression/ anxiety parameters in adulthood. Good or poor levels of maternal care received during the neonatal period evidently determine low and high levels of HPA axis responsiveness and anxiety, mediated by elevated or reduced oxytocin receptor binding in the amygdala, respectively (Boccia and Pedersen, 2001; Francis et al., 2002). Thus, the plasticity of the oxytocin system in optimising brain and body responses to stress is not only important in adulthood but also in the neonatal period.

Effect of physiological circumstances on oxytocin responses to stress

As for vasopressin, the secretion of oxytocin both centrally and peripherally in relation to stress alters under a variety of physiological conditions. These include aging and reproduction, both cases exhibiting a reduction in oxytocin neuron responsiveness and secretion in response to stress.

Aging

In old age when plasma corticosterone is elevated, hypothalamic oxytocin content decreases (Ozawa et al., 2001). Basal oxytocin release is not altered, but intra-PVN release in response to stress in rats is attenuated, and this mirrors an attenuated peripheral oxytocin secretory response (Keck et al., 2000). The lack of inhibition of the HPA axis by oxytocin may at least partly explain the hypercortisolemia. However, the enhancing effect of corticosterone on oxytocin release and oxytocin receptor expression seems lost and thus is apparently uncoupled with age.

Reproduction

Changes in the patterns of oxytocin neuron activity and secretion are important in parturition and lactation and underlie the normal processes of birth (The Ferguson Reflex), maternal behaviour and suckling (the milk-ejection reflex). Thus oxytocin neurons behave differently at these times compared to virgin rats, and the process of optimising their responses begins in mid-late pregnancy. This has consequences for secretory stress responses, and the converse is also true, stress exposure seriously disrupts birth, maternal behaviour and suckling (Cross, 1955; Newton et al., 1968; Lincoln et al., 1973; Leng et al., 1987; Pardon et al., 2000; Lau, 2001; Douglas et al., 2002b).

Pregnancy
Oxytocin peripheral secretory responses to stress, like those of the HPA axis, become attenuated in late pregnancy (novel environment and forced swimming; Neumann et al., 1998; Johnstone et al., 2000). This hyporesponsiveness is caused by a variety of adaptations within the brain. Endogenous opioids regulate oxytocin secretory responses to stress (Carter and Lightman, 1987a), and the nature of this control alters in late pregnancy with the emergence of a strong opioid inhibition, restraining oxytocin secretory responses to stress, as well as to other stimuli (Douglas et al., 1995, 1998). The opioid inhibition of oxytocin secretory responses can be reproduced in virgin rats by artificially mimicking the estrogen and progesterone profiles of pregnancy (Douglas et al., 2000), and thus is sex steroid dependent. Sex steroids additionally have effects on anxiety-related behaviours, and estrogen enhances the anxiolytic effect of oxytocin, accompanied by increased oxytocin receptor binding in the lateral septum (McCarthy et al., 1996). In contrast to peripheral secretion, stress-induced increases in oxytocin release within the PVN are not attenuated in late pregnancy, although restraint by endogenous opioids of somato-dendritic oxytocin release emerges (Wigger and Neumann, 2002). Inhibition of central oxytocin release would prevent premature depletion of oxytocin peptide from the magnocellular neurons, which is required later during parturition when oxytocin autoregulation (by positive feedback) of oxytocin neurons promotes birth (Neumann et al., 1996).

A role for oxytocin in the attenuation of HPA axis responses to stress in pregnancy has also been sought, since oxytocin in the PVN inhibits HPA axis secretion. However, central oxytocin antagonist administration has no effect on plasma ACTH or corticosterone responses to novel environment or forced swimming in late pregnant rats (Neumann et al., 2000b), and so the inhibitory effect of central oxytocin in virgins is lost in pregnancy. This is likely to be a result of opioid inhibition of oxytocin release in the PVN at this time, but also will be partly due to changes in other neurotransmitters/neuromodulators that are known to influence oxytocin neurons in late pregnancy (e.g. noradrenaline, nitric oxide, GABA). In late pregnancy, before estrogen levels dominate the secretory profile of pregnancy hormones, anxiety-related behaviours are enhanced (Neumann et al.,1998). Anxiety in these rats is further enhanced by central oxytocin antagonist administration either icv or directly into the amygdala (Neumann et al., 2000a; Neumann, 2002), indicating that oxytocin restrains mechanisms that underlie anxiety in pregnancy to a certain extent. Although oxytocin receptor expression also increases in the brain, including the PVN and limbic system, during mid-late pregnancy (Insel, 1990, 1992; Ingram et al., 1995; Young et al., 1997), the effect of oxytocin on anxiety is region-dependent (not effective when given icv, Neumann, 2002). Thus in pregnancy there is a dissociation of the effects of oxytocin on HPA axis responsiveness and anxiety-related behaviour.

Parturition

Exposure to stress severely disrupts parturition in many species (mouse, Newton et al., 1968; rat, Leng et al., 1987; Douglas et al., 2002b; pig, Lawrence et al., 1995; human Hedegaard et al., 1993, 1996; Wadhwa et al., 1998, 2001); this occurs in the rat even if the stressor is of a non-intervention type, within the home environment (using air puff startle; Douglas et al., 2002a; Neumann et al., 2003). Delay in rodent birth due to stress involves a reduction in uterine contractility, which in the mouse has been shown to be mediated by a beta-adrenergic mechanism (Douglas et al., 2002b) and in both rat and mouse is accompanied by a reduction in plasma oxytocin concentration compared to that at birth (Leng et al., 1987, 1988; Douglas et al., 2002b). Although at first sight this appears contrary to the usual increase in oxytocin secretion in response to stress, it has to be seen in the light of the patterns of oxytocin secretion during birth. Oxytocin neuron firing rate becomes intermittent with high-frequency bursts, resulting in pulsatile secretion coincident with individual births (Higuchi et al., 1986; Summerlee, 1989; Gilbert et al., 1994). The Ferguson Reflex, initiated by the stretch of the uterus and cervix, mediates this unique pattern and thus when uterine contraction is inhibited and births are absent the positive feedback to oxytocin neuron activity and secretion is disrupted, resulting in low plasma oxytocin concentrations. Oxytocin administration effectively restarts stress-delayed labour and birth (Douglas et al., 2002b) and oxytocin is commonly utilised in the labour ward to promote birth in women. As in pregnancy, opioids continue to be inhibitory to oxytocin secretion during birth and opioid antagonist administration accelerates the birth profile in rats (Bicknell et al., 1988), including during birth delayed by stress (Leng et al., 1987, 1988) and birth delayed by opioid administration is advanced by exogenous oxytocin (Luckman et al., 1993; Douglas et al., 1993). Endogenous opioids also inhibit birth and oxytocin secretion in the pig, and stress-delayed birth is restored by opioid antagonist (Lawrence et al., 1992). However, in the human and mouse opioid antagonist has no effect on normal or stress-delayed birth (Lindow et al., 1992; Douglas et al., 2002b).

HPA axis secretion is reduced during parturition in the rat (Wigger et al., 1999), so despite a common perception that labour is painful and/or stressful it is not processed as such in the rat. Furthermore, our preliminary data show that exposure to stress during birth also does not activate HPA axis secretory responses in rats (Douglas et al., 2002a; Neumann et al., 2003), thus stress hyporesponsiveness is greater in parturition than in late pregnancy. We have found a similar effect of birth and stress during birth on ACTH secretion for the mouse (preliminary data, not shown), but in pigs and humans plasma glucocorticoids are elevated during late pregnancy and birth (Jarvis et al., 1999; Wadhwa et al., 2001), although whether secretion is further elevated with stress during birth is not clear (Jarvis et al., 1997, 1999). Again, opioids strongly restrain secretory stress responsiveness in the rat since opioid antagonist increases ACTH and corticosterone concentrations (Wigger et al., 1999).

Oxytocin exerts positive feedback on magnocellular oxytocin neurons during parturition, as part of the burst firing mechanism promoting a speedy birth (Neumann et al., 1996) and oxytocin release in the PVN is elevated during parturition (Neumann et al., 1993). However, similar to the reports in late pregnancy, central oxytocin antagonist administration does not increase HPA axis secretion in basal or stress conditions during parturition as it does in virgins (Douglas et al., 2002a; Neumann et al., 2003), and so effects of oxytocin alone cannot explain the hyporesponsiveness of the HPA axis in parturition.

Lactation

In lactation, as in parturition, oxytocin is secreted in a pulsatile pattern, following burst firing of oxytocin neurons in response to suckling which is known as the milk-ejection reflex (Wakerley and Lincoln, 1973; Lincoln and Wakerley, 1974). Stress exposure, or even minor disruption, during lactation interrupts the reflex, and a calm state and safe environment for humans is important to permit the reflex to occur; in the rat a state of 'slow-wave activity' of cortical neurons is required for the milk-ejection reflex. Oxytocin secretion into the blood in response to stress is almost completely abolished in lactation

(Carter and Lightman, 1987b; Higuchi et al., 1988; Lightman, 1992; Neumann et al., 1998). This arises since with stress the interruption of suckling, and thus a loss of the intermittent oxytocin neuron activity and secretion, results in low blood oxytocin levels. Opioid inhibition can partly explain these reduced oxytocin secretory responses since naloxone restores the stress-induced increased in oxytocin secretion (Saridaki and Lightman, 2003), but also CRF-stimulated oxytocin secretion as seen in virgin rats is lost in lactation (Bruhn et al., 1986; Patel et al., 1991; Lightman, 1992) perhaps as a consequence of the reduced CRF expression (Lightman, 1992). Prolactin, which is also a stress hormone and is hypersecreted in lactation, was investigated as a modulator of the oxytocin secretory responses but artificially induced hyperprolactinemia in virgin female rats does not cause the attenuation of oxytocin responses (Carter and Lightman, 1987b).

Central oxytocin action is also not responsible for the attenuation of HPA axis secretory responses in lactation since oxytocin antagonist has no effect on ACTH and corticosterone secretion (Neumann et al., 2000b; Neumann, 2002), similar to the lack of effect in late pregnancy and parturition, although whether oxytocin release within the PVN during stress changes during lactation is not known. Anxiety remains dissociated from HPA axis secretory responses in lactation and oxytocin antagonist enhances anxiety, although to a lesser extent than in late pregnancy (Neumann et al., 2000b; Neumann, 2002). Therefore, throughout reproduction an important role for central oxytocin may be in restraining anxiety-related behaviours.

Maternal aggression is one aspect of the gamut of maternal behaviours expressed in lactating rat mothers and is directed towards intruders into their home cages. As this is a stressful situation for the mother the HPA axis becomes activated, although to a reduced degree (Neumann et al., 2001), as part of the hyporesponsiveness of stress responses observed throughout the reproductive cycle. Oxytocin peripheral secretion remains unchanged in the mothers during the time the intruder is in her cage and a lack of an oxytocin secretory response to stress is to be expected in lactation. Centrally, basal oxytocin release within the amygdala is elevated in lactation, perhaps tonically preventing anxiety. However, oxytocin release here does not further increase during this stress and it has no effect on the offensive/attack behaviours exhibited (Neumann et al., 2001; Neumann, 2002). Thus, central oxytocin may not be involved in maternal aggressive behaviours.

Summary

Oxytocin is released into the blood in response to stress and within the brain in the PVN and SON, as well as in limbic regions. Although the role of oxytocin secreted into the blood is uncertain, central oxytocin has clear anti-stress and anti-anxiety effects: increased central oxytocin in pregnancy and lactation restrains anxiety-related behaviour, but does not explain the attenuated HPA axis secretory responses then or in parturition. Oxytocin is a key neuropeptide mediating parturition and lactation, and stress exposure of a mother disrupts both processes, at least partly by interfering with the unique pattern of oxytocin neuron activity and secretion at these times. There is a progressive development of hyporesponsiveness of oxytocin secretion in response to stress through pregnancy, parturition and lactation; perhaps to ensure retention of oxytocin stores for its role in parturition and lactation.

Conclusion

While vasopressin and oxytocin are similar neuropeptides with a highly overlapping location of their source neurons and their release sites, they retain differential, though related, effects in terms of mediation of stress responses. Although some of their functions in stress are clearly defined, for example the secretagogue role of vasopressin at the anterior pituitary and the central inhibition of the HPA axis secretory responses by both vasopressin and oxytocin, there are other related brain and periperhal roles that remain to be precisely elucidated. The complex network of vasopressin and oxytocin pathways and patterns of release within the brain undoubtedly are involved in the perception of and responses to stress and anxiety and, with the growing literature on their involvement in human behaviour, mood and disease, are potential future therapeutic targets.

Abbreviations

ACTH	adrenocorticotrophic hormone
cAMP	cyclic adenosine monophosphate
CCK	cholecystokinin
CREB	cAMP response-element binding protein
CRF	corticotropin-releasing factor
GABA	gamma aminobutyric acid
HPA	hypothalamo–pituitary–adrenal
icv	intracerebroventricular
NGFI-B	nerve growth factor inducing factor B
RNA	ribonucleic acid, (m = messenger, hn = heterogeneous nuclear)
PVN	paraventricular nucleus
SON	supraoptic nucleus
V1b	vasopressin 1b receptor

Acknowledgements

The author's research was supported by The BBSRC, The Wellcome Trust, The Royal Society of London and The British Council. I would like to thank my colleagues in the Laboratory of Neuroendocrinolgy at the University of Edinburgh, in particular Dr. Mike Ludwig for critical reading of the manuscript, and Professor John Russell, Professor Gareth Leng and Dr. Paula Brunton for discussions on the experiments and techniques reported. I also thank Professor Inga Neumann for constructive comments on an early draft of the manuscript.

References

Adan, R.A., van Leeuwen, F.W., Sonnemans, M.A., Brouns, M., Hoffman, G., Verbalis, J.G. and Burbach, J.P. (1995) Rat oxytocin receptor in brain, pituitary, mammary gland, and uterus: partial sequence and immunocytochemical localization. Endocrinology, 136: 4022–4028.

Aguilera, G. (1994) Regulation of pituitary ACTH secretion during chronic stress. Front. Neuroendocrinol., 15: 321–350.

Aguilera, G. and Rabadan-Diehl, C. (2000a) Vasopressinergic regulation of the hypothalamic-pituitary-adrenal axis: implications for stress adaptation. Reg. Peptides, 96: 23–29.

Aguilera, G. and Rabadan-Diehl, C. (2000b) Regulation of vasopressin V-1b receptors in the anterior pituitary gland of the rat. Exp. Physiol., 85: 19S–26S.

Altemus, M., Redwine, L.S., Leong, Y.M., Frye, C.A., Porges, S.W. and Carter, C.S. (2001) Responses to laboratory psychosocial stressors in postpartum women. Psychosom. Med., 63: 814–821.

Antoni, F.A. (1993) Vasopressinergic control of pituitary adrenocorticotropin secretion comes of age. Front. Neuroendocrinol., 14: 76–122.

Aravich, P.F. and Sladek, J.R. (1987) Aging of rodentvasopressin systems: morphometric and functional considerations. In: Gash, D.M. and Boer, G.J. (Eds.), Vasopressin: Principles and Properties. Plenum Press, New York, pp. 579–610.

Arletti, R. and Bertolini, A. (1987) Oxytocin acts as an antidepressant in two animal models of depression. Life Sci., 41: 1725–1730.

Bartanusz, V., Aubry, J.M., Jezova, D., Baffi, J. and Kiss, J.Z. (1993a) Up-regulation of vasopressin mRNA in paraventricular hypophysiotrophic neurons after acute immobilization stress. Neuroendocrinology, 58: 625–629.

Bartanusz, V., Jezova, D., Bertini, L.T., Tilders, F.J.H., Aubry, J. and Kiss, J.Z. (1993b) Stress-induced increases in vasopressin and corticotropin-releasing factor expression in hypophysiotropic paraventricular neurons. Endocrinology, 132: 895–902.

Bertini, L.T. and Kiss, J.Z. (1991) Hypophysiotropic neurons are capable of altering the ratio of co-packaged neurohormones. Neuroscience, 42: 237–244.

Bicknell, R.J., Leng, G., Russell, J.A., Dyer, R.G., Mansfield, S. and Zhao, B.G. (1988) Hypothalamic opioid mechanisms controlling oxytocin neurones during parturition. Brain Res. Bull., 20: 743–749.

Bilezikjian, L.M., Woodgett, J.R., Hunter, T. and Vale, W.W. (1987) Phorbol ester-induced down-regulation of protein kinase C abolishes vasopressin-mediated responses in rat anterior pituitary cells. Mol. Endocrinol., 1: 555–560.

Boccia, M.L. and Pedersen, C.A. (2001) Brief vs. long maternal separations in infancy: contrasting relationships with adult maternal behavior and lactation levels of aggression and anxiety. Psychoneuroendocrinology, 26: 657–672.

Brinton, R.E., Wamsley, J.K., Gee, K.W., Wan, Y.-P. and Yamamura, H.I. (1984) [^{3}H] Oxytocin binding sites in the rat brain demonstrated by quantitative light microscopic autoradiography. Eur. J. Pharmacol., 102: 365–367.

Broad, K.D., Levy, F., Evans, G., Kimura, T., Keverne, E.B. and Kendrick, K.M. (1999) Previous maternal experience potentiates the effect of parturition on oxytocin receptor mRNA expression in the paraventricular nucleus. Eur. J. Neurosci., 11: 3725–3737.

Brownstein, M.J., Russel, J.T. and Gainer, H. (1980) Synthesis, transport and release of posterior pituitary hormones. Science, 207: 373–378.

Bruhn, T.O., Sutton, S.W., Plotsky, P.M. and Vale, W.W. (1986) Central administration of corticosterone-releasing factor modulates oxytocin secretion in the rat. Endocrinology, 119: 1558–1563.

Buijs, R.M., Kalsbeek, A., Vanderwoude, T.P., Vanheerikhuize, J.J. and Shinn, S. (1993a) Suprachiasmatic nucleus lesion increases corticosterone secretion. Am. J. Physiol., 264: R1186–R1192.

Buijs, R.M., Markman, M., Nunescardoso, B., Hou, Y.X. and Shinn, S. (1993b) Projections of the suprachiasmatic nucleus to stress-related areas in the rat hypothalamus – a light and electron microscopic study. J. Comp. Neurol., 335: 42–54.

Buma, P. and Nieuwenhuys, R. (1987) Ultrastructural demonstration of oxytocin and vasopressin release sites in the neural lobe and median eminence of the rat by tannic acid and immunogold methods. Neurosci. Lett., 74: 151–157.

Burke, Z.D., Ho, M.Y., Morgan, H., Smith, M., Murphy, D. and Carter, D. (1997) Repression of vasopressin gene expression by glucocorticoids in transgenic mice: evidence of a direct mechanism mediated by proximal 5′ flanking sequence. Neuroscience, 78: 1177–1185.

Caffe, A.R., Van Leeuwen, F.W. and Luiten, P.G.M. (1987) Vasopressin cells of the medial amygdala project to the lateral septum and ventral hippocampus. J. Comp. Neurol., 261: 237–252.

Callahan, M.F., Thore, C.R., Sundberg, D.K., Gruber, K.A., O'Steen, K. and Morris, M. (1992) Excitotoxin paraventricular nucleus lesions: stress and endocrine reactivity and oxytocin mRNA levels. Brain Res., 597: 8–15.

Carter, C.S. and Altemus, M. (1997) Integrative functions of lactational hormones in social behavior and stress management. Ann. NY Acad. Sci., 807: 164–174.

Carter, D.A. and Lightman, S.L. (1987a) Opioid control of oxytocin secretion: evidence of distinct regulatory actions of two opiate receptor types. Life Sci., 40: 2289–2296.

Carter, D.A. and Lightman, S.L. (1987b) Oxytocin responses to stress in lactating and hyperprolactinaemic rats. Neuroendocrinology, 46: 532–537.

Carvallo, P. and Aguilera, G. (1989) Protein kinase C mediates the effects of vasopressin in pituitary corticotrophs. Mol. Endocrinol., 3: 1935–1943.

Chan, R.K.W., Brown, E.R., Ericsson, A., Kovacs, K.J. and Sawchenko, P.E. (1993) A comparison of 2 immediate-early genes, c-fos and NGFIB, as markers for functional activation in stress-related neuroendocrine circuitry. J. Neurosci., 13: 5126–5138.

Chini, B. and Fanelli, F. (2000) Molecular basis of ligand binding and receptor activation in the oxytocin and vasoressin receptor family. Exp. Physiol., 85S: 59–66.

Crine, A.F. and Buijs, R.M. (1987) Electric footshocks differentially affect plasma and spinal cord vasopressin and oxytocin levels. Peptides, 8: 243–246.

Cross, B.A. (1955) Neurohormonal mechanisms in emotional inhibiton of milk ejection. J. Endocrinol., 12: 29–37.

da Costa, A.P., Wood, S., Ingram, C.D. and Lightman, S.L. (1996) Region-specific reduction in stress-induced c-fos mRNA expression during pregnancy and lactation. Brain Res., 742: 177–844.

da Costa, A.P.C., Kampa, R.J., Windle, R.J., Ingram, C.D., and Lightman, S.L. (1997) Region-specific immediate-early gene expression following the administration of corticotropin-releasing hormone in virgin and lactating rats. Brain Res., 770: 151–162.

da Costa, A.P.C., Ma, X.M., Ingram, C.D., Lightman, S.L. and Aguilera, G. (2001) Hypothalamic and amygdaloid corticotropin-releasing hormone (CRH) and CRH receptor-1 mRNA expression in the stress- hyporesponsive late pregnant and early lactating rat. Mol. Brain Res., 91: 119–130.

Dallman, M.F., Bhatnagar, S. and Viau, V. (2000) Hypothalamo-pituitary-adrenal axis. In: Fink, G. (Ed.), Enclyclopedia of Stress, Vol. 2. Academic Press, San Diego, pp. 468–476.

de Groeij, D.C.E., Dijkstra, H. and Tilders, F.J.H. (1992) Stress-induced increases in vasopressin and corticotropin-releasing factor expression in hypophysiotrophic paraventricular neurons. Endocrinology, 131: 847–853.

DeVries, A.C., Young, W.S. and Nelson, R.J. (1997) Reduced aggressive behaviour in mice with targeted disruption of the oxytocin gene. J. Neuroendocrinol., 9: 363–368.

Diamond, M.I., Miner, N.J., Yoshinaga, S.K. and Yamamoto, K.R. (1990) Transcription factor regulation interactions; selector of positive and negative regulation from a single DNA element. Science, 249: 1266–1272.

Douglas, A.J., Clarke, G., MacMillan, S.J.A., Bull, P.M., Neumann, I., Way, S.A., Wright, D.M., McGrory, B.G. and Russell, J.A. (1993) Effects of the kapp-opioid agonist U50,488 on parturition in rats. Br. J. Pharm., 109: 251–258.

Douglas, A.J., Neumann, I., Meeren, H.K.M., Leng, G., Johnstone, L.E., Munro, G. and Russell, J.A. (1995) Central endogenous opioid inhibition of supraoptic oxytocin neurons in pregnant rats. J. Neurosci., 15: 5049–5057.

Douglas, A.J., Johnstone, H.A., Wigger, A., Landgraf, R., Russell, J.A. and Neumann, I.D. (1998) The role of endogenous opioids in neurohypophysial and hypothalamo-pituitary-adrenal axis hormone secretory responses to stress in pregnant rats. J. Endocrinol., 158: 285–293.

Douglas, A.J., Johnstone, H., Brunton, P. and Russell, J.A. (2000) Sex-steroid induction of endogenous opioid inhibition on oxytocin secretory responses to stress. J. Neuroendocrinol., 12: 343–350.

Douglas, A.J., Brunton, P.J., Russell, J.A. and Neumann, I.D. (2001a) Greater attenuation of hypothalamo-pituitary-adrenal (HPA) axis stress responses in pregnant vs lactating mice. BNA Abstracts, 16: 30.02.

Douglas, A.J., Russell, J.A. and Neumann, I.D. (2001b) Investigation of mechanisms regulating the attenuated secretory stress responses in pregnant mice. Reproduction Abstract Series, 27: 6.

Douglas, A.J., Bosch, O., Toschi, N. and Neumann, I.D. (2002a) Stress hyporesponsiveness during parturition in the rat: antagonising brain oxytocin does not reveal an inhibitory influence. Stress, 5 (Suppl.): P025.

Douglas, A.J., Leng, G. and Russell, J.A. (2002b) The importance of oxytocin mechanisms in the control of mouse parturition. Reproduction, 123: 543–552.

Ebner, K., Wotjak, C.T., Holsboer, F., Landgraf, R. and Engelmann, M. (1999) Vasopressin released within the septal brain area during swim stress modulates the behavioural stress response in rats. Eur. J. Neurosci., 11: 997–1002.

Ebner, K., Wotjak, C.T., Landgraf, R. and Engelmann, M. (2000) A single social defeat experience selectively stimulates the release of oxytocin, but not vasopressin, within the septal brain area of male rats. Brain Res., 872: 87–92.

Ebner, K., Wotjak, C.T., Landgraf, R. and Engelmann, M. (2002) Forced swimming triggers vasopressin release within the amygdala to modulate stress-coping strategies in rats. Eur. J. Neurosci., 15: 384–388.

Engelmann, M., Ebner, K., Landgraf, R., Holsboer, F. and Wotjak, C.T. (1999) emotional stress triggers intrahypothalamic but not peripheral release of oxytocin in male rats. J. Neuroendocrinol., 11: 867–872.

Engelmann, M., Ebner, K., Landgraf, R. and Wotjak, C.T. (1998) Swim stress triggers the release of vasopressin within the suprachiasmatic nucleus of male rats. Brain Res., 792: 343–347.

Engelmann, M., Ludwig, M., Singewald, N., Ebner, K., Sabatier, N., Lubec, G., Landgraf, R. and Wotjak, C.T. (2001) Taurine selectively modulates the secretory activity of vasopressin neurons in conscious rats. Eur. J. Neurosci., 14: 1047–1055.

Engelmann, M., Wotjak, C.T., Ebner, K. and Landgraf, R. (2000) Behavioural impact of intraseptally released vasopressin and oxytocin in rats. Exp. Physiol., 85: 125S–130S.

Everts, H.G. and Koolhaas, J.M. (1999) Differential modulation of lateral septal vasopressin receptor blockade in spatial learning, social recognition, and anxiety-related behaviors in rats. Behav. Brain Res., 99: 7–16.

Feldman, S. and Weidenfeld, J. (1995) Neural mechanisms involved in the corticosteroid feedback effects on the hypothalamo-pituitary-adrenocortical axis. Prog. Neurobiol., 45: 129–141.

Ferguson, J.N., Aldag, J.M., Insel, T.R. and Young, L.J. (2001) Oxytocin in the medial amygdala is essential for social recognition in the mouse. J. Neurosci., 21: 8278–8285.

Francis, D.D., Young, L.J., Meaney, M.J. and Insel, T.R. (2002) Naturally occurring differences in maternal care are associated with the expression of oxytocin and vasopressin (V1a) receptors: gender differences. J. Neuroendocrinol., 14: 349–353.

Geelen, G. and Corman, B. (1992) Relationshiup between vasopressin and renal concentrating ability in aging rats. Am. J. Physiol., 262: R826–833.

Geenen, V, Kecha, O. and Martens, H. (1998) Thymic expression of neuroendocrine self-peptide precursors: role in T cell survival and self-tolerance. J. Neuroendocrinol., 10: 811–822.

Gibbs, D.M. (1984) Dissociation of oxytocin, vasopressin and corticotropin secretion during different types of stress. Life Sci., 35: 487–491.

Gilbert, C.L., Goode, J.A. and McGrath, T.J. (1994) Pulsatile secretion of oxytocin during parturition in the pig: temporal relationship with fetal expulsion. J. Physiol., 475: 129–137.

Gillies, G.E., Linton, E.A. and Lowry, P.J. (1982) Corticotropin-releasing activity of the new CRH is potentiated several times by vasopressin. Nature, 229: 355–357.

Gutkowska, J., Jankowski, M., Lambert, C., Mukaddam-Daher, S., Zingg, H.H. and McCann, S.M. (1997) Oxytocin releases atrial natriuretic peptide by combining with oxytocin receptors in the heart. Proc. Natl. Acad. Sci. USA, 94: 11704–11709.

Haanwinckel, M.A., Elias, L., Favoretto, A.L., Gutkowska, J., McCann, S.M. and Antunes-Rodrigues, J. (1995) Oxytocin mediates atrial natriuretic peptide release and natriuresis after volume expansion in the heart. Proc. Natl. Acad. Sci. USA, 92: 7902–7906.

Hashiguchi, H., Ye, S.H., Morris, M. and Alexander, N. (1997) Single and repeated environmental stress: effect on plasma oxytocin, corticosterone, catecholamines, and behavior. Physiol. Behav., 61: 731–736.

Hatzinger, M., Wotjak, C.T., Naruo, T., Simchen, R., Keck, M.E., Landgraf, R., Holsboer, F. and Neumann, I.D. (2000) Endogenous vasopressin contributes to hypothalamic-pituitary-adrenocortical alterations in aged rats. J. Endocrinol., 164: 197–205.

Hedegaard, M., Henriksen, T.B., Sabroe, S. and Secher, N.J. (1993) Psychological distress in pregnancy and preterm delivery. Brit. Med. J., 307: 234–239.

Hedegaard, M., Henriksen, T.B., Secher, N.J., Hatch, M.C. and Sabroe, S. (1996) Do stressful life events affect duration of gestation and risk of preterm delivery? Epidemiology, 7: 339–345.

Heinrichs, M., Neumann, I. and Ehlert, U. (2002) Lactation and stress: protective effects of breast-feeding in humans. Stress, 5: 195–203.

Herman, J.P. (1995) In situ hybridization analysis of vasopressin gene transcription in the paraventricular and supraoptic nuclei of the rat: regulation by stress and glucocorticoids. J. Comp. Neurol., 363: 15–27.

Herman, J.P., Cullinan, W.E., Young, E.A., Akil, H. and Watson, S.J. (1992) Selective forebrain fiber tract lesions implicate ventral hippocampal structures in tonic regulation of paraventricular nucleus corticotropin-releasing hormone (CRH) and arginine vasopressin (AVP) messenger RNA expression. Brain Res., 592: 228–238.

Herman, J.P., Adams, D. and Prewitt, C. (1995) Regulatory changes in neuroendocrine stress integrative ciruitry produced by a variable stress paradigm. Neuroendocrinology, 61: 180–190.

Higuchi, T., Tadokoro, Y., Honda, K. and Negoro, H. (1986) Detailed analysis of blood oxytocin levels during suckling and parturition in the rat. J. Endocrinol., 110: 251–256.

Higuchi, T., Honda, S., Takano, S. and Negoro, H. (1988) Reduced oxytocin response to osmotic stimulus and immobilisation stress in lactating rats. J. Endocrinol., 116: 225–230.

Hisano, S., Li, S., Kagotani, Y. and Daikoku, S. (1992) Synaptic associations between oxytocin-containing magnocellular neurons and neurons containing corticotropin-releasing factor in the rat magnocellular paraventricular nucleus. Brain Res., 576: 311–318.

Holsboer, F. (1995) Neuroendocrinology of mood disorders. In: Bloom, F.E. and Kupfer, D.J. (Eds.), Psycholpharmcaology: the fourth generation of progress. Raven Press, New York, pp. 957–968.

Hoogendijk, J.E., Fliers, E., Swaab, D.F. and Verwer, R.W.H. (1985) Activation of vasopresin neurons in the human supraoptic and paraventricular nucleus in senescence and senile dementia. J. Neurol. Sci., 69: 291–299.

Ingram, C.D., Adams, T.S., Jiang, Q.B., Terenzi, M.G., Lambert, R.C., Wakerley, J.B. and Moos, F. (1995) Limbic regions mediating central actions of oxytocin on the milk-ejection reflex in the rat. J. Neuroendocrinol., 7: 1–13.

Insel, T.R. (1990) Regional changes in brain oxytocin receptors post-partum: time-course and relationship to maternal behaviour. J. Neuroendocrinol., 2: 539–545.

Insel, T.R. (1992) Oxytocin – a neuropeptide for affiliation: evidence from behavioral, receptor autoradiographic, and comparative studies. Psychoneuroendocrinology, 17: 3–35.

Jankowski, M., Hajjar, F., Kawas, S.A., Mukaddam-Daher, S., Hoffman, G., McCann, S.M. and Gutkowska, J. (1998) Rat heart: a site of oxytocin production and action. Proc. Natl. Acad. Sci. USA, 95: 14558–14563.

Jarvis, S., Lawrence, A.B., McLean, K.A., Deans, L.A., Chirnside, J. and Calvert, S.K. (1997) The effect of environment on behavioural activity, ACTH, beta-endorphin and cortisol in pre-farrowing gilts. Animal Sci., 65: 465–472.

Jarvis, S., Lawrence, A.B., McLean, K.A., Deans, L.A., Chirnside, J. and Calvert, S.K. (1999) The effect of piglet expulsion in the sow on plasma cortisol, adrenocorticotropic hormone and beta-endorphin. Reprod. Dom. Animals, 34: 89–94.

Jia, L.-G., Canny, B.J. and Leong, D.A. (1992) Paracrine communication regulates adrenocorticotropin secretion. Endocrinology, 130: 534–539.

Jingami, H., Matsukura, S., Numa, S. and Imura, H. (1985) Effects of adrenalectomy and dexanmethasone administration on the level of preprocorticotropin-releasing factor messenger ribonucleic acid (mRNA) in the hypothalamus and adrenocorticotropin/b-lipotropin precursor mRNA in the pituitary in rats. Endocrinology, 117: 1314–1320.

Jirikowski, G.F., McGimsey, W.C., Caldwell, J.D. and Sar, M. (1993) Distribution of oxytocinergic glucocorticoid target neurons in the rat hypothalamus. Horm. Metab. Res., 25: 543–544.

Johnson, A.G., Crawford, G.A., Kelly, D., Nguyen, T.V. and Gyory, A.Z. (1994) Arginine vasopressin and osmolality in the elderly. J. Am. Geriatr. Soc., 42: 399–404.

Johnstone, H.A., Wigger, A., Douglas, A.J., Neumann, I.D., Landgraf, R., Seckl, J.R. and Russell, J.A. (2000) Attenuation of hypothalamic-pituitary-adrenal axis stress responses in late pregnancy: changes in feedforward and feedback mechanisms. J. Neuroendocrinol., 12: 811–822.

Kalin, N.H., Gibbs, D.M., Barksdale, C.M., Shelton, S.E. and Carnes, M. (1985) Behavioral stress decreases plasma oxytocin concentrations in primates. Life Sci., 36: 1275–1280.

Kalsbeek, A., Palm, I.F. and Buijs, R.M. (2002) Central vasopressin systems and steroid hormones. Prog. Br. Res., 139: 57–73.

Kalsbeek, A., vanderVliet, J. and Buijs, R.M. (1996) Decrease of endogenous vasopressin release necessary for expression of the circadian rise in plasma corticosterone: a reverse microdialysis study. J. Neuroendocrinol., 8: 299–307.

Kasting, N.W. (1988) Simultaneous and independent release of vasopressin and oxytocin in the rat. Can. J. Physiol. Pharmacol., 66: 22–26.

Kc, P., Haxhiu, M.A., Trouth, C.O., Balan, K.V., Anderson, W.A. and Mack, S.O. (2002) CO_2-induced c-Fos expression in hypothalamic vasopressin containing neurons. Resp. Physiol., 129: 289–296.

Keck, M.E., Hatzinger, M., Wotjak, C.T., Landgraf, R., Holsboer, F. and Neumann, I.D. (2000) Ageing alters intrahypothalamic release patterns of vasopressin and oxytocin in rats. Eur. J. Neurosci., 12: 1487–1494.

Keck, M.E., Wigger, A., Welt, T., Muller, M.B., Gesing, A., Reul, J.M.H.M., Holsboer, F., Landgraf, R. and Neumann, I.D. (2002) Vasopressin mediates the response of the combined dexamethasone/CRH test in hyper-anxious rats: implications for pathogenesis of affective disorders. Neuropsychopharmacology, 26: 94–105.

Kim, J.K., Summer, S.N., Wood, W.M. and Schrier, R.W. (2001) Role of glucocorticoid hormones in arginine vasopressin gene regulation. Biochem. Biophys. Res. Comm., 289: 1252–1256.

King, M.G., Brown, R. and Kusnecov, A. (1985) An increase in startle response in rats administered oxytocin. Peptides, 6: 567–568.

Kiss, J.Z., Mezey, E. and Skirboll, L. (1984) Corticotropin-releasing factor-immunoreactive neurons of the paraventricular nucleus become vasopressin positive after adrenalectomy. Proc. Natl. Acad. Sci. USA, 81: 1854–1858.

Klavdieva, M.M. (1996) The history of neuropeptides II. Front. Neuroendocrinol., 17: 126–153.

Koolhaas, J.M., Everts, H., De Ruiter, A.J., de Boer, S.F. and Bohus, B. (1998) Coping with stress in rats and mice: differential peptidergic modulation of the amygdala-lateral septum complex. Prog. Br. Res., 119: 437–448.

Kovacs, K. and Mezey, E. (1987) Dexamethasone inhibits corticotropin-releasing factor gene expression in the rat paraventricular nucleus. Neuroendocrinology, 46: 365–368.

Kovacs, K., Kiss, J.Z. and Makara, G.B. (1986) Glucocorticoid implants around the hypothalamic paraventricular nucleus prevent the increase of corticotropin-releasing factor and arginine vasopressin immunostaining induced by adrenalectomy. Neuroendocrinology, 44: 229–234.

Kovacs, K.J. and Sawchenko, P.E. (1996a) Sequence of stress-induced alterations in indices of synaptic and transcriptional activation in parvocellular neurosecretory neurons. J. Neurosci., 16: 262–273.

Kovacs, K.J. and Sawchenko, P.E. (1996b) Regulation of stress-induced transcriptional changes in the hypothalamic neurosecretory neurons. J. Mol. Neurosci., 7: 125–133.

Kovacs, K.J., Foldes, A. and Sawchenko, P.E. (2000) Glucocorticoid negative feedback selectively targets vasopressin transcription in parvocellular neurosecretory neurons. J. Neurosci., 20: 3843–3852.

Krukoff, T.L., MacTavish, D. and Jhamandas, J.H. (1997) Activation by hypotension of neurons in the hypothalamic paraventricular nucleus that project to the brainstem. J. Comp. Neurol., 385: 285–296.

Landgraf, R. (2001) Neuropeptides and anxiety-related behavior. Endocr. J., 48: 517–533.

Landgraf, R., Gerstberger, R., Montkowski, A., Probst, J.C., Wotjak, C.T., Holsboer, F. and Engelmann, M. (1995) V1 vasopressin receptor antisense oligodeoxynucleotide into septum reduces vasopressin binding, social discrimination abilities, and anxiety-related behavior in rats. J. Neurosci., 15: 4250–4258.

Landgraf, R., Wigger, A., Holsboer, F. and Neumann, I.D. (1999) Hyper-reactive hypothalamo-pituitary-adrenocortical axis in rats bred for high anxiety-related behaviour. J. Neuroendocrinol., 11: 405–407.

Lang, R.E., Heil, J.W., Ganten, D., Hermann, K., Unger, T. and Rascher, W. (1983) Oxytocin unlike vasopressin is a stress hormone in the rat. Neuroendocrinology, 37: 314–316.

Larsen, P.J., Vrang, N., Moller, M., Jessop, D.S., Lightman, S.L., Chowdrey, H.S. and Mikkelsen, J.D. (1994) The diurnal expression of genes encoding vasopressin and vasoactive intestinal peptide within the rat suprachiasmatic nucleus is influenced by circulating glucocorticoids. Mol. Brain Res., 27: 342–346.

Lau, C. (2001) Effects of stress on lactation. Pediatric Clinics of North America, 48: 221–234.

Lawrence, A.B., Petherick, J.C., McLean, K., Gilbert, C.L., Chapman and Russell, J.A. (1992) Naloxone prevents interruption of parturition and increases plasma oxytocin following environmental disturbance in parturient sows. Physiol. Behav., 52: 917–923.

Lawrence, A.B., Petherick, J.C., McLean, K.A., Deans, L., Chirnside, J., Vaughan, A., Gilbert, C.L., Forsling, M.L. and Russell, J.A. (1995) The effects of chronic environmental stress on parturition and on oxytocin and vasopressin secretion in the pig. Animal Reprod. Sci., 38: 251–264.

Leng, G. (2000) Oxytocin. In: Fink, G. (Ed.), Enclyclopedia of Stress, Vol. 3. Academic Press, San Diego, pp. 109–114.

Leng, G., Mansfield, S., Bicknell, R.J., Blackburn, R.E., Brown, D., Chapman, C., Dyer, R.G., Hollingsworth, S., Shibuki, K., Yates, J.O. and Way, S. (1988) Endogenous opioid actions and effects of environmental disturbance on parturition and oxytocin secretion in rats. J. Reprod. Fertil., 84: 345–356.

Leng, G., Mansfield, S., Bicknell, R.J., Brown, D., Chapman, C., Hollingsworth, S., Ingram, C.D., Marsh, M.I., Yates, J.O. and Dyer, R.G. (1987) Stress-induced disruption of parturition in the rat may be mediated by endogenous opioids. J. Endocrinol., 114: 247–252.

Levis, M.C. and Sawchenko, P.E. (1993) Neuropeptide co-expressionin the magnocellular neurosecretory system of the female rat: evidence for differential modulation by estrogen. Neuroscience, 54: 1001–1018.

Liberzon, I. and Young, E.A. (1997) Effects of stress and glucocorticoids on CNS oxytocin receptor binding. Psychoneuroendocrinology, 22: 411–422.

Liberzon, I., Chalmers, D.T., Mansour, A., Lopez, J.F., Watson, S.J. and Young, E.A. (1994) Glucocoticoid regulation of hippocampal oxytocin receptor binding. Brain Res., 650: 317–322.

Liebsch, G., Montkowski, A., Holsboer, F. and Landgraf, R. (1998) Behavioral profiles of two wistar rat lines selectively bred for high or low anxiety-related behaviour. Behav. Brain Res., 94: 301–310.

Liebsch, G., Wotjak, C.T., Landgraf, R. and Engelmann, M. (1996) Septal vasopressin modulates anxiety-related behaviour in rats. Neurosci. Lett., 217: 101–104.

Light, K.C., Smith, T.E., Johns, J.M., Brownley, K.A., Hofheimer, J.A. and Amico, J.A. (2000) Oxytocin responsivity in mothers of infants: a preliminary study of relationships with blood pressure during laboratory stress and normal ambulatory activity. Health Psychol., 19: 560–567.

Lightman, S.L. (1992) Alterations in hypothalamic-pituitary responsiveness during lactation. Ann. N. Y. Acad. Sci., 652: 340–346.

Lightman, S.L. (1993) Alterations in hypothalamic-pituitary-adrenal responsiveness during lactation. Ann. N.Y. Acad. Sci., 689: 340–346.

Lightman, S.L. and Young, W.S. (1987) Vasopressin, oxytocin, dynorphin, enkephalin and corticotrophin-releasing factor mRNA stimulation in the rat. J. Physiol., 394: 23–39.

Lightman, S.L. and Young, W.S. (1988) Corticotrophin-releasing factor, vasopressin and pro-opiomelanocortin mRNA responses to stress and opiates in the rat. J. Physiol., 403: 511–523.

Lightman, S.L., Windle, R.J., Wood, S.A., Kershaw, Y.M., Shanks, N. and Ingram, C.D. (2001) Peripartum plasticity within the hypothalamo-pituitary-adrenal axis. Prog. Br. Res., 133: 111–129.

Lincoln, D.W. and Wakerley, J.B. (1974) Electrophysiological evidence for the activation of supraoptic neurones during the release of oxytocin. J. Physiol., 242: 553–544.

Lincoln, D.W., Hill, A. and Wakerley, J.B. (1973) The milk-ejection reflex of the rat: an intermittent function not abolished by surgical levels of anaesthesia. J. Endocrinol., 57: 459–467.

Lindow, S.W., van der Spuy, Z.M., Hendricks, M.S., Rosselli, A.P., Lombard, C. and Leng, G. (1992) The effect of morphine and naloxone administration on plasma oxytocin concentrations in the first stage of labour. Clin. Endocrinol., 37: 349–353.

Lucassen, P.J., Ravid, R., Gonatas, N.K. and Swaab, D.F. (1993) Activation of the human supraoptic and paraventricular nucleus neurons with aging and in Alzheimer's disease as judged from increasing size of the golgi apparatus. Brain Res., 632: 105–113.

Luckman, S.M., Antonijevic, I., Leng, G., Dye, S., Douglas, A.J., Russell, J.A. and Bicknell, R.J. (1993) The maintenance of normal parturition in the rat requires neurohypophysial oxytocin. J. Neuroendocrinol., 5: 7–12.

Ludwig, M. (1998) Dendritic release of vasopressin and oxytocin. J. Neuroendocrinol., 10: 881–895.

Ma, S., Shipston, M. and Russell, J.A. (2001) Vasopressin and the attenuated ACTH responses to stressors during pregnancy in rats. WCNH proceedings Bordeaux, P2–4.

Ma, X.M. and Aguilera, G. (1999a) Transcriptional responses of the vasopressin and corticotropin-releasing hormone genes to acute and repeated intraperitoneal hypertonic saline injection in rats. Mol. Brain Res., 68: 129–140.

Ma, X.M. and Aguilera, G. (1999b) Differential regulation of corticotropin-releasing hormone and vasopressin transcription by glucocorticoids. Endocrinology, 140: 5642–5650.

Ma, X.-M., Levy, A. and Lightman, S.L. (1997a) Rapid changes in heteronuclear RNA for corticotrophin-releasing hormone and arginine vasopressin in response to acute stress. J. Endocrinol., 152: 81–89.

Ma, X.-M., Levy, A. and Lightman, S.L. (1997b) Rapid changes of heteronuclear RNA for arginine vasopressin but not for corticotropin releasing hormone in response to acute corticosterone administration. J. Neuroendocrinol., 9: 723–728.

Ma, X.-M., Levy, A. and Lightman, S.L. (1997c) Emergence of an isolated arginine vasopressin (AVP) response to stress after repeated restraint: a study of both AVP and corticotropin-releasing hormone messenger ribonucleic acid (RNA) and heteronuclear RNA. Endocrinology, 138: 4351–4357.

Makino, S., Schulkin, J., Smith, M.A., Pacak, K., Palkovits, M. and Gold, P.W. (1995a) Regulation of corticotropin-releasing hormone receptor messenger ribonucleic acid in the rat brain and pituitary by glucocorticoids and stress. Endocrinology, 136: 4517–4525.

Makino, S., Smith, M.A. and Gold, P.W. (1995b) Increased expression of corticotropin-releasing hormone and vasopressin messenger ribonucleic acid (messenger RNA) in the hypothalamic paraventricular nucleus during repeated stress-association with reduction in glucocorticoid receptor messenger RNA levels. Endocrinology, 136: 3299–3309.

McCarthy, M.M., McDonald, C.H., Brooks, P.J. and Goldman, D. (1996) An anxiolytic action of oxytocin is enhanced by estrogen in the mouse. Physiol. Behav., 60: 1209–1215.

Meaney, M.J., O'Donnell, D., Rowe, W., Tannenbaum, B., Steverman, A., Walker, M., Nair, N.P. and Lupien, S. (1995) Individual differences in hypothalamic-pituitary-adrenal acitivity in later life and hippocampal aging. Exp. Gerontol., 30: 229–251.

Miller, M. (1987) Increased vasopressin secretion: an early life manifestation of aging in the rat. J. Gerontol., 42: 3–7.

Miyata, S., Itoh, T., Lin, S.H., Ishiyama, M., Nakashima, T. and Kiyohara, T. (1995) Temporal changes of c-fos expression in oxytocinergic magnocellular neuroendocrine cells of the rat hypothalamus with restraint stress. Brain Res. Bull., 37: 391–395.

Mohr, E. and Schmitz, E. (1991) Functional characterization of estrogen and glucocorticoid responsive elements in the rat oxytocin gene. Brain Res. Mol., Brain Res., 9: 293–298.

Muller, M.B., Landgraf, R., Preil, J., Sillaber, I., Kresse, A.E., Keck, M.E., Zimmermann, S., Holsboer, F. and Wurst, W. (2000) Selective activation of the hypothalamic vasopressinergic system in mice deficient for the corticotropin-releasing hormone receptor 1 is dependent on glucocorticoids. Endocrinology, 141: 4262–4269.

Murphy, H.M., Wideman, C.H. and Nadzam, G.R. (1998) The role of vasopressin in modulating circadian rhythm responses to phase shifts. Peptides, 19: 1191–1208.

Neumann, I.D. (2000) Antisense oligodeoxynucleotide effects on the hypothalamic-neurohypophysial system and the hypothalamic-pituitary-adrenal axis. Methods-A Companion to Methods in Enzymology, 22: 227–237.

Neumann, I.D. (2002) Involvement of the brain oxytocin system in stress coping: interactions with the hypothalamo-pituitary-adrenal axis. Prog. Br. Res., 139: 147–162.

Neumann, I., Douglas, A.J., Pittman, Q.J., Russell, J.A. and Landgraf, R. (1996) Oxytocin released within the supraoptic nucleus of the rat brain by positive feedback action is involved in parturition-related events. J. Neuroendocrinol., 8: 227–233.

Neumann, I.D., Johnstone, H.A., Hatzinger, M., Liebsch, G., Shipston, M., Russell, J.A., Landgraf, R. and Douglas, A.J. (1998) Attenuated neuroendocrine responses to emotional and physical stressors in pregnant rats involve adenohypophysial changes. J. Physiol., 508: 289–300.

Neumann, I.D., Kromer, S.A., Toschi, N. and Ebner, K. (2000a) Brain oxytocin inhibits the (re)activity of the hypothalamo-pituitary-adrenal axis in male rats: involvement of hypothalamic and limbic brain regions. Reg. Peptides, 96: 31–38.

Neumann, I., Russell, J.A. and Landgraf, R. (1993) Oxytocin and vasopressin release within the supraoptic and paraventricular nuclei of pregnant, parturient and lactating rats: a microdialysis study. Neuroscience, 53: 65–75.

Neumann, I.D., Torner, L. and Wigger, A. (2000b) Brain oxytocin: differential inhibition of neuroendocrine stress responses and anxiety-related behaviour in virgin, pregnant and lactating rats. Neuroscience, 95: 567–575.

Neumann, I.D., Toschi, N. and Douglas, A.J. (2003) No response of the hypothalamo-pituitary-adrenal axis to airpuff stress in parturient rats: lack of involvement of brain oxytocin. Endocrinology, 144(6): 2473–2479.

Neumann, I.D., Toschi, N., Ohl, F., Torner, L. and Kromer, S.A. (2001) Maternal defence as an emotional stressor in female rats: correlation of neuroendocrine and behavioural parameters and involvement of brain oxytocin. Eur. J. Neurosci., 13: 1016–1024.

Neumann, I.D., Wigger, A., Torner, L., Holsboer, F. and Landgraf, R. (2000c) Brain oxytocin inhibits basal and stress-induced activity of the hypothalamo-pituitary-adrenal axis in male and female rats: partial action within the paraventricular nucleus. J. Neuroendocrinol., 12: 235–243.

Newton, N., Peeler, D. and Newton, M. (1968) Effect of disturbance on labor. an experiment with 100 mice with dated pregnancies. Am. J. Obstet. Gynecol., 101: 1096–1102.

Nishioka, T., Anselmo-Franci, J.A., Li, P., Callahan, M.F. and Morris, M. (1998) Stress increases oxytocin release within the hypothalamic paraventricular nucleus. Brain Res., 781: 57–61.

Onaka, T. and Yagi, K. (1993) Effects of novelty stress on vasopressin and oxytocin secretion by the pituitary in the rat. J. Neuroendocrinol., 5: 365–369.

Ozawa, H., Nishimura, A., Xie, C.X. Akiyama, K. and Kawata, M. (2001) Influence of aging on the vasopressin and oxytocin neurons of the rat hypothalamus. WCNH proceedings Bordeaux: P2–14.

Pardon, M.C., Gerardin, P., Joubert, C., Perez-Diaz, F. and Cohen-Salmon, C. (2000) Influence of prepartum chronic ultramild stress on maternal pup care behavior in mice. Biol. Psychiatry, 47: 858–863.

Patchev, V.K., Schlosser, S.F., Hassan, A.H. and Almeida, O.F. (1993) Oxytocin binding sites in rat limbic and hypothalamic structures: site-specific modulation by adrenal and gonadal steroids. Neuroscience, 57: 537–543.

Patel, H., Chowdrey, H.S. and Lightman, S.L. (1991) Lactation abolishes corticotropin-releasing factor-induced oxytocin secretion in the conscious rat. Endocrinology, 128: 725–727.

Pearce, D., Matsui, W., Miner, J.N. and Yamamoto, M. (1998) Glucocorticoid receptor transcriptional activity determined by spacing of receptor and non-receptor DNA sites. J. Biol. Chem., 273: 30081–30085.

Petersson, M. (2002) Cardiovascular effects of oxytocin. Prog. Br. Res., 139: 281–288.

Petersson, M., Alster, P., Lundeberg, T. and Uvnas-Moberg, K. (1996) Oxytocin increases nociceptive thresholds in a long term perspective in female and male rats. Neurosci. Lett., 212: 87–90.

Petersson, M., Hulting, A.-L. and Uvnas-Moberg, K. (1999) Oxytocin causes a sustained decrease in plasma levels of corticosterone in rats. Neurosci. Lett., 264: 41–44.

Pinnock, S.B. and Herbert, J. (2001) Corticosterone differentially modulates expression of corticotropin releasing factor and arginine vasopressin mRNA in the hypothalamic paraventricular nucleus following either acute or repeated restraint stress. Eur. J. Neurosci., 13: 576–584.

Plotsky, P.M., Bruhn, T.O. and Vale, W. (1984) Central modulation of immunoreactive corticotropin-releasing factor secretion by arginine vasopressin. Endocrinology, 115: 1639–1641.

Purba, J.S., Hoogendijk, W.J.G., Hofman, M.A. and Swaab, D.F. (1996) Increased number of vasopressin and oxytocin-expressing neurons in the paraventricular nucleus of the hypothalamus in depression. Arch. Gen. Psychiatry, 53: 137–143.

Quinones-Jenab, V., Jenab, S., Ogawa, S., Adan, R.A.M., Burbach, J.P.H. and Pfaff, D.W. (1997) Effects of estrogen on oxytocin receptor messenger ribonucleic acid expression in the uterus, pituitary, and forebrain of the female rat. Neuroendocrinology, 65: 9–17.

Raadsheer, F.C., Sluiter, A.A., Ravid, R., Tilders, F.J. and Swaab, D.F. (1993) Localization of corticotropin-releasing hormone (CRH) neurons in the paraventricular nucleus of the human hypothalamus: age-dependent colocalization with vasopressin. Brain Res., 615: 50–62.

Rabadan-Diehl, C. and Aguilera, G. (1998) Glucocorticoids increase vasopressin V1b receptor coupling to phospholipase C. Endocrinology, 139: 3220–3226.

Renaud, L.P. (2000) Vasopressin. In: Fink, G. (Ed.), Enclyclopedia of Stress, Vol. 3. Academic Press, San Diego, pp. 650–656.

Rene, P. and De Keyzer, Y. (2002) The vasopressin receptor of cortioctroph pituitary cells. Prog. Br. Res., 139: 346–357.

Rivier, C., Rivier, J., Mormede, P. and Vale, W. (1984) Studies of the nature of the interaction between vasopressin and corticotropin-releasing factor on adrenocorticotropin release in the rat. Endocrinology, 115: 882–886.

Roberts, M.M., Robinson, A.G., Fitzsimmons, M.D., Grant, F., Lee, W.S. and Hoffman, G.E. (1993) c-fos expression in vasopressin and oxytocin neurons reveals functional heterogeneity within magnocellular neurons. Neuroendocrinology, 57: 388–400.

Roozendaal, B., Wiersma, A., Driscoll, P., Koolhaas, J.M., Bohus, B. and Wiersma, A. (1992) Vasopressinergic modulation of stress responses in the central amygdala of the roman high-avoidance and low-avoidance rat. Brain Res., 596: 35–40.

Russell, J.A., Ma, S., Shipston, M.J., Landgraf, R., Wigger, A., Neumann, I.D., Douglas, A.J. and Brunton, P.J. (2000) Global reduction in hypothalamo-pituitary-adrenal (HPA) axis stress responses in pregnant rats: attenuated feedforward mechanisms. ICE 2000 Abstracts: P7.

Saameli, K. (1968) The circulatory actions of the neurohypophysial hormones. In: Berde, B. (Ed.), Neurohypophysial hormones and similar polypeptides. Springer-Verlag, Berlin, pp. 545–612.

Sapolsky, R.M. (1991) Do glucocorticoid concentrations rise with age in the rat? Neurobiol. Aging, 13: 171–174.

Saridaki, E. and Lightman, S.L. (2003) Naloxone increases the oxytocin response to stress in lactating rats. In: Thorn, N.A. Vilhardt, H. and Treiman, M. (Eds.), Proceedings of the 4th International Conference on the Neurohypophysis. Oxford University Press, Oxford, pp. 32–34.

Sawchenko, P.E., Swanson, L.W. and Vale, W.W. (1984a) Co-expression of corticotropin-releasing factor and vasopressin immunoreactivity in parvocellular neurosecretory neurons of the adrenalectomized rat. Proc. Natl. Acad. Sci. USA, 81: 1883–1887.

Sawchenko, P.E., Swanson, L.W. and Vale, W.W. (1984b) Corticotropin-releasing factor: co-expression within distinct subsets of oxytocin-, vasopressin-, and neurotensin-immunoreactive neurons in the hypothalamus of male rats. J. Neurosci., 4: 1118–1129.

Sawchenko, P.E., Arias, C.A. and Mortrund, M.T. (1993) Local tetrodotoxin blocks chronic stress effects on corticotrophin releasing factor and vasopressin messenger ribonucleic acid in hypophysiotrophic neurons. J. Neuroendocrinol., 5: 341–348.

Scaccianoce, S., di Sciullo, A. and Angelicci, L. (1990) Age-related changes in hypothalamo-pituitary-adrenal axis activity in the rat. Neuroendocrinology, 52: 150–155.

Scaccianoce, S., Muscolo, L.A., Cigliana, G., Navarra, D. Nicolai, R. and Angelucci, L. (1991) Evidence for a specific role of vasopressin in sustaining pituitary-adrenocortical stress response in the rat. Endocrinology, 128: 3138–3143.

Schule, R., Rangarajan, P., Kliewer, S., Ransone, L.J., Bolado, J., Yang, N., Verma, I.M. and Evans, R.M. (1990) Functional antagonism between oncoprotein c-jun and the glucocorticoid receptor. Cell, 62: 1217–1226.

Seeman, T.E. and Robbins, J. (1994) Aging and hypothalamo-pituitary-adrenal response to challenge in humans. Endocr. Rev., 15: 233–260.

Shen, E., Dun, S.L., Ren, C., Bennettclarke, C. and Dun, N.J. (1992) Hypotension preferentially induces c-fos immunoreactivity in supraoptic vasopressin neurons. Brain Res., 593: 136–139.

Silverman, W.F., Aravich, P.F., Sladek, J.R. and Sladek, C. (1990) Physiological and biochemical indices of neurohypophyseal function in aging the Fischer rat. Neuroendocrinology, 52: 181–190.

Stachowiak, A., Macchi, C., Nussdorfer, G.G. and Malendowicz, L.K. (1995) Effects of oxytocin on the function and morphology of the rat adrenal cortex:in vitro and in vivo investigations. Res. Exp. Med., 195: 265–274.

Stauber, C., Altschmeid, J., Akerblom, I.E., Marron, J.L. and Mellon, P.L. (1990) Mutual cross-reference between glucocorticoid receptor and CREB inhibits transactivation in placental cells. New Biologist, 4: 527–540.

Summerlee, A.J.S. (1989) Relaxin, opioids and the timing of birth in rats. In: Dyer, R.G. and Bicknell, R.J. (Eds.), Brain Opioid Systems in Reproduction. Oxford University Press, pp. 257–270.

Swanson, L.W. and Simmons, D.M. (1989) Differential steroid hormone and neural influences on peptide mRNA levels in CRH cells of the paraventricular nucleus: a hybridization histochemical study in the rat. J. Comp. Neurol., 285: 413–435.

Terwel, D., Markerink, M. and Jolles, J. (1992) Age-related changes in concentrations of vasopressin in the central nervous system and plasma of the male Wistar rat. mech. Ageing Devel., 65: 127–136.

Torner, L., Thrivikraman, K.V., Toschi, N., Holsboer, F., Plotsky, P.M. and Neumann, I.D. (2000) Effect of adrenalectomy and corticosterone on hypothalamic and peripheral release of oxytocin and behavioral stress coping in male rats. Soc. Neuroscience Abstracts, 30: P 153.12.

Toufexis, D.J., Tesolin, S., Huang, N. and Walker, C.D. (1999) Altered pituitary sensitivity to corticotropin-releasing factor and arginine vasopressin participates in the stress hyporesponsiveness of lactation in the rat. J. Neuroendocrinol., 11: 757–764.

Tribollet, E., Barberis, C. and Jard, S. (1988) Localization and pharmacological characterization of high affinity binding sites for vasopressin and oxytocin in the rat brain by light microscopic autoradiography. Brain Res., 442: 105–118.

Turner, R.A., Altemus, M., Yip, D.N., Kupferman, E., Fletcher, D., Bostrom, A., Lyons, D.M. and Amico, J.A. (2002) Effects of emotion on oxytocin, prolactin and ACTH in women. Stress, 5: 269–276.

Uht, R.M., McKelvy, J.F., Harrison, R.W. and Bohn, M.C. (1988) Demonstration of glucocorticoid receptor-like immunoreactivity in glucocorticoid-sensitive vasopressin and corticotropin releasing factor neurons in the hypothalamic paraventricular nucleus. J. Neurosci. Res., 19: 405–411.

Unlap, T. and Jope, R.S. (1994) Dexamethasone attenuates kainate-induced AP-1 activation in rat brain. Mol. Brain Res., 24: 275-282.

Ur, E. and Grossman, A. (1994) The neuroregulation of corticotropin secretion. In: Imura, H. (Ed.), The Pitutiary Gland. Raven Press, New York, pp. 309–330.

Uvnas-Moberg, K. (1998a) Oxytocin may mediate the benefits of positive social interaction and emotions. Psychoneuroendocrinology, 23: 819–836.

Uvnas-Moberg, K. (1998b) Anti-stress pattern induced by oxytocin. News Physiol. Sci., 13: 22–26.

Uvnas-Moberg, K., Ahlenius, S., Hillegaart, V. and Alster, P. (1994) High doses of oxytocin cause sedation and low doses cause an anxiolytic-like effect in male rats. Pharmacol. Biochem. Behav., 49: 101–106.

van Bardeleben, U. and Holsboer, F. (1991) Effect of age on the cortisol response to human corticotropin-releasing hormone ion depressed patients pretreated with dexamethasone. Biol. Psychiatry, 29: 1042–1050.

Van der Woude, P.F., Goudsmit, E., Wierda, M., Purba, J.S., Hofman, M.A., Bogte, H. and Swaab, D.F. (1995) No vasopressin cell loss in the human hypothalamus in aging and Alzheimer's disease. Neurobiol. Aging, 16: 11–18.

van Eekelen, J.A.M., Rots, N.Y., Sutanto, W. and De Kloet, E.R. (1991) The effect of aging on stress responsiveness and central corticosteroid receptors in the Brown Norway rat. Neurobiol. Aging, 13: 159–170.

van Londen, L., Goekoop, J.G., van Kempen, G.M., Frankhuijzen-Sierevogel, A.C., Wiegant, V.M., van, d., V and De Wied, D. (1997) Plasma levels of arginine vasopressin elevated in patients with major depression. Neuropsychopharmacology, 17: 284–292.

Vellucci, S.V. and Parrott, R.F. (1997) Vasopressin and oxytocin gene expression in the porcine forebrain under basal conditions and following acute stress. Neuropeptides, 31: 431–438.

Vellucci, S.V. and Parrott, R.F. (1998) Expression of mRNAs for vasopressin, oxytocin and corticotrophin releasing hormone in the hypothalamus and of cyclooxygenases-1 and -2 in the cerebral vasculature, of endotoxin challenged pigs. Neuropeptides, 32: 439–446.

Verbalis, J.G., Baldwin, E.F., Ronnekleiv, O.K. and Robinson, A.G. (1986) In vitro release of vasopressin and oxytocin from rat median eminence tissue. Neuroendocrinology, 42.

Verbalis, J.G., Mangione, M.P. and Stricker, E.M. (1991) Oxytocin produces natriuresis in rats at physiological plasma concentrations. Endocrinology, 128: 1317–1322.

Viau, V. (2002) Functional cross-talk between the hypothalamic-pituitary-gonadal and -adrenal axes. J. Neuroendocrinol., 14: 506–513.

Viau, V., Chu, A., Soriano, L. and Dallman, M.F. (1999) Independent and overlapping effects of corticosterone and testosterone on corticotropin-releasing hormone and arginine vasopressin mrna expression in the paraventricular nucleus of the hypothalamus and stress-induced adrenocorticotropic hormone release. J. Neurosci., 19: 6684–6693.

Viau, V., Soriano, L. and Dallman, M.F. (2001) Androgens alter corticotropin releasing hormone and arginine vasopressin mRNA within forebrain sites known to regulate activity in the hypothalamic-pituitary-adrenal axis. J. Neuroendocrinol., 13: 442–452.

Wadhwa, P.D., Porto, M., Garite, T.J., Chicz-DeMet, A. and Sandman, C.A. (1998) Maternal corticotropin-releasing hormone levels in the early third trimester predict length of gestation in human pregnancy. Am. J. Obstet. Gynecol., 179: 1079–1085.

Wadhwa, P.D., Sandman, C.A. and Garite, T.J. (2001) The neurobiology of stress in human pregnancy:implications for prematurity and development of the fetal central nervous system. Prog. Br. Res., 133: 131–142.

Wakerley, J.B. and Lincoln, D.W. (1973) The milk-ejection reflex of the rat: a 20- to 40-fold acceleration in the firing of paraventricular neurones during oxytocin release. J. Endocrinol., 57: 477–493.

Walker, C.D. (1995) Neuroendocrine changes associated with the adrenocortical system during lactation in the rat. Annal. Endocrinol., 56: 169–172.

Walker, C.-D., Toufexis, D.J. and Burlet, A. (2001a) Hypothalamic and limbic expression of CRH and vasopressin during lactation:implications for the control of ACTH secretion and stress hyporesponsiveness. Prog. Br. Res., 133: 99–110.

Walker, C.D., Tilders, F.J.H. and Burlet, A. (2001b) Increased colocalization of corticotropin-releasing factor and arginine vasopressin in paraventricular neurones of the hypothalamus in lactating rats:evidence from immunotargeted lesions and immunohistochemistry. J. Neuroendocrinol., 13: 74–85.

Watanabe, T., Oki, Y. and Orth, D.N. (1989) Kinetic actions and interactions of arginine vasopressin, angiotensin-II and oxytocin on adrenocorticotropin secretion by rat anterior pituitary cells in the microperifusion system. Endocrinology, 125: 1921–1931.

Webster, J.C. and Cidlowski, J.A. (1999) Mechanisms of glucocorticoid-receptor-mediated repression of gene expression. Trends Endocrinol. Metab., 1: 396–402.

Weinstock, M. (1997) Does prenatal stress impair coping and regulation of hypothalamic- pituitary-adrenal axis? Neurosci. Biobehav. Rev., 21: 1–10.

Whitnall, M.H. (1993) Regulation of the hypothalamic corticotropin-releasing hormone neurosecretory system. Prog. Neurobiol., 40: 573–629.

Wigger, A. and Neumann, I.D. (2002) Endogenous opioid regulation of stress-induced oxytocin release within the hypothalamic paraventricular nucleus is reversed in late pregnancy: a microdialysis study. Neuroscience, 112: 121–129.

Wigger, A., Lorscher, P., Oehler, I., Keck, M.E., Naruo, T. and Neumann, I.D. (1999) Nonresponsiveness of the rat hypothalamo-pituitary-adrenocortical axis to parturition-related events: inhibitory action of endogenous opioids. Endocrinology, 140: 2843–2849.

Windle, R.J., Herlihy, Y.M., Wood, S.A., Shanks, N. and Lightman, S.L. (1997a) Central actions of oxytocin and vasopressin on behavioural and endocrine responses to stress in rats. J. Physiol., 505: 11–12.

Windle, R.J., Shanks, N., Lightman, S.L. and Ingram, C.D. (1997b) Central oxytocin administration reduces stress-induced corticosterone release and anxiety behavior in rats. Endocrinology, 138: 2829–2834.

Winslow, J.T. and Insel, T.R. (2002) The social deficits of the oxytocin knockout mouse. Neuropeptides, 36: 221–229.

Winslow, J.T., Hearn, E.F., Ferguson, J., Young, L.J., Matzuk, M.M. and Insel, T.R. (2000) Infant vocalization, adult aggression and fear behavior of an oxytocin null mutant mouse. Hormones Behav., 37: 145–155.

Wolfson, B., Manning, R.W., Davis, L.G., Arentzen, R. and Baldino, J.F. (1985) Co-localization of corticotrophin-releasing factor and vasopressin mrna in neurons after adrenalectomy. Nature, 315: 59–61.

Wotjak, C.T., Ganster, J., Kohl, G., Holsboer, F., Landgraf, R. and Engelmann, M. (1998a) Dissociated central and peripheral release of vasopressin, but not oxytocin, in response to repeated swim stress: new insights into the secretory capacities of peptidergic neurons. Neuroscience, 85: 1209–1222.

Wotjak, C.T., Kubota, M., Engelmann, M., Neumann, I. and Landgraf, R. (1995) Physical and emotional stressors stimulate the release of vasopressin within the hypothalamic paraventricular nucleus of rats: a microdialysis study. Soc. Neuroscience Abstracts, 21: 806.7.

Wotjak, C.T., Kubota, M., Kohl, G. and Landgraf, R. (1996a) Release of vasopressin from supraoptic neurons within the median eminence in vivo. a combined microdialysis and push-pull perfusion study in the rat. Brain Res., 726: 237–241.

Wotjak, C.T., Kubota, M., Liebsch, G., Montkowski, A., Holsboer, F., Neumann, I. and Landgraf, R. (1996b) Release of vasopressin within the rat paraventricular nucleus in response to emotional stress: a novel mechanism of regulating adrenocorticotropic hormone secretion? J. Neurosci., 16: 7725–7732.

Wotjak, C.T., Ludwig, M., Ebner, K., Russell, J.A., Singewald, N., Landgraft, R. and Engelmann, M. (2002) Vasopressin from hypothalamic magnocelleular neurons has opposite actions at the adenohypophysis and in the supraoptic nucleus on ACTH secretion. Eur. J. Neurosci., 16: 477–485.

Wotjak, C.T., Naruo, T., Muraoka, S., Landgraf, R. and Engelmann, M. (1998b) Swim stress alters the expression of vasopressin and oxytocin mRNAs in the rat paraventricular nucleus. Eur. J. Neurosci., 10: 11625.

Wotjak, C.T., Naruo, T., Muraoka, S., Simchen, R., Landgraf, R. and Engelmann, M. (2001) Forced swimming stimulates the expression of vasopressin and oxytocin in magnocellular neurons of the rat hypothalamic paraventricular nucleus. Eur. J. Neurosci., 13: 2273–2281.

Yang-Yen, H.-F., Chambard, J.-C., Sun, Y.-L., Smeal, T., Schmidt, T.J., Drouin, J. and Karin, M. (1990) Transcriptional interference between c-jun and the glucocorticoid receptor:mutual inhibition of DNA binding due to direct protein-protein interaction. Cell, 62: 1205–1215.

Young, L.J., Muns, S., Wang, Z.X. and Insel, T.R. (1997) Changes in oxytocin receptor mRNA in rat brain during pregnancy and the effects of estrogen and interleukin-6. J. Neuroendocrinol., 9: 859–865.

Young, W.S., Mezey, E. and Siegel, R.E. (1986a) Vasopressin and oxytocin mRNAs in adrenalectomized and brattleboro rats: analysis by quantitative in situ hybridization histochemistry. Brain Res., 387: 231–241.

Young, W.S., Mezey, E. and Siegel, R.E. (1986b) Quantitative in situ hybridization histochemistry reveals increased levels of corticotropin-releasing factor mRNA after adrenalectomy in rats. Neurosci. Lett., 70: 198–203.

T. Steckler, N.H. Kalin and J.M.H.M. Reul (Eds.)
Handbook of Stress and the Brain, Vol. 15
ISBN 0-444-51173-3

CHAPTER 2.6

The role of vasopressin in behaviors associated with aversive stimuli

Kathleen C. Chambers[1,*] and UnJa L. Hayes[2]

[1]*University of Southern California, Dept. Psychol-SGM 501, Los Angeles, CA 90089, USA*
[2]*University of Massachusetts, Dept. Psychol – Tobin Hall, Amherst, MA 01003, USA*

Abstract: When an individual encounters stressors such as physical pain, internal malaise, or drought conditions, a constellation of physiological and behavioral responses are evoked that aid the individual in coping with the stressors and learning strategies to minimize exposure to these stressors. Vasopressin (VP) plays an important role in modulating these responses and this role is complex. Acute painful stimuli such as electric footshock stimulate hormonal and neural release of VP and this in turn can contribute to analgesia and facilitate maintenance of behaviors that allow the individual to avoid future painful encounters in the same environmental context. The effect of VP on avoidance maintenance is consistent no matter when during the stressful situation it is released. Two hypotheses have emerged to account for the action of VP on shock avoidance maintenance. One states that central VP is necessary for optimal mnemonic processing of aversive environmental stimuli and the other states that VP produces nonspecific effects on arousal systems that modulate memory processes. There is evidence to support both positions. Internal malaise induced by consumption of toxic substances also stimulates hormonal release of VP and this in turn facilitates a reduction in exposure to the toxic substance and modulates maintenance of behaviors that avoid future contact with the toxic substance. The hormonal and neural release of low levels of VP promotes relinquishing avoidance behavior when it is present after the first exposure to the stressor but the release of high levels of VP facilitates maintenance. It is suggested that low doses of VP reduce the sensory impact of toxins while high doses of VP have aversive stimulus properties that are additive with the properties of toxins. Finally, scarcity of water stimulates hormonal release of VP, which directly promotes water retention. The neural changes in VP release during the conditions of water restriction can affect avoidance behavior maintenance when an individual encounters other stressors. It facilitates maintaining avoidance behavior for physical pain stressors and relinquishing avoidance behavior for internal malaise stressors. The picture emerging is that VP has variable effects on learning and memory processes. Some of these effects are due to its direct action on the neural systems mediating these processes while other effects are due to its action on modulatory neural systems, such as those mediating the aversive properties of VP and those mediating arousal.

Introduction

Stressors trigger the hormonal release of vasopressin (VP) from the posterior pituitary and stimulate or inhibit the neuromodulator release in a number of different nuclei in the brain. As a hormone, VP is synthesized in the supraoptic and paraventricular hypothalamic nuclei and released into the general circulatory system from axons terminating near capillaries in the posterior pituitary (Swanson and Sawchenko, 1980). As a neuromodulator, it is synthesized in and released from neurons in the suprachiasmatic hypothalamus, bed nucleus of the stria terminalis (BST), and medial amygdala as well as the supraoptic and paraventricular (PVN) hypothalamic nuclei (Bargmann and Scharrer, 1951;

*Corresponding author. Fax: +1(213) 746 9082;
E-mail: Kchamber@usc.edu

Buijs, 1978; van Leeuwen and Caffe, 1983; Sofroniew, 1985; Caffe et al., 1987; Jolkkonen et al., 1988; Miller et al., 1988). These synthesis sites project either directly or indirectly to a number of different areas of the brain known to be critical for the expression of various behaviors.

The pioneering work of DeWied and his colleagues revealed that increased hormonal and neuromodulator release of VP has broader effects than its putative role in fluid balance (Verney, 1947). It is involved in a wide range of behaviors, which include defensive, eating, and parental behaviors (Bamshad et al., 1993; Wang et al., 1998; Bray, 2000; McCann et al., 2000), and it modulates performance in a number of different kinds of learning situations, which include appetitive, spatial, social, and aversive situations (van Wimersma Greidanus, et al., 1986; DeWied et al., 1993; Dietrich and Allen, 1997; Dantzer, 1998; Koolhaas et al., 1998; Popik and van Ree, 1998; Alescio-Lautier et al., 2000). Because a detailed review of the extensive literature on the influence of VP on behavior is beyond the scope of this chapter, the discussion is focused on behaviors that are part of the defensive system and are associated with pain and malaise. Particular emphasis is placed on shock and taste avoidance learning paradigms.

Behaviors associated with pain

Pain-elicited responses

Effects of pain on vasopressin release

Various pain stressors have been reported to increase plasma VP levels in a number of different species. For example, plasma VP levels increase in humans after exposure to pain stressors such as surgical stress and they are elevated in individuals with chronic pain disorder (Furyua et al., 1993; Wahlbeck et al., 1996). In rats, acute and continuous footshock (3 mA for 60 s) produces an immediate increase in VP levels (Onaka et al., 1986a,b). However, certain painful experiences produce a type of emotional stress that actually reduces the release of VP. Increases in plasma VP levels are either reduced or suppressed when rats are exposed to intermittently applied footshock (3 mA at 2.0 s durations every 15 s or 1 mA at 0.5 s durations every 5 s; Onaka et al., 1986a; Callahan et al., 1992). The attenuation in VP release is not due to a depletion of VP in the pituitary, since subsequent intraperitoneal injection of hypertonic saline or urethane 10 min after termination of the intermittent footshock produces a significant increase in VP levels. Other types of emotionally stressful situations also have been found to suppress VP release, e.g., social defeat, forced swimming, and exposure to learned fear stimuli, that is, environmental stimuli that previously had been paired with footshock (Landgraf et al., 1998; Yagi et al., 1998; Engelmann et al., 2000). The failure of emotional stressors to trigger the hormonal release of VP does not mean that there is no involvement of VP. A number of stressors, including social defeat and forced swimming, trigger the release of VP within the hypothalamus (paraventricular, supraoptic and suprachiasmatic nuclei) and limbic areas (septum and amygdala) but fail to cause the release of VP into the general circulation (Landgraf et al., 1998; Engelmann et al., 2000).

Evidence supports the involvement of noradrenergic projections to the BST in the hormonal suppression induced by intermittent footshock (Onaka and Yagi, 1998). When these projections are destroyed by infusion of a selective neurotoxin into the BST, the suppression of hormonal VP is abolished. However, the picture for the suppression of hormonal VP that occurs in response to learned fear stimuli is more complicated. Exposure to these stimuli suppresses the hormonal release of VP and increases noradrenergic activity in the BST (Yagi et al., 1998). However, the hormonal suppression is not abolished when the noradrenergic fibers are destroyed (Onaka and Yagi, 1998). This suggests that either the noradrenergic fibers in the BST are not involved in the suppression of hormonal VP in response to learned fear stimuli or suppression is mediated by the BST as well as some other neural pathway.

Effects of vasopressin on pain-elicited behaviors

Immunocytochemical, radioimmunoassay, and retrograde tracing procedures have revealed that vasopressinergic neurons project to and receive afferent connections from various areas known to be involved in nociception (Buijs, 1978; Swanson and Sawchenko, 1980; Pittman et al., 1981; Sawchenko and Swanson, 1982; Hallbeck and Blomqvivt, 1999). Studies of the possible involvement of VP in the behavioral response to pain have employed a number of different kinds of measures. These have included tail-flick latency, which is the amount of time a rat keeps its tail under a high level of radiant heat, and flinch-jump threshold, which is the lowest shock intensity needed to elicit the withdrawal of a paw from an electrified grid. The results from the studies using these two measures have been inconsistent, but when VP does have an effect, it is in the direction of increased analgesia, that is, longer tail-flick latencies and higher flinch-jump thresholds.

Both systemic and central injections increase analgesia as measured by tail-flick latency. Subcutaneous injections of VP and intracerebroventricular infusion of VP or a VP analogue produce analgesia, which can be reduced by central infusions of VP antiserum (Berntson and Berson, 1980; Bodnar et al., 1982; Kordower et al., 1982). On the other hand, neither systemic nor central injections of VP are effective in producing analgesia as measured by flinch-jump threshold (Kordower et al., 1982). However, there is evidence that normal circulating levels of VP contribute to analgesia. Homozygous Brattleboro rats, which are derived from the Long-Evans strain, have an almost complete lack of detectable VP in the brain and bloodstream and suffer from diabetes insipidus. They also have lower flinch-jump thresholds than their Long-Evans counterparts both before and after a forced swim in cold water, a procedure that typically reduces pain sensitivity (Sokol et al., 1976; Vandersande and Dierickx, 1976; Bodnar et al., 1980, 1982). This impairment can be partially restored to levels exhibited by Long-Evans rats after systemic administration of VP. Perhaps more consistent results across the different behavioral measures of analgesia will be obtained when VP is infused directly into specific central sites. Recent studies using microinjection techniques have identified the central nucleus of the amygdala as a site for at least some VP-mediated analgesic effects (Ahn et al., 2001).

Shock avoidance learning

Studies of the influence of VP on avoidance learning have employed two categories of shock avoidance conditioning: passive and active. Passive shock avoidance learning situations are usually employed in a dynamic way, that is, animals are first exposed to an environment under one kind of stimulus condition and then the stimulus condition of the environment is changed. For example, an animal is placed on a raised and lighted platform above a grid floor or in a lighted chamber with an adjoining darkened chamber. For the first few trials, it is allowed to step down to the grid floor or step into the darkened chamber without aversive consequences. After obtaining information that the adjoining environment is safe, it is then shocked. In subsequent trials, the animal again does not receive a footshock when it moves to the adjoining environment and the animal eventually extinguishes its avoidance of this environment. The timing of the first post-acquisition test ranges from a few hours after the acquisition trial to a couple of days.

In the typical active shock avoidance learning situation, an auditory or visual stimulus is presented briefly and then an aversive stimulus such as electric footshock is presented while the auditory or visual stimulus is still present. In order to terminate exposure to the shock, an animal must learn a route of escape to another safe environment that is provided by the experimenter. Examples of these routes include jumping over a partition to an adjoining compartment and jumping onto a pole that extends out of the center of a grid floor emitting the shock. The dangerous and safe environments are distinctly different, and which environment is dangerous and which is safe remains consistent. An animal makes the transition from escape to avoidance when it learns that the auditory or light stimulus predicts the occurrence of footshock and it takes the route of escape before the footshock is activated. In another active avoidance situation, the two-way shuttle box, the animal must learn to move back and forth between two compartments to avoid shock when an auditory or light stimulus is presented. During the transition from escape to avoidance, the animal experiences shock in both chambers, which means that both chambers are unsafe. The animal must learn that if it moves to the unoccupied chamber when the auditory or light stimulus is presented, it can avoid being shocked. All of these learned avoidance behaviors can be extinguished when shock is no longer given after presentation of the auditory or visual stimulus. During both the acquisition and extinction phases, animals are given a number of conditioning sessions (e.g., 3 sessions), each of which has a specified number of trials (e.g., 10 trials/session), across a number of days (e.g., 3 days).

Effects of vasopressin on shock avoidance learning

Effects of peripheral administration of vasopressin

Most of the evidence shows that VP facilitates maintenance of the shock avoidance memory. Although VP does not influence the acquisition of passive or active shock avoidance learning, it increases the likelihood that animals will show avoidance behavior when tested 24 and 48 h after acquisition of a passive avoidance task and it prolongs extinction of both passive and active shock avoidance. These effects have been found when systemic injections of VP are given before or after the one acquisition trial in passive shock avoidance and the last acquisition session in active shock avoidance or before or after the first extinction trial in both passive and active shock avoidance (DeWied, 1971, 1976; Bohus et al., 1972, 1973; Koob, et al., 1981; Lebrun et al., 1985a; Gaffori and DeWied, 1985; Gunther et al., 1988; DeWied et al., 1991; Faiman et al., 1992).

We have replicated the prolonged extinction results using a one-trial step-through passive avoidance situation (see Ader and DeWied, 1972; Bohus et al., 1972; DeWied et al., 1991). Male rats were injected subcutaneously with saline or 2 different doses of VP (6 and 18 μg/kg body weight) immediately after the acquisition trial and they were given 3 post-acquisition tests 0, 24, and 48 h later. The first post-acquisition test is a measure of acquisition and the ability to maintain the avoidance behavior. In this test, a significant difference between saline and VP males was not evident (see Fig. 1). Because the animals did not receive shock after the first post-acquisition test, the two post-acquisition tests given 24 and 48 h after acquisition were measures of extinction. The results clearly showed that by the second extinction test, the reentry latencies were shorter in the saline than VP males. Similar results have been obtained for VP doses of 3 and 9 ug/kg doses (Hayes and Chambers, in press, unpublished data).

Effects of central administration of vasopressin

Central administration of VP has similar effects on maintenance of shock avoidance as systemic administration. VP facilitates maintenance of the

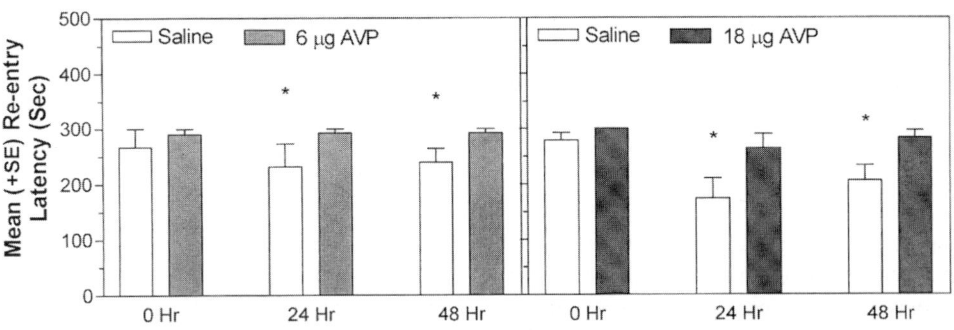

Fig. 1. Mean (+SE) latency to reenter a chamber in male rats tested 0, 24, and 48 h after experiencing electric footshock in the chamber. The males were given subcutaneous injections of 6 or 18 μg/kg of arginine vasopressin (AVP) or saline immediately after experiencing the footshock. *Significantly shorter latency than AVP during the given post-acquisition test, $p \leq 0.04$.

avoidance behavior when it is infused into the lateral ventricles shortly before the first post-acquisition test in passive shock avoidance situations and immediately after each acquisition session or immediately after the first extinction session in active shock avoidance tasks (Bohus et al., 1978; Koob et al., 1981, 1986; DeWied et al., 1984; DeJong et al., 1985, 1986; Kovács et al., 1986; Bernstein et al., 1994; Winnicka and Wisniewski, 1999). Antagonists of the V_1 receptor can block the facilitation of passive avoidance maintenance induced by VP, which suggests that the central effect of VP on passive avoidance is mediated by the V_1 receptor (Boccia et al., 1998; Klimkiewicz, 2001). However, a novel type of receptor may mediate the effects of VP fragments (Reijmers et al., 1998; Nakayama et al., 2000). Because VP fragments apparently have no peripheral effects and their function seems to be tied specifically to memory-associated behaviors, it has been proposed that the VP fragment receptor is a link between the neuropeptide VP and its effect on these behaviors (Reijmers et al., 1998).

Intracerebroventricular administration of VP has been found to be 440 times more potent than when given subcutaneously. The ED50, or the amount necessary to produce a behavioral effect in 50% of the test subjects, has been found to be 0.57 μg for subcutaneous injections and 1.3 ng for intracerebroventricular infusions (DeWied et al., 1984). DeWied et al. (1984) have concluded that the potency of intracerebroventricular injections, as reflected in the small amount of VP needed to affect avoidance maintenance, limits the site of action to the central nervous system. This conclusion is supported by antiserum, lesion, and local infusion studies. Microinjection of VP antiserum into the dorsal hippocampus and lesions of the rostral septal nuclei and anterodorsal hippocampus attenuate the effects of VP on passive and active shock avoidance (DeWied et al., 1976; van Wimersma Greidanus and DeWied, 1976a; Kovács et al., 1982). There is facilitation of passive shock avoidance maintenance after local infusion of VP into the hippocampal dentate gyrus, dorsal septal area, and dorsal raphe nucleus immediately after the single acquisition trial or 1 h before the first retention test (Kovács et al., 1979, 1986). In addition, direct administration of VP or its fragments into specific regions of the brain have been found to facilitate maintenance of other learning tasks (Engelmann et al., 1996; Alescio-Laurier and Soumireu-Mourat, 1998; Orowaska-Majdak et al., 2001).

Because both peripheral and central administration of VP can maintain avoidance behavior, the question of how peripherally administered VP influences neural areas mediating avoidance behavior arises. Although VP can cross the cerebrospinal blood-brain barrier, its entry is limited and it is degraded quickly, which suggests that it does not act directly on the neural areas that modulate avoidance behavior (Zaidi and Heller, 1974; Ang and Jenkins, 1982; Mens et al., 1983). Peripheral VP may affect these neural areas by acting via peripheral entry routes, such as the vagus or splanchnic nerves or via a circumventricular route such as the area postrema, which contains V_1 receptors (Phillips et al., 1988; Gerstberger and Fahrenholz, 1989; Raggenbass et al., 1989; Barberis et al., 1995; Tribollet et al., 1998). It should be noted, however, that VP could be degraded into more potent fragments and these fragments could act directly on the neural areas modulating avoidance behavior (Burbach et al., 1984; DeWied et al., 1987).

Vasopressin and compromised memory models

Retrograde amnesia is the inability to recall information or events acquired before neurological damage or trauma. In experimental settings, retrograde amnesia is induced by a number of different procedures, e.g., chronic alcohol consumption prior to acquisition of the learned avoidance and electroconvulsive shock, hypoxia, intercisternal injection of glutamic acid or NMDA, cycloheximide or beta-endorphin injection, and intracerebroventricular infusion of beta-amyloid protein-(1–40) after acquisition of the learned avoidance. Amnesia is evidenced by a shortened latency to respond in passive avoidance tasks and a failure to display avoidance in active avoidance tasks. Administration of VP can reduce or reverse these effects. Amnesic animals treated with VP or VP fragments exhibit longer passive avoidance latencies and a higher number of active avoidance responses than untreated amnesics; acquisition is not affected (Rigter et al., 1974; Car et al., 1993, 1994; Hirate et al., 1997; Izquierdo et al., 1997; Tanaka et al., 1998; Sato et al., 1999).

Role of endogenous vasopressin

Although most studies have focused on the effects of increases in the peripheral and central release of VP, the possible involvement of the normal presence of VP in learning situations also has been addressed. Two approaches have been used in pursuing this issue, Brattleboro rats and antivasopressin serum. The results from studies of Brattleboro rats suggest that the normal presence of VP modulates both learning and memory processes. In a one-trial step-through passive avoidance task, homozygous Brattleboro rats showed reduced ability to maintain the avoidance than heterozygous Brattleboro rats, a variant with levels of VP intermediate between the homozygous variant and the Long-Evans parent strain and with relatively normal water metabolism (Bohus et al., 1975; DeWied et al., 1975). Treating the homozygous Brattleboro rats with arginine VP or desglycinamide lysine-8-vasopressin (DG-LVP) significantly improved retention. Because DG-LVP has minimal peripheral activity, the improved performance was not due to a correction in water metabolism. In both the shuttle box and pole jump active shock avoidance tasks, homozygous Brattleboro rats acquired the avoidance at a slower rate than Wistar rats and once learned, they extinguished it at a faster rate (Bohus et al., 1975; Baily and Weiss, 1981; Liard, 1988). These differences were not based on differences in shock sensitivity or activity levels because the homozygous Brattleboro and Wistar rats did not differ in the appearance of flinch and jerk responses when given inescapable footshock or in ambulation, grooming, rearing, and defecation responses when placed in an open field (DeWied et al., 1975).

One of the problems with using Brattleboro rats is that it is unclear whether the deficits found in shock avoidance are the direct result of the absence of VP or due to some other effect of the genetic abnormality. Studies using antivasopressin serum in rats with an intact VP system have provided additional evidence that the normal presence of VP modulates memory processes but they have failed to support the role of VP in learning processes (van Wimersma Greidanus et al., 1975a,b; van Wimersma Greidanus and DeWied, 1976b; Leccese and Isenhour, 1983; Croiset et al., 1990). In these studies, deficits were found in maintenance of passive and active shock avoidance, but not in acquisition. Furthermore, impaired avoidance behavior was found only when central levels of VP were depleted.

Interpretation of the effects of vasopressin on shock avoidance learning

Examination of numerous studies across different investigators reveals that for the majority of animals tested, VP facilitates maintenance of passive and active shock avoidance. Many investigators have concluded that VP facilitates consolidation of the learned avoidance memory by acting directly on neural areas mediating mnemonic processes and that central VP is necessary for optimal mnemonic processing of aversive environmental events (van Wimersma Greidanus et al., 1975a, 1986; Leshner et al., 1978; Leccese and Isenhour, 1983; DeWied et al., 1988). But not all investigators agree with this interpretation, suggesting instead that VP produces nonspecific effects that modulate memory processes (Sahgal et al., 1982; Sahgal and Wright, 1983; Lebrun, et al., 1985b; Skopkova et al., 1991). These suggested effects have included state dependency, general activity, and arousal state. The evidence suggests that state dependency and general activity changes do not contribute to the effects of VP on shock avoidance maintenance but arousal state during the time of VP administration may be a factor that determines the effectiveness of VP.

It is well known that a task learned in one physiological state may not be retrievable under another physiological state (Overton, 1966; Patel et al., 1979; Rabin, et al., 1982). Thus state dependency is an issue when VP is given only before acquisition tests or when it is given only before avoidance maintenance or extinction tests. That VP produces similar effects on shock avoidance when given before acquisition, after acquisition, before extinction, and before acquisition and extinction suggests that state dependency does not play a role (DeWied, 1971; Ader and DeWied, 1972; Bohus et al., 1972, 1978; King and DeWied, 1974; Rigter et al., 1974; DeWied et al., 1984; DeJong et al., 1985; Gaffori and DeWied, 1985; Koob et al., 1986; Faiman et al., 1992).

A drug that alters general activity can affect performance of both passive and active shock avoidance tasks. If the drug reduces movement, it could be concluded falsely that the drug had a facilitatory effect on avoidance maintenance in a passive avoidance situation and detrimental effects on avoidance maintenance in an active avoidance task. The opposite would be true for a drug that increases movement. The effects of VP on activity are dependent on when it is given with respect to behavioral testing and the dose that is given. Locomotion in an open field test is reduced when VP is administered 0–20 min before testing but not when given 60 min before testing (Krejčí et al., 1979; Andrews et al., 1983; Gaffori and DeWied, 1985; Alescio-Lautier and Soumirev-Mourat, 1990). Bar pressing is reduced when high doses of VP are given but not when animals are treated with low doses (Andrews et al., 1983). This suggests that alterations in general activity do not contribute to the effects of VP under appropriate timing and dose conditions.

The arousal state, however, may be an important determinant of how effective VP is in facilitating shock avoidance maintenance. Several studies have reported that VP produces bimodal effects, that is, it maintains the avoidance behavior in most animals but disrupts it in some (Sahgal, et al., 1982; Sahgal and Wright, 1983). Sahgal and his colleagues have drawn a parallel between these effects of VP and those of amphetamine, which increases arousal state, and chlordiazepoxide, which decreases arousal state. In a one-trial step-though active shock avoidance task, all three agents produced prolonged as well as accelerated retention latencies (Sahgal and Wright, 1983). It has been suggested that VP heightens arousal and that the initial arousal state of the animal may determine whether VP will enhance or disrupt avoidance maintenance. If the initial arousal levels are too high, VP will disrupt maintenance and if they are too low, VP will enhance it.

One source of arousal may be the peripheral effects of VP. It has been suggested that VP maintains avoidance because its vascular effects serve as arousal stimuli (Lebrun, et al., 1985). Increases in systemic VP levels produce peripheral increases in blood pressure and bradycardia (Malayan et al., 1980; Pittman et al., 1982; DeWied et al., 1984). Central administration of VP also alters peripheral blood pressure and heart rate. An increase in blood pressure has been observed after intracerebroventricular infusions of VP or local infusions into the nucleus of the solitary tract, anterior hypothalamus, locus coeruleus, and nucleus reticularis rostroventrolateralis (Matsaguchi et al., 1982; Pittman et al., 1982; Wang et al., 1982; Berecek et al., 1984; Feuerstein et al., 1984; Gomez et al., 1993) and tachycardia has been found after intracerebral infusions of VP into the nucleus of the solitary tract and locus coeruleus (Matsuguchi et al., 1982; Berecek et al., 1984). It has been suggested that these effects might in some way be the basis for the effects of VP on shock avoidance maintenance (Ettenberg et al., 1983; Bluthe et al., 1985a; Lebrun et al., 1985b). There are active shock avoidance studies that have found an association between the pressor properties of VP and the ability of VP to produce memory effects. Intracerebroventricular pretreatment with an AP antagonist analog prevents the prolongation of extinction induced by subcutaneous injections of VP only when the dose used blocks systolic blood pressure (Lebrun et al., 1985). Similar results have been obtained for the effects of VP on an appetitive food-finding task (Ettenberg, 1984). However, fragments of VP have been shown to influence shock avoidance without the confounding cardiovascular effects. This is evident when the fragments are administered either peripherally or centrally (DeWied, 1976; Krejčí et al., 1979; DeWied et al., 1984, 1987, 1991; DeJong et al., 1985; Gaffori and DeWied, 1985). For example, the VP fragment, [pGlu^4, Cyt^6] AVP$_{4-8}$, influences retention and extinction of avoidance behaviors without producing any of the classic endocrine effects (DeWied et al., 1984; DeJong et al., 1985). Like VP, intracerebroventricular infusions of [pGlu^4, Cyt^6] AVP$_{4-8}$ are more effective than when given subcutaneously, in this case, 20,000 times more effective. Taken together, the data suggest that activation of vascular changes is not necessary for VP to maintain avoidance behavior.

Failure to support the hypothesis that vascular effects serve as arousal stimuli does not eliminate arousal as the basis for the effects of VP on avoidance maintenance. The problem with studying an arousal hypothesis has been finding an operational measure of arousal. The two most common measures used in the study of arousal have been behavioral activity levels and EEG signs of cortical activation. However, behavioral activity is a poor indicator of arousal state. Clearly, in the external defense system, an animal that freezes or exhibits behavioral immobility after sensing the presence of a predator is as aroused as when it changes from immobility to a running response after assessing the location of the predator. This issue has been alluded to in disagreements about the interpretation of reduced behavioral activity following VP treatment, some reporting it as sedation and others pointing to the immobile crouching and staring as indicative of elevated arousal (Krejčí et al., 1979; Sahgal, 1984). It leads us back to the problem of whether one can say that failure to find alterations in open field behavior 60 min after administration of VP means that the arousal state of the animal is not elevated (Gaffori and DeWied, 1985). Clearly, simple measures of activity are not sufficient. There is general consensus that EEGs with predominant beta activity are indicative of arousal, gamma activity of hyperarousal, and alpha activity of relaxation. Unfortunately, EEG measurements in unanesthetized, freely moving animals have not been applied to the study of the effects of VP on avoidance maintenance. However, human studies have shown that the VP produces EEG activity indicative of arousal (Gais et al., 2002). Future work will need to determine whether EEG measurements during behavioral testing can adequately address the arousal issue.

Taking all of the data into consideration, there is nothing that will allow elimination of either the mnemonic or arousal hypothesis. One could argue that because VP can maintain avoidance behavior when applied directly to areas of the brain known to be critically involved in learning and memory processes, namely, the dentate gyrus of the hippocampus, then VP must be having direct mnemonic effects (Kovács et al., 1979, 1986; O'Connell et al., 2000; Chang et al., 2001; Jones et al., 2001; Babar et al., 2002). However, this does not preclude the possibility that VP also acts on mnemonic areas via its action on one of the arousal circuits. The study of arousal neural systems has mushroomed over the past 15 years and a number of different neuromodulator systems have been implicated in arousal. These include the basal forebrain and acetylcholine, the locus coeruleus and noradrenaline, the tuberomammillary nucleus and histamine, and the raphe nucleus and serotonin (Aston-Jones and Bloom, 1981; Peck and Vanderwolf, 1991; Vanderwolf, 1992; Rasmusson et al., 1994; Khateb et al., 1995; Brown et al., 2001). It is interesting that VP also maintains avoidance when it is infused into the lateral septum and the dorsal raphe nucleus (Kovács et al., 1979, 1986). Distinctions have been made between general arousal and tense arousal and these neural areas appear to be associated with at least some tense arousal states (Thayer, 1989). Responses in situations that are associated with distress, such as shock avoidance in a shuttle-box, performance in an elevated plus maze, and vocalization from an air puff, increase c-Fos-like immunoreactivity (c-FLI) in the lateral septum while vigorous bar pressing for sweetened condensed milk reinforcement does not (Duncan et al., 1996). Although extracellular levels of serotonin in the dorsal raphe nucleus do not change when rats are given escapable shocks, they increase in an inescapable shock situation (Maswood et al., 1998). It has been suggested that this increase contributes to the behavioral effects of inescapable shock. One of these effects is maintaining a failure to respond even in new situations that allow escape.

Although the majority of studies have found that VP maintains avoidance behavior, there are some negative findings. As indicated above, VP can either have no effect or it can disrupt maintenance of avoidance in some animals (Alliot and Alexinsky, 1982; Sahgal et al., 1982; Andrews et al., 1983; Sahgal and Wright, 1983). These failures to find avoidance maintenance may be based on genetic differences. Examination of six different inbred strains of mice during acquisition of a two-way shuttle box task have revealed that VP accelerated acquisition in one strain, disrupted acquisition in another strain, and had no effect in the remaining strains (Hamburger et al., 1985). The genetic issue has been raised to account for some failures to find that VP maintains avoidance responses in Brattleboro rats (Celestian et al., 1975; Bailey and Weiss, 1981; Bohus and DeWied, 1998). Since many of the colonies started with a limited number of rats for mating, it has been suggested that genetic drift occurred among the different breeding colonies. Differences in the ability to find deficits are associated with the colony source. Genetic factors may also interact with experiential factors to influence responsiveness to the effects of VP and avoidance maintenance. Studies have reported such interactions in the display of stress responses. For example, BALB/cByJ mice are more reactive to stressors than

C57BL/6ByJ mice. Exposure of the BALB/cByJ mice to early life handling and allowing them to be raised by C57 BL/6ByJ mothers reduced their reactions to stress. Allowing C57BL/6ByJ mice to be raised by BALB/cByJ mothers had no effect (Anisman et al., 1998).

Genetic issues probably can be tied to two other findings. First, there are individual differences in pain threshold to footshock and the intensity and duration of shock can determine if an animal will reliably display an avoidance behavior (Ader et al., 1972; Seliger, 1977; Sahgal, 1990). Second, there are good and poor learners and the effects of VP in these animals can be different. In many learning studies, experimenter generated criterion are set for deciding whether an animal has learned a task and thus whether that animal will be included in further tests. For example, in one study, only those rats that displayed at least seven out of 10 pole-jump responses during the second acquisition test were tested further (DeWied, 1976). Yet, exclusion of poor learners can alter the conclusions that are made. It has been reported that VP injections do not influence the acquisition of a discriminative appetitive task when good learners are tested but it impairs the performance of poor learners (Alliot and Alexinsky, 1982). All of these findings raise the question of whether genetics alters the neural pathways associated with avoidance memory processes or arousal.

Effects of shock avoidance learning on neural vasopressin

Clearly, alterations in brain VP levels can influence maintenance of learned shock avoidance. The opposite is true as well. The experience of acquiring shock avoidance can alter brain VP levels. Rats that have acquired one-trial passive shock avoidance, show decreased levels of VP in the PVN, suprachiasmatic nucleus, lateral septum, and hippocampus when tested for avoidance maintenance 24 or 120 h after acquisition (Laczi et al., 1983a,b). They also show increased VP levels in the central amygdala, subfornical organ, and locus coeruleus when measured after a 24-h retention test (Laczi et al., 1983b) and in the CSF when measured after a 120-h retention test (Kovács et al., 1979, 1986; Laczi et al., 1984). At least some of these changes are tied to the facilitation of avoidance maintenance. Decreases in VP levels in the hippocampus are found in good avoiders but not poor avoiders and as mentioned in the *Effects of Central Administration of Vasopressin* section, maintenance of passive shock avoidance is facilitated after local infusion of VP into the hippocampal dentate gyrus (Laczi et al., 1983a). The effect of learning on brain VP also has been found in other learning situations. For example, the pattern of neuronal activation after intracerebroventricular infusion of VP, as measured by c-FLI, is less pervasive after learning a visual discrimination task than it is in unconditioned mice (Paban et al., 1999).

Behaviors associated with malaise

Toxin-triggered responses

Agents such as LiCl are toxic when administered in high doses and they will trigger a constellation of behavioral and physiological responses that are indicative of a general state of malaise. These responses include the following: decreases in motor activity and increases in prostration or lying-on-belly (Ladowsky and Ossenkopp, 1986; Meachum and Bernstein, 1990), increases in EEG slow waves and sleepiness (Schou, 1968; Karniol et al., 1978), suppression of food intake (Bernstein and Goehler, 1983), triggering of diarrhea and vomiting in species with this capability (Schou, 1968; Stricker et al., 1988; Rabin and Hunt, 1992), hypothermia (Taukulis, 1982; Batson, 1983; Cunningham and Niehus, 1993), decreases in gastric motility and emptying (Flanagan et al., 1989; McCann et al., 1989), inhibition of basal and food-stimulated gastrin secretion (Lauritsen et al., 1978), increases in mean arterial pressure (O'Connor et al., 1987), and decreases in heart rate (Wilkin and Cunningham, 1982; O'Connor et al., 1987).

Plasma VP levels in nonhuman primates have been reported to increase after exposure to LiCl (Verbalis et al., 1987). Although no change in plasma VP levels has been reported when measured shortly after an injection of LiCl in rats, changes in peripheral VP levels are evident when measurements are taken 1 h later (Cheng et al., 1986; Verbalis et al., 1986). In addition, LiCl induces c-FLI expression in VP neurons in the PVN and supraoptic nucleus (Olsewski et al., 2001). It has been suggested that VP functions to reduce intestinal blood flow through constriction of splanchnic blood vessels and thereby reduces exposure to the toxic effects of LiCl (Stricker and Verbalis, 1991).

Conditioned consumption reduction and conditioned taste aversion

When illness-inducing agents such as LiCl are administered after an animal has consumed a novel food or drink, the animal learns an association between the sensory properties of the novel food or drink and the sensory properties of the toxic agent. As a result of this learned association, there is a change in the behaviors expressed when the animal again encounters the food or drink (see Chambers 1990, Chambers and Bernstein, 1995, 2003). These behaviors can be divided into three categories. First, individuals will show conditioned responses based on the illness-inducing agent, that is, the conditioned taste will elicit responses that are similar to the unconditioned responses elicited by the agent. In humans, simply hearing or thinking about the conditioned taste elicits reports of nausea and in rats, conditioned tastes will elicit some of the unconditioned responses to LiCl, e.g., lying-on-belly and decreased heart rate (Garcia, 1989; Kosten and Contreras, 1989; Meachum and Bernstein, 1990). These learned responses to the novel taste are termed conditioned illness responses or conditioned agent responses. Second, animals will show consumption reduction responses. These responses are measured as decreases in consumption, decreases in preference for the conditioned taste when it is made available along with another taste substance, or decreases in ingestive orofacial responses (a series of rhythmical mouth movements and alternations between tongue protrusions and tongue retractions that result in swallowing). These learned responses to the novel taste are termed conditioned taste avoidances or conditioned consumption reductions. Finally, animals will show aversive responses to the novel taste. When rats are poisoned after consuming a sweet taste such as sucrose, which appears to be innately liked, their subsequent behavioral responses to sucrose resemble those exhibited after consumption of a bitter taste such as quinine, which appears to be innately disliked. These behavioral responses include increases in aversive orofacial responses (mouth gaping with tongue retraction followed by long-duration tongue protrusion and mouth closure, which results in a reduction in swallowing), spillage of food from eating containers, and defensive burying of drinking spouts (Carpenter, 1956; Garcia and Koelling, 1966; Rozin, 1967; Wilkie et al., 1979; Berridge et al., 1981; Parker, 1988; Bowers et al., 1992). In studies of humans, an avoided food is identified as distasteful if consumption of that food previously had been followed by emesis (Steiner, 1979; Garcia, 1989). These learned responses to the novel taste are termed conditioned taste aversions.

The specific sensory properties that allow an agent to induce condition consumption reduction and taste aversion remain unidentified. But unlike LiCl, not all agents condition both types of responses. Agents such as morphine induce conditioned consumption reduction but do not induce conditioned aversive reactions to the conditioned taste (Parker and Brosseau, 1990). In addition, these types of agents are rewarding at the same doses that induce conditioned consumption reduction (Best et al., 1973; Wise et al., 1976; van der Kooy et al., 1983; Parker, 1995). They produce conditioned place preferences and are effective reinforcers in a drug self-administration paradigm (Baxter et al., 1974; Wise et al., 1976; Spyraki et al., 1982; van der Kooy et al., 1983). On the basis of these results, it has been suggested that the physiological stimuli mediating conditioned consumption reduction induced by reinforcing agents are qualitatively different than those induced by the illness agent LiCl (Hunt and Amit, 1987; Parker, 1988, 1995). A distinction is now made between those agents that induce both conditioned consumption reduction and conditioned aversive responses and those that only induce conditioned consumption reduction. Left out of this picture is whether a taste substance can elicit conditioned agent responses after it has been paired with any agent capable of inducing conditioned consumption reduction, simply because most studies have not measured these responses when animals are exposed to a conditioned taste substance.

Booth (1977) suggested a number of years ago that learned reductions in consumption could be a conditioned reaction based on states other than illness. A number of studies have shown that pairing a food with satiating properties can produce a learned reduction in the consumption of that food (Booth, 1985). Learned reductions in consumption also can occur under conditions of reward comparison (Flaherty and Checke, 1982; Lucas et al., 1990). In this type of situation, animals are given access to a saccharin solution, which is less preferred, before they are given access to a sucrose solution, which is more preferred. They learn that the presence of the saccharin solution predicts the future availability of the sucrose solution, and they consequently reduce their consumption of the saccharin solution. It has been suggested that rewarding drugs serve a preferred reward role rather than an illness-inducing role (Grigson, 1997). The implication of all of these data is that the neural mechanisms mediating conditioned consumption reduction based on satiety, reward comparison, and illness are different.

Vasopressin induces conditioned consumption reduction

Intracerebroventricular infusion of VP does not induce conditioned consumption reduction, as long as infusion does not trigger seizures, but when peripheral injections of VP are given after an animal has consumed a novel taste solution, the animal exhibits conditioned consumption reduction to the novel taste (Ettenberg, 1984; Tam et al., 1985; Bluthe et al., 1985a,b; Hayes and Chambers, 2002, in press; Koob et al., 1986; Langhans et al., 1991). In our conditioned consumption reduction studies, non-deprived male rats were allowed access to a 10% sucrose solution for 1 h. Either immediately or 50 min after this consumption period, they were injected with saline or VP. This acquisition procedure was followed three times. When VP was administered immediately after sucrose consumption, the highest doses (4.5, 6, 9, and 18 μg/kg body weight) induced CCR after only one pairing but the lowest doses (1.5 and 3 μg/kg body weight) required two pairings before they induced reduction. When VP was administered 50 min after sucrose consumption, an 18 μg/kg dose induced conditioned consumption reduction after one pairing but a 6 μg/kg dose failed to induce reduction even after three pairings (see Fig. 2).

Vasopressin-triggered responses

Peripheral administration of VP increases blood pressure, decreases heart rate, inhibits gastric emptying, and disrupts spontaneous locomotor activity (Malayan et al., 1980; Pittman et al., 1982; Ettenberg et al., 1983; De Wied et al., 1984; Langhans et al., 1991). In addition, VP has been found to suppress eating even in food-deprived animals and it does so

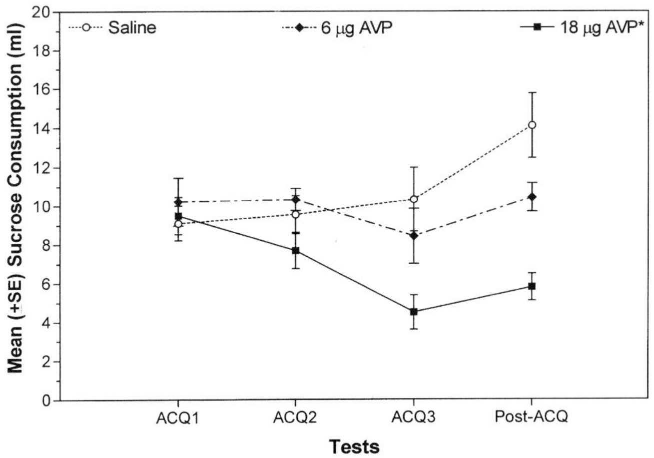

Fig. 2. Mean (±SE) amount of sucrose consumed by male rats before (ACQ1) and after one (ACQ2), two (ACQ3), and three (Post-ACQ) pairings of sucrose consumption with subcutaneous injections of 6 or 18 μg/kg of AVP or saline given 50 min after consumption. *Significant reduction in consumption across tests, $p < 0.02$.

with doses that are within physiological levels (Meyer et al., 1989; Langhans et al., 1991). This hypophagic effect is based on a delay in meal onset in rats and a reduction in meal size and first intermeal interval in African Pygmy goats (Meyer et al., 1989; Langhans et al., 1991). It is reversed by a V_1 receptor antagonist, an α-adrenergic receptor antagonist, and the Ca^{++} channel blocker verapamil, which serves as an intracellular messenger for both V_1 and α-adrenergic receptors. However, neither the α-adrenergic receptor antagonist nor verapamil reverse the decrease in gastric emptying produced by VP. On the basis of all of these results, it has been suggested that VP inhibits eating through a V_1 receptor mediated activation of an α-adrenergic mechanism and that this inhibition is not a consequence of the inhibited gastric emptying.

It has been suggested that it is the pressor properties of VP that allow it to induce conditioned consumption reduction. This hypothesis is supported by three pieces of evidence. First, doses of VP that do not alter systemic blood pressure or trigger seizures also do not induce conditioned consumption reduction when VP is infused intracerebroventricularly (DeWied et al., 1984; Hayes and Chambers, 2002). Second, the VP analog desglycinamide arginine vasopressin, which has weak pressor-agonist activity, does not induce conditioned consumption reduction (Ettenberg et al., 1983; Bluthe et al., 1985b). Third, VP-induced conditioned consumption reduction can be blocked by subcutaneous and intracerebroventricular administration of the V_1 pressor antagonist dPTyr(Me)AVP (Ettenberg, 1984; Bluthe et al., 1985b). The amount required to block a VP-induced consumption reduction also blocks peripheral changes in blood pressure (Le Moal et al., 1981; Lebrun et al., 1985b). Despite the strength of this evidence, it does not allow the exclusion of satiety as the sensory property that allows VP to induce conditioned consumption reduction.

Aversive properties of vasopressin

As indicated above, finding that VP induces conditioned consumption reduction does not mean that it has aversive properties. However, there is some evidence to suggest that high doses of peripherally administered VP can induce conditioned taste aversion. First, the properties of VP that allow it to induce conditioned consumption reduction are additive with those of LiCl (Hayes and Chambers, unpublished data). When a low dose of LiCl (0.075 mEq/kg) was administered immediately after access to a novel sucrose solution and a high dose of VP (18 μg/kg) was administered 50 min later, the rate of acquisition of the conditioned consumption reduction was faster than if only one of these two agents is given (see Fig. 3). Second, intermediate doses of VP induce

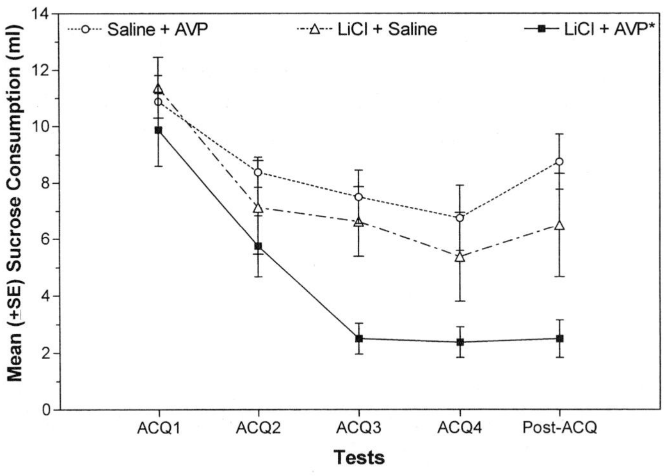

Fig. 3. Mean (±SE) amount of sucrose consumed by male rats before (ACQ1) and after one (ACQ2), two (ACQ3), three (ACQ4), and four (Post-ACQ) pairings of sucrose consumption with intraperitoneal injections of 0.075 mEq/kg of 0.15 M LiCl or saline given immediately after consumption and subcutaneous injections of 18 μg/kg of arginine vasopressin (AVP) or saline given 50 min after consumption. *Significantly greater reduction in consumption than each of the other two groups, $p \leq 0.04$.

conditioned place avoidance. In this procedure, rats are repeatedly injected with a drug and placed in one of two adjoining, yet contextually distinct, chambers. On test day, the rats are allowed to move freely between the chambers. Preference and avoidance are inferred from the percentage of time spent in each of the chambers. It has been found that the rats spend less time in the chamber associated with injections of VP (6 μg/kg; Ettenberg, 1983; Ettenberg et al., 1984). This effect is attenuated by injections of a VP antagonist (Ettenberg, 1984). Likewise, pairing a higher dose of VP with the dark compartment of the step-through apparatus results in a greater latency to reenter the dark compartment when placed in the illuminated compartment (10 μg/kg; Ebenezer, 1988). Finally, preliminary evidence in our lab indicates that pairing a high dose of VP (18 μg/kg) can induce the aversive taste reaction, mouth gaping (Hayes et al., unpublished data).

Neural mechanisms

Conditioned consumption reductions induced by low doses of VP may be based on one sensory property or they may be the end result of multiple sensory properties, each of which are associable with taste, and conditioned taste aversions induced by high doses of VP may be the result of excessive stimulation of the same sensory properties that induce conditioned consumption reduction or the addition of another sensory property that is inherently aversive. Two peripheral sensory pathways have been identified as critical for acquisition of conditioned consumption reduction. The brain entry route for one of these pathways is the area postrema, which projects to the parabrachial nucleus and the access route for the other pathways is the vagus, which projects to the NST and this structure in turn projects to the parabrachial nucleus (Shapiro and Miscelis, 1985; Borison, 1989; Papas and Ferguson, 1990; Miller and Leslie, 1994).

For illness-inducing agents such as LiCl, the critical access route is the area postrema. Intraperitoneal administration of LiCl increases the expression of c-FLI and the electrical activity of neurons in the area postrema (Adachi et al., 1991; Yamamoto et al., 1992a,b; Swank et al., 1995). In addition, the learning of consumption reduction and aversive taste reaction is prevented by permanent and reversible cooling lesions of the area postrema (Ritter et al., 1980; Eckel and Ossenkopp, 1996; Wang et al., 1997a,b). On the other hand, the positive reinforcement agent morphine does not require the area postrema to induce conditioned consumption reduction but acts via the vagus nerve (van der Kooy, 1984; Bechara and van der Kooy, 1985).

It has been suggested that pressor sensory properties of VP induce conditioned consumption reduction by acting via baroreceptor vagal connections to the brainstem (Bluthe et al., 1985; Hreash et al., 1990). If satiety is a sensory property of VP in conditioned consumption reduction learning, a vagal access route is unlikely to be involved because hepatic branch vagotomy does not block the hypophagic effect of VP (Langhans et al., 1991). The sympathetic nervous system has been suggested as an access route by which VP produces hypophagia (Langhans et al., 1991). It is possible, however, that the area postrema mediates whatever sensory properties are critical for induction of conditioned consumption reduction and conditioned taste aversion by VP and that the clusters of neurons activated determines whether only conditioned consumption reduction or both conditioned consumption reduction and conditioned taste aversion will be expressed. This circumventricular structure has been implicated in pressor regulation, it is part of the neural pathway that mediates food intake, it contains V_1 receptors, and it is directly activated by VP (Fink et al., 1987a,b; Carpenter, et al., 1988; Phillips et al., 1988; Skoog and Mangiapane, 1988; Borison, 1989; Gerstberger and Fahrenholz, 1989; Mangiapane et al., 1989; Raggenbass et al., 1989; Ritter and Taylor, 1990; Skoog et al., 1990; Barberis et al., 1995; Tribollet et al., 1998;).

Effects of vasopressin on extinction of LiCl-induced conditioned consumption reduction

Vasopressin after acquisition

As indicated in the *Shock Avoidance Learning* section, animals receiving a wide range of doses of VP after acquisition show prolonged extinction of shock avoidance (Ader and DeWied, 1972; Bohus et al., 1978; DeWied et al., 1984; DeJong et al., 1985). However, low doses of VP accelerate the completion of extinction of a LiCl-induced conditioned consumption reduction when administered after LiCl injection, but the onset of extinction is delayed when high doses of VP are administered (Hayes and Chambers, 2002, in press, unpublished data).

In our conditioned consumption reduction studies, nondeprived male rats were injected with LiCl (1.5 mEq/kg) after they had been allowed access to a 10% sucrose solution for 1 h (Hayes and Chambers, 2002, in press, unpublished data). Subcutaneous injections of VP (6 and 18 μg/kg body weight) were given 0 (immediately), 25, or 50 min after the LiCl injection and intracerebroventricular infusions (1 ng/rat) were given 50 min after LiCl. In shock avoidance procedures, the application of shock produces an immediate response in animals. Thus, if VP is injected immediately after the presentation of the shock, it is acting when the animal is no longer experiencing the stimulus effects of the shock. With conditioned consumption reduction, however, the effects of LiCl gradually increase in intensity, peaking approximately 30 min after injection, and then slowly dissipating thereafter (Morrison et al., 1971; Sterner, 1990). Therefore, in order to assure that temporal parameters were not a factor in any differences in the effects of VP on shock avoidance and LiCl-induced conditioned consumption reduction, VP was injected at a time when the animal was experiencing essentially no aversive effects of LiCl (0 min) or varying levels of aversive effects (25 and 50 min). In the central study, VP did not influence the onset of extinction of the LiCl-induced conditioned

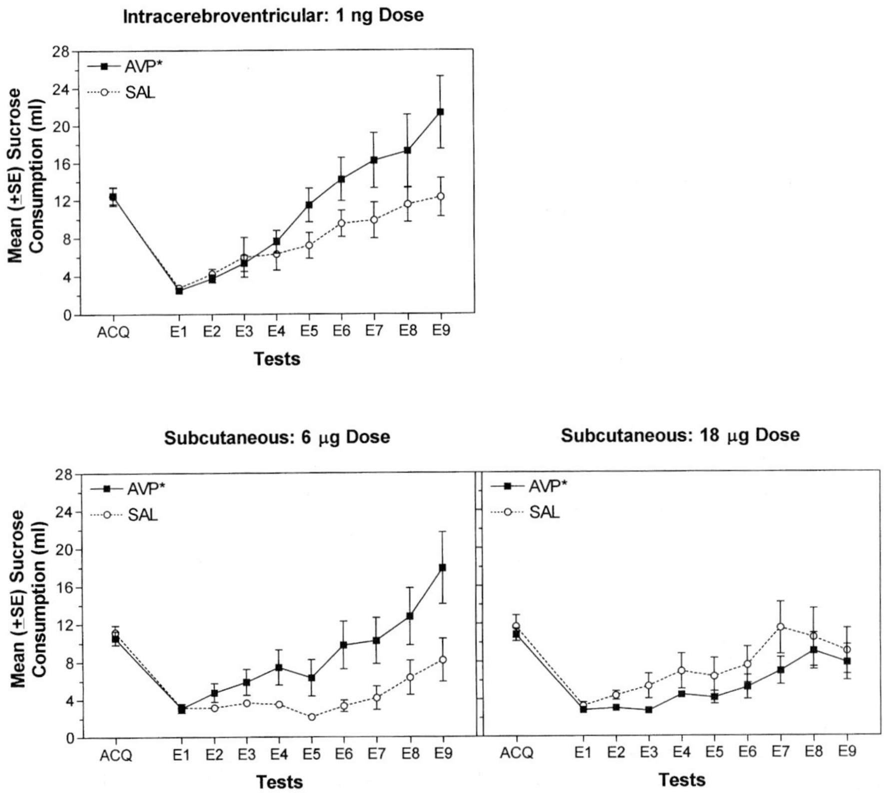

Fig. 4. Mean (± SE) amount of sucrose consumed by male rats before (ACQ) and after (E1–E9) pairing sucrose consumption with intraperitoneal injections of 1.5 mEq/kg LiCl given immediately after consumption. Subcutaneous injections of 6 or 18 μg/kg of AVP or saline and intracerebroventricular infusions of 1 ng/rat of AVP or saline were given 50 min after LiCl injection. *Significantly faster completion of extinction than the saline group for 1 ng and the 6 μg doses and significantly slower initiation of extinction than the saline group for the 18 μg dose, $p < 0.05$. The intracerebroventricular data were adapted from Hayes and Chambers (2002).

consumption reduction but facilitated the completion of extinction (see Fig. 4). In the peripheral study, the 6-μg/kg dose had no effect on the initiation of extinction but it accelerated the completion of extinction when given 50 min after LiCl injection. The 18 μg/kg dose of VP had no effect on the completion of extinction but it delayed the onset of extinction when injected 25 and 50 min after LiCl.

Vasopressin before acquisition and extinction

There are a few studies that have found prolonged extinction of a LiCl-induced conditioned consumption reduction after peripheral administration of VP. In these studies, VP was administered before acquisition and/or extinction (Cooper et al., 1980; Vawter and Green, 1980). Finding a differential effect of VP on learning and memory processes depending on the timing of administration would not be unique. Although VP consistently prolongs extinction of shock avoidance no matter when in the learning process it is given, consistent effects are not always found in other learning situations (Ader et al., 1972; Bohus et al., 1972, 1978; King and De Wied, 1974; Rigter et al., 1974; DeWied et al., 1984, 1991; DeJong et al., 1985; Gaffori et al., 1985; Car et al., 1994).

Despite several attempts to replicate the prolonged extinction findings by administering VP either before acquisition or before both acquisition and extinction, we have failed to do so. First, central administration of 1 ng/rat of VP into the right lateral ventricle 20 min before presentation of sucrose during the acquisition test has no effect on extinction of LiCl-induced conditioned consumption reduction. Second, central administration of 1ng/rat of VP into the right lateral ventricle 20 min before presentation of sucrose during the acquisition test and each extinction test has no effect on extinction of LiCl-induced conditioned consumption reduction. Third, subcutaneous injections of VP in the amounts of 9 and 18 μg/kg fail to influence behavior when injected 1 h before acquisition and extinction tests. The lack of effect is found with both fluid-restricted (23 h schedule) and non-restricted animals. Finally, subcutaneous administration of DG-LVP (3 and 9 ug/kg) 1 h before acquisition and extinction testing has no effect on extinction in fluid-restricted male rats (23 h schedule). It should also be noted that inconsistent results were found in the studies of Cooper et al. (1980). In one study, both young and old fluid deprived males were injected with lysine VP before acquisition of a LiCl-induced conditioned consumption reduction. Although the old males consumed less saccharin during extinction, the young males did not and neither age exhibited a slower rate of extinction. The reason for the inconsistencies in the results remains unsolved. However, it cannot be tied simply to strain, sex, fluid balance state, form of VP, type of bottle test (1 vs. 2), kind of taste solution or conditioning agent, time interval between injection of VP and pairing of the taste solution with the conditioning agent, or time interval between extinction tests. In some studies, vasopressin delayed extinction and in others, it had no effect even though these factors were similar (see Table 1). The only apparent procedural difference between studies finding a delay and those failing to find this effect was the number of

Table 1. Procedures used when vasopressin delayed extinction of conditioned consumption reduction and when it had no effect on extinction

	Delay	No effect
Strain	Sprague-Dawley[a] Wistar[b]	Sprague-Dawley[a,c]
Sex	Male[a,b]	Male[a,c]
Fluid balance state	Fluid deprived[a,b]	Fluid deprived[a,c] Nondeprived[c]
Form of vaspressin	LVP[a] DG-LVP[b]	LVP[a] DG-LVP[c] AVP[c]
Bottle test	1-Bottle[b] 2-Bottle[a]	 2-Bottle[a,c]
Taste solution	Saccharin[a,b]	Saccharin[a] Sucrose[c]
Conditioning agent	LiCl[b] Amphetamine[a]	LiCl[a,c]
Time of vasopressin administration	Before ACQ[b] Before E[a,b] Before ACQ and E[b]	Before ACQ[a,c] Before ACQ and E[c]
Interval: vaopressin-conditioning agent	80 min[b]	80 min[c] 90 min[b]
Interval: extinction tests	Every 4th day[a] Daily[b]	 Daily[c]
Number of ACQ tests	3 Tests[a,b]	1 Test[a,c]

[a] Cooper et al. (1980).
[b] Vawter amd Green (1980).
[c] Hayes and Chambers, unpublished data.

acquisition tests given. VP prolonged extinction when three acquisition tests were given and had no effect when one test was given. Even if this is the critical differentiating factor, the data clearly indicate that the ability of VP to prolong extinction of conditioned consumption reduction is very limited.

What does it mean that vasopressin accelerates or delays extinction?

There is often an inclination to view extinction simply as a mnemonic process, so that when extinction is fast, memory is weak and when extinction is slow, memory is strong. The value judgment under this view is that the effects of VP are negative when it produces faster extinction and it is positive when it produces slower extinction. However, extinction is a reflection of the integration of many different processes. In a conditioned consumption reduction situation, extinction is a reflection of the strength of the taste–malaise association, the ability to retain the association in memory, the ability to retrieve the memory of the association, and the ability to unlearn that consumption of the taste substance is followed by malaise and/or relearn that consumption of the taste substance is followed by neutral or positive consequences. If VP modulates relearning, then the effects of VP would be positive when it produced faster extinction and negative when slower extinction was produced.

The role of VP in conditioned consumption reduction situations has yet to be determined but there are suggestive data. It is likely that the delay in the onset of extinction with the high doses of VP is due to a strengthening of the acquisition process because of the aversive properties of this dose. As indicated in the Vasopressin Induces Conditioned Consumption Reduction section, an 18 μg/kg dose of VP can induce both conditioned consumption reduction and conditioned taste aversion and the addition of this dose of VP 50 min after LiCl injections accelerates the rate of acquisition (Hayes and Chambers, unpublished data; Hayes et al. unpublished data). These results are similar to what has been found when the second injection after LiCl is another LiCl injection. A stronger conditioned consumption reduction is produced when two separate doses of 1.5 mEq/kg of LiCl are injected immediately and 35 or 70 min after sucrose consumption than when a single dose of 3.0 mEq/kg of LiCl is injected immediately after consumption (Domjan et al., 1979).

The reason for the acceleration of the completion of extinction of LiCl-induced conditioned consumption reduction when low doses of vasopressin are used is not as clear. The modulatory effect of low doses of VP cannot be attributed to any associative properties that these doses have. Although peripheral administration of 6-μg/kg dose of VP induced conditioned consumption reduction when injected immediately after sucrose consumption and produced conditioned place avoidance after repeated pairings, this dose did not induce conditioned consumption reduction when injected 50 min after sucrose consumption (Ettenberg et al., 1983; Ettenberg, 1984; Hayes and Chambers, in press). In addition, the dose of vasopressin infused into the lateral ventricle did not induce conditioned consumption reduction when injected immediately after sucrose consumption (Hayes and Chambers, 2002). One explanation, for which there is support, is that VP reduces the aversiveness of LiCl. Studies have shown that VP can attenuate the unconditioned effects of illness-inducing agents. The reduction in social exploration of conspecific juveniles that occurs when adult animals are injected with the cytokine interleukin-1 is partially reversed by administration of VP (Dantzer et al., 1991). Recently, we have found that lowering a dose of LiCl from 1.5–0.3 mEq/kg accelerated the completion but not the initiation of extinction of a conditioned consumption reduction (Chambers and Wang, 2004). These are the same kinds of effects that VP had on extinction of a LiCl-induced conditioned consumption reduction. Taken together, these findings are consistent with the hypothesis that VP effectively reduces the dose of LiCl. If future work substantiates this hypothesis, this would mean that VP does not act on taste avoidance memory processes but rather acts to disrupt sensory input critical for acquisition.

One way in which VP could diminish the effectiveness of LiCl is to reduce the input into neural areas mediating extinction. Two neural areas that have been implicated in extinction of conditioned consumption reduction are the BST and the lateral septum (LS). We have found that VP levels are elevated in the BST during extinction testing, which suggests an involvement of the vasopressinergic system in this neural area in the extinction process (see the next section; Brownson et al., 2002). The BST is an extrahypothalamic site of VP biosynthesis and the LS is one of its major efferents (DeVries and Buijs, 1983; van Leeuwen and Caffe, 1983; Sofroniew, 1985; Miller et al., 1988). Electrolytic lesions of the stria terminalis after acquisition of a conditioned consumption reduction abolish the acquired reduction and lesions of the LS before acquisition facilitate extinction (McGowan et al., 1969; Yamamoto, et al., 1981). Taken together, these data suggest that a BST–LS system is involved in the extinction process. In a recent study, we compared c-FLI expression in males given only LiCl after sucrose consumption and those given LiCl and intracerebroventricular administration of VP 50 min later. Those males given only LiCl had significantly higher expression in the dorsal and ventral LS and the lateral division of the BST (see Fig. 5). These data raise the possibility that the effect of VP on extinction of a LiCl-induced conditioned consumption reduction is mediated by a BST–LS system. If this is the case, then peripherally administered VP could be acting at the sensory level, that is, on neurons in the area postrema, to reduce input into the BST and intracerebroventricularly administered VP could be acting directly on neurons in the BST.

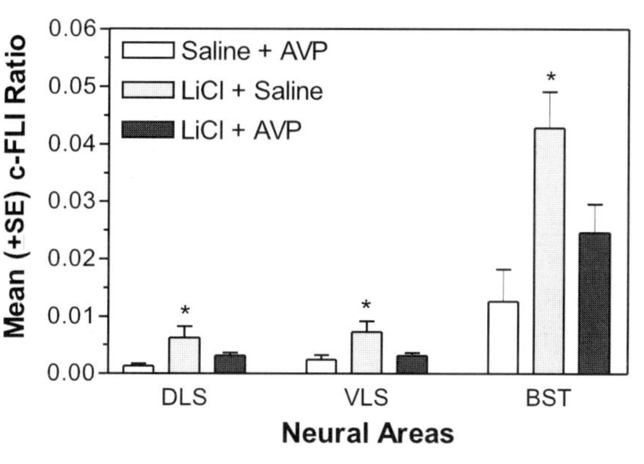

Fig. 5. Mean (+SE) c-FLI ratio (area of c-FLI in the neural structure divided by total area of the neural structure) in the dorsal (DLS) and ventral (VLS) lateral septum and the BST of male rats 2 h after pairing sucrose consumption with intraperitoneal injections of 1.5 mEq/kg LiCl or saline given immediately after consumption and intracerebroventricular infusions of 1 ng/rat of AVP or saline given 50 min after consumption. *Significantly greater than each of the other two groups, $p \leq 0.04$.

Effects of LiCl-induced conditioned consumption reduction on neural vasopressin levels

As was true for shock avoidance, the experience of acquisition and extinction of conditioned consumption reduction can alter VP levels in the brain (Brownson et al., 2002). In our study, three groups of male rats were kept on *ad libitum* access to food and water while receiving different amounts of conditioning and extinction testing. The first group (Ext-1) was exposed to seven days of preconditioning (two cylinders of chilled water given for 1 h every day), acquisition of the conditioned consumption reduction (one cylinder of chilled sucrose and another cylinder of chilled water given for 1 h followed by an injection of 1.5 mEq/kg LiCl), and one extinction test. The second group (Ext-9+) was given the same experiences as the first group and at least 8 more extinction tests. The third group (No-CCR) was not exposed to behavioral testing, that is, no preconditioning or acquisition and extinction testing. Because the unconditioned males had been adapted to their housing environment for two weeks, had been handled only for the purposes of cage cleaning, and had not been exposed to any conditioning procedures, it is reasonable to assume that the VP levels in these males are representative of a homeostatic state.

VP levels in the PVN and the BST were higher in the two groups given behavioral testing than in the unconditioned group (see Fig. 6). There are at least two explanations for this effect. One possibility is that the elevation in VP is an anticipatory response based on a circadian rhythm, especially in light of the involvement of VP in circadian time keeping (Ingram et al., 1998). These animals had been adapted for 10 days (Ext-1) or longer (Ext-9) to a schedule in which they were given chilled fluids every 24 h at the beginning of the dark portion of the light/dark cycle. The VP levels were measured just before the next scheduled exposure. It is well known that animals display anticipatory responses based on a circadian rhythm of drinking or feeding (Bolles, 1963, 1968; Bolles and Moot, 1973). The neural site of the

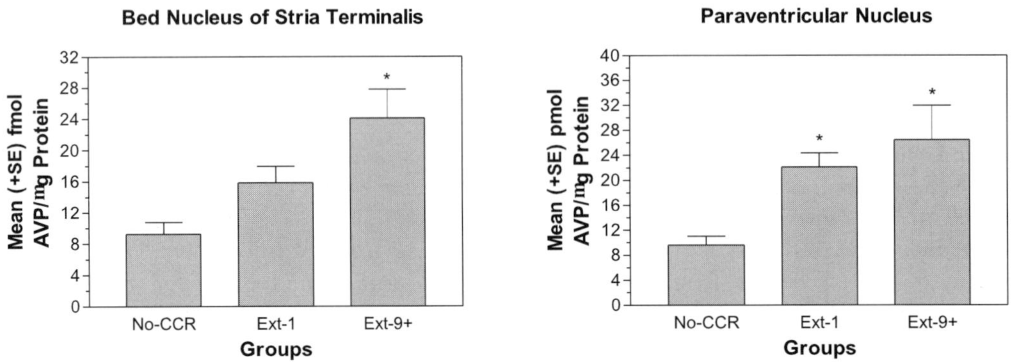

Fig. 6. Mean (+SE) fmol of AVP/μg of protein in the BST and pmol of AVP/μg of protein in the paraventricular nucleus of unconditioned male rats (no conditioned consumption reduction; No-CCR) and of conditioned male rats terminated 23 h after the first extinction test (Ext-1) or after at least nine extinction tests (Ext-9+). None of the conditioned rats had begun extinguishing at the time of termination. *Significantly higher than the No-CCR group, $p \leq 0.05$. These data were adapted from Brownson et al., 2002.

pacemaker for this circadian rhythm remains unknown, but it apparently is not the suprachiasmatic nucleus (Boulos et al., 1998). Another explanation for the elevated VP level is that it mediates the continued expression of consumption suppression during extinction. None of the males had begun extinguishing at the time of termination. As mentioned in the previous section, there is some evidence for an involvement of the BST in extinction of conditioned consumption reduction (Yamamoto, et al., 1981; Brownson et al., 2002). Whether the PVN plays a role in extinction remains unexplored. It should be noted that one cannot assume that the elevated VP levels in the conditioned males represent an increase in VP release and vasopressinergic neuromodulation. The interpretation of increases in VP content is not straightforward and does not necessarily reflect increased release (Landgraf et al., 1998).

Fluid imbalance and behaviors associated with pain and malaise

Effects of fluid imbalance on vasopressin release

An initial physiological response to fluid imbalance, either by fluid deprivation or salt loading, is an increase in the synthesis and release of VP and the consequent increase in retention of water. Various effects of fluid deprivation and salt loading on systemic VP levels have been found in a number of different species, including rats, dogs, hoofed animals, and humans (Ramsay et al., 1977; Wang et al., 1982; Doris and Bell, 1984; Geelen et al., 1984; Szczepanska-Sadowska et al., 1984; Houpt et al., 1989; Carvalho et al., 1996; Thornton et al., 1987; Huch et al., 1998). In rats, plasma VP increases in a linear manner across 24–72 h of total fluid deprivation and when drinking water is replaced with a 2% saline solution, plasma VP reaches a peak after 24 h, which is maintained until at least 72 h (Mens et al., 1980).

Investigators have attempted to assess the effects of fluid deprivation and salt loads on the neuromodulator release of VP by measuring levels in various regions of the brain. However, interpretation of this work is limited by inconsistent findings and the inability to state whether increased or decreased content reflects increased or decreased vasopressinergic transmission (Landgraf et al., 1998). These qualifiers aside, there is a picture that has emerged. VP content in the PVN is elevated in rats that have been fluid deprived for 23 h and induction of c-FLI has been observed in this nucleus after 5, 16, and 24 h of fluid deprivation (see Fig. 7; Morien et al., 1999; Brownson et al., 2002). Because systemic levels of VP are elevated after 24 h of fluid deprivation, the activation of cells projecting to the neurohypophysis and releasing VP into the general circulation are the source of at least some of the elevated VP content.

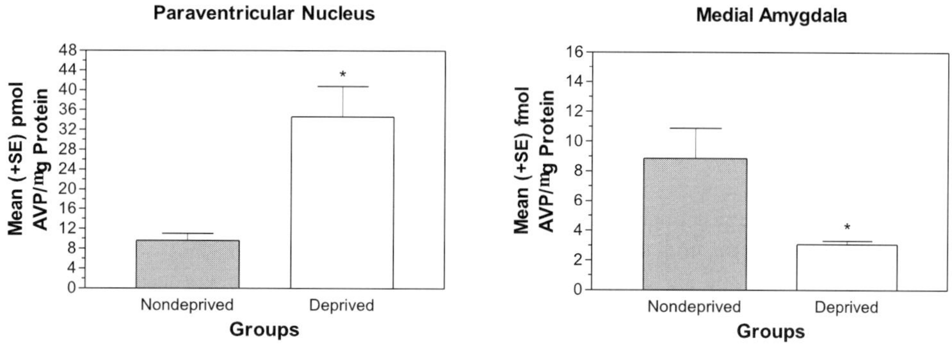

Fig. 7. Mean (+SE) pmol of AVP/μg of protein in the paraventricular nucleus and fmol of AVP/μg of protein in the medial amygdala of nondeprived male rats and fluid-deprived male rats terminated 23 h after the initiation of fluid deprivation. *Significantly different than the nondeprived group, $p \leq 0.03$. These data were adapted from Brownson et al., 2002.

VP content in the PVN is depressed after 3, 4, and 7 days of fluid deprivation (Negro-Vilar and Samson, 1979; Rougon-Rapuzzi et al., 1978; Epstein et al., 1983). Because systemic VP levels and VP messenger RNA levels in the PVN are elevated after 3 days of fluid deprivation (Mens et al., 1980; Arima et al., 1998), the suppressed content within this nucleus indicates that there is facilitation in both synthesis and release. Recent neural lesion studies suggest that the area postrema contributes to the regulation of VP synthesis and release during fluid imbalance (Arima et al., 1998). The increase in plasma VP levels after salt loading is attenuated in area postrema-lesioned rats and VP messenger RNA levels in the PVN are lower in area postrema lesioned than sham rats after 3 days of water deprivation. Although elevation in VP content in the PVN can be tied to hormonal release of VP, this does not preclude the possibility that some of the elevation represents activation of neurons that release VP as a neuromodulator.

The medial amygdala is a VP synthesis site that has two major projections. One travels via the ventral amygdalofugal pathway to the dorsal and ventral lateral septum and medial preoptic area and the other travels through the amygdalohippocampal transition zone to the ventral hippocampus (Caffe et al., 1987). In a recent study, we failed to find changes in VP levels in the medial amygdala when an overall analysis involving a number of different groups was computed. Direct comparison of VP levels in the medial amygdala and lateral septum of only nondeprived and 23-h fluid deprived males shows that levels are lower in the medial amygdala but not the lateral septum of the fluid-deprived males (see Fig. 7; Brownson et al., 2002). However, decreased VP content has been found in the amygdala as well as the lateral septum and hippocampus after 3 days of fluid deprivation (Epstein et al., 1983). Taken together, these data suggest that fluid deprivation reduces vasopressinergic synthesis in and release from the medial amygdala. The medial amygdala is implicated in reproductive behaviors, such as sexual behavior (De Jonge et al., 1992; Wood and Newman, 1993). Reduced release may reflect an inhibition of these behaviors so that the animal will not be distracted from the more urgent search for water.

Effects of fluid imbalance on learned behaviors

Shock avoidance

Because fluid imbalance induced by fluid deprivation or salt loading increases endogenous levels of VP, one would expect either of these procedures to facilitate shock avoidance maintenance. Evidence for this suggestion was found when intraperitoneal injections of hypertonic saline were given immediately after rats experienced footshock in a one-trial step-through passive shock avoidance situation. These animals showed longer avoidance latencies than untreated rats when tested 24 h later (Lebrun et al., 1985a). This improvement in avoidance maintenance was comparable to rats that had received peripheral administration of 2 μg of VP and it could be blocked by the subcutaneous administration of the VP antagonist, dPtyr(Me)AVP.

Conditioned consumption reduction

Extinction of conditioned consumption reduction is accelerated under fluid deprivation and fluid restriction schedules in a number of different strains and stocks of rats (Grote and Brown, 1973; Peck and Ader, 1974; Sengstake et al., 1978). In our laboratory, a faster extinction rate in fluid restricted (22–23 h schedule) as compared to nonrestricted Sprague-Dawley and Fischer 344 male rats has been replicated numerous times (Sengstake et al., 1978; Sengstake and Chambers, 1979; Chambers et al., 1993; Brownson et al., 1994).

Sprague-Dawley rats

We have suggested that fluid restriction accelerates extinction of conditioned consumption reduction by decreasing testosterone availability through a reduction in serum testosterone levels (Sengstake et al., 1978; Sengstake and Chambers, 1979; Chambers, 1985; Chambers et al., 1993, 1997; Brownson et al., 1994). There is a considerable amount of evidence to support this hypothesis in Sprague-Dawley male rats. First, reducing testosterone levels in nonrestricted male rats through gonadectomy accelerates extinction (Chambers, 1976, 1980; Sengstake and Chambers, 1991). Second, fluid restriction accelerates extinction in a choice situation in males but not females (Sengstake et al., 1978). Third, serum testosterone levels are lower in fluid restricted male rats than in nondeprived males and administering testosterone to fluid restricted males restores extinction rates to those of untreated nonrestricted males (Sengstake et al., 1978; Chambers et al., 1993, 1997). Finally, testosterone and fluid restriction have the same pattern of behavioral effects, that is, they affect extinction by acting during extinction but not during acquisition (Chambers and Sengstake, 1979; Sengstake and Chambers, 1979).

We have suggested that the reduced serum levels of testosterone in fluid restricted males are the direct result of elevations in circulating VP (Chambers et al., 1993). In vitro studies have demonstrated that VP exerts a dose-dependent inhibition of androgen biosynthesis in the Leydig cells of the testes. Only those VP agonists that selectively exert pressor, but not antidiuretic, effects are capable of inhibiting androgen biosynthesis. This suggests that the antigonadal activity of VP is mediated by V_1 receptors within the Leydig cells (Adashi and Hsueh, 1981a,b; Meidan and Hsueh, 1985; Kasson and Hsueh, 1986). Because higher levels of VP are required to inhibit androgen biosynthesis in vitro than circulates endogenously, it has been hypothesized that androgen biosynthesis is modulated by local production of VP in the testes rather than by fluctuations in the general circulation (Kasson et al., 1985; Kasson and Hsueh, 1986). Additional support for this hypothesis includes the presence of VP-like peptide, functional receptors, and a degradative enzyme in the testes of homozygous Brattleboro rats. However, there is evidence that the decreases in plasma testosterone levels found after exposure to another type of stressor, immobilization, is due to increased release of VP into the general circulation (Keil and Severs, 1977; Husain et al., 1979; Collu et al., 1984). Immobilization imposed for 2 h increases plasma VP and decreases plasma testosterone levels in Long-Evans but not homozygous Brattleboro rats (Collu et al., 1984).

Fischer 344 rats

Circulating testosterone appears to play a less important role in Fischer 344 males than it does in Sprague-Dawley males. Gonadectomy has no effect on extinction in nonrestricted Fisher 344 males (Chambers et al., 1997). Fluid restriction accelerates extinction in a choice situation in both females and males and it does not decrease testosterone levels in males. Although exogenous testosterone treatment can prolong extinction in fluid restricted males, it takes extremely high doses to do so (Brownson, 1992; Brownson et al., 1994). This suggests that there is a testosterone-independent mechanism that contributes to the acceleration of extinction in fluid restricted Fischer 344 males. A similar mechanism also may exist along with the testosterone-dependent mechanism in Sprague-Dawley rats. When circulating levels of testosterone are controlled via exogenous treatment in gonadectomized Sprague-Dawley males, testosterone is less effective in prolonging extinction in fluid restricted males than nonrestricted males even

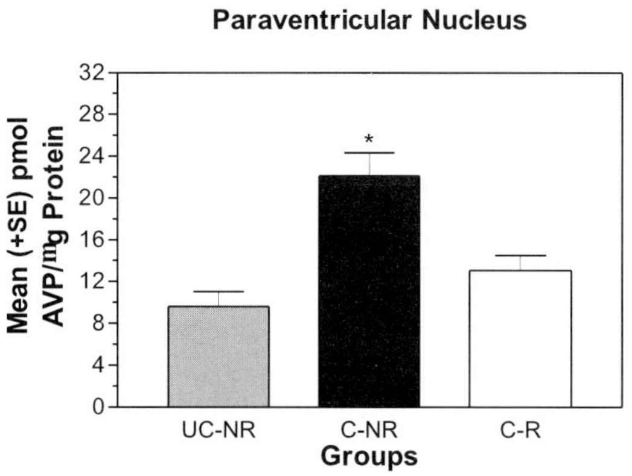

Fig. 8. Mean (+SE) pmol of AVP/μg of protein in the paraventricular nucleus of unconditioned male rats and of conditioned male rats terminated 23 h after the first extinction test. The conditioned males were either on a daily 23 h fluid-restriction schedule and were terminated just before their scheduled 1-h access to fluids (C-R) or they were not fluid restricted (C-NR). The unconditioned rats were not fluid restricted (UC-NR). All of the rats showed maximum consumption reduction during the first extinction test. *Significantly higher than each of the other two groups, $p < 0.003$. These data were adapted from Brownson et al. (2002).

though testosterone levels are similar (Chambers et al., 1993). In a recent study, we measured the VP levels in a number of different neural areas of fluid restricted (23 h schedule) and nonrestricted Fischer 344 males that were given a LiCl-induced conditioned consumption reduction or were unconditioned (Brownson et al., 2002). When these data were reanalyzed comparing VP levels of only the nonrestricted unconditioned male rats and both the fluid restricted and nonrestricted conditioned males 23 h after the first extinction test, VP levels were higher in the PVN of the nonrestricted conditioned male rats (see Fig. 8). These results raise the possibility that a testosterone-independent vasopressinergic system in the PVN plays a critical role in the differential extinction rate of fluid restricted and nonrestricted males, which will need to be verified by manipulating VP levels in this brain site during extinction of conditioned consumption reduction.

Conclusion

After the initial studies showing that VP maintained shock avoidance learning, VP was launched as a mnemonic hormone/neuromodulator. A flurry of studies across a wide range of learning tasks was conducted and many of these revealed a similar maintenance effect, which solidified the hypothesis. Although this hypothesis has been attractive, it has proven to be unable to handle a vast and growing amount of data. In consumption reduction learning studies, VP accelerates or delays extinction depending on the dose and whether it is administered before acquisition and extinction tests or after acquisition tests. Studies employing other learned tasks associated with the eating system also have found vaiability in the kinds of effects VP exerts. It has detrimental or facilitatory effects depending on the particular task and behavioral measure or the time during the learning process that it is given. These studies have used food deprivation to motivate animals to learn how to obtain food. VP slowed acquisition of a continuously reinforced and fixed interval bar press response (Alliot and Alexinsky, 1982; Andrews et al., 1983). However, it had no effect on acquisition and extinction of a light–dark discrimination requiring bar press responses (Alliot and Alexinsky, 1982). In another type of discrimination task two distinctive runways were employed; food was always present at the end of one of the runways and it was never present at the end of the other runway. Infusion of VP into the ventral hippocampus had detrimental effects on performance at the beginning of the learning process and facilatatory effects toward the end of acquisition (Paban et al., 1997; Alescio-Lautier and Soumireu-Mourat, 1998; Alescio-Lautier et al., 2000). In a third discrimination task, one of two retractable levers was presented (the sample) and the rat had to press the lever to obtain food. Then after a delay, both levers were presented and the rats were required to remember the lever that had most recently been presented (the choice). During the delay, the rats were required to approach and operate a magazine flap with a nose-poke in order to release the retracted levers. This prevented the rat from staying at the side of the sample lever and eliminated a simple response strategy. VP-treated rats selected a particular side and restricted their sample and choice responses to that side, which resulted in few errors of commission on the alternative lever (incorrect responses) but an increase in errors of omission on the alternative lever (Sahgal, 1987, 1990).

Because of the variability in the effects of VP in appetitive tasks, the mnemonic hypothesis has been challenged for these tasks as well as for shock avoidance tasks (Andrews et al., 1983; Sahgal, 1984, 1987, 1988, 1990; van Haaren et al., 1986). VP facilitates maintenance of shock avoidance in most animals and this effect is found over a wide range of doses and at different times of administration. However, even for these types of tasks, there are individuals within a strain and strains within a species that show deleterious effects when VP is administered. The question all of the results raise is whether any one hypothesis can handle all of the data. Should we abandon the mnemonic hypothesis for another? We think the answer to that question is no. The data for shock avoidance suggests that VP acts directly on neural areas mediating learning and memory. The data for consumption reduction learning suggests VP acts on an illness sensory system, reducing its activation when low doses are given after pairing sucrose consumption with LiCl illness and increasing its activation when high doses are given. The data for the delayed matching to position food-appetitive task suggests that VP acts on an arousal–motivational system. This conclusion is based on application of the theory of signal detection, a theory proposing that performance in choice working memory tasks is determined by two independent factors, sensitivity of the neural pathways mediating learning and memory and those mediating motivational states (Pontecorvo, et al., 1996). We suggest that some of the effects of VP are due to its direct action on the neural systems mediating learning and memory processes while other effects are due to its action on modulatory neural systems, including systems that mediate the aversive properties of VP and systems that mediate arousal.

References

Adachi, A., Kobashi, M., Miyoshi, N. and Tsukamoto, G. (1991) Chemosensitive neurons in the area postrema of the rat and their possible functions. Brain Res. Bull., 26: 137–140.

Adashi, E.Y. and Hsueh, J.W. (1981a) Direct inhibition of testicular androgen biosynthesis by arginine-vasopressin: mediation through pressor-selective testicular recognition sites. Endocrinol., 109: 1793–1795.

Adashi, E.Y. and Hsueh, J.W. (1981b) Direct inhibition of testicular androgen biosynthesis revealing antigonadal activity of neurohypohysial hormones. Nature 293: 650–652.

Ader, R., Weijnen, J.A.W.M. and Moleman, P. (1972) Retention of a passive avoidance response as a function of the intensity and duration of electric shock. Psychon. Sci., 26: 125–128.

Ader, R. and DeWied, D. (1972) Effects of lysine vasopressin on passive avoidance learning. Psychon. Sci., 29: 46–84.

Ahn, D.K., Kim, K.H., Ju, J.S., Kwon, S. and Park J.S. (2001) Microinjection of arginine vasopressin into the central nucleus of amygdala suppressed nociceptive jaw opening reflex in freely moving rats. Brain Res. Bull., 55: 117–121.

Alescio-Laurier, B. and Soumireu-Mourat, B. (1990) Effects of peripheral administered arginine-vasopressin on learning, retention and forgetting in mice. Behav. Brain Res., 41: 117–128.

Alescio-Laurier, B. and Soumireu-Mourat, B. (1998) Role of vasopressin in learning and memory in the hippocampus. Prog. Brain Res., 119: 501–521.

Alescio-Lautier, B., Paban, V. and Soumireu-Mourat, B. (2000) Neuromodulation of memory in the hippocampus by vasopressin. Eur. J. Pharmacol., 405: 63–72.

Alliot, J. and Alexinsky, T. (1982) Effects of posttrial vasopressin injections on appetitively motivated learning in rats. Physiol. Behav., 28: 525–530.

Andrews, J.S., Newton, B.A. and Sahgal, A. (1983) The effects of vasopressin on positively rewarded responding and on locomotor activity in rats. Neuropeptides, 4: 17–29.

Ang, V.T.Y. and Jenkins, J.S. (1982) Blood-cerebrospinal fluid barrier to arginine-vasopressin, desmopressin and desglycinamide arginine-vasopressin in the dog. J. Endocrinol., 93: 319–325.

Anisman, H., Zaharia, M.D., Meaney, M.J., and Merali, Z. (1998) Do early-life events permanently alter behavioral and hormonal responses to stressors? Internat. J. Devel. Neurosci. 16: 149–164.

Arima, H., Konido, K., Murase, T., Yokol, H., Iwasaki, Y., Saito, H. and Oiso, Y. (1998) Regulation of vasopressin synthesis and release by area postrema in rats. Endocrinol., 139: 1481–1486.

Aston-Jones, G. and Bloom, F.E. (1981) Activity of norepinephrine-containing locus coeruleus neurons in behaving rats anticipates fluctuations in the sleep-waking cycle. J. Neurosci., 1: 876–886.

Bailey, W.H. and Weiss, J.M. (1981) Avoidance conditioning and endocrine function in brattleboro rats. In: Martinez Jr., J.L., Jensen, R.A., Messing, R.B., Rigter, H., McGaugh, J.L. (Eds.), Endogenous Peptides and Learning and Memory Processes. Academic Press, New York, pp. 371–395.

Bamshad, M., Novak, M.A. and DeVries, G. (1993) Sex and species differences in the vasopressin innervation of sexually naïve and parental prairie voles, *Microtus ochrogaster* and meadow voles, *Microtus pennsylvanicus*. J. Neuroendocrinol., 5: 247–255.

Babar, E., Melik, E., Ozgunen, T. and Polat, S. (2002) Effects of excitotoxic median raphe lesion on working memory deficits produced by the dorsal hippocampal muscarinic receptor blockade in the inhibitory avoidance in rats. Brain Res. Bull., 57: 683–688.

Barberis, C., Balestre, M.N., Jard, S., Tribollet, E., Arsenijevic, Y., Dreifuss, J.J., Bankowski, K., Manning, M., Chan, W.Y., Schlosser, S.S., Holsboer, F. and Elands, J. (1995) Characterization of a novel, linear radioiodinated vasopressin antagonist: an excellent radioligand for vasopressin V1a receptors. Neuroendocrinol., 62: 135–146.

Bargmann, W. and Scharrer, E. (1951) The site of origin of the hormones of the posterior pituitary. Am. Sci., 39: 255–259.

Batson, J.D. (1983) Effects of repeated lithium injections on temperature, activity and flavor conditioning in rats. Anim. Behav., 11: 199–204.

Baxter, B.L., Gluckman, M.I., Stein, L. and Scerni, R.A. (1974) Self-injection of apomorphine in the rat: positive reinforcement by a dopamine receptor stimulant. Pharmacol. Biochem. Behav., 2: 387–391.

Bechara, A. and van der Kooy, D. (1985) Opposite motivational effects of endogenous opioids in brain and periphery. Nature, 314: 533–534.

Berecek, K.H., Olpe, H.R., Jones, R.S. and Hofbauer, K.G. (1984) Microinjection of vasopressin into the locus coeruleus of conscious rats. Am. J. Physiol., 247(4, Pt. 2): H675–681.

Berman, E.E. and Dudai, Y. (2001) Memory extinction, learning anew, and learning the new: dissociations in the molecular machinery of learning in cortex. Sci., 291: 2417–2419.

Bernstein, I.L. and Goehler, L.E. (1983) Chronic LiCl infusions: conditioned suppression of food intake and preference. Behav. Neurosci., 97: 290–298.

Bernstein, I.L., Schwarzberg, H., Poeggel, G. and Reiser, M. (1994) Intracerebroventricular but not intraperitoneal administration of aluminum attenuates vasopressin-enhanced retrieval of a passive avoidance tasks in rats. Pharmacol. Biochem. Behav., 47: 587–590.

Berntson, G.G. and Berson, B.S. (1980) Antinociceptive effects of intraventricular or systemic administration of vasopressin in the rat. Life Sci., 26: 455–459.

Berridge, K., Grill, H.J. and Norgren, R. (1981) Relation of consummatory responses and preabsorptive insulin release to palatability and learned taste aversions. J. Comp. Physiol. Psychol., 95: 363–382.

Best, M., Best, P. and Mickley, G. (1973) Conditioned aversion to distinct environmental stimuli resulting from gastrointestinal distress. J. Comp. Physiol. Psychol., 85: 250–257.

Bluthe, R.M., Dantzer, R. and LeMoal, M. (1985a) Peripheral injections of vasopressin control behavior by way of interoceptive signals for hypertension. Behav. Brain Res., 18: 31–39.

Bluthe, R.M., Dantzer, R., Mormede, P. and LeMoal, M. (1985b) Specificity of aversive stimulus properties of vasopressin. Psychopharmacol., 87: 238–241.

Boccia, M.M., Kopf, S.R. and Baratti, C.M. (1998) Effects of a single administration of oxytocin or vasopressin and their interactions with two selective receptor antagonists on memory storage in mice. Neurobiol. Learn. Mem., 69: 136–146.

Bodnar, R.J., Wallace, M.M., Kordower, J.H., Nilaver, G., Cort, J. and Zimmerman, E.A. (1982) Modulation of nociceptive thresholds by vasopressin in the brattleboro and normal rat. Ann. NY Acad. Sci., 394: 735–739.

Bodnar, R.J., Zimmerman, E.A., Nilaver, G., Mansour, A., Thomas, L.W., Kelly, D.D. and Glusman, M. (1980) Dissociation of cold-water swim and morphine analgesia in Brattleboro rats with diabetes insipidus. Life Sci., 26: 1581–1590.

Bohus, B., Ader, R. and DeWied, D. (1972) Effects of vasopressin on active and passive avoidance behavior. Horm. Behav., 3: 191–197.

Bohus, B. and DeWied, D. (1998) The vasopressin deficient Brattleboro rats: A natural knockout model used in the search for CNS effects of vasopressin. Prog. Brain Res., 119: 555–573.

Bohus, B., Gispen, W.H. and DeWied, D. (1973) Effect of lysine vasopressin and $ACTH_{4-10}$ on conditioned avoidance behavior of hypophysectomized rats. Neuroendocrinol., 11: 137–143.

Bohus, B., Urban, I. van Wimersma Greidanus, T.B. and DeWied, D. (1978) Opposite effects oxytocin and vasopressin on avoidance behaviour and hippocampal theta rhythm in the rat. Neuropharmacol., 17: 239–247.

Bohus, B., van Wimersma Greidanus, T.J.B. and DeWied, D. (1975) Behavioral and endocrine responses of rats with hereditary hypothalamic diabetes insipidus (Brattleboro strain). Physiol. Behav., 14: 609–615.

Bolles, R.C. (1963) Effect of food deprivation upon the rat's behavior in its home cage. J. Comp. Physiol. Psychol., 56: 456–460.

Bolles, R.C. (1968) Anticipatory general activity in thirsty rats. J. Comp. Physiol. Psychol., 65: 511–513.

Bolles, R.C. and Moot, S.A. (1973) The rat's anticipation of two meals a day. J. Comp. Physiol. Psychol., 83: 510–514.

Booth, D.A. (1977) Satiety and appetite are conditioned reactions. Psychosom. Med., 39: 76–81.

Booth, D.A. (1985) Food-conditioned eating preferences and aversion with interoceptive elements: conditioned appetites and satieties. Ann. NY Acad. Sci., 443: 22–41.

Borison, H.L. (1989) Area postrema: Chemoreceptor circumventricular organ of the medulla oblongata. Prog. Neurobiol., 32: 351–390.

Boulos, Z., Rosenwasser, A.M. and Terman, M. (1998) Feeding schedules and the circadian organization of behavior in the rat. Behav. Brain Res., 1: 39–65.

Bowers, R.L., Herzog, C.D., Stone, E.H. and Dionne, T.J. (1992) Defensive burying following injections of cholecystokinin, bombesin, and LiCl in rats. Physiol. Behav., 41: 969–972.

Bray, G.A. (2000) Afferent signals regulating food intake. Proc. Nutri. Soc., 59: 373–384.

Brown, R.E., Stevens, D.R. and Haas, H.L. (2001) The physiology of brain histamine. Prog. Neurobiol., 63: 637–672.

Brownson, E.A. (1992) Hormonal mechanisms for fast extinction of a conditioned taste aversion in fluid deprived rats. Unpublished doctoral dissertation, University of Southern California, Los Angeles.

Brownson, E.A., Sengstake, C.B. and Chambers, K.C. (1994) The role of serum testosterone in the accelerated extinction of a conditioned taste aversion in fluid deprived male rats. Physiol. Behav., 55: 273–278.

Brownson, E.A., Chambers, K.C. and Brinton, R.D. (2002) Vasopressin content increases in select brain regions during extinction of a conditioned taste aversion. Brain Res. Bull., 59: 125–134.

Buijs, R.M. (1978) Intra- and extrahypothalamic vasopressin and oxytocin pathways in the rat: pathways to the limbic system, medulla oblongata and spinal cord. Cell Tiss. Res., 192: 423–435.

Burbach, J.P.H., Wang, X.C., Ten Haaf, J.A. and De Wied, D. (1984) Substances resembling C-terminal vasopressin fragments are present in the brain but not in the pituitary gland. Brain Res., 306: 384–387.

Bures, J. and Buresova, O. (1979) Neurophysiological analysis of conditioned taste aversion. In: Brazier, M.A. (Ed.), Brain Mechanisms in Memory and Learning: From the Single Neuron to Man. Raven Press, New York, pp. 127–138.

Caffe, A.R., van Leeuwen, F.W. and Luiten, P.G.M. (1987) Vasopressin cells in the medial amygdala of the rat project to the lateral septum and ventral hippocampus. J. Comp. Neurol., 261: 237–252.

Callahan, M.F., Thore, C.R., Sunberg, D.K., Gruber, K.A., O'Steen K. and Morris, M. (1992) Excitoxin paraventricular nucleus lesions: Stress and endocrine reactivity and oxytocin mRNA levels. Brain Res., 597: 8–15.

Car, H., Borawska, M. and Wisniewski, K. (1993) The effects of vasopressin analog: [d(CH_2)$_5$, Tyr(Me)^2, Δ^3Pro^7]AVP on learning and memory processes in rats with experimental amnesia. Pol. J. Pharmacol., 45: 11–22

Car, H., Borawska, M. and Wisniewski, K. (1994) Effects of vasopressin and vasopressin analogue [d(CH_2)$_5$, Tyr(Me)^2, Val^4, Δ^3Pro^7] AVP on learning and memory in rats chronically treated with ethanol. J. Physiol. Pharmacol., 45: 163–171.

Carpenter, D.O., Briggs, D.B., Knox, A.P. and Strominger, N. (1988) Excitation of area postrema neurons by transmitters, peptides and cyclic nucleotides. J. Neurophysiol., 59: 358–369.

Carpenter, J.A. (1956) Species difference in taste preferences. J. Comp. Physiol. Psychol., 49: 139–144.

Carvalho, A.C., Botelho, L.M., Green, L.J. and Santos, R.A. (1996) Effect of acute volume expansion associated with salt load on the profile of plasma angiotensins in rats. Immunopharmacol., 33: 143–145.

Celestian, J.F., Carey, R.J. and Miller, M. (1975) Unimpaired maintenance of a conditioned avoidance response in the rat with diabetes insipidus. Physiol. Behav., 15: 707–711.

Chambers, K.C. (1976) Hormonal influences on sexual dimorphism in the rate of extinction of a conditioned taste aversion in rats. J. Comp. Physiol. Psychol., 90: 851–856.

Chambers, K.C. (1980) Progesterone, estradiol, testosterone and dihydrotestosterone: effects on rate of extinction of a conditioned taste aversion in rats. Physiol. Behav., 24: 1061–1065.

Chambers, K.C. (1985) Sexual dimorphisms as an index of hormonal influences on conditioned food aversions. Ann. NY Acad. Sci., 443: 110–125.

Chambers, K.C. (1990) A neural model for conditioned taste aversions. Ann. Rev. Neurosci., 13: 373–385.

Chambers, K.C. and Bernstein, I.L. (1995) Conditioned flavor aversions. In: Doty, R.L. (Ed.), Handbook of Clinical Olfaction and Gustation, 1st Edn. Marcel Dekker, New York, pp. 745–773.

Chambers, K.C. and Bernstein, I.L. (2003) Conditioned flavor aversions. In: Doty, R.L. (Ed.), Handbook of Clinical Olfaction and Gustation, 2nd Edn. Marcel Dekker, New York.

Chambers, K.C. and Sengstake, C.B. (1979) Temporal aspects of the dependency of a dimorphic rate of extinction on testosterone. Physiol. Behav., 22: 53–56.

Chambers, K.C., Sengstake, C.B., Brownson, E.A. and Westfahl, P.K. (1993) Decreased testosterone levels and accelerated extinction of a conditioned taste aversion in fluid deprived male rats. Behav. Neurosci., 107: 299–305.

Chambers, K.C. and Wang, Y. (2004) Role of the lateral parabrachial nucleus in apomorphine-induced conditioned consumption reduction: cooling lesions and relationship of c-Fos-like-immunoreactivity to strength of conditioning. Behav. Neurosci., 118: 199–213.

Chambers, K.C., Yuan, D.L., Brownson, E.A. and Wang, Y. (1997) Sexual dimorphism in conditioned taste aversions: mechanism and function. In: Bouton, M.E. and Fanselow, M.S. (Eds.), Learning, Motivation, and Cognition: The Functional Behaviorism of Robert C. Bolles. American Psychological Association, Washington D.C., pp. 195–224.

Chang, H.P., Ma, Y.L., Wan, F.J., Tsai, L.Y., Lindberg, F.P. and Lee, E.H. (2001) Functional blocking of integrin-associated protein impairs memory retention and decreases glutamate release from the hippocampus. Neurosci., 102: 289–296.

Cheng, S.W.T., O'Connor, E.F. and North, W.G. (1986) Acute infusion of lithium chloride raises blood pressure in the conscious rat. Neuropharmacol., 25: 1185–1190.

Collu, R., Gibb, W., Bichet, D.G. and Ducharme, J.R. (1984) Role of arginine-vasopressin (AVP) in stress-induced inhibition of testicular steroidogenesis in normal and in AVP-deficient rats. Endocrinol., 115: 1609–1615.

Cooper, R.L., McNamara, M.C. and Thompson, W.G. (1980) Vasopressin and conditioned flavor aversion in aged rats. Neurobiol. Aging, 1: 53–57.

Croiset, G., Heijnen, C.J. and DeWied, D. (1990) Passive avoidance behavior, vasopressin and the immune system. Neuroendocrinol., 51: 156–161.

Cunningham, C.L. and Niehus, J.S. (1993) Drug-induced hypothermia and conditioned place aversion. Behav. Neurosci., 107: 468–479.

Dantzer, R. (1998) Vasopressin, gonadal steroids and social recognition. Prog. Brain Res., 119: 409–414.

Dantzer, R., Bluthe, R.M. and Kelly, K.W. (1991) Androgen-dependent vasopressinergic neruotransmission attenuates interleukin-1-induced sickness behavior. Brain Res., 557: 115–120.

DeJong, W., Gaffori, O., van Ree, J.M. and DeWied, D. (1985) Differentiation of behavioral and peripheral actions of neurpeptides generated from vasopressin in the brain. In: Schrier, R.W. (Ed.), Vasopressin. Raven Press, New York, pp. 189–194.

De Jonge, F.H., Oldenburger, W.P., Louwerse, A.L., and van de Poll, N.E. (1992) Changes in male copulatory behaviour after sexual exciting stimuli: Effects of medial amygdala lesions. Physiol. Behav. 52: 327–332.

DeVries, G.J. and Buijs, R.M. (1983) The origin of the vasopressinergic and oxytocinergic innervation of the rat brain with special reference to the lateral septum. Brain Res., 273: 307–317.

DeWied, D. (1971) Long term effect of vasopressin on the maintenance of a conditioned avoidance response in rats. Nature, 232: 58–60.

DeWied, D. (1976) Behavioral effects of intraventricularly administered vasopressin and vasopressin fragments. Life Sci., 19: 685–690.

DeWied, D., Bohus, B. and van Wimersma Greidanus, T.J.B. (1975) Memory deficit in rats with hereditary diabetes insipidus. Brain Res., 85: 152–156.

DeWied, D., Diamant, M. and Fodor, M. (1993) Central nervous system effects of the neurohypophyseal hormones and related peptides. Front. Neuroendocrinol., 14: 251–302.

DeWied, D., Elands, J. and Kovács, G. (1991) Interactive effects of neurohypophyseal neuropeptides with receptor antagonists on passive avoidance behavior: mediation by a cerebral neurohypophyseal hormone receptor? Proc. Nat. Acad. Sci., 88: 1494–1498.

DeWied, D., Gaffori, O., Burbach, J.P.H., Kovács, G.L. and van Ree, J.M. (1987) Structure activity relationship studies with C-terminal fragments of vasopressin and oxytocin on avoidance behaviors in rats. J. Pharmacol. Exp. Ther., 241: 268–274.

DeWied, D., Gaffori, O., van Ree, J.M. and DeJong, W. (1984) Central target for the behavioural effects of vasopressin neuropeptides. Nature, 308: 276–278.

DeWied, D., Joels, M., Burbach, J.P.H., DeJong, W., De Kloet, E.R., Gaffori, O.W.J., Urban, I.J.A., van Ree, J.M., van Wimersma Greidanus, T.B., Veldhuis, H.D., Versteeg, D.H.G. and Wiegant, V.M. (1988) Vasopressin effects on the central nervous system. In: Negro-Vilar, A. and Conn, P.M. (Eds.), Peptide Hormones: Effects and Mechanisms of Action, Vol. 1. CRC Press, Inc., Boca Raton, Fl, pp. 97–140.

DeWied, D., van Wimersma Greidanus, T.B., Bohus, B., Urban, I. and Gispen, W.H. (1976) Vasopressin and memory consolidation. Prog. Brain Res., 45: 181–194.

Dietrich, A. and Allen, J.D. (1997) Vasopressin and Memory. I. The vasopressin analogue AVP4-9 enhances working memory as well as reference memory in the radial arm maze. Behav. Brain Res., 87: 195–200.

Domjan, M., Foster, K. and Gillan, D.J. (1979) Effects of distribution of the drug unconditioned stimulus on taste-aversion learning. Physiol. Behav., 23: 931–938.

Doris, P.A. and Bell, F.R. (1984) Vasopressin in plasma and cerebrospinal fluid of hydrated and dehydrated steers. Neuroendocrinol, 38: 290–296.

Duncan, G.E., Knapp, D.J. and Breese, G.R. (1996) Neuroanatomical characterization of Fos induction in rat behavioral models of anxiety. Brain Res., 713: 79–91.

Ebenezer, I. (1988) Can vasopressin alone act as an unconditioned stimulus to produce passive avoidance behaviour in rats in a typical memory experiment? Neuropharmacol., 27: 903–907.

Eckel, L.A. and Ossenkopp, K.-P. (1996) Area postrema mediates the formation of rapid, conditioned palatability shifts in lithium-treated rats. Behav. Neurosci., 110: 202–212.

Engelmann, M. Wotjak, C.T., Ebner, K. and Landgraf, R. (2000) Behavioural impact of intraseptally released vasopressin and oxytocin in rats. Exp. Physiol., 85: 125S–130S.

Engelmann, M., Wotjak, C.T., Neumann, I., Ludwig, M. and Landgraf, R. (1996) Behavioral consequences of intracerebral vasopressin and oxytocin: focus on learning and memory. Neurosci. Biobehav. Rev., 20: 341–358.

Epstein, Y., Castel, M., Glick, S.M., Sivan, N. and Ravid, R. (1983) Changes in hypothalamic and extra-hypothalamic vasopressin content of water deprived rats. Cell Tiss. Res., 233: 99–111.

Ettenberg, A. (1984) ICV Application of a vasopressin antagonist peptide prevents the behavioral actions of vasopressin. Behav. Brain Res., 14: 201–211

Ettenberg, A., van der Kooy, D., Le Moal, M., Koob, G.F. and Bloom, F.E. (1983) Can aversive properties of peripherally injected vasopressin account for its putative role in memory? Behav. Brain Res., 7: 331–350.

Faiman, C.P., deErausquin, G.A. and Baratti, C.M. (1992) Modulation of memory retrieval by pre-testing vasopressin: Involvement of a central cholinergic nicotinic mechanism. Methods Findings Exp. Clin. Pharmacol., 14: 607–613.

Feuerstein, G., Zerbe, R.L. and Faden, A.L. (1984) Central cardiovascular effects of vasotocin, oxytocin and vasopressin in conscious rats. J. Pharmacol. Exp. Ther., 228: 348–353.

Fink, G.D., Bruner, C.A. and Mangiapane, M.L. (1987a) Area postrema is critical for angiotensin-induced hypertension in rats. Hypertension, 9: 355–361.

Fink, G.D., Pawloski, C.M., Blair, M.L. and Mangiapane, M.L. (1987b) The area postrema in deoxycorticosterone-salt hypertension in rats. Hypertension, 9: 206–209.

Flaherty, C.F. and Checke, S. (1982) Anticipation of incentive gain. Anim. Learn. Behav., 10: 177–182.

Flanagan, L.M., Verbalis, J.G. and Stricker, E.M. (1989) Effects of anorexigenic treatments on gastric motility in rats. Amer. J. Physiol., 256: R955–R961.

Furyua, K., Shimizu, R, Hirabayashi, Y, Ishii, R. and Fukuda, H. (1993) Stress hormone responses to major intra-abdominal surgery during and immediately after sevoflurane-nitrous oxide anaesthesia in elderly patients. Can. J. Anaesth., 40: 435–439.

Gaffori, O. and DeWied, D. (1985) Further evidence for a dissociation of peripheral and central effects of vasopressin. Psychoneuroendocrinol., 10: 439–444.

Gaffori, O., van Ree, J.M. and DeWied, D. (1985) Effect of Desglycinamide (Arg^8)vasopressin (DGAVP) and ACTH (4-10) on memory and attentional processes as assessed with food search behavior in rats. Neurosci. Lett. Suppl., 22: S494.

Gais, S., Sommer, M., Fischer, S., Perras, B. and Born, J. (2002) Post-trial administration of vasopressin in humans does not enhance memory formation (vasopressin and memory consolidation). Peptides, 23: 581–583.

Garcia, J. (1989) Food for Tolman: Cognition and Cathexis in Concert. In: Archer, T. and Nilson, L.-G. (Eds.), Aversion, Avoidance and Anxiety. Lawrence Erlbaum, Hillsdale, NJ, pp. 45–85.

Garcia, J. and Koelling, R.A. (1966) Relation of cue to consequence in avoidance learning. Pychon. Sci., 4: 123–124.

Geelen, G., Keil, L.C., Kravik, S.E., Wade, C.E., Thrasher, T.N., Barnes, P.R., Pyka, G., Nesvig, C. and Greenleaf, J.E. (1984) inhibition of plasma vasopressin after drinking in dehydrated humans. Am. J. Physiol., 247(6, pt.2): R968–R971.

Gerstberger, R. and Fahrenholz, F. (1989) Autoradiographic localization of V1 vasopressin binding sites in rat brain and kidney. Eur. J. Pharmacol., 167: 105–116.

Gomez, R.E., Cannata, M.A., Milner, T.A., Anwar, M., Reis, D.J. and Ruggiero, D.A. (1993) Vasopressinergic mechanism in the nucleus reticularis in blood pressure control. Brain Res., 604: 90–105.

Grigson, P.S. (1997) Conditioned taste aversions and drugs of abuse: a reinterpretation. Behav. Neurosci., 111: 129–136.

Grote, F.W., Jr. and Brown, R.T. (1973) Deprivation level affects extinction of a conditioned taste aversion. Learn. Motivat., 4: 314–329.

Gunther, O., Kovács, G.L., Szabo, G., Schwarzberg, H. and Telegdy, G. (1988) Differential effect of vasopressin on open-field activity and passive avoidance behaviour following intracerebroventricular versus intracisternal administration in rats. Acta Physiologica Hungarica, 71: 203–206.

Hallbeck, M. and Blomqvist, A. (1999) Spinal cord-projecting vasopressinergic neurons in the rat paraventricular hypothalamus. J. Comp. Neurol., 411: 201–211.

Hamburger, R., Sela, A. and Belmaker, R.H. (1985) Differences in learning and extinction in response to vasopressin in sex inbred mouse strains. Psychopharmacol., 87: 124–125.

Hayes, U.L. and Chambers, K.C. (2002) Central infusion of vasopressin in male rats accelerates extinction of conditioned taste avoidance induced by LiCl. Brain Res. Bull., 57: 727–733.

Hayes, U.L. and Chambers, K.C. (in press) Peripheral vasopressin accelerates extinction of conditioned taste avoidance. Physiol. Behavior.

Hirate, K., Hirano, M., Nakajima, Y., Hiyama, A., Maeda, O. and Asakura, W. (1997) No. 302, a newly synthesized [pGlu4,Cyt6]AVP(4-9) analogue, prevents the disruption of avoidance behavior. Behav. Brain Res., 83: 205–208.

Houpt, K.A., Thornton S.N. and Allen W.R. (1989) Vasopressin in dehydrated and rehydrated ponies. Physiol. Behav., 45: 659–661.

Hreash, R., Keil, L.C., Chou, L. and Reid, I.A. (1990) Effects of carotid occlusion and angiotensin II on vasopressin secretion in intact and vagotomized conscious rabbits. Endocrinol., 127: 1160–1166.

Huch, K.M., Wall, B.M., Mangold, T.A., Bobal, M.A. and Cooke, C.R. (1998) Hemodynamic response to vasopressin in dehydrated human subjects. J. Investig Med., 46: 312–318.

Hunt, T. and Amit, Z. (1987) Conditioned taste aversion induced by self-administered drugs: Paradox revisited. Neurosci. Biobehav. Rev., 1: 107–130.

Husain, M.K., Manger, W.M., Rock, T.W., Weiss, R.J. and Frantz, A.G. (1979) Vasopressin release due to manual restraint in the rat: role of body compression and comparison with other stressful stimuli. Endocrinol., 10: 641–644.

Ingram, C.D., Ciobanu, R., Coculescu, I.L., Tanasescu, R., Coculescu, M. and Mihai, R. (1998) Vasopressin neurotransmission and the control of circadian rhythms in the suprachiasmatic nucleus. Prog. Brain Res., 119: 351–364.

Izquierdo, L.A., Schroder, N., Ardenghi, P., Quevedo, J., Netto, C.A., Medina, J.H. and Izquierdo, I. (1997) Systemic administration of ACTH or vasopressin reverses the amnestic effect of posttraining beta-endorphin or electroconvulsinve shock but not that of intrahippocampal infusion of protein kinase inhibitors. Neurobiol. Learn. Mem., 68: 197–202.

Jolkkonen, J., Tuomisto, L., van Wimersma Greidanus, T.B. and Riekkinen, P.J. (1988) Vasopressin levels in the cerebrospinal fluid of rats with lesions of the paraventricular and suprachiasmatic nuclei. Neurosci. Letters, 86: 184–188.

Jones, M.W., Errington, M.L., French, P.J., Fine, A., Bliss, T.V., Garel, S., Charnay, P., Bozon, B. Laroche, S. and Davis, S. (2001) A requirement for the early gene Zif268 in the expression of late LTP and long-term memories. Nature Neurosci., 4: 289–296.

Karniol, I.G., Dalton, J. and Lader, M.H. (1978) Acute and chronic effects of lithium chloride on physiological and psychological measures in normals. Psychopharmacol., 57: 289–294.

Kasson, B.G. and Hsueh, J.W. (1986) Arginine vasopressin as an intragonadal hormone in Brattleboro rats: presence of a testicular vasopressin-like peptide and functional vasopressin receptors. Endocrinol., 118: 23–31.

Kasson, B.G., Meidan, R. and Hsueh, A.J.W. (1985) Identification and characterization of arginine vasopressin-like substances in the rat testis. J. Biol. Chem., 260(10): 5302–5307.

Keil, L.C. and Severs, W.B. (1977) Reduction in plasma vasopressin levels of dehydrated rats following acute stress. Endocrinol., 100: 30–38.

Khateb, A., Fort, P., Pegna, A., Jones, B.E. and Muhlethaler, M. (1995) Cholinergic nucleus basalis neurons are excited by histamine in vitro. Neurosci., 69: 495–506.

King, A.R. and DeWied, D. (1974) Localized behavioral effects of vasopressin on maintenance of an active avoidance response in rats. J. Comp. Physiol. Psychol., 86: 1008–1018.

Klimkiewicz, T. (2001) Memory effects of arginine vasopressin (AVP) and [7–9] fragment of its peptide chain in rats. Acta Neurobiol. Exp., 61: 267–276.

Koob, G.F., Dantzer, R., Bluthe, R., Lebrun, C., Bloom, F.E. and Le Moal, M. (1986) Central injections of arginine vasopressin prolong extinction of active avoidance. Peptides, 7: 213–218.

Koob, G.F., LeMoal, M., Gaffori, O., Manning, M., Sawyer, W.H., Rivier, J. and Bloom, F.E. (1981) Arginine vasopressin and a vasopressin antagonist peptide: opposite effects on extinction of active avoidance in rats. Regul. Pep. 2: 153–163.

Koolhaas, J.M., Everts, H., de Ruiter, A.J., de Boer, S.F. and Bohus, B. (1998) Coping with stress in rats and mice: differential peptidergic modulation of the amygdala-lateral septum complex. Prog. Brain Res., 119: 437–448.

Kordower, J.H., Kikorszy, V. and Bodnar, R.J. (1982) Central antinociceptive effects of lysine-vasopressin and an analogue. Peptides, 3: 613–617.

Kosten, T. and Contreras, R.J. (1989) Deficits in conditioned heart rate and taste aversion in area postrema-lesioned rats. Behav. Brain Res., 35: 9–21.

Kovács, G.L., Bohus, B., Versteeg, D.H.G., De Kloet, E.R. and DeWied, D. (1979) Effect of oxytocin and vasopressin on memory consolidation: sites of action and catecholaminergic correlates after local microinjection into limbic-midbrain structures. Brain Res., 175(2): 303–314.

Kovács, G.L. Buijs, R.M., Bohus, B. and van Wimersma Greidanus, T.B. (1982) Microinjection of arginine 8-vasopressin antiserum into the dorsal hippocampus attenuates passive avoidance behavior in rats. Physiol. Behav., 28: 45–48.

Kovács, G.L., Veldhuis, H.D., Versteeg, D.H.G. and DeWied, D. (1986) Facilitation of avoidance behavior by vasopressin fragments microinjected into limbic-midbrain structures. Brain Res., 371: 17–24.

Krejčí, I., Kupkova, B., Metys, J., Barth, T. and Jost, K. (1979) Vasopressin analogs: sedative properties and passive avoidance behavior in rats. Eur. J. Pharmacol., 56: 347–353.

Laczi, F., Gaffori, O, De Kloet, E.R. and De Wied, D. (1983a) Arginine vasopressin content of hippocampus and amygdala during passive avoidance behavior in rats. Brain Res., 280: 309–315.

Laczi, F., Gaffori, O., De Kloet, E.R. and De Wied, D. (1983b) Differential responses in immunoreactive arginine-vasopressin content of microdissected brain regions during passive avoidance behavior. Brain Res., 26: 342–346.

Laczi, F., Mens, Gaffori, O., Fekete and DeWied, D. (1984) Levels of arginine-vasopressin in cerebrospinal fluid during passive avoidance behavior in rats. Life Sci., 34: 2385–2391.

Ladowsky, R.L. and Ossenkopp, K.-P. (1986) Conditioned taste aversions and changes in motor activity in lithium-treated rats: mediating role of the area postrema. Neuropharmacol., 25: 71–77.

Landgraf, R., Wotjak, C.T., Neumann, I.D. and Engelmann, M. (1998) Release of vasopressin within the brain contributes to neuroendocrine and behavioral regulation. Prog. Brain Res., 119: 201–220.

Langhans, W., Delprete, E. and Scharrer, E. (1991) Mechanisms of vasopressin's anorectic effect. Physiol. Behav., 49: 169–176.

Lauritsen, K.B., Heltberg, J., Hornum, I. and Rehfeld, J.F. (1978) Lithium inhibits basal and food-stimulated gastrin secretion. Gastroenterol., 75: 59–60.

Le Moal, M., Koob, G.F., Koda, L.Y., Bloom, F.E., Manning, M., Sawyer, W.H. and Rivier, J. (1981) Vasopressor receptor antagonist prevents behavioral effects of vasopressin. Nature, 291: 491–493.

Lebrun, C., Le Moal, M., Dantzer, R., Bloom, F.E. and Koob, G.F. (1985a) Hypertonic saline mimics the effects of vasopressin on inhibitory avoidance in the rat. Behav. Neural Biol., 47: 130–137.

Lebrun, C., Le Moal, M., Koob, G.F. and Bloom, F.E. (1985b) Vasopressin pressor antagonist injected centrally reverses behavioral effects of peripheral injection of vasopressin, but only at doses that reverse increase in blood pressure. Regul. Pept., 11: 173–181.

Leccese, A.P. and Isenhour, J.L. (1983) Intraventicular administration of antivasopressin serum inhibits retention in mice. Peptides, 4: 265–267.

Leshner, A.I., Hofstein, R., Samuel, D. and van Wimersma Greidanus, T.B. (1978) Intraventricular injection of anti-vasopressin serum blocks learned helplessness in rats. Pharmacol. Biochem. Behav., 9: 889–892.

Liard, J.F. (1988) Characterization of acute hemodynamic effects of antidiuretic agonists in conscious dogs. J. Cardiovas. Pharmacol., 11: 174–180.

Lucas, G.A., Timberlake, W., Gawley, D.J. and Drew, J. (1990) Anticipation of future food: suppression and facilitation of saccharin intake depending on the delay and type of future food. J. Exp. Psychol.: Anim. Behav. Proc. 16: 169–177.

Malayan, S.A., Ramsay, D.J., Keil, L.C. and Reid, I.A. (1980) Effects of increases in plasma vasopressin concentration on plasma rennin activity, blood pressure, heart rate, and plasma corticosteroid concentration in conscious dogs. Endocrinol., 107: 1899–1904.

Mangiapane, M.L., Skoog, K.M., Rittenhouse, P.A., Blair, M.L. and Sladek, C.D. (1989) Lesions of the area postrema region attenuates hypertension in spontaneously hypertensive rats. Circ. Res., 64: 129–135.

Maswood, S., Barter, J.E., Watkins, L.R. and Maier, S.F. (1998) Exposure to inescapable but not escapable shock increases extracellular levels of 5-HT in the dorsal raphe nucleus of the rat. Brain Res., 783: 115–120.

Matsuguchi, H., Sharabi, R.M., Gordon, F.J., Johnson, A.K. and Schmid, P.G. (1982) Blood pressure and heart rate responses to microinjections of vasopressin into the nucleus of the solitarius region of the rat. Neuropharmacol., 21: 687–693.

McCann, M.J., Verbalis, J.G. and Stricker, E.M. (1989) LiCl and CCK inhibit gastric emptying and feeding and stimulate OT secretion in rats. Amer. J. Physiol., 256: R463–R468.

McCann, S.M., Antunes-Rodrigues, J., Franci, C.R., Anselmo-Franci, J.A., Karanth, S. and Rettori, V. (2000) Role of the hypothalamic pituitary adrenal axis in the control of the response to stress and infection. Braz. J. Med. Biol. Res., 33: 1121–1131.

McGowan, B.K., Garcia, J., Ervin, F.R. and Schwartz, J. (1969) Effects of septal lesions on bait-shyness in the rat. Physiol. Behav., 4: 907–909.

Meachum, C.L. and Bernstein, I.L. (1990) Conditioned responses to a taste CS paired with LiCl administration. Behav. Neurosci., 104: 711–715.

Meidan, R. and Hsueh, A.J.W. (1985) Identification and characterization of arginine vasopressin receptors in the rat testis. Endocrinol., 116: 416–423

Mens, W.B.J., Bouman, H.J., Bakker, E.A.D. and van Wimersma Greidanus, T.B. (1980) Differential effects of various stimuli on AVP levels in blood and cerebrospinal fluid. Eur. J. Pharmacol., 68: 89–92.

Mens, W.B.J., Witter, A. and van Wimersma Greidanus, T.B. (1983) Penetration of neurohypophyseal hormones from plasma into cerebrospinal fluid (CSF): half-times of disappearance of these neuropeptides from CSF. Brain Res., 262: 143–149.

Meyer, A.H., Langhans, W. and Scharrer, E. (1989) Vasopressin reduces food intake in goats. Quarterly J. Exp. Physiol., 74: 465–473.

Miller, A.D. and Leslie, R.A. (1994) The area postrema and vomiting. Front. Neuroendocrinol., 15: 301–320.

Miller, M.A., Zoeller, R.T. and Dorsa, D.M. (1988) Detection of vasopressin messenger RNA in cells within the bed nucleus of the stria terminalis by *in situ* hybridization histochemistry. Neurosci. Lett., 94: 264–268.

Morien, A., Garrard, L. and Rowland, N.E. (1999) Expression of Fos immunoreactivity in rat brain during dehydration: effect of duration and timing of water deprivation. Brain Res., 816: 1–7.

Morrison, J.M., Pritchard, H.D., Braude, M.L. and D'Aguanno, W. (1971) Plasma and brain lithium levels after lithium carbonate and lithium chloride administration by different routes in rats. Proc. Soc. Exp. Bio. Med., 137: 889–892.

Nakayama, Y., Takano, Y., Shimohigashi, Y., Tanabe, S., Fujita, T., Kamiya, H. and Tsujimoto, G. (2000) Pharmacological characterization of a novel AVP(4-9) binding site in rat hippocampus. Brain Res., 858: 416–423.

Negro-Vilar, A. and Samson, W.K. (1979) Dehydration-induced changes in immunoreactive vasopressin levels in specific hypothalamic structures. Brain Res., 169: 585–589.

Nozaki, K., Boccalini, P. and Moskowitz, M.A. (1992) Expression of c-fos-like immunoreactivity in brain stem after meningeal irritation by blood in the subarachnoid space. Neurosci., 49: 669–680.

O'Connell, C., Gallagher, H.C., O'Malley, A., Bourke, M. and Regan, C.M. (2000) CREB phosphorylation coincides with transient synapse formation in the rat hippocampal dentate gyrus following avoidance learning. Neural Plast., 7: 279–289.

O'Connor, E.F., Cheng, S.W.T. and North, W.G. (1987) Effects of intraperitoneal injection of lithium chloride on neurohypophyseal activity: implication for behavioral studies. Physiol. Behav., 40: 91–95.

Olsewski, P.K., Wirth, M.M., Shaw, T.J., Grace, M.K., Billington, C.J., Giraudo, S.Q. and Levine, A.S. (2001) Role of alpha-MSH in the regulation of consummatory behavior: immunohistochemical evidence. Amer. J. Physiol.-Reg. Integr. Comp. Physiol., 281: R673–R680.

Onaka, T. and Yagi, K. (1998) Role of noradrenergic projections to the bed nucleus of the stria terminalis in neuroendocrine and behavioral responses to fear-related stimuli in rats. Brain Res., 788: 287–293.

Onaka, T., Hamamura, M. and Yagi, K. (1986a) Potentiation of vasopressin secretion by footshocks in rats. Jap. J. Physiol., 36: 1253–1260.

Onaka, T., Hamamura, M. and Yagi, K. (1986b) Suppression of vasopressin secretion by classically conditioned stimuli in rats. Jap. J. Physiol., 36: 1253–1260.

Orowaska-Majdak, M., Koodziejski, P. and Traczyk, W.Z. (2001) Hippocampal vasopressin (AVP) dialysis and the conditioned eyelid reflex in rabbits. J. Physiol. Pharmacol., 52: 767–780.

Overton, D.A. (1974) Experimental methods for the study of state-dependent learning. Fed. Proc., 33: 1800–1813.

Paban, V., Alescio-Lautier, B., Devigne, C. and Soumireu-Mourat, B. (1999) Fos protein expression induced by intracerebroventricular injection of vasopressin in unconditioned and conditioned mice. Brain Res., 825: 115–131.

Paban, V., Alescio-Lautier, B., Devigne, C. and Soumireu-Mourat, B. (1997) Effects of arginine8-vasopressin administered at different times in the learning of an appetitive visual discriminative task in mice. Behav. Brain Res., 87: 149–157.

Papas, S. and Ferguson, A.V. (1990) Electrophysiological characterization of reciprocal connections between the parabrachial nucleus and the area postrema in the rat. Brain Res. Bull., 24: 577–582.

Parker, L.A. (1988) Defensive burying of flavors paired with lithium but not amphetamine. Psychopharmacol., 96: 250–252.

Parker, L.A. (1995) Rewarding drugs produce taste avoidance, but not taste aversion. Neurosci. Biobehav. Rev., 19: 143–151.

Parker, L.A. and Brosseau, B. (1990) Apomorphine-induced flavor-drug associations: a dose-response analysis by the taste reactivity test and the conditioned taste avoidance test. Pharmacol. Biochem. Behav., 35: 583–587.

Patel, J.B., Ciofalo, V.B. and Iorio, L.C. (1979) Benzodiazepine blockade of passive avoidance task in mice: a state-dependent phenomenon. Psychopharmacol., 61: 25–28.

Peck, B.K. and Vanderwolf, C.H. (1991) Effects of raphe stimulation on hippocampal and neocortical activity and behaviour. Brain Res., 568: 244–252.

Peck, J.H. and Ader, R. (1974) Illness-induced taste aversion under states of deprivation and satiation. Anim. Learn. Behav., 2: 6–8.

Phillips, P.A., Abrahams, J.M., Kelly, J., Paxinos, G., Grzonka, Z., Mendelsohn, F.O.A. and Johnston, C.I. (1988) Localization of vasopressin binding sites in rat brain by *in vitro* autoradiography using a radioiodinated V_1 receptor antagonist. Neurosci., 27: 749–761.

Pittman, Q.J., Blume, H.W. and Renaud, L.P. (1981) Connections of the hypothalamic paraventricular nucleus with the neurohypophysis, median eminence, amygdala, lateral septum and midbrain periacqueductal gray: an electrophysiological study in the rat. Brain Res., 215: 15–28.

Pittman, Q.J., Lawrence, D. and McLean, L. (1982) Central effects of arginine vasopressin on blood pressure in rats. Endocrinol., 110: 1058–1060.

Pontecoro, M.J., Sahgal, A. and Steckler, T. (1996) Further developments in the measurement of working memory in rodents. Cog. Brain Res., 3: 205–213.

Popik, P. and van Ree, J.M. (1998) Neurohypophyseal peptides and social recognition in rats. Prog. Brain Res., 119: 415–436.

Rabin, B.M. and Hunt, W.A. (1992) Relationship between vomiting and taste aversion learning in the ferret: studies with ionizing radiation, lithium chloride, and amphetamine. Behav. Neural Biol., 58: 83–93.

Rabin, B.M., Hunt, W.A. and Lee, J. (1982) Studies on the role of central histamine in the acquisition of radiation-induced conditioned taste aversion. Radiat. Res., 90: 609–620.

Raggenbass, M., Tribollet, E., Dubois-Dauphin, M. and Dreifuss, J.J. (1989) Vasopressin receptor of the vasopressor (V1) type in the nucleus of the solitary tract of the rat mediate direct neuronal excitation. J. Neurosci., 9: 3929–3936.

Ramsay, D.J., Rolls, B.J. and Wood, R.J. (1977) Thirst following water deprivation in dogs. Am. J. Physiol., 232: R93–R100.

Rasmusson, D.D., Clow, K. and Szerb, J.C. (1994) Modification of neocortical acetylcholine release and electroencephalogram desynchronization due to brain stem stimulation by drugs applied to the basal forebrain. Neurosci., 60: 665–677.

Reijmers, L.G., van Ree, J.M., Spruijt, B.M., Burbach, J.P. and DeWied, D. (1998) Vasopressin metabolites: a link between vasopressin and memory? Prog. Brain Res., 119: 523–535.

Rigter, H., van Riezen, H. and DeWied, D. (1974) The effects of ACTH- and vasopressin-analogues on CO_2-induced retrograde amnesia in rats. Physiol. Behav., 13: 381–388.

Ritter, S. and Taylor, J.S. (1990) Vagal sensory neurons are required for lipoprivic but not glucoprivic feeding in rats. Amer. J. Physiol., 258: R1395–R1401.

Ritter, S., McGlone, J.J. and Kelley, K.W. (1980) Absence of lithium-induced taste aversion after area postrema lesion. Brain Res., 201: 501–506.

Rougon-Rapuzzi, G., Cau, P., Boudier, J.A. and Cupo, A. (1978) Evolution of vasopressin levels in the hypothalamo-posthypophysial system of the rat during rehydration following water deprivation. Neuroendocrinol., 27: 46–62.

Rozin, P. (1967) Specific aversions as a component of specific hungers. J. Comp. Physiol. Psychol., 64: 237–242.

Sahgal, A. (1984) A critique of the vasopressin-memory hypothesis. Psychopharmacol., 83: 215–228.

Sahgal, A. (1987) Contrasting effects of vasopressin, desglycinamide-vasopressin and amphetamine on a delayed matching to position task in rats. Psychopharmacol., 93: 243–249.

Sahgal, A. (1988) Vasopressin and amphetamine, but not desglycinaminde vasopressin, impair positively reinforced visual attention performance in rats. Behav. Brain Res., 29: 35–42.

Sahgal, A. (1990) How do neuropeptides alter cognitive performance? Pol. J. Pharmacol. Pharm., 42: 527–535.

Sahgal, A. and Wright, C. (1983) A comparison of the effects of vasopressin and oxytocin with amphetamine and chlordiazepoxide on passive avoidance behavior in rats. Psychopharmacol., 80: 88–92.

Sahgal, A., Keith, A.B., Wright, C. and Edwardson, J.A. (1982) Failure of vasopressin to enhance memory in a passive avoidance task in rats. Neurosci. Lett., 28: 87–92

Sato, T., Tanaka, K., Ohhishi, Y., Teramoto, T., Hirate, K. and Nishikawa, T. (1999) The ameliorating effects of a novel NC-1900 on impairments of learning/memory caused by glutamic acid. Folia Pharamcol. Jap., 114 (Suppl 1): 198P–203P.

Sawchenko, P.E. and Swanson, L.W. (1982) Immunohistochemical identification of neurons in the paraventricular nucleus of the hypothalamus that project to the medulla or to the spinal cord in the rat. J. Comp. Neurol., 205: 260–272.

Schou, M. (1968) Lithium in psychiatry – a review. In: Efron, E.H. (Ed.), Pharmacology: A Review of Progress 1957–1967, Public Health Service Publication No 1836, pp. 701–718.

Seliger, D.L. (1977) Passive avoidance learning in the rat as functions of d-amphetamine dosage and shock intensity. Psychopharmacol., 54: 241–242.

Sengstake, C.B. and Chambers, K.C. (1991) Sensitivity of male, female, and androgenized female rats to testosterone during extinction of a conditioned taste aversion. Behav. Neurosci., 105: 120–125.

Sengstake, C.B. and Chambers, K.C. (1979) Differential effects of fluid deprivation on the acquisition and extinction phases of a conditioned taste aversion. Bull. Psychon. Soc., 14: 85–87.

Sengstake, C.B., Chambers, K.C. and Thrower, J.H. (1978) Interactive effects of fluid deprivation and testosterone on the expression of a sexually dimorphic conditioned taste aversion. J. Comp. Physiol. Psychol., 92: 1150–1155.

Shapiro, R.E. and Miselis, R.R. (1985) The central neural connections of the area postrema of the rat. J. Comp. Neurol., 234: 344–364.

Siegel, J.L. (1976) Effect of medial septal lesions on conditioned taste aversion in the rat. Physiol. Behav., 17: 761–765.

Skoog, K.M., Blair, M.L., Sladek, C.D., Williams, W.M. and Mangiapane, M.L. (1990) Area postrema: essential for support of arterial pressure after hemorrhage in rats. Am. J. Physiol., 258: R1472–R1478.

Skoog, K.M. and Mangiapane, M.L. (1988) Area postrema and cardiovascular regulation in rats. Am. J. Physiol., 254: H963–H969.

Skopkova, J., Croiset, G. and De Wied, D. (1991) Differential effects of DGAVP on acquisition and extinction of active avoidance behavior. Peptides, 12: 471–475.

Sofroniew, M.V. (1985) Vasopressin- and neurophysin-immunoreactive neurons in the septal region, medial amygdala and locus coeruleus in colchicine-treated rats. Neurosci., 15: 347–358.

Sokol, H.W., Zimmerman, E.A., Sawyer, W.H. and Robinson, A.G. (1976) The hypothalamic neurohypophysial system of the rat: localization and quantitation of neurophysin by light microscopic immunocytochemistry in normal rats and in Brattleboro rats deficient in vasopressin and a neurophysin. Endocrinol., 98: 176–188.

Spyraki, C., Fibiger, H.C. and Phillips, A.G. (1982) Dopaminergic substrates of amphetamine-induced place preference conditioning. Brain Res., 253: 185–193.

Steiner, J.E. (1979) Human facial expressions in response to taste and smell stimulation. In: Reese, H.W. and Lipsitt, L.P. (Eds.), Advances in Child Development and Behavior. Academic Press, New York, pp. 257–295.

Sterner, R.T. (1990) Plasma lithium as a marker of lithium chloride in wild Norway rats (*R. norvegicus*). Physiol. Behav., 47: 1013–1015.

Stricker, E.M. and Verbalis, J.G. (1991) Caloric and noncaloric controls of food intake. Brain Res. Bull., 27: 299–303.

Stricker, E.M., McCann, M.J., Flanagan, L.M. and Verbalis, J.G. (1988) Neurohypophyseal secretion and gastric function: biological correlates of nausea. In: Takagi, H., Oomura, Y., Ito, M. and Otsuka, M. (Eds.), Biowarning System in the Brain. University Tokyo Press, Tokyo, pp. 295–307.

Swank, M.W., Schafe, G.E. and Bernstein, I.L. (1995) c-Fos induction in response to taste stimuli previously paired with amphetamine or LiCl during taste aversion learning. Brain Res., 673: 251–261.

Swanson, L.W. and Sawchenko, P.E. (1980) Paraventricular nucleus: A site for integration of neuroendocrine and autonomic mechanism. Neuroendocrinol., 31: 410–417.

Syme, L.A. and Syme, G.J. (1973) Effects of lithium chloride on the activity of rats tested alone or in pairs. Psychopharmacol., 29: 85–89.

Syme, L.A. and Syme, G.J. (1974) The role of sex and novelty in determining the social response to lithium chloride. Psychopharmacol., 40: 91–100.

Szczepanska-Sadowska, E., Simon-Oppermann, C., Gray, D. and Simon, E. (1984) Plasma and cerebrospinal fluid vasopressin and osmolality in relation to thirst. Pfluger Arch., 400: 294–299.

Tam, F.W.K., Chen, C., Alpert, J.E. and Iversen, S.D. (1985) aversive effects of subcutaneously injected vasopressin in the rat: independence of the ascending dorsal noradrenergic bundle. Brain Res., 337: 133–137.

Tanaka, T., Yamada, K., Senzaki, K., Narimatsu, H., Nishimura, K., Kameyama, T. and Nabeshima, T. (1998) NC-1900, an active fragment analog of arginine vasopressin, improves learning and memory deficits induced by beta-amyloid protein in rats. Eur. J. Pharmacol., 352: 135–142.

Taukulis, H. (1982) Attenuation of pentobarbital-elicited hypothermia in rats with a history of pentobarbital-LiCl pairings. Pharmacol. Biochem. Behav., 17: 695–697.

Thayer, R.E. (1989) The Psychobiology of Mood and Arousal. Oxford University Press, New York, pp. 88–109.

Thornton, S.N., Parrott, R.F. and Delaney, C.E. (1987) Differential responses of plasma oxytocin and vasopressin to dehydration in non-stressed sheep. Acta Endocrinol., 114: 519–523.

Tribollet, E., Arsenijevic, Y. and Barberis, C. (1998) Vasopressin binding sites in the central nervous system: distribution and regulation. Prog. Brain Res., 119: 45–55.

van der Kooy, D. (1984) Area postrema: site where cholecystokinin acts to decrease food intake. Brain Res., 295: 345–347.

van der Kooy, D., Swerdlow, N.R. and Koob, G.F. (1983) Paradoxical reinforcing properties of apomorphine: effects of nucleus accumbens and area postrema lesions. Brain Res., 259: 111–118.

van Haaren, F., van Zanten, S., and van de Poll, N.E. (1986) Vasopressin disrupts radial-maze performance in rats. Behav. Neural Biol. 45(3): 350–357.

van Leeuwen, F.W. and Caffe, A.R. (1983) Vasopressin-immunoreactive cell bodies in the bed nucleus of the stria terminalis of the rat. Cell Tiss. Res., 228: 525–534.

van Wimersma Greidanus, T.B. and DeWied, D. (1976a) Dorsal hippocampus: a site of action of neuropeptides on avoidance behavior? Pharmacol. Biochem. Behav., 5 (Suppl 1): 29–33.

van Wimersma Greidanus, T.B. and DeWied, D. (1976b) Modulation of passive-avoidance behavior of rats by intracerebroventricular administration of antivasopressin serum. Behav. Biol., 18: 325–333.

van Wimersma Greidanus, T.B., Bohus, B. and DeWied, D. (1975a) The role of vasopressin in memory processes. Prog. Brain Res., 42: 135–141.

van Wimersma Greidanus, T.B., Dogterom, J. and DeWied, D. (1975b) Intraventricular administration of anti-vasopressin serum inhibits memory consolidation in rats. Life Sci., 16: 637–644.

van Wimersma Greidanus, T.B., van Ree, J.M. and DeWied, D. (1986) Vasopressin and Memory. In: DeWied, D., Gispen, W.H. and van Wimersma Greidanus, T.J.B. (Eds.), Neuropeptides and Behavior, Vol. 2. Pergamon Press, Oxford, pp. 99–122.

Vanderwolf, C.H. (1992) The electrocorticogram in relation to physiology and behavior: A new analysis. Electrophysiol. Clin. Neurophysiol., 82: 165–175.

Vandesande, F. and Dierickx, K. (1976) Immunocytochemical demonstration of separate vasotocinergic and mesotocinergic neurons in the amphibians hypothalamic magnocellular neurosecretory system. Cell Tiss. Res., 175: 289–296.

Vawter, M.P. and Green, K.F. (1980) Effects of desglycinamide-lysine vasopressin on a conditioned taste aversion in rats. Physiol. Behav., 25: 851–854.

Verbalis, J.G., McHale, C.M., Gardiner, T.W. and Stricker, E.M. (1986) Oxytocin and vasopressin secretion in response to stimuli producing learned taste aversions in rats. Behav. Neurosci., 100: 466–475.

Verbalis, J.G., Richardson, D.W. and Stricker, E.M. (1987) Vasopressin release in response to nausea-producing agents and cholecystokinin in monkeys. Amer. J. Physiol., 252: R749–R753.

Verney, E.B. (1947) The antidiuretic hormone and the factors which determine its release. Proc. Royal Soc. Series B, 135: 25–106.

Wahlbeck, K., Sundblom, M., Kalso, E., Tigerstedt, I. and Rimon, R. (1996) Elevated plasma vasopressin and normal cerebrospinal fluid angiotensin-converting enzyme in chronic pain disorder. Biol. Psychiatry, 40: 994–999.

Wang, B.C., Share, L., Crofton, J.T. and Kimura, T. (1982) Effects of intravenous and intracerebroventricular infusion of hypertonic solutions on plasma and cerebrospinal fluid vasopressin concentrations. Neuroendocrinol., 34: 215–221.

Wang, Y., Lavond, D.L. and Chambers, K.C. (1997a) The effects of cooling the area postrema of male rats on conditioned taste aversions induced by LiCl and apomorphine. Behav. Brain Res., 82: 149–158.

Wang, Y., Lavond, D.L. and Chambers, K.C. (1997b) Cooling the area postrema induces conditioned taste aversions and blocks acquisition of LiCl-induced aversions. Behav. Neurosci. 111: 768–776.

Wang, Z., Young, L.J., De Vries, G.J. and Insel, T.R. (1998) Voles and vasopressin: a review of molecular, cellular and behavioral studies of pair bonding and paternal behaviors. Prog. Brain Res., pp. 483–499.

Wilkie, D.M., Maclennan, A.J. and Pinel, J.P. (1979) Rat defensive behavior: burying noxious food. J. Exp. Anal. Behav., 31: 299–306.

Wilkin, L.D. and Cunningham, C.L. (1982) Pavlovian conditioning with ethanol and lithium: effects on heart rate and taste aversion in rats. J. Comp. Physiol. Psychol., 96: 781–790.

Winnicka, M.M. and Wisniewski, K. (1999) Bilateral transection of temporoentorhinal connections attenuate vasopressin improvement of memory in rats. Pharmacol. Res., 37: 61–66.

Wise, R., Yokel, P. and Dewitt, H. (1976) Both positive reinforcement and conditioned aversion from amphetamine and from apomorphine in rats. Sci., 191: 1273–1274.

Wood, R.I., and Newman. S.W. (1993) Mating activates androgen receptor-containing neurons in chemosensory pathways of the male Syrian hamster brain. Brain Res. 614: 65–77.

Yagi, K., Onaka, T. and Yoshida, A. (1998) Role of NMDA receptors in the emotional memory associated with neuroendocrine responses to conditioned fear stimuli in the rat. Neurosci. Res., 30: 279–286.

Yamamoto, T. (1994) A neural model for taste aversion learning. In: Kurihara, K., Suzuki, N. and Ogawa, H. (Eds.), Olfaction and Taste XI. Springer-Verlag, Tokyo, pp. 471–474.

Yamamoto, T., Azuma, S. and Kawamura, Y. (1981) Significance of cortical-amygdalar-hypothalamic connections in retention of conditioned taste aversion in rats. Exp. Neurol., 74: 758–768.

Yamamoto, T., Fujimoto, Y., Shimura, T. and Sakai, N. (1995) Conditioned taste aversion in rats with excitotoxic brain lesions. Neurosci. Res., 22: 31–49.

Yamamoto, T., Shimura, T., Azuma, S., Bai, W.-Zh., Fujimoto, Y. and Wakisaka, S. (1992a) Induction of *c-fos*-like protein in the central nervous system of the rat following internal malaise. In: Inoki, R., Shigenaga, Y. and Tohyama, M. (Eds.), Processing and Inhibition of Nociceptive Information. Elsevier, Amsterdam, pp. 165–168.

Yamamoto, T., Shimura, T., Sako, N., Azuma, S., Bai, W.-Zh. and Wakisaka, S. (1992b) C-fos expression in the rat brain after intraperitoneal injection of lithium chloride. Neuroreport, 3: 1049–1052.

Zaidi, S.M.A. and Heller, H. (1974) Can neurohypophysial hormones cross the blood-cerebrospinal fluid barrier? J. Endocrinol., 60: 195–196.

SECTION 3

Stress and the HPA Axis

T. Steckler, N.H. Kalin and J.M.H.M. Reul (Eds.)
Handbook of Stress and the Brain, Vol. 15
ISBN 0-444-51173-3

CHAPTER 3.1

Corticosteroid receptors and HPA-axis regulation

E. Ronald de Kloet*, Mathias Schmidt and Onno C. Meijer

Division of Medical Pharmacology/LACDR-LUMC, University of Leiden, P.O. Box 9502, 2300 RA Leiden, The Netherlands

Abstract: The activity of the hypothalamic–pituitary–adrenal (HPA) axis has three modes of operation: (1) pulsatility with intervals of approximately 60 min; (2) circadian variation with peak activity prior to the onset of the active period; and (3) a profound activation in response to physical and psychological stressors. End products of the HPA axis are cortisol and corticosterone, which co-ordinate, in concert with the other components of the HPA axis, the body and brain responses to the stressor and thereby facilitate adaptive processes. The actions of the corticosteroids are mediated by receptors of which the mineralocorticoid (MR) and glucocorticoid (GR) receptors acting as gene transcription factors are best investigated. This chapter addresses the following aspects of corticosteroid action on HPA-axis activity:

First, MR binds cortisol and corticosterone with a ten-fold higher affinity than GR. This differential affinity has led to the concept that via MR homeostasis and HPA activity is maintained, while GR-mediated signalling facilitates their recovery from stress and promotes adaptive processes in preparation for future events.

Second, the MR- and GR-mediated effects target the core of the HPA axis as well as its afferents. This provides an enormous diversity to corticosteroid action.

Third, corticosteroid feedback varies as a function of the phase in HPA pulsatility, the nature and intensity of the stressor and depends on genetic determinants expressed as receptor variants and polymorphisms.

Fourth, during development HPA activity is low and stable due to enhanced corticosteroid feedback and adrenal hyporesponsiveness. Early life experience at that time can program HPA reactivity and corticosteroid feedback for life.

Fifth, the molecular mechanism of corticosteroid action proceeds along two fundamentally different pathways. MR and GR bind to specific DNA motifs (glucocorticoid response elements – GREs) in regulatory regions like promoters and exert direct control over the transcription machinery for which recruitment of co-regulators is often indispensable. This mode of operation mostly leads to transactivation or occasionally to repression of transcription. The other mode involves interaction of the receptor with transcription factors (i.e. NFκB and AP-1) to prevent them from transcription regulation and is selective for GR. The outcome is mostly transrepression of gene transcription aimed to dampen stress-induced processes.

The chapter is concluded with the thesis that imbalance in MR- and GR-mediated actions may lead to neuroendocrine dysregulation and behavioural impairments, which after passing a certain threshold enhances the vulnerability to stress-related disorders for which the individual is genetically pre-disposed.

Introduction

Since the discovery of corticosteroid receptors in 1968 by Bruce McEwen (McEwen et al., 1968), several highlights have marked leaps in our understanding of the role these receptors play in HPA-axis regulation. These highlights coincided with breakthroughs of new methodologies and discoveries.

First of all, the hallmark discovery is that corticosteroid receptors are abundant in the limbic system beyond the core of the hypothalamic–pituitary adrenal (HPA) axis. Back in the sixties it was thought that the glucocorticoid feedback should take place exclusively in the hypothalamic paraventricular nucleus (PVN) and the pituitary corticotrophs. When sites could be visualised with radiolabelled

*Corresponding author. Tel.: +31-71-527-6210/6290;
Fax: +31-71-527-4715; E-mail: e.kloet@lacdr.leidenuniv.nl

corticosterone of sufficient specific activity, much to the surprise of the established endocrine society the corticosterone receptors were particularly abundant in extra-hypothalamic limbic regions, notably the hippocampus. These limbic sites were only indirectly implicated in the HPA regulation and are distinct from the core of the HPA axis.

Second, the discovery by Reul and De Kloet (1985) that corticosteroids act through high ($K_d \sim 0.5\,nM$, 4°C) and ten-fold lower-affinity ($K_d \sim 5\,nM$, 4°C) nuclear receptors (i.e. the mineralocorticoid receptors [MRs] and glucocorticoid receptors [GRs]) in brain and pituitary. This discovery was possible by the synthesis of 'pure' glucocorticoids by Roussel–Uclaf researchers (Moguilevski and Philibert, 1984). It occurred at the time that the GR was cloned by Ron Evans (Hollenberg et al., 1985) soon followed by the cloning of the MR (Arriza et al., 1987). Of great importance was also the application of the first antibody to the GR for immunocytochemistry (Fuxe et al., 1985). The discovery of MR and GR has had a large impact on the field, because it represents a binary receptor system controlling gene networks underlying the onset and the termination of the stress response (De Kloet and Reul, 1987; Arriza et al., 1988).

The third revolution came with the finding that the corticosteroid receptors can affect signalling pathways beyond activation of glucocorticoid response elements (GREs) by interaction with other transcription factors (Karin et al., 1993). MR and GR both bind to these GREs, but only GR is capable to interact with transcription factors such as activating protein (AP-1) and nuclear factor κB (NFκB) to attenuate stress-induced activity in specific pathways. This finding provided a firm mechanistic underpinning to the concept advanced by Tausk (1951) and Munck et al. (1984) that glucocorticoids actually dampen primary stress reactions. Also, the GRE site became complex. Thus, co-activator and co-repressor molecules were identified that appeared powerful modulators of nuclear receptor function (Meijer, 2002; Xu and O'Malley 2002).

Right now corticosteroid receptor science has arrived at a critical point. On the one hand, the corticosteroid receptor is a sound starting point for in-depth studies in the genomics of the stress response. Such studies open up a bewildering array of stress-responsive genes (Datson et al., 2001; Feldker et al., 2003) that need to be subjected in minute analysis to answer questions like how, where and when these genes become active in stress-induced signalling pathways, and foremost what their precise function is. The other side of the coin is the awareness that the stress system, in particular the corticosteroids, through their receptors orchestrate body and brain responses to changing environments, individuals and social contexts. The challenge today is therefore to combine molecular techniques piecing together the relationships between all the gene products with the analysis of higher brain functions, i.e. emotions and cognitive processes. The ultimate goal is to understand how stress hormones help to preserve health and how these hormones may become damaging under chronic adverse conditions and cause diseases like depression.

This contribution contains three parts. In the first part, approaches and methodologies are described to the study of corticosteroid receptor function in the regulation of the HPA axis and its afferent pathways (see Fig. 1). The second part describes the role of the corticosteroid receptors in the organisation of the HPA axis, which is a prime example of how corticosteroids may change HPA responsiveness and behavioural adaptation for months and years, way beyond the minute to hour dimensions characteristic for their physiological regulations. The third part describes the current state-of-the-art in approaches and concepts on the molecular underpinnings of the HPA-axis regulation via corticosteroid receptors with the focus on transcription factors and the recently discovered co-activators and repressors. The chapter is concluded with future directions on the main avenues of corticosteroid stress hormone research.

Corticosteroid receptors in HPA-axis regulation: system level

The distinction between MR and GR domains of corticosteroid action is now a gold mine in the exploration of valuable new data on regulation of the HPA axis for more than 15 years. Initially, the receptor properties and the precise neuroanatomical localisation were the landmarks for digging with endocrine and pharmacological tools. These

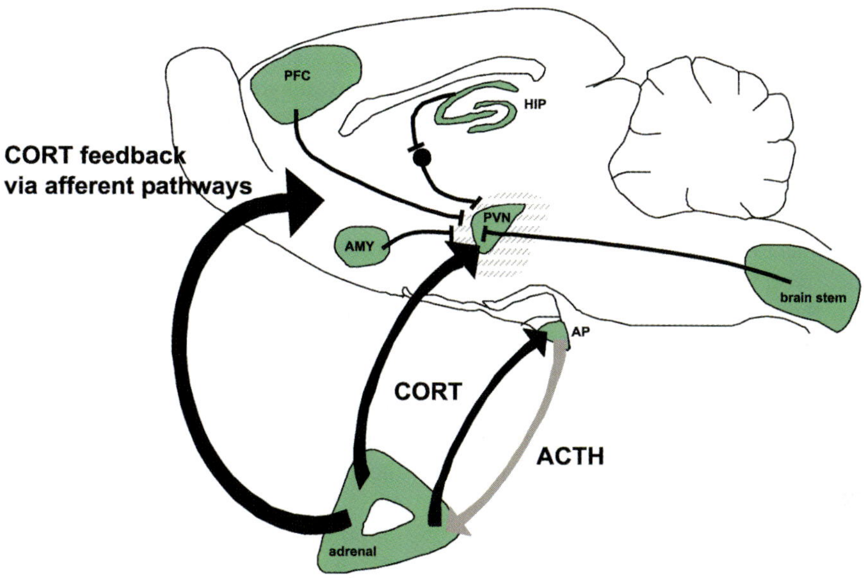

Fig. 1. Schematic overview of corticosterone feedback action on the HPA axis and afferent inputs. AMY, amygdala; AP, anterior pituitary; HIP, hippocampus; PFC, prefrontal cortex; PVN, paraventricular nucleus.

explorations became more and more successful when the awareness grew that the outcome of hormone action depends on the context in which the hormones operate. Right now we are in the fortunate position that these context-dependent aspects of corticosteroid action potentially can be 'translated' in molecular events, and vice versa.

Properties and localisation

In McEwen's initial observation tracer amounts of ^{3}H-corticosterone were administered and, by using cell fractionation or autoradiography, one hour later a pronounced labelling of cell nuclei in hippocampal pyramidal neurons and dentate gyrus neurons was noticed. Cell nuclei in dorsal septum and amygdala, cortical areas and circumventricular regions were labelled as well. The animals were adrenalectomized (ADX) to deplete the receptors of endogenous hormone (McEwen et al., 1968; Gerlach and McEwen, 1972). The tracer doses (0.7 μg) produced, after systemic administration, very low circulating corticosterone levels, yet the receptors were occupied near saturation with radiolabelled corticosterone. Uptake and retention could be suppressed with aldosterone, but not with the potent synthetic glucocorticoid dexamethasone (De Kloet et al., 1975). Given the pulsatile nature of corticosterone secretion in vivo (Windle et al., 1998) one can assume that under 'normal' conditions extensive MR occupancy is maintained with endogenous corticosterone in rat and with cortisol in man. Some debate remains to what extent precisely MR is occupied (Kalman and Spencer, 2002). The answer to this question must await better methods to measure in unbiased fashion available and occupied MR under in vivo conditions.

In vitro cytosol-binding studies of tritiated corticosterone, in the absence and presence of the pure glucocorticoid RU 28362, revealed two types of soluble corticosteroid receptors that did bind corticosterone with ten-fold difference in affinity (Veldhuis et al., 1982; Reul and De Kloet, 1985). One site did bind also the pure glucocorticoid and was therefore designated GR. The other site bound either corticosterone or aldosterone with very high affinity and was initially dubbed 'corticosterone' receptor (Veldhuis et al., 1982) or 'aldosterone' receptor (Krozowski and Funder, 1983). In 1983 Jan Åke

Gustafsson showed at a steroid meeting in Marseille, the very first immunohistochemical staining of GR in the brain. To the surprise of one of us present at that meeting (ERdK), immunoreactive GR was present in CA1 and CA2 (Fuxe et al., 1985), but not in the hippocampal CA3 region, although that region was labelled in vivo with the tracer corticosterone. Then we realised that the tracer dose perhaps labelled only the "corticosterone" receptor, but was insufficient to occupy the GR. In the subsequent studies this idea was tested. With a post-hoc in vitro cytosol binding we demonstrated subsequently that in vivo much higher amounts of corticosterone beyond tracer level were needed to occupy the GR (Reul and de Kloet, 1985). Thus, the high-affinity corticosterone receptor, initially identified by McEwen, was not the classical GR.

At that time we still indicated the high-affinity corticosterone receptor 'CR' as type I, and the lower-affinity GR as type II. Immunocytochemical and in situ hybridisation patterns of GR appeared identical to the *anatomical* distribution with in vitro autoradiography of sites labelled with tritiated RU28362, the pure GR ligand (Fuxe et al., 1985; Reul and de Kloet, 1986; Van Eekelen et al., 1988). The MR was cloned from the kidney (Arriza et al., 1987) and it appeared that in vitro the ligand-binding profile of the cloned MR was similar to that of the high affinity CR/type 1 receptor previously shown in hippocampus of the rat (Veldhuis et al., 1982; Reul and de Kloet, 1985; Arriza et al., 1987). Finally, the corticosterone receptor (or type 1 receptor) in hippocampal neurons detected with in vitro autoradiography with tritiated corticosterone and aldosterone in the presence of excess RU28362 (Reul and de Kloet, 1986) was proven to be the MR using in situ hybridisation (Van Eekelen et al., 1988). This MR appeared co-localised with the GR in discrete clusters in nuclear domains with confocal laser microscopy (Van Steensel et al., 1996). The MR was found restricted to limbic structures, e.g. hippocampus, lateral septum, amygdaloid and cortical neurons, while the GR was detected widely in brain, neurons and glial cells. GR was particularly abundant in the PVN, limbic regions and brainstem aminergic neurons.

The mystery that an MR did bind with very high affinity ($K_d \sim 0.5\,nM$, 4°C), the naturally occurring glucocorticoid corticosterone in rat and mouse was soon solved by the identification of the 11β-hydroxysteroid dehydrogenase (11β-HSD type 2) (Edwards et al., 1988; Funder et al., 1988). This oxydase inactivates corticosterone in kidney leaving the MR open for binding to aldosterone. The brain lacks this oxydase except for the aldosterone targets involved in the regulation of salt homeostasis in periventricular brain regions. Instead, the larger part of the brain contains the reductase isoform, the 11β-HSD type 1, which generates corticosterone from inactive 11β-dehydrocorticosterone. This enzyme is present in the limbic brain and thus may function as a corticosterone trap (Seckl, 1997). However, administration of the 11β-HSD blocker carbexonolone i.c.v. does not affect retention of tracer amounts of corticosterone in hippocampus under conditions that it did block the enzyme (Van Haarst et al., 1996). Accordingly, 11β-HSD type 2 confers pre-receptor-binding specificity in typical MR target tissues, but for the brain the jury is still out for a conclusive role of the type 1 form in retention of corticosterone. This question will be answered after further functional analysis of the different 11β-HSD mutant lines and the analysis of carbenoloxone on brain function and behaviour (see Holmes, this volume).

Administration of the dexamethasone tracer revealed one-hour post-injection intense labelling of the ventricular system, and of the pituitary corticotrophs (De Kloet et al., 1975). The intense pituitary labelling matched the evidence for a pituitary site of synthetic glucocorticoids in suppression of stress-induced HPA activation (De Kloet et al., 1974; Miller et al., 1992). In this way dexamethasone action is distinct from that of corticosterone. The latter steroid does not reach the pituitary GR in sufficient quantities because of excess of competing transcortin-like molecules that are present in abundance in the same pituitary cells (De Kloet et al., 1977). Dexamethasone did not label bona fide GR in the brain in comparable density as in pituitary, in spite of its ventricular localisation. We now know that dexamethasone is hampered in its penetration in the brain by a multidrug resistance (mdr) 1a-P-glycoprotein (Pgp) in the blood–brain barrier (Meijer et al., 1998; Karssen et al., 2001). Hence, administration of a low dose of dexamethasone blocks the stress-induced HPA axis, depletes the brain of endogenous corticosterone and does not replace for the depleted hormone. This condition is called brain-selective chemical ADX (see the chapter by Karssen et al., this volume).

In summary, the low amounts of corticosterone secreted in a pulsatile fashion during the circadian trough substantially occupy the MR, while higher corticosterone concentrations towards the circadian peak and after stress progressively occupy GR over MR. The MR is abundant in limbic brain regions, while GR is ubiquitous, but unevenly distributed with abundance in brain stress centres and very low levels in suprachiasmatic nucleus and hippocampal CA3 (Van Eekelen et al., 1988). There is some debate over GR expression in the primate hippocampus, but this issue was resolved when the proper antibodies were used (Patel et al., 2000; Sanchez et al., 2000).

Corticosteroid antagonist studies

Compelling evidence for the role of MR and GR in HPA-axis regulation came from the studies in which rather selective antagonists were infused in the cerebroventricular system, or in discrete brain regions. Central administration of doses in the ng range appeared necessary, while systemic administration required mg amounts of the antagonist to affect central mechanisms. The reason for this million times higher dose difference appears due to the rapid clearance of the antagonists from the circulation, the mdr1A Pgp in the blood–brain barrier for which the antagonists are substrates and the counter-regulatory effects on the HPA axis that increase the circulating corticosterone level (see Karssen et al., this volume).

Mineralocorticoid antagonists

Ratka et al. (1989) showed that the MR antagonist RU 28318 icv in a bolus dose of 100 ng/1 µl caused, under basal morning conditions, a transient rise in circulating corticosterone level. The MR antagonist icv enhanced the evening rise in ACTH and corticosterone secretion (Oitzl et al., 1995) irrespective of the site of administration (100 ng) in the ventricular system or in the dorsal hippocampus (10 ng) (Van Haarst et al., 1997). MR blockade also facilitates the corticosterone response to novelty (Ratka et al., 1989) (Fig. 2). Systemic administration of mg amounts (50 mg/kg) also triggered an am corticosterone response (Spencer et al., 1998). Likewise, icv MR antisense infusion increased circulating corticosterone levels (Reul et al., 1997). Chronic RU28318 icv (100 ng/h), administered via an Alzet minipump, gave an initial rise in ACTH and corticosterone in the pm phase of the first infusion

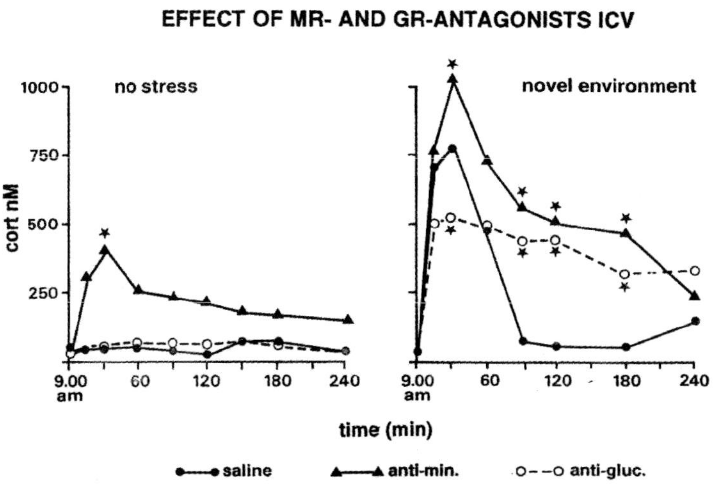

Fig. 2. Effect of mineralocorticoid antagonist RU 28318 icv and glucocorticoid antagonist RU 486 icv on plasma corticosterone levels under basal am resting conditions and after exposure to a novel environment. Adapted from Rakta et al. (1989), Fig. 2 and De Kloet (1991), Fig. 6.

day. Subsequently, HPA activity normalised, but adrenal sensitivity to ACTH was two-fold increased (Van Haarst, 1996).

In man the MR antagonists spironolactone (Young et al., 1998) or canrenoate (Dodt et al., 1993; Grottoli et al., 2002) p.o. induced a rise in cortisol. The latter effect could be blocked by the benzodiazepine agonist alprazolam. In some cases the antagonist was effective in an acute injection in the morning (Kellner et al., 2002; Young et al., 2003) and in others in the evening (Grottoli et al., 2002). Repeated daily injections over 2 days (Young et al., 1998) or 8 days (Heuser et al., 2000) increased cortisol levels but not ACTH, which points to adaptive changes resulting in enhanced adrenal sensitivity. The effect of spironolactone on cortisol was further enhanced during aging (Heuser et al., 2000). The MR antagonist also has been tested in psychiatric disorders. In depressed patients the basal and CRF-induced ACTH and cortisol levels were elevated by spironolactone (Deuschle et al., 1998; Grottoli et al., 2002; Kellner et al., 2002; Young et al., 2003). In patients suffering from post-traumatic stress syndrome (PTSD) the response to MR antagonist was not different from controls (Kellner et al., 2002).

In summary, brain MR controls the HPA tone. The data demonstrate that acute blockade of hippocampal MR disinhibits HPA-axis activity under basal resting conditions and enhances the daily corticosterone surge. The effect is more pronounced under conditions that the central drive is enhanced, such as during aging and depression. MR blockade also enhances the HPA effect of a stressor that taps on hippocampus function, i.e. exposure to novelty. The HPA axis slowly adapts to repeated MR blockade ultimately producing hypercorticism and enhanced adrenocortical sensitivity to ACTH.

Site of MR antagonist action

MR antagonists administered icv or in the dorsal hippocampus triggered similar ACTH and corticosterone responses under basal conditions during the circadian rise (Van Haarst et al., 1997). Furthermore, in most, but not all, studies the expression and binding of MR in hippocampus correlates with HPA activity under basal and stressful conditions. Interestingly, the cyclic increase of HPA activity in female rats on the evening of pro-estrus occurs when the high estrogen and progesterone levels impair MR function; estrogens lower hippocampal MR mRNA levels and binding capacity, while progesterone causes a profound decrease in MR-binding affinity (Carey et al., 1995). Additional evidence on stressor- and gender-specific effects was provided by Karandrea et al. (2002). Rat strains with high levels of hippocampal MR expression and lower pituitary GR expression (e.g. male Lewis rats) show lower basal and stress-induced HPA activity as compared to Wistar rats from which they are derived (Oitzl et al., 1995). However, Jongen-Relo et al. (2002) did find a higher GR rather than MR mRNA level in hippocampus of Lewis versus Fisher rats, demonstrating again that mRNA cannot be always extrapolated to the protein level. Aged rats and dogs generally have reduced MR (and GR) expression and an increased basal HPA activity, and prolonged stress-induced ACTH release (Van Eekelen et al., 1991; Rothuizen et al., 1993; Morano et al., 1994; 1995; Herman et al., 2001), but in rats these effects are strain dependent. Icv administered endotoxin impairs MR function and causes a chronically elevated basal HPA activity (Schöbitz et al., 1994). Tricyclic antidepressants increase expression of hippocampal MR and decrease basal and stress-induced HPA activity (Brady et al., 1991; Seckl and Fink, 1992).

In a careful parametric study, Reul et al. (1993) established that after an initial down-regulation MR was induced by amitryptiline over a period of weeks (GR was also slightly upregulated at that time), while basal and stress-induced HPA activity were suppressed and this effect persisted in transgenic mice with reduced GR expression (Holsboer and Barden, 1996). A recent study by Gesing et al. (2001) points to yet another level of complexity involving MR. If rats were exposed to a forced swim stressor the rise in stress- or CRF-induced hippocampal MR occurred, which required the presence of corticosterone. The CRF–MR link is functional because MR antagonists produce, at the time of MR induction, a markedly enhanced HPA response. The CRF–hippocampal MR link may explain the enhanced responses to MR antagonists under conditions that hypercorticism can be assumed from animal experiments (Cole et al., 1998) or during depression and aging (Heuser et al., 2000; Young et al., 2002).

In summary, the data clearly demonstrate the importance of MR function in hippocampus for the regulation of HPA tone, i.e. the basal levels and the onset of stress-induced HPA activity. The role of MR in other limbic structures (amygdala, septum), in frontal cortex and in brainstem nuclei, e.g. A6 and A2, awaits further study.

Glucocorticoid antagonists

The GR (and progesterone) antagonist mifepristone (RU 486, C 1073) does not activate the HPA axis when administered acutely under basal morning resting conditions in the rat, neither 1 h nor 24 h after injection (Ratka et al., 1989; Van Haarst, 1996; Van Haarst et al., 1997). In the pm phase mifepristone icv enhanced the circadian rise in ACTH and corticosterone release one-hour post-icv injection. The explanation probably is that the GR occupancy is too low for blockade during nadir, because the circulating plasma corticosterone levels are too low.

A condition of central glucocorticoid resistance was evoked by chronic icv infusion of the anti-glucocorticoid mifepristone through Alzet mini-pumps (100 ng/h) over a period of 14 days (Van Haarst et al., 1996). The experiments occurred in parallel with the MR antagonist infusion mentioned in the previous paragraph. Controls received the vehicle icv. Mifepristone administration triggered the first day an enhanced pm rise in both ACTH and corticosterone. In the following days the HPA response, including CRF mRNA expression, was not different from controls, but at day 4 the enhanced pm corticosterone surge reappeared. Also, adrenal weight and sensitivity to ACTH were markedly enhanced (Van Haarst et al., 1996). This finding indicates that the HPA axis slowly adapts to the condition of central glucocorticoid resistance induced by chronic mifepristone with an increased corticosterone output during the circadian peak. Since basal trough levels are not affected, the result of chronic central GR blockade is an enhanced circadian amplitude in corticosterone (Fig. 3).

The response to stressors was also affected differentially by mifepristone treatment (Fig. 2). One hour after am GR antagonist administration the initial corticosterone response to novelty was suppressed, an effect opposite to that of MR antagonist (Ratka et al., 1989). The duration of the corticosterone response was prolonged after both the MR and the GR antagonist, but the underlying reason is different. The GR antagonist interferes with negative feedback after the novelty stressor, but the MR antagonist produces higher peak values and therefore, the response lasts longer. However, 24 h after the single administration of mifepristone novelty-induced ACTH and corticosterone responses was enhanced, but there was no interference with negative feedback. Yet, chronic icv infusion resulting in the actual presence of the antagonist at the time of the novelty stressor interfered again with negative feedback and as a consequence the response to the novelty stressor then remained elevated for prolonged periods of time.

Mifepristone was also applied as therapeutic agent in man. Several years ago mifepristone appeared to ameliorate fast and effective the depressive and psychotic symptoms in Cushing patients (Van Der Lely et al., 1991). More recently, patients suffering from psychotic depression also rapidly improved (Belanoff et al., 2002b). In these patients mifepristone enhanced the amplitude of the flattened corticol rhythm (Belanoff et al., 2001), a result reminiscent to the effect of chronic treatment of rats, as mentioned

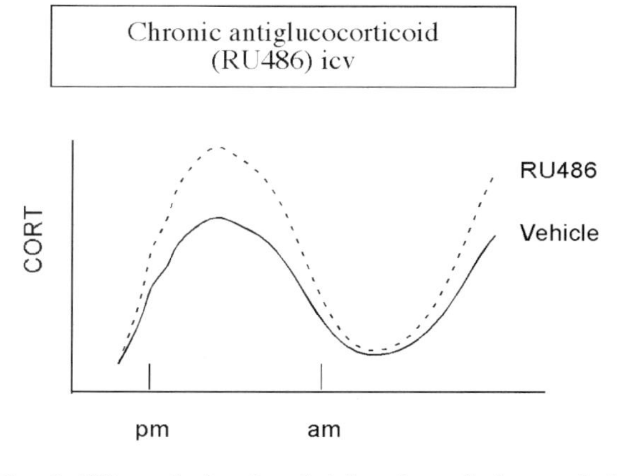

Fig. 3. Effect of chronic administration of glucocorticoid antagonist RU 486 icv on circadian rhythm of plasma corticosterone levels (modified from Van Haarst et al., 1996).

in the previous paragraph. Other researchers (see Wolkowitz and Reuss, 2002 for a review) also reported positive results in affect and cognitive performance. Although mifepristone is a mixed glucocorticoid and progesterone antagonist, its effect on cognition, mood, affect and HPA regulation in animals and man are due to blockade of the GR.

In summary, brain GR blockade interferes with the termination of the stress response. In a longer timeframe, beyond the presence of the antagonist the circadian surge and the stress responsiveness is enhanced. Glucocorticoid feedback resistance induced by chronic central GR blockade produces a sequelae of adaptations, ultimately resulting in an enhanced HPA reactivity as exemplified by a larger circadian corticosterone amplitude and enhanced stress responsiveness.

Site of GR antagonist action

As pointed out previously acute GR antagonist am administration is effective neither after icv nor after intrahippocampal administration. During the p.m. phase ACTH and corticosterone were increased one hour after the GR antagonist icv (Van Haarst et al., 1997) and the same stimulatory response occurred acutely after GR blockade at the level of the PVN (De Kloet et al., 1988). ACTH, but not corticosterone concentrations were suppressed after intrahippocampal administration (De Kloet et al., 1988; Van Haarst et al., 1997). A suppression of novelty-induced HPA responses was observed after icv administration (Ratka et al., 1989). Implants of glucocorticoids near the PVN act similarly (Kovács et al., 1986; Kovács and Makara, 1988; Kovács and Sawchenko, 1996), while local application of the antagonist RU 38486 has the opposite effect (De Kloet et al., 1988). Corticosterone implants were also effective in suppressing ACTH release in the medial prefrontal cingulate cortex (Diorio et al., 1993; Akana et al., 2001). Corticosterone implants in the central amygdala had no immediate effects on ACTH release, but increase expression of CRF mRNA in the amygdala. This effect likely accounts for the increased stress responsiveness of limbic circuitry and the enhanced autonomic outflow under conditions of chronic stress and high levels of corticosteroids (Akana et al., 2001; Cook, 2002).

In summary, the data show that, as expected, GR mediates negative-feedback action in the hypothalamic PVN. The feedback action in the other areas is ambiguous and seems to be context and state dependent. In general, chronic high corticosteroid concentrations acting on frontal cortex, hippocampus and amygdala disinhibit the HPA axis.

Adrenalectomy and agonist replacement

ADX, and hormone replacement also have been used to explore the role of corticosteroid receptors in HPA control. ADX triggers a profound increase in CRF and particularly AVP mRNA and peptide levels in the PVN and in the external layer of the median eminence. Basal ACTH levels are dramatically elevated, while circadian changes and stress responses of the hormone show a large amplitude in excursions (Dallman et al., 1987). The rise in basal trough levels of ACTH after ADX is prevented by chronic replacement with very low amounts of exogenous corticosterone. Corticosterone also suppresses the ADX-induced synthesis of AVP, while CRF was not affected by either treatment (Bradbury et al., 1994). The IC_{50} of corticosterone suppression was about 0.5 nM in terms of circulating free corticosterone, in the range of the MR K_d value. At the circadian peak much higher levels of exogenous corticosteroids were required, and half maximal suppression was achieved by a free concentration of about 5 nM, close to the K_d of GR (Dallman et al., 1989). However, exclusive activation of GR was insufficient to suppress the circadian peak, and MR activation appeared to be indispensable (Bradbury et al., 1994). The corticosteroid concentration does not need to be continuously high, in that an episodic rise in corticosterone levels by injection or ingestion via the normal evening drink is sufficient to occupy both receptor types, and to maintain ACTH levels with small amplitude changes over the 24-h period (Bradbury et al., 1991). Support for a co-operative MR- and GR-mediated action in control of HPA axis was provided by Spencer et al. (1998), following peripheral administration of the antagonists to adrenally intact animals in a dose range of 30–50 mg/kg.

An interesting twist in the story around the ADX model recently came forward from the work of the Dallman group (Dallman et al., 2002). They observed that the ingestion of sucrose and saline in the ADX rats also normalised CRF and ACTH, as does corticosterone replacement (Laugero et al., 2001). Both sucrose and corticosterone restored parameters for cell metabolism. This observation led to the notion that in ADX rats corticosterone replacement acts to restore the HPA axis through recovery of energy metabolism. Alternatively, corticosterone may act on the brain, through a sucrose-responsive pathway that interacts with central mechanisms underlying metabolism. A subsequent study showed that the latter possibility is less likely because icv corticosterone in ADX rats caused weight loss (Laugero et al., 2002), under conditions that systemic corticosterone restored endocrine and metabolic parameters to the level of the Sham-ADX rats (Kamara et al., 1992). This finding was reproduced recently by icv infusion of corticosterone hemisuccinate (100 ng/24 h) in ADX animals. Intriguingly, they also showed that icv corticosterone blocked recovery by sucrose of ADX effects on metabolism and HPA axis (Laugero et al., 2002). Moreover, corticosterone icv enhanced CRF expression, while basal and stress-induced ACTH of ADX rats was also enhanced.

As explanation of this intriguing series of observations, Dallman et al. (2002) state that in the periphery corticosterone restores the HPA axis through recovery of the metabolic disturbance. In the brain corticosterone icv infusion mimics the effect of stress-induced corticosterone activating GR-specific brain functions. In a hypothetical model (Laugero et al., 2002) corticosterone icv is assumed to activate sympathetic outflow and its concomitant metabolic changes by induction of amygdala CRF mRNA. This model is in agreement with the thesis of Cook (2002). Naturally, many studies need to be done to explore the interactions between the adrenocortical steroids and medullar hormones, since both are removed in the ADX animals, and both are powerful modulators of the stress response and energy metabolism.

These findings have at least four obvious implications. First, it emphasizes that stimulus-specific afferent pathways mediate a great variety of stress reactions activating the core of the HPA axis. These stress reactions can be evoked by the metabolic crises evoked by ADX, as discussed in the previous paragraphs, by tissue damage, infection, pain, hemodynamics, toxic agents or psychological processes in response to other individuals and the environment, either real or imaginary. Second, corticosterone rises, feeds back and dampens these primary stress responses of various types in the periphery and the brain and thus prevents them from overshooting (Tausk, 1951; Munck et al., 1984). The corticosteroids therefore eliminate the response to the stressor, and thus the source of the HPA activation. Third, the stress-induced rise in corticosterone therefore has an enormous diversity in action. It dampens the primary stress reactions in the specific stressor-stimulated afferent pathway, but at the same time acts via MR and GR on many targets including those in the core of the HPA axis. In this way the hormone co-ordinates and synchronises body and brain reactions to a stressor.

In summary, ADX presents an interesting model to probe selective pathways in the stress response with specific MR and GR agonists in the face of conditions that restore the metabolic disturbance induced by adrenal ablation. Stressors may activate a stimulus-specific afferent to the HPA axis (Kovács et al., this volume) and evoke, at the same time, a common signal triggered by depletion of energy resources (Selye, 1952; Ingle, 1954). It follows that corticosterone acts to restore stressor-depleted energy resources, where it restrains the stimulus-specific stress reactions, while it acts on the brain to bias the most opportune behavioural adaptations.

Function of glucocorticoid feedback sites in stressor-specific pathways

The role of glucocorticoids and sympathetic nervous system in the integrated control of energy deposition and caloric intake, involving the hypothalamic and amygdala CRF systems, is extensively reviewed by Dallman et al. (2002). In addition, several pathways have been identified that convey stressor-specific information in the brain (Palkovits, 2000). These include ascending aminergic neurons via direct, monosynaptic inputs that excite the PVN, an action potentiated in the acute phase of the stress response by corticosteroids (Marinelli and Piazza, 2002), but attenuated by excess of the same hormones if stress persists (Karten et al., 1999). Secondly, the SCN conveys excitatory and inhibitory circadian pacemaker activity to the PVN, an activity that is modulated by corticosteroids. One such SCN output regulates–via spinal projections–pronounced daily shifts in adrenal sensitivity and corticosterone secretion (Kalsbeek et al., 1996). Thirdly, GABA-ergic neurons in the hypothalamus and pre-optic area form a network surrounding the PVN. Corticosteroids have been reported to enhance GABA turnover in the hypothalamus, suggesting that an enhanced GABA-ergic tone may govern inhibitory control over the PVN (Herman and Cullinan, 1997; Cole and Sawchenko, 2002; Herman et al., 2002). All inputs to the PVN mentioned above express GRs. They also contain numerous co-localised neuropeptides, which also regulate PVN activity in their own right (Swanson, 1991).

The activation of a particular afferent neuronal network innervating the PVN area is stressor specific and depends on the nature of the stimulus (Herman et al., 2002, this volume). If it constitutes a direct threat to survival through physical stressors (e.g. respiratory distress, haemorrhage, inflammation, infection and trauma), the ascending aminergic pathways promptly activate the autonomic and neuroendocrine centres in the hypothalamus. If sensory stimuli are subject to appraisal and interpretation processing in higher brain regions is required, this may subsequently lead to modulation of GABA-ergic tone and change in synthesis of CRF, AVP and other neuropeptides of the PVN secretagogue cocktail. Activation of brainstem and limbic circuitry is not separated, but, in fact, mutually interactive and stress-induced corticosteroids readily enter the brain and feed back on all components of the neural stress circuitry, but in a context-dependent manner.

Limbic inputs impinging on the PVN and the hypothalamic GABA-ergic neurons express high levels of MR in addition to GR, suggesting dual regulation of these inputs by corticosteroids. Moreover, these inputs can be either excitatory from hippocampus (Joëls and De Kloet, 1993; Joëls, 2001, this volume), enhancing GABA-ergic tone, or inhibitory (e.g. from amygdala) and reducing GABA-ergic tone. This implies that, with enhanced hippocampal input, the HPA axis is predicted to be relatively more suppressed, and that enhanced amygdaloid input would lead to disinhibition of GABA-ergic input to the PVN and enhanced HPA activity. Using the selective antagonists and agonists this is precisely what has been observed. The MR antagonist attenuates the excitatory hippocampal output (Joëls and De Kloet, 1994) and leads to an enhanced HPA response to novelty (Ratka et al., 1989).

Glucocorticoids activate the central amydala leading to disinhibition of the HPA axis and an enhanced sympathetic outflow (Dallman et al., 2002). Roles for the amygdala in fear, anxiety and activation of HPA axis are well documented (Roozendaal et al., 2001). The amygdala expresses CRF, part of an extra-hypothalamic CRF network mediating the behavioural expressions of stress, fear and anxiety. Corticosteroids activate the central amygdala and enhance the expression of CRF, suggesting a positive feedback in this network as opposed to the negative-feedback role in the hypothalamic PVN. The activation of the central amygdala by corticosteroids leads to disinhibition of the HPA response to stress and an enhanced sympathetic outflow. The significance of MR and GR in regulation of CRF expression and function in the amygdaloid nuclei requires further study. The frontal cortex pathway that inhibits the PVN through GABA is enhanced by corticosterone (Diorio et al., 1995; Akana et al., 2001).

In summary, observations on HPA regulation have often been made without consideration of the corticosteroid effects on higher brain functions involved in arousal and processing of information. Two features of the behavioural effects of corticosteroids need to be addressed, which are of relevance for the neuroendocrine focus of this review. First, MR- and GR-mediated effects on neuronal excitability and aspects of behaviour can be discriminated. Hippocampal MR mediates effects of corticosterone on appraisal of information and response selection (Oitzl and De Kloet 1992, 1994; De Kloet et al., 1999), while GR function does not modify these aspects of sensory integration but rather promotes processes underlying consolidation of acquired information. Second, MR- and GR-mediated effects on information processing facilitate behavioural adaptation, which promotes the inhibitory control exerted by the higher brain circuits over HPA activity (De Kloet et al., 1998, 1999).

Modulation of glucocorticoid feedback

Different time domains of negative feedback of glucocorticoids can be distinguished (Dallman, 2000). For each time domain there are different mechanisms and sites of action in the core and extended HPA axis, and at least two classical corticosteroid receptor types, MR and GR, are involved.

The most rapid effects of glucocorticoids on their own secretion occur within minutes at the pituitary level. These are independent of gene transcription or de novo synthesis of proteins. The receptor that mediates these kinds of rapid effects is not known – nor are the precise mechanisms that are used to shut off secretion of ACTH. There are a number of possible candidate types of receptors for these effects. First, there may be a dedicated membrane receptor for glucocorticoids, but such a receptor has not been identified in mammals as yet (Orchinik, 1998). Second, corticosteroids or rapidly formed metabolites may act as neuroactive steroids (Rupprecht et al., 2001). Third, in analogy to the estrogen receptor (Levin, 2002), corticosteroids may bind to classical intracellular receptors, which may interact in a non-classical way with membrane proteins or second messenger pathways. Although there is a substantial number of central corticosteroid effects, which occur in a similarly rapid timeframe (Haller et al., 1998), mechanistic understanding of such processes awaits an easily manipulative experimental system with a robust readout.

Corticosteroid feedback depends on the balance between GR function on the one hand, and stress-induced activation of CRF neurons and the HPA axis on the other (De Kloet et al., 1997). One way in which this balance can be disturbed is under conditions of a local GR deficit. This can be congenital, as in the recent transgenic mouse line with brain-selective reduced expression of GR. Such mice display hypercorticoidism, cognitive impairment and metabolic disturbances, which in many ways resemble the symptoms of Cushing's syndrome (Holsboer and Barden, 1996). Feedback resistance can also be acquired, as in administration of the antiglucocorticoid RU 38486 (Lamberts et al., 1992; Van Haarst et al., 1996).

Reset of feedback sensitivity occurs when the input from the multiple sensory signalling pathways converging on CRF neurons becomes disproportionate. This may occur due to environmental changes, emotion, arousal or cognitive stimuli, which may become particularly potent chronic stressors under conditions of uncertainty, lack of control or poor predictability of upcoming events. Such conditions can be created in models of psychosocial stress in rats housed in mixed-gender groups in a complex environment resulting in a sustained HPA activation in subordinates (Blanchard et al., 1993), or in the sensory contact model, where mice can smell and see, but are not attacked by an aggressive opponent (Veenema et al., 2003). The elevated glucocorticoid levels caused by such chronic psychological stressors produce resistance to elevated glucocorticoids, through downregulation of GR in the CRF/AVP neurons (Makino et al., 1995) and increased activity of stressor-driven transcription factors.

Resistance to corticosteroid feedback in CRF neurons causes increased HPA activity and produces hypercorticoidism. As a consequence, the rest of the body including the brain and its neural stress response circuitry suffer from glucocorticoid overexposure. Importantly, glucocorticoid elevation initially synergizes with stress-induced activation of serotonergic, dopaminergic and noradrenergic neurons in the brainstem, and thus increases the sensitivity of limbic–forebrain areas to aminergic inputs (Marinelli and Piazza, 2002). These include direct aminergic input to the CRF/AVP neurons as well as indirect afferent inputs to these CRF neurons via the hippocampus. Moreover, chronic stress and corticosterone also activate the amygdaloid CRF system involved in stress-related behaviours. By these mechanisms the feedback resistance at the level of the CRF neurons is increasingly reinforced.

Resistance or increased sensitivity to glucocorticoid feedback can also be caused by mutations in the GR (and MR) (see, for review, De Rijk et al., 2002). In humans, the GR gene variants and some GR polymorphisms are common and sometimes associated with impaired control of stress reactions. For instance, the variant GRβ may modulate the interaction of the cognate GRα form with NFκB. One study demonstrated in rheumatoid arthritis a polymorphism in GRβ that increased its stability and hence may cause reduced GR activity. Unfortunately there were no HPA data in this patient group available (De Rijk et al., 2001). Other polymorphisms affect various aspects of the life cycle of the GR. This may include either selective repression of gene activation through interaction with other transcription factors or interaction of the GR with GREs (Lamberts, 2002). Data on the linkage of GR polymorphisms with disease are emerging and the preliminary data are fascinating. For instance, the Arg22Lys polymorphism decreased sensitivity to glucocorticoids in vivo, resulting in a better metabolic health profile (Van Rossum et al., 2002). The BclI restriction fragment polymorphism and an Asp363Ser may not only influence the regulation of the HPA axis, but are also associated with energy metabolism and cardiovascular control (Diblasio et al., 2003; Van Rossum et al., 2003). A GR polymorphism biasing the GRE pathway rather than interaction with transcription factors may be expected to enhance vulnerability to stress-related metabolic disease, cognitive decline and perhaps a decreased life expectancy during aging.

In summary, overt steroid resistance due to mutations impairing GR produce high cortisol levels. As a consequence, MR becomes overstimulated resulting in a compromised Na/K balance. However, more subtle genetic polymorphisms exist, that bias either GRE interaction or the interaction of GR with transcription factors and other receptor regulatory proteins. Such mutations may be unnoticed under "normal" conditions, but inflict imbalance in metabolism and stress regulation. This opens up new *vistas* towards understanding the co-morbidity between depression and co-morbid metabolic and cardiovascular diseases.

The issue of corticosteroid feedback and HPA pulsatility

The activity of the HPA axis not only shows a circadian rhythm, but also an ultradian rhythm, which results in the pulsatile bursts of corticosteroid hormone secretion every hour. The ultradian rhythm occurs in humans, monkey's and other mammals including rodents (Windle et al., 1998; Lightman et al., 2000). There is evidence from the monkey that the pulse generator exists within the hypothalamus (Mershon et al., 1992). The synchrony between ACTH and corticosteroid pulses is, however, poor suggesting that non-ACTH mechanisms are also involved. Indeed removal of the splanchnic nerve input to the adrenal increases the number of corticosterone pulses during the diurnal trough by increasing the adrenal sensitivity to ACTH (Jasper and Engeland, 1997). Moreover, the amplitude of the ultradian rhythm shows profound changes during the day and during a variety of (patho)physiological states (Windle et al., 2001). The nature of the HPA pulse generator is unknown, in particular how the alternating phases of excitation and inhibition are regulated. Unresolved is also if disruption of the ultradian rhythm results in abnormal tissue and cell responses.

Corticosteroid receptors and feedback: developmental level

During the first two weeks after birth the core of the HPA axis of the rat is characterised by a very low activity (Schapiro et al., 1962; Levine et al., 1967; Levine, 1970; Sapolsky and Meaney, 1986). During this so-called stress hyporesponsive period (SHRP), basal corticosterone output from the adrenal is much lower than in adult animals and mild stressors do not elicit a marked increase of corticosterone release. Following the discovery of the SHRP, it soon became clear that the quiescence of the HPA axis during this time is dependent on the sort of stimulus used. In 1980 Schoenfeld et al. could show, for the first time, that the HPA axis of the developing rat is capable to respond under specific circumstances, such as stimulation by ether fumes (Schoenfeld et al., 1980). Additional studies by Walker and others supported the study of Schoenfeld, demonstrating that the HPA axis of the developing rat responds in a time- and stressor-dependent manner (Witek-Janusek, 1988; Walker et al., 1990, 1991; Walker and Dallman, 1993). Already at this stage of development an overall theme can be recognised for the function of the stress system in adulthood, namely the stressor specificity of the response. It is now evident that the low postnatal activity of the HPA axis is dependent on the afferent pathway, via which a stressful stimulus is communicated. If the stressor is relayed via limbic system pathways (e.g. hippocampus or amygdala), the HPA system indeed responds only poorly. This has been shown in a number of studies, where mild stressors such as a saline injection or exposure to novelty have been used (Rosenfeld et al., 1993a; Smith et al., 1997; van Oers et al., 1998b Schmidt et al., 2002a). Yet, if the stimulus is relayed via alternative afferent pathways, e.g. norepinephric pathways from the brainstem, the suppression of HPA-axis activity in young rats or mice is overcome and the system does respond with a strong activation (Schoenfeld et al., 1980; Witek-Janusek, 1988; Walker et al., 1990; Walker et al., 1991; Walker and Dallman, 1993). Thus, the response of the HPA system during development is stimulus and context dependent.

There are two main questions in regard to the function of the corticosteroid receptors and negative feedback that arise from the phenomenon of the SHRP. First, how is negative feedback involved in the maintenance of the quiescence of the SHRP? Secondly, if via some afferent pathways an activation of the HPA axis occurs during the SHRP, how and via which receptors and pathways is this activation terminated?

SHRP and negative feedback

Both questions have of course to do with the function of the MR and the GR during development. It has been suggested that the function of both receptors in mediating different forms of negative feedback is dependent on the developmental state of the animal. In the adult, the MR is thought to mainly mediate the basal tone of the HPA axis via a proactive mode. The GR, on the other hand, is thought to be predominantly activated under conditions of high corticosteroid secretion (stress), terminating the stress response and restoring homeostasis (reactive mode). A number of studies indicated that the function of both receptors during development correlates with their ontogeny in different brain regions (Sapolsky et al., 1985; Meaney et al., 1996; van Oers et al., 1998a). In rats and mice, MR expression is already very high at birth and remains at adult levels throughout development (Rosenfeld et al., 1990). This developmental pattern correlates with the strong basal inhibition of the HPA system during postnatal development. Several studies have suggested that the MR during development plays an important role in maintaining the quiescence of the HPA axis (Ratka et al., 1989; van Oers et al., 1998a). Using adrenalectomy, in combination with either an implanted corticosterone pellet (proactive feedback mode) or an acute corticosterone injection (reactive feedback mode), van Oers et al., could demonstrate that the MR-mediated actions are functional in the neonate. However, it is also possible that hippocampal MR regulate the activity of the neonatal HPA axis in a more subtle way and that these receptors are not primarily responsible for the suppression of the HPA axis during the SHRP.

In contrast, the evidence is mounting that the main inhibition of the HPA activity in the neonate is occurring via the GR, specifically on the peripheral level of the pituitary. The answer to the question of why the pituitary of the neonate is not responding to mild stressors with an adult-like ACTH response seems to lie in the negative-feedback control of the POMC gene via the pituitary GRs. This hypothesis was first proposed by Schapiro (1965) and has since been supported by a number of findings. First, GR is expressed in the pituitary at adult-like levels during postnatal development (Sakly and Koch, 1981). Second, CBG levels are very low during the SHRP, resulting in high levels of free, biological active corticosterone (Henning, 1978). Third, elimination of circulating corticosterone levels by adrenalectomy greatly enhanced POMC gene expression in the pituitary, but had no or only little effect on CRF expression in the PVN (Grino et al., 1989). Fourth, treatment of mouse pups during the SHRP with GR antagonists greatly enhanced ACTH and corticosterone levels (Schmidt, 2004). This effect was most likely due to the blockade of pituitary GR rather than brain GRs, as it has been reported that the chronic blockade of GR in the PVN in neonatal rats only slightly enhanced CRF expression and corticosterone secretion (Yi et al., 1993). Thus, it seems likely that a high inhibitory tone on ACTH production via pituitary GR is the most proximal cause for the low responsiveness of the pituitary during the SHRP.

A number of arguments have also been brought forward which do not support a strong GR-mediated feedback at the pituitary during development. First, POMC gene expression has been shown to increase steadily during postnatal development, while GR expression in the pituitary is relatively high (Vazquez and Akil, 1992). This finding is indeed not easily explained. However, it is possible that other transcription factors increase the drive of POMC transcription during the SHRP, even though POMC expression is under a strong (and constant) GR feedback. Secondly, anogenital stroking during maternal deprivation has been shown to normalise ACTH release from the pituitary (Van Oers et al., 1998b). However, stroking might modulate the central effects of maternal deprivation rather than the HPA regulation in the periphery. Thus, the finding that central modulation by stroking is capable to alter ACTH release from the pituitary, following HPA axis activation by maternal deprivation, does not contradict a strong GR-mediated feedback signal on POMC expression during the SHRP.

But what about the concept that GR mediate the effect of high corticosterone concentrations and stress-induced HPA activity, both conditions that do not occur during the SHRP? The answer lies in the balance of free and bound corticosterone. As mentioned earlier, CBG levels are low during the SHRP, resulting in relatively high levels of free corticosterone, even though the total concentration of the hormone is very low. As a consequence, pituitary GR should be largely occupied by corticosterone even under basal conditions. In contrast, brain GR will not be activated by corticosterone, as in the brain the concentration of free corticosterone is much lower than in the blood. In addition, while GR expression has been reported at adult-like levels in the pituitary during the SHRP, hippocampal GR expression is very low at birth and only increases slowly afterwards (Kalinyak et al., 1989; Lawson et al., 1991; Rosenfeld et al., 1990; 1993b; Bohn et al., 1994). Even at the time around weaning (postnatal day 20) the concentration of the GR in the hippocampus is still lower compared to the adult situation, especially in the dentate gyrus (Van Eekelen et al., 1991a; 1991b; Rosenfeld et al., 1993b). Concordantly, the reactive negative-feedback mode does not function efficiently in the neonate. Numerous studies have demonstrated that neonates do not terminate an activation of the HPA axis as quickly as in the adult. Elevated corticosterone concentrations can still be found even 2 h after a mild stressor has been given (Suchecki et al., 1995; Smith et al., 1997). In correlation with the ontogeny of the GR in the hippocampus, a fully functional negative feedback in the rat is only developing after weaning (Vázquez and Akil, 1993; Schmidt et al., 2002b).

In addition, the function of negative feedback via the MR or the GR may be dependent on the specific stimulus. It can be hypothesised that if a stimulus activates the HPA axis via afferent inputs from, e.g. the brainstem, glucocorticoid feedback will also act mainly on this afferent pathway. Other structures that were not directly involved in the activation of the HPA axis, as in this example the hippocampus, will also be less affected by negative feedback. This hypothesis is supported by the finding that the efficiency of negative feedback during development is dependent on the nature of the stressor (Schoenfeld et al., 1980; Walker et al., 1991; Levine et al., 1994; Shanks and Meaney, 1994; Dent et al., 1999).

It can be concluded that the glucocorticoid feedback plays an important role during HPA-axis development. Numerous arguments point to the pituitary as the most proximal site of feedback inhibition of HPA activity during the SHRP via GR. As a consequence the adrenal is during the SHRP hyporesponsive to stressors and ACTH stimulation (Rosenfeld et al., 1992; Levine et al., 2000). Proactive control via the MR may contribute to the quiescence of the HPA axis postnatally, but in a subtler manner. Reactive feedback via the brain GR on stress-induced HPA activation, on the other hand, is a late developmental process, but may function specific to the region or afferent pathway activated by a certain stimulus.

Disruption of HPA-axis development

The importance of negative-feedback control during postnatal development becomes evident, when the normal function of the HPA axis is disturbed. Most methods that are used to activate the HPA system during development are based on an alteration (disruption or enhancement) of normal mother–infant interaction. Frequently used methods (among others) are daily handling repeated maternal separation for 3 or 8 h on consecutive days during the SHRP or a single interval of prolonged maternal deprivation for 24 h (Levine, 2001; Meaney, 2001; Pryce and Feldon, 2003). Most of these methods have in common that they activate the HPA axis during the SHRP, thereby increasing circulating corticosterone levels. It has been found that the corticosterone treatment or a disruption of the low corticosteroid secretion during development results in a number of short- and long-term consequences related to MR and GR control of HPA function.

Long-term Consequences

The predominant short-term effect of a separation of mother and pups for a prolonged period is an increased basal corticosterone secretion. In addition, it has been shown that maternal deprivation results in a decrease of MR and GR transcription (Smith et al., 1997; van Oers et al., 1998b; Schmidt et al., 2002a). It is not yet clear whether a possible change of function of either one of the both receptors is a cause or a consequence of maternal deprivation effects on corticosterone secretion.

Alterations of the circulating corticosterone levels either by corticosterone treatment or by manipulations of the mother–infant interaction result in an altered HPA function during adulthood. Corticosterone-enriched drinking water of the mother, resulting in increased corticosterone levels in suckling pups, has been shown to result in neurochemical and behavioural consequences, such as a reduced glucocorticoid stress response and an increased learning ability (Angelucci et al., 1985; Casolini et al., 1997). Also a disruption of mother–pup interaction and subsequently HPA development does not only alter GR/MR function during ontogeny, it also affects negative-feedback function throughout life. A number of studies had already indicated quite early that handling of pups during postnatal development could influence the corticosterone response of these animals during adulthood (Levine et al., 1967; Zarrow et al., 1972). In handling, animals are separated from their mother daily for about 10–15 min, a procedure which increases the overall licking and grooming behaviour of the mother. Subsequent studies could show a long-term effect of early handling that led to a reduced CRF expression and enhanced negative feedback in the brain (Meaney et al., 1989; Plotsky et al., 1993; Viau et al., 1993; Caldji et al., 1998).

In contrast to the seemingly beneficial effects of early handling, longer absence of the mother (e.g. maternal separation) resulted in an increased pituitary stress response in adult animals and in a decreased negative feedback (Plotsky and Meaney, 1993; Liu et al., 2000). Furthermore, 24 h of maternal deprivation at postnatal day 3 resulted in enhanced pituitary–adrenal basal levels, increased CRF expression and decreased MR and GR expression in adult animals (Ladd et al., 1996; Rots et al., 1996; Sutanto et al., 1996). Sutanto et al. (1996) could demonstrate that the binding capacity of both corticosteroid receptors in adulthood is sensitive to early maternal deprivation. Interestingly, these effects were also gender specific. Long-term effects of brief manipulations occurring early in development (e.g. maternal deprivation) on HPA function and behaviour were also shown by other studies (Suchecki and Tufik, 1997; Workel et al., 1997, 2001; Oitzl et al., 2000; Suchecki et al., 2000; Jimenez-Vasquez et al., 2001; Penke et al., 2001; Husum et al., 2002). Thus, the function of the HPA system and especially the negative-feedback system during adulthood are in strong correlation with the postnatal development of the animal. Increased maternal care results in a generally lower responsiveness of the HPA axis (enhanced negative feedback), while decreased maternal care results in an increased activity of the stress system (decreased negative feedback) (Caldji et al., 1998).

In summary, it becomes clear that the negative feedback during ontogeny is a very dynamic process. The function of negative feedback is constantly adapted in correlation to environmental stimuli and the vulnerability of an individual to stressful events during adulthood is determined by its development. It is therefore not sufficient to study the state of the stress system at a given point in time. It is necessary to include the whole developmental process in order to increase our understanding of the regulation or dysregulation of the stress system in a certain individual.

Corticosteroid receptors in HPA-axis regulation: molecular level

The MR and GR which mediate glucocorticoid effects, are present at multiple levels in the HPA axis, as well as in brain sites projecting to the PVN. Ligand-activated MR and GR act as transcription factors in the cell nucleus. MR and GR can modulate gene transcription in several ways, and these effects may show a substantial degree of dependence on cell type and cellular state (e.g. in case of neurons: depending on afferent inputs). The high degree of cellular specificity is, for example, clear from the PNMT gene (responsible for adrenalin synthesis). GR is necessary for its expression in the adrenal medulla and in this tissue a strong regulator of mRNA abundance, while in the brain the gene seems to be unresponsive to GR activation.

The MR and GR can act as transcription factors in basically two different ways. First, the receptors may bind independently (Fig. 4a), or in conjunction with other transcription factors (Fig. 4b), to DNA motifs in regulatory regions like promoters (such as the consensus glucocorticoid response element or GRE). This may either lead to transactivation, as is mostly the case when the receptors bind to consensus palindromic response elements, or to repression of transcription in case of 'negative GREs'. The latter elements typically deviate substantially from the classical GRE consensus.

Second, MR and GR may influence transcription by interacting with other, non-receptor transcription factors, which may or may not be bound to the DNA (Fig. 4c). The latter mode is mostly referred to as 'transrepression' as it is assumed to lead to mostly repressive effects of glucocorticoids on gene expression. The repression, as observed in test systems, tends to be mutual, in that GR- and MR-mediated responses are inhibited by activity of their cross-talk partners. The best-characterised cross-talk partners of glucocorticoid signalling are AP-1 and NF-κB (Gottlicher et al., 1998), but a number of other interesting factors exist, such as Stat 5 (Stoecklin et al., 1999).

The stoichiometry of cross-talk partners is critical to the outcome of interactions with corticosteroid receptors. This was first shown at a composite response element, where GR could either stimulate or repress AP-1 activity, depending on whether the AP-1 activity was constituted of c-Jun homodimers or c-jun/c-fos heterodimers (Diamond et al., 1990). At the same DNA element it was shown that GR, but

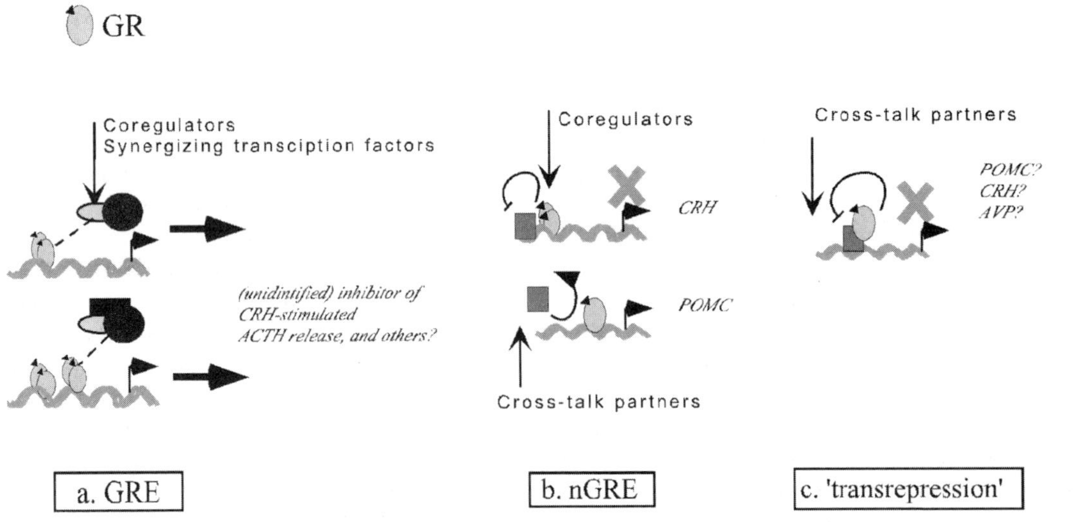

Fig. 4. In the core of the HPA axis, GR mediates the transcriptional responses to corticosterone in different way (a–c). The relevant promoters are indicated at the right of each cartoon. Factors, which determine the magnitude of the response, are indicated with arrows. Outside the core of the axis, MR influence gene expression in similar ways. See text for details.

not MR, transrepresses AP-1 activity (Pearce and Yamamoto, 1993). This finding offered a first transcriptional basis for differential effects mediated by MR and GR. However, in vivo MR is also capable of mediating negative effects on transcription, for example in case of the 5-HT1A receptor gene (Meijer and De Kloet, 1995). Accordingly, MR is able to mediate transrepression at the 5-HT1A receptor promoter, critically depending on the transcription factors which drive the expression of the gene (Meijer et al., 2000a).

The presence of cross-talk partners is highly variable, depending on the cell type and cellular state of activity induced by other signals. c-Fos expression, for example, is a function of neuronal activity, and its abundance in a glucocorticoid target area like the paraventricular nucleus of the hypothalamus (PVN) differs dramatically as depending on the stress state of an animal (Ceccatelli et al., 1989). Elegant studies have shown that in hypothalamic extracts of stressed rats the binding to AP-1 response elements on the DNA is indeed impaired, suggesting in vivo relevance of GR–AP-1 interactions in negative feedback of glucocorticoids on the HPA axis (Kovács et al., 2000). Thus, GR would repress transcription in activated PVN neurons, but at the same time be less efficient at transactivating from its response elements.

Repression of gene transcription in negative feedback at the core of the HPA axis

In the core of the HPA axis, there are two clear targets for glucocorticoid negative feedback through transcription: the proopiomelanocortin gene in the anterior pituitary, from which ACTH is derived, and the CRF gene in the parvocellular part of the PVN.

POMC expression and ACTH release

The promoter of the POMC gene has been studied extensively with regard to its negative regulation by glucocorticoids. This is by merit of the availability of the AtT-20 cell line (Gumbiner and Kelly, 1981), which synthesises and secretes ACTH in response to physiological stimuli, and thus forms an excellent model for the corticotropic cells from the anterior pituitary. The POMC promoter is driven by amongst others, AP-1 and CREB, which are activated by stimulation with CRF (Boutillier et al., 1995). These proteins are potential targets for transrepression by GR via protein–protein interactions. However, at least part of the negative regulation of this promoter takes place via a negative GRE (Drouin et al., 1993). This site also binds the positive CRF-driven transcription factor Nurr77 (Philips et al., 1997). The repressive effect of GR may be brought about by competition for DNA binding at the same DNA element (Murphy and Conneely, 1997). Because repression via GR seems to depend on which of the multiple-stimulating pathways is activated (both CRF dependent and CRF independent), the gene may have ways to 'escape' from glucocorticoid negative feedback under some conditions (Bousquet et al., 2000).

CRF-stimulated ACTH secretion (as opposed to POMC expression) is sensitive to glucocorticoid feedback through induction of an, as yet, unidentified gene which counteracts the effects of the CRF-induced second messenger pathway (Tian et al., 2001). This is an example of the 'intermediate' time-frame of negative feedback. Interestingly, AVP, which strongly potentiates the effect of CRF on ACTH release, reduces the sensitivity to this glucocorticoid feedback in isolated corticotropes. This may explain why (chronic) stress conditions with a strong AVP-ergic drive to the pituitary are relatively stress resistant (Lim et al., 2002).

The importance of the negative GRE in the POMC promoter was confirmed by the *dim/dim* mouse (Reichardt et al., 1998). This elegant knock-in mouse model expresses a GR, which is severely compromised in binding to the DNA (Heck et al., 1994). Although not all GRE-dependent promoters are affected by this mutation (Adams et al., 2003), these mice have clear phenotypical disturbances in GR-mediated effects, which depend on DNA binding of the receptor. The point in case is the expression of the POMC mRNA and ACTH peptide in corticotropes, which are substantially increased in the *dim/dim* mice. The distinct mechanism of glucocorticoid negative feedback on ACTH release from the pituitary is revealed by these mice, as ACTH levels are not markedly elevated in *dim/dim* mice (Reichardt et al., 1998). Interestingly, this is in spite of the aforementioned transactivation-dependent effect of dexamethasone on CRF-stimulated ACTH release.

Although ligand-binding studies consistently demonstrate that MR is also present in the pituitary (Spencer et al., 1993), the role of these receptors in gene regulation of e.g. POMC is not clear. It may be that, if MR is indeed expressed in the corticotropes, this receptor type mediates effects similar to GR, and that co-expression of the receptors serves to broaden the range of sensitivity to the hormone.

CRF and AVP in the PVN

The next step up in control of the HPA axis is the PVN, where the ACTH secretagogues CRF and AVP are synthesised. Expression of both peptides is subject to negative feedback by corticosterone in vivo (Swanson and Simmons, 1989) but, in part, because cell lines representative of parvocellular neurons are lacking, the molecular mechanism of CRF and AVP gene regulation is understood less well than that of POMC. GR are expressed at high levels in the PVN, and the promoters of both genes have been shown to be directly regulated via GR in heterologous cell lines (Guardiola-Diaz et al., 1996; Iwasaki et al., 1997; Malkoski and Dorin, 1999).

However, both MR and GR are abundantly present in neurons, which project to the PVN. Hence, it is difficult to distinguish in vivo direct transcriptional effects of GR in the parvocellular neurons from transsynaptic effects which are the consequence of corticosteroids acting at an afferent site (Herman et al., 1990). In particular, the feedback effects on the activity of the parvocellular neurons (as opposed to expression levels of CRF and AVP) may be direct or indirect via a transsynaptic mechanism. Blockade of (hippocampal) MR by antagonist infusion, for example, leads to increased activity of the HPA axis (Ratka et al., 1989), via unknown molecular targets. Delineation of the different receptor populations in the brain that are involved in control of PVN factors under basal and stress conditions may await the generation of site-specific knockout mice, or other efficient local knockdown techniques.

The promoter of the human CRF gene contains, among other putative glucocorticoid-sensitive *cis*-elements (Guardiola-Diaz et al., 1996), a negative GRE (Malkoski and Dorin, 1999). This is a compound element to which both the GR as well as the transcription factor AP-1 can bind, much like the element from the proliferin gene which has been subject to intense investigation (e.g. Diamond et al., 1990; Pearce and Yamamoto, 1993). The mechanism of DNA binding to these types of elements differs from that to consensus GRE, and the normal CRF expression in the *dim/dim* mouse does not preclude in vivo relevance of this repressive DNA element (Reichardt et al., 1998 – see under 'POMC expression and ACTH release'). The strong involvement of GR in control of CRF expression levels is, for example clear from newborn GR knockout mice, which show dramatically increased CRF expression (Kretz et al., 1999).

Like CRF, AVP expression in vivo is negatively regulated by corticosteroids, but the genes do not react in parallel under every condition. As with CRF, negative transcriptional regulation of AVP is dependent on nature and duration of the inputs into the PVN (Pinnock and Herbert, 2001). The promoter of the gene is transrepressed via GR in heterologous cells, but no responsible cross-talk partner or *cis*-element was described in these studies (Iwasaki et al., 1997). Disparate regulation of CRF and AVP expression does occur in vivo, as under conditions of chronic stress the drive on ACTH secretion is thought to shift towards AVP dependence.

Other targets in the brain: multiple targets and multiple mechanisms

MR and GR in many brain regions can influence the HPA axis (and many other processes) through transsynaptic pathways and via unidentified molecular targets. Whether or not gene transcription is influenced by corticosteroids depends on the cellular activity state. Genome-wide analysis using tissue from adrenalectomized and corticosterone-replaced animals has identified hundreds of regulated transcripts in hippocampus as a whole under resting conditions (Datson et al., 2001). A bottleneck in such studies is that only medium to high abundantly (differentially) expressed genes can be reliably detected using the current methodology, be it SAGE or microarrays (Evans et al., 2002). Also, in general, the magnitude of corticosteroid-induced changes in mRNA expression in the brain is smaller than in tissues like liver, kidney (Chen et al., 1999) or adrenal medulla (Sabban and Kvetnánsky, 2001). This limits the power of genomic approaches considerably.

While negative feedback at the core of the HPA axis involves repressive effects on transcription, in many other regulatory mechanisms transactivation can also be involved. In fact, recent studies on the glucocorticoid-induced leucine zipper protein, or GILZ, demonstrated for the immune system that glucocorticoids may use parallel mechanisms to achieve a single goal. While activated GR may interfere by protein–protein interactions with other transcription factors such as AP-1, corticosteroids also induce the expression of GILZ (D'Adamio et al., 1997), which in turn may inhibit AP-1 and/or NF-κB (Mittelstadt and Ashwell, 2001). In fact, as GILZ seems to be corticosteroid regulated in many tissues (Robert-Nicoud et al., 2001), such mechanisms could be relevant for regulation in the core of the HPA axis.

Receptor Co-regulators

The effects of MR and GR (like those of other steroid receptors) on transcription after binding to the DNA are mediated and modulated by a large number of proteins, referred to as co-regulators. These consist of multiple protein families with co-activating and co-repressing effects. As these co-regulators differ in their specific interactions with steroid receptor types, their downstream effects and in their cellular expression patterns, the co-regulator stoichiometry is thought to determine the magnitude and nature of steroid responses in a given cell (McKenna et al., 1999; Rosenfeld and Glass, 2001).

Prominent examples of co-regulators are the p160 Steroid Receptor Co-activators (SRCs). The SRC family consists of three genes, coding for the structurally related proteins SRC-1 (NCoA-1), SRC-2 (NCoA-2, TIF2, GRIP-1) and the somewhat more distant SRC-3 (NCoA-3, pCIP/ACTR/AIB1/RAC3/TRAM1). The SRC family members have different interactions with steroid receptors and unique expression patterns, and therefore are a good example of possible determinants of cellular specificity of glucocorticoid actions in vivo (Meijer, 2002; Nishihara et al., 2003). Also, splice variants of the SRC-1 gene have been shown to differentially affect steroid receptor signalling (Ding et al., 1998; Kalkhoven et al., 1998), and to have highly specific expression patterns in the brain, e.g. in the PVN (Meijer, 2000). The fact that the signalling for all nuclear receptor types seems to converge at the level of the SRCs makes these proteins interesting as determinants for cross-talk between multiple steroid pathways.

Regulation of the expression or activity levels of co-activators is expected to change steroid responses. The expression of SRC-1 can be subject to hormonal regulation in the pituitary (Misiti et al., 1998), and also shows variation in the brain (Bousios et al., 2001). Thus, SRCs may well be factors involved in physiological modulation of steroid responsiveness. As a last fascinating aspect of SRCs it may be mentioned that their interactions with steroid receptors can be ligand dependent. For the vitamin D receptor, a synthetic ligand was shown to preferentially interact with SRC-2, suggesting relative resistance to that particular ligand in tissues with low expression of that particular co-activator (Takeyama et al., 1999).

Many criteria for determinants of inherent or acquired differences in glucocorticoid sensitivity apply to the SRC family members: there is proof of principle for specific interactions with steroid receptors, there are specific expression patterns in the brain and other tissues, and their activity and abundance can be regulated. However, the precise roles of the SRCs and other relevant co-regulators, such as the co-repressors NCoR and SMRT, for glucocorticoid sensitivity of different organ systems in vivo, remains to be resolved.

In conclusion, the molecular biology of corticosteroid receptor action in negative feedback in brain and pituitary has been elucidated to a considerable extent with respect to regulation of expression of AVP, CRF and POMC genes. In addition, there is a lot known about *possible* mechanisms of general MR- and GR-mediated transcription regulation. However, a lot remains unknown. Issues such as which exact mechanisms are at work in the relevant brain areas in vivo, which other primary transcriptional and non-transcriptional (rapid feedback) targets are of importance and which other cellular components are involved in differences and changes in steroid sensitivity remain to be resolved.

Future directions

The role of brain MR and GR in the control of the HPA axis was reviewed here along with criteria prescribed by their localisation and properties. GR appears as the predominant receptor in the core of the HPA axis. GR is also abundant in discrete extrahypothalamic brain regions, while co-localised with the high-affinity MR. Accordingly, corticosteroids exert feedback control in the core of the HPA-axis proper as well as in the afferent pathways involved in a wide variety of cognitive, emotional and vegetative functions. The steroids exert these actions in concert with the other components of the HPA axis, and together these signals of the neuroendocrine system orchestrate behavioural adaptations to changing environments.

Today it is believed that the stress system is organised in two anti-parallel modes (see Table 1): CRF drives via CRF-1 receptors the rapid sympathetic and HPA-axis reactions to stress. MR affects the cognitive input by facilitating processes underlying interpretation of environmental changes and response selection. According to Aaron Hsueh (Hsu and Hsueh, 2001), urocortin 2 and 3 drive via the CRF-2 receptors the slower parasympathetic responses aimed towards recovery, coping and adaptation. Implicated in this slower mode of action is the GR, which facilitates recovery and adaptation upon activation by rising corticosterone concentrations. GR helps to prepare for future events by promoting the storage of energy resources and the storage of information how to deal with a challenge. MR also operates under high corticosteroid concentrations during stress and appears to be under control of CRF (Gesing et al., 2001).

Table 1. Anti-parallel organized stress systems. MR controls neuronal networks underlying the sensitivity or threshold of the immediate CRF-1 receptor-driven stress reactions. GR controls the termination of the immediate stress reactions, facilitates recovery and prepares for future events

Stress	Adaptation
CRH	Stresscopin
CRH-1 receptor	CRH-2 receptor
Sympathetic	Para-sympathetic
Immediate	Late sustained
Fight/flight	Coping
MR	GR

Some time ago we have proposed that the MR- and GR-mediated effects operate in balance to ensure homeostasis and health (De Kloet and Reul, 1987; De Kloet, 1991; De Kloet et al., 1998). This hypothesis has its roots in the cellular studies that clearly demonstrate the opposing MR- versus GR-mediated effects on almost every possible neurophysiological endpoint (Joëls and De Kloet, 1994). The consensus from these studies was that MR maintains stability in the hippocampal circuitry, while via GR excitability is restored that is transiently raised by excitatory stimuli. It follows that MR/GR imbalance destabilises the stress circuitry, causes neuroendocrine dysregulation and impairs behavioural adaptation. If the adverse condition persists and a certain threshold is passed susceptibility to stress-related diseases is increased. In other words, the MR- and GR-mediated effects are crucial for allostasis (McEwen, 2002), i.e. the dynamic processes underlying the maintenance of homeostasis and health, and its cost (allostatic load).

The concepts outlined in this review summarised in the previous paragraphs raise a number of questions for future research:

1. How does the GR-mediated negative feedback in the HPA core operate in the face of corticosteroid action on the afferent inputs to the PVN? Are these feedback sites in the HPA core and in its afferents part of a co-ordinated check and balance system via which corticosteroid hormones exert control over the maintenance of homeostasis?
2. How do the MR- and GR-mediated actions operate in the pulsatile mode of HPA-axis activity?
3. What is the role of the corticosteroid receptors in the development of the stress system?
4. What is the molecular basis for the differential MR- and GR-mediated effects, and which other transcriptionally active proteins are determinants and modulators of their actions?
5. What are the receptor mechanisms underlying the rapid feedback actions?
6. How does dysregulation of MR- and GR-mediated processes occur, and how does it contribute to the pathogenesis of stress-related brain disorders, and possibly, co-morbid metabolic, cardiovascular and immunological diseases?

Today the approaches to address these questions are often inspired by the new technologies that can be used for molecular dissection of the MR- and GR-dependent signalling pathways. One approach is to study the phenotype occurring after (transient) ablation of receptor function via selective antagonists, antisense DNA and si-RNA, or conditional knockout procedures (Reichardt et al., 1998; Karst et al., 2000; Oitzl et al., 2001). In other cases disruption of corticosteroid target genes may cause a complete reversal in cognitive performance (Grootendorst, 2001). Using genome-wide screening corticosteroid-responsive signalling pathways are being identified (Datson et al., 2001; Feldker et al., 2003). These new findings are very promising, but it will be a tall order to identify the prime mover activated by MR and GR imbalance in the cascade that ultimately may lead to a stress-related disorder.

Acknowledgments

We thank the Netherlands Organisation for Scientific Research (NWO) for their continuous support of our research. The editorial assistance of Ms. Ellen M. Heidema is gratefully acknowledged.

References

Adams, M., Meijer, O.C., Wang, J., Bhargava, A. and Pearce, D. (2003) Homodimerization of the glucocorticoid receptor is not necessary for DNA binding: activation of the Phenylethanolamine-N-Methyltransferase gene by dimerization-defective mutants. Mol. Endocrinol., (e publication ahead of print).

Akana, S.F., Chu, A., Soriano, L. and Dallman, M.F. (2001) Corticosterone exerts site-specific and state-dependent effects in prefrontal cortex and amygdala on regulation of adrenocorticotropic hormone, insulin and fat depots. J. Neuroendocrinol., 13: 625–637.

Angelucci, L., Patacchioli, F.R., Scaccianoce, S., Di Sciullo, A., Cardillo, A. and Maccari, S. (1985) A model for later-life effects of perinatal drug exposure: maternal hormone mediation. Neurobehav. Toxicol. Teratol., 7: 511–517.

Arriza, J.L., Weinberger, C., Cerelli, G., Glaser, T.M., Handelin, B.L., Housman, D.E. and Evans, R.M. (1987) Cloning of human mineralocorticoid receptor complementary DNA: structural and functional kinship with the glucocorticoid receptor. Science, 237: 268–275.

Arriza, J.L., Simerly, R.B., Swanson, L.W. and Evans, R.M. (1988) The neuronal mineralocorticoid receptor as a mediator of glucocorticoid response. Neuron, 1: 887–900.

Belanoff, J.K., Flores, B.H., Kalezhan, M., Sund, B. and Schatzberg, A.F. (2001) Rapid reversal of psychotic depression using mifepristone. J. Clin. Psychopharmacol., 21: 516–521.

Belanoff, J.K., Jurik, J., Schatzberg, L.D., DeBattista, C. and Schatzberg, A.F. (2002a) Slowing the progression of cognitive decline in Alzheimer's disease using mifepristone. J. Mol. Neurosci., 19: 201–206.

Belanoff, J.K., Rothschild, A.J., Cassidy, F., DeBattista, C., Baulieu, E.E., Schold, C. and Schatzberg, A.F. (2002b) An open label trial of C-1073 (mifepristone) for psychotic major depression. Biol. Psychiat., 52: 386–392.

Blanchard, D.C., Sakai, R.R., McEwen, B.S., Weiss, S.M. and Blanchard, R.J. (1993) Subordination stress: behavioral, brain, and neuroendocrine correlates. Behav. Brain Res., 58: 113–121.

Bohn, M.C., Dean, D., Hussain, S. and Giuliano, R. (1994) Development of mRNAs for glucocorticoid and mineralocorticoid receptors in rat hippocampus. Brain Res. Dev. Brain Res., 77: 157–162.

Bousios, S., Karandrea, D., Kittas, C. and Kitraki, E. (2001) Effects of gender and stress on the regulation of steroid receptor coactivator-1 expression in the rat brain and pituitary. J. Steroid Biochem. Mol. Biol., 78: 401–407.

Bousquet, C., Zatelli, M.C. and Melmed, S. (2000) Direct regulation of pituitary proopiomelanocortin by STAT3 provides a novel mechanism for immuno-neuroendocrine interfacing. J. Clin. Invest., 106: 1417–1425.

Boutillier, A.L., Monnier, D., Lorang, D., Lundblad, J.R., Roberts, J.L. and Loeffler, J.P. (1995) Corticotropin-releasing hormone stimulates proopiomelanocortin transcription by cFos-dependent and -independent pathways: characterization of an AP1 site in exon 1. Mol. Endocrinol., 9: 745–755.

Bradbury, M.J., Akana, S.F., Cascio, C.S., Levin, N. and Dallman, M.F. (1991) Regulation of basal ACTH secretion by corticosterone is mediated by both type I (MR) and type II (GR) receptors in rat brain. J. Steroid Biochem. Mol. Biol., 40: 133–142.

Bradbury, M.J., Akana, S.F. and Dallman, M.F. (1994) Roles of type I and II corticosteroid receptors in regulation of basal activity in the hypothalamo–pituitary–adrenal axis during the diurnal trough and the peak: evidence for a non-additive effect of combined receptor occupation. Endocrinology, 134: 1286–1296.

Brady, L.S., Whitfield, H.J., Jr., Fox, R.J., Gold, P.W. and Herkenham, M. (1991) Long-term antidepressant administration alters corticotropin-releasing hormone, tyrosine hydroxylase, and mineralocorticoid receptor gene expression in rat brain. Therapeutic implications. J. Clin. Invest., 87: 831–837.

Caldji, C., Tannenbaum, B., Sharma, S., Francis, D., Plotsky, P.M. and Meaney, M.J. (1998) Maternal care during infancy regulates the development of neural systems mediating the expression of fearfulness in the rat. Proc. Natl. Acad. Sci. USA, 95: 5335–5340.

Carey, M.P., Deterd, C.H., De Koning, J., Helmerhorst, F.M. and De Kloet, E.R. (1995) The influence of ovarian steroids on hypothalamic–pituitary–adrenal regulation in the female rat. J. Endocrinol., 144: 311–321.

Casolini, P., Cigliana, G., Alema, G.S., Ruggieri, V., Angelucci, L. and Catalani, A. (1997) Effect of increased maternal corticosterone during lactation on hippocampal corticosteroid receptors, stress response and learning in offspring in the early stages of life. Neurosci., 79: 1005–1012.

Ceccatelli, S., Villar, M.J., Goldstein, M. and Hokfelt, T. (1989) Expression of c-Fos immunoreactivity in transmitter-characterized neurons after stress. Proc. Natl. Acad. Sci. USA, 86: 9569–9573.

Chen, S.Y., Bhargava, A., Mastroberardino, L., Meijer, O.C., Wang, J., Buse, P., Firestone, G.L., Verrey, F. and Pearce, D. (1999) Epithelial sodium channel regulated by aldosterone-induced protein sgk. Proc. Natl. Acad. Sci. USA, 96: 2514–2519.

Cole, R.L. and Sawchenko, P.E. (2002) Neurotransmitter regulation of cellular activation and neuropeptide gene expression in the paraventricular nucleus of the hypothalamus. J. Neurosci., 22: 959–969.

Cook, C.J. (2002) Glucocorticoid feedback increases the sensitivity of the limbic system to stress. Physiol. Behav., 75: 455–464.

D'Adamio, F., Zollo, O., Moraca, R., Ayroldi, E., Bruscoli, S., Bartoli, A., Cannarile, L., Migliorati, G. and Riccardi, C. (1997) A new dexamethasone-induced gene of the leucine zipper family protects T lymphocytes from TCR/CD3-activated cell death. Immunity, 7: 803–812.

Dallman, M.F., Akana, S.F., Cascio, C.S., Darlington, D.N., Jacobson, L. and Levin, N. (1987) Regulation of ACTH secretion: variations on a theme of B. Recent Prog. Horm. Res., 43: 113–173.

Dallman, M.F., Levin, N., Cascio, C.S., Akana, S.F., Jacobson, L. and Kuhn, R.W. (1989) Pharmacological evidence that the inhibition of diurnal adrenocorticotropin secretion by corticosteroids is mediated via type I corticosterone-preferring receptors. Endocrinology, 124: 2844–2850.

Dallman, M.F., Viau, V.G., Bhatnagar, S., Gomez, F., Laugero, K. and Bell, M.E. (2002) Corticotropin-releasing factor, corticosteroids, stress, and sugar: energy balance, the brain, and behavior. Horm. Brain Behav., 1: 571–631.

Datson, N.A., Van der Perk, J., De Kloet, E.R. and Vreugdenhil, E. (2001) Identification of corticosteroid-responsive genes in rat hippocampus using serial analysis of gene expression. Eur. J. Neurosci., 14: 675–689.

De Kloet, E.R. (1991) Brain corticosteroid receptor balance and homeostatic control. Front Neuroendocrinol., 12: 95–164.

De Kloet, E.R., Burbach, P. and Mulder, G.H. (1977) Localization and role of transcortin-like molecules in the anterior pituitary. Mol. Cell. Endocrinol., 7: 261–273.

De Kloet, E.R., Oitzl, M.S. and Joëls, M. (1999) Stress and cognition: are corticosteroids good or bad guys? Trends Neurosci., 22: 422–426.

De Kloet, E.R. and Reul, J.M.H.M. (1987) Feedback action and tonic influence of corticosteroids on brain function: a concept arising from the heterogeneity of brain receptor systems. Psychoneuroendocrinol., 12: 83–105.

De Kloet, E.R., van der Vies, J. and de Wied, D. (1974) The site of suppressive action of dexamethasone on pituitary-adrenal activity. Endocrinology, 94: 61–73.

De Kloet, E.R., Vreugdenhil, E., Oitzl, M.S. and Joëls, M. (1998) Brain corticosteroid receptor balance in health and disease. Endocr. Rev., 19: 269–301.

De Kloet, E.R., Wallach, G. and McEwen, B.S. (1975) Differences in corticosterone and dexamethasone binding to rat brain and pituitary. Endocrinology, 96: 598–609.

De Kloet, E.R., De Kock, S., Schild, V. and Veldhuis, H.D. (1988) Antiglucocorticoid RU 38486 attenuates retention of a behaviour and disinhibits the hypothalamic–pituitary–adrenal axis at different brain sites. Neuroendocrinol., 47: 109–115.

De Kloet, E.R., Vreugdenhil, E., Oitzl, M.S. and Joëls, M. (1997) Glucocorticoid feedback resistance. Trends Endocrinol. Metabol., 8: 26–33.

Dent, G.W., Smith, M.A. and Levine, S. (1999) The ontogeny of the neuroendocrine response to endotoxin. Brain Res. Dev. Brain Res., 117: 21–29.

De Rijk, R.H., Schaaf, M.J., Turner, G., Datson, N.A., Vreugdenhil, E., Cidlowski, J., de Kloet, E.R., Emery, P., Sternberg, E.M. and Detera-Wadleigh, S.D. (2001) A human glucocorticoid receptor gene variant that increases the stability of the glucocorticoid receptor beta-isoform mRNA is associated with rheumatoid arthritis. J. Rheumatol., 28: 2383–2388.

De Rijk, R.H., Schaaf, M. and de Kloet, E.R. (2002) Glucocorticoid receptor variants: clinical implications. J. Steroid Biochem. Mol. Biol., 81: 103–122.

Deuschle, M., Weber, B., Colla, M., Müller, M., Kniest, A. and Heuser, I. (1998) Mineralocorticoid receptor also modulates basal activity of hypothalamus–pituitary–adrenocortical system in humans. Neuroendocrinol., 68: 355–360.

Diamond, M.I., Miner, J.N., Yoshinaga, S.K. and Yamamoto, K.R. (1990) Transcription factor interactions: selectors of positive or negative regulation from a single DNA element. Science, 249: 1266–1272.

Di Blasio, A.M., van Rossum, E.F., Maestrini, S., Berselli, M.E., Tagliaferri, M., Podesta, F., Koper, J.W., Liuzzi, A. and Lamberts, S.W. (2003) The relation between two polymorphisms in the glucocorticoid receptor gene and body mass index, blood pressure and cholesterol in obese patients. Clin. Endocrinol., 59: 68–74.

Ding, X.F., Anderson, C.M., Ma, H., Hong, H., Uht, R.M., Kushner, P.J. and Stallcup, M.R. (1998) Nuclear receptor-binding sites of coactivators glucocorticoid interacting protein 1 (GRIP1) and steroid receptor coactivator 1 (SRC-1): multiple motifs with different binding specificities. Mol. Endocrinol., 12: 302–313.

Diorio, D., Viau, V. and Meaney, M.J. (1993a) The role of the medical prefrontal cortex (cingulate gyrus) in the regulation of hypothalamic–pituitary–adrenal responses to stress. J. Neurosci., 13: 3839–3847.

Diorio, D., Viau, V. and Meaney, M.J. (1993b) The role of the medical prefrontal cortex (cingulate gyrus) in the regulation of hypothalamic–pituitary–adrenal responses to stress. J. Neurosci., 13: 3829–3847.

Dodt, C., Kern, W., Fehm, H.L. and Born, J. (1993) Antimineralocorticoid canrenoate enhances secretory activity of the hypothalamus–pituitary–adrenocortical (HPA) axis in humans. Neuroendocrinol., 58: 570–574.

Drouin, J., Sun, Y.L., Chamberland, M., Gauthier, Y., De Lean, A., Nemer, M. and Schmidt, T.J. (1993) Novel glucocorticoid receptor complex with DNA element of the hormone-repressed POMC gene. EMBO J., 12: 145–156.

Edwards, C.R.W., Stewart, P.M., Burt, D., Brett, L., McIntyre, M.A., Sutanto, W., De Kloet, E.R. and Monder, C. (1988) Localisation of 11β-hydroxysteroid dehydrogenase-tissue specific protector of the mineralocorticoid receptor. Lancet, 2(8618): 986–989.

Evans, S.J., Datson, N.A., Kabbaj, M., Thompson, R.C., Vreugdenhil, E., De Kloet, E.R., Watson, S.J. and Akil, H. (2002) Evaluation of Affymetrix Gene Chip sensitivity in rat hippocampal tissue using SAGE analysis. Serial Analysis of Gene Expression. Eur. J. Neurosci., 16: 409–413.

Feldker, D.E., Datson, N.A., Veenema, A.H., Meulmeester, E., De Kloet, E.R. and Vreugdenhil, E. (2003) Serial analysis of gene expression predicts structural differences in hippocampus of long attack latency and short attack latency mice. Eur. J. Neurosci., 17: 379–387.

Funder, J.W., Pearce, P.T., Smith, R. and Smith, A.I. (1988) Mineralocorticoid action: target tissue specificity is enzyme, not receptor, mediated. Science, 242: 583–585.

Fuxe, K., Wikström, A.C., Ökret, Agnati, L.F., Harfstrand, A., Yu, Z.Y., Granholm, L., Zoli, M., Vale, W. and Gustafsson, J. (1985) Mapping of the glucocorticoid receptor immunoreactive neurons in the rat tel- and diencephalons using a monoclonal antibody against rat liver glucocorticoid receptors. Endocrinology, 117: 1803–1812.

Gerlach, J. and McEwen, B.S. (1972) Rat brain binds adrenal steroid hormone: radioautography of hippocampus. Science, 175: 1133–1136.

Gesing, A., Bilang-Bleuel, A., Droste, S.K., Linthorst, A.C., Holsboer, F. and Reul, J.M. (2001) Psychological stress increases hippocampal mineralocorticoid receptor levels: involvement of corticotropin-releasing hormone. J. Neurosci., 21: 4822–4829.

Gottlicher, M., Heck, S. and Herrlich, P. (1998) Transcriptional cross-talk, the second mode of steroid hormone receptor action [see comments]. J. Mol. Med., 76: 480–489.

Grino, M., Burgunder, J.M., Eskay, R.L. and Eiden, L.E. (1989). Onset of glucocorticoid responsiveness of anterior pituitary corticotrophs during development is scheduled by corticotropin-releasing factor. Endocrinology 124: 2686–2692.

Grootendorst, J., De Kloet, E.R., Vossen, C., Dalm, S. and Oitzl, M.S. (2001) Repeated exposure to rats has persistent genotype-dependent effects on learning and locomotor activity of apolipoprotein E knockout and C57Bl/6 mice. Behav. Brain Res., 125: 249–259.

Grottoli, S., Giordano, R., Maccagno, B., Pellegrino, M., Ghigo, E. and Arvat, E. (2002) The stimulatory effect of canrenoate, a mineralocorticoid antagonist, on the activity of the hypothalamus–pituitary–adrenal axis is abolished by alprazolam, a benzodiazepine, in humans. J. Clin. Endocrinol. Metab., 87: 4616–4620.

Guardiola-Diaz, H.M., Kolinske, J.S., Gates, L.H. and Seasholtz, A.F. (1996) Negative glucorticoid regulation of cyclic adenosine 3′, 5′-monophosphate-stimulated corticotropin-releasing hormone-reporter expression in AtT-20 cells. Mol. Endocrinol., 10: 317–329.

Gumbiner, B. and Kelly, R.B. (1981) Secretory granules of an anterior pituitary cell line, AtT-20, contain only mature forms of corticotropin and beta-lipotropin. Proc. Natl. Acad. Sci. USA, 78: 318–322.

Haller, J., Halasz, J., Makara, G.B. and Kruk, M.R. (1998) Acute effects of glucocorticoids: behavioral and pharmacological perspectives. Neurosci. Biobehav. Rev., 23: 337–344.

Heck, S., Kullmann, M., Gast, A., Ponta, H., Rahmsdorf, H.J, Herrlich, P. and Cato, A.C.B. (1994) A distinct modulating domain in glucocorticoid receptor monomers in the repression of activity of the transcription factor AP-1. EMBO J., 13: 4087–4095.

Henning, S.J. (1978) Plasma concentrations of total and free corticosterone during development in the rat. Am. J. Physiol., 235: E451–E456.

Herman, J.P. and Cullinan, W.E. (1997) Neurocircuitry of stress: central control of the hypothalamo–pituitary–adrenocortical axis. Trends Neurosci., 20: 78–84.

Herman, J.P., Wiegand, S.J. and Watson, S.J. (1990) Regulation of basal corticotropin-releasing hormone and arginine vasopressin messenger ribonucleic acid expression in the paraventricular nucleus: effects of selective hypothalamic deafferentations. Endocrinology, 127: 2408–2417.

Herman, J.P. and Larson, B.R. (2001) Differential regulation of forebrain glutamic acid decarboxylase mRNA expression by aging and stress. Brain Res., 912: 60–66.

Herman, J.P., Tasker, J.G., Ziegler, D.R. and Cullinan, W.E. (2002) Local circuit regulation of paraventricular nucleus stress integration: glutamate-GABA connections. Pharmacol. Biochem. Behav., 71: 457–468

Heuser, I., Deuschle, M., Weber, A., Kniest, A., Ziegler, C., Weber, B. and Colla, M. (2000a) The role of mineralocorticoid receptors in the circadian activity of the human hypothalamus–adrenal system: effect of age. Neurobiol. Aging, 21: 585–589.

Heuser, I., Deuschle, M., Weber, B., Stalla, G.K. and Holsboer, F. (2000b) Increased activity of the hypothalamus–pituitary–adrenal system after treatment with the mineralocorticoid receptor antagonist spironolactone. Psychoneuroendocrinol., 25: 513–518.

Hollenberg, S.M., Weinberger, C., Ong, E.S., Cerelli, G., Oro, A., Lebo, R., Thompson, E.B., Rosenfeld, M.G. and Evans, R.M. (1985) Primary structure and expression of a functional human glucocorticoid receptor cDNA. Nature, 318: 635–641.

Holsboer, F. and Barden, N. (1996) Antidepressants and hypothalamic–pituitary–adrenocortical regulation. Endocr. Rev., 17: 187–205.

Hsu, S.Y. and Hsueh, A.J. (2001) Human stresscopin and stresscopin-related peptide are selective ligands for the type 2 corticotropin-releasing hormone receptor. Nat. Med., 7: 605–611.

Husum, H., Termeer, E., Mathe, A.A., Bolwig, T.G. and Ellenbroek, B.A. (2002) Early maternal deprivation alters hippocampal levels of neuropeptide Y and calcitonin-gene related peptide in adult rats. Neuropharmacol., 42: 798–806.

Ingle, D.J. (1954) Permissibility of hormone action: a review. Acta Endocrinol., 17: 172–194.

Iwasaki, Y., Oiso, Y., Saito, H. and Majzoub, J.A. (1997) Positive and negative regulation of the rat vasopressin gene promoter. Endocrinology, 138: 5266–5274.

Jasper, M.S. and Engeland, W.C. (1997) Splanchnicotomy increases adrenal sensitivity to ACTH in non-stressed rats. Am. J. Physiol., 273: E363–E368.

Jimenez-Vasquez, P.A., Mathe, A.A., Thomas, J.D., Riley, E.P. and Ehlers, C.L. (2001) Early maternal separation alters neuropeptide Y concentrations in selected brain regions in adult rats. Brain Res. Dev. Brain Res., 131: 149–152.

Joëls, M. and De Kloet, E.R. (1993) Corticosteroid actions on amino acid-mediated on peptide mRNA levels in CRH cells of the paraventricular nucleus: a hybridization histochemical study in the rat. J. Comp. Neurol., 285: 413–435.

Joëls, M. and De Kloet, E.R. (1994) Mineralocorticoid and glucocorticoid receptors in the brain. Implications for ion permeability and transmitter systems. Prog. Neurobiol., 43: 1–36.

Jongen-Relo, A.L., Pothuizen, H.H., Feldon, J. and Pryce, C.R. (2002) Comparison of central corticosteroid receptor expression in male Lewis and Fisher rats. Brain Res., 953: 223–231.

Kalinyak, J.E., Griffin, C.A., Hamilton, R.W., Bradshaw, J.G., Perlman, A.J. and Hoffman, A.R. (1989) Developmental and hormonal regulation of glucocorticoid receptor messenger RNA in the rat. J. Clin. Invest., 84: 1843–1848.

Kalkhoven, E., Valentine, J.E., Heery, D.M. and Parker, M.G. (1998) Isoforms of steroid receptor co-activator 1 differ in their ability to potentiate transcription by the oestrogen receptor. EMBO J., 17: 232–243.

Kalman, B.A. and Spencer, R.L. (2002) Rapid corticosteroid-dependent regulation of mineralocorticoid receptor protein expression in rat brain. Endocrinology, 143: 4184–4195.

Kalsbeek, A., Van Heerikhuize, J.J., Wortel, J. and Buijs, R.M. (1996) A diurnal rhythm of stimulatory input to the hypothalamo–pituitary–adrenal system as revealed by timed intrahypothalamic administration of the vasopressin V1 antagonist. J. Neurosci., 16: 5555–5565.

Kamara, K.S., Kamara, A.K. and Castonguay, T.W. (1992) A reexamination of the effects of intracerebroventricular glucocorticoids in adrenalectomized rats. Brain Res. Bull., 29: 355–358.

Karandrea, D., Kittas, C. and Kitraki, E. (2002) Forced swimming differentially affects male and female brain corticosteroid receptors. Neuroendocrinol., 75: 217–226.

Karin, M., Yang-Yen, H.F., Chambard, J.C., Deng, T. and Saatcioglu, F. (1993) Various modes of gene regulation by nuclear receptors for steroid and thyroid hormones. Eur. J. Clin. Pharmacol., 45(Suppl 1): 9–5; discussion 43–44.

Karssen, A.M., Meijer, O.C., van der Sandt, I.C., Lucassen, P.J., de Lange, E.C. de Boer, A.G. and De Kloet, E.R. (2001) Multidrug resistance P-glycoprotein hampers the access of cortisol but not of corticosterone to mouse and human brain. Endocrinology, 142: 2686–2694.

Karst, H., Karten, Y.J., Reichardt, H.M., De Kloet, E.R., Schütz, G. and Joëls, M. (2000) Corticosteroid actions in hippocampus require DNA binding of glucocorticoid receptor homodimers. Nat. Neurosci., 3: 977–978.

Karten, Y.J., Nair, S.M., van Essen, L., Sibug, R. and Joels, M. (1999) Long-term exposure to high corticosterone levels attenuates serotonin responses in rat hippocampal CA1 neurons. Proc. Natl. Acad Sci. USA, 96: 13456–13461.

Kellner, M., Baker, D.G., Yassouridis, A., Bettinger, S., Otte, C., Naber, D. and Wiedemann, K. (2002) Mineralocorticoid receptor function in patients with posttraumatic stress disorder. Am. J. Psychiat., 159: 1938–1940.

Kovács, K.J. and Makara, G.B. (1988) Corticosterone and dexamethasone act at different brain sites to inhibit adrenalectomy-induced adrenocorticotropin hypersecretion. Brain Res., 474: 205–210.

Kovács, K.J. and Sawchenko, P. (1996) Sequence of stress-induced alterations in indices of synaptic and transcriptional activation in parvocellular neurosecretory neurons. J. Neurosci., 16: 262–273.

Kovács, K.J., Kiss, J.Z. and Makara, G.B. (1986) Glucocorticoid implants around the hypothalamic paraventricular nucleus prevent the increase of corticotropin-releasing factor and arginine vasopressin immunostaining induced by adrenalectomy. Neuroendocrinol., 44: 229–234.

Kovács, K.J., Foldes, A. and Sawchenko, P.E. (2000) Glucocorticoid negative feedback selectively targets vasopressin transcription in parvocellular neurosecretory neurons. J. Neurosci., 20: 3843–3852.

Kretz, O., Reichardt, H.M., Schütz, G. and Bock, R. (1999) Corticotropin-releasing hormone expression is the major target for glucocorticoid feedback-control at the hypothalamic level. Brain Res., 818: 488–491.

Krozowski, Z.S. and Funder, J.W. (1983) Renal mineralocorticoid receptors and hippocampal corticosterone-binding species have identical intrinsic steroid specificity. Proc. Natl. Acad. Sci. USA, 80: 6056–6060.

Ladd, C.O., Owens, M.J. and Nemeroff, C.B. (1996) Persistent changes in corticotropin-releasing factor neuronal systems induced by maternal deprivation. Endocrinology, 137: 1212–1218.

Lamberts, S.W. (2002) Glucocorticoid receptors and Cushing's disease. Mol. Cell. Endocrinol., 29: 69–72.

Lamberts, S.W.J., Koper, J.W., Biemond, P., Den Holder, F.H. and De Jong, F.H. (1992) Cortisol receptor resistance: the variability of its clinical presentation and response to treatment. J. Clin. Endocrinol. Metabol., 74: 313–321.

Laugero, K.D., Bell, M.E., Bhatnagar, S., Soriano, L. and Dallman, M.F. (2001) Sucrose ingestion normalizes central expression of corticotropin-releasing-factor messenger ribonucleic acid and energy balance in adrenalectomized rats: a glucocorticoid-metabolic-brain axis? Endocrinology, 142: 2796–2804.

Laugero, K.D., Gomez, F., Manalo, S. and Dallman, M.F. (2002) Corticosterone infused intracerebroventricularly inhibits energy storage and stimulates the hypothalamo–pituitary axis in adrenalectomized rats drinking sucrose. Endocrinology, 143: 4552–4562.

Lawson, A., Ahima, R., Krozowski, Z. and Harlan, R. (1991) Postnatal development of corticosteroid receptor immunoreactivity in the rat hippocampus. Brain Res. Dev. Brain Res., 62: 69–79.

Levine, S. (1970) The pituitary–adrenal system and the developing brain. Progr. Brain Res., 32: 79–85.

Levine, S. (2001) Primary social relationships influence the development of the hypothalamic–pituitary–adrenal axis in the rat. Physiol. Behav., 73: 255–260.

Levin, E.R. (2002) Cellular functions of plasma membrane estrogen receptors. Steroids, 67: 471–475.

Levine, S., Glick, D. and Nakane, P.K. (1967) Adrenal and plasma corticosterone and vitamin A in rat adrenal glands during postnatal development. Endocrinology, 80: 910–914.

Levine, S., Berkenbosch, F., Suchecki, D. and Tilders, F.J. (1994) Pituitary–adrenal and interleukin-6 responses to recombinant interleukin-1 in neonatal rats. Psychoneuroendocrinol., 19: 143–153.

Levine, S., Dent, G. and De Kloet, E.R. (2000) Stress-hyporesponsive period. In: Fink, G. (Ed.), Encyclopaedia of Stress, Vol. 3. Academic Press, pp. 518–526.

Lightman, S.L., Windle, R.J., Julian, M.D., Harbuz, M.S., Shanks, N., Wood, S.A., Kershaw, Y.M. and Ingram, C.D. (2000) Significance of pulsatility in the HPA axis. Novartis Foundation Symposium, 277: 244–260.

Lim, M.C., Shipston, M.J. and Antoni, F.A. (2002) Posttranslational modulation of glucocorticoid feedback inhibition at the pituitary level. Endocrinology, 143: 3769–3801.

Liu, D., Caldji, C., Sharma, S., Plotsky, P.M. and Meaney, M.J. (2000) Influence of neonatal rearing conditions on stress-induced adrenocorticotropin responses and norepinepherine release in the hypothalamic paraventricular nucleus. J. Neuroendocrinol., 12: 5–12.

Makino, S., Smith, M.A. and Gold, P.W. (1995) Increased expression of corticotropin-releasing hormone and vasopressin messenger ribonucleic acid (mRNA) in the hypothalamic paraventricular nucleus during repeated stress: association with reduction in glucocorticoid receptor mRNA levels. Endocrinology, 136: 3299–3309.

Malkoski, S.P. and Dorin, R.I. (1999) Composite glucocorticoid regulation at a functionally defined negative glucocorticoid response element of the human corticotropin-releasing hormone gene. Mol. Endocrinol., 130: 1629–1644.

Marinelli, M. and Piazza, P.V. (2002) Interaction between glucocorticoid hormones, stress and psychostimulant drugs. Eur. J. Neurosci., 16: 387–394.

McEwen, B.S., Weiss, J.M. and Schwartz, L.S. (1968) Selective retention of corticosterone by limbic structures in rat brain. Nature, 220: 911–912.

McKenna, N.J., Lanz, R.B. and O'Malley, B.W. (1999) Nuclear receptor coregulators: cellular and molecular biology. Endocr. Rev., 20: 321–344.

Meaney, M.J. (2001) Maternal care, gene expression, and the transmission of individual differences in stress reactivity across generations. Ann. Rev. Neurosci., 24: 1161–1192.

Meaney, M.J., Aitken, D.H., Viau, V., Sharma, S. and Sarrieau, A. (1989) Neonatal handling alters adrenocortical negative feedback sensitivity and hippocampal type II glucocorticoid receptor binding in the rat. Neuroendocrinol., 50: 597–604.

Meaney, M.J., Diorio, J., Francis, D., Widdowson, J., LaPlante, P., Caldji, C., Sharma, S., Seckl, J.R. and Plotsky, P.M. (1996) Early environmental regulation of forebrain glucocorticoid receptor gene expression: implications for adrenocortical responses to stress. Dev. Neurosci., 18: 49–72.

Meijer, O.C. (2002) Coregulator proteins and corticosteroid action in the brain. J. Neuroendocrinol., 14: 499–505.

Meijer, O.C. and De Kloet, E.R. (1995) A role for the mineralocorticoid receptor in a rapid and transient suppression of 5-HT1A receptor mRNA by corticosterone. J. Neuroendocrinol., 7: 653–657.

Meijer, O.C., De Lange, E.C.M., De Boer, A.G., Breimer, D.D., Workel, J.O. and De Kloet, E.R. (1998) Penetration of dexamethasone in brain glucocorticoid targets is enhanced in mdr1a P glycoprotein knockout mice. Endocrinology, 139: 1789–1793.

Meijer, O.C., Steenbergen, P.J. and De Kloet, E.R. (2000) Differential expression and regional distribution of steroid receptor coactivators SRC-1 and SRC-2 in brain and pituitary. Endocrinology, 141: 2192–2199.

Meijer, O.C., Williamson, A., Dallman, M.F. and Pearce, D. (2000) Transcriptional repression of the 5-HT1A receptor promoter by corticosterone via mineralocorticoid receptors depends on the cellular context. J. Neuroendocrinol., 12: 245–254.

Meijer, O.C., Williamson, A., Dallman, M.F. and Pearce, D. (2000) Transcriptional rrpression of the 5-HT1A receptor promoter by corticosterone via mineralocorticoid receptors depends on the cellula context. J. Neuroendocrinol., 12: 245–254.

Mershon, J.L., Sehlhorst, E.S. and Rebar, R.W., Liew, J.H. (1992) Evidence of a corticotrophin-releasing hormone pulse generator in the macaque hypothalamus. Endocrinology, 130: 2991–2996.

Miller, A.H., Spencer, R.L., Pulera, M., Kang, S., McEwen, B.S. and Stein, M. (1992) Adrenal steroid receptor activation in rat brain and pituitary following dexamethasone: implications for the dexamethasone suppression test. Biol. Psychiatry, 32: 850–869.

Misiti, S., Schomburg, L., Yen, P.M. and Chin, W.W. (1998) Expression and hormonal regulation of coactivator and corepressor genes. Endocrinology, 139: 2493–2500.

Mittelstadt, P.R. and Ashwell, J.D. (2001) Inhibition of AP-1 by the glucocorticoid-inducible protein GILZ. J. Biol. Chem., 276: 29603–29610.

Moguilewsky, M. and Philibert, D. (1984) RU 38486: potent antiglucocorticoid activity correlated with strong binding to the cytosolic glucocorticoid receptor followed by an impaired activation. J. Steroid Biochem., 20: 271–276.

Morano, M.I., Vázquez, D.M. and Akil, H. (1994) The role of the hippocampal mineralocorticoid and glucocorticoid receptors in the hypothalamo–pituitary–adrenal axis of the aged Fisher rat. Mol. Cell. Neurosci., 5: 400–412.

Munck, A., Guyre, P.M. and Holbrook, N.J. (1984) Physiological functions of glucocorticoids in stress and their relation to pharmacological actions. Endocr. Rev., 5: 25–44.

Murphy, E.P. and Conneely, O.M. (1997) Neuroendocrine regulation of the hypothalamic pituitary adrenal axis by the nurr1/nur77 subfamily of nuclear receptors. Mol. Endocrinol., 11: 39–47.

Nishihara, E., Yoshida-Komiya, H., Chan, C.S., Liao, L., Davis, R.L., O'Malley, B.W. and Xu, J. (2003) SRC-1 null mice exhibit moderate motor dysfunction and delayed development of cerebellar purkinje cells. J. Neurosci., 23: 213–222.

Oitzl, M. and De Kloet, E.R. (1992) Selective corticosteroid-antagonists modulate specific aspects of spatial orientation learning. Behav. Neurosci., 106: 62–71.

Oitzl, M.S., Fluttert, M. and De Kloet, E.R. (1994) The effect of corticosterone on reactivity to spatial novelty is mediated by central mineralocorticosteroid receptors. Eur. J. Neuroscience, 6: 1072–1079.

Oitzl, M.S., Van Haarst, A.D., Sutanto, W. and De Kloet, E.R. (1995) Corticosterone, brain mineralocorticoid receptors (MR) and the activity of the hypothalamic-pituitary-adrenal (HPA) axis: the Lewis rat as an example of increased central MR capacity and a hyporesponsive HPA axis. Psychoneuroendocrinol., 20: 655–675.

Oitzl, M.S., Workel, J.O., Fluttert, M., Frosch, F. and De Kloet, E.R. (2000) Maternal deprivation affects behaviour from youth to senescence: amplification of individual differences in spatial learning and memory in senescent Brown Norway rats. Eur. J. Neurosci., 12: 3771–3780.

Oitzl, M.S., Reichardt, H.M., Joëls, M. and De Kloet, E.R. (2001) Point mutation in the mouse glucocorticoid receptor preventing DNA binding impairs spatial memory. Proc. Natl. Acad. Sci. USA, 98: 12790–12795.

Orchinik, M. (1998) Glucocorticoids, stress, and behavior: shifting the timeframe. Horm. Behav., 34: 320–327.

Palkovits, M. (2000) Stress-induced expression of co-localized neuropeptides in hypothalamic and amygdaloid neurons. Eur. J. Pharmacol., 405: 161–166.

Patel, P.D., Lopez, J.F., Lyons, D.M., Burke, S., Wallace, M. and Schatzberg, A.F. (2000) Glucocorticoid and mineralocorticoid receptor mRNA expression in squirrel monkey brain. J. Psychiatr. Res., 34: 383–392.

Pearce, D. and Yamamoto, K.R. (1993) Mineralocorticoid and glucocorticoid receptor activities distinguished by nonreceptor factors at a composite response element. Science, 259: 1161–1165.

Penke, Z., Felszeghy, K., Fernette, B., Sage, D., Nyakas, C. and Burlet, A. (2001) Postnatal maternal deprivation produces long-lasting modifications of the stress response, feeding and stress-related behaviour in the rat. Eur. J. Neurosci., 14: 747–755.

Philips, A., Maira, M., Mullick, A., Chamberland, M., Lesage, S., Hugo, P. and Drouin, J. (1997) Antagonism between Nur77 and glucocorticoid receptor for control of transcription. Mol. Cell. Biol., 17: 5952–5959.

Pinnock, S.B. and Herbert, J. (2001) Corticosterone differentially modulates expression of corticotropin releasing factor and arginine vasopressin mRNA in the hypothalamic paraventricular nucleus following either acute or repeated restraint stress. Eur. J. Neuroscience, 13: 576–584.

Plotsky, P.M. and Meaney, M.J. (1993) Early, postnatal experience alters hypothalamic corticotropin-releasing factor (CRF) mRNA, median eminence CRF content and stress-induced release in adult rats. Brain Res. Mol. Brain Res., 18: 195–200.

Plotsky, P.M., Thrivikraman, K.V. and Meaney, M.J. (1993) Central and feedback regulation of hypothalamic corticotropin-releasing factor secretion. Ciba Found. Symp., 172: 59–75; discussion 75–84: 59–75.

Pryce, C.R. and Feldon, J. (2003) Long-term neurobehavioural impact of the postnatal environment in rats: manipulations, effects and mediating mechanisms. Neurosci. Biobehav. Rev., 27: 57–71.

Ratka, A., Sutanto, W., Bloemers, M. and De Kloet, E.R. (1989) On the role of brain mineralocorticoid (type I) and glucocorticoid (type II) receptors in neuroendocrine regulation. Neuroendocrinol., 50: 117–123.

Reichardt, H.M., Kaestner, K.H., Tuckermann, J., Kretz, O., Wessely, O., Bock, R., Gass, P., Schmid, W., Herrlich, P., Angel, P. and Schütz, G. (1998) DNA binding of the glucocorticoid receptor is not essential for survival [see comments]. Cell, 93: 531–541.

Reul, J.M.H.M. and De Kloet, E.R. (1985) Two receptor systems for corticosterone in rat brain: microdistribution and differential occupation. Endocrinology, 117: 2505–2511.

Reul, J.M.H.M. and De Kloet, E.R. (1986) Anatomical resolution of two types of corticosterone receptors in rat brain with in vitro autoradiography and computerized image analysis. J. Ster. Biochem., 24: 269–272.

Reul, J.M.H.M., Stec, I., Soder, M. and Holsboer, F. (1993) Chronic treatment of rats with the antidepressant amitriptyline attenuates the activity of the hypothalamic–pituitary–adrenocortical system. Endocrinology, 133: 312–320.

Reul, J.M.H.M., Probst, J.C., Skutella, T., Hirschmann, M., Stec, I.S., Montkowski, A., Landgraf, R. and Holsboer, F. (1997) Increased stress-induced adrenocorticotropin response after long-term intracerebroventricular treatment of rats with antisense mineralocorticoid receptor oligodeoxynucleotides. Neuroendocrinol., 65: 189–199.

Robert-Nicoud, M., Flahaut, M., Elalouf, J.M., Nicod, M., Salinas, M., Bens, M., Doucet, A., Wincker, P., Artiguenave, F., Horisberger, J.D., Vandewalle, A., Rossier, B.C. and Firsov, D. (2001) Transcriptome of a mouse kidney cortical collecting duct cell line: effects of aldosterone and vasopressin. Proc. Natl. Acad. Sci. USA, 98: 2712–2716.

Roozendaal, B., de Quervain, D.J., Ferry, B., Setlow, B. and McGaugh, J.L. (2001) Basolateral amygdala-nucleus accumbens interactions in mediating glucocorticoid enhancement of memory consolidation. J. Neurosci., 21: 2518–2525.

Rosenfeld, M.G. and Glass, C.K. (2001) Coregulator codes of transcriptional regulation by nuclear receptors. J. Biol. Chem., 276: 36865–36868.

Rosenfeld, P., Sutanto, W., Levine, S. and De Kloet, E.R. (1990) Ontogeny of mineralocorticoid (type 1) receptors in brain and pituitary: an in vivo autoradiographical study. Brain Res. Dev. Brain Res., 52: 57–62.

Rosenfeld, P., Suchecki, D. and Levine, S. (1992) Multifactorial regulation of the hypothalamic–pituitary–adrenal axis during development. Neurosci. Biobehav. Rev., 16, 553–568.

Rosenfeld, P., Ekstrand, J., Olson, E., Suchecki, D. and Levine, S. (1993a) Maternal regulation of adrenocortical activity in the infant rat: effects of feeding. Dev. Psychobiol., 26: 261–277.

Rosenfeld, P., Van Eekelen, J.A.M., Levine, S. and De Kloet, E.R. (1993b) Ontogeny of corticosteroid receptors in the brain. Cell. Mol. Neurobiol., 13: 295–319.

Rothuizen, J., Reul, J.M., Van Sluijs, F.J., Mol, J.A., Rijnberk, A. and De Kloet, E.R. (1993) Increased neuroendocrine reactivity and decreased brain mineralocorticoid receptor-binding capacity in aged dogs. Endocrinology, 132: 161–168.

Rots, N.Y., de Jong, J., Workel, J.O., Levine, S., Cools, A.R. and De Kloet, E.R. (1996) Neonatal maternally deprived rats have as adults elevated basal pituitary–adrenal activity and enhanced susceptibility to apomorphine. J. Neuroendocrinol., 8: 501–506.

Rupprecht, R., Di Michele, F., Hermann, B., Strohle, A., Lancel, M., Romeo, E. and Holsboer, F. (2001) Neuroactive steroids: molecular mechanisms of action and implications for neuropsychopharmacology. Brain Res. Brain Res. Rev., 37: 59–67.

Sabban, E.L. and Kvetnánsky, R. (2001) Stress-triggered activation of gene expression in catecholaminergic systems: dynamics of transcriptional events. Trends Neurosci., 24: 91–98.

Sakly, M. and Koch, B. (1981) Ontogenesis of glucocorticoid receptors in anterior pituitary gland: transient dissociation among cytroplasmic receptor density, nuclear uptake, and regulation of corticotropic activity. Endocrinology, 108: 591–596.

Sakly, M. and Koch, B. (1983) Ontogenetical variations of transcortin modulate glucocorticoid receptor function and corticotropic activity in the pituitary gland. Horm. Metab. Res., 15: 92–96.

Sanchez, M.M., Young, L.J., Plotsky, P.M. and Insel, T.R. (2000) Distribution of corticosteroid receptors in the rhesus brain: relative absence of glucocorticoid receptors in the hippocampal formation. J. Neurosci., 20: 4657–4668.

Sapolsky, R.M. and Meaney, M.J. (1986) Maturation of the adrenocortical stress response: neuroendocrine control mechanisms and the stress hyporesponsive period. Brain Res., 396: 64–76.

Sapolsky, R.M., Meaney, M.J. and McEwen, B.S. (1985) The development of the glucocorticoid receptor system in the rat limbic brain. III. Negative-feedback regulation. Brain Res., 350: 169–173.

Schapiro, S., Geller, E. and Eiduson, S. (1962) Neonatal adrenal cortical response to stress and vasopressin. Proc. Soc. Exp. Biol. Med., 109: 937–941.

Schapiro, S. (1965) Androgen treatment in early infancy: effect upon adult adrenal cortical response to stress and adrenal and ovarian compensatory hypertrophy. Endocrinology, 77: 585–587.

Schmidt, M. (2004) Stress system development: molecular mechanism of inhibition and activation. PhD Thesis, Leiden University, The Netherlands.

Schmidt, M., Oitzl, M.S., Levine, S. and De Kloet, E.R. (2002a) The HPA system during the postnatal development of CD1 mice and the effects of maternal deprivation. Brain Res. Dev. Brain Res., 139: 39–49.

Schmidt, M., Okimoto, D.K., Dent, G.W., Gordon, M.K. and Levine, S. (2002b) Maternal regulation of the hypothalamic–pituitary–adrenal axis in the 20-day-old rat: consequences of laboratory weaning. J. Neuroendocrinol., 14: 450–457.

Schöbitz, B., Sutanto, W., Carey, M.P., Holsboer, F. and De Kloet, E.R. (1994) Endotoxin and interleukin 1 decrease the affinity of hippocampal mineralocorticoid (type 1) receptor in parallel to activation of the hypothalamic–pituitary–adrenal axis. Neuroendocrinol., 60: 124–133.

Schoenfeld, N.M., Leathem, J.H. and Rabii, J. (1980) Maturation of adrenal stress responsiveness in the rat. Neuroendocrinol., 31: 101–105.

Seckl, J.R. (1997) 11β-Hydroxysteroid dehydrogenase in the brain: a neural regulator of glucocorticoid action? Front. Neuroendocrinol., 18: 49–99.

Seckl, J.R. and Fink, G. (1992) Antidepressants increase glucocorticoid and mineralocorticoid receptor mRNA expression in rat hippocampus in vivo. Neuroendocrinol., 55: 621–626.

Selye, H. (1952) The Story of the Adaptation Syndrome. Acta, Montreal.

Shanks, N. and Meaney, M.J. (1994) Hypothalamic–pituitary–adrenal activation following endotoxin administration in the developing rat: a CRH-mediated effect. J. Neuroendocrinol., 6: 375–383.

Smith, M.A., Kim, S.Y., Van Oers, H.J. and Levine, S. (1997) Maternal deprivation and stress induce immediate early genes in the infant rat brain. Endocrinology, 138: 4622–4628.

Spencer, R.L., Miller, A.H., Moday, H., Stein, M. and McEwen, B.S. (1993) Diurnal differences in basal and acute stress levels of type I and type II adrenal steroid receptor activition in neural and immune tissues. Endocrinology, 133: 1941–1950.

Spencer, R.L., Kim, P.J., Kalman, B.A. and Cole, M.A. (1998) Evidence for mineralocorticoid receptor facilitation of glucocorticoid receptor-dependent regulation of hypothalamic–pituitary–adrenal axis activity. Endocrinology, 139: 2718–2726.

Stoecklin, E., Wissler, M., Schaetzle, D., Pfitzner, E. and Groner, B. (1999) Interactions in the transcriptional regulation exerted by Stat5 and by members of the steroid hormone receptor family. J. Steroid Biochem. Mol. Biol., 69: 195–204.

Suchecki, D. and Tufik, S. (1997) Long-term effects of maternal deprivation on the corticosterone response to stress in rats. Am. J. Physiol., 273: R1332–R1338.

Suchecki, D., Nelson, D.Y., Van Oers, H. and Levine, S. (1995) Activation and inhibition of the hypothalamic–pituitary–adrenal axis of the neonatal rat: effects of maternal deprivation [published erratum appears in Psychoneuroendocrinol., 20(6): 677]; Psychoneuroendocrinol., 20: 169–182.

Suchecki, D., Duarte, P.B. and Tufik, S. (2000) Pituitary–adrenal axis and behavioural responses of maternally deprived juvenile rats to the open field. Behav. Brain Res., 111: 99–106.

Sutanto, W., Rosenfeld, P., De Kloet, E.R. and Levine, S. (1996) Long-term effects of neonatal maternal deprivation and ACTH on hippocampal mineralocorticoid and glucocorticoid receptors. Brain Res. Dev. Brain Res., 92: 156–163.

Swanson, L.W. (1991) Biochemical switching in hypothalamic circuits mediating responses to stress. Progr. Brain Res., 87: 181–200.

Swanson, L.W. and Simmons, D.M. (1989) Differential steroid hormone and neural influences on peptide mRNA levels in CRH cells of the paraventricular nuclues: a hybridization histochemical study in the rat. J. Comp. Neurol., 285: 413–435.

Takeyama, K., Masuhiro, Y., Fuse, H., Endoh, H., Murayama, A., Kitanaka, S., Suzawa, M., Yanagisawa, J. and Kato, S. (1999) Selective interaction of vitamin D receptor with transcriptional coactivators by a vitamin D analog. Mol. Cell. Biol., 19: 1049–1055.

Tausk, M. (1951) Das Hormon: Hat die Nebenniere tatsächlich eine Verteidigungsfunktion? Vol. 3. Organon International BV, The Netherlands.

Tian, L., Hammond, M.S., Florance, H., Antoni, F.A. and Shipston, M.J. (2001) Alternative splicing determines sensitivity of murine calcium-activated potassium channels to glucocorticoids. J. Physiol., 537: 57–68.

Van Der Lely, A.J., Foeken, K., Van der Mast, R.C. and Lamberts, S.W. (1991) Rapid reversal of acute psychosis in the Cushing syndrome with the cortisol-receptor antagonist mifepristone (RU 486). Ann. Intern. Med., 114: 143–144.

Van Eekelen, J.A.M., Jiang, W., De Kloet, E.R. and Bohn, M.C. (1988) Distribution of the mineralocorticoid and the glucocorticoid receptor mRNAs in the rat hippocampus. J. Neurosci. Res., 21: 88–94.

Van Eekelen, J.A.M., Kiss, J.Z., Westphal, H.M. and De Kloet, E.R. (1987b) Immunocytochemical study on the intracellular localization of the type 2 glucocorticoid receptor in the rat brain. Brain Res., 436: 120–128.

Van Eekelen, J.A.M., Rosenfeld, P., Levine, S., Westphal, H.M. and De Kloet, E.R. (1987b) Post-natal disappearance of glucocorticoid receptor immunoreactivity. Neurosci. Res. Comm., 1: 129–133.

Van Eekelen, J.A.M., Bohn, M.C. and De Kloet, E.R. (1991a) Postnatal ontogeny of mineralocorticoid and glucocorticoid receptor gene expression in regions of the rat tel- and diencephalon. Brain Res. Dev. Brain Res., 61: 33–43.

Van Eekelen, J.A.M., Rots, N.Y., Sutanto, W., Oitzl, M.S. and De Kloet, E.R. (1991b) Brain corticosteroid receptor gene expression and neuroendocrine dynamics during aging. J. Steroid. Biochem. Mol. Biol., 40: 679–683.

Van Eekelen, J.A.M., Oitzl, M.S. and De Kloet, E.R. (1995) Adrenocortical hyporesponsiveness and glucocorticoid feedback resistance in old male Brown Norway rats. J. Gerontol. A Biol. Sci. Med. Sci., 50: B83–B89.

Van Haarst, A.D., Oitzl, M.S., Workel, J.O. and De Kloet, E.R. (1996) Chronic brain glucocorticoid receptor blockade enhances the rise in circadian and stress-induced pituitary-adrenal activity. Endocrinology, 137: 4935–4943.

Van Haarst, A.D., Oitzl, M.S., Workel, J.O. and De Kloet, E.R. (1996a) Chronic brain glucocorticoid receptor blockade enhances the rise in circadian and stress-induced pituitary–adrenal activity. Endocrinology, 137: 4935–4943.

Van Haarst, A.D., Welberg, L.A.M., Sutanto, W., Oitzl, M.S. and De Kloet, E.R. (1996a) 11β-hydroxysteroid dehydroxygenase activity in the hippocampus: implications for in vivo corticosterone receptor binding and cell nuclear retention. J. Neuroendocrinol., 8: 595–600.

Van Haarst, A.D., Oitzl, M.S. and De Kloet, E.R. (1997) Facilitation of feedback inhibition through blockade of glucocorticoid receptors in the hippocampus. Neurochem. Res., 22: 1323–1328.

Van Oers, H.J., De Kloet, E.R., Li, C. and Levine, S. (1998a) The ontogeny of glucocorticoid negative feedback: influence of maternal deprivation. Endocrinology, 139: 2838–2846.

Van Oers, H.J., De Kloet, E.R., Whelan, T. and Levine, S. (1998b) Maternal deprivation effect on the infant's neural stress markers is reversed by tactile stimulation and feeding but not by suppressing corticosterone. J. Neurosci., 18: 10171–10179.

Van Rossum, E.F., Koper, J.W., Huizenga, N.A., Uitterlinden, A.G., Janssen, J.A., Brinkmann, A.O., Grobbee, D.E., de Jong, F.H., van Duyn, C.M., Pols, H.A. and Lamberts, S.W. (2002) A polymorphism in the glucocorticoid receptor gene, which decreases sensitivity to glucocorticoids in vivo, is associated with low insulin and cholesterol levels. Diabetes, 51: 3128–3134.

Van Rossum, E.F., Koper, J.W., Van Den Beld, A.W., Uitterlinden, A.G., Arp, P., Ester, W., Janssen, J.A., Brinkmann, A.O., De Jong, F.H., Grobbee, D.E., Pols, H.A.J. and Lamberts, S.W. (2003) Identification of the BclI polymorphism in the glucocorticoid receptor gene: association with sensitivity to glucocorticoids in vivo and body mass index. Clin. Endocrinol. (Oxf)., 59: 585–592.

Van Steensel, B., Van Binnendijk, E., Hornsby, C.D., Van der Voort, H.T.M., De Kloet, E.R. and Van Driel, R. (1996) Partial colocalization of glucocorticoid and mineralocorticoid receptors in discrete compartments in nuclei of rat hippocampus neurons. J. Cell. Sci., 109: 787–792.

Vázquez, D.M. and Akil, H. (1992) Development of pituitary pro-opiomelanocortin gene and peptide expression: characterization and effect of repeated intermittent maternal isolation. Neuroendocrinol., 56: 320–330.

Vázquez, D.M. and Akil, H. (1993) Pituitary–adrenal response to ether vapor in the weanling animal: characterization of the inhibitory effect of glucocorticoids on adrenocorticotropin secretion. Pediatr. Res., 34: 646–653.

Veenema, A.H., Meijer, O.C., De Kloet, E.R. and Koolhaas, J.M. (2003) Genetic selection for coping style predicts stressor susceptibility. J. Neuroendocrinol., 15: 256–267.

Veldhuis, H.D., Van Koppen, C., Van Ittersum, M. and De Kloet, E.R. (1982) Specificity of the adrenal steroid receptor system in the rat hippocampus. Endocrinology, 110: 2044–2051.

Viau, V., Sharma, S., Plotsky, P.M. and Meaney, M.J. (1993) Increased plasma ACTH responses to stress in nonhandled compared with handled rats require basal levels of corticosterone and are associated with increased levels of ACTH secretagogues in the median eminence. J. Neurosci., 13: 1097–1105.

Walker, C.D. and Dallman, M.F. (1993) Neonatal facilitation of stress-induced adrenocorticotropin secretion by prior stress: evidence for increased central drive to the pituitary. Endocrinology, 132: 1101–1107.

Walker, C.D., Akana, S.F., Cascio, C.S. and Dallman, M.F. (1990) Adrenalectomy in the neonate: adult-like adrenocortical system responses to both removal and replacement of corticosterone. Endocrinology, 127: 832–842.

Walker, C.D., Scribner, K.A., Cascio, C.S. and Dallman, M.F. (1991) The pituitary–adrenocortical system of neonatal rats is responsive to stress throughout development in a time-dependent and stressor-specific fashion. Endocrinology, 128: 1385–1395.

Windle, R.J., Wood, S.A., Lightman, S.L. and Ingram, C.D. (1998) The pulsatile characteristics of hypothalamo–pituitary–adrenal activity in female Lewis and Fischer rats and its relationship to differential stress responses. Endocrinology, 139: 4044–4052.

Windle, R.J., Wood, S.A., Kershaw, Y.M., Lightman, S.L., Ingram, C.D. and Harbuz, M.S. (2001) Increased corticosterone pulse frequency during adjuvant-induced arthritis and its relationship to alterations in stress responsiveness. J. Neuroendocrinol., 13: 905–911.

Witek-Janusek, L. (1988) Pituitary–adrenal response to bacterial endotoxin in developing rats. Am. J. Physiol., 255: E525–E530.

Wolkowitz, O.M., Epel, E.S. and Reus, V.I. (2001) Stress hormone-related psychopathology: pathophysiological and treatment implications. World J. Biol. Psychiatry, 2: 115–143.

Workel, J.O., Oitzl, M.S., Ledeboer, A. and De Kloet, E.R. (1997) The Brown Norway rat displays enhanced stress-induced ACTH reactivity at day 18 after 24-h maternal deprivation at day 3. Brain Res. Dev. Brain Res., 103: 199–203.

Workel, J.O., Oitzl, M.S., Fluttert, M., Lesscher, H., Karssen, A. and De Kloet, E.R. (2001) Differential and age-dependent effects of maternal deprivation on the hypothalamic–pituitary–adrenal axis of Brown Norway rats from youth to senescence. J. Neuroendocrinol., 13: 569–580.

Xu, J. and O'Malley, B.W. (2002) Molecular mechanisms and cellular biology of the steroid receptor coactivator (SRC) family in steroid receptor function. Rev. Endocr. Metab. Disord., 3: 185–192.

Yi, S.J., Masters, J.N., and Baram, T.Z. (1993) Effects of a specific glucocorticoid receptor antagonist on corticotropin releasing hormone gene expression in the paraventricular nucleus of the neonatal rat. Brain Res. Dev. Brain Res. 73: 253–259.

Young, E.A., Lopez, J.F., Murphy-Weinberg, V., Watson, S.J. and Akil, H. (1998) The role of mineralocorticoid receptors in hypothalamic–pituitary–adrenal axis regulation in humans. J. Clin. Endrinol. Metab., 83: 3339–3345.

Young, E.A., Lopez, J.F., Murphy-Weinberg, V., Watson, S.J. and Akil, H. (2003) Mineralocorticoid receptor function in major depression. Arch. Gen. Psychiatry, 60: 24–28.

Zarrow, M.X., Campbell, P.S. and Denenberg, V.H. (1972) Handling in infancy: increased levels of the hypothalamic corticotropin releasing factor (CRF) following exposure to a novel situation. Proc. Soc. Exp. Biol. Med., 141: 356–358.

T. Steckler, N.H. Kalin and J.M.H.M. Reul (Eds.)
Handbook of Stress and the Brain, Vol. 15
ISBN 0-444-51173-3
Published by Elsevier B.V.

CHAPTER 3.2

Glucocorticoid effects on gene expression

Tomoshige Kino* and George P. Chrousos

Pediatric and Reproductive Endocrinology Branch, National Institute of Child Health and Human Development, National Institutes of Health, Bethesda, MD 20892, USA

Abstract: Glucocorticoids secreted from the adrenal glands have extremely diverse effects on many physiological functions. They exert most of their actions by changing the transcription rates of responsive genes through intracellular glucocorticoid receptor (GR). Glucocorticoid-bound GR interacts with specific DNA sequences called glucocorticoid response elements located in the promoter regions of many responsive genes, directly stimulating or suppressing their transcriptional activities. Glucocorticoid-bound GR also indirectly modulates the transcriptional activity of numerous transcription factors by associating with them via mutual protein–protein interactions. Though these mechanisms of transcriptional regulation, glucocorticoids exert their effects in different, distinct cells, tissues and organs, including the central nervous, metabolic, musculo-skeletal, connective tissue and cardiovascular systems.

Introduction

Glucocorticoids are steroid hormones secreted by the adrenal glands, important for maintenance of basal and stress-related homeostasis (Bamberger et al., 1996). These hormones regulate a variety of biological processes and exert profound influences on many physiological functions. In pharmacological amounts, glucocorticoids are used as potent immunosuppressive agents in the management of many inflammatory, autoimmune, and lymphoproliferative diseases. At the cellular level, the actions of glucocorticoids are mediated by an intracellular receptor protein, the glucocorticoid receptor (GR), which functions as a hormone-activated transcription factor that regulates the expression of glucocorticoid-responsive genes (Kino et al., 2003). These genes probably represent up to 20% of the human genome and are influenced by the ligand-activated GR directly or indirectly (Galon et al., 2002).

The GR is ubiquitously expressed in almost all human tissues and organs. The human (h) GR, a single polypeptide chain of 777 amino acid residues, belongs to the steroid/sterol/thyroid/retinoid/orphan receptor superfamily of nuclear transcription factors, with over 150 members currently cloned and characterized across species (Mangelsdorf et al., 1995). Through this receptor, glucocorticoids mainly change the transcription rates of glucocorticoid-responsive genes, modulating thus the expression of downstream protein molecules. Glucocorticoids exert their biological activities in almost all organs/tissues of the organism. In the following section we review the recent progress in the understanding of glucocorticoid actions, primarily focusing on how they regulate gene expression of the organism.

Glucocorticoid receptor

Structure and actions of the glucocorticoid receptor

The GR, together with the mineralocorticoid, progesterone, estrogen, and androgen receptors, form the steroid hormone receptor subfamily. Steroid receptors

*Corresponding author. Pediatric and Reproductive Endocrinology Branch, National Institute of Child Health and Human Development, National Institutes of Health, Bldg. 10, Rm. 9D42, 10 Center Drive MSC 1583, Bethesda, MD 20892-1583, USA. Tel.: +1301-496-6417;
Fax: +1301-480-2024; E-mail: kinot@mail.nih.gov

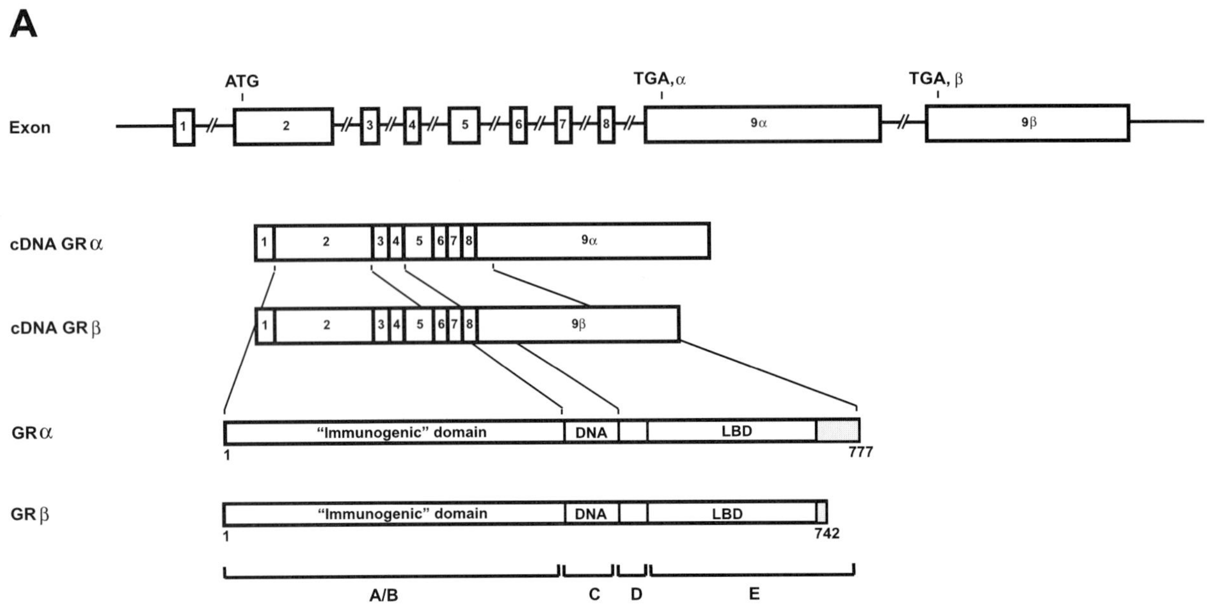

Fig. 1. Genomic and complementary DNA, and protein structures of the human (h) glucocorticoid receptor (GR). (A): The hGR gene consists of 10 exons. Exon 1 is untranslated region, exon 2 codes for the immunogenic domain (A/B), exons 3 and 4 for the DNA-binding domain (C), and exons 5–9 for the hinge region (D) and the ligand-binding domain (E). The GR gene contains two terminal exons 9 (exon 9α and 9β) alternatively spliced to produce the classic GRα and the nonligand-binding GRβ. C-terminal gray-colored domains in GRα and GRβ show their specific portions (from [149]).

display a modular structure comprised of five to six regions (A–F), with the N-terminal A/B region harboring an autonomous activation function (activation function 1, AF1), and the C and E regions corresponding to the DNA- and ligand-binding domains (Kino et al., 2001). The GR cDNA was isolated by the expression cloning in 1985 (Hollenberg et al., 1985).

The GR gene consists of nine exons and is located on chromosome 5. Alternative splicing of the GR gene in exon 9 generates two highly homologous receptor isoforms, termed α and β. These are identical through amino acid 727, but then diverge, with GRα having an additional 50 amino acids and GRβ having an additional, nonhomologous 15 amino acids (Fig. 1). GRα resides primarily in the cytoplasm, and represents the classic glucocorticoid receptor, which functions as a ligand-dependent transcription factor. In contradistinction to GRα, GRβ resides primarily in the nucleus of cells independently of the presence of ligand, does not bind glucocorticoids or antiglucocorticoids, does not activate glucocorticoid-responsive genes, and is transcriptionally inactive (Bamberger et al., 1996).

The GR in its unliganded state is located primarily in the cytoplasm, as part of heterooligomeric complexes containing heat-shock proteins 90, 70, and 50, and, possibly, other proteins (Bamberger et al., 1996). After binding to its ligand, the GR undergoes conformational changes, dissociates from the heat-shock proteins, homodimerizes, and translocates into the nucleus, where it interacts with hormone-responsive elements and/or other transcription factors in the promoter regions of target genes.

Mechanisms of direct transcriptional activation by GRα

GRα exerts its classic transcriptional activity on its responsive genes after binding to glucocorticoid response elements (GREs), small sequences of DNA, in the promoters of these genes (Bamberger et al., 1996). These include genes encoding methallothionein, the growth hormone receptor, the interleukin 6 receptor, microsomal cytochrome P450 monooxygenase (CYP3A), rat tyrosine aminotransferase, and

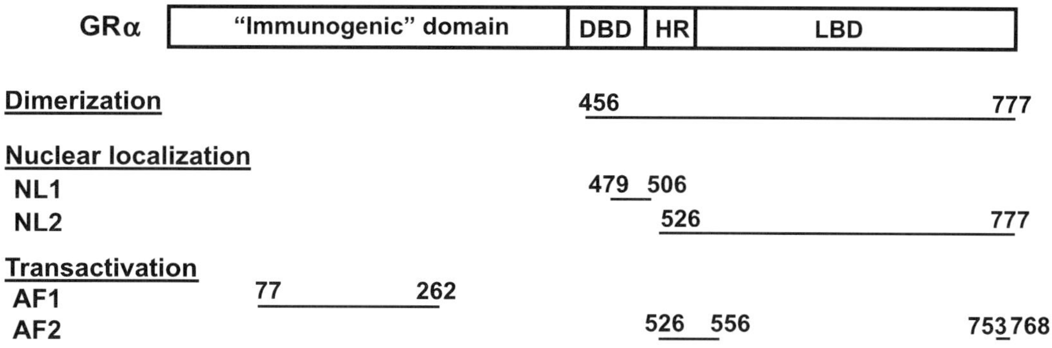

Fig. 2. Functional domains of hGRα. Functional domains and subdomains are indicated (from (Kino and Chrousos, 2002)). DBD, DNA-binding domain; LBD, ligand-binding domain; NL1 and 2, nuclear translocation signal 1 and 2; AF-1 and 2, activation function 1 and 2.

the mitochondrial sterol 27-hydroxylase (CYP27A1) (Karin et al., 1984). In addition, the long-terminal repeat promoter of the murine mammary tumor virus contains four classic GREs, which allow this promoter to fully respond to glucocorticoids (Bresnick et al., 1990). The optimal recognition site of GREs is an inverted hexameric palindrome separated by three basepairs, PuGNACANNNTGTNCPy, with each GR molecule binding to one of the palindromes (Lieberman et al., 1993). The GRE-bound GRα stimulates the transcription rate of responsive genes by facilitating the formation of the transcription initiation complex, including RNA polymerase II and its ancillary components via its AF-1 and AF-2 domains, which are located in the immunogenic and the ligand-binding domains, respectively (Beato and Sanchez-Pacheco, 1996) (Fig. 2). The former is ligand independent while the latter is ligand dependent (Giguere et al., 1986).

Research studies aimed to identify molecules that interact with the AF-2 of the GR, have led to identification of several protein and protein complexes, called coactivators, that form a bridge between the DNA-bound GRα and the transcription initiation complex and assist enzymatically with the transmission of the glucocorticoid signal to the RNA polymerase II (McKenna et al., 1999). These include: (1) p300 and the homologous cAMP-responsive element-binding protein (CREB)-binding protein (CBP), which also serve as macromolecular docking "platforms" for transcription factors from several signal transduction cascades, including nuclear receptors, CREB, AP-1, NFκB, p53, Ras-dependent growth factor, and STATs (Goodman and Smolik, 2000). Because of their central position in many signal-transduction cascades, the p300/CBP coactivators are also called cointegrators; (2) p300/CBP-associated factor (p/CAF), originally reported as a human homologue of yeast Gcn5, which interacts with p300/CBP and is also a broad transcription coactivator (Yang et al., 1996); and (3) the p160 family of coactivators, which preferentially interact with the steroid hormone receptors (Leo and Chen, 2000). These include the steroid receptor coactivator-1 (SRC-1), SRC-2, which consists of transcription intermediate factor-II (TIF-II) and the glucocorticoid receptor-interacting protein-1 (GRIP-1), and SRC-3, which consists of the p300/CBP/cointegrator-associated protein (p/CIP), activator of thyroid receptor (ACTR), and the receptor-associated coactivator-3 (RAC3) (McKenna et al., 1999; Leo and Chen, 2000).

The p160 coactivators are the first to be attracted to the DNA-bound steroid hormone receptor and help accumulate p300/CBP and p/CAF proteins to the promoter region, indicating that the p160 proteins play an important role in steroid hormone receptor-mediated transactivation. These coactivators also have intrinsic histone acetyltransferase (HAT) activity through which they loosen the tightly assembled chromatin structure and facilitate access of transcriptional complexes to the promoter regions (McKenna et al., 1999). HAT activity also modulates the binding of transcription factors to specific elements on their responsive promoters (Gu and Roeder, 1997; Boyes et al., 1998), as well as the dissociation of coactivators from nuclear receptors or other transcription factors (Chen et al., 1999). The p160 family of coactivators and p300/CBP proteins contain one or more copies of the coactivator signature motif sequence LXXLL, where L is leucine and X is any amino acid (Heery et al., 1997; Leo and Chen, 2000). LXXLL forms an α helical structure aligning leucine residues in such a way as to form hydrophobic bonds with the AF-2 surface of the receptor. The latter is formed by helices 3, 4, and 12 of the ligand-binding domain (LBD) of GRα in response to ligand binding.

The AF-2 transactivation domain of GRα also attracts several other distinct chromatin modulators, such as the mating-type switching/sucrose non-fermenting (SWI/SNF) complex and components of the vitamin D receptor-interacting protein/thyroid hormone receptor-associated protein (DRIP/TRAP) complex (McKenna et al., 1999). The SWI/SNF complex is an ATP-dependent chromatin-remodeling factor with a multi-subunit structure in which an ATPase functions as the catalytic center (Fry and Peterson, 2001). Depending on the energy of ATP hydrolysis, it introduces superhelical torsion into DNA. One of its components, SNF2 binds to AF-2 of GRα directly, functioning as an interface between the GR and the SWI/SNF complex (Yoshinaga et al., 1992). The DRIP/TRAP complex is also a multi-protein conglomerate, which consists of over 10 different proteins, including DRIP205/TRAP220/PBP and components of SMCC (McKenna et al., 1999). The DRIP/TRAP complex may modulate transcription through interaction and modification of general transcription factors, such as TFIIH or the C-terminal tail of the RNA polymerase II. DRIP205/TRAP220 contains two LXXLL motifs through which it binds to the ligand-activated AF-2 directly (Rachez and Freedman, 2001). Since the DRIP/TRAP complex and p160 coactivators use the same motif for interaction with the steroid hormone receptors, they may bind to the same site of these receptors and sequentially interact with them for full activation of transcription. Alternatively, they may interact with a particular set of steroid hormone receptors, or have a different effect on different tissues (McKenna et al., 1999; Glass and Rosenfeld, 2000).

In contrast to the mechanisms of transactivation via the AF-2, those of AF-1 are not as well elucidated yet. Using the yeast system, the Ada complex may act on AF-1 transactivation through direct interaction (Henriksson et al., 1997). The SWI/SNF complex and the HAT coactivators, such as p160 and p300/CBP, also physically interact with AF-1 and mediate its transcriptional activity (Webb et al., 1998; Ma et al., 1999; Wallberg et al., 2000). In addition, DRIP150, a component of the DRIP/TRAP complex, and the tumor susceptibility gene 101 (TSG101) interact with the AF-1 of GRα in a modified yeast two-hybrid screening (Hittelman et al., 1999). An RNA coactivator, the steroid RNA activator (SRA) also interacts with AF-1 (Lanz et al., 1999). Given that any of these molecules and complexes interact with both AF-1 and AF-2, it is likely the that concurrent activation of AF-1 and AF-2 by their common and/or distinct binding partners may be necessary for optimal activation of GRα-induced transcription (Fig. 3) (Benecke et al., 2000).

Interaction of GR with and modification of the activity of other transcription factors

Glucocorticoids exert most of their extremely diverse effects through only one protein molecule, the classic GR. This becomes possible not only by direct interaction of the GR with the promoter of its responsive genes, but also because the hormone–receptor complexes affect other signal transduction cascades through mutual protein–protein interactions with the transcription factors regulating these cascades (Bemberger et al., 1996). A mouse model harboring a mutant GR, which is active in protein–protein interaction but unable to dimerize and bind to GREs, survive relatively free of major clinical problems, in contrast to mice in which the entire GR gene is deleted, which die at birth from severe respiratory distress syndrome (Cark et al., 1992; Reichardt et al., 1998). The former mouse model and additional in vitro data indicate that GR interacts with and influences the effect of other transcription factors as a monomer (Reichardt et al., 1998; Reichardt et al., 2001). The protein–protein interactions of GRα with other transcription factors take place on promoters that do not contain GREs, as well as on the promoters that have both GRE(s) and responsive element(s) of transcription factors that interact with

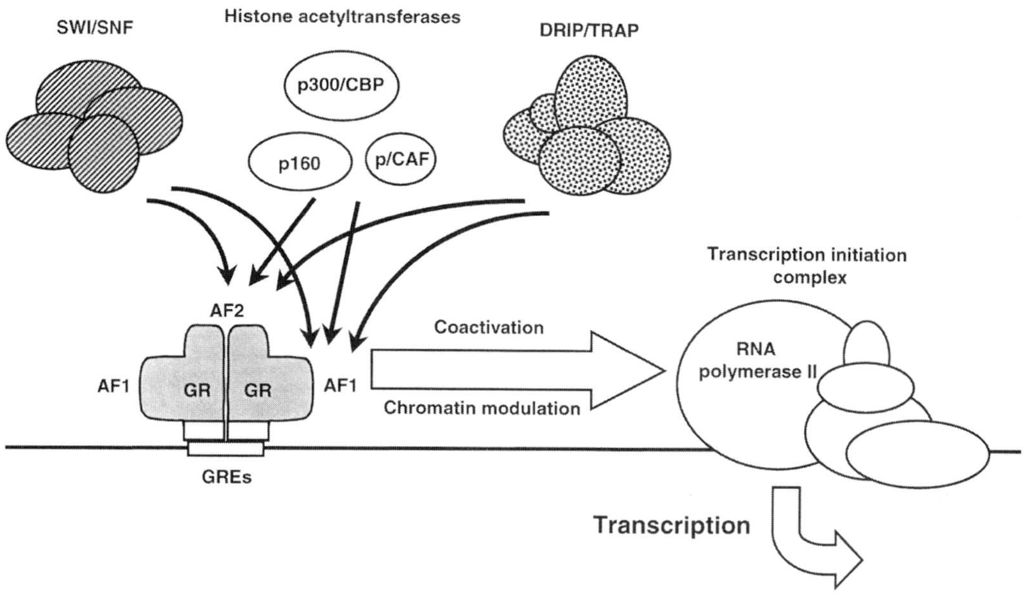

Fig. 3. Schematic model showing the interaction and activity of coactivators and other chromatin modulators, attracted by GR to the promoter region of glucocorticoid-responsive genes. AF-1 and 2, activation function 1 and 2; p/CAF, p300/CBP-associated factor; CBP: cAMP-responsive element-binding protein (CREB)-binding protein; TRAP, thyroid hormone receptor-associated protein; DRIP, vitamin D receptor-interacting protein; SWI, mating-type switching; SNF, sucrose nonfermenting.

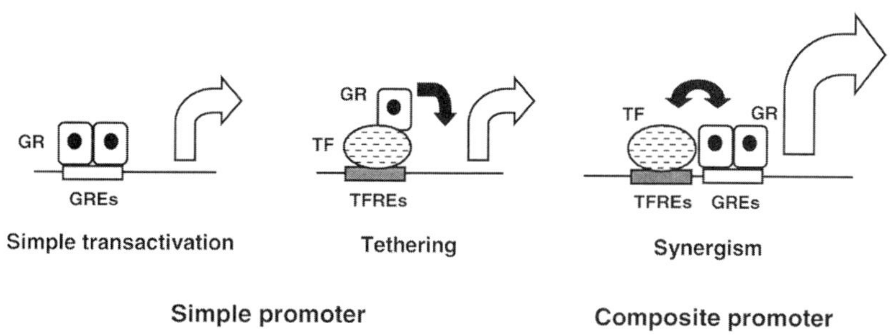

Fig. 4. Three different modes of transcriptional regulation of the glucocorticoid-responsive promoters by GR. GR may interact with other transcription factors directly or indirectly. A protein or a protein complex may intermediate their association in the latter case. GR, glucocorticoid receptor; TF, transcription factor; GRE, glucocorticoid-responsive element; TFRE, transcription factor-responsive element.

GRα ("composite promoters") (Fig. 4) (Miner and Yamamoto, 1991).

Glucocorticoid-induced suppression of transactivation through protein–protein interaction may be particularly important in the immunosuppressive and anti-inflammatory actions of glucocorticoids (Reichardt et al., 1998; Reichardt et al., 2001). Substantial part of glucocorticoid effect on the immune system may be explained by its interaction with NF-κB, AP-1, and probably several STATs (Didonato et al., 1996; Barnes and Karin, 1997; Karin and Chang, 2001). In addition, GR also mutually influences the transcriptional activity of other transcription factors, such as CREB, CAAT/enhancer-binding protein (C/EBP), chicken ovalbumin upstream promoter-transcription factor (COUP-TF) II, Nur77, p53, hepatocyte nuclear factor (HNF)-6, GATA-1, Oct-1 and -2, nuclear factor (NF)-1, and Sp-1 (Kino et al., 2003). The following section further discusses the interactions of NF-κB,

AP-1, and with STATs and their effects on GRα transcriptional activity as examples.

Nuclear factor-κB (NF-κB)

NF-κB is one of the most important transcription factors that regulate immune function and the inflammatory reaction. NF-κB is stimulated by many inflammatory signals and cytokines (Barnes and Karin, 1997; Perkin, 2000). It is a dimer of various members of the NF-κB/Rel, family, including p50 (and its precursor p105), p52 (and its precursor p100), c-Rel, RelA, and RelB, in mammalian organisms. The heterodimer p65/p50 is the major and most abundant form of NF-κB. In its inactive form, NF-κB forms a trimer with an additional regulatory protein, IκB in the cytoplasm. A variety of extracellular signals, such as bacterial and viral products like lipopolysaccharide (LPS), released intracellular components, such as heat-shock proteins after physical or chemical stress, and several proinflammatory cytokines induce phosphorylation of IκB by activating a cascade of kinases. The phosphorylated IκB then dissociates from NF-κB and is catabolized, while the liberated NF-κB translocates into the nucleus, where it binds to the κB-responsive elements in the promoter regions of its responsive genes.

Ligand-activated GRα directly binds NF-κB p65 at its Rel homology domain through its DNA-binding domain (DBD) and suppresses the transcriptional activity of NF-κB, while NF-κB suppresses GRα-induced transactivation through GREs. Interaction with GRα inhibits binding of NF-κB to its responsive elements or neutralizes its ability to transmit an effective signal (Caldenhoven et al., 1995; Liden et al., 1997; Wissink et al., 1997; McKay and Cidlowski, 1999). The LBD of GRα is necessary for this suppressive action (McKay and Cidlowski, 1998). GRα also suppresses NF-κB-induced transactivation by an additional mechanism, in which GRα tethered to the κB-responsive promoters attracts histone deacetylases and/or modulates the phosphorylation of the RNA polymerase II C-terminal tail (Ito et al., 2000; Nissen and Yamamoto, 2000). In addition, the ligand-activated GRα increases the synthesis of IκB by stimulating its promoter activity through classic GREs, thus segregating active NF-κB away from the nucleus by forming inactive heterocomplexes with IκB in the cytoplasm (Auphan et al., 1995).

Activator protein-1 (AP-1)

AP-1 is a transcription factor that regulates the expression of diverse genes involved in cell proliferation and differentiation (Herrlich, 2001; Karin and Chang, 2001; van Dam and Castellazzi, 2001). It acts as a dimer of members of the bZip protein family, in which c-Fos and c-Jun heterodimers are the most abundant. AP-1 transduces the signal of phorbol esters, growth factors, and proinflammatory cytokines, such as IL-1 and TNFα. These molecules stimulate different members of the mitogen-activated protein kinase family, e.g., extracellular signal-regulated kinase p38, and Jun N-terminal kinase (JNK). Once these kinases are activated, they stimulate the synthesis of specific transcription factors involved in the induction of *fos* and *jun* gene transcription, as well as enhance the transcriptional activity of both preexisting and newly synthesized c-Fos/c-Jun proteins. AP-1 and GRα mutually interact and repress each other's transcriptional activities on their respective responsive promoters. The LBD and DBD of GRα and the leucine zipper domain of c-Jun are necessary for this interaction (Schule et al., 1990). Inhibition of binding of AP-1 to DNA may be one of the underlying mechanisms of GRα-induced suppression of AP-1-mediated transactivation. Furthermore, GRα competes with AP-1 for the p300/CBP coactivators, which have limited reserves; therefore, AP-1 may not have access to adequate amounts of these coactivators to exert its transcriptional activity fully (Kamei et al., 1996).

Signal transducers and activators of transcription (STATs)

This class of transcription factors forms a distinct family, which regulates transcription induced by many extracellular signals of receptors that are members of the cytokine receptor superfamily (Leonard and O'Shea et al., 1998; Ihle, 2001). These include many cytokines, interferons, granulocyte macrophage colony-stimulation factor, erythropoietin, growth hormone, prolactin, and leptin. So far, seven STAT proteins (STAT1, 2, 3, 4, 5a, 5b, and 6) and their isoforms have been reported. Extracellular signals induce receptor oligomerization and subsequently activate Janus kinases (JAKs) that phosphorylate their own tyrosine residues, as well as the cytoplasmic domains of the membrane receptors that induced this activity. STATs are then attracted to phosphorylated residues of the receptors through their SH2 domains and are phosphorylated by the JAKs. Phosphorylated STATs then dimerize and translocate into the nucleus where they bind specific DNA sequences on their responsive promoters.

GRα mutually interacts with one of the STATs, STAT5, which functions downstream of growth hormone, prolactin, and several cytokines (Grimely et al., 1999). In an artificial overexpression system, ligand-activated GRα and STAT5 mutually affect each other's transcriptional activities on their simple responsive promoters; GRα enhances STAT5-induced transactivation, while STAT5 represses the transcriptional activity of GRα on GRE-driven promoters (Stocklin et al., 1996; Stoecklin et al., 1997). However, in physiological conditions, the GRα-STAT5 interaction may be seen only on the promoters that have both GRα- and STAT5-binding sites (Doppler et al., 2001). Although GRα and STAT5 can be coprecipitated with each other's specific antibodies, their direct binding has not been unequivocally shown. The N-terminal transactivation domain of GRα is necessary for the enhancement of STAT5 transactivation, while phosphorylation and subsequent activation of STAT5 is not necessary for its repression of GR transactivation (Stocklin et al., 1996; Groner et al., 2000). Similar functional interaction of GRα has also been reported with STAT1 and 3 (Zhang et al., 1997; Takeda et al., 1998; Aittomaki et al., 2000).

Glucocorticoid effects on tissue-specific gene expression

Through the above described mechanisms of GR actions on transcription, and/or through possibly other, as yet undefined actions, such as direct effects on molecules localized in the cytoplasmic membrane or the cytoplasm and/or posttranscriptional effects on the facilitation of degradation/stabilization of mRNAs, glucocorticoids exert their specific complex effects in different, distinct cells, tissues, organs, and systems. In the following sections, the biological actions of glucocorticoids on several systems are reviewed briefly.

Central nervous system (CNS)

Glucocorticoids exert many effects on the central nervous system (CNS) in vivo, influencing cognition, memory, mood, and sleep (Belanoff et al., 2001). Glucocorticoids may even change the anatomical structure of the CNS, causing a reduction of hippocampal volume, ventricular enlargement, and reversible cortical atrophy (McEwen et al., 1997; Sapolsky, 2000; Fuchs et al., 2001). The pathophysiology of these in vivo effects has not been well elucidated, even though there are many reports demonstrating that glucocorticoids change several functions of neuronal cells, including their electrical activity, synthesis/secretion of neurotransmitters and/or neuropeptides, and cell degeneration or death (Holsboer and Barden, 1996; Falkenstein et al., 2000). For example, glucocorticoids may induce neuronal death in the hippocampus by reducing energy supply and/or by increasing cytoplasmic calcium concentrations. The former may be mediated by reduction of glucose supply via inhibition of its transport, while the latter may be mediated by a decrease of the transcription rates of the calcium-ATPase pump in the plasma membrane; the latter plays a key role in extruding cytoplasmic calcium into the extracellular space (Bhargava et al., 2000; Lee et al., 2002). Glucocorticoids also modulate the expression of glutamine synthetase, fibrillary acidic protein, nerve growth factor, fibroblast growth factor, and lipocortin-1 in glial cells (Vardimon et al., 1999). Among them, glutamate synthetase plays a pivotal role in the circulation of the excitotoxic neurotransmitter glutamate between neurons and glial cells. Since its substrate glutamate acts as a toxic agent for neurons, glucocorticoids may further regulate neuronal death by changing the expression levels of this enzyme.

The CNS contains two types of "glucocorticoid" receptors, which can bind glucocorticoids and mediate their actions. One is the mineralocorticoid receptor (MR) (type I) and the other is the classic GR (type II) (de Kloet et al., 2000; Reul et al., 2000). The former is distributed predominantly in the limbic system, including the hippocampus, septum, septohippocampal nucleus, and amygdala, while the latter is widely distributed in the entire CNS. Since the CNS, particularly the hippocampus, does not contain a significant amount of 11β-hydroxysteroid dehydrogenase (HSD) type 2 that inactivates glucocorticoids and causes the MR to bind mineralocorticoids in mineralocorticoid-responsive tissues, MR can act as a functional high-affinity receptor for glucocorticoids in the hippocampus (de Kloet et al., 2000). Therefore, this portion of the brain employs two systems for sensing circulating glucocorticoids, distinguishing different secretion patterns and evoking different responses. Limited information indicates that the MR mediates a tonic effect of glucocorticoids on circadian fluctuations, the sensitivity of the stress response, and organization of the behavioral response to stress, while the GR transduces the negative effect of glucocorticoids on stress-induced HPA-axis activation and facilitates memory storage (Schobitz et al., 1994; de Kloet, 1995).

Intermediary metabolism

Glucocorticoids modulate diverse metabolic activities in humans including carbohydrate, lipid and protein metabolism, influencing the flows between the liver and peripheral tissues, such as muscle, fat, and connective tissue (Orth et al., 1992). The direction of changes induced by stress or pharmacological doses of glucocorticoids is to increase the circulating levels of energy fuel, such as glucose, free fatty acids, amino acids, and glycerol, by facilitating catabolism of glycogen, fat, and protein in peripheral tissues. These compounds are then taken up by the liver where they are used as substrates for gluconeogenesis and subsequent glycogen synthesis and glucose secretion. Glycogen is accumulated in the liver by glucocorticoids through activation of glucokinase, glycogen synthase (conversion from the inactive form to the active form), and inactivation of glycogen phosphorylase, which helps catalyze glycogen hydrolysis (Stalmans and Laloux, 1979; Miller and Chrousos, 2001).

Glucose metabolism

Glucocorticoids stimulate gluconeogenesis by increasing the production of its rate-limiting enzyme phosphoenol-pyruvate carboxykinase (PEPCK), and by activating the enzyme glucose-6-phosphatase (Coufalik and Monder, 1981; van de Werve et al., 2000; Miller and Chrousos, 2001). Glucocorticoids directly stimulate PEPCK promoter activity through the GR that is attracted to the promoter region of the PEPCK gene via atypical GREs (Wang et al., 1999). The protein–protein interactions of GR with the promoter-bound chicken ovalbumin upstream promoter-transcription factor II (COUP-TFII) may be also required for maximal activation of this promoter (De Martino et al., 2004). Peroxisome proliferator-activated receptor-γ coactivator-1 (PGC-1), which has a major role in cellular respiration and adaptive thermogenesis in muscle and brown fat, was recently shown to play an important role in the activation of the PEPCK promoter interphasing with the activated GR (Herzig et al., 2001; Yoon et al., 2001). Glucocorticoids also decrease the insulin-stimulated rate of glucose uptake and utilization by peripheral tissues (Olefsky, 1975; Weinstein et al., 1995; Sakoda et al., 2000). There are several conflicting reports demonstrating induction of glucose transporters in different tissues by glucocorticoids (Hajduch et al., 1995; Dimitriadis et al., 1997).

Fat metabolism

Glucocorticoids exert lipolytic activity in adipose tissue, thereby increasing the levels of plasma-free fatty acids and glycerol (Orth et al., 1992). Although the exact molecular mechanism of this action has not been elucidated yet, glucocorticoids seem to act by potentiating the effects of several lipolytic hormones, e.g., growth hormone and the catecholamines, on hormone-sensitive lipase in adipose tissue (Orth et al., 1992). Physiological levels of glucocorticoids are also necessary to maintain normal levels of circulating free fatty acids. Chronic exposure with excess amount of glucocorticoids causes visceral obesity, a known feature of Cushing's syndrome (Chrousos, 2001). The centripedal obesity phenotype suggests that sensitivity and/or responsiveness of adipocytes to glucocorticoids are different in tissues located in different regions of the human body; however, the clear molecular mechanisms underlying this phenomenon are not known.

In in vitro experimental conditions, glucocorticoids stimulate differentiation of NIH3T3 preadipocyte cells into mature, fat-containing adipocytes. An epidermal growth factor-related protein termed the preadipocyte factor-1 (pref-1) was discovered in these cells, whose expression negatively regulates their differentiation; glucocorticoids suppress the expression of this polypeptide (Smas and Sul, 1993). Therefore, differential expression/induction of pref-1 by glucocorticoids in visceral and subcutaneous fat might explain the development of the particular fat redistribution caused by glucocorticoids (Smas et al., 1999; Moon et al., 2002). It is also known that visceral adipocytes from omental fat express higher 11β-HSD type 1 than those from subcutaneous fat (Bujalska et al., 1997). Exposure of these cells to glucocorticoids further enhances this enzyme activity in omental fat, but not in subcutaneous fat. Since this enzyme converts inactive cortisone to active cortisol, these results suggest that visceral fat is exposed to higher concentrations of glucocorticoids than subcutaneous fat. Thus, the differential expression of 11β-HSD type 1 in visceral and subcutaneous fat might also contribute to the characteristic redistribution of fat tissues by glucocorticoids.

Skeletal muscle and connective tissues

Excessive amounts of glucocorticoids act as a catabolic agent on the protein turnover that supplies amino acids from mainly skeletal muscles and connective tissues to the liver for gluconeogenesis (Miller and Chrousos, 2001). Physiological levels of glucocorticoids, on the other hand, may be necessary to maintain proper turnover of protein. The overall molecular mechanisms underlying glucocorticoid actions on protein turnover has not been fully elucidated yet.

Effect on skeletal muscle

Excessive amounts of glucocorticoids induce muscle wasting, i.e., marked reduction in muscle volume, particularly in fast-twitch muscle fibers (Christy, 1971; Askari et al., 1976). The onset of these changes is usually gradual and often requires weeks or months to develop. Glucocorticoids preferentially affect the proximal muscles; however, the distal muscles are also affected with prolonged exposure to high doses of glucocorticoids. Differential effect of glucocorticoids on synthesis and degradation of muscle-contractile proteins, such as the myosin light and heavy chains, were reported (Seene, 1994; Almon and Dubois, 1990; Rooyackers and Nair, 1997).

Effects on connective tissues

Glucocorticoid treatment causes thinning and atrophy of skin and formation of striae, which are characterized histologically by loss of elastic fibers (Oikarinen and Autio, 1991; Oikarinen, 1992; Autio et al., 1993). Glucocorticoids also impair wound healing (Anstead, 1998). Glucocorticoids suppress fibroblast proliferation mainly through inhibiting DNA and RNA synthesis (Jung-Testas and Baulieu, 1985; Durant et al., 1986). They also suppress production of all types of collagen fibers both by inhibiting their mRNA synthesis and by stimulating their degradation (Nacht and Garzon, 1974). Glucocorticoids reduce the activity of enzymes needed for the biosynthesis of collagen. Glucocorticoids inhibit the synthesis of glycosaminoglycans, which may explain the rapid thinning of the dermis following topical glucocorticoid treatment (Greene and Kochhar, 1975).

Bones

Physiological amounts of glucocorticoids seem to be necessary to maintain normal function of bone, its turnover, and growth (Raisz and Bingham, 1972; Canalis and Delany, 2002). On the other hand, excess amounts and/or prolonged usage of glucocorticoids frequently cause osteopenia and osteoporosis, which is one of the major side effects of long-term glucocorticoid therapy (Clowes et al., 2001).

Glucocorticoid effects are more obvious in cancellous bone, such as vertebral bodies and ribs, than in compact bone, which turns over more slowly (Baylink, 1983). Thus, the most pronounced bone loss occurs in the axial skeleton, rather than in the extremities. Excess glucocorticoids mainly suppress bone formation through modulating osteoblast activity (Reid, 2000; Canalis and Delany, 2002). They decrease proliferation of osteoblasts and increase their apoptosis, and reduce their protein synthesis, including production of type I collagen, osteocalcin, and insulin-like growth factors. On the other hand, their effect on bone resorption through the osteoclasts is a controversial issue with reports showing both suppression and activation of these cells (Reid, 2000). Glucocorticoids also reduce the absorption of calcium and phosphate in the intestine and renal tubules (Reid, 2000). This effect is not mediated by vitamin D but rather by direct inhibition of the absorption of these electrolytes.

Cardiovascular homeostasis

Physiological amounts of glucocorticoids are essential to maintain vascular tone, preserving proper peripheral resistance, and are probably necessary for normal renal tubular functions, which determine reabsorption of water and electrolytes (Brem, 2001). Excess amounts of glucocorticoids increase peripheral vascular resistance (Ullian, 1999; Brem, 2001). Their effect is exerted mainly on the vascular smooth muscle cells of peripheral blood vessels. Glucocorticoids do not seem to directly affect vascular tone, but mainly increase it by potentiating the action of vasoconstrictor and by suppressing the activity of vasodilator hormones, in part through upregulation or downregulation of the expression of their receptors or signaling systems, respectively (Brem, 2001). For example, glucocorticoids increase the expression of the angiotensin II receptor, probably through stimulating its transcription, while they potentiate the action of α-adrenergic agonists via a postreceptor mechanism, as well as by upregulating their binding sites (Haigh and Jones, 1990; Russo et al., 1990; Sato et al., 1994). On the other hand, glucocorticoids suppress production of prostaglandin E_2 and nitric oxide (NO), both of which are potent vasodilators (Handa et al., 1984). Glucocorticoids suppress production of the latter by inhibiting the enzyme NO synthase, the transmembrane transport of arginine, its precursor, and the synthesis of NO synthase cofactor tetrahydrobiopterin (Kelly et al., 1998; Ullian, 1999; Wallerath et al., 1999; Johns et al., 2001). Glucocorticoid effects on ion transport in vascular smooth-muscle cells have also been reported (Kato et al., 1992). These hormones increase calcium uptake, possibly by potentiating de novo synthesis of membrane calcium channels.

Immune system

Physiological concentrations of glucocorticoids are necessary to maintain homeostasis and proper function of the immune system, while pharmacological amounts and/or concentrations obtained in the stress state strongly suppress activity of its components (Chrousos, 1995; Wilckens and De Rijk, 1997). The latter activity may be beneficial pharmacologically; indeed, glucocorticoids are one of the most efficient therapeutic compounds for allergic, inflammatory, and autoimmune diseases, as well as for suppressing the host rejection reaction in organ transplantation. High doses or levels of glucocorticoids, however, increase the risk of several types of infection, including those from bacteria, viruses, fungi, and parasites; infections by gram-negative bacteria and fungi appear to be particularly prevalent in patients exposed to very high levels of glucocorticoids (Kass and Finland, 1958; Dale and Petersdorf, 1973).

Glucocorticoids exert quite diverse activities on nearly all steps of the immune reaction, including the innate, cellular, and humoral immunities (Fauci et al., 1976; Boumpas et al., 1993). Glucocorticoids produce complex final effects on the immune system depending on the situation in which particular components of the immune system are differently involved. In general, glucocorticoids cause (a) suppression of inflammatory mediator secretion, (b) inhibition of monocyte/granulocyte migration at inflammed tissues, (c) suppression of cellular immunity, and (d) stimulation of secondary humoral immunity.

Suppression of proinflammatory mediators

Inflammed tissues and local immunocompetent cells produce many bioactive substances that cause vasodilation, increase vascular permeability, and attract circulating leukocytes to the local inflammation sites. These include the prostanoids, histamine, bradykinin, serotonin, many interleukins, nitric oxide, tumor necrosis factor (TNF) α, interferons, plasminogen activators, several growth factors for immunocompetent cells, and inflammation-related enzymes, such as several proteases (Tanaka et al., 1997; Bazan et al., 1990; O'Sullivan, 1999; Venarske and deShazo, 2003). Through modulation of these bioactive compounds, glucocorticoids suppress acute inflammatory reactions, such as redness, edema, pain, and loss of function of inflammed tissues/organs.

Glucocorticoids strongly downregulate the arachidonic acid cascade, which produces bioactive prostaglandins, thromboxanes, hydroperoxyeicosatetraenoic acids (HPETE), hydroxyeicosatetraenoic acids (HETE), and leukotrienes, by suppressing de novo synthesis of phospholipase A_2, cyclooxygenase 2, prostaglandin synthase, and lipooxygenase, all produced in response to inflammatory stimuli, such as interleukin-1β and TNFα (Herschman et al., 1995). Glucocorticoids also inhibit production of nitric oxide and oxygen radicals, and block the action and/or release of histamine, bradykinin, serotonin, chemokines, interleukins (IL) 1, 2, 6, and 12, and leukotrienes (Charlesworth et al., 1991; Cuzzocrea et al., 2001). The exact mechanisms of glucocorticoid action on these bioactive agents are still not well elucidated, and dependent on the tissue and experimental system employed.

Inhibition of monocyte/granulocyte migration

Glucocorticoids have a strong effect on the trafficking/distribution of leukocytes during inflammation (Parrillo and Fauci, 1979; Charlesworth et al., 1991). First, they increase blood neutrophil concentration and cause a profound depletion of eosinophils and basophils in circulation. Glucocorticoids also promote survival and proliferation of neutrophils, while they induce apoptosis of eosinophils and basophils (Meagher et al., 1996). Glucocorticoids induce neutrophilia by increasing release of polymorphonuclear cells from the bone marrow and by inhibiting the transmigration of neutrophils to inflammatory sites. The latter occurs through (1) suppression of leukocyte adhesion molecule leukocyte factor adhesion-1 (LFA-1) and *L*-secretin, (2) suppression of the endothelial adhesion molecule intercellular adhesion molecule-1 (ICAM-1), and inhibition of chemokine release (IL-8 and other CXC chemokines) by immune accessory cells (Cronstein et al., 1992; Cronstein and Weissmann, 1995; Nakagawa et al., 1999). Similarly, glucocorticoids block eosinophil transmigration through inhibition of adhesion molecule ICAM-3, and the release of chemokines (Jahnsen et al., 1999). Macrophage transmigration is also suppressed through similar mechanisms (Cronstein and Weissmann, 1995).

Glucocorticoids act on the innate immune system by modulating secretion of both proinflammatory and/or antiinflammatory mediators. Glucocorticoids suppress inflammation by enhancement of clearance of foreign antigens, toxins, microorganisms, and dead cells through enhancing opsonization of scavenger systems and by stimulating the phagocytotic ability of macrophages (Liu et al., 1999).

Suppression of cellular and stimulation of humoral immunity

In general, glucocorticoids suppress cellular immunity. They do this by reducing IL-12 and TNFα production by monocytes/macrophages and dendritic cells and by shifting differentiation of CD4+ T cells from T lymphocytes helper (Th) 1 into Th2, which respectively drive cellular and humoral immunity (Elenkov et al., 2000a,b). Glucocorticoids suppress differentiation into Th1 lymphocytes by inhibiting secretion of IL-12, while they promote Th2 lymphocytes by stimulating production of IL-10. Moreover, glucocorticoids suppress cellular immunity by downregulating antigen presentation of dendritic cells to T lymphocytes, which is known as a trigger of the cellular immune response. Thus, glucocorticoids suppress the expression of the major histocompatibility complex (MHC) II antigens and its costimulatory molecules in monocytes/dendritic cells, which are necessary for T lymphocytes to recognize presented antigens (Piemonti et al., 1999). In addition, glucocorticoids induce apoptosis in specific fractions of T lymphocytes in some conditions (Ashwell et al., 2000).

Glucocorticoids enhance actions of IL-4 on B cell differentiation and isotype switching, thereby helping to induce development of immunoglobin (Ig) E- and IgG4-secreting B cells (Wu et al., 1991; Kimata et al., 1995). In parallel, however, glucocorticoids prevent the IgE-induced allergic immune response mediated by mast cells and eosinophils, suppressing this component of humoral immunity (Barnes, 2001a,b).

Summary

This chapter describes the diverse actions of glucocorticoids on the expression of responsive genes. All these actions are mediated by a single intracellular receptor molecule, the GR. Glucocorticoid-bound GR modulates the transcription rates of glucocorticoid-responsive genes by (1) directly binding to the DNA of their promoter regions, and by (2) modulating the transcriptional activity of many other transcription factors through mutual protein–protein interactions. Glucocorticoids reach and influence the functions of almost all organs/tissues of the organism, including those of the CNS. Glucocorticoids have major roles in behavioral, and cardiovascular homeostasis, intermediary metabolism, bone turnover, and immune function.

References

Aittomaki, S., Pesu, M., Groner, B., Janne, O.A., Palvimo, J.J. and Silvennoinen, O. (2000) Cooperation among Stat1, glucocorticoid receptor, and PU.1 in transcriptional activation of the high-affinity Fc gamma receptor I in monocytes. J. Immunol., 164: 5689–5697.

Almon, R.R. and Dubois, D.C. (1990) Fiber-type discrimination in disuse and glucocorticoid-induced atrophy. Med. Sci. Sports Exerc., 22: 304–311.

Anstead, G.M. (1998) Steroids, retinoids, and wound healing. Adv. Wound Care, 11: 277–285.

Ashwell, J.D., Lu, F.W. and Vacchio, M.S. (2000) Glucocorticoids in T cell development and function. Annu. Rev. Immunol., 18: 309–345.

Askari, A., Vignos, P.J., Jr. and Moskowitz, R.W. (1976) Steroid myopathy in connective tissue disease. Am. J. Med., 61: 485–492.

Auphan, N., DiDonato, J.A., Rosette, C., Helmberg, A. and Karin, M. (1995) Immunosuppression by glucocorticoids: inhibition of NF-kappa B activity through induction of I kappa B synthesis. Science, 270: 286–290.

Autio, P., Risteli, J., Kiistala, U., Risteli, L., Karvonen, J. and Oikarinen, A. (1993) Serum markers of collagen synthesis and degradation in skin diseases. Altered levels in diseases with systemic manifestation and during systemic glucocorticoid treatment. Arch. Dermatol. Res., 285: 322–327.

Bamberger, C.M., Schulte, H.M. and Chrousos, G.P. (1996) Molecular determinants of glucocorticoid receptor function and tissue sensitivity to glucocorticoids. Endocr. Rev., 17: 245–261.

Barnes, P.J. (2001a) Corticosteroids, IgE, and atopy. J. Clin. Invest., 107: 265–266.

Barnes, P.J. (2001b) Molecular mechanisms of corticosteroids in allergic diseases. Allergy, 56: 928–936.

Barnes, P.J. and Karin, M. (1997) Nuclear factor-kappaB: a pivotal transcription factor in chronic inflammatory diseases. N. Engl. J. Med., 336: 1066–1071.

Baylink, D.J. (1983) Glucocorticoid-induced osteoporosis. N. Engl. J. Med, 309: 306–308.

Bazan, N.G., de Abreu, M.T., Bazan, H.E. and Belfort, R.J. (1990) Arachidonic acid cascade and platelet-activating factor in the network of eye inflammatory mediators: therapeutic implications in uveitis. Int. Ophthalmol., 14: 335–344.

Beato, M. and Sanchez-Pacheco, A. (1996) Interaction of steroid hormone receptors with the transcription initiation complex. Endocr. Rev., 17: 587–609.

Belanoff, J.K., Gross, K., Yager, A. and Schatzberg, A.F. (2001) Corticosteroids and cognition. J. Psychiatr. Res., 35: 127–145.

Benecke, A., Chambon, P. and Gronemeyer, H. (2000) Synergy between estrogen receptor alpha activation functions AF1 and AF2 mediated by transcription intermediary factor TIF2. EMBO Rep., 1: 151–157.

Bhargava, A., Meijer, O.C., Dallman, M.F. and Pearce, D. (2000) Plasma membrane calcium pump isoform 1 gene expression is repressed by corticosterone and stress in rat hippocampus. J. Neurosci., 20: 3129–3138.

Boumpas, D.T., Chrousos, G.P., Wilder, R.L., Cupps, T.R. and Balow, J.E. (1993) Glucocorticoid therapy for immune-mediated diseases: basic and clinical correlates. Ann. Intern. Med., 119: 1198–1208.

Boyes, J., Byfield, P., Nakatani, Y. and Ogryzko, V. (1998) Regulation of activity of the transcription factor GATA-1 by acetylation. Nature, 396: 594–598.

Brem, A.S. (2001) Insights into glucocorticoid-associated hypertension. Am. J. Kidney Dis., 37: 1–10.

Bresnick, E.H., John, S., Berard, D.S., LeFebvre, P. and Hager, G.L. (1990) Glucocorticoid receptor-dependent disruption of a specific nucleosome on the mouse mammary tumor virus promoter is prevented by sodium butyrate. Proc. Natl. Acad. Sci. USA, 87: 3977–3981.

Bujalska, I.J., Kumar, S. and Stewart, P.M. (1997) Does central obesity reflect "Cushing's disease of the omentum"? Lancet, 349: 1210–1213.

Caldenhoven, E., Liden, J., Wissink, S., Van de Stolpe, A., Raaijmakers, J., Koenderman, L., Okret, S., Gustafsson, J.A. and Van der Saag, P.T. (1995) Negative cross-talk between RelA and the glucocorticoid receptor: a possible mechanism for the antiinflammatory action of glucocorticoids. Mol. Endocrinol., 9: 401–412.

Canalis, E. and Delany, A.M. (2002) Mechanisms of glucocorticoid action in bone. Ann. NY Acad. Sci., 966: 73–81.

Charlesworth, E.N., Kagey-Sobotka, A., Schleimer, R.P., Norman, P.S. and Lichtenstein, L.M. (1991) Prednisone inhibits the appearance of inflammatory mediators and the influx of eosinophils and basophils associated with the cutaneous late-phase response to allergen. J. Immunol., 146: 671–676.

Chen, H., Lin, R.J., Xie, W., Wilpitz, D. and Evans, R.M. (1999) Regulation of hormone-induced histone hyperacetylation and gene activation via acetylation of an acetylase. Cell, 98: 675–686.

Christy, N.P. (1971) Iatrogenic Cushing's syndrome. In: Christy, N.P. (Ed.), The Human Adrenal Cortex. Harper & Row, New York, pp. 395.

Chrousos, G.P. (1995) The hypothalamic–pituitary–drenal axis and immune-mediated inflammation. N. Engl. J. Med., 332: 1351–1362.

Chrousos, G.P. (2001) Glucocorticoid therapy. In: Felig, P. and Frohman, L.A. (Eds.), Endocrinology and Metabolism. McGraw-Hill, New York, pp. 609–632.

Clark, J.K., Schrader, W.T. and O'Malley, B.W. (1992) Mechanism of steroid hormones. In: Wilson, J.D. and Foster, D.W. (Eds.), Williams Textbook of Endocrinology. W.B. Sanders Co., Philadelphia, pp. 35–90.

Clowes, J.A., Peel, N. and Eastell, R. (2001) Glucocorticoid-induced osteoporosis. Curr. Opin. Rheumatol., 13: 326–332.

Coufalik, A.H. and Monder, C. (1981) Stimulation of gluconeogenesis by cortisol in fetal rat liver in organ culture. Endocrinology, 108: 1132–1137.

Cronstein, B.N., Kimmel, S.C., Levin, R.I., Martiniuk, F. and Weissmann, G. (1992) A mechanism for the antiinflammatory effects of corticosteroids: the glucocorticoid receptor regulates leukocyte adhesion to endothelial cells and expression of endothelial-leukocyte adhesion molecule 1 and intercellular adhesion molecule 1. Proc. Natl. Acad. Sci. USA, 89: 9991–9995.

Cronstein, B.N. and Weissmann, G. (1995) Targets for antiinflammatory drugs. Annu. Rev. Pharmacol. Toxicol., 35: 449–462.

Cuzzocrea, S., Riley, D.P., Caputi, A.P. and Salvemini, D. (2001) Antioxidant therapy: a new pharmacological approach in shock, inflammation, and ischemia/reperfusion injury. Pharmacol. Rev., 53: 135–159.

Dale, D.C. and Petersdorf, R.G. (1973) Corticosteroids and infectious diseases. Med. Clin. North Am., 57: 1277–1287.

de Kloet, E.R. (1995) Steroids, stability and stress. Front Neuroendocrinol., 16: 416–425.

de Kloet, E.R., Van Acker, S.A., Sibug, R.M., Oitzl, M.S., Meijer, O.C., Rahmouni, K. and de Jong, W. (2000) Brain mineralocorticoid receptors and centrally regulated functions. Kidney Int., 57: 1329–1336.

De Martino, M.U., Bhattachryya, N., Alesci, S., Ichijo, T., Chrousos, G.P. and Kino, T. (2004) The glucocorticoid receptor (GR) and the orphan nuclear receptor chicken ovalbumin upstream promoter-transcription factor II (COUP-TFII) interact with and mutually affect each other's transcriptional activities: implications for intermediary metabolism. Mol. Endocrinol., 18: 820–833.

Didonato, J.A., Saatcioglu, F. and Karin, M. (1996) Molecular mechanisms of immunosuppression and anti-inflammatory activities by glucocorticoids. Am. J. Respir. Crit. Care Med., 154: S11–S15.

Dimitriadis, G., Leighton, B., Parry-Billings, M., Sasson, S., Young, M., Krause, U., Bevan, S., Piva, T., Wegener, G. and Newsholme, E.A. (1997) Effects of glucocorticoid excess on the sensitivity of glucose transport and metabolism to insulin in rat skeletal muscle. Biochem. J., 321(Pt 3): 707–712.

Doppler, W., Windegger, M., Soratroi, C., Tomasi, J., Lechner, J., Rusconi, S., Cato, A.C., Almlof, T., Liden, J., Okret, S., Gustafsson, J.A., Richard-Foy, H., Starr, D.B., Klocker, H., Edwards, D. and Geymayer, S. (2001) Expression level-dependent contribution of glucocorticoid receptor domains for functional interaction with STAT5. Mol. Cell Biol., 21: 3266–3279.

Durant, S., Duval, D. and Homo-Delarche, F. (1986) Factors involved in the control of fibroblast proliferation by glucocorticoids: a review. Endocr. Rev., 7: 254–269.

Elenkov, I.J., Chrousos, G.P. and Wilder, R.L. (2000a) Neuroendocrine regulation of IL-12 and TNF-alpha/IL-10 balance. Clinical implications. Ann. NY Acad. Sci., 917: 94–105.

Elenkov, I.J., Wilder, R.L., Chrousos, G.P. and Vizi, E.S. (2000b) The sympathetic nerve – an integrative interface between two supersystems: the brain and the immune system. Pharmacol. Rev., 52: 595–638.

Falkenstein, E., Tillmann, H.C., Christ, M., Feuring, M. and Wehling, M. (2000) Multiple actions of steroid hormones – a focus on rapid, nongenomic effects. Pharmacol. Rev., 52: 513–556.

Fauci, A.S., Dale, D.C. and Balow, J.E. (1976) Glucocorticosteroid therapy: mechanisms of action and clinical considerations. Ann. Intern. Med., 84: 304–315.

Fry, C.J. and Peterson, C.L. (2001) Chromatin remodeling enzymes: who's on first? Curr. Biol., 11: R185–R197.

Fuchs, E., Flugge, G., Ohl, F., Lucassen, P., Vollmann-Honsdorf, G.K. and Michaelis, T. (2001) Psychosocial stress, glucocorticoids, and structural alterations in the tree shrew hippocampus. Physiol. Behav., 73: 285–291.

Galon, J., Franchimont, D., Hiroi, N., Frey, G., Boettner, A., Ehrhart-Bornstein, M., O'Shea, J.J., Chrousos, G.P. and Bornstein, S.R. (2002) Gene profiling reveals unknown enhancing and suppressive actions of glucocorticoids on immune cells. FASEB J., 16: 61–71.

Giguere, V., Hollenberg, S.M., Rosenfeld, M.G. and Evans, R.M. (1986) Functional domains of the human glucocorticoid receptor. Cell, 46: 645–652.

Glass, C.K. and Rosenfeld, M.G. (2000) The coregulator exchange in transcriptional functions of nuclear receptors. Genes Dev., 14: 121–141.

Goodman, R.H. and Smolik, S. (2000) CBP/p300 in cell growth, transformation, and development. Genes Dev., 14: 1553–1577.

Greene, R.M. and Kochhar, D.M. (1975) Some aspects of corticosteroid-induced cleft palate: a review. Teratology, 11: 47–55.

Grimley, P.M., Dong, F. and Rui, H. (1999) Stat5a and Stat5b: fraternal twins of signal transduction and transcriptional activation. Cytokine Growth Factor Rev., 10: 131–157.

Groner, B., Fritsche, M., Stocklin, E., Berchtold, S., Merkle, C., Moriggl, R. and Pfitzner, E. (2000) Regulation of the transactivation potential of STAT5 through its DNA-binding activity and interactions with heterologous transcription factors. Growth Horm. IGF Res., 10: S15–S20.

Gu, W. and Roeder, R.G. (1997) Activation of p53 sequence-specific DNA binding by acetylation of the p53 C-terminal domain. Cell, 90: 595–606.

Haigh, R.M. and Jones, C.T. (1990) Effect of glucocorticoids on alpha 1-adrenergic receptor binding in rat vascular smooth muscle. J. Mol. Endocrinol., 5: 41–48.

Hajduch, E., Hainault, I., Meunier, C., Jardel, C., Hainque, B., Guerre-Millo, M. and Lavau, M. (1995) Regulation of glucose transporters in cultured rat adipocytes: synergistic effect of insulin and dexamethasone on GLUT4 gene expression through promoter activation. Endocrinology, 136: 4782–4789.

Handa, M., Kondo, K., Suzuki, H. and Saruta, T. (1984) Dexamethasone hypertension in rats: role of prostaglandins and pressor sensitivity to norepinephrine. Hypertension, 6: 236–241.

Heery, D.M., Kalkhoven, E., Hoare, S. and Parker, M.G. (1997) A signature motif in transcriptional co-activators mediates binding to nuclear receptors. Nature, 387: 733–736.

Henriksson, A., Almlof, T., Ford, J., McEwan, I.J., Gustafsson, J.A. and Wright, A.P. (1997) Role of the Ada adaptor complex in gene activation by the glucocorticoid receptor. Mol. Cell. Biol., 17: 3065–3073.

Herrlich, P. (2001) Cross-talk between glucocorticoid receptor and AP-1. Oncogene, 20: 2465–2475.

Herschman, H.R., Xie, W. and Reddy, S. (1995) Inflammation, reproduction, cancer and all that.... The regulation and role of the inducible prostaglandin synthase. Bioessays, 17: 1031–1037.

Herzig, S., Long, F., Jhala, U.S., Hedrick, S., Quinn, R., Bauer, A., Rudolph, D., Schutz, G., Yoon, C., Puigserver, P., Spiegelman, B. and Montminy, M. (2001) CREB regulates hepatic gluconeogenesis through the coactivator PGC-1. Nature, 413: 179–183.

Hittelman, A.B., Burakov, D., Iniguez-Lluhi, J.A., Freedman, L.P. and Garabedian, M.J. (1999) Differential regulation of glucocorticoid receptor transcriptional activation via AF-1-associated proteins. EMBO J., 18: 5380–5388.

Hollenberg, S.M., Weinberger, C., Ong, E.S., Cerelli, G., Oro, A., Lebo, R., Thompson, E.B., Rosenfeld, M.G. and Evans, R.M. (1985) Primary structure and expression of a functional human glucocorticoid receptor cDNA. Nature, 318: 635–641.

Holsboer, F. and Barden, N. (1996) Antidepressants and hypothalamic–pituitary–adrenocortical regulation. Endocr. Rev., 17: 187–205.

Ihle, J.N. (2001) The Stat family in cytokine signaling. Curr. Opin. Cell Biol., 13: 211–217.

Ito, K., Barnes, P.J. and Adcock, I.M. (2000) Glucocorticoid receptor recruitment of histone deacetylase 2 inhibits interleukin-1beta-induced histone H4 acetylation on lysines 8 and 12. Mol. Cell. Biol., 20: 6891–6903.

Jahnsen, F.L., Haye, R., Gran, E., Brandtzaeg, P. and Johansen, F.E. (1999) Glucocorticosteroids inhibit mRNA expression for eotaxin, eotaxin-2, and monocyte-chemotactic protein-4 in human airway inflammation with eosinophilia. J. Immunol., 163: 1545–1551.

Johns, D.G., Dorrance, A.M., Tramontini, N.L. and Webb, R.C. (2001) Glucocorticoids inhibit tetrahydrobiopterin-dependent endothelial function. Exp. Biol. Med. (Maywood), 226: 27–31.

Jung-Testas, I. and Baulieu, E.E. (1985) Effects of steroid hormones and antihormones in cultured cells. Exp. Clin. Endocrinol., 86: 151–164.

Kamei, Y., Xu, L., Heinzel, T., Torchia, J., Kurokawa, R., Gloss, B., Lin, S.C., Heyman, R.A., Rose, D.W., Glass, C.K. and Rosenfeld, M.G. (1996) A CBP integrator complex mediates transcriptional activation and AP-1 inhibition by nuclear receptors. Cell, 85: 403–414.

Karin, M. and Chang, L. (2001) AP-1-glucocorticoid receptor crosstalk taken to a higher level. J. Endocrinol., 169: 447–451.

Karin, M., Haslinger, A., Holtgreve, H., Richards, R.I., Krauter, P., Westphal, H.M. and Beato, M. (1984) Characterization of DNA sequences through which cadmium and glucocorticoid hormones induce human metallothionein-IIA gene. Nature, 308: 513–519.

Kass, E.H. and Finland, M. (1958) Corticosteroids and infections. Adv. Intern. Med., 9: 45–80.

Kato, H., Hayashi, T., Koshino, Y., Kutsumi, Y., Nakai, T. and Miyabo, S. (1992) Glucocorticoids increase Ca2+ influx through dihydropyridine-sensitive channels linked to activation of protein kinase C in vascular smooth muscle cells. Biochem. Biophys. Res. Commun., 188: 934–941.

Kelly, J.J., Tam, S.H., Williamson, P.M., Lawson, J. and Whitworth, J.A. (1998) The nitric oxide system and cortisol-induced hypertension in humans. Clin. Exp. Pharmacol. Physiol., 25: 945–946.

Kimata, H., Lindley, I. and Furusho, K. (1995) Effect of hydrocortisone on spontaneous IgE and IgG4 production in atopic patients. J. Immunol., 154: 3557–3566.

Kino, T. and Chrousos, G.P. (2002) Tissue-specific glucocorticoid resistance-hypersensitivity syndromes: multifactorial states of clinical importance. J. Allergy Clin. Immunol., 109: 609–613.

Kino, T., De Martino, M.U., Charmandari, E., Mirani, M. and Chrousos, G.P. (2003) Tissue glucocorticoid resistance/hypersensitivity syndromes. J. Steroid Biochem. Mol. Biol., 85: 457–467.

Kino, T., Vottero, A. and Chrousos, G.P. (2001) Mineralocorticoid and glucocorticoid receptors In: Burris, T.P. and McCabe, E.R.B. (Eds.), Nuclear Receptors and Genetic Disease. Academic Press, San Diego, pp. 297–308.

Lanz, R.B., McKenna, N.J., Onate, S.A., Albrecht, U., Wong, J., Tsai, S.Y., Tsai, M.J. and O'Malley, B.W. (1999) A steroid receptor coactivator, SRA, functions as an RNA and is present in an SRC-1 complex. Cell, 97: 17–27.

Lee, A.L., Ogle, W.O. and Sapolsky, R.M. (2002) Stress and depression: possible links to neuron death in the hippocampus. Bipolar Disord., 4: 117–128.

Leo, C. and Chen, J.D. (2000) The SRC family of nuclear receptor coactivators. Gene, 245: 1–11.

Leonard, W.J. and O'Shea, J.J. (1998) Jaks and STATs: biological implications. Annu. Rev. Immunol., 16: 293–322.

Liden, J., Delaunay, F., Rafter, I., Gustafsson, J. and Okret, S. (1997) A new function for the C-terminal zinc finger of the glucocorticoid receptor. Repression of RelA transactivation. J. Biol. Chem., 272: 21467–21472.

Lieberman, B.A., Bona, B.J., Edwards, D.P. and Nordeen, S.K. (1993) The constitution of a progesterone response element. Mol. Endocrinol., 7: 515–527.

Liu, Y., Cousin, J.M., Hughes, J., Van Damme, J., Seckl, J.R., Haslett, C., Dransfield, I., Savill, J. and Rossi, A.G. (1999) Glucocorticoids promote nonphlogistic phagocytosis of apoptotic leukocytes. J. Immunol., 162: 3639–3646.

Ma, H., Hong, H., Huang, S.M., Irvine, R.A., Webb, P., Kushner, P.J., Coetzee, G.A. and Stallcup, M.R. (1999) Multiple signal input and output domains of the 160-kilodalton nuclear receptor coactivator proteins. Mol. Cell. Biol., 19: 6164–6173.

Mangelsdorf, D.J., Thummel, C., Beato, M., Herrlich, P., Schutz, G., Umesono, K., Blumberg, B., Kastner, P., Mark, M., Chambon, P. et al. (1995) The nuclear receptor superfamily: the second decade. Cell, 83: 835–839.

McEwen, B.S., Conrad, C.D., Kuroda, Y., Frankfurt, M., Magarinos, A.M. and McKittrick, C. (1997) Prevention of stress-induced morphological and cognitive consequences. Eur. Neuropsychopharmacol., 7(Suppl. 3): S323–S328.

McKay, L.I. and Cidlowski, J.A. (1998) Cross-talk between nuclear factor-kappa B and the steroid hormone receptors: mechanisms of mutual antagonism. Mol. Endocrinol., 12: 45–56.

McKay, L.I. and Cidlowski, J.A. (1999) Molecular control of immune/inflammatory responses: interactions between nuclear factor-kappa B and steroid receptor-signaling pathways. Endocr. Rev., 20: 435–459.

McKenna, N.J., Lanz, R.B. and O'Malley, B.W. (1999) Nuclear receptor coregulators: cellular and molecular biology. Endocr. Rev., 20: 321–344.

Meagher, L.C., Cousin, J.M., Seckl, J.R. and Haslett, C. (1996) Opposing effects of glucocorticoids on the rate of apoptosis in neutrophilic and eosinophilic granulocytes. J. Immunol., 156: 4422–4428.

Miller, W.L. and Chrousos, G.P. (2001) The adrenal cortex. In: Felig, P. and Frohman, L.A. (Eds.), Endocrinology and Metabolism. McGraw-Hill, New York, pp. 387–524.

Miner, J.N. and Yamamoto, K.R. (1991) Regulatory crosstalk at composite response elements. Trends Biochem. Sci., 16: 423–426.

Moon, Y.S., Smas, C.M., Lee, K., Villena, J.A., Kim, K.H., Yun, E.J. and Sul, H.S. (2002) Mice lacking paternally expressed Pref-1/Dlk1 display growth retardation and accelerated adiposity. Mol. Cell. Biol., 22: 5585–5592.

Nacht, S. and Garzon, P. (1974) Effects of corticosteroids on connective tissue and fibroblasts. Adv. Steroid. Biochem. Pharmacol., 4: 157–187.

Nakagawa, M., Bondy, G.P., Waisman, D., Minshall, D., Hogg, J.C. and van Eeden, S.F. (1999) The effect of glucocorticoids on the expression of L-selectin on polymorphonuclear leukocyte. Blood, 93: 2730–2737.

Nissen, R.M. and Yamamoto, K.R. (2000) The glucocorticoid receptor inhibits NFkappaB by interfering with serine-2 phosphorylation of the RNA polymerase II carboxy-terminal domain. Genes Dev., 14: 2314–2329.

O'Sullivan, S. (1999) On the role of PGD2 metabolites as markers of mast cell activation in asthma. Acta Physiol. Scand. Suppl., 644: 1–74.

Oikarinen, A. (1992) Dermal connective tissue modulated by pharmacologic agents. Int. J. Dermatol., 31: 149–156.

Oikarinen, A. and Autio, P. (1991) New aspects of the mechanism of corticosteroid-induced dermal atrophy. Clin. Exp. Dermatol., 16: 416–419.

Olefsky, J.M. (1975) Effect of dexamethasone on insulin binding, glucose transport, and glucose oxidation of isolated rat adipocytes. J. Clin. Invest., 56: 1499–1508.

Orth, D.N., Kovacs, W.J. and DeBold, C.R. (1998) The adrenal cortex. In: Wilson, J.D. and Foster, D.W. (Eds.), Textbook of Endocrinology. W.B. Saunders Co., Philadelphia., pp. 517–664

Parrillo, J.E. and Fauci, A.S. (1979) Mechanisms of glucocorticoid action on immune processes. Annu. Rev. Pharmacol. Toxicol., 19: 179–201.

Perkins, N.D. (2000) The Rel/NF-kappa B family: friend and foe. Trends Biochem. Sci., 25: 434–440.

Piemonti, L., Monti, P., Allavena, P., Sironi, M., Soldini, L., Leone, B.E., Socci, C. and Di Carlo, V. (1999) Glucocorticoids affect human dendritic cell differentiation and maturation. J. Immunol., 162: 6473–6481.

Rachez, C. and Freedman, L.P. (2001) Mediator complexes and transcription. Curr. Opin. Cell Biol., 13: 274–280.

Raisz, L.G. and Bingham, P.J. (1972) Effect of hormones on bone development. Annu. Rev. Pharmacol., 12: 337–352.

Reichardt, H.M., Kaestner, K.H., Tuckermann, J., Kretz, O., Wessely, O., Bock, R., Gass, P., Schmid, W., Herrlich, P., Angel, P. and Schutz, G. (1998) DNA binding of the glucocorticoid receptor is not essential for survival. Cell, 93, 531–541.

Reichardt, H.M., Tuckermann, J.P., Gottlicher, M., Vujic, M., Weih, F., Angel, P., Herrlich, P. and Schutz, G. (2001) Repression of inflammatory responses in the absence of DNA binding by the glucocorticoid receptor. EMBO J., 20: 7168–7173.

Reid, I.R. (2000) Glucocorticoid-induced osteoporosis. Baillieres Best Pract. Res. Clin. Endocrinol. Metab., 14: 279–298.

Reul, J.M., Gesing, A., Droste, S., Stec, I.S., Weber, A., Bachmann, C., Bilang-Bleuel, A., Holsboer, F. and Linthorst, A.C. (2000) The brain mineralocorticoid receptor: greedy for ligand, mysterious in function. Eur. J. Pharmacol., 405: 235–249.

Rooyackers, O.E. and Nair, K.S. (1997) Hormonal regulation of human muscle protein metabolism. Annu. Rev. Nutr., 17: 457–485.

Russo, D., Fraser, R. and Kenyon, C.J. (1990) Increased sensitivity to noradrenaline in glucocorticoid-treated rats: the effects of indomethacin and desipramine. J. Hypertens., 8: 827–833.

Sakoda, H., Ogihara, T., Anai, M., Funaki, M., Inukai, K., Katagiri, H., Fukushima, Y., Onishi, Y., Ono, H., Fujishiro, M., Kikuchi, M., Oka, Y. and Asano, T. (2000) Dexamethasone-induced insulin resistance in 3T3-L1 adipocytes is due to inhibition of glucose transport rather than insulin signal transduction. Diabetes, 49: 1700–1708.

Sapolsky, R.M. (2000) Glucocorticoids and hippocampal atrophy in neuropsychiatric disorders. Arch. Gen. Psychiatry, 57: 925–935.

Sato, A., Suzuki, H., Murakami, M., Nakazato, Y., Iwaita, Y. and Saruta, T. (1994) Glucocorticoid increases angiotensin II type 1 receptor and its gene expression. Hypertension, 23: 25–30.

Schobitz, B., De Kloet, E.R. and Holsboer, F. (1994) Gene expression and function of interleukin 1, interleukin 6 and tumor necrosis factor in the brain. Prog. Neurobiol., 44: 397–432.

Schule, R., Rangarajan, P., Kliewer, S., Ransone, L.J., Bolado, J., Yang, N., Verma, I.M. and Evans, R.M. (1990) Functional antagonism between oncoprotein c-Jun and the glucocorticoid receptor. Cell, 62: 1217–1226.

Seene, T. (1994) Turnover of skeletal muscle contractile proteins in glucocorticoid myopathy. J. Steroid. Biochem. Mol. Biol., 50: 1–4.

Smas, C.M. and Sul, H.S. (1993) Pref-1, a protein containing EGF-like repeats, inhibits adipocyte differentiation. Cell, 73: 725–734.

Smas, C.M., Chen, L., Zhao, L., Latasa, M.J. and Sul, H.S. (1999) Transcriptional repression of pref-1 by glucocorticoids promotes 3T3-L1 adipocyte differentiation. J. Biol. Chem., 274: 12632–12641.

Stalmans, W. and Laloux, M. (1979) Glucocorticoids and hepatic glycogen metabolism. Monogr. Endocrinol., 12: 517–533.

Stocklin, E., Wissler, M., Gouilleux, F. and Groner, B. (1996) Functional interactions between Stat5 and the glucocorticoid receptor. Nature, 383: 726–728.

Stoecklin, E., Wissler, M., Moriggl, R. and Groner, B. (1997) Specific DNA binding of Stat5, but not of glucocorticoid receptor, is required for their functional cooperation in the regulation of gene transcription. Mol. Cell Biol., 17: 6708–6716.

Takeda, T., Kurachi, H., Yamamoto, T., Nishio, Y., Nakatsuji, Y., Morishige, K., Miyake, A. and Murata, Y. (1998) Crosstalk between the interleukin-6 (IL-6)-JAK-STAT and the glucocorticoid-nuclear receptor pathway: synergistic activation of IL-6 response element by IL-6 and glucocorticoid. J. Endocrinol., 159, 323–330.

Tanaka, S., Sohen, S. and Fukuda, K. (1997) A role for histamine receptors in rheumatoid arthritis. Semin. Arthritis Rheum.,26: 824–833.

Ullian, M.E. (1999) The role of corticosteriods in the regulation of vascular tone. Cardiovasc. Res., 41: 55–64.

van Dam, H. and Castellazzi, M. (2001) Distinct roles of Jun: Fos and Jun:ATF dimers in oncogenesis. Oncogene, 20: 2453–2464.

van de Werve, G., Lange, A., Newgard, C., Mechin, M.C., Li, Y. and Berteloot, A. (2000) New lessons in the regulation of glucose metabolism taught by the glucose 6-phosphatase system. Eur. J. Biochem., 267: 1533–1549.

Vardimon, L., Ben-Dror, I., Avisar, N., Oren, A. and Shiftan, L. (1999) Glucocorticoid control of glial gene expression. J. Neurobiol., 40: 513–527.

Venarske, D. and deShazo, R.D. (2003) Molecular mechanisms of allergic disease. South Med. J., 96: 1049–1054.

Wallberg, A.E., Neely, K.E., Hassan, A.H., Gustafsson, J.A., Workman, J.L. and Wright, A.P. (2000) Recruitment of the SWI-SNF chromatin remodeling complex as a mechanism of gene activation by the glucocorticoid receptor tau1 activation domain. Mol. Cell. Biol., 20: 2004–2013.

Wallerath, T., Witte, K., Schafer, S.C., Schwarz, P.M., Prellwitz, W., Wohlfart, P., Kleinert, H., Lehr, H.A., Lemmer, B. and Forstermann, U. (1999) Down-regulation of the expression of endothelial NO synthase is likely to contribute to glucocorticoid-mediated hypertension. Proc. Natl. Acad. Sci. USA, 96: 13357–13362.

Wang, J.C., Stromstedt, P.E., Sugiyama, T. and Granner, D.K. (1999) The phosphoenolpyruvate carboxykinase gene glucocorticoid response unit: identification of the functional domains of accessory factors HNF3 beta (hepatic nuclear factor-3 beta) and HNF4 and the necessity of proper alignment of their cognate binding sites. Mol. Endocrinol., 13: 604–618.

Webb, P., Nguyen, P., Shinsako, J., Anderson, C., Feng, W., Nguyen, M.P., Chen, D., Huang, S.M., Subramanian, S., McKinerney, E., Katzenellenbogen, B.S., Stallcup, M.R. and Kushner, P.J. (1998) Estrogen receptor activation function 1 works by binding p160 coactivator proteins. Mol. Endocrinol., 12: 1605–1618.

Weinstein, S.P., Paquin, T., Pritsker, A. and Haber, R.S. (1995) Glucocorticoid-induced insulin resistance: deiegelman BAxamethasone inhibits the activation of glucose transport in rat skeletal muscle by both insulin- and non-insulin-related stimuli. Diabetes, 44: 441–445.

Wilckens, T. and De Rijk, R. (1997) Glucocorticoids and immune function: unknown dimensions and new frontiers. Immunol. Today, 18: 418–424.

Wissink, S., van Heerde, E.C., Schmitz, M.L., Kalkhoven, E., van der Burg, B., Baeuerle, P.A. and van der Saag, P.T. (1997) Distinct domains of the RelA NF-kappaB subunit are required for negative cross-talk and direct interaction with the glucocorticoid receptor. J. Biol. Chem., 272: 22278–22284.

Wu, C.Y., Sarfati, M., Heusser, C., Fournier, S., Rubio-Trujillo, M., Peleman, R. and Delespesse, G. (1991) Glucocorticoids increase the synthesis of immunoglobulin E by interleukin 4-stimulated human lymphocytes. J. Clin. Invest., 87: 870–877.

Yang, X.J., Ogryzko, V.V., Nishikawa, J., Howard, B.H. and Nakatani, Y. (1996) A p300/CBP-associated factor that competes with the adenoviral oncoprotein E1A. Nature, 382: 319–324.

Yoon, J.C., Puigserver, P., Chen, G., Donovan, J., Wu, Z., Rhee, J., Adelmant, G., Stafford, J., Kahn, C.R., Granner, D.K., Newgard, C.B. and Spiegelman, B.M. (2001) Control of hepatic gluconeogenesis through the transcriptional coactivator PGC-1. Nature, 413: 131–138.

Yoshinaga, S.K., Peterson, C.L., Herskowitz, I. and Yamamoto, K.R. (1992) Roles of SWI1, SWI2, and SWI3 proteins for transcriptional enhancement by steroid receptors. Science, 258: 1598–1604.

Zhang, Z., Jones, S., Hagood, J.S., Fuentes, N.L. and Fuller, G.M. (1997) STAT3 acts as a co-activator of glucocorticoid receptor signaling. J. Biol. Chem., 272: 30607–30610.

T. Steckler, N.H. Kalin and J.M.H.M. Reul (Eds.)
Handbook of Stress and the Brain, Vol. 15
ISBN 0-444-51173-3

CHAPTER 3.3

The role of 11β-hydroxysteroid dehydrogenases in the regulation of corticosteroid activity in the brain

Jonathan R. Seckl*, Joyce L.W. Yau and Megan C. Holmes

Endocrinology Unit, Molecular Medicine Centre, Edinburgh University, Western General Hospital, Edinburgh EH4 2XU, UK

Abstract: Glucocorticoids influence a broad range of central nervous system (CNS) processes, altering neurotransmission, electrophysiological activity, metabolism, cell division and survival. These actions are mediated by nuclear glucocorticoid and mineralocorticoid receptors, which regulate transcription of target genes. The amount of steroid available to activate these receptors is not only dependent on circulating hormone levels but also, crucially, on intracellular, pre-receptor metabolism of glucocorticoids by 11β-hydroxysteroid dehydrogenases (11β-HSDs). There are two distinct 11β-HSD isozymes, the products of distantly related genes. 11β-HSD type 2 is a high-affinity dehydrogenase that rapidly inactivates glucocorticoids to inert 11-keto-derivatives, thus 'protecting' co-localised receptors from the ligand. In contrast, in many intact cells and whole tissues, 11β-HSD type 1 catalyses the reverse reaction, regenerating active glucocorticoids from circulating inert forms, locally amplifying glucocorticoid actions. This review highlights recent results that have illuminated the important and very different roles that the two isozymes play in the brain.

Introduction: glucocorticoids and the CNS

Glucocorticoids (cortisol and corticosterone in most mammals, corticosterone alone in rats and mice) have a plethora of biological effects, influencing most systems of the body including, of course, the brain. Very many neural and glial systems are modified by glucocorticoids and their target genes include a host of neurotransmitter synthetic and degradative enzymes, receptors and ion channels (e.g. NMDA and serotonin receptor subtypes, K^+ channels), cytoskeletal proteins, enzymes involved in calcium activation and key enzymes underpinning neuronal and glial metabolism. Acute elevations of glucocorticoids are adaptive in response to diurnal activation cues or stress, facilitating survival pathways, such as fear conditioning, and inhibiting immediately unnecessary or unhelpful processes, such as inflammation. In contrast, chronically high levels of glucocorticoids (e.g. in Cushing's syndrome) are deleterious to homeostasis at all stages of life from embryonic development to senescence (de Kloet, 1991). In the brain, chronic glucocorticoid excess exerts profound adverse effects, producing neuropsychiatric dysfunction (depression, psychosis), cognitive impairments, structural deterioration and neuro endocrine dysfunction. Such effects are underpinned by altered neuronal metabolism, electrical activity and the pruning of dendritic structure at least in the hippocampus, a region especially sensitive to glucocorticoids (McEwen, 1999).

*Corresponding author. Tel.: +144-131-651-1035; Fax: +144-131-651-1085; E-mail: J.Seckl@ed.ac.uk

The HPA axis

It is obviously crucial that normally glucocorticoid levels are strictly controlled and this occurs primarily by their negative feedback upon the hypothalamic–pituitary–adrenal (HPA) axis. This is achieved, as are glucocorticoid actions throughout the central nervous system (CNS) and the periphery, by binding to intracellular receptors which are ligand-activated members of the nuclear hormone receptor superfamily of transcription factors (Glass, 1994). For glucocorticoids there are two types, lower-affinity, widely distributed glucocorticoid receptors (GR) and the higher-affinity, more narrowly expressed mineralocorticoid receptor (MR) (de Kloet, 1991). When activated by binding ligand, both translocate from cytoplasm to the nucleus and, either directly or indirectly with other transcription factors, bind to specific DNA sequences in the promoter regions of target genes. In the CNS, gene-profiling studies have suggested that the genes regulated by GR and MR are largely distinct (Vreugdenhil and deKloet, 1998).

Tissue actions of glucocorticoids

It has been a long-held dogma in endocrinology that glucocorticoid actions are determined by the concentration of active steroid hormones in the circulation (modulated by hormone binding to plasma proteins, notably corticosteroid-binding globulin; CBG) and by the density of intracellular GR and MR in target tissues. However, while such a system may be sufficient for some purposes, corticosteroids are lipophilic and access most cells readily and their nuclear receptors, especially GR, are very widely distributed indeed, rather limiting the differential signalling possibilities of this system in different tissues. And there, until recently, it all lay, with little interest in any physiological importance of metabolism of steroids in the CNS and elsewhere by a complexity of arcane tissue enzymes. Recently, however, it has become apparent that the enzymic pre-receptor metabolism of glucocorticoids by one enzyme complex at least, 11β-hydroxysteroid dehydrogenase (11β-HSD), also potently regulates corticosteroid access to receptors within specific tissues, and thus physiology. Here we review the emerging biology of these fascinating enzymes and highlight their roles in the brain.

11β-Hydroxysteroid dehydrogenases

11β-Hydroxysteroid dehydrogenases (11β-HSDs) were first identified over 50 years ago (Amelung et al., 1953). They are microsomal enzymes that catalyse the interconversion of active 11-hydroxy glucocorticoids (cortisol, corticosterone) and their inert 11-keto forms (cortisone, 11-dehydrocorticosterone, 11-DHC) which do not bind to GR or MR. There are two isozymes of 11β-HSD which are the products of distinct genes and differ considerably in tissue distribution and function (Stewart and Krozowski, 1999).

11β-HSD type 2

11β-HSD type 2, though more recently characterised and cloned (Brown et al., 1993; Albiston et al., 1994), was the first isozyme to have a biological function ascribed to it. 11β-HSD2 is highly expressed in classical aldosterone-selective target tissues (distal nephron, colon, sweat gland) where it acts as a potent NAD-dependent, 11β-dehydrogenase, rapidly inactivating glucocorticoids to their inert keto forms. 11β-HSD2 action allows selective access of aldosterone (a non-substrate) to intrinsically non-selective MR in vivo, in the face of a substantial excess of glucocorticoids in the circulation. Humans (Mune et al., 1995) or mice (Kotelevtsev et al., 1999) homozygous for deleterious mutations of the 1β-HSD2 gene exhibit the syndrome of apparent mineralocorticoid excess in which cortisol and corticosterone illicitly occupy renal MR causing sodium retention, severe hypertension and hypokalaemia. An analogous syndrome is produced by liquorice, the active component glycyrrhetinic acid and its hemisuccinate derivative carbenoxolone (CBX), all of which potently inhibit 11β-HSDs with a nanomolar Ki in vitro (Stewart et al., 1987, 1990). By analogy, any 11β-HSD2 in the CNS might be anticipated to relate to putative aldosterone-selective central effects.

11β-HSD type 1

In contrast to 11β-HSD2, the function of original 11β-HSD type 1 isozyme, though purified and characterised from rat liver almost two decades ago by Monder and his colleagues (Lakshmi and Monder, 1988; Agarwal et al., 1989), has been much less readily discerned (reviewed by Seckl and Walker, 2001). However, it is the 11β-HSD1 isoform that is widely expressed in the adult CNS (Moisan et al., 1990c). 11β-HSD1 expression is found in both neurons and glia. High levels occur in the hippocampus, cerebellum and neocortex (Moisan et al., 1990c; Lakshmi et al., 1991), suggesting that it plays a role in modulating the effects of glucocorticoids on mood, learning and memory. Lower 11β-HSD1 expression is found in the hypothalamus and anterior pituitary, indicating a potential role in neuroendocrine control (Moisan et al., 1990c; Sakai et al., 1992).

The key to understanding the functions of 11β-HSD1 in the brain and most, if not all, other tissues came from studies of its reaction direction in *intact* cells cultured in vitro. Although 11β-HSD1 shows bi-directional activity in tissue homogenates and purified microsomal preparations (Lakshmi and Monder, 1988), it acts predominantly as a *reductase* in intact cells transfected with 11β-HSD1 cDNA. Such cells convert 11-dehydrocorticosterone in the medium to corticosterone and not vice versa (Duperrex et al., 1993; Low et al., 1994a). This is not merely due to differential access of 11-hydroxy and 11-keto steroids through the cell membrane, since both steroids can activate reporter genes driven by GR or MR in cells transfected with 11β-HSD1 cDNA, but only corticosterone activates transcription in the same cell line lacking 11β-HSD1 (Low et al., 1994a). This is also not merely an effect confined to clonal cells, since intact primary cultures of hepatocytes (Jamieson et al., 1995), lung cells (Hundertmark et al., 1995) and adipocytes (Bujalska et al., 1997; Napolitano et al., 1998) also show predominant 11β-reduction of inert 11-keto steroids to active forms, though the same cells as soon as they are homogenised readily exhibit fully bi-directional activity. This suggests that it is the subcellular context that is likely to determine 11β-HSD1 reaction direction in intact cells; presumably the environment in the inner leaflet of the endoplasmic reticulum, where the enzyme is presumed to lie (Ozols, 1995), favours 11β-reduction. Certainly, there are several potent NADPH-generating enzymes located there which are assumed to provide ample co-substrate for reduction by 11β-HSD1. This 11β-reductase reaction direction, far from inactivating glucocorticoids, regenerates active glucocorticoids in target cells from the substantial levels of inert 11-keto steroids in the blood (Walker et al., 1992). Circulating 11-ketosteroids are produced largely by the actions of renal 11β-HSD2 (Fig. 1).

Recent data suggest that 11β-reductase activity of 11β-HSD1 also predominates in intact organs and in vivo. Thus the intact liver perfused in situ converts ~50% of 11-dehydrocorticosterone to active corticosterone on a single pass, whereas dehydrogenation of corticosterone is 4–5 times less (Jamieson et al., 2000). Moreover, the liver can activate large amounts of 11-dehydrocorticosterone, suggesting a large turnover of this reaction direction in vivo. In humans, reduction also predominates across the liver (Walker et al., 1992).

To dissect this in vivo, mice homozygous for targeted disruption of the 11β-HSD1 gene have been generated (Kotelevtsev et al., 1997). These animals have clearly shown that 11β-HSD1 is the sole 11β-reductase, at least in mice. Moreover, the reaction appears important since the animals, despite normal or modestly elevated plasma corticosterone

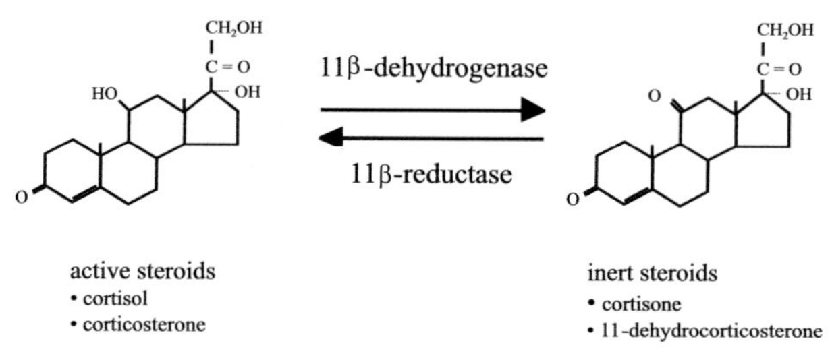

Fig. 1. 11β-Hydroxysteroid dehydrogenase: the glucocorticoid 'shuttle'.

levels, resist the hyperglycaemia associated with stress or high-fat diets. This is due, apparently, to attenuated responses of hepatic glucocorticoid-sensitive gluconeogenic enzymes implying a reduction in glucocorticoid levels within the hepatocyte. This contention is supported by lower levels of other glucocorticoid-sensitive enzymes of hepatic metabolism of lipids and a 'cardioprotective' lipid profile (Morton et al., 2001). In contrast, transgenic mice overexpressing 11β-HSD1 in adipose tissue, at levels chosen to mimic the 2–3-fold elevation of enzyme activity observed in adipose tissue in human and rodent models of obesity/metabolic syndrome, have elevated intra-adipose corticosterone

	11β-HSD1		**11β-HSD2**
purified	1985		1996
cloned	1989		1994
Protein	34kD 287-292 aa	20% homology	44kD 405-427
cDNA	1200 bp	14% homology	1900 bp
Human chr	1		16
Cofactor	NADP(H)		NAD
Affinities (Km)			
corticosterone	1.8 μM		5 nM
cortisol	17 μM		50 nM
cortisone	100 nM		>μM
Reaction	*Reductase*		*Dehydrogenase*
Peripheral sites	Liver Adipose ovary and testis (some species) vasculature		Distal nephron colon Placenta & fetus
CNS sites	**widespread** including:- cerebellum hippocampus cortex brain stem hypothalamus pituitary		**restricted in adult (rat)** subcommissural organ nucl. tractus solitarius (NTS is only site in adult mouse CNS) central nucl. amygdala lateral hypothalamus **widespread in midgestation fetus**

Fig. 2. 11β-HSD isozymes.

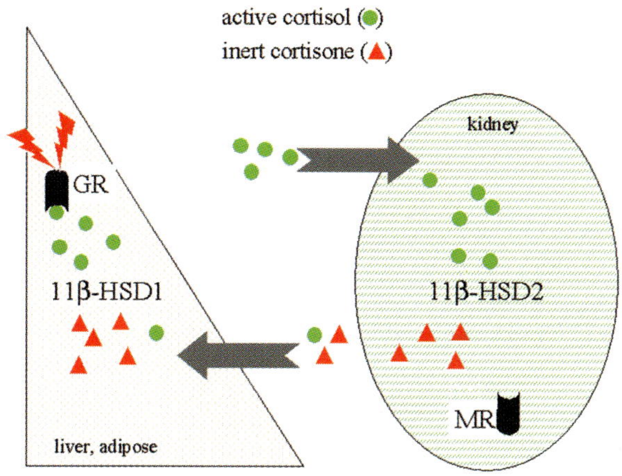

Fig. 3. The peripheral 11β-HSD shuttle. In kidney, 11β-HSD2 acting as a dehydrogenase inactivates cortisol and corticosterone, preventing their illicit occupation of non-selective MR. In the liver, adipose and other tissues, 11β-HSD1, acting as a reductase, takes the inert cortisone (generated in kidney by 11β-HSD2) and regenerates active glucocorticoids, amplifying their action on local GR.

levels, with normal plasma steroid concentrations (Masuzaki et al., 2001). The mice faithfully model the metabolic syndrome and show visceral obesity, hyperglycaemia, insulin resistance, dyslipidaemia and hypertension. Clearly this enzyme can potently modulate glucocorticoid action in target tissues in the periphery. But is this relevant in the CNS?

11β-HSDs in the CNS

Early studies

In the 1960s, a series of studies, using histochemical and biochemical techniques, examined the activity of 11β-HSD in a range of tissues, including the brain. 11-Keto oxidation of steroids was found in mouse, rat, dog and primate whole brain extracts, as well as fetal brain and the C6 glioma cell line (Peterson et al., 1965; Sholiton et al., 1965; Grosser, 1966; Grosser and Axelrod, 1968; Miyabo et al., 1973). Studies of the interconversion of radiolabelled cortisol and cortisone in vivo and in vitro in murine tissues also documented 11β-HSD activity in brain (Burton and Tufnell, 1967), though at lower levels than found in liver and kidney. In contrast, the key studies in the late 1980s, which demonstrated the crucial renal role of 11β-HSD (later shown to be type 2) in preventing glucocorticoids from binding to renal MR in vivo, did not find 11β-HSD activity in the hippocampus (Edwards et al., 1988; Funder et al., 1988). These data were interpreted as demonstrating that the clear non-selectivity of hippocampal MR for corticosteroid ligands in vivo (de Kloet et al., 1975; McEwen et al., 1986a) reflected the absence of 11β-HSD.

However, several studies suggested aldosterone-specific central effects on salt appetite and blood pressure, and in several periventricular regions aldosterone is selectively concentrated in vivo in the face of corticosterone (Yongue and Roy, 1987) and cannot be readily displaced by excess glucocorticoid (McEwen et al., 1986b). Thus, the presence of 11β-HSD in the brain was re-examined.

Studies, a decade ago, clearly demonstrated 11β-HSD activity in homogenates first of rat cerebellum (Moisan et al., 1990b) and then from a broad range of CNS regions, including the hippocampus (Moisan et al., 1990c; Lakshmi et al., 1991), though the high density of MR here are clearly occupied by corticosterone rather than aldosterone in vivo (de Kloet et al., 1975). These studies also showed that 11β-HSD activity is highest in the cerebellum, hippocampus and neocortex, with levels some 10–30% of kidney and liver (Moisan et al., 1990c; Lakshmi et al., 1991). 11β-HSD is also clearly detectable in most other brain subregions, including the hypothalamus, amygdala and brainstem (Moisan et al., 1990c; Lakshmi et al., 1991; Sakai et al., 1992; Seckl et al., 1993). So which isozymes are present and what is their possible function?

11β-HSD2 in the CNS

Although the original studies up until the mid-1990s were performed before the clear realisation that there were two distinct isoforms of 11β-HSD, here we will assume that knowledge and review each isozyme in turn. We begin with 11β-HSD2 because, though the second identified and characterised, it was the first isozyme for which a clear function was determined, namely protection of MR from occupation by glucocorticoids.

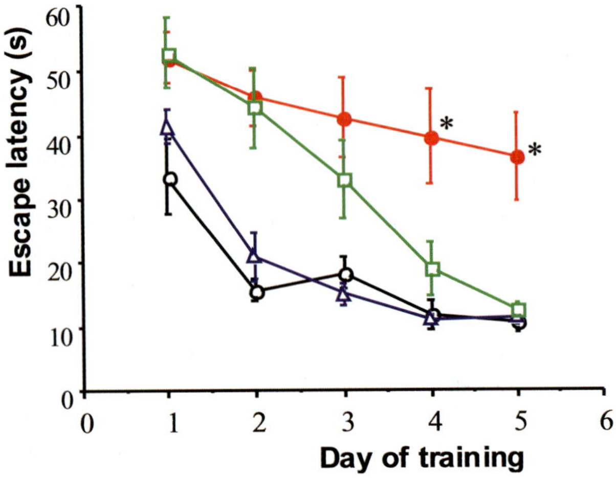

Fig. 4. Impaired learning of aged wild-type, but not 11β-HSD-$1^{-/-}$ mice in the watermaze. Mice were trained for 5 consecutive days (4 trials/day) to find and escape onto a randomly located flagged hidden platform. Young (4 m) wild type (open black circles), young 11β-HSD-$1^{-/-}$ mice (open blue triangle), aged 18–20 m wild-type (closed red circle) and 11β-HSD-$1^{-/-}$ (open green square), all on an isogenic 129/Ola strain background. The aged 11β-HSD-$1^{-/-}$ mice perform, as well as young mice, despite their age and elevated plasma corticosterone levels. * $P < 0.05$ compared to all other groups.

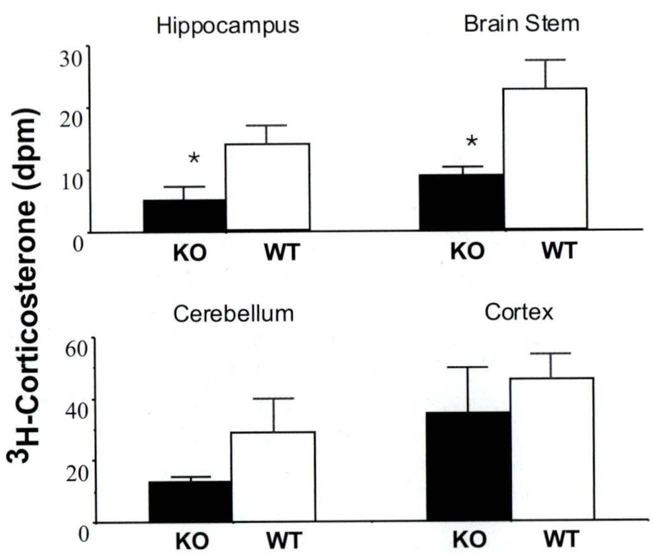

Fig. 5. Intracerebral corticosterone levels are lower in 11β-HSD1 knockout mice. Distribution of chronically infused [^{3}H]-corticosterone in brain tissues of aged wild-type (WT) and 11β-HSD-$1^{-/-}$ (KO) mice.

In vivo, several effects of aldosterone on the brain are specific to this steroid and are not mimicked by corticosterone. The clearest aldosterone-specific actions in the brain relate to the central control of blood pressure and salt appetite (Gomez-Sanchez, 1986; McEwen et al., 1986b; Gomez-Sanchez et al., 1990). Thus, aldosterone given intracerebroventricularly in tiny nanogram doses elevates blood pressure in the rat, whilst the same dose given peripherally is ineffective, implying a specific central action rather than effects via leakage to the periphery. Corticosterone icv does not reproduce this action and indeed antagonises the effect of aldosterone (Gomez-Sanchez et al., 1990). Similarly, aldosterone exerts central actions on salt appetite which cannot be reproduced by a range of doses of corticosterone (McEwen et al., 1986b). Since CBG, another major candidate to confer ligand specificity upon MR, is absent from most brain regions, the presence of 11β-HSD in the CNS is inferred.

Following the cloning of 11β-HSD2, a series of studies examined the distribution of its mRNA in the brain. These showed that 11β-HSD2 is only expressed in a few discrete sites in the adult rat brain (Roland et al., 1995; Robson et al., 1998). These loci also express MR, albeit not in strikingly high density. However, these regions are appropriate to produce the selective central effects of aldosterone to increase blood pressure (Gomez-Sanchez, 1986), including the nucleus tractus solitarius, and salt appetite (Nitabach et al., 1989), notably in the ventromedial hypothalamus, subcommissural organ and central nucleus of the amygdala. When the 11β-HSD inhibitor carbenoxolone is infused icv in rats, at a dose that again has no effect when given peripherally, it increases blood pressure, suggesting that central 11β-HSD2 gates access of glucocorticoids to central receptors regulating blood pressure, likely MR as the effect is lost with MR antagonists (Gomez-Sanchez and Gomez-Sanchez, 1992). Interestingly, in the mouse there is an even more limited expression of 11β-HSD2, with no expression in the forebrain or even the subcommissural organ (Robson and Holmes, unpublished observation), although this is the site of highest expression in the rat. This species difference may perhaps be compatible with the different salt regulation in the mouse, which is less mineralocorticoid dependent than in the rat. However, overall, the picture is of strikingly little 11β-HSD2 in the adult CNS.

11β-HSD2 in the developing CNS

In contrast, 11β-HSD2 is highly expressed in the fetal and early postnatal rodent CNS (Diaz et al., 1998; Robson et al., 1998), perhaps preventing premature exposure to glucocorticoids which have potent effects on neuronal and glial maturation and survival during prenatal and early postnatal development (Meyer, 1983). High levels of glucocorticoids have rather deleterious actions in CNS development, inhibiting cell proliferation and differentiation. 11β-HSD2 acts as a major barrier to glucocorticoids reaching the fetus and the neonate (Lopez-Bernal et al., 1980). It is highly expressed in the placenta, inactivating corticosterone prior to reaching the fetus (Benediktsson et al., 1997). Perhaps as a further layer of defence, 11β-HSD2 is expressed in the developing brain and other fetal tissues (Brown et al., 1996; Diaz et al., 1998). CNS expression is observed from midgestation in a wide range of sites, but the expression is somewhat dramatically switched-off as each nucleus develops (Brown et al., 1996; Diaz et al., 1998). At birth in the rat, the main areas of 11β-HSD2 expression are in the thalamus and cerebellum, areas exhibiting substantial postnatal development. By weaning at postnatal day 21, CNS 11β-HSD2 expression is confined to those few areas seen in the adult (Robson et al., 1998). So, there appears to be an exquisitely timed system of protection and then exposure of developing brain regions to circulating glucocorticoids. What might this determine?

Glucocorticoid effects on postnatal cerebellar development

Several previous studies in the rat have shown that postnatal treatment with glucocorticoids (hydrocortisone, dexamethasone, corticosterone) results in a delayed and malformed development of the cerebellum (Bohn and Lauder, 1978; Bohn and Lauder, 1980). The cerebellum is still proliferating postnatally and a considerable amount of cell migration, differentiation as well as cerebellar secondary folding, is taking place between P1–28. Neonatal adrenalectomy prolongs mitosis and delays disappearance of the external granule layer (EGL), a secondary germinal zone present on the surface of the developing cerebellum. Exogenously administered glucocorticoids accelerate the disappearance of the EGL due to premature cessation of granule cell division and an inhibition of cell proliferation, which results in a decrease in cell numbers within the internal granule layer (IGL) and abnormalities in the secondary folding of the lobes (most marked in lobes VIII and IX of the cerebellar cortex) (Bohn and Lauder, 1978, 1980). The EGL has particularly high expression of 11β-HSD2 (Robson et al., 1998), so does loss of this GC protective enzyme mimic the effects of exogenously elevated glucocorticoids? The mice used for these studies have the 11β-HSD2 gene deletion isogenic on an inbred strain (C57BL/6) to facilitate investigation of the phenotype. In preliminary studies cerebellar growth was noted to be significantly impaired in 11β-HSD2$^{-/-}$ mice (Holmes et al., unpublished), consistent with a neuroprotective role for 11β-HSD2 in postnatal cerebellar development. Further details are being determined.

Fetal programming: prenatal glucocorticoids and 11β-HSD2

The early life effects of glucocorticoids in the rodent have been studied in some detail in the rat (Seckl, 1998). Other species have also been studied, such as the guinea-pig and sheep, and correlations have been made with human data (Weinstock, 2001; Welberg and Seckl, 2001; Matthews, 2002). In the rat, exogenous administration of corticosterone or the synthetic glucocorticoid, dexamethasone, during pregnancy can have life-long effects on the offspring, making them more susceptible to pathology in adult life. These effects are called glucocorticoid 'programming' and the brain is a prime site of action. Prenatal dexamethasone, a poor 11β-HSD2 substrate that therefore crosses the placenta intact, results in offspring with increased HPA-axis activity demonstrated by elevated basal corticosterone levels (Levitt et al., 1996). This is thought to be the consequence of altered expression of corticosteroid receptors at sites of glucocorticoid feedback such as the PVN, hippocampus and/or amygdala (Welberg et al., 2001). The precise pattern of GR/MR changes is dependent on the time-window of the glucocorticoid exposure and the species involved (Dean and Matthews, 1999). Furthermore, as adults these prenatal dexamethasone-exposed rats develop behavioural abnormalities consistent with a more anxious phenotype. Similar data have been obtained in rats which have been exposed to prenatal stress (a consequence of which will be elevation of endogenous glucocorticoids) (Maccari et al., 1995), as well as offspring of pregnant rats treated with carbenoxolone, which inhibits placental 11β-HSD2 activity to elevate glucocorticoid exposure to the fetus (Welberg et al., 2000).

The next question is 'are the 11β-HSD2$^{-/-}$ mice behaving as a glucocorticoid-programmed animal?'. Before addressing this question, some of the key differences in glucocorticoid regulation and secretion during pregnancy of the rat versus the mouse need to be explored, as these could potentially modify maternal environment and programming. The pregnant rat maintains an environment of low levels of glucocorticoids by becoming less responsive to stress (Johnstone et al., 2000). Consequently, the placental 11β-HSD2 barrier is less likely to be breached. In the mouse, in contrast, basal levels of plasma corticosterone rise dramatically throughout pregnancy to reach a plateau by day 16 when they are approximately 50–100 times higher than non-pregnant levels (Holmes et al., unpublished). Induction of CBG occurs in an attempt to buffer this astronomical rise in plasma corticosterone, but although CBG levels increase over 10-fold, full buffering of corticosterone is not accomplished. Perhaps unsurprisingly, restraint stress cannot induce any further increases in corticosterone. However, throughout pregnancy a circadian rhythm of plasma corticosterone is preserved and only peak levels of corticosterone penetrate the 11β-HSD2 barrier (Venihaki et al., 2000), presumably adding to the provision of glucocorticoids for normal key developmental processes such as maturation of the lung.

In the rat, central programming by prenatal glucocorticoid overexposure or 11β-HSD inhibition, alters subsequent adult HPA activity and behaviour such that the offspring appear anxious (Welberg et al., 2000, 2001). At present there is little evidence for programming of the HPA axis in the 11β-HSD2$^{-/-}$ mouse, in which both basal and peak stress corticosterone responses appear normal. However, more detailed work is necessary to determine if there is altered 'switch-off hairline' characteristics analogous to the effects seen in the, pre-natally stressed animals. Furthermore, 11β-HSD2$^{-/-}$ mice have significantly smaller adrenals (C.J. Kenyon and M.C. Holmes, unpublished), which although presumably due to the reduced turnover of corticosterone, may mask any subtle hyperactivity of the HPA axis induced by glucocorticoid programming. Alternatively, the mouse may be more resistant to glucocorticoid programming of the HPA axis than the rat, perhaps reflecting the difference in the maternal glucocorticoid levels. However, our preliminary experiments suggest that the 11β-HSD2$^{-/-}$ mice exhibit increased anxiety, at least consistent with the phenotype of prenatal glucocorticoid exposure (Holmes, unpublished observations).

11β-HSD1 in the brain

11β-HSD activity was not uniformly detected in CNS samples until it was realised that in brain homogenates enzyme activity is markedly potentiated by addition of dinucleotide co-substrate (Moisan et al., 1990b, c; Lakshmi et al., 1991). These effects presumably reflect lower levels of endogenous NADP(H) in the brain than in liver and kidney (Glock and McLean, 1955), leading to the notion that variations in co-substrate levels may potently determine enzyme activity in the brain in vivo (Moisan et al., 1990b). These data also strongly suggest that 11β-HSD1 is responsible for this brain activity, since purified 11β-HSD2 is potentiated only by NAD (Brown et al., 1993), albeit various tissues can interconvert the dinucleotides in vivo. 11β-HSD activity has been reported in rat, mouse, dog and primate brain (Grosser, 1966; Grosser and Axelrod, 1968; Miyabo et al., 1973). Biochemical, immunological and molecular studies suggest that it is 11β-HSD1 that is responsible for widespread CNS activity. Thus, western blots detect the 34 kD 'liver' 11β-HSD1 species in rat brain extracts (Monder and Lakshmi, 1990). Northern analysis confirms a single 11β-HSD1 mRNA species expressed in a range of brain subregions (Moisan et al., 1990b, c), transcribed from the same predominant gene promoter employed in the liver (Low et al., 1993). In general, the level of 11β-HSD1 mRNA expression parallels enzyme activity (Low et al., 1994b).

Localisation

Immunocytochemical studies suggest that 11β-HSD1-like immunoreactivity is widespread in brain cells, reportedly in neurons more than glia (Moisan et al., 1990b; Sakai et al., 1992). However, 11β-HSD activity has also been found in clonal glioma lines (Grosser, 1966) and the expression in glia merits more careful examination. Immunostaining is not only found in neuronal cell bodies, but also in axons and dendritic processes, suggesting a possible membrane action. Indeed, cortisone and cortisol appear to exert distinct effects on cell membrane $GABA_A$ receptors, at least in the guinea-pig ileum model (Ong et al., 1990). In situ hybridisation studies demonstrate expression of 11β-HSD1 mRNA is also predominantly neuronal (Moisan et al., 1990b, c) with most neurons expressing at least some 11β-HSD1 mRNA, although with considerable variation. 11β-HSD1 mRNA is highly expressed in cerebellar granule and Purkinje cells, some brainstem nuclei, a subset of hippocampal cells, neurons in layers II and IV/V of the neocortex, some hypothalamic nuclei and the anterior pituitary (Moisan et al., 1990b, c). In general, immunohistochemistry has paralleled these findings (Moisan et al., 1990b, c; Sakai et al., 1990, 1992).

Ontogeny of 11β-HSD1 in the CNS

11β-HSD bioactivity has been demonstrated in fetal brain (Tye and Burton, 1980; Moisan et al., 1990a). 11β-HSD1 mRNA expression is first observed in the fetal rat around embryonal day 15.5, and gradually increases with age, localised to the neuronal layers of the hippocampus, neocortex, cerebellum, dorsal brainstem, midbrain, hypothalamus and pituitary (Diaz et al., 1998). 11β-HSD1 bioactivity is also detectable in late gestation in rat hippocampus, cerebellum and parietal cortex suggesting the mRNA is translated to protein. Postnatally, there are complex tissue-specific patterns of ontogeny of 11β-HSD activity and 11β-HSD1 mRNA, with a nadir of enzyme activity in hippocampus and cortex at postnatal day 10, followed by a gradual rise to adult values (Moisan et al., 1990a). In contrast, 11β-HSD activity in the cerebellum peaks at postnatal day 10 and then falls to adult levels by postnatal day 15.

Reaction direction of 11β-HSD1 in intact CNS cells

So which reaction direction predominates in the brain? Primary cultures of fetal rat hippocampal cells express 11β-HSD1, but no 11β-HSD2 (Rajan et al., 1996). Whilst homogenates of hippocampal cells show bi-directional activity the intact cells show only reductase activity, in line with intact cells from other organs and with the much greater concentrations of NADPH than NADP in brain. 11β-reduction ensures that, otherwise, inert 11-dehydrocorticosterone is equipotent with corticosterone in hippocampal cells at least in potentiating excitatory amino acid neurotoxicity (Rajan et al., 1996). But does this matter in vivo where such amplification of neurotoxicity by glucocorticoids is only erratically observed?

The CNS phenotype of 11β-HSD1 knockout mice

11β-HSD1 expression is switched on fairly late in gestation and loss of expression does not lead to manifestation of a gross developmental phenotype, the mice appear healthy and robust (Kotelevtsev et al., 1997). However, some caution should be observed as the enzyme is expressed in liver, lung and CNS at the end of gestation (Hundertmark et al., 2002a) and 11β-HSD1 null mice have subtle but significant perinatal developmental changes in their lungs, compatible with modest immaturity of the surfactant system (Hundertmark et al., 2002b).

11β-HSD1 and HPA-axis regulation

As detailed above, 11β-HSD1 is expressed in the hippocampus, hypothalamic paraventricular nucleus (PVN) and pituitary, key loci of glucocorticoid negative feedback upon the HPA axis. So an obvious question is to determine the effect of 11β-HSD1 knockout on the HPA axis. 11β-HSD1$^{-/-}$ mice have hypertrophied adrenals and elevated basal ACTH and corticosterone levels at the nadir of the rhythm (Kotelevtsev et al., 1997; Harris et al., 2001). The increased ACTH level may underpin the adrenal hypertrophy. Indeed, in vitro, 11β-HSD1$^{-/-}$ adrenals are hyper-responsive to ACTH. These effects are compatible with the decreased half-life (increased metabolic clearance) of corticosterone in this model due to lack of regeneration in the liver and other sites. However, the unexpected finding is that the basal (morning) plasma corticosterone levels are also elevated. This would not be expected unless there was altered central drive to or feedback upon the HPA axis, which should otherwise act to maintain normal basal levels of glucocorticoids. Altered central drive is rather unlikely since CRF mRNA levels in the PVN are unaltered. Moreover, this is unlikely to be an artefact of altered plasma binding as CBG is not affected by loss of 11β-HSD1 (Harris et al., 2001).

Interestingly, in 11β-HSD1 null mice, restraint stress leads to an exaggerated peak corticosterone response, though peak ACTH levels are unaltered (Kotelevtsev et al., 1997; Harris et al., 2001). This suggests that the corticosterone hypersecretion is mainly due to adrenal hypersensitivity. However, the turn-off phase of the ACTH response, which correlates with glucocorticoid negative-feedback efficiency, is delayed. Indeed, exogenous cortisol, 2 h prior to restraint, produces less inhibition of the corticosterone stress response in 11β-HSD1$^{-/-}$ mice, indicating attenuated negative feedback. There is no change in central GR or MR density to offer an alternative explanation. This subtle feedback deficit might explain the elevated basal levels of glucocorticoids in the null mice (Kotelevtsev et al., 1997; Harris et al., 2001). This is also consistent with the finding that the brain accumulates significantly less ^{3}H-corticosterone after a 7-day infusion in 11β-HSD1$^{-/-}$ mice, emphasising that 11β-HSD1 activity is an important determinant of intracellular glucocorticoid levels in the brain (Yau et al., 2001). Interestingly, recent data suggest that 11β-HSD1 mRNA expression is reduced in a subset of hippocampal cells in obese Zucker rats (Mattsson et al., 2003). This may in part explain why this strain lacks full HPA feedback sensitivity and hence has increased corticosterone levels. Again subtle changes in intracellular steroid metabolism appear to make surprisingly important contributions to phenotype.

The circadian periodicity of plasma corticosterone is also modified in 11β-HSD1 null mice. The evening elevation in corticosterone is shifted much earlier, producing an extended period of hypersecretion. 11β-HSD1 activity is not thought to be regulated in a circadian pattern, as 11β-HSD1 mRNA expression is unaltered at 8 am versus 8 pm (Harris et al., 2001). It therefore is possible that 11β-HSD1 may modify central glucocorticoid signalling on circadian pathways. The mechanisms and loci of these interesting effects remain to be determined.

Regulation of 11β-HSD1 in the CNS

Given the importance of 11β-HSD in determining glucocorticoid action, many studies have addressed the regulation of enzyme activity. Dexamethasone induces 11β-HSD1 gene expression and activity in primary hippocampal cells in vitro and in rat hippocampus, cortex, cerebellum and hypothalamus in vivo (Low et al., 1994b; Rajan et al., 1996). However, the longer-term regulation of 11β-HSD1 by adrenal steroids is more complex (Jamieson et al., 1999). Arthritis stress for 15 days, which persistently and markedly elevates plasma corticosterone levels, also induces hippocampal 11β-HSD (Low et al., 1994b). This induction of 11β-HSD1 to inflammation occurs in other tissues (Thieringer et al., 2001), and in the CNS may be rather specific to this stress, since induction of 11β-reductase would be predicted to *increase* cellular exposure to glucocorticoids and thus amplify any deleterious effects. Indeed, in the tree shrew, chronic psychosocial stress (for 28 days) attenuates hippocampal 11β-HSD activity (Jamieson et al., 1997). It is conceivable that the rise in 11β-HSD1 activity in response to some acute stressors may amplify the glucocorticoid signal during its acute adaptive phase, whereas the later downregulation of the enzyme with longer-term stress ameliorates the adverse effects of chronic glucocorticoid excess.

An intriguing physiological correlate of altered brain 11β-HSD1 has recently emerged. In late pregnancy in rats, the HPA axis becomes rather refractory to stressful stimulation, perhaps to obviate premature effects of glucocorticoids upon late fetal development and the induction of labour. Interestingly, 11β-HSD1 activity is selectively increased in the hypothalamic paraventricular nucleus at this time (Johnstone et al., 2000), providing the potential to further amplify glucocorticoid feedback and hence flatten HPA responses. Such assertions remain to be examined in the knockout mice.

11β-HSD1 and brain aging

Although the hippocampus requires glucocorticoids for neuronal function and survival, it is also particularly vulnerable to the adverse effects of chronic glucocorticoid excess, which produces atrophy of dendrites, neuronal and cognitive dysfunction and occasionally, at least in some strains of rat, neuronal loss (McEwen, 1999). The damaging effects of chronic glucocorticoid excess on neuronal structure and function become more marked with ageing. Indeed, chronic glucocorticoid overexposure has been implicated in the pathogenesis of age-related cognitive decline and Alzheimer's disease (Meaney et al., 1995). In a subgroup of aged rodents, there is an association of cognitive decline with a chronic elevation of plasma corticosterone levels and a loss of circadian periodicity. If glucocorticoids are maintained at low levels, by adrenalectomising the rats and replacing with low-dose corticosterone, they are less susceptible to age-related impairments (Landfield et al., 1978). Maintenance of low glucocorticoid levels throughout life can also be achieved by neonatal manipulations (neonatal handling) that permanently facilitate glucocorticoid feedback tone upon the HPA axis and this also prevents the decline in hippocampal structure and cognitive function with age (Meaney et al., 1988). So what does 11β-HSD1 in the hippocampus and perhaps other loci contribute to the wear and tear effects of stress with ageing?

As outlined above, in primary cultures of hippocampal cells, 11β-HSD1 potentiates kainic acid-induced neurotoxicity not only by corticosterone, but equally by intrinsically inert 11-dehydrocorticosterone, an effect prevented by carbenoxolone (Rajan et al., 1996). 11-Dehydrocorticosterone in vivo also increases kainic acid toxicity in adrenalectomised rats, an action also blocked by carbenoxolone

(Ajilore and Sapolsky, 1999). This work indicates that 11β-HSD is working as a reductase, but does not reveal where this reaction is occurring (CNS or periphery). Such studies are hampered by the non-selectivity of CBX (it inhibits both 11β-HSD isozymes as well as some other short-chain dehydrogenases) and its variable access to brain regions in vivo, even when given intracerebroventricularly (Jellinck et al., 1993). This non-selectivity might explain why carbenoxolone only appears to block 11-dehydrocorticosterone potentiation of kainate neurotoxicity when glucocorticoid levels are low. The balance of effects of carbenoxolone, which inhibits both 11β-HSDs and gains rather patchy tissue access, is difficult to predict. So what about ageing in 11β-HSD1 null mice which have many less variable compared with the non-selective inhibitors?

Old (24-month) wild-type mice, as aged rats, show glucocorticoid-associated impairments in hippocampus-dependent learning and memory tasks in the water maze (Yau et al., 2001). Young 11β-HSD1$^{-/-}$ mice, despite elevated plasma corticosterone levels, perform as well as young wild-type, suggesting that they are relatively 'blind' to the tissue effects of elevated plasma glucocorticoids. Strikingly, aged 11β-HSD1 null mice, as well as young mice also learn and avoid the cognitive decline seen in the majority of aged wild-type mice. This cognitive protection in aged 11β-HSD1 null mice associates with substantially *reduced* intrahippocampal corticosterone levels indicating the potency of intracellular metabolism by 11β-HSDs in determining effective glucocorticoid action upon target receptors. Whether 11β-HSD1 deficiency also protects against the emergence of structural pathology in the hippocampus of aged mice remains to be determined.

Interestingly, 11β-HSD1 mRNA levels appear to decrease within the hippocampus in aged rats, which inversely correlates with their increase in plasma corticosterone levels (Yau and Seckl, 2001). Possibly reduction in 11β-HSD1 activity will impair glucocorticoid feedback and hence may underpin HPA-axis hyperactivity with ageing. Reduced 11β-HSD1 may also help attenuate tissue exposure to glucocorticoids and thus protect the vulnerable hippocampus. In general, reduction of 11β-HSD1 activity results in beneficial consequences to the animal. Although 11β-HSD1$^{-/-}$ mice have elevated circulating corticosterone levels and impaired HPA regulation, the lower *intracellular* corticosterone levels in brain (and other tissues) results in a neuroprotective phenotype. Very recently, a class of selective 11β-HSD1 inhibitors have been found to reduce blood glucose levels in hyperglycaemic mice (Alberts et al., 2002; Barf et al., 2002). It will be interesting to see if these compounds are able to access the CNS and if they also produce the beneficial cognitive effects seen in 11β-HSD1$^{-/-}$ mice.

Relevance of 11β-HSD1 in the human brain

In situ hybridisation studies in post-mortem human brain material have shown high expression of 11β-HSD1 mRNA in the hippocampus, prefrontal cortex and the cerebellum (J. Noble, J.L.W. Yau and J.R. Seckl, unpublished data) mirroring the findings in rodents. It will be important to determine whether this mRNA is paralleled by 11β-HSD1 activity, as it is in the rat brain, and if 11β-HSD1 expression is dysregulated in specific brain regions with normal aging and in disease (e.g. Alzheimer's, depression). In vivo studies in humans are hampered by the uncertain CNS penetration and non-selectivity of the currently available liquorice-based inhibitors. Thus, despite reports implicating liquorice abuse in a range of psychiatric disorders, it has been difficult to dissect the marked peripheral actions (hypokalaemia) from any central effects (reviewed by Seckl, 1997). The development of centrally active selective 11β-HSD1 inhibitors will facilitate the study of whether 11β-reductase activity plays a role in mood and cognitive function in humans. Such drugs may prevent cognitive impairments with ageing and ameliorate other psychopathologies associated with glucocorticoid excess, such as depression.

Summary

11β-HSDs are microsomal enzymes which interconvert active glucocorticoids and inert 11-keto forms. This, otherwise, arcane reaction is surprisingly important in peripheral tissues and the CNS, determining steroid access to nuclear receptors and hence biological effects in health and disease. 11β-HSD1 is a reductase that amplifies glucocorticoid levels. Studies in null mice show a role in feedback. Deficiency of this isozyme reduces intracerebral glucocorticoid levels and prevents the emergence of cognitive deficits with ageing. 11β-HSD2 is a potent dehydrogenase, little expressed in the adult CNS, but which plays an emerging role in gating glucocorticoid actions on the developing nervous system. The roles of these isozymes in CNS biology is under scrutiny and the importance of manipulations of the type 1 enzyme has therapeutic promise.

Acknowledgements

Work in the authors' laboratory was supported by a Programme grant from the Wellcome Trust (J.R.S.), a Wellcome Trust Senior Clinical Fellowship (J.R.S.) and project grants from the Wellcome Trust (J.R.S., J.L.Y. and M.C.H.). J.L.Y. is a Carter Fellow of the Alzheimer's Research Trust.

References

Agarwal, A.K., Monder, C., Eckstein, B. and White, P.C. (1989) Cloning and expression of rat cDNA encoding corticosteroid 11dehydrogenase. J. Biol. Chem., 264: 18939–18943.

Ajilore, O.A. and Sapolsky, R.M. (1999) In vivo characterization of 11 beta-hydroxysteroid dehydrogenase in rat hippocampus using glucocorticoid neuroendangerment as an endpoint. Neuroendocrinology, 69(2): 138–144.

Alberts, P., Engblom, L., Edling, N., Forsgren, M., Klingstrom, G., Larsson, C., Ronquist-Nii, Y., Ohman, B., Abrahmsen, L. (2002) Selective inhibition of 11 beta-hydroxysteroid dehydrogenase type 1 decreases blood glucose concentrations in hyperglycaemic mice. Diabetologia, 45: 1528–1532.

Albiston, A.L., Obeyesekere, V.R., Smith, R.E. and Krozowski, Z.S. (1994) Cloning and tissue distribution of the human 11β-hydroxysteroid dehydrogenase type 2 enzyme. Mol. Cell. Endo., 105(2): R11–R17.

Amelung, D., Huebener, H.J., Roka, L. and Meyerheim, G. (1953) Conversion of cortisone to compound F. Journal of Clinical Endocrinology and Metabolism, 13: 1125–1129.

Barf, T., Vallgarda, J., Emond, R., Haggstrom, C., Kurz, G., Nygren, A., Larwood, V., Mosialou, E., Axelsson, K., Olsson, R., Engblom, L., Edling, N., Ronquist-Nii, Y., Ohman, B., Alberts, P. and Abrahmsen, L. (2002) Arylsulfonamidothiazoles as a new class of potential antidiabetic drugs. Discovery of potent and selective inhibitors of the 11 beta-hydroxysteroid dehydrogenase type 1. Journal of Medicinal Chemistry, 45(18): 3813–3815.

Benediktsson, R., Calder, A.A., Edwards, C.R.W. and Seckl, J.R. (1997) Placental 11β-hydroxysteroid dehydrogenase type 2 is the placental barrier to maternal glucocorticoids: ex vivo studies. Clin. Endocrinol., 46: 161–166.

Bohn, M.C. and Lauder, J.M. (1978) The effects of neonatal hydrocortisone on rat cerebellar development: an autoradiographic and light microscopic study. Dev. Neurosci., 1: 250–266.

Bohn, M.C. and Lauder, J.M. (1980) Cerebellar granule cell genesis in hydrocortisone-treated rat. Dev. Neurosci., 3: 81–89.

Brown, R.W., Chapman, K.E., Edwards, C.R.W. and Seckl, J.R. (1993) Human placental 11β-hydroxysteroid dehydrogenase: partial purification of and evidence for a distinct NAD-dependent isoform. Endocrinol., 132: 2614–2621.

Brown, R.W., Diaz, R., Robson, A.C., Kotelevtsev, Y., Mullins, J.J., Kaufman, M.H. and Seckl, J.R. (1996) The ontogeny of 11β-hydroxysteroid dehydrogenase type 2 and mineralocorticoid receptor gene expression reveal intricate control of glucocorticoid action in development. Endocrinol., 137: 794–797.

Bujalska, I., Kumar, S. and Stewart, P.M. (1997) Central obesity: "Cushing's disease of the omentum". Lancet, 349: 1210–1213.

Burton, A.F. and Tufnell, R.W. (1967) 11-dehydrocorticosteroids in tissues of mice. Canadian Journal of Biochemistry, 46: 497–502.

de Kloet, E.R. (1991) Brain corticosteroid receptor balance and homeostatic control. Frontiers in Neuroendocrinology, 12: 95–164.

de Kloet, E.R., Wallach, G. and McEwen, B.S. (1975) Differences in corticosterone and dexamethasone binding to rat brain and pituitary. Endocrinol., 96: 598–609.

Dean, F. and Matthews, S. (1999) Maternal dexamethasone treatment in late gestation alters glucocorticoid and mineralocorticoid receptor mRNA in the fetal guinea pig brain. Brain Res., 846: 253–259.

Diaz, R., Brown, R.W. and Seckl, J.R. (1998) Ontogeny of mRNAs encoding glucocorticoid and mineralocorticoid receptors and 11β-hydroxysteroid dehydrogenases in prenatal rat brain development reveal complex control of glucocorticoid action. J. Neurosci., 18: 2570–2580.

Duperrex, H., Kenouch, S., Gaeggeler, H.-P., Seckl, J.R., Edwards, C.R.W., Farman, N. and Rossier, B.C. (1993) Rat liver 11β-hydoxysteroid dehydrogenase cDNA encodes oxoreductase activity in a mineralocorticoid-responsive toad bladder cell line. Endocrinology, 132: 612–619.

Edwards, C.R.W., Stewart, P.M., Burt, D., Brett, L., McIntyre, M.A., Sutanto, W.S., de Kloet, E.R. and Monder, C. (1988) Localisation of 11β-hydroxysteroid dehydrogenase-tissue specific protector of the mineralocorticoid receptor. Lancet, ii: 986–989.

Funder, J.W., Pearce, P.T., Smith, R. and Smith, A.I. (1988) Mineralocorticoid action: target tissue specificity is enzyme, not receptor, mediated. Science, 242: 583–585.

Glass, C.K. (1994) Differential recognition of target genes by nuclear receptor monomers, dimers and heterodimers. Endocrine Reviews, 15: 391–407.

Glock, G.E. and McLean, P. (1955) Levels of oxidized and reduced diphosphopyridine nucleotide and triphosphopyridine nucleotide in animal tissues. Biochem. J., 61: 388–390.

Gomez-Sanchez, E.P. (1986) Intracerebroventricular infusion of aldosterone induces hypertension in rats. Endocrinol., 118: 819–823.

Gomez-Sanchez, E.P., Fort, C.M. and Thwaites, D. (1990) Icv infusion of corticosterone antagonises icv-aldosterone hypertension. Am. J. Physiol., E649–E653.

Gomez-Sanchez, E.P. and Gomez-Sanchez, C.E. (1992) Central hypertensinogenic effects of glycyrrhizic acid and carbenoxolone. Am. J. Physiol., 263: E1125–E1130.

Grosser, B.I. (1966) 11β-hydroxysteroid metabolism by mouse brain and glioma 261. J. Neurochem., 13: 475–478.

Grosser, B.I. and Axelrod, L.R. (1968) Conversion of cortisol to cortisol acetate, cortisone acetate and cortisone by the developing primate brain. Steroids, 11: 827–836.

Harris, H.J., Kotelevtsev, Y., Mullins, J.J., Seckl, J.R. and Holmes, M.C. (2001) 11β-hydroxysteroid dehydrogenase type 1 null mice have altered hypothalamic–pituitary–adrenal axis activity: a novel control of glucocorticoid feedback. Endocrinology, 142: 114–120.

Hundertmark, S., Buhler, H., Ragosch, V., Dinkelborg, L., Arabin, B. and Weitzel, H.K. (1995) Correlation of surfactant phosphatidylcholine synthesis and 11beta-hydroxysteroid dehydrogenase in the fetal lung. Endocrinology, 136: 2573–2578.

Hundertmark, S., Dill, A., Buhler, H., Stevens, P., Looman, K., Ragosch, V., Seckl, J.R. and Lipka, C. (2002a) 11 beta-hydroxysteroid dehydrogenase type 1: a new regulator of fetal lung maturation. Hormone and Metabolic Research, 34(10): 537–544.

Hundertmark, S., Dill, A., Ebert, A., Zimmermann, B., Kotelevtsev, Y.V., Mullins, J.J. and Seckl, J.R. (2002b) Foetal lung maturation in 11 beta-hydroxysteroid dehydrogenase type 1 knockout mice. Hormone and Metabolic Research, 34(10): 545–549.

Jamieson, P.M., Chapman, K.E., Edwards, C.R.W. and Seckl, J.R. (1995) 11β-hydroxysteroid dehydrogenase is an exclusive 11β-reductase in primary cultured rat hepatocytes: effect of physicochemical and hormonal manipulations. Endocrinology, 136: 4754–4761.

Jamieson, P.M., Chapman, K.E. and Seckl, J.R. (1999) Tissue- and temporal-specific regulation of 11β-hydroxysteroid dehydrogenase type 1 by glucocorticoids in vivo. J. Steroid Biochem. Molec. Biol., 68: 245–250.

Jamieson, P.M., Chapman, K.E., Walker, B.R. and Seckl, J.R. (2000) 11β-hydroxysteroid dehydrogenase type 1 is a predominant 11β-reductase in the intact perfused rat liver. J. Endocrinol., 165: 685–692.

Jamieson, P., Fuchs, E. and Seckl, J. (1997) Chronic psychosocial stress attenuates 11β-hydroxysteroid dehydrogenase activity in the hippocampus and liver in the tree-shrew. Stress, 2: 123–132.

Jellinck, P.H., Monder, C., McEwen, B.S. and Sakai, R.R. (1993) Differential inhibition of 11β-hydroxysteroid dehydrogenase by carbenoxolone in rat brain regions and peripheral tissues. J. Steroid Biochem. Molec. Biol., 46: 209–213.

Johnstone, H.A., Wigger, A., Douglas, A.J., Neumann, I.D., Landgraf, R., Seckl, J.R. and Russell, J.A. (2000) Attenuation of hypothalamic–pituitary–adrenal axis stress responses in late pregnancy: changes in feedforward and feedback mechanisms. Journal of Neuroendocrinology, 12(8): 811–822.

Kotelevtsev, Y., Brown, R.W., Fleming, S., Kenyon, C.J., Edwards, C.R.W., Seckl, J.R. and Mullins, J.J. (1999) Hypertension in mice lacking 11β-hydroxysteroid dehydrogenase type 2. J. Clin. Invest., 103: 683–689.

Kotelevtsev, Y., Holmes, M.C., Burchell, A., Houston, P.M., Schmoll, D., Jamieson, P.M., Best, R., Brown, R., Edwards, C.R.W. et al. (1997) 11β-hydroxysteroid dehydrogenase type 1 knockout mice show attenuated glucocorticoid inducible responses and resist hyperglycaemia on obesity or stress. Proc. Natl. Acad. Sci. USA, 94: 14924–14929.

Lakshmi, V. and Monder, C. (1988) Purification and characterisation of the corticosteroid 11β-dehydrogenase component of the rat liver 11β-hydroxysteroid dehydrogenase complex. Endocrinol., 123: 2390–2398.

Lakshmi, V., Sakai, R.R., McEwen, B.S. and Monder, C. (1991) Regional distribution of 11β-hydroxysteroid dehydrogenase in rat brain. Endocrinol., 128: 1741–1748.

Landfield, P.W., Waymire, J. and Lynch, G. (1978) Hippocampal aging and adrenocorticoids: a quantitative correlation. Science, 202: 1098–1102.

Levitt, N., Lindsay, R.S., Holmes, M.C. and Seckl, J.R. (1996) Dexamethasone in the last week of pregnancy attenuates hippocampal glucocorticoid receptor gene expression and elevates blood pressure in the adult offspring in the rat. Neuroendocrinology, 64: 412–418.

Lopez-Bernal, A., Flint, A.P.F., Anderson, A.B.M. and Turnbull, A.C. (1980) 11β-hydroxysteroid dehydrogenase activity (E.C.1.1.1.146) in human placenta and decidua. J. Steroid Biochem., 13: 1081–1087.

Low, S.C., Assaad, S.N., Rajan, V., Chapman, K.E., Edwards, C.R.W. and Seckl, J.R. (1993) Regulation of 11β-hydroxysteroid dehydrogenase by sex steroids in vivo: further evidence for the existence of a second dehydrogenase in rat kidney. Journal of Endocrinology, 139: 27–35.

Low, S.C., Chapman, K.E., Edwards, C.R.W. and Seckl, J.R. (1994a) Liver-type 11β-hydroxysteroid dehydrogenase cDNA encodes reductase not dehydrogenase activity in intact mammalian COS-7 cells. Journal of Molecular Endocrinology, 13: 167–174.

Low, S.C., Moisan, M.-P., Edwards, C.R.W. and Seckl, J.R. (1994b) Glucocorticoids and chronic stress up-regulate 11β-hydroxysteroid dehydrogenase activity and gene expression in the hippocampus. J. Neuroendocrinol., 6: 285–290.

Maccari, S., Piazza, P.V., Kabbaj, M., Barbazanges, A., Simon, H. and Le Moal, M. (1995) Adoption reverses the long-term impairment in glucocorticoid feedback induced by prenatal stress. Journal of Neuroscience, 15: 110–116.

Masuzaki, H., Paterson, J., Shinyama, H., Morton, N.M., Mullins, J.J., Seckl, J.R. and Flier, J.S. (2001) A transgenic model of visceral obesity and the Metabolic Syndrome. Science, 294: 2166–2170.

Matthews, S.G. (2002). Early programming of the hypothalamo–pituitary–adrenal axis. Trends Endocrinol. Metab., 130: 373–380.

Mattsson, C., Lai, M., Noble, J., McKinney, E., Yau, J.L.W., Seckl, J.R. and Walker, B.R. (2003) Reduced mineralocorticoid receptor and 11β-hydroxysteroid dehydrogenase type 1 expression in hippocampus of Zucker fat rats; mechanisms for altered feedback control of the hypothalamic–pituitary–adrenal axis in obesity. Endocrinology. 144: 2997–3003.

McEwen, B.S. (1999). Stress and hippocampal plasticity. Annual Review of Neuroscience, 22: 105–122.

McEwen, B.S., de Kloet, E.R. and Rostene, W. (1986a) Adrenal steroid receptors and action in the nervous system. Physiol. Rev., 66: 1121–1188.

McEwen, B.S., Lambdin, L.T., Rainbow, T.C. and De Nicola, A.F. (1986b) Aldosterone effects on salt appetite in adrenalectomised rats. Neuroendocrinol., 43: 38–43.

Meaney, M.J., Aitken, D.H., van Berkel, C., Bhatnagar, S. and Sapolsky, R.M. (1988) Effect of neonatal handling on age-related impairments associated with the hippocampus. Science, 239: 766–768.

Meaney, M.J., O'Donnell, D., Rowe, W., Tannenbaum, B., Steverman, A., Walker, M., Nair, N.P.V. and Lupien, S. (1995) Individual differences in hypothalamic–pituitary–adrenal activity in later life and hippocampal aging. Experimental Gerontology, 30: 229–251.

Meyer, J.S. (1983) Early adrenalectomy stimulates subsequent growth and development of the rat brain. Exp. Neurol., 82: 432–446.

Miyabo, S., Kishida, S. and Hisada, T. (1973) Metabolism and conjugation of cortisol by various dog tissues in vitro. Journal of Steroid Biochemistry, 4: 567–576.

Moisan, M.-P., Edwards, C.R.W. and Seckl, J.R. (1990a) Ontogeny of 11β-hydroxysteroid dehydrogenase in rat brain. Neuroendocrinol., 52(Suppl. I): 137.

Moisan, M.-P., Seckl, J.R., Brett, L.P., Monder, C., Agarwal, A.K., White, P.C. and Edwards, C.R.W. (1990b) 11β-hydroxysteroid dehydrogenase messenger ribonucleic acid expression, bioactivity and immunoreactivity in rat cerebellum. J. Neuroendocrinol., 2: 853–858.

Moisan, M.-P., Seckl, J.R. and Edwards, C.R.W. (1990c) 11β-hydroxysteroid dehydrogenase bioactivity and messenger RNA expression in rat forebrain: localization in hypothalamus, hippocampus and cortex. Endocrinology, 127: 1450–1455.

Monder, C. and Lakshmi, V. (1990) Corticosteroid 11β-dehydrogenase of rat tissues: immunological studies. Endocrinology, 126: 2435–2443.

Morton, N.M., Holmes, M.C., Fievet, C., Staels, B., Mullins, J.J. and Seckl, J.R. (2001) Improved lipid and lipoprotein profile, hepatic insulin sensitivity and glucose tolerance in 11beta-hydroxysteroid dehydrogenase type 1 null mice. J. Biol. Chem., 276: 41293–41300.

Mune, T., Rogerson, F.M., Nikkilä, H., Agarwal, A.K. and White, P.C. (1995) Human hypertension caused by mutations in the kidney isozyme of 11β-hydroxysteroid dehydrogenase. Nature Genetics, 10: 394–399.

Napolitano, A., Voice, M., Edwards, C.R.W., Seckl, J.R. and Chapman, K.E. (1998) 11β-hydroxysteroid dehydrogenase type 1 in adipocytes: expression is differentiation-dependent and hormonally-regulated. J. Steroid Biochem. Molec. Biol., 64: 251–260.

Nitabach, M.N., Schulkin, J. and Epstein, A. (1989) The medial amygdala is part of a mineralocorticoid-sensitive circuit controlling NaCl ingestion in the rat. Behav. Brain Res., 35: 159–161.

Ong, J., Kerr, D.I.B., Capper, H.R. and Johnston, G.A.R. (1990) Cortisone: a potent $GABA_A$ antagonist in the guinea pig isolated ileum. J. Pharm. Pharmacol., 42: 662–664.

Ozols, J. (1995) Lumenal orientation and post-translational modification of the liver microsomal 11β-hydroxysteroid dehydrogenase. J. Biol. Chem., 270: 2305–2312.

Peterson, N.A., Chaikoff, I.L. and Jones, C. (1965) The in vitro conversion of cortisol to cortisone by subcellular brain fractions of young and adult rats. J. Neurochem., 12: 273–278.

Rajan, V., Edwards, C.R.W. and Seckl, J.R. (1996) 11β-hydroxysteroid dehydrogenase in cultured hippocampal cells reactivates inert 11-dehydrocorticosterone, potentiating neurotoxicity. J. Neuroscience, 16: 65–70.

Robson, A.C., Leckie, C., Seckl, J.R. and Holmes, M.C. (1998) Expression of 11β-hydroxysteroid dehydrogenase type 2 in the postnatal and adult rat brain. Molec. Brain Res., 61: 1–10.

Roland, B.L., Li, K.X.Z. and Funder, J.W. (1995) Hybridization histochemical localization of 11β-hydroxysteroid dehydrogenase type 2 in rat brain. Endocrinol., 136: 4697–4700.

Sakai, R.R., Lakshmi, V., Monder, C., Funder, J.W., Krozowski, Z. and McEwen, B.S. (1990) Colocalization of 11β-hydroxysteroid dehydrogenase and mineralocorticoid receptor in rat brain. Soc. Neurosci. Abstr. # 7773.

Sakai, R.R., Lakshmi, V., Monder, C. and McEwen, B.S. (1992) Immunocytochemical localisation of 11beta-hydroxysteroid dehydrogenase in hippocampus and other brain regions of the rat. J. Neuroendocrinol., 4: 101–106.

Seckl, J.R. (1997) 11β-hydroxysteroid dehydrogenase in the brain: a novel regulator of glucocorticoid action? Front. Neuroendocrinol., 18: 49–99.

Seckl, J.R. (1998) Physiologic programming of the fetus. Clinics Perinatol., 25: 939–964.

Seckl, J.R. and Walker, B.R. (2001) 11β-hydroxysteroid dehydrogenase type 1: a tissue-specific amplifier of glucocorticoid action. Endocrinology, 142: 1371–1376.

Seckl, J.R., Dow, R.C., Low, S.C., Edwards, C.R.W. and Fink, G. (1993) The 11β-hydroxysteroid dehydrogenase inhibitor glycyrrhetinic acid affects corticosteroid feedback regulation of hypothalamic corticotrophin-releasing peptides. J. Endocrinol., 136: 471–477.

Sholiton, L.J., Werk, E.E. Jr., and MacGee, J. (1965) Metabolism of cortisol-4-^{14}C and cortisone-4-^{14}C by rat brain homogenates. Metab. Clin. Exp., 14: 1122–1127.

Stewart, P.M. and Krozowski, Z.S. (1999) 11 beta-hydroxysteroid dehydrogenase. Vitamins and Hormones, 57: 249–324.

Stewart, P.M., Valentino, R., Wallace, A.M., Burt, D., Shackleton, C.H.L. and Edwards, C.R.W. (1987) Mineralocorticoid activity of liquorice: 11β-hydroxysteroid dehydrogenase deficiency comes of age. Lancet, ii: 821–824.

Stewart, P.M., Wallace, A.M., Atherden, S.M., Shearing, C.H. and Edwards, C.R.W. (1990) Mineralocorticoid activity of carbenoxolone: contrasting effects of carbenoxolone and liquorice on 11β-hydroxysteroid dehydrogenase activity in man. Clin. Sci., 78: 49–54.

Thieringer, R., Le Grand, C., Carbin, L., Cai, T.-Q., Wong, B., Wright, S.D. and Hermanowski-Vosatka, A. (2001) Hydroxysteroid dehydrogenase type 1 is induced in human monocytes upon differentiation to macrophages. J. Immunol., 167: 30–35.

Tye, L.M. and Burton, A.F. (1980) Variation in the pattern of metabolism of corticosteroids in fetal mouse tissues. Life Sci., 26: 35–39.

Venihaki, M.A., Carrigan, P., Dikkes, P. and Majzoub, A. (2000) Circadian rise in maternal glucocorticoid prevents pulmonary dysplasia in fetal mice with adrenal insufficiency. Proc. Natl. Acad. Sci. USA, 97: 7336–7341.

Vreugdenhil, E. and deKloet, E.R. (1998) Corticosteroid hormones and neuronal vulnerability: towards identification of candidate vulnerability genes. Progress in Brain Research, 117: 9–22.

Walker, B.R., Campbell, J.C., Fraser, R., Stewart, P.M. and Edwards, C.R.W. (1992) Mineralocorticoid excess and inhibition of 11β-hydroxysteroid dehydrogenase in patients with ectopic ACTH syndrome. Clinical Endocrinology, 37: 483–492.

Weinstock, M. (2001). Alterations induced by gestational stress in brain morphology and behaviour of the offspring. Prog. Neurobiol., 65: 427–451.

Welberg, L.A.M. and Seckl, J.R. (2001). Prenatal stress, glucocorticoids and the programming of the brain. J. Neuroendocrinol., 13: 113–128.

Welberg, L.A.M., Seckl, J.R. and Holmes, M.C. (2000) Inhibition of 11β-hydroxysteroid dehydrogenase, the feto-placental barrier to maternal glucocorticoids, permanently programs amygdala glucocorticoid receptor mRNA expression and anxiety-like behavior in the offspring. Europ. J. Neurosci., 12: 1047–1054.

Welberg, L.A.M., Seckl, J.R. and Holmes, M.C. (2001) Prenatal glucocorticoid programming of brain corticosteroid receptors and corticotrophin-releasing hormone: possible implications for behaviour. Neuroscience, 104: 71–79.

Yau, J.L.W. and Seckl, J.R. (2001) 11 beta-hydroxysteroid dehydrogenase type I in the brain; thickening the glucocorticoid soup. Molecular Psychiatry, 6(6): 611–614.

Yau, J.L.W., Noble, J., Kenyon, C.J., Hibberd, C., Kotelevtsev, Y., Mullins, J.J. and Seckl, J.R. (2001) Lack of tissue glucocorticoid reactivation in 11β-hydroxysteroid dehydrogenase type 1 knockout mice ameliorates age-related learning impairments. Proc. Natl. Acad. Sci. USA, 98: 4716–4721.

Yongue, B.G. and Roy, E.J. (1987) Endogenous aldosterone and corticosterone in brain cell nuclei of adrenal-intact rats: regional distribution and effects of physiological variations in serum steroids. Brain Res., 436: 49–61.

T. Steckler, N.H. Kalin and J.M.H.M. Reul (Eds.)
Handbook of Stress and the Brain, Vol. 15
ISBN 0-444-51173-3

CHAPTER 3.4

Corticosteroids and the blood–brain barrier

A.M. Karssen*[,1], O.C. Meijer and E.R. de Kloet

Division of Medical Pharmacology, Leiden/Amsterdam Center for Drug Research, Leiden University Medical Center, P.O. Box 9502, 2300 RA Leiden, The Netherlands

Abstract: To reach their central target areas corticosteroid hormones have to enter the brain by passing the blood–brain barrier (BBB), a dynamic barrier that protects the brain from peripheral influences. At this barrier, the efflux transporter P-glycoprotein (Pgp) is expressed, which hampers the uptake into the brain of a wide range of compounds including several glucocorticoids such as dexamethasone, prednisolone, and cortisol. Remarkably, corticosterone uptake is not affected by this Pgp.

P-glycoprotein may play a crucial role as an intermediate between brain and periphery by controlling transport of corticosteroids at the BBB and thus exposure of corticosteroid receptors in brain. Indeed, low-dose dexamethasone treatment leads to dissimilarity in the potency of direct central and peripheral glucocorticoid actions. It suppresses pituitary–adrenal secretion at the pituitary level without replacing endogenous glucocorticoids in the brain.

The lack of Pgp-mediated transport of corticosterone, in contrast to cortisol, suggests a more important role for corticosterone in modulating human brain function than hitherto recognized. In post-mortem human brain samples the ratio of corticosterone over cortisol is six times higher compared with the ratio in plasma.

In the next years, modulation of glucocorticoid access by efflux transporters like Pgp will increasingly emerge as a new level of regulation of glucocorticoid actions in brain.

Introduction

Glucocorticoids are important modulators of brain function and hypothalamus–pituitary–adrenal (HPA)-axis activity. To reach their central target areas they have to enter the brain by passing the blood–brain barrier (BBB), a dynamic barrier that protects the brain from peripheral influences. Although the importance of the actions of glucocorticoids in brain is commonly accepted, modulation of glucocorticoid access at the BBB level has hardly been a subject of research as these compounds are considered to readily pass this barrier. Now that it has been demonstrated that transmembrane proteins are able to transport these hormones (Bourgeois et al., 1993; Thompson, 1995; Ueda et al., 1996), this issue becomes an increasingly interesting subject to study. Any process at the BBB that can influence the endothelial crossing of glucocorticoids would directly affect central corticosteroid receptor occupancy and the magnitude of the central response to corticosteroids.

Glucocorticoid transport at the blood–brain barrier

Hampered brain uptake of dexamethasone

The naturally occurring glucocorticoid, corticosterone, readily gains access to the brain and accumulates in limbic brain areas expressing mineralocorticoid receptor (MR) (McEwen et al., 1968; De Kloet et al.,

*Corresponding author. Tel.: +1-650-498-5187; Fax: +1-650-498-7761;
[1]Present address: School of Medicine, Department of Psychiatry and Behavioral Sciences, Stanford University, MSLS, P124, 1201 Welch Road, Palo Alto, CA 94304-5485, USA. Fax: +1-650-498-7761. E-mail: akarssen@stanford.edu

1975; De Kloet, 1991). In contrast, the synthetic glucocorticoid dexamethasone, when administered in tracer doses to adrenalectomised rats or mice, is poorly retained in glucocorticoid receptor (GR) containing areas in brain (De Kloet et al., 1974; Rees et al., 1975). Their diversity in receptor affinity can only partially explain this differential retention pattern (Reul and De Kloet, 1985; Reul et al., 2000b), as the anterior pituitary which also expresses high amounts of GR, retains high amounts of dexamethasone. To explain this puzzling phenomenon the existence of a BBB limiting the uptake of dexamethasone into the brain was postulated (De Kloet et al., 1975; Rees et al., 1975). Recently, it was indeed demonstrated that penetration of dexamethasone into the brain is hampered because the efflux transporter P-glycoprotein (Pgp) excludes this exogenous compound from glucocorticoid receptors in mouse brain (Schinkel et al., 1995; Meijer et al., 1998). Pgp is expressed at the luminal membranes of endothelial cells of the BBB (Schinkel, 1999).

Blood–brain barrier and P-glycoprotein

The BBB is a dynamic physical and metabolic barrier consisting of specialized endothelial cells that protects the brain from blood-borne compounds, and plays a role in maintaining brain homeostasis (Bradbury, 1993) (Fig. 1). Just like the pituitary, some brain areas such as the circumventricular organs lie outside this barrier, but most of the brain is shielded from the periphery by the tight junctions between brain capillary endothelial cells and other barrier properties of these cells. The latter features comprise the lack of fenestrations and pinocytotic vesicles and the presence of metabolic enzymes and special transporter proteins, which can facilitate the uptake or impede the entry of compounds (Lee et al., 2001b). The BBB can strongly interfere with the access to and the distribution to the brain of endogenous and exogenous compounds. Generally, hydrophilic and large lipophilic compounds are not able to penetrate the brain, as they are not able to pass cell membranes, whereas small lipophilic compounds can easily cross the BBB by passive diffusion through the endothelial cells. For a number of highly lipophilic compounds BBB permeability is, however, unexpectedly low.

The multidrug transporter Pgp is an important functional component of the BBB and various other tissues with a barrier function (Gottesman and Pastan, 1993; Schinkel, 1999) (Fig. 1). It acts like a "gatekeeper" at the BBB actively keeping a wide variety of lipophilic, potentially neurotoxic substances

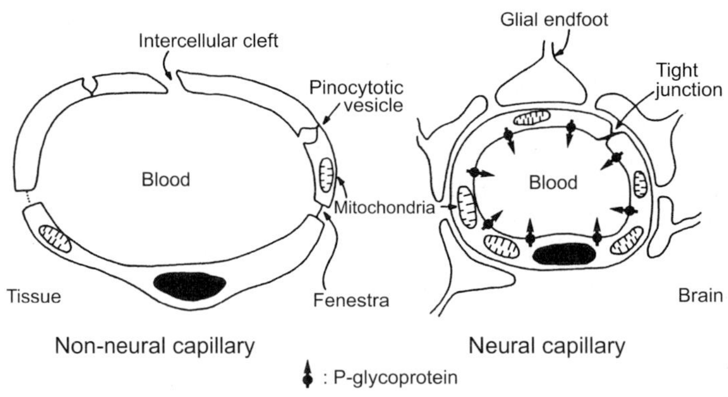

Fig. 1. Schematic representations of the anatomy of a typical blood vessel in peripheral and brain tissue, respectively. Unlike peripheral capillary endothelial cells, brain capillary endothelial cells are closely sealed by tight junctions, they display no intercellular clefts and little fenestration or pinocytosis, and they have a relatively high number of mitochrondia. Some of these characteristics are induced and maintained by astrocyte foot processes that are closely attached to and extensively envelop the brain endothelium. For simplicity, the supporting pericytes and the basal lamina, structural connective tissue surrounding the blood capillaries and separating the glial endfoots from the brain endothelial cells are not shown. The various BBB-specific transporters are not shown except for the efflux transporter P-glycoprotein (indicated by the balls and arrows). This transmembrane protein is localized at the luminal membrane of the endothelial cells and transports its substrates (back) into the blood in an energy-dependent manner. Reprinted from Schinkel (1999) with copyright permission from Elsevier Science.

out of the brain. Pgp belongs to the adenosine triphosphate (ATP)-binding cassette (ABC) transporter proteins (Borst and Elferink, 2002). This transmembrane protein is encoded by the multidrug resistance (MDR) genes, mdr1a in rodents and the highly homologous MDR1 in humans. Studies using mdr1a (−/−) knockout mice, which lack functional Pgp at the BBB, have shown that this efflux transporter is responsible for the apparent low permeation of dexamethasone and a wide range of other unrelated compounds that should easily penetrate the BBB as expected on the basis of their size and their sufficient high lipid solubility (Schinkel et al., 1994; Meijer et al., 1998). Pgp likely extrudes its substrates directly from the lipid bilayer even before they can enter the cytoplasm (Gottesman and Pastan, 1993; Sharom, 1997). Corticosteroids are commonly believed to cross endothelial barriers with relative ease by virtue of their highly lipophilic nature and their small size. Dexamethasone has thus long been considered an exception.

P-glycoprotein-mediated steroid transport

Interestingly, in vitro studies have revealed that not only dexamethasone is a substrate of Pgp, but also several other steroids (Ueda et al., 1996). Human Pgp expressed in MDR1 cDNA-transfected pig kidney epithelial cell lines transports dexamethasone, but also cortisol, cortisone, prednisolone and, to a lesser extent, aldosterone (Ueda et al., 1992; Karssen et al., 2001; Karssen et al., 2002) (Table 1). When seeded on filters these cells form monolayers with Pgp expressed at the apical side and show polar translocation of Pgp substrates reminiscent of the BBB Pgp function (Fig. 2). In addition, Pgp-expressing cells show reduced accumulation of dexamethasone and cortisol (Barnes et al., 1996). These findings suggested that Pgp might play a role in modulating exposure of the brain to various glucocorticoids including prednisolone and cortisol. Autoradiographic studies in adrenalectomised mice with a disrupted mdr1a gene have indeed confirmed that both cortisol and prednisolone like dexamethasone are hampered to enter the mouse brain and to reach glucocorticoid receptors (Karssen et al., 2001, 2002) (Fig. 3A and B).

Remarkably, Pgp does not affect corticosterone uptake in mouse brain (Fig. 3C and D), nor does it affect transport of corticosterone in stably MDR1-transfected monolayers (Karssen et al., 2001). Corticosterone transport by Pgp has been reported, but these studies mostly rely on Pgp encoded by mdr1b (Wolf and Horwitz, 1992; Bourgeois et al., 1993). This second rodent mdr1 gene has been found to express another Pgp with very similar substrate specificity (Devault and Gros, 1990), but this gene is not expressed at the BBB (Jette et al., 1995; Demeule et al., 2001). It may have some capacity to transport corticosterone in addition to its large capacity to transport dexamethasone, prednisolone, and cortisol (Bourgeois et al., 1993). MDR1 and mdr1b Pgp might play a presently unestablished role in steroid transport out of adrenocortical cells (Ueda et al., 1996; Ambudkar et al., 1999), although the rat, in contrast to mouse and human, lacks any Pgp in the adrenal, and mice with a disrupted mdr1b gene do not show gross disturbances in corticosteroid handling (Borst and Elferink, 2002).

Molecular structure

It is remarkable that, in spite of its broad spectrum of substrates (Schinkel et al., 1994), Pgp distinguishes subtle differences in corticosteroid structure. Comparison of the molecular structures of steroids with reference to their ability to be transported by Pgp reveals that the 17-hydroxyl moiety, in combination

Table 1. Steroid transport capability of P-glycoprotein

Corticosteroid	Transport by Pgp	References
Cortisol/hydrocortisone	+	1,2,3
Corticosterone	−*	1,3*
Dexamethasone	+	2,4,5
Prednisolone	+	3,6
Methylprednisolone	+	3,7
Triamcinolone	+	3
Betamethasone	+	3
Cortisone	+	1
Cortexolone/deoxycortisol	−	3,6
Deoxycorticosterone	−	3
Aldosterone	+/−	2,3,6
RU486/mifepristone**	−/inhibitor	8

*Weakly transported by mdr1b Pgp.
**GR antagonist.
1, Karssen et al., 2001; 2, Ueda et al., 1992, Bourgeois et al., 1993, 4, Meijer et al., 1998; 5, Schinkel et al., 1995; 6, Karssen et al., 2002; 7, Koszdin, et al., 2000; 8, Gruol et al., 1994.

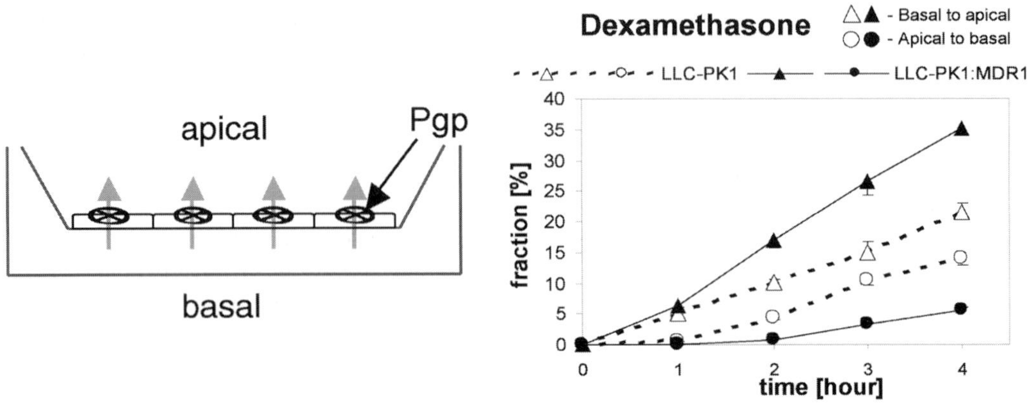

Fig. 2. Pgp-mediated transepithelial transport of dexamethasone. (A) In monolayers of pig kidney epithelial (LLC-PK1) cells stably transfected with human MDR1 cDNA Pgp is localized at the apical side, transporting its substrates from the basal compartment to the apical compartment. (B) Transepithelial transport of 7 nM ^{3}H-dexamethasone from basal to apical (triangles) and from apical to basal (circles) was measured in monolayers of MDR1 transfected LLC-PK1 (solid line) or host cells (broken line). Dexamethasone was transported in a polarized fashion in MDR1 monolayers, but not in untransfected monolayers. Presented is the fraction of the dose of radioactivity added to the opposite compartment at $t = 0$ that is present in medium at different time points. Each point represents the mean of three monolayers ± SEM.

with an 11-hydroxyl or 11-oxo moiety, might determine the ability of Pgp to transport steroids (Bourgeois et al., 1993; Karssen et al., 2001, 2002). Steroids having both these hydroxyl-groups (such as dexamethasone, cortisol, (methyl)prednisolone, and cortisone) are effectively transported by Pgp. Steroids lacking one of these groups (like aldosterone, corticosterone, and cortexolone) and steroids without any of these groups are minimally if at all transported. The high-affinity MR-ligand deoxycorticosterone belongs to the latter group and therefore it should easily be retained in brain. However, although it readily enters the brain (Kraulis et al., 1975), deoxycorticosterone is poorly retained by MR in different brain areas and pituitary of adrenalectomised rats (McEwen et al., 1976). This suggests that there are additional factors, e.g. local metabolism, determining the retention of this mineralocorticoid in potential target areas.

Cerebrospinal fluid

The hampered uptake into the brain implies that plasma levels of cortisol, even "free," non-CBG-bound cortisol, or of dexamethasone may not directly represent brain levels. Unfortunately, determination of cerebrospinal fluid (CSF) levels may be of limited value, as Pgp is not expressed at the blood side of the blood–CSF barrier (BCB) (Rao et al., 1999). Indeed, autoradiography film data showed that in wild-type mice radioactive labelling was restricted to the ventricles (Fig. 3), indicating free access of glucocorticoids through the BCB. Cortisol, dexamethasone, and other Pgp substrates like prednisolone may slowly gain access to the brain through the cerebroventricular system (Rees et al., 1975) or circumventricular organs and may reach brain areas in the immediate vicinity of the ventricles. However, since the surface of the BBB is approximately 5000 times greater than the surface of the BCB (Pardridge et al., 1981), the uptake in brain tissue will remain considerably reduced even in presence of high plasma levels. Brain and CSF likely constitutes different compartments not reflecting linear relationships in glucocorticoid concentrations.

Direct effects of glucocorticoids on BBB

Glucocorticoids may be able to tighten the BBB decreasing its permeability to other compounds. Adrenalectomy increases permeability of the BBB to macromolecules, which was restored by corticosterone replacement (Long and Holaday, 1985). Since the GR-antagonist mifepristone (RU486) blocks the

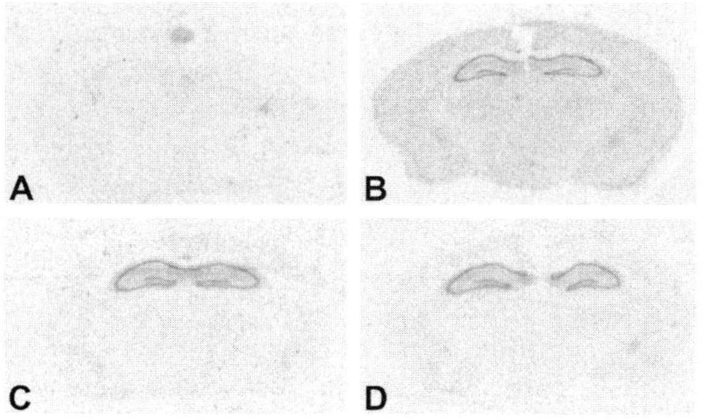

Fig. 3. Differential uptake of corticosterone and cortisol in presence of P-glycoprotein. Representative autoradiograms of 12-μm coronal sections of the brain of wild-type (A,C) and mdr1a knockout mice (B,D) at hippocampus level. Autoradiograms show radioactive labeling at 1 h after systemic treatment with tracer doses of ^{3}H-cortisol (A,B) or ^{3}H-corticosterone (C,D). Note that cortisol rather than corticosterone is hampered in penetration into the mouse brain by P-glycoprotein present at the BBB. The dark spots in (A) represent transverse sectioning of the cerebroventricular space and chorioid plexus. Data are from Karssen et al. 2001), reprinted with permission, © The Endocrine Society. Dexamethasone also poorly enters wild-type, but not mdr1a knockout brain (Meijer et al., 1998).

gradual tightening of an in vitro BBB induced by glucocorticoids, the GR is probably involved in the physiological regulation of BBB permeability (Gaillard et al., 2001). As various stressors may increase permeability of the BBB, glucocorticoids may mediate a protective response to potentially damaging effects of these stressors at the level of the BBB. It is not known whether glucocorticoids may also directly alter Pgp functionality at the BBB. Unlike the mdr1b promoter, the mdr1a, and MDR1 promoters do not have a glucocorticoid responsive-element (Cohen et al., 1991; Labialle et al., 2002), although regulation of expression through protein–protein interactions of glucocorticoid receptors and other transcription factors might be possible.

Functional implications

Differential uptake of cortisol and corticosterone

The differential uptake of corticosterone and cortisol may have important implications for glucocorticoid feedback to the human brain. In contrast to rodents, both cortisol and corticosterone are circulating in human plasma, although corticosterone is present at 10–20 fold lower levels than cortisol. Relevance of Pgp-mediated transport of cortisol for endogenous glucocorticoid exposure of the human brain is suggested by the increased corticosterone : cortisol ratio in dichloromethane extracts of post-mortem brain as compared to human plasma (Karssen et al., 2001). While in human plasma corticosterone concentrations are only 5% of cortisol levels, in the brain corticosterone levels are 30% of those of cortisol as determined by liquid chromatography–mass spectrometry. Although no direct comparison can be made between absolute plasma and brain glucocorticoid levels from the same subjects, these findings do suggest that cortisol levels in human brain are six times lower than those in human blood. Thus, in contrast to corticosterone, which readily enters rodent and human brain, the main endogenous glucocorticoid in human appears to be partially excluded from the human brain. However, this notion should not be misconstrued, as, in spite of its hampered access, cortisol will still reach the brain in sufficient amounts to activate corticosteroid receptors. Indeed, cortisol exerts some feedback actions in the human brain and influences brain functioning. On the other hand, peripheral actions probably are relatively more potent, provided that these target cells do not express Pgp.

The preferential uptake of corticosterone in human brain suggests that this endogenous glucocorticoid may play a more prominent role in human brain function than hitherto recognized. Owing to the differential uptake of cortisol and corticosterone, the human glucocorticoid feedback system might be more complex than the rodent system. The resultant species difference may impede extrapolation of data regarding central glucocorticoid action from rodent to human. The question arises whether cortisol and corticosterone might actually affect brain function differently. The differential interaction of Pgp with both hormones is exceptional in the sense that neither pharmacological nor physiological differences between both hormones have been acknowledged thus far.

However, a more thorough examination of literature data reveals small but consistent differences between both corticosteroids in affinity and transactivation properties of corticosteroid receptors, particularly the MR. Different research groups have consistently shown that corticosterone has a slightly higher affinity for rat, dog, and human MR (Krozowski and Funder, 1983; De Kloet et al., 1984; Arriza et al., 1987; Reul et al., 1990), which may underlie the reported higher effectiveness of corticosterone in promoting human MR transactivation (Lombes et al., 1994; Hellal-Levy et al., 1999). Species differences do exist however, since hamster MR appears to be a cortisol-preferring receptor (Sutanto et al., 1988). With regard to GR, cortisol might be the more potent one in transactivation (Hellal-Levy et al., 1999), but more thorough studies are needed to verify this difference.

Tentatively, corticosterone may be the more effective glucocorticoid at the MR in human brain with a potentially different role than cortisol. In light of this, it may be of relevance that the distribution of MR in humans seems to be broader than in rodents, with relatively high levels found in the prefrontal cortex (Lopez et al., 1999). This structure is essential for mood and cognitive processing and may be particularly sensitive to glucocorticoid feedback in humans (Lupien and Lepage, 2001). Furthermore, a recent study reported that GR levels are relatively low in hippocampus of rhesus monkey in contrast to MR levels (Sanchez et al., 2000). Although this latter finding should be confirmed for human hippocampus, the species-specific distribution of MR suggests that this receptor might have a more pronounced role in mediating glucocorticoid actions in human brain than in rodent brain.

It would be interesting to know how much either corticosterone and cortisol contributes to stabilization of neuronal excitability, maintenance of neuronal integrity, suppression of HPA activity, and facilitation of behavioral adaptation (De Kloet et al., 1998).

Consequences of impaired access of dexamethasone

Owing to its hampered uptake dexamethasone may create a low-corticosteroid condition selectively in the brain reminiscent of a central adrenalectomy-like state (Fig. 4). This is the case because, when present at low to moderate plasma concentrations, the high-affinity GR ligand dexamethasone predominantly act on the anterior pituitary rather than on the brain to suppress pituitary–adrenal secretion (De Kloet et al., 1974; Miller et al., 1992). Under these conditions dexamethasone will replace corticosterone at peripheral glucocorticoid targets, but in the brain it would poorly substitute for the depleted endogenous hormone, particularly at the MR. Since these low concentrations of dexamethasone already strongly suppress ACTH and corticosterone secretion, dexamethasone-induced effects on brain function should probably be ascribed to decreased rather than increased central glucocorticoid action. We have indeed demonstrated that dexamethasone circulating at low concentrations – after a three-week treatment with less than 1 μg/ml through drinking water – does not feed back at several glucocorticoid-responsive genes expressed in the hypothalamic paraventricular nucleus (PVN) (Karssen et al., 2003). In contrast, these concentrations of dexamethasone potently act on various peripheral glucocorticoid targets including the pituitary.

At higher plasma levels, direct central action might progressively emerge, as after high-dose treatment dexamethasone suppresses expression of glucocorticoid-responsive genes indicating that under these conditions it is able to enter the brain in sufficient amounts to activate central GR.

The low-corticosteroid state created by low levels of dexamethasone in the brain may even increase the stress responsivity of the paraventricular corticotropin-releasing factor/vasopressin (CRF/AVP)-secreting neurons due to removal of glucocorticoid negative feedback to the PVN itself and to brain areas involved in activation of the PVN. Consequently, the central drive on the pituitary in response to stress will be augmented, which may override the dexamethasone-mediated inhibition of pituitary ACTH production and secretion.

As dexamethasone does not bind to MR in vivo, prolonged treatment with dexamethasone might, irrespective of the dose, result in a reduced occupation of MR through suppression of the endogenous MR ligands cortisol/corticosterone. Although acute administration of dexamethasone did not lead to a noticeable depletion of hippocampal MR

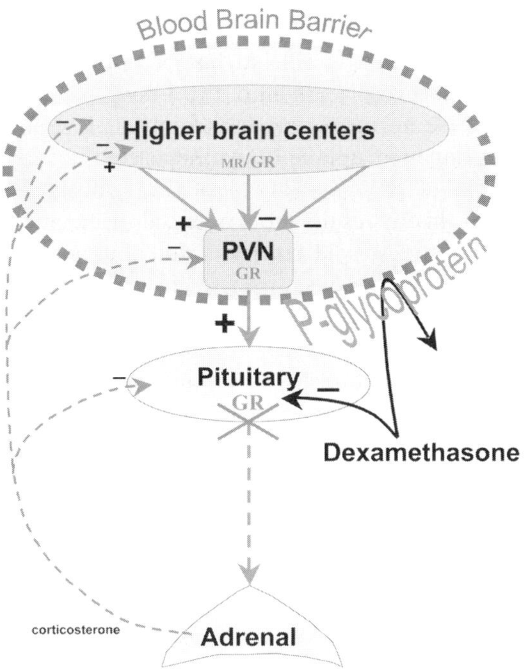

Fig. 4. Creation of brain-selective low-corticosteroid condition by low-dose dexamethasone. Low plasma concentrations of dexamethasone primarily act on the anterior pituitary to suppress pituitary–adrenal secretion. Dexamethasone replaces corticosterone at peripheral glucocorticoid targets. However, due to the presence of P-glycoprotein at the BBB dexamethasone cannot replace corticosterone in the brain. The ensuing low-corticosteroid state in the brain is reminiscent of a brain-selective adrenalectomy condition, and will likely affect glucocorticoid actions on brain function.

(Reul et al., 1987), prolonged administration did (Spencer et al., 1990).

This will lead to a shift in activation of MR relative to GR. As both receptor types mediate distinct but co-ordinated actions on neuronal excitability, synaptic plasticity, and learning and memory, this shift in MR/GR balance may disturb brain function and may thus further impair the ability to maintain brain homeostasis.

Other corticosteroids that are substrates of Pgp are less likely to create this central adrenalectomy-like condition, as the high affinity to GR, lack of plasma binding, and long-lasting activity favors the potency of dexamethasone to strongly inhibit pituitary–adrenal secretion. To suppress pituitary–adrenal secretion to the same extent high doses of cortisol (about 70-fold those of dexamethasone) are needed, doses that will likely reach the brain activating central MR and GR.

Glucocorticoid feedback in depression

The dexamethasone suppression test (DST) is widely used in the clinic often in combination with a CRF challenge, to evaluate the dysregulation of the HPA axis in depressive patients (Holsboer, 2000). In many of these patients excessive release of CRF and vasopressin (AVP) from PVN neurons into the portal circulation leads to a hyperactivity of the HPA axis (Gold and Chrousos, 2002). A majority of depressed patients show escape from suppression of baseline or CRF-induced cortisol levels after administration of a low dose of dexamethasone the night before. Although nonsuppression in the DST is often ascribed to decreased negative feedback through corticosteroid receptors, an imbalance between central drive and dexamethasone feedback inhibition at the pituitary level may also underlie the HPA abnormalities. As dexamethasone is administered at a low dose it will mainly act on the pituitary. At this level, the excessively increased drive of CRF/AVP may be strong enough to override even properly functioning negative feedback. On the other hand, it cannot be excluded that the excessively increased drive itself is caused by an aberrant glucocorticoid feedback mechanism in the brain, a level of glucocorticoid action that is most likely not tested in the DST.

The dexamethasone-induced reduction of central feedback may even further aggravate the hyperactive central drive of both CRF and AVP in depressed patients leading to the escape from the suppressive effect of dexamethasone on cortisol plasma levels. Increased AVP release may particularly result in an apparent glucocorticoid feedback resistance at the pituitary level, as AVP-stimulated ACTH secretion is refractory to glucocorticoid feedback (Aguilera and Rabadan-Diehl, 2000). Dexamethasone suppression in healthy subjects can indeed be overcome by concurrent infusion of CRF and AVP (Holsboer, 1999).

A better way to explore the efficacy of the central corticosteroid receptor feedback system may be provided by the application of corticosterone. As corticosterone easily enters the brain, it likely acts predominantly on the central feedback target areas. It would therefore be able to directly suppress the CRF/AVP drive in the PVN. Besides, as corticosterone also activates MR, a corticosterone suppression test would also probe the function of this receptor. Regulation of HPA axis is partly mediated by MR (De Kloet, 1991; Reul et al., 2000b) and may even be the more relevant receptor type involved in dysregulation of HPA-axis seen in stress-related disorders (Reul et al., 2000a; Makino et al., 2002).

Antidepressants and P-glycoprotein

An intriguing alternative mechanism of action of antidepressants involving steroid membrane transporters like Pgp was recently postulated by Pariante et al. (2001, 2003, 2004) based on in vitro findings. They showed that coincubation of dexamethasone or cortisol with various antidepressants that are Pgp substrates resulted in enhanced GR function without affecting GR levels in L929 fibroblast cells. Coincubation with dexamethasone in presence of a Pgp inhibitor or with corticosterone does not lead to a facilitation of GR function, suggesting involvement of transport mediated by a steroid membrane transporter like Pgp. This may imply that antidepressants may inhibit the function of Pgp at the BBB, thus increasing uptake of cortisol into the brain and enhancing GR-mediated negative feedback on the HPA axis, decreasing HPA-axis hyperactivity in depression. Acute facilitation of GR function may precede upregulation of GR levels, implying that GR upregulation may be the consequence of facilitated GR function rather than the cause (Pariante and Miller, 2001). However, upregulation of MR preceding GR upregulation may also make up the primary cause of restoring normal HPA-axis activity in major depression (Seckl and Fink, 1992; Reul et al., 1993).

Conclusions and perspectives

In conclusion, the importance of corticosteroid transport at the BBB in controlling corticosteroid access to the brain has become increasingly evident in the last decade. Efflux transporters like Pgp may play a crucial role as an intermediate between brain and periphery by controlling transport of corticosteroids at the BBB. Pgp is able to hamper penetration of various corticosteroids into the brain, particularly when these hormones are circulating at low plasma levels. Impaired uptake of synthetic glucocorticoids, but also of the naturally occurring glucocorticoid cortisol, likely results in a reduced occupation of central corticosteroid receptors and thus in a diminished central response to these glucocorticoids.

Intriguingly, both mouse and human Pgp do not transport corticosterone in contrast to cortisol. Future investigations will reveal whether corticosterone rather than cortisol may be the major endogenous corticosteroid in mediating corticosteroid actions, particularly via MR, on human brain function, as suggested by the preferential uptake of endogenous corticosterone into human brain.

The brain-selective low-corticosteroid state created by administration of low-dose dexamethasone to rats might be used as a novel animal model to specifically study central roles of corticosterone without the potentially confounding effects of reduced peripheral glucocorticoid effects.

At the BBB several other efflux transporters are expressed besides Pgp (Lee et al., 2001a; Borst and Elferink, 2002). For instance, several members of the multidrug resistance-associated proteins (MRPs) have been detected at the brain capillaries. Whether any of these or any yet unknown transporter may transport corticosteroids as well remains to be resolved. Furthermore, steroid membrane transporters might be present at neuronal cells as well, affecting uptake of glucocorticoids directly at the neuronal level (Pariante et al., 2003). Efflux transporters may even co-exist with a membrane uptake system for corticosterone, as recently suggested (Pariante et al., 2004).

Modulation of Pgp-mediated transport of corticosteroids may influence central glucocorticoid actions, as exemplified by the inhibition of corticosteroid transport by antidepressants. Altered uptake of glucocorticoids may reset MR/GR balance affecting neuronal function and HPA-axis activity. Therefore, Pgp may provide an interesting new target to regulate glucocorticoid feedback to the brain in disorders with disturbed central glucocorticoid signaling such as major depression and posttraumatic stress disorder, and possibly also chronic fatigue syndrome and fibromyalgia.

P-glycoprotein plays a key role in protection against a wide variety of drugs including anticancer and antiepileptic drugs and HIV protease inhibitors. These drugs may influence transport of endogenous substrates including corticosteroid hormones, but also centrally and peripherally acting compounds (King et al., 2001; Lam et al., 2001). Particularly, inhibitors of Pgp transport may interfere with physiological Pgp function. Some steroids (e.g. progesterone, RU486) may inhibit Pgp function (Gruol and Bourgeois, 1994). As Pgp impairs the efficacy of treatment of (brain) cancer, much effort has been put in finding reversal agents to bypass Pgp by inhibiting its transport function. Although some of these Pgp modulators are now in clinical trials, the clinical efficacy remains to be established particularly with regard to their potential side effects (Van Zuylen et al., 2000). Inhibition of Pgp may have undesired side effects by increasing the uptake of cortisol into the brain potentially endangering neuronal survival.

Polymorphisms in the MDR1 gene resulting in altered levels and functionality of Pgp have been shown to affect efficiency of this transporter (Hoffmeyer et al., 2000). Regulation of expression and post-translational modification of Pgp presumably also affect Pgp-efflux transport function. With regard to BBB Pgp, little is known about these features. Much work is still to be done to resolve when and how Pgp operates as a dynamic regulator of the central access of Pgp substrates including glucocorticoids.

In addition, it should be resolved whether corticosteroids can directly affect Pgp functionality. Preliminary data suggest that glucocorticoids might be able to increase Pgp expression in brain (Sérée et al., 1998; Aquilante et al., 2000). This would not only affect glucocorticoid access to the brain but also access of other Pgp substrates.

Abbreviations

ABC transporter	adenosine triphosphate (ATP)-binding cassette transporter
ACTH	adrenocorticotropic hormone
AVP	arginine vasopressin
BBB	blood–brain barrier
BCB	blood–CSF barrier
CRF	corticotropin-releasing factor
CSF	cerebrospinal fluid
DST	dexamethasone suppression test
GR	glucocorticoid receptor
HPA axis	hypothalamus–pituitary–adrenal axis
MDR	multidrug resistance
MR	mineralocorticoid receptor
MRP	multidrug resistance-associated protein
Pgp	P-glycoprotein
PVN	hypothalamic paraventricular nucleus

References

Aguilera, G. and Rabadan-Diehl, C. (2000) Vasopressinergic regulation of the hypothalamic–pituitary–adrenal axis: implications for stress adaptation. Regul. Pept., 96: 23–29.

Ambudkar, S.V., Dey, S., Hrycyna, C.A., Ramachandra, M., Pastan, I. and Gottesman, M.M. (1999) Biochemical, cellular, and pharmacological aspects of the multidrug transporter. Annu. Rev. Pharmacol. Toxicol., 39: 361–398.

Aquilante, C.L., Letrent, S.P., Pollack, G.M. and Brouwer, K.L.R. (2000) Increased brain P-glycoprotein in morphine tolerant rats. Life Sci., 66: L47–L51.

Arriza, J.L., Weinberger, C., Cerelli, G., Glaser, T.M., Handelin, B.L., Housman, D.E. and Evans, R.M. (1987) Cloning of human mineralocorticoid receptor complementary DNA: structural and functional kinship with the glucocorticoid receptor. Science, 237: 268–275.

Barnes, K.M., Dickstein, B., Cutler, G.B., Fojo, T. and Bates, S.E. (1996) Steroid transport, accumulation, and antagonism of P-glycoprotein in multidrug-resistant cells. Biochemistry, 35: 4820–4827.

Borst, P. and Elferink, R.O. (2002) Mammalian abc transporters in health and disease. Annu. Rev. Biochem., 71: 537–592.

Bourgeois, S., Gruol, D.J., Newby, R.F. and Rajah, F.M. (1993) Expression of an mdr gene is associated with a new form of resistance to dexamethasone-induced apoptosis. Mol. Endocrinol., 7: 840–851.

Bradbury, M.W. (1993) The blood–brain barrier. Exp. Physiol., 78: 453–472.

Cohen, D., Piekarz, R.L., Hsu, S.I., DePinho, R.A., Carrasco, N. and Horwitz, S.B. (1991) Structural and functional analysis of the mouse mdr1b gene promoter. J. Biol. Chem., 266: 2239–2244.

De Kloet, E.R. (1991) Brain corticosteroid receptor balance and homeostatic control. Front. Neuroendocrinol., 12: 95–164.

De Kloet, E.R., Van der Vies, J. and De Wied, D. (1974) The site of the suppressive action of dexamethasone on pituitary–adrenal activity. Endocrinology, 94: 61–73.

De Kloet, E.R., Wallach, G. and McEwen, B.S. (1975) Differences in corticosterone and dexamethasone binding to rat brain and pituitary. Endocrinology, 96: 598–609.

De Kloet, E.R., Veldhuis, H.D., Wagenaars, J.L. and Bergink, E.W. (1984) Relative binding affinity of steroids for the corticosterone receptor system in rat hippocampus. J. Steroid Biochem., 21: 173–178.

De Kloet, E.R., Vreugdenhil, E., Oitzl, M.S. and Joëls, M. (1998) Brain corticosteroid receptor balance in health and disease. Endocr. Rev., 19: 269–301.

Demeule, M., Labelle, M., Regina, A., Berthelet, F. and Beliveau, R. (2001) Isolation of endothelial cells from brain, lung, and kidney: expression of the multidrug resistance P-glycoprotein isoforms. Biochem. Biophys. Res. Commun., 281: 827–834.

Devault, A. and Gros, P. (1990) Two members of the mouse mdr gene family confer multidrug resistance with overlapping but distinct drug specificities. Mol. Cell. Biol., 10: 1652–1663.

Gaillard, P.J., van der Meide, P.H., De Boer, A.G. and Breimer, D.D. (2001) Glucocorticoid and type I interferon interactions at the blood–brain barrier: relevance for drug therapies for multiple sclerosis. Neuroreport, 12: 2189–2193.

Gold, P.W. and Chrousos, G.P. (2002) Organization of the stress system and its dysregulation in melancholic and atypical depression: high vs low CRH/NE states. Mol. Psychiatry, 7: 254–275.

Gottesman, M.M. and Pastan, I. (1993) Biochemistry of multidrug resistance mediated by the multidrug transporter. Annu. Rev. Biochem., 62: 385–427.

Gruol, D.J. and Bourgeois, S. (1994) Expression of the mdr1 P-glycoprotein gene: a mechanism of escape from glucocorticoid-induced apoptosis. Biochem. Cell Biol., 72: 561–571.

Gruol, D.J., Zee, M.C., Trotter, J. and Bourgeois, S. (1994) Reversal of multidrug resistance by RU 486. Cancer Res., 54: 3088–3091.

Hellal-Levy, C., Couette, B., Fagart, J., Souque, A., Gómez-Sánchez, C.E. and Rafestin-Oblin, M. (1999) Specific hydroxylations determine selective corticosteroid recognition by human glucocorticoid and mineralocorticoid receptors. FEBS Lett., 464: 9–13.

Hoffmeyer, S., Burk, O., Von Richter, O., Arnold, H.P., Brockmoller, J., Johne, A., Cascorbi, I., Gerloff, T., Roots, I., Eichelbaum, M. and Brinkmann, U. (2000) Functional polymorphisms of the human multidrug-resistance gene: multiple sequence variations and correlation of one allele with P-glycoprotein expression and activity in vivo. Proc. Natl. Acad. Sci. USA, 97: 3473–3478.

Holsboer, F. (1999) The rationale for corticotropin-releasing hormone receptor (CRH-R) antagonists to treat depression and anxiety. J. Psychiatr. Res., 33: 181–214.

Holsboer, F. (2000) The corticosteroid receptor hypothesis of depression. Neuropsychopharmacology, 23: 477–501.

Jette, L., Pouliot, J.F., Murphy, G.F. and Béliveau, R. (1995) Isoform I (mdr3) is the major form of P-glycoprotein expressed in mouse brain capillaries. Evidence for cross-reactivity of antibody C219 with an unrelated protein. Biochem. J., 305: 761–766.

Karssen, A.M., Meijer, O.C., Van der Sandt, I.C.J., Lucassen, P.J., De Lange, E.C.M., De Boer, A.G. and De Kloet, E.R. (2001) Multidrug resistance P-glycoprotein hampers the access of cortisol but not of corticosterone to mouse and human brain. Endocrinology, 142: 2686–2694.

Karssen, A.M., Meijer, O.C., Van der Sandt, I.C.J., De Boer, A.G., De Lange, E.C.M. and De Kloet, E.R. (2002) The role of the efflux transporter P-glycoprotein in brain penetration of prednisolone. J. Endocrinol., 175: 251–260.

Karssen, A.M., Meijer, O.C., Visser, M., Sanjuan Piñol, R. and De Kloet, E.R. (2003) Exclusion of dexamethasone from the brain leads to a selective central hypocorticosteroid state. In: Karssen, A.M. (Ed.), Thesis, Ch. 5. Leiden University, Leiden, pp. 95–114. http://www.etcl.nl/diss/med/2003/002/Karssen.htm.

King, M., Su, W., Chang, A., Zuckerman, A. and Pasternak, G.W. (2001) Transport of opioids from the brain to the periphery by P-glycoprotein: peripheral actions of central drugs. Nat. Neurosci., 4: 268–274.

Koszdin, K.L., Shen, D.D. and Bernards, C.M. (2000) Spinal cord bioavailability of methylprednisolone after intravenous and intrathecal administration – the role of P-glycoprotein. Anesthesiology, 92: 156–163.

Kraulis, I., Foldes, G., Traikov, H., Dubrovsky, B. and Birmingham, M.K. (1975) Distribution, metabolism and biological activity of deoxycorticosterone in the central nervous system. Brain Res., 88: 1–14.

Krozowski, Z.S. and Funder, J.W. (1983) Renal mineralocorticoid receptors and hippocampal corticosterone-binding species have identical intrinsic steroid specificity. Proc. Natl. Acad. Sci. USA, 80: 6056–6060.

Labialle, S., Gayet, L., Marthinet, E., Rigal, D. and Baggetto, L.G. (2002) Transcriptional regulators of the human multidrug resistance 1 gene: recent views. Biochem. Pharmacol., 64: 943–948.

Lam, F.C., Liu, R., Lu, P., Shapiro, A.B., Renoir, J.M., Sharom, F.J. and Reiner, P.B. (2001) beta-Amyloid efflux mediated by p-glycoprotein. J. Neurochem., 76: 1121–1128.

Lee, G., Dallas, S., Hong, M. and Bendayan, R. (2001a) Drug transporters in the central nervous system: brain barriers and brain parenchyma considerations. Pharmacol. Rev., 53: 569–596.

Lee, G., Schlichter, L., Bendayan, M. and Bendayan, R. (2001b) Functional expression of P-glycoprotein in rat brain microglia. J. Pharmacol. Exp. Ther., 299: 204–212.

Lombes, M., Kenouch, S., Souque, A., Farman, N. and Rafestin-Oblin, M.E. (1994) The mineralocorticoid receptor discriminates aldosterone from glucocorticoids independently of the 11β-hydroxysteroid dehydrogenase. Endocrinology, 135: 834–840.

Long, J.B. and Holaday, J.W. (1985) Blood–brain barrier: endogenous modulation by adrenal–cortical function. Science, 227: 1580–1583.

Lopez, J.F., Akil, H. and Watson, S.J. (1999) Neural circuits mediating stress. Biol. Psychiatry, 46: 1461–1471.

Lupien, S.J. and Lepage, M. (2001) Stress, memory, and the hippocampus: can't live with it, can't live without it. Behav. Brain Res., 127: 137–158.

Makino, S., Hashimoto, K. and Gold, P.W. (2002) Multiple feedback mechanisms activating corticotropin-releasing hormone system in the brain during stress. Pharmacol. Biochem. Behav., 73: 147–158.

McEwen, B.S., Weiss, J.M. and Schwartz, L.S. (1968) Selective retention of corticosterone by limbic structures in rat brain. Nature, 220: 911–912.

McEwen, B.S., De Kloet, E.R. and Wallach, G. (1976) Interactions in vivo and in vitro of corticoids and progesterone with cell nuclei and soluble macromolecules from rat brain regions and pituitary. Brain Res., 105: 129–136.

Meijer, O.C., De Lange, E.C.M., Breimer, D.D., De Boer, A.G., Workel, J.O. and De Kloet, E.R. (1998) Penetration of dexamethasone into brain glucocorticoid targets is enhanced in mdr1a P-glycoprotein knockout mice. Endocrinology, 139: 1789–1793.

Miller, A.H., Spencer, R.L., Pulera, M., Kang, S., McEwen, B.S. and Stein, M. (1992) Adrenal steroid receptor activation in rat brain and pituitary following dexamethasone: implications for the dexamethasone suppression test. Biol. Psychiatry, 32: 850–869.

Pardridge, W.M., Frank, H.J., Cornford, E.M., Braun, L.D., Crane, P.D. and Oldendorf, W.H. (1981) Neuropeptides and the blood–brain barrier. Adv. Biochem. Psychopharmacol., 28: 321–328.

Pariante, C.M. and Miller, A.H. (2001) Glucocorticoid receptors in major depression: relevance to pathophysiology and treatment. Biol. Psychiatry, 49: 391–404.

Pariante, C.M., Makoff, A., Lovestone, S., Feroli, S., Heyden, A., Miller, A.H. and Kerwin, R.W. (2001) Antidepressants enhance glucocorticoid receptor function in vitro by modulating the membrane steroid transporters. Br. J. Pharmacol., 134: 1335–1343.

Pariante, C.M., Hye, A., Williamson, R., Makoff, A., Lovestone, S. and Kerwin, R.W. (2003) The antidepressant clomipramine regulates cortisol intracellular concentrations and glucocorticoid receptor expression in fibroblasts and rat primary neurones. Neuropsychopharmacology, 28: 1553–1561.

Pariante, C.M., Thomas, S.A., Lovestone, S., Makoff, A. and Kerwin, R.W. (2004) Do antidepressants regulate how cortisol affects the brain. Psychoneuroendocrinology, 29: 423–447.

Rao, V.V., Dahlheimer, J.L., Bardgett, M.E., Snyder, A.Z., Finch, R.A., Sartorelli, A.C. and Piwnica-Worms, D. (1999) Choroid plexus epithelial expression of MDR1 P glycoprotein and multidrug resistance-associated protein contribute to the blood–cerebrospinal-fluid drug-permeability barrier. Proc. Natl. Acad. Sci. USA, 96: 3900–3905.

Rees, H.D., Stumpf, W.E. and Sar, M. (1975) Autoradiographic studies with ^{3}H-dexamethasone in the rat brain and pituitary. In Stumpf, W.E. and Grant, L. (Eds.), Anatomical Neuroendocrinology. S. Karger, Basel, pp. 262–269.

Reul, J.M. and De Kloet, E.R. (1985) Two receptor systems for corticosterone in rat brain: microdistribution and differential occupation. Endocrinology, 117: 2505–2511.

Reul, J.M., Van den Bosch, F.R. and De Kloet, E.R. (1987) Relative occupation of type-I and type-II corticosteroid receptors in rat brain following stress and dexamethasone treatment: functional implications. J. Endocrinol., 115: 459–467.

Reul, J.M., De Kloet, E.R., Van Sluijs, F.J., Rijnberk, A. and Rothuizen, J. (1990) Binding characteristics of mineralocorticoid and glucocorticoid receptors in dog brain and pituitary. Endocrinology, 127: 907–915.

Reul, J.M., Stec, I., Soder, M. and Holsboer, F. (1993) Chronic treatment of rats with the antidepressant amitriptyline attenuates the activity of the hypothalamic–pituitary–adrenocortical system. Endocrinology, 133: 312–320.

Reul, J.M., Bilang-Bleuel, A., Droste, S., Linthorst, A.C.E., Holsboer, F. and Gesing, A. (2000a) New mode of hypothalamic–pituitary–adrenocortical axis regulation: significance for stress-related disorders. Z. Rheumatol., 59: 22–25.

Reul, J.M., Gesing, A., Droste, S., Stec, I.S.M., Weber, A., Bachmann, C., Bilang-Bleuel, A., Holsboer, F. and Linthorst, A.C.E. (2000b) The brain mineralocorticoid receptor: greedy for ligand, mysterious in function. Eur. J. Pharmacol., 405: 235–249.

Sanchez, M.M., Young, L.J., Plotsky, P.M. and Insel, T.R. (2000) Distribution of corticosteroid receptors in the rhesus brain: relative absence of glucocorticoid receptors in the hippocampal formation. J. Neurosci., 20: 4657–4668.

Schinkel, A.H. (1999) P-glycoprotein, a gatekeeper in the blood–brain barrier. Adv. Drug Delivery Rev., 36: 179–194.

Schinkel, A.H., Smit, J.J., Van Tellingen, O., Beijnen, J.H., Wagenaar, E., Van Deemter, L., Mol, C.A.A.M., Van der Valk, M.A., Robanus Maandag, E.C., Te Riele, H.P., Berns, A.J.M. and Borst, P. (1994) Disruption of the mouse mdr1a P-glycoprotein gene leads to a deficiency in the blood–brain barrier and to increased sensitivity to drugs. Cell, 77: 491–502.

Schinkel, A.H., Wagenaar, E., Van Deemter, L., Mol, C.A.A.M. and Borst, P. (1995) Absence of the mdr1a P-glycoprotein in mice affects tissue distribution and pharmacokinetics of dexamethasone, digoxin, and cyclosporin A. J. Clin. Invest., 96: 1698–1705.

Seckl, J.R. and Fink, G. (1992) Antidepressants increase glucocorticoid and mineralocorticoid receptor mRNA expression in rat hippocampus in vivo. Neuroendocrinology, 55: 621–626.

Sérée, E., Villard, P.H., Hevér, A., Guigal, N., Puyoou, F., Charvet, B., Point-Scomma, H., Lechevalier, E., Lacarelle, B. and Barra, Y. (1998) Modulation of MDR1 and CYP3A expression by dexamethasone: evidence for an inverse regulation in adrenals. Biochem. Biophys. Res. Commun., 252: 392–395.

Sharom, F.J. (1997) The P-glycoprotein efflux pump: how does it transport drugs? J. Membr. Biol., 160: 161–175.

Spencer, R.L., Young, E.A., Choo, P.H. and McEwen, B.S. (1990) Adrenal steroid type I and type II receptor binding: estimates of in vivo receptor number, occupancy, and activation with varying level of steroid. Brain Res., 514: 37–48.

Sutanto, W., Reul, J.M., Van Eekelen, J.A.M. and De Kloet, E.R. (1988) Corticosteroid receptor analyses in rat and hamster brains reveal species specificity in the type I and type II receptors. J. Steroid Biochem., 30: 417–420.

Thompson, E.B. (1995) Membrane transporters of steroid-hormones. Curr. Biol., 5: 730–732.

Ueda, K., Okamura, N., Hirai, M., Tanigawara, Y., Saeki, T., Kioka, N., Komano, T. and Hori, R. (1992) Human P-glycoprotein transports cortisol, aldosterone, and dexamethasone, but not progesterone. J. Biol. Chem., 267: 24248–24252.

Ueda, K., Kino, K., Taguchi, Y., Yamada, K., Saeki, T., Tanigawara, Y. and Komano, T. (1996) Role of P-glycoprotein in the transport of hormones and peptides. In: Gupta, S. and Tsuruo, T. (Eds.), Multidrug Resistance in Cancer Cells: Molecular, Biochemical, Physiological, and Biological Aspects, Ch. 20. John Wiley, Chichester, pp. 303–319.

Van Zuylen, L., Nooter, K., Sparreboom, A. and Verweij, J. (2000) Development of multidrug-resistance convertors: sense or nonsense? Invest. New Drugs, 18: 205–220.

Wolf, D.C. and Horwitz, S.B. (1992) P-glycoprotein transports corticosterone and is photoaffinity-labeled by the steroid. Int. J. Cancer, 52: 141–146.

T. Steckler, N.H. Kalin and J.M.H.M. Reul (Eds.)
Handbook of Stress and the Brain, Vol. 15
ISBN 0-444-51173-3

CHAPTER 3.5

Glucocorticoids and motivated behaviour

V. Lemaire, P.V. Piazza and M. Le Moal*

INSERM U588, Institut François Magendie, 1 rue Camille St Saëns, 33077 Bordeaux, France

Abstract: In this chapter the modulatory roles of corticosteroid hormones in motivated behaviours are reviewed. On the one side, motivation is considered as a complex construct central to adaptive behaviours and related to cognition and emotion. On the other side, these hormones act upon many brain regions through glucocorticoid and mineralocorticoid receptors. These central actions have been shown to participate in the regulatory processes leading to adaptation. Pathologies of the adrenal gland provide insights into the role of these hormones, in general. Both in Addison's disease and Cushing's syndrome many symptoms relate on motivational–emotional deficits. Examples will be provided that evidence a role of glucocorticoids on reinforcement processes. These hormones act on the reward system and are thought to facilitate dopamine utilization. Through a direct reward process, the hormones, like drugs of abuse, are self-administered orally or intravenously. Some individuals are vulnerable to drug and characterized by an increase in dopamine utilization in the mesocorticolimbic system during environmental challenges and are prone to self-administer the hormone and drug of abuse. It will be shown that the stress hormones participate in pathophysiological processes making the organism more vulnerable to various pathologies.

Introduction

Motivation is a definable property of behaviour and behaviour is the outcome of activity in the brain that emerged in evolution (Stellar and Stellar, 1985). It is the operation of many brain mechanisms. Motivation refers to arousal, intention, inner drive and direction of behaviours. Historically, motivation has been assimilated to instinct, then to drive, then to drive theory of reinforcement and learning in which bodily need led to drive and to new learned behaviour reducing the drive. Other complementary perspectives propose rather than a need-reduction, a drive-reduction theory of learning. In this sense, motivation is largely purposive, with reinforcement feedbacks, negative or positive. In higher animals, particularly in vertebrates, behaviour will be heavily dependent on hormones. With increasing age, behaviour became more dependent on sensory and cerebral processes, knowing and control about experience (see Stellar and Stellar, 1985). As the nervous system evolved more complex, sensory and internal environments, control were more and more added and motivated behaviour was more variable and modifiable. Emotion, mood, affects and hedonic experience became important accompaniments.

When we consider the action of a hormone or a transmitter in functional terms, the analysis has to be placed in a structure–function relationship. The constructs generally defined from psychological or neuropsychological approaches do not fit into a segregated modular neuronal organization. Terms such as memory, arousal, attention, pleasure, reinforcement and reward are not "localized" in selective brain regions such as those described for a long time by anatomists. In fact, these functional entities are distributed along several interconnected structures at different levels of the encephalon. Moreover, even if integrated functional entities have been proven to exist from neuropsychological studies,

*Corresponding author. Tel.: +33 557 57 36 60;
Fax: +33 557 57 36 69; E-mail: lemoal@bordeaux.inserm.fr

it is nevertheless observed that a given psychological ability is not insulated and influences many others. A classic example is the natural relation between cognition, emotion, mood and motivation, something experienced in everyday life by the brain as it works in an integrated manner for adaptive behaviours (Le Moal, 1995). Bearing these considerations in mind, it will not be surprising to observe the ubiquitous functional effects of glucocorticoids to which receptors have been found in many regions of the brain (Reul and De Kloet, 1985; De Kloet, 1991). Pathologies of the adrenal glands that induce hyper- or hypo-corticosteroid secretion produce complex syndromes with many functional abilities involved and disturbed either directly or indirectly. It is in some way artificial to separate motivation, reinforcement, incentive and reward processes from cognition, learning and memory, the one mediating the others and considering that mood and related cognitive states are in human so deeply influenced by the same neurotransmitters and hormones, especially stress hormones (Plihal et al., 1996). Corticosteroids facilitate cognitive abilities that tend to be better for emotionally arousing information than for mental information (Buchanan and Lovallo, 2001); this has implications for the understanding of memories for stressful or traumatic experiences.

Many psychological theories have considered the subjective phenomenon of positive motivation, pleasure and reward at the basis of the development and stabilization of goal-directed behaviours. Dysregulation of pleasure-related processes has been suggested to be involved in the appearance of different psychiatric pathologies such as depression and addictive disorders. Because animal research cannot deal directly with such subjective phenomena, psychobiologists have primarily focused on the neural bases of behaviours directed towards stimuli generally generating pleasure in humans, and defined as incentives, positive reinforcers or rewards in animals. Research on the biological substrates of reward, not surprisingly, has essentially focused primarily on the central nervous system (CNS). Starting with the discovery of intra-cranial self-stimulation (Olds and Milner, 1954), it has been possible to progressively develop, first, an anatomical chart of reward-related brain sites, and, later on, to discover the neurochemical systems involved in such a process. Many good reviews have been written in the past on this issue now largely neglected (Fibiger and Phillips, 1988; Wise and Rompre, 1989; Robbins and Everitt, 1996). It is generally agreed that the release of the neurotransmitter dopamine in brain structures, such as the nucleus accumbens, plays a central role in motivation and in reward-related process (Fibiger and Phillips, 1988; Wise and Rompre, 1989; Robbins and Everitt, 1996).

Glucocorticoids, cortisol in humans and corticosterone in rodents, are the final step of the activation of the hypothalamo–pituitary–adrenal (HPA) axis. These hormones have numerous effects in the periphery, where they modify energy metabolism and the activity of the immune system (Munck et al., 1984). Glucocorticoids also act at the level of the CNS. These hormones easily cross the blood–brain barrier, and bind to two types of specific intracellular receptors, which are hormone-activated transcription factors (McEwen et al., 1986). Glucocorticoids also seem to have direct membrane effects, though the substrate of this action remains unclear (Joëls and de Kloet, 1994). Fast and slow effects of glucocorticoids on neural activity and behaviour have been largely documented and reviewed (McEwen et al., 1986; Joëls and de Kloet, 1994). This guideline has been used to demonstrate the role of peptides and hormones on reinforcement-related behaviours and, in particular, for corticoid hormones (Piazza and Le Moal, 1996, 1997). First, it should be shown that there exists a correlation between the activity of the candidate neural system and reward-related behaviours. These studies have been principally based on the development of in vivo neurochemical techniques in freely behaving animals. Second, it should be shown that the functional activity of the candidate neural system influences behavioural responses to rewarding stimuli. These studies have principally analyzed changes in motivation-related behaviours after excitotoxic lesions of brain structures or specific neurotoxic lesions of neurotransmitter systems. Additionally, experimental evidences on this issue have also been collected using pharmacological tools that more or less specifically modify the activity of reward-related brain systems. Finally, it should be shown that the activation of the candidate neuronal system has positive-reinforcing effects on its own. These studies have used intracranial self-stimulation and also the self-administration of pharmacological stimuli, such as direct or indirect agonists, mimicking or inducing the activation of a given neuronal system. The latter approach has been particularly useful in the field of drug abuse and has generated considerable knowledge on the biological bases of the reinforcing properties of drugs (Koob and Bloom, 1988).

Clinical observations and interpretations

Increased or decreased levels of circulating corticosteroids due to direct or indirect adrenal dysfunctions of various origins lead to two opposite syndromes, Cushing's syndrome on one hand and Addison's disease on the other hand. Both present a variety of complex, mental and behavioural manifestations that are frequently the first complaints for which the patient consults in clinic. These manifestations can be a source of diagnostic difficulties in the absence of physical symptoms (see Kaplan and Sadock, 1995).

Addison's disease

Diminished functioning of the adrenal glands, with decreased steroids (cortisol, aldosterone, adrenal androgens), results in a complex clinical syndrome whose development is frequently insidious and slowly progressive. Here again, significant changes in personality and behaviour almost always occur in the early course of Addison's disease. Abulia, apathy, fatigue, lack of initiative, reclusiveness, irritability, negativism, depression and poverty of thought may be evident (Kaplan and Sadock, 1995). Often, the patient becomes reclusive and irritable as the disease progresses; in advanced stages, depressive mood and psychomotor retardation may be marked. Some patients develop mild organic mental disorders with recent memory deficits. Acute exacerbations of adrenal–cortical insufficiency – Addisonian crises – may occur, during which an acute brain syndrome may develop, with hallucinations, delusions and other signs of frank psychosis that may later proceed to stupor and coma. Physiological signs, such as weakness, fatigability, anorexia and weight loss, hypotension, perceptual dysfunctions with hyperactivity of taste and smell, to which the patient are often not aware, add to a general picture and it is not surprising that Addison's disease is often misdiagnosed as depression, hypochondriasis or chronic anxiety. It is often claimed that patients suffering of Addison's disease can pretend to almost normal life. In fact, the replacement therapy is not a mechanistic one and practice evidences how difficult, at the individual level, the therapeutic adjustment is. Patients often complain of difficulties to maintain efforts, hours of work or everyday duties, of increased reactivity – or reduced tolerance – to life events and stressors, of fatigue and weariness (Stoffer, 1993; Oelkers, 1996). These symptoms suggest a limitation of incentives motivation and well-being. The subjective quality of health status is difficult to evaluate. In spite of clinical assessments and questionnaires developed to measure health-related well-being, a few is still known about motivation and quality of life in patients with adrenocortical failure.

Cushing's syndrome

Cushing's syndrome results from chronic excess of circulating cortisol. Here again, mental manifestations often precede the physical signs and symptoms and a wide variety of behavioural symptoms are observed and may differ according to the origins of glucocorticoid excess, either exogenous or endogenous. Psychiatric complaints are present in almost 60% of the patients with spontaneously occurring Cushing's syndrome due to endogenous hypercorticolism (Starkman and Schteingart, 1981). Most of the problems are on the motivational and mood sides. The most common is depression. Suicide risk is high. Moreover, patients complain of loss of libido (almost always, but increased in women). Mood problems are not unidirectional and there may be brief episodes of disturbed behaviour characterized by excitement, acute anxiety and emotional lability (Kaplan and Sadock, 1995). Elated mood or signs of manic excitement occur, but they are not as common as depression. Mixtures of affective and organic syndromes with memory deficits, especially loss of recent memory and difficulty in concentrating, may be seen. It is classic to describe severe manifestations that include a non-psychotic organic mental disorder with confusion and disorientation that may mimic other organic, toxic or metabolic conditions. Indeed, disturbances caused by the secondary metabolic effects of Cushing's syndrome – electrolyte imbalance, diabetic ketosis and hypertensive encephalopathy – themselves cause mental syndromes. More rare – found in 15–25% of patients – is a psychotic organic brain syndrome with paranoia and hallucinations that may be mistaken for schizophrenia. These psychiatric disturbances tend to vary in severity and type, perhaps related to the underlying psychological state.

Interpretation of human data

Whatever the direction of the syndrome, in spite of overlaps in symptoms and no specific valence, incentive motivation, mood and arousal problems dominate and are undoubted at the roots of dysfunction in general activation and cognitive abilities. The interest of clinical observations is that they refer to slowly chronic, increasing modifications of the circulating hormone. In man, acute administration of dexamethasone (DEX) that activates glucocorticoid receptors have an energizing influence that can lead to dysphoric influence on mood. Activation of mineralocorticoid receptors that dominate after cortisol administration – and during a pre-treatment with DEX – induces changes towards euphoric mood (Plihal et al., 1996). Corticosteroids are routinely prescribed for a variety of common illnesses both on an outpatient and inpatient basis and psychiatric symptoms often occur within the first two weeks of therapy, are dose dependent, include mood and motivational problems, mania or depression, may worsen by use of classic antidepressants, are improved with concomitant use of lithium and antipsychotics (Brown et al., 1999). This later observation suggests a role for dopamine transmission. It remains to know the part of previous inherent acquired specific individual vulnerability in the deleterious effects of such acute treatments (Brown et al., 2002).

Glucocorticoids as a biological substrate of motivation and reward-related behaviours

The secretion of glucocorticoids increases in response to rewarding stimuli

The physiological secretion of glucocorticoids by the adrenal gland can be divided into two distinct ranges: a low physiological range, that in the rodent corresponds to plasmatic concentrations of corticosterone that are around 1 µg/100 ml, and a high physiological range, that in the rodent ranges from around 5 to 20 µg/100 ml of corticosterone. A shift between the two ranges is observed during the circadian cycle. Corticosterone concentrations are in the low range during the rest period and increase up to the high physiological range during the activity period (Joëls and de Kloet, 1994; McEwen et al., 1986). Corticosterone secretion also increases, shifting at various degrees into the high physiological range, in response to several types of environmental stimulations. Among these, the most largely studied are probably exposures to experimental models of stress. However, a similar increase in the concentrations of glucocorticoids is also observed in response to positive reinforcers, such as food and a receptive sexual partner.

The relationship between the secretion of glucocorticoids and the consumption of food has been well established. For example, in the rodent, the daily increase in corticosterone secretion is at least in part related to food consumption (Krieger, 1974; Honma et al., 1984). Indeed, in restricted food conditions, a shift in the circadian time at which the food is presented is accompanied by a parallel shift in the daily corticosterone peak. Moreover, corticosterone levels increase shortly before food consumption and decrease afterwards (Krieger, 1974; Honma et al., 1984). A positive correlation between the secretion of glucocorticoids and the behaviours elicited by a receptive sexual partner has also been reported in different species. For example, in amphibians (Orchinik et al., 1988) or in lizards (Manzo et al., 1994), an increase in glucocorticoids accompanies courtship and sexual behaviours. It is noteworthy that the burst in sexual activity observed in the toad after heavy rain is selectively associated with an increase in corticosterone concentrations without changes in testosterone (Orchinik et al., 1988). In male mice, an increase in corticosterone secretion is observed briefly after the exposure to a receptive female if the male is sexually rested, but not if it is sexually sated (Bronson and Desjardins, 1982). In parallel, in male rats, the levels of glucocorticoids are increased by the presence of females in the colony (Taylor et al., 1987). In this case, an interesting dissociation with stress is observed. Indeed, when the stress levels of mixed-sex colonies are enhanced, glucocorticoids do not further increase but actually decrease (Taylor et al., 1987).

Not only natural reinforcers but also drugs of abuse increase corticosterone secretion. For example, drug-induced stimulation of the secretion of glucocorticoids has been shown for psychostimulants (Fuller and Snody, 1981), nicotine (Caggiula et al., 1991) and ethanol (Trudeau et al., 1990). It should be considered that drugs of abuse might activate the glucocorticoid message also by a direct action on corticosteroid receptors. In conclusion, though only indirect evidences exist until now, it is reasonable to suggest that drugs of abuse can induce the activation and translocation of corticosteroid receptors with an hormone-independent mechanism (Lowy, 1990).

Glucocorticoids influence reward-related behaviours

Not very much is known about the effects of glucocorticoids on food-motivated behaviour. It has been reported that the suppression of glucocorticoids by adrenalectomy accelerates the extinction of an operant schedule reinforced by food, whereas administration of glucocorticoids has opposite effects (Micco et al., 1979). Interestingly, repeated blockade of mineralocorticoid receptors, but not of glucocorticoid receptors, impairs food-rewarded learning (Douma et al., 1998). Decrease of responding during extinction may be interpreted as the result of a learning disturbance. In this case, it should be expected that the learning disturbance is found no matter the type of reinforcer used, either positive or negative. However, suppression of glucocorticoids accelerates the extinction of food reinforcement but also delays the extinction of a negatively reinforced task, such as active avoidance (Weiss et al., 1970). It is then probable that the effect of glucocorticoids on the extinction of food-reinforced behaviour is not mediated by a learning disturbance, but results from a decrease in the reinforcing effect of this reward. The fact that glucocorticoids enhance food-directed behaviours is also supported by a much more complete series of experiments on the effects of these hormones on spontaneous food consumption (Kumar and Leibowitz, 1988). Suppression of glucocorticoids by adrenalectomy reduces the burst of food intake that accompanies the beginning of the activity period, a glucocorticoid-dependent effect since in adrenalectomized rats food intake is normalized by the administration of corticosterone (Kumar and Leibowitz, 1988). However, it should be noted that corticosterone itself is not able to trigger food intake since it will induce eating in adrenalectomized rats only if it is administered at the beginning of the activity period, when the highest levels of food consumption naturally occur in the rodent, but not at other times (Kumar and Leibowitz, 1988).

A much larger body of evidence indicates that glucocorticoids also facilitate motivational and rewarding effects of pharmacological positive reinforcers, such as drugs of abuse (for a review, see Piazza and Le Moal, 1996). The interaction between glucocorticoids and drugs of abuse is relevant to the issue discussed here because the rewarding effects of drugs and the ones of natural reinforcers share common neurobiological substrates (Koob and Bloom, 1988; Wise, 1996). Several lines of evidence suggest that glucocorticoids increase the reinforcing effects of psychostimulants. Early works have shown that administration of corticosterone facilitates the acquisition of amphetamine self-administration (Piazza et al., 1991). More recently, it has been shown (Deroche et al., 1997) that suppression of glucocorticoids by adrenalectomy induces a downward vertical shift in the dose–response curve for cocaine self-administration, decreasing the intake of cocaine for all the doses tested. This effect of adrenalectomy is dose dependently reversed by corticosterone (Deroche et al., 1997), and suggests that glucocorticoids increase the efficacy of cocaine to act as a positive reinforcer (Fig. 1). Metyrapone, a blocker of cortisone synthesis, has been shown to induce withdrawal symptoms in opioid-dependent patients (Kennedy et al., 1990).

The idea that corticosterone increases the reinforcing effects of psychostimulants is also supported by the effects of this hormone on the reinstatement of self-administration, which is considered as a model of relapse to drug use. Administration of corticosterone to animals first trained to self-administer cocaine and then submitted to an extinction procedure dose dependently induces the reinstatement of the

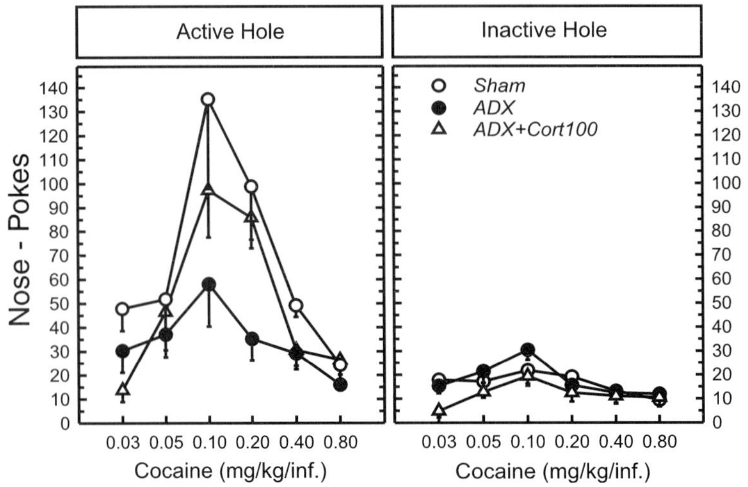

Fig. 1. Effects of adrenalectomy (ADX) and corticosterone replacement (ADX + CORT) with corticosterone subcutaneous pellet (50 mg) + corticosterone in drinking water (100 μg/ml) on cocaine self-administration. Suppression of endogenous corticosteroids by adrenalectomy modified the dose–response function for cocaine self-administration. ADX rats showed a lower number of active nose pokes, but a similar number of inactive responses (inactive hole). Thus, ADX prevents the acquisition of cocaine self-administration. This can be reversed by a therapy of substitution with a chronic corticosterone treatment. Thus, corticosterone facilitates rewarding effects of positive reinforcers like cocaine. Adapted from Deroche et al. (1997).

responding for the drug (Deroche et al., 1997). Furthermore, a chronic treatment with metyrapone, during 8 days of withdrawal from cocaine self-administration, decreases the intake of cocaine during relapse, i.e. when animals have the opportunity to self-administer cocaine again (Piazza et al., 1994). Evidence also exists that glucocorticoids increase the reinforcing effects of other drugs of abuse and in particular alcohol. Adrenalectomy decreases the intake of alcohol in alcohol-preferring rats and glucocorticoids reverse this effect and can increase consumption above baseline (Fahlke et al., 1995).

Self-administered glucocorticoids have positive reinforcing effects

Affective properties of stressful experiences: a paradox

Avoidance is the usual response to stressful situations. However, certain individuals appear to seek situations involving a strong activation accompanied by a degree of stress that are generally avoided by others. "Stress-seeking" behaviour has been described in various animal species. For example, in the monkey, Spealman and collaborators (1978) have shown that high and constant rates of responding may be maintained on a lever that delivers electric shocks. In rats, it has been reported that a mild stress such as intense handling can induce place preference (Bozarth, 1987), a behavioural response commonly seen with drugs of abuse, and that certain subjects electrically self-stimulate aversive brain regions, inducing behavioural and autonomic disturbances similar to those of physiological stress (Cazala et al., 1985). Seeking activating or stressful situations, like exposure to novelty in the rodent, has an interesting adaptive correlate. Some rats exhibit a high locomotor reactivity when forced to a novel environment (Piazza et al., 1989) or a high preference for novelty when given the choice between a familiar and a novel environment (Dellu et al., 1993). These animals, defined as high responders or high reactives (HRs) as opposed to low responders or low reactives (LRs), also show a higher sensitivity to the behavioural and neurochemical effects of drugs of abuse (Piazza et al., 1989; Hooks et al., 1991) and a higher predisposition to self-administer these drugs intravenously (Fig. 2) (Piazza et al., 1989).

To account for the appetitive properties of stressful and stimulating experiences, it may be postulated

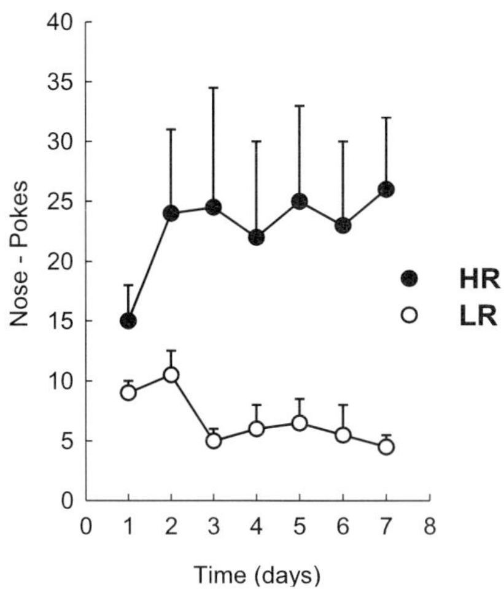

Fig. 2. Cocaine intravenous self-administration acquisition in HRs and LRs. These two groups were constructed according to the locomotor reactivity in a novel environment. Only HRs, which are more sensitive to behavioural and neurochemical effects of drugs of abuse, acquired self-administration of cocaine as shown by the higher number of visits to the active hole. Interestingly, HPA-axis response to stress is more important in HRs, suggesting appetitive properties of stressful experiences in specific subjects. Adapted from Piazza et al. (2000).

that some of the biological responses to stressful and activating situations have reinforcing effects. The HPA axis is activated by environmental stimuli and in particular by stress (Selye, 1950; McEwen et al., 1986), and in humans, administration of glucocorticoids has been reported to produce euphoric effects in some individuals (Van Zersen, 1976; Hall et al., 1979; Ling et al., 1981). Furthermore, chronic treatment with glucocorticoids can induce either physical or psychological dependence, in the absence of any abnormality in HPA function or reappearance of the disease for which glucocorticoids were administered (Dixon and Christy, 1980). The involvement of corticosterone in the appetitive properties of stress in some individuals has been intensively studied (Piazza et al., 1993) and it was of interest to examine whether corticoids, in the range of stress-induced levels, possess reinforcing properties, and if so, whether there were individual differences in sensitivity to its reinforcing properties.

Convincing evidence of the involvement of glucocorticoids in reward-related processes has been provided by the discovery that corticosterone has positive reinforcing effects (Deroche et al., 1993; Piazza et al., 1993), which has been shown by using oral and intravenous self-administration.

Oral self-administration

The effects of glucocorticoids on drugs of abuse may be explained by their interactions with reinforcing processes. Administration of corticoids has been reported to produce euphoric effects in some individuals (Von Zerssen, 1976), and chronic corticoid treatment can induce physical or psychological dependence (Dixon and Christy, 1980) in the absence of any recurrence of the underlying disease or abnormality in HPA-axis function. It has been shown that the animals orally self-administer corticosterone (Deroche et al., 1993).

Two procedures widely employed to test the reinforcing properties of orally delivered drugs (Meisch and Caroll, 1987) were used. In the first procedure, originally used to establish ethanol as a reinforcer in rats (Meisch and Beardsley, 1975), rats were offered a choice between tap water and a solution of corticosterone during the dark phase (active and feeding period) of the circadian cycle. The second procedure, first used to establish ethanol as a reinforcer in the rhesus monkey (Henningfield and Meisch, 1976), dissociates drinking of a given drug solution from eating (Meisch and Caroll, 1987). In this procedure, animals only have access to food during a fixed period of the day. Initially, corticosterone solution or tap water was presented in association with the daily ration of food, while water was available for the rest of the day. During the test period access to food was shifted forward in time, while the availability of the corticosterone solution did not change. Since the corticosterone solution was no longer available at the same time as the food, intake of the corticosterone solution was tested independently of food intake.

The two paradigms provide complementary information about the motivational properties of corticosterone as an orally administered substance. The first, and most important experiment, will determine if corticosterone has reinforcing properties when orally administered. The second experiment will test if the reinforcing properties of the hormone are strong enough to induce drinking independently of its regulatory function. Thus, in rats, drinking is strictly associated with eating such that these animals drink very small amounts of fluid when food is not available. Rats prefer a corticosterone solution to tap water, developing self-administration in a dose-dependent manner. This preference could be extinguished, but was regained during the reversal phase. Moreover, animals that had access to the corticosterone solution drank more than rats that had access to water in the absence of food. These results indicate that corticosterone has reinforcing properties after oral administration. Furthermore, they suggest that corticosterone may exert its effects on the reinforcing properties of addictive drugs by a direct action on reinforcement systems. This could account for some of the side effects of chronic steroid treatment. In human, corticoids are generally administered via the oral route.

Intravenous self-administration

Rats will also develop intravenous self-administration of corticosterone in a dose-dependent manner (Piazza et al., 1993), and animals will adapt their responses in order to maintain a constant intake of corticosterone in a single session (Fig. 3).

A strict relationship between the rate of responding and the dose delivered per infusion is classically observed for all pharmacological compounds acting as positive reinforcers. This phenomenon has been interpreted as an attempt of the subject to maintain an ideal level of reinforcement. Measurement of the plasma levels of corticosterone during its intravenous self-administration has shown that animals adjust their responding in order to obtain plasmatic levels of the hormone that are in the high physiological range, as for example those observed after intense stress. This finding clearly shows that the increase in corticosterone levels during stress has no aversive

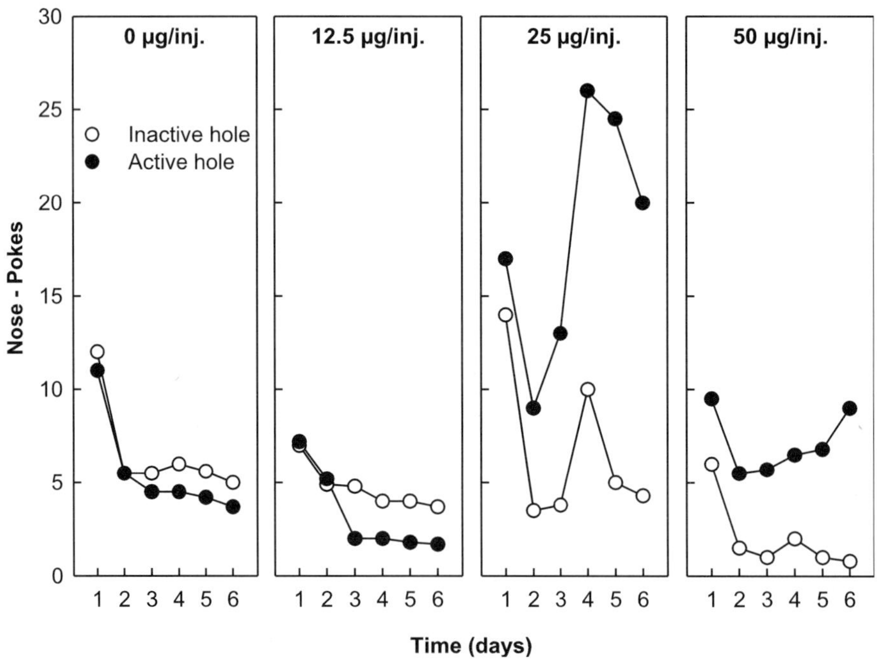

Fig. 3. Dose-dependent corticosterone intravenous self-administration during 6 days: corticosterone induced self-administration at the doses of 25 and 50 µg/injection as showed by the higher number of nose pokes in the active hole (eliciting corticosterone injection) compared to the inactive hole (control). Those results indicate that corticosterone itself has reinforcing properties. The dose–response curve resembled that of other reinforcing drugs, namely, a decreasing number of injections per session with the increasing dose per injection. This is assumed to be the animal's attempt to obtain an optimal level of reinforcement. Adapted from Piazza et al. (1993).

effects but it is instead rewarding. Furthermore, animals with different reactivities to novelty and propensities to self-administer drugs of abuse (HRs and LRs) differ in corticosterone self-administration (Fig. 4) (Piazza et al., 1993).

Positive reinforcing effects could be thus part of the physiological role of corticosterone secretion during stress. Glucocorticoids are thought to prevent an overreaction of physiological mechanisms designed to protect the organism from the body's own responses to stressors (Munck et al., 1984). This protective role of glucocorticoids in adaptation to stress is generally attributed to their peripheral actions and rather overlooks the central effects of these hormones. The positive reinforcing effect of glucocorticoids could extend protection to the CNS, helping defend the individual from the highly aversive effects of stress, thereby enabling him to cope better with it.

Positive reinforcing effects of glucocorticoids in rats are further evidenced by some observations in humans. As shown before, administration of synthetic glucocorticoids has been reported to produce euphoric effects in some individuals (Von Zerssen, 1976). Furthermore, a long-lasting increase in plasma glucocorticoids, either from exogenous administration (Von Zerssen, 1976; Ling et al., 1981) or in hypothalamo–pituitary–adrenal axis dysfunctions such as Cushing's syndrome (Krieger, 1983; Holsboer, 1989), may be associated with certain psychotic reactions resembling those observed in chronic users of psychostimulant drugs (Antelman, 1988). Finally, a general anhedonic state is one of the major symptoms of pathologies, such as Addison's disease, in which hypocortisolism is observed (Reiser and Reiser, 1995).

Neural substrates of reward-related effects of glucocorticoids

Glucocorticoids participate in the mediation of reinforcement and reward. This means, in neurobiological terms, that those hormones act on the biological substrate reinforcement.

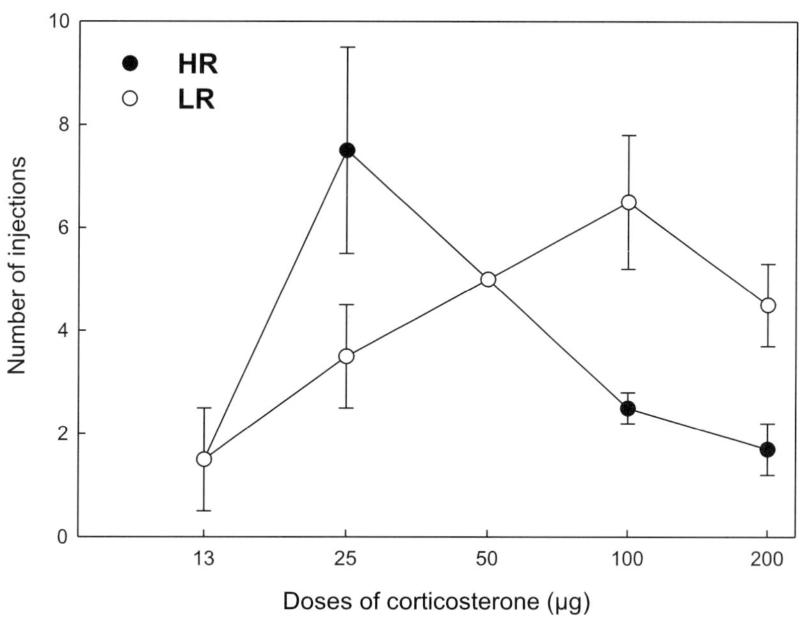

Fig. 4. Corticosterone self-administration (number of injections) in HRs and LRs. HRs and LRs had different dose–response curves for corticosterone self-administration. LRs showed a shift to the right in the dose–response function compared to HRs. Neither HRs nor LRs developed self-administration at 12.5 μg dose. HRs had a highest rate of self-administration at the 25 μg dose, whereas LRs only exhibited a comparable rate at a dose of 100 μg. This suggests that HRs are more sensitive to the reinforcing effect of corticosterone. Adapted from Piazza et al. (1993).

Indeed, to accept these hormones as one of the biological systems mediating reward, it is necessary to show that glucocorticoids act on the neuronal substrates of reward in such a way as to explain their reward-related effects.

The reinforcing properties of corticosteroids may be mediated by a central effect such as an action on mesencephalic dopaminergic neurons (Fig. 5). Several observations support this hypothesis. First, mesencephalic dopaminergic neurons are thought to be an anatomical substrate of the reinforcing properties of drugs such as psychostimulants and opioids (Koob and Bloom, 1988; Wise and Rompre, 1989; Le Moal and Simon, 1991). Second, dopaminergic neurons possess glucocorticoid receptors (Harfstrand et al., 1986), and glucocorticoids may alter dopaminergic activity (Hall and McGinley, 1982; Rothschild et al., 1985). Furthermore, synthetic glucocorticoids, such as dexamethasone, and corticosterone have been shown to induce dopamine release within this system, i.e. administration of corticosterone in drinking water at the dose of 100 μg/ml significantly increased the extracellular concentration of dopamine in the nucleus accumbens (Piazza et al., 1996). Studies from our laboratory reinforced the hypothesis of an action on dopaminergic neurons. Differences in dopaminergic activity and in dopamine response to corticosterone between HRs and LRs could underlie the differences in sensitivity to corticosterone. It has been shown (Piazza et al., 1991; Rouge-Pont et al., 1993) that HRs have a higher dopaminergic activity in the nucleus accumbens. Furthermore, in response to corticosterone administration, the rise in extracellular dopamine concentration in the nucleus accumbens of HRs is twice that observed in the same brain region of the LRs.

Although an action of corticosterone on dopaminergic neurons may explain the reinforcing effects of this hormone, the wide variety of actions of glucocorticoids on the CNS (McEwen et al., 1986) suggests that other neurochemical systems may be involved. Glucocorticoids may act on GABA (Bormann and Clapham, 1985; Lambert and Peters, 1989), glutamate (Tischler et al., 1988; Sapolsky, 1990), opioid (Chao and McEwen, 1990, 1991) or serotonin (De Kloet et al., 1982; Dickinson et al., 1985) neurons, all of which influence dopaminergic activity

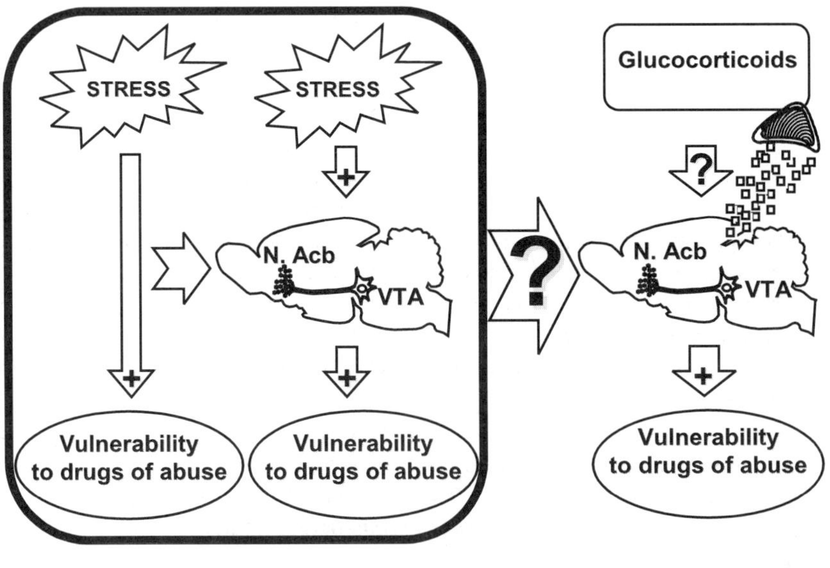

Fig. 5. Stressful experiences stimulate drug of abuse self-administration by acting on the mesencephalic dopaminergic neurons, precisely on the nucleus accumbens (Acb) which is known to be implicated in the positive reinforcing effects of drug of abuse. Indeed, stress-induced dopamine release in the nucleus accumbens has been demonstrated. Many results suggest that glucocorticoids participate to the mediation of this effect, i.e. these neurons possess glucocorticoid receptors and corticosterone has been shown to induce dopamine release in this system.

(Le Moal and Simon, 1991) and modulate the reinforcing properties of drugs of abuse.

Chronic hormone elevation: detrimental behavioural effects

Corticosterone actions in the brain are mediated by glucocorticoid (GR) and mineralocorticoid (MR) receptors. The GRs occur everywhere in the brain, while most abundant in CRF neurons of the hypothalamic paraventricular nucleus and pituitary corticotropes. MRs are expressed in hippocampus without aldosterone selectivity. MRs bind corticosterone with high affinity, 10-fold higher than co-localized GRs. These two signalling pathways correspond to two influences, tonic influence of the hormone exerted via hippocampal MRs, while the additional occupancy of GRs with higher levels of corticosterone (or cortisone) mediates feedback actions aimed to restore disturbances in homeostasis (De Kloet and Reul, 1987; De Kloet et al., 1998). MRs are involved in maintenance of stress system activity and GRs mediate steroid control for recovery. A balance between these two influences is critical; it is genetically programmed and limits vulnerability to stress-related disease in genetically predisposed individuals to anxiety and negative motivation (De Kloet and Reul, 1987; De Kloet et al., 1998; Korte, 2001).

From readiness to behave to pathological emotional states

Glucocorticoids increase activation in a site-specific manner, in amygdala, hippocampus, nucleus accumbens, medial prefrontal and orbitofrontal regions and consequently are implicated in emotional and cognitive functions and reinforcement processes. Hormonal secretion acting on receptor imbalance dysfunctions contributes to the pathophysiology of mood and cognitive disorders. In other words, corticosterone or cortisol serve a wide range of functions in contexts that are not necessarily stressful in an aversive sense of the term, such as territoriality, attachment, food intake, predation, social behaviours, effortful demands or in general, a "readiness to behave" (see Erickson et al., 2003). It is important to mention that the hormone exerts normally its effects through genomic actions, relatively slow and that rapid effects exist, from reward-reinforcement-related processes, as it has been shown, to fast cognitive appraisals suggesting that some of these effects might be supported by fast membranes-associated receptor proteins (Orchinik et al., 1991; Brann et al., 1995; Lupien et al., 1997; Moore and Evans, 1999). At the roots of most of the classic cognitive actions of glucocorticoids are basic motivational and emotional primary effects. Subjective reports and behavioural observations of arousal and energy levels correlate with cortisol measures in human or after administration of the hormone (Schmidt et al., 1999); the "wake-up" energy-enhancing role of cortisol is a correlation of its morning–evening cycle and is accompanied by enhanced glucose metabolism (Adam and Gunnar, 2001). Adults reported as having high self-esteem had higher cortisol level, and lower psychological distress than their lower self-esteem counterparts (Zorilla et al., 1995), a temperamental trait that has been described in children (Gunnar, 1994; Granger et al., 1994) as extraverted at the start of a new-school year, showed larger increases of control than did more introverted children (Kagan et al., 1987). These data reflect a state of heightened arousal and attention, while more increase may lead to enhanced anticipatory anxiety (Kagan et al., 1987; Schmidt et al., 1997). Most of these subjective effects follow a classic inverted u-shape curve. Chronic cortisol elevations have detrimental effects on attention, especially towards an emotionally arousing stimulus; however, short-term high levels allow mobilization of cognitive resources and enhance memory for these emotionally arousing events (Corodimas et al., 1994; Lupien et al., 1999).

When emotionally arousing stimuli are processed, the amygdala, temporal and prefrontal cortex regions appear to be important for memories. The amygdala, in particular, plays a major role to evaluate varieties of emotionally and anxiogenic salient stimuli. During states of fear and anxiety, glucocorticoids are elevated. Moreover, glucocorticoid administration increases the response of fear and its memory. As the experience of fear precedes the rise of cortisol, it is supposed that CRF is one of the mechanisms mediating corticosteroid release for organizing behavioural responses (Butler et al., 1990; Davis et al., 1997; Shepard et al., 2000); in consequence, elevated cortisol concentration promotes the facilitation of CRF gene expression, in amygdala in particular, and finally enhances perception of fear and anxiety. The arousal produced by these effects enhances memory for fearful events, and for the detection of uncertainty and discrepancy in the environment. It might be hypothesized that the induction of these central states may be long lasting and the experience of anxiety, fear, negative emotional states may persist as learning processes, even if cortisol is not elevated anymore (Coplan et al., 1996; Baker et al., 1999). In another pathological dimension, mood disorders – unipolar and bipolar depression –, glucose metabolism is elevated in the amygdala and stressed plasma cortisol concentration were correlated positively with this activation in the same depressed patients.

Fear and anxiety

It is now clear that corticosteroids play extremely important roles in these negative motivations, such as fear and anxiety (see Korte, 2001). Corticosteroids exert their effects on behaviour most often indirectly, because they do not regulate behaviour but they induce chemical changes in particular sets of neurons which control particular behavioural outcomes. Furthermore, according to the context and the phase of the stress response, corticosteroids have different, even opposite, effects on behaviour. Corticosteroids, at low circulating levels, exert a permissive action via brain MRs on the mediation of acute freezing behaviour and acute fear-related plus-maze behaviour (Korte et al., 1995). Corticosteroids, at high circulating levels, enhance acquisition, conditioning and consolidation of an inescapable stressful experience via GR mechanisms. Brain GR occupation also promotes processes underlying fear potentiation (Roozendaal et al., 1996). Fear potentiation can be seen as an adjustment in anticipation of changing demands. However, such feed-forward regulation may be particularly vulnerable to dysfunction. MR and/or GR mechanisms are involved in fear extinction (Korte, 2001). Brain MRs may be involved in the extinction of passive avoidance, and GRs may be involved in mediating the extinction of active avoidance (Bohus, 1987). In the developing brain, corticosteroids play a facilitatory role in the ontogeny of freezing behaviour, probably via GRs in the dorsal hippocampus, and their influence on the development of the septo-hippocampal cholinergic system (Takahashi, 1996).

In chronic stress conditions, corticosteroid actions, via MRs and GRs, can become maladaptive. Mental health of human and animal welfare are likely to be seriously threatened after psychosocial stress, prolonged stress, prenatal stress or postnatal stress, because central MR/GR balance is chronically deregulated (Korte, 2001).

Chronic stress and anxiety participate to numerous bio-behavioural disorders, such as for example, abdominal obesity syndrome. Recently, Björntorp (2001) has shown that the perceived stress-related cortisol secretion, measured under ordinary living conditions during a random working day, is related to various disadvantageous psychological factors. When the HPA axis has been adapted to blunted activity with poor coping and feedback control, the association between these factors and hormone secretion becomes stronger (Rosmond et al., 1998). This means that stress with subsequent HPA axis activation is associated with problems in the homeostasis of several somatic systems. This might also result in a sensitization of the axis to the perceived stress, with the consequence of pronounced stress-related cortisol secretion (Dallman, 1993).

Long-term effects of perinatal stress

One of the most important questions raised by modern psychiatry and experimental psychopathology is the origin and pathology of mental diseases. More concisely, clinical and experimental neurosciences are increasingly concerned with the factors that render one individual more vulnerable than another to a given pathological outcome. Animal models are now available to understand the sources of individual differences for specific phenotypes prone to behavioural disadaptations and bio-behavioural disorders (Dellu et al., 1996).

To summarize, over the last 10 years, we have explored the consequences of environmental perinatal manipulations in the rat (see Koehl et al., 2002).

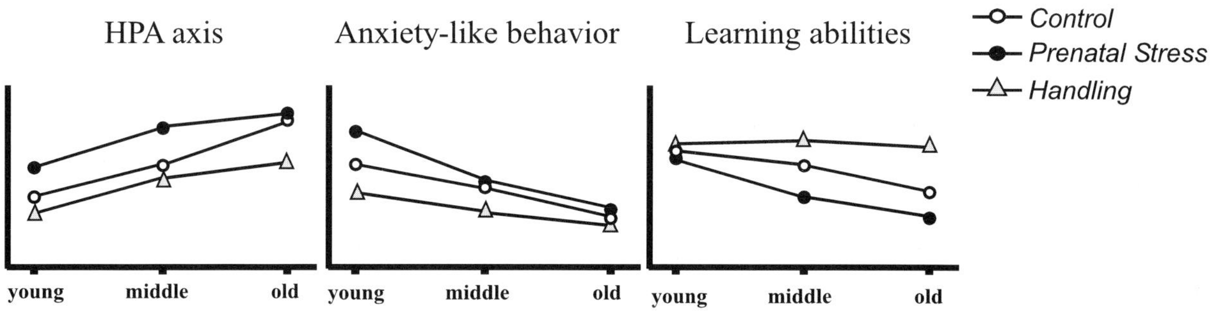

Fig. 6. Age-related evolution of the HPA-axis response to stress (left), anxiety-like behaviour (middle) and learning abilities (right) in control, prenatal stress and handling animals throughout life. Prenatal stress and postnatal handling affect the response of the HPA axis throughout the life of the animals, while they alter anxiety-like behaviour only in young animals and learning abilities later in life. Adapted from Koehl et al. (2001).

We have shown that prenatal stress is at the origin of a wide range of physiological and behavioural aberrances which define a bio-behavioural syndrome. Furthermore, the cognitive disabilities observed in prenatally stressed rats were recently related to an alteration of neurogenesis in the dentate gyrus, confirming the impact of early life events on brain morphology (Lemaire et al., 2000). A second model (handling model) has also been developed in which pups are briefly separated from their mothers during early postnatal life. Handling rats exhibit opposite behavioural and physiological modifications compared to prenatally stressed rats (Vallée et al., 1997, 1999).

Taken together, the results of these investigations show that the bio-behavioural phenotype that characterizes each individual is strongly linked to the nature and timing of perinatal experience. Longitudinal studies are of heuristic value. There are long-term relationships between stress-axis deregulation, anxiety-like behaviours that appear early in life and decrease with age, and cognitive impairments that appear later (see Fig. 6).

Conclusion

Corticosteroid hormones have been isolated many decades ago. By 1943, no less than 25 corticosteroids were identified – and the history of their physiological and behavioural roles is certainly one of the most complex and controversial of biological sciences, as wrote Angelucci (2000), "a moving from pedestal to dust and back". Considering the research interests were oriented towards physiology, homeostasis and adaptation or towards pathology, clinical neurosciences and therapeutics, the discourses were operating in different worlds. The concept of stress and the discovery of the selective role of various stressors, making corticosteroids – the end product of the HPA axis – as "stress hormones" made the picture even more complex and confusing. Few biological products seem to have such abundant ubiquitous roles, linking peripheral organs and brain, beneficial at certain concentration, deleterious at others depending of the situations and of the coping and control abilities of the organism.

Emerging from recent analysis is a picture of extraordinary diversity, whether viewed in terms of the target cells, peripheral or central, the metabolic pathways, the physiological functions regulated but the main question remains to understand how these diverse actions are coordinated and integrate, because these biological products are supposed to participate to homeostasis and adaptation, and to protect the organism from various challenges. An example is their role in carbohydrate metabolism, in day-to-day regulation of food disposal and blood glucose levels or in a prototypical challenge of a chase. The permissive or proactive actions are rooted in basal glucocorticoid levels, i.e. of the HPA axis, promoting coordination of circadian events, sleep–wake and food-intake cycles, processes underlying basic motivation, selective attention, integration of sensory information response selection and emotional stability. Suppressive or stimulating reactivities are the consequences of challenge-induced levels, but the final goal is to terminate stress-induced HPA axis activation and the surge in hormone, to facilitate the organism ability to cope, to adapt and recover, in other words, a CNS control of the secretion (De Kloet, 1991). More and more it appears that marked heterogeneity of neuroendocrine responses to various stressors exist and that each stressor has a neurochemical signature (Pacak and Palkovits, 2001). With evolution, these hormones have been harnessed to protection against stressors and aid in recovery from the subsequent responses. There are differences between the physiological role of glucocorticoids, i.e., salutary responses, for survival in natural environment, and the pathological effects of prolonged hormone elevations, when the natural recovery phase to the deleterious stimuli is prevented from occurring.

References

Adam, E.K. and Gunnar, M.R. (2001) Relationship functioning and home and work demands predict individual differences in diurnal cortisol patterns in women. Psychoneuroendocrinology, 26: 189–208.

Angelucci, L. (2000) The glucocorticoid hormone: from pedestal to dust and back. Eur. J. Pharmacol., 405: 139–147.

Antelman, S.M. (1988) Stressor-induced sensitization to subsequent stress: implications for the development and treatment of clinical disorders. In: Kalivas, P.W. (Ed.), Sensitization in the Central Nervous System. Academic Press, New York, pp. 227–259.

Baker, D.G., West, S.A., Nicholson, W.E., Ekhator, N.N., Kasckow, J.W., Hill, K.K., Bruce, A.B., Orth, D.N. and Geracioti, T.D., Jr. (1999) Serial CSF corticotropin-releasing hormone levels and adrenocortical activity in combat veterans with posttraumatic stress disorder. Am. J. Psychiatry, 156: 585–588.

Björntorp, P. (2001) Do stress reactions cause abdominal obesity and comorbidities? Obesity Reviews, 2: 73–86.

Bohus, B. (1987) Limbic-midbrain mechanisms and behavioural physiology interactions with CRF, ACTH and adrenal hormones. In: Helhammer, D., Florin, I. and Weiner, H. (Eds.), Neurobiological Approaches to Human Disease. Hans Huber Publishers, Toronto, pp. 267–285.

Bormann, J. and Clapham, D.E. (1985) Gamma-amino-butyric acid receptor channels in adrenal chromaffin cells: a patch-clamp study. Proc. Natl. Acad. Sci. USA, 82: 2168–2172.

Born, J., Hitzler, V., Pietrowsky, R., Pauschinger, P. and Fehm, H.L. (1989) Influences of cortisol on auditory evoked potentials (AEPs) and mood in humans. Neuropsychobiology, 20: 145–151.

Bozarth, M.A. (1987) In: Bozarth, M. (Ed.), Methods of Assessing the Reinforcing Properties of Abused Drugs. Springer, New York, pp. 241–274.

Brann, D.W., Hendry, L.B. and Mahesh, V.B. (1995) Emerging diversities in the mechanism of action of steroid hormones. J. Steroid Biochem. Mol. Biol., 52: 113–133.

Bronson, F.H. and Desjardins, C. (1982) Endocrine response to sexual arousal in mice. Endocrinology, 111: 1286–1291.

Brown, E.S., Khan, D.A. and Nejteck, V.A. (1999) The psychiatric side effects of corticosteroids. Ann. Allergy Asthma Immunol., 83: 495–504.

Brown, E.S., Suppes, T., Khan, D. and Carmody, T.J. (2002) Mood changes during prednisone bursts in outpatients with asthma. J. Clin. Psychopharmacology, 22: 55–61.

Buchanan, T.W. and Lovallo, W.R. (2001) Enhanced memory for emotional material following stress-level cortisol treatment in humans. Psychoneuroendocrinology, 26: 307–317.

Butler, P.D., Weiss, J.M., Stout, J.C. and Nemeroff, C.B. (1990) Corticotropin-releasing factor produces fear-enhancing and behavioral activating effects following infusion into the locus coeruleus. J. Neurosci., 10: 176–183.

Caggiula, A.R., Epstein, L.H., Antelman, S.M., Saylor, S.S., Perkins, K.A., Knopf, S. and Stiller, R. (1991) Conditioned tolerance to the anorectic and corticosterone-elevating effects of nicotine. Pharmacol. Biochem. Behav., 40: 53–59.

Cazala, P., Bendani, T. and Zielinski, A. (1985) Self-stimulation of an 'aversive' brain structure: the mesencephalic central gray area. Brain Res., 327: 53–60.

Chao, H.M. and McEwen, B.S. (1990) Glucocorticoid regulation of preproenkephalin messenger ribonucleic acid in the rat striatum. Endocrinology, 126: 3124–3130.

Chao, H.M. and McEwen, B.S. (1991) Glucocorticoid regulation of neuropeptide mRNAs in the rat striatum. Mol. Brain Res., 9: 307–311.

Coplan, J.D., Andrews, M.W., Rosenblum, L.A., Owens, M.J., Friedman, S., Gorman, J.M. and Nemeroff, C.B. (1996) Persistent elevations of cerebrospinal fluid concentrations of corticotropin-releasing factor in adult nonhuman primates exposed to early-life stressors: implications for the pathophysiology of mood and anxiety disorders. Proc. Natl. Acad. Sci. USA, 93: 1619–1623.

Corodimas, K.P., LeDoux, J.E., Gold, P.W. and Schulkin, J. (1994) Corticosterone potentiation of conditioned fear in rats. Ann. N.Y. Acad. Sci., 746: 392–393.

Dallman, M.F. (1993) Stress update. Adaptation of the hypothalamic–pituitary–adrenal axis to chronic stress. Trends Endocrinol. Metab., 4: 62–69.

Davis, M., Walker, D.L. and Lee, Y. (1997) Roles of the amygdala and bed nucleus of the stria terminalis in fear and anxiety measured with the acoustic startle reflex. Possible relevance to PTSD. Ann. NY. Acad. Sci., 821: 305–331.

De Kloet, E.R. (1991) Brain corticosteroid receptor balance and homeostatic control. Frontiers Neuroendocrinol., 12: 95–164.

De Kloet, E.R., Kovacs, G.L., Szabo, G., Telegdy, G., Bohus, B. and Versteeg, D.H.G. (1982) Decreased serotonin turnover in the dorsal hippocampus of the rat brain shortly after adrenalectomy: selective normalization after corticosterone substitution. Brain Res., 239: 659–663.

De Kloet, E.R. and Reul, J.M.H.M. (1987) Feedback action and tonic influence of corticosteroids on brain function: a concept arising from the heterogeneity of brain receptor systems. Psychoneuroendocrinology, 12: 83–105.

De Kloet, E.R., Vreugdenhil, E., Oitzl, M, and Joëls, M. (1998) Brain corticosteroid receptors balance in health and disease. Endocrine Rev., 19: 269–301.

Dellu, F., Mayo, W., Piazza, P.V., Le Moal, M. and Simon, H. (1993) Individual differences in behavioural responses to novelty. Possible relationship with sensation-seeking trait in man. Pers. Individual Differences, 15: 411–418.

Dellu, F., Mayo, W., Vallée, M., Maccari, S., Piazza, P.V., Le Moal, M. and Simon, H. (1996) Behavioral reactivity to novelty during youth as a predictive factor of stress-induced corticosterone secretion in the elderly: a life-span study in rats. Psychoneuroendocrinology, 21: 441–453.

Deroche, V., Marinelli, M., Le Moal, M. and Piazza, P.V. (1997) Glucocorticoids and behavioral effects of psychostimulants. II: Cocaine intravenous self-administration and reinstatement depend on glucocorticoid levels. J. Pharm. Exp. Ther., 281: 1401–1407.

Deroche, V., Piazza, P.V., Deminière, J.M., Le Moal, M. and Simon, H. (1993) Rats orally self-administer corticosterone. Brain Res., 622: 315–320.

Dickinson, S.L., Kennett, G.A. and Curzon, G. (1985) Reduced 5-hydroxytryptamine-dependent behaviour in rats following chronic corticosterone treatment. Brain Res., 345: 10–18.

Dixon, R.B. and Christy, N.P. (1980) On the various forms of corticosteroid withdrawal syndrome. Am. J. Med., 68: 224–230.

Douma, B.R., Korte, S.M., Buwalda, B., La Fleur, S.E., Bohus, B. and Luiten, P.G. (1998) Repeated blockade of mineralocorticoid receptors, but not of glucocorticoid receptors impairs food rewarded spatial learning. Psychoneuroendocrinology, 23: 33–44.

Erickson, K., Drevets, W. and Schulkin, J. (2003) Glucocorticoid regulation of diverse cognitive functions in normal and pathological emotional states. Neurosci. Biobehav. Rev., 27: 233–246.

Fahlke, C., Hard, E., Eriksson, C.J.P., Engel, J.A. and Hansen, S. (1995) Consequences of long term exposure to corticosterone or dexamethasone on ethanol consumption in the adrenalectomized rat, and the effect of type I and type II corticosteroid receptor antagonists. Psychopharmacology, 117: 216–224.

Fibiger, H.C. and Phillips, A.G. (1988) Mesocorticolimbic dopamine systems and reward. Ann. NY Acad. Sci., 537: 206–215.

Fuller, R.W. and Snody, H.D. (1981) Elevation of serum corticosterone by pergolide and other dopaminergic agonists. Endocrinology, 109: 1026–1032.

Granger, D.A., Weisz, J.R. and Kauneckis, D. (1994) Neuroendocrine reactivity, internalizing behavior problems, and control-related cognitions in clinic-referred children and adolescents. J. Abnorm. Psychol., 103: 267–276.

Gunnar, M. (1994) Psychoendocrine studies of temperament and stress in early childhood: expanding current models. In: Bates, J. and Wachs, T. (Eds.), Temperament: Individual Differences at the Interface of Biology and Behavior. American Psychological Association, Washington, DC, pp. 175–198.

Hall, E.D. and McGinley, P.A. (1982) Effects of single intravenous glucocorticoid dose on biogenic amine levels in cat lumbar spinal cord. Neurochem., 38: 1787–1790.

Hall, R.C.W., Popkin, M.K., Stickney, S.K. and Gardner, E.R. (1979) Presentation of the steroid psychoses. J. Nerv. Ment. Dis., 167: 229–236.

Harfstrand, A., Fuxe, K., Cintra, A., Agnati, L.F., Zini, I., Wilkstrom, A.C., Okret, S., Zhao-Ying, Y., Goldstein, M., Steinbusch, H., Verhofstad, A. and Gustafsson, J.A. (1986) Glucocorticoid receptor immunoreactivity in monoaminergic neurons in the rat brain. Proc. Natl. Acad. Sci. USA, 83: 9776–9783.

Henningfield, J.E. and Meisch, R.A. (1976) Establishment of orally delivered ethanol as a positive reinforcer for rhesus monkeys. Reports from the Research Laboratories of the Department of Psychiatry, University of Minnesota, Minneapolis, PR: 77-4.

Holsboer, F. (1989) Psychiatric implications of altered limbic-hypothalamic–pituitary–adrenocortical activity. Eur. Arch. Psychiatr. Neurol. Sci., 238: 302–322.

Honma, K., Honma, S. and Hiroshige, T. (1984) Feeding-associated corticosterone peak in rats under various feeding cycles. Am. J. Physiol., 246: R721–R726.

Hooks, M.S., Jones, G.H., Smith, A.D., Neill, D.B. and Justice, J.B., Jr. (1991) Response to novelty predicts the locomotor and nucleus accumbens dopamine response to cocaine. Synapse, 9: 121–128.

Joëls, M. and De Kloet, E.R. (1994) Mineralocorticoid and glucocorticoid receptors in the brain. Implications for ion permeability and transmitter systems. Prog. Neurobiol., 43: 1–36.

Kagan, J.S., Reznick, J.S. and Snidman, N. (1987) The physiology and psychology of behavioral inhibition in children. Child Dev., 58: 1459–1473. 65.

Kaplan, H.I. and Sadock, B.J. (1995) Comprehensive Textbook of Psychiatry, 4th ed. Williams and Wilkins, Baltimore/London.

Kennedy, J.A., Hartman, N., Sbriglio, R., Khuri, E., Kreek, M.J. (1990) Metyrapone-induced withdrawal symptoms. Br. J. Addict., 85: 1133–1140.

Koehl, M., Lemaire, V., Mayo, W., Abrous, D.N., Maccari, S., Piazza, P.V., Le Moal, M. and Vallée, M. (2002) Individual vulnerability to substance abuse and affective disorders: role of early environmental influences. Neurotoxicity Res., 4: 281–296.

Koob, G.F. and Bloom, F.E. (1989) Cellular and molecular mechanisms of drug dependence. Science, 242: 715–723.

Korte, S.M. (2001) Corticosteroids in relation to fear, anxiety and psychopathology. Neurosci. Biobehav. Rev., 25: 117–142.

Korte, S.M., De Boer, S.F., De Kloet, E.R. and Bohus, B. (1995) Anxiolitic-like effects of selective mineralocorticoid and glucocorticoid antagonists on fear-enhanced behavior in elavated plus-maze. Psychoneuroendocrinology, 20: 385–394.

Krieger, D.T. (1974) Food and water restriction shifts corticosterone temperature activity and brain amine periodicity. Endocrinology, 95: 1195–1201.

Krieger, D.T. (1983) Physiopathology of Cushing's disease. Endocrinol. Rev., 4: 22–53.

Kumar, B.A. and Leibowitz, S.F. (1988) Impact of acute corticosterone administration on feeding and macronutrient self-selection patterns. Am. J. Physiol., 254: R222–R228.

Lambert, J.J. and Peters, J.A. (1989) Steroidal modulation of the GABAA benzodiazepine receptor complex: an electrophysiological investigation. In: Barnard, E.A. and Costa, E. (Eds.), Allosteric Modulation of Amino Acid Receptors: Therapeutic Implications. Raven, New York, pp. 139–154.

Le Moal, M. (1995) Mesocorticolimbic dopaminergic neurons. Functional and regulatory roles. In: Bloom, F.E. and Kupfer, D.J. (Eds.), Psychopharmaocology: The Fourth Generation of Progress. Raven Press, Ltd, New York, pp. 283–294.

Le Moal, M. and Simon, H. (1991) Mesocorticolimbic dopaminergic network: functional and regulatory roles. Physiol. Rev., 71: 155–234.

Lemaire, V., Koehl, M., Le Moal, M. and Abrous, D.N. (2000) Prenatal stress produces learning deficits associated with an inhibition of neurogenesis in the hippocampus. Proc. Natl. Acad. Sci. USA, 97: 11032–11037.

Ling, M.H.M., Perry, P.J. and Tsuang, M.T. (1981) Side effects of corticosteroid therapy. Arch. Gen. Psychiatry, 38: 471–477.

Lowy, T.M. (1990) MK-801 antagonizes methamphetamine induced decreases in hippocampal and striatal corticosteroid receptors. Brain Res., 533: 348–352.

Lupien, S.J., Gillin, C.J. and Hauger, R.L. (1999) Working memory is more sensitive than declarative memory to the acute effects of corticosteroids: a dose–response study in humans. Behav. Neurosci., 113: 420–430.

Lupien, S.J., Gaudreau, S., Tchiteya, B.M., Maheu, F., Sharma, S., Nair, N.P., Hauger, R.L., McEwen, B.S. and Meaney, M.J. (1997) Stress-induced declarative memory impairment in healthy elderly subjects: relationship to cortisol reactivity. J. Clin. Endocrinol. Metab., 82: 2070–2075.

Manzo, C., Zerani, M., Gobbetti, A., Di Fiore, M.M. and Angelini, F. (1994) Is corticosterone involved in the reproductive process of the male lizard, Podarcis sicula sicula? Horm. Behav., 28: 117–129.

McEwen, B.S. (1993) Endocrine effects on the brain and their relationship to behaviour. In: Seigel, S.J. (Ed.), Basic Neurochemistry: Molecular, Cellular, and Medical Aspects. Raven Press, New York, pp. 1003–1023.

McEwen, B.S., De Kloet, E.R. and Rostene, W. (1986) Adrenal steroid receptors and actions in the nervous systems. Physiol. Rev., 66: 1121–1188.

Meisch, R.A. and Beardsley, P. (1975) Ethanol as a reinforcer for rats: effects of concurrent access to water and alternate positions of water and ethanol. Psychopharmacologia, 43: 19–23.

Meisch, R.A. and Carroll, M.E. (1987) Oral drug self-administration: drugs as reinforcers. In: Bozarth, M.A. (Ed.), Methods of Assessing the Reinforcing Properties of Abused Drugs. Springer Verlag, pp. 143–160.

Micco, D.J. Jr., McEwen, B.S. and Shein, W. (1979) Modulation of behavioral inhibition in appetitive extinction following manipulation of adrenal steroids in rats: implications for the involvement of the hippocampus. J. Comp. Physiol. Psychol., 93: 323–329.

Moore, F. and Evans, S. (1999) Steroid hormones use non-genomic mechanisms to control brain functions and behaviors: a review of evidence. Brain Behav. Evol., 54: 41–50.

Munck, A., Guyre, P.M. and Holbrook, N.J. (1984) Physiological function of glucocorticoids in stress and their relation to pharmacological actions. Endocrine Rev., 5: 25–44.

Oelkers, W. (1996) Adrenal insufficiency. N. Engl. J. Med., 335: 1206–1212.

Olds, J. and Milner, P.M. (1954) Positive reinforcement produced by electrical stimulation of septal area and other regions of the brain. J. Comp. Physiol. Psychol., 47: 419–427.

Orchinik, M., Licht, P. and Crews, D. (1988) Plasma steroid concentrations change in response to sexual behavior in Bufo marinus. Horm. Behav., 22: 338–350.

Orchinik, M., Murray, T.F. and Moore, F.L. (1991) A corticosteroid receptor in neuronal membranes. Science, 252: 1848–1851.

Pacak, K. and Palkovits, M. (2001) Stressor specificity of central neuroendocrine responses: implications for stress-related disorders. Endocrine Rev., 22: 502–548.

Piazza, P.V., Deroche-Gamonent, V., Rouge-Pont, F. and Le Moal, M. (2000) Vertical shifts in self-administration dose-response functions predict a drug-vulnerable phenotype predisposed to addiction. J. Neurosci., 20: 4226–4232.

Piazza, P.V. and Le Moal, M. (1996) Pathophysiological basis of vulnerability to drug abuse: role of an interaction between stress, glucocorticoids and dopaminergic neurons. Annu. Rev. Pharmacol. Toxicol., 36: 359–378.

Piazza, P.V. and Le Moal, M. (1997) Glucocorticoids as a biological substrate of reward: physiological and pathophysiological implications. Brain Res. Rev., 25: 359–372.

Piazza, P.V., Deminière, J.M., Le Moal, M., and Simon, H. (1989) Factors that predict individual vulnerability to amphetamine self-administration. Science, 245: 1511–1513.

Piazza, P.V., Deroche, V., Deminière, J.M., Maccari, S., Le Moal, M. and Simon, H. (1993) Reinforcing properties of corticosterone demonstrated by intravenous self-administration. Possible biological basis of sensation-seeking. Proc. Natl. Acad. Sci. USA, 90: 11738–11742.

Piazza, P.V., Maccari, S., Deminière, J.M., Le Moal, M., Mormède, P. and Simon, H. (1991) Corticosterone levels determine individual vulnerability to amphetamine self-administration. Proc. Natl. Acad. Sci. USA, 88: 2088–2092.

Piazza, P.V., Marinelli, M., Jodogne, C., Deroche, V., Rougè-Pont, F., Maccari, S., Le Moal, M. and Simon, H. (1994) Inhibition of corticosterone synthesis by metyrapone decreases cocaine-induced locomotion and relapse of cocaine self-administration. Brain Res., 65: 259–264.

Piazza, P.V., Rouge-Pont, F., Deroche, V., Maccari, S., Simon, H. and Le Moal, M. (1996) Glucocorticoids have state-dependent stimulant effects on the mesencephalic dopaminergic transmission. Proc. Natl. Acad. Sci. USA, 93: 8716–8720.

Plihal, W., Krug, R., Pietrowsky, R., Fehm, H.L. and Born, J. (1996) Corticosteroid receptor mediated effects on mood in humans. Psychoneuroendocrinology, 21: 515–523.

Reiser, L.W. and Reiser, M.F. (1995) Endocrine disorders. In: Kaplan, H.I. and Sadock, B.J. (Eds.), Comprehensive Textbook of Psychiatry, 4th ed. William and Wilkins, Baltimore, pp. 1170–1171.

Reul, J.M.H.M. and De Kloet, E.R. (1985) Two receptor systems for corticosterone in rat brain; microdistribution and differential occupation. Endocrinology, 117: 2505–2512.

Robbins, T.W. and Everitt, B.J. (1996) Neurobehavioral mechanisms of reward and motivation. Curr. Opin. Neurobiol., 6: 228–236.

Roozendaal, B., Bohus, B. and McGaugh, J.L. (1996) Dose-dependent suppression of adrenocortical activity with metyrapone: effects on emotion and memory. Psychoneuroendocrinology, 21: 681–693.

Rosmond, R., Dallman, M.F. and Björntorp, P. (1998) Stress-related cortisol secretion in men: relationships with abdominal obesity and endocrine, metabolic and hemodynamic abnormalities. J. Clin. Encocr. Metab., 83: 1853–1859.

Rothschild, A.J., Langlaii, P.J., Schatzberg, A.F., Miller, M.M., Saloman, M.S., Lerbinger, I.E., Cole, J.O. and Bird, E.D. (1985) The effect of a single acute dose of dexamethasone on monoamine and metabolite levels in the rat brain. Life Sci., 36: 2491–2505.

Rouge-Pont, F., Piazza, P.V., Kharouby, M., Le Moal, M. and Simon, H. (1993) Higher and longer stress-induced increase in dopamine concentrations in the nucleus accumbens of animals predisposed to amphetamine self-administration. A microdialysis study. Brain Res. 602: 169–174.

Sapolsky, R.M. (1990) Glucocorticoids, hippocampal damage and the glutamatergic synapse. Prog. Brain Res., 6: 13–23.

Schmidt, L.A., Fox, N.A., Goldberg, M.C., Smith, C.C. and Schulkin, J. (1999) Effects of acute prednisone administration on memory, attention and emotion in healthy human adults. Psychoneuroendocrinology, 24: 461–483.

Schmidt, L.A., Fox, N.A., Rubin, K.H., Sternberg, E.M., Gold, P.W., Smith, C.C. and Schulkin, J. (1997) Behavioral and neuroendocrine responses in shy children. Dev. Psychobiol., 30: 127–140.

Selye, H. (1950) Stress: The Physiology and the Pathology of Exposure to Stress. Acta Medica, Montreal.

Shepard, J.D., Barron, K.W. and Myers, D.A. (2000) Corticosterone delivery to the amygdala increases corticotropin-releasing factor mRNA in the central amygdaloid nucleus and anxiety-like behavior. Brain Res., 861: 288–295.

Spealman, R.D., Katz, J.L. and Witkin, J.M. (1978) Drug effects on responding maintained by stimulus-reinforcer and response-reinforcer contingencies. J. Exp. Anal. Behav., 30: 187–962.

Starkman, M.N. and Schteingart, D.E. (1981) Neuropsychiatric manifestations of patients with Cushing's syndrome. Arch. Intern. Med., 141: 215–219.

Stellar, J.R. and Stellar, E. (1985) The Neurobiology of Motivation and Reward. Springer Verlag, New York.

Stoffer, S.S. (1993) Addison's disease. How to improve patients' quality of life? Postgrad. Med., 93: 265–278.

Takahashi, L.K. and Rubin, W.W. (1993) Corticosteroid induction of threat-induced behavioural inhibition in preweanling rat. Behav. Neurosci., 107: 860–866.

Takahashi, L.K. (1996) Glucocorticoids and the hippocampus. Developmental interactions facilitating the expression of behavioral inhibition. Mol. Neurobiol., 13(3): 213–226.

Taylor, G.T., Weis, S.J. and Rupich, R. (1987) Male rat behavior, endocrinology and reproductive physiology in a mixed-sex, socially stressful colony. Physiol. Behav., 39: 429–433.

Tischler, M.E., Henriksen, E.J. and Cook, P.H. (1988) Role of glucocorticoids in increased muscle glutamine production in starvation. Muscle Nerve, 11: 752–756.

Trudeau, L.E., Aragon, C.M.G. and Amit, Z. (1990) Effects of ethanol on locomotor depression and corticosterone release induced by restraint-stress: support for a stress ethanol interaction. Pharmacol. Biochem. Behav., 36: 273–278.

Vallée, M., Maccari, S., Dellu, F., Simon, H., Le Moal, M. and Mayo, W. (1999) Long-term effects of prenatal stress and postnatal handling on age-related glucocorticoid secretion and cognitive performance: a longitudinal study in the rat. Eur. J. Neurosci., 11: 2906–2916.

Vallée, M., Mayo, W., Dellu, F., Le Moal, M., Simon, H. and Maccari, S. (1997) Prenatal stress induces high anxiety and postnatal handling induces low anxiety in adult offspring: correlation with stress-induced corticosterone secretion. J. Neurosci., 17: 2626–2636.

Von Zerssen, U. (1976) Mood and behavioural changes under corticosteroid therapy. In: Itel, T.M., Laudahn, G. and Hermann, W.M. (Eds.), Psychotropic Action of Hormones. Spectrum, New York, pp. 195–214.

Weiss, J.M., McEwen, B.S., Silva, M.T. and Kalkut, M. (1970) Pituitary–adrenal alterations and fear responding. Am. J. Physiol., 218: 864–868.

Wise, R.A. (1996) Neurobiology of addiction. Curr. Opin. Neurobiol., 6: 243–251.

Wise, R.A. and Rompre, P.P. (1989) Brain dopamine and reward. Annu. Rev. Psychol., 40: 191–225.

Zorrilla, E.P., De Rubeis, R.J. and Redei, E. (1995) High self-esteem, hardiness and affective stability are associated with higher basal pituitary–adrenal hormone levels. Psychoneuroendocrinology, 20: 591–601.

T. Steckler, N.H. Kalin and J.M.H.M. Reul (Eds.)
Handbook of Stress and the Brain, Vol. 15
ISBN 0-444-51173-3

CHAPTER 3.6

Effects of glucocorticoids on emotion and cognitive processes in animals

Jos Prickaerts* and Thomas Steckler

CNS-Affective Spectrum Disorders, Johnson & Johnson Pharmaceutical Research & Development, Turnhoutseweg 30, B-2340 Beerse, Belgium

Abstract: Glucocorticoids are released from the adrenal glands in response to both physiological and psychological stress exposure. Two nuclear receptors can be distinguished, the mineralocorticoid receptor (MR) and the glucocorticoid receptor (GR). Amongst a variety of actions, glucocorticoids play an important role in the modulation of cognitive processes, which will be the focus of this chapter. In particular, the acute and chronic effects of glucocorticoids on learning, memory, and attention will be discussed in relation to relevant brain regions where glucocorticoids may mediate these effects (in particular the hippocampus, amygdala, and prefrontal cortex). Acute activation of the MR seems to facilitate both acquisition and extinction processes, either indirectly via changes in behavioral reactivity or directly, as might be the case during extinction. The GR plays an important role in consolidation and retrieval processes of declarative and/or spatial/contextual memory. There seems to be a bell-shaped effect of GR activation on consolidation, with lower level of GR activation facilitating consolidation, while strong activation of the GR inhibits consolidation. The effects of GR activation on retrieval seem to be primarily detrimental. Glucocorticoids modulate hippocampus-dependent consolidation and retrieval processes via the basolateral complex of the amygdala, partly at the level of the nucleus accumbens, which is a site of convergence for both hippocampus and the basolateral complex, through interactions with the noradrenergic system. Chronic glucocorticoid excess, on the other hand, leads to cognitive deficits, presumable via hippocampal, amygdaloidal, and/or prefrontal impairment. However, chronic lack of GR activity, as can be induced in, for example, mouse mutants, also impairs learning and memory function. Both acute and chronic changes in glucocorticoid level may have effects on arousal and/or attention, but the animal literature is sparse when it comes to the investigation of these processes. The molecular basis of these effects gained increasing interest over recent years. Although it is known that glucocorticoids can affect gene transcription in several ways, it is not exactly clear yet how these effects can be translated into the effects of glucocorticoids on cognitive processes. For memory processes it has been suggested that dimerization of the GR as well as protein–protein interactions are involved.

Acute effects of glucocorticoids

Stress and memory

The glucocorticoids corticosterone and its human equivalent cortisol are produced after stimulation of the hypothalamo–pituitary–adrenal (HPA) axis and are amongst the main players in the response toward a stressor. There are intricate relationships between stress responsivity and cognitive processes. On the one hand, cognitive processes are necessary to cope adequately, both actively and passively, with a stressor in that a subject has to be aware that there is a stressor and at the same time it has to learn that the stressor can be controlled by an appropriate response (Steckler, 2004). Adaptation to stress occurs

*Corresponding author. Tel.: +32 14 607925;
Fax: +32 14 603753; E-mail: jprickae@prdbe.jnj.com

when the acquired response is successful in reducing the impact of a stressor. If not, maladaptation may occur. Maladaptation can be induced in models of chronic stress, such as in the inescapable shock paradigm (Sherman et al., 1982). On the other hand, there is strong evidence for an important role of stress and stress hormones in the modulation of cognitive processes. Here we will focus on the role of glucocorticoids in animal learning, memory, and attention.

Before starting a discussion on the role of glucocorticoids in the modulation of animal cognition, it appears prudent to point out that stress plays an extraordinary, and often neglected, role in *all* cognitive paradigms. Most learning and memory tasks involve a clear aversive or appetitive stimulus, possibly linked to a conditioning stimulus (Pavlovian or instrumental learning), in order to motivate an animal to make the desired response. In all of these paradigms, stress will play a role at some stage, although the degree and the nature of the stressor will vary by task and in time. It still appears plausible to most that any animal exposed to an aversive situation, e.g., a foot shock or a predator, will perceive a certain degree of stress, but this might be less obvious when it comes to appetitive tasks. However, even here, subjects are exposed to some level of stress at some point in time. For instance, in a food-motivated learning task the subjects are usually food deprived, which induces stress. To give yet another example, novelty stress will play a role in every paradigm where subjects are exposed to a novel situation, be it a novel object or the novel test situation itself. Thus, there appears to be some overlap between motivational processes and stress in that an animal will try to alter its exposure to a stressor, be it the drive to avoid a foot shock, the reduction of a food-deprivation state, or even the exposure to a mildly stressful situation, as would be the case under conditions where a novel environment is explored, albeit it should be clear that motivation and stress are not the same.

One important difference between the situations described above, e.g., the aversive situation of foot-shock exposure versus the appetitive situation of food reward in food-deprived animals, seems to be the emotional valence of the task. Avoidance tasks are based on the principle to induce fear in an animal and hence incorporate a substantial fear component. It has been suggested that memory involving such a component might differ from memory where the fear component is less pronounced, i.e., different brain regions might be involved in the mediation of these two types of memory and the two types of memory might also be differentially susceptibility to the effects of glucocorticoids, although we will argue that this is a gradual process and not an absolute distinction. Nevertheless, it is useful to briefly introduce the concept of what is called emotional memory.

Mostly, the term emotional memory is used when there is an obvious contribution of fear to learning, which is thought to be mediated via the amygdala. Passive avoidance learning or fear conditioning would be considered emotional learning since aversive electrical shocks are used. Theoretically, nonemotional memory would be activated in tasks involving an appetitive component, such as in appetitively reinforced spatial-learning tasks, e.g., in mazes where animals try to locate a food source on the basis of spatial cues. Spatial memory depends heavily on hippocampal integrity. Such nonemotional memory would not necessarily require the involvement of the same amygdaloidal nuclei as emotional memory. However, care must be taken to clearly define the task at hand. For example, spatial learning in a water maze, which is hippocampus dependent, also involves an aversive stimulus, i.e., water. In fact, selective amygdala lesions have clearly been shown to affect spatial water-maze learning (Roozendaal et al., 1996b), which would be in line with the suggestion that emotional memory is involved. It could even be argued that it remains questionable whether pure nonemotional memory exists (cf. Newman, 1990) and the distinction between emotional and nonemotional memory might be more gradual, depending on task and experience of the animal, since every task will involve a certain aversive component. Thus, even in an appetitively reinforced maze task, there will be an emotional memory component contributing to performance. This component is likely to be of greater impact during early training stages, when the test apparatus is unfamiliar and hence fear provoking, than during later stages when animals are familiar with the maze.

Glucocorticoids and memory

Glucocorticoid receptors

A common outcome of foot shock, food deprivation, and the exposure to a novel environment is an increase in plasma glucocorticoid levels. Glucocorticoids are known to bind to two corticosteroid receptor subtypes, the glucocorticoid receptor (GR) and the mineralocorticoid receptor (MR; Veldhuis et al., 1982; Reul and De Kloet, 1985). These receptors differ in both their affinity for glucocorticoids and their localization within the brain. The MR binds cortisol and corticosterone with an affinity that is about 10-fold higher than that of the GR (Reul and De Kloet, 1985; McEwen et al., 1986). Consequently, during basal conditions, at the trough of the circadian rhythm, it is primarily the MRs that is occupied. However, during the circadian peak, during stress, or after external administration of glucocorticoids, MRs become saturated and the occupancy of the GRs increases. In the rat, MRs are most exclusively expressed in the hippocampus (Reul and De Kloet, 1985), although lower levels of expression can also be observed in some other limbic structures such as the septum, medial amygdala, the olfactory nucleus, and some hypothalamic nuclei (Van Eekelen et al., 1988; Ahima et al., 1990). In contrast, GRs are more widely expressed in most rat brain regions, but also particularly dense in the hippocampus and amygdala, and with a substantial amount in the hypothalamic paraventricular nucleus (PVN). Species differences in the expression of these receptors can be observed. In the rhesus monkey it was found that GR mRNA expression and immunocytochemical GR staining was high within the PVN, cerebellum, and prefrontal cortex (Sanchez et al., 2000). However, the density of GR mRNA expression and GR-like immunoreactive cells in the hippocampus was rather low, especially in contrast to the rat brain. MR mRNA and protein expression were abundant in the rhesus monkey hippocampal formation. Of note, human probes and antibodies were used. In a squirrel monkey study it was found, using squirrel monkey-specific probes, that MR mRNA is strongly expressed in the hippocampus, but high levels of MR mRNA expression were also found in the amygdala and to a lesser extent in the cerebellum and caudate putamen complex (Patel et al., 2000). GR mRNA in the squirrel monkey brain was expressed in the hippocampus, amygdala, PVN, and cerebellum. Remarkably, although mRNA expression of MR and GR in deep brain structures and the hippocampal formation was similar to that reported in rodents, MR and, in particular, GR mRNA were expressed at higher levels in the squirrel monkey prefrontal cortex. These findings suggest that in primates, including humans, the prefrontal cortex is an important target for glucocorticoids besides the hippocampus, and that the effects of glucocorticoids on cognitive processes mediated by the prefrontal cortex may be species specific.

Are the brain areas rich in MR and GR involved in mediating memory processes?

Some further definitions

Besides the distinction between emotional and non-emotional memory, several other operational definitions of memory exist. Mostly, memory is subdivided into procedural (or implicit) and declarative (or explicit) memory (Eichenbaum et al., 1992; Squire, 1992; Cohen and Eichenbaum, 1993; Clark et al., 2002). Declarative memory is thought to be dependent on conscious awareness and has to do with learning of events and facts. Procedural memory is related to automatic and reflexive learning (priming, nonassociative learning, simple conditioning, learning of skills, and habits).

To our knowledge there are only few studies about procedural learning and glucocorticoids. In humans, procedural learning seems to be unaffected by cortisol (Kirschbaum et al., 1996). However, in rats it has recently been found that corticosterone is involved in procedural learning as corticosterone treatment improved simple (delay eyeblink) conditioning (Beylin and Shors, 2003; see Table 1 for a brief explanation of the various paradigms used to evaluate the effects of glucocorticoids on cognitive function in animals).

In contrast to procedural learning, the contribution of corticosterone to declarative memory has been intensively studied in experimental animals (see below) and humans (Lupien and McEwen, 1997;

Table 1. Overview of several behavioral tests assessing learning and memory or attention in rodents, with a brief description

Eyeblink conditioning: delay and trace eyeblink conditioning

Usually, electrodes are implanted to deliver a periorbital shock unconditioned stimulus (US) and to record eyelid electromyographic responses. In the delay model, animals are exposed to a cued (tone, light) conditioning stimulus (CS) that overlaps and coterminates with the US. In the trace model, the CS is separated for less than half a second from the US. Animals are exposed to about 200 trials of paired stimuli a day until the task is learned, i.e., there is no further increase in the number of conditioned responses, which are eye blinks occurring after termination of the CS. Trace eyeblink conditioning is hippocampus dependent, i.e., hippocampal lesions prevent acquisition of this task. Delay eyeblink conditioning is not dependent on the hippocampus (Beylin et al., 2001).

Fear conditioning: cued and contextual conditioning

Animals are placed in a test chamber. After an acclimatization baseline period, the animals are exposed to, e.g., about 10 trials of paired stimuli (can also be used in combination with a trace paradigm; see eyeblink conditioning). For each trial, a cued (tone, light) CS is presented for about 30 s immediately followed by a short footshock US. In most studies, this training is used for both contextual fear conditioning and cued fear conditioning in one combined study. Note that for contextual fear conditioning only the administration of footshocks is already sufficient, with the contextual information of the test chamber being the CS. Twenty-four hours after training, the animals are placed in the test chamber for about 5 min and no conditioning stimuli are presented. During this contextual test phase, freezing (no observable movement except breathing) is measured. Approximately 1 h later, animals are returned to the test chamber, which has been altered to reduce associations with the context. After an acclimatization baseline period, the conditioned stimulus is presented in the cued test phase and freezing is measured. The learned fear response is calculated as the increase in freezing expressed as a percentage of one of the two baselines, either during acclimatization to the context or in the novel context. The amygdala is essential for acquisition and expression of fear conditioning. The hippocampus mediates the conditioning to the context (LeDoux, 2000; Sanders et al., 2003).

Avoidance learning: passive- and active-avoidance learning

Passive (or inhibitory) avoidance can be measured as step through, step down, or other variant. In each case, the animal is refraining from making the measured response. In step through passive avoidance, the apparatus usually consists of a light and dark compartment. A guillotine door separates the two compartments. On day one (training trial) animals are placed in the light compartment and the latency to enter the dark compartment is measured. The guillotine door is lowered and a short foot shock is delivered. Twenty-four hours later, a retention (test) trial is given. Learning is reflected by an increase in the latency to reenter the dark compartment. Passive-avoidance learning is also assessed in one-day-old chicks (Sandi and Rose, 1994). In the training trial, the chick is offered a bright chrome bead coated with methylanthranilate. After pecking the bead, the bird displays a peck and shake response (the disgust response to methylanthranilate). At 24 h after training, each chick is tested by offering a dry chrome bead for 30 s, and its response (pecking or avoidance) is noted.

Active avoidance learning can be assessed in an apparatus consisting of two equal compartments, separated from each other via a small barrier or wall with an open door. Animals are trained in one session of about 50 trials of paired stimuli. A trial starts with the presentation of a cued signal (tone, light) as the CS of a certain duration of x seconds (e.g., 8 s). The signal is terminated when an animal crosses from one compartment to the other within this timeframe. In this case, the animal makes an avoidance response. If the animal does not respond within this timeframe, the CS continues for some more time (y seconds, e.g., 2 s), but now a foot shock is also given as the US. If the animal responds within the y seconds, this is called an escape response. The CS and US are terminated when the animal escapes to the other compartment or when, in this example, $x+y$ seconds (e.g., 10 s) had elapsed. Note that instead of a step-through response, a lever response or chain pulling might be alternatively required. Learning is reflected in an increase in avoidance responses during acquisition. Both hippocampus and amygdala are involved in avoidance learning (Roozendaal and McGaugh, 1996, 1997a).

Water-maze spatial navigation

This paradigm was originally described by Morris (1981). A circular tank (with a diameter of about 1.5–2.5 m for rats; 1 m for mice) is filled with water. The animal is placed in the water and has to swim toward an escape platform, which is hidden underneath the water surface and remains in a stable position relatively to extramaze spatial cues. Several daily trials are given until the task is learned. Parameters as swimming distance and swimming velocity are measured. The place version of this test (hidden platform below the water surface) measures spatial learning since the animal has to use spatial extramaze cues, and requires an intact hippocampus (Morris et al., 1982). Twenty-four hours after learning, a retention test (one trial) can be done in which the platform is removed and the time spent in the quadrant, where the platform was located, can be measured to assess spatial memory retrieval. The cue version of the water maze, in which the platform is visible above the water surface, assesses sensorimotor function, and is dependent on caudate nucleus function. The amygdala has been found to have a modulatory influence on both the hippocampal and caudate nucleus memory systems involved in spatial and cued water-maze learning, respectively (Packard and Teather, 1998).

(*continued*)

Table 1. Continued.

Object recognition task
Originally described by Ennaceur and Delacour (1988). In the training trial, the animal is allowed to explore two identical objects in an arena for a fixed amount of time. After a delay interval, the animal is put back in the arena for the test trial, but now with two dissimilar objects, a familiar one and a new one. The times spent exploring each object are recorded. When an animal recognizes the familiar object, it spends significantly more time on exploring the new one. When the delay is long enough, there is no more discrimination between the objects. The rhinal cortex is essential to perform this task (Mumby, 2001), although there is also evidence for a role of the hippocampus (Clark et al., 2000).
Latent inhibition
Latent inhibition is apparent when prior exposure to a CS (e.g., tone) in the absence of reinforcement retards subsequent conditioning to the stimulus when it is paired directly with an US (e.g., footshock). This test requires attentional abilities. A deficit in selective attention results in better learning in this task. Note that changes in selective attention may be influenced by associative learning. Although the medial prefrontal cortex is considered to play a role in attentional processes (Robbins, 2000), lesioning the medial prefrontal cortex of rats has no clear effect on latent inhibition (Weiner and Feldon, 1997). The role of the hippocampus in selective attention processes, as measured in latent inhibition, is still a matter of debate (Buhusi et al., 1998). The nucleus accumbens is involved latent inhibition (Weiner and Feldon, 1997).

Belanoff et al., 2001; Alderson and Novack, 2002; Lupien et al., 2004).

Declarative memory – role of the hippocampus

Declarative memory is thought to depend on the integrity of the hippocampus, whereas procedural memory is probably hippocampus independent (Eichenbaum et al., 1992; Squire, 1992; Eichenbaum, 2000; but see Poldrack and Rodriguez, 2003). According to the declarative memory theory, the hippocampus is regarded as a more general learning system important for encoding relationships between environmental stimuli and creating episodic memories. Examples of such associative learning are trace eyeblink conditioning, object recognition memory, and cued (e.g., tone or light) fear conditioning (e.g., Clark et al., 2000, 2002; Prickaerts et al., 2002; Moita et al., 2003; Ahi et al., 2004) (see Table 1).

Besides different *types* of memory, there are different *processes* of memory, i.e., acquisition, consolidation, and retrieval of information (D'Mello and Steckler, 1996; Abel and Lattal, 2001). In general, the hippocampus is assumed to be necessary for the acquisition and consolidation of new information (e.g., Baddeley, 1995). However, recent evidence is accumulating that the hippocampus is also involved in (certain aspects) of retrieval of information (Moser and Moser, 1998; Izquierdo et al., 2000; Maren and Holt, 2000; Roozendaal et al., 2001b; Steffenach et al., 2002).

Spatial/contextual memory – role of the hippocampus

An alternative hypothesis proposes that the hippocampus is the substrate of a cognitive map (O'Keefe and Nadel, 1978), which plays an important role in processing of spatial information (allocentric learning). This can be demonstrated in various spatial tasks, scheduled in, for example, water mazes (e.g., Morris et al., 1982) or radial mazes (e.g., Jarrard, 1995). In fact, performance in nonspatial associative learning tasks, such as object recognition and cued fear conditioning, appears to be independent of hippocampal function (for reviews, see Steckler et al., 1998; Mumby, 2001; Sanders et al., 2003).

Although the hippocampus plays an important role in the mediation of declarative memory, other structures such as the rhinal cortex (object information; e.g., Steckler et al., 1998; Bussey et al., 2000) and the amygdala (information about tone-shock associations, which plays a role in, for example, fear conditioning; e.g., LeDoux, 2000) are important for the mediation of declarative memory as well. Obviously, both spatial and nonspatial stimuli need to be integrated to have an optimal information processing.

It has been suggested that hippocampal neurons encode the combinations of stimuli and locations (cf. Knierim, 2003), i.e., the hippocampus itself may be the site where the contextual representation is assembled. Alternatively, it has also been proposed that the particular brain structure integrating the different types of information depends on the situation, i.e., on the specificity of the trait(s) and intensity of a stimulus (cf. Bussey et al., 2000; Richter-Levin and Akirav, 2000).

Amygdala and memory

With respect to memory performance and in the context of the present chapter, two regions of the amygdala are of importance: the central nucleus of the amygdala (CeA) and the complex of the basolateral nuclei of the amygdala (BLA). The BLA is the major input area of the amygdala, and the CeA is the major output relay (Pitkanen et al., 1997). As already mentioned above, the amygdala seems to be responsible for the emotional/fear component which is associated with the context during learning (LeDoux, 2000; Knierim, 2003; Sanders et al., 2003; Zald, 2003). Lesions of the CeA affect passive-coping behavior toward a stressor (Roozendaal et al., 1997) and those learning processes which involve a passive-coping strategy, such as cued (tone) fear conditioning (Phillips and LeDoux, 1992; LeDoux, 2000) and passive-avoidance learning (Roozendaal and McGaugh, 1997a), and impair spatial water-maze learning (Roozendaal et al., 1996). Lesions of the BLA have no major effect on passive avoidance or water-maze performance (Roozendaal and McGaugh, 1996; Roozendaal et al., 1996; Roozendaal and McGaugh, 1997a).

In contrast to contextual fear conditioning, it has been shown that cued (tone) fear conditioning is hippocampus independent, i.e., it depends on the CeA and not on both the hippocampus and the CeA as in contextual fear conditioning (LeDoux, 2000; but see Knierim, 2003; Ahi et al., 2004). Interestingly, cued (tone) fear conditioning seems to be also independent of glucocorticoids (Pugh et al., 1997a; but see Hui et al., 2004), while corticosterone seems to be involved in the consolidation of contextual fear conditioning (Pugh et al., 1997b). This would, to a certain degree, argue in favor of a role of glucocorticoids in the modulation of contextual memory, rather than of a noncontextual or nonmnemonic effect on the fear component of the task, although it can of course not be excluded that the degrees of fear exhibited or the stimulus salience in cued and contextual fear conditioning differ, and that these potential differences in the degree of fear or stimulus salience are, at least in part, responsible for the different effects of glucocorticoids in these two variants of the task.

Alternatively, it could be *emotionally relevant* contextual memory, rather than contextual memory per se, which is affected by glucocorticoids. Indeed, a fear component will also be associated with other contextual tasks, such as in spatial navigation in the water maze (water avoidance/novelty) and radial maze (novelty component), and glucocorticoids have been found to be involved in the modulation of the acquisition of spatial-learning water- and radial-maze tasks (e.g., Oitzl et al., 1997b). Indeed, it could be argued that glucocorticoids might be of special relevance for the remembrance of contexts in which emotionally relevant events took place, i.e., glucocorticoids are involved in processes leading to storage of information, which may be needed in the future to decide whether a situation is threatening. Both the hippocampus and amygdala seem to play a role in mediating this effect.

However, glucocorticoids do not only affect hippocampal and amygdaloidal function, but these areas in turn influence the activity of the HPA axis, hence lending feedback to glucocorticoid release from the adrenal glands. Lesion studies suggest an inhibitory role of the hippocampus on HPA-axis regulation (Moberg et al., 1971; Fischette et al., 1980), primarily via a relay at the level of the bed nucleus of the stria terminalis (Cullinan et al., 1993; Herman and Cullinan, 1997). The CeA seems to have a direct influence on plasma corticosterone levels (and noradrenaline levels) during stress (Roozendaal et al., 1997). This then would suggest that the neuroendocrine state is of importance in learning about a stressful situation and in the consolidation of an emotional experience and that this experience in turn will affect the neuroendocrine state (Roozendaal et al., 1997).

Given that the association of an emotional stimulus with a context seems to require both an intact hippocampus and the amygdala, in particular the CeA (LeDoux, 2000), and that both GRs and MRs are highly expressed in the hippocampus and amygdala (albeit expression of the MR is less in the amygdala), it could be argued that glucocorticoids might play an important role in processing and integrating emotional information with contextual information and in the storage of this information.

Other brain areas involved
It should of course not be forgotten that a number of other structures are of importance for consolidation of memory information besides the hippocampus and amygdala. The involvement of these brain structures seem to be, at least in part, determined by the stimulus modality (depending upon, for example, whether stimuli are of acoustic or visual nature) and task requirements (for example, whether spatial/contextual learning or cue learning is required) (Packard and Teather, 1998; Avanzi et al., 2003). Although the hippocampus and the amygdala are of prime importance for the mediation of emotionally relevant information (see above), the hippocampus and amygdala are generally believed not to be the brain structures in which information about emotionally significant events or parts of information are stored. However, it has been suggested that the hippocampus might store certain aspects of the context (O'Reilly and Rudy, 2001), while context-independent representations of simple stimulus associations have been suggested to be stored in the amygdala, as seems to be the case with information obtained in tone fear conditioning (LeDoux, 2000).

Interestingly, glucocorticoids are also involved in the modulation of delay eyeblink conditioning (Beylin and Shors, 2003). Delay eyeblink conditioning is mediated by the cerebellum, opening the possibility that glucocorticoids modulate these processes at this level (Medina et al., 2002). Likewise, glucocorticoids have been implicated in the modulation of object memory (McCormick et al., 1997), which might be mediated via actions at rhinal cortical level. Moreover, GRs are highly expressed in dopaminergic neurons projecting to the rodent prefrontal cortex (Harfstrand et al., 1986) and glucocorticoids have been reported to regulate dopaminergic neurotransmission at the level of the prefrontal cortex (as well as the amygdala) (Thomas et al., 1994). Thus, even though direct effects of glucocorticoids on prefrontal function may differ across species, prefrontally mediated behavioral processes could still be indirectly affected by glucocorticoids via interactions with the dopaminergic system. In addition, elevated levels of cortisol, at least in humans and monkeys, are likely to affect prefrontal functions also directly via GR and MR activation. These direct and indirect effects on prefrontal activity would be expected to lead to the observed changes in attentional function (Forget et al., 2000) and response inhibition (Lyons et al., 2000). Consitent with these findings is the recent observation that corticosterone also affects medial prefrontal cortex-dependent working memory of rats in a T-maze alternation task (Roozendaal et al., 2004). However, the corticosterone-induced modulation of working memory involved an intact BLA and noradrenergic activation, possibly at the level of the BLA and/or medial PFC.

Acute effects of glucocorticoids on learning and memory

Having delineated the different memory processes and the main brain areas involved in the mediation of declarative and contextual/spatial memory processes, and also knowing that all these brain areas are rich in glucocorticoid receptors, we will next focus on the acute effects of glucocorticoids on declarative and contextual/spatial memory performance in animals.

In general, glucocorticoids have been studied using both adrenalectomized (ADX) or intact animals. More specifically, the effects of increased glucocorticoid levels on cognitive function have been investigated by simple administration of glucocorticoids or by stress exposure, whereby the latter has been assumed to be equivalent to increased glucocorticoid activity. Studies on the reduction in glucocorticoid levels, on the other hand, often focused on the effects of selective GR or MR antagonists in intact animals or on ADX. After ADX, animals are deprived of glucocorticoids, but of course also of other hormones produced in the adrenal glands.

Adrenalectomized animals
Adrenalectomized rats display impaired learning in different behavioral paradigms, including the water maze (Oitzl and De Kloet, 1992; Roozendaal et al., 1996a; McCormick et al., 1997), object recognition (McCormick et al., 1997), and contextual conditioning (Pugh et al., 1997b). The ADX-induced training impairment in the water maze was reversed by

posttraining systemic injection with the synthetic glucocorticoid dexamethasone (Roozendaal et al., 1996) or corticosterone replacement prior to and during testing (McCormick et al., 1997). These findings suggest that the ADX-induced impairment is primarily due to lack of glucocorticoids. Likewise, corticosterone replacement prior to and during acquisition of an object recognition task restored learning (McCormick et al., 1997) and corticosterone replacement after a contextual fear-conditioning episode attenuated the deleterious effects seen in ADX rats (Pugh et al., 1997b).

GR and MR antagonism

The use of selective GR and MR antagonists allows a further dissociation of the role of glucocorticoids in the modulation of cognitive processes. Numerous studies suggest that the GR is heavily involved in these processes. Systemic injection of GR antagonists (RU 38486 and RU 40555) prior or immediately after contextual fear conditioning impaired retention of the fear response of rats (Pugh et al., 1997a). Intracerebroventricular (ICV) administration of the GR antagonist RU 38486 before conditioning also attenuated contextual fear expression (Cordero and Sandi, 1998), which would argue that this GR effect is centrally mediated. In contrast, systemic GR blockade prior or immediately after cued (tone) fear conditioning was ineffective (Pugh et al., 1997a). This indicates that glucocorticoids are not involved in cued (tone) fear conditioning (but see Hui et al., 2004).

Along similar lines, systemic injection of the GR antagonist RU 555 prior to training impaired spatial learning of rats in a Y-maze spatial discrimination task (Conrad et al., 1999). Further evidence for a role of GR in spatial learning and memory comes from findings that the GR antagonist RU 38486 impaired consolidation when administered ICV shortly after the first training session in the water maze (Oitzl and De Kloet, 1992; Roozendaal et al., 1996).

In contrast, administration of the MR antagonist spironolactone was ineffective in the water-maze learning task (Oitzl and De Kloet, 1992). However, when spironolactone was given before training it changed the search pattern, which led to the suggestion that MR activation is essential for interpretation of environmental stimuli and selection of a behavioral response (Oitzl and De Kloet, 1992; Oitzl et al., 1994). Of course, altered stimulus processing would affect subsequent performance. Hence, not surprising, spironolactone administration resulted in impaired performance during the next water-maze session (Oitzl and De Kloet, 1992). Thus, GR, but not MR, blockade impairs consolidation of spatial memory, while the MR is involved in stimulus processing and/or response selection.

ICV injection of the MR antagonist RU 28318 was also ineffective on passive avoidance performance when given after training (Oitzl et al., 1993). However, when administered before training, RU 28318 impaired retention performance. Again, this would be in line with an effect secondary to the alterations in behavioral reactivity induced by MR blockade during acquisition (Oitzl and De Kloet, 1992; Oitzl et al., 1994).

A number of studies also investigated the effects of GR and MR blockade on passive-avoidance learning in the one-day-old chick: post-training intracerebral injections of the GR antagonists RU 28318 or RU 38486 impaired retention of a passive-avoidance response in these experiments (Sandi and Rose, 1997). Again, MR blockade (RU 28318) was ineffective when given after training (Sandi and Rose, 1994a), but impaired retention performance when administered before training. Moreover, the MR antagonist altered the birds' pecking pattern, suggesting altered reactivity to nonspecific aspects of the training task (Sandi and Rose, 1994a), which is also in agreement with a role of MRs in behavioral reactivity (Oitzl and De Kloet, 1992; Oitzl et al., 1994). An alteration in behavioral reactivity could also explain the deficit in retention performance seen in passive-avoidance performance in chicks. Of note, this is an indirect effect due to changes in nonspecific factors and not a direct memory effect.

GR and MR Agonism

Posttraining intracerebral injections of corticosterone, on the other hand, enhanced retention of a passive-avoidance response in one-day-old chicks (Sandi and Rose, 1994b, 1997). Likewise, systemic administration of corticosterone immediately after training has been reported to improve consolidation in the water maze in rats (Sandi et al., 1997). Further, it has recently been found that systemic injection of corticosterone immediately after cued (tone) fear conditioning improved cued fear expression also (Hui et al., 2004). Of note, this finding is in contrast with earlier observations that ADX or posttraining administration of a GR antagonist failed to impair cued (tone) fear conditioning (Pugh et al., 1997a). Finally and more specifically, posttraining systemic injections with the GR agonist dexamethasone (Flood et al., 1978; Roozendaal and McGaugh, 1996) or infusions of the GR agonist RU 28362 directly into the rat hippocampus (Roozendaal and McGaugh, 1997a) enhanced passive-avoidance retention.

Although memory-improving actions of glucocorticoid agonists have been reported in a wide variety of animal models, several studies also reported that glucocorticoids impaired memory performance (Lupien and McEwen, 1997; De Kloet et al., 1999). Of note, these memory impairments only become apparent at relative high doses of glucocorticoids (Bohus and Lissak, 1968; Sandi and Rose, 1997). Given that the MR is already almost completely occupied at low corticosterone levels, these findings suggest that the detrimental effects of glucocorticoids on memory are mediated via the GR. Additional support for this suggestion comes from a study showing impaired spatial Y-maze performance following systemic administration of corticosterone or the GR agonist RU 362, both of which resulted in strong stimulation of the GR (Conrad et al., 1999).

Thus, there is a bell-shaped relationship between consolidation processes and glucocorticoid levels. In further support of this suggestion, it has been shown that plasma corticosterone levels are biphasically related to memory consolidation following stress exposure (Sandi and Rose, 1997; Sandi et al., 1997; Cordero and Sandi, 1998).

Are GRs involved in consolidation or retrieval?

We have seen that glucocorticoids affect certain types of memory, i.e., declarative and spatial/contextual memory processes of high emotional significance. Next, we will address the question of which processes are influenced by GR stimulation or blockade, as there is still a debate on whether GR activation affects consolidation and/or retrieval.

The evidence discussed above suggests that glucocorticoids seem to be clearly involved in consolidation processes. However, systemic administration of corticosterone before retention testing also impaired performance in a study by De Quervain et al. (1998). In that study, rats were first trained in a water maze, followed 24 h after the last daily training by retention testing in a probe trial. During the probe trial the animals were swimming in the maze in the absence of an escape platform. In general, animals remembering the previous platform location will spend more time in the vicinity of that location. Corticosterone given before the probe trial impaired water-maze performance, suggesting that GRs also affect memory retrieval.

However, it has been argued by De Kloet and colleagues that the effects of corticosterone on memory retrieval are not specifically GR mediated, but are due to a disrupted MR function, which would induce nonspecific effects via impaired responsivity to novelty (in the water-maze probe trial, novelty could be induced during retention testing by the absence of the escape platform; De Kloet et al., 1999, 2002). Interestingly, it has been shown that ICV administration of spironolactone on two consecutive water-maze training days did not change the latencies to find the platform (Oitzl and De Kloet, 1992). However, retention testing immediately after training on day two revealed that although the rats headed directly for the former platform position, they started exploring the remaining parts of the water maze more than controls (Oitzl and De Kloet, 1992). Such "random" search pattern naturally would also impair probe-trial performance and reflects an altered behavioral response to novelty. However, it is unlikely that retrieval processes are directly impaired by MR blockade since latencies during training were unaffected in the study by Oitzl and De Kloet (1992). Rather, a retrieval impairment would be secondary to an altered behavioral reactivity after MR blockade.

In addition, others have argued that the effects on retrieval are directly mediated via GRs based on human data, where retrieval processes are also impaired following glucocorticoid administration (Roozendaal, 2002). This hypothesis is further supported by a recent study, showing that the GR agonist RU 28362 impaired water-maze retention when administered directly into the hippocampus (Roozendaal et al., 2003).

Given that the GR facilitates consolidation of novel information while at the same time it impairs retrieval of previously stored information, how could such a discrepancy be explained? It has been suggested by Roozendaal (2002) that a temporary disruption of memory retrieval during stressful conditions induced by GR activation may diminish proactive interference, i.e., an impaired retrieval of previously stored information will interfere less with the consolidation of newly learned information. This would result in facilitated performance under conditions where consolidation is measured, while performance would be impaired under conditions assessing retrieval, i.e., the effects of GR activation on consolidation might be indirect. Clearly, more studies are needed to further clarify the role of GRs in consolidation and retrieval.

Extinction learning

A specific form of learning is extinction learning, which is characterized by a decrease in the amplitude and frequency of a conditioned response when the conditioned stimulus that elicits it is repeatedly nonreinforced (for example, behavioral testing of animals previously trained under footshock or food-reward conditions in the absence of electrical-shock punishment or food reward) (Myers and Davis, 2002). Elevated brain corticosterone levels during active avoidance extinction have been shown to facilitate extinction (Bohus, 1970). Thus, glucocorticoids seem to be involved in extinction learning. In particular, the MR may have a permissive action on the acquisition of extinction. In support of this idea, very low doses of corticosterone normalized the extinction behavior in a passive-avoidance task in ADX rats (Bohus and De Kloet, 1981). Given the expression pattern of the MR in the rat brain and the fact that the hippocampus plays a role in extinction learning, it was suggested that this MR effect is most likely mediated at the hippocampal level (Korte, 2001). Alternatively, it could of course be argued that this may reflect a recovered contribution of the MR's function toward behavioral reactivity in a new situation during passive-avoidance extinction learning, rather than being a direct effect on extinction performance per se.

Thus, activation of the MR seems to facilitate both acquisition and extinction processes, either indirectly via changes in behavioral reactivity or directly, as might be the case during extinction. The GR plays an important role in consolidation and retrieval processes. Again, it remains a matter of debate at the moment whether these are direct or indirect effects. There seems to be a bell-shaped effect of GR activation on consolidation, with lower level of GR activation facilitating consolidation, while strong activation of the GR inhibits consolidation. The effects of GR activation on retrieval seem to be primarily detrimental. In other words, lower levels of glucocorticoids are beneficial to learn and remember information related to a stressful event – and hence facilitate stress coping – while acute high levels of glucocorticoids have a negative effect on learning. On a speculative note, it might be argued that one of the physiological roles of the latter process could be to facilitate forgetting of very traumatic events.

Amygdala and hippocampus – two brain areas mediating the acute effects of glucocorticoids on memory

We have by now seen that the hippocampus is an important structure mediating the acute effects of both MR and GR activation and blockade on learning and memory. This is supported by the similarity between the effects of glucocorticoids and hippocampal manipulations on the various types of memories, but even more clearly by studies showing that direct, intrahippocampal administration of glucocorticoids affect mnemonic processes (e.g., Roozendaal and McGaugh, 1997a; Roozendaal et al., 2003). However, the hippocampus does not stand in isolation, but interacts with other brain areas to mediate these glucocorticoid-induced effects. Interestingly, the BLA has to be intact in order for intrahippocampal GR agonists to affect consolidation (Roozendaal and McGaugh, 1997a) or retention (Roozendaal et al., 2003), suggesting complex interactions between these two areas.

How do the amygdala and the hippocampus interact in the mediation of the glucocorticoid-induced effects on mnemonic performance?

Consolidation of contextual/spatial information in tasks, such as contextual fear conditioning, passive avoidance, or water-maze spatial navigation, requires both an intact hippocampus and appropriate glucocorticoid activity (see above). Both systemic (Roozendaal and McGaugh, 1996; Roozendaal et al., 1996) and intrahippocampal injections (Roozendaal and McGaugh, 1997a) of GR agonists and GR antagonists affect performance in passive-avoidance or spatial water-maze learning, but only if the BLA is intact. Interestingly, BLA lesions on their own fail to affect performance in these tasks (Roozendaal and McGaugh, 1996; Roozendaal et al., 1996; Roozendaal and McGaugh, 1997a). However, GR ligands were effective in both passive-avoidance and spatial-learning tasks when injected directly into the BLA (Roozendaal and McGaugh, 1997b; Roozendaal et al., 2001a), suggesting that the BLA is an important site where glucocorticoids affect this type of behavior.

How could the BLA influence the effects of GR activation at the hippocampal level? The BLA projects to the entorhinal cortex, to the dentate gyrus, and hippocampal areas CA1–CA3 (Thomas et al., 1984; Pitkanen et al., 1997), suggesting that the BLA could directly influence GR effects at the level of the hippocampus. Moreover, both hippocampus and BLA could have a common site of convergence. Both the BLA (through the stria terminalis) and the hippocampus project to the nucleus accumbens (NAc), which could be such a site of convergence. In support of this view, it was found that lesions of either the NAc or the stria terminalis blocked the effects of intra-BLA or intra-hippocampal infusions with the GR agonist RU 28362 on consolidation in a rat passive-avoidance task, while having no effect on their own (Roozendaal et al., 2001a).

To make matters even more complex, it has been shown that the memory-modulating effects of glucocorticoids in the hippocampus depend on an intact noradrenergic neurotransmission within the BLA (Roozendaal et al., 1999; McGaugh et al., 2002; McIntyre et al., 2003). Thus, we have a situation where GR activation at the hippocampal level is under influence of the BLA, either directly or at the level of a hippocampal output station, the NAc. GR activation at the BLA itself also affects consolidation processes via the NAc. On top of that, there is a noradrenergic input modulating this control function of the BLA. Glucocorticoids in turn seem to have a permissive action on the noradrenergic system in the BLA (Roozendaal et al., 2002b), which would allow for a fine tuning of this system.

Since glucocorticoids do not only affect consolidation, but high levels of corticosterone have detrimental effects on memory-retention processes as well (De Quervain et al., 1998; Roozendaal et al., 2001b, 2003), it is of interest to investigate whether this complex interaction also applies for retrieval processes. As is the case with learning, BLA lesions fail to affect retrieval in passive-avoidance or water-maze tasks (see McGaugh et al., 1996; Roozendaal et al., 2003), but again the BLA influences the effects of intrahippocampal injection with a GR agonist on spatial memory retention (Roozendaal et al., 2003). However, while GR agonism affects spatial-learning tasks when injected directly into the BLA, intra-BLA injection of a GR agonist fails to affect water-maze retention (Roozendaal and McGaugh, 1997b), i.e., there is a dissociation between effects on consolidation, which are influenced by GR action at the BLA–NAc pathway, and effects on retrieval, which are independent of direct glucocorticoid-mediated effects at the level of the BLA.

In contrast, lesions of the CeA impair memory performance, but do not influence glucocorticoid-induced behavioral changes (Roozendaal and McGaugh, 1996; Roozendaal et al., 1996). Moreover, GR ligands failed to affect either passive avoidance or spatial learning when injected directly into the CeA (Roozendaal and McGaugh, 1997b). The CeA also lacks direct projections to the hippocampus (Thomas et al., 1984; Pitkanen et al., 1997). However, it should be noted that lesions of the CeA result in robust behavioral impairments in the same tasks which are susceptible to modulation by glucocorticoids (Phillips and LeDoux, 1992; Roozendaal and McGaugh, 1996, 1997a; Roozendaal et al., 1996) and that the CeA affects HPA-axis function. Thus, we cannot exclude a more general modulatory role of the CeA on the glucocorticoid/noradrenaline-induced BLA–hippocampal interplay (cf. Roozendaal et al., 1997).

Taken together, it may be argued that glucocorticoids modulate hippocampus-dependent memory storage via the BLA at NAc level, through interactions with the noradrenergic system. The interaction between BLA and hippocampus is also of relevance for retrieval processes, although a direct effect of glucocorticoids at BLA level seems to be lacking. Moreover, it is also likely that there are other neurotransmitter systems influencing this interaction, besides corticosterone and noradrenaline. Possible candidates are, for example, acetylcholine (Power et al., 2000; McGaugh et al., 2002; McIntyre et al., 2003; Power et al., 2003) and corticotropin-releasing factor (CRF) (e.g., Roozendaal et al., 2002a). Although the exact nature of the interaction between the BLA and the hippocampus remains unclear at present, it is tempting to suggest that emotional information from the BLA is transmitted to the NAc (and possibly also directly to the hippocampus), thereby providing an emotional tone to the mnemonic information from the hippocampus.

The interplay between hippocampus, amygdala, and the HPA axis in the recognition of novelty and familiarity

In this section we will try to put the effects of glucocorticoids on consolidation and retrieval into a physiological context and show their importance for stress reactivity, as expressed by activation of the HPA axis. This is of relevance as there are clear differences between plasma corticosterone and adrenocorticotrophic hormone (ACTH) levels following stress exposure, depending on whether the stressor is novel or familiar.

The impact of a novel stressor

The PVN is the area that is central for the regulation of the HPA axis (see Fig. 1A). There are intimate links between the hippocampus and the amygdala, i.e., those brain areas mediating emotional memory processes, and the HPA axis (Fig. 1B). In general, it is believed that a new and thus unfamiliar (unconditioned) stressor results in an increase of glucocorticoids by stimulating the HPA axis via activation of the amygdala (McGaugh et al., 1996; Herman and Cullinan, 1997; Roozendaal et al., 1997).

As discussed before, glucocorticoids are involved in the consolidation of contextual/spatial information at the hippocampal level (Roozendaal and McGaugh, 1997a; see Fig. 1B, solid orange pathway) and play a role in memory formation at the level of the BLA, i.e., which in turn influences the effects of glucocorticoids on consolidation processes in the hippocampus, at least in part via its projection to the NAc through the stria terminalis (Roozendaal and McGaugh, 1997a, b; Roozendaal et al., 2001a; McGaugh et al., 2002; McIntyre et al., 2003; see Fig. 1B, dashed green lines).

Besides stimulating the HPA axis, the CeA is involved in the expression of certain fear behaviors including passive coping and tone-fear conditioning (Roozendaal et al., 1997; Ledoux, 2000). Tone fear conditioning is hippocampus and glucocorticoid independent (Pugh et al., 1997a; Ledoux, 2000; but see Ahi et al., 2004; Hui et al., 2004) and according to some authors the tone-shock association is stored within the CeA (Ledoux, 2000). Moreover, it has been suggested that even in contextual fear learning, which is assumed to be hippocampus and glucocorticoid dependent, the CeA is the central output relay of behavior instead of the hippocampus (Ledoux, 2000). There are connections between the EC of the hippocampus and the CeA, both direct and indirect via the BLA (McDonald and Mascagni, 1997; Pitkanen et al., 1997, 2000). Thus, it is assumed that the CeA receives contextual information necessary to associate with tone-shock information (see Fig. 1B, dashed orange pathway).

In contrast to the facilitatory effect of the amygdala on the HPA-axis function, it has been suggested that the hippocampus mediates the negative feedback on the HPA axis (see Herman and Cullinan, 1997). However, recent data indicate that the hippocampus does not inhibit the HPA axis if an unconditioned stressor is presented (Tuvnes et al., 2003). This intuitively makes sense, as an inhibited HPA axis is not favorable in the presence of an unknown stressor, where an increased level of stress reactivity is required to cope successfully with that stressor.

The impact of familiar stressors

Although the amygdala seems to play a role in HPA-axis regulation under conditions of novelty, this may change if a stressor becomes familiar. In fact, lesions of the CeA have only limited effects on conditioned stress responses, both at the endocrine and behavioral level (Roozendaal et al., 1997; note that in case of retention of tone-fear information, lesioning the CeA after conditioning might yet be detrimental since the CeA is necessary for its behavioral expression). Likewise, BLA lesions made after passive-avoidance water-maze learning did not impair memory retention (McGaugh et al., 1996; Roozendaal et al., 2003) and had no effect on the corticosterone response after retention testing in the water maze (Roozendaal et al., 2003). These findings are in agreement with the assumption that the amygdala is not involved in stimulation of the HPA axis if a stressor is familiar. It might be argued that in this case the relevant information (for example, about context or shock presentation) is already stored somewhere else and the appropriate behavioral response can be selected.

In contrast, lesions of the hippocampus increase the corticosterone response after presentation of a familiar stressor, for example, in animals previously exposed to an aversive retention task (Roozendaal et al., 2001b). Thus, the hippocampus may inhibit the HPA axis after exposure to a familiar stressor, but not after presentation of a novel stressor (Tuvnes et al., 2003). Again, this makes sense as it might be assumed that inhibition of the HPA axis is desirable when the stressor is already known and an appropriate adaptive response is possible.

Hippocampal lesions have been demonstrated to impair the retention performance in, for instance, the water maze (Moser and Moser, 1998; Roozendaal et al., 2001; Steffenach et al., 2002). It has been suggested that this can be related to the elevated levels of corticosterone due to an impaired hippocampal-dependent inhibition of the HPA axis (Roozendaal et al., 2001). Accordingly, as mentioned before, injection of a GR agonist into the dorsal hippocampus affected the retention performance in the water maze (Roozendaal et al., 2003). Interestingly, injection of the GR agonist had no effect on the corticosterone response of the HPA axis after the retention test, i.e., there is an increase in corticosterone levels between control and GR agonist-treated rats. Thus, the hippocampal inhibiting effect on HPA-axis activity seems to be glucocorticoid independent. Yet a distinction in processes of memory retrieval and HPA-axis inhibition might not be anticipated since it intuitively seems not necessary to still inhibit the HPA axis to an old (familiar) stimulus when it cannot be remembered due to the GR agonist treatment, i.e., it is in fact a new stressor. Apparently, processes of memory retrieval and HPA-axis inhibition are far more complex involving more mediators and brain structures.

Taken together, an unfamiliar stressor is processed via the amygdala, which leads to stimulation of the HPA axis. Once the animal has learned about the stressor and the stressor becomes familiar, the hippocampus comes into play and inhibits HPA-axis function. Thus, glucocorticoids affect learning processes and cognitive processes in turn influencing the release of glucocorticoids.

Synaptic plasticity

A proposed physiological correlate of learning and memory at the hippocampal level is long-term potentiation (LTP) (Bliss and Collingridge, 1993; Ishihara et al., 1997). LTP has been suggested to reflect a form of synaptic plasticity and can be operationally defined as an increase in the efficiency of synaptic transmission following repeated high-electrical stimulation. Another form of synaptic plasticity following low-frequency stimulation is long-term depression (LTD). LTD is a decrease in synaptic efficiency and assumed to also play an active role in learning and memory (Jaffard et al., 1996; Nakao et al., 2002). However, there is also considerable evidence indicating that LTP and LTD as the mechanisms underlying learning and memory can be questioned (e.g., Keith and Rudy, 1990; Cain et al., 1997; Izquierdo et al., 2002).

Given that LTP/LTD and learning and memory are linked, do glucocorticoids affect these electrophysiological correlates of synaptic plasticity in the hippocampus? The role of glucocorticoids in LTP and LTD has been widely investigated. In slice preparations in vitro, a bell-shaped dose–response relationship has been found between corticosterone and hippocampal LTP, which is impaired by both too low and too high corticosterone levels, i.e., hippocampal LTP is optimal when glucocorticoid levels are mildly elevated, leading to complete occupation of MRs and partial occupation of GRs (Diamond et al., 1992). In vivo, the BLA was found to modulate hippocampal LTP, and this modulation

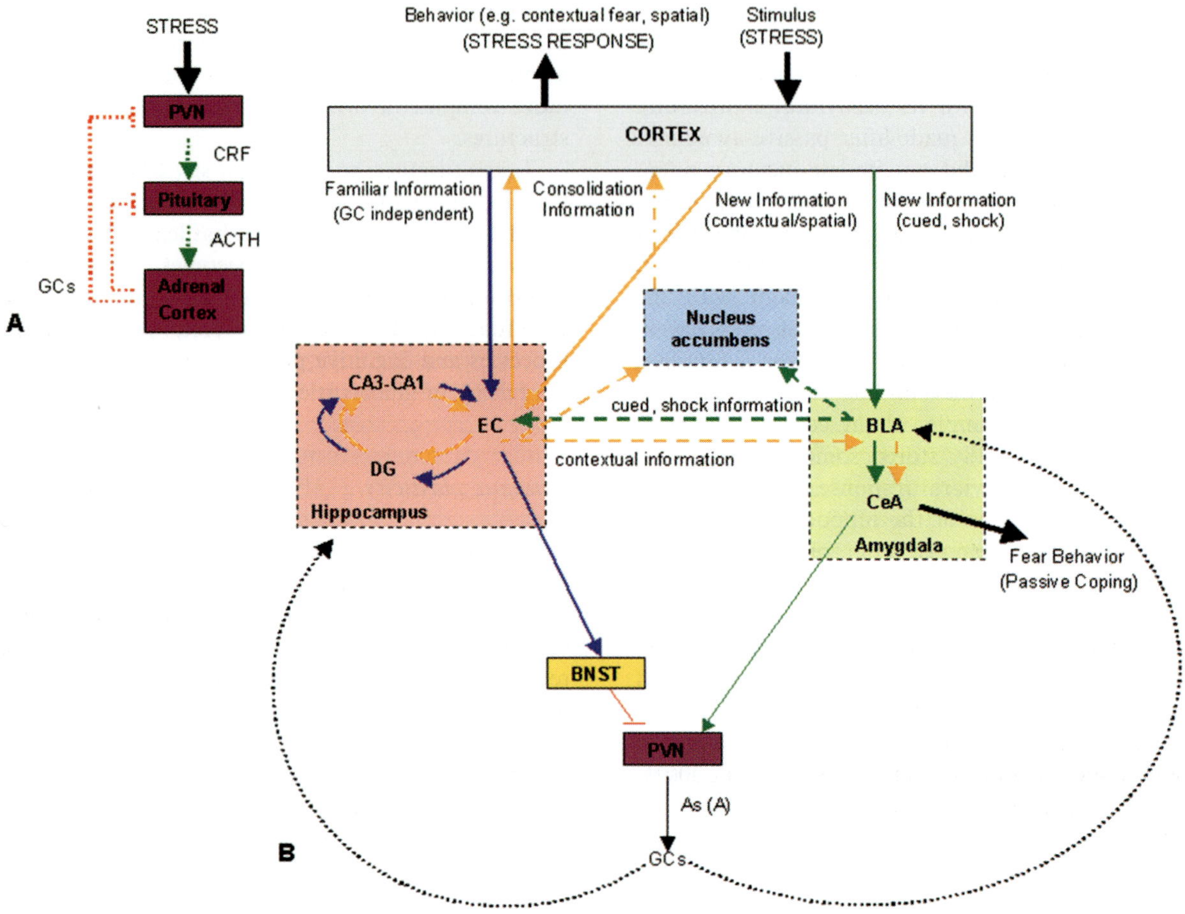

Fig. 1. Diagram showing functional connections between the amygdala and hippocampus involved in conditioned stress responses and the involvement of glucocorticoids (GCs: cortisol in humans and corticosterone in rodents). Only major pathways have been depicted. Arrows indicate facilitatory projections but do not imply monosynaptic connections. Dotted lines represent effects after secretion in plasma. Blunt-ended lines denote inhibitory feedback loops or projections. (A) The paraventricular nucleus (PVN) is a critical relay station for the central regulation of the hypothalamo–pituitary–adrenal (HPA) axis. In case of physiological stressors, which do not require higher-order processing, the PVN is stimulated via brainstem nuclei. This causes the secretion of corticotropin-releasing factor (CRF) from the PVN. CRF induces the release of adrenocorticotropic hormone (ACTH) from the pituitary, and ACTH elicits secretion of GCs from the cortex of the adrenal gland. GCs cross the blood–brain barrier and shut-off the neuroendocrine stress response at the level of the PVN and pituitary. It has been suggested that this also applies to the hippocampus, which has an indirect inhibitory input to the PVN via a GABAergic projection from the bed nucleus of the stria terminalis (BNST). However, recent data suggest that the hippocampus is not necessary for inhibition of the HPA axis to a new stressor. (B) Psychological and physical stressors require processing of signals from multiple-sensory modalities of a stimulus prior to initiation of a stress response. It is assumed that comparison of a stimulus with previous experience takes place at cortical level. The central nucleus of the amygdala (CeA) receives information about unfamiliar stimuli (e.g., about an electrical shock) from the cortex via the complex of the basolateral nuclei of the amygdala (BLA), although the BLA may be bypassed. Subsequently, the CeA has a stimulating effect on the HPA axis, thus resulting in increased levels of GCs (solid green pathway). The entorhinal cortex (EC), which is the main interface between the hippocampus and other brain areas, receives contextual/spatial information of the stimulus from the cortex. The hippocampus is involved in storage of this information (solid orange pathway). GCs have a mediating effect on memory storage through influences on both BLA and hippocampus (see dotted GC lines). The memory-modulating effects of GCs in the hippocampus depend on an intact BLA, either via direct input to the hippocampus or through interaction with the nucleus accumbens (dashed green lines), although it may be argued that the BLA itself does not project the relevant emotional/fear information toward these output structures. Thus, the nucleus

seems corticosterone dependent (Akirav and Richter-Levin, 2002). This is in agreement with the modulating effect of the BLA on glucocorticoid-dependent hippocampal memory processes (Roozendaal and McGaugh, 1997b; Roozendaal et al., 1999; McGaugh et al., 2002). Under conditions of high levels of GR occupation, hippocampal LTP induction is impaired in brain slices, but at the same time LTD induction is facilitated (Xu et al., 1998).

An attempt to directly relate the effects of glucocorticoids on LTP and LTD to those on memory processes as has been made in two interesting studies. In the first study it was found that although elevations of corticosterone during novelty acquisition may contribute to the expression of hippocampal LTD, corticosterone elevations alone do not appear to be sufficient for LTD expression (Manahan-Vaughan and Braunewell, 1999). This would suggest that factors other than glucocorticoids are involved under in vivo conditions, which differs from the in vitro findings discussed above. In the second study, it was shown that the amygdala can modulate the effects of a stressor on hippocampal LTP and spatial water-maze learning, but unexpectedly corticosterone appeared to play no major role in these effects (Kim et al., 2001). Again, this suggests that the picture is more complex and that these processes are clearly under the control of more than just glucocorticoids.

Molecular mechanisms

Given that glucocorticoids affect synaptic plasticity, what could be the underlying molecular mechanisms? A number of possibilities have been proposed, such as alterations in cell adhesion molecules (CAMs), interactions with the mitogen-activated protein (MAP) kinase pathway, and/or effects on brain-derived neurotrophic factor (BDNF) activity.

GRs and MRs are nuclear receptors that can affect gene transcription either through homodimerization of the GR or MR and direct binding of these homodimers to the DNA, or through protein–protein interactions of the receptor monomers with other transcription factors (Beato et al., 1995; Whitfield et al., 1999).

It has been reported that a point mutation in the mouse glucocorticoid receptor, which prevented dimerization and DNA binding, resulted in impaired spatial learning in the water maze (Oitzl et al., 2001). This would suggest a role for direct DNA binding of the GR receptor in the GR-mediated effects on spatial memory. Nevertheless, the participation of protein–protein interactions of the receptor with factors that are part of intracellular-signaling cascades, or with other transcription factors, cannot be ruled out.

One possible candidate affected by glucocorticoids is the MAP kinase pathway. Activation of this pathway can lead to phosphorylation of the transcription factor cAMP response element-binding protein (CREB), a process thought to be of key importance for learning and memory at amygdala, hippocampal, and NAc levels. Stress exposure seems to interact with learning experience by activating this pathway in certain brain areas. Thus, an increase in phosphorylation of the hippocampal MAP kinase ERK2 was observed in rats that had learned a water-maze spatial navigation paradigm. If these animals were trained under conditions of high stress (i.e., in cold water), ERK2 was also phosphorylated in the amygdala, but only in animals that had learned the task. This was accompanied by high levels of plasma corticosterone. However, no ERK2 phosphorylation was seen in the amygdala if animals had learned the task under low stress conditions in warm water (Akirav et al., 2001). These findings suggest that the

accumbens is also a site of convergence for the memory–modulatory information from the BLA and hippocampus and this information may subsequently feed back to cortical areas (including the hippocampus) influencing consolidation processes (irregularly dashed orange line). The CeA itself is thought to be involved in the expression of certain fear behaviors (passive coping, tone-fear conditioning). Some studies even suggest an involvement of the CeA in contextual fear conditioning and as such the relevant contextual information from the hippocampus would be conveyed to the CeA (dashed orange pathway, although the BLA may be bypassed). Recently, it has been found that the hippocampus does not inhibit the HPA axis in case of a new (unconditioned) stressor. In fact, a stimulated HPA axis is desirable in case of a new stressor. However, when a stressor becomes familiar (conditioned), it might be desirable to inhibit the HPA axis (blue pathway toward BNST). The hippocampal inhibition of the HPA axis is corticosterone independent. There is no stimulation of the HPA axis by the amygdala in case of a familiar stressor and it can be argued that this is also not necessary since the stressor has become less stressful.

amygdala and the hippocampus are differentially activated following spatial learning, depending on the level of stress involved. Likewise, stress exposure has been shown to lead to CREB phosphorylation in various brain areas, including the hippocampus and amygdala, which has been linked to processes of stress-related learning (Bilang-Bleuel et al., 2002).

However, although phosphorylation of ERK2 or CREB was seen in the amygdala under conditions of elevated plasma corticosterone levels, it is questionable whether this is a direct glucocorticoid-mediated effect as glucocorticoids in general reduce ERK2 and CREB phosphorylation, and inhibit the MAP kinase pathway in various cell types (Son et al., 2001; Stockand and Meszaros, 2003). This opens the possibility that factors other than glucocorticoids could be involved in the mediation of these phosphorylation events, e.g., noradrenaline.

Glucocorticoids do not only suppress the activity of the MAP kinase pathway and CREB phosphorylation, but also the expression of mRNA and protein levels of hippocampal BDNF both in vitro and in vivo (Schaaf et al., 1998). The expression of the neurotrophic factor BDNF is enhanced via phosphorylation of transcription factors, such as CREB. However, BDNF has been strongly implicated in spatial memory formation. Another possibility to explain the discrepancy between the in vitro results from studies investigating the effects of glucocorticoids on the activation of these signaling cascades and the in vivo results following stress/learning-task exposure, besides the involvement of other components of the stress system, could be that the acute effects of glucocorticoids depend on whether glucocorticoids were administered at high concentrations or are present under physiological conditions. In fact, high levels of corticosterone observed during water-maze learning did not affect hippocampal BDNF mRNA expression (Schaaf et al., 1999). This led to the suggestion that BDNF is relatively resistant to regulation by endogenous corticosterone and that this constellation might contribute to an optimal emotional memory performance in combination with MR and/or moderate GR activation (Schaaf et al., 1999, 2000; see also Scaccianoce et al., 2003).

The mechanism of the protection of BDNF against the actions of glucocorticoids is unclear at present. Recently, it has even been hypothesized that normal levels and low increases of glucocorticoids might even produce increases in BDNF, which would optimize certain memory functions (Garcia, 2002), although the mechanism of action for this effect remains elusive. However, if this is correct, one could speculate that similar differences also apply for the intracellular-signaling cascades and possible CREB phosphorylation in the brain (cf. (Bilang-Bleuel et al., 2002). In fact, it has been shown that selective GR activation by dexamethasone inhibits the MAP kinase pathway and CREB phosphorylation in hippocampal progenitor cells (Son et al., 2001). Stimulation of the MR by aldosterone, on the other hand, has been reported to activate the MAP kinase cascade in cardiac fibroblasts in a spironolactone-sensitive, RU 486-insensitive manner (Stockand and Meszaros, 2003). If that is also true in brain, it could explain why low levels of glucocorticoids could have facilitatory effects on memory and the molecular mechanisms underlying this effect, i.e., activation of the MAP kinase pathway via MR stimulation. In agreement with this idea, it is evident that too high levels of glucocorticoids, presumably via GR activation, especially for longer durations, should overcome these protective mechanisms, resulting in a strong decrease in BDNF and in memory impairments. This hypothesis, however, would fail to explain the beneficial effects of acute GR agonism on memory performance or it has also to be assumed that a moderate GR activation is beneficial for memory performance.

The third group of molecules claimed to play a role in glucocorticoid-mediated memory processes are the CAMs. CAMs are members of the immunoglobulin superfamily. In particular, the CAMs NCAM and L1 have been implicated in the neural mechanisms underlying memory formation. There are indications that glucocorticoids play a role in the regulation of CAMs as, for instance, expression of NCAM and L1 in the hippocampus is dependent on stress intensity during contextual fear conditioning (Merino et al., 2000). In fact, bidirectional effects of acute and chronic (21 days) administration of corticosterone on NCAM expression have been reported, with acute corticosterone resulting in enhanced NCAM levels and chronic treatment decreasing its expression in the frontal cortex in rats

(Sandi and Loscertales, 1999). This suggests a time course comparable to the effects of acute and chronic glucocorticoids on learning and memory performance. However, no changes were observed in hippocampal NCAM expression in that study.

Arousal and attention

Are there cognitive functions other than learning and memory, which are affected by acute changes in glucocorticoid activity? The level of arousal of a subject determines its ability to direct its attention toward information from the environment. This enables the subject to distinguish between relevant and irrelevant cues (selective attention), and thereby affects memory functions (Yerkes and Dodson, 1908). The effects of glucocorticoids on arousal and/or attention have been extensively investigated in humans, but only very sparse in animals. In fact, there is only one study reporting that acute administration of corticosterone disrupted latent inhibition in rats (Shalev et al., 1998). There are many theories of latent inhibition, amongst them that it critically depends on selective attention. This would suggest that selective attention processes are under control of glucocorticoids, although further studies are needed before firm conclusions can be drawn.

In humans, it has been suggested that glucocorticoids affect arousal in a bell-shaped manner (Beckwith et al., 1983; Born et al., 1989). In addition, there is evidence for glucocorticoids affecting selective attention in humans (Kopell et al., 1970; Skosnik et al., 2000), although conflicting data have also been reported (Born et al., 1987; Newcomer et al., 1999; Monk and Nelson, 2002).

Given that the MR is involved in the process of interpretation of environmental stimuli and response selection (Oitzl and De Kloet, 1992; Sandi and Rose, 1994a) or sensory integration (Lupien and McEwen, 1997), it might be suggested that this receptor plays a role in attentional processes as well. The definitions of these MR-mediated processes are very similar to the definition of the process of selective attention of humans, i.e., a selection of appropriate stimuli in the environment. Whether these definitions refer to one and the same process or are of relevance for one and the same brain structure remains to be determined. Consequently, a necessity in both animal and human research is to investigate whether a direct effect of glucocorticoids on learning and memory can be influenced or explained by effects on arousal and attention.

Chronic effects of glucocorticoids

As was already briefly discussed above, the acute effects of glucocorticoids on cognitive function may be substantially different from their chronic effects. This will be reviewed in more detail in the following section.

Stress and memory

Hippocampal damage

When levels of glucocorticoids are too high for too long, their effects on learning and memory will become pathological. In patients with Cushing's syndrome, who have a chronically elevated cortisol level, or in patients with major depression, of whom about 50% exhibit chronic hypersecretion of cortisol, deficits in declarative memory can be observed (Brown et al., 1999; Steckler et al., 1999a). Magnetic resonance imaging (MRI) studies revealed that the hippocampus undergoes selective volume reduction in Cushing's syndrome (Starkman et al., 1992) and major depression (Sheline et al., 1996). It has been suggested that it is the excess of glucocorticoids that leads to damage of the hippocampus, which in turn might explain some of the learning and memory impairments seen in these disorders (Brown et al., 1999). Concomitantly, many animal studies have demonstrated that chronic stress, as well as chronic corticosterone treatment, results in learning and memory deficits in several behavioral tasks, including water-maze and passive-avoidance tests, and that this impairment was accompanied by hippocampal damage (Luine et al., 1994; Bodnoff et al., 1995; Krugers et al., 1997; McLay et al., 1998; Bisagno et al., 2000; Sousa et al., 2000; Coburn-Litvak et al., 2003).

Several reasons could account for the damage of the hippocampus and, subsequently, its reduction in volume. It has been shown that chronic stress, just as high levels of glucocorticoid administration, causes atrophy of dendrites of pyramidal neurons in the CA3 region of the hippocampus of mammals through a mechanism involving both glucocorticoids and excitatory amino acids (for a review, see McEwen, 1999). In addition, decreased neurogenesis could contribute to the reduced hippocampal volume since chronic (but also acute) stress, as well as glucocorticoid administration, can decrease the generation of new granular neurons in the dentate gyrus of the hippocampus (Gould and Tanapat, 1999; Czeh et al., 2002; Pham et al., 2003). This effect could be mediated via a reduction in BDNF considering its possible role in the regulation of cell proliferation in the hippocampus (Katoh-Semba et al., 2002; Lee et al., 2002). Indeed, corticosterone administration (e.g. Schaaf et al., 1998) as well as stress exposure (Nibuya et al., 1999; Xu et al., 2002) decrease both the expression of mRNA and protein levels of BDNF in the hippocampus.

The glucocorticoid cascade hypothesis proposes that hippocampal damage, which results in downregulation of GRs in hippocampal neurons, due to long periods of stress or excess levels of glucocorticoids could result in an impaired negative feedback to the HPA axis (Sapolsky et al., 1986). Consequently, this would lead to further elevations of cortisol/corticosterone. Subsequently, the latter could provide further injury to the hippocampus, and even more substantial cognitive impairments. Ultimately, there would be permanent destruction of the hippocampal neurons. However, more and more evidence is accumulating that hippocampal neuronal cell death, i.e., apoptosis, is not always observed in humans and animals after chronic stress or glucocorticoid treatment (for a review, see Belanoff et al., 2001). In addition, rat studies showed that hippocampal damage, i.e., dendritic CA3 atrophy, and resulting cognitive impairments are reversible following rehabilitation from stress or corticosterone treatments (Luine et al., 1994; Sousa et al., 2000). Thus, in this perspective neuronal function and structural plasticity are probably of more importance than neuronal death.

However, more recently, it has been suggested that hippocampal CA3 damage per se is not responsible for the deficits in memory retrieval after chronic stress (Roozendaal, 2002). This suggestion is based on the finding that the 11β-hydroxylase inhibitor metyrapone, which leads to reduced corticosterone levels, attenuated an excitotoxic CA3 lesion-induced water-maze retrieval deficit in rats (Roozendaal et al., 2001b). As mentioned above, it is the hippocampal CA3 area that is most susceptible to chronic glucocorticoid excess. In addition, corticosterone levels were elevated after the lesion, probably due to an impaired hippocampal-dependent inhibition of the HPA axis. This would suggest that metyrapone blocked a corticosterone-induced retrieval deficit that was independent on the integrity of the hippocampal CA3 area. However, a contribution of acquisition and consolidation deficits to the memory impairments after chronic stress, as a result of glucocorticoid-induced hippocampal damage, cannot be ruled out yet.

Indeed, it should be noted that the hippocampus is probably not the only structure that is subject to excess glucocorticoid-induced damage (cf. Wellman, 2001). In patients with major depression, MRI scans revealed volume reductions in core nuclei of the amygdala (i.e., in BLA and the accessory basal nucleus; Sheline et al., 1998) and medial and orbital prefrontal cortical areas (Drevets, 2001). Thus, it may be suggested that chronic glucocorticoids lead to learning and memory deficits (possibly induced by damage of the hippocampus and/or amygdala) and/or impaired attention and response inhibition (possibly induced by prefrontal damage).

Chronic alterations in GR and MR function

In contrast to hippocampal damage as the cause of a decrease in HPA-axis feedback mechanism, a decreased HPA-axis feedback has been suggested to be due to the reduced function of the GR in major depression, most likely at the level of the PVN and pituitary (for a review, see Pariante and Miller, 2001). In addition, it might be speculated that an initial reduction in function of the GR, and possibly also the MR, at the level of the hippocampus contributes directly to the learning and memory deficits in patients with major depression. The possible contribution of a decreased GR function and hippocampal damage to chronic stress-induced memory deficits is illustrated in Fig. 2.

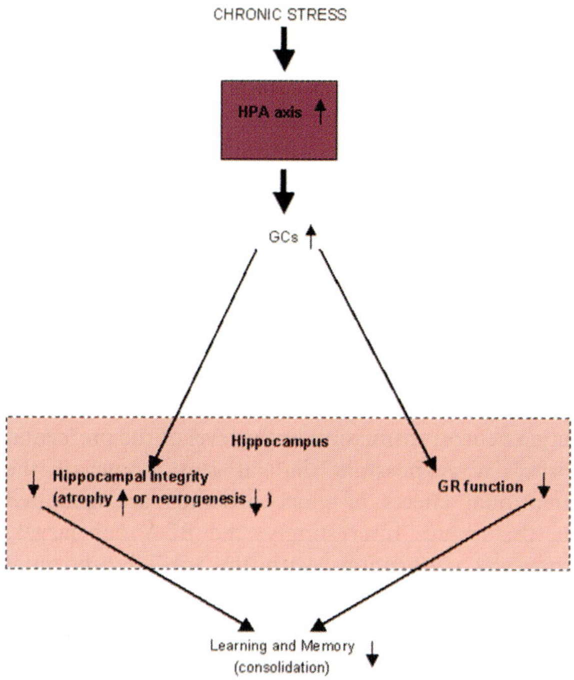

Fig. 2. Diagram illustrating the contribution of hippocampal damage and a decreased glucocorticoid receptor (GR) function to chronic stress-induced memory impairments. During chronic stress the activated HPA axis results in elevated basal levels of glucocorticoids (GCs). Excess levels of glucocorticoids (GCs) are assumed to damage the hippocampus either by causing atrophy of dendrites of pyramidal neurons in the CA3 region or by decreasing the generation of new granular neurons in the dentate gyrus. This damage to the hippocampus might explain some of the memory impairments seen after chronic stress or in diseases with excess levels of GCs, including major depression. In patients with major depression, excess levels of GCs are also assumed to decrease the function of the GR. It might be speculated that an initial reduction in function of the GR, and possibly also the mineralocorticoid receptor (MR), at the level

To study the function of chronic alterations of GR and MR activity in the brain, several mouse mutants have been generated with genetic modifications of the GR or MR function (Gass et al., 2001; Steckler, 2001; Kellendonk et al., 2002; Muller et al., 2002). The following mouse mutants are of particular interest in relation to memory functions: transgenic mice with decreased GR expression due to overexpression of an antisense directed against the GR (AGR), homo- and heterozygous knockout mice with inactivation of the GR gene ($GR^{-/-}$ and $GR^{+/-}$), conditional knockout mice with brain-specific inactivation of the GR gene ($GR^{Nes/Cre}$), mouse mutants with a point mutation that abolishes GR dimerization ($GR^{dim/dim}$), and knockout mice with inactivation of the MR gene ($MR^{-/-}$). $GR^{-/-}$ and $MR^{-/-}$ knockouts are problematic to use for behavioral testing due to reduced viability, since most $GR^{-/-}$ mice die immediately after birth due to atelectasis of the lungs and homozygous $MR^{-/-}$ mice develop pseudo-hypoaldosteronism after birth and die between postnatal days 8 and 13 as a result of dehydration caused by renal sodium and water loss. Some $GR^{-/-}$ mice will survive and both $GR^{-/-}$ and $GR^{+/-}$ mice have been used in tests of learning and memory (Oitzl et al., 1997a), even though nonspecific effects due to hypoxia early in life cannot be ruled out. Furthermore, $MR^{-/-}$ mice can be rescued by exogenous salt supply, thereby allowing the study of certain aspects of brain function in adult mice (Gass et al., 2000), although again it is likely that there are some nonspecific effects due to gross homeostatic alterations.

AGR, $GR^{dim/dim}$, $GR^{-/-}$, $GR^{+/-}$, and $GR^{Nes/Cre}$ mice all have impaired spatial learning in the water maze (Oitzl et al., 1997a; Steckler et al., 1999b; Oitzl et al., 2001; Kellendonk et al., 2002), which would suggest a role of hippocampal GRs in specific learning and memory processes. Of course, it is unlikely that the behavioral alterations in these mice are restricted to hippocampal dysfunction. AGR mice, for example, showed response disinhibition characterized by decreased response latencies and increased premature responding in a five-choice simultaneous discrimination task, opening the possibility of an involvement of the prefrontal cortex (Steckler et al., 2000).

Since AGR, $GR^{dim/dim}$, and $GR^{+/-}$ mice still have residual GR activity, as well as increased plasma corticosterone levels, either under basal conditions or under conditions of stress exposure, it is also possible that the residual glucocorticoid activity contributes to the learning and memory deficits seen in these mice. In fact, these mutants also suffer from the risk that compensatory processes during development took place and a relative overactivity of the GR in AGR mice has even been proposed to explain some of the paradoxical behavioral alterations seen in these animals (Steckler and Holsboer, 1999). In this respect, it is interesting to note that $GR^{Nes/Cre}$ mice, which completely lack neuronal and glial GR, also show an initial learning impairment in the water maze, suggesting that the spatial memory deficit of $GR^{Nes/Cre}$ mice is due to the absence of GR. However, further training in the water maze overcomes this initial learning impairment, indicating that the GR itself is not essential for spatial memory (Kellendonk et al., 2002).

Arousal and attention

Not only acute changes in corticosterone level may have effects on arousal and/or attention, but chronic changes might also have consequences on these processes. Again, the animal literature is sparse when it comes to the investigation of these processes. A simultaneous visual discrimination deficit has been reported in AGR mice (Steckler et al., 2000), but this seemed to be due to alterations in impulsive responding. Since glucocorticoids have been reported to regulate dopaminergic neurotransmission at the level of the prefrontal cortex and altered dopaminergic activity has been documented in these mice (Sillaber et al., 1998), it is possible that this effect is secondary to alterations in dopaminergic activity. Further, Shalev et al. (1998) showed that the long-term repeated administration of corticosterone disrupted latent inhibition.

In the human literature, it has been reported that chronic exposure to elevated levels of cortisol in patients with Cushing's syndrome is associated with deficits in selective attention (Forget et al., 2000). Likewise, in patients with major depression selective attention was impaired (Schatzberg et al., 2000). These findings suggest that the anatomy and function of the prefrontal cortex may be disrupted in these disorders (see also Austin et al., 2001). However, further animal studies are required to further delineate the exact nature of a glucocorticoid-induced attentional deficit.

Conclusions

From the studies reviewed above, it can be concluded that acute glucocorticoid elevations at physiological level enhance memory consolidation and memory retrieval via the GR. This suggests that glucocorticoids released after a stressful event enhance memory for that event. The hippocampus is the brain region that mediates the effects of glucocorticoids on memory, although an intact amygdala and accumbens function with an intact noradrenergic system is also required for optimal consolidation. The MR is thought to be involved in mediating "behavioral reactivity to novelty," possibly also mediated at the hippocampal level.

There is limited evidence for an involvement of glucocorticoids in attentional processes. Since glucocorticoids have been reported to regulate dopaminergic neurotransmission at the level of the prefrontal cortex, it is possible that at least some of the attentional effects of glucocorticoids are mediated via this route. Interestingly, the BLA has largely reciprocal connections with the medial prefrontal cortex (Ongur and Price, 2000) and it is tempting to speculate that an intact BLA is also required to modulate the glucocorticoid effects on attention.

In contrast, sustained elevated levels of glucocorticoids impair learning and memory and seem to disrupt attentional mechanisms in animals and humans. This is particularly evident in human disorders with chronically elevated cortisol levels, including Cushing's syndrome or major depression.

Although it is known that glucocorticoids can affect gene transcription in several ways, it is not exactly clear yet how these effects can be translated into the effects of glucocorticoids on cognitive processes. For memory processes it has been suggested that dimerization of the GR as well as protein–protein interactions are involved.

Clearly, much progress has been made over recent years in understanding the involvement of glucocorticoids in the modulation of cognitive processes in animals, including the dissociation between the functions of the two receptor subtypes, the dissociation between acute and chronic effects, the delineation of the brain areas involved, and a more clearcut description of the different memory processes affected by glucocorticoids. Although many questions still need to be addressed, it is clear that the knowledge obtained from these animal experiments will also foster our understanding of the cognitive processes in patients suffering from abnormalities in the activity of the HPA axis, as can be seen, for example, in patients with Cushing's syndrome and with major depressive disorder.

Abbreviations

ACTH	adrenocorticotrophic hormone
ADX	adrenalectomized
BDNF	brain-derived neurotrophic factor
BLA	basolateral nuclei complex of the amygdala
BNST	bed nucleus of the stria terminalis
CAM	cell adhesion molecule
CeA	central nucleus of the amygdala
CREB	cAMP response-binding element
CRF	corticotropin-releasing factor
CS	conditioning stimulus
EC	entorhinal cortex
GC	glucocorticoid
GR	glucocorticoid receptor
HPA	hypothalamo–pituitary–adrenal
ICV	intracerebroventricular
LTD	long-term depression
LTP	long-term potentiation
MAP	mitogen-activated protein
MR	mineralocorticoid receptor
MRI	magnetic resonance imaging
NAc	nucleus accumbens
US	unconditioned stimulus
PVN	hypothalamic paraventricular nucleus

References

Ahi, J., Radulovic, J. and Spiess, J. (2004) The role of hippocampal signaling cascades in consolidation of fear memory. Behav. Brain Res., 149: 17–31.

Ahima, R., Krozowski, Z. and Harlan, R. (1991) Type I corticosteroid receptor-like immunoreactivity in the rat CNS: distribution and regulation by corticosteroids. J. Comp. Neurol., 313: 522–538.

Akirav, I. and Richter-Levin, G. (2002) Mechanisms of amygdala modulation of hippocampal plasticity. J. Neurosci., 22: 9912–9921.

Akirav, I., Sandi, C. and Richter-Levin, G. (2001) Differential activation of hippocampus and amygdala following spatial learning under stress. Eur. J. Neurosci., 14: 719–725.

Alderson, A.L. and Novack, T.A. (2002) Neurophysiological and clinical aspects of glucocorticoids and memory: a review. J. Clin. Exp. Neuropsychol., 24: 335–355.

Austin, M.P., Mitchell, P. and Goodwin, G.M. (2001) Cognitive deficits in depression: possible implications for functional neuropathology. Br. J. Psychiatry, 178: 200–206.

Avanzi, V., Silva, R.C., Macedo, C.E. and Brandao, M.L. (2003) 5-HT mechanisms of median raphe nucleus in the conditioned freezing caused by light/foot-shock association. Physiol. Behav., 78: 471–477.

Baddeley, A.D., Wilson, B.A. and Watts, F.N. (1995) Handbook of Memory Disorders. John Wiley & Sons, Chichester, West Sussex.

Beato, M., Herrlich, P. and Schutz, G. (1995) Steroid hormone receptors: many actors in search of a plot. Cell, 83: 851–857.

Beckwith, B.E., Lerud, K., Antes, J.R. and Reynolds, B.W. (1983) Hydrocortisone reduces auditory sensitivity at high tonal frequencies in adult males. Pharmacol. Biochem. Behav., 19: 431–433.

Belanoff, J.K., Gross, K., Yager, A. and Schatzberg, A.F. (2001) Corticosteroids and cognition. J. Psychiatr. Res., 35: 127–145.

Beylin, A.V. and Shors, T.J. (2003) Glucocorticoids are necessary for enhancing the acquisition of associative memories after acute stressful experience. Horm. Behav., 43: 124–131.

Beylin, A.V., Gandhi, C.C., Wood, G.E., Talk, A.C., Matzel, L.D. and Shors, T.J. (2001) The role of the hippocampus in trace conditioning: temporal discontinuity or task difficulty? Neurobiol. Learn. Mem., 76: 447–461.

Bilang-Bleuel, A., Rech, J., De Carli, S., Holsboer, F. and Reul, J.M. (2002) Forced swimming evokes a biphasic response in CREB phosphorylation in extrahypothalamic limbic and neocortical brain structures in the rat. Eur. J. Neurosci., 15: 1048–1060.

Bisagno, V., Ferrini, M., Rios, H., Zieher, L.M. and Wikinski, S.I. (2000) Chronic corticosterone impairs inhibitory avoidance in rats: possible link with atrophy of hippocampal CA3 neurons. Pharmacol. Biochem. Behav., 66: 235–240.

Bliss, T.V. and Collingridge, G.L. (1993) A synaptic model of memory: long-term potentiation in the hippocampus. Nature, 361: 31–39.

Bodnoff, S.R., Humphreys, A.G., Lehman, J.C., Diamond, D.M., Rose, G.M. and Meaney, M.J. (1995) Enduring effects of chronic corticosterone treatment on spatial learning, synaptic plasticity, and hippocampal neuropathology in young and mid-aged rats. J. Neurosci., 15: 61–69.

Bohus, B. (1970) Central nervous structures and the effect of ACTH and corticosteroids on avoidance behavior: a study with intracerebral implantation of corticosteroids in the rat. Prog. Brain Res., 32: 171–184.

Bohus, B. and Lissak, K. (1968) Adrenocortical hormones and avoidance behaviour of rats. Int. J. Neuropharmacol., 7: 301–306.

Bohus, B. and De Kloet, E.R. (1981) Adrenal steroids and extinction behavior: antagonism by progesterone, deoxycorticosterone and dexamethasone of a specific effect of corticosterone. Life Sci., 28: 433–440.

Born, J., Kern, W., Fehm-Wolfsdorf, G. and Fehm, H.L. (1987) Cortisol effects on attentional processes in man as indicated by event-related potentials. Psychophysiology, 24: 286–292.

Born, J., Hitzler, V., Pietrowsky, R., Pauschinger, P. and Fehm, H.L. (1989) Influences of cortisol on auditory evoked potentials (AEPs) and mood in humans. Neuropsychobiology, 20: 145–151.

Brown, E.S., Rush, A.J. and McEwen, B.S. (1999) Hippocampal remodeling and damage by corticosteroids: implications for mood disorders. Neuropsychopharmacology, 21: 474–484.

Buhusi, C.V., Gray, J.A. and Schmajuk, N.A. (1998) Perplexing effects of hippocampal lesions on latent inhibition: a neural network solution. Behav. Neurosci., 112: 316–351.

Bussey, T.J., Duck, J., Muir, J.L. and Aggleton, J.P. (2000) Distinct patterns of behavioural impairments resulting from fornix transection or neurotoxic lesions of the perirhinal and postrhinal cortices in the rat. Behav. Brain Res., 111: 187–202.

Cain, D.P., Saucier, D. and Boon, F. (1997) Testing hypotheses of spatial learning: the role of NMDA receptors and NMDA-mediated long-term potentiation. Behav. Brain Res., 84: 179–193.

Clark, R.E., Zola, S.M. and Squire, L.R. (2000) Impaired recognition memory in rats after damage to the hippocampus. J. Neurosci., 20: 8853–8860.

Clark, R.E., Manns, J.R. and Squire, L.R. (2002) Classical conditioning, awareness, and brain systems. Trends Cogn. Sci., 6: 524–531.

Coburn-Litvak, P.S., Pothakos, K., Tata, D.A., McCloskey, D.P. and Anderson, B.J. (2003) Chronic administration of corticosterone impairs spatial reference memory before spatial working memory in rats. Neurobiol. Learn. Mem., 80: 11–23.

Cohen, H. and Eichenbaum, H.B. (1993) Memory, Amnesia, and the Hippocampal System. MIT Press, Cambridge, MA.

Conrad, C.D., Lupien, S.J. and McEwen, B.S. (1999) Support for a bimodal role for type II adrenal steroid receptors in spatial memory. Neurobiol. Learn. Mem., 72: 39–46.

Cordero, M.I. and Sandi, C. (1998) A role for brain glucocorticoid receptors in contextual fear conditioning: dependence upon training intensity. Brain Res., 786: 11–17.

Cullinan, W.E., Herman, J.P. and Watson, S.J. (1993) Ventral subicular interaction with the hypothalamic paraventricular nucleus: evidence for a relay in the bed nucleus of the stria terminalis. J. Comp. Neurol., 332: 1–20.

Czeh, B., Welt, T., Fischer, A.K., Erhardt, A., Schmitt, W., Muller, M.B., Toschi, N., Fuchs, E. and Keck, M.E. (2002) Chronic psychosocial stress and concomitant repetitive transcranial magnetic stimulation: effects on stress hormone levels and adult hippocampal neurogenesis. Biol. Psychiatry, 52: 1057–1065.

De Kloet, E.R., Oitzl, M.S. and Joels, M. (1999) Stress and cognition: are corticosteroids good or bad guys? Trends Neurosci., 22: 422–426.

De Kloet, E.R., Grootendorst, J., Karssen, A.M. and Oitzl, M.S. (2002) Gene x environment interaction and cognitive performance: animal studies on the role of corticosterone. Neurobiol. Learn. Mem., 78: 570–577.

De Quervain, D.J., Roozendaal, B. and McGaugh, J.L. (1998) Stress and glucocorticoids impair retrieval of long-term spatial memory. Nature, 394: 787–790.

Diamond, D.M., Bennett, M.C., Fleshner, M. and Rose, G.M. (1992) Inverted-U relationship between the level of peripheral corticosterone and the magnitude of hippocampal primed burst potentiation. Hippocampus, 2: 421–430.

D'Mello, G.D. and Steckler, T. (1996) Animal models in cognitive behavioural pharmacology: an overview. Cogn. Brain Res., 3: 345–352.

Drevets, W.C. (2001) Neuroimaging and neuropathological studies of depression: implications for the cognitive-emotional features of mood disorders. Curr. Opin. Neurobiol., 11: 240–249.

Eichenbaum, H. (2000) A cortical–hippocampal system for declarative memory. Nat. Rev. Neurosci., 1: 41–50.

Eichenbaum, H., Otto, T. and Cohen, N.J. (1992) The hippocampus – what does it do? Behav. Neural Biol., 57: 2–36.

Ennaceur, A. and Delacour, J. (1988) A new one-trial test for neurobiological studies of memory in rats. 1: Behavioral data. Behav. Brain Res., 31: 47–59.

Fischette, C.T., Komisaruk, B.R., Edinger, H.M., Feder, H.H. and Siegel, A. (1980) Differential fornix ablations and the circadian rhythmicity of adrenal corticosteroid secretion. Brain Res., 195: 373–387.

Flood, J.F., Vidal, D., Bennett, E.L., Orme, A.E., Vasquez, S. and Jarvik, M.E. (1978) Memory facilitating and anti-amnesic effects of corticosteroids. Pharmacol. Biochem. Behav., 8: 81–87.

Forget, H., Lacroix, A., Somma, M. and Cohen, H. (2000) Cognitive decline in patients with Cushing's syndrome. J. Int. Neuropsychol. Soc., 6: 20–29.

Garcia, R. (2002) Stress, metaplasticity, and antidepressants. Curr. Mol. Med., 2: 629–638.

Gass, P., Kretz, O., Wolfer, D.P., Berger, S., Tronche, F., Reichardt, H.M., Kellendonk, C., Lipp, H.P., Schmid, W. and Schutz, G. (2000) Genetic disruption of mineralocorticoid receptor leads to impaired neurogenesis and granule cell degeneration in the hippocampus of adult mice. EMBO Rep., 1: 447–451.

Gass, P., Reichardt, H.M., Strekalova, T., Henn, F. and Tronche, F. (2001) Mice with targeted mutations of glucocorticoid and mineralocorticoid receptors: models for depression and anxiety? Physiol. Behav., 73: 811–825.

Gould, E. and Tanapat, P. (1999) Stress and hippocampal neurogenesis. Biol. Psychiatry, 46: 1472–1479.

Harfstrand, A., Fuxe, K., Cintra, A., Agnati, L.F., Zini, I., Wikstrom, A.C., Okret, S., Yu, Z.Y., Goldstein, F., Steinbusch, H., Verhofstad, A. and Gustafsson, J.A. (1986) Glucocorticoid receptor immunoreactivity in monoaminergic neurons of rat brain. Proc. Natl. Acad. Sci. USA, 82: 9779–9783.

Herman, J.P. and Cullinan, W.E. (1997) Neurocircuitry of stress: central control of the hypothalamo–pituitary–adrenocortical axis. Trends Neurosci., 20: 78–84.

Hui, G.K., Figueroa, I.R., Poytress, B.S., Roozendaal, B., McGaugh, J.L. and Weinberger, N.M. (2004) Memory enhancement of classical fear conditioning by post-training injections of corticosterone in rats. Neurobiol. Learn. Mem., 81: 67–74.

Ishihara, K., Mitsuno, K., Ishikawa, M. and Sasa, M. (1997) Behavioral LTP during learning in rat hippocampal CA3. Behav. Brain Res., 83: 235–238.

Izquierdo, L.A., Barros, D.M., Ardenghi, P.G., Pereira, P., Rodrigues, C., Choi, H., Medina, J.H. and Izquierdo, I. (2000) Different hippocampal molecular requirements for short- and long-term retrieval of one-trial avoidance learning. Behav. Brain Res., 111: 93–98.

Izquierdo, I., Vianna, M.R., Medina, J.H. and Viola, H. (2002) Repetition of memories lost or never acquired. Trends Neurosci., 25: 77–78.

Jarrard, L.E. (1995) What does the hippocampus really do? Behav. Brain Res., 71: 1–10.

Jaffard, R., Vouimba, R.M., Marighetto, A. and Garcia, R. (1996) Long-term potentiation and long-term depression in the lateral septum in spatial working and reference memory. J. Physiol. Paris, 90: 339–341.

Katoh-Semba, R., Asano, T., Ueda, H., Morishita, R., Takeuchi, I.K., Inaguma, Y. and Kato, K. (2002) Riluzole enhances expression of brain-derived neurotrophic factor with consequent proliferation of granule precursor cells in the rat hippocampus. FASEB J., 16: 1328–1330.

Keith, J.R. and Rudy, J.W. (1990) Why NMDA-receptor-dependent long-term potentiation may not be a mechanism of learning and memory: reappraisal of the NMDA-receptor blockade strategy. Psychobiology, 18: 251–257.

Kellendonk, C., Gass, P., Kretz, O., Schutz, G. and Tronche, F. (2002) Corticosteroid receptors in the brain: gene targeting studies. Brain Res. Bull., 57: 73–83.

Kim, J.J., Lee, H.J., Han, J.S. and Packard, M.G. (2001) Amygdala is critical for stress-induced modulation of hippocampal long-term potentiation and learning. J. Neurosci., 21: 5222–5228.

Kirschbaum, C., Wolf, O.T., May, M., Wippich, W. and Hellhammer, D.H. (1996) Stress- and treatment-induced elevations of cortisol levels associated with impaired declarative memory in healthy adults. Life Sci., 58: 1475–1483.

Knierim, J.J. (2003) Hippocampus and memory. Can we have our place and fear it too? Neuron, 37: 372–374.

Kopell, B.S., Wittner, W.K., Lunde, D., Warrick, G. and Edwards, D. (1970) Cortisol effects on averaged evoked potentials, alpha-rhythm, time estimation, and two-flash fusion threshold. Psychosom. Med., 32: 39–49.

Korte, S.M. (2001) Corticosteroids in relation to fear, anxiety and psychopathology. Neurosci. Biobehav. Rev., 25: 117–142.

Krugers, H.J., Douma, B.R., Andringa, G., Bohus, B., Korf, J. and Luiten, P.G. (1997) Exposure to chronic psychosocial stress and corticosterone in the rat: effects on spatial discrimination learning and hippocampal protein kinase Cgamma immunoreactivity. Hippocampus, 7: 427–436.

LeDoux, J.E. (2000) Emotion circuits in the brain. Annu. Rev. Neurosci., 23: 155–184.

Lee, J., Duan, W. and Mattson, M.P. (2002) Evidence that brain-derived neurotrophic factor is required for basal neurogenesis and mediates, in part, the enhancement of neurogenesis by dietary restriction in the hippocampus of adult mice. J. Neurochem., 82: 1367–1375.

Luine, V., Villegas, M., Martinez, C. and McEwen, B.S. (1994) Repeated stress causes reversible impairments of spatial memory performance. Brain Res., 639: 167–170.

Lupien, S.J. and McEwen, B.S. (1997) The acute effects of corticosteroids on cognition: integration of animal and human model studies. Brain Res. Brain Res. Rev., 24: 1–27.

Lupien, S.J., Maheu, F.S. and Weekes, N. (2005) Glucocorticoids: effects on human cognition. In: Steckler, T., Kalin, N.H. and Reul, J.M.H.M. (Eds.), Handbook of Stress and the Brain, Part 1, Elsevier, Amsterdam, pp. 387–402.

Lyons, D.M., Lopez, J.M., Yang, C. and Schatzberg, A.F. (2000) Stress-level cortisol treatment impairs inhibitory control of behavior in monkeys. J. Neurosci., 20: 7816–7821.

Manahan-Vaughan, D. and Braunewell, K.H. (1999) Novelty acquisition is associated with induction of hippocampal long-term depression. Proc. Natl. Acad. Sci. USA, 96: 8739–8744.

Maren, S. and Holt, W. (2000) The hippocampus and contextual memory retrieval in Pavlovian conditioning. Behav. Brain Res., 110: 97–108.

McCormick, C.M., McNamara, M., Mukhopadhyay, S. and Kelsey, J.E. (1997) Acute corticosterone replacement reinstates performance on spatial and nonspatial memory tasks 3 months after adrenalectomy despite degeneration in the dentate gyrus. Behav. Neurosci., 111: 518–531.

McDonald, A.J. and Mascagni, F. (1997) Projections of the lateral entorhinal cortex to the amygdala: a Phaseolus vulgaris leucoagglutinin study in the rat. Neuroscience, 77: 445–459.

McEwen, B.S. (1999) Stress and hippocampal plasticity. Annu. Rev. Neurosci., 22: 105–122.

McEwen, B.S., De Kloet, E.R. and Rostene, W. (1986) Adrenal steroid receptors and actions in the nervous system. Physiol. Rev., 66: 1121–1188.

McGaugh, J.L., Cahill, L. and Roozendaal, B. (1996) Involvement of the amygdala in memory storage: interaction with other brain systems. Proc. Natl. Acad. Sci. USA, 93: 13508–13514.

McGaugh, J.L., McIntyre, C.K. and Power, A.E. (2002) Amygdala modulation of memory consolidation: interaction with other brain systems. Neurobiol. Learn. Mem., 78: 539–552.

McIntyre, C.K., Power, A.E., Roozendaal, B. and McGaugh, J.L. (2003) Role of the basolateral amygdala in memory consolidation. Ann. N.Y. Acad. Sci., 985: 273–293.

McLay, R.N., Freeman, S.M. and Zadina, J.E. (1998) Chronic corticosterone impairs memory performance in the Barnes maze. Physiol. Behav., 63: 933–937.

Medina, J.F., Christopher, R.J., Mauk, M.D. and LeDoux, J.E. (2002) Parallels between cerebellum- and amygdala-dependent conditioning. Nat. Rev. Neurosci., 3: 122–131.

Merino, J.J., Cordero, M.I. and Sandi, C. (2000) Regulation of hippocampal cell adhesion molecules NCAM and L1 by contextual fear conditioning is dependent upon time and stressor intensity. Eur. J. Neurosci., 12: 3283–3290.

Moberg, G.P., Scapagnini, U., Groot, J.D. and Ganong, W.F. (1971) Effect of sectioning of the fornix on diurnal fluctuation in plasma corticosterone levels in the rat. Neuroendocrinology, 7: 11–15.

Moita, M.A., Rosis, S., Zhou, Y., LeDoux, J.E. and Blair, H.T. (2003) Hippocampal place cells acquire location-specific responses to the conditioned stimulus during auditory fear conditioning. Neuron, 37: 485–497.

Monk, C.S. and Nelson, C.A. (2002) The effects of hydrocortisone on cognitive and neural function: a behavioral and event-related potential investigation. Neuropsychopharmacology, 26: 505–519.

Morris, R.G.M. (1981) Spatial localization does not require the presence of local cues. Learn. Motiv., 12: 239–260.

Morris, R.G., Garrud, P., Rawlins, J.N. and O'Keefe, J. (1982) Place navigation impaired in rats with hippocampal lesions. Nature, 297: 681–683.

Moser, M.B. and Moser, E.I. (1998) Distributed encoding and retrieval of spatial memory in the hippocampus. J. Neurosci., 18: 7535–7542.

Muller, M.B., Keck, M.E., Steckler, T. and Holsboer, F. (2002) Genetics of endocrine–behavior interactions. In: Pfaff, Arnold, Etgen, Fahrbach and Moss, Rubin, (Eds.), Hormones, Brain and Behavior, Vol. 5. Academic Press, San Diego, pp. 263–301.

Mumby, D.G. (2001) Perspectives on object-recognition memory following hippocampal damage: lessons from studies in rats. Behav. Brain Res., 127: 159–181.

Myers, K.M. and Davis, M. (2002) Behavioral and neural analysis of extinction. Neuron, 36: 567–584.

Nakao, K., Ikegaya, Y., Yamada, M.K., Nishiyama, N. and Matsuki, N. (2002) Hippocampal long-term depression as an index of spatial working memory. Eur. J. Neurosci., 16: 970–974.

Newman, L.S. (1990) Intentional and unintentional memory in young children: remembering vs. playing. J. Exp. Child Psychol., 50: 243–258.

Newcomer, J.W., Selke, G., Melson, A.K., Hershey, T., Craft, S., Richards, K. and Alderson, A.L. (1999) Decreased memory performance in healthy humans induced by stress-level cortisol treatment. Arch. Gen. Psychiatry, 56: 527–533.

Nibuya, M., Takahashi, M., Russell, D.S. and Duman, R.S. (1999) Repeated stress increases catalytic TrkB mRNA in rat hippocampus. Neuroscience Letters, 267: 81–84.

O'Keefe, J. and Nadel, L. (1978) The Hippocampus as a Cognitive Map. Clarendon Press, Oxford.

Oitzl, M.S. and De Kloet, E.R. (1992) Selective corticosteroid antagonists modulate specific aspects of spatial orientation learning. Behav. Neurosci., 106: 62–71.

Oitzl, M.S., De Kloet, E.R., Joels, M., Schmid, W. and Cole, T.J. (1997a). Spatial learning deficits in mice with a targeted glucocorticoid receptor gene disruption. Eur. J. Neurosci., 9: 2284–2296.

Oitzl, M.S., van Haarst, A.D. and De Kloet, E.R. (1997b). Behavioral and neuroendocrine responses controlled by the concerted action of central mineralocorticoid (MRS) and glucocorticoid receptors (GRS). Psychoneuroendocrinology, 22(Suppl. 1) 1: S87–S93.

Oitzl, M.S., Josephy, M. and Spruijt, B.M. (1993) An ACTH/MSH(4–9) analog counteracts the behavioral effects of a mineralocorticoid receptor antagonist. Pharmacol. Biochem. Behav., 44: 447–450.

Oitzl, M.S., Fluttert, M. and De Kloet, E.R. (1994) The effect of corticosterone on reactivity to spatial novelty is mediated by central mineralocorticosteroid receptors. Eur. J. Neurosci., 6: 1072–1079.

Oitzl, M.S., Reichardt, H.M., Joels, M. and De Kloet, E.R. (2001) Point mutation in the mouse glucocorticoid receptor preventing DNA binding impairs spatial memory. Proc. Natl. Acad. Sci. USA, 98: 12790–12795.

Ongur, D. and Price, J.L. (2000) The organization of networks within the orbital and medial prefrontal cortex of rats, monkeys and humans. Cereb. Cortex, 10: 206–219.

O'Reilly, R.C. and Rudy, J.W. (2001) Conjunctive representations in learning and memory: principles of cortical and hippocampal function. Psychol. Rev., 108: 311–345.

Packard, M.G. and Teather, L.A. (1998) Amygdala modulation of multiple memory systems: hippocampus and caudate-putamen. Neurobiol. Learn. Mem., 69: 163–203.

Pariante, C.M. and Miller, A.H. (2001) Glucocorticoid receptors in major depression: relevance to pathophysiology and treatment. Biol. Psychiatry, 49: 391–404.

Patel, P.D., Lopez, J.F., Lyons, D.M., Burke, S., Wallace, M. and Schatzberg, A.F. (2000) Glucocorticoid and mineralocorticoid receptor mRNA expression in squirrel monkey brain. J. Psychiatr. Res., 34: 383–392.

Pham, K., Nacher, J., Hof, P.R. and McEwen, B.S. (2003) Repeated restraint stress suppresses neurogenesis and induces biphasic PSA-NCAM expression in the adult rat dentate gyrus. Eur. J. Neurosci., 17: 879–886.

Phillips, R.G. and LeDoux, J.E. (1992) Differential contribution of amygdala and hippocampus to cued and contextual fear conditioning. Behav. Neurosci., 106: 274–285.

Pitkanen, A., Savander, V. and LeDoux, J.E. (1997) Organization of intra-amygdaloid circuitries in the rat: an emerging framework for understanding functions of the amygdala. Trends Neurosci., 20: 517–523.

Pitkanen, A., Pikkarainen, M., Nurminen, N. and Ylinen, A. (2000) Reciprocal connections between the amygdala and the hippocampal formation, perirhinal cortex, and postrhinal cortex in rat. A review. Ann. N.Y. Acad. Sci., 911: 369–391.

Poldrack, R.A. and Rodriguez, P. (2003) Sequence learning. what's the hippocampus to do? Neuron, 37: 891–893.

Porsolt, R.D., Le Pichon, M. and Jalfre, M. (1977) Depression: a new animal model sensitive to antidepressant treatments. Nature, 266: 730–732.

Power, A.E., Roozendaal, B. and McGaugh, J.L. (2000) Glucocorticoid enhancement of memory consolidation in the rat is blocked by muscarinic receptor antagonism in the basolateral amygdala. Eur. J. Neurosci., 12: 3481–3487.

Power, A.E., McIntyre, C.K., Litmanovich, A. and McGaugh, J.L. (2003) Cholinergic modulation of memory in the basolateral amygdala involves activation of both m1 and m2 receptors. Behav. Pharmacol., 14: 207–213.

Prickaerts, J., de Vente, J., Honig, W., Steinbusch, H.W. and Blokland, A. (2002) cGMP, but not cAMP, in rat hippocampus is involved in early stages of object memory consolidation. Eur. J. Pharmacol., 436: 83–87.

Pugh, C.R., Fleshner, M. and Rudy, J.W. (1997a) Type II glucocorticoid receptor antagonists impair contextual but not auditory-cue fear conditioning in juvenile rats. Neurobiol. Learn. Mem., 67: 75–79.

Pugh, C.R., Tremblay, D., Fleshner, M. and Rudy, J.W. (1997b). A selective role for corticosterone in contextual-fear conditioning. Behav. Neurosci., 111: 503–511.

Reul, J.M. and De Kloet, E.R. (1985) Two receptor systems for corticosterone in rat brain: microdistribution and differential occupation. Endocrinology, 117: 2505–2511.

Richter-Levin, G. and Akirav, I. (2000) Amygdala–hippocampus dynamic interaction in relation to memory. Mol. Neurobiol., 22: 11–20.

Robbins, T.W. (2000) From arousal to cognition: the integrative position of the prefrontal cortex. Prog. Brain Res., 126: 469–483.

Roozendaal, B. (2002) Stress and memory: opposing effects of glucocorticoids on memory consolidation and memory retrieval. Neurobiol. Learn. Mem., 78: 578–595.

Roozendaal, B. and McGaugh, J.L. (1996) Amygdaloid nuclei lesions differentially affect glucocorticoid-induced memory enhancement in an inhibitory avoidance task. Neurobiol. Learn. Mem., 65: 1–8.

Roozendaal, B. and McGaugh, J.L. (1997a). Basolateral amygdala lesions block the memory-enhancing effect of glucocorticoid administration in the dorsal hippocampus of rats. Eur. J. Neurosci., 9: 76–83.

Roozendaal, B. and McGaugh, J.L. (1997b) Glucocorticoid receptor agonist and antagonist administration into the basolateral but not central amygdala modulates memory storage. Neurobiol. Learn. Mem., 67: 176–179.

Roozendaal, B., Portillo-Marquez, G. and McGaugh, J.L. (1996) Basolateral amygdala lesions block glucocorticoid-induced modulation of memory for spatial learning. Behav. Neurosci., 110: 1074–1083.

Roozendaal, B., Koolhaas, J.M. and Bohus, B. (1997) The role of the central amygdala in stress and adaption. Acta Physiol. Scand. Suppl., 640: 51–54.

Roozendaal, B., Nguyen, B.T., Power, A.E. and McGaugh, J.L. (1999) Basolateral amygdala noradrenergic influence enables enhancement of memory consolidation induced by hippocampal glucocorticoid receptor activation. Proc. Natl. Acad. Sci. USA, 96: 11642–11647.

Roozendaal, B., De Quervain, D.J., Ferry, B., Setlow, B. and McGaugh, J.L. (2001a). Basolateral amygdala–nucleus accumbens interactions in mediating glucocorticoid enhancement of memory consolidation. J. Neurosci., 21: 2518–2525.

Roozendaal, B., Phillips, R.G., Power, A.E., Brooke, S.M., Sapolsky, R.M. and McGaugh, J.L. (2001b). Memory retrieval impairment induced by hippocampal CA3 lesions is blocked by adrenocortical suppression. Nat. Neurosci., 4: 1169–1171.

Roozendaal, B., Brunson, K.L., Holloway, B.L., McGaugh, J.L. and Baram, T.Z. (2002a) Involvement of stress-released corticotropin-releasing hormone in the basolateral amygdala in regulating memory consolidation. Proc. Natl. Acad. Sci. USA, 99: 13908–13913.

Roozendaal, B., Quirarte, G.L. and McGaugh, J.L. (2002b). Glucocorticoids interact with the basolateral amygdala beta-adrenoceptor – cAMP/cAMP/PKA system in influencing memory consolidation. Eur. J. Neurosci., 15: 553–560.

Roozendaal, B., Griffith, Q.K., Buranday, J., De Quervain, D.J. and McGaugh, J.L. (2003) The hippocampus mediates glucocorticoid-induced impairment of spatial memory retrieval: dependence on the basolateral amygdala. Proc. Natl. Acad. Sci. USA, 100: 1328–1333.

Roozendaal, B., McReynolds, J.R. and McGaugh, J.L. (2004) The basolateral amygdala interacts with the medial prefrontal cortex in regulating glucocorticoid effects on working memory impairment. J Neurosci., 24: 1385–1392.

Sanchez, M.M., Young, L.J., Plotsky, P.M. and Insel, T.R. (2000) Distribution of corticosteroid receptors in the rhesus brain: relative absence of glucocorticoid receptors in the hippocampal formation. J. Neurosci., 20: 4657–4668.

Sanders, M.J., Wiltgen, B.J. and Fanselow, M.S. (2003) The place of the hippocampus in fear conditioning. Eur. J. Pharmacol., 463: 217–223.

Sandi, C. and Rose, S.P. (1994a) Corticosteroid receptor antagonists are amnestic for passive avoidance learning in day-old chicks. Eur. J. Neurosci., 6: 1292–1297.

Sandi, C. and Rose, S.P. (1994b) Corticosterone enhances long-term retention in one-day-old chicks trained in a weak passive avoidance learning paradigm. Brain Res., 647: 106–112.

Sandi, C. and Rose, S.P. (1997) Training-dependent biphasic effects of corticosterone in memory formation for a passive avoidance task in chicks. Psychopharmacology (Berl), 133: 152–160.

Sandi, C. and Loscertales, M. (1999). Opposite effects on NCAM expression in the rat frontal cortex induced by acute vs. chronic corticosterone treatments. Brain Res., 828: 127–134.

Sandi, C., Loscertales, M. and Guaza, C. (1997) Experience-dependent facilitating effect of corticosterone on spatial memory formation in the water maze. Eur. J. Neurosci., 9: 637–642.

Sapolsky, R.M., Krey, L.C. and McEwen, B.S. (1986) The neuroendocrinology of stress and aging: the glucocorticoid cascade hypothesis. Endocr. Rev., 7: 284–301.

Scaccianoce, S., Del Bianco, P., Caricasole, A., Nicoletti, F. and Catalani, A. (2003) Relationship between learning, stress and hippocampal brain-derived neurotrophic factor. Neuroscience, 121: 825–828.

Schaaf, M.J., de Jong, J., De Kloet, E.R. and Vreugdenhil, E. (1998) Downregulation of BDNF mRNA and protein in the rat hippocampus by corticosterone. Brain Res., 813: 112–120.

Schaaf, M.J., Sibug, R.M., Duurland, R., Fluttert, M.F., Oitzl, M.S., De Kloet, E.R. and Vreugdenhil, E. (1999) Corticosterone effects on BDNF mRNA expression in the rat hippocampus during morris water maze training. Stress, 3: 173–183.

Schaaf, M.J., De Kloet, E.R. and Vreugdenhil, E. (2000) Corticosterone effects on BDNF expression in the hippocampus. Implications for memory formation. Stress, 3: 201–208.

Schatzberg, A.F., Posener, J.A., DeBattista, C., Kalehzan, B.M., Rothschild, A.J. and Shear, P.K. (2000). Neuropsychological deficits in psychotic versus nonpsychotic major depression and no mental illness. Am. J. Psychiatry, 157: 1095–1100.

Shalev, U., Feldon, J. and Weiner, I. (1998) Latent inhibition is disrupted by acute and repeated administration of corticosterone. Int. J. Neuropsychopharmcol., 1: 103–113.

Sheline, Y.I., Gado, M.H. and Price, J.L. (1998) Amygdala core nuclei volumes are decreased in recurrent major depression. Neuroreport, 9: 2023–2028.

Sheline, Y.I., Wang, P.W., Gado, M.H., Csernansky, J.G. and Vannier, M.W. (1996) Hippocampal atrophy in recurrent major depression. Proc. Natl. Acad. Sci. USA, 93: 3908–3913.

Sherman, A.D., Sacquitne, J.L. and Petty, F. (1982) Specificity of the learned helplessness model of depression. Pharmacol. Biochem. Behav., 16: 449–454.

Sillaber, I., Montkowski, A., Landgraf, R., Barden, N., Holsboer, F. and Spanagel, R. (1998) Enhanced morphine-induced behavioural effects and dopamine release in the nucleus accumbens in a transgenic mouse model of impaired glucocorticoid (type II) receptor function: influence of long-term treatment with the antidepressant moclobemide. Neuroscience, 85: 415–425.

Skosnik, P.D., Chatterton, R.T., Jr., Swisher, T. and Park, S. (2000) Modulation of attentional inhibition by norepinephrine and cortisol after psychological stress. Int. J. Psychophysiol., 36: 59–68.

Son, G.H., Geum, D., Jung, H. and Kim, K. (2001) Glucocorticoid inhibits growth factor-induced differentiation of hippocampal progenitor HiB5 cells. J. Neurochem., 79: 1013–1021.

Sousa, N., Lukoyanov, N.V., Madeira, M.D., Almeida, O.F. and Paula-Barbosa, M.M. (2000) Reorganization of the morphology of hippocampal neurites and synapses after stress-induced damage correlates with behavioral improvement. Neuroscience, 97: 253–266.

Squire, L.R. (1992) Memory and the hippocampus: a synthesis from findings with rats, monkeys, and humans. Psychol. Rev., 99: 195–231.

Starkman, M.N., Gebarski, S.S., Berent, S. and Schteingart, D.E. (1992) Hippocampal formation volume, memory dysfunction, and cortisol levels in patients with Cushing's syndrome. Biol. Psychiatry, 32: 756–765.

Steckler, T. (2001) The molecular neurobiology of stress – evidence from genetic and epigenetic models. Behav. Pharmacol., 12: 381–427.

Steckler, T. (2005) Psychological concepts of stress. In: Steckler, T., Kalin, N.H. and Reul, J.M.H.M. (Eds.), Handbook of Stress and the Brain, Part 1, Elsevier, Amsterdam, pp. 25–42.

Steckler, T. and Holsboer, F. (1999) Enhanced conditioned approach responses in transgenic mice with impaired glucocorticoid receptor function. Behav. Brain Res., 102: 151–163.

Steckler, T., Drinkenburg, W.H., Sahgal, A. and Aggleton, J.P. (1998) Recognition memory in rats – II. Neuroanatomical substrates. Prog. Neurobiol., 54: 313–332.

Steckler, T., Holsboer, F. and Reul, J.M.H.M. (1999a) Glucocorticoids and depression. Bailliere's Clin. Endocrinol. Metabol., 13: 597–614.

Steckler, T., Weis, C., Sauvage, M., Mederer, A. and Holsboer, F. (1999b). Disrupted allocentric but preserved egocentric spatial learning in transgenic mice with impaired glucocorticoid receptor function. Behav. Brain Res., 100: 77–89.

Steckler, T., Sauvage, M. and Holsboer, F. (2000) Glucocorticoid receptor impairment enhances impulsive responding in transgenic mice performing on a simultaneous visual discrimination task. Eur. J. Neurosci., 12: 2559–2569.

Steffenach, H.A., Sloviter, R.S., Moser, E.I. and Moser, M.B. (2002) Impaired retention of spatial memory after transection of longitudinally oriented axons of hippocampal CA3 pyramidal cells. Proc. Natl. Acad. Sci. USA, 99: 3194–3198.

Stockand, J.D. and Meszaros, J.G. (2003) Aldosterone stimulates proliferation of cardiac fibroblasts by activating Ki-RasA and MAPK1/2 signaling. Am. J. Physiol., 284: H176–H184.

Thomas, S.R., Assaf, S.Y. and Iversen, S.D. (1984) Amygdaloid complex modulates neurotransmission from the entorhinal cortex to the dentate gyrus of the rat. Brain Res., 307: 363–365.

Thomas, D.N., Post, R.M. and Pert, A. (1994) Central and systemic corticosterone differentially affect dopamine and norepinephrine in the frontal cortex of the awake freely moving rat. Ann. N.Y. Acad. Sci., 746: 467–469.

Tuvnes, F.A., Steffenach, H.A., Murison, R., Moser, M.B. and Moser, E.I. (2003) Selective hippocampal lesions do not increase adrenocortical activity. J. Neurosci., 23: 4345–4354.

Van Eekelen, J.A., Jiang, W., De Kloet, E.R. and Bohn, M.C. (1988) Distribution of the mineralocorticoid and the glucocorticoid receptor mRNAs in the rat hippocampus. J. Neurosci. Res., 21: 88–94.

Veldhuis, H.D., Van Koppen, C., Van Ittersum, M. and De Kloet, E.R. (1982) Specificity of the adrenal steroid receptor system in rat hippocampus. Endocrinology, 110: 2044–2051.

Weiner, I. and Feldon, J. (1997) The switching model of latent inhibition: an update of neural substrates. Behav. Brain Res., 88: 11–25.

Wellman, C.L. (2001) Dendritic reorganization in pyramidal neurons in medial prefrontal cortex after chronic corticosterone administration. J. Neurobiol., 49: 245–253.

Whitfield, G.K., Jurutka, P.W., Haussler, C.A. and Haussler, M.R. (1999) Steroid hormone receptors: evolution, ligands, and molecular basis of biologic function. J. Cell Biochem., Suppl. 32–33: 110–122.

Xu, L., Holscher, C., Anwyl, R. and Rowan, M.J. (1998) Glucocorticoid receptor and protein/RNA synthesis-dependent mechanisms underlie the control of synaptic plasticity by stress. Proc. Natl. Acad. Sci. USA, 95: 3204–3208.

Xu, H., Qing, H., Lu, W., Keegan, D., Richardson, J.S., Chlan-Fourney, J. and Li, X.M. (2002) Quetiapine attenuates the immobilization stress-induced decrease of brain-derived neurotrophic factor expression in rat hippocampus. Neurosci. Lett., 321: 65–68.

Yerkes, R.M. and Dodson, J.D. (1908) The relation of strength of stimulus to rapidity of habit formation. J. Comp. Neurol. Psychol., 18: 459–482.

Zald, D.H. (2003) The human amygdala and the emotional evaluation of sensory stimuli. Brain Res. Brain Res. Rev., 41: 88–123.

T. Steckler, N.H. Kalin and J.M.H.M. Reul (Eds.)
Handbook of Stress and the Brain, Vol. 15
ISBN 0-444-51173-3

CHAPTER 3.7

Glucocorticoids: effects on human cognition

Sonia J. Lupien[1,2,*], Françoise S. Maheu[1,3] and Nicole Weekes[4]

[1]*Laboratory of Human Psychoneuroendocrine Research, Douglas Hospital, Montreal, QC, Canada*
[2]*Department of Psychiatry, McGill University, 6875 Bld. Lasalle, Montreal, QC H4H 1R3, Canada*
[3]*Department of Psychology, University of Montreal, 6875 Bld. Lasalle, Montreal, QC, H4H 1R3, Canada*
[4]*Department of Psychology, Pomona College, 550 N. Harvard Avenue, Claremont, California 91711, USA*

Abstract: In this chapter, we review the literature on the effects of glucocorticoids on human cognition with a particular emphasis on the results of studies which have measured the effects of exogenous administrations of glucocorticoids on human cognitive function. In order to describe this field of research, we use history as a background and we present the various models of glucocorticoid effects on human cognition. We use this background because history has taught us a lot of things. First, that these steroids could access the brain and lead to steroid psychosis. Second, that glucocorticoids' access to the hippocampus could lead to impairments in declarative memory. Third, that glucocorticoids' access to the frontal lobes and the amygdala could lead to working memory impairments and dysregulation of emotional cognition. Finally, that glucocorticoids' action on cognitive process can also be positive, and that the positive or negative effects of glucocorticoids depend on the balance of the two glucocorticoid receptor types known to exist today.

Introduction

Many will keep stringent, long-lasting memories of 11 September 2001 attacks on New York and Washington. This *Flashbulb Memory phenomenon* is defined as the vivid recollection one has of striking events because of the high level of surprise and emotional arousal these events create (Brown and Kulik, 1977). Human studies demonstrated that witnesses to violent crimes have enhanced memory for the arousing aspects of the traumatic event, although they tend to forget the neutral information concerning the event (Loftus, 1979; Christianson, 1992). A second characteristic of human memory for traumatic events concerns retention intervals. It has been shown that emotional and/or traumatic information are better recalled after a delay than immediately after learning, whereas memory for neutral events decreases over longer intervals (Kleinsmith and Kaplan, 1963; LaBar and Phelps, 1998; Adolphs et al., 2000; Quevedo et al., 2003). These data show that stressful experiences differentially modulate human memory for the emotional and neutral component of the learning situation.

Glucocorticoids (corticosterone in rat, cortisol in human) secreted during stress are among the factors thought to be responsible for these modulatory effects on cognitive functions (Lupien and McEwen, 1997). In young (Kirschbaum et al., 1996; Wolf et al., 2001a) and elderly (Lupien et al., 1997) healthy controls, glucocorticoid elevation in response to a psychological stressor (TSST test, i.e., public speaking and arithmetic task; Kirschbaum et al., 1993) has been correlated with deficits in memory for neutral words. To mimic the physiological effects of stress, studies measuring the effects of exogenous administration of synthetic glucocorticoids on cognitive function were

*Corresponding author. Laboratory of Human Psychoneuroendocrine Research, Douglas Hospital Research Center, 6875 Bld. Lasalle, Verdun, QC H4H 1R3, Canada.
Tel.: +1(514) 762-3028; Fax: +1(514) 888-4064;
E-mail: sonia.lupien@mcgill.ca

undertaken. Important glucocorticoid elevations following the acute or chronic administration of high doses of synthetic glucocorticoids led to very inconsistent results when compared to the stress literature (for reviews, see Lupien and McEwen, 1997; Belanoff et al., 2001; Lupien and Lepage, 2001).

One very important reason for such a discrepancy is that "stress" is not equal to "glucocorticoids" in neuroendocrinological terms. During a stress, there is the release of catecholamines and glucocorticoids. It is thus unclear at this point whether the effects of psychological stress on human cognitive function are due to catecholamines and/or glucocorticoids variations alone, or to their combined activity on the brain (for a review, see Maheu and Lupien, 2003). Second, there are major neuroendocrine differences between stress-induced glucocorticoid elevations, and psychopharmacologically induced glucocorticoids elevations, and this difference lies directly in the activity of the *hypothalamic–pituitary–adrenal* (*HPA*) *axis*.

Under most conditions, the HPA axis lies under the dominance of specific releasing factors secreted by neurons located in paraventricular nucleus (PVN) of the hypothalamus. Most notable among these releasing factors is corticotropin-releasing factor (CRF). CRF neurons from the parvocellular cells of the PVN project extensively to the portal capillary zone of the median eminence, furnishing a potent excitatory signal for the synthesis and release of ACTH from the anterior pituitary. CRF, along with cosecretagogues such as arginine vasopressin (AVP) and oxytocin provide the signals by which neural inputs are coded into endocrine signals through their actions on pituitary corticotrophes. Dynamic changes in ACTH levels occurring during the circadian peak or in response to stress are associated with changes in the secretion of one or more of these secretagogues from hypothalamic neurons into the portal system of the anterior pituitary, enhancing the synthesis and release of adrenocorticotropin (ACTH). Elevated ACTH levels, in turn, increase the synthesis and release of glucocorticoids from the adrenal. The responsivity of the HPA axis to stress is, in part, determined by the ability of the glucocorticoids to regulate ACTH release (i.e., glucocorticoid negative feedback). Circulating glucocorticoids feedback onto the pituitary and specific brain regions to inhibit the secretion of releasing factors from hypothalamic neurons and pituitary ACTH.

Given the dynamic nature of the HPA axis, endogenous increases of glucocorticoids as induced by stress will have very different effects on HPA activity, when compared to exogenous administration of synthetic glucocorticoids. As presented in Fig. 1, perception of a stressor (psychological stress protocols) will lead to increased circulating levels of

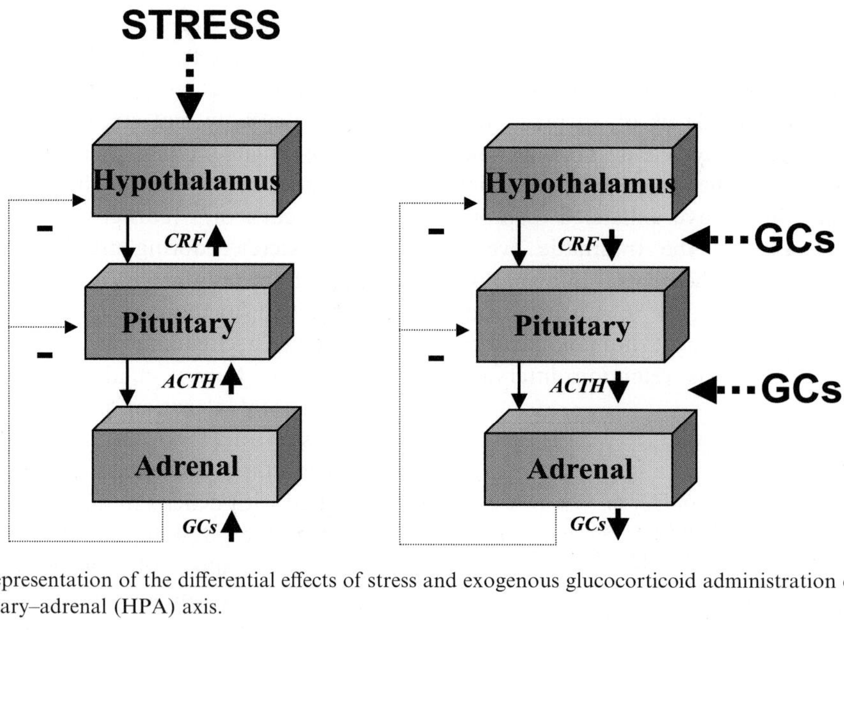

Fig. 1. Schematic representation of the differential effects of stress and exogenous glucocorticoid administration on the activity of the hypothalamic–pituitary–adrenal (HPA) axis.

CRF and ACTH, while exogenous administrations of glucocorticoids (pharmacological protocols) will lead to suppression of CRF and ACTH, and thus, decreased circulating levels of both CRF and ACTH at high doses. Given that both CRF (Swerdlow et al., 1986; Dunn and Berridge, 1990; Cole and Koob, 1991) and ACTH (Molle et al., 1997) have been involved in learning and memory, it is unclear at this point whether the effects of stress on human memory are sustained solely by glucocorticoids, catecholamines, CRF, or ACTH alone, or by a combination of some of these hormones and peptides for their actions on the brain.

In an attempt to present the reader with a clearer view of the effects of glucocorticoids on human cognition, we have decided to discuss exclusively the results of studies which have measured the effects of exogenous administrations of glucocorticoids on human cognitive function. In doing so, our background will be history, as we will present the various models of glucocorticoid effects on human cognition that have been described for the last 60 years.

Glucocorticoids and steroid psychosis

The first human evidence suggesting that exogenous glucocorticoids might affect cognitive function comes from clinical studies of cases of steroid psychosis. In 1949, Hench and collaborators discovered the therapeutic effects of glucocorticoids on inflammatory diseases, such as rheumatoid arthritis and asthma. Regarded by many scientists and clinicians as the "wonder drugs," glucocorticoids soon became very popular for the treatment of various diseases. Besides being employed in hormone-replacement therapy in Addison's disease or after adrenalectomy, they have also been used in treating rheumatoid arthritis, ulcerative colitis, asthma, Hodgkins' disease, systemic lupus erythematosus, and various dermatological disorders. However, no more than two years after their introduction as *antiinflammatory* drugs, the enthusiasm engendered by glucocorticoids was dampened by the finding that the therapeutic use of glucocorticoids was followed by several side effects, particularly on affect and cognition. The first case was published in 1951 by Borman and Schmallenberg who reported suicide following cortisone treatment. A year later, three papers were published reporting severe mental disturbances in patients under *glucocorticoid therapy* (Brody, 1952; Clark et al., 1952; Rome and Braceland, 1952).

The mental side effects of glucocorticoid therapy constituted a full spectrum of a psychotic disorder. Most often, a "vitalization effect" was observed, particularly in the aged patient (Kountz et al., 1953) or in individuals with reduced vitality due to the underlying illness (Von Zerssen, 1957 cited in Von Zerssen, 1976). In these cases, the patient displayed an elevation of mood, varying in degree from a feeling of well-being to abnormal degrees of euphoria, psychomotor activation, increased appetite, and reduced sleep. However, these changes of mood and behavior were generally observed in the first days or weeks of glucocorticoid therapy (Brody, 1952; Rees, 1953; Dordick and Gluck, 1955), and it was not clear whether these signs were related to glucocorticoid actions or to the relief from severe physical complaints and handicap related to the underlying disease treated with glucocorticoids (Lidz et al., 1952; Rees, 1953).

In a large proportion of patients, euphoria was not present in the first days or weeks of treatment. Rather, tension, irritability, and sleeplessness were present. Both the euphoric and dysphoric states could gradually increase to a full-blown manic episode in which the patient would present marked euphoria or dysphoria, pronounced self-assertiveness, hyperactivity, logorrhea, and flight of ideas (Cobb et al., 1954). These are the mental symptoms that have led some scientists and clinicians to name the side effects of glucocorticoid therapy a "*steroid psychosis*" (Clark et al., 1952; Rome and Braceland, 1952).

There were other types of aberrant behaviors that could appear with glucocorticoid therapy and these later behaviors closely resembled those observed in Cushing's disease patients (Trethowan and Cobb, 1952). These mood and behavioral change covered a wide range of symptoms and included feeling of weakness, fatigue, and drowsiness, lack of concentration, apathy, anxiety, depression, and sometimes suicide as reported by Borman and Schmallenberg (1951). These behavioral changes were also associated with changes in brain activity as measured by *electroencephalogram* (EEG). In man, glucocorticoid therapy sometimes led to epileptic seizures and even to status epilepticus (Stephen and Noad, 1951; Loewenberg, 1954). Similarly, some EEG abnormalities were also found in patients with Cushing's disease (Plotz et al., 1952; Glaser et al., 1955). On the whole, the clinical descriptions of cases of steroid psychosis following glucocorticoid therapy were strikingly similar to the mental disturbances associated with Cushing's disease, leading to the idea that glucocorticoids might be the underlying causes of steroid psychosis.

Throughout history, certain authors have questioned whether glucocorticoids really cause psychiatric adverse effects (Mitchell and Collins, 1984). However, a recent metaanalysis of randomized controlled trials has provided firm confirmation that they can (Conn and Poynard, 1994). In general, *prednisone* has been most frequently implicated in causing psychiatric side effects (for a review, see Hall et al., 1979), but other less widely used steroids, such as *methylprednisone* (Greeves, 1984; Perry et al., 1984), *dexamethasone* (Bick, 1983), and even inhaled *beclomethasone* (Kreus et al., 1975), have been reported to induce mental disturbances. Most patients exhibiting side effects are between 21 and 60 years (Ling et al., 1981), but adverse mental changes are also being reported in children (De la Riva, 1958; Sullivan and Dickerman, 1979; Bender and Milgrom, 1995) and elderly individuals (Varney et al., 1984). Finally, females appear to be at somewhat greater risk than males, even after controlling for diseases that are more common to women (Lewis and Smith, 1983). Today, glucocorticoid-induced mood disturbances are classified in the DSM-IV as substance-induced mood disorders, with an associated specification of depressive, manic, or mixed features.

Glucocorticoids, the hippocampus, and declarative memory

The observation that glucocorticoid treatment could lead to steroid psychosis suggested that the excessive concentrations of the steroid may access the brain and bind to receptors able to recognize the glucocorticoid molecule. The search for brain receptors able to recognize peripheral hormones was then opened. In 1968, it culminated with Bruce McEwen's seminal Nature paper (McEwen et al., 1968), showing that the rodent brain was indeed able to recognize glucocorticoids. The story then took a very important detour when McEwen and collaborators reported that the brain region showing the highest density of receptors for glucocorticoids was the *hippocampus*.

Scientists from the field of animal studies were excited by the findings of McEwen and collaborators (1968), showing that the largest quantity of *glucocorticoid receptors* were found in the hippocampus. The reason for this interest came from an earlier study published by Scoville and Milner (1957), which reported that the hippocampus is essential for learning and memory. Given the presence of glucocorticoid receptors in the rodent hippocampus, it became clear that glucocorticoids might be related to some aspects of learning and memory in rodents (for a complete review of animal data, see Lupien and McEwen, 1997; and the chapters from Prickaerts and Steckler, dealing with animal cognitive function and effects of glucocorticoids in this book).

About 20 years after the discovery of glucocorticoid receptors in the rodent hippocampus, and supported by a significant amount of animal data showing that glucocorticoids have a significant impact on learning and memory, human studies started to appear in the literature. Here, scientists used another important characteristic of hippocampal function originally described by Scoville and Milner (1957), and confirmed in rodents by the group of Squire (for a review, see Squire, 1992), i.e., its specific role in declarative memory function. *Declarative memory* refers to conscious or voluntary recollection of previous information (such as remembering what one had for breakfast), whereas *nondeclarative memory* refers to the facilitation of recollection of previous information without a conscious and deliberate intention to retrieve this information (such as measured in priming). Declarative memory function is usually measured by tasks in which the subject has to explicitly recall some information learned previously, while nondeclarative memory function relates to an unconscious type of memory for which recall can take place without necessary consciousness from the subject. This somewhat specialized role of the hippocampus served as the basis for specific hypotheses regarding the effects of glucocorticoids on human learning and memory.

Now, remember that it is the observation of severe memory impairments in patients suffering from steroid psychosis that led scientists to search for potential detrimental effects of exogenous administration of glucocorticoids in human populations. What is important to note here is that in cases of steroid psychosis, chronic administration of glucocorticoids led to memory deficits. Interestingly, although some earlier studies measured the effects of chronic administration of synthetic glucocorticoids on memory function in normal subjects, most of the human studies performed in this field of research used acute administration of glucocorticoids in order to assess the effects of these stress hormones on human cognitive function, and in many instances, different effects of acute and chronic administration of synthetic glucocorticoids have been reported.

Effects of chronic administration of glucocorticoids on declarative memory

In 1990, Wolkowitz and collaborators observed impaired memory performance in normal adults following five days administration of high doses of prednisone (80 mg p.o. daily). However, the authors also reported the induction of mental disturbances in the subjects treated with this regimen, so it was not clear at this point whether the observed memory impairments in the subjects were due to glucocorticoids, or to the psychic disturbances induced by glucocorticoid treatment.

However, in 1994, Newcomer and collaborators, using a four-day administration procedure of 0.5 and 1 mg dexamethasone in normal controls reported impaired declarative memory performance on the fourth day of treatment only, with no psychic disturbances. This result suggested that the impaired declarative memory performance was related to the glucocorticoid treatment. Similar results were obtained by the same group using four days administration of 40 and 160 mg cortisol (Newcomer et al., 1999). In both studies, no immediate or delayed effects of the synthetic glucocorticoid were observed on nondeclarative memory, again suggesting a specific effect of glucocorticoids on nondeclarative, hippocampal-dependent type of memory.

In 1999, Schmidt and collaborators assessed the effects of four days administration of 160 mg prednisone on memory, attention, and emotion in 24 young men. Mood, regional brain electrical activity (EEG), the startle eyeblink response, memory recall, and performance on an attention task was evaluated after four days of treatment. As with previous data obtained by Wolkowitz and collaborators (1990), results showed that subjects treated with prednisone showed a significantly greater increase in self-reported negative emotion. They also presented a greater relative right-frontal EEG alpha activity, and they recalled fewer objects on the memory task following treatment. No significant group differences were found on posterior EEG activity, the startle eyeblink measure, or the attention measure. These findings confirmed thee facilitation of negative emotion in subjects receiving high doses of synthetic glucocorticoids for at least four days, and further showed that glucocorticoids, in isolation or in combination with impaired mood, have negative effects on memory function in humans.

In a more recent study, McAllister-Williams and Rugg (2002) examined the effects of one-week administration of hydrocortisone (40 mg/day) on the neural correlates of declarative memory in healthy subjects as measured using evoked-related potentials (ERPs). They found that response times on the declarative memory task were significantly speeded by cortisol, but that recognition accuracy was reduced after hydrocortisone administration, suggesting the presence of a speed accuracy trade-off between speed and performance after chronic administration of synthetic glucocorticoids.

Effects of acute administration of glucocorticoids on declarative memory

The first study performed on the acute effects of glucocorticoids on human memory process was a dose–response. In 1986, Beckwith and collaborators showed that the effects of 5, 10, 20, and 40 mg hydrocortisone on human memory performance depend upon the dose administered. Only the highest doses of glucocorticoids enhanced the recall of previously presented lists of words, leading to the suggestion that glucocorticoids could have beneficial effects on learning and memory.

In 1993, Fehm-Wolfsdorf and collaborators reported that glucocorticoid administration in the morning impaired declarative memory function, while it had no effect on cognitive performance when administered at night. This chapter introduced a very important methodological point to take into consideration in studies of the effects of glucocorticoids on cognitive function, i.e., time of day. Indeed, glucocorticoids follow a *circadian rhythm* with highest circulating levels observed in the morning, with slowly decreasing levels through the afternoon and evening periods in humans. Administration of synthetic glucocorticoids to an organism with high (AM phase in humans) or low (PM phase in humans) circulating levels of glucocorticoids should thus have very different effects on the total levels of glucocorticoids that can reach the brain. The results reported by Fehm-Worlsdorf and collaborators (1993) confirmed that the effects of glucocorticoids on human cognitive function depend on the baseline levels of circulating glucocorticoids at the time of drug administration. When administered at the time of glucocorticoid peak, exogenous glucocorticoids impaired memory, while when administered at the time of the glucocorticoid trough, exogenous glucocorticoid had no impact on memory. This result suggested the presence of an inverted-U shape function between circulating levels of glucocorticoids (endogenous + exogenous) and memory function, and went along with animal studies showing the existence of an inverted-U shape curve between circulating levels of glucocorticoids and cognitive performance (for a complete review, see Lupien and McEwen, 1997). The possibility that an inverted-U shape function relates circulating levels of glucocorticoids and memory performance in humans would also explain the positive effects of a high dose of hydrocortisone previously reported by Beckwith et al. (1986), who administered the drug in the afternoon.

In 1996, Kirschbaum and collaborators took advantage of the declarative/nondeclarative memory dissociation of the hippocampus in order to assess whether glucocorticoids would have a specific impact on declarative memory function in humans. They reported that administration of 10 mg hydrocortisone led to a significant decrease in declarative memory performance, while it had no effect on nondeclarative memory performance. These results suggested that glucocorticoids interact with hippocampal neurons to induce cognitive deficits in humans.

More recently, de Quervain and collaborators (2000) tested the impact of an acute increase of glucocorticoids as a function of the nature of memory processing. They administered 25 mg cortisone either before the acquisition of a list of words, immediately after, or just before the retrieval of the list. The results revealed significant impairments in memory when the drug was administered just before retrieval, thus suggesting specific effects of glucocorticoids on the retrieval of a previously learned information.

A specific effect of acute cortisol elevations on retrieval process in humans has recently been replicated by Wolf and collaborators (2001b). Young and aged men were administered with 0.5 mg/kg cortisol *after* having learned a list of 10 words. A second word list was learned and recalled after drug administration. Results showed that cortisol impaired recall from the word list learned before treatment in both groups, but did not influence recall of the list learned after treatment. These results go along with previous data obtained by de Quervain et al. (2000) showing that acute exogenous administration of glucocorticoids have an impairing effects on retrieval process, and they also confirm other studies showing that acute administration of synthetic glucocorticoids in young controls has no effect on acquisition of declarative memory tasks (Lupien et al., 1999; Hsu et al., 2003).

The in vivo demonstration of these glucocorticoid effects on memory-retrieval process was recently performed by the group of de Quervain and collaborators (2003) using positron emission tomography (PET). Young subjects were administered 25 mg cortisone 24 h after learning various declarative memory tasks. Brain activation was measured by PET 1 h after drug administration. Results showed that cortisone induced a large decrease in regional cerebral blood flow in the right-posterior medial temporal lobe, that was associated with impaired cued recall of word pairs learned 24 h earlier. These results were the first to provide an in vivo demonstration that acutely elevated glucocorticoid levels can impair declarative memory-retrieval processes that are related to a disturbance of medial temporal lobe function.

In summary, although chronic (at least four days) administration of glucocorticoids has been shown to impair declarative memory function in humans, acute elevations of glucocorticoids on memory led to inconsistent results. To this day, however, a specific effect on memory-retrieval process seems to best characterize the acute effects of exogenous administration of glucocorticoids on human declarative memory (see Fig. 2).

Interim: lessons from history

Now, let us stop time and let us go back to Bruce McEwen's 1968 discovery of glucocorticoid receptors in the rodent hippocampus. It is clear from the literature summarized above that the presence of glucocorticoid receptors in the rodent hippocampus served as the basis for the hypothesis that glucocorticoids should significantly, and specifically impair declarative memory function in humans. However, and although the hypothesis was simple, economical, and easy to test, a closer look at history teaches us that the *glucocorticoid–hippocampus link* might not be the best hypothesis to fully explain acute and/or chronic glucocorticoid-induced cognitive changes in humans (for a complete review on this topic, see Lupien and Lepage, 2001). Here is what history has to tell us.

In their 1968 paper, McEwen and collaborators described the retention of corticosterone – a naturally occurring glucocorticoid in the rodent brain – in the adrenalectomized rat brain. The rats were first adrenalectomized in order to deplete the rat's system of any endogenous circulating glucocorticoids, and then corticosterone was injected and retention of this naturally occurring glucocorticoid was assessed. Using this method, McEwen and collaborators (1968) showed that corticosterone was highly retained by the hippocampus.

In 1975, de Kloet and collaborators, assuming that a synthetic glucocorticoid called dexamethasone would be even better retained by the hippocampus than the naturally occurring corticosterone, duplicated McEwen's study using dexamethasone and found out that this assumption was in fact incorrect. Indeed, dexamethasone was very poorly retained by the hippocampus of the rodent brain, and this, irrespective of the route of administration (peripheral vs. intracerebroventricular). Later, these authors, in collaboration with Bruce McEwen, showed that the small amounts of dexamethasone that prenetrated the brain was retained in a regional pattern that was distinctively different from corticosterone (McEwen et al., 1976). In the same study, another

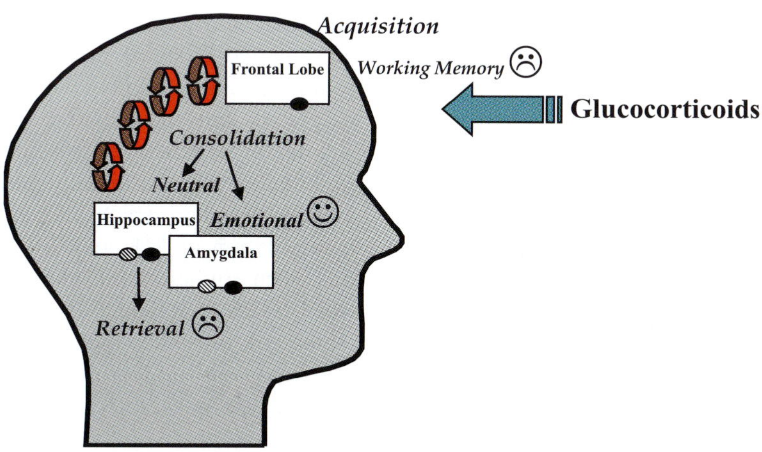

Fig. 2. Schematic representation of the effects of glucocorticoids on human cognitive function. Black dots refer to glucocorticoid receptors (GRs), while pattern dots refer to mineralocorticoid receptors (MRs). The types of cognitive processes known to be impaired after acute and/or chronic administration of glucocorticoids are represented with a symbol ☹, while those that benefit from glucocorticoids are represented with a symbol ☺.

steroid (cortisol) that is not a natural glucocorticoid in the rat (although it is in humans) was very poorly retained in the rodent brain. On the contrary, corticosterone as well as aldosterone (a mineralocorticoid) were highly retained by the rodent hippocampus and surrounding limbic structures.

The different modes of action of dexamethasone and corticosterone on the rodent brain gave indications that these two compounds might bind to *different* types of glucocorticoid receptors. This idea was confirmed six years later by Veldhuis and collaborators (1982), who observed the presence of *mineralocorticoid receptors* (MRs) and *glucocorticoid receptors* (GRs) in the rodent hippocampus.

In 1985, Reul and DeKloet showed that the tracer amounts of corticosterone that were previously retained so abundantly by the rodent hippocampus were actually bound to MRs, and not to GRs. In fact, these authors showed that affinity of hippocampal GRs for corticosterone in the rat brain was actually too low for any signal to be detected. At this point in time, history taught us one important fact about MRs and GRs, i.e., there exists a tremendous difference in the two receptor types in terms of affinity.

Following the work performed by Reul and De Kloet (1985), we now know that MRs bind glucocorticoids with an affinity that is about 6–10 times higher than that of GRs. This *differential affinity* results in a striking difference in occupation of the two receptor types under different conditions and time of day. Thus, during the circadian trough (the PM phase in humans), the endogenous hormone occupies more than 90% of MRs, but only 10% of GRs. However, during stress and/or the circadian peak of GC secretion (the AM phase in humans), MRs are saturated, and there is occupation of approximately 67–74% of GRs (Reul and de Kloet, 1985). This differential affinity of MRs and GRs first permitted to explain the well-known inverted-U shape curve known to exist between circulating levels of glucocorticoids and cognitive function (for a review, see Lupien and McEwen, 1997), and second, opened the door to a brand new hypothesis about glucocorticoid effects on cognitive function, i.e., the fact that glucocorticoids could *also* have *positive effects* on cognitive function.

Positive effects of glucocorticoids on human cognition

In contrast to human studies in which glucocorticoids were consistently shown to have detrimental effects on declarative memory function (see above), many studies performed in rodents reported that the ratio of MR/GR occupation is a major determinant of the direction of glucocorticoid-induced cognitive changes (for a review, see De Kloet et al., 1999). For example, long-term potentiation (LTP), a proposed neurobiological substrate of memory formation, has been shown to be optimal when gluocorticoid levels are mildly elevated, i.e., when the ratio of MR/GR occupation is high (see Diamond et al., 1992). In contrast, significant decreases in LTP are observed after adrenalectomy, when MR occupancy is very low (Dubrovsky et al., 1987; Filipini et al., 1991), or after exogenous administration of synthetic glucocorticoids (Bennett et al., 1991; Pavlides et al., 1993), which activate GRs.

In their recent paper, De Kloet and collaborators (1999) have reinterpreted the well-known inverted-U shape function between circulating levels of glucocorticoids and cognitive performance in line with the *MR/GR ratio hypothesis*. In this view, cognitive function can be enhanced when most of the MRs and only part of the GRs are activated (top of the interted-U shape function; increased MR/GR ratio). However, when circulating levels of glucocorticoids are significantly decreased or increased (extremes of the inverted-U shape function; low MR/GR ratio), cognitive impairments will result. The authors suggested that the negative view of glucocorticoid actions on human cognitive function could be partly explained by limitations in previous human experimental designs, which did not allow differential manipulation of MR and GR levels. In order to do this, such studies should measure cognitive function when GRS occupancy is decreased (rather than increased), thus allowing functional measures of MR/GR occupancy on learning and memory. One way to test such a hypothesis is the use of a *hormone removal-replacement protocol.*

In a hormone removal-replacement protocol, the behavior resulting from the absence of the hormone of interest is first measured, and then, baseline hormonal levels are restored to normal values and the same behavior is measured once again. It is postulated that if the hormone of interest has a real impact in the behavior tested, then this behavior should be restored to normal value after hormonal replacement (see Brown, 1998).

In order to test this suggestion, our group performed a hormone replacement study in a population of young normal controls (Lupien et al., 2002). In this protocol, we used a within-subject double-blind experimental protocol in which we first induced a chemical lowering of glucocorticoid levels by administration of *metyrapone*, a potent inhibitor of glucocorticoid synthesis, and then restored baseline-circulating glucocorticoid levels with subsequent infusion of hydrocortisone. Memory performance of participants under each of these conditions was compared to that measured on a placebo day. The results showed that, when compared to placebo, the pharmacological decrease of circulating levels of glucocorticoids induced by metyrapone significantly impaired delayed memory performance. Most importantly, we showed that this impairment was completely restored after hydrocortisone replacement. These results showed that glucocorticoids can modulate memory function, and most importantly, they showed that the absence of circulating glucocorticoid levels is as detrimental for human memory function, as is a significant increase of glucocorticoids.

We have suggested that this modulation can happen through a differential activation of MRs and GRs. Indeed, during the metyrapone condition, MR occupancy was low, given the significant decrease of glucocorticoids secretion induced by metyrapone. At this point, impairment in delayed memory was observed. On the contrary, during the hydrocortisone-replacement condition, glucocorticoids levels were restored to the levels typical of those measured in the AM phase, i.e., leading to a saturation of MRs, with partial occupancy of GRs. This differential occupation thus led to an increased MR/GR ratio, and a restoration of baseline cognitive performance.

In a second study (also in Lupien et al., 2002), we took advantage of the circadian variation in circulating levels of glucocorticoids and tested the impact of a bolus injection of glucocorticoids in the late afternoon, at a time of very low glucocorticoid concentrations. In a previous study with young normal controls, we injected a similar dose of glucocorticoids in the morning, at the time of the circadian peak, and reported detrimental effects of glucocorticoids on memory (Lupien et al., 1999). Here, when we injected a similar dose of hydrocortisone in the afternoon, at the time of the circadian trough, we observed that glucocorticoids had a positive impact on cognitive efficiency by significantly decreasing reaction times on a recognition memory task when compared to a placebo condition.

In the later study, administration of glucocorticoids at the time of circadian trough might have led to partial activation of GRs, thus increasing cognitive efficiency in the group of participants who received hydrocortisone, when compared to placebo. This later finding is interesting in line with data obtained by Oitzl and de Kloet (1992) and recently reviewed by de Kloet et al. (1999), suggesting that MRs and GRs mediate different effects of glucocorticoids in different time domains. According to this view, MR activation is involved in behavioral reactivity in response to environmental cues, while GR-mediated effects promote consolidation of acquired information. The significant decrease in reaction times observed after glucocorticoid administration in the PM phase are in line with a MR-mediated effect of behavioral reactivity, while the delayed memory impairment observed after metyrapone administration is in line with a glucocorticoid-mediated effect of memory consolidation. However, it is clear that only a careful study manipulating activation of MRs and GRs in humans could permit confirmation of the above suggestion.

Be this as it may, the results obtained in the PM versus AM phase in humans are in accordance with the previous data obtained by Fehm-Wolfsdorf and collaborators (1993) showing that (1) recall performance on a memory test is better in the AM phase, compared to the PM phase in young normal controls, and (2) administration of glucocorticoids suppresses this circadian variation in cognitive performance. These results, along with others by Newcomer et al. (1999) showing that four days administration of glucocorticoids in the PM phase impairs performance on the fourth day, show that the impact of glucocorticoids on cognitive are relative rather than absolute, i.e., the direction and magnitude of the effects of glucocorticoids on cognition depend both on the time of day, as well as on the time course of the administration.

Interim: lessons from history

Now, let us stop time again and go back to Reul and DeKloet's 1985 work on MRs and GRs. The study of MRs and GRs activity in the brain showed that there exists large differences in terms of affinity for MRs and GRs, a finding that led to the view that glucocorticoids are not only bad guys (DeKloet et al., 1999).

However, the search for MRs and GRs in the brain led to another important discovery, i.e., the *differential distribution of MRs and GRs* in the rodent brain (see Fig. 3). Following the work by Reul and DeKloet (1985), it was established that in the rodent brain the MR is present exclusively in the limbic system, with a preferential distribution in the hippocampus, parahippocampal gyrus, entorhinal, and insular cortices. On the contrary, the GR is present in both subcortical (PVN and other hypothalamic nuclei, the hippocampus and parahippocampal gyrus) and cortical structures, with a preferential distribution in the *prefrontal cortex* (McEwen et al., 1968, 1986; Meaney and Aitken, 1985; Diorio et al., 1993). Still, in the rodent brain, the largest concentration of both MRs and GRs was found in the hippocampus, which again led to the glucocorticoid–hippocampus link described before.

However, in 2000, two papers were published which described the distribution of MRs and GRs in the primate brain, more closely related to the *human brain* in terms of neocortex development. These two recent studies mapping both MRs and GRs distribution revealed that in the primate's brain, there are less GR then originally proposed in the hippocampus, but there are more GR in the frontal lobes than the levels originally described in the rodent literature. These results strongly suggested that extrapolation from rat brain to primate brain may be misleading when discussing the impact of glucocorticoids on the hippocampus.

The first study was published by Sanchez and collaborators (2000) who reported that, in contrast to its well-established distribution in the rat brain, GR mRNA is only weakly detected in the dentate gyrus and Cornu Ammonis of the macaque hippocampus. In contrast, GR mRNA is strongly detected in the pituitary, cerebellum, hypothalamic PVN, and prefrontal cortices. In a second study published by Patel and collaborators (2000), it was reported on an experiment where the authors used a specific squirrel monkey antibody and found that GR receptors were well expressed in the hippocampus, but were more prominently found in the prefrontal cortex.

These recent evidences in the primate brain showed that MRs are present in large quantities in the hippocampus and limbic structures, while GRs are present in all these structures and additionally in the frontal regions. This latter finding suggested that in humans, glucocorticoids should not only affect the hippocampus, but also the frontal lobes. Such results have recently been obtained.

Glucocorticoids, the frontal lobe, and working memory

Studies in nonhuman primates (Goldman-Rakic, 1987, 1995) and humans (Petrides and Milner, 1982; Owen et al., 1990) showed that lesions of the dorsolateral prefrontal cortex (DLPFC) give rise to impairments in *working memory*. Working memory is the cognitive mechanism that allows us to keep a limited amount of information active for a limited period of time (see Baddeley, 1995). Thus, working memory impairments have been found in several experiments using a variety of delay-task procedure. In these tasks, a temporal gap is introduced between a stimulus and a response, which creates the need to maintain the stimulus in temporary memory storage. Data obtained in monkeys showed that cells in the lateral prefrontal cortex become particularly active during delayed response tasks, suggesting that these cells are actively involved in holding on to the information during the delay (Goldman-Rakic, 1990, 1995).

Neuropsychological evidence suggests that humans with prefrontal damage are impaired in working memory (Luria, 1966; Fuster, 1980). These patients are also highly susceptible to cognitive interference and they perform poorly on neuropsychological tests that require response inhibition such as the Wisconsin Card Sorting Test (Stuss et al., 1982;

Shimamura, 1995). Moreover, recent neuroimaging data, summarized and reviewed by Smith et al. (1998; see also Dolan and Fletcher, 1997; Ungerleider et al., 1998), show a significant relationship between working memory processing and activation observed in the prefrontal cortex (Smith et al., 1998; Ungerleider et al., 1998).

In 1999, two studies performed in humans reported that working memory is more sensitive than declarative memory to acute and short-term administration of glucocorticoids. Young and collaborators (1999) administered 20 mg hydrocortisone for 10 days to young normal male volunteers and measured various cognitive functions in a randomized, placebo control, crossover, within-subject design. They showed that this regimen of glucocorticoids led to deficits in cognitive function sensitive to frontal-lobe dysfunction (working memory), while it did not impact on cognitive function sensitive to hippocampal damage.

Similar results were obtained by our group (Lupien et al., 1999) using an acute dose–response protocol. In this study, 40 young subjects were infused for 100 min with either glucocorticoids or placebo and declarative and working memory function was tested during the infusion period. The results revealed that performance on the working memory task decreased significantly whereas performance on the declarative memory task remained the same, following an acute elevation of glucocorticoids. Curve-fit estimations revealed the existence of a significant quadratic function (U-shape curve) between performance on the working memory task and changes in glucocorticoid levels after hydrocortisone infusion. The results of these two studies suggested that in young individuals, working memory is more sensitive than declarative memory to an acute elevation of glucocorticoids, which goes along with the suggestion that glucocorticoids have a significant impact on frontal-lobe functions in humans (see Fig. 2).

Glucocorticoids, the amygdala, and emotional memory

Although most of the literature on the acute effects of glucocorticoids on animal and human cognitive process was reported using the hippocampus and the frontal lobes as models for glucocorticoid-induced cognitive changes, there is now evidence showing that glucocorticoids also act as modulators of the formation of emotional memory in the amygdala.

The role of the amygdala in the modulation and/or storage of emotional memory has been demonstrated in various animal models. The amygdala contains both MRs and GRs (Allen and Allen, 1974; Honkaniemi et al., 1992), and the interaction between glucocorticoids and the amygdala has recently been demonstrated in humans by the presence of a significantly smaller amygdala volume in children with congenital adrenal hyperplasia, which is a genetic disease where there is a block in cortisol production (Merke et al., 2003).

Glucocorticoid receptors, in particular nuclei of the amygdala (particularly the central and medial), have been implicated in emotional expression and in neuroendocrine control of emotions (for a recent review, see Roozendaal, 2002). In rodents, Roozendaal et al. (1996a,b) demonstrated that posttraining injections of dexamethasone enhances inhibitory-avoidance retention, while inhibition of glucocorticoid synthesis by administration of metyrapone impairs performance on this same task.

To this day, there are two studies which measured the effects of exogenous admnistration of glucocorticoids on emotional memory. In 2001, Buchanan and Lovallo exposed young participants to pictures varying in emotional arousal after they received 20 mg cortisol. During acquisition, subjects were not aware that their memory for the pictures would be tested a week later (incidental memory). Results revealed that elevated cortisol levels during memory encoding enhanced the long-term recall performance of emotionally arousing pictures while it had no impact on the delayed recall of the neutral pictures.

In a more recent study, Adercrombie and collaborators (2003) tested the effects of exogenous administration of synthetic glucocorticoids on emotional memory using a dose–response study. Young men were presented with emotionally arousing and neutral stimuli after receiving either 20 or 40 mg cortisol. Free recall of the stimuli was performed 1 h after drug administration and recognition memory of the stimuli was performed two evenings later. Results showed that cortisol elevations decreased the number of errors committed on the free-recall tasks. More importantly, the authors showed that when tested for recognition two evenings later, when cortisol levels were no longer manipulated, recognition performance presented an inverted-U quadratic curve, with recognition memory for both emotionally arousing and neutral stimuli being facilitated at the 20 mg dose. In contrast to the data obtained by Buchanan and Lovallo (2001), these results showed beneficial effects of synthetic cortisol on both emotionally arousing and neutral material.

Although there are no data at this point that can help explain these discrepant data, it is important to note that presentation of neutral stimuli that are embedded in a context of emotionally arousing material can have a different effect on recall of the neutral material, due to a phenomenon that one could name a *proactive facilitatory effect*. Research has shown that memorizing elements that are interrelated promotes the establishment of associations between the different aspects of the information to be learned, leading to a deeper encoding and consolidation of the information in memory (Craik and Lockhart, 1972). The presentation of emotionally arousing material in a memory task could potentially serve as a contextual cue for remembering neutral material, and promote the enhanced recall of both types of stimuli. In order to assess the potential impact of proactive facilitatory effects of emotional material over neutral stimuli, it will be important in future studies assessing the effects of exogenous glucocorticoids on emotional memory to compare drug actions on neutral materials when the neutral stimuli are presented with and without emotionally arousing stimuli.

Still, in both studies, positive effects of exogenous glucocorticoids were found for emotionally arousing material (see Fig. 2). These results, obtained in humans, are similar to what has been found in animal studies where positive effects of glucocorticoids on emotional memory are reported (Roozendaal and McGaugh, 1996, 1997a,b; Roozendaal et al., 1996a,b, 1997; also see Roozendaal, 2000, 2002 for reviews).

Conclusion

Altogether, the findings on the impact of exogenous glucocorticoids on human congnitive function now show that not only the *direction* of glucocorticoid-induced cognitive changes should vary as a function of MRs and GRs occupancy, but also the *nature* and/or *anatomical substrate* of these cognitive changes.

Based on the MR/GR ratio hypothesis and the distribution of these two types of receptors in the primate (Sanchez et al., 2000; Patel et al., 2000) and human (Sarrieau et al., 1988) brain, one has to come to the conclusion that the anatomical substrate of the cognitive changes induced by the absence of MR and GR activation should be very different from that induced by a saturation of MRs and GRs. The reason for this lies in the fact that the absence of MR/GR activation would preferentially impact on the hippocampus, while the saturation of MR/GR would recruit additional frontal regions, given the exclusive presence of GRs in this region. Although GRs are also present in the hippocampus, many recent reports suggest that the presence of MRs in the hippocampus acts by creating a physiological balance of both types of receptors for their action on the HPA axis [called the "Binary Hormone Response System" by Evans and Arriza (1989) and the "MR/GR balance hypothesis" by Oitzl et al., 1995)]. This suggests that the presence of MRs within a structure acts by decreasing GRs responsivity to glucocorticoids because of the tonic influence of MRs on the HPA axis (Oitzl et al., 1995). This also implies that the absence of the tonic influence of MRs in the prefrontal regions of the human brain would increase GRs sensitivity to glucocorticoids and lead to increased sensitivity of prefrontal regions to acute increases in glucocorticoid levels, when compared to the hippocampus. This new model of glucocorticoid actions on the brain opens the door to new and exciting studies on the impact of glucocorticoids on human cognitive function.

After more than 60 years of research on the effects of exogenous glucocorticoids on human cognitive function, we have come to view the actions of this steroid as being modulatory rather than direct. History has taught us a lot of things. First, that these steroids could access the brain and lead to steroid psychosis. Second, that glucocorticoids' access to the hippocampus could lead to impairments in learning and memory function. Third, that glucocorticoids' access to the frontal lobes and the amygdala could lead to working memory impairments and dysregulation of emotional cognition. Finally, that glucocorticoids' action on cognitive process can also be positive, and that the positive or negative effects of glucocorticoids depend on the balance of the two glucocorticoid receptor types known to exist today.

The story was supposed to become simpler as new discoveries were being made about the actions of glucocorticoids on the animal and human brain. Yet, it became more complex, dynamic, and difficult to grasp. However, this should not be taken as a sign that one will not be eventually able to describe and predict the effects of glucocorticoids on cognitive function in any particular individual. On the contrary, it should tell us that the glucocorticoid hormone is a very important player in the interface between the brain and the environment. This renders the study of its action more exciting than ever.

Acknowledgments

S.J.L. research summarized in this paper was funded by a Scientist Research Award from Fonds de la recherche en santé du Québec (FRSQ), by an operating grant from Canadian Institute of Health Research (CIHR), and by a Research Scholar Award from EJLB Foundation. F.S.M. research is funded by a studentship from the Canadian Institutes of Health Research. The Douglas Hospital Longitudinal Study of Normal and Pathological Aging is funded by a grant for the Canadian Alzheimer Society.

References

Abercrombie, H.C., Kalin, N.H., Thurow, M.E., Rosenkranz, M.A. and Davidson, R.J. (2003) Cortisol variation in human affects memory for emotionally laden and neutral information. Behavioral Neuroscience, 117: 505–516.

Adolphs, R., Tranel, D. and Denburg, N. (2000) Impaired emotional declarative memory following unilateral amygdala damage. Learn. Mem., 7: 180–186.

Allen, J.P. and Allen, C.F. (1974) Role of the amygdaloid complexes in the stress-induced release of ACTH. Neuroendocrinology, 15: 220–230.

Baddeley, A. (1995) Working memory: the interface between memory and cognition. In: Schacter, D.L. and Tulving, E. (Eds.), Memory Systems. The MIT Press, Cambridge, pp. 351–368.

Beckwith, B.E., Petros, T.V., Scaglione, C. and Nelson, J. (1986) Dose-dependent effects of hydrocortisone on memory in human males. Physiology and Behavior, 36: 283–286.

Belanoff, J.K., Gross, K., Yager, A. and Schatzberg, A.F. (2001) Corticosteroids and cognition. J. Psychiatry Res., 35: 127–145.

Bender, B. and Milgrom, H. (1995) Neuropsychiatric effects of medications for allergic diseases. Journal of Allergy and Immunology, 95: 523–528.

Bennett, M.C., Diamond, D.M., Fleshner, M. and Rose, G.M. (1991) Serum corticosterone level predicts the magnitude of hippocampal primed burst potentiation and depression in urethane-anesthetized rats. Psychobiology, 19: 301–307.

Bick, P.A. (1983) Obsessive–compulsive behavior associated with dexamethasone treatment. Journal of Nervous and Mental Disorders, 171: 253–254.

Borman, M.C. and Schmallenberg, H.C. (1951) Suicide following cortisone treatment. JAMA, 146: 337–338.

Brody, S. (1952) Psychiatric observations in patients treated with cortisone and ACTH. Psychosomatic Medicine, 14: 94–103.

Brown, N. (1998) An Introduction to Neuroendocrinology. Cambridge University Press, Cambridge, pp. 408.

Brown, R. and Kulik, J. (1977) Flashbulb memories. Cognition, 5: 73–99.

Buchanan, T.W. and Lovallo, W.R. (2001) Enhanced memory for emotional material following stress-level cortisol treatment in humans. Psychoneuroendocrinology, 26: 307–317.

Christianson, S.A. (1992) Emotional stress and eyewitness memory: a critical review. Psychol. Bull., 112: 284–309.

Clark, L.D., Bauer, W. and Cobb, S. (1952) Preliminary observations on mental disturbances occurring in patients under therapy with cortisone and ACTH. New England Journal of Medicine, 246: 205–216.

Cobb, S., Quarton, G.C. and Clark, L.D. (1954) Neuropsychiatric Disorders in Lukens FDW, Medical Uses of Cortisone. Blakiston, Toronto, pp. 506–518.

Cole, B. and Koob, G.F. (1991) Corticotropin-releasing factor, stress, and animal behavior. In: McCubbin, J.A., Kaufman, P.G. and Nemeroff, C.B. (Eds.), Stress, Neuropeptides and Systemic Diseases. Academic Press Inc., New York, NY.

Conn, H.O. and Poynard, T. (1994) Corticosteroids and peptic ulcer: meta-analysis of adverse events during steroid therapy. Journal of Internal Medicine, 236: 619–632.

Craik, F.I.M. and Lockhart, R.S. (1972) Levels of processing: a framework for memory research. J. Verb. Learn. Verb. Behav., 11: 671–684.

De Kloet, E.R., Wallach, G. and McEwen, B.S. (1975) Differences in corticosterone and dexamethasone binding to rat brain and pituitary. Endocrinology, 96: 598–609.

De Kloet, E.R., Oitzl, M.S. and Joels, M. (1999) Stress and cognition: are corticosteroids good or bad guys? Trends in Neurosciences, 22: 422–426.

De la Riva, G.E. (1958) Psychic and somatic changes observed in allergic children after prolonged steroid therapy. South Medicine Journal, 5: 865–868.

de Quervain, D.J.F., Roozendaal, B., Nitsch, R.M., McGaugh, J.L. and Hock, C. (2000) Acute cortisone administration impairs retrieval of long-term declarative memory in humans. Nature Neuroscience, 3: 313–314.

de Quervain, D.J., Henke, K., Aerni, A., Treyer, V., McGaugh, J.L., Berthold, T., Nitsch, R.M., Buck, A., Roozendaal, B., Hock, C. (2003). Glucocorticoid-induced impairment of declarative memory retrieval is associated with reduced blood flow in the medial temporal lobe. European Journal of Neuroscience, 17: 1296–1302.

Diamond, D.M., Bennett, M.C., Fleshner, M. and Rose, G.M. (1992) Inverted-U relationship between the level of peripheral corticosterone and the magnitude of hippocampal primed burst potentiation. Hippocampus, 2: 421–430.

Diorio, D., Viau, V. and Meaney, M.J. (1993) The role of the medial prefrontal cortex (cingulate gyrus) in the regulation of hypothalamic–adrenal responses to stress. J. Neurosci., 13: 3839–3847.

Dolan, R.J. and Fletcher, P.C. (1997) Dissociating prefrontal and hippocampal function in episodic memory encoding. Nature, 388: 582–585.

Dordick, J.R. and Gluck, E.J. (1955) Preliminary clinical trials with prednisone (meticorten) in rheumatic diseases. JAMA, 158: 166–170.

Dubrovsky, B.O., Liquornik, M.S., Noble, P. and Gijsbers, K. (1987) Effects of 5alpha-dihydrocorticosterone on evoked responses and long-term potentiation. Brain Research Bulletin, 19: 635–638.

Dunn, A.J. and Berridge, C.W. (1990) Physiological and behavioral responses to corticotropin-releasing factor administration: is CRF a mediator of anxiety or stress responses? Brain Res. Rev., 15: 71–100.

Evans, R.M. and Arriza, J.L. (1989) A molecular framework for the actions of glucocorticoid hormones in the nervous system. Neuron, 2: 1105–1112.

Fehm-Wolfsdorf, G., Reutter, D., Zenz, H., Born, J. and Lorenz-Fehm, H. (1993) Are circadian variations in taste thresholds cortisol-dependent? Journal of Psychophysiology, 7: 5–72.

Filipini, D., Gijsbers, K., Birmingham, M.K. and Dubrovsky, B. (1991) Effects of adrenal steroids and their reduced metabolites on hippocampal long-term potentiation. Journal of Steroid Biochemistry and Molecular Biology, 40: 87–92.

Fuster, J.M. (1980) The Prefrontal Cortex. Raven Press, New York, pp. 419.

Glaser, G.H., Kornfeld, D.S. and Knight, R.P. (1955) Intravenous hydrocortisone, corticotrophin, and the electroencephalogram. Arch. Neurol. Psychiatry, 73: 338–344.

Goldman-Rakic, P.S. (1987) Circuitry of primate prefrontal cortex and regulation of behavior by representational memory. In: Plum, F. (Ed.), Handbook of Physiology. Section 1: The Nervous System, Vol. 5. American Psychological Society, Bethesda, pp. 373–417.

Goldman-Rakic, P.S. (1990) Cellular and circuit basis of working memory in prefrontal cortex of nonhuman primates. Progress in Brain Research, 85: 325–335.

Goldman-Rakic, P.S. (1995) Cellular basis of working memory. Neuron, 14: 477–485.

Greeves, J.A. (1984) Rapid-onset steroid psychosis with very low dosage of prenisolone (letter). Lancet, 1: 1119–1120.

Hall, R.C.W., Popkin, M.K., Stickney, S.K. and Gardner, E.R. (1979) Presentation of the steroid psychoses. Journal of Nervous and Mental Disease, 167: 229–236.

Hench, P.S., Kendall, E.C., Slocumb, C.H. and Polley, H.F. (1949) The Effect of a Hormone of the Adrenal Cortex (17-Hydroxy-11-dehydrocorticosterone: Compound E) and of Pituitary Adrenocorticotropic Hormone on Rheumatoid Arthritis. Procof Staff Meeting Mayo Clinic, Rochester, pp. 181.

Honkaniemi, J., Pelto-Huikko, M., Rechardt, L., Isola, J., Lammi, A., Fuxe, K., Gustaffson, J.A., Wikstrom, A.C., Hokfelt, T. (1992) Colocalization of peptide and glucocorticoid receptor immunoreactivities in the rat central amygdaloid nucleus. Neuroendocrinology, 55: 451–459.

Hsu, F.C., Garside, M.J., Massey, A.E., McAllister-Williams, R.H. (2003). Effects of a single dose of cortisol on the neural correlates of episodic memory and error processing in healthy volunteers. Psychopharmacology (Berl), 164: 431–442.

Kirschbaum, C., Pirke, K.M. and Hellhammer, D.H. (1993) The "Trier Social Stress Test" – a tool for investigating psychobiological stress responses in a laboratory setting. Neuropsychobiology, 28: 76–81.

Kirschbaum, C., Wolf, O.T., May, M., Wippich, W. and Hellhammer, D.H. (1996) Stress and drug-induced elevation of cortisol levels impair explicit memory in healthy adults. Life Sciences, 58: 1475–1483.

Kleinsmith, L.J. and Kaplan, S. (1963) Paired-associate learning as a function of arousal and interpolated interval. J. Exp. Psychol., 65: 190–193.

Kountz, W.B., Ackermann, P.G. and Kheim, T. (1953) The influence of some hormonal substances on the nitrogen balance and clinical state of elderly patients. Journal of Clinical Endocrinology and Metabolism, 13: 534–547.

Kreus, K.E., Viljanen, A.A. and Kujale, E. (1975) Treatment of steroid-dependent asthma patients with beclomethasone dipropionate aerosol. Scandinavian Journal of Respiratory Disease, 56: 47–57.

LaBar, K.S. and Phelps, E.A. (1998) Arousal-mediated memory consolidation: role of the medial temporal lobe in humans. Psychol. Sci., 9: 490–493.

Lewis, D.A. and Smith, R.E. (1983) Steroid-induced psychiatric syndromes: a report of 14 cases and a review of the literature. Journal of Affective Disorders, 5: 319–322.

Lidz, T., Carter, J.D. and Lewis, B.I. (1952) Effects of ACTH and cortisone on mood and mentation. Psychosomatic Medicine, 14: 363–377.

Ling, M.H.M., Perry, P.J. and Tsuang, M.T. (1981) Side effects of corticosteroid therapy. Archive of General Psychiatry, 38: 471–477.

Loewenberg, W.H. (1954) Convulsive seizures complicating cortisone and ACTH therapy: clinical and electroencephalographic observations. JCEM, 13: 1039–1045.

Loftus, E.F. (1979) Eyewitness Testimony. Harvard University Press, London, pp. 256.

Lupien, S.J. and McEwen, B.S. (1997) The acute effects of corticosteroids on cognition: integration of animal and human model studies. Brain Research Reviews, 24: 1–27.

Lupien, S.J. and Lepage, M. (2001) Stress, memory, and the hippocampus: can't live with it, can't live without it. Behavioral and Brain Research, 127: 137–158.

Lupien, S.J., Gaudreau, S., Tchiteya, B.M., Maheu, F., Sharma, S., Nair, N.P.V., Hauger, R.L., McEwen, B.S. and Meaney, M.J. (1997) Stress-induced declarative memory impairment in healthy elderly subjects: relationship to cortisol reactivity. J. Clin. Endocrinol. Metab., 82: 2070–2075.

Lupien, S.J., Gillin, C. and Hauger, R.L. (1999) Working memory is more sensitive than declarative memory to the acute effects of corticosteroids: a dose-response study. Behavioral Neuroscience, 113: 1–11.

Lupien, S.J., Wilkinson, C.W., Briere, S., Menard, C., Ng Ying Kin, N.M.K. and Nair, N.P.V. (2002) The modulatory effects of corticosteroids on cognition: studies in young human populations. Psychoneuroendocrinology, 27: 401–416.

Luria, A.R. (1966) Higher Cortical Functions in Man. Basic Books, New York, pp. 312.

Maheu, F. and Lupien, S. (2003). La mémoire aux prises avec les émotions et le stress: un impact nécessairement dommageable? Médecine/Sciences, 19: 118–124.

McAllister-Williams, R.H. and Rugg, M.D. (2002) Effects of repeated cortisol administration on brain potential correlates of episodic memory retrieval. Psychopharmacology, 160: 74–83.

McEwen, B.S., Weiss, J.M., Schwartz, L.S. (1968) Selective retention of corticosterone by limbic structure in rat brain. Nature, 220: 911–912.

McEwen, B.S., Wallach, G. and de Kloet, E.R. (1976) Interaction in vivo and in vitro of corticoids and progesterone with cell nuclei and soluble macromolecules from rat brain regions and pituitary. Brain Res., 105: 129–136.

McEwen, B.S., DeKloet, E.R. and Rostene, W.H. (1986) Adrenal steroid receptors and actions in the nervous system. Physiol. Rev., 66: 1121–1150.

Meaney, M.J. and Aitken, D.H. (1985) [^{3}H]dexamethasone binding in rat frontal cortex. Brain Res., 328: 176–180.

Merke, D.P., Fields, J.D., Keil, M.F., Vaituzis, A.C., Chrousos, G.P. and Giedd, J.N. (2003) Children with classic congenital adrenal hyperplasia have decreased amygdala volume: potential prenatal and postnatal hormonal effects. JCEM, 88: 1760–1765.

Mitchell, D.M. and Collins, J.V. (1984) Do corticosteroids really alter mood? Postgraduate Medicine Journal, 60: 467–470.

Molle, M., Pietrowsky, R., Fehm, H.L. and Born, J. (1997) Regulation of human thought by neuropeptide ACTH 4–10: an analysis of the EEG's dimensional complexity. Neuroreport, 8: 2715–2720.

Newcomer, J.W., Craft, S., Hershey, T., Haskins, K. and Bardgett, M.E. (1994) Glucocorticoid-induced impairment in declarative memory performance in adult humans. Journal of Neuroscience, 14: 2047–2053.

Newcomer, J.W., Selke, G., Melson, A.K., Hershey, T., Craft, S., Richards, K. and Alderson, A.L. (1999) Decreased memory performance in healthy humans induced by stress-level cortisol treatment. Archives of General Psychiatry, 56: 527–533.

Oitzl, M.S. and De Kloet, E.R. (1992) Selective corticosteroid antagonists modulate specific aspects of spatial orientation learning. Beh. Neurosci., 106: 62–71.

Oitzl, M.S., vanHaarst, A.D., Sutanto, W. and De Kloet, E.R. (1995) Corticosterone, brain mineralocorticoid receptors (MRs) and the activity of the hypothalamic–pituitary–adrenal (HPA) axis: the Lewis rat as an example of increased central MR capacity and a hyporesponsive HPA axis. Psychoneuroendocrinology, 6: 655–675.

Owen, A.M., Downes, J.J., Sahakian, V.J., Polkey, C.E. and Robbins, T.W. (1990) Planning and spatial working memory following frontal lobe lesions in man. Neuropsychologia, 28: 1201–1034.

Patel, P.D., Lopez, J.F., Lyons, D.M., Burke, S., Wallace, M. and Schatzberg, A.F. (2000) Glucocorticoid and mineralocorticoid receptor mRNA expression in squirrel monkey brain. J. Psychiatr. Res., 34: 383–392.

Pavlides, C., Watanabe, Y. and McEwen, B.S. (1993) Effects of glucocorticoids on hippocampal long-term potentiation. Hippocampus, 3: 183–192.

Perry, P.J., Tsuang, M.T. and Hwang, M.H. (1984) Prednisolone psychosis: clinical observations. Drug Intell. Clin. Pharm., 18: 603–609.

Petrides, M. and Milner, B. (1982) Deficits on subject-ordered tasks after frontal- and temporal-lobe lesions in man. Neuropsychologia, 31: 1–15.

Plotz, C.M., Knowlton, A.I. and Ragan, C. (1952) The natural history of Cushing's syndrome. American Journal of Medicine, 13: 597–614.

Quevedo, J., Sant'Anna, M.K., Madruga, M., Lovato, I., de-Paris, F., Kapczinski, F., Izquierdo, I. and Cahill, L. (2003) Differential effects of emotional arousal in short- and long-term memory in healthy adults. Neurobiol. Learn. Mem., 79: 132–135.

Rees, L. (1953) Psychological concomitants of cortisone and ACTH therapy. Journal of Mental Sciences, 99: 497–504.

Reul, J.M.H.M. and De Kloet, E.R. (1985) Two receptor systems for corticosterone in rat brain: microdistribution and differential occupation. Endocrinology, 117: 2505–2512.

Rome, H.P. and Braceland, F.J. (1952) Psychological response to corticotrophin, cortisone, and related steroid substances. JAMA, 148: 27–30.

Roozendaal, B. (2000) Glucocorticoids and the regulation of memory consolidation. Psychoneuroendocrinology, 25: 213–38.

Roozendaal, B. (2002) Stress and memory: opposing effects of glucocorticoids on memory consolidation and memory retrieval. Neurobiol. Learn. Mem., 78: 578–595.

Roozendaal, B. and McGaugh, J.L. (1996) Amygdaloid nuclei lesions differentially affect glucocorticoid-induced memory enhancement in an inhibitory avoidance task. Neurobiol. Learn. Mem., 65: 1–8.

Roozendaal, B. and McGaugh, J.L. (1997a) Basolateral amygdala lesions block the memory-enhancing effect of glucocorticoid administration in the dorsal hippocampus of rats. Eur. J. Neurosci., 9: 76–83.

Roozendaal, B. and McGaugh, J.L. (1997b) Glucocorticoid receptor agonist and antagonist administration into the basolateral but not central amygdala modulates memory storage. Neurobiol. Learn. Mem., 67: 176–179.

Roozendaal, B., Bohus, B. and McGaugh, J.L. (1996a) Dose-dependent suppression of adrenocortical activity with metyrapone: effects on emotion and memory. Psychoneuroendocrinology, 21: 681–693.

Roozendaal, B., Cahill, L. and McGaugh, J.L. (1996b) Interaction of emotionally activated neuromodulatory systems in regulating memory storage. In: Ishikawa, K., McGaugh, J.L. and Sakata, H. (Eds.), Brain Processes and Memory. Elsevier, Amsterdam, pp. 39–54.

Sanchez, M.M., Young, L.J., Plotsky, P.M. and Insel, T.R. (2000) Distribution of corticosteroid receptors in the rhesus brain: relative absence of glucocorticoid receptors in the hippocampal formation. Journal of Neuroscience, 20: 4657–4668.

Sarrieau, A., Dussaillant, M., Sapolsky, R.M., Aitken, D.H., Olivier, A., Lal, S., Rostene, W.H., Quirion, R. and Meaney, M.J. (1988) Glucocorticoid binding sites in human temporal cortex. Brain Research, 442: 159–163.

Schmidt, L.A., Fox, N.A., Goldbert, M.C., Smith, C.C. and Shulkin, J. (1999) Effects of acute prednisone administration on memory, attention and emotion in healthy human subjects. Psychoneuroendocrinology, 24: 461–483.

Scoville, W.B. and Milner, B. (1957) Loss of recent memory after bilateral hippocampal lesions. J. Neurol., Neurosurg. Psychiatry, 20: 11–21.

Shimamura, A.P. (1995) Memory and frontal lobe function. In: Gazzaniga, M.S. (Ed.), The Cognitive Neurosciences. MIT Press, Cambridge, pp. 803–813.

Smith, E.E., Jonides, J., Marshuetz, C. and Koeppe, R.A. (1998) Components of verbal working memory: evidence from neuroimaging. Proceedings of the National Academy of Science, 95: 876–882.

Squire, L.R. (1992) Memory and the hippocampus: a synthesis from findings with rats, monkeys, and humans. Psychol. Rev., 99: 195–231.

Stephen, E.H.M. and Noad, K.B. (1951) Status epilepticus occurring during cortisone therapy. Medical Journal of Austria, 2: 334–335.

Stuss, D.T., Kaplan, E.F., Benson, D.F., Weir, W.S., Chiulli, S. and Sarazin, F.F. (1982) Evidence for the involvement of orbitofrontal cortex in memory functions: an interference effects. Journal of Comparative Physiological Psychology, 96: 913–925.

Sullivan, B.J. and Dickerman, J.D. (1979) Steroid-associated catatonia: report of a case. Pediatrics, 63: 677–678.

Swerdlow, N.R., Geyer, M.A., Vale, W.W. and Koob, G.F. (1986) Corticotropin-releasing factor potentiates acoustic startle in rats: blockade by chlordiazepoxide. Psychopharmacology, 88: 147–152.

Trethowan, W.H. and Cobb, S. (1952) Neuropsychiatric aspects of Cushing's syndrome. Arch. Neurol. Psychiat., 67: 283–309.

Ungerleider, L.G., Courtney, S.M. and Haxby, J. (1998) A neural system for human visual working memory. Proceedings of the National Academy of Science, 95: 883–890.

Varney, N.R., Alexander, B. and MacIndoe, J.H. (1984) Reversible steroid dementia in patients with steroid psychosis. American Journal of Psychiatry, 141: 369–384.

Veldhuis, H.D., Van Koppen, C., Van Ittersum, M. and de Kloet, E.R. (1982) Specificity of the adrenal steroid receptor system in rat hippocampus. Endocrinology, 110: 2044–2051.

Von Zerssen, D. (1976) Mood and behavioral changes under corticosteroid therapy. In: Ifil, T.M., Laudahn, G. and Herrmann, W.M. (Eds.), Psychotropic Action of Hormones. Spectrum, New York, pp. 195–222.

Wolf, O.T., Schommer, N.C., Hellhammer, D.H., McEwen, B.S. and Kirschbaum, C. (2001a) The relationship between stress induced cortisol levels and memory differs between men and women. Psychoneuroendocrinology, 26: 711–720.

Wolf, O.T., Convit, A., McHugh, P.F., Kandil, E., Thorn, E.L., De Santi, S., McEwen, B.S. and de Leon, M.J. (2001b). Cortisol differentially affects memory in young and elderly men. Behavioral Neuroscience, 115: 1002–1011.

Wolkowitz, O.M., Reus, V.I., Weingartner, H., Thompson, K., Breier, A., Doran, A., Rubinow, D. and Pickar, D. (1990) Cognitive effects of corticosteroids. American Journal of Psychiatry, 147: 1297–1303.

Young, A.H., Sahakian, B.J., Robbins, T.W. and Cowen, P.J. (1999) The effects of chronic administration of hydrocortisone on cognitive function in normal male volunteers. Psychopharmacology, 145: 260–266.

SECTION 4

Neurotransmitter Systems Involved in Stress Responsivity

T. Steckler, N.H. Kalin and J.M.H.M. Reul (Eds.)
Handbook of Stress and the Brain, Vol. 15
ISBN 0-444-51173-3

CHAPTER 4.1

Neurocircuit regulation of the hypothalamo–pituitary–adrenocortical stress response – an overview

James P. Herman[1,*], Nancy K. Mueller[1], Helmer Figueiredo[1] and William E. Cullinan[2]

[1]Department of Psychiatry, University of Cincinnati, Cincinnati, OH 45267-0559, USA
[2]Department of Biomedical Sciences, Marquette University, Milwaukee, WI 53233, USA

Abstract: Central integration of the hypothalamo–pituitary–adrenocortical (HPA) axis stress response is controlled by neurosecretory neurons in the medial parvocellular paraventricular nucleus (PVN). Activation of the PVN is a complex process regulated by both direct and indirect neuronal connections, as well as communication with blood-borne messengers. Ascending brainstem pathways from the nucleus of the solitary tract (both catecholaminergic and noncatecholaminergic systems) and serotonergic midbrain raphe nuclei provide direct neuronal excitation of PVN neurons. Stimulation of the HPA axis is also mediated by transsynaptic inputs from the medial and central amygdaloid nuclei, which disinhibit the PVN by way of GABAergic relay neurons in the hypothalamus and bed nucleus of the stria terminalis (BST), and perhaps enhance excitatory input from the brainstem. Inhibition of the HPA axis is controlled by PVN-projecting GABA neurons in the hypothalamus and BST, which are driven in part by descending stimulatory inputs from ventral subiculum and infralimbic cortex. The PVN is profoundly affected by blood-borne factors (including peptides and cytokines); these messengers communicate with the PVN through interactions with circumventricular organs or by induction of perivascular prostaglandin synthesis. Finally, glucocorticoids can directly inhibit PVN neurons, by diffusion from the dense vascular beds localized in this region. Thus, activation of the HPA axis is controlled by a wide variety of signals from both brain and periphery, which are thence effectively integrated into a net secretory signal at the level of the hypophysiotrophic PVN neuron.

The hypothalamo–pituitary–adrenocortical (HPA) axis is vital for adaptation of the organism to an ever-changing environment. Glucocorticoids represent the end product of the HPA axis cascade, and act at multiple loci to redistribute resources to meet a real or perceived challenge. A rich literature indicates that control of glucocorticoid secretion is essential to health. Dysregulation of the HPA axis has multiple deleterious actions on the brain and body, and has been implicated as a causal or complicating factor in numerous affective disease states, including depression and posttraumatic stress disorder (McEwen and Stellar, 1993; Sapolsky et al., 1986).

The principle integrators of the HPA stress response are neurosecretory neurons localized to the medial parvocellular subdivision of the paraventricular nucleus (PVN) (Antoni, 1986; Whitnall, 1993). These neurons synthesize and secrete a number of hypophysiotrophic peptides, the most prominent of which are corticotropin-releasing factor (CRF) and arginine vasopressin (AVP) (Antoni, 1986; Kiss, 1988; Whitnall, 1993). Corticotropin-releasing factor

*Corresponding author. Tel.: +1513-558-4813;
Fax: +1513-558-9104; E-mail: james.herman@uc.edu

stimulates adrenocorticotrophic hormone (ACTH) release from anterior pituitary corticotropes under both basal and stressed conditions. In contrast, AVP has little ACTH-releasing activity alone, but potentiates the action of CRF in a synergistic fashion (Gillies et al., 1982), thus acting to adjust the "gain" of the HPA axis. As the executors of the HPA response, hypophysiotropic neurons are in receipt of information from a wide variety of sources, and have the daunting task of summating this vast array of information into a meaningful signal.

The complex anatomy of the PVN

Prior to discussion of neurocircuit integration of PVN function, it is critical to examine the rather unique context of PVN neurons in the CNS. There are several aspects of the PVN in general, and the parvocellular PVN in particular, that give it privileged access to a variety of inputs. First and foremost, it is critical to consider that the parvocellular hypophysiotrophic PVN cells are endocrine "motoneurons." These cells project to the hypophysial portal vasculature in a blood-brain barrier deficient region of the brain (median eminence) (Antoni, 1986; Whitnall, 1993). Thus, the terminals of these cells are privy to information from vascular targets that may be of relevance to growth of CRF axons and neurons (and may play a role in modulating hormone release). Second, the somata of these neurons reside in one of the richest vascular environments of the brain (Fig. 1). While there is no evidence that these capillary beds are fenestrated, it is clear that the ability of these cells to sample blood-brain barrier permeable molecules (e.g., steroids) is substantial. Third, Golgi impregnations (van den Pol, 1982) and intracellular fills of parvocellular PVN neurons (Rho and Swanson, 1989) indicate that the dendritic trees of PVN neurons are largely confined within the boundaries of the nucleus. Thus, direct synaptic modulation of ACTH secretagogue release likely occurs within the nucleus. Finally, the PVN receives a restricted set of direct inputs, but is surrounded by a shell of fibers and terminals emanating from a variety of sources, including cholinergic inputs from brainstem (Ruggiero et al., 1990) and descending limbic input from the prefrontal cortex, ventral

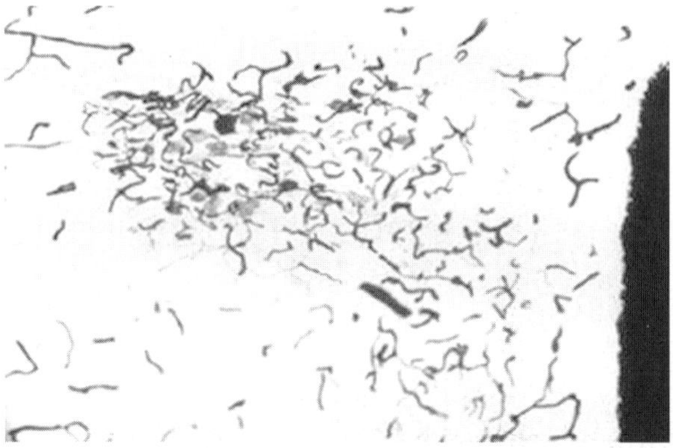

Fig. 1. Vascular configuration of the paraventricular nucleus. Section through the PVN of a rat perfused with blue gelatin-ink solution and processed for immuncytochemistry for vasopressin-specific neurophysin (light-gray cell bodies). Note the rich vascular plexus in the PVN proper compared with the surrounding brain parenchyma.

subiculum, lateral septum, and medial amygdaloid nucleus (Hurley et al., 1991; Cullinan et al., 1993; Canteras et al., 1995; Risold and Swanson, 1997; Prewitt and Herman, 1998). An example of this arrangement can be appreciated in Fig. 2B, illustrating dense innervation of the peri-PVN region using an antibody staining vesicular glutamate transporter 1 (VGlut1), a marker for forebrain glutamatergic neurons (Bellocchio et al., 2000; Takamori et al., 2000). This outer shell is populated with PVN-projecting GABAergic neurons (Roland and Sawchenko, 1993); as PVN dendrites do not ramify outside the PVN, the current data suggest that inputs innervating the shell may be processed locally prior to accessing the PVN itself. The latter observations also present the opportunity for modulating PVN function via "volume transmission," whereby neurotransmitters/neuromodulators affect postsynaptic neurons via diffusion; the latter process has been proposed (van den Pol, 1982), though this remains controversial.

The "direct line"– ascending inputs from the brainstem

A wealth of evidence indicates that neurons in the brainstem play a major role in integrating HPA responses to stress (see schematic in Fig. 3). It is clear that there are sets of direct inputs to the medial

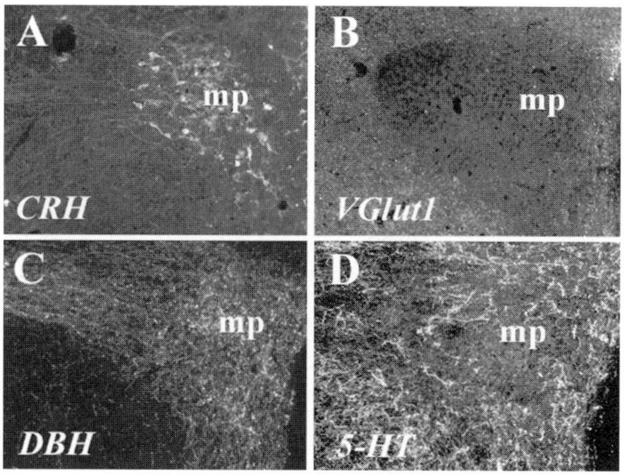

Fig. 2. Various configurations of neurotransmitter innervation to the paraventricular nucleus. Diversity of excitatory neurotransmitter innervation of the PVN. A: Localization of CRH-immunoreactive neurons in the medial parvocellular (mp) division of the PVN. B: Glutamate innervation (visualized using an antibody to vesicular glutamate transporter 1 (VGlut1), which specifically labels terminals from cortical and limbic glutamate neurons) forms a plexus of terminal boutons immediately outside the PVN, but does not innervate the PVN proper to any substantial extent. C: Noradrenaline/epinephrine terminals (visualized using an antibody to dopamine beta-hydroxylase (DBH)) are largely confined to the PVN proper, with particularly dense aggregations in the mp PVN. D: Serotonin (5-HT) immunoreactivity is located both in the PVN and its surround.

parvocellular PVN from noradrenergic and adrenergic neurons in the region of the nucleus of the solitary tract (NTS) and to a lesser extent, the A1/C1 region of the ventrolateral medulla (Sawchenko and Swanson, 1981; Cunningham and Sawchenko, 1988; Cunningham et al., 1990). Accordingly, the medial parvocellular PVN is heavily innervated by noradrenaline and epinephrine fibers (Sawchenko and Swanson, 1981; Mezey et al., 1984; Cunningham and Sawchenko, 1988; Cunningham et al., 1990) (see Fig. 2C). These direct pathways are integral to excitation of the HPA axis; deafferentation and 6-OHDA lesion studies indicate that removal of ascending input to the PVN disrupts ACTH and corticosterone release and blocks intrinsic cFos activation following exposure to a variety of stressors, including interleukin-1 beta administration, hemorrhage, and ether (Szafarczyk et al., 1985; Alonso et al., 1986; Sawchenko, 1988; Li et al., 1996). Pharmacological analyses indicate that the catecholaminergic component of this input is transduced by way of alpha-1 adrenergic receptors (Szafarczyk et al., 1987; Plotsky et al., 1989).

Evidence is emerging to indicate that neuromodulators other than catecholamines can contribute to HPA axis stimulation from the NTS region. For example, work from our group indicates that glucagon-like peptide-1 (GLP-1) neurons of the NTS modulate HPA responses to both visceral illness and elevated plus maze exposure (Kinzig et al., 2002). GLP-1 does not colocalize with tyrosine hydroxylase in the NTS (Larsen et al., 1997), indicating that effects are mediated by way of noncatecholaminergic neurons. Indeed, combined tract tracing-Fos studies suggest that the catecholaminergic neurons do not account for all stress-activated cells in the NTS (c.f., Dayas et al., 1999), implying a role for other cell types in stress integration.

The prominent position of the NTS in stress integration is consistent with its role in coordinating responses to perturbations of the internal milieu. The role of the NTS in mediating reflex control of the cardiovascular system is well documented (c.f., Loewy and McKellar, 1980; Lawrence and Jarrott, 1996). Similarly, the NTS is involved in relay of information mediating visceral illness (Seeley et al., 2000) and systemic infection (Ericsson et al., 1994). All of the above processes are accompanied by activation of the HPA axis, and provide a means through which visceral signals can recruit the support of glucocorticoids en route to restoration of homeostasis. However, recent findings (Dayas et al., 2001b) suggest that the NTS system may also be recruited to relay stressful information from forebrain regions, as removal of both catecholaminergic and noncatecholaminergic inputs from the NTS can dampen responses to nonvisceral (i.e., psychogenic) stimuli (see discussion on limbic inputs, below).

While the NTS plays a prominent role in mediating ACTH release, it is by no means the only brainstem system involved in HPA axis function. The PVN is innervated by 5-HT fibers emanating from the median and dorsal raphe cell groups (B7–B9) (Sawchenko et al., 1983) (Fig. 2D). In situ hybridization and immunohistochemical studies indicate that serotonin receptors (e.g., 5-HT1A, 5-HT2C, 5-HT2A, 5HT7) are present in the parvocellular region of the PVN (Lovenberg et al., 1993;

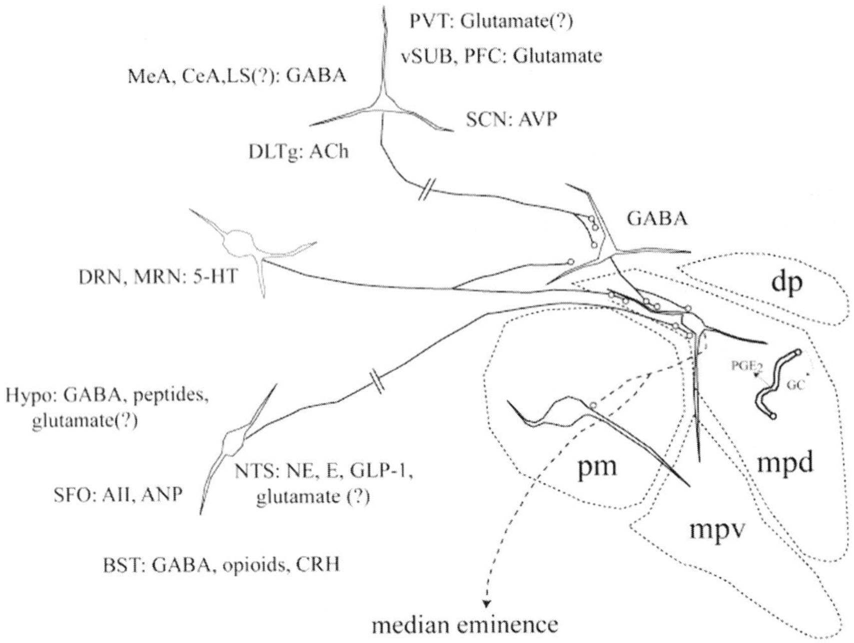

Fig. 3. Schematic diagram of stress-circuit inputs to the PVN. Hypophysiotrophic neurons of the dorsolateral medial parvocellular PVN (mpd) receive direct inputs from noradrenaline, epinephrine (E) and peptidergic neurons from the nucleus of the solitary tract (NTS), peptidergic and GABAergic neurons in several hypothalamic nuclei (Hypo), and GABAergic and peptidergic neurons in the BST. The subfornical organ sends excitatory angiotensin II (AII) projections to the PVN. Neurons in the peri-PVN region send direct (probably GABAergic) projections into the mpd. These neurons play a role in gating descending information from the limbic system (ventral subiculum (vSub), medial prefrontal cortex (PFC), medial amygdala (MeA) and lateral septum (LS)), paraventricular thalamus (PVT), and suprachiasmatic nucleus (SCN), and likely relay cholinergic (ACh) input from the dorsolateral tegmental nucleus (DLTg). Serotonergic (5-HT) input from the median (MRN) and dorsal raphe (DRN) nuclei access both the PVN and its surround. It is not clear which cell group mediates serotonergic excitation of CRF release. The mpd neurons project to the median eminence (dotted line), and have intranuclear axon collaterals that ramify in both magnocellular and parvocellular zones. Dendrites of parvocellular neurons extend across subdivision boundaries of the nucleus, but rarely extend outside the nucleus proper. Neurons can also receive blood-borne glucocorticoid signals (GCs) and cytokine-mediated signals from vascular epithelial cells (PGE_2). Modified from Herman et al. (2002), with permission.

Wright et al., 1995; Zhang et al., 2002), supporting an action of serotonin at the CRF neuron. Pharmacological studies generally ascribe a stimulatory action of serotonin on the HPA axis that is mediated by activation of PVN CRF neurons and CRF release (see Van de Kar and Blair, 1999). These stimulatory actions appear to be mediated by PVN 5-HT2A and perhaps 5-HT1A receptors (Pan and Gilbert, 1992; Van de Kar et al., 2001). However, the serotonin effect is somewhat complicated by the fact that the immediate surround of the PVN is also extensively innervated by serotonergic fibers (Sawchenko et al., 1983), (Fig. 2D); thus, it is unclear whether serotonin effects reflect direct activation of PVN neurons, the PVN surround, or both regions.

Numerous studies using both in vivo and in vitro methods indicate that acetylcholine may also elicit ACTH release (c.f., Jones and Gillham, 1988). Microinjection of acetylcholine into the PVN can promote ACTH release and CRF gene transcription via muscarinic receptors (Ohmori et al., 1995). Whereas nicotine also powerfully activates the HPA axis, it does so indirectly by stimulation of neurons in the NTS region (Matta et al., 1998). Despite the evidence for direct stimulatory actions of acetylcholine, anatomical studies indicate that direct innervation of the PVN proper by cholinergic neurons is very sparse; indeed, the overwhelming majority of choline acetyltransferase positive fibers are found in its immediate surround (Ruggiero et al., 1990). Thus, it is possible that muscarinic stimulation of CRF neurons may be mediated by local interneurons in the immediate vicinity of the PVN.

The parvocellular PVN is also heavily innervated by the lateral parabrachial nucleus and periaqueductal grey (Swanson and Sawchenko, 1983). Both regions are well-positioned to relay "stressful" information to the PVN; the parabrachial nucleus is involved in cardiovascular integration (Saper, 1995), and the periaqueductal gray in mediating pain responses (Behbehani, 1995). The exact contribution of these regions to HPA regulation remains to be elucidated.

Registering blood-borne factors – circumventricular organs and intrinsic vasculature

The medial parvocellular PVN retains a rich innervation from the subfornical organ (SFO) (Fig. 3) and attendant structures of the lamina terminalis pathway, including the median preoptic nucleus and organum vasculosum of the lamina terminalis (Sawchenko and Swanson, 1983; Swanson and Sawchenko, 1983). These structures have a strong parallel innervation of magnocellular subnuclei (Sawchenko and Swanson, 1983; Swanson and Sawchenko, 1983), whereby they are actively involved in fluid and electrolyte regulation (Ganong, 2000; McKinley et al., 2001).

The organization of the SFO–PVN pathway presents an interesting link between the peripheral circulation and HPA activation. The blood–brain barrier in the SFO is deficient (Oldfield and McKinley, 1995), likely allowing access of resident neurons to circulating peptide factors such as angiotensin II and atrial natriuretic peptide (ANP). Intrinsic SFO neurons possess receptors to these species and project directly to the parvocellular PVN (Langub et al., 1995; Herman et al., 1996a; Ferguson et al., 2001) affording a rapid (1-2 synapse) relay between peripheral peptide levels and initiation of the stress response. Importantly, SFO neurons also express angiotensin II and atrial natriuretic peptide (Lind et al., 1984; Langub et al., 1995), suggesting that peripheral and central effectors may use the same peptide messenger. This is likely the case for angiotensin II, which excites the HPA axis by way of angiotensinergic SFO neurons (Plotsky et al., 1988); it remains to be determined whether the inhibitory effects of ANP on the HPA axis are mediated by a parallel pathway. Nonetheless, it is clear that the SFO is positioned to integrate short-lived peptide signals from kidney and perhaps heart into modulation of CRF-neuronal activity.

The area postrema may also play a role in integrating blood-borne information, particularly with respect to circulating cytokines. Lesion studies indicate that damage to the area postrema attenuates PVN cFos induction and noradrenaline release following systemic interleukin-1 beta injection (Ishizuka et al., 1997; Lee et al., 1998). This blood-brain barrier deficient region does not directly innervate the PVN, but is positioned to modulate activity of CRF neurons transsynaptically via connections in the NTS.

Emerging evidence suggests that parvocellular PVN neurons may also be susceptible to information from blood-brain barrier-intact vasculature as well (illustrated in Fig. 3). Recent studies indicate that cytokines (such as TNF-alpha, IL-1beta, and IL-6) activate prostaglandin E2 synthesis in perivascular cells in the medulla, which in turn activate PVN neurons (Ericsson et al., 1997; Rivest, 2001). Moreover, work from Rivest's group indicates that prostaglandin receptors (EP4) are expressed in dendrites of CRF neurons, and thereby allow direct signaling at the CRF neuron itself (Zhang and Rivest, 1999).

Given the ready transit of steroids across the blood-brain barrier, the fact that CRF neurons live in a particularly rich vascular environment is likely to be highly significant. Parvocellular CRF neurons express high levels of glucocorticoid receptors (Uht et al., 1988) (as well as very low amounts of mineralocorticoid receptor), affording them the opportunity to integrate glucocorticoid signals directly. Accordingly, there is good evidence to indicate that CRF neurons are inhibited by local action of glucocorticoids (Kovacs et al., 1986), and exogenous corticosterone and dexamethasone appear to suppress CRF and vasopressin gene transcription in parvocellular neurons (Wolfson et al., 1985; Harbuz and Lightman, 1989; Herman, 1995). However, it is important to note that intrinsic PVN glucocorticoid receptors do not account for all feedback effects of glucocorticoids on the HPA system (Kovacs and Makara, 1988; Herman et al., 1990), implying an important contribution of synaptic inputs.

Naturally, the vascularity of the PVN also places it in position to be influenced by other vascular and lipophilic factors, including endothelial nitric oxide, gonadal steroids, and related compounds. The contributions of these elements to PVN signaling have yet to be delineated.

Intrahypothalamic circuits

The parvocellular PVN receives rich innervation from numerous hypothalamic nuclei. Prominent input is derived from the anteroventral preoptic nucleus, medial preoptic area, dorsomedial hypothalamic nucleus, lateral hypothalamus, arcuate nucleus, and ventral premammillary nucleus (Sawchenko and Swanson, 1983). These structures are known to be critical effector systems for numerous homeostatic functions, including reproduction, energy balance, fluid and electrolyte balance, and thermogenesis.

Lesion and stimulation studies indicate that the majority of intrahypothalamic PVN projections inhibit HPA activation. These observations are consistent with the largely GABAergic phenotype of PVN-projecting cells in these regions (Okamura et al., 1990; Cullinan et al., 1993) (see Fig. 3), and with the prominent innervation of parvocellular CRF neurons by GABA (Decavel and Van Den Pol, 1990, 1992). However, the presence of rich populations of peptidergic neurons in these regions suggests that GABA does not act in isolation. Indeed, it is highly likely that a substantial proportion of GABA neurons in all of these regions co-express peptidergic species, including dynorphin, enkephalin, proopiomelanocortin, atrial natriuretic peptide, and C-type natriuretic peptide, all of which are HPA-inhibitory, and neuropeptide Y and even CRF, which enhance ACTH secretion (see Herman et al., 1996b). In addition, despite the fact that the majority of PVN-projecting cells are GABAergic, the hypothalamus is rich in glutamate-expressing neurons, many of which are interspersed with GABA neurons in PVN regulatory regions (Ziegler et al., 2002). Thus, hypothalamic PVN-projecting neurons appear to have the opportunity to activate or inhibit hypophysiotropic CRF cells, and in view of the extensive interconnectivity among such hypothalamic subnuclei, it may well prove to be the case that these systems act as "tuners" of PVN outflow, rather than monolithic excitatory or inhibitory influences.

In addition to inputs emanating from distinct hypothalamic subnuclei, the parvocellular PVN also receives an extremely rich innervation from neurons in its immediate surround, including the subparaventricular zone, perifornical region, and neurons situated in the immediate vicinity of its nuclear boundaries (Roland and Sawchenko, 1993) (see Fig. 3). These regions are rich in GABAergic neurons (Roland and Sawchenko, 1993, Fig. 4A) and, as noted below, receive inputs from upstream forebrain structures. Stimulation of these neurons with local glutamate causes GABA-mediated inhibition of PVN neurons (Boudaba et al., 1996), consistent with an active role for these neurons in inhibiting CRF neurons. Recent work from our group indicates that these neurons are actively involved in HPA inhibition, as microinjection of a broad-spectrum ionotropic glutamate receptor antagonist in this region enhances the corticosterone response to restraint (Ziegler and Herman, 2000). Accordingly, neurons in this region are c-Fos activated following stress exposure (Cullinan et al., 1996; Cole and Sawchenko, 2002; Fig. 4B), consistent with their involvement in mitigating PVN activation.

The importance of local-circuit GABA innervation of the PVN is underscored by the fact that dendrites of PVN neurons rarely extend beyond the nuclear boundaries (Rho and Swanson, 1989; van den Pol, 1982). As such, it is unlikely that the rich fiber plexus located around the PVN directly innervates the CRF neurons. Rather, it appears that these local neurons may serve as a gating mechanism integrating inputs that project to the shell of the PVN, but not to the CRF neurons themselves. These observations therefore imply that the peri-PVN region plays a major role in integrating various forebrain and brainstem signals into a net inhibitory input to neuronal populations controlling the HPA stress response.

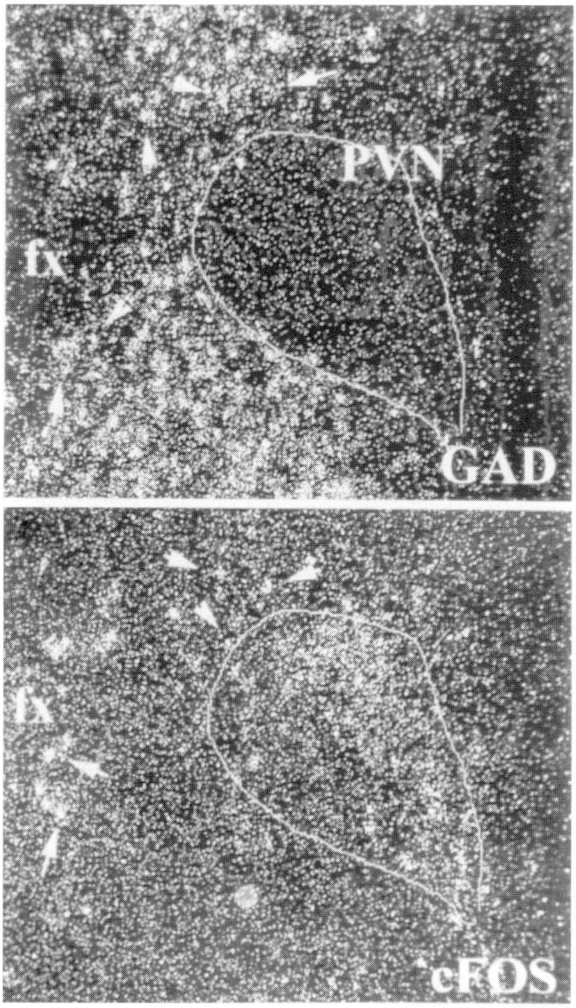

Fig. 4. Neurons in the peri-PVN region. A: A sizable population of GABAergic neurons can be localized to the surround of the PVN, by using in situ hybridization for glutamic acid decarboxylase (GAD) 65 mRNA. B: Numerous c-Fos mRNA positive neurons can be seen in the PVN surround following 30 min of restraint stress. Note that c-Fos mRNA is also expressed in the mpPVN, reflecting activation of CRH neurons.

Do all roads lead to Rome? Limbic interactions with the PVN

The parvocellular PVN receives substantial direct input from only two limbic-related structures: the bed nucleus of the stria terminalis (BST) and the substantia innominata (Sawchenko and Swanson, 1983). The PVN is heavily innervated by the fusiform, anterodorsal, and interfasicular subnuclei of the BST, with additional, less extensive projections from the dorsomedial, principle, and transverse subnuclei (Cullinan et al., 1993; Dong et al., 2001). Dual-label analysis indicates that the vast majority of PVN-projecting neurons are GABAergic (Cullinan et al., 1993), predicting that BST inputs strongly inhibit PVN neurons. In support of these data, lesion and stimulation studies indicate that PVN-projecting regions of the BST inhibit corticosterone release and PVN CRF mRNA levels (Dunn, 1987; Herman et al., 1994). However, it should be noted that lesions of the anterior and lateral regions of the BST attenuate ACTH secretion (Dunn, 1987; Gray et al., 1993; Herman et al., 1994), suggesting that some BST areas may be involved in excitation of PVN stress responses.

The role of the substantia innominata projection to the PVN is ill-defined. It should be noted that this region receives input from the amygdala, bed nucleus of the stria terminalis, as well as the nucleus accumbens and ventral striatal structures (Grove, 1988; Spooren et al., 1991). Interaction with the nucleus accumbens is of relevance to the interface between "reward systems" and HPA regulation. In general, this link is poorly understood; there is one report indicating that stimulation of the nucleus accumbens inhibits corticosterone secretion (Saphier and Feldman, 1985), consistent with inhibitory actions at the PVN. Nonetheless, the nucleus accumbens is exquisitely sensitive to the effects of stress (c.f., Marinelli and Piazza, 2002), and is thus in a logical position to modulate HPA responses in concert with processing of information signaling reward value.

By far, the vast majority of limbic influences on HPA function require an intervening synapse (see Fig. 3). This is clearly the case for the ventral subiculum (hippocampus), infralimbic (prefrontal) cortex, medial amygdala, central amygdala, and lateral septum, all of which modulate HPA activity (c.f., Feldman et al., 1995, Herman and Cullinan, 1997) without substantial direct input to the PVN (Hurley et al., 1991; Cullinan et al., 1993; Canteras et al., 1995; Risold and Swanson, 1997; Prewitt and Herman, 1998). Notably, in all cases there is impressive innervation of PVN-projecting regions of the peri-PVN zone, hypothalamus, and/or the

BST. Dual-tracing studies indicate the potential for synaptic interactions between PVN-projecting cells in these regions and projection neurons from the ventral subiculum, medial amygdala, and central amygdala (Cullinan et al., 1993; Prewitt and Herman, 1998; Dong et al., 2001). Thus, these regions are poised to influence the HPA axis through modulating the activity of intervening (predominantly GABAergic) projections to the parvocellular PVN.

An integrative role for GABAergic PVN-projecting peri-PVN, BST, and hypothalamic cell populations is bolstered by the observation that hippocampal, cortical, and amygdaloid influences on HPA tone differ substantially. Notably, ventral subiculum and infralimbic cortex inhibit HPA responses to stressful stimuli (Figueiredo et al., 2003; Diorio et al., 1993; Herman et al., 1998). Output neurons of both regions are predominantly glutamatergic (Walaas and Fonnum, 1980), and thus set up a monosynaptic relay that converts excitatory cortical/hippocampal outflow to inhibition at the PVN (see Fig. 5A-B). In contrast, medial and central amygdaloid projections to these regions are predominantly GABAergic (Swanson and Petrovich, 1998) (Fig. 5C-D); thus, interactions with GABAergic PVN populations would result in disinhibition, resulting in a net increase in HPA activation. The intense overlap of limbic inputs into PVN-projecting regions implies a convergence of input, and by extension, an area where excitatory and inhibitory inputs can be summated into excitation or inhibition of ACTH release.

Importantly, the nature of limbic interaction with stress responsiveness is stimulus specific. Ventral subiculum and prefrontal cortex inhibit stress responses to stimuli that are psychogenic in origin, involving multi-modal sensory stimuli without obvious homeostatic disruption. Lesions of these areas result in hyperresponsiveness to putative psychogenic stimuli such as restraint or novelty, but not ether inhalation or hypoxia (Figueiredo et al., 2003; Bradbury et al., 1993; Herman et al., 1998). In contrast, both medial and central amygdaloid nuclei appear stress excitatory. However, there are considerable differences in stimulus specificity across the amygdaloid nuclei; lesion and c-fos mapping studies indicate the medial nucleus is selectively involved in psychogenic responses, such as restraint, whereas the central nucleus mediates responses

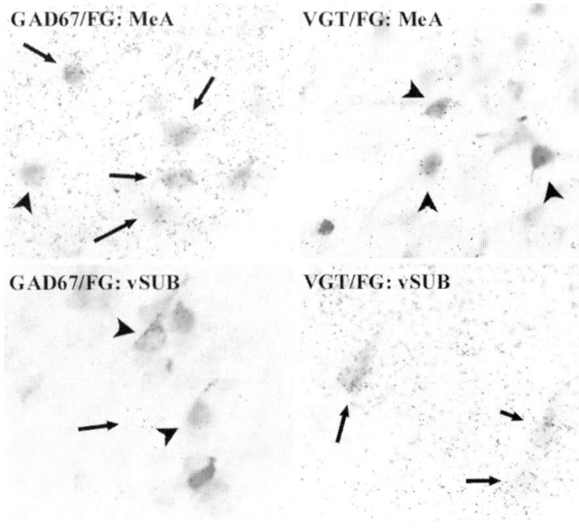

Fig. 5. Innervation of the PVN-projecting region of the BST by GABAergic and glutamatergic limbic regions. Combined tract tracing/immunohistochemistry (Fluorogold (FG)) and in situ hybridization was used to identify the phenotype of medial amygdaloid (MeA)(A, B) and ventral subicular (vSUB)(C, D) neurons projecting to the PVN-projecting interfascicular region of the BST. Following BST injection of FG, the vast majority of dual-labeled neurons in the MeA were dual labeled with FG and GAD67 mRNA (A, arrows), indicating a GABAergic phenotype. No vesicular glutamate transporter 1 (VGT)/GAD67 colocalization was observed (B). In contrast, BST-projecting neurons in the ventral subiculum did not express significant levels of GAD67 (C, arrowheads), but virtually all expressed VGT (D). Note that unlabeled cells (arrow) in the vSUB did express GAD67, probably reflecting non-pyramidal interneurons localized in this region. Thus, MeA afferents to this PVN-projecting region of the BST are likely to be GABAergic, whereas vSUB afferents appear to be glutamate containing.

to interleukin-1beta and perhaps hemorrhage (Thrivikraman et al., 1997; Dayas et al., 1999, 2001a; Xu et al., 1999). Thus, it is clear that programming the stress response is a distributed process, involving a summation of inputs encoding various internally and externally perceived stimuli.

Importantly, additional HPA-modulatory regions interact with PVN-projecting regions, but not the PVN proper. The hypothalamic suprachiasmatic nucleus, which entrains diurnal corticosterone rhythms, connects with the PVN through relays in the subparaventricular zone (Watts et al., 1987). This pathway apparently involves vasopressinergic activation of intermediary GABA neurons, consistent with the known anatomy of the peri-PVN region (Buijs et al., 1998). In addition, the paraventricular thalamus also interacts with the PVN in a transsynaptic manner, perhaps by the way of the dorsomedial hypothalamic nucleus (Bhatnagar and Dallman, 1998). The paraventricular thalamus is critical to processing information regarding repeated stress, as lesion of the posterior aspect of this region prohibits sensitization of HPA responses to novel stimuli following chronic homotypic stress exposure (Bhatnagar and Dallman, 1998). Subsequent studies also implicate this region in stress habituation, as lesions retard the decrement in HPA reactivity seen with repeated exposure to the same stressor (Bhatnagar et al., 2002). Finally, a number of hyphalamic regions appear to connect with the parvocellular PVN through intermediary hypothalamic synapses. Most notable among these is the ventromedial nucleus, which communicates with the PVN through the dorsomedial nucleus and peri-PVN zone (ter Horst and Luiten, 1987). The ventromedial nucleus is critical to integration of neural and bloodborne orexigenic and anorectic signals, and may thereby tune the HPA response to perturbations of energy balance.

It is also evident that limbic regions may use long-loop pathways to influence HPA axis activity. Numerous limbic regions, including the infralimbic cortex and central amygdaloid nucleus, send descending projections to the NTS (Schwaber et al., 1982; van der Kooy et al., 1984). Thus, it is probable that stressful information relayed to the brainstem may modulate subsequent HPA responses secondary to changes in NTS outflow (Xu et al., 1999). Indeed, emerging evidence suggests that such relays play a major role in forebrain stress integration. Lesions of medullary A1/A2 noradrenergic neurons reduce PVN c-fos activation following restraint (Dayas et al., 2001b), consistent with a role for ascending systems in mediating responses to nonvisceral stressors. In addition, intra-PVN injection of a GLP-1 antagonist inhibits both ACTH and corticosterone responses to novelty stress in an elevated plus maze; as GLP-1 is synthesized only in the NTS (Larsen et al., 1997), the data indicate that this noncatecholaminergic system also regulates responses to psychogenic stressors.

Perspectives on PVN integration: A multifaceted effector of stress responsiveness?

The bulk of the current discussion has centered on hypophysiotropic PVN neurons; however, it is important to note that much of this micro- and macroanatomy is relevant to magnocellular and preautonomic cell populations in the PVN. Thus, the same classes of input also modulate activity of posterior pituitary vasopressin and oxytocin neurons (e.g., Boudaba et al., 1996). Furthermore, dendrites of parvocellular, magnocellular, and preautonomic neurons ramify freely within the PVN (van den Pol, 1982; Rho and Swanson, 1989), and there is evidence for communication among PVN cell populations via intranuclear axon collaterals as well (van den Pol, 1982). As such, it is reasonable to suggest that the HPA component of PVN integration is a microcosm of the functional role of the nucleus as a central motoneuron for the organismic stress response.

The convergence of limbic inputs onto BST, hypothalamic, and peri-PVN cell populations indicates that an active filtering of information is taking place in these regions. We propose that these cell populations are in essence "weighing" information from the respective limbic sites prior to sending an integrated signal to the PVN. The hippocampus likely communicates learned information, including spatial aspects of the stimuli, whereas the amygdala, which processes fearful stimuli, may relay input on perceived danger. (It is notable that the hippocampus and amygdala are interconnected and may subserve complementary roles in assessing the relative safety of an environment.) By converging at key nodal points prior to the PVN, the net result of activation of these two systems can be efficiently integrated into an appropriate HPA signal.

The placement of potential pre-PVN integrators may also reveal an additional checkpoint. For instance, hypothalamic PVN-projecting populations are centered around key integrative sites (lateral hypothalamus, preoptic area, and dorsomedial nucleus) that are also involved in organizing behavioral and autonomic responses to homeostatic challenges. Processing in these regions may integrate the net limbic signal with respect to what the organism perceives about the internal milieu.

Thus, environmental familiarity and prospective danger may be filtered with respect to the organism's internal state – allowing an override to the net limbic signal. For example, an animal may need to engage in risky behavior if it is hungry, with an attendant glucocorticoid response generated to anticipate the need to mobilize energy should the perceived danger come to fruition.

Overall inputs to the PVN present ample opportunity to coordinate a stress response in accordance with internal state, external environment and memory. This triad of functional inputs appear to be inextricably linked – all classes of information need to be weighed prior to response. Whereas internal state (communicated by ascending brainstem systems, circumventricular organs, and perhaps local trans-vascular signaling within the PVN itself) has direct input to the CRF neuron, this system may be tempered by the integrated inhibitory signal coming from the BST and hypothalamus. Conversely, inhibition of the HPA system upon habituation or learned safety may be overridden by homeostatic challenges. As such, stress circuits offer a fine-tuned HPA response that accounts for the exquisite control of glucocorticoid balance in the healthy organism.

Acknowledgments

This work was supported by NIH grants MH49698 to JPH, MH56577 to WEC, MH60819 to JPH and WEC, and a NIH NRSA grant to NKM. We would also like to thank Mark Dolgas, Garrett Bowers, and Melanie Parrish for their fine technical assistance.

References

Alonso, G., Szafarczyk, A., Balmefrezol, M. and Assenmacher, I. (1986) Immunocytochemical evidence for stimulatory control by the ventral noradrenergic bundle of parvocellular neurons of the paraventricular nucleus secreting corticotropin releasing hormone and vasopressin in rats. Brain Res., 397: 297–307.

Antoni, F.A. (1986) Hypothalamic control of adrenocorticotropin secretion: advances since the discovery of 41-residue corticotropin-releasing factor. Endocrine Rev., 7: 351–378.

Behbehani, M.M. (1995) Functional characteristics of the midbrain periaqueductal gray. Prog. Neurobiol., 46: 575–605.

Bellocchio, E.E., Reimer, R.J., Fremeau, R.T., Jr. and Edwards, R.H. (2000) Uptake of glutamate into synaptic vesicles by an inorganic phosphate transporter. Science, 289: 957–960.

Bhatnagar, S. and Dallman, M. (1998) Neuroanatomical basis for facilitation of hypothalamic-pituitary-adrenal responses to a novel stressor after chronic stress. Neuroscience, 84: 1025–1039.

Bhatnagar, S., Huber, R., Nowak, N. and Trotter, P. (2002) Lesions of the posterior paraventricular thalamus block habituation of hypothalamic-pituitary-adrenal responses to repeated restraint. J. Neuroendocrinol., 14: 403–410.

Boudaba, C., Szabo, K. and Tasker, J.G. (1996) Physiological mapping of local inhibitory inputs to the hypothalamic paraventricular nucleus. J. Neurosci., 16: 7151–7160.

Bradbury, M.J., Strack, A.M. and Dallman, M.F. (1993) Lesions of the hippocampal efferent pathway (fimbria-fornix) do not alter sensitivity of adrenocorticotropin to feedback inhibition by corticosterone in rats. Neuroendocrinol., 58: 396–407.

Buijs, R.M., Hermes, M.H. and Kalsbeek, A. (1998) The suprachiasmatic nucleus-paraventricular nucleus interactions: a bridge to the neuroendocrine and autonomic nervous system. Prog. Brain Res., 119: 365–382.

Canteras, N.S., Simerly, R.B. and Swanson, L.W. (1995) Organization of projections from the medial nucleus of the amygdala: a PHAL study in the rat. J. Comp. Neurol., 360: 213–245.

Cole, R.L. and Sawchenko, P.E. (2002) Neurotransmitter regulation of cellular activation and neuropeptide gene expression in the paraventricular nucleus of the hypothalamus. J. Neurosci., 22: 959–969.

Cullinan, W.E., Helmreich, D.L. and Watson, S.J. (1996) Fos expression in forebrain afferents to the hypothalamic paraventricular nucleus following swim stress. J. Comp. Neurol., 368: 88–99.

Cullinan, W.E., Herman, J.P. and Watson, S.J. (1993) Ventral subicular interaction with the hypothalamic paraventricular nucleus: evidence for a relay in the bed nucleus of the stria terminalis. J. Comp. Neurol., 332: 1–20.

Cunningham, E.T., Jr and Sawchenko, P.E. (1988) Anatomical specificity of noradrenergic inputs to the paraventricular and supraoptic nuclei of the rat hypothalamus. J. Comp. Neurol., 274: 60–76.

Cunningham, E.T., Jr., Bohn, M.C. and Sawchenko, P.E. (1990) Organization of adrenergic inputs to the paraventricular and supraoptic nuclei of the hypothalamus in the rat. J. Comp. Neurol., 292: 651–667.

Dayas, C.V., Buller, K.M., Crane, J.W., Xu, Y. and Day, T.A. (2001a) Stressor categorization: acute physical and psychological stressors elicit distinctive recruitment patterns in the amygdala and in medullary noradrenergic cell groups. Eur. J. Neurosci., 14: 1143–1152.

Dayas, C.V., Buller, K.M. and Day, T.A. (1999) Neuroendocrine responses to an emotional stressor: evidence for involvement of the medial but not the central amygdala. Eur. J. Neurosci., 11: 2312–2322.

Dayas, C.V., Buller, K.M. and Day, T.A. (2001b) Medullary neurones regulate hypothalamic corticotropin-releasing factor cell responses to an emotional stressor. Neuroscience, 105: 707–719.

Decavel, C. and Van Den Pol, A.N. (1990) GABA: a dominant neurotransmitter in the hypothalamus. J. Comp. Neurol., 302: 1019–1037.

Decavel, C. and Van Den Pol, A.N. (1992) Converging GABA and glutamate-immunoreactive axons make synaptic contact with identified hypothalamic neurosecretory axons. J. Comp. Neurol., 316: 104–116.

Diorio, D., Viau, V. and Meaney, M.J. (1993) The role of the medial prefrontal cortex (cingulate gyrus) in the regulation of hypothalamo-pituitary-adrenal responses to stress. J. Neurosci., 13: 3839–3847.

Dong, H.W., Petrovich, G.D., Watts, A.G. and Swanson, L.W. (2001) Basic organization of projections from the oval and fusiform nuclei of the bed nuclei of the stria terminalis in adult rat brain. J. Comp. Neurol., 436: 430–455.

Dunn, J.D. (1987) Plasma corticosterone responses to electrical stimulation of the bed nucleus of the stria terminalis. Brain Res., 407: 327–331.

Ericsson, A., Arias, C. and Sawchenko, P.E. (1997) Evidence for an intramedullary prostaglandin-dependent mechanism in the activation of stress-related neuroendocrine circuitry by intravenous interleukin-1. J. Neurosci., 17: 7166–7179.

Ericsson, A., Kovacs, K.J. and Sawchenko, P.E. (1994) A functional anatomical analysis of central pathways subserving the effects of interleukin-1 on stress-related neuroendocrine neurons. J. Neurosci., 14: 897–913.

Feldman, S., Conforti, N. and Weidenfeld, J. (1995) Limbic pathways and hypothalamic neurotransmitters mediating adrenocortical responses to neural stimuli. Neurosci. Biobehav. Rev., 19: 235–240.

Ferguson, A.V., Washburn, D.L. and Latchford, K.J. (2001) Hormonal and neurotransmitter roles for angiotensin in the regulation of central autonomic function. Exp. Biol. Med. (Maywood), 226: 85–96.

Figueiredo, H.F., Bodie, B.L., Tauchi, M., Dolgas, C.M. and Herman, J.P. (2003) Stress integration following acute and chronic predator stress: Differential activation of central stress circuitry and sensitization of the hypothalamo-pituitary adrenocortical axis. Endocrinology, 144: 5249–5258.

Ganong, W.F. (2000) Circumventricular organs: definition and role in the regulation of endocrine and autonomic function. Clin. Exp. Pharmacol. Physiol., 27: 422–427.

Gillies, G.E., Linton, E.A. and Lowry, P.J. (1982) Corticotropin-releasing activity of the new CRF is potentiated several times by vasopressin. Nature, 299: 355–357.

Gray, T.S., Piechowski, R.A., Yracheta, J.M., Rittenhouse, P.A., Bethea, C.L. and Van de Kar, L.D. (1993) Ibotenic acid lesions in the bed nucleus of the stria terminalis attenuate conditioned stress-induced increases in prolactin, ACTH and corticosterone. Neuroendocrinology, 57: 517–524.

Grove, E.A. (1988) Neural associations of the substantia innominata in the rat: afferent connections. J. Comp. Neurol., 277: 315–346.

Harbuz, M.S. and Lightman, S.L. (1989) Glucocorticoid inhibition of stress-induced changes in hypothalamic corticotrophin-releasing factor messenger RNA and proenkephalin A messenger RNA. Neuropeptides, 14: 17–20.

Herman, J.P. (1995) In situ hybridization analysis of vasopressin gene transcription in the paraventricular and supraoptic nuclei of the rat: regulation by stress and glucocorticoids. J. Comp. Neurol., 363: 15–27.

Herman, J.P. and Cullinan, W.E. (1997) Neurocircuitry of stress: central control of the hypothalamo-pituitary-adrenocortical axis. TINS, 20: 78–83.

Herman, J.P., Cullinan, W.E. and Watson, S.J. (1994) Involvement of the bed nucleus of the stria terminalis in tonic regulation of paraventricular hypothalamic CRH and AVP mRNA expression. J. Neuroendocrinol., 6: 433–442.

Herman, J.P., Cullinan, W.E., Ziegler, D.R. and Tasker, J.G. (2002) Role of the paraventricular nucleus microenvironment in stress integration. Eur. J. Neurosci., 16: 381–385.

Herman, J.P., Dolgas, C.M. and Carlson, S.C. (1998) Ventral subiculum co-ordinates situation-specific neuroendocrine and behavioral stress responses. Neurosci., 86: 449–459.

Herman, J.P., Dolgas, C.M., Rucker, D. and Langub, M.C., Jr. (1996a) Localization of natriuretic peptide-activated guanylate cyclase mRNAs in the rat brain. J. Comp. Neurol., 369: 165–187.

Herman, J.P., Prewitt, C.M.-F. and Cullinan, W.E. (1996b) Neuronal circuit regulation of the hypothalamo-pituitary-adrenocortical stress axis. Crit. Rev. Neurobiol., 11: 371–394.

Herman, J.P., Wiegand, S.J. and Watson, S.J. (1990) Regulation of basal corticotropin-releasing hormone and arginine vasopressin messenger ribonucleic acid expression in the paraventricular nucleus: effects of selective hypothalamic deafferentations. Endocrinology, 127: 2408–2417.

Hurley, K.M., Herbert, H., Moga, M.M. and Saper, C.B. (1991) Efferent projections of the infralimbic cortex of the rat. J. Comp. Neurol., 308: 249–276.

Ishizuka, Y., Ishida, Y., Kunitake, T., Kato, K., Hanamori, T., Mitsuyama, Y. and Kannan, H. (1997) Effects of area postrema lesion and abdominal vagotomy on interleukin-1 beta-induced norepinephrine release in the hypothalamic paraventricular nucleus region in the rat. Neurosci. Lett., 223: 57–60.

Jones, M.T. and Gillham, B. (1988) Factors involved in the regulation of adrenocorticotropic hormone/beta-lipotropic hormone. Physiol. Rev., 68: 743–818.

Kinzig, K.P., Figueiredo, H.F., Murphy, E.K., Herman, J.P., Sakai, R.R., D'Alessio, D.A. and Seeley, R.J. (2002) Multiple role of glucagon-like peptide-1 in response to interoceptive and exteroceptive stressors. 2002 Abstract Viewer/Itinerary Planner, Society for Neuroscience, Washington, DC, 865.15, (CD-ROM).

Kiss, J.Z. (1988) Dynamism of chemoarchitecture in the hypothalamic paraventricular nucleus. Brain Res. Bull., 20: 699–708.

Kovacs, K., Kiss, J.Z. and Makara, G.B. (1986) Glucocorticoid implants around the hypothalamic paraventricular nucleus prevent the increase of corticotropin-releasing factor and arginine vasopressin immunostaining induced by adrenalectomy. Neuroendocrinology, 44: 229–234.

Kovacs, K.J. and Makara, G.B. (1988) Corticosterone and dexamethasone act at different brain sites to inhibit adrenalectomy-induced adrenocorticotropin hypersecretion. Brain Res., 474: 205–210.

Langub, M.C., Jr., Watson, R.E., Jr. and Herman, J.P. (1995) Distribution of natriuretic peptide precursor mRNAs in the rat brain. J. Comp. Neurol., 356: 183–199.

Larsen, P.J., Tang-Christensen, M., Holst, J.J. and Orskov, C. (1997) Distribution of glucagon-like peptide-1 and other preproglucagon-derived peptides in the rat hypothalamus and brainstem. Neuroscience, 77: 257–270.

Lawrence, A.J. and Jarrott, B. (1996) Neurochemical modulation of cardiovascular control in the nucleus tractus solitarius. Prog. Neurobiol., 48: 21–53.

Lee, H.Y., Whiteside, M.B. and Herkenham, M. (1998) Area postrema removal abolishes stimulatory effects of intravenous interleukin-1beta on hypothalamic-pituitary-adrenal axis activity and c-fos mRNA in the hypothalamic paraventricular nucleus. Brain Res. Bull., 46: 495–503.

Li, H.-Y., Ericsson, A. and Sawchenko, P.E. (1996) Distinct mechanisms underlie activation of hypothalamic neurosecretory neurons and their medullary catecholaminergic afferents in categorically different stress paradigms. Proc. Natl. Acad. Sci., 93: 2359–2364.

Lind, R.W., Swanson, L.W. and Ganten, D. (1984) Angiotensin II immunoreactive pathways in the central nervous system of the rat: evidence for a projection from the subfornical organ to the paraventricular nucleus of the hypothalamus. Clin. Exp. Hypertension [A], 6: 1915–1920.

Loewy, A.D. and McKellar, S. (1980) The neuroanatomical basis of central cardiovascular control. Fed. Proc., 39: 2495–2503.

Lovenberg, T.W., Baron, B.M., de Lecea, L., Miller, J.D., Prosser, R.A., Rea, M.A., Foye, P.E., Racke, M., Slone, A.L., Siegel, B.W. Danielson, P.E., Sutcliffe, J.G. and Erlander, M.G. (1993) A novel adenylyl cyclase-activating serotonin receptor (5-HT7) implicated in the regulation of mammalian circadian rhythms. Neuron, 11: 449–458.

Marinelli, M. and Piazza, P.V. (2002) interaction between glucocorticoid hormones, stress and psychostimulant drugs. Eur. J. Neurosci., 16: 387–394.

Matta, S.G., Fu, Y., Valentine, J.D. and Sharp, B.M. (1998) Response of the hypothalamo-pituitary-adrenal axis to nicotine. Psychoneuroendocrinology, 23: 103–113.

McEwen, B.S. and Stellar, E. (1993) Stress and the individual. Mechanisms leading to disease. Arch. Intern. Med., 153: 2093–2101.

McKinley, M.J., Allen, A.M., May, C.N., McAllen, R.M., Oldfield, B.J., Sly, D. and Mendelsohn, F.A. (2001) Neural pathways from the lamina terminalis influencing cardiovascular and body fluid homeostasis. Clin. Exp. Pharmacol. Physiol., 28: 990–992.

Mezey, E., Kiss, J.Z., Skirboll, L.R., Goldstein, M. and Axelrod, J. (1984) Increase of corticotropin-releasing factor staining in rat paraventricular nucleus neurones by depletion of hypothalamic adrenaline. Nature, 310: 140–141.

Ohmori, N., Itoi, K., Tozawa, F., Sakai, Y., Sakai, K., Horiba, N., Demura, H. and Suda, T. (1995) Effect of acetylcholine on corticotropin-releasing factor gene expression in the hypothalamic paraventricular nucleus of conscious rats. Endocrinology, 136: 4858–4863.

Okamura, H., Abitbol, M., Julien, J.F., Dumas, S., Berod, A., Geffard, M., Kitahama, K., Bobillier, P., Mallet, J. and Wiklund, L. (1990) Neurons containing messenger RNA encoding glutamate decarboxylase in rat hypothalamus demonstrated by in situ hybridization, with special emphasis on cell groups in medial preoptic area, anterior hypothalamic area and dorsomedial hypothalamic nucleus. Neuroscience, 39: 675–699.

Oldfield, B.J. and McKinley, M.J. (1995) Circumventricular Organs. In: Paxinos, G. (Ed.), The Rat Nervous System, 2nd Edn. Academic Press, San Diego, pp. 391–403.

Pan, L. and Gilbert, F. (1992) Activation of 5-HT1A receptor subtype in the paraventricular nuclei of the hypothalamus induces CRH and ACTH release in the rat. Neuroendocrinology, 56: 797–802.

Plotsky, P.M., Cunningham, E.T., Jr. and Widmaier, E.P. (1989) Catecholaminergic modulation of corticotropin-releasing factor and adrenocorticotropin secretion. Endocrine Rev., 10: 437–458.

Plotsky, P.M., Sutton, S.W., Bruhn, T.O. and Ferguson, A.V. (1988) Analysis of the role of angiotensin II in mediation of adrenocorticotropin secretion. Endocrinology, 122: 538–545.

Prewitt, C.M. and Herman, J.P. (1998) Anatomical interactions between the central amygdaloid nucleus and the hypothalamic paraventricular nucleus of the rat: a dual tract-tracing analysis. J. Chem. Neuroanat., 15: 173–185.

Rho, J.H. and Swanson, L.W. (1989) A morphometric analysis of functionally defined subpopulations of neurons in the

paraventricular nucleus of the rat with observations on the effects of colchicine. J. Neurosci., 9: 1375–1388.
Risold, P.Y. and Swanson, L.W. (1997) Connections of the rat lateral septal complex. Brain Res. Brain Res. Rev., 24: 115–195.
Rivest, S. (2001) How circulating cytokines trigger the neural circuits that control the hypothalamic-pituitary-adrenal axis. psychoneuroendocrinology, 26: 761–788.
Roland, B.L. and Sawchenko, P.E. (1993) Local origins of some GABAergic projections to the paraventricular and supraoptic nuclei of the hypothalamus of the rat. J. Comp. Neurol., 332: 123–143.
Ruggiero, D.A., Giuliano, R., Anwar, M., Stornetta, R. and Reis, D.J. (1990) Anatomical substrate of cholinergic-autonomic regulation in the rat. J. Comp. Neurol., 292: 1–53.
Saper, C.B. (1995) Central autonomic control. In: Paxinos, G. (Ed.), The Rat Nervous System, 2nd Edn. Academic Press, San Diego, pp. 107–135.
Saphier, D. and Feldman, S. (1985) Effects of neural stimuli on paraventricular nucleus neurones. Brain Res. Bull., 14: 401–417.
Sapolsky, R.M., Krey, L.C. and McEwen, B.S. (1986) The neuroendocrinology of stress and aging: the glucocorticoid cascade hypothesis. Endocr. Rev., 7: 284–301.
Sawchenko, P.E. (1988) Effects of catecholamine-depleting medullary knife cuts on corticotropin-releasing factor and vasopressin immunoreactivity in the hypothalamus of normal and steroid-manipulated rats. Neuroendocrinology, 48: 459–470.
Sawchenko, P.E. and Swanson, L.W. (1981) Central noradrenergic pathways for the integration of hypothalamic neuroendocrine and autonomic responses. Science, 214: 685–687.
Sawchenko, P.E. and Swanson, L.W. (1983) The organization of forebrain afferents to the paraventricular and supraoptic nuclei of the rat. J. Comp. Neurol., 218: 121–144.
Sawchenko, P.E., Swanson, L.W., Steinbusch, H.W. and Verhofstad, A.A. (1983) The distribution and cells of origin of serotonergic inputs to the paraventricular and supraoptic nuclei of the rat. Brain Res., 277: 355–360.
Schwaber, J.S., Kapp, B.S., Higgins, G.A. and Rapp, P.R. (1982) Amygdala and basal forebrain direct connections with the nucleus of the solitary tract and the dorsal motor nucleus. J. Neurosci., 2: 1424–1438.
Seeley, R.J., Blake, K., Rushing, P.A., Benoit, S., Eng, J., Woods, S.C. and D'Alessio, D. (2000) The role of CNS glucagon-like peptide-1 (7-36) amide receptors in mediating the visceral illness effects of lithium chloride. J. Neurosci., 20: 1616–1621.
Spooren, W.P., Veening, J.G., Groenewegen, H.J. and Cools, A.R. (1991) Efferent connections of the striatopallidal and amygdaloid components of the substantia innominata in the cat: projections to the nucleus accumbens and caudate nucleus. Neuroscience, 44: 431–447.
Swanson, L.W. and Petrovich, G.D. (1998) What is the amygdala?. Trends Neurosci, 21: 323–331.
Swanson, L.W. and Sawchenko, P.E. (1983) Hypothalamic integration: organization of the paraventricular and supraoptic nuclei. Ann. Rev. Neurosci., 6: 269–324.
Szafarczyk, A., Alonso, G., Ixart, G., Malaval, F. and Assenmacher, I. (1985) Diurnal-stimulated and stress-induced ACTH release in rats is mediated by ventral noradrenergic bundle. Am. J. Physiol., 249: E219–E226.
Szafarczyk, A., Malaval, F., Laurent, A., Gibaud, R. and Assenmacher, I. (1987) Further evidence for a central stimulatory actions of catecholamines on adrenocorticotropin release in the rat. Endocrinology, 121: 883–892.
Takamori, S., Rhee, J.S., Rosenmund, C. and Jahn, R. (2000) Identification of a vesicular glutamate transporter that defines a glutamatergic phenotype in neurons. Nature, 407: 189–194.
ter Horst, G.J. and Luiten, P.G.M. (1987) Phaseolus vulgaris leuco-agglutinin tracing of intrahypothalamic connections of the lateral, ventromedial, dorsomedial and paraventricular hypothalamic nuclei in the rat. Brain Res. Bull., 18: 191–203.
Thrivikraman, K.V., Su, Y. and Plotsky, P.M. (1997) Patterns of Fos-immunoreactivity in the CNS induced by repeated hemorrhage in conscious rats: correlations with pituitary-adrenal axis activity. Stress, 2: 145–158.
Uht, R.M., McKelvy, J.F., Harrison, R.W. and Bohn, M.C. (1988) Demonstration of glucocorticoid receptor-like immunoreactivity in glucocorticoid-sensitive vasopressin and corticotropin-releasing factor neurons in the hypothalamic paraventricular nucleus. J. Neurosci. Res., 19: 405–411.
Van de Kar, L.D. and Blair, M.L. (1999) Forebrain pathways mediating stress-induced hormone secretion. Front Neuroendocrinol, 20: 1–48.
Van de Kar, L.D., Javed, A., Zhang, Y., Serres, F., Raap, D.K. and Gray, T.S. (2001) 5-HT2A Receptors stimulate ACTH, corticosterone, oxytocin, renin, and prolactin release and activate hypothalamic CRF and oxytocin-expressing cells. J. Neurosci., 21: 3572–3579.
van den Pol, A.N. (1982) The magnocellular and parvocellular paraventricular nucleus of rat: intrinsic organization. J. Comp. Neurol., 206: 317–345.
van der Kooy, D., Koda, L.Y., McGinty, J.F., Gerfen, C.R. and Bloom, F.E. (1984) The organization of projections from the cortex, amygdala, and hypothalamus to the nucleus of the solitary tract in rat. J. Comp. Neurol., 224: 1–24.
Walaas, I. and Fonnum, F. (1980) Biochemical evidence for glutamate as a transmitter in hippocampal efferents to the basal forebrain and hypothalamus in the rat brain. Neuroscience, 5: 1691–1698.
Watts, A.G., Swanson, L.W. and Sanchez-Watts, G. (1987) Efferent projections of the suprachiasmatic nucleus: studies using anterograde transport of phaseolus vulgaris-leucoagglutinin in the rat. J. Comp. Neurol., 258: 204–229.

Whitnall, M.H. (1993) Regulation of the hypothalamic corticotropin-releasing hormone neurosecretory system. Prog. Neurobiol., 40: 573–629.

Wolfson, B., Manning, R.W., Davis, L.G., Arentzen, R. and Baldino, F., Jr. (1985) Co-localization of corticotropin releasing factor and vasopressin mRNA in neurones after adrenalectomy. Nature, 315: 59–61.

Wright, D.E., Seroogy, K.B., Lundgren, K.H., Davis, B.M. and Jennes, L. (1995) Comparative localization of serotonin1A, 1C, and 2 receptor subtype mRNAs in rat brain. J. Comp. Neurol., 351: 357–373.

Xu, Y., Day, T.A. and Buller, K.M. (1999) The central amygdala modulates hypothalamic-pituitary-adrenal axis responses to systemic interleukin-1beta administration. Neuroscience, 94: 175–183.

Zhang, J. and Rivest, S. (1999) Distribution, regulation and colocalization of the genes encoding the EP2- and EP4-PGE2 receptors in the rat brain and neuronal responses to systemic inflammation. Eur. J. Neurosci., 11: 2651–2668.

Zhang, Y., Damjanoska, K.J., Carrasco, G.A., Dudas, B., D'Souza, D.N., Tetzlaff, J., Garcia, F., Hanley, N.R., Scripathirathan, K., Petersen, B.R., Gray, T.S., Battaglia, G., Muma, N.A. and Van de Kar, L.D. (2002) Evidence that 5-HT2A receptors in the hypothalamic paraventricular nucleus mediate neuroendocrine responses to (-)DOI. J. Neurosci., 22: 9635–9642.

Ziegler, D.R., Cullinan, W.E. and Herman, J.P. (2002) Distribution of vesicular glutamate transporter mRNA in rat hypothalamus. J. Comp. Neurol., 448: 217–229.

Ziegler, D.R. and Herman, J.P. (2000) Local integration of glutamate signaling in the hypothalamic paraventricular region: regulation of glucocorticoid stress responses. Endocrinology, 141: 4801–4804.

T. Steckler, N.H. Kalin and J.M.H.M. Reul (Eds.)
Handbook of Stress and the Brain, Vol. 15
ISBN 0-444-51173-3

CHAPTER 4.2

Sympatho-adrenal activity and hypothalamic–pituitary–adrenal axis regulation

Yvonne M. Ulrich-Lai and William C. Engeland*

Departments of Surgery and Neuroscience, University of Minnesota, Minneapolis, MN 55455, USA

Abstract: The secretion of glucocorticoids from the adrenal cortex is considered an essential adaptive component of the response to stress. Although adrenocorticotropic hormone (ACTH) regulates adrenal steroidogenesis, the dissociation between changes in plasma ACTH and glucocorticoids under nonstress and stress conditions has prompted the search for alternative non-ACTH mechanisms. The innervation of the adrenal cortex and the intermingling of adrenal cortical and medullary cells form the anatomical substrate for regulation of adrenocortical secretion by neural elements. In vitro studies demonstrate the effectiveness of a variety of neurotransmitters to affect adrenal steroidogenesis. In vivo experiments show that adrenal denervation alters glucocorticoid secretion under nonstress and stress conditions. Further investigations are required to reveal the full extent of neural control of adrenal steroidogenesis and to delineate the mechanisms that underlie it. Nonetheless, the collective results suggest that the brain can control glucocorticoid production by at least two complimentary mechanisms, by stimulation of the neuroendocrine hypothalamic–pituitary–adrenal axis and through direct activation of adrenal neural elements.

Survival of an organism depends on the proper maintenance of ones internal milieu despite ever-changing surroundings, an adaptive process known as homeostasis (Canon, 1929). Stress is generally defined as a physical or psychological stimulus that threatens homeostasis (Chrousos and Gold, 1992). The primary systems employed by vertebrates to cope with stress are the sympatho-adrenal medullary axis and the hypothalamic–pituitary–adrenal (HPA) axis (Chrousos and Gold, 1992). Early studies by Hans Selye (1936) showed that the adrenal medulla and cortex provided hormonal support that has complementary effects essential for survival following stress. The adrenal medulla is viewed as a response element of the sympathetic nervous system, whereas the adrenal cortex is thought to be regulated exclusively by pituitary-derived adrenocorticotropic hormone (ACTH). Subsequent work has shown clearly that the adrenal medulla and cortex are intertwined anatomically and functionally (Ehrhart-Bornstein et al., 1998) and has provided support for the concept that adrenal innervation contributes to the function of the adrenal cortex (reviewed in Engeland, 1998). These studies have extended our understanding of sympatho-adrenal control to include not only regulation of adrenal medullary secretion of catecholamines, but also regulation of cortical function by adrenal innervation. There is substantial experimental support for the hypothesis that adrenal cortical secretion of steroids is regulated both by the sympatho-adrenal and the hypothalamic–pituitary systems.

*Corresponding author. Mayo Mail Code 120, 420 Delaware St. SE University of Minnesota, Minneapolis, MN 55455, USA. Tel: +1(612) 625-4976; Fax: +1(612) 624-8909; E-mail: engel002@tc.umn.edu.

Dissociation between ACTH and glucocorticoids

The HPA axis is activated when afferent information specific to a particular stressor impinges upon hypophysiotropic neurons in the medial parvocellular division of the paraventricular nucleus of the hypothalamus. The axons from the paraventricular nucleus neurons extend into the median eminence at the base of the pituitary stalk. Once activated, the neurons secrete releasing hormones, including corticotropin-releasing factor (CRF) and vasopressin, into the portal blood system of the median eminence. The CRF and vasopressin are transported via the portal blood to the anterior pituitary, where the pituitary corticotrophs are stimulated to release ACTH into the systemic circulation. Circulating ACTH then stimulates the zona fasciculata of the adrenal cortex to synthesize and secrete glucocorticoids (i.e. cortisol in humans and corticosterone in rats) into the circulation.

Experimental observations

If adrenal cortical secretion is controlled exclusively by plasma ACTH, then the changes in plasma glucocorticoids should occur in parallel with changes in plasma ACTH. However, there are several reports of dissociation between changes in plasma ACTH and glucocorticoids, suggesting that factors in addition to ACTH could contribute to adrenal glucocorticoid production. Examples in experimental animals include dogs after small hemorrhage (Dempsher and Gann, 1983), cats after tooth pulp stimulation (Bereiter et al., 1982), mice and rats after colon inflammation (Franchimont et al., 2000; Kojima et al., 2002), and rats after water deprivation (Aguilera et al., 1993; Kiss et al., 1994). In humans, there have been reports of dissociation between plasma ACTH and cortisol after major abdominal surgery (Naito et al., 1991), experimentally induced ischemic pain (Gullner et al., 1982), and brain injury (Feibel et al., 1983). In these studies, stress-induced changes in plasma ACTH and glucocorticoids are temporally uncoupled and/or the changes in plasma ACTH and glucocorticoids differ markedly in magnitude.

The dissociation between plasma ACTH and glucocorticoids might result from changes in adrenal responsiveness to ACTH induced by stress. For example, chronic intermittent stimulation produced by immobilization results in augmented corticosterone responses compared to acute stimulation (Makino et al., 1995); the enhanced plasma corticosterone response occurs without an enhanced plasma ACTH response, suggesting that the adrenals are hyperresponsive to ACTH. Chronic stress results in hyperresponsive adrenals that have been attributed to adrenal hyperplasia induced by ACTH (Makino et al., 1995). Although stimulation by ACTH has been suggested to account for the adrenal response, chronic elevation in plasma ACTH has not been observed consistently (Chappell et al., 1986; Gomez et al., 1996). For example, plasma corticosterone is increased from 12 h to 48 h of water deprivation in rats, whereas plasma ACTH is elevated at 12 h, but returns to baseline by 24 h (Aguilera et al., 1993). Although it is possible that the dissociation between plasma ACTH and corticosterone may result in part from the inability to detect small but biologically relevant changes in plasma ACTH (Wood et al., 1982; Engeland et al., 1989), it is also possible that chronic stress induces changes in adrenocortical structure and function that are mediated in part by non-ACTH factors, such as the autonomic neural input to the adrenal.

Clinical observations

Abnormal functioning of the HPA system has been reported in a number of neurological disorders including major depression and Alzheimer's disease. In depression, plasma ACTH responses to stress and to CRF administration are blunted, whereas plasma cortisol responses are normal or augmented (Gold et al., 1986; Charlton and Ferrier, 1989). A similar pattern of responses have been reported for patients with Alzheimer's disease (O'Brien et al., 1994). The differential responsiveness of the pituitary and the adrenal in these disorders has led to the hypothesis that augmented plasma cortisol responses result from increases in adrenal responsiveness to ACTH (Amsterdam et al., 1983; Jaeckle et al., 1987; Nasman et al., 1996). Although it is likely that changes in adrenal responsivity are involved, the mechanism responsible for changing adrenal responsiveness to ACTH has not been defined. Interestingly, increased basal sympathetic neural activity has been associated with augmented plasma cortisol responses to stress in Alzheimer's patients (Pascualy et al., 2000). These findings are consistent with a possible role for the sympathetic nervous system to modulate adrenal glucocorticoid production via adrenal innervation.

Adrenal neural anatomy

The adrenal is composed of two glands: an inner medulla that is composed primarily of chromaffin cells and an outer cortex that is composed of steroidogenic tissue, all of which is surrounded by the adrenal capsule. Innervation of the adrenal medulla has been well-recognized, but more recently it has been shown that the adrenal cortex also receives abundant innervation by nerve fibers of both extrinsic and intrinsic origin (Fig. 1). The extrinsic innervation of the rat adrenal gland includes primary afferent, preganglionic sympathetic and postganglionic sympathetic innervation; there is little evidence for direct parasympathetic innervation of the rat adrenal gland (Coupland et al., 1989; Holgert et al., 1998). Adrenal primary afferent fibers are largely calcitonin gene-related peptide (CGRP) positive, with a subset colabeling for substance P (Kuramoto et al., 1987; Mohamed et al., 1988; Zhou et al., 1991). The adrenal afferents have cell bodies in the ipsilateral dorsal root ganglion from thoracic (T) segments T3–T13, and in the nodose ganglia (Kuramoto et al., 1987; Coupland et al., 1989; Zhou et al., 1991). Adrenal preganglionic sympathetic innervation is cholinergic, and thus contains the vesicular acetylcholine transporter (VAChT), with a subset of fibers colabeling for neuronal nitric oxide synthase (nNOS) (Holgert et al., 1995; Arvidsson et al., 1997). The adrenal preganglionic sympathetic innervation originates in the ipsilateral intermediolateral cell column of the spinal cord in segments T4–T13 (Schramm et al., 1975; Kesse et al., 1988; Strack et al., 1989). Lastly, the adrenal postganglionic sympathetic innervation is noradrenergic and thus contains tyrosine hydroxylase (TH); the adrenal postganglionic innervation

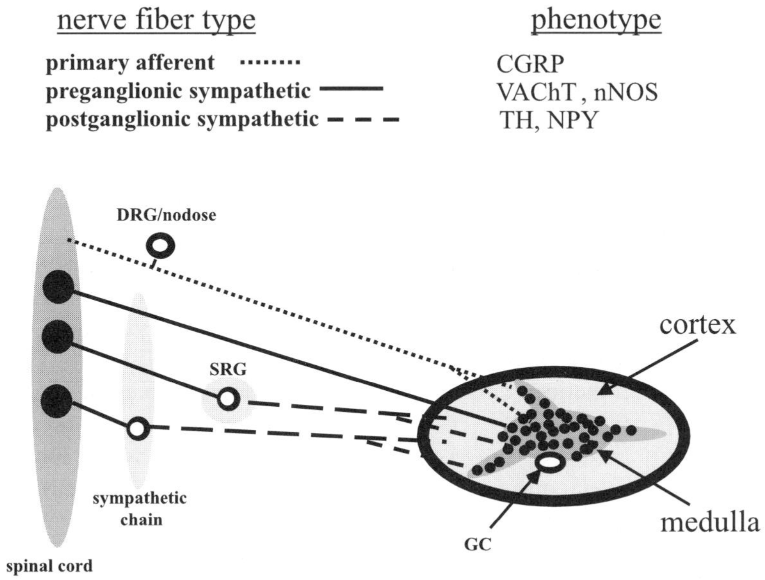

Fig. 1. Schematic of innervation of the rat adrenal gland. Extrinsic innervation includes primary afferent neurons, preganglionic sympathetic fibers, and postganglionic sympathetic neurons. The intrinsic innervation originates from ganglion cells located in the medulla. CGRP, calcitonin gene-related peptide; VAChT, vesicular acetylcholine transporter; nNOS, neuronal nitric oxide synthase, TH, tyrosine hydroxylase, NPY, neuropeptide Y; VIP, vasoactive intestinal polypeptide; DRG/nodose, dorsal root ganglia and nodose ganglion; SRG, suprarenal ganglion; GC, adrenal ganglion cell.

is also neuropeptide Y (NPY) positive (Kondo, 1985; Holgert et al., 1998). Adrenal postganglionic innervation is derived from cell bodies in the sympathetic chain (levels T5–T12) and the suprarenal ganglia (Kesse et al., 1988). Importantly, the thoracic splanchnic nerve constitutes a primary conduit for the extrinsic innervation to the adrenal (Holgert et al., 1998), in addition to the innervation that enters the gland along the blood vessels (Kleitman and Holzwarth, 1985). The intrinsic innervation originates from two types of medullary ganglion cells that have been observed in the rat adrenal: type I cells are noradrenergic and NPY positive, whereas type II cells produce nNOS and vasoactive intestinal peptide (VIP) (Holgert et al., 1998).

In addition to direct innervation of cortical cells, cortical function could be affected indirectly by catecholamines or neuropeptides secreted by the adrenal medulla. Chromaffin cells not only secrete hormones into the systemic circulation, but also may secrete locally to influence cortical cells in a paracrine fashion. This latter possibility had received little support until recently, since adrenal blood flow is directed centripetally from the outer cortex through the medulla (Sparrow and Coupland, 1987), limiting exposure of cortical cells to medullary factors. However, recent work has shown that cortical cells are interspersed within the adrenal medulla and clusters of chromaffin cells lie adjacent to cortical cells in the outer cortex in rat (Gallo-Payet et al., 1987) and human (Bornstein et al., 1994) adrenals. These studies show that cortical-medullary cell contacts can occur throughout the adrenal. Since secretion by chromaffin cells is regulated predominantly by sympathetic preganglionic innervation (Holgert et al., 1998), sympathetic innervation of the adrenal could regulate secretion of medullary factors that act locally to influence cortical cell function (Ehrhart-Bornstein et al., 1998).

Experimental evidence for neural control

In vitro studies: cells and tissue fragments

The evidence supporting neural modulation of adrenocortical function includes a large body of in vitro work demonstrating that the application of neurotransmitters to adrenal cortical tissue alters steroidogenesis. Most neurotransmitters identified in adrenal nerve fibers or chromaffin cells have been found to have some effect in vitro. However, the response to specific neurotransmitters has been variable, depending in part on the species studied. Since a comprehensive review of this subject has recently been published (Ehrhart-Bornstein et al., 1998), only a few examples of these experiments will be given here (summarized in Table 1). Cultures of bovine zona fasciculata cells release cortisol in response to catecholamines, an effect blocked by ß-adrenergic antagonists (Walker et al., 1988, 1991). In contrast, corticosterone responses from rat zona fasciculata cells are not affected by adrenergic agonists, despite detection of adrenergic receptor binding (Shima et al., 1984). Acetylcholine stimulates cortisol secretion from bovine zona fasciculata cells (Walker et al., 1991) and from frog adrenal tissue (Benyamina et al., 1987), and these responses are blocked by muscarinic receptor antagonists. There are no studies reporting cholinergic effects on corticosteroid release from rat adrenocortical cells. There are numerous studies evaluating neuropeptide effects; discussion will be limited to VIP and NPY. In rat adrenals, VIP receptor binding is localized to the outer adrenal cortex (Cunningham and Holzwarth, 1989), and VIP stimulates corticosterone and aldosterone release in perifused rat capsular/glomerulosa tissue (Cunningham and Holzwarth, 1988). In rat fasciculata cells, VIP is ineffective in stimulating corticosterone (Hinson et al., 1992). In contrast, human adrenal fasciculata cells release cortisol in response to VIP (Bornstein et al., 1996). Interestingly, the effect of VIP on steroidogenesis may be indirect, since the response in human (Bornstein et al., 1996) and rat (Hinson et al., 1992) cortical cells is reduced by ß-adrenergic receptor antagonists. These results have been interpreted indicating that the effects of VIP are mediated by the local release of catecholamines from chromaffin cells present in the cortical cell preparations. Using similar in vitro approaches, evidence has been presented for modulation of corticosteroid secretion by NPY (Renshaw and Hinson, 2001). In vitro approaches have been informative in defining cellular mechanisms, but they preclude investigation of a neurotransmitter's role in a physiological setting.

Table 1. Neurotransmitters in the adrenal gland that affect steroidogenesis in the inner zona fasciculata as reflected by changes in glucocorticoid secretion

Transmitter	Species/Preparation	Effect	Reference
Epinephrine	Bovine/ZF cells	Increase	Walker et al. (1988)
	Porcine/perfused	Increase	Bornstein et al. (1990)
	Bovine/ZF cells	Increase	Walker et al. (1991)
	Guinea pig/perfused	Increase	Mokuda et al. (1992)
Noradrenaline	Porcine/perfused	Increase	Bornstein et al. (1990)
	Porcine/perfused	Increase	Holst et al. (1991)
Acetylcholine	Frog/ZF cells	Increase	Benyamina et al. (1991)
	Bovine/ZF cells	Increase	Walker et al. (1991)
VIP	Frog/ZF cells	Increase	Leboulenger et al. (1984)
	Porcine/perfused	Increase	Ehrhart-Bornstein et al. (1991)
	Rat/perfused	Increase	Hinson et al. (1994)
	Human/ZF cells	Increase	Bornstein et al. (1996)
NPY	Rat/ZF cells	Decrease	Malendowicz et al. (1990)
	Rat/perfused	Increase	Renshaw and Hinson (2001)

The effect of neurotransmitters using in situ perfused adrenals may result in part from indirect vasoactive effects that alter perfusion flow rate. zona fasciculata, ZF; vasoactive intestinal peptide, VIP; neuropeptide Y, NPY.

In situ perfused adrenals

Since isolation of adrenal cells disrupts adrenal zonation, studies have used in situ perfused adrenals in order to assess neurotransmitter effects in a preparation that maintains in vivo anatomical relationships; media is pumped through the vasculature of the adrenal gland while collecting the effluent for assay of steroid levels. Using perfused porcine adrenals, cortisol secretion is increased in a dose-related fashion by noradrenaline, epinephrine, VIP and galanin (Bornstein et al., 1990; Ehrhart-Bornstein et al., 1991; Holst et al., 1991); perfused guinea pig adrenals also secrete cortisol in response to epinephrine (Mokuda et al., 1992). Using perfused rat adrenals, application of VIP, substance P, NPY, and neurotensin increased corticosterone production (Hinson et al., 1994). Since each of the neuropeptides examined is vasoactive, it is unclear whether the neuropeptides stimulated corticosterone release directly or affected release indirectly by changing perfusion flow rate. For most peptides, changes in perfusion flow rate and corticosterone release were well correlated. For example, VIP increased perfusion flow rate and corticosterone release, whereas NPY reduced flow rate and had minimal effects on corticosterone release (Hinson et al., 1994). These results support the possibility that adrenal neurotransmitters produce effects through direct activation of receptors on adrenocortical parenchymal cells and through changing adrenal vascular dynamics (see section on Adrenal Blood Flow).

In addition to measuring corticosteroid responses to administration of putative adrenal neurotransmitters, perfused adrenal preparations have been useful in assessing responses to peripheral nerve stimulation. If adrenal innervation contributes to the control of adrenal steroid secretion, electrical stimulation of adrenal nerves should influence corticosteroid secretion. This approach permits activation of adrenal neural input to assess whether a more physiological administration of adrenal neurotransmitters is effective. Splanchnic nerve stimulation of perfused pig adrenals results in increases in catecholamines, VIP, galanin, and cortisol secretion (Bornstein et al., 1990; Ehrhart-Bornstein et al., 1991; Holst et al., 1991); infusion of catecholamines, galanin, or VIP stimulate cortisol secretion suggesting that each could contribute to neurally-induced changes in cortical function.

In vivo studies: splanchnic nerve stimulation

In vivo work studying neural modulation of adrenocortical function has been performed in anesthetized and/or hypophysectomized animals. For example, stimulation of the splanchnic nerve in anesthetized calves (Edwards and Jones, 1987a) and in hypophysectomized dogs replaced by constant infusion of ACTH (Engeland and Gann, 1989) increases cortisol secretion. The response occurs by modulating adrenal responsiveness to ACTH. These studies have established that neural modulation of adrenocortical steroidogenesis is possible. However, the physiological relevance of neural control is not clear, since these studies involve hypophysectomy, a procedure that profoundly disrupts many neuroendocrine systems, and/or the acute stress of anesthesia and surgical preparation.

In vivo studies: splanchnic nerve section

In order to assess the effects of adrenal innervation in awake behaving animals, responses of the HPA axis have been studied in animals after transection of the thoracic splanchnic nerve. Cutting the thoracic splanchnic nerve affects cholinergic and peptidergic innervation, but not the noradrenergic innervation of the gland (Kesse et al., 1988). For example, immunolabeling of the adrenal innervation showed that splanchnic nerve transection removed the CGRP-positive primary afferent nerve fibers, and the VAChT-positive preganglionic sympathetic innervation (Figs. 2 and 3). In contrast, splanchnic nerve transection did not affect immunolabeling for the NPY- and TH-positive postganglionic sympathetic fibers and Type I ganglion cells, and the VIP- and nNOS-positive Type II ganglion cells (Ulrich-Lai and Engeland, 2002). These findings corroborate reports that the thoracic splanchnic nerve provides afferent and preganglionic sympathetic innervation to the adrenal, whereas postganglionic sympathetic innervation enters the adrenal primarily along the vasculature (Kleitman and Holzwarth, 1985; Holgert et al., 1998; Ulrich-Lai and Engeland, 2000). After recovery from the surgical stress induced by splanchnic denervation, animals have been studied under nonstress and stress conditions.

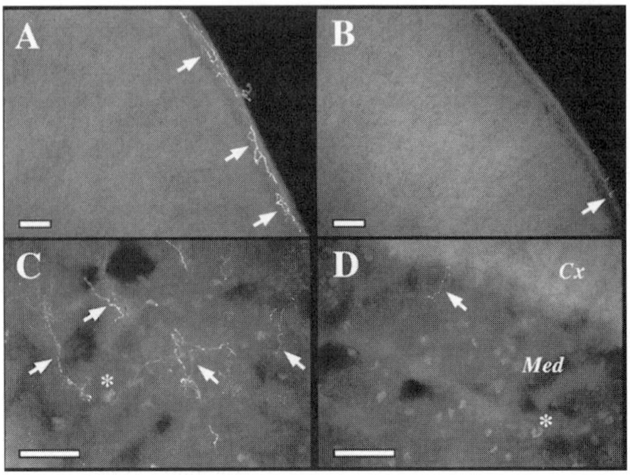

Fig. 2. Splanchnic nerve transection reduced the CGRP-positive innervation of the adrenal gland. Sham nerve-cut adrenals had several CGRP-positive fibers (*arrows*) in the capsule and outer cortex (A) and medulla (C). In contrast, CGRP-positive fibers (*arrows*) were rarely observed in the capsule (B) and medulla (D) of splanchnic nerve-cut adrenals. Also note the presence of CGRP-positive chromaffin cells (*) in the medulla of both the sham nerve-cut (C) and nerve-cut (D) adrenals. Cx = cortex, Med = medulla. Bar = 100 μm. (Ulrich-Lai and Engeland, 2002).

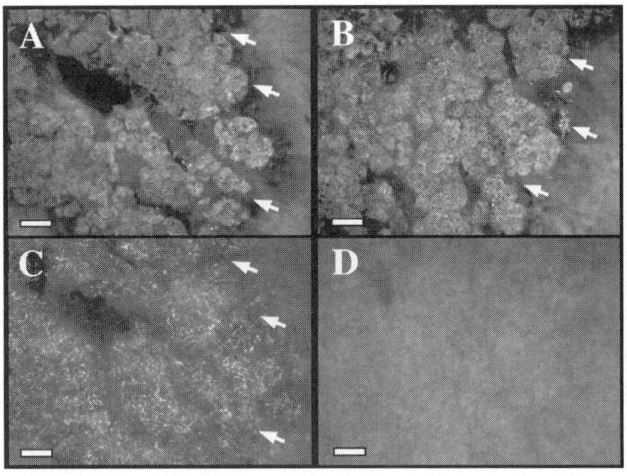

Fig. 3. Splanchnic nerve transection reduced the VAChT-positive innervation of the adrenal medulla. The medulla of sham nerve-cut adrenals (A) and nerve-cut adrenals (B) was largely composed of TH-positive chromaffin cells (*arrows*). Double labeling showed that the medulla of sham nerve-cut adrenals was innervated by VAChT-positive nerve fibers (C, *arrows*), whereas the medulla of nerve-cut adrenals was not innervated by VAChT-positive nerve fibers (D). Bar = 50 μm (Ulrich-Lai and Engeland, 2002).

Control of nonstress corticosteroids

There is a circadian rhythm in HPA axis activity. Nonstress levels of plasma ACTH and corticosterone vary throughout the day, with a nadir at the onset of the inactive period (i.e. in the morning (AM) for nocturnal animals like the rat) and a peak at the onset of the active period (i.e. in the late afternoon (PM)). There is a high-amplitude rhythm in plasma corticosterone, typically of 5- to 10-fold from the trough (AM) to peak (PM) levels, whereas there is a low-amplitude rhythm in plasma ACTH (up to 2-fold) that frequently is not significant throughout the day (Dallman et al., 1978; Kaneko et al., 1980; Wilkinson et al., 1981; Akana et al., 1986; Kalsbeek et al., 1996). The robust rhythm in plasma corticosterone despite a modest rhythm in plasma ACTH is largely due to a coincident rhythm in adrenal sensitivity to ACTH (Dallman et al., 1978; Kaneko et al., 1980, 1981).

The circadian rhythm in plasma corticosterone persists in hypophysectomized rats that have been implanted with ACTH pellets, suggesting that the corticosterone rhythm does not depend on the rhythmic release of ACTH (Meier, 1976). In addition, dexamethasone eliminates the plasma ACTH rhythm without affecting the adrenal responsiveness rhythm (Dallman et al., 1978), suggesting that the diurnal rhythm in adrenal responsiveness occurs independent of changes in plasma ACTH. Spinal cord transection at T-7, a cut that should disrupt adrenal sympathetic innervation, abolishes the plasma corticosterone rhythm (Ottenweller and Meier, 1982). There is evidence demonstrating that transection of the splanchnic nerve in rats disrupts the circadian rhythm in nonstress plasma corticosterone in rats (Jasper and Engeland, 1994; Dijkstra et al., 1996); however, there is disagreement as to whether this effect is via increasing AM plasma corticosterone levels (Jasper and Engeland, 1994) or via decreasing PM plasma corticosterone levels (Dijkstra et al., 1996). Moreover, it is not clear whether the disruption in the circadian rhythm results from alterations in adrenal sensitivity to ACTH (Dijkstra et al., 1996; Jasper and Engeland, 1997). We have recently reexamined these questions and showed that splanchnic nerve transection reduces the PM rise in plasma corticosterone, without affecting plasma ACTH (Ulrich-Lai and Engeland, 2001). Moreover, the splanchnic nerve-mediated effects were associated with decreased adrenal cAMP content and decreased adrenal sensitivity to ACTH in dexamethasone-blocked rats treated with exogenous ACTH. The finding that splanchnic nerve section did not affect plasma corticosterone in the AM is contradictory to our previous studies showing that splanchnicectomy increased pulsatile release of corticosterone in the AM (Jasper and Engeland, 1994). In this work, corticosterone pulsatility was examined using adrenal microdialysis that enabled high-frequency sampling of adrenal extracellular fluid (Jasper and Engeland, 1994); based on microdialysis sampling, it was hypothesized that splanchnic innervation provided inhibitory input to the adrenal during the nadir of the circadian rhythm. However, since coincident blood and adrenal dialysate sampling was not done in these experiments, it is unclear whether the presence of the adrenal probe altered corticosterone secretion and affected the response to splanchnicectomy. Additional experiments will be needed to resolve these differences.

Control of stress-induced corticosteroids

The contribution of adrenal innervation to the control of steroidogenesis under conditions of stress has been reported in a variety of species. Bilateral transection of the splanchnic nerve decreases the glucocorticoid response to noradrenaline injection in conscious lambs and decreases adrenal sensitivity to ACTH in conscious calves (Edwards et al., 1986; Edwards and Jones, 1987b). In the ovine fetus, splanchnicectomy reduced plasma cortisol responses to hypotension without changing plasma ACTH responses (Myers et al., 1990). In neonatal rats, maternal separation-induced increases in adrenal sensitivity to ACTH are suppressed by chemical sympathectomy supporting a facilitatory role for sympathetic activity in mediating the response (Walker, 1995).

In adult rats, splanchnic innervation has been implicated in controlling plasma corticosterone responses to dehydration stress (Brooks et al., 1997). The possibility that adrenal innervation contributed to this response was suggested by work demonstrating a dissociation between plasma ACTH and corticosterone following dehydration stress in rats (Aguilera et al., 1993; Kiss et al., 1994). Subsequent experiments (Ulrich-Lai and Engeland, 2002) have shown that splanchnic nerve transection attenuates the plasma corticosterone and the adrenal corticosterone response to dehydration stress (Fig. 4). In addition, reduced corticosterone responses to dehydration stress are associated with decreases in adrenal sensitivity to ACTH, assessed by measuring corticosterone responses to exogenous ACTH after dexamethasone treatment (Fig. 5).

Transection of the thoracic splanchnic nerve removes the VAChT-positive fibers and the CGRP-positive fibers, presumably reflecting the loss of preganglionic sympathetic nerves and of primary afferent nerve fibers, respectively. Both of these fiber types could mediate the alterations in adrenocortical function that occur after splanchnicectomy. The preganglionic sympathetic fibers predominantly innervate the adrenal medulla (Holgert et al., 1998). Adrenal medullectomy or enucleation (i.e. surgical removal of the adrenal medulla) was used to determine whether the splanchnic nerve-mediated effects are dependent on the preganglionic sympathetic innervation and the adrenal medulla. This approach has been used previously to assess the influence of the medulla on cortical function (Wilkinson et al., 1981). Adrenal demedullation prevented the effect of splanchnic nerve transection on the plasma corticosterone and adrenal cAMP response to dehydration stress (Fig. 6), suggesting that the adrenal medulla is required for splanchnic nerve-mediated facilitation of the plasma corticosterone response. However, a limitation of medullectomy is that the removal of the medulla occurs at the expense of inner cortical cell loss. Although the cortex regenerates, there is a marked difference in the pattern and density of cortical innervation; for example, the regenerating adrenal gland becomes hyperinnervated by nerve fibers positive for CGRP, NPY, and TH (Ulrich-Lai and Engeland, 2000). It is difficult to assess whether changes in cortical function after medullectomy result from loss of the medulla or from differences in cortical mass, innervation or function.

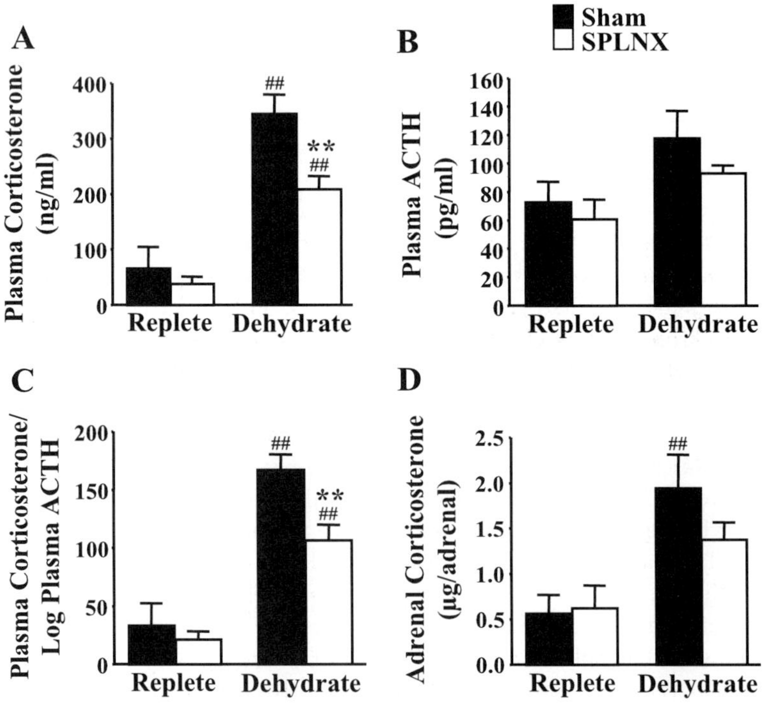

Fig. 4. Splanchnic nerve transection reduced the plasma and adrenal corticosterone response to dehydration stress. Plasma hormone concentrations and adrenal corticosterone content were assessed after 48 h of dehydration (*Dehydrate*) or ad lib water (*Replete*) in rats with transected (*open bars*) or sham-transected (*solid bars*) splanchnic nerves. (A) plasma corticosterone, (B) plasma ACTH, (C) plasma corticosterone/log of plasma ACTH, and (D) adrenal corticosterone content. Data are shown as mean ± SEM. $n = 4$–6/group. ##$p < 0.01$ versus replete. **$p < 0.01$ versus sham nerve cut (Ulrich-Lai and Engeland, 2002).

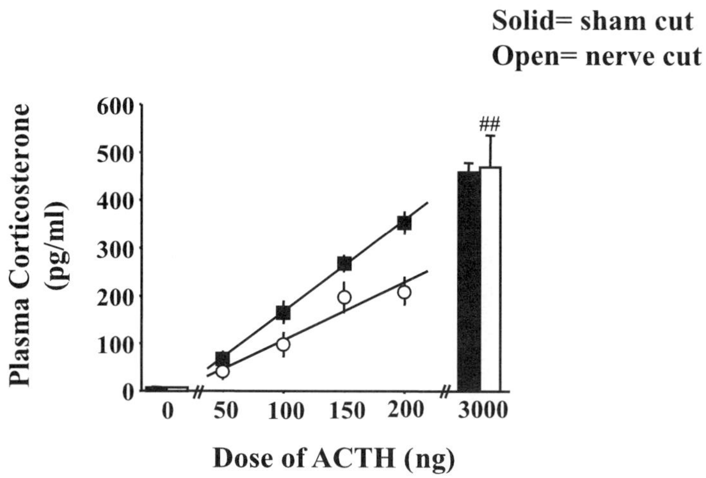

Fig. 5. Splanchnic nerve transection reduced postdehydration adrenal sensitivity to submaximal doses of ACTH. Plasma corticosterone responses to exogenous ACTH (0–3000 ng; s.c.) were assessed in dexamethasone-blocked, 48 h dehydrated rats with transected (*open symbols*) or sham-transected (*solid symbols*) splanchnic nerves. The slope of the corticosterone response in sham-transected rats was 1.78 ± 0.10, whereas that in transected rats was 1.13 ± 0.15 ($p = 0.013$). Data are shown as mean $\pm$ SEM. $n = 4$–6/group. ##$p < 0.01$ versus 200 ng dose (Ulrich-Lai and Engeland, 2002).

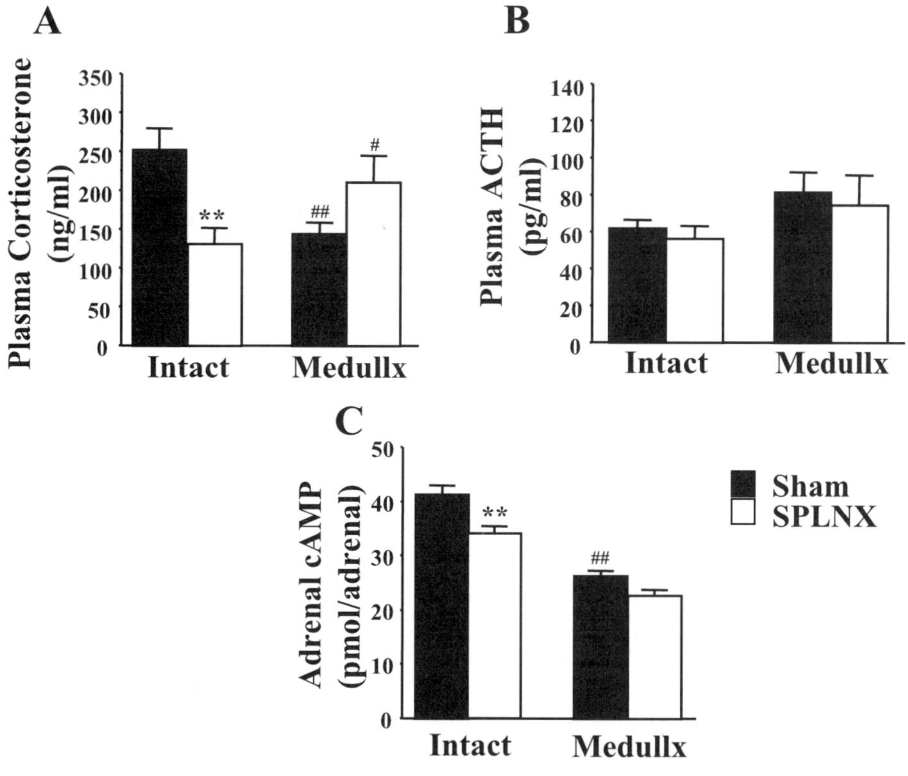

Fig. 6. Adrenal demedullation prevented the splanchnic nerve-mediated effect on postdehydration plasma corticosterone and adrenal cAMP. (A) Plasma corticosterone, (B) plasma ACTH, and (C) adrenal cAMP after 48 h dehydration stress in rats with intact adrenals (*Intact*) or demedullated adrenals (*Medullx*). $n = 10$–11/group. **$p < 0.01$ vs. sham nerve cut, #$p < 0.05$, ##$p < 0.01$ vs. intact.

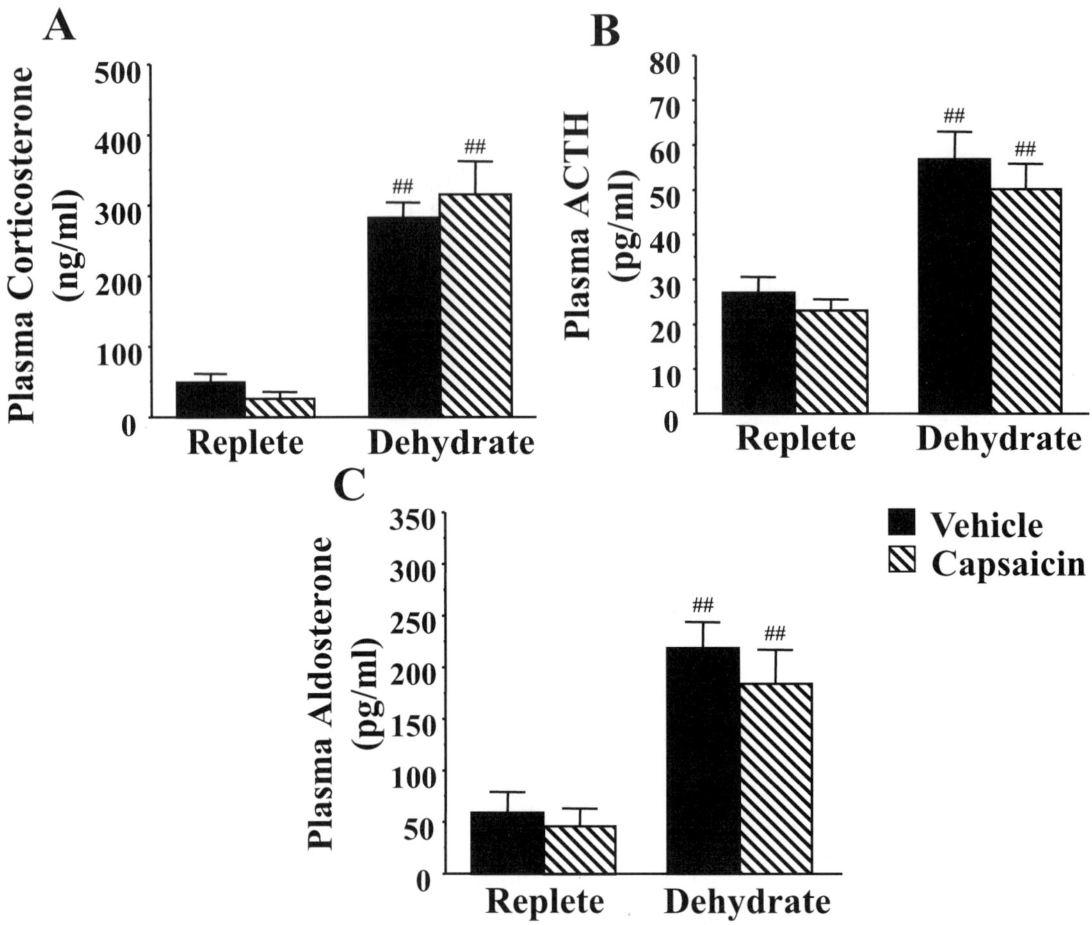

Fig. 7. (A) Plasma corticosterone , (B) ACTH, and (C) aldosterone were increased by 48 h dehydration stress (*Dehydrate*) relative to water replete controls (*Replete*), and were not affected by capsaicin treatment of the splanchnic nerve. $n = 8$–12/group. ##$p < 0.01$ vs. replete.

To address a possible effect of adrenal primary afferent fibers in dehydration-induced increases in plasma corticosterone, rats were treated with capsaicin, a neurotoxin selective for a subset of afferent nerves (Caterina et al., 1997; Holzer, 1998). Capsaicin applied locally to the splanchnic nerve removes a subset of primary afferent adrenal nerves and inhibits compensatory adrenal growth (Ulrich-Lai et al., 2002). Topical treatment of the splanchnic nerve with capsaicin did not affect plasma ACTH or corticosterone responses to dehydration stress (Fig. 7). These results suggest that adrenal capsaicin-sensitive primary afferents are not involved in mediating the splanchnic nerve effect on plasma corticosterone responses to dehydration stress. It is possible that a more complete removal of primary afferents would be effective, since capsaicin does not produce a complete deafferentation of the adrenal gland (Ulrich-Lai et al., 2002).

Potential adrenal mechanisms for splanchnic neural effects

Neurotransmitter-induced change in steroidogenesis

Neurotransmitters binding to receptors on adrenal cortical cells could directly alter glucocorticoid

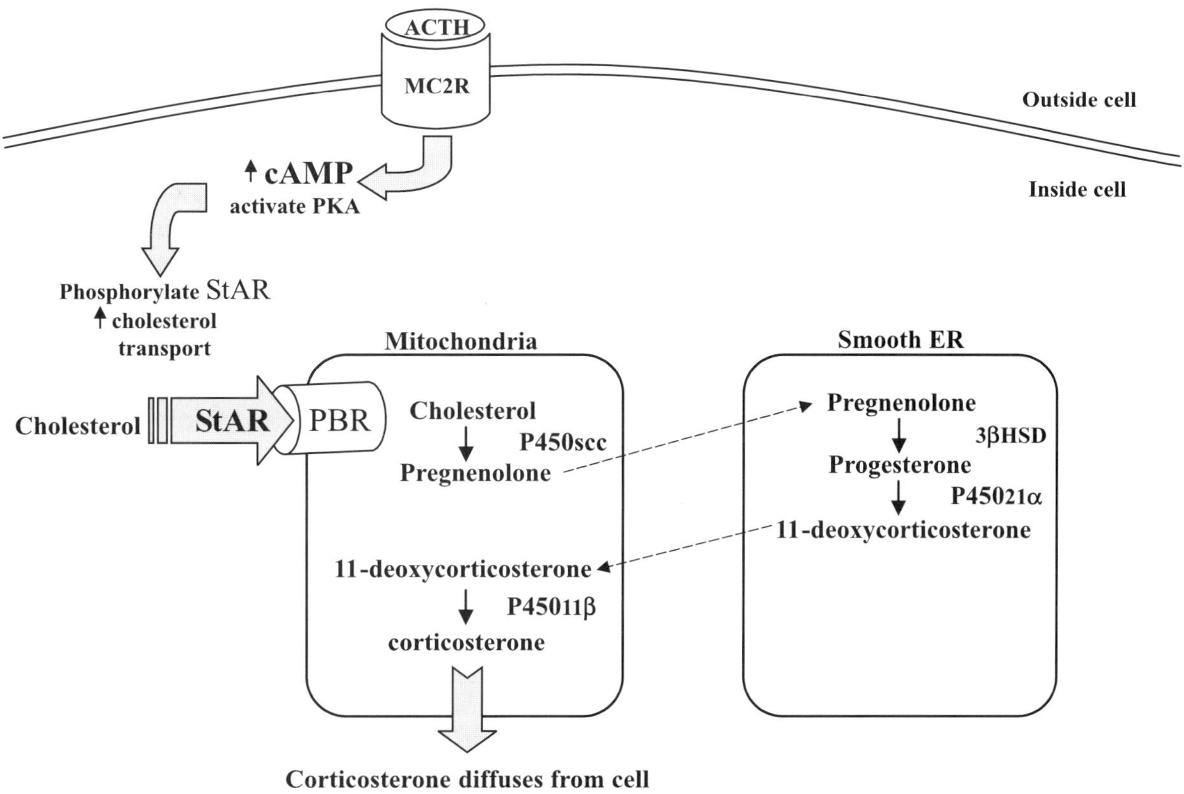

Fig. 8. Schematic of signal transduction pathways in the rat adrenal fasciculata cell. ACTH, adrenocorticotropin; MC2R, melanocortin-2 receptor; cAMP, cyclic AMP; PKA, protein kinase A; StAR, steroidogenic acute regulatory protein; PBR, peripheral benzodiazepine receptor; P450scc, cytochrome P450 side chain cleavage enzyme; 3ßHSD, 3ßhydroxysteroid dehydrogenase; P45021α, cytochrome P450 21 α-hydroxylase; P45011ß, cytochrome P450 11ß-hydroxylase; ER, endoplasmic reticulum.

secretion at different steps in the steroidogenic pathway. Glucocorticoid steroidogenesis is dependent on activation of the zona fasciculata by ACTH (Fig. 8). ACTH binds to the melanocortin 2 receptor (MC2R), a G-protein-coupled receptor that activates adenylyl cyclase, thereby increasing intracellular cyclic 3′5′-adenosine monophosphate (cAMP) levels (Mountjoy et al., 1992). Increases in intracellular cAMP stimulate protein kinase A (PKA) which stimulates the rapid mobilization of intracellular stores of cholesterol to the outer mitochondrial membrane and facilitates cholesterol transport to the inner mitochondrial membrane (Hall, 2001). The delivery of cholesterol to the inner mitochondrial membrane, where it is converted to pregnenolone by the cytochrome P450 side chain cleavage complex (P450scc), is the rate-limiting step in steroidogenesis. At least two molecules are involved in cholesterol transport, steroidogenic acute regulatory protein (StAR), and peripheral benzodiazepine receptor (PBR) (reviewed in Hall, 2001). StAR is a 30 kDa protein that is phosphorylated by PKA and transports cholesterol from the outer to inner mitochondrial membranes (Stucco and Clark, 1996; Arakane et al., 1997). StAR is rapidly turned over and stimulation with ACTH increases the amount of StAR protein (Ariyoshi et al., 1998; Lehoux et al., 1999). Increases in StAR are associated with chronic elevations of plasma corticosterone (Ariyoshi et al., 1998; Zilz et al., 1999), and defects in the StAR gene cause congenital lipoid adrenal hyperplasia in humans, a disorder characterized by a lack of adrenal and gonadal steroid hormones (Bose et al., 2000). PBR is a 18 kDa protein that also facilitates the transport of cholesterol to the inner mitochondrial membrane

(Papadopoulos and Brown, 1995), perhaps by creating contact points between the inner and outer mitochondrial membranes or by forming a pore between the membranes (Papadopoulos and Brown, 1995). It is not clear whether StAR and PBR interact, or act independently, to facilitate mitochondrial transport of cholesterol. Following the conversion of cholesterol to pregnenolone by P450scc, a series of other steroidogenic enzymes in fasciculata cells catalyze subsequent steps that result in the synthesis of cortisol (humans) or corticosterone (rodents) (Hall, 2001).

Neurotransmitters could alter steroidogenesis through receptor coupling to adenylyl cyclase and changing intracellular cAMP levels. For example, activation of ß-adrenergic receptors in bovine fasciculata cells (Guse-Behling et al., 1992) or VIP receptors in adrenocortical tumor cells (Birnbaum et al., 1980) increases adrenal cAMP. This possibility is supported by experiments demonstrating that splanchnicectomy decreases adrenal cAMP responses to dehydration stress (Ulrich-Lai and Engeland, 2002) and reduces adrenal cAMP observed at the peak of the diurnal rhythm in rats (Ulrich-Lai and Engeland, 2001). These results suggest that changes in cAMP may be a common mechanism for splanchnic neural effects, and that the effects of adrenal splanchnicectomy occur in part at a step proximal to the generation of cAMP. However, it is also possible that some neurotransmitters could have additional actions distal to the generation of cAMP. To address this issue, experiments have been done to determine whether splanchnic denervation affects the amount of adrenal StAR or PBR protein. Although adrenal StAR increases in the PM in nonstressed rats and in response to dehydration stress, splanchnicectomy had no effect on adrenal StAR or PBR protein levels for either response (Ulrich-Lai and Engeland, 2001, 2002). Moreover, some neurotransmitters may affect steroidogenic enzyme levels; in bovine adrenocortical cells, epinephrine increases the level of mRNA expression for multiple steroidogenic enzymes including P450scc, P45017α-hydroxylase, P45011ß-hydroxylase, and P450 21 α-hydroxylase (Guse-Behling et al., 1992). Thus, splanchnic neurotransmitters could affect adrenal responsiveness by changing steroidogenic enzyme protein and activity.

Adrenal blood flow

Vasoactive neurotransmitters are present in nerve fibers in the adrenal capsule and subcapsular region; this innervation likely acts to change arteriolar diameter and alter blood flow within the sinusoids traversing the cortex. Changes in cortical blood flow would alter corticosterone secretion by changing the presentation rate of ACTH (Urquhart, 1965; L'Age et al., 1970). There is evidence that noradrenergic innervation of the rat adrenal provided by postganglionic nerves is involved in the regulation of adrenocortical blood flow (Carlsson et al., 1993). However, it is unclear which neurotransmitters contribute to physiological changes in cortical blood flow and under which physiological conditions blood flow is changed. Studies have shown that hemorrhage in dogs increases adrenal medullary flow without changing cortical blood flow (Breslow et al., 1986) and this response is neurally mediated (Breslow et al., 1987). Attempts have been made to monitor changes in intracortical blood flow using fluorescent microspheres (Jasper et al., 1990). However, the ability to monitor differences in flow between the outer capsule/glomerulosa and the inner fasciculata using microspheres was prevented by the close anatomical arrangement of blood vessels in this region. Moreover, the regulation of adrenal blood flow has been assessed primarily in anesthetized acutely prepared animals. Since anesthesia affects adrenal blood flow (Faraci et al., 1989), studies in awake animals are required to determine how adrenal innervation affects cortical and medullary blood flow under different physiological conditions.

Conclusion

This chapter evaluates the hypothesis that adrenal innervation represents a physiologically relevant, extra-ACTH mechanism for regulating adrenocortical function. Anatomical studies demonstrate that the adrenal cortex receives neural input either through direct innervation or through interaction with medullary chromaffin cells. In vitro data show that neurotransmitters can affect glucocorticoid secretion. In addition, surgical denervation of the adrenal alters steroidogenic responses under nonstress and stress conditions. The adrenal cellular mechanism for this effect has not been completely delineated. Additional studies are also required to determine the range of physiological conditions under which adrenal innervation plays a role. However, the literature surveyed demonstrates that ACTH is not the exclusive physiological regulator of the adrenal cortex and that adrenal innervation also plays a significant role.

Abbreviations

ACTH	adrenocorticotropin
HPA	hypothalamic–pituitary–adrenal
CRF	corticotropin-releasing factor
CGRP	calcitonin gene-related peptide
T	thoracic
VAChT	vesicular acetylcholine transporter
nNOS	neuronal nitric oxide synthase
TH	tyrosine hydroxylase
NPY	neuropeptide Y
VIP	vasoactive intestinal polypeptide
MC2R	melanocortin-2 receptor
cAMP	cyclic 3′5′-adenosine monophosphate
PKA	protein kinase A
P450scc	cytochrome P450 side chain cleavage complex
StAR	steroidogenic acute regulatory protein
PBR	peripheral benzodiazepine receptor
3ßHSD	3ß Hydroxysteroid dehydrogenase
P45021α	cytochrome P45021 α-hydroxylase
P45011ß	cytochrome P45011ß-hydroxylase
ER	endoplasmic reticulum

Acknowledgments

The authors wish to thank Michelle Bland, Ada Fraticelli, Debra Hebert, Carolyn Morris, and Cheryl Wotus for their technical assistance and Dr. Brett Levay-Young for help with methods development. This work was supported by NSF grants IBN-9728132 and IBN-0112543, a Howard Hughes Medical Institute Predoctoral Fellowship (YMU), and a University of Minnesota Graduate School Doctoral Dissertation Fellowship (YMU).

References

Aguilera, G., Lightman, S.L. and Kiss, A. (1993) Regulation of hypothalamic–pituitary–adrenal axis during water deprivation. Endocrinology, 132: 241–248.

Akana, S.F., Cascio, C.S., Du, J.Z., Levin, N. and Dallman, M. F. (1986) Reset of feedback in the adrenocortical system: an apparent shift in sensitivity of adrenocorticotropin to inhibition by corticosterone between morning and evening. Endocrinology, 119: 2325–2332.

Amsterdam, J.D., Winkour, A., Abelman, E., Lucki, I. and Rickels, K. (1983) Cosyntropin (ACTH 1–24) stimulation test in depressed patients and healthy subjects. Am. J. Psychiatry, 140: 907–909.

Arakane, F., King, S.R., Du, Y., Kallen, C.B., Walsh, L.P., Watari, H., Stocco, D.M. and Strauss, J.F. (1997) Phosphorylation of steroidogenic acute regulatory protein (StAR) modulates its steroidogenic activity. J. Biol. Chem., 272: 32656–32662.

Ariyoshi, N., Kim, Y.C., Artemenko, I., Bhattacharyya, K.K. and Jefcoate, C.R. (1998) Characterization of the rat StAR gene that encodes the predominant 3.5-kilobase pair mRNA. ACTH stimulation of adrenal steroids in vivo precedes elevation of StAR mRNA and protein. J. Biol. Chem., 273: 7610–7619.

Arvidsson, U., Riedl, M., Elde, R. and Meister, B. (1997) Vesicular acetylcholine transporter (VAChT) protein: a novel and unique marker for cholinergic neurons in the central and peripheral nervous systems. J. Comp. Neurol., 378: 454–467.

Benyamina, M., Leboulenger, F., Lirhmann, I., Delarue, C., Feuilloley, M. and Vaudry, H. (1987) Acetylcholine stimulates steroidogenesis in isolated frog adrenal gland through muscarinic receptors: evidence for a desensitization mechanism. J.Endocr., 113: 339–348.

Bereiter, D.A., Plotsky, P.M. and Gann, D.S. (1982) Tooth pulp stimulation potentiates the adrenocorticotropic response to hemorrhage in cats. Endocrinology, 111: 1127–1132.

Birnbaum, R.S., Alfonzo, M. and Kowal, J. (1980) Vasoactive intestinal peptide- and adrenocorticotropin-stimulated adenyl cyclase in cultured adrenal tumor cells: evidence for a specific vasoactive intestinal peptide receptor. Endocrinology, 106: 1270–1275.

Bornstein, S.R., Ehrhart-Bornstein, M., Scherbaum, W.A., Pfeiffer, E.F. and Holst, J.J. (1990) Effects of splanchnic nerve stimulation on the adrenal cortex may be mediated by chromaffin cells in a paracrine manner. Endocrinology, 127: 900–906.

Bornstein, S.R., Gonzales-Hernandez, J.A., Ehrhart-Bornstein, M., Adler, G. and Scherbaum, W.A. (1994) Intimate contact of chromaffin and cortical cells within the human adrenal gland forms the cellular basis for important intraadrenal interactions. J. Clin. Endocrinol. Metab., 78: 225–232.

Bornstein, S.R., Haidan, A. and Ehrhart-Bornstein, M. (1996) Cellular communication in the neuroadrenocortical axis: role of vasoactive polypeptide. Endocrine Res., 22: 819–829.

Bose, H.S., Sato, S., Aisenberg, J., Shalev, S.A., Matsuo, N. and Miller, W.L. (2000) Mutations in the steroidogenic acute regulatory protein (StAR) in six patients with congenital lipoid adrenal hyperplasia. J. Clin. Endocrinol. Metab,. 85: 3636–3639.

Breslow, M.J., Jordan, D.A., Thellman, S.T. and Traystman, R.J. (1987) Neural control of adrenal medullary and cortical blood flow during hemorrhage. Am. J. Physiol., 252: H521–H528.

Breslow, M.J., Mennen, A., Koehler, R.C. and Traystman, R.J. (1986) Adrenal medullary and cortical blood flow during hemorrhage. Am. J. Physiol. 250: H954–H960.

Brooks, V.L., Huhtala, T.A., Silliman, T.L. and Engeland, W.C. (1997) Differential effect of water deprivation on mRNAs encoding tyrosine hydroxylase and the norepinephrine transporter in rat adrenal. Am. J. Physiol (Reg. Int. Compar.) 272: R1897–R1903.

Canon, W. (1929) Organization for physiological homeostasis. Physiol. Rev., 9: 399–431.

Carlsson, S., Jonsdottir, I.H., Skarphedinsson, J.O. and Thoren, P. (1993) Evidence for an adrenergic innervation of the adrenal cortical blood vessels in rats. Acta Physiologica Scandinavica, 149: 23–30.

Caterina, M.J., Schumacher, M.A., Tominaga, M., Rosen, T.A., Levine, J.D. and Julius, D. (1997) The capsaicin receptor: a heat-activated ion channel in the pain pathway. Nature, 389: 816–824.

Chappell, P.B., Smith, M.A., Kilts, C.D., Bissette, G., Ritchie, J., Anderson, C. and Nemeroff, C.B. (1986) Alterations in corticotropin-releasing factor-like immunoreactivity in discrete rat brain regions after acute and chronic stress. J. Neuroscience, 6: 2908–2914.

Charlton, B.G. and Ferrier, I.N. (1989) Hypothalamo–pituitary–adrenal axis abnormalities in depression: a review and a model. Psych. Med., 19: 331–336.

Chrousos, G.P. and Gold, P.W. (1992) The concepts of stress and stress system disorders. Overview of physical and behavioral homeostasis. J. Amer. Med. Assoc., 267: 1244–1252.

Coupland, R.E., Parker, T.L., Kesse, W.K. and Mohamed, A.A. (1989) The innervation of the adrenal gland. III. Vagal innervation. J. Anatomy, 163: 173–181.

Cunningham, L.A. and Holzwarth, M.A. (1988) Vasoactive intestinal peptide stimulates adrenal aldosterone and corticosterone secretion. Endocrinology, 122: 2090–2097.

Cunningham, L.A. and Holzwarth, M.A. (1989) Autoradiographic distribution of 125 I-VIP binding in the rat adrenal cortex. Peptides, 10: 1105–1108.

Dallman, M.F., Engeland, W.C., Rose, J.C., Wilkinson, C.W., Shinsako, J. and Siedenburg, F. (1978) Nychthemeral rhythm in adrenal responsiveness to ACTH. Am. J. Physiol., 235: R210–R218.

Dempsher, D.P. and Gann, D.S. (1983) Increased cortisol secretion after small hemorrhage is not attributable to changes in adrenocorticotropin. Endocrinology, 113: 86–93.

Dijkstra, I., Binnekade, R. and Tilders, F.J.H. (1996) Diurnal variation in resting levels of corticosterone is not mediated by variation in adrenal responsiveness to ACTH but involves splanchnic nerve integrity. Endocrinology, 137: 540–547.

Edwards, A.V. and Jones, C.T. (1987a) The effect of splanchnic nerve stimulation on adrenocortical activity in conscious calves. J. Physiol., 382: 385–396.

Edwards, A.V. and Jones, C.T. (1987b) The effect of splanchnic nerve section on the sensitivity of the adrenal cortex to ACTH in the calf. J. Physiol., 390: 23–31.

Edwards, A.V., Jones, C.T. and Bloom, S.R. (1986) Reduced adrenal cortical sensitivity to ACTH in lambs with cut splanchnic nerves. J. Endocr., 110: 81–85.

Ehrhart-Bornstein, M., Bornstein, S.R., Scherbaum, W.A., Pfeiffer, E.F. and Holst, J.J. (1991) Role of vasoactive intestinal peptide in a neuroendocrine regulation of the adrenal cortex. Neuroendocrinology, 54: 623–628.

Ehrhart-Bornstein, M., Hinson, J.P., Bornstein, S.R., Scherbaum, W.A. and Vinson, G.P. (1998) Intraadrenal interactions in the regulation of adrenocortical steroidogenesis. Endocrine Rev., 19: 101–143.

Engeland, W.C. (1998) Functional innervation of the adrenal cortex by the splanchnic nerve. Horm. Metab. Res., 30: 311–314.

Engeland, W.C. and Gann, D.S. (1989) Splanchnic nerve stimulation modulates steroid secretion in hypophysectomized dogs. Neuroendocrinology, 50: 124–131.

Engeland, W.C., Miller, P. and Gann, D.S. (1989) Dissociation between changes in plasma bioactive and immunoreactive adrenocorticotropin after hemorrhage in awake dogs. Endocrinology, 124: 2978–2985.

Faraci, F.M., Chilian, W.M., Williams, J.K. and Heistad, D.D. (1989) Effects of reflex stimuli on blood flow to the adrenal medulla. Am. J. Physiol., 257: H590–H596.

Feibel, J., Kelly, M., Lee, L. and Woolf, P. (1983) Loss of adrenocortical suppression after acute brain injury: role of increased intracranial pressure and brain stem function. J. Clin. Endocrinol. Metab., 57: 1245–1250.

Franchimont, D., Bouma, G., Galon, J., Wolkersdorfer, G.W., Haidan, A., Chrousos, G.P. and Bornstein, S.R. (2000) Adrenal cortical activation in murine colitis. Gastroenterology, 119: 1560–1568.

Gallo-Payet, N., Pothier, P. and Isler, H. (1987) On the presence of chromaffin cells in the adrenal cortex: their possible role in adrenocortical function. Biochem. Cell Biol., 65: 588–592.

Gold, P.W., Loriaux, D.L., Roy, A., Kling, M.A., Calabrese, J.R., Kellner, C.H., Nieman, L.K., Post, R.M., Pickar, D., Gallucci, W., Avgerinos, P., Paul, S., Oldfield, E.H., Cutler,

G.B. and Chrousos, G.P. (1986) Responses to corticotropin-releasing hormone in the hypercortisolism of depression and Cushing's disease. New Eng. J. Med., 314: 1329–1334.

Gomez, F., Lahmame, A., Kloet, E.R.D. and Armario, A. (1996) Hypothalamic–pituitary–adrenal response to chronic stress in five inbred strains: differential responses are mainly located at the adrenocortical level. Neuroendocrinology, 63: 327–337.

Gullner, H.-G., Nicholson, W., Wilson, M., Bartter, F. and Orth, D. (1982) The response of plasma immunoreactive adrenocorticotropin beta-endorphin/beta-lipotropin, delta-lipotropin and cortisol to experimentally induced pain in normal subjects. Clinical Science, 63: 397–400.

Guse-Behling, H., Ehrhart-Bornstein, M., Bornstein, S.R., Waterman, M.R., Scherbaum, W.A. and Adler, G. (1992) Regulation of adrenal steroidogenesis by adrenaline expression of cytochrome P450 genes. J. Endocr., 135: 229–237.

Hall, P.F. (2001) Actions of corticotropin on the adrenal cortex: biochemistry and cell biology. In: McEwen. B.S. (Ed.), Handbook of Physiology. IV. Coping with the Environment: Neural and Endocrine Mechanisms. Oxford University Press, New York, pp. 61–101.

Hinson, J.P., Cameron, L.A., Purbrick, A. and Kapas, S. (1994) The role of neuropeptides in the regulation of adrenal vascular tone: effects of vasoactive intestinal polypeptide, substance P, neuropeptide Y, neurotensin, Met-enkephalin, and Leu-enkephalin on perfusion medium flow rate in the intact perfused rat adrenal. Regul. Pept., 51: 55–61.

Hinson, J.P., Kapas, S., Orford, C.D. and Vinson, G.P. (1992) Vasoactive intestinal peptide stimulation of aldosterone secretion by the rat adrenal cortex may be mediated by the local release of catecholamines. J. Endocr., 133: 253–258.

Hinson, J.P., Purbick, A., Cameron, L.A. and Kapas, S. (1994) The role of neuropeptides in the regulation of adrenal zona fasciculata/reticularis function. Effects of vasoactive intestinal peptide, substance P, neuropeptide Y, met- and leu-enkephalin and neurotensin on corticosterone secretion in the intact perfused rat adrenal gland in situ. Neuropeptides, 26: 391–397.

Holgert, H., Aman, K., Cozzari, C., Hartman, B. K., Brimijoin, S., Emson, P., Goldstein, M. and Hokfelt, T. (1995) The cholinergic innervation of the adrenal gland and its relation to enkephalin and nitric oxide synthase. Neuroreport, 6: 2576–2580.

Holgert, H., Dagerlind, A. and Hokfelt, T. (1998) Immunohistochemical characterization of the peptidergic innervation of the rat adrenal gland. Horm. Metab. Res., 30: 315–322.

Holst, J.J., Ehrhart-Bornstein, M., Messel, T., Poulsen, S.S. and Harling, H. (1991) Release of galanin from isolated perfused porcine adrenal glands: role of splanchnic nerves. Am. J. Physiol., 261: E31–E40.

Holzer, P. (1998) Neural injury, repair and adaptation in the GI tract. II. The elusive action of capsaicin on the vagus nerve. Am. J. Physiol., (Gastrointest. Liver Physiol.) 275: G8–G13.

Jaeckle, R.S., Kathol, R.G., Lopez, J.F., Meller, W.H. and Krummel, S.J. (1987) Enhanced adrenal sensitivity to exogenous cosyntropin (ACTH 1–24) stimulation in major depression. Arch. Gen. Psychiatry, 44: 233–240.

Jasper, M.S. and Engeland, W.C. (1994) Splanchnic neural activity modulates ultradian and circadian rhythms in adrenocortical secretion in awake rats. Neuroendocrinology, 59: 97–109.

Jasper, M.S. and Engeland, W.C. (1997) Splanchnicotomy alters adrenal sensitivity to ACTH in non-stressed rats. Am. J. Physiol. (Endo. Metab.), 273: E363–368.

Jasper, M.S., McDermott, P., Gann, D.S. and Engeland, W.C. (1990) Measurement of blood flow to the adrenal capsule, cortex and medulla in dogs after hemorrhage by fluorescent microspheres. J. Auton. Nerv. Syst., 30: 159–167.

Kalsbeek, A., Heerikhuize, J.J.V., Wortel, J. and Buijs, R.M. (1996) A diurnal rhythm of stimulatory input to the hypothalamo–pituitary–adrenal system as revealed by timed intrahypothalamic administration of the vasopressin V1 antagonist. J. Neuroscience, 16: 5555–5565.

Kaneko, M., Hiroshige, T., Shinsako, J. and Dallman, M.F. (1980) Diurnal changes in amplification of hormone rhythms in the adrenocortical system. Am. J. Physiol., 239: R309–R316.

Kaneko, M., Kaneko, K., Shinsako, J. and Dallman, M.F. (1981) Adrenal sensitivity to ACTH varies diurnally. Endocrinology, 109: 70–75.

Kesse, W.K., Parker, T.L. and Coupland, R.E. (1988) The innervation of the adrenal gland I. The source of pre- and postganglionic nerve fibers to the rat adrenal gland. J. Anatomy, 157: 33–41.

Kiss, A., Jezova, D. and Aquilera, G. (1994) Activity of the hypothalamic pituitary adrenal axis and sympathoadrenal system during food and water deprivation in the rat. Brain Research, 663: 84–92.

Kleitman, N. and Holzwarth, M.A. (1985) Catecholaminergic innervation of the rat adrenal cortex. Cell Tissue Res., 241: 139–147.

Kojima, K., Naruse, Y., Iijima, N., Wakabayashi, N., Mitsufuji, S., Ibata, Y. and Tanaka, M. (2002) HPA-axis responses during experimental colitis in the rat. Am. J. Physiol. (Regul Integr Comp Physiol), 282: R1348–R1355.

Kondo, H. (1985) Immunohistochemical analysis of the localization of neuropeptides in the adrenal gland. Arch. Histo. Japon., 48: 453–481.

Kuramoto, H., Kondo, H. and Fujita, T. (1987) Calcitonin gene-related peptide immunoreactivity in scattered chromaffin cells and nerve fibers in the adrenal gland of rats. Cell Tissue Res., 247: 309–315.

L'Age, M., Gonzalez-Luque, A. and Yates, F.E. (1970) Adrenal blood flow dependence of cortisol secretion rate in unanesthetized dogs. J. Physiol,. 219: 281–287.

Leboulenger, F., Perroteau, I., Netchitailo, P., Lihrmann, I., Leroux, P., Delarue, C., Coy, D.H. and Vaudry, H. (1984)

Action of vasoactive intestinal peptide (VIP) on amphibian adrenocortical function, in vitro. Peptides, 4: 299–303.

Lehoux, J.-G., Hales, D.B., Fleury, A., Briere, N., Martel, D. and Ducharme, L. (1999) The in vivo effects of ACTH and sodium restriction on the formation of different species of steroidogenic acute regulatory protein in rat adrenal. Endocrinology, 140: 5154–5164.

Makino, S., Smith, M.A. and Gold, P.W. (1995) Increased expression of corticotropin-releasing hormone and vasopressin messenger ribonucleic acid (mRNA) in the hypothalamic paraventricular nucleus during repeated stress: association with reduction in glucocorticoid receptor mRNA levels. Endocrinology, 136: 3299–3309.

Malendowicz, L.K., Lesniewska, B. and Miskowiak, B. (1990) Neuropeptide Y inhibits corticosterone by isolated rat adrenocortical cells. Experientia, 46: 721–722.

Meier, A.H. (1976) Daily variation in concentration of plasma corticosteroid in hypophysectomized rats. Endocrinology, 98: 1475–1479.

Mohamed, A.A., Parker, T.L. and Coupland, R.E. (1988) The innervation of the adrenal gland. II. The source of spinal afferent nerve fibers to the guinea pig adrenal. J. Anatomy, 160: 51–58.

Mokuda, O., Sakamoto, Y., Kawagoe, R., Ubukata, E. and Shimizu, N. (1992) Epinephrine augments cortisol secretion from isolated perfused adrenal glands of guinea pigs. Am. J. Physiol., 262: E806–E809.

Mountjoy, K.G., Robbins, L.S., Mortrud, M.T. and Cone, R.D. (1992) The cloning of a family of genes that encode the melanocortin receptor. Science, 257: 1248–1251.

Myers, D.A., Robertshaw, D. and Nathanielsz, P.W. (1990) Effect of bilateral splanchnic nerve section on adrenal function in the ovine fetus. Endocrinology, 127: 2328–2335.

Naito, Y., Fukata, J., Tamai, S., Seo, N., Nakai, Y., Mori, K. and Imura, H. (1991) Biphasic changes in hypothalamo–pituitary–adrenal function during the early recovery period after major abdominal surgery. J. Clin. Endo. Metab., 73: 111–117.

Nasman, B., Olsson, T., Fagerlund, M., Eriksson, S., Viitanen, M. and Carlstrom, K. (1996) Blunted adrenocorticotropin and increased adrenal steroid response to human corticotropin-releasing hormone in Alzheimer's disease. Biol. Psychiatry, 39: 311–318.

O'Brien, J.T., Schweitzer, I., Ames, D., Mastwyk, M. and Colman, P. (1994) The function of the hypothalamic–pituitary–adrenal axis in Alzheimer's disease: response to insulin hypoglycemia. Br. J. Psychiatry, 165: 650–657.

Ottenweller, J. and Meier, A. (1982) Adrenal innervation may be an extrapituitary mechanism able to regulate adrenocortical rhythmicity in rats. Endocrinology, 111: 1334–1338.

Papadopoulos, V. and Brown, A.S. (1995) Role of the peripheral-type benzodiazepine receptor and the polypeptide diazepam binding inhibitor in steroidogenesis. J. Steroid Biochem. Mol. Biol., 53: 103–110.

Pascualy, M., Petrie, E.C., Brodkin, K., Piskind, E.R., Wilkinson, C.W. and Raskind, M.A. (2000) Hypothalamic pituitary adrenocortical and sympathetic nervous system responses to the cold pressor test in Alzheimer's disease. Biol. Psychiatry, 48: 247–254.

Renshaw, D. and Hinson, J.P. (2001) Neuropeptide Y and the adrenal gland: a review. Peptides, 22: 429–438.

Schramm, L.P., Adair, J.R., Stribling, J.M. and Gray, L.P. (1975) Preganglionic innervation of the adrenal gland of the rat: a study using horseradish peroxidase. Exp. Neurol., 49: 540–553.

Selye, H. (1936) A syndrome produced by diverse nocuous agents. Nature, 138: 32.

Shima, S., Komoriyama, K., Hirai, M. and Kouyama, H. (1984) Studies on cyclic nucleotides in the adrenal gland. XI. Adrenergic regulation of adenylate cyclase activity in the adrenal cortex. Endocrinology, 114: 325–329.

Sparrow, R.A. and Coupland, R.E. (1987) Blood flow to the adrenal gland of the rat: its distribution between the cortex and the medulla before and after haemorrhage. J. Anat., 155: 51–61.

Strack, A.M., Sawyer, W.B., Platt, K.B. and Loewy, A.D. (1989) CNS cell groups regulating the sympathetic outflow to adrenal gland as revealed by transneuronal cell body labeling with pseudorabies virus. Brain Res., 491: 274–296.

Stucco, D.M. and Clark, B.J. (1996) Role of steroidogenic acute regulatory protein (StAR) in steroidogenesis. Biochem. Pharmacol., 5: 197–205.

Ulrich-Lai, Y.M. and Engeland, W.C. (2000) Hyperinnervation during adrenal regeneration influences the rate of functional recovery. Neuroendocrinology, 71: 107–123.

Ulrich-Lai, Y.M. and Engeland, W.C. (2001) Splanchnicectomy-induced attenuation of corticosterone diurnal rhythmicity is prevented by adrenal medullectomy in rats. Exper. Biol. Mtg.: Abstract 127.2.

Ulrich-Lai, Y.M. and Engeland, W.C. (2002) Adrenal splanchnic innervation modulates adrenal cortical responses to dehydration stress in rats. Neuroendocrinology, 76: 79–92.

Ulrich-Lai, Y.M., Marek, D.M. and Engeland, W.C. (2002) Capsaicin-sensitive adrenal sensory fibers participate in compensatory adrenal growth in rats. Am. J. Physiol., (Reg. Integr. Compar.) 283: R877–R884.

Urquhart, J. (1965) Adrenal blood flow and the adrenocortical response to corticotropin. Am. J. Physiol., 209: 1162–1168.

Walker, C.-D. (1995) Chemical sympathectomy and maternal separation affect neonatal stress responses and adrenal sensitivity to ACTH. Amer. J. Physiol., 268: R1281–R1288.

Walker, S.W., Lightly, E.R.T., Milner, S.W. and Williams, B.C. (1988) Catecholamine stimulation of cortisol secretion by 3-day primary cultures of purified zona fasciculata/reticularis cells isolated from bovine adrenal cortex. Mol. Cell Endocrinol., 57: 139–147

Walker, S.W., Lightly, E.R.T., Clyne, C., Williams, B.C. and Bird, I.M. (1991) Adrenergic and cholinergic regulation of cortisol secretion from the zona fasciculata/reticularis of bovine adrenal cortex. Endocrine Res., 17: 237–264.

Wilkinson, C.W., Shinsako, J. and Dallman, M.F. (1981) Return of pituitary–adrenal function after adrenal enucleation or transplantation: diurnal rhythms and responses to ether. Endocrinology, 109: 162–169.

Wood, C., Shinsako, J. and Dallman, M. (1982) Comparison of canine corticosteroid responses to mean and phasic increases in ACTH. Am. J. Physiol., 242: E102–E108.

Zhou, X.-F., Oldfield, B.J. and Livett, B.G. (1991) Substance P-containing sensory neurons in the rat dorsal root ganglia innervate the adrenal medulla. J. Auton. Nerv. Syst., 33: 247–254.

Zilz, A., Li, H., Castello, R., Papadoupolos, V. and Widmaier, E.P. (1999) Developmental expression of the peripheral-type benzodiazepine receptor and the advent of steroidogenesis in rat adrenal glands. Endocrinology, 140: 859–864.

T. Steckler, N.H. Kalin and J.M.H.M. Reul (Eds.)
Handbook of Stress and the Brain, Vol. 15
ISBN 0-444-51173-3

CHAPTER 4.3

The locus coeruleus–noradrenergic system and stress: modulation of arousal state and state-dependent behavioral processes

Craig W. Berridge*

Departments of Psychology and Psychiatry, University of Wisconsin, 1202 W. Johnson St., Madison, WI 53706, USA

The locus coeruleus–noradrenergic system and stress

The current conceptualization of stress as a behavioral state, elicited by challenging or threatening events, arises from nearly a century of research starting with the seminal work of Cannon (1914) and Selye (1946). In these studies, various physiological systems were similarly affected by disparate environmental events, which have in common a potential to disrupt homeostasis or threaten animal well-being. Initially, emphasis was placed on stressor-induced activation of peripheral systems, primarily endocrine systems. This work identified the activation of both peripheral catecholamine systems and the pituitary–adrenal axis as hallmark features of the state of stress. The activation of these systems results in enhanced ability of the animal to physically contend with a challenging situation. More recently, emphasis has been placed on the neurobiology of the affective, cognitive, and behavioral components of stress.

This raises the long-standing issue of what are the psychological defining features of stress. In contrast to the well-delineated physiological indices of stress, the affective and cognitive features of stress remain less clear. The terms stress and anxiety are frequently used interchangeably. However, the precise definition of these terms and the relationship between stress and anxiety are poorly understood. Typically, anxiety is viewed as having both anticipatory and affective components. In contrast, as defined above, stress is most commonly viewed as a response to a present challenge. Moreover, the extent to which stress has an affective component, which can or cannot be dissociated from anxiety, is unclear. Regardless of the exact configuration of cognitive and affective responses associated with stress, it appears a heightened level of readiness for action is paramount to a state of stress. A prominent component of this preparatory state is an elevated level of arousal defined, for the purposes of this review, by a heightened awareness of, and sensitivity to, environmental stimuli. Associated with stressor-induced alterations in arousal level are alterations in specific state-dependent processes, including attention, memory, and sensory information processing.

Candidate neural systems that participate in stress-related alterations in arousal state and state-dependent behavioral processes include the monoaminergic neurotransmitter systems. For example, it has long been known that stress is associated with a robust activation of cerebral noradrenergic systems, resulting in an increase in rates of noradrenaline release widely throughout the brain.

*Tel.: +1608-265-5938; Fax: +1608-262-4029; E-mail: berridge@wisc.edu

Relatively, recent evidence suggests that central noradrenergic systems modulate behavioral and forebrain neuronal activity states as well as state-dependent cognitive and physiological processes. In many cases, actions of noradrenaline have been demonstrated to play a critical role in stressor-induced alterations in cognitive function. Combined, these observations suggest a prominent role of central noradrenergic systems in both adaptive and possibly maladaptive stressor-induced alterations in cognition and affect.

Substantial evidence indicates that the activation of the prototypical stress systems (hypothalamo–pituitary–adrenal axis as well as peripheral and central catecholaminergic systems) occurs across both aversive and appetitive conditions (see below; for review see, Marinelli and Piazza, 2002). This suggests the working hypothesis that at least a subset of the physiological indices of stress may be independent of affective valence (pleasant vs. unpleasant) and more closely aligned with arousal level, motivational state, and/or the need for action. In this context, noradrenergic modulation of behavioral state and state-dependent cognitive processes may serve a critical function in stress-related behavior, while not being limited to the state of stress (see Clinical Implications below for further discussion).

Anatomical features of the locus coeruleus–noradrenergic system

Noradrenaline-containing axons are distributed widely throughout the central nervous system (CNS). A majority of brain noradrenergic neurons are concentrated in the brainstem nucleus, locus coeruleus (LC). The LC is a well-delineated cluster of noradrenaline-containing neurons, located adjacent to the fourth ventricle in the pontine brainstem. It is composed of a small number of neurons: approximately 1500 per nucleus in rat, several thousand in monkey, and 10,000–15,000 in human. Despite these relatively small numbers, these neurons possess immensely ramified axons permitting the nucleus to project broadly throughout the neuraxis, from spinal cord to neocortex (Swanson and Hartman, 1975; Foote et al., 1983; Lewis, 2001), excluding the basal ganglia. Importantly, the LC provides the sole source of noradrenaline to hippocampus and neocortex, regions critical for higher cognitive and affective function.

Despite the widespread distribution of LC efferent fibers within the brain, there is substantial regional specificity of noradrenergic fiber distribution across cortical and subcortical structures. For example, within neocortex, there is both regional and laminar variability in noradrenergic fiber density (Morrison and Foote, 1986; Lewis et al., 1987; Lewis, 2001; reviewed in Foote and Morrison, 1987; Lewis, 2001). An exception to the rule that the basal ganglia does not receive a noradrenergic innervation is the shell subregion of the nucleus accumbens. The nucleus accumbens can be divided into distinct subfields, delineated based on histochemical markers as well as efferent and afferent projection patterns. In contrast to that of the core subregion, the shell subregion has extensive and reciprocal connections with a variety of limbic and brainstem autonomic structures. Thus, it is of interest that this subregion of the nucleus accumbens is the only striatal subfield to receive a moderately dense noradrenergic innervation (Berridge et al., 1997), although the majority of this arises from non-LC noradrenergic sources (Delfs et al., 1998).

Similar to other neurotransmitter systems, noradrenaline acts at multiple receptors in target tissues. Traditionally three noradrenergic receptor subtypes have been recognized: α_1, α_2, and β. α_1- and β-receptors are thought to exist primarily at postsynaptic sites, whereas α_2-receptors exist both pre- and postsynaptically. The distribution and second messenger coupling of these receptor subtypes varies within and across brain regions. For example, within neocortex, β-receptors appear to be more broadly distributed across laminae and are positively coupled to the Gs/cAMP second messenger system, whereas α_1- and α_2-receptors are concentrated in the superficial layers and are coupled to the phosphoinositol and Gi/cAMP systems, respectively (Dohlman et al., 1991). Recently, molecular, biological, and pharmacological studies have revealed an even greater diversity of adrenergic receptors, with multiple subtypes each of β-, α_1-, and α_2-receptors identified (Boyajian and Leslie, 1987; Jones and Palacios, 1991; Bylund et al., 1992). Currently, three β- (β1–3), three α_1 (α_1a, α_1b, α_1d) and four α_2- (α_2A–D) receptor subtypes are recognized. These newly identified subtypes appear to be distributed differentially across cortical laminae (Young, III and Kuhar, 1980; Rainbow and Biegon, 1983; Rainbow et al., 1984; Goldman-Rakic et al., 1990; McCune et al., 1993; Nicholas et al., 1993a,b; Pieribone et al., 1994). These observations suggest that different adrenoceptor subtypes mediate distinct actions within noradrenergic-innervated circuits by virtue of their differential distribution both within and across functionally-distinct terminal fields. Interestingly, adrenergic receptors also reside on glial cells, prompting speculation regarding the influence of the LC–noradrenergic system on glial function and the impact of such actions on neighboring neurons (Stone and Ariano, 1989; Stone et al., 1990; Stone and John, 1991; Aoki, 1992).

Electrophysiological features of LC neurons

Locus coeruleus neurons fire in at least two distinct activity modes: *tonic* and *phasic*. *Tonic* activity is characterized by relatively low-frequency, sustained, and highly regular discharge patterns. The pioneering work of Hobson and McCarley and colleagues demonstrated that the tonic discharge activity is state dependent: LC neurons display highest discharge rates during waking (quiet waking < 2 Hz; active waking > 2 Hz), slower rates during slow-wave sleep (< 1 Hz), and minimal activity during REM or paradoxical sleep (Hobson et al., 1975; Foote et al., 1980). Of particular interest is the fact that, in general, changes in LC discharge rates anticipate changes in behavioral state (Hobson et al., 1975; Foote et al., 1980; Aston-Jones and Bloom, 1981a). Within waking, sustained increases in tonic discharge rates are elicited by environmental stimuli that elicit sustained increases in EEG and behavioral indices of arousal or attentiveness (Foote et al., 1980; Aston-Jones and Bloom, 1981a). Tonic discharge rates as high as 15 Hz have been reported for brief periods under high-arousal conditions associated with specific appetitive stimuli (e.g. preferred food, Foote et al., 1980). However, the extent to which these rates can be sustained for prolonged periods under high-arousal conditions, including stress, remains unclear. This appears to be a substantial lacuna in our understanding of both the neurobiology of stress and the LC–noradrenergic system.

Within waking, LC neurons also display *phasic* alterations in discharge rates in response to both unconditioned and conditioned salient sensory stimuli (Foote et al., 1980; Aston-Jones and Bloom, 1981b; Aston-Jones et al., 1994). These phasic responses are observed with a relatively short latency (15–70 ms in rat) and are comprised of a brief burst of 2–3 action potentials followed by a more prolonged period of suppression of discharge activity (approximately 300–700 ms). Phasic responses are observed in association with overt attending to a novel stimulus within a particular environmental location (e.g. an orienting response). Phasic LC responses habituate with repeated stimulus presentation. This habituation of phasic discharge is accompanied by habituation of the behavioral response to that stimulus. Further, phasic responses are less robust during lower levels of vigilance, including those associated with sleep, grooming, and eating, which are associated with lower tonic discharge rates (Aston-Jones and Bloom, 1981b). Moreover, higher levels of tonic discharge activity, are also associated with less robust phasic discharge activity. For example, both hypotension stress and corticotropin-releasing factor elevate tonic discharge activity and reduce sensory driven phasic discharge (Valentino and Foote, 1987, 1988; Valentino and Wehby, 1988). The functional impact of stressor-induced disruption of phasic discharge remains to be determined.

Relationship between rates of noradrenaline efflux to rates of LC discharge activity

Extracellular (extrasynaptic) levels of noradrenaline are linearly related to tonic LC discharge rates across the range of LC-firing rates typically observed across the sleep–wake cycle (Florin-Lechner et al., 1996; Berridge and Abercrombie, 1999). These observations indicate that relatively small fluctuations in absolute LC discharge rates within the range typically observed in normal sleep and waking (i.e. a change from 1.5 to 3.0 Hz) result in pronounced changes in noradrenaline efflux. At higher levels of tonic discharge, the relationship between firing rate and noradrenaline efflux appears to be of a nonmonotonic nature, with smaller increases in noradrenaline efflux associated with increased discharge rates beyond 3–4 Hz (Berridge and Abercrombie, 1999). Dopaminergic neurons can display a sustained burst-type mode of firing. The burst mode is associated with greater rates of DA efflux relative to similar frequency tonic discharge activity (Bean and Roth, 1991; Manley et al., 1992). In contrast to dopaminergic neurons, LC neurons do not display prolonged burst-type firing activity. Instead, LC neurons display quite brief phasic discharge superimposed upon tonic discharge activity. Importantly, the 2–3 action potential burst comprising phasic discharge is followed by a sustained period (200–500 ms) of suppressed firing. The net effect of 2–3 action potentials followed by a prolonged period of inhibition on synaptic levels of noradrenaline and neuronal function within LC terminal fields remains to be determined.

Plasticity of the LC–noradrenergic system in stress

Central noradrenergic (as well as dopaminergic) systems possess robust compensatory mechanisms that permit adjustment to long-term alterations in activity. These alterations are observed in response to damage as well as environmental- (e.g. stress) and pharmacological-based (e.g. antidepressant) manipulations. This plasticity may well be a key feature of catecholaminergic systems and likely contributes to at least some of the cognitive and affective consequences of certain environmental conditions and certain classes of psychoactive drugs. Of particular relevance to the current discussion is stressor-induced plasticity within the LC–noradrenergic system.

It has long been known that prolonged, or repeated, exposure to stressors, such as footshock, cold, or restraint, elicit a decrease in β-receptor-driven accumulation of cAMP (Stone, 1979; Stone, 1981). The stressor-induced downregulation of the β-dependent cAMP response appears to result largely from a reduction in α_1-receptor potentiation of the β-receptor cAMP response (Stone et al., 1984, 1985; Stone, 1987). Repeated exposure to a stressor also attenuates LC neuronal responsivity and noradrenaline release to the same (homotypic) stressor (Abercrombie and Jacobs, 1987; Nisenbaum et al., 1991). Although repeated presentation of certain stressors results in tolerance to the LC-activating actions of those stressors, enhanced responsivity of LC neurons to repeated immobilization stress has been observed (Pavcovich et al., 1990), indicating that tolerance to a given stressor is not obligatory.

The above-described development of tolerance to stressor-induced LC activation is in contrast to the well-documented ability of both acute and chronic stressors to increase activity and/or quantity of the rate-limiting enzyme in noradrenaline biosynthesis, tyrosine hydroxylase (TH; Stone et al., 1978; Kramarcy et al., 1984). Thus, although chronic/repeated stressors do not tend to produce elevated levels of LC neuronal discharge, they do result in increased capacity of the system to release noradrenaline, due to elevated rates of noradrenaline synthesis (for review, see Dunn and Kramarcy, 1984). These observations raise the question under what conditions would increased synthetic capacity be utilized? Insight into this issue is provided by the observation that, in contrast to homotypic stressors, repeated/chronic stress results in an increased responsiveness of the LC–noradrenergic system to presentation of a *different* (heterotypic) stressor. For example, following exposure to chronic cold-stress, greater increases in extracellular noradrenaline levels are observed within hippocampus in response to tail-shock (Nisenbaum et al., 1991) or tail-pinch (Finlay et al., 1995).

Thus, during prolonged exposure to a particular stressor the LC–noradrenergic system develops an increased capacity to respond to additional challenges. The ability of this system to modulate output as a consequence of prior stress has important implications for understanding stressor-induced alterations in cognitive and affective function. However, in general, the contribution of plasticity within the LC–noradrenergic system to resilience and/or dysregulation of cognition and affect in stress remains poorly understood.

Sensitivity of the LC–noradrenergic system to appetitive and aversive stimuli

As reviewed above and elsewhere in this volume, extensive evidence indicates a robust activation of the

LC–noradrenergic system by a variety of stressors (Thierry et al., 1968; Weiss et al., 1970; Zigmond and Harvey, 1970; Korf et al., 1973; Stone, 1973a, 1975; Weiss et al., 1975, 1980; Anisman, 1978; Redmond, and Huang, 1979; Grant and Redmond, 1984; Dunn and Kramarcy, 1984; Nisenbaum et al., 1991; Finlay et al., 1995). The early demonstration of a sensitivity of LC neurons to stressors suggested a possibly selective role of the LC in stress and led to a number of hypotheses concerning alarm- or anxiety-specific functions of these neurons. However, it is important to note that electrophysiological studies in unanesthetized animals demonstrate a sensitivity of tonic and phasic LC discharge to both appetitive as well as aversive stimuli (Foote et al., 1980; Aston-Jones and Bloom, 1981b; Aston-Jones et al., 1994, 1997, 1998). Further, recent microdialysis studies demonstrate elevated extracellular levels of noradrenaline in response to appetitively conditioned stimuli, presumably reflecting elevation in tonic discharge rates (Feenstra et al., 1999, 2001; Feenstra, 2000). These observations combined, suggest that both tonic and phasic LC discharge activity are more closely related to the overall salience, arousing and/or motivating nature of a given stimulus rather than affective valence. As reviewed in the following sections, evidence indicates the LC–noradrenergic system modulates a variety of physiological and behavioral processes related to the acquisition, processing, and responding to salient sensory information.

Modulatory actions of LC–noradrenergic efferents on forebrain neuronal and behavioral activity states

Alterations in arousal level may be a primary component of the state of stress. Increased arousal, including that associated with the transition from sleep to waking, is characterized by an enhanced ability to detect, process, and respond to information arising from the environment (Steriade, 1969; Livingstone and Hubel, 1981). Accompanying changes in sleep–wake state are changes in activity patterns of cortical and thalamic neurons. These changes in neuronal activity are, in turn, reflected in EEG measures (Timo-Iaria et al., 1970; Vanderwolf and Robinson, 1981). For example, during non-REM sleep, cortical EEG is characterized by the presence of large-amplitude, slow-wave activity, reflecting slow, synchronous firing of cortical and thalamic neurons. During active waking and REM sleep, these neurons no longer fire slowly in synchrony, and this is reflected by the presence of high-frequency, low-amplitude activity in cortical EEG recordings (desynchronized activity; EEG activation). At the neuronal level, waking and REM sleep are associated with increased excitability of thalamic and neocortical neurons (for review, Steriade and Buzsaki, 1990). An obvious distinction between REM sleep and waking is the extent to which an animal is aware of and responsive to environmental stimuli. Differentiation between quiet and active waking, slow-wave sleep and REM sleep can be made based on combined EEG and electromyographic (EMG) recordings: from REM to active waking, progressively larger amplitude EMG activity is observed, corresponding to progressively greater muscle tone (Timo-Iaria et al., 1970).

Within waking, fluctuations in attention to the environment occur that are similarly associated with changes in cortical/thalamic activity patterns. Thus, animals engaged in self-directed behaviors (such as grooming) display lower frequency, larger amplitude activity in cortical EEG than animals actively attending to the environmental stimuli (for review, see Vanderwolf and Robinson, 1981). In humans, the degree of cortical EEG activation is highly associated with the ability to sustain focused attention (e.g. vigilance), indicating a close association between cortical activity patterns and higher cognitive processes (Makeig and Inlow, 1992; Makeig and Jung, 1996; Jung et al., 1997).

The fact that LC neurons increase firing rates in anticipation of waking and waking-associated forebrain activation (as measured by EEG) suggests the hypothesis that LC efferents participate in the induction of either the awake state and/or cortical/thalamic activity patterns associated with waking and enhanced arousal. Additionally, given LC neuronal discharge rate is positively related to EEG and behavioral measures of arousal within the waking state, the LC–noradrenergic system may contribute to state-dependent alterations in cognitive and affective processes. These actions of the LC–noradrenergic system may contribute to stressor-induced alterations in forebrain neuronal and behavioral activity states. Data collected over the past decade provide strong support for these hypotheses.

Noradrenergic modulation of cortical and thalamic neuronal activity state, in vitro

Cortical and thalamic neurons display distinct activity modes during sleeping and waking. Thus, during slow-wave sleep, these neurons are hyperpolarized relative to waking and display a burst-type activity mode. This activity mode is associated with a relative insensitivity to incoming sensory information. In contrast, during waking these neurons display a single-spike mode associated with the efficient and accurate processing of sensory information (Mukhametov et al., 1970; McCarley et al., 1983; Domich et al., 1986; Steriade et al., 1986; McCormick and Bal, 1997). The above-described electrophysiological observations indicate increased rates of noradrenaline release during conditions associated with the single-spike mode, suggesting that LC efferents contribute to the induction of this activity state (Foote et al., 1980; Aston-Jones and Bloom, 1981a). Consistent with this hypothesis, McCormick and colleagues have demonstrated that, in vitro, noradrenaline induces a shift in the firing pattern of cortical and thalamic neurons from a burst mode to a single-spike mode (McCormick and Prince, 1988; McCormick, 1989; Pape and McCormick, 1989). The ability of noradrenaline to induce the single-spike activity mode involves actions of both α_1-receptors and β-receptors (McCormick et al., 1991).

LC modulation of EEG and behavioral indices of arousal

Supporting a role of the LC in behavioral and EEG indices of waking are the well-documented sedative effects of systemic, ICV, or intra-brainstem administration of α_2-agonists, which acutely suppress LC neuronal discharge activity and noradrenaline release (De Sarro et al., 1987, 1988, 1989; Gatti et al., 1988; Waterman et al., 1988). In fact, due to their ability to reduce the effective anesthetic dose, α_2-agonists are commonly used as adjuncts to surgical anesthesia (Kaukinen and Pyykko, 1979; Bloor and Flacke, 1982). These sedative effects of α_2-agonists are opposite to those observed following intrabrainstem infusions of α_2-antagonists (De Sarro et al., 1988, 1989).

The small size of the LC, situated in close proximity to a variety of brainstem structures, presents a substantial challenge for the *selective* manipulation of LC neuronal discharge rates. A combined recording/infusion probe has been described that permits a greater degree of anatomical localization of intratissue infusions (Adams and Foote, 1988). Using this approach, it was demonstrated that *unilateral* LC activation, produced by a small infusion of the cholinergic agonist bethanechol, elicits a robust *bilateral* activation of cortical and hippocampal EEG (see Fig. 1; Berridge and Foote, 1991). In contrast, bilateral suppression of LC neuronal discharge activity via infusion of an α_2-agonist produces a robust increase in slow-wave activity in cortical and hippocampal EEG. Importantly, even minimal (i.e. 10% of basal levels) neuronal discharge activity unilaterally is sufficient to maintain bilateral

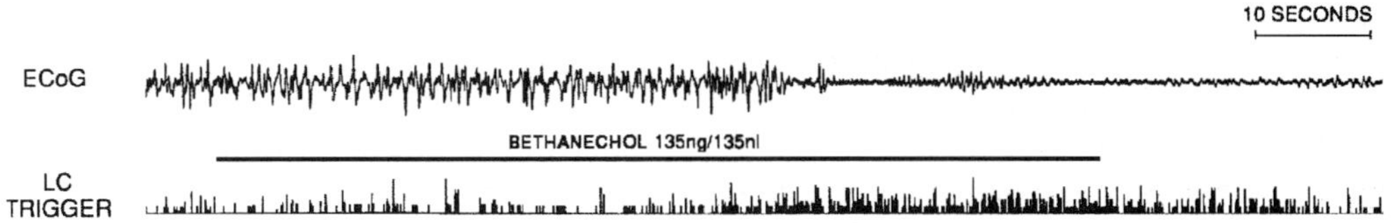

Fig. 1. Relationship of LC activity to cortical (ECoG) before, during and after peri-LC infusions of the cholinergic agonist, bethanechol. Cholinergic agonists exert potent excitatory effects on LC neuronal discharge rates. Bethanechol was infused at a constant rate throughout the interval indicated. EEG activity is shown in the *top trace* and the raw trigger output from LC activity in the *bottom trace*. LC activity is seen to increase during the latter part of the infusion. Several seconds later, reduced amplitude and increased frequency becomes evident in the ECoG. Simultaneous alterations in hippocampal EEG activity were observed with changes in ECoG (data not shown). The EEG activating effect of LC stimulation was prevented by ICV pretreatment with a β-antagonist (data not shown). Modified from Berridge and Foote (1991).

forebrain activation (Berridge et al., 1993). These and other observations demonstrate that minimal LC neuronal discharge activity within one hemisphere is sufficient to maintain bilateral activation of the forebrain. This observation is consistent with previous observations demonstrating diminished, yet not absent, LC discharge activity in animals that are clearly awake, yet engaged in consummatory or grooming behavior (Aston-Jones and Bloom, 1981a). Combined, these observations indicate the LC is a potent modulator of forebrain EEG state, with unilateral LC neuronal discharge activity causally related to the bilateral maintenance of EEG activity patterns associated with arousal.

Noradrenaline acts within the basal forebrain to modulate forebrain EEG and behavioral activity states

The complete array of sites within which noradrenergic efferents act to modulate behavioral state remain to be elucidated. Potential sites include cortical, thalamic, basal forebrain, and brainstem regions. Basal forebrain structures implicated in the regulation of cortical and hippocampal activity state include the general region of the basal forebrain encompassing the medial septal area/diagonal band of Broca (MS; Buzsaki et al., 1983; Smythe et al., 1991; Berridge et al., 1996; Berridge and Foote, 1996), the general region of the anterior-medial hypothalamus, encompassing the medial preoptic area (MPOA; McGinty and Sterman, 1968; Findlay and Hayward, 1969; Mallick and Alam, 1992; Sherin et al., 1996; Berridge et al., 1999), and the substantia innominata (SI; Buzsaki et al., 1988; Metherate et al., 1992). Each of these regions receives a relatively dense noradrenergic innervation, the preponderance of which arises from LC and thus could be involved in LC-dependent alterations in forebrain activity state (Swanson and Hartman, 1975; Zaborsky et al., 1991). Results from a series of studies conducted in our laboratory demonstrate potent EEG activating and wake-promoting actions of noradrenaline via actions at both β- and α_1-receptor subtypes located within MS and MPOA, but not SI.

Thus, in the halothane-anesthetized rat, unilateral infusions of the β-agonist, isoproterenol, into MS elicit a robust and sustained bilateral activation of cortical and hippocampal EEG. These EEG responses are observed with a latency of approximately 3–8 min. Conversely, bilateral, but not unilateral, infusions of the β-antagonist, timolol, increase cortical and hippocampal slow-wave activity (Berridge et al., 1996).

In sleeping, unanesthetized animals in which remote-controlled infusions are used to avoid waking of the animal, infusions of either a β- or α_1-agonist into MS or MPOA elicits a robust and sustained increase in time spent awake while suppressing REM sleep. Both β- and α_1-agonist-induced waking resembles spontaneous waking and was not associated with excessive locomotor activity or stereotypy (Berridge and Foote, 1996; Berridge and O'Neill, 2001). In both MS and MPOA, additive wake-promoting actions of simultaneous stimulation of β- and α_1-receptors is observed (Fig. 2; Berridge et al., 2002).

Interestingly, in contrast to that observed with infusions into MS and MPOA, neither infusions of noradrenaline, a β-agonist, an α_1-agonist, or the indirect noradrenergic agonist, amphetamine, into SI exert wake-promoting actions (Berridge et al., 1996, 1999; Berridge and O'Neill, 2001). The only exception to this is observed with the highest dose of noradrenaline tested, in which case the magnitude and duration of waking is substantially less than that observed with MPOA infusions (Berridge and O'Neill, 2001). Based on these and other observations, it is concluded that waking observed following high-dose noradrenaline infusion into SI likely results from diffusion of noradrenaline into MPOA. In vitro, noradrenaline depolarizes cholinergic basal forebrain neurons (Fort et al., 1995), indicating a neuromodulatory role of noradrenaline within SI. Combined with previous anatomical observations, the current observations indicate that although noradrenaline acts within SI to modulate neuronal activity, these actions do not elicit a transition from sleep to waking. To date, the actions of noradrenaline within SI on behavior, cognition, and affect remain to be elucidated.

Noradrenaline is necessary for EEG and behavioral indices of alert waking: synergistic sedative actions of α_1- and β-receptor blockade

As described above, α_1- and β-receptors exert unique (non-redundant, additive) wake-promoting actions,

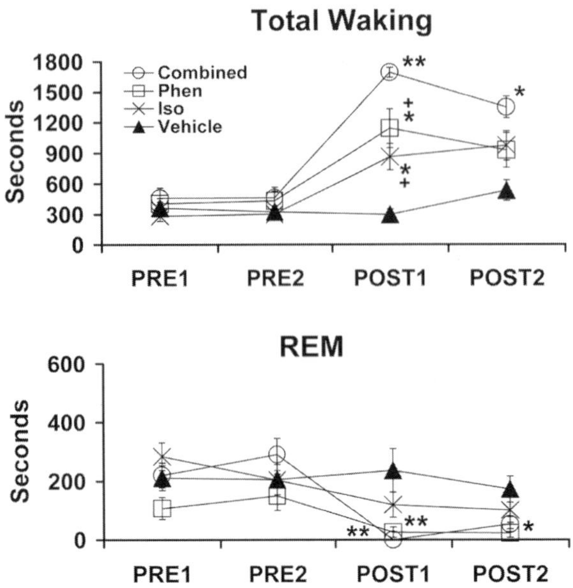

Fig. 2. Additive wake-promoting effects of combined α_1- and β-receptor stimulation within MS. Shown are the effects of individual and combined α_1- and β-receptor stimulation within MS on total time spent awake as determined from EEG/EMG measures. Animals received 250 nl infusions unilaterally into MS of either: (1) vehicle; (2) 4 nmol of the β-agonist, isoproterenol (*Iso*); (3) 10 nmol of the α_1-agonist, phenylephrine (*Phen*), or; (4) 4 nmol isoproterenol + 10 nmol phenylephrine (*Combined*). Marginally effective doses of isoproterenol and phenylephrine were used to better observe any additive effects of combined receptor stimulation. Symbols represent mean (± SEM) time (seconds) spent in a given behavioral state per 30-min testing epoch. PRE1 and PRE2 represent preinfusion epochs. POST1 and POST2 represent postinfusion epochs, beginning 15-min following infusion. Prior to infusion, animals spent the majority of time in slow-wave sleep. Combined stimulation of α_1- and β-receptors elicited substantially larger increases in measures of waking than that observed with stimulation of either receptor subtype individually. During POST2, the only significant alteration in time spent awake was observed in the combined treated animals. Associated with infusion-induced increases in time spent awake were decreases in slow-wave and REM sleep. $^*P < 0.05$, $^{**}P < 0.01$ compared with vehicle-treated controls. $^+P < 0.05$ compared with combined treated animals. Modified from Berridge et al. (2003).

which contribute to the overall arousal state of the animal. Given this, it might be proposed that combined blockade of α_1- and β-receptors would be necessary to impact noticeably EEG and behavioral indices of arousal. In support of this hypothesis, combined administration of a β-antagonist

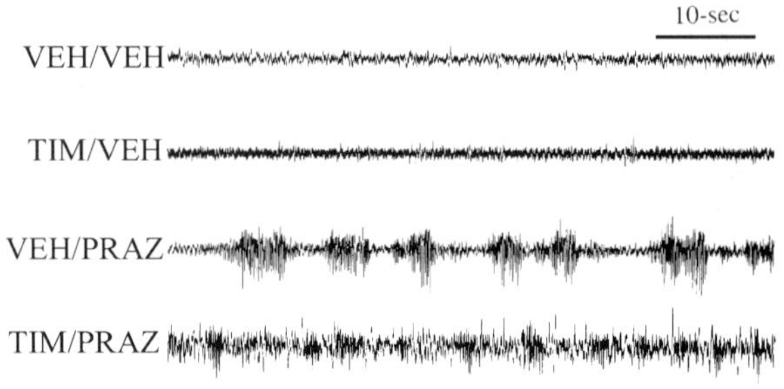

Fig. 3. Effects of α_1-receptor blockade, β-receptor blockade, and combined α_1/β-receptor blockade on ECoG activity. The β-antagonist, timolol, was infused into the lateral ventricles (ICV, 150 μg/2 μl) whereas, the α_1-antagonist, prazosin, was administered intraperitonally (IP, 500 μg/kg). 30-min prior to testing, animals were treated with either: (1) ICV vehicle + IP vehicle (VEH/VEH); (2) ICV timolol + IP vehicle (TIM/VEH); (3) ICV vehicle + prazosin (VEH/PRAZ), and; (4) combined timolol + prazosin (TIM/PRAZ). At the time of testing, animals were placed in a brightly-lit novel environment. (A) Shown are approximately 1-min ECoG traces (printed at 5 mm/s) from the second 5-min epoch of testing. Vehicle-treated controls displayed behavioral and ECoG indices of waking throughout much of the recording session. This is reflected in sustained ECoG desynchronization (low amplitude, high frequency). Treatment with the β-antagonist, timolol, alone had no effects on ECoG activity. Treatment with the α_1-antagonist alone elicited substantial increases in the frequency, amplitude, and duration of HVS. In contrast to either α_1- or β-receptor blockade alone, combined α_1/β-receptor blockade a substantial increase in large-amplitude, slow-wave activity. Modified from Berridge and España (2000).

(timolol, ICV) and an α_1-antagonist (prazosin, IP) results in a profound increase in large-amplitude, slow-wave activity in cortical EEG in animals exposed to an arousal-increasing and stress-inducing, brightly-lit novel environment (see Fig. 3; Berridge and España, 2000). This increase in slow-wave activity is in contrast to the minimal EEG effects observed following administration of a β-antagonist or the high-voltage spindles elicited by α_1-antagonist administration (Buzsaki et al., 1991). The extent to which slow-wave activity is observed with combined α_1/β-receptor blockade is dependent on the time spent in the testing apparatus. Thus, although α_1-antagonist-induced high-voltage spindles or α_1/β-antagonist-induced slow-wave activity is usually observed to a small degree during the first 5 min of testing, the magnitude of these responses increases over the subsequent testing period. This suggests that although prevention of noradrenergic neurotransmission at β- and α_1-receptors has a profound impact on behavioral state even under high-arousal/stressful conditions, under certain conditions this effect can be minimal.

Enhanced LC discharge activity contributes to stressor-induced activation of the forebrain

The above-described observations suggest a critical role of LC neuronal activity in stressor-induced increases in arousal. To test this hypothesis, Page et al. (1993) examined the effects of suppression of LC discharge on hypotension-stress-induced activation of cortical and hippocampal EEG in the halothane-anesthetized rat. These studies demonstrated that bilateral suppression of LC neuronal activity via peri-LC infusions of an α_2-agonist (clonidine) prevent EEG activation elicited by hypotension stress. Thus, these results support the hypothesis that the LC–noradrenergic system participates in stressor-induced alterations in arousal level. Additionally, these actions would be predicted to have a substantial impact on a variety of state-dependent cognitive and/or affective processes in stress.

Summary: LC modulation of arousal in stress

A large body of information demonstrates a clear role of the LC system in the modulation of EEG and behavioral indices of arousal. Additional, though limited, data demonstrate a critical role of the LC in stressor-induced activation of the forebrain, under certain conditions. Combined, these observations suggest the prominent participation of this neurotransmitter system in stressor-induced increases in arousal. Forebrain neuronal and behavioral activity states are influenced by a variety of ascending modulatory systems, including serotonergic, cholinergic, histaminergic, and dopaminergic systems. The above-described observations do not address the relative contribution of these systems to stressor-induced alterations in arousal. The fact that the combined blockade of α_1- and β-receptors had only moderate effects on EEG indices of arousal during the first 5 min of testing under high-arousal/stressful conditions suggests noradrenergic systems may provide only a minimal contribution to stressor-induced EEG activation under certain conditions. Of course, these studies did not completely suppress noradrenergic neurotransmission given postsynaptic α_2-receptors were not targeted. Therefore, the extent to which noradrenergic systems are necessary for stressor-induced alterations across a variety of stressors and testing conditions remains to be fully characterized.

Modulatory actions of the LC–noradrenergic system on sensory processing within cortical and thalamic circuits

During periods of environmental demand (e.g. stress), information collection and processing is critical for guidance of appropriate behavior and potentially survival. Sensory information processing is highly dependent on the arousal/behavioral state of the animal (Evarts, 1960; Steriade, 1969; Pfingst et al., 1977; Hyvarinen et al., 1980; Bushnell et al., 1981; Goldberg and Bushnell, 1981; Livingstone and Hubel, 1981; Mountcastle et al., 1981). These early studies indicate that neuronal responses are facilitated under conditions of elevated arousal and/or enhanced selective attention. Given the above-described state-dependent nature of the LC–noradrenergic system, this system may well contribute to state-dependent modulation of sensory information processing during stress. Additionally, substantial evidence demonstrates direct actions of noradrenaline within cortical and thalamic sensory processing regions, which likely contributes to information processing under conditions of stress and elevated arousal levels.

Initially, evidence suggested the LC–noradrenergic system exerted primarily an inhibitory action on cortical neuronal activity (Stone, 1973b; Armstrong-James and Fox, 1983). Subsequent work indicated noradrenaline exerted a greater inhibition of spontaneous firing rate relative to stimulus-evoked discharge, resulting in a net increase in the "signal-to-noise" ratio (Foote et al., 1975; Woodward et al., 1979; Moises et al., 1981). More recent work has indicated a more complicated array of electrophysiological actions of noradrenaline on sensory neuronal activity (Rogawski and Aghajanian, 1980a,b;

Collins et al., 1984; Videen et al., 1984; Mooney et al., 1990; Manunta and Edeline, 1997; Ciombor et al., 1999), including the facilitation of both inhibitory and excitatory responses to afferent information (Woodward et al., 1979). Outside hippocampus, noradrenergic enhancement of neuronal responses to excitatory synaptic stimuli involves actions of α_1-receptors (Rogawski and Aghajanian, 1980a; Marwaha and Aghajanian, 1982; Waterhouse et al., 1983; Mouradian et al., 1991), whereas the facilitation of inhibitory responses involves β-receptors (Segal and Bloom, 1974; Waterhouse et al., 1982; Sessler et al., 1989). Within hippocampus, noradrenaline enhancement of excitatory synaptic transmission is dependent upon β-receptors, whereas augmentation of synaptic inhibition is mediated by α_1-receptor mechanisms (Mueller et al., 1982; Harley, 1991; Heginbotham and Dunwiddie, 1991).

More recent work suggests that noradrenaline may exert more specific actions on information processing than simply modulating responsiveness to excitatory and inhibitory input, including modulation of feature extraction properties of sensory neurons. For example, iontophoretically applied noradrenaline has been demonstrated to alter specific receptive field properties (e.g. direction selectivity, velocity tuning, response threshold) of visually responsive neurons in cat (McLean and Waterhouse, 1994) and rat (Waterhouse et al., 1990) primary visual cortex. Noradrenaline-induced alterations in the receptive field properties of sensory neurons have also been observed within somatosensory and auditory regions (Kossl and Vater, 1989; George, 1992; Manunta and Edeline, 1997).

Adding further complexity to the multiplicity of electrophysiological actions of noradrenaline, the actions of noradrenaline on sensory cortical neuronal activity are highly dose dependent (Devilbiss and Waterhouse, 2000). For example, the magnitude of glutamate-induced neuronal activity is maximally enhanced at a specific concentration of noradrenaline, and only moderately or minimally affected by concentrations above and below this level (see Fig. 4). Recent work in unanesthetized rat supports non-monotonic, dose-dependent actions of noradrenaline on sensory information processing within neocortex and thalamus.

Combined, these observations indicate that within neocortex noradrenaline exerts a complicated array

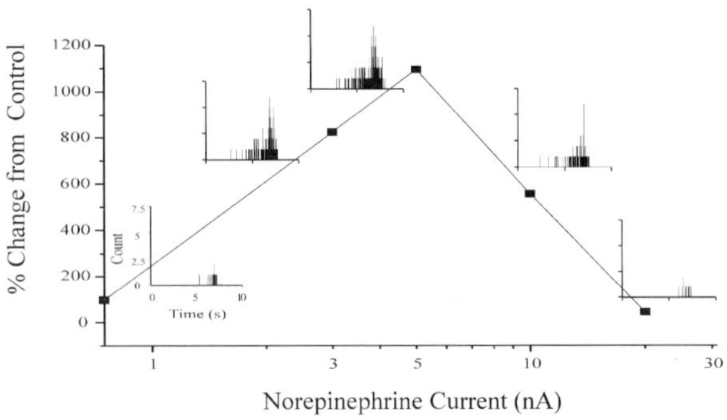

Fig. 4. Biphasic dose–response relationship for noradrenaline modulation of glutamate-evoked responses in rat sensory cortical neurons. Each point on the graph and its associated perievent histogram represent the extracellularly recorded response of a single-layer V somatosensory cortical neuron (in vitro tissue slice preparation) to uniform iontophoretic pulses (20 nA, 8-s duration, 40-s cycle) of glutamate before (control, at left) and during administration of different tonic levels of iontophoretic noradrenaline (1–20 nA). Noradrenaline increases the glutamate-evoked neuronal response maximally when administered at an intermediate dose. Above and below this concentration, noradrenaline does not facilitate glutamate-evoked neuronal responses to a comparable extent. Each histogram sums unit activity during six consecutive glutamate applications. From Waterhouse et al. (1998).

of modulatory actions, which are dependent on cell type, activity state of the neuron, cellular expression of receptors, and levels of noradrenaline available for interaction with those receptors. Such actions are likely of particular importance under conditions of threat when a rapid and accurate behavioral response is required. Given stress is associated with elevated levels of noradrenaline release, it is likely that actions of noradrenaline on sensory information processing plays an important role in the ability of an animal to collect, process, and respond to salient information in stress. It remains for future work to identify the degree to which these actions of the LC–noradrenergic system contribute to cognitive, affective, and behavioral responding in stress.

Modulatory actions of the LC–noradrenergic system on neuronal plasticity

Long-term survival may require long-term alterations in behavior to either more effectively deal with, or avoid, subsequent exposure to a challenging/threatening situation. As described above, the LC–noradrenergic system displays stressor-induced longterm alterations in a variety of cellular processes. In addition to plasticity within the LC system, this system has been demonstrated to elicit long-term alterations in synaptic efficacy, which may underlie learning and memory. Additionally, noradrenaline modulates rates of transcription of immediate-early genes (IEGs), including during stress. Combined, these actions may contribute to long-term alterations in cognition and behavior that may facilitate contending with repeated exposure to challenging situations.

Long-term modulatory actions of the LC–noradrenergic system on synaptic efficacy within neuronal ensembles

Long-lasting, LC-dependent alterations in responsiveness to afferent information are observed at the level of large-population neuronal ensembles. For example, potent modulatory actions of noradrenaline have been observed in long-term potentiation (LTP), a cellular model of memory. LTP refers to a use-dependent, long-lasting increase in synaptic strength or efficacy. Thus, when excitatory synapses are rapidly and repetitively stimulated for brief periods (tetanic stimulation), the postsynaptic neurons generate action potentials more readily upon subsequent stimulation. This effect is manifested as an enhancement of the hippocampal population spike. Three forms of LTP have been described in the hippocampal formation; one involving mossy-fiber input to CA3 of the hippocampus (from the dentate gyrus), one involving the Schaffer collateral input to region CA1 (from the entorhinal cortex), and one involving perforant path input to the dentate gyrus. That LTP is readily observed in a structure critical for memory function has further stimulated interest in LTP as a possible mechanism underlying memory.

Noradrenaline has been demonstrated to influence LTP within CA3 and the dentate gyrus. For example, depletion of noradrenaline decreases the population spike observed in dentate gyrus (Stanton and Sarvey, 1985), whereas noradrenaline application elicits a frequency-dependent enhancement of LTP in the CA3 subfield (Hopkins and Johnston, 1984). These effects appear to be dependent on actions of noradrenaline at β-receptors (Hopkins and Johnston, 1988).

Noradrenaline also elicits a long-lasting enhancement of synaptic efficacy in both the dentate gyrus and CA1 region of the hippocampus in vitro in the absence of tetanic stimulation (Stanton and Sarvey, 1987; Heginbotham and Dunwiddie, 1991). These effects also appear to involve actions of noradrenaline at β-receptors (Heginbotham and Dunwiddie, 1991). In vivo, a similar facilitation of the population spike is observed in the dentate gyrus following enhanced noradrenaline neurotransmission induced by LC activation (Harley and Sara, 1992), α_2-antagonist administration (Segal et al., 1991), or direct application of noradrenaline (Stanton and Sarvey, 1987). The potentiating effect of noradrenaline in the dentate gyrus involves actions of both β- and α_1-receptors (Chaulk and Harley, 1998).

Importantly, enhancement of synaptic strength within the dentate gyrus is observed following behaviorally relevant, sensory-driven increases in LC discharge rate. For example, LC neurons display a short duration (approximately 1–2 s) increase in discharge activity when a rat encounters a novel stimulus within a familiar environment (Vankov et al., 1995). This novelty-induced activation of LC is accompanied by a short duration (approximately 20 s) enhancement in the population spike within the dentate gyrus in response to electrical stimulation of the perforant path, which is attenuated by pretreatment with a β-antagonist (Kitchigina et al., 1997).

Taken together, these studies demonstrate substantial and coordinated enhancement of synaptic strength within the hippocampal formation following activation of the LC–noradrenaline system. An additional form of noradrenaline-dependent plasticity has also been described in neocortex (Kirkwood et al., 1999). In these studies, it was observed that noradrenaline elicits a long-term synaptic depression (LTD) of the population response recorded from layer III of visual cortex. These actions of noradrenaline appear to derive from actions at α_1-receptors and are dependent on neurotransmission at NMDA receptors (Kirkwood et al., 1999).

Overall, these observations indicate a potentially prominent role of the LC–noradrenaline system in mediating long-lasting modifications in neurotransmission within large populations of forebrain neurons. It is particularly intriguing that these actions are observed in structures that support higher-level cognitive processes that are dependent on neuronal plasticity (e.g. learning and memory). The actions of the LC–noradrenaline system within these structures may serve an essential role in the ability of these structures to maintain optimal plasticity and, by doing so, optimize behavioral plasticity under varying environmental conditions. Such actions may be particularly critical in permitting an animal to deal rapidly and effectively when environmental situations that pose a threat to the animal (e.g. stress) are reencountered.

Facilitatory actions of the LC–noradrenergic system on transcription rates of immediate-early and other plasticity-related genes

Long-term alterations in CNS function involve, at least in part, alterations in rates of gene transcription and protein production. A set of "immediate-early genes" (IEGs) has been identified that are activated rapidly by a variety of neuromodulators. The IEG's in this group include c-fos, c-jun, nur77, tis-1, tis-7, tis-21, NGFI-A and -B, and zif-268. Many of these genes regulate transcription rates of a variety of additional genes. Through these actions, IEGs may provide an intervening step through which relatively short-term alterations in neuronal discharge patterns are transduced into long-term biochemical events that underlie plasticity, including learning and memory (Milbrandt, 1987; Dragunow et al., 1989).

Recent work demonstrates a prominent role of the LC–noradrenaline system in the regulation of IEGs, suggesting a potentially prominent role of this neurotransmitter system in the regulation of behavioral processes dependent on plasticity of forebrain circuits. For example, increased noradrenaline release elicited by systemic administration of an α_2-antagonist increases c-fos, nur77, tis-7, zif-268, and tis-21 mRNA (Gubits et al., 1989; Bing et al., 1991) and protein levels (Bing et al., 1992) in rat cerebral cortex. Direct infusion of noradrenaline into cortex (Stone et al., 1991) or amygdala (Stone et al., 1997) elicited a similar increase in c-fos mRNA.

Interestingly, stress is associated with similar activating effects on IEG expression (Gubits et al., 1989; Bing et al., 1991, 1992). Importantly, the activating effects of pharmacologically or stressor-induced increases in noradrenaline neurotransmission on IEG expression are attenuated with pretreatment of either β- or α_1-antagonists (Gubits et al., 1989; Stone et al., 1991, 1997; Bing et al., 1991) as well as (in the case of stress) LC lesions (Stone et al., 1993). These observations indicate a dependence of stressor-induced alterations in IEG expression on stressor-induced increases in noradrenaline release.

More recent work demonstrates a similar critical role of the LC system in waking-related c-fos and other IEG expression. For example, unilateral LC lesion prevented waking-related increases in Fos and nerve-growth factor-induced A (NGFI-A) as well as levels of phosphorylated cyclic AMP response element-binding protein (p-CREB) in the ipsilateral cortex (Cirelli et al., 1996; Cirelli and Tononi, 1998). Other plasticity-related cellular signals that are increased during waking and decreased by LC lesions are Arc and BDNF (Cirelli and Tononi, 2000).

In general, the functional impact of noradrenaline-depenent rates of IEG transcription has remained unexplored. However, recent observations indicate inhibition of *Arc* protein expression impairs the maintenance phase of hippocampal LTP (Guzowski et al., 2000). Thus, noradrenaline-dependent alterations in rates of IEG expression likely impact a variety of physiological systems that support higher-cognitive processes, including learning and memory. Again, such actions are likely important in learning how best to contend with challenging environmental conditions.

Facilitatory actions of the LC–noradrenergic system on neuronal energy availability

Neuronal activity is tightly linked to energy utilization. Astrocytes are a repository of stored glucose in the form of glycogen, which appears to be highly metabolically active and linked to neuronal activity (Phelps, 1972; Watanabe and Passonneau, 1973; Ibrahim, 1975). For example, glycogen levels are increased during slow-wave sleep, suggesting increased glucose utilization during waking

(Karnovsky et al., 1983). Evidence suggests that noradrenaline exerts a robust modulatory effect on astrocytic glycogen levels. Thus, application of noradrenaline to cortical slice or astrocyte cultures induces a rapid-onset glycogenolysis (Magistretti et al., 1981; Magistretti, 1986; Hof et al., 1988; Sorg and Magistretti, 1991). This action of noradrenaline is mimicked by application of either an α_1-agonist or a β-agonist (Sorg and Magistretti, 1992). This initial glycogenolysis is followed by an increase in glycogen synthesis to a point where glycogen levels exceed baseline conditions. This noradrenaline-induced resynthesis of glycogen is dependent on actions of β-receptors and permits a more sustained level of glycogenolysis (Sorg and Magistretti, 1992). Thus, these observations suggest that under conditions of increased LC discharge activity, noradrenaline acts to increase glucose availability. Gold and colleagues have demonstrated a substantial impact of glucose availability on memory processes via actions within multiple LC terminal fields (Lee et al., 1988; Gold, 1995). Combined, these observations suggest noradrenaline-dependent modulation of glucose availability may have a substantial impact on learning and memory and possibly other cognitive and/or affective processes. These actions may be of particular importance under conditions of environmental demand associated with enhanced rates of noradrenaline release.

Modulatory actions of the LC–noradrenergic system on cognitive processes

The above-described actions of the LC efferent system suggest a widespread influence of this monoaminergic pathway on information processing at both the single neuron and neuronal network levels across a variety of LC terminal fields. These actions likely have a critical impact on cognitive and behavioral processes and may be of particular importance under threatening and/or stressful conditions requiring accurate collection and processing of information and appropriate behavioral response selection. Consistent with these observations, evidence suggests that the LC–noradrenergic system plays a prominent role in a variety of cognitive processes related to the collection, processing, retention, and utilization of sensory information. Of particular relevance to this chapter, noradrenaline appears to play a prominent role in stressor-induced alterations in these processes.

Relationship between LC electrophysiological activity and vigilance

The waking state is associated with enhanced attention and sensitivity to environmental stimuli. Within this state, the extent to which an animal engages in focused versus scanning attention varies with the environmental configuration. The ability to regulate focused attention can be essential for both normal behavior and, ultimately, survival. This may be particularly true under stressful conditions that pose a threat to the animal. Vigilance, a measure of sustained attention (Mackworth, 1970; Parasuraman, 1984; Warm and Jerison 1984), is highly correlated with forebrain neuronal activity patterns, as measured by EEG (Makeig and Inlow, 1992). In support of a role of the LC–noradrenergic system in vigilance, Aston-Jones and colleagues have demonstrated a high correlation between *phasic* fluctuations in LC discharge activity and performance on a vigilance task in monkeys (Rajkowski et al., 1994; Aston-Jones et al., 1997, 1998). In this task, animals were required to press a lever when they detected a visual target stimulus embedded in a series of nontarget stimuli. Phasic LC responses were elicited preferentially by target stimuli that evoked correct responses from the animal. Importantly, these LC responses preceded the behavioral response. Thus, in this task, LC neurons display phasic discharge activity in response to salient sensory stimuli that contain information relevant to goal attainment.

The relationship between phasic activity and target detection is dependent on *tonic* discharge activity: phasic responses to correct target detection are observed only when tonic discharge activity falls within a relatively narrow range. Specifically, moderately elevated tonic discharge rates (approximately 2.0 Hz) are associated with optimal target detection and maximal phasic LC response magnitude. At discharge levels below and above this, vigilance is impaired due to sedation of the animal and a failure to detect target stimuli. Under these conditions, target stimuli fail to elicit phasic discharge. At moderately higher levels of tonic discharge (3.0 Hz), increased eye movements, decreased ability to foveate on a pretest eye fixation stimulus (e.g. enhanced scanning), and more frequent false alarm errors occur, accompanied by diminished phasic responses to target stimuli (Rajkowski et al., 1994). The increase in false alarms results from a decreased ability to discriminate target from distractor (decreased d') and an increase in nontarget responding (decreased β factor; Aston-Jones et al., 1994; Usher et al., 1999).

These observations indicate a three-way relationship between tonic discharge, phasic discharge, and behavior (vigilance). As part of this relationship, there is an inverted U-shaped relationship between tonic discharge rates and vigilance with poor performance on vigilance tasks observed at both low and high rates of tonic discharge. It is interesting to speculate that the nonmonotonic relationship between LC discharge rate and vigilance may involve, at least in part, nonmonotonic dose-dependent effects of noradrenaline observed at the cellular level in neocortex, as described above. Of particular relevance to the current discussion, the decrease in phasic discharge observed at higher rates of tonic discharge activity is similar to that observed with stressor-induced increases in tonic discharge activity described above. Given acute stress is most likely associated with tonic discharge levels that approach or exceed 4–5 Hz, it is predicted that within this range of tonic discharge activity both phasic discharge and performance in a vigilance task would be disrupted. To date, the degree to which stressors impact vigilance, and the role of noradrenergic systems in stressor-induced alterations in vigilance, remains to be determined.

Modulatory actions of the LC–noradrenergic system on attention

The observations described above suggest a prominent role of the LC in at least a subset of attentional processes. Additionally, the ability of noradrenaline to enhance cortical function by reducing "noise" and/or facilitating processing of relevant sensory signals suggests that the LC–noradrenergic system might enhance cognitive function under "noisy" conditions, where irrelevant stimuli could impair performance. Results from studies conducted in rodents, monkeys, and humans in which rates of noradrenergic neurotransmission were manipulated largely support these hypotheses. For example, noradrenaline depletion produces deficits in the performance of nonaged, otherwise intact animals on a variety of tasks when irrelevant stimuli are presented during testing (for review, see Mehta et al., 2001). Thus, the addition of distracting visual stimuli at the choice point in a T-maze produces a greater disruption of performance in noradrenaline-depleted rats than in sham-treated animals (Roberts et al., 1975; Oke and Adams, 1978). Similarly, the presentation of irrelevant, auditory stimuli impairs sustained attention in rats with forebrain noradrenaline depletion, although these animals perform normally under nondistracting conditions (Carli et al., 1983). Further, noradrenaline depletion increases conditioned responses to irrelevant stimuli, while decreasing responses to relevant stimuli (Lorden et al., 1980; Selden et al., 1990, 1991). Thus, overall, impairment of noradrenergic neurotransmission impacts attentional and other cognitive tasks under conditions associated with high-demand and/or increased arousal.

The LC–noradrenergic system may be particularly sensitive to novel environmental stimuli. For example, enhanced LC discharge rates are observed when rats encounter novel stimuli within a familiar environment (Vankov et al., 1995). Further, pharmacological manipulations that enhance noradrenaline release increase physical contact/interaction with a novel stimulus located within a familiar environment (Devauges and Sara, 1990). In contrast, when examined in a novel environment, enhanced noradrenergic neurotransmission decreases attention to an individual object, possibly reflecting enhanced scanning of the environment (Arnsten et al., 1981; Berridge and Dunn, 1989). These observations indicate that not only is the LC particularly responsive to novelty, but that this system exerts a strong modulatory influence on exploration of, and interaction with, novel aspects of the environment. Novel stimuli, by their very nature, may be particularly salient given the unknown potential of these stimuli to harm or help the animal.

Combined, the above-described evidence suggests a role of the LC–noradrenaline system in the attention to salient environmental stimuli (including novel stimuli), particularly under challenging, distracting conditions, including stress. Consistent with this conclusion, prior exposure to a stressor decreases the degree to which an animal attends to a specific stimulus within the environment (Arnsten et al., 1985), an effect reversed by treatment with a noradrenergic α_1-antagonist (Berridge and Dunn, 1989).

Modulatory actions of the LC–noradrenergic system on working memory

In primates, prefrontal cortex (PFC) serves a critical role in inhibition of processing of irrelevant stimuli (Knight et al., 1981; Woods and Knight, 1986). In monkeys, this region can be functionally subdivided, with the dorsolateral PFC associated with performance in the spatial delayed-response task, a test of spatial working memory. Working memory is highly sensitive to stress (Arnsten and Goldman-Rakic, 1998). Substantial evidence indicates noradrenaline exerts a potent modulatory influence on working memory via actions within the PFC and that these actions contribute to stressor-induced alterations in working memory. Importantly, and similar to that observed with vigilance, working memory performance displays an inverted U-type relationship with rates of noradrenaline release such that both low rates (associated with lesions or aging) and high rates (associated with stress) are associated with impaired performance in tests of working memory.

Early observations indicated that damage to the catecholaminergic innervation of the PFC impairs working memory (Brozoski et al., 1979; Simon, 1981; Collins et al., 1998). Additional work indicated a facilitatory role of noradrenaline in the modulation of working memory performance through actions at postsynaptic α_2-noradrenergic receptors. For example, α_2-agonists improve performance in dorsolateral PFC-dependent tasks in monkeys with neurotoxin-induced catecholamine depletion (Arnsten and Goldman-Rakic, 1985; Schneider and Kovelowski, 1990; Cai et al., 1993). The ability of α_2-agonists to improve working memory is most readily observed under conditions that challenge PFC function, such as during the presentation of distracting stimuli (Jackson and Buccafusco, 1991; Arnsten and Contant, 1992). These beneficial effects of α_2-agonists on working memory involve drug actions directly within the PFC. For example, direct infusion of an α_2-antagonist into PFC impairs working memory performance (Li and Mei, 1994), whereas direct infusion of the α_2-agonists into PFC improves performance in both monkey and rat (Tanila et al., 1996; Arnsten, 1997; Mao et al., 1999).

In contrast to that of α_2-receptors, α_1-receptor stimulation within PFC exerts a debilitating effect on working memory performance. Thus, infusions of an α_1-receptor agonist into the PFC of rats (Arnsten et al., 1999) or monkeys (Mao et al., 1999) impairs working memory performance, an effect reversed by coadministration of an α_1-antagonist (Arnsten et al., 1999). Further, stressors also impair working memory performance (Arnsten and Goldman-Rakic, 1998). Importantly, this stressor-induced impairment in working memory is blocked by infusion of an α_1-antagonist directly within PFC (Birnbaum et al., 1999).

Based on these and other observations, Arnsten (2000) has proposed that under moderate rates of release associated with quiet, alert waking, noradrenaline facilitates working memory performance via actions at α_{2A}-receptors located within PFC. Under conditions associated with low PFC noradrenaline levels (e.g. aging), performance is impaired due to insufficient stimulation of α_{2A}-receptors. α_{2A}-receptors have a higher affinity for noradrenaline than do α_1-receptors (O'Rourke et al., 1994) and thus under normal conditions (moderate arousal levels), minimal α_1-mediated neurotransmission occurs. Under conditions of stress and higher arousal levels, which are associated with increased rates of noradrenaline release, stimulation of α_1-receptors impairs working memory.

It should be noted that impaired working memory and/or vigilance associated with stress is not necessarily detrimental to the animal. Presumably, increased labile attention associated with higher-arousal levels, which occurs at the expense of working memory, serves an important survival benefit under appropriate environmental conditions. Of course, these environmental conditions are rarely encountered at home, work, or school, and thus under these conditions, elevated levels of noradrenaline release and subsequent noradrenaline-dependent alterations in working memory and/or vigilance may impair cognitive and behavioral processes appropriate for these conditions.

Modulatory actions of the LC–noradrenergic system on arousal-related memory

Memory strength can be enhanced by stressful, emotionally-arousing conditions (Christianson, 1992). Steroid (e.g. glucocorticoids) and catecholamine (e.g. epinephrine) hormones participate in this arousal-induced enhancement of memory (Kety, 1972; Gold et al., 1975; Liang et al., 1985; Introini-Collison and McGaugh, 1986; Ferry et al., 1999c). Circulating epinephrine stimulates release of central noradrenaline via β-receptors located on vagal afferents (see McGaugh, 2000). Noradrenaline release within the amygdala plays a critical role in the memory-enhancing actions of both arousing stimuli and circulating epinephrine. For example, epinephrine-induced memory enhancement is blocked by intra-amygdala infusions of a β-antagonist (Liang et al., 1986; McGaugh et al., 1996). Recent evidence indicates that the basolateral nucleus of the amygdala is a critical site within the amygdala in the memory-modulating effects of noradrenaline. Thus, posttraining infusions of noradrenaline into the basolateral nucleus of the amygdala enhances spatial learning, whereas β-antagonist infusions have an opposite effect on performance in this (Hatfield and McGaugh, 1999) and an inhibitory avoidance task (Ferry and McGaugh, 1999). Further, glucocorticoid-induced enhancement of performance in an inhibitory avoidance task is blocked by intra-basolateral amygdala infusion of β-antagonists. This effect is observed with either β_1- or β_2-blockade and is not observed when infusions are placed within the central nucleus of the amygdala (Quirarte et al., 1997).

Similar to that observed with β-receptor-dependent neurotransmission, α_1-agonists and antagonists also facilitate or impair, respectively, performance in an inhibitory avoidance task when infused directly within the basolateral amygdala (Ferry et al., 1999b). This facilitatory action of α_1-receptors on memory appears to result from the α_1-dependent enhancement of β-receptor-mediated cAMP production (Ferry et al., 1999a). For example, β-antagonists block the α_1-agonist-induced enhancement of memory (Ferry et al., 1999a). In contrast, α_1-antagonist infusions into basolateral amygdala shift the dose–response curve for β-agonist-induced memory enhancement to the right (Ferry et al., 1999a).

These observations indicate a prominent role of noradrenaline (via actions within the basolateral amygdala) in the regulation of memory under conditions of high-arousal conditions, particularly those associated with stress. In support of a role of noradrenaline in emotionally-related memory in humans, Cahill et al. (1994) demonstrated β-receptor blockade in human subjects blocks the typically observed enhanced memory for emotionally activating images relative to emotionally-neutral images.

Noradrenergic modulation of motor function

Currently much of the information regarding the actions of noradrenaline on neuronal activity concerns systems involved in the acquisition and processing of sensory information. Less is known about the actions of noradrenaline on central motor systems. However, available information suggests, as in sensory neuronal circuits, LC efferent pathway stimulation or noradrenaline application enhances motoneuron responsiveness to excitatory synaptic inputs (VanderMaelen and Aghajanian, 1980; White and Neuman, 1980; Fung and Barnes, 1981; Fung and Barnes, 1987; Foehring et al., 1989; Rasmussen and Aghajanian, 1990). Further, and similar to that observed within sensory systems, noradrenaline appears to facilitate both excitatory and inhibitory responses of cerebellar perkinje cells (Freedman et al., 1977; Moises et al., 1979, 1981, 1983; Moises and Woodward, 1980). Finally, descending LC efferent pathways have been demonstrated to exert multiple actions on spinal interneurons, motoneurons, spinal reflex activity, and postural mechanisms, associated with motor performance (Fung et al., 1991; Pompeiano et al., 1991). Thus, the LC efferent pathway appears to facilitate neuronal function at multiple levels of the motor system. In conjunction with its proposed role in facilitating sensory signal transmission according to the behavioral demands of the organism, LC output may also regulate the speed and efficiency of motor responses to salient stimuli. As such, an overall outcome of LC activation may both increased detection and response to stimuli that have survival value for the organism.

Clinical implications

The above-described observations indicate the LC–noradrenergic system impacts widespread neural circuits involved in the collection and processing of sensory information and that these actions may play a prominent role in physiological and behavioral responding in stress. As such, dysregulation of LC–noradrenergic neurotransmission might impact any number of stress-related cognitive and affective processes. In keeping with this, the dysregulation of noradrenergic neurotransmission has been proposed to contribute to a large number of cognitive and affective disorders, including stress-related disorders. Much of the evidence suggesting a potential role of the LC in behavioral disorders derives from the therapeutic actions of pharmacological treatments that target noradrenergic neurotransmission. However, it is essential to note that a pharmacological intervention can be therapeutic while not targeting the specific biological deficit causing the symptoms. In most cases, there is little evidence indicating a direct, or causal, relationship between dysregulation of noradrenergic neurotransmission and a particular behavioral disorder. This may simply reflect the difficulty of assessing CNS processes in vivo in humans and the limitations of indirect measures that are necessarily used to assess central noradrenergic neurotransmission in humans. Alternatively, this may suggest that the dysregulation of noradrenergic systems is not a primary etiological factor in cognitive and/or affective dysfunction associated with these psychiatric/behavioral disorders. In this latter view, the LC–noradrenergic system may well represent an appropriate target for pharmacological intervention because this system innervates and modulates a dysfunctional neural circuit, and not because this neurotransmitter system itself is dysfunctional. The extent to which noradrenergic systems are causally involved in cognitive and affective symptoms associated with a variety of behavioral disorders remains a critical question for future work.

Much has been written about the potential involvement of noradrenergic systems in a variety of stress- and anxiety-related disorders (for review, see Southwick et al., 1999; Sullivan et al., 1999; Anand and Charney, 2000). Discussion of this broad and complicated topic exceeds the scope of the current review. However, it is worth noting that much of the original impetus behind speculation of an anxiogenic action of noradrenaline was the observation that stressors were particularly potent at activating the LC–noradrenaline system. As reviewed above, recent work demonstrates a similar sensitivity/responsivity of LC neurons to appetitive stimuli. These observations indicate that enhanced rates of noradrenaline release per se are not sufficient to induce a negative affective state, such as anxiety. Thus, rather than conveying aversive content, the LC–noradrenaline system may convey more general information regarding stimulus attributes, such as salience. Further, as mentioned above, the terms stress and anxiety have not been well defined operationally. Most animal work addressing the role of noradrenergic systems in stress and anxiety examines motoric responses under conditions that are stressful, as evidenced by physiological indices described above. The extent to which motor responses under these conditions also reflect specific emotional states such as anxiety or fear is difficult to ascertain in animals. As such, it is difficult to make definitive conclusions regarding the role of noradrenergic systems in a particular emotional state from these types of studies. Thus, despite substantial speculation and research, the extent to which noradrenaline exerts an anxiogenic action, under certain conditions, remains unclear. Adding to the confusion, it has been suggested recently that in contrast to an anxiogenic action of the LC–noradrenergic system, this system might in fact serve as an anxiolytic function under stressful conditions (Weiss et al., 1994). Studies in humans indicate increased anxiety following peripheral manipulations that increase noradrenaline neurotransmission (Charney et al., 1983).

However, the relationship between generalized arousal, which may well be sensitive to pheripheral manipulations of noradrenergic neurotransmission, and anxiety has not been fully explored in humans. Thus, the extent to which results obtained in humans indicate direct versus indirect actions of central noradrenergic systems on anxiety-related circuits remains unclear.

Finally, noradrenergic efferents target a variety of CNS regions, many presumably without a role in affect. This being said, it is certainly feasible that through actions on stress and/or anxiety-specific circuits, noradrenaline could well participate in anxiety and/or stress-related processes. In the context of the current discussion, it is suggested that at the very least, certain *affect-independent* components of stress- and anxiety-related disorders may be modulated by the LC–noradrenaline system, including arousal, attention, and learning and memory.

Among stress-related disorders, substantial evidence indicates a hyperreactivity of noradrenergic systems in panic disorder as well as posttraumatic stress disorder (PTSD; see Southwick et al., 1999). This is consistent with the above-described ability of prolonged or intense stressors to sensitize noradrenergic systems to heterotypic stressors in animals. Excessive reactivity of noradrenergic systems in PTSD suggests a causal relationship between noradrenaline release and panic in patients suffering from these disorders. In support of this hypothesis is the ability of increased noradrenergic neurotransmission, via systemic administration of α_2-antagonists, to elicit episodes of panic in these patient populations, but not normal controls (Southwick et al., 1993; Bremner et al., 1996). In the case of PTSD, panic attacks can be associated with memories of traumatic events (see Southwick et al., 1999). As reviewed above, noradrenaline modulates strongly emotional memory via actions within the amygdala and synaptic strength of neuronal ensembles in hippocampus and neocortex. These actions could provide the neural substrates for intrusive memories associated with PTSD. Aside from a possible contributory role of noradrenaline to panic and/or anxiety associated with PTSD, excessive reactivity of central noradrenergic systems in this patient population could lead to dysregulation of arousal and state-dependent memory and/or attentional processes also associated with PTSD, as well as other stress-related disorders.

Summary

A defining feature of stressful conditions is the need to confront challenging, or threatening, conditions. Associated with this is the need to acquire and process sensory information rapidly and efficiently prior to making an accurate response selection. Long-term survival may be dependent on behavioral plasticity to better contend with, or avoid, a given threatening environmental stimulus. Evidence to date argues for a prominent role of the LC–noradrenergic system in a variety of physiological, cognitive, and behavioral processes associated with information processing, response selection, and behavioral plasticity. Thus, results from a variety of studies reveal a surprising degree of cohesion: whether at the level of the single cell, populations of neurons, or behavior, noradrenaline increases the organism's ability to process relevant or salient stimuli while suppressing responding to irrelevant stimuli. This involves two basic categories of action. First, the system contributes to the initiation of behavioral and forebrain neuronal activity states appropriate for the collection of sensory information (e.g. waking). Second, within waking the LC–noradrenergic system modulates a variety of state-dependent processes including sensory information processing, attention, and memory. Noradrenaline-dependent modulation of long-term changes in synaptic strength, gene transcription, and other processes suggest a potentially critical role of this system in experience-dependent alterations in neural function and behavior.

Stress is associated with elevated rates of noradrenaline release. Thus, it is expected that under stressful conditions, noradrenaline-dependent modulation of the above-described physiological and behavioral process will occur. In fact, evidence demonstrates an involvement of the LC–noradrenergic system in stressor-induced alterations in many of these processes (e.g. arousal, working memory, high arousal-related memory, IEG expression). These actions of the LC–noradrenergic system are likely independent of affective valence (e.g. appetitive vs. aversive) and are dependent only on whether a stimulus is salient (relevant) to ongoing and/or future behavioral action. In this capacity noradrenaline facilitates the collection and processing of information most critical to survival. The extent to which a stimulus is deemed salient will be dependent on environmental, homeostatic, and experiential factors: water may be highly salient only when in a water-deprived state. Under certain conditions, the collection of salient information may require attending to a single stimulus or set of stimuli for relatively prolonged periods (e.g. sustained focused attention). Under other conditions, typically associated with higher-arousal levels (e.g. threat/stress, but also certain appetitive conditions), it may be necessary to scan the environment for rapid detection of multiple stimuli. Novelty may be particularly salient, given novel stimuli need to be assessed quickly and efficiently to determine the extent to which they pose a threat. Threatening stimuli are similarly salient, often requiring rapid response selection for survival. The actions of noradrenaline on energy availability may support neural circuits under conditions associated with increased demand, including stress. Moreover, the long-term actions of noradrenaline, whether at the level of the gene (IEGs), neural ensembles (LTP/LTD), or behavior (memory), may facilitate rapid and accurate response selection when a stimulus is reencountered or an environment suggests a particular level of preparedness is warranted (e.g. repeated or chronic stress). The actions of the LC–noradrenergic system on sensory information acquisition and processing likely occur in conjunction with similar facilitatory actions on motor responses.

Combined, these observations suggest that the LC–noradrenergic system is a critical component of the neural architecture that allows an organism to contend with complex and challenging environments. As such, it appears reasonable to propose dysregulation of this system might contribute to the dysregulation of a variety of attention and/or arousal-related processes associated with stress-related disorders. Available evidence suggests a potentially prominent role of the LC–noradrenergic system in PTSD. Independent of whether noradrenergic systems contribute to the etiology of stress- related disorders, noradrenergic systems may well be an appropriate target for pharmacological intervention in the treatment of specific attention, memory, and/or arousal dysfunction associated with these disorders. It remains for future research to delineate completely the multiplicity of cognitive and affective actions of noradrenergic systems and identify the terminal fields and receptor subtypes associated with these actions. The better understanding of the wide range of behavioral actions of this neural system may well provide insight into the development of better pharmacological treatments for a variety of cognitive and affective disorders.

Acknowledgments

The author gratefully acknowledges support by the University of Wisconsin Graduate School and PHS grants DA10681, DA00389, and MH62359.

References

Abercrombie, E.D. and Jacobs, B.L. (1987) Single-unit response of noradrenergic neurons in the locus coeruleus of freely moving cats. II. Adaptation to chronically presented stressful stimuli. J. Neurosci., 7: 2844–2848.

Adams L.M. and Foote, S.L. (1988) Effects of locally infused pharmacological agents on spontaneous and sensory-evoked activity of locus coeruleus neurons. Brain Res. Bull., 21: 395–400.

Anand, A. and Charney, D.S. (2000) Norepinephrine dysfunction in depression. J. Clin. Psychiatry, 61(Suppl. 10): 16–24.

Anisman, H. (1978) Neurochemical changes elicited by stress. In: psychopharmacology of aversively motivated behavior (Anisman H, Bignami G, Eds.), pp 119–172. New York, Plenum Press.

Aoki, C. (1992) Beta-adrenergic receptors: astrocytic localization in the adult visual cortex and their relation to catecholamine axon terminals as revealed by electron microscopic immunocytochemistry. J. Neurosci., 12: 781–792.

Armstrong-James, M. and Fox, K. (1983) Effects of ionophoresed noradrenaline on the spontaneous activity of neurones in rat primary somatosensory cortex. J. Physiol., 335: 427–447.

Arnsten, A.F. (1997) Catecholamine regulation of the prefrontal cortex. J. Psychopharmaco, 11: 151–162.

Arnsten, A.F. (2000) Through the looking glass: differential noradenergic modulation of prefrontal cortical function. Neural Plast., 7: 133–146.

Arnsten, A.F., Berridge, C. and Segal, D.S. (1985) Stress produces opioid-like effects on investigatory behavior. Pharmacol. Biochem. Behav., 22: 803–809.

Arnsten, A.F. and Contant, T.A. (1992) Alpha-2 adrenergic agonists decrease distractibility in aged monkeys performing the delayed response task. Psychopharmacology (Berl), 108: 159–169.

Arnsten, A.F. and Goldman-Rakic, P.S. (1985) Alpha 2-adrenergic mechanisms in prefrontal cortex associated with cognitive decline in aged nonhuman primates. Science, 230: 1273–1276.

Arnsten, A.F. and Goldman-Rakic, P.S. (1998) Noise stress impairs prefrontal cortical cognitive function in monkeys: evidence for a hyperdopaminergic mechanism. Arch. Gen. Psychiatry, 55: 362–368.

Arnsten, A.F., Mathew, R., Ubriani, R., Taylor, J.R. and Li, B.M. (1999) Alpha-1 noradrenergic receptor stimulation impairs prefrontal cortical cognitive function. Biol. Psychiatry, 45: 26–31.

Arnsten, A.F., Segal, D.S., Loughlin, S.E. and Roberts, D.C. (1981) Evidence for an interaction of opioid and noradrenergic locus coeruleus systems in the regulation of environmental stimulus-directed behavior. Brain Res., 222: 351–363.

Aston-Jones, G. and Bloom, F.E. (1981a) Activity of norepinephrine-containing locus coeruleus neurons in behaving rats anticipates fluctuations in the sleep-waking cycle. J. Neurosci., 1: 876–886.

Aston-Jones G. and Bloom FE (1981b) Norepinephrine-containing locus coeruleus neurons in behaving rats exhibit pronounced responses to non-noxious environmental stimuli. J. Neurosci., 1: 887–900.

Aston-Jones, G., Rajkowski, J., Ivanova, S., Usher, M. and Cohen, J. (1998) Neuromodulation and cognitive performance: recent studies of noradrenergic locus ceruleus neurons in behaving monkeys. Adv. Pharmacol., 42: 755–759.

Aston-Jones, G., Rajkowski, J. and Kubiak, P. (1997) Conditioned responses of monkey locus coeruleus neurons anticipate acquisition of discriminative behavior in a vigilance task. Neuroscience, 80: 697–715.

Aston-Jones, G., Rajkowski, J., Kubiak, P. and Alexinsky, T. (1994) Locus coeruleus neurons in monkey are selectively activated by attended cues in a vigilance task. J. Neurosci., 14: 4467–4480.

Bean, A.J. and Roth, R.H. (1991) Extracellular dopamine and neurotensin in rat prefrontal cortex in vivo: effects of median forebrain bundle stimulation frequency, stimulation pattern, and dopamine autoreceptors. J. Neurosci., 11: 2694–2702.

Berridge, C.W. and Abercrombie, E.D. (1999) Relationship between locus coeruleus discharge rates and rates of norepinephrine release within neocortex as assessed by in vivo microdialysis. Neuroscience, 93: 1263–1270.

Berridge, C.W., Bolen, S.J., Manley, M.S. and Foote, S.L. (1996) Modulation of forebrain electroencephalographic activity in halothane-anesthetized rat via actions of noradrenergic beta-receptors within the medial septal region. J. Neurosci., 16: 7010–7020.

Berridge, C.W. and Dunn, A.J. (1989) Restraint-stress-induced changes in exploratory behavior appear to be mediated by norepinephrine-stimulated release of CRF. J. Neurosci., 9: 3513–3521.

Berridge, C.W. and España, R.A. (2000) Synergistic sedative effects of noradrenergic alpha(1)- and beta-receptor blockade on forebrain electroencephalographic and behavioral indices. Neuroscience, 99: 495–505.

Berridge, C.W. and Foote, S.L. (1991) Effects of locus coeruleus activation on electroencephalographic activity in neocortex and hippocampus. J. Neurosci., 11: 3135–3145.

Berridge, C.W. and Foote, S.L. (1996) Enhancement of behavioral and electroencephalographic indices of waking following stimulation of noradrenergic beta-receptors within the medial septal region of the basal forebrain. J. Neurosci., 16: 6999–7009.

Berridge, C.W., Isaac, S.O. and España, R.A. (2002) Additive Wake-Promoting Actions of Medial Basal Forebrain Noradrenergic α_1- and β-Receptor Stimulation. Behav Neurosci.,

Berridge, C.W., O'Neil, J. and Wifler, K. (1999) Amphetamine acts within the medial basal forebrain to initiate and maintain alert waking. Neuroscience, 93: 885–896.

Berridge, C.W. and O'Neill, J. (2001) Differential sensitivity to the wake-promoting actions of norepinephrine within the medial preoptic area and the substantia innominata. Behav. Neurosci., 115: 165–174.

Berridge, C.W., Page, M.E., Valentino, R.J. and Foote, S.L. (1993) Effects of locus coeruleus inactivation on electroencephalographic activity in neocortex and hippocampus. Neuroscience, 55: 381–393.

Berridge, C.W., Stratford, T.L., Foote, S.L. and Kelley, A.E. (1997) Distribution of dopamine beta-hydroxylase-like immunoreactive fibers within the shell subregion of the nucleus accumbens. Synapse, 27: 230–241.

Bing, G., Stone, E.A., Zhang, Y. and Filer, D. (1992) Immunohistochemical studies of noradrenergic-induced expression of c-fos in the rat CNS. Brain Res., 592: 57–62.

Bing, G.Y., Filer, D., Miller, J.C. and Stone, E.A. (1991) Noradrenergic activation of immediate early genes in rat cerebral cortex. Brain Res. Mol. Brain Res., 11: 43–46.

Birnbaum, S., Gobeske, K.T., Auerbach, J., Taylor, J.R. and Arnsten, A.F. (1999) A role for norepinephrine in stress-induced cognitive deficits: alpha-1-adrenoceptor mediation in the prefrontal cortex. Biol. Psychiatry, 46: 1266–1274.

Bloor, B.C. and Flacke, W.E. (1982) Reduction in halothane anesthetic requirement by clonidine, an alpha-adrenergic agonist. Anesth. Analg., 61: 741–745.

Boyajian, C.L. and Leslie, F.M. (1987) Pharmacological evidence for alpha-2 adrenoceptor heterogeneity: differential binding properties of [3H]rauwolscine and [3H]idazoxan in rat brain. J. Pharmacol. Exp. Ther., 241: 1092–1098.

Bremner, J.D., Krystal, J.H., Southwick, S.M. and Charney, D.S. (1996) Noradrenergic mechanisms in stress and anxiety: II. Clinical studies. Synapse, 23: 39-51.

Brozoski, T.J., Brown, R.M., Rosvold, H.E. and Goldman, P.S. (1979) Cognitive deficit caused by regional depletion of dopamine in prefrontal cortex of rhesus monkey. Science, 205: 929–932.

Bushnell, M.C., Goldberg, M.E. and Robinson, D.L. (1981) Behavioral enhancement of visual responses in monkey cerebral cortex. I. Modulation in posterior parietal cortex related to selective visual attention. J. Neurophysiol., 46: 755–772.

Buzsaki, G., Bickford, R.G., Ponomareff, G., Thal, L.J., Mandel, R. and Gage, F.H. (1988) Nucleus basalis and thalamic control of neocortical activity in the freely moving rat. J. Neurosci., 8: 4007–4026.

Buzsaki, G., Kennedy, B., Solt, B.V. and Ziegler, M. (1991) Noradrenergic control of thalamic oscillation: the role of α-2 receptors. Eur. J. Neurosci., 3: 222–229.

Buzsaki, G., Leung, L.W. and Vanderwolf., C.H. (1983) Cellular bases of hippocampal EEG in the behaving rat. Brain Res., 287: 139–171.

Bylund, D.B., Blaxall, H.S., Iversen, L.J., Caron, M.G., Lefkowitz, R.J. and Lomasney, J.W. (1992) Pharmacological characteristics of alpha 2-adrenergic receptors: comparison of pharmacologically defined subtypes with subtypes identified by molecular cloning. Mol. Pharmacol., 42: 1–5.

Cahill, L., Prins, B., Weber, M. and McGaugh, J.L. (1994) Beta-adrenergic activation and memory for emotional events. Nature, 371: 702–704.

Cai, J.X., Ma, Y.Y., Xu, L. and Hu, X.T. (1993) Reserpine impairs spatial working memory performance in monkeys: reversal by the alpha 2-adrenergic agonist clonidine. Brain Res., 614: 191–196.

Cannon, W.B. (1914) The emergency function of the adrenal medulla in pain and the major emotions. Am. J. Physiol., 33: 356–372.

Carli, M., Robbins, T.W., Evenden, J.L. and Everitt, B.J. (1983) Effects of lesions to ascending noradrenergic neurones on performance of a 5-choice serial reaction task in rats; implications for theories of dorsal noradrenergic bundle function based on selective attention and arousal. Behav. Brain Res., 9: 361–380.

Charney, D.S., Heninger, G.R. and Redmond, D.E., Jr. (1983) Yohimbine induced anxiety and increased noradrenergic function in humans: effects of diazepam and clonidine. Life Sci., 33: 19–29.

Chaulk, P.C. and Harley, C.W. (1998) Intracerebroventricular norepinephrine potentiation of the perforant path-evoked potential in dentate gyrus of anesthetized and awake rats: A role for both alpha- and beta-adrenoceptor activation. Brain Res., 787: 59–70.

Christianson, S.A. (1992) Emotional stress and eyewitness memory: a critical review. Psychol. Bull., 112: 284–309.

Ciombor, K.J., Ennis, M. and Shipley, M.T. (1999) Norepinephrine increases rat mitral cell excitatory responses to weak olfactory nerve input via alpha-1 receptors in vitro. Neuroscience, 90: 595–606.

Cirelli, C., Pompeiano, M., d'Ascanio, P., Arrighi, P. and Pompeiano, O. (1996) c-fos Expression in the rat brain after unilateral labyrinthectomy and its relation to the uncompensated and compensated stages. Neuroscience, 70: 515–546.

Cirelli, C. and Tononi, G. (1998) Changes in anti-phosphoserine and anti-phosphothreonine antibody binding during the sleep-waking cycle and after lesions of the locus coeruleus. Sleep Res. Online, 1: 11–18.

Cirelli, C. and Tononi, G. (2000) Differential expression of plasticity-related genes in waking and sleep and their regulation by the noradrenergic system. J. Neurosci., 20: 9187–9194.

Collins, G.G., Probett, G.A., Anson, J. and McLaughlin, N.J. (1984) Excitatory and inhibitory effects of noradrenaline on synaptic transmission in the rat olfactory cortex slice. Brain Res., 294: 211–223.

Collins, P., Roberts, A.C., Dias, R., Everitt, B.J. and Robbins, T.W. (1998) Perseveration and strategy in a novel spatial self-ordered sequencing task for nonhuman primates. Effects Of excitotoxic lesions and dopamine depletions of the prefrontal cortex. J. Cogn. Neurosci., 10: 332–354.

De Sarro, G.B., Ascioti, C., Froio, F., Libri, V. and Nistico, G. (1987) Evidence that locus coeruleus is the site where clonidine and drugs acting at alpha 1- and alpha 2-adrenoceptors affect sleep and arousal mechanisms. Br. J. Pharmacol., 90: 675–685.

De Sarro, G.B., Bagetta, G., Ascioti, C., Libri, V. and Nistico, G. (1988) Microinfusion of clonidine and yohimbine into locus coeruleus alters EEG power spectrum: effects of aging and reversal by phosphatidylserine. Br. J. Pharmacol., 95: 1278–1286.

De Sarro, G.B., Bagetta, G., Ascioti, C., Libri, V. and Nistico, G. (1989) Effects of pertussis toxin on the behavioural and ECoG spectrum changes induced by clonidine and yohimbine after their microinfusion into the locus coeruleus. Br. J. Pharmacol., 96: 59–64.

Delfs, J.M., Zhu, Y., Druhan, J.P. and Aston-Jones, G.S. (1998) Origin of noradrenergic afferents to the shell subregion of the nucleus accumbens: anterograde and retrograde tract-tracing studies in the rat. Brain Res., 806: 127–140.

Devauges, V. and Sara, S.J. (1990) Activation of the noradrenergic system facilitates an attentional shift in the rat. Behav. Brain Res., 39: 19–28.

Devilbiss, D.M. and Waterhouse, B.D. (2000) Norepinephrine exhibits two distinct profiles of action on sensory cortical neuron responses to excitatory synaptic stimuli. Synapse, 37: 273–282.

Dohlman, H.G., Thorner, J., Caron, M.G. and Lefkowitz, R.J. (1991) Model systems for the study of seven-transmembrane-segment receptors. Annu. Rev. Biochem., 60: 653–688.

Domich, L., Oakson, G. and Steriade, M. (1986) Thalamic burst patterns in the naturally sleeping cat: a comparison between cortically projecting and reticularis neurones. J. Physiol., 379: 429–449.

Dragunow, M., Currie, R.W., Faull, R.L., Robertson, H.A. and Jansen, K. (1989) Immediate-early genes, kindling and long-term potentiation. Neurosci. Biobehav. Rev., 13: 301–313.

Dunn, A.J. and Kramarcy, N.R. (1984) Neurochemical responses in stress: relationships between the hypothalamic–pituitary–adrenal and catecholamine systems. In: Iversen, L.L., Iversen, S.D., Snyder, S.H., (Eds.), Handbook of Psychopharmacology. Plenum Publishing Corporation, New York, pp. 455–515.

Evarts, E.V. (1960) Effects of sleep and waking on spontaneous and evoked discharge of single units in visual cortex. Fed. Proc., 19: 828–837.

Feenstra, M.G. (2000) Dopamine and noradrenaline release in the prefrontal cortex in relation to unconditioned and conditioned stress and reward. Prog. Brain Res., 126: 133–163.

Feenstra M.G., Teske G., Botterblom M.H. and de Bruin J.P. (1999) Dopamine and noradrenaline release in the prefrontal cortex of rats during classical aversive and appetitive conditioning to a contextual stimulus: interference by novelty effects. Neurosci. Lett., 272: 179–182.

Feenstra, M.G., Vogel, M., Botterblom, M.H. and Joosten, R.N. and de Bruin, J.P. (2001) Dopamine and noradrenaline efflux in the rat prefrontal cortex after classical aversive conditioning to an auditory cue. Eur. J. Neurosci., 13: 1051–1054.

Ferry, B. and McGaugh, J.L. (1999) Clenbuterol administration into the basolateral amygdala post-training enhances retention in an inhibitory avoidance task. Neurobiol. Learn Mem., 72: 8–12.

Ferry, B., Roozendaal, B. and McGaugh, J.L. (1999a) Basolateral amygdala noradrenergic influences on memory storage are mediated by an interaction between beta- and alpha1-adrenoceptors. J. Neurosci., 19: 5119–5123.

Ferry, B., Roozendaal, B. and McGaugh, J.L. (1999b) Involvement of alpha1-adrenoceptors in the basolateral amygdala in modulation of memory storage. Eur. J. Pharmacol., 372: 9–16.

Ferry, B., Roozendaal, B. and McGaugh, J.L. (1999c) Role of norepinephrine in mediating stress hormone regulation of long-term memory storage: a critical involvement of the amygdala. Biol. Psychiatry., 46: 1140–1152.

Findlay, A.L. and Hayward, J.N. (1969) Spontaneous activity of single neurones in the hypothalamus of rabbits during sleep and waking. J. Physiol., 201: 237–258.

Finlay, J.M., Zigmond, M.J. and Abercrombie, E.D. (1995) Increased dopamine and norepinephrine release in medial prefrontal cortex induced by acute and chronic stress: effects of diazepam. Neuroscience, 64: 619–628.

Florin-Lechner, S.M., Druhan, J.P., Aston-Jones, G. and Valentino, R.J. (1996) Enhanced norepinephrine release in prefrontal cortex with burst stimulation of the locus coeruleus. Brain Res., 742: 89–97.

Foehring, R.C., Schwindt, P.C. and Crill, W.E. (1989) Norepinephrine selectively reduces slow Ca2+- and Na+-mediated K+ currents in cat neocortical neurons. J. Neurophysiol., 61: 245–256.

Foote, S.L., Aston-Jones, G. and Bloom, F.E. (1980) Impulse activity of locus coeruleus neurons in awake rats and monkeys is a function of sensory stimulation and arousal. Proc. Natl. Acad. Sci. USA, 77: 3033–3037.

Foote, S.L., Bloom, F.E. and Aston-Jones, G. (1983) Nucleus locus ceruleus: new evidence of anatomical and physiological specificity. Physiol. Rev., 63: 844–914.

Foote, S.L., Freedman, R. and Oliver, A.P. (1975) Effects of putative neurotransmitters on neuronal activity in monkey auditory cortex. Brain Res., 86: 229–242.

Foote, S.L. and Morrison, J.H. (1987) Extrathalamic modulation of cortical function. Annu. Rev. Neurosci., 10: 67–95.

Fort, P., Khateb, A., Pegna, A., Muhlethaler, M. and Jones, B.E. (1995) Noradrenergic modulation of cholinergic nucleus basalis neurons demonstrated by in vitro pharmacological and immunohistochemical evidence in the guinea-pig brain. Eur. J. Neurosci., 7: 1502–1511.

Freedman, R., Hoffer, B.J., Woodward, D.J. and Puro, D. (1977) Interaction of norepinephrine with cerebellar activity evoked by mossy and climbing fibers. Exp. Neurol., 55: 269–288.

Fung, S.J. and Barnes, C.D. (1981) Evidence of facilitatory coerulospinal action in lumbar motoneurons of cats. Brain Res., 216: 299–311.

Fung, S.J. and Barnes, C.D. (1987) Membrane excitability changes in hindlimb motoneurons induced by stimulation of the locus coeruleus in cats. Brain Res., 402: 230–242.

Fung, S.J., Manzoni, D., Chan, J.Y., Pompeiano, O. and Barnes, C.D. (1991) Locus coeruleus control of spinal motor output. Prog. Brain Res., 88: 395–409.

Gatti, P.J., Hill, K.J., Da Silva, A.M., Norman, W.P. and Gillis, R.A. (1988) Central nervous system site of action for the hypotensive effect of clonidine in the cat. J. Pharmacol. Exp. Ther., 245: 373–380.

George, M.J. (1992) Modification of receptive fields of posteriomedial barrel subfield neocortical single units by known concentrations of iontophoresed noradrenaline in the rat. Int. J. Neurosci., 65: 69–81.

Gold, P.E. (1995) Role of glucose in regulating the brain and cognition. Am. J. Clin. Nutr., 61: 987S–995S.

Gold, P.E., van Buskirk. R.B. and McGaugh, J.L. (1975) Effects of hormones on time-dependent memory storage processes. Prog. Brain Res., 42: 210–211.

Goldberg, M.E. and Bushnell, M.C. (1981) Behavioral enhancement of visual responses in monkey cerebral cortex.

II. Modulation in frontal eye fields specifically related to saccades. J. Neurophysiol., 46: 773–787.

Goldman-Rakic, P.S., Lidow, M.S. and Gallager, D.W. (1990) Overlap of dopaminergic, adrenergic, and serotoninergic receptors and complementarity of their subtypes in primate prefrontal cortex. J. Neurosci., 10: 2125–2138.

Grant, S.J. and Redmond, D.E., Jr. (1984) Neuronal activity of the locus ceruleus in awake Macaca arctoides. Exp. Neurol., 84: 701–708.

Gubits, R.M., Smith, T.M., Fairhurst, J.L. and Yu, H. (1989) Adrenergic receptors mediate changes in c-fos mRNA levels in brain. Brain Res. Mol. Brain Res., 6: 39–45.

Guzowski, J.F., Lyford, G.L., Stevenson, G.D., Houston, F.P., McGaugh, J.L., Worley, P.F. and Barnes, C.A. (2000) Inhibition of activity-dependent arc protein expression in the rat hippocampus impairs the maintenance of long-term potentiation and the consolidation of long-term memory. J. Neurosci., 20: 3993–4001.

Harley, C. (1991) Noradrenergic and locus coeruleus modulation of the perforant path-evoked potential in rat dentate gyrus supports a role for the locus coeruleus in attentional and memorial processes. Prog. Brain Res., 88: 307–321.

Harley, C.W. and Sara, S.J. (1992) Locus coeruleus bursts induced by glutamate trigger delayed perforant path spike amplitude potentiation in the dentate gyrus. Exp. Brain Res., 89: 581–587.

Hatfield, T. and McGaugh, J.L. (1999) Norepinephrine infused into the basolateral amygdala posttraining enhances retention in a spatial water maze task. Neurobiol. Learn Mem., 71: 232–239.

Heginbotham, L.R. and Dunwiddie, T.V. (1991) Long-term increases in the evoked population spike in the CA1 region of rat hippocampus induced by beta-adrenergic receptor activation. J. Neurosci., 11: 2519–2527.

Hobson, J.A., McCarley, R.W. and Wyzinski, P.W. (1975) Sleep cycle oscillation: reciprocal discharge by two brainstem neuronal groups. Science, 189: 55–58.

Hof, P.R., Pascale, E. and Magistretti, P.J. (1988) K+ at concentrations reached in the extracellular space during neuronal activity promotes a Ca2+-dependent glycogen hydrolysis in mouse cerebral cortex. J. Neurosci., 8: 1922–1928.

Hopkins, W.F. and Johnston, D. (1984) Frequency-dependent noradrenergic modulation of long-term potentiation in the hippocampus. Science, 226: 350–352.

Hopkins, W.F. and Johnston, D. (1988) Noradrenergic enhancement of long-term potentiation at mossy fiber synapses in the hippocampus. J. Neurophysiol., 59: 667–687.

Hyvarinen, J., Poranen A. and Jokinen, Y. (1980) Influence of attentive behavior on neuronal responses to vibration in primary somatosensory cortex of the monkey. J. Neurophysiol., 43: 870–882.

Ibrahim, M.Z. (1975) Glycogen and its related enzymes of metabolism in the central nervous system. Adv. Anat. Embryol. Cell Biol., 52: 3–89.

Introini-Collison, I.B. and McGaugh, J.L. (1986) Epinephrine modulates long-term retention of an aversively motivated discrimination. Behav. Neural. Biol., 45: 358–365.

Jackson, W.J. and Buccafusco, J.J. (1991) Clonidine enhances delayed matching-to-sample performance by young and aged monkeys. Pharmacol. Biochem. Behav., 39: 79–84.

Jones, C.R. and Palacios, J.M. (1991) Autoradiography of adrenoceptors in rat and human brain: alpha-adrenoceptor and idazoxan binding sites. Prog. Brain Res., 88: 271–291.

Jung, T.P., Makeig, S., Stensmo, M. and Sejnowski, T.J. (1997) Estimating alertness from the EEG power spectrum. IEEE Trans. Biomed. Eng., 44: 60–69.

Karnovsky, M.L., Reich, P., Anchors, J.M. and Burrows, B.L. (1983) Changes in brain glycogen during slow-wave sleep in the rat. J. Neurochem., 41: 1498–1501.

Kaukinen, S. and Pyykko, K. (1979) The potentiation of halothane anaesthesia by clonidine. Acta Anaesthesiol Scand, 23: 107–111.

Kety, S.S. (1972) The possible role of the adrenergic systems of the cortex in learning. Res. Publ. Assoc. Res. Nerv. Ment. Dis., 50: 376–389.

Kirkwood, A., Rozas, C., Kirkwood, J., Perez, F. and Bear, M.F. (1999) Modulation of long-term synaptic depression in visual cortex by acetylcholine and norepinephrine. J. Neurosci., 19: 1599–1609.

Kitchigina, V., Vankov, A., Harley, C. and Sara, S.J. (1997) Novelty-elicited, noradrenaline-dependent enhancement of excitability in the dentate gyrus. Eur. J. Neurosci., 9: 41–47.

Knight, R.T., Hillyard, S.A., Woods, D.L. and Neville, H.J. (1981) The effects of frontal cortex lesions on event-related potentials during auditory selective attention. Electroencephalogr. Clin. Neurophysiol., 52: 571–582.

Korf, J., Aghajanian, G.K. and Roth, R.H. (1973) Increased turnover of norepinephrine in the rat cerebral cortex during stress: role of the locus coeruleus. Neuropharmacology, 12: 933-938.

Kossl, M., and Vater, M. (1989) Noradrenaline enhances temporal auditory contrast and neuronal timing precision in the cochlear nucleus of the mustached bat. J. Neurosci., 9: 4169–4178.

Kramarcy, N.R., Delanoy, R.L. and Dunn, A.J. (1984) Footshock treatment activates catecholamine synthesis in slices of mouse brain regions. Brain Res., 290: 311–319.

Lee, M.K., Graham, S.N. and Gold, P.E. (1988) Memory enhancement with posttraining intraventricular glucose injections in rats. Behav. Neurosci., 102: 591–595.

Lewis, D.A. (2001) The catecholamine innervation of primate cerebral cortex. In: Solanto, M.V, Arnsten, A.F.T. and Castellanos, F.X., (Eds.), Stimulant Drugs and ADHD: Basic and Clinical Neuroscience. Oxford University Press, New York, pp. 77–103.

Lewis, D.A., Campbell, M.J., Foote, S.L., Goldstein, M. and Morrison, J.H. (1987) The distribution of tyrosine hydroxylase-immunoreactive fibers in primate neocortex is widespread but regionally specific. J. Neurosci., 7: 279–290.

Li, B.M. and Mei, Z.T. (1994) Delayed-response deficit induced by local injection of the alpha 2-adrenergic antagonist yohimbine into the dorsolateral prefrontal cortex in young adult monkeys. Behav. Neural. Biol., 62: 134–139.

Liang, K.C., Bennett, C. and McGaugh, J.L. (1985) Peripheral epinephrine modulates the effects of post-training amygdala stimulation on memory. Behav. Brain Res., 15: 93–100.

Liang, K.C., Juler, R.G. and McGaugh, J.L. (1986) Modulating effects of posttraining epinephrine on memory: involvement of the amygdala noradrenergic system. Brain Res., 368: 125–133.

Livingstone, M.S. and Hubel, D.H. (1981) Effects of sleep and arousal on the processing of visual information in the cat. Nature, 291: 554–561.

Lorden, J.F., Callahan, M. and Dawson, R., Jr. (1980) Depletion of central catecholamines alters amphetamine- and fenfluramine-induced taste aversions in the rat. J. Comp. Physiol. Psychol., 94: 99–114.

Mackworth, J.F. (1970) Vigilance and attention. Penguin, Baltimore.

Magistretti, P.J. (1986) Intercellular communication mediated by VIP in the cerebral cortex. Peptides, 7(1): 169–173.

Magistretti, P.J., Morrison, J.H., Shoemaker, W.J., Sapin, V. and Bloom, F.E. (1981) Vasoactive intestinal polypeptide induces glycogenolysis in mouse cortical slices: a possible regulatory mechanism for the local control of energy metabolism. Proc. Natl. Acad. Sci. USA, 78: 6535–6539.

Makeig, S. and Inlow, M. (1992) Lapses in alertness: coherence of fluctuations in performance and EEG spectrum. Electroencephalogr Clin. Neurophysiol., 86: 23–35.

Makeig, S. and Jung, T.P. (1996) Tonic, phasic, and transient EEG correlates of auditory awareness in drowsiness. Brain Res. Cogn. Brain Res., 4: 15–25.

Mallick, B.N. and Alam, M.N. (1992) Different types of norepinephrinergic receptors are involved in preoptic area mediated independent modulation of sleep-wakefulness and body temperature. Brain Res., 591: 8–19.

Manley, L.D., Kuczenski, R., Segal, D.S., Young, S.J. and Groves, P.M. (1992) Effects of frequency and pattern of medial forebrain bundle stimulation on caudate dialysate dopamine and serotonin. J. Neurochem., 58: 1491–1498.

Manunta, Y. and Edeline, J.M. (1997) Effects of noradrenaline on frequency tuning of rat auditory cortex neurons. Eur. J. Neurosci., 9: 833–847.

Mao, Z.M., Arnsten, A.F. and Li, B.M. (1999) Local infusion of an alpha-1 adrenergic agonist into the prefrontal cortex impairs spatial working memory performance in monkeys. Biol. Psychiatry, 46: 1259–1265.

Marinelli, M. and Piazza, P.V. (2002) Interaction between glucocorticoid hormones, stress and psychostimulant drugs. Eur. J. Neurosci., 16: 387–394.

Marwaha, J. and Aghajanian, G.K. (1982) Relative potencies of alpha-1 and alpha-2 antagonists in the locus ceruleus, dorsal raphe and dorsal lateral geniculate nuclei: an electrophysiological study. J. Pharmacol. Exp. Ther., 222: 287–293.

McCarley, R.W., Benoit, O. and Barrionuevo, G. (1983) Lateral geniculate nucleus unitary discharge in sleep and waking: state- and rate-specific aspects. J. Neurophysiol., 50: 798–818.

McCormick, D.A. (1989) Cholinergic and noradrenergic modulation of thalamocortical processing. Trends Neurosci., 12: 215–221.

McCormick, D.A. and Bal, T. (1997) Sleep and arousal: thalamocortical mechanisms. Annu. Rev. Neurosci., 20: 185–215.

McCormick, D.A., Pape, H.C. and Williamson, A. (1991) Actions of norepinephrine in the cerebral cortex and thalamus: implications for function of the central noradrenergic system. Prog. Brain Res., 88: 293–305.

McCormick, D.A. and Prince, D.A. (1988) Noradrenergic modulation of firing pattern in guinea pig and cat thalamic neurons, in vitro. J. Neurophysiol., 59: 978–996.

McCune, S.K., Voigt, M.M. and Hill, J.M. (1993) Expression of multiple alpha adrenergic receptor subtype messenger RNAs in the adult rat brain. Neuroscience, 57: 143–151.

McGaugh, J.L. (2000) Memory – a century of consolidation. Science, 287: 248–251.

McGaugh, J.L., Cahill, L. and Roozendaal, B. (1996) Involvement of the amygdala in memory storage: interaction with other brain systems. Proc. Natl. Acad Sci. USA, 93: 13508–13514.

McGinty, D.J. and Sterman, M.B. (1968) Sleep suppression after basal forebrain lesions in the cat. Science, 160: 1253–1255.

McLean, J. and Waterhouse, B.D. (1994) Noradrenergic modulation of cat area 17 neuronal responses to moving visual stimuli. Brain Res., 667: 83–97.

Mehta, M.A., Sahakian, B.J. and Robbins, T.W. (2001) Comparative psycholpharmacology of methylphenidate and related drugs in human volunteers, patients with ADHD, and experimental animals. In: Solanto, M.V., Arnsten, A.F.T., Castellanos, F.X., (Eds.), Stimulant Drugs and ADHD: Basic and Clinical Neuroscience. Oxford University Press, New York, pp. 303–331.

Metherate, R., Cox, C.L. and Ashe, J.H. (1992) Cellular bases of neocortical activation: modulation of neural oscillations by the nucleus basalis and endogenous acetylcholine. J. Neurosci., 12: 4701–4711.

Milbrandt, J. (1987) A nerve growth factor-induced gene encodes a possible transcriptional regulatory factor. Science, 238: 797–799.

Moises, H.C., Waterhouse, B.D. and Woodward, D.J. (1981) Locus coeruleus stimulation potentiates Purkinje cell responses to afferent input: the climbing fiber system. Brain Res., 222: 43–64.

Moises, H.C., Waterhouse, B.D. and Woodward, D.J. (1983) Locus coeruleus stimulation potentiates local inhibitory processes in rat cerebellum. Brain Res. Bull., 10: 795–804.

Moises, H.C. and Woodward, D.J. (1980) Potentiation of GABA inhibitory action in cerebrllum by locus coeruleus stimulation. Brain Res., 182: 327–344.

Moises, H.C., Woodward, D.J., Hoffer, B.J. and Freedman, R. (1979) Interactions of norepinephrine with purkinje cell responses to putative amino acid neurotransmitters applied by microiontophoresis. Exp. Neurol., 64: 493–515.

Mooney, R.D., Bennett-Clarke, C., Chiaia, N.L., Sahibzada, N. and Rhoades, R.W. (1990) Organization and actions of the noradrenergic input to the hamster's superior colliculus. J. Comp. Neurol., 292: 214–230.

Morrison, J.H. and Foote, S.L. (1986) Noradrenergic and serotoninergic innervation of cortical, thalamic, and tectal visual structures in old and new world monkeys. J. Comp. Neurol., 243: 117–138.

Mountcastle, V.B., Andersen, R.A. and Motter, B.C. (1981) The influence of attentive fixation upon the excitability of the light-sensitive neurons of the posterior parietal cortex. J. Neurosci., 1: 1218–1225.

Mouradian, R.D., Sessler, F.M. and Waterhouse, B.D. (1991) Noradrenergic potentiation of excitatory transmitter action in cerebrocortical slices: evidence for mediation by an alpha 1 receptor-linked second messenger pathway. Brain Res., 546: 83–95.

Mueller, A.L., Palmer, M.R., Hoffer, B.J. and Dunwiddie, T.V. (1982) Hippocampal noradrenergic responses in vivo and in vitro. Characterization of alpha and beta components. Naunyn Schmiedebergs Arch. Pharmacol., 318: 259–266.

Mukhametov, L.M., Rizzolatti, G. and Seitun, A. (1970) An analysis of the spontaneous activity of lateral geniculate neurons and of optic tract fibers in free moving cats. Arch. Ital. Biol., 108: 325–347.

Nicholas, A.P., Pieribone, V. and Hokfelt, T. (1993a) Distributions of mRNAs for alpha-2 adrenergic receptor subtypes in rat brain: an in situ hybridization study. J. Comp. Neurol., 328: 575–594.

Nicholas, A.P., Pieribone, V.A. and Hokfelt, T. (1993b) Cellular localization of messenger RNA for beta-1 and beta-2 adrenergic receptors in rat brain: an in situ hybridization study. Neuroscience, 56: 1023–1039.

Nisenbaum, L.K., Zigmond, M.J., Sved, A.F. and Abercrombie, E.D. (1991) Prior exposure to chronic stress results in enhanced synthesis and release of hippocampal norepinephrine in response to a novel stressor. J. Neurosci., 11: 1478–1484.

O'Rourke, M.F., Blaxall, H.S., Iversen, L.J. and Bylund, D.B. (1994) Characterization of [3H]RX821002 binding to alpha-2 adrenergic receptor subtypes. J. Pharmacol. Exp. Ther., 268: 1362–1367.

Oke, A.F. and Adams, R.N. (1978) Selective attention dysfunctions in adult rats neonatally treated with 6-hydoxydopamine. Pharmacol. Biochem. Behav., 9: 429–432.

Page, M.E., Berridge, C.W., Foote, S.L. and Valentino, R.J. (1993) Corticotropin-releasing factor in the locus coeruleus mediates EEG activation associated with hypotensive stress. Neurosci. Lett., 164: 81–84.

Pape, H.C. and McCormick, D.A. (1989) Noradrenaline and serotonin selectively modulate thalamic burst firing by enhancing a hyperpolarization-activated cation current. Nature, 340: 715–718.

Parasuraman, R. (1984) The psychobiology of sustained attention. In: Warm, J.S. (ed.), Sustained Attention in Human Performance. John Wiley & Sons, New York, pp. 61–101.

Pavcovich, L.A., Cancela, L.M., Volosin, M., Molina, V.A. and Ramirez, O.A. (1990) Chronic stress-induced changes in locus coeruleus neuronal activity. Brain Res. Bull., 24: 293–296.

Pfingstm, B.E., O'Connor, T.A. and Miller, J.M. (1977) Response plasticity of neurons in auditory cortex of the rhesus monkey. Exp. Brain Res., 29: 393–404.

Phelps, C.H. (1972) Barbiturate-induced glycogen accumulation in brain. An electron microscopic study. Brain Res., 39: 225–234.

Pieribone, V.A., Nicholas, A.P., Dagerlind, A. and Hokfelt, T. (1994) Distribution of alpha 1 adrenoceptors in rat brain revealed by in situ hybridization experiments utilizing subtype-specific probes. J. Neurosci., 14: 4252–4268.

Pompeiano, O., Horn, E. and d'Ascanio, P. (1991) Locus coeruleus and dorsal pontine reticular influences on the gain of vestibulospinal reflexes. Prog. Brain Res., 88: 435–462.

Quirarte, G.L., Roozendaal, B. and McGaugh, J.L. (1997) Glucocorticoid enhancement of memory storage involves noradrenergic activation in the basolateral amygdala. Proc. Natl. Acad Sci. USA, 94: 14048–14053.

Rainbow, T.C. and Biegon, A. (1983) Quantitative autoradiography of [3H]prazosin binding sites in rat forebrain. Neurosci Lett., 40: 221–226.

Rainbow, T.C., Parsons, B. and Wolfe, B.B. (1984) Quantitative autoradiography of beta 1- and beta 2-adrenergic receptors in rat brain. Proc. Natl. Acad Sci. USA, 81: 1585–1589.

Rajkowski, J., Kubiak, P. and Aston-Jones, G. (1994) Locus coeruleus activity in monkey: phasic and tonic changes are associated with altered vigilance. Brain Res. Bull., 35: 607–616.

Rasmussen, K. and Aghajanian, G.K. (1990) Serotonin excitation of facial motoneurons: receptor subtype characterization. Synapse, 5: 324–332.

Redmond, D.E., Jr. and Huang, Y.H. (1979) Current concepts. II. New evidence for a locus coeruleus-norepinephrine connection with anxiety. Life Sci., 25: 2149–2162.

Roberts, D.C., Zis, A.P. and Fibiger, H.C. (1975) Ascending catecholamine pathways and amphetamine-induced locomotor activity: importance of dopamine and apparent non-involvement of norepinephrine. Brain Res., 93: 441–454.

Rogawski, M.A. and Aghajanian, G.K. (1980a) Activation of lateral geniculate neurons by norepinephrine: mediation by an alpha-adrenergic receptor. Brain Res., 182: 345–359.

Rogawski, M.A. and Aghajanian, G.K. (1980b) Modulation of lateral geniculate neurone excitability by noradrenaline microiontophoresis or locus coeruleus stimulation. Nature, 287: 731–734.

Schneider, J.S. and Kovelowski, C.J. (1990) Chronic exposure to low doses of MPTP. I. Cognitive deficits in motor asymptomatic monkeys. Brain Res., 519: 122–128.

Segal, M. and Bloom, F.E. (1974) The action of norepinephrine in the rat hippocampus. II. Activation of the input pathway. Brain Res., 72: 99–114.

Segal, M., Markram, H. and Richter-Levin, G. (1991) Actions of norepinephrine in the rat hippocampus. Prog. Brain Res., 88: 323–330.

Selden, N.R., Everitt, B.J. and Robbins, T.W. (1991) Telencephalic but not diencephalic noradrenaline depletion enhances behavioural but not endocrine measures of fear conditioning to contextual stimuli. Behav. Brain Res., 43: 139–154.

Selden, N.R., Robbins, T.W. and Everitt, B.J. (1990) Enhanced behavioral conditioning to context and impaired behavioral and neuroendocrine responses to conditioned stimuli following ceruleocortical noradrenergic lesions: support for an attentional hypothesis of central noradrenergic function. J. Neurosci., 10: 531–539.

Selye, H. (1946) The general adaptation syndrome and the diseases of adaptation. J. Clin. Endocrinol. Metab., 6: 117–230.

Sessler, F.M., Mouradian, R.D., Cheng, J.T., Yeh, H.H., Liu, W.M. and Waterhouse, B.D. (1989) Noradrenergic potentiation of cerebellar Purkinje cell responses to GABA: evidence for mediation through the beta-adrenoceptor-coupled cyclic AMP system. Brain Res., 499: 27–38.

Sherin, J.E., Shiromani, P.J., McCarley, R.W. and Saper, C.B. (1996) Activation of ventrolateral preoptic neurons during sleep. Science, 271: 216–219.

Simon H (1981) [Dopaminergic A10 neurons and frontal system (author's transl)]. J. Physiol. (Paris), 77: 81–95.

Smythe, J.W., Cristie, B.R., Colom, L.V., Lawson, V.H. and Bland, B.H. (1991) Hippocampal theta field activity and theta-on/theta-off cell discharges are controlled by an ascending hypothalamo-septal pathway. J. Neurosci., 11: 2241–2248.

Sorg, O. and Magistretti, P.J. (1991) Characterization of the glycogenolysis elicited by vasoactive intestinal peptide, noradrenaline and adenosine in primary cultures of mouse cerebral cortical astrocytes. Brain Res., 563: 227–233.

Sorg, O. and Magistretti, P.J. (1992) Vasoactive intestinal peptide and noradrenaline exert long-term control on glycogen levels in astrocytes: blockade by protein synthesis inhibition. J. Neurosci., 12: 4923–4931.

Southwick, S.M., Bremner, J.D., Rasmusson, A., Morgan, C.A., III, Arnsten, A. and Charney D.S. (1999) Role of norepinephrine in the pathophysiology and treatment of posttraumatic stress disorder. Biol. Psychiatry, 46: 1192–1204.

Southwick, S.M., Krystal, J.H., Morgan, C.A., Johnson, D., Nagy, L.M., Nicolaou, A., Heninger, G.R. and Charney, D.S. (1993) Abnormal noradrenergic function in posttraumatic stress disorder. Arch. Gen. Psychiatry, 50: 266–274.

Stanton, P.K. and Sarvey, J.M. (1985) Depletion of norepinephrine, but not serotonin, reduces long-term potentiation in the dentate gyrus of rat hippocampal slices. J. Neurosci., 5: 2169–2176.

Stanton, P.K. and Sarvey, J.M. (1987) Norepinephrine regulates long-term potentiation of both the population spike and dendritic EPSP in hippocampal dentate gyrus. Brain Res. Bull., 18: 115–119.

Steriade, M. (1969) Alterations of evoked neuronal discharges in motor and somesthetic cortices during the sleep-wakefulness cycles in the monkey. Electroencephalogr. Clin. Neurophysiol., 27: 678.

Steriade, M and Buzsaki, G. (1990) Parallel activation of thalamic and cortical neurons by brainstem and basal forebrain cholinergic systems. Oxford University Press, Oxford.

Steriade, M, Domich, L. and Oakson, G. (1986) Reticularis thalami neurons revisited: activity changes during shifts in states of vigilance. J. Neurosci., 6: 68–81.

Stone, E.A. (1973a) Accumulation and metabolism of norepinephrine in rat hypothalamus after exhaustive stress. J. Neurochem., 21: 589–601.

Stone, E.A. (1975) Neurochemical and behavioral effects of severe stress. Psychopharmacol. Bull., 11: 71–72.

Stone, E.A. (1979) Reduction by stress of norepinephrine-stimulated accumulation of cyclic AMP in rat cerebral cortex. J. Neurochem., 32: 1335–1337.

Stone, E.A. (1981) Mechanism of stress-induced subsensitivity to norepinephrine. Pharmacol. Biochem. Behav., 14: 719–723.

Stone, E.A. (1987) Central cyclic-AMP-linked noradrenergic receptors: new findings on properties as related to the actions of stress. Neurosci. Biobehav. Rev., 11: 391–398.

Stone, E.A. and Ariano, M.A. (1989) Are glial cells targets of the central noradrenergic system? A review of the evidence. Brain Res. Brain Res. Rev., 14: 297–309.

Stone, E.A., Freedman, L.S. and Morgano, L.E. (1978) Brain and adrenal tyrosine hydroxylase activity after chronic footshock stress. Pharmacol. Biochem. Behav., 9: 551–553.

Stone, E.A. and John, S.M. (1991) Further evidence for a glial localization of rat cortical beta-adrenoceptors: studies of in vivo cyclic AMP responses to catecholamines. Brain Res., 549: 78–82.

Stone, E.A., Platt, J.E., Trullas, R. and Slucky, A.V. (1984) Reduction of the cAMP response to norepinephrine in rat cerebral cortex following repeated restraint stress. Psychopharmacology (Berl), 82: 403–405.

Stone, E.A., Sessler, F.M. and Liu, W.M. (1990) Glial localization of adenylate-cyclase-coupled beta-adrenoceptors in rat forebrain slices. Brain Res., 530: 295–300.

Stone, E.A., Slucky, A.V., Platt, J.E. and Trullas, R. (1985) Reduction of the cyclic adenosine 3′,5′-monophosphate response to catecholamines in rat brain slices after repeated restraint stress. J. Pharmacol. Exp. Ther., 233: 382–388.

Stone, E.A., Zhang, Y., Hiller, J.M., Simon, E.J. and Hillman, D.E. (1997) Activation of fos in mouse amygdala by local infusion of norepinephrine or atipamezole. Brain Res., 778: 1–5.

Stone, E.A., Zhang, Y., John, S., Filer, D. and Bing, G. (1993) Effect of locus coeruleus lesion on c-fos expression in the cerebral cortex caused by yohimbine injection or stress. Brain Res., 603: 181–185.

Stone, E.A., Zhang, Y., John, S.M. and Bing, G. (1991) c-Fos response to administration of catecholamines into brain by microdialysis. Neurosci. Lett., 133: 33–35.

Stone, T.W. (1973b) Pharmacology of pyramidal tract cells in the cerebral cortex. Naunyn Schmiedebergs Arch. Pharmacol., 278: 333–346.

Sullivan, G.M., Coplan, J.D., Kent, J.M. and Gorman, J.M. (1999) The noradrenergic system in pathological anxiety: a focus on panic with relevance to generalized anxiety and phobias. Biol. Psychiatry, 46: 1205–1218.

Swanson, L.W. and Hartman, B.K. (1975) The central adrenergic system. An immunofluorescence study of the location of cell bodies and their efferent connections in the rat utilizing dopamine-beta-hydroxylase as a marker. J. Comp. Neurol., 163: 467–505.

Tanila, H., Rama, P. and Carlson, S. (1996) The effects of prefrontal intracortical microinjections of an alpha-2 agonist, alpha-2 antagonist and lidocaine on the delayed alternation performance of aged rats. Brain Res. Bull., 40: 117–119.

Thierry, A.M., Javoy, F., Glowinski, J. and Kety, S.S. (1968) Effects of stress on the metabolism of norepinephrine, dopamine and serotonin in the central nervous system of the rat. I. Modifications of norepinephrine turnover. J. Pharmacol. Exp. Ther., 163: 163–171.

Timo-Iaria, C., Negrao, N., Schmidek, W.R., Hoshino, K., Lobato de Menezes, C.E. and Leme, D.R. (1970) Phases and states of sleep in the rat. Physiol. Behav., 5: 1057–1062.

Usher, M., Cohen, J.D., Servan-Schreiber, D., Rajkowski, J. and Aston-Jones, G. (1999) The role of locus coeruleus in the regulation of cognitive performance. Science, 283: 549–554.

Valentino, R.J. and Foote, S.L. (1987) Corticotropin-releasing factor disrupts sensory responses of brain noradrenergic neurons. Neuroendocrinology, 45: 28–36.

Valentino, R.J. and Foote, S.L. (1988) Corticotropin-releasing hormone increases tonic but not sensory-evoked activity of noradrenergic locus coeruleus neurons in unanesthetized rats. J. Neurosci., 8: 1016–1025.

Valentino, R.J. and Wehby, R.G. (1988) Corticotropin-releasing factor: evidence for a neurotransmitter role in the locus ceruleus during hemodynamic stress. Neuroendocrinology, 48: 674–677.

VanderMaelen, C.P. and Aghajanian, G.K. (1980) Intracellular studies showing modulation of facial motoneurone excitability by serotonin. Nature, 287: 346–347.

Vanderwolf, C.H. and Robinson, T.E. (1981) Reticulo-cortical activity and behavior: A critique of the arousal theory and a new synthesis. Behavioral Brain Science, 4: 459–514.

Vankov, A., Herve-Minvielle, A. and Sara, S.J. (1995) Response to novelty and its rapid habituation in locus coeruleus neurons of the freely exploring rat. Eur. J. Neurosci., 7: 1180–1187.

Videen, T.O., Daw, N.W. and Rader R.K. (1984) The effect of norepinephrine on visual cortical neurons in kittens and adult cats. J. Neurosci., 4: 1607–1617.

Warm, J.S. and Jerison, H.J. (1984) The psychophysics of vigilance. In: Warm, J.S. (Ed.), Sustained Attention in Human Performance. John Wiley & Sonspp, New York, 15–59.

Watanabe, H. and Passonneau, J.V. (1973) Factors affecting the turnover of cerebral glycogen and limit dextrin in vivo. J. Neurochem., 20: 1543–1554.

Waterhouse, B.D., Azizi, S.A., Burne, R.A. and Woodward, D.J. (1990) Modulation of rat cortical area 17 neuronal responses to moving visual stimuli during norepinephrine and serotonin microiontophoresis. Brain Res., 514: 276–292.

Waterhouse, B.D., Devilbiss, D., Fleischer, D., Sessler, F.M. and Simpson, K.L. (1998) New perspectives on the functional organization and postsynaptic influences of the locus ceruleus efferent projection system. Adv. Pharmacol., 42: 749–754.

Waterhouse, B.D., Lin, C.S., Burne, R.A. and Woodward, D.J. (1983) The distribution of neocortical projection neurons in the locus coeruleus. J. Comp. Neurol., 217: 418–431.

Waterhouse, B.D., Moises, H.C., Yeh, H.H. and Woodward, D.J. (1982) Norepinephrine enhancement of inhibitory synaptic mechanisms in cerebellum and cerebral cortex: mediation by beta adrenergic receptors. J. Pharmacol. Exp. Ther., 221: 495–506.

Waterman, A, Livingston, A. and Bouchenafa, O. (1988) Analgesic effects of intrathecally-applied alpha 2-adrenoceptor agonists in conscious, unrestrained sheep. Neuropharmacology, 27: 213–216.

Weiss, J.M., Bailey, W.H., Pohorecky, L.A., Korzeniowski, D. and Grillione, G. (1980) Stress-induced depression of motor activity correlates with regional changes in brain norepinephrine but not in dopamine. Neurochem. Res., 5: 9–22.

Weiss, J.M., Glazer, H.I., Pohorecky, L.A. Brick, J. and Miller, N.E. (1975) Effects of chronic exposure to stressors on avoidance-escape behavior and on brain norepinephrine. Psychosom. Med., 37: 522–534.

Weiss, J.M., Stone, E.A. and Harrell, N. (1970) Coping behavior and brain norepinephrine level in rats. J. Comp. Physiol. Psychol., 72: 153–160.

Weiss, J.M., Stout, J.C., Aaron, M.F., Quan, N., Owens, M.J., Butler, P.D. and Nemeroff, C.B. (1994) Depression and anxiety: role of the locus coeruleus and corticotropin-releasing factor. Brain Res. Bull., 35: 561–572.

White, S.R. and Neuman R.S. (1980) Facilitation of spinal motoneurone excitability by 5-hydroxytryptamine and noradrenaline. Brain Res., 188: 119–127.

Woods, D.L. and Knight, R.T. (1986) Electrophysiologic evidence of increased distractibility after dorsolateral prefrontal lesions. Neurology, 36: 212–216.

Woodward, D.J., Moises, H.C., Waterhouse, B.D., Hoffer, B.J. and Freedman, R. (1979) Modulatory actions of norepinephrine in the central nervous system. Fed. Proc., 38: 2109–2116.

Young, W.S., III. and Kuhar, M.J. (1980) Noradrenergic alpha 1 and alpha 2 receptors: light microscopic autoradiographic localization. Proc. Natl. Acad. Sci. USA, 77: 1696–1700.

Zaborsky, L., Cullinan, W.E. and Braun, A. (1991) Afferent to basal forebrain cholinergic projection neurons: an update. In: Napier, T.C. (Ed.), The Basal Forebrain. Plenum, New York, pp. 43–100.

Zigmond, M.J. and Harvey, J.A. (1970) Resistance to central norepinephrine depletion and decreased mortality in rats chronically exposed to electric foot shock. J. Neurovisc. Relat., 31: 373–381.

T. Steckler, N.H. Kalin and J.M.H.M. Reul (Eds.)
Handbook of Stress and the Brain, Vol. 15
ISBN 0-444-51173-3

CHAPTER 4.4

Functional interactions between stress neuromediators and the locus coeruleus–norepinephrine system

Rita J. Valentino[1,*] and Elisabeth J. Van Bockstaele[2]

[1]*The Children's Hospital of Philadelphia, 402C Abramson Pediatric Research Center, 34th and Civic Center Blvd., Philadelphia, PA 19104, USA*
[2]*Department of Pathology, Anatomy and Cell Biology, Thomas Jefferson University, 1020 Locust St., Philadelphia, PA 19107, USA*

Abstract: Activation of the hypothalamic–pituitary–adrenal axis is considered a hallmark of the stress response. Coincident with this, peripheral and central noradrenaline systems are activated. By engaging medullary noradrenaline projections to the paraventricular nucleus of the hypothalamus, stressors regulate the endocrine limb of the stress response. Engaging noradrenaline projections from the pontine nucleus, locus coeruleus (LC), to the forebrain may serve as a behavioral or cognitive limb of the stress response. This chapter will focus on the impact of stress on the LC–noradrenaline system and its forebrain targets. The body of evidence supporting the view that diverse stressors consistently activate this system will be briefly reviewed. Convergent lines of evidence suggest that corticotropin-releasing factor (CRF) is a critical substrate of the stress response that regulates activity of the LC–noradrenaline system during stress. This chapter will discuss the impact of CRF and stressors on the LC–noradrenaline system from the level of individual LC neurons to the distributed network of LC projections. The concept that stress regulates activity of the LC–noradrenaline system by integrating the actions of CRF with other neuromodulators will be developed and a circuitry underlying these interactions will be proposed. These findings will be integrated with physiological and behavioral studies that have suggested specific roles for the LC–noradrenaline system in the modulation of arousal and attention into the hypothesis that activation of this system is an important component of a cognitive limb of the stress response. Finally, as the endocrine limb of the stress response exhibits plasticity, which is expressed with manipulations such as chronic stress or adrenalectomy, evidence will be reviewed which suggests that the limb comprising the LC–noradrenaline system exhibits similar plasticity, which may have both adaptive and pathological consequences.

Stress-elicited activation of the locus coeruleus–noradrenaline system

Early studies that characterized the physiological properties of locus coeruleus (LC) neurons in unanesthetized rats, cats, and primates demonstrated a general increase in discharge frequency in response to all modes of nonnoxious sensory stimuli (Foote et al., 1980; Aston-Jones and Bloom, 1981b; Rasmussen et al., 1986). In some cases this was found to precede cortical electroencephalographic (EEG) indices of arousal and orientation toward the source of the stimulus (Aston-Jones and Bloom, 1981a). Subsequently, it was observed that a variety of nonnoxious visceral challenges including hypotension (Svensson and Thoren, 1979; Elam et al., 1984, 1985; Morilak et al., 1987; Svensson, 1987), hypoxia (Elam et al., 1981), hypoglycemia, and nonnoxious

*Corresponding author. Tel.: +1215-590-0650; Fax: +1215-590-4107; E-mail: valentino@email.chop.edu

distention of viscera (Elam et al., 1986) also increased LC discharge and in cases in which it was simultaneously recorded, forebrain EEG activation was temporally correlated with LC activation (Page et al., 1992, 1993; Lechner et al., 1997). These studies suggested that LC activation may facilitate arousal and behavioral responses to stimuli.

The initial studies that examined the effects of stressors on the LC–noradrenaline system used measurements of noradrenaline metabolites and turnover in forebrain targets of the LC as endpoints of activation (Thierry et al., 1968; Korf et al., 1973; Cassens et al., 1980, 1981). These studies demonstrated that shock increased noradrenaline turnover and that the effects could be abolished by LC lesions, underscoring the necessity of the LC in these effects. Importantly, unpredictable shock resulted in a greater response compared to shock that was signaled by stimuli (Tsuda et al., 1989). Similarly, immunological challenge elicited by injection of interleukin-2 increased noradrenaline utilization in regions targeted by LC projections (Lacosta et al., 2000).

More direct methods for measuring noradrenaline release in LC target regions, such as microdialysis, have yielded results consistent with turnover studies. Thus, restraint, tailshock, and tailpinch increase extracellular noradrenaline levels in hippocampus, which receives its sole source of noradrenaline from the LC (Abercrombie et al., 1988). Auditory stress also increased noradrenaline release in the cortex (Britton et al., 1992) and the same hypotensive stress that was previously demonstrated to increase LC discharge rates also increased noradrenaline release (measured by microdialysis) in frontal cortex and hypothalamus (Smagin et al., 1994).

Additional endpoints of activation of the LC–noradrenaline system, including expression of tyrosine hydroxylase mRNA (measured by in situ hybridization) and protein (measured by immunohistochemistry) or c-fos mRNA or protein, consistently suggest that this system is activated by a variety of challenges that would be considered stressors, including restraint, shock, hypotension, swim, immune challenge, water avoidance stress, and social stress (Smith et al., 1991; 1992; Duncan et al., 1993; Imaki et al., 1993; Bonaz and Tache, 1994; Beck and Fibiger, 1995; Chan and Sawchenko, 1995; Dun et al., 1995; Graham et al., 1995; Campeau and Watson, 1997; Kollack-Walker et al., 1997; Chang et al., 2000; Funk and Amir, 2000; Rusnak et al., 2001; Sabban and Kvetnansky, 2001; Ishida et al., 2002; Makino et al., 2002). Interestingly, stress-induced increases in tyrosine hydroxylase mRNA were observed selectively in LC neurons and not in dopaminergic nuclei, indicating selective effects on the noradrenergic system (Smith et al., 1991).

The LC–noradrenaline system is activated by stressors is perhaps not surprising, given that it is also activated by nonnoxious sensory stimuli. Questions that arise are whether different stimuli (sensory vs. stressors) produce a qualitatively distinct activation, whether they engage distinct circuits to activate the LC–noradrenaline system and whether LC activation by sensory stimuli versus stressors have different functional consequences.

Glucocorticoids as potential neuromediators of LC activation

As LC neurons robustly express glucocorticoid receptors (Towle et al., 1982; Harfstrand et al., 1986; Cintra et al., 1994; Morimoto et al., 1996), corticosteroids might be considered a potential mediator of LC activation during stress. Indeed, iontophoretic application of glucocorticoids into the LC has been reported to increase discharge rates of LC neurons (Avanzino et al., 1987). However, it is unlikely that glucocorticoids mediate activation of the LC–noradrenaline system by stress. Neither adrenalectomy nor corticosterone administration alter tyrosine hydroxylase expression in LC neurons (Smith et al., 1991). Moreover, adrenalectomy results in tonically elevated LC discharge rates (Pavcovich and Valentino, 1997) (see below). Rather, glucocorticoids may serve to limit LC activation during stress. In support of this, acute immobilization stress increased tyrosine hydroxylase expression in LC neurons to a greater extent in adrenalectomized compared to sham rats (Makino et al., 2002). The presence of glucocorticoid receptor mRNA in LC neurons during pre- and postnatal development suggests an additional role for glucocorticoids in modulating the development of this system (Cintra et al., 1991, 1993).

Corticotropin-releasing factor (CRF) as a mediator of LC activation

Following the identification and characterization of CRF as the neurohormone that initiates the endocrine limb of the stress response (Vale et al., 1981), convergent findings from multiple laboratories implicated CRF as a neuromodulator/neurotransmitter that engaged extrahypophyseal neuronal systems to regulate behavioral, autonomic, and cognitive limbs of the stress response. These findings included the widespread distribution of CRF-immunoreactive nerve terminals (Swanson et al., 1983; Sakanaka et al., 1987) and CRF receptor binding sites (De Souza, 1987) throughout the brain. Particularly convincing were studies demonstrating that centrally administered CRF could mimic certain autonomic, behavioral, and immunological effects of stress even in hypophysectomized animals and that central administration of CRF antagonists could attenuate nonendocrine aspects of the stress response (for review see Dunn and Berridge, 1990; Owens and Nemeroff, 1991). These studies led to the speculation that CRF may have a dual role in the stress response, serving as a neurohormone to initiate the endocrine limb and a neurotransmitter to mediate nonendocrine (e.g., behavioral, autonomic, immunological, cognitive responses) limbs of the stress.

The initial finding of CRF-immunoreactive terminals within the LC region, although relatively moderate, suggested that this noradrenaline-containing nucleus was a target of extrahypophyseal CRF and that CRF might mediate LC activation by stressors. This hypothesis has now been tested by numerous electrophysiological, neurochemical, and behavioral approaches.

CRF effects on LC-neuronal activity and noradrenaline release

When administered intracerebroventricularly (i.c.v.), CRF increases tonic discharge rates of LC neurons of anesthetized and unanesthetized rats in a dose-dependent manner (Valentino et al., 1983; Valentino and Foote, 1988). The magnitude and duration of this response is somewhat greater in unanesthetized rats. CRF is approximately 300 times more potent when administered into the LC versus i.c.v., pointing to a site of action within the LC region (Curtis et al., 1997). Additionally, intracoerulear administration of CRF antagonists almost completely abolished LC activation by i.c.v. administered CRF, suggesting that LC activation by i.c.v. administered CRF is primarily due to actions within the LC (Curtis et al., 1997). Inhibition of LC discharge by CRF has been reported by one group and the reasons for the discrepancy have not been reconciled (Borsody and Weiss, 1996). Nonetheless, activation of LC neurons by CRF is consistent with microdialysis and c-fos studies described below.

LC activation by CRF differs qualitatively from that produced by other agents such as excitatory amino acids or muscarinic agonists in that the maximum magnitude of activation is less (approximately a twofold increase in discharge rate), the onset slower, and duration longer. This may reflect either indirect actions or different intracellular signaling systems. The qualitatively differential response of LC neurons to CRF versus excitatory amino acids is of interest given that the latter may mediate responses to sensory stimuli, while CRF may selectively mediate LC activation by stressors. This concept is discussed in greater detail below.

Studies of the effects of CRF on LC neurons in vitro in a slice preparation are consistent with in vivo findings that CRF increases LC-neuronal activity, although the cellular mechanisms underlying this have not been investigated in detail (Jedema and Grace, 2000). Activation of LC-neurons by CRF is mediated by CRF-R1 receptors (Schulz et al., 1996). As CRF-R1 receptors are coupled via the Gs protein to adenylate cyclase (Chalmers et al., 1996), induction of pathways linked to cyclic AMP are potential mediators of LC activation. Consistent with this, CRF increased cyclic AMP accumulation in an LC-like cell line, CATH.a (Bundey and Kendall, 1999). This mechanism of action of CRF may be relevant for conditions that result in heterologous sensitization of Gs-coupled processes (Watts, 2002) (see below).

Consistent with electrophysiological evidence that CRF activates LC neurons, direct CRF administration into the LC increases fos expression in LC neurons (Rassnick et al., 1994, 1998). The pattern of c-fos expression in brain following CRF administration into the LC partially resembled the pattern produced by footshock, suggesting that CRF in the LC may engage many of the same circuits that are engaged by footshock. Notably, c-fos expression in paraventricular hypothalamic neurons was produced by footshock, but not by intra-LC CRF, suggesting that LC activation may mediate non-endocrine, but not endocrine responses to footshock.

As the magnitude of LC activation by CRF is significantly less than that produced by excitatory amino acids, muscarinic agonists, or vasoactive intestinal peptide (Curtis et al., 1999), the question arises as to whether this activation is sufficient to impact on LC targets. This has been addressed in part by microdialysis studies measuring noradrenaline release in LC targets following CRF administration. Both i.c.v. and intracoerulear administration of CRF increase noradrenaline extracellular levels in hippocampus and cortex indicating that CRF-elicited LC activation is translated to enhanced forebrain noradrenaline release (Smagin et al., 1995; Schulz and Lehnert, 1996; Curtis et al., 1997; Page and Abercrombie, 1999). Studies described below demonstrating that LC activation by locally administered CRF is temporally correlated to forebrain EEG activation support the notion that these effects are translated across the synapse to cells targeted by the LC–noradrenaline system (Curtis et al., 1997).

As discussed above, LC neurons respond to multimodal sensory stimuli and this is typically characterized by a brief robust increase in discharge rate, followed by a longer period of relative inactivity (see Fig. 1). In addition to increasing tonic LC discharge rate, CRF alters the LC sensory response by elevating baseline discharge and attenuating evoked discharge, thereby reducing the signal-to-noise ratio of the LC sensory response (Fig. 1) (Valentino and Foote, 1987, 1988). This effect of CRF has been observed in anesthetized rats, using sciatic nerve stimulation to evoke LC discharge, and in unanesthetized rats using auditory stimuli. A similar disruption of the LC sensory response is apparent during hypotensive stress (Fig. 1) (Valentino and Wehby, 1988; Aston-Jones et al., 1991). That this effect on LC neurons may translate to network effects on sensory processing is suggested by the finding that CRF injection into the LC mimics the effect of immobilization to reduce auditory evoked cortical responses (Miyazato et al., 2000).

As glutamate afferents to the LC have been proposed to mediate LC responses to sensory stimuli (Aston-Jones et al., 1991), CRF effects on LC sensory responses suggest important interactions with glutamate. Potential anatomical substrates for such interactions between CRF and glutamate are described below (see Fig. 2). Given that the effects of CRF and glutamate on LC neurons are qualitatively different, the interaction of these two neuromodulators may serve to fine-tune LC discharge and may play a role in shifts in attentional state (see below).

Regulation of the LC–noradrenaline system by endogenous CRF

It is unlikely that the LC–noradrenaline system is under tonic regulation by CRF because CRF antagonists administered either i.c.v. or directly into the LC have no effect on LC discharge rate or noradrenaline release in targets in unstressed rats (Curtis et al., 1994; Page and Abercrombie, 1999). In contrast, evidence for tonic CRF release in the LC is apparent in adrenalectomized rats. LC discharge rates are typically higher in adrenalectomized rats compared to sham-operated rats and administration of CRF antagonists into the LC normalizes this difference (Pavcovich and Valentino, 1997). These findings suggest that under nonstressed conditions, CRF release within the LC is restrained by basal levels of corticosteroids, perhaps acting at mineralocorticoid receptors.

CRF is released to impact on the LC during specific challenges and subsequently increases tonic activity and disrupts sensory-evoked responses. Evidence for this has been best demonstrated during hypotensive stress (Valentino and Wehby, 1988; Valentino et al., 1991; Page et al., 1993; Curtis et al., 1997). Similar to the effects of exogenously administered CRF, hypotensive stress increases LC discharge rate and decreases the signal-to-noise ratio of LC sensory responses (Figs. 1 and 4). CRF antagonists (administered i.c.v.) blocked LC activation by hypotensive challenge in a dose-dependent manner and the IC50s for an individual antagonist were identical against CRF and hypotensive challenge, suggesting that similar receptor binding sites are involved in LC activation by these two stimuli

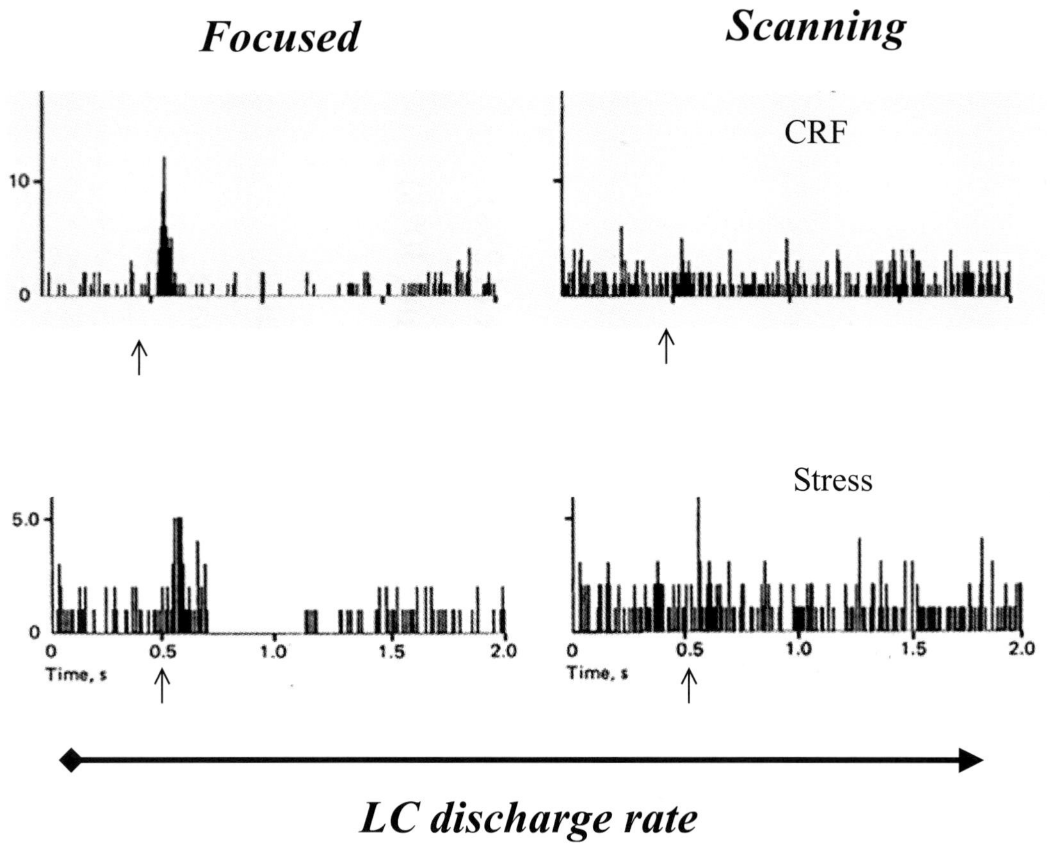

Fig. 1. Schematic depicting the effect on LC sensory responses as a function of tonic discharge rate. Shown are poststimulus time histograms generated during presentation of repeated sciatic nerve stimulation (small arrows). The top pair of histograms was generated before (left) and after (right) i.c.v administration of CRF. The bottom pair of histograms was generated before and during a hypotensive challenge. Note that in the absence of stress or CRF, the stimulus elicits LC discharge and this is followed by a brief period of relative inactivity. In contrast, when LC neurons are exposed to CRF (or during a challenge that releases CRF in the LC), discharge activity is somewhat greater and the magnitude of the evoked response is greatly attenuated. It is proposed that the loss of LC responses to selective stimuli is associated with a shift from focused to scanning attention (modified from Valentino and Wehby, 1988).

(Curtis et al., 1994). Importantly, intracoerulear microinjection of CRF antagonists prevented LC activation by hypotensive stress, and the antagonists were approximately 300–1000 times more potent when administered directly into the LC compared to i.c.v. (Curtis et al., 1994). The selectivity of the CRF antagonists in blocking this response was supported by the finding that excitatory amino acid antagonists, in doses that prevent LC activation by certain other stimuli (i.e., sciatic nerve stimulation, bladder distention), did not alter LC activation by hypotensive challenge (Page et al., 1992; Curtis et al., 1994). Conversely, doses of CRF antagonists that prevented LC activation by hypotensive challenge did not alter LC activation by sciatic nerve stimulation or bladder distention. LC activation by CRF during hypotensive stress is translated to increased noradrenaline release in forebrain regions and cortical and hippocampal EEG activation (Page et al., 1993). LC activation by hypotensive stress is also apparent as increased c-fos expression in LC neurons and functional–anatomical studies using this endpoint, as well as expression of the phosphorylated cyclic AMP response-element binding protein (PCREB) implicate the central nucleus of the amygdala (CNA) as a potential source of CRF that mediates LC activation by hypotensive challenge (Curtis et al., 2002) (see below).

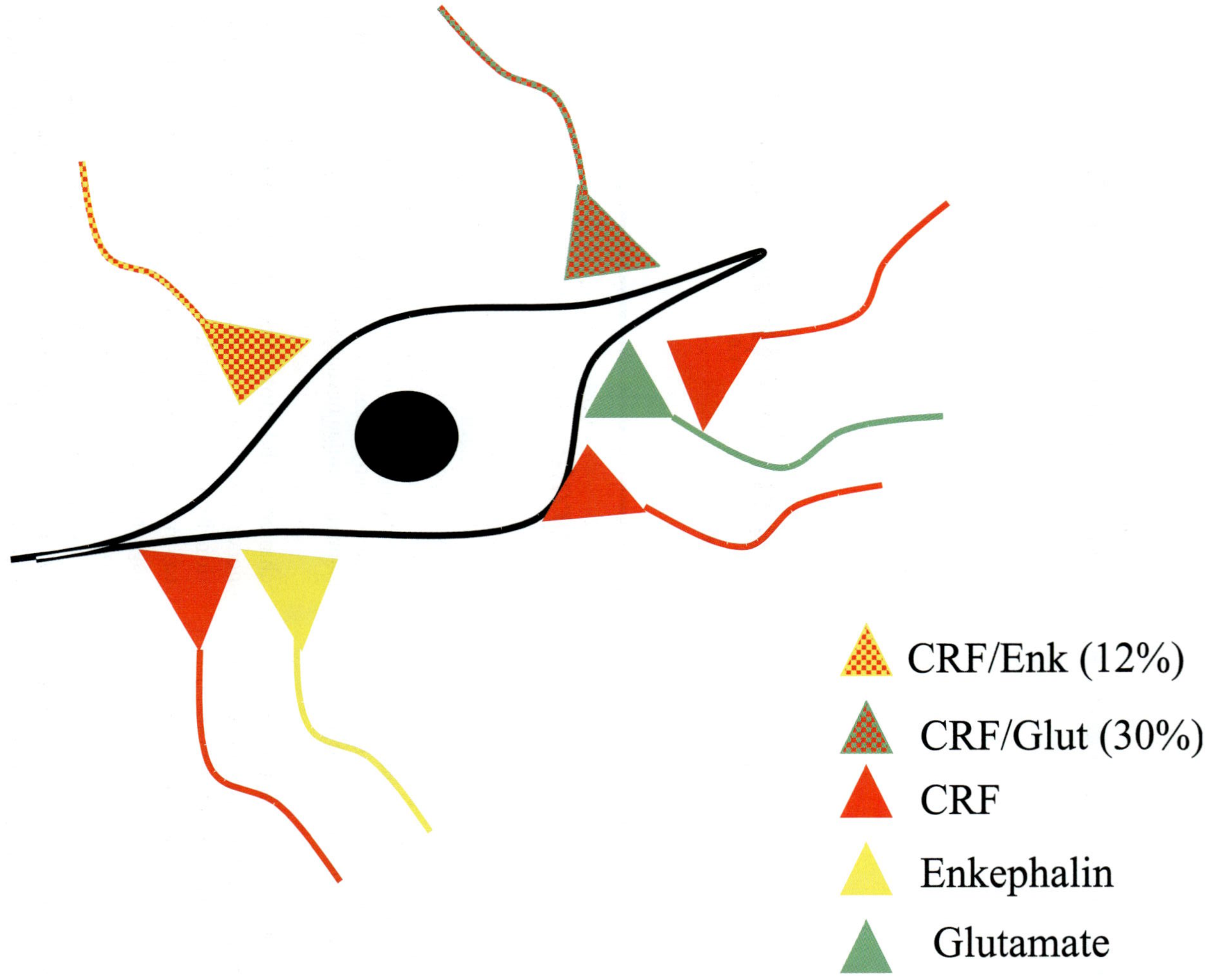

Fig. 2. Schematic depicting potential modes of interaction between CRF axon terminals in the LC region and either glutamate or enkephalin based upon ultrastructural studies. A substantial number of CRF axon terminals also contain glutamate and fewer colocalize enkephalin. CRF coreleased with glutamate from common terminals could impact on glutamate release or its postsynaptic effect. CRF axon terminals were also apposed to axon terminals containing glutamate (18%) and contacted dendrites that received convergent input from glutamate-containing axon terminals (13%). Through these interactions, CRF could also influence glutamate release or its postsynaptic effect. In contrast, only occassional CRF terminals were found to be apposed to enkephalin-containing axon terminals. However, 22% of CRF terminals contacted dendrites that received convergent input from enkephalin-containing nerve terminals, suggesting they may comodulate the same LC neurons.

When hypotensive stress is terminated, LC activation is also abruptly terminated. This is apparent as a rapid decrease in LC-neuronal discharge, which falls below prestress rates for several minutes (See Fig. 4) (Curtis et al., 2001). This rebound inhibition may serve as a counter-regulatory measure to restore activity of the LC–noradrenaline system to baseline when the stress is no longer present. This rebound inhibition was recently demonstrated to be mediated by endogenous opioid afferents to the LC because it was prevented by prior administration of naloxone into the LC (Curtis et al., 2001). These results suggest important interactions between CRF and endogenous opioids in regulating the activity of the LC–noradrenaline system in response to stress. Whereas activation of the LC–noradrenaline system by CRF during stress may be important in maintaining arousal and facilitating specific behaviors or attentional processes in response to the stress, termination of this response when the stressor is no longer present is adaptive. Engaging endogenous opioid control over LC activity during the termination of stress may serve this role. Anatomical substrates for CRF-opioid coregulation of the LC are described below (Fig. 2).

Like hypotension, low magnitudes of colon distention also engage CRF release to activate LC neurons (Elam et al., 1986; Lechner et al., 1997; Rouzade-Dominguez et al., 2001). This is prevented by microinfusion of CRF antagonists, but not excitatory amino acid antagonists into the LC (Lechner et al., 1997). Similar to LC activation by hypotensive stress, this effect is temporally correlated to cortical EEG activation and may be important for facilitating arousal and behaviors that are compatible with the visceral response to this stimulus. This response may play a role in the hypersensitivity to colonic stimuli that is a feature of irritable bowel disorder. The intersection between CRF circuits and the LC–noradrenaline system that underlie this effect are described below.

Given the constraints of recording single unit LC activity in rats during stress presentation, it has been difficult to use this electrophysiological endpoint to elucidate the role of CRF in LC activation by other stressors. However, microinfusion of CRF antagonists into the LC has also been shown to interfere with other endpoints of LC activation, including restraint stress-induced increases in tyrosine hydroxylase expression in LC neurons (Melia and Duman, 1991) and restraint or handling stress-induced increases in noradrenaline release in the medial prefrontal cortex (Smagin et al., 1996b; Kawahara et al., 2000). These results are consistent with an impact of endogenous CRF in the LC that is released under certain conditions. However, it is unlikely that CRF is the sole mediator of LC activation by stimuli that may be considered stressors. For example, LC activation by opiate withdrawal is insensitive to CRF antagonists (Valentino and Wehby, 1989) and evidence suggests that this is mediated by excitatory amino acid neurotransmission in the LC (Rasmussen and Aghajanian, 1989; Akaoka and Aston-Jones, 1991). Additionally, recent findings suggest that swim stress-induced activation of LC neurons (as indicated by c-fos expression) is insensitive to CRF antagonists (Roche et al., in press). Thus, CRF release in the LC is engaged only by specific stimuli and may not occur in response to all stimuli that may be considered stressors.

Anatomical substrates for stress-elicited LC–noradrenaline activation

At the light microscopic level, CRF innervation of the nuclear region of the rat LC appears relatively moderate (Valentino et al., 1992). However, CRF-immunoreactive fibers densely innervate surrounding regions into which LC dendrites extend, including the ventromedial peri-LC at mid to caudal levels and the dorsolateral peri-LC at the rostral pole of the LC (Valentino et al., 1992). Additionally, CRF fibers appear more numerous in the periventricular region medial to the LC where LC dendrites extend. CRF innervation of the primate LC is more dense than rat and exhibits a similar topographical pattern of innervation. The primate LC is divided into medial and lateral aspects by the tract of the mesencephalic trigeminal nerve and although CRF fibers can be seen throughout the LC, they are most dense rostrolaterally (Foote and Cha, 1988; Austin et al., 1995). The significance of the topographical organization of CRF terminal fields within the LC region remains to be established. Evidence suggests that different sources of CRF innervate different LC/peri-LC regions (Fig. 4) and the extent of colocalization with other neurotransmitters may differ in the core versus extranuclear dendritic zones (discussed below). However, because LC neurons exhibit extensive collateralization and innervate multiple targets (Foote et al., 1983), it is not clear whether LC neurons that are targeted by different sources of CRF have different projection fields.

Ultrastructural studies suggest multiple substrates by which endogenous CRF can impact on the LC–noradrenaline system. To date two different regions have been analyzed, the ventromedial peri-LC and rostrolateral peri-LC (Van Bockstaele et al., 1996, 1998, 1999). In both regions, synaptic specializations between CRF-immunoreactive terminals and LC dendrites are numerous, with 20–25% of CRF-immunoreactive terminals contacting tyrosine hydroxylase-immunoreactive dendrites in these regions. Although both asymmetric (excitatory)- and symmetric (inhibitory)-type synapses were observed, asymmetric synapses were approximately twice as common. As asymmetric synapses are generally associated with excitatory neurotransmission, this is consistent with the electrophysiological effects of CRF and the finding of a substantial colocalization of CRF with glutamate in terminals in the LC region (described below). In addition to direct contacts, CRF terminals were also observed to be apposed to unlabeled terminals that formed synaptic specializations with LC dendrites. This is suggestive of indirect modulation of LC activity via presynaptic interactions and may be a substrate for regulation of LC sensory responses.

CRF-containing nerve terminals in the LC region contain other nonlabeled dense core vesicles and smooth clear vesicles, indicative of colocalization of CRF with other neuropeptides and/or classical neurotransmitters (Van Bockstaele et al., 1996). Because the impact of CRF may be a result of modulation of the colocalized transmitter, the identity of the colocalized neurotransmitter is of interest. Approximately 30% of the CRF-immunoreactive terminals in the rostrolateral dendritic region colocalize glutamate (Valentino et al., 2001). This stands in contrast to a minor percentage (5%) of CRF-containing terminals in the same region that colocalize GABA (Valentino et al., 2001). The high degree of colocalization of CRF and glutamate is consistent with the finding that majority of synapses between CRF terminals and LC dendrites are asymmetric. In addition to colocalization, individual CRF- and glutamate-immunoreactive terminals were often apposed each other or found to converge onto common dendrites in the rostrolateral LC dendritic zone (Fig. 2). The frequency of these interactions with glutamate terminals was substantially greater than with GABA-containing terminals, suggesting that CRF interacts with glutamate, but not GABA, in this region.

Modulation of glutamate transmission in the LC has important functional implications. Because glutamate is a primary neuromediator of LC activation by sensory stimuli (Ennis and Aston-Jones, 1988; Chiang et al., 1991; Ennis et al., 1992), these interactions may mediate the disruption of LC sensory responses by stress. For example, glutamate release could be influenced by corelease of CRF from common terminals or from an apposed terminal (Fig. 2). The postsynaptic effects of glutamate could also be altered by CRF that is coreleased with glutamate or that exists in separate terminals that converge on the same LC neuron. As discussed above, although both glutamate and CRF increase LC discharge rates, the mode of activation is qualitatively different in both magnitude and time course. As excitatory amino acid neurotransmission in the LC, which produces a robust short-lasting activation, has been linked to sensory responses, it is tempting to speculate that glutamate inputs are engaged by sensory stimuli to facilitate focusing of attention toward the stimulus. In contrast, tonic increases in LC activity, such as that produced by CRF, have been linked to a disruption of focused attention and a shift to scanning modes of attention (Rajkowski et al., 1994). By engaging CRF afferents to the LC, stressors may disable the influence of glutamate and selective responses to sensory stimuli. The net effect of this may be to shift the mode of attention from being focused on an individual stimulus to scanning of multiple stimuli in the environment, an effect which would be adaptive in a threatening environment (Fig. 1).

The physiological studies discussed above suggest that the onset and termination of stress may signal CRF and opioid modulation of the same LC neuron, respectively (Curtis et al., 2001). In this regard, it is of interest that some CRF terminals also colocalize enkephalin, although these are less abundant than those that colocalize glutamate (Valentino and Van Bockstaele, 2001). Colocalization of the two peptides is more often observed in the nuclear part of the LC, as opposed to the extranuclear dendritic zone. In addition to colocalization, individual enkephalin- and CRF-containing nerve terminals were found to converge onto common dendrites (Fig. 2). This finding suggests that parallel CRF and opioid afferent terminals are positioned to regulate discharge activity of the same LC neurons in an opposing manner, depending on the onset or termination of stress.

CRF receptors in the LC

The receptor mediating LC activation by CRF is thought to be the CRF-R1 subtype because this activation is sensitive to selective CRF-R1 antagonists (Schulz et al., 1996). However, CRF-R1 receptor binding (as demonstrated by receptor autoradiography) is of relatively low density in the LC (De Souza, 1987) and the CRF-R1 mRNA (measured by in situ hybridization) is not present in rat LC neurons (Van Pett et al., 2000), although it is abundant in primate LC (Sanchez et al., 1999). The lack of evidence for CRF receptors in rat LC is surprising, given the physiological responses of LC neurons to CRF and the ultrastructural findings of synapses between CRF-containing axon terminals and LC dendrites. One explanation for this discrepancy is that CRF acts presynaptically with receptors on terminals within the LC region, whose cell bodies are distant from the LC. However, CRF-R1 mRNA is not highly expressed in regions containing LC afferents (see below). In contrast to in situ hybridization studies, immunohistochemical studies (at the light microscopic and electron microscopic level) using antisera directed against the CRF-R1 receptor (but also cross reacting with CRF-R2) support a robust presence of CRF receptor in rat and mouse LC neurons (Fig. 3) (Chen et al., 2000; Sauvage and Steckler, 2001). At the ultrastructural level, CRF-R immunoreactivity was present within LC dendrites and non labeled axon terminals within the LC, consistent with both pre- and post synaptic actions of CRF (Fig. 3). Although these findings provide evidence for the existence of CRF-R1 receptors on LC neurons, the discrepancy with in situ hybridization studies has not been resolved.

Circuitry linking CRF to the LC

Based on retrograde tracing from the nuclear core of the LC, it has been proposed that regulation of the LC–noradrenaline system occurs through a restricted set of afferents (Aston-Jones et al., 1986). The sources of these afferent fibers included the nucleus paragigantocellularis (PGi) in the ventrolateral medulla, nucleus prepositus, and the dorsal cap of the PVN, regions which contain CRF–immunoreactive neurons in colchicine-treated rats (Valentino et al., 1992). Double labeling of neurons that are retrogradely labeled from the nuclear LC for CRF immunoreactivity implicated the PGi, the dorsal cap of the PVN, and the Barrington's nucleus as sources of CRF that innervate the nuclear LC (Valentino et al., 1992, 1996; Lechner and Valentino, 1999). In contrast, CRF neurons in the nucleus prepositus appeared to be a different population lying lateral to those that were retrogradely labeled from the LC (Valentino et al., 1992).

LC dendrites extend for several hundred microns outside the nucleus (Shipley et al., 1996). Retrograde tracing from the nucleus does not label afferents that may impact on the LC through contacts with extranuclear dendrites. Notably, CRF terminals are more dense in extranuclear dendritic regions, particularly in the rostrolateral peri-LC. Injection of retrograde tracers into the rostrolateral peri-LC where CRF-immunoreactive fibers are dense labeled a distinct set of CRF afferents including the CNA, bed nucleus of the stria terminalis (BNST) and the PVN (Van Bockstaele et al., 1998, 1999). Projections from the CNA and BNST have been further examined at an ultrastructural level (Van Bockstaele et al., 1998, 1999). Terminals arising from both the CNA and BNST form synaptic specializations with LC dendrites in the rostrolateral peri-LC. A semi-quantitative analysis revealed that the percentage of terminals from each nucleus contacting LC dendrites in this region was comparable (approximately 20%). However, a greater percentage (35% vs. 13%)

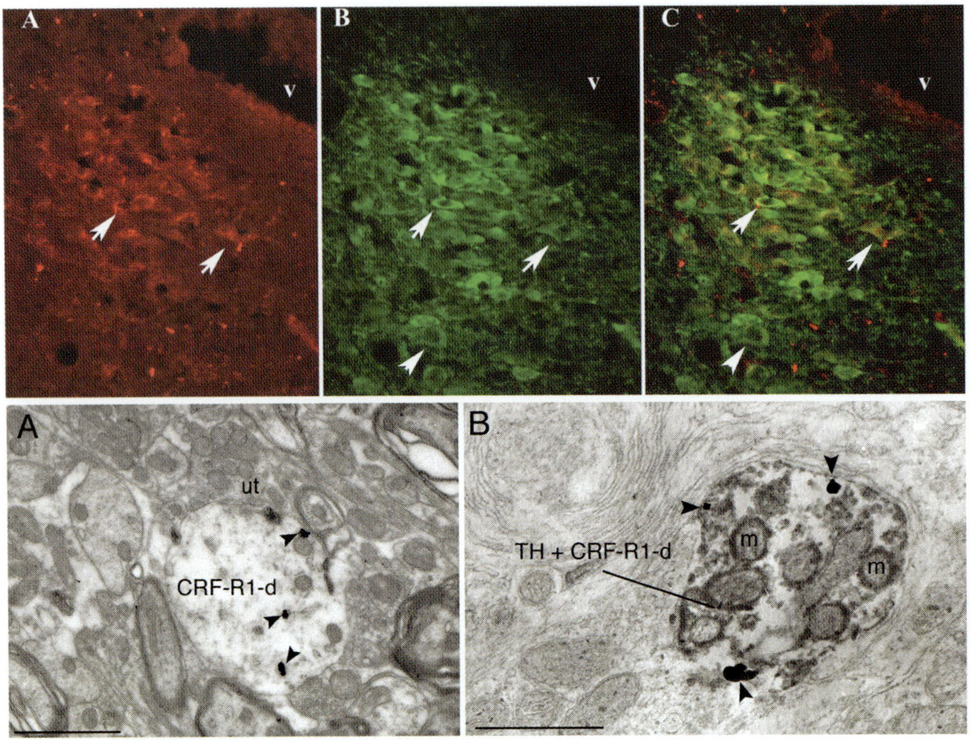

Fig. 3. CRF-R1-receptor immunoreactivity in LC neurons. The top panel shows fluorescent photomicrographs of a 15-μm thick coronal section through the LC immunolabeled for-A) CRF-R1 (red), (B) tyrosine hydroxylase immunoreactivity (green), and (C) the resultant merged image. Note the presence of CRF-R1 immunoreactivity in almost all LC neurons which sometimes appears yellow in the merged image. The arrows point to examples of dual-labeled neurons; the arrowheads point to a neuron that is tyrosine hydroxylase immunoreactive only. In the bottom panel electron photomicrographs show immunogold-silver detection of a rabbit antibody directed against CRF-R1 alone (A) or with tyrosine hydroxylase (TH; B) in LC. In (A), a dendrite (d) contains gold-silver deposits (arrowheads) for CRF-R1 along the cytoplasmic portion of its plasma membrane. An unlabeled terminal (ut) is apposed to the CRF-R1-labeled dendrite (CRF-R1-d). Bar = 0.45 μm. In (B) a dendrite (d) containing peroxidase labeling for TH exhibits gold-silver labeling for CRF-R1 (arrowheads). The dendrite contains several mitochondria (m). Bar = 0.35 μm.

of amygdalar terminals were CRF immunoreactive. These results suggested that CRF is a major neuromediator in the amygdalar–LC pathway, while it may play less of a role in the BNST–LC projection. Distinct roles for the CNA and BNST have been postulated in fear responses, with the former being engaged during conditioned fear and the latter during unconditioned fear (Walker and Davis, 1997). The anatomical studies described above would suggest that both nuclei (and therefore both conditions) impact on the LC, but that the qualitative nature of the impact may differ because of the different neurochemical nature of the projections.

The ultrastructural studies also implicated the CNA as a primary, although not the sole, contributor to CRF in the rostrolateral peri-LC. Consistent with this, CNA lesions dramatically reduce CRF immunoreactivity in the rostrolateral peri-LC (Sakanaka et al., 1986). The finding that different CRF afferents which target LC dendrites terminate in distinct regions leads to the speculation that different stressors engage specific CRF circuits to activate the LC–noradrenaline system. Recent studies designed to identify the sources of CRF that are engaged by different stimuli support this hypothesis. For example, LC neurons are activated by nonnoxious levels of bladder and colon distention and CRF is involved in the latter response (Elam et al., 1986; Page et al., 1992; Lechner et al., 1997). Ibotenic acid lesions of Barrington's nucleus, but not surrounding

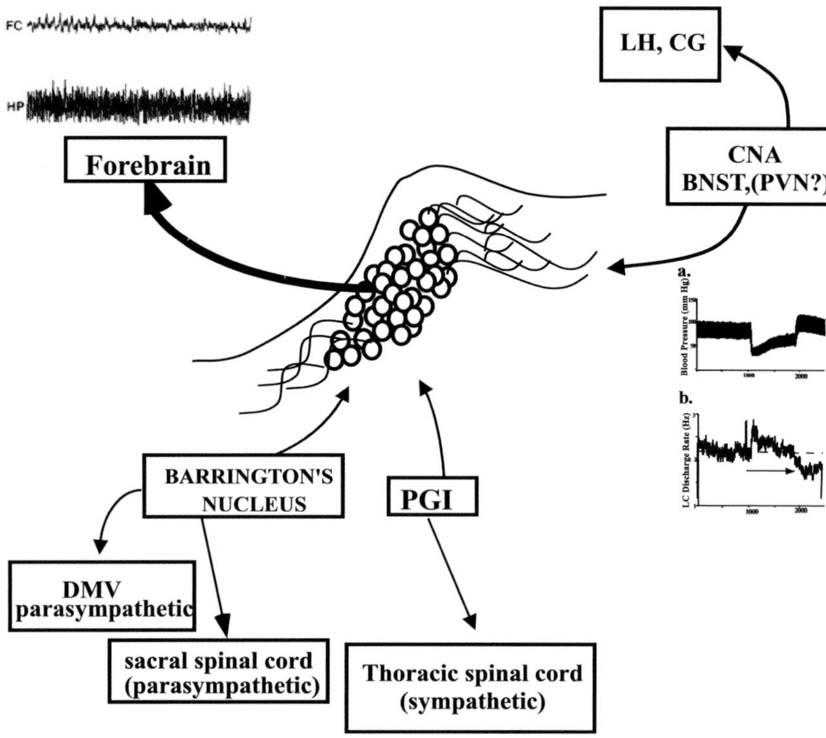

Fig. 4. Schematic depicting potential CRF afferents to the LC as determined from anatomical studies described in the text. Barrington's nucleus and the PGi project to the nuclear core of the LC. LC projections from Barrington's nucleus and the PGi may be coordinated with their projections to preganglionic parasympathetic or sympathetic neurons, respectively. Thus, activation of the LC–noradrenaline system may be coordinated with autonomic responses to stimuli. Sources of CRF terminating in the rostrolateral LC include the BNST, CNA, and possibly the PVN. Of these, the CNA appears to be the major source of CRF to this region. Projections to the lateral hypothalamus (LH) and central gray (CG) are thought to mediate autonomic and behavioral components of a fear response, respectively. Simultaneous activation of the LC–noradrenaline system may serve as a cognitive limb of this reponse. LC activation is associated with forebrain EEG activation. The traces shown (above Forebrain) are of frontal cortical (FC) and hippocampal (HP), EEG recorded during hypotensive stress. The cortical EEG became desynchronized and hippocampal EEG exhibited theta rhythm indicative of activation. (a) The traces on the right show the blood pressure response to infusion of nitroprusside and (b) simultaneously recorded LC discharge rate. The time of the infusion is indicated by the arrow. Note the increase in LC discharge rate during the infusion and the inhibition when the stress is terminated. The CNA has been implicated as the source of CRF that mediates LC activation by this challenge (Modified from Van Bockstacle et al. 1988).

regions, were found to prevent LC activation by colon distention, implicating Barrington's nucleus as the source of CRF responsible for LC activation by colon distention (Rouzade-Dominguez et al., 2001). Importantly, these lesions did not alter LC activation by hypotensive stress, indicating that this CRF pathway is selectively activated by certain stressors. Barrington's nucleus projects to the column of preganglionic parasympathetic neurons in the lumbosacral spinal cord that gives rise to the pelvic nerve and regulates pelvic visceral functions (Loewy et al., 1979; De Groat et al., 1993; Valentino et al., 1999). Dual retrograde tracing from the spinal cord and the nuclear LC demonstrated that CRF-containing LC-projecting neurons in Barrington's nucleus also project to the spinal cord (Valentino et al., 1996). Thus, Barrington's nucleus is positioned to coregulate the sacral parasympathetic system and the brain noradrenergic system in response to colon distention, perhaps in an effort to coordinate pelvic visceral responses with arousal or compatible behaviors (Fig. 4).

In contrast to colon distention, LC activation by hypotensive stress is prevented by lesion of the CNA, which targets the rostrolateral extranuclear dendritic zone, but not Barrington's nucleus or the BNST (Curtis et al., 2002). Moreover, hypotensive stress increases PCREB expression in CNA–CRF neurons that project to the LC (Curtis et al., 2002). The CNA coordinates autonomic and behavioral responses to emotional stimuli (eg., conditioned fear) via projections to the lateral hypothalamus and central gray, respectively (LeDoux et al., 1988). CRF-containing projections from the CNA to LC dendrites in the rostrolateral peri-LC may serve as a cognitive limb of this fear response that is expressed as enhanced arousal and altered attention. In addition to its role in fear responses, the CNA coordinates information from the cardiovascular system and other viscera with behavior (Fig. 4). Thus, it is not surprising that this plays a critical role in activation of the LC–noradrenaline system by hypotension.

Together, the anatomical findings suggest that different stimuli may engage selective CRF afferents to the LC that terminate in specific subfields of the LC (Fig. 4). The significance of these specific termination patterns has yet to be determined and as discussed above, the LC is not a clearly topographically organized nucleus. However, one may speculate that projections to the rostrolateral peri-LC may be more highly processed as this is a more heterogeneous region compared to the nuclear LC.

Functional consequences of stress-induced activation of the LC–noradrenaline system

The electrophysiological characteristics of LC neurons in rats and monkeys recorded during behavioral tasks have implicated the LC–noradrenaline system in arousal and attentional states (Foote et al., 1980; Aston-Jones and Bloom, 1981a,b). Thus, LC discharge rate is positively correlated to EEG indices of arousal (Foote et al., 1980; Aston-Jones and Bloom, 1981a). Moreover, selective manipulation of LC activity in halothane-anesthetized rats is sufficient to alter forebrain electroencephalographic (EEG) activity recorded in the cortex or hippocampus (Berridge and Foote, 1991; Berridge et al., 1993). For example, intracoerulear microinfusion of agents that increase LC discharge rate, such as muscarinic agonists, produce a temporally linked forebrain EEG activation which resembles that seen during behavioral arousal (e.g., cortical activity becomes desynchronized and shifts from high amplitude slow frequency to higher frequency; hippocampal EEG converts to theta rhythm) (Berridge and Foote, 1991; Berridge et al., 1993). Conversely, selective inhibition of LC discharge produced by local microinfusion of α_2-adrenergic agonists is sufficient to increase the incidence of slow-wave synchronous activity, similar to that seen during sleep or quiet waking (Berridge et al., 1993). Although the maximum magnitude of LC activation produced by CRF microinfusion is substantially less than that produced by muscarinic agonists, this is sufficient to activate forebrain EEG in anesthetized rats (Curtis et al., 1997). Likewise, physiological challenges that activate the LC by a magnitude of 20–50% above baseline (e.g., hypotension, bladder distention, and colon distention) produce a simultaneous forebrain EEG activation, suggesting that relatively small increases in LC discharge are sufficient for activation of the forebrain EEG (Valentino et al., 1991; Page et al., 1992, 1993; Lechner et al., 1997). CRF release within the LC is necessary for both LC and forebrain EEG activation during hypotensive challenge. Thus, microinfusion of a CRF antagonist into the LC, which prevents its activation by hypotensive challenge, also prevents the cortical EEG activation produced by this stress (Page et al., 1993). These results suggest that a consequence of CRF-induced activation of the LC–noradrenaline system may be to increase or maintain arousal. As a central response to visceral stimuli (e.g., bladder or colonic distention), forebrain arousal may serve to shift attention and facilitate behaviors that are compatible with the visceral responses (elimination). It should be noted that increased arousal would be predicted to be a common consequence of LC activation that could be elicited by neurochemically distinct afferents that are engaged by other challenges.

The robust response of LC neurons to sensory stimuli and the observation that this electrophysiological response often preceded behavioral orientation toward the stimulus suggested that LC activation may be important in directing attention toward salient stimuli in the environment. Studies by Aston-Jones and colleagues in which LC activity was recorded in monkeys during performance on an odd-ball visual discrimination task requiring focused attention have provided a more detailed view of the relationship between LC–noradrenaline activity and attention (Rajkowski et al., 1992, 1994; Aston-Jones et al., 1999; Usher et al., 1999). In these animals, low frequencies of LC discharge were associated with inattention, drowsiness, and poor task performance, whereas optimal performance was associated with relatively higher discharge rates. However, increases in discharge that exceeded this optimal rate were associated with a decrement in attention to the target stimuli and poor task performance. Thus, the relationship between LC discharge rate and task performance resembled an inverted U-shaped function. During the period that LC discharge was elevated and task performance was poor, the monkeys attended to task-irrelevant stimuli in the behavioral chamber and LC responses to target stimuli were greatly attenuated, an effect reminiscent of the effects of CRF on LC sensory responses. This group hypothesized that increases in LC discharge rate (above some optimal level) are associated with a shift from focused to scanning or labile attention (Fig. 1). Based on these studies, one would predict that the impact of CRF on the LC–noradrenergic system is dependent on the existing state of arousal. In a state of low arousal and attentiveness, CRF release may serve to arouse and mobilize forebrain targets to facilitate adaptive behaviors in response to a stimulus. However, during active waking, release of CRF would be predicted to shift the mode of attention from being focused on discrete stimuli to scanning of broad stimuli within the environment, an effect that would be adaptive in a life-threatening situation (Fig. 1). This may be a cognitive limb of a fear response that is set into motion when CRF projections from the amygdala to the LC are engaged.

The consequences of CRF release in the LC may not be limited to the cognitive effects described above. A number of behavioral, autonomic, and immunological effects have been reported following microinfusion of CRF or CRF antagonists into the LC, which have implicated CRF regulation of LC activity in other functions. For example, administration of CRF antagonists into the LC attenuate shock-induced freezing responses and immobilization-induced defensive behavior suggesting a role for stress-elicited CRF release in the LC in anxiogenic behaviors (Swiergel et al., 1992; Smagin et al., 1996a). Consistent with this, administration of propranolol decreased conditioned suppression produced by CRF administration (Cole and Koob, 1988). CRF in the LC may also be a target of anxiolytic drugs, such as alprazolam and adinazolam, as these agents decrease CRF levels in the LC region (Owens et al., 1989, 1993). Administration of CRF or the α_2-adrenergic antagonist, yohimbine, enhanced retention of passive avoidance behavior, implicating activation of the LC–noradrenaline system in memory (Chen et al., 1992). Finally, CRF in the LC in concentrations sufficient to induce c-fos expression in LC neurons, decreased T-lymphocyte mitogenic responses, implicating the LC–noradrenaline system in certain immunological aspects of the stress response (Rassnick et al., 1994).

Plasticity of CRF–LC interactions

The neurohormone CRF (in the PVN) is highly regulated by both corticosteroids and prior stress. Corticosteroids, the product of the endocrine limb of the stress response, exert a negative feedback regulation, in part by decreasing CRF mRNA in PVN hypothalamic neurons and tonic and evoked release of CRF into the median eminence (Plotsky, 1985; Plotsky et al., 1986; Plotsky and Sawchenko, 1987; Dallman et al., 1992). In contrast, prior stress (either repeated, chronic, or a single severe stress) generally increases CRF mRNA in PVN hypothalamic neurons and stress-evoked release of CRF (Imaki et al., 1991; Dallman et al., 1992; De Goeij et al., 1992; Bartanusz et al., 1993). CRF in different nuclei is differentially regulated by corticosteroids and stress (Swanson and Simmons, 1989; Imaki et al., 1991, 1992; Makino et al., 1994, 1995; Palkovits et al., 1998). For example, adrenalectomy and corticosteroids have reciprocal effects on CRF mRNA in PVN neurons, which are opposite to those on CRF mRNA in the CNA. Differential regulation is also seen with a single session of footshock or hypoosmotic challenge, which selectively increase CRF expression in neurons of Barrington's nucleus (Imaki et al., 1991, 1992).

The mechanisms underlying regulation of the neurohormone CRF are of interest as they may play a role in HPA dysfunctions that accompany certain stress-related disorders, such as depression and posttraumatic stress disorder (PTSD). Because the regulation of the LC–noradrenaline system by CRF may be part of a behavioral or cognitive limb of the stress response, it is important to know whether this action is similarly regulated. Recent studies of the regulation of CRF neurotransmitter actions in the LC by corticosteroids and prior stress support this idea.

As discussed above, adrenalectomy results in a tonic secretion of CRF in the LC that is observed 14 days after surgery (Pavcovich and Valentino, 1997), corresponding to the same timeframe during which increased CRF mRNA and protein levels in PVN hypothalamic neurons, as well as enhanced tonic and stress-evoked release of CRF in the median eminence are apparent (Sawchenko et al., 1984; Plotsky and Sawchenko, 1987). Because CRF antagonists increase LC discharge rates only in adrenalectomized animals, tonic CRF release in the LC must be normally restrained by corticosteroids. This has important implications for stress-related disorders, such as depression, in which hypersecretion of CRF has been postulated (Nemeroff et al., 1984; Gold et al., 1988a,b; Holsboer, 1999). Corticosteroid receptor dysfunction would be predicted to result in hypersecretion of neurotransmitter CRF in the LC and tonic hyperactivity of the LC–noradrenergic system. This would be expressed as increased arousal, sleep disturbances, and inability to concentrate. Consistent with this, decreases in slow frequency EEG activity and increases in high frequency EEG have been described in adrenalectomized rats (Bradbury et al., 1998).

In contrast to corticosteroids, which regulate CRF neurotransmission in the LC at a presynaptic level, affecting CRF release, prior stress regulates CRF neurotransmission in the LC at the postsynaptic level, altering the sensitivity of LC neurons to CRF (Fig. 5). The manner in which postsynaptic sensitivity is altered may depend upon the severity of the stimulus and the time between the last stress and the challenge dose of CRF. Immediately following an acute (30 min) session of footshock, LC neurons become desensitized to CRF as indicated by a shift to the right of the CRF dose–response curve for LC activation (Curtis et al., 1995). The physiological impact of this change in sensitivity to CRF is apparent as a loss of LC responses to hypotensive challenge (Curtis et al., 1995) and colon distention (Lechner et al., 1997) (i.e., cross desensitization). This change in LC sensitivity appears

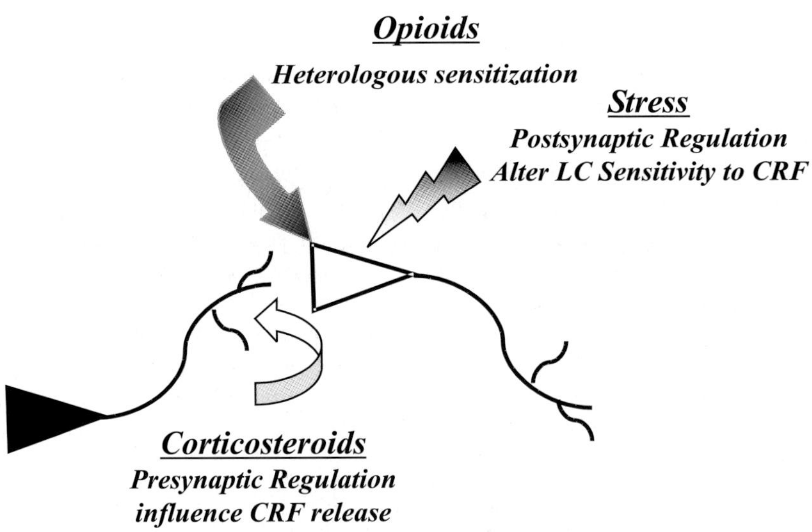

Fig. 5. Schematic depicting potential mechanisms by which CRF neurotransmission in the LC may be regulated. The solid triangle represents a CRF neuron that synapses with an LC neuron (open triangle). Evidence suggests that corticosteroids regulate CRF release in the synapse. Prior stress alters postsynaptic sensitivity of LC neurons to CRF. Prior exposure to opioids which interact with Gi-coupled receptors on LC neurons may result in heterologous sensitization to Gs-coupled processes through which CRF acts.

to be selective to CRF as LC spontaneous discharge and sensitivity to other inputs is not altered. Because these effects are observed immediately (within 1 h) following the shock session, desensitization may be an acute result of receptor internalization. A similar desensitization has been reported following exposure to prior CRF administration or auditory stress (Conti and Foote, 1995, 1996).

LC sensitivity to CRF is also altered following repeated sessions of footshock (once daily for 5 days), repeated daily i.p. injections of saline (21 days), and 24 h after a single exposure to swim stress (Curtis et al., 1995, 1999). Like acute footshock and unlike adrenalectomy, none of these conditions altered LC basal discharge rate or sensitivity to nonCRF inputs (excitatory amino acid, muscarinic, vasoactive intestinal peptide). Each of these conditions produced a similar complex shift of the CRF doseresponse curve for LC activation such that LC neurons were more sensitive to low (previously inactive) doses of CRF and less sensitive to higher doses, suggesting complex changes in CRF receptor kinetics, although the mechanisms underlying stress-induced changes in LC sensitivity to CRF have yet to be determined. Stress-induced sensitization of the LC–noradrenaline system is also expressed at the level of noradrenaline release in target regions. Thus, stress-elicited noradrenaline release in hippocampus (measured by microdialysis) was enhanced in animals with a history of chronic stress (Nisenbaum et al., 1991). Parallel to the electrophysiological findings, increased sensitivity to CRF-induced increases in noradrenaline release in previously stressed rats has also been demonstrated (Finlay et al., 1997).

A relevant feature of stress-induced sensitization of the LC is that the magnitude of LC activation produced by previously subthreshold doses is comparable to that necessary to increase forebrain EEG activity, i.e., sufficient to impact on forebrain targets. This implies that in animals with a history of severe or repeated stress the consequences of activation of the LC–noradrenergic system (i.e., increased arousal, loss of focused attention) may be elicited inappropriately by stimuli that would not typically activate the system. Because the sensitization appears to be selective to CRF, only stimuli that engage CRF release in the LC would be expected to produce an exaggerated response. This selectivity may account for some features of posttraumatic stress disorder (PTSD), a condition where exposure to a severe trauma selectively sensitizes to stimuli previously associated with the trauma. Subjects with PTSD exhibit hyperarousal, hypervigilance, and sleep disturbances, symptoms that would be associated with a hyperactive LC–noradrenergic system (Charney et al., 1993). LC sensitization to CRF could also underlie symptoms of stress-related bowel disorders, such as irritable bowel syndrome (Valentino et al., 1999). This disorder is characterized by increased cognitive awareness of colonic distention (Lynn and Friedman, 1993), which activates the LC via CRF-dependent mechanisms.

Stress-induced plasticity of the neurohormone and neurotransmitter effects of CRF in the LC may underlie symptoms of stress-related disorders that are characterized by concomitant hyperactivity of the HPA axis and the brain-noradrenergic system. This leads to the prediction that pharmacological manipulation of CRF would be a therapeutic target for these disorders. Several selective antagonists for the CRF-R1 receptor subtype have been synthesized for therapeutic use (Mansbach et al., 1997; Holsboer, 1999; Zobel et al., 2000). These agents have been demonstrated to be effective in certain behavioral models of depression, such as learned helplessness (Mansbach et al., 1997). A recent clinical trial in a small group of patients supports the efficacy of these agents in depression (Zobel et al., 2000). Although CRF antagonists are useful at preventing the acute effects of CRF, a more appropriate therapeutic intervention would involve reversing stress-induced plasticity. In this regard, antidepressants, which are effective in several stress-related disorders, have been shown to increase glucocorticoid receptor mRNA, an action that would be predicted to reverse or attenuate stress-induced "upregulation" of CRF (Reul et al., 1993, 1994; Barden et al., 1995). Whether this class of drugs reverses stress-induced changes in LC sensitivity to CRF is of interest and remains to be established.

Exposure to opioids represents another potential mechanism of plasticity of the LC–noradrenaline system to stress. Opioids activate Gi-coupled receptors on LC neurons (Duman et al., 1988; Nestler et al., 1999). In numerous preparations, activation of Gi-coupled receptors results in subsequent sensitization to Gs-coupled processes, a phenomenon known as heterologous sensitization (Watts, 2002). Opioid-induced heterologous sensitization of Gs-coupled receptors would be predicted to sensitize LC neurons to CRF and stressors that release CRF. This has important implications for subjects that abuse opiates or are chronically administered opiates for clinical reasons. Sensitization of LC neurons to CRF, coupled with a loss of the opioid counter-regulatory mechanism would result in an stress-induced activation the LC–noradrenaline system that is exaggerated and a more prolonged. Thus, exposure to opiates may make these individuals vulnerable to the adverse consequences of stress.

Summary and conclusions

The LC–noradrenaline system is generally activated by stressors and through its widespread projections, this activation is translated to forebrain targets resulting in enhanced arousal and alterations in sensory processing that may facilitate scanning of diverse stimuli in the environment. This may represent a cognitive limb of the stress response. One mediator of this activation is CRF, which impacts on the LC from diverse nuclei, including Barrington's nucleus and the CNA. The net effect of CRF on LC neurons is integrated with other neuromodulators of this system, including glutamate and opioids. Opioid modulation of the LC–noradrenaline system may serve to counterbalance the excitatory effects of CRF and play a role in rapidly terminating LC–noradrenaline activation when the stressor is no longer present. The neurotransmitter effects of CRF on LC neurons are regulated by corticosteroids, prior stress, and potentially by exposure to opiates. Dysregulation of CRF neurotransmission in the LC may play a role in cognitive symptoms (e.g., hyperarousal, sleep disorders, attentional dysfunction) of stress-related disorders. Targeting mechanisms underlying plasticity of CRF–LC interactions may be important novel treatments for these disorders.

Abbreviations

BNST	bed nucleus of the stria terminalis
CNA	central nucleus of the amygdala
CRF	corticotropin-releasing factor
EEG	electroencephalographic
i.c.v.	intracerebroventricular
LC	locus coeruleus
mRNA	messenger ribonucleic acid
PGi	nucleus paragigantocellularis
PTSD	posttraumatic stress disorder
PVN	paraventricular nucleus of the hypothalamus

Acknowledgments

This work is supported by PHS Grants MH40008, MH02006 and a NARSAD Distinguished Investigator Award (R.J.V.).

References

Abercrombie, E.D., Keller, R.W. and Zigmond, M.J. (1988) Characterization of hippocampal norepinephrine release as measured by microdialysis perfusion: pharmacological and behavioral studies. Neuroscience, 27: 897–904.

Akaoka, H. and Aston-Jones, G. (1991) Opiate withdrawal-induced hyperactivity of locus coeruleus neurons is substantially mediated by augmented excitatory amino acid input. J. Neurosci., 11: 3830–3839.

Aston-Jones, G. and Bloom, F.E. (1981a) Activity of norepinephrine-containing locus coeruleus neurons in behaving rats anticipates fluctuations in the sleep-waking cycle. J. Neurosci., 1: 876–886.

Aston-Jones, G. and Bloom, F.E. (1981b) Norepinephrine-containing locus coeruleus neurons in behaving rats exhibit pronounced responses to non-noxious environmental stimuli. J. Neurosci., 1: 887–900.

Aston-Jones, G., Ennis, M., Pieribone, V.A., Nickell, W.T. and Shipley, M.T. (1986) The brain nucleus locus coeruleus: restricted afferent control of a broad efferent network. Science, 234: 734–737.

Aston-Jones, G., Rajkowski, J. and Cohen, J. (1999) Role of locus coeruleus in attention and behavioral flexibility. Biological Psychiatry, 46: 1309–1320.

Aston-Jones, G., Shipley, M.T., Chouvet, G., Ennis, M., Van Bockstaele, E.J., Pieribone, V., Shiekhattar, R., Akaoka, H., Drolet, G., Astier, B., Charlety, P., Valentino, R. and Williams, J.T. (1991) Afferent regulation of locus coeruleus neurons: anatomy, physiology and pharmacology. Prog. Brain Res., 85: 47–75.

Austin, M.C., Rice, P.M., Mann, J.J. and Arango, V. (1995) Localization of corticotropin-releasing hormone in the human locus coeruleus and pedunculopontine tegmental nucleus: an immunocytochemical and in situ hybridization study. Neuroscience, 64: 713–727.

Avanzino, G.L., Ermirio, R., Cogo, C.E., Ruggeri, P. and Molinari, C. (1987) Effects of corticosterone on neurones of the locus coeruleus, in the rat. Neurosci. Lett., 80: 85–88.

Barden, N., Reul, J.M.H.M. and Holsboer, F. (1995) Do antidepressants stabilize mood through actions on the hypothalamic-pituitary-adrenocortical system? Trends in Neuroscience, 18: 6–11.

Bartanusz, V., Jezova, D., Bertini, L.T., Tilders, F.J.H., Aubry, J.M. and Kiss, J.Z. (1993) Stress-induced increase in vasopressin and corticotropin-releasing factor expression in hypophysiotropic paraventricular neurons. Endocrinology, 132: 895–902.

Beck, C.H.M. and Fibiger, H.C. (1995) Conditioned fear-induced changes in behavior and in the expression of the immediate early gene c-fos: with and without diazepam pretreatment. J. Neurosci., 15: 709–720.

Berridge, C.W. and Foote, S.L. (1991) Effects of locus coeruleus activation on electroencephalographic activity in the neocortex and hippocampus. J. Neurosci., 11: 3135–3145.

Berridge, C.W., Page, M.E. Valentino, R.J. and Foote, S.L. (1993) Effects of locus coeruleus inactivation on electroencephalographic activity in neocortex and hippocampus. Neuroscience, 55: 381–383.

Bonaz, B. and Tache, Y. (1994) Water-avoidance stress-induced c-fos expression in the rat brain and stimulation of fecal output: role of corticotropin-releasing factor. Brain Res., 641: 21–28.

Borsody, M.K. and Weiss, J.M. (1996) Influence of corticotropin-releasing hormone on electrophysiological activity of locus coeruleus neurons. Brain Res., 724: 149–168.

Bradbury, M.J., Dement, W.C. and Edgar, D.M. (1998) Effects of adrenalectomy and subsequent corticosterone replacement on rat sleep state and EEG power spectra. Am. J. Physiol., 275: R555–R565.

Britton, K.T., Segal, D.S., Kuczenski, R. and Hauger, R. (1992) Dissociation between in vivo hippocampal norepinephrine response and behavioral/neuroendocrine responses to noise stress in rats. Brain Res., 574: 125–130.

Bundey, R.A. and Kendall, D.A. (1999) Inhibition of receptor-mediated calcium responses by corticotropin-releasing hormone in the CATH.a cell line. Neuropharmacology, 38: 39–47.

Campeau, S. and Watson, S.J. (1997) Neuroendocrine and behavioral response and brain pattern of c-fos induction associated with audiogenic stress. J. Neuroendocrinol., 9: 577–588.

Cassens, G., Kuruc, A., Roffman, M., Orsulak, P.J. and Schildkraut, J.J. (1981) Alterations in brain norepinephrine metabolism and behavior induced by environmental stimuli previously paired with inescapable shock. Brain Res., 2: 387–407.

Cassens, G., Roffman, M., Kuruc, A., Orsulak, P.J. and Schildkraut, J.J. (1980) Alterations in brain norepinephrine metabolism induced by environmental stimuli previously paired with inescapable shock. Science, 209: 1138–1139.

Chalmers, D.T., Lovenberg, T.W., Grigoriadis, D.E., Behan, D.P. and De Souza, E.B. (1996) Corticotropin-releasing factor receptors: from molecular biology to drug design. Trends in Pharmacological Science, 17: 166–172.

Chan, R.K.W. and Sawchenko, P.E. (1995) Hemodynamic regulation of tyrosine hydroxylase messenger RNA in medullary catecholamine neurons: a c-fos-guided hybridization histochemical study. Neuroscience, 66: 377–390.

Chang, M.S., Sved, A.F., Zigmond, M.J. and Austin, M.C. (2000) Increased transcription of the tyrosine hydroxylase gene in individual locus coeruleus neurons following footshock stress. Neuroscience, 101: 131–139.

Charney, D.S., Deutch, A.Y., Krystal, J.H., Southwick, S.M. and Davis, M. (1993) Psychobiologic mechanisms of posttraumatic stress disorder. Arch. Gen. Psychiatry, 50: 294–305.

Chen, M.F., Chiu, T.H. and Lee, E.H.Y. (1992) Noradrenergic mediation of the memory-enhancing effect of corticotropin-releasing factor in the locus coeruleus. Psychoneuroendocrinology, 17: 113–124.

Chen, Y., Brunson, K.L., Muller, M.B., Cariaga, W. and Baram, T.Z. (2000) Immunocytochemical distribution of corticotropin-releasing hormone receptor type-1 (CRF(1))-like immunoreactivity in the mouse brain: light microscopy analysis using an antibody directed against the C-terminus. J. Comp. Neurol., 420: 305–323.

Chiang, C., Curtis, A.L., Drolet, G., Valentino, R.J. and Aston-Jones, G. (1991) Auditory responses of locus coeruleus neurons are attenuated by excitatory amino acid receptor antagonists in the awake rat. Soc. Neurosci. Abstr., 17: 1540.

Cintra, A., Solfrini, V., Agnati, L.F., Gustafsson J.A. and Fuxe, K. (1991) Strongly glucocorticoid receptor immunoreactive neurons in the neonatal rat brain. Neuroreport, 2: 85–88.

Cintra, A., Solfrini, V., Bunnemann, B., Okret, S., Bortolotti, F., Gustafsson, J.A. and Fuxe, K. (1993) Prenatal development of glucocorticoid receptor gene expression and immunoreactivity in the rat brain and pituitary gland: a combined in situ hybridization and immunocytochemical analysis. Neuroendocrinology, 57: 1133–1147.

Cintra, A., Zoli, M., Rosen, L., Agnati, L.F., Okret, S. Wikstrom, A.-C., Gustafsson, J.-A. and Fuxe, K. (1994) Mapping and computer assisted morphometry and microdensitometry of glucocorticoid receptor immunoreactive neurons and glial cells in the rat cental nervous system. Neuroscience, 62: 843–897.

Cole, B.J. and Koob, G.F. (1988) Propranolol antagonizes the enhanced conditioned fear produced by corticotropin-releasing factor. J. Pharmacol. Exp. Ther., 247: 902–910.

Conti, L.H. and Foote, S.L. (1995) Effects of pretreatment with corticotropin-releasing factor (CRF) on the electrophysiological responsivity of the locus coeruleus to subsequent CRF challenge. Neuroscience, 69: 209–219.

Conti, L.H. and Foote, S.L. (1996) Reciprocal cross-desensitization of locus coeruleus electrophysiological responsivity to corticotropin-releasing factor and stress. Brain Res., 722: 19–29.

Curtis, A.L., Bello, N.T., Connally, K.R. and Valentino, R.J. (2002) Corticotropin-releasing factor neurons of the central nucleus of the amygdala mediate locus coeruleus activation by cardiovascular stress. J. Neuroendocrinol., 14: 667–682.

Curtis, A.L., Bello, N.T., and Valentino, R.J. (2001) Endogenous opioids in the locus coeruleus function to limit the noradrenergic response to stress. J. Neurosci., 21: RC152.

Curtis, A.L., Florin-Lechner, S.M., Pavcovich, L.A. and Valentino, R.J. (1997) Activation of the locus coeruleus noradrenergic system by intracoerulear microinfusion of corticotropin-releasing factor: effects on discharge rate, cortical norepinephrine levels and cortical electroencephalographic activity. J. Pharmacol. Exp. Ther., 281: 163–172.

Curtis, A.L., Grigoradis, D., Page, M.E., Rivier, J. and Valentino, R.J. (1994) Pharmacological comparison of two corticotropin-releasing factor antagonists: in vivo and in vitro studies. J. Pharmacol. Exp. Ther., 268: 359–365.

Curtis, A.L., Pavcovich, L.A. and Valentino, R.J. (1995) Previous stress alters corticotropin-releasing factor neurotransmission in the locus coeruleus. Neuroscience, 65: 541–550.

Curtis, A.L., Pavcovich, L.A. and Valentino, R.J. (1999) Long term regulation of locus coeruleus sensitivity to corticotropin-releasing factor by swim stress. J. Pharmacol. Exp. Ther., 289: 1211–1219.

Dallman, M.F., Akana, S.F., Scriber, K.A., Bradbury, M.J., Walker, C.D., Strack, A.M., and Cascio, C.S. (1992) Stress, feedback and facillitation in the hypothalamo-pituitary-adrenal axis. J. Neruoendocrinol., 4: 517–526.

De Goeij, D.C.E., Jezova, D. and Tilders, F.J.H. (1992) Repeated stress enhances vasopressin synthesis in corticotorpin releasing factor neurons in the paraventricular nucleus. Brain Res., 557: 265–268.

De Groat, W.C., Booth, A.M. and Yoshimura, N. (1993) Neurophysiology of micturition and its modification in animal models of human disease. In: Maggi, C.A. (Eds.), The Autonomic Nervous System. Harwood Academic Publishers, London, pp. 227–290.

De Souza, E.B. (1987) Corticotropin-releasing factor receptors in the rat central nervous system: Characterization and regional distribution. J. Neurosci., 7: 88–100.

Duman, R.S., Tallman, J.F. and Nestler, E.J. (1988) Acute and chronic opiate-regulation of adenylate cyclase in brain: specific effects in locus coeruleus. J. Pharmacol. Exp. Ther., 246: 1033–1039.

Dun, N.J., Dun, S.L., Shen, E., Tang, H., Huang, R. and Chiu, T.H. (1995) c-fos expression as a marker of central cardiovascular neurons. Biol. Signals, 4: 117–123.

Duncan, G.E., Johnson, K.B., and Breese, G.R. (1993) Topographic patterns of brain activity in response to swim stress: assessment by 2-deoxyglucose uptake and expression of fos-like immunoreactivity. J. Neurosci., 13: 3932–3943.

Dunn, A.J. and Berridge, C.W. (1990) Physiological and behavioral responses to corticotropin-releasing factor administration: is CRF a mediator of anxiety or stress. Brain Res. Rev., 15: 71–100.

Elam, M., Svensson, T.H. and Thoren, P. (1985) Differentiated cardiovascular afferent regulation of locus coeruleus neurons and sympathetic nerves. Brain Res., 358: 77–84.

Elam, M., Thoren, T. and Svensson, T.H. (1986) Locus coeruleus neurons and sympathetic nerves: activation by visceral afferents. Brain Res., 375: 117–125.

Elam, M., Yao, T., Svensson, T.H. and Thoren, P. (1984) Regulation of locus coeruleus neurons and splanchnic, sympathetic nerves. Brain Res., 290: 281–287.

Elam, M., Yao, T., Thoren, P. and Svensson, T.H. (1981) Hypercapnia and hypoxia: chemoreceptor-mediated control of locus coeruleus neurons and splanchnic, sympathetic nerves. Brain Res., 222: 373–381.

Ennis, M. and Aston-Jones, G. (1988) Activation of locus coeruleus from nucleus paragigantocellularis: a new excitatory amino acid pathwway in brain. J. Neurosci., 8: 3644–3657.

Ennis, M., Aston-Jones, G. and Shiekhattar, R. (1992) Activation of locus coeruleus neurons by nucleus paragigantocellularis or noxious sensory stimulation is mediated by intracoerulear excitatory amino acid neurotransmission. Brain Res., 598: 185–195.

Finlay, J.M., Jedema, H.P., Ravinovic, A.D., Mana, M.J., Zigmond, M.J. and Sved, A.F. (1997) Impact of corticotropin-releasing hormone on extracellular norepinephrine in prefrontal cortex after cold stress. J. Neurochem., 69: 144–150.

Foote, S.L., Aston-Jones, G. and Bloom, F.E. (1980) Impulse activity of locus coeruleus neurons in awake rats and monkeys is a function of sensory stimulation and arousal. Proc. Natl. Acad. Sci. USA, 77: 3033–3037.

Foote, S.L., Bloom, F.E. and Aston-Jones, G. (1983) Nucleus locus coeruleus: new evidence of anatomical and physiological specificity. Physiol. Rev., 63: 844–914.

Foote, S.L. and Cha, C.I. (1988) Distribution of corticotropin-releasing factor-like immunoreactivity in brainstem of two monkey species (saimiri sciureus and macaca fasicularis): an immunohistochemical study. J. Comp. Neurol., 276: 239–264.

Funk, D. and Amir, S. (2000) Circadian modulation of fos resonses to odor of the red fox, a rodent predator, in the rat olfactory system. Brain Res., 866: 262–267.

Gold, P.W., Goodwin, F.K. and Chrousos, G.P. (1988a) Clinical and biochemical manifestations of depression: relationship to the neurobiology of stress (Part 1). New Eng. J. Med., 319: 348–353.

Gold, P.W., Goodwin, F.K. and Chrousos, G.P. (1988b) Clinical and biochemical manifestations of depression: relationship to the neurobiology of stress (Part 2). New Eng. J. Med., 319: 413–420.

Graham, J.C., Hoffman, G.E. and Sved, A.F. (1995) c-fos expression in brain in response to hypotension and hypertension in conscious rats. J. Auton. Nerv. Sys., 55: 92–104.

Harfstrand, A., Fuxe, K., Cintra, A., Agnati, L.F., Zini, I. Wikstrom, A.-C., Okret, S., Yu, Z.-Y., Goldstein, M., Steinbusch, H., Verhofstad, A. and Gustafsson, H.-A. (1986) Glucocorticoid receptor immunoreactivity in monoaminergic neurons of rat brain. Proc. Natl. Acad. Sci. USA, 83: 9779–9783.

Holsboer, F. (1999) The rationale for corticotropin-releasing hormone receptor (CRH-R) antagonists to treat depression and anxiety. J. Psychiatr. Res., 33: 181–214.

Imaki, T., Nahan, J.L., Sawachenko, P.E. and Vale, W. (1991) Differential regulation of corticotropin-releasing factor mRNA in rat brain regions by glucocorticoids and stress. J. Neuroscience, 11: 585–599.

Imaki, T., Shibasaki, T., Hotta, M. and Demura, H. (1993) Intracerebroventricular administration of corticotropin-releasing factor induces c-fos mRNA expression in brain regions related to stress responses: comparison with pattern of c-fos mRNA induction after stress. Brain Res., 616: 114–125.

Imaki, T., Vale, W. and Sawchenko, P.E. (1992) Regulation of corticotropin-releasing hormone mRNA in neuroendocrine and autonomic neurons by osmotic stimuli and volume loading. Neuroendocrinology, 56: 633–640.

Ishida, Y., Hashiguchi, H., Takeda, R., Ishizuka, Y., Mitsuyama, Y., Kannan, H., Nishimori, T. and Nakahara, D. (2002) Conditioned-fear stress increases Fos expression in monoaminergic and GABAergic neurons of the locus coeruleus and dorsal raphe nuclei. Synapse, 45: 46–51.

Jedema, H.P. and Grace, A.A. (2000) Effect of corticotropin-releasing hormone on electrophysiological properties of locus coeruleus neurons recorded in vitro. Soc. Neurosci. Abstr., 26: 1433.

Kawahara, H., Kawahara, Y. and Westerink, B.H. (2000) The role of afferents to the locus coeruleus in the handling stress-induced increase in the release of norepinephrine in the medial prefrontal cortex: a dual-probe microdialysis study in the rat brain. Eur. J. Pharmacol., 387: 279–286.

Kollack-Walker, S., Watson, S.J. and Akil, H. (1997) Social stress in hamsters: defeat activates specific neurocircuits within the brain. J. Neurosci., 17: 8842–8865.

Korf, J., Aghajanian, G.K. and Roth, R.H. (1973) Increased turnover of norepinephrine in the rat cerebral cortex during stress: role of the locus coeruleus. Neuropharmacology, 12: 933–938.

Lacosta, S., Merali, Z. and Anisman, H. (2000) Central monoamine activity following acute and repeated systemic interleukin-2 administration. Neuroimmunomodulation, 8: 83–90.

Lechner, S., Curtis, A., Brons, R. and Valentino, R. (1997) Locus coeruleus activation by colon distention: role of corticotropin-releasing factor and excitatory amino acids. Brain Res., 756: 114–124.

Lechner, S.M. and Valentino, R.J. (1999) Glucocorticoid receptor-immunoreactivity in corticotropin-releasing factor afferents to the locus coeruleus. Brain Res., 816: 17–28.

LeDoux, J.E., Iwata, J., Cicchetti, P. and Reis, D.J. (1988) Different projections of the central amygdaloid nucleus mediate autonomic and behavioral correlates of conditioned fear. J. Neurosci., 8: 2517–2529.

Loewy, A.D., Saper, C.B. and Baker, R.P. (1979) Descending Projections from the pontine micturition center. Brain Res., 172: 533–538.

Lynn, R.B. and Friedman, L.S. (1993) Irritable bowel syndrome. New Eng. J. Med., 329: 1940–1945.

Makino, S., Gold, P.W. and Schulkin, J. (1994) Corticosterone effects on corticotropin-releasing hormone mRNA in the central nucleus of the amygdala and the parvocellular region of the paraventricular neucleus of the hypothalamus. Brain Res., 640: 105–112.

Makino, S., Schulkin, J., Smith, M.A., Pacak, K., Palkovits, M. and Gold, P.W. (1995) Regulation of corticotropin-releasing hormone receptor messenger ribonucleic acid in the rat brain and pituitary by glucocorticoids and stress. Endocrinology, 136: 4517–4525.

Makino, S., Smith, M.A. and Gold, P.W. (2002) Regulatory role of glucocorticoids and glucocorticoid receptor mRNA levels on tyrosine hydroxylase gene expression in the locus coeruleus during repeated immobilization stress. Brain Res., 943: 216–223.

Mansbach, R.S., Brooks, E.N. and Chen, Y.L. (1997) Antidepressant-like effects of CP154526, a selective CRF1 receptor antagonist. Eur. J. Pharmacol., 323: 21–26.

Melia, K.R. and Duman, R.S. (1991) Involvement of corticotropin-releasing factor in chronic stress regulation of the brain noradrenergic system. Proc. Natl. Acad. Sci., 88: 8382–8386.

Miyazato, H., Skinner, R.D. and Garcia-Rill, E. (2000) Locus coeruleus involvement in the effects of immobilization stress on the p13 midlatency auditory evoked potential in the rat. Prog Neuropsychopharmacol. Biol. Psychiatry, 24: 1177–1201.

Morilak, D.A., Fornal, C. and Jacobs, B.L. (1987) Effects of physiological manipulations on locus coeruleus neuronal activity in freely moving cats. II. Cardiovascular challenge. Brain Res., 422: 24–31.

Morimoto, M., Morita, N., Ozawa, H., Yokoyama, K. and Kawata, M. (1996) Distribution of glucocorticoid receptor immunoreactivity and mRNA in the rat brain: an immunohistochemical and in situ hybridization study. Neurosci Res., 26: 235–269.

Nemeroff, C., Widerlov, E., Bissette, G.T., Walleus, H., Karlson, I., Eklund, K., Kilts, C., Loosen, P. and Vale, W. (1984) Elevated concentrations of CSF corticotropin-releasing factor-like immunoreactivity in depressed patients. Science, 226: 1342–1344.

Nestler, E.J., Alreja, M. and Aghajanian, G.K. (1999) Molecular control of locus coeruleus neurotransmission. Biol. Psychiatry, 46: 1131–1139.

Nisenbaum, L.K., Zigmond, M.J., Sved, A.F. and Abercrombie, E.D. (1991) Prior exposure to chronic stress results in enhanced synthesis and release of hippocampal norepinephrine in response to a novel stressor. J. Neurosci., 11: 1478–1484.

Owens, M.J., Bissette, G. and Nemeroff, C.B. (1989) Acute effects of alprazolam and adinazolam on the concentrations of corticotropin-releasing factor in the rat brain. Synapse, 4: 196–202.

Owens, M.J. and Nemeroff, C.B. (1991) Physiology and pharmacology of corticotropin-releasing factor. Pharmacol. Revs., 43: 425–474.

Owens, M.J., Vargas, M.A. and Nemeroff, C.B. (1993) The effects of alprazolam on corticotropin-releasing factor neurons in the rat brain: implications for a role for CRF in the pathogenesis of anxiety disorders. J. Psychiatr. Res., 27: 209–220.

Page, M.E. and Abercrombie, E.D. (1999) Discrete local application of corticotropin-releasing factor increases locus coeruleus discharge and extracellular norepinephrine in rat hippocampus. Synapse, 33: 304–313.

Page, M.E., Akaoka, H., Aston-Jones, G. and Valentino, R.J. (1992) Bladder distention activates locus coeruleus neurons by an excitatory amino acid mechanism. Neuroscience, 51: 555–563.

Page, M.E., Berridge, C.W., Foote, S.L. and Valentino, R.J. (1993) Corticotropin-releasing factor in the locus coeruleus mediates EEG activation associated with hypotensive stress. Neurosci. Lett., 164: 81–84.

Palkovits, M., Young, W.S. 3rd, Kovacs, K., Toth, Z. and Makara, G.B. (1998) Alterations in corticotropin-releasing hormone gene expression of central amygdaloid neurons following long-term paraventricular lesions and adrenalectomy. Neuroscience, 85: 135–147.

Pavcovich, L.A. and Valentino, R.J. (1997) Regulation of a putative neurotransmitter effect of corticotropin-releasing factor: effects of adrenalectomy. J. Neurosci., 17: 401–408.

Plotsky, P.M. (1985) Hypophysiotropic regulation of adenohypophyseal adrenocorticotropin secretion. Fed. Proc., 344: 207–214.

Plotsky, P.M., Otto, S. and Sapolsky, R.M. (1986) Inhibition of immunoreactive corticotropin-releasing factor secretion into the hypophysial-portal circulation by delayed glucocorticoid feedback. Endocrinology, 119: 1126–1130.

Plotsky, P.M. and Sawchenko, P.E. (1987) Hypophysial-portal plasma levels, median eminence content, and immunohistochemical staining of corticotropin-releasing factor, vasopressin and oxytocin after pharmacological adrenalectomy. Endocrinology, 120: 1361–1369.

Rajkowski, J., Kubiak, P. and Aston-Jones, G. (1992) Activity of locus coeruleus neurons in behaving monkeys varies with focused attention: short- and long-term changes. Soc. Neurosci. Abstr., 18: 538.

Rajkowski, J., Kubiak, P. and Aston-Jones, G. (1994) Activity of locus coeruleus neurons in monkey: phasic and tonic changes correspond to altered vigilance. Brain Res. Bull., 35: 607–616.

Rasmussen, K. and Aghajanian, G.K. (1989) Withdrawal-induced activation of locus coeruleus neurons in opiate-dependent rats: attenuation by lesions of the nucleus paragigantocellularis. Brain Res., 505: 346–350.

Rasmussen, K., Morilak, D.A. and Jacobs, B.L. (1986) Single unit activity of locus coeruleus neurons in the freely moving cat. I. During naturalistic behaviors and in response to simple and complex stimuli. Brain Res., 371: 324.

Rassnick, S., Hoffman, G.E., Rabin, B.S. and Sved, A.F. (1998) Injection of corticotropin-releasing hormone into the locus coeruleus or foot shock increases neuronal Fos expression. Neuroscience, 85: 259–268.

Rassnick, S., Sved, A.F. and Rabin, B.S. (1994) Locus coeruleus stimulation by corticotropin-releasing hormone suppresses in vitro cellular immune responses. J. Neurosci., 14: 6033–6040.

Reul, J.M.H.M., Labeur, M.S., Grigoriadis, D.E., DeSouza, E.B. and Holsboer, F. (1994) Hypothalamic-pituitary-adrenocortical axis changes in the rat after long-term treatment with the reversible monoamine oxidase-A inhibitor moclobemide. Neuroendocrinology, 60: 509–519.

Reul, J.M.H.M., Stec, I., Soder, M. and Holsboer, F. (1993) Chronic treatment of rats with the antidepressant amitriptyline attenuates the activity of the hypothalamic-pituitary-adrenocortical system. Endocrinology, 133: 312–320.

Roche, M., Commons, K.G., Peoples, A. and Valentino, R.J. (2003) Circuitry underlying regulation of the serotonergic system by swim stress. J. Neurosci, 23: 970–977.

Rouzade-Dominguez, M.-L., Curtis, A.L. and Valentino, R.J. (2001) Role of Barrington's nucleus in the activation of rat locus coeruleus neurons by colonic distension. Brain Res., 917: 206–218.

Rusnak, M., Kvetnansky, R., Jelokova, J. and Palkovits, M. (2001) Effect of novel stressors on gene expression of tyrosine hydroxylase and monoamine transporters in brainstem noradrenergic neurons of long-term repeatedly immobilized rats. Brain Res., 899: 20–35.

Sabban, E.L. and Kvetnansky, R. (2001) Stress-triggered activation of gene expression in catecholaminergic systems: dynamics of transcriptional events. Trends Neurosci., 24: 91–98.

Sakanaka, M., Shibasaki, T. and Lederes, K. (1987) Corticotropin-releasing factor-like immunoreactivity in the rat brain as revealed by a modified cobalt-glucose oxide-diaminobenzidene method. J. Comp. Neurol., 260: 256–298.

Sakanaka, M., Shibasaki, T. and Lederis, K. (1986) Distribution and efferent projections of corticotropin-releasing factor-like immunoreactivity in the rat amygdaloid complex. Brain Res., 382: 213–238.

Sanchez, M.M., Young, L.J., Plotsky, P.M. and Insel, T.R. (1999) Autoradiographic and in situ hybridization localization of corticotropin-releasing factor 1 and 2 receptors in nonhuman primate brain. J. Comp. Neurol., 408: 365–377.

Sauvage, M. and Steckler, T. (2001) Detection of corticotropin-releasing hormone receptor 1 immunoreactivity in cholinergic, dopaminergic and noradrenergic neurons of the murine basal forebrain and brainstem nuclei-potential implication for arousal and attention. Neuroscience, 104: 643–652.

Sawchenko, P.E., Swanson, L.W. and Vale, W.W. (1984) Co-expression of corticotropin-releasing factor and vasopressin immunoreactivity in parvocellular neurosecretry neurons of the adrenalectomized rat. Proc. Natl. Acad. Sci. USA, 81: 1883–1887.

Schulz, C. and Lehnert, H. (1996) Activation of noradrenergic neurons in the locus coeruleus by corticotropin-releasing factor, a microdialysis study. Neuroendocrinology, 63: 454–458.

Schulz, D.W., Mansbach, R.S., Sprouse, J., Braselton, J.P., Collins, J., Corman, M., Dunaiskis, A., Faraci, S., Schmidt, A.W., Seeger, T., Seymour, P., Tingley, F.D.I., Winston, E.N., Chen, Y.L. and Heym, J. (1996) CP-154,526: a potent and selective nonpeptide antagonist of corticotropin releasing factor receptors. Proc. Natl. Acad. Sci. USA, 93: 10477–10482.

Shipley, M.T., Fu, L., Ennis, M., Liu, W. and Aston-Jones, G. (1996) Dendrites of locus coeruleus neurons extend preferentially into two pericoerulear zones. J. Comp. Neurol., 365: 56–68.

Smagin, G.N., Harris, R.B. and Ryan, D.H. (1996a) Corticotropin-releasing factor antagonist infused into the locus coeruleus attenuates immobilization stress-induced defensive withdrawal in rats. Neurosci. Lett., 220: 167–170.

Smagin, G.N., Swiergiel, A.H. and Dunn, A.J. (1994) Sodium nitroprusside infusions activate cortical and hypothalamic noradrenergic systems in rats. Neurosci. Res. Comm., 14: 85–91.

Smagin, G.N., Swiergiel, A.H. and Dunn, A.J. (1995) Corticotropin-releasing factor administered into the locus coeruleus, but not the parabrachial nucleus, stimulates norepinephrine release in the prefrontal cortex. Brain Res. Bull., 36: 71–76.

Smagin, G.N., Zhou, J., Harris, R.B.S. and Ryan, D.H. (1997) Corticotropin-releasing factor antagonist attenuates immobilization stress-induced norepinephrine release in the prefrontal cortex in rats. Brain Res. Bull. 42: 431–434.

Smith, M.A., Banerjee, S., Gold, P.W. and Glowa, J. (1992) Induction of c-fos mRNA in rat brain by conditioned and unconditioned stressors. Brain Res., 578: 135–141.

Smith, M.A., Brady, L.S., Glowa, J., Gold, P.W. and Herkenham, M. (1991) Effects of stress and adrenalectomy on tyrosine hydroxylase mRNA levels in the locus ceruleus by in situ hybridization. Brain Res., 544: 26–32.

Svensson, T.H. (1987) Peripheral, autonomic regulation of locus coeruleus noradrenergic neurons in brain: putative implications for psychiatry and psychopharmacology. Psychopharmacology, 92: 1–7.

Svensson, T.H. and Thoren, P. (1979) Brain noradrenergic neurons in the locus coeruleus: inhibition by blood volume load through vagal afferens. Brain Res., 172: 174.

Swanson, L.W., Sawchenko, P.E., Rivier, J. and Vale, W. (1983) Organization of ovine corticotropin-releasing factor immunoreactive cells and fibers in rat brain: an immunohistochemical study. Neuroendocrinology, 36: 165–186.

Swanson, L.W. and Simmons, D.M. (1989) Differential steroid hormone and neural influences on peptide mRNA levels in CRH cells of the paraventricular nucleus: a hybridization histochemical study in the rat. J. Comp. Neurol., 285: 413–435.

Swiergel, A.H., Takahashi, L.K. Ruben, W.W. and Kalin, N.H. (1992) Antagonism of corticotropin-releasing factor receptors in the locus coeruleus attenuates shock-induced freezing in rats. Brain Res., 587: 263–268.

Thierry, A.M., Javoy, F., Glowinski, J. and Kety, S.S. (1968) Effects of stress on the metabolism of norepinephrine, dopamine and serotonin in the central nervous system of the rat: modification of norpinephrine turnover. J. Pharmacol. Exp. Ther., 163: 163–171.

Towle, A.C., Sze, P.Y. and Lauder, J.M. (1982) Cytosol glucocorticoid binding in monoaminergic cell groups. Dev. Neurosci., 5: 458–464.

Tsuda, A., Ida, Y., Satoh, H., Tsujimaru, S. and Tanaka, M. (1989) Stressor predictability and rat brain noradrenaline metabolism. Pharmacol. Biochem. Behav., 32: 569–572.

Usher, M., Cohen, J.D., Servan-Schreiber, D., Rajkowski, J. and Aston-Jones, G. (1999) The role of locus coeruleus in the regulation of cognitive performance. Science, 283: 549–554.

Vale, W., Spiess, J., Rivier, C. and Rivier, J. (1981) Characterization of a 41-residue ovine hypothalamic peptide that stimulates secretion of corticotropin and beta-endorphin. Science, 213: 1394–1397.

Valentino, R.J., Chen, S., Zhu, Y. and Aston-Jones, G. (1996) Evidence for divergent projections of corticotropin-releasing hormone neurons of Barrington's nucleus to the locus coeruleus and spinal cord. Brain Res., 732: 1–15.

Valentino, R.J. and Foote, S.L. (1987) Corticotropin-releasing factor disrupts sensory responses of brain noradrenergic neurons. Neuroendocrinology, 45: 28–36.

Valentino, R.J. and Foote, S.L. (1988) Corticotropin-releasing factor increases tonic but not sensory-evoked activity of noradrenergic locus coeruleus neurons in unanesthetized rats. J. Neurosci., 8: 1016–1025.

Valentino, R.J., Foote, S.L. and Aston-Jones, G. (1983) Corticotropin-releasing factor activates noradrenergic neurons of the locus coeruleus. Brain Res., 270: 363–367.

Valentino, R.J., Miselis, R.R. and Pavcovich, L.A. (1999) Pontine regulation of pelvic viscera: pharmacological target for pelvic visceral dysfunctions. Trends Pharmacol. Sci., 20: 253–266.

Valentino, R.J., Page, M.E. and Curtis, A.L. (1991) Activation of noradrenergic locus coeruleus neurons by hemodynamic stress is due to local release of corticotropin-releasing factor. Brain Res., 555: 25–34.

Valentino, R.J., Page, M.E., Van Bockstaele, E. and Aston-Jones, G. (1992) Corticotropin-releasing factor innervation of the locus coeruleus region: distribution of fibers and sources of input. Neuroscience, 48: 689–705.

Valentino, R.J., Rudoy, C., Saunders, A. and Van Bocstaele, E.J. (2001) Corticotropin-releasing factor is preferentially colocalized with glutamate in terminals in the locus coeruleus region. Neuroscience, 106: 375–384.

Valentino, R.J. and Van Bockstaele, E.J. (2001) Opposing regulation of the locus coeruleus by corticotropin-releasing factor and opioids: potential for reciprocal interactions between stress and opioid sensitivity. Psychopharmacology, 158: 331–342.

Valentino, R.J. and Wehby, R.G. (1988) Corticotropin-releasing factor: evidence for a neurotransmitter role in the locus coeruleus during hemodynamic stress. Neuroendocrinology, 48: 674–677.

Valentino, R.J. and Wehby, R.G. (1989) Locus coeruleus discharge characteristics of morphine-dependent rats: Effects of naltrexone. Brain Res., 488: 126–134.

Van Bockstaele, E.J., Colago, E.E.O. and Valentino, R.J. (1996) Corticotropin-releasing factor-containing axon terminals synapse onto catecholamine dendrites and may presynaptically modulate other afferents in the rostral pole of the nucleus locus coeruleus in the rat brain. J. Comp. Neurol., 364: 523–534.

Van Bockstaele, E.J., Colago, E.E.O. and Valentino, R.J. (1998) Amygdaloid corticotropin-releasing factor targets locus coeruleus dendrites: substrate for the coordination of emotional and cognitive limbs of the stress response. J. Neuroendocrinol., 10: 743–757.

Van Bockstaele, E.J., Peoples, J. and Valentino, R.J. (1999) Differential regulation of the rostrolateral peri-locus coeruleus region by limbic afferents. Biol. Psych., 46: 1352–1363.

Van Pett, K., Viau, V., Bittencourt, J.C., Chan, R.K., Li, H.Y., Arias, C., Prins, G.S., Perrin, M., Vale, W. and Sawchenko, P.E. (2000) Distribution of mRNAs encoding CRF receptors in brain and pituitary of rat and mouse. J. Comp. Neurol., 428: 191–212.

Walker, D.L. and Davis, M. (1997) Double dissociation between the involvement of the bed nucleus of the stria terminalis and the central nucleus of the amygdala in startle increases produced by conditioned versus unconditioned fear. J. Neurosci., 17: 9375–9383.

Watts, V.J. (2002) Molecular mechanisms for heterologous sensitization of adenylate cyclase. J. Pharmacol. Exp. Ther., 302: 1–7.

Zobel, A.W., Nickel, T., Kunzel, H.E., Ackl, N., Sonntag, A., Ising, M. and Holsboer, F. (2000) Effects of the high-affinity corticotropin-releasing hormone receptor 1 antagonist R121919 in major depression: the first 20 patients treated. J. Psychiatr. Res., 34: 171–181.

T. Steckler, N.H. Kalin and J.M.H.M. Reul (Eds.)
Handbook of Stress and the Brain, Vol. 15
ISBN 0-444-51173-3

CHAPTER 4.5

Regional specialisation in the central noradrenergic response to unconditioned and conditioned environmental stimuli

S.C. Stanford[1,*] and C.A. Marsden[2]

[1]*Department of Pharmacology, University College London, Gower Street, London WC1E 6BT, UK*
[2]*School of Biomedical Sciences, Institute of Neuroscience, University of Nottingham Medical School, Queen's Medical Centre, Nottingham NG7 2UH, UK*

Abstract: Many studies, using a wide range of techniques, have indicated that the central noradrenergic system of rats is extremely sensitive to non-noxious, but aversive environmental stimuli. The effects of stress experienced in early life (social isolation) are particularly interesting because the ensuing changes in the central noradrenergic system seem to be relevant to the causes of psychiatric disorders in humans that are exacerbated by stress (e.g. depression and schizophrenia). However, noradrenergic neurones projecting to different regions of the brain make different contributions to the response and adaptation to stress. This is especially apparent when comparing the noradrenergic response to conditioned and unconditioned environmental stimuli. Finally, evidence from studies of inbred rat strains points to differences in noradrenergic transmission, which influence the stress response and behaviour, that are determined by genotype. This chapter draws on evidence from all these experimental approaches and discusses the proposal that noradrenergic transmission makes an important contribution to coping with mild stress but that, when the stress is intense, it exacerbates the impact of the stressor on mood and behaviour.

Introduction

Walter Cannon was the first to note that there were individual differences in the behavioural responses of cats to the psychological stress of hearing dogs barking. Further, he proposed that these differences were related to the concentration of chemical factor(s) in the circulation ('excited blood'). Nearly 100 years later, we are still trying to explain the neurochemical coding of the stress response. This chapter outlines recent work aimed at gaining more insight into this question and concentrates on the role of noradrenaline, one of the most studied neuronal factors. Evidence gathered over the last three decades has consistently confirmed that stressful stimuli activate noradrenergic neurones in the brain. What is less clear is whether, and under what circumstances, recruitment of a noradrenergic response is beneficial in terms of coping with stress.

There are several reasons why it is difficult to define the effects of stress on central noradrenergic neurones and, conversely, the role of these neurones in the stress response. First, the extent to which central noradrenergic neurones are activated during stress depends on the type of stress being experienced. Indeed, it is important to note that noradrenergic neurones are also activated by hedonic stimuli (Cenci et al., 1992) and so it is not necessarily the aversive

*Corresponding author. Tel.: +44 7679 3731;
Fax: +44 7679 7298; E-mail: c.stanford@ucl.ac.uk

(stressful) features of a given stimulus that determine the noradrenergic response. Second, the noradrenergic response can depend on an individual's history of stress and, third, the response to a given form of stress differs across brain regions and can depend on genetic background.

Stimuli that activate central noradrenergic neurones

Electrophysiological studies have revealed that typical laboratory stressors, such as restraint (Abercrombie and Jacobs, 1987) or footshock (Hirata and Aston-Jones, 1994) increase the firing rate of noradrenergic neurones that project from the locus coeruleus, which is located in the brainstem, to the forebrain. Changes in the noradrenaline content of peripheral tissues and brain regions can also be detected after exposure of animals to intense forms of stress, especially if they are unpredictable and uncontrollable. Such stimuli generally cause depletion of neuronal stores of noradrenaline, but the extent of the depletion varies across different brain regions: it is typically greater and more rapid in the hypothalamus and locus coeruleus than in the frontal cortex (reviewed by Stanford, 1995). These findings could reflect differences in the activation of noradrenergic neurones projecting to different brain regions. They could also be explained by regional differences in the extent to which replenishment of noradrenaline stores, by neuronal reuptake and synthesis, keeps pace with the rate of transmitter release.

When defining the role of central noradrenergic transmission in the stress response, it should be borne in mind that many different types of sensory stimuli activate these neurones, including those that do not cause overt discomfort (e.g. light flashes and tonal pips, Table 1). Even handling of animals has this effect and not only increases the concentration of extracellular noradrenaline in the cortex (Dalley and Stanford, 1995; Feenstra et al., 1998) but, when it is repeated on a daily basis, eventually causes downregulation of α_2- and β-adrenoceptors and an increase in the activity of tyrosine hydroxylase in the hippocampus (and possibly other brain regions too) (Stanford et al., 1984). If the handling incorporates an intraperitoneal injection of saline,

Table 1. Common laboratory stressors

Interoceptive cues		
	Hypotension	rat
	Bladder distension	rat
	Hypoglycaemia	cat
Somatosensory		
	Tail pinch	rat
	Footshock	rat
	Restraint	rat
Sensory		
	Light flash	rat and cat
	Fur contact	rat
	Olfaction	rat
	Tonal pips and clicks	rat, cat and monkey
	Auditory click	cat
	Air puff (on hair)	monkey
Complex environmental stimuli		
	Threatening cues	cat and monkey
	Tapping on cage	monkey

the neurochemical changes are even larger (Stanford et al., 1984). Such changes can obviously be confounding factors in stress research (see Stanford, 1996).

Noradrenergic transmission and resistance to stress ('coping')

The impact of noradrenergic transmission on the process of coping with stress could well depend on the nature of the stress, also. Radioligand binding studies, using [^{3}H] dihydroalprenolol to measure the density of cortical β-adrenoceptors in the rat, investigated whether there was a correlation between individual differences in the density of these receptors and animals' behavioural resistance to stress. Notwithstanding the limitation that this radioligand is now known to bind to 5-HT_{1A} receptors to some extent, general principles emerging from this study are still relevant. The density of [^{3}H] dihydroalprenolol binding sites within a group of rats correlated positively with individuals' behavioural resistance to the stress of exposure to an open field (Salmon and Stanford, 1989). There was also a positive correlation between radioligand binding and resistance to the stress of frustrative non-reward, when animals experienced one extinction trial per day (Stanford and Salmon, 1989). These findings imply that the greater the density of [^{3}H] dihydroalprenolol binding sites (of which the majority comprise β–adrenoceptors) the greater an individual's resistance to stress. Such a change is reminiscent of early findings that it was human subjects with a higher rate of urinary excretion of noradrenaline who performed best in an audiovisual conflict task (the Stroop test; Frankenhaeuser, 1971).

However, when the stress of frustrative non-reward was intensified, by carrying out all the extinction trials in one day ('massed extinction'), there was a negative correlation between the density of [^{3}H] dihydroalprenolol binding sites and behaviour. This suggests that, with more intense forms of stress, a greater density of [^{3}H] dihydroalprenolol binding sites is associated with reduced resistance to stress (Salmon and Stanford, 1992). In both cases, these 'within group' correlations were independent of animals' locomotor activity, although in a subsequent microdialysis study, comparing noradrenaline efflux across different groups of rats, the increase in noradrenaline efflux induced by somatosensory stimuli was greatest in the group that showed the greatest locomotor activity in an open field (Rosario and Abercrombie, 1999).

Even setting aside the welfare implications of studying the effects of noxious stimuli, such findings underline the need to ensure that the types and intensity of stressors being studied experimentally are realistic analogues of those commonly encountered by humans. Since the neurochemical effects of stimuli that impose physical discomfort cannot be assumed to generalize to those of non-noxious, stressors, the effects of naturalistic (psychological), environmental stimuli are of particular interest.

Noradrenergic response to naturalistic stress

Relatively few studies have investigated the noradrenergic response to naturalistic aversive stimuli. Yet, it has been known for many years that the concentration of the noradrenaline metabolite, methylhydroxyphenylglycol (MHPG) is increased in the amygdala, hypothalamus and locus coeruleus of rats that have watched conspecifics experiencing footshock (Tanaka et al., 1991). In macaque monkeys, even the approach of the experimenter increases the firing rate of neurones in the locus coeruleus (Grant et al., 1988). In a study in which rats were exposed to ultrasound, at a frequency known to be associated with aversive events (20 kHz), there was an aversive response in the rat characterised by escape behaviours followed by freezing. A lesion of the locus coeruleus attentuated the freezing, but not the escape behaviour. A similar reduction in freezing after the lesion was observed using a conditioned emotional response (CER) paradigm (Neophytou et al., 2001). These findings support evidence, described below, that the noradrenergic innervation derived from the locus coeruleus is involved in the modification of fear-related behaviours in the rat in both unconditioned and conditioned situations.

Recently, in vivo microdialysis has been used to investigate the effects of stress on the concentration of extracellular noradrenaline in the brains of freely moving rats. As might be expected, a wide range of aversive stimuli increases the amount of noradrenaline reaching a microdialysis probe ('efflux') implanted in the hippocampus (e.g. Abercrombie et al., 1988) or paraventricular nucleus of the hypothalamus (Pacak et al., 1992). We have used this technique to investigate whether changes in noradrenaline efflux can be detected when animals are exposed to a novel environment.

The first experiments explored whether the magnitude or duration of the noradrenaline response was influenced by the quality or intensity of the environmental stimulus. Transfer of rats from their home cage to an identical novel cage that was illuminated to the same intensity (20 lux) produced only a small and transient ($<$ 60 min) increase in noradrenaline efflux. Most of this increase could be attributed to the handling required to transfer the rats from one cage to another (Dalley and Stanford, 1995) (Fig. 1). However, transfer of the rats from their home cage to a brightly lit novel cage (1500 lux) produced a larger and longer-lasting increase. In fact, when monitoring noradrenaline efflux in rats exposed to different configurations of novel environment, there were clearly graded changes in noradrenaline efflux in environments that differed in respect of their illumination and whether or not they contained an unfamiliar rat (Dalley and Stanford, 1994, 1995) (Fig. 1). Other groups have similarly reported an increase in noradrenaline efflux on placing rats in a

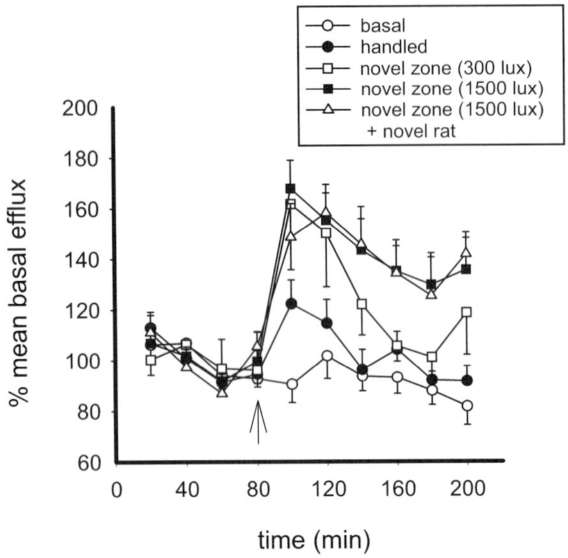

Fig. 1. Effect of environment on noradrenaline efflux in the frontal cortex of rats. Points show mean ± s.e. mean efflux, expressed as a percentage of the last basal value. This was taken at time '80 min' (indicated by the arrow), immediately before transfer of the rats from their home cage to the test environment. The test zone was configured in one of the ways indicated. $n = 6$, except for the group transferred to the novel zone (1500 lux) and those maintained in the home cage throughout the experiment ('basals') ($n = 10$) and the group transferred to the novel zone (1500 lux) ($n = 12$). See: Dalley and Stanford (1995).

novel environment (Feenstra et al., 2000). It is interesting that, although these environmental features (e.g. novelty, bright light, the presence of a conspecific) are used routinely in behavioural screens for anti-anxiety drugs, neither buspirone nor diazepam affected the incremental increase in noradrenaline efflux caused by exposure to the novel environment (Dalley et al., 1996). Similarly, drugs modifying the function of $GABA_A$-receptor inputs to the locus coeruleus have no effect on the noradrenaline response to handling (Kawahara et al., 2000).

Although all these findings indicate that environmental stimuli are extremely effective at causing an increase in noradrenaline efflux, they give no indication of how the change in noradrenergic transmission relates to the animals' emotional response to the novel environment. This question can be explored by first characterising animals' behavioural response to different configurations of test environment and then monitoring noradrenaline efflux when the rats are confined within each type of test environment. In this way, it should be possible to distinguish whether or not the changes in noradrenaline efflux parallel the extent to which animals perceive each type of test environment to be aversive. As described below, a modified place preference test can be used to investigate this question.

Grading rats' behavioural response to environmental stimuli

The test we have used to investigate the relationship between noradrenergic transmission and animals' behavioural reaction to a test environment is based on the assumption that rats will avoid an environment they perceive to be aversive and express a preference ('approach') for one they find less aversive, or more interesting, or rewarding. A light/dark exploration box (Crawley and Goodwin, 1980) that has been modified for use in these particular experiments, is ideal for this type of experiment because this apparatus can be used for both behavioural and microdialysis studies lasting several hours. Details of the test box are given in detail elsewhere (McQuade et al., 1999) but briefly the apparatus comprises three chambers arranged in series. The walls and floor of the central chamber (the 'neutral' zone) are constructed of black, opaque Perspex and this zone is illuminated to the same intensity as the rats' home cages (30 lux). One of the two end chambers is made of black Perspex (for the 'dark' test condition) while the other has transparent walls and floor and can be illuminated separately (the 'light' arena). The test and neutral zones are separated by partitions that can be raised and which incorporate a guillotine-style door. In the behavioural experiments, one of these doors is raised to enable the rat to commute between the neutral arena and one of the test chambers. However, in microdialysis experiments, these doors remain closed so as to confine the rat in the test arena.

When testing the behavioural response of naïve (unoperated) outbred Sprague-Dawley rats during exposure to these different types of novel environment, there was no difference in their locomotor activity in the light and dark arena. However, the

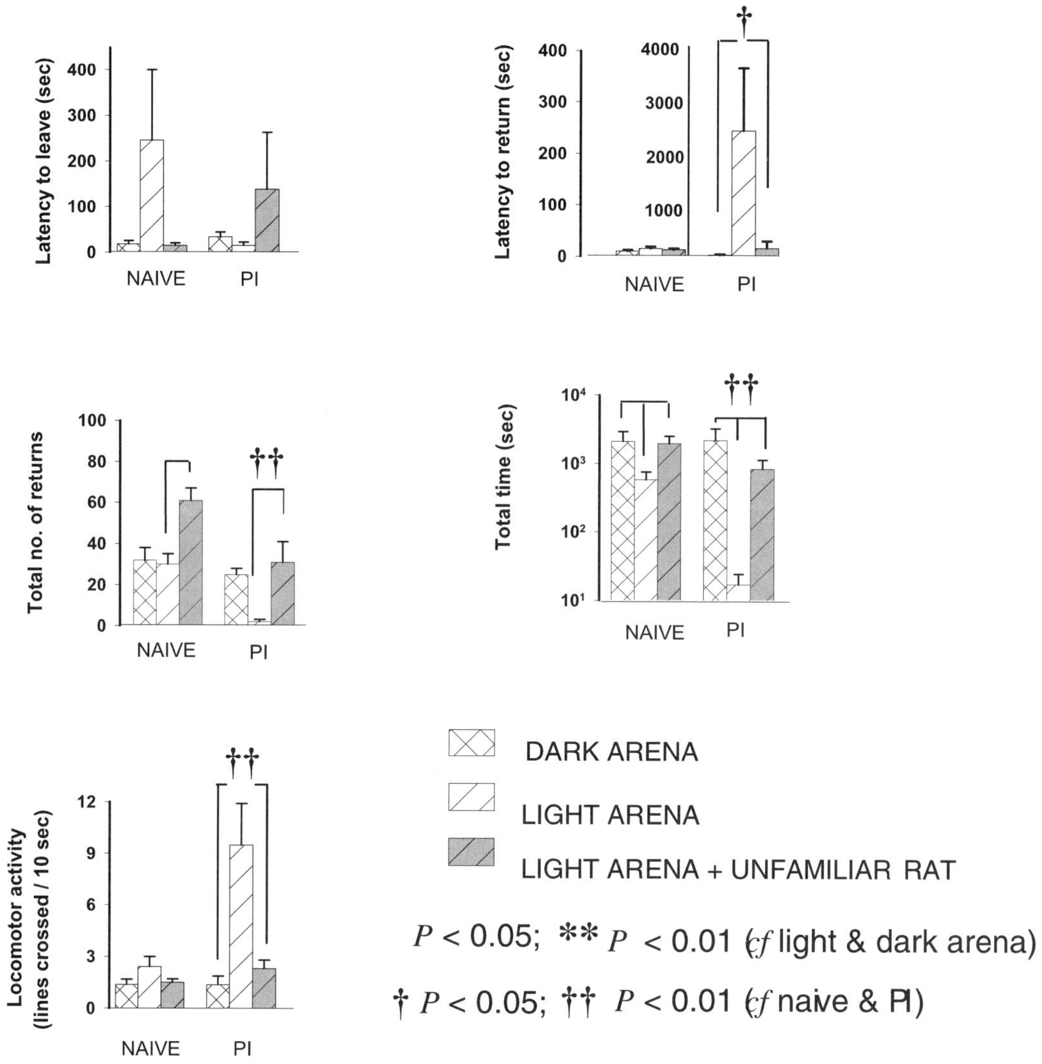

Fig. 2. The behaviour of rats after transfer from the neutral zone of a light/dark exploration box to an adjacent, test arena (configured in one of the ways indicated). Behaviour was scored for a 2-h period during which rats were able to commute freely between the two compartments. The columns show behavioural scores of naïve rats and of rats that had experienced surgery with implantation of a microdialysis probe in the frontal cortex and hypothalamus ('PI'). Bars show mean ± s.e. mean. $n = 7$ or 8. See: McQuade et al. (1999).

rats spent less time in the light arena than in the 'dark' one, suggesting that they found the former to be the more aversive (McQuade et al., 1999) (Fig. 2). Interestingly, the presence of an unfamiliar rat in the light arena abolished many of the differences in behaviour in the light and dark zones and so seemed to ameliorate the aversive impact of a brightly lit novel arena (McQuade et al., 1999) (Fig. 2).

An important factor to consider in these experiments is that implantation of a microdialysis probe might itself modify animals' behaviour. Changes in the rhythm of body temperature and locomotor activity following implantation of a microdialysis probe are thought to last for at least 7 days (Drijfhout et al., 1995). To see whether animals' response to novel stimuli is similarly affected, we investigated whether implantation of microdialysis probes in the frontal cortex and hypothalamus alters rats' behaviour in the light/dark exploration box. In these experiments, the probes were not connected to the perfusion system for collection of dialysis samples and so rats' movement within the light/dark exploration box was unimpeded.

Implantation of the microdialysis probes modified several aspects of the rats' behaviour. They spent considerably less time in the light arena than did rats that had not undergone surgery (Fig. 2). They also showed a greater latency to return and made fewer returns to the test arena. These findings suggest that probe implantation increases avoidance of an aversive novel environment. However, in another study using microdialysis of the ventral hippocampus to investigate the effects of social interaction on 5-hydroxytryptamine (5-HT) and cAMP efflux, there was no effect of probe implantation on either normal social interaction or the effects of diazepam on the behaviour (Cadogan et al., 1994). Nevertheless, all these findings highlight the need for appropriate controls in such experiments, otherwise the behavioural response to different types of test environment (especially aversive ones) could easily be misinterpreted.

Noradrenergic responses to a novel environment

Having characterised animals' behavioural reaction to each type of novel environment, the next step is to establish whether these responses might be related to any changes in noradrenaline efflux.

One advantage of using in vivo microdialysis is that it is possible to implant two probes into different brain regions and to compare local changes in noradrenaline efflux induced by the same stimulus. We have used this approach to compare, within each rat, changes in noradrenaline efflux in the frontal cortex and hypothalamus. These two brain regions were chosen for several reasons, not least of which was the evidence, cited above, that the noradrenergic response in the hypothalamus differs from that in the frontal cortex. In addition to this, the frontal cortex is one of the few brain areas to be innervated exclusively by noradrenergic neurones projecting from the locus coeruleus, whereas the majority of hypothalamic nuclei are mainly innervated by noradrenergic neurones in the lateral tegmental nuclei. There is a good deal of indirect evidence to suggest that there are differences in the function and neurochemical control of these two groups of neurones. For instance, their morphology differs considerably (Fallon et al., 1978) as does their sensitivity to the neurotoxin, N-(2-chloroethyl)-N-ethyl-2-bromobenzylamine (DSP-4, Fritschy and Grzanna, 1989).

When rats were transferred from the neutral zone of the light/dark exploration box and confined within a 'dark' test environment, illuminated to the same intensity as the neutral zone, there was no change in noradrenaline efflux in either the frontal cortex or the hypothalamus (McQuade et al., 1999). However, confinement in the brightly lit test zone caused a large increase in noradrenaline efflux in both brain regions. These findings suggest that novelty alone is not a sufficient stimulus to increase efflux of noradrenaline, i.e. the dark test arena was just as 'novel' as the brightly lit one and yet only the latter increased noradrenaline efflux in either brain region. It is also evident that the extent to which animals find the arena aversive is not the key factor. Thus, the presence of another rat in the brightly lit arena greatly diminished its impact on behaviour and yet noradrenaline efflux was still increased in both the frontal cortex and the hypothalamus.

Overall, the noradrenaline response seems to be determined by a combination of factors that cannot obviously be classified as either 'aversive' or 'novel'. 'Emotional significance' or salience (i.e. its significance or relevance to the individual) has been suggested (Feenstra, 2000) whereas electrophysiological studies indicate that a 'change in salience' of a stimulus could be a key factor (Sara and Segal, 1991). The question of whether a change in noradrenergic transmission serves as an alarm system or to increase vigilance and govern selective attention (Aston-Jones et al., 2000), or has all these functions, is unresolved. Regardless of these questions, the experiments described here confirm that, assuming that rats' behaviour gives any indication of their emotional reaction to a novel environment, they perceive a brightly lit novel environment to be aversive and that noradrenaline efflux is increased when rats are confined in such an environment. Moreover, in all the configurations of the light/dark exploration box, the noradrenaline response in the frontal cortex was similar to that in the hypothalamus. It seems that, under these experimental conditions, there is no clear distinction in the cortical and hypothalamic noradrenergic responses in outbred Sprague-Dawley rats when they experience unconditioned, naturalistic environmental stimuli.

Noradrenergic response to conditioned stimuli

Landmark electrophysiological studies indicated that neurones in the rat locus coeruleus respond to formerly neutral stimuli, which, through conditioning, assume the role of a cue predicting an aversive somatosensory stimulus (such as footshock or an air-puff: Rasmussen and Jacobs, 1986; Sara and Segal, 1992). Microdialysis studies have also shown conditioned noradrenaline responses to a conditioned environmental cue for footshock in the frontal cortex (Feenstra et al., 1999) and the hypothalamus (Yokoo et al., 1990). Furthermore, there is cytochemical evidence that c-Fos (a marker for activated neurones) is increased in noradrenergic neurones of the locus coeruleus, as well as the A5 and A7 nuclei, following exposure of rats to a conditioned cue associated with footshock (Pezzone et al., 1993).

In these experiments, the conditioned cue was either contextual (the environment) or explicit (e.g. a tone) but, in all cases, the unconditioned stimulus was a noxious stimulus: footshock. A key question is whether cues for non-noxious but aversive unconditioned stimuli can have the same effect? Indeed, if noradrenaline has a key role in the stress response, such changes must be anticipated: humans often express a physiological stress response on exposure to an apparently harmless environment that, through either innate fear or a previous (aversive) experience in that environment, is interpreted as threatening. The light/dark exploration box can be used to investigate this question by testing whether a noradrenaline response is evoked by a (formerly) neutral stimulus (the sound of a 10-s, 70 dB tone) which, after a series of conditioning trials, rats associated with imminent transfer to the aversive, brightly lit zone of the light/dark exploration box (McQuade and Stanford, 2000).

With animals placed in the neutral zone of the light/dark exploration box, the tone alone had no effect on noradrenaline efflux after either a single exposure, or after a series of seven twice-daily exposures over 3.5 days. However, in a series of conditioning trials (seven over 3.5 days) rats were exposed to the tone and, 10 s later, transferred to the brightly lit arena of the light/dark exploration box. Microdialysis probes were implanted in the frontal cortex and hypothalamus immediately after the last conditioning trial. In the final phase of the experiment (the test trial), rats were exposed to the tone but, on this occasion, they remained in the neutral zone of the light/dark exploration box. There was a clear, transient increase in noradrenaline efflux in the frontal cortex of the rats indicating the development of a noradrenergic response to the conditioned cue. However, there was no response in the hypothalamus (McQuade and Stanford, 2000) (Fig. 3).

These findings could suggest that the hypothalamus does not develop adaptive changes on repeated presentation of a stressful stimulus. Indeed, the hypothalamus does seem to be less able to recruit adaptive neurochemical changes than other brain regions, such as the cortex and hippocampus (Shanks et al., 1994; reviewed by Stanford, 1995). However, the basal efflux of noradrenaline in both the frontal cortex and the hypothalamus was reduced by approximately 50% in animals that had undergone repeated conditioning trials (Table 2). Moreover, when animals were transferred to the brightly lit arena (i.e. the unconditioned stimulus) after the conditioning trials, the increase in noradrenaline efflux in both the frontal cortex and the hypothalamus, was somewhat prolonged compared with that in rats experiencing this stimulus for the first time. Both these changes suggest that some form of neurochemical adaptation has occurred that modifies transmission from noradrenergic neurones projecting to the hypothalamus as well as those innervating the frontal cortex.

Obviously, it is important to establish that after repeated exposure to the unconditioned stimulus the animals still find it aversive. Indeed, on monitoring the behaviour of rats that had had repeated

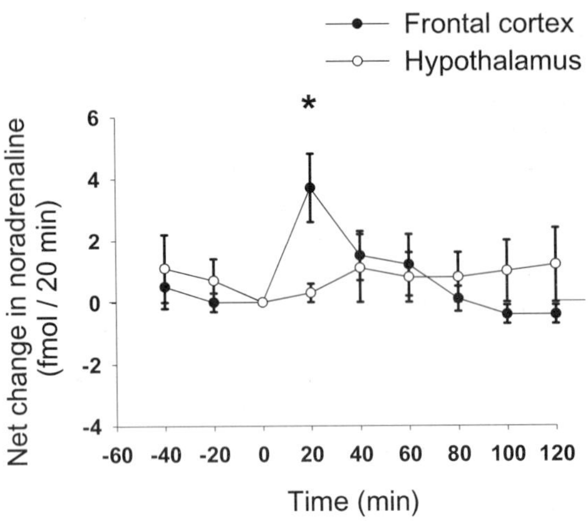

Fig. 3. The net change in the concentration of noradrenaline in microdialysis samples of the frontal cortex and hypothalamus in rats on exposure to a 10-s conditioned cue (400 Hz tone; 70 dB at source) (marked by the arrow). '0' (abscissa) indicates the time at which rats were exposed to the tone (i.e. the last basal sample). Points show mean ± SE mean ($n = 7$ or 8). $*P < 0.05$, compared with final basal sample. See: McQuade and Stanford (2000).

Table 2. Basal efflux of noradrenaline in the rat during the first and last of a series of conditioning trials in the light/dark exploration box

	Before first exposure to light/dark exploration box ($n = 11$)	After repeated conditioning trials ($n = 15$)
Frontal cortex	11.0 ± 1.1	7.7 ± 0.6*
Hypothalamus	10.4 ± 1.1	8.7 ± 0.7*

Values show mean ± SE basal noradrenaline efflux (fmol/20 min). Sample sizes indicated in brackets. $*P < 0.05$ (cf first exposure to the light/dark exploration box). Data from McQuade and Stanford (2000).

experience of the brightly lit arena, some behavioural habituation was evident: animals spent more time in the brightly lit test zone than did those experiencing this stimulus for the first time. However, other behavioural measures were unchanged and the animals still spent the majority of their time in the neutral zone suggesting that they perceived the brightly lit zone as the more aversive.

Although there is still no clear explanation of how the response to stress is regulated in different components of the central noradrenergic system, the experiments described here suggest that noradrenergic neurones projecting to different brain regions respond in different ways to conditioned aversive cues. By analogy with the serotonergic system (Greaff et al., 1996) this suggests that there could be (at least) two major components of this system. One could consist mainly of brainstem circuits that directly influence the function of the HPA axis and which govern the 'flight/fight' response to unconditioned stimuli. A second, activated by contextual (conditioned) stimuli could recruit higher limbic circuits (see also, Herman and Cullinan, 1997) that indirectly modulate monoamine function in the brainstem. Such a circuit would undoubtedly incorporate noradrenergic inputs into the amygdala (see e.g. Dayas and Day, 2001).

Noradrenergic response to prolonged environmental stress during post-weaning development

There is substantial evidence that early life events influence brain development and the response to stress in the adult, and that such factors may play an important causative role in various psychiatric disorders, such as schizophrenia and depression. Understanding the underlying developmental abnormalities that occur in response to various early life events could provide important new information on the mechanisms of specific psychiatric disorders of which stress is a component. One model that has attracted considerable attention is isolation rearing, in which rats are reared in social isolation, from weaning for at least four weeks, and compared with group reared rats (4–6 per cage). Isolation rearing produces a range of persistent behavioural effects in the young adult that include a very robust finding of hyperactivity in response to novelty (e.g. Jones et al., 1992). There are also changes in more specific behaviours, which include deficits in pre-pulse inhibition (PPI) (Varty et al., 1999) and altered sensitivity to reward; these observations suggest that isolation rearing may model some of the neurodevelopmental aspects associated with schizophrenia. In the case of the response to rewarding stimuli, the results indicate that the animals show an enhanced incentive motivational response (Hall et al., 1997) but, interestingly, there is a reduced response to reward-producing drugs measured either by self-administration (Phillips et al., 1994) or conditioned place preference (Wongwitdecha and Marsden, 1996).

Importantly for the present discussion, isolation rearing has also shown to produce an anxiogenic behavioural profile on the elevated plus-maze (Stanford et al., 1988; Wright et al., 1991) and increased aversion in a CER paradigm. The change in CER-induced behaviour may indicate that sensory deprivation affects the development of neural pathways, resulting in an altered response to cognitive information, with abnormalities in attentional and emotional functioning. The changes in cognition have focussed attention on hippocampal and cortical regions as primary sites for the neurodevelopmental effects of isolation rearing. This view is supported by evidence of decreased synaptophysin immmunoreactivity in the molecular layer of the dentate gyrus of the isolates (Varty et al., 1999), as well as the CA3 region of the hippocampus (Lapiz et al., in preparation), indicating a reduction in the number of synaptic contacts and thus, a reduction in the neural input to the hippocampus.

A major question is whether the alterations in stress-related behaviours in the isolates and the reduction in synaptic activity are associated with changes in the function of specific neurotransmitters. Particular attention has been given to the monoamine neurotransmitters, including noradrenaline. Isolation rearing reduces the release of noradrenaline in response to a stressor (tail pinch) in the hippocampus in vivo. This change parallels the reduced release of serotonin in response to stress, which is also seen in the hippocampus (Muchimapura et al., 2002). The decrease in noradrenergic function is associated with enhanced sensitivity of the terminal presynaptic α_2-adrenoceptors (Table 3) with a similar situation occurring with the terminal $5HT_{2B/D}$ autoreceptors (Fulford and Marsden, 1997a,b; Muchimapura et al., 2003). These findings support the idea that a suitable noradrenergic (and serotonergic) response is required to cope appropriately with stressful stimuli. The increase in noradrenaline function in the hippocampus needs to be appropriate and is part of a complex neuronal response essential for coping with stress; too great a noradrenergic response may be as adverse as too small a response. Interestingly depletion of brain noradrenaline in the isolation-reared rats, using the neurotoxin, DSP-4, enhanced memory in the water-maze but not through an increase in acquisition. The isolates with noradrenaline depletion also showed a decrease in inspective exploration (i.e. objective-specific exploration) but not general exploration (Lapiz et al., 2001). Whether these alterations in sensory perception are involved in the altered response to stress in the animals remains to be determined.

Table 3. Effect of an α_2-adrenoceptor agonist (clonidine) and antagonist (idazoxan) and tail pinch on extracellular noradrenaline in the dorsal hippocampus in group and isolation-reared rats

	Group Reared (%)	Isolation Reared (%)
Clonidine (0.3 mg/kg)	42 ± 6	62 ± 4*
Idazoxan (1 mg/kg)	115 ± 18	197 ± 30*
Tail pinch (5 min)	278 ± 30	124 ± 16*

Results presented as the maximum increase or decrease given as a percentage of the pretreatment basal values. $^*p < 0.05$ (ANOVA with post hoc Dunnett's test on non normalised data) $n = 4$–6 per group. Noradrenaline was measured using in vivo microdialysis. Data from Fulford and Marsden (1997a) and unpublished.

Overall, the isolation-reared rat can provide clues as to how the noradrenergic system is involved in response to long-term social stress and the importance of the hippocampus. However, it is clear that noradrenaline is only one of several factors involved in this process as similar changes are seen with the serotonergic system in the hippocampus. Furthermore, while the cortex and the hippocampus may be the primary targets for isolation-induced changes in neuronal development, there are also marked effects on amygdala/accumbens and striatal function with increased dopamine and serotonin responses to CER in these regions (Fulford and Marsden, 1997c, 1998) (Fig. 4). Models such as the isolation-reared rat may help us to understand the important inter-relationships between different brain areas and neurotransmitter systems in the response to stress and mechanisms involved with coping. However, it should be remembered that, in humans, there is substantial individual variation in the response thus implicating genetic factors.

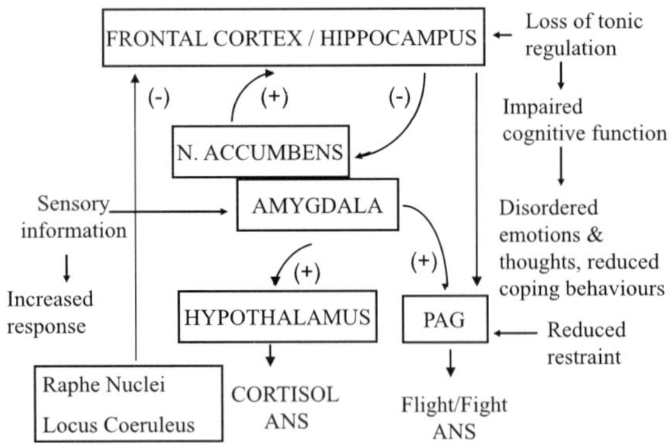

Fig. 4. Schematic diagram showing the suggested effects of isolation rearing that results in increased response to aversion. Early environmental social deprivation results in altered development of the frontal cortex and hippocampus causing reduced input from the noradrenergic pathways from the locus coeruleus and serotonergic pathways from the medial/dorsal raphe involved in coping behaviour. This leads to changes in the functional state of the accumbens/amygdala (increased dopamine and serotonin responsivity) with altered interpretation of incoming sensory information. The net effect is increased output from the main motor pathways involved in response to aversion from the hypothalamus and the periqueductal grey (PAG). ANS: autonomic nervous system.

Genetic background and the noradrenergic response to stress

To establish that noradrenaline efflux is (or is not) increased on exposure to aversive environmental stimuli does not help to ascertain whether the response transmits or ameliorates its impact on behaviour. One way of investigating that question is to manipulate noradrenergic transmission by administering drugs and to characterise any ensuing changes in animals' behavioural response to the test stimulus. However, this approach assumes that test compounds target only a single population of receptors, which is rarely the case (if ever). Alternatively, chemical lesions of central noradrenergic pathways can be induced by selective neurotoxins. Again, selectivity can be a problem. Further, it cannot be assumed that a decline in the tissue transmitter content reflects a reduction in its extracellular concentration. Indeed, despite a substantial depletion (> 70%) of noradrenaline in the brain, following treatment with the neurotoxin, DSP-4, there was a twofold increase in the concentration of extracellular noradrenaline in the frontal cortex (Hughes and Stanford, 1998). Another limitation of both these approaches, as experience with humans confirms, is that they ignore the factors that determine vulnerability to stress, and its long-term impact, which leads to physical and mental illnesses in some individuals but not others.

In order to avoid these problems, and yet investigate how any difference in neurochemical responses might parallel behaviour (or vice versa) we have studied two inbred strains of rats, which express innate differences in their behavioural responses to stress: the Maudsley Reactive (MR) and Nonreactive (Wistar, MNRA) rats. These animals are ideal for experiments investigating the relationship between noradrenergic transmission and the behavioural response to aversive environmental stimuli because the criterion for their inbreeding is their behavioural response to a novel environment (the 'open field'). The behaviour of these rats has been studied a good deal in the past (see Broadhurst, 1975) and, albeit controversially, was deemed to hold the key to innate differences in 'emotionality'.

We carried out behavioural and microdialysis studies in these two strains of rat, using the light/dark exploration box, as described above. Only in the MR strain were time spent/visit and activity/visit in the light arena less than the dark one (McQuade and Stanford, 2001). This is consistent with evidence that MR and MNRA rats differ in their exploratory behaviour (Broadhurst, 1975) and that these differences are seen most clearly under high-stress conditions (Sara et al., 1994).

In parallel microdialysis experiments, there was no difference in the concentration of extracellular noradrenaline in MR and MNRA rats while they were in the neutral zone. As in outbred Sprague-Dawley rats, noradrenaline efflux in the frontal cortex was transiently increased in both MR and MNRA rats when they were confined within the brightly lit but not the dark, test arena of the light/dark exploration box. Noradrenaline efflux was also increased in the brightly lit arena in the hypothalamus of both strains of Maudsley rats. However, in MNRA but not MR rats, the noradrenaline response was maintained for at least 2 h (i.e. until the animals were returned to the neutral zone of the light/dark exploration box) (McQuade and Stanford, 2001).

It is striking that it is the MNRA rats, which are regarded as being the more resistant to stress, that showed a prolonged hypothalamic noradrenaline response. Whether this means that the increased efflux had a stress-protective effect is as yet unknown. However, it is notable that a prolongation of the noradrenaline response was also evident in outbred Sprague-Dawley rats that had been repeatedly exposed to, and showed some behavioural habituation to, the light zone of the light/dark exploration box (see above; McQuade and Stanford, 2000). Further evidence to support the possibility that augmentation or prolongation of a noradrenaline response increases behavioural resistance to an aversive environment comes from reports that basal firing rate of neurones in the locus coeruleus and incidence of burst firing was greater in the MNRA than the MR strain (Verbanac et al., 1994), albeit in anaesthetised rats. Consistent with this, there was a greater increase in tyrosine hydroxylase activity in the locus coeruleus of MNRA rats after a bout of footshock (Blizard et al., 1983). These changes are possibly explained by a lower sensitivity of α_2-autoreceptors to activation by agonists (Verbanac et al., 1994). Finally, a recent comparative study of outbred Sprague-Dawleys and the Lewis (low stress responsivity) and Wistar Kyoto (high stress responsivity) strains of rats showed that stress-induced increases in mRNA for tyrosine hydroxylase and noradrenaline efflux were greatest in the Lewis rats. Such findings have led to the conclusion that increased noradrenergic transmission is an essential component of coping with stress (Pardon et al., 2002).

Despite these promising leads, the relationship between noradrenergic transmission and the stress response in the Maudsley inbred strains is not straightforward (reviewed by Blizard and Adams, 2002). There are reports of a greater increase in the concentration of noradrenaline in the plasma of MR rats after immobilisation (Blizard et al., 1983). Also, the increase in DOPAC concentration (used as an index of noradrenaline synthesis) in the locus coeruleus and ventrolateral medulla (which contains the A1/C1 nucleus) after a bout of immobilisation was greater in MR than in MNRA rats (Buda et al., 1994). Clearly, more research is needed to explain these apparently disparate findings. Also, it is important to establish whether a difference in the cortical noradrenaline response is apparent when MR and MNRA rats are exposed to conditioned environmental stimuli, as described above.

Noradrenergic coding of coping with stress

It is now clear that there is no stereotypical noradrenaline response to stress, but that the role and consequences of central noradrenergic transmission depend on the type or severity of the stimulus, the brain region to which these neurones project and individual differences in the neurobiological coding of behaviour. On this basis, it is likely that the optimal behavioural response to a given environmental stimulus requires a specific noradrenergic response (or combination of responses).

Evidence for such a proposal first emerged from a series of studies of peripheral catecholamine responses to psychological stress, carried out by Frankenhaeuser's group (e.g. Frankenhaeuser, 1968; reviewed by Stanford, 1993). In these experiments, subjects with low basal urinary noradrenaline excretion perceived themselves to cope better with the stress of an audiovisual conflict test, the Stroop test, and actual task control was better, in those subjects that rallied the greatest noradrenaline response. Conversely, high baseline noradrenaline secretors performed best in an unstressed condition and it was the subjects who showed the smallest increase in plasma noradrenaline who actually coped better with the test stress.

A caveat of this proposal is that it appears that changes in plasma catecholamines during psychological stress are evident only under carefully controlled conditions. In a study of three different groups of cardiac patients, plasma catecholamines exceeded those commonly reported in studies of relaxed, young, healthy individuals, and in none of these patients was it possible to demonstrate an increase in plasma catecholamines during a psychological stress (the Stroop test: Stanford et al., 1997). Moreover, in heart transplant patients, haemodynamic responses to the Stroop test were clearly disrupted and there was a greater reliance on hormonal secretion of catecholamines. Yet, this seemed to have no effect on their emotional response to the stress (Salmon et al., 2001).

Nevertheless, based on Frankenhaeuser's findings, and those describing correlations between individual differences in neurochemical markers and rats' behavioural responses during stress (see above), it was suggested that the relationship between central noradrenergic transmission and coping with stress is described by a bell-shaped curve (Stanford, 1993). Accordingly, optimal coping with stress rests on rallying an appropriate noradrenergic response; this could be determined genetically and/or by the individuals' previous experience of that stimulus.

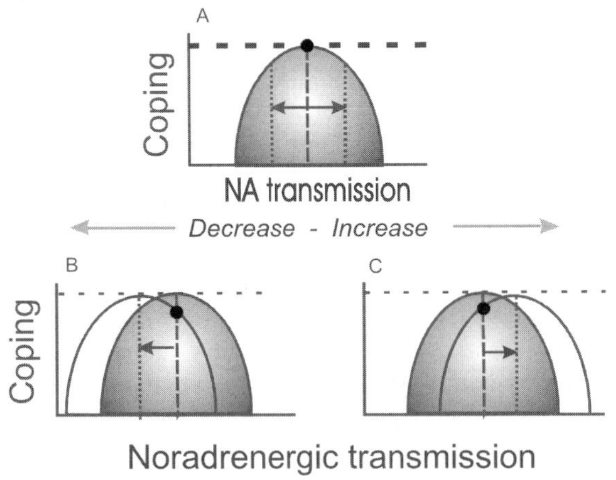

Fig. 5. Schematic diagram showing the hypothetical relationship between noradrenergic transmission and an individual's resistance to stress. A: Optimal coping is attained when the brain rallies a specific noradrenergic response (•); this could be determined genetically and/or by previous experience of the stress. Either a reduction or an increase in noradrenergic transmission diminishes coping. B: The shaded area depicts the relationship between noradrenergic transmission and stress resistance in normal individuals (as shown in 'A'). If there is a leftward shift of the curve then the (predetermined) noradrenergic response that would be optimal in normal individuals, now produces suboptimal stress resistance (•). One remedy for such a dysfunction is to reduce noradrenergic transmission so as to restore optimal coping. C: In the case of a rightward shift of the curve, a predetermined noradrenergic response to a specific stimulus, that would be optimal in normal individuals, will again produce suboptimal coping (•). This time, the remedy is to increase noradrenergic transmission. In both B and C, an alternative way to restore optimal coping would be to reverse the shift in the noradrenergic transmission–coping curve. This could explain the changes in mood that occur after chronic administration of drugs or behavioural strategies that cause long-latency changes in neurochemical factors that influence noradrenergic transmission.

According to such a scheme, either a deficit or overactivity of noradrenergic transmission would impair the ability to cope with the stimulus (Fig. 5A). It is also possible that the neurochemical coding of coping with stress can be disrupted by a shift, to either the right or the left, of the curve defining the relationship between noradrenergic transmission and coping (Fig. 5B and C). Under these circumstances, a given noradrenergic response, which would be optimal in normal subjects, would now produce a suboptimal coping response (see also Stanford, 1996). The remedy would be to adjust the extent to which noradrenergic neurones are activated by an acute stress, or to recruit long-term adaptive changes that restore the appropriate relationship between the neuronal response and coping.

Obviously, noradrenergic transmission is not the only factor that determines the behavioural response to even simple environmental stimuli. Indeed, a bell-shaped dose–response curve itself confirms the intervention of one or more factors that govern noradrenergic transmission. Ultimately, it is these interactions between noradrenergic neurones and other neurotransmitters that determine the role of noradrenergic transmission in the neurochemical coding of coping with stress.

References

Abercrombie, E.D. and Jacobs, B.L. (1987) Single-unit response of noradrenergic neurons in the locus coeruleus of freely moving cats. I. Acutely presented stressful and nonstressful stimuli. J. Neurosci., 7: 2837–2843.

Abercrombie, E.D., Keller, R.W. and Zigmond, M.J. (1988) Characterization of hippocampal norepinephrine release as measured by microdialysis perfusion: pharmacological and behavioral studies. Neuroscience, 27: 897–904.

Aston-Jones, G., Rajkowski, J. and Cohen, J. (2000) Locus coeruleus and regulation of behavioral flexibility and attention. Prog. Brain Res., 126: 165–182.

Blizard, D.A. and Adams, N. (2002) The Maudsley reactive and nonreactive strains: a new perspective. Behav. Genet., 32: 277–299.

Blizard, D.A., Freedman, L.S. and Liang, B. (1983) Genetic variation, chronic stress and the central and peripheral noradrenergic systems. Am. J. Physiol., 245: R600–R605.

Broadhurst, P.I. (1975) The Maudsley reactive and nonreactive strains of rats: a survey. Behav. Genet., 5: 288–319.

Buda, M., Lachuer, J., Devauges, V., Barbagli, B., Blizard, D. and Sara, S.J. (1994) Central noradrenergic reactivity to stress in Maudsley rat strains. Neurosci. Lett., 167: 33–36

Cadogan, A.K., Kendall, D.A., Fink, H. and Marsden, C.A. (1994) Social interaction increases 5-HT release and cAMP efflux in the rat ventral hippocampus in vivo. Behav. Pharmacol., 5: 299–305.

Cenci, M.A., Kalen, P., Mandel, R.J. and Bjorkland, A. (1992) Regional differences in the regulation of dopamine and noradrenaline release in medial frontal cortex, nucleus accumbens and caudate-putamen: a microdialysis study in the rat. Brain Res., 581: 217–228.

Crawley, J. and Goodwin, F.K. (1980) Preliminary report of a simple animal behaviour model for the anxiolytic effects of benzodiazepines. Pharmacol. Biochem. Behav., 13: 167–170.

Dalley, J.W. and Stanford, S.C. (1994) Graded changes in noradrenaline (NA) efflux in dialysates of the rat cortex evoked by non-noxious stressors. J. Psychopharmacol., 8 (suppl.): A47.

Dalley, J.W. and Stanford, S.C. (1995) Incremental changes in extracellular noradrenaline availability in the frontal cortex induced by naturalistic environmental stimuli: a microdialysis study in the freely moving rat. J. Neurochem., 65: 2644–2651.

Dalley, J.W., Mason, K. and Stanford, S.C. (1996) Increased levels of extracellular noradrenaline in the frontal cortex of rats exposed to naturalistic environmental stimuli: modulation by acute systemic administration of diazepam or buspirone. Psychopharmacol., 127: 47–54.

Dayas, C.V. and Day, T.A. (2001) Opposing roles for medial and central amygdala in the initiation of noradrenergic cell responses to a psychological stressor. Eur. J. Neuroscience, 15: 1712–1718.

Drijfhout, W.J., Kemper, R.H., Meerlo, P., Koolhaas, J.M. and Westerink, B.H. (1995) A telemetry study of the chronic effects of microdialysis probe implantation on the activity pattern and temperature rhythm of the rat. J. Neurosci. Methods, 61: 191–196.

Fallon, J.H., Koziell, D.A. and Moore, R.Y. (1978) Catecholamine innervation of the basal forebrain. J. Comp. Neurol., 180: 509–532.

Feenstra, M.G. (2000) Dopamine and noradrenaline release in the prefrontal cortex in relation to unconditioned and conditioned stress and reward. Prog. Brain Res. 126: 133–163.

Feenstra, M.G.P., Botterblom, M.H.A. and Uum, J.F.M. (1998) Local activation of metabotropic glutamate receptors inhibits the handling-induced increased release of dopamine in the nuclues accumbens but not that of dopamine or noradrenaline in the prefrontal cortex: comparison with inhibition of ionotropic receptors. J. Neurochem., 70: 1104–1113.

Feenstra, M.G.P., Teske, G., Botterblom, M.H.A., and De Bruin, J.P.C. (1999) Dopamine and noradrenaline release in the prefrontal cortex of rats during classical aversive and appetitive conditioning to a contextual stimulus: interference by novelty effects. Neurosci. Lett., 272: 179–182.

Feenstra, M.G.P., Botterblom, M.H.A. and Masterbroek, S. (2000) Dopamine and noradrenaline efflux in the prefrontal cortex in the light and dark period: effects of novelty and handling and comparison to the nucleus accumbens. Neuroscience, 100: 741–748.

Frankenhaeuser, M. (1968) Catecholamine excretion as related to cognitive and emotion reaction patterns. Psychosomatic Med., 30: 109–120.

Frankenhaeuser, M. (1971) Behavior and circulating catecholamines. Brain Res., 31: 241–262.

Fritschy, J.-M. and Grzanna, R. (1989) Immunohistochemical analysis of the neurotoxic effects of DSP-4 identifies two populations of noradrenergic axon terminals. Neuroscience, 30: 181–197.

Fulford, A.J. and Marsden, C.A. (1997a) Social isolation in the rat enhances α_2-autoreceptor function in the hippocampus *in vivo*. Neuroscience, 77: 57–64.

Fulford, A.J. and Marsden, C.A. (1997b) Conditioned release of 5-hyroxytryptamine *in vivo* in the nucleus accumbens following isolation-rearing in the rat. Neuroscience, 83: 481–487.

Fulford, A.J. and Marsden, C.A. (1997c) Effect of isolation-rearing on noradrenaline release in rat hypothalamus and hippocampus *in vitro*. Brain Res., 748: 93–99.

Fulford, A.J. and Marsden, C.A. (1998) Effect of isolation-rearing on conditioned dopamine release *in vivo* in the nucleus accumbens of the rat. J. Neurochem., 70: 384–390.

Graeff, F.G., Guimaraes, F.S., de Andrade, T.G. and Deakin, W.F. (1996) Role of 5-HT in stress, anxiety and depression. Pharmacol. Biochem. Behav., 54: 129–141.

Grant, S.J., Aston-Jones, G. and Redmond, D.E. (1988) Responses of primate locus coeruleus neurons to simple and complex sensory stimuli. Brain Res. Bull., 21: 401–410.

Hall, F.S., Humby, T., Wilkinson, L.S. and Robbins, T.W. (1997) The effects of isolation-rearing of rats on the behavioural response to food and environmental novelty. Physiol. Behav., 62: 281–290.

Herman, J.P. and Cullinan, W.E. (1997) Neurocircuitry of stress: central control of the hypothalmo–pituitary–adrenocortical axis. Trends Neurosci., 20: 74–84.

Hirata, H. and Aston-Jones, G. (1994) A novel long-latency response of locus coeruleus neurons to noxious stimuli: mediation by peripheral C-fibres. J. Neurophysiol., 71: 1752–1761.

Hughes, Z.A. and Stanford, S.C. (1998) A partial noradrenergic lesion induced by DSP-4 increases extracellular noradrenaline concentration in rat frontal cortex: a microdialysis study in vivo. Psychopharmacol., 136: 299–303.

Jones, G.H., Hernandez, T.D., Kendall, D.A., Marsden, C.A. and Robbins, T.W. (1992) Dopaminergic and serotonergic function following isolation rearing in rats, a study of behavioural responses and post-mortem and *in vivo* neurochemistry. Pharmacol. Biochem. Behav., 43: 17–35.

Kawahara, H., Kawahara, Y. and Westerink, B.H.C. (2000) The role of afferents to the locus coeruleus in the handling stress-induced increase in the release of noradrenaline in the medial prefrontal cortex: a dual-probe microdialysis study in the rat brain. Eur. J. Pharmacol., 387: 279–286.

Lapiz, M.D.S., Mateo, Y., Durkin, S., Parker, T. and Marsden, C.A. (2001) Effects of central noradrenaline depletion by the selective neurotoxin DSP-4 on the behaviour of the isolated rat in the elevated plus maze and water maze. Psychopharmacology, 155: 251–259.

McQuade, R. and Stanford, S.C. (2000) A microdialysis study of the noradrenergic response in rat frontal cortex and hypothalamus to a conditioned cue for aversive, naturalistic environmental stimuli. Psychopharmacol., 148: 201–208.

McQuade, R. and Stanford, S.C. (2001) Differences in central noradrenergic and behavioural responses of Maudsley non-reactive and Maudsley reactive inbred rats on exposure to an aversive novel environment. J. Neurochem., 76: 21–28.

McQuade, R., Creton, D. and Stanford, S.C. (1999) Effect of novel environmental stimuli on rat behaviour and central noradrenaline function measured by *in vivo* microdialysis. Psychopharmacol., 145: 393–400.

Muchimapura, S., Mason, R. and Marsden, C.A. (2002) Isolation rearing in the rat disrupts the hippocampal response to stress. Neuroscience, 112: 697–706.

Muchimapura, S., Mason, R. and Marsden, C.A. (2003) The effect of isolation rearing on pre- and post-synaptic serotonergic function in the rat dorsal hippocampus. Synapse, 47: 209–217.

Neophytou, S.I., Aspley, S., Butler, S., Beckett, S. and Marsden, C.A. (2001) Effects of lesioning noradrenergic neurones in the locus coeruleus on conditioned and unconditioned aversive behaviour in the rat. Prog. Neuropsychopharmacol. Biol. Psychiatry, 25: 1307–1321.

Pacak, K., Armando, I., Fukuhara, K., Kvetnansky, R., Palkovits, M., Kopin, I.J. and Goldstein, D.S. (1992) Noradrenergic activation in the paraventricular nucleus during acute and chronic immobilization stress in rats: an *in vivo* microdialysis study. Brain Res., 589: 91–96

Pardon, M.-C., Gould, G.G., Garcia, A., Phillips, L., Cook, M.C., Miller, S.A., Mason, P.A. and Morilak, D.A. (2002) Stress reactivity of the brain noradrenergic system in three rat strains differing in their neuroendocrine and behavioural responses to stress: implications for susceptibility to stress-related neuropsychiatric disorders. Neuroscience. 115: 229–242.

Pezzone, M.A., Lee, W.-S., Hoffman, G.E., Pezzone, K.M. and Rabin, B.S. (1993) Activation of brainstem catecholaminergic neurons by conditioned and unconditioned aversive stimuli as revealed by c-Fos immunoreactivity. Brain Res., 608: 310–318.

Phillips, G.D., Howes, S.R., Whitelaw, R.B., Wilkinson, L.S. and Robbins, T.W. (1994) Isolation rearing enhances the locomotor response to cocaine and a novel environment, but impairs the intravenous administration of cocaine. Psychopharmacology, 115: 407–418.

Rasmussen, K. and Jacobs, B.L. (1986) Single unit activity of locus coeruleus neurons in the freely moving cat. II. Conditioning and pharmacologic studies. Brain Res., 371: 335–344

Rosario, L.A. and Abercrombie, E.D. (1999) Individual differences in behavioural reactivity: correlation with stress-induced norepinephrine efflux in the hippocampus of Sprague-Dawley rats. Brain Res. Bull., 48: 595–602.

Salmon, P. and Stanford, S.C. (1989) Beta-adrenoceptor density correlates with behaviour of rats in the open field. Psychopharmacology, 98: 412–416.

Salmon, P. and Stanford, S.C. (1992) Research strategies for decoding the neurochemical basis of resistance to stress. J. Psychopharmacol., 6: 1–7.

Salmon, P., Stanford, S.C., Mikhail, G., Zielinski, S. and Pepper, J.R. (2001) Hemodynamic and emotional responses to a psychological stressor after cardiac transplantation. Psychosomatic. Med., 63: 289–299.

Sara, S.J. and Segal, M. (1991) Plasticity of sensory responses of locus coeruleus neurons in the behaving rat: implications for cognition. Prog. Brain. Res., 88: 571–585.

Sara, S.J., Devauges, V., Biegon, A. and Blizard, D.A. (1994) The Maudsley rat strains as a probe to investigate noradrenaline-cholinergic interaction in cognitive function. J. Physiol., 88: 337–345.

Shanks, N., Griffiths, J. and Anisman, H. (1994) Norepinephrine and serotonin alterations following stressor exposure: mouse strain differences. Pharmacol. Biochem. Behav., 49: 57–65.

Stanford, S.C. (1993) Monoamines in response and adaptation to stress. In: Stanford, S.C and Salmon, P (Eds.), Stress: From Synapse to Syndrome. Academic Press, London, pp. 281–331.

Stanford, S.C. (1995) Central noradrenergic neurones and stress. Pharmac. Ther., 68: 297–343.

Stanford, S.C. (1996) Stress: a major variable in the psychopharmacological response. Pharmacol. Biochem. Behav., 54: 211–217.

Stanford, S.C. and Salmon, P. (1989) Neurochemical correlates of behavioural responses to frustrative nonreward in the rat: implications for the role of central noradrenergic neurones in behavioural adaptation to stress. Exp. Brain. Res., 75: 133–138.

Stanford, S.C., Fillenz, M. and Ryan, E. (1984) The effect of repeated mild stress on cerebral cortical adrenoceptors and noradrenaline synthesis in the rat. Neurosci. Lett., 45: 163–167.

Stanford, S.C., Parker, V. and Morinan, A. (1988) Deficits in exploratory behaviour in socially-isolated rats are not accompanied by changes in cerebral cortical adrenoceptor binding. J. Affect. Disord., 15: 175–180.

Stanford, S.C., Mikhail, G., Salmon, P., Gettins, D., Zielinski, S. and Pepper, J.R. (1997) Psychological stress does not affect plasma catecholamines in subjects with cardiovascular disorder. Pharmacol. Biochem. Behav., 58: 1167–1174.

Tanaka, M., Tsuda, A., Yokoo, H., Yoshida, M., Mizoguchi, K. and Shimizu, T. (1991) Psychological stress-induced increase in noradrenaline release in rat brain regions are attenuated by diazepam but not by morphine. Pharmacol. Biochem. Behav., 39: 191–195.

Varty, G.B., Marsden, C.A. and Higgins, G.A. (1999) Reduced synaptophysin immunoreactivity in the dentate gyrus of prepulse inhibition-impaired isolation-reared rats. Brain Res., 824: 197–203.

Verbanac, J.S., Commissaris, R.L., Altman, H.J. and Pitts, D.K. (1994) Electrophysiological characteristics of locus coeruleus neurons in the Maudsley reactive (MR) and non-reactive (MNRA) rat strains, Neurosci. Lett., 179: 137–140.

Wongwitdecha, N. and Marsden, C.A. (1996) Effect of social isolation on the reinforcing properties of morphine in the conditioned place preference test. Pharmacol. Biochem. Behav., 53: 531–534.

Wright, I.K., Ismail, H., Upton, N. and Marsden, C.A. (1991) Resocialisation of isolation-reared rats does not alter their anxiogenic profile in the elevated X-maze model of anxiety. Physiol. Behav., 50: 1129–1132.

Yokoo, H., Tanaka, M., Yoshida, M., Tsuda A., Takahiki, T. and Mizoguchi, K. (1990) Direct evidence of conditioned fear-elicited enhancement of noradrenaline release in the rat hypothalamus assessed by intracranial microdialysis. Brain Res., 536: 305–308.

T. Steckler, N.H. Kalin and J.M.H.M. Reul (Eds.)
Handbook of Stress and the Brain, Vol. 15
ISBN 0-444-51173-3

CHAPTER 4.6

Stress, corticotropin-releasing factor and serotonergic neurotransmission

Astrid C.E. Linthorst*

The Henry Wellcome Laboratories for Integrative Neuroscience and Endocrinology, Dorothy Hodgkin Building, University of Bristol, Whitson Street, Bristol BS1 3NY, UK

Abstract: An intricate interplay between different brain neurotransmitter and neuropeptide systems coordinates the neuroendocrine, autonomic and behavioural responses to stress. Under normal conditions, the body has appropriate mechanisms to respond to acute stressful challenges, but chronic stress may evolve in maladaptive coping processes, resulting in an enhanced risk for illness. Disturbances at the level of both corticotropin-releasing factor (CRF) and serotonin (5-HT) have been implicated in the etiology of stress-related psychiatric disorders, such as major depression and anxiety. Therefore, this chapter is dedicated to the effects of stress on serotonergic neurotransmission and the role of CRF herein. The first part of this chapter covers the neuroanatomical evidence for interactions between CRF and 5-HT at the level of the raphe nuclei. Further, the effects of stress and CRF on different aspects of serotonergic neurotransmission will be addressed, including expression of *c-fos* in the raphe nuclei, serotonin synthesis and firing rate of serotonergic neurons. Moreover, the effects of five selected stressors (immune stress, forced swimming, tail pinch, electric shock/fear conditioned stress and predator stress), differing in their physical and psychological impact, on the levels of 5-HT and its metabolite 5-hydroxyindoleacetic acid in forebrain regions will be discussed. The data presented here underscore the concept that stress, possibly via activation of the CRF system, affects serotonergic neurotransmission. Stress seems to exert predominantly stimulatory effects on the 5-HT system, at least in higher brain structures such as the hippocampus. Long-term changes in the CRF system, e.g. chronic elevation of brain levels of CRF or dysfunctioning of CRF receptor type 1, have profound consequences for the stress responsiveness of the hippocampal 5-HT system. Notwithstanding the vast amount of data on the effects of stress on serotonergic neurotransmission, the wide variety of experimental protocols used until now has uncovered the clear need for more systematic and neuroanatomically detailed approaches to further elucidate the interactions between stress, CRF and 5-HT. Such strategies will increase our knowledge on 'healthy' stress processing and may consequently lead to a better understanding of the putative maladaptive processes involved in the etiology of stress-related psychiatric and other disorders.

Introduction

The increasing load of psychological and physical stress weighs heavily on a vast number of people in our society. Under normal circumstances the body will exert appropriate responses to overcome acute stressful situations. Chronic stress may, however, lead to maladaptive stress coping processes and, consequently, to an enhanced risk for illness. In this respect, it is of interest that the prevalence of stress-related psychiatric disorders, such as major depression and anxiety, seems to increase considerably. It is, therefore, of utmost importance to develop and propagate programmes to improve stress coping strategies in (young) individuals. However, to tackle stress-related health problems and to improve coping

*Tel.: +44 117 33 13140;
E-mail: Astrid.Linthorst@bristol.ac.uk

strategies, it will be essential to increase our understanding of the processing of stress at the level of the central nervous system.

Interactions between different brain neurotransmitter and neuropeptide systems coordinate the neuroendocrine, autonomic and behavioural responses to stress. In this chapter, I will discuss the present knowledge on the effects of stress on the serotonin (5-HT) system in animals, because this neurotransmitter is known to regulate hypothalamic–pituitary–adrenocortical (HPA) axis activity (Dinan, 1996; Lowry, 2002; Carrasco and Van de Kar, 2003) and behaviour (Lucki, 1998) during stressful challenges. Moreover, disturbed functioning of serotonergic neurotransmission seems to play a key role in the etiology of depression, as evidenced by the decreased levels of 5-hydroxyindoleacetic acid (5-HIAA, the metabolite of 5-HT) in the cerebrospinal fluid, the specific changes in the expression of 5-HT receptors and the therapeutical efficacy of selective serotonin reuptake inhibitors (SSRIs) observed in depressed patients (for review see Mann, 1998; Maes and Meltzer, 1995; Ressler and Nemeroff, 2000).

As has been excellently described in other chapters in this Handbook (see Section 2. Hypothalamic Hormones involved in stress responsivity), corticotropin-releasing factor (CRF) is a key mediator of the stress-induced activation of the HPA axis and the autonomic nervous system, and of behavioural responses to stress. During the past two decades evidence has been accumulating for a role of a hyperactive CRF system in major depression. In depressed patients, enhanced levels of CRF in the cerebrospinal fluid (Nemeroff et al., 1984), decreased numbers of CRF binding sites in the frontal cortex (Nemeroff et al., 1988), an elevated level of CRF mRNA expression in the hypothalamic paraventricular nucleus (Raadsheer et al., 1995) and an increased number of CRF-expressing neurons in this nucleus (Raadsheer et al., 1994) have been found. Hence, because both CRF and 5-HT are involved in the regulation of the stress response and both systems seem to be malfunctioning in mood disorders, this chapter will focus on the present evidence for a role of CRF in the regulation of serotonergic neurotransmission under basal and stress conditions. Furthermore, observations showing that chronic changes in the CRF system have profound consequences for the responsiveness of serotonergic neurotransmission to stress will also be discussed.

The information presented in this chapter underscores the concept that stress, via activation of the CRF system, affects serotonergic neurotransmission. Long-term changes in the status of the CRF system, for instance due to the experience of chronic stress, may lead to changes in the responsiveness of 5-HT to stressful challenges. Whether such changes in serotonergic neurotransmission eventually precipitate in psychiatric illness or mental imbalance should be subject of further investigations.

Relationship between serotonin and corticotropin-releasing factor at the neuroanatomical level

Serotonergic neurons, located in the brainstem, are mainly confined to the raphe nuclei (Dahlström and Fuxe, 1964; Steinbusch, 1981). A rostral and a caudal 5-HT system can be distinguished (Törk, 1990; Jacobs and Azmitia, 1992). The rostral 5-HT system consists of neurons located in the caudal linear nucleus, the dorsal (DRN) and median raphe nucleus (MRN) and the supralemniscal region (B9 group). This system provides a dense serotonergic innervation of the whole forebrain. On the other hand, the caudal 5-HT system, encompassing the raphe pallidus nucleus, the raphe obscurus nucleus, the raphe magnus nucleus and serotonergic cell bodies in the ventral lateral medulla, projects to different targets in the spinal cord. Connections between the raphe nuclei and with other brainstem structures have also been described.

A neuroanatomical basis for a relationship between CRF and 5-HT is now emerging. CRF-immunoreactive cell bodies and fibres have been demonstrated in the rostral and caudal raphe nuclei (Cummings et al., 1983; Swanson et al., 1983; Sakanaka et al., 1987; Ruggiero et al., 1999; Kirby et al., 2000; Valentino et al., 2001). Cell bodies immunoreactive for the CRF-like neuropeptide urocortin 1 (Vaughan et al., 1995) have also been detected in the DRN and MRN, and in the raphe magnus nucleus after colchicine treatment (Kozicz et al., 1998; Bittencourt et al., 1999). Moreover, a dense distribution of urocortin 1-immunoreactive fibres has been observed in the DRN together with a low to moderate innervation of the MRN and the caudal raphe nuclei (Bittencourt et al., 1999). In contrast to the clear presence of CRF and urocortin 1 fibres in the DRN, no (or very few) immunoreactive fibres (or mRNA levels) for the recently discovered CRF-related neuropeptide urocortin 3 (Lewis et al., 2001) have been found in this nucleus and in the other serotonergic cell body regions (Li et al., 2002). Unfortunately, no detailed information on the localization of urocortin 2, another member of the CRF family (Reyes et al., 2001), in the raphe nuclei is available yet.

Some recent studies have begun to focus on the topographical distribution of CRF-immunoreactive fibres in the DRN. Although CRF-immunoreactive fibres are present throughout the DRN, a distinct rostral to caudal innervation pattern has been described. Dense CRF innervation has been found in the interfascicular and ventromedial regions at rostral to medial levels, which shifts to dorsal and dorsolateral regions in the caudal DRN (Kirby et al., 2000; Lowry et al., 2000; Valentino et al., 2001). CRF-immunoreactive fibres are found in close association with neurons immunopositive for tryptophan hydroxylase (TPH; the rate-limiting enzyme for the synthesis of serotonin) (Lowry et al., 2000) or 5-HT (Kirby et al., 2000), but also with non-serotonergic neurons. A detailed electron microscopy study revealed that CRF axons in the DRN make preferentially synaptic contacts with dendrites and (non-CRF) axon terminals (Valentino et al., 2001). In the dorsal and dorsolateral DRN the dendritic contacts are in majority asymmetric (excitatory) and in the ventromedial/interfascicular regions largely symmetric (inhibitory). Moreover, the percentage of contacts of CRF fibres with axon terminals is higher in the rostral ventromedial/interfascicular DRN as compared to the caudal dorsolateral DRN (Valentino et al., 2001). Of interest to note is that the distribution of CRF fibres and 5-HT neurons in the DRN is not identical. For instance, a high number of 5-HT neurons is found in the ventromedial/interfascicular region of the caudal DRN, whereas the CRF innervation is relatively low in this region as compared to the dorsolateral aspects at the same rostral–caudal level. This observation suggests that CRF fibres may form synaptic contacts also with non-serotonergic neurons. Indeed, a recent study has demonstrated immunoreactivity for CRF receptors associated with GABA-immunopositive neurons (Roche et al., 2003).

CRF and the urocortins show characteristic binding affinities for the two types of CRF receptors presently known (for review see Reul and Holsboer, 2002). Whereas CRF binds relatively selectively to CRF receptor type 1 (CRF_1) over CRF receptor type 2 (CRF_2), urocortin 1 shows high-binding affinity for both the receptor types (Chen et al., 1993; Vaughan et al., 1995; Lovenberg et al., 1995). In contrast, urocortin 2 and urocortin 3 seem to represent selective ligands for CRF_2 (Lewis et al., 2001; Reyes et al., 2001). A low to moderate CRF_1 mRNA expression has been found in both the rostral and caudal raphe nuclei (Chalmers et al., 1995; Bittencourt and Sawchenko, 2000; Van Pett et al., 2000). The MRN contains moderate and the DRN and the caudal linear nucleus higher levels of CRF_2 mRNA (Chalmers et al., 1995; Bittencourt and Sawchenko, 2000; Van Pett et al., 2000), but expression of CRF_2 mRNA seems to be absent in the caudal serotonergic cell groups. However, CRF_2-*immunoreactive* neuronal profiles have been demonstrated recently in the rostral as well as in the caudal raphe nuclei (Lowry et al., 2002). Immunocytochemistry studies applying a double-labeling approach have shown the presence of CRF_1 in TPH-positive neurons in the raphe nuclei (Lowry et al., 2002). Of interest is also the observation that CRF-R immunoreactivity can be found in GABAergic neurons in the dorsolateral region of the DRN (Roche et al., 2003). According to the manufacturer's specifications, the antibody for CRF_1 used in the latter study may, however, show cross reactivity with CRF_2.

Based on the above-described findings, it is evident that the CRF system is in a unique position to influence serotonergic neurotransmission by direct and indirect (GABAergic interneurons) mechanisms at the level of the raphe nuclei. However, the exact interplay between the different members of the CRF family, the role of the two types of CRF-R, and their functional implications is subject of intensive research. It should be emphasized that further detailed studies addressing the topographical specificity of the relationship between the CRF and 5-HT systems are needed. Such studies should also include more elaborated investigations in the MRN, because this raphe nucleus plays an important role in the regulation of the output of the hippocampus and the cortex, brain regions fundamentally involved in stress coping strategies.

Raphe nuclei: stress- and corticotropin-releasing factor-induced alterations in immediate early gene expression and firing rate of serotonergic neurons

c-fos

Several studies, implementing the immediate early gene product c-fos as a functional marker for neuronal activation (for review see Kovacs, 1998), have shown that stress may result in activation of raphe neurons. Different psychological stress models such as the exposure to an elevated plus maze (Silveira et al., 1993) or a conditioned fear stress paradigm (Pezzone et al., 1993; Beck and Fibiger, 1995; Ishida et al., 2002), social defeat stress (Martinez et al., 1998), restraint (Watanabe et al., 1994; Cullinan et al., 1995) and inescapable tailshock (as compared to escapable tailshock; Grahn et al., 1999) increase c-fos expression within the DRN and/or the MRN. The mixed psychological and physical stressor forced swimming also results in elevated levels of c-fos mRNA in the DRN and MRN (Cullinan et al., 1995). In contrast, immune stress (for instance induced by intraperitoneal injection of the bacterial endotoxin lipopolysaccharide (LPS) or intravenous administration of interleukin-1) seems to have no effect on c-fos expression in the DRN or MRN (Ericsson et al., 1994; Elmquist et al., 1996; Laflamme et al., 1999).

Until now little attention has been paid to the detailed neuroanatomical topography of stress-induced expression of c-fos in the different raphe nuclei. A recent study by the group of Rita Valentino, however, shows that 15 min of forced swimming in water of 25°C increases c-fos immunoreactivity especially in the dorsolateral DRN (Roche et al., 2003). A high percentage of the c-fos-positive neurons were doubly labeled for GABA and enveloped by CRF fibres. Interestingly, these GABA neurons also contained CRF-R (Roche et al., 2003). Hence, these authors have speculated that, at least in this subregion of the DRN, stress results (via CRF-R) in stimulation of inhibitory interneurons leading to inhibition of 5-HT neurons and consequently to lower levels of 5-HT in specific terminal regions. This possibility is underscored by the observation that conditioned fear stress results in c-fos immunostaining of not only 5-HT, but also of GABA-immunopositive neurons in the DRN (Ishida et al., 2002). The effect of stress on c-fos expression in the raphe nuclei seems to be mimicked by i.c.v. administration of CRF and urocortin 1. Both the neuropeptides increase c-fos immunoreactivity profoundly in the DRN but only moderately in the MRN (Bittencourt and Sawchenko, 2000). Moderate effects of these neuropeptides were also demonstrated in the caudal raphe nuclei, i.e. raphe magnus, pallidus and obscurus nuclei (Bittencourt and Sawchenko, 2000). In contrast, urocortin 2 has no effects on c-fos expression in the raphe nuclei (Reyes et al., 2001).

Firing rate of serotonergic neurons

Different types of serotonergic neurons have been described in the rostral raphe nuclei. The majority of 5-HT neurons fire solitary spikes in a slow (0.3–5 Hz) and regular discharge pattern (Aghajanian et al., 1968; Aghajanian and Vandermaelen, 1982), which is closely related to the sleep–wake cycle. The highest levels of discharge are observed during active waking. The discharge rate, however, decreases during quiet waking and even further during non-rapid eye movement sleep. Finally, these 5-HT cell bodies are almost salient during rapid eye movement sleep (see Jacobs and Azmitia, 1992). A substantial number of 5-HT neurons (about 25%) in the cat DRN and MRN has been found to be tonically activated specifically during oral–buccal movements, such as chewing, licking and grooming. These neurons are not activated during other behaviours and even decrease their firing rate during locomotion and orienting behaviours (Fornal et al., 1996). In contrast, a small subpopulation of serotonergic neurons has been described in a region between the medial longitudinal fasciculi at the caudal interface of the DRN and the MRN. The firing frequency of these 5-HT neurons is not clearly related to the vigilance state of the animal (Rasmussen et al., 1984). Finally, recent studies report on 5-HT neurons in the DRN (and to a lesser extent in the MRN) that fire pairs of spikes or bursts of spikes in a very short (< 20 ms) time interval but with a highly regular pattern (Hajós et al., 1995; Morzorati and Johnson, 1999). Mimicking the burst-firing pattern of 5-HT neurons by electrical stimulation of the DRN produces a greater increase in the extracellular levels of 5-HT in the medial prefrontal cortex than observed after single pulse stimulation (Gartside et al., 2000).

Barry Jacobs and colleagues have carefully analysed the effects of stress on the firing patterns of 5-HT neurons in the raphe nuclei of the cat. During different types of stress, including thermoregulatory and glucoregulatory challenges as well as noise stress and painful stimuli, the firing rate of 5-HT neurons in the DRN and MRN is not different from discharge rates found during a normal phase of active waking (for review see Jacobs and Azmitia, 1992). These observations have led to the proposition that 5-HT plays no specific role during stress, but that this neurotransmitter merely facilitates motor output and coordinates the concurrent neuroendocrine and autonomic responses, together with the suppression of sensory processing (Jacobs and Fornal, 1999). More recent studies on the effects of CRF on discharge rate suggest, however, that there are, at least in the DRN, distinct subsets of 5-HT neurons. Intracerebroventricular (i.c.v.) and intraraphe administration of low doses of CRF decrease the in vivo firing rate of 5-HT neurons in rats under halothane anaesthesia in the rostral and medial aspects of the DRN without an apparent dorsoventral or mediolateral organization (Price et al., 1998; Kirby et al., 2000). In contrast, an in vitro study by Lowry and colleagues showed a CRF-induced rise in the firing rate of a specific set of neurons located in the ventral and interfascicular region of the caudal DRN (Lowry et al., 2000). No effect was found on 5-HT neurons in the dorsomedial region of the DRN at the same rostral–caudal level in the latter study. Although the respective influence of anaesthesia and of the absence of afferent input in the in vivo and in vitro studies cannot be excluded at this moment, the data suggest the presence of discrete subsets of CRF-sensitive 5-HT neurons in the DRN. Together with the distinct efferent innervation pattern of the forebrain by the different subregions of the DRN and the MRN (there is evidence for a rostrocaudal, mediolateral and dorsoventral encephalotopy within the DRN; see discussion in Valentino et al., 2001), these data put forward that 5-HT neurons may display a topographically specific response to stress. The effects of stress and CRF on the subpopulation of 5-HT neurons that fire in a burst-like pattern are not known yet. Studies on stress-induced changes in the firing rate of these neurons will be of high relevance, given the observation that electrical stimulation of the DRN in a burst-like pattern causes a greater rise in extracellular 5-HT in the prefrontal cortex than single pulse stimulation (Gartside et al., 2000).

Effects of stress and corticotropin-releasing factor on the synthesis of serotonin

Serotonin is synthesized from the essential amino acid L-tryptophan, which is transported from the blood into the brain via an L-type neutral amino acid transporter in the blood-brain barrier. Proteins in the diet are the principal source of L-tryptophan. In serotonergic neurons L-tryptophan is hydroxylated into 5-hydroxytryptophan (5-HTP) by the enzyme TPH, forming the rate-limiting step in the synthesis of 5-HT. Next, 5-HTP is decarboxylated into 5-HT by the enzyme aromatic L-amino acid decarboxylase (AADC). A recent report indicates that two isoforms of TPH may exist, TPH1 and TPH2, responsible for the synthesis of 5-HT in the periphery and the brain respectively (Walther et al., 2003). TPH becomes activated during situations where a replenishment of the stores of 5-HT is needed. For instance, during electrical stimulation of serotonergic cell bodies a (frequency-dependent) rise in the synthesis of 5-HT from L-tryptophan can be observed (Shields and Eccleston, 1972; Herr et al., 1975; Boadle-Biber et al., 1986). The activation of TPH involves protein phosphorylation by calcium/calmodulin-dependent protein kinase II (Lysz and Sze, 1978; Hamon et al., 1981; Kuhn and Lovenberg, 1982; Ehret et al., 1989) and/or cAMP-dependent protein kinase A (Johansen et al., 1995; see also Stenfors and Ross, 2002; see also Mockus and Vrana, 1998). There is also evidence that the availability of (free) L-tryptophan to the brain can affect the level of synthesis of 5-HT under certain circumstances (Fernstrom and Wurtman, 1971; Knott and Curzon, 1972; for further discussion see Boadle-Biber, 1993).

As described above, distinct effects of stress on neuronal activation and on the firing rate of 5-HT neurons have been found. Therefore, the question arises whether stress also affects the synthesis of 5-HT and whether it does so in a stressor and neuroanatomically specific manner. Initial work by Azmitia and McEwen has shown that prolonged cold and electric footshock stress increase TPH activity in the midbrain and forebrain of rats; an effect which is prevented by adrenalectomy (Azmitia and McEwen, 1974). The group of Margaret Boadle-Biber has collected interesting and comprehensive data on the effects of loud sound stress on 5-HT synthesis, using an ex vivo TPH activity assay and an in vivo 5-HTP accumulation paradigm, with emphasis on dissecting the different effects of acute versus chronic stress. One session of randomly presented sound stress results in a short-lasting increase in TPH activity in supernatants of the cortex and midbrain of Fischer 344 rats (Boadle-Biber et al., 1989). Repeated exposure of the animals to the sound stress paradigm induces a persistent elevation of TPH activity which is still present 24 h after the last stress session (Boadle-Biber et al., 1989). The effects of sound stress on TPH activity are related to increases in firing rate, because no effects of sound stress were found after pretreatment of the animals with the 5-HT_{1A} receptor agonist gepirone, which inhibits the discharge of 5-HT neurons (Corley et al., 1992). Interestingly, the effects of sound stress on TPH activity are blocked by adrenalectomy (Singh et al., 1990) and are mimicked by i.c.v. administration of CRF (Singh et al., 1992). Moreover, glucocorticoid receptors (GRs) seem to play a permissive role in the CRF- and sound stress-induced activation of TPH (Singh et al., 1990, 1992). Studies, in which the accumulation of 5-HTP after inhibition of AADC by m-hydroxylbenzylamine (NSD 1015) was assessed, revealed that sound stress increases 5-HT synthesis in the MRN but not in the DRN (Dilts and Boadle-Biber, 1995; Daugherty et al., 2001). A similar specific activation only in the MRN is observed after exposure to forced swim stress or tailshock (Corley et al., 2002). The notion that stress may cause a neuroanatomically specific effect on 5-HT synthesis is further underscored by the observation that i.c.v. administration of CRF has no effect on 5-HTP accumulation after AADC inhibition in the mediobasal hypothalamus (Van Loon et al., 1982). Moreover, also repeated immobilization stress seems to exert neuroanatomically distinct effects on TPH activity. Whereas Miklos Palkovits and colleagues found no changes in TPH activity in various brain regions, such as the hypothalamus, amygdala, hippocampus and DRN, after repeated (5 times) immobilization stress (Palkovits et al., 1976), Culman et al. observed increased TPH activity in the locus coeruleus, decreased activity in the suprachiasmatic nucleus together with no effects in the DRN after seven immobilization sessions (Culman et al., 1984). In contrast, this stress paradigm has been found to increase TPH *mRNA levels* 10-fold in the DRN and 6-fold in the MRN, without affecting TPH mRNA levels in the pineal gland (Chamas et al., 1999). Interestingly, measurement of 5-HTP accumulation by microdialysis revealed that 1 h of immobilization increases 5-HT synthesis in the medial prefrontal cortex and the nucleus accumbens (Nakahara and Nakamura, 1999).

The effects of stress, and especially chronic stress, on 5-HT synthesis are highly relevant also with respect to the pathophysiology of affective and anxiety disorders, in which a deficit in 5-HT neurotransmission has been suggested. Indeed depressive symptoms may exacerbate following a low-tryptophan diet (Delgado et al., 1990). Clearly more detailed studies will be needed, also implementing the effects of stress on the phosphorylation of TPH. From the presently available data, however, the picture is emerging that stress can influence the synthesis of 5-HT, in a neuroanatomical and stress duration specific manner, by changing the activity of the rate-limiting enzyme TPH.

Effects of stress and corticotropin-releasing factor on the levels of serotonin and 5-hydroxyindoleacetic acid in forebrain terminal regions

The previous sections of this chapter have delineated specialized interactions between the CRF and 5-HT systems at the neuroanatomical level, together with stressor- and site-specific effects of stress on the firing rate of raphe neurons and the synthesis of 5-HT. The impact of stress-induced alterations in serotonergic neurotransmission for behaviour and neuroendocrine regulation will, however, highly depend on the ultimate changes in the extracellular levels of 5-HT in the terminal regions. The present section will, therefore, be devoted to the question whether stress in a stressor- and brain structure-specific manner influences terminal levels of 5-HT, with an emphasis on its extracellular level as assessed by in vivo microdialysis. The effects of five stressors, varying in intensity and their physical versus psychological character, and of CRF and urocortin 1 will be discussed. A summary of the effects of these stressors on the levels of 5-HT and 5-HIAA in terminal regions is presented in Table 1.

Immune stress

Immune stress is often considered to represent a so-called systemic stressor (Li et al., 1996; Herman and Cullinan, 1997), indicating its immediate physiologic threat. During an infectious challenge the organism responds with a variety of defence mechanisms, among which are the activation of the HPA axis and the development of fever. The behavioral changes frequently observed during infection, i.e. fatigue, loss of appetite, immobility and social disinterest, are collectively termed as 'sickness behaviour' (Hart, 1988; Dantzer et al., 1991), and indicate that immune stress may also possess a psychological element, involving higher brain structures. We have provided extensive evidence showing that an infectious challenge results in a highly specific answer of the brain 5-HT system. Intraperitoneal (i.p.) injection of LPS results in a prolonged rise in extracellular levels of 5-HT (Fig. 1A) and 5-HIAA in the hippocampus of rats (Linthorst et al., 1995b). This effect can be mimicked by i.p. (Merali et al., 1997) and i.c.v. injection of interleukin-1ß (Linthorst et al., 1995b) and interleukin-2, but not tumor necrosis factor-α (Pauli et al., 1998). The stimulatory effects of immune stress on hippocampal serotonergic neurotransmission are underscored by studies on the turnover of 5-HT, as indicated by an increase in the ratio between the post-mortem levels of 5-HIAA and 5-HT (Kabiersch et al., 1988; Zalcman et al., 1994; Lacosta et al., 1999). Moreover, studies on tissue levels of 5-HT and on tissue and dialysate levels of 5-HIAA have shown that the stimulatory effects of immune activation on serotonergic neurotransmission are not confined to the hippocampus, but can also be found in other higher brain structures such as the frontal cortex (Dunn and Welch, 1991; Dunn, 1992; Zalcman et al., 1994; Lavicky and Dunn, 1995). In contrast, we observed no effect of i.p. injection of LPS on extracellular levels of 5-HT (Fig. 1B), and only a minute rise in 5-HIAA, in the preoptic area of the rat (Linthorst et al., 1995a). A similar observation in this brain region has been made after administration of the pyrogen muramyl dipeptide in cats (Wilkinson et al., 1991). At the level of the hypothalamus predominantly stimulatory effects of systemic immune stimulation and administration of interleukin-1 on 5-HT turnover and dialysate levels of 5-HIAA have been described (Mefford and Heyes, 1990; Dunn, 1992; Mohankumar et al., 1993; Shintani et al., 1993; Lavicky and Dunn, 1995; Lacosta et al., 1999), although also a decrease in tissue levels of 5-HIAA after i.c.v. injection of interleukin-1 and tumor-necrosis factor-α has been observed (Connor et al., 1998).

The effects of immune stress on serotonergic neurotransmission are also interesting from a neuroanatomical point of view. As described above, administration of LPS and interleukin-1 do not induce c-fos activation in the raphe nuclei (Ericsson et al., 1994; Elmquist et al., 1996; Laflamme et al., 1999). Hence, it has been suggested that the immune stress-induced alterations in 5-HT levels are generated locally in the terminal regions. This possibility is underlined by our observation that local infusion of interleukin-1ß into the hippocampus by reversed dialysis indeed results in a rise in extracellular 5-HT in this brain region (Linthorst et al., 1994).

Forced swim stress

Forced swim stress is a widely used stress paradigm in rats and mice, mainly applied to assess putative antidepressant properties of drugs in the so-called Porsolt forced swim test paradigm (Porsolt et al., 1977;

Table 1. Summary of the effects of stress on 5-HT and 5-HIAA levels in terminal regions as reported in the literature

Stressor	Brain structure	Observations	References
Immune stress (lipopolysaccharide; cytokines)	Hippocampus	↑ 5-HT and 5-HIAA (m)	Linthorst et al., 1994, 1995b; Merali et al., 1997; Pauli et al., 1998
		↑ turnover (p)	Kabiersch et al., 1988; Zalcman et al., 1994
		↑ 5-HT and 5-HIAA (p)	Connor et al., 1998; Lacosta et al., 1999
	Frontal cortex	↑ turnover (p)	Dunn and Welch, 1991; Dunn, 1992; Zalcman et al., 1994
		↑ 5-HIAA (m)	Lavicky and Dunn, 1995
	Preoptic area	↔ 5-HT (m), minute rise in 5-HIAA	Wilkinson et al., 1991; Linthorst et al., 1995a
	Hypothalamus	↔ turnover (p)	Dunn and Welch, 1991
		↑ turnover (p)	Mefford and Heyes, 1990; Dunn, 1992
		↑ 5-HT and 5-HIAA (m)	Mohankumar et al., 1993; Shintani et al., 1993; Lavicky and Dunn, 1995
		↓ 5-HIAA (p)	Connor et al., 1998
		↑ 5-HIAA (p)	Lacosta et al., 1999
Forced swim stress	Hippocampus	↑ 5-HT, 5-HIAA biphasic (m)	Rueter and Jacobs, 1996; Peñalva et al., 2002; Linthorst et al., 2002; Fujino et al., 2002
		↓ 5-HT (m)	Adell et al., 1997
		↔ 5-HT, ↓ 5-HIAA (m)	Kirby et al., 1995
	Frontal cortex	↑ 5-HT (m)	Rueter and Jacobs, 1996; Fujino et al., 2002
		↓ 5-HT (m)	Adell et al., 1997
		↔ 5-HT, ↓ 5-HIAA (m)	Kirby et al., 1995
		↑ turnover (p)	Connor et al., 2000
	Amygdala	↑ 5-HT (m)	Rueter and Jacobs, 1996; Adell et al., 1997
		↓ 5-HT and 5-HIAA (m)	Kirby et al., 1995
		↑ turnover (p)	Connor et al., 2000
	Striatum	↑ 5-HT (m)	Rueter and Jacobs, 1996
		↑ 5-HT, ↓ 5-HIAA (m)	Kirby et al., 1995
	Lateral septum	↓ 5-HT and 5-HIAA (m)	Kirby et al., 1995
Tail pinch	Hippocampus	↑ 5-HT and 5-HIAA (m)	Kalén et al., 1989; Pei et al., 1990; Vahabzadeh and Fillenz, 1994; Rueter and Jacobs, 1996; Fujino et al., 2002
		↑ turnover (p)	Pei et al., 1990
	Frontal cortex	↑ 5-HT (m)	Rueter and Jacobs, 1996; Fujino et al., 2002
		↑ turnover (p)	Pei et al., 1990
	Amygdala	↑ 5-HT (m)	Rueter and Jacobs, 1996
	Striatum	↑ 5-HT (m)	Rueter and Jacobs, 1996
		↔ turnover (p)	Pei et al., 1990
	Hypothalamus	↔ turnover (p)	Pei et al., 1990

(*continued*)

Table 1. Continued

Stressor	Brain structure	Observations	References
Electric shock/conditioned fear stress	Hippocampus	↑ 5-HT (m) footshock	Hajos-Korcsok et al., 2003
		↑ turnover (p) after footshock	Dunn, 1988
		↔ turnover (p) after footshock and conditioned fear stress	Inoue et al., 1993, 1994
		↑ 5-HT and 5-HIAA (m) inescapable vs. escapable tailshock	Amat et al., 1998
		↑ 5-HT (m) during conditioned fear stress	Wilkinson et al., 1996
	Frontal cortex	↑ turnover (p) after footshock	Inoue et al., 1993, 1994
		↓ turnover (p) after footshock	Dunn, 1988
		↑ 5-HT (p) inescapable vs. escapable shock	Heinsbroek et al., 1991
		↑ 5-HT (m) during conditioned fear stress	Yoshioka et al., 1995; Hashimoto et al., 1999
		↑ turnover (p) after conditioned fear stress	Inoue et al., 1993, 1994
	Amygdala	↑ turnover (p) after footshock and conditioned fear stress	Dunn, 1988; Inoue et al., 1993, 1994
	Hypothalamus	↑ turnover (p) after footshock (lateral hypothalamus)	Inoue et al., 1994
		↔ turnover (p) after footshock (paraventricular nucleus)	Inoue et al., 1993, 1994
		↓ turnover (p)	Dunn, 1988
Predator stress	Hippocampus	↑ 5-HT and 5-HIAA (m)	Rueter and Jacobs, 1996; Linthorst et al., 2000
		↑ turnover or 5-HIAA (p)	Hayley et al., 2001; Belzung et al., 2001
		↔ turnover (p) (fox odour)	Dias Soares et al., 2003
	Frontal cortex	↑ 5-HT and 5-HIAA (m)	Rueter and Jacobs, 1996
		↑ 5-HIAA (p)	Hayley et al., 2001
		↑ 5-HIAA (p) (fox odour)	Hayley et al., 2001
	Amygdala	↑ 5-HT and 5-HIAA (m)	Rueter and Jacobs, 1996
		↑5-HIAA (p) (fox odour)	Hayley et al., 2001
	Striatum	↑ 5-HT (m)	Rueter and Jacobs, 1996
		↔ turnover (p)	Belzung et al., 2001
	Hypothalamus	↔ turnover (p)	Belzung et al., 2001
		↑ 5-HIAA (p) (paraventricular nucleus; fox odour)	Hayley et al., 2001
i.c.v. administration of CRF	Hippocampus	↑ 5-HT and 5-HIAA (m) 0.03–10 μg CRF (and urocortin 1)	Linthorst et al., 2002
		↔ 5-HT and 5-HIAA basal levels (m) after 7 days of infusion	Linthorst et al., 1997
	Frontal cortex	↑ 5-HIAA (m) 17, 330 pmol	Lavicky and Dunn, 1993
	Striatum	↓ 5-HT (m) 0.1 and 0.3 μg	Price et al., 1998
		↔ 5-HT (m) 1.0 μg	
		↑ 5-HT and 5-HIAA (m) 3.0 μg	
	Lateral septum	↓ 5-HT (m) 0.3 and 1.0 μg	Price and Lucki, 2001
		↔ 5-HT (m) 0.1 and 3.0 μg	
	Hypothalamus	↑ 5-HIAA (m) 17, 330 pmol	Lavicky and Dunn, 1993

↑, increase; ↓, decrease; ↔, no effect. m, as assessed by in vivo microdialysis (Mohankumar et al., 1993: push–pull method); p, post-mortem measurement of tissue levels. Turnover is defined as the ratio between tissue levels of 5-HIAA and 5-HT. For more details, see text.

Cryan et al., 2002), but also to study neurotransmitter responses to stress. Forced swim stress is characterized by a clear psychological stress element, because the animal has to cope with a situation from which it cannot escape. However, the physical component of this stressor is often underestimated. Not only do the animals spend a significant portion of their time in the water on escape attempts and swimming, hence resulting in a high amount of motor activity, but, depending on the temperature of the water and duration of the stressor, a dramatic drop in body temperature can be observed (for instance the body temperature of rats reaches about 30°C after swimming for 15 min in water of 25°C; Linthorst et al., unpublished observations; see also Stone, 1970).

In a recent set of microdialysis studies in rats and mice, we have aimed to elucidate the effects of swim stress on 5-HT in the hippocampus and the role of CRF herein. We collected unexpected, but exciting data. Forced swimming for 15 min in water of 25°C results in an enormous (about 900% of baseline) increase in hippocampal extracellular levels of 5-HT in rats (Linthorst et al., 2002; Figs. 2A,B). This dramatic response is, however, only observed in animals that dive during the swim session (Fig. 2C; maximum is about 1500% of baseline). Pretreatment of the rats with the CRF-R antagonist D-Phe-CRF$_{12-41}$ prevented the diving-associated rise in hippocampal 5-HT, pointing to an essential role of CRF or the urocortins in this effect (Figs. 2B,C). Because the animals are connected to a swivel system via a plastic collar around their neck, we have hypothesized that the exaggerated 5-HT response in diving animals

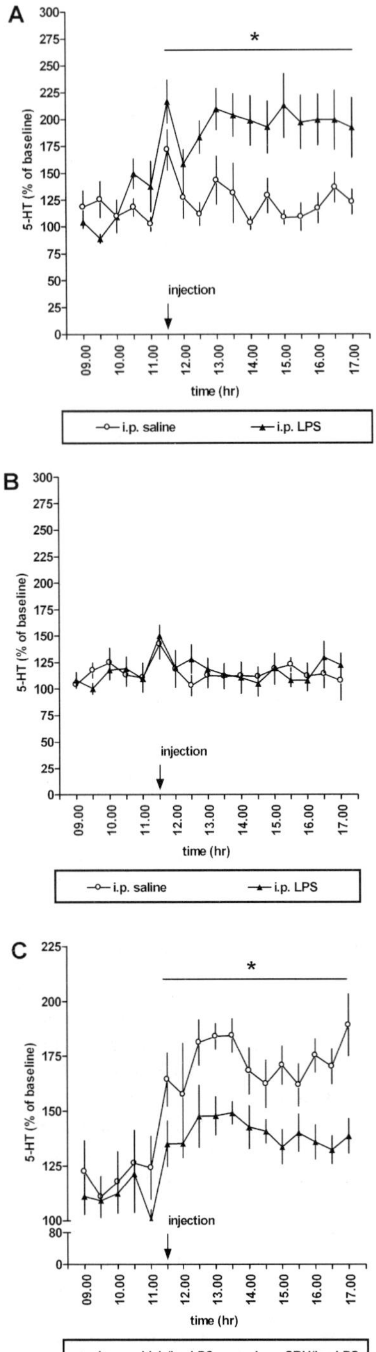

Fig. 1. Effects of intraperitoneal administration of LPS (100 μg/kg body weight) on extracellular levels of 5-HT in the hippocampus (A and C) and preoptic area (B), as assessed by in vivo microdialysis in conscious, freely moving Wistar rats. 5-HT levels are expressed as percentage of baseline. The time points on the x-axis correspond to the time of the day at which the collection of the respective microdialysis sample was started. Intraperitoneal administration of LPS significantly increases 5-HT levels in the hippocampus (A), but not in the preoptic area (B). Figure C shows the results in rats long-term i.c.v. infused with CRF (1 μg/μl/h) via a miniosmotic pump. LPS was injected on day 7 of the infusion treatment. As compared to control-infused rats, long-term CRF-treated animals show a significantly reduced responsiveness of hippocampal 5-HT to systemic injection of LPS. Values represent mean ± SEM. *, $P < 0.05$ (analysis of variance with repeated measures design). Data are taken from Linthorst et al., 1995b, (A) Linthorst et al., 1995a (B) and Linthorst et al., 1997 (C), with the permission of the Journal of Neuroscience (A,C) and the European Journal of Neuroscience (B).

may be related to a different appreciation of the stressful situation, putatively leading to a panic-like response (Linthorst et al., 2002). This postulate is underscored by the observation that the panic-inducing substance m-chloro-phenylpiperazine (mCPP) is also able to increase hippocampal extracellular levels of 5-HT to 300–1400% of baseline, depending on the dose applied (Eriksson et al., 1999). Moreover, this extreme response is absent in animals connected to the swivel system via a peg on their head (Linthorst, unpublished observations). The exceptional situation of diving in rats with a plastic collar is further underscored by the observation that forced swimming in mice, which do not dive during the swim stress, causes an increase in hippocampal 5-HT levels of not more than 170–240% of baseline (Peñalva et al., 2002; Fujino et al., 2002; Oshima et al., 2003). In both rats and mice a small and transient swim stress-induced decrease in hippocampal levels of 5-HIAA has been observed during or immediately after the swim procedure, followed by a rise to levels of about 130% of baseline (Peñalva et al., 2002; Linthorst et al., 2002).

The complex nature of forced swimming is also evident from the results of other research groups. Whereas forced swimming for 30 min in water of 30–35°C has been reported to enhance extracellular 5-HT levels to about 120–170% of baseline in the hippocampus, frontal cortex, amygdala and striatum of rats during the dark period of the diurnal cycle

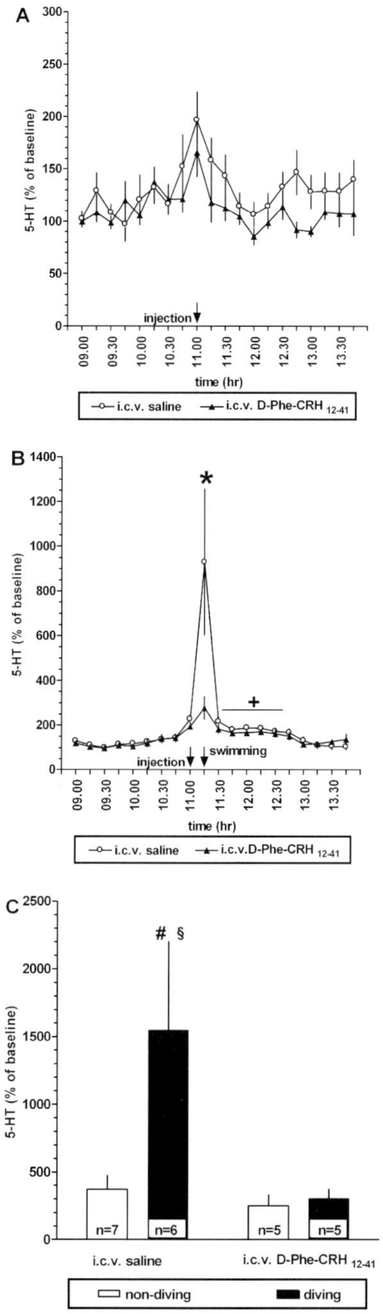

Fig. 2. Effects of forced swim stress (15 min, water temperature 25°C) on extracellular levels of 5-HT in the hippocampus, as assessed by in vivo microdialysis in conscious, freely moving Wistar rats. The rats were i.c.v. pretreated with saline or the CRF-R antagonist D-Phe-CRF_{12-41} (5 µg). 5-HT levels are expressed as percentage of baseline. The time points on the x-axis correspond to the time of the day at which the collection of the respective microdialysis sample was started. Figure A shows that i.c.v. administration of D-Phe-CRF_{12-41} slightly decreases hippocampal extracellular levels of 5-HT in non-stressed rats ($F(1,10) = 6.04$, $P < 0.05$ (analysis of variance with repeated measures)). Forced swim stress causes, as depicted in Figure B, an enormous rise in hippocampal 5-HT concentrations, which was inhibited by i.c.v. pretreatment with D-Phe-CRF_{12-41}. Dividing the rats into two subgroups, i.e. animals that had dived during the swim session and non-diving animals, revealed that the immense increase in 5-HT during forced swimming was only observed in saline-pretreated diving rats (C). This effect could be prevented by i.c.v. pretreatment with D-Phe-CRF_{12-41}. Values represent mean ± SEM. *, $P < 0.05$, stressed versus nonstressed saline-pretreated controls; +, $P < 0.05$, stressed versus respective nonstressed control groups; #, $P < 0.05$, versus nondiving saline-pretreated rats; §, $P < 0.05$, versus diving D-Phe-CRF_{12-41} pretreated rats (all posthoc tests: Scheffé). Data are taken from Linthorst et al., 2002, with the permission of the European Journal of Neuroscience.

(Rueter and Jacobs, 1996), 5 min of swimming in water of 25°C results in a decrease in hippocampal and prefrontal cortical 5-HT and in an elevation in the amygdala (Adell et al., 1997) (in the latter study, however, a SSRI has been added to the perfusion medium). Sacrifice of rats after a second exposure to forced swimming (classical Porsolt forced swim test) revealed an enhanced 5-HIAA/5-HT ratio in the frontal cortex and amygdala (Connor et al., 2000). Furthermore, work of Irwin Lucki and colleagues shows that forced swimming for 30 min in water of 21–22°C causes a region-specific effect on 5-HT and 5-HIAA in rats. A decrease in 5-HIAA levels is observed in the lateral septum, amygdala, hippocampus and striatum. In contrast, whereas no effect on 5-HT is found in the hippocampus and frontal cortex, extracellular levels of the neurotransmitter increase in the striatum and decrease in the lateral septum and amygdala as a consequence of the forced swim procedure (Kirby et al., 1995, 1997). Interestingly, in a follow-up study, the swim stress-induced decrease in 5-HT in the lateral septum (20 min, water temperature 24–25°C) could be prevented by pretreatment of the animals with D-Phe-CRF_{12-41}, indicating the involvement of the CRF system in this effect (Price et al., 2002).

The inconsistency in the results on swim stress may at first sight seem somewhat disappointing. It is, however, conceivable that the divergent results published so far are related to the different experimental protocols, especially concerning the duration of swimming and the temperature of the water. Our preliminary data show that the hippocampal 5-HT and 5-HIAA responses to swim stress highly depend on the water temperature and the concomitant decrease in body temperature (Linthorst et al., 2001). Given the role of 5-HT in thermoregulation, mainly involving the preoptic area and anterior hypothalamus but also the raphe pallidus nucleus, these data suggest that during forced swimming counteracting effects generated by the psychological and thermoregulatory aspects of the stressor may determine the final outcome of the (hippocampal) 5-HT response. Moreover, our data in diving rats, i.e. the dramatic CRF-R-dependent rise in hippocampal 5-HT, may have uncovered a new phenomenon in the relationship between the CRF and 5-HT systems, with putative relevance for the pathophysiology of panic disorder.

Tail pinch

Tail pinch is performed by applying a paper clip on the tail of a rat or mouse. Apart from arousing the animal, tail pinch also stimulates pain sensory pathways and elicits consummatory behaviours, such as eating, gnawing and chewing. Microdialysis and turnover studies from various research groups show consistently that tail pinch increases serotonergic neurotransmission in the hippocampus, prefrontal cortex and amygdala (Kalén et al., 1989; Pei et al., 1990; Vahabzadeh and Fillenz, 1994; Rueter and Jacobs, 1996; Fujino et al., 2002). Interestingly, in three studies the maximum extracellular levels of 5-HT in the hippocampus were found with a delay, i.e. after termination of the stressor (Kalén et al., 1989; Pei et al., 1990; Vahabzadeh and Fillenz, 1994). This observation may be related to consummatory behaviours and grooming displayed by the animals after the removal of the paper clip. This possibility is underscored by the description of a subset of serotonergic neurons in the DRN of the cat that is highly active during oral buccal movements (Fornal et al., 1996). Tail pinch has also been reported to increase levels of 5-HT in the locus coeruleus (Singewald et al., 1997) and in the striatum (Rueter and Jacobs, 1996). In contrast, Pei and colleagues observed no effect of this stressor on the turnover of 5-HT in the striatum and the hypothalamus (Pei et al., 1990).

Electric foot- or tailshock and conditioned fear stress

Electric foot- or tailshock is an intense stressor, combining physical and psychological aspects. In contrast, conditioned fear stress, i.e. confronting an animal with the context during which previously a shock was experienced, represents a psychological stress paradigm. Unfortunately, data on the effects of these stress models on serotonergic neurotransmission are sparse. In addition, to our knowledge, there are no in vivo microdialysis studies available monitoring 5-HT levels during both the presentation of the shock and the conditioned fear context in the same animals. Electric footshock increases serotonergic neurotransmission in various brain regions, including the hippocampus, the nucleus accumbens, the amygdala, the prefrontal cortex and the lateral hypothalamus (Dunn, 1988; Inoue et al., 1993, 1994; Hajos-Korcsok et al., 2003). However, the exact outcome of footshock stress on neurotransmitter systems seems to depend on the shock intensity (Inoue et al., 1994) and the coping possibilities. Hence, there is some evidence that *inescapable* foot- and tailshocks in rats result in augmented 5-HT responses in the frontal cortex (Heinsbroek et al., 1991) and the ventral hippocampus (Amat et al., 1998) as compared to *escapable* shocks. Interestingly, in the dorsal periaqueductal gray opposite results were found, i.e. an augmented increase in dialysate 5-HT in animals that could escape from the shock as compared to rats experiencing a yoked inescapable shock (Amat et al., 1998). Conditioned fear stress leads, as assessed by microdialysis, to an elevation of extracellular levels of 5-HT in the hippocampus (Wilkinson et al., 1996) and the prefrontal cortex (Yoshioka et al., 1995; Hashimoto et al., 1999). An increased turnover of 5-HT in the prefrontal cortex was also observed in post-mortem level studies in conditioned fear stressed rats (Inoue et al., 1994). Interestingly, in the hippocampus the effects of conditioned fear stress seem to be related to the contextual aversive cues and not to the conditioned discrete stimulus (Wilkinson et al., 1996).

Predator stress

A few research groups have assessed the consequences of predator stress for the extracellular levels or turnover of 5-HT. Predator stress is an interesting stress model, because it represents a psychological stressor, with apart from increased motor activity during risk assessment strategies, no major physical stress component (see also Blanchard et al., 1998). As may be taken from Table 1, exposure to a predator (i.e. exposing a rat to a cat, a mouse to a rat or cat) results invariably in elevated serotonergic neurotransmission in the hippocampus, frontal cortex and amygdala (Rueter and Jacobs, 1996; Linthorst et al., 2000; Belzung et al., 2001; Hayley et al., 2001; Beekman and Linthorst, unpublished observations). Interestingly, the hippocampal 5-HT response was highly augmented in GR-impaired mice confronted with a rat as compared to control mice (Linthorst et al., 2000). This mutant mouse line shows no activation of the HPA axis and a different coping strategy (i.e., significantly more investigatory behavior along the separation wall) during predator stress. It is, however, at present not clear whether the different responses of GR-impaired mice are related to GR dysfunction itself, or to changes at the level of the CRF system also reported in these mice (Dijkstra et al., 1998). The data, however, clearly suggest that activation of GR is not (under all circumstances) required for activation of the 5-HT system, despite the permissive action of GR activation on 5-HT turnover and TPH activity described in the literature (Singh et al., 1990; De Kloet, 1991). This notion is underscored by the absence of an effect of adrenalectomy on CRF- and sleep deprivation-induced elevations in hippocampal serotonergic neurotransmission (Linthorst et al., 2002; Peñalva et al., 2003).

Exposing mice to fox odour causes also a rise in 5-HT turnover in the prefrontal cortex of BALB/cByJ and C57Bl/6ByJ mice, whereas this stimulation increases 5-HIAA levels in the central amygdala of the C57Bl/6ByJ strain only (Hayley et al., 2001). However, no effect on hippocampal 5-HT has been observed in rats exposed to fox odour (Dias Soares et al., 2003). Similarly as for other stressors, the effects of predator exposure seem to be more variable in the striatum (Rueter and Jacobs, 1996; Belzung et al., 2001). Moreover, the data available at present suggest that exposure to a predator exerts no effects on 5-HT at the level of the hypothalamus (Belzung et al., 2001). The presentation of fox odour increases, however, 5-HIAA concentrations in the paraventricular nucleus of the hypothalamus in mice (Hayley et al., 2001). Interestingly, the effects of predator stress on brain serotonergic neurotransmission may be phylogenetically very old, because similar observations as in rodents have been made in bicolour damselfish (*Pomacentrus partitus*) (Winberg et al., 1993).

Corticotropin-releasing factor and urocortin 1

Because of its key role in the coordination of various responses to stress, it is of certain interest to look into the effects of CRF, and CRF-related peptides, on 5- HT levels in terminal regions. Intracerebroventricular administration of CRF and urocortin 1 (0.03–10 μg) increases extracellular levels of 5-HT and 5-HIAA dose dependently in the hippocampus as assessed by in vivo microdialysis in rats (Linthorst et al., 2002). The CRF_1 antagonist CP-154,526 has been found to reduce extracellular levels of 5-HT in the rat hippocampus, but not in the prefrontal cortex (Isogawa et al., 2000). Long-term infusion of CRF (7 days via a miniosmotic pump) has no effect on the basal levels of hippocampal 5-HT and 5-HIAA as measured on the last day of the treatment. However, a significantly decreased immune stress-induced rise in 5-HT in the hippocampus was found in long-term CRF-treated rats (Linthorst et al., 1997; Fig. 1C), putatively involving desensitization of CRF-R. Based on this observation we hypothesize that long-term disturbances in the CRF system may have profound consequences for the functioning of (hippocampal) serotonergic neurotransmission. This postulate is further underscored by our recent observations that mice with a life-long deficiency of CRF_1 (Timpl et al., 1998) show elevated hippocampal levels of 5-HIAA (Fig. 3B), but not 5-HT (Fig. 3A), over the diurnal rhythm (Peñalva et al., 2002). Moreover, the forced swim stress-induced increase in hippocampal 5-HT is augmented in CRF_1-deficient mice (Fig. 3C). Chronic oral treatment with the CRF_1 antagonist NBI30775 also evolves in an altered responsiveness of hippocampal 5-HT and 5-HIAA to stress, although no effect of this treatment on diurnal levels was found (Oshima et al., 2003). These data clearly underscore a role for CRF in the regulation of extracellular levels of 5-HT in the hippocampus of rats and mice under basal and stress conditions. However, at present, the exact interplay between the two types of CRF-R and the possible role of urocortin 2 and urocortin 3 are not clear yet and should be further elucidated.

Biphasic effects of i.c.v. administration of CRF were observed in the striatum of rats. Whereas low doses of CRF (0.1 and 0.3 μg) decrease extracellular levels of 5-HT, no effect after injection of 1 μg and a rise after injection of 3 μg CRF was found (Price et al., 1998). No decreases in 5-HIAA were found at the lower doses of CRF, but an increase in extracellular 5-HIAA levels was observed after administration of 3 μg CRF. A biphasic response of extracellular 5-HT was also shown in the lateral septum (Price and Lucki, 2001). In this regard, it is of special interest that injection of CRF into the dorsal raphe nucleus also decreases extracellular levels of 5-HT in both the lateral septum and striatum (Price and Lucki, 2001).

Unfortunately, information on the effects of CRF on serotonergic neurotransmission in other brain regions is sparse. Intracerebroventricular administration of CRF has been found to increase extracellular levels of 5-HIAA in the medial hypothalamus and medial prefrontal cortex (5-HT was not measurable in this study) (Lavicky and Dunn, 1993). In contrast, direct injection of CRF into the lateral hypothalamus decreases dialysate concentrations of 5-HIAA levels in both fed and fasted animals (Shimizu and Bray, 1989).

Some remarks on the effects of stress on serotonin in terminal regions

After having read the previous sections, the reader may have been left with the impression that the effects of stress on the levels of 5-HT and 5-HIAA in different terminal regions are not extremely consistent. This may in part indeed be true. The data presented here are collected by various research groups, all using their own stress protocols and microdialysis or post-mortem techniques in rats or mice. The influence of differences in experimental protocols is strikingly demonstrated by the effects of swim stress on hippocampal extracellular levels of 5-HT and 5-HIAA. However, as can also be taken from Table 1, some general conclusions may be drawn. First, 5-HT levels in the hippocampus and in the frontal cortex seem to be increased by diverse stressors, varying in intensity and physical versus psychological impact. During the past 12 years, we have found that different stressors, including immune stress, novelty, sleep deprivation, forced swimming and predator stress, all increase consistently hippocampal 5-HT levels as assessed by in vivo microdialysis in rats and mice. The effects of stress on hippocampal 5-HT are mimicked by i.c.v. administration of CRF and urocortin 1, and the 5-HT response in diving rats can be inhibited by a CRF-R antagonist, pointing to the involvement of the CRF system, at least under defined circumstances, in the effects of stress on hippocampal serotonergic neurotransmission. However, the inhibitory effects of CRF on extracellular levels of 5-HT demonstrated in the striatum and lateral septum suggest additional interactions between the CRF and 5-HT systems. Second, the effects of stress on 5-HT levels in the striatum, amygdala and lateral septum also seem to be in general stimulatory. Forced swim stress may represent a special situation regarding these brain regions. The decreased levels of 5-HT and/or 5-HIAA as reported by Irwin Lucki and colleagues, and the evident role of CRF herein, suggest that during swim stress additional neuroanatomical circuits may be stimulated, for instance involved in the regulation of body temperature. Third, the data available so far suggest that 5-HT at the level of the hypothalamus/preoptic area is not responding consistently to stressful stimuli. However, more coherent and, with respect to the highly specialized organization of the different hypothalamic nuclei, neuroanatomically more detailed studies will be needed to verify whether this conclusion is justified. Fourth, because of the general correlation between 5-HT levels (and firing

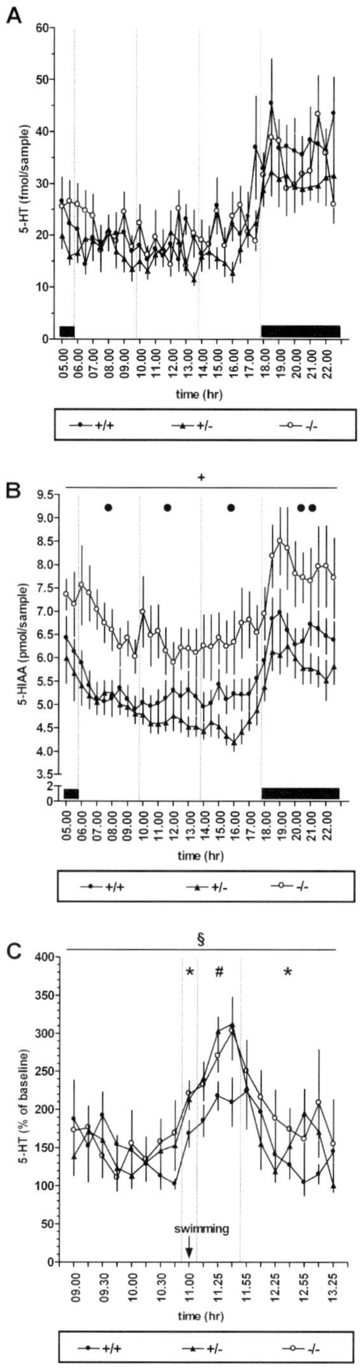

Fig. 3. Serotonergic neurotransmission in conscious, freely moving CRF_1-deficient mutant mice under basal (A,B) and forced swim stress (10 min, water temperature 25°C; C) conditions as assessed by in vivo microdialysis. 5-HT levels are expressed as fmol/30 min sample (A; diurnal rhythm) or percentage of baseline (C; stress). 5-HIAA levels are expressed as pmol/30 min sample (B). The time points on the x-axis correspond to the time of the day at which the collection of the respective microdialysis sample was started. The dotted lines indicate the time periods used for the statistical analyses. Hippocampal 5-HT and 5-HIAA levels were lower during the light phase than during the dark phase of the light–dark cycle (A,B). Whereas for 5-HT no apparent differences between the genotypes were observed (A), significantly elevated levels of 5-HIAA were found in the homozygous CRF_1 mutant mice (B) over the whole diurnal rhythm. Figure C shows that forced swim stress increases hippocampal extracellular levels of 5-HT, followed by a further increase after termination of the stressor. The forced swim stress-induced rise in 5-HT was significantly enhanced in heterozygous and homozygous CRF_1 mutant mice as compared to wild-type controls. Values represent mean ± SEM. +, $P < 0.05$, effect of genotype (analysis of variance with repeated measures). §, $P < 0.05$, interaction between time period and genotype (analysis of variance with repeated measures). •, $P < 0.05$, homozygous mutant versus wild-type and heterozygous mutant mice. ••, $P < 0.05$, homozygous versus heterozygous mutant mice. *, $P < 0.05$, homozygous mutant versus wild-type mice. . , $P < 0.05$, heterozygous mutant versus wild-type mice (all posthoc tests: Duncan multiple range test for mean of period). Data are taken from Peñalva et al., 2002, with the permission of Neuroscience.

rate of serotonergic neurons) and the behavioural activity status of an animal, it has been proposed that stress-induced increases in the activity of the 5-HT system subserve the facilitation of motor output and coordination of concurrent adaptations in autonomic and neuroendocrine functioning and in sensory information processing (Jacobs and Fornal, 1999). In our experience, increases in extracellular levels of 5-HT in different brain structures (hippocampus, cortex, preoptic area, lateral septum, caudate nucleus, DRN) under normal and stress conditions are indeed often accompanied by enhanced behavioral activity. However, our data clearly suggest that the 5-HT system may play, via alternative mechanisms, additional roles in the regulation of homeostasis and in stress coping strategies. For instance, during an infectious challenge the animals display sickness behaviour and are immobile, notwithstanding a vast increase in extracellular levels of 5-HT and 5-HIAA in the hippocampus. Furthermore, the dramatic increase in hippocampal 5-HT in diving rats with a plastic collar around their neck cannot be explained on the sole basis of increased behavioural activity.

Conclusion

The studies described in this chapter underscore the concept that stress impacts on serotonergic neurotransmission at different levels, including neuronal firing rate, synthesis, release and metabolism of 5-HT. Although not discussed here, stress also affects serotonergic neurotransmission by specifically changing 5-HT pre- and postsynaptic receptors (for discussion the reader is referred to Chaouloff, 2000). CRF is in a unique neuroanatomical position to coordinate the effects of stress on the 5-HT system. Indeed, a role for CRF in stress-induced changes in 5-HT neurotransmission could be determined under defined circumstances. Moreover, we have provided evidence that long-term changes in the status of the CRF system can lead to alterations in the responsiveness of 5-HT to stressful challenges. However, the wide variety of stress models and the non-standardized experimental protocols used by the different research groups have frequently resulted in variable or conflicting results. Moreover, the neuroanatomical specificity (also with respect to the raphe nuclei) of the effects of different types of stress and of CRF has drawn relatively little attention. Hence, given the important role of 5-HT in a broad range of physiological processes and the putative defect of this system in stress-related psychiatric illness, future experiments should address the effects of stress on this neurotransmitter system in a greater and more systematic stressor- and neuroanatomical detail. Such studies will increase our knowledge on the role of 5-HT in 'healthy' stress processing and in maladaptive processes possibly underlying the development of stress-related psychiatric disorders. Moreover, a better insight into the relationship between stress, CRF and 5-HT may stimulate discussions on the improvement of strategies to cope with (chronic) stress in everyday private, school and professional life.

Abbreviations

5-HT	serotonin (5-hydroxytryptamine)
5-HIAA	5-hydroxyindoleacetic acid
5-HTP	5-hydroxytryptophan
AADC	L-amino acid decarboxylase
CRF	corticotropin-releasing factor
CRF-R	corticotropin-releasing factor receptor
CRF_1	corticotropin-releasing factor receptor type 1
CRF_2	corticotropin-releasing factor receptor type 2
DRN	dorsal raphe nucleus
GR	glucocorticoid receptor
HPA	hypothalamic–pituitary–adrenocortical
i.p.	intraperitoneal
i.c.v.	intracerebroventricular
LPS	lipopolysaccharide
MRN	median raphe nucleus
SSRI	selective serotonin reuptake inhibitor
TPH	tryptophan hydroxylase

Acknowledgements

I am indebted to the present and former members of my group for their continuous dedication to the work on serotonergic neurotransmission and stress. I am especially grateful to Ms. Cornelia Flachskamm for her unsurpassed technical assistance. Ms. Monika

Mailinger is thanked for her superb secretarial assistance. The work of the author described in this chapter has been supported by the Max Planck Society and the Volkswagen Foundation.

References

Adell, A., Casanovas, J.M. and Artigas, F. (1997) Comparative study in the rat of the actions of different types of stress on the release of 5-HT in raphe nuclei and forebrain areas. Neuropharmacology, 36: 735–741.

Aghajanian, G.K., Foote, W.E. and Sheard, M.H. (1968) Lysergic acid diethylamide: sensitive neuronal units in the midbrain raphe. Science, 161: 706–708.

Aghajanian, G.K. and Vandermaelen, C.P. (1982) Intracellular identification of central noradrenergic and serotonergic neurons by a new double labeling procedure. J. Neurosci., 2: 1786–1792.

Amat, J., Matusamat, P., Watkins, L.R. and Maier, S.F. (1998) Escapable and inescapable stress differentially and selectively alter extracellular levels of 5-HT in the ventral hippocampus and dorsal periaqueductal gray of the rat. Brain Res., 797: 12–22.

Azmitia, E.C., Jr. and McEwen, B.S. (1974) Adrenalcortical influence on rat brain tryptophan hydroxylase activity. Brain Res., 78: 291–302.

Beck, C.H. and Fibiger, H.C. (1995) Conditioned fear-induced changes in behavior and in the expression of the immediate early gene c-fos: with and without diazepam pretreatment. J. Neurosci., 15: 709–720.

Belzung, C., El Hage, W., Moindrot, N. and Griebel, G. (2001) Behavioral and neurochemical changes following predatory stress in mice. Neuropharmacology, 41: 400–408.

Bittencourt, J.C. and Sawchenko, P.E. (2000) Do centrally administered neuropeptides access cognate receptors? An analysis in the central corticotropin-releasing factor system. J. Neurosci., 20: 1142–1156.

Bittencourt, J.C., Vaughan, J., Arias, C., Rissman, R.A., Vale, W.W. and Sawchenko, P.E. (1999) Urocortin expression in rat brain: Evidence against a pervasive relationship of urocortin-containing projections with targets bearing type 2 CRF receptors. J. Comp. Neurol., 415: 285–312.

Blanchard, D.C., Griebel, G., Rodgers, R.J. and Blanchard, R.J. (1998) Benzodiazepine and serotonergic modulation of antipredator and conspecific defense. Neurosci. Biobehav. Rev., 22: 597–612.

Boadle-Biber, M.C. (1993) Regulation of serotonin synthesis. Prog. Biophys. Mol. Biol., 60: 1–15.

Boadle-Biber, M.C., Corley, K.C., Graves, L., Phan, T.H. and Rosecrans, J. (1989) Increase in the activity of tryptophan hydroxylase from cortex and midbrain of male Fischer 344 rats in response to acute or repeated sound stress. Brain Res., 482: 306–316.

Boadle-Biber, M.C., Johannessen, J.N., Narasimhachari, N. and Phan, T.H. (1986) Tryptophan hydroxylase: increase in activity by electrical stimulation of serotonergic neurons. Neurochem. Int., 8: 83–92.

Carrasco, G.A. and Van de Kar, L.D. (2003) Neuroendocrine pharmacology of stress. Eur. J. Pharmacol., 463: 235–272.

Chalmers, D.T., Lovenberg, T.W. and De Souza, E.B. (1995) Localization of novel corticotropin-releasing factor receptor (CRF_2) mRNA expression to specific subcortical nuclei in rat brain: comparison with CRF_1 receptor mRNA expression. J. Neurosci., 15: 6340–6350.

Chamas, F., Serova, L. and Sabban, E.L. (1999) Tryptophan hydroxylase mRNA levels are elevated by repeated immobilization stress in rat raphe nuclei but not in pineal gland. Neuroscience Letters, 267: 157–160.

Chaouloff, F. (2000) Serotonin, stress and corticoids. J. Psychopharmacol., 14: 139–151.

Chen, R., Lewis, K.A., Perrin, M.H. and Vale, W.W. (1993) Expression cloning of a human corticotropin-releasing-factor receptor. Proc. Natl. Acad. Sci. USA, 90: 8967–8971.

Connor, T.J., Kelliher, P., Shen, Y., Harkin, A., Kelly, J.P. and Leonard, B.E. (2000) Effect of subchronic antidepressant treatments on behavioral, neurochemical, and endocrine changes in the forced-swim test. Pharmacol. Biochem. Behav., 65: 591–597.

Connor, T.J., Song, C., Leonard, B.E., Merali, Z. and Anisman, H. (1998) An assessement of the effects of central interleukin-1-beta, -2, -6, and tumor necrosis factor-alpha administration on some behavioural, neurochemical, endocrine and immune parameters in the rat. Neuroscience, 84: 923–933.

Corley, K.C., Phan, T.H., Daugherty, W.P. and Boadle-Biber, M.C. (2002) Stress-induced activation of median raphe serotonergic neurons in rats is potentiated by the neurotensin antagonist, SR 48692. Neurosci. Lett., 319: 1–4.

Corley, K.C., Singh, V.B., Phan, T.H. and Boadle-Biber, M.C. (1992) Effect of gepirone on increases in tryptophan hydroxylase in response to sound stress. Eur. J. Pharmacol., 213: 417–425.

Cryan, J.F., Markou, A. and Lucki, I. (2002) Assessing antidepressant activity in rodents: recent developments and future needs. Trends Pharmacol. Sci., 23: 238–245.

Cullinan, W.E., Herman, J.P., Battaglia, D.F., Akil, H. and Watson, S.J. (1995) Pattern and time course of immediate early gene expression in rat brain following acute stress. Neuroscience, 64: 477–505.

Culman, J., Kiss, A. and Kvetnansky, R. (1984) Serotonin and tryptophan hydroxylase in isolated hypothalamic and brain stem nuclei of rats exposed to acute and repeated immobilization stress. Exp. Clin. Endocrinol., 83: 28–36.

Cummings, S., Elde, R., Ells, J. and Lindall, A. (1983) Corticotropin-releasing factor immunoreactivity is widely distributed within the central nervous system of the rat: an immunohistochemical study. J. Neurosci., 3: 1355–1368.

Dahlström, A. and Fuxe, K. (1964) Evidence for the existence of monoamine-containing neurons in the central nervous system. Acta Physiol. Scand., 62: 2–55.

Dantzer, R., Bluthé, R.M., Kent, S. and Kelley, K.W. (1991) Behavioural effects of cytokines. In: Rothwell, N. and Dantzer, R. (Eds.), Interleukin-1 in the brain, Pergamon Press, Oxford, pp. 135–150.

Daugherty, W.P., Corley, K.C., Phan, T.H. and Boadle-Biber, M.C. (2001) Further studies on the activation of rat median raphe serotonergic neurons by inescapable sound stress. Brain Res., 923: 103–111.

De Kloet, E.R. (1991) Brain corticosteroid receptor balance and homeostatic control. Front Neuroendocrinol., 12: 95–164.

Delgado, P.L., Charney, D.S., Price, L.H., Aghajanian, G.K., Landis, H. and Heninger, G.R. (1990) Serotonin function and the mechanism of antidepressant action. Reversal of antidepressant-induced remission by rapid depletion of plasma tryptophan. Arch. Gen. Psychiatry, 47: 411–418.

Dias Soares, D., Fernandez, F., Aguerre, S., Foury, A., Mormede, P. and Chaouloff, F. (2003) Fox odour affects corticosterone release but not hippocampal serotonin reuptake and open field behaviour in rats. Brain Res., 961: 166–170.

Dijkstra, I., Tilders, F.J.H., Aguilera, G., Kiss, A., Rabadandiehl, C., Barden, N., Karanth, S., Holsboer, F. and Reul, J.M.H.M. (1998) Reduced activity of hypothalamic corticotropin-releasing hormone neurons in transgenic mice with impaired glucocorticoid receptor function. J. Neurosci., 18: 3909–3918.

Dilts, R.P. and Boadle-Biber, M.C. (1995) Differential activation of the 5-hydroxytryptamine-containing neurons of the midbrain raphe of the rat in response to randomly presented inescapable sound. Neurosci. Lett., 199: 78–80.

Dinan, T.G. (1996) Serotonin and the regulation of hypothalamic-pituitary-adrenal axis function. Life Sci., 58: 1683–1694.

Dunn, A.J. (1992) Endotoxin-induced activation of cerebral catecholamine and serotonin metabolism: comparison with interleukin-1. J. Pharmacol. Exp. Therapeut., 261: 964–969.

Dunn, A.J. (1988) Stress-related changes in cerebral catecholamine and indoleamine metabolism: lack of effect of adrenalectomy and corticosterone. J. Neurochem., 51: 406–412.

Dunn, A.J. and Welch, J. (1991) Stress- and endotoxin-induced increases in brain tryptophan and serotonin metabolism depend on sympathetic nervous system activity. J. Neurochem., 57: 1615–1622.

Ehret, M., Cash, C.D., Hamon, M. and Maitre, M. (1989) Formal demonstration of the phosphorylation of rat brain tryptophan hydroxylase by Ca2+/calmodulin-dependent protein kinase. J. Neurochem., 52: 1886–1891.

Elmquist, J.K., Scammell, T.E., Jacobson, C.D. and Safer, C.B. (1996) Distribution of Fos-like immunoreactivity in the rat brain following intravenous lipopolysaccharide administration. J. Comp. Neurol., 371: 85–103.

Ericsson, A., Kovacs, K.J. and Sawchenko, P.E. (1994) A functional anatomical analysis of central pathways subserving the effects of interleukin-1 on stress-related neuroendocrine neurons. J. Neurosci., 14: 897–913.

Eriksson, E., Engberg, G., Bing, O. and Nissbrandt, H. (1999) Effects of mCPP on the extracellular concentrations of serotonin and dopamine in rat brain. Neuropsychopharmacology, 20: 287–296.

Fernstrom, J.D. and Wurtman, R.J. (1971) Brain serotonin content: physiological dependence on plasma tryptophan levels. Science, 173: 149–152.

Fornal, C.A., Metzler, C.W., Marrosu, F., Ribiero-do-Valle, L.E. and Jacobs, B.L. (1996) A subgroup of dorsal raphe serotonergic neurons in the cat is strongly activated during oral-buccal movements. Brain Res., 716: 123–133.

Fujino, K., Yoshitake, T., Inoue, O., Ibii, N., Kehr, J., Ishida, J., Nohta, H. and Yamaguchi, M. (2002) Increased serotonin release in mice frontal cortex and hippocampus induced by acute physiological stressors. Neurosci. Lett., 320: 91–95.

Gartside, S.E., Hajós-Korcsok, E., Bagdy, E., Harsing, L.G., Sharp, T. and Hajós, M. (2000) Neurochemical and electrophysiological studies on the functional significance of burst firing in serotonergic neurons. Neuroscience, 98: 295–300.

Grahn, R.E., Will, M.J., Hammack, S.E., Maswood, S., McQueen, M.B., Watkins, L.R. and Maier, S.F. (1999) Activation of serotonin-immunoreactive cells in the dorsal raphe nucleus in rats exposed to an uncontrollable stressor. Brain Res., 826: 35–43.

Hajós, M., Gartside, S.E., Villa, A.E.P. and Sharp, T. (1995) Evidence for a repetitive (burst) firing pattern in a subpopulation of 5-hydroxytryptamine neurons in the dorsal and median raphe nuclei of the rat. Neuroscience, 69: 189–197.

Hajos-Korcsok, E., Robinson, D.D., Yu, J.H., Fitch, C.S., Walker, E. and Merchant, K.M. (2003) Rapid habituation of hippocampal serotonin and norepinephrine release and anxiety-related behaviors, but not plasma corticosterone levels, to repeated footshock stress in rats. Pharmacol Biochem. Behav., 74: 609–616.

Hamon, M., Bourgoin, S., Artaud, F. and El Mestikawy, S. (1981) The respective roles of tryptophan uptake and tryptophan hydroxylase in the regulation of serotonin synthesis in the central nervous system. J. Physiol. (Paris), 77: 269–279.

Hart, B.L. (1988) Biological basis of the behavior of sick animals. Neurosci. Biobehav. Rev., 12: 123–137.

Hashimoto, S., Inoue, T. and Koyama, T. (1999) Effects of conditioned fear stress on serotonin neurotransmission and freezing behavior in rats. Eur. J. Pharm., 378: 23–30.

Hayley, S., Borowski, T., Merali, Z. and Anisman, H. (2001) Central monoamine activity in genetically distinct strains of mice following a psychogenic stressor: effects of predator exposure. Brain Res., 892: 293–300.

Heinsbroek, R.P., van Haaren, F., Feenstra, M.G., Boon, P. and van de Poll, N.E. (1991) Controllable and uncontrollable footshock and monoaminergic activity in the frontal cortex of male and female rats. Brain Res., 551: 247–255.

Herman, J.P. and Cullinan, W.E. (1997) Neurocircuitry of stress: central control of the hypothalamo-pituitary-adrenocortical axis. Trends Neurosci., 20: 78–84.

Herr, B.E., Gallager, D.W. and Roth, R.H. (1975) Tryptophan hydroxylase: activation in vivo following stimulation of central serotonergic neurons. Biochem. Pharmacol., 24: 2019–2023.

Inoue, T., Koyama, T. and Yamashita, I. (1993) Effect of conditioned fear stress on serotonin metabolism in the rat brain. Pharmacol. Biochem. Behav., 44: 371–374.

Inoue, T., Tsuchiya, K. and Koyama, T. (1994) Regional changes in dopamine and serotonin activation with various intensity of physical and psychological stress in the rat brain. Pharmacol. Biochem. Behav., 49: 911–920.

Ishida, Y., Hashiguchi, H., Takeda, R., Ishizuka, Y., Mitsuyama, Y., Kannan, H., Nishimori, T. and Nakahara, D. (2002) Conditioned-fear stress increases Fos expression in monoaminergic and GABAergic neurons of the locus coeruleus and dorsal raphe nuclei. Synapse, 45: 46–51.

Isogawa, K., Akiyoshi, J., Hikichi, T., Yamamoto, Y., Tsutsumi, T. and Nagayama, H. (2000) Effect of corticotropin releasing factor receptor 1 antagonist on extracellular norepinephrine, dopamine and serotonin in hippocampus and prefrontal cortex of rats in vivo. Neuropeptides, 34: 234–239.

Jacobs, B.L. and Azmitia, E.C. (1992) Structure and function of the brain serotonin system. Physiol. Rev., 72: 165–229.

Jacobs, B.L. and Fornal, C.A. (1999) Activity of serotonergic neurons in behaving animals. Neuropsychopharmacology, 21: S9–S15.

Johansen, P.A., Jennings, I., Cotton, R.G. and Kuhn, D.M. (1995) Tryptophan hydroxylase is phosphorylated by protein kinase A. J. Neurochem., 65: 882–888.

Kabiersch, A., del Rey, A., Honegger, C.G. and Besedovsky, H.O. (1988) Interleukin-1 induces changes in norepinephrine metabolism in the rat brain. Brain Behavior and Immunity, 2: 267–274.

Kalén, P., Rosegren, E., Lindvall, O. and Björklund, A. (1989) Hippocampal noradrenaline and serotonin release over 24 hours as measured by the dialysis technique in freely moving rats: correlation to behavioural activity state, effect of handling and tail-pinch. Eur. J. Neurosci., 1: 181–188.

Kirby, L.G., Allen, A.R. and Lucki, I. (1995) Regional differences in the effects of forced swimming on extracellular levels of 5-hydroxytryptamine and 5- hydroxyindoleacetic acid. Brain Res., 682: 189–196.

Kirby, L.G., Chou-Green, J.M., Davis, K. and Lucki, I. (1997) The effects of different stressors on extracellular 5-hydroxytryptamine and 5-hydroxyindoleacetic acid. Brain Res., 760: 218–230.

Kirby, L.G., Rice, K.C. and Valentino, R.J. (2000) Effects of corticotropin-releasing factor on neuronal activity in the serotonergic dorsal raphe nucleus. Neuropsychopharmacology, 22: 148–162.

Knott, P.J. and Curzon, G. (1972) Free tryptophan in plasma and brain tryptophan metabolism. Nature, 239: 452–453.

Kovacs, K.J. (1998) C-Fos as a transcription factor – a stressful (re)view from a functional map. Neurochem. Int., 33: 287–297.

Kozicz, T., Yanaihara, H. and Arimura, A. (1998) Distribution of urocortin-like immunoreactivity in the central nervous system of the rat. J. Comp. Neurol., 391: 1–10.

Kuhn, D.M. and Lovenberg, W. (1982) Role of calmodulin in the activation of tryptophan hydroxylase. Fed. Proc., 41: 2258–2264.

Lacosta, S., Merali, Z. and Anisman, H. (1999) Behavioral and neurochemical consequences of lipopolysaccharide in mice: anxiogenic-like effects. Brain Res., 818: 291–303.

Laflamme, N., Feuvrier, E., Richard, D. and Rivest, S. (1999) Involvement of serotonergic pathways in mediating the neuronal activity and genetic transcription of neuroendocrine corticotropin-releasing factor in the brain of systemically endotoxin-challenged rats. Neuroscience, 88: 223–240.

Lavicky, J. and Dunn, A.J. (1993) Corticotropin-releasing factor stimulates catecholamine release in hypothalamus and prefrontal cortex in freely moving rats as assessed by microdialysis. J. Neurochem., 60: 602–612.

Lavicky, J. and Dunn, A.J. (1995) Endotoxin administration stimulates cerebral catecholamine release in freely moving rats as assessed by microdialysis. J. Neurosci. Res., 40: 407–413.

Lewis, K., Li, C., Perrin, M.H., Blount, A., Kunitake, K., Donaldson, C., Vaughan, J., Reyes, T.M., Gulyas, J., Fischer, W., Bilezikjian, L., Rivier, J., Sawchenko, P.E. and Vale, W.W. (2001) Identification of urocortin III, an additional member of the corticotropin-releasing factor (CRF) family with high affinity for the CRF2 receptor. Proc. Natl. Acad. Sci. USA, 98: 7570–7575.

Li, C., Vaughan, J., Sawchenko, P.E. and Vale, W.W. (2002) Urocortin III-immunoreactive projections in rat brain: partial overlap with sites of type 2 corticotrophin-releasing factor receptor expression. J. Neurosci., 22: 991–1001.

Li, H.Y., Ericsson, A. and Sawchenko, P.E. (1996) Distinct mechanisms underlie activation of hypothalamic neurosecretory neurons and their medullary catecholaminergic afferents in categorically different stress paradigms. Proc. Natl. Acad. Sci. USA, 93: 2359–2364.

Linthorst, A.C.E., Flachskamm, C., Barden, N., Holsboer, F. and Reul, J.M.H.M. (2000) Glucocorticoid receptor impairment alters CNS responses to a psychological stressor: an in vivo microdialysis study in transgenic mice. Eur. J. Neurosci., 12: 283–291.

Linthorst, A.C.E., Flachskamm, C., Holsboer, F. and Reul, J.M.H.M. (1994) Local administration of recombinant human interleukin-1 beta in the rat hippocampus increases serotonergic neurotransmission, hypothalamic-pituitary-adrenocortical axis activity, and body temperature. Endocrinology, 135: 520–532.

Linthorst, A.C.E., Flachskamm, C., Holsboer, F. and Reul, J.M.H.M. (1995a) Intraperitoneal administration of bacterial endotoxin enhances noradrenergic neurotransmission in the rat preoptic area: relationship with body temperature and hypothalamic-pituitary-adrenocortical axis activity. Eur. J. Neurosci., 7: 2418–2430.

Linthorst, A.C.E., Flachskamm, C., Hopkins, S.J., Hoadley, M.E., Labeur, M.S., Holsboer, F. and Reul, J.M.H.M. (1997) Long-term intracerebroventricular infusion of corticotropin-releasing hormone alters neuroendocrine, neurochemical, autonomic, behavioral, and cytokine responses to a systemic inflammatory challenge. J. Neurosci., 17: 4448–4460.

Linthorst, A.C.E., Flachskamm, C., Müller-Preuss, P., Holsboer, F. and Reul, J.M.H.M. (1995b) Effect of bacterial endotoxin and interleukin-1 beta on hippocampal serotonergic neurotransmission, behavioral activity, and free corticosterone levels: an in vivo microdialysis study. J. Neurosci., 15: 2920–2934.

Linthorst, A.C.E., Flachskamm, C. and Reul, J.M.H.M. (2001) Hippocampal serotonin responses to forced swim stress in rats: influence of water temperature. In: Proceedings of the 31st Annual Meeting of the Society for Neuroscience. San Diego, CA, USA.

Linthorst, A.C.E., Peñalva, R.G., Flachskamm, C., Holsboer, F. and Reul, J.M.H.M. (2002) Forced swim stress activates rat hippocampal serotonergic neurotransmission involving a corticotropin-releasing hormone receptor-dependent mechanism. Eur. J. Neurosci., 16: 2441–2452.

Lovenberg, T.W., Liaw, C.W., Grigoriadis, D.E., Clevenger, W., Chalmers, D.T., De Souza, E.B. and Oltersdorf, T. (1995) Cloning and characterization of a functionally distinct corticotropin-releasing factor receptor subtype from rat brain. Proc. Natl. Acad. Sci. USA, 92: 836–840.

Lowry, C.A. (2002) Functional subsets of serotonergic neurones: implications for control of the hypothalamic-pituitary-adrenal axis. J. Neuroendocrinol., 14: 911–923.

Lowry, C.A., Johnson, P.L., Hathway, N.J.A. and Lightman, S.L. (2002) Distribution of corticotropin-releasing factor receptor 1-(CRFR1) and CRFR2-immunoreactivity in limbic and caudal brainstem raphe nuclei. Program No.867.6. 2002 Abstract Viewer/Itinerary Planner. Washington, DC, Society for Neuroscience, 2002. CD-ROM.

Lowry, C.A., Rodda, J.E., Lightman, S.L. and Ingram, C.D. (2000) Corticotropin-releasing factor increases in vitro firing rates of serotonergic neurons in the rat dorsal raphe nucleus: evidence for activation of a topographically organized mesolimbocortical serotonergic system. J. Neurosci., 20: 7728–7736.

Lucki, I. (1998) The spectrum of behaviors influenced by serotonin. Biol. Psychiatry, 44: 151–162.

Lysz, T.W. and Sze, P.Y. (1978) Activation of brain tryptophan hydroxylase by a phosphorylating system. J. Neurosci. Res., 3: 411–418.

Martinez, M., Phillips, P.J. and Herbert, J. (1998) Adaptation in patterns of c-fos expression in the brain associated with exposure to either single or repeated social stress in male rats. Eur. J. Neurosci., 10: 20–31.

Maes, M. and Meltzer, H.Y. (1995) The serotonin hypothesis of major depression. In: Bloom, F.E. and Kupfer, D.J. (Eds.), Psychopharmacology: The fourth Generation of Progress. pp. 933–944.

Mann, J.J. (1998) The neurobiology of suicide. Nature Medicine, 4: 25–30.

Mefford, I.N. and Heyes, M.P. (1990) Increased biogenic amine release in mouse hypothalamus following immunological challenge: antagonism by indomethacin. J. Neuroimmunol., 27: 55–61.

Merali, Z., Lacosta, S. and Anisman, H. (1997) Effects of interleukin-1-beta and mild stress on alterations of norepinephrine, dopamine and serotonin neurotransmission – a regional microdialysis study. Brain Res., 761: 225–235.

Mockus, S.M. and Vrana, K.E. (1998) Advances in the molecular characterization of tryptophan hydroxylase. J. Mol. Neurosci., 10: 163–179.

Mohankumar, P.S., Thyagarajan, S. and Quadri, S.K. (1993) Interleukin-1 beta increases 5-hydroxyindoleacetic acid release in the hypothalamus in vivo. Brain Res. Bulletin, 31: 745–748.

Morzorati, S.L. and Johnson, T.B. (1999) Serotonergic neuronal activity in the dorsal raphe nucleus of selectively bred alcohol-preferring and alcohol-nonpreferring rats and unselected wistar rats. Alcohol Clin. Exp. Res., 23: 1362–1367.

Nakahara, D. and Nakamura, M. (1999) Differential effect of immobilization stress on in vivo synthesis rate of monoamines in medial prefrontal cortex and nucleus accumbens of conscious rats. Synapse, 32: 238–242.

Nemeroff, C.B., Owens, M.J., Bissette, G., Andorn, A.C. and Stanley, M. (1988) Reduced corticotropin releasing factor binding sites in the frontal cortex of suicide victims. Arch. Gen. Psychiatry, 45: 577–579.

Nemeroff, C.B., Widerlov, E., Bissette, G., Walleus, H., Karlsson, I., Eklund, K., Kilts, C.D., Loosen, P.T. and Vale, W. (1984) Elevated concentrations of CSF corticotropin-releasing factor-like immunoreactivity in depressed patients. Science, 226: 1342–1344.

Oshima, A., Flachskamm, C., Reul, J.M.H.M., Holsboer, F. and Linthorst, A.C.E. (2003) Altered serotonergic neurotransmission but normal hypothalamic-pituitary-adrenocortical axis activity in mice chronically treated with the corticotropin-releasing hormone receptor type 1 antagonist NBI 30775. Neuropsychopharmacology, 28: 2148–2159.

Palkovits, M., Brownstein, M., Kizer, J.S., Saavedra, J.M. and Kopin, I.J. (1976) Effect of stress on serotonin concentration and tryptophan hydroxylase activity of brain nuclei. Neuroendocrinology, 22: 298–304.

Pauli, S., Linthorst, A.C.E. and Reul, J.M.H.M. (1998) Tumour necrosis factor-alpha and interleukin-2 differentially affect hippocampal serotonergic neurotransmission, behavioural activity, body temperature and hypothalamic-pituitary-adrenocortical axis activity in the rat. European J. Neurosci., 10: 868–878.

Pei, Q., Zetterstrom, T. and Fillenz, M. (1990) Tail pinch-induced changes in the turnover and release of dopamine and 5-hydroxytryptamine in different brain regions of the rat. Neuroscience, 35: 133–138.

Peñalva, R.G., Flachskamm, C., Zimmermann, S., Wurst, W., Holsboer, F., Reul, J.M.H.M. and Linthorst, A.C.E. (2002) Corticotropin-releasing hormone receptor type 1-deficiency enhances hippocampal serotonergic neurotransmission: an in vivo microdialysis study in mutant mice. Neuroscience, 109: 253–266.

Peñalva, R.G., Lancel, M., Flachskamm, C., Reul, J.M.H.M., Holsboer, F. and Linthorst, A.C.E. (2003) Effect of sleep and sleep depriviation on serotonergic neurotransmission in the hippocampus: a combined in vivo microdialysis/EEG study in rats. Eur. J. Neurosci., 17: 1896–1906.

Pezzone, M.A., Lee, W.S., Hoffman, G.E., Pezzone, K.M. and Rabin, B.S. (1993) Activation of brainstem catecholaminergic neurons by conditioned and unconditioned aversive stimuli as revealed by c-Fos immunoreactivity. Brain Res., 608: 310–318.

Porsolt, R.D., Le Pichon, M. and Jalfre, M. (1977) Depression: a new animal model sensitive to antidepressant treatments. Nature, 266: 730–732.

Price, M.L., Curtis, A.L., Kirby, L.G., Valentino, R.J. and Lucki, I. (1998) Effects of corticotropin-releasing factor on brain serotonergic activity. Neuropsychopharmacology, 18: 492–502.

Price, M.L., Kirby, L.G., Valentino, R.J. and Lucki, I. (2002) Evidence for corticotropin-releasing factor regulation of serotonin in the lateral septum during acute swim stress: adaptation produced by repeated swimming. Psychopharmacology (Berl), 162: 406–414.

Price, M.L. and Lucki, I. (2001) Regulation of serotonin release in the lateral septum and striatum by corticotropin-releasing factor. J. Neurosci., 21: 2833–2841.

Raadsheer, F.C., Hoogendijk, W.J.G., Stam, F.C., Tilders F.J.H. and Swaab, D.F. (1994) Increased numbers of corticotropin-releasing hormone expressing neurons in the hypothalamic paraventricular nucleus of depressed patients. Neuroendocrinology, 60: 436–444.

Raadsheer, F.C., Van Heerikhuize, J.J., Lucassen, P.J., Hoogendijk, W.J.G., Tilders, F.J.H. and Swaab, D.F. (1995) Corticotropin-releasing hormone mRNA levels in the paraventricular nucleus of patients with Alzheimer's disease and depression. Am. J. Psychiatry, 152: 1372–1376.

Rasmussen, K., Heym, J. and Jacobs, B.L. (1984) Activity of serotonin-containing neurons in nucleus centralis superior of freely moving cats. Exp. Neurol., 83: 302–317.

Ressler, K.J. and Nemeroff, C.B. (2000) Role of serotonergic and noradrenergic systems in the pathophysiology of depression and anxiety disorders. Depression & Anxiety, 12: 2–19.

Reul, J.M.H.M. and Holsboer, F. (2002) Corticotropin-releasing hormone receptors 1 and 2 in anxiety and depression. Curr. Opin. Pharmacol., 2: 23–33.

Reyes, T.M., Lewis, K., Perrin, M.H., Kunitake, K.S., Vaughan, J., Arias, C.A., Hogenesch, J.B., Gulyas, J., Rivier, J., Vale, W.W. and Sawchenko, P.E. (2001) Urocortin II: a member of the corticotropin-releasing factor (CRF) neuropeptide family that is selectively bound by type 2 CRF receptors. Proc. Natl. Acad. Sci. USA, 98: 2843–2848.

Roche, M., Commons, K.G., Peoples, A. and Valentino, R.J. (2003) Circuitry underlying regulation of the serotonergic system by swim stress. J. Neurosci., 23: 970–977.

Rueter, L.E. and Jacobs, B.L. (1996) A microdialysis examination of serotonin release in the rat forebrain induced by behavioral environmental manipulations. Brain Res., 739: 57–69.

Ruggiero, D.A., Underwood, M.D., Rice, P.M., Mann, J.J. and Arango, V. (1999) Corticotropic-releasing hormone and serotonin interact in the human brainstem: behavioral implications. Neuroscience, 91: 1343–1354.

Sakanaka, M., Shibasaki, T. and Lederis, K. (1987) Corticotropin releasing factor-like immunoreactivity in the rat brain as revealed by a modified cobalt-glucose oxidase-diaminobenzidine method. J. Comp. Neurol., 260: 256–298.

Shields, P.J. and Eccleston, D. (1972) Effects of electrical stimulation of rat midbrain on 5-hydroxytryptamine synthesis as determined by a sensitive radioisotope method. J. Neurochem., 19: 265–272.

Shimizu, H. and Bray, G.A. (1989) Modulation by corticotropin-releasing factor of monoamine metabolism in the lateral hypothalamus. Neurosci. Lett., 103: 74–80.

Shintani, F., Kanba, S., Nakaki, T., Nibuya, M., Kinoshita, N., Suzuki, E., Yagi, G., Kato, R. and Asai, M. (1993) Interleukin-1 beta augments release of norepinephrine, dopamine, and serotonin in the rat anterior hypothalamus. J. Neurosci., 13: 3574–3581.

Silveira, M.C., Sandner, G. and Graeff, F.G. (1993) Induction of Fos immunoreactivity in the brain by exposure to the elevated plus-maze. Behav. Brain Res., 56: 115–118.

Singewald, N., Kaehler, S., Hemeida, R. and Philippu, A. (1997) Release of serotonin in the rat locus coeruleus: effects of cardiovascular, stressful and noxious stimuli. Eur. J. Neurosci., 9: 556–562.

Singh, V.B., Corley, K.C., Phan, T.H. and Boadle-Biber, M.C. (1990) Increases in the activity of tryptophan hydroxylase from rat cortex and midbrain in response to acute or repeated sound stress are blocked by adrenalectomy and restored by dexamethasone treatment. Brain Res., 516: 66–76.

Singh, V.B., Hao-Phan, T., Corley, K.C. and Boadle-Biber, M.C. (1992) Increase in cortical and midbrain tryptophan hydroxylase activity by intracerebroventricular administration of corticotropin releasing factor: block by adrenalectomy, by RU 38486 and by bilateral lesions to the central nucleus of the amygdala. Neurochem. Int., 20: 81–92.

Steinbusch, H.W.M (1981) Distribution of serotonin-immunoreactivity in the central nervous system of the rat-cell bodies and terminals. Neuroscience, 6: 557–618.

Stenfors, C. and Ross, S.B. (2002) Evidence for involvement of protein kinases in the regulation of serotonin synthesis and turnover in the mouse brain in vivo. J. Neural Transm., 109: 1353–1363.

Stone, E.A. (1970) Behavioral and neurochemical effects of acute swim stress are due to hypothermia. Life Sciences – Part 1 – Physiology & Pharmacology, 9: 877–888.

Swanson, L.W., Sawchenko, P.E.., Rivier, J. and Vale, W.W. (1983) Organization of ovine corticotropin-releasing factor immunoreactive cells and fibres in the rat brain: an immunohistochemical study. Neuroendocrinology, 36: 165–186.

Timpl, P., Spanagel, R., Sillaber, I., Kresse, A., Reul, J.M.H.M, Stalla, G.K., Blanquet, V., Steckler, T., Holsboer, F. and Wurst, W. (1998) Impaired stress response and reduced anxiety in mice lacking a functional corticotropin-releasing hormone receptor 1. Nature Genetics, 19: 162–166.

Törk, I. (1990) Anatomy of the serotonergic system. Ann. N.Y. Acad. Sci., 600: 9–35.

Vahabzadeh, A. and Fillenz, M. (1994) Comparison of stress-induced changes in noradrenergic and serotonergic neurons in the rat hippocampus using microdialysis. Eur. J. Neurosci., 6: 1205–1212.

Valentino, R.J., Liouterman, L., Van Bockstaele, E.J. (2001) Evidence for regional heterogeneity in corticotropin-releasing factor interactions in the dorsal raphe nucleus. J. Comp. Neurol., 435: 450–463.

Van Loon, G.R., Shum, A. and Ho, D. (1982) Lack of effect of corticotropin releasing factor on hypothalamic dopamine and serotonin synthesis turnover rates in rats. Peptides, 3: 799–803.

Van Pett, K., Viau, V., Bittencourt, J.C., Chan, R.K.W., Li, H.Y., Arias, C., Prins, G.S., Perrin, M., Vale, W. and Sawchenko, P.E. (2000) Distribution of mRNAs encoding CRF receptors in brain and pituitary of rat and mouse. J. Comp. Neurol., 428: 191–212.

Vaughan, J., Donaldson, C., Bittencourt, J., Perrin, M.H., Lewis, K., Sutton, S., Chan, R., Turnbull, A.V., Lovejoy, D., Rivier, C., Rivier, J., Sawchenko, P.E. and Vale, W. (1995) Urocortin, a mammalian neuropeptide related to fish urotensin I and to corticotropin-releasing factor. Nature, 378: 287–292.

Walther, D.J., Peter, J.U., Bashammakh, S., Hörtnagl, H., Voits, M., Fink, H. and Bader, M. (2003) Synthesis of serotonin by a second tryptophan hydroxylase isoform. Science, 299: 76.

Watanabe, Y., Stone, E. and Mcewen, B.S. (1994) Induction and habituation of c-fos and zif/268 by acute and repeated stressors. Neuroreport, 5: 1321–1324.

Wilkinson, L.O., Auerbach, S.B. and Jacobs, B.L. (1991) Extracellular serotonin levels change with behavioral state but not with pyrogen-induced hyperthermia. J. Neurosci., 11: 2732–2741.

Wilkinson, L.S., Humby, T., Killcross, S., Robbins, T.W. and Everitt, B.J. (1996) Dissociations in hippocampal 5-hydroxytryptamine release in the rat following pavlovian aversive conditioning to discrete and contextual stimuli. Eur. J. Neurosci., 8: 1479–1487.

Winberg, S., Myrberg, A.A., Jr. and Nilsson, G.E. (1993) Predator exposure alters brain serotonin metabolism in bicolour damselfish. Neuroreport, 4: 399–402.

Yoshioka M., Matsumoto M., Togashi H. and Saito H. (1995) Effects of conditioned fear stress on 5-HT release in the rat prefrontal cortex. Pharmacol. Biochem. Behav., 51: 515–519.

Zalcman, S., Green-Johnson, J.M., Murray, L., Nance, D.M., Dyck, D., Anisman, H. and Greenberg, A.H. (1994) Cytokine-specific central monoamine alterations induced by interleukin-1, -2 and -6. Brain Res., 643: 40–49.

T. Steckler, N.H. Kalin and J.M.H.M. Reul (Eds.)
Handbook of Stress and the Brain, Vol. 15
ISBN 0-444-51173-3

CHAPTER 4.7

Modulation of glutamatergic and GABAergic neurotransmission by corticosteroid hormones and stress

Marian Joëls*, Harmen J. Krugers and J. Martin Verkuyl

Swammerdam Institute for Life Sciences, Section Neurobiology, University of Amsterdam, The Netherlands

Abstract: This chapter highlights the current views on corticosteroid modulation of amino acid-mediated transmission. Amino acids such as glutamate or γ-amino butyric acid (GABA) are responsible for most of the excitatory and inhibitory transmission in brain. By targeting these systems, corticosteroid hormones could alter information processing in a very effective way.

So far, data were mostly collected in the rodent hippocampus, which expresses mineralo- as well as glucocorticoid receptors. In general, occupation of mineralocorticoid receptors is found to be necessary to maintain neuronal integrity, stable amino acid-mediated transmission and synaptic potentiation. In the CA1 area, additional activation of glucocorticoid receptors suppresses amino acid-mediated responses to synaptic stimulation, administered either with low-frequency (basal transmission) or with high-frequency patterns. Probably, glutamate as well as GABA-mediated responses are suppressed. Effects were usually reported for high doses of the hormone, found to be relatively fast in onset and are thus possibly caused by a nongenomic pathway. Mild stressors affected amino acid transmission in the hippocampus in vivo similarly as corticosteroid administration in vitro, although it is not clear (i) whether the in vivo effects are partly due to an altered availability of the amino acids – e.g., by changes in release or uptake –, (ii) whether these effects can be exclusively ascribed to stress-induced rise in corticosteroid hormones, and (iii) to what extent relay of areas other than the hippocampus is involved. Chronic stress was found to raise the glutamate-mediated transmission in all hippocampal subareas, albeit caused by regionally different mechanisms. Functional changes in amino acid-mediated pathways in other brain regions have been much less investigated, although recent data indicate that moderately high doses of corticosterone suppress GABAergic innervation of parvocellular neurons in the hypothalamic paraventricular nucleus, the key site for regulation of hypothalamo–pituitary–adrenal activity.

Although amino acid projections (and functional processes depending on them) clearly seem to be modulated by corticosteroid hormones and stress, the mechanism underlying this modulation is virtually unknown. Future studies will need to clarify the signaling pathways underlying corticosteroid effects on amino acid transmission and will have to address the functional consequences for behavioral processes.

*Corresponding author. Tel.: +31-20-5257626;
Fax: +31-20-5257709; E-mail: joels@science.uva.nl

Introduction

Binding of corticosteroid hormones to their receptors in brain, i.e. the high-affinity mineralocorticoid receptor (MR) and the glucocorticoid receptor (GR) with approximately tenfold lower affinity (review by De Kloet, 1991), results in transcriptional regulation of responsive genes (Beato and Sanchez-Pacheco, 1996). Many studies over the past decade have shown that these transcriptional effects strongly modulate the efficacy of neurotransmitter systems in brain (Joëls, 1997). In particular the central serotonergic system but also the cholinergic and (nor)adrenergic systems (see elsewhere, this volume) were found to be sensitive to effects exerted by stress, more specifically by corticosteroid hormones. Far less is known about the sensitivity of central amino acid-mediated projections to corticosteroids and stress. This is surprising, since glutamate and γ-amino butyric acid (GABA) are by far the most prevalent excitatory and inhibitory transmitters, respectively, in brain. Therefore, gene-mediated modulation of these neurotransmitter systems would be an extremely powerful way to modulate central information processing. As it is, only bits and pieces are known about the effect of corticosteroids and stress on amino acid-mediated transmission and at this moment it is not even clear whether the effects observed so far are due to transcriptional regulation or mediated by another, nongenomic mechanism. A nongenomic pathway was, for instance, found to be involved in the modulatory effects of 5-α-reduced metabolites of classical steroids on ionotropic receptors for GABA and glutamate. These so-called 'neurosteroid' actions are not discussed here; for this subject we refer to dedicated reviews (Baulieu et al., 1996; Lambert et al., 2001; see also elsewhere in this volume).

Several aspects of amino acid transmission could form a target for corticosteroid modulation: (i) enzymes involved in the formation or conversion of glutamate/GABA; (ii) the release of amino acids; (iii) local reuptake mechanisms; (iv) amino acid receptors; and (v) signaling pathways involved in the functional response to amino acids. The former three together determine the extracellular concentrations that can be determined e.g., with microdialysis. Interpretation of these levels is complicated by the fact that amino acids are also part of the Kreb's cycle, and their levels in part depend on local metabolic activity. Studies on these three aspects of amino acid transmission are briefly reviewed in the next section. The latter two aspects (amino acid receptors and the signaling pathways linked to the receptors) have been investigated mostly at the RNA level with in situ hybridization and at the protein level with either biochemical or electrophysiological methods. Most of this chapter is dedicated to these studies.

Corticosteroid modulation of amino acid-mediated transmission has been primarily investigated in the hippocampal formation. The hippocampus is very suitable for electrophysiological recording due to its laminar structure and trisynaptic circuit involving three major glutamatergic pathways. Amino acid-mediated projections in the hippocampus are involved in basal information flow in all hippocampal subfields but also play a key role in long-term potentiation and depression, two forms of synaptic plasticity that are presumed to contribute to learning and memory formation (Bliss and Collingridge, 1993; Bear and Abraham, 1996). Also, the hippocampus expresses high amounts of MRs as well as GRs in principal neurons (Reul and de Kloet, 1985). A second area in which amino acid transmission in relation to stress and corticosteroid actions has been investigated is the paraventricular nucleus (PVN) of the hypothalamus. The parvocellular neurons in this nucleus are key regulators of the hypothalamo–pituitary–adrenal (HPA) axis and receive a dense GABAergic inhibition, which (amongst other things) relays input from higher brain regions including the hippocampus (Herman and Cullinan, 1997; see also elsewhere, this volume). Investigations in other brain regions are relatively limited and are not reviewed here. Interestingly though, recent studies suggest that corticosteroid hormones effectively modulate long-term potentiation in amygdala nuclei (Akirav and Richter-Levin, 2002) and may thus exert important influences on activity in the hippocampus (Akirav and Richter-Levin, 1999; Kim et al., 2001).

The final part of this review is dedicated to a summary of the present state of affairs and a discussion about the mechanism that may underlie the observed phenomena. Open questions and the experiments necessary to answer these questions are outlined.

Levels of amino acids: modulation by corticosteroids and stress

Functional studies examining the effect of corticosteroids and/or stress on amino acid-mediated transmission (next section) usually address the efficacy of the postsynaptic machinery (receptors, signaling pathways), especially when these studies are performed in vitro with continuous perfusion. However, to fully appreciate how corticosteroids or stress modulate amino acid-mediated transmission in vivo it is also important to monitor the availability of the transmitters.

The availability of amino acids is first determined by the enzymes that synthesize or convert the transmitters. Of these, two important enzymes have been studied in some detail, i.e., glutamine synthetase and glutamic acid decarboxylase (GAD; see Table 1). Glutamine synthetase protein or activity was found to be either unaffected (Tombaugh and Sapolsky, 1990) or increased by corticosteroids (Patel et al., 1983; Jackson et al., 1995). When tested in an isolated preparation – rat astrocyte cultures – the latter turned out to be a nongenomic effect (Jackson et al., 1995). Results with GAD were more equivocal, although often an increased GAD activity or mRNA expression was observed (Yoneda et al., 1983; Otero Losada, 1988; Maroulakou and Stylianopoulou, 1991; Bowers et al., 1998; Herman and Larson, 2001; Stone et al., 2001). However, decreases in GAD activity or mRNA expression were also reported (Otero Losada, 1988; Acosta et al., 1993; Herman and Larson, 2001; Stone et al., 2001). Variation in the results can be ascribed to the area in which the measurements were carried out, the stress paradigm that was applied (e.g., acute versus chronic stress) and the GAD isoform that was investigated. Collectively, the data suggest that glutamate is more effectively converted to glutamine by corticosterone, while stress or corticosterone generally stimulate the conversion of glutamate into GABA. If so, stress-induced rises in corticosteroids would result in more GABA and less glutamate.

Extracellular levels of amino acids, however, are also largely determined by high-affinity uptake in glial cells or presynaptic terminals (overview in Table 2). Uptake of GABA was found to be decreased or unaffected (e.g., in hippocampus) by acute or chronic stress (Otero Losada, 1989; Acosta et al., 1993). Only in striatum an increase in GABA uptake was reported (Otero Losada, 1989). Glutamate uptake was generally increased after corticosteroid treatment or stress in cortex and limbic areas but not striatum (Gilad et al., 1990; Zhu et al., 1998). The effects of stress/corticosteroid on GABA and glutamate uptake were in most cases fast in onset and probably nongenomic in nature (Zhu et al., 1998). A slower decrease in glutamate uptake was found in one study (Virgin et al., 1991); this effect may have occurred secondary to a glucocorticoid-dependent downregulation of glucose-transporters. Since high-affinity uptake of amino acids is energy dependent, glucose availability may be a rate-limiting factor in the uptake process (see for extensive literature on this view: Sapolsky, 1992). Nevertheless, the majority of the studies point to a decreased GABA uptake and increased glutamate uptake after stress or corticosteroid treatment.

Extracellular concentrations of the amino acids also depend on release of the transmitters. After acute exposure to stressful events increased glutamate release was observed in synaptosomes prepared from hippocampus, frontal cortex, and septum but not striatum (Gilad et al., 1990). Effects were visible after 30 min and peaked after 1 h. Possibly, stress-induced enhancements of neurotransmitter release can explain the rise in extracellular levels of glutamate seen in several studies (Table 3). Most of these studies employed push–pull cannulas or more recently microdialysis in freely moving rats. Thus, corticosterone and dexamethasone, given either intraperitoneally or intrahippocampally, were found to increase glutamate and aspartate but not GABA levels in hippocampus through a nongenomic pathway (Venero and Borrell, 1999). In agreement, stress evoked a corticosterone-dependent and relatively fast increase in the hippocampal glutamate concentration and a somewhat slower rise in aspartate concentration (Gilad et al., 1990; Abraham et al., 1998). Similar stress-dependent rises in extracellular excitatory amino acid concentrations were also found for the locus coeruleus (Singewald et al., 1995) and the frontal cortex (Gilad et al., 1990; Moghaddam, 1993). In general, results on extracellular GABA concentrations are extremely variable: a stress-induced rise, decrease as well as no change were observed,

Table 1. Effect of corticosteroids, and acute or chronic stress on amino acid synthesizing and degrading enzymes

Experiment design	Effect	Study
DEX (10^{-7}–10^{-5} M); neonatal rat atrocyte culture	Glutamine synthetase mRNA and activity increased; nongenomic	Jackson et al. (1995)
CORT treatment in vivo; 11 and 20 day old rats	Glutamine synthetase protein levels increased. Most clearly in young animals, particularly in cerebellum	Patel et al. (1983)
CORT injections: prenatally and/or at PND48 and PND60; study 24 h or 5 days later, in hippocampus	After 24 h: GAD 65 mRNA decreased, only when CORT was given postnatally. No effect on GAD 67 mRNA. After 5 d: GAD 67 mRNA increased	Stone et al. (2001)
ADX, ADX + CORT, stress	In hippocampus: no change in glutamine synthetase activity	Tombaugh and Sapolsky (1990)
ADX, control and thermal stress; measurements in synaptosomes	GAD activity decreased by ADX in striatum. Thermal stress: increased GAD activity in cortex, hypothalamus, hippocampus and striatum. Same effects in control and ADX: not via CORT?	Maroulakou and Stylianpoulou (1991)
Immobilization stress for 3 h	GAD activity in striatum and hypothalamus increased; effects disappeared after 12 h	Yoneda et al. (1983)
Chronic intermittent stress; young middle aged, and aged Fisher/Brown Norway hybrid rats	Basal GAD 65 mRNA increased in medical preoptic area and pBNST, in aged rats. Chronic stress decreased GAD 65 mRNA in hypothalamus of aged rats	Herman and Larson (2001)
Acute or chronic intermittent stress in rats	Acute stress: GAD 67 mRNA increased in arcuate nucleus, dorsomedial nucleus hypothalamus BST, hippocampus; GAD 65 mRNA increased in BST and dentate gryus Peak change: 1 h; back to normal at 2 h Chronic stress: GAD65 mRNA increased in dorsomedical nucleus, medical preoptic area, SCN, aBST; GAD 67 mRNA: increased in medical preoptic area, aBST, hippocampus	Bowers et al. (1998)
Acute and chronic cold stress	Acute stress: GAD activity decreased in olfactory bulb Chronic stress: GAD activity decreased in olfactory bulb and striatum	Acosta et al. (1993)
Acute stress (cold or immobilization); chronic immobilization stress	Acute stress: GAD activity decreased in striatum Chronic stress: GAD activity increased in frontal cortex	Otero Losada (1988)

sometimes even within the same area (Yoneda et al., 1983; Otero Losada, 1988, 1989; Acosta et al., 1993; Singewald et al., 1995; Venero and Borrell, 1999).

In summary: Although corticosteroids/stress were generally seen to increase the extracellular concentration of glutamate and not GABA, it is not quite clear how this rise is accomplished. Studies on synthesizing and degrading enzymes rather favor a rise in GABA at the cost of glutamate, while uptake studies also support this. All of these effects develop rapidly and are probably nongenomic. Possibly, stress stimulates synaptic release of glutamate. Alternatively, release from nonvesicular sources could contribute to the rise in glutamate. This could

Table 2. Effect of corticosteroids, and acute and chronic stress on amino acid uptake

Experimental design	Effect	Study
CORT: 10^{-6} M; DEX: 10^{-9}–10^{-7} M; Synaptosomes from cortex, neuroblastoma cells	DEX and CORT: increased GLU uptake; rapid effects, via G-protein	Zhu et al. (1998)
High dose of CORT hippocampal astrocyte culture; secondary to glucose problem	In culture from hippocampus, but not cortex or cerebellum: high dose of GC gave decreased GLU-uptake; seen after 4 h	Virgin et al. (1991)
Acute immobilization stress	GABA uptake decreased in olf. bulb, but increased in striatum; no change in frontal cortex, hippocampus, hypothalamus; effects after 5 min	Otero Losada (1989)
Acute restraint stress; synaptosomes	GLU uptake increased in frontal cortex, septum and hippocampus; not in striatum 30 min after stress, plateau after 1 h	Gilad et al. (1990)
Acute and chronic cold stress	Acute stress; decreased GABA uptake in frontal cortex, hypothalamus, olf. bulb. Chronic stress: decreased uptake in hypothalamus	Acosta et al. (1993)

concern amino acids that gain access to the extracellular space through reversal of high-affinity transporters. It could also involve amino acids that are not related to transmitter pools but are a by-product of the Kreb's cycle and rather reflect the local metabolic status. Most studies did not discriminate between these two pools of amino acids. In this view, changes in amino acid levels develop secondary to stress/corticosteroid-dependent effects on cell metabolism (Horner et al., 1990). It is not at all clear, though, whether such extrasynaptic rises in amino acid levels will result in altered postsynaptic function of the transmitters.

Hippocampus

Corticosteroid hormones clearly affect basal (low-frequency) amino acid-mediated synaptic transmission in several hippocampal areas. Recent evidence also demonstrates that long-term potentiation as well as depression – which involve glutamatergic transmission (Bliss and Collingridge, 1993; Bear and Abraham, 1996) – are strongly modulated by corticosteroids (reviewed by Kim and Diamond, 2002). Below we give an overview of studies that have appeared over the past two decades. For each hippocampal subfield, we start by describing the effects exerted by exogenously administered corticosteroid hormones and by acute stress – giving rise to elevated levels of the endogenous hormone. This comparison allows a first impression to what extent effects of acute stress are caused by corticosteroids or by other compounds such as corticotropin-releasing factors (CRF) or factors related to activation of the sympathetic system. Next, we give an overview of the effects exerted by chronic stress. Very often these effects are not merely an extrapolation of acute stress effects, but rather illustrate the aberrations that develop once an organism is chronically exposed to stressful events.

Basal neurotransmission

CA1 region

In the CA1 area, corticosteroid effects were determined with regard to Schaffer collateral fibers. Schaffer collateral stimulation activates glutamatergic afferent projections, but in addition also feedforward and feedback inhibitory (GABAergic) projections. This results in a sequential excitatory postsynaptic potential (EPSP), a fast GABAa recepor-mediated inhibitory postsynaptic potential (IPSP) and a slow GABAb receptor-mediated IPSP. Adrenalectomy

Table 3. Effect of corticosteroids, and acute and chronic stress on extracellular levels of glutamate and GABA

Experimental design	Effect	Study
CORT, DEX ip or intrahippocampal: microdialysis in freely moving rats; measured in CA1 hippocampus	[ASP], [GLU] increased (15–45 min). [GABA] not changed; nongenomic	Venero and Borrell (1999)
Acute immobilization stress	[GABA] decreased in olfactory bulb, no changes in frontal cortex, hippocampus, mediobasal hypothalamus; effects after 5 min.	Otero Losada (1989)
Immobilization stress for 3 h	[GABA] increased in striatum and hypothalamus, [GLU] decreased; effects disappeared after 12 hrs	Yoneda et al. (1983)
Ether stress: sham or ADX + CORT animals; microdialysis in hippocampus	[GLU] increased after 30 min, [ASP] after 120 min; parallels CORT peak; effect not seen in CORT replaced ADX rats; depends on CORT peak after stress	Abraham et al. (1998)
Tail pinch or immobilization stress–push pull cannula in locus coeruleus	Tail pinch: GLU], [ASP] after 120 min; parallels CORT peak; effect not seen in CORT replaced ADX rats; depends on CORT peak after stress	Singewald et al. (1995)
20 min restraint or swim stress; microdialysis in prefrontal cortex, hippocampus, striatum, n. accumbens	[GLU] increased after both stressors, particularly in prefrontal cortex; similar though less marked result for [ASP]; less clear effects in striatum	Moghaddam (1993)
Restraint stress: synaptosomes	[GLU] release increased in frontal cortex, hippocampus, septum, not in striatum; effect after 30 min., plateau after 1 hr	Gilad et al. (1990)
Acute and chronic cold stress	Acute stress: [GABA] decreased in striatum; chronic stress: [GABA] decreased in frontal cortex, hypothalamus, olf. Bulb	Acosta et al. (1993)
Acute and chronic immobilization	Acute stress: [GABA] decreased in striatum; turnover down. Chronic stress: [GABA] decreased in frontal cortex; turnover up.	Otero Losada (1988)

causes a decline in synaptic responses evoked by stimulation of Schaffer collaterals, particularly with repeated low-frequency stimulation (Joëls and de Kloet, 1993; Birnstiel et al., 1995). It was proposed that the latter may be due to general metabolic disturbances rather than a specific modulation of glutamatergic and GABAergic transmission (Joëls and de Kloet, 1993). The attenuation was fully prevented by activation of MRs. In general, corticosterone application in doses that supposedly activate predominantly MRs resulted in relatively large and stable responses to synaptic activation (Reiheld, 1984; Rey et al., 1987, 1989; Joëls and de Kloet, 1993). By contrast, additional GR activation by high doses of corticosterone in most studies suppressed both GABA- and glutamate-mediated transmission, in field potential as well as single-cell recording studies (Vidal et al., 1986; Rey et al., 1987, 1989; Zeise et al., 1992; Joëls and de Kloet, 1993; Birnstiel et al., 1995).

The mechanism underlying corticosteroid modulation of amino acid transmission in the CA1 area is unknown. First of all, glutamate and GABA receptor-mediated components of the synaptic responses were not always pharmacologically separated; since their timecourse partially overlaps, effects on glutamatergic transmission may have been confounded by effects on GABAergic transmission (or vice versa). With regard to the excitatory responses, pharmacological distinction between AMPA and NMDA receptor-mediated events was only made in one preliminary study, showing that corticosterone rapidly and reversibly enhances the frequency of miniature postsynaptic excitatory currents, whereas at the longterm it changes the amplitude rather than frequency of these miniature synaptic currents (Karst et al., 2003). Most studies employed high doses of corticosterone and also reported relatively fast (within 20 min) and sometimes reversible actions. This favors a nongenomic effect. In one study, fast effects on GABAergic responses, induced by high doses of corticosterone, were found to depend on cytosolic factors (Teschemacher et al., 1996). In situ hybridization and binding studies also indicate that transcriptional regulation of glutamate and GABA receptors probably does not underlie the observed functional modulations (Orchinik et al., 1994; Watanabe et al., 1995; Weiland et al., 1997), although some of the NMDA receptor subunits and kainate receptor binding were modulated by corticosterone and/or adrenalectomy (Watanabe et al., 1995; Weiland et al., 1997). Interestingly, GR-mediated suppression of field responses was already seen at quite low doses of corticosterone when the extracellular [Ca^{2+}] was high (Talmi et al., 1992). This raises the possibility that intracellular Ca^{2+} is involved in the GR effects on glutamatergic transmission.

Although most studies agree that GR activation by administration of corticosterone results in suppression of excitatory transmission in the CA1 area, exposure to a single stressor (also resulting in GR activation) was not found to have marked effects on basal excitatory transmission (Alfarez et al., 2002). It cannot be excluded that stress activates compounds (e.g., CRF) that functionally oppose the effect of corticosteroid hormones. Nevertheless, it seems likely that glutamatergic processes are affected by acute stress in some way, since very clear changes have been reported with regard to synaptic plasticity, which involves amino acid-mediated transmission (see below).

It is feasible that following chronic stress, atrophy of neurons in the CA3 area (Magarinos and McEwen, 1995) – providing a dense innervation of the CA1 region – will evoke adaptive changes in synaptic transmission in the CA1 area. This is indeed supported by a study showing that 3 weeks following a 6-month exposure of rats to a shuttle escape task, the threshold for synaptically evoked field responses in the CA1 area was reduced (Kerr et al., 1991). Although the maximal field response amplitude was unaffected, a change in the afferent fiber properties was found, particularly in midaged rats. Indications for adaptive changes in the CA1 area were also inferred from a study in which animals received high doses of corticosterone in vivo, either once daily, for one week or for three weeks. Thus, in these animals a single injection with corticosterone suppressed the field potential recorded in subsequently prepared slices (Karten et al., 1999). After three weeks of overexposure to corticosterone, this suppression was no longer observed. GABAergic transmission was not specifically investigated but that may be of interest since expression of some GABAa receptor subunits was altered after prolonged corticosterone administration (Orchinik et al., 1995).

CA3 region

Relatively little is known about corticosteroid effects on amino acid-mediated transmission in the CA3 area. Electrophysiologically, the effect of adrenalectomy on synaptic responses of CA3 neurons has not been investigated. Also, it is unknown whether temporary rises in corticosterone, e.g., after an acute stressor, affect amino acid-mediated transmission in CA3 neurons. CA3 pyramidal neurons are quite unique in brain, as they contain relatively low levels of GRs compared to MRs (Reul and de Kloet, 1985). It is therefore interesting to see whether acute stressors alter synaptic transmission in this area, perhaps through a presynaptic mechanism.

Recently, glutamate transmission was studied in the CA3 region from chronically stressed rats (Kole et al., 2002). A day after the last stressor, the decay-constant and amplitude of the NMDA receptor-mediated synaptic component was increased compared to the control group (Fig. 1). No changes were observed in the nonNMDA receptor-mediated component of the synaptic response. The functional increase in NMDA receptor-mediated responses is not paralleled by changes in NMDA receptor binding (Watanabe et al., 1995), although chronic corticosterone treatment did increase the expression level of some of the NMDA receptor subunits (Weiland et al., 1997). Given the dissociation between effects on the NMDA and AMPA receptor-mediated synaptic components a presynaptic effect does not seem likely. In agreement,

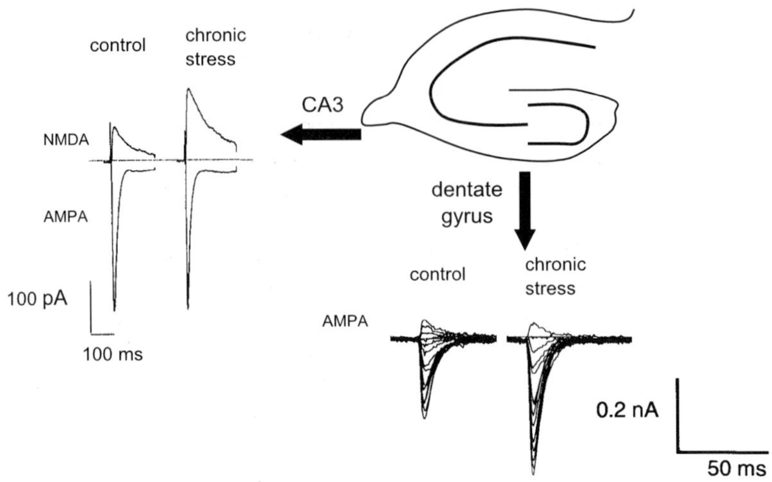

Figure 1. Increased responsiveness to excitatory amino acids in chronically stressed rats. Traces (left) from CA3 pyramidal neurons showing that NMDA-receptor mediated, synaptically evoked excitatory postsynaptic currents (EPSCs, top) are enhanced in amplitude in rats subjected for 3 weeks to restraint stress compared to control. AMPA receptor-mediated responses (bottom) are unchanged. Adapted from Kole et al. (2002). Traces on the lower right show typical EPSCs in dentate granule cells from a rat subjected to 21 days of unpredictable chronic stress (right) and a control rat (left). EPSCs were evoked by voltage steps ranging from -80 to $+40$ mV. Experiments were performed in the presence of the NMDA receptor blocker APV, so that the depicted traces represent nonNMDA (AMPA) receptor-mediated responses. AMPA responses of dentate cells are significantly enhanced after chronic stress, while NMDA responses (not shown) are unchanged. Adapted from Karst and Joëls (2003).

paired pulse facilitation was unaltered (Kole et al., 2002; but see Pavlides et al. 2002), so that chronic stress probably does not alter the release probability of glutamate. This is somewhat surprising, since electronmicroscopical analysis of mossy fiber projections to CA3 neurons showed a marked redistribution of synaptic vesicles toward the active zone (Magarinos et al., 1997). The latter is expected to result in increased release probability of glutamate-containing vesicles. The discrepancy may be due to spatial segregation: in the electrophysiological study, effects were most clearly seen at sites that are innervated by commissural/associational fibers rather than mossy fibers (Kole et al., 2002).

Dentate gyrus

Depletion of endogenous corticosterone by adrenalectomy was found to significantly reduce synaptically evoked excitatory postsynaptic potentials (EPSPs) in identified rat dentate granule cells (Joëls et al., 2001). This reduction was also seen in the field response evoked by perforant path stimulation (Stienstra et al., 1998). The reduction of both the EPSP amplitude and the field response could be normalized by in vivo treatment of rats with a low dose of corticosterone, sufficient to activate MRs but not GRs (Stienstra and Joëls, 2000). Further analysis of the EPSP with the use of pharmacological tools revealed that responses mediated by the AMPA as well as the NMDA receptors were reduced after adrenalectomy, to an approximately equal extent (Joëls et al., 2001). The latter suggests a presynaptic rather than postsynaptic site of corticosteroid actions. Since paired pulse stimulation was not largely changed after adrenalectomy (or corticosterone replacement; Stienstra et al., 1998, Stienstra and Joëls, 2000), it seems likely that synaptic innervation rather than release probability is decreased in the absence of corticosterone.

Interpretation of the electrophysiological data, however, is confounded by the fact that parallel to the electrophysiological changes, the viability of dentate granule is affected by adrenalectomy (reviewed by Gould and Tapanat, 1999; see also elsewhere in this volume). Thus, corticosterone depletion was shown to enhance apoptosis, preferentially of "aged" granule cells. At the same time, neurogenesis in the dentate gyrus is stimulated. Apoptosis as well as neurogenesis could be restored by activation of MRs in the dentate. Since "aged" cells die within three days of adrenalectomy by apoptosis while young new cells are only functional after several weeks, the adrenalectomy-induced effect on cell turnover temporarily results in a shift toward younger age of the surviving dentate cells. Several observations, though, suggest that the effects of adrenalectomy and corticosterone replacement on electrophysiological properties of the perforant path projection are not caused by the changes in turnover of the cells. First, young dentate cells are known to display basal cell characteristics that differ from those of 'aged' cells (Liu et al., 1996). Electrophysiological recording from (surviving) cells three days after adrenalectomy, however, did not demonstrate large changes in basal membrane properties (Joëls et al., 2001), indicating that the changes in cell turnover concern only a small part of the cells. Second, in vitro corticosterone treatment of slices from adrenalectomized rats (displaying apoptosis) was able to fully restore electrophysiological signals after several hours (Stienstra and Joëls, 2000); effects seen with aldosterone (preferentially activating MRs) resembled those of 30 nM corticosterone (activating MRs as well as GRs) so that opposed to the CA1 area GR activation in the dentate does not alter field potentials compared to slices with predominant MR activation. Finally, reduced field responses were already observed one day after adrenalectomy, when apoptosis has not yet occurred (Stienstra et al., 1998). Together these data suggest that shortly after adrenalectomy, glutamatergic innervation of dentate granule cells becomes impaired. This causes a reduced synaptic response and may also result in atrophy of dentate cells (Wossink et al., 2001). Possibly as part of the entire synaptic reorganization in this area vulnerable cells may die, triggering the proliferation of new cells from progenitor cells in the subgranular zone and hilus. In agreement, lesion of the entorhinal projection was found to interfere with corticosteroid effects on dentate cell turnover or vice versa (Scheff et al., 1986; Cameron et al., 1995).

Apparently, corticosteroid hormones are necessary to maintain steady glutamatergic innervation of dentate granule cells. Brief elevations of corticosterone level (as may e.g., occur after an acute stressor) did not affect the amplitude of the response to perforant path input, as shown with field potential (Stienstra and Joëls, 2000; Alfarez et al., 2003) and whole cell recording (Karst and Joëls, 2003); the risetime but not decay of glutamate-mediated responses were slightly though significantly faster after activation of GRs (Karst and Joëls, 2003). This may signify GR-dependent shifts in glutamate receptor subunit composition, e.g., of the NMDA receptor (Weiland et al., 1997), or redistribution of existing glutamate receptors. The amplitude of glutamate-mediated responses of dentate cells, however, was largely potentiated when animals had been exposed to variable stressors for three weeks prior to the electrophysiological experiment (Karst and Joëls, 2003). Thus, AMPA but not NMDA receptor-mediated currents were increased in amplitude after chronic stress. Kinetic properties as well as voltage dependency were not affected by chronic stress. It is presently unclear how this facilitated response through AMPA receptors after chronic stress is accomplished. It may point to a potentiated GR-induced transcriptional regulation of AMPA receptor subunits after chronic stress, as was indeed demonstrated with single cell RNA amplification (Qin et al., 2004). In accordance, RT-PCR showed that GLU-R1 subunits are enhanced after chronic stress, while effects on the NMDA-R1 subunit expression and binding were less clear (Schwendt and Jezova, 2000; but see also Weiland et al., 1997). Yet, other studies employing binding essays yielded no effect of acute corticosterone treatment or chronic stress (Watanabe et al., 1995). Therefore, posttranslational modifications of AMPA receptors cannot be ruled out.

In summary: In the dentate gyrus as well as the CA1 region, corticosteroid hormones are indispensible to maintain amino acid-mediated transmission. In both cases, occupation of the MR seems to be crucial. The effect of additional GR activation, however, displays regional differences. In the dentate gyrus GR activation increases glutamatergic transmission, but only in chronically stressed and not in naive rats. In the CA1 region, acute rises in corticosterone level resulting in GR (in addition to MR) activation generally suppress glutamate mediated field responses, probably through a nongenomic pathway. After chronic elevation of corticosterone levels adaptive changes are seen. Notwithstanding these differences, persistently raised corticosteroid levels result in enhanced responsiveness to excitatory amino acid, both in the dentate gyrus and CA1 area. This was also seen in the CA3 area. The enhanced responsiveness to excitatory amino acids may contribute to the hippocampal atrophy observed in morphological studies. Many issues are still unresolved, most importantly the mechanism underlying the corticosteroid modulation of amino acid transmission.

Long-term potentiation and long-term depression

While corticosteroids and stress thus alter the responsiveness of hippocampal cells to low-frequency single stimulation of glutamatergic and GABAergic pathways, steroid-dependent alterations may also occur with repetitive stimulation causing long-term depression or potentiation (depending on the stimulus frequency; Bliss and Collingridge, 1993; Bear and Abraham, 1996). These effects on low frequency and repetitive stimulation may involve different modes of action by the hormone. However, it is also possible that effects of corticosteroids on low-frequency-evoked amino acid-mediated transmission are in some respects comparable to the effects normally evoked by high-frequency stimulation, so that the latter is occluded by the former. Effects of corticosteroid hormones and stressful events on synaptic efficacy are extremely interesting from a functional point of view, since alterations in synaptic efficacy are widely believed to underlie learning and memory processes, which in turn are profoundly modulated by exposure to stressful events and corticosteroid hormones (see reviews by De Kloet et al., 1999; Lupien and Lepage, 2001; McGaugh and Roozendaal, 2002). Thus, emotionally arousing events tend to be remembered very well on the one hand (see elsewhere in this volume). On the other hand, stressful situations have also been reported to hamper retrieval of learned information. Considerable evidence now indicates that the effects of emotionally arousing events on cognition are mediated by the release of corticosteroid hormones during these events, and subsequent receptor activation (e.g., De Kloet et al., 1999; McGaugh and Roozendaal, 2002). Understanding how corticosteroid hormones affect cognitive performance therefore, requires knowledge of how these hormones modulate synaptic efficacy and the underlying cellular mechanisms. Below we review recent studies showing that corticosteroid hormones modulate long-term potentiation (LTP) and long-term depression (LTD).

CA1 area

A large number of studies have addressed the issue whether exposure to stressful events, and the concomitant release of corticosteroid hormones, modulates hippocampal synaptic plasticity. These studies reveal that exposure to relatively severe stressful events impairs hippocampal LTP in the CA1 area of hippocampal slices (Foy et al., 1987; Shors et al., 1989,1990a,b; Shors and Thompson, 1992; Kim et al., 1996; Garcia et al., 1997). Milder stressors too impair LTP induction, but this requires more subtle stimulation paradigms. Thus, Mesches et al. (1999) and Alfarez et al. (2002) recently showed that psychosocial stress impairs primed burst potentiation in vitro, whereas LTP was not affected (Fig. 2). The effects of exposure to stressful events on hippocampal synaptic efficacy are supported by in vivo studies of the hippocampal CA1 area: exposure to relatively mild stressful events impairs hippocampal LTP (Xu et al., 1997, 1998). Moreover, placing animals in a novel cage prevents the induction of primed burst potentiation (Diamond et al., 1990, 1994). Importantly, exposure to stressful experiences and novelty not only suppresses LTP, but also facilitates the induction of LTD (Kim et al., 1996; Xu et al., 1997, 1998a; Manahan-Vaughan and Braunewell, 1999; Manahan Vaughan, 2000).

These studies raise the question whether the effects of exposure to stressful events on hippocampal synaptic plasticity are mediated by the concomitant release of corticosteroid hormones. Indeed, corticosteroid-hormones have profound effects on synaptic plasticity in the hippocampal CA1 area. Rey et al. (1994) showed that treatment of mouse slices with low dosages of corticosterone or aldosterone (both activate mainly MRs) facilitates LTP, while application of aldosterone enhances synaptic potentiation in the rat (Pavlides et al., 1996). With respect to elevated corticosteroid levels, in vitro studies have revealed

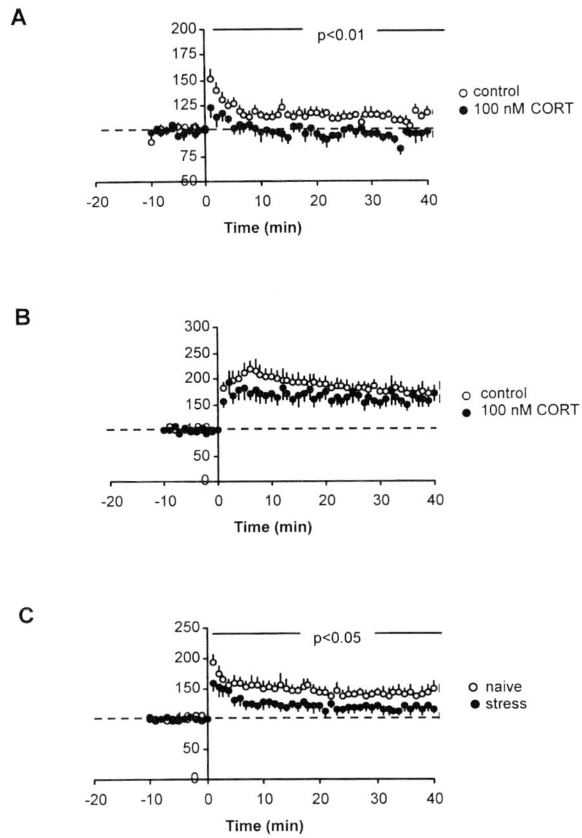

Figure 2. Primed burst potentiation (A) but not theta burst potentiation (B) in the mouse CA1 hippocampal area is suppressed by 100 nM corticosterone, administered to a hippocampal slice for 20 min in vitro, 1–4 h before recording. Comparable effects on primed burst potentiation (C) were obtained in mice that were subjected to a predator stress 1 h before preparation of the slices. Adapted from Alfarez et al. (2002).

that application of corticosterone (at dosages high enough to activate the GR in addition to the MR) hampers hippocampal CA1 synaptic efficacy later on (Rey et al., 1994; Alfarez et al., 2002, 2003). Importantly, these studies indicate that an acute elevation of plasma corticosterone levels impairs LTP by direct activation of steroid receptors in the hippocampal formation, not requiring the involvement of other brain regions. The selective GR-antagonist RU 38486 blocked the effect of corticosterone on LTP, indicating that the effect of corticosterone on hippocampal synaptic plasticity is indeed GR mediated (Rey et al., 1994). In addition, a role for elevated plasma corticosteroid hormones and the GR in modulating hippocampal synaptic efficacy is substantiated by studies examining hippocampal synaptic efficacy in vitro after in vivo treatment with corticosterone. Thus, treatment of ADX-animals with the GR agonist RU 28362 impairs synaptic potentiation (Pavlides et al., 1996). In addition to the effects of exposure to stressful events on LTP, Coussens et al. (1997) showed that very high doses of the GR agonist RU 28362 facilitate NMDA receptor-dependent LTD.

A critical question is whether the release of corticosteroid hormones during and after a stressful experience and the subsequent binding to GRs mediates the effects of stressful events on synaptic function. Such a role for glucocorticoids was substantiated by Xu et al. (1998b), demonstrating that RU 38486 prevents the effects of stressful events on LTP and LTD in a protein synthesis-dependent and RNA-dependent way. However, the influence of other compounds released after stress cannot be ruled out, as was e.g., recently demonstrated for CRF (Blank et al., 2002). Taken together, from these studies a picture emerges that under (resting) conditions, i.e., when MR activation is predominant, synaptic potentiation in the hippocampal CA1 area is facilitated. By contrast, exposure to stressful situations and the concomitantly released corticosteroid hormones hamper hippocampal CA1 LTP, but facilitate LTD via a GR-mediated mechanism.

In addition to the effects of exposure to acute stressful events and elevated corticosteroid hormone levels on synaptic efficacy, there is also evidence that prolonged exposure to stress and elevated corticosteroid levels hampers hippocampal synaptic function (Kerr et al., 1991; Gerges et al., 2001; Alfarez et al., 2003). Initially, synaptic plasticity was examined in anaesthetized animals (Kerr et al., 1991; Gerges et al., 2001), leaving open the possibility that synaptic plasticity was modulated not by exposure to prolonged stress per se, but by acute elevations in plasma corticosterone levels. Alfarez et al. (2003) recently addressed this issue showing that prolonged exposure to stress hampers hippocampal CA1 synaptic plasticity when studied in vitro under conditions where basal plasma corticosterone levels are low.

CA3 area

In the hippocampal CA3 area, Pavlides and McEwen (1999) reported that MR activation enhanced LTP evoked by stimulating the commissural/associational input to CA3, while GR activation reduced it. By contrast, neither MR nor GR activation modulated LTP depending on mossy fiber input to CA3. Importantly, prolonged stress also reduced LTP in the commissural/associational input to CA3, but not in the mossy fiber input to CA3 (Pavlides et al., 2002). Since commissural/associational synaptic potentiation is NMDA receptor-dependent, while mossy fiber LTP is NMDA receptor independent, these studies suggest that corticosterone and chronic stress only affect LTP that critically depends on NMDA receptor function.

Dentate gyrus

Stress and corticosteroids modulate synaptic responsiveness not only in the hippocampal CA1 and CA3 area, but also in dentate gyrus. Thus, Shors and Dryver (1994) reported that acute exposure to relatively severe stress hampers synaptic plasticity in the rat dentate gyrus. A potential role for corticosteroid hormone receptors in these effects was further elaborated by Pavlides et al. (1995a,b). Thus, these authors demonstrated through in vivo studies that selective occupation of the GR hampers hippocampal synaptic plasticity in the dentate gyrus and facilitates LTD. Moreover, MR activation enhanced synaptic potentiation (Pavlides and McEwen, 1999).

Chronic exposure to stress and corticosteroid hormones have also been reported to modulate synaptic efficacy in the dentate gyrus. Thus, three weeks of repeated unpredictable stressors or restraint stress hampers synaptic efficacy in the dentate gyrus in vitro (Alfarez et al., 2003) as well as in vivo (Pavlides et al., 2002). Moreover, three weeks of exposure to elevated corticosterone levels hampers hippocampal synaptic plasticity in the dentate gyrus (Pavlides et al., 1993).

In summary: There is considerable evidence that acute exposure to stressful experiences reduces synaptic potentiation and facilitates LTD in the different hippocampal subfields via a GR-mediated mechanism (Diamond et al., 1992). By contrast, under resting conditions, activation of MRs rather facilitates LTP. Prolonged exposure to stressful events and elevated corticosteroid levels hamper synaptic potentiation in the hippocampal CA1 area and dentate gyrus.

Paraventricular nucleus of the hypothalamus (PVN)

The PVN receives both excitatory and inhibitory input from brainstem structures, local hypothalamic areas, and higher brain areas such as limbic regions and the prefrontal cortex (see elsewhere in this volume). Important in understanding the effects of innervation to the PVN is the concept of the peri-PVN. Surrounding the PVN there is a shell of interneurons projecting onto the neuroendocrine cells of the nucleus (Roland and Sawchenko, 1993). Innervation descending to the PVN can either directly innervate the neuroendocrine cells, or relay via these local GABAergic interneurons. If input is relayed via these interneurons this will switch the "sign" of the input. For instance, glutamatergic projections from the ventral subiculum stimulate GABAergic interneurons, resulting in inhibition of the PVN. By contrast, GABAergic innervation from the amygdala also terminating on GABAergic interneurons promotes activity of PVN neurons. Studies concerning glutamatergic and GABAergic inputs are reviewed below.

Glutamate

The importance of glutamate in the control of excitability of parvocellular neurons was already suggested by observations with electron microscopy and in situ hybridization. High numbers of glutamatergic synapses were found to contact dendrites and cell somata of parvocellular neurons (Decavel and van den Pol, 1992). Postsynaptically, glutamatergic receptors are abundantly expressed. A specific set of receptors is expressed in the parvocellular part of PVN, containing the CRF-producing neurons. With respect to the NMDA receptor complex, subunit NR1, and NR2A and B, but not NR2C and D are expressed. For the AMPA receptor, especially the GluR1, and for the kainate the GluR5 and KA2 receptor subunits are being expressed in the PVN

(Van den Pol et al., 1994; Herman et al., 2000; Eyigor et al., 2001). A more detailed study found that neurons positive for CRF express at least NMDA receptor subunit NR1 mRNA, AMPA receptor subunits GluR1 and GluR2, and the high-affinity kainate receptor subunit KA2 (Aubry et al., 1996). Glutamate was indeed found to affect the synaptic responses of parvocellular neurons (Boudaba et al., 1996, 1997), although the interpretation of these data is hampered by the fact that the effects of glutamate may have been accomplished directly or indirectly. Very recently, Di et al. (2003) showed that high doses of corticosterone can suppress excitatory glutamatergic synaptic inputs to parvocellular neurosecretory neurons in the PVN, through a retrograde pathway involving cannabinoid receptors. This mechanism may contribute to the rapid feedback effects of corticosterone.

Despite this histological evidence for the presence of glutamate and its receptors in the PVN, studies on the functional role of this transmitter in HPA activation are not conclusive. Some in vivo studies find activation of the HPA axis by glutamate (Ziegler and Herman, 2000), while others find no effect (Cole and Sawchenko, 2002). Regarding in vitro studies, conflicting results have been reported. One study showed that a glutamate agonist increases the amount of ACTH secretagogues but others found a decrease or no effect (Costa et al., 1992; Patchev et al., 1994; Joanny et al., 1997, 2000).

Furthermore, so far little information exists regarding modulation of glutamate transmission by stress. One study showed subtle changes in NR1 mRNA expression 24 h after a single immobilization stress (Bartanusz et al., 1995). Clearly the role of glumate in HPA activation and the effects of stress and corticosteroids on glutamatergic transmission in the PVN need further investigation.

GABA

The GABAergic innervation of the hypothalamus is very dense: about 50% of the hypothalamic synapses are GABAergic (Decavel and Van den Pol, 1990). Many subunits of the GABAergic receptor complex are expressed in the PVN (Fritschy and Mohler, 1995), of which at least α1 and α2, and β1–3 are expressed in CRF-producing neurons (Cullinan, 2000).

In vivo studies showed strong responses of the HPA axis to modulation of GABAergic transmission. Injection of the GABAa receptor antagonist bicuculline close to the PVN caused an increased cellular activity as measured by c-FOS expression, increased CRF and vasopressin levels in the parvocellular subregion of the PVN and increased corticosterone levels (Cole and Sawchenko, 2002). Also, stress-induced increases in ACTH level could be suppressed by potentiation of GABAergic transmission, using muscimol injection into the PVN (Stotz-Potter et al., 1996).

Corticosteroids and stress on the other hand do not leave the GABAergic transmission of the PVN undisturbed. In the mid-eighties pharmacological experiments showed that corticosteroids affect benzodiazepine binding, indicative for effects on the GABAa receptor complex. Removal of the corticosteroids by adrenalectomy led to increased benzodiazepine binding, as measured in whole hypothalamus preparations of the rat. This effect was reversed by corticosteroid substitution (Majewska et al., 1985; De Souza et al., 1986; Goeders et al., 1986; Smith et al., 1992). Similar findings were reported in mice (Miller et al., 1988). Stress too altered benzodiazepine binding. In mice, depending on the strain and stressor, either an increase (social stress) or a decrease (single/repeated forced swim) of benzodiazepine binding was found (Miller et al., 1987; Weizman et al., 1989, 1990). Corticosterone seems to be directly responsible for these changes since the reduction of benzodiazepine binding after forced swim stress could be prevented by adrenalectomy of the mice (Weizman et al., 1990). A detailed in situ hybridization study of the PVN of chronically stressed rats revealed reduced levels of the β2 subunit of the GABAergic complex specifically in parvocellular neurons, leaving other paraventricular cell types (not directly involved in the HPA axis) unaffected (Cullinan and Wolfe, 2000).

Recently, detailed electrophysiological analysis of GABAergic synapses demonstrated that corticosteroids seem to affect GABAergic function primarily presynaptically. By electrophysiological measurements, monitoring so-called miniature postsynaptic currents, individual synaptic responses of vesicle release can be analyzed. Without apparent change in receptor characteristics such as the amplitude or decay of the response, the release of GABAergic vesicles was increased after adrenalectomy (Verkuyl and Joëls, 2003; see Fig. 3). Counts of GABA positive terminals onto CRF-producing neurons with electron microscopy revealed that the reduced vesicle release could be explained by an increase in the number of GABAergic synapses (Miklos and Kovacs, 2002). Restoring the corticosteroid levels in adrenalectomized rats normalized vesicle release to control levels indicating that the effects are indeed mediated by corticosteroids. Such regulation of GABAergic synapses is not unprecedented in the hypothalamus: in the arcuate nucleus of ovariectomized rats estrogen specifically decreases the number of GABAergic synapses, probably by regulating the interaction between glia and neurons (Garcia-Segura et al., 1994).

In summary: Parvocellular neurons in the PVN receive both glutamatergic and GABAergic input. At present, it is not clear to what extent the glutamatergic input is modulated by corticosteroids and/or stress. However, corticosteroids were quite consistently found to suppress GABAergic transmission.

Concluding Remarks

Studies so far indicate that amino acid-mediated transmission can be altered by corticosteroid hormones. The data, however, are incomplete and therefore difficult to generalize.

Most studies carried out so far involve corticosteroid modulation of synaptically evoked responses. Often this was done without pharmacological tools, so that multiple transmitter systems were activated at the same time. This precludes for instance distinction between glutamatergic and GABAergic pathways, which usually are both activated with stimulation. Those studies that allowed dinstinction between these two amino acids indicate that in the hippocampus at least glutamatergic pathways are functionally

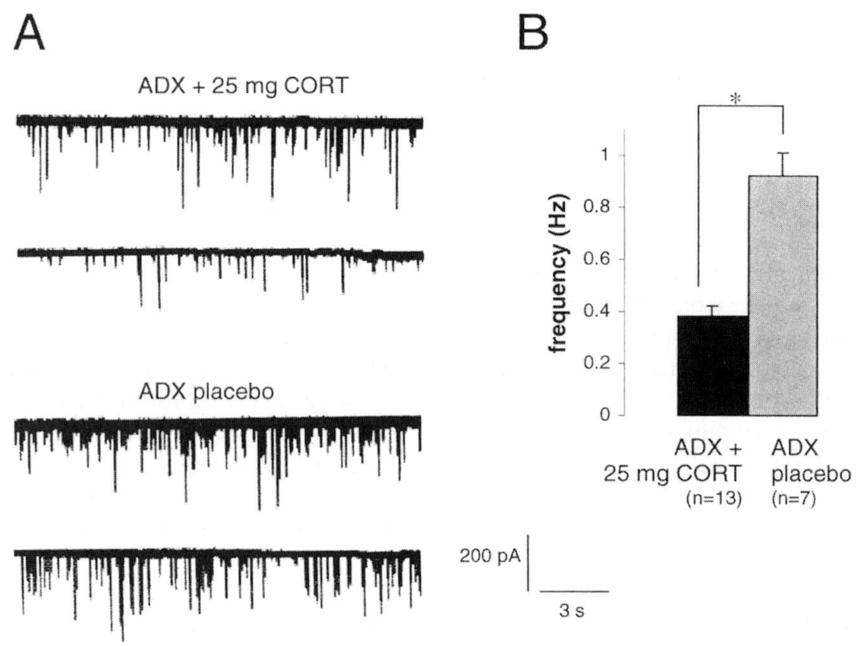

Figure 3. A. Typical traces showing that the frequency of GABAa receptor-mediated miniature inhibitory postsynaptic potentials (mIPSC) is suppressed when circulating levels of corticosterone are moderately high (ADX-25 mg CORT), compared to animals that were deprived of their endogenous hormone (ADX-pla). Other mIPSC properties, such as amplitude or decay-time were not affected by the hormone. B. Bar histogram showing the averaged mIPSC frequency in ADX rats with and without corticosterone replacement. Adapted from Verkuyl and Joëls (2003).

affected by corticosteroid hormones; GABAergic pathways were not as extensively studied.

In the hippocampus, corticosteroids act through two receptor subtypes. In the CA1 region and the dentate gyrus, MR activation was found to be necessary to maintain stable glutamatergic transmission; GABAergic transmission was not specifically investigated. Similarly, information about the CA3 area lacks to date. Higher doses of corticosterone, resulting in additional GR activation, generally suppress glutamatergic and probably also GABAergic transmission in the CA1 area, but seem ineffective in the dentate gyrus. Little information is available on the effects of an acute stressor (also resulting in GR activation), but so far stress exposure does not seem to be very effective in modulating the basal synaptic transmission. Postsynaptically this may be due to the fact that more compounds are involved in the stress response, such as CRF, which may induce effects opposite to those of corticosterone. It should also be realized that in vivo responsiveness to stimulation of glutamatergic and GABAergic pathways depends on the availability of the transmitter. In the hippocampus, extracellular levels of glutamate but not GABA were found to be enhanced by acute stress. It is not quite clear how this is caused, since studies on synthesizing enzymes and uptake rather favor an increase in GABA levels, at the cost of glutamate. Possibly, these effects develop secondary to GR-mediated suppression of glucose transporters (Virgin et al., 1991). The extracellularly measured glutamate is not necessarily of vesicular origin. It is thus not certain whether the enhanced glutamate levels will indeed result in increased responses due to activation of glutamate receptors in the postsynaptic density. Clearly, recording of synaptic responses in freely moving animals subjected to appropriate stress paradigms is necessary, resolving at the same time the stress factors and receptors involved in this process. Studies on chronic stress so far point to an enhanced responsiveness to glutamate, although the glutamate receptor involved differs for the hippocampal subregions. GABAergic transmission after chronic stress has not been investigated functionally.

In the PVN, GRs predominate. Studies so far indicate that local GABAergic innervation is considerable when corticosteroid levels are very low, thus providing a tonic inhibition of the HPA axis. When corticosteroid levels rise to moderate or high amounts GABAergic inhibition was found to be diminished (Verkuyl and Joëls, 2003). Normally, this is overriden by feedback actions of corticosterone on CRF production and release, and possibly by effects on other transmitter systems such as glutamatergic pathways (Di et al., 2003). Only under conditions that corticosteroids are less effective in their feedback action, e.g., with partial GR resistance or when reduced GABAergic tone is present already with relatively low corticosterone levels, disinhibition of the HPA axis is expected to occur.

Although experimental evidence thus supports that amino acid transmission is indeed affected by the hormones, many issues remain open for further investigation. As mentioned above, there is a great need for recording of amino acid-mediated transmission in behaviorally relevant stressful situations, in a way that also allows resolution of the stress factors involved. The latter could involve the use of pharmacological tools, which specifically block one of the corticosteroid receptor subtypes or continuous sampling of circulating corticosterone. Furthermore, the mechanism by which corticosteroid hormones affect amino acid-mediated transmission is virtually unknown. There is at present no proof that the corticosteroid effects are gene-mediated. In fact, several studies support a nongenomic pathway. For instance, especially in the CA1 area, often effects (both on the pre- and postsynaptic site) were observed within twenty minutes or required high doses of corticosterone. Also, the site of action is not very clear. Some studies indicate corticosteroid effects on presynaptic aspects of transmission, like synaptic innervation (Di et al., 2003; Kars et al., 2003). This can only be resolved by experimental designs that allow dissociation of the two aspects, e.g., by studying the responsiveness to exogenously applied amino acid receptor agonists. It may well be possible that corticosteroids do not specifically alter amino acid release, uptake, or receptor function, but merely induce synaptic reorganization – either by promoting the formation of new synapses or by rearranging the existing amino acid receptors. Clearly, these possibilities need to be examined through carefully designed experiments.

The functional consequences of corticosteroid actions on amino acid-mediated transmission can be considerable. After all, amino acids mediate most of the (fast) excitatory and inhibitory pathways in the brain: Modulation of these pathways by corticosteroids would be a tremendously effective way of adapting information processing at times of stress. In the PVN, manipulations with amino acids directly change the functionality of the HPA axis (Stotz-Potter et al., 1996; Ziegler and Herman, 2000; Bailey and Dimicco, 2001; Cole and Sawchenko, 2002). But in the hippocampus too, amino acids play a crucial role. A vast amount of data support that LTP in the hippocampus is suppressed and LTD facilitated by corticosteroids. It is tempting to speculate that this stress-induced reduction in synaptic efficacy underlies cognitive deficits as found after exposure to stressful situations. However, a number of important questions have to be addressed in the future: What are the mechanisms via which stress and corticosteroid hormones hamper synaptic communication? Can exposure to stress and elevated corticosteroid levels also facilitate synaptic efficacy? And do such alterations underlie stress- and corticosteroid-modulation of learning and memory processes?

All in all, it is clear that corticosteroid hormones and stress alter extracellular levels of amino acids as well as pre- and postsynaptic aspects of amino acid-mediated transmission. Data so far suggest that this may involve both nongenomic, fast effects, and genomic, delayed actions, although the precise mechanism awaits further unraveling. Whatever the precise mechanism, corticosteroid hormones and stress apparently can alter amino acid-mediated transmission over a wide range of time. Since amino acid-mediated transmission is prevalent in mammalian brain and involved in important physiological and pathological processes, these effects of the hormone are extremely powerful tools to alter brain function in health and disease.

References

Abraham, I., Juhasz, G., Kekesi, K.A. and Kovacs, K.J. (1998) Corticosterone peak is responsible for stress-induced elevation of glutamate in the hippocampus. Stress, 2: 171–181.

Acosta, G.B., Otero Losada, M.E. and Rubio, M.C. (1993) Area-dependent changes in GABAergic function after acute and chronic cold stress. Neurosci. Lett., 154: 175–178.

Akirav, I. and Richter-Levin, G. (1999) Biphasic modulation of hippocampal plasticity by behavioral stress and basolateral amygdala stimulation in the rat. J. Neurosci., 19: 10530–10535.

Akirav, I., and Richter-Levin, G. (2002) Mechanisms of amygdala modulation of hippocampal plasticity. J. Neurosci., 22: 9912–9921.

Alfarez, D.N., Wiegert, O., Joëls, M. and Krugers, H.J. (2002) Corticosterone and stress reduce synaptic potentiation in mouse hippocampal slices with mild stimulation. Neuroscience, 115: 1119–1126.

Alfarez, D.N., Joëls, M. and Krugers, H.J. (2003) Chronic unpredictable stress impairs Long-Term Potentiation in hippocampal CA1 area and dentate gyrus *in vitro*. Eur. J. Neurosci., 17: 1928–1934.

Aubry, J.M., Bartanusz, V., Pagliusi, S., Schulz, P. and Kiss, J.Z. (1996) Expression of ionotropic glutamate receptor subunit mRNAs by paraventricular corticotropin-releasing factor (CRF) neurons. Neurosci. Lett., 205: 95–98.

Bailey, T.W. and Dimicco, J.A. (2001) Chemical stimulation of the dorsomedial hypothalamus elevates plasma ACTH in conscious rats. Am. J. Physiol. Regul. Integr. Comp. Physiol., 280: R8–15.

Bartanusz, V., Aubry, J.M., Pagliusi, S., Jezova, D., Baffi, J. and Kiss, J.Z. (1995) Stress-induced changes in messenger RNA levels of N-methyl-D-aspartate and AMPA receptor subunits in selected regions of the rat hippocampus and hypothalamus. Neuroscience, 66: 247–252.

Baulieu, E.E., Schumacher, M., Koenig, H., Jung-Testas, I. and Akwa, Y. (1996) Progesterone as a neurosteroid: actions within the nervous system. Cell. Mol. Neurobiol., 16: 143–154.

Bear, M.F. and Abraham, W.C. (1996) Long-term depression in hippocampus. Annu. Rev. Neurosci., 19: 437–462.

Beato, M. and Sanchez-Pacheco, A. (1996) Interaction of steroid hormone receptors with the transcription initiation complex. Endocr. Rev., 17: 587–609.

Birnstiel, S., List, T.J. and Beck, S.G. (1995) Chronic corticosterone treatment maintains synaptic activity of CA1 hippocampal pyramidal cells: acute high corticosterone administration increases action potential number. Synapse, 20: 117–124.

Blank, T., Nijholt, I., Eckart, K. and Spiess, J. (2002) Priming of long-term potentiation in mouse hippocampus by corticotropin-releasing factor and acute stress: implications for hippocampus-dependent learning. J. Neurosci., 22: 3788–3794.

Bliss, T.V.P. and Collingridge, G.L. (1993) A synaptic model of memory: long-term potentiation in the hippocampus. Nature, 361: 31–40.

Boudaba, C., Szabo, K. and Tasker, J. G. (1996) Physiological mapping of local inhibitory inputs to the hypothalamic paraventricular nucleus. J. Neurosci., 16: 7151–7160.

Boudaba, C., Schrader, L.A. and Tasker, J.G. (1997) Physiological evidence for local excitatory synaptic circuits in the rat hypothalamus. J. Neurophysiol., 77: 3396–3400.

Bowers, G., Cullinan, W.E. and Herman, J.P. (1998) Region-specific regulation of glutamic acid decarboxylase (GAD) mRNA expression in central stress circuits. J. Neurosci., 18: 5938–5947.

Cameron, H.A., McEwen, B.S. and Gould, E. (1995) Regulation of adult neurogenesis by excitatory input and NMDA receptor activation in the dentate gyrus. J. Neurosci., 15: 4687–4692.

Cole, R.L. and Sawchenko, P.E. (2002) Neurotransmitter regulation of cellular activation and neuropeptide gene expression in the paraventricular nucleus of the hypothalamus. J. Neurosci., 22: 959–969.

Costa, A., Yasin, S.A., Hucks, D., Forsling, M.L., Besser, G.M. and Grossman, A. (1992) Differential effects of neuroexcitatory amino acids on corticotropin-releasing hormone-41 and vasopressin release from rat hypothalamic explants. Endocrinology, 131: 2595–2602.

Coussens, C.M., Kerr, D.S. and Abraham, W.C. (1997) Glucocorticoid receptor activation lowers the threshold for NMDA-receptor-dependent homosynaptic long-term depression in the hippocampus through activation of voltage-dependent calcium channels. J. Neurophysiol., 78: 1–9.

Cullinan, W.E. (2000) GABA(A) receptor subunit expression within hypophysiotropic CRH neurons: a dual hybridization histochemical study. J. Comp. Neurol., 419: 344–351.

Cullinan, W.E. and Wolfe, T.J. (2000) Chronic stress regulates levels of mRNA transcripts encoding beta subunits of the GABA(A) receptor in the rat stress axis. Brain Res., 887: 118–124.

De Kloet, E.R. (1991) Brain corticosteroid receptor balance and homeostatic control. Front. Neuroendocrinol., 12: 95–164.

De Kloet, E.R., Oitzl, M.S. and Joëls, M. (1999) Stress and cognition: are corticosteroids good or bad guys? Trends Neurosci., 22: 422–426.

De Souza, E.B., Goeders, N.E. and Kuhar, M.J. (1986) Benzodiazepine receptors in rat brain are altered by adrenalectomy. Brain Res., 381: 176–181.

Decavel, C. and Van den Pol, A.N. (1990) GABA: a dominant neurotransmitter in the hypothalamus. J. Comp. Neurol., 302: 1019–1037.

Decavel, C. and van den Pol, A.N. (1992) Converging GABA- and glutamate-immunoreactive axons make synaptic contact with identified hypothalamic neurosecretory neurons. J. Comp. Neurol., 316: 104–116.

Di, S., Malcher-Lopes, R., Halmos, K.C. and Tasker, J.G. (2003) Nongenomic glucocorticoid inhibition via endocannabinoid release in the hypothalamus: a fast feedback mechanism. J. Neurosci. 23: 4850–4857.

Diamond, D.M., Bennett, M.C., Stevens, K.E., Wilson, R.L. and Rose, G.M. (1990) Exposure to a novel environment interferes with the induction of hippocampal primed burst potentiation in the behaving rat. Psychobiology, 18: 273–281.

Diamond, D.M., Bennett, M.C., Fleshner, M. and Rose, G.M. (1992) Inverted-U relationship between the level of peripheral corticosterone and the magnitude of hippocampal primed burst potentiation. Hippocampus, 2: 421–430.

Diamond, D.M., Fleshner, M. and Rose, G.M. (1994). Psychological stress repeatedly blocks hippocampal primed burst potentiation in behaving rats. Behav. Brain Res., 62: 1–9.

Eyigor, O., Centers, A. and Jennes, L. (2001) Distribution of ionotropic glutamate receptor subunit mRNAs in the rat hypothalamus. J. Comp. Neurol., 434: 101–124.

Fritschy, J.M. and Mohler, H. (1995) GABAA-receptor heterogeneity in the adult rat brain: differential regional and cellular distribution of seven major subunits. J. Comp. Neurol., 359: 154–194.

Foy, M.R., Stanton, M.E., Levine, S. and Thompson, R.F. (1987) Behavioral stress impairs long-term potentiation in rodent hippocampus. Behav. Neural. Biol., 48: 138–149.

Garcia, R., Musleh, W., Tocco, G., Thompson, R.F. and Baudry, M. (1997) Time-dependent blockade of STP and LTP in hippocampal slices following acute stress in mice. Neurosci. Lett., 233: 41–44.

Garcia-Segura, L.M., Chowen, J.A., Parducz, A. and Naftolin, F. (1994) Gonadal hormones as promoters of structural synaptic plasticity: cellular mechanisms. Prog. Neurobiol., 44: 279–307.

Gerges, N.Z., Stringer, J.L. and Alkadhi, K.A. (2001) Combination of hypothyroidism and stress abolishes early LTP in the CA1 but not dentate gyrus of hippocampus of adult rats. Brain Res., 922: 250–260.

Gilad, G.M., Gilad, V.H., Wyatt, R.J. and Tizabi, Y. (1990) Region-selective stress-induced increase of glutamate uptake and release in rat forebrain. Brain Res., 525: 335–338.

Goeders, N.E., De Souza, E.B. and Kuhar, M.J. (1986) Benzodiazepine receptor GABA ratios: regional differences in rat brain and modulation by adrenalectomy. Eur. J. Pharmacol., 129: 363–366.

Gould, E. and Tanapat, P. (1999) Stress and hippocampal neurogenesis. Biol. Psychiatry, 46: 1472–1479.

Herman, J.P. and Cullinan, W.E. (1997) Neurocircuitry of stress: central control of the hypothalamo-pituitary- adrenocortical axis. Trends Neurosci., 20: 78–84.

Herman, J.P. and Larson, B.R. (2001) Differential regulation of forebrain glutamic acid decarboxylase mRNA expression by aging and stress. Brain Res., 912: 60–66.

Herman, J.P., Eyigor, O., Ziegler, D.R. and Jennes, L. (2000) Expression of ionotropic glutamate receptor subunit mRNAs in the hypothalamic paraventricular nucleus of the rat. J. Comp. Neurol., 422: 352–362.

Horner, H.C., Packan, D.R. and Sapolsky, R.M. (1990) Glucocorticoids inhibit glucose transport in cultured hippocampal neurons and glia. Neuroendocrinology, 52: 57–64.

Jackson, M.J., Zielke, H.R. and Max, S.R. (1995) Effect of dibutyryl cyclic AMP and dexamethasone on glutamine synthetase gene expression in rat astrocytes in culture. Neurochem. Res., 20: 201–207.

Joanny, P., Steinberg, J., Oliver, C. and Grino, M. (1997) Glutamate and N-methyl-D-aspartate stimulate rat hypothalamic corticotropin-releasing factor secretion in vitro. J. Neuroendocrinol., 9: 93–97.

Joanny, P., Steinberg, J., Guerrero, F., Sauze, N., Oliver, C. and Grino, M. (2000) The effects of ionotropic agonists of excitatory amino acids on the release of arginine vasopressin in rat hypothalamic slices. J. Neuroendocrinol., 12: 970–976.

Joëls, M. (1997) Steroid hormones and excitability in the mammalian brain. Front. Neuroendocrinol., 18: 2–48.

Joëls, M. and de Kloet, E.R. (1993) Corticosteroid actions on amino acid-mediated transmission in rat CA1 hippocampal cells. J. Neurosci., 13: 4082–4090.

Joëls, M., Stienstra, C. and Karten, Y. (2001) Effect of adrenalectomy on membrane properties and synaptic potentials in rat dentate granule. J. Neurophysiol., 85: 699–707.

Karst, H. and Joëls, M. (2003) Effect of chronic stress on synaptic currents in rat hippocampal dentate gyrus neurons. J. Neurophysiol., 89: 625–633.

Karst, H., Krugers, H.J. and Joëls, M. (2003) Slow and rapid effects of corticosterone on spontaneous AMPA miniature EPSCs in mouse hippocampal CA1 neurons. Soc. Neurosci Abstr. 192: 4.

Karten, Y.J., Slagter, E. and Joëls, M. (1999) Effect of long-term elevated corticosteroid levels on field responses to synaptic stimulation, in the rat CA1 hippocampal area. Neurosci. Lett., 265: 41–44.

Kerr, D.S., Campbell, L.W., Apllegate, M.D., Brodish, A. and Landfield, P.W. (1991) Chronic stress-induced acceleration of electrophysiologic and morphometric biomarkers of hippocampal aging. J. Neurosci., 11: 1316–1324.

Kim, J.J. and Diamond, D.M. (2002) The stressed hippocampus, synaptic plasticity and lost memories. Nature Rev. Neurosci., 3: 453–462.

Kim, J.J., Foy, M.R. and Thompson, R.F. (1996) Behavioral stress modifies hippocampal plasticity through N-methyl-D-aspartate receptor activation. Proc. Natl. Acad. Sci. U.S.A., 93: 4750–4753.

Kim, J.J., Lee, H.J., Han, J.S. and Packard, M.G. (2001) Amygdala is critical for stress-induced modulation of hippocampal long-term potentiation and learning. J. Neurosci., 21: 5222–5228.

Kole, M.H., Swan, L. and Fuchs, E. (2002) The antidepressant tianeptine persistently modulates glutamate receptor currents of the hippocampal CA3 commissural associational synapse in chronically stressed rats. Eur. J. Neurosci., 16: 807–816.

Lambert, J.J., Belelli, D., Harney, S.C., Peters, J.A. and Frenguelli, B.G. (2001) Modulation of native and recombinant GABA(A) receptors by endogenous and synthetic neuroactive steroids. Brain Res. Brain Res. Rev., 37: 68–80.

Liu, Y.B., Lio, P.A., Pasternak, J.F. and Trommer, B.L. (1996) Developmental changes in membrane properties and post-synaptic currents of granule cells in rat dentate gyrus. J. Neurophysiol., 76: 1074–1088.

Lupien, S.J. and Lepage, M. (2001) Stress, memory, and the hippocampus: can't live with it, can't live without it. Behav. Brain Res., 127: 137–158.

Magarinos, A.M. and McEwen, B.S. (1995) Stress-induced atrophy of apical dendrites of hippocampal CA3c neurons: comparison of stressors. Neuroscience, 69: 83–88.

Magarinos, A.M., Garcia Verdugo, J.M. and McEwen, B.S. (1997) Chronic stress alters synaptic terminal structure in hippocampus. Proc. Natl. Acad. Sci. USA, 94: 14002–14008.

Majewska, M.D., Bisserbe, J.C. and Eskay, R.L. (1985) Glucocorticoids are modulators of GABAA receptors in brain. Brain Res., 339: 178–182.

Manahan-Vaughan, D. (2000) Long-term depression in freely moving rats is dependent upon strain variation, induction protocol and behavioral state. Cereb. Cortex, 10: 482–487.

Manahan-Vaughan, D. and Braunewell, H.-K. (1999) Novelty acquisition is associated with induction of hippocampal long-term depression. Proc. Natl. Acad. Sci. USA, 96: 8739–8744.

Maroulakou, I.G. and Stylianopoulou, F. (1991) The effects of adrenalectomy and thermal stress on glutamic acid decarboxylase activity in different regions of the rat brain. Neurochem. Res., 16: 1265–1268.

McGaugh, J.L. and Roozendaal, B. (2002) Role of adrenal stress hormones in forming lasting memories in the brain. Curr. Opin. Neurobiol., 12: 205–210.

Mesches, M.H., Fleshner, M., Heman, K.L., Rose, G.M. and Diamond, D.M. (1999) Exposing rats to a predator blocks primed burst potentiation in the hippocampus *in vitro*. J. Neurosci., 19: RC18.

Miklos, I.H. and Kovacs, K.J. (2002) GABAergic innervation of corticotropin-releasing hormone (CRH)- secreting parvocellular neurons and its plasticity as demonstrated by quantitative immunoelectron microscopy. Neuroscience, 113: 581–592.

Miller, L.G., Thompson, M.L., Greenblatt, D.J., Deutsch, S.I., Shader, R.I. and Paul, S.M. (1987) Rapid increase in brain benzodiazepine receptor binding following defeat stress in mice. Brain Res., 414: 395–400.

Miller, L.G., Greenblatt, D.J., Barnhill, J.G., Thompson, M.L. and Shaderh, R.I. (1988) Modulation of benzodiazepine receptor binding in mouse brain by adrenalectomy and steroid replacement. Brain Res., 446: 314–320.

Moghaddam, B. (1993) Stress preferentially increases extra-neuronal levels of excitatory amino acids in the prefrontal cortex: comparison to hippocampus and basal ganglia. J. Neurochem., 60: 1650–1657.

Orchinik, M., Weiland, N.G. and McEwen, B.S. (1994) Adrenalectomy selectively regulates GABAA receptor subunit expression in the hippocampus. Mol. Cell. Neurosci., 5: 451–458.

Orchinik, M., Weiland, N.G. and McEwen, B.S. (1995) Chronic exposure to stress levels of corticosterone alters GABAA receptor subunit mRNA levels in rat hippocampus. Brain Res. Mol. Brain Res,. 34: 29–37.

Otero Losada, M.E. (1988) Changes in central GABA ergic function following acute and repeated stress. Br. J. Pharmacol., 93: 483–490.

Otero Losada, M.E. (1989) Acute stress and GABAergic function in rat brain. Br. J. Pharmacol., 96: 507–512.

Patchev, V.K., Karalis, K. and Chrousos, G.P. (1994) Effects of excitatory amino acid transmitters on hypothalamic corticotropin-releasing hormone (CRH) and arginine-vasopressin (AVP) release in vitro: implications in pituitary-adrenal regulation. Brain Res., 633: 312–316.

Patel, A.J., Hunt, A. and Tahourdin, C.S. (1983) Regulation of in vivo glutamine synthetase activity by glucocorticoids in the developing rat brain. Brain Res., 12: 83–91.

Pavlides, C. and McEwen, B.S. (1999) Effects of mineralocorticoid and glucocorticoid receptors on long-term potentiation in the CA3 hippocampal field. Brain Res., 851: 204–214.

Pavlides, C., Watanabe, Y. and McEwen, B.S. (1993) Effects of glucocorticoids on hippocampal long-term potentiation. Hippocampus, 3: 183–192.

Pavlides, C., Kimura, A., Magarinos, A.M. and McEwen, B.S. (1995a) Hippocampal homosynaptic long-term depression/depotentiation induced by adrenal steroids. Neuroscience, 68: 379–385.

Pavlides, C., Watanabe, Y., Magarinos, A.M. and McEwen, B.S. (1995b) Opposing roles of type I and type II adrenal steroid receptors in hippocampal long-term potentiation. Neuroscience, 68: 387–394.

Pavlides, C., Ogawa, S., Kimura, A. and McEwen, B.S. (1996) Role of adrenal steroid mineralocorticoid and glucocorticoid receptors in long-term potentiation in the CA1 field of hippocampal slices. Brain Res., 738: 229–235.

Pavlides, C., Nivon, L.G. and McEwen, B.S. (2002) Effects of chronic stress on hippocampal long-term potentiation. Hippocampus, 12: 245–257.

Qin, Y., Karst, H. and Joëls, M. (2004) Chronic unpredictable stress alters gene expression in rat single dentate granule cells. J. Neurochem. 89: 364–374.

Reiheld, C.T., Teyler, T.J. and Vardaris, R.M. (1984) Effects of corticosterone on the electro-physiology of hippocampal CA1 pyramidal cells in vitro. Brain Res. Bull., 12: 349–353.

Reul, J.M. and de Kloet, E.R. (1985) Two receptor systems for corticosterone in rat brain: microdistribution and differential occupation. Endocrinology, 117: 2505–2511.

Rey, M., Carlier, E. and Soumireu-Mourat, B. (1987) Effects of corticosterone on hippocampal slice electrophysiology in normal and adrenalectomized BALB/c mice. Neuroendocrinology, 46: 424–429.

Rey, M., Carlier, E. and Soumireu-Mourat, B. (1989) Effects of RU 486 on hippocampal slice electrophysiology in normal and adrenalectomized BALB/c mice. Neuroendocrinology, 49: 120–124.

Rey, M., Carlier, E., Talmi, M. and Soumireu-Mourat, B. (1994) Corticosterone effects on long-term potentiation in mouse hippocampal slices. Neuroendocrinology, 60: 36–41.

Roland, B.L. and Sawchenko, P.E. (1993) Local origins of some GABAergic projections to the paraventricular and supraoptic nuclei of the hypothalamus in the rat. J. Comp. Neurol., 332: 123–143.

Sapolsky, R.M. (1992) Stress, The Aging Brain and the Mechanisms of Neuron Death. The MIT Press, Cambridge, USA.

Scheff, S.W., Hoff, S.F. and Anderson, K.J. (1986) Altered regulation of lesion-induced synaptogenesis by adrenalectomy and corticosterone in young adult rats. Exp. Neurol., 93: 456–470.

Schwendt, M. and Jezova, D. (2000) Gene expression of two receptor subunits in response to repeated stress exposure in rat hippocampus. Cell. Mol. Neurobiol., 20: 319–329.

Shors, T.J. and Dryver, E. (1994) Effects of stress and long-term potentiation (LTP) on subsequent LTP and the theta burst response in the dentate gyrus. Brain Res., 666: 232–238.

Shors, T.J. and Thompson, R.F. (1992) Acute stress impairs (or induces) synaptic long-term potentiation (LTP) but does not affect paired-pulse facilitation in the stratum radiatum of rat hippocampus. Synapse, 11: 262–265.

Shors, T.J., Seib, T.B., Levine, S. and Thompson, R.F. (1989) Inescapable versus escapable shock modulates long-term potentiation in the rat hippocampus. Science, 244: 224–226.

Shors, T.J., Levine, S. and Thompson, R.F. (1990a) Opioid antagonist eliminates the stress-induced impairment of long-term potentiation (LTP). Brain Res., 506: 316–318.

Shors, T.J., Levine, S. and Thompson, R.F. (1990b) Effect of adrenalectomy and demedullation on the stress-induced impairment of long-term potentiation. Neuroendocrinology, 51: 70–75.

Singewald, N., Zhou, G.Y. and Schneider, C. (1995) Release of excitatory and inhibitory amino acids from the locus coeruleus of conscious rats by cardiovascular stimuli and various forms of acute stress. Brain Res., 704: 42–50.

Smith, C.C., Hauser, E., Renaud, N.K., Leff, A., Aksentijevich, S., Chrousos, G.P., Wilder, R.L., Gold, P.W. and Sternberg, E.M. (1992) Increased hypothalamic [3H]flunitrazepam binding in hypothalamic-pituitary-adrenal axis hyporesponsive Lewis rats. Brain Res., 569: 295–299.

Stienstra, C.M. and Joëls, M. (2000) Effect of corticosteroid treatment in vitro on adrenalectomy-induced impairment of synaptic transmission in the rat dentate gyrus. J. Neuroendocrinol., 12: 199–205.

Stienstra, C.M., Van Der Graaf, F., Bosma, A., Karten, Y.J., Hesen, W. and Joëls, M. (1998) Synaptic transmission in the rat dentate gyrus after adrenalectomy. Neuroscience, 85: 1061–1071.

Stone, D.J., Walsh, J.P., Sebro, R., Stevens, R., Pantazopoulos, H. and Benes, F.M. (2001) Effects of pre- and postnatal corticosterone exposure on the rat hippocampal GABA system. Hippocampus, 11: 492–507.

Stotz-Potter, E.H., Morin, S.M. and DiMicco, J.A. (1996) Effect of microinjection of muscimol into the dorsomedial or paraventricular hypothalamic nucleus on air stress-induced neuroendocrine and cardiovascular changes in rats. Brain Res., 742: 219–224.

Talmi, M., Carlier, E., Rey, M. and Soumireu-Mourat, B. (1992) Modulation of the in vitro electrophysiological effect of corticosterone by extracellular calcium in the hippocampus. Neuroendocrinology, 55: 257–263.

Teschemacher, A., Zeise, M.L. and Zieglgansberger, W. (1996) Corticosterone-induced decrease of inhibitory postsynaptic potentials in rat hippocampal pyramidal neurons in vitro depends on cytosolic factors. Neurosci. Lett., 215: 83–86.

Tombaugh, G.C. and Sapolsky, R.M. (1990) Hippocampal glutamine synthetase: insensitivity to glucocorticoids and stress. Am. J. Physiol., 258: E894–897.

Van den Pol, A., Hermans-Borgmeyer, I., Hofer, M., Ghosh, P. and Heinemann, S. (1994) Ionotropic glutamate-receptor gene expression in hypothalamus: localization of AMPA, kainate, and NMDA receptor RNA with in situ hybridization. J. Comp. Neurol., 343: 428–444.

Venero, C. and Borrell, J. (1999) Rapid glucocorticoid effects on excitatory amino acid levels in the hippocampus: a microdialysis study in freely moving rats. Eur. J. Neurosci., 11: 2465–2473.

Verkuyl, J.M. and Joëls, M. (2003) Effect of adrenalectomy on miniature inhibitory postsynaptic currents in the paraventricular nucleus of the hypothalamus. J. Neurophysiol., 89: 237–245.

Vidal, C., Jordan, W. and Zieglgansberger, W. (1986) Corticosterone reduces the excitability of hippocampal pyramidal cells in vitro. Brain Res., 383: 54–59.

Virgin Jr., C.E., Ha, T.P., Packan, D.R., Tombaugh, G.C., Yang, S.H., Horner, H.C. and Sapolsky, R.M. (1991) Glucocorticoids inhibit glucose transport and glutamate uptake in hippocampal astrocytes: implications for glucocorticoid neurotoxicity. J. Neurochem., 57: 1422–1428.

Watanabe, Y., Weiland, N.G. and McEwen, B.S. (1995) Effect of adrenal steroid manipulations and repeated restraint stress on dynorphin mRNA levels and excitatory amino acid receptor binding in hippocampus. Brain Res., 680: 217–225.

Weiland, N.G., Orchinik, M. and Tanapat, P. (1997) Chronic corticosterone treatment induces parallel changes in N-methyl-D-aspartate receptor subunit messenger RNA levels and antagonist binding sites in the hippocampus. Neuroscience, 78: 653–662.

Weizman, R., Weizman, A., Kook, K.A., Vocci, F., Deutsch, S.I. and Paul, S.M. (1989) Repeated swim stress alters brain benzodiazepine receptors measured in vivo. *J. Pharmacol.* Exp. Ther., 249: 701–707.

Weizman, A., Weizman, R., Kook, K.A., Vocci, F., Deutsch, S.I. and Paul, S.M. (1990) Adrenalectomy prevents the stress-induced decrease in in vivo [3H]Ro15- 1788 binding to GABAA benzodiazepine receptors in the mouse. Brain Res., 519: 347–350.

Wossink, J., Karst, H., Mayboroda, O. and Joëls, M. (2001) Morphological and functional properties of rat dentate granule cells after adrenalectomy. Neuroscience, 108: 263–272.

Xu, L., Anwyl, R. and Rowan, M.J. (1997) Behavioral stress facilitates the induction of long-term depression in the hippocampus. Nature, 387: 497–500.

Xu, L., Anwyl, R. and Rowan, M.J. (1998a) Spatial exploration induces a persistent reversal of long-term potentiation in rat hippocampus. Nature, 394: 891–894.

Xu, L., Holscher, C., Anwyl, R. and Rowan, M.J. (1998b) Glucocorticoid receptor and protein/RNA synthesis-dependent mechanisms underlie the control of synaptic plasticity by stress. Proc. Natl. Acad. Sci. USA, 95: 3204–3208.

Yoneda, Y., Kanmori, K., Ida, S. and Kuriyama, K. (1983) Stress-induced alterations in metabolism of gamma-aminobutyric acid in rat brain. J. Neurochem., 40: 350–356.

Zeise, M.L., Teschemacher, A., Arriagada, J. and Zieglgänsberger, W. (1992) Corticosterone reduces synaptic inhibition in rat hippocampal and neocortical neurons in vitro. J. Neuroendocrinology, 4: 107–112.

Ziegler, D.R. and Herman, J.P. (2000) Local integration of glutamate signaling in the hypothalamic paraventricular region: regulation of glucocorticoid stress responses. Endocrinology, 141: 4801–4804.

Zhu, B.G., Zhu, D.H. and Chen, Y.Z. (1998) Rapid enhancement of high affinity glutamate uptake by glucocorticoids in rat cerebral cortex synaptosomes and human neuroblastoma clone SK-N-SH: possible involvement of G-protein. Biochem. Biophys. Res. Commun., 247: 261–265.

T. Steckler, N.H. Kalin and J.M.H.M. Reul (Eds.)
Handbook of Stress and the Brain, Vol. 15
ISBN 0-444-51173-3

CHAPTER 4.8

Neuroactive steroids

Rainer Rupprecht[1,2]*

[1]*Department of Psychiatry, Ludwig Maximilian University, Nußbaumstr. 7, Munich, Germany*
[2]*Max Planck Institute of Psychiatry, 80336 Munich, Germany*

Abstract: Steroids influence neuronal function through binding to cognate intracellular receptors which may act as transcription factors in the regulation of gene expression. In addition, certain so called neuroactive steroids modulate ligand-gated ion channels via non-genomic mechanisms. Especially distinct 3α-reduced metabolites of progesterone and deoxycorticosterone are potent positive allosteric modulators of γ-aminobutyric acid type A ($GABA_A$) receptors. However, also classical steroid hormones such as 17β-estradiol, testosterone and progesterone are neuroactive steroids because they may act as functional antagonists at ligand-gated ion channels such as the 5-hydroxytryptamine type 3 ($5\text{-}HT_3$) receptor or certain glutamate receptors. On the other hand, also 3α-reduced neuroactive steroids may regulate gene expression via the progesterone receptor after intracellular oxidation into 5α-pregnane steroids. Animal studies showed that progesterone is converted rapidly into GABAergic neuroactive steroids in vivo. Moreover, progesterone and 3α-reduced neuroactive steroids produce a benzodiazepine-like sleep EEG profile in rats and humans. During major depression, there is an altered composition of such 3α-reduced neuroactive steroids which is corrected by successful treatment with antidepressant drugs. Studies in patients with panic disorder suggest that neuroactive steroids may also play a role in modulating human anxiety. Neuroactive steroids affect a broad spectrum of behavioral functions through their unique molecular properties and may represent a new treatment strategy for neuropsychiatric disorders.

Keywords: Neurosteroids, neuroactive steroids, ligand-gated ion channels, depression, anxiety, $GABA_A$ receptor

Introduction

Steroid hormone action involves binding of the steroids to their respective intracellular receptors (Evans, 1988; Truss and Beato, 1993; Rupprecht and Holsboer, 1999). These receptors subsequently change their conformation by dissociation from chaperone molecules, e.g. the heat shock proteins, and translocate to the nucleus where they bind as homo- or heterodimers to the respective response elements that are located in the regulatory regions of target promoters. Thus, steroid hormone receptors act as transcription factors in the regulation of gene expression (Evans, 1988; Truss and Beato, 1993; Rupprecht and Holsboer, 1999). Meanwhile, there is increasing evidence that certain steroids may alter neuronal excitability via the cell surface through interaction with certain neurotransmitter receptors (Majewska et al., 1986; Paul and Purdy, 1992; Lambert et al., 1995; Rupprecht and Holsboer, 1999). The term "neuroactive steroids" has been coined for steroids with these particular properties (Paul and Purdy, 1992). While the action of steroids at the genome requires a time period from minutes to hours that is limited by the rate of protein biosynthesis (McEwen, 1991), the modulatory effects of neuroactive steroids are fast occurring events requiring only milliseconds to seconds (McEwen, 1991).

*Corresponding author. Tel.: +49 89 5160 2770; Fax: +49 89 5160 5524, E-mail: Rainer.Rupprecht@psy.med.uni-muenchen.de

Initially, it has been believed that steroid hormones act exclusively through the classical genomic pathway, whereas certain neuroactive steroids that do not bind to either known steroid hormone receptor, e.g. 3α-reduced metabolites of progesterone and deoxycorticosterone such as 3α, 5α-tetrahydroprogesterone (3α, 5α-THP; 3α-hydroxy-5α-pregnan-20-one; allopregnanolone) and 3α, 5α-tetrahydrodeoxycorticosterone (3α, 5α-THDOC; 3α, 21-dihydroxy-5α-pregnan-20-one; allotetrahydrodeoxycorticosterone), pregnenolone sulfate (PS) or dehydroepiandrosterone sulfate (DHEA-S) are allosteric modulators of specific neurotransmitter receptors such as γ-aminobutyric acid type A ($GABA_A$) receptors (Evans, 1988; Paul and Purdy, 1992). This concept, however, has been challenged by the identification of binding sites for classical steroid hormones, e.g. progesterone (Ramirez and Zheng, 1996), estradiol (Pappas et al., 1995; Ramirez and Zheng, 1996), testosterone (Ramirez and Zheng, 1996), glucocorticoids (Orchinik et al., 1991) or aldosterone (Wehling, 1997), at membranes of cells or tissues and of a large number of signal transduction pathways involved in steroid hormone action (Wehling, 1997). Moreover, the modulation of ligand-gated ion channels or G-protein coupled receptors by steroids may alter the activity of intracellular kinases, which consequently affects the expression patterns of downstream genes, e.g. via the cyclic AMP-protein kinase A-cyclic AMP reponsive element binding protein (CREB) pathway (Wehling, 1997; Zakon, 1998). It is important to emphasize that a variety of steroid hormones have been identified that interact with different neurotransmitter receptors and thus also need to be defined as neuroactive steroids.

Biosynthesis of neuroactive steroids

Due to their lipophilic nature steroids that are produced in various endocrine organs can easily cross the blood-brain barrier. However, a variety of neuroactive steroids may be synthesized in the brain itself without the aid of peripheral sources (Baulieu, 1991, 1998; Akwa et al., 1992). These steroids that are formed within the brain from cholesterol have been defined also as "neurosteroids" (Baulieu, 1998). As excellent reviews are available elsewhere on the biosynthesis and metabolism of steroids in the brain (Mellon, 1994; Baulieu, 1998; Compagnone and Mellon, 2000), only a short description of some major pathways involved in the synthesis of neuroactive steroids that are important for neuropsychopharmacology is given (Fig. 1). Progesterone may be formed from pregnenolone by the 3β-hydroxysteroid dehydrogenase/Δ^5-Δ^4-isomerase. The 5α-reductase catalyzes the reduction of progesterone and deoxycorticosterone into the 5α-pregnane steroids 5α-dihydroprogesterone (5α-DHP) and 5α-dihydrodeoxycorticosterone (5α-DHDOC), respectively, the 5β-reductase reduces progesterone to 5β-dihydroprogesterone (5β-DHP). These are irreversible reactions in mammalian cells (Celotti et al., 1992). These pregnane steroids may be further reduced to the neuroactive steroids 3α, 5α-THP, 3α, 5β-tetrahydroprogesterone (3α, 5β-THP; 3α-hydroxy-5β-pregnan-20-one; pregnanolone) and 3α, 5α-THDOC by the 3α-hydroxysteroid oxidoreductase. This reaction may work both in the reductive and in the oxidative direction depending on the cofactors present in the environment (Rupprecht et al., 1993). Both the 5α-reductase and the 3α-hydroxysteroid oxidoreductase exist in various isoforms that are expressed in a tissue-specific manner (Compagnone and Mellon, 2000). Pregnenolone is also a precursor for dehydroepiandrosterone (DHEA). These two steroids exist also as conjugated sulfate esters, e.g. PS and DHEA-S, and fatty acid esters at concentrations frequently exceeding those of the free steroids (Baulieu, 1998). Both progesterone and DHEA are converted to androstenedione, which is a precursor of testosterone. Estradiol is formed by the aromatase either from testosterone or from androstenedione via estrone.

Modulation of neurotransmitter receptors by neuroactive steroids

Steroid modulation of γ-aminobutyric acid type A ($GABA_A$) receptors

The 3α-reduced metabolites of progesterone and deoxycorticosterone 3α, 5α-tetrahydroprogesterone (3α, 5α-THP; 3α-hydroxy-5α-pregnan-20-one; allopregnanolone) and 3α, 5α-tetrahydrodeoxycorticosterone

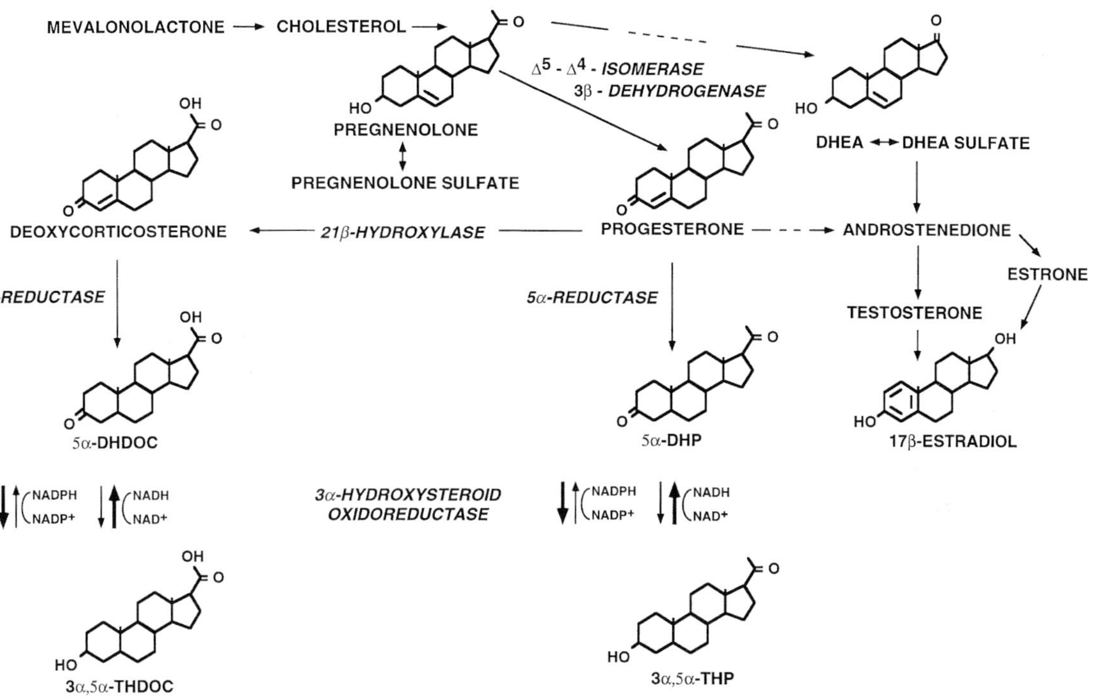

Fig. 1. Biosynthesis of neuroactive steroids. Reproduced with permission from Rupprecht and Holsboer (1999).

(3α, 5α-THDOC; 3α, 21-dihydroxy-5α-pregnan-20-one; allotetrahydrodeoxycorticosterone) were the first steroids that have been shown to modulate neuronal excitability via their interaction with γ-aminobutyric acid type A ($GABA_A$) receptors (Majewska et al., 1986). $GABA_A$ receptors consist of various subunits that form ligand-gated ion channels with considerable homology to glycine, nicotinic acetylcholine and serotonin type 3 ($5\text{-}HT_3$) receptors (Paul and Purdy, 1992; Lambert et al., 1995; Wetzel et al., 1998) (Fig. 2). A variety of different classes of drugs act through $GABA_A$ receptors (Fig. 2): agonists for the GABA binding site, benzodiazepines, and also barbiturates, clomethiazol, neuroactive steroids, alcohols and anesthetics. While a specific and saturable binding site at $GABA_A$ receptors has been clearly identified for GABA and benzodiazepines, the occurrence for such a binding site has not yet been proven for the latter compounds. The assumption of a steroid-binding site at this ligand-gated ion channel is based on pharmacological studies concerning the strong stereoselectivity and structure–activity relationship of the action of neuroactive steroids at this neurotransmitter receptor (Lambert et al., 1995). However, no direct binding of steroids to the receptor protein has been demonstrated so far by biochemical methods. At a functional level, the steroids 3α, 5α-THP and 3α, 5α-THDOC may displace *t*-butylbicyclophosphorothionate (TBPS) from the chloride channel and enhance the binding of muscimol and benzodiazepines to $GABA_A$ receptors (Paul and Purdy, 1992). These neuroactive steroids are potent positive allosteric modulators of $GABA_A$ receptors because they enhance the GABA-evoked chloride current (Fig. 3) through increasing the frequency and/or duration of openings of the GABA-gated chloride channel (Majewska et al., 1986; Paul and Purdy, 1992; Lambert et al., 1995). However, similar to barbiturates but in contrast to benzodiazepines, high concentrations in the micromolar range of these neuroactive steroids have been shown to exert a certain intrinsic agonistic activity in the absence of GABA (Puia et al., 1990). Within the steroid molecule, the presence of a 3α-hydroxy group within the A-ring of these molecules is an absolute requirement

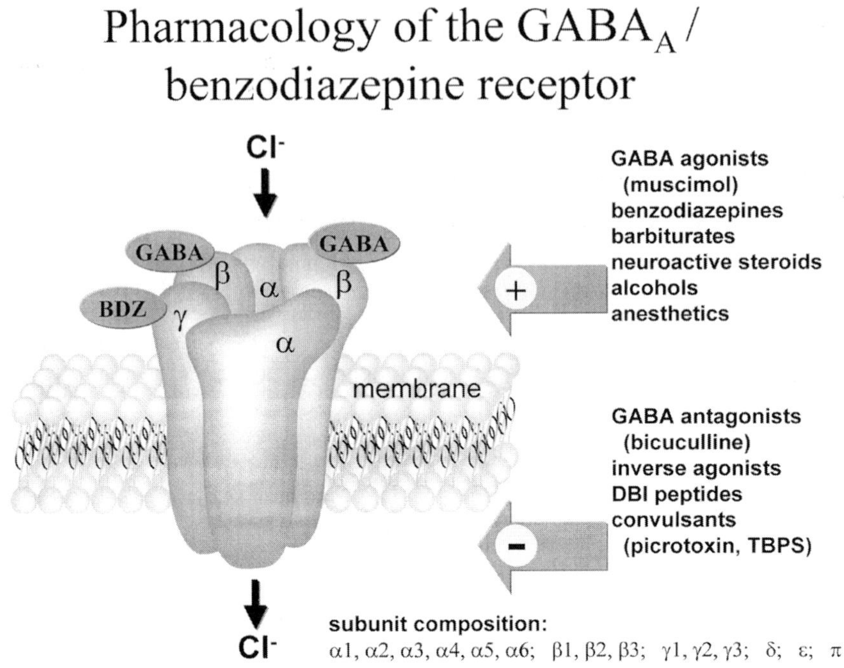

Fig. 2. Pharmacology of the GABA$_A$/benzodiazepine receptor complex.

for a positive allosteric activity at GABA$_A$ receptors (Gee et al., 1988; Paul and Purdy, 1992; Lambert et al., 1995) as 5α-pregnane steroids such as 5α-dihydroprogesterone (5α-DHP) are inactive (Fig. 3) while 3β, 5α-tetrahydroprogesterone (3β, 5α-THP) may even act as a functional antagonist for GABA-agonistic steroids (Prince and Simmonds, 1992; Maitra and Reynolds, 1998). Moreover, pregnenolone sulfate (PS) or dehydroepiandrosterone sulfate (DHEA-S) display GABA-antagonistic properties (Paul and Purdy, 1992; Lambert et al., 1995; Rupprecht and Holsboer, 1999). Further details on the modulation of GABA$_A$ receptors by neuroactive steroids are reviewed elsewhere (Paul and Purdy, 1992; Lambert et al., 1995).

Steroid modulation of various neurotransmitter receptors

Using whole-cell voltage-clamp recordings of human embyonic kidney cells stably expressing the 5-HT$_3$ type A receptor (Maricq et al., 1991), we could show that the gonadal steroids 17β-estradiol and progesterone may also act as non-competitive functional antagonists at the 5-HT$_3$ receptor (Wetzel et al., 1998). There is a distinct structure–activity relationship for the actions of steroids at the 5-HT$_3$ receptor that, however, differed considerably from that known for GABA$_A$ receptors. Our data and recent findings using spin labelling techniques with the nicotinic acetycholine receptor (Barrantes et al., 2000) are in favor of the view that the steroids insert into the membrane at the receptor–membrane interface and thereby allosterically modulate the function of these neurotransmitter receptors in a structure-specific manner. Although future research is needed to definitely clarify the issue how steroids interact with ligand-gated ion channels at the molecular level, it appears that the allosteric modulation of neurotransmitter receptors by steroids is a highly complex phenomenon that is dependent on the molecule structure of the respective steroid, the amino acid composition of the individual receptor and the physicochemical properties of the cell membrane (Rupprecht and Holsboer, 1999) and

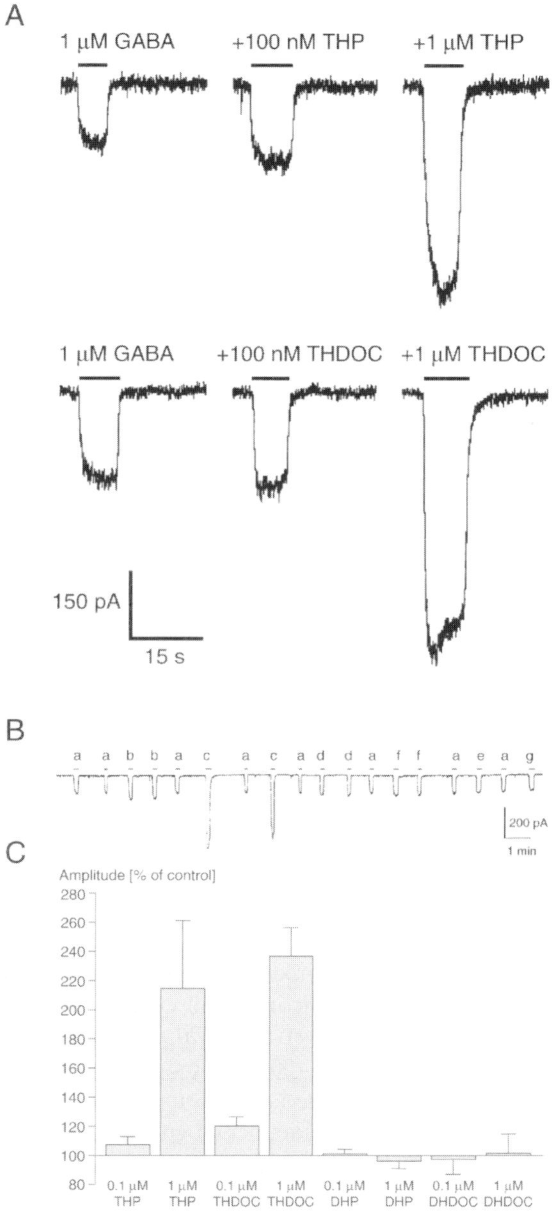

Fig. 3. Allosteric modulation of the GABA-evoked chloride current by neuroactive steroids and 5α-pregnane steroids. Rat hypothalamic neurons were recorded in the whole-cell voltage-clamp configuration. Positive allosteric modulation of the GABA-evoked chloride current by neuroactive steroids (A). The bar indicates the presence of 1 μM GABA. Modulatory properties of neuroactive steroids and of 5α-pregnane steroids at the $GABA_A$ receptor in form of a representative experiment (B) and as the mean ± SD several independent experiments (C). Reproduced with permission from Rupprecht and Holsboer (1999).

is not merely determined by a putative steroid binding site.

In addition to $GABA_A$ receptors or $5\text{-}HT_3$ receptors, the other members within the family of ligand-gated ion channels, e.g. nicotinic acetycholine receptors (Valera et al., 1992; Bullock et al., 1997) or glycine receptors (Wu et al., 1990) have been shown to be steroid-sensitive (Rupprecht and Holsboer, 1999). Within the glutamate receptor family, *N*-methyl-D-aspartate (NMDA) receptors, α-amino-3-hydroxy-5-methyl-4-isoxazolepropionic acid (AMPA) receptors and kainate receptors have also been demonstrated to be a target for steroid modulation (Wu et al., 1991; Park-Chung et al., 1994; Weaver et al., 1997a; Weaver et al., 1997b; Rupprecht and Holsboer, 1999). Again, the structure–activity requirements for the modulation of ligand-gated ion channels by steroids apparently differ considerably between members of these neurotransmitter receptor families. The oxytocin receptor was recently identified as the first G-protein coupled receptor for which steroids can be ligands (Grazzini et al., 1998). Also sigma receptors, which still lack detailed molecular characterization, may bind steroids (Su et al., 1988) and are sensitive to steroid modulation (Monnet et al., 1995).

Effects of neuroactive steroids on gene expression

For a long time it was assumed that neuroactive steroids modulating $GABA_A$ receptors do not regulate gene expression via intracellular steroid receptors because they do not bind to either known steroid hormone receptor (Paul and Purdy, 1992). Using a cotransfection system with recombinant progesterone receptors and the mouse mammary tumor virus (MTV) promoter upstream of the luciferase gene as a reporter gene, we could show that the neuroactive steroids 3α, 5α-THP and 3α, 5α-THDOC effectively activate gene expression via progesterone receptors after intracellular oxidation into 5α-dihydroprogesterone (5α-DHP) and 5α-dihydrodeoxycorticosterone (5α-DHDOC). These steroids, in contrast to 3α, 5α-THP and 3α, 5α-THDOC, bind to progesterone receptors of different species (Rupprecht et al., 1993). Synthetic analogues of naturally occurring neuroactive steroids should therefore avoid genomic effects through progesterone receptors (Rupprecht et al., 1996; Gasior et al., 1999) in order to prevent gynecological side effects. This, for example, has been achieved with ganaxolone (Carter et al., 1997; Monaghan et al., 1997; Gasior et al., 1999), where a 3β-methyl group has been introduced into the A ring of the steroid molecule to block the intracellular oxidation. In vivo, it has recently been shown that 3α, 5α-THP may enhance (Smith et al., 1998a) or decrease (Grobin and Morrow, 2000) the expression of the gene encoding for the α_4 subunit of the $GABA_A$ receptor, which is responsible for the sensitivity to benzodiazepines. In addition, both 3α, 5α-THP and 3α, 5α-THDOC, when administered to rats, suppress the expression of vasopressin and corticotropin-releasing factor (CRF) (Patchev et al., 1994; Patchev et al., 1997). Therefore, neuroactive steroids affect also the activity of the hypothalamic–pituitary–adrenal (HPA) system. In conclusion, both genomic and non-genomic effects of neuroactive steroids have to be considered in the further exploitation of this new class of drugs in neuropsychopharmacology, either with regard to putative clinical effects or potential side effects.

Systemic effects of neuroactive steroids

Sleep

Already early investigations suggested a sleep promoting and hypnotic effect of the 3α-reduced neuroactive steroid 3α, 5α-THDOC (Mendelson et al., 1987). Early studies with synthetic analogues of 3α-reduced neuroactive steroids showed that such steroids may shorten sleep latency and thus might be suitable for treatment of sleep disturbances (Edgar et al., 1997). However, these studies did not include data on EEG power spectra which allows only limited conclusions on a pharmacological profile of such compounds. Therefore, we performed detailed studies with progesterone (Lancel et al., 1996) and 3α, 5α-THP (Lancel et al., 1997) in rats using spectral analysis of sleep. These investigations revealed a sleep EEG profile for such steroids that was quite similar to that observed with benzodiazepines. These steroids shortened the latency to non-REM sleep and promoted pre-REM sleep. Both within non-REM and REM sleep they decreased the EEG power within the delta frequency range, whereas there was a pronounced increase in EEG activity within the spindle frequencies and the beta frequencies (Fig. 4). Also in humans, progesterone exerted benzodiazepine-like effects on sleep (Friess et al., 1997). Therefore, particular attention has to be drawn to putative side effects of such steroids on sleep such as development of tolerance and withdrawal symptoms, e.g. rebound insomnia. However, a recent study investigating the effects of a subchronic administration of 3α, 5α-THP on sleep did not show any tolerance development or withdrawal symptoms (Damianisch et al., 2001) (Fig. 4). Thus, in spite of their benzodiazepine-like effects on sleep, 3α -reduced neuroactive steroids appear to differ from benzodiazepines when administered over longer time periods and might therefore be suitable as a treatment strategy for sleep disturbances.

Anticonvulsant properties

Drugs that enhance the function of $GABA_A$ receptors such as benzodiazepines and barbiturates as well as drugs targeting the GABA-binding site of the $GABA_A$ receptor are commonly used as effective antiepileptic agents. 3α-reduced neuroactive steroids exerted pronounced anticonvulsant effects in various animal models (Belelli et al., 1990; Devaud et al., 1995; Frye and Scalise, 2000). First clinical experiences using progesterone as a precursor molecule in women suffering from catamenial epilepsy, which is related to the menstrual cycle, reported a decrease in epileptiform discharges following administration of progesterone (Bäckström et al., 1984; Herzog, 1995). Currently, first synthetic analogues of 3α-reduced neuroactive steroids, e.g. ganaxolone, are under investigation for antiepileptic activity (Gasior et al., 1997; Monaghan et al., 1997; Beekman et al., 1998; Gasior et al., 1999). Although first animal studies with subchronic adminstration of ganaxolone suggest that this steroid induces anticonvulsant tolerance to benzodiazepines but not to itself (Reddy and Rogawski, 2000), putative side effects such as

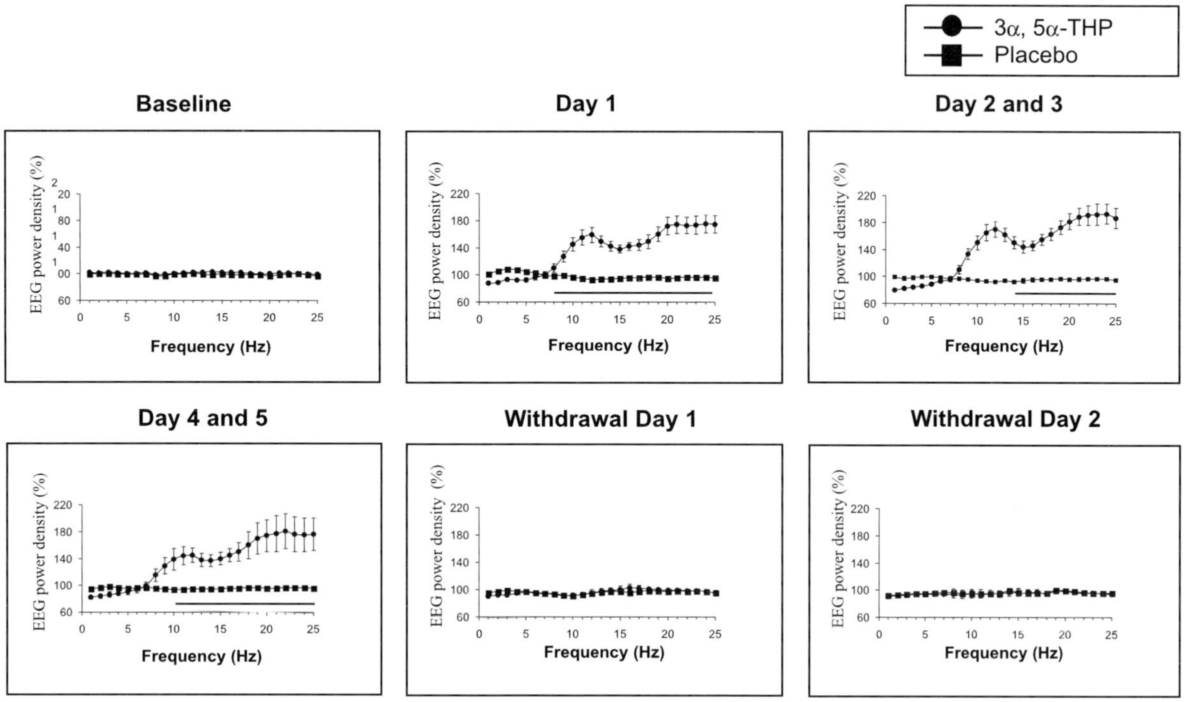

Fig. 4. Effets of 3α, 5α-THP (15 mg/kg body weight) and placebo on EEG power densitity within non-REM sleep in male Wistar rats. Data represent mean ± SEM ($n = 8$ per group) and are shown as percentage of the average power density during the baseline condition. The lines in the panels day 1, day 2 and 3 and day 4 and 5 indicate the frequency bands with a significant difference between both treatment groups. Modified according to Damianisch et al. (2001).

sedation, alteration of sleep architecture and development of tolerance have to be taken into account, especially when considering long-term treatment with this new class of drugs.

Anesthesia

3α-reduced neuroactive steroids may also exert antinociceptive (Frye and Duncan, 1994; Nadeson and Goodchild, 2001) and anesthetic (Korneyev and Costa, 1996) effects in various animal models. 3α-reduced neuroactive steroids such as 3α, 5β-THP (pregnanolone) may exert strong sedative (Schulz et al., 1996) or even anesthetic (Carl et al., 1990) effects in humans. A mixture of alphaxolone and alphadolone has also been developed (Althesin®) as an anesthetic in humans (Gyermek and Soyka, 1975). However, solubility problems and a hypersensitivity to the respective solvent have led to the withdrawal of this compound (Zorumski et al., 2000). It remains to be determined whether 3α-reduced neuroactive steroids will receive further consideration as putative anesthetics in humans.

Nootropic properties, cognition and dementia disorders

Pregnenolone sulfate (PS) and DHEA sulfate (DHEA-S) display GABA antagonistic properties and exert complex effects at NMDA receptors (Zorumski et al., 2000). Moreover, sulfate derivatives of pregnanolone have been shown to exert neuroprotective effects via inhibition of NMDA receptor function (Weaver et al., 1997b). In addition to the effects of DHEA via the cell membrane, potential antiglucocorticoid effects of DHEA have been reported in vivo (Browne et al., 1993; Araneo and Daynes, 1995). Therefore, such steroids might possess nootropic properties. Indeed, early studies have suggested that intracerebroventricular administration of pregnenolone and pregnenolone sulfate leads to an amelioration in various memory tasks in rodents (Flood et al., 1992). Moreover, also DHEA has been suggested to enhance memory retention in mice (Flood et al., 1988). In aged rats, low PS levels have been found in the hippocampus and were correlated with memory deficits that could be transiently corrected by PS injection (Vallee et al., 1997). Also prolonged intracerebroventricular infusion of PS enhanced cognitive performance in mice (Ladurelle et al., 2000). However, valid clinical data concerning the memory-enhancing properties of pregnenolone in dementia disorders are lacking to date. There is evidence that DHEA levels decrease with age (Thomas et al., 1994) and decreased concentrations of DHEA have been reported in patients suffering from Alzheimer's disease and multi-infarct dementia (Sunderland et al., 1989; Näsman et al., 1991). Decreased DHEA-S concentrations may constitute an enhanced risk for the development of Alzheimer's disease (Hillen et al., 2000). Thus, an interplay between neuroactive steroids and the HPA system may be of importance for the pathophysiology of dementia disorders. Meanwhile, DHEA is sold as an antiaging drug especially in the USA. However, systematic research as to whether DHEA supplementation may enhance cognitive performance in normal aging people or in dementia disorders is scarcely available. One open study suggested beneficial effects of DHEA-S on daily living in patients with multiinfarct dementia (Azuma et al., 1999). However, controlled studies with DHEA in Alzheimer's diesase or multi-infarct dementia are not available to date.

Antipsychotic properties

Epidemiological studies suggest that the onset of psychiatric symptoms may be related to changes in the secretion of gonadal hormones (Hallonquist et al., 1993; Häfner et al., 1993). Moreover, there is a difference between pre- and post-menopausal women with an increased vulnerability for the onset of schizophrenic episodes after the menopause (Häfner et al., 1993). Thus, it may be hypothesized that a sudden drop of steroid concentrations may contribute to the development of such disorders and a steroid replacement might be of therapeutic value. In contrast to haloperidol, progesterone neither produced catalepsy nor antagonized amphetamine-induced stereotypy. However, both progesterone and haloperidol but not $3\alpha,5\alpha$-THP (Rupprecht et al., 1999) effectively restored the disruption of the prepulse inhibition (PPI) of the acoustic startle response that was evoked by apomorphine. This behavioral profile of progesterone is compatible with the possibility that progesterone itself shares some properties with atypical antipsychotics, which may be relevant for the development and treatment of psychotic disturbances, e.g. postpartum psychosis. It has recently been demonstrated that the atypical neuroleptic agent olanzapine may increase the concentrations of $3\alpha,5\alpha$-THP in rat brain (Marx et al., 2000). Also clozapine, in contrast to haloperidol, may enhance the concentrations of both $3\alpha,5\alpha$-THP and of progesterone in rat brain in a time and dose dependent fashion (Barbaccia et al., 2001). Thus, neuroactive steroids might also contribute to the pharmacological profile of atypical antipsychotic drugs.

Premenstrual dysphoric disorder, pregnancy and postpartum period

Concentrations of neuroactive steroids vary throughout the menstrual cycle and throughout pregnancy, which is accompanied by changes in $GABA_A$ receptor plasticity (Concas et al., 1998). In patients with premenstrual dysphoric disorder (PMDD), decreased levels of 3α, 5α-THP have been reported during the luteal phase (Rapkin et al., 1997; Bicikova et al., 1998; Monteleone et al., 2000) which might contribute to the development of mood symptoms and irritability. Both at baseline and after mental stress, an enhanced ratio $3\alpha,5\alpha$-THP/cortisol has been observed (Girdler et al., 2001). Interestingly, PMDD patients with a high symptom score had lower levels of $3\alpha,5\alpha$-THP when compared with less symptomatic patients (Girdler et al., 2001). Treatment with citalopram may enhance the sensitivity of $GABA_A$ receptors to modulation by $3\alpha,5\beta$-THP in women suffering from PMDD (Sundström and Bäckström, 1998) and the importance of neuroactive steroid-serotonergic interactions is further underlined by an increased response of 3α, 5α-THP to challenge with L-tryptophan in patients with PMDD (Rasgon et al., 2001). Thus, the interplay between the serotonergic system and neuroactive steroids may contribute to the efficacy of selective serotonin reuptake inhibitors in the treatment of PMDD (Steiner et al., 1995). During pregnancy, there is a rise in the concentrations of progesterone and of an array of neuroactive steroids (Pearson Murphy et al., 2001). While progesterone concentrations decline rapidly after delivery, neuroactive steroids are still elevated several weeks postpartum (Pearson Murphy et al., 2001). There was a tendency for increased concentrations of neuroactive steroids in depressed women during the latter half of pregnancy when compared with nondepressed women (Pearson Murphy et al., 2001). Thus, neuroactive steroids might also contribute to psychiatric complaints during pregnancy and the postpartum period.

Antidepressant properties and major depression and stress

Stress is a key factor of major depression which is accompanied by an overdrive of the HPA-system. 3α-reduced neuroactive steroids may suppress the expression of vasopressin and CRF in rats (Patchev et al., 1994; Patchev et al., 1997). As endogenous 3α-reduced neuroactive steroids rise during stress, e.g. during a forced swimming procedure (Paul and Purdy, 1992), such an increase of endogenous neuroactive steroids might contribute to the termination of a stress period as a counterregulatory mechanism.

The selective serotonin reuptake inhibitor (SSRI) fluoxetine may enhance the concentrations of 3α, 5α-THP in different areas of the rat brain (Uzunov et al., 1996; Serra et al., 2001). At the molecular level, it has recently been demonstrated that SSRIs may shift in the activity of the 3α-hydroxysteroid oxidreductase, which catalyzes the conversion of 5α-DHP into 3α, 5α-THP, towards the reductive direction thereby enhancing the formation of 3α, 5α-THP (Griffin and Mellon, 1999). In addition, 3α, 5α-THP has been suggested to possess antidepressant-like effects in mice using the Porsolt swim test (Khisti et al., 2000). These preclinical findings suggest that 3α-reduced neuroactive steroids such as 3α, 5α-THP may play a role for the treatment of depression with antidepressant drugs. Indeed, the concentrations of the GABA agonistic neuroactive steroids 3α, 5α-THP and pregnanolone were reduced in the plasma of depressed patients, while there was an increase in 3β, 5α-THP, an antagonistic isomer for 3α, 5α-THP (Romeo et al., 1998). This disequilibrium of neuroactive steroids could be corrected by treatment with fluoxetine throughout several weeks (Romeo et al., 1998; Uzunova et al., 1998) (Fig. 5). In contrast to the preclinical data, also tri- and tetracyclic antidepressants interfered with the composition of neuroactive steroids in a similar way as did SSRIs (Romeo et al., 1998). However, treatment of depressed patients with repetitive transcranial magnetic stimulation did not affect the concentrations of neuroactive steroids neither in responders nor in nonresponders to this nonpharmacological treatment (Padberg et al., 2002).

Studies investigating DHEA or DHEA-S concentrations in depression have yielded divergent results. While one study noted a decrease in DHEA-S concentrations associated with depression (Barrett-Connor et al., 1999), other studies reported an increase during major depression (Heuser et al., 1998; Fabian et al., 2001). Elevated baseline DHEA-S concentrations have even been suggested to predict non-response to electroconvulsive therapy (ECT) (Maayan et al., 2000). Nevertheless, treatment with DHEA either as the only medication or as an adjunct to stable antidepressant medication may exert beneficial effects on depressed mood (Wolkowitz et al., 1997; Wolkowitz et al., 1999). Therefore, the role of DHEA in depression and antidepressant therapy should receive further consideration in the future.

Ethanol tolerance and withdrawal

Animal studies have shown that systemic ethanol administration may elevate the concentrations of 3α, 5α-THP in rat brain (Janis et al., 1998) and that 3α, 5α-THP might contribute to the pharmacological actions of ethanol (van Doren et al., 2000). On the other hand, 3α, 5α-THP protects against

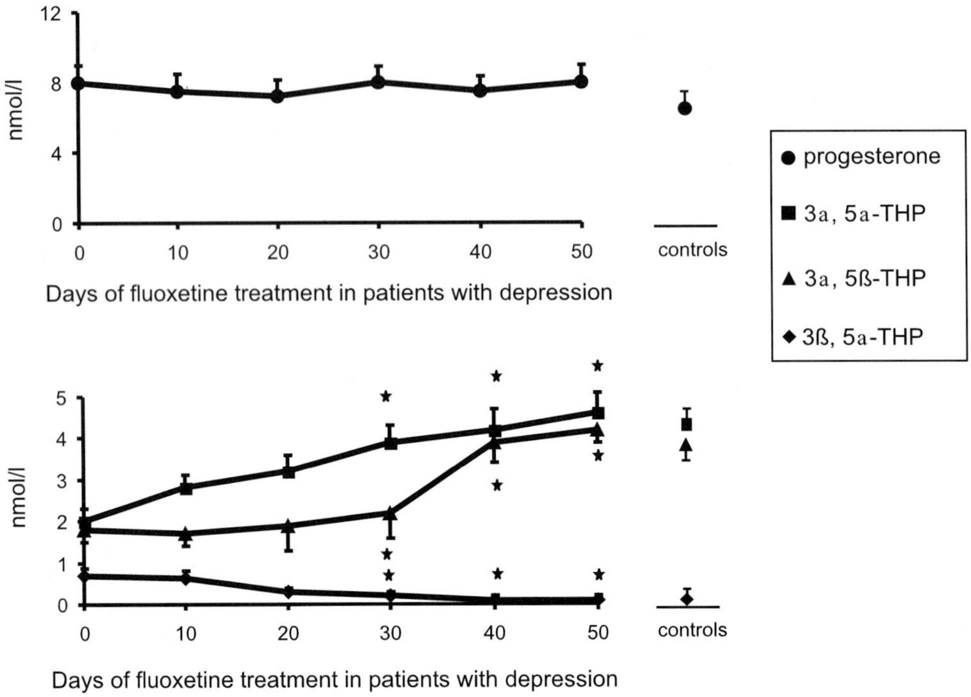

Fig. 5. Plasma concentrations of neuroactive steroids in depressed patients during treatment with 20 mg fluoxetine. The asterisks indicate significant differences from baseline values. Modified according Romeo et al. (1998).

bicuculline-induced seizures during ethanol withdrawal (Devaud et al., 1995). Interestingly, an abstinence syndrome with increased seizure liability may also occur after discontinuation of GABAergic steroids (Janis et al., 1998; Smith et al., 1998b) which may be related to changes in the kinetics of $GABA_A$ receptor channels. Moreover, rats selectively bred for high sensitivity to ethanol exhibit also an enhanced sensitivity to 3α-reduced neuroactive steroids (Korpi et al., 2001). However, also pregnenolone sulfate (PS) and DHEA-S have been suggested to be involved in the development of tolerance to ethanol in mice (Barbosa and Morato, 2001). In patients suffering from ethanol abuse, the concentrations of 3α, 5α-THP are markedly reduced during ethanol withdrawal (Romeo et al., 1996; Romeo et al., 2000) and normalized within four weeks (Romeo et al., 2000). The reduced concentrations of 3α, 5α-THP might contribute to the enhanced seizure liability of such patients during ethanol withdrawal. Treatment with fluoxetine results in an earlier rise in 3α, 5α-THP concentrations and is accompanied by a decrease of depression and anxiety during ethanol withdrawal.

Anxiolytic properties and anxiety disorders

Positive allosteric modulators of $GABA_A$ receptors, e.g. benzodiazepines, are effective anxiolytic substances. Thus, also 3α-reduced neuroactive steroids should exert anxiolytic effects. Indeed, such steroids were effective anxiolytics in different animal models, e.g. the elevated plus maze test (Crawley et al., 1986; Bitran et al., 1991; Wieland et al., 1991). Also progesterone as a precursor molecule for 3α-reduced neuroactive steroids may act as an anxiolytic via $GABA_A$ receptors (Bitran et al., 1995). Meanwhile, anxiolytic properties have also been demonstrated for synthetic analogues of 3α-reduced neuroactive steroids (Vanover et al., 2000). 3α-reduced neuroactive steroids may further counteract the anxiogenic effects of CRF and reduce the expression of the

CRF gene (Patchev et al., 1994). Although the anxiolytic effects of 3α-reduced neuroactive steroids in animal models are promising, putative side effects such as toxicity, sedation and withdrawal effects have to be taken into consideration and no firm conclusion can be drawn at the moment whether such steroids are superior to benzodiazepines as anxiolytics. First studies in patients with panic disorder from our research group suggest that 3α-reduced neuroactive steroids may play a pivotal role in human anxiety in that they may serve as a counterregulatory mechanism against the occurrence of spontaneous panic attacks (Ströhle et al., 2002). Studies of neuroactive steroids during experimentally induced panic attacks in patients with panic disorder and healthy controls in our research group showed that there is a pronounced decrease in 3α, 5α-THP and 3α, 5β-THP together with an increase 3β, 5α-THP following both cholecystokinin tetrapeptide (CCK-4) and sodium lactate administration in patients with panic disorder (Ströhle et al., in press). However, such changes do not occur in healthy controls (Ströhle et al., in press). These changes in neuroactive steroid composition might result in a decreased GABAergic tone that may be related to the pathophysiology of panic attacks in patients with panic disorder. Although treatment with paroxetine did not further increase the concentrations of GABAergic neuroactive steroids, antidepressants such as SSRIs might be effective as antipanic agents also through stabilizing the equilibrium of endogenous neuroactive steroids (Ströhle et al., 2002).

Conclusions

Neuroactive steroids may modulate neuronal function through their concurrent influence on neuronal excitability and gene expression (Fig. 6). This intracellular cross-talk between genomic and nongenomic steroid effects provides the molecular basis for steroid action in the brain and the future development of such compounds in neuropsychopharmacology, both with regard to putative clinical

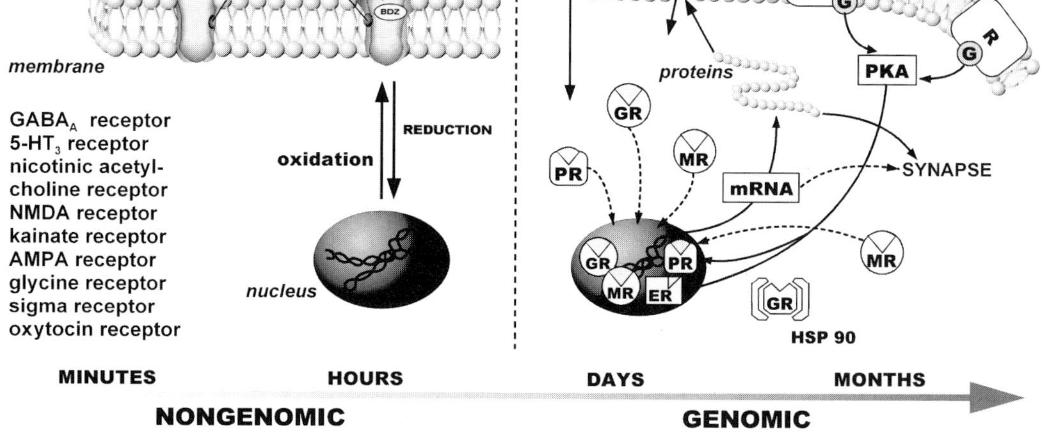

Fig. 6. Nongenomic and genomic effects of neuroactive steroids. Abbreviations: BDZ, benzodiazepines; R, receptor; G, G-protein; PKA, protein kinase A; HSP 90, heat shock protein 90; GR, glucocorticoid receptor; MR, mineralocorticoid receptor; PR, progesterone receptor; ER, estrogen receptor. Reproduced with permission from Rupprecht and Holsboer (1999).

effects and side effects. One important issue is specificity. As yet, no naturally occurring steroid with a really specific and selective action at a distinct steroid receptor or neurotransmitter receptor has been identified. Another issue that deserves further consideration is treatment duration. While behavioural properties of neuroactive steroids are quite well characterized in numerous paradigms after acute administration, studies on the consequences of long-term administration of such compounds are widely lacking to date. As a prerequisite for further clinical exploitation of such steroids in neuropsychopharmacology especially these types of studies are particularly needed in the future. In conclusion, endogenous or exogenous neuroactive steroids offer a considerable potential in the treatment of neuropsychiatric disorders. Future studies addressing the effects of neuroactive steroids on multiple neurotransmitter receptors and the behavioural consequences of long-term administration will be crucial to explore the neuropsychopharmacological potential of this yet unexploited class of drugs.

Acknowledgements

The studies on neuroactive steroids at the Max-Planck-Institute of Psychiatry and the Department of Psychiatry, Ludwig Maximilian University, Munich are supported by the Gerhard Heβ Programm of the Deutsche Forschungsgemeinschaft and the German Federal Research Ministry within the promotional emphasis "Competence Nets in Medicine".

References

Akwa, Y., Morfin, R.F., Robel, P. and Baulieu, E.E. (1992) Neurosteroid metabolism. 7 alpha-hydroxylation of dehydroepiandrosterone and pregnenolone by rat brain microsomes. Biochem. J., 288: 959–964.

Araneo, B. and Daynes, R. (1995) Dehydroepiandrosterone functions as more than an antiglucocorticoid in preserving immunocompetence after thermal injury. Endocrinology, 136: 393–401.

Azuma, T., Nagai, Y., Saito, T., Funauchi, M., Matsubara, T. and Sakoda, S. (1999) The effect of dehydroepiandrosterone sulfate administration to patients with multi-infarct dementia. J. Neurol. Sci., 162: 69–73.

Barbaccia, M.L., Affricano, D., Purdy, R.H., Maciocco, E., Spiga, F. and Biggio, G. (2001) Clozapine, but not haloperidol, increases brain concentrations of neuroactive steroids in the rat. Neuropsychopharmacology, 25: 489–497.

Barbosa, A.D. and Morato, G.S. (2001) Influence of neurosteroids on the development of rapid tolerance to ethanol in mice. Eur. J. Pharmacol., 431: 179–188.

Barrantes, F.J., Antollini, S.S., Bouzat, C.B., Garbus, I. and Massol, R.H. (2000) Nongenomic effects of steroids on the nicotinic acetylcholine receptor. Kidney Int., 57: 1382–1389.

Barrett-Connor, E., von Muhlen, D., Laughlin, G.A. and Kripke, A. (1999) Endogenous levels of dehydroepiandrosterone sulfate, but not other sex hormones, are associated with depressed mood in older women: the Rancho Bernardo Study. J. Am. Geriatr. Soc., 47: 685–691.

Baulieu, E.E. (1991) Neurosteroids: A new function in the brain. Biol. Cell., 71: 3–10.

Baulieu, E.E. (1998) Neurosteroids: a novel function of the brain. Psychoneuroendocrinology, 23: 963–987.

Bäckström, T., Zetterlund, B., Blom, S. and Romano, M. (1984) Effects of intravenous progesterone infusions on the epileptic discharge frequency in women with partial epilepsy. Acta Neurol. Scand., 69: 240–248.

Beekman, M., Ungard, J.T., Gasior, M., Carter, R.B., Dijkstra, D., Goldberg, S.R. and Witkin, J.M. (1998) Reversal of behavioral effects of pentylenetetrazol by the neuroactive steroid ganaxolone. J. Pharmacol. Exp. Ther., 284: 868–877.

Belelli, D., Lan, N.C. and Gee, K.W. (1990) Anticonvulsant steroids and the GABA/benzodiazepine receptor-chloride ionophore complex. Neurosci. Biobehav. Rev., 14: 315–322.

Bicikova, M., Dibbelt, L., Hill, M., Hampl, R. and Starka, L. (1998) Allopregnanolone in women with premenstrual syndrome. Hormon. Metab. Res., 30: 227–230.

Bicikova, M., Tallová, J., Hill, M., Krausova, Z. and Hampl, R. (2000) Serum concentrations of some neuroactive steroids in women suffering from mixed anxiety-depressive disorder. Neurochem. Res., 25: 1623–1627.

Bitran, D., Hilvers, R.J. and Kellogg, C. K. (1991) Anxiolytic effects of 3α-hydroxy-5α [β]-pregnan-20-one: endogenous metabolites of progesterone that are active at the $GABA_A$ receptor. Brain Res., 561: 157–161.

Bitran, D., Shiekh, M. and McLeod, M. (1995) Anxiolytic effect of progesterone is mediated by the neurosteroid allopregnanolone at brain $GABA_A$ receptors. Journal of Neuroendocrinology, 7: 171–177.

Browne, E.S., Porter, J.R., Correa, G., Abadie, J. and Svec, F. (1993) Dehydroepiandrosterone regulation of the hepatic glucocorticoid receptor in the Zucker rat. The obesity research program. J. Steroid Biochem. Molec. Biol., 45: 517–524.

Bullock, A.E., Clark, A.L., Grady, S.R., Robinson, S.F., Slobe, B.S., Marks, M.J. and Collins, A.C. (1997) Neurosteroids modulate nicotinic receptor function in mouse striatal and thalamic synaptosomes. J. Neurochem., 68: 2412–2423.

Carl, P., Högskilde, S., Nielsen, J.W., Sörensen, M.B., Lindholm, M., Karlen, B. and Bäckström, T. (1990) Pregnanolone emulsion: a preliminary pharmacokinetic and pharmacodynamic study of a new intravenous agent. Anaesthesia, 45: 189–197.

Carter, R.B., Wood, P.L., Wieland, S., Hawkinson, J.E., Belelli, D., Lambert, J.J., White, H.S., Wolf, H.H., Mirsadeghi, S., Tahir, S.H., Bolger, M.B., Lan, N.C. and Gee, K.W. (1997) Characterization of the anticonvulsant properties of ganaxolone (CCD 1042; 3α-hydroxy-3-methyl-5α-pregnan-20-one), a selective high-affinity steroid modulator of the $GABA_A$ receptor. J. Pharmacol. Exp. Ther., 280: 1284–1295.

Celotti, F., Melcangi, R.C. and Martini, L. (1992) The 5α-reductase in the brain: molecular aspects and relation to brain function. Front. Neuroendocrinol., 13: 163–215.

Compagnone, N.A. and Mellon, S.H. (2000) Neurosteroids: Biosynthesis and function of these novel neuromodulators. Front. Neuroendocrinol. 21, 1–56.

Concas, A., Mostallino, M.C., Porcu, P., Follesa, P., Barbaccia, M.L., Trabucchi, M., Purdy, R.H., Grisenti, P. and Biggio, G. (1998) Role of brain allopregnanolone in the plasticity of γ-aminobutyric acid type A receptor in rat brain during pregnancy and after delivery. Proc. Natl. Acad. Sci. USA, 95: 13284–13289.

Crawley, J.N., Glowa, J.R., Majewska, M.D. and Paul, S.M. (1986) Anxiolytic activity of an endogenous adrenal steroid. Brain Res., 398: 382–385.

Damianisch, K., Rupprecht, R. and Lancel, M. (2001) The influence of subchronic administration of the neurosteroid allopregnanolone on sleep in the rat. Neuropsychopharmacology, 25: 576–584.

Devaud, L.L., Purdy, R.H. and Morrow, A.L. (1995) The neurosteroid 3α-hydroxy-5α-pregnan-20-one, protects against bicuculline-induced seizures during ethanol withdrawal in rats. Alcoholism: Clinical and Experimental Research, 19: 350–355.

Edgar, D.M., Seidel, W.F., Gee, K.W., Lan, N.C., Field, G., Xia, H., Hawkinson, J.E., Wieland, S., Carter, R.B. and Wood, P.L. (1997) CCD-3693: An orally bioavailable analog of the endogenous neuroactive steroid, pregnanolone, demonstrates potent sedative hypnotic actions in the rat. J. Pharmacol. Exp. Ther., 282: 420–429.

Evans, R.M. (1988) The steroid and thyroid hormone receptor superfamily. Science, 240: 889–895.

Fabian, T.J., Dew, M.A., Pollock, B.G., Reynolds, C.F., Mulsant, B.H., Butters, M.A., Zmuda, M.D., Linares, A.M., Trottini, M. and Kroboth, P.D. (2001) Endogenous concentrations of DHEA and DHEA-S decrease with remission of depression in older adults. Biol. Psychiatry, 50: 767–774.

Flood, J.F., Morley, J.E. and Roberts, E. (1992) Memory-enhancing effects in male mice of pregnenolone and of steroids metabolically derived from it. Proc. Natl. Acad. Sci. USA, 89: 1567–1571.

Flood, J.F., Smith, G.E. and Roberts, E. (1988) Dehydroepiandrosterone and its sulfate enhance memory retention in mice. Brain Res., 447: 269–278.

Friess, E., Tagaya, H., Trachsel, L., Holsboer, F. and Rupprecht, R. (1997) Progesterone-induced changes in sleep in male subjects. Am. Journal Physiol., 272: E885–E891.

Frye, C.A. and Duncan, J.E. (1994) Progesterone metatolites effective at the $GABA_A$ receptor complex modulate pain sensitivity in rats. Brain Res., 643: 194–203.

Frye, C.A. and Scalise, T.J. (2000) Anti-seizure effects of progesterone and 3α,5α–THP in kainic acid and perforant pathway models of epilepsy. Psychoneuroendocrinology, 25: 407–420.

Gasior, M., Carter, R.B., Goldber, S.R. and Witkin, J.M. (1997) Anticonvulsant and behavioral effects of neuroactive steroids alone and in conjuction with diazepam. J. Pharmacol. Exp. Ther., 282: 543–553.

Gasior, M., Carter, R.B. and Witkin, J.M. (1999) Neuroactive steroids: potential therapeutic use in neurological and psychiatric disorders. Trends Pharmacol. Sci., 20: 107–112.

Gee, K.W., Bolger, M.B., Brinton, R.E., Coirini, H. and McEwen, B.S. (1988) Steroid modulation of the chloride ionophore in rat brain: Structure-activity requirements, regional dependance and mechanism of action. J. Pharmacol. Exp. Ther., 246: 803–812.

Girdler, S.S., Straneva, P.A., Light, K.C., Pedersen, C.A. and Morrow, A.L. (2001) Allopregnanolone levels and reactivity to mental stress in premenstrual dysphoric disorder. Biol. Psychiatry., 49: 788–797.

Grazzini, E., Guillon, G., Mouillac, B. and Zingg, H.H. (1998) Inhibition of oxytocin receptor function by direct binding of progesterone. Nature, 392: 509–512.

Griffin, L.D. and Mellon, S.H. (1999) Selective serotonin reuptake inhibitors directly alter activity of neurosteroidogenic enzymes. Proc. Natl. Acad. Sci. USA, 96: 13512–13517.

Grobin, A.C. and Morrow, A.L. (2000) 3α-hydroxy-5α-pregnan-20-one exposure reduces GABA(A) receptor α4 subunit mRNA levels. Eur. J. Pharmacol., 409: R1–R2.

Gyermek, L. and Soyka, L.F. (1975) Steroid anesthetics. Anesthesiology, 42: 116–121.

Hallonquist, J.D., Seeman, M.V., Lang, M. and Rector, N.A. (1993) Variation in symptom severity over the menstrual cycle of schizophrenics. Biol. Psychiatry, 33: 207–209.

Häfner, H., Riecher-Rössler, A., an der Heiden, W., Maurer, K., Fätkenheuer, B. and Löffler, W. (1993) Generating and testing a causal explanation of the gender difference in age at first onset of schizophrenia. Psychological Med., 23: 925–940.

Herzog, A.G. (1995) Progesterone therapy in women with complex partial and secondary generalized seizures. Neurology, 45: 1660–1662.

Heuser, I., Deuschle, M., Luppa, P., Schweiger, U., Standhardt, H. and Weber, B. (1998) Increased diurnal plasma concentrations of dehydroepiandrosterone in depressed patients. J. Clin. Endocrinol. Metab., 83: 3130–3133.

Hillen, T., Lun, A., Reischies, F.M., Borchelt, M., Steinhagen-Thiessen, E. and Schaub, R.T. (2000) DHEA-S plasma levels and incidence of Alzheimer's Disease. Biol. Psychiatry, 47: 161–163.

Janis, G.C., Devaud, L.L., Mitsuyama, H. and Morrow, A.L. (1998) Effects of chronic ethanol consumption and withdrawal on the neuroactive steroid 3α-hydroxy-5α-pregnan-20-one in male and female rats. Alcohol. Clin. Exp. Res., 22: 2055–2061.

Khisti, R.T., Chopde, C.T. and Jain, S.P. (2000) Antidepressant-like effect of the neurosteroid 3α-hydroxy-5α-pregnan-20-one in mice forced swim test. Pharmacol. Biochem. Behav., 67: 137–143.

Korneyev, A. and Costa, E. (1996) Allopregnanolone (THP) mediates anesthetic effects of progesterone in rats. Horm. Behav., 30: 37–43.

Korpi, E.R., Makela, R., Romeo, E., Guidotti, A., Uusi-Oukari, M., Furnari, C., di Michele, F., Sarviharju, M., Xu, M. and Rosenberg, P.H. (2001) Increased behavioral neurosteroid sensitivity in a rat line selectively bred for high alcohol sensitivity. Eur. J. Pharmacol., 421: 31–38.

Ladurelle, N., Eychenne, B., Denton, D., Blair-West, J., Schumacher, M., Robel, P. and Baulieu, E. (2000) Prolonged intracerebroventricular infusion of neurosteroids affects cognitive performances in the mouse. Brain Res., 858: 371–379.

Lambert, J.J., Belelli, D., Hill-Venning, C. and Peters, J.A. (1995) Neurosteroids and $GABA_A$ receptor function. Trends Pharmacol. Sci., 16: 295–303.

Lancel, M., Faulhaber, J., Holsboer, F. and Rupprecht, R. (1996) Progesterone induces changes in sleep EEG comparable to those of agonistic $GABA_A$ receptor modulators. Am. J. Physiol., 271: E763–E772.

Lancel, M., Faulhaber, J., Schiffelholz, T., Romeo, E., di Michele, F., Holsboer, F. and Rupprecht, R. (1997) Allopregnanolone affects sleep in a benzodiazepine-like fashion. J. Pharmacol. Exp. Ther., 282: 1213–1218.

Maayan, R., Yagorowski, Y., Grupper, D., Weiss, M., Shtaif, B., Kaoud, M.A. and Weizman, A. (2000) Basal plasma dehydroepiandrosterone sulfate level: a possible predictor for response to electroconvulsive therapy in depressed psychotic inpatients. Biol. Psychiatry, 48: 693–701.

Maitra, R. and Reynolds, J.N. (1998) Modulation of $GABA_A$ receptor function by neuroactive steroids: evidence for heterogeneity of steroid sensitivity of recombinant $GABA_A$ receptor isoforms. Can. J. Physiol. Pharmacol., 76: 909–920.

Majewska, M.D., Harrison, N.L., Schwartz, R.D., Barker, J.L. and Paul, S.M. (1986) Steroid hormone metabolites are barbiturate-like modulators of the GABA receptor. Science, 232: 1004–1007.

Maricq, A.V., Peterson, A.S., Brake, A.J., Myers, R.M. and Julius, D. (1991) Primary structure and functional expression of the $5HT_3$ receptor, a serotonin-gated ion channel. Science, 254: 432–437.

Marx, C.E., Duncan, G.E., Gilmore, J.H., Lieberman, J.A. and Morrow, A.L. (2000) Olanzapine increases allopregnanolone in the rat cerebral cortex. Biol. Psychiatry, 47: 1000–1004.

McEwen, B.S. (1991) Non-genomic and genomic effects of steroids on neural activity. Trends Pharmacol. Sci., 12: 141–147.

Mellon, S.H. (1994) Neurosteroids: Biochemistry, modes of action and clinical relevance. J. Clin. Endocrinol. Metab., 78: 1003–1008.

Mendelson, W.B., Martin, J.V., Perlis, M., Wagner, R., Majewska, M.D and Paul, S.M. (1987) Sleep induction by an adrenal steroid in the rat. Psychopharmacology, 93: 226–229.

Monaghan, E.P., McAuley, J.W. and Data, J.L. (1999) Ganaxolone: a novel positive allosteric modulator of the GABA(A) receptor complex for the treatment of epilepsy. Expert Opin. Investig. Drugs, 8: 1663–1671.

Monaghan, E.P., Navalta, L.A., Shum, L., Ashbrock, D.W. and Lee, D.A. (1997) Initial human experience with ganaxolone, a neuroactive steroid with antiepileptic activity. Epilepsia, 38: 1026–1031.

Monnet, F.P., Mahé, V., Robel, P. and Baulieu, E.E. (1995) Neurosteroids, via sigma receptors, modulate the $[^3H]$norepinephrine release evoked by N-methyl-D-aspartate in the rat hippocampus. Proc. Natl. Acad. Sci. USA, 92: 3774–3778.

Nadeson, R. and Goodchild, C.S. (2001) Antinociceptive properties of neurosteroids III: experiments with alphadolone given intravenously, intraperitoneally, and intragastrically. Br. J. Anaesth., 86: 704–708.

Näsman, B., Olsson, B., Bäckström, T., Eriksson, S., Grankvist, K., Viitanen, M. and Bucht, G. (1991) Serum dehydroepiandrosterone sulfate in Alzheimer's disease and in multiinfarct dementia. Biol. Psychiatry, 30: 684–690.

Orchinik, M., Murray, T.F. and Moore, F.L. (1991) A corticosteroid receptor in neuronal membranes. Science, 252: 1848–1851.

Padberg, F., di Michele, F., Zwanzger, P., Romeo, E., Bernardi, G., Schüle, C., Baghai, T.C., Ella, R., Pasini, A. and Rupprecht, R. (2002) Plasma concentrations of neuroactive steroids before and after repetitive transcranial magnetic stimulation (rTMS) in major depression. Neuropsychopharmacology, 27: 874–878.

Pappas, T.D., Gametchu, B. and Watson, C.S. (1995) Membrane estrogen receptors identified by multiple antibody labelling and impeded-ligand binding. FASEB J., 9: 404–410.

Park-Chung, M., Wu, F.S. and Farb, D.H. (1994) 3α-hydroxy-5beta-pregnan-20-one sulfate: a negative modulator of the NMDA-induced current in cultured neurons. Mol. Pharmacol., 46: 146–150.

Patchev, V.K., Montkowski, A., Rouskova, D., Koranyi, L., Holsboer, F. and Almeida, O.F.X. (1997) Neonatal treatment of rats with the neuroactive steroid tetrahydrodeoxycorticosterone (THDOC) abolishes the behavioral and neurondocrine consequences of adverse early life events. J. Clin. Invest., 99: 962–966.

Patchev, V.K., Shoaib, M., Holsboer, F. and Almeida, O.F.X. (1994) The neurosteroid tetrahydroprogesterone counteracts corticotropin-releasing hormone-induced anxiety and alters the release and gene expression of corticotropin-releasing hormone in the rat hypothalamus. Neurosci., 62: 265–271.

Paul, S.M. and Purdy, R.H. (1992) Neuroactive steroids. FASEB J., 6: 2311–2322.

Pearson Murphy, B.E., Steinberg, S.I., Hu, F.-Y. and Allison, C.M. (2001) Neuroactive ring A-reduced metabolites of progesterone in human plasma during pregnancy: elevated levels of 5alpha-dihydroprogesterone in depressed patients during the latter half of pregnancy. J. Clin. Endocrinol. Metab., 86: 5981–5987.

Prince, R.J. and Simmonds, M.A. (1992) 5β-pregnan-3β-ol-20-one, a specific antagonist at the neurosteroid site of the $GABA_A$ receptor-complex. Neurosci. Lett.,135: 273–275.

Ramirez, V.D. and Zheng, J. (1996) Membrane sex-steroid receptors in the brain. Front. Neuroendocrinol., 17: 402–439.

Rapkin, A.J., Morgan, M., Goldman, L., Brann, D.W., Simone, D. and Mahesh, V.B. (1997) Progesterone metabolite allopregnanolone in women with premenstrual syndrome. Obstet. Gynecol., 90: 709–714.

Rasgon, N., Serra, M., Biggio, G., Pisu, M. G., Fairbanks, L., Tanavoli, S. and Rapkin, A. (2001) Neuroactive steroid-serotonergic interaction: responses to an intravenous L-tryptophan challenge in women with premenstrual syndrome. Eur. J. Endocrinol., 145: 25–33.

Reddy, D.S. and Rogawski, M.A. (2000) Chronic treatment with the neuroactive steroid ganaxolone in the rat induces anticonvulsant tolerance to diazepam but not to itself. J. Pharmacol. Exp. Ther., 295: 1241–1248.

Romeo, E., Brancati, A., De Lorenzo, A., Fucci, P., Furnari, C., Pompili, E., Sasso, G. F., Spalletta, G., Troisi, A. and Pasini, A. (1996) Marked decrease of plasma neuroactive steroids during alcohol withdrawal. Clin. Neuropharmacol., 19: 366–369.

Romeo, E., Pompili, E., di Michele, F., Pace, M., Rupprecht, R., Bernardi, G. and Pasini, A. (2000) Effects of fluoxetine, indomethacine and placebo on 3α, 5α tetrahydroprogesterone (THP) plasma levels in uncomplicated alcohol withdrawal. World J. Biol. Psychiatry, 1: 101–104.

Romeo, E., Ströhle, A., Spalletta, G., di Michele, F., Hermann, B., Holsboer, F., Pasini, A. and Rupprecht, R. (1998) Effects of antidepressant treatment on neuroactive steroids in major depression. Am. J. Psychiatry, 155: 910–913.

Rupprecht, R., Berning, B., Hauser, C.A.E., Holsboer, F. and Reul, J.M.H.M. (1996) Steroid receptor mediated effects of neuroactive steroids: characterization of structure-activity relationship. Eur. J. Pharmacol., 303: 227–234.

Rupprecht, R. and Holsboer, F. (1999) Neuroactive steroids: mechanisms of action and neuropsychopharmacological perspectives. Trends Neurosci., 22: 410–416.

Rupprecht, R., Koch, M., Montkowski, A., Lancel, M., Faulhaber, J., Harting, J. and Spanagel, R. (1999) Assessment of neuroleptic-like properties of progesterone. Psychopharmacology, 143: 29–38.

Rupprecht, R., Reul, J.M.H.M., Trapp, T., van Steensel, B., Wetzel, C., Damm, K., Zieglgänsberger, W. and Holsboer, F. (1993) Progesterone receptor-mediated effects of neuroactive steroids. Neuron, 11: 523–530.

Schulz, H., Jobert, M., Gee, K.W. and Ashbrook, D.W. (1996) Soporific effect of the neurosteroid pregnanolone in relation to the substance's plasma level: a pilot study. Neuropsychobiology, 34: 106–112.

Serra, M., Pisu, M.G., Muggironi, M., Parodo, V., Papi, G., Sari, R., Dazzi, L., Spiga, F., Purdy, R.H. and Biggio, G. (2001) Opposite effects of short- versus long-term administration of fluoxetine on the concentrations of neuroactive steroids in rat plasma and brain. Psychopharmacology, 158: 48–54.

Smith, S.S., Gong, Q.H., Hsu, F.-C., Markowitz, R.S., ffrench-Mullen, J.M.H. and Li, X. (1998a) $GABA_A$ receptor α4 subunit suppression prevents withdrawal properties of an endogenous steroid. Nature, 392: 926–930.

Smith, S.S., Gong, Q.H., Li, X., Moran, M.H., Bitran, D., Frye, C.A. and Hsu, F.C. (1998b) Withdrawal from 3α-OH-5α-pregnan-20-one using a pseudopregnancy model alters the kinetics of hippocampal $GABA_A$-gated current and increases the $GABA_A$ receptor alpha4 subunit in association with increased anxiety. J. Neurosci., 18: 5275–5284.

Steiner, M., Steinberg, S., Stewart, D., Carter, D., Berger, C., Reid, R., Grover, D. and Streiner, D. (1995) Fluoxetine in the treatment of premenstrual dysphoria. N. Engl. J. Med., 332, 1529–1534.

Ströhle, A., Romeo, E., di Michele, F., Pasini, A., Yassouridis, A., Holsboer, F. and Rupprecht, R. (2002) $GABA_A$ receptor modulatory neuroactive steroid composition in panic disoder and during paroxetine treatment. Am. J. Psychiatry, 159: 145–147.

Ströhle, A., Romeo, E., di Michele, F., Pasini, A., Hermann, B., Gajewski, G., Holsboer, F. and Rupprecht, R. (2003) Induced panic attacks shift $GABA_A$ receptor modulatory steroid composition in patients with panic disorder: preliminary results. Archives of General Psychiatry, 60: 161–168.

Su, T.-P., London, E.D. and Jaffe, J.H. (1988) Steroid binding at σ receptors suggests a link between endocrine, nervous, and immune systems. Science, 240: 219–221.

Sunderland, T., Merril, C.R., Harrington, M.G., Lawlor, B. A., Molchan, S.E., Martinez, R. and Murphy, D.L. (1989) Reduced plasma dehydroepiandrosterone concentrations in Alzheimer's disesase. Lancet, II: 570–572.

Sundström, I. and Bäckström, T. (1998) Citalopram increases pregnanolone sensitivity in patients with premenstrual syndrome: an open trial. Psychoneuroendocrinology, 23: 73–88.

Thomas, G., Frenoy, N., Legrain, S., Sebag-Lanoe, R., Baulieu, E.E. and Debuire, B. (1994) Serum dehydroepiandrosterone sulfate levels as an individual marker. J. Clin. Endocrinol. Metab., 79: 1273–1276.

Truss, M. and Beato, M. (1993) Steroid hormone receptors: interaction with deoxyribonucleic acid and transcription factors. Endocrine Rev., 14: 459–479.

Uzunov, D.P., Cooper, T.B., Costa, E. and Guidotti, A. (1996) Fluoxetine-elicited changes in brain neurosteroid content measured by negative ion mass fragmentography. Proc. Natl. Acad. Sci., USA 93: 12599–12604.

Uzunova, V., Sheline, Y., Davis, J.M., Rasmusson, A., Uzunov, D.P., Costa, E. and Guidotti, A. (1998) Increase in the cerebrospinal fluid content of neurosteroids in patients with unipolar major depression who are receiving fluoxetine or fluvoxamine. Proc. Natl. Acad. Sci.USA, 95: 3239–3244.

Valera, S., Ballivet, M. and Bertrand, D. (1992) Progesterone modulates a neuronal nicotinic acetylcholine receptor. Proc. Natl. Acad. Sci.USA, 89: 9949–9953.

Vallee, M., Mayo, W., Darnaudery, M., Corpechot, C., Young, J., Koehl, M., Le Moal, M., Baulieu, E.E., Robel, P. and Simon, H. (1997) Neurosteroids: deficient cognitive performance in aged rats depends on low pregnenolone sulfate levels in the hippocampus. Proc. Natl. Acad. Sci. USA, 94: 14865–14870.

VanDoren, M.J., Matthews, D.B., Janis, G.C., Grobin, A.C., Devaud, L.L. and Morrow, A.L. (2000) Neuroactive steroid 3alpha-hydroxy-5alpha-pregnan-20-one modulates electrophysiological and behavioral actions of ethanol. J. Neuroscience, 20: 1982–1989.

Vanover, K.E., Rosenzweig-Lipson, S., Hawkinson, J.E., Lan, N.C., Belluzzi, J.D., Stein, L., Barrett, J.E., Wood, P.L. and Carter, R.B. (2000) Characterization of the anxiolytic properties of a novel neuroactive steroid, Co 2-6749 (GMA-839; WAY-141839; 3α, 21-Dihydroxy-3β-trifluoromethyl-19-nor-5β-pregnan-20-one), a selective modulator of γ-aminobutyric $acid_A$ receptors. J. Pharmacol. Exp. Ther., 295: 337–345.

Weaver, C.E. Jr., Marek, P., Park-Chung, M., Tam, S.W. and Farb, D.H. (1997b). Neuroprotective activity of a new class of steroidal inhibitors of the N–methyl–D–aspartate receptor. Proc. Natl. Acad. Sci. USA, 94: 10450–10454.

Weaver, C.E. Jr., Park-Chung, M., Gibbs, T.T. and Farb, D.H. (1997a) 17β-estradiol protects against NMDA-induced excitotoxicity by direct inhibition of NMDA receptors. Brain Res., 761: 338–341.

Wehling, M. (1997) Specific, nongenomic actions of steroid hormones. Annu. Rev. Physiol., 59: 365–393.

Wetzel, C.H.R., Hermann, B., Behl, C., Pestel, E., Rammes, G., Zieglgänsberger, W., Holsboer, F. and Rupprecht, R. (1998) Functional antagonism of gonadal steroids at the 5-HT_3 receptor. Mol. Endocrinol., 12: 1441–1451.

Wieland, S., Lan, N.C., Mirasedeghi, S., Gee, K.W. (1991) Anxiolytic activity of the progesterone metabolite 5α-pregnan-3α-ol-20-one. Brain Res., 565: 263–268.

Wolkowitz, O.M., Reus, V.I., Keebler, A., Nelson, N., Friedland, M., Brizendine, L. and Roberts, E. (1999) Double-blind treatment of major depression with dehydroepiandrosterone. Am. J. Psychiatry, 156: 646–649.

Wolkowitz, O.M., Reus, V.I., Roberts, E., Manfredi, F., Chan, T., Raum, W.J., Ormiston, S., Johnson, R., Canick, J., Brizendine, L. and Weingartner, H. (1997) Dehydroepiandrosterone (DHEA) treatment of depression. Biol. Psychiatry, 41: 311–318.

Wu, F.-S., Gibbs, T.T. and Farb, D.H. (1990) Inverse modulation of γ–aminobutyric acid- and glycine-induced currents by progesterone. Mol. Pharmacol., 37: 597–602.

Wu, F.-S., Gibbs, T.T. and Farb, D.H. (1991) Pregnenolone sulfate: A positive allosteric modulator at the N-Methyl-D-aspartate receptor. Mol. Pharmacol., 40: 333–336.

Zakon, H.H. (1998) The effects of steroid hormones on electrical activity of excitable cells. Trends Neurosci., 21: 202–207.

Zorumski, C.F., Mennerick, S., Isenberg, K.E. and Covey, D.F. (2000) Potential clinical uses of neuroactive steroids. Curr. Opin. Investig. Drugs, 1: 360–369.

T. Steckler, N.H. Kalin and J.M.H.M. Reul (Eds.)
Handbook of Stress and the Brain, Vol. 15
ISBN 0-444-51173-3

CHAPTER 4.9

Endogenous opioids, stress, and psychopathology

Andrea L.O. Hebb, Sylvie Laforest and Guy Drolet*

Centre de Recherche en Neurosciences, CHUL, Université Laval, QC, Canada

Abstract: The maintenance of homeostasis during stressful conditions is mediated through a complex, highly interactive organization of neuroanatomical pathways in the central nervous system (CNS). Specific neurotransmitter systems initiate immune, endocrine, metabolic, cardiovascular, respiratory, and behavioral changes in response to stress. Among these neurotransmitter systems, endogenous opioids (enkephalin, β-endorphin, and dynorphin) represent major modulatory systems in responding and adapting to stress. The endogenous opioids are important neuromodulators, acting at multiple levels of the CNS, participating in the mediation, modulation, and regulation of the stress response, including neuroendocrine (hypothalamo–pituitary–adrenal, HPA axis), autonomic (sympathoadrenal axis), and behavioral (fear, anxiety, memory, locomotor activity, mood, perception of reward) responses. Endogenous opioids also play a fundamental role in stressor-induced analgesia and the modulation of organismic defensive repertoires. Although being part of the same peptide family, the endogenous opioids do not act as a single functional unit in stress regulation. Indeed, there is growing evidence demonstrating the existence of functionally opposed opioid systems affecting emotional and perceptual experiences. Therefore, conflicting reports on the functional roles of endogenous opioids in stress regulation are likely to reflect the fact that different opioid systems may have opposite actions depending on the site of action, the subtype of opioid receptor involved, or the stressor conditions.

The present review is certainly not an exhaustive evaluation of the past years on endogenous opioids and stress. Although in this review we have focused our attention mainly on the traditional opioid peptides (enkephalin, dynorphin, β-endorphin) and receptors (mu, μ-OP; delta, δ-OP; kappa, κ-OR) and their roles in stress adaptation, we will also introduce new endogenous opioid members (endomorphin and nociceptin/orphanin FQ). We will also review how psychological stressors induced alterations in endogenous opioid peptides and how this could have important implications for anxiety, depression, and panic disorder.

Opioid and stress: endogenous opioids

The endogenous opioids and their receptor systems are ubiquitous, located with varying densities throughout the central, peripheral, and autonomic nervous systems as well as in several endocrine tissues and target organs. This widespread distribution is consistent with the involvement of endogenous opioids in a broad range of functions and behavior, including regulation of pain, reinforcement, and reward, release of neurotransmitters, as well as autonomic and neuroendocrine modulation (for review, see Bolles and Fanselow, 1982; Holaday, 1983; Akil et al., 1984; Howlett and Rees, 1986; Przewlocki et al., 1991; Hayden-Hixson and Nemeroff, 1993; Yamada and Nabeshima, 1995; Akil et al., 1998; Russell and Douglas, 2000). In fact, these anatomical and functional opioid characteristics argue against the conceptualization of the endogenous opioid peptides as a single, functional unit.

Three distinct (classical) families of endogenous opioid peptides have been identified to date; the enkephalins, dynorphins, and β-endorphin. Each family is derived from different multifunctional precursor polypeptides: proenkephalin (proenkephalin A), prodynorphin (proenkephalin B), and proopioimelanocortin (POMC), respectively. The

*Corresponding author. Tel.: +1(418) 654 2152; Fax: +1(418) 654 2753; E-mail: gdrolet@crchul.ulaval.ca

endogenous opioid peptides produce their biological effects through three main types of receptors (all being members of the family of seven transmembrane G-protein-coupled receptors) referred to as mu (μ-OP), delta (δ-OP), and kappa (κ-OP). The different types of opioid receptors have been defined based on pharmacological- and radioligand-binding experiments (Mansour et al., 1987), and more recently by their recent cloning (Mansour et al., 1994; Uhl et al., 1994). Although in vitro studies indicated that each opioid peptide has affinity for more than one type of opioid receptor (enkephalins and δ-OP, dynorphins and κ-OP, endorphin and μ-OP), proenkephalin and prodynorphin cleavage products are able to bind μ-OP, δ-OP and κ-OP receptors (Quirion and Pert, 1981; Wuster et al., 1981).

More recently, additional endogenous opioid peptides have been characterized; endomorphin-1 and endomorphin-2 appear to have properties consistent with neurotransmitter/neuromodulator actions in mammals. Endomorphin-1 (Tyr-Pro-Trp-Phe-NH2, EM-1) and endomorphin-2 (Tyr-Pro-Phe-Phe-NH2, EM-2) are peptides recently isolated from brain that show the highest affinity and selectivity for the μ-OR of all the known endogenous opioids (Zadina et al., 1997). The distribution of endomorphins is consistent with a role for the peptides in the modulation of diverse functions, including perception of pain, autonomic, neuroendocrine, and some homeostatic functions as well as modulation of responses to stress (Martin-Schild et al., 1999; Horvath, 2000). However, the physiological significance of endomorphins in the stress response still has to be determined.

A novel group of a related but nonopioid system, the nociceptin/orphanin FQ (N/OFQ) and its receptor, has been characterized in the brain (Meunier et al., 1995; Reinscheid et al., 1995). The endogenous ligand of the N/OFQ system is a heptadecapeptide that binds with high affinity to the nociceptin–orphanin peptide (NOP) receptor (formerly know as ORL1). The N/OFQ has a high degree of sequence identity to the other endogenous opioids (especially with dynorphin A), but does not bind to traditional opioid receptors. In the same vein, the NOP receptor, which is also G-protein-coupled receptor, shows low binding affinities for selective opioid agonists and antagonists. The functions and distribution of N/OFQ and NOP receptors have been described and reviewed in detail elsewhere (Mogil and Pasternak, 2001; Reinscheid and Civelli, 2002).

There is a very important bibliographic source of information on endogenous opioids published each year since 1979. This interesting series of 24 papers constitute an annual source on research concerning the opioid system, including stress and behavior (Olson et al., 1979, 1980, 1981, 1982; Olson et al., 1983, 1984, 1985, 1986, 1987, 1989a,b, 1990, 1991, 1992, 1993, 1994, 1995, 1996, 1997, 1998; Vaccarino et al., 1999; Vaccarino and Kastin, 2000, 2001; Bodnar and Hadjimarkou, 2002).

Opioid and stress: anatomical distribution

All traditional endogenous opioid systems are widely represented in regions which are heavily involved in the stress response, including the hypothalamus, pituitary, and adrenal gland. Similarly opioid neurons are found in central autonomic centers which receive important opioidergic innervation. More specifically, central endorphinergic neurons originate from two nuclei, the arcuate nucleus in the posterior hypothalamus and the nucleus tractus solitarius in the brainstem. Both endorphin groups have extensive projections to other areas providing a rich network of POMC fibers throughout the brain (Palkovits et al., 1987; Schafer et al., 1991). POMC fibers originating from the arcuate nucleus innervate hypothalamic nuclei, including the paraventricular nucleus of the hypothalamus (PVH), the median eminence, and limbic structures including the septum, bed nucleus of the stria terminalis, amygdala, and the preoptic area (Palkovits et al., 1987; Schafer et al., 1991). In the brainstem, endorphinergic fibers are observed in the parabrachial nucleus, the ventrolateral medulla, the nucleus tractus solitarius, the dorsal motor nucleus of the vagus, and the nucleus ambiguus, regions involved in autonomic regulation.

The neuronal distribution of the enkephalin-derived and dynorphin-derived peptides is far more complex than the neuronal distribution of the POMC family. Therefore, the description of the structures containing enkephalin or dynorphin perikarya and fibers has been restricted to the nuclei that may be involved in the neurobiology of stress. The complete distribution of enkephalin and dynorphin neurons (as seen by immunohistochemistry and in situ hybridization histochemistry) and fibers throughout the brain have been extensively described by previous studies (Watson et al., 1982; Khachaturian et al., 1983; Guthrie and Basbaum, 1984; Petrusz et al., 1985; Fallon and Leslie, 1986; Harlan et al., 1987; Menetrey and Basbaum, 1987; Hurd, 1996). In brief, neurons containing enkephalin- or dynorphin-derived peptides can be found virtually at all levels of the brain, from the telencephalon to the spinal cord. Enkephalinergic neurons are present in most regions of the telencephalon, including the cerebral and piriform cortex, amygdala, septum, bed nucleus of the stria terminalis, and the preoptic area. In the hypothalamus, perikarya are seen in most nuclei (including the PVH). At the level of the brainstem, enkephalin neurons are identified in the parabrachial nucleus, ventrolateral medulla (nucleus paragigantocellularis, lateral reticular nucleus), and nucleus tractus solitarius. The distribution of enkephalin fibers and terminals is roughly similar to that of neuronal perikarya. Dense enkephalin innervation is found in the parabrachial nucleus, the nucleus tractus solitarius, and the lateral reticular nucleus which are nuclei involved in autonomic regulation. A moderate innervation is also found in PVH, median eminence, amygdala, LC, and ventrolateral medulla. Dynorphin-derived peptide neurons are present in the cerebral cortex, amygdala, the dentate gyrus of the hippocampal formation, PVH, supraoptic nucleus, parabrachial nucleus, nucleus tractus solitarius, and ventrolateral medulla. Areas containing a relatively high density of dynorphin fibers are the hippocampus, septum, median eminence, LC, dorsal motor nucleus of the vagus, and the lateral reticular nucleus.

Each opioid receptor type demonstrates a distinct anatomical distribution which was previously determined by binding studies (Mansour et al., 1987; Blackburn et al., 1988) and, more recently, by in situ hybridization histochemistry (Mansour et al., 1994, 1995) to visualize the expression of their mRNAs. In relation with the structures that are involved in the neurobiology of stress, cells expressing μ-OR are distributed in such regions as the septum, bed nucleus of the stria terminalis, hippocampus (dentate gyrus), amygdala, medial preoptic area, locus coeruleus, parabrachial nucleus, nucleus tractus solitarius, and dorsal motor nucleus of the vagus. Neurons expressing δ-OR are located in hippocampus, amygdala, and lateral reticular nucleus. Finally, the neurons expressing κ-OR are localized in regions such as piriform cortex, medial preoptic area, bed nucleus of the stria terminalis, amygdala, PVH, supraoptic nucleus, median eminence, and the nucleus tractus solitarius.

Opioid and stress: adaptation to stress

The relevance of the role(s) played by endogenous opioids in adaptation to stress may be considered critical in the etiology and pathology of certain physiological disorders associated with repeated or prolonged stress, such as cardiovascular diseases, affective, and behavior disorders. Although, several attempts have been made to assess stress-induced alterations of opioid receptors, the results of the studies are varied, even contradictory (Akil et al., 1984; Grossman, 1986; Howlett and Rees, 1986; Szekely, 1990; Przewlocki et al., 1991). Many possibilities have been advanced to explain discrepancies in the involvement of endogenous opioids in stress processes. The obvious technical considerations, such as species, agonist, and antagonist specificity, dose selectivity, route of administration, time course of the response, state and nature of the anesthetic, and stress levels, have all been offered to account for these differences. Moreover, the existence of contradictory results may also reflect the fact that opioid pathways in the brain do not act as a single entity. Furthermore, endogenous opioid peptides are involved in many physiological processes which are not directly related to the stress response.

There is solid evidence indicating major involvement for enkephalin, β-endorphin, and dynorphin in both stress-induced physiological and behavioral responses (Howlett and Rees, 1986; Katoh et al., 1990; Szekely, 1990; Przewlocki et al., 1991; Katoh et al., 1992; Pechnick, 1993). Endogenous opioids may exert their action on the HPA axis and the autonomic nervous system which are major effector systems that serve to maintain homeostasis during exposure to stressors. Indeed, considerable neuroanatomical and pharmacological data exist suggesting an involvement of opioid-derived peptides in the regulation of the sympathetic nervous system and cardiovascular system (Holaday, 1983; Morilak et al., 1990a,b; Drolet et al., 1991a,b; McCubbin, 1993) as well as on the role of opioids in neuroendocrine regulation, particularly at the level of the paraventricular nucleus of the hypothalamus. The importance of the PVH in the central regulation and coordination of the stress response is well recognized. Indeed, the PVH is one important coordinating center of the stress system having virtually all the corticotropin-releasing factor (CRF) neurons that control the release of adrenocorticotropic hormone (ACTH) at the level of the median eminence (Sawchenko, 1986; Swanson et al., 1986, 1987; Palkovits, 1987; Sawchenko et al., 1993). Furthermore, the PVH is one of the few structures that project directly onto preganglionic sympathetic neurons (Strack et al., 1989a,b), as well as onto the preganglionic neurons of the parasympathetic nervous system (Lawrence and Pittman, 1985).

Many studies have examined the expression of enkephalin mRNA (and dynorphin mRNA) within the PVH. It was demonstrated that stress caused adaptive increases in Enkephalin gene expression in these neurons. enkephalin mRNA levels are increased in the PVH after acute or chronic stress including intraperitoneal injection of hypertonic saline (Lightman and Young, 1987a,b, 1988, 1989; Harbuz and Lightman, 1989; Watts, 1992; Young and Lightman, 1992), ether stress (Watts, 1991; Ceccatelli and Orazzo, 1993), restraint (Ceccatelli and Orazzo, 1993), morphine withdrawal (Lightman and Young, 1987a, 1988; Harbuz et al., 1991), or colchicine injection (Ceccatelli and Orazzo, 1993). These findings all suggest a role for opioids in the PVH with respect to some aspects of adaptation of the organism to stress. We have recently investigated the effects of acute and chronic exposure to psychological stress on enkephalin–neuron activation (enkephalin mRNA and Fos immunoreactivity) in the PVH (Dumont et al., 2000) and in the ventrolateral medulla (Mansi et al., 2000), which provides the highest density of enkephalinergic parvocellular PVH afferents (Beaulieu et al., 1996). Acute immobilization caused a marked increase in both the number of Fos-ir and Fos-enkephalin double-labeled cells in all the parvocellular subdivisions of the paraventricular nucleus of the hypothalamus as well as in the caudal and rostral ventrolateral medulla. Chronic immobilization had no effect on basal Fos labeling of both regions, but had opposite effects on the basal number of enkephalin cells (PVH: 43% increase and ventrolateral medulla: 50% decrease). Moreover, chronically stressed rats displayed an attenuated Fos response (PVH: 67% decrease and ventrolateral medulla: 100% decrease) to subsequent immobilization exposure. Conversely, there was no significant attenuation of the activation of PVH–enkephalin neurons. By contrast, the stress-induced activation of enkephalin neurons in the ventrolateral medulla was completely abolished following chronic immobilization. These results indicate that chronic psychological stress induced a differential, apparently region-specific adaptation response of the enkephalin system. The influence of enkephalin neurons in the PVH appears to be increased following chronic stress as suggested by the increased number of basal enkephalin neurons and their sustained activation (Dumont et al., 2000). Conversely, the enkephalinergic influence originating from the ventrolateral medulla is virtually removed following exposure to chronic psychological stress (Mansi et al., 2000). These observations therefore imply that during acute stress exposure, activation of enkephalin neurons from PVH and ventrolateral medulla regulates some aspects of the stress response. During chronic stress conditions, this decrease of activation of the enkephalinergic input from ventrolateral medulla may be translated into a decrease inhibition in the PVH, while the resistance to habituation of the activation of enkephalin neurons within the PVH could contribute to buffer the potentially detrimental effects of a physiological response (i.e., HPA and autonomic axis) to stress.

A role for opioids in the amygdaloid complex in adaptation to stress

There is strong evidence indicating that endogenous opioids (and particularly enkephalin) are involved in attenuating or terminating stress responses (i.e., defensive reactions of the organism) (Tanaka et al., 1988, 1989, 2000; McCubbin, 1993; Janssens et al., 1995; Mansi et al., 2000; Curtis et al., 2001). For instance, naloxone (a nonselective opioid receptor antagonist) induced a greater increment in the HPA responses (ACTH and cortisol) in chronically stressed animals as compared to unstressed control animals, suggesting that the impact of opioid systems had increased due to chronic stress (Janssens et al., 1995). In the same vein, Tanaka's group reported that central administration of opioid agonists (including enkephalin itself) can attenuate not only stress-induced increases in norepinephrine release in cortex and limbic structures, including the amygdala, but also emotional responses shown during stress exposure (i.e., 60 min immobilization stress) (Tanaka et al., 2000).

A substantial literature exists implicating opioids in the amygdala in modulation of the stress response, and specifically in attenuating the impact of psychological stressors. Direct injection of opioid agonists within the amygdala diminishes anxiety-like behavior (File and Rodgers, 1979; Rodgers and File, 1979; Good and Westbrook, 1995). Injection of an enkephalin analog into the central nucleus of the amygdala produced attenuation of cold restraint-induced gastric mucosal lesions in rats while intra-amygdala (central nucleus) naloxone administration potentiated restraint-induced gastric pathology (Ray et al., 1988; Ray and Henke, 1990). An enkephalinergic pathway from the amygdala to the periacquaductal gray nucleus has also been shown to be involved in the suppression of anxiety-related behaviors (Shaikh et al., 1991a,b; Siegel et al., 1997). Further, in addition to modulating overt physiological and behavioral expressions of anxiety elicited in direct response to stressful stimuli, opioids in the amygdala have also been shown to impair fear conditioning, an aspect of the stress response that may be related to a more general role for opioids in processes such as learning and memory (Westbrook et al., 1997). Morphine injected within the amygdala (central nucleus) impaired the acquisition of fear in rats exposed to a hot-plate apparatus (Good and Westbrook, 1995). Activation of opioid receptors in the central nucleus of the amygdala during exposure of rats to the hot plate may have prevented the formation of the excitatory connections between representations of the apparatus cues and the noxious thermal stimulation, thereby reducing the ability of these cues to provoke fear (Good and Westbrook, 1995). This is consistent with a previous study (Gallagher et al., 1982) that showed injection of an enkephalin analogue within the central nucleus of the amygdala attenuated the acquisition of classically conditioned heart-rate responding in rabbits.

The advent of knockout mice for opioid genes brought further evidence for a major role of enkephalin and δ-OR in anxiety-like behavior. Enkephalin knockout mice exhibit an elevated level of anxiety-like behavior (Konig et al., 1996; Kieffer, 1999; Ragnauth et al., 2001). Consistent with that, the δ-OR knockout mice also showed higher anxiety in the elevated plus maze and the light–dark box, suggesting that the activity of δ-OR may diminish anxiety (Filliol et al., 2000). The amygdala could represent a major site for these effects since overexpression of proenkephalin in the amygdala potentiates the anxiolytic effects of benzodiazepines. Indeed, Kang et al. (2000) addressed the role of enkephalin in the control of anxiety-like behavior and anxiety-reducing actions of benzodiazepines, using a recombinant, replication-defective herpes virus carrying human preproenkephalin cDNA that was delivered to rat amygdala. While enkephalin gene infection alone did not reduce anxiety-like behavior, rats infected with enkephalin gene exhibited a greater response to the anxiolytic effect of diazepam when compared to rats infected with a control virus containing the lacZ gene. The enhancement of diazepam action by enkephalin transfection was naloxone-reversible, region-specific, and correlated with the time course of preproenkephalin expression. These findings implicate amygdala opioid peptides in regulating the anxiolytic effects of benzodiazepines (Kang et al., 2000).

Considerably fewer studies have addressed the physiological functions of dynorphin (and κ-OR) within the amygdaloid complex. The existence of functionally opposed opioid systems affecting emotional and perceptual experiences has been proposed on the basis that the μ-OR and/or δ-OR mediate the euphorigenic properties (rewarding effect) of opioid agonists, while κ-OR seem to mediate aversive effects (Pan, 1998; Matsuzawa et al., 1999; Nobre et al., 2000; Sante et al., 2000; Ge et al., 2002). However, recent research, using local microinjections within the infralimbic cortex (Wall and Messier, 2000a,b) or systemic injections (Privette and Terrian, 1995; Agmo and Belzung, 1998) revealed an anxiolytic role for κ-OP agonists as well. Therefore, amygdaloid dynorphin and enkephalin neurons could potentially exert either different or similar effects in response to stressful stimuli. Activation of the enkephalinergic system may attenuate, whereas activation of the dynorphinergic system may potentiate the response to stressors. Alternatively, it is also justified to postulate that dynorphin as well as enkephalin may contribute to attenuating the impact of psychological stressors in the amygdaloid complex. As a result, changes in the relative activity of dynorphinergic compared to enkephalinergic neuronal systems within the amygdala invoked by stressors may be important in behavioral and physiological changes induced by stress. More research is needed to elucidate and understand the role(s) played by each endogenous peptide and their receptors within the amygdaloid complex in the stress response.

Stressor-induced alterations in endogenous opioid peptides: implications for anxiety, depression, and panic

The putative influence of aversive life events to the provocation, maintenance, and exacerbation of psychological disturbance is well documented (Breier, 1989; Anisman and Zacharko, 1990, 1992; Masure, 1994; Loas, 1996; Cui and Vaillant, 1997; Risch, 1997; Kessing et al., 1998; Kim and Yoon, 1998; Bremner, 1999; Weiss et al., 1999). The notion that stressful life events provoke or exacerbate psychopathology in humans is appealing. The severity of any stressor experience may be defined by the release kinetics of neurotransmitters associated with mood, affect, and motivation and putative neurotransmitters, which modulate individual experiences associated with anxiety and cognitive appraisal of environmental stimuli (see Zacharko et al., 1995). The following discussion outlines the mesolimbic opioid alterations, specifically enkephalin, associated with stressor imposition and the propensity of such neural variations to influence anxiety and motivation.

Human studies

Among clinicians it is well versed that uncontrollable aversive life events result in the formation of attributions concerning the stressor, leading to negative expectancies and alterations in mood which ultimately provoke feelings of anxiety as well as depression (Abramson et al., 1978; Maier, 1984; Breier et al., 1987; Brown and Siegel, 1988; Metalsky and Joiner, 1992; Porteous and Tyndall, 1994). Increased perceptions of stressful life events, a concomitant of anxiety disorders, impairs the anatomy, physiology, and behavioral functions of the prefrontal cortex (Diorio et al., 1993) and hippocampus (McEwen and Sapolsky, 1995; Sapolsky, 1996), damage thought to be related to an increase in cortisol levels and glucocorticoids associated with stress (Van Dijken et al., 1992; Bremner, 1999). In humans, the hippocampus, prefrontal cortex, and amygdala are believed to contribute to emotion, in particular fear-related negative affect and affective working memory (Morgan et al., 1993; Morgan and LeDoux, 1995; Jinks and McGregor, 1997; Davidson and Irwin, 1999). In a recent magnetic resonance imaging study of mood disorders, experimenters demonstrated a reduction in the mean gray matter volume of the prefrontal cortex (Drevets et al., 1998) and amygdala (Sheline et al., 1998, 1999) in subjects with recurrent episodes of major depressive disorder. Similarly, functional neuroimaging studies have identified abnormalities of resting blood flow and glucose metabolism in the medial prefrontal cortex and amygdala of depressed individuals which are correlated with depression severity (Drevets, 1999). Liberzon et al. (2002) employed positron-emission tomography to measure cerebral blood flow and μ-OR binding in limbic and cortical structures among subjects during presentation of emotionally salient visual pictures. It was found that emotionally charged stimuli induced an increase in cerebral blood flow to the prefrontal cortex and lateral amygdala and was associated with reduced μ-OP binding in these identical sites. Immunohistochemical and retrograde tracing has identified μ-OR on parabrachial neurons projecting to the amygdala in the rat (Chamberlin et al., 1999). Furthermore, D-Ala^2, N-Me-Phe^4, Gly-Ol^5-enkephalin (DAMGO) induced μ-OP activation in the parabrachial nucleus attenuated the impact of relatively severe psychological stressors as revealed by cardiovascular attenuation (Kiritsy-Roy et al., 1986; Marson et al., 1989; Sun et al., 1996; Wisniewska and Wisniewski, 1996). Moreover, the central and basolateral amygdaloid nuclei (Gelsema et al., 1987; Soltis et al., 1997) and the ventral tegmental area (VTA) (van den Buuse, 1998) provide prominent parabrachial innervation and accordingly may influence cardiovascular responsivity to environmental challenge. Enkephalinergic neurons having ascending projections into the A10 region (VTA) are located in the dorsal raphe, dorsal tegmental nucleus, and brainstem nuclei associated with respiration and cardiovascular control. These data suggest an anatomical basis for the abnormal hedonic, motivational, neuroendocrine, and autonomic manifestations evident in clinical cases as well as provide a neural model for the increased sensitivity of individuals with psychological dysfunction to aversive life events and to recurrent and severe episodes of illness.

Exposure to life stressors provoke increased CSF β-endorphin levels among individuals with mild anxiety disorders (Eriksson et al., 1989; Darko et al., 1992; Goodwin et al., 1993; Baker et al., 1997; Westrin et al., 1999), while panic patients and individuals with major depression display reduced CSF β-endorphin levels (Zis et al., 1985). Severe anxiety syndromes, including panic and posttraumatic stress disorder/individuals with panic disorder or posttraumatic stress disorder, also display augmented levels of aggression (Korn et al., 1997; Southwick et al., 1999). For example, some panic patients report an increased incidence of panic-associated suicidal ideation, engage in property destruction, initiate physical assaults, and may exhibit homicidal tendencies (Korn et al., 1997). Moreover, the incidence of aggressive episodes among patients with posttraumatic stress disorder appeared to increase with attending symptom severity (McFall et al., 1999). Interestingly, mice lacking δ-OP (Filliol et al., 2000) or preproenkephalin-derived peptides display increased fear and anxiety in novel environments and increased aggression toward conspecifics (Konig et al., 1996; McFall et al., 1999). In a similar vein, naltrexone has been reported to increase aggression and blood pressure in an individual with posttraumatic stress disorder (Ibarra et al., 1994) and induce panic attacks in individuals with panic disorder (Maremmani et al., 1998). Aggression in animal subjects, individuals with panic disorder, and posttraumatic stress disorder has also been associated with increased sensitivity of 5-HT_{1A} receptors in the raphe nuclei and hypothalamus (Korn et al., 1997; Southwick et al., 1999). Met- and leu-enkephalin and β-endorphin are colocalized with 5-HT in the caudal raphe nuclei (Korn et al., 1997; Southwick et al., 1999). These data suggest that increased anxiety-like behavior may be associated with a dysinhibition of 5-HT in raphe nuclei (Albert and Walsh, 1982; Ferris et al., 1997, 1999). In conclusion, it would seem appropriate to target pharmacological manipulations which effect μ-OP and δ-OP to the treatment of psychological disturbance in which enhanced vulnerability to stressors is a prominent feature of psychopathology (e.g., severe depression, anxiety).

Animal studies: psychological stressors

Exposure of animals to species-specific predatory cues may provide a relevant simulation of clinical psychopathology (Ferris et al., 1999). For example, exposure of rats to predators and predator odor incites anxiety-like behaviour in the elevated plus maze (McGregor and Dielenberg, 1999), increased acoustic startle (Adamec et al., 1997; Plata-Salaman et al., 2000), increased freezing in novel environments (Hotsenpiller and Williams, 1997), and decreased sucrose consumption (Calvo-Torrent et al., 1999). Exposure of rats to a cat is associated with an enhanced acoustic startle response for up to seven days following predator exposure (Adamec et al., 1998). Drugs that decrease anxiety, including diazepam (Bitsios et al., 1999; Cannizzaro et al., 2001) and enkephalin μ-OR and δ-OR agonists (Tilson et al., 1986; Vivian and Miczek, 1998), attenuate startle. Rats which are exposed to predator cues display enhanced risk-assessment activity, including defensive burying, stretch attend, and freezing behaviors (Kemble and Bolwahnn, 1997; Dielenberg et al., 2001) (for a review, see Blanchard et al., 1998) which are attenuated by diazepam and imipramine pretreatment (Blanchard et al., 1993;

Molewijk et al., 1995; Grewal et al., 1997). Exposure of rats or mice to fox odor (TMT), weasel odor (PT), or cat odor is associated with increased opioid-dependent analgesia and freezing in animals relative to control odors (Lester and Fanselow, 1985; Kavaliers, 1988; Lichtman and Fanselow, 1990; Hotsenpiller and Williams, 1997; Kavaliers and Choleris, 1997; Kavaliers et al., 1997). Among nonhuman primates, naloxone increased and morphine decreased vocalizations among infants separated from their mother (Kalin et al., 1988). These findings were replicated in rats exposed to a predator. In particular, morphine decreased (Shepherd et al., 1992) while naloxone increased (Blanchard et al., 1991) ultrasonic emissions in rats exposed to a cat.

It has recently been demonstrated in our laboratory (Hebb et al., 2002a,b) that a 10-min predator odor exposure (fox odor) increased anxiety-like behavior in the light–dark box and enhanced acoustic startle in mice relative to mice exposed to saline or the pungent control odor butyric acid. Mice exposed to TMT displayed enhanced freezing relative to control mice which was associated with reduced time spent in the light of the test apparatus. Anxiety-like behavior in the light–dark box following predator odor exposure was accompanied by decreased proenkephalin gene expression in the core of the nucleus accumbens and increased proenkephalin gene expression in the basolateral, central, and medial amygdala. In particular, among mice exposed to fox odor, enhanced levels of anxiety-like behavior in mice was associated with increased activation of enkephalin neurons in the basolateral, medial, and central amygdaloid nuclei gene expression while the overall level of enkephalin transcript was decreased in these animals. However, enhanced levels of freezing among mice not assessed for anxiety in the light–dark test was associated with decreased activation of enkephalin neurons in the central amygdala relative to mice which exhibited less freezing to predator odor. It is well established that the central amygdala plays a prominent role in the acquisition and mediation of fear responses (Roozendaal et al., 1991) and is the major output pathway to the basolateral amygdala and many subcortical nuclei that mediate fear-related behaviors, including freezing and startle (Rosen et al., 1998). The medial amygdala has been associated with a variety of neuroendocrine as well as behavioral responses, including mating and aggression. The induction of enkephalin in the medial amygdala of mice is consistent with a role of this nucleus in behavioral arousal and social memory (Vochteloo and Koolhaas, 1987; Kollack-Walker and Newman, 1995). Interestingly, it has recently been reported that the male rats display an increase in delta-opioid receptor immunoreactivity in the medial nucleus of the amygdala compared to female rats (Wilson et al., 2002), which may underlie sex differences in response to stressor experiences (Heinsbroek et al., 1988). Klein et al. (1998) reported that in male rats opioid blockade by peripheral naloxone administration enhanced unconditioned freezing in the home cage following footshock imposition. However, conditioned freezing was associated with increased enkephalin mRNA from the central amygdala, suggesting that enkephalin within this nucleus is associated with learning and memory (Petrovich et al., 2000). It will be recalled that preproenkephalin-deficient mice display increased anxiety in the open field and elevated plus-maze tests (Petrovich et al., 2000; Ragnauth et al., 2001). Psychological stressors increase amygdaloid δ-OR receptor binding (Pohorecky et al., 1999) and μ-OR in the ventral tegmental area (Nikulina et al., 1999), paralleling anxiety induction in rats and mice (MacNeil et al., 1997; Blanchard et al., 1998; Moynihan et al., 2000).

In other investigations, employing various psychological stressors, evidence exists to support a role of enkephalin in the stress response of animals. For example, the social interaction and defeat paradigm indices correlate with extracellular met-enkephalin availability in the rostral nucleus accumbens (Bertrand et al., 1997) and μ-OR mRNA density in the VTA (Nikulina et al., 1999), respectively. Indeed, naltrexone increased initial contact latency and decreased active interaction duration among rats in a social-interaction paradigm (Zhang et al., 1996). Pretreatment with naloxone (1 mg/kg s.c.) antagonized increased behavioral activity (e.g., ambulation, rearing, sniffing) of rats exposed to a psychological stressor (perception of another rat receiving footshocks) while having no effect on behavior of control animals or animals subjected to footshock (VandenBerg et al., 1998). In rats, social isolation (e.g., seven days) was associated with decreased proenkephalin mRNA in the nucleus accumbens (Angulo et al., 1991) and increased δ-OR binding in the prefrontal cortex, the amygdala, nucleus accumbens, and hypothalamus (Pohorecky et al., 1999) compared to group-housed rats. Interestingly, prior stressor history of mice influences locomotor activity to subsequent administration of enkephalin agonists (Calenco-Choukroun et al., 1991). For example, intra-VTA administration of the δ-OR agonists in the rat induced hyperactivity in a familiar (i.e., home cage), unfamiliar (four-hole box), and an open-field environment, while DAGO enhanced locomotion in a familiar but not in the unfamiliar environment or the open-field paradigm 24 h following a 3-day habituation baseline schedule. The environment-μ-OP and δ-OP associated increase in locomotor activity was blocked by systemic naloxone (0.3 mg/kg) in mice (Calenco-Choukroun et al., 1991). Moreover, such stressor scenarios may define the profile of symptoms and determine vulnerability (e.g., latency to the emergence of psychological dysfunction) to anxiety disorders. Such predisposing neurochemical variations may (a) identify region- specific neural sequelae contributing to behavioral sensitization and conditioning among subjects; (b) provide a parallel for the development of psychological disturbance in human subjects; and (c) suggest therapeutic-intervention strategies.

Mesolimbic opioid availability underlying coping and therapeutic efficacy of antidepressant regimens

It is well documented that endogenous opioids contribute to the expression of affect and motivation in human (Cohen et al., 1984; Zis et al., 1985; Castilla-Cortazar et al., 1998) and animal subjects (Hernandez et al., 1997). Indeed, major affective disorder has been associated with decreased cerebrospinal (CSF) endorphin concentrations (Djurovic et al., 1999). Endogenous opioids have been implicated in the mechanism of action of antidepressant therapies (de Felipe et al., 1985, 1989). In fact, the involvement of opioid systems in clinical depression is based on clinical and animal studies which report that inhibitors of enkephalin-degrading enzymes (e.g., RB 101) have antidepressant properties in various paradigms (Tejedor-Real et al., 1998), chronic imipramine treatment promotes the expression of the μ-OR in the hippocampus and frontal cortex of the rat (de Gandarias et al., 1998), chronic imipramine and desipramine, inhibit the enkephalin-degrading aminopeptidase in a concentration-dependent manner in rat brain (Gallego et al., 1998; de Gandarias et al., 1999), and endogenous opioids have been implicated in the mechanism of action of antidepressant therapies (de Felipe et al., 1985, 1989). Inhibitors of enkephalin-degrading enzymes (de Felipe et al., 1985, 1989; Tejedor-Real et al., 1998) demonstrate antidepressant properties which are linked to specific opioid and nonopioid receptor subtypes (e.g., δ-OR; Tejedor-Real et al., 1998) in various animal paradigms. For example, in rodents the antidepressant effects of RB 101 in conditioned suppression of motility (CSM) test were reversed by the selective δ opioid receptor antagonist, naltrindole (Baamonde et al., 1992). Met-enkephalin administration attenuated suppression of motility among mice placed in an environment previously paired with electric shock, an effect that was reversed by naloxone and 6-OHDA lesions of nucleus accumbens neurons (Katoh et al., 1991). Acute administration of imipramine (10 mg/kg, i.p.) decreased proenkephalin mRNA (20%) and prodynorphin mRNA (25%) in the nucleus accumbens and striatum, while chronic administration (10 mg/kg i.p. twice daily for 10 days) increased nucleus accumbens met-enkephalin and proenkephalin gene levels (Dziedzicka-Wasylewska and Rogoz, 1995; Dziedzicka-Wasylewska and Papp, 1996) as well as prodynorphin mRNA 24 h following cessation of treatment (Kurumaji et al., 1988; Przewlocki et al., 1997). Similar changes in the levels of met-enkephalin in the rat hippocampus, striatum, hypothalamus, and pituitary were reported following chronic administration of the antidepressant drugs, tianeptine and fluoxetine (Dziedzicka-Wasylewska et al., 2002). Treatment-resistant depression in clinical populations sometimes responds to electroconvulsive shock-therapy treatment which promotes increased enkephalin in brain (Hong et al., 1979; Holaday et al., 1986; Lason et al., 1987; He et al., 1989), including hippocampus (He et al., 1989). Typically, chronic antidepressant regimes (4–6 weeks) are necessary to restore mood in clinical depression (Stimmel, 1995).

It may be that chronic antidepressant therapy is necessary to promote stable levels of enkephalin in central sites before any noticeable change in mood and affect occurs. Taken together, coping and antidepressant agents favor mesolimbic enkephalin release among subjects which may blunt the impact of the stressor or stressor-associated cues.

It has been reported that the release of enkephalin within mesolimbic sites reduces the effect of the stress response by decreasing numerous physiological responses, including emotional and affective states (Dziedzicka-Wasylewska and Papp, 1996). It should be considered that while mild stressors or the cues associated with aversive events may elicit cognitions which reenlist neurochemical variations associated with the original stressor experience (Ahmed and Koob, 1997; Erb et al., 1998), coping may prompt anxiolytic agent release. In humans, hyperactivity of endogenous opioids has been postulated to underlie defensive coping styles, characterized by orientation away from threatening stimuli thereby minimizing distress and negative emotions (Janssen and Arntz, 1996, 1997; Jamner and Leigh, 1999). The exact relationship between increased endogenous opioids and coping has not been defined (Goodwin and Barr, 1997; Jamner and Leigh, 1999). Cardiac patients demonstrated enhanced blood pressure and heart rate accompanied by increased plasma levels of met-enkephalin, dynorphin, and β-endorphin in response to a psychological stressor (mental arithmetic test) (Fontana et al., 1998). Systemic administration of the opioid antagonists, naloxone and naltrexone, enhanced cardiovascular indices induced by a cognitive stressor in normal and cardiac patients (Fontana et al., 1997, 1998; McCubbin et al., 1998). Interestingly, naloxone and naltrexone also decreased approach behaviour among phobic individuals (Arntz, 1993; Janssen and Arntz, 1996, 1997) and attenuated the anxiolytic influence of diazepam among patients awaiting surgery (Duka et al., 1982). Interestingly, naloxone can also reverse placebo analgesia (Benedetti and Amanzio, 1997), suggesting that there is a modulatory effect of opioid function on subject expectations which may be extended to anticipatory anxiety. In mice, naloxone attenuated the anxiolytic properties of diazepam and chlordiazepoxide in mice in the light–dark and elevated plus-maze paradigms (Agmo et al., 1995; Belzung and Agmo, 1997). In contrast to benzodiazepine administration, chronic benzodiazepine withdrawal elicits significant anxiety (Fontaine et al., 1984; Otto et al., 1993). Such behavioural reactivity to anxiolytic withdrawal has been linked to decreased met-enkephalin immunoreactivity in the nucleus accumbens (Kurumaji et al., 1988; Przewlocki et al., 1997). It should be considered that acute and chronic administration of the anti-depressant/anxiolytic agent imipramine effects neurochemically distinct profiles of met and leu enkephalin release within the ventral tegmental area and nucleus accumbens (Dziedzicka-Wasylewska and Rogoz, 1995). The variable sensitivity of central mesocorticolimbic sites to antidepressant regimens is consistent with data outlining differential responsivity of mesocorticolimbic sites to stressor imposition (Zacharko and Anisman, 1991).

It should be emphasized that cognitive and performance-based therapeutic interventions among anxiety patients frequently introduce coping strategies to reduce anxiety during pharmacotherapy and eventual drug taper (Shear et al., 1991; Nagy et al., 1993; Spiegel et al., 1994). Such interventions attempt to reduce the saliency of associational cues which precipitate anxiety. In fact, the efficacy of performance-associated phobic treatment, for example, can be reliably traced to cognitive appraisal of performance self-adequacy and coping repertoires (Williams et al., 1989). Perceived control over footshock enhances cerebrospinal (CSF) release of benzodiazepine-like agent among rats (Piva et al., 1991; Drugan et al., 1994, 1997), which may detract from the severity of the experience. Yet, the source of such neurochemical activity may be associated with mesocorticolimbic sensitivity to stressor controllability. Indeed, the release of "anxiolytic" agents following stressor imposition has been detected in the amygdala (Kang et al., 1999), mesocortex (Cabib and Puglisi-Allegra, 1996), the ventral tegmental area, and the nucleus accumbens (Dziedzicka-Wasylewska and Papp, 1996). Indeed, it has been suggested that endogenous enkephalin release among animal subjects may, in some instances, diminish the propensity of restraint to sustain protracted alterations (e.g., one week) of mesocortical dopamine release (Cuadra et al., 1999) and anxiety-like behavior in animals in the light–dark paradigm

(Cancela et al., 1995). Central administration of μ-OR and δ-OR opioid receptor agonists, before or after footshock, reduced stressor-associated deficits in locomotor activity (Hebb et al., 1997) and VTA brain stimulation deficits (Zacharko et al., 1998) among mice. Systemic administration of met-enkephalin (10 mg/kg) 30 min before a 6-h restraint stress reduced the stressor-induced increase in plasma corticosterone among mice (Sverko et al., 1997), while daily naltrindole (1 mg/kg) from birth to postnatal day 19 inhibited subsequent increased corticosterone release in 25-day-old female rats in response to a 3-min swim session (Fernandez et al., 1999). Moreover, systemic administration of RB-101 or naltrexone reduced and exaggerated freezing to shock-associated environmental cues, respectively, in mice (Baamonde et al., 1992; Calcagnetti and Schechter, 1994). The enkephalin-inhibitor RB-101, met-enkephalin, leu-enkephalin as well as the specific δ-OP agonist, BUBU, also reversed escape deficits among rats previously subjected to uncontrollable footshock (Tejedor-Real et al., 1993, 1995, 1998). The effects of RB-101 on escape deficits were attenuated by prior administration of the DA receptor antagonist, SCH-23390, or the selective δ opioid receptor antagonist, naltrindole (Tejedor-Real et al., 1998). In contrast, naloxone potentiated the effect of inescapable shock among rats (Tejedor-Real et al., 1998). Taken together, acute exposure of animal and human subjects to stressors promotes enkephalin release within the mesocortex, nucleus accumbens, and ventral tegmental area. Such a neurochemical signature within the mesolimbic system may be influenced by the severity of the environmental experience and determine behavioral responsivity to stressor applications (Siegel et al., 1984; Imperato et al., 1992; Izumi, 1998; Giardino et al., 1999).

Mesolimbic opioid receptor density appears to coincide with somatodendritic dopamine, cholecystokinin, and GABA interneuron density (Mansour et al., 1988; Kalivas, 1993). Notably, central and basolateral nuclei amygdaloid neurochemical perturbations associated with benzodiazepine-GABA receptor variations including alterations in GABA and glutamate (Davis et al., 1994; Kunos and Varga, 1995; Soltis et al., 1997; Walker and Davis, 1997) most likely potentiate the efficacy of anxiolytic agents, including diazepam. It should be emphasized that μ-OP have been identified on both GABA- and DA-containing neurons and δ-OP have been detected on afferent inputs to GABA neurons in the VTA (Kalyuzhny and Wessendorf, 1997). Such anatomical data provide a putative framework for a μ/δ-OR interaction in the VTA that may impact on stressor-associated behavioral deficits. The effects of opioid activity on the anxiolytic properties of benzodiazepine-like substances have been attributed to specific opioid receptor subtypes (Tsuda et al., 1996; Tejedor-Real et al., 1998; Abbadie et al., 2000). In summary, increased enkephalin availability may represent a modulatory system in stress adaptation in both animal and human subjects and avert exaggerated behavioral and neurochemical cascades associated with reexposure to stressors or stressor-associated cues previously associated with initial stressor presentations. In effect, behavioral-coping strategies and antidepressant regimens favor mesolimbic enkephalin release which may blunt the impact of the stressor or stressor-associated cues.

In animals, a variety of environmental stressors, both nocioceptive and nonpainful threat stimuli (Lightman and Young, 1987a; Rodgers and Randall, 1987; Sribanditmongkol et al., 1994; Yamada and Nabeshima, 1995; Beaulieu et al., 1996), elicit opioid release and changes in opioid receptor binding (Stuckey et al., 1989) which may mediate specific types of stress analgesia (Zadina et al., 1999) which is blocked by opioid antagonists and interacts with anxiolytic (Kelly et al., 1986; Watkins and Mayer, 1986) and antidepressant (Zurita et al., 1999) agents. For example, chronic footshock induced a 40–50% reduction of Met and Leu enkephalin levels in whole brain of rats (McGivern et al., 1983) and a time-dependent decrease in the level of immunoreactive met-enkephalin in the medial A10 region (VTA), which may be indicative of a footshock-induced increase in release and metabolism of met-enkephalin (Kalivas et al., 1988). Exposure of young rats to a male intruder increased preproenkephalin mRNA in the central amygdaloid nucleus and enhanced heat analgesia (Wiedenmayer et al., 2002). Stressor-induced analgesia is thought to be an integral and adaptive component of the defensive repertoire (Amit and Galina, 1986; Fanselow, 1986; Rodgers and Randall, 1987; Vaccarino and Kastin, 2001).

It has also been proposed that endogenous opioid mechanisms, specifically leu-enkephalin availability and delta receptor activation, are involved in stimulus processing and early stages of learning in conditioning tasks among animals (Schulteis and Martinez, 1992; Hernandez and Watson, 1997; Martinez et al., 1998). Indeed, salient environmental experiences modulate peripheral leu-enkephalin availability and may increase the associated memory strength and preparedness of the animal to future challenges (Martinez and Weinberger, 1987, 1988; Derrick et al., 1991; Schulteis and Martinez, 1993). One of the mechanisms associated with the expression of anxiety incorporates an amygdaloid-opioid neurochemical mosaic which may encode the saliency of the environmental experience, consistent with an amygdaloid-dependent modulation of emotional memory (Ferry et al., 1999). Central administration of the δ_1-OP agonist D-Pen^2, L-Pen^5 (DDLPE) decreased passive-avoidance learning in rats trained to avoid footshock (Ukai et al., 1997). It has been demonstrated that central δ-OP activation following uncontrollable footshock and novelty stress in CD-1 mice precipitated an increase in locomotor activity with reexposure to only the identical, milder form of the original stressor. The behavioral response of δ-OR activation following stressor imposition were also dose and time dependent (Hebb et al., 1997). Indeed, endogenous δ-OR activation may increase the saliency of particular attributes or characteristics of environmental stimuli (Hernandez and Watson, 1997). Clearly, the distribution of μ-OP and δ-OR in mesolimbic sites contributes to coping (Goodwin and Barr, 1997). Increased preproenkephalin gene expression in central sites, including the amygdala, nucleus accumbens, locus coeruleus, among others, associated with fear and anxiety, presumably underlies compensatory physiological responses that attenuate the deleterious effects of uncontrollable stressor applications (Dumont et al., 2000; Curtis et al., 2001). Conceptually, it is conceivable that central μ-OP and δ-OR activation interferes with the encoding of stressor-associated events. Such processes may alter long-term responsivity of organisms to future stressor encounters.

In human and animal subjects increased preproenkephalin availability is associated with mood elevation and coping ability which may detract from the aversiveness of the stressor experience. In animal and human subjects, activation of μ-OP and δ-OP in the basolateral amygdala, ventral tegmental area, and mesocortex attenuates the aversiveness of psychological stressors. The temporal expression of enkephalin in mesolimbic sites associated with specific indices of anxiety and fear would delineate specific behavioral subsets associated with stressor exposure and predict vulnerability to future-like events. In particular, activation of μ-OP and δ-OR within the ventral tegmental area, nucleus accumbens, and amygdala may diminish anxiety and motivational loss accompanying stressor exposure, permit expression of coping behavior, detract from the saliency of the stressor, and alter long-term responsivity to ensuing stressor encounters. The relative contributions of δ-OP and μ-OP activation to motivated behavior is predicated on mesocorticolimbic site-specific differences in opioid receptor density as well as environmental stimuli which may modulate the release of variable neurotransmitter and neuropeptide systems with different affinities for the δ-OP and μ-OR sub-types.

Conclusion

Clearly, prominent behavioral correlates of central opioid activation are in evidence following stressor manipulations. In particular, opioid availability in diverse brain areas associated with anxiety-like behaviors (i.e., enkephalin within the amygdala) may contribute to affective aspects of an anxiety-provoking situation following stressor imposition. Opioid peptide availability is linked to colocalization of other neurotransmitters in distinct central sites which suggests that opioids may modulate (a) different aspects of anxiety, including anticipatory reactions to anxiogenic stimuli, (b) variations in cognitive arousal and vigilance, and (c) modulation of behavior and central neurochemical activity. This seemingly overlap of function may supply an organism with many stress responses, but more importantly the ability to manufacture assorted behavioral reactions in response to specific stressful situations. Likewise, multiple anxiogenic agents and putative neurotransmitters or neuromodulators in the mesolimbic system as well as prefrontal cortex and brainstem would appear to participate in the promotion or alleviation of stress-induced symptoms and mood alterations. In any event, these data suggest that pharmacological management of stress-induced pathologies should be directed toward specific symptoms characterizing the psychological pathology.

Acknowledgments

G.D. held a scholarship from le Fonds de la Recherche en Santé du Québec (FRSQ). A.L.O.H. held a fellowship from The Canadian Institute of Health Research.

References

Abbadie, C., Pan, Y., Drake, C. and Pasternak, G. (2000) Comparative immunohistochemical distributions of carboxy terminus epitopes from the mu-opioid receptor splice variants MOR-1D, MOR-1 and MOR-1C in the mouse and rat CNS. Neuroscience, 100: 141–153.

Abramson, L.Y., Garber, J. and Edwards, N.B. (1978) Expectancy changes in depression and schizophrenia. J. Abnorm. Psychol., 87: 102–109.

Adamec, R.E., Shallow, T. and Budgell, J. (1997) Blockade of CCK(B) but not CCK(A) receptors before and after the stress of predator exposure prevents lasting increases in anxiety-like behavior: implications for anxiety associated with posttraumatic stress disorder. Behav. Neurosci., 111: 435–449.

Adamec, R., Kent, P., Anisman, H., Shallow, T. and Merali, Z. (1998) Neural plasticity, neuropeptides and anxiety in animals: implications for understanding and treating affective disorder following traumatic stress in humans. Neurosci. Biobehav. Rev., 23: 301–318.

Agmo, A. and Belzung, C. (1998) The role of subtypes of the opioid receptor in the anxiolytic action of chlordiazepoxide. Neuropharmacology, 37: 223–232.

Agmo, A., Galvan, A., Heredia, A. and Morales, M. (1995) Naloxone blocks the antianxiety but not the motor effects of benzodiazepines and pentobarbital: experimental studies and literature review. Psychopharmacology (Berl), 120: 186–194.

Ahmed, S.H. and Koob, G.F. (1997) Cocaine – but not food-seeking behavior is reinstated by stress after extinction. Psychopharmacology (Berl), 132: 289–295.

Akil, H., Watson, S.J., Young, E., Lewis, M.E., Khachaturian, H. and Walker, J.M. (1984) Endogenous opioids: biology and functions. Ann. Rev. Neurosci., 7: 223–255.

Akil, H., Owens, C., Gutstein, H., Taylor, L., Curran, E. and Watson, S. (1998) Endogenous opioids: overview and current issues. Drug Alcohol Depend., 51: 127–140.

Albert, D.J. and Walsh, M.L. (1982) The inhibitory modulation of agonistic behavior in the rat brain: a review. Neurosci. Biobehav. Rev., 6: 125–143.

Amit, Z. and Galina, Z.H. (1986) Stress-induced analgesia: adaptive pain suppression. Physiol. Rev., 66: 1091–1120.

Angulo, J.A., Printz, D., Ledoux, M. and McEwen, B.S. (1991) Isolation stress increases tyrosine hydroxylase mRNA in the locus coeruleus and midbrain and decreases proenkephalin mRNA in the striatum and nucleus accumbens. Mol. Brain Res., 11: 301–308.

Anisman, H. and Zacharko, R.M. (1990) Multiple neurochemical and behavioral consequences of stressors: implications for depression. Pharmacol. Ther., 46: 119–136.

Anisman, H. and Zacharko, R.M. (1992) Depression as a consequence of inadequate neurochemical adaptation in response to stressors. Br. J. Psychiatry, Suppl, 36–43.

Arntz, A. (1993) Endorphins stimulate approach behaviour, but do not reduce subjective fear. A pilot study. Behav. Res. Ther., 31: 403–405.

Baamonde, A., Dauge, V., Ruiz-Gayo, M., Fulga, I.G., Turcaud, S., Fournie-Zaluski, M.C. and Roques, B.P. (1992) Antidepressant-type effects of endogenous enkephalins protected by systemic RB 101 are mediated by opioid delta and dopamine D1 receptor stimulation. Eur. J. Pharmacol., 216: 157–166.

Baker, D.G., West, S.A., Orth, D.N., Hill, K.K., Nicholson, W.E., Ekhator, N.N., Bruce, A.B., Wortman, M.D., Keck, P.E., Jr. and Geracioti, T.D., Jr. (1997) Cerebrospinal fluid and plasma beta-endorphin in combat veterans with post-traumatic stress disorder. Psychoneuroendocrinology, 22: 517–529.

Beaulieu, J., Champagne, D. and Drolet, G. (1996) Enkephalin innervation of the paraventricular nucleus of the hypothalamus: distribution of fibers and origins of input. J. Chem. Neuroanat., 10: 79–92.

Belzung, C. and Agmo, A. (1997) Naloxone blocks anxiolytic-like effects of benzodiazepines in Swiss but not in Balb/c mice. Psychopharmacology (Berl), 132: 195–201.

Benedetti, F. and Amanzio, M. (1997) The neurobiology of placebo analgesia: from endogenous opioids to cholecystokinin. Prog. Neurobiol., 52: 109–125.

Bertrand, E., Smadja, C., Mauborgne, A., Roques, B.P. and Dauge, V. (1997) Social interaction increases the extracellular levels of [Met]enkephalin in the nucleus accumbens of control but not of chronic mild stressed rats. Neuroscience, 80: 17–20.

Bitsios, P., Philpott, A., Langley, R.W., Bradshaw, C.M. and Szabadi, E. (1999) Comparison of the effects of diazepam on the fear-potentiated startle reflex and the fear-inhibited light reflex in man. J. Psychopharmacol., 13: 226–234.

Blackburn, T.P., Cross, A.J., Hille, C. and Slater, P. (1988) Autoradiographic localization of delta opiate receptors in rat and human brain. Neuroscience, 27: 497–506.

Blanchard, D.C., Weatherspoon, A., Shepherd, J., Rodgers, R.J., Weiss, S.M. and Blanchard, R.J. (1991) "Paradoxical" effects of morphine on antipredator defense reactions in wild and laboratory rats. Pharmacol. Biochem. Behav., 40: 819–828.

Blanchard, R.J., Shepherd, J.K., Rodgers, R.J., Magee, L. and Blanchard, D.C. (1993) Attenuation of antipredator defensive behavior in rats following chronic treatment with imipramine. Psychopharmacology (Berl), 110: 245–253.

Blanchard, R.J., Nikulina, J.N., Sakai, R.R., McKittrick, C., McEwen, B. and Blanchard, D.C. (1998) Behavioral and endocrine change following chronic predatory stress. Physiol. Behav., 63: 561–569.

Bodnar, R.J. and Hadjimarkou, M.M. (2002) Endogenous opiates and behavior: 2001. Peptides, 23: 2307–2365.

Bolles, R.C. and Fanselow, M.S. (1982) Endorphins and behavior. Annu. Rev. Psychol., 33: 87–101.

Breier, A. (1989) A.E. Bennett award paper. Experimental approaches to human stress research: assessment of neurobiological mechanisms of stress in volunteers and psychiatric patients. Biol. Psychiatry, 26: 438–462.

Breier, A., Albus, M., Pickar, D., Zahn, T.P., Wolkowitz, O.M. and Paul, S.M. (1987) Controllable and uncontrollable stress in humans: alterations in mood and neuroendocrine and psychophysiological function. Am. J. Psychiatry, 144: 1419–1425.

Bremner, J.D. (1999) Does stress damage the brain? Biol. Psychiatry, 45: 797–805.

Brown, J.D. and Siegel, J.M. (1988) Attributions for negative life events and depression: the role of perceived control. J. Pers. Soc. Psychol., 54: 316–322.

Cabib, S. and Puglisi-Allegra, S. (1996) Stress, depression and the mesolimbic dopamine system. Psychopharmacology (Berl), 128: 331–342.

Calcagnetti, D.J. and Schechter, M.D. (1994) Deficits in shock-induced freezing and naltrexone enhancement of freezing in fawn hooded rats. Brain Res. Bull., 35: 37–40.

Calenco-Choukroun, G., Dauge, V., Gacel, G., Feger, J. and Roques, B.P. (1991) Opioid delta agonists and endogenous enkephalins induce different emotional reactivity than mu agonists after injection in the rat ventral tegmental area. Psychopharmacology (Berl), 103: 493–502.

Calvo-Torrent, A., Brain, P.F. and Martinez, M. (1999) Effect of predatory stress on sucrose intake and behavior on the plus-maze in male mice. Physiol. Behav., 67: 189–196.

Cancela, L.M., Bregonzio, C. and Molina, V.A. (1995) Anxiolytic-like effect induced by chronic stress is reversed by naloxone pretreatment. Brain Res. Bull., 36: 209–213.

Cannizzaro, C., Martire, M., Cannizzaro, E., Provenzano, G., Gagliano, M., Carollo, A., Mineo, A. and Steardo, L. (2001) Long-lasting handling affects behavioural reactivity in adult rats of both sexes prenatally exposed to diazepam. Brain Res., 904: 225–233.

Castilla-Cortazar, I., Castilla, A. and Gurpegui, M. (1998) Opioid peptides and immunodysfunction in patients with major depression and anxiety disorders. J. Physiol. Biochem., 54: 203–215.

Ceccatelli, S. and Orazzo, C. (1993) Effect of different types of stressors on peptide messenger ribonucleic acids in the hypothalamic paraventricular nucleus. Acta. Endocrinol., 128: 485–492.

Chamberlin, N.L., Mansour, A., Watson, S.J. and Saper, C.B. (1999) Localization of mu-opioid receptors on amygdaloid projection neurons in the parabrachial nucleus of the rat. Brain Res., 827: 198–204.

Cohen, M.R., Pickar, D., Cohen, R.M., Wise, T.N. and Cooper, J.N. (1984) Plasma cortisol and beta-endorphin immunoreactivity in human obesity. Psychosom. Med., 46:454–462.

Cuadra, G., Zurita, A., Lacerra, C. and Molina, V. (1999) Chronic stress sensitizes frontal cortex dopamine release in response to a subsequent novel stressor: reversal by naloxone. Brain Res. Bull., 48: 303–308.

Cui, X.J. and Vaillant, G.E. (1997) Does depression generate negative life events? J. Nerv. Ment. Dis., 185: 145–150.

Curtis, A.L., Bello, N.T. and Valentino, R.J. (2001) Evidence for functional release of endogenous opioids in the locus ceruleus during stress termination. J. Neurosci., 21: RC152.

Darko, D.F., Risch, S.C., Gillin, J.C. and Golshan, S. (1992) Association of beta-endorphin with specific clinical symptoms of depression. Am. J. Psychiatry, 149: 1162–1167.

Davidson, R.J. and Irwin, W. (1999) The functional neuroanatomy of emotion and affective style. Trends Cogn. Sci., 3: 11–21.

Davis, M., Hitchcock, J.M., Bowers, M.B., Berridge, C.W., Melia, K.R. and Roth, R.H. (1994) Stress-induced activation of prefrontal cortex dopamine turnover: blockade by lesions of the amygdala. Brain Res., 664: 207–210.

de Felipe, M.C., De Ceballos, M.L., Gil, C. and Fuentes, J.A. (1985) Chronic antidepressant treatment increases enkephalin levels in n. accumbens and striatum of the rat. Eur. J. Pharmacol., 112: 119–122.

de Felipe, M.C., Jimenez, I., Castro, A. and Fuentes, J.A. (1989) Antidepressant action of imipramine and iprindole in mice is enhanced by inhibitors of enkephalin-degrading peptidases. Eur. J. Pharmacol., 159: 175–180.

de Gandarias, J.M., Echevarria, E., Acebes, I., Silio, M. and Casis, L. (1998) Effects of imipramine administration on mu-opioid receptor immunostaining in the rat forebrain. Arzneimittelforschung, 48: 717–719.

de Gandarias, J.M., Irazusta, J., Varona, A., Gil, J., Fernandez, D. and Casis, L. (1999) Effect of imipramine on enkephalin-degrading peptidases. Eur. Neuropsychopharmacol., 9: 493–499.

Derrick, B.E., Weinberger, S.B. and Martinez, J.L., Jr. (1991) Opioid receptors are involved in an NMDA receptor-independent mechanism of LTP induction at hippocampal mossy fiber-CA3 synapses. Brain Res. Bull., 27: 219–223.

Dielenberg, R.A., Carrive, P. and McGregor, I.S. (2001) The cardiovascular and behavioral response to cat odor in rats: unconditioned and conditioned effects. Brain Res., 897: 228–237.

Diorio, D., Viau, V. and Meaney, M.J. (1993) The role of the medial prefrontal cortex (cingulate gyrus) in the regulation of hypothalamic–pituitary–adrenal responses to stress. J. Neurosci., 13: 3839–3847.

Djurovic, D., Milic-Askrabic, J. and Majkic-Singh, N. (1999) Serum beta-endorphin level in patients with depression on fluvoxamine. Farmaco, 54: 130–133.

Drevets, W.C. (1999) Prefrontal cortical-amygdalar metabolism in major depression. Ann. N.Y. Acad. Sci., 877: 614–637.

Drevets, W.C., Ongur, D. and Price, J.L. (1998) Neuroimaging abnormalities in the subgenual prefrontal cortex: implications for the pathophysiology of familial mood disorders. Mol. Psychiatry, 3: 220–226, 190–221.

Drolet, G., Morilak, D.A. and Chalmers, J. (1991a) Endogenous opioid tonically inhibit the depressor neurons in the caudal ventrolateral medulla of rabbits: mediation through delta and kappa receptors. Neuropharmacology, 30: 383–390.

Drolet, G., Morilak, D.A. and Chalmers, J. (1991b) Opioid innervation of the caudal ventrolateral medulla is not critical for the expression of the aortic depressor nerve response in the rabbit. J. Auton. Nerv. Syst., 32: 37–46.

Drugan, R.C., Basile, A.S., Ha, J.H. and Ferland, R.J. (1994) The protective effects of stress control may be mediated by increased brain levels of benzodiazepine receptor agonists. Brain Res., 661: 127–136.

Drugan, R.C., Basile, A.S., Ha, J.H., Healy, D. and Ferland, R.J. (1997) Analysis of the importance of controllable versus uncontrollable stress on subsequent behavioral and physiological functioning. Brain Res. Protoc., 2: 69–74.

Duka, T., Millan, M.J., Ulsamer, B. and Doenicke, A. (1982) Naloxone attenuates the anxiolytic action of diazepam in man. Life Sci., 31: 1833–1836.

Dumont, E.C., Kinkead, R., Trottier, J., Gosselin, I. and Drolet, G. (2000) Effect of chronic psychogenic stress exposure on enkephalin neuronal activity and expression in the rat hypothalamic paraventricular nucleus. J. Neurochem., 75: 2200–2211.

Dziedzicka-Wasylewska, M. and Rogoz, R. (1995) The effect of prolonged treatment with imipramine and electroconvulsive shock on the levels of endogenous enkephalins in the nucleus accumbens and the ventral tegmentum of the rat. J. Neural. Transm. Gen. Sect., 102: 221–228.

Dziedzicka-Wasylewska, M. and Papp, M. (1996) Effect of chronic mild stress and prolonged treatment with imipramine on the levels of endogenous Met-enkephalin in the rat dopaminergic mesolimbic system. Pol. J. Pharmacol., 48: 53–56.

Dziedzicka-Wasylewska, M., Dlaboga, D., Pierzchala-Koziec, K. and Rogoz, Z. (2002) Effect of tianeptine and fluoxetine on the levels of Met-enkephalin and mRNA encoding proenkephalin in the rat. J. Physiol. Pharmacol., 53: 117–125.

Erb, S., Shaham, Y. and Stewart, J. (1998) The role of corticotropin-releasing factor and corticosterone in stress- and cocaine-induced relapse to cocaine seeking in rats. J. Neurosci., 18: 5529–5536.

Eriksson, E., Westberg, P., Thuresson, K., Modigh, K., Ekman, R. and Widerlov, E. (1989) Increased cerebrospinal fluid levels of endorphin immunoreactivity in panic disorder. Neuropsychopharmacology, 2: 225–228.

Fallon, J.H. and Leslie, F.M. (1986) Distribution of dynorphin and enkephalin peptides in the rat brain. J. Comp. Neurol., 249: 293–336.

Fanselow, M.S. (1986) Conditioned fear-induced opiate analgesia: a competing motivational state theory of stress analgesia. Ann. N.Y. Acad. Sci., 467: 40–54.

Fernandez, B., Alberti, I., Kitchen, I. and Viveros, M.P. (1999) Neonatal naltrindole and handling differently affect morphine antinociception in male and female rats. Pharmacol. Biochem. Behav., 64: 851–855.

Ferris, C.F., Melloni, R.H., Jr., Koppel, G., Perry, K.W., Fuller, R.W. and Delville, Y. (1997) Vasopressin/serotonin interactions in the anterior hypothalamus control aggressive behavior in golden hamsters. J. Neurosci., 17: 4331–4340.

Ferris, C.F., Stolberg, T. and Delville, Y. (1999) Serotonin regulation of aggressive behavior in male golden hamsters (Mesocricetus auratus). Behav. Neurosci., 113: 804–815.

Ferry, B., Roozendaal, B. and McGaugh, J. (1999) Role of norepinephrine in mediating stress hormone regulation of long-term memory storage: a critical involvement of the amygdala. Biological Psychiatry, 46: 1140–1152.

File, S.E. and Rodgers, R.J. (1979) Partial anxiolytic action of morphine sulphate following microinjection into the central nucleus of the amygdala in rats. Pharmacol. Biochem. Behav., 11: 313–318.

Filliol, D., Ghozland, S., Chluba, J., Martin, M., Matthes, H.W., Simonin, F., Befort, K., Gaveriaux-Ruff, C., Dierich, A., LeMeur, M., Valverde, O., Maldonado, R. and Kieffer, B.L. (2000) Mice deficient for delta- and mu-opioid receptors exhibit opposing alterations of emotional responses. Nat. Genet., 25: 195–200.

Fontaine, R., Chouinard, G. and Annable, L. (1984) Rebound anxiety in anxious patients after abrupt withdrawal of benzodiazepine treatment. Am. J. Psychiatry, 141: 848–852.

Fontana, F., Bernardi, P., Pich, E.M., Boschi, S., De Iasio, R., Spampinato, S. and Grossi, G. (1997) Opioid peptide modulation of circulatory and endocrine response to mental stress in humans. Peptides, 18: 169–175.

Fontana, F., Bernardi, P., Pich, E.M., Boschi, S., De Iasio, R. and Spampinato, S. (1998) Endogenous opioid peptides and mental stress in congestive heart failure patients. Peptides, 19: 21–26.

Gallagher, M., Kapp, B.S. and Pascoe, J.P. (1982) Enkephalin analogue effects in the amygdala central nucleus on conditioned heart rate. Pharmacol. Biochem. Behav., 17: 217–222.

Gallego, M., Casis, L. and Casis, O. (1998) Imipramine inhibits soluble enkephalin-degrading aminopeptidase activity in vitro. Eur. J. Pharmacol., 360: 113–116.

Ge, Y., Lundeberg, T. and Yu, L.C. (2002) Blockade effect of mu and kappa opioid antagonists on the anti-nociception induced by intra-periaqueductal grey injection of oxytocin in rats. Brain Res., 927: 204–207.

Gelsema, A.J., McKitrick, D.J. and Calaresu, F.R. (1987) Cardiovascular responses to chemical and electrical stimulation of amygdala in rats. Am. J. Physiol., 253: R712–R718.

Giardino, L., Bettelli, C., Pozza, M. and Calza, L. (1999) Regulation of CCK mRNA expression in the rat brain by stress and treatment with sertraline, a selective serotonin re-uptake inhibitor. Brain Res., 824: 304–307.

Good, A.J. and Westbrook, R.F. (1995) Effects of a microinjection of morphine into the amygdala on the acquisition and expression of conditioned fear and hypoalgesia in rats. Behav. Neurosci., 109: 631–641.

Goodwin, G.A. and Barr, G.A. (1997) Evidence for opioid and nonopioid processes mediating adaptive responses of infant rats that are repeatedly isolated. Dev. Psychobiol., 31: 217–227.

Goodwin, G.M., Austin, M.P., Curran, S.M., Ross, M., Murray, C., Prentice, N., Ebmeier, K.P., Bennie, J., Carroll, S. and Dick, H. et al. (1993) The elevation of plasma beta-endorphin levels in major depression. J. Affect. Disord., 29: 281–289.

Grewal, S.S., Shepherd, J.K., Bill, D.J., Fletcher, A. and Dourish, C.T. (1997) Behavioural and pharmacological characterisation of the canopy stretched attend posture test as a model of anxiety in mice and rats. Psychopharmacology (Berl), 133: 29–38.

Grossman, A. (1986) Opioids and stress: the role of ACTH and epinephrine. In: Taché, Y. and Cantin, M. (Eds.), Neuropeptides and Stress. Springer-Verlag, New York, pp. 313–324.

Guthrie, J. and Basbaum, A. (1984) Colocalization of immunoreactive proenkephalin and prodynorphin products in medullary neurons of the rat. Neuropeptides, 4: 437–445.

Harbuz, M. and Lightman, S. (1989) Responses of hypothalamic and pituitary mRNA to physical and psychological stress in the rat. J. Endocrinol., 122: 705–711.

Harbuz, M., Russell, J., Summer, B., Kawata, M. and Lightman, S. (1991) Rapid changes in the content of proenkephalin A and corticotrophin releasing hormone mRNAs in the paraventricular nucleus during morphine withdrawal in urethane-anaesthetized rats. Mol. Brain Res., 9: 285–291.

Harlan, R.E., Shivers, B.D., Romano, G.J., Howells, R.D. and Pfaff, D.W. (1987) Localization of proenkephalin mRNA in the rat brain and spinal cord by in situ hybridization. J. Comp. Neurol., 258: 159–184.

Hayden-Hixson, D.M. and Nemeroff, C.B. (1993) Role(s) of neuropeptides in responding and adaptation to stress: a focus on corticotropin-releasing factor and opioid peptides. In: Stanford, S.C. and Salmon, P. (Eds.), Stress, from Stress to Syndrome. Academic Press, London, pp. 356–391.

He, X.P., Chen, B.Y., Zhu, J.M. and Cao, X.D. (1989) Change of Leu-enkephalin- and B-endorphin-like immunoreactivity in the hippocampus after electroconvulsive shock and electroacupuncture. Acupunct. Electrother. Res., 14: 131–139.

Hebb, A.L.O., Mendella, P.D. and Zacharko, R.M. (1997) Immediate delta-receptor activation following stressor imposition influences behavioral reactivity to a subsequent stressful encounter. Soc. Neurosci. Abst., 23: 1857.

Hebb, A.L., Zacharko, R.M., Dominguez, H., Trudel, F., Laforest, S. and Drolet, G. (2002a) Odor-induced variation in anxiety-like behavior in mice is associated with discrete and differential effects on mesocorticolimbic cholecystokinin mRNA expression. Neuropsychopharmacology, 27: 744–755.

Hebb, A.L.O., Lemerrer, J., Laforest, S., Zacharko, R.M. and Drolet, G. (2002b) Exposure to the predator odor (TMT) enhances mesocorticolimbic neuronal activation and enkephalin mRNA expression associated with anxiety in mice. In: Program No, 670.15. 2002 Abstract Viewer/Itinerary Planner. Society for Neuroscience, Washington, DC, Online.

Heinsbroek, R.P., van Haaren, F. and van de Poll, N.E. (1988) Sex differences in passive avoidance behavior of rats: sex-dependent susceptibility to shock-induced behavioral depression. Physiol. Behav., 43: 201–206.

Hernandez, L.L. and Watson, K.L. (1997) Opioid modulation of attention-related responses: delta-receptors modulate habituation and conditioned bradycardia. Psychopharmacology (Berl), 131: 140–147.

Hernandez, L.L., Watson, K.L., Fowler, B.M., Bair, K.D. and Singha, A.K. (1997) Opioid modulation of attention-related responses: peripheral-to-central progression and development of mu influence as learning occurs. Psychopharmacology (Berl), 132: 50–60.

Holaday, J.W. (1983) Cardiovascular effects of endogenous opiate systems. Ann. Rev. Pharmacol. Toxicol., 23: 541–594.

Holaday, J.W., Tortella, F.C., Long, J.B., Belenky, G.L. and Hitzemann, R.J. (1986) Endogenous opioids and their receptors. Evidence for involvement in the postictal effects of electroconvulsive shock. Ann. N.Y. Acad. Sci., 462: 124–139.

Hong, J.S., Gillin, J.C., Yang, H.Y. and Costa, E. (1979) Repeated electroconvulsive shocks and the brain content of endorphins. Brain Res., 177: 273–278.

Horvath, G. (2000) Endomorphin-1 and endomorphin-2: pharmacology of the selective endogenous mu-opioid receptor agonists. Pharmacol. Ther., 88: 437–463.

Hotsenpiller, G. and Williams, J.L. (1997) A synthetic predator odor (TMT) enhances condiotioned analgesia and fear when paired with a benzodiazepine receptor inverse agonist (FG-7142). Psychobiology, 25: 83–88.

Howlett, T.A. and Rees, L.H. (1986) Endogenous opioid peptides and hypothalamo–pituitary function. Ann. Rev. Physiol., 48: 527–536.

Hurd, Y.L. (1996) Differential messenger RNA expression of prodynorphin and proenkephalin in the humain brain. Neuroscience, 72: 767–783.

Ibarra, P., Bruehl, S.P., McCubbin, J.A., Carlson, C.R., Wilson, J.F., Norton, J.A. and Montgomery, T.B. (1994) An unusual reaction to opioid blockade with naltrexone in a case of post-traumatic stress disorder. J. Trauma. Stress, 7: 303–309.

Imperato, A., Puglisi-Allegra, S., Grazia Scrocco, M., Casolini, P., Bacchi, S. and Angelucci, L. (1992) Cortical and limbic dopamine and acetylcholine release as neurochemical correlates of emotional arousal in both aversive and non-aversive environmental changes. Neurochem. Int., 20 Suppl.: 265S–270S.

Izumi, T. (1998) Behavioral and neurochemical study on the role of the brain cholecystokinin system in anxiety. Hokkaido Igaku Zasshi, 73: 463–473.

Jamner, L.D. and Leigh, H. (1999) Repressive/defensive coping, endogenous opioids and health: how a life so perfect can make you sick. Psychiatry Res., 85: 17–31.

Janssen, S.A. and Arntz, A. (1996) Anxiety and pain: attentional and endorphinergic influences. Pain, 66: 145–150.

Janssen, S.A. and Arntz, A. (1997) No evidence for opioid-mediated analgesia induced by phobic fear. Behav. Res. Ther., 35: 823–830.

Janssens, C.J.J.G., Helmond, F.A., Loyens, L.W.S., Schouten, W.G.P. and Wiegant, V.M. (1995) Chronic stress increases the opioid-mediated inhibition of the pituitary–adrenocortical response to acute stress in pigs. Endocrinology, 136: 1468–1473.

Jinks, A.L. and McGregor, I.S. (1997) Modulation of anxiety-related behaviours following lesions of the prelimbic or infralimbic cortex in the rat. Brain Res., 772: 181–190.

Kalin, N.H., Shelton, S.E. and Barksdale, C.M. (1988) Opiate modulation of separation-induced distress in non-human primates. Brain Res., 440: 285–292.

Kalivas, P.W. (1993) Neurotransmitter regulation of dopamine neurons in the ventral tegmental area. Brain Res. Rev., 18: 75–113.

Kalivas, P.W., Duffy, P., Dilts, R. and Abhold, R. (1988) Enkephalin modulation of A10 dopamine neurons: a role in dopamine sensitization. Ann. N.Y. Acad. Sci., 537: 405–414.

Kalyuzhny, A.E. and Wessendorf, M.W. (1997) CNS GABA neurons express the mu-opioid receptor: immunocytochemical studies. Neuroreport., 8: 3367–3372.

Kang, W., Wilson, S.P. and Wilson, M.A. (1999) Changes in nociceptive and anxiolytic responses following herpes virus-mediated preproenkephalin overexpression in rat amygdala are naloxone-reversible and transient. Ann. N.Y. Acad. Sci., 877: 751–755.

Kang, W., Wilson, S.P. and Wilson, M.A. (2000) Overexpression of proenkephalin in the amygdala potentiates the anxiolytic effects of benzodiazepines. Neuropsychopharmacology, 22: 77–88.

Katoh, A., Nabeshima, T. and Kameyama, T. (1990) Behavioral changes induced by stressful situations: effects of enkephalins, dynorphin, and their interactions. J. Pharmacol. Exp. Ther., 253: 600–607.

Katoh, A., Nabeshima, T. and Kameyama, T. (1991) Interaction between enkephalinergic and dopaminergic systems in stressful situations. Eur. J. Pharmacol., 193: 95–99.

Katoh, A., Nabeshima, T., Ukai, R. and Kameyama, T. (1992) Endorphins: do not affect behavioral stress responses in mice. Peptides, 13: 737–739.

Kavaliers, M. (1988) Brief exposure to a natural predator, the short-tailed weasel, induces benzodiazepine-sensitive analgesia in white-footed mice. Physiol. Behav., 43: 187–193.

Kavaliers, M. and Choleris, E. (1997) Sex differences in N-methyl-D-aspartate involvement in kappa opioid and non-opioid predator-induced analgesia in mice. Brain Res., 768: 30–36.

Kavaliers, M., Colwell, D.D. and Perrot-Sinal, T.S. (1997) Opioid and non-opioid NMDA-mediated predator-induced analgesia in mice and the effects of parasitic infection. Brain Res., 766: 11–18.

Kelly, P.A., Ford, I. and McCulloch, J. (1986) The effect of diazepam upon local cerebral glucose use in the conscious rat. Neuroscience, 19: 257–265.

Kemble, E.D. and Bolwahnn, B.L. (1997) Immediate and long-term effects of novel odors on risk assessment in mice. Physiol. Behav., 61: 543–549.

Kessing, L.V., Mortensen, P.B. and Bolwig, T.G. (1998) Clinical definitions of sensitisation in affective disorder: a case register study of prevalence and prediction. J. Affect. Disord., 47: 31–39.

Khachaturian, H., Lewis, M. and Watson, S.J. (1983) Enkephalin systems in the diencephalon and brainstem of the rat. J. Comp. Neurol., 220: 310–320.

Kieffer, B.L. (1999) Opioids: first lessons from knockout mice. Trends Pharmacol. Sci., 20: 19–26.

Kim, J.J. and Yoon, K.S. (1998) Stress: metaplastic effects in the hippocampus. Trends Neurosci., 21: 505–509.

Kiritsy-Roy, J., Appel, N., Bobbitt, F. and Van Loon, G. (1986) Effects of mu-opioid receptor stimulation in the hypothalamic paraventricular nucleus on basal and stress-induced catecholamine secretion and cardiovascular responses. J. Pharmacol. Exp. Toxicol., 239: 814–822.

Klein, L.C., Popke, E.J. and Grunberg, N.E. (1998) Sex differences in effects of opioid blockade on stress-induced freezing behavior. Pharmacol. Biochem. Behav., 61: 413–417.

Kollack-Walker, S. and Newman, S.W. (1995) Mating and agonistic behavior produce different patterns of Fos immunolabeling in the male Syrian hamster brain. Neuroscience, 66: 721–736.

Konig, M., Zimmer, A.M., Steiner, H., Holmes, P.V., Crawley, J.N., Brownstein, M.J. and Zimmer, A. (1996) Pain responses, anxiety and aggression in mice deficient in pre- proenkephalin. Nature, 383: 535–538.

Korn, M.L., Plutchik, R. and Van Praag, H.M. (1997) Panic-associated suicidal and aggressive ideation and behavior. J. Psychiatr. Res., 31: 481–487.

Kunos, G. and Varga, K. (1995) The tachycardia associated with the defense reaction involves activation of both GABAA and GABAB receptors in the nucleus tractus solitarii. Clin. Exp. Hypertens., 17: 91–100.

Kurumaji, A., Mitsushio, H. and Takashima, M. (1988) Chronic dietary treatment with antidepressants decrease brain Met-enkephalin-like immunoreactivity in the rat. Psychopharmacology (Berl), 94: 188–192.

Lason, W., Przewlocka, B. and Przewlocki, R. (1987) Single and repeated electroconvulsive shock differentially affects the prodynorphin and pro-opiomelanocortin system in the rat. Brain Res., 403: 301–307.

Lawrence, D. and Pittman, Q.J. (1985) Interaction between descending paraventricular neurons and vagal motor neurons. Brain Res., 332: 158–160.

Lester, L.S. and Fanselow, M.S. (1985) Exposure to a cat produces opioid analgesia in rats. Behav. Neurosci., 99: 756–759.

Liberzon, I., Zubieta, J.K., Fig, L.M., Phan, K.L., Koeppe, R.A. and Taylor, S.F. (2002) mu-Opioid receptors and limbic responses to aversive emotional stimuli. Proc. Natl. Acad. Sci. USA, 99: 7084–7089.

Lichtman, A.H. and Fanselow, M.S. (1990) Cats produce analgesia in rats on the tail-flick test: naltrexone sensitivity is determined by the nociceptive test stimulus. Brain Res., 533: 91–94.

Lightman, S.L. and Young, W.S.I. (1987a) Changes in hypothalamic preproenkephalin A mRNA following stress and opiate withdrawal. Nature, 328: 643–645.

Lightman, S.L. and Young, W.S.I. (1987b) Vasopressin, oxytocin, dynorphin, enkephalin and corticotrophin-releasing factor mRNA stimulation in the rat. J. Physiol. (London), 394: 23–39.

Lightman, S.L. and Young, W.S.I. (1988) Corticotrophin-releasing factor, vasopressin and pro-opiomelanocortin mRNA responses to stress and opiates in the rat. J. Physiol. (London), 403: 511–523.

Lightman, S.L. and Young, W.S.I. (1989) Influence of steroids on the hypothalamic corticotropin-releasing factor and preproenkephalin mRNA responses to stress. Proc. Natl. Acta. Sci. USA, 86: 4306–4310.

Loas, G. (1996) Vulnerability to depression: a model centered on anhedonia. J. Affect. Disord., 41: 39–53.

MacNeil, G., Sela, Y., McIntosh, J. and Zacharko, R.M. (1997) Anxiogenic behavior in the light-dark paradigm following intraventricular administration of cholecystokinin-8S, restraint stress, or uncontrollable footshock in the CD-1 mouse. Pharmacol. Biochem. Behav., 58: 737–746.

Maier, S.F. (1984) Learned helplessness and animal models of depression. Prog. Neuropsychopharmacol. Biol. Psychiatry, 8: 435–446.

Mansi, J.A., Laforest, S. and Drolet, G. (2000) Effect of stress exposure on the activation pattern of enkephalin-containing perikarya in the rat ventral medulla. J. Neurochem., 74: 2568–2575.

Mansour, A., Khachaturian, H., Lewis, M., Akil, H. and Watson, S. (1987) Autoradiographic differentiation of mu, delta, and kappa opioid receptors in the rat forebrain and midbrain. J. Neurosci., 7: 2445–2464.

Mansour, A., Khachaturian, H., Lewis, M.E., Akil, H. and Watson, S.J. (1988) Anatomy of CNS opioid receptors. Trends Neurosci., 11: 308–314.

Mansour, A., Fox, C.A., Burke, S., Meng, F., Thompson, R.C., Akil, H. and Watson, S.J. (1994) Mu, delta and kappa opioid receptor mRNA expression in the rat CNS: an in situ hybridization study. J. Comp. Neurol., 350: 412–438.

Mansour, A., Fox, C.A., Akil, H. and Watson, S.J. (1995) Opioid-receptor mRNA expression in the rat CNS: anatomical and functional implications. Trends Neurosci., 18: 22–29.

Maremmani, I., Marini, G. and Fornai, F. (1998) Naltrexone-induced panic attacks. Am. J. Psychiatry, 155: 447.

Marson, L., JA, K.-R. and Van, L.G. (1989) μ-Opioid peptide modulation of cardiovascular and sympathoadrenal responses to stress. Am. J. Physiol., 257: R901–R908.

Martin-Schild, S., Gerall, A.A., Kastin, A.J. and Zadina, J.E. (1999) Differential distribution of endomorphin 1- and endomorphin 2-like immunoreactivities in the CNS of the rodent. J. Comp. Neurol., 405: 450–471.

Martinez, J.L., Jr. and Weinberger, S.B. (1987) Memory traces: how to increase and decrease their strength. Int. J. Neurol., 21–22: 33–50.

Martinez, J.L., Jr., Weinberger, S.B. and Schulteis, G. (1988) Enkephalins and learning and memory: a review of evidence for a site of action outside the blood–brain barrier. Behav. Neural Biol., 49: 192–221.

Martinez, M., Phillips, P.J. and Herbert, J. (1998) Adaptation in patterns of c-fos expression in the brain associated with exposure to either single or repeated social stress in male rats. Eur. J. Neurosci., 10: 20–33.

Masure, C.M. (1994) Stress and Psychiatric Disorders. American Psychiatric Press, Washington.

Matsuzawa, S., Suzuki, T., Misawa, M. and Nagase, H. (1999) Different roles of mu-, delta- and kappa-opioid receptors in ethanol-associated place preference in rats exposed to conditioned fear stress. Eur. J. Pharmacol., 368: 9–16.

McCubbin, J.A. (1993) Stress and endogenous opioids: behavioral and circulatory interactions. Biol. Psychol., 35: 91–122.

McCubbin, J., Bruehl, S., Wilson, J., Sherman, J., Norton, J. and Colclough, G. (1998) Endogenous opioids inhibit ambulatory blood pressure during naturally occurring stress. Psychosomatic Medicine, 60: 227–231.

McEwen, B.S. and Sapolsky, R.M. (1995) Stress and cognitive function. Curr. Opin. Neurobiol., 5: 205–216.

McFall, M., Fontana, A., Raskind, M. and Rosenheck, R. (1999) Analysis of violent behavior in Vietnam combat veteran psychiatric inpatients with posttraumatic stress disorder. J. Trauma Stress, 12: 501–517.

McGivern, R.F., Mousa, S., Couri, D. and Berntson, G.G. (1983) Prolonged intermittent footshock stress decreases Met and Leu enkephalin levels in brain with concomitant decreases in pain threshold. Life Sci., 33: 47–54.

McGregor, I.S. and Dielenberg, R.A. (1999) Differential anxiolytic efficacy of a benzodiazepine on first versus second exposure to a predatory odor in rats. Psychopharmacology (Berl), 147: 174–181.

Menetrey, D. and Basbaum, A.I. (1987) The distribution of substance P-, enkephalin- and dynorphin-immunoreactive neurons in the medulla of the rat and their contribution to bulbospinal pathways. Neuroscience, 23: 173–187.

Metalsky, G.I. and Joiner, T.E., Jr. (1992) Vulnerability to depressive symptomatology: a prospective test of the diathesis-stress and causal mediation components of the hopelessness theory of depression. J. Pers. Soc. Psychol., 63: 667–675.

Meunier, J.C., Mollereau, C., Toll, L., Suaudeau, C., Moisand, C., Alvinerie, P., Butour, J.L., Guillemot, J.C., Ferrara, P. and Monsarrat, B. et al. (1995) Isolation and structure of the endogenous agonist of opioid receptor-like ORL1 receptor. Nature, 377: 532–535.

Mogil, J.S. and Pasternak, G.W. (2001) The molecular and behavioral pharmacology of the orphanin FQ/nociceptin peptide and receptor family. Pharmacol. Rev., 53: 381–415.

Molewijk, H.E., van der Poel, A.M. and Olivier, B. (1995) The ambivalent behaviour "stretched approach posture" in the rat as a paradigm to characterize anxiolytic drugs. Psychopharmacology (Berl), 121: 81–90.

Morgan, M.A. and LeDoux, J.E. (1995) Differential contribution of dorsal and ventral medial prefrontal cortex to the acquisition and extinction of conditioned fear in rats. Behav. Neurosci., 109: 681–688.

Morgan, M.A., Romanski, L.M. and LeDoux, J.E. (1993) Extinction of emotional learning: contribution of medial prefrontal cortex. Neurosci. Lett., 163: 109–113.

Morilak, D.A., Drolet, G. and Chalmers, J. (1990a) Cardiovascular effects of the opioid antagonist naloxone in the rostral ventrolateral medulla of rabbits. Am. J. Physiol., 258: R325–R331.

Morilak, D.A., Drolet, G. and Chalmers, J. (1990b) Tonic opioid inhibition of the pressor region of the rostral ventrolateral medulla of rabbits is mediated by delta receptors. J. Pharmacol. Exp. Toxicol., 254: 671–676.

Moynihan, J.A., Karp, J.D., Cohen, N. and Ader, R. (2000) Immune deviation following stress odor exposure: role of endogenous opioids. J. Neuroimmunol., 102: 145–153.

Nagy, L.M., Krystal, J.H., Charney, D.S., Merikangas, K.R. and Woods, S.W. (1993) Long-term outcome of panic disorder after short-term imipramine and behavioral group treatment: 2.9-year naturalistic follow-up study. J. Clin. Psychopharmacol., 13: 16–24.

Nikulina, E.M., Hammer, R.P., Jr., Miczek, K.A. and Kream, R.M. (1999) Social defeat stress increases expression of mu-opioid receptor mRNA in rat ventral tegmental area. Neuroreport, 10: 3015–3019.

Nobre, M.J., Ribeiro dos Santos, N., Aguiar, M.S. and Brandao, M.L. (2000) Blockade of mu- and activation of kappa-opioid receptors in the dorsal periaqueductal gray matter produce defensive behavior in rats tested in the elevated plus-maze. Eur. J. Pharmacol., 404: 145–151.

Olson, G.A., Olson, R.D., Kastin, A.J. and Coy, D.H. (1979) Endogenous opiates: through 1978. Neurosci. Biobehav. Rev., 3: 285–299.

Olson, G.A., Olson, R.D., Kastin, A.J. and Coy, D.H. (1980) Endogenous opiates: 1979. Peptides, 1: 365–379.

Olson, G.A., Olson, R.D., Kastin, A.J. and Coy, D.H. (1981) Endogenous opiates: 1980. Peptides, 2: 349–369.

Olson, G.A., Olson, R.D., Kastin, A.J. and Coy, D.H. (1982) Endogenous opiates: 1981. Peptides, 3: 1039–1072.

Olson, G.A., Olson, R.D. and Kastin, A.J. (1983) Endogenous opiates: 1982. Peptides, 4: 563–576.

Olson, G.A., Olson, R.D. and Kastin, A.J. (1984) Endogenous opiates: 1983. Peptides, 5: 975–992.

Olson, G.A., Olson, R.D. and Kastin, A.J. (1985) Endogenous opiates: 1984. Peptides, 6: 769–791.

Olson, G.A., Olson, R.D. and Kastin, A.J. (1986) Endogenous opiates: 1985. Peptides, 7: 907–933.

Olson, G.A., Olson, R.D. and Kastin, A.J. (1987) Endogenous opiates: 1986. Peptides, 8: 1135–1164.

Olson, G.A., Olson, R.D. and Kastin, A.J. (1989a) Endogenous opiates: 1987. Peptides, 10: 205–236.

Olson, G.A., Olson, R.D. and Kastin, A.J. (1989b) Endogenous opiates: 1988. Peptides, 10: 1253–1280.

Olson, G.A., Olson, R.D. and Kastin, A.J. (1990) Endogenous opiates: 1989. Peptides, 11: 1277–1304.

Olson, G.A., Olson, R.D. and Kastin, A.J. (1991) Endogenous opiates: 1990. Peptides, 12: 1407–1432.

Olson, G.A., Olson, R.D. and Kastin, A.J. (1992) Endogenous opiates: 1991. Peptides, 13: 1247–1287.

Olson, G.A., Olson, R.D. and Kastin, A.J. (1993) Endogenous opiates: 1992. Peptides, 14: 1339–1378.

Olson, G.A., Olson, R.D. and Kastin, A.J. (1994) Endogenous opiates: 1993. Peptides, 15: 1513–1556.

Olson, G.A., Olson, R.D. and Kastin, A.J. (1995) Endogenous opiates: 1994. Peptides, 16: 1517–1555.

Olson, G.A., Olson, R.D. and Kastin, A.J. (1996) Endogenous opiates: 1995. Peptides, 17: 1421–1466.

Olson, G., Olson, R. and Kastin, A. (1997) Endogenous opiates: 1996. Peptides, 18: 1651–1688.

Olson, G.A., Olson, R.D., Vaccarino, A.L. and Kastin, A.J. (1998) Endogenous opiates: 1997. Peptides, 19: 1791–1843.

Otto, M.W., Pollack, M.H., Sachs, G.S., Reiter, S.R., Meltzer-Brody, S. and Rosenbaum, J.F. (1993) Discontinuation of benzodiazepine treatment: efficacy of cognitive-behavioral therapy for patients with panic disorder. Am. J. Psychiatry., 150: 1485–1490.

Palkovits, M. (1987) Organization of the stress response at the anatomical level. Prog. Brain Res., 72: 47–55.

Palkovits, M., Mezey, E. and Eskay, R.L. (1987) Distribution and possible origin of b-endorphin and ACTH in discrete brainstem nuclei of rats. Neuropeptides, 9: 123–137.

Pan, Z.Z. (1998) mu-Opposing actions of the kappa-opioid receptor. Trends Pharmacol. Sci., 19: 94–98.

Pechnick, R.N. (1993) Effects of opioids on the hypothalamo-pituitary–adrenal axis. Ann. Rev. Pharmacol. Toxicol., 33: 353–382.

Petrovich, G.D., Scicli, A.P., Thompson, R.F. and Swanson, L.W. (2000) Associative fear conditioning of enkephalin mRNA levels in central amygdalar neurons. Behav. Neurosci., 114: 681–686.

Petrusz, P., Merchenthaler, I. and Maderdrut, J.L. (1985) Distribution of enkephalin-containing neurons in the central nervous system. In: Björklund, A. and Hökfelt, T., (Eds.), Handbook of Chemical Neuroanatomy. Elsevier Science, Amsterdam, pp. 273–334.

Piva, M.A., Medina, J.H., de Blas, A.L. and Pena, C. (1991) Formation of benzodiazepine-like molecules in rat brain. Biochem. Biophys. Res. Commun., 180: 972–981.

Plata-Salaman, C.R., Ilyin, S.E., Turrin, N.P., Gayle, D., Flynn, M.C., Bedard, T., Merali, Z. and Anisman, H. (2000) Neither acute nor chronic exposure to a naturalistic (predator) stressor influences the interleukin-1beta system, tumor necrosis factor-alpha, transforming growth factor-beta1, and neuropeptide mRNAs in specific brain regions. Brain Res. Bull., 51: 187–193.

Pohorecky, L.A., Skiandos, A., Zhang, X., Rice, K.C. and Benjamin, D. (1999) Effect of chronic social stress on delta-opioid receptor function in the rat. J. Pharmacol. Exp. Ther., 290: 196–206.

Porteous, A. and Tyndall, J. (1994) Yes, I want to walk to the OR. Can. Oper. Room Nurs. J., 12: 15–25.

Privette, T.H. and Terrian, D.M. (1995) Kappa opioid agonists produce anxiolytic-like behavior on the elevated plus-maze. Psychopharmacology (Berl), 118: 444–450.

Przewlocki, R., Przewlocka, B. and Lason, W. (1991) Adaptation of opioid systems to stress. In: Almeida, O.F.X. and Shippenberg, T.S. (Eds.), The Neurobiology of Opioid Systems to Stress. Springer-Verlag, Berlin, pp. 229–243.

Przewlocki, R., Lason, W., Turchan, J. and Przewlocka, B. (1997) Imipramine induces alterations in proenkephalin and prodynorphin mRNAs level in the nucleus accumbens and striatum in the rat. Pol. J. Pharmacol., 49: 351–355.

Quirion, R. and Pert, C.B. (1981) Dynorphins: similar relative potencies on mu, delta- and kappa-opiate receptors. Eur. J. Pharmacol., 76: 467–468.

Ragnauth, A., Schuller, A., Morgan, M., Chan, J., Ogawa, S., Pintar, J., Bodnar, R.J. and Pfaff, D.W. (2001) Female preproenkephalin-knockout mice display altered emotional responses. Proc. Natl. Acad. Sci. USA, 98: 1958–1963.

Ray, A. and Henke, P.G. (1990) Enkephalin-dopamine interactions in the central amygdalar nucleus during gastric stress ulcer formation in rats. Behav. Brain Res., 36: 179–183.

Ray, A., Henke, P.G. and Sullivan, R.M. (1988) Opiate mechanisms in the central amygdala and gastric stress pathology in rats. Brain Research, 23: 195–198.

Reinscheid, R.K. and Civelli, O. (2002) The orphanin FQ/nociceptin knockout mouse: a behavioral model for stress responses. Neuropeptides, 36: 72–76.

Reinscheid, R.K., Nothacker, H.P., Bourson, A., Ardati, A., Henningsen, R.A., Bunzow, J.R., Grandy, D.K., Langen, H., Monsma, F.J., Jr. and Civelli, O. (1995) Orphanin FQ: a neuropeptide that activates an opioidlike G protein-coupled receptor. Science, 270: 792–794.

Risch, S.C. (1997) Recent advances in depression research: from stress to molecular biology and brain imaging. J. Clin. Psychiatry, 58 Suppl. 5: 3–6.

Rodgers, R.J. and File, S.E. (1979) Exploratory behaviour and aversive thresholds following intra-amygdaloid application of opiates in rats. Pharmacol. Biochem. Behav., 11: 505–511.

Rodgers, R.J. and Randall, J.I. (1987) Social conflict analgesia: inhibition of early non-opioid component by diazepam or flumazepil fails to affect appearance of late opioid component. Brain Res. Bull., 19: 141–144.

Roozendaal, B., Koolhaas, J.M. and Bohus, B. (1991) Attenuated cardiovascular, neuroendocrine, and behavioral responses after a single footshock in central amygdaloid lesioned male rats. Physiol. Behav., 50: 771–775.

Rosen, J.B., Fanselow, M.S., Young, S.L., Sitcoske, M. and Maren, S. (1998) Immediate-early gene expression in the amygdala following footshock stress and contextual fear conditioning. Brain Res., 796: 132–142.

Russell, J.A. and Douglas, A.J. (2000) Opioids. In: Fink, G., (Ed.), Encyclopedia of Stress. Academic Press, San Diego, pp. 86–98

Sante, A.B., Nobre, M.J. and Brandao, M.L. (2000) Place aversion induced by blockade of mu or activation of kappa opioid receptors in the dorsal periaqueductal gray matter. Behav. Pharmacol., 11: 583–589.

Sapolsky, R.M. (1996) Why stress is bad for your brain. Science, 273: 749–750.

Sawchenko, P.E. (1986) Neuropeptides, the paraventricular nucleus and the integration of hypothalamic neuroendocrine and autonomic function. In: Taché, Y. and Cantin, M. (Eds.), Neuropeptides and Stress. Springer-Verlag, New York, pp. 73–91.

Sawchenko, P.E., Imaki, T., Potter, E., Kovacs, K., Imaki, J. and Vale, W. (1993) The functional neuroanatomy of corticotropin-releasing factor. In: Chadwick, D.J., Marsh, J. and Ackrill, K. (Eds.), Corticotropin-Releasing Factor. John Wiley & Sons Ltd., Chichester, pp. 5–29.

Schafer, M.K.-H., Day, R., Watson, S.J. and Akil, H. (1991) In: Almeida, O.F.X. and Shippenberg, T.S. (Eds.), Neurobiology of Opioids. Springer-Verlag, Berlin, pp. 53–71.

Schulteis, G. and Martinez, J.L., Jr. (1992) Peripheral modulation of learning and memory: enkephalins as a model system. Psychopharmacology (Berl), 109: 347–364.

Schulteis, G. and Martinez, J.L., Jr. (1993) Experience-dependent regulation of Leu-enkephalin hydrolysis in rat plasma. Peptides, 14: 161–167.

Shaikh, M.B., Lu, C.L. and Siegel, A. (1991a) An enkephalinergic mechanism involved in amygdaloid suppression of affective defence behavior elicited from the midbrain periaqueductal gray in the cat. Brain Res., 559: 109–117.

Shaikh, M.B., Lu, C.L. and Siegel, A. (1991b) Affective defense behavior elicited from the feline midbrain periqueductal gray is regulated by mu and delta opioid receptors. Brain Res., 557: 344–348.

Shear, M.K., Ball, G., Fitzpatrick, M., Josephson, S., Klosko, J. and Frances, A. (1991) Cognitive-behavioral therapy for panic: an open study. J. Nerv. Ment. Dis., 179: 468–472.

Sheline, Y.I., Gado, M.H. and Price, J.L. (1998) Amygdala core nuclei volumes are decreased in recurrent major depression. Neuroreport, 9: 2023–2028.

Sheline, Y.I., Sanghavi, M., Mintun, M.A. and Gado, M.H. (1999) Depression duration but not age predicts hippocampal volume loss in medically healthy women with recurrent major depression. J. Neurosci., 19: 5034–5043.

Shepherd, J.K., Blanchard, D.C., Weiss, S.M., Rodgers, R.J. and Blanchard, R.J. (1992) Morphine attenuates antipredator ultrasonic vocalizations in mixed-sex rat colonies. Pharmacol. Biochem. Behav., 41: 551–558.

Siegel, R.A., Duker, E.M., Fuchs, E., Pahnke, U. and Wutte, W. (1984) Responsiveness of mesolimbic, mesocortical, septal, and hippocampal cholecystokinin and substance P neuronal systems to stress in the male rat. Neurochem. Int., 6: 783–789.

Siegel, A., Schubert, K.L. and Shaikh, M.B. (1997) Neurotransmitters regulating defensive rage behavior in the cat. Neurosci. Biobehav. Rev., 21: 733–742.

Soltis, R.P., Cook, J.C., Gregg, A.E. and Sanders, B.J. (1997) Interaction of GABA and excitatory amino acids in the basolateral amygdala: role in cardiovascular regulation. J. Neurosci., 17: 9367–9374.

Southwick, S.M., Paige, S., Morgan, C.A., 3rd, Bremner, J.D., Krystal, J.H. and Charney, D.S. (1999) Neurotransmitter alterations in PTSD: catecholamines and serotonin. Semin Clin. Neuropsychiatry, 4: 242–248.

Spiegel, D.A., Bruce, T.J., Gregg, S.F. and Nuzzarello, A. (1994) Does cognitive behavior therapy assist slow-taper alprazolam discontinuation in panic disorder? Am. J. Psychiatry, 151: 876–881.

Sribanditmongkol, P., Sheu, M.J. and Tejwani, G.A. (1994) Inhibition of morphine tolerance and dependence by diazepam and its relation to the CNS Met-enkephalin levels. Brain Res., 645: 1–12.

Stimmel, G.L. (1995) How to counsel patients about depression and its treatment. Pharmacotherapy, 15: 100S–104S.

Strack, A.M., Sawyer, W.B., Platt, K.B. and Loewy, A.D. (1989a) CNS cell groups regulating the sympathetic outflow to adrenal gland as revealed by transneuronal cell body labeling with pseudorabies virus. Brain Res., 491: 274–296.

Strack, A.M., Sawyer, W.B., Hughes, J.H., Platt, K.B. and Loewy, A.D. (1989b) A general pattern of CNS innervation of the sympathetic outflow demonstrated by transneuronal pseudorabies viral infections. Brain Res., 491: 156–162.

Stuckey, J., Marra, S., Minor, T. and Insel, T.R. (1989) Changes in mu opiate receptors following inescapable shock. Brain Res., 476: 167–169.

Sun, S.Y., Liu, Z., Li, P. and Ingenito, A.J. (1996) Central effects of opioid agonists and naloxone on blood pressure and heart rate in normotensive and hypertensive rats. Gen. Pharmacol., 27: 1187–1194.

Sverko, V., Marotti, T., Rocic, B., Rabatic, S. and Hrsak, I. (1997) Influence of methionine-enkephalin on stress-induced parameters. Int. J. Immunopharmacology, 19: 691–698.

Swanson, L.W., Sawchenko, P.E. and Lind, R.W. (1986) Regulation of multiple peptides in CRF parvocellular neurosecretory neurons: implications for the stress response. Prog. Brain Res., 169–190.

Swanson, L.W., Sawchenko, P.E., Lind, R.W. and Rho, J.-H. (1987) The CRH motoneuron: differential peptide regulation in neurons with possible synaptic, paracrine and endocrine outputs. Ann. N.Y. Acad. Sci., 512: 12–23.

Szekely, J.I. (1990) Opioid peptides and stress. Crit. Rev. Neurobiol., 6: 1–12.

Tanaka, M., Ida, Y. and Tsuda, A. (1988) Naloxone, given before but not after stress exposure, enhances stress-induced increases in noradrenaline release in rats brain regions. Pharmacol. Biochem. Behav., 29: 613–619.

Tanaka, M., Ida, Y., Tsuda, A., Tsujimaru, S., Shiarao, I. and Oguchi, M. (1989) Met-enkephalin, injected during the early phase of stress, attenuates stress-induced increases in noradrenaline release in rat brain regions. Pharmacol. Biochem. Behav., 32: 791–795.

Tanaka, M., Yoshida, M., Emoto, H. and Ishii, H. (2000) Noradrenaline systems in the hypothalamus, amygdala and locus coeruleus are involved in the provocation of anxiety: basic studies. Eur. J. Pharmacol., 405: 397–406.

Tejedor-Real, P., Mico, J.A., Maldonado, R., Roques, B.P. and Gibert-Rahola, J. (1993) Effect of mixed (RB 38A) and selective (RB 38B) inhibitors of enkephalin degrading enzymes on a model of depression in the rat. Biol. Psychiatry, 34: 100–107.

Tejedor-Real, P., Mico, J.A., Maldonado, R., Roques, B.P. and Gibert-Rahola, J. (1995) Implication of endogenous opioid system in the learned helplessness model of depression. Pharmacol. Biochem. Behav., 52: 145–152.

Tejedor-Real, P., Mico, J.A., Smadja, C., Maldonado, R., Roques, B.P. and Gilbert-Rahola, J. (1998) Involvement of delta-opioid receptors in the effects induced by endogenous enkephalins on learned helplessness model. Eur. J. Pharmacol., 354: 1–7.

Tilson, H., McLamb, R. and Hong, J. (1986) Behavioral effects of centrally administered dynorphin and [D-ala2-D-leu] enkephalin (DADLE) in rats. Neuropeptides, 8: 193–206.

Tsuda, M., Suzuki, T., Misawa, M. and Nagase, H. (1996) Involvement of the opioid system in the anxiolytic effect of diazepam in mice. Eur. J. Pharmacol., 307: 7–14.

Uhl, G.R., Childers, S. and Pasternak, G. (1994) An opiate-receptor gene family reunion. Trends Neurosci., 17: 89–93.

Ukai, M., Takada, A., Sasaki, Y. and Kameyama, T. (1997) Stimulation of delta1- and delta2-opioid receptors produces amnesia in mice. Eur. J. Pharmacol., 338: 1–6.

Vaccarino, A.L. and Kastin, A.J. (2000) Endogenous opiates: 1999. Peptides, 21: 1975–2034.

Vaccarino, A.L. and Kastin, A.J. (2001) Endogenous opiates: 2000. Peptides, 22: 2257–2328.

Vaccarino, A.L., Olson, G.A., Olson, R.D. and Kastin, A.J. (1999) Endogenous opiates: 1998. Peptides, 20: 1527–1574.

VandenBerg, C., Lamberts, R., Wolterink, G., Wiegant, V. and VanRee, J. (1998) Emotional and footshock stimuli induce differential long-lasting behavioural effects in rats; involvement of opioids. Brain Research, 799: 6–15.

van den Buuse, M. (1998) Role of the mesolimbic dopamine system in cardiovascular homeostasis. Stimulation of the ventral tegmental area modulates the effect of vasopressin on blood pressure in conscious rats. Clin. Exp. Pharmacol. Physiol., 25: 661–668.

Van Dijken, H.H., Van der Heyden, J.A., Mos, J. and Tilders, F.J. (1992) Inescapable footshocks induce progressive and long-lasting behavioural changes in male rats. Physiol. Behav., 51: 787–794.

Vivian, J.A. and Miczek, K.A. (1998) Effects of mu and delta opioid agonists and antagonists on affective vocal and reflexive pain responses during social stress in rats. Psychopharmacology (Berl), 139: 364–375.

Vochteloo, J.D. and Koolhaas, J.M. (1987) Medial amygdala lesions in male rats reduce aggressive behavior: interference with experience. Physiol. Behav., 41: 99–102.

Walker, D.L. and Davis, M. (1997) Double dissociation between the involvement of the bed nucleus of the stria terminalis and the central nucleus of the amygdala in startle increases produced by conditioned versus unconditioned fear. J. Neurosci., 17: 9375–9383.

Wall, P.M. and Messier, C. (2000a) Concurrent modulation of anxiety and memory. Behav. Brain Res., 109: 229–241.

Wall, P.M. and Messier, C. (2000b) U-69,593 microinjection in the infralimbic cortex reduces anxiety and enhances spontaneous alternation memory in mice. Brain Res., 856: 259–280.

Watkins, L.R. and Mayer, D.J. (1986) Multiple endogenous opiate and non-opiate analgesia systems: evidence of their existence and clinical implications. Ann. N.Y. Acad. Sci., 467: 273–299.

Watson, S.J., Khachaturian, H., Akil, H., Coy, D.H. and Goldstein, A. (1982) Comparaison of the distribution of dynorphin systems and enkephalin systems in brain. Science, 218: 1134–1136.

Watts, A.G. (1991) Ether anesthesia differentially affects the content of prepro-corticotropin-releasing hormone, prepro-neurotensin/neuromedin N and prepro-enkephalin mRNAs in the hypothalamic paraventricular nucleus of the rat. Brain Res., 544: 353–357.

Watts, A.G. (1992) Disturbance of fluid homeostasis leads to temporally and anatomically distinct responses in neuropeptide and tyrosine hydroxylase mRNA levels in the paraventricular and supraoptic nuclei of the rat. Neuroscience, 46: 859–879.

Weiss, E.L., Longhurst, J.G. and Mazure, C.M. (1999) Childhood sexual abuse as a risk factor for depression in women: psychosocial and neurobiological correlates. Am. J. Psychiatry, 156: 816–828.

Westbrook, R.F., Good, A.J. and Kiernan, M.J. (1997) Microinjection of morphine into the nucleus accumbens impairs contextual learning in rats. Behav. Neurosci., 111: 996–1013.

Westrin, A., Ekman, R. and Traskman-Bendz, L. (1999) Alterations of corticotropin releasing hormone (CRH) and neuropeptide Y (NPY) plasma levels in mood disorder patients with a recent suicide attempt. Eur. Neuropsychopharmacol., 9: 205–211.

Wiedenmayer, C.P., Noailles, P.A., Angulo, J.A. and Barr, G.A. (2002) Stress-induced preproenkephalin mRNA expression in the amygdala changes during early ontogeny in the rat. Neuroscience, 114: 7–11.

Williams, S.L., Kinney, P.J. and Falbo, J. (1989) Generalization of therapeutic changes in agoraphobia: the role of perceived self-efficacy. J. Consult. Clin. Psychol., 57: 436–442.

Wilson, M.A., Mascagni, F. and McDonald, A.J. (2002) Sex differences in delta opioid receptor immunoreactivity in rat medial amygdala. Neurosci. Lett., 328: 160–164.

Wisniewska, R.J. and Wisniewski, K. (1996) The effect of cholecystokinin (CCK-33) and C-terminal fragments of cholecystokinin: CCK-8 and CCK-4 on the cardiovascular system in rats. Gen. Pharmacol., 27: 159–163.

Wuster, M., Rubini, P. and Schultz, R. (1981) The preference of putative pro-enkephalins for different types of opiate receptors. Life Sci., 29: 1219–1227.

Yamada, K. and Nabeshima, T. (1995) Stress-induced behavioral responses and multiple opioid systems in the brain. Beh. Brain Res., 67: 133–145.

Young, W.S.I. and Lightman, S.L. (1992) Chronic stress elevates enkephalin expression in the rat paraventricular and supraoptic nuclei. Mol. Brain Res., 13: 111–117.

Zacharko, R.M. and Anisman, H. (1991) Stressor-induced anhedonia in the mesocorticolimbic system. Neurosci Biobehav. Rev., 15: 391–405.

Zacharko, R.M., Koszycki, D., Mendella, P.D. and Bradwejn, J. (1995) Behavioral, neurochemical, anatomical and electrophysiological correlates of panic disorder: multiple transmitter interaction and neuropeptide colocalization. Prog. Neurobiol., 47: 371–423.

Zacharko, R.M., Maddeaux, C., Hebb, A.L., Mendella, P.D. and Marsh, N.J. (1998) Vulnerability to stressor-induced disturbances in self-stimulation from the dorsal and ventral A10 area: differential effects of intraventricular D-Ala2-Met5-enkephalinamide, D-Ala2, N-Me-Phe4, Gly-Ol5-enkephalin, and D-Pen2, D-Pen5-enkephalin administration. Brain Res. Bull., 47: 237–248.

Zadina, J.E., Hackler, L., Ge, L.J. and Kastin, A.J. (1997) A potent and selective endogenous agonist for the mu-opiate receptor. Nature, 386: 499–502.

Zadina, J.E., Martin-Schild, S., Gerall, A.A., Kastin, A.J., Hackler, L., Ge, L.J. and Zhang, X. (1999) Endomorphins: novel endogenous mu-opiate receptor agonists in regions of high mu-opiate receptor density. Ann. N.Y. Acad. Sci., 897: 136–144.

Zhang, H.T., Xu, Z.M., Luo, Z.P. and Qin, B.Y. (1996) Anxiogenic effect of naltrexone in social interaction test in rats. Zhongguo Yao Li Xue Bao, 17: 314–317.

Zis, A.P., Haskett, R.F., Albala, A.A., Carroll, B.J. and Lohr, N.E. (1985) Opioid regulation of hypothalamic–pituitary–adrenal function in depression. Arch. Gen. Psychiatry, 42: 383–386.

Zurita, A., Cuadra, G. and Molina, V.A. (1999) The involvement of an opiate mechanism in the sensitized behavioral deficit induced by early chronic variable stress: influence of desipramine. Behav. Brain Res., 100: 153–159.

T. Steckler, N.H. Kalin and J.M.H.M. Reul (Eds.)
Handbook of Stress and the Brain, Vol. 15
ISBN 0-444-51173-3

CHAPTER 4.10

Acetylcholinesterase as a window onto stress responses

Hermona Soreq[1,*], Raz Yirmiya[2], Osnat Cohen[3] and David Glick[1]

[1]*Department of Biological Chemistry, The Hebrew University of Jerusalem, Jerusalem, Israel*
[2]*Department of Psychology, The Hebrew University of Jerusalem, Jerusalem, Israel*
[3]*Departments of Biological Chemistry and Psychology, The Hebrew University of Jerusalem, Jerusalem, Israel*

Abstract: It has long been known that cholinergic neurotransmission is intimately associated with mammalian stress responses. Inhibition of acetylcholinesterase (AChE), like stress, elevates the levels of acetylcholine (ACh) in the short term, and both conditions induce some common long-lasting behavioral symptoms. Therefore, AChE manipulations provide an interesting window onto stress responses. Like many other stimuli, both stress and inhibition of AChE cause an increase in AChE gene expression that is also associated with a shift in its pre-mRNA splicing pattern. Of the several variants of AChE that arise due to alternative splicing, it is specifically the usually rare, soluble AChE-R variant that is up-regulated. Transgenic mice that over-express AChE also show many of the same symptoms as stress: erratic behavior following circadian light/dark shift, progressive failure of learning and memory, intensified long-term potentiation (LTP), development of neuropathologies, progressive muscle fatigue and degeneration of neuromuscular junctions. Altered expression of other cholinergic proteins in these mice, e.g. protein kinase CβII choline acetyltransferase and the high affinity choline transporter, suggests chronic feedback responses to the cholinergic imbalance. Stress-associated characteristics can be ameliorated in mice and humans by treatment with antisense agents that induce selective destruction of AChE-R, which provides further support for changes in alternative splicing, and in particular the accumulation of this variant, having a role in the etiology of stress responses.

Introduction

Acetylcholine (ACh), the modulator of numerous physiological reactions, also modulates mammalian stress responses. In particular, a striking similarity exists between reported long-term responses to anti-cholinesterase exposure and traumatic experiences (Fullerton and Ursano, 1990; Somani et al., 2000). Furthermore, anti-cholinesterases and stress both induce expression of the acetylcholinesterase gene, *ACHE* (Kaufer et al., 1998; Kaufer et al., 1999). The realization of this similarity has drawn us into stress studies from backgrounds in molecular biology and neuroimmunology. The melding of our molecular, biological and behavioral approaches to cholinergic neurotransmission has contributed, we feel, to a deeper appreciation of the role of cholinergic neurotransmission, and of acetylcholinesterase (AChE) in particular, in stress responses. This chapter, therefore, presents long-term cellular and behavioral stress responses as molecular and biochemical processes that are amenable to further study, and, we hope, to therapeutic intervention.

*Corresponding author. Tel.: +972 2 658 5109;
Fax: +972 2 652 0258; E-mail: soreq@cc.huji.ac.il

Cholinergic responses to stress

It is well accepted that stress in mammals is rapidly followed by a pronounced activation of central cholinergic pathways that is correlated with transiently enhanced release of ACh (Imperato et al.,

1991; Dazzi et al., 1995). In addition to ACh release, exposure to stress induces short- and long-term alterations in various other components of the cholinergic system. These include the density and composition of cholinergic receptors, choline uptake and expression and activity of ACh-related enzymes.

In rodents, exposure to acute stressors, such as restraint, foot-shock or swimming, produces a rapid but transient activation of central cholinergic systems in general and the septo-hippocampal system in particular. These effects of stress were measured by various methods, including the release of newly synthesized [^{3}H]-ACh from hippocampal synaptosomal preparations (Finkelstein et al., 1985) or in vivo microdialysis (Tajima et al., 1996; Mizuno and Kimura, 1997). The changes in ACh release depend on the parameters of the stressor. For example, in several studies enhanced hippocampal ACh release could be observed as early as 10–15 min following the initiation of restraint stress. The cholinergic response declined after 50–60 min of restraint, but when the animals were freed from the restraining tubes after 2 h, ACh levels increased again, possibly because of the additional stress (Imperato et al., 1989; Tajima et al., 1996). In another study, rats that were restrained and immersed in a 37°C water bath also showed stimulated release of hippocampal ACh, but only following removal from restraint. In contrast, rats that were restrained and immersed in a 20°C water bath displayed decreased ACh levels during cold exposure, which returned to near-baseline values following removal from cold water (Stillman et al., 1997). Other stressors produce similar effects, e.g., exposure to foot-shock stress, delivered for 8 min, induced a rapid and marked (75%) increase in hippocampal ACh output, which persisted for about 40 min (Dazzi et al., 1995).

Changes in number and affinity of cholinergic receptors were also demonstrated following exposure to stress. These changes depend on the parameters of the stressful stimulus, particularly its intensity and duration. For example, in one study, immobilization stress for 2 hr, but not 10 min, caused a significant increase in the affinity of the muscarinic receptor in several brain areas, including the hippocampus and basal ganglia (Gonzalez and Pazos, 1992, but see Finkelstein et al., 1985). Similarly, chronic exposure to short periods of immobilization (10 min/day for 3–21 days) had no effect on cholinergic receptors, whereas exposure to 2 hr/day for 3–21 days resulted in significantly increased number (Gonzalez and Pazos, 1992) and affinity (Finkelstein et al., 1985) of muscarinic receptors. Changes in the sensitivity of cholinergic receptors have also been demonstrated pharmacologically: following chronic food-restriction stress, rats exhibited a greater behavioral response to a muscarinic agonist and a lower behavioral response to a muscarinic antagonist in modulation of passive avoidance learning (Orsini et al., 2001).

In laboratory animals both stress and inhibition of AChE indirectly cause elevation of AChE mRNA and protein levels in the brain (Fullerton and Ursano, 1990; Somani et al., 2000), while suppressing levels of mRNAs that encode the ACh-synthesizing enzyme choline acetyltransferase (ChAT) and the vesicular acetylcholine transporter (vAChT) (Friedman et al., 1996; Kaufer et al., 1998). These observations indicate that a decrease in ACh, due to increased hydrolysis is re-enforced by reduced ACh-synthesis in cholinergic pathways. These complementary modulations of cholinergic gene expression ensure that transient acute cholinergic hyperactivation is followed by a persistent suppression of cholinergic neurotransmission in the mammalian brain. These clues have led us to the initiation of a systematic investigation of the role of AChE in stress responses in several experimental paradigms, including the human hematopoietic system (Grisaru et al., 2001), mouse hippocampal slices (Friedman et al., 1998) and intestinal epithelium (Shapira et al., 2000b), myasthenic rat muscle (Brenner et al., 2003) and testicular tubules (Mor et al., 2001). In all of these systems we found that various stressors induce expression of AChE-R mRNA, one of the AChE pre-mRNA splicing variants (Grisaru et al., 1999; Soreq and Glick, 2000). These findings highlight the role of the normally rare AChE-R in mammalian stress. The observations, in five major organ systems, suggest a central role for cholinergic pathways in mediating long-term stress responses. The effects of chronic stress exposure on cholinergic neurotransmission were examined by others, revealing long-term changes that were present even many days after termination of the exposure. For example, 10 days following the termination of chronic stress (daily exposure to water immersion and 2-hr restraint for 4 weeks), rats exhibited markedly enhanced KCl-evoked cholinergic response within the hippocampus, although basal ACh release was not changed (Mizoguchi et al., 2001). Several lines of evidence suggest that these initial events contribute to the behavioral consequences of stress. For example, scopolamine, a muscarinic antagonist, affects rodent responses to foot-shock stress (Kaneto, 1997), and long-term psychological disturbances, strikingly reminiscent of those that characterize post-traumatic stress disorder (PTSD), are associated with both acute (Burchfiel and Duffy, 1982; Rosenstock et al., 1991) and chronic (Li et al., 2000; Stephens et al., 1995) exposure to inhibitors of AChE. The scope of these observations could have been expanded to include panic disorders, which have an important cholinergic component (Battaglia, 2002), and depression (Lupien et al., 1999), but this would take us beyond the mandate of this chapter.

The relationships between stress-induced pituitary-adrenocortical activation and ACh release

The effects of the hypothalamo–pituitary–adrenal (HPA) axis on the cholinergic system and its role in stress-induced cholinergic activation have been intensively studied. The hypothalamus is the source of corticotropin-releasing factor (CRF); the pituitary of ACTH, and the adrenal of glucocorticoids. Initial studies demonstrated that large doses of either adrenocorticotropic hormone (ACTH) or corticosterone activate the septo-hippocampal cholinergic system similarly to the activation of this system by short-term stress (Gilad, 1987; Gilad et al., 1985). Choline uptake and ACh release were also elevated 2 days after adrenalectomy. The effects of adrenalectomy were attenuated by corticosterone, which reduces ACTH levels, but not by ACTH administration (Gilad et al., 1985). This effect may be the result of reduced feedback inhibition, which results in elevated CRF and ACTH levels. A role for CRF in cholinergic activation is also suggested by the presence of CRF receptors on cholinergic cells (Chen et al., 2000; Sauvage and Steckler, 2001; Stillman et al., 1997) and the finding that exogenous administration of CRF activates cholinergic neurons and induces hippocampal ACh release (Day et al., 1998a; Day et al., 1998b; Stillman et al., 1997). Further studies demonstrated that the relationships between stress, adrenocortical activation and ACh release are not direct, because the time course of stress-induced plasma corticosterone elevation does not parallel that of ACh release. For example, during continuous restraint for 2 hr, ACh release gradually decreases following initial elevation, but it increases again when animals are released from the restraining tubes. In contrast, corticosterone levels remain elevated for the whole restraining period and decrease when animals are released (Imperato et al., 1991). Furthermore, stress-induced ACh release was found to be similar in adrenalectomized and sham-operated rats, and was not influenced by pre-treatment with the glucocorticoid antagonist RU38486 (Imperato et al., 1991), providing additional support for the hypothesis that the effects of stress on cholinergic systems are not mediated solely by adrenocortical activation.

Gulf War syndrome and penetrance of the blood–brain barrier under stress

Cholinergic neurotransmission is probably involved in most of the known arousal-related sensory stimulation effects. For example, ACh release increases dramatically in the frontal cortex and hippocampus of rats introduced into a novel environment (Inglis et al., 1994), as well as following presentation of auditory, visual, olfactory and tactile sensory stimuli (Acquas et al., 1996). The cholinergic activation is restricted to presentation of novel stimuli, which may be regarded as mildly stressful. If a novel stimulus, such as tone or light, is followed by an unconditioned strong stressful stimulus (such as foot-shock), it will continue to elicit ACh release. However, presentation of a tone or light to habituated animals, i.e., following repetitive presentations of the same stimulus, will fail to elicit a cholinergic response (Acquas et al., 1996). These findings indicate that ACh release in the cortex and hippocampus are reliably activated only by behaviorally relevant stimuli. During stressful conditions, the activation of the cholinergic system by novel, conditioned or the stressful stimulus itself provides an important mechanism for adaptive enhancement of arousal and/or attentional processes.

Stress-induced ACh release is probably also involved in other cognitive functions, particularly learning and memory. The relationship between memory processes and cholinergic neurotransmission is well established in the literature. Early reports of the role of ACh in learning and memory (Drachman and Leavitt, 1974), and evidence of substantial reductions in neocortical cholinergic function in Alzheimer's disease (Bartus et al., 1982; Nilsson et al., 1986) provided evidence for the cholinergic hypothesis of memory. In particular, cholinergic stimulation is known to facilitate working memory performance (Ellis and Nathan, 2001), supposedly by influencing the signal-to-noise ratio for the response of single neurons in the cortex (Furey et al., 2000). This increases the selectivity of perceptual responses, possibly enlarging the response to afferent input and reducing background activity. Consistent with this notion, cholinergic antagonists, such as scopolamine, selectively interfere with encoding of new information, but not with its retrieval, whereas cholinesterase inhibitors, such as physostigmine or Alzheimer's disease drugs (Giacobini, 2002) improve working memory. It can be speculated that under conditions of acute stress, working memory is an important asset for survival. Thus, cholinergic-mediated improvement in working memory performance might be adaptive. Stress-induced cholinergic activation may also play a role in other forms of memory, such as aversive conditioning. For example, enhancement of inescapable foot-shock stress-induced ACh transmission by pretreatment with an anticholinesterase (eserine sulfate) was found to increase the learned helplessness produced in this paradigm, reflected by increased escape deficits observed 3 days later (Kelsey, 1983). The mechanisms that mediate the effects of stress-induced cholinergic activation on learning and memory are not clear. It has been suggested that the CRF system may interact with cholinergic activation in memory modulation, but this hypothesis has not received much experimental support (Steckler and Holsboer, 2001).

Gulf War syndrome and penetrance of the blood–brain barrier under stress

Cholinergic pathways are involved in controlling numerous peripheral functions, e.g. neuromuscular activity, salivation, intestinal functions and lacrymation. Over the years, this led to the development of drugs aimed at controlling these functions. To avoid unwanted effects over higher brain functions, these drugs were designed to remain in the circulation, where they are held, thanks to the blood-brain barrier (BBB) (Rubin and Staddon, 1999). A notable example is pyridostigmine bromide (PB), used for the past 40 years for treating patients with myasthenia gravis, an autoimmune syndrome that causes debilitating muscle fatigue (Berrouschot et al., 1997). During the 1991 Gulf War, injection of PB was a prophylaxic treatment of soldiers, in anticipation of chemical warfare, to transiently block AChE and protect it from permanent inhibition by nerve agents. This treatment was based on clinical studies that demonstrated its effectiveness and showed side effects that were strictly limited to peripheral symptoms (Beck et al., 2001). However, impairments in higher functions of treated soldiers and subsequent symptoms that appeared to originate in the CNS led to questions regarding the ability of PB to remain peripheral under stresses associated with war (Golomb, 1999). This, in turn, provoked further studies that explored behavioral changes under PB administration. It was thus found that PB modified the acoustic startle response (Servatius et al., 1998), locomotor activity in an open field (Kant et al., 2001), hand and eye coordination (Wolthuis et al., 1995), the capacity to press a lever in a delayed reinforced manner (van Haaren et al., 1999) and visual discrimination properties (Liu, 1992). A relatively simple explanation of these behavioral effects would be that the stress associated with PB injection modifies the properties of the BBB and enabled PB penetrance of the brain, in spite of the cationic group that had been introduced into this carbamate to prevent penetrance through the BBB. This was the working hypothesis of a study (Friedman et al., 1996) that reported efficient inhibition of brain AChE by intraperitoneally injected PB into forced swim-stressed, but not naïve, mice. However, the complete picture is far from simple, as other studies – with other stressors and other rodent models – reported no inhibition of brain AChE under systemic administration of PB (Cook et al., 1988; Stitcher et al., 1978). The later-reported transcriptional feedback response, including AChE overproduction under exposure to anti-cholinesterases (Kaufer et al., 1998) offers a tentative explanation of this inconsistency. Yet more recently, the lethality of PB in animals was reported to be increased (Chaney et al., 1999; Chaney et al., 2000) and its capacity to inhibit brain AChE facilitated, when co-injected with permethrin or *N*,*N*-diethyl-*m*-tolumide (Abou-Donia et al., 2001; Abou-Donia et al., 1996). While it is difficult to compare these studies to one another, because they used different stressors or animals, it seems possible that under some circumstances stress can cause permeabilization of the BBB.

Corticotropin-releasing factor, a 41-residue peptide originating in the hypothalamus, apparently has a central role in response to stress. It has long been known to cause release of ACTH from the pituitary into the bloodstream, where it travels to the adrenal cortex to promote synthesis and secretion of cortisol, the source of many other stress responses, among them activation of *ACHE* gene expression. More recently, CRF, acting with mast cells, has been shown to increase permeability of the BBB (Esposito et al., 2001; Esposito et al., 2002). In a dramatic demonstration of the effect of stress, Relyea and Mills found that the anti-AChE pesticide, carbaryl, was 2 to 4 times more lethal to amphibians when they were exposed to the stress of predatory cues (Relyea and Mills, 2001). This may be an additional demonstration of the effect of stress on the BBB.

mRNA splicing variations

One of the strategies by which Nature has greatly expanded the expression potential of the genome is by alternative splicing of pre-mRNAs. In approximately 20% of the expressed genes, the original transcript can generate a number of different mature mRNAs by selection of only some of the open reading frames and elimination of others. Thus, subtly or substantially different variant proteins may be generated from a single gene. The process of splicing is accomplished by complex molecular machines, under the direction of a number of sometimes competing small splicing factor proteins (Stamm, 2002; Zhou et al., 2002). The limited number of these factors, and the balance among them, would allow a concerted response to an external change by the generation of characteristic splicing variants of a potentially large number of genes. Thus, a stress, starvation for example, may initially bring about the up-regulation of a single splicing factor, which because of its involvement in splicing of a wide variety of pre-mRNAs, results in the biosynthesis of the proteins that can minimize the effects of starvation. The integration of the response is built into the participation of that one (or those few) splicing factor(s) in the synthesis of physiologically appropriate proteins by splicing of their gene transcripts. Similarly, a defect, inherited or acquired, in one of these splicing factors would affect a correspondingly wide variety of such events. Although the extent of alternative splicing as an integrating (or disintegrating) principle is still being explored, it is a fruitful source of hypotheses that may help to explain the many forms that responses to stress may take in different tissues, and the similarities among responses to various kinds of stresses. One example, taken from the neurosciences, is the stress-induced modification of K^+ channels (Xie and McCobb, 1998; Xie and Black, 2001), which is regulated by neuronal activity-dependent transcriptional changes in a number of splicing regulatory proteins (Daoud et al., 1999). Another example, is the stress-directed splicing of AChE pre-mRNA (Meshorer et al., 2002; Soreq and Glick, 2000; Meshorer and Soreq, 2002), which we explored.

Acetylcholinesterase and stress responses

The short-term stress response is typified by increases of epinephrine and glucocorticoids, which facilitate a mobilization of energy stores and a suppression of competing activities, such as growth, reproduction, immune response and tissue repair. Prolonged elevation of glucocorticoids by stress results in central nervous system (CNS) abnormalities, among them learning and memory defects, perhaps caused by decreased branching of dendrites, and a variety of psychiatric problems, including those associated with

PTSD. However, short-lived traumatic effects, which do not result in long-term activation of the HPA axis, can also result in PTSD (Sapolsky, 2002). It appears, therefore, that some stress responses are mediated by other, more persistent factors. While it is highly unlikely that AChE is the only mediator of HPA activation, the process leading to AChE-R production appears to fit the description of a prime suspect: the *ACHE* promoter contains glucocorticoid response elements, which increase AChE expression following stress (Grisaru et al., 1999; Shapira et al., 2000b). Following stress, the major newly-synthesized neuronal AChE variant is AChE-R. This implies that glucocorticoids and/or other stress-induced modulator(s) must also influence the splicing of the AChE pre-mRNA. AChE-R mRNA and AChE-R itself remain in neurons for weeks following stress (Meshorer et al., 2002) and the increased presence of (transgenic) AChE-R exerts a protective effect from the microanatomical neuropathologies that accumulate under excess of neuronal AChE-S (Sternfeld et al., 2000). Moreover, the suppression of AChE-R by a specific antisense agent (Galyam et al., 2001) alters neurite extension (Grifman et al., 1998), improves the outcome of closed head injuries (Shohami et al., 2000) and reverses myasthenic symptoms, which are associated with increased levels of circulating AChE-R (Argov et al., 2003; Brenner et al., 2003).

A significant source of the variety of AChE isoforms is 3'-alternative pre-mRNA splicing (Fig. 1) which confers different C-terminal sequences on a 543-residue core protein.

A single human *ACHE* gene gives rise to a wide variety of protein variants. The 543-residue core of human AChE is encoded by 3 exons, 2, 3 and 4, and by itself is catalytically competent; 3'-alternative splicing of the pre-mRNA result in additional C-terminal sequences of the S (synaptic), R (readthrough) and E (erythrocytic) variants. Human AChE-S terminates with 40 residues (DTLDEAERQWKAEFHRWSSYMVHWKNQF-DHYSKQDRCSDL); AChE-E with 14 residues (ASEAPSTCPGFTHG); and AChE-R with 26 residues (GMQGPAGSGWEEGSGSPPGVTPL FSP). The translation start codon is in exon 2, which encodes a leader sequence that does not appear in any of these mature proteins. In addition to a proximal promoter next to exon 1 (Atanasova et al., 1999; Camp and Taylor, 1998; Chan et al., 1998), a far upstream distal enhancer region is rich in potential regulatory sequences. For instance, the *ACHE* transcriptional activation by cortisol (Grisaru et al., 2001) is likely due to the distal glucocorticoid response element (Meshorer et al., 2002). A deletion mutation in this region disrupts one of two HNF3 binding sites, which consequently activates transcription (Shapira et al., 2000b). Normally, much more AChE-S than AChE-R mRNA is produced, but under stress or inhibition of AChE, alternative splicing produces much more of the AChE-R mRNA, likely under c-fos regulation (Kaufer et al., 1998). The C-terminal sequence of AChE-S enables it to form multimers that can then be joined to anchoring proteins (Ohno et al., 2000; Perrier et al., 2002) that attach them to the synaptic membrane; and AChE-E dimerizes and undergoes a transesterification reaction to replace its C-terminal sequence with a glycophosphatidylinositol group that can be embedded in the erythrocyte membrane (Silman and Futerman, 1987). AChE-R, however, cannot form multimers, and when secreted, it remains

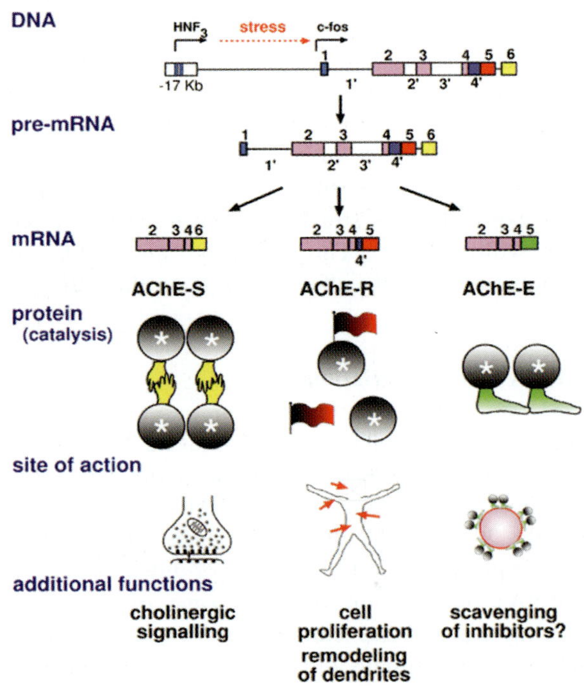

Fig. 1. The molecular biology of human AChE.

soluble. Additional variation arises from post-translational changes: glycosylation, which affects turnover (Kronman et al., 2000), formation of intermolecular disulfide bridges and attachment to a phospholipid or collagen-like protein anchor to synapse membranes (as mentioned), or intracellular interaction with partner proteins (Birikh et al., 2003).

Note on nomenclature

Another nomenclature is based on physical rather than molecular biological relations, and names the synaptic and erythropoietic variants T (tailed), and H (hydrophobic), respectively (Massoulie, 2000). The many guises in which AChE-T occurs are further termed G, for globular, or A for asymmetric and G_1, G_2 etc. for monomer, dimer, etc.

Transcriptional feedback response to stress

The functioning of the *ACHE* gene is subject to massive developmental pressures, yet retains a certain level of plasticity also in the adult. In both neocortical and hippocampal neurons, various external stimuli induce rapid, yet long-lasting *ACHE* gene expression. In fact, the *ACHE* gene responds with increased transcription to psychological stress (Kaufer et al., 1998), anti-AChE intoxication (Shapira et al., 2000b), closed head injury (Shohami et al., 2000) and autoimmune impairments of neuromuscular function (Brenner et al., 2003) and immobilization stress (Nijholt et al., 2003). It is presumed that psychological or physical stress induces cholinergic excitation via release of ACh. Elevated cortisol and the consequent feedback over-expression of AChE then act to dampen excessive neurotransmission towards its normal level (Kaufer et al., 1998). This is important both for cholinergic neurotransmission and for other neurotransmission circuits modulated by ACh, for example hippocampal glutamatergic activity (Gray et al., 1996; Meshorer et al., 2002), and dopaminergic circuits in the substantia nigra (Llinas and Greenfield, 1987). That over-produced AChE can also protect the organism from the toxicity of anti-AChEs, has been demonstrated in laboratory animals (Ashani et al., 1991; Wolfe et al., 1992; Doctor et al., 1993; Raveh et al., 1989).

Transcriptional activation is common to many genetically determined responses to pharmaceuticals, e.g. by cytochrome P-450 proteins (Evans and Relling, 1999). However, in addition to transcriptional activation, AChE mRNA transcripts in nerve, muscle and blood cells are subject to calcineurin-controlled differentiation-induced stabilization (Chan et al., 1998; Luo et al., 1999). Both these processes increase the amount of AChE when and where it is needed. As AChE-R mRNA is significantly less stable than AChE-S mRNA (Chan et al., 1998; Luo et al., 1999), any stabilizing effect should significantly favor the R-variant. This should be of particular interest in the context of the routine use of anti-AChEs in the treatment of Alzheimer's disease (AD) patients (Palmer, 2002). Indeed, AChE-R was recently shown to accumulate in the cerebrospinal fluid of AD patients treated with anti-AChEs (Darreh-Shori et al., 2002, 2003). The variable efficacy of such agents among individuals may therefore reflect differential capacities to induce transcriptional activation and/or stabilization of AChE mRNA.

Transgenic overexpression of neuronal AChE-S recapitulates chronic stress effects

We created a transgenic mouse model (TgS), in which overexpression of human synaptic AChE-S, limited to CNS neurons (Beeri et al., 1995), promotes a late-onset and progressive impairment in learning and memory that in humans is associated with PTSD (Kaufer and Soreq, 1999). The cognitive defects observed in these mice were assumed to reflect the physiological state induced by an excess of AChE, but we cannot rule out the defects being secondary effects, resulting from adaptations to the increased levels of CNS AChE. The possibility that modified regulation of *ACHE* gene expression may have imposed profound disruption of both central and peripheral cholinergic systems is strengthened by our more recent findings of multi-levelled impairments in these mice in neuromuscular junction (NMJ) structure and function (Farchi et al., 2003), which eventually lead to amyotrophy in TgS mice (Andres et al., 1997). However, there is a temporal gap between over-expression of AChE in young TgS mice and the delayed onset of neurodegenerative and cognitive processes, perhaps reflecting the fact that excess AChE causes damage only when present for a long time.

Compensatory mechanisms as suppressors of stress symptoms

If, indeed, AChE over-production is generally associated with long-term stress responses, the question arises, how does the brain handle the resultant state that is induced by elevated ACh hydrolysis? The obvious answer is, by initiating compensatory mechanisms that would elevate the cholinergic state to retrieve functional balance. Similar compensatory mechanisms are generally assumed to enable the extended pre-symptomatic stages that accompany neurodegenerative disease, e.g. AD and Parkinson's disease (Zigmond,1997). Neurodeterioration, in this view, would be due to inadequate compensation for consequences of both inherited defects and stress. Elevation of the cholinergic status would, however, be possible only if sufficient numbers of viable cholinergic neurons were available. Indeed, animal models are known that develop cognitive impairments at relatively early ages due to inherited loss of cholinergic forebrain neurons (Zigmond, 1997; Sago et al., 1998).

Primary stress-related function subserved by the cholinergic system, particularly within the hypothalamus, is in regulation of the HPA axis. Indeed, the secretion of CRF can be also altered by ACh, although the direction of the cholinergic influence may vary in different brain areas: e.g. ACh stimulated CRF secretion from the hypothalamus, but inhibited its release from cortical tissues (Tizabi and Calogero, 1992). Moreover, microinjection of ACh into the hypothalamus induced the expression of CRF and pro-opiomelanocortin mRNA within the hypothalamus and pituitary, respectively (Ohmori et al., 1995). Both nicotinic and muscarinic receptors may be involved in HPA axis regulation. Nicotine was found to activate CRF neurons in various brain areas, including the hypothalamic paraventricular nucleus (Nilsson et al., 1986) and to induce CRF secretion in vitro (Calogero et al., 1988). The effects of nicotine on the release of ACTH from the pituitary may reflect indirect noradrenergic pathways mediated via the brainstem (Fu et al., 1997; Matta et al., 1990). Muscarinic mechanisms are also involved: administration of the muscarinic agonist arecoline stimulates the HPA axis in the rat and this effect was found to be mediated mainly by the release of endogenous CRF (Calogero et al., 1989). Furthermore, blockade of hippocampal muscarinic receptors augmented ACTH and corticosterone responses to restraint without altering basal HPA activity (Bhatnagar et al., 1997). Together, these findings strongly suggest that cholinergic systems regulate stress-induced HPA activity and may serve to coordinate behavioral and neuroendocrine responses to stress.

Cholinesterase genetics and stress responses

AChE levels depend on multiple inherited and acquired elements, so that in some humans, there is a higher than usual basal level of *ACHE* expression (Silver, 1974). As in most circumstances the individual shows no ill effects, there is apparently an adaptation to this state. This may involve an increased level of ACh receptors, similar to TgS mice (Perry et al., 2000), or increased high affinity choline transporter, also shown in TgS mice (Erb et al., 2001). It was hypothesized that similar to TgS mice, individuals with constitutive AChE over-expression would be unable to respond appropriately to stress and that their *ACHE* gene would contain some clues as to the cause. Therefore the genomic DNA from 340 subjects was analyzed, with special attention to a region of the promoter sequence that was rich in transcription factor binding elements and which includes a glucocorticoid response element. Two adjacent mutations in this distal upstream enhancer domain of the human *ACHE* gene were discovered in heterozygous carriers (Shapira et al., 2000a): a 4-bp deletion and a single nucleotide substitution. The deletion, identified in a woman who presented acute hypersensitivity to pyridostigmine, was found in transfected cells to constitutively increase AChE expression by abolishing 1 of 2 adjacent HNF3 binding sites. Because the deletion confers a gain of function of *ACHE*, the trait is dominant; the substitution impairs the glucocorticoid receptor binding site in this region. Further studies will be required to find whether this trait is also associated with increased risk for exaggerated stress responses.

Environmental stress on AChE mitigated by butyrylcholinesterase

Since its discovery, AChE has been known as the enzyme that hydrolyzes the neurotransmitter, ACh. The biological role of the AChE-homologous enzyme, butyrylcholinesterase (BuChE), has long remained elusive. It has been postulated that BuChE is a back-up for AChE, and in the very special case of the AChE-knockout mouse, it may be BuChE that in fact performs ACh hydrolysis. Nevertheless, in the real world of Nature, there are no known cases of human AChE mutations that abolish its activity, which is a powerful message that AChE serves an irreplaceable function. It must be mentioned, however, that there are mutations of the AChE-S-anchoring protein, which result in end-plate AChE deficiencies and cause major neuromuscular defects (Ohno et al., 2000). Another role proposed for BuChE is as a molecular decoy that absorbs anti-AChEs that may find their ways into the body and minimize this source of stress on the organ systems that depend on a functional AChE. AChE and BuChE have overlapping specificities for substrates and inhibitors, with BuChE being somewhat more promiscuous. In consequence, just about every anti-AChE is also an anti-BuChE. The environment contains many and varied anti-AChEs, ranging from the anatoxins, natural organophosphate poisons of blue-green algae (Carmichael, 1994) and the abundant glycoalkaloids of *Solanaceae* (potatoes, tomatoes, aubergines) (Friedman, 2002; McGehee et al., 2000; Roddick, 1989) to the toxins of snake venoms, e.g. fasciculin (Marchot et al., 1997) and natural medicines of plant origin, e.g. huperzine and physostigmine (Giacobini, 2000).

An anti-AChE entering the body will react with serum BuChE (and, for that matter, AChE-E on erythrocyte membranes) before even coming into contact with AChE-S at neuromuscular junctions or brain synapses. The individual is thus protected by the ability of BuChE to adsorb AChE inhibitors. Consistent with its being a molecular decoy for AChE, are the prominence of BuChE in the serum and its capacity to react quickly with a wide spectrum of compounds. Furthermore, some polymorphisms of the *BCHE* gene render carriers increasingly susceptible to the ill effects of anti-AChE exposure. Polymorphism of *BCHE* has been extensively surveyed, originally by study of the variant-characteristic susceptibility to inhibitors of the serum activity (Kalow and Genest, 1957), more recently by molecular genotyping (La Du et al., 1990; Loewenstein et al., 1995). *BCHE* mutations are very unevenly distributed around the world, with dramatically high or low frequencies found especially in historically isolated ethnic groups, possibly reflecting genetic founder effects. The different BuChE variants and their frequencies may also reflect an evolutionary adaptation to local environmental factors. Because BuChE mutants offer varying protection against anti-AChEs, carriers of these mutations may be more vulnerable than non-carriers to anti-AChEs and may show exaggerated responses when exposed. An extension of this idea, in conjunction with the similarity of chemical and other stressors, is increased vulnerability of BuChE mutation carriers to late-onset diseases. In the case of AD, some have found such an association (Lehmann et al., 1997; Lehmann et al., 2000), but others have not (Brindle et al., 1998).

Non-classical biological roles of acetylcholinesterase

Because stress events induce AChE overexpression, the concentration of the AChE protein will be higher following stress. This will effect both the hydrolytic and the non-classical actions of AChE, both of which may be relevant to the organismal response to stress. Non-classical actions of AChE, i.e. those not associated with hydrolysis of ACh at the synapse or NMJ, have been reported by a number of laboratories, among them, those of M.E. Appleyard (Appleyard, 1992), John Bigbee (Bigbee et al., 1999), W. Stephen Brimijoin (Koenigsberger et al., 1997), Susan Greenfield (Greenfield, 1996), Paul Layer (Layer, 1996) and David Small (Small et al., 1996). It is still a challenge to integrate the disparate, but not contradictory, findings of these research groups into a coherent view of the biology of AChE. Recent discoveries may indicate a biochemical basis of at least some of these seemingly anomalous effects. There is a homology among AChE and other members of the α/β fold family of proteins, including the neurotactins, which are involved in cell–cell adhesion. The conserved domain of neurotactins may be exchanged for AChE and still retain cell–cell interaction (Darboux et al., 1996). Moreover, genetic inactivation of AChE does not prevent its neurite growth-promoting activity (Grifman et al., 1998; Sternfeld et al., 1998a). The mammalian non-catalytic AChE-homologous proteins, the neuroligins, reside in excitatory synapses (Song et al., 1999), and are known to bind neuronal cell surface proteins, the neurexins (Nguyen and Sudhof, 1997); neurexins, neurotactins and neuroligins are transmembrane proteins with C-terminal cytoplasmic tails which could enable signal transduction through the binding of PDZ domain proteins, e.g. membrane-associated guanylate kinases (MAGUKs; (Ponting et al., 1997), and thus provide the molecular basis of the complex consequences that are characteristic of these cell–cell interactions. Many in the research community are now open to the recognition of non-cholinergic roles of AChE (Botti et al., 1998), and more such functions will doubtlessly be identified in the near future.

Cell–cell interactions are mediated by the interaction of membrane-bound neuroligin-1 with membrane-bound β-neurexin. If neuroligin-1 is displaced from β-neurexin by the homologous (Tsigelny et al., 2000), soluble, AChE-R, the cell–cell interaction is broken. This is likely to modify the properties and/or intracellular signalling activities of PDZ domain proteins that interact with both neuroligin-1 and β-neurexin, with predictable effects on excitatory synapse activities. When fully understood, this version (Fig. 2) may prove to have been too simplistic, as it cannot be readily reproduced in cultured neurons (Scheiffele et al., 2000), but it's elements seem safely established.

The behavioral implications of neuroligins' expression have recently been demonstrated in that genetic polymorphisms in the human neuroligin gene were found to increase the risk for behavioral impairments and autism (Zoghbi, 2003; Jamain et al., 2003). This, in turn, highlights the potential importance that displacement of neuroligin with AChE-R may have on stereotypic behavior.

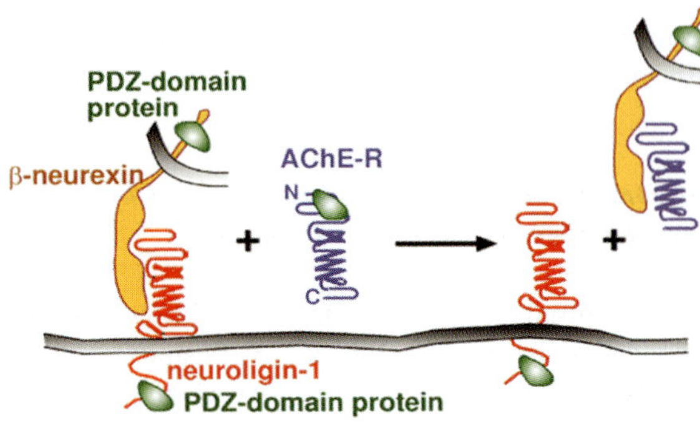

Fig. 2. Proposed molecular basis of non-catalytic properties of AChE.

Compensation for neuron loss by increased neurite outgrowth can delay the symptoms of neurodegeneration. Various studies support the notion of AChE's participation in such processes. Studies of the morphogenic roles of AChE have been facilitated by the construction of transgenic mouse lines that overexpress a specific AChE variant, AChE-S (TgS) (Beeri et al., 1995) or AChE-R (TgR) (Sternfeld et al., 1998b), by the use of stably transfected cell lines (Koenigsberger et al., 1997; Grifman et al., 1998; Bigbee et al., 1999) and in primary neurons that express and produce small quantities of a recombinant variant (Sternfeld et al., 1998a). In several of these model systems, human AChE-R emerged as having effects distinct from those of AChE-S. In cultured glioblastoma cells, over-expressed AChE-R confers a phenotype of small, round, rapidly dividing cells as opposed to the AChE-S phenotype of process growth (Karpel et al., 1996; Perry and Soreq, 2002). Antisense suppression of AChE mRNA in neuroblastoma cells was associated with complete loss of neuritogenesis, which was retrieved by re-transfection with AChE-S (Koenigsberger et al., 1997). Similar results were obtained in PC12 cells where either AChE-R or the non-catalytic homolog, neuroligin, retrieved neurite growth following antisense suppression of AChE-R (Grifman et al., 1998).

The changes under stress in the levels of AChE variants further imply an altered ratio between AChE-R and AChE-S in the stressed nervous system. This highlights one of the key challenges in this field, namely the search for the physiological functions of the different splice variants. While previous theories of AChE's involvement in neurophysiological activities were largely limited to cholinergic neurotransmission, its non-catalytic activities likely span many more circuits and brain regions. Moreover, the soluble AChE-R monomers secreted under stress were recently shown to modulate glutamatergic neurotransmission and affect the stress-induced changes in long-term potentiation (Vereker et al., 2000; Nijholt et al., 2003), a cellular function related to memory, or the facilitation of long-term depression (Xu et al., 1997), its opposite.

Acetylcholinesterase-R is overproduced under the influence of several stressors

Facilitation of the capacity for ACh hydrolysis provides useful short-term suppression of the cholinergic hyperexcitation that is associated with stress responses. This can prevent epileptic seizures, a known consequence of anti-cholinesterase exposure (Blanchet et al., 1994; Shih and McDonough, 1997), and head injury (Shaw, 2002). Moreover, in the long run, these forms of stress or trauma – acute psychological stress, exposure to anti-AChEs, head injury – all can lead to delayed neurodeterioration. Further studies will be required to determine whether the association of AChE-R with these physiological conditions reflects a causal relationship to neurodeterioration, whether the expression of AChE-R is a protective mechanism that is not always sufficient to prevent the deterioration, or if both assumptions are correct, with the AChE-R concentration determining its effects.

To begin to address these questions we examined stress-induced and stress-related neuropathological and behavioral parameters in mice with transgenic overexpression of various AChE isoforms (Table 1). As described below, TgS mice exhibit impairments in spatial learning and memory (Beeri et al., 1995), which may be partially accounted for by the biochemical alterations in their brains. TgS mice adapt to the high levels of AChE by increased synthesis of high-affinity choline transporter (elevating pre-synaptic choline uptake) and acetyl cholinetransferase (facilitating ACh synthesis). The rather counter-intuitive result is an unchanged level of ACh in conscious mice; under halothane anesthesia the TgS ACh levels were lower than in parent strain mice, attesting to the transient nature of these compensatory effects (Erb et al., 2001). The modified AChE-R/AChE-S ratio may induce persistent changes in the CNS. Exposure of TgS mice to acute levels of the anti-AChE, diisopropylfluorophosphonate (DFP), failed to induce AChE-R overproduction in their intestinal endothelium, an exposure response that occurred readily in the parent FVB/N strain (Shapira et al., 2000b). The high level of AChE in brains of TgS or the altered AChE-S/AChE-R ratio may render these mice particularly vulnerable to the long-term consequences of acute stress. The TgS mice make an intriguing model in which to study the roles of AChE and cholinergic signalling in mammalian stress responses, but because several components of the cholinergic system have been perturbed, a model with which it is necessary to perform a large number of control experiments.

Behavioral manifestations of stress in transgenic mice with cholinergic imbalances

Behavioral differences between TgS and control mice were first sought by telemetric recording of locomotion patterns, both under basal conditions, as well as following the stress of a switch in the circadian light/dark cycle. Under normal conditions, naïve TgS mice displayed close to normal locomotion behavior,

Table 1. Stress-inducing physiological phenotypes associated with cholinergic impairments

Insult	Outcome	Reference
Forced swim (psychological)	Hippocampal hyperexcitation	Friedman et al. (1998)
Light/dark switch (physiological)	Enhanced locomotion	Cohen et al. (2002)
Anti-cholinesterase exposure	Impaired working memory	Kaufer et al. (1999)
Head injury	Impaired motor coordination	Shohami et al. (2000)

Table 2. Behavioral stress correlates in transgenic mice overexpressing distinct AChE variants.

Transgene	Behavioral test	Stress-relevant phenotype	Reference
S	Morris water maze	Impaired acquisition of spatial information	Beeri et al. (1995); Beeri et al. (1997)
S	Social exploration	Working memory deficit (extended sniffing time)	Cohen et al. (2002)
S	Elevated plus maze	Reduced anxiety (prolonged open arm time)	Erb et al. (2001)
S	Locomotion	Aimless hyperactivity after light/dark switch	Cohen et al. (2002)
R	Emergence into a new field	Extended conflict behavior (delayed emergence)	Birikh et al. (2003)

similar to that of naïve mice from the non-transgenic strain. However, their locomotor response to stress, either in their cages or in an elevated plus maze, was distinct from that of the parent strain. We also assessed the effects of AChE overexpression in these mice on social memory functioning, which is known to be markedly affected by exposure to various physical and psychological stressors. This issue is summarized in Table 2 and detailed below.

Response to circadian shift

Mammalian stress responses are known to be subject to circadian regulation. Circadian differences were reported in the stress-induced pressor activity in mice (Bernatova et al., 2002) and the circadian regulation of cortisol levels in human infants was found to include a significant genetic component (Bartels et al., 2003). Moreover, women with and without sexual abuse and post-traumatic stress disorder differ substantially in the diurnal pattern of their HPA axis activity as well as in their response to neuroendocrine challenges (Bremner et al., 2003). Cholinergic neurotransmission circuits are also known to be subject to circadian changes (Carlson, 1994) and control the sensorimotor cortical regions regulating such activity (Fibiger, 1991). Conversely, the stressful shift in circadian cycle is known to produce considerable impairment of the locomotor behavior of both animals and humans (Weibel et al., 1999; Lightman et al., 2000). To assess the general responsiveness of normal and TgS mice to this stressor, we recorded their locomotor activity, using telemetric transmitters, implanted in the peritoneal cavity. Movement was monitored by a sensor to estimate the distance the animal moved over several days.

Under routine conditions, both the parental strain FVB/N mice and TgS mice displayed similar home cage activity. Their circadian rhythms included, as expected, significantly more frequent and pronounced locomotor activity during the early part of the dark phase of the circadian cycle. Seventy-two hours following reversal of the light/dark phases, both genotypes lost most of the circadian rhythm in their locomotor activity. This response is common to most rodents, as reported by others (Hillegaart and Ahlenius, 1994) with a strong contribution of their background genotypes. However, FVB/N and TgS mice subjected to a circadian light/dark shift presented distinctive behavioral patterns. The parent strain, similarly to other mammals, presented general reduction in post-shift locomotor activity. In sharp contrast, TgS mice showed a general increase in post-shift activity, both during the dark and the light phases. In particular, activity in the dark phase of the reversed cycle was significantly greater in TgS compared with control mice. These findings indicate that adjustment to the circadian insult was markedly impaired in TgS mice, suggesting that these mice display a genetic predisposition to abnormal responses to changes in the circadian rhythm, and perhaps other stresses, as well.

Impaired social interactions

Decrements in social investigation following repeated contact between mice were reported by others to be subject to cholinergic regulation (Winslow and Camacho, 1995). The intensified response of TgS mice to the stress of a circadian switch therefore suggested that the excess of AChE is the cause. However, in spite of their inbred genotype, individual TgS mice presented distinct patterns of social interaction, attesting to the acquired nature of much of the phenotype. Intriguingly, the inbred TgS mice presented relatively high variability in their locomotion activity, suggesting experience-derived contribution to this phenotype. The variable nature of the excessive locomotor activity in individual TgS mice indicated daily stress origin(s), variable in its extent and duration. The molecular origin of such heterogeneity could be the increased amount of neuronal vAChT, ChAT and mAChE-R mRNAs in the sensorimotor cortex and hippocampal neurons. While both psychological and physical stressors induce neuronal AChE-R over-production, the acquired feedback responses to their cholinergic imbalance may depend on individual experiences, explaining this variability. This conclusion further implies that transcriptional activation and shifted alternative splicing in the brain of mammals are most valuable, in that they prevent excess accumulation of AChE-S and its consequent behavioral damage. Recently, it was reported that the C-terminus of AChE-S includes a region with neurotoxic properties (Greenfield and Vaux, 2002), perhaps through its capacity to promote nuclear translocation of AChE-S (Perry et al., 2002). While the exaggerated stress responses, such as the intense locomotor response to the mild stress of a circadian switch can be expected to exacerbate the hypocholinergic state of these already compromised animals, it cannot compensate for the yet-unknown nuclear effects and/or neurotoxicity of the excess AChE-S in neurons.

Transient antisense suppression of transgenics' locomotor activity further substantiates the involvement of AChE-R in the circadian light/dark shift-associated hyperactive response. To assess this assumption, TgS mice were injected intraperitoneally with 50 g/kg of a 3′-terminally 2′-*O*-methyl protected antisense oligonucleotide, EN101, that suppresses de novo production of mouse AChE-R (Shohami et al., 2000; Cohen et al., 2002). The difference between pre- and post-treatment locomotor activity was calculated for each animal, following EN101 or saline injection. TgS mice presented only a transient decrease in locomotor activity following EN101 treatment. In myasthenic rats with impaired locomotion, and massive AChE-R excess in their circulation, 500 mg/kg EN101 sufficed to improve locomotion for over 24 h, demonstrating dose and time dependence for this effect (Brenner et al., 2003).

AChE overexpression and memory functioning

Stress insults notably intensify fear memory (Nijholt et al., 2003). To explore the possibility that the modified expression pattern of AChE is causally involved in this phenomenon, immobilization-stressed wild-type mice were subjected to antisense suppression of their hippocampal AChE-R. Reduction of immunochemically detected AChE-R by 25% prevented the elevation of LTP and the characteristic freezing response to a foot-shock insult (ibid.). The finding that AD is associated with premature death of cholinergic neurons, initially within thc basal forebrain, and the limited efficacy of anti-AChE drugs in ameliorating these deficits, raised the question whether cholinergic imbalance may produce cognitive impairments. Indeed, stress is also associated with cognitive disturbances (Lupien and Lepage, 2001). If so, the stress-induced increases in AChE expression and catalytic activity, may be causally involved with the stress component of cognitive impairment in the demented brain. In addition, AChE-S overexpression was found to enhance the formation of amyloid plaques in the brains of double transgenics, which express both human AChE-S and the Swedish double mutated amyloid protein, as compared with the parent strain, with amyloid mutation alone (Rees et al., 2003). This change in the properties of another neuroactive protein emphasizes the complexity of the feedback responses that may be induced under ACE-S over-expression and complicates the interpretation of the stress-induced changes.

Progressive decline in dendritic arbors

Concomitant with the effects of stress on memory are remodelling changes in the dendritic arbors of neurons in various brain regions. These depend both on the type of stress and the brain region where they occur; for example, chronic immobilization stress induces dendritic atrophy and debranching in CA3 pyramidal neurons of the hippocampus, while pyramidal and stellate neurons in the amygdala exhibit enhanced dendritic arborization in response to the same stress (Vyas et al., 2002). Chronic, unpredictable stress, however, had little effect on CA3 neurons, but induced atrophy in bipolar neurons of the basolateral complex in the amygdala (Vyas et al., 2002). Thus, the cellular and molecular differences among the responses to different kinds of stress are still largely unexplained.

To examine their neurodeterioration, we compared the dendritic arbors in cortical neurons of TgS mice and wild-type controls and tested them in parallel in the Morris water maze. This test assesses spatial memory, which is known to deteriorate with age; however, the age-dependent deterioration in spatial memory was rapidly accelerated in TgS mice (Beeri et al., 1995), reaching a level of total failure in young adults (Beeri et al., 1997).

The progressively reduced dendritic fields in the brains of these mice (Beeri et al., 1997) might represent the structural correlate of their spatial memory impairment. This supports the above summarized studies that have frequently revealed neuronal damage or atrophy, mirroring cognitive dysfunction in demented animals and humans. While it is not yet clear whether the limited dendritic arbors in TgS mice reflect enhanced pruning or a suppressed growth of neurites, the thinned neural network that this entails mimics a parallel phenomenon in the brain of AD patients (Ruppin and Reggia, 1995). The impairments in the Morris water maze performance in TgS mice thus suggest that prolonged AChE overexpression is associated with a progressive decline in learning and memory.

AChE-R contribution to adverse stress responses is highly variable

The delayed consequences of stress and neurodegeneration are both known to involve impairment of other aspects of learning and memory, in particular those that related to individual relationships. To test whether this aspect of behavior, as well, is associated with AChE regulation, we assessed the memory functioning of TgS mice in the social recognition test, which is based on olfactory perception (Dantzer et al., 1990). To examine the effect of AChE-R overexpression on social recognition, and its correlation with pre-treatment symptom severity, TgS mice were divided into three groups, and those exhibiting short or long exploration of a familiar juvenile were selected based on a baseline social recognition test. AChE-R mRNA was then suppressed by EN101 (AS3) treatment. Social recognition was tested again 1, 3 and 6 days following two intracerebroventricular injections of EN101, separated by 24 h. As expected, there was a significant overall difference between the short and the long groups in exploration time. However, post-hoc tests revealed that these groups differed significantly only during the pre-treatment day, and not after the EN101 treatment. Furthermore, within the long, but not the short explorers group, social exploration of the 'same' juvenile was significantly reduced 1 day after the EN101 injection, (at the same dose as before), with progressive increases in social exploration time during the 5 subsequent days. This experiment thus demonstrated both the causal involvement and the reversibility of the AChE-R effect, especially in animals with severe pre-treatment impairments and in comparison to the short-term efficacy of tacrine (Cohen et al., 2002).

Antisense AChE-R mRNA suppression selectively reduces brain AChE-R protein

To conceptually prove the putative role of AChE-R in mediating the impaired social recognition, we conducted a second experiment, in which mice completed a social recognition test before and after two injections with either EN101 or a sequence specificity control. The tested EN101 effect was central (due to the intracerebroventricular administration). Twenty-four hours following the second test, their brains were removed and AChE-R expression was assessed. As found in the first experiment, Tg mice with long pre-treatment explorative behavior displayed a significant improvement in social exploration of the 'same' juvenile 24 h following the second treatment with EN101, but not with the irrelevant AS-ON targeted to BuChE mRNA (ASB). Control mice with either long or short pre-treatment social exploration showed no response to either EN101 or the irrelevant antisense oligonucleotide, perhaps reflecting a limitation in the resolution power of these behavioral tests.

Immunodetected AChE-R protein levels were significantly lower in EN101 treated mice as compared with ASB-treated mice, regardless of their genotype or pre-treatment behavior pattern. In contrast, densitometric analysis of immunodetected total AChE protein (detected by an antibody targeted to the N-terminus, common to both isoforms) revealed essentially unchanged signals. Together, these findings attest to the selectivity of the antisense treatment for treating AChE-R over-expressing animals and its sequence-specificity in reversing the AChE-R induced impairment of behavior. The outcome of this second experiment has also provided a tentative explanation of the long duration of the antisense effect, in that disrupted function appears to be associated with higher levels of AChE-R. So long as AChE-R remains below a threshold level, function remains normal, even if the antisense agent is no longer present. AChE-R expression, thus, presents a relatively wide safety margins, above which it causes deleterious effects.

Peripheral/central interactions: transfer of stress signals through cholinergic pathways

Recently, a "cholinergic anti-inflammatory pathway" has been identified in which cholinergic signalling through the efferent vagus nerve modulates the mammalian inflammatory response (Bernik et al., 2002; Borovikova et al., 2000; Tracey et al., 2001). ACh, the principal vagal neurotransmitter, significantly attenuated the release of the proinflammatory cytokines, tumor necrosis factor-α (TNFα), interleukin (IL)-1β, IL-6 and IL-18 (but not the anti-inflammatory cytokine IL-10), in lipopolysaccharide-stimulated human macrophage cultures and in live rats. The signalling pathway involves the α7 nicotinic receptor (Tracey, 2002). A parallel process in the brain, or indeed in other leukocytes, has not yet been explored. An interesting point of this

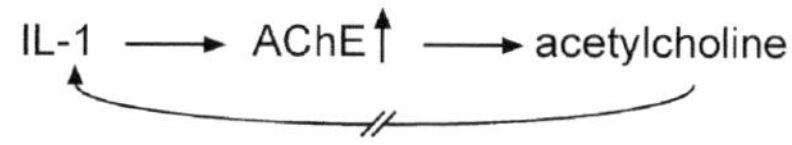

Scheme 1

observation is that ACh in the blood is given a physiological role, and as a corollary, the AChE of blood, notably the AChE-E on the surface of erythrocytes is placed in the spotlight. However, the amount of blood AChE-E changes only very slowly, which seems to disqualify it from a regulatory role in inflammatory responses; AChE-S, which is expressed in multiple leukocyte lineages (Deutsch et al., 2002), as well as AChE-R, being soluble and having a short half-life, are more likely regulators of responses to inflammatory challenges, as well as controllers of the heightened anxiety that accompanies the inflammatory responses (Danzer, 1999; Reichenberg et al., 2001).

In light of reports that stress, both in humans and in animals, involves increased production of pro-inflammatory cytokines, e.g. IL-1 (Maes et al., 1998; Nguyen et al., 1998; Spivak et al., 1997), and that IL-1 causes AChE over-production in PC12 cells (Li et al., 2000), we postulate the following relationship (Scheme 1): IL-1 induction of AChE over-expression suppresses ACh, ablating the interference by ACh in IL-1 production, a cycle that may explain prolongation of both the stress responses and the over-expression of AChE and cytokines (Scheme 1). Further experiments will be required for establishing the complete circle and demonstrating the causal involvement of AChE in regulating the brain levels of pro-inflammatory cytokines.

A logical extension of this concept is that AChE-S and/or AChE-R levels should be pivotal for regulation of proinflammatory cytokines in both the peripheral and central nervous systems. Thus, stress-induced elevation of cortisol levels results in elevated neuronal AChE production (Meshorer et al., 2002). This would reduce ACh and elevate production of pro-inflammatory cytokines (Scheme 2). According to this model, in addition to the direct suppression of blood cytokine production by cortisol (Marx, 1995), cortisol would also activate brain cytokine production, by an indirect cholinergic-mediated mechanism. This indirect route may explain some of the permissive and pro-inflammatory actions of

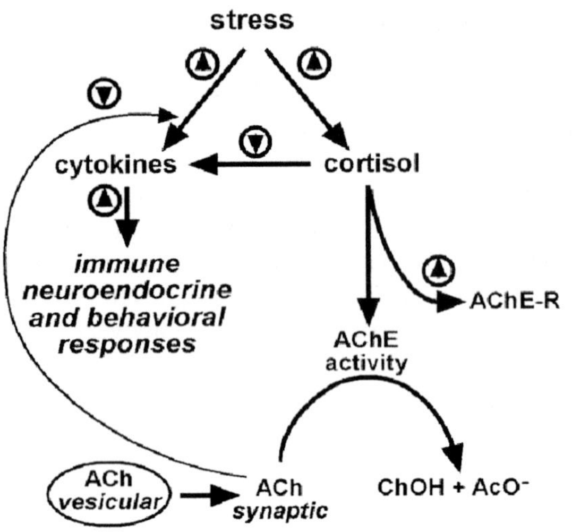

Scheme 2 The cellular and biochemical events to which we attribute predicted stress-associated changes.

glucocorticoids (Brooke and Sapolsky, 2002; Munck and Naray-Fejes-Toth, 1994; Wilckens and De Rijk, 1997). In support of this hypothesis, we have recently shown that the endotoxin-induced changes in human working and declarative memory associate both with degradation of plasma AChE-R and with the circulation levels of pro-inflammatory cytokines (Cohen et al., 2003).

Stress induces the release of cytokines and cortisol. Cytokines elevation is associated with immune, neuroendocrine and behavioral responses (Konsman et al., 2002). Cortisol induces AChE-R production, which should elevate (▲) AChE plasma activity. In the periphery, ACh released from neuronal vesicles into the synaptic cleft, acts to suppress (▼) cytokines production in macrophages (upward curved arrow). If this is also the case within the brain, stress-induced increase in AChE-R expression and activity will result in lower ACh levels, which may further enhance neuronal cytokine production.

AChE-R accumulation is transient because this enzyme and its mRNA are relatively unstable. Therefore, the cognitive effects of inflammation would be expected to be limited to the duration of cortisol's presence in the circulation and perhaps several hours after, but not much longer. This theory is compatible with previous findings in humans (Yirmiya et al., 2000) and with our on-going studies.

The future of acetylcholinesterase in stress studies

Anti-AChE therapies

As the role of AChE-R in stress responses is brought to light, it is becoming apparent that a key to regulating these responses is manipulation of the levels of this protein. It will be a challenge to devise therapeutic strategies that will regulate it, while leaving normal cholinergic neurotransmission unaffected. One such strategy uses an antisense reagent to specifically destroy the mRNA that encodes that variant. Although the antisense approach has been around for several decades, the mechanism of action of antisense oligonucleotides is incompletely understood (Opalinska and Gewirtz, 2002). Nevertheless, whatever the molecular mechanisms there have been experimental and even clinical successes in using this approach (Orr, 2001). In our hands, such an agent, EN101, has successfully aided recovery from closed head injury in mice (Shohami et al., 2000), retrieved functional working memory in TgS mice (Cohen et al., 2002) and reversed the symptoms of experimental autoimmune myasthenia gravis in rats (Brenner et al., 2003), EN101 is now being successfully tested in the clinic for treatment of human myasthenia gravis (Argov et al., 2003).

The cholinergic component of stress may confer bidirectional effects on cognitive functions

Behind the conventional view of stress responses reflecting a complex syndrome or disease, lies the fact that acute stress represents an extreme example of a natural process, enabling the adjustment to changing environment through neuronal plasticity. This is consistent with the fact that acute glucocorticoid administration, while impairing retrieval of long-term declarative memory (Kirschbaum et al., 1996), improves the working memory in humans (de Quervain et al., 2000). It is also compatible with the reports that cholinergic enhancement facilitates the increased selectivity of perceptual processing during working memory (Furey et al., 2000), and that higher cortisol values facilitate spatial memory in toddlers (Stansbury et al., 2000). Thus, the stress-induced effects on memory functions are complex, and may reflect both the intensity of the stressful experience and the type of memory function that was measured.

This chapter would be incomplete without mentioning the complex signalling cascades which are pivotal for the described stress responses. These likely involve changes in, especially PCKβII, which has been found essential for the contextual fear response, a characteristic consequence of acute stress (Weeber et al., 2000). PCKβII is an alternative splicing product of the PCKβ gene. A tentative link between this signaling cascade and the cholinergic feedback response to stress was indicated in a recent two-hybrid search for cellular-binding partners of AChE-R, which revealed association of AChE-R with the PKCβII scaffold protein, RACK1 (Birikh et al., 2003). Compatible with this finding, inherited AChE-R overexpression in transgenic mice resulted in perikaryal clusters enriched with PKCβII, accompanied by PKC-augmented LTP enhancement (Nijholt et al., 2003). The decreases in RACK1 in the brain of AD patients (Battaini et al., 1999) further supports the yet elusive link between the stress load with which one is confronted and the onset of neurodegeneration.

Abbreviations

ACh	acetylcholine
ACTH	adrenocorticotropic hormone; a.k.a. corticotropin
ACHE	acetylcholinesterase gene
AChE	acetylcholinesterase protein
AD	Alzheimer's disease
BuChE	butyrylcholinesterase
ChAT	choline acetyltransferase
CNS	central nervous system
CRF	corticotrophin releasing factor
DFP	diisopropylfluorophosphonate
HPA	hypothalamo–pituitary–adrenal (axis)
NMJ	neuromuscular junction
PB	pyridostigmine bromide
PTSD	post-traumatic stress disorder
TgR	a transgenic mouse strain that expresses human AChE-R
TgS	a transgenic mouse strain that expresses human AChE-S
vAChT	vesicular acetylcholine transporter

Acknowledgements

Some of the work reported here was supported by the US Army Medical Research and Materiel Command under grant No. DAMD17-99-1-9547, the Israel Science Foundation (618/02-1), the US-Israel Binational Science Foundation (1999-115), European Community Grant (LSHM-CT-2003-503330) and Ester Neuroscience.

References

Abou-Donia, M.B., Goldstein, L.B., Jones, K.H., Abdel-Rahman, A.A., Damodaran, T. Ca^{++}-dependent protein kinase (PKC) activity and intracellular translocation (Young et al., 2002). V., Dechkovskaia, A.M., Bullman, S.L., Amir, B.E. and Khan, W.A. (2001) Locomotor and sensorimotor performance deficit in rats following exposure to pyridostigmine bromide, DEET, and permethrin, alone and in combination. Toxicol Sci, 60: 305–314.

Abou-Donia, M.B., Wilmarth, K.R., Jensen, K.F., Oehme, F.W. and Kurt, T.L. (1996) Neurotoxicity resulting from coexposure to pyridostigmine bromide, deet, and permethrin: implications of Gulf War chemical exposures. J. Toxicol. Environ. Health, 48: 35–56.

Acquas, E., Wilson, C. and Fibiger, H.C. (1996) Conditioned and unconditioned stimuli increase frontal cortical and hippocampal acetylcholine release: effects of novelty, habituation, and fear. J. Neurosci., 16: 3089–3096.

Andres, C., Beeri, R., Friedman, A., Lev-Lehman, E., Henis, S., Timberg, R., Shani, M. and Soreq, H. (1997) Acetylcholinesterase-transgenic mice display embryonic modulations in spinal cord choline acetyltransferase and neurexin Iß gene expression followed by late-onset neuromotor deterioration. Proc. Natl. Acad. Sci. USA, 94: 8173–8178.

Appleyard, M.E. (1992) Secreted acetylcholinesterase: non-classical aspects of a classical enzyme. Trends Neurosci., 15: 485–490.

Argov, Z., McKee, D., Agus, S., Soreq, H., Ben-Yoseph, O., Brawer, S. and Sussman, J. (2003) EN101: a novel antisense therapeutic strategy for myasthenia gravis. Paper presented at: Fifty-fifth Annual Meeting of the American Academy of Neurology, 2003 (Honolulu).

Ashani, Y., Shapira, S., Levy, D., Wolfe, A.D., Doctor, B.P. and Raveh, L. (1991) Butyrylcholinesterase and acetylcholinesterase prophylaxis against soman poisoning in mice. Biochem. Pharmacol., 41: 37–41.

Atanasova, E., Chiappa, S., Wieben, E. and Brimijoin, S. (1999) Novel messenger RNA and alternative promoter for murine acetylcholinesterase. J. Biol. Chem., 274: 21078–21084.

Bartels, M., de Geus, E.J., Kirschbaum, C., Sluyter, F., and Boomsma, D.I. (2003) Heritability of daytime cortisol levels in children. Behav. Genet., 33: 421–433.

Bartus, R.T., Dean, R.L., 3rd, Beer, B. and Lippa, A.S. (1982) The cholinergic hypothesis of geriatric memory dysfunction. Science, 217: 408–414.

Battaglia, M. (2002) Beyond the usual suspects: a cholinergic route for panic attacks. Mol Psychiatry, 7: 239–246.

Battaini, F., Pascale, A., Lucchi, L., Pasinetti, G.M. and Govoni, S. (1999) Protein kinase C anchoring deficit in postmortem brains of Alzheimer's disease patients. Exp. Neurol., 159: 559–564.

Beck, K.D., Zhu, G., Beldowicz, D., Brennan, F.X., Ottenweller, J.E., Moldow, R.L. and Servatius, R.J. (2001) Central nervous system effects from a peripherally acting cholinesterase inhibiting agent: interaction with stress or genetics. Ann. N. Y. Acad. Sci., 933: 310–314.

Beeri, R., Andres, C., Lev-Lehman, E., Timberg, R., Huberman, T., Shani, M. and Soreq, H. (1995) Transgenic expression of human acetylcholinesterase induces progressive cognitive deterioration in mice. Curr. Biol., 5: 1063–1071.

Beeri, R., Le Novere, N., Mervis, R., Huberman, T., Grauer, E., Changeux, J.P. and Soreq, H. (1997) Enhanced hemicholinium binding and attenuated dendrite branching in cognitively impaired acetylcholinesterase-transgenic mice. J. Neurochem., 69: 2441–2451.

Bernatova, I., Key, M.P. Lucot, J.B. and Morris, M. (2002) Circadian differences in stress-induced pressor reactivity in mice. Hypertension, 40: 768–773.

Bernik, T.R., Friedman, S.G., Ochani, M., DiRaimo, R., Ulloa, L., Yang, H., Sudan, S., Czura, C.J., Ivanova, S.M. and Tracey, K.J. (2002) Pharmacological stimulation of the cholinergic antiinflammatory pathway. J. Exp. Med., 195: 781–788.

Berrouschot, J., Baumann, I., Kalischewski, P., Sterker, M. and Schneider, D. (1997) Therapy of myasthenic crisis. Crit. Care Med., 25: 1228–1235.

Bhatnagar, S., Costall, B. and Smythe, J.W. (1997) Hippocampal cholinergic blockade enhances hypothalamic-pituitary-adrenal responses to stress. Brain Res., 766: 244–248.

Bigbee, J.W., Sharma, K.V., Gupta, J.J. and Dupree, J.L. (1999) Morphogenic role for acetylcholinesterase in axonal outgrowth during neural development. Environ. Health Perspect., 107: Suppl 1, 81–87.

Birikh, K., Sklan, E., Shoham, S. and Soreq, H. (2003) Interaction of "readthrough" acetylcholinesterase with RACK1 and PKCbetaII correlates with intensified fear-induced conflict behavior. Proc. Natl. Acad. Sci. USA, 100: 283–288.

Blanchet, G., Carpentier, P., Lallement, G. and Sentenac-Roumanou, H. (1994) Prevention and treatment of status epilepticus induced by soman. Ann. Pharm. Fr., 52: 11–24.

Borovikova, L.V., Ivanova, S., Zhang, M., Yang, H., Botchkina, G.I., Watkins, L.R., Wang, H., Abumrad, N., Eaton, J.W. and Tracey, K.J. (2000) Vagus nerve stimulation attenuates the systemic inflammatory response to endotoxin. Nature, 405: 458–462.

Botti, S.A., Felder, C.E., Sussman, J.L. and Silman, I. (1998) Electrotactins: a class of adhesion proteins with conserved electrostatic and structural motifs. Protein Eng., 11: 415–420.

Bremner, J.D., Vythilingam, M., Anderson, G., Vermetten, E., McGlashan, T., Heninger, G., Rasmusson, A., Southwick, S.M. and Charney, D.S. (2003) Assessment of the hypothalamic-pituitary-adrenal axis over a 24-hour diurnal period and in response to neuroendocrine challenges in women with and without childhood sexual abuse and posttraumatic stress disorder. Biol. Psychiatry, 54: 710–718.

Brenner, T., Hamra-Amitay, Y., Evron, T., Boneva, N., Seidman, S. and Soreq, H. (2003) The role of readthrough acetylcholinesterase in the pathophysiology of myasthenia gravis. FASEB J., 17: 214–222.

Brindle, N., Song, Y., Rogaeva, E., Premkumar, S., Levesque, G., Yu, G., Ikeda, M., Nishimura, M., Paterson, A., Sorbi, S. et al. (1998) Analysis of the butyrylcholinesterase gene and nearby chromosome 3 markers in Alzheimer disease. Hum. Mol. Genet., 7: 933–935.

Brooke, S.M. and Sapolsky, R.M. (2002) Glucocorticoid Exacerbation of gp120 Neurotoxicity: Role of Microglia. Exp. Neurol., 177: 151–158.

Burchfiel, J.L. and Duffy, F.H. (1982) Organophosphate neurotoxicity: chronic effects of sarin on the electroencephalogram of monkey and man. Neurobehav. Toxicol. Teratol., 4: 767–778.

Calogero, A.E., Gallucci, W.T., Bernardini, R., Saoutis, C., Gold, P.W. and Chrousos, G.P. (1988) Effect of cholinergic agonists and antagonists on rat hypothalamic corticotropin-releasing hormone secretion in vitro. Neuroendocrinology, 47: 303–308.

Calogero, A.E., Kamilaris, T.C., Gomez, M.T., Johnson, E.O., Tartaglia, M.E., Gold, P.W. and Chrousos, G.P. (1989) The muscarinic cholinergic agonist arecoline stimulates the rat hypothalamic-pituitary-adrenal axis through a centrally-mediated corticotropin-releasing hormone-dependent mechanism. Endocrinology, 125: 2445–2453.

Camp, S. and Taylor, P. (1998) Intronic elements appear essential for the differentiation-specific expression of acetylcholinesterase in C2C12 myotubes. In: Doctor, B.P., Taylor, P., Quinn, D.M., Rotundo, R.L. and Gentry, M.K. (Eds.), Structure and Function of Cholinesterases and Related Proteins. Plenum Press, New York, pp. 51–55.

Carlson, N.R. (1994) Physiology of Behavior, Allyn and Bacon. Needham Heights, MA.

Carmichael, W. (1994) The toxins of cyanobacteria. Sci. Amer., 270: 78–86.

Chan, R.Y., Adatia, F.A., Krupa, A.M. and Jasmin, B.J. (1998) Increased expression of acetylcholinesterase T and R transcripts during hematopoietic differentiation is accompanied by parallel elevations in the levels of their respective molecular forms. J. Biol. Chem., 273: 9727–9733.

Chaney, L.A., Rockhold, R.W., Wineman, R.W., and Hume, A.S. (1999) Anticonvulsant-resistant seizures following pyridostigmine bromide (PB) and N,N-diethyl-m-toluamide (DEET) Toxicol. Sci., 49: 306–311.

Chaney, L.A., Wineman, R.W., Rockhold, R.W. and Hume, A.S. (2000) Acute effects of an insect repellent, N,N-diethyl-m-toluamide, on cholinesterase inhibition induced by pyridostigmine bromide in rats. Toxicol. Appl. Pharmacol., 165: 107–114.

Chen, Y., Brunson, K.L., Muller, M.B., Cariaga, W. and Baram, T.Z. (2000) Immunocytochemical distribution of corticotropin-releasing hormone receptor type-1 (CRF(1))-like immunoreactivity in the mouse brain: light microscopy analysis using an antibody directed against the C-terminus. J. Comp. Neurol., 420: 305–323.

Cohen, O., Erb, C., Ginzberg, D., Pollak, Y., Seidman, S., Shoham, S., Yirmiya, R. and Soreq, H. (2002) Neuronal overexpression of 'readthrough' acetylcholinesterase is associated with antisense-suppressible behavioral impairments. Mol. Psychiatry, 7: 874–885.

Cohen, O., Reichenberg, A., Perry, C., Ginzberg, D., Pollmächer, T., Soreq, H. and Yirmiya, R. (2003) Endotoxin-induced changes in human working and declarative memory are associated with C – terminal cleavage of plasma "readthrough" acetylcholinesterase. J. Molec. Neurosci., 21: 199–212.

Cook, W.O., Beasley, V.R., Dahlem, A.M., Dellinger, J.A., Harlin, K.S. and Carmichael, W.W. (1988) Comparison of effects of anatoxin-a(s) and paraoxon, physostigmine and pyridostigmine on mouse brain cholinesterase activity. Toxicon., 26: 750–753.

Dantzer, R., Tazi, A. and Bluthe, R.M. (1990) Cerebral lateralization of olfactory-mediated affective processes in rats. Behav. Brain Res., 40: 53–60.

Danzer, R., Whollman, E.E. and Yirmiya, R. (1999) Cytokines, Stress and Depression. Kluwer Academic/Plenum, New York.

Daoud, R., Da Penha Berzaghi, M., Siedler, F., Hubener, M. and Stamm, S. (1999) Activity-dependent regulation of alternative splicing patterns in the rat brain. Eur. J. Neurosci., 11: 788–802.

Darboux, I., Barthalay, Y., Piovant, M. and Hipeau Jacquotte, R. (1996) The structure-function relationships in Drosophila neurotactin show that cholinesterasic domains may have adhesive properties. EMBO J., 15: 4835–4843.

Darreh-Shori, T., Almkvist, O., Guan, Z.Z., Garlind, A., Strandberg, B., Svensson, A.L., Soreq, H., Hellstrom-Lindahl, E. and Nordberg, A. (2002) Sustained cholinesterase inhibition in AD patients receiving rivastigmine for 12 months. Neurology, 59: 563–572.

Darreh-Shori, T., Hellström-Lindahl, E., Flores-Flores, C, Guan, Z., Soreq, H. and Nordberg, A. (2004) Long-lasting acetylcholinesterase splice variations in anticholinesterase-treated Alzheimer's disease patients. J. Neurochem., 88: 1102–1113.

Day, J.C., Koehl, M., Deroche, V., Le Moal, M. and Maccari, S. (1998a) Prenatal stress enhances stress- and corticotropin-releasing factor-induced stimulation of hippocampal acetylcholine release in adult rats. J. Neurosci., 18: 1886–1892.

Day, J.C., Koehl, M., Le Moal, M. and Maccari, S. (1998b) Corticotropin-releasing factor administered centrally, but not peripherally, stimulates hippocampal acetylcholine release. J. Neurochem., 71: 622–629.

Dazzi, L., Motzo, C., Imperato, A., Serra, M., Gessa, G.L. and Biggio, G. (1995) Modulation of basal and stress-induced release of acetylcholine and dopamine in rat brain by abecarnil and imidazenil, two anxioselective gamma-aminobutyric acidA receptor modulators. J. Pharmacol. Exp. Ther., 273: 241–247.

de Quervain, D.J., Roozendaal, B., Nitsch, R.M., McGaugh, J.L., and Hock, C. (2000) Acute cortisone administration impairs retrieval of long-term declarative memory in humans. Nat. Neurosci., 3: 313–314.

Deutsch, V.R., Pick, M., Perry, C., Grisaru, D., Hemo, Y., Golan-Hadari, D., Grant, A., Eldor, A. and Soreq, H. (2002) The stress-associated acetylcholinesterase variant AChE-R is expressed in human CD34(+) hematopoietic progenitors and its C-terminal peptide ARP promotes their proliferation. Exp. Hematol., 30: 1153–1161.

Doctor, B.P., Blick, D.W., Caranto, G., Castro, C.A., Gentry, M.K., Larrison, R., Maxwell, D.M., Murphy, M.R., Schutz, M., Waibel, K. et al. (1993) Cholinesterases as scavengers for organophosphorus compounds: protection of primate performance against soman toxicity. Chem. Biol. Interact, 87: 285–293.

Drachman, D.A. and Leavitt, J. (1974) Human memory and the cholinergic system. A relationship to aging? Arch. Neurol., 30: 113–121.

Ellis, K.A. and Nathan, P.J. (2001) The pharmacology of human working memory. Int. J. Neuropsychopharmacol., 4: 299–313.

Erb, C., Troost, J., Kopf, S., Schmitt, U., Loffelholz, K., Soreq, H. and Klein, J. (2001) Compensatory mechanisms enhance hippocampal acetylcholine release in transgenic mice expressing human acetylcholinesterase. J. Neurochem., 77: 638–646.

Esposito, P., Chandler, N., Kandere, K., Basu, S., Jacobson, S., Connolly, R., Tutor, D. and Theoharides, T.C. (2002) Corticotropin-releasing hormone and brain mast cells regulate blood-brain-barrier permeability induced by acute stress. J. Pharmacol. Exp. Ther., 303: 1061–1066.

Esposito, P., Gheorghe, D., Kandere, K., Pang, X., Connolly, R., Jacobson, S. and Theoharides, T.C. (2001) Acute stress increases permeability of the blood-brain-barrier through activation of brain mast cells. Brain Res., 888: 117–127.

Evans, W.E. and Relling, M.V. (1999) Pharmacogenomics: translating functional genomics into rational therapeutics. Science, 286: 487–491.

Farchi, N., Soreq, H. and Hochner, B. (2003) Chronic acetylcholinesterase overexpression induces multilevelled aberrations in mouse neuromuscular physiology. J. Physiol., 546: 165–173.

Fibiger, H.C. (1991) Cholinergic mechanisms in learning, memory and dementia: a review of recent evidence. Trends Neurosci., 14: 220–223.

Finkelstein, Y., Koffler, B., Rabey, J.M. and Gilad, G.M. (1985) Dynamics of cholinergic synaptic mechanisms in rat hippocampus after stress. Brain Res., 343: 314–319.

Friedman, A., Kaufer, D., Pavlovsky, L. and Soreq, H. (1998) Cholinergic excitation induces activity-dependent electrophysiological and transcriptional responses in hippocampal slices. J. Physiol. Paris, 92: 329–335.

Friedman, A., Kaufer, D., Shemer, J., Hendler, I., Soreq, H. and Tur-Kaspa, I. (1996) Pyridostigmine brain penetration under stress enhances neuronal excitability and induces early immediate transcriptional response. Nat. Med., 2: 1382–1385.

Friedman, M. (2002) Tomato glycoalkaloids: role in the plant and in the diet. J. Agric. Food Chem., 50: 5751–5780.

Fu, Y., Matta, S.G., Valentine, J.D., and Sharp, B.M. (1997) Adrenocorticotropin response and nicotine-induced norepinephrine secretion in the rat paraventricular nucleus are mediated through brainstem receptors. Endocrinology, 138: 1935–1943.

Fullerton, C.S., and Ursano, R.J. (1990) Behavioral and psychological responses to chemical and biological warfare. Mil. Med., 155: 54–59.

Furey, M.L., Pietrini, P., and Haxby, J.V. (2000) Cholinergic enhancement and increased selectivity of perceptual processing during working memory. Science, 290: 2315–2319.

Galyam, N., Grisaru, D., Grifman, M., Melamed-Book, N., Eckstein, F., Seidman, S., Eldor, A. and Soreq, H. (2001) Complex host cell responses to antisense suppression of ACHE gene expression. Antisense Nucl. Acid Drug Dev., 11: 51–57.

Giacobini, E. (2000) Cholinesterase inhibitors: from the Calabar bean to Alzheimer therapy. In: Giacobini, E. (ed.), Cholinesterases and Cholinesterase Inhibitors, Martin Dunitz, London, pp. 181–226.

Giacobini, E. (2002) Long-term stabilizing effect of cholinesterase inhibitors in the therapy of Alzheimer' disease. J. Neural. Transm. Suppl., 62: 181–187.

Gilad, G.M. (1987) The stress-induced response of the septo-hippocampal cholinergic system. A vectorial outcome of psychoneuroendocrinological interactions. Psychoneuroendocrinology, 12: 167–184.

Gilad, G.M., Mahon, B.D., Finkelstein, Y., Koffler, B. and Gilad, V.H. (1985) Stress-induced activation of the hippocampal cholinergic system and the pituitary-adrenocortical axis. Brain Res., 347: 404–408.

Golomb, B. (1999) A Review of the Scientific Literature as it Pertains to Gulf War Illnesses, Vol 2 Santa Monica, Rand Corp.

Gonzalez, A.M., and Pazos, A. (1992) Modification of muscarinic acetylcholine receptors in the rat brain following chronic immobilization stress: an autoradiographic study. Eur. J. Pharmacol., 223: 25–31.

Gray, R., Rajan, A.S., Radcliffe, K.A., Yakehiro, M., and Dani, J.A. (1996) Hippocampal synaptic transmission enhanced by low concentrations of nicotine. Nature, 383: 713–716.

Greenfield, S. (1996) Non-classical actions of cholinesterases: role in cellular differentiation, tumorigenesis and Alzheimer's disease. Neurochem. Int., 28: 485–490.

Greenfield, S. and Vaux, D.J. (2002) Parkinson's disease, Alzheimer's disease and motor neurone disease: identifying a common mechanism. Neuroscience, 113: 485–492.

Grifman, M., Galyam, N., Seidman, S. and Soreq, H. (1998) Functional redundancy of acetylcholinesterase and neuroligin in mammalian neuritogenesis. Proc. Natl. Acad. Sci. USA, 95: 13935–13940.

Grisaru, D., Deutch, V., Shapira, M., Galyam, N., Lessing, B., Eldor, A. and Soreq, H. (2001) ARP, a peptide derived from the stress-associated acetylcholinesterase variant, has hematopoietic growth promoting activities. Mol. Med., 7: 93–105.

Grisaru, D., Sternfeld, M., Eldor, A., Glick, D. and Soreq, H. (1999) Structural roles of acetylcholinesterase variants in biology and pathology. Eur. J. Biochem., 264: 672–686.

Hillegaart, V. and Ahlenius, S. (1994) Time course for synchronization of spontaneous locomotor activity in the rat following reversal of the daylight (12:12 h) cycle. Physiol. Behav., 55: 73–75.

Imperato, A., Puglisi-Allegra, S., Casolini, P. and Angelucci, L. (1991) Changes in brain dopamine and acetylcholine release during and following stress are independent of the pituitary-adrenocortical axis. Brain Res., 538: 111–117.

Imperato, A., Puglisi-Allegra, S., Casolini, P., Zocchi, A. and Angelucci, L. (1989) Stress-induced enhancement of dopamine and acetylcholine release in limbic structures: role of corticosterone. Eur. J. Pharmacol., 165: 337–338.

Inglis, F.M., Day, J.C. and Fibiger, H.C. (1994) Enhanced acetylcholine release in hippocampus and cortex during the anticipation and consumption of a palatable meal. Neuroscience, 62: 1049–1056.

Jamain, S., Quach, H., Betancur, C., Rastam, M., Colineaux, C., Gillberg, I.C., Soderstrom, H., Giros, B., Leboyer, M., Gillberg, C., et al. (2003) Mutations of the X-linked genes encoding neuroligins NLGN3 and NLGN4 are associated with autism. Nat. Genet., 34: 27–29.

Kalow, W. and Genest, K. (1957) A method for detection of atypical forms of human serum cholinesterase: Determination of dibucaine numbers. Can. J. Biochem., 35: 339–346.

Kaneto, H. (1997) Learning/memory processes under stress conditions. Behav. Brain Res., 83: 71–74.

Kant, G.J., Bauman, R.A., Feaster, S.R., Anderson, S.M., Saviolakis, G.A. and Garcia, G.E. (2001) The combined effects of pyridostigmine and chronic stress on brain cortical and blood acetylcholinesterase, corticosterone, prolactin and alternation performance in rats. Pharmacol. Biochem. Behav., 70: 209–218.

Karpel, R., Sternfeld, M., Ginzberg, D., Guhl, E., Graessmann, A. and Soreq, H. (1996) Overexpression of alternative human acetylcholinesterase forms modulates process extensions in cultured glioma cells. J. Neurochem., 66: 114–123.

Kaufer, D., Friedman, A., Seidman, S. and Soreq, H. (1998) Acute stress facilitates long-lasting changes in cholinergic gene expression. Nature, 393: 373–377.

Kaufer, D., Friedman, A., Seidman, S. and Soreq, H. (1999) Anticholinesterases induce multigenic transcriptional feedback response suppressing cholinergic neurotransmission. Chem. Biol. Interact., 119–120: 349–360.

Kaufer, D. and Soreq, H. (1999) Tracking cholinergic pathways from psychological and chemical stressors to variable neurodeterioration paradigms. Curr. Opin. Neurol., 12: 739–743.

Kelsey, J.E. (1983) The role of norepinephrine and acetylcholine in mediating escape deficits produced by inescapable shocks. Behav. Neural. Biol., 37: 326–331.

Kirschbaum, C., Wolf, O.T., May, M., Wippich, W. and Hellhammer, D.H. (1996) Stress- and treatment-induced elevations of cortisol levels associated with impaired declarative memory in healthy adults. Life Sci., 58: 1475–1483.

Koenigsberger, C., Chiappa, S. and Brimijoin, S. (1997) Neurite differentiation is modulated in neuroblastoma cells engineered for altered acetylcholinesterase expression. J. Neurochem., 69: 1389–1397.

Konsman, J.P., Parnet, P. and Dantzer, R. (2002) Cytokine-induced sickness behaviour: mechanisms and implications. Trends Neurosci., 25: 154–159.

Kronman, C., Chitlaru, T., Elhanany, E., Velan, B. and Shafferman, A. (2000) Hierarchy of post-translation modifications involved in the circulatory longevity of glycoproteins. Demonstration of concerted contributions of glycan sialylation and subunit assembly to the pharmacokinetic behavior of acetylcholinesterase. J. Biol. Chem., 275: 29488–29502.

La Du, B.N., Bartels, C.F., Nogueira, C.P., Hajra, A., Lightstone, H., Van der Spek, A. and Lockridge, O. (1990) Phenotypic and molecular biological analysis of human butyrylcholinesterase variants. Clin. Biochem., 23: 423–431.

Layer, P.G. (1996) Non-classical actions of cholinesterases: role in cellular differentiation, tumorigenesis and Alzheimer's disease. Neurochem. Int., 28: 491–495.

Lehmann, D.J., Johnston, C. and Smith, A.D. (1997) Synergy between the genes for butyrylcholinesterase K variant and apolipoprotein E4 in late-onset confirmed Alzheimer's disease. Hum. Mol. Genet., 6: 1933–1936.

Lehmann, D.J., Nagy, Z., Litchfield, S., Borja, M.C. and Smith, A.D. (2000) Association of butyrylcholinesterase K variant with cholinesterase-positive neuritic plaques in the temporal cortex in late-onset Alzheimer's disease. Hum. Genet., 106: 447–452.

Li, Y., Liu, L., Kang, J., Sheng, J.G., Barger, S.W., Mrak, R.E. and Griffin, W.S. (2000) Neuronal-glial interactions mediated by interleukin-1 enhance neuronal acetylcholinesterase activity and mRNA expression. J. Neurosci., 20: 149–155.

Lightman, S.L., Windle, R.J., Julian, M.D., Harbuz, M.S., Shanks, N., Wood, S.A., Kershaw, Y.M. and Ingram, C.D. (2000) Significance of pulsatility in the HPA axis. Novartis Found Symp., 227: 244–257.

Liu, W.F. (1992) Acute effects of oral low doses of pyridostigmine on simple visual discrimination and unconditioned consummatory acts in rats. Pharmacol. Biochem. Behav., 41: 251–254.

Llinas, R.R. and Greenfield, S.A. (1987) On-line visualization of dendritic release of acetylcholinesterase from mammalian substantia nigra neurons. Proc. Natl. Acad. Sci. USA, 84: 3047–3050.

Luo, Z.D., Wang, Y., Werlen, G., Camp, S., Chien, K.R. and Taylor, P. (1999) Calcineurin enhances acetylcholinesterase mRNA stability during C2-C12 muscle cell differentiation. Mol. Pharmacol., 56: 886–894.

Lupien, S.J. and Lepage, M. (2001) Stress, memory, and the hippocampus: can't live with it, can't live without it. Behav. Brain Res., 127: 137–158.

Lupien, S.J., Maheu, F., Tu, M.T., Lemay, M., McEwen, B.S. and Meaney, M.J. (1999) Increased cortisol levels and impaired cognition in human aging: implication for depression and dementia in later life. Rev. Neurosci., 10: 117–139.

Maes, M., Song, C., Lin, A., De Jongh, R., Van Gastel, A., Kenis, G., Bosmans, E., De Meester, I., Benoy, I., Neels, H., Demedts, P., Janca, A., Scharpe, S. and Smith, R.S. (1998) The effects of psychological stress on humans: increased production of pro-inflammatory cytokines and a Th1-like response in stress-induced anxiety. Cytokine, 10: 313–318.

Marchot, P., Prowse, C.N., Kanter, J., Camp, S., Ackermann, E.J., Radic, Z., Bougis, P.E., and Taylor, P. (1997) Expression and activity of mutants of fasciculin, a peptidic acetylcholinesterase inhibitor from mamba venom. J. Biol. Chem., 272: 3502–3510.

Marx, J. (1995) How the glucocorticoids suppress immunity. Science, 270: 232.

Massoulie, J. (2000) Molecular forms and anchoring of acetylcholinesterase. In: E. Giacobini, (Ed.), Cholinesterases and Cholinesterase Inhibitors. Martin Dunitz, London, pp. 81–101.

Matta, S.G., Singh, J. and Sharp, B.M. (1990) Catecholamines mediate nicotine-induced adrenocorticotropin secretion via alpha-adrenergic receptors. Endocrinology, 127: 1646–1655.

McGehee, D.S., Krasowski, M.D., Fung, D.L., Wilson, B., Gronert, G.A. and Moss, J. (2000) Cholinesterase inhibition by potato glycoalkaloids slows mivacurium metabolism. Anesthesiology, 93: 510–519.

Meshorer, E., Erb, C., Gazit, R., Pavlovsky, L., Kaufer, D., Friedman, A., Glick, D., Ben-Arie, N. and Soreq, H. (2002) Alternative splicing and neuritic mRNA translocation under long-term neuronal hypersensitivity. Science, 295: 508–512.

Meshorer, E. and Soreq, H. (2002) Pre-mRNA splicing modulations in senescence Aging Cell, 1(1): 10–16.

Mizoguchi, K., Yuzurihara, M., Ishige, A., Sasaki, H. and Tabira, T. (2001) Effect of chronic stress on cholinergic transmission in rat hippocampus. Brain Res., 915: 108–111.

Mizuno, T. and Kimura, F. (1997) Attenuated stress response of hippocampal acetylcholine release and adrenocortical secretion in aged rats. Neurosci. Lett., 222: 49–52.

Mor, I., Grisaru, D., Titelbaum, L., Evron, T., Richler, C., Wahrman, J., Sternfeld, M., Yogev, L., Meiri, N., Seidman, S. and Soreq, H. (2001) Modified testicular expression of stress-associated "readthrough" acetylcholiensterase predicts male infertility. FASEB J., 15: 2039–2041.

Munck, A. and Naray-Fejes-Toth, A. (1994) Glucocorticoids and stress: permissive and suppressive actions. Ann. N. Y. Acad. Sci., 746: 115–130.

Nguyen, K.T., Deak, T., Owens, S.M., Kohno, T., Fleshner, M., Watkins, L.R., and Maier, S.F. (1998) Exposure to acute stress induces brain interleukin-1beta protein in the rat. J. Neurosci., 18: 2239–2246.

Nguyen, T. and Sudhof, T.C. (1997) Binding properties of neuroligin 1 and neurexin 1ß reveal function as heterophilic cell adhesion molecules. J Biol Chem, 272: 26032–26039.

Nijholt, I., Farchi, N., Kye, M.-J., Sklan, E., Shoham, S., Verbeurre, B., Owen D., Hochner, B., Spiess, J., Soreq, H. and Blank, T. (2003) Alternative splicing modulation of hippocampal long-term potentiation and fear memory. Mol. Psych., in press.

Nilsson, L., Nordberg, A., Hardy, J., Wester, P. and Winblad, B. (1986) Physostigmine restores 3H-acetylcholine efflux from Alzheimer brain slices to normal level. J. Neural. Transm., 67: 275–285.

Ohmori, N., Itoi, K., Tozawa, F., Sakai, Y., Sakai, K., Horiba, N., Demura, H. and Suda, T. (1995) Effect of acetylcholine on corticotropin-releasing factor gene expression in the hypothalamic paraventricular nucleus of conscious rats. Endocrinology, 136: 4858–4863.

Ohno, K., Engel, A.G., Brengman, J.M., Shen, X.M., Heidenreich, F., Vincent, A., Milone, M., Tan, E., Demirci, M., Walsh, P., et al. (2000) The spectrum of mutations causing end-plate acetylcholinesterase deficiency. Ann. Neurol., 47: 162-170.

Opalinska, J.B. and Gewirtz, A.M. (2002) Nucleic-acid therapeutics: basic principles and recent applications. Nat. Rev. Drug Discov., 1: 503–514.

Orr, R.M. (2001) Technology evaluation: fomivirsen, Isis Pharmaceuticals Inc/CIBA vision. Curr. Opin. Mol. Ther., 3: 288–294.

Orsini, C., Castellano, C. and Cabib, S. (2001) Pharmacological evidence of muscarinic-cholinergic sensitization following chronic stress. Psychopharmacology (Berl), 155: 144–147.

Palmer, A.M. (2002) Pharmacotherapy for Alzheimer's disease: progress and prospects. Trends Pharmacol. Sci., 23: 426–433.

Perrier, A.L., Massoulie, J. and Krejci, E. (2002) PRiMA: the membrane anchor of acetylcholinesterase in the brain. Neuron, 33: 275–285.

Perry, C., Sklan, E.H., Birikh, K., Shapira, M., Trejo, L., Eldor, A. and Soreq, H. (2002) Complex regulation of acetylcholinesterase gene expression in human brain tumors. Oncogene, 21: 8428–8441.

Perry, C. and Soreq, H. (2002) Transcriptional regulation of erythropoiesis. Fine tuning of combinatorial multi-domain elements. Eur. J. Biochem., 269: 3607–3618.

Perry, E., Martin-Ruiz, C., Lee, M., Griffiths, M., Johnson, M., Piggott, M., Haroutunian, V., Buxbaum, J.D., Nasland, J., Davis, K., Gotti, C., Clementi, F., Tzartos, S., Cohen, O., Soreq, H., Jaros, E., Perry, R., Ballard, C., McKeith, I. and Court, J. (2000) Nicotinic receptor subtypes in human brain ageing, Alzheimer and Lewy body diseases. Eur. J. Pharmacol., 393: 215–222.

Ponting, C., Phillips, C., Davies, K. and Blake, D. (1997) PDZ domains: targeting signalling molecules to sub-membranous sites. Bioessays., 19: 469–479.

Raveh, L., Ashani, Y., Levy, D., De La Hoz, D., Wolfe, A.D. and Doctor, B.P. (1989) Acetylcholinesterase prophylaxis against organophosphate poisoning. Quantitative correlation between protection and blood-enzyme level in mice. Biochem. Pharmacol., 38: 529–534.

Rees, T., Hammond, P.I., Soreq, H., Younkin, S. and Brimijoin, S. (2003) Acetylcholinesterase promotes beta-amyloid plaques in cerebral cortex. Neurobiol. Aging., 24: 777–787.

Rees, T., Hammond, P.I., Soreq, H., Younkin, S. and Brimijoin, S. (2003) Acetylcholinesterase promotes beta-amyloid plaques in cerebral cortex. Neurobiol. Aging, 24: 777–787

Reichenberg, A., Yirmiya, R., Schuld, A., Kraus, T., Haack, M., Morag, A. and Pollmacher, T. (2001) Cytokine-associated emotional and cognitive disturbances in humans. Arch. Gen. Psychiatry., 58: 445–452.

Relyea, R.A. and Mills, N. (2001) Predator-induced stress makes the pesticide carbaryl more deadly to gray treefrog tadpoles (Hyla versicolor) Proc. Natl. Acad. Sci. USA, 98: 2491–2496.

Roddick, J.G. (1989) The acetylcholinesterase-inhibitory activity of steroidal glycoalkaloids and their aglycones. Phytochemistry, 28: 2631–2634.

Rosenstock, L., Keifer, M., Daniell, W., McConnell, R. and Claypoole, K. (1991) Chronic central nervous system effects of acute organophosphate pesticide intoxication. The Pesticide Health Effects Study Group. Lancet., 338: 223–227.

Rubin, L.L. and Staddon, J.M. (1999) The cell biology of the blood-brain barrier. Annu. Rev. Neurosci., 22: 11–28.

Ruppin, E. and Reggia, J.A. (1995) Patterns of functional damage in neural network models of associative memory. Neural Comput., 7: 1105–1127.

Sago, H., Carlson, E.J., Smith, D.J., Kilbridge, J., Rubin, E.M., Mobley, W.C., Epstein, C.J. and Huang, T.T. (1998) Ts1Cje, a partial trisomy 16 mouse model for Down syndrome, exhibits learning and behavioral abnormalities. Proc. Natl. Acad. Sci. USA, 95: 6256–6261.

Sapolsky, R.M. (2002) Chickens, eggs and hippocampal atrophy. Nat. Neurosci., 5: 1111–1113.

Sauvage, M. and Steckler, T. (2001) Detection of corticotropin-releasing hormone receptor 1 immunoreactivity in cholinergic, dopaminergic and noradrenergic neurons of the murine basal forebrain and brainstem nuclei–potential implication for arousal and attention. Neurosci., 104: 643–652.

Scheiffele, P., Fan, J., Choih, J., Fetter, R. and Serafini, T. (2000) Neuroligin expressed in nonneuronal cells triggers presynaptic development in contacting axons. Cell, 101: 657–669.

Schwarz, M., Glick, D., Loewenstein, Y. and Soreq, H. (1995) Engineering of human cholinesterases explains and predicts diverse consequences of administration of various drugs and poisons. Pharmacol. Ther., 67: 283–322.

Servatius, R.J., Ottenweller, J.E., Beldowicz, D., Guo, W., Zhu, G. and Natelson, B.H. (1998) Persistently exaggerated startle responses in rats treated with pyridostigmine bromide. J. Pharmacol. Exp. Ther., 287: 1020–1028.

Shapira, M., Grant, A., Korner, M. and Soreq, H. (2000a) Genomic and transcriptional characterization of the human ACHE locus: complex involvement with acquired and inherited diseases. Isr. Med. Assoc. J., 2: 470–473.

Shapira, M., Tur-Kaspa, I., Bosgraaf, L., Livni, N., Grant, A.D., Grisaru, D., Korner, M., Ebstein, R.P. and Soreq, H. (2000b) A transcription-activating polymorphism in the ACHE promoter associated with acute sensitivity to anti-acetylcholinesterases. Hum. Mol. Genet., 9: 1273–1281.

Shaw, N. (2002) The neurophysiology of concussion. Prog. Neurobiol., 67, 281.

Shih, T.M. and McDonough, J.H., Jr. (1997) Neurochemical mechanisms in soman-induced seizures. J. Appl. Toxicol., 17: 255–264.

Shohami, E., Kaufer, D., Chen, Y., Seidman, S., Cohen, O., Ginzberg, D., Melamed-Book, N., Yirmiya, R. and Soreq, H. (2000) Antisense prevention of neuronal damages following head injury in mice. J. Mol. Med., 78: 228–236.

Silman, I. and Futerman, A.H. (1987) Modes of attachment of acetylcholinesterase to the surface membrane. Eur. J. Biochem., 170: 11–22.

Silver, A. (1974) The Biology of Cholinesterases. Amsterdam, North Holland.

Small, D.H., Michaelson, S. and Sberna, G. (1996) Non-classical actions of cholinesterases: role in cellular differentiation, tumorigenesis and Alzheimer's disease. Neurochem. Int., 28: 453–483.

Somani, S.M., Husain, K., Asha, T. and Helfert, R. (2000) Interactive and delayed effects of pyridostigmine and physical stress on biochemical and histological changes in peripheral tissues of mice. J. Appl. Toxicol., 20: 327–334.

Song, J.Y., Ichtchenko, K., Sudhof, T.C. and Brose, N. (1999) Neuroligin 1 is a postsynaptic cell-adhesion molecule of excitatory synapses. Proc. Natl. Acad. Sci. USA, 96: 1100–1105.

Soreq, H. and Glick, D. (2000) Novel roles for cholinesterases in stress and inhibitor responses. In: Giacobini, E. (Ed.), Cholinesterases and Cholinesterase Inhibitors: Basic, Preclinical and Clinical Aspects. Martin Dunitz, London, pp. 47–61.

Spivak, B., Shohat, B., Mester, R., Avraham, S., Gil-Ad, I., Bleich, A., Valevski, A. and Weizman, A. (1997) Elevated levels of serum interleukin-1 beta in combat-related post-traumatic stress disorder. Biol. Psychiatry., 42: 345–348.

Stamm, S. (2002) Signals and their transduction pathways regulating alternative splicing: a new dimension of the human genome. Hum. Mol. Genet., 11: 2409–2416.

Stansbury, K., Haley, D. and Koeneker, A. (2000) Higher cortisol values facilitate spatial memory in toddlers. Brief report. Ann. N. Y. Acad. Sci., 911: 456–458.

Steckler, T. and Holsboer, F. (2001) Interaction between the cholinergic system and CRF in the modulation of spatial discrimination learning in mice. Brain Res., 906: 46–59.

Stephens, R., Spurgeon, A., Calvert, I.A., Beach, J., Levy, L.S., Berry, H. and Harrington, J.M. (1995) Neuropsychological effects of long-term exposure to organophosphates in sheep dip. Lancet, 345: 1135–1139.

Sternfeld, M., Ming, G., Song, H., Sela, K., Timberg, R., Poo, M. and Soreq, H. (1998a) Acetylcholinesterase enhances neurite growth and synapse development through alternative contributions of its hydrolytic capacity, core protein, and variable C termini. J. Neurosci., 18: 1240–1249.

Sternfeld, M., Patrick, J.D. and Soreq, H. (1998b) Position effect variegations and brain-specific silencing in transgenic mice overexpressing human acetylcholinesterase variants. J. Physiol. Paris, 92: 249–255.

Sternfeld, M., Shoham, S., Klein, O., Flores-Flores, C., Evron, T., Idelson, G.H., Kitsberg, D., Patrick, J.W. and Soreq, H. (2000) Excess "readthrough" acetylcholinesterase attenuates but the "synaptic" variant intensifies neurodeterioration correlates. Proc. Natl. Acad. Sci. USA, 97: 8647–8652.

Stillman, M.J., Shukitt-Hale, B., Coffey, B.P., Levy, A. and Lieberman, H.R. (1997) In vivo hippocampal acetylcholine release during exposure to acute stress. Stress 1, 191–200.

Stitcher, D.L., Harris, L.W., Heyl, W.C., and Alter, S.C. (1978) Effects of pyridostigmine and cholinolytics on cholinesterase and acetylcholine in Soman poisoned rats. Drug Chem. Toxicol., 1: 355–362.

Tajima, T., Endo, H., Suzuki, Y., Ikari, H., Gotoh, M. and Iguchi, A. (1996) Immobilization stress-induced increase of hippocampal acetylcholine and of plasma epinephrine, norepinephrine and glucose in rats. Brain Res., 720: 155–158.

Tizabi, Y. and Calogero, A.E. (1992) Effect of various neurotransmitters and neuropeptides on the release of corticotropin-releasing hormone from the rat cortex in vitro. Synapse, 10: 341–348.

Tracey, K.J. (2002) The inflammatory reflex. Nature, 420: 853–859.

Tracey, K.J., Czura, C.J. and Ivanova, S. (2001) Mind over immunity. FASEB J., 15: 1575–1576.

Tsigelny, I., Shindyalov, I.N., Bourne, P.E., Sudhof, T.C. and Taylor, P. (2000) Common EF-hand motifs in cholinesterases and neuroligins suggest a role for $Ca2^{+}$ binding in cell surface associations. Protein Sci., 9: 180–185.

van Haaren, F., De Jongh, R., Hoy, J.B., Karlix, J.L., Schmidt, C.J., Tebbett, I.R. and Wielbo, D. (1999) The effects of acute and repeated pyridostigmine bromide administration on response acquisition with immediate and delayed reinforcement. Pharmacol. Biochem. Behav., 62: 389–394.

Vereker, E., O'Donnell, E. and Lynch, M.A. (2000) The inhibitory effect of interleukin-1beta on long-term potentiation is coupled with increased activity of stress-activated protein kinases. J. Neurosci., 20: 6811–6819.

Vyas, A., Mitra, R., Shankaranarayana Rao, B.S. and Chattarji, S. (2002) Chronic stress induces contrasting patterns of dendritic remodeling in hippocampal and amygdaloid neurons. J. Neurosci., 22: 6810–6818.

Weeber, E.J., Atkins, C.M., Selcher, J.C., Varga, A.W., Mirnikjoo, B., Paylor, R., Leitges, M. and Sweatt, J.D. (2000) A role for the beta isoform of protein kinase C in fear conditioning. J. Neurosci., 20: 5906–5914.

Weibel, L., Follenius, M. and Brandenberger, G. (1999) Biologic rhythms: their changes in night-shift workers. Presse Med., 28: 252–258.

Wilckens, T. and De Rijk, R. (1997) Glucocorticoids and immune function: unknown dimensions and new frontiers. Immunol. Today, 18: 418–424.

Winslow, J.T. and Camacho, F. (1995) Cholinergic modulation of a decrement in social investigation following repeated contacts between mice. Psychopharmacology (Berl), 121: 164–172.

Wolfe, A.D., Blick, D.W., Murphy, M.R., Miller, S.A., Gentry, M.K., Hartgraves, S.L., and Doctor, B.P. (1992) Use of cholinesterases as pretreatment drugs for the protection of rhesus monkeys against soman toxicity. Toxicol. Appl. Pharmacol., 117: 189–193.

Wolthuis, O.L., Groen, B., Busker, R.W. and van Helden, H.P. (1995) Effects of low doses of cholinesterase inhibitors on behavioral performance of robot-tested marmosets. Pharmacol. Biochem. Behav., 51: 443–456.

Xie, J. and Black, D.L. (2001) A CaMK IV responsive RNA element mediates depolarization-induced alternative splicing of ion channels. Nature, 410: 936–939.

Xie, J. and McCobb, D.P. (1998) Control of alternative splicing of potassium channels by stress hormones. Science, 280: 443–446.

Xu, L., Anwyl, R. and Rowan, M.J. (1997) Behavioural stress facilitates the induction of long-term depression in the hippocampus. Nature, 387: 497–500.

Yirmiya, R., Pollak, Y., Morag, M., Reichenberg, A., Barak, O., Avitsur, R., Shavit, Y., Ovadia, H., Weidenfeld, J., Morag, A., Newman, M.E. and Pollmacher, T. (2000) Illness, cytokines, and depression. Ann. N. Y. Acad. Sci., 917: 478–487.

Zhou, Z., Licklider, L.J., Gygi, S.P. and Reed, R. (2002) Comprehensive proteomic analysis of the human spliceosome. Nature, 419: 182–185.

Zigmond, M.J. (1997) Do compensatory processes underlie the preclinical phase of neurodegenerative disease? Insights from an animal model of parkinsonism. Neurobiol. Dis., 4: 247–253.

Zoghbi H.Y. (2003) Postnatal neurodevelopmental disorders: meeting at the synapse? Science, 302: 826–830.

T. Steckler, N.H. Kalin and J.M.H.M. Reul (Eds.)
Handbook of Stress and the Brain, Vol. 15
ISBN 0-444-51173-3

CHAPTER 4.11

Pathways and transmitter interactions mediating an integrated stress response

Colin D. Ingram*

Psychobiology Research Group, School of Neurology, Neurobiology and Psychiatry, University of Newcastle, Royal Victoria Infirmary, Newcastle NE1 4LP, UK

Abstract: Stress evokes a number of coincident neuroendocrine, autonomic, and behavioral responses which serve to avoid the damaging effects of the threat and restore homeostasis. Various techniques, notably those using immediate-early gene expression (IEG), have helped to define the neural networks which are activated by different types of stress stimuli and which serve to coordinate the different elements of the response. Although different stress modalities utilize distinct neural pathways each can be generalized to comprise: (i) a cognitive or sensory system which responds to introceptive or exteroceptive signals and includes a "threshold detector mechanism" which, when exceeded, will activate, (ii) a "response activating network" which distributes a stress signal to the various output systems to generate an appropriate response. Diversity and integration within this network is achieved through a hierarchy of overlapping pathways, with some having the capacity for direct, rapid activation in response to an immediate threat to homeostasis, while higher order pathways provide a distributed signal to several output systems. Certain transmitters have a preeminent role in regulating diverse aspects of the stress response (notable corticotropin-releasing factor) and it is suggested that the organization of the response activating network enables these transmitters to fulfil important roles in integrating the stress response.

This book is testament to the fact that in recent years the application of new pharmacological, neurochemical, anatomical, and molecular technologies to systems level neuroscience has dramatically advanced our understanding of the field of stress neurobiology. The availability of pharmacological tools to selectively modulate synaptic transmission and the ability to identify specific neurons activated by stressors, combined with our growing knowledge of neuroanatomical relations, has led to the point where it is possible to begin to provide an overall description of the neural circuits which serve to integrate the stress response. However, with understanding has also come complexity. The multimodal nature of the stress response and the increasing number of factors shown to modulate this response have added to the diversity of processes that need to be accommodated by any integrated system. In addition, the list of neurotransmitters and neuromodulators for which there is evidence for involvement in stress grows ever longer. The foregoing chapters in this section have provided in-depth reviews of the evidence for specific transmitter involvement in the response to stress, and a number of other recent reviews have provided insight into the overall organization of stress-related systems (Lopez et al., 1999; Sawchenko et al., 2000; Pacák and Palkovits, 2001; Vermetten and Bremner, 2002; Carrasco and van de Kar, 2003; Herman et al., 2003; Millan, 2003; Phillips et al., 2003). However, whilst these have considered the evidence that one or other particular pathway or transmitter is either affected by or affects the stress response, there remains a need to provide a conceptual framework for how stress systems operate and how different transmitters and pathways might contribute to an

* Tel.: +44 191 282 5678; Fax: +44 191 282 5108;
E-mail: c.d.ingram@ncl.ac.uk

overall integrated response. In the introductory chapter, Herman et al. (Chapter 4.1) provide an excellent overview of the current status of neural circuitry relevant to the stress activation of the hypothalamo–pituitary–adrenal (HPA) axis. Because of the well understood target for this circuitry (the paraventricular nucleus (PVN)) and the easily measurable index of response (plasma ACTH or corticosterone), this is perhaps the stress-activated system which is currently best understood. Once similar understanding is available for the other behavioral and physiological responses to stress it may be possible to map out the complete integrated system, which generates a coordinated response to stressful stimuli. Since we are not yet at this stage, this chapter will not adopt a strictly hodological (brain connectivity) approach to integration between the various neural systems subserving the response to stress, but will discuss some generalized models for integration. Furthermore, this consideration will be restricted to the response to acute stress stimuli, rather than chronic or repeated stresses which can lead to adaptive processes in stress circuits. Whilst this approach runs the risk of being too simplistic, it serves to highlight some basic attributes of stress–responsive circuits and the role that particular transmitters may have. These models will consider the hierarchical nature of different circuits and will propose the possible organization of a network that would account for the apparent coordinating role of one particular transmitter, corticotropin-releasing factor (CRF).

Defining the stress response and determining the underlying circuits

Stress–response systems

At a systems level the central stress response can be considered in terms of a classical reflex arc: the afferent limb comprising a variety of sensory modalities (nociceptors, chemoreceptors, auditory and visual systems, etc.), as well as cognitive neural circuits that underlie emotional processes, while the efferent or effector limb comprises appropriate behavioral, physiological, and autonomic responses that subserve a number of defensive, protective, and adaptive functions (Fig. 1). In this respect, stress may be considered an extension of homeostasis and all homeostatic pathways potentially can contribute to the integrated stress system. Whilst this view touches on the much debated boundary between Cannon's concepts of homeostasis and Selye's "stress syndrome" considered earlier in this volume (Chapters 1.1 and 1.2) and elsewhere (Pacák and Palkovits, 2001), it does serve to highlight two important features of an integrated stress–response system. Firstly, there are no specific "stress receptors" and thus the neural origin of any stress response are normal sensory and emotional pathways. However, within these pathways there may be thresholds that determine whether a specific level of activity, or condition that is perceived to endanger the wellbeing of the individual, alters the stimulus so that it takes on a "stressful" quality. Secondly, the output is not through a single effector system. Therefore, an integrated stress–response system should not be defined on the basis of either a generalised, non-specific activation (Selye), or of one particular behavioral, autonomic, or neuroendocrine measure (often determined by the specialism of the researcher). With this in mind it is possible to construct a simple integrated stress–response system in which sensory and cognitive/emotional inputs feed into a central "response activating network" that, in turn, provides the appropriate signals to generate a coordinated output (Fig. 1).

The output or effector limb of the stress response involves widespread effects on endocrine systems which can signal a whole body reaction. These include changes in the hypothalamo–pituitary–gonadal axis, oxytocin, prolactin, growth hormone, and renin-angiotensin, as well as HPA activation (van de Kar and Blair, 1999; Tsigos and Chrousos, 2002; and see Chapter 1.4). In this respect there is considerable participation of hypothalamic tubero-infundibular neurons, which synthesize and secrete hormones or regulatory factors (e.g. CRF, GnRH, TRH, GHRH, dopamine). Stress also modulates autonomic outflow, including sympathetic changes to cardiovascular, respiratory, metabolic, renal, and sweat gland function, and vagal and sacral para-sympathetic efferents, which control gut motility. The sympathoexcitatory response will involve activation of the supraspinal innervation of the

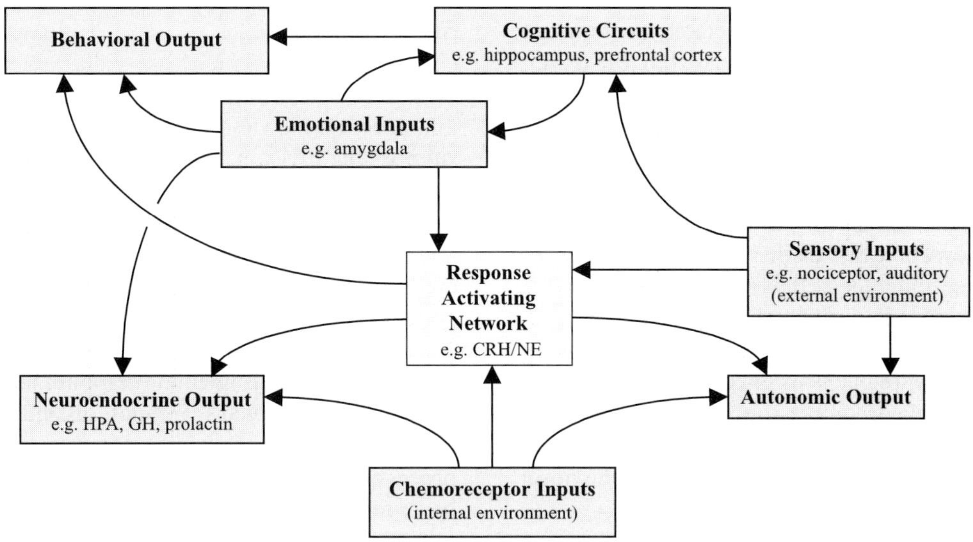

Fig. 1. Model of the basic organization of the central stress–responsive network. Three main input systems provide the trigger for stress responses: emotional/cognitive inputs reflect the psychological state of the individual, chemoreceptor/mechanoreceptor inputs provide signals about the internal environment, and sensory inputs respond to various external stimuli. Three major output systems encompass those regulating behavior, neuroendocrine, and autonomic activity. The inputs can interact directly with the outputs as part of normal function in the form of a reflex arc. However, when the inputs pass a threshold level to trigger a stress reaction a central response activating network activates the different output systems to achieve a coordinated response.

cholinergic preganglionic neurons which principally arises from the ventrolateral and ventromedial areas of the rostral medulla, the spinal projections of the raphe nuclei, the A5 (ventrolateral pons) noradrenergic neurons, and a population of dorsomedial parvocellular neurons in the PVN (Senba et al., 1993; Pacák and Palkovits, 2001; Sved et al., 2001). Lesser hypothalamic pathways also arise in the lateral hypothalamus and arcuate, perifornical, and dorsomedial hypothalamic (DMH) nuclei. Many of these descending projections are peptidergic in phenotype (CRF, vasopressin (AVP), oxytocin, somatostatin, enkephalin, and atrial natriuretic peptide) and relay via the catecholaminergic neurons in the medulla or terminate directly on the preganglionic neurons in the thoracolumbar intermediolateral cell column (Palkovits, 1999). Although stressors may cause simultaneous activation of many of these descending pathways, anatomical tract tracing and immediate-early gene (IEG) expression studies indicate that functional subgroups exist within this network, allowing for variations in the response.

The final effector system of the stress response is that of behavior. Of all the components of the stress response, behavior is the most species specific and, like the majority of the data in this review, what is considered here are data mainly obtained from the rat. However, although the expression of stress behavior may vary in other species, the underlying circuitry is likely to be common. This circuitry generates an appropriate emotional coping strategy. These may be divided into "active" coping strategies (confrontation, flight, flight) that are particularly adaptive if the stress is escapable and are characterized by coincident sympathoexcitation (hypertension, tachycardia), and "passive" coping strategies (immobility, hyporeactivity) appropriate for inescapable stress and characterized by coincident sympathoinhibition. Either strategy also encompasses a number of displacement activities which, in the rat, involve considerable orofacial activity (grooming, chewing). The medial hypothalamus, amygdala, and midbrain periaqueductal gray (PAG) constitute the main neural substrates for the integration of aversive states, and distinct divisions mediate the different forms of emotional coping and their associated patterns of autonomic activity. For example, within the PAG, excitation of the lateral region generates active coping, while the ventrolateral region generates passive coping (Bandler et al., 2000). Recent evidence has also shown that both these regions of the PAG communicate with the medial prefrontal cortex (mPFC) and parabrachial area in topographically distinct parallel circuits to generate these different behavioral coping strategies (Bernard and Bandler, 1998). Importantly the parabrachial area is also a major relay for viscerosensory information from the nucleus of the tractus solitarius (NTS) and spinal cord going to the central amygdala (CeA) and bed nuclei of the stria terminalis (BNST) and, thus, may provide the link between behavioral and autonomic responses.

The appropriate combination of these varied adaptive outputs allows the animal to resist the effects of stress. In this respect although activation of the HPA axis outside the normal daily pattern has been considered to be the principal characteristic of a stress response, it should not be forgotten that there are instances in which stress does not result in increased HPA activity. For example, certain acute stressors like hypercapnia can cause acute anxiety and sympathoactivation with no HPA response, and repeated restraint stress results in dissociation of the various outputs, with habituation of both the HPA response and associated the expression of Fos in the PVN, while the autonomic response (tachycardia) persists, indicating the continued stressful quality of the stimulus (Stamp and Herbert, 1999). Furthermore, any stress–responsive system needs to be sufficiently flexible to accommodate the varying patterns of neuroendocrine activity in response to different stimuli; e.g. ether stress will increase ACTH, AVP, and oxytocin, while restraint will increase ACTH and oxytocin but not AVP, and cold stress affects ACTH neither oxytocin nor AVP (Gibbs, 1984). Thus, whilst the response activating network which orchestrates the appropriate output may be illustrated by a single box in a model of integration (Fig. 1), this does not mean that it exists as a unitary mechanism. Indeed, considerable evidence argues against the concept of a unitary stress system, most notable being the evidence of different stress categories or modalities (Li, et al., 1996; Dayas et al., 2001a; Pacák and Palkovits, 2001; Herman et al., 2003).

Stress modalities

As well as the separation of stressors into escapable and inescapable described above, a wealth of data support the division of stress stimuli into two broad categories. The first category has been variously termed physical, systemic, introceptive, or reactive, and covers stimuli such as hemorrhage or other cardiovascular challenges, challenges to the immune system (infection), pain, respiratory stimuli (e.g. CO_2 or ether), and heat stress. Broadly speaking these stimuli require an immediate response and the underlying neural pathways are not dissimilar to a fast reflex arc, with little dependence on higher order processing. The second category of stimuli has been termed emotional, psychological/psychogenic, processive, exteroceptive, or anticipatory, and includes restraint, fear (conditioned avoidance), novel environment (e.g. open field or plus maze), swim stress, noise, motion stress, visual, predator stimuli, psychosocial stress, and conflict. The important feature of these stressors is the dependence on information processing and on prior experience, and thus the underlying neural pathways incorporate several cognitive areas of the telencephalon. Obviously stressors may convey varying proportions of these different modalities (e.g. footshock not only has a painful physical component but also an emotional reaction to the unpredictable stimulus from the animal's environment) and so the boundary may not appear precise. This spectrum has led others to include further subdivisions of stress responses (e.g. physical, psychological, social and cardiovascular/metabolic: Pacák and Palkovits, 2001), however, a binary division remains the best validated. Until recently this categorization was based on subjective and anthropomorphic views as to the attributes of the stress. However, an increasing body of objective measures confirm this distinction, including neurotransmitter responses to different stressors (e.g. Inoue et al., 1994; Adell et al., 1997), the regional pattern of activation of different brain areas (e.g. Li et al., 1996; Emmert and Herman, 1999; Dayas et al., 2001a; Herman et al., 2003), and the selective effects of different stressors on physiological and neuroendocrine responses (e.g. Gibbs, 1984; Sawchenko et al., 2000; Pacák and Palkovits, 2001). The following sections will consider the identity of the neural systems which may integrate responses to these different stress modalities.

Immediate-early gene (IEG) mapping of stress–responsive circuits

Since the early 1990s, studies of the patterns of cellular activation using the expression of IEGs (e.g. *c-fos*, *fos-b*, *jun b*, NGFI-A, NGFI-B, *fra-2*) and their protein products have provided considerable insight into the pathways activated by a wide range of stressful stimuli. However, before considering how these data have contributed to understanding an integrated stress–responsive circuit, it is important to note a few technical limitations. Firstly, for practical reasons, the areas of the brain analyzed are frequently preselected. Thus, the majority of studies have focused on the activation of areas with well-established roles in stress neurobiology (e.g. PVN, amygdala, locus coeruleus (LC)), although a few have mapped wider regions of the brain, allowing for the identification of novel responsive areas (e.g. Smith et al., 1992; Senba et al., 1993; Cullinan et al., 1995; Elmquist et al., 1996; Campeau and Watson, 1997; Emmert and Herman, 1999; Palkovits, 1999; Windle et al., 2004). Secondly, most studies have examined the *c-fos* gene or its product Fos. Whilst comparisons with other IEGs have shown similar patterns of activation (e.g. Stamp and Herbert, 1999), the *c-fos* gene may not be always activated by stress. For example, swim stress increases NGFI-A (*zif/268*) but not *c-fos* mRNA in the lateral BNST and peripeduncular nucleus (Cullinan et al., 1995). Thirdly, only a minority of studies have determined either the neurochemical identity or afferent connections of the neurons activated by stress, and where this has happened most studies have focussed on a limited number of transmitters, such as CRF, noradrenaline and serotonin (5-HT) (e.g. Ceccatelli et al., 1989; Pezzone et al., 1993; Grahn et al., 1999; Ishida et al., 2002; Helfferich and Palkovits, 2003; see also Chapters 4.1, 4.4, and 4.6). However, accepting these limitations, studies of IEG expression have established a number of important characteristics of the stress–responsive network.

One of the most important features of stress circuitry established by IEG mapping is the existence of modality-specific pathways. Thus, while some areas exhibit consistent activation by stress and may be considered to be the signature of a stressful stimulus (e.g. PVN), others show dependence on the two major categories of stressors described above. This is particularly the case for telencephalic regions and reflects the varying contribution of cognitive processing to the response. Therefore, while two psychological stressors, swim stress and restraint, evoke largely similar regional patterns of gene expression (Cullinan et al., 1995), exposure to an open field (psychological stress) evokes much greater activation of the medial amygdala (MeA) and lateral septum than ether vapour (physical stress), with the reverse being the case in the parietal cortex (Emmert and Herman, 1999). Indeed the amygdala displays a high degree of modality dependence, in that physical stressors (hemorrhage and immune challenge) elicit Fos expression in neurons of the CeA while psychological stressors (noise, restraint and forced swim) primarily have effects on neurons of the MeA (Dayas et al., 2001a; Herman et al., 2003). Thus, it has been well established that the psychological stress of restraint (as distinct from complete immobilization) will induce Fos protein or *c-fos* mRNA expression in the MeA rather than CeA (e.g. da Costa et al., 1996; Dayas et al., 1999), and ibotenic acid lesions of the MeA (but not CeA) will reduce restraint-induced activation of the PVN (Dayas et al., 1999). However, this dichotomous circuitry is not limited to the amygdala. For example, although all stressors recruit both A1 and A2 (noradrenergic) and C1 and C2 (adrenergic) neurons, physical stressors appear to target a more rostral population of A1 and A2 neurons than psychological stressors (Dayas et al., 2001a). Likewise, within the PAG physical stressors (hemorrhage and pain) increase Fos in the ventrolateral area, consistent with involvement in a passive coping strategy, while psychological stressors (restraint and swim stress) increase Fos in the lateral area, consistent with its role in active coping (Bernard and Bandler, 1998). These differential patterns confirm the plurality of the circuits which serve to integrate the appropriate stress response.

A second feature of stress circuits revealed by IEG mapping is the differentiation between those areas which are activated by the sensory stimulus (i.e. the afferent pathway), and those which are only activated when there is a stress response (i.e. the response activating network) (Fig. 1). In this respect it should be remembered that there are no specific stress receptors and it is necessary to consider the mechanisms by which a normal sensory (or cognitive) pathway may lead to the activation of the stress response. This is likely to involve a threshold detection mechanism. For example, auditory stimuli will not activate a stress response until the intensity (volume), frequency, and/or duration of the stimulus reaches a threshold when it is considered to be stressful. The pathways underlying this transition have been elegantly illustrated by Campeau and Watson (1997) using measures of plasma corticosterone and regional *c-fos* mRNA expression following exposure of rats to differing intensities of white noise. Areas of the brain in which *c-fos* mRNA indicated activation could be classified into three categories: (i) those which were part of a general arousal system and which were similarly activated by the experimental conditions, irrespective of the intensity of the noise (e.g. anterior cortical amygdala, basolateral amygdala (BLA), and MeA, anteroventral and mediodorsal thalamus, cingulate and piriform cortices); (ii) those which were part of the normal auditory pathway and which showed graded increases in gene expression as a function of the noise intensity, irrespective of whether the intensity was below or above the threshold required for HPA activation (e.g. cochlear nuclei, nuclei of the inferior colliculus; medial geniculate nucleus, the superior olivary complex, auditory cortex); and (iii) those areas considered to be part of the stress processing pathway which were only excited at intensities of 90 dB and above, at which point plasma corticosterone levels indicated that the HPA axis had become activated. These areas included the medial and ventral nuclei of the BNST and adjacent septohypothalamic area/ventrolateral septum (SHy/VLS), the ventral dentate gyrus, lateral and medial preoptic areas (MPOA), median raphe nucleus (MRN), pedunculopontine tegmentum (PPTg), and PVN. Although the neurochemical identities of the neurons comprising this latter category were not determined, it is likely that the PVN neurons will have included CRF neurons driving the HPA response. Furthermore, it is interesting that in this and studies of other stressors (e.g. Arnold et al., 1992; Senba et al., 1993; Chen and Herbert, 1995; Cullinan et al., 1995; da Costa et al., 1996; Stamp and Herbert, 1999; Windle et al., 2004) the BNST and SHy/VLS regions have been shown to be activated, since these regions are the origin of GABAergic relays to the PVN (Sawchenko and Swanson, 1983; Herman et al., 2002) and may serve as important stress processing areas, integrating inputs from the amygdala (Weller and Smith, 1982). The BNST is also a site that projects to autonomic centers in the brainstem and may be important for coordinating limbic control of cardiovascular function (Loewy, 1991). Activation of the MRN suggests the involvement of serotonergic neurons, while the PPTg neurons are likely to be cholinergic, both of which have been implicated in regulating arousal. Thus, each of these areas may constitute a part of the network coordinating the different components of the stress response. Although Campeau and Watson (1997) provided a schematic of the connections between these different auditory-, arousal-, and stress-related systems, it was not possible to conclude at what point in this putative stress circuit the normal response to an auditory signal had been converted to a stress response and, therefore, the identity of the threshold detector mechanism remains to be determined.

Elements of the response activating network

Evidence from a large number of anatomical, recording, and lesioning studies have identified areas which may be considered to be the principal constituents of the response activating network. Each of these areas contributes to the network function by integrating specific types of signals, by distributing signals throughout the network, or by contributing to the output function. The following sections briefly summarize these different integrative functions.

Medial hypothalamus: PVN and DMH

The PVN is often considered to be the principal nucleus mediating efferent responses to stress, as both the HPA axis and autonomic responses originate here (see Chapter 4.1 for detailed review). Indeed activation of the stress-related neurons of PVN may be regarded as the signature without which a stimulus cannot be considered stressful. Within the PVN CRF-containing parvocellular neurons are activated by both physical and psychological stress stimuli, reflecting the neuroendocrine component of the response (Ceccatelli et al., 1989; Helfferich and Palkovits, 2003). In addition, efferents containing oxytocin, AVP, and possibly neuropeptide Y (NPY) arising from the dorsal parvocellular region project both to the brainstem and directly to the spinal preganglionic neurons in the intermediolateral cell column, where they are involved in integrating autonomic (sympathetic) responses (Coote et al., 1998). Further integration of autonomic, neuroendocrine, and behavioral responses to emotional stress may also occur in the adjacent area of the DMH (DiMicco et al., 2002), and both the PVN and DMH constitute part of the hypothalamic defense area which interconnects with other medial hypothalamic areas to generate appropriate responses to threatening stimuli (Canteras, 2002). Many of the afferents to the PVN (at least those subserving HPA activation) relay in a ring of GABAergic neurons in the sub-PVN region, peri-PVN, BNST, MPOA, ventromedial hypothalamus, and DMH (Herman et al., 2002). However, the PVN also receives direct inputs from circulating factors in the blood and cerebrospinal fluid and in this respect some factors which induce stress responses (e.g. immune factors or glucoprivation) may involve a very short neural pathway.

Extended amygdala

As noted above and reviewed in Chapter 6.2, the CeA and MeA are involved in processing different stress modalities, and for relaying stress signals to other limbic and brainstem areas (Davis et al., 1994). Electrical stimulation of the CeA evokes both signs of arousal and sympathetic activation, and lesions of the CeA decreases HPA activation during immobilization and blocks fear conditioning (Hitchcock and Davis, 1986). While the MeA and CeA are involved in amygdaloid output, the BLA is involved in processing afferent sensory information and plays an important role in integrating inputs from various cortical and thalamic sites which are relayed via intra-amygdaloid connections to other subnuclei. The BLA may also play an important role in associative learning which underlies fear conditioning and anxiety. This integrative role is illustrated by the ability of neurotoxic lesions of the BLA and CeA to block the effect of a conditioned fear stress on dopamine, noradrenaline, and 5-HT transmission in the mPFC and stress-induced freezing, defecation, and HPA activation (Goldstein et al., 1996). The BNST is considered to be part of the extended amygdala and, as such, there are marked nucleus-specific associations: i.e. the CeA is connected to lateral BNST (oval nucleus), while the MeA is connected to the anterodorsal/medial nuclei of the BNST, both through the stria terminalis and ventral amygdalofugal pathways. Importantly the BNST appears to be the relay for inputs to the PVN-controlling tonic HPA activity (Herman et al., 1994) and in this respect plays an important role in determining basal HPA activity on which stress acts. The anterolateral BNST is also involved both in higher order activation of stress–responsive circuits and in the generation of coping behaviors.

Hippocampus

The precise functions of the hippocampus in stress responding are somewhat controversial, but are principally twofold. Firstly, it is an area important for learning and memory, and thereby contributes to the processing of stressful information, The hippocampus has a particularly important role in episodic memory for the emotional context of memories and plays a role in regulating behavioral responses to threatening environmental contexts (Phillips and LeDoux, 1992). In this respect, although Fos expression indicates that the hippocampus is activated by various stress conditions, it appears to be principally reactive to exteroceptive (psychogenic) stress and not the interoceptive (physical) stresses. Secondly, the hippocampus provides inhibitory control over the HPA axis and may mediate its effects on the PVN via inhibitory connections to subcortical areas (Herman et al., 2002, 2003). There has also been considerable emphasis placed on the hippocampus as a site for glucocorticoid negative feedback (see Chapter 3 of this volume). This inhibition involves both basal (tonic) control and the termination of responses to stress. Indeed, lesions of the hippocampus result in prolonged HPA responses to psychological (processive) stressors, but not physical (reactive) stressors (Herman et al., 2003), consistent with its involvement in cognitive processing. However, the role of the hippocampus and subiculum in regulating basal and stress-induced HPA activity has been questioned (e.g. Tuvnes et al., 2003). In part this may relate to misperceptions of the hippocampus being a unitary structure with little regional specialization. In this respect, within the septohippocampal complex it is an area in the ventral subiculum that appears to provide the main output, at least in respect of regulating the PVN and HPA axis activity (Herman et al., 1998). The outflow from this ventral subicular area is glutamatergic and innervates a number of areas including the mPFC and the GABAergic neurons in the peri-PVN region, BNST, and hypothalamus (MPOA and DMH).

Septum

The septal area has two major functions in stress responding. Firstly, it regulates hippocampal function through the septohippocampal pathway and this connection is important in controlling the termination of stress responses. Secondly, it functions as a relay for limbic afferents that pass through the SHy/VLS, an area which is consistently activated by the psychological stress of restraint, noise, or novelty (da Costa et al., 1996; Campeau and Watson, 1997; Emmert and Herman, 1999), but not by physical stressors.

Prefrontal cortex (cingulate cortex)

Various cortical areas display increased Fos expression following psychological stress, but not after physical stress, most important among which is the mPFC (see Chapter 6.3). Thus, expression of Fos in this region occurs in response to restraint (Chen and Herbert, 1995; Cullinan et al., 1995) and noise (Campeau and Watson, 1997), but not in response to other stressors, like lipopolysaccharide (LPS) (Elmquist et al., 1996; Yokoyama and Sasaki, 1999) or osmotic challenge (Sharp et al., 1991). There is increasing interest in the PFC as an area integrating both the stress-induced activation of the HPA axis and glucocorticoid negative feedback with executive function and cognitive processing (Sullivan and Gratton, 2002). Furthermore, the PFC plays an important role in generating an active coping strategy in response to a stressful stimulus (Giorgi et al., 2003), and lesions of the mPFC markedly alter stress responses, primarily affecting responses to emotional stressors (Diorio et al., 1993; Sullivan and Gratton, 1999). Indeed lesions of the mPFC increase both the HPA response to restraint stress and the expression of *c-fos* mRNA in various areas including PVN, MeA, ventrolateral BNST, and piriform and dorsal endopiriform cortices (Figueiredo et al., 2003), consistent with the mPFC being a higher order area providing inhibitory control over pathways mediating psychological stress. In this respect the PFC has major connections to several areas involved in stress responding, including the amygdala, paraventricular thalamic nucleus, raphe nuclei, NTS, and the GABAergic interneurons in the BNST and hypothalamus that project to the PVN. It may, therefore, have a higher order function to regulate the activity of much of the response activating network, perhaps setting an emotional tone to information processing.

Locus coeruleus (LC) and A1/A2 medullary nuclei

Numerous studies have demonstrated the activation of brainstem catecholaminergic neurons in the integrated response to stress and this is extensively reviewed in Chapters 4.3, 4.4, and 4.5 of this volume. The ventrolateral medulla (A1), dorsomedial medulla/NTS (A2), ventrolateral pons (A5), and LC (A6) all display increased IEG expression following stress. Increased Fos immunoreactivity or *c-fos* mRNA is observed in response to psychological stressors, such as restraint, noise, and forced swim (Cullinan et al., 1995; Campeau and Watson, 1997; Stamp and Herbert, 1999; Dayas et al., 2001b; Helfferich and Palkovits, 2003), or physical stressors, such as intraperitoneal injection of hypertonic saline, hypercapnia, and hemorrhage (Ceccatelli et al., 1989; Haxhiu et al., 1996; Dayas et al., 2001b), as well as following the mixed stimulus of footshock

(Pezzone et al., 1993) or the fear conditioned by exposure to footshock (Ishida et al., 2002). Combined immunocytochemistry for tyrosine hydroxylase has confirmed the noradrenergic phenotype of many of these neurons, although other noncatecholaminergic neurons are also activated, including GABA interneurons (Ishida et al., 2002). The known connectivity of these nuclei have implicated them in several integrative functions, particularly in relation to relaying primary viscerosensory information (A1/A2) and in sympathetic activation (A1, A2, and A5). The LC–noradrenaline system also has major roles in arousal and in the initiation and maintenance of forebrain neuronal activity appropriate for the collection and processing of salient stress-related information through diverse sensory, attention, and memory circuits (Berridge and Waterhouse, 2003).

Raphe Nuclei

As reviewed extensively in Chapter 4.6, both the MRN and dorsal raphe nucleus (DRN) show responses to stress stimuli and, since these nuclei provide the majority of 5-HT innervation to the brain, increased serotonergic neurotransmission is implicit in their activation. However, the attributes of the stress stimuli have major effects on the pattern of responses in the two nuclei. Stressors, which are uncontrollable or which evoke conditioned fear, cause activation of DRN neurons. For example, inescapable tailshock or footshock leads to an increased expression of Fos in 5-HT-immunoreactive DRN neurons (Pezzone et al., 1993; Grahn et al., 1999), and fear conditioned by exposure to footshock induced Fos expression in both 5-HT and GABAergic neurons in the DRN (Ishida et al., 2002). In contrast, swim stress may reduce forebrain 5-HT release through inhibition of raphe neurons, and this is consistent with the fact that the increased Fos expression in the DRN occurs primarily in neurons immunoreactive for GABA in the dorsolateral subdivision and few have a 5-HT phenotype (Roche et al., 2003). Thus, while psychological stressors are effective in increasing Fos expression in both the raphe and mPFC, an immune stress (LPS) which has little or no psychological component, has no effect in either structure (Elmquist et al., 1996). However, despite this apparent lack of Fos expression in the raphe, LPS does cause a marked increase in 5-HT neurotransmission in the hippocampus (Linthorst et al., 1995), suggesting either that the increase in transmission arises from an activity independent mechanism, or that the *c-fos* gene is not induced by this particular stimulus, or that a selective subpopulation of raphe neurons is affected. Indeed different stress modalities may selectively affect median versus dorsal raphe, or specific subpopulations of DRN neurons. For example, inescapable sound stress activates serotonergic neurons in the MRN, but not DRN, and selectively increases 5-HT turnover in MRN projection areas (hippocampus, nucleus accumbens (NAc), and cortex) but not in the caudate nucleus which receives input from the DRN (Daugherty et al., 2001). This functional separation may be important for integrating appropriate behavioral responses. It has been suggested that activation of DRN in response to acute threatening situations facilitate cognitive processes in the amygdala that evaluate the threat, whilst inhibiting escape behavior mediated via the PAG (Graeff et al., 1996). In contrast, the MRN–hippocampal system may promote resistance to unavoidable stress by disconnecting aversive events from other cognitive and behavioral processes. A recent review of these functional subpopulations suggested that 5-HT neurons within the middle and caudal regions of the DRN may project to multiple components of the extended amygdala to facilitate various components of the stress response, while connections of the MRN with structures involved in the efferent pathway from the ventral subiculum may provide the control over the adaptation or coping in response to stress (Lowry, 2002).

A hierarchical model of integration within the stress–responsive network

With increasing knowledge of areas activated by stress it has become possible to define the specific pathways underlying the response to individual stressors, including immobilization, cold, hemorrhage, hypoglycemia, and pain (Pacák and Palkovits, 2001), and auditory stress (Campeau and Watson, 1997). However, from the foregoing description of the major elements of the stress–responsive network, it is clear that many areas fulfil several functions or are only recruited under particular conditions. Therefore, simple static pathways, akin to fixed electrical circuits, may not be appropriate to explain the organization of a network which can respond to different stressors and which generates an integrated stress response. It is perhaps more appropriate to consider stress responses as generated by a number of parallel or overlapping pathways. Figure 2 illustrates a hypothetical model of this form of organization. Within this model, intrinsic and extrinsic signals are processed by cognitive and sensory systems, but input to the response activating network does not occur until they reach a threshold for activation. Output from the network is generated by well-defined motor neurons that generate specific behavioral and/or physiological effects (e.g. CRF neurons driving HPA activity or sympathetic cholinergic neurons regulating cardiovascular activity). Processing between these two occurs by at least three parallel streams:

- Pathway (1) represents a simple, direct pathway where output is principally focussed on one (or a few) specific output system(s) that has high homeostatic value, although other outputs may play a minor role. These pathways are similar to reflex arcs and are particularly common for rapid responses to physical stressors.
- Pathway (2) represents a more complex distributed network which provides coordinate activation of several output systems. This organization is common for the higher order pathways subserving psychological stressors.
- Pathway (3) represent the highest order pathway which has important modulatory effects on the output systems, but may not transmit the primary stress signal. The overlapping of these higher order pathways generates the diversity of potential responses, and explains the existence of convergence on output systems which has been extensively reported by studies of retrograde labelling of stress-activated pathways.

This model, in which integration is achieved through a combination of divergent and convergent streams, is compatible with the principles of integration, which Herman and colleagues have suggested might regulate the HPA axis. In this they propose that circuits subserving "anticipatory" responses are superimposed on the direct "reactive" (reflex) pathways to converge on the PVN, and have applied the term "hierarchical" to describe this pattern of organization (Herman et al., 2003; Chapter 4.1). This is a highly appropriate term, since,

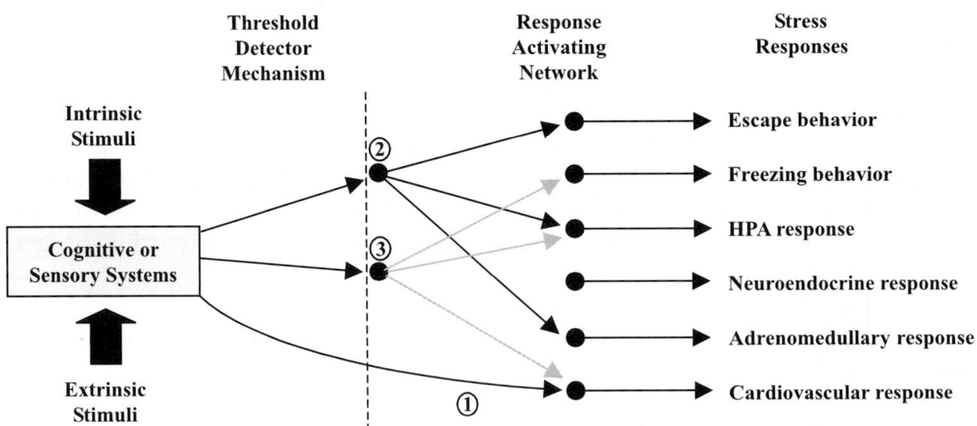

Fig. 2. Hierarchical model for integrated responses to stressful stimuli. Cognitive and sensory systems function normally to respond to intrinsic and extrinsic signals but without generating a stress response. However, a detector mechanism operates to establish a threshold above which a signal is passed to a response activating network to generate an appropriate stress response. The response activating network comprises proximal components which provide fast input to systems that generate specific physiological or behavioral responses, and distributed networks which function to integrate a number of different responses. Depending on the nature of the stressful stimulus the threshold detector mechanism may either directly activate the proximal system (e.g. (1)) or may have a more indirect effect, activating overlapping higher order systems that generate the appropriate repertoire of responses (e.g. (2) and (3)).

while the model illustrated in Fig. 2 shows only a relatively small number of layers, the situation in reality may involve a more complex hierarchy. For example, it has been suggested that emotional perception of stress is processed through at least two parallel systems: a ventral system, comprising amygdala, insula, cingulate gyrus, and PFC, which is important for identifying the emotional significance of stimuli and the production of affective states and the accompanying autonomic responses; and a dorsal system, comprising the hippocampus and dorsal areas of the cingulate and PFC, which is responsible for executive functions, selective attention, and effortful regulation of affect (Phillips et al., 2003). Importantly, within this model the overall integrated response to a stress is not predetermined but is a function of the number and intensity of the different overlapping systems which are recruited. Furthermore, convergence may alter the pattern of stress responding at a molecular level. In this respect, it is interesting to note that recent data using oligonucleotide microarrays to compare the gene expression profiles in the PVN evoked by LPS (physiological stress) and restraint (psychological stress) have shown that, while several neuropeptides (e.g. orexin, preproenkephalin, NPY) show common changes in expression, many of the transcription factors induced were stimulus specific (Reyes et al., 2003). Thus, while the distinct pathways which convey the different stress modalities may evoke apparently similar HPA and sympathoadrenal activation, differences in the postreceptor signalling cascade evoked by the different convergent pathways generates separate molecular responses.

As previously mentioned, a critical regulatory point in any model of stress responding is the gating or threshold detector mechanism that allows signals from normal sensory or cognitive pathways to enter the response activating network. This mechanism is likely to differ for the different processing streams. For responses to physical stressors, this gating could be a simple electrophysiological or signalling threshold which determines whether or not a neuron receiving the sensory information will fire and transmit the signal into the stress circuit. However, it is important to note that this threshold is likely to: (i) vary between pathways allowing for differential recruitment; (ii) show graded signal transduction allowing for the variable levels of output observed for many stressors; and (iii) vary between individuals and between different circumstances, leading to observed differences in responses. For pathways that respond to psychological stressors, threshold detection will involve more extensive circuitry that engages cognitive processing which determines whether or not a stimulus constitutes a threat to the wellbeing of the individual and which ensures the response is contextually appropriate (Phillips et al., 2003). Therefore, gating cannot be a simple threshold but involves appraisal and identification of the stimulus salience. In respect of many sensory stimuli the threshold detector is likely to reside either in the thalamus or in the brainstem reticular activating system, which act as relays for pathways either to the sensory cortex for normal processing, or to the amygdala for emotional responses.

Transmitter involvement in stress-reactive circuits

The areal (or regional) approach to determining integration within stress reactive circuits, considered above, is based primarily on known connectivity between and functionality of various activated brain nuclei. However, an alternative view to integration within stress circuits is based on the multiple functions of specific transmitters. Whilst these complementary hodological (connectivity) and neurochemical approaches have considerable overlap, there are a number of reasons for considering them separately. This is particularly the case where a transmitter shows widespread and synchronous release, or where a transmitter participates in several components of the stress response and, therefore, it is not possible to ascribe a specific site of action. Furthermore, for many transmitters their precise anatomical role in the pathways between stress stimulus and stress response remains to be determined.

At least four approaches have been successfully used to define transmitter involvement in the integration of stress responses. These are: (i) pharmacological approaches based primarily on whether agonists or antagonists of particular transmitter pathways selectively modify the various measures of stress; (ii) neurochemical approaches, which include measurement of changes in transmitter release either by in vivo or ex vivo methodology during the course of stress. This approach has particularly benefited from the development of real-time measurements that can be made simultaneously with behavior or other physiological responses, including microdialysis, voltammetry, and immunosensors; (iii) neuroanatomical approaches that determine the chemical identity of neurons activated by stressful stimuli, such as combined IEG and immunocytochemistry; and (iv) genetic approaches, which include identifying genes which show changes in expression during stress, or demonstrating the effect that transgenic methods of under- or over-expression have on the stress response.

The ways in which a particular transmitter may be considered to be "involved" in the functioning of an integrated stress–responsive network are varied (Fig. 3). Whilst the involvement usually implies a contribution to direct transduction pathways, transmitters may also fulfil important modulatory or permissive functions (see 6 and 7 in Fig. 3). Furthermore, it is important to distinguish between neurotransmitter responses which are integral to driving the stress response and those which arise from the effects of stress (i.e. secondary to autonomic, behavioral, or neuroendocrine activation; transmitter 8 in Fig. 3). This is particularly important in respect of the effects of glucocorticoids which themselves may have an integrating role (see Chapter 3). Thus, while the

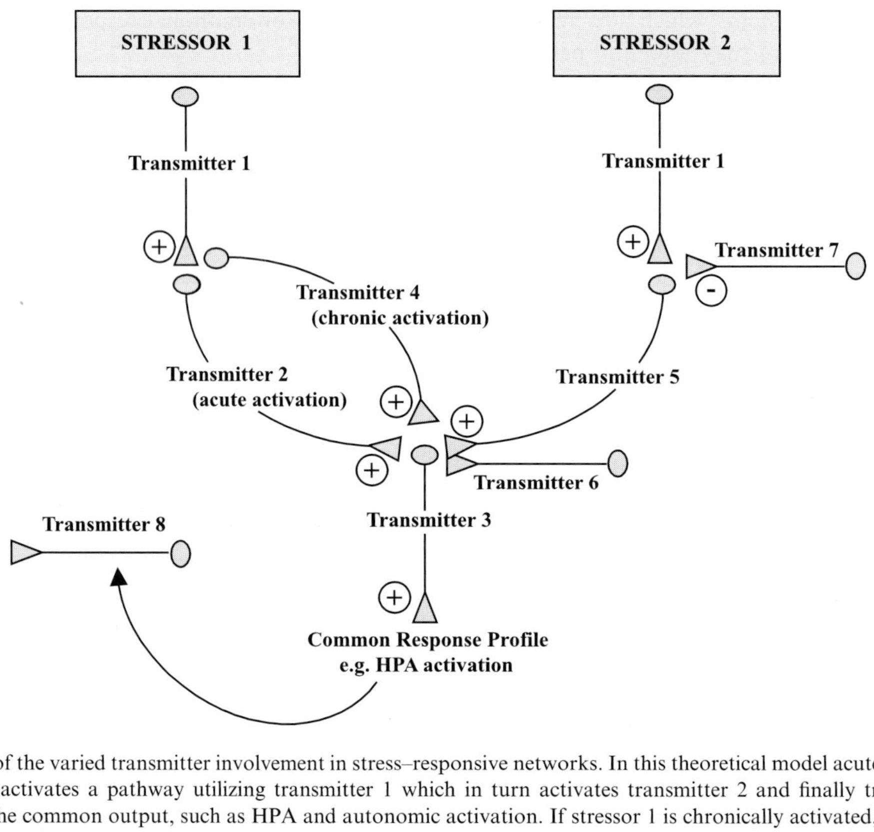

Fig. 3. Examples of the varied transmitter involvement in stress–responsive networks. In this theoretical model acute exposure to a first stress (stressor 1) activates a pathway utilizing transmitter 1 which in turn activates transmitter 2 and finally transmitter 3 which controls some of the common output, such as HPA and autonomic activation. If stressor 1 is chronically activated, then transmitter 4 becomes activated and may assume some of the function of transmitter 2. The effects of a second stressor may also be mediated through transmitter 1 but in a separate neuronal population. In this case it is able to activate transmitter 3 and the common response profile through transmitter 5, the function of which has a dependence on the activity of transmitter 6. Neurons producing transmitter 6 are not themselves activated by the stress but have a permissive role. This second stress pathway is suppressed by the activity of transmitter 7 (an endogenous antistress factor). Finally, the activation of the common stress response leads to activation of transmitter 8 which has no function in either pathway.

majority of secondary effects of corticosteroids are long term and contribute to the adaptive phenomena of habituation, cross-sensitization, and priming which modify subsequent responses to stimuli, some corticosteroid effects are sufficiently rapid to be considered part of the acute response to stress. Examples of this are the rise in hippocampal glutamate release following ether stress which parallel the rise in corticosterone and is abolished in adrenalectomized, corticosterone-replaced animals (Abraham et al., 1998), or the secondary increase in amygdaloid CRF and GABA release, which parallels the rise in corticosterone (Fig. 4; Cook, 2001).

In addition to the effects of steroids, a wide variety of other factors may have an impact on the way a transmitter is involved in stress pathways. These include:

Stress history

Whether previously exposed to the same (homotypic) or a different (heterotypic) stress stimulus, the processes of habituation and sensitization, and the factor of chronicity will all alter stress pathways by adaptation of central transmitter, receptor, and/or transporter involvement (e.g. Fuchs and Flügge, 2003). Indeed a single stressful event can have persistent effects that alter several neuroendocrine systems for days and weeks (Servatius et al., 2000; Valles et al., 2002), and this most likely involves reorganization of the neurocircuitry or transmitter expression levels. Furthermore, early life stress leads to life-long programming of transmitter involvement (see Part II: Chapters 1.1 and 1.2). This plasticity of transmitter function is important for ensuring maintained or appropriate responses to potentially life threatening stimuli.

Genotype and gender

Aside from the obvious differences between species, marked differences in stress responses occur between strains (see Part II: Chapter 1.4) and this may have a

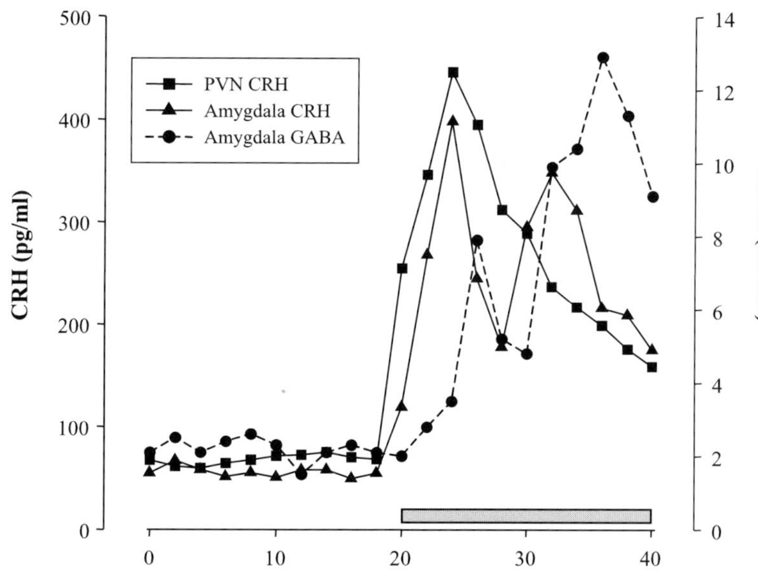

Fig. 4. Simultaneous real-time measurement of dynamic changes in extracellular CRF and GABA in the PVN and amygdala in response to psychological stress in a female sheep. The shaded bar represents 20 min during which the animal was placed in a padded cradle off the ground and exposed to the sound of a dog sporadically barking. CRF was measured from a single chamber microdialysis immunosensor located in the PVN and sampled at 2 min intervals. A second probe located in the amygdala had a double chamber allowing simultaneous measurement of both GABA and CRF. Data kindly provided by Dr. C.J. Cook, HortResearch, Hamilton, New Zealand. For methodology see Cook (1998, 2001).

transmitter basis. For example, the Maudsley Reactive and Nonreactive strains display marked differences in emotionality, possibly as a result of altered noradrenergic responses to stress (Blizard and Adams, 2002), and the High and Low anxiety-related (HAB and LAB) rats differ in their stress coping strategies (active vs. passive), possibly as a result of the differential expression and release of AVP in the PVN arising from single nucleotide polymorphisms in the gene promotor (Landgraf and Wigger, 2002). Furthermore, sexually diergic responses to stress may have their basis either in the effects of ovarian steroids on transmitter expression (Young, 1998; McCormick et al., 2002) or in sexually dimorphic transmitter pathways (Rhodes and Rubin, 1999; Figueiredo et al., 2002).

Physiological or pathological status

Various physiological and pathological conditions will also alter the contribution a transmitter plays in stress pathways. For example, reproductive status has a major effect on stress responses, such that lactating rats display less anxious behavior and both pregnant and lactating rats show reduced responsiveness of the HPA axis to various physical or emotional stimuli (Lightman et al., 2001; Neumann, 2001), which correlates with attenuated activation of central stress-reactive nuclei (da Costa et al., 1996). This reduced emotionality may be controlled, in part, by the increased central expression of oxytocin and prolactin at this time.

Transmitter systems mediating integrated stress responses

While the diversity of transmitters implicated in one or other aspect of stress-reactive pathways continues to expand, relatively few hold pre-eminent roles in integrating multiple aspects of the response. These are primarily the amines, noradrenaline, 5-HT, and dopamine, and the peptides, CRF, and vasopressin. Other sections of this book provide comprehensive reviews of many of these transmitters, but here a brief overview of their organization and functions is considered in respect of their possible contribution toward an integrated stress system.

Noradrenaline

The involvement of central noradrenaline pathways in stress is well documented (see Chapters 4.3, 4.4, and 4.5 for reviews) and many areas implicated in stress processing receive noradrenaline innervation. This innervation arises either from the ventrolateral (A1) and dorsomedial (A2) medullary regions, which innervate the hypothalamus and amygdala via the ventral noradrenergic bundle, or from the LC (A6), which innervates mainly limbic and cortical areas via the dorsal noradrenergic bundle (Cunningham and Sawchenko, 1988). Furthermore, noradrenaline neurons of both the A5 and LC/subcoeruleus areas send descending projections to the intermediolateral cell column to regulate sympathetic activity. As a result of this widespread innervation many sensory stimuli, whether aversive or not, will stimulate noradrenergic activity in stress-related areas (Pacák et al., 1995). For example, noradrenaline release occurs in the hippocampus in response to both psychological stress (restraint; Vahabzadeh and Fillenz, 1992) and physical stress (intraperitoneal LPS; Linthorst et al., 1996), and several stress modalities will increase noradrenaline release in the mPFC, including immobilization, novel environment, or conditioned fear (McQuade and Stanford, 2000; Swanson et al., 2004). The dorsal noradrenergic bundle is a major source of these mPFC afferents (Nakane et al., 1994) and simultaneous microdialysis in the mPFC and LC has shown synchronous release of noradrenaline in response to hypotension or handling stress (Kawahara et al., 1999), consistent with its origin in the LC. Furthermore, consistent with a widely distributed signalling function, any one particular stressor will cause noradrenaline release across several stress-related areas. For example, microdialysis studies have shown that electric footshock induces noradrenaline release in the PVN (Yokoo et al., 1990; Ishizuka et al., 2000), amygdala (Quirarte et al., 1998; Williams et al., 1998), hippocampus (Hajos-Korcsok et al., 2003), and mPFC (Ishizuka et al., 2000). However, noradrenaline release does show a certain degree of regional specificity, as application of mild stressors (handling and tail pinch) will stimulate noradrenaline release in the mPFC and NAc, but not in the caudate-putamen (Cenci et al., 1992). Interestingly, the magnitude of this noradrenaline release has been shown to be directly proportional to the stimulus intensity (Pacák, 2000) suggesting that, unlike other transmitters, the pathway mediating this response may not incorporate a threshold detection mechanism.

Coupled with its widespread release, noradrenaline appears to play a role in many aspects of the stress response and, therefore, can be considered to subserve an important integrative function. However, this role may be largely restricted to that of activating or initiating a response rather than sustaining it, as measures of hippocampal noradrenaline release have shown that release does not persist during a sustained stress while behavioral and HPA components of the response do (Britton et al., 1992). In respect of innervation of the PVN, noradrenaline appears to have involvement in HPA regulation as α_{1b} receptor mRNA colocalizes with CRF mRNA in the PVN (Day et al., 1999) and the α_1 antagonist prazosin will attenuate ACTH response to ether stress (Szafarczyk et al., 1987). However, noradrenaline involvement in mediating HPA responses to stress is strongly modality dependent. Using a wide variety of introceptive and exteroceptive stressors, Pacák (2000) showed that noradrenaline release in the PVN varied in a way that did not correlate with the plasma ACTH response, suggesting that the relative contribution of noradrenaline to HPA activation was stressor specific. Indeed evidence from a range of lesion and retrograde labelling studies has indicated that direct noradrenaline projections to the PVN contribute to the HPA response to physical stressors, while responses to psychological stress may involve indirect noradrenaline connections with higher centers, possibly including relays in the MeA (Dayas et al., 2001a). Thus, 6-hydroxydopamine lesion of the ventral noradrenergic bundle, which carries noradrenaline afferents from the A1/A2 to the PVN, does not affect the HPA response to novel environment (Castagné et al., 1990), but attenuates the response to ether stress (Szafarczyk et al., 1985). Furthermore, recent evidence has shown that injection of an anti-dopamine-β-hydroxylase-saporin conjugate into the PVN, which causes almost complete ablation of noradrenaline terminals in the PVN and their associated cell bodies, profoundly impaired HPA activation and PVN Fos expression in response to glucoprivation (insulin-induced hypoglycemia) but was without effect on the corticosterone response to forced swim (Ritter et al., 2003). This modality-specific differentiation is consistent with the involvement of different subgroups of A1/A2 neurons in the response to emotional and physical stressors (Dayas et al., 2001a).

As well as affecting neuroendocrine function, noradrenaline release in the amygdala may be involved in regulating memory storage and behavior. Evidence from region-specific administration of the α_1-adrenoceptor antagonist benoxathian has suggested that the noradrenaline innervation of the lateral BNST and CeA are respectively involved in aspects of anxiety-like behavior measured by the elevated plus maze and social interaction test (Cecchi et al., 2002). Furthermore, within cortical areas noradrenaline regulates variations in arousal during the sleep–wake cycle and may facilitate arousal in response to a number of physiological stimuli that have the potential to take on stressful properties (e.g. sensory stimuli from the viscera and cardiovascular system). In this respect noradrenaline pathways may play an important role in orientating attention toward salient environmental stimuli and facilitation of the cognitive processes that underlie stress responding.

The control of ascending noradrenaline systems under both stressful and nonstressful conditions arises from a limited number of limbic and brainstem regions (Singewald and Philippu, 1998; Berridge and Waterhouse, 2003). Inputs to the A1, A2, and LC arise in the dorsal PVN, CeA, and BNST, while the A1/A2 also receive major innervation from viscerosensory areas in the NTS. Importantly many of the forebrain descending afferents arise from CRF-containing regions. In this respect, Valentino and van Bockstaele suggest that stimulus evoked bursts of LC activity, possibly mediated by glutamate, may be important for selective attention, while sustained LC activation mediated by CRF may cause a shift from stimulus specific responding to a general arousal state which is important for scanning of multiple stimuli appropriate for the response to threatening situations (see Chapter 4.4).

Serotonin (5-HT)

In a similar arrangement to noradrenaline, 5-HT is present in a widely distributed network. This arises from the midbrain raphe (DRN and MRN) and appears to contribute both to direct pathways activating components of the stress response and to higher order coordinated activation of a number of different areas. Direct serotonergic projections to the PVN have been demonstrated (Carasco and van der Kar, 2003; Herman et al., 2003), where they form synapses with CRF, oxytocin, and AVP neurons. In addition, 5-HT_{1A} and 5-HT_{2A} receptors are present on neurons of the PVN (Wright et al., 1995; Zhang et al., 2002), where they may contribute to activation of the HPA axis. Indeed the $5HT_{2A}$ agonist DOI causes neuroendocrine responses similar to stress and induces Fos expression in CRF and oxytocin neurons of the PVN (van de Kar et al., 2001). 5-HT projections also go to most other nuclei activated by stress and a variety of stressors can evoke release or turnover of 5-HT in the PFC, hippocampus, amygdala, NAc, and LC (Pei et al., 1990; Shimizu et al., 1992; Kawahara et al., 1993; Inoue et al., 1994; Linthorst et al., 1995; Adell et al., 1997; Amat et al., 1998; Kaehler et al., 2000; Funada and Hara, 2001). However, whilst the serotonergic system provides widespread projections to the forebrain, it is perhaps a misconception that 5-HT generates a diffuse and nonspecific stress signal, as recent evidence has demonstrated the existence of functional subsets of raphe 5-HT neurons, which respond to stress in a selective manner (see above). In addition, microdialysis data obtained from forced swim or conditioned fear, suggest that 5-HT responses may require very specific behavioral reactivity and/or stimulus conditions, with responses dependent on the escape potential of the model. In this respect serotonergic antagonists will potentiate and SSRIs attenuate the expression of learned helplessness, suggesting that 5-HT plays an important part in stress coping strategies.

Within the hierarchical model of stress responding (Fig. 2) 5-HT may have a relatively minor role in direct activation of output neurons (such as HPA activation), but primarily function to modulate the appropriate level of neuroendocrine and autonomic activity and coping strategy, whether it be through facilitation of motor output or facilitation of freezing behavior (equivalent to pathway (3)). Regional differences in the levels of 5-HT attained during specific stressful stimuli may alter this pattern of coping. In this respect it should not be forgotten that the serotonergic system is under important control by noradrenaline afferents and there is accumulating evidence for an important innervation by CRF (see Linthorst, Chapter 4.6). Differential sensitivity of subpopulations of serotonergic neurons to these inputs will alter the balance of transmission to the respective projection regions.

Dopamine

The role of dopamine in responses to stress are largely restricted to a subset of neurons in the ventral tegmental area (VTA) projecting to the PFC and NAc, with the former projection being the more stress sensitive (Roth et al., 1988; Pani et al., 2000). In this respect, handing stress and hypotension strongly stimulate the release of dopamine in the mPFC (Kawahara et al., 1999), while the acute stress of restraint or footshock will activate the mesolimbic dopamine system (Puglisi-Allegra et al., 1991). Conditioned stress causes selective increase in dopamine utilization in medial and lateral PFC and NAc, but not in perirhinal or cingulate cortices, BLA, or caudate-putamen (Goldstein et al., 1994), and likewise handling and tail pinch stimulates DA release in the mPFC, but has only small effects in NAc and no effects in caudate-putamen (Cenci et al., 1992). Consistent with this pattern of release, footshock increases Fos expression in prelimbic and infralimbic cortices and in tyrosine hydroxylase-labeled neurons of the VTA (Morrow et al., 2000), and restraint stress increases the concentrations of the dopamine metabolite DOPAC in the PFC and NAc, and induces Fos immunoreactivity in dopamine neurons of the VTA but not the substantia nigra (Deutch et al., 1991). Furthermore, retrograde tracer studies confirm the activation of a distinct subset of VTA dopamine neurons, which project to the PFC. Thus, from a large number of studies it has been established that acute stress evokes a greater increase in dopamine metabolism and release within the PFC than other subcortical areas (Finlay and

Zigmond, 1997). Indeed the mesoprefrontal system appears to be particularly responsive to low intensity stresses that do not affect other ascending dopaminergic systems (Horger and Roth, 1996), and the basis of this differential sensitivity may relate to the relative lack of D2 inhibitory autoreceptors in the mesoprefrontal pathway, coupled to the presence of extensive excitatory inputs to the VTA. Interestingly the effect of stress on dopamine-dependent behaviors and on the activation of afferents to the NAc appears to depend on the chronicity and degree of control that the animal can exert on the stress stimulus (Cabib and Puglisi-Allegra, 1996). In this respect dopamine in the mPFC may normally act to suppress mesolimbic dopamine transmission, but this fails in conditions of extreme or unpredictable stress.

Dopamine innervation also appears to be important for stress-induced activation of neurons in the anterolateral BNST (Kozicz, 2002), which are involved both in higher order activation of stress–responsive circuits and in the generation of coping behaviors (see above). Dopamine plays a role in the hedonic and reward aspects of stress, and the effects of stress on sexual activity and appetite and on the sensitivity to drugs of abuse are possibly mediated through the dopamine system. Dopamine also has the capacity to increase the gain of neuronal information processing and, therefore, can affect the learning and information processing associated with future stress. Finally it is notable that the CeA plays an important role in dopamine neurotransmission in the PFC, such that lesions of CeA block stress-induced dopamine release in the PFC and infusion of AMPA into the CeA evokes both a rapid increase in PFC dopamine release and an increase in arousal state (Stalnaker and Berridge, 2003). This is consistent with CeA involvement in coordination of neural systems regulating the behavioral state during stress.

Acetylcholine (ACh)

Acetylcholine is released in the hippocampus and cortex in response to a number of stressful stimuli, including footshock (Dazzi et al., 1995) and immobilization (Tajima et al., 1996), and this response may have dependence on serotonergic inputs from the DRN. However, it is unclear whether it is stress itself which triggers ACh release or whether it is novelty, since the responses are transient and appear to occur with a variety of novel stimuli (Acquas et al., 1996). In this regard cholinergic neurons in the PPTg have an important role in controlling arousal and attention and, since forebrain cholinergic innervation is known to facilitate working memory performance by increasing the discriminatory capacity of cortical neurons, the role of acute changes in ACh may have particular relevance to the acquisition of memories. However, there is evidence that acute hippocampal release of ACh may also be involved in the termination of the effects of stress. Blocking hippocampal muscarinic receptors has the effect of augmenting restraint-induced release of ACTH and corticosterone (Bhatnagar et al., 1997), and septohippocampal cholinergic inputs may be essential for the hippocampus to express its normal inhibitory effect on the HPA axis (Han et al., 2002). Interestingly, the expression of certain variants of the metabolic enzyme, acetylcholinesterase, is highly sensitive to stress, and differences in both basal and stress-induced levels of this enzyme may underlie inter-individual and adaptive responses to stress (see Soreq et al., Chapter 4.10),

Amino acids (glutamate and GABA)

Despite (or possibly due to) the importance of glutamate as a central transmitter, its role in mediating the various aspects of the stress response have received relatively scant attention. Both restraint and swim stress evoke widespread release of glutamate in mPFC, hippocampus, striatum, and NAc, with the largest increase in mPFC (Moghaddam, 1993). However, whether glutamate released in response to various noxious and nonnoxious stimuli is functional has been questioned (Timmerman et al., 1999). Nevertheless, because of its well-established role in cognitive processes, it is likely that glutamate will modulate responses to psychological stressors which utilize cognitive circuits. Indeed hippocampal (ventral subicular) efferents are glutamatergic, and glutamate appears to have an important role in modifying stress-induced activation of the mPFC, particularly the mesoprefrontal dopamine system. For example, conditioned stress (tone paired with footshock) causes increased mPFC dopamine and 5-HT turnover, and increased serum corticosterone and freezing behavior. Freezing behavior and dopamine utilization in the PFC could be blocked by an NMDA glycine site antagonist, but HPA responses and other neurochemical changes were unaffected, indicating functional selectivity (Goldstein et al., 1994). Handling stress also increases extracellular dopamine and its metabolites in the PFC, but does not modify glutamate. Retrodialysis of ionotropic glutamate receptor agonists, NMDA and AMPA, significantly attenuates this handling-induced dopamine release (del Arco and Mora, 2001), while retrodialysis of a metabotropic glutamate (mGlu2/3) receptor agonist significantly attenuated immobilization-induced noradrenaline release (but not DA) (Swanson et al., 2004).

The near ubiquitous inhibitory effect of GABA on neural activity means that it is hard to attribute the effects of GABAergic agonists and antagonists to specific aspects of the stress response. Furthermore, for technical reasons relatively few studies have measured regional stress-induced GABA release. However, increased GABA release has been measured in the LC in response to immobilization or hypovolemia (Singewald et al., 1995), and in the amygdala in response to noise (Singewald et al., 2000) or psychogenic stress (Fig. 4). This release may modulate transmission through the activated response network. Furthermore, many PVN afferents are immunoreactive for glutamic acid decarboxylase and these GABAergic afferents appear to play a major role in regulating CRF neurons and activity of the HPA axis (Boudaba et al., 1996; Herman et al., 2002; see Chapter 4.1). Evidence from organotypic culture has shown that tonic GABAergic inhibition suppresses CRF neurons and withdrawal of this tone (either by antagonist or appropriate afferent input) unmasks a glutamatergic AMPA/kainate receptor mediated excitation, which drives CRF neuron activity (Bartanusz et al., 2004). The presence of a GABAergic tone has been demonstrated in vivo by the fact that local application of bicuculline into the PVN increases Fos expression and HPA activity (Cole and Sawchenko, 2002). However, while glutamate has a role in regulating PVN CRF neurons (see Chapter 4.7), no study has yet directly addressed its involvement in stress-induced HPA activation. Finally, one should remember that corticosteroids have major effects on glutamate and GABA neurotransmission and, therefore, one cannot exclude possible indirect effects during stress (see Chapter 4.7).

Nitric oxide (NO)

There is increasing evidence that NO plays an important regulatory role in many parts of the stress activating network. Restraint stress activates putative NO-producing neurons in many autonomic centers (including medial septum, amygdala, PVN, raphe nuclei, NTS, and ventrolateral medulla) and may function to decrease sympathetic output (Krukoff, 1998). NO appears to be particularly important for regulating the PVN outflow which regulates sympathetic activity (Kenney et al., 2003). Indeed many PVN neurons that express Fos following immobilization stress also contain NO synthase, and treatment with competitive NO synthase inhibitors blocks stress-induced Fos expression in the PVN (Amir et al., 1997). The levels of NO metabolites in the PVN region also increase after intraperitoneal administration of interleukin-1β, but not after footshock (Ishizuka et al., 2000). However, footshock will increase NO in the mPFC, indicating regional selectivity. Interestingly, many (if not all) cholinergic neurons in the PPTg colocalize NO synthase (Sugaya and McKinney, 1995), suggesting NO may also contribute to the arousal and attentional aspects of the stress response.

Vasopressin (AVP) and oxytocin

Along with CRF, AVP is released from parvocellular PVN neurons during a stress response and plays a key role in the synergistic activation of pituitary corticotrophs (see Chapter 1.3). However, in addition to this peripheral action, AVP may have important central effects. Microdialysis within the PVN shows that AVP is released in response to the emotional stress of social defeat, but not by exposure to a novel cage (Wotjak et al., 1996). In contrast, within the septum and amygdala, social defeat (a purely emotional stress) does not evoke AVP release (Engelmann et al., 2000) despite the ability of this stress to evoke powerful behavioral responses and activation of the HPA axis (Ebner et al., 2000). However, forced swim stress (a combined physical and emotional stress) will induce the release of AVP (Ebner et al., 2002). The function of this centrally released AVP may relate to acute stress-coping strategies, anxiety-related behavior, and/or learning and memory processes (see Chapter 2.6). Retrodialysis of a V1 antagonist in the amygdala demonstrates that AVP regulates coping strategy during forced swim, reducing active coping (struggling) and increasing passive coping (floating) (Ebner et al., 2002). Furthermore, recent data suggests that differences in central AVP pathways correlates with anxiety-like behavior (Landgraf and Wigger, 2002). AVP is also important in the adaptive responses to persistent or repeated stress, since the increased expression by the hypophysiotrophic neurons of the PVN may play a role in driving corticotroph activity when the levels of CRF are reduced by the feedback effects of glucocorticoids (Ma and Lightman, 1998).

Oxytocin is also released into the circulation from magnocellular neurons of the supraoptic nucleus and PVN in response to stressful stimuli, such as ether, restraint, swim stress, motion stress, novel environment, or fear conditioned by foot-shocks (Gibbs, 1984; Nishioka et al., 1998; Zou et al., 1998; Neumann, 2002). The function of this neuroendocrine response is unclear, but may contribute to hepatic glycogenolysis or cardiovascular regulation. In addition to this endocrine function, oxytocin may also be released centrally. This can occur either coincident with peripheral release, as for the release into the PVN during motion stress (Nishioka et al., 1998), or in the absence of peripheral release, as for the release into the supraoptic nucleus and septum following social defeat (Ebner et al., 2000; Engelmann et al., 2000). This centrally released oxytocin may play an important integrative function. For example, oxytocin is expressed by a number of descending PVN efferents and is implicated in the regulation of sympathetic outflow (Coote et al., 1998). In addition, intracerebroventricular administration of oxytocin will attenuate the HPA responses to noise or restraint, and will attenuate stress-induced *c-fos* mRNA expression in the PVN and SHy/VLS (Windle et al., 1997, 2004). While infusion of an antagonist indicates that endogenous oxytocin modulates the HPA response to a novel environment (Neumann, 2002), it appears not to regulate coping strategy during the conflict test (Ebner et al., 2000).

Other neuropeptides with potential integrative roles in stress responding

It would be beyond the scope of this review to consider all the transmitters which have the potential to contribute to the integrated stress response, however, it is worth mentioning a few for which recent evidence has suggested an important role.

Substance P

Central substance P pathways may be important in the process of terminating the activation of the stress-activated circuits, particularly those controlling the HPA axis (Jessop et al., 2000). They may also contribute to the activation of the sympathetic system during acute stress response (Ku et al., 1998), e.g. pressor responses involve the CeA, ventromedial hypothalamus, PVN, and rostral ventrolateral medulla (A1). Microinjection of substance P into these areas will induce pressor responses and substance P has been suggested to mediate the pressor response to CeA activation (Wu et al., 1999). Furthermore there is increasing evidence that substance P plays a role in modulating anxiety-like behavior (Aguiar and Brandao, 1996; Boyce et al., 2001; Santarelli et al., 2001) and studies of the effects of NK-1 antagonists have demonstrate important behavioral functions (see Part II: Chapter 4.3).

Galanin

Galanin may be involved in the stress coping generated by the BNST–amygdala complex. Injection of galanin into the CeA has an anxiolytic effect and recent evidence has suggested that noradrenaline neurotransmission can stimulate stress-induced release of galanin which acts to attenuate anxiety behavior during acute stress (Khoshbouei et al., 2002). This galanin may arise from the closely associated lateral BNST, where dense noradrenergic and dopaminergic fibers contact galanin neurons (Kozicz, 2001). However, galanin administered into the lateral BNST appears to facilitate behavioral responses to stress (Morilak et al., 2003). Whether this is due to an autoinhibitory effect on efferents to the CeA is not clear.

Neuropeptide Y (NPY)

Stress has marked effects on regional levels of expression of NPY (Thorsell et al., 1998; Krukoff et al., 1999; Makino et al., 2000). Interestingly, intraamygdaloid injections of NPY have anxiolytic effects (Heilig et al., 1993) and microinjection of NPY receptor ligands has suggested that NPY Y1 receptors may mediate these effects. Furthermore, recent evidence has shown that increased emotionality is observed in mice with an homologous recombination knockout of the preproNPY gene, while rats overexpressing NPY in the hippocampus display no overt behavioral responses to stress (Heilig and Thorsell, 2002). However, it is not clear whether this is a direct effect on stress circuits or on the acquisition and processing of memories necessary for responses to emotional stressors.

Integration through the coordinating effect of specific transmitters: role for CRF

The large body of evidence, provided above and elsewhere in this volume, has shown that individual transmitters can have multiple involvement in generating the behavioral and physiological responses to stress, or can evoke responses which have a superficial resemblance to those occurring during stress. This has led to the popular concept that specific transmitters may fulfil a function to integrate or coordinate the stress response. This integration may occur either early in the process of perception, at the point at which the sensory or cognitive process passes a threshold to become a stress response, or at a later stage in the pathway, where a distributed network synchronously contacts several output systems (cf. Pathways (2) and (3) in Fig. 2). However, while either of these arrangements would confer integration on a stress–response network, it is important that this should be distinguished from the situation where the same transmitter contributes to different stress pathways which does not confer integration (cf. Transmitter 1 in Fig. 3). Indeed, evidence based on global changes in transmitter function (e.g. agonist/antagonist administration or knockout/overexpression studies) may not make this important distinction. Nevertheless, it is clear that certain transmitters not only possess the appropriate anatomical arrangement to generate integration within a stress–responsive network, but to do so in a way that is selective for stress responding. One transmitter which has been widely accepted to fulfil this role is CRF, which is both the most important neuroendocrine factor controlling HPA activity, and acts as a central transmitter controlling aspects of behavior and autonomic/physiological activation (Brown and Fisher, 1985; Dunn and Berridge, 1990; Owens and Nemeroff, 1991; de Souza, 1995; Bakshi and Kalin, 2000; Smagin and Dunn, 2000; Smagin et al., 2001). Some of the features of these roles are briefly addressed here.

Role of CRF in mediating the stress response

The functions of central CRF and related peptides (e.g. urocortins) are extensively reviewed elsewhere in this volume and it is unnecessary to repeat the evidence here (see Chapters 2.3, 2.4, 4.4, 4.6, Part II: 1.3, and 4.1). However, it is worth noting that there is an extensive network of CRF-containing neurons that originate in several stress–responsive areas of the brain (PVN, BNST, CeA, LC, parabrachial area, hippocampus, and NAc) (Sawchenko and Swanson, 1985). Furthermore, the increase in CRF mRNA in response to a wide variety of stressors of different modalities (Bakshi and Kalin, 2000) indicates that these CRF neurons form part of a common responsive system. However, consistent with a more selective role in regulating transmission, intracerebral injection of the CRF antagonist, α-helical $CRF_{9\text{-}41}$, has been shown to decrease restraint-induced *c-fos* mRNA expression in the PVN without affecting IEG induction in the VLS or LC (Imaki et al., 1995), suggesting that not all of the network activated by stress is dependent upon CRF neurotransmission.

Shortly after its identification as a regulator of the HPA axis, CRF was shown to induce behavioral activation which resembled some of the features seen during stress (Sutton et al., 1982; Dunn and Berridge, 1990). Importantly the behavioral effects of intracerebrally administered CRF depend upon context: testing in a familiar environment or low arousal state evoked locomotor activity, exploration (rearing), displacement (grooming), and attenuated feeding, while prominent behavioral suppression occurred when CRF was tested in an unfamiliar state under conditions of arousal (Koob, 1999; Smagin et al., 2001). This suggests that CRF acts to modulate the response of a pathway rather than having a predefined effect. Lesions of the CeA will block some of the behavioral effects of centrally administered CRF, and this may be mediated through the reciprocal excitatory connections with LC–noradrenaline system (see Valentino and van Bockstaele, Chapter 4.4). Beyond this pharmacological effect of exogenous CRF, the role of endogenous CRF in response to stress is demonstrated by the ability of antagonists to attenuate the behavioral (e.g. Hotta et al., 1999; Smagin et al., 2001; see Part II: Chapters 1.3 and 4.1 for review) and sympathoexcitatory responses (Jezova et al., 1999) to stress, as well as the stress-induced increase in Fos expression in the PVN (Imaki et al., 1995). Furthermore, genetic modification of the CRF system evokes marked differences in stress- or anxiety-like behaviors consistent with its involvement (Bakshi and Kalin, 2000; Coste et al., 2001; Müller et al., 2003; and see Part II: Chapter 1.3 for review). The pharmacology of these responses and the relative contribution of CRF and urocortins has been reviewed elsewhere (Smagin et al., 2001), but CRF-R1 has been particularly associated with the cognitive aspects of stress behavior, including emotionality, attention, and executive function (Steckler and Holsboer, 1999; Bakshi and Kalin, 2000).

Consistent with its potential role in coordinating multiple aspects of the stress response, CRF receptors are distributed in many of the areas of the brain which exhibit stress-induced IEG expression. CRF-R1 receptors are located in the pituitary, PVN, hippocampus, amygdala, and neocortex (Wong et al., 1994; Chalmers et al., 1995; Sauvage and Steckler, 2001), while CRF-R2α receptors are found in the lateral septum, amygdala, and ventromedial hypothalamus (Lovenberg et al., 1995; Sanchez et al., 1999). These two receptors appear to have opposing actions, since mice deficient for CRF-R1 display markedly attenuated responses to stress (Smith et al., 1998; Coste et al., 2001), while deficiency for CRF-R2 leads to anxiety-like behavior and exaggerated stress responses (Bale et al., 2000). However, recent evidence suggests that the coordinating role of CRF is not straightforward, since, while CRF-R1 knockout mice show reduced stress and anxiety levels, conditional knockout mutants, in which CRF-R1 function is selectively inactivated in the anterior forebrain and limbic brain structures but which leave HPA activity is intact, display hypersensitivity to stress (Müller et al., 2003).

CRF is able to generate these responses through interactions with other transmitter systems. One of the best characterized transmitter interactions is the ability of CRF to regulate the activity of the noradrenergic system arising from the LC (Koob, 1999; Valentino and van Bockstaele, Chapter 4.4). CRF also evokes the release of 5-HT (or its metabolite) in the hippocampus (Linthorst et al., 2002), striatum (Price et al., 1998), and medial hypothalamic area (Lavicky and Dunn, 1993; for review see Linthorst, Chapter 4.6). This release is consistent with excitatory effects on the neurons of the raphe (Price et al., 1998; Lowry et al., 2000), although it may also be driven by increased afferent effects of noradrenaline. Intracerebroventricular administration of CRF will evoke release of hippocampal ACh similar to that evoked by stress (Day et al., 1998), and this may arise from the presence of CRF-R1 on the cholinergic neurons of the PPTg (Sauvage and Steckler, 2001). Exogenous CRF also increases dopamine turnover (measured by tissue DOPAC levels) in a number of hypothalamic and limbic areas, including the NAc, PVN, and periventricular nuclei (Pan et al., 1995). Thus, these widespread transmitter effects, coupled to the distribution of receptors in stress-related areas, are consistent with an integrative function for CRF.

CRF involvement in an integrated stress–response

Despite the evidence for CRF involvement in multiple aspects of the stress response, it is not clear that this supports an integrative function within a hierarchical model of stress pathways (Fig. 2). While it may be conceptually useful to consider CRF as a single integrating system underlying the stress response, several lines of evidence suggest a more complex organization. Firstly, the distribution of CRF-containing neurons is such that they cannot operate as a single network which distributes a stress signal to coordinate several downstream response systems (i.e. similar to Pathways (2) or (3) in Fig. 2). Although it is possible that this network of CRF neurons may be innervated by a common transmitter system, such as noradrenaline, to evoke coordinated activation, in this case CRF is not acting as the integrating factor. Secondly, while it is clear that CRF is released into both the PVN and amygdala in response to a stressful stimulus and thereby might coordinate neuroendocrine and behavioral responses, recent technical advances have shown that the temporal profiles may be quite distinct (Fig. 4), arguing against a single system coordinating the responses. Thirdly, CRF administration, either intracerebrally (Arnold et al., 1992; Imaki et al., 1993; da Costa et al., 1997) or directly into the LC (Rassnick et al., 1998), produces a regional pattern of Fos expression that does not mirror the pattern induced by stress, and probably is a reflection of the distribution of the CRF-R1 receptor and downstream innervated areas (Bittencourt and Sawchenko, 2000). While there may be significant overlap between these two stimuli, it does not prove the integrative function of CRF. For example, both intracerebral CRF and immobilization stress increase Fos in the PVN, MeA, and LC, but only CRF increases Fos in the BNST, CeA, and NTS, while only stress has an effect in the SHy (Imaki et al., 1993).

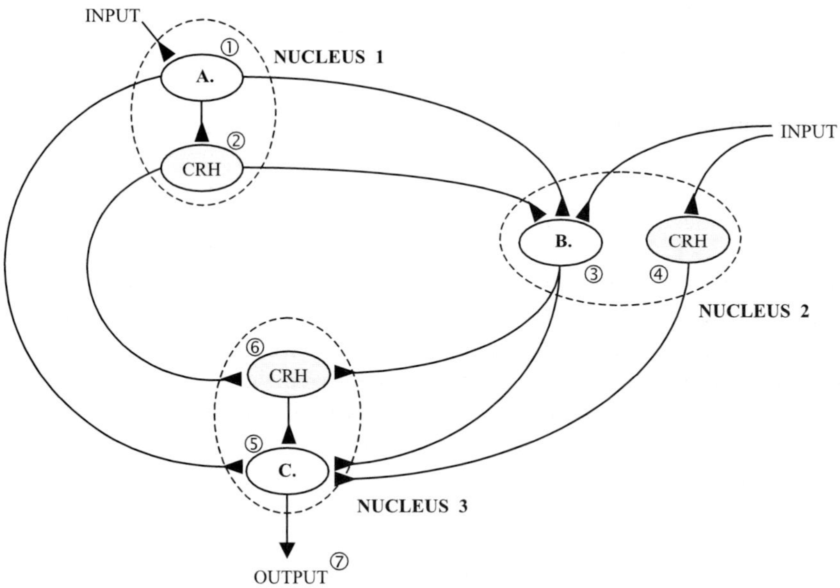

Fig. 5. Possible organization of CRF neurons within a distributed stress–responsive circuit. Neurons (1) in nucleus 1 receive inputs which may potentially be considered stressful. Output from these neurons goes to neurons (3) in nucleus 2 and also to local CRF neurons (2). Above a particular threshold for activation these CRF neurons are also activated and the stimulus takes on a stressful quality, signalling this to nucleus 2 by a convergent pathway. A population of CRF neurons (4) in nucleus 2 receives an input which is direct and does not rely on intranuclear connections. Once again a threshold for activation determines whether or not these neurones are activated and pass a stress signal to neurons (5) in nucleus 3. CRF neurons (6) in or close to nucleus 3 receive their inputs via collaterals from nucleus 2. When this input exceeds a threshold for activation, intrinsic connections within or close to nucleus 3 signal this as stressful. Output (7) from nucleus 3 is regulated by activity within the whole network. CRF may influence the activity of the network at several points through gain control and/or parallel processing. The overall contribution of CRF to transmission within this network may depend upon the relative number of CRF neurons and their different thresholds for activation.

Indeed since IEG expression displays at least two modality-specific patterns this unlikely to arise from a single CRF system. So how might stress pathways be organized in order for CRF to fulfil a pivotal role in integrating the activation of several transmitter systems and the expression of various aspects of the stress response?

In the models described above, one or more discrete regulatory center(s) was considered to project to the various effector systems. However, while this distributed organization may be appropriate for transmitters that have a relatively restricted origin (e.g. noradrenaline in the LC or 5-HT in the DRN) this does not hold for CRF, which is both synthesized by neurons within many stress–responsive areas and has receptors within the same areas (see above and Chapter 4.4). However, transmitters with discrete origins but distributed projection sites, such as noradrenaline and 5HT, fulfil many functions in addition to contributing to the stress response and so cannot be considered to carry the stress signal alone. Thus, it may be more appropriate to consider the stress response as generated by a number of sensory and cognitive networks which variably employ noradrenaline, 5-HT, dopamine, ACh, and glutamate, but that, throughout these networks, CRF functions to modulate the signal in order to attribute a "stressful" characteristic. Using this concept of parallel processing, CRF neurons might function both as a threshold detector mechanism and as a mechanism to modulate the nature of signals passing through the network by changes in gain control (Fig. 5). This may operate through a number of different connections of CRF neurons which are known to exist within the network of stress–responsive areas. This model allows for the circuit to fulfil functions unrelated to stress, but to switch to a stress–responsive mode through the action of CRF at any of a number of different sites. This model incorporates the concept of threshold detection and distributed activation, and the integrative role of CRF can be conceived not simply as its contribution to one specific pathway, but by its ability to change the nature of the signal passing through the network. Importantly the CRF system may be activated by a wide range of stress stimuli but may only modulate transmission through pathways which are activated by specific stimuli, and therefore the model also incorporates aspects of the modality selective responding.

Conclusions

So, from the foregoing account is it possible to provide an integrated view of pathway and transmitter involvement in stress? The cautious response to this is, "to some extent." Stress is a term which encompasses diverse range of complex responses and while technical advances have helped to identify many of the key areas and transmitters that contribute to generating these responses, we are still a long way from fully determining the way these join up to generate an integrated behavioral, neuroendocrine, and autonomic response. Nevertheless, certain principles have emerged which allow the construction of theoretical models. The future challenge will be to refine these models in order to eventually determine the neural basis of stress.

Abbreviations

5-HT	5-hydroxytryptamine (serotonin)
ACh	acetylcholine
AVP	vasopressin
BLA	basolateral amygdala
BNST	bed nuclei of the stria terminalis
CeA	central nucleus of the amygdala
CRF	corticotropin-releasing factor
CRF-R1/ CRF-R2	type 1/2 CRF receptor
DMH	dorsomedial hypothalamus
DRN	dorsal raphe nucleus
HPA	hypothalamo–pituitary–adrenal
IEG	immediate-early gene
LC	locus coeruleus
LPS	lipopolysaccharide
MeA	medial nucleus of the amygdala
(m)PFC	(medial) Prefrontal cortex
MPOA	medial preoptic area
MRN	median raphe nucleus
NAc	nucleus accumbens
NO	nitric oxide
NPY	neuropeptide Y
NTS	nucleus tractus solitarius
PAG	periqueductal gray of the midbrain

PPTg	pedunculopontine tegmentum
PVN	paraventricular nucleus of the hypothalamus
SHy	septohypothalamic area
VLS	ventrolateral septum
VTA	ventral tegmental area

Acknowledgments

The author is grateful to the members of his group and to colleagues, with whom discussions have helped to shape these ideas. He is especially grateful to Dr. Christian Cook for providing Fig. 4.

References

Abraham, I., Juhasz, G., Kekesi, K.A. and Kovacs, K.J. (1998) Corticosterone peak is responsible for stress-induced elevation of glutamate in the hippocampus. Stress, 2: 171–181.

Acquas, E., Wilson, C. and Fibiger, H.C. (1996) Conditioned and unconditioned stimuli increase frontal cortical and hippocampal acetylcholine release: effects of novelty, habituation, and fear. J. Neurosci., 16: 3089–3096.

Adell, A., Casanovas, J.M. and Artigas, F. (1997) Comparative study in the rat of the actions of different types of stress on the release of 5-HT in raphe nuclei and forebrain areas. Neuropharmacology, 36: 735–741.

Aguiar, M.S. and Brandao, M.L. (1996) Effects of microinjections of the neuropeptides substance P in the dorsal periaqueductal gray on the behaviour of rats in the plus-maze test. Physiol. Behav., 60: 1183–1186.

Amat, L., Matus-Amat, P., Watkins, L.R. and Maier, S.F. (1998) Escapable and inescapable stress differentially and selectively alter extracellular levels of 5-HT in the ventral hippocampus and dorsal periaqueductal gray of the rat. Brain Res., 797: 12–22.

Amir, S., Rackover, M. and Funk, D. (1997) Blockers of nitric oxide synthase inhibit stress activation of c-fos expression in neurons of the hypothalamic paraventricular nucleus in the rat. Neuroscience, 77: 623–627.

Arnold, F.J.L., de Lucas Bueno, M., Shiers, H., Hancock, D.C., Evans, G.I. and Herbert, J. (1992) Expression of *c-fos* in regions of the basal limbic forebrain following intracerebroventricular corticotropin-releasing factor (CRF) in unstressed and stressed male rats. Neuroscience, 51: 377–390.

Bakshi, V.P. and Kalin, N.H. (2000) Corticotropin-releasing hormone and animal models of anxiety: Gene-environment interactions. Biol. Psychiatry, 48: 1175–1198.

Bale, T.L., Contarino, A., Smith, G.W., Chan, R., Gold, L.H., Sawchenko, P.E., Koob, G.F., Vale, W.W. and Lee, K.F. (2000) Mice deficient for corticotropin-releasing hormone, receptor-2 display anxiety-like behaviour and are hypersensitive to stress. Nat. Genet., 24: 410–414.

Bandler, R., Keay, K.A., Floyd, N. and Price, J. (2000) Central circuits mediating patterned autonomic activity during active vs. passive emotional coping. Brain Res. Bull., 53: 95–104.

Bartanusz, V., Muller, D., Gaillard, R.C., Streit, P., Vutskits, L. and Kiss, J.Z. (2004) Local γ-aminobutyric acid and glutamate circuit control of hypophyseotrophic corticotropin-releasing factor neuron activity in the paraventricular nucleus of the hypothalamus. Eur. J. Neurosci., 19: 777–782.

Bernard, J.F. and Bandler, R. (1998) Parallel circuits for emotional coping behaviour: new pieces in the puzzle. J. Comp. Neurol., 401: 429–436.

Berridge, C.W. and Waterhouse, B.D. (2003) The locus coeruleus-noradrenergic system: modulation of behavioral state and state-dependent cognitive processes. Brain Res. Rev., 42: 33–84.

Bhatnagar, S., Costall, B. and Smythe, J.W. (1997) Hippocampal cholinergic blockade enhances hypothalamic-pituitary-adrenal responses to stress. Brain Res., 766: 244–248.

Bittencourt, J.C. and Sawchenko, P.E. (2000) Do centrally administered neuropeptides access cognate receptors?: an analysis in the central corticotropin-releasing factor system. J. Neurosci., 20: 1142–1156.

Blizard, D.A. and Adams, N. (2002) The Maudsley Reactive and Nonreactive strains: a new perspective. Behav. Genet., 32: 277–299.

Boudaba, C., Szabo, K. and Tasker, J.G. (1996) Physiological mapping of local inhibitory inputs to the hypothalamic paraventricular nucleus. J. Neurosci., 16: 7151–7160.

Boyce, S., Smith D., Carlson, E., Hewson, L., Rigby, M., O'Donnell, R., Harrison, T. and Rupniak, N.M. (2001) Intra-amygdala injection of the substance P [NK(1) receptor] antagonist L-760735 inhibits neonatal vocalisations in guinea pigs. Neuropharmacology, 41: 130–137.

Britton, K.T., Segal, D.S., Kuczenski, R. and Hauger, R. (1992) Dissociation between in vivo hippocampal norepinephrine response and behavioral/neuroendocrine responses to noise stress in rats. Brain Res., 574: 125–130.

Brown, M.R. and Fisher, L.A. (1985) Corticotropin-releasing factor: effects on the autonomic nervous system and visceral systems. Fed. Proc., 44: 243–248.

Cabib, S. and Puglisi-Allegra, S. (1996) Stress, depression and the mesolimbic dopamine system. Psychopharmacology, 128: 331–342.

Campeau, S. and Watson, S.J. (1997) Neuroendocrine and behavioral responses and brain pattern of *c-fos* induction associated with audiogenic stress. J. Neuroendocrinol., 9: 577–588.

Canteras, N.S. (2002) The medial hypothalamic defense system: hodological organization and functional implications. Pharmacol. Biochem. Behav., 71: 481–491.

Carrasco, G.A. and van de Kar, L.D. (2003) Neuroendocrine pharmacology of stress. Eur. J. Pharmacol., 463: 235–272.

Castagné, V., Rivet, J.-M. and Mormède, P. (1990) The integrity of the ventral noradrenergic bundle (VNAB) is not necessary for neuroendocrine stress response. Brain Res., 511: 349–352.

Ceccatelli, S., Villar, M.J., Goldstein, M. and Hökfelt, T. (1989) Expression of c-fos immunoreactivity in transmitter-characterized neurons after stress. Proc. Natl. Acad. Sci. USA, 86: 9569–9573.

Cecchi, M., Khoshbouei, H., Javors, M. and Morilak, D.A. (2002) Modulatory effects of norepinephrine in the lateral bed nucleus of the stria terminalis on behavioral and neuroendocrine responses to acute stress. Neuroscience, 112: 13–21.

Cenci, M.A., Kalen, P., Mandel, R.J. and Björklund, A. (1992) Regional differences in the regulation of dopamine and noradrenaline release in medial frontal cortex, nucleus accumbens and caudate-putamen: a microdialysis study in the rat. Brain Res., 581: 217–228.

Chalmers, D.T., Lovenberg, T.W. and de Souza, E.B. (1995) Localization of novel corticotropin-releasing factor 2 (CRF(2)) mRNA expression to specific subcortical nuclei in rat brain – comparison with CRF(1) receptor mRNA expression. J. Neurosci., 15: 6340–6350.

Chen, X. and Herbert, J. (1995) Regional changes in c-fos expression in the basal forebrain and brainstem during adaptation to repeated stress: correlations with cardiovascular, hypothermic and endocrine responses. Neuroscience, 64: 675–685.

Cole, R.L. and Sawchenko, P.E. (2002) Neurotransmitter regulation of cellular activation and neuropeptide gene expression in the paraventricular nucleus of the hypothalamus. J. Neurosci., 22: 959–969.

Cook, C.J. (1998) Monitoring on-line of extracellular gamma-amino-4-butyric acid using microdialysis coupled to immunosensor analysis. J. Neurosci. Meth., 82: 145–150.

Cook, C.J. (2001) Measuring of extracellular cortisol and corticotropin-releasing hormone in the amygdala using immunosensor coupled microdialysis. J. Neurosci. Meth., 110: 95–101.

Coote, J.H., Yang, Z., Pyner, S. and Deering, J. (1998) Control of sympathetic outflows by the hypothalamic paraventricular nucleus. Clin. Exp. Pharmacol. Physiol., 25: 461–463.

Coste, S.C., Murray, S.E. and Stenzel-Poore, M.P. (2001) Animal models of CRH excess and CRH receptor deficiency display altered adaptations to stress. Peptides, 22: 733–741.

Cullinan, W.E., Herman, J.P., Battaglia, D.F., Akil, H. and Watson, S.J. (1995) Pattern and time course of immediate early gene expression in the rat brain following acute stress. Neuroscience, 64: 477–505.

Cunningham, E.T. and Sawchenko, P.E. (1988) Anatomical specificity of noradrenergic inputs to the paraventricular and supraoptic nuclei of the rat hypothalamus. J Comp. Neurol., 274: 60–76.

da Costa, A.P.C., Wood, S., Ingram, C.D. and Lightman, S.L. (1996) Region-specific reduction in stress-induced *c-fos* mRNA expression during pregnancy and lactation. Brain Res., 743: 177–184.

da Costa, A.P.C., Kampa, R.J., Windle, R.J., Ingram, C.D. and Lightman, S.L. (1997) Region-specific immediate-early gene expression following the administration of corticotropin-releasing hormone in virgin and lactating rats. Brain Res., 770: 151–162.

Daugherty, W.P., Corley, K.C., Phan, T.H. and Boadle-Biber, M.C. (2001) Further studies on the activation of rat median raphe serotonergic neurons by inescapable sound stress. Brain Res., 923: 103–111.

Davis, M., Rainnie, D. and Cassell, M. (1994) Neurotransmission in the rat amygdala related to fear and anxiety. Trends Neurosci., 17: 208–214.

Day, H.E.W., Campeau, S., Watson, S.J. Jr and Akil, H. (1999) Expression of α_{1b} adrenoceptor mRNA in corticotropin-releasing hormone-containing cells of the rat hypothalamus and its regulation by corticosterone. J. Neurosci., 19: 10098–10106.

Day, J.C., Koehl, M., Le Moal, M. and Maccari, S. (1998) Corticotropin-releasing factor administered centrally, but not peripherally, stimulates hippocampal acetylcholine release. J. Neurochem., 71: 622–629.

Dayas, C.V., Buller, K.M., Crane, J.W., Xu, Y. and Day, T.A. (2001a) Stressor categorization: acute physical and psychological stressors elicit distinctive recruitment patterns in the amygdala and in medullary noradrenergic cell groups. Eur. J. Neurosci., 14: 1143–1152.

Dayas, C.V., Buller, K.M. and Day, T.A. (1999) Neuroendocrine responses to an emotional stressor: evidence for involvement of the medial but not the central amygdala. Eur. J. Neurosci., 11: 2312–2322.

Dayas, C.V., Buller, K.M. and Day, T.A. (2001b) Medullary neurones regulate hypothalamic corticotropin-releasing factor cell responses to an emotional stressor. Neuroscience, 105: 707–719.

Dazzi, L., Motzo, C., Imperato, A., Serra, M., Gessa, G.L. and Biggio, G. (1995) Modulation of basal and stress-induced release of acetylcholine and dopamine in rat brain by abecarnil and imidazenil, two anxioselective gamma-aminobutyric acid A receptor modulators. J. Pharmacol. Exp. Ther., 273: 241–247.

de Souza, E.B. (1995) Corticotropin-releasing factor receptors: physiology, pharmacology, biochemistry and role in central nervous system and immune disorders. Psychoneuroendocrinology, 20: 789–819.

del Arco, A. and Mora, F. (2001) Dopamine release in the prefrontal cortex during stress is reduced by the local activation of glutamate receptors. Brain Res. Bull., 56: 125–130.

Deutch, A.Y., Lee, M.C., Gillham, M.H., Cameron, D.A., Goldstein, M. and Iadarola, M.J. (1991) Stress selectively increases fos protein in dopamine neurons innervating the prefrontal cortex. Cereb. Cortex, 1: 273–292.

DiMicco, J.A., Samuels, B.C., Zaretskaia, M.V. and Zaretsky, D.V. (2002) The dorsomedial hypothalamus and the response to stress: part renaissance, part revolution. Pharmacol. Biochem. Behav., 71: 469–480.

Diorio, D., Viau, V. and Meaney, M.J. (1993) The role of the medial prefrontal cortex (cingulate gyrus) in the regulation of hypothalamic-pituitary-adrenal responses to stress. J. Neurosci., 13: 3839–3847.

Dunn, A.J. and Berridge, C.W. (1990) Physiological and behavioral responses to corticotropin-releasing factor administration: is CRF a mediator of anxiety or stress responses? Brain Res. Rev., 15: 71–100.

Ebner, K., Wotjak, C.T., Landgraf, R. and Engelmann, M. (2000) A single social defeat experience selectively stimulates the release of oxytocin, but not vasopressin, within the septal brain area of male rats. Brain Res., 872: 87–92.

Ebner, K., Wotjak, C.T., Landgraf, R. and Engelmann, M. (2002) Forced swimming triggers vasopressin release within the amygdala to modulate stress-coping strategies in rats. Eur. J. Neurosci., 15: 384–388.

Elmquist, J.K., Scammell, T.E., Jacobson, C.D. and Saper, C.B. (1996) Distribution of Fos-like immunoreactivity in the rat brain following intravenous lipopolysaccharide administration. J. Comp. Neurol., 371: 85–103.

Emmert, M.H. and Herman, J.P. (1999) Differential forebrain *c-fos* mRNA induction by ether inhalation and novelty: evidence for distinctive stress pathways. Brain Res., 845: 60–67.

Engelmann, M., Wotjak, C.T., Ebner, K. and Landgraf, R. (2000) Behavioural impact of intraseptally released vasopressin and oxytocin in rats. Exp. Physiol., 85: 125S–130S.

Figueiredo, H.F., Bruestle, A., Bodie, B., Dolgas, C.M. and Herman, J.P. (2003) The medial prefrontal cortex differentially regulates stress-induced *c-fos* expression in the forebrain depending on type of stressor. Eur. J. Neurosci., 18: 2357–2364.

Figueiredo, H.F., Dolgas, C.M. and Herman, J.P. (2002) Stress activation of cortex and hippocampus is modulated by sex and stage of estrus. Endocrinology, 143: 2534–2540.

Finlay, J.M. and Zigmond, M.J. (1997) The effects of stress on central dopaminergic neurons: possible clinical implications. Neurochem. Res., 22: 1387–1394.

Fuchs, E. and Flügge, G. (2003) Chronic social stress: effects on limbic brain structures. Physiol. Behav., 79: 417–427.

Funada, M. and Hara, C. (2001) Differential effects of psychological stress on activation of the 5-hydroxytrptamine- and dopamine-containing neurons in the brain of freely moving rats. Brain Res., 901: 247–251.

Gibbs, D.M. (1984) Dissociation of oxytocin, vasopressin and corticotropin secretion during different types of stress. Life Sci., 35: 487–491.

Giorgi, O., Lecca, D., Piras, G., Driscoll, P. and Corda, M.G. (2003) Dissociation between mesocortical dopamine release and fear-related behaviours in two psychogenetically selected lines of rats that differ in coping strategies to aversive conditions. Eur. J. Neurosci., 17: 2716–2726.

Goldstein, L.E., Rasmusson, A.M., Bunney, B.S. and Roth, R.H. (1994) The NMDA glycine site antagonist (+)-HA-966 selectively regulates conditioned stress-induced metabolic activation of the mesoprefrontal cortical dopamine but not serotonin systems: a behavioral, neuroendocrine, and neurochemical study in the rat. J. Neurosci., 14: 4937–4950.

Goldstein, L.E., Rasmusson, A.M., Bunney, B.S. and Roth, R.H. (1996) Role of the amygdala in the coordination of behavioral, neuroendocrine, and prefrontal cortical monoamine responses to psychological stress in the rat. J. Neurosci., 16: 4787–4798.

Graeff, F.G., Guimarães, F.S., de Andrade, T.G.C.S. and Deakin, J.F.W. (1996) Role of 5-HT in stress, anxiety, and depression. Pharmacol. Biochem. Behav., 54: 129–141.

Grahn, R.E., Will, M.J., Hammack, S.E., Maswood, S., McQueen, M.B., Watkins, L.R. and Maier, S.F. (1999) Activation of serotonin-immunoreactive cells in the dorsal raphe nucleus in rats exposed to an uncontrollable stressor. Brain Res., 826: 35–43.

Hajos-Korcsok, E., Robinson, D.D., Yu, J.H., Fitch, C.S., Walker, E. and Merchant, K.M. (2003) Rapid habituation of hippocampal serotonin and norepinephrine release and anxiety-related behaviors, but not plasma corticosterone levels, to repeated footshock stress in rats. Pharmacol. Biochem. Behav., 74: 609–616.

Han, J.S., Bizon, J.L., Chun, H.J., Maus, C.E. and Gallagher, M. (2002) Decreased glucocorticoid receptor mRNA and dysfunction of HPA axis in rats after removal of the cholinergic innervation to the hippocampus. Eur. J. Neurosci., 16: 1399–1404.

Haxhiu, M.A., Yung, K., Erokwu, B. and Cherniack, N.S. (1996) CO_2-induced c-fos expression in the CNS catecholaminergic neurons. Respir. Physiol., 105: 35–45.

Heilig, M., McLeod, S., Brot, M., Heinrichs, S.C., Menzaghi, F., Koob, G.F. and Britton, K.T. (1993) Anxiolytic-like action of neuropeptide Y: mediation by Y1 receptors in the amygdala and dissociation from food intake effects. Neuropsychopharmacology, 8: 357–363.

Heilig, M. and Thorsell, A. (2002) Brain neuropeptide Y (NPY) in stress and alcohol dependence. Rev. Neurosci., 13: 85–94.

Helfferich, F. and Palkovits, M. (2003) Acute audiogenic stress-induced activation of CRH neurons in the hypothalamic paraventricular nucleus and catecholaminergic neurons in the medulla oblongata. Brain Res., 975: 1–9.

Herman, J.P., Cullinan, W.E. and Watson, S.J. (1994) Involvement of the bed nucleus of the stria terminalis in tonic regulation of paraventricular hypothalamic CRH and AVP mRNA expression. J. Neuroendocrinol., 6: 433–442.

Herman, J.P., Dolgas, C.M. and Carlson, S.C. (1998) Ventral subiculum coodinates situation-specific neuroendocrine and behavioral stress responses. Neuroscience, 86: 449–459.

Herman, J.P., Figueiredo, H., Mueller, N.K., Ulrich-Lai, Y., Ostrander, M.M., Choi, D.C. and Cullinan, W.E. (2003) Central mechanisms of stress integration: hierarchical circuitry controlling hypothalamo-pituitary-adrenocortical responsiveness. Front. Neuroendocrinol., 24: 151–180.

Herman, J.P., Tasker, J.G., Ziegler, D.R. and Cullinan, W.E. (2002) Local circuit regulation of paraventricular nucleus stress integration: Glutamate-GABA connections. Pharmacol. Biochem. Behav., 71: 457–468.

Hitchcock, J. and Davis, M. (1986) Lesions of the amygdala, but not of the cerebellum or red nucleus, block conditioned fear as measured with the potentiated startle paradigm. Behav. Neurosci., 100: 11–22.

Horger, B.A. and Roth, R.H. (1996) The role of mesoprefrontal dopamine neurons in stress. Crit. Rev. Neurobiol., 10: 395–418.

Hotta, M., Shibasaki, T., Arai, K. and Demura, H. (1999) Corticotropin-releasing factor receptor type 1 mediates emotional stress-induced inhibition of food intake and behavioral changes in rats. Brain Res., 823: 221–225.

Imaki, T., Shibasaki, T., Hotta, M. and Demura, H. (1993) Intracerebroventricular administration of corticotropin-releasing factor induces *c-fos* mRNA expression in brain regions related to stress responses: comparison with pattern of *c-fos* mRNA induction after stress. Brain Res., 616: 114–125.

Imaki, T., Shibasaki, T., Wang, X.Q. and Demura, H. (1995) Intracerebroventricular administration of corticotropin-releasing factor antagonist attenuates *c-fos* mRNA expression in the paraventricular nucleus after stress. Neuroendocrinology, 61: 445–452.

Inoue, T., Tsucgiya, K. and Koyama, T. (1994) Regional changes in dopamine and serotonin activation with various intensity of physical and psychological stress in the rat brain. Pharmacol. Biochem. Behav., 49: 911–920.

Ishida, Y., Hashiguchi, H., Takeda, R., Ishizuka, Y., Mitsuyama, Y., Kannan, H., Nishimori, T. and Nakahara, D. (2002) Conditioned-fear stress increases Fos expression in monoaminergic and GABAergic neurons of the locus coeruleus and dorsal raphe nuclei. Synapse, 45: 46–51.

Ishizuka, Y., Ishida, Y., Jin, Q., Kato, K., Kunitake, T., Mitsuyama, Y. and Kannan, H. (2000) Differential profiles of nitric oxide and norepinephrine releases in the paraventricular nucleus region in response to mild footshock in rats. Brain Res., 862: 17–25.

Jessop, D.S., Renshaw, D., Larsen, P.J., Chowdrey, H.S. and Harbuz, M.S. (2000) Substance P is involved in terminating the hypothalamic-pituitary-adrenal axis response to acute stress through centrally located neurokinin-1 receptors. Stress, 3: 209–220.

Jezova, D., Ochedalski, T., Glickman, M., Kiss, A. and Aguilera, G. (1999) Central corticotropin-releasing hormone receptors modulate hypothalamic-pituitary-adrenocortical and sympathoadrenal activity during stress. Neuroscience, 94: 797–802.

Kaehler, S.T., Singewald, N., Sinner, C., Thurnher, C. and Philippu, A. (2000) Conditioned fear and inescapable shock modify the release of serotonin in the locus coeruleus. Brain Res., 859: 249–254.

Kawahara, Y., Kawahara, H. and Westerink, B.H. (1999) Comparison of effects of hypotension and handling stress on the release of noradrenaline and dopamine in the locus coeruleus and medial prefrontal cortex of the rat. Naunyn Schmiedebergs Arch Pharmacol., 360: 42–49.

Kawahara, H., Yoshida, M., Yokoo, H., Nishi, M. and Tanaka, M. (1993) Psychological stress increases serotonin release in the rat amygdala and prefrontal cortex assessed by in vivo microdialysis. Neurosci. Lett., 162: 81–84.

Kenney, M.J., Weiss, M.L. and Haywood, J.R. (2003) The paraventricular nucleus: an important component of the central neurocircuitry regulating sympathetic nerve outflow. Acta Physiol. Scand., 177: 7–15.

Khoshbouei, H., Cecchi, M., Dove, S., Javors, M. and Morilak, D.A. (2002) Behavioral reactivity to stress: amplification of stress-induced noradrenergic activation elicits a galanin-mediated anxiolytic effect in central amygdala. Pharmacol. Biochem. Behav., 71: 407–417.

Koob, G.F. (1999) Corticotropin-releasing factor, norepinephrine, and stress. Biol. Psychiatry, 46: 1167–1180.

Kozicz, T. (2001) Axon terminals containing tyrosine hydroxylase- and dopamine-β-hydroxylase immunoreactivity form synapses with galanin immunoreactive neurons in the lateral division of the bed nucleus of the stria terminalis in the rat. Brain Res., 914: 23–33.

Kozicz, T. (2002) Met-enkephalin immunoreactive neurons recruited by acute stress are innervated by axon terminals immunopositive for tyrosine hydroxylase and dopamine-alpha-hydroxylase in the anterolateral division of bed nuclei of the stria terminalis in the rat. Eur. J. Neurosci., 16: 823–835.

Krukoff, T.L. (1998) Central regulation of autonomic function: no brakes? Clin. Exp. Pharmacol. Physiol., 25: 474–478.

Krukoff, T.L., MacTavish, D. and Jhamandas, J.H. (1999) Effects of restraint stress and spontaneous hypertension on neuropeptide Y neurones in the brainstem and arcuate nucleus. J. Neuroendocrinol., 11: 715–723.

Ku, Y.-H., Tan, L., Li, L.-S. and Ding, X. (1998) Role of corticotropin-releasing factor and substance P in pressor responses of nuclei controlling emotion and stress. Peptides, 19: 677–682.

Landgraf, R. and Wigger, A. (2002) High vs low anxiety-related behavior rats: an animal model of extremes in trait anxiety. Behav. Genet., 32: 301–314.

Lavicky, J. and Dunn, A.J. (1993) Corticotropin-releasing factor stimulates catecholamine release in hypothalamus and prefrontal cortex in freely moving rats as assessed by microdialysis. J. Neurochem., 60: 602–612.

Li, H.Y., Ericsson, A. and Sawchenko, P.E. (1996) Distinct mechanisms underlie activation of hypothalamic neurosecretory neurons and their medullary catecholaminergic afferents in categorically different stress paradigms. Proc. Natl. Acad. Sci. USA, 93: 2359–2364.

Lightman, S.L., Windle, R.J., Wood, S.A., Kershaw, Y.M., Shanks, N. and Ingram, C.D. (2001) Peripartum plasticity within the hypothalamo-pituitary-adrenal axis. Prog. Brain Res., 133: 111–129.

Linthorst, A.C.E., Flachskamm, C., Holsboer, F. and Reul, J.M.H.M. (1996) Activation of serotonergic and noradrenergic neurotransmission in the rat hippocampus after peripheral administration of bacterial endotoxin: involvement of the cyclo-oxygenase pathway. Neuroscience, 72: 989–997.

Linthorst, A.C.E., Flachskamm, C., Müller-Preuss, P., Holsboer, F. and Reul, J.M.H.M. (1995) Effect of bacterial endotoxin and interleukin-1β on hippocampal serotonergic neurotransmission, behavioral activity, and free corticosterone levels: an *in vivo* microdialysis study. J. Neurosci., 15: 2920–2934.

Linthorst, A.C.E., Panalva, R.G., Flachskamm, C., Holsboer, F. and Reul, J.M.H.M. (2002) Forced swim activates rat hippocampal serotonergic neurotransmission involving a corticotropin-releasing hormone-dependent mechanism. Eur. J. Neurosci., 16: 2441–2452.

Loewy, A.D. (1991) Forebrain nuclei involved in autonomic control. Prog. Brain Res., 87: 253–268.

Lopez, J.F., Akil, H. and Watson, S.J. (1999) Neural circuits mediating stress. Biol. Psychiat., 46: 1461–1471.

Lovenberg, T.W., Liaw, C.W., Grigoriadis, D.E., Clevenger, W., Chalmers, D.T., de Souza, E.B. and Oltersdorf, T. (1995) Cloning and characterization of a functionally distinct corticotropin-releasing factor receptor subtype from rat brain. Proc. Natl. Acad. Sci. USA, 92: 836–840.

Lowry, C.A. (2002) Functional subsets of serotonergic neurones: implications for control of the hypothalamic-pituitary-adrenal axis. J. Neuroendocrinol., 14: 911–923.

Lowry, C.A., Rodda, J.E., Lightman, S.L. and Ingram, C.D. (2000) Corticotropin-releasing factor increases *in vitro* firing rates of serotonergic neurons in the rat dorsal raphe nucleus: evidence for activation of a topographically organized mesolimbocortical serotonergic system. J. Neurosci., 20: 7728–7736.

Ma, X.M. and Lightman, S.L. (1998) The arginine vasopressin and corticotrophin-releasing hormone gene transcription responses to varied frequencies of repeated stress in rats. J. Physiol., 510: 605–614.

Makino, S., Baker, R.A., Smith, M.A. and Gold, P.W. (2000) Differential regulation of neuropeptide Y mRNA expression in the arcuate nucleus and locus coeruleus by stress and antidepressants. J. Neuroendocrinol., 12: 387–395.

McCormick, C.M., Linkroum, W., Sallinen, B.J. and Miller, N.W. (2002) Peripheral and central sex steroids have differential effects on the HPA axis of male and female rats. Stress, 5: 235–247.

McQuade, R. and Stanford, S.C. (2000) A microdialysis study of the noradrenergic response in rat frontal cortex and hypothalamus to a conditioned cue for aversive, naturalistic environmental stimuli. Psychopharmacology, 148: 201–208.

Millan, M.J. (2003) The neurobiology and control of anxious states. Prog. Neurobiol., 70: 83–244.

Moghaddam, B. (1993) Stress preferentially increases extraneuronal levels of excitatory amino acids in the prefrontal cortex: comparison to hippocampus and basal ganglia. J. Neurochem., 60: 1650–1657.

Morilak, D.A., Cecchi, M. and Khoshbouei, H. (2003) Interactions of norepinephrine and galanin in the central amygdala and lateral bed nucleus of the stria terminalis modulate the behavioral response to acute stress. Life Sci., 73: 715–726.

Morrow, B.A., Elsworth, J.D., Lee, E.J. and Roth, R.H. (2000) Divergent effects of putative anxiolytics on stress-induced Fos expression in the mesoprefrontal system of the rat. Synapse, 36: 143–154.

Müller, M.B., Zimmermann, S., Sillaber, I., Hagemeyer, T.P., Deussing, J.M., Timpl, P., Kormann, M.S.D., Droste, S.K., Kühn, R., Reul, J.M.H.M., Holsboer, F. and Wurst, W. (2003) Limbic corticotropin-releasing hormone receptor 1 mediates anxiety-related behavior and hormonal adaptation to stress. Nat. Neurosci., 6: 1100–1107.

Nakane, H., Shimizu, N. and Hori, T. (1994) Stress-induced norepinephrine release in the rat prefrontal cortex measured by microdialysis. Am. J. Physiol., 267: R1559–R1566.

Neumann, I.D. (2001) Alterations in behavioral and neuroendocrine stress coping strategies in pregnant, parturient and lactating rats. Prog. Brain Res., 133: 143–152.

Neumann, I.D. (2002) Involvement of the brain oxytocin system in stress coping: interactions with the hypothalamo-pituitary-adrenal axis. Prog. Brain Res., 139: 147–162.

Nishioka, T., Anselmo-Franci, J.A., Li, P., Callahan, M.F. and Morris, M. (1998) Stress increases oxytocin release within the hypothalamic paraventricular nucleus. Brain Res., 781: 57–61.

Owens, M.J. and Nemeroff, C.B. (1991) Physiology and pharmacology of corticotropin-releasing factor. Pharmacol. Rev., 43: 425–473.

Pacák, K. (2000) Stressor-specific activation of the hypothalamic-pituitary-adrenocortical axis. Physiol. Res., 49: S11–S17.

Pacák, K. and Palkovits, M. (2001) Stressor specificity of central neuroendocrine responses: implications for stress-related disorders. Endocr. Rev., 22: 502–548.

Pacák, K., Palkovits, M., Kopin, I.J. and Goldstein, D.S. (1995) Stress-induced norepinephrine release in the hypothalamic paraventricular nucleus and pituitary-adrenocortical and sympathoadrenal activity: in vivo microdialysis studies. Front. Neuroendocrinol., 16: 89–150.

Palkovits, M. (1999) Interconnections between the neuroendocrine hypothalamus and the central autonomic system. Front Neuroendocrinol., 20: 270–295.

Pan, J.T., Lookingland, K.J. and Moore, K.E. (1995) Differential effects of corticotropin-releasing hormone on central dopaminergic and noradrenergic neurons. J. Biomed. Sci., 2: 50–56.

Pani, L., Porcella, A. and Gessa, G.L. (2000) The role of stress in the pathophysiology of the dopaminergic system. Mol. Psychiatry, 5: 133–138.

Pei, Q., Zetterström, T. and Fillenz, M. (1990) Tail pinch-induced changes in the turnover and release of dopamine and 5-hydroxytryptamine in different brain regions of the rat. Neuroscience, 35: 133–138.

Pezzone, M.A., Lee, W.S., Hoffman, G.E., Pezzone, K.M. and Rabin, B.S. (1993) Activation of brainstem catecholaminergic neurons by conditioned and unconditioned aversive stimuli as revealed by c-Fos immunoreactivity. Brain Res., 608: 310–318.

Phillips, M.L., Drevets, W.C., Rauch, S.L. and Lane, R. (2003) Neurobiology of emotion perception I: the neural basis of normal emotion perception. Biol. Psychiat., 54: 504–514.

Phillips, R.G. and LeDoux, J.E. (1992) Differential contribution of amygdala and hippocampus to cued and contextual fear conditioning. Behav. Neurosci., 106: 274–285.

Price, M.L., Curtis, A.L., Kirby, L.G., Valentino, R.J. and Lucki, I. (1998) Effect of corticotropin-releasing factor on brain serotonergic activity. Neuropsychopharmacology, 18: 492–502.

Puglisi-Allegra, S., Imperato, A., Angelucci, L. and Cabib, S. (1991) Acute stress induces time-dependent responses in dopamine mesolimbic system. Brain Res., 217–222.

Quirarte, G.L., Galvez, R., Roozendaal, B. and McGaugh, J.L. (1998) Norepinephrine release in the amygdala in response to footshock and opioid peptidergic drugs. Brain Res., 808: 134–140.

Rassnick, S., Hoffman, G.E., Rabin, B.S. and Sved, A.F. (1998) Injection of corticotropin-releasing hormone into the locus coeruleus or footshock increases neuronal Fos expression. Neuroscience, 85: 259–268.

Reyes, T.M., Walker, J.R., DeCino, C., Hogenesch, J.B. and Sawchenko, P.E. (2003) Categorically distinct acute stressors elicit dissimilar transcriptional profiles in the paraventricular nucleus of the hypothalamus. J. Neurosci., 23: 5607–5616.

Rhodes, M.E. and Rubin, R.T. (1999) Functional sex differences ('sexual diergism') of central nervous system cholinergic systems, vasopressin, and hypothalamic-pituitary-adrenal axis activity in mammals: a selective review. Brain Res. Rev., 30: 135–152.

Ritter, S., Watts, A.G., Dinh, T.T., Sanchez-Watts, G. and Pedrow, C. (2003) Immunotoxin lesion of hypothalamically projecting norepinephrine and epinephrine neurons differentially affects circadian and stressor-stimulated corticosterone secretion. Endocrinology, 144: 1357–1367.

Roche, M., Commons, K.G., Peoples, A. and Valentino, R.J. (2003) Circuitry underlying regulation of the serotonergic system by swim stress. J. Neurosci., 23: 970–977.

Roth, R.H., Tam, S.Y., Ida, Y., Yang, J.X. and Deutch, A.Y. (1988) Stress and the mesocorticolimbic dopamine systems. Ann. N.Y. Acad. Sci., 537: 138–147.

Sanchez, M.M., Young, L.J., Plotsky, P.M. and Insel, T.R. (1999) Autoradiographic and in situ hybridization localization of corticotropin-releasing factor 1 and 2 receptors in nonhuman primate brain. J. Comp. Neurol., 408: 365–377.

Santarelli, L., Gobbi, G., Debs, P.C., Sibille, E.L., Blier, P., Hen, R. and Heath, M.J.S. (2001) Genetic and pharmacological disruption of neurokinin 1 receptor function decreases anxiety-related behaviors and increases serotonergic function. Proc. Natl. Acad. Sci. USA, 98: 1912–1917.

Sauvage, M. and Steckler, T. (2001) Detection of corticotropin-releasing hormone receptor 1 immunoreactivity in cholinergic, dopaminergic, and noradrenergic neurons of the murine basal forebrain and brainstem nuclei. Potential implications for arousal and attention. Neuroscience, 104: 643–652.

Sawchenko, P.E., Li, H.Y. and Ericsson, A. (2000) Circuits and mechanisms governing hypothalamic responses to stress: a tale of two paradigms. Prog. Brain Res., 122: 61–78.

Sawchenko, P.E. and Swanson, L.W. (1983) The organization of the forebrain afferents to the paraventricular and supraoptic nuclei. J. Comp. Neurol., 186: 621–656.

Sawchenko, P.E. and Swanson, L.W. (1985) Localization, colocalization, and plasticity of corticotropin-releasing factor immunoreactivity in rat brain. Fed. Proc., 44: 221–227.

Senba, E., Matsunaga, K., Tohyama, M. and Noguchi, K. (1993) Stress-induced *c-fos* expression in the rat brain: activation mechanism of sympathetic pathway. Brain Res. Bull., 31: 329–344.

Servatius, R.J., Natelson, B.H., Moldow, R., Pogach, L., Brennan, F.X. and Ottenweller, J.E. (2000) Persistent neuroendocrine changes in multiple hormonal axes after a single or repeated stressor exposures. Stress, 3: 263–274.

Sharp, F.R., Sagar, S.M., Hicks, K., Lowenstein, D. and Hisanga, K. (1991) C-fos mRNA, Fos, and Fos-related antigen induction by hypertonic saline and stress. J. Neurosci., 11: 2321–2331.

Shimizu, N., Take, S., Hori, T. and Oomura, Y. (1992) In vivo measurement of hypothalamic serotonin release by intracerebral microdialysis: significant enhancement by immobilization stress in rats. Brain Res. Bull., 28: 727–734.

Singewald, N., Kouvelas, D., Mostafa, A., Sinner, C. and Philippu, A. (2000) Release of glutamate and GABA in the amygdala of conscious rats by acute stress and baroreceptor activation: differences between SHR and WKY rats. Brain Res., 864: 138–141.

Singewald, N., Zhou, G.Y. and Schneider, C. (1995) Release of excitatory and inhibitory amino acids from the locus coeruleus of conscious rats by cardiovascular stimuli and various forms of acute stress. Brain Res., 704: 42–50.

Singewald, N. and Philippu, A. (1998) Release of neurotransmitters in the locus coeruleus. Prog. Neurobiol., 56: 237–267.

Smagin, G.N. and Dunn, A.J. (2000) The role of CRF receptor subtypes in stress-induced behavioural responses. Eur. J. Pharmacol., 405: 199–206.

Smagin, G.N., Heinrichs, S.C. and Dunn, A.J. (2001) The role of CRH in behavioral responses to stress. Peptides, 22: 713–724.

Smith, G.W., Aubry, J.M., Dellu, F., Contarino, A., Bilezigian, L.M., Gold, L.H., Chen, R., Marchuk, Y., Hauser, C., Bentley, C.A., Sawchenko, P.E., Koob, G.F., Vale, W. and Lee, K.F. (1998) Corticotropin releasing factor receptor 1-deficient mice display decreased anxiety, impaired stress response, and aberrant neuroendocrine development. Neuron, 20: 1093–1102.

Smith, M.A., Banerjee, S., Gold, P.W. and Glowa, J. (1992) Induction of *c-fos* mRNA in rat brain by conditioned and unconditioned stressors. Brain Res., 578: 135–141.

Stalnaker, T.A. and Berridge, C.W. (2003) AMPA receptor stimulation within the central nucleus of the amygdala elicits a differential activation of central dopaminergic systems. Neuropsychopharmacology, 28: 1923–1934.

Stamp, J.A. and Herbert, J. (1999) Multiple immediate-early gene expression during physiological and endocrine adaptation to repeated stress. Neuroscience, 94: 1313–1322.

Steckler, T. and Holsboer, F. (1999) Corticotropin-releasing hormone receptor subtypes and emotion. Biol. Psychiatry, 46: 1480–1508.

Sugaya, K. and McKinney, M. (1994) Nitric oxide synthase gene expression in cholinergic neurons in the rat brain examined by combined immunocytochemistry and *in situ* hybridization histochemistry. Mol. Brain Res., 23: 111–125.

Sullivan, R.M. and Gratton, A. (1999) Lateralized effects of medial prefrontal cortex lesions on neuroendocrine and autonomic responses in rats. J. Neurosci., 19: 2834–2840.

Sullivan, R.M. and Gratton, A. (2002) Prefrontal cortical regulation of hypothalamic-pituitary-adrenal function in the rat and implications for psychopathology: side matters. Psychoneuroendocrinology, 27: 99–114.

Sutton, R.E., Koob, G.F., Le Moal, M., Rivier, C. and Vale, W.W. (1982) Corticotropin releasing factor produces behavioural activation in rats. Nature, 297: 331–333.

Sved, A.F., Cano, G. and Card, J.P. (2001) Neuroanatomical specificity of the circuits controlling sympathetic outflow to different targets. Clin. Exp. Pharmacol. Physiol., 28: 115–119.

Swanson, C.J., Perry, K.W. and Schoepp, D.D. (2004) The mGlu2/3 receptor agonist, LY354740, blocks immobilization-induced increases in noradrenaline and dopamine release in the rat medial prefrontal cortex. J. Neurochem., 88: 194–202.

Szafarczyk, A., Alonso, G., Ixart, G., Malaval, F. and Assenmacher, I. (1985) Diurnal stimulated and stress-induced ACTH release in rats is mediated by ventral noradrenergic bundle. Am. J. Physiol., 249: E219–E226.

Szafarczyk, A., Malaval, F., Laurent, A., Gibaud, R. and Assenmacher, I. (1987) Further evidence for a central stimulatory action of catecholamines on adrenocorticotropin release in the rat. Endocrinology, 121: 883–892.

Tajima, T., Endo, H., Suzuki, Y., Ikari, H., Gotoh, M. and Iguchi, A. (1996) Immobilization stress-induced increase of hippocampal acetylcholine and of plasma epinephrine, norepinephrine and glucose in rats. Brain Res., 720: 155–158.

Thorsell, A., Svensson, P., Wiklund, L., Sommer, W., Ekman, R. and Heilig, M. (1998) Suppressed neuropeptide Y (NPY) mRNA in rat amygdala following restraint stress. Regul. Pept., 75/76: 247–254.

Timmerman, W., Cisci, G., Nap, A., de Vries, J.B. and Westerink, B.H. (1999) Effects of handling on extracellular levels of glutamate and other amino acids in various areas of the brain measured by microdialysis. Brain Res., 833: 150–160.

Tsigos, C. and Chrousos, G. (2002) Hypothalamic-pituitary-adrenal axis, neuroendocrine factors and stress. J. Psychosom. Res., 53: 865–871.

Tuvnes, F.A., Steffenach, H.A., Murison, R., Moser, M.B. and Moser, E.I. (2003) Selective hippocampal lesions do not increase adrenocortical activity. J. Neurosci., 23: 4345–4354.

Vahabzadeh, A. and Fillenz, M. (1994) Comparison of stress-induced changes in noradrenergic and serotonergic neurons in the rat hippocampus using microdialysis. Eur. J. Neurosci., 6: 1205–1212.

Valles, A., Marti, O., Harbuz, M.S. and Armario, A. (2002) A single lipopolysaccharide administration is sufficient to induce a long-term desensitization of the hypothalamic-pituitary-adrenal axis. Neuroscience, 112: 383–389.

van de Kar, L.D. and Blair, M.L. (1999) Forebrain pathways mediating stress-induced hormone secretion. Front. Neuroendocrinol., 20: 1–48.

van de Kar, L.D., Javed, A., Zhang, Y.H., Serres, F., Raap, D.K. and Gray, T.S. (2001) 5-HT_{2A} receptors stimulate ACTH, corticosterone, oxytocin, renin, and prolactin release and activate hypothalamic CRF and oxytocin-expressing cells. J. Neurosci., 21: 3572–3579.

Vermetten, E. and Bremner, J.D. (2002) Circuits and systems in stress. I. Preclinical studies. Depress. Anxiety, 15: 126–147.

Weller, K.L. and Smith, D.A. (1982) Afferent connections to the bed nucleus of the stria terminalis. Brain Res., 232: 255–270.

Williams, C.L., Men, D., Clayton, E.C. and Gold, P.E. (1998) Norepinephrine release in the amygdala after systemic injection of epinephrine or escapable footshock: contribution of the nucleus of the solitary tract. Behav. Neurosci., 112: 1414–1422.

Windle, R.J., Kershaw, Y.M., Shanks, N., Wood, S.A., Lightman, S.L. and Ingram, C.D. (2004) Oxytocin attenuates stress-induced *c-fos* mRNA expression in specific forebrain regions associated with modulation of hypothalamo-pituitary-adrenal activity. J. Neurosci., 24: 2974–2982.

Windle, R.J., Shanks, N., Lightman, S.L. and Ingram, C.D. (1997) Central oxytocin administration reduces stress-induced corticosterone release and anxiety behavior in rats. Endocrinology, 138: 2829–2834.

Wong, M.L., Licinio, J., Pasternak, K.I. and Gold, P.W. (1994) Localization of corticotropin-releasing hormone (CRH) receptor mRNA in adult rat brain by in situ hybridization histochemistry. Endocrinology, 135: 2275–2278.

Wotjak, C.T., Kubota, M., Liebsch, G., Montkowski, A., Holsboer, F., Neumann, I. and Landgraf, R. (1996) Release of vasopressin within the rat paraventricular nucleus in response to emotional stress: a novel mechanism of regulating adrenocorticotropic hormone secretion? J. Neurosci., 16: 7725–7732.

Wright, D.E., Seroogy, K.B., Lundgren, K.H., Davis, B.M. and Jennes, L. (1995) Comparative localization of serotonin 1A, 1C, and 2 receptor subtype mRNAs in rat brain. J. Comp. Neurol., 351: 357–373.

Wu, J.S., Ku, Y.H., Li, L.S., Lu, Y.C., Ding, X. and Wang, Y.G. (1999) Corticotropin releasing factor and substance P mediate the nucleus amygdaloideus centralis-nucleus ventromedialis-nucleus dorsomedialis pressor system. Brain Res., 842: 392–398.

Yokoo, H., Tanaka, M., Yoshida, M., Tsuda, A., Tanaka, T. and Mizoguchi, K. (1990) Direct evidence of conditioned fear-elicited enhancement of noradrenaline release in the rat hypothalamus assessed by intracranial microdialysis. Brain Res., 536: 305–308.

Yokoyama, C. and Sasaki, K. (1999) Regional expressions of Fos-like immunoreactivity in rat cerebral cortex after stress; restraint and intraperitoneal lipopolysaccharide. Brain Res., 816: 267–275.

Young, E.A. (1998) Sex differences and the HPA axis: implications for psychiatric disease. J. Gend. Specif. Med., 1: 21–27.

Zhang, Y., Damjanoska, K.J., Carrasco, G.A., Dudas, B., D'Souza, D.N., Tetzlaff, J., Garcia, F., Hanley, N.R., Scripathirathan, K., Petersen, B.R., Gray, T.S., Battaglia, G., Muma, N.A. and van de Kar, L.D. (2002) Evidence that 5-HT_{2A} receptors in the hypothalamic paraventricular nucleus mediate neuroendocrine responses to (-)DOI. J. Neurosci., 22: 9635–9642.

Zou, C.J., Onaka, T. and Yagi, K. (1998) Role of adrenoceptors in vasopressin, oxytocin, and prolactin responses to conditioned fear stimuli in the rat. J. Neuroendocrinol., 10: 905–910.

SECTION 5

Neuroplasticity and Stress

T. Steckler, N.H. Kalin and J.M.H.M. Reul (Eds.)
Handbook of Stress and the Brain, Vol. 15
ISBN 0-444-51173-3

CHAPTER 5.1

The intracellular signaling cascade and stress

Yogesh Dwivedi* and Ghanshyam N. Pandey

Psychiatric Institute, Department of Psychiatry, University of Illinois at Chicago, 1601 W. Taylor St., Chicago IL, 60612, USA

Abstract: Exposure of stress precipitates a coordinated series of responses which may lead to changes in brain functions. Most of the changes in brain function are mediated by stress-induced activation of neurotransmitter systems in the brain, which include neurotransmitter receptors and these receptors-mediated signal transduction pathways. The signal transduction pathways are ultimately responsible for stress-induced changes in neuronal functions. Some of these changes are beneficial, however, depending upon the nature and duration of stressful stimuli, the changes in neuronal functions may be harmful, which may lead to many disorders. In the present chapter, we describe comprehensively the effects of stress on neurotransmitter receptors and various molecules of signal transduction pathways in brain.

Introduction

It is well established that several types of stressors, including psychosocial, social, or environmental stress, may lead to changes in brain function. Some of the changes due to acute stress are beneficial for the ability to deal with everyday challenges; however, failure to cope with the response to acute stress may lead to the syndrome of chronic stress, which is associated with alterations in the hypothalamic-pituitary-adrenal (HPA) axis and with elevated secretion of glucocorticoids, with consequent illnesses including immunosuppression and psychiatric disorders. One of the mechanisms by which these changes occur could be by influencing the activation of neurotransmitters, their receptors, and these receptor-mediated intracellular signaling. Several studies demonstrate that chronic stress may cause abnormalities in intracellular signaling that may lead to changes in the functional properties of various important proteins or in the expression of certain genes, which may in turn participate in stress-related disorders such as depression and other psychiatric disorders.

Intracellular signaling coordinates the behavior of individual cells within the brain in various physiological processes. This signaling requires three essential components: a molecular signal, also known as a neurotransmitter, that sends the information from one cell to another, a receptor that receives the signal and transmits the information provided by the signal, and a target molecule that mediates the cellular responses.

A large number of receptors have been identified and grouped into three different families depending upon the mechanism used to transduce signal binding into a cellular response. These families are (1) receptors linked to ligand-gated ion channels, (2) receptors that have intrinsic enzymatic properties, and (3) G protein-coupled receptors. The rapid action of stress is mediated by neurotransmitters that regulate neuronal activity via the gating of ion channels, which causes ion influx and changes the membrane potential or can lead to the entry of Ca^{2+} ions, which serves as a second messenger signal within the cell. On the other hand, most of the chronic effects of stress are mediated by G protein-coupled receptors or receptors that contain intrinsic enzymatic properties. Most neurotransmitter receptors belong to the

*Corresponding author. Tel.: (312) 413-4557;
Fax: (312) 355-3857; Email: ydwivedi@psych.uic.edu

G protein-coupled receptor family. These receptors regulate intracellular reactions by an indirect mechanism involving an intermediate transducing molecule known as guanosine 5′-triphosphate (GTP) binding protein or G protein. Activated G proteins alter the functions of many downstream effectors. Most of these effectors are enzymes that produce intracellular second messengers. The second messengers then trigger the signaling cascade further downstream and cause a cellular response. Stress may affect intracellular signaling at multiple levels. In the present chapter, the focus is primarily on the stress modulation of G protein-coupled neurotransmitter receptors, G proteins, second messengers, and second messenger-dependent and -independent effector molecules.

Stress and G protein-coupled receptors

The two most studied G protein-coupled receptors with respect to stress regulation are the serotonin (5HT) and adrenergic receptors. We will briefly describe the current state of knowledge of the effects of stress on these receptors.

Stress is known to alter the brain monoaminergic system. Several studies indicate that serotonergic neurons are especially affected by stress and related disorders (reviewed by Chaouloff, 1993). A marked increase in 5HT synthesis and release has been observed in various brain areas of experimental animals in response to different stressful conditions such as electrical footshock, tail pinches, immobilization stress, or a cold environment (Curzon et al., 1972; Joseph and Kennett, 1983; Pei et al., 1990; Clement et al., 1993). Since corticosterone (or cortisol in humans) is released after stress-induced activation of HPA axis, many studies have demonstrated that adrenalectomy and replacement of corticosterone or the administration of corticosterone exogenously influences tryptophan hydroxylase activity and 5HT turnover in the brain (Azmitia and McEwen, 1974; Singh et al., 1990).

Serotonin receptors

The neurotransmitter activity of 5HT has been attributed to its binding to 5HT receptors. 5HT receptors have been classified into three different families on the basis of their signal transduction mechanism association (Barnes and Sharp, 1999; Albert and Tiberi, 2001; Hoyer et al., 2002). Class I are those linked to the adenylyl cyclase-cyclic AMP signaling system, such as $5HT_{1A}$, $5HT_{1B}$, $5HT_{1D}$, and $5HT_4$; class II are those linked to the phosphoinositide (PI) hydrolysis signaling system, such as $5HT_{2A}$, $5HT_{2B}$, and $5HT_{2C}$; and class III are those linked to ion channels such as $5HT_3$. Cloning studies have also revealed the existence of $5HT_{1E}$, $5HT_{1F}$, $5HT_{2F}$, $5HT_5$, $5HT_6$, and $5HT_7$ receptor subtypes. Of these $5HT_{1E}$, $5HT_{1F}$, $5HT_5$, $5HT_6$, and $5HT_7$ are linked to the adenylyl cyclase-cAMP and $5HT_{2F}$ is linked to the PI signaling system. The specificity of the serotonergic responses to stress is made evident by the interactions with pre- and postsynaptic 5HT receptor subtypes.

Since $5HT_{1A}$ receptors are inhibitory somatodendritic receptor in raphe serotonergic cells and a postsynaptic receptor in selective serotonergic terminal fields, and since they play a key role in nerve firing activity and/or release of 5HT, many studies have focused on the role of $5HT_{1A}$ in stress and stress-related disorders. For example, chronic unpredictable stress significantly decreases the expression and number of binding sites for $5HT_{1A}$ receptors across all hippocampal subfields (López et al., 1998). Similar effects have been found following chronic restraint stress (Watanabe et al., 1993). Several studies also have demonstrated that administration of corticosterone or adrenalectomy can cause changes in the number or expression of $5HT_{1A}$ receptors (reviewed by Chaouloff, 1993; 1995). These receptors show differences in their degree of sensitivity in various brain regions toward lower-versus higher-circulating corticosterone levels. For example, continuous exposure to corticosterone decreases $5HT_{1A}$ receptor binding in the dentate gyrus and the CA4 areas of the hippocampus (Mendelson and McEwen, 1992a, 1992b). On the other hand, upon removal of corticosterone 3 weeks prior to sacrifice of the rat, $5HT_{1A}$ receptors increase in these areas of the hippocampus. A negative correlation between corticosterone treatment and $5HT_{1A}$ receptor mRNA in rat dentate gyrus has also been reported (Meijer and de Kloet, 1994). Cortical areas are also sensitive to glucocorticoid treatment toward $5HT_{1A}$ receptors.

This is evident from a study showing that corticosterone treatment to rats significantly decreases the number of $5HT_{1A}$ binding sites in frontal cortex (Crayton et al., 1996). Studies of the effects of adrenalectomy and replacement with corticosterone show that adrenalectomy increases $5HT_{1A}$ receptors in the CA1 area of the hippocampus, with parallel trends in CA4 and the dentate gyrus. These changes are reversed by the administration of corticosterone (Biegon et al., 1985; Rostene et al., 1985; Huang and Azmitia, 1999). Similar effects have been observed by many investigators who found an increase in $5HT_{1A}$ receptor binding in CA1, 2, 3, and the dentate gyrus after adrenalectomy, which was restored to control levels 7–28 days after corticosterone replacement (Burnet et al., 1992; Chalmers et al., 1994; Kuroda et al., 1994; Tijani-Butt and Labow, 1994; Zong and Ciarnello, 1995; Holmes et al., 1995a).

Another 5HT receptor subtype which has been studied the most with respect to stress and the HPA axis is $5HT_{2A}$. $5HT_{2A}$ receptors have been widely implicated in the pathophysiology of depression and other psychiatric disorders (Graeff, 1997; Bell and Nutt, 1998; van Veelen and Kahn, 1999; Mann, 1999; 2003). Initial studies on the effects of corticosterone or ACTH treatment revealed a significant increase in $5HT_{2A}$ receptor density in rat forebrain and adrenalectomy prevented this increase (Kuroda et al., 1992). Dexamethasone, an agonist to glucocorticoid receptors, also causes similar effects on $5HT_{2A}$ receptors in rat brain (Kuroda et al., 1993). Consistent with these observations, several other studies have also demonstrated that corticosterone treatment to rats increases $5HT_{2A}$ receptors in the frontal and parietal cortices (Fernandes et al., 1997; Takao et al., 1997). In our own studies, we observed that corticosterone treatment increases $5HT_{2A}$ receptors in both cerebral cortex and hippocampus (Pandey and Dwivedi, 1999). The effects of adrenalectomy alone on $5HT_{2A}$ receptors, on the other hand, have been inconsistent. Few studies demonstrate that adrenalectomy, like corticosterone treatment, increases $5HT_{2A}$ receptors in the hippocampus (Martire et al., 1989; Jitsuiki et al., 2000; Katagiri et al., 2001), whereas several other studies show that adrenalectomy does not affect $5HT_{2A}$ receptor density in frontal cortex of rats (Kuroda et al., 1992, 1994; Chaouloff, 1993).

Besides $5HT_{1A}$ and $5HT_2$ receptors several other 5HT receptor subtypes are also regulated by stress and glucocorticoids. $5HT_{2C}$ receptors, which have been involved in anxiety and various autonomic and neuroendocrine functions, have been shown to be increased after chronic stress in the CA2 area of the rat hippocampus. Depletion of cortisol by adrenalectomy also increases $5HT_{2C}$ receptor mRNA in posterior CA1 and CA3 areas, and this effect is reversed by glucocorticoid replacement (Holmes et al., 1995b). The newly discovered $5HT_6$ and $5HT_7$ receptors have also been the subjects of study in the regulation of stress. This is mainly because of their similarities with $5HT_{1A}$ receptors. Both $5HT_6$ and $5HT_7$ receptors are expressed in hippocampal and other limbic brain regions where $5HT_{1A}$ receptors are highly enriched (Ward et al., 1995; Gerard et al., 1996; Gustafson et al., 1996; Stowe and Barnes, 1998). But $5HT_6$ and $5HT_7$ receptors are coupled to adenylyl cyclase-cAMP signal transduction system in a positive manner (Bard et al., 1993; Plassat et al., 1993; Kohen et al., 1996; Hirst et al., 1997; Sleight et al., 1998), in contrast to $5HT_{1A}$ receptors, which are linked to this signaling system in a negative manner. The pharmacological properties of $5HT_6$ and $5HT_7$ receptors are also quite similar to those of $5HT_{1A}$ receptor (Barnes and Sharp, 1999). Using in situ hybridization studies, it has been shown that mRNA expression of $5HT_7$ receptors is increased in CA1 and CA3 areas of the hippocampus and in the retrosplenal cortex after adrenalectomy (Le Corre et al., 1997). Another study suggests that adrenalectomy not only increases the mRNA expression of $5HT_7$ receptors, but also increases the mRNA expression of $5HT_6$ receptor in hippocampus and these increases are partially reversed by corticosterone replacement (Yau et al., 1997).

Adrenergic receptors

Stress-induced disruption of central nervous noradrenergic activity is a well known phenomenon. An extensive body of literature has documented that there is increased sensitivity of the noradrenergic system in people under stress or suffering from depression (Bremner et al., 1996; Heninger et al.,

1998; Sullivan et al., 1999; Southwick et al., 1999a, 1999b, 1999c), in stressed nonhuman primates (Rosenblum et al., 1994), and in rats (Pacak et al., 1995; Dalley et al., 1996; Hellriegel and D'Mello, 1997). Many studies have also shown enhanced release of noradrenaline during stress exposure (Finlay et al., 1995; Goldstein et al., 1996; Birnbaum et al., 1999).

The activity of the noradrenergic and adrenergic system is regulated via adrenoreceptors. Various responses to stress reactions are known to involve changes in the sensitivity and number of adrenoreceptors. Among the various subtypes of adrenergic receptors, modulation in α_1, α_2, and β-adrenoreceptors plays an important role in stressful conditions (Stone et al., 1985; Weiss et al., 1994; Stone et al., 1996; Stone and Quartermain, 1999). Single-stress challenges cause downregulation of β-adrenergic receptors in the hypothalamus after footshock (Cohen et al., 1986) and in the cerebellum after immobilization stress (U'Prichard and Kvetnansky, 1980). The effect of repeated stress on β-adrenergic receptors, on the other hand, depends upon the paradigm used. In stress-induced behavioral depression (also known as learned helplessness) higher β-adrenergic receptor density has been reported in the hippocampus but not in the cortex or the hypothalamus of rats (Martin et al., 1990). In our study, we found higher β-adrenergic receptor density in hippocampus of learned helpless rats as compared with tested controls (Pandey et al., 1995). However, studies using other behavioral paradigm did not find significant differences in β-adrenergic receptor density in frontal cortex, hippocampus, or hypothalamus of learned helpless rats (Brannan et al., 1995; Gurguis et al., 1996). In contrast to these studies, immobilization stress combined with immersion of rats in cold water cause decrease in β-adrenergic receptor in the neurohypophysis and intermediate lobes of pituitary gland (Klenerova and Sida, 1994). Effect of a previous experience of stress by brief swim stress or daily injections of saline although do not affect β-adrenergic receptors; however, swim stress followed by a course of saline injections causes downregulation of β-adrenergic receptors (Davis et al., 1994).

As far as α adrenergic receptors are concerned, most studies show that a single-stress challenge does not cause any significant change in the number of binding sites for α_1 adrenergic receptors in rat brain (U'Prichard and Kvetnansky, 1980; Lynch et al., 1983; Cohen et al., 1986). However, a recent study suggests that restraint stress may alter expression of α_1 adrenergic receptors in a brain-region and time-dependent manner. In hypothalamus, 30 and 60 min restraint stress results in a significant decrease in α_1 adrenergic receptor mRNA. On the other hand, a significant increase in α_1 adrenergic receptor mRNA has been noted after 60, 120, and 240 min of restraint stress in mid brain (Miyahara, 1999). Additional studies have shown that the mRNA expression of a subtype of α_1 adrenergic receptor, i.e., α_{1b} is increased in the paraventricular nucleus of the hypothalamus after adrenalectomy and that corticosterone replacement reduces this increase (Day et al., 1999). Similar to α_1 adrenergic receptors, many studies have shown that α_2 adrenergic receptors are resistant to change after single-stress challenge (reviewed by Stanford, 1995). However, downregulation in the mid-brain and the brain stem after immobilization stress (U'Prichand and Kvetnansky, 1980; Weiss, 1994) and upregulation in the hypothalamus (Nukina et al., 1987) and the cortex (U'Prichand and Kvetnansky, 1980; Cohen et al., 1986) of α_2 adrenergic receptors after several types of single-stress challenge have also been reported.

When the same stress is given repeatedly, α_1 adrenergic receptor binding does not change (reviewed by Stanford, 1995) but the effect of repeated stress on α_2 adrenergic receptors is more labile, however, is quite inconsistent. For example, there are reports which suggest upregulation (U'Prichand and Kvetnansky, 1980), no change (Yamanaka et al., 1987), or even downregulation (Torda et al., 1981; Lynch et al., 1983) of α_2 adrenergic receptors in cerebral cortex after repeated stress. Chronic psychosocial stress, on the other hand, induces dynamic changes in region-specific up- or downregulation of α_2 adrenergic receptors (Flugge, 1996). A recent study suggests that chronic stress as well as short-term treatment with corticosterone downregulate α_2 adrenergic receptors in several brain regions of male tree shrews whereas long-term treatment with corticosterone upregulates α_2 adrenergic receptors in these animals (Flugge, 1999).

Stress and G proteins

Guanine nucleotide binding proteins (G proteins) occupy a central position and play a critical role in transducing extracellular signals to cellular targets, thus transmitting messages from cell surface receptors to cellular effectors (Neer, 1995; Clapham and Neer, 1997; Hamm, 1998; Freissmuth et al., 1999; Neves et al., 2002). About 80% of the receptors for neurotransmitters, hormones, and neuromodulators have been shown to elicit their responses through G proteins. G proteins are heterotrimers consisting of 3 subunits, α, β, and γ, each encoded by a specific gene. β and γ subunits bind tightly to each other, whereas the β subunit also contains a common binding site for α subunit recognition. The α subunit binds to guanosine triphosphate (GTP) and confers receptor–effector specificity to G proteins. The γ subunit has been reported to have a G protein-specific recognition site. In the resting state, guanosine diphosphate (GDP) is bound to the α subunit, and the three subunits (α, β, and γ) are associated as a trimer. Receptor-mediated activation of G proteins causes the release of GDP from the α subunit, allowing GTP to bind and induce the dissociation of the G protein α subunit from the $\beta\gamma$ subunits. The α and $\beta\gamma$ subunits can then activate various effectors to modulate cellular responses. The free α subunit has intrinsic GTPase activity, which causes the hydrolysis of bound GTP to GDP. This causes the reassociation of the α and the $\beta\gamma$ subunits with the G protein complex and thus termination of their activity. The lifetime of the activated G protein subunit depends upon the GTPase activating proteins (GAPs), which determine the rate at which the GTP bound to the α subunit is hydrolyzed (Clapham and Neer, 1997; Hamm, 1998; Freissmuth et al., 1999). The responsiveness of G protein-coupled receptors may desensitize in response to persistent stimulation. This may occur through the phosphorylation of G protein-coupled receptors by second messenger-regulated kinases, blocking their interaction with G proteins. A second mechanism is through the selective phosphorylation of activated receptors by specific G protein-coupled receptor kinases. This is followed by binding of a protein of the arrestin family to the phosphorylated receptors, blocking their ability to activate G proteins, and thus results in the termination of the response (Cabrera-Vera et al., 2003).

On the basis of cDNA sequencing and the similarity of amino acid sequences, Gα subunits have been classified into four major classes: G_s, G_i, G_q, and G_{12}. More than 16 distinct genes encode the G protein α subunits and there is a splice variant in at least two genes (Gilman, 1987; Simon et al., 1991; Neer, 1995; Hamm, 1998). Five distinct β subunit genes and 12 γ subunit genes have also been identified (Neer, 1995; Hildebrandt, 1997). Once activated, both α and $\beta\gamma$ subunits can activate or inhibit multiple effectors to modulate cellular responses. $G_s\alpha$ stimulates adenylyl cyclase, whereas $G_i\alpha$ mediates the inhibition of adenylyl cyclase. $G_q\alpha$ is coupled to phospholipase C, an enzyme that is involved in PI hydrolysis. $G_s\alpha$ and $G_i\alpha$ have also been shown to be operative in gating L-type Ca^{2+} channels and K^+ channels, respectively. The G_i family includes G_t (transducin), which activates cyclic gualyl monophosphate (cGMP) phosphodiesterase and two G_o isoforms. The role of $G_{12}\alpha$ is not clear. Recent studies suggest that $G_{12}\alpha$ can stimulate GAPs and nonreceptor tyrosine kinases. $G_o\alpha$ is believed to be associated with phosphoinositide hydrolysis and ion channels. In addition, G proteins are also linked to effectors such as phospholipase A_2, phosphodiesterases, and phosphoinositide 3 kinase, and to second messengers such as cyclic GMP, arachidonic acid, and phosphatidic acid (Marinissen and Gutkind, 2001). A particular G protein can couple to more than one effector. Recent studies also demonstrate that different classes of G proteins can couple to a single effector. This occurs through the molecular switching of one G protein-effector system to a different G protein-effector system, triggered by agonist-mediated phosphorylation of receptors (Daaka et al., 1997; Lefkowitz, 1998).

Several evidences show that G proteins may play an important role in stress. These evidences are based on either the effects of stress on the expression of G protein subunits or effects of glucocorticoids treatment on the expression or functions of G protein subunits. For example, rats subjected to restraint stress have been shown to exhibit increase in levels of $G_s\alpha$, $G_o\alpha$ and Gβ subunits in brain (Wolfgang et al., 1994). Corticosterone treatment to rats, on the one hand, increases the levels of $G_s\alpha$, on the other hand, decreases the levels of $G_i\alpha$ in the cerebral cortex. Corticosterone treatment to adrenalectomized rats increases the levels of $G_o\alpha$ and decreases the levels of $G_i\alpha$ and adrenalectomy produces the opposite changes in the levels of these G protein subunits rat cerebral cortex (Saito et al., 1989). In another study, it has been shown that a low dose of corticosterone treatment to adrenalectomized rats increases $G_s\alpha$ level whereas high dose of corticosterone increases levels of not only $G_s\alpha$, but also of $G_i\alpha$ and $G_o\alpha$ in hippocampus (Okuhara et al., 1997). Cell culture studies show that dexamethasone treatment of CH, UMR-106-01 or R0S17/2-8 cells increases the levels of $G_{q/11}\alpha$, $G_s\alpha$, and $G\beta$ subunits (Rodan and Rodan, 1986; Chang and Bourne, 1987; Mitchell and Bansal, 1997). Other studies demonstrate that glucocorticoids not only alter the levels of G proteins but also alter their functional response. For example, $5HT_{1A}$ receptor-mediated activation of an inward rectifying K^+ current, linked to pertussis toxin-sensitive G protein, is significantly altered in the CA3 field of corticosterone-treated rats, suggesting the possible role of G protein-mediated functions in $5HT_{1A}$ receptor-mediated neuronal activity (Okuhara and Beck, 1998).

Stress can also affect another class of G proteins, known as monomeric G proteins or small G proteins. These monomeric G proteins relay signals from activated cell surface receptors to intracellular targets such as the cytoskeleton and the vesicular trafficking apparatus of cells. A large number of small G proteins have been identified, including Ras, which helps regulate cell differentiation and proliferation by relaying signals from receptor kinases to the nucleus (Barbacid, 1987; Campbell et al., 1998; Downward, 1998, Huang and Reichardt, 2001; Downward, 2003). Several of the small GTPases are involved in vesicle trafficking in the presynaptic terminal, while others play an important role in protein and RNA trafficking in and out of the nucleus.

Regulators of G protein signaling, also known as RGS protein, negatively modulate G protein functions by accelerating the GTPase activity of G protein α subunits (Hepler, 1999; Ross and Wilkie, 2000; Ishii and Kurachi, 2003). Recent studies suggest that RGS may play an important role in corticosterone regulation of G protein functions. It has been found that the expression of RGS type 4, which is excessively expressed in the brain, is decreased in neuroendocrine AtT 20 cells and increased in locus coeruleus type cells CATHa after dexamethasone treatment (Ni et al., 1999). In a similar fashion chronic administration of corticosterone to rats decreases RGS type 4 mRNA level in the paraventricular nucleus of the hypothalamus but increases its level in the locus coeruleus. This differential regulation of RGS4 in these brain areas may contribute to the brain's region-specific and long-term adaptation to stress (Ni et al., 1999).

Stress and the adenylyl cyclase-cAMP signaling system

In the adenylyl cyclase-cAMP signaling system, agonist binding to receptors causes the activation of G proteins such as $G_s\alpha$, $G_i\alpha$, or $G\beta\gamma$ subunits. These G protein subunits then bind to the enzyme adenylyl cyclase and modulate its activity. Activation of adenylyl cyclase causes the conversion of ATP to cAMP, which serves as a second messenger. cAMP then activates the phosphorylation enzyme protein kinase A (PKA). Once activated, PKA phosphorylates various intracellular proteins and thereby modifies hormonal and neurotransmitter responses, including receptor downregulation or desensitization, alteration of neurotransmitter release, and activation or repression of gene expression (Fig. 1) (Nestler and Greengard, 1984; Borrelli et al., 1992; Spaulding, 1993). Since regulation of the intracellular level of a second messenger is critical to the transduction of a cellular response, many studies have examined cAMP generation after stress or after alterations of the HPA function. The intracellular level of cAMP is determined by the rate of its synthesis from ATP by adenylyl cyclase and its catabolic conversion to 5′AMP by a cyclic nucleotide, phosphodiesterase. It has been shown that repeated stress attenuate adenylyl cyclase activity in response to Ca^{2+}-calmodulin in cerebral cortex, without affecting the basal or forskolin-stimulated adenylyl cyclase activity (Gannon and McEwen, 1990). These changes are not present in hippocampus. Corticosterone treatment shows the similar response in cerebral cortex (Gannon and McEwen, 1990). On the other hand, adrenalectomy attenuates hippocampal, but not cortical, calmodulin-dependent adenylyl cyclase activity (Gannon et al., 1991). Downregulation of β-adrenergic receptors, which are linked to the

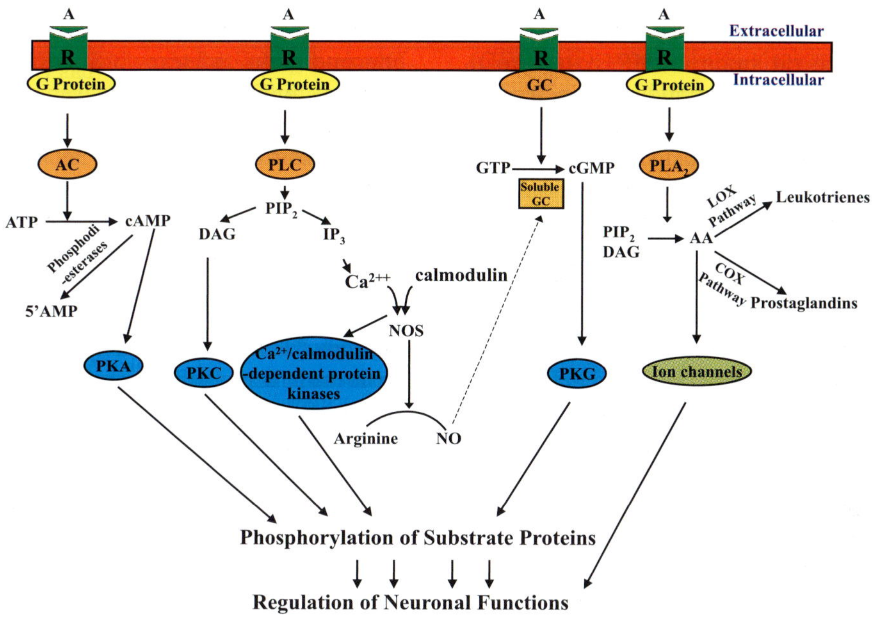

Fig. 1. Schematic representation of major intracellular signaling mechanisms regulated by stress. Agonist (A) binding to neurotransmitter receptors (R) causes activation of G proteins. There are several subtypes of G proteins. Some of the G proteins are stimulatory in nature (G_s), whereas others are inhibitory in nature (G_i). G_s protein stimulates whereas G_i protein inhibits adenylyl cyclase (AC) in adenylyl cyclase-cAMP signaling pathway. Another subtype of G protein, i.e. G_q activates phospholipase C (PLC) enzyme in phosphoinositide signaling pathway. Certain G proteins such as G_o are linked to ion channels. Activation of adenylyl cyclase causes formation of cAMP from ATP. The response of cAMP is terminated by hydrolysis of cAMP to 5′AMP by phosphodiesterases. Activation of PLC leads to the hydrolysis of phosphatidylinositol 4,5-biphosphate into diacylglycerol (DAG) and inositol triphosphate (IP_3). IP_3 mobilizes Ca^{2+} from intracellular sources. Signaling induced by release of nitric oxide (NO) involves activation of nitric oxide synthase (NOS) activated by Ca^{2+} released from intracellular sources by IP_3. NOS then converts arginine to nitric oxide (NO). NO activates cGMP. In the phospholipase A_2 (PLA_2)-mediated signaling pathway, binding of agonist to the receptor activates, through a G protein, the enzyme PLA_2. This enzyme then hydrolyzes PIP_2, forming arachidonic acid (AA). AA is metabolized into leukotrienes or prostaglandins through lipooxygenase (LOX) or cyclooxygenase (COX) pathway. The second messengers such as cAMP, DAG, IP_3, Ca^{2+}, cGMP or AA mediate the neuronal functions either by activating certain protein kinases or by activating ion channels. For example, DAG activates protein kinase C (PKC), whereas, cAMP activates protein kinase A (PKA). Ca^{2+} activates Ca^{2+}-dependent protein kinases whereas cGMP activates cGMP-dependent protein kinase (PKG). Formation of AA leads to the modulation of ion channels such as potassium channels. The protein kinases then phosphorylate specific proteins in the cytoplasm or transcription factors in the nucleus, which results in the alterations in functions of proteins or modulation of the transcription of specific genes. This leads to the generation of specific biological responses.

adenylyl cyclase-cAMP signaling system in a positive manner, is associated with the desensitization of cAMP synthesis after corticosterone treatment (Stone, 1979, Stone et al., 1984). Repeated footshock, although does not downregulate β-adrenergic receptors, it causes desensitization of cAMP synthesis (Stone, 1979). Repeated stress can also attenuate noradrenergic-isoprenaline-induced cAMP formation (Stone et al., 1984, 1985). Adrenalectomy, on the other hand, not only increases noradrenergic-stimulated cAMP accumulation in the limbic forebrain and the frontal cortex (Mobley and Sulser, 1980; Mobley et al., 1983) but also in hippocampus without altering β-adrenergic receptor number (Harrelson et al., 1987; Harlesson and McEwen, 1987). The administration of corticosterone to adrenalectomized rats reverses the effects caused by adrenalectomy in both the hippocampus (Roberts et al., 1984) and the frontal cortex (Mobley et al., 1983). In another study, it has been shown that high levels of corticosterone can attenuate noradrenaline response even below control levels in rat hippocampus (Gannon and McEwen,

1990). Dexamethasone or ACTH can enchance the sensitivity of the rat brain cAMP system to a variety of agents known to stimulate this second messenger system. These studies suggest a bimodal effect of glucocorticoids, in that an initial response to continuous treatment with dexamethasone or ACTH is a decrease in α-adrenergic receptor-stimulated cAMP production but more prolonged administration of these substances also increases β-adrenergic-stimulated cAMP formation. Furthermore, dexamethasone and corticosterone both enhance the sensitivity of the rat brain cAMP-generating system to forskolin (Duman et al., 1989; Czyrak, 1996).

Besides the effects of stress or of glucocorticoid administration on cAMP formation, alterations in the expression of adenylyl cyclase have also been reported. Adenylyl cyclase exists in multiple forms, each a unique gene product (Tang and Gilman, 1992). Adenylyl cyclase can be grouped into three main classes: (1) Types II, IV, and VII are activated synergistically by $G_s\alpha$ and $G\beta\gamma$ subunits; (2) Types V and VI are inhibited by $G_i\alpha$ and Ca^{2+} ; and (3) Types VIII, I, and III are activated synergistically by $G_s\alpha$ together with Ca^{2+}-calmodulin. Adrenalectomy causes a decrease in the expression of Type II and Type I adenylyl cyclase mRNA in rat hippocampus. When sham or adrenalectomized rats are exposed to 1 hr restraint stress for 4 days, the stressed rats show increased expression of both Type I and Type II adenylyl cyclase in hippocampus. Stressed adrenalectomized rats also show an increase in the expression of Type II adenylyl cyclase without any change in the expression of Type I adenylyl cyclase in the hippocampus (Wolfgang et al., 1994).

As mentioned earlier, cAMP generated in response to adenylyl cyclase activates phosphorylating enzyme protein kinase A (PKA). Activation of this enzyme is crucial for various cellular functions. PKA is a holoenzyme composed of two homodimeric regulatory (R) and two catalytic (C) subunits. In the absence of cAMP, PKA is inactive and exists as a stable tetramer. After an increase in intracellular cAMP, the regulatory PKA subunits bind to cAMP in a cooperative manner, which results in the disassociation of the holoenzyme into a dimeric regulatory unit and two monomers of catalytic subunits. The free catalytic subunits then phosphorylate substrates in the cytosol or translocate into the nucleus and phosphorylate nuclear proteins. Thus both the catalytic and the regulatory subunits are important in facilitating PKA-mediated functions. The response generated by cAMP can be terminated by the hydrolysis of cAMP into 5′AMP by phosphodiesterases or by removal of the phosphate group by protein phosphatases. The activity of phosphodiesterases can be enhanced through phosphorylation by PKA, thus causing an abrupt termination of the signal.

To examine whether PKA is regulated by endogenous glucocorticoids, we have extensively studied various aspects of PKA in rat brain at different time intervals after bilateral adrenalectomy, i.e., after days 1, 4, and 14. We have also studied if the effects on PKA caused by adrenalectomy are reversed by exogenous treatment with corticosterone and if corticosterone treatment to normal rats causes changes opposite to those observed with adrenalectomized rats (Dwivedi and Pandey, 2000a). We observed that 1 day of corticosterone treatment has no significant effects, but 4 days of corticosterone treatment decreases both $[^3H]$cAMP binding to the regulatory subunit of PKA and PKA catalytic activity in the rat cortex and hippocampus. These changes are much more profound after 14 days of corticosterone. These effects are also dose-dependent. The higher dose of corticosterone is much more effective in causing changes in PKA than the lower dose. Adrenalectomy, opposite to corticosterone treatment, increases PKA activity and $[^3H]$cAMP binding in both cortex and hippocampus in a time-dependent manner. These changes are reversed by corticosterone treatment in a dose-dependent manner. The higher dose of corticosterone completely reverses the changes in PKA after adrenalectomy (Dwivedi and Pandey, 2000a).

To examine whether the changes in regulatory and catalytic properties of PKA due to adrenalectomy or corticosterone treatment are related to expression of regulatory and/or catalytic subunits of PKA, we examined the mRNA and protein expression of individual subunits of PKA. Multiple isoforms of regulatory (RIα, RIIα, RIβ, and RIIβ) and catalytic (Cα, Cβ, and Cγ) subunits exist and are encoded by separate genes (Scott et al., 1987; Clegg and McKnight, 1988). The tissue distribution of regulatory subunits is such that RIα and RIIα are present ubiquitously, whereas RIβ is present in the brain and in developing sperms. However, RIIβ is the predominant isoform and the principal mediator of cAMP-mediated activity in the CNS. The catalytic subunit isoforms Cα and Cβ are ubiquitously expressed, although Cβ is the predominant isoform in the brain and Cγ is a testis-specific isoform. In our study, we found that the mRNA and protein expressions of RIα, RIIβ, and Cβ isoforms are significantly decreased in cortex and hippocampus after corticosterone treatment. Removal of adrenal glands increases the expression of these subunits, and corticosterone treatment of adrenalectomized rats reverses the adrenalectomy-induced changes in PKA RIα, RIIβ, and Cβ subunits (Dwivedi and Pandey, 2000a). Our studies thus suggest that the expression of specific isoforms of PKA regulatory and catalytic subunits is under the regulation of glucocorticoids and these changes may be relevant in stress-related disorders.

Stress and the phosphoinositide signaling pathway

In this lipid signaling pathway, agonist binding to neurotransmitter receptors such as $5HT_{2A}$, $5HT_{2C}$, or α_1 adrenergic causes activation of G proteins, which then activate the enzyme phosphoinositide-specific phospholipase C (PI-PLC). PI-PLC then hydrolyzes phosphatidylinositol 4,5 biphosphate (PIP_2) and generates two important intracellular second messengers: inositol 1,4,5-triphosphate (IP_3) and diacylglycerol (DAG) (Berridge and Irvine, 1989; Joseph and Williamson, 1989). DAG remains within the membrane and activates the phosphorylating enzyme protein kinase C (PKC), which phosphorylates substrate proteins in the plasma membrane and cytosol. DAG is short-lived and is rapidly phosphorylated to form phosphatidate by diacylglycerol kinase, thus causing the termination of PKC activation. The other second messenger, IP_3, leaves the cell membrane and diffuses within the cytosol. IP_3 then binds to IP_3 receptors and mobilizes Ca^{2+} from intracellular sources such as the endoplasmic reticulum (Fig. 1). These events mediate cellular activation and subsequent biological responses, such as neurotransmitter release, cell growth, differentiation, neuronal development, and gene expression (Nishizuka, 1998; Berridge and Irvine, 1989).

Several reports suggest that stress or glucocorticoids can modulate the overall PI signaling system or alter specific components of this signaling system. For example, greater induction of PI hydrolysis induced by noradrenaline, or the excitatory amino acid agonists quisqualate and *trans*- amino-1,3-cyclopentanedicarboxylic acid, in rat hippocampal slices from adrenalectomized rats has been reported (Kolasa et al., 1992). It has been suggested that these enhanced responses may be due to the removal of the inhibitory influence of glucocorticoids on phosphoinositide metabolism. Administration of dexamethasone every second day for 14 days on the other hand markedly reduces noradrenaline-stimulated PI metabolism in the rat hippocampus. In the cortex, this effect is smaller than in the hippocampus (Takahashi et al., 1996). In a further attempt to examine whether the changes in PI metabolism by glucocorticoids are related to the effects on the components of the PI signaling pathway, we and other investigators have examined effector proteins such as PI-PLC, the second messenger IP_3, and the downstream phosphorylating enzyme PKC after glucocorticoid treatment or after adrenalectomy. On the basis of sequence homology and the localization of structural domains, PLC has been characterized into three major families: PLC β_1, δ_1, and γ_1 (Rhee and Choi, 1992; Exton, 1994). All PLC isozymes recognize PIP_2 as a substrate and carry out Ca^{2+}-dependent hydrolysis of inositol lipids; however, these isozymes are differentially regulated and expressed (Cockcroft and Thomas, 1992). It has been shown that PLC β is activated by receptors that activate the $G_{q/11}$, G_i, or G_o family of G proteins. Also, $G\beta_1$ in combination with $\gamma_{2,3}$ or γ_5, or β_2 with γ_2 G proteins also activates the PLC β isozyme. PLC γ is regulated by receptor and nonreceptor tyrosine kinases. Little is known about the regulation of PLC δ.

In a recent study, we observed that chronic but not acute treatment of rats with dexamethasone increases the catalytic activity of PI-PLC in cortex and hippocampus. This increase is associated with a selective increase in expression of the PLC β_1 isozyme, without any change in γ_1 or δ_1 isozyme of PLC (Dwivedi and Pandey, 1999a). To further examine whether the endogenous glucocorticoid corticosterone causes similar or dissimilar effects to that of dexamethasone on PI-PLC, normal rats as well as adrenalectomized rats were treated with low or high doses of corticosterone. These experiments were performed after single or repeated administrations of corticosterone (14 days). Interestingly, 14-days duration of corticosterone treatment to normal rats dose-dependently decreases PI-PLC activity and selectively the mRNA and protein expression of the PLC β_1 isozyme in cortex and hippocampus. Bilateral adrenalectomy causes the opposite changes in these measures, and corticosterone treatment into adrenalectomized rats reverses these effects. The effects of corticosterone, thus, are opposite to that of dexamethasone. One of the possibilities could be poor penetration of dexamethasone in the brain. It has been reported that dexamethasone, when administered to adrenalectomized rats, is poorly retained by nuclear receptors in brain whereas at the same time a large amount of this synthetic glucocorticoid is retained in the pituitary (de Kloet, 2000). This poor retention is due to presence of multidrug resistance P-glycoprotein (mdr 1a-) in the blood–brain barrier, which extrudes dexamethasone along with other xenobiotics (Schinkel et al., 1996). Meijer et al. (1998) showed that adrenalectomized mice with a genetic disruption of mdr 1a gene had ten-fold increase in dexamethasone retention in brain, reaching levels observed in the pituitary, confirming that mdr 1a-P-glycoproteins are involved in extrusion of dexamethasone from brain. These studies support the concept of a pituitary site of action of dexamethasone, in which on one hand, dexamethasone blocks stress-induced release of ACTH from pituitary, on the other hand, it does not replace the endogenous corticosteroids in brain. This creates a state of 'chemical adrenalectomy' (reviewed by de Kloet, 2000). Our findings of the opposite effects of dexamethasone and corticosterone on PI-PLC in brain could be due to this so-called 'chemical adrenalectomy' caused by dexamethasone.

Another component in the PI pathway that plays an important role in mediating the functional response to this signaling mechanism is IP_3. IP_3 is generated by hydrolysis of PIP_2 by PI-PLC. The IP_3 signal is physiologically effective only on IP_3 receptors, which transduce this intermediate signal to a Ca^{2+} signal by mobilizing Ca^{2+} from intracellular sources. Ca^{2+} is required for the activation of many regulatory enzymes, including Ca^{2+}/calmodulin-dependent protein kinases and PKC. We have examined the status of IP_3 levels and of IP_3 receptors both in response to corticosterone treatment and after adrenalectomy and supplementation of corticosterone to adrenalectomized rats (Dwivedi et al., 2000b). One day of treatment with corticosterone has no significant effects on IP_3 levels in cortex, hippocampus, or cerebellum of rat. However, 14 days' duration of adrenalectomy causes a substantial and significant increase in the levels of IP_3 in all these brain areas. Treatment of low or high dose of corticosterone to adrenalectomized rats decreases IP_3 level in a dose-dependent manner. Furthermore, although acute corticosterone treatment to normal rats (1 day) does not change the IP_3 level, 14 days of treatment with corticosterone decreases IP_3 level in a dose-dependent manner. The binding characteristics of IP_3 in the brains of adrenalectomized and corticosterone-treated rats reveals that adrenalectomy decreases the number of IP_3 binding sites in cortex, hippocampus, and cerebellum and that corticosterone treatment of adrenalectomized rats reverses these changes. Furthermore, corticosterone treatment of normal rats produces changes opposite to those elicited by adrenalectomy. Because IP_3 binds to all the IP_3 receptor isoforms and because the specificity and binding affinity of IP_3 do not differ in different classes of IP_3 receptors, we have examined whether the decrease in number of IP_3 binding sites is due to an altered expression of IP_3 receptor isoforms. Three distinct types of the IP_3 receptor family have been molecularly cloned: IP_3 RI, IP_3 RII, and IP_3 RIII. Corticosterone treatment does not affect the expression of any of the IP_3 receptor isoforms. It is possible that the decrease in IP_3 levels caused by corticosterone is a consequence of a decrease in PI-PLC activity, and that the number of binding sites for IP_3 increases after corticosterone treatment as a compensatory event in response to the decreased IP_3 levels (Dwivedi et al., 2000b).

The other second messenger generated in response to PLC is diacylglycerol. Diacylglycerol activates the phosphorylating enzyme PKC. The isoforms of PKC constitute a family of Ca^{2+}-/or phospholipid-dependent serine/threonine kinases. Twelve subtypes of PKC have been identified and grouped into four major classes: conventional, Ca^{2+}-dependent (α, βI, βII, and γ); novel, Ca^{2+}-independent (δ, ε, η, and θ); atypical (ξ, λ, and τ); and PKC μ (Mellor and Parker, 1998). Novel PKCs differ from the conventional PKCs because they do not require Ca^{2+} for activation, and they can be differentiated from the atypical PKCs, which are also Ca^{2+}-dependent but are insensitive to activation of diacylglycerol and to phorbol esters (Nishizuka, 1998; Newton, 2001). Of these types, the conventional PKCs are involved in the cellular regulation of hormonal stimulation. Structurally, PKC consists of a regulatory domain (containing a pseudosubstrate sequence cystine-rich region and a NH_2 terminus) and a catalytic domain (containing a carboxyl terminus and binding sites for ATP and substrate proteins). Activation of diacylglycerol by phorbol esters requires the translocation of the enzyme from the cytoplasm to the membrane. Diacylglycerol increases the affinity of PKC for phospholipid binding and reduces the requirement of Ca^{2+}. Phorbol dibutyrate (PDBU) binds to the regulatory domain of PKC and thus can provide the status of regulatory activity of PKC, whereas the catalytic activity of PKC can be determined by measuring the transfer of the γ phosphate group of ATP to a PKC substrate.

In our study, we have shown that repeated administration of dexamethasone to rats increases both the catalytic activity of PKC and [^{3}H] PDBU binding to its regulatory subunits in cortex and hippocampus (Dwivedi and Pandey, 1999a). Furthermore, dexamethasone selectively increases the expression of PKC γ and ε in cortex and hippocampus. Similarly as with dexamethasone treatment, depletion of corticosterone by adrenalectomy causes significant increases in [^{3}H]PDBU binding to PKC regulatory subunits and in the catalytic activity of PKC in rat cortex and hippocampus. This is associated with increases in the expression of PKC γ and ε isozymes in these brain areas (Dwivedi and Pandey, 1999b). Treatment of adrenalectomized rats with corticosterone dose-dependently reverses the adrenalectomy-induced increase in [^{3}H]PDBU binding, catalytic activity, and expression of PKCγ and ε isozymes. Exogenous corticosterone treatment causes changes opposite to those observed with dexamethasone or adrenalectomy (Pandey and Pandey, 1999). These results thus suggest that PKCs are under the regulation of glucocorticoids, which in turn may play an important role in stress and in altered HPA function. Studies from other investigators have shown that chronic psychosocial stress to rats specifically increases PKC γ levels in the dentate gyrus and CA2 at anterior hippocampal levels, but corticosterone treatment does not affect PKC γ levels in these areas of the hippocampus. Furthermore, inhibition of PKC attenuates immobilization stress-induced plasma corticosterone levels in mice (Kim et al., 2000).

Stress and calcium regulation

Calcium is an important modulator of many signaling systems, including phosphoinositide. Several acute effects of corticosterone can be influenced by rapid changes in extracellular Ca^{2+}. For example, corticosterone increases depolarization-induced Ca^{2+} influx and potentiates calmodulin-induced activation of voltage-sensitive calcium channels (Sze and Iqbal, 1994). In cell culture studies, it has been shown that corticosterone inhibits nicotine-induced Ca^{2+} release in PC12 cells. This release is blocked by pertussis toxin and PKC inhibitors (Qiu et al., 1998). Furthermore, dexamethasone potentiates $5HT_2$ receptor-mediated Ca^{2+} mobilization in C6 glioma cells (Muraoka et al., 1993). In rat hippocampal neurons, corticosterone inhibits the N- and L-type Ca currents^{2+} which is blocked by pertussis toxin (Ffrench-Mullen, 1995). Extracellular Ca^{2+} decreases or increases the corticosterone-induced decrease in the population spike amplitude in hippocampal neurons. Antiglucocorticoids inhibit the electrophysiological effects of corticosterone (Talmi et al., 1992).

Stress and phospholipase A_2

Another target of G protein action is phospholipase A_2. This enzyme acts on DAG and on certain membrane phospholipids, such as PIP_2, to release the fatty acid arachidonic acid and lysophospholipids (Farooqui et al., 1997; Kudo and Murakami, 2002;

Brown et al., 2003). These lipid products may serve as second messengers. The arachidonic acid is then metabolized, either by the lipooxygenase pathway, forming products like leukotrienes or the cyclooxygenase pathway, leading to the formation of prostaglandins and thromboxanes. Arachidonic acid modulates neuronal signaling by direct effects on ion channels or indirectly through the activation of PKC and through the activation of metabolites (Fig. 1). Many glucocorticoids have been shown to inhibit prostaglandin formation by preventing arachidonic acid release from phospholipids as well as the induction of cyclooxygenase (Blackwell et al., 1980; Croxtall et al., 2002) suggesting that elevation of glucocorticoids in response to stress can affect signaling mechanism mediated by phospholipase A_2.

Stress and nitric oxide

Another important G protein-coupled receptor signaling pathway is mediated through nitric oxide (Bredt et al., 1990; Forstermann et al., 1995). Nitric oxide is produced from arginine by nitric oxide synthase and acts as a transmitter by diffusing from the cytosol and activating guanylyl cyclase. Receptors such as glutamate or *N*-methyl-D-aspartate (NMDA) receptors lead to the activation of PLC, which results in the formation of IP_3 and the release of Ca^{2+} from intracellular sources. Ca^{2+} combines with calmodulin and activates nitric oxide synthase, producing nitric oxide. Nitric oxide then stimulates guanylyl cyclase, causing an increase in cyclic guanine monophosphate (GMP), which in turn activates cGMP-dependent protein kinases. This results in the phosphorylation of proteins and subsequently in the modulation of K^+ and Ca^{2+} channels (Fig. 1).

Numerous studies show that nitric oxide participates in the control of many neurosecretory processes, including the corticotropin-releasing factor (CRF) neurosecretory system (Costa et al., 1993; Sandi and Guaza, 1995; Mcleod et al., 2001). Stress may influence neuronal excitability, in which nitric oxide functions as an amplifier or as a feedback regulator of neuronal excitation or inhibition, thus altering the homeostasis of the neurosecretory system (reviewed by Riedel, 2000). Long-term stress has been shown to increase the activity of Ca^{2+}-dependent nitric oxide synthase and induces the expression of inducible nitric oxide synthase in the rat cortex (Lorenzo Fernandez, 1999). It has been shown that inducible nitric oxide synthase production may be responsible for the neurodegenerative diseases caused by stress. Recently it has been reported that nitric oxide synthase may be responsible for the glucocorticoid-induced atrophy of hippocampal neurons (Regan et al., 1999). Several other studies also demonstrate that stress and corticosterone can regulate nitric oxide synthase expression in the brain (Ceccatelli and Orazzo, 1993; Weber et al., 1994). Removal of glucocorticoids by adrenalectomy increases the expression of nitric oxide synthase in the hippocampus, and administration of glucocorticoids attenuates the adrenalectomy-induced increase in nitric oxide synthase expression (Lopez-Figueroa et al., 1998). More evidence for the role of nitric oxide in stress comes from studies showing the prevention of behavioral effects of corticosterone (Sandi et al., 1996) and of the stress-induced adrenocorticotropic hormone and corticosterone response (Tsuchiya et al., 1997) by nitric oxide synthase inhibitors.

Stress and protein tyrosine kinase-mediated signaling

There are two types of protein tyrosine kinases: (1) those that contain a receptor domain as well as a kinase domain, and thus cause self-phosphorylation and activation, and (2) those that do not contain a receptor domain but can phosphorylate substrate proteins. The category 1 protein tyrosine kinases are known as tyrosine receptor kinase or Trk (Patapoutian and Reichardt, 2001). Trks mediate the signaling systems of growth factors and neurotrophins and have been divided into three families: TrkA, TrkB, and TrkC. TrkA mediates the signaling of nerve growth factors, whereas TrkB is responsible for brain derived neurotrophic factor (BDNF)- and neurotrophin (NT) 4-induced signaling. TrkC mediates the functions of NT3. NT3 can also bind to and mediate functions through TrkA and TrkB. The binding of neurotrophins/growth factors to the appropriate Trk receptor leads to the dimerization and transphosphorylation of tyrosine residues in the intracellular domain of Trk receptors and the subsequent activation of cytoplasmic signaling pathways (Huang and Reichardt, 2003).

The signaling systems that are activated by Trk are extracellular-signaling-regulated mitogen-activated protein kinase (ERK-MAP kinase) and phosphoinositide (PI) 3-kinase/Akt. The ERK-mediated signaling cascade is an evolutionary conserved signaling mechanism that is involved in signal transduction from cell surface receptors to the nucleus (Lewis et al., 1998; Cobb, 1999). In this pathway, dimerization and activation of Trk receptors causes the recruitment of coupling proteins such as ShC, growth factor receptor bound protein (GRB_2), and coupling protein son of sevenless (SOS). This in turn causes the activation of a small G protein, Ras (Nimnual et al., 1998). Ras then recruits another kinase, Raf to the membrane, where it is phosphorylated and activated by serine/threonine/tyrosine kinases. This leads to the activation of another kinase known as ERK kinase (also known as MEK). MEK then phosphorylates and activates ERK on serine/threonine residues (English et al., 1999) (Fig. 2). Once activated, ERK can phosphorylate many substrate proteins in the cytosol or translocate to the nucleus, where they regulate the activity of transcription factors (Chen et al., 1992; Lin et al., 1993; Marais et al., 1993). This leads to specific biological responses, including cell proliferation, differentiation, survival, gene expression, and cell cycle response of neurons to neural activity (Segar and Krebs, 1995; Lewis et al., 1998; Grewal et al., 1999; Kolkova et al., 2000). Because of the role played by ERK signaling in cell survival, recent studies are focusing on the role of the ERK signaling pathway in the stress response. For example, recently, it has been reported that chronic stress initiated by mild footshock for 21 days causes pronounced and persistent hyperphosphorylation and therefore activation of ERK isoforms, i.e., ERK1 and ERK2, in dendrites of the rat prefrontal cortex (Trentani et al., 2002). This hyperphosphorylation of ERK1 and 2 may represent a specific path by which chronic stress may affect the functioning of cortical structures and cause defects in the neural networks involved in elaboration of sensory stimuli. Cell culture studies also show that corticosterone activates ERK1 and ERK2 in PC 12 cells, which suggests that glucocorticoids mediate the activation of ERK1 and ERK2 signaling (Qiu et al., 2001).

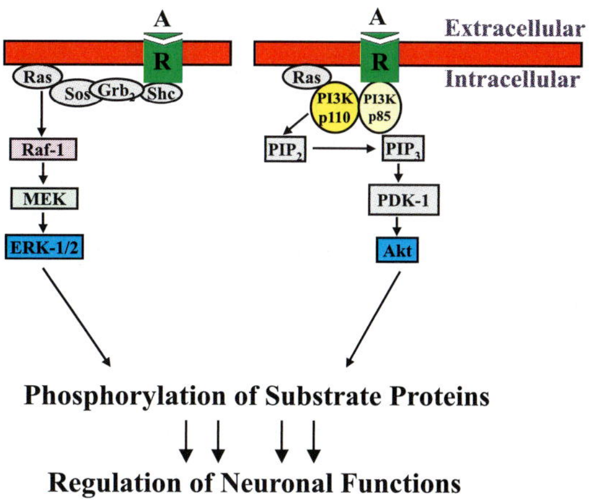

Fig. 2. Schematic representation of signal transduction mechanisms of extracellular signal-regulated (ERK) and phosphoinositide (PI) 3-kinase pathways. In the ERK pathway, the binding of trophic factors/growth factors to the tyrosine receptor kinase (Trk) causes its dimerization and activation. This results in the activation of Raf-1 through Shc, Grb_2, Sos, and Ras. Activation of Raf-1 leads to the activation of ERK kinase (MEK) and of ERK-1/2 in a sequential manner. ERK-1 and ERK-2 then phosphorylate substrate proteins in cytosol or transcription factors in nucleus after translocation. In the PI3-kinase pathway, binding of growth factors to Trk receptors causes dimerization and activation of Trk receptors. PI3-kinase is activated following binding to Trk receptors through the p85 regulatory domain. This causes the recruitment of PI3-kinase (PI3K) from cytosol to the plasma membrane, where the catalytic domain of PI3-kinase, i.e., p110, phosphorylates PIP_2 into PIP_3. PIP_3 then activates phosphoinositide-dependent kinase (PDK)-1, which phosphorylates and activates Akt. Akt then phosphorylates and activates or inactivates substrate proteins in cytosol or transcription factors in nucleus after translocation.

Other members of the MAP kinase family are c-Jun N-terminal kinase (JNK) and p38. Stress responses may also lead to the activation of these two kinases.

Another signal transduction system that is activated by growth factors or neurotrophins and can be affected by stress is PI3 kinase/Akt. In this signaling pathway, activation of Trk receptors elicits the recruitment of the catalytic subunit of PI3 kinase to the vicinity of the plasma membrane. The catalytic subunit of PI3 kinase then phosphorylates phosphatidylinositol 4,5 biphosphate to phosphatidylinositol (3,4,5)-triphosphate. Phosphatidylinositol (3,4,5)-triphosphate then activates the phosphorylating enzyme Akt through PI-dependent protein kinase-1

(Downward, 1998; Duronio et al., 1998). Akt then phosphorylates several proteins in the cytosol or transcription factors in the nucleus after translocation (Holgado-Madruga et al., 1997; Vaillant et al., 1999) (Fig. 2). Akt plays an important role in cell survival, which is mainly achieved by phosphorylating and therefore inhibiting a variety of substrates that affect apoptosis directly (Yao and Cooper, 1995; Dudek et al., 1997; Vanhaesebroeck et al., 1997; Eves et al., 1998; Datta et al., 1999; Hetman et al., 1999, Yuan and Yanker, 2000). In addition, the signaling pathways that are activated by protein tyrosine kinase are PLC γ-mediated signaling and STAT (signal transducers and activators of transcription) (Darnell, 1997; Pellegrini and Dusanter-Fourt, 1997; Leonard and O'shea, 1998; Kaplan and Miller, 2000; Huang and Reichardt, 2003). The elevation of cytosolic Ca^{2+} following activation of PLC γ results in a widespread protein phosphorylation by serine/threonine protein kinases. These include Ca^{2+}-calmodulin dependent kinase, phosphorylase kinase, and elongation factor-2.

The nonreceptor protein tyrosine kinases are a large group of signaling proteins that have diverse roles in cell proliferation, differentiation, and cell death (Gomperts et al., 2002). This family of proteins posseses no intrinsic catalytic activity, although they produce a response similar to that of receptors containing intrinsic catalytic activity. These protein tyrosine kinases recruit catalytic subunits within the cell in the form of one or more nonreceptor protein tyrosine kinases. This family of receptors is divided into ten different subfamilies: Src, Syk, Btk, Csk, Abl, JAK, Fak, Brk, Fes, and Ack.

Conclusion

The data reviewed in this chapter lead one to conclude that stress may influence neurotransmitter receptors and a receptor-mediated signal transduction systems in the brain. Since altered HPA function and release of cortisol are well known phenomena during stress, most of the studies have focused on the regulation of neurotransmitter receptors and intracellular signaling by glucocorticoids. Some of the effects of stress on intracellular signaling may be receptor mediated, and some could be related to direct effects on individual components of signaling pathways. Furthermore, some of the effects could be indirect, as many of the effects of glucocorticoids on PKA and PKC are evident only after chronic administration of glucocorticoids. These effects could be secondary to the changes in neurotransmitter receptors or components of intracellular signaling, or as a compensatory response to such changes. Nonetheless, stress alters the intracellular cascade at multiple levels. The final outcome of the alterations in the intracellular cascade is alterations in the functional properties of certain proteins, either through genomic or nongenomic action. Most of the protein kinases alter the functional properties of various proteins by phosphorylating them or by altering the transcription of genes, thereby modifying the expression of the proteins. The regulation of the gene expression of critical proteins by stress is described comprehensively in another chapter of this book ('Transcription factors as modulators of stress responsivity'). As stated earlier in this chapter, some of the acute effects of stress are beneficial; however, the chronic effects of stress may lead to abnormalities in the brain, including altered brain structures. Several studies indicate atrophy of the hippocampus and loss of neurons and/or glial cells in several structures of the brain. These deleterious effects of stress may be associated with psychiatric disorders such as post-traumatic stress disorder (PTSD) or depression. The findings of stress-induced modulation of neurotransmitter receptors and of the intracellular signaling cascade may be quite relevant in the pathophysiology of these psychiatric disorders, as similar findings, particularly those associated with $5HT_{2A}$, PKA, PKC, IP_3, and G proteins, have been reported in the blood cells of patients with affective disorders and in postmortem brain of depressed or suicide victims.

References

Albert, P.R. and Tiberi, M. (2001) Receptor signaling and structure: insights from $serotonin_1$ receptors. TRENDS in Endocrol. Metab., 12: 453–460.

Azmitia, E.C. and McEwen, B.S. (1974) Adrenocortical influence on rat brain tryptophan hydroxylase activity. Brain Res., 78: 291–302.

Barbacid, M. (1987) RAS genes. Annu. Rev. Biochem., 56: 779–827.

Bard, J.A., Zgombick, J., Adham, N., Vaysse, P., Branchek, T.A. and Weinshank, R.L. (1993) Cloning of a novel human serotonin receptor ($5HT_7$) positively linked to adenylate cyclase. J. Biol. Chem., 268: 23422–23426.

Barnes, N.M. and Sharp, T. (1999) A review of central 5HT receptors and their function. Neuropharmacol., 38: 1083–1152.

Bell, C.J. and Nutt, D.J. (1998) Serotonin and panic. Br. J. Psychiatry., 172: 465–471.

Berridge, M.J. and Irvine, R.F. (1989) Inositol phosphates and cell signaling. Nature, 341: 197–205.

Biegon, A., Rainbow, T.C. and McEwen, B.S. (1985) Corticosterone modulation of neurotransmitter receptors in rat hippocampus: A quantitative autoradiographic study. Brain Res., 332: 309–314.

Birnbaum, S., Gobeske, K.T., Auerbach, J., Taylor, J.R. and Arnsten, A.F.T. (1999) A role for norepinephrine in stress-induced cognitive deficits: α-1-adrenoceptor mediation in the prefrontal cortex. Biol. Psychiatry, 46: 1266–1274.

Blackwell, G.J., Carnucio, R., Di Rosa, M., Flower, R.J., Parente, L. and Persico, P. (1980) Macrocortin: A polypeptide causing the anti-phospholipase effect of glucocorticoids. Nature, 287: 147–149.

Borrelli, E., Montmayeur, J.P. and Foulkes, N.S. (1992) Sassone-Corsi P. Signal transduction and gene control: The cAMP pathway. Crit. Rev. Oncog., 3: 321–338.

Brannan, S.K., Miller, A., Jones, D.J., Kramer, G.L. and Petty, F. (1995) Beta-adrenergic receptor changes in learned helplessness may depend on stress and test parameters. Pharmacol. Biochem. Behav., 51: 553–556.

Bredt, D.S., Hwang, P.M. and Snyder, S.H. (1990) Localization of nitric oxide synthase indicating a neural role for nitric oxide. Nature, 347: 768–770.

Bremner, J.D., Krystal, J.H., Southwick, S.M. and Charney, D.S. (1996) Noradrenergic mechanism in stress and anxiety: II. Clinical studies. Synapse, 23: 39–51.

Brown, W.J., Chambers, K. and Doody, A. (2003) $Phospholipase_{A2}$ (PLA_2) enzymes in membrane trafficking: mediators of membrane shape and function. Traffic, 4: 214–221.

Burnet, P.W.J., Mefford, I.N., Smith, C.C., Gold, P.W. and Sternberg, E.M. (1992) Hippocampal 8-[^{3}H]hydroxy-2-(di-n-propylamino) tetralin binding site densities, serotonin receptor ($5HT_{1A}$) messenger ribonucleic acid abundance, and serotonin levels parallel the activity of the hypothalamopituitary-adrenal axis in rat. J. Neurchem., 59: 1062–1070.

Cabrera-Vera, T.M., Vanhauwe, J., Thomas, T.O., Medkova, M., Preininger, A., Mazzoni, M.R. and Hamm, E. (2003) Insights into G protein structure, function, and regulation. Endocrine Rev., 24: 765–781.

Campbell, S.L., Khosravi-Far, R., Rossman, K.L., Clark, G.J. and Der, C.J. (1998) Increasing complexity of RAS signaling. Oncogene, 17: 1395–1413.

Ceccatelli, S. and Orazzo, C. (1993) Effect of different types of stressors on peptide messenger ribonucleic acids in the hypothalamic paraventricular nucleus. Acta Endocrinol., 128: 485–492.

Chalmers, D.T., López, J.F., Vázquez, D.M., Akil, H. and Watson, S.J. (1994) Regulation of hippocampal $5HT_{1A}$ receptor gene expression by dexamethasone. Neuropsychopharmacology, 10: 215–222.

Chang, F.-H. and Bourne, H.R. (1987) Dexamethasone increases adenylyl cyclase activity and expression of the α subunit of G_S in GH_3 cells. Endocrinology, 121: 1711–1715.

Chaouloff, F. (1993) Physiopharmacological interactions between stress hormones and central serotonergic systems. Brain Res. Rev., 18: 1–32.

Chaouloff, F. (1995) Regulation of 5HT receptors by corticosteroids: where do we stand? Fundam. Clin. Pharmacol., 9: 219–333.

Chen, R.H., Sarnecki, C. and Blenis, J. (1992) Nuclear localization and regulation of erk- and rsk-encoded protein kinases. Mol. Cell. Biol., 12: 915–927.

Clapham, D.E. and Neer, E.J. (1997) G protein beta gamma subunits. Annu. Rev. Pharmacol. Toxicol., 37: 167–203.

Clegg, C.H. and McKnight, G.S. (1988) Genetic characterization of a brain-specific form of the type I regulatory subunit of cAMP-dependent protein kinase. Proc. Natl. Acad. Sci. USA, 85: 3703–3707.

Clement, H.W., Schäfer, F., Rowe, C., Gensa, D. and Wesemann, W. (1993) Stress-induced changes of extracellular 5-hydroxyindoleacetic acid concentrations followed in the nucleus raphe dorsalis and the frontal cortex of the rat. Brain Res., 64: 117–124.

Cobb, M.H. (1999) MAP kinase pathways. Prog. Biophys. Mol. Biol., 71: 479–500.

Cockcroft, S. and Thomas, G.M.H. (1992) Inositol-lipid-specific phospholipase C isoenzymes and their differential regulation by receptors. Biochem. J., 288: 1–14.

Cohen, R.M., Cohen, M.R. and McLellan, C.A. (1986) Foot shock induces time and region specific adrenergic receptor changes in rat brain. Pharmacol. Biochem. Behav., 24: 1587–1592.

Costa, A., Trainer, P., Besser, M. and Grossman, A. (1993) Nitric oxide modulates the release of corticotropin-releasing hormone from the rat hypothalamus in vitro. Brain Res., 605: 187–192.

Crayton, J.W., Joshi, I., Gulati, A., Arora, R.C. and Wolf, W.A. (1996) Effect of corticosterone on serotonin and catecholamine receptors and uptake sites in rat frontal cortex. Brain Res., 728: 260–262.

Croxtall, J.D., van Hal, P.T., Choudhury, Q., Gilroy, D.W. and Flower, R.J. (2002) Different glucocorticoids vary in their genomic and non-genomic mechanism of action in A549 cells. Br. J. Pharmacology., 135: 511–519.

Curzon, G., Joseph, M.H. and Knott, P.J. (1972) Effects of immobilization and food deprivation on rat brain tryptophan metabolism. J. Neurochem., 19: 1967–1974.

Czyrak, A. (1996) Modulation of the forskolin-induced cyclic AMP accumulation by corticosterone. Pol. J. Pharmacol., 48: 595–599.

Daaka, Y., Luttrell, L.M. and Lefkowtiz, R.J. (1997) Switching of the coupling of the beta2-adrenergic receptor to different G proteins by protein kinase A. Nature, 390: 88–91.

Dalley, J.W., Mason, K. and Stanford, S.C. (1996) Increased levels of extracellular noradrenaline in the frontal cortex of rats exposed to naturalistic environmental stimuli: Modulation by acute systemic administration of diazepam or buspirone. Psychopharmacology, 127: 47–54.

Darnell, J.E. Jr. (1997) STATs and gene regulation. Science, 227: 1630–1635.

Datta, S.R., Brunet, A. and Greenberg, M.E. (1999) Cellular survival: a play in three Akts. Genes. Dev., 13: 2905–2927.

Davis, S., Heal, D.J. and Stanford, S.C. (1994) Sensitization to β-adrenoceptor downregulation in mouse cerebral cortex in response to a brief swim stress. Br. J. Pharmacol., 12: 157P.

Day, H.E.W., Campeau, S., Watson, S.J. Jr. and Akil, H. (1999) Expression of α_{1B} adrenoceptor mRNA in corticotropin-releasing hormone-containing cells of the rat hypothalamus and its regulation by corticosterone. J. Neurosci., 19: 10098–10106.

de Kloet, E.R. (2000) Stress in the brain. Eur. J. Pharmacol., 405: 187–198.

Downward, J. (1998) Mechanisms and consequences of activation of protein kinase B/AKT. Curr. Opin. Neurobiol., 10: 262–267.

Downward, J. (1998) RAS signaling and apoptosis. Curr. Opin. Genet. Dev., 8: 49–54.

Downward, J. (2003) Targeting RAS signaling pathways in cancer therapy. Nat. Rev. Cancer, 3: 11–22.

Dudek, H., Datta, S.R., Franke, T.F., Bimbaum, M.J., Yao, R., Cooper, G.M., Segal, R.A., Kaplan, D.R. and Greenberg, M.E. (1997) Regulation of neuronal survival by the serine-threonine protein kinase Akt. Science., 275: 661–665.

Duman, R.S., Strada, S.J. and Enna, S.J. (1989) Glucocorticoid administration increases receptor-mediated and forskolin-stimulated cyclic AMP accumulation in rat brain cerebral cortical slices. Brain Res., 477: 166–171.

Duronio, V., Scheid, M.P. and Ettinger, S. (1998) Downstream signaling events regulated by phosphatidylinositol-3 kinase activity. Cell Signal, 10: 233–239.

Dwivedi, Y. and Pandey, G.N. (1999a) Administration of dexamethasone upregulates protein kinase C activity and the expression of γ and ε protein kinase C isozymes in the rat brain. J. Neurochem., 72: 380–387.

Dwivedi, Y. and Pandey, G.N. (1999b) Repeated administration of dexamethasone increases phosphoinositide-specific phospholipase C activity and mRNA and protein expression of the phospholipase C β_1 isozyme in rat brain. J. Neurochem., 73: 780–790.

Dwivedi, Y. and Pandey, G.N. (2000a) Adrenal glucocorticoids modulate [^{3}H]cyclic AMP binding to protein kinase A (PKA), cyclic AMP-dependent PKA activity, and protein levels of selective regulatory and catalytic subunit isoforms of PKA in rat brain. J. Pharmacol. Exp. Ther., 294: 103–116.

Dwivedi, Y., Rizavi, H.S., Rao, J.S. and Pandey, G.N. (2000b) Modifications in the phosphoinositide signaling pathway by adrenal glucocorticoids in rat brain: Focus on phosphoinositide-specific phospholipase C and inositol 1,4,5-trisphosphate. J. Pharmacol. Exp. Ther., 295: 244–254.

English, J., Pearson, G., Wilsbacher, J., Swantek, J., Karandikar, M., Xu, S. and Cobb, M.H. (1999) New insights into the control of MAP kinase pathways. Exp. Cell. Res., 253: 255–270.

Eves, E.M., Xiong, W., Bellacosa, A., Kennedy, S.G., Tsichlis, P.N., Rosner, M.R. and Hay, N. (1998) Akt, a target of phosphatidylinositol-3 kinase, inhibits apoptosis in a differentiating neuronal cell line. Mol. Cell. Biol., 18: 2143–2152.

Exton, J.H. (1994) Phosphoinositide phospholipases and G proteins in hormone action. Annu Rev Physiol, 56: 349–369.

Farooqui, A.A., Yang, H.C., Rosenberger, T.A. and Horrocks, L.A. (1997) Phospholipase A_2 and its role in brain tissue. J. Neurochem., 69: 889–901.

Fernandes, C., McKittrick, C.R., File, S.E. and McEwen, B.S. (1997) Decreased $5HT_{1A}$ and increased $5HT_{2A}$ receptor binding after chronic corticosterone associated with a behavioral indication of depression but not anxiety. Psychoneuroendocrinology, 22: 477–491.

Ffrench-Mullen, J.M. (1995) Cortisol inhibition of calcium currents in guinea pig hippocampal CA1 neurons via G-protein-coupled activation of protein kinase C. J. Neurosci., 15: 903–911.

Finlay, J.M., Zigmond, M.J. and Abercrombie, E.D. (1995) Increased dopamine and norepinephrine release in medial prefrontal cortex induced by acute and chronic stress: Effects of diazepam. Neuroscience, 64: 619–628.

Flugge, G. (1996) Alterations in the central nervous $alpha_2$-adrenoceptor system under chronic psychosocial stress. Neuroscience, 75: 187–196.

Flugge, G. (1999) Effects of cortisol on brain alpha2-adrenoceptors: Potential role in stress. Neurosci. Biobehav. Rev., 23: 949–956.

Forstemann, U., Gath, I., Schwartz, P., Closs, E.I. and Kleiner, H. (1995) Isoforms of nitric oxide synthase. Properties, cellular distribution and expressional control. Biochem. Pharmacol., 50: 1321–1332.

Freissmuth, M., Waldhoer, M., Bofill-Cardoba, E. and Nanoff, C. (1999) G protein antagonists. Trends Pharmacol. Sci., 20: 237–245.

Gannon, M.N., Brinton, R.E., Sakai, R.R. and McEwen, B.S. (1991) Diurnal differences and adrenal involvement in calmodulin stimulation of hippocampal adenylyl cyclase activity. J. Neuroendocrinol., 3: 37–41.

Gannon, M.N. and McEwen, B.S. (1990) Calmodulin involvement in stress- and corticosterone-induced regulation of cyclic AMP-generating systems in brain. J. Neurochem., 55: 276–284.

Gerard, C., el Mestikawy, S., Lebrand, C., Adrien, J., Ruat, M., Traiffort, E., Hamon, M. and Martres, M.P. (1996) Quantitative RT-PCR distribution of serotonin $5HT_6$ receptor mRNA in the central nervous system of control or 5,7-dihydroxytryptamine-treated rats. Synapse, 23: 164–173.

Gilman, A.G. (1987) G proteins: transducers of receptor-generated signals. Annu. Rev. Biochem., 56: 515–649.

Goldstein, L.E., Rasmusson, A.M., Bunney, S.B. and Roth, R.H. (1996) Role of the amygdala in the coordination of behavioral, neuroendocrine and prefrontal cortical monoamine responses to psychological stress in the rat. J. Neurosci., 16: 4787–4798.

Gomperts, B.D., Tatham, P.E.R. and Kramer, I.M. (2002) Signal Transduction. Signaling pathways operated by non-receptor protein tyrosine kinases. Academic Press, pp. 283–297

Graeff, F.G. (1997) Serotonergic systems. Psychiatr. Clin. North Am., 20: 723–739.

Grewal, S.S., York, R.D. and Stork, P.J. (1999) Extracellular signal-regulated kinase signaling in neurons. Curr. Opin. Neurobiol., 9: 544–553.

Gurguis, G.N.M., Kramer, G. and Petty, F. (1996) Indices of brain beta-adrenergic receptor signal transduction in the learned helplessness animal model of depression. J. Psychiatr. Res., 30: 135–146.

Gustafson, E.L., Durkin, M.M., Bard, J.A., Zgombick, J. and Branchek, T.A. (1996) A receptor autoradiographic and in situ hybridization analysis of the distribution of the $5HT_7$ receptor in rat brain. Br. J. Pharmacol., 117: 657–666.

Hamm, H.E. (1998) The many faces of G protein signaling. J. Biol. Chem., 273: 669–672.

Harrelson, A. and McEwen, B.S. (1987) Steroid hormone influences on cyclic AMP-generating systems. Curr. Topics Membr. Transport., 31: 217–247.

Harrelson, A.L., Rostene. W. and McEwen. B.S. (1987) Adrenocortical steroids modify neurotransmitter-stimulated cyclic AMP accumulation in the hippocampus and limbic brain of the rat. J. Neurochem., 48: 1648–1655.

Hellriegel, E.T. and D'Mello, A.P. (1997) The effect of acute, chronic and chronic intermittent stress on the central noradrenergic system. Pharmacol. Biochem. Behav., 57: 207–214.

Heninger, G.R., Charney, D.S. and Price, L.H. (1988) Noradrenergic and serotonergic receptor system function in panic disorder and depression. Acta Psychiatr. Scand. Suppl., 341: 138–150.

Hepler, J.R. (1999) Emerging roles for RGS proteins in cell signaling. Trends in Pharmacol. Sci., 20: 376–382.

Hetman, M., Kanning, K., Cavanaugh, J.E. and Xia, Z. (1999) Neuroprotection by brain-derived neurotrophic factor is mediated by extracellular signal-regulated kinase and phosphatidylinositol-3 kinase. J. Biol. Chem., 274: 22569–22580.

Hildebrandt, J.D. (1997) Role of subunit diversity in signaling by heterotrimeric G proteins. Biochem. Pharmacol., 54: 325–339.

Hirst, W.D., Price, G.W., Rattray, M. and Wilkin, G.P. (1997) Identification of 5-hydroxytryptamine receptors positively coupled to adenylyl cyclase in rat cultured astrocytes. Br. J. Pharamcol., 120: 509–515.

Holgado-Madruga, M., Moscatello, D.K., Emlet, D.R., Dieterich, R. and Wong, A.J. (1997) Grb2-associated binder-1 mediates phosphatidylinositol 3-kinase activation and the promotion of cell survival by nerve growth factor. Proc. Natl. Acad. Sci. USA, 94: 12419–12424.

Holmes, M.C., French, K.L. and Seckl, J.R. (1995a) Modulation of serotonin and corticosteroid receptor gene expression in the rat hippocampus with circadian rhythm and stress. Brain Res. Mol. Brain Res., 28: 186–192.

Holmes, M.C., Yau, J.L.W., French, K.L. and Seckl, J.R. (1995b) The effect of adrenalectomy on 5HT and corticosteroid receptor subtype mRNA expression in rat hippocampus. Neuroscience, 64: 327–337.

Hoyer, D., Hannon, J.P. and Martin, G.R. (2002) Molecular, pharmacological and functional diversity of 5HT receptors. Pharmacol. Biochem. Behav., 71: 533–554.

Huang, E.J. and Reichardt, L.F. (2001) Neurotrophins: roles in neuronal development and function. Annu. Rev. Neurosci. 24: 677–736.

Huang, E.J. and Reichardt, L.F. (2003) Trk receptors: roles in neuronal signal transduction. Annu. Rev. Biochem., 72: 609–642.

Huang, J. and Azmitia, E.C. (1999) Homologous regulation of $5HT_{1A}$ receptor mRNA in adult rat hippocampal dentate gyrus. Neurosci. Lett., 270: 5–8.

Ishii, M. and Kurachi, Y. (2003) Physiological actions of regulators of G protein signaling (RGS) proteins. Life Sci., 74: 163–171.

Jitsuiki, H., Kagaya, A., Goto, S., Horiguchi, J. and Yamawaki, S. (2000) Effect of lithium carbonate on the enhancement of $serotonin_{2A}$ receptor elicited by dexamethasone. Neuropsychobiology, 41: 55–61.

Joseph, M.H. and Kennett, G.A. (1983) Stress-induced release of 5HT in hippocampus and its dependence on increased tryptophan availability: an in vivo electrochemical study. Brain Res., 270: 251–257.

Joseph, S.K. and Williamson, J.R. (1989) Inositol polyphosphates and intracellular calcium release. Arch. Biochem. Biophys. 273: 1–15.

Kaplan, D.R. and Miller, F.D. (2000) Neurotrophin signal transduction in the nervous system. Curr. Opin. Neurobiol., 10: 381–391.

Katagiri, H., Kagaya, A., Nakae, S., Morinobu, S. and Yamawaki, S. (2001) Modulation of $serotonin_{2A}$ receptor function in rats after repeated treatment with dexamethasone and L-type calcium channel antagonist nimodipine. Prog. NeuroPsychopharmacol. Biol. Psychiatry, 25: 1269–1281.

Kim, D.-H., Jung, J.-S., Kim, H.-S., Suh, H.-W., Son, B.-K., Kim, Y.-H. and Song, D.-K. (2000) Inhibition of brain protein kinase C attenuates immobilization stress-induced plasma corticosterone levels in mice. Neurosci. Lett., 291: 69–72.

Klenerova, V. and Sida, P. (1994) Changes in beta-adrenergic receptors in the neurohypophysis and intermediate lobe or rat hypophysis exposed to stress. Physiol. Res., 43: 289–292.

Kohen, R., Metcalf, M.A., Khan, N., Druck, T., Huebner, K., Lachowicz, J.E., Meltzer, H.Y., Sibley, D.R., Roth, B.L. and Hamblin, M.W. (1996) Cloning, characterization, and chromosomal localization of a human $5HT_6$ serotonin receptor. Neurochem., 66: 47–56.

Kolasa, K., Song, L. and Jope, R.S. (1992) Adrenalectomy increases phosphoinositide hydrolysis induced by norepinephrine or excitatory amino acids in rat hippocampal slices. Brain Res., 579: 128–134.

Kolkova, K., Novitskaya, V., Pedersen, N., Berezin, V. and Bock, E. (2000) Neural cell adhesion molecule-stimulated neurite outgrowth depends on activation of protein kinase C and the RAS-mitogen-activated protein kinase pathway. J. Neurosci.., 20: 2238–2246.

Kudo, I. and Murakami, M. (2002) Phospholipase A_2 enzymes. Prostaglandis Other Lipid Mediat., 68–69: 3–58.

Kuroda, Y., Mikuni, M., Nomura, N. and Takahashi, K. (1993) Differential effect of subchronic dexamethasone treatment on $serotonin_2$ and β adrenergic receptors in the rat cortex and hippocampus. Neurosci. Lett., 195–198.

Kuroda, Y., Mikuni, M., Ogawa, T. and Takahashi, K. (1992) Effect of ACTH, adrenalectomy and the combination treatment on the density of $5HT_2$ receptor binding sites in neocortex of rat forebrain and $5HT_2$ receptor-mediated wet-dog shake behaviors. Psychopharmacology, 108: 27–32.

Kuroda, Y., Watanabe, Y., Albeck, D.S., Hastings, N.B. and McEwen, B.S. (1994) Effects of adrenalectomy and Type I or Type II glucocorticoid receptor activation on $5HT_{1A}$ and $5HT_2$ receptor binding and 5HT transporter mRNA expression in rat brain. Brain Res., 648: 157–161.

Le Corre, S., Sharp, T., Young, A.H. and Harrison, P.J. (1997) Increase of $5HT_7$ (serotonin-7) and $5HT_{1A}$ (serotonin-1A) receptor mRNA expression in rat hippocampus after adrenalectomy. Psychopharmacology, 130: 368–374.

Lefkowitz, R.J. (1998) G protein-coupled receptors. III. New roles for receptor kinases and beta-arrestins in receptor signaling and desensitization. J. Biol. Chem., 273: 18677–18680.

Leonard, W.J. and O'Shea, J.J. (1998) JAKs and STATs: biological implications. Annu. Rev. Immunol., 16: 293–322.

Lewis, T.S., Shapiro, P.S. and Ahn, N.G. (1998) Signal transduction through MAP kinase cascades. Adv. Cancer Res., 49–139.

Lewis, T.S., Shapiro, P.S. and Ahn, N.G. (1998) Signal transduction through MAP kinase cascades. Adv. Cancer Res., 74: 49–139.

Lin, L.L., Watermann, M., Lin, A.Y., Knopf, J.L., Seth, A., Davis, R.J. (1993) cPLA2 is phosphorylated and activated by MAP kinase. Cell 72: 269–278.

López, J.F., Chalmers, D.T., Little, K.Y. and Watson, S.J. (1998) Regulation of $serotonin_{1A}$, glucocorticoid, and mineralocorticoid receptor in rat and human hippocampus: implications for the neurobiology of depression. Biol. Psychiatry., 43: 547–573.

Lopez-Figueroa, M.O., Itoi, K. and Watson, S.J. (1998) Regulation of nitric oxide synthase messenger RNA expression in the rat hippocampus by glucocorticoids. Neuroscience, 87: 439–446.

Lorenzo Fernandez, P. (1999) Stress and neurodegeneration: pharmacologic strategies. Ann. Rev. Acad. Nac. Med. (Madr)., 116: 417–428.

Lynch, M., Littleton, J., McKernan, R.M., Durcan, M.J., McMillian, T. and Campbell, I.C. (1983) α-adrenoceptor number and function in rat cortex after ethanol and immobilization stress. Brain Res., 288: 145–149.

Mann, J.J. (1999) Role of the serotonergic system in the pathogenesis of major depression and suicidal behavior. Neuropsychopharmacology, Suppl. 1: S99–S105.

Mann, J.J. (2003) Neurobiology of suicidal behavior. Nat. Rev. Neurosci., 4: 819–828.

Marais, R., Wynne, J. and Treisman, R. (1993) The SRF accessory protein E lk-1 contains a growth factor-regulated transcriptional activation domain. Cell, 73: 381–393.

Marinissen, M.J. and Gutkind, J.S. (2001) G-protein-coupled receptors and signaling networks: Emerging paradigms. Trends Pharmacol. Sci., 22: 368–376.

Martin, J.V., Edwards, E., Johnson, J.O. and Henn, F.A. (1990) Monoamine receptors in an animal model of affective disorder. J. Neurochem., 55: 1142–1148.

Martire, M., Pistritto, G. and Prexiosi, P. (1989) Different regulation of serotonin receptors following adrenal hormone imbalance in the rat hippocampus and hypothalamus. Neural Transm., 78: 109–120.

McEwen, B.S. (1991) Stress and hippocampus. An update on current knowledge. Presse. Med., 20: 1801–1806.

McLeod, T.M., Lopez-Figueroa, A.L. and Lopez-Figueroa, M.O. (2001) Nitric oxide, stress, and depression. Psychopharmacol. Bull., 35: 24–41.

Meijer, O.C. and de Kloet, E.R. (1994) Corticosterone suppresses the expression of $5HT_{1A}$ receptor mRNA in rat dentate gyrus. Eur. J. Pharmacol., 266: 255–261.

Meijer, O.C., de Lange, E.C., Breimer, D.D., de Boer, A.G., Workel, J.O. and de Kloet, E.R. (1998) Penetration of dexamethasone into brain glucocorticoid targets is enhanced in mdrlA P-glycoprotein knockout mice. Endocrinology, 139: 1789–1793.

Mellor, H. and Parker, P.J. (1998) The extended protein kinase C superfamily. Biochem. J., 332: 281–292.

Mendelson, S.D. and McEwen, B.S. (1992a) Autoradiographic analyses of the effects of adrenalectomy and corticosterone on $5HT_{1A}$ and $5HT_{1B}$ receptors in the dorsal hippocampus and cortex of the rat. Neuroendocrinology, 55: 444–450.

Mendelson, S.D. and McEwen, B.S. (1992b) Quantitative autoradiographic analyses of the time course and reversibility of corticosterone-induced decreases in binding at $5HT_{1A}$ receptors in rat forebrain. Neuroendocrinology, 56: 881–888.

Mitchell, J. and Bansal, A. (1997) Dexamethasone increases $G\alpha_{q/11}$ expression and hormone-stimulated phospholipase C activity in UMR-106-01 cells. Am. J. Physiol., 273: E528–E535.

Miyahara, S., Komori, T., Fujiwara, R., Shizuya, K., Yamamoto, M., Ohmori, M. and Okazaki, Y. (1999) Effects of restraint stress on α_1 adrenoceptor mRNA expression in the hypothalamus and midbrain of the rat. Brain Res., 843: 130–135.

Mobley, P.L., Manier, D.H. and Sulser, F. (1983) Norepinephrine-sensitive adenylate cyclase system in rat brain: Role of adrenal corticosteroids. J. Pharmacol. Exp. Ther., 226: 71–77.

Mobley, P.L. and Sulser, F. (1980) Adrenal corticoids regulate sensitivity of noradrenaline receptor-coupled adenylate cyclase in brain. Nature, 286: 608–609.

Mobley, P.L., Sulser, F. (1980) Adrenal steroids affect the norepinephrine sensitive adenylate cyclase system in the rat limbic forebrain. Eur. J. Pharmacol., 65: 321–322.

Muraoka, S.-I., Mikuni, M., Kagaya, A., Saitoh, K. and Takahashi, K. (1993) Dexamethasone potentiates $serotonin_2$ receptor-mediated intracellular Ca^{2+} mobilization in C6 glioma cells. Neuroendocrinology, 57: 322–329.

Neer, E.J. (1995) Heterotrimeric G proteins: organizers of transmembrane signals. Cell, 80: 249–257.

Nestler, E.J. and Greengard, P. (1984) Protein phosphorylation and the regulation of neuronal function. In: Siegel, G.H., Albers, R.W., Agranoff, B.W. and Molioff, P. (Eds.), Basic Neurochemistry: Molecular, Cellular, and Medical Aspects, Little Brown, Boston, pp. 449–474.

Nestler, E.J., Terwilliger, R.Z. and Halm, E. (1989) Corticosterone increases protein tyrosine kinase activity in the locus coeruleus and other monoaminergic nuclei of rat brain. Mol. Pharmacol., 35: 265–270.

Neves, S.R., Ram, P.T. and Iyengar, R. (2002) G protein pathways. Science, 296: 1636–1639.

Newton, A.C. (2001) Protein kinase C: structural and spatial regulation by phosphorylation, cofactors, and macromolecular interactions. Chem. Rev., 101: 2353–2364.

Ni, Y.G., Gold, S.J., Iredale, P.A., Terwilliger, R.Z., Duman, R.S. and Nestler, E.J. (1999) Region-specific regulation of RGS4 (Regulator of G protein-signaling protein Type 4) in brain by stress and glucocorticoids: In vivo and in vitro studies. J. Neurosci., 19: 3674–3680.

Nimnual, A.S., Yatsula, B.A. and Bar-Sagi, D. (1998) Coupling of RAS and RAC guanosine triphosphatases through the RAS exchanger SOS. Science, 279: 560–563.

Nishizuka, Y. (1988) The molecular heterogeneity of protein kinase C and its implications for cellular regulation. Nature, 334: 661–665.

Nukina, I., Glavin, G.B. and LaBella, F.S. (1987) Acute cold-restraint stress affects alpha 2-adrenoceptors in specific brain regions of the rat. Brain Res., 401: 30–33.

Okuhara, D.Y., Beck, S.G. and Muma, N.A. (1997) Corticosterone alters G protein α-subunit levels in the rat hippocampus. Brain Res., 745: 144–151.

Okuhara, D.Y. and Beck, S.G. (1998) Corticosteroids alter $5\text{-hydroxytryptamine}_{1A}$ receptor-effector pathway in hippocampal subfield CA3 pyramidal cells. J. Pharmacol. Exp. Ther., 284: 1227–1233.

Pacak, K., McCarty, R., Palkovits, M., Kopin, I.J. and Goldstein, D.S. (1995) Effects of immobilization on in vivo release of norepinephrine in the bed nucleus of the stria terminalis in conscious rats. Brain Res., 688: 242–246.

Pandey, G.N. and Dwivedi, Y. (1999) Effects of adrenal glucocorticoids on protein kinase C (PKC) binding sites, PKC activity, and expression of PKC isozymes in the rat brain. Soc. Neuroscience Abstr., 25: 1812.

Pandey, S.C., Ren, X., Sagen, J. and Pandey, G.N. (1995) Beta-adrenergic receptor subtypes in stress-induced behavioral depression. Pharmacol. Biochem. Behav., 51: 339–344.

Patapoutian, A. and Reichardt, L.F. (2001) Trk receptors: mediators of neurotrophin action. Curr. Opin. Neurobiol., 11: 272–280.

Pei, Q., Zetterström, T. and Fillenz, M. (1990) Tail pinch-induced changes in the turnover and release of dopamine and 5-hydroxytryptamine in different brain regions of the rat. Neuroscience, 35: 133–138.

Pellegrini, S. and Dusanter-Fourt, I. (1997) The structure, regulation and function of the Janus kinases (JAKs) and the signal transducers and activators of transcription (STATs). Eur. J. Biochem., 248: 615–633.

Plassat, J.-L., Amlaiky, N. and Hen, R. (1993) Molecular cloning of a mammalian serotonin receptor that activates adenylate cyclase. Mol. Pharmacol., 44: 229–236.

Qiu, J., Lou, L.G., Huang, X.Y., Lou, S.J., Pei, G. and Chen, Y.Z. (1998) Nongenomic mechanisms of glucocorticoid inhibition of nicotine-induced calcium influx in PC12 cells: involvement of protein kinase C. Endocrinology, 139: 5103–5108.

Qiu, J., Wang, P., Jing, Q., Zhang, W., Lix, X., Zhong, Y., Sun, G., Pei, G. and Chen, Y. (2001) Rapid activation of ERK1/2 mitogen-activated protein kinase by corticosterone in PC12 cells. Biochem. Biphys. Res. Commun. 287: 1017–1024.

Reagan, L.P., McKittrick, C.R. and McEwen, B.S. (1999) Corticosterone and phenytoin reduce neuronal nitric oxide synthase messenger RNA expression in rat hippocampus. Neuroscience, 91(1): 211–219.

Rhee, S.G. and Choi, K.D. (1992) Regulation of inositol phospholipid-specific phospholipase C isozymes. J. Biol. Chem., 267: 12393–12396.

Riedel, W. (2000) Role of nitric oxide in the control of the hypothalamic-pituitary-adrenocortical axis. Z Rheumatol., 59 (Suppl. 2): 36–42.

Roberts, V.J., Singhal, R.L. and Roberts, D.C. (1984) Corticosterone prevents the increase in noradrenaline-stimulated adenylyl cyclase activity in rat hippocampus following adrenalectomy or metopirone. Eur. J. Pharmacol., 103: 235–240.

Rodan, S. and Rodan, G. (1986) Dexamethasone effects of β-adrenergic receptors and adenylate cyclase regulatory proteins G_S and G_I in ROS 17/2.8 cells. Endocrinology, 118: 2510–2518.

Rosenblum, L.A., Coplan, J.D., Friedman, S., Bassoff, T., Gorman, J.M. and Andrews, M.W. (1994) Adverse early experiences affect noradrenergic and serotonergic functioning in adult primates. Biol. Psychiatry., 35: 221–227.

Ross, E.M. and Wilkie, T.M. (2000) GTPase-activating proteins for heterotrimeric G proteins: regulators of G protein signaling (RGS) and RGS-like proteins. Annu. Rev. Biochem., 69: 795–827.

Rostene, W.H., Rischette, C.T., Dussaillant, M. and McEwen, B.S. (1985) Adrenal steroid modulation of vasactive intestinal peptide effect on $serotonin_1$ binding sites in the rat brain shown by in vitro quantitative autoradiography. Neuroendocrinology, 40: 129–134.

Saito, N., Guitart, X., Hayward, M., Tallman, J.F., Duman, R.S. and Nestler, E.J. (1989) Coritcosterone differentially regulates the expression of $G_S\alpha$ and $G_i\alpha$ messenger RNA and protein in rat cerebral cortex. Proc. Natl. Acad. Sci. USA, 86: 3906–3910.

Sandi, C. and Guaza, C. (1995) Evidence for a role of nitric oxide in the corticotrophin-releasing factor release induced by interleukin-1β. Eur. J. Pharmacol., 274: 17–23.

Sandi, C., Venero, C. and Guaza, C. (1996) Nitric oxide synthase inhibitors prevent rapid behavioral effects of corticosterone in rats. Neuroendocrinology, 63: 446–453.

Sapolsky, R.M. (1987) Glucocorticoids and hippocampal damage. Trends Neurosci., 10: 346–349.

Schinkel, A.H., Wagenaar, E., Mol, C.A. and van Deemter, L. (1996) P-glycoprotein in the blood-brain barrier of mice influces the brain pentration and pharmacological activity of many drugs. J. Clin. Invest., 97: 2517–2524.

Scott, J.D., Glaccum, M.B., Zoller, M.J., Uhler, M.D., Helfman, D.M., McKnight, G.S. and Krebs, E.G. (1987) The molecular cloning of a type II regulatory subunit of the cAMP-dependent protein kinase from rat skeletal muscle and mouse brain. Proc. Natl. Acad. Sci. USA, 84: 5192–5196.

Seger, R. and Krebs, E.G. (1995) The MAPK signaling cascade. FASEB J., 9: 726–735.

Simon, M.I., Strathmann, M.P. and Gautam, N. (1991) Diversity of G proteins in signal transduction. Science, 252: 802–808.

Singh, V.B., Onaivi, E.S., Phan, T.H. and Boadle-Biber, M.C. (1990) The increases in rat cortical and midbrain tryptophan hydroxylase activity in response to acute or repeated sound stress are blocked by biolateral lesions to the central nucleus of the amygdala. Brain Res., 530: 49–53.

Sleight, A.J., Boess, F.G., Bos, M. and Bourson, A. (1998) The putative $5HT_6$ receptor: localization and function. Ann. NY Acad. Sci., 861: 91–96.

Southwick, S.M., Bremner, J.D., Rasmusson, A., Morgan, C.A. III, Arnsten, A. and Charney, D.S. (1999a) Role of norepinephrine in the pathophysiology and treatment of posttraumatic stress disorder. Biol. Psychiatry., 46: 1192–1204.

Southwick, S.M., Morgan, C.A. III, Charney, D.S. and High, J.R. (1999b) Yohimbine use in a natural setting: Effects on posttraumatic stress disorder. Biol. Psychiatry., 46: 442–444.

Southwick, S.M., Paige, S., Morgan, C.A. III, Bremner, J.D., Krystal, J.H. and Charney, D.S. (1999c) Neurotransmitter alterations in PTSD: Catecholamines and serotonin. Semin. Clin. Neuropsychiatry, 4: 242–248.

Spaulding, S.W. (1993) The ways in which hormones change cyclic adenosine 3′5′-monophosphate-dependent protein kinase subunits, and how such changes affect cell behavior. Endocr. Rev., 14: 632–560.

Stanford, S.C. (1995) Central noradrenergic neurones and stress. Pharmacol. Ther., 68: 297–342.

Stone, E.A., Platt, J.E., Trullas, R. and Slucky, A.V. (1984) Reduction of the cAMP response to norepinephrine in rat cerebral cortex following repeated restraint stress. Psychopharmacology, 82: 403–405.

Stone, E.A. and Quartermain, D. (1999) Alpha-1-noradrenergic neurotransmission, corticosterone, and behavioral depression. Biol. Psychiatry., 46: 1287–300.

Stone, E.A., Rhee, J. and Quartermain, D. (1996) Blockade of effect of stress on risk assessment behavior in mice by a beta-1 adrenoceptor antagonist. Pharmacol. Bechem. Behav., 55: 215–217.

Stone, E.A., Slucky, A.V., Platt, J.E. and Trullas, R. (1985) Reduction of the cyclic adenosine 3′5′-monophosphate response to catecholamines in rat brain slices after repeated restraint stress. J. Pharmacol. Exp. Ther., 233: 382–388.

Stone, E.A. (1979) Reduction by stress of norepinephrine-stimulated accumulation of cyclic AMP in rat cerebral cortex. J. Neurochem., 32: 1335–1337.

Stowe, R.L. and Barnes, N.M. (1998) Cellular distribution of $5HT_7$ receptor mRNA in rat brain. Br. J. Pharmacol., 123: 229P.

Sullivan, G.M., Coplan, J.D., Kent, J.M. and Gorman, J.M. (1999) The noradrengeric system in pathological anxiety: A focus on panic with relevance to generalized anxiety and phobias. Biol. Psychiatry, 46: 1205–1218.

Sze, P.Y. and Iqbal, Z. (1994) Glucocorticoid action on depolarization-dependent calcium influx in brain synaptosomes. Neuroendocrinology, 59: 457–465.

Takahashi, M., Morinobu, S., Totsuka, S. and Endoh, M. (1996) Chronic dexamethasone administration decreases noradrenaline-stimulated, but not serotonin-stimulated, phosphoinositide metabolism in the rat brain. Naunyn. Schmiedebergs Arch. Pharmacol., 353: 616–620.

Takao, K., Nagatani, T., Kitamura, Y. and Yamawaki, S. (1997) Effects of corticosterone on $5HT_{1A}$ and $5HT_2$ receptor binding and on the receptor-mediated behavioral responses of rats. Eur. J. Pharmacol., 123–128.

Talmi, M., Carlier, E., Rey, M. and Soumireu-Mourat, B. (1992) Modulation of the vitro electrophysiological effect of corticosterone by extracellular calcium in the hippocampus. Neuroendocrinology, 55: 257–263.

Tang, W.J. and Gilman, A.G. (1992) Adenylyl cyclase. Cell, 70: 869–872.

Tijani-Butt, S.M. and Labow, D.M. (1994) Time course of the effects of adrenalectomy and corticosterone replacement on $5HT_{1A}$ receptors and 5HT uptake sites in the hippocampus and dorsal raphe nucleus of the rat brain: an autoradiographic analysis. Psychopharmacology, 113: 481–486.

Torda, T., Yamaguchi, I., Hirata, F., Kopin, I.J. and Axelrod, J. (1981) Mepacrine treatment prevents immobilization-induced desensitization of beta-adrenergic receptors in rat hypothalamus and brain stem. Brain Res., 205: 441–444.

Trentani, A., Kuipers, S.D., Ter Horst, G.J. and Den Boer, J.A. (2002) Selective chronic stress-induced in vivo ER1/2 hyperphosphorylation in medial prefrontocortical dendrites: Implications for stress-related cortical pathology? Eur. J. Neurosci.., 15: 1681–1691.

Tsuchiya, T., Kishimoto, J., Koyama, J. and Ozawa, T. (1997) Modulatory effect of L-NAME, a specific nitric oxide synthase (NOS) inhibitor, on stress-induced changes in plasma adrenocorticotropic hormone (ACTH) and corticosterone levels in rats: physiological significance of stress-induced NOS activation in hypothalamic-pituitary-adrenal axis. Brian Res., 776: 68–74.

U'Prichard, D.C. and Kvetnansky, R. (1980) Central and peripheral adrenergic receptors in acute and repeated immobilization stress. In: Usdin, E., Kvetnansky, R., Kopin, I.J. (Eds.), Catecholamines and stress: Recent Advances, Elsevier, Amsterdam, pp. 293–308.

Vaillant, A.R., Mazzoni, I., Tudan, C., Boudreau, M., Kaplan, D.R. and Miller, F.D. (1999) Depolarization and neurotrophins converge on the phosphatidylinositol-3 kinase-Akt pathway to synergistically regulate neuronal survival. J. Cell. Biol., 146: 955–966.

van Veelen, N.M. and Kahn, R.S. (1999) Dopamine, serotonin, and schizophrenia. Adv. Neurol., 80: 425–429.

Vanhaesebroeck, B., Leevers, S.J., Panayaton, G. and Waterfield, M.D. (1997) The phosphoinositide-3 kinase: a conserved family of signal transducers. Trend Biochem. Sci., 22: 267–272.

Ward, R.P., Hamblin, M.W., Lachowicz, J.E., Hoffman, B.J., Sibley, D.R. and Dorsa, D.M. (1995) Localization of serotonin subtype 6 receptor messenger RNA in the rat brain by in situ hybridization histochemistry. Neurosci., 64: 1105–1111.

Watanabe, Y., Sakai, R.R., MeEwen, B.S. and Mendelson, S. (1993) Stress and antidepressant effects on hippocampal and cortical $5HT_{1A}$ and $5HT_2$ receptors and transport sites for serotonin. Brain Res., 615: 87–94.

Weber, C.M., Eke, B.C. and Maines, M.D. (1994) Corticosterone regulates heme oxygenase-2 and NO synthase transcription and protein expression in rat brain. J. Neurochem., 63: 953–962.

Weiss, G.K., Ratner, A., Voltura, A., Savage, D., Lucero, K. and Castillo, N. (1994) The effect of two different types of stress on locus coeruleus $alpha_2$ receptor binding. Brain Res. Bull., 33: 219–221.

Wolfgang, D., Chen, I. and Wand, G.S. (1995) Effects of restraint stress on components of adenylyl cyclase signal tranduction in the rat hippocampus. Neuropsychopharmacology, 11: 187–193.

Yamanaka, K., Muramatsu, I. and Kigoshi, S. (1987) Effect of chronic nicotine treatment against repeated immobilization stress. Pharamcol. Biochem. Behav., 26: 259–263.

Yao, R. and Cooper, G.M. (1995) Requirement for phosphatidylinositol-3 kinase in the prevention of apoptosis by nerve growth factor. Science, 267: 2003–2006.

Yau, J.L.W., Noble, J., Widdowson, J. and Seckl, J.R. (1997) Impact of adrenalectomy on $5HT_6$ and $5HT_7$ receptor gene expression in the rat hippocampus. Mol. Brain Res., 45: 182–186.

Yuan, J. and Yankner, B.A. (2000) Apoptosis in the nervous system. Nature, 407: 802–809.

Zhong, P. and Ciarnello, R.D. (1995) Transcriptional regulation of hippocampal $5HT_{1A}$ receptors by corticosteroid hormones. Mol. Brain Res., 29: 23–34.

T. Steckler, N.H. Kalin and J.M.H.M. Reul (Eds.)
Handbook of Stress and the Brain, Vol. 15
ISBN 0-444-51173-3

CHAPTER 5.2

The role of neurotrophic factors in the stress response

Marco A. Riva*

Center for Neuropharmacology, Department of Pharmacological Sciences, University of Milan, Milan 20133, Italy

Abstract: Neurotrophic factors are important regulators of cell function and participate in activity-dependent synaptic plasticity. Herein these proteins may represent an ideal candidate to mediate short- and long-term effects of stress on brain function and structure. The modulation of different neurotrophic molecules, including neurotrophins and fibroblast growth factors, is rather specific from the anatomical and temporal point of view. Acute stress can upregulate the expression of NGF, NT-3 as well as FGF-2, in different brain structures including the hippocampal formation and hypothalamus, suggesting the possibility that such a rapid effect may represent a protective mechanism to cope with stressful situations. Conversely the regulation of brain-derived neurotrophic factor (BDNF) expression appears to be more complex and region specific. Acute or chronic exposure to different types of stress downregulates BDNF expression in the hippocampus, an effect that may contribute to stress-induced impairment in brain function leading to cellular atrophy and cognitive deterioration. It is believed that pharmacological intervention aimed at normalizing the biosynthesis of neurotrophic factors may represent a novel and valuable strategy to ameliorate defects associated with psychiatric disorders where stress represents a major element of vulnerability.

Introduction

The stress response can be considered a general reaction of an organism, which can monitor internal and external conditions in order to develop proper coping strategies that will allow survival. The stress response occurs through multiple pathways involving hormones and neurotransmitters. The hypothalamic–pituitary–adrenal (HPA) system operates to control glucocorticoid hormones secreted by the adrenal glands, which are the most important steroid hormones secreted during stress. Glucocorticoids mobilize energy, increase cardiovascular tone, reinforce aspects of the immune system and modulate different systems of the body. However glucocorticoid hormones exert multiple effects on brain function and neuronal viability: excessive glucocorticoids are deleterious for neuronal cells and may play a relevant role in several psychiatric disorders, such as depression, schizophrenia, and posttraumatic stress disorders (PTSD), which are characterized by complex changes in brain plasticity (Sapolsky, 2000; Kim and Diamond, 2002).

The brain can be considered a major player for the effects exerted by stress on body function. In fact the brain is not only responsible for the interpretation of the stressful experience and the development of proper coping strategies, but it is crucial for the memory of the stressful event. Depending on the type of stress, it has been postulated that the brain can set in motion different coping strategies. First of all, we should bear in mind that the response to psychological stressors is determined by the way an organism perceives and reacts to the stimulus. This individual susceptibility represents a major challenge in the identification of molecular substrate for the effects determined by stress on brain function.

*Tel.: +39-02-50318334; Fax: +39-02-50318278;
E-mail: M.Riva@unimi.it

Second, the response of the brain to stress depends very much upon the timing and the duration of stress exposure. It has been well documented that stress during prenatal and early postnatal development may leave permanent traces on brain function determining changes in the activity of the HPA axis and interfering with the process of neuronal maturation (Ladd et al., 2000; McEwen, 2000a; Meaney, 2001).

Later in life, the effects exerted by stress on brain function will depend very much on the type of stress and the length of exposure. On the short-term (acute effects) stress may impact neurotransmission and alter the performance in specific brain functions, including learning abilities (McEwen and Sapolsky, 1995; de Kloet et al., 1999): it is expected that such changes will be finalized to coping and survival. Conversely, prolonged exposure to stress can alter brain activity and function through changes, which usually lead to reduced cellular plasticity and enhanced neuronal vulnerability (Sapolsky, 2000; Kim and Diamond, 2002). Indeed structural alteration and reduced neurogenesis in selected brain regions have been reported to occur following prolonged stress exposure (Gould and Tanapat, 1999; McEwen, 2000a).

During aging, a progressive deterioration of the mechanisms governing the function of the HPA axis can be observed and this, together with a "physiological" reduction in neuroplasticity, may lead to a higher susceptibility to stressful events.

On this basis, during the last few years, several investigators have begun to address the question of how stress can alter brain function and what are the cellular and molecular correlates of the changes observed following stress exposure. Such strategy would ultimately lead to the identification of "vulnerability" factors, which might become the target of drugs aimed at reinstating normal brain activity. This scenario appears to be of great interest for several neuropsychiatric disorders where stress represents, if not the primary cause, an important vulnerability element. It is becoming widely accepted that the interaction between genetic and environmental factors, stress among the others, is a prerequisite for disease onset. For example, gene polymorphisms have been associated to several neuropsychiatric disorders without necessarily yielding a pathological phenotype, which instead can be revealed following stress exposure at a certain stage of life.

Neurotrophic factors appear to be good candidates for mediating short- and long-term effects of stress on brain function. In fact, this large and heterogeneous class of proteins is not only important in neuronal maturation and survival but plays a more complex role in controlling neuronal function and cellular resilience (Huang and Reichardt, 2001; Poo, 2001).

I will therefore review and discuss the data supporting a role of different neurotrophic molecules in the response of the brain to stress, with a specific focus on the anatomical and temporal pattern of these responses. I will deal with two families of neurotrophic molecules, namely the neurotrophins and fibroblast growth factors, since their modulation by stress has been characterized in detail in more recent years.

Neurotrophic factors and brain plasticity

Neurotrophic factors represent a wide and heterogeneous class of polypeptides exerting a complex array of activities on different cellular phenotypes. They were originally discovered for their role in development, as target-derived factors, but it is now generally agreed that neurotrophic factors, such as neurotrophins, are molecular effectors of several aspects of brain plasticity: they promote growth and health of neurons, modulate neurotransmitter function and are integral to the modifiability of the central nervous system.

Several classes of neurotrophic factors for central and peripheral nervous system have been identified including the neurotrophins, the transforming growth factor ß superfamily and the fibroblast growth factors.

The neurotrophin family of neurotrophic factors comprises several polypeptides including nerve growth factor (NGF), brain-derived neurotrophic factor (BDNF), neurotrophin-3 (NT-3), NT-4/5, and NT-6. NGF was the first trophic factor to be discovered more than 50 years ago as target-derived trophic molecules that regulates the survival and maturation of developing neurons in the peripheral nervous system (Levi-Montalcini, 1987). However during the last 10–15 years it has become clear that the neurotrophins not only support the survival of postmitotic neurons (Lewin and Barde, 1996), but also regulate other neuronal functions, including axon growth and synaptic plasticity (Black, 1999; McAllister, 1999; Lu and Gottschalk, 2000; Thoenen, 2000).

Neurotrophins are synthesized as precursors (proneurotrophins) that are proteolytically cleaved to mature, biologically active neurotrophins (Edwards et al., 1988). It is assumed that they are packaged into vesicles in the soma in direct proportion to the level of their mRNA's, and that they are then transported to either presynaptic axon terminals or postsynaptic dendrites for local secretion. Proneurotrophins, including proNGF and proBDNF, can also be secreted and appear to have an opposite function with respect to their mature forms (Chao and Bothwell, 2002).

Neurotrophins activate two different classes of receptors, the tropomyosin-related kinase (trk) family of receptor tyrosine kinases and the p75 receptor, a member of the tumor necrosis factor (TNF) receptor superfamily. Trk receptors appear to mediate almost all the survival-promoting activities of neurotrophins, with NGF activating TrkA, BDNF and NT4/5 activating TrkB, and NT-3 activating TrkC. In addition, NT-3 can also activate, although with lower affinity, TrkA and TrkB. These receptors exist as different splicing variants in their extracellular domain, which may affect ligand interaction (Patapoutian and Reichardt, 2001).

The binding of the neurotrophin to Trk dimers initiates trans-autophosphorylation of specific tyrosine residues on the intracellular domain of the receptor (Kaplan and Miller, 2000). These phosphotyrosine residues serve as docking sites for elements of intracellular signaling cascades. Activated receptors initiate several signal transduction cascades, including mitogen-activated protein kinase (MAPK) pathways, the phosphatidylinositol 3-kinase (PI3K) pathway, and the phospholipase C-γ pathway (Kaplan and Miller, 2000; Huang and Reichardt, 2001).

There is an extensive literature on the role of neurotrophin in synaptic plasticity. (Thoenen, 1995; Schinder and Poo, 2000; Poo, 2001). High levels of neurotrophins may induce modifications of synaptic function and the formation of new synaptic contacts. The neurotrophin BDNF has been shown to regulate synaptic transmission across a broad temporal spectrum ranging from short-term modulation, which occurs in the order of seconds to minutes (Kang and Schuman, 1995), to prolonged effects that persist for many hours, such as the long-term potentiation (LTP) (Korte et al., 1995; Patterson et al., 1996) or long-term depression (LTD) response (Linden and Connor, 1995). Furthermore, the activity-dependent secretion of neurotrophins and their acute modulatory effects on synaptic efficacy suggest that they may be responsible for activity-induced LTP or LTD. Genetic deletion of BDNF in mice disrupted normal induction of LTP in the CA1 region of the hippocampus, which was rescued by exogenous administration of the neurotrophin (Korte et al., 1995; Patterson et al., 1996).

In cultures of hippocampal neurones, the application of BDNF induces a rapid potentiation of glutamate-mediated synaptic transmission (Schinder et al., 2000), but reduces inhibitory transmission. Most synaptic effects of neurotrophins are accounted for by presynaptic modification of transmitter secretion, which may be triggered by a BDNF-induced increase in cytosolic calcium (Berninger et al., 1993). A recent report has demonstrated that the activity of BDNF at presynaptic sites may involve the phosphorylation of synapsin I (Jovanovic et al., 2000). Further evidence has indicated that BDNF may regulate the number of docked vesicle at CA1 synapses as well as the expression of proteins associated with synaptic vesicles (synapsins, synaptobrevin, synaptophysin and synaptotagmin) (Pozzo-Miller et al., 1999).

The regulation of neurotrophin can occur at several levels. As we will discuss further in detail, a major mechanism is the regulation of their biosynthesis, which mainly occurs through changes in gene transcription. The modulation of neurotrophin biosynthesis is activity-dependent and can involve a complex interplay between neurotransmitters and hormones.

However, in the case of BDNF, neuronal activity may also modulate the subcellular targeting of the neurotrophin to distal dendrites (Tongiorgi et al., 1997) as well as its secretion from pre- and postsynaptic compartments (Canossa et al., 1997).

Based on these observations neurotrophins, among other trophic molecules, may represent a key factor to link neuronal activity and synaptic plasticity. It is expected that changes in different aspects of the neurotrophin "system" represent one of the mechanisms through which environmental manipulation, such as stress, can alter cellular resilience and modify brain structure. Since the threshold of cellular vulnerability can be determined by the expression and activity of neurotrophic factors, a functional reduction of these elements would be harmful for specific cellular population and may increase the vulnerability of selected cellular phenotypes under challenging situations.

The modulation of neurotrophic factors in response to stress

When analyzing the role of neurotrophic factors in the stress response, different aspects must be taken into consideration.

The first element is the type of stressor employed. In fact, depending on the experimental paradigm, activation of different brain regions may occur. Although some brain regions, such as the hippocampus, express high levels of glucocorticoid and mineralocorticoid receptors, these hormones do not represent the only player in the stress response, which may also involve enhanced release of several neurotransmitters including noradrenaline, dopamine, and glutamate (Jedema and Moghaddam, 1994; Finlay et al., 1995; Moghaddam, 2002). Hence the modulation of neurotrophic factors following stress may represent one of the mechanisms through which different extracellular signals can integrate to regulate cell function. It can be hypothesized that these responses will depend upon the "state" of these systems and may be altered by pathologic situations, which affect the function of the HPA axis as well as neurotransmitter responses.

The second important element is the duration of stress. As we will discuss later on, the functional implication of acute changes may be quite different, even opposed, to the changes observed after prolonged exposure to stress. Acutely, stress can suppress specific functions via a rapid modulation of neurotrophic factors, but it may also contribute to the activation of protective/defensive mechanisms, which tend to counteract the adverse effects deriving from stress exposure. On the contrary the significance of the changes occurring after chronic stress points toward a reduction and long-term impairment in brain function, characterized by neuronal atrophy and decreased neurogenesis (Gould and Tanapat, 1999; McEwen, 2000a), and may be relevant for diseases that are characterized by enhanced vulnerability to stress.

A third aspect is the timing of stress exposure: increasing evidences suggest that adverse life events during brain maturation can increase the vulnerability to psychiatric disorders (Meaney, 2001) and that one important feature of these conditions is a reduced cellular resilience (Duman et al., 2000; Manji et al., 2000). Stressful events during pregnancy or early postnatal life can alter the normal program of neuronal maturation, a highly regulated process that requires the timely expression of different neurotrophic molecules. Hence, stressful experience during specific developmental periods may alter the expression and function of neurotrophic factors, an event that will affect the proper maturation of specific neuronal circuitry.

Finally stress-related changes may be qualitatively and quantitatively different in aging, when normal homeostatic mechanisms may be less functional.

Neurotrophins

Nerve growth factor

Nerve growth factor (NGF) was discovered 50 years ago as a molecule regulating the development and differentiation of peripheral nervous system (PNS) neurons. It has been classically considered a target-derived trophic molecule although this does not appear to be the only mode of action for this neurotrophin (Levi-Montalcini, 1987; Thoenen et al., 1987; Thoenen et al., 1988). NGF is expressed in the CNS, mainly in the hippocampus, cortex and olfactory bulb, where it is developmentally regulated and represents a major trophic support for cholinergic neurons of the basal forebrain (Hefti and

Weiner, 1986). NGF can be upregulated in response to injury within peripheral and central nervous system, and endogenous NGF signaling in neuronal and nonneuronal cells is believed to subserve neuroprotective function and facilitate neural repair (Sofroniew et al., 2001).

In rodents, an increase in circulating and central levels of NGF has been reported to occur following stress exposure. Social stress, including intermale fighting and maternal aggression, results in an increase of NGF in the blood stream and in the CNS (Lakshmanan, 1986; Spillantini et al., 1989). The raise in serum levels of NGF is rapid, reaching a peak during the first 2–3 h following the fighting session. The source of NGF appears to be the submaxillary salivary gland where the trophic factor is present at very high levels. Interestingly, increased levels of NGF are not necessarily linked to fighting but appear to be related to the social status of the animals. Serum NGF levels of mice with repeated defeat and submission experience showed an increase double that is seen in dominant, attacking mice (Aloe et al., 1986). Moreover, a subordinate-like profile can be induced by systemic NGF administration (Bigi et al., 1992). It has been hypothesized (Alleva and Santucci, 2001) that NGF may exert a physiological role via the adrenal glands and could affect behavior also via changes in the production of adrenal hormones. In fact exogenous administration of NGF results in marked adrenal gland hypertrophy (Aloe et al., 1986; Bigi et al., 1992), an effect more pronounced in the medulla rather in the cortex.

A stress-related increase of plasma NGF levels was also reported to occur in humans: an elevation of circulating neurotrophin was indeed observed during the first parachute jump from an airplane by young soldiers (Aloe et al., 1994). This effect appears to be due to the anticipation of the jump, but does not correlate with the activity of the HPA axis. Moreover, an increase was also observed in the high (Trk-A) and low (p75) affinity receptors for NGF, present in peripheral blood lymphocytes of the soldiers. Since NGF induces the expression of interleukin-2 receptors in these cells, it could be hypothesized that the neurotrophin might act on peripheral targets: more specifically, NGF released under a stressful situation may contribute to the reinforcement of the immune responses of the organism.

Stress can also modulate the expression of brain NGF, although the anatomical specificity of this effect depends upon the type of stressor. Aggressive behavior rapidly upregulates NGF mRNA in the hypothalamus, an effect that is not evident in other CNS or peripheral areas such as cerebral cortex, hippocampus and heart (Spillantini et al., 1989). The elevation of hypothalamic NGF mRNA was accompanied by an increase of biologically active NGF (Spillantini et al., 1989), which was not abolished by sialoadenectomy suggesting that the NGF found in the hypothalamus is not of salivary origin (Aloe et al., 1986). Several findings support the hypothesis that hypothalamic NGF might be involved in neuroendocrine function, thereby regulating behavioral outcome. Furthermore, NGF can facilitate structural changes and might affect the levels of other peptides or hormones present in the hypothalamus.

Conversely, changes in hippocampal NGF appears to be less reproducible: a long immobilization (8 h) reduces NGF mRNA levels (Ueyama et al., 1997), but a shorter exposure to restraint stress does not alter its expression, although it reduces NGF mRNA levels in basal forebrain (Scaccianoce et al., 2000).

An open question is the mechanism whereby stress regulates brain NGF. There is no direct relationship between these changes and variations in plasma corticosterone. Nevertheless, adrenal hormones appear to modulate the expression of the neurotrophin in several brain regions and within selected cellular phenotypes. Removal of adrenal glands reduces the levels of NGF in hippocampus and cerebral cortex (Barbany and Persson, 1992), which is in line with the observation that administration of the synthetic glucocorticoid dexamethasone increase the expression of the neurotrophin in these brain regions, without affecting its levels in hypothalamus, striatum, and cerebellum (Mocchetti et al., 1996). This regulation appears to occur mainly in neurons, although direct exposure of cultured astroglial cells to the synthetic glucocorticoid dexamethasone produces a marked reduction of NGF mRNA contents (Riva et al., 1995b).

Furthermore, adrenalectomy reduces, at least in part, the upregulation of NGF expression following kainic acid injection, suggesting that stress and the enhanced release of adrenal hormones, occurring during seizure activity, can also contribute to adaptive changes involving the modulation of this neurotrophin (Barbany and Persson, 1993).

The stress sensitivity to NGF is also demonstrated by the effects produced by environmental manipulation occurring during development. An elevation of NGF expression has been observed in hippocampus and hypothalamus following brief or prolonged maternal deprivation during the first week of life (Cirulli, 2001). A similar effect was also reported to occur at postnatal day-11 in pups reared with hypercorticosteronemic mothers (Scaccianoce et al., 2001). These changes may represent a protective coping mechanism to limit the impact of a stressful situation, but could also interfere with the program of brain maturation and contribute to long-lasting changes in brain function as a consequence of adverse life events during development.

Brain-derived neurotrophic factor

The regulation of another neurotrophin, BDNF, appears to be somewhat different from NGF and has been investigated by several researchers. Since BDNF is an important player in neuronal plasticity, it may represent an ideal mediator for short- and long-term changes taking place following exposure to stress. Single or repeated immobilization stress decreases BDNF expression in the hippocampus (Smith et al., 1995c; Vaidya et al., 1997; Butterweck et al., 2001; Roceri et al., 2002): the reduction of neurotrophin levels following immobilization is rapid, occurring as early as 1 h after the beginning of the stress, and mainly affects dentate gyrus cells, whereas the effects on CA3 and CA1 layers appears to be less pronounced and consistent (Butterweck et al., 2001). A decrease of hippocampal neurotrophin levels is observed following swim stress (Russo-Neustadt et al., 2001) and it has also been associated with a psychological stress, which consists in the reexposure to cues previously associated with footshock (Rasmusson et al., 2002). The magnitude of stress-induced downregulation of BDNF expression in the hippocampus is quite variable (and not reproduced by all authors) depending upon the type and duration of stress as well as the time elapsed between the end of stress and the sacrifice of the animals.

Corticosterone injection reduces BDNF mRNA levels in hippocampal dentate gyrus, suggesting the contribution of adrenal steroids in stress-induced regulation of BDNF mRNA levels (Smith et al., 1995c; Hansson et al., 2000; Schaaf et al., 2000). Despite the fact that adrenalectomy does not produce any consistent change in BDNF gene expression (Barbany and Persson, 1992; Smith et al., 1995c; Hansson et al., 2000), the downregulation of the neurotrophin following acute restraint stress was less pronounced in adrenalectomized rats, thus suggesting that adrenal hormones contribute, to a certain extent, to stress-induced changes in BDNF expression (Smith et al., 1995c). Interestingly it has been demonstrated that 5HT2A, but not 5HT2C or 5HT1A, receptor antagonists can effectively antagonize the reduction of hippocampal BDNF mRNA following stress, suggesting that an enhanced release of serotonin may play a role in these regulatory mechanisms (Vaidya et al., 1997). It remains to be established if other neurotransmitters such as dopamine and glutamate, which are known to affect BDNF expression and are released during stress, may participate in the modulation of the neurotrophin under adverse conditions. However it is difficult to reconcile an enhanced release of glutamate following stress (Moghaddam, 2002) with hippocampal BDNF downregulation, since it is known that glutamate increases, and not decreases, its expression (Zafra et al., 1990; Kokaia et al., 1993). Conversely, stimulation of dopamine D2 receptors reduces the mRNA levels for BDNF in the hippocampus (Fumagalli et al., 2001), although it is not well established if stress can effectively increase dopamine efflux in the hippocampal region, as it does in other brain regions such as the prefrontal cortex and nucleus accumbens (Finlay et al., 1995).

Since stress is an important component of many neuropsychiatric disorders, a reduced production of BDNF, mainly within the dentate gyrus, can contribute to some of the defects that characterize stress-related disorders, such as depression and PTSD. Prolonged exposure to stress or glucocorticoids may determine structural changes within the hippocampus, characterized by atrophy of apical dendrites in CA3 pyramidal neurons, which is accompanied by specific cognitive deficits in spatial learning and memory (Magarinos et al., 1997;

McEwen, 2000a,b; Sapolsky, 2000). Such effect may originate from a loss of neuropil volume, an inhibition of the genesis of new neurons and glia or the loss of pre-existing neuronal or glial cells. Even though it is known that glutamate can play a role in these events (McEwen, 2000b), we may speculate that the dendritic atrophy, that is observed following stress, can be a consequence of reduced trophic supports provided by BDNF (and other trophic factors) in this structure. Although withdrawal of trophic factor support may not necessarily lead, per se, to cellular damage or neuronal cell death, it may render specific structures or cellular phenotypes more vulnerable to toxic or noxious stimuli. With this regard, it may be hypothesized that, under stressful situations, the vulnerability threshold of a specific cellular phenotype can be reduced because the protective "supply" of neurotrophin is diminished.

Furthermore stress can alter brain function with impairment in learning and memory: it must be kept in mind that BDNF, beyond its important neurotrophic activity, exerts a relevant role in neuronal plasticity, regulates neurotransmitter release and has a role in memory and cognition (Schinder and Poo, 2000; Poo, 2001). Although the role of stress and corticosteroids on cognition depends upon the context of the learning task (de Kloet et al., 1999), the rapid reduction of BDNF mRNA levels following stress may prevent the initial encoding of contextual memory associated with a traumatic experience, thus representing a sort of defensive mechanism. However, a reduced expression of the neurotrophin after prolonged severe stress, may be one of the mechanisms through which stress and glucocorticoid hormones will impair brain function leading to cognitive deterioration and cellular atrophy (Magarinos et al., 1997; de Kloet et al., 1999; McEwen, 2000b). If this is the case pharmacological intervention aimed at reinstating the correct expression of BDNF may represent a valuable strategy to ameliorate functional defects associated with stress.

Accordingly electroconvulsive seizure therapy and chronic antidepressant treatment blocked the down-regulation of BDNF mRNA in the hippocampus in response to restraint stress (Nibuya et al., 1995) although, in another report, chronic exposure to imipramine failed to counteract the changes occurring in hippocampal BDNF after acute or chronic immobilization (Butterweck et al., 2001).

A further note is that stress may not directly alter the basal expression of a specific trophic factor, but it may interfere with its activity-dependent regulation. For example, BDNF expression is upregulated during learning and exercise, as well as in other experimental paradigms, and these mechanisms may be impaired following stress exposure. If this is the case, the alteration of BDNF activity can be "unmasked" during specific tasks that require the proper modulation of the neurotrophin expression.

Another aspect that deserves attention is the anatomical specificity of stress-related changes in BDNF expression. Acute or repeated immobilization stress increases BDNF mRNA levels in the hypothalamus and pituitary (Smith et al., 1995b), suggesting that, similar to NGF, this neurotrophin may be important in brain regions that are primarily involved in the function of the HPA axis. This finding was confirmed by other investigators, who demonstrated not only a rapid upregulation of the neurotrophin but a differential modulation of its transcripts, which originates by specific 5' exons, in response to immobilization (Rage et al., 2002). The rat BDNF gene has a complex structure, with four 5' noncoding exons (exons I to IV) and one 3' exon (exon V) encoding the mature BDNF protein (Timmusk et al., 1993). Transcription of these exons is regulated by multiple promoters that yield ten different BDNF mRNAs: although each BDNF mRNA encodes an identical BDNF protein, these transcripts are differentially expressed and regulated throughout the brain (Timmusk et al., 1993; Lauterborn et al., 1998). This may affect RNA stability and trafficking, as well as translation, suggesting that a differential usage of each promoter may represent an important mechanism in the regulation of BDNF function.

Following stress it was shown that the increase of exon III mRNA levels was rapid (within 15 min) but transient, whereas exon I expression was increased only after 3 h of immobilization (Rage et al., 2002). These changes are localized in the paraventricular and supraoptic nuclei suggesting a possible role of the neurotrophin in autocrine and paracrine regulation of arginin-vasopressin secretory process, as well as in the neuroplastic and functional changes of the hypothalamus following stress.

Reducing glucocorticoid feedback through adrenalectomy upregulates BDNF expression in the paraventricular nuclei of the hypothalamus, as well as in the anterior pituitary, an effect that was normalized by corticosterone replacement therapy (Smith et al., 1995b). Furthermore the elevation of hypothalamic BDNF levels after immobilization stress appears to be more pronounced in adrenalectomized rats. These data suggest that adrenal hormones exert an inhibitory role on BDNF expression in these regions reminiscent of their negative feedback on corticotrophin-releasing factor (CRF) in the paraventricular nuclei, but not in extrahypothalamic areas. The mechanism by which stress can increase the neurotrophin expression in the hypothalamus and pituitary gland has not been established. Glutamate is a likely player because it is released during stress and it is able to stimulate BDNF transcription (Zafra et al., 1991), although other mediators, such as IL-1ß may also contribute to the observed effects.

The role of BNDF in the hypothalamus is a matter of investigation. Apart from its ability to modulate dendritic branching and synaptic strength (Poo, 2001), BDNF may mediate the synthesis of specific neuropeptides (Carnahan and Nawa, 1995) and appears to be involved in the control of body weight and food consumption. Exogenous administration of the neurotrophin can determine a reduction in weight, whereas mice with a conditional knockout of the BDNF gene show an increase in body weight as well as in daily food intake, effects that can in part be mediated by elevated levels of leptin and insulin (Rios et al., 2001). Interestingly, weight loss is observed during prolonged stress, suggesting a possible association of this effect with the upregulation of hypothalamic BDNF expression.

Another brain structure where BDNF expression is regulated by stress is the prefrontal-cingulate cortex, whose function appears to be altered in schizophrenia and depression. A single immobilization stress increases BDNF mRNA levels both in prefrontal and cingulate cortex (Molteni et al., 2001b), an effect that, similar to the hypothalamus, may be attributed to an enhanced release of dopamine and glutamate (Finlay et al., 1995; Moghaddam, 2002) rather than being related to glucocorticoids. Interestingly, chronic stress reduces the expression of the neurotrophin in prefrontal cortex (Roceri et al., 2004), suggesting that adaptive changes taking place following chronic stress may recapitulate defects observed in psychiatric disorders, including reduced activity of dopamine and glutamate within the prefrontal cortex, which may ultimately contribute to BDNF downregulation. On this basis, although a rapid elevation of BDNF mRNA levels may represent a compensatory-protective mechanism, prolonged exposure to stress will have a negative impact on the neuroplastic capacity of prefrontal cortex function through a downregulation of the neurotrophin expression.

Stress-induced changes in BDNF expression are attenuated in aged animals. It has been reported that aging per se decreases the hippocampal levels of the neurotrophin in rats as well as in monkeys (Smith and Cizza, 1996; Hayashi et al., 2001). Exposure of old rats to an immobilization stress determined a less pronounced reduction of hippocampal expression and failed to alter the levels of BDNF mRNA in the hypothalamus, suggesting that different molecular mechanisms may subserve the modulation of this neurotrophin in specific brain structures. Since the acute modulation of NGF under the same experimental conditions was not affected by aging, it may be inferred that the age-dependent impairment in these regulatory mechanisms may be specific for BDNF (Smith and Cizza, 1996).

There is evidence that the stress-induced regulation of BDNF expression may represent a mechanism contributing to long-term changes in brain plasticity following stress exposure during development. It is known that stressful events occurring early in development may alter the normal program of brain maturation and produce long-lasting effects on brain function, which alter stress responsiveness and may increase the vulnerability to psychiatric disorders (Meaney, 2001). Accordingly, several animal models that can mimic such events have been developed in order to characterize the molecular changes contributing to long-term impairment in brain function (Lipska and Weinberger, 2000). A reduction of hippocampal BDNF mRNA and protein levels occurred in adult rats that were exposed to a single period of 24 maternal deprivation on postnatal day 9 (Roceri et al., 2002), suggesting that changes in neurotrophin expression may represent one of the mechanisms through which stress during development produce long-lasting effects on brain function.

Although the regulation of BDNF biosynthesis following stress exposure has been investigated in great detail, limited data are available on the changes produced by stress on BDNF signaling machinery.

The expression of Trk-B, the high-affinity receptor for BDNF is reduced after an acute immobilization stress in the hypothalamus (Givalois et al., 2001), whereas the catalytic form of the receptor appears to be upregulated in the hippocampus following chronic stress (Nibuya et al., 1999). These changes may oppose those observed for BDNF suggesting the possibility that they may represent a compensatory mechanism to stress exposure.

Neurotrophin-3

The regulation of neurotrophin-3 following stress is opposite with respect to BDNF. In fact, its mRNA levels are upregulated in the hippocampus by chronic, but not acute, stress or following prolonged exposure to corticosterone (Barbany and Persson, 1992; Smith et al., 1995c). A similar upregulation of NT-3 expression has also been observed in the locus coeruleus (Smith et al., 1995a). Surgical removal of the adrenal glands reduces NT-3 mRNA levels in hippocampal dentate gyrus and within the CA2 pyramidal layer (Barbany and Persson, 1992; Hansson et al., 2000) and prevents stress-induced changes of the neurotrophin (Smith et al., 1995c), thus suggesting the important role exerted by circulating glucocorticoids in these mechanisms.

In analogy to what has been observed with other trophic molecules, such as FGF-2 (see later), it may be hypothesized that the upregulation of NT-3 biosynthesis may represent a compensatory response aimed at preventing or limiting stress-induced damage in the hippocampus.

Fibroblast growth factors

FGF-2 is the prototype member of a large family of neurotrophic molecules existing in different protein isoforms, which display selective subcellular localization thus implying different functional roles (Bikfalvi et al., 1997). FGF-2 binds to four related tyrosine kinase receptors (FGFR1-4), which exist in different splice variants and, in many respects, appear similar to other growth factor receptors (Klint and Claesson-Welsh, 1999). FGF-2, a potent angiogenic factor, may stimulate hematopoiesis and play an important role in the differentiation and function of the central nervous system. Its mRNA and protein are found at relatively high concentrations in several regions of the embryonic, postnatal, and adult central nervous system, where it is mainly expressed in astroglial cells. FGF-2 can stimulate neonatal and adult brain neurogenesis (Palmer et al., 1999), plays an important role in regeneration after CNS injury and may participate in a cascade of events to facilitate neuronal repair and survival. Its neuroprotective activity has been shown over a wide range of neuronal phenotypes in vitro (Mattson et al., 1989) as well as in vivo (Anderson et al., 1988; Otto and Unsicker, 1990).

The expression of FGF-2 in the brain is regulated by different neurotransmitter and hormonal pathways. In accordance with the possibility that its modulation may represent a rapid protective mechanism to maintain cell homeostasis and reduce neuronal damage under challenging or adverse situations, an acute immobilization stress determines an upregulation of FGF-2 in several brain regions of adult animals (Molteni et al., 2001a). Within some structures, such as the hippocampal formation, the effect is probably mediated by glucocorticoid hormones: in fact exogenous administration of adrenal steroids increases FGF-2 mRNA and protein levels (Riva et al., 1995b; Hansson et al., 2000), whereas adrenalectomy reduces its expression in hippocampus and frontal cortex (Riva et al., 1995a; Hansson et al., 2000). However we cannot rule out the possibility that neurotransmitters, such as dopamine, glutamate or norephinephrine, which are released upon stress exposure, can contribute to stress-related changes in FGF-2 expression. In fact, it has been previously shown that all these mediators can increase the biosynthesis of FGF-2 in specific brain structures (Follesa and Mocchetti, 1993; Riva et al., 1996; Roceri et al., 2001).

The modulation of FGF-2 by stress is rapid and transient and, to some extent, resembles the modulation of FGF-2 described in other experimental paradigms (Riva et al., 1992; Riva et al., 1994).

Being that FGF-2 is neuroprotective (Anderson et al., 1988; Otto and Unsicker, 1990) and that its endogenous production may be relevant to preserve neuronal function under adverse situations (Rowntree and Kolb, 1997), it may be inferred that the prompt upregulation after stress exposure represents a possible strategy to preserve neuronal viability under challenging situations. It is interesting to notice that the expression of FGF-2 is induced by stress in dopamine producing regions (ventral tegmental area and substantia nigra) as well as in dopaminergic target regions, including striatum and prefrontal cortex, to underline a close relation between stress, dopamine and FGF-2. In this regard, we may speculate that the upregulation of the trophic factor may contribute to the sensitizing effects of stress and glucocorticoids in analogy to what has been previously described for amphetamine (Flores et al., 1998; Flores et al., 2000).

Although the regulation of FGF-1 (acidic FGF), a close congener of FGF-2, has not been investigated following stress exposure, surgical removal of adrenal glands reduces its mRNA levels in frontal cortex and prevents its upregulation elicited by kainic acid (Riva et al., 1995a), suggesting a functional role of adrenal steroids also in the control of brain FGF-1 expression.

Conclusions

The data accumulated over the last few years demonstrate that profound alterations in the expression profile of trophic molecules, mainly neurotrophins and fibroblast growth factors, can take place in response to stress. These changes can occur at different developmental stages and with a specific anatomical profile, thus suggesting that they depend upon different molecular mechanisms. Most of the studies have focussed on changes in mRNA and protein levels of different neurotrophic factors, but limited information is still available on their signaling pathways. In particular it will be crucial to assess if stress exposure may affect not only the biosynthesis of these proteins within selected brain regions, but could also regulate their subcellular distribution, local translation and receptor interaction. For example, in the case of BDNF, it will be important to analyze different isoforms and investigate whether the changes in the transcription of 5' exons will interfere with the localization and targeting of the neurotrophin within specific cellular populations. Furthermore, changes in signaling pathways will ultimately influence post-receptor mechanisms involving the modulation of different intracellular cascades that, for example, might lead to changes in proteins involved in cellular survival and vulnerability (Manji et al., 2000).

In summary, changes in the expression and function of neurotrophic factors following stress may represent a major component of the plastic changes set in motion within the CNS in order to cope with challenging situations. However repetitive stress or prolonged adverse situations may lead to a protracted reduction of neurotrophic factor production that could enhance cellular vulnerability. If this is indeed the case, pharmacological strategies aimed at reinstating the normal expression and function of neurotrophic molecules within selected brain regions might prove useful for the treatment of neuropsychiatric disorders where stress represent a major vulnerability factor.

References

Alleva, E. and Santucci, D. (2001) Psychosocial vs. "physical" stress situations in rodents and humans: role of neurotrophins. Physiol. Behav., 73: 313–320.

Aloe, L., Alleva, E., Bohm, A. and Levi-Montalcini, R. (1986) Aggressive behavior induces release of nerve growth factor from mouse salivary gland into the bloodstream. Proc. Natl. Acad. Sci. U S A, 83: 6184–6187.

Aloe, L., Bracci-Laudiero, L., Alleva, E., Lambiase, A., Micera, A. and Tirassa, P. (1994) Emotional stress induced by parachute jumping enhances blood nerve growth factor levels and the distribution of nerve growth factor receptors in lymphocytes. Proc. Natl. Acad. Sci. U S A, 91: 10440–10444.

Anderson, K.J., Dam, D., Lee, S. and Cotman, C.W. (1988) Basic fibroblast growth factor prevents death of lesioned cholinergic neurons in vivo. Nature, 332: 360–361.

Barbany, G. and Persson, H. (1992) Regulation of Neurotrophin mRNA Expression in the Rat Brain by Glucocorticoids. Eur. J. Neurosci., 4: 396–403.

Barbany, G. and Persson, H. (1993) Adrenalectomy attenuates kainic acid-elicited increases of messenger RNAs for neurotrophins and their receptors in the rat brain. Neuroscience, 54: 909–922.

Berninger, B., Garcia, D.E., Inagaki, N., Hahnel, C. and Lindholm, D. (1993) BDNF and NT-3 induce intracellular Ca2+ elevation in hippocampal neurones. Neuroreport, 4: 1303–1306.

Bigi, S., Maestripieri, D., Aloe, L. and Alleva, E. (1992) NGF decreases isolation-induced aggressive behavior, while increasing adrenal volume, in adult male mice. Physiol. Behav., 51: 337–343.

Bikfalvi, A., Klein, S., Pintucci, G. and Rifkin, D.B. (1997) Biological roles of fibroblast growth factor-2. Endocr. Rev., 18: 26–45.

Black, I.B. (1999) Trophic regulation of synaptic plasticity. J. Neurobiol., 41: 108–118.

Butterweck, V., Winterhoff, H. and Herkenham, M. (2001) St John's wort, hypericin, and imipramine: a comparative analysis of mRNA levels in brain areas involved in HPA axis control following short-term and long-term administration in normal and stressed rats. Mol. Psychiatry, 6: 547–564.

Canossa, M., Griesbeck, O., Berninger, B., Campana, G., Kolbeck, R. and Thoenen, H. (1997) Neurotrophin release by neurotrophins: implications for activity-dependent neuronal plasticity. Proc. Natl. Acad. Sci. U S A, 94: 13279–13286.

Carnahan, J. and Nawa, H. (1995) Regulation of neuropeptide expression in the brain by neurotrophins. Potential role in vivo. Mol. Neurobiol., 10: 135–149.

Chao, M.V. and Bothwell, M. (2002) Neurotrophins: to cleave or not to cleave. Neuron, 33: 9–12.

Cirulli, F. (2001) Role of environmental factors on brain development and nerve growth factor expression. Physiol. Behav., 73: 321–330.

de Kloet, E.R., Oitzl, M.S. and Joels, M. (1999) Stress and cognition: are corticosteroids good or bad guys? Trends Neurosci., 22: 422–426.

Duman, R.S., Malberg, J., Nakagawa, S. and D'Sa, C. (2000) Neuronal plasticity and survival in mood disorders. Biol. Psychiatry, 48: 732–739.

Edwards, R.H., Selby, M.J., Garcia, P.D. and Rutter, W.J. (1988) Processing of the native nerve growth factor precursor to form biologically active nerve growth factor. J. Biol. Chem., 263: 6810–6815.

Eide, F.F., Vining, E.R., Eide, B.L., Zang, K., Wang, X.Y. and Reichardt, L.F. (1996) Naturally occurring truncated trkB receptors have dominant inhibitory effects on brain-derived neurotrophic factor signaling. J. Neurosci., 16: 3123–3129.

Finlay, J.M., Zigmond, M.J. and Abercrombie, E.D. (1995) Increased dopamine and norepinephrine release in medial prefrontal cortex induced by acute and chronic stress: effects of diazepam. Neuroscience, 64: 619–628.

Flores, C., Rodaros, D. and Stewart, J. (1998) Long-lasting induction of astrocytic basic fibroblast growth factor by repeated injections of amphetamine: blockade by concurrent treatment with a glutamate antagonist. J. Neurosci., 18: 9547–9555.

Flores, C., Samaha, A.N. and Stewart, J. (2000) Requirement of endogenous basic fibroblast growth factor for sensitization to amphetamine. J. Neurosci., 20: RC55.

Follesa, P. and Mocchetti, I. (1993) Regulation of basic fibroblast growth factor and nerve growth factor mRNA by beta-adrenergic receptor activation and adrenal steroids in rat central nervous system. Mol. Pharmacol., 43: 132–138.

Fumagalli, F., Santero, R., Gennarelli, M., Giorgio, R. and Andrea Riva, M. (2001) Decreased hippocampal BDNF expression after acute systemic injection of quinpirole. Neuropharmacology, 40: 954–957.

Givalois, L., Marmigere, F., Rage, F., Ixart, G., Arancibia, S. and Tapia-Arancibia, L. (2001) Immobilization stress rapidly and differentially modulates BDNF and TrkB mRNA expression in the pituitary gland of adult male rats. Neuroendocrinology, 74: 148–159.

Gould, E. and Tanapat, P. (1999) Stress and hippocampal neurogenesis. Biol. Psychiatry, 46: 1472–1479.

Hansson, A.C., Cintra, A., Belluardo, N., Sommer, W., Bhatnagar, M., Bader, M., Ganten, D. and Fuxe, K. (2000) Gluco- and mineralocorticoid receptor-mediated regulation of neurotrophic factor gene expression in the dorsal hippocampus and the neocortex of the rat. Eur. J. Neurosci., 12: 2918–2934.

Hayashi, M., Mistunaga, F., Ohira, K. and Shimizu, K. (2001) Changes in BDNF-immunoreactive structures in the hippocampal formation of the aged macaque monkey. Brain Res., 918: 191–196.

Hefti, F. and Weiner, W.J. (1986) Nerve growth factor and Alzheimer's disease. Ann. Neurol., 20: 275–281.

Huang, E.J. and Reichardt, L.F. (2001) Neurotrophins: roles in neuronal development and function. Annu. Rev. Neurosci., 24: 677–736.

Jedema, H.P. and Moghaddam, B. (1994) Glutamatergic control of dopamine release during stress in the rat prefrontal cortex. J. Neurochem., 63: 785–788.

Jovanovic, J.N., Czernik, A.J., Fienberg, A.A., Greengard, P. and Sihra, T.S. (2000) Synapsins as mediators of BDNF-enhanced neurotransmitter release. Nat. Neurosci., 3: 323–329.

Kang, H.J. and Schuman, E.M. (1995) Neurotrophin-induced modulation of synaptic transmission in the adult hippocampus. J. Physiol. Paris, 89: 11–22.

Kaplan, D.R. and Miller, F.D. (2000) Neurotrophin signal transduction in the nervous system. Curr. Opin. Neurobiol., 10: 381–391.

Kim, J.J. and Diamond, D.M. (2002) The stressed hippocampus, synaptic plasticity and lost memories. Nat. Rev. Neurosci., 3: 453–462.

Klein, R., Conway, D., Parada, L.F. and Barbacid, M. (1990) The trkB tyrosine protein kinase gene codes for a second neurogenic receptor that lacks the catalytic kinase domain. Cell, 61: 647–656.

Klint, P. and Claesson-Welsh, L. (1999) Signal transduction by fibroblast growth factor receptors. Frontiers in Bioscience, 4: D165–177.

Knipper, M., da Penha Berzaghi, M., Blochl, A., Breer, H., Thoenen, H. and Lindholm, D. (1994) Positive feedback between acetylcholine and the neurotrophins nerve growth factor and brain-derived neurotrophic factor in the rat hippocampus. Eur. J. Neurosci., 6: 668–671.

Kokaia, Z., Gido, G., Ringstedt, T., Bengzon, J., Kokaia, M., Siesjo, B.K., Persson, H. and Lindvall, O. (1993) Rapid increase of BDNF mRNA levels in cortical neurons following spreading depression: regulation by glutamatergic mechanisms independent of seizure activity. Brain Res. Mol. Brain Res., 19: 277–286.

Korte, M., Carroll, P., Wolf, E., Brem, G., Thoenen, H. and Bonhoeffer, T. (1995) Hippocampal long-term potentiation is impaired in mice lacking brain-derived neurotrophic factor. Proc. Natl. Acad. Sci. U S A, 92: 8856–8860.

Ladd, C.O., Huot, R.L., Thrivikraman, K.V., Nemeroff, C.B., Meaney, M.J. and Plotsky, P.M. (2000) Long-term behavioral and neuroendocrine adaptations to adverse early experience. Prog. Brain Res., 122: 81–103.

Lakshmanan, J. (1986) Beta-nerve growth factor measurements in mouse serum. J. Neurochem., 46: 882–891.

Lauterborn, J.C., Poulsen, F.R., Stinis, C.T., Isackson, P.J. and Gall, C.M. (1998) Transcript-specific effects of adrenalectomy on seizure-induced BDNF expression in rat hippocampus. Brain Res. Mol. Brain Res., 55: 81–91.

Levi-Montalcini, R. (1987) The nerve growth factor 35 years later. Science, 237: 1154–1162.

Lewin, G.R. and Barde, Y.A. (1996) Physiology of the neurotrophins. Annu. Rev. Neurosci., 19: 289–317.

Linden, D.J. and Connor, J.A. (1995) Long-term synaptic depression. Annu. Rev. Neurosci., 18: 319–357.

Lipska, B.K. and Weinberger, D.R. (2000) To model a psychiatric disorder in animals: schizophrenia as a reality test. Neuropsychopharmacology, 23: 223–239.

Lu, B. and Gottschalk, W. (2000) Modulation of hippocampal synaptic transmission and plasticity by neurotrophins. Prog. Brain Res., 128: 231–241.

Magarinos, A.M., Verdugo, J.M. and McEwen, B.S. (1997) Chronic stress alters synaptic terminal structure in hippocampus. Proc. Natl. Acad. Sci. USA, 94: 14002–14008.

Manji, H.K., Moore, G.J., Rajkowska, G. and Chen, G. (2000) Neuroplasticity and cellular resilience in mood disorders. Mol. Psychiatry, 5: 578–593.

Mattson, M.P., Murrain, M., Guthrie, P.B. and Kater, S.B. (1989) Fibroblast growth factor and glutamate: opposing roles in the generation and degeneration of hippocampal neuroarchitecture. J. Neurosci., 9: 3728–3740.

McAllister, A.K. (1999) Subplate neurons: a missing link among neurotrophins, activity, and ocular dominance plasticity? Proc. Natl. Acad. Sci. U S A, 96: 13600–13602.

McEwen, B.S. (2000a) Effects of adverse experiences for brain structure and function. Biol. Psychiatry, 48: 721–731.

McEwen, B.S. (2000b) The neurobiology of stress: from serendipity to clinical relevance. Brain Res., 886: 172–189.

McEwen, B.S. and Sapolsky, R.M. (1995) Stress and cognitive function. Curr. Opin. Neurobiol., 5: 205–216.

Meaney, M.J. (2001) Maternal care, gene expression, and the transmission of individual differences in stress reactivity across generations. Annu. Rev. Neurosci., 24: 1161–1192.

Mocchetti, I., Spiga, G., Hayes, V.Y., Isackson, P.J. and Colangelo, A. (1996) Glucocorticoids differentially increase nerve growth factor and basic fibroblast growth factor expression in the rat brain. J. Neurosci., 16: 2141–2148.

Moghaddam, B. (2002) Stress activation of glutamate neurotransmission in the prefrontal cortex: implications for dopamine-associated psychiatric disorders. Biol. Psychiatry, 51: 775–787.

Molteni, R., Fumagalli, F., Magnaghi, V., Roceri, M., Gennarelli, M., Racagni, G., Melcangi, R.C. and Riva, M.A. (2001a) Modulation of fibroblast growth factor-2 by stress and corticosteroids: from developmental events to adult brain plasticity. Brain Res. Brain Res. Rev., 37: 249–258.

Molteni, R., Lipska, B.K., Weinberger, D.R., Racagni, G. and Riva, M.A. (2001b) Developmental and stress-related changes of neurotrophic factor gene expression in an animal model of schizophrenia. Mol. Psychiatry, 6: 285–292.

Nibuya, M., Morinobu, S. and Duman, R.S. (1995) Regulation of BDNF and trkB mRNA in rat brain by chronic electroconvulsive seizure and antidepressant drug treatments. J. Neurosci., 15: 7539–7547.

Nibuya, M., Takahashi, M., Russell, D.S. and Duman, R.S. (1999) Repeated stress increases catalytic TrkB mRNA in rat hippocampus. Neurosci. Lett., 267: 81–84.

Ninkina, N., Adu, J., Fischer, A., Pinon, L.G., Buchman, V.L. and Davies, A.M. (1996) Expression and function of TrkB variants in developing sensory neurons. EMBO J., 15: 6385–6393.

Otto, D. and Unsicker, K. (1990) Basic FGF reverses chemical and morphological deficits in the nigrostriatal system of MPTP-treated mice. J. Neurosci., 10: 1912–1921.

Palmer, T.D., Markakis, E.A., Willhoite, A.R., Safar, F. and Gage, F.H. (1999) Fibroblast growth factor-2 activates a latent neurogenic program in neural stem cells from diverse regions of the adult CNS. J. Neurosci., 19: 8487–8497.

Patapoutian, A. and Reichardt, L.F. (2001) Trk receptors: mediators of neurotrophin action. Curr. Opin. Neurobiol., 11: 272–280.

Patterson, S.L., Abel, T., Deuel, T.A., Martin, K.C., Rose, J.C. and Kandel, E.R. (1996) Recombinant BDNF rescues deficits in basal synaptic transmission and hippocampal LTP in BDNF knockout mice. Neuron, 16: 1137–1145.

Poo, M.M. (2001) Neurotrophins as synaptic modulators. Nature Rev. Neurosci., 2: 24–32.

Pozzo-Miller, L.D., Gottschalk, W., Zhang, L., McDermott, K., Du, J., Gopalakrishnan, R., Oho, C., Sheng, Z.H. and Lu, B. (1999) Impairments in high-frequency transmission, synaptic vesicle docking, and synaptic protein distribution in the hippocampus of BDNF knockout mice. J. Neurosci., 19: 4972–4983.

Rage, F., Givalois, L., Marmigere, F., Tapia-Arancibia, L. and Arancibia, S. (2002) Immobilization stress rapidly modulates BDNF mRNA expression in the hypothalamus of adult male rats. Neuroscience, 112: 309–318.

Rasmusson, A.M., Shi, L. and Duman, R. (2002) Downregulation of BDNF mRNA in the hippocampal dentate gyrus after re-exposure to cues previously associated with footshock. Neuropsychopharmacology, 27: 133–142.

Rios, M., Fan, G., Fekete, C., Kelly, J., Bates, B., Kuehn, R., Lechan, R.M. and Jaenisch, R. (2001) Conditional deletion of brain-derived neurotrophic factor in the postnatal brain leads to obesity and hyperactivity. Mol. Endocrinol., 15: 1748–1757.

Riva, M.A., Donati, E., Tascedda, F., Zolli, M. and Racagni, G. (1994) Short- and long-term induction of basic fibroblast growth factor gene expression in rat central nervous system following kainate injection. Neuroscience, 59: 55–65.

Riva, M.A., Fumagalli, F., Blom, J.M., Donati, E. and Racagni, G. (1995a) Adrenalectomy reduces FGF-1 and FGF-2 gene expression in specific rat brain regions and differently affects their induction by seizures. Mol. Brain Res., 34: 190–196.

Riva, M.A., Fumagalli, F. and Racagni, G. (1995b) Opposite regulation of basic fibroblast growth factor and nerve growth factor gene expression in rat cortical astrocytes following dexamethasone treatment. J. Neurochem., 64: 2526–2533.

Riva, M.A., Gale, K. and Mocchetti, I. (1992) Basic fibroblast growth factor mRNA increases in specific brain regions following convulsive seizures. Mol. Brain Res., 15: 311–318.

Riva, M.A., Molteni, R., Lovati, E., Fumagalli, F., Rusnati, M. and Racagni, G. (1996) Cyclic AMP-dependent regulation of fibroblast growth factor-2 messenger RNA levels in rat cortical astrocytes: comparison with fibroblast growth factor-1 and ciliary neurotrophic factor. Mol. Pharmacol., 49: 699–706.

Roceri, M., Molteni, R., Fumagalli, F., Gennarelli, M., Racagni, G., Corsini, G.U., Maggio, R. and Riva, M.A. (2001) Stimulatory role of dopamine on FGF-2 expression in rat striatum. J. Neurochem., 76: 990–997.

Roceri, M., W., H., Racagni, G., B.A., E. and Riva, M.A. (2002) Early maternal deprivation reduces the expression of BDNF and NMDA receptor subunits in rat hippocampus. Mol. Psychiatry, 7: 609–616.

Roceri, M., Cirulli, F., Pessina, C., Peretto, P., Racagni, G. and Riva, M.A. (2004) Postnatal repeated maternal deprivation produces age dependent changes of BDNF expression in selected rat brain regions. Biological Psychiatry, 55: 708–714.

Rowntree, S. and Kolb, B. (1997) Blockade of basic fibroblast growth factor retards recovery from motor cortex injury in rats. Eur. J. Neurosci., 9: 2432–2441.

Russo-Neustadt, A., Ha, T., Ramirez, R. and Kesslak, J.P. (2001) Physical activity-antidepressant treatment combination: impact on brain-derived neurotrophic factor and behavior in an animal model. Behav. Brain Res., 120: 87–95.

Sapolsky, R.M. (2000) Glucocorticoids and hippocampal atrophy in neuropsychiatric disorders. Archives of General Psychiatry, 57: 925–935.

Scaccianoce, S., Catalani, A., Lombardo, K., Consoli, C. and Angelucci, L. (2001) Maternal glucocorticoid hormone influences nerve growth factor expression in the developing rat brain. Neuroreport, 12: 2881–2884.

Scaccianoce, S., Lombardo, K. and Angelucci, L. (2000) Nerve growth factor brain concentration and stress: changes depend on type of stressor and age. Int. J. Dev. Neurosci., 18: 469–479.

Schaaf, M.J., De Kloet, E.R. and Vreugdenhil, E. (2000) Corticosterone effects on BDNF expression in the hippocampus. Implications for memory formation. Stress, 3: 201–208.

Schinder, A.F., Berninger, B. and Poo, M. (2000) Postsynaptic target specificity of neurotrophin-induced presynaptic potentiation. Neuron, 25: 151–163.

Schinder, A.F. and Poo, M.M. (2000) The neurotrophin hypothesis for synaptic plasticity. Trends Neurosci., 23: 639–645.

Smith, M.A. and Cizza, G. (1996) Stress-induced changes in brain-derived neurotrophic factor expression are attenuated in aged Fischer 344/N rats. Neurobiol. Aging, 17: 859–864.

Smith, M.A., Makino, S., Altemus, M., Michelson, D., Hong, S.K., Kvetnansky, R. and Post, R.M. (1995a) Stress and antidepressants differentially regulate neurotrophin 3 mRNA expression in the locus coeruleus. Proc. Natl. Acad. Sci. U S A, 92: 8788–8792.

Smith, M.A., Makino, S., Kim, S.Y. and Kvetnansky, R. (1995b) Stress increases brain-derived neurotropic factor messenger ribonucleic acid in the hypothalamus and pituitary. Endocrinology, 136: 3743–3750.

Smith, M.A., Makino, S., Kvetnansky, R. and Post, R.M. (1995c) Stress and glucocorticoids affect the expression of brain-derived neurotrophic factor and neurotrophin-3 mRNAs in the hippocampus. J. Neurosci., 15: 1768–1777.

Sofroniew, M.V., Howe, C.L. and Mobley, W.C. (2001) Nerve growth factor signaling, neuroprotection, and neural repair. Annu. Rev. Neurosci., 24: 1217–1281.

Spillantini, M.G., Aloe, L., Alleva, E., De Simone, R., Goedert, M. and Levi-Montalcini, R. (1989) Nerve growth factor mRNA and protein increase in hypothalamus in a mouse model of aggression. Proc. Natl. Acad. Sci. U S A, 86: 8555–8559.

Thoenen, H. (1995) Neurotrophins and neuronal plasticity. Science, 270: 593–598.

Thoenen, H. (2000) Neurotrophins and activity-dependent plasticity. Prog. Brain Res., 128: 183–191.

Thoenen, H., Bandtlow, C. and Heumann, R. (1987) The physiological function of nerve growth factor in the central nervous system: comparison with the periphery. Rev. Physiol. Biochem. Pharmacol., 109: 145–178.

Thoenen, H., Bandtlow, C., Heumann, R., Lindholm, D., Meyer, M. and Rohrer, H. (1988) Nerve growth factor: cellular localization and regulation of synthesis. Cell Mol. Neurobiol., 8: 35–40.

Timmusk, T., Palm, K., Metsis, M., Reintam, T., Paalme, V., Saarma, M. and Persson, H. (1993) Multiple promoters direct tissue-specific expression of the rat BDNF gene. Neuron, 10: 475–489.

Tongiorgi, E., Righi, M. and Cattaneo, A. (1997) Activity-dependent dendritic targeting of BDNF and TrkB mRNAs in hippocampal neurons. J. Neurosci., 17: 9492–9505.

Ueyama, T., Kawai, Y., Nemoto, K., Sekimoto, M., Tone, S. and Senba, E. (1997) Immobilization stress reduced the expression of neurotrophins and their receptors in the rat brain. Neurosci. Res., 28: 103–110.

Vaidya, V.A., Marek, G.J., Aghajanian, G.K. and Duman, R.S. (1997) 5-HT2A receptor-mediated regulation of brain-derived neurotrophic factor mRNA in the hippocampus and the neocortex. J. Neurosci., 17: 2785–2795.

Zafra, F., Castren, E., Thoenen, H. and Lindholm, D. (1991) Interplay between glutamate and gamma-aminobutyric acid transmitter systems in the physiological regulation of brain-derived neurotrophic factor and nerve growth factor synthesis in hippocampal neurons. Proc. Natl. Acad. Sci. U S A, 88: 10037–10041.

Zafra, F., Hengerer, B., Leibrock, J., Thoenen, H. and Lindholm, D. (1990) Activity dependent regulation of BDNF and NGF mRNAs in the rat hippocampus is mediated by non-NMDA glutamate receptors. EMBO J., 9: 3545–3550.

T. Steckler, N.H. Kalin and J.M.H.M. Reul (Eds.)
Handbook of Stress and the Brain, Vol. 15
ISBN 0-444-51173-3

CHAPTER 5.3

Transcription factors as modulators of stress responsivity

Ronald S. Duman*, David H. Adams and Birgitte B. Simen

Departments of Psychiatry and Pharmacology, Laboratory of Molecular Psychiatry, Yale University School of Medicine, 34 Park Street, New Haven, CT 06508, USA

Abstract: Exposure to stress can lead to both adaptive and maladaptive changes that control neuronal function and behavior. The mechanisms underlying these adaptive changes include regulation of synaptic transmission, intracellular signal transduction, gene expression, and even structural alterations. Because acute and chronic exposure to stress can lead to long-term changes in neuronal function, special emphasis has been placed on regulation of transcription factors and patterns of gene expression that could underlie these changes. In this chapter, the basic mechanisms that regulate transcription factors and gene expression, including *cis*- and *trans*-acting factors, are discussed. The influence of stress on three major classes of transcription factors, activating transcription factor (ATF) (e.g., the cAMP response element-binding protein or CREB), activator protein-1 (AP-1) (e.g., c-Fos and c-Jun), and nuclear factor kappa B (NF-κB) is examined. Characterization of the transcription factors and target genes underlying the actions of stress will provide critical information for understanding stress-related neurobiological disorders and ultimately better treatment interventions.

Introduction

The identification and characterization of genes that control neuronal function and behavior have been the focus of intense research in recent years. The goal of this work is to understand the genes that control complex behavior and thereby identify gene mutations that contribute to neurobiological disorders. The completion of the human genome project, as well as the sequencing of the DNA of other species, has provided the coding information that is needed to study the molecular basis of neuronal function and behavior. However, DNA sequence alone does not tell us which genes are turned on and off in response to different endocrine, immune, environmental, and behavioral conditions. This information is critical because it represents one of the primary mechanisms by which the brain processes information and allows an organism to make appropriate adaptive responses to the same or related stimuli. The expression of genes in the brain is regulated by multiple internal and external stimuli that can induce discrete and unique patterns of gene expression in different brain regions and even within a single neuron. These unique patterns of expression are involved in shaping the function of the brain and its ability to adapt and generate long-term and informed responses to subsequent stimuli.

Exposure to short- or long-term stress produces significant and sometimes profound effects on neuronal function and behavior, and many of these effects occur at the level of gene transcription. This chapter will provide an overview of the fundamental concepts of gene transcription, and will then focus on several classes of transcription factors that have been extensively studied in the context of stress, anxiety, and depression. These include the cAMP response element-binding protein (CREB), the c-Fos/activator protein-1 (AP-1), and the NFκB systems. This is not meant to be an exhaustive list of the transcription factors regulated by stress, but a more focused overview of a few significant factors that have been examined in this context. The glucocorticoid receptors represent another major class of transcription factors that are influenced by elevated levels of glucocorticoids that are activated by stress. Glucocorticoid receptor regulation of gene expression will be covered in another chapter.

DNA and gene transcription

Gene expression is controlled by a complex interaction of transcription factors (referred to as *trans*-regulatory elements) with specific sequences of DNA in the promoter elements of genes (*cis* elements). The process of gene transcription and the expression of cassettes of genes under different conditions are of primary importance because these processes control all aspects of cellular/neuronal function (for reviews of gene transcription, see Armstrong and Montminy, 1993; Nestler et al., 2001). In 1953, Watson and Crick reported on the structure of DNA. They postulated that single strands of nucleotides, made up of adenine (A), thymine (T), guanine (G), and cytosine (C), are able to form a double helix by pairing of complementary strands (A with T and G with C). Double-stranded DNA has the characteristics required for duplication of genetic material during cell division and reproduction and for expression of genes that are necessary for cell survival and function.

Gene transcription and expression are tightly controlled and most of the nuclear DNA is in an inactive state, tightly coiled around nucleosomes that are the major components of chromosomes. This form of DNA is inaccessible to transcription factors and must be uncoiled for initiation of gene transcription to occur. Relaxation of the nucleosomes, which are made up largely of histones, requires enzymatic processing and remodeling of the nucleosomes. The histone acetylases, one of the major classes of enzymes responsible for remodeling of nucleosomes, modify core DNA-binding proteins and allow for recruitment of transcription factors and RNA polymerase. This is referred to as the initiation phase of gene transcription, which is then followed by elongation and termination. Histone deacetylases return DNA to the coiled/bound nucleosome state and contribute to inactivation of gene transcription. Interestingly, histone deacetylases are inhibited by valproic acid, an anticonvulsant drug used for the treatment of bipolar disorder (Phiel et al., 2001). Inhibition of histone deacetylase leads to increased expression of genes that are thought to contribute to the therapeutic action of valproic acid and other mood-stabilizing drugs, as well as its side effects.

The initiation of gene transcription is controlled by interactions of transcription factors with specific DNA sequences in the gene promoter. The elements that determine where initiation occurs and control basal rates of transcription are referred to as core elements. The association of RNA polymerase to the core promoter, referred to as a TATA box because of the high number of T and A nucleotides, is required for initiation (Fig. 1). Genes that do not contain a TATA box have a poorly conserved element referred to as an initiator. RNA polymerase binds to the core promoter upon remodeling of a nucleosome and this step is required for transcriptional initiation. Type II RNA polymerase (pol II) is utilized for transcription of mRNA, while other forms are used for very large RNA (I) or for small nuclear RNA (snRNA)

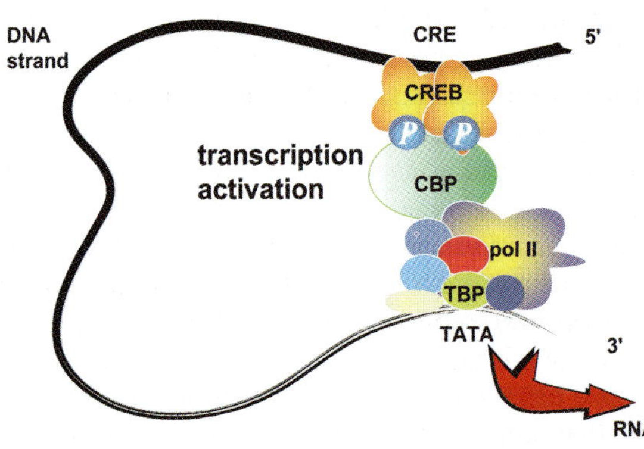

Fig. 1. Schematic model of *cis*- and *trans*-acting DNA elements involved in regulation of gene transcription. This model depicts a complex of transcriptional proteins associated with a TATA box or transcription initiation site. The TATA-binding protein (TBP) and RNA polymerase II (pol II), as well as several additional binding proteins, make up this complex. Transcription activation requires association with a promoter element, such as a cAMP response element (CRE) and the CRE-binding protein (CREB). Such elements are often located several hundreds of basepairs upstream from the TATA box but they can associate with the TATA complex by folding of the DNA strand upon itself. CREB binds to the CRE, but does not become fully active until it is phosphorylated (P). This allows for binding with the CREB-binding protein (CBP) that in turn couples to and activates the pol II transcription complex, resulting in the synthesis of RNA.

(I and III). The TATA-binding protein is responsible for binding of pol II as well as other transcription factors and cofactors and is therefore critical for initiation of basal transcription. Promoter elements located further upstream from the initiator help to recruit activator and/or repressor proteins and control higher rates of RNA synthesis and gene expression.

The mRNA formed in the nucleus is transported to the cytoplasm, where it is translated to form cellular proteins. It is important to point out that posttranscriptional modifications of mRNA also play a critical role and are important for further regulation and fine tuning of specific patterns and types of genes expressed. For example, certain genes can encode multiple splice variants, each of which may have different regulatory and functional domains. Another example is RNA editing whereby a single nucleotide can be altered and result in an amino acid switch that changes the function of a protein. The stability and cellular localization of mRNA are additional mechanisms for regulation. The half-life of mRNA is controlled by sequences found in the 3′ untranslated region (or 3′ UTR) that influence stability and degradation. Sequences in the 3′UTR are also thought to act as signals for transport of mRNA to distal sites, such as axon and dendrite terminals, for local protein expression. The combination of these and other mechanisms provides many sites for fine tuning the translation and expression of proteins in a unique fashion and thereby control of neuronal function.

Transcription factor families

Basal levels of gene transcription occur when RNA polymerase binds to the TATA box/initiation site, but efficient and higher rates of transcription require additional gene transcription factors. Transcription factors bind relatively close to the initiation site or bind to promoter elements that can be located several kilobases upstream of the initiation site (Nestler et al., 2001). Transcription factors are referred to as either promoters or enhancers depending on the distance from the initiation site, although functionally there is little difference because the distal enhancers can be brought into close proximity to the TATA box/initiation site when the DNA loops upon itself (Fig. 1). In addition to a DNA-binding domain, transcription factors generally have activation and regulatory domains. Activation often requires an interaction with another subunit of the same or a different transcription factor for complete activation. These dimers may act in concert to enhance transcription, or a heterodimer may repress activity compared to that in response to a homodimer, or vice versa. The requirement for dimers that enhance or repress transcription is another mechanism for fine-tuning gene transcription.

In this chapter we will discuss three major families of transcription factors that are regulated by stress and the immune system. These are the CREB-like transcription factors, the AP-1, and the NFκB family. Each of these classes of transcription factors is regulated by a different mechanism. The primary means for activation of most CREB family members is by phosphorylation of specific amino acid residues in the regulatory domain. NFκB is also activated by phosphorylation, including phosphorylation of an NFκB-binding protein that then releases NFκB. Members of the AP-1 family are regulated primarily by induction of the total amount of the transcription factor protein, although these factors also have sites for regulation by phosphorylation. Another major class of transcription factor that plays a major role in the stress response is the glucocorticoid receptor, a member of the steroid hormone receptor superfamily. Steroid hormone receptors are cytoplasmic proteins that translocate to the nucleus upon binding to a specific class of steroid. Other members of this class include receptor/transcription factors for thyroid hormones, sex steroids (estrogen, progesterone, and testosterone), retinoic acid, and vitamin D.

Stress and the transcription factor CREB

CREB and related transcription factors were initially studied and identified to contribute to the long-term adaptations that underlie learning and memory (Silva et al., 1998). However, CREB is expressed throughout the brain and is now known to be regulated by a variety of stimuli and in turn to regulate several classes of genes that influence many different aspects of CNS function. This includes acute and long-term responses to stress. This section will provide a description of the structural and functional aspects of CREB and then will describe studies demonstrating the regulation and function of CREB in stress and related conditions.

CREB structure and function

CREB belongs to the activating transcription factor family (ATF), and as the name of this family implies most members are activators of gene transcription. There are three different splice variants of CREB found in the brain, α, β, and Δ. CREB was originally identified as a transcriptional mediator of cAMP-signaling and cAMP-dependent protein kinase. However, CREB is also phosphorylated and activated by other protein kinases, including Ca^{2+}-dependent protein kinases and ribosomal S6 kinase (RSK) (Fig. 2). The phosphorylation of CREB (pCREB) by multiple signal transduction pathways makes this transcription factor a point of convergence for rapid as well as long-term adaptive signaling pathways. This could play an important

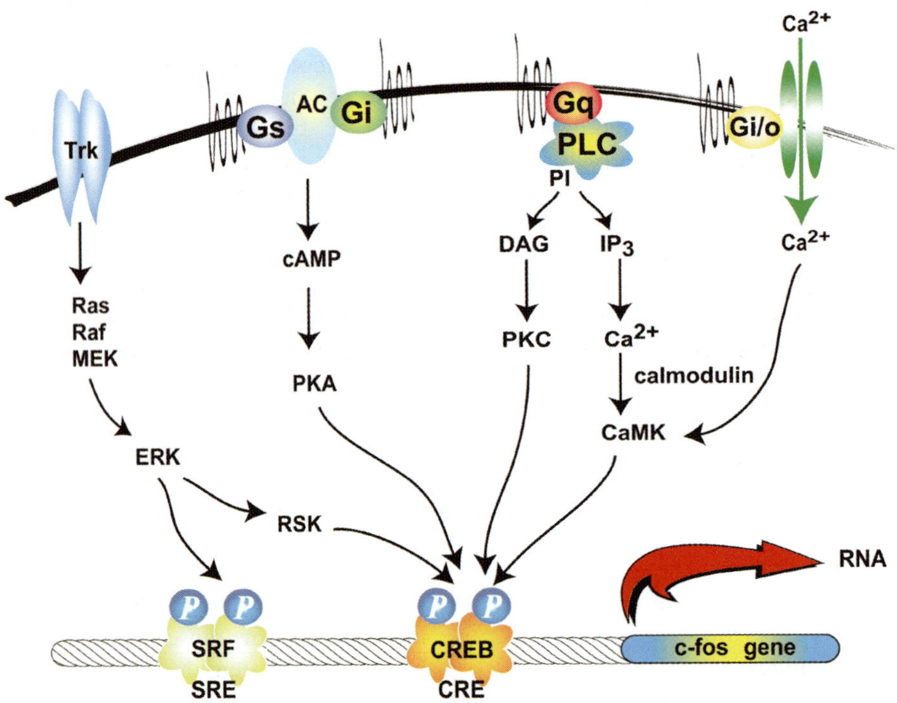

Fig. 2. Diagram of signal-transduction pathways that can activate CREB and that increase c-Fos gene expression. G protein receptor-coupled second messenger, neurotrophic factor, and Ca^{2+} activated pathways activate protein kinases that subsequently regulate CREB and c-Fos gene expression. Regulation of cAMP levels is controlled by receptors that either stimulate or inhibit adenylyl cyclase (AC) via coupling to Gs or Gi, respectively. Activation of adenylyl cyclase and elevation of cAMP levels stimulates cAMP-dependent protein kinase (PKA), one of several protein kinases that can phosphorylate CREB at Ser133. Activation of phospholipase C (PLC) and the phosphotidyl inositol (PI) pathway occurs via receptor coupling with Gq. Activation of PLC leads to the formation of inositol trisphosphate (IP_3) and diacylglycerol (DAG), which stimulates protein kinase C (PKC). Ca^{2+} can be released from intracellular stores by IP3, or by influx through ion channels (e.g., voltage-dependent or glutamate-*N*-methyl D aspartate). Ca^{2+}, in conjunction with calmodulin, activates Ca^{2+}/calmodulin-dependent protein kinase (CaMK). PKC and CaMK can also phosphorylate and activate CREB. In addition to these second messenger-dependent pathways, neurotrophic factor activation of Trk receptors and the Ras-Raf-MEK-ERK pathway can lead to regulation of gene expression. ERK can activate ribosomal S6 kinase (RSK), another CREB kinase. Phosphorylation of CREB leads to transcriptional activation of the c-Fos promoter which contains a CRE. In addition, ERK can directly activate c-Fos gene expression via phosphorylation of a serum response factor (SRF) which binds to a serum response element (SRE) in the promoter of the c-Fos gene.

role in the function of CREB in response to acute and chronic stress exposures. These and other pathways may be differentially regulated by stress in subsets of neurons and glia and result in a unique gene-expression signature. A number of technological approaches, from single-cell real-time polymerase chain reaction and genechip analysis, to double and triple labeling of proteins in cells, are being used to identify these unique signatures.

CREB and related family members bind to a specific sequence of DNA, referred to as the cAMP response element (CRE) that is located in the promoter regions of genes. CREB binds with highest affinity to the consensus CRE, an eight base-pair, palindromic sequence, TGACGTCA. CREB also binds to CRE sequences that have one or two substitutions, although affinity of CREB for these elements is reduced. The binding of CREB to DNA requires the interaction of two molecules of CREB or related family members to form either homo- or heterodimers. Most dimers of ATF factors result in transcriptional activation, although there are a few examples of heterodimers that result in repression. The ability of CREB to form a dimer is dependent on the leucine zipper domain (Fig. 3). This region has a leucine repeat every seven amino acids and hydrophobic interactions of the leucine residues of two transcription factors results in the formation of a "leucine zipper" that stabilizes the dimer. The leucine zipper domain is adjacent to a highly basic DNA binding domain. The name basic leucine zipper, or bZIP, refers to transcription factors that utilize these structural and functional motifs.

Homo- or heterodimers of CREB and related proteins are capable of binding to CRE sites, but transcriptional activation predominantly occurs when CREB is phosphorylated (see Figs. 1–3). There are several phosphorylation sites in the CREB molecule, but it is phosphorylation of a conserved amino acid residue, Ser^{133}, that is necessary for activation of gene transcription. Phosphorylation of Ser^{133} is necessary for the interaction of CREB with the CREB-binding protein (CBP), which in turn interacts with the basal transcription complex at the TATA box/initiator of a gene and thereby results in transcriptional activation (Fig. 1).

Another closely related subgroup of the ATF family is referred to as the CRE modulators or CREMs. The CREM transcription factors include members that are regulated by either phosphorylation or expression. Most of the CREM isoforms, like CREB, are phosphorylated and this leads to activation of gene transcription. However, there is also a transcriptional inhibitor, referred to as

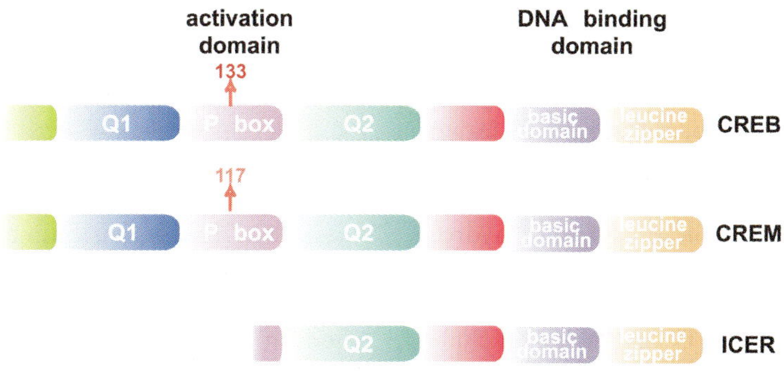

Fig. 3. Schematic structure of the CREB and CREM transcription factors. These models depict the primary functional domains of CREB and CREM. This includes the activation domain, also referred to as the P box or kinase-inducible domain, which contains a phosphorylation site for transcriptional activation. The amino terminus of the proteins contains the basic domain that binds to DNA and the leucine zipper domain that is responsible for dimerization of CREB and related proteins. Also shown at the bottom is the structure of ICER, or the inducible cAMP early response factor. ICER is generated by use of an alternative start site in the CREM gene that is activated by cAMP/Ca^{2+} signaling and a CRE site. This results in the synthesis of a truncated form of the transcription factor that lacks the kinase-inducible domain. ICER is able to dimerize with CREB, CREM and related proteins, but cannot be activated by phosphorylation. ICER thereby serves as a negative-feedback transcription factor that shuts down CRE-mediated gene expression.

inducible cAMP early repressor (ICER). ICER is especially interesting because it is induced by activation of a CRE in an alternative promoter of the CREM gene. Initiation from this alternative site results in the formation of a truncated form of CREM that lacks the kinase-inducible domain (Fig. 3). This is an example of a transcription factor isoform that can dimerize with an activator like CREM or CREB, but because it cannot be phosphorylated it represses gene transcription. In this way ICER serves as a negative-feedback mechanism for counterregulation of CRE-mediated gene expression.

Regulation of CREB by acute physical stress

The potential role of CREB in responses to stress has been a topic of interest in recent years. These studies demonstrate that CREB is activated in several brain regions by exposure to acute stress, suggesting that this transcription factor could underlie early gene expression responses. The activation of CREB varies with the type of stress exposure (e.g., physical vs. psychological stress) and the brain regions examined.

Acute exposure to a physical stressor, such as an electrified prod, ether, hyperosmolar solutions, or hypothermia, increases pCREB in the paraventricular nucleus (PVN) of the hypothalamus and the locus coeruleus (LC) (Borsok et al., 1994; Kovacs and Sawchenkno, 1997; Legradi et al., 1997; Bruijnzeel et al., 2001). The activation of CREB has been studied with regard to regulation of target genes in these regions. In LC, one of the suggested target genes of CREB is tyrosine hydroxylase, the rate-limiting enzyme for the synthesis of noradrenaline. Stress increases the expression of tyrosine hydroxylase in the LC and the promoter of this gene contains a CRE (Lewis et al., 1987). Other genes that could be regulated by CREB and stress include certain isoforms of adenylyl cyclase, which could contribute to an elevation of cAMP-CREB signaling (Chao et al., 2002).

In the PVN, one of the key target genes that is regulated by stress and could be a target of CREB is corticotrophin-releasing factor (CRF). CRF is one of the key endocrine factors in the hypothalamic–pituitary–adrenal (HPA) axis. This neuropeptide is expressed in the parvocellular neurons in the PVN of the hypothalamus. Exposure to stress results in a rapid stimulation of the HPA axis and the release of existing CRF, as well as activation of CRF gene expression (Kovacs and Sawchenko, 1997). The CRF gene contains a CRE in the promoter region that could underlie its regulation by CREB and stress. Studies to directly examine the regulation of CRF in vivo by CREB have not been conducted. For example, the influence of expression of a dominant negative mutant of CREB in the PVN or null mutation of CREB on the expression of CRF in PVN could be examined to determine if CREB is necessary for the induction of CRF by stress.

Regulation of CREB by psychological stress

Another type of acute stress paradigm that has been examined is the forced swim test, which is suggested to be a model of psychological stress (Bilang-Bleuel et al., 2002). In this model, animals are exposed to swim stress for short periods of time (e.g., 5–15 min) and eventually become immobile. The forced swim test was originally designed as a model of antidepressant action as treatment with most classes of antidepressants decreases immobility time (Porsolt et al., 1977). Exposure to forced swim results in a biphasic induction of pCREB in the dentate gyrus of the hippocampus, many subregions of cerebral cortex, and amygdala, but not in the PVN, dorsal raphe, or LC. Levels of pCREB are dramatically increased between 15 and 30 min and then return to basal levels, or below by 1 h (Bilang-Bleuel et al., 2002). Interestingly, there is a second surge in levels of pCREB by 6 h and this effect is sustained for at least 48 h. The biphasic time course has similarities with the early and late phases of long-term potentiation and induction of pCREB in this cellular model (CaMKII and then PKA mediated, respectively), raising the possibility that pCREB regulates genes that contribute to long-term changes in neuronal function and behavior. Levels of Fos-immunoreactivity are increased in many of the same regions, but are also increased in other regions, including PVN, LC, dorsal raphe, and bed nucleus of the stria terminalis (Bilang-Bleuel et al., 2002).

The induction of pCREB in higher limbic forebrain structures in response to forced swim stress has been discussed by Bilang-Bleuel and colleagues (2002) with regard to the psychological stress associated with this model. Forced swim stress also includes a physical component and activates deep subcortical structures as well. The higher limbic structures where pCREB is induced are involved in processing sensory information, as well as cognitive and emotional inputs compared to those regions where induction of pCREB is observed in response to physical stress (i.e., PVN and LC). This indicates that different neuronal circuits are activated depending on the type of stress exposure (psychological vs. physical stress) (Bilang-Bleuel et al., 2002). A similar pattern was observed for induction of mineralocorticoid receptors (Gesing et al., 2001). The induction of pCREB is observed in populations of neurons that represent the initial input subregions within these circuits (dentate gyrus in hippocampus; lateral nucleus in the amygdala). This suggests that modulation of pCREB may be involved in adjusting neuronal function at a proximal level and thereby affecting processing of information entering the circuit.

Nucleus accumbens CREB: an emotional gating switch

The influence of forced swim stress, as well as other types of stress, on pCREB immunoreactivity in the nucleus accumbens has also been examined (Pliakas et al., 2001; Barrot et al., 2002). These studies demonstrate that forced swim, footshock, restraint, or social stress increase pCREB immunoreactivity in the nucleus accumbens shell and to a lesser extent in core, but not dorsal striatum. Repeated unpredictable stress also increases pCREB in the nucleus accumbens (Barrot et al., 2002). In addition, Barrot and colleagues have utilized a CRE transgenic reporter mouse to demonstrate that CRE-mediated gene expression is also increased under these conditions. Chronic morphine administration also increases pCREB- and CRE-mediated gene expression in NAc (Barrot et al., 2002). These studies demonstrate that the rewarding stimuli as well as stressful stimuli increase pCREB- and CRE-mediated gene expression in the nucleus accumbens shell. This has led to the hypothesis by Barrot and colleagues that emotional stimuli, regardless of the valence, increase pCREB- and CRE-mediated gene expression in the nucleus accumbens.

The possible role of CREB in the functional response to anxiogenic, aversive, and nociceptive stimuli has also been examined to directly test the hypothesis that CREB acts as an emotional gating sensor in the nucleus accumbens (Barrot et al., 2002). In general, when CREB in nucleus accumbens is increased the responses in behavioral tests of these endpoints are decreased: (1) viral expression of CREB in the nucleus accumbens reduces the preference for opiates or sucrose; (2) viral expression of CREB in the nucleus accumbens decreases anxiety-related behavior (i.e., plus maze, open field) and expression of dominant negative CREB (mCREB) has the opposite effect; (3) studies of aversive and/or nociceptive responses demonstrate that viral expression of CREB reduces the sensitivity to naloxone in a conditioned aversion paradigm and mCREB has the opposite effect; (4) analysis of unconditioned responses demonstrate the expression of CREB in the nucleus accumbens increases the threshold footshock intensities required to elicit vocalization or jumping (increased threshold), while mCREB decreases threshold (i.e., animals vocalize and jump at lower intensity). Finally, expression of CREB in the nucleus accumbens decreases swimming/struggling in the forced swim test and decreases active avoidance in the learned helplessness paradigm, while mCREB expression has the opposite effects (Pliakas et al., 2001; Newton et al., 2002).

Influence of repeated stress on CREB

The influence of chronic or repeated stress on activation of CREB has not been studied as extensively. Bruijnzeel and colleagues (2001) have investigated the influence of repeated stress on pCREB in the PVN to determine if CREB could contribute to the increased expression of CRF in response to repeated stress that has also been observed (Bruijnzeel et al., 1999). Prior exposure to footshock stress increases the induction of CRF upon challenge with another stress (Bruijnzeel et al., 2001). However, this paradigm does not lead to increased expression of pCREB upon the second challenge stress. This suggests that the elevated induction of CRF is not due to increased activation of pCREB. However, one potential problem with the analysis is that only the total number of pCREB immunoreactive cells was determined. Although this is a typical analysis it is possible that there are different amounts of pCREB per cell and a small increase in cells that already express pCREB could be missed. It is also possible that other proteins that make up the transcriptional complex, such as the CREB-binding protein, are upregulated and more efficiently recruit pCREB. It is also possible that other transcription factors regulate the expression of CRF in PVN. Previous studies have demonstrated that prior footshock stress increases the induction of Fos-immunoreactivity in the PVN in response to a challenge stress (Rivest and Rivier, 1994; Bruijnzeel et al., 1999).

Another study has examined the influence of repeated, chronic footshock stress (21 days) on pCREB immunoreactivity (Trentani et al., 2002). The results of this study are interesting and demonstrate that in contrast to the acute stress paradigms, chronic stress decreases levels of pCREB in subregions of cerebral cortex (prefrontal, cingulate, and perirhinal) and paraventricular thalamic nucleus. There was no effect on pCREB in somatosensory cortex, dentate gyrus, or PVN in this report. A study of maternal isolation stress (1 h per day from D2 to D9) also demonstrates that levels of pCREB are decreased in several brain regions when examined at a later developmental time point (postnatal day 80) (Huang et al., 2002). These studies demonstrate that chronic stress can result in downregulation of pCREB in certain brain regions. It is interesting to speculate that activation of pCREB by acute stress may represent a normal adaptive response and that this effect is reversed with chronic exposure to stress in certain brain regions.

Role of CREB in conditioned responses

Owing to the well-characterized role of CREB in learning and memory, the regulation and function of CREB in conditioned fear has been extensively examined. These studies have reported biphasic regulation of pCREB in parietal cortex, hippocampus, and amygdala. The early phase (0–30 min) correlates with the unconditioned stimulus (footshock stress), while the late phase (3–6 h) is associated with the conditioned response (freezing in response to context associated with footshock) (Stanciu et al., 2001). A recent study found a greater pCREB response in animals exposed to context and then footshock versus animals receiving footshock at the same time as context (Stanciu et al., 2001). This suggests a role for pCREB in memory consolidation, in agreement with regulation of hippocampal pCREB during the late phase of memory consolidation (Bernabeu et al., 1997). The time course of pCREB expression significantly varies depending on the type and duration of the stimuli. This suggests that the early phase induction of pCREB may be more closely related to the arousal and anxiety induced by the experimental conditions than the process related to memory consolidation. In contrast to pCREB, induction of Fos is observed between 60 and 90 min after footshock, and this appears to be independent of the stimulus type or duration (see Stanciu et al., 2001). This suggests a highly constrained regulation of the Fos gene.

CREB in the etiology and treatment of depression

Mood disorders are often associated with stress, which can precipitate an episode of depression or can worsen an existing situation. Therefore, stress or stress-related behavioral paradigms are often used as models of depression and/or antidepressant actions. The regulation of CREB by acute and chronic stress suggests that this transcription factor could also be involved in depression. This possibility is supported by both basic and preclinical studies of CREB. Basic research studies first demonstrated a role for CREB in the actions of antidepressant treatments. These studies report that chronic antidepressant treatment increases CREB in the hippocampus and cerebral cortex of rodents (Fig. 4) (Frechilla et al., 1998; Nibuya et al., 1996; Thome et al., 2000). Upregulation of CREB expression and function is dependent on chronic antidepressant treatment, consistent with the time course for the therapeutic action of antidepressants. In addition, several different classes of antidepressants, including

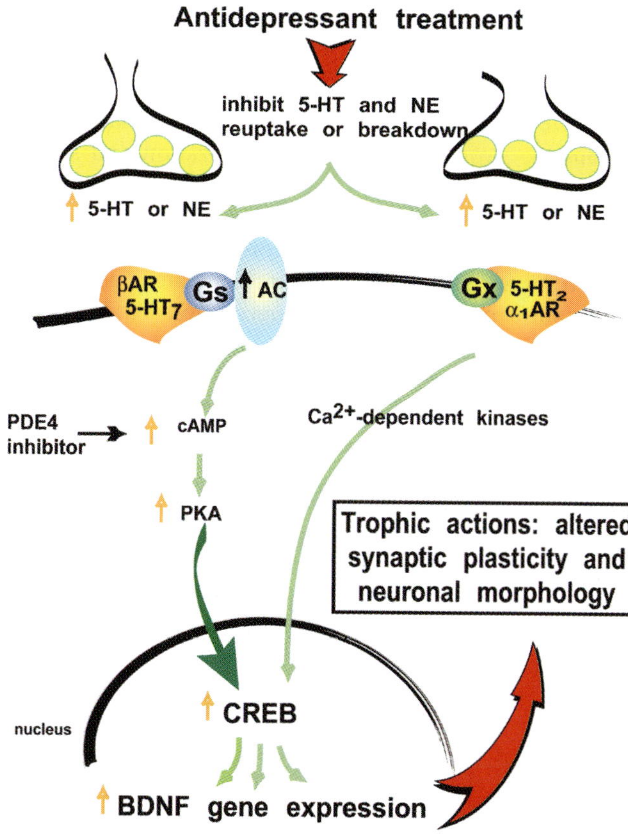

Fig. 4. Model depicting antidepressant regulation of the cAMP–CREB cascade and gene targets. Antidepressants block the reuptake or metabolism of NE, as well as 5-HT, and increase synaptic levels of these monoamines. Chronic antidepressant administration results in adaptations of the receptor-coupled signal transduction pathways and regulation of gene expression. One pathway that is regulated by antidepressant treatment is the cAMP–CREB cascade. Chronic antidepressant treatment leads to upregulation of cAMP-dependent protein kinase (PKA), increased levels of this kinase in nuclear fractions, and increased function and expression of the cAMP response element-binding protein (CREB). Inhibitors of phosphodiesterase type IV (PDE4) increase levels of cAMP and are known to have antidepressant efficacy in behavioral models and in clinical trials. CREB can also be regulated by Ca^{2+}-stimulated protein kinases, as well as ribosomal S6 kinase (see Fig. 2) and thereby acts as a common transcription factor target for multiple signal transduction pathways. Chronic antidepressant treatment and the cAMP–CREB cascade also increase the expression of specific gene targets in limbic brain structures, most notably brain-derived neurotrophic factor (BDNF) and its receptor TrkB. Upregulation of the cAMP–CREB cascade and increased expression of BDNF/TrkB are thought to produce antidepressant effects, in part, by blocking or reversing the atrophy and decreased neurogenesis resulting from stress.

5-HT and noradrenaline-selective reuptake inhibitors increase CREB, suggesting that this transcription factor is a common downstream target of antidepressants.

The influence of antidepressant treatment on the function and expression of CREB has been examined using several different approaches. Antidepressant treatment increases the expression of CREB mRNA and immunoreactivity, indicating that the total amount of CREB protein is increased (Nibuya et al., 1996). Levels of CREB binding to synthetic DNA containing a consensus CRE (i.e., determined in a gel mobility shift assay) and levels of pCREB determined by immunohistochemistry are upregulated by antidepressant treatment, suggesting that the function of CREB is increased (Nibuya et al., 1996; Thome et al., 2000). This possibility has been confirmed by studies demonstrating that antidepressant treatment increases CRE-mediated gene transcription in a line of transgenic mice that expresses a CRE-regulated reporter gene, β-galactosidase (Thome et al., 2000). These mice can be used to visualize CRE-mediated gene expression in vivo in response to various types of pharmacological or behavioral stimuli.

The transcriptional activity of CREB can also be assessed by the expression of target genes that contain CRE sites. Two genes of interest that have been identified are brain-derived neurotrophic factor (BDNF) and its receptor TrkB. Chronic, but not acute antidepressant treatment increases the expression of BDNF and TrkB mRNA in the hippocampus and this effect is blocked in CREB null mutant mice (Nibuya et al., 1995; 1996; Conti et al., 2002). In contrast to antidepressant treatment, stress decreases the expression of BDNF in the hippocampus (Nibuya et al., 1999; Smith et al., 1995). Decreased expression of BDNF could contribute to the atrophy and loss of cells resulting from chronic stress exposure (see Duman et al., 2000). The exon III-specific promoter of BDNF contains a Ca^{2+}/CRE site, referred to as a CaRE that is responsible for the induction of BDNF expression in response to activation of the cAMP-PKA cascade, as well as in response to neuronal depolarization and stimulation of Ca^{2+}-calmodulin-dependent protein kinase (Tao et al., 1998).

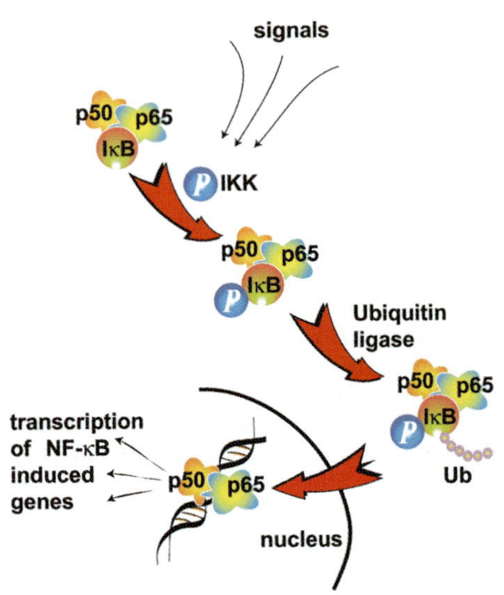

Fig. 5. Model depicting NFκB signaling to the nucleus. Signals transduced from cell-surface receptors converge on IκB kinase (IKK), which is activated by phosphorylation. IKK in turn phosphorylates IκB, causing it to become a substrate for an ubiquitin ligase. After ubiquitination, IκB is rapidly degraded to release the active NFκB dimer (here shown as a p50/p65 heterodimer). The active NFκB complex partitions preferentially to the nucleus where it activates transcription from κB-containing promotors.

The function of CREB in the actions of antidepressant treatment has also been examined using viral-mediated gene transfer and mutant mouse approaches. Viral-mediated expression of CREB in the hippocampus produces an antidepressant-like effect in the forced swim and learned helplessness paradigms (Chen et al., 2001). Preliminary studies demonstrate that viral expression of CREB in the amygdala also produces an antidepressant response in the learned helplessness model of depression (Wallace et al., 2002). However, as mentioned above, viral or transgenic expression of CREB in the nucleus accumbens results in a prodepressive phenotype in both the forced swim and learned helplessness models (Pliakas et al., 2001; Newton et al., 2002). These studies demonstrate that CREB influences behavior in models of depression, but that the effect depends on the brain region examined. This is not surprising because CREB is known to influence different target genes in these brain structures. In the hippocampus and amygdala, one of the targets of CREB is BDNF, which is capable of producing an antidepressant effect (Shirayama et al., 2002). In the nucleus accumbens, one of the key target genes of CREB is prodynorphin, which can produce aversive effects that could contribute to the prodepressive effects of CREB expression in this brain region (Carlezon et al., 1998; Pliakas et al., 2001; Newton et al., 2002).

Clinical postmortem studies also demonstrate that CREB is altered in depressed patients. There is one report demonstrating that levels of CREB immunoreactivity are decreased in temporal cerebral cortex of patients not on antidepressant medication at the time of death (Dowlatshahi et al., 1998). In contrast, levels of CREB immunoreactivity are significantly increased in temporal cortex of patients receiving an antidepressant at the time of death. Further studies of the levels of CREB expression, as well as function, in postmortem brains of depressed patients must be conducted to determine if altered CREB is a marker of depression. These preliminary studies are consistent with the hypothesis that CREB may be involved in the pathophysiology and treatment of depression.

Stress and the AP-1 family transcription factors

AP-1 transcription factors represent another family of proteins that regulate gene expression. The primary mechanism for regulation of these proteins is via induction of gene expression. The AP-1 proteins bind to the AP-1 promoter element, which is a sequence of seven nucleotides and is similar to the consensus CRE sequence, TGACTCA. Although this sequence only differs from the CRE sequence by one nucleotide, it is sufficient to confer a relatively high degree of selective binding of the AP-1 transcription factors over the CREB-like factors. There are many genes in the brain that contain AP-1 elements, including those for neuropeptides, neurotransmitter synthetic enzymes, receptors, and neurotrophic factors.

AP-1 structure and function

AP-1 transcription factors, like CREB and ATF family members, belong to the superfamily of basic-leucine zipper DNA-binding proteins. AP-1 transcription factors form dimers via the leucine zipper domain and bind to DNA via the adjacent basic amino acid-rich domain. The two major families of AP-1 factors are the Fos and Jun transcription factors. Members of the Fos family include c-Fos, Fos-related antigen-1 and -2 (FRA-1 and FRA-2), FosB and the FosB splice variant △FosB. The Jun family includes c-Jun, JunB, and JunD. A functional dimer is made up of one Fos and one Jun family member. Homodimers of Jun or Fos are not uncommon, but the DNA-binding affinity of these complexes is lower than for the Fos/Jun heterodimers.

Under basal, unstimulated conditions cellular levels of most Fos and Jun transcription factors are low, and in some cells undetectable. Activation of many signal-transduction pathways increases the expression levels of Fos and Jun resulting in the formation of AP-1 complexes. JunD is one exception in that it is constitutively expressed. The rapid induction of most AP-1 transcription factors has led to their classification as immediate early genes (IEGs). The protypical IEG is c-Fos, which can be induced rapidly, within minutes of the presentation of an extracellular stimulus. The ability of c-Fos to be turned on so rapidly is due to the presence of multiple response elements in the promoter, which includes three CREs. Activation of cAMP and/or Ca^{2+}-dependent protein kinases and CREB thereby result in rapid induction of c-Fos expression (Fig. 2). Induction of c-Fos gene expression also occurs via additional signaling cascades, including the MAP kinase–ERK signaling pathway. Increased activity of this cascade can led to phosphorylation of CREB via activation of RSK. ERK signaling also results in activation of another transcription factor, Elk-1, which forms a complex with the serum response factor (SRF) that binds to serum response element (SRE) in the c-Fos promoter (Fig. 2). AP-1 transcription factors can also be regulated by phosphorylation, which in many cases functions as a negative-feedback mechanism to turn off AP-1-mediated gene expression.

AP-1-mediated transcription is further complicated by formation of heterodimers with other transcription factors such as p-CREB and nuclear steroid receptors. These cross-family dimers bind to different consensus sequences with differing affinities. Glucocorticoid receptors and Ap-1 proteins can antagonize one another both in vitro and in vivo (Karin and Chang, 2001). The mechanisms of this interference are not completely understood but may include direct protein–protein interactions as well as inhibition of upstream signaling molecules involved in the activation of AP-1 transcription. Jun N-terminal kinase (JNK) is stimulated by growth factors and cytokines and enhances AP-1 activity by induction of c-Fos and c-Jun transcription and phosphorylation of Jun proteins. Glucocorticoids inhibit JNK activity, so this is one possible mechanism of glucocorticoid-mediated inhibition of AP-1 activity.

Stress and regulation of fos transcription factors

A brief overview of the effects of stress on c-Fos and IEG expression will be presented in this section. Stress-induced alterations in the levels or activity of c-Fos and other IEGs are likely important for regulating both the adaptive and detrimental responses to stress. Stress increases the expression of c-Fos and other AP-1 family members in both endocrine organs and various brain regions. However, it should be noted that further studies are needed to determine if these AP-1 transcription factors underlie the actions of stress on gene expression and function. This must be done using approaches similar to those described for CREB, particularly inducible transgenic and knockout mice combined to determine the role of c-Fos/AP-1 transcription factors in stress-stimulated gene expression.

c-Fos mapping

Although c-Fos and other IEGs ultimately function as transcription factors and regulate downstream target genes, c-Fos has most widely been used as a functional marker of activity in neurons and neuronal circuitries after a variety of stimuli. Expression of c-Fos and related IEG transcription factors is very rapid and robust and occurs in response to a variety of receptor-coupled signal-transduction pathways. Very low basal levels (i.e., unstimulated conditions) of c-Fos/IEG expression provides a high signal to noise ratio. This approach is not without limitations, which include a lack of expression in certain cell types and nonspecific induction.

Expression of c-Fos has been used to determine the neuronal circuits underlying the neuroendocrine, autonomical, and behavioral responses induced by stress. c-Fos mapping indicates that acute stress challenges activate the HPA axis, specifically the CRF-containing parvocellular neurons of the hypothalamus. The induction of c-Fos in CRF-containing neurons is intensity dependent and correlates with the increases in plasma corticosterone (Ericsson et al., 1994; Campeau and Watson, 1997). In addition to the hypothalamic PVN, acute stressors increase c-Fos mRNA and immunoreactivity in various brain regions. Stress can be classified as physical stress such as hemorrhage, ether, hyperosmolarity, and hyperthermia and psychological stress such as footshock, restraint, and immobilization as discussed above in relation to CREB activation. The neurocircuitry of both physical and psychological stressors as determined by c-Fos mapping generally includes the effector neurons in the PVN, cingulate cortex, lateral septum, septohypothalamic nucleus, medial preoptic area, bed nucleus of stria terminalis, central amygdala, dorsal raphe, and locus coeruleus (Kovacs, 1998). In addition, psychological stressors such as swim or restraint stress activate neocortex, allocortex, hippocampus, nucleus accumbens, and medial amygdala brain regions involved in processing sensory, cognitive, and emotional input (Kovacs, 1998; Lopez et al., 1999; Bilang-Bleuel et al., 2002). However, there is some degree of specificity to the c-Fos response depending upon the stress that is administered. For instance, swim stress causes a relatively greater induction of c-Fos in the hippocampus than does restraint stress (Lopez et al., 1999). In addition, even when similar brain regions are activated by stress, the afferent inputs that mediate this activation are stressor specific.

Fos expression and chronic stress

Repeated stimulation of the c-Fos gene eventually leads to a refractory state where further transcription is limited. This is likely due to counterregulatory mechanisms, including the induction of inhibitory transcription factors such as ICER, that block CRE-meditated gene expression. However, under conditions of repeated and long-term activation of the c-Fos gene the accumulation an alternative splice variant known as ΔFosB is observed. Unlike Fos, which degrades very rapidly (i.e., half-life of minutes to hours), ΔFosB is relatively stable and has a very long half-life of days to weeks (see Nestler et al., 2001). This different pattern of expression is thought to contribute to the long-term adaptations to repeated stimulation. Such differences in the temporal stability of AP-1 transcription factors, combined with different patterns of stimulation, provide mechanisms for discrete regulation of gene expression and cellular function.

The activation of c-Fos following acute stress provides a map of the circuitry involved in the stress response. However, the pattern of c-Fos induction after repeated or chronic stress may reveal pathological processes. Repeated stress or chronic glucocorticoid administration attenuates the subsequent induction of c-Fos, Fos B, and Jun B expression in the PVN by acute immoblization stress (Umemoto et al., 1997). Similarly, repeated restraint stress reduces c-Fos expression compared to acute restraint in the medial amygdala, hippocampus, septum, and brainstem. However, the degree of habituation of c-Fos and as well as other immediate early genes after repeated stress varies between brain regions and specific stressors (Chen and Hebert, 1995; Stamp and Herbert, 1999).

The decreased activation of c-Fos after repeated stressors suggests that other members of the AP-1 family may mediate the effects of chronic stress. Sustained elevations of other members of the c-Fos family have been reported after chronic stimuli as discussed above. Protein products of the ΔFosB splice variant, termed chronic FRAs, are induced in a region-specific manner by a variety of chronic stimuli including electroconvulsive seizures, chronic drug of abuse administration, and lesions (Nestler et al., 1999). However, repeated immobilization stress does not alter the expression of the chronic FRA gene products of the FosB gene in the adrenal medulla (Nankova et al., 2000). In contrast, FRA-2 expression is increased by acute stress and to even greater levels by chronic stress in the adrenal medulla (Nankova et al., 2000). Other FRAs are also increased in the adrenal medulla, but these appear to be distinct from the chronic FRAs induced in the brain after ECS and the ΔFosB products. FRA-2 is induced in the PVN after capsaicin-induced stress (Honkaniemi et al., 1994). Additional studies are needed to completely characterize the effects of chronic stress on the large family of c-Fos transcription factors in brain regions that mediate the stress response. Inducible transgenic and knockout mice will be useful for linking the expression of c-Fos/AP-1 transcription factors with gene expression and behaviors produced by chronic stress.

c-Fos target genes

As discussed above, c-Fos and other related family members ultimately function as transcription factors and regulate downstream target genes through interaction with an AP-1 promoter element to either stimulate or repress transcription. There are a number of genes that contain the AP-1 consensus sequence within their promoter, including vasopressin, enkephalin, dynorphin, somatostatin, cholecystokinin, luteinizing hormone-releasing hormone, tyrosine hydroxylase (TH), and glutamic acid decarboxylase (GAD). However, direct evidence for gene regulation by AP-1 transcription factors is limited.

In the adrenal medulla acute and chronic stress-induced increases in the catecholamine synthesis enzymes, TH, and dopamine β-hydroxylase (DBH) are correlated with increased binding of AP-1 factors (Sabban and Kvetnansky, 2001). A single immobilization stress increases binding of c-Fos and c-Jun to the AP-1 transcription site and the binding is not further elevated with repeated stress (Nankova et al., 1994). Studies of c-Fos knockout mice suggest that c-Fos is necessary for chronic stress-induced DBH expression in the adrenals of female rats and in the brainstem, independent of gender (Serova et al., 1998). However, TH and phenyl-ethanolamine *N*-methyltransferase (PNMT) levels are still increased by chronic stress in c-Fos knockout mice (Serova et al., 1998). This could result from the actions of other members of the c-Fos family that are driving expression of TH and DBH after chronic stress. FRA-2 activates the TH and DBH promoters in a cell-expression system in agreement with this possibility (Nankova et al., 2000).

AP-1 transcription factors also play a role in regulating genes involved in cell survival (Shaulian and Karin, 2002). Depending upon the composition of the AP-1 proteins, the complex can either positively or negatively regulate cell proliferation. Fibroblasts derived from c-Fos or FosB single knockout mice proliferate normally, but fibroblasts from double c-Fos/FosB knockout mice have reduced proliferation. JunD and c-Jun knockout mice show similar deficits in cell proliferation. The induction of cyclin D1 is repressed in these knockout mice compared to wildtype. AP-1 proteins bind and activate the cyclin D1 promoter. Induction of cyclin D1 is critical for cell-cycle progression. Cyclin D1 binds to and activates G1 phase cyclin-dependent kinases (CDKs) that in turn activate and inhibit proteins that facilitate the transition from G1 phase to S phase of the cell cycle. These data suggest that the induction of cyclin D1 is one mechanism by which AP-1 proteins stimulate cell proliferation. However, not all AP-1 proteins promote proliferation. Studies with transgenic mice overexpressing JunB suggest that JunB antagonizes the effects of c-Jun and inhibits proliferation in a cell-type specific manner. Consistent with its ability to antagonize c-Jun, JunB inhibits the cyclin D1 promoter. There is also evidence that AP-1 proteins are involved in regulating genes that induce apoptosis (Shaulian and Karin, 2002). c-Fos is induced by kainic acid which induces apoptosis in the hippocampus through glutamate receptor stimulation. In addition, overexpression of c-Jun or c-Fos in various cell lines can be proapoptotic. Inhibition of c-Jun through the expression of a dominant negative mutant protects neurons from apoptosis induced by NGF withdrawal or chronic depolarization. One established AP-1 target gene that is proapoptotic is the Fas-ligand (FasL), as c-Jun has been shown to induce FasL. However, c-Jun is not proapoptotic in all cell or tissue types as demonstrated by c-Jun knockout mice from which some cell types undergo massive apoptosis. It is likely the balance between proapoptotic and antiapoptotic target genes that ultimately determines whether AP-1 activity leads to cell death or increases cell survival.

NF-κB-mediated gene transcription

Nuclear factor kappa B (NF-κB) is a ubiquitous transcription factor activated by a variety of cellular stresses (Sen and Baltimore, 1986). Originally identified as an inducer of immunoglobulin κ light-chain expression, it has since been implicated in protective responses to mutagenic factors, inflammation, infections, redox stress, radiation, and CNS injury as well as many homeostatic functions. In this section, we will give an overview of NF-κB's composition and regulation, with particular emphasis on its protective stress-responsive functions, both in the immune system and in the CNS.

Molecular composition and activation of NF-κB

NF-κB is composed of hetero- and homodimers of five Rel family proteins, NF-κB1 (p50), NF-κB2 (p52), RelA (p65), RelB, and c-Rel, all of which can bind to DNA. The three latter subunits also function as transcriptional activators, and one of these must therefore be present in an active NF-κB complex (Ghosh et al., 1998). The most common and therefore most intensely studied ("prototypical") NF-κB is the p50-RelA heterodimer. Under basal conditions, most of the NF-κB in a cell exists in an inactive complex with an inhibitor, one of the IκB-family proteins (Baeuerle and Baltimore, 1988). When IκB is phosphorylated by IκB kinase (IKK), it becomes a proteasomal substrate and is rapidly degraded, thereby releasing the active NF-κB complex, which is now preferentially located in the nucleus. IKK consists of two kinase subunits, IKKα and IKKβ, and a noncatalytic subunit IKKγ (Zandi, et al., 1997; Rothuort et al., 1998). NF-κB can also be activated through a noncanonical pathway, wherein p52 precursor in association with RelB is phosphorylated by NIK kinase in an IKKα-dependent manner and cleaved to release the active p52-RelB dimer. This pathway is utilized during development of lymphoid organs and adaptive immunity (Senftleben et al., 2001). After activation, the nuclear NF-κB complex binds to κB promoter elements and increases expression of target genes (Sen and Baltimore, 1986).

Conditions that lead to activation of NF-κB are often linked to stress, be it an infection, tissue damage, or a toxic chemical compound that threatens the homeostasis of the organism. Thus, NF-κB activation is driven by such diverse stressors as proinflammatory cytokines, acute-phase proteins, and cigarette smoke. NF-κB also responds to DNA damage from irradiation, shear stress, and backup of unfolded proteins in the endoplasmic reticulum (Pahl, 1999). In response to the activation of NF-κB, genes containing the κB promoter element are rapidly transcribed. Among these are many molecules central to the immune system, adhesion molecules, acute-phase proteins, and a large variety of transcription factors (Pahl, 1999).

In addition to relief of NF-κB inhibition through the canonical IKK–IκB pathway, the activity of NF-κB can be modulated by several factors. For instance, PKA-mediated phosphorylation of p65 renders NF-κB competent to bind to p300/CBP, a cofactor that appears to be necessary not only for CREB-mediated transcription, but also for NF-κB activity (Zhong et al., 1998). ξPKC, casein kinase II and IKKα-mediated phosphorylation events enhance the transcriptional activity of NF-κB (Leitges et al., 2001). In addition, p300/CBP-dependent acetylation of RelA at one site enhances activity, and further acetylation increases NF-κB's affinity for κB sites (Chen et al., 2002).

Antiapoptotic effects of NF-κB

NF-κB protects against apoptosis by induction of a number of antiapoptotic target genes. Targeted deletion of RelA in mice causes lethality around embryonic days 15–16 due to liver apoptosis, and other cell types lacking NF-κB are also more prone to apoptosis (Beg et al., 1995; Beg and Baltimore, 1996; Van Antwerp et al., 1996). Moreover, embryonic liver apoptosis phenotypes are observed in both IKKβ and IKKγ knockouts (Li and Verma, 2002).

Apoptosis is in many cases initiated by signaling through members of the tumor necrosis factor (TNF) family of receptors. Activation of TNF receptor I by the proinflammatory cytokine TNF-α initiates several signaling cascades including JNK and a pro-apoptotic pathway leading to activation of caspase-8, but the outcome is attenuated by the simultaneous activation of NF-κB (Liu et al., 1996). By comparison, activation of another TNF receptor family member, Fas, which leads to similar proapoptotic events, does not efficiently induce NF-κB and therefore causes full-blown apoptosis (Karin and Lin, 2002). Antiapoptotic genes transcribed upon NF-κB activation include IAPs (inhibitors of apoptosis), TRAF1 and 2 (TNF receptor-associated factors), GADD45β, and c-FLIP (Karin and Lin, 2002). IAPs directly associate with caspases and prevent activation of initiator caspase 9 and inhibit effector caspases 3 and 7 (Deveraux and Reed, 1999). c-FLIP mimics caspases in its composition, but contains a catalytically inactive effector domain and appears to interfere with caspase-8 recruitment and activation (Krueger et al., 2001). Members of the Bcl-2 protein family are also expressed upon NF-κB regulation, including A1, Bcl-X_L, and Bcl-2 itself (all antiapoptotic), whereas the proapoptotic Bax protein is downregulated (Bentires-Alj et al., 2001; Karin and Lin et al., 2002). GADD45β and XIAP (X-linked inhibitor of apoptosis) may protect against apoptosis by interfering with proapoptotic JNK signaling (De Smaele et al., 2001; Tang et al., 2001). Finally, TRAF1 and 2 are critical components of the TNF receptor super family signaling pathway, and their induction by NF-κB may play a role in modulating the balance between pro- and antiapoptotic signals (Karin and Lin, 2002).

Pulling in the opposite direction is the fact that many of the components of the signaling pathways leading to NF-κB activation are caspase targets (including NF-κB itself), and cleavage of these attenuates the antiapoptotic outcome. Additionally, IκB can be cleaved to render the molecule resistant to the phosphorylation step necessary for its degradation to release active NF-κB (Karin and Lin, 2002). NF-κB may also in certain situations promote apoptosis, rather that fight it, but there are only a few examples of this. The most interesting may be that MEK1-rsk-dependent activation of NF-κB by tumor supressor p53 is necessary for p53-induced apoptosis (Ryan et al., 2000).

The generally cell-protective effect of NF-κB activation is, not surprisingly, a double-edged sword with regard to oncogenesis. For instance, as many as 10% of lymphocyte neoplasias have, in addition to other abnormalities, mutations in their Rel or IκB sequences (Foo and Nolan, 1999). Several members of the NF-κB pathway induce transformation in vivo or in vitro when mutated or dysregulated, and many common oncogenes affect NF-κB activation. Additionally, activation of NF-κB in response to chemotherapy and radiation counteracts the ability of the treatment to induce apoptosis of the cancer cells (Baldwin, 2001). It is also of concern that NF-κB regulates the expression of adhesion molecules, some cell-cycle proteins, and COX-2, and therefore may be involved in proliferation, inhibition of differentiation, angiogenesis, and metastasis (Baldwin, 2001). Therefore, inhibition of NF-κB may be useful in treatment of certain cancers, and some experimental tumors are indeed responsive to such treatment. However, considering the sometimes opposing effects of NF-κB actions, more studies are needed to determine the overall benefit of such treatment in vivo.

Role of NF-κB in immune system activation

The NF-κB pathway is intimately involved in maturation and activation of most immune cells by controlling the expression of a broad spectrum of immune system effectors, including MHC proteins, proinflammatory cytokines, chemokines, interferons, and adhesion molecules. Individual actions of the pathway effectors have been elucidated partly by transgenic techniques. For instance, c-Rel knockout mice have defects in immune cell activation leading to decreased production of cytokines and immunoglobulins (Kontgen et al., 1995). RelB knockouts show dendritic cell defects and die postnatally from T-cell-dependent multiorgan inflammation (Weih et al., 1996). Removal of p50 or p52 also causes immune defects, especially involving B cells (Li and Verma, 2002).

During an immune response, NF-κB is activated rapidly to induce an inflammatory response. The major pathways of activation include signaling by proinflammatory cytokines, especially TNF-α and IL-1 (interleukin-1). In addition, the pattern recognition receptors called Toll-like receptors induce NF-κB in response to bacterial and viral products, such as lipopolysaccharide, dsRNA, and lipopeptides as part of the innate immune response. Antigen-specific T cells that are activated both through their T-cell receptors and costimulatory signals also induce NF-κB. This in turn promotes production of IL-2, an essential component of a humoral immune response. NF-κB induction in other immune cells leads to massive secretion of more proinflammatory molecules, including IL-1, IL-6, IL-8, TNF-α, COX-2, matrix metalloproteinases, and inducible nitric oxide synthase (Li and Verma, 2002). Several of these molecules lead to further NF-κB activation, propagating an inflammatory response to protect the host against the infectious agent. However, the analogy of the double-edged sword also applies here, since many viruses have acquired κB sites in their viral promotors. Thus, when the host's immune response to the virus activates NF-κB to fight against the viral infection, transcription of viral proteins is enhanced. This may partly explain how certain viruses, including EBV and HIV-1, maintain chronic infections (Pahl, 1999).

Several commonly used antiinflammatory drugs inhibit NF-κB activation (in addition to their other effects). These include corticosteroids and nonsteroidal antiinflammatory drugs such as aspirin, at least some of which inhibit the phosphorylation of IκB proteins. The antiinflammatory cytokines IL-10 and IL-13 suppress nuclear localization of NF-κB and upregulate IκB expression. Specific blockade of NF-κB activation continues to be a focus for new antiinflammatory drug development (Epinat and Gilmore, 1999; Yamamoto and Gaynor, 2001).

Role of NF-κB in CNS pathology and stress

Neuronal NF-κB has important neuroprotective roles in response to CNS injury such as that following ischemia and seizures. It also appears to attenuate neuronal death in ALS, Alzheimer's, Parkinson's, and Huntington's diseases. In contrast, microglial NF-κB activation appears to promote neuronal degradation, and the end result of general NF-κB activation is therefore dependent on the balance of neuronal and glial responses (Mattson and Camandola, 2001). For example, data from studies with p50 knockout mice suggests that ischemic neuronal death is enhanced by NF-κB activation, whereas NF-κB protects neurons against seizure-related excitotoxicity (Schneider et al., 1999; Yu et al., 1999). In addition to proinflammatory cytokines, potent CNS inducers of NF-κB include bradykinin, glutamate, increases in intracellular Ca^{2+}, and reactive oxygen species, all of which are important mediators of CNS pathology (Mattson and Camandola, 2001).

CNS injury-responsive NF-κB-induced molecules include the proinflammatory cytokines TNF-α and IL-6, βAPP (β-amyloid protein precursor), Mn-SOD (manganese superoxide dismutase), calbindin, ICAM-1, GFAP, and NAIP-1 (neuronal apoptosis inhibitory protein-1), in addition to the antiapoptotic factors described above (Mattson and Camandola, 2001). βAPP has neurotrophic properties and also further induces NF-κB activation. Mn-SOD is a neuroprotective mitochondrial antioxidant molecule, and calbindin is involved in calcium-mediated neuronal signaling and cell death. ICAM-1 and GFAP mediate intracellular interactions and structural stability of glial cells, respectively, but the functional outcomes of their upregulation are unclear.

In addition to its role in CNS injury, NF-κB is also involved in regulation of synaptic function. In this context, NF-κB is activated in response to membrane depolarization, low-frequency stimulation, and during long-term potentiation of synaptic transmission. In fact, pretreatment with κB decoy DNA (to "soak up" activated NF-κB) in hippocampal slices abolishes the ability to induce long-term depression and significantly decreases the amplitude of long-term potentiation (Albensi and Mattson, 2000). The effects of NF-κB activation in this phenomenon are not well understood, but induced genes may include NMDA and AMPA glutamate receptor subunits (Furukawa and Mattson, 1998).

NF-κB activation has also been suggested to be involved in stress and some psychiatric diseases. Emotional distress in women scheduled for breast biopsies led to decreased levels of NF-κB that returned to baseline when stress was relieved, providing a possible explanation for the phenomenon of stress-induced immunosuppression (Nagabhushan et al., 2001). Gene expression analysis in frontal cortex from patients with bipolar disorder showed a significant increase in NF-κB transcription factor complex components, and increased levels of the same gene transcripts were found in some patients with schizophrenia and depression (Sun et al., 2001). However, it is not clear whether these increases reflect a protective response or an integral part of the pathology.

Conclusions

Significant progress has been made in elucidating the molecular mechanisms that control gene transcription and genetic code of humans and other species. This work provides the tools necessary to study the genes that are differentially expressed in response to stress and other stimuli and that ultimately control neuronal function and behavior. This is an exciting and interesting time as we begin to unravel the fundamental molecular basis for complex behaviors and the genetic basis that underlie differences between individuals. The studies outlined in this chapter regarding a few classes of transcription factors and target genes provide a framework for future studies that will provide a complete analysis of the signal transduction and gene expression patterns that are critical to stress responses. Using well-defined approaches that clearly define the intensity and type of stress (e.g., physical vs. psychological stress) will eventually elucidate the complex cellular responses that underlie adaptive changes as well as those that contribute to maladaptive alterations in response to stress.

Acknowledgments

This work is supported by USPHS grants MH45481 and 2 PO1 MH25642, a Veterans Administration National Center Grant for PTSD, and by the Connecticut Mental Health Center.

References

Albensi, B.C. and Mattson, M.P. (2000) Evidence for the involvement of TNF and NF-kappaB in hippocampal synaptic plasticity. Synapse, 35: 151–159.

Armstrong, R. and Montminy, M.R. (1993) Transsynaptic control of gene expression. Ann. Rev. Neurosci., 16: 17–29.

Baeuerle, P.A. and Baltimore, D. (1988) I kappa B: a specific inhibitor of the NF-kappa B transcription factor. Science, 242: 540–546.

Baldwin, A.S. (2001) Control of oncogenesis and cancer therapy resistance by the transcription factor NF-kappaB. J. Clin. Invest., 107: 241–246.

Barrot, M., Olivier, J.D.A., Perrotti, L.I., DiLeone, R.J., Berton, O., Eisch, A.J., Impey, S., Storm, D.R., Neve, R.L., Yin, J.C., Zachariou, V. and Nestler, E.J. (2002) CREB activity in the nucleus accumbens shell controls gating of behavioral responses to emotional stimuli. PNAS, 99: 11435–11440.

Beg, A.A. and Baltimore, D. (1996) An essential role for NF-kappaB in preventing TNF-alpha-induced cell death. Science, 274: 782–784.

Beg, A.A., Sha, W.C., Bronson, R.T., Ghosh, S. and Baltimore, D. (1995) Embryonic lethality and liver degeneration in mice lacking the RelA component of NF-kappa B. Nature, 376: 167–170.

Bentires-Alj, M., Dejardin, E., Viatour, P., Van Lint, C., Froesch, B., Reed, J.C., Merville, M.P. and Bours, V. (2001) Inhibition of the NF-kappa B transcription factor increases Bax expression in cancer cell lines. Oncogene, 20: 2805–2813.

Bernabeu, R., Bevilaqua, L., Ardenghi, P., Bromberg, E., Schmitz, P., Bianchin, M., Izquierdo, I. and Medina, J.H. (1997) Involvement of hippocampal cAMP/cAMP-dependent protein kinase signaling pathways in a late memory consolidation phase of aversively motivated learning in rats. PNAS, 94: 7041–7046.

Bilang-Bleuel, A., Rech, J., De Carli, S., Holsboer, F. and Reul, J.M.H.M. (2002) Forced swimming evokes a biphasic response in CREB phosphorylation in extrahypothalamic limbic and neocortical brain structures in the rat. Eur. J. Neurosci., 15: 1048–1060.

Borsok, D., Konradi, C., Falkowski, O., Comb, M. and Hyman, S. (1994) Molecular mechanisms of stress-induced proenkephalin gene regulation: CREB interacts with the proenkephalin gene in the mouse hypothalamus and is phosphorylated in response to hyperosmolar stress. Mol. Endocrinol., 8: 240–248.

Bruijnzeel, A., Stam, R., Compaan, J.C., Croiset, G., Akkermans, L.M., Olivier, B. and Wiegant, V.M. (1999) Long-term sensitization of Fos-responsivity in the rat central nervous system after a single stressful experience. Br. Res., 819: 15–22.

Bruijnzeel, A., Stam, R., Compaan, J.C. and Wiegant, V.M. (2001) Stress-induced sensitization of CRH-ir but not P-CREB-ir responsivity in the rat central nervous system. Brain Res., 908: 187–196.

Campeau, S. and Watson, S.J. (1997) Neuroendocrine and behavioral responses and brain pattern of c-fos induction associated with audiogenic stress. J. Neuroendocrinol., 9: 577–588.

Carlezon, W.J., Thome, J., Olson, V.G., Lane-Ladd, S.B., Brodkin, E.S., Hiroi, N., Duman, R.S., Neve, R.L. and Nestler, E.J. (1998) Regulation of cocaine reward by CREB. Science, 282: 2272–2275.

Chao, J., Ni, Y.G., Bolanos, C.A., Rahman, Z., DiLeone, R.J. and Nestler, E.J. (2002) Characterization of the mouse adenylyl cyclase type VIII gene promoter: regulation by cAMP and CREB. Eur. J. Neurosci., 16: 1284–1294.

Chen, X. and Herbert, J. (1995) Regional changes in c-fos expression in the basal forebrain and brainstem during adaptation to repeated stress: correlations with cardiovascular, hypothermic and endocrine responses. Neurosci., 64: 675–685.

Chen, A.-H., Shirayama, Y., Shin, K.-H., Neve, R.L. and Duman, R.S. (2001) Expression of the cAMP response element binding protein (CREB) in hippocampus produces antidepressant effect. Biol. Psychiat., 49: 753–762.

Chen, L.F., Mu, Y. and Greene, W.C. (2002) Acetylation of RelA at discrete sites regulates distinct nuclear functions of NF-kappaB. EMBO J., 21: 6539–6548.

Conti, A., Cryan, J.F., Dalvi, A., Lucki, L. and Blendy, J.A. (2002) CREB is essential for the upregulation of BDNF transcription, but not the behavioral or endocrine responses to antidepressant drugs. J. Neurosci., 22: 3262–3268.

De Smaele, E., Zazzeroni, F., Papa, S., Nguyen, D.U., Jin, R., Jones, J., Cong, R. and Franzoso, G. (2001) Induction of gadd45beta by NF-kappaB downregulates pro-apoptotic JNK signalling. Nature, 414: 308–313.

Deveraux, Q.L. and Reed, J.C. (1999) IAP family proteins – suppressors of apoptosis. Genes Dev., 13: 239–252.

Dowlatshahi, D., MacQueen, G.M., Wang, J.F. and Young, L.T. (1998) Increased temporal cortex CREB concentrations and antidepressant treatment in major depression. The Lancet, 352: 1754–1755.

Duman, R.S., Malberg, J., Nakagawa, S. and D'Sa, C. (2000) Neuronal plasticity and survival in mood disorders. Biol. Psychiatry. 48: 732–739.

Epinat, J.C. and Gilmore, T.D. (1999) Diverse agents act at multiple levels to inhibit the Rel/NF-kappaB signal transduction pathway. Oncogene, 18: 6896–6909.

Ericsson, A., Kovacs, K.J. and Sawchenko, P.E. (1994) A functional anatomical analysis of central pathways subserving the effects of interleukin-1 on stress-related neuroendocrine neurons. J. Neurosci., 14: 897–913.

Foo, S.Y. and Nolan, G.P. (1999) NF-kappaB to the rescue: RELs, apoptosis and cellular transformation. Trends Genet., 15: 229–235.

Frechilla, D., Otano, A. and Del Rio, J. (1998) Effect of chronic antidepressant treatment on transcription factor binding activity in rat hippocampus and frontal cortex. Progr. Neuro. Psychopharm. Biol. Psychiatr., 22: 787–802.

Furukawa, K. and Mattson, M.P. (1998) The transcription factor NF-kappaB mediates increases in calcium currents and decreases in NMDA- and AMPA/kainate-induced currents induced by tumor necrosis factor-alpha in hippocampal neurons. J. Neurochem., 70: 1876–1886.

Gesing, A., Bilang-Bleuel, Droste A., Linthorst, S.K., Holsboer, F. and Reul, J.M.H.M. (2001) Psychological stress increased hippocampal mineralocorticoid receptor levels: involvement of corticotropin-releasing hormone. J. Neurosci., 21: 4822–4829.

Ghosh, S., May, M.J. and Kopp, E.B. (1998) NF-kappa B and Rel proteins: evolutionarily conserved mediators of immune responses. Annu. Rev. Immunol., 16: 225–260.

Honkaniemi, J., Kononen, J., Kainu, T. and Pelto-Hiukko, M. (1994) Induction of multiple immediate early genes in rat hypothalamic paraventricular nucleus after stress. Br. Res. Mol. Br. Res., 25: 234–241.

Huang, L., Holmes, G.L., Lai, M.C., Hung, P.L., Wang, C.L., Wang, T.J., Yang, C.H., Liou, C.W. and Yang, S.N. (2002) Maternal deprivation stress exacerbates cognitive deficits in immature rats with recurrent seizures. Epilepsia, 43: 1141–1148.

Karin, M. and Chang, L. (2001) AP-1–glucocorticoid receptor crosstalk taken to a higher level. J. Endocrinol., 169: 447–451.

Karin, M. and Lin, A. (2002) NF-kappaB at the crossroads of life and death. Nat. Immunol., 3: 221–227.

Kontgen, F., Grumont, R.J., Strasser, A., Metcalf, D., Li, R., Tarlinton, D. and Gerondakis, S. (1995) Mice lacking the c-rel proto-oncogene exhibit defects in lymphocyte proliferation, humoral immunity, and interleukin-2 expression. Genes Dev., 9: 1965–1977.

Kovacs, K. (1998) c-Fos as a transcription factor: a stressful (re) view from a functional map. Neurochem. Int., 33: 287–297.

Kovacs, K. and Sawchenko, P.E. (1997) Sequence of stress-induced alterations in indices of synaptic and transcriptional activation in parvocellular neurosecretory neurons. J. Neurosci., 16: 262–273.

Krueger, A., Baumann, S., Krammer, P.H. and Kirchhoff, S. (2001) FLICE-inhibitory proteins: regulators of death receptor-mediated apoptosis. Mol. Cell. Biol., 21: 8247–8254.

Legradi, G., Holzer, D., Kapcala, L.P. and Lechan, R.M. (1997) Glucocorticoids inhibit stress-induced phosphorylation of CREB in corticotropin-releasing hormone neurons of the hypothalamic paraventricular nucleus. Neuroendocrinology, 66: 3550–3554.

Leitges, M., Sanz, L., Martin, P., Duran, A., Braun, U., Garcia, J.F., Camacho, F., Diaz-Meco, M.T., Rennert, P.D. and Moscat, J. (2001) Targeted disruption of the zetaPKC gene results in the impairment of the NF-kappaB pathway. Mol. Cell., 8: 771–780.

Lewis, E., Harrington, C.A. and Chikaraishi, D.M. (1987) Transcriptional regulation of the tyrosine hydroxylase gene by glucorticoid and cyclic AMP. PNAS, 84: 3550–3554.

Li, Q. and Verma, I.M. (2002) NF-kappaB regulation in the immune system. Nat. Rev. Immunol., 2: 725–734.

Liu, Z.G., Hsu, H., Goeddel, D.V. and Karin, M. (1996) Dissection of TNF receptor 1 effector functions: JNK activation is not linked to apoptosis while NF-kappaB activation prevents cell death. Cell 87: 565–576.

Lopez, J., Akil, H. and Watson, S.J. (1999) Neural circuits mediating stress. Biol. Psychiat., 45: 1461–1471.

Mattson, M.P. and Camandola, S. (2001) NF-kappaB in neuronal plasticity and neurodegenerative disorders. J. Clin. Invest., 107: 247–254.

Nagabhushan, M., Mathews, H.L. and Witek-Janusek, L. (2001) Aberrant nuclear expression of AP-1 and NFkappaB in lymphocytes of women stressed by the experience of breast biopsy. Brain Behav. Immun., 15: 78–84.

Nankova, B., Kvetnansky, R., McMahon, A., Viskupic, E., Hiremagalur, B., Frankle, G., Fukuhara, K., Kopin, I.J. and Sabban, E.L. (1994) Induction of tyrosine hydroxylase gene expression by a nonneuronal nonpituitary-mediated mechanism in immobilization stress. PNAS, 91: 5937–5941.

Nankova, B., Rivkin, M., Kelz, M., Nestler, E.J. and Sabban, E.L. (2000) Fos-related antigen 2: potential mediator of the transcriptional activation in rat adrenal medulla evoked by repeated immobilization stress. J. Neurosci., 20: 5647–5653.

Nestler, E., Kelz, M.B. and Chen, J. (1999) DeltaFosB: a molecular mediator of long-term neural and behavioral plasticity. Brain Res., 835: 10–17.

Nestler, E.J., Hyman, S.E. and Malenka, R.C. (2001) Signaling to the nucleus. In: Molecular Neuropharmacology (eds. E.J. Nestler, S.E. Hyman and R.C. Malenka). McGraw Hill, New York, NY, pp. 115–137.

Newton, S., Thome, J., Wallace, T.L., Shirayama, Y., Schlesinger, L., Sakai, N., Chen, N., Neve, R., Nestler, E.J. and Duman, R.S. (2002) Inhibition of cAMP response element-binding protein or dynorphin in the nucleus accumbens produces an antidepressant-like effect. J Neurosci., 24: 10883–10890.

Nibuya, M., Morinobu, S. and Duman, R.S. (1995) Regulation of BDNF and trkB mRNA in rat brain by chronic electroconvulsive seizure and antidepressant drug treatments. J. Neurosci., 15: 7539–7547.

Nibuya, M., Nestler, E.J. and Duman, R.S. (1996) Chronic antidepressant administration increases the expression of cAMP response element binding protein (CREB) in rat hippocampus. J. Neurosci., 16: 2365–2372.

Nibuya, M., Takahashi, M., Russell, D.S. and Duman, R.S. (1999) Chronic stress increases catalytic TrkB mRNA in rat hippocampus. Neurosci. Letts., 267: 81–84.

Pahl, H.L. (1999) Activators and target genes of Rel/NF-kappaB transcription factors. Oncogene, 18: 6853–6866.

Phiel, C., Zhang, F., Huang, E.Y., Guenther, M.G., Lazzar, M.A. and Klein, P.S. (2001) Histone deacetylase is a direct target of valproic acid, a potent anticonvulsant, mood stabilizer, and teratogen. J. Biol. Chem., 276: 36734–36741.

Pliakas, A., Carlson, R.R., Neve, R.L., Konradi, C., Nestler, E.J. and Carlezon, W.A. (2001) Altered responsiveness to cocaine and increased immobility in the forced swim test associated with elevated CREB expression in the nucleus accumbens. J. Neurosci., 21: 7397–7403.

Porsolt, R., Le Pichon, M. and Jalfre, M. (1977) Depression: a new animal model sensitive to antidepressant treatments. Nature, 266: 730–732.

Rivest, S. and Rivier, C. (1994) Stress and interleukin-1 beta-induced activation of c-fos, NGF1-B and CRF gene expression in the hypothalamic PVN: comparison between Sprague-Dawley, Fisher-344 and Lewis rats. J. Neuroendocrinol., 6: 101–117.

Rothwarf, D.M., Zandi, E., Natoli, G. and Karin, M. (1998) IKK-gamma is an essential regulatory subunit of the IkappaB kinase complex. Nature 395: 297–300.

Ryan, K.M., Ernst, M.K., Rice, N.R. and Vousden, K.H. (2000) Role of NF-kappaB in p53-mediated programmed cell death. Nature, 404: 892–897.

Sabban, E. and Kvetnansky, R. (2001) Stress-triggered activation of gene expression in catecholaminergic systems: dynamics of transcriptional events. Trends Neurosci., 24: 91–98.

Schneider, A., Martin-Villalba, A., Weih, F., Vogel, J., Wirth, T. and Schwaninger, M. (1999) NF-kappaB is activated and promotes cell death in focal cerebral ischemia. Nat. Med., 5: 554–559.

Sen, R. and Baltimore, D. (1986) Multiple nuclear factors interact with the immunoglobulin enhancer sequences. Cell, 46: 705–716.

Senftleben, U., Cao, Y., Xiao, G., Greten, F.R., Krahn, G., Bonizzi, G., Chen, Y., Hu, Y., Fong, A., Sun, S.C. and Karin, M. (2001) Activation by IKKalpha of a second, evolutionary conserved, NF-kappa B signaling pathway. Science, 293: 1495–1499.

Serova, L., Saez, E., Spiegelman, B.M. and Sabban, E.L. (1998) c-Fos deficiency inhibits induction of mRNA for some, but not all, neurotransmitter biosynthetic enzymes by immobilization stress. J. Neurochem., 70: 1935–1940.

Shaulian, E. and Karin, M. (2002) AP-1 as a regulator of cell life and death. Nat. Cell. Biol., 4: E131–E136.

Shirayama, Y., Chen, A.C.-H., Nakagawa, S., Russell, R.S. and Duman, R.S. (2002) Brain derived neurotrophic factor produces antidepressant effects in behavioral models of depression. J. Neurosci., 22: 3251–3261.

Silva, A., Kogan, J.H., Frankland, P.W. and Kida, S. (1998) CREB and memory. Ann. Rev. Neurosci., 21: 127–148.

Smith, M.A., Makino, S., Kvetnansky, R. and Post, R.M. (1995) Stress alters the express of brain-derived neurotrophic factor and neurotrophin-3 mRNAs in the hippocampus. J. Neurosci., 15: 1768–1777.

Stamp, J. and Herbert, J. (1999) Multiple immediate-early gene expression during physiological and endocrine adaptation to repeated stress. Neurosci., 94: 1313–1322.

Stanciu, M., Radulovic, J. and Spiess, J. (2001) Phosphorylated cAMP response element binding protein in the mouse brain after fear conditioning: relationship to Fos production. Mol. Brain Res., 94: 15–24.

Sun, Y., Zhang, L., Johnston, N.L., Torrey, E.F. and Yolken, R.H. (2001) Serial analysis of gene expression in the frontal cortex of patients with bipolar disorder. Br. J. Psychiatry, Suppl 41: s137–s141.

Tang, G., Minemoto, Y., Dibling, B., Purcell, N.H., Li, Z., Karin, M. and Lin, A. (2001) Inhibition of JNK activation through NF-kappaB target genes. Nature, 414: 313–317.

Tao, X., Finkbeiner, S., Arnold, D.B., Shaywitz, A.J. and Greenberg, M.E. (1998) Ca2+ influx regulates BDNF transcription by a CREB family transcription factor-dependent mechanism. Neuron, 20: 709–726.

Thome, J., Sakai, N., Shin, K.H., Steffen, C., Zhang, Y.-J., Impey, S., Storm, D.R. and Duman, R.S. (2000) cAMP response element-mediated gene transcription is upregulated by chronic antidepressant treatment. J. Neurosci., 20: 4030–4036.

Trentani, A., Kuipers, S.D., Ter Horst, G.J. and Den Boer, J.A. (2002) Selective chronic stress-induced in vivo ERK1/2 hyperphosphorylation in medial prefrontocortical dendrites: implications for stress-related cortical pathology? Eur. J. Neurosci., 15: 1681–1691.

Umemoto, S., Kawai, Y., Ueyama, T. and Senba, E. (1997) Chronic glucocorticoid administration as well as repeated stress affects the subsequent acute immobilization stress-induced expression of immediate early genes but not that of NGFI-A. Neurosci., 80: 763–773.

Van Antwerp, D.J., Martin, S.J., Kafri, T., Green, D.R. and Verma, I.M. (1996) Suppression of TNF-alpha-induced apoptosis by NF-kappaB. Science, 274: 787–789.

Wallace, T., Stellitano, K.E., Neve, R.L. and Duman, R.S. (2002) Effects of CREB expression in the basolateral amygdala on behavioral models of depression and anxiety. Soc. for Neuros. Sci. Abstract.

Weih, F., Durham, S.K., Barton, D.S., Sha, W.C., Baltimore, D. and Bravo, R. (1996) Both multiorgan inflammation and myeloid hyperplasia in RelB-deficient mice are T cell dependent. J. Immunol., 157: 3974–3979.

Yamamoto, Y. and Gaynor, R.B. (2001) Therapeutic potential of inhibition of the NF-kappaB pathway in the treatment of inflammation and cancer. J. Clin. Invest., 107: 135–142.

Yu, Z., Zhou, D., Bruce-Keller, A.J., Kindy, M.S. and Mattson, M.P. (1999) Lack of the p50 subunit of nuclear factor-kappaB increases the vulnerability of hippocampal neurons to excitotoxic injury. J. Neurosci., 19: 8856–8865.

Zandi, E., Rothwarf, D.M., Delhase, M., Hayakawa, M. and Karin, M. (1997) The IkappaB kinase complex (IKK) contains two kinase subunits, IKKalpha and IKKbeta, necessary for IkappaB phosphorylation and NF-kappaB activation. Cell 91: 243–252.

Zhong, H., Voll, R.E. and Ghosh, S. (1998) Phosphorylation of NF-kappa B p65 by PKA stimulates transcriptional activity by promoting a novel bivalent interaction with the coactivator CBP/p300. Mol. Cell, 1: 661–671.

T. Steckler, N.H. Kalin and J.M.H.M. Reul (Eds.)
Handbook of Stress and the Brain, Vol. 15
ISBN 0-444-51173-3

CHAPTER 5.4

Experience, structural plasticity and neurogenesis

Jennifer D. Peters* and Elizabeth Gould

Department of Psychology, Princeton University, Princeton, NJ 08544, USA

Abstract: The adult hippocampus is a structurally plastic region in which dendritic remodeling, synaptogenesis, neurogenesis, and cell death have all been reported to occur in the absence of damage or pathology. These ongoing structural changes are modulated by both hormones and experience, and may contribute to aspects of hippocampal function, such as learning and memory, response to novelty, and regulation of the hypothalamic–pituitary–adrenal (HPA) axis. In turn, environmental complexity, learning, stress, and aversive experience appear to shape the hippocampus by regulating its structural plasticity. This chapter focuses on structural changes in the adult hippocampus, including dendritic remodeling and neurogenesis, and how these processes are affected by hormones and experience throughout life.

The adult hippocampus exhibits a variety of dynamic structural phenomena previously believed to occur only during development. This structural plasticity appears to be modulated by both hormones and experience, suggesting that these changes may represent mechanisms that underlie hippocampal function. In this regard, the hippocampus has been linked to several functions that are affected by stress, including learning, response to novelty, and regulation of the hypothalamic–pituitary–adrenal (HPA) axis.

The first section of this chapter examines hormones and experiences that appear to contribute to ongoing anabolic processes in the unstressed adult hippocampus. The second section focuses on how stress hormones and aversive experience affect these processes, and discusses some implications that this high-degree of plasticity may have for understanding how the adult brain normally functions and what goes wrong in cases of psychiatric illness.

*Corresponding author. Tel.: (609) 258-5625, (609) 258-4483; Fax: (609) 258-1113; E-mail: peters@princeton.edu

Hormones and experience: hippocampal plasticity in the unstressed brain

The adult hippocampus has been shown to undergo significant structural change in the absence of damage or pathology. Hormones and experience appear to induce fluctuations in the number and complexity of synapses and dendritic spines, the production and survival of new neurons, and overall hippocampal volume throughout life.

Synaptic and dendritic remodeling

Considerable transformations of hippocampal synapses and dendrites have been observed in response to changing behavioral states, hormonal levels, or environmental conditions. For example, during periods of torpor in the ground squirrel, mossy fiber synapses on CA3 pyramidal cells become smaller and show fewer postsynaptic densities and dendritic spine infoldings, but within two hours of arousal, their structures are completely restored (Popov and Bocharova, 1992). These rapid morphological changes parallel those seen in the apical dendrites of hippocampal pyramidal cells, which appear much shorter, less-branched, and have fewer dendritic spines prior to arousal (Popov et al., 1992).

Excitatory synapses on pyramidal cells in the CA1 region of female rodents are continuously formed and retracted in response to fluctuations in ovarian steroid levels via an NMDA receptor-mediated mechanism (McEwen et al., 2001). These estrous cycle oscillations in synapse number are coincident with changes in dendritic spine density on CA1 pyramidal neurons, such that during proestrus, rats have approximately 30% more spines than during late estrus (Woolley et al., 1990a).

Environmental enrichment has long been known to produce pronounced structural changes in the brains of laboratory animals (Diamond et al., 1964). These paradigms usually give animals access to larger cages, different toys and opportunities for sensory stimulation and physical exercise, and a chance for increased social interaction and learning experiences. Exposure to laboratory enrichment early in life enhances the growth of granule and pyramidal cell dendrites in the developing rat hippocampus (Fiala et al., 1978). Other studies have noted increases in cell volumes and the number of dendritic branches in pyramidal cells, and increased dendrite length in the granule cells of the dentate gyrus in mice raised to adulthood in enriched environments. These effects appeared to be independent of physical activity, as mice with running wheels, but not other aspects of enrichment, did not show these changes (Faherty et al., 2003). Although the timescale and persistence of the effects of enrichment are not as well-characterized in the adult hippocampus as they are in other brain regions, enriched environment living for as little as 4 days increases spine number and length in the adult rat cortex (Wallace et al., 1992), and rats returned to isolated laboratory cages show persistent dendritic changes for several weeks following 30 days or more experience in enriched environments (Camel et al., 1986).

One parameter that may potentially contribute to the environmental enrichment effect is learning, or the opportunity for the animals to engage in a greater variety of behaviors that involve the hippocampus. Some investigators have observed an increase in spine density on granule cells in the dentate gyrus of the hippocampus six hours after training on a passive avoidance paradigm, an effect that persists for fewer than 3 days (O'Malley et al., 1998). A similar learning-associated effect was seen in granule cells after water maze training (O'Malley et al., 2000). An increase in basal spine density on the pyramidal cells of the CA1 region of the hippocampus was measured 24 h post-training on either a trace eyeblink (hippocampus-dependent) or delay eyeblink (hippocampus-independent) conditioning task. No changes were observed in the dentate gyrus in this experiment. When given an NMDA receptor antagonist, animals neither acquired the conditioned responses nor showed increases in CA1 pyramidal spine density (Leuner et al., 2003). Taken together, these results suggest that learning alters dendritic spines, and it seems likely that these dendritic changes (which are generally accepted to indicate changes in the number of excitatory synapses as well) contribute to the learning process itself.

Adult neurogenesis

Although it was previously assumed that neurogenesis was restricted to early developmental periods, the adult brain of a variety of mammalian species, including humans, continues to generate significant numbers of neurons throughout life (Gould and Gross, 2002). One brain region where this phenomenon is particularly robust is the hippocampus. Approximately 9000 cells are produced each day in the dentate gyrus of the adult rat, the majority of which appear to differentiate into neurons (Cameron and McKay, 2001). Like the resident population of granule neurons, these adult-generated cells extend axons into the CA3 region (Stanfield and Trice, 1988; Hastings and Gould, 1999; Markakis and Gage, 1999) and form synapses (Kaplan and Hinds, 1977). Many of the new neurons die after several weeks (Cameron et al., 1993b; Biebl et al., 2000; Gould et al., 2001), but despite their transient existence, these cells appear to contribute to some aspects of hippocampal function, such as learning (Gould and Gross, 2002), reaction to novelty (Lemaire et al., 1999), and mood regulation (Santarelli et al., 2003).

Adult neurogenesis, like dendritic remodeling, is modulated by hormones and the environment. Ovarian steroid manipulations demonstrate that transiently high levels of estrogen result in increases in the rate of cell proliferation in the dentate gyrus. Ovariectomy diminishes the number of new neurons, whereas estrogen replacement restores neuron production to normal. A natural fluctuation in cell proliferation was also observed across the estrous cycle, such that female rats in proestrus produced more neurons than in other phases when estrogen levels are lower (Tanapat et al., 1999).

Numerous studies suggest that environmental complexity can enhance the number of new neurons in the hippocampus over the number observed in control animals. The first study to report this examined black-capped chickadees and found that birds living in the wild had more new neurons than those living in captivity. A seasonal difference in neurogenesis was observed as well, such that during the season of maximal seed caching and retrieval, more new neurons were maintained as compared to other times of year (Barnea and Nottebohm, 1994).

These findings suggested that activation of the hippocampus, by engagement in spatial navigation learning (or some other seasonal activity), enhanced the survival of new neurons. Another study showed more directly that food storage and retrieval stimulates hippocampal neurogenesis, since marsh tits that were allowed to forage for food showed considerable increases in neurogenesis over age-matched controls (Patel et al., 1997). Subsequent work demonstrated that living in a laboratory enriched environment setting also enhanced the survival of newly generated granule cells in the dentate gyrus of adult rats and mice (Kempermann et al., 1997; Nilsson et al., 1999). Some of this enrichment effect may also be due to increased motor activity, as mice given access to a running wheel have been shown to have increased neurogenesis over controls (see Brown et al., 2003).

Another possibility is that various learning tasks involving the hippocampus increase the number of new cells either by upregulating cell proliferation or enhancing the survival of cells produced in adulthood. Support for this hypothesis comes from studies that show twice the number of adult-generated neurons in the dentate gyri of rats trained in hippocampus-dependent associative learning tasks over controls or rats trained in analogous tasks that do not require the hippocampus (Gould et al., 1999a). Subsequent research has indicated that these new neurons actually participate in the learning process, since the toxin methylazoxymethanol acetate (MAM), which kills proliferating cells, not only blocked ongoing hippocampal neurogenesis but also the formation of trace memories. When rats were allowed to recover and neurogenesis resumed, the ability to acquire new trace memories returned (Shors et al., 2001b). It appears that only certain hippocampus-dependent learning tasks require ongoing neurogenesis, as MAM treatment does not disrupt all kinds of hippocampus-dependent learning (Shors et al., 2002). The continuous addition of new neurons with their immature (and thus unique) properties suggests that adult-generated neurons may play a crucial role in some forms of associative learning (Gould et al., 1999b).

Hippocampal volume

Some evidence suggests that the overall volume of the hippocampus is related to some of its learning and memory functions. Vertebrate species with behavioral adaptations involving a need for extensive spatial navigation learning exhibit larger hippocampi than similar species without these adaptations. For example, birds and small mammals that cover wide territories in search of food or mates have significantly larger hippocampal volumes than other related species living in more restricted areas (see Sherry et al., 1992). Furthermore, avian species that presumably engage the hippocampus to distribute and relocate seeds in disparate locations show variations in the hippocampal formation across seasons, such that periods of intense caching and retrieval are correlated with increases in hippocampal volume (Smulders et al., 1995). These seasonal differences are not found in similar species of non-caching birds (Lee et al., 2001).

Sex differences in hippocampal volume have also been reported in birds and mammals. These dimorphisms may be related to sex differences in hippocampus-dependent behavior. For example, female parasitic brooding cowbirds spend considerable amounts of time searching for and remembering the location of appropriate nests in which to lay eggs. They have larger hippocampi than male cowbirds who do not engage in such behaviors (Sherry et al., 1993). In addition, polygamous male voles who traverse wide territories in search of mates have larger hippocampi than the more sedentary females (Jacobs et al., 1990).

Differences in hippocampal volume have also been reported within same sex conspecifics, including in humans. Cab drivers in London exhibit larger posterior hippocampi, and smaller anterior hippocampi, than controls (Maguire et al., 2000). It is possible that an individual's hippocampal structure prior to experience might impose a predisposition to engage in certain behaviors or alter the likelihood of succeeding in activities that require certain types of learning, as becoming a cab driver inevitably involves extensive spatial navigation learning, but it may also be that the actual process of engaging in behaviors that require the hippocampus alters the size and shape of this brain region. Considerable evidence suggests that experience is, in fact, capable of modulating hippocampal structure, even in adulthood.

Laboratory-based experimental manipulations in environment suggest that more complex experiences lead to gross changes in the hippocampus. When compared to control animals living in standard laboratory housing, animals living in a more complex environment had larger dentate gyri (Kempermann et al., 1997). While it is unclear whether enriched environment studies actually reflect the influence of experience on the hippocampus, as opposed to removing the effects of deprivation, these findings do suggest that conditions capable of activating the hippocampus alter its size.

Changes in hippocampal volume might reflect any number of mechanisms, including alterations in total cell number, the length or complexity of dendritic arborizations, or the density of connections between cells. Since structural plasticity at each of these levels has been reported in the intact hippocampus of a variety of species under a variety of conditions that involve that brain region, it seems likely that the behaviors of the hippocampus in turn contribute to hippocampal function in a more general sense.

Stress hormones and aversive experience: hippocampal plasticity and psychopathology

Owing to its high density of glucocorticoid receptors (Van Eekelen and De Kloet, 1992) and putative role in regulating shut-off of the HPA axis (Feldman and Conforti, 1980), the hippocampus has been the focus of intensive investigation regarding the influence of stress on the brain. A number of studies have shown that aversive experiences and stress hormones mediate several aspects of structural plasticity in the adult hippocampus (Fig. 1).

Dendritic remodeling and cell survival

Stress appears to affect hippocampal dendritic spines differently depending upon the sex of the animal, the cell type examined, and the duration and intensity of the stressor. Male rats grow more pyramidal cell spines 24 h following exposure to an acute stressor. This increase in spine density appears to correlate with stress-induced enhancements in hippocampus-dependent learning. An opposite effect of stress is observed in female rats during diestrus; acute stress decreases the number of dendritic spines on CA1 pyramidal cells, a change that correlates with a stress-induced decrement in learning (Shors et al., 1998, 2001a; Wood and Shors, 1998, Wood et al., 2001).

Chronic, in contrast to acute, exposure to stress or stress hormones (glucocorticoids such as corticosterone), causes reversible dendritic atrophy in the apical dendrites of pyramidal cells in the CA3 region of the adult hippocampus. For instance, 21 days of 6-h-per-day restraint stress or 21 days of exogenous corticosterone treatment leads to significant atrophy of these dendrites (Woolley et al., 1990b; Watanabe et al., 1992c; Magarinos and McEwen, 1995a). Glucocorticoids are implicated in this change, since treatment with cyanoketone, an adrenal steroid synthesis blocker, prevents this stress-induced dendritic atrophy (Magarinos and McEwen, 1995b). Social stress in rats and tree shrews was found to have a similar effect on CA3 pyramidal cell dendrites (Magarinos et al., 1996; McKittrick et al., 2000).

Additional components of the stress response, in concert with or downstream from glucocorticoids, likely contribute to dendritic remodeling. Stress is thought to increase the release of glutamate in the hippocampus, leading to the activation of NMDA receptors. Treatment with phenytoin, which blocks glutamate release, or blockade of NMDA receptors, preserved CA3 dendrites exposed to stress or stress

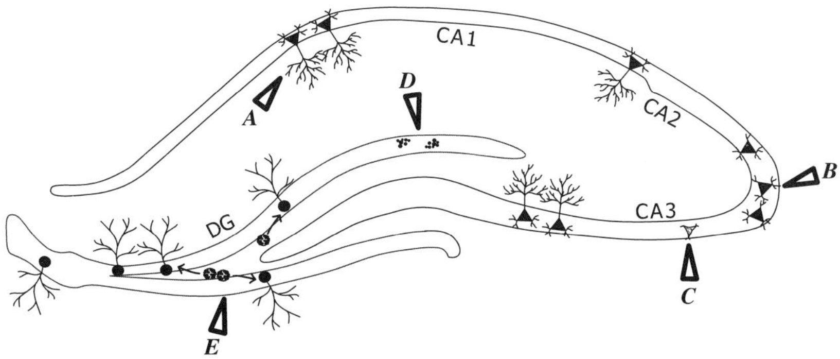

Fig. 1. Diverse actions of stress and adrenal steroids on the adult hippocampus. (*A*) Dendritic spine density increases on pyramidal cells of the CA1 region of male rats within 24 h following acute stress. The opposite finding, a decrease in spine density, was demonstrated in female rats stressed during diestrus (Shors et al., 2001a). (*B*) Chronic exposure to stress or glucocorticoids causes apical dendritic atrophy in pyramidal cells of the CA3 region (Woolley et al., 1990b; Watanabe et al., 1992c; Magarinos et al., 1996; McKittrick et al., 2000). (*C*) Chronic exposure to extreme stress or high levels of glucocorticoids may result in CA3 pyramidal cell death (Uno et al., 1989; Sapolsky et al., 1990). (*D*) Basal levels of glucocorticoids prevent apoptosis in granule cells, but adrenalectomy results in an increase in pyknotic cells in the dentate gyrus (Gould et al., 1991b; Cameron and Gould, 1996). (*E*) Stressors or elevated glucocorticoids suppress the proliferation of granule cell precursors in the adult dentate gyrus (Gould et al., 1992, 1997, 1998; Cameron and Gould, 1994; Tanapat et al., 2001).

hormones (Watanabe et al., 1992a; Magarinos and McEwen, 1995b). NMDA receptor activity is enhanced by serotonin (Rahmann and Neumann, 1993), which is released in response to many stressors as well (see Chaouloff, 2000). Giving tianeptine, an atypical tricyclic antidepressant that reduces extracellular serotonin, serves to prevent stress- or corticosterone-induced atrophy of dendrites in the hippocampus (Watanabe et al., 1992b).

Dendritic atrophy may ultimately result from glutamate excitotoxicity, or increased levels of intracellular calcium following activation of NMDA receptors. Excessive amounts of calcium can be harmful to cells by causing depolymerization or proteolysis of the cytoskeleton, allowing degradation of dendrites. Shrinkage of dendrites may serve a protective function, reducing the number of excitatory synapses and subsequent risk of excitotoxicity, or such a process might herald cell death.

Chronic, long-term treatment of rodents and monkeys with high doses of glucocorticoids has been demonstrated to lead to degeneration even beyond dendritic atrophy in the hippocampus, as evidenced by irregularities in the CA3 and CA2 cell layers, shrinkage and condensation of the soma, and sometimes nuclear pyknosis in pyramidal neurons (Sapolsky et al., 1985, 1990). This pattern of damage was also observed in monkeys subjected to severe or fatal stress during their lifetimes, further suggesting that stress can seriously compromise the hippocampus (Uno et al., 1989).

As is postulated to be the case for dendritic atrophy, this extreme cell damage and eventual death is hypothesized to be mediated by the glucocorticoid receptor (GR) (Packan and Sapolsky, 1990) along with the accompanying excessive excitatory neurotransmission resulting in oxidative damage, compromised energy utilization, and possibly the induction of programmed cell death cascades associated with excess intracellular calcium (Reagan and McEwen, 1997; Sapolsky, 2000). Whether stress and glucocorticoids directly kill cells or simply lower their capacities to withstand injury (even bouts of hypoglycemia or ischemia that would be tolerated by healthy cells) remains debatable (see Lee et al., 2002). It is important to note, however, that cell death is not always observed when high levels of glucocorticoids are sustained over time, or even if electrophysiological or cognitive impairments are evident (Kerr et al., 1991; Bodnoff et al., 1995).

Adult neurogenesis

Stress and glucocorticoids are potent modulators of hippocampal neurogenesis. Exogenous applications of supraphysiological levels of corticosterone during development or later in life decreases the rate of proliferation of granule cell precursors in the rat (Gould et al., 1991a, 1992; Cameron and Gould, 1994; Gould, 1994). Alternatively adrenalectomy stimulates cell proliferation in the dentate gyrus during development and adulthood (Gould et al., 1992; Cameron and Gould, 1994). Still further evidence linking adrenal steroids to the modulation of hippocampal neurogenesis stems from the fact that conditions that elevate glucocorticoids, such as natural aging, have been shown to correlate with lower rates of granule cell production, and adrenalectomy in aged rats reverses this trend (Kuhn et al., 1996; Cameron and McKay, 1999).

Stressful situations, which involve elevations in glucocorticoid levels and often increased glutamatergic transmission as well (Moghaddam et al., 1994), have been demonstrated to inhibit hippocampal neurogenesis in a variety of mammalian species. Acute exposure of rats to the odor of their natural predator, the fox, rapidly suppresses precursor proliferation, and ultimately the production of immature neurons, in the dentate gyrus. This effect involves adrenal steroids since normalizing corticosterone levels, by adrenalectomizing animals and replacing with low dose corticosterone in the drinking water, prevents the stress-induced decrease in cell proliferation (Tanapat et al., 2001).

Tree shrews respond similarly to acute subordination stress, and chronic one-hour exposures to dominant animals for 28 days results not only in significant reductions in cell proliferation, but also decreased dentate gyrus volumes in the subordinates (Gould et al., 1997; Lucassen et al., 2001). A single brief exposure to an aggressive resident marmoset, an experience known to elevate cortisol levels, significantly suppresses cell proliferation in the dentate gyrus of intruder marmosets (Gould et al., 1998). These findings suggest that stress-induced inhibition of neurogenesis is a phenomenon that is common to many mammalian species.

Since most progenitor cells in the dentate gyrus of adult animals do not appear to express mineralocorticoid or glucocorticoid receptors, adrenal steroids probably affect proliferation via an indirect mechanism (Cameron et al., 1993a). Recent evidence suggests that this mechanism involves NMDA receptor activation, since blocking the activation of NMDA receptors prevents the stress-induced suppression of cell proliferation, whereas activating NMDA receptors prevents the adrenalectomy-induced increase in proliferation (Cameron et al., 1998). Independent of adrenal steroids, competitive and noncompetitive NMDA receptor antagonists enhance proliferation, as do lesions of the entorhinal cortex, which normally provides excitatory input to the dentate gyrus (Cameron et al., 1995). Experimentally induced seizures (which involve extreme levels of excitatory neurotransmission) have been shown to increase hippocampal neurogenesis (Parent et al., 1997; Nakagawa et al., 2000), which may seem to contradict the above findings. However, it is important to note that such seizures cause cell death, a condition that stimulates hippocampal neurogenesis (Gould and Tanapat, 1997; Kelsey et al., 2000).

Although several studies have demonstrated a role for glucocorticoids in controlling the production of new neurons in the dentate gyrus both during development and in adulthood, many questions remain unanswered. First, is there a rebound of cell proliferation after glucocorticoid treatment (or stress) ceases? A recent report suggests that proliferation remains dampened for at least nine days following inescapable shock training (Malberg and Duman, 2003), but whether this suppression is common to different stressors and how long it may continue is unclear. Second, if cell proliferation does undergo a compensatory increase, it is uncertain whether such a rebound could suffice to make up for the deficits accumulated over long periods of intense stress. Third, and perhaps most importantly, the functional consequences of these stress-induced changes remain unknown. This issue is particularly difficult to address because stress and glucocorticoids are known to affect so many processes within the hippocampus that any measurable behavioral, cognitive, or affective changes are difficult to link to one cellular mechanism.

Hippocampal structure and psychiatric disorders

In general, a wide array of psychiatric and endocrine disorders appear to result in reduced hippocampal volume in humans. Decreased hippocampal size has been observed in all of the following conditions: Cushing's syndrome (hypercortisolemia) (Starkman et al., 1992), recurrent depression (Sheline et al., 1996), post-traumatic stress disorder (PTSD) (Bremner et al., 1995; Gurvits et al., 1996), and schizophrenia (Bogerts et al., 1993; Fukuzako et al., 1996).

It is not yet clear whether these disorders cause the hippocampus to become smaller, or whether patients who are born with smaller hippocampi are then predisposed to develop these disorders. On the one hand, some studies of patients with major depression suggest that damage to the hippocampus accumulates over time, since patients experiencing their first episode have hippocampi of similar volume to age-matched controls, whereas patients who have had several recurrences show decreases in volume that correlate with the amount of time spent depressed (Sheline et al., 1999; MacQueen et al., 2003). On the other hand, some recent evidence suggests that smaller hippocampi in PTSD are present prior to the development of the condition, however the degree of atrophy has also been related to the duration of trauma (Gurvits et al., 1996; Gilbertson et al., 2002).

Since stress often precedes or appears to be linked to the development of psychiatric disorders, and since these conditions are often associated with hippocampus-related cognitive deficits, it is tempting to speculate that differences in hippocampal volume in these human conditions are the result of hormonal or experiential modulation of synapses, dendrites, and the production and survival of new neurons. Many investigators have shown that antidepressant treatment in animals reverses stress-induced changes in neurogenesis, mitigates reductions in hippocampal volume, and prevents dendritic atrophy (Czeh et al., 2001; Duman et al., 2001; Malberg and Duman, 2003). However, a stronger link between these animal studies and clinical reports will require the development of convincing animal models of these disorders and higher resolution neuroimaging. Furthermore, much work is needed to link these structural changes to symptoms associated with the specific psychopathology that accompanies different conditions (McEwen and Magarinos, 2001).

In summary, the hippocampus appears to play a critical role in learning and memory, affect, and HPA axis regulation. In turn, enrichment and learning, as well as stress and aversive experience, seem to modulate hippocampal structure and function. Ongoing remodeling of synapses and dendrites, continued cell birth and death, and gross volume fluctuations may be important for normal functioning. However, this plasticity may also make the hippocampus particularly vulnerable to structural abnormalities that can lead to pathology.

References

Barnea, A. and Nottebohm, F. (1994) Seasonal recruitment of hippocampal neurons in adult free-ranging black-capped chickadees. Proc. Natl. Acad. Sci. USA, 91: 11217–11221.

Biebl, M., Cooper, C.M., Winkler, J. and Kuhn, H.G. (2000) Analysis of neurogenesis and programmed cell death reveals a self-renewing capacity in the adult rat brain. Neurosci. Lett., 291: 17–20.

Bodnoff, S.R., Humphreys, A.G., Lehman, J.C., Diamond, D.M., Rose, G.M. and Meaney, M.J. (1995) Enduring effects of chronic corticosterone treatment on spatial learning, synaptic plasticity, and hippocampal neuropathology in young and mid-aged rats. J. Neurosci., 15: 61–69.

Bogerts, B., Lieberman, J.A., Ashtari, M., Bilder, R.M., Degreef, G., Lerner, G., Johns, C. and Masiar, S. (1993) Hippocampus-amygdala volumes and psychopathology in chronic schizophrenia. Biol. Psychiatry., 33: 236–246.

Bremner, J.D., Randall, P., Scott, T.M., Bronen, R.A., Seibyl, J.P., Southwick, S.M., Delaney, R.C., McCarthy, G., Charney, D.S. and Innis, R.B. (1995) MRI-based measurement of hippocampal volume in patients with combat-related posttraumatic stress disorder. Am. J. Psychiatry, 152: 973–981.

Brown, J., Cooper-Kuhn, C.M., Kempermann, G., Van Praag, H., Winkler, J., Gage, F.H. and Kuhn, H.G. (2003) Enriched environment and physical activity stimulate hippocampal but not olfactory bulb neurogenesis. Eur. J. Neurosci., 17(10): 2042–2046.

Camel, J.E., Withers, G.S. and Greenough, W.T. (1986) Persistence of visual cortex dendritic alterations induced by postweaning exposure to a "superenriched" environment in rats. Behav. Neurosci., 100: 810–813.

Cameron, H.A. and Gould, E. (1994) Adult neurogenesis is regulated by adrenal steroids in the dentate gyrus. Neuroscience, 61: 203–209.

Cameron, H.A. and Gould, E. (1996) Distinct populations of cells in the adult dentate gyrus undergo mitosis or apoptosis in response to adrenalectomy. J. Comp. Neurol., 369: 56–63.

Cameron, H.A. and McKay, R.D. (1999) Restoring production of hippocampal neurons in old age. Nat. Neurosci., 2: 894–897.

Cameron, H.A. and McKay, R.D. (2001) Adult neurogenesis produces a large pool of new granule cells in the dentate gyrus. J. Comp. Neurol., 435: 406–417.

Cameron, H.A., Woolley, C.S. and Gould, E. (1993a) Adrenal steroid receptor immunoreactivity in cells born in the adult rat dentate gyrus. Brain Res., 611: 342–346.

Cameron, H.A., Woolley, C.S., McEwen, B.S. and Gould, E. (1993b) Differentiation of newly born neurons and glia in the dentate gyrus of the adult rat. Neuroscience, 56: 337–344.

Cameron, H.A., McEwen, B.S. and Gould, E. (1995) Regulation of adult neurogenesis by excitatory input and NMDA receptor activation in the dentate gyrus. J. Neurosci., 15: 4687–4692.

Cameron, H.A., Tanapat, P. and Gould, E. (1998) Adrenal steroids and *N*-methyl-D-aspartate receptor activation regulate neurogenesis in the dentate gyrus of adult rats through a common pathway. Neuroscience, 82: 349–354.

Chaouloff, F. (2000) Serotonin, stress and corticoids. J. Psychopharmacol., 14: 139–151.

Czeh, B., Michaelis, T., Watanabe, T., Frahm, J., de Biurrun, G., van Kampen, M., Bartolomucci, A. and Fuchs, E. (2001) Stress-induced changes in cerebral metabolites, hippocampal volume, and cell proliferation are prevented by antidepressant treatment with tianeptine. Proc. Natl. Acad. Sci. USA, 98(22): 12796–12801.

Diamond, M.C., Drech, D. and Rosenzweig, M.R. (1964) The effects of an enriched environment on the histology of the rat cerebral cortex. J. Comp. Neurol., 123: 111–120.

Duman, R.S., Nakagawa, S. and Malberg, J. (2001) Regulation of adult neurogenesis by antidepressant treatment. Neuropsychopharmacology, 25(6): 836–844.

Faherty, C.J., Kerley, D. and Smeyne, R.J. (2003) A Golgi-Cox morphological analysis of neuronal changes induced by environmental enrichment. Brain Res. Dev. Brain Res., 141(1–2): 55–61.

Feldman, S. and Conforti, N. (1980) Participation of the dorsal hippocampus in the glucocorticoid feedback effect on adrenocortical activity. Neuroendocrinology, 30: 52–55.

Fiala, B.A., Joyce, J.N. and Greenough, W.T. (1978) Environmental complexity modulates growth of granule cell dendrites in developing but not adult hippocampus of rats. Exp. Neurol., 59: 372–383.

Fukuzako, H., Fukazako, T., Hashiguchi, T., Hokazono, Y., Takeuchi, K., Hirakawa, K., Ueyama, K., Takigawa, M., Kajiya, Y., Nakajo, M. and Fujimoto, T. (1996) Reduction in hippocampal formation volume is caused mainly by its shortening in chronic schizophrenia: assessment by MRI. Biol. Psychiatry, 39: 938–945.

Gilbertson, M.W., Shenton, M.E., Ciszewski, A., Kasai, K., Lasko, N.B., Orr, S.P. and Pitman, R.K. (2002) Smaller hippocampal volume predicts pathologic vulnerability to psychological trauma. Nat. Neurosci., 5: 1242–1247.

Gould, E. (1994) The effects of adrenal steroids and excitatory input on neuronal birth and survival. Ann. N. Y. Acad. Sci., 743: 73–92; discussion 92–93.

Gould, E. and Gross, C.G. (2002) Neurogenesis in adult mammals: some progress and problems. J. Neurosci., 22(3): 619–623.

Gould, E. and Tanapat, P. (1997) Lesion-induced proliferation of neuronal progenitors in the dentate gyrus of the adult rat. Neuroscience, 80: 427–436.

Gould, E., Woolley, C.S., Cameron, H.A., Daniels, D.C. and McEwen, B.S. (1991a) Adrenal steroids regulate postnatal development of the rat dentate gyrus: II. Effects of glucocorticoids and mineralocorticoids on cell birth. J. Comp. Neurol., 313: 486–493.

Gould, E., Woolley, C.S. and McEwen, B.S. (1991b) Adrenal steroids regulate postnatal development of the rat dentate gyrus: I. Effects of glucocorticoids on cell death. J. Comp. Neurol., 313: 479–485.

Gould, E., Cameron, H.A., Daniels, D.C., Woolley, C.S. and McEwen, B.S. (1992) Adrenal hormones suppress cell division in the adult rat dentate gyrus. J. Neurosci., 12: 3642–3650.

Gould, E., McEwen, B.S., Tanapat, P., Galea, L.A. and Fuchs, E. (1997) Neurogenesis in the dentate gyrus of the adult tree shrew is regulated by psychosocial stress and NMDA receptor activation. J. Neurosci., 17: 2492–2498.

Gould, E., Tanapat, P., McEwen, B.S., Flugge, G. and Fuchs, E. (1998) Proliferation of granule cell precursors in the dentate gyrus of adult monkeys is diminished by stress. Proc. Natl. Acad. Sci. USA, 95: 3168–3171.

Gould, E., Beylin, A., Tanapat, P., Reeves, A. and Shors, T.J. (1999a) Learning enhances adult neurogenesis in the hippocampal formation. Nat. Neurosci., 2: 260–265.

Gould, E., Tanapat, P., Hastings, N.B. and Shors, T.J. (1999b) Neurogenesis in adulthood: a possible role in learning. Trends Cogn. Sci., 3: 186–192.

Gould, E., Vail, N., Wagers, M. and Gross, C.G. (2001) Adult-generated hippocampal and neocortical neurons in macaques have a transient existence. Proc. Natl. Acad. Sci. USA, 95: 3168–3171.

Gurvits, T.V., Shenton, M.E., Hokama, H., Ohta, H., Lasko, N.B., Gilbertson, M.W., Orr, S.P., Kikinis, R., Jolesz, F.A., McCarley, R.W. and Pitman, R.K. (1996) Magnetic resonance imaging study of hippocampal volume in chronic, combat-related posttraumatic stress disorder. Biol. Psychiatry, 40: 1091–1099.

Hastings, N.B. and Gould, E. (1999) Rapid extension of axons into the CA3 region by adult-generated granule cells. J. Comp. Neurol., 413: 146–154.

Jacobs, L.F., Gaulin, S.J., Sherry, D.F. and Hoffman, G.E. (1990) Evolution of spatial cognition: sex-specific patterns of spatial behavior predict hippocampal size. Proc. Natl. Acad. Sci. USA, 87: 6349–6352.

Kaplan, M.S. and Hinds, J.W. (1977) Neurogenesis in the adult rat: electron microscopic analysis of light radioautographs. Science, 197: 1092–1094.

Kelsey, J.E., Sanderson, K.L. and Frye, C.A. (2000) Perforant path stimulation in rats produces seizures, loss of hippocampal neurons, and a deficit in spatial mapping which are reduced by prior MK-801. Behav. Brain Res., 107: 59–69.

Kempermann, G., Kuhn, H.G. and Gage, F.H. (1997) More hippocampal neurons in adult mice living in an enriched environment. Nature, 386: 493–495.

Kerr, D.S., Campbell, L.W., Applegate, M.D., Brodish, A. and Landfield, P.W. (1991) Chronic stress-induced acceleration of electrophysiologic and morphometric biomarkers of hippocampal aging. J. Neurosci., 11: 1316–1324.

Kuhn, H.G., Dickinson-Anson, H. and Gage, F.H. (1996) Neurogenesis in the dentate gyrus of the adult rat: age-related decrease of neuronal progenitor proliferation. J. Neurosci., 16: 2027–2033.

Lee, A.L., Ogle, W.O. and Sapolsky, R.M. (2002) Stress and depression: possible links to neuron death in the hippocampus. Bipolar. Disord., 4: 117–128.

Lee, D.W., Smith, G.T., Tramontin, A.D., Soma, K.K., Brenowitz, E.A. and Clayton, N.S. (2001) Hippocampal volume does not change seasonally in a non food-storing songbird. Neuroreport, 12: 1925–1928.

Lemaire, V., Aurousseau, C., Le Moal, M. and Abrous, D.N. (1999) Behavioral trait of reactivity to novelty is related to hippocampal neurogenesis. Eur. J. Neurosci., 11(11): 4006–4014.

Leuner, B., Falduto, J. and Shors, T.J. (2003) Associative memory formation increases the observation of dendritic spines in the hippocampus. J. Neurosci., 23(2): 659–665.

Lucassen, P.J., Vollmann-Honsdorf, G.K., Gleisberg, M., Czeh, B., De Kloet, E.R. and Fuchs, E. (2001) Chronic psychosocial stress differentially affects apoptosis in hippocampal subregions and cortex of the adult tree shrew. Eur. J. Neurosci., 14: 161–166.

MacQueen, G.M., Campbell, S., McEwen, B.S., Macdonald, K., Amano, S., Joffe, R.T., Nahmias, C. and Young, L.T. (2003) Course of illness, hippocampal function, and hippocampal volume in major depression. Proc. Natl. Acad. Sci. USA, 100: 1387–1392.

Magarinos, A.M. and McEwen, B.S. (1995a) Stress-induced atrophy of apical dendrites of hippocampal CA3c neurons: comparison of stressors. Neuroscience, 69: 83–88.

Magarinos, A.M. and McEwen, B.S. (1995b) Stress-induced atrophy of apical dendrites of hippocampal CA3c neurons: involvement of glucocorticoid secretion and excitatory amino acid receptors. Neuroscience, 69: 89–98.

Magarinos, A.M., McEwen, B.S., Flugge, G. and Fuchs, E. (1996) Chronic psychosocial stress causes apical dendritic atrophy of hippocampal CA3 pyramidal neurons in subordinate tree shrews. J. Neurosci., 16: 3534–3540.

Maguire, E.A., Gadian, D.G., Johnsrude, I.S., Good, C.D., Ashburner, J., Frackowiak, R.S. and Frith, C.D. (2000) Navigation-related structural change in the hippocampi of taxi drivers. Proc. Natl. Acad. Sci. USA, 97: 4398–4403.

Malberg, J.E. and Duman, R.S. (2003) Cell proliferation in adult hippocampus is decreased by inescapable stress: reversal by fluoxetine treatment. Neuropsychopharmacology, 28(9): 1562–1571.

Markakis, E.A. and Gage, F.H. (1999) Adult-generated neurons in the dentate gyrus send axonal projections to field CA3 and are surrounded by synaptic vesicles. J. Comp. Neurol., 406: 449–460.

McEwen, B. and Magarinos, A.M. (2001) Stress and hippocampal plasticity: implications for the pathophysiology of affective disorders. Hum. Psychopharmacol., 16(S1): S7–S19.

McEwen, B., Akama, K., Alves, S., Brake, W.G., Bulloch, K., Lee, S., Li, C., Yuen, G. and Milner, T.A. (2001) Tracking the estrogen receptor in neurons: implications for estrogen-induced synapse formation. Proc. Natl. Acad. Sci. USA, 98: 7093–7100.

McKittrick, C.R., Magarinos, A.M., Blanchard, D.C., Blanchard, R.J., McEwen, B.S. and Sakai, R.R. (2000) Chronic social stress reduces dendritic arbors in CA3 of hippocampus and decreases binding to serotonin transporter sites. Synapse, 36: 85–94.

Moghaddam, B., Bolinao, M.L., Stein-Behrens, B. and Sapolsky, R. (1994) Glucocorticoids mediate the stress-induced extracellular accumulation of glutamate. Brain Res., 655: 251–254.

Nakagawa, E., Aimi, Y., Yasuhara, O., Tooyama, I., Shimada, M., McGeer, P.L. and Kimura, H. (2000) Enhancement of progenitor cell division in the dentate gyrus triggered by initial limbic seizures in rat models of epilepsy. Epilepsia, 41: 10–18.

Nilsson, M., Perfilieva, E., Johansson, U., Orwar, O. and Eriksson, P.S. (1999) Enriched environment increases neurogenesis in the adult rat dentate gyrus and improves spatial memory. J. Neurobiol., 39: 569–578.

O'Malley, A., O'Connell, C. and Regan, C.M. (1998) Ultrastructural analysis reveals avoidance conditioning to induce a transient increase in hippocampal dentate spine density in the 6 hour post-training period of consolidation. Neuroscience, 87(3): 607–613.

O'Malley, A., O'Connell, C., Murphy, K.J. and Regan, C.M. (2000) Transient spine density increases in the mid-molecular layer of hippocampal dentate gyrus accompany consolidation of a spatial learning task in the rodent. Neuroscience, 99(2): 229–232.

Packan, D.R. and Sapolsky, R.M. (1990) Glucocorticoid endangerment of the hippocampus: tissue, steroid and receptor specificity. Neuroendocrinology, 51: 613–618.

Parent, J.M., Yu, T.W., Leibowitz, R.T., Geschwind, D.H., Sloviter, R.S. and Lowenstein, D.H. (1997) Dentate granule cell neurogenesis is increased by seizures and contributes to aberrant network reorganization in the adult rat hippocampus. J. Neurosci., 17: 3727–3738.

Patel, S.N., Clayton, N.S. and Krebs, J.R. (1997) Spatial learning induces neurogenesis in the avian brain. Behav. Brain Res., 89: 115–128.

Popov, V.I. and Bocharova, L.S. (1992) Hibernation-induced structural changes in synaptic contacts between mossy fibres and hippocampal pyramidal neurons. Neuroscience, 48: 53–62.

Popov, V.I., Bocharova, L.S. and Bragin, A.G. (1992) Repeated changes of dendritic morphology in the hippocampus of ground squirrels in the course of hibernation. Neuroscience, 48: 45–51.

Rahmann, S. and Neumann, R.S. (1993) Activation of 5-HT2 receptors facilitates depolarization of neocortical neurons by *N*-methyl-D-aspartate. European Journal of Pharmacology, 231: 347–354.

Reagan, L.P. and McEwen, B.S. (1997) Controversies surrounding glucocorticoid-mediated cell death in the hippocampus. J. Chem. Neuroanat., 13: 149–167.

Santarelli, L., Saxe, M., Gross, C., Surget, A., Battaglia, F., Dulawa, S., Weisstaub, N., Lee, J., Duman, R., Arancio, O., Belzung, C. and Hen, R. (2003) Requirement of hippocampal neurogenesis for the behavioral effects of antidepressants. Science, 301(5634): 805–809.

Sapolsky, R.M. (2000) The possibility of neurotoxicity in the hippocampus in major depression: a primer on neuron death. Biol. Psychiatry, 48: 755–765.

Sapolsky, R.M., Krey, L.C. and McEwen, B.S. (1985) Prolonged glucocorticoid exposure reduces hippocampal neuron number: implications for aging. J. Neurosci., 5: 1222–1227.

Sapolsky, R.M., Uno, H., Rebert, C.S. and Finch, C.E. (1990) Hippocampal damage associated with prolonged glucocorticoid exposure in primates. J. Neurosci., 10: 2897–2902.

Sheline, Y.I., Wang, P.W., Gado, M.H., Csernansky, J.G. and Vannier, M.W. (1996) Hippocampal atrophy in recurrent major depression. Proc. Natl. Acad. Sci. USA, 93: 3908–3913.

Sheline, Y.I., Sanghavi, M., Mintun, M.A. and Gado, M.H. (1999) Depression duration but not age predicts hippocampal volume loss in medically healthy women with recurrent major depression. J. Neurosci., 19: 5034–5043.

Sherry, D.F., Jacobs, L.F. and Gaulin, S.J. (1992) Spatial memory and adaptive specialization of the hippocampus. Trends Neurosci., 15: 298–303.

Sherry, D.F., Forbes, M.R., Khurgel, M. and Ivy, G.O. (1993) Females have a larger hippocampus than males in the brood-parasitic brown-headed cowbird. Proc. Natl. Acad. Sci. USA, 90: 7839–7843.

Shors, T.J., Lewczyk, C., Pacynski, M., Mathew, P.R. and Pickett, J. (1998) Stages of estrous mediate the stress-induced impairment of associative learning in the female rat. Neuroreport, 9: 419–423.

Shors, T.J., Chua, C. and Falduto, J. (2001a) Sex differences and opposite effects of stress on dendritic spine density in the male versus female hippocampus. J. Neurosci., 21: 6292–6297.

Shors, T.J., Miesegaes, G., Beylin, A., Zhao, M., Rydel, T. and Gould, E. (2001b) Neurogenesis in the adult is involved in the formation of trace memories. Nature, 410: 372–376.

Shors, T.J., Townsend, D.A., Zhao, M., Kozorovitskiy, Y. and Gould, E. (2002) Neurogenesis may relate to some but not all types of hippocampal-dependent learning. Hippocampus, 12: 578–584.

Smulders, T.V., Sasson, A.D. and DeVoogd, T.J. (1995) Seasonal variation in hippocampal volume in a food-storing bird, the black-capped chickadee. J. Neurobiol., 27: 15–25.

Stanfield, B.B. and Trice, J.E. (1988) Evidence that granule cells generated in the dentate gyrus of adult rats extend axonal projections. Exp. Brain Res., 72: 399–406.

Starkman, M.N., Gebarski, S.S., Berent, S. and Schteingart, D.E. (1992) Hippocampal formation volume, memory dysfunction, and cortisol levels in patients with Cushing's syndrome. Biol. Psychiatry, 32: 756–765.

Tanapat, P., Hastings, N.B., Reeves, A.J. and Gould, E. (1999) Estrogen stimulates a transient increase in the number of new neurons in the dentate gyrus of the adult female rat. J. Neurosci., 19(14): 5792–5801.

Tanapat, P., Hastings, N.B., Rydel, T.A., Galea, L.A. and Gould, E. (2001) Exposure to fox odor inhibits cell proliferation in the hippocampus of adult rats via an adrenal hormone-dependent mechanism. J. Comp. Neurol., 437: 496–504.

Uno, H., Tarara, R., Else, J.G., Suleman, M.A. and Sapolsky, R.M. (1989) Hippocampal damage associated with prolonged and fatal stress in primates. J. Neurosci., 9: 1705–1711.

Van Eekelen, J.A. and De Kloet, E.R. (1992) Co-localization of brain corticosteroid receptors in the rat hippocampus. Prog. Histochem. Cytochem., 26: 250–258.

Wallace, C.S., Kilman, V.L., Withers, G.S. and Greenough, W.T. (1992) Increases in dendritic length in occipital cortex after 4 days of differential housing in weanling rats. Behav. Neural Biol., 58: 64–68.

Watanabe, Y., Gould, E., Cameron, H.A., Daniels, D.C. and McEwen, B.S. (1992a) Phenytoin prevents stress- and corticosterone-induced atrophy of CA3 pyramidal neurons. Hippocampus, 2: 431–435.

Watanabe, Y., Gould, E., Daniels, D.C., Cameron, H. and McEwen, B.S. (1992b) Tianeptine attenuates stress-induced morphological changes in the hippocampus. Eur. J. Pharmacol., 222: 157–162.

Watanabe, Y., Gould, E. and McEwen, B.S. (1992c) Stress induces atrophy of apical dendrites of hippocampal CA3 pyramidal neurons. Brain Res., 588: 341–345.

Wood, G.E. and Shors, T.J. (1998) Stress facilitates classical conditioning in males, but impairs classical conditioning in females through activational effects of ovarian hormones. Proc. Natl. Acad. Sci. USA, 95: 4066–4071.

Wood, G.E., Beylin, A.V. and Shors, T.J. (2001) The contribution of adrenal and reproductive hormones to the opposing effects of stress on trace conditioning in males versus females. Behav. Neurosci., 115: 175–187.

Woolley, C.S., Gould, E., Frankfurt, M. and McEwen, B.S. (1990a) Naturally occurring fluctuation in dendritic spine density on adult hippocampal pyramidal neurons. J. Neurosci., 10: 4035–4039.

Woolley, C.S., Gould, E. and McEwen, B.S. (1990b) Exposure to excess glucocorticoids alters dendritic morphology of adult hippocampal pyramidal neurons. Brain Res., 531: 225–231.

T. Steckler, N.H. Kalin and J.M.H.M. Reul (Eds.)
Handbook of Stress and the Brain, Vol. 15
ISBN 0-444-51173-3

CHAPTER 5.5

Adult neurogenesis in rodents and primates: functional implications

Eberhard Fuchs* and Boldizsár Czéh

Clinical Neurobiology Laboratory, German Primate Center, Kellnerweg 4, 37077, Göttingen, Germany

Abstract: Within the last two decades, our view of the mature mammalian brain has been changed. It is far from being fixed and immutable, as a number of factors such as environmental stimulation, learning, growth factors, glucocorticoid, and sexual hormones, stress, aging, neurotransmitters such as glutamate and serotonin, and a number of drugs regulate the production of new neurons in the adult dentate gyrus. This newfound capacity has forced a new look at plasticity of the brain, an organ previously considered to have an anatomically stable structure. The presence of neuronal precursors in the adult mammalian brain suggests a new level of plasticity for brain regions, whereby neurons are constantly replaced. Moreover, it raises questions regarding the underlying mechanisms of the newly generated neurons, and how they influence the functioning of the differentiated adult brain. The evolutionary conservation of adult neurogenesis and the persistently high level of neuron production throughout adulthood suggest that this process is of fundamental biological importance. The functional implications of adult neurogenesis remain unknown, but several studies have suggested a link between the formation of new neurons in the dentate gyrus and learning.

New neurons for the adult mammalian brain? The story of a controversy

Unlike most body cells such as those in the skin, the gut or the blood, which are constantly renewed and easily regenerated, the brain, and in particular the mammalian brain, has always been regarded as a nonrenewable organ. Most neurons of the adult central nervous system are terminally differentiated. Sometimes the adult brain can compensate for damage by making new connections among surviving neurons. However, it cannot repair itself because it lacks the stem cells that are necessary for neuronal regeneration. This limitation and unclear evolutionary benefit were first described by Santiago Ramón y Cajal, who stated: "In adult centers the nerve paths are something fixed, ended, immutable. Everything may die, nothing may be regenerated. It is for science of the future to change, if possible, this harsh decree." (Cajal, 1928). This dogma of neurobiology was initially challenged almost four decades ago. Using ^{3}H-thymidine autoradiography, Altman and Das (1965) discovered the production of new granular neurons in the dentate gyrus and olfactory bulb of adult rat brains. Although only limited work followed this initial finding, it confirmed and further substantiated the neuronal character of the newly generated hippocampal cells by demonstrating that they receive synaptic input and extend axons into the mossy fiber pathway that projects to the CA3 subfield (Kaplan and Hinds, 1977; Stanfield and Trice, 1988). The next landmark was in the early 1980s, when a substantial neurogenesis in the vocal control nucleus of the adult canary brain was demonstrated (Goldman and Nottebohm, 1983) and a functional link between behavior, song learning, and the production of new neurons could be established (Alvarez-Buylla et al., 1988). Neuronal turnover in the high vocal control centers is thought

*Corresponding author. Tel.: +49-551-3851 130;
Fax: +49-551-3851 307; E-mail: efuchs@gwdg.de

to play a role in the modification of perceptual memories or motor programs for song production in these animals (Goldman, 1998).

Despite several attempts, neurogenesis could not be demonstrated in the brains of adult nonhuman primates such as rhesus monkeys, thereby leading to the assumption that neuronal replication could not be tolerated in primates – including humans – because it might interfere with learning and memory. In his initial study, Rakic (1985) investigated neurogenesis in adult rhesus monkeys using ^{3}H-thymidine, examining major structures and subdivisions of the brain including the visual, motor, and association neocortex, hippocampus, and olfactory bulb. Rakic found "not a single heavily labeled cell with the morphological characteristics of a neuron in any brain in any adult animal" and concluded that "all neurons of the rhesus monkey brain are generated during prenatal and early postnatal life" (Rakic, 1985; Eckenhoff and Rakic, 1988). Furthermore, Rakic argued that "a stable population of neurons may be a biological necessity in an organism whose survival relies on learned behavior acquired over a long period of time." These reports had a profound influence on the development of the field. Presumably, these negative results formed the basis for researchers of the time to show little interest in neurogenesis in the adult mammalian brain.

This view was clearly changed, with the field consequently reviving and "exploding" when the thymidine analog 5-bromo-2′-deoxyuridine (BrdU) was introduced for labeling newborn neurons (Miller and Nowakowski, 1988). Using this new technique, it became clear that adult hippocampal neurogenesis in mammals is not restricted only to rodents but has been conserved throughout mammalian evolution. The formation of new granule neurons was demonstrated in tree shrews (Gould et al., 1997), animals considered to be phylogenetically located between insectivores and primates. This evidence was followed by initial reports of neurogenesis in the brains of marmoset monkeys (Gould et al., 1998), a small nonhuman primate from South America, and in the rhesus monkey, a typical representative of the nonhuman Old-world primates (Gould et al., 1999b; Kornack and Rakic, 1999).

Work with terminally ill patients has confirmed that humans also generate new neurons. The thymidine analog BrdU was injected into patients to monitor tumor cell proliferation. Some of these individuals subsequently died from their illness and small samples of hippocampus were evaluated for the presence of BrdU-labeled neurons. Since BrdU was systemically administered, all dividing cells would have been labeled and, indeed, newborn neurons were detected in the granule cell layer of all individuals (Eriksson et al., 1998). These data unequivocally show that neurogenesis is a common phenomenon across mammalian species.

The BrdU method

Like ^{3}H-thymidine, which was the label of choice in the initial experiments, BrdU is incorporated into a dividing progenitor cell during the S-phase. Fully differentiated neurons do not divide and cannot integrate the label. The BrdU immunolabel is not quantifiable in the same way as ^{3}H-thymidine, which allows the counting of silver grains, thereby enabling objective criteria for strong or weak labeling. Depending on the survival time after injection, BrdU is a marker of proliferating cells (short survival time) and their progeny (longer survival time). BrdU labeling is a nonisotopic method and is visualized with immunocytochemical methods. Importantly, the BrdU technique has two major advantages. First, it allows stereological estimation of the total number of newborn cells and by this the effects of different treatments can be evaluated and compared. Second, combining BrdU labeling with other immunocytochemical labels permits the determination of the phenotypes of the new cells. Markers for immature neurons are class III ß-tubulin (Tuj1), doublecortin, turned-off-after-division 64 KDa (TOAD-64), or polysialated neural cell adhesion molecule (PSA-NCAM). Markers for mature neurons are antibodies directed against neuron-specific nuclear protein (NeuN), microtubule-associated-protein 2 (MAP-2), or the calcium-binding protein, calbindin. Double-staining with markers for nonneuronal cells can demonstrate a glial phenotype: glial fibrillary acidic protein (anti-GFAP) or anti-S-100ß can visualize astrocytes, whereas oligodendrocytes can be labeled by anti-O4 or anti-CNP (2′,3′-cyclic nucleotide 3′-phosphodiesterase).

It should be noted, however, that there is an ongoing debate about the accuracy of the BrdU technique (Gould and Gross, 2002; Hayes and Nowakowski, 2002). Some groups argue that application of an inappropriately high dose of BrdU may give false positive results (Rakic, 2002a).

Restriction of adult neurogenesis in the mammalian brain

It has been reported that within vertebrates, at least for all bony fish investigated so far, there is an enormous potential for the production of new neurons in the adult brain. In gymnotiform fish, an average of 10^5 cells, corresponding to approximately 0.2% of the total population of the adult brain, are in the S-phase within any 2-h period (Zupanc, 1999). Comparing the massive neurogenesis observed throughout the life span of fish, amphibians and even birds with that observed in mammals, it is clear that the extent and the capacity for adult neurogenesis has diminished during the course of evolution. Furthermore, a local restriction is observed. In contrast to the widespread parenchymal neuronal migration by new neurons in the adult avian brain, there is a clear spatial restriction in the mammalian brain. Here, two regions of active proliferation generate neurons continuously throughout life, namely the subependymal zone of the lateral ventricle and the dentate gyrus of the hippocampal formation (see sections: "The dentate gyrus of the hippocampal formation" and "The subependymal zone," Fig. 1).

In the rat dentate gyrus, approximately 9000 new cells are generated daily, i.e. 0.75% of the existing granule cell population or one new cell per 130 mature granule neurons each day (Cameron and

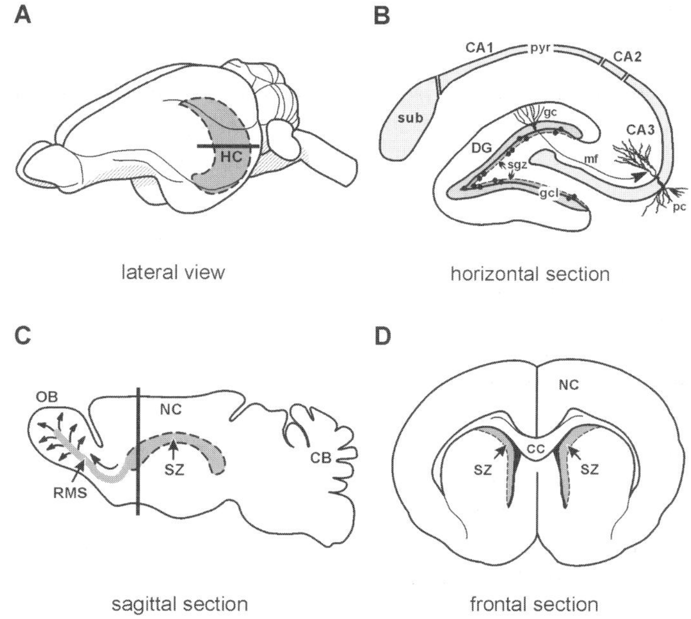

Fig. 1. Areas of neurogenesis in the adult mammalian brain. (A) Lateral view of a tree shrew brain displaying the localization of the hippocampus (HC). (B) Schematic drawing of horizontal section of the hippocampal formation, displaying the localization of neural progenitor cells within the germinative subgranular zone (sgz) adjacent to the granule cell layer (gcl) of the dentate gyrus (DG). Newly generated granule cells (gc) develop synapses on their cell bodies and dendrites extend axons and contact CA3 pyramidal cells (pc). (C) Schematic view of the ventricular subependymal zone (SZ), where multipotential self-renewing stem cells reside and spontaneously proliferate. Newly generated cells then migrate along the rostral migratory stream (RMS) toward the olfactory bulb (OB), into which they finally incorporate. (**D**) In the frontal section, the subependymal (or subventricular) zone appears adjacent to the wall of the lateral ventricle. NC: neocortex; CB: cerebellum; CC: corpus callosum; mf: mossy fibers.

McKay, 2001). This is a relatively small percentage when compared to the total number of granular neurons, which is estimated to be in the order of 10^6, depending on the strain and age of the animal and the counting method used (Amaral and Witter, 1989; West et al., 1991). Since only about 30% of the BrdU-positive cells survive permanently and differentiate into neurons (Kempermann et al., 2003), it was therefore estimated that approximately 6% of the total granule cell population is renewed within one month (Cameron and McKay, 2001).

Because of their presumed ability to form new synapses rapidly, newly generated neurons might be responsible for a greater proportion of new connections than the resident neurons, thereby having a proportionately larger influence on the physiology and functioning of the hippocampal formation. Morphologically indistinguishable from their neighbors with synapses on cell bodies and dendrites (Markakis and Gage, 1999), adult-generated granule neurons are thought to be integrated into the neuronal network. However, it is not clear whether the new neurons receive the same inputs as the ones produced during development.

Kornack and Rakic (1999) estimated that in the adult macaque hippocampus approximately 0.004% of the total granule cell population is generated per day, i.e. one new neuron per 24,000 existing granule neurons each day.

In the adult human hippocampus, cells do proliferate, yielding 50–400 BrdU-positive cell/mm^3 within the germinative zone (Erikson et al., 1998). Another study examining cell formation in the human hippocampal formation of newborn infants came to a much more cautious conclusion, stating that "neurogenesis in the adult primate brain may possibly be very limited or absent" (Seress et al., 2001). In their study, Seress et al. (2001) investigated postmortem hippocampal tissue of human infants from the 24th gestational week until the end of the first postnatal year. To detect cell proliferation, they used an endogenous mitotic marker Ki-67, instead of the exogenous marker BrdU. The authors found that newly generated cells represent $\sim 1\%$ of the granule cell population in the newborn infant hippocampus, but they claimed that the majority (80%) of them had glial morphology, while a maximum of 20% of the labeled cells were granule cells. Thus, it appears that in primates newborn neurons represent a much smaller fraction of the mature granule cells compared to rodents. Whether these newly generated cells in the human hippocampus become functional and, if so, their role remains unknown.

The dentate gyrus of the hippocampal formation

Adult neurogenesis is found in the dentate gyrus of the hippocampal formation (Fig. 1A, B), a unique brain structure that contains a high degree of structural plasticity and is intimately involved in the processing and storage of new information (Suzuki and Eichenbaum, 2000). During development of the hippocampus, a secondary germinal zone, separate from the ventricle wall, is formed along the border between the hilus and the granular cell layer, i.e. the subgranular zone (Altman and Bayer, 1990). Developmental studies have revealed that in both rodents and primates, the majority of neurons in the entorhinal cortex, subicular complex, and Ammon's horn are generated before birth. The major difference in hippocampal development between primates and rodents is the formation of the dentate gyrus. In rodents, approximately 85% of the dentate granule cells are formed postnatally, whereas similar numbers of cells are formed already prenatally in primates (Bayer, 1980; Rakic and Nowakowski, 1981).

In rhesus monkeys the formation of granule cells starts as early as week E38, but the majority of them are generated between E60–E120 (Rakic and Nowakowski, 1981). In humans, the granule cell layer first appears during the 13–14th gestational week, and similar to rhesus monkeys, the majority of the granule cells are formed throughout the second trimester (Humphrey, 1967; Seress et al., 2001). At term, about 50–60% of the adult number of granule cells are present in the dentate gyrus of the rhesus monkey (Keuker et al., 2003), and this estimate is approximately 70–85% in case of humans (L. Seress, personal communication).

Neurogenesis in the dentate gyrus continues throughout life, but displays a steady decline from early postnatal days (Kuhn et al., 1996). Nevertheless, neurogenesis is still detectable into very old age in both rodents and nonhuman primates (Kuhn et al., 1996; Gould, 1999b).

Importantly, neural stem cells that exhibit long-term self-renewal and multipotentiality can be isolated from the adult ventricular subependyma, whereas the adult dentate gyrus does not contain resident stem cells. Instead, separate neuronal and glial progenitors with only limited self-renewal capacity are the source of newly generated dentate neurons throughout adulthood (Seaberg and van der Kooy, 2002).

In the last few years, a theory has been developed that describes this population of cells (Gage, 1998, 2000). According to this theory, the cells are dividing asymmetrically resulting in daughter cells that enter the pathway leading to differentiation. The remaining cells are thought to form a sustaining pool of proliferating cells. From studies in rodents, it was estimated that about 50% of the BrdU-positive cells develop a neuronal phenotype, migrate into the granular cell layer, become morphologically indistinguishable from the other surrounding granule cells, and are capable of extending axonal projections along the mossy fiber tract to their natural target area, the hippocampal CA3 region (Cameron et al., 1993b; Markakis and Gage, 1999). About 15% of the BrdU-positive cells differentiate into glia cells. The remaining 35% do not show a clear neuronal or glial phenotype up to four weeks after cell division. Thus, the new neuron is the end product of a series of steps consisting of proliferation, survival, migration, differentiation, and establishment of functional connections with other neurons.

The subependymal zone

The subependymal zone of the adult lateral ventricle (Fig. 1C, D) gives rise to new neurons and is seen as a residual proliferative zone left over from the embryonic neural tube. Multipotential, self-renewing stem cells in the adult subependymal zone are the source of newly generated cells that pass through the rostral migratory stream, complete their last divisions, and incorporate into the olfactory bulb where they differentiate into interneurons (Luskin, 1993; Lois and Alvarez-Buylla, 1994; Menezes et al., 1995). Furthermore, Kornack and Rakic (2001a) provided evidence for the presence of spontaneous adult neurogenesis in the primate subependymal zone, which raises the possibility that an active subependymal zone/rostral migratory stream system is also present in humans.

The migration into the olfactory bulb has been described as a tangential chain migration, which does not require radial glia as a stationary partner (Lois et al., 1996). It is important to note that two neuronal phenotypes are generated in the olfactory bulb, GABAergic granule cells, and dopaminergic periglomerular interneurons (McLean and Shipley, 1988; Betarbet et al., 1996). Moreover, retroviral fate mapping studies confirmed that multipotential neural stem cells in the ventricle wall also generate glial cells (Goldman, 1995; Morshead et al., 1998). Therefore, within specific regions of the adult brain, all signals are present for instructing stem cells to generate glia and neurons with specific neurotransmitter phenotypes.

Neurogenesis in forebrain areas

In an unperturbed mammalian brain, neurogenesis is thought to be strictly limited to the dentate gyrus and the subependymal ventricle wall. Other areas of the brain contain an abundant population of proliferative precursors, but these cells generate only glia. The most common assumption is that neurogenic zones are defined by the location of the neural stem/progenitor cells in the adult. However, this may not be true. Immature progenitors in white matter generate oligodendrocytes in vivo and in vitro (Wolswijk and Noble, 1992; Horner et al., 2000), but recent studies show that these "glial" progenitors can actually make neurons in culture if treated with the appropriate growth factors (Palmer et al., 1999; Kondo and Raff, 2000; Nunes et al., 2003). This implies that the local environment, not the distribution of cells, is the key factor in defining where neurons are made.

A progenitor cell may "see" quite specific local environments depending on where it resides, and it would react appropriately to the local signals being produced by neighboring cells. In the dentate gyrus, the progenitor cell neighbors include other precursors, glia, granule cell neurons and, surprisingly, vascular endothelium. Recent studies show that the neural progenitor cells in the hippocampal subgranular zone proliferate in small clusters and that these clusters are located around the periphery of small capillaries (Palmer et al., 2000). Dividing endothelia are found within the core of many proliferating clusters and this angiogenic microenvironment appears to be relatively unique to the hippocampal subgranular zone and the ventricular subependymal zone. Precursors in white matter do not associate with vessels and it is possible, though not yet shown, that this vascular environment provides some of the cues necessary for stem cells to generate neurons.

There have been a few reports on spontaneous adult neurogenesis in neocortical structures. Kaplan (1981) reported on continuous neurogenesis in the three-month-old rat visual cortex. Recently, Gould and coworkers demonstrated that in adult nonhuman primates new neurons originating from the subependymal zone migrate through the white matter to the prefrontal, inferior temporal, and posterior parietal cortices, where they extend axons (Gould et al., 1999c, 2001). Furthermore, Bernier et al. (2002) reported on adult neurogenesis in the amygdala, piriform cortex, and adjoining inferior temporal cortex in squirrel monkeys and macaques. These findings invoked great skepticism in the scientific community (Nowakowski and Hayes, 2000; Rakic, 2002b). Clearly, because of the considerable conceptual and biomedical implications of this claim, it is essential to validate the reliability and robustness of this putative phenomenon. Accordingly, Kornack and Rakic (2001b) examined the proliferation and phenotypic differentiation of cells in the cerebrum of adult rhesus monkeys, but their findings do not substantiate the claim of neurogenesis in adult nonhuman primate neocortex. Further, similar negative findings were reported recently by Koketsu et al. (2003). Today there is an agreement that in nonhuman primates continuous adult neurogenesis takes place in the hippocampal dentate gyrus and in the subependymal zone, with cells from the latter region migrating to the olfactory bulb. However, additional studies are needed to clarify the extent of adult neurogenesis in other regions of the primate brain.

Neuromodulatory factors regulating adult hippocampal neurogenesis

Steroid hormones

Glucocorticoid hormones, such as cortisol (in primates) and corticosterone (in rats) secreted by the adrenal cortex, have been shown to inhibit the production of new granule neurons by suppressing the proliferation of granule cell precursors. The suppressive action of glucocorticoids seems to have biological relevance. Conditions associated with hypercortisolism, such as stress and ageing, are also associated with reduced granular cell production in rodents, tree shrews, and nonhuman primates (Gould et al., 1997, 1998; Cameron and McKay, 1999). On the other hand, removal of circulating glucocorticoids by adrenalectomy results in a clear increase in neurogenesis in the dentate gyrus of young adult rats (Cameron and Gould, 1994). The mechanism by which glucocorticoids inhibit adult neurogenesis is unknown. Glucocorticoids may act in part via mineralocorticoid receptors (Gass et al., 2000; Fischer et al., 2002). However, newborn cells in the subgranular zone do not express detectable levels of glucocorticoid or mineralocorticoid receptors (Cameron et al., 1993a).

In contrast to the suppressive effects of glucocorticoids on cell proliferation, the ovarian steroid estrogen has been shown to stimulate the proliferation of granule cell precursors in the dentate gyrus of adult female rats. This increase in the rate of cell proliferation occurs naturally across the rat estrous cycle, with maximal levels of cell production during proestrus, a time when estrogen levels are highest (Tanapat et al., 1999). This finding raises the interesting question regarding which mechanisms mediate the survival of newly generated cells in females as well as in males.

Several studies suggest that glutamate receptors may mediate the effects of glucocorticoids on hippocampal proliferation. *N*-methyl-aspartate (NMDA) receptor blockade or entorhinal cortical lesions (which deprive the hippocampus of a major glutamatergic input) prevent glucocorticoid-mediated decreases in adult neurogenesis (Cameron et al., 1995). NMDA receptor blockade also prevents adrenalectomy mediated increases in adult neurogenesis (Cameron et al., 1998). Alternatively, stress- and glucocorticoid-induced changes in other steroids, such as estrogen or testosterone, or in neurotransmitter systems (see following section), could conceivably be involved, especially because plasma stress hormone levels do not always correlate with the rate of adult hippocampal neurogenesis (Czeh et al., 2001, 2002).

Neurotransmitters

As indicated previously, another strong regulator of proliferation is the glutamatergic input to the granule layer of the dentate gyrus. NMDA receptor agonists and antagonists decrease or increase cell proliferation respectively (Cameron et al., 1995). Nitric oxide (NO) seems to be another negative regulator of cell proliferation in the hippocampal dentate gyrus (Packer et al., 2003).

Recent evidence supports the view that other neurotransmitter systems influence the production of new granule neurons in the dentate gyrus. For example, serotonin may stimulate granule cell production (Brezun and Daszuta, 2000a,b), whereas depletion of serotonin reduces neurogenesis (Brezun and Daszuta, 1999). In line with the growing body of evidence suggesting that mood stabilizers and antidepressants exert neurotrophic effects (Chen et al., 2000; Malberg et al., 2000), recent reports showed that different classes of antidepressant drugs, as well as the NMDA receptor antagonist MK-801, may prevent stress-induced suppression of adult hippocampal cell proliferation (Gould et al., 1997; Czeh et al., 2001; Lee et al., 2001; van der Hart et al., 2002; Malberg and Duman, 2003). It should be noted that to date, receptors neither for steroid hormones nor for glutamate or other neurotransmitters have been found on hippocampal progenitor cells. It will be interesting in future studies to identify the types of receptors expressed by these progenitor cells; such receptors should provide clues as to the intercellular signals that are critical in regulating adult neurogenesis.

Growth factors

Although the neuromodulatory signals discussed in the previous section trigger proliferation, the direct mitogenic stimulus to the progenitor cells appears to be mediated via growth factors, such as epidermal growth factor (EGF). Dentate precursor cells are known to express EGF receptors (Okano et al., 1996) and direct infusion of the growth factor into the dentate gyrus stimulates proliferation (Tanapat and Gould, 1997). Chronic infusion of EGF into the ventricular system of adult rats triggered neurogenesis, predominantly in the subependymal zone, and was nearly ineffective in stimulating proliferation in the subgranular zone (Kuhn et al., 1997). Moreover, when using this route of administration, EGF induced a prominent phenotypic shift that led to more astrocytes and fewer neurons. Intracerebroventricular administration of the vascular endothelial growth factor (VEGF) stimulates neurogenesis both in the subependymal zone and in the subgranular zone of the dentate gyrus, where the VEGFR2/Flk-1 receptor was colocalized with the immature neurons (Jin et al., 2002). Via peripheral application, selective induction of neurogenesis has also been achieved using FGF-2 (Wagner et al., 1999) or IGF-1 (Aberg et al., 2000). It is presently unclear whether growth factors can pass directly through the blood-brain barrier or whether other mechanisms such as angiogenesis are triggered, which have a secondary positive effect on neurogenesis.

Systemic influences on neurogenesis

Strain, age, and environment

Regulation of adult hippocampal neurogenesis has different regulatory levels, including cell proliferation, survival, and differentiation. Several studies have examined whether strain differences could possibly affect adult hippocampal neurogenesis, revealing that different aspects of adult hippocampal neurogenesis are differentially influenced by the genetic background (Kempermann et al., 1997a; Kempermann and Gage, 2002a,b).

Another internal factor that has a strong effect on adult hippocampal neurogenesis is age. With increasing age, the proliferation rate and ratio of cells differentiating into neurons decreases, whereas the percentage of surviving newborn cells remains constant (Kuhn et al., 1996; Gould et al., 1999b).

Interestingly, reduction of corticosteroid levels in aged rats could restore the naturally low rate of cell proliferation (Cameron and McKay, 1999). This may suggest that the neuronal progenitor population in the dentate gyrus in fact remains stable during ageing, but the spontaneous rate of neurogenesis is suppressed by elevated levels of corticosteroids.

External stimuli and environmental complexity enhance the survival of new neurons in the adult brain. This was first demonstrated in black-capped chickadees (Barnea and Nottebohm, 1994, 1996). Newly formed neurons were shown to survive longer in birds living in their natural environment compared to those animals living in captivity. Moreover, seasonal differences in the adult-generated hippocampal neurons correlated with seasonal changes in food storage and retrieval, behaviors that require spatial learning (Barnea and Nottebohm, 1994, 1996).

This line of research has been extended to the mammals by studies that analyzed the stimulatory effect of the environment on adult dentate neurogenesis (Kempermann et al., 1997b; van Praag et al., 2000). "Enriched environment" paradigms, where female mice are placed into housing conditions that are more similar to their natural surroundings, have been shown to increase neurogenesis by stimulating a better survival of the newly generated cells. The "enriched" animals also showed improved motor skills and better performance in learning tasks. Most importantly, the stimulatory effect on neurogenesis occurred at all ages, including senescence, even when the animals were housed under enriched conditions for only a few weeks (Kempermann et al., 1998).

Among the stimulatory factors within an enriched environment, voluntary physical activity appears to be a very strong activator of the proliferation of hippocampal progenitor cells (van Praag et al., 1999). The sole introduction of a running wheel into a standard laboratory home cage doubled hippocampal neurogenesis, suggesting that physiological parameters, such as blood flow, glucose uptake, and neovascularization, could be mediators of this effect.

Today it is evident that the mammalian hippocampus does indeed show an activity-dependent regulation of adult neurogenesis. These findings imply that data related to the number, regulation, and longevity of newly generated cells must be interpreted in light of the manner in which animals are housed. Relatively impoverished environments may also account for the detection of low numbers of new neurons in some regions, or the inability to find new neurons in some areas.

Environmental complexity is not only related to structural enrichment. Exposing mice to an odor-enriched environment can markedly increase the survival of newborn cells in the olfactory bulb (Rochefort et al., 2002). Furthermore, this effect is region specific, because enriched odor exposure does not influence hippocampal neurogenesis and the mice living in an odor-enriched environment display improved olfactory memory without a change in spatial learning performance (Rochefort et al., 2002). Another study using mutant mice demonstrated that deficits in the migration of olfactory-bulb neuron precursors result in an impairment of discrimination between odors, whereas general olfactory functions are unaltered (Gheusi et al., 2000).

Learning

Probably the most convincing evidence for a functional role of newborn neurons comes from studies of songbirds. For instance, in canaries, song is specific to males, which modify their repertoire by listening to congeners, and by adding, dropping, or altering song syllables. Some studies have shown that rates of neuron turnover in the high-vocal center (HVC), a nucleus involved in song production, are highest at times of year when canaries modify their songs, whereas song stability is greatest when canaries breed and recruitment of new neurons is at its lowest (Kirn et al., 1994; Alvarez-Buylla and Kirn, 1997). It has been hypothesized that neuronal replacement in the HVC provides a cellular basis for song plasticity in adult canaries (Alvarez-Buylla et al., 1992; Kirn and Nottebohm, 1993). A similar correlation has been reported among food-storing birds, which cache food and retrieve caches based on the spatial memory of their locations. In black-capped chickadees, recruitment of hippocampal neurons increases in the autumn at the peak of caching behavior (Barnea and Nottebohm, 1994, 1996). A study on mountain chickadees, however, could not replicate this correlation

(Pravosudov et al., 2002). Such results suggest that seasonal patterns of hippocampus-dependent learning (spatial memory for cache locations) might also correlate with differential patterns of neurogenesis, thereby affecting the number of neurons in the mammalian dentate gyrus.

In rodents, Gould et al. (1999a) were the first to demonstrate that learning can increase the survival of the newly generated granule cells. In naïve laboratory animals, the majority of the newborn neurons degenerate within two weeks of their production (Cameron et al., 1993b; Kempermann et al., 2003). It was demonstrated that if rats were trained to learn hippocampal-dependent tasks like place learning in a Morris water maze (Morris et al., 1982) or trace eyeblink conditioning (Solomon et al., 1986), then these learning experiences could significantly increase the survival of the newly generated granule cells (Gould et al., 1999a). This stimulating effect was attributed to learning, and not mere general experience, because exposure of animals to the same environment and conditions in the absence of learning had no effect on the number of new neurons. Furthermore, learning tasks that do not require the hippocampus, like delay eyeblink conditioning and cue learning in a Morris water maze (Schmaltz and Theios, 1972; Morris et al., 1982), did not alter the number of new neurons. These data clearly demonstrate that an enriched environment and certain types of learning are sufficient to enhance the number of new neurons in the dentate gyrus of adult rats and suggest the possibility that these new cells may play a role in learning.

In a seminal study, Shors et al. (2001) used a toxin, the DNA-methylating agent methylazoxymethanol acetate (MAM), to block cell proliferation. MAM dramatically diminished the number of adult-generated cells in the dentate gyrus of rats, which resulted in a significant impairment of hippocampal dependent, but not hippocampal-independent forms of associative memory formation. In this study, adult rats were trained with delay or trace eyeblink conditioning. This task requires the animals to learn to associate two stimuli, a conditioned stimulus (white noise) with an unconditioned stimulus (shock to the eyelid), and as the animal learns, it blinks in response to the conditioned stimulus. During delay conditioning, the conditioned stimulus and unconditioned stimulus overlap, and acquisition does not require an intact hippocampus (Schmaltz and Theios, 1972). During trace eyeblink conditioning, the animals have to associate the two stimuli separated temporally (during the trace interval), and acquisition of the trace conditioning requires intact hippocampus (Solomon et al., 1986; Weiss et al., 1999). Treating the rats with MAM for two weeks reduced the number of newly generated cells in the hippocampal dentate gyrus by 75–84%, yet the animals could rapidly acquire the hippocampal-independent task of delay conditioning, whereas their conditioned responses during trace conditioning were reduced by ~60% (Shors et al., 2001). Until now, this study is probably the most compelling evidence indicating the role of adult-generated neurons in mammalian learning.

Contrasting somewhat to the findings of Shors et al. (2001), in a recent experiment conducted in our laboratory we could not demonstrate any learning impairment after stress-induced suppression of dentate cytogenesis in a hippocampal-dependent task (Bartolomucci et al., 2002). We exposed adult male tree shrews to five weeks of daily psychosocial stress and tested them repeatedly on a holeboard apparatus using two different learning tasks devised to evaluate hippocampal-dependent and hippocampal-independent cognitive function. We could show that despite the fact that stress significantly suppressed hippocampal neurogenesis, learning was enhanced in a hippocampal-dependent task in which animals had to learn the spatial distribution of hidden food rewards. Importantly, this stress-induced improvement of learning was found only for the number of reference memory errors (opening of an unbaited hole), whereas working memory performance, as measured by the number of repeated choices, was not affected by the stress. However, it should be emphasized that in the study by Shors et al. (2001), the disruption of dentate cell proliferation was much more pronounced than in our experiment and, more importantly, the memory tasks employed in the two experiments account for different memory types – namely, the accurate timing of a learned response, and the acquisition and retrieval of a complex spatial distribution. These conflicting results seemed to be resolved by a more recent study from the same group (Shors et al., 2002). As in their previous study, they used again the antimitotic agent MAM, and tested whether reduction of cytogenesis affected learning and performance associated with different hippocampal-dependent tasks: spatial navigation learning in a Morris water maze, contextual fear conditioning, and trace fear conditioning. Reduction of new neurons in the adult hippocampus was associated with impaired performance in the trace fear conditioning paradigm but affected neither spatial navigation learning in the Morris water maze nor contextual fear conditioning. It is possible that spatial navigation learning can still occur with a very small percentage of new neurons. Another possibility is that only certain types of hippocampal-dependent learning require new cells, and spatial navigation learning is not one of those types. Nevertheless, a study using genetically different mouse strains did show a significant correlation between adult hippocampal neurogenesis and parameters describing the acquisition of the Morris water maze task (Kempermann and Gage, 2002a).

A recent study used fractionated brain irradiation to block the formation of new neurons in the dentate gyrus (Madsen et al., 2003). Blockade of dentate neurogenesis induced significant impairment in a place recognition task using a T-maze, but lesioned animals performed equally to controls when tested in the Morris water maze (Madsen et al., 2003).

Furthermore, several recent studies made an attempt to correlate hippocampal cell proliferation with water maze performance in aged rats, but majority of them failed to support a direct relationship of adult neurogenesis with spatial learning and memory capability in the Morris water maze (Merrill et al., 2003; Bizon and Gallagher, 2003), except one (Drapeau et al., 2003).

Yet another hypothesis suggested that the functional role of these adult-generated granule cells does not relate to the acquisition of new memories. Instead, they may play a role in the periodic clearance of outdated hippocampal memory traces after cortical memory consolidation, thereby ensuring that the hippocampus is continuously available to process new memories (Feng et al., 2001).

Apparently, the outcome of experiments investigating the role of adult hippocampal neurogenesis in learning is largely dependent on the approach, i.e. which specific learning paradigm is used for evaluating hippocampal-dependent performance. The functional role of the hippocampus is broad (Suzuki and Clayton, 2000) and a great variety of learning tests can be used for assessment, each of which may be specific to a certain type of associative learning. For example, some groups put extra effort into their experimental paradigm by not using animals that were housed and tested in laboratory conditions; rather, they investigated wild animals living in their natural environment. A well-designed study on wild eastern gray squirrels (a long-lived mammal that scatter-hoards food) found negative results when testing whether seasonal variations in spatial memory processing (i.e. increased processing during caching season in the autumn) correlate with changes in neurogenesis and total granule cell number in the hippocampal dentate gyrus (Lavenex et al., 2000).

Acute and chronic stressful experience

Collectively, the above-discussed observations demonstrate that cell proliferation in the dentate gyrus can be modulated by environmental signals and experience. However, environmental signals can also be detrimental to the functioning of neurogenesis. Stressful experiences are known to activate the hypothalamic–pituitary–adrenal (HPA) axis and increase levels of circulating adrenocortical steroid hormones, cortisol or corticosterone.

There is compelling evidence demonstrating that both acute and chronic stressful experience can affect the production of new hippocampal granule cells by suppressing both the proliferation rate of precursor cells as well as the survival rate of the daughter cells. In collaboration with Elizabeth Gould, we were the first to demonstrate that acute psychosocial stress can dramatically suppress cell proliferation in the hippocampal dentate gyrus of adult tree shrews (Gould et al., 1997). In nonhuman primates, we found a similar effect of acute stressful experience (Gould et al., 1998) suggesting that these characteristics may be common to most mammalian species. In rats, a single exposure to a predator odor resulted in a marked reduction of cell proliferation (Tanapat et al., 2001). Chronic psychosocial stress seems to have a relatively mild suppressive effect on neurogenesis.

Subjecting tree shrews to five weeks of daily psychosocial stress resulted in a ~30% decrease of dentate cell proliferation (Czeh et al., 2001), and similar moderate effects were observed in rats (Czeh et al., 2002). Therefore, one may speculate that acute stress seems to cause a more robust inhibition of dentate cell proliferation, and proliferating cells may "habituate" after chronic stress and therefore are not as responsive to the inhibitory effects of stress hormones. However, once again the effect of stress on adult hippocampal neurogenesis may largely depend on the applied experimental stress paradigm. This could explain the findings of a recent study in which acute restraint stress of rats did not affect the proliferation rate of dentate precursor cells, whereas three weeks of daily restraint stress suppressed proliferation by ~20%, and six weeks of chronic stress resulted in a substantial ~50% reduction of neurogenesis, producing a significant decrement in the total number of granule cells by 13% (Pham et al., 2003). It should be emphasized that comparisons across studies are often difficult, because of the different BrdU-labeling schedules.

Importantly, stress exposure suppresses proliferation in the dentate gyrus, while proliferation in the subependymal zone is unaffected, indicating that the stress-induced decrease in cytogenesis is not due to a nonspecific effect, for example, decreased bioavailability of the marker molecule BrdU. Furthermore, there is compelling evidence that not only cell proliferation but also the survival of the newborn granule neurons are suppressed by stress (Czeh et al., 2002; Pham et al., 2003).

Long-term effect of prenatal stress

It has become increasingly evident that the antecedents of many illnesses begin in fetal life, and further, that prenatal conditions can bias us toward either health or disease in the postpartum period. For example, several retrospective studies have confirmed that chronic maternal stress during pregnancy significantly increases the likelihood of disturbed physical and/or psychological development of the child (Jones and Tauscher, 1978; Meijer, 1985; Lou et al., 1994). An association has also been described between maternal stress and an increased probability of schizophrenia (Huttunen and Niskanen, 1978; Myhrman et al., 1996; van Os and Selten, 1998) and depressive symptomatology (Watson et al., 1999; Brown et al., 2000) in prenatally stressed offspring.

Studies investigating the long-term effect of prenatal stress on adult hippocampal neurogenesis are scarce. Lemaire et al. (2000) performed an extensive study on the effect of prenatal stress in rats. They restrained pregnant female rats during late pregnancy for 45 min three times a day, at the same time exposing them to bright light; adult dentate neurogenesis was later evaluated at different ages in the offspring. They reported that prenatal stress induced a life span reduction of hippocampal neurogenesis and decline in total granule cell numbers, accompanied by learning impairment in hippocampal-related spatial tasks (Lemaire et al., 2000).

In a recent study that investigated whether prenatal stress can alter neural, hormonal, and behavioral status in nonhuman primates, pregnant rhesus monkeys were acutely stressed on a daily basis for 25% of their 24-week gestation using an acoustic startle protocol (Coe et al., 2003). At two-to-three years of age, hippocampal volume, cytogenesis in the dentate gyrus, and cortisol levels were evaluated in the offspring generated from stressed and control pregnancies. Prenatal stress, both early and late in pregnancy, resulted in reduced hippocampal volume and an inhibition of dentate gyrus neurogenesis. These changes were associated with higher cortisol levels, a more rapid escape from dexamethasone suppression, lower levels of exploration, and higher levels of motor behavior. These findings indicate that the prenatal environment can alter behavior, deregulate neuroendocrine systems, and affect the hippocampal structure of primates in a persistent manner. Moreover, these data strengthen pathophysiological hypotheses that propose an early neurodevelopmental origin for psychopathological vulnerability in adulthood.

Injury-induced adult neurogenesis

Pathological conditions, such as ischemia or epileptic seizures, result in a marked increase of neurogenesis in the adult hippocampus. Enhanced dentate neurogenesis after experimental stroke may represent a natural form of neural self-repair (Liu et al., 1998; Kee et al., 2001; Jin et al., 2001; Yagita et al., 2001), whereas seizures-induced neurogenesis may have either a reparative role or alternatively it can promote abnormal hyperexcitability (Parent et al., 1997, 1998; Scott et al., 1998; Blumcke et al., 2001; Parent, 2002). Ethanol exposure also seems to affect neurogenesis in the adult dentate gyrus, although so far the experimental results are somewhat conflicting. Some studies report decreased neurogenesis after ethanol exposure (Nixon and Crews, 2002; Herrera et al., 2003), whereas others demonstrate the opposite effect (Pawlak et al., 2002; Zharkovsky et al., 2003).

Functional role of adult-generated neurons

The evolutionary conservation of adult neurogenesis in the mammalian brain suggests that it is of fundamental biological importance. Newly generated dentate granule cells become incorporated into the granule cell layer, attain the morphological and biochemical characteristics of neurons (Cameron et al., 1993b; Okano et al., 1993), develop synapses on their cell bodies, and dendrites (Kaplan and Hinds, 1977; Kaplan and Bell, 1984), extend axons into the CA3 region (Stanfield and Trice, 1988; Markakis and Gage, 1999), and generate action potentials (van Praag et al., 2002). They show distinct morphological and electrophysiological properties compared to mature granule cells (Liu et al., 2000), present a lower threshold for induction of long-term potentiation (LTP) and display robust LTP (Wang et al., 2000). Furthermore, the fact that continuous neurogenesis takes place in the hippocampal formation raises the possibility that newborn neurons could participate in learning. Indeed, a continually rejuvenating population of new neurons seems well suited for the proposed transient role of the hippocampal formation in information storage (Squire, 1992). The question of what the individual new neurons are used for is difficult to address experimentally. The major difficulty is that to date, no specific agent is available that can selectively block neurogenesis. Any of the currently available treatments that suppress adult neurogenesis may affect other brain regions or other processes in the same region. Probably, an elegant way to resolve this exciting question would be to develop a mutant animal in which neurogenesis could be blocked in a specific region at particular times.

Acknowledgment

We are grateful to J. Keuker for her help in the preparation of the figures.

References

Aberg, M.A., Aberg, N.D., Hedbacker, H., Oscarsson, J. and Eriksson, P.S. (2000) Peripheral infusion of IGF-I selectively induces neurogenesis in the adult rat hippocampus. J. Neurosci., 20: 2896–2903.

Altman, J. and Bayer, S.A. (1990) Migration and distribution of two populations of hippocampal granule cell precursors during the perinatal and postnatal periods. J. Comp. Neurol., 301: 365–381.

Altman, J. and Das, G.D. (1965) Autoradiographic and histological evidence of postnatal hippocampal neurogenesis in rats. J. Comp. Neurol., 124: 319–335.

Alvarez-Buylla, A. and Kirn, J.R. (1997) Birth, migration, incorporation, and death of vocal control neurons in adult songbirds. J. Neurobiol., 33: 585–601.

Alvarez-Buylla, A., Theelen, M. and Nottebohm, F. (1988) Birth of projection neurons in the higher vocal center of the canary forebrain before, during, and after song learning. Proc. Natl. Acad. Sci. USA, 85: 8722–8726.

Alvarez-Buylla, A., Ling, C.Y. and Nottebohm, F. (1992) High vocal center growth and its relation to neurogenesis, neuronal replacement and song acquisition in juvenile canaries. J. Neurobiol., 23: 396–406.

Amaral, D.G. and Witter, M.P. (1989) The three-dimensional organization of the hippocampal formation: a review of anatomical data. Neuroscience, 31: 571–591.

Barnea, A. and Nottebohm, F. (1994) Seasonal recruitment of hippocampal neurons in adult free-ranging black-capped chickadees. Proc. Natl. Acad. Sci. USA, 91: 11217–11221.

Barnea, A. and Nottebohm, F. (1996) Recruitment and replacement of hippocampal neurons in young and adult chickadees: an addition to the theory of hippocampal learning. Proc. Natl. Acad. Sci. USA, 93: 714–718.

Bartolomucci, A., de Biurrun, G., Czeh, B., van Kampen, M. and Fuchs, E. (2002) Selective enhancement of spatial learning under chronic psychosocial stress. Eur. J. Neurosci., 15: 1863–1866.

Bayer, S.A. (1980) Development of the hippocampal region in the rat. I. Neurogenesis examined with 3H-thymidine autoradiography. J. Comp. Neurol., 190: 87–114.

Bernier, P.J., Bedard, A., Vinet, J., Levesque, M. and Parent, A. (2002) Newly generated neurons in the amygdala and adjoining cortex of adult primates. Proc. Natl. Acad. Sci. USA, 99: 11464–11469.

Betarbet, R., Zigova, T., Bakay, R.A. and Luskin, M.B. (1996) Dopaminergic and GABAergic interneurons of the olfactory bulb are derived from the neonatal subventricular zone. Int. J. Dev. Neurosci., 14: 921–930.

Bizon, J.L. and Gallagher, M. (2003) Production of new cells in the rat dentate gyrus over the lifespan: relation to cognitive decline. Eur. J. Neurosci., 18: 215–219.

Blumcke, I., Schewe, J.C., Normann, S., Brustle, O., Schramm, J., Elger, C.E. and Wiestler, O.D. (2001) Increase of nestin-immunoreactive neural precursor cells in the dentate gyrus of pediatric patients with early-onset temporal lobe epilepsy. Hippocampus, 11: 311–321.

Brezun, J.M. and Daszuta, A. (1999) Depletion in serotonin decreases neurogenesis in the dentate gyrus and the subventricular zone of adult rats. Neuroscience, 89: 999–1002.

Brezun, J.M. and Daszuta, A. (2000a) Serotonin may stimulate granule cell proliferation in the adult hippocampus, as observed in rats grafted with foetal raphe neurons. Eur. J. Neurosci., 12: 391–396.

Brezun, J.M. and Daszuta, A. (2000b) Serotonergic reinnervation reverses lesion-induced decreases in PSA-NCAM labeling and proliferation of hippocampal cells in adult rats. Hippocampus, 10: 37–46.

Brown, A.S., van Os, J., Driessens, C., Hoek, H.W. and Susser, E.S. (2000) Further evidence of relation between prenatal famine and major affective disorder. Am. J. Psychiatry, 157: 190–195.

Cajal, R.S. (1928) Degeneration and regeneration of the nervous system. May, R.M. trans., Vol. II. Oxford University Press, London: Humphrey Milford, p. 750.

Cameron, H.A. and Gould, E. (1994) Adult neurogenesis is regulated by adrenal steroids in the dentate gyrus. Neuroscience, 61: 203–209.

Cameron, H.A. and McKay, R.D. (1999) Restoring production of hippocampal neurons in old age. Nat. Neurosci., 2: 894–897.

Cameron, H.A. and McKay, R.D. (2001) Adult neurogenesis produces a large pool of new granule cells in the dentate gyrus. J. Comp. Neurol., 435: 406–417.

Cameron, H.A., Woolley, C.S. and Gould, E. (1993a) Adrenal steroid receptor immunoreactivity in cells born in the adult rat dentate gyrus. Brain Res., 611: 342–346.

Cameron, H.A., Woolley, C.S., McEwen, B.S. and Gould, E. (1993b) Differentiation of newly born neurons and glia in the dentate gyrus of the adult rat. Neuroscience, 56: 337–344.

Cameron, H.A., McEwen, B.S. and Gould, E. (1995) Regulation of adult neurogenesis by excitatory input and NMDA receptor activation in the dentate gyrus. J. Neurosci., 15: 4687–4692.

Cameron, H.A., Tanapat, P. and Gould, E. (1998) Adrenal steroids and N-methyl-D-aspartate receptor activation regulate neurogenesis in the dentate gyrus of adult rats through a common pathway. Neuroscience, 82: 349–354.

Chen, G., Rajkowska, G., Du, F., Seraji-Bozorgzad, N. and Manji, H.K. (2000) Enhancement of hippocampal neurogenesis by lithium. J. Neurochem., 75: 1729–1734.

Coe, C.L., Kramer, M., Czeh, B., Gould, E., Reeves, A.J., Kirschbaum, C. and Fuchs, E. (2003) Prenatal stress diminishes neurogenesis in the dentate gyrus of juvenile rhesus monkeys. Biol. Psychiatry, 54: 1025–1034.

Czeh, B., Michaelis, T., Watanabe, T., Frahm, J., de Biurrun, G., van Kampen, M., Bartolomucci, A. and Fuchs, E. (2001) Stress-induced changes in cerebral metabolites, hippocampal volume, and cell proliferation are prevented by antidepressant treatment with tianeptine. Proc. Natl. Acad. Sci. USA, 98: 12796–12801.

Czeh, B., Welt, T., Fischer, A.K., Erhardt, A., Schmitt, W., Muller, M.B., Toschi, N., Fuchs, E. and Keck, M.E. (2002) Chronic psychosocial stress and concomitant repetitive transcranial magnetic stimulation: effects on stress hormone levels and adult hippocampal neurogenesis. Biol. Psychiatry, 52: 1057–1065.

Drapeau, E., Mayo, W., Aurousseau, C., Le Moal, M., Piazza, P.V. and Abrous, D.N. (2003) Spatial memory performances of aged rats in the water maze predict levels of hippocampal neurogenesis. Proc. Natl. Acad. Sci. USA, 100: 14385–14390.

Eckenhoff, M.F. and Rakic, P. (1988) Nature and fate of proliferative cells in the hippocampal dentate gyrus during the life span of the rhesus monkey. J. Neurosci., 8: 2729–2747.

Eriksson, P.S., Perfilieva, E., Bjork-Eriksson, T., Alborn, A.M., Nordborg, C., Peterson, D.A. and Gage, F.H. (1998) Neurogenesis in the adult human hippocampus. Nat. Med., 4: 1313–1317.

Feng, R., Rampon, C., Tang, Y.P., Shrom, D., Jin, J., Kyin, M., Sopher, B., Miller, M.W., Ware, C.B., Martin, G.M., Kim, S.H., Langdon, R.B., Sisodia, S.S. and Tsien, J.Z. (2001) Deficient neurogenesis in forebrain-specific presenilin-1 knockout mice is associated with reduced clearance of hippocampal memory traces. Neuron, 32: 911–926.

Fischer, A.K., von Rosenstiel, P., Fuchs, E., Goula, D., Almeida, O.F. and Czeh, B. (2002) The prototypic mineralocorticoid receptor agonist aldosterone influences neurogenesis in the dentate gyrus of the adrenalectomized rat. Brain Res., 947: 290–293.

Gage, F.H. (1998) Stem cells of the central nervous system. Curr. Opin. Neurobiol., 8: 671–676.

Gage, F.H. (2000) Mammalian neural stem cells. Science, 287: 1433–1438.

Gass, P., Kretz, O., Wolfe, D.P., Berger, S., Tronche, F., Reichardt, H.M., Kellendonk, C., Lipp, H.P., Schmid, W. and Schutz, G. (2000) Genetic disruption of mineralocorticoid receptor leads to impaired neurogenesis and granule cell degeneration in the hippocampus of adult mice. EMBO Rep., 1: 447–451.

Gheusi, G., Cremer, H., McLean, H., Chazal, G., Vincent, J.D. and Lledo, P.M. (2000) Importance of newly generated neurons in the adult olfactory bulb for odor discrimination. Proc. Natl. Acad. Sci. USA, 97: 1823–1828.

Goldman, J.E. (1995) Lineage, migration, and fate determination of postnatal subventricular zone cells in the mammalian CNS. J. Neurooncol., 24: 61–64.

Goldman, S.A. (1998) Adult neurogenesis: from canaries to the clinic. J. Neurobiol., 36: 267–286.

Goldman, S.A. and Nottebohm, F. (1983) Neuronal production, migration, and differentiation in a vocal control nucleus of the adult female canary brain. Proc. Natl. Acad. Sci. USA, 80: 2390–2394.

Gould, E. and Gross, C.G. (2002) Neurogenesis in adult mammals: some progress and problems. J. Neurosci., 22: 619–623.

Gould, E., McEwen, B.S., Tanapat, P., Galea, L.A. and Fuchs, E. (1997) Neurogenesis in the dentate gyrus of the adult tree shrew is regulated by psychosocial stress and NMDA receptor activation. J. Neurosci., 17: 2492–2498.

Gould, E., Tanapat, P., McEwen, B.S., Flügge, G. and Fuchs, E. (1998) Proliferation of granule cell precursors in the dentate gyrus of adult monkeys is diminished by stress. Proc. Natl. Acad. Sci. USA, 95: 3168–3171.

Gould, E., Beylin, A., Tanapat, P., Reeves, A. and Shors, T.J. (1999a) Learning enhances adult neurogenesis in the hippocampal formation. Nat. Neurosci., 2: 260–265.

Gould, E., Reeves, A.J., Fallah, M., Tanapat, P., Gross, C.G. and Fuchs, E. (1999b) Hippocampal neurogenesis in adult Old World primates. Proc. Natl. Acad. Sci. USA, 96: 5263–5267.

Gould, E., Reeves, A.J., Graziano, M.S. and Gross, C.G. (1999c) Neurogenesis in the neocortex of adult primates. Science, 286: 548–552.

Gould, E., Vail, N., Wagers, M. and Gross, C.G. (2001) Adult-generated hippocampal and neocortical neurons in macaques have a transient existence. Proc. Natl. Acad. Sci. USA, 98: 10910–10917.

Hayes, N.L. and Nowakowski, R.S. (2002) Dynamics of cell proliferation in the adult dentate gyrus of two inbred strains of mice. Brain Res. Dev. Brain Res., 134: 77–85.

Herrera, D.G., Yague, A.G., Johnsen-Soriano, S., Bosch-Morell, F., Collado-Morente, L., Muriach, M., Romero, F.J. and Garcia-Verdugo, J.M. (2003) Selective impairment of hippocampal neurogenesis by chronic alcoholism: protective effects of an antioxidant. Proc. Natl. Acad. Sci. USA, 100: 7919–7924.

Horner, P.J., Power, A.E., Kempermann, G., Kuhn, H.G., Palmer, T.D., Winkler, J., Thal, L.J. and Gage, F.H. (2000) Proliferation and differentiation of progenitor cells throughout the intact adult rat spinal cord. J. Neurosci., 20: 2218–2228.

Humphrey, T. (1967) The development of the human hippocampal fissure. J. Anat., 101: 655–676.

Huttunen, M.O. and Niskanen, P. (1978) Prenatal loss of father and psychiatric disorders. Arch. Gen. Psychiatry, 35: 429–431.

Jin, K., Minami, M., Lan, J.Q., Mao, X.O., Batteur, S., Simon, R.P. and Greenberg, D.A. (2001) Neurogenesis in dentate subgranular zone and rostral subventricular zone after focal cerebral ischemia in the rat. Proc. Natl. Acad. Sci. USA, 98: 4710–4715.

Jin, K., Zhu, Y., Sun, Y., Mao, X.O., Xie, L. and Greenberg, D.A. (2002) Vascular endothelial growth factor (VEGF) stimulates neurogenesis in vitro and in vivo. Proc. Natl. Acad. Sci. USA, 99: 11946–11950.

Jones, F. and Tauscher, J. (1978) Residence under an airport-landing pattern as a factor in teratism. Arch. Environ. Health, 33: 10–12.

Kaplan, M.S. (1981) Neurogenesis in the 3-month-old rat visual cortex. J. Comp. Neurol., 195: 323–338.

Kaplan, M.S. and Bell, D.H. (1984) Mitotic neuroblasts in the 9-day-old and 11-month-old rodent hippocampus. J. Neurosci., 4: 1429–1441.

Kaplan, M.S. and Hinds, J.W. (1977) Neurogenesis in the adult rat: electron microscopic analysis of light radioautographs. Science, 197: 1092–1094.

Kee, N.J., Preston, E. and Wojtowicz, J.M. (2001) Enhanced neurogenesis after transient global ischemia in the dentate gyrus of the rat. Exp. Brain Res., 136: 313–320.

Kempermann, G. and Gage, F.H. (2002a) Genetic determinants of adult hippocampal neurogenesis correlate with acquisition, but not probe trial performance, in the water maze task. Eur. J. Neurosci., 16: 129–136.

Kempermann, G. and Gage, F.H. (2002b) Genetic influence on phenotypic differentiation in adult hippocampal neurogenesis. Brain Res. Dev. Brain Res., 134: 1–12.

Kempermann, G., Kuhn, H.G. and Gage, F.H. (1997a) Genetic influence on neurogenesis in the dentate gyrus of adult mice. Proc. Natl. Acad. Sci. USA, 94: 10409–10414.

Kempermann, G., Kuhn, H.G. and Gage, F.H. (1997b) More hippocampal neurons in adult mice living in an enriched environment. Nature, 386: 493–495.

Kempermann, G., Kuhn, H.G. and Gage, F.H. (1998) Experience-induced neurogenesis in the senescent dentate gyrus. J. Neurosci., 18: 3206–3212.

Kempermann, G., Gast, D., Kronenberg, G., Yamaguchi, M. and Gage, F.H. (2003) Early determination and long-term persistence of adult-generated new neurons in the hippocampus of mice. Development, 130: 391–399.

Keuker, J.I., Luiten, P.G. and Fuchs, E. (2003) Preservation of hippocampal neuron numbers in aged rhesus monkeys. Neurobiol. Aging, 24: 157–165.

Kirn, J.R. and Nottebohm, F. (1993) Direct evidence for loss and replacement of projection neurons in adult canary brain. J. Neurosci., 13: 1654–1663.

Kirn, J.R., O'Loughlin, B., Kasparian, S. and Nottebohm, F. (1994) Cell death and neuronal recruitment in the high vocal center of adult male canaries are temporally related to changes in song. Proc. Natl. Acad. Sci. USA, 91: 7844–7848.

Koketsu, D., Mikami, A., Miyamoto, Y. and Hisatsune, T. (2003) Nonrenewal of neurons in the cerebral neocortex of adult macaque monkeys. J. Neurosci., 23: 937–942.

Kondo, T. and Raff, M. (2000) Oligodendrocyte precursor cells reprogrammed to become multipotential CNS stem cells. Science, 289: 1754–1757.

Kornack, D.R. and Rakic, P. (1999) Continuation of neurogenesis in the hippocampus of the adult macaque monkey. Proc. Natl. Acad. Sci. USA, 96: 5768–5773.

Kornack, D.R. and Rakic, P. (2001a) The generation, migration, and differentiation of olfactory neurons in the adult primate brain. Proc. Natl. Acad. Sci. USA, 98: 4752–4757.

Kornack, D.R. and Rakic, P. (2001b) Cell proliferation without neurogenesis in adult primate neocortex. Science, 294: 2127–2130.

Kuhn, H.G., Dickinson-Anson, H. and Gage, F.H. (1996) Neurogenesis in the dentate gyrus of the adult rat: age-related decrease of neuronal progenitor proliferation. J. Neurosci., 16: 2027–2033.

Kuhn, H.G., Winkler, J., Kempermann, G., Thal, L.J. and Gage, F.H. (1997) Epidermal growth factor and fibroblast growth factor-2 have different effects on neural progenitors in the adult rat brain. J. Neurosci., 17: 5820–5829.

Lavenex, P., Steele, M.A. and Jacobs, L.F. (2000) The seasonal pattern of cell proliferation and neuron number in the dentate gyrus of wild adult eastern grey squirrels. Eur. J. Neurosci., 12: 643–648.

Lee, H.J., Kim, J.W., Yim, S.V., Kim, M.J., Kim, S.A., Kim, Y.J., Kim, C.J. and Chung, J.H. (2001) Fluoxetine enhances cell proliferation and prevents apoptosis in dentate gyrus of maternally separated rats. Mol. Psychiatry, 6: 610, 725–728.

Lemaire, V., Koehl, M., Le Moal, M. and Abrous, D.N. (2000) Prenatal stress produces learning deficits associated with an inhibition of neurogenesis in the hippocampus. Proc. Natl. Acad. Sci. USA, 97: 11032–11037.

Liu, J., Solway, K., Messing, R.O. and Sharp, F.R. (1998) Increased neurogenesis in the dentate gyrus after transient global ischemia in gerbils. J. Neurosci., 18: 7768–7778.

Liu, X., Tilwalli, S., Ye, G., Lio, P.A., Pasternak, J.F. and Trommer, B.L. (2000) Morphologic and electrophysiologic maturation in developing dentate gyrus granule cells. Brain Res., 856: 202–212.

Lois, C. and Alvarez-Buylla, A. (1994) Long-distance neuronal migration in the adult mammalian brain. Science, 264: 1145–1148.

Lois, C., Garcia-Verdugo, J.M. and Alvarez-Buylla, A. (1996) Chain migration of neuronal precursors. Science, 271: 978–981.

Lou, H.C., Hansen, D., Nordentoft, M., Pryds, O., Jensen, F., Nim, J. and Hemmingsen, R. (1994) Prenatal stressors of human life affect fetal brain development. Dev. Med. Child Neurol., 36: 826–832.

Luskin, M.B. (1993) Restricted proliferation and migration of postnatally generated neurons derived from the forebrain subventricular zone. Neuron, 11: 173–189.

Madsen, T.M., Kristjansen, P.E., Bolwig, T.G. and Wortwein, G. (2003) Arrested neuronal proliferation and impaired hippocampal function following fractionated brain irradiation in the adult rat. Neuroscience, 119: 635–642.

Malberg, J.E., Duman, R.S. (2003) Cell proliferation in adult hippocampus is decreased by inescapable stress: reversal by fluoxetine treatment. Neuropsychopharmacology, 28: 1562–1571.

Malberg, J.E., Eisch, A.J., Nestler, E.J. and Duman, R.S. (2000) Chronic antidepressant treatment increases neurogenesis in adult rat hippocampus. J. Neurosci., 20: 9104–9110.

Markakis, E.A. and Gage, F.H. (1999) Adult-generated neurons in the dentate gyrus send axonal projections to field CA3 and are surrounded by synaptic vesicles. J. Comp. Neurol., 406: 449–460.

McLean, J.H. and Shipley, M.T. (1988) Postmitotic, postmigrational expression of tyrosine hydroxylase in olfactory bulb dopaminergic neurons. J. Neurosci., 8: 3658–3669.

Meijer, A. (1985) Child psychiatric sequelae of maternal war stress. Acta Psychiatr. Scand., 72: 505–511.

Menezes, J.R., Smith, C.M., Nelson, K.C. and Luskin, M.B. (1995) The division of neuronal progenitor cells during migration in the neonatal mammalian forebrain. Mol. Cell Neurosci., 6: 496–508.

Merrill, D.A., Karim, R., Darraq, M., Chiba, A.A. and Tuszynski, M.H. (2003) Hippocampal cell genesis does not correlate with spatial learning ability in aged rats. J. Comp. Neurol., 459: 201–207.

Miller, M.W. and Nowakowski, R.S. (1988) Use of bromodeoxyuridine-immunohistochemistry to examine the proliferation, migration and time of origin of cells in the central nervous system. Brain Res., 457: 44–52.

Morris, R.G., Garrud, P., Rawlins, J.N. and O'Keefe, J. (1982) Place navigation impaired in rats with hippocampal lesions. Nature, 297: 681–683.

Morshead, C.M., Craig, C.G. and van der Kooy, D. (1998) In vivo clonal analyses reveal the properties of endogenous neural stem cell proliferation in the adult mammalian forebrain. Development, 125: 2251–2261.

Myhrman, A., Rantakallio, P., Isohanni, M., Jones, P. and Partanen, U. (1996) Unwantedness of a pregnancy and schizophrenia in the child. Br. J. Psychiatry, 169: 637–640.

Nixon, K. and Crews, F.T. (2002) Binge ethanol exposure decreases neurogenesis in adult rat hippocampus. J. Neurochem., 83: 1087–1093.

Nowakowski, R.S. and Hayes, N.L. (2000) New neurons: extraordinary evidence or extraordinary conclusion? Science, 288: 771.

Nunes, M.C., Roy, N.S., Keyoung, H.M., Goodman, R.R., McKhann, G., Jiang, L., Kang, J., Nedergaard, M. and Goldman, S.A. (2003) Identification and isolation of multipotential neural progenitor cells from the subcortical white matter of the adult human brain. Nat. Med., 9: 439–447.

Okano, H.J., Pfaff, D.W. and Gibbs, R.B. (1993) RB and Cdc2 expression in brain: correlations with 3H-thymidine incorporation and neurogenesis. J. Neurosci., 13: 2930–2938.

Okano, H.J., Pfaff, D.W. and Gibbs, R.B. (1996) Expression of EGFR-, p75NGFR-, and PSTAIR (cdc2)-like immunoreactivity by proliferating cells in the adult rat hippocampal formation and forebrain. Dev. Neurosci., 18: 199–209.

Packer, M.A., Stasiv, Y., Benraiss, A., Chmielnicki, E., Grinberg, A., Westphal, H., Goldman, S.A. and Enikolopov, G. (2003) Nitric oxide negatively regulates mammalian adult neurogenesis. Proc. Natl. Acad. Sci. USA, 100: 9566–9571.

Palmer, T.D., Markakis, E.A., Willhoite, A.R., Safar, F. and Gage, F.H. (1999) Fibroblast growth factor-2 activates a latent neurogenic program in neural stem cells from diverse regions of the adult CNS. J. Neurosci., 19: 8487–8497.

Palmer, T.D., Willhoite, A.R. and Gage, F.H. (2000) Vascular niche for adult hippocampal neurogenesis. J. Comp. Neurol., 425: 479–494.

Parent, J.M. (2002) The role of seizure-induced neurogenesis in epileptogenesis and brain repair. Epilepsy Res., 50: 179–189.

Parent, J.M., Yu, T.W., Leibowitz, R.T., Geschwind, D.H., Sloviter, R.S. and Lowenstein, D.H. (1997) Dentate granule cell neurogenesis is increased by seizures and contributes to aberrant network reorganization in the adult rat hippocampus. J. Neurosci., 17: 3727–3738.

Parent, J.M., Janumpalli, S., McNamara, J.O. and Lowenstein, D.H. (1998) Increased dentate granule cell neurogenesis following amygdala kindling in the adult rat. Neurosci. Lett., 247: 9–12.

Pawlak, R., Skrzypiec, A., Sulkowski, S. and Buczko, W. (2002) Ethanol-induced neurotoxicity is counterbalanced by increased cell proliferation in mouse dentate gyrus. Neurosci. Lett., 327: 83–86.

Pham, K., Nacher, J., Hof, P.R. and McEwen, B.S. (2003) Repeated restraint stress suppresses neurogenesis and induces biphasic PSA-NCAM expression in the adult rat dentate gyrus. Eur. J. Neurosci., 17: 879–886.

Pravosudov, V.V., Lavenex, P. and Clayton, N.S. (2002) Changes in spatial memory mediated by experimental variation in food supply do not affect hippocampal anatomy in mountain chickadees (*Poecile gambeli*). J. Neurobiol., 51: 142–148.

Rakic, P. (1985) Limits of neurogenesis in primates. Science, 227: 1054–1056.

Rakic, P. (2002a) Adult neurogenesis in mammals: an identity crisis. J. Neurosci., 22: 614–618.

Rakic, P. (2002b) Neurogenesis in adult primate neocortex: an evaluation of the evidence. Nat. Rev. Neurosci., 3: 65–71.

Rakic, P. and Nowakowski, R.S. (1981) The time of origin of neurons in the hippocampal region of the rhesus monkey. J. Comp. Neurol., 196: 99–128.

Rochefort, C., Gheusi, G., Vincent, J.D. and Lledo, P.M. (2002) Enriched odor exposure increases the number of newborn neurons in the adult olfactory bulb and improves odor memory. J. Neurosci., 22: 2679–2689.

Schmaltz, L.W. and Theios, J. (1972) Acquisition and extinction of a classically conditioned response in hippocampectomized rabbits (*Oryctolagus cuniculus*). J. Comp. Physiol. Psychol., 79: 328–333.

Scott, B.W., Wang, S., Burnham, W.M., De Boni, U. and Wojtowicz, J.M. (1998) Kindling-induced neurogenesis in the dentate gyrus of the rat. Neurosci. Lett., 248: 73–76.

Seaberg, R.M. and van der Kooy, D. (2002) Adult rodent neurogenic regions: the ventricular subependyma contains neural stem cells, but the dentate gyrus contains restricted progenitors. J. Neurosci., 22: 1784–1793.

Seress, L., Abraham, H., Tornoczky, T. and Kosztolanyi, G. (2001) Cell formation in the human hippocampal formation from mid-gestation to the late postnatal period. Neuroscience, 105: 831–843.

Shors, T.J., Miesegaes, G., Beylin, A., Zhao, M., Rydel, T. and Gould, E. (2001) Neurogenesis in the adult is involved in the formation of trace memories. Nature, 410: 372–376.

Shors, T.J., Townsend, D.A., Zhao, M., Kozorovitskiy, Y. and Gould, E. (2002) Neurogenesis may relate to some but not all types of hippocampal-dependent learning. Hippocampus, 12: 578–584.

Solomon, P.R., Vander Schaaf, E.R., Thompson, R.F. and Weisz, D.J. (1986) Hippocampus and trace conditioning of the rabbit's classically conditioned nictitating membrane response. Behav. Neurosci., 100: 729–744.

Squire, L.R. (1992) Memory and the hippocampus: a synthesis from findings with rats, monkeys, and humans. Psychol. Rev., 99: 195–231.

Stanfield, B.B. and Trice, J.E (1988) Evidence that granule cells generated in the dentate gyrus of adult rats extend axonal projections. Exp. Brain Res., 72: 399–406.

Suzuki, W.A. and Clayton, N.S. (2000) The hippocampus and memory: a comparative and ethological perspective. Curr. Opin. Neurobiol., 10: 768–773.

Suzuki, W.A. and Eichenbaum, H. (2000) The neurophysiology of memory. Ann. N.Y. Acad. Sci., 911: 175–191.

Tanapat, P. and Gould, E. (1997) EGF stimulates proliferation of granule cell precursors in the dentate gyrus of adult rats. Soc. Neurosci. Abstr., 23: 317.

Tanapat, P., Hastings, N.B., Reeves, A.J. and Gould, E. (1999) Estrogen stimulates a transient increase in the number of new neurons in the dentate gyrus of the adult female rat. J. Neurosci., 19: 5792–5801.

Tanapat, P., Hastings, N.B., Rydel, T.A., Galea, L.A. and Gould, E. (2001) Exposure to fox odor inhibits cell proliferation in the hippocampus of adult rats via an adrenal hormone-dependent mechanism. J. Comp. Neurol., 437: 496–504.

van der Hart, M.G., Czeh, B., de Biurrun, G., Michaelis, T., Watanabe, T., Natt, O., Frahm, J. and Fuchs, E. (2002) Substance P receptor antagonist and clomipramine prevent stress-induced alterations in cerebral metabolites, cytogenesis in the dentate gyrus and hippocampal volume. Mol. Psychiatry, 7: 933–941.

van Os, J. and Selten, J.P. (1998) Prenatal exposure to maternal stress and subsequent schizophrenia. The May 1940 invasion of The Netherlands. Br. J. Psychiatry, 172: 324–326.

van Praag, H., Kempermann, G. and Gage, F.H. (1999) Running increases cell proliferation and neurogenesis in the adult mouse dentate gyrus. Nat. Neurosci., 2: 266–270.

van Praag, H., Kempermann, G. and Gage, F.H. (2000) Neural consequences of environmental enrichment. Nat. Rev. Neurosci., 1: 191–198.

van Praag, H., Schinder, A.F., Christie, B.R., Toni, N., Palmer, T.D. and Gage, F.H. (2002) Functional neurogenesis in the adult hippocampus. Nature, 415: 1030–1034.

Wagner, J.P., Black, I.B. and DiCicco-Bloom, E. (1999) Stimulation of neonatal and adult brain neurogenesis by subcutaneous injection of basic fibroblast growth factor. J. Neurosci., 19: 6006–6016.

Wang, S., Scott, B.W. and Wojtowicz, J.M. (2000) Heterogenous properties of dentate granule neurons in the adult rat. J. Neurobiol., 42: 248–257.

Watson, J.B., Mednick, S.A., Huttunen, M. and Wang, X. (1999) Prenatal teratogens and the development of adult mental illness. Dev. Psychopathol., 11: 457–466.

Weiss, C., Bouwmeester, H., Power, J.M. and Disterhoft, J.F. (1999) Hippocampal lesions prevent trace eyeblink conditioning in the freely moving rat. Behav. Brain Res., 99: 123–132.

West, M.J., Slomianka, L. and Gundersen, H.J. (1991) Unbiased stereological estimation of the total number of neurons in the subdivisions of the rat hippocampus using the optical fractionator. Anat. Rec., 231: 482–497.

Wolswijk, G. and Noble, M. (1992) Cooperation between PDGF and FGF converts slowly dividing O-2A adult progenitor cells to rapidly dividing cells with characteristics of O-2A perinatal progenitor cells. J. Cell Biol., 118: 889–900.

Yagita, Y., Kitagawa, K., Ohtsuki, T., Takasawa, K., Miyata, T., Okano, H., Hori, M. and Matsumoto, M. (2001) Neurogenesis by progenitor cells in the ischemic adult rat hippocampus. Stroke, 32: 1890–1896.

Zharkovsky, T., Kaasik, A., Jaako, K. and Zharkovsky, A. (2003) Neurodegeneration and production of the new cells in the dentate gyrus of juvenile rat hippocampus after a single administration of ethanol. Brain Res., 978: 115–123.

Zupanc, G.K. (1999) Neurogenesis, cell death and regeneration in the adult gymnotiform brain. J. Exp. Biol., 202: 1435–1446.

T. Steckler, N.H. Kalin and J.M.H.M. Reul (Eds.)
Handbook of Stress and the Brain, Vol. 15
ISSN 0-444-51173-3

CHAPTER 5.6

Cellular and molecular analysis of stress-induced neurodegeneration – methodological considerations

J. Lu[1], Z. Némethy[1], J.M. Pego[2], J.J. Cerqueira[2], N. Sousa[2] and O.F.X. Almeida[1,*]

[1]*Max-Planck Institute of Psychiatry, Kraepelinstrasse 2-10, D-80804 Munich, Germany*
[2]*Life and Health Science Research Institute, Health Science School, University of Minho, Campus de Gualtar, 4710-057 Braga, Portugal*

Abstract: Evidence that chronic hypercorticalism induces a broad spectrum of deleterious cellular effects in the brain has accumulated over the last two decades. These principal effects of hypercorticalism include neuronal atrophy, neuronal death and glial responses. Importantly, these changes, which may occur interdependently and/or concomitantly, lead to neurodegeneration. While there has been a significant expansion of the number of techniques available for examining effects of chronic stress in the brain, the cellular and molecular mechanisms underpinning stress-induced neurodegeneration are still only partially known. This article appraises the major current methodologies available for analyzing stress-induced neurodegeneration, and considers the advantages and limitations of each of these methods.

What do we understand by stress?

Stress refers to the organism's attempt to mount an 'adaptive' (beneficial) response to aversive stimuli in order to maintain or restore homeostasis. Different sensory and motor systems are differentially activated depending on the quality and intensity of the stressful (aversive) stimulus, and the magnitude and duration of the response are influenced by the "context" of the stimulus (experience, mood, age, environmental factors) (Herman and Cullinan, 1997). Thus, extreme caution is necessary before generalizing about the effects of one particular "stress"; it is fair to say that much of the confusion existing in the field is a consequence of the false presumption that elevated corticosteroids mimic stress and/or one stressor is a representative of every stressor. Another important point to be noted is that prolonged elevations in corticosteroid secretion are a crucial accompaniment of the chronic response to stress (Sapolsky et al., 2000). Briefly, in the event that adequate adaptive mechanisms cannot be recruited, chronic stress will result in a state of chronic hypercorticalism and, as a consequence, deleterious effects, including immune suppression and a variety of mental disturbances will emerge (Sapolsky et al., 2000).

Forms of neurodegeneration

Numerous studies have demonstrated that chronic hypercorticalism induces a broad spectrum of deleterious cellular effects in the brain, which can be conveniently categorized as neuronal atrophy, neuronal death, and glial responses. These changes, which may occur interdependently and/or concomitantly constitute neurodegeneration. The neurodegenerative changes associated with hypercorticalism are by far less-marked than the damage seen in the so-called

*Corresponding author: Tel.: 9 89 30622216;
Fax: 9 89 306 22461; E-mail: osa@mpipsykl.mpg.de.

neurodegenerative diseases (Parkinson's, Alzheimer's, etc.) and it would therefore, probably be more prudent to consider them as representative of the selective vulnerability of given brain regions. Most of the examples given in this review relate to the responses of hippocampal cells to stress (and pharmacological hypercorticalism). The hippocampus has been the most extensively studied brain region in this respect; its particular vulnerability to corticosteroids most probably reflects its high concentrations of corticosteroid (mineralocorticoid and glucocorticoid) receptors.

Neuronal atrophy

An important notion to be kept in mind when referring to neuronal degeneration is that it does not necessarily imply the death of neurons. Indeed, most events perceived by a living organism (either positive or negative) are believed to modulate the structure of neuronal networks rather than lead to changes in neuronal number (Segal, 2002; Erickson et al., 2003).

To evaluate dendritic arborizations one can use the Golgi technique, which selectively impregnates single neurons with silver chromate (Camillo Golgi, 1843–1924). This method has provided indispensable information about the way in which sets of neuronal elements contribute to the gross structure of the neuropil and tracts. Impregnations show up as black, purple or reddish-brown against a pale yellow background; it is essentially a stochastic technique, the exact chemical mechanism of which remains unclear. This approach allows impregnated neurons and boundaries in any region of interest to be traced and reconstructed from successive serial sections. Two-dimensional (2D) analysis can be performed from traces obtained using a drawing tube attached to a light microscope. This type of analysis does not require any sophisticated equipment and has been widely used in the past. However, it has one major disadvantage: converting a 3D probe into a 2D probe results in a loss of information on the suppressed dimension. To achieve 3D reconstructions, cell bodies, dendritic arborizations, and boundaries of the region of interest should be drawn (under 25–100 × oil immersion objectives) and plotted in 3D using a video computer system (e.g., Neurolucida from MicroBrightField, Inc.). Three-dimensional models of neurons can be visualized using appropriate software. Three-dimensional reconstructions of neurons can be rotated around any of the x-, y-, and z-axes to allow the best visualization of the dendritic trees. Total dendritic lengths, number of segments/bifurcations, Sholl analysis (which provides an estimate of dendritic densities, based on the number of intersections between concentric circles centered in cell soma and the dendritic segments) and spine densities are just some of the parameters this analysis provides.

Importantly, the use of these techniques allowed the pioneers of neuroanatomy to recognize the organization of neuronal networks, and to eventually demonstrate the occurrence of remarkable alterations in dendritic trees and synaptic contacts following neuronal insults. Indeed, such knowledge existed long before the description of different forms of neuronal death.

Several studies in the 1990s demonstrated that hypercorticalism (pharmacological or stress-induced) induces alterations in cytoplasm organelles (Miller et al., 1989) and, ultimately, atrophy of CA3 pyramidal cell dendrites in the hippocampal formation; (Watanabe et al., 1992; Magarinos and McEwen 1995a); subsequent work confirmed these results in this neuronal population but also observed similar alterations in all the other major subdivisions of the hippocampal formation (Sousa et al., 2000). Furthermore, the later studies noted a marked loss of synapes in at least one of the links of the intrinsic hippocampal circuitry (the mossy fiber–CA3 connection). It may therefore be concluded that elevated corticosteroids trigger structural responses within cytoplasmic organelles, dendrites, axons, and their synaptic contacts; importantly, such changes do not necessarily involve the irreversible loss of neurons (Sousa et al., 2000).

Neuritic alterations of the type described above correlate with behavioral deficits and would appear to serve as the neuroanatomical basis of adaptive mechanisms underlying learning and memory (Erickson et al., 2003). The cellular basis of learning and memory has long been believed to include alterations in dendrites (mainly in spines) and in the number and structure of synapses (Cajal, 1893). The validity of this notion was explored in a number of quantitative light and electron microscopic studies, which, in the majority of cases, showed that the richness of dendritic arborizations and the numerical density of synapses increases as a consequence of learning of novel behaviors. More recent studies have also shown that, despite numerical changes in dendritic spines and synapses, the cellular mechanisms of hippocampus-dependent associative learning include the remodeling of existing hippocampal synapses; these changes most likely reflect an involvement of signal transduction proteins and the transformation of silent postsynaptic synapses into active ones (Rusakov et al., 1997; Stewart et al., 2000). In light of these robust correlations between neuritic (dendritic spine and synapse) changes and cognitive performance, it seems warranted to conclude that perturbations of the former will result in impaired performance in hippocampus-dependent learning tasks.

Most interestingly, although the neuritic atrophy and synaptic loss referred to above would be expected to provoke some degree of functional impairment, together with the fact that these paradigms are not necessarily associated with neuronal cell loss, it seems more than likely that neuronal reorganization (regrowth of dendrites and axons and establishment of new synapses) of damaged neuronal circuits is an important mechanism allowing recovery from insults (McEwen, 1999). The above proposition appears to be valid insofar that studies in rats have shown that, whereas no significant structural reorganization occurs during or immediately after the termination of elevated corticosteroid levels (by pharmacological means or after the imposition of stressors), significant reorganization does occur within one month of withdrawal from the damaging stimulus (Sousa et al., 2000). This so-called "reactive synaptogenesis" occurs throughout the hippocampal formation and is commensurate with restoration of spatial learning and memory to levels found in control animals. Thus, the more recent findings match well with older observations that hypercorticalism-induced cognitive impairment is a reversible phenomenon.

Importantly, regeneration of dendritic, axonal, and synaptic elements does not seem to be compromised in conditions when profound neuronal loss has occurred, e.g., in the dentate granule cell layer after adrenalectomy, a manipulation accompanied by marked collapse of the mossy fiber inputs to the CA3 pyramidal layer. Administration of low doses of corticosterone to adrenalectomized rats can, at least partially, restore the total dendritic length of granule cells and the volume and surface area of the mossy fiber terminals (Sousa et al., 1999a). In addition, substitution therapy with corticosterone results in complete recovery of the volume of the suprapyramidal bundle, number, and surface area of mossy fiber–CA3 synapses, and the surface area of dendritic excrescences (Sousa et al., 1999a). These observations on the fine structural adjustments fit with results of other work showing that behavioral functions impaired by adrenalectomy can be partially reinstated by the administration of corticosterone (McCormick et al., 1997).

The evidence summarized above firmly indicates that alterations of the corticosteroid milieu can induce profound, but largely reversible, changes in the ultrastructural organization of the hippocampal formation; these bidirectional alterations, more than changes in neuron viability, may represent the neuroanatomical correlation of hippocampus-dependent learning and memory. Presently, there is no clear data available as to what neurochemical mechanisms might underlie the fine structural observations described above. However, NMDA and serotonin (5-HT) receptors appear to be key players since the administration of either NMDA antagonists or serotonin reuptake inhibitors have been shown to abrogate CA3 dendritic atrophy (Watanabe et al., 1992; Magarinos and McEwen, 1995b). Growth factors also seem to be likely mediators, as suggested by data showing that elevated corticosteroid levels (including those produced in response to stress) attenuate hippocampal brain-derived growth factor (BDNF) and nerve growth factor (NGF) levels (Smith et al., 1995; Hansson et al., 2000) and that adrenalectomy results in significant alterations in the levels of neurotrophin-3 and fibroblast growth factor-2 (FGF-2) (Barbany and Persson, 1992; Hansson et al., 2000). Finally, it seems highly plausible that neurotrophins play a major role in the neuritic regrowth seen after recovery from exposure to high corticosteroid levels because the recovery phase is characterized by an increase in their synthesis (Smith et al., 1995).

Neuronal death

Neuronal death can occur through one of the two basic mechanisms – necrosis or apoptosis (see Fig. 1 and for review Majno and Joris, 1995). Necrosis is the unexpected death of cells resulting from "external damage," usually mediated via destruction of the integrity of plasma membrane and/or the trophic support of the cell. Morphologically, there is lysis of the plasma membrane of the swollen necrotic cell, which leads to release of cytoplasmic components into the surrounding tissue spaces. Inflammatory cells, attracted by the necrotic debris, trigger tissue destruction. Necrosis of isolated cells can occur, although necrosis usually affects large clusters. Consequently, there is significant tissue inflammation (with subsequent repair and scarring), with permanent alteration of architecture and function. Since necrosis usually ensues from cytotoxins, the process is completed rapidly within seconds-to-minutes.

Apoptosis differs from necrosis in that it involves the triggering of specific, sequentially occurring, events. Although the term apoptosis was originally coined to describe a specific morphological sequel, it is now known that apoptosis depends on activation of a genomic program; as such, the term apoptosis is frequently used synonymously with the term programmed cell death (Fig. 2). It should be mentioned that most current methods for the detection of apoptosis can only detect late stages of the process, and that some programmed cell death may not involve the mechanisms of apoptosis (e.g., oncosis – the term oncosis (derived from onkos, meaning swelling) was proposed in 1910 by von Recklinghausen precisely to mean cell death with swelling; oncosis leads to necrosis with karyolysis and stands in contrast to apoptosis, which leads to necrosis with karyorhexis and cell shrinkage). In contrast to necrosis, apoptosis is a much slower process; depending on the initiating stimulus, apoptosis requires from a few hours to several days for its complete manifestation. Conceptually, this form of cell death is analogous to "suicide," inasmuch as death results from the activation of the dying cell's own death machinery. Apoptosis, first recognized by embryologists, has now come to be recognized as being important for maintaining tissue homeostasis and to constitute a major component of many pathological responses, including neurodegenerative diseases. It is important to note that the genetic program for apoptosis can be triggered by both intrinsic (e.g., during histogenesis) and extrinsic

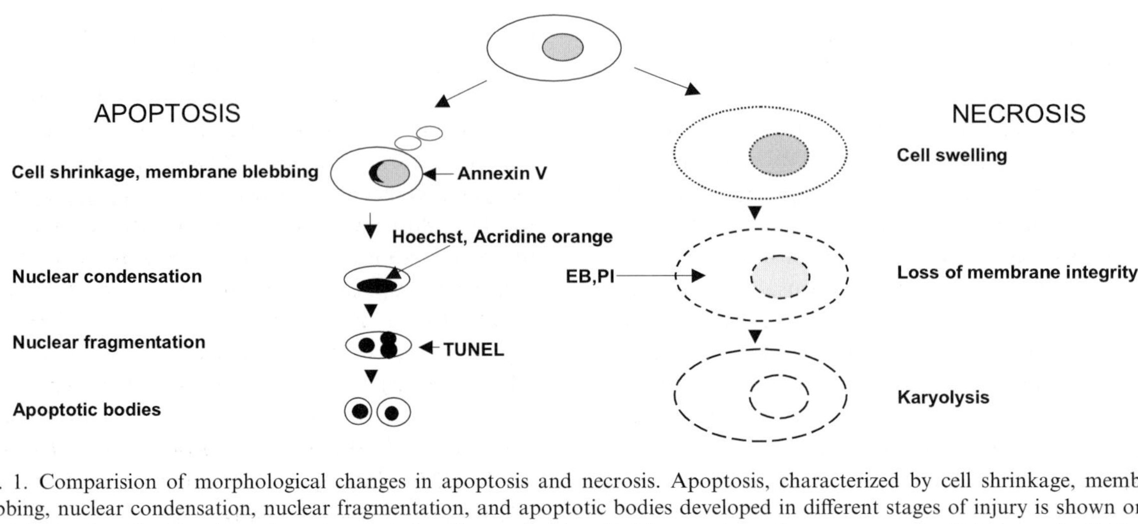

Fig. 1. Comparision of morphological changes in apoptosis and necrosis. Apoptosis, characterized by cell shrinkage, membrane blebbing, nuclear condensation, nuclear fragmentation, and apoptotic bodies developed in different stages of injury is shown on the left-side. As described in the main text, apoptotic cells can be identified in a variety of ways, some of which (annexin V-binding, Hoechst, acridine orange, and TUNEL staining) are indicated here. Note that the majority of methods are based primarily on changes in the properties of the cell membrane and nucleus. Necrosis, characterized by cell swelling, loss of membrane integrity, and karyolysis is shown on the right. Membrane-impermeable markers such as ethidium bromide (EB) and propidium iodide (PI) can be used to identify necrosis.

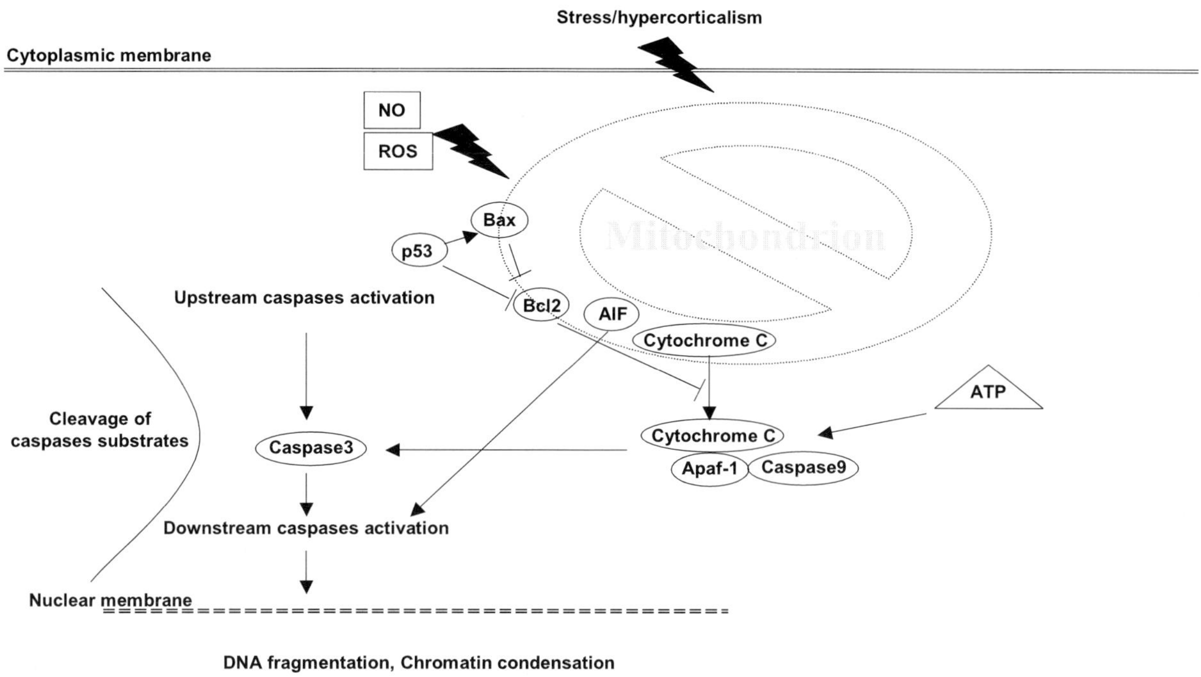

Fig. 2. Signal pathways in apoptosis. The mitochondrion as the integrator of apoptotic signals from stress or other factors such as nitric oxide (NO) or reactive oxygen species (ROS), can release cytochrome *c* and apoptosis-inducing factor (AIF); cytochrome *c*, together with Apaf1, Caspase9, and ATP activate caspase3, which, in turn, activates downstream caspases for DNA cleavage. Note that Bcl2 from the mitochondrial membrane can prevent mitochondrial pore formation, which antagonizes cytochrome *c* activity; Bax can increase mitochondrial permeability. Separation of living/apoptotic cells by flow cytometry. For flow cytometric analysis, ethanol-fixed cells are washed in phosphate–citrate buffer and stained with propidium iodide. As cells pass in front of a laser, they absorb, diffract, refract and reflect incident light, and emit fluorescence. The scattered light is focused by a lens into a photomultiplier, the emitted fluorescent signal is optically filtered through dichroic mirrors, and subsequently processed by wide bandpass filters selected to optimize the various fluorescent emissions; signals are detected by photomultiplier tubes, and based on fluorescence intensity profiles, living and apoptotic cells can be distinguished.

factors, including stressful stimuli and exogenous corticosteroids, although the intracellular signaling cascades and morphological changes are essentially the same in both situations.

As already mentioned, apoptosis is now known to be important during embryogenesis/histogenesis but also in the course of normal tissue turnover. Of course, the mature brain is traditionally not regarded as an organ where cell and tissue turnover occurs, but with the increasing number of reports that, besides glial cells, neurons can also be generated in certain regions, the original concept does not seem to be strictly correct. Furthermore, it is being increasingly recognized that apoptosis makes a significant contribution to neural cell loss in pathological conditions, e.g., in neurodegenerative diseases such as Alzheimer's and Parkinson's disease (Honig and Rosenberg, 2000; Friedlander, 2003). It is also pertinent to mention that a revisionist view with respect to the distinctive roles and mechanisms of necrosis and apoptosis has emerged since the late 1990s; according to these authors it is now accepted that virtually any insult just below the threshold to induce necrosis results in an apoptotic response (McConkey, 1998). The cellular response becomes relevant in this process in that, in contrast to the situation in necrosis, apoptosis involves active processes within the dying cell and does not merely depend on the insult itself.

In contrast to most other cells, neurons have elaborate morphologies with complex neuritic arborizations that often extend long distances from the perikarya. It is in fact the richness of complex contacts between neurons that results in the establishment of functional networks. With this concept in mind, it is not difficult to accept that neuronal degeneration does not necessarily imply neuronal death; neuronal atrophy and synaptic loss also represent forms of nervous tissue degeneration. It was recently shown that the biochemical cascades leading to apoptosis can be activated locally in synapses and dendrites (Mattson, 2000), indicating a much more complex role for apoptosis than previously envisaged, i.e., in synaptic loss and dendritic remodeling.

Glial response

Glia mediate neuroendocrine and neuroimmune functions that are altered in the face of a number of neuronal insults, including prolonged stress. The biological functions of glia involve changes in shape, interactions with neurons and other glia, and gene expression. Glia cells become activated in the presence of ongoing neurodegeneration and progress to produce what is termed "reactive gliosis" (Nichols, 1999; Liu and Hong, 2003). Since good markers to distinguish normal from reactive glia are not commonly available, most researchers currently depend on well-defined morphological criteria. In several neurodegenerative conditions, astrocytes exhibit hypertrophy and signs of metabolic activation, and astrocytic processes begin to entwine neurons. Microglia also become activated and subsets of these cells increase in number and may enter the phagocytic or reactive stage. Glial markers of brain aging and glial activation include glial fibrillary acidic protein (GFAP) and transforming growth factor (TGF)-β1, which are increased in astrocytes and microglia, respectively (Nichols, 1999). Interestingly, steroids (Laping et al., 1994), such as those produced in the adrenals (Melcangi et al., 1997), regulate the interactions between glia and neurons, and glial gene expression, including GFAP and TGF-β1.

Despite the recognized relevance of the biological functions of glia, little is known about the effect of stress on hippocampal glial cells. Anecdotal evidence suggests an increase in glial cell number and signs of cytoplasmic transformation of astrocytes and microglia in areas of the brain implicated in stress-induced disorders (namely the prefrontal cortex and the hippocampus) (Ramos-Remus et al., 2002). Based on these findings, it appears that the hippocampal glial response to chronic stress may be similar to that found in endangered or challenged hippocampal environments, such as in ischemia.

A different line of evidence on the glial response to imbalances in the corticosteroid milieu has come from studies in surgically lesioned animals (Vijayan and Cotman, 1987). Animals with surgical entorhinal lesions concomitantly treated with hydrocortisone demonstrated more astrocytes and fewer nonastrocytes in the dentate outer molecular layer compared with untreated animals. Glia in the treated animals also showed a decrease in average optical density of cytoplasmic acid phosphatase staining. These findings suggest that hydrocortisone treatment prior to, and following, an entorhinal lesion accelerates lesion-induced migration of astrocytes to the outer molecular layer, and reduces the increase in microglial number resulting from the lesion. The observed effect on microglia may result from direct hormonal inhibition of local proliferation of microglia or from the well-known systemic anti-inflammatory action of glucocorticoids on monocytes, the putative precursors of brain microglia. In light of these findings it has been suggested that glucocorticoid hormones significantly alter the response of nonneuronal cells to neural tissue damage. Lending support to this view is the observation that adrenalectomized animals show induction of GFAP immunoreactivity, which occurs contemporaneously with neurodegeneration (Trejo et al., 1998). Although no variation in the total number of glial cells is found, signs of astroglial activation can be observed in the adrenalectomized group: astroglial cells change in size and shape, and their processes in the molecular layer, which normally show unipolarity become randomly organized (Sousa et al., 1997). Both effects are confined to the dentate gyrus and mossy fiber zone. The degeneration and astroglial reaction become more pronounced with increasing duration after adrenalectomy, and both can be prevented by placing animals on corticosterone replacement therapy. Results such as these illustrate the close relationship between the glial response and neuronal degeneration in the dentate gyrus following adrenalectomy, in terms of both, time and space (Sousa et al., 1997).

What are the neural targets of stress-mediated degeneration?

Corticosteroids are secreted distal to their brain targets but distribution maps of their receptors serve as reliable indicators of their sites of action. In a landmark study on the rat brain, Reul and de Kloet (1985) reported that radioactively labeled corticosterone binds with differing affinities to two distinct receptors, and that the hippocampus showed the highest signal retention for both receptors; subsequent cloning studies revealed significant homologies between the high-affinity and low-affinity central and peripheral corticosteroid receptors: mineralocorticoid (MR) and glucocorticoid (GR) receptors, respectively. In vitro studies showed that the high-affinity binding site in brain can also bind aldosterone; in practice however, the endogenous production of aldosterone only reaches concentrations sufficient to activate renal mineralocorticoid receptors (Funder, 1996); thus, cerebral MR show promiscuity in that, like GR, they bind corticosterone; however, since they have a ca. 10-fold greater affinity for corticosterone as compared to GR, MR are predominantly occupied during periods when corticosteroid levels are low, whereas GR only become occupied when corticosteroid secretion increases above a certain threshold (e.g., during stress or in pathological conditions). Further, the presence of two isoforms of the pre-receptor enzyme 11ß-hydroxysteroid dehydrogenase, involved in the interconversion of corticosteroids to active and inactive forms, contribute to the selective access to intracellular receptors (Yau and Seckl, 2001).

While GR are widely distributed throughout the brain, but are particularly concentrated in the hippocampus, hypothalamus, and lower brainstem, MR are almost exclusively confined to the hippocampus and other limbic structures such as the septum, central nucleus of the amygdala, the olfactory nucleus, and some hypothalamic nuclei (Van Eekelen et al., 1988; Ahima and Harlan, 1990; Ahima et al., 1991). Within the hippocampal formation, subfield-specific differences in MR and GR concentration profiles have been described: MR levels are high in CA1 pyramidal layer $\approx$ granule cell layer (dentate gyrus) > CA3 pyramidal layer, and GR are concentrated in the CA1 $\approx$ dentate gyrus $\gg$ CA3 (Van Eekelen et al., 1988). The functional significance, if any, of these differential patterns of receptor distribution may be inferred from the known functions of the particular brain nuclei displaying high levels of MR and GR expression and/or ligand binding. At this juncture, it is important to point out that while the described patterns of MR and GR occurrence in the various hippocampal subdivisions may serve as eventual predictors of function, they do not necessarily reflect the receptor repertoire of individual cells in any region; further, it is still not known to what extent receptor composition (concentration of individual receptors or co-localization of MR and GR in the same cell) determines the fate of a particular cell (e.g., survival vs. death) in response to changes in the corticosteroid milieu.

Experimental paradigms for examining stress-mediated degeneration

Designing models of stress implies a clear definition of the question under study; more specifically, if one wants to determine the effect of stress upon a specific region of the brain, several issues need to be considered. One of them is adaptation; if a single stressor is applied for a prolonged period, then the organism tends to adapt to that stressor and the stress response gets blunted. A second issue to consider is unpredictability; even when applying different stressors, care must be taken to avoid adaptation, e.g., by applying a battery of stressors at different clock times and in random order. A final point to consider is that stressors vary in quality; for example, physical and psychological stressors activate different regions of the brain, with the former depending on perception by brain stem centers as opposed to the latter, which depends on the activation of higher regions of the brain (in particular the limbic system). Obviously, comparisons between different experimental paradigms (and the results therefrom) must also take into account factors such as intensity and duration/chronicity.

A commonly used approach in evaluating the cellular effects of stress involves decomposition of the effectors of these actions, e.g., by mimicking the endocrine response to stress by administering high doses of corticosteroids, a paradigm that does not exactly reproduce the physical, behavioral/emotional, and neurochemical manifestations of stress. Nevertheless, our current understanding of the actions mediated by the two corticosteroid receptors has largely benefited from the exploitation of the high selectivity of aldosterone (the prototypic MR agonist) and dexamethasone or RU28362 (prototypic GR agonists) as well as the antagonists spironolactone and RU28318 (for blocking MR effects) and RU38486 (for blocking GR effects). Further insights into the biological actions of MR and GR are now being gained from MR and GR gain- and loss-of-function mouse models (Muller et al., 2002). The use of such models has proved particularly useful in proving and understanding the importance and role of these receptors in stress-mediated neuronal damage, and neuronal disorders influenced by stress such as anxiety, depression, and dementia.

While in vivo models are necessary for the evaluation of stress effects, in vitro models are indispensable for understanding the cellular and molecular mechanisms underlying those effects. The latter approach is particularly amenable to analysis at the molecular level, but the major caveat here is that in vitro observations do not necessarily apply to the whole organism whose ultimate response to the same stimulus reflects an integration of a plethora of adaptive and signaling pathways emanating from cells with diverse properties, e.g., the liver can substantially influence the response of the brain to endogenous and exogenous stress hormones. As a result, a neurotoxic stimulus in vitro might just happen to be protective or to have no effect in the living organism.

Use of stereology in analyzing neurodegeneration

Another extremely relevant issue to consider when designing an experiment is the sensitivity and specificity of the methodological procedures employed to test the hypothesis. Obviously, the analysis of stress-induced neurodegeneration also follows this rule. A common first analytical approach is to make observations on histological sections. Histological sections define the normal appearance of tissue and organs, and detect natural or induced alterations in structure. Histological descriptions often include terms such as "large," "small," "many," "few," "absent," or "present." Helpful as these terms are for the description of basic features, they are often open to subjectivity and, being qualitative, they do not allow statistical evaluation of the effects of a particular treatment, e.g., stress exposure. Quantitative data can take several forms, but all basically depend on counting cells in a section. One modern approach, superior to previous methods (Abercrombie, 1946; Weibel et al., 1966) is that of stereology, which is given detailed consideration below. Using stereology, one can obtain estimates of object volumes and derive numbers of objects from this data increasing the precision and relevance of data (Gundersen et al., 1988; West, 1999).

The principle behind stereology is to recreate or estimate the properties of geometrical objects in space. Its application to tissue or organ sections allows relatively precise estimation of the geometrical properties of the objects in a given section. As space has three dimensions, objects within it have properties for each possible number of dimensions, and objects within a given space can be defined in terms of their volumes (3 dimensions), surfaces (2 dimensions), lengths (1 dimension) and numbers. Each of these properties can be estimated by stereological methods, usually a two-step procedure involving: sampling and subsequently measuring.

A characteristic of many tissues and organs is that they contain a large number of the objects of interest, but too many to be measured individually. Producing a good sample is an essential step in stereological methods. Errors incurred during sampling can result in difficulties in obtaining meaningful stereological estimates later. Sampling usually starts before the investigator has any predictions as to the study outcome and even before the investigator has thought about applying stereological methods. To avoid later regrets, it is advisable to sample correctly from the very beginning; however, the researcher can be consoled by the fact that stereology-based sampling methods are compatible with all other types of analysis.

One usually wants to make statements about a structure (e.g., the hippocampus) or a cellular population (neurons) by sampling only a part of the structure or population. If such statements are to be valid for the entire structure or population, the sample must be a representative one. Selecting representative samples requires: (i) access to the entire structure or population; (ii) ability to recognize and/or define the entire structure or population; and (iii) that all parts of the structure or population contribute equally to the sample.

These pre-requisites can be met by random sampling in one of two ways: (i) Random independent samples – This is the most obvious approach in which one selects an initial location at random. After measuring the objects of interest, subsequent locations for measurement are chosen independent of the first. When a sufficient number of locations has been sampled, the individual measurements are averaged. Despite providing reliable and reproducible estimates, this method is, however, not an efficient sampling procedure. (ii) Uniform random systematic (URS) samples – In URS sampling, a random starting point is selected and samples are drawn at regular (or uniform systematic) intervals. Choosing a random starting point means that all areas to be analyzed have an equal chance to contribute to the final sample measure. By eliminating sample clustering, the URS sampling procedure, on average, yields a more accurate estimate than the random independent sampling approach, and is the recommended method of choice. In practice, sections are selected using the URS sampling procedure at the time of tissue sectioning. If, for example, every fifth section is collected, the only requirement is to assure compliance with the need to randomly select the initial section in the series.

The next step in stereology is 'measuring' which involves relatively easy, routine work depending on identification of the object of interest and based on a simple set of rules. Curiously, the volume, surface, length, and number of objects are such basic parameters that it may be surprising to realize that methods for their accurate measurement only became available in the 1980s and did not enter widespread use until the 1990s. Stereological tools have now virtually replaced the earlier error-prone methods, which all suffered from the assumption that histological sections are two-dimensional images from which three-dimensional measurements were nevertheless attempted. The traditional methods involved certain well-grounded assumptions about the "missing dimension" in two-dimensional images. Inherently, the proximity to the true values achieved using such assumption-based methods, depended largely on how good the assumptions were.

In modern stereology, design-based methods have replaced assumption-based ones. These newer approaches involve measurements on a series of sections which in fact do have three dimensions. Therefore, information about the third dimension is based on fact, rather than assumption; obtaining precise measurements then depends on one other factor – the availability of good probes to apply to the sample. The selection of the adequate probe ultimately determines the precision of the estimation (West, 1999).

Estimation of volumes

Estimating volumes using points is conceptually the easiest of all stereological methods and was first described by the Italian mathematician Bonaventura Cavalieri (1598–1647). The "Cavalieri Principle" holds that, if one places a grid of regularly spaced points over an object of interest, the measured surface area will be a function of the number of points falling within it. To calculate the volume of an object (in our case, section), one simply has to multiply the average areas of different sections by the thickness of each section. The point-counting principle can, theoretically, also be applied to very small objects like cells, but this would require very thin sections in order to reduce error and the ability to identify the object in consecutive sections. Other approaches, like the nucleator – in which a point associated with a small particle (e.g., a nucleolus within a cell) is identified and from which rays are extended until the intersection of particle's boundaries to allow the estimation of its profile area and, subsequently, the absolute volume of the particle – have been developed for this purpose (Gundersen et al., 1988).

Estimation of total cell numbers

Measuring neuronal loss has preoccupied many neuroscientists interested in the effects of stress and glucocorticoids on the brain, in particular, the hippocampus. Such information can be generated by simply counting the number of cells in a given section and comparing the values obtained with those for sections from an anatomically matched area in control (e.g., nonstressed) subjects. This procedure will yield an estimate of cell number per unit area (N_A); to date, this is probably the most widely-used method to count neurons. The precision (relevance) of such an estimate relies entirely on how similar the sections being compared are. A serious (but common) error of such estimates is the "reference trap," which refers to how variation in the volume of reference can affect the final result. This can be illustrated by considering the fact that because the N_A of granule cells in the hippocampus in two different sections from different experimental groups is similar, it does not necessarily follow that the total number of cells in each section is the same; this is because the volume (derived from the third dimension, which is not taken into account in deriving the N_A value) of the dentate granule cell layer might differ significantly between individual sections and subjects. Design-based methods (e.g., estimating neurons using volumes within a probe) help avoid the introduction of such biases (West, 1999).

Essentially, estimating numbers within a volume is just the corollary of estimating volumes with points. One takes two adjacent sections that are thinner than the diameter of the object (e.g., nuclei) to be counted; the objects visible in the second, but not first, section are counted. Then, the number of objects in the volume represented by the two sections will, on average, correspond to the number of objects counted in the second. This approach is called the (physical) dissector because the counting principle is based on a comparison of two sections. Application of this technique provides the numerical density (N_v) of objects (e.g., neurons) within a region of interest. Now, if the total volume of the structure (e.g., hippocampus) is known, the total number of objects (e.g., neurons) can be derived from the product of N_v and the total volume.

The optical fractionator is another means for obtaining the total number of objects in a given 3D structure. It is based on the combination of systematic sampling (which yields an estimate of the fraction of the tissue – fractionator) and the dissector in thick optical sections that intrinsically have three dimensions.

Detection of cell death

The application of stereological methods to histological sections (e.g., stained with Nissl, Giemsa) can certainly provide information of neuronal loss based on the comparison of total number of surviving neurons between experimental groups (West et al., 1991; Sousa et al., 1999b). However, this approach can also be applied in combination with markers of neuronal death to directly determine the number of dying cells at a particular time-point. Several specific staining methods for detecting neurodegeneration have been developed but their use has not yet been generalized. The earliest markers of neuronal degeneration were based on silver-impregnation methods that provide unspecific indications of degeneration in neuronal soma and neurites.

More recently, the use of two anionic fluorescein derivatives have proved very useful for the simple and definitive localization of neuronal degeneration in brain tissue sections. Initial work on the first generation fluorochrome, Fluoro-Jade, demonstrated the utility of this compound for the detection of neuronal degeneration induced by a variety of well-characterized neurotoxicants, including kainic acid, 3-nitropropionic acid, isoniazid, ibogaine, domoic acid, and high doses of dizocilpine maleate (MK-801) (Schmued et al., 1997). After validation, the tracer was used to reveal previously unreported sites of neuronal degeneration associated with other neurotoxicants. Preliminary findings with a second-generation fluorescein derivative, Fluoro-Jade B, suggest that this tracer is a specific and selective marker for the identification of neurons undergoing degeneration (both apoptotic and necrotic) (Eyupoglu et al., 2003); Fluoro-Jade B also provides improved staining and can stain the distal portion as well as the proximal portion of the dissected axon (the so-called anterograde and retrograde degeneration after axotomy). Furthermore, Fluoro-Jade tracers can be combined with other histologic methods, including immunofluoresence that can help in discriminating different types of neurodegeneration to obtain information on the neurochemical identity of the affected cells (Schmued and Hopkins, 2000); recent preliminary findings on a number of specialized applications of Fluoro-Jade include the detection of apoptosis, amyloid plaques, astrocytes, and dead cells in tissue culture.

An early observation concerning apoptosis was that cells entering apoptosis showed dramatic and characteristic changes in nuclear shape and organization (Fig. 1) (see for review Kerr et al., 1972; Wyllie, 1980; Ucker, 1991). It is still probably correct to say that the characteristic change in nuclear morphology is the most accurate indicator of the involvement of apoptosis in the death of a cell. This is true even in light of the apparently paradoxical observation that nuclear fragmentation per se is not essential for apoptosis; enucleated cells can still undergo other changes characteristic of apoptosis. This unequivocally demonstrates that the effectors of the apoptotic machinery are located in the cytoplasm. However, under normal conditions, changes in nuclear morphology remain an early and relatively unequivocal hallmark of apoptosis, with such changes occurring at an early point in the series of morphological events, usually soon after the onset of surface blebbing.

Apoptosis is an ATP (energy)-dependent process (Reed and Green, 2002). Since ATP levels fall to a point where the cell can no longer perform basic metabolic functions, the cell will die. Apoptotic cells exhibit significant reductions in their ATP levels, which can serve as an early marker of cell death. Depletion of energy pools is, however, not specific to apoptosis. Either exposure to toxic agents (secondary necrosis) or metabolic damage (primary necrosis) can also induce drops in ATP levels, albeit rapid ones (Leist et al., 1999), followed by necrotic cell death. The change in both ATP and ADP levels (ADP/ATP ratio) has been used to differentiate apoptosis from necrosis (Bradbury et al., 2000). In contrast, cell proliferation and growth arrest can both be recognized by increased levels of ATP and decreased levels of ADP. Determination of the ADP/ATP ratio offers highly consistent results and with excellent correlation to other markers of apoptosis (e.g., TUNEL-based techniques and caspase assays) (Bradbury et al., 2000).

c-Jun N-terminal kinase (JNK) is one of the main MAP kinase groups identified in mammals. Recent evidence suggests that activation of JNK plays an important role in neuronal apoptosis and other physiological and pathological processes (Ham et al., 2000). For measuring JNK activity easily in a large number of samples, one can use an assay that utilizes an N-terminal c-Jun fusion protein bead to selectively "pull down" JNK from cell lysate; c-Jun phosphorylation is then measured using a phospho-c-Jun-specific antibody. Alternatively, one might analyze JNK-specific activity by determining the phosphorylation of c-Jun by Western blotting using a phospho-c-Jun-specific antibody. Given the involvement of JNK in signaling pathways, which may not be directly related to apoptosis, care needs to be exercised in interpreting results obtained with such methods.

One of the first questions to resolve whenever searching for neurodegeneration, whether necrotic or apoptotic, is the ability to distinguish if the cells undergoing degeneration are neurons or glial cells. For this, immunohistochemistry is the most convenient and commonly used approach. Using specific antibodies for each cell population, one can easily identify the lineage of dying cells. Numerous neural cell type-specific (neurons, astroglia, oligodendrocytes, etc.) markers are currently available. For example, one may use antiGFAP to label astrocytes, antidoublecortin to identify neuroblasts (stem cells), antiNeuN to mark mature, differentiated neurons, or antiTuJ1 to study fibers.

As mentioned already, apoptosis is a genetically programmed phenomenon. A complex network of genes (Steller, 1995; Lossi and Merighi, 2003), in particular encoding members of the Bcl-2 family of proteins, play a central role in the regulation of apoptosis. Here, we focus on Bcl-2 family members as these have received most attention in the context of this article. The Bcl-2 family of proteins comprises death-inducer (proapoptotic) molecules such as Bax and Bcl-x_S and death-repressor (antiapoptotic) molecules such as Bcl-2 and Bcl-x_L. These various proteins, which can homo- or heterodimerize with each other, are activated by physiological or injurious stimuli, and appear to operate upstream of events leading to the final execution phase of the apoptotic process; the latter results from the activation of cysteine proteases, the caspases.

Caspases convey the apoptotic signal in a proteolytic cascade, with caspases cleaving and activating other caspases that then degrade other cellular targets that lead to cell death (Friedlander, 2003). Caspase activation can directly initiate the permeability transition of the mitochondrial membrane, resulting in the release of several mitochondrial proteins (see Fig. 2 for a simplified scheme). The large number of products developed to study caspases and their substrates is indirect testimony to their importance; because of space constraints, the authors here only review a few of these. Caspase-3 is a key protease that becomes activated during the early stages of apoptosis. Synthesized as an inactive proenzyme, the activated form cleaves and activates other caspases, in addition to cleaving specific targets in the cytoplasm and nucleus (e.g., DNA and nuclear membrane fragmentation). Once activated, caspase-3 serves as a marker for cells undergoing apoptosis. Several biotin- or FITC-tagged anti-active caspase 3 antibodies are available, facilitating their routine use (Gown and Willingham, 2002). Because caspase activity is likely to be the most specific indicator of the apoptotic process, the assay of caspase activity through the detection of specific cleavage products in target proteins represents a theoretically valid approach for measuring apoptosis. Recently, antibodies to the caspase-generated cleavage products of cytokeratin 18 have appeared on the market, with several studies demonstrating their utility, especially in cell culture, but probably also in fixed tissue sections (Leers et al., 1999). However, cytokeratin 18 is expressed only in certain cell types and this antibody is not broadly applicable to all cell types. The use of antibodies specific for more generally distributed cleaved substrates of caspases, such as the cleaved form of caspase 3, would have more general applicability (Srinivasan et al., 1998). Owing to their cell-permeable nature, a new line of cell-permeable fluorogenic caspase substrates enables the visualization of intracellular protease activities by standard fluorescence microscopy or multiparameter flow cytometry (see below) in living cells. The substrates, designed for caspase-1, caspase-6, caspase-8 (the caspase-3 processing enzyme), and caspase-9, detect early events in the apoptotic pathway before DNA degradation has started (Davis et al., 1998; Komoriya et al., 2000). Recently, these caspase substrates have been used to demonstrate that the pattern of caspase activation is not only dependent on the apoptosis-inducing agent employed, but also on the cell type (Komoriya et al., 2000). Events occurring downstream of caspase-3 activation include cleavage of poly(ADP-ribose) polymerase (PARP), an enzyme implicated in DNA damage and repair mechanisms. Cleavage of PARP from the native 116 kDa to 85 kDa is considered a hallmark of apoptosis (Sallmann et al., 1997). The availability of FITC-tagged anti-PARP antibodies therefore, provide another useful marker of apoptosis.

In healthy cells, cytochrome *c* is located in the space between the inner and outer mitochondrial membranes. An apoptotic stimulus triggers the release of cytochrome *c* from the mitochondria into the cytosol where it binds to Apaf-1. The cytochrome *c*/Apaf-1 complex activates caspase-9, which then activates caspase-3 and other downstream caspases. Cytochrome *c* released from the mitochondria into the cytosol can be detected by Western blotting using antibodies directed against cytochrome *c*. The procedure is simple, straightforward, and provides an effective means for detecting cytochrome *c* translocation from mitochondria into cytosol during apoptosis (Jemmerson et al., 1999).

As already alluded to, Bcl-2 family proteins form complexes, these complexes can enter the mitochondrial membrane where they regulate the release of cytochrome *c* and other proteins. When Bax, for example, localizes to the mitochondrial membrane, it acts to increase mitochondrial permeability, induces the release of cytochrome *c* and other mitochondrial proteins, leading to apoptosis ultimately. In contrast, Bcl-2 and Bcl-x_L prevent mitochondrial pore formation and therefore, block apoptosis. Antibodies (applications include immunocytochemistry, Western blotting) and gene probes (for Northern blotting, in situ hybridization histochemistry, and polymerase chain reaction analysis) are now available for measuring most key members of the Bcl-2 family in a variety of species, including humans, rats, and mice. Such studies have shown that Bcl-2 levels in the brain decline rapidly after birth, except for those areas displaying postnatal neurogenesis such as the dentate gyrus of the hippocampus. Further, numerous studies have shown that Bcl-2 expression can be induced in the adult brain, including the hippocampus, upon experience of various noxious stimuli. Unlike that of Bcl-2, the expression of Bcl-X_L occurs in neurons from early development through to senescence. The proapoptotic protein Bax is expressed through all life stages whereas the smaller proapoptotic splice variant of Bcl-2, Bcl-X_S is only barely detectable in the mature brain. To date, there is no evidence that corticosteroids, which represent the endocrine response to stress, can directly regulate or interact with any members of the *bcl-2* gene family. Rather, corticosteroids appear to influence the pro- and antiapoptotic gene expression and activity by interacting with p53, a ubiquitously distributed tumor suppressor protein, which has been shown to induce and repress the transcription of *bax* and *bcl-2*; glucocorticoids were recently shown to enhance the transactivation potential of p53 (Crochemore et al., 2002). Although measurements of gene or protein expression of Bcl-2 family members might be reasonably expected to correlate with apoptosis, recent studies have shown that absolute levels of these molecules do not reflect the actual viability of neurons in situ. Rather, the ratio of expression of pro-apoptotic (e.g., Bax) to antiapoptotic (e.g., Bcl-2) molecules factor has proven to be the factor determining neuronal survival (Almeida et al., 2000) insofar that this derivative correlates with the incidence of apoptosis measured by histochemical techniques such as TUNEL (see below).

Disruption of the mitochondrial transmembrane potential is one of the earliest events after apoptosis induction. Normally, cellular energy generated by mitochondrial respiration accumulates in the transmembrane space as an electron gradient called the mitochondrial transmembrane potential $\Delta\Psi_m$. Disruption of the mitochondrial transmembrane potential occurs following the onset of apoptosis. Using a fluorescent lipophilic cation as a mitochondrial activity marker, one can measure differences in the fluorescence displayed by healthy cells versus apoptotic cells: in healthy cells, the dye accumulates and aggregates in the mitochondria, producing a bright red fluorescence, while in apoptotic cells the fluorescence cation cannot aggregate in the mitochondria because of the altered transmembrane potential, thus remaining in its monomeric (green fluorescent) form within the cytoplasm. These fluorescent signals are analyzed by flow cytometry using the FITC channel for green monomers and the propidium iodide (PI) channel for red aggregates. Additionally, apoptotic and healthy cells can be viewed simultaneously by fluorescence microscopy using a wide-band pass filter. Some kits combine detection of disrupted mitochondrial transmembrane potential with changes in the composition of the plasma membrane.

Flow cytometry can be summarized as a method for measuring physical and biochemical features of cell components on a cell-by-cell basis, primarily by optical means. Fluorescent dyes or fluorophore-conjugated antibodies are used to report the quantities of specific cellular components, density of cellular markers and receptors – or even activation state of various enzymes. Put simpler, flow cytometers are highly sophisticated fluorescence microscopes, where fixed or living cells are not attached to a well-defined surface, but rather travel one by one, by continuous flow of a stream of the suspension past a sensor. Each cell scatters some of the excitation laser light, and the labeled cells emit fluorescent signals from the dye. These two parameters are sensed by photodetectors, data are collected, and processed by a computer.

Thc tcrm 'FACS' is Becton-Dickinson's registered trademark and is an acronym for "Fluorescence-Activated Cell Sorter." FACS is therefore, a machine that can rapidly separate cells in a suspension, based on the size and color of their fluorescence. (Note: not all flow cytometers are necessarily able to separate cells into different vials, but all can analyze the distribution of cell size and/or physical or biochemical cellular properties). These particular features of flow cytometric methods allow the identification and quantification of apoptotic cells as well as possible mechanisms of cell death.

The main flow cytometric approaches that can be used to identify apoptotic cells may be summarized as follows: (1) Apoptosis-associated changes in cell size and granularity can be detected by analysis of laser light scattered by the cell. (2) Using annexin V in combination with propidium iodide (PI), it is possible to differentiate between healthy, early apoptotic, and necrotic cells on the basis of the distribution of plasma membrane phospholipids as well as changes in membrane integrity. (3) Fluorochromes like Rhodamine 123 (Rhod123) or 3,3′-dihexiloxa-dicarbocyanine ($DiDOC_6$) reveal decreases in the mitochondrial transmembrane potential ($\Delta\Psi_m$) that occurs early during apoptosis. (4) Apoptotic cells can be recognized by their fractional DNA content, or by the presence of DNA strand breaks using fluorochrome-labeled nucleotides attached to the 3′-OH termini in a reaction catalyzed by exogenous terminal deoxynucleotidyl transferase (TdT). As regards the identification of putative mechanisms of cell death, after labeling with primary and fluorescent secondary antibodies, one can detect and measure: (a) cellular levels of death-related proteins (members of Bcl-2 family, proto-oncogenes like c-myc and ras, tumor-suppressor genes such as p53, etc.) or (b) study particular cell functions, such as mitochondrial metabolism, in the context of cell sensitivity to apoptosis.

The main virtue of flow cytometry lies in the possibility of multiparametric, correlated analysis of a multitude of cell attributes and markers, thus addressing problems of cellular heterogeneity. Flow cytometry also provides more effective data acquisition as compared to fluorescence microscopy (which has similar capabilities, but in which the sample size is limited up to a few hundred cells. Flow cytometry can easily measure 10,000–100,000 cells per sample!). There are, however, certain difficulties associated with this technique, which have to be taken into consideration when using flow cytometry in general. With respect to cell death detection, a major problem is that the single parameter on which the identification of apoptotic or necrotic cells relies on, may be absent, when apoptosis is atypical. Moreover, in the case of nonfixed, living cells, the dissociation procedure may damage the plasma membrane, resulting in PI (a widely used cell viability dye) to enter and label the cells as if they were dead. Clumping of cells may also pose technical difficulties. Since high-speed FACS machines use high pressure to achieve rapid acquisition rate, limitations such as cell type and viability must also be considered.

Externalization of phosphatidylserine (PS) and phosphatidylethanolamine are hallmarks of changes in the cell surface during apoptosis. Annexin V binds to PS with strong avidity and can be used as a marker of PS externalization using either microscopy or flow cytometry (when fluorescent-labeled annexin V is applied). Importantly, annexin V-binding cannot be applied to tissue sections or adherent cells, and when flow cytometric analysis is used on cell suspensions. PS are phospholipids only present on the cytoplasmic face of the plasma membrane and other internal membranes, and it remains unclear as to why certain subpopulations of cells ($< 30\%$) entering apoptosis externalize PS at a very early stage of the process, just after the fragmentation of the nucleus begins. However, since inhibition of caspase activity blocks PS externalization, a role for caspases is indicated. It is important to keep in mind that, during necrosis or the terminal lytic steps of apoptosis, PS that are actually localized on the inner face of the membrane might be accessed by annexin V, giving rise to false positives. When combined with a vital dye such as the red fluorescent DNA-binding compound PI, FITC-annexin V labeling can be used to distinguish necrotic from apoptotic cells; this is because PI does not penetrate live cell membranes or cells in the early phases of apoptosis but only cells that have lost membrane integrity as a result of necrosis or very late apoptosis.

The permeability of the plasma membrane is substantially different in necrotic and apoptotic cells, a fact that can be taken advantage of in the distinction between these forms of cell death. Thus, it is possible to distinguish between stages of apoptosis mainly on the basis of the plasma membrane permeability changes. Large-molecular weight DNA-binding dyes, such as PI or the homodimer of ethidium bromide (EB), cannot enter intact cells because of their large size and, without permeabilization treatments, do not label apoptotic cells until the final stage of cell lysis. On the other hand, in ethanol-fixed cells, which have been subsequently washed in phosphate–citrate buffer, extraction of the low-molecular weight DNA from apoptotic cells takes place, and apoptotic cells appear to the left of the normal G1 peak after PI staining (Fig. 3). This is a fast, simple, but not very specific, method for detecting apoptosis, and has the disadvantage that apoptosis of cells in late S phase or from G2 may be missed. Smaller dyes that can attach to DNA (such as DAPI, Hoechst 33342 or 33258), are furthermore able to enter, and differentially label apoptotic and healthy cells based on the condensation and subsequent fragmentation of the chromatin, which occurs early during apoptosis. Using flow cytometry, for instance, one can distinguish between healthy, apoptotic and dead cells by the simultaneous use of the blue-fluorescent Hoechst 33342 dye (which stains the condensed chromatin of apoptotic cells more brightly than that of normal cells) and PI, which

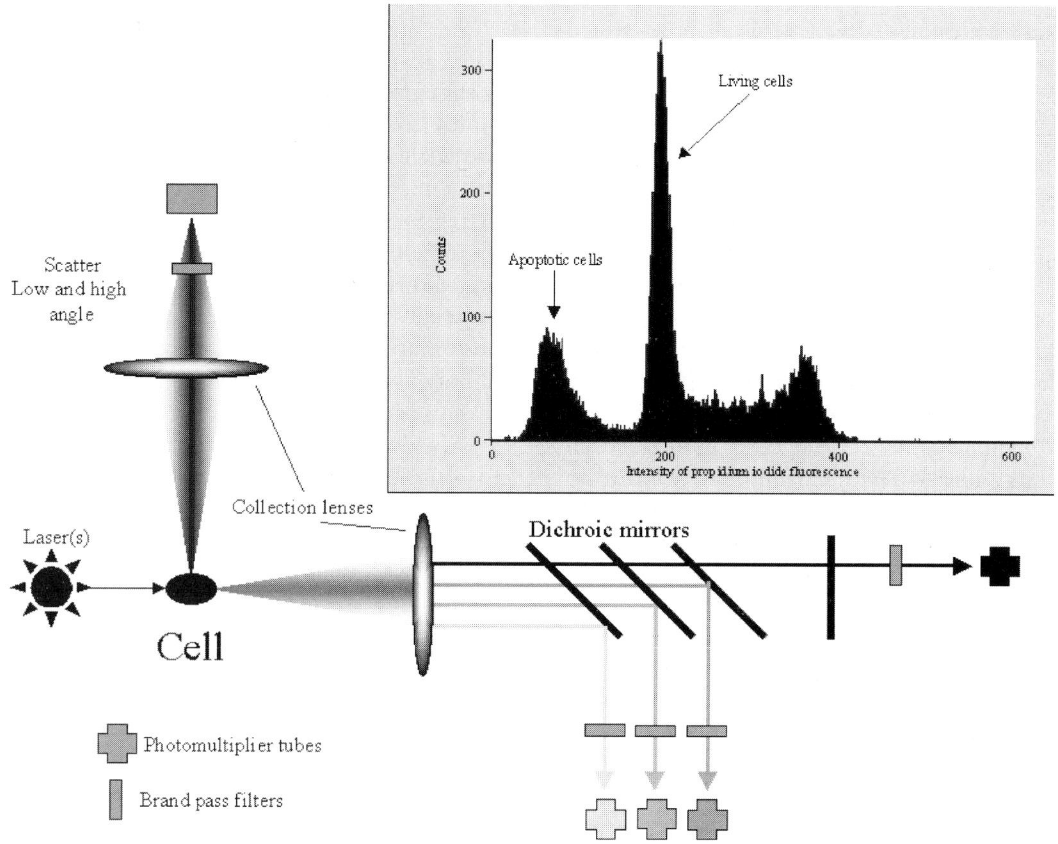

Fig. 3. Separation of living/apoptotic cells by flow cytometry. For flow cytometric analysis, ethanol-fixed cells are washed in phosphate–citrate buffer and stained with propidium iodide. As cells pass in front of a laser, they absorb, diffract, refract, and reflect incident light, and emit fluorescence. The scattered light is focused by a lens into a photomultiplier, the emitted fluorescent signal is optically filtered through dichroic mirrors, and subsequently processed by wide-band pass filters selected to optimize the various fluorescent emissions; signals are detected by photomultiplier tubes, and based on fluorescence intensity profiles, living and apoptotic cells can be distinguished.

labels dead cells (Pollack and Ciancio, 1991). Acridine orange (AO) (another cell-permeant nucleic acid-binding dye that emits green fluorescence when bound to double-stranded DNA, and red fluorescence when bound to single-stranded DNA or RNA) is another useful probe for identifying apoptotic cells, because its metachromatic fluorescence is sensitive to DNA conformation. Careful combination of fluorescent dyes, furthermore, allows even more accurate determination of different stages of the apoptotic process: for example, 7-aminoactinomycin D (7-AAD) can be used alone or in combination with Hoechst 33342 to separate populations of live, early apoptotic, and late apoptotic, cells (Schmid et al., 1994). In mixed cell populations, however, identification of cell types is necessary, and has also to be taken into consideration. For instance, the combination of acridine orange and ethidium bromide (AO/EB) is useful to accurately differentiate between healthy, early apoptotic, late apoptotic, and necrotic cells (Liegler et al., 1995), but cannot be used for phenotypic analyses due to the broad emission spectrum of AO and EB. Certain techniques, like DNA strand-break labeling by terminal deoxynucleotidyl transferase (TdT) can overcome this problem, but are technically very demanding. TdT adds biotinylated or digoxigenin-labeled nucleotides to the strand-breaks in the DNA of apoptotic cells – apoptotic cells therefore, can be detected by using fluorochrome-labeled streptavidin conjugates or fluorochrome-labeled anti-digoxygenin antibodies in flow cytometric analysis. By combining this procedure with phenotypic markers tagged to other dyes, it is even possible to obtain cell cycle profiles in cells of a given phenotype (Li et al., 1996).

As mentioned earlier, changes in plasma membrane permeability are signs of late phases of cell lysis. Changes in mitochondrial membrane permeability, however, occur much earlier during apoptosis, and are considered to be a distinctive feature of early programmed cell death. The mitochondrial permeability transition (MPT) is intimately linked to the opening of a "megachannel," the permeability transition pore (PTP). Ionic equilibration through the PTP results in disruption of the mitochondrial transmembrane potential ($\Delta\Psi_m$), uncoupling of the respiratory chain, and release of cytochrome *c* into the cytoplasm. Of all these features, changes in the mitochondrial permeability can be relatively easily followed by application of fluorescent dyes (e.g., $DiOC_6$), while the subsequent ionic and electrical fluctuations can be investigated by patch-clamp techniques or certain fluorophores. While certain drugs, like the green-fluorescent calcein (which is produced from the nonfluorescent calcein-AM form within the cell itself) are used to indicate PTP opening, and, subsequently, the taking up of the dye into the mitochondrial matrix, others (like JC-1, JC-9, or $DiOC_6$) do not just simply accumulate in the mitochondria, but also indicate changes in $\Delta\Psi_m$ in single-cell imaging or flow cytometric assays. Other dyes, like MitoTracker® Red CMXRos can be fixed by aldehyde-based fixatives and can thus be used for other subsequent analytical procedures such as immunocytochemistry, DNA end-labeling, in situ hybridization, or counterstaining. Ionic concentrations in the mitochondria can be monitored using patch-clamp techniques or fluorescent dyes like the Ca^{2+}-sensor Rhod-2.

Loss of DNA integrity is characteristic of apoptosis (Collins et al., 1997). When DNA extracted from apoptotic cells is analyzed using gel electrophoresis, a characteristic internucleosomal "ladder" of DNA fragments (typically, 180–200 bp in length) is revealed (Compton and Cidlowski, 1986; Walker et al., 1999); larger DNA fragments have also been seen at earlier stages in apoptotic cell cultures. Although these electrophoretic methods are commonly used in apoptosis detection, the results they provide can present interpretational difficulties. Also, these methods cannot be easily applied, requiring extraction of DNA from large numbers of cells undergoing apoptosis in a relatively synchronous way; however, such synchrony is not always present, especially in tissues (Collins et al., 1997), and as noted above, apoptosis is a relatively rare event, occurring in only a subset of cells within a given structure, thus raising problems of sensitivity.

A widely used method that has contributed much to our knowledge of stress- and corticosteroid-induced apoptosis in the brain is also based on the detection of DNA strand-breaks. This approach detects 3′-OH ends of single-stranded DNA after the addition of labeled nucleotides to the open ends of DNA in a procedure known as in situ end-labeling (ISEL). The latter may be achieved using either *E. coli* polymerase (or its Klenow fragment) in a method called in situ nick-translation (ISNT), or terminal transferase in a method referred to as terminal deoxynucleotidyltransferase-mediated dUTP nick-end labeling (TUNEL) (Modak and Bollum, 1972; Gavrieli et al., 1992; Jin et al., 1999). These methods allow the cytochemical demonstration of free DNA strand openings.

TUNEL staining is now widely used for the detection of apoptotic cells in tissue sections and cells in culture. Despite its apparent simplicity, unless used optimally, this technique may lack sensitivity and, worse, specificity. For example, TUNEL can reportedly label both apoptotic and necrotic cells, and potentially, proliferating cells also, although these problems are less-frequently encountered in tissue sections than when cultured cells are stained. Moreover, as already noted in the main text, apoptotic cells can be easily recognized on the basis of their unambiguous morphological characteristics. With regard to mitotic cells, it deserves mentioning that although chromatin condensation at telophase may mimic apoptosis, the greatest analogy between mitotic and apoptotic aspects occurs in abortive mitosis, a form of cell division that leads to active cell death (sometimes named "mitotic catastrophe").

The major problems associated with the TUNEL technique, especially in tissues, can be summarized as follows: (i) without pretreatment, TUNEL sensitivity is poor and can lead to false negatives; (ii) established pretreatments (proteinase K, microwaves) can easily result in labeling of morphologically normal nuclei; and (iii) the method depends on good fixation and can prove problematic when large tissue blocks are used (outside–inside gradients of penetration of fixative). Other considerations include: (i) the DNA breaks, which are targeted by TUNEL, are less accessible than intact DNA; (ii) besides apoptosis, DNA recombination, replication, repair or compaction–relaxation during mitosis, tissue electrocoagulation, autolysis, fixation, paraffin embedding, cutting, and pretreatments with H2O2, detergent, proteinase K, and microwaves can all result in DNA breaks; and (iii) DNA compaction (a hallmark of apoptosis) and protein cross-linking and precipitation induced by fixation can mask the 3-OH recessed ends.

Despite the above caveats, TUNEL is still regarded as a reliable marker of the DNA fragmentation, which typically occurs in apoptosis. The key to distinguishing between apoptotic and nonapoptotic DNA is the cautious use of "break disclosure" reagents (detergents, proteases, microwaves). Extensive tests have led some authors to propose that optimal staining results from qualitative adaptations of retrieval techniques rather than retrieval reinforcement; for example, quite different pHs are necessary to obtain specific labeling in formalin- versus Bouin-fixed tissues. When fixation is controlled (e.g., homogeneous and light) as is usually the case in prospective studies, proteinase K alone may be sufficient for all cross-linking aldehyde fixatives (paraformaldehyde, formalin, B5). Proteinase K and microwave treatment may be necessary when tissues are fixed for too long and/or in precipitating solutions (Bouin's).

Table 1. Dyes commonly used for quantifying apoptosis by flow cytometry

Marker dye	MW	Absorption max.	Emission max.
DAPI	350.25	358	461
PI	668.4	535	617
$DiOC_6$	572.73	484	501
Rhod 123	380.83	507	529
JC-1	652.23	514	529
Annexin V conjugates	Depends on fluorescent conjugate		
AO	301.82	500	526
7-AAD	1270	546	647
MitoTracker Red® CMXRos	531.52	578	599
Rhod-2, AM	1123.96	550	571
Hoechst 33342, 33258	623.96	352	461

Nonspecific (background) staining can also present a problem, even when optimal pretreatments are applied. This can be overcome by optimizing the detection system, e.g., dilution of the enzyme-coupled antibody, choice of enzyme, careful monitoring of color development. Absence of standardization of color reaction implies suboptimal quantification of those cells which might otherwise show morphological signs of apoptosis. Also to be remembered is that all labeled cells, irrespective of intensity of labeling, should to be counted as long as they show morphological features of apoptosis.

Another method for detecting these single-strand ends is the use of monoclonal antibody reactive with single-stranded DNA (Naruse et al., 1994; Frankfurt et al., 1996). Since preservation/fixation procedures can have dramatic effects on the detection of single-stranded DNA (Labat-Moleur et al., 1998; Tateyama et al., 1998), careful consideration must be given to this issue and optimized for each cell type or tissue. The investigator should also keep in mind that in cases of overfixation, for example, open DNA strands will be inaccessible to assay reagents (Nakamura et al., 1997). This can be overcome by introducing protease treatments prior to ISNT or TUNEL procedures. Proteases must be used cautiously – protease treatments can mimic the actions of endogenous caspases, thus leading to artefactual DNA strand-breaks. Here, it is also important to note that, depending on permeabilization and fixation protocols, some methods detect so-called preapoptotic nuclei in which strand breaks are detected in the absence of apoptosis-like changes in the morphology of the nucleus. Alternatively, positively labeled strand-breaks may not correlate with nuclear fragmentation in individual cells, or DNA strand-breaks may only become detectable at relatively late stages of the apoptotic process (Collins et al., 1997). Recently, a number of authors have indicated reservations about the use of the TUNEL and ISNT assays for detecting apoptosis. It has become apparent that single-stranded DNA ends are not necessarily specific for apoptosis since they may also occur in necrotic cells (Kockx et al., 1998; Mizoguchi et al., 1998). Therefore, although these methods have been, and remain, very useful (their major advantage being that they can be applied directly to intact tissue sections, thus providing good anatomical resolution), the results they yield must be treated with extreme care; for example, in our studies (e.g. Hassan et al., 1996), we only consider TUNEL-positive cells as apoptotic if they simultaneously display the typical morphological characteristics of apoptotic cells; positively stained cells, which have a clearly defined nucleus and cell body are excluded, as are cell fragments and endothelial cells; further stringency is added by ensuring that the person performing the cell counts is unaware of the treatments.

Concluding remarks

The main objective of this article was to provide a brief overview of the methodologies available to study the cellular and molecular basis of stress-induced neurodegeneration. While our coverage is by no means exhaustive, we aimed to review each of the major approaches in current use, both in brain tissue and cell culture, and to discuss each of the methods in terms of their advantages and inherent drawbacks; it should become obvious to the reader that no single method can be considered to be definitive by itself, and investigators are encouraged to confirm results obtained one method with that from an alternative technique whenever feasible, in order to avoid from misinterpretation of results. We also attempted to discuss certain important aspects of experimental design in the hope that the use of standardized procedures will contribute to our increased understanding of stress-induced neuronal damage.

List of abbreviations

5-HT	serotonin
7-ADD	7-aminoactinomycin D
AIF	apoptosis-inducing factor
AO	acridine orange
APAF-1	apoptotic protease activating factor 1
BDNF	brain-derived nerve factor
CA	field of hippocampus
DAPI	4′-6-diamidino-2-phenylindole
$DiDOC_6$	3,3′-dihexiloxa-dicarbocyanine
EB	ethidium bromide
FACS	fluorescence-activated cell sorter
FGF-2	fibroblast growth factor
FITC	fluorescein isothiocyanate
GFAP	glial fibrillary acidic protein
GR	glucocorticoid receptor
ISEL	in situ end-labeling
ISNT	in situ nick-translation
JC-1	5,5′,6,6′-tetra-chloro-1,1′,3,3′-tetraethyl benzimidazolyl-carbocyanine iodide
JNK	c-Jun N-terminal kinase
MK-801	dizocilpine maleate
MPT	mitochondrial permeability transition
MR	mineralocorticoid receptor
N_A	number per unit area
NeuN	neuronal-specific nuclear protein
NGF	nerve growth factor
NMDA	*N*-methyl-*D*-aspartic acid
NO	nitric oxide
N_V	number per unit volume
PARP	poly(ADP-ribose) polymerase
PI	propidium iodide
PS	phosphatidylserine
PTP	permeability transition pore
Rhod	rhodamine
ROS	reactive oxygen species
TdT	terminal deoxynucleotidyl transferase
TGF	transforming growth factor
TuJ1	neuron-specific class III beta-tubulin
TUNEL	Terminal deoxynucleotidyl transferase-mediated dUTP nick end-labeling
URS	uniform random systematic
$\Delta\Psi_m$	mitochondrial transmembrane potential

Acknowledgements

This article was written under the auspices of the German–Portuguese cooperation – Gabinete de Relações Internacionais da Ciência e do Ensino

Superior (GRICES) and the German Academic Exchange Service (DAAD). Zs. Némethy was supported by Marie Curie Indivdual Fellowship from the European Commission (QLK6-CT-2001-51072). Many of the methods described here were established in the authors' laboratories by past and present colleagues who are duly thanked.

References

Abercrombie, M. (1946) Estimation of nuclear population from microtome sections. Anat. Rec., 94: 239–247.

Ahima, R.S. and Harlan, R.E. (1990) Charting of type II glucocorticoid receptor-like immunoreactivity in the rat central nervous system. Neuroscience, 39: 579–604.

Ahima, R., Krozowski, Z. and Harlan, R. (1991) Type I corticosteroid receptor-like immunoreactivity in the rat CNS: distribution and regulation by corticosteroids. J. Comp. Neurol., 313: 522–538.

Almeida, O.F.X., Conde, G.L., Crochemore, C., Demeneix, B.A., Fischer, D., Hassan, A.H., Meyer, M., Holsboer, F. and Michaelidis, T.M. (2000) Subtle shifts in the ratio between pro- and antiapoptotic molecules after activation of corticosteroid receptors decide neuronal fate. FASEB J., 14: 779–790.

Barbany, G. and Persson, H. (1992) Regulation of Neurotrophin mRNA expression in the rat brain by glucocorticoids. Eur. J. Neurosci., 4: 396–403.

Bradbury, D.A., Simmons, T.D., Slater, K.J. and Crouch, S.P. (2000) Measurement of the ADP: ATP ratio in human leukaemic cell lines can be used as an indicator of cell viability, necrosis and apoptosis. J. Immunol. Methods, 240: 79–92.

Cajal, R. (1893) Neue darstellung vom histologischen bau des centralnervensystem. Arch. Anat. Physiol., 319: 428.

Collins, J.A., Schandi, C.A., Young, K.K., Vesely, J. and Willingham, M.C. (1997) Major DNA fragmentation is a late event in apoptosis. J. Histochem. Cytochem., 45: 923–934.

Compton, M.M. and Cidlowski, J.A. (1986) Rapid in vivo effects of glucocorticoids on the integrity of rat lymphocyte genomic deoxyribonucleic acid. Endocrinology, 118: 38–45.

Crochemore, C., Michaelidis, T.M., Fischer, D., Loeffler, J.P. and Almeida, O.F.X. (2002) Enhancement of p53 activity and inhibition of neural cell proliferation by glucocorticoid receptor activation. FASEB J., 16: 761–770.

Davis, R.E., Mysore, V., Browning, J.C., Hsieh, J.C., Lu, Q.A. and Katsikis, P.D. (1998) In situ staining for poly(ADP-ribose) polymerase activity using an NAD analogue. J. Histochem. Cytochem., 46: 1279–1289.

Erickson, K., Drevets, W. and Schulkin, J. (2003) Glucocorticoid regulation of diverse cognitive functions in normal and pathological emotional states. Neurosci. Biobehav. Rev., 27: 233–246.

Eyupoglu, I.Y., Savaskan, N.E., Brauer, A.U., Nitsch, R. and Heimrich, B. (2003) Identification of neuronal cell death in a model of degeneration in the hippocampus. Brain Res. Protoc., 11: 1–8.

Frankfurt, O.S., Robb, J.A., Sugarbaker, E.V. and Villa, L. (1996) Monoclonal antibody to single-stranded DNA is a specific and sensitive cellular marker of apoptosis. Exp. Cell Res., 226: 387–397.

Friedlander, R.M. (2003) Apoptosis and caspases in neurodegenerative diseases. New Engl. J. Med., 348: 1365–1375.

Funder, J.W. (1996) Mineralocorticoid receptors in the central nervous system. J. Steroid Biochem. Mol. Biol., 56: 179–183.

Gavrieli, Y., Sherman, Y. and Ben-Sasson, S.A. (1992) Identification of programmed cell death in situ via specific labeling of nuclear DNA fragmentation. J. Cell Biol., 119: 493–501.

Gown, A.M. and Willingham, M.C. (2002) Improved detection of apoptotic cells in archival paraffin sections: immunohistochemistry using antibodies to cleaved caspase 3. J. Histochem. Cytochem., 50: 449–454.

Gundersen, H.J.G., Bagger, P., Bendtsen, T.F., Evans, S.M., Korbo, L., Marcussen, N., Moller, A., Nielsen, K., Nyengaard, J.R., Pakkenberg, B., Sorensen, F.B., Vesterby, A. and West, M.J. (1988) The new stereological tools: disector, fractionator, nucleator and point sampled intercepts and their use in pathological research and diagnosis. APMIS, 96: 379–394.

Ham, J., Eilers, A., Whitfield, J., Neame, S.J. and Shah, B. (2000) c-Jun and the transcriptional control of neuronal apoptosis. Biochem. Pharmacol., 60: 1015–1021.

Hansson, A.C., Cintra, A., Belluardo, N., Sommer, W., Bhatnagar, M., Bader, M., Ganten, D. and Fuxe, K. (2000) Gluco- and mineralocorticoid receptor-mediated regulation of neurotrophic factor gene expression in the dorsal hippocampus and the neocortex of the rat. Eur. J. Neurosci., 12: 2918–2934.

Hassan, A.H., von Rosenstiel, P., Patchev, V.K., Holsboer, F. and Almeida, O.F.X. (1996) Exacerbation of apoptosis in the dentate gyrus of the aged rat by dexamethasone and the protective role of corticosterone. Exp. Neurol., 140: 43–52.

Herman, J.P. and Cullinan, W.E. (1997) Neurocircuitry of stress: central control of the hypothalamo–pituitary–adrenocortical axis. Trends Neurosci., 20: 78–84.

Honig, L.S. and Rosenberg, R.N. (2000) Apoptosis and neurologic disease. Am. J. Med., 108: 317–330.

Jemmerson, R., Liu, J., Hausauer, D., Lam, K.P., Mondino, A. and Nelson, R.D. (1999) A conformational change in cytochrome c of apoptotic and necrotic cells is detected by monoclonal antibody binding and mimicked by association of the native antigen with synthetic phospholipid vesicles. Biochemistry, 38: 3599–3609.

Jin, K., Chen, J., Nagayama, T., Chen, M., Sinclair, J., Graham, S.H. and Simon, R.P. (1999) In situ detection of neuronal DNA strand breaks using the Klenow fragment of DNA polymerase I reveals different mechanisms of neuron death after global cerebral ischemia. J. Neurochem., 72: 1204–1214.

Kerr, J.F., Wyllie, A.H. and Currie, A.R. (1972) Apoptosis: a basic biological phenomenon with wide-ranging implications in tissue kinetics. Br. J. Cancer, 26: 239–257.

Kockx, M.M., Muhring, J., Knaapen, M.W. and de Meyer, G.R. (1998) RNA synthesis and splicing interferes with DNA in situ end labeling techniques used to detect apoptosis. Am. J. Pathol., 152: 885–888.

Komoriya, A., Packard, B.Z., Brown, M.J., Wu, M.L. and Henkart, P.A. (2000) Assessment of caspase activities in intact apoptotic thymocytes using cell-permeable fluorogenic caspase substrates. J. Exp. Med., 191: 1819–1828.

Labat-Moleur, F., Guillermet, C., Lorimier, P., Robert, C., Lantuejoul, S., Brambilla, E. and Negoescu, A. (1998) TUNEL apoptotic cell detection in tissue sections: critical evaluation and improvement critical evaluation and improvement. J. Histochem. Cytochem., 46: 327–334.

Laping, N.J., Teter, B., Nichols, N.R., Rozovsky, I. and Finch, C.E. (1994) Glial fibrillary acidic protein: regulation by hormones, cytokines, and growth factors. Brain Pathol., 4: 259–75.

Leers, M.P., Kolgen, W., Bjorklund, V., Bergman, T., Tribbick, G., Persson, B., Bjorklund, P., Ramaekers, F.C., Bjorklund, B., Nap, M., Jornvall, H. and Schutte, B. (1999) Immunocytochemical detection and mapping of a cytokeratin 18 neo-epitope exposed during early apoptosis. J. Pathol., 187: 567–572.

Leist, M., Single, B., Naumann, H., Fava, E., Simon, B., Kuhnle, S. and Nicotera, P. (1999) Inhibition of mitochondrial ATP generation by nitric oxide switches apoptosis to necrosis. Exp. Cell Res., 249: 396–403.

Li, X., Melamed, M.R. and Darzynkiewicz, Z. (1996) Detection of apoptosis and DNA replication by differential labeling of DNA strand breaks with fluorochromes of different color. Exp. Cell Res., 222: 28–37.

Liegler, T.J., Hyun, W., Yen, T.S.B. and Stittes, D.P. (1995) Detection and quantification of live, apoptotic, and necrotic human peripheral lymphocytes by single-laser flow cytometry. Clin. Diagnostic Lab. Immunol., 2: 369–376.

Liu, B. and Hong, J.S. (2003) Role of microglia in inflammation-mediated neurodegenerative diseases: mechanisms and strategies for therapeutic intervention. J. Pharmacol. Exp. Ther., 304: 1–7.

Lossi, L. and Merighi, A. (2003) In vivo cellular and molecular mechanisms of neuronal apoptosis in the mammalian CNS. Prog. Neurobiol., 69: 287–312.

Magarinos, A.M. and McEwen, B.S. (1995a) Stress-induced atrophy of apical dendrites of hippocampal CA3c neurons: comparison of stressors. Neuroscience, 69: 83–88.

Magarinos, A.M. and McEwen, B.S. (1995b) Stress-induced atrophy of apical dendrites of hippocampal CA3c neurons: involvement of glucocorticoid secretion and excitatory amino acid receptors. Neuroscience, 69: 89–98.

Majno, G. and Joris, I. (1995) Apoptosis, oncosis, and necrosis. An overview of cell death. Am. J. Pathol., 146: 3–15.

Mattson, M.P. (2000) Apoptotic and anti-apoptotic synaptic signaling mechanisms. Brain Pathol., 10: 300–312.

McConkey, D.J. (1998) Biochemical determinants of apoptosis and necrosis. Toxicol. Lett., 99: 157–168.

McCormick, C.M., McNamara, M., Mukhopadhyay, S. and Kelsey, J.E. (1997) Acute corticosterone replacement reinstates performance on spatial and nonspatial memory tasks 3 months after adrenalectomy despite degeneration in the dentate gyrus. Behav. Neurosci., 111: 518–531.

McEwen, B.S. (1999) Stress and the aging hippocampus. Front. Neuroendocrinol., 20: 49–70.

Melcangi, R.C., Magnaghi, V., Cavarretta, I., Riva, M.A. and Martini, L. (1997) Corticosteroid effects on gene expression of myelin basic protein in oligodendrocytes and of glial fibrillary acidic protein in type 1 astrocytes. J. Neuroendocrinol., 9: 729–733.

Miller, M.M., Antecka, E. and Sapolsky, R. (1989) Short term effects of glucocorticoids upon hippocampal ultrastructure. Exp. Brain Res., 77: 309–314.

Mizoguchi, M., Manabe, M., Kawamura, Y., Kondo, Y., Ishidoh, K., Kominami, E., Watanabe, K., Asaga, H., Senshu, T. and Ogawa, H. (1998) Deimination of 70-kD nuclear protein during epidermal apoptotic events in vitro. J. Histochem. Cytochem., 46: 1303–1309.

Modak, S.P. and Bollum, F.J. (1972) Detection and measurement of single-strand breaks in nuclear DNA in fixed lens sections. Exp. Cell Res., 75: 307–313.

Muller, M., Holsboer, F. and Keck, M.E. (2002) Genetic modification of corticosteroid receptor signalling: novel insights into pathophysiology and treatment strategies of human affective disorders. Neuropeptides, 36: 117–131.

Nakamura, M., Yagi, H., Ishii, T., Kayaba, S., Soga, H., Gotoh, T., Ohtsu, S., Ogata, M. and Itoh, T. (1997) DNA fragmentation is not the primary event in glucocorticoid-induced thymocyte death in vivo. Eur. J. Immunol., 27: 999–1004.

Naruse, I., Keino, H. and Kawarada, Y. (1994) Antibody against single-stranded DNA detects both programmed cell death and drug-induced apoptosis. Histochemistry, 101: 73–78.

Nichols, N.R. (1999) Glial responses to steroids as markers of brain aging. J. Neurobiol., 40: 585–601.

Pollack, A. and Ciancio, G. (1991) Cell-cycle phase-specific analysis of cells viability using Hoechst 33342 and propidium iodide after ethanol preservation. In: Darzynkiewicz, Z. and Crissman, H.A. (Eds.), Flow Cytometry. Academic Press, San Diego, pp. 19–24.

Ramos-Remus, C., Gonzalez-Castaneda, R.E., Gonzalez-Perez, O., Luquin, S. and Garcia-Estrada, J. (2002) Prednisone induces cognitive dysfunction, neuronal degeneration, and reactive gliosis in rats. J. Investig. Med., 50: 458–464.

Reed, J.C. and Green, D.R. (2002) Remodeling for demolition: changes in mitochondrial ultrastructure during apoptosis. Mol. Cell, 9: 1–3.

Reul, J.M. and de Kloet. E.R. (1985) Two receptor systems for corticosterone in rat brain: microdistribution and differential occupation. Endocrinology, 117: 2505–2511.

Rusakov, D.A., Davies, H.A., Harrison, E., Diana, G., Richter-Levin, G., Bliss, T.V. and Stewart, M.G. (1997) Ultrastructural synaptic correlates of spatial learning in rat hippocampus. Neuroscience, 80: 69–77.

Sallmann, F.R., Bourassa, S., Saint-Cyr, J. and Poirier, G.G. (1997) Characterization of antibodies specific for the caspase cleavage site on poly(ADP-ribose) polymerase: specific detection of apoptotic fragments and mapping of the necrotic fragments of poly(ADP-ribose) polymerase. Biochem. Cell Biol., 75: 451–456.

Sapolsky, R.M., Romero, L.M. and Munck, A.U. (2000) How do glucocorticoids influence stress responses? Integrating permissive, suppressive, stimulatory, and preparative actions. Endocr. Rev., 21: 55–89.

Schmid, I., Uittenbogaart, C.H. and Giorgi, J.V. (1994) Sensitive method for measuring apoptosis and cell surface phenotype in human thymocytes by flow cytometry. Cytometry, 15: 12–20.

Schmued, L.C. and Hopkins, K.J. (2000) Fluoro-Jade B: a high affinity fluorescent marker for the localization of neuronal degeneration. Brain Res., 874: 123–130.

Schmued, L.C., Albertson, C. and Slikker Jr., W. (1997) Fluoro-Jade: a novel fluorochrome for the sensitive and reliable histochemical localization of neuronal degeneration. Brain Res., 751: 37–46.

Segal, M. (2002) Changing views of Cajal's neuron: the case of the dendritic spine. Prog. Brain Res., 136: 101–107.

Smith, M.A., Makino, S., Kvetnansky, R. and Post, R.M. (1995) Stress and glucocorticoids affect the expression of brain-derived neurotrophic factor and neurotrophin-3 mRNAs in the hippocampus. J. Neurosci., 15: 1768–1777.

Sousa, N., Madeira, M.D. and Paula-Barbosa, M.M. (1997) Structural alterations of the hippocampal formation of adrenalectomized rats: an unbiased stereological study. J. Neurocytol., 26: 423–438.

Sousa, N., Madeira, M.D. and Paula-Barbosa, M.M. (1999a) Corticosterone replacement restores normal morphological features to the hippocampal dendrites, axons and synapses of adrenalectomized rats. J. Neurocytol., 28: 541–558.

Sousa, N., Paula-Barbosa, M.M. and Almeida, O.F.X. (1999b) Ligand and subfield specificity of corticoid-induced neuronal loss in the rat hippocampal formation. Neuroscience, 89: 1079–1087.

Sousa, N., Lukoyanov, N.V., Madeira, M.D., Almeida, O.F.X. and Paula-Barbosa, M.M. (2000) Reorganization of the morphology of hippocampal neurites and synapses after stress-induced damage correlates with behavioral improvement. Neuroscience, 97: 253–266.

Srinivasan, A., Roth, K.A., Sayers, R.O., Shindler, K.S., Wong, A.M., Fritz, L.C. and Tomaselli, K.J. (1998) In situ immunodetection of activated caspase-3 in apoptotic neurons in the developing nervous system. Cell Death Differ., 5: 1004–1016.

Steller, H. (1995) Mechanisms and genes of cellular suicide. Science, 267: 1445–1449.

Stewart, M.G., Harrison, E., Rusakov, D.A., Richter-Levin, G. and Maroun, M. (2000) Re-structuring of synapses 24 hours after induction of long-term potentiation in the dentate gyrus of the rat hippocampus in vivo. Neuroscience, 100: 221–227.

Tateyama, H., Tada, T., Hattori, H., Murase, T., Li, W.X. and Eimoto, T. (1998) Effects of prefixation and fixation times on apoptosis detection by in situ end-labeling of fragmented DNA. Arch. Pathol. Lab. Med., 122: 252–255.

Trejo, J.L., Rua, C., Cuchillo, I. and Machin, C. (1998) Calbindin-D28k- and astroglial protein-immunoreactivities, and ultrastructural differentiation in the prenatal rat cerebral cortex and hippocampus are affected by maternal adrenalectomy. Dev. Brain Res., 108: 161–177.

Ucker, D.S. (1991) Death by suicide: one way to go in mammalian cellular development? New Biol., 3: 103–109.

Van Eekelen, J.A., Jiang, W., De Kloet, E.R. and Bohn, M.C. (1988) Distribution of the mineralocorticoid and the glucocorticoid receptor mRNAs in the rat hippocampus. J. Neurosci. Res., 21: 88–94.

Vijayan, V.K. and Cotman, C.W. (1987) Hydrocortisone administration alters glial reaction to entorhinal lesion in the rat dentate gyrus. Exp. Neurol., 96: 307–320.

Walker, P.R., Leblanc, J., Smith, B., Pandey, S. and Sikorska, M. (1999) Detection of DNA fragmentation and endonucleases in apoptosis. Methods, 17: 329–338.

Watanabe, Y., Gould, E., Cameron, H.A., Daniels, D.C. and McEwen, B.S. (1992) Phenytoin prevents stress- and corticosterone-induced atrophy of CA3 pyramidal neurons. Hippocampus, 2: 431–435.

Weibel, E.R., Kistler, G.S. and Scherle, W.F. (1966) Practical stereological methods for morphometric cytology. J. Cell Biol., 30: 23–38.

West, M.J. (1999) Stereological methods for estimating the total number of neurons and synapses: issues of precision and bias. Trends Neurosci., 22: 51–61.

West, M.J., Slomianka, L. and Gundersen, H.J.G. (1991) Unbiased stereological estimation of the total number of neurons in the subdivisions of the rat hippocampus using the optical fractionator. Anat. Rec., 231: 482–497.

Wyllie, A.H. (1980) Glucocorticoid-induced thymocyte apoptosis is associated with endogenous endonuclease activation. Nature, 284: 555–556.

Yau, J.L. and Seckl, J.R. (2001) 11beta-hydroxysteroid dehydrogenase type I in the brain; thickening the glucocorticoid soup. Mol. Psychiatry, 6: 611–614.

T. Steckler, N.H. Kalin and J.M.H.M. Reul (Eds.)
Handbook of Stress and the Brain, Vol. 15
ISBN 0-444-51173-3

CHAPTER 5.7

Enhancing resilience to stress: the role of signaling cascades

Pei-Xiong Yuan*, Rulun Zhou, Neda Farzad, Todd D. Gould, Neil A. Gray, Jing Du and Husseini K. Manji

Laboratory of Molecular Pathophysiology, National Institute of Mental Health, Building 49, Room B1EE16, 49 Convent Drive, Bethesda, MD 20892-4405, USA

Abstract: A central role of the brain involves both the perception and response to stressful stimuli. The mechanisms by which the brain responds to stress are of critical importance to the appropriate function of an organism. In this regard cellular resilience in the brain and responses at the neuronal level to stress have become an intriguing area of research. The results of stress in the brain appear to include atrophy of hippocampal neurons, other morphometric and structural brain changes, a decrease in neurotrophic support, and changes in behavior in preclinical models. The hypothalamic–pituitary–adrenal (HPA) axis appears to play a critical role in mediating these effects. Increasing recent data implicate a critical role for glucocorticoids, and corticotrophic hormone-releasing factor (CRF), in long-term effects of early-life stress on hippocampal integrity and function. Clinical evidence is consistent with the preclinical evidence including structural and morphometric brain changes, and the finding that a significant percentage of patients with mood disorders display some form of HPA axis activation. Stress is a critical factor in the development of some psychiatric disorders. Some antidepressants, electroconvulsive shock therapy (ECT), and mood stabilizers (lithium) appear to modulate glucocorticoid receptor number and/or function, components of the HPA axis, and neurotrophic pathways and molecules in preclinical models. The possibility arises that regulation of these factors may be a principle component to the susceptibility to develop and the treatment of psychiatric disorders. We also discuss the role that epigenetic factors, perhaps mediated by stress, may have on behavior, and response to future stressors.

Introduction

After many decades of stress research, beginning with the pioneering work of Hans Selye (Selye, 1956), the term "stress" is still defined today in a variety of ways. However, the term is generally accepted to include a disruptive force, whether good or bad, which affects the homeostatic balance of an organism. Stress may result from a disparate array of psychological or physiological stressors, or disruptive stimuli. Psychological stress can be defined as involving a reaction to an aversive stimulus in an individual's external environment (Kollack-Walker et al., 2000), for example, novelty stress and social defeat; physiological stress can be defined as disturbing an individual's internal milieu, leading to activation of regulatory mechanisms that serve to restore homeostasis (Kollack-Walker et al., 2000), for example, starvation, noise, cold exposure, or hemorrhage. Coping with stressful stimuli is often described as a conscious cognitive effort. However, it may also be physiological in nature, whereby the body activates a series of counteractive measures in response to a stressor. For example, changes in gene expression, neurotransmitter or receptor levels, or synaptic plasticity may all be adaptive responses to stress. Whether psychological or physiological, the brain plays a key role in perceiving, adapting and responding to stress. There is a vast array of neuronal

*Corresponding author. Tel.: +1(301) 496 9802; Fax: +1(301) 480-0123; E-mail: yuanp@intra.nimh.nih.gov

cellular adaptive mechanisms that are of central importance in understanding the brain's response to stressful stimuli. Failure of these adaptive mechanisms may antedate illness; for example, much evidence suggest stress as an antecedent of psychiatric disorders (Heim and Nemeroff, 2001). For example, in clinical populations environmental events (for example, early childhood stressors) correlate with the development of psychiatric disorders in adults (Heim and Nemeroff, 2001). Particularly strong is the epidemiological evidence in regard to the development of depression and anxiety disorders (Heim and Nemeroff, 2001). Indeed, accumulated data suggests that various stressors such as physical abuse, sexual abuse, parental loss, and even prenatal stress are all correlated with the development of severe mood and anxiety disorders in adulthood (Heim and Nemeroff, 2001). Furthermore, the clinical presentation of psychiatric disorders is often associated with acute life events or ongoing stress in adulthood (Heim and Nemeroff, 2001). In total, these data lead to a hypothesis whereby early life stressors lead to a state of enhanced vulnerability, and that adult mood and anxiety disorders present when later life events have effects on the vulnerable brain of these individuals (Heim and Nemeroff, 2001).

Evidence implicates the hypothalamic–pituitary–adrenal (HPA) axis (corticotrophic hormone-releasing factor (CRF) and glucocorticoids in particular) in long-term effects of early-life stress on hippocampal integrity and function. Most well-studied is the hippocampus, where high levels of glucocorticoid receptors are thought to mediate stress effects including cellular morphological changes, and changes in regional circuits and gross behavioral differences. A growing body of data is specifically implicating glutamatergic neurotransmission in stress-induced hippocampal atrophy and death (McEwen, 1999; Sapolsky, 2000b). As we discuss, the abundant data for the critical roles of CRF and glucocorticoids are noteworthy in regard to the pathophysiology of mood disorders.

In addition to directly causing neuronal atrophy, stress and glucocorticoids also appear to reduce cellular resilience, thereby making certain neurons more vulnerable to other insults, such as ischemia, hypoglycemia, and excitatory amino acid toxicity (Sapolsky, 2000a). The long-term effects of stress on behavior may be mediated by epigenetic gene regulatory factors. This chapter focuses primarily on the adaptive neuronal resilience to stress and the mechanisms underlying this phenomenon. However, as we discuss at the end of the chapter, these data and their implications may have relevance for the treatment of severe psychiatric disorders. Thus, recent efforts to understand the cellular and molecular actions of mood stabilizers and antidepressants have focused efforts on understanding what enzymes, signaling pathways, and gene expression profiles are altered in the brain (Duman, 1998; Duman et al., 2000; Gould and Manji, 2002). This evidence suggests that the downstream targets of both antidepressants and mood stabilizers are critical intracellular signaling pathways (Gould et al., 2002b); many of the same pathways are those involved in modulation of stress and the stress response. Accumulating evidence suggests that these medications may target pathways involved in mediating cellular resilience and neuroplasticity within the brain (Duman et al., 2000; Manji et al., 2000b). It is possible that the effects of these medications are to counteract – at least in part – the deleterious effects of stress that antedate the development of psychiatric illnesses.

Stress modulates neural plasticity

"Neuroplasticity" subsumes diverse processes of vital importance by which the brain perceives, adapts and responds to a variety of internal and external stimuli. The manifestations of neuroplasticity in the adult central nervous system (CNS) have been characterized as including alterations of dendritic function, synaptic remodeling, long-term potentiation (LTP), axonal sprouting, neurite extension, synaptogenesis, and even neurogenesis (see (Mesulam, 1999) for an excellent overview).

The adult brain is more plastic than previously believed. The alterations in cellular morphology resulting from various stressors have been the focus of considerable recent research (D'Sa and Duman, 2002) (Fig. 1). Remodeling of synaptic contacts and dendrites in the hypothalamus coinciding with the onset of lactation (Michaloudi et al., 1997; Stern and Armstrong, 1998) and branching of dendrites of cerebral–cortical neurons in an enriched environment

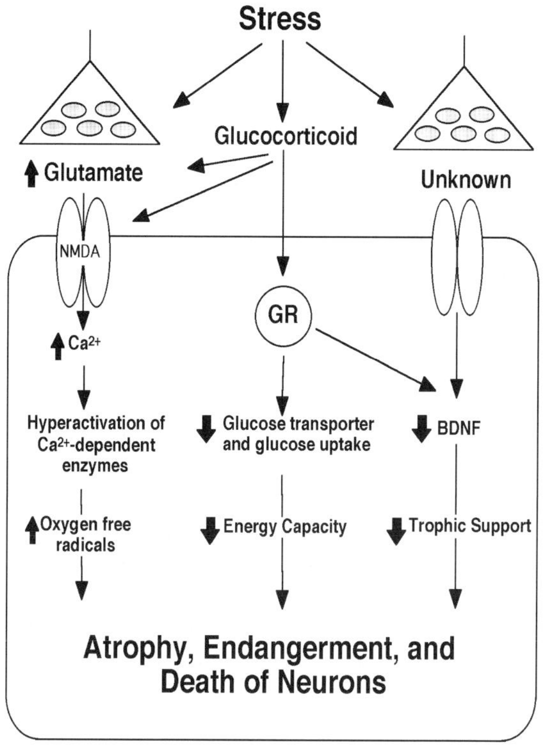

Fig. 1. Cellular mechanisms by which stress and mood disorders may bring about impairments of structural plasticity. This figure depicts the multiple mechanisms by which stress and potentially affective episodes may attenuate cellular resiliency, thereby resulting in atrophy, death, and endangerment of hippocampal neurons. The primary mechanisms appear to be: (i) excessive NMDA and nonNMDA glutamatergic throughput; (ii) downregulation of cell surface glucose transporters, which are involved in bringing glucose into the cell. Reduced levels of glucose transporters thus reduce the neurons' energy reservoir, making them susceptible to energy failure when faced with excessive demands; (iii) reduction in the levels of BDNF, which is essential for the neuron's normal trophic support and synaptic plasticity. The well-documented reduction in glial cells may contribute to impairments of neuronal structural plasticity by reducing the neuron's energy supply and reduced glial-mediated clearing of excessive synaptic glutamate. NMDA, *N*-methyl-D-aspartate glutamate receptor; GR, glucocorticoid receptor; BDNF, brain-derived neurotrophic factor. Modified and reproduced, with permission (Manji and Duman, 2001).

and after training (Withers and Greenough, 1989), are just two examples of adaptive plasticity.

Most studies of adaptive plasticity of neurons in response to stress, as well as hormones of the HPA axis, have focused on the hippocampus. This is due, in part, to the well-defined and easily studied neuronal populations of the limbic brain regions, including the dentate gyrus granule cell layer, and the CA1 and CA3 pyramidal cell layers. These cell layers and their connections (mossy fiber pathway and Schaffer collateral) have also been used as cellular models of learning and memory (i.e., LTP). Another major reason that the hippocampus has been the focus of stress research is that the highest levels of glucocorticoid receptors (GR) are expressed in this brain region (Lopez et al., 1998). However, it is clear that stress and glucocorticoids also influence the survival and plasticity of neurons in other brain regions (e.g., prefrontal cortex, vide infra) that have not yet been studied in the same detail as the hippocampus.

One of the most consistent effects of stress on cellular morphology is dendritic remodeling of hippocampal neurons (for reviews see (McEwen, 1999; Sapolsky, 2000a). The remodeling of dendrites is observed profoundly in the CA3 pyramidal neurons as atrophy-decreased number and length of the apical dendritic branches. The stress-induced atrophy of CA3 neurons occurs after two to three weeks of exposure to restraint stress or more long-term social stress, and has been observed in rodents and tree shrews (McEwen, 1999; Sapolsky, 2000a). Although the effects of chronic stress in the CA3 tend to be the greatest, subtle structural changes are also found in the CA1 and dentate gyrus after a one-month multiple stress paradigm (Sousa et al., 2000). Additionally, profound changes in the morphology of the mossy fiber terminals and significant loss of synapses were also observed. These alterations were partially reversible following rehabilitation from stress. Moreover, these fine structural changes also occur upon exposure to high levels of glucocorticoids, and were accompanied by impairments in spatial learning and memory. The latter were undetectable following rehabilitation, suggesting that activation of the HPA axis likely plays a major role in mediating the stress-induced neural plasticity and hippocampal-dependent learning and memory (Sapolsky, 1996, 2000a; McEwen, 1999). The hippocampus has a very high concentration of glutamate and expresses both mineralcorticoid receptor (MR) and glucocorticoid receptor (GR), though GR may be relatively scarce in the hippocampus of primates (Patel et al., 2000; Sanchez et al., 2000), and more abundant in cortical regions. MR activation in the hippocampus (CA1) is associated with reduced calcium currents, whereas activation of GR causes increased calcium currents and enhanced responses to excitatory amino acids. Very high levels of GR activation markedly increases calcium currents and leads to increased *N*-methyl-D-aspartate (NMDA) receptor throughput that could predispose to neurotoxicity. Indeed, as we discuss in greater detail below, a growing body of data has implicated glutamatergic neurotransmission in stress-induced hippocampal atrophy and death ((McEwen, 1999); see Fig. 1).

These observations are noteworthy with respect to the pathophysiology of mood disorders, since a significant percentage of patients with mood disorders display some form of HPA axis activation, and the subtypes of depression most frequently associated with HPA activation are those most likely to be associated with hippocampal volume reductions (Sapolsky, 2000a). A significant percentage of patients with Cushing's disease, in which pituitary gland adenomas result in cortisol hypersecretion, are also known to manifest prominent depressive symptoms, as well as hippocampal atrophy. Additionally, some patients with Cushing's disease show a reduction in hippocampal volume that correlates inversely with plasma cortisol concentrations; following corrective surgical treatment, enlargement of hippocampal volume in proportion to the treatment-associated decrement in urinary free-cortisol concentrations is observed (Starkman et al., 1999; Simmons et al., 2001).

The HPA axis hyperactivity in mood disorder patients is most commonly measured by increased cortisol levels in plasma (especially at the circadian nadir), urine and CSF, increased cortisol response to adrenocorticotropin hormone (ACTH), blunted ACTH response to CRF challenge, enlarged pituitary and adrenal glands, and downregulation at post-mortem examination of frontal cortical CRF. Reduced corticosteroid receptor feedback is implicated in this process by challenge studies with dexamethasone and dexamethasone plus CRF in bipolar and unipolar subjects (McQuade and Young, 2000; Reul and Holsboer, 2002).

Hippocampal changes and HPA axis over-activation is also a finding in anxiety disorders where a number of clinical studies suggest that patients with post-traumatic stress disorder (PTSD) have a smaller hippocampal volume than matched control subjects (Bremner et al., 1995, 1997; Gurvits et al., 1996; Stein et al., 1997; Charney and Bremner, 1999). To date, no quantitative neuroimaging studies have been performed in other anxiety disorders such as panic disorder, phobic disorder, or generalized anxiety disorder – but a single study does suggest the presence of abnormalities in the temporal lobe in panic disorder (Ontiveros et al., 1989; Charney and Bremner, 1999). Also, some studies suggest that there is an increase in cortisol release in response to stress in PTSD and panic disorder. Furthermore, it is generally consistent that there is a chronic increase of CRF in patients with anxiety. A cautionary note in the interpretation of the clinical studies is suggested by the results of the recent longitudinal studies undertaken to investigate the effects of early-life stress and inherited variation in monkey hippocampal volumes (Lyons et al., 2001). In these studies, paternal half-siblings raised apart from one another by different mothers in the absence of fathers were randomized to one of three postnatal conditions that disrupted diverse aspects of early maternal care. These researchers found that all paternal half-siblings, with small adult hippocampal volumes, responded to the removal of mothers after weaning with initially larger relative increases in cortisol levels (Lyons et al., 2001). Plasma cortisol levels 3 and 7 days later, and measures of cortisol-negative feedback in adulthood were not, however, correlated with hippocampal size. Thus, these studies suggest that small hippocampi also reflect an inherited characteristic of the brain, and highlight the need for caution in attribution of causality in the cross-sectional human morphometric studies of the hippocampus.

A recent study by Gilbertson et al. also supports the hypothesis that smaller hippocampal volume is associated with susceptibility to stress (Gilbertson et al., 2002). The brains of monozygotic twin pairs, in which one twin experienced combat in Vietnam and the other did not, were imaged by MRI. As reported in prior studies, veterans who had developed PTSD displayed reduced hippocampal volume in comparison to those who did not. However, it was also observed that the combat-naive cotwins of the PTSD-sufferers also had reduced hippocampal volume, in comparison to cotwins of veterans who never developed PTSD. Likewise, the hippocampal volume of the combat-naive cotwins was inversely correlated with the severity of PTSD symptoms in their veteran counterparts. These data suggest that a genetic contributor to PTSD susceptibility is associated with reduced hippocampal volume (and/or a propensity to sustain PTSD-unrelated hippocampal shrinkage).

Although not as extensively studied as the hippocampus, recent research has also demonstrated histopathological changes in rat prefrontal cortex after corticosterone administration (Wellman, 2001). Thus, using a Golgi-Cox procedure, Wellman (2001) investigated pyramidal neurons in layer II-III of medial prefrontal cortex, and quantified dendritic morphology in three dimensions. This study demonstrated a significant redistribution of apical dendrites in corticosterone-treated animals, with the amount of dendritic material proximal to the soma being increased and distal dendritic material being decreased. These findings suggest that stress may produce a significant reorganization of the apical dendritic arbor from medial prefrontal cortex in rats. Most recently, Lyons (2002) demonstrated that four years after a brief stressor (intermittent postnatal separations from maternal availability) young adult squirrel monkeys showed significantly larger right ventral medial prefrontal volumes. Neither overall brain volumes nor left prefrontal measures were altered, suggesting selective (rather than nonspecific) effects. An intriguing observation of this study was that, similar to their hippocampal findings (vide supra), these investigators found a strong heritability of the right ventral medial prefrontal volume. Thus, in this study, certain fathers produced offspring with large right ventral medial prefrontal volumes, whereas others produced offspring with small right ventral medial prefrontal volumes (Lyons, 2002). Since the paternal half-siblings were raised apart by different mothers in the absence of fathers, the phenotypic similarities in right ventral medial prefrontal volume likely represent a major genetic contribution, effects which were not seen for other prefrontal regions (Lyons, 2002).

Effects of stress, glucocorticoids, and psychotrophic medications on hippocampal neurogenesis

The demonstration that neurogenesis occurs in the adult human brain has reinvigorated research into the cellular mechanisms by which the birth of new neurons is regulated in the mammalian brain (Eriksson et al., 1998). The localization of pluripotent progenitor cells and neurogenesis occurs in restricted brain regions. The greatest density of new cell birth is observed in the subventricular zone and the subgranular layer of the hippocampus. Cells born in the subventricular zone migrate largely to the olfactory bulb and those in the subgranular zone into the granule cell layer. The newly generated neurons send out axons and appear to make connections with surrounding neurons, indicating that they are capable of integrating into the appropriate neuronal circuitry in hippocampus and cerebral cortex. Neurogenesis in the hippocampus is increased by enriched environment, exercise, and hippocampal-dependent learning (Kempermann et al., 1997; van Praag et al., 1999; Gould et al., 2000). Upregulation of neurogenesis in response to these behavioral stimuli and the localization of this process to hippocampus has led to the proposal that new cell birth is involved in learning and memory (Gould et al., 2000).

Chronic, but not acute, antidepressant treatment also increases the neurogenesis of dentate gyrus granule cells (Jacobs et al., 2000; Manev et al., 2001; D'Sa and Duman, 2002). These studies demonstrate that chronic administration of different classes of antidepressant treatment, including noradrenaline (NA), selective serotonin reuptake inhibitors (SSRIs), and electroconvulsive seizure, increases the proliferation and survival of new neurons. Lithium also increases neurogenesis in the dentate gyrus (Chen et al., 2000b). In contrast, increased neurogenesis is not observed in response to chronic administration of nonantidepressant psychotropic drugs. Studies demonstrating that neurogenesis is increased by conditions that stimulate neuronal activity (e.g., enriched environment, learning, exercise) suggest that this process is also positively regulated by, and may even be dependent on, neuronal plasticity (Kempermann, 2002). It is clear that the enhancement of hippocampal neurogenesis by antidepressants serves to highlight the degree to which these effective treatments are capable of regulating long-term neuroplastic events in the brain. At this point, it is not completely clear precisely what the clinical significance of enhancement of adult hippocampal neurogenesis by antidepressants truly represents. In view of the opposite effects of stress and antidepressants on hippocampal neurogenesis, it is quite plausible that alterations in hippocampal neurogenesis are fundamental to the clinical syndrome of depression (Jacobs et al., 2000; Manev et al., 2001; D'Sa and Duman, 2002; Kempermann, 2002). However, as Kempermann (2002) has clearly articulated, much more research is required in order to adequately link changes in adult hippocampal neurogenesis to the pathophysiology and treatment of depression.

Recent studies have shown that decreased neurogenesis occurs in response to both acute and chronic stress (see (Gould et al., 2000)). Removal of adrenal steroids (i.e., adrenalectomy) increases neurogenesis and treatment with high levels of glucocorticoids reproduces the downregulation of neurogenesis that occurs in response to stress. Aging also influences the rate of neurogenesis. Although neurogenesis continues into late life, the rate is significantly reduced (Cameron and McKay, 1999). The decreased rate of cell birth could result from upregulation of the HPA axis and higher levels of adrenal steroids that occur in later life. Lowering glucocorticoid levels in aged animals restores neurogenesis to levels observed in younger animals, indicating that the population of progenitor cells remains stable but that it is inhibited by glucocorticoids (Cameron and McKay, 1999). Interestingly, studies in glucocorticoid receptor knockout mice showed significant alterations in hippocampal neurogenesis (Gass et al., 2000). A reduction of granule cell neurogenesis (up to a 65% of control levels) was found in MR−/− mice, whereas GR−/− mice did not show neurogenic disruption, eventually relating the MR receptor in the pathogenesis of hippocampal changes observed in chronic stress and affective disorders (Gass et al., 2000). These preclinical observations raise the interesting possibility that CRF and GR antagonists, currently being developed for the treatment of mood and anxiety disorders, may have particular utility in the treatment of elderly depressed patients. Also, of potential relevance (noting the effect of hormonal fluctuations on mood disorders) for our understanding of the neurobiology and treatment of mood disorders, ovariectomy decreases the proliferation of new cells in the hippocampus, effects which are reversed by estrogen replacement. The rate of neurogenesis fluctuates over the course of the estrus cycle in rodents, and the total rate of cell birth is higher in female rodents relative to males. In addition to potentially playing a role in the beneficial cognitive effects of estrogen, the regulation of neurogenesis by this gonadal steroid may also provide important clues about certain sexually dimorphic characteristics of mood disorders.

Mechanisms underlying stress-induced morphometric changes

Glutamate, calcium, and apoptosis

Microdialysis studies have shown that stress increases extracellular levels of glutamate in hippocampus, and NMDA glutamate receptor antagonists attenuate stress-induced atrophy of CA3 pyramidal neurons (McEwen, 1999; Sapolsky, 2000b). Although a variety of methodological issues remain to be fully resolved, the preponderence of the evidence to date suggests that the atrophy, and possibly death, of CA3 pyramidal neurons arises, at least in part, from increased glutamate neurotransmission (McEwen, 1999; Sapolsky, 2000b). It should be noted, however, that although NMDA antagonists block stress-induced hippocampal atrophy, there have not been any studies demonstrating that they are able to block the cell death induced by severe stress. This suggests that the mechanisms underlying atrophy and death may lie on a continuum, with severe (or prolonged) stresses "recruiting" additional pathogenic pathways in addition to enhanced NMDA-mediated neurotransmission. As discussed, stress increases extracellular levels of glutamate and sustained activation of NMDA, as well as nonNMDA ionotropic receptors could result in high intracellular levels of calcium. Overactivation of the glutamate ionotropic receptors is known to contribute to the neurotoxic effects of a variety of insults, including repeated seizures and ischemia (Fig. 1). Neurotoxicity follows as a response to overactivation of calcium-dependent enzymes and the generation of oxygen-free radicals. Stress or glucocorticoid exposure also compromises the metabolic capacity of neurons, thereby increasing the vulnerability to other types of neuronal insults.

Hypothalamic–pituitary–adrenal (HPA) axis

Activation of the HPA axis appears to play a critical role in mediating these effects, since stress-induced neuronal atrophy is prevented by adrenalectomy, and duplicated by exposure to high concentrations of glucocorticoids (reviewed in Sapolsky, 1996, 2000a; McEwen, 1999). Increasing recent data also suggests a critical role for CRF in long-term effects of early-life stress on hippocampal integrity and function. Thus, the administration of CRF to the brains of immature rats has been demonstrated to reduce memory function throughout life; these deficits are associated with progressive loss of hippocampal CA3 neurons and chronic upregulation of hippocampal CRF expression, effects that do not require the presence of stress levels of glucocorticoids (Brunson et al., 2001). The CRF_1 receptor, which binds CRF with higher affinity than CRF_2 receptor, plays a major role in regulating adrenocorticotropin hormone (ACTH) release, and has been implicated in animal models of anxiety. Indeed, the central administration of CRF_1 antisense oligodeoxynucleotides has been demonstrated to have anxiolytic effects against both CRF and psychological stressors. Although CRF_2 receptors appear to act in an antagonistic manner, i.e., CRF_1 activates and CRF_2 attenuates the stress response, its precise role is still being characterized (reviewed in (Reul and Holsboer, 2002). Interestingly, pretreatment with a CRF antagonist also attenuates the stress-induced increases in MR levels in hippocampus, neocortex, frontal cortex, and amygdala (Gesing et al., 2001). Rats that underwent a stressor also showed increased ACTH and cortisol levels following the administration of an MR antagonist, suggesting that the upregulation of MR in the stressed group is associated with increased inhibitory tone of the HPA axis.

Together, the abundant data for the critical roles of CRF and glucocorticoids are noteworthy with respect to the pathophysiology of mood disorders, since a significant percentage of patients with mood disorders display some form of HPA axis activation, and the subtypes of depression most frequently associated with HPA activation are those most likely to be associated with hippocampal volume reductions (Sapolsky, 2000a). A significant percentage of patients with Cushing's disease, in which pituitary gland adenomas result in cortisol hypersecretion, are also known to manifest prominent depressive symptoms, as well as hippocampal atrophy. Additionally, some patients with Cushing's disease show a reduction in hippocampal volume that correlates inversely with plasma cortisol concentrations; following corrective surgical treatment, enlargement of hippocampal volume in proportion to the treatment-associated decrement in urinary free-cortisol concentrations is observed (Starkman et al., 1999; Simmons et al., 2001).

The HPA axis hyperactivity in mood disorder patients is generally manifest by increased cortisol levels in plasma (especially at the circadian nadir), urine, and CSF, increased cortisol response to ACTH, blunted ACTH response to CRF challenge, enlarged pituitary and adrenal glands, and down-regulation at postmortem of frontal cortical CRF. Reduced corticosteroid receptor feedback is implicated in this process by challenge studies with dexamethasone and dexamethasone plus CRF in bipolar and unipolar subjects (McQuade and Young, 2000; Reul and Holsboer, 2002).

Evidence in humans suggests that decreased corticosteroid receptor number (postmortem reduction of GR messenger ribonucleic acid (mRNA) (Webster et al., 1999)) may be present in the hippocampus of individuals with bipolar and unipolar disorder, and some antidepressants (tricyclics), electroconvulsive shock therapy (ECT), and mood stabilizers (lithium) may modulate GR number and/or function (reviewed in (Holsboer, 2000). Furthermore, transgenic mice with reduced GR have HPA and cognitive disturbance that may parallel depression in humans and that normalizes with antidepressant exposure (Steckler et al., 2001). Finally, antisense oligonucleotides targeted to GR reduced immobility on the forced swim test (suggesting an antidepressant-like effect), as does the antiglucocorticoid drug mifepristone (RU-486) (Korte et al., 1996).

In addition to directly causing neuronal atrophy, stress and glucocorticoids also appear to reduce cellular resilience, thereby making certain neurons more vulnerable to other insults, such as ischemia, hypoglycemia, and excitatory amino acid toxicity (Sapolsky, 2000a). Thus, recurrent stress (and presumably recurrent mood disorders episodes, which are often associated with hypercortisolemia) may lower the threshold for cellular death/atrophy in response to a variety of physiological (e.g., aging) and pathological (e.g., ischemia) events. The potential functional significance of these effects is supported by the demonstration that overexpression of the glucose transporter blocks the neurotoxic effects of neuronal insults (Fig. 1) (Sapolsky, 2000a,b; Manji and Duman, 2001b). Such processes may conceivably also play a role in the relationship between mood disorders and cerebrovascular events, considering that individuals who develop their first depressive episode in late-life have an increased likelihood of showing magnetic resonance imaging (MRI) evidence of cerebrovascular disease (Kumar et al., 1997; Murray and Lopez, 1997; Steffens and Krishnan, 1998; Steffens et al., 1999; Chemerinski and Robinson, 2000; Drevets, 2000). The precise mechanisms by which glucocorticoids exert these deleterious effects remain to be fully elucidated, but likely involve the inhibition of glucose transport (thereby diminishing capability of energy production and augmenting susceptibility to hypoglycemic conditions), and the aberrant, excessive facilitation of glutamatergic signaling (Sapolsky, 2000a).

The reduction in the resilience of discrete brain regions including hippocampus and potentially prefrontal cortex, may also reflect the propensity for various stressors to decrease the expression of brain-derived neurotrophic factor (BDNF) in this region (Smith et al., 1995; Nibuya et al., 1999). The mechanisms underlying the downregulation of BNDF by stress have not been fully elucidated. However, as we discuss in a later section recent evidence suggests that mood stabilizers and antidepressants activate neurotrophic signaling pathways, and in particular have effects on BDNF-mediated signaling.

Neurotrophic signaling cascades: a focus on brain-derived neurotrophic factor (BDNF)

Neurotrophins are a family of regulatory factors that mediate the differentiation and survival of neurons, as well as the modulation of synaptic transmission and synaptic plasticity (Patapoutian and Reichardt, 2001; Poo, 2001). The neurotrophin family now include – among others – nerve growth factor (NGF), brain-derived neurotrophic factor (BDNF), neurotrophin (NT)-3, NT-4/5 and NT-6 (Patapoutian and Reichardt, 2001). These various proteins are closely related in terms of sequence homology and receptor specificity. They bind to and activate specific receptor tyrosine kinases belonging to the Trk family of receptors, including TrkA, TrkB, TrkC and a pan-neurotrophin receptor P75 (Patapoutian and Reichardt, 2001; Poo, 2001). Additionally, there are two isoforms of TrkB receptors: the full length TrkB and the truncated form of TrkB, which does not contain the intracellular tyrosine kinase domain (Fryer et al., 1996). The truncated form of TrkB can thus function as a dominant-negative inhibitor for the TrkB receptor tyrosine kinase, thereby providing another mechanism to regulate BDNF signaling in the CNS (Gonzalez et al., 1999). Neurotrophins can be secreted constitutively or transiently, and often in an activity-dependent manner. Recent observations support a model wherein neurotrophins are secreted from the dendrite and act retrogradely at presynaptic terminals where they act to induce long-lasting modifications (Poo, 2001). Within the neurotrophin family, BDNF is a potent physiological survival factor, which has also been implicated in a variety of pathophysiological conditions. The cellular actions of BDNF are mediated through two types of receptors: a high-affinity tyrosine receptor kinase (TrkB) and a low-affinity pan-neurotrophin receptor (p75). TrkB is preferentially activated by BDNF and NT4/5, and appears to mediate most of the cellular responses to these neurotrophins.

BDNF and other neurotrophic factors are necessary for the survival and function of neurons (Mamounas et al., 1995), implying that a sustained reduction of these factors could affect neuronal viability. As discussed already, BDNF is best known for its long-term neurotrophic and neuroprotective effects, which may be very important for its putative role in the pathophysiology and treatment of mood disorders, and its putative role in the effects of stress (vide infra). In this context, it is noteworthy that although endogenous neurotrophic factors have traditionally been viewed as increasing cell survival by providing necessary trophic support, it is now clear that their survival-promoting effects are mediated in large part by an inhibition of cell death cascades (Riccio et al., 1999). Increasing evidence suggests that neurotrophic factors inhibit cell death cascades by activating the mitogen-activated protein (MAP) kinase signaling pathway and the phosphotidylinositol-3 kinase (PI-3K)/Akt pathway (Fig. 2).

Signaling through the mitogen-activated protein (MAP) kinase cascade

Shc (a protein that recognizes specific phosphotyrosine residues on receptors) recruitment and phosphorylation results in recruitment to the membrane

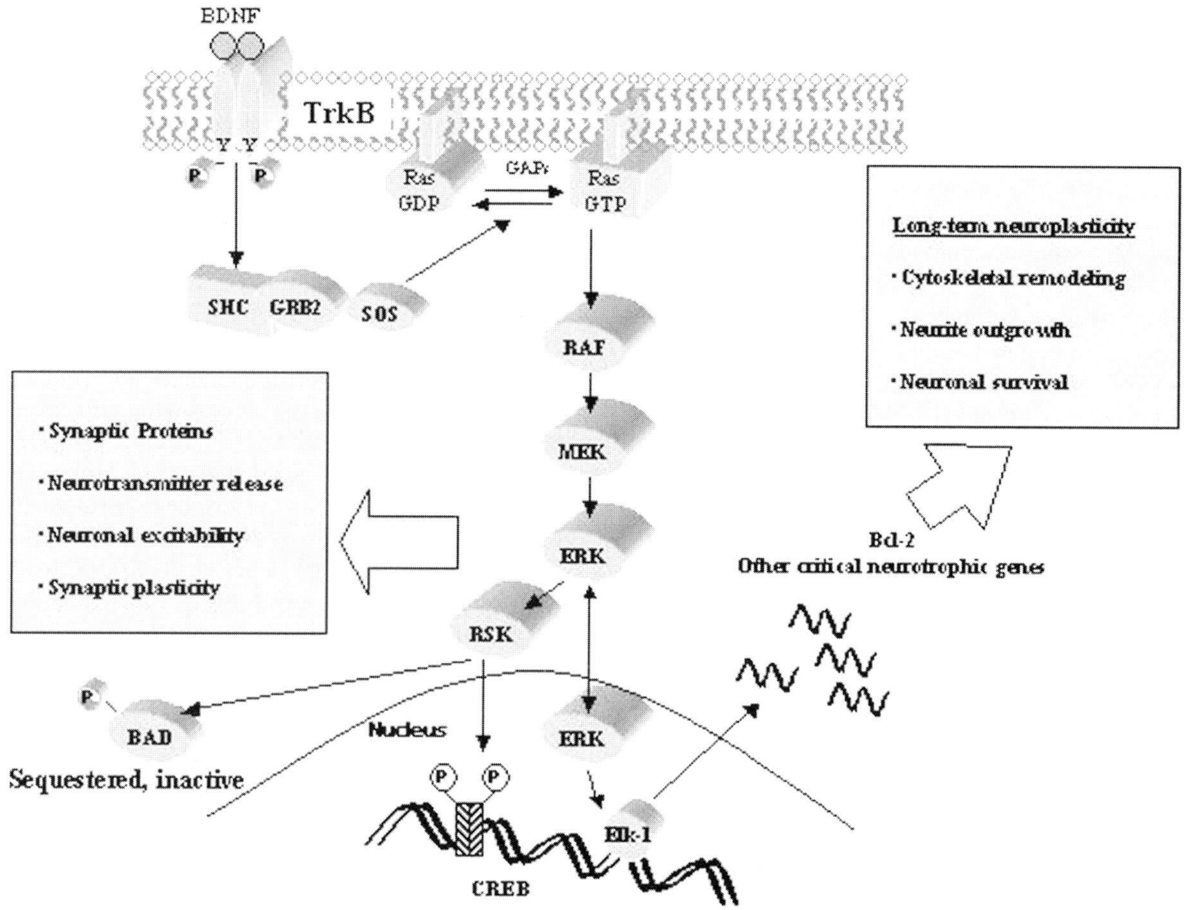

Fig. 2. Neurotrophins and the ERK MAP minase signaling cascade. Cell survival is dependent on neurotrophic factors, such as brain-derived neurotrophic factor (BDNF) and nerve growth factor (NGF), and the expression of these factors can be induced by synaptic activity. The influence of neurotrophic factors on cell survival is mediated by activation of the mitogen-activated protein (MAP) kinase cascade. Activation of neurotrophic factor receptors, also referred to as Trks, results in activation of the MAP kinase cascade via several intermediate steps, including phosphorylation of the adaptor protein SHC and recruitment of the guanine nucleotide exchange factor son of sevenless (Sos). This results in activation of the small guanosine triphosphate – binding protein Ras, which leads to activation of a cascade of serine/threonine kinases. This includes Raf, MAP kinase (MEK), and MAP kinase also referred to as extracellular response kinase, or Erk). One target of the MAP kinase cascade is Rsk, which influences cell survival in at least two ways. Rsk phosphorylates and inactivates the proapoptotic factor BAD. Rsk also phosphorylates cyclic adenosine monophosphate response element-binding protein (CREB) and thereby increases the expression of the antiapoptotic factor bcl-2. These mechanisms underlie many of the long-term effects of neurotrophins, including neurite outgrowth, cytoskeletal remodeling, and cell survival. It is now clear, however, that BDNF also exerts many acute synaptic effects, including regulating the release of a number of neurotransmitters.

of a complex of the adaptor Grb-2 and the Ras exchange factor, son of sevenless (SOS), thereby stimulating transient activation of Ras. Ras, in turn, activates PI3K, the p38 MAPK/MAPK-activating protein kinase 2 pathway and the c-Raf/ERK pathway. Among the targets of ERK are the ribosomal S6 kinases (RSKs). Both RSK and MAPK-activating protein kinase 2 phosphorylate CRE-binding protein (CREB) and other transcription factors. Recent studies have demonstrated that the activation of the MAP kinase pathway can inhibit apoptosis by inducing the phosphorylation of BAD (Bcl-xl/Bcl-2 Associated Death promoter), and increasing the expression of the antiapoptotic protein Bcl-2, the latter effect likely involves the cAMP (cyclic adenosine monophosphate) response element binding protein (CREB) (Riccio et al., 1997; Bonni et al., 1999). Phosphorylation of BAD occurs via activation of Rsk. Rsk phosphorylates BAD and thereby promotes its inactivation. Activation of Rsk also mediates the actions of the MAP kinase cascade and neurotrophic factors on the expression of Bcl-2. Rsk can phosphorylate CREB, leading to induction of Bcl-2 gene expression (Fig. 2).

MAP kinases are abundantly present in the brain, and in recent years have been postulated to play a major role in a variety of long-term CNS functions, both in the developing and mature CNS (Suzuki et al., 1995; Matsubara et al., 1996; Kornhauser and Greenberg, 1997; Fukunaga and Miyamoto, 1998; Robinson et al., 1998). With respect to their actions in the mature CNS, MAP kinases have been implicated in mediating neurochemical processes associated with long-term facilitation (Martin et al., 1997), long-term potentiation (English and Sweatt, 1996, 1997), associative learning (Atkins et al., 1998), one-trial and multi-trial classical conditioning (Crow et al., 1998), long-term spatial memory (Blum et al., 1999), and have also been postulated to integrate information from multiple, infrequent bursts of synaptic activity (Murphy et al., 1994). Important for the present discussion, MAP kinase pathways have recently been demonstrated to regulate the responses to environmental stimuli and stressors in rodents (Xu et al., 1997), and to couple protein kinase A (PKA) and protein kinase C (PKC) to CREB protein phosphorylation in area CA1 of hippocampus (Roberson et al., 1996, 1999). These recent studies suggest the possibility of a broad role for the MAPK cascade in regulating gene expression in long-term forms of synaptic plasticity (Roberson et al., 1999). Overall, the data suggests that MAP kinases play important physiological roles in the mature CNS, and furthermore, may represent important targets for the actions of CNS-active agents (Nestler, 1998; Yuan et al., 1998).

One important mechanism by which the MAP kinase signaling cascade inhibits cell death is by increasing the expression of the antiapoptotic protein Bcl-2 (Bonni et al., 1999; Finkbeiner, 2000). Bcl-2 is now recognized as a major neuroprotective protein, since Bcl-2 overexpression protects neurons against diverse insults including ischemia, MPTP (1-methyl-4-phenyl-1,2,3,6-tetrahydropyridine), β-amyloid, free radicals, excessive glutamate, and growth factor deprivation (Manji et al., 2001c). Accumulating data suggests that not only is Bcl-2 neuroprotective, but that it also exerts neurotrophic effects and promotes neurite sprouting, neurite outgrowth, and axonal regeneration (Oh et al., 1996; Chen et al., 1997; Chen and Tonegawa, 1998; Chierzi et al., 1999; Holm et al., 2001). Moreover, a recent study demonstrated that severe stress exacerbates stroke outcome by suppressing Bcl-2 expression (DeVries et al., 2001). In this study, the stressed mice expressed ~70% less Bcl-2 mRNA than unstressed mice after ischemia. Furthermore, stress greatly exacerbated infarct in control mice but not in transgenic mice that constitutively express increased neuronal Bcl-2. Finally, high corticosterone concentrations were significantly correlated with larger infarcts in wild-type mice but not in Bcl-2 overexpressing transgenic mice. Thus, enhanced Bcl-2 expression appears to be capable of offsetting the potentially deleterious consequences of stress-induced neuronal endangerment, and suggests that pharmacologically induced upregulation of Bcl-2 may have considerable utility in the treatment of a variety of disorders associated with endogenous or acquired impairments of cellular resilience (vide infra). Overall, it is clear that the neurotrophic factor/MAP kinase/Bcl-2 signaling cascade plays a critical role in cell survival in the CNS, and that there is a fine balance maintained between the levels and activities of cell survival and cell death factors; modest changes in this signaling cascade or in the levels of the Bcl-2 family of proteins (potentially due to genetic, illness, or insult-related factors) may therefore profoundly affect cellular viability.

Recently, it was reported that chronic stress (21 days footshock) induced a pronounced and persistent extracellular response kinase 1/2 (ERK1/2) hyperphosphorylation in dendrites of the higher prefrontocortical layers, while phospho-CREB was reduced in several cortical regions including frontal cortex (Trentani et al., 2002). Since CREB is phosphorylated and activated by phospho-ERK1/2 directly, this reduction indicate that chronic stress could downregulate CREB phosphorylation indirectly, and subsequently downregulate the transcription of some genes such as Bcl-2 and BDNF. As mentioned previously, severe stressors decrease the expression of Bcl-2 (DeVries et al., 2001)and BDNF (Smith et al., 1995; Nibuya et al., 1999) in brain regions, but the mechanisms that mediate this effect still remain unclear. An intriguing possibility holds that this effect may be attributable to dysregulation of the BDNF–ERK–CREB coordination. This disruption of the coordination may be a key mechanism by which prolonged stress induces atrophy of selective subpopulations of vulnerable neurons and/or distal dendrites. Conceivably, the precise kinetics of ERK and CREB activation ultimately dictate whether the activated kinases participate in a cell survival or death-promoting pathway.

Phosphotidylinositol-3 kinase (PI3K-Akt) pathway: a major pathway mediating neuronal survival

The PI3K-Akt pathway is also particularly important for mediating neuronal survival under a wide variety of circumstances. Trk receptors can activate PI3K through at least two distinct pathways, the relative importance of which differs between neuronal subpopulations. In many neurons, Ras-dependent activation of PI3K is the most important pathway through which neurotrophins promote cell survival. In some cells, however, PI3K can also be activated through three adaptor proteins, Shc, Grb-2, and Gab-1. Binding to phosphorylated tyrosine 490 of Shc results in recruitment of Grb-2. Phosphorylated Grb-2 provides a docking site for Gab-1, which in turn is bound by PI3K (Brunet et al., 2001).

PI3 kinase directly regulates certain cytoplasmic apoptotic pathways. Akt has been proposed to act both prior to the release of cytochrome-c by pro-apoptotic Bcl-2 family members, and subsequent to the release of cytochrome-c, by regulating components of the apoptosome. Akt phosphorylates the proapoptotic Bcl-2 family member BAD, thereby inhibiting BAD's proapoptotic functions (Datta et al., 1997).

Akt may also promote survival in an indirect fashion by regulating another major signaling enzyme–glycogen synthase kinase 3 (GSK-3) (Woodgett, 2001). Elevated GSK-3 has been shown to promote apoptosis in cultured neurons (Bijur et al., 2000). Furthermore, neurotrophin withdrawal increases, whereas phosphorylation by Akt decreases GSK-3 activity (Hetman et al., 2000). Moreover, a series of studies indicates that Akt controls a major class of transcriptional factors – the Forkhead box transcription factor, class O (FOXO) subfamily of Forkhead transcriptional regulators (FKHR, FKHRL1, and AFX). Several groups have independently shown that Akt directly phosphorylates FOXOs and inhibits their ability to induce the death genes (Brunet et al., 1999; Dijkers et al., 2000). Finally, activation of Akt also results in phosphorylation of nuclear factor-κB (NF-κB). Transcription activated by NF-κB has been shown recently to promote neuronal survival (Maggirwar et al., 1998). Thus, PI3K acting through Akt may promote survival by variety of mechanisms; precisely which of these mechanisms is operative in the actions of neurotrophins, and under what circumstances is the focus of extensive current research (Fig. 2).

Epigenetic gene regulatory factors likely regulate how organisms adapt to and respond to stress: regulation of neurotrophic factors may be important

The discussion in the chapter have centered upon the fact that the stress response, mood disorders, and other behaviors, while traditionally being conceptualized as neurochemical disorders, are now being thought of in a more dynamic sense, wherein changes in the underlying pathophysiology of these processes are being reconceptualized as being due primarily to changes in synaptic plasticity and cellular resilience.

In the next section of this chapter we discuss the emerging evidence suggesting that antidepressant medications and mood stabilizers (valproic acid and lithium) upregulate BDNF (Duman et al., 2000) and other proteins with neurotrophic/neuroprotective functions in the brain (Manji and Chen, 2002). Furthermore, there exists a great deal of evidence – both clinical and preclinical – showing that activating these pathways has functional effects (increased neurogenesis, neuroprotection, etc.).

As discussed, it is well-established that acute stress – or the stress hormone corticosterone – can regulate the expression of neurotrophic molecules in rodents. Specifically, a number of studies have shown that acute stress and/or administration of exogenous corticosterone administration to rodents decreases the mRNA and protein levels of BDNF in the hippocampus (Smith et al., 1995). In addition to the broadly replicated BDNF results, studies have looked at other neurotrophic molecules, with variable results dependent upon the experimental conditions. Thus, both stress and thymoleptic medications regulate neurotrophic factors, and their regulation of these molecules appears to be opposite to one another. These diametric changes in the regulation of neurotrophic molecules give rise to the notion that regulation of these factors may be a principal component to the susceptibility and treatment of psychiatric disorders (Manji and Duman, 2001).

In addition to the well-documented acute effects of stress, increasing evidence – both preclinical and clinical – suggests that stress can have behavioral, biochemical, and cellular effects far temporally removed from the initial stressful event. These data – by correlation and association – suggest major importance of these pathways in both the pathophysiology and treatment of mood disorders (Duman et al., 2000). In this brief section we discuss, some of the ways – in particular epigenetic mechanisms – by which stress may have lasting effects on behavior, and response to future stressors.

Stress – and the individual response to stress – appears to be a causal factor in the development of psychiatric diseases. In clinical populations environmental events (for example, early childhood stressors) correlate with the development of psychiatric disorders in adults (Heim and Nemeroff, 2001). Particularly strong is the epidemiological evidence in regard to the development of depression and anxiety disorders (Heim and Nemeroff, 2001). Indeed, accumulated data suggests that various stressors such as physical abuse, sexual abuse, parental loss, and even prenatal stress are all correlated with the development of severe mood and anxiety disorders in adulthood (Heim and Nemeroff, 2001). Furthermore, the clinical presentation of psychiatric disorders is often associated with acute life events or ongoing stress in adulthood (Heim and Nemeroff, 2001). In toto, these data lead to a hypothesis whereby early-life stressors lead to a state of enhanced vulnerability, and that adult mood and anxiety disorders present when later-life events have effects on the vulnerable brain of these individuals (Heim and Nemeroff, 2001). As witnessed by multiple studies of discordant monozygotic twins – in addition to the aforementioned studies – where one has the disorder and the other does not – nonmendelian mechanisms must be operative to control behavior in genetically identical populations (Gottesman et al., 1982). A critical question thus becomes what are the mechanism by which early-life events regulate long-term changes in behavior and sustained differences in gene expression. Genetic vulnerability factors undoubtedly play a critical role, but these abundant data – from multiple studies of discordant monozygotic twins – have shown that nongenetic (and more specifically nonmendelian genetic factors) are also critical mediators for the phenotypic expression of psychiatric diseases.

Similarly, preclinical studies utilizing animals have found that early-life stress produces long-lasting and sometimes profound biochemical and behavioral changes that persist for a lifetime. While there are undoubtedly mendelian genetic contributions (both susceptibility and protective), which effect the impact of neonatal stressors on brain development, it is noteworthy that recent studies have also demonstrated nongenomic transmission across generations of not only maternal behavior, but also stress responses in rodents (Francis et al., 1999). Additional studies utilizing genetically identical inbred animals have identified specific environmental events, which can result in permanent alterations in behavior, gene expression, and subsequent responses to stimuli such as stress (Heim and Nemeroff, 2001; Sanchez et al., 2001). Specifically, a growing body of data has demonstrated that neonatal stress can have a major impact on brain development, in particular by bringing about persistent changes in CRF-containing neurons, the HPA axis, the serotonergic system, the noradrenergic system, and the sympathetic nervous system (Graham et al., 1999). One well-studied model relies upon a maternal separation paradigm, wherein neonatal rodents are separated from their mother for defined periods. Behavioral, gene expression, and hormone production, changes have been noted to result from this paradigm, suggesting that there are epigenetic mechanisms that regulate these differential responses (Meaney, 2001). An additional line of studies by Meaney and colleagues has even documented that the degree of parental care that a newborn rat pup receives has lasting effects. This group has accumulated a large amount of data based on identifying litters in which pups were analyzed based upon the degree of maternal licking and grooming, followed by separation of groups receiving a standard deviation above and below the mean. Specific findings in these animals include lowered glucocorticoid receptor levels, increased HPA activity, and differential reactivity to stress (Ladd et al., 2000; Meaney, 2001; Weaver et al., 2001).

What are the mechanisms by which early-life stress – both in humans and animals – leads to a state of enhanced vulnerability?

The evidence discussed so far describe the clinical and preclinical evidence that nongenetic factors may play a role in behavioral and biochemical responses in the brain. In most areas of the brain neurons are not replaced. Thus, permanent and semipermanent modifications to deoxyribonucleic acid (DNA) that occur in early life, which affect gene transcription, could have downstream effects that persist for a lifetime – thus being temporally distant from the initial event. Mechanisms of gene regulation that are not determined solely by DNA base pair sequence are termed epigenetic and likely play a role in the formation of cellular memory and the modulation of neural circuitry in a manner that alters lifetime cellular and behavioral responses in an organism. Mendelian (DNA base pair) genetics involve inheritance patterns of nucleotide (cytosine, guanine, thymine, and adenine) inheritance. This is determined almost exclusively by the gametes from the parents (except for spontaneous mutations, trinucleotide repeats, etc.). Whereas epigenetics (or the epigenome) refers to factors such as cytosine methylation and the affinity of DNA regions to nucleosomes (made up of histones). This type of inheritance can dramatically influence gene expression. The true extent of the dynamic mechanisms responsible is unknown, and is an active area of research. However, it is known to involve the interplay of transcription factors interacting with covalent DNA modifications, such as cytosine methylation, and the accessibility of DNA that is regulated by histone acetylation (Petronis, 2001; Geiman and Robertson, 2002).

These epigenetic mechanisms are likely involved in modulating how previous experience may regulate subsequent behavioral responses (Meaney, 2001). Clearly, environmental factors play a role, but much of this may be primarily through epigenetic modifications. Thus, epigenetic interactions are likely prominent in gene expression, and formation of cellular memory. Indeed, work by Meaney and colleagues (with their grooming and licking model of parental care discussed above) has recently begun to document that some of the resultant changes – in particular GR regulation – may be mediated by epigenetic gene regulatory events (Weaver et al., 2002). The regulation of GR levels appears to be under the control of differential methylation of regions of the promoter, and the degree of parental care appears to regulate the methylation of this region – and thus long-term gene expression (Weaver et al., 2002). A potential interesting avenue of research may be the role that early-life stress plays in regulating changes in neurotrophic signaling cascades. Epigenetic factors may play a role in regulating these cascades thereby contributing to a long-term state of reduced cellular resilience, and ultimately to phenotypic expression of disease. Future experiments will help to further delineate the effect of stress on neurotrophic molecules, the regulation of BDNF by stress, and provide a cellular model for studying these effects. They will additionally lay groundwork for potential future experiments addressing specific mechanisms of epigenetic gene regulation (such as cytosine methylation and histone acetylation) with specific relevance for studying gene–environment interactions in the pathophysiology of mood disorders, and suggest mechanisms by which behavior can be regulated by epigenetic "marks," which are stable and permanent for the life of the organism.

Evidence that mood stabilizing and antidepressant medications regulate intracellular signaling pathways that exert critical neurotrophic/ neuroprotective effects

Noting the effects of stress on the brain, brain function, and behavior in preclinical models it is notable that the brains of patients with mood disorders show macroscopic and microscopic alterations that suggest impairments in neuroplasticity and cellular resilience. It is beyond the scope of this chapter to review these findings in depth; however, the evidence derives from postmortem and in vivo imaging. Studies utilizing neuroimaging report a decrease in frontal and temporal gray matter and an increase in ventricular size in patients with mood disorders (Drevets et al., 1997, 1999). Many studies also report a decrease in the size of some neuronal structures, including the hippocampus and portions of the basal ganglia (Drevets et al., 1999; Rajkowska, 2002). Functional imaging additionally reveals multiple abnormalities of regional cerebral blood flow and glucose metabolism in limbic and prefrontal cortical structures in mood disorders (Drevets et al., 1999; Drevets, 2000; Manji et al. 2001a).

Recent evidence using stereology techniques has identified decreased size and/or density of neurons, and decreased number and density of glial cells in portions of the anterior cingulate cortex, orbital frontal cortex, and dorsal lateral prefrontal cortex, as well changes in the hippocampus and amygdala, in patients with mood disorders (Rajkowska, 2000, 2002). Thus, both neuroimaging and postmortem findings suggest that the pathology of mood disorders may involve structural, as well as functional, changes in the brain. While there is not total reproducibility among studies, differences likely represent variations of experimental design, and as would be expected in heterogenous disorders such as mood disorders, in patient populations. Thus, research is required in order to understand if subtypes of depression, or mood disorders, are associated with any particular abnormality (Lenox et al., 2002).

As discussed more extensively earlier, endogenous neurotrophic factors are necessary for survival and functioning of neurons (Mamounas et al., 1995). They increase cell survival by providing necessary trophic support for growth, but also by exerting inhibitory effects on cell death cascades (Riccio et al., 1999). Evidence suggests that mood stabilizers and antidepressants may regulate these pathways. Antidepressant treatment in rats increases CREB phosphorylation and CREB-mediated gene expression in mouse limbic brain regions (Thome et al., 2000). Different classes of chronic antidepressant treatments, including NA and SSRIs and electroconvulsive seizure, upregulates CREB and BDNF expression, indicating that the CREB cascade and BDNF are common post-receptor targets of these therapeutic agents (Nibuya et al., 1995, 1996); furthermore the increase was only seen with chronic use, thus corresponding to the onset of action of these medications. More evidence that relates upregulation of these pathways and antidepressant utilization comes from antidepressant-like performance in behavioral models (Duman et al., 1999). Thus, it was observed that CREB overexpression in the dentate gyrus or BDNF injection results in an antidepressant-like effect in the learned–helplessness paradigm and the forced swim test model of antidepressant efficacy in rats (Siuciak et al., 1997; Chen et al., 2001; Shirayama et al., 2002). Chronic antidepressant treatment also results in an increase in the number of newly formed neurons (i.e., neurogenesis) in the hippocampus of rats.

Another series of studies involving mood stabilizing medications lithium and valproic acid have demonstrated that these medications activate the ERK MAP kinase cascade (Yuan et al., 2001; Gould et al., 2002a). Chronic lithium or valproic acid robustly increases the levels of activated ERK in the frontal cortex and hippocampus of rats (G. Chen and H. K. Manji, unpublished observations). As described earlier, this pathway regulates the expression of the neuroprotective protein Bcl-2, and chronic treatment of rats with either lithium or valproic acid produces a twofold increase of Bcl-2 levels in the frontal cortex (Chen et al., 1999). Furthermore, chronic lithium treatment increases Bcl-2 levels in the mouse hippocampus (Chen et al., 2000a), and in cerebellar granule cells in culture

(Chen and Chuang, 1999). Valproic acid also increases Bcl-2 levels in human cells of neuronal origin (Yuan et al., 2001).

It has also been observed that lithium is neuroprotective in animal models of ischemia, Huntington's disease, promotes neurogenesis in the hippocampus of rats, and is neuroprotective in many cell culture models (Manji et al., 1999; Chuang et al., 2002). Valproic acid also exerts neuroprotective actions in a number of cellular models including glutamate toxicity, β-amyloid toxicity, and following exposure to other toxins (Bruno et al., 1995; Mark et al., 1995; Manji et al., 2000a; Hashimoto et al., 2002).

Thus, regulation of neurotrophic pathways is seen with both classes of medications (i.e., antidepressants and mood stabilizers) in limbic and frontal brain regions (Duman et al., 2000; Manji et al., 2000b). However, it is likely that therapeutic responses are generated by changes in specific cell types, and in neural pathways specific to each disorder. Proof from clinical studies are required to validate whether these actions are clinically relevant for the treatment of mood disorders.

Based upon the above evidence longitudinal clinical studies were undertaken to address whether lithium produces effects consistent with changes in neuroplasticity. It was found using proton magnetic resonance spectroscopy (MRS) that chronic (4 week) lithium administration at therapeutic doses increases *N*-acetyl-aspartate (NAA, generally considered a marker of neuronal viability) concentration in the human brain in vivo (Moore et al., 2000a). These findings provide intriguing indirect support for the contention that: similar to the findings observed in the rodent brain and in human neuronal cells in culture, chronic lithium increases neuronal viability/function in the human brain. As a follow up of the NAA findings, brain tissue volumes were examined using high-resolution three-dimensional MRI and validated quantitative brain tissue segmentation methodology to identify and quantify the various components by volume, including total brain white and gray matter content. This study revealed that 4 weeks of lithium at therapeutic doses significantly increases total gray matter content in the human brain of patients with bipolar disorder (Moore et al., 2000b). No significant changes were observed in white matter volume or regional cerebral water content, thereby providing strong evidence that the observed increases in gray matter content are likely due to neurotrophic effects.

Conclusions

Despite the collective logic of the adaptive mechanisms and impaired cellular resilience mentioned in this chapter, there are heterogeneous arrays of processes, involving intra- and intercellular actions, adaptations in neurons, glia, vasculature and peripheral organs. This chapter only represents a partial review of the cellular adaptive responses during stress. By understanding both the adaptations of the nervous system and the limits it faces in defending itself, efficacious clinical therapies may well emerge. Most importantly, impairments of structural plasticity and cellular resilience have also been implicated in the preclinic and clinic studies of mood disorders. Furthermore, antidepressants and mood stabilizers exert major effects on signaling pathways, which regulate neuroplasticity and cell survival (see review by (Manji et al., 2000c)). All of these findings have generated considerable excitement among the clinical neuroscience community, and are reshaping views about the neurobiological underpinnings of these stress-related disorders (Manji et al., 2000d, 2001b; D'Sa and Duman 2002; Nestler et al., 2002; Young, 2002).

Abbreviations

NMDA	*N*-methyl-*D*-aspartate
CRF	corticotrophic hormone-releasing factor; also called corticotrophin-releasing factor
HPA	hypothalamic–pituitary–adrenal
GR	glucocorticoid receptor
ECT	electroconvulsive shock therapy
mRNA	messenger ribonucleic acid
BDNF	brain-derived neurotrophic factor
MAPK	mitogen-activated protein kinase
(PI-3K)	phosphotidylinositol-3 kinase
CNS	central nervous system
LTP	long-term potentiation
MR	mineralocorticoid receptor
ACTH	adrenocorticotrophic hormone

NGF	nerve growth factor
NT	neurotrophin
Trk	tyrosine kinase
P75	pan-neurotrophin receptor
PKA	protein kinase A
PKC	protein kinase C
SOS	son of sevenless
RSK	ribosomal S6 kinases
CREB	cAMP response element binding protein
cAMP	cyclic adenosine monophosphate
BAD	Bcl-xl/Bcl-2 Associated Death promoter
Bcl-xl	B-cell lymphoma/leukemia-xl
Bcl-2	B-cell lymphoma/leukemia-2
MPTP	1-methyl-4-phenyl-1,2,3,6-tetrahydropyridine
GSK	glycogen synthase kinase
FOX	Forkhead box transcription factor
FKHR	Forkhead transcriptional regulators?
AFX	ALL1 fused gene from chromosome X
NF-κB	nuclear factor κB
GABA	gamma-aminobutyric acid
AMPA	alpha-amino-3-hydroxy-5-methylisoxazole-4-propionate
mEPSC	miniature excitatory postsynaptic currents
mIPSC	miniature inhibitory postsynaptic currents
IR	immunoreactivity
ION	isthmo-optic nucleus
MEK	mitogen-activated/ERK-activating kinase
ERK	extracellular response kinase
MRS	magnetic resonance spectroscopy
NAA	*N*-acetyl-aspartate
PTSD	post-traumatic stress disorder
SSRI	selective serotonin reuptake inhibitor
NA	noradrenaline

Acknowledgments

We would like to acknowledge the support of the Intramural Research Program of the National Institute of Mental Heath and the Stanley Medical Research Institute. Due to space limitations we often cited review papers and apologize to those authors whose original data was not included.

References

Atkins, C.M., Selcher, J.C., Petraitis, J.J., Trzaskos, J.M. and Sweatt, J.D. (1998) The MAPK cascade is required for mammalian associative learning. Nat. Neurosci., 1: 602–609.

Bijur, G.N., De Sarno, P. and Jope, R.S. (2000) Glycogen synthase kinase-3beta facilitates staurosporine- and heat shock-induced apoptosis. Protection by lithium. J. Biol. Chem., 275: 7583–7590.

Blum, S., Moore, A.N., Adams. F. and Dash, P.K. (1999) A mitogen-activated protein kinase cascade in the CA1/CA2 subfield of the dorsal hippocampus is essential for long-term spatial memory. J. Neurosci., 19: 3535–3544.

Bonni, A., Brunet, A., West, A.E., Datta, S.R., Takasu, M.A. and Greenberg, M.E. (1999) Cell survival promoted by the Ras-MAPK signaling pathway by transcription-dependent and -independent mechanisms. Science, 286: 1358–1362.

Bremner, J.D., Randall, P., Scott, T.M., et al. (1995) MRI-based measurement of hippocampal volume in patients with combat-related posttraumatic stress disorder. Am. J. Psychiatry, 152: 973–981.

Bremner, J.D., Randall, P., Vermetten, E., et al. (1997) Magnetic resonance imaging-based measurement of hippocampal volume in posttraumatic stress disorder related to childhood physical and sexual abuse – a preliminary report. Biol. Psychiatry, 41: 23–32.

Brunet, A., Bonni, A., Zigmond, M.J., et al. (1999) Akt promotes cell survival by phosphorylating and inhibiting a Forkhead transcription factor. Cell, 96: 857–868.

Brunet, A., Datta, S.R. and Greenberg, M.E. (2001) Transcription-dependent and -independent control of neuronal survival by the PI3K-Akt signaling pathway. Curr. Opin. Neurobiol., 11: 297–305.

Bruno, V., Sortino, M.A., Scapagnini, U., Nicoletti, F. and Canonico, P.L. (1995) Antidegenerative effects of Mg(2+)-valproate in cultured cerebellar neurons. Funct. Neurol., 10: 121–130.

Brunson, K.L., Eghbal-Ahmadi, M., Bender, R., Chen, Y. and Baram, T.Z. (2001) Long-term, progressive hippocampal cell loss and dysfunction induced by early-life administration of corticotropin-releasing hormone reproduce the effects of early-life stress. Proc. Natl. Acad. Sci. USA, 98: 8856–8861.

Cameron, H.A. and McKay, R.D. (1999) Restoring production of hippocampal neurons in old age. Nat. Neurosci., 2: 894–897.

Charney, D.S. and Bremner, J.D. (1999) The neurobiology of anxiety disorders. In: Bunney, B.S. (Ed.), Neurobiology of Mental Illness. Oxford University Press, New York, pp. 494–517.

Chemerinski, E. and Robinson, R.G. (2000) The neuropsychiatry of stroke. Psychosomatics, 41: 5–14.

Chen, A.C., Shirayama, Y., Shin, K.H., Neve, R.L. and Duman, R.S. (2001) Expression of the cAMP response element binding protein (CREB) in hippocampus produces an antidepressant effect. Biol. Psychiatry, 49: 753–762.

Chen, D.F. and Tonegawa, S. (1998) Why do mature CNS neurons of mammals fail to re-establish connections following injury – functions of bcl-2. Cell Death Differ., 5: 816–822.

Chen, D.F., Schneider, G.E., Martinou, J.C. and Tonegawa, S. (1997) Bcl-2 promotes regeneration of severed axons in mammalian CNS. Nature, 385: 434–439.

Chen, G., Zeng, W.Z., Yuan, P.X., et al. (1999) The mood-stabilizing agents lithium and valproate robustly increase the levels of the neuroprotective protein bcl-2 in the CNS. J. Neurochem., 72: 879–882.

Chen, G., Rajkowska, G., Du, F., Seraji-Bozorgzad, N. and Manji, H.K. (2000a) Enhancement of hippocampal neurogenesis by lithium. J. Neurochem., 75: 1729–1734.

Chen, G., Masana, M.I. and Manji, H.K. (2000b) Lithium regulates PKC-mediated intracellular cross-talk and gene expression in the CNS in vivo. Bipolar Disorder 2000 Sep; 2(3 Pt 2)217–236. Review.

Chen, R.W. and Chuang, D.M. (1999) Long term lithium treatment suppresses p53 and Bax expression but increases Bcl-2 expression. A prominent role in neuroprotection against excitotoxicity. J. Biol. Chem., 274: 6039–6042.

Chierzi, S., Strettoi, E., Cenni, M.C. and Maffei, L. (1999) Optic nerve crush: axonal responses in wild-type and bcl-2 transgenic mice. J. Neurosci., 19: 8367–8376.

Chuang, D.M., Chen, R., Chalecka-Franaszek, E., et al. (2002) Neuroprotective effects of lithium in cultured cells and animal model of diseases. Bipolar Disorders, 4: 129–136.

Crow, T., Xue-Bian, J.J., Siddiqi, V., Kang, Y. and Neary, J.T. (1998) Phosphorylation of mitogen-activated protein kinase by one-trial and multi-trial classical conditioning. J. Neurosci., 18: 3480–3487.

Datta, S.R., Dudek, H., Tao, X., et al. (1997) Akt phosphorylation of BAD couples survival signals to the cell – intrinsic death machinery. Cell, 91: 231–241.

DeVries, A.C., Joh, H.D., Bernard, O., et al. (2001) Social stress exacerbates stroke outcome by suppressing Bcl-2 expression. Proc. Natl. Acad. Sci. USA, 98: 11824–11828.

Dijkers, P.F., Medema, R.H., Lammers, J.W., Koenderman, L. and Coffer, P.J. (2000) Expression of the pro-apoptotic Bcl-2 family member Bim is regulated by the forkhead transcription factor FKHR-L1. Curr. Biol., 10: 1201–1204.

Drevets, W.C. (2000) Neuroimaging studies of mood disorders. Biol. Psychiatry, 48: 813–829.

Drevets, W.C., Price, J.L., Simpson Jr., J.R., et al. (1997) Subgenual prefrontal cortex abnormalities in mood disorders. Nature, 386: 824–827.

Drevets, W.C., Gadde, K. and Krishnan, R. (1999) Neuroimaging studies of depression. In: Bunney, B.S. (Ed.), Neurobiology of Mental Illness. Oxford University Press, New York, NY, pp 394–418.

D'Sa, C. and Duman, R. (2002) Antidepressants and neuroplasticity. Bipolar Disorder, 4: 183.

Duman, R.S. (1998) Novel therapeutic approaches beyond the serotonin receptor. Biol. Psychiatry, 44: 324–335.

Duman, R.S., Malberg, J. and Thome, J. (1999) Neural plasticity to stress and antidepressant treatment. Biol. Psychiatry, 46: 1181–1191.

Duman, R.S., Malberg, J., Nakagawa, S. and D'Sa, C. (2000) Neuronal plasticity and survival in mood disorders. Biol. Psychiatry, 48: 732–739.

English, J.D. and Sweatt, J.D. (1996) Activation of p42 mitogen-activated protein kinase in hippocampal long term potentiation. J. Biol. Chem., 271: 24329–24332.

English, J.D. and Sweatt, J.D. (1997) A requirement for the mitogen-activated protein kinase cascade in hippocampal long term potentiation. J. Biol. Chem., 272: 19103–19106.

Eriksson, P.S., Perfilieva, E., Bjork-Eriksson, T., et al. (1998) Neurogenesis in the adult human hippocampus. Nat. Med., 4: 1313–1317.

Finkbeiner, S. (2000) CREB couples neurotrophin signals to survival messages. Neuron, 25: 11–14.

Francis, D., Diorio, J., Liu, D. and Meaney, M.J. (1999) Nongenomic transmission across generations of maternal behavior and stress responses in the rat. Science, 286: 1155–1158.

Fryer, R.H., Kaplan, D.R., Feinstein, S.C., Radeke, M.J., Grayson, D.R. and Kromer, L.F. (1996) Developmental and mature expression of full-length and truncated TrkB receptors in the rat forebrain. J. Comp. Neurol., 374: 21–40.

Fukunaga, K. and Miyamoto, E. (1998) Role of MAP kinase in neurons. Mol. Neurobiol., 16: 79–95.

Gass, P., Kretz, O., Wolfer, D.P., et al. (2000) Genetic disruption of mineralocorticoid receptor leads to impaired neurogenesis and granule cell degeneration in the hippocampus of adult mice. EMBO Rep., 1: 447–451.

Geiman, T.M. and Robertson, K.D. (2002) Chromatin remodeling, histone modifications, and DNA methylation – how does it all fit together? J. Cell. Biochem., 87: 117–125.

Gesing, A., Bilang-Bleuel, A., Droste, S.K., Linthorst, A.C., Holsboer, F. and Reul, J.M. (2001) Psychological stress increases hippocampal mineralocorticoid receptor levels: involvement of corticotropin-releasing hormone. J. Neurosci., 21: 4822–4829.

Gilbertson, M.W., Shenton, M.E., Ciszewski, A., et al. (2002) Smaller hippocampal volume predicts pathologic vulnerability to psychological trauma. Nat. Neurosci., 5: 1242–1247.

Gonzalez, M., Ruggiero, F.P., Chang, Q., et al. (1999) Disruption of Trkb-mediated signaling induces disassembly of postsynaptic receptor clusters at neuromuscular junctions. Neuron, 24: 567–583.

Gottesman, I.I., Shields, J. and Hanson, D.R. (1982) Schizophrenia, the epigenetic puzzle. Cambridge University Press, Cambridge, New York.

Gould, E., Tanapat, P., Rydel, T. and Hastings, N. (2000) Regulation of hippocampal neurogenesis in adulthood. Biol. Psychiatry, 48: 715–720.

Gould, T., Chen, G. and Manji, H. (2002a) Mood stabilizer psychopharmacology. Clinical Neuroscience Research, 2: 193–212.

Gould, T., Du, J., Yuan, P., Chen, G. and Manji, H. (2002b) Evidence suggesting activation of the Wnt signaling pathway in rat brain after lithium and valproic acid administration. Biol. Psychiatry, 51: 39S.

Gould, T.D. and Manji, H.K. (2002) The Wnt signaling pathway in bipolar disorder. Neuroscientist, 8.

Graham, Y.P., Heim, C., Goodman, S.H., Miller, A.H. and Nemeroff, C.B. (1999) The effects of neonatal stress on brain development: implications for psychopathology. Dev. Psychopathol., 11: 545–565.

Gurvits, T.V., Shenton, M.E., Hokama, H., et al. (1996) Magnetic resonance imaging study of hippocampal volume in chronic, combat-related posttraumatic stress disorder. Biol. Psychiatry, 40: 1091–1099.

Hashimoto, R., Hough, C., Nakazawa, T., Yamamoto, T. and Chuang, D.M. (2002) Lithium protection against glutamate excitotoxicity in rat cerebral cortical neurons: involvement of NMDA receptor inhibition possibly by decreasing NR2B tyrosine phosphorylation. J. Neurochem., 80: 589–597.

Heim, C. and Nemeroff, C.B. (2001) The role of childhood trauma in the neurobiology of mood and anxiety disorders: preclinical and clinical studies. Biol. Psychiatry, 49: 1023–1039.

Hetman, M., Cavanaugh, J.E., Kimelman, D. and Xia, Z. (2000) Role of glycogen synthase kinase-3beta in neuronal apoptosis induced by trophic withdrawal. J. Neurosci., 20: 2567–2574.

Holm, K.H., Cicchetti, F., Bjorklund, L., et al. (2001) Enhanced axonal growth from fetal human bcl-2 transgenic mouse dopamine neurons transplanted to the adult rat striatum. Neuroscience, 104: 397–405.

Holsboer, F. (2000) The corticosteroid receptor hypothesis of depression. Neuropsychopharmacology, 23: 477–501.

Jacobs, B.L., Praag, H. and Gage, F.H. (2000) Adult brain neurogenesis and psychiatry: a novel theory of depression. Mol. Psychiatry, 5: 262–269.

Kempermann, G. (2002) Regulation of adult hippocampal neurogenesis – implications for novel theories of major depression. Bipolar Disord., 4: 17–33.

Kempermann, G., Kuhn, H.G. and Gage, F.H. (1997) More hippocampal neurons in adult mice living in an enriched environment. Nature, 386: 493–495.

Kollack-Walker, S., Day, H.E.W. and Akil, H. (2000) Central stress neurocircuits. In: Encyclopedia of Stress. Ed. G. Fink, Academic Press, San Diego.

Kornhauser, J.M. and Greenberg, M.E. (1997) A kinase to remember: dual roles for MAP kinase in long-term memory. Neuron, 18: 839–842.

Korte, S.M., De Kloet, E.R., Buwalda, B., Bouman, S.D. and Bohus, B. (1996) Antisense to the glucocorticoid receptor in hippocampal dentate gyrus reduces immobility in forced swim test. Eur. J. Pharmacol., 301: 19–25.

Kumar, A., Miller, D., Ewbank, D., et al. (1997) Quantitative anatomic measures and comorbid medical illness in late-life major depression. Am. J. Geriatr. Psychiatry, 5: 15–25.

Ladd, C.O., Huot, R.L., Thrivikraman, K.V., Nemeroff, C.B., Meaney, M.J. and Plotsky, P.M. (2000) Long-term behavioral and neuroendocrine adaptations to adverse early experience. Prog. Brain Res., 122: 81–103.

Lenox, R.H., Gould, T.D. and Manji, H.K. (2002) Endophenotype in bipolar disorder. Am. J. Med. Genet., 114: 391–406.

Lopez, J.F., Chalmers, D.T., Little, K.Y. and Watson, S.J. (1998) A.E. Bennett Research Award. Regulation of serotonin1A, glucocorticoid, and mineralocorticoid receptor in rat and human hippocampus: implications for the neurobiology of depression. Biol. Psychiatry, 43: 547–573.

Lyons, D.M. (2002) Stress, depression, and inherited variation in primate hippocampal and prefrontal brain development. Psychopharmacology Bulletin, 36: 27–43.

Lyons, D.M., Yang, C., Sawyer-Glover, A.M., Moseley. M.E. and Schatzberg, A.F. (2001) Early life stress and inherited variation in monkey hippocampal volumes. Arch. Gen. Psychiatry, 58: 1145–1151.

Maggirwar, S.B., Sarmiere, P.D., Dewhurst, S. and Freeman, R.S. (1998) Nerve growth factor-dependent activation of NF-kappaB contributes to survival of sympathetic neurons. J. Neurosci., 18: 10356–10365.

Mamounas, L.A., Blue, M.E., Siuciak, J.A. and Altar, C.A. (1995) Brain-derived neurotrophic factor promotes the survival and sprouting of serotonergic axons in rat brain. J. Neurosci., 15: 7929–7939.

Manev, H., Uz, T., Smalheiser, N.R. and Manev, R. (2001) Antidepressants alter cell proliferation in the adult brain in vivo and in neural cultures in vitro. Eur. J. Pharmacol., 411: 67–70.

Manji, H.K. and Chen, G. (2002) PKC, MAP kinases and the bcl-2 family of proteins as long-term targets for mood stabilizers. Mol. Psychiatry, 7(Suppl. 1): S46–56.

Manji, H. and Duman, R. (2001a) Impairments of neuroplasticity and cellular resilience in severe mood disorder: implications for the development of novel therapeutics. Psychopharmacology Bulletin, 35: 5–49.

Manji, H.K., Moore, G.J. and Chen, G. (1999) Lithium at 50: have the neuroprotective effects of this unique cation been overlooked? Biol. Psychiatry, 46: 929–940.

Manji, H.K., Moore, G.J. and Chen, G. (2000a) Clinical and preclinical evidence for the neurotrophic effects of mood stabilizers: implications for the pathophysiology and treatment of manic-depressive illness. Biol. Psychiatry, 48: 740–754.

Manji, H.K., Moore, G.J., Rajkowska, G. and Chen, G. (2000b) Neuroplasticity and cellular resilience in mood disorders. Mol. Psychiatry, 5: 578–593.

Manji, H.K., Moore, G.J. and Chen, G. (2000d) Lithium up-regulates the cytoprotective protein Bcl-2 in the CNS in vivo: a role for neurotrophic and neuroprotective effects in manic depressive illness. J. Clin. Psychiatry, 61: 82–96.

Manji, H.K., Moore, G.J. and Chen, G. (2001a) Bipolar disorder: leads from the molecular and cellular mechanisms of action of mood stabilizers. Br. J. Psychiatry, 178: s107–119.

Manji, H.K., Drevets, W.C. and Charney, D.S. (2001b) The cellular neurobiology of depression. Nat. Med., 7: 541–547.

Manji, H.K., Moore, G.J. and Chen, G. (2001c) Bipolar disorder: leads from the molecular and cellular mechanisms of action of mood stabilizers. Br. J. Psychiatry, Suppl 41: s107–119.

Mark, R.J., Ashford, J.W., Goodman, Y. and Mattson, M.P. (1995) Anticonvulsants attenuate amyloid beta-peptide neurotoxicity, Ca2+ deregulation, and cytoskeletal pathology. Neurobiol. Aging, 16: 187–198.

Martin, K.C., Michael, D., Rose, J.C., et al. (1997) MAP kinase translocates into the nucleus of the presynaptic cell and is required for long-term facilitation in Aplysia. Neuron, 18: 899–912.

Matsubara, M., Kusubata, M,. Ishiguro, K., Uchida, T., Titani, K. and Taniguchi, H. (1996) Site-specific phosphorylation of synapsin I by mitogen-activated protein kinase and Cdk5 and its effects on physiological functions. J. Biol. Chem., 271: 21108–21113.

McEwen, B.S. (1999) Stress and hippocampal plasticity. Annu. Rev. Neurosci., 22: 105–122.

McQuade, R. and Young, A.H. (2000) Future therapeutic targets in mood disorders: the glucocorticoid receptor. Br. J. Psychiatry, 177: 390–395.

Meaney, M.J. (2001) Maternal care, gene expression, and the transmission of individual differences in stress reactivity across generations. Annu. Rev. Neurosci., 24: 1161–92.

Mesulam, M.M. (1999) Neuroplasticity failure in Alzheimer's disease: bridging the gap between plaques and tangles. Neuron, 24: 521–529.

Michaloudi, H.C., el Majdoubi, M., Poulain, D.A., Papadopoulos, G.C. and Theodosis, D.T. (1997) The noradrenergic innervation of identified hypothalamic magnocellular somata and its contribution to lactation-induced synaptic plasticity. J. Neuroendocrinol., 9: 17–23.

Moore, G.J., Bebchuk, J.M., Hasanat, K., et al. (2000a) Lithium increases N-acetyl-aspartate in the human brain: in vivo evidence in support of bcl-2's neurotrophic effects? Biol. Psychiatry, 48: 1–8.

Moore, G.J., Bebchuk, J.M., Wilds, I.B., Chen, G., Manji, H.K. and Menji, H.K. (2000b) Lithium-induced increase in human brain grey matter. Lancet, 356: 1241–1242.

Murphy, T.H., Blatter, L.A., Bhat, R.V., Fiore, R.S., Wier, W.G. and Baraban, J.M. (1994) Differential regulation of calcium/calmodulin-dependent protein kinase II and p42 MAP kinase activity by synaptic transmission. J. Neurosci., 14: 1320–1331.

Murray, C.J. and Lopez, A.D. (1997) Global mortality, disability, and the contribution of risk factors: global burden of disease study. Lancet, 349: 1436–1442.

Nestler, E.J. (1998) Antidepressant treatments in the 21st century. Biol. Psychiatry, 44: 526–533.

Nestler, E.J., Barrot, M., DiLeone, R.J., Eisch, A.J., Gold, S.J. and Monteggia, L.M. (2002) Neurobiology of depression. Neuron, 34: 13–25.

Nibuya, M., Morinobu, S. and Duman, R.S. (1995) Regulation of BDNF and trkB mRNA in rat brain by chronic electroconvulsive seizure and antidepressant drug treatments. J. Neurosci., 15: 7539–7547.

Nibuya, M., Nestler, E.J. and Duman, R.S. (1996) Chronic antidepressant administration increases the expression of cAMP response element binding protein (CREB) in rat hippocampus. J. Neurosci., 16: 2365–2372.

Nibuya, M., Takahashi, M., Russell, D.S. and Duman, R.S. (1999) Repeated stress increases catalytic TrkB mRNA in rat hippocampus. Neurosci. Lett., 267: 81–84.

Oh, Y.J., Swarzenski, B.C. and O'Malley, K.L. (1996) Overexpression of Bcl-2 in a murine dopaminergic neuronal cell line leads to neurite outgrowth. Neurosci. Lett., 202: 161–164.

Ontiveros, A., Fontaine, R., Breton, G., Elie, R., Fontaine, S. and Dery, R. (1989) Correlation of severity of panic disorder and neuroanatomical changes on magnetic resonance imaging. J. Neuropsychiatry Clin. Neurosci., 1: 404–408.

Patapoutian, A. and Reichardt, L.F. (2001) Trk receptors: mediators of neurotrophin action. Curr. Opin. Neurobiol., 11: 272–280.

Patel, P.D., Lopez, J.F., Lyons, D.M., Burke, S., Wallace, M. and Schatzberg, A.F. (2000) Glucocorticoid and mineralocorticoid receptor mRNA expression in squirrel monkey brain. J. Psychiatr. Res., 34: 383–392.

Petronis, A. (2001) Human morbid genetics revisited: relevance of epigenetics. Trends Genet., 17: 142–146.

Poo, M.M. (2001) Neurotrophins as synaptic modulators. Nat. Rev. Neurosci., 2: 24–32.

Rajkowska, G. (2000) Postmortem studies in mood disorders indicate altered numbers of neurons and glial cells. Biol. Psychiatry, 48: 766–777.

Rajkowska, G. (2002) Cell pathology in bipolar disorder. Bipolar Disord., 4: 105–116.

Reul, J.M. and Holsboer, F. (2002) Corticotropin-releasing factor receptors 1 and 2 in anxiety and depression. Curr. Opin. Pharmacol., 2: 23–33.

Riccio, A., Pierchala, B.A., Ciarallo, C.L. and Ginty, D.D. (1997) An NGF-TrkA-mediated retrograde signal to transcription factor CREB in sympathetic neurons. Science, 277: 1097–1100.

Riccio, A., Ahn, S., Davenport, C.M., Blendy, J.A. and Ginty, D.D. (1999) Mediation by a CREB family transcription factor of NGF-dependent survival of sympathetic neurons. Science, 286: 2358–2361.

Roberson, E.D., English, J.D. and Sweatt, J.D. (1996) A biochemist's view of long-term potentiation. Learn Mem., 3: 1–24.

Roberson, E.D., English, J.D., Adams, J.P., Selcher, J.C., Kondratick, C. and Sweatt, J.D. (1999) The mitogen-activated protein kinase cascade couples PKA and PKC to cAMP response element binding protein phosphorylation in area CA1 of hippocampus. J. Neurosci., 19: 4337–4348.

Robinson, M.J., Stippec, S.A., Goldsmith, E., White, M.A. and Cobb, M.H. (1998) A constitutively active and nuclear form of the MAP kinase ERK2 is sufficient for neurite outgrowth and cell transformation. Curr. Biol., 8: 1141–1150.

Sanchez, M.M., Young, L.J., Plotsky, P.M. and Insel, T.R. (2000) Distribution of corticosteroid receptors in the rhesus brain: relative absence of glucocorticoid receptors in the hippocampal formation. J. Neurosci., 20: 4657–4668.

Sanchez, M.M., Ladd, C.O. and Plotsky, P.M. (2001) Early adverse experience as a developmental risk factor for later psychopathology: evidence from rodent and primate models. Dev. Psychopathol., 13: 419–449.

Sapolsky, R.M. (1996) Stress, glucocorticoids, and damage to the nervous system: the current state of confusion. Stress, 1: 1–19.

Sapolsky, R.M. (2000a) Glucocorticoids and hippocampal atrophy in neuropsychiatric disorders. Arch. Gen. Psychiatry, 57: 925–935.

Sapolsky, R.M. (2000b) The possibility of neurotoxicity in the hippocampus in major depression: a primer on neuron death. Biol. Psychiatry, 48: 755–765.

Selye, H. (1956) The Stress of Life. McGraw-Hill, New York.

Shirayama, Y., Chen, A.C., Nakagawa, S., Russell, D.S. and Duman, R.S. (2002) Brain-derived neurotrophic factor produces antidepressant effects in behavioral models of depression. J. Neurosci., 22: 3251–3261.

Simmons, N.E., Alden, T.D., Thorner, M.O. and Laws Jr., E.R. (2001) Serum cortisol response to transsphenoidal surgery for Cushing disease. J. Neurosurg., 95: 1–8.

Siuciak, J.A., Lewis, D.R., Wiegand, S.J. and Lindsay, R.M. (1997) Antidepressant-like effect of brain-derived neurotrophic factor (BDNF). Pharmacol. Biochem. Behav., 56: 131–137.

Smith, M.A., Makino, S., Kvetnansky, R. and Post, R.M. (1995) Stress and glucocorticoids affect the expression of brain-derived neurotrophic factor and neurotrophin-3 mRNAs in the hippocampus. J. Neurosci., 15: 1768–1777.

Sousa, N., Lukoyanov, N.V., Madeira, M.D., Almeida, O.F. and Paula-Barbosa, M.M. (2000) Reorganization of the morphology of hippocampal neurites and synapses after stress-induced damage correlates with behavioral improvement. Neuroscience, 97: 253–266.

Starkman, M.N., Giordani, B., Gebarski, S.S., Berent, S., Schork, M.A. and Schteingart, D.E. (1999) Decrease in cortisol reverses human hippocampal atrophy following treatment of Cushing's disease. Biol. Psychiatry, 46: 1595–1602.

Steckler, T., Rammes, G., Sauvage, M., van Gaalen, M.M., Weis, C., Zieglgansberger, W. and Holsboer, F. (2001) Effects of the monoamine oxidase A inhibitor moclobemide on hippocampal plasticity in GR-impaired transgenic mice. J. Psychiat. Res., 35(1): 29–42.

Steffens, D.C. and Krishnan, K.R. (1998) Structural neuroimaging and mood disorders: recent findings, implications for classification, and future directions. Biol. Psychiatry, 43: 705–712.

Steffens, D.C., Helms, M.J., Krishnan, K.R. and Burke, G.L. (1999) Cerebrovascular disease and depression symptoms in the cardiovascular health study. Stroke, 30: 2159–2166.

Stein, M.B., Koverola, C., Hanna, C., Torchia, M.G. and McClarty, B. (1997) Hippocampal volume in women victimized by childhood sexual abuse. Psychol. Med., 27: 951–959.

Stern, J.E. and Armstrong, W.E. (1998) Reorganization of the dendritic trees of oxytocin and vasopressin neurons of the rat supraoptic nucleus during lactation. J. Neurosci., 18: 841–853.

Suzuki, T., Okumura-Noji, K. and Nishida, E. (1995) ERK2-type mitogen-activated protein kinase (MAPK) and its substrates in postsynaptic density fractions from the rat brain. Neurosci. Res., 22: 277-285

Thome, J., Sakai, N., Shin, K., et al. (2000) cAMP response element-mediated gene transcription is upregulated by chronic antidepressant treatment. J. Neurosci., 20: 4030–4036.

Trentani, A., Kuipers, S.D., Ter Horst, G.J. and Den Boer, J.A. (2002) Selective chronic stress-induced in vivo ERK1/2 hyperphosphorylation in medial prefrontocortical dendrites: implications for stress-related cortical pathology? Eur. J. Neurosci., 15(10): 1681–1691.

van Praag, H., Kempermann, G. and Gage, F.H. (1999) Running increases cell proliferation and neurogenesis in the adult mouse dentate gyrus. Nat. Neurosci., 2: 266–270.

Weaver, I.C., La Plante, P., Weaver, S., et al. (2001) Early environmental regulation of hippocampal glucocorticoid receptor gene expression: characterization of intracellular mediators and potential genomic target sites. Mol. Cell. Endocrinol., 185: 205–218.

Weaver, I.C.G., Cervoni, N., Szyf, M. and Meaney, M.J. (2002) Maternal behavior in infancy regulates methylation of hippocampal glucocorticoid receptor promoter. Abstract Soc. Neuroscience Annual Meeting, Vol. abstract number 866.9.

Webster, O'Grady, Orthmann, et al. (1999) Decreased glucocorticoid receptor mRNA levels in individuals with depression, bipolar disorder and schizophrenia. Schizophrenia Research, 41: 111.

Wellman, C.L. (2001) Dendritic reorganization in pyramidal neurons in medial prefrontal cortex after chronic corticosterone administration. J. Neurobiol., 49: 245–253.
Withers, G.S. and Greenough, W.T. (1989) Reach training selectively alters dendritic branching in subpopulations of layer II-III pyramids in rat motor-somatosensory forelimb cortex. Neuropsychologia, 27: 61–69.
Woodgett, J.R. (2001) Judging a protein by more than its name: gsk-3. Sci STKE, 2001: RE12.
Xu, Q., Fawcett, T.W., Gorospe, M., Guyton, K.Z., Liu, Y. and Holbrook, N.J. (1997) Induction of mitogen-activated protein kinase phosphatase-1 during acute hypertension. Hypertension, 30: 106–111.
Young, L.T. (2002) Neuroprotective effects of antidepressant and mood stabilizing drugs. J. Psychiatry Neurosci., 27: 8–9.
Yuan, D., Komatsu, K., Tani, H., Cui, Z. and Kano, Y. (1998) Pharmacological properties of traditional medicines. XXIV. Classification of antiasthmatics based on constitutional predispositions. Biol. Pharm. Bull., 21: 1169–1173.
Yuan, P.X., Huang, L.D., Jiang, Y.M., Gutkind, J.S., Manji, H.K. and Chen, G. (2001) The mood stabilizer valproic acid activates mitogen-activated protein kinases and promotes neurite growth. J. Biol. Chem., 276: 31674–31683.

SECTION 6

The Stressed Brain

T. Steckler, N.H. Kalin and J.M.H.M. Reul (Eds.)
Handbook of Stress and the Brain, Vol. 15
ISBN 0-444-51173-3

CHAPTER 6.1

Psychological and physiological stressors

Krisztina J. Kovács*, Ildikó H. Miklós and Balázs Bali

Laboratory of Molecular Neuroendocrinology, Institute of Experimental Medicine, Hungarian Academy of Science, Szigony u. 43, Budapest H-1083, Hungary

Abstract: Stressors posed by the external and internal environment are divided into two main categories: physiological and psychological. These categories are not mutually exclusive and their classification is based on the differences in perception and registration of relevant stimuli and the afferent pathways that mediate and/or may modulate stress responses. This chapter summarizes and compares afferent mechanisms, activated circuits, as well as transcriptional, hormonal, and autonomic responses to physiological and psychological challenges. While both type of stressors initiate complex adaptive responses physiological stressors directly target homeostatic parameters, transduced via viscerosensory pathways, psychological stressors recruit various somatosensory and nociceptive afferentations, the information is processed through complex cortical and limbic circuits to include cognitive, learned, and emotional components. Functional anatomical mapping strategies that are based on the induction of immediate-early genes (i.e. c-fos) identified corticotropin-releasing factor (CRF)-synthesizing parvocellular neurons as a common target of both type of stressors. At the cellular level, double imaging technique revealed that even parvocellular neurosecretory neurons may respond differentially to categorically distinct paradigms. Significant differences have also been found at various medullary (nucleus of the solitary tract), subcortical, and limbic areas (amygdala and bed nucleus of stria terminalis) that involved in mediation of relevant information to the stress-related hypophysiotropic neurons. In addition, the chapter contains a brief summary on laboratory stressors used on human subjects.

Introduction

Stressors have been defined by Selye as exogenous or endogenous challenges that threaten homeostasis (Selye, 1936; Selye, 1955). Indeed, vital parameters such as serum osmolarity and pH, extracellular sodium concentration, blood glucose levels, oxygen tension and core temperature should be kept in a narrow range around a *fixed setpoint* under tight homeostatic control. In addition to stressors that directly affect these physiological variables and require immediate corrections, there are challenges posed by the external or internal environment that initiate centrally mediated responses to serve adaptation. These situations involve failures to satisfy internal drives, aversive environmental stimuli, presence of predators or changes in the individual's real or anticipated state in a social context. The concept of allostasis, introduced by Sterling and Eyer (1988), emphasizes the *change of setpoints*. McEwen broadened this concept and defined allostasis as a process maintaining stability through change and promoting adaptation and coping (McEwen and Stellar, 1993; McEwen, 1998a,b, 2000). At the time of birth of the stress concept, stressors have been regarded as what are now referred to as physiological stressors. Stressful stimuli provoke centrally mediated coordinated responses including neuroendocrine [i.e. activation of the hypothalamo–pituitary–adrenocortical (HPA) axis], autonomic (i.e. activation of the sympatho-adrenal system), and behavioral (i.e. flight or fight) changes. However, it became clear, that psychological challenges we, human beings, are experiencing every day are among the most powerful stimuli to induce these responses.

*Corresponding author Tel.: +36-1-210-9952;
Fax: +36-1-210-9423; E-mail: kovacs@koki.hu

Stressor categories

Stress paradigms used in the *experimental stress research* can be divided into two main categories:

1. *Physiological stressors* often referred to as homeostatic, systemic or physical stressors.

These challenges target physiological parameters, their effect is mediated through well-defined internal or external receptor systems, the afferentation include viscerosensory pathways, activate subcortical autonomic circuits and directly affect stress-related motoneurons without significant cortical contribution. Examples include osmotic, metabolic, cardiovascular and immune stressors.

Blood-borne signals [plasma osmolarity, immune mediators and certain hormones (i.e. angiotensin II and insulin)] are registered at the sensory circumventricular organs that lack the blood brain barrier (Katsuura et al., 1990; McKinley et al., 2003) and project to the paraventricular nucleus (PVN) (Kovacs and Sawchenko, 1993; Ericsson et al., 1994; Sawchenko et al., 2000). Blood pressure and volume-related information is mediated through viscerosensory vagal and glossopharingeal inputs to the nucleus of the solitary tract (NTS) (Chan and Sawchenko, 1994; Chan and Sawchenko, 1998; Dampney and Horiuchi, 2003). Glucose-sensitive vagal inputs also reach NTS neurons that give rise ascending catecholaminergic pathways to the stress-related neurosecretory cells of the hypothalamus (Adachi et al., 1984; Nagase et al., 1993). Temperature- (Scammell et al., 1993) and glucose (Williams et al., 2001) sensitive neurons were identified in the hypothalamus that may also provide direct inputs to the neurosecretory motoneurons (Fig. 1A). Thus, homeostatic challenges launch relative simple neuroendocrine reflexes by direct activation of the parvocellular CRF secreting neurons in the paraventricular nucleus to initiate the hormonal stress cascade. In addition to neuroendocrine responses, autonomic efferent projections to preganglionic cells groups in the medulla and spinal cord became also activated to induce adaptive cardiovascular, respiratory and other vegetative responses and include the activation of the sympatoadrenal system as well (Fig. 1B).

2. *Psychological stressors* are also called neurogenic, emotional, or processive stimuli. Registered and initiated by complex somatosensory and nociceptive mechanisms, include less-specific exteroceptive and/or somatic inputs, processed through higher order brain circuits and involve learned, emotional, and cognitive components. Most often used emotional stressors are exposure to novel environment, restraint, immobilization, or footshock.

The mechanisms that initiate a psychological stress response are less easily specified compared to physiological stress category. These probably involve spinal and trigeminal somatosensory and nociceptive pathways (spinothalamic-, trigeminothalamic-, medial leminiscus pathway, spinoreticulothalamic-, spinohypothalamic-, spinoreticular pathways). Scarce relevant viscerosensory information travel through the vagus and glossopharingeal nerves to the NTS (for review, see Sawchenko et al., 2000 and Pacak and Palkovits, 2001). Some of these pathways give collaterals to the hypothalamic effector neurons, but a great deal of relevant activational inputs is relayed at extrahypothalamic sites. Inputs from the extrahypothalamic areas do not reach directly the stress-related motorneurons, their effect is integrated at the local GABA and glutamatergic interneuronal population in the periparaventricular region (Boudaba et al., 1996; Boudaba et al., 1997; Herman et al., 2002). Psychological stressors result in c-fos induction at cortical- (prefrontal) and subcortical limbic areas (septum, basolateral, and medial amygdala and bed nucleus of stria terminalis), midline thalamic nuclei, periaqueductal gray, locus coeruleus and catecholaminergic neurons in NTS and ventrolateral medulla (Cullinan et al., 1995; Li and Sawchenko, 1998; Dayas et al., 2001). Although these medullary cell groups are acknowledged as the major source of ascending aminergic pathways that stimulate CRF-secreting neurons in the PVN, their activation in response to psychogen stressors such as intermittent footshock is rather downstream to the recruitment of parvocellular stress-related neurons (Li et al., 1996; Sawchenko et al., 2000).

Like physiological stressors, psychological and emotional stimuli also activate parvocellular neurosecretory motoneurons in the PVN and result in upregulation of neuropeptide gene expression in these cells (Ma et al., 1997; Viau and Sawchenko, 2002). Visceral responses to psychological stressors also include cardiovascular, respiratory, gastrointestinal

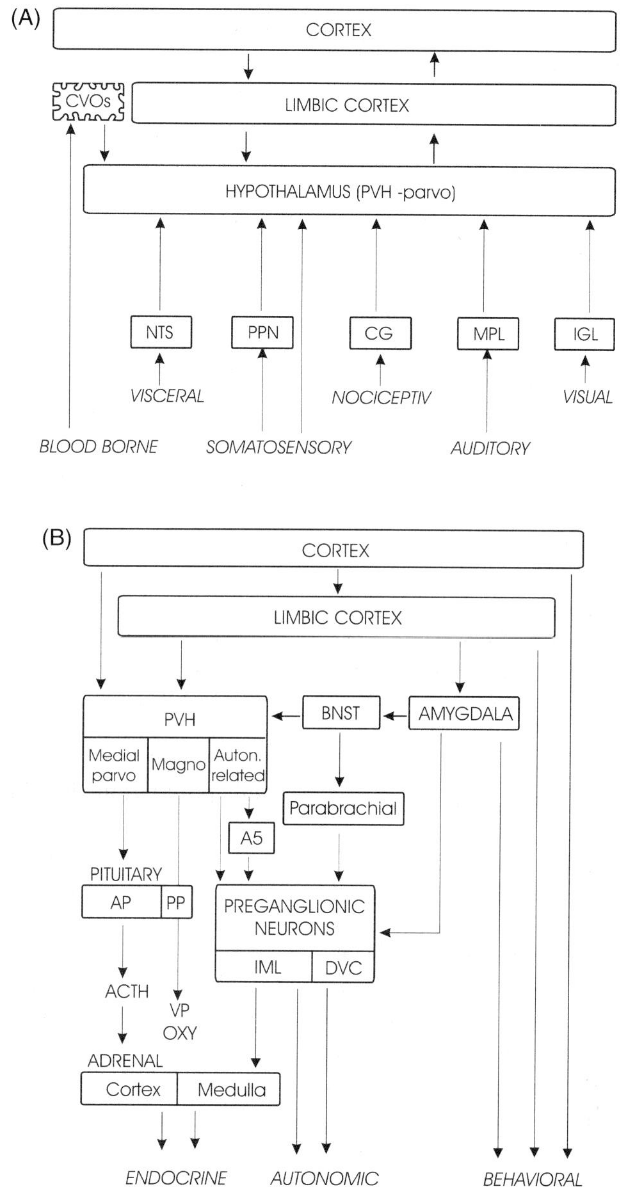

Fig. 1. Schematic summary of afferent (A) and efferent (B) stress pathways. Abbreviations (A)- CG: central gray; CVOs: circumventricular organs; IGL: intergeniculate leaflet; MPL: medial paralemniscal nucleus; NTS: nucleus of the solitary tract; PPN: pedunculopontine nucleus; PVH-parvo: medial dorsal parvocellular subdivision of the hypothalamic paraventricular nucleus. Abbreviations (B)- AP: anterior pituitary; BNST: bed nucleus of stria terminalis; DVC: dorsal vagal complex; IML: intermediolateral cell column of the spinal cord; OXY:oxytocin; PP: posterior pituitary; VP: vasopressin.

and thermoregulatory changes and are stressor-specific. Efferent pathways originate from cortical, limbic, hypothalamic and brainstem areas to activate motor and autonomic responses (Fig. 1B).

These stressor categories however, are not mutually exclusive. Experimentally used stressors as well as those posed by the environment should be viewed in their complexity. Immobilization or footshock, for example, may cause pain that initiate well-defined stress reflex responses. Ether inhalation includes physiological stress components such as hypoxia and respiratory distress as well as emotional challenges such as restraint, aversive smell and loss of consciousness. Immune stimuli is often accompanied by hypotension and/or initiate sickness behavior (Berkenbosch et al., 1989). Stress responses can be conditioned and cross-sensitized among these categories (Goldstein et al., 1996; Morrow et al., 1999).

In addition to stressors used in experimental biology, there are *behaviorally* more *relevant, naturalistic stressors* such as predator exposure or social stressors such as isolation, agonistic encounter (defeat) or changes in social hierarchy.

Acute or chronic exposure of rats to predators (cats, ferrets) or their odors (synthetic fox fecal odor, trimethylthiazoline, TMT) evoke significant activation of the HPA axis (Blanchard et al., 1998; Morrow et al., 2000; Figueiredo et al., 2003). In most of these studies there is no direct contact between the subjects, indicating that visual, olfactory and auditory cues associated with the presence of the predator are able to start innate response programs, including the stress response. Brain regions, recruited by predator exposure correspond fairly well with those activated by emotional stressor (restraint), however, it is interesting to note that cat-exposed rats do not express c-fos mRNA in their PVNs in spite of the increase of CRF mRNA levels (Figueiredo et al., 2003).

Agonistic encounter in males results in differential c-fos activation patterns in dominant and subordinate hamsters (Kollack-Walker et al., 1997) or in the resident and intruder rats (Halasz et al., 2002; Martinez et al., 2002). Fighting itself activates c-fos expression in the medial amygdala in both partners, however the number of areas showing increased neuronal activation is much higher in subordinates than in dominant males (Kollack-Walker et al., 1997; Martinez et al., 2002). Repeated exposure to social defeat results in a selective pattern of habituation of immediate-early gene expression in stress-releated pathways, however, there are brain areas showing persistent activation (Kollack-Walker et al., 1999). We refer to other chapters of this book for detailed neurobiological analysis of social stress effects.

Stressor characteristics

Intensity of the stressors can be controlled experimentally in some cases: different doses of bacterial lipopolysaccharide (LPS) or cytokines (Ericsson et al., 1994) as immune stressors, different doses of insulin to induce hypoglycemia, amount of blood withdrawn during hemorrhage (Pacak and Palkovits, 2001), different intensity of noise to induce acoustic stress (Campeau and Watson, 1997) etc.

Most stressors can also be categorized according to the *duration* of the challenge or the *frequency* of exposure to stressful situations.

1. *Acute stress* means single exposure to a single challenge. In case of some stressors the duration of the challenge can be experimentally manipulated (footshock, ether, restraint, immobilization, swim, cold), in other cases, acute stress responses can only be initiated (hyperosmotic challenge, insulin-induced hypoglycemia, formaline injection).
2. *Repeated stress* (may also be referred to as chronic intermittent stress) covers a situation when single exposure to one acute stressor is repeated once or couple of times daily for a prolonged time. Although most acute stressors can be repeated, emotional (restraint, immobilization) or immune (bacterial LPS or cytokine exposure) stressors are more often repeated than homeostatic challenges (hyperosmotic challenge, hypoglycemia etc.)

Chronic repeated variable stress paradigm (Chappell et al., 1986; Willner et al., 1992; Herman et al., 1995; Stout et al., 2002) includes different stressors from different classes, randomly distributed and repeated over a long time period.

3. The term *chronic stress* refers to continuos exposure to a stressor. Certain experimentally induced certain disease states, such as arthritis (Sternberg et al., 1989; Harbuz et al., 1992), experimental allergic encephalomyelitis (EAE) (Harbuz et al., 1997), streptozotocin-induced diabetes (Akana et al., 1996; Chan et al., 2002), drug abuse (Sarnyai et al., 2001), as well as continuos social conflict or subordinate status (Albeck et al., 1997) are among the examples.

Adrenalectomy can also be regarded as a chronic stressor because of the metabolic imbalance caused by the lack of glucocorticoid hormones (Laugero et al., 2001).

Activity-dependent modifications of the stress responses

During repeated or chronic exposure to homo- or heterotypic stressors, two general types of activity-dependent changes can occur in the stress-related neuronal circuitry. Both involve changes in response produced by previous inputs: in other words the present response depends on the history of preceding inputs that can be regarded as a special form of memory. The first is a gradual decrement in response elicited by repeated or chronic application of the identical stressor referred to as habituation. The second is a progressive increase of the response: the phenomenon of sensitization or facilitation. In general, habituation occurs in response to nondamaging stimuli, which may enable an animal to remain in a relatively safe environment without eliciting a response constantly to the same innocuous stimulus. Sensitization has also an adaptive value to detect or to react to any stimulus earlier or in a more exaggerated way having been exposed to a harmful stimulus.

Habituation

Rats habituate their hormonal and transcriptional responses to repeated exposure to the same type (homotypic) of stressors. Habituation refers to the decrement in HPA activity that occurs with repeated exposure to the same or homotypic stressor.

Basal measures of HPA activity remain unchanged or slightly elevated (Aguilera, 1994; Ma et al., 1997), however, even a few exposures to a stressor result in decreased or attenuated response to the *last* challenge of the same kind.

Plasma CORT and hypothalamic CRF hetronuclear (hn)RNA responses are decreasing with increasing frequency of exposure to stress (Ma and Lightman, 1998). However, parvocellular AVP hnRNA levels seems to be facilitated by frequent exposure to restraint (Ma and Lightman, 1998). In line with these changes, chronic intermittent stress differentially affects CRF and AVP stores and neuropeptide release from the median eminence (De Goeij et al., 1992a,b). Repeated hypoglycemia resulted in a decrease of CRF and an increase of AVP content and a shift in peptide release towards AVP (De Goeij et al., 1992). Neural or humoral mechanisms underlying stress habituation remains to be fully characterized. Theoretically, decreased drives or increased inhibitory mechanisms posed on the stress-related hypothalamic motorneurons could be involved. Recent findings implicate reduced afferent drive to the hypothalamic PVN in habituation. PVN-projecting cell groups that display attenuated activation to repeated stimuli included central autonomic (nucleus of the solitary tract, ventrolateral medulla, parabrachial nucleus) and limbic forebrain structures (septum, amygdala/bed nucleus of stria terminalis, prefrontal cortex) (Gosselink, 2002). However, no changes of c-fos expression are detected in hypothalamic areas involved in local inhibitory control of PVN. Involvement of stressor-specific "upstream" mechanisms in habituation is further supported by the increased responsibility of the axis to heterotypic stimuli that are mediated through distinct afferent inputs (Fernandes et al., 2002).

In addition to the neural mechanisms, impaired negative feedback control of hypothalamic neurosecretory neurons that occur upon repeated exposure to stress may also be involved in habituation. Decreased negative feedback favors the synthesis and release of parvocellular AVP that may be responsible for exaggerated response to heterotypic challenges, which gain access to releasable neuropeptide pool through pathways that bypass the habituated ones. Decreased sensitivity of the glucocorticoid feedback, probably due to decreased number of glucocorticoid receptors and their interaction with transcription factors induced by CRF and AVP, is critical for the maintenance of ACTH responses in the presence of elevated plasma glucocorticoid levels during chronic stress (Aguilera, 1994; Herman et al., 1995).

Stress-induced changes in hippocampal mineralocorticoid receptors (MR) might also contribute to altered regulation of HPA activity. Recent studies revealed CRF-dependent increase of MR in the hippocampus after psychologic stressors, which was associated with an increased MR-mediated tonic inhibition of HPA axis (Gesing et al., 2001). In line with these studies Cole et al. suggest that mineralocorticoid receptors may also play an important role in constraining the HPA axis response to homotypic stressors in habituated rats (Cole et al., 2000).

Posterior division of the paraventricular nucleus of the thalamus (pPVTh) has been also implicated in habituation: lesion of this structure prevented habituation to repeated restraint without altering HPA responses to the first challenge (Bhatnagar et al., 2002). It should be noted however, that other reports did not support this hypothesis and showed no effect of PVTh or BNST on habituation to restraint (Fernandes et al., 2002).

Facilitation, sensitization, cross-sensitization

Prior exposure to repeated stressors results in habituation to the same stressor, however the responsiveness to heterotypic stressors is maintained or even facilitated (Scribner et al., 1993; Ma and Lightman, 1998). Rats exposed to repeated restraint and challenged with hypertonic saline show elevated CORT levels compared to naïve animals. CRF transcription is induced normally in response to hypertonic saline injection in repeatedly restrained animals, while parvocellular AVP hnRNA levels rise more rapidly and to higher levels than in controls (Ma and Aguilera, 1999).

Facilitated response to a novel challenge in rats habituated to another type of stressor may implicate differential effect of chronic/repeated stressor on synthesis and secretion of ACTH secretagogues. Increased synthesis of AVP in the parvocellular neurons and its accumulation in the external zone of the median eminence over that of CRF provide the neuropeptide basis for potentiation of ACTH release in response to a subsequent, heterotypic stressor (Schmidt et al., 1996; Schmidt et al., 2001).

The posterior division of the paraventricular nucleus of the thalamus (pPVTh) seems to play an essential role in facilitation. pPVTh has been shown to inhibit the enhanced or facilitated HPA responses to novel, heterotypic restraint in previously chronically cold stressed rats (Bhatnagar et al., 2002).

Sensitization occurs even after a single exposure to a stressor, which rather referred to as priming.

Priming

A single exposure to one stressor induces delayed and long-lasting hyperresponsiveness in all indices of HPA axis activity to subsequent to homotypic and heterotypic stressors (Schmidt et al., 1995; Schmidt et al., 1996; Schmidt et al., 2001). Increased production, storage, and secretion of AVP from CRF-synthesizing hypophyseotropic neurons play a dominant role in this phenomenon (Aubry et al., 1999).

It should be emphasized, however, that acute, repeated, chronic and primed stress-response categories are operational only in a certain experimental time domain and do not take into account the stress or allostatic load "history" of the subjects.

Strain and individual differences

All aspects of the stress reactivity to various stressful challenges including neuroendocrine, autonomic, and behavioral responses show significant individual differences (Kabbaj and Akil, 2001) and depend on the strain (Harbuz, 1994; Dhabhar, 1997), pre- and postnatal experiences, and maternal care (Levine et al., 1991; Meaney et al., 1993; Reul et al., 1994; Abraham and Kovacs, 2000). We refer to Part II: Chapters 1.1; 1.5 and 1.6 of this book for details.

Selye's concept of stress emphasized the nonspecific nature of the response to wide range of "nocuous agents" (i.e. stressors) and identified the stress triad (adrenal enlargement, gastric ulceration, and thymocolymphatic involution) as the hallmarks of the stress reaction. Although the stereotypic activation of stress-effector mechanisms has been confirmed even at molecular level, with the advent of the technical achievements in the analysis, data are accumulating supporting the heterogeneity of the responses, including differences in afferent and efferent pathways recruited during stress responses, plasma catecholamine profiles, and neurochemical measures.

Functional anatomical mapping of stress-activated circuits

Adaptation of the functional activity mapping strategy that is based on the inducibility of different immediate-early genes provided a useful tool in stress research to identify stress-related cells and extended circuits activated in response to various stressors. This approach confirmed the activation of CRF-secreting parvocellular neurosecretory neurons that initiate the neuroendocrine stress cascade, but also revealed the heterogeneity of afferent pathways and efferent mechanisms in response to different challenges.

Immediate-early gene markers of stress-induced neuronal activation

Stereotypic inducibility of c-fos proto-oncogene rendered this immediate-early gene (IEG) to be the most widely used functional anatomical mapping tool to identify cells and extended circuitries that became activated in response to various stressful stimuli (Greenberg and Ziff, 1984; Ceccatelli et al., 1989; Morgan and Curran, 1991). In addition to c-fos, nerve growth factor-induced protein B (NGFI-B) and Fos-related antigens (FRAs) are also frequently used markers of neuronal activation in stress research (Chan et al., 1993; Hoffman et al., 1993; Kovacs, 1998).

Pros of IEG-based mapping strategy

- Baseline expression of IEGs is low,
- Induced stereotypically in response to wide range of extracellular stimuli,
- Phenotype of activated neurons can easily be identified,
- Number of activated profiles can be quantitatively analyzed,
- Differential induction of mRNA and protein product of IEGs allows identification of profiles responsive to two different challenges (Chaudhuri, 1997; Chaudhuri et al., 1997; Kovacs et al., 2001).

Cons of IEG-based mapping strategy

- Induction is transient, timing of the analysis is important,
- May not equally label all activated neurons in a given circuit,
- May not identify neurons that gain net inhibitory input after challenge,
- Relation of IEG induction with release from tonic inhibition is not completely known,
- Nuclear localization of the protein product does not allow to reveal connectivity and morphology of activated neurons.

Timing of immediate-early gene (IEG) induction following acute challenges

Acute physiological, psychological, and immune challenges induce transient expression of IEG in the hypothalamic PVN (Honkaniemi et al., 1994; Hughes and Dragunow, 1995). Generally, c-fos mRNA cannot be detected under stress-free conditions, it's transcription is rapidly upregulated showing maximum 30 min after challenge, is diminished by 1 hour and is not detectable by 2–3 h poststress (Ericsson et al., 1994; Cullinan et al., 1995). c-Fos protein is also undetectable under basal conditions, first c-Fos immunoreactive cell nuclei are revealed 15–30 min after rapid stress, show maximum at 90–180 min and decline thereafter (Morgan and Curran, 1991; Giovannelli et al., 1992; Kovacs and Sawchenko, 1996).

Markers of cellular activation in chronic situations

One of the significant drawbacks of c-fos-based activational mapping strategy is the transient induction of the marker, which does not allow detection of activated profiles under chronic situations (Hoffman et al., 1993). There are however, IEG markers, such as FRAs (Fos-related antigens, Fra1 and Fra2) that are induced in response to acute stimuli and, due to their long half-life, gradually accumulate in response to chronic and/or repeated stimuli (Sharp et al., 1991). Fos-related antigens but not c-fos are detected in the parvocellular neurosecretory neurons following adrenalectomy (Jacobson et al., 1990; Brown and Sawchenko, 1997).

Neuronal activation in the paraventricular hypophyseotropic neurons

Most of acute stressful stimuli result in an induction of c-fos or NGFI-B in the dorsal medial part of the parvocellular subdivision of the PVN (mpPVN), where hypophyseotropic CRF-secreting neurons are concentrated (Ceccatelli et al., 1989; Chan et al., 1993). However, there is a significant heterogeneity of IEG induction patterns seen in other functional domains of the PVN (see Fig. 2). Stressors that selectively activate CRF-synthesizing neurons in the PVN include novel environment (Handa et al., 1993), saline injection (Sharp et al., 1991), restraint (Ceccatelli et al., 1989; Abraham and Kovacs, 2000), ether (Kovacs and Sawchenko, 1996), footshock (Rivest and Rivier, 1994), swim (Duncan et al., 1993; Cullinan et al., 1995), and IL-1β (Ericsson et al., 1994). The level of recruitment of mpPVN CRF cell population following restraint, swim, hemorrhage and IL-1β seems to be submaximal, as c-Fos positive CRF cells account only for one-third of hypophyseotropic CRF cells in this subdivision (Dayas et al., 2001).

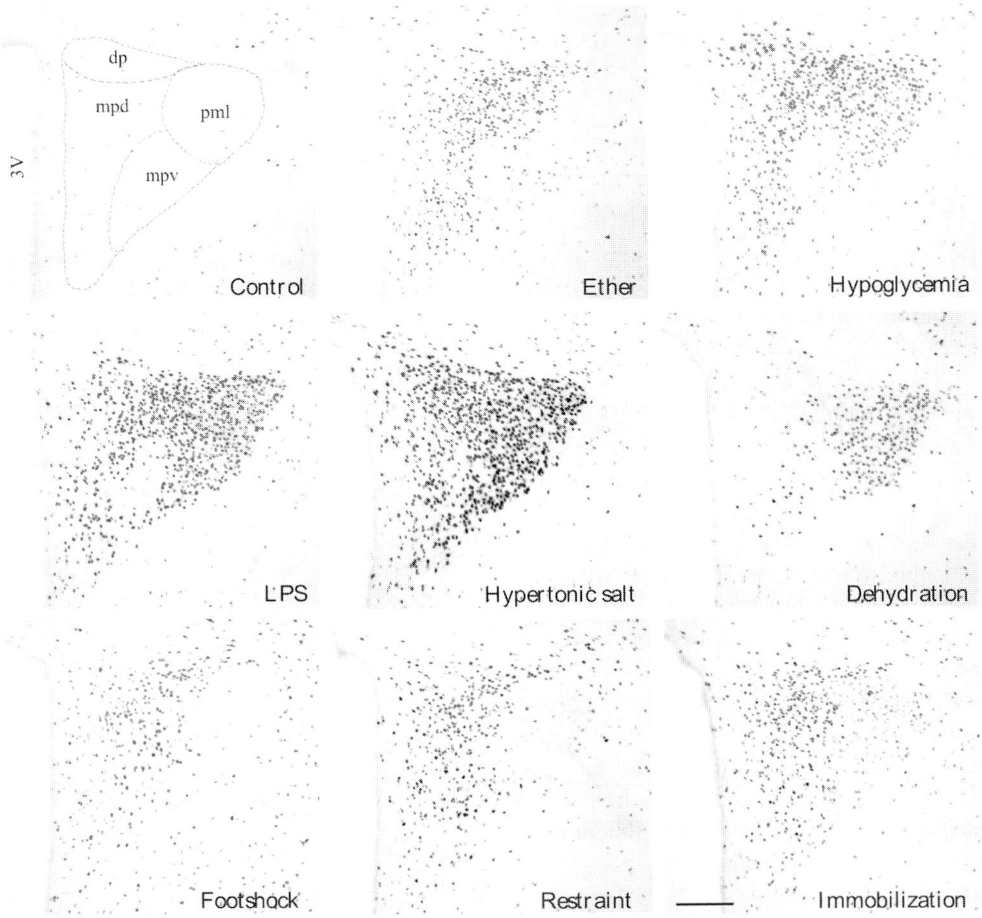

Fig. 2. Challenge-specific patterns of c-Fos protein expression in the hypothalamic paraventricular nucleus (PVN). Microphotographs through a common level of the PVN to compare neuronal activation patterns in response to various physiological (ether, insulin-induced hypoglycemia, systemic LPS challenge, hypertonic salt injection and chronic dehydration) or psychological (footshock, restraint, immobilization) stressors. All acute challenges result in immediate-early gene expression in the medial dorsal parvocellular subdivision (mpd), the acknowledged seat of hypophysiotropic CRF neurons. Ether, footshock and restraint-induced neuronal activation patterns are restricted to the parvocellular subdivision, while other stressors recruit autonomic-related [dorsal parvocellular, (dp) and ventral medial parvocellular (mpv)] and magnocellular (pm) portions of the nucleus as well. Chronic dehydration yields nearly pure magnocellular activation. 3V, third ventricle; Bar, 150 μm.

Immediate-early gene induction in other visceromotor neurons of PVN

In contrast to the common and robust induction of cellular markers of neuronal activation in the medial dorsal parvocellular subdivision, the intensity and spatial distribution of IEG expression in the other two functionally distinct subdivisions of the PVN (i.e. magnocellular and autonomic projection neurons) varies markedly as a function of the nature of the stimulus and/or of the response (Fig. 2.) (Swanson and Sawchenko, 1980). Systemic IL-1 injections (Ericsson et al., 1994) resulted in a significant level of c-fos expression in the oxytocinergic magnocellular neurosecretory neurons, while LPS injections (Rivest and Rivier, 1994; Rivest and Laflamme, 1995; Elmquist and Saper, 1996), anaphylactoid reaction (Foldes et al., 2000), insulin-induced hypoglycemia (Brown and Sawchenko, 1997) activated vasopressin-synthesizing magnocellular neurons as well. Following hemorrhage or acute hyperosmotic challenge IEG markers were distributed uniformly throughout the paraventricular nucleus (Chan et al., 1993; Pacak and Palkovits, 2001). Chronic salt loading results in a qualitatively distinct activation pattern in the PVN with exclusive IEG induction in the magnocellular compartment and lack of c-fos or NGFI-B mRNA expression in the CRF-secreting parvocellular neurons (Sharp et al., 1991; Kovacs and Sawchenko, 1993). Autonomic projection neurons that give rise long ascending efferents to the sympathetic and parasympathetic centers in the brainstem and spinal cord also display IEG induction in response to various stressful stimuli (Li and Sawchenko, 1998; Zhang et al., 2000).

Stressor-specific activation of afferent pathways

Although stressors of each class commonly activate certain PVN effector neurons, the registration of the stimuli and afferent inputs that may relay these effects to hypothalamic effector neurons are clearly differential (Sawchenko et al., 1996).

Depending on the challenge (i.e. salt loading, acustic stimuli, smell or electric footshock and restraint etc.) specific cell populations are activated. Systemic stressors provoke simple reflex responses initiated by neuronal cell groups and associated circumventricular organs that relay *visceral sensory* information. In case of psychological stressors it is assumed that pathways conveying *somatosensory/ nociceptive* information mediate visceromotor responses (Li et al., 1996).

Comparison of neuronal activation patterns induced by physiological and psychological stressors

Immediate-early gene expression-based mapping strategy provides a unique tool to compare brain areas and pathways specifically recruited by categorically different challenges. Comprehensive analyses revealed certain limbic and medullary areas as major structures in which c-fos patterns are to differentiate between physiological and emotional stressors.

Patterns induced by physiological challenges such as hemorrhage, IL-1 injection, or ether involve primarily autonomic-associated aspects of the amygdala (the central nucleus, CeA) and bed nucleus of stria terminalis (oval subnucleus) with little if any neocortical representation (Li et al., 1996; Sawchenko et al., 1996; Abraham and Kovacs, 2000; Dayas et al., 2001). In contrast, psychological stressors such as immobilization, restraint, footshock or forced swim specifically recruited cortical and subcortical limbic structures including prelimbic, infralimbic and cingulate cortices, lateral septal nucleus, medial, and basolateral amygdala.

Both psychological and physiological challenges activate A1 and A2 catecholaminergic areas in the ventrolateral and dorsomedial medulla respectively (Smith et al., 1995; Sawchenko et al., 1996; Dayas et al., 2001; Pacak and Palkovits, 2001). Detailed spatial analysis of activated profiles revealed, however, that systemic stress paradigms such as hemorrhage and IL-1 injection induced c-fos expression in a distinct noradrenergic subpopulation that are more rostral to those recruited by emotional stressors (Dayas et al., 2001). The dependence of activation of paraventricular effector neurons on ascending catecholaminergic cell groups is also a hallmark that differentiates between stress paradigms. While hypothalamic responses to systemic stressor (IL-1) are attenuated following knife cuts that separate

CRF-secreting paraventricular neurons from their inputs arising in the medulla, such lesions do not abolish PVN transcriptional responses to psychological stressor, footshock. Moreover, medullary c-fos activation to footshock is dependent on rather than to initiate stress responses seen in "upstream" structures.

Stressor-specific recruitment of histaminergic neuron populations of the posterior hypothalamus was also found recently (Miklos and Kovacs, 2003). Histaminergic fibers were revealed in close contact with CRF-synthesizing neurons in the PVN and histaminergic mechanisms have been thought to have a high impact on stress-induced hormonal and transcriptional responses (Miklós et al., 2000). Dual localization of activation marker c-fos with histidine decarboxylase (HDC) mRNA, revealed that only a small subset of tuberomammillary histaminergic neurons is responsive to stress. Histaminergic activational patterns do not, however, differentiate between categorically distinct stressors, as both physiological (insulin-induced hypoglycemia) and psychological challenges (restraint and footshock) were among those that selectively activated histaminergic cell groups (Miklos and Kovacs, 2003).

Cellular targets of different stressors can be directly compared in the same animal using *double activity imaging approach* (Chaudhuri, 1997). This technique exploits the transient and distinct temporal inducibility of *c-fos* mRNA and c-Fos protein to reveal and compare cellular targets of different stimuli by concomitant localization of *c-fos* mRNA and c-Fos immunoreactivity in brain sections of animals that were timely challenged with two different stressors.

Double activity imaging confirmed the heterogeneity of cellular targets of physiological and psychological stressors in limbic and brainstem areas. Within the parvocellular subdivision, most of the neurons were responsive to both systemic and emotional stimuli, however, some heterogeneity was also revealed (Fig. 3).

Stress-induced transcriptional changes in the paraventricular nucleus

Various acute stressful stimuli that are conveyed to the visceromotor neurons of the PVN evoke transcriptional activation of CRF gene. Hypophyseotropic parvocellular neurons display positive

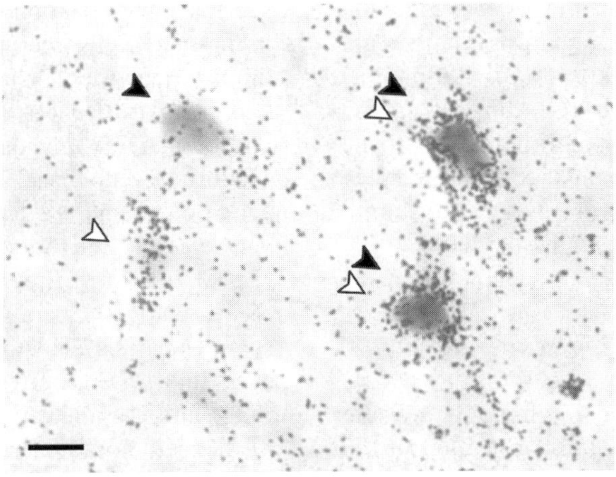

Fig. 3. Double localization of c-fos mRNA and c-Fos protein in the rat hypothalamus. To compare the heterogeneity of the response of parvocellular neurosecretory cells to categorically distinct paradigms, rats were challenged with two stressors consecutively: first it was exposed to ether vapor for 5 min 2 hours before the experiment then restrained for 15 min 1.5 h later. Profiles showing DAB reaction in their cell nuclei correspond to c-Fos protein positive, ether-responsive neurons in the parvocellular subdivision of the hypothalamic paraventricular nucleus (black arrowheads). Cells with small autoradiographic grains over their cytoplasm are c-fos mRNA expressing neurons that are responsive to restraint only (white arrowheads). Double-labeled cells are targets of both stimuli. Bar, 20 μm.

hybridization signals corresponding to CRF mRNA under basal conditions (Young et al., 1986; Kovacs and Mezey, 1987). Various acute physiological [hypoglycemia (Suda, 1988 p. 692), intraperitoneal hypertonic saline (Lightman and Young, 1989)] and psychological insults [restraint (Harbuz and Lightman, 1989; Kalin et al., 1994), immobilization (Makino et al., 1995; Aubry et al., 1999), swim (Harbuz et al., 1994), naloxone-precipitated morphine withdrawal (Lightman and Young, 1988)] upregulate steady-state levels of CRF mRNA in the stress-related parvocellular neurons. Depending on the challenge, maximal responses were detected at 2–4 h poststress. However, in other acute cases [ether (Watts, 1991; Kovacs and Sawchenko, 1996), footshock (Imaki et al., 1991) or cold exposure (Harbuz and Lightman, 1989; Ceccatelli and Orazzo, 1993)] hybridization signals corresponding to CRF mRNA were not significantly elevated over the high basal level of mature transcript.

Intron-specific cRNA probes hybridize to sequences in hnRNA before being processed to mRNA and, therefore, provide a direct index of transcriptional activation. In situ hybridization using such probes revealed rapid, stress-induced upregulation of CRF transcription in the parvocellular neurosecretory neurons. Under stress-free, baseline conditions, CRF hnRNA expression is low (Herman et al., 1992) but rapidly increased in response to ether (Kovacs and Sawchenko, 1996) acute hypertonic saline (Ma et al., 1999) or restraint (Imaki et al., 1995). The CRF transcriptional response is generally transient, showing maximum 5–15 min after stress and return to the baseline between 30 min and 1 h poststress. Hypovolemia, however, results in a sustained activation of CRF hnRNA and RNA levels in the parvocellular neurons that were detected 5 h after stress.

In addition to CRF, stress-related parvocellular neurons can express multiple bioactive peptides, several of which act as ACTH secretagogues. Foremost among these is arginine vasopressin (AVP) that potentiates CRF action at the pituitary and generally held as the dominant determinant of the challenge-induced drive on HPA axis. Acute stressors induce AVP transcription in the CRF-secreting parvocellular neurons, where its expression (hnRNA and mRNA) remains undetectable in nonstressed animals (Kiss et al., 1984; Sawchenko et al., 1984; Lightman and Young, 1988; Herman et al., 1991; Aubry et al., 1999).

Transcriptional activation of AVP gene in the parvocellular neurosecretory neurons either follows similar time course (in case of hemorrhage, anaphylaxis, restraint) or is significantly delayed (following ether and hypertonic saline injection) compared to that of CRF hnRNA. Although the molecular mechanisms underlying these differences are not fully explored, recent data suggest that the delayed AVP response during ether inhalation is likely due to the fast feedback effect of stress-induced corticosterone (Ma et al., 1997; Kovacs et al., 2000), while this aspect of steroid feedback is essentially not detectable in the hemorrhage paradigm (Thrivikraman and Plotsky, 1993).

Hormonal responses

Both physiological and psychological stressors commonly activate the pituitary-adrenocortical hormone secretion. There are however, significant differences in the timing of the maximum, the peak amplitude and in the decay of the hormone response. Short duration acute stressors such as ether, result in ACTH maximum within 5–10 min after challenge, ACTH peak occurs 15 min after formalin-induced pain, while restraint or immobilization-induced maximal ACTH levels are detected at 15 and 30 min respectively. On the contrary, acute cold exposure does not provoke ACTH elevation at all (for review, see Pacak and Palkovits, 2001). The time required for maximal activation of pituitary hormone secretion is independent of a stressor being physiological or emotional (processed), however seems to correlate with the maximal level reached during the response (Pacak and Palkovits, 2001). Plasma ACTH levels decline to the baseline shortly after the response, because of (1) the cessation of the stimulus, (2) negative glucocorticoid feedback or (3) due to inhibitory neural mechanisms activated during stress. In case of long duration stressors such as immobilization or restraint, ACTH secretion is gradually decreasing after the peak, or kept on a plateau but remains significantly elevated over the baseline until the end of the stress.

Adrenocortical corticosterone or cortisol secretion follows that of ACTH, being maximal between 15–60 min after challenge with strong heterogeneity between the stressors. Adrenomedullary epinephrine secretion is increased within the first 15 min after challenge and peaks at different times and values. It is also interesting to note that cold stress does not result in increase of epinephrine plasma levels (Pacak and Palkovits, 2001). Plasma noradrenaline concentration that referred to as a measure of sympathoneural activation also shows stressor-specificity. Hemorrhage for example does not elevate plasma noradrenaline levels, while cold stress and immobilization are strong inducers of this measure.

Because of the heterogeneity of the responses outlined above we emphasize the importance of time course studies in measuring the hormonal secretory responses to different stressors.

In addition to activation of HPA axis, stress also increases the secretion of stress-responsive hormones such as prolactin and β-endorphin and may inhibit hypothalamo-pituitary-gonadal and hypothalamo-pituitary-thyroid axes and growth hormone secretion. Significant elevations of PRL plasma levels have been reported following novelty, ether, footshock, restraint, immobilization, social conflict etc. (Kjaer et al., 1991; Herman et al., 1995; Zelena et al., 1999), however, the hypothalamic mechanisms that regulate stress-induced PRL release seem to be independent of the paraventricular nucleus (Makara and Kovacs, 1997). Plasma PRL levels peak within 5 min after ether inhalation and decrease to the baseline within 30 min after exposure (Makara and Kovacs, 1997).

Taken together, physiological and psychological stressors provoke common and distinct transcriptional, neurochemical and hormonal efferent mechanisms corresponding to stereotypical and specific aspects of the general alarm- and adaptation reaction.

Human stressors

When considering stressors in the human context, one of the first impressions is how freely and easily people use the term "stress" in everyday situations, but how challenging it is to define and classify human stressors from a scientific perspective. In contrast to the definition problems and often lack of evidence for stress-induced pathologies, many fields of medicine including psychiatry, endocrinology, immunology, cardiology etc. describe conditions related to stress. Psychiatry is the only field that classifies stressors from its own point of view: many psychiatric disorders are considered to be stress-related, the 4th edition of the Diagnostic and Statistical Manual of Mental Disorders (DSM IV, 1994) of the American Psychiatric Association call for clinicans to evaluate individuals on five levels or axes. Two out of the five axes could be considered as stress-associated conditions that affect individuals' functioning. Axis III identifies relevant physical diseases and conditions, and furthermore, Axis IV includes "Psychosocial and Environmental Factors". Among the latter, there seem to be reasonable to separate a distinct subset of stressors. Besides *ordinary/everyday stressors* that "constitute the normal vicissitudes of life" and potentially include death of a loved person, divorce, rejection, financial reverses, homelessness, caretaking of elderly parents etc., there is another type of events "involving actual or threatened death or serious injury, or a threat to the physical integrity of others" even being subjected, witnessed, or confronted by, and causes "intense fear, helplessness, or horror". This *catastrophic/traumatic event* is "outside the range of usual human experience" such as war, torture, airplane crash, car accident, terror attack, factory explosion, military combat, assault, homicide, sexual abuse of child, or natural disasters (earthquakes, hurricanes, volcano eruptions). In contrast to the first group – often referred to as "major life events" – , a catastrophic stressor does not allow the individual to respond with an active "fight-or-flight" reaction, and often resulted in a posttraumatic stress disorder. However, there is a wide variety of potential challenges in vivo, experimental human data are based only on a very limited number of laboratory stressors. *Experimental stressors* involve (1) *psychological tasks* (interview, arithmetic tasks e.g. Trier Social Stress Test, public speaking, university exam, Stroop Color Word Test (Jensen and Rohwer, 1966; Ray, 1979), films or videotapes, mild electric shock), and (2) *physiological tasks* (cold pressor test). Sport and space medicine are also a rich source of human data related to different type of physical challenges e.g. excessive physical exercise during marathon running, or the combined vestibular and cardiovascular demand during weightlessness and space flight. Finally, *chronic diseases* could be also considered as chronic stressors, as a consequence of sustained metabolic (diabetes), immune (rheumatoid arthritis), painful (fibromyalgia) stimuli.

Abbreviations

ACTH	adrenocorticotrophic hormone
AVP	arginine vasopressin
BNST	bed nucleus of stria terminalis
CeA	central nucleus of amygdala
CORT	corticosterone
CRF	corticotropin-releasing factor
EAE	experimental allergic encephalomyelitis
Fra	Fos-related antigen
HDC	histidine decarboxylase
hnRNA	heteronuclear ribonucleic acid
HPA	hypothalamo–pituitary–adrenocortical
IEG	immediate-early gene
ip	intraperitoneal
-ir	immunoreactive
IL-1	interleukin-1
LPS	bacterial lipopolysaccharide
mpPVN	medial parvocellular subdivision of the hypothalamic paraventricular nucleus
mRNA	messenger RNA
NGFI-B	nerve growth factor-induced B protein
NTS	nucleus of the solitary tract
pPVTh	posterior region of the thalamic paraventricular nucleus
PVN	hypothalamic paraventricular nucleus

Appendix

Ethical issues & resources

Research on animal subjects

Guidelines for Ethical Conduct in the Care and Use of Animals
American Psychological Association's Committee on Animal Research and Ethics (CARE) http://www.apa.org/science/anguide.html

The Biomedical Investigator's Handbook for Researchers Using Animal Models
Foundation for Biomedical Research. Washington, D.C.: FBR, 1987. http://www.fbresearch.org/

Guide for the Care and Use of Laboratory Animals
National Academy of Sciences. 7th ed. Washington, D.C.: National Research Council, Institute for Laboratory Animal Research, NAS, 1996. http://www.nap.edu/catalog/5140.html

Research on human subjects

Declaration of Helsinki. Recommendations Guiding Physicians in Biomedical Research Involving Human Subjects
Adopted by 18th World Medical Assembly, Helsinki, 1964; revised by 29th World Medical Assembly, Tokyo, 1975; Venice, 1983; and Hong Kong, 1989. http://www.opt.auckland.ac.nz/public/staffpgs/myap/helsinki.html; http://www.wma.net/e/policy/17-c_e.html

Federal Policy for the Protection of Human Subjects: Notes and Rules.
Department of Health and Human Services, National Institutes of Health, Office for Protection from Research Risks. *Federal Register. (June 18, 1991) 56: 28001–32.* http://ohrp.osophs.dhhs.gov/humansubjects/guidance/45cfr46.htm

American Psychological Association's Ethics Committee Rules and Procedures
Ethical principles of psychologist and code of conduct http://www.apa.org/ethics/code.html

General readings

UNESCO International Bioethics Committee
http://bioethics.net/

NIH's Bioethics Resources on the Web
http://www.nih.gov/sigs/bioethics/

National Reference Center for Bioethics Literature
http://www.georgetown.edu/research/nrcbl/

References

Abraham, I.M. and Kovacs, K.J. (2000) Postnatal handling alters the activation of stress-related neuronal circuitries. Eur. J. Neurosci., 12(8): 3003–3014.

Adachi, A., Shimizu, N., Oomura, Y. and Kobashi, M. (1984) Convergence of hepatoportal glucose-sensitive afferent signals to glucose-sensitive units within the nucleus of the solitary tract. Neurosci. Lett., 46(2): 215–218.

Aguilera, G. (1994) Regulation of pituitary ACTH secretion during chronic stress. Front Neuroendocrinol, 15(4): 321–50.

Akana, S.F., Hanson, E.S., Horsley, C.J., Strack, A.M., Bhatnagar, S., Bradbury, M.J., Milligan, E.D. and Dallman, M.F. (1996) Clamped corticosterone (B) reveals the effect of endogenous B on both facilitated responsivity to acute restraint and metabolic responses to chronic stress. Stress, 1(1): 33–49.

Albeck, D.S., McKittrick, C.R., Blanchard, D.C., Blanchard, R.J., Nikulina, J., McEwen, B.S. and Sakai, R.R. (1997) Chronic social stress alters levels of corticotropin-releasing factor and arginine vasopressin mRNA in rat brain. J. Neurosci., 17(12): 4895–4903.

Aubry, J.M., Bartanusz, V., Jezova, D., Belin, D. and Kiss, J.Z. (1999) Single stress induces long-lasting elevations in vasopressin mRNA levels in CRF hypophysiotrophic neurones, but repeated stress is required to modify AVP immunoreactivity. J. Neuroendocrinol., 11(5): 377–384.

Berkenbosch, F., de Goeij, D.E., Rey, A.D. and Besedovsky, H.O. (1989) Neuroendocrine, sympathetic and metabolic responses induced by interleukin-1. Neuroendocrinology, 50(5): 570–576.

Bhatnagar, S., Huber, R., Nowak, N. and Trotter, P. (2002) Lesions of the posterior paraventricular thalamus block habituation of hypothalamic–pituitary–adrenal responses to repeated restraint. J. Neuroendocrinol., 14(5): 403–410.

Blanchard, R.J., Nikulina, J.N., Sakai, R.R., McKittrick, C., McEwen, B. and Blanchard, D.C. (1998) Behavioral and endocrine change following chronic predatory stress. Physiol. Behav., 63(4): 561–569.

Boudaba, C., Schrader, L.A. and Tasker, J.G. (1997) Physiological evidence for local excitatory synaptic circuits in the rat hypothalamus. J. Neurophysiol., 77(6): 3396–3400.

Boudaba, C., Szabo, K. and Tasker, J.G. (1996) Physiological mapping of local inhibitory inputs to the hypothalamic paraventricular nucleus. J. Neurosci., 16(22): 7151–7160.

Brown, E.R. and Sawchenko, P.E. (1997) Hypophysiotropic CRF neurons display a sustained immediate-early gene response to chronic stress but not to adrenalectomy. J. Neuroendocrinol., 9(4): 307–316.

Campeau, S. and Watson, S.J. (1997) Neuroendocrine and behavioral responses and brain pattern of c-fos induction associated with audiogenic stress. J. Neuroendocrinol., 9(8): 577–588.

Ceccatelli, S. and Orazzo, C. (1993) Effect of different types of stressors on peptide messenger ribonucleic acids in the hypothalamic paraventricular nucleus. Acta Endocrinol. (Copenh), 128(6): 485–92.

Ceccatelli, S., Villar, M.J., Goldstein, M. and Hokfelt, T. (1989) Expression of c-Fos immunoreactivity in transmitter-characterized neurons after stress. Proc. Natl. Acad. Sci. USA, 86(23): 9569–9573.

Chan, O., Chan, S., Inouye, K., Shum, K., Matthews, S.G. and Vranic, M. (2002) Diabetes impairs hypothalamo-pituitary-adrenal (HPA) responses to hypoglycemia, and insulin treatment normalizes HPA but not epinephrine responses. Diabetes, 51(6): 1681–1689.

Chan, R.K., Brown, E.R., Ericsson, A., Kovacs, K.J. and Sawchenko, P.E. (1993) A comparison of two immediate-early genes, c-fos and NGFI-B, as markers for functional activation in stress-related neuroendocrine circuitry. J. Neurosci., 13(12): 5126–38.

Chan, R.K. and Sawchenko, P.E. (1994) Spatially and temporally differentiated patterns of c-fos expression in brainstem catecholaminergic cell groups induced by cardiovascular challenges in the rat. J. Comp. Neurol., 348(3): 433–460.

Chan, R.K. and Sawchenko, P.E. (1998) Organization and transmitter specificity of medullary neurons activated by sustained hypertension: implications for understanding baroreceptor reflex circuitry. J. Neurosci., 18(1): 371–387.

Chappell, P.B., Smith, M.A., Kilts, C.D., Bissette, G., Ritchie, J., Anderson, C. and Nemeroff, C.B. (1986) Alterations in corticotropin-releasing factor-like immunoreactivity in discrete rat brain regions after acute and chronic stress. J. Neurosci., 6(10): 2908–2914.

Chaudhuri, A. (1997) Neural activity mapping with inducible transcription factors. Neuroreport, 8(16): 5–9.

Chaudhuri, A., Nissanov, J., Larocque, S. and Rioux, L. (1997) Dual activity maps in primate visual cortex produced by different temporal patterns of zif268 mRNA and protein expression. Proc. Natl. Acad. Sci. USA, 94(6): 2671–2675.

Cole, M.A., Kalman, B.A., Pace, T.W., Topczewski, F., Lowrey, M.J. and Spencer, R.L. (2000) Selective blockade of the mineralocorticoid receptor impairs hypothalamic-pituitary-adrenal axis expression of habituation. J. Neuroendocrinol., 12(10): 1034–1042.

Cullinan, W.E., Herman, J.P., Battaglia, D.F., Akil, H. and Watson, S.J. (1995) Pattern and time course of immediate early gene expression in rat brain following acute stress. Neuroscience, 64(2): 477–505.

Dampney, R.A. and Horiuchi, J. (2003) Functional organisation of central cardiovascular pathways: studies using c-fos gene expression. Prog. Neurobiol., 71(5): 359–384.

Dayas, C.V., Buller, K.M., Crane, J.W., Xu, Y. and Day, T.A. (2001a) Stressor categorization: acute physical and psychological stressors elicit distinctive recruitment patterns in the amygdala and in medullary noradrenergic cell groups. Eur. J. Neurosci., 14(7): 1143–1152.

Dayas, C.V., Buller, K.M. and Day, T.A. (2001b) Medullary neurones regulate hypothalamic corticotropin-releasing factor cell responses to an emotional stressor. Neuroscience, 105(3): 707–719.

De Goeij, D.C., Binnekade, R. and Tilders, F.J. (1992a) Chronic stress enhances vasopressin but not corticotropin-releasing factor secretion during hypoglycemia. Am. J. Physiol., 263(2 Pt 1): E394–E399.

De Goeij, D.C., Dijkstra, H. and Tilders, F.J. (1992b) Chronic psychosocial stress enhances vasopressin, but not corticotropin-releasing factor, in the external zone of the median eminence of male rats: relationship to subordinate status. Endocrinology, 131(2): 847–853.

Dhabhar, F.S., McEwen, B.S., Spencer, R.L. (1997) Adaptation to prolonged or repeated stress-comparison between rat strains showing intrinsic differences in reactivity to acute stress. Neuroendocrinology 65(5): 360–368.

Duncan, G.E., Johnson, K.B. and Breese, G.R. (1993) Topographic patterns of brain activity in response to swim stress: assessment by 2-deoxyglucose uptake and expression of Fos-like immunoreactivity. J. Neurosci., 13(9): 3932–3943.

Elmquist, J.K. and Saper, C.B. (1996). Activation of neurons projecting to the paraventricular hypothalamic nucleus by intravenous lipopolysaccharide. J. Comp. Neurol., 374(3): 315–331.

Ericsson, A., Kovacs, K.J. and Sawchenko, P.E. (1994) A functional anatomical analysis of central pathways subserving the effects of interleukin-1 on stress-related neuroendocrine neurons. J. Neurosci., 14(2): 897–913.

Fernandes, G.A., Perks, P., Cox, N.K., Lightman, S.L., Ingram, C.D. and Shanks, N. (2002) Habituation and cross-sensitization of stress-induced hypothalamic- pituitary-adrenal activity: effect of lesions in the paraventricular nucleus of the thalamus or bed nuclei of the stria terminalis. J. Neuroendocrinol., 14(7): 593–602.

Figueiredo, H.F., Bodie, B.L., Tauchi, M., Dolgas, C.M. and Herman, J.P. (2003) Stress integration after acute and chronic predator stress: differential activation of central stress circuitry and sensitization of the hypothalamo-pituitary-adrenocortical axis. Endocrinology, 144(12): 5249–5158.

Foldes, A., Nemethy, Z., Szalay, O. and Kovacs, K.J. (2000) Anaphylactoid reactions activate hypothalamo-pituitary-adrenocortical axis: comparison with endotoxic reactions. Brain Res. Bull., 52(6): 573–579.

Gesing, A., Bilang-Bleuel, A., Droste, S.K., Linthorst, A.C., Holsboer, F. and Reul, J.M. (2001) Psychological stress increases hippocampal mineralocorticoid receptor levels: involvement of corticotropin-releasing hormone. J. Neurosci., 21(13): 4822–4829.

Giovannelli, L., Shiromani, P.J., Jirikowski, G.F. and Bloom, F.E. (1992) Expression of c-fos protein by immunohistochemically identified oxytocin neurons in the rat hypothalamus upon osmotic stimulation. Brain Res., 588(1): 41–48.

Goldstein, L.E., Rasmusson, A.M., Bunney, B.S. and Roth, R.H. (1996) Role of the amygdala in the coordination of behavioral, neuroendocrine, and prefrontal cortical monoamine responses to psychological stress in the rat. J. Neurosci., 16(15): 4787–4798.

Gosselink, K.L.S.P. (2002) *Differential resistance to habituation among identified paraventricular nucleus afferents in rats subjected to repeated restraint.* Society for Neuroscience Abstracts.

Greenberg, M.E. and Ziff, E.B. (1984) Stimulation of 3T3 cells induces transcription of the c-fos proto-oncogene. Nature, 311: 433–438.

Halasz, J., Liposits, Z., Kruk, M.R. and Haller, J. (2002) Neural background of glucocorticoid dysfunction-induced abnormal aggression in rats: involvement of fear- and stress-related structures. Eur. J. Neurosci., 15(3): 561–569.

Handa, R.J., Nunley, K.M. and Bollnow, M.R. (1993) Induction of c-fos mRNA in the brain and anterior pituitary gland by a novel environment. Neuroreport, 4(9): 1079–1082.

Harbuz, M.S., Conde, G.L., Marti, O., Lightman, S.L. and Jessop, D.S. (1997) The hypothalamic-pituitary-adrenal axis in autoimmunity. Ann. NY Acad. Sci., 823: 214–224.

Harbuz, M.S., Jessop, D.S., Lightman, S.L. and Chowdrey, H.S. (1994) The effects of restraint or hypertonic saline stress on corticotrophin- releasing factor, arginine vasopressin, and proenkephalin A mRNAs in the CFY, Sprague-Dawley and Wistar strains of rat. Brain Res., 667(1): 6–12.

Harbuz, M.S. and Lightman, S.L. (1989) Responses of hypothalamic and pituitary mRNA to physical and psychological stress in the rat. J. Endocrinol., 122(3): 705–711.

Harbuz, M.S., Rees, R.G., Eckland, D., Jessop, D.S., Brewerton, D. and Lightman, S.L. (1992) Paradoxical responses of hypothalamic corticotropin-releasing factor (CRF) messenger ribonucleic acid (mRNA) and CRF-41 peptide and adenohypophysial proopiomelanocortin mRNA during chronic inflammatory stress. Endocrinology, 130(3): 1394–1400.

Herman, J.P., Adams, D. and Prewitt, C. (1995) Regulatory changes in neuroendocrine stress-integrative circuitry produced by a variable stress paradigm. Neuroendocrinology, 61(2): 180–190.

Herman, J.P., Cullinan, W.E., Ziegler, D.R. and Tasker, J.G. (2002) Role of the paraventricular nucleus microenvironment in stress integration. Eur. J. Neurosci., 16(3): 381–385.

Herman, J.P., Schafer, M.K., Thompson, R.C. and Watson, S.J. (1992) Rapid regulation of corticotropin-releasing hormone gene transcription in vivo. Mol. Endocrinol., 6(7): 1061–1069.

Herman, J.P., Schafer, M.K., Watson, S.J. and Sherman, T.G. (1991) In situ hybridization analysis of arginine vasopressin gene transcription using intron-specific probes. Mol. Endocrinol., 5(10): 1447–1456.

Hoffman, G.E., Smith, M.S. and Verbalis, J.G. (1993) c-Fos and related immediate early gene products as markers of activity in neuroendocrine systems. Front. Neuroendocrinol., 14(3): 173–213.

Honkaniemi, J., Kononen, J., Kainu, T., Pyykonen, I. and Pelto-Huikko, M. (1994) Induction of multiple immediate early genes in rat hypothalamic paraventricular nucleus after stress. Brain Res. Mol. Brain Res., 25(3–4): 234–241.

Hughes, P. and Dragunow, M. (1995) Induction of immediate-early genes and the control of neurotransmitter-regulated gene expression within the nervous system. Pharmacol. Rev., 47(1): 133–178.

Imaki, T., Nahan, J.L., Rivier, C., Sawchenko, P.E. and Vale, W. (1991) Differential regulation of corticotropin-releasing factor mRNA in rat brain regions by glucocorticoids and stress. J. Neurosci., 11(3): 585–599.

Imaki, T., Xiao-Quan, W., Shibasaki, T., Yamada, K., Harada, S., Chikada, N., Naruse, M. and Demura, H. (1995) Stress-induced activation of neuronal activity and corticotropin-releasing factor gene expression in the paraventricular nucleus is modulated by glucocorticoids in rats. J. Clin. Invest., 96(1): 231–238.

Jacobson, L., Sharp, F.R. and Dallman, M.F. (1990) Induction of fos-like immunoreactivity in hypothalamic corticotropin-releasing factor neurons after adrenalectomy in the rat. Endocrinology, 126(3): 1709–1719.

Jensen, A.R. and Rohwer, W.D., Jr. (1966) The Stroop color-word test: a review. Acta. Psychol., 25(1): 36–93.

Kabbaj, M. and Akil, H. (2001) Individual differences in novelty-seeking behavior in rats: a c-fos study. Neuroscience, 106(3): 535–545.

Kalin, N.H., Takahashi, L.K. and Chen, F.L. (1994) Restraint stress increases corticotropin-releasing hormone mRNA content in the amygdala and paraventricular nucleus. Brain Res., 656(1): 182–186.

Katsuura, G., Arimura, A., Koves, K. and Gottschall, P.E. (1990) Involvement of organum vasculosum of lamina terminalis and preoptic area in interleukin 1 beta-induced ACTH release. Am. J. Physiol., 258(1 Pt 1): E163–E171.

Kiss, J.Z., Mezey, E. and Skirboll, L. (1984) Corticotropin-releasing factor-immunoreactive neurons of the paraventricular nucleus become vasopressin positive after adrenalectomy. Proc. Natl. Acad. Sci. USA, 81(6): 1854–1858.

Kjaer, A., Knigge, U., Olsen, L., Vilhardt, H. and Warberg, J. (1991) Mediation of the stress-induced prolactin release by hypothalamic histaminergic neurons and the possible involvement of vasopressin in this response. Endocrinology, 128(1): 103–110.

Kollack-Walker, S., Don, C., Watson, S.J. and Akil, H. (1999) Differential expression of c-fos mRNA within neurocircuits of male hamsters exposed to acute or chronic defeat. J. Neuroendocrinol., 11(7): 547–559.

Kollack-Walker, S., Watson, S.J. and Akil, H. (1997) Social stress in hamsters: defeat activates specific neurocircuits within the brain. J. Neurosci., 17(22): 8842–8855.

Kovacs, K.J. (1998) c-Fos as a transcription factor: a stressful (re)view from a functional map. Neurochem. Int., 33(4): 287–297.

Kovacs, K.J., Csejtei, M. and Laszlovszky, I. (2001) Double activity imaging reveals distinct cellular targets of haloperidol, clozapine and dopamine D(3) receptor selective RGH-1756. Neuropharmacology, 40(3): 383–393.

Kovacs, K.J., Foldes, A. and Sawchenko, P.E. (2000) Glucocorticoid negative feedback selectively targets vasopressin transcription in parvocellular neurosecretory neurons. J. Neurosci., 20(10): 3843–3852.

Kovacs, K.J. and Mezey, E. (1987) Dexamethasone inhibits corticotropin-releasing factor gene expression in the rat paraventricular nucleus. Neuroendocrinology, 46(4): 365–358.

Kovacs, K.J. and Sawchenko, P.E. (1993) Mediation of osmoregulatory influences on neuroendocrine corticotropin-releasing factor expression by the ventral lamina terminalis. Proc. Natl. Acad. Sci. USA, 90(16): 7681–7685.

Kovacs, K.J. and Sawchenko, P.E. (1996) Sequence of stress-induced alterations in indices of synaptic and transcriptional activation in parvocellular neurosecretory neurons. J. Neurosci., 16(1): 262–273.

Laugero, K.D., Bell, M.E., Bhatnagar, S., Soriano, L. and Dallman, M.F. (2001) Sucrose ingestion normalizes central expression of corticotropin-releasing-factor messenger ribonucleic acid and energy balance in adrenalectomized rats: a glucocorticoid-metabolic-brain axis? Endocrinology, 142(7): 2796–2804.

Levine, S., Huchton, D.M., Wiener, S.G. and Rosenfeld, P. (1991) Time course of the effect of maternal deprivation on the hypothalamic–pituitary–adrenal axis in the infant rat. Dev. Psychobiol., 24(8): 547–558.

Li, H.Y., Ericsson, A. and Sawchenko, P.E. (1996) Distinct mechanisms underlie activation of hypothalamic neurosecretory neurons and their medullary catecholaminergic afferents in categorically different stress paradigms. Proc. Natl. Acad. Sci. USA, 93(6): 2359–2364.

Li, H.Y. and Sawchenko, P.E. (1998) Hypothalamic effector neurons and extended circuitries activated in 'neurogenic' stress: a comparison of footshock effects exerted acutely, chronically, and in animals with controlled glucocorticoid levels. J. Comp. Neurol., 393(2): 244–266.

Lightman, S.L. and Young, W.S., 3rd (1988) Corticotrophin-releasing factor, vasopressin and pro-opiomelanocortin mRNA responses to stress and opiates in the rat. J. Physiol., 403: 511–523.

Lightman, S.L. and Young, W.S., 3rd (1989) Influence of steroids on the hypothalamic corticotropin-releasing factor and preproenkephalin mRNA responses to stress. Proc. Natl. Acad. Sci. USA, 86(11): 4306–4310.

Ma, X.M. and Aguilera, G. (1999) Differential regulation of corticotropin-releasing hormone and vasopressin transcription by glucocorticoids. Endocrinology, 140(12): 5642–5650.

Ma, X.M., Levy, A. and Lightman, S.L. (1997) Emergence of an isolated arginine vasopressin (AVP) response to stress after repeated restraint: a study of both AVP and corticotropin-releasing hormone messenger ribonucleic acid (RNA) and heteronuclear RNA. Endocrinology, 138(10): 4351–4357.

Ma, X.M., Levy, A. and Lightman, S.L. (1997a) Rapid changes in heteronuclear RNA for corticotrophin-releasing hormone and arginine vasopressin in response to acute stress. J. Endocrinol., 152(1): 81–89.

Ma, X.M., Levy, A. and Lightman, S.L. (1997b) Rapid changes of heteronuclear RNA for arginine vasopressin but not for corticotropin releasing hormone in response to acute corticosterone administration. J. Neuroendocrinol., 9(10): 723–728.

Ma, X.M. and Lightman, S.L. (1998) The arginine vasopressin and corticotrophin-releasing hormone gene transcription responses to varied frequencies of repeated stress in rats. J. Physiol., 510(Pt 2): 605–614.

Ma, X.M., Lightman, S.L. and Aguilera, G. (1999) Vasopressin and corticotropin-releasing hormone gene responses to novel stress in rats adapted to repeated restraint. Endocrinology, 140(8): 3623–3632.

Makara, G.B. and Kovacs, K.J. (1997) Lesioning of the hypothalamic paraventricular nucleus inhibits ether-induced ACTH but not prolactin release. Neurobiology, 5(3): 403–411.

Makino, S., Smith, M.A. and Gold, P.W. (1995) Increased expression of corticotropin-releasing hormone and vasopressin messenger ribonucleic acid (mRNA) in the hypothalamic paraventricular nucleus during repeated stress: association with reduction in glucocorticoid receptor mRNA levels. Endocrinology, 136(8): 3299–3309.

Martinez, M., Calvo-Torrent, A. and Herbert, J. (2002) Mapping brain response to social stress in rodents with c-fos expression: a review. Stress, 5(1): 3–13.

McEwen, B.S. (1998) Stress, adaptation, and disease. Allostasis and allostatic load. Ann. NY Acad. Sci., 840: 33–44.

McEwen, B.S. (2000a) Allostasis and allostatic load: implications for neuropsychopharmacology. Neuropsychopharmacology, 22(2): 108–24.

McEwen, B.S. (2000b). The neurobiology of stress: from serendipity to clinical relevance. Brain Res., 886(1–2): 172–189.

McEwen, B.S. and Stellar, E. (1993) Stress and the individual. Mechanisms leading to disease. Arch. Intern. Med., 153(18): 2093–2101.

McKinley, M.J., McAllen, R.M., Davern, P., Giles, M.E., Penschow, J., Sunn, N., Uschakov, A. and Oldfield, B.J. (2003) The sensory circumventricular organs of the mammalian brain. Adv. Anat. Embryol. Cell Biol., 172: III–XII, 1–122, back cover.

Meaney, M.J., Bhatnagar, S., Larocque, S., McCormick, C., Shanks, N., Sharma, S., Smythe, J., Viau, V. and Plotsky, P.M. (1993) Individual differences in the hypothalamic-pituitary-adrenal stress response and the hypothalamic CRF system. Ann. NY Acad. Sci., 697: 70–85.

Miklós, I.H., Fekete, C., Szabó, G., Kovács, K.J. and Liposits, Z. (2000) *Histaminergic innervation of the corticotropin-releasing hormone neurons in the paraventricular nucleus.* Society for Neuroscience Abstracts.

Miklos, I.H. and Kovacs, K.J. (2003) Functional heterogeneity of the responses of histaminergic neuron subpopulations to various stress challenges. Eur. J. Neurosci., 18(11): 3069–3079.

Morgan, J.I. and Curran, T. (1991) Stimulus-transcription coupling in the nervous system: involvement of the inducible proto-oncogenes fos and jun. Annu. Rev. Neurosci., 14: 421–451.

Morrow, B.A., Elsworth, J.D., Inglis, F.M. and Roth, R.H. (1999) An antisense oligonucleotide reverses the footshock-induced expression of fos in the rat medial prefrontal cortex and the subsequent expression of conditioned fear-induced immobility. J. Neurosci., 19(13): 5666–7563.

Morrow, B.A., Redmond, A.J., Roth, R.H. and Elsworth, J.D. (2000) The predator odor, TMT, displays a unique, stress-like pattern of dopaminergic and endocrinological activation in the rat. Brain Res., 864(1): 146–151.

Nagase, H., Inoue, S., Tanaka, K., Takamura, Y. and Niijima, A. (1993) Hepatic glucose-sensitive unit regulation of glucose-induced insulin secretion in rats. Physiol. Behav., 53(1): 139–143.

Pacak, K. and Palkovits, M. (2001) Stressor specificity of central neuroendocrine responses: implications for stress-related disorders. Endocr. Rev., 22(4): 502–548.

Ray, C. (1979). Examination stress and performance on a color-word interference test. Percept. Mot. Skills, 49(2): 400–402.

Reul, J.M., Stec, I., Wiegers, G.J., Labeur, M.S., Linthorst, A.C., Arzt, E. and Holsboer, F. (1994) Prenatal immune challenge alters the hypothalamic-pituitary- adrenocortical axis in adult rats. J. Clin. Invest., 93(6): 2600–2607.

Rivest, S. and Laflamme, N. (1995) Neuronal activity and neuropeptide gene transcription in the brains of immune-challenged rats. J. Neuroendocrinol., 7(7): 501–525.

Rivest, S. and Rivier, C. (1994) Stress and interleukin-1 beta-induced activation of c-fos, NGFI-B and CRF gene expression in the hypothalamic PVN: comparison between Sprague-Dawley, Fisher-344 and Lewis rats. J. Neuroendocrinol., 6(1): 101–117.

Sarnyai, Z., Shaham, Y. and Heinrichs, S.C. (2001) The role of corticotropin-releasing factor in drug addiction. Pharmacol. Rev., 53(2): 209–243.

Sawchenko, P.E., Brown, E.R., Chan, R.K., Ericsson, A., Li, H.Y., Roland, B.L. and Kovacs, K.J. (1996) The paraventricular nucleus of the hypothalamus and the functional neuroanatomy of visceromotor responses to stress. Prog. Brain Res., 107: 201–222.

Sawchenko, P.E., Li, H.Y. and Ericsson, A. (2000) Circuits and mechanisms governing hypothalamic responses to stress: a tale of two paradigms. Prog. Brain Res., 122: 61–78.

Sawchenko, P.E., Swanson, L.W. and Vale, W.W. (1984) Co-expression of corticotropin-releasing factor and vasopressin immunoreactivity in parvocellular neurosecretory neurons of the adrenalectomized rat. Proc. Natl. Acad. Sci. USA, 81(6): 1883–1887.

Scammell, T.E., Price, K.J. and Sagar, S.M. (1993) Hyperthermia induces c-fos expression in the preoptic area. Brain Res., 618(2): 303–307.

Schmidt, E.D., Binnekade, R., Janszen, A.W. and Tilders, F.J. (1996) Short stressor induced long-lasting increases of vasopressin stores in hypothalamic corticotropin-releasing hormone (CRH) neurons in adult rats. J. Neuroendocrinol., 8(9): 703–712.

Schmidt, E.D., Janszen, A.W., Wouterlood, F.G. and Tilders, F.J. (1995) Interleukin-1-induced long-lasting changes in hypothalamic corticotropin-releasing hormone (CRH)–neurons and hyperresponsiveness of the hypothalamus-pituitary-adrenal axis. J. Neurosci., 15(11): 7417–7426.

Schmidt, E.D., Schoffelmeer, A.N., De Vries, T.J., Wardeh, G., Dogterom, G., Bol, J.G., Binnekade, R. and Tilders, F.J. (2001) A single administration of interleukin-1 or amphetamine induces long-lasting increases in evoked noradrenaline release in the hypothalamus and sensitization of ACTH and corticosterone responses in rats. Eur. J. Neurosci., 13(10): 1923–1930.

Scribner, K.A., Akana, S.F., Walker, C.D. and Dallman, M.F. (1993) Streptozotocin-diabetic rats exhibit facilitated adrenocorticotropin responses to acute stress, but normal sensitivity to feedback by corticosteroids. Endocrinology, 133(6): 2667–2674.

Selye, H. (1936) A syndrome produced by diverse nocuous agents. Nature, 138: 32.

Selye, H. (1955) Stress and disease. Science, 122: 625–631.

Sharp, F.R., Sagar, S.M., Hicks, K., Lowenstein, D. and Hisanaga, K. (1991) c-fos mRNA, Fos, and Fos-related antigen induction by hypertonic saline and stress. J. Neurosci., 11(8): 2321–2331.

Smith, D.W., Buller, K.M. and Day, T.A. (1995) Role of ventrolateral medulla catecholamine cells in hypothalamic neuroendocrine cell responses to systemic hypoxia. J. Neurosci., 15(12): 7979–7988.

Sterling, P. and Eyer, J. (1988) Allostasis: a new paradigm to explain arousal pathology. In: Fisher, S., Reason, J. (Eds.), Handbook of Life, Stress, Cognition and Health. John Wiley & Sons, New York, pp. 629–649.

Sternberg, E.M., Hill, J.M., Chrousos, G.P., Kamilaris, T., Listwak, S.J., Gold, P.W. and Wilder, R.L. (1989) Inflammatory mediator-induced hypothalamic-pituitary-adrenal axis activation is defective in streptococcal cell wall arthritis-susceptible Lewis rats. Proc. Natl. Acad. Sci. USA, 86(7): 2374–2378.

Stout, S.C., Owens, M.J. and Nemeroff, C.B. (2002) Regulation of corticotropin-releasing factor neuronal systems and hypothalamic-pituitary-adrenal axis activity by stress and chronic antidepressant treatment. J. Pharmacol. Exp. Ther., 300(3): 1085–1092.

Swanson, L.W. and Sawchenko, P.E. (1980) Paraventricular nucleus: a site for the integration of neuroendocrine and autonomic mechanisms. Neuroendocrinology, 31(6): 410–417.

Thrivikraman, K.V. and Plotsky, P.M. (1993) Absence of glucocorticoid negative feedback to moderate hemorrhage in conscious rats. Am. J. Physiol., 264(4 Pt 1): E497–E503.

Viau, V. and Sawchenko, P.E. (2002) Hypophysiotropic neurons of the paraventricular nucleus respond in spatially, temporally, and phenotypically differentiated manners to acute vs. repeated restraint stress: rapid publication. J. Comp. Neurol., 445(4): 293–307.

Watts, A.G. (1991) Ether anesthesia differentially affects the content of prepro- corticotropin-releasing hormone, prepro-neurotensin/neuromedin N and prepro-enkephalin mRNAs in the hypothalamic paraventricular nucleus of the rat. Brain Res., 544(2): 353–357.

Williams, G., Bing, C., Cai, X.J., Harrold, J.A., King, P.J. and Liu, X.H. (2001) The hypothalamus and the control of energy homeostasis: different circuits, different purposes. Physiol. Behav., 74(4–5): 683–701.

Willner, P., Muscat, R. and Papp, M. (1992) Chronic mild stress-induced anhedonia: a realistic animal model of depression. Neurosci. Biobehav. Rev, 16(4): 525–534.

Young, W.S., 3rd, Mezey, E. and Siegel, R.E. (1986) Quantitative in situ hybridization histochemistry reveals increased levels of corticotropin-releasing factor mRNA after adrenalectomy in rats. Neurosci. Lett., 70(2): 198–203.

Zelena, D., Makara, G.B. and Jezova, D. (1999) Simultaneous blockade of two glutamate receptor subtypes (NMDA and AMPA) results in stressor-specific inhibition of prolactin and corticotropin release. Neuroendocrinology, 69(5): 316–323.

Zhang, Y.H., Lu, J., Elmquist, J.K. and Saper, C.B. (2000) Lipopolysaccharide activates specific populations of hypothalamic and brainstem neurons that project to the spinal cord. J. Neurosci., 20(17): 6578–6586.

T. Steckler, N.H. Kalin and J.M.H.M. Reul (Eds.)
Handbook of Stress and the Brain, Vol. 15
ISBN 0-444-51173-3

CHAPTER 6.2

Involvement of the amygdala in the neuroendocrine and behavioral consequences of stress

Irit Akirav and Gal Richter-Levin*

Department of Psychology, and The Brain and Behavior Research Center University of Haifa, Haifa 31905, Israel

The amygdala is long known to be involved in the response to stress. There is extensive evidence indicating that a major contribution of the amygdala is to mediate neuroendocrine influences of emotional arousal and stress on learning and memory. Specifically, the basolateral amygdala (BLA) is involved in modulating the formation of long-term memory associated with emotionally arousing events in other brain structures (e.g., the hippocampus) where the memories are actually consolidated. By facilitating or impairing memory consolidation, the amygdala serves a highly adaptive process allowing organisms to select from recent experiences those that should be stored.

Considerable evidence suggests that the stress hormones, noradrenaline, and glucocorticoids, are the main mediators of the amygdala's modulation effects on the hippocampus, and presumably other memory-related brain structures.

This chapter reviews evidence from animal studies investigating the involvement of the amygdala in mediating the influences of noradrenaline and glucocorticoids on stress-related modifications of hippocampal memory functions.

The exposure to a threatening or stressful experience may result in excitement and increased motivation which supports the ability of an organism to respond in an adaptive manner to threat. However, stress may also have a deleterious influence on the organism and enhance its vulnerability to disease.

It is known that aversive or stressful events release stress hormones and that the amount of hormones released increases with the aversiveness of the stimulus.

The response to stress can either facilitate or impair cognition; under certain conditions, despite substantially high levels of stress hormones release, animals may display enhanced cognitive performance in hippocampus-dependent tasks whereas certain types of learning are impaired by stress (Bartolomucci et al., 2002).

One critical question regarding stress and its influence on cognitive function is when the effect of stress switches from cognitive improvement to cognitive impairment (Oitzl et al., 1998). Animal studies using highly aversive motivated training situations suggest that the amygdala critically influences both the acquisition and the expression of emotional memories (Jellestad and Bakke, 1985; Hitchcock and Davis, 1987; Dunn and Everitt, 1988; LeDoux et al., 1988; Davis, 1992; Cahill and McGaugh, 1998; LeDoux, 2000). In these experiments, clear evidence has been obtained that the behavioral expression of emotionally motivated learning depends on the integrity of the amygdala during a narrow and highly specific window of time (Cahill and McGaugh, 1998; Bianchin et al., 1999). Moreover, amygdala activity is apparently more important for the translation of

*Corresponding author. Tel.: +1972-4-8240962; Fax: +1972-4-8240966; E-mail: gal.r-l@psy.haifa.ac.il

an emotional reaction into heightened or impaired recall than it is for the generation of an emotional reaction per se.

Notably, the amygdala appears to play a critical role immediately after acquisition of the behavior and for some duration of time thereafter (Adolphs et al., 2000). These findings have been interpreted as evidence for the amygdala's role specifically in the consolidation processes of motivated learning that are influenced by emotional arousal, an interpretation consistent with studies in humans (Adolphs et al., 1997, 2000; Cahill and McGaugh, 1998; Hamann et al., 1999; McGaugh, 2000). There are memory-related changes in amygdala activity that appear selective for emotional (aversive) material and are consistent with the view that the amygdala modulates memory storage for emotionally arousing events. Direct manipulations of neural activity in the amygdala (for example, with pharmacological inactivation or electrophysiological stimulation) have shown that the amygdala, and specifically the basolateral amygdala (BLA), exerts its modulatory effect on secondary structures, such as the hippocampus, that are directly involved in memory consolidation (Packard et al., 1994; McGaugh et al., 1996; Packard and Teather, 1998; Roozendaal et al., 1999; Richter-Levin and Akirav, 2000). The hippocampus is believed to play a critical role in the acquisition and consolidation of episodic declarative memories in humans (Squire, 1992; 1998; Eichenbaum, 1999; Maguire et al., 1999). Complementary observations from animal studies have reported that the hippocampus is necessary for generating spatio-temporal and contextual representations (Morris et al., 1982; Eichenbaum, 1992; Squire, 1992; Moser et al., 1993; Alvarez et al., 1995; Bunsey and Eichenbaum, 1996; O'Keefe, 1999).

In the context of sufficiently arousing stimuli the amygdala and the hippocampus may interact to intensify or harm memory (Layton and Krikoriani, 2002).

It has been shown that hippocampal memory-related processes are potently influenced by the adrenal stress hormones noradrenaline and glucocorticoids (GLUC) (Gold et al., 1984; Bennett et al., 1991; Diamond et al., 1992; Pavlides et al., 1993; Rey et al., 1994; McEwen, 1995).

Naturally, multiple factors mediate stress effects on hippocampal functioning and behavioral stress leads to the activation of a wide variety of neurotransmitters and neuroendocrine systems that can potentially affect learning and memory. These include corticotropin-releasing factor (CRF), opioid peptides, acetylcholine, neurosteroids, etc. (de Wied and Croiset, 1991). However, the adrenal stress hormones appear to play an important role in enabling the significance of an experience to regulate the strength of memory of the experience (Diamond et al., 1992; Pavlides et al., 1993; Rey et al., 1994). The participation of the amygdala in learning depends on the degree to which the training conditions induce phasic increases in arousal associated with the release of the adrenal stress hormones. The activation of the amygdala is critical for mediating the influences of noradrenaline and GLUC on hippocampal function because amygdala lesions block the effects of these modulators on consolidation (Liang et al., 1990; Roozendaal and McGaugh, 1996; McGaugh, 2000; Kim et al., 2001).

Here, we will describe a memory system composed of the amygdala, the hippocampus and the stress hormones, which is activated following an arousing event. Specifically, we will focus on the involvement of the amygdala in mediating the influences of noradrenaline and GLUC on stress-related modifications of hippocampal memory functions. Under extreme conditions, these modifications are likely to be associated with cognitive memory disturbances or pathological augmentation following stressful and traumatic life events.

Emotional arousal leads to the secretion of the catecholamines epinephrine and noradrenaline by the sympathetic system and to the activation of the hypothalamic-pituitary adrenocortical (HPA) axis and its final product glucocorticoids [corticosterone (CORT) in rats, cortisol in humans]. There is strong evidence that hippocampal functioning is highly susceptible to disruption by increased emotionality and the resulting secretion of the stress hormones. Since CORT levels are reliably elevated under stressful conditions it has become widely accepted that an elevation in GLUC levels is indicative of a stress state (Kim and Diamond, 2002). There are two types of receptors for glucocorticoids [Type I, also known as the mineralcorticoid receptor (MR), and type II, also called the glucocorticoid receptor (GR)], and the hippocampus is one of the few sites in the body where there are substantial concentrations of both types of receptors (de Kloet et al., 1990). The effect of CORT on memory performance should take into account the specific roles of the two types of receptors. It has been suggested that MRs have a role in behavioral reactivity during novel situations, whereas GRs are involved in consolidation of learned information (de Kloet et al., 1999).

The effects of stress and/or GLUC on hippocampal plasticity and on memory consolidation of hippocampus-dependent learning were reported to follow an inverted U-shaped dose–response relationship: extremely low and high levels may impair consolidation or plasticity whereas moderate activation seems to be a prerequisite for the long-term storage of information or to its reinforcement. Thus, removal of circulating corticosteroids (via adrenalectomy) or selective MR or GR antagonist injections impaired acquisition and retention in hippocampus-dependent tasks, such as spatial learning, avoidance and contextual fear conditioning. At the same time exposure to a stressor or experimentally induced high levels of CORT were reported to impair acquisition and retention of those tasks (Oitzl and de Kloet, 1992; Cahill et al., 1994; Conrad et al., 1996, 1997, 1999; Diamond et al., 1994, 1996, 1999; Diamond and Rose, 1994; Oitzl et al., 1994; Sandi and Rose, 1994b; Vaher et al., 1994; Pugh et al., 1997; de Quervain et al., 1998). In contrast, intermediate increase in circulating corticosteroids has been shown to facilitate memory in different learning paradigms (Akirav et al., in press; Flood et al., 1978; Oitzl and de Kloet, 1992; Sandi and Rose, 1994a; Pugh et al., 1997; Sandi et al., 1997; Cordero and Sandi, 1998; Liu et al., 1999; Akirav et al., 2001).

Similarly, post-training administration of noradrenaline influences retention with an inverted U-shaped curve; retention is enhanced at moderate doses in a variety of training tasks, including inhibitory avoidance, active avoidance, discrimination learning and appetitively motivated tasks, and retention is impaired at high doses (Gold and van Buskirk, 1975; Izequirdo and Dias, 1985; Introini-Collison and McGaugh, 1986; Liang et al., 1986; McGaugh et al., 1990).

This inverted U-shaped function of the stress hormones may be related to motivational and attentional processes.

The lower performance associated with very low levels of stress is usually explained by the low motivation that accompanies the low stress and the ease with which the subject is therefore diverted from the problem by extraneous factors (Vroom, 1964; Anderson, 1976). During intermediate stress levels, the level of motivation to solve the problem reaches the optimum zone in which the stress broadens the span of attention so that the subject is more amenable to relevant information (for example, spatial cues in the environment) that may actually aid its performance (Easterbrook, 1959; Selden et al., 1990). Under high levels of stress, on the other hand, it may be that the motivation to solve the task is so high that the subject's perception narrows to only very obvious cues, and it ignores other relevant information. Thus, impaired attentional processes may prevent successful acquisition of information and by this the successful consolidation and recall of it (Conrad et al., 1997). However, it may be that under such conditions subjects will perform well on an easy task and will fail to acquire more demanding tasks (Selden et al., 1990). Under extreme conditions, however, the high anxiety that is associated with the high stress may lead to physiological involuntary autonomic responses that interfere with performance (Vroom, 1964).

Several lines of evidence suggest that the emotional arousal that activates the amygdala, and specifically the BLA (composed of the lateral nucleus, the basal nucleus and the accessory basal nucleus), results in modulation (enhancement or impairment) of memory-related processes in the hippocampus (for review, see Cahill and McGaugh, 1998; McGaugh, 2000; Richter-Levin and Akirav, 2000; Roozendaal, 2000; Packard and Cahill, 2001; Pare, 2003).

The hippocampus puts a specific event into its proper context, it binds together multiple events that co-occur during an experience, organizes and categorizes them, and through this kind of rich processing it converts short-term into long-term memories, enabling accurate episodic memories to be formed (Chiba et al., 1994; Kesner et al., 1996). However, since the hippocampus is constantly receiving a vast amount of information, what could help filter what is important and what is less relevant? In emotionally arousing conditions, activating the amygdala may have a key role in modulating hippocampal consolidation; by facilitating or dampening memory consolidation, this modulatory system would serve a highly adaptive process allowing organisms to select from recent experiences those that should be permanently stored (Gold and McGaugh, 1975).

Generally, emotional arousal improves memory, thus emotionally arousing events are better remembered than neutral events that are normally weakly retained or require repetition to endure (Cahill and McGaugh, 1998). We suggested the "emotional tagging" concept (Richter-Levin and Akirav, 2003), according to which the activation of the amygdala in emotionally arousing events mark the experience as important presumably by strengthening of synapses located on neurons that have just been activated in another brain-memory system that is engaged in the learning situation. Accordingly, adding to the general arousing influence of an emotional experience, there is also a more specific impact on memory processes, i.e. in potentiating the consolidation of emotionally loaded aspects of an experience into enhanced long-term memory.

However, this is not necessarily a linear correlation, i.e. more intense emotional valence is not always associated with stronger memory. Similar to the effects of stress on hippocampal learning processes, the effects of amygdala activation follow an inverted U-shaped dose–response relationship; extreme low and high levels of amygdala activation may impair consolidation whereas moderate activation seems to be a prerequisite for the long-term storage of information (Diamond et al., 1992; Yau et al., 1995).

Accordingly, under conditions of high stress and amygdala activation, subjects tend to show impaired attentional processes, for example, problems discriminating between relevant and irrelevant stimuli and such impairments will easily prevent successful acquisition of useful information (Lupien and McEwen, 1997). Thus, very high levels of emotional arousal may prevent the proper evaluation and categorization of experience by interfering with hippocampal function (van der Kolk, 1997), and because the hippocampus is prevented from fulfilling its integrative function, some aspects of the experience may be consolidated while others may be impaired.

Extremely high arousal and stress levels may lead to pathological conditions. During excessive cases of stress, the augmentation of the stress-hormones activation or their long-term presence in the system may underlie the high-anxiety levels that possibly will cause subjects to over-concentrate on emotional and defensive coping mechanisms instead of paying attention to problem-solving strategies. Likewise, it may underlie the repetitive reliving of the stress experience (or traumatic event), which may well result in disturbances such as post-traumatic stress disorder (PTSD). PTSD is conceived as a condition that involves the coupling of the amplification of memory for a traumatic stimulus with the decrement in memory for surrounding contextual material. It has been suggested that the failure to consolidate material proximate to the traumatic stimulus (for example, the context of the trauma) is due to amygdala suppression of hippocampal function and that the enhancement of memory at the highest levels of emotional arousal is because the amygdala then becomes the exclusive locus of consolidation of the traumatic event (Layton and Krikorian, 2002). Accordingly, under certain stressful conditions hippocampal functioning may be impaired while amygdala processing will be facilitated at the expanse of hippocampus-dependent spatio-temporal processing (Diamond et al., 2001).

Memory formation is considered to involve long-lasting alterations in synaptic efficacy known as synaptic plasticity. It was demonstrated that specific patterns of activation, such as brief high-frequency stimulation of afferent fibers to the hippocampus, can result in these long-lasting alterations of synaptic efficacy. The most widely studied cellular model for synaptic plasticity is long-term potentiation (LTP; Bliss and Collingridge, 1993; Martin et al., 2000).

Long-term potentiation in the hippocampal dentate gyrus (DG) area of freely moving rats was reinforced after its induction by appetitive and aversive stimuli (which are known to activate the amygdala). The efficacy of these stimuli terminated about 1 h after tetanization, probably reflecting time constants of the mechanisms underlying consolidation. The appetitive and aversive stimuli-induced reinforcement was blocked by the β-adrenergic antagonist propranolol, implicating noradrenaline in the underlying cellular processes (Seidenbecher et al., 1997).

We have shown (Akirav and Richter-Levin, 1999; 2002) that activating the BLA prior to perforant path

(PP) tetanization has a biphasic effect on hippocampal plasticity; priming the BLA immediately before PP tetanization results in the enhancement of DG LTP, whereas stimulation in a spaced interval (1 or 2 h before PP tetanization) results in its suppression. Moreover, we found that whereas a stressful experience suppresses hippocampal LTP, priming the BLA in stressed animals relieves the depressant effect of behavioral stress on hippocampal LTP (Akirav and Ricter-Levin, 1999). This study strongly supports the notion that the amygdala and the hippocampus may act synergistically to form long-term memories of significantly emotional events, and that the temporal relationship between the activation of the amygdala and the hippocampus may be critical to the outcome. We further found that the effects of both the priming and the spaced activation of the BLA on hippocampal plasticity were mediated by the stress hormones noradrenaline and CORT. Because both hormones seem to be involved in the enhancing as well as the inhibitory effects of the BLA, a third factor must be postulated that will define whether an enhancement or inhibition of plasticity will take place. One factor may be time, i.e. the effects of a brief exposure to these hormones are excitatory, whereas their prolonged presence in the spaced phase may lead to the inhibitory effect. Thus, it may be that at the onset of an emotional event the stress hormones permissively mediate plasticity and lead to its facilitation whereas their prolonged presence in the system may suppress the cognitive response to stress.

Memory is generally better for emotionally arousing events than for neutral events and it has been suggested that the amygdala may be more extensively involved in training situations that are highly arousing, and that those stimuli invoking weaker emotional responses are probably less effective at consistently or robustly activating the amygdala (Cahill and McGaugh, 1990; 1998). We have demonstrated that a highly arousing learning experience that significantly activated the amygdala, led to better hippocampus-dependent memory and that CORT is critically involved in this effect (Akirav et al., 2001; Akirav et al., in press). We examined the activation of the memory-related biochemical marker ERK2 (ERK/MAPK; extracellular-signal regulated kinase/mitogen-activated protein kinase) in the hippocampus and the amygdala following a spatial learning task performed under different stressful conditions (Akirav et al., 2001). ERK2 is an essential component of the signal transduction mechanisms subserving some forms of learning (Atkins et al., 1998; Berman et al., 1998; Blum et al., 1999; Schafe et al., 1999; Selcher et al., 1999). Animals trained for a massed spatial task in a water-maze under cold-water conditions (moderate level of stress) showed better performance in the spatial task and higher levels of CORT compared with animals trained in warm water (mild level of stress). Significant activation of ERK2 in the hippocampus was found in all the animals that had acquired the spatial task (irrespective of the level of stress involved) whereas ERK2 activation in the amygdala was found only in animals that acquired the task in cold water. Moreover, animals that were exposed to the cold water with no escape platform in the maze (and thus with no specific task to learn) and showed the highest CORT levels did not show ERK2 activation in the amygdala, indicating that ERK2 activation in the amygdala was learning specific. Thus, it is likely that the activation of the amygdala (as seen by the activation of ERK2) following an emotionally charged hippocampal-dependent learning experience led to the better performance of the cold water trained rats in the spatial task (Akirav et al., 2001).

In a following work we examined whether CORT is a main candidate to determine the strength of the hippocampal-dependent spatial task consolidation by exogenously manipulating the levels of CORT (Akirav et al., in press). Rats were injected with metyrapone [which reduces CORT synthesis; 25, 50, 75 mg/kg, intraperitoneal (i.p.)] or with CORT (10, 25 mg/kg, i.p.) and trained in a massed spatial task in either cold (19°C) or warm (25°C) water. We found that animals injected with metyrapone showed impairment in performance in cold water, whereas rats injected with CORT showed dose-dependent improvement in performance in warm water. These two mirror experiments of CORT increase and decrease strongly suggest that, under the conditions described, CORT is instrumental in the acquisition of the spatial learning task. These ideas are further supported from another study in which rats trained in 19°C showed a quicker rate of acquisition and better long-term retention than rats trained in 25°C water. In addition, post-training corticosterone levels, on the first day of training, were significantly higher in rats in the 19°C group than in the 25°C group. Performance of rats trained in 25°C, but not in 19°C, was improved by injecting them i.p. with corticosterone immediately after each training session. Thus, the effect of exogenously administered corticosterone on the neural mechanisms determining the strength of spatial information consolidation may be facilitatory (Sandi et al., 1997).

Our spatial learning experiments described above (Akirav et al., 2001; Akirav et al, in press) show that training in cold water (a stressful experience) significantly activates the amygdala and that the good performance of these stressed animals is mediated by corticosterone. Since the amygdala is involved in regulating the effects of stress and stress hormones on hippocampal-dependent memory processes, including spatial learning (McGaugh, 2000), it is plausible that the effects of corticosterone on the hippocampal dependent task are mediated via influences involving the amygdala. These ideas are strongly supported by a recent study (Kim et al., 2001) in which amygdala lesions effectively blocked stress effects on hippocampal LTP and spatial memory without significantly affecting the increase in corticosterone secretion in response to stress. Thus, the increase in corticosterone levels is probably not a sufficient condition to mediate stress effects on hippocampal plasticity and learning. The critical role of the amygdala in regulating corticosterone effects on hippocampus-dependent learning is further discussed and supported by various experiments described in the next section.

In any case, the data collectively suggests that the stress hormones and specifically CORT may be necessary for the establishment of an enduring memory (de Kloet et al., 1998), though the outcome may depend on whether the stressor is within or out of the learning context. In fact, the cold water-trained rats showed considerably high levels of CORT (Sandi et al., 1997; Akirav et al., 2001; Akirav et al., in press), however, animals still performed the task well. The release of CORT was within the context of the learning experiment and seemed to have a positive effect in the formation of the spatial information.

Others have also shown that administration of exogenous CORT in the appropriate temporal context, i.e. in close relation to training, may potentiate memory in a dose-related fashion (Flood et al., 1978; Bohus et al., 1982; McEwen et al., 1986; Sandi and Rose, 1994a; Lupien and McEwen, 1997). For example, using a passive avoidance task in day-old chicks, intra-cerebral corticosterone administration at either 15 min pre-training or at 5, 30, 60 min (but not 120, 180, or 360 min) post-training retained the passive avoidance response when tested 24 h post-training [which is otherwise retained only for a few hours (< 10) after training] (Sandi and Rose, 1994a). In rats, an i.p. corticosterone injection, given immediately after training at low-shock intensity in contextual fear conditioning, enhanced long-term expression of the fear response (Cordero and Sandi, 1998). Similarly, the GR agonist dexamethasone (0.3 mg/kg) administered immediately after training in a one-trial inhibitory avoidance task, enhanced retention that was tested 48 h later (Roozendaal and McGaugh, 1996). In the chapter by Lupien et al., the positive effects of exogenous glucocorticoids for emotionally arousing material in humans are depicted.

However, GR activation triggered by a distracting stressor that is out of the context of the learning task disrupts ongoing consolidation (de Kloet et al., 1990) as seen in different hippocampus-dependent tasks in which the exposure to an unrelated stressor (e.g., footshock, exposure to an unfamiliar environment, etc.) interrupted the performance in the learning task (Diamond et al., 1996; de Quervain et al., 1998). Accordingly, when a low dose of corticosterone was injected prior to a retrieval spatial test in the water-maze, performance in the task declined (de Quervain et al., 1998). Also see the chapter by Lupien in which an acute exogenous administration of glucocorticoids impaired retrieval processes in humans.

In another study, rats foraged for food in 7 arms of a 14-arm radial maze. After they ate the food in 4 of the 7 baited arms, they were placed in an unfamiliar environment (stress) for a 4-h delay. At the end of the delay they were returned to the maze to locate the food in the 3 remaining baited arms. Stress impaired their performance in the task, probably by way of transiently disrupting hippocampal function (Diamond et al., 1996).

Collectively, these studies suggest that the temporal relationship between GR activation (or the exposure to stress) and the behavioral task is important, thus an exposure to the stressor that is out of the context of the original learning task may disrupt ongoing consolidation and influence the retrieval process (Oitzl et al., 1997; 1998; de Kloet et al., 1999; 2002).

Studies with animals have shown that the BLA is necessary for the expression of the modulatory effects of stress and stress-related hormones on hippocampal learning and memory function (Packard et al., 1994; Cahill and McGaugh, 1996; McGaugh et al., 1996; Packard and Chen, 1999; Roozendaal et al., 1999; Kim et al., 2001; Kim and Diamond, 2002).

It has been demonstrated that an intact amygdala is necessary for the stress hormones to exert their influence on hippocampal memory function. Lesions in the amygdala block the modulatory effects of systemic and post-training intra-hippocampal injections of stress hormones on long-term memory assessed in a variety of learning tasks, including inhibitory avoidance, Y-maze discrimination and water-maze tasks (Cahill and McGaugh, 1990; Roozendaal and McGaugh, 1996; Roozendaal et al., 1996, 1998; Packard and Chen, 1999).

Accordingly, post-training injections of the synthetic glucocorticoid dexamethasone or the GR agonist RU 28362 enhanced the performance in a 48-h inhibitory avoidance retention test, and a selective neurochemically induced lesion of the BLA blocked this enhancement (Roozendaal and McGaugh, 1996; 1997). In another task, intra-hippocampal infusions of RU 28362 given 60 min before a spatial task retention test impaired retrieval and a selective *N*-methyl-D-aspartate (NMDA)-induced lesion of the BLA 1 week before training blocked this impairment (Roozendaal et al., 2003). Likewise, BLA lesions reversed the impairing effects of GR antagonists on retention of a spatial task (Roozendaal et al., 1996).

In adrenalectomized rats lesions of the BLA blocked the impairing effects of adrenalectomy (ADX) on spatial learning and memory even though these lesions did not affect the neurodegenerative changes in the DG (Roozendaal et al., 1998).

Similar results have been obtained with noradrenaline; lesions of the stria terminalis, a major pathway carrying projections to and from the amygdala, blocked the memory-enhancing effects of a low dose of noradrenaline as well as the memory-impairing effects of a high dose of noradrenaline infused into the amygdala post-training in a inhibitory avoidance task (Liang et al., 1990). Other experiments have shown that selective activation of adrenoceptors in the BLA modulates memory consolidation in other brain areas (for review, see McGaugh, 2002; McGaugh and Roozendaal, 2002; McGaugh et al., 2002). For example, post-training microinfusions of noradrenaline or the β-noradrenergic antagonist propranolol into the BLA immediately following training in a spatial version of the water-maze task was shown to modulate spatial performance. Retention latencies obtained on the second day of training revealed that noradrenaline dose-dependently enhanced retention for the location of the hidden platform whereas propranolol significantly impaired retention (Hatfield and McGaugh, 1999). Similarly, post-training intra-BLA infusions of the β-adrenoceptor agonist clenbuterol enhanced memory for inhibitory avoidance (Ferry and McGaugh, 1999).

Taken together, the data suggests that the effects of the stress hormones on hippocampal learning and memory process may be mediated via influences involving the amygdala.

Stress may exert an inhibitory or facilitatory influence on cognitive and electrophysiological measures of hippocampal functioning.

The full expression of stress effects on the hippocampus requires co-activation of the amygdala and hippocampus, in concert with the actions of neuromodulators (mainly of CORT) directly on the hippocampus. We suggest that the BLA co-ordinates the interactions of hormonal systems in their influences on the storage of information in other brain regions and specifically the hippocampus. There are wide projections from the BLA to the hippocampus; the basal nucleus projects substantially to the EC, CA3 and CA1 fields of the hippocampus, the subiculum and the parasubiculum. The accessory basal nucleus projects substantially to the EC, the CA1 field and the parasubiculum, and the main projections from the lateral nucleus are directed to the EC and the parasubiculum (Pikkarainen et al., 1999).

Although the stress hormones may affect hippocampal functioning directly, additional influence from the amygdala is required for these hormones to influence memory consolidation (also see the chapter by Prickaerts and Steckler regarding glucocorticoids modulation of hippocampal-dependent memory storage via the BLA at the nucleus accumbens level, through interactions with the noradrenergic system).

There are several possible ways in which the BLA may be involved in mediating the effects of the stress hormones on memory storage. The stress hormones can act in parallel to affect memory function via binding to receptors in the amygdala, in the hippocampus or in other brain structures. This will activate the amygdala and the hippocampus and thereby mediate the effects of the stress hormones on memory formation. Noradrenaline may be released directly into the hippocampus and the amygdala from ascending terminals of the locus coeruleus (LC) upon arousal and stress, which in turn may induce changes (enhancement or impairment) on the neural activity engaged in memory processes (Seidenbecher et al., 1997). Stress produces increases in noradrenaline turnover in the LC, the hippocampus and the amygdala, as well as in the hypothalamus and the cerebral cortex (Charney et al., 1995). GLUC released by an arousing experience bind to steroid receptors in the hippocampus, in the BLA, and other parts of the brain. For example, it has been suggested that GLUC binds directly to GRs in the BLA and their effects may be mediated via an interaction with β-adrenergic mechanisms in the BLA (Roozendaal, 2000). However, these receptors are most abundant in the hippocampus (de Kloet et al, 1999) and thus GLUC released following an emotional experience also binds to GRs and MRs in the hippocampus and exert facilitative or damaging influence directly there. A schematic model of BLA modulation of hippocampal memory processes with the neuromodulation of the stress hormones, noradrenaline and CORT, is represented in Fig. 1.

The amygdala is critically involved in modulating the formation of long-term memory associated with emotionally arousing events (Gold et al., 1975; Cahill and McGaugh, 1990). Its modulatory role on hippocampal memory is mediated by the function of the stress hormones. During acquisition, the amygdala influences memory formation by selecting stimuli of a highly arousing nature that are likely to be of relevance to the organism. Accordingly, the amygdala is probably important in the acquisition, but less so in the retrieval, of declarative knowledge about emotions (Adolphs et al., 1997; Hamann et al., 1997; Phelps and Anderson, 1997).

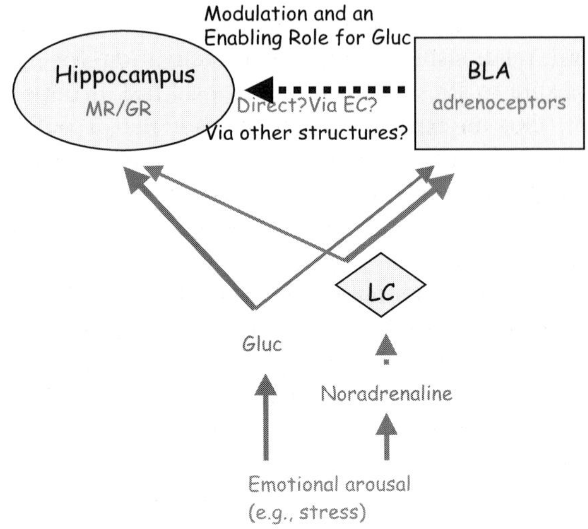

Fig. 1. A schematic model of the involvement of the BLA in mediating the influences of norepinephrine and glucocorticoids (Gluc) on stress-related modifications of hippocampal memory functions. noradrenaline and Gluc affect both the amygdala and the hippocampus. However, evidence suggests that the major influence of noradrenaline is on the amygdala while the main effects of Gluc are in the hippocampus. In addition, the amygdala has a role in enabling the modulating function of Gluc in the hippocampus. BLA – basolateral amygdala; CORT – corticosterone; EC – entorhinal cortex; Gluc – glucocorticoids; GR – glucocorticoid receptor; LC – locus coeruleus; MR – mineralcorticoid receptor.

However, in addition to modulating memory-related processes in other brain regions, the amygdala may be the site of some aspects of the memory of the experience. LTP, and more importantly, learning-induced LTP, was described in the amygdala (McKernan and Shinnick-Gallagher, 1997; Rogan et al., 1997; Yaniv et al., 2001). Furthermore, memory-related gene expression and activation of biochemical pathways known to be involved in the formation of long-term memory in the amygdala were also described (Akirav et al., 2001; Hall et al., 2001; Stork et al., 2001; Radwanska et al., 2002; Ressler et al., 2002). It is compelling to suggest that under stressful conditions, when plasticity is suppressed in the hippocampus (Shors et al., 1990; Kim et al., 1996; Diamond and Park, 2000;

Pavlides et al., 2000; Wang et al., 2000), the contribution of the amygdala to the formation of the memory of the stressful event increases. This shift may be at the heart of the formation of traumatic memories under extremely stressful conditions. Because the amygdala may lack the complexity of hippocampal and cortical structures, the result may be a very intense but simplified and inaccurate memory of the traumatic events. The role of the stress hormones specifically in these traumatic memories is not yet clear. It has been suggested that noradrenaline has a key role in the activation of the amygdala and in enabling its modulation of memory-related processes in other brain regions (McGaugh, 2000). Whether noradrenaline is also instrumental in the establishment of amygdala-based aspects of traumatic memories is yet to be established. Clearly, understanding the mechanisms underlying the formation of these memories will more thoroughly enable the development of more efficient approaches to the treatment of anxiety disorders and, in particular, of PTSD.

References

Adolphs, R., Cahill, L., Schul, R. and Babinsky, R. (1997) Impaired declarative memory for emotional material following bilateral amygdala damage in humans. Learn. Mem., 4: 291–300.

Adolphs, R., Tranel, D. and Denburg, N. (2000) Impaired emotional declarative memory following unilateral amygdala damage. Learn Mem., 73: 180–186.

Akirav, I., Kozenicky, M., Tal, D., Sandi, C., Venero, C. and Richter-Levin, G. A facilitative role for corticosterone in the acquisition of a spatial task under moderate stress. Learn and Mem., in press.

Akirav, I. and Richter-Levin, G. (1999) Biphasic modulation of hippocampal plasticity by behavioral stress and basolateral amygdala stimulation in the rat. J. Neurosci., 1: 10530–10535.

Akirav, I. and Richter-Levin, G. (2002) Mechanisms of amygdala modulation of hippocampal plasticity. J. Neurosci., 15: 9912–9921.

Akirav, I., Sandi, C. and Richter-Levin, G. (2001) Differential activation of hippocampus and amygdala following spatial learning under stress. Eur. J. Neurosci., 14: 719–725.

Alvarez, P., Zola-Morgan, S. and Squire, L.R. (1995) Damage limited to the hippocampal region produces long-lasting memory impairment in monkeys. J. Neurosci., 15: 3796–807.

Anderson, C.R. (1976) Coping behaviors as intervening mechanisms in the inverted U stress-performance relationship. J. Appl. Psychol., 61: 30–34.

Atkins, C.M., Selcher, J.C., Petraitis, J.J., Trzaskos, J.M. and Sweatt, J.D. (1998) The MAPK cascade is required for mammalian associative learning. Nat. Neurosci., 1: 602–609.

Bartolomucci, A., de Biurrun, G., Czeh, B., van Kampen, M. and Fuchs, E. (2002) Selective enhancement of spatial learning under chronic psychosocial stress. Eur. J. Neurosci., 15: 1863–1866.

Bennett, M.C., Diamond, D.M., Fleshner, M. and Rose, G.M. (1991) Serum corticosterone level predicts the magnitude of hippocampal primed burst potentiation and depression in urethane-anesthetized rats. Psychobiol., 19: 301–307.

Berman, D.E., Hazvi, S., Rosenblum, K., Seger, R. and Dudai, Y. (1998) Specific and differential activation of mitogen-activated protein kinase cascades by unfamiliar taste in the insular cortex of the behaving rat. J. Neurosci., 1: 10037–10044.

Bianchin, M., Mello e Souza, T., Medina, J.H. and Izquierdo, I. (1999) The amygdala is involved in the modulation of long-term memory, but not in working or short-term memory. Neurobiol. Learn. Mem., 71: 127–131.

Bliss, T.V.P. and Collingridge, G.L. (1993) A synaptic model of memory: long-term potentiation in the hippocampus. Nature, 361: 31–39.

Blum, S., Moore, A.N., Adams, F. and Dash, P.K. (1999) A mitogen-activated protein kinase cascade in the CA1/CA2 subfield of the dorsal hippocampus is essential for long-term spatial memory. J. Neurosci., 1: 3535–3544.

Bohus, B., De Kloet, E.R. and Veldhuis, H.D. (1982) Adrenal steroids and behavioral adaptation: relationship to brain corticoid receptors. In: Ganten, D. and Pfaff, D. (Eds.), Current Topics in Neuroendocrinology: Adrenal Actions on Brain, Vol 2. Springer, Berlin, 107–148.

Bunsey, M. and Eichenbaum, H. (1996) Conservation of hippocampal memory function in rats and humans. Nature, 18: 255–257.

Cahill, L. and McGaugh, J.L. (1990) Amygdaloid complex lesions differentially affect retention of tasks using appetitive and aversive reinforcement. Behav. Neurosci., 104: 532–543.

Cahill, L. and McGaugh, J.L. (1996) Modulation of memory storage. Curr. Opin. Neurobiol., 6: 237–242.

Cahill, L. and McGaugh, J.L. (1998) Mechanisms of emotional arousal and lasting declarative memory. Trends Neurosci., 21: 294–299.

Cahill, L., Prins, B., Weber, M. and McGaugh, J.L. (1994) Beta-adrenergic activation and memory for emotional events. Nature, 20: 702–4.

Charney, D.S., Deutch, A.Y., Southwick, S.M. and Krystal, J.H. (1995) Neural circuits and mechanisms of post traumatic stress disorder. In: Freiedman, M.J., Charney, D.S. and Deutch, A.Y. (Eds.), Neurobiological and Clinical

Consequences of Stress: from Normal Adaptation to PTSD. Lipoincott-Raven publishers, Philadelpia. pp: 271–290.

Chiba, A.A., Kesner, R.P. and Reynolds, A.M. (1994) Memory for spatial location as a function of temporal lag in rats: role of hippocampus and medial prefrontal cortex. Behav. Neural. Biol., 61: 123–131.

Conrad, C.D., Galea, L.A., Kuroda, Y. and McEwen, B.S. (1996) Chronic stress impairs rat spatial memory on the Y maze, and this effect is blocked by tianeptine pretreatment. Behav. Neurosci., 110: 1321–1334.

Conrad, C.D., Lupien, S.J. and McEwen, B.S. (1999) Support for a bimodal role for type II adrenal steroid receptors in spatial memory. Neurobiol. Learn. Mem., 72: 39–46.

Conrad, C.D., Lupien, S.J., Thanasoulis, L.C. and McEwen, B.S. (1997) The effects of type I and type II corticosteroid receptor agonists on exploratory behavior and spatial memory in the Y-maze. Brain Res., 6: 76–83.

Cordero, M.I. and Sandi, C. (1998) A role for brain glucocorticoid receptors in contextual fear conditioning: dependence upon training intensity. Brain Res., 9: 11–17.

Davis, M. (1992) The role of the amygdala in fear and anxiety. Annu. Rev. Neurosci., 15: 353–375.

Diamond, D.M., Bennett, M.C., Fleshner, M. and Rose, G.M. (1992) Inverted-U relationship between the level of peripheral corticosterone and the magnitude of hippocampal primed burst potentiation. Hippocampus, 2: 421–430.

Diamond, D.M., Fleshner. M., Ingersoll. N. and Rose, G.M. (1996) Psychological stress impairs spatial working memory: relevance to electrophysiological studies of hippocampal function. Behav. Neurosci., 110: 661–772.

Diamond, D.M., Fleshner, M. and Rose, G.M. (1994) Psychological stress repeatedly blocks hippocampal primed burst potentiation in behaving rats. Behav. Brain Res., 30: 1–9.

Diamond, D.M. and Park, C.R. (2000) Predator exposure produces retrograde amnesia and blocks synaptic plasticity. Progress toward understanding how the hippocampus is affected by stress. Ann. NY Acad. Sci. USA, 911: 453–455.

Diamond, D.M., Park, C.R., Heman, K.L. and Rose, G.M. (1999) Exposing rats to a predator impairs spatial working memory in the radial arm water maze. Hippocampus., 9: 542–552.

Diamond, D.M., Park, C.R., Puls, M.J. and Rose, G.M. (2001) Differential effects of stress on hippocampal and amygdaloid LTP. In: Holscher, C. (Ed.), Neuronal Mechanism of Memory Formation. Cambridge UP, UK, pp. 379–403.

Diamond, D.M. and Rose, G.M. (1994) Stress impairs LTP and hippocampal-dependent memory. Ann. NY Acad. Sci., 30: 411–414.

Dunn, L.T. and Everitt, B.J. (1988) Double dissociations of the effects of amygdala and insular cortex lesions on conditioned taste aversion, passive avoidance, and neophobia in the rat using the excitotoxin ibotenic acid. Behav. Neurosci., 102: 3–23.

Easterbrook, J.A. (1959) The effect of emotion on cue utilization and the organization of behavior. Psychol. Rev., 66: 183–201.

Eichenbaum, H. (1992) The hippocampal formation and declarative memory in animals. J. Cog. Neurosci., 4: 217–231.

Eichenbaum, H. (1999) The hippocampus and mechanisms of declarative memory. Behav. Brain Res., 103: 123–133.

Ferry, B. and McGaugh, J.L. (1999) Clenbuterol administration into the basolateral amygdala post-training enhances retention in an inhibitory avoidance task. Neurobiol. Learn. Mem., 72: 8–12.

Flood, J.F., Vidal, D., Bennett, E.L., Orme, A.E., Vasquez, S. and Jarvik, M.E. (1978) Memory facilitating and anti-amnesic effects of corticosteroids. Pharmacol. Biochem. Behav., 8: 81–87.

Gold, P.E. and van Buskirk, R. (1975) Facilitation of time-dependent memory processes with posttrail epinephrine injections. Behav. Biol., 13: 145–153.

Gold, P.E., Delanoy, R.L. and Merrin, J. (1984) Modulation of long-term potentiation by peripherally administered amphetamine and epinephrine. Brain Res., 2: 103–107.

Gold, P.E., Hankins, L., Edwards, R.M., Chester, J. and McGaugh, J.L. (1975) Memory interference and facilitation with posttrial amygdala stimulation: effect on memory varies with footshock level. Brain Res., 28: 509–513.

Gold, P.E. and McGaugh, J.L. (1975) A single-trace, two process view of memory storage processes. In: Deutsch, D. and Deutsch, J.A. (Eds.), Short-term Memory. Academic Press, NY, pp. 355–378.

Hall, J., Thomas, K.L. and Everitt, B.J. (2001) Fear memory retrieval induces CREB phosphorylation and Fos expression within the amygdala. Eur. J. Neurosci., 13: 1453–1458.

Hamann, S.B., Cahill, L., McGaugh, J.L. and Squire, L.R. (1997) Intact enhancement of declarative memory for emotional material in amnesia. Learn Mem., 4: 301–309.

Hamann, S.B., Ely, T.D., Grafton, S.T. and Kilts, C.D. (1999) Amygdala activity related to enhanced memory for pleasant and aversive stimuli. Nat. Neurosci., 2: 289–293.

Hatfield, T. and McGaugh, J.L. (1999) Norepinephrine infused into the basolateral amygdala posttraining enhances retention in a spatial water maze task. Neurobiol. Learn. Mem., 71: 232–239.

Hitchcock, J.M. and Davis, M. (1987) Fear-potentiated startle using an auditory conditioned stimulus: effect of lesions of the amygdala. Physiol. Behav., 39: 403–408.

Introini-Collison, I.B. and McGaugh, J.L. (1986) Epinephrine modulates long-term retention of an aversively motivated discrimination. Behav. Neural Biol., 45: 358–365.

Izquierdo, I. and Dias, R.D. (1985) Influence on memory of posttraining or pre-test injections of ACTH, vasopressin, epinephrine, and beta-endorphin, and their interaction with naloxone. Psychoneuroendocrinol., 10: 165–172.

Jellestad, F.K. and Bakke, H.K. (1985) Passive avoidance after ibotenic acid and radio frequency lesions in the rat amygdala. Physiol. Behav., 34: 299–305.

Kesner, R.P., Hunt, M.E., Williams, J.M. and Long, J.M. (1996) Prefrontal cortex and working memory for spatial response, spatial location, and visual object information in the rat. Cereb Cortex., 6: 311–318.

Kim, J.J. and Diamond, D.M. (2002) The stressed hippocampus, synaptic plasticity and lost memories. Nat. Rev. Neurosci., 3: 453–462.

Kim, J.J., Foy, M.R. and Thompson, R.F. (1996) Behavioral stress modifies hippocampal plasticity through NMDA receptor activation. Proc. Natl. Acad. Sci. USA, 93: 4750–4753.

Kim, J.J., Lee, H.J., Han, J.S. and Packard, M.G. (2001) Amygdala is critical for stress-induced modulation of hippocampal long-term potentiation and learning. J. Neurosci., 15: 5222–5228.

de Kloet, E.R., Grootendorst, J., Karssen, A.M. and Oitzl, M.S. (2002) Gene x environment interaction and cognitive performance: animal studies on the role of corticosterone. Neurobiol. Learn Mem., 78: 570–577.

de Kloet, E.R., Oitzl, M.S. and Joels, M. (1999) Stress and cognition: are corticosteroids good or bad guys? Trends Neurosci., 22: 422–426.

de Kloet, E.R., Reul, J.M. and Sutanto, W. (1990) Corticosteroids and the brain. J. Steroid. Biochem. Mol. Biol., 20: 387–394.

de Kloet, E.R., Vreugdenhil, E., Oitzl, M.S. and Joels, M. (1998) Brain corticosteroid receptor balance in health and disease. Endocr. Rev., 19: 269–301.

Van der Kolk, B.A. (1997) The psychobiology of posttraumatic stress disorder. J. Clin. Psychiatry, 58 Suppl 9: 16–24.

Layton, B. and Krikorian, R. (2002) Memory mechanisms in posttraumatic stress disorder. J. Neuropsychiatry Clin. Neurosci., 14: 254–261.

LeDoux, J.E. (2000) Emotion circuits in the brain. Annu. Rev. Neurosci., 23: 155–184.

LeDoux, J.E., Iwata, J., Cicchetti, P. and Reis, D.J. (1988) Different projections of the central amygdaloid nucleus mediate autonomic and behavioral correlates of conditioned fear. J. Neurosci., 8: 2517–2529.

Liang, K.C., Juler, R.G. and McGaugh, J.L. (1986) Modulating effects of posttraining epinephrine on memory: involvement of the amygdala noradrenergic system. Brain Res., 12: 125–133.

Liang, K.C., McGaugh, J.L. and Yao, H.Y. (1990) Involvement of amygdala pathways in the influence of post-training intra-amygdala norepinephrine and peripheral epinephrine on memory storage. Brain Res., 5: 225–233.

Liu, L., Tsuji, M., Takeda, H., Takada, K. and Matsumiya, T. (1999) Adrenocortical suppression blocks the enhancement of memory storage produced by exposure to psychological stress in rats. Brain Res., 6: 134–140.

Lupien, S.J. and McEwen, B.S. (1997) The acute effects of corticosteroids on cognition: integration of animal and human model studies. Brain Res. Rev., 24: 1–27.

Maguire, E.A., Burgess, N. and O'Keefe, J. (1999) Human spatial navigation: cognitive maps, sexual dimorphism, and neural substrates. Curr. Opin. Neurobiol., 9: 171–177.

Martin, S.J., Grimwood, P.D. and Morris, R.G. (2000) Synaptic plasticity and memory: an evaluation of the hypothesis. Annu. Rev. Neurosci., 23: 649–711.

McEwen, B.S. (1995) Stressful experience, brain, and emotions: developmental, genetic, and hormonal influences. In: Gazzagnia, M. (Ed.), The Cognitive Neurosciences. MIT Press, Cambridge, MA, pp. 1117–1135.

McEwen, B.S., De Kloet, E.R. and Rostene, W. (1986) Adrenal steroid receptors and actions in the nervous system. Physiol. Rev., 66:1121–1188.

McGaugh, J.L. (2000a) Memory—a century of consolidation. Science, 14: 248–251.

McGaugh, J.L. (2000b) Memory consolidation and the amygdala: a systems perspective. Trends Neurosci., 25: 456.

McGaugh, J.L., Cahill, L. and Roozendaal, B. (1996) Involvement of the amygdala in memory storage: interaction with other brain systems. Proc. Natl. Acad. Sci. USA, 26: 13508–13514.

McGaugh, J.L., Introini-Collison, I.B., Nagahara, A.H., Cahill, L., Brioni, J.D. and Castellano, C. (1990) Involvement of the amygdaloid complex in neuromodulatory influences on memory storage. Neurosci. Biobehav. Rev., 14: 425–431.

McGaugh, J.L., McIntyre, C.K. and Power, A.E. (2002) Amygdala modulation of memory consolidation: interaction with other brain systems. Neurobiol. Learn. Mem., 78: 539–552.

McGaugh, J.L. and Roozendaal, B. (2002) Role of adrenal stress hormones in forming lasting memories in the brain. Curr. Opin. Neurobiol., 12: 205–210.

McKernan, M.G. and Shinnick-Gallagher, P. (1997) Fear conditioning induces a lasting potentiation of synaptic currents in vitro. Nature, 390: 607–611.

Morris, R.G., Garrud, P., Rawlins, J.N. and O'Keefe, J. (1982) Place navigation impaired in rats with hippocampal lesions. Nature, 24: 681–683.

Moser, E., Moser, M.B. and Andersen, P. (1993) Spatial learning impairment parallels the magnitude of dorsal hippocampal lesions, but is hardly present following ventral lesions. J. Neurosci., 13: 3916–3925.

Oitzl, M.S., Fluttert, M., Sutanto, W. and de Kloet, E.R. (1998) Continuous blockade of brain glucocorticoid receptors facilitates spatial learning and memory in rats. Eur. J. Neurosci., 10: 3759–3766.

Oitzl, M.S., Fluttert, M. and de Kloet, E.R. (1994) The effect of corticosterone on reactivity to spatial novelty is mediated by central mineralocorticosteroid receptors. Eur. J. Neurosci., 1: 1072–1079.

Oitzl, M.S., van Haarst, A.D. and de Kloet, E.R. (1997) Behavioral and neuroendocrine responses controlled by the concerted action of central mineralocorticoid (MRS) and glucocorticoid receptors (GRS). Psychoneuroendocrinology, 22: S87–S93.

Oitzl, M.S. and de Kloet, E.R. (1992) Selective corticosteroid antagonists modulate specific aspects of spatial orientation learning. Behav. Neurosci., 106: 62–71.

O'Keefe, J. (1999) Do hippocampal pyramidal cells signal non-spatial as well as spatial information? Hippocampus, 9: 352–364.

Packard, M.G. and Cahill, L. (2001) Affective modulation of multiple memory systems. Curr. Opin. Neurobiol., 11: 752–756.

Packard, M.G., Cahill, L. and McGaugh, J.L. (1994) Amygdala modulation of hippocampal-dependent and caudate nucleus-dependent memory processes. Proc. Natl. Acad. Sci. USA, 30: 8477–8481.

Packard, M.G. and Chen, S.A. (1999) The basolateral amygdala is a co-factor in memory enhancement produced by intra-hippocampal glutamate injections. Psychobiol., 27: 377–385.

Packard, M.G. and Teather, L.A. (1998) Amygdala modulation of multiple memory systems: hippocampus and caudate-putamen. Neurobiol. Learn. Mem., 69: 163–203.

Pare, D. (2003) Role of the basolateral amygdala in memory consolidation. Prog. Neurobiol., 70: 409–420.

Pavlides, C., Nivon, L.G. and McEwen, B.S. (2000) Effects of chronic stress on hippocampal long-term potentiation. Hippocampus, 12: 245–257.

Pavlides, C., Watanabe, Y. and McEwen, B.S. (1993) Effects of glucocorticoids on hippocampal long-term potentiation. Hippocampus, 3: 183–192.

Phelps, E.A. and Anderson, A.K. (1997) Emotional memory: what does the amygdala do? Curr. Biol., 1: R311–R314.

Pikkarainen, M., Ronkko, S., Savander, V., Insausti, R. and Pitkanen, A. (1999) Projections from the lateral, basal, and accessory basal nuclei of the amygdala to the hippocampal formation in rat. J. Comp. Neurol., 11: 229–260.

Pugh, C.R., Tremblay, D., Fleshner, M. and Rudy, J.W. (1997) A selective role for corticosterone in contextual-fear conditioning. Behav. Neurosci., 111: 503–511.

de Quervain, D.J., Roozendaal, B. and McGaugh, J.L. (1998) Stress and glucocorticoids impair retrieval of long-term spatial memory. Nature, 20: 787–790.

Radwanska, K., Nikolaev, E., Knapska, E. and Kaczmarek, L. (2002) Differential response of two subdivisions of lateral amygdala to aversive conditioning as revealed by c-Fos and P-ERK mapping. Neuroreport, 3: 2241–2246.

Ressler, K.J., Paschall, G., Zhou, X.L. and Davis, M. (2002) Regulation of synaptic plasticity genes during consolidation of fear conditioning. J. Neurosci., 15: 7892–7902

Rey, M., Carlier, E., Talmi, M. and Soumireu-Mourat, B. (1994) Corticosterone effects on long-term potentiation in mouse hippocampal slices. Neuroendocrinol., 60: 36–41.

Richter-Levin, G. and Akirav, I. (2000) Amygdala-hippocampus dynamic interaction in relation to memory. Mol. Neurobiol., 22: 11–20.

Richter-Levin, G. and Akirav, I. (2003) Emotional tagging of memory formation—in the search for neural mechanisms. Brain Res. Rev., 43: 247–256.

Rogan, M.T., Staubli, U.V. and LeDoux, J.E. (1997) Fear conditioning induces associative long-term potentiation in the amygdala. Nature, 390: 604–607.

Roozendaal, B. (2000) Glucocorticoids and the regulation of memory consolidation. Psychoneuroendocrinol, 25: 213–38.

Roozendaal, B., Griffith, Q.K., Buranday, J., De Quervain, D.J. and McGaugh, J.L. (2003) The hippocampus mediates glucocorticoid-induced impairment of spatial memory retrieval: dependence on the basolateral amygdala. Proc. Natl. Acad. Sci. USA, 4: 1328–1333.

Roozendaal, B. and McGaugh, J.L. (1996) Amygdaloid nuclei lesions differentially affect glucocorticoid-induced memory enhancement in an inhibitory avoidance task. Neurobiol. Learn. Mem., 65: 1–8.

Roozendaal, B. and McGaugh, J.L. (1997) Basolateral amygdala lesions block the memory-enhancing effect of glucocorticoid administration in the dorsal hippocampus of rats. Eur. J. Neurosci., 9: 76–83.

Roozendaal, B., Nguyen, B.T., Power, A.E. and McGaugh, J.L. (1999) Basolateral amygdala noradrenergic influence enables enhancement of memory consolidation induced by hippocampal glucocorticoid receptor activation. Proc. Natl. Acad. Sci. USA, 28: 11642–11647.

Roozendaal, B., Portillo-Marquez, G. and McGaugh, J.L. (1996) Basolateral amygdala lesions block glucocorticoid-induced modulation of memory for spatial learning. Behav. Neurosci., 110: 1074–1083.

Roozendaal, B., Sapolsky, R.M. and McGaugh, J.L. (1998) Basolateral amygdala lesions block the disruptive effects of long-term adrenalectomy on spatial memory. Neurosci., 84: 453–465.

Sandi, C., Loscertales, M. and Guaza, C. (1997) Experience-dependent facilitating effect of corticosterone on spatial memory formation in the water maze. Eur. J. Neurosci., 9: 637–642.

Sandi, C. and Rose, S.P. (1994a) Corticosterone enhances long-term retention in one-day-old chicks trained in a weak passive avoidance learning paradigm. Brain Res., 30: 106–112.

Sandi, C. and Rose, S.P. (1994b) Corticosteroid receptor antagonists are amnestic for passive avoidance learning in day-old chicks. Eur. J. Neurosci., 1: 1292–1297.

Schafe, G.E., Atkins, C.M., Swank, M.W., Bauer, E.P., Sweatt, J.D. and LeDoux, J.E. (2000) Activation of ERK/MAP kinase in the amygdala is required for memory consolidation of pavlovian fear conditioning. J. Neurosci., 1: 8177–8187.

Seidenbecher, T., Reymann, K.G. and Balschun, D. (1997) A post-tetanic time window for the reinforcement of long-term potentiation by appetitive and aversive stimuli. Proc. Natl. Acad. Sci. USA, 18: 1494–1499.

Selcher, J.C., Atkins, C.M., Trzaskos, J.M., Paylor, R. and Sweatt, J.D. (1999) A necessity for MAP kinase activation in mammalian spatial learning. Learn. Mem., 6: 478–470.

Selden, N.R., Cole, B.J., Everitt, B.J. and Robbins, T.W. (1990) Damage to ceruleo-cortical noradrenergic projections impairs locally cued but enhances spatially cued water maze acquisition. Behav. Brain Res., 18: 29–51.

Shors, T.J., Foy, M.R., Levine, S. and Thompson, R.F. (1990) Unpredictable and uncontrollable stress impairs neuronal plasticity in the rat hippocampus. Brain Res. Bull., 24: 663–667.

Squire, L.R. (1992) Memory and the hippocampus: a synthesis from findings with rats, monkeys, and humans. Psychol. Rev., 99: 195–231.

Squire, L.R. (1998) Memory systems. C. R. Acad. Sci. III, 321: 153–156.

Stork, O., Stork, S., Pape, H.C. and Obata, K. (2001) Identification of genes expressed in the amygdala during the formation of fear memory. Learn. Mem., 8: 209–219.

Vaher, P.R., Luine, V.N., Gould, E. and McEwen, B.S. (1994) Effects of adrenalectomy on spatial memory performance and dentate gyrus morphology. Brain Res., 5: 71–78.

Vroom, V.H. (1964) Work and Motivation. Wiley, New York.

Yaniv, D., Schafe, G.E., LeDoux, J.E. and Richter-Levin, G. (2001) A gradient of plasticity in the amygdala revealed by cortical and subcortical stimulation, in vivo. Neuroscience, 106: 613–620.

Yau, J.L., Olsson, T., Morris, R.G., Meaney, M.J. and Seckl, J.R. (1995) Glucocorticoids, hippocampal corticosteroid receptor gene expression and antidepressant treatment: relationship with spatial learning in young and aged rats. Neuroscience, 66: 571–581.

de Wied, D. and Croiset, G. (1991) Stress modulation of learning and memory processes. Methods Achiev. Exp. Pathol., 15: 167–199.

Wang, J., Akirav, I. and Richter-Levin, G. (2000) Short-term behavioral and electrophysiological consequences of underwater trauma. Physiol. Behav., 70: 327–332.

T. Steckler, N.H. Kalin and J.M.H.M. Reul (Eds.)
Handbook of Stress and the Brain, Vol. 15
ISBN 0-444-51173-3

CHAPTER 6.3

Role of prefrontal cortex in stress responsivity

Alain Gratton[1,*] and Ronald M. Sullivan[2]

[1]McGill University, Douglas Hospital Research Center, 6875 LaSalle Bvld., Montreal (Verdun), QC H4H 1R3, Canada
[2]Unviversité de Montréal, Centre de Recherche Fernand-Séguin, Montreal, QC, Canada

Abstract: In this chapter, we review evidence derived from studies in animals and humans implicating the medial prefrontal cortex in the regulation of the autonomic, neuroendocrine, and behavioral responses to stress. The main focus of the review is on the functional differences that have been uncovered between the prelimbic and infralimbic subregions of the prefrontal cortex. We also discuss at some length, the mounting evidence of hemispheric specialization of medial prefrontal cortical function as it relates to the regulation of the autonomic and neuroendocrine responses to stress as well as of emotional reactivity and stress-related behaviors. The last section of the chapter deals with some of the long-term consequences of early developmental perturbations on medial prefrontal cortical function and their implications for stress-related psychopathologies. The discussion here centers primarily on the lasting effects of early postnatal maternal separation and handling on medial prefrontal cortex-mediated stress reactivity and their potential role in the development of major depression.

Introduction

The importance of the prefrontal cortex (PFC), particularly the medial PFC, in the brain circuitry that evolved to regulate stress reactivity is well documented. Exposure to a wide variety of stressors markedly increases neuronal and genomic activation within the medial PFC, as reflected in the pronounced release of the excitatory neurotransmitter glutamate in this area (Moghaddam, 1993) and the expression of the immediate early gene c-fos (e.g. Handa et al., 1993; Beck and Fibiger, 1995; Morrow et al., 2000). Various monoamine pathways modulate medial PFC activity in times of stress, most notably and selectively, the mesocortical DA system originating in the ventral tegmental area or VTA (e.g. Thierry et al., 1976; Deutch and Roth, 1990; Finlay et al.,1995; Yoshioka et al., 1995; Sullivan and Gratton, 1998; Jedema et al., 1999; Hayley et al., 2001). While the responsivity of these and other neurochemical systems to various stressors and pharmacological challenges have received a great deal of attention, the role of the PFC per se in regulating the (downstream) physiological and behavioral responses to stress is less understood. It is the purpose of the present chapter to review the available information on this aspect of PFC function.

The PFC, in keeping with its well-recognized role as the 'executive' control center of the brain, is well-positioned to regulate physiological and behavioral responses to stressors, particularly the orbital and medial PFC networks. The orbital network receives abundant sensory, limbic, and other cortical inputs and is regarded as a substrate for integrating viscerosensory information with affective signals (Price, 1999). The ventromedial PFC, particularly the infralimbic (IL) cortex, functions in close association with the orbital network and is regarded as a visceromotor system (Cechetto and Saper, 1990; Price, 1999). This region of the PFC has a diverse afferentation, especially from limbic regions such as amygdala and ventral hippocampus/subiculum from which it receives heavy glutamatergic inputs. Indeed, in the case of the amygdala, there is increasing evidence indicating that the PFC-mediated responses to stressors are regulated by a dopamine-sensitive

*Corresponding author. E-mail: gratal@douglas.mcgill.ca

mechanism in the basolateral amygdala (BLA) which is known to also receive a stress-responsive dopamine input from the ventral tegmental area (Stevenson and Gratton, 2003; Stevenson et al., 2003). The ventromedial PFC also provides direct outputs to hypothalamic and numerous brainstem areas involved in emotion and stress regulation, including the periaqueductal gray, the nucleus of the solitary tract, the dorsal motor nucleus of the vagus, the parabrachial nucleus and the nucleus ambiguus (Terreberry and Neafsey, 1987; Sesack et al., 1989; Hurley et al., 1991; Takagishi and Chiba, 1991; Bacon and Smith, 1993; Jodo et al., 1998). The ventromedial system is, therefore, in a pivotal position to regulate autonomic and neuroendocrine activity in conjunction with the appropriate emotional states, particularly in times of stress.

The PFC and autonomic function

Numerous early studies have described a variety of effects on autonomic function (primarily cardiovascular and respiratory) following stimulation or lesioning of PFC sites across a number of species (for reviews, see Cechetto and Saper, 1990; Van Eden and Buijs, 2000). In rats, lesions of the ventromedial PFC significantly alter the patterns of respiratory and cardiovascular changes normally associated with conditional emotional responses (Frysztak and Neafsey, 1991; 1994). These authors concluded that this region is necessary for the complete sympathetic activation of cardiovascular responses to both mild or severe stressors. Similarly, excitotoxic lesions of medial PFC output neurons greatly reduce gastric stress ulcer formation, an autonomically mediated form of stress pathology (Sullivan and Gratton, 1999). Conversely, local depletion of the predominantly inhibitory DA input to medial PFC exacerbates stress-induced gastric pathology (Sullivan and Szechtman, 1995).

Animal studies have also revealed important functional distinctions within subdivisions of the medial PFC, such that stimulation of the ventral zone (IL cortex or area 25) typically elicits sympathetic responses, while stimulation of more dorsal zones (prelimbic/anterior cingulate or areas 32 and 24, respectively) results in a parasympathetic profile of responses (Powell et al., 1994). Lesion studies of autonomic activity during classically conditioned emotional responses (CERs) support this distinction. In situations where the CER normally involves bradycardia, typically when the animal is restrained, lesions to the prelimbic cortex greatly attenuate this conditioned cardiac response, while lesions to the IL cortex are ineffective (Powell et al., 1994). However, in situations where the CER involves tachycardia, typically in freely moving animals, lesions to the IL cortex greatly decrease this autonomic response, in many cases eliciting a bradycardic response instead (Frysztak and Neafsey, 1991). As will be seen later, such functional distinctions between the dorsal and ventral extents of the PFC appear to span numerous other aspects of PFC-mediated regulation of stress responses and emotional processing.

Patients with damage to the ventromedial PFC fail to mount an autonomic response to emotionally charged stimuli, and exhibit considerable impairments in emotional and social functioning, decision-making, and risk assessment (Damasio et al., 1990; Damasio, 1994). More recent brain imaging studies have confirmed an association between PFC activity and autonomic function, reporting evidence of increased neuronal activity in medial orbital/medial PFC in response to a variety of procedures designed to alter blood pressure and/or heart rate (King et al., 1999; Harper et al., 2000).

The PFC and neuroendocrine function

There is now considerable evidence that the PFC is significantly involved in the regulation of the hypothalamic-pituitary-adrenal (HPA) axis as well. Negative feedback control of HPA activity by circulating corticosteroids is achieved via both mineralocorticoid (MR or type I) and glucocorticoid (GR or type II) receptors, with the latter being more important in limiting or terminating HPA activation in times of stress (McEwen et al., 1986; Ratka et al., 1989; Bradbury et al., 1994; De Kloet et al., 1998). Given that it has such a high density of GR, the hippocampus has received a great deal of attention as an important extrahypothalamic site in the regulation of HPA function (Sapolsky et al., 1984). Yet, one should not forget that high levels of GR are also found in the PFC (Cintra et al., 1994; McEwen et al., 1986). In rat, PFC levels of GR have been estimated to be 75–80% of those seen in hippocampus (Meaney and Aitken, 1985). Indeed, chronic glucocorticoid treatment in rats leads to dramatic dendritic reorganization of medial PFC neurons, not unlike that seen in hippocampus (Wellman, 2001). In monkey, much higher levels of GR binding are observed in PFC than in hippocampus (Sanchez et al., 2000), suggesting that, in humans, the PFC may play an important, if not a dominant role in modulating HPA activity during or following stressful experiences.

In their study, Diorio et al. (1993) reported evidence that the medial PFC is indeed involved in HPA negative-feedback regulation, showing that corticosterone (CORT) implanted in anterior cingulate/prelimbic cortex, reduces the peak plasma increases in adrenocorticotrophic hormone (ACTH) and CORT elicited by restraint stress, without affecting basal HPA function. Akana et al. (2001) recently reported the same effect in both acutely and chronically stressed rats. Diorio et al. (1993) also reported that lesioning this PFC feedback region, resulted in exaggerated stress-induced ACTH and CORT release. Prefrontal cortical regulation of stress-induced activation of HPA function was found to be specific to restraint stress, as no such effects were observed when animals were exposed to ether, a systemic stressor that, presumably, does not require cortical processing. Similar findings were reported in a study by Brake et al. (2000a) who showed that neonatal ibotenate lesions to dorsomedial PFC enhances the adult plasma CORT responses to restraint stress.

However, ibotenate lesions have been reported to *suppress* restraint stress-induced plasma CORT responses (Sullivan and Gratton, 1999) when more ventral regions of medial PFC (IL cortex) are targeted, suggesting that the ventromedial PFC serves to promote stress-induced activation of HPA function. Such an activational role of ventromedial PFC on HPA function can be seen as being parallel to the autonomic (sympathetic) activation mediated by this same cortical region in times of stress or emotional arousal in contrast to the primarily parasympathetic involvement of more dorsal regions of PFC. It also raises the possibility that the effects of damage to the ventromedial PFC may supercede those resulting from damage to more dorsal sites, by virtue of the direct anatomical links from IL cortex to neuroendocrine and autonomic control centers. An HPA-activating role for ventromedial PFC is consistent with the increases in plasma CORT seen following electrical stimulation of this area (Feldman and Conforti, 1985), an effect mediated by the prefrontal-hypothalamic pathway. It would also be in general agreement with early findings in human patients undergoing limbic leucotomy, where electrical stimulation of the ventral (orbital) frontal cortex, but not of the cingulate cortex, selectively increased plasma ACTH (Frankel and Jenkins, 1975) without affecting levels of other hormones.

The PFC and stress-related behaviors

Similar functional dissociations between dorsal and ventral PFC may help explain the widely varying effects of lesions to this region on stressful or anxiety-provoking situations. Lesions of the prelimbic cortex and/or the anterior cingulate cortex are reported to increase timidity or fear reactivity (Holson, 1986; Morgan and LeDoux, 1995). For example, in a classical conditioned-fear paradigm, dorsomedial PFC lesions increased fear behavior (freezing) both during the acquisition phase and in response to the conditioned stimulus, indicating such lesions results in a general increase in fearfulness (Morgan and LeDoux, 1995). Slightly more ventral lesions, however, had little effect during acquistion but did significantly impair extinction of the conditioned-fear response (Morgan et al., 1993). In another lesion study, however, in which the IL cortex was specifically targeted, freezing behavior was dramatically reduced in response to the conditioned-fear stimulus, as was the frequency of ultrasonic vocalizations which were a typical component of the conditioned-fear response of control animals (Frysztak and Neafsey, 1991).

A number of studies have reported on the effects of medial PFC lesions on measures of anxiety, as reflected in the animals' performance on the elevated plus maze test. In cases where damage to IL cortex is substantial, if not complete, anxiolytic effects have been reported such that animals spend significantly more time exploring the open arms of the plus maze (Gonzalez et al., 2000; Lacroix et al., 2000; Sullivan and Gratton, 2002a). In contrast, 6-OHDA-induced DA depletion of IL cortex results in an anxiogenic effect with rats exploring the open arms significantly less than controls (Espejo, 1999). One study not in keeping with this pattern, reported an anxiogenic effect (reduced open arm exploration) following electrolytic lesions to IL cortex (Jinks and McGregor, 1997); these lesions, however, were particularly small, in no case destroying the entire target area. Thus, it is possible that the anxiogenic effect reported in this study reflected a compensatory response by the surviving IL cortex neurons.

Lesions restricted to prelimbic cortex have resulted in an intermediate effect. While animals with such lesions tend to spend more time in the open arm of the elevated plus maze, they also are generally more active; thus lesions to the prelimbic cortex should perhaps not be considered to produce a selective anxiolytic effect (Maaswinkel et al., 1996). The essential role of IL cortex in the expression of anxiety and fearfulness was highlighted by two studies from the same group. When excitotoxic lesions of medial PFC clearly destroyed the IL cortex, an anxiolytic effect was observed on performance in the plus maze (Lacroix et al., 2000), but when the excitotoxin was applied to more dorsal sites, thus sparing the IL cortex, no such effect was found (Lacroix et al., 1998).

Another behavioral response thought to be mediated by the IL cortex is the simple taste aversion. It was reported that, while rats with excitotoxic IL cortex lesions do not differ from controls in their consumption of a novel palatable food (sweetened condensed milk), they fail to show the expected decrease in consumption when the same palatable food is adulterated with bitter tasting quinine (Sullivan and Gratton, 2002a). Thus, rats with such lesions to the IL cortex were either made insensitive to the bitterness of the quinine-laced food, or were impaired in their ability to produce the expected 'emotional' response to the aversive stimulus. From the evidence reviewed thus far, it appears that the ventromedial (IL) region of PFC plays a crucial role in the integration or coordination of emotional and stress-related physiological states, optimizing adaptive behavioral responses to situations of perceived threat or conflict.

Recent high-resolution brain-imaging studies in humans have linked this same region of PFC with negative emotional states. Subjects with high ratings of negative affect (which incorporate unpleasant mood states such as anxiety, irritability and anger) are at significantly higher risk of developing clinical depression and anxiety disorders. Individual differences in reported negative affect have been shown to correlate significantly with resting cerebral blood flow specifically in a focal region of posterior ventromedial PFC, corresponding to IL cortex in the rat (Zald et al., 2002). As well, when normal subjects are anticipating an electric shock, those who reported experiencing low levels of anxiety were able to suppress activity in this region, while highly anxious individuals were unable to do so (Simpson et al., 2001).

Hemispheric specialization and PFC stress regulatory systems

One of the more intriguing aspects of PFC-mediated regulation of stress reactivity and emotional processing is the extent to which it is lateralized to the right hemisphere, not only in humans, but in rodents as well. In humans, numerous electroencephalographic (EEG) studies of frontal brain asymmetries have established that right frontal biases are associated with negative emotional states and affective styles (see review by Davidson, 1998). Even from the first year of life, left-biased frontal EEG asymmetry is associated with approach behaviors and positive affect, whereas right-sided biases are linked to withdrawal and defensive behavior. Brain imaging data also indicate that the ventromedial PFC region that most closely correlated with negative affect was also localized to the right hemisphere (Zald et al., 2002). Moreover, the deficits linked to autonomic dysfunction and emotional processing resulting from ventromedial PFC damage, appear to be accounted for almost exclusively by the damage in the right hemisphere (Tranel et al., 2002). In addition to its involvement in processing negative emotion, the right PFC is known to play a dominant role in the activation of neuroendocrine (HPA axis) and autonomic (especially sympathetic) systems regulating cardiovascular and electrodermal responses (Wittling and Roschmann, 1993; Meadows and Kaplan, 1994; Henry, 1997; Wittling, 1997; Yoon et al., 1997).

In the rhesus monkey, additional studies of frontal brain asymmetries in EEG activity have demonstrated that individuals with strong right-biased asymmetries also have persistently elevated levels of plasma cortisol and very high cerebrospinal fluid levels of corticotropin-releasing factor (Kalin et al., 1998; 2000). Such animals are typically very fearful and adopt highly defensive behaviors, suggesting that strongly lateralized right PFC activity is a trait marker for highly anxious individuals, the consequence of which may be a greater risk for various pathologies that have been linked to chronically high levels of circulating glucocorticoids.

Similar findings of prefrontal asymmetry have been found in the rat. Using a left/right asymmetry index in the ventromedial PFC (IL cortex) DA response to a species-typical threat (predator odor), it was found that right-biased responses were significantly related to high stress-induced increases in plasma CORT levels (Sullivan and Gratton, 1998). As well, suppression of restraint stress-induced plasma corticosterone responses can be produced by unilateral excitotoxic lesions to the right, but not to the left IL cortex (Sullivan and Gratton, 1999). In that same study, unilateral lesions to the right IL cortex were also found to greatly reduce stress-induced gastric ulcer formation (as much as bilateral lesions), with left IL cortex lesions again being ineffective. The same right lateralized IL cortex lesion effects are seen also on the animals' performance in the elevated plus maze and simple taste aversion test (Sullivan and Gratton, 2002a).

The majority of animal studies of asymmetries in PFC function have focused on the role of the DA input to this region, which is highly asymmetrical in nature, both under basal conditions and under conditions of increased activation in response to stress (Slopsema et al., 1982; Carlson et al., 1991; 1993; 1996; Sullivan and Szechtman, 1995; Sullivan et al., 1998; Sullivan and Gratton, 1998; Anderson and Teicher, 1999; Berridge et al., 1999; Thiel and Schwarting, 2001). Right-sided increases in PFC DA transmission have been specifically associated with responses to stressful novel environments (Berridge et al., 1999), anxiolytic behavior in the plus maze (Anderson and Teicher, 1999), protection from stress ulcer pathology (Sullivan and Szechtman, 1995), and escape performance following exposure to uncontrollable shock (Carlson et al., 1993). Taken together, these findings indicate that not only is the medial PFC intrinsically lateralized with respect to the regulation of stress and emotional reactivity, but at least one of its major afferent systems is similarly lateralized to optimally modulate this cortical regulation.

The reasons for such right-lateralized PFC regulation of stress responses and emotional reactivity are not clear, but it is clear that this phenomenon transcends species and is not restricted to the PFC (e.g. Denenberg, 1981; Adamec and Morgan, 1994; Coleman-Mesches and McGaugh, 1995), although the PFC may represent the top of a hierarchy of structures, orchestrating the activity of these 'stress networks'. It has been suggested that cerebral lateralization of this sort may reflect, in part, the central representation of peripheral (visceral) asymmetries. Visceral inputs to brain, especially from heart and stomach, are processed primarily in the right brainstem (Geschwind and Galaburda, 1987). For example, the nucleus ambiguus (directly linked with IL cortex) regulates the changes in heart rate and vocalization frequencies in response to stress, but it is the right nucleus ambiguus that is critical for these functions (Porges, 1995). Right-hemispheric structures are thus more directly linked with basic autonomic and neuroendocrine life-sustaining functions and mature earlier in development than their left-hemisphere counterparts (Geschwind and Galaburda, 1987). As the associated cortical networks mature, a right-dominant cortical system may then follow from these more basic asymmetries, as part of optimally efficient and tightly linked regulatory systems of stress reactivity. One implication of this is that the more slowly maturing cortical systems would also be more vulnerable to a number of pre- and early postnatal insults. The consequences of this may not only be to alter the development of normal PFC asymmetry, but also the development of stress-regulatory systems in general.

Early development and PFC function: implications for stress-related pathology

Numerous early developmental perturbations have been shown to have long-lasting if not permanent effects on PFC function. In rats, prenatal stress is known to lead to enhanced anxiety in the offspring which is associated with lateralized changes in mesocortical DA function (Fride and Weinstock, 1988). Similarly, cocaine exposure in utero results in offspring with dramatically increased levels of Fos expression selectively in the ventromedial PFC (Morrow et al., 2002). The same prenatal treatment also greatly enhances the stress-induced activation of DA transmission in the ventromedial PFC (Elsworth et al., 2001). In contrast, a blunted PFC (IL cortex) DA response to stress that is lateralized to the right hemishere is observed in adult rats that had sustained a brief episode of global anoxia at the time of birth (Brake et al., 2000b). As adults, animals that have sustained this type of early-life insult also display a right-sided increase in PFC DA transporter levels (Brake et al., 2000b) and a significant impairment of stress-induced activation of the HPA axis (Boksa et al., 1996).

Early postnatal maternal separation and social isolation, have been shown to result in abnormally high synaptic densities within the IL cortex (Ovtscharoff and Braun, 2001), as well as significantly altered densities of DA and 5-HT terminals throughout the medial PFC (Braun et al., 2000). Early social isolation has also been reported to result in decreases in basal DA turnover, selectively within the IL cortex (Heidbreder et al., 2000). Early maternal separation results in greatly increased HPA activity that persists throughout adulthood and this is reflected in elevated hypothalamic CRF mRNA levels, and exaggerated stress-induced plasma corticosterone responses (Plotsky and Meaney, 1993; Meaney et al., 1996). This treatment also results in decreases in GR mRNA levels within the frontal cortex and hippocampus (Avishai-Eliner et al., 1999), thus diminishing the brain's capacity for feedback regulation of the HPA axis. A recent magnetic resonance imaging study in the monkey, has confirmed the impact of maternal separation on PFC, as this condition resulted in a significant *enlargement* of the ventromedial PFC, specific to the right hemisphere (Lyons et al., 2002). Together with the reported increase in synaptic density of this region, such findings could explain the heightened sensitivity to stressors and greater fearfulness of these animals.

Whereas the above developmental insults appear to have long-lasting detrimental consequences, one early-life manipulation which appears to promote efficient regulation of stress and emotiomal reactivity is early postnatal handling, which promotes maternal behaviors such as the licking and grooming of pups and arch-back nursing (Levine, 1975; Smotherman, 1983; Liu et al., 1997). In contrast to the effects of maternal separation, early postnatal handling increases frontal cortex and hippocampal GR mRNA levels, while decreasing hypothalamic CRF mRNA levels and attenuating stress-induced increases in plasma ACTH and corticosterone levels. In addition, adult rats that were handled as pups habituate to the acute effects of repeated mild stress on the HPA axis (Meaney et al., 1985; 1996; Sullivan and Gratton, 2003). As adults, handled animals also show significantly *reduced* synaptic density in IL cortex compared to controls (Ovtscharoff and Braun, 2001), which may in part account for the reduced drive of the HPA axis and the lower fearfulness and higher levels of exploratory behaviors of these animals.

An additional effect of early postnatal handling is to stimulate the 'normal' development of cerebral lateralization. It has been shown that handled rats exhibit right-cortical lateralization for the regulation of a variety of emotion-related, species typical behaviors, while nonhandled rats do not (Denenberg, 1981; Denenberg et al., 1986). Similarly, handling has been shown to induce a rightward shift in benzodiazepine receptor binding in IL cortex and hippocampus, the extent of which was correlated with reductions in anxiety-related behaviors (Sullivan and Gratton, 2003). As well, it was recently observed that the potentiating effect of DA receptor blockade in IL cortex on stress-induced HPA activation, is lateralized to the right hemisphere among handled rats, but not among nonhandled rats (Sullivan and Dufresne, 2002). Taken together, these findings suggest that right hemispheric lateralization of PFC function is necessary for optimal regulation of stress and emotional reactivity.

It is intuitively obvious that any perturbations in the development of PFC stress regulatory systems could have serious consequences, not only in terms of behavior and stress-related psychopathologies, but also of somatic manifestations that would be expected to result from chronic neuroendocrine and/or autonomic dysfunction. Either excessive, deficient, or abnormally lateralized development of these prefrontal cortical networks may contribute in greater or lesser part, to numerous pathologies.

Clinical depression and anxiety disorders have received considerable attention in this regard, being associated with not only neuroendocrine and/or autonomic dysfunction, but functional abnormalities of PFC function as well. The implications for disturbed patterns of PFC lateralization on a variety of stress-related psychopathologies have been recently reviewed (Sullivan and Gratton, 2002b). Variations in the normal asymmetrical (right-biased) regulation of neuroendocrine function, have as well been found to be associated with a number of somatic complaints (Wittling and Schweiger, 1993). Frontal brain asymmetries in EEG activity have even been shown to predict individual differences in immune function (natural killer cell activity), which could provide a vital link in our understanding of the relationship between stress and disease or psychosomatic illness generally (Davidson et al., 1999).

The experience-dependent maturation of PFC systems, particularly as derived from social (maternal) relationships, has formed the basis of theories of human development (Henry, 1997). The proper development of these ventral prefrontal cortical systems, particularly in the right hemisphere, dependent importantly on the DA projection to this area. As such these represent higher-level systems that play a critical role in the development of successful coping strategies and optimal emotional self-regulation (Schore, 1996; 1997). Variations in the quality of such early postnatal experiences is suggested to account for individual differences in vulnerability to a wide range of pathologies. The importance of the early development of these systems is highlighted by cases of early trauma to ventral PFC regions, where not only emotional and social functioning are impaired, but the development of moral reasoning itself is gravely affected (Anderson et al., 1999).

Within the normal spectrum of development however, there is considerable room for optimism in overcoming the consequences of adverse conditions during early life. In rats, it has been shown that the neuroendocrine and behavioral consequences of early maternal separation can be reversed by environmental enrichment during the peripubertal period, or by various pharmacological interventions (Kaufman et al., 2000; Francis et al., 2002). While many of the molecular changes initially induced by maternal separation, particularly at the level of the hypothalamus, appear to persist throughout adult life, it appears that such subsequent interventions can override the influence exerted by these neuroendocrine and behavioral regulators. Such a capacity may well reflect adaptive changes in hippocampal and PFC circuitry, such that the plasticity which originally makes such systems vulnerable, may also allow the restoration of normal stress and emotional reactivity, through alternate, compensatory mechanisms.

References

Adamec, R.E. and Morgan, H.D. (1994) The effect of kindling of different nuclei in the left and right amygdala on anxiety in the rat. Neuroscience, 55: 1–12.

Akana, S.F., Chu, A., Soriano, L. and Dallman, M.F. (2001) Corticosterone exerts site-specific and state-dependent effects in prefrontal cortex and amygdala on regulation of adrenocorticotropic hormone, insulin and fat depots. J. Neuroendocrinol., 13(7): 625–637.

Anderson, S.L. and Teicher, M.H. (1999) Serotonin laterality in amygdala predicts performance in the elevated plus maze. Neuroreport, 10: 3497–3500.

Anderson, S.W., Bechara, A., Damasio, H., Tranel, D. and Damasio, A.R. (1999) Impairment of social and moral behavior related to early damage in human prefrontal cortex. Nat. Neurosci., 2(11): 1032–1037.

Avishai-Eliner, S., Hatalski, C.G., Tabachnik, E., Eghbal-Ahmadi, M. and Baram, T.Z. (1999) Differential regulation of glucocorticoid receptor messenger RNA (GR–mRNA) by maternal deprivation in immature rat hypothalamus and limbic regions. Brain Res. Dev. Brain Res., 114(2): 265–268.

Bacon, S.J. and Smith, A.D. (1993) A monosynaptic pathway from an identified vasomotor centre in the medial prefrontal cortex to an autonomic area in the thoracic spinal cord. Neuroscience, 54: 719–728.

Beck, C.H. and Fibiger, H.C. (1995) Conditioned fear-induced changes in behavior and in the expression of the immediate early gene c-fos: with and without diazepam pretreatment. J. Neurosci., 15(1): 709–720.

Berridge, C.W., Mitton, E., Clark, W. and Roth, R.H. (1999) Engagement in a non-escape (displacement) behavior elicits a selective and lateralized suppression of frontal cortical dopaminergic utilization in stress. Synapse, 32: 187–197.

Boksa, P., Krishnamurthy, A. and Sharma, S. (1996). Hippocampal and hypothalamic type I corticosteroid receptor affinities are reduced in adult rats born by a caesarean procedure with or without an added period of anoxia. Neuroendocrinology, 64: 25–34.

Bradbury, M.J., Akana, S.F. and Dallman, M.F. (1994) Roles of type I and II corticosteroid receptors in regulation of basal activity in the hypothalamo–pituitary–adrenal axis during the diurnal trough and the peak: evidence for a nonadditive effect of combined receptor occupation. Endocrinology, 134(3): 1286–1296.

Brake, W.G., Flores, G., Francis, D., Meaney, M.J., Srivastava, L.K. and Gratton, A. (2000a) Enhanced nucleus accumbens dopamine and plasma corticosterone stress responses in adult rats with neonatal excitotoxic lesions to the medial prefrontal cortex. Neuroscience, 96(4):687–695.

Brake, W.G., Sullivan, R.M. and Gratton, A. (2000b) Perinatal distress leads to lateralized medial prefrontal cortical dopamine hypofunction in adult rats. J. Neurosci., 20(14): 5538–5543.

Braun, K., Lange, E., Metzger, M. and Poeggel, G. (2000) Maternal separation followed by early social deprivation affects the development of monoaminergic fiber systems in the medial prefrontal cortex of Octodon degus. Neuroscience, 95: 309–318.

Carlson, J.N., Fitzgerald, L.W., Keller, R.W. and Glick, S.D. (1991) Side and region dependent changes in dopamine activation with various durations of restraint stress. Brain Res., 550: 313–318.

Carlson, J.N., Fitzgerald, L.W., Keller, R.W. and Glick, S.D. (1993) Lateralized changes in prefrontal cortical dopamine activity induced by controllable and uncontrollable stress in the rat. Brain Res., 630: 178–187.

Carlson, J.N., Visker, K.E., Keller, Jr., R.W. and Glick, S.D. (1996) Left and right 6-hydroxydopamine lesions of the medial prefrontal cortex differentially alter subcortical dopamine utilization and the behavioral response to stress. Brain Res., 711: 1–9.

Cechetto, D.F. and Saper, C.B. (1990) Role of the cerebral cortex in autonomic function. In: Loewy, A.D. and Spyer, K.M. (Eds.), Central Regulation of Autonomic Functions, Oxford University Press, Oxford, pp. 208–223.

Cintra, A., Zoli, M., Rosen, L., Agnati, L.F., Okret, S., Wikstrom, A.-C., Gustafsson, J.-A. and Fuxe, K. (1994) Mapping and computer densitometry of glucocorticoid receptor immunoreactive neurons and glial cells in the rat central nervous system. Neuroscience, 62(3): 843–897.

Coleman-Mesches, K. and McGaugh, J.L. (1995) Differential effects of pretraining inactivation of the right or left amygdala on retention of inhibitory avoidance training. Behav. Neurosci., 109: 642–647.

Damasio, A.R. (1994) Descartes' Error, Grosset/Putnam, New York.

Damasio, A.R., Tranel, D. and Damasio, H. (1990) Individuals with sociopathic behavior caused by frontal damage fail to respond autonomically to social stimuli. Behav. Brain Res., 41: 81–94.

Davidson, R.J. (1998) Cerebral asymmetry, emotion and affective style. In: Davidson, R.J. and Hughdahl, K. (Eds.), Brain Asymmetry, MIT Press, Cambridge, pp. 361–387.

Davidson, R.J., Coe, C.C., Dolski, I. and Donzella, B. (1999) Individual differences in prefrontal activation asymmetry predict natural killer cell activity at rest and in response to challenge. Brain Behav. Immun., 13: 93–108.

De Kloet, E.R., Vreugdenhil, E., Oitzl, M.S. and Joels, M. (1998) Brain corticosteroid receptor balance in health and disease. Endocrine Rev., 19(3): 269–301.

Denenberg, V.H. (1981) Hemispheric laterality in animals and the effects of early experience. Behav. Brain Sci., 4: 1–49.

Denenberg, V.H., Gall, J.S., Berrebi, A. and Yutzey, D.A. (1986) Callosal mediation of cortical inhibition in the lateralized rat brain. Brain Res., 397: 327–332.

Deutch, A.Y. and Roth, R.H. (1990) The determinants of stress-induced activation of the prefrontal cortical dopamine system. In: Uylings, H.B.M., Van Eden, C.G., De Bruin, J.P.C., Corner, M.A. and Feenstra, M.G.P. (Eds.), Progress in Brain Research, Vol 85, The prefrontal cortex: Its structure, function and pathology, Elsevier, Amsterdam, pp. 367–403.

Diorio, D., Viau, V. and Meaney, M.J. (1993) The role of the medial prefrontal cortex (cingulate cortex) in the regulation of hypothalamic–pituitary–adrenal responses to stress. J. Neurosci., 13: 3839–3847.

Elsworth, J.D., Morrow, B.A. and Roth, R.H. (2001) Prenatal cocaine exposure increases mesoprefrontal dopamine neuron responsivity to mild stress. Synapse, 42: 80–83.

Espejo, E.F. (1999) Selective dopamine depletion within the medial prefrontal cortex induces anxiogenic-like effects in rats placed on the elevated plus maze. Brain Res., 762: 281–284.

Feldman, S. and Conforti, N. (1985) Modifications of adrenocortical responses following frontal cortex stimulation in rats with hypothalamic deafferentations and medial forebrain bundle lesions. Neuroscience, 15: 1045–1047.

Finlay, J.M., Zigmond, M.J. and Abercrombie, E.D. (1995) Increased dopamine and norepinephrine release in medial prefrontal cortex induced by acute and chronic stress: effects of diazepam. Neuroscience, 64(3): 619–628.

Francis, D.D., Diorio, J., Plotsky, P.M. and Meaney, M.J. (2002) Environmental enrichment reverses the effects of maternal separation on stress reactivity. J. Neurosci., 22(18): 7840–7843.

Frankel, R.J. and Jenkins, J.S. (1975) Pituitary hormone response to brain stimulation in man. J. Endocrinol., 67(1): 113–117.

Fride, E. and Weinstock, M. (1988) Prenatal stress increases anxiety related behavior and alters cerebral lateralization of dopamine activity. Life Sci., 42(10): 1059–1065.

Frysztak, R.J. and Neafsey, E.J. (1991) The effect of medial frontal cortex lesions on respiration, "freezing," and ultrasonic vocalizations during conditioned emotional responses in rats. Cerebr. Cortex, 1: 418–425.

Frysztak, R.J. and Neafsey, E.J. (1994) The effect of medial frontal cortex lesions on cardiovascular conditioned emotional responses in the rat. Brain Res., 643: 181–193.

Geschwind, N. and Galaburda, A.M. (1987) Cerebral lateralization: biological mechanisms, associations, and pathology. The MIT Press, Cambridge, MA.

Gonzalez, L.E., Rujano, M., Tucci, S., Paredes, D., Silva, E., Alba, G. and Hernandez, L. (2000) Medial prefrontal transection enhances social interaction. I: behavioral studies. Brain Res., 887: 7–15.

Handa, R.J., Nunley, K.M. and Bollnow, M.R. (1993) Induction of c-fos mRNA in the brain and anterior pituitary gland by a novel environment. Neuroreport, 4(9): 1079–1082.

Harper, R.M., Bandler, R., Spriggs, D. and Alger, J.R. (2000) Lateralized and widespread brain activation during transient blood pressure elevation revealed by magnetic resonance imaging. J. Comp. Neurol., 417(2): 195–204.

Hayley, S., Borowski, T., Merali, Z. and Anisman, H. (2001) Central monoamine activity in genetically distinct strains of mice following a psychogenic stressor: effects of predator exposure. Brain Res., 892(2): 293–300.

Heidbreder, C.A., Weiss, I.C., Domeney, A.M., Pryce, C., Homberg, J., Hedou, G., Feldon, J., Moran, M.C. and Nelson, P. (2000) Behavioral, neurochemical and endocrinological characterization of the early social isolation syndrome. Neuroscience, 100(4): 749–768.

Henry, J.P. (1997) Psychological and physiological responses to stress: The right hemisphere and the hypothalamic–pituitary–adrenal axis, an inquiry into problems of human bonding. Acta Physiol. Scand., (Suppl. 640): 10–25.

Holson, R.R. (1986) Mesial prefrontal cortical lesions and timidity in rats. I. Reactivity to aversive stimuli. Physiol. Behav., 37: 221–230.

Hurley, K.M., Herbert, H., Moga, M.M. and Saper, C.B. (1991) Efferent projections of the infralimbic cortex of the rat. J. Comp. Neurol., 308: 249–276.

Jedema, H.P., Sved, A.F., Zigmond, M.J. and Finlay, J.M. (1999) Sensitization of norepinephrine release in medial prefrontal cortex: effect of different chronic stress protocols. Brain Res., 830(2): 211–217.

Jinks, A.L. and McGregor, I.S. (1997) Modulation of anxiety-related behaviours following lesions of the prelimbic or infralimbic cortex in the rat. Brain Res., 772: 181–190.

Jodo, E., Chiang, C. and Aston-Jones, G. (1998) Potent excitatory influence of prefrontal cortex activity on noradrenergic locus coeruleus neurons. Neuroscience, 83(1): 63–79.

Kalin, N.H., Larson, C., Shelton, S.E. and Davidson, R.J. (1998) Asymmetric frontal brain activity, cortisol, and behavior associated with fearful temperament in rhesus monkeys. Behav. Neurosci., 112: 286–292.

Kalin, N.H., Shelton, S.E. and Davidson, R.J. (2000) Cerebrospinal fluid corticotropin-releasing hormone levels are elevated in monkeys with patterns of brain activity associated with fearful temperament. Biol. Psychiat., 47: 579–585.

Kaufman, J., Plotsky, P.M., Nemeroff, C.B. and Charney, D.S. (2000) Effects of early adverse experiences on brain structure and function: clinical implications. Biol. Psychiat., 48(8): 778–790.

King, A.B., Menon, R.S., Hachinski, V. and Cechetto, D.F. (1999) Human forebrain activation by visceral stimuli. J. Comp. Neurol., 413(4): 572–582.

Lacroix, L., Broersen, L.M., Feldon, J. and Weiner, I. (2000) Effects of local infusions of dopaminergic drugs into the medial prefrontal cortex of rats on latent inhibition, prepulse inhibition and amphetamine induced activity. Behav. Brain Res., 107: 111–121.

Lacroix, L., Broersen, L.M., Weiner, I. and Feldon, J. (1998) The effects of excitotoxic lesion of the medial prefrontal cortex on latent inhibition, prepulse inhibition, food hoarding, elevated plus maze, active avoidance and locomotor activity in the rat. Neuroscience, 84: 431–442.

Levine, S. (1975) Infantile experience and resistance to physiological stress. Science, 126: 405–406.

Liu, D., Diorio, J., Tannenbaum, B., Caldji, C., Francis, D., Freedman, A., Sharma, S., Pearson, D., Plotsky, P.M. and Meaney, M.J. (1997) Maternal care, hippocampal glucocorticoid receptors, and hypothalamic–pituitary–adrenal responses to stress. Science, 277: 1659–1662.

Lyons, D.M., Afarian, H., Schatzberg, A.F., Sawyer-Glover, A. and Moseley, M.E. (2002) Experience-dependent asymmetric variation in primate prefrontal morphology. Behav. Brain Res., 136: 51–59.

Maaswinkel, H., Gispen, W.H. and Suruijt, B.M. (1996) Effects of an electrolytic lesion of the prelimbic area on anxiety-related and cognitive tasks in the rat. Behav. Brain Res., 79: 51–59.

McEwen, B.S., De Kloet, E.R. and Rostene, W.H. (1986) Adrenal steroid receptors and actions in the nervous system. Physiol. Rev., 66: 1121–1150.

Meadows, M.-E. and Kaplan, R.F. (1994) Dissociation of autonomic and subjective responses to emotional slides in right hemisphere damaged patients. Neuropsychologia, 32: 847–856.

Meaney, M.J. and Aitken, D.H. (1985) [3H]Dexamethasone binding in rat frontal cortex. Brain Res., 328: 176–180.

Meaney, M.J., Aitken, D.H., Bodnoff, S.R., Iny, L.J., Tatarewicz, J.E. and Sapolsky, R.M. (1985) Early postnatal handling alters glucocorticoid receptor concentrations in selected brain regions. Behav. Neurosci., 99(4): 765–770.

Meaney, M.J., Diorio, J., Francis, D., Widdowson, J., LaPlante, P., Caldji, C., Sharma, S., Seckl, J.R. and Plotsky, P.M. (1996) Early environmental regulation of forebrain glucocorticoid receptor gene expression: implications for adrenocortical responses to stress. Dev. Neurosci., 18(1–2): 49–72.

Moghaddam, B. (1993) Stress preferentially increases extraneuronal levels of excitatory amino acids in the prefrontal cortex: comparison to hippocampus and basal ganglia. J. Neurochem., 60(5): 1650–1657.

Morgan, M.E. and LeDoux, J.E. (1995) Differential contribution of dorsal and ventral medial prefrontal cortex to the acquisition and extinction of conditioned fear in rats. Behav. Neurosci., 109: 681–688.

Morgan, M.A., Romanski, L.M. and LeDoux, J.E. (1993) Extinction of emotional learning: contribution of medial prefrontal cortex. Neurosci. Lett., 163: 109–113.

Morrow, B.A., Elsworth, J.D., Lee, E.J. and Roth, R.H. (2000). Divergent effects of putative anxiolyticson stress-induced fos expression in the mesoprefrontal system of the rat. Synapse, 36(2): 143–154.

Morrow, B.A., Elsworth, J.D. and Roth, R.H. (2002) Male rats exposed to cocaine in utero demonstrate elevated expression of Fos in the prefrontal cortex in response to environment. Neuropsychopharmacology, 26(3): 275–285.

Ovtscharoff, W. Jr. and Braun, K. (2001) Maternal separation and social isolation modulate the postnatal development of synaptic composition in the infralimbic cortex of Octodon degus. Neuroscience, 104: 33–40.

Plotsky, P.M. and Meaney, M.J. (1993) Early, postnatal experience alters hypothalamic corticotropin-releasing factor (CRF) mRNA, median eminence CRF content and stress-induced release in adult rats. Brain Res. Mol. Brain Res., 18(3): 195–200.

Porges, S.W. (1995) Cardiac vagal tone: A physiological index of stress. Neurosci. Biobehav. Rev., 19: 225–233.

Powell, D.A., Watson, K. and Maxwell, B. (1994) Involvement of subdivisions of the medial prefrontal cortex in learned cardiac adjustments in rabbits. Behav. Neurosci., 108: 294–307.

Price, J.L. (1999) Prefrontal cortical networks related to visceral function and mood. Ann. NY Acad. Sci., 877: 383–396.

Ratka, A., Sutanto, W., Bloemers, M. and de Kloet, E.R. (1989) On the role of brain mineralocorticoid (type I) and glucocorticoid (type II) receptors in neuroendocrine regulation. Neuroendocrinology, 50(2): 117–123.

Sanchez, M.M., Young, L.J., Plotsky, P.M. and Insel, T.R. (2000) Distribution of corticosteroid receptors in the rhesus brain: relative absence of glucocorticoid receptors in the hippocampal formation. J. Neurosci., 20(12): 4657–4668.

Sapolsky, R.M., Krey, L.C. and McEwen, B.S. (1984) Glucocorticoid-sensitive hippocampal neurons are involved in terminating the adrenocortical stress response. Proc. Natl. Acad. Sci. USA, 81: 6174–6177.

Schore, A.N. (1996) The experience-dependent maturation of a regulatory system in the orbital prefrontal cortex and the origin of developmental psychopathology. Dev. Psychopathol., 8: 59–87.

Schore, A.N. (1997) Early organization of the nonlinear right brain and development of a predisposition to psychiatric disorders. Dev. Psychopathol., 9: 595–631.

Sesack, S.R., Deutch, A.Y., Roth, R.H. and Bunney, B.S. (1989) Topographical organization of the efferent projections of the medial prefrontal cortex in the rat: an anterograde tract-tracing study with phaseolus vulgaris leucoagglutinin. J. Comp. Neurol., 290: 213–242.

Simpson, J.R. Jr., Drevets, W.C., Snyder, A.Z., Gusnard, D.A. and Raichle, M.E. (2001) Emotion-induced changes in human medial prefrontal cortex: II. During anticipatory anxiety. Proc. Natl. Acad. Sci. USA, 98(2): 688–693.

Slopsema, J.S., VanDerGugten, J. and DeBruin, J.P.C. (1982) Regional concentrations of noradrenaline and dopamine in the frontal cortex of the rat: dopaminergic innervation of the prefrontal subareas and lateralization of prefrontal dopamine. Brain Res., 250: 197–200.

Smotherman, W.P. (1983) Mother-infant interaction and the modulation of pituitary-adrenal activity in rat pups after early stimulation. Dev. Psychobiol., 16: 169–176.

Stevenson, C. and Gratton, A. (2003) Basolateral amygdala modulation of the nucleus accumbens dopamine response to stress: role of medial prefrontal cortical dopamine. Eur. J. Neurosci., 17: 1287–1295.

Stevenson, C., Sullivan, R.M. and Gratton, A. (2003) Effects of basolateral amygdala dopamine depletion on the nucleus accumbens and medial prefrontal cortical dopamine responses to stress. Neurosci., 116: 285–293.

Sullivan, R.M. and Gratton, A. (1998) Relationships between stress-induced increases in medial prefrontal cortical dopamine and plasma corticosterone levels in rats: role of cerebral laterality. Neuroscience, 83: 81–91.

Sullivan, R.M. and Gratton, A. (1999) Lateralized effects of medial prefrontal cortex lesions on neuroendocrine and autonomic stress responses in rats. J. Neurosci., 19: 2834–2840.

Sullivan, R.M. and Gratton, A. (2002a) Behavioral effects of excitotoxic lesions of ventral medial prefrontal cortex in the rat are hemisphere-dependent. Brain Res., 927: 69–79.

Sullivan, R.M. and Gratton, A. (2002b) Prefrontal cortical regulation of hypothalamic–pituitary–adrenal function in the rat and implications for psychopathology: side matters. Psychoneuroendocrinology, 27: 99–114.

Sullivan, R.M. and Gratton, A. (2003) Behavioral and neuroendocrine relevance of hemispheric asymmetries in benzodiazepine receptor binding induced by postnatal handling in the rat. Brain & Cogn., 51: 218–222.

Sullivan, R.M. and Szechtman, H. (1995) Asymmetrical influence of mesocortical dopamine depletion on stress ulcer development and subcortical dopamine systems in rats: implications for psychopathology. Neuroscience, 65: 757–766.

Takagishi, M. and Chiba, T. (1991) Efferent projections of the infralimbic (area 25) region of the medial prefrontal cortex in the rat: an anterograde tracer PHA-L study. Brain Res., 566(1–2): 26–39.

Terreberry, R.R. and Neafsey, E.F. (1987) The rat medial frontal cortex projects directly to autonomic regions of the brainstem. Brain Res. Bull., 19: 639–649.

Thiel, C.M. and Schwarting, R.K. (2001) Dopaminergic lateralisation in the forebrain: relations to behavioral asymmetries and anxiety in male Wistar rats. Neuropsychobiology, 43: 192–199.

Thierry, A.M., Tassin, J.P., Blanc, G. and Glowinski, J. (1976) Selective activation of mesocortical DA systems by stress. Nature, 263: 242–243.

Tranel, D., Bechara, A. and Denburg, N.L. (2002) Asymmetric functional roles of right and left ventromedial prefrontal cortices in social conduct, decision-making, and emotional processing. Cortex, 38(4): 589–612.

Van Eden, C.G. and Buijs, R.M. (2000) Functional neuroanatomy of the prefrontal cortex: autonomic interactions. Prog. Brain Res., 126: 49–62.

Wellman, C.L. (2001) Dendritic reorganization in pyramidal neurons in medial prefrontal cortex after chronic corticosterone administration. J. Neurobiol., 49(3): 245–253.

Wittling, W. (1997) The right hemisphere and the human stress response. Acta Physiol. Scand., (Suppl. 640): 55–59.

Wittling, W. and Roschmann, R. (1993) Emotion-related hemisphere asymmetry: subjective emotional responses to laterally presented films. Cortex, 29: 431–448.

Wittling, W. and Schweiger, E. (1993) Alterations of neuroendocrine brain asymmetry: a neural risk factor affecting physical health. Neuropsychobiology, 28: 25–29.

Yoon, B.W., Morillo, C.A., Cechetto, D.F. and Hachinski, V. (1997) Cerebral lateralization in cardiac autonomic control. Arch. Neurol., 54(6): 741–744.

Yoshioka, M., Matsumoto, M., Togashi, H. and Saito, H. (1995) Effects of conditioned fear stress on 5-HT release in the rat prefrontal cortex. Pharmacol. Biochem. Behav., 51(2): 515–519.

Zald, D.H., Mattson, D.L. and Pardo, J.V. (2002) Brain activity in ventromedial prefrontal cortex correlates with individual differences in negative affect. Proc. Natl. Acad. Sci. USA, 99: 2450–2454.

Index

1: Refers to Part 1; **2:** Refers to Part 2